10 0260886 4

UNIVERSITY OF NOTTINGHAM

N
RY

D1625984

# Handbook of Plant and Crop Physiology

# BOOKS IN SOILS, PLANTS, AND THE ENVIRONMENT

*Editorial Board*

*Agricultural Engineering*   Robert M. Peart, University of Florida, Gainesville

*Animal Science*   Harold Hafs, Rutgers University, New Brunswick, New Jersey

*Crops*   Mohammad Pessarakli, University of Arizona, Tucson

*Irrigation and Hydrology*   Donald R. Nielsen, University of California, Davis

*Microbiology*   Jan Dirk van Elsas, Research Institute for Plant Protection, Wageningen, The Netherlands

*Plants*   L. David Kuykendall, U.S. Department of Agriculture, Beltsville, Maryland

*Soils*   Jean-Marc Bollag, Pennsylvania State University, University Park, Pennsylvania

Tsuyoshi Miyazaki, University of Tokyo

*Soil Biochemistry, Volume 1,* edited by A. D. McLaren and G. H. Peterson
*Soil Biochemistry, Volume 2,* edited by A. D. McLaren and J. Skujiņš
*Soil Biochemistry, Volume 3,* edited by E. A. Paul and A. D. McLaren
*Soil Biochemistry, Volume 4,* edited by E. A. Paul and A. D. McLaren
*Soil Biochemistry, Volume 5,* edited by E. A. Paul and J. N. Ladd
*Soil Biochemistry, Volume 6,* edited by Jean-Marc Bollag and G. Stotzky
*Soil Biochemistry, Volume 7,* edited by G. Stotzky and Jean-Marc Bollag
*Soil Biochemistry, Volume 8,* edited by Jean-Marc Bollag and G. Stotzky
*Soil Biochemistry, Volume 9,* edited by G. Stotzky and Jean-Marc Bollag

*Organic Chemicals in the Soil Environment, Volumes 1 and 2,* edited by C. A. I. Goring and J. W. Hamaker
*Humic Substances in the Environment,* M. Schnitzer and S. U. Khan
*Microbial Life in the Soil: An Introduction,* T. Hattori
*Principles of Soil Chemistry,* Kim H. Tan
*Soil Analysis: Instrumental Techniques and Related Procedures,* edited by Keith A. Smith
*Soil Reclamation Processes: Microbiological Analyses and Applications,* edited by Robert L. Tate III and Donald A. Klein
*Symbiotic Nitrogen Fixation Technology,* edited by Gerald H. Elkan
*Soil–Water Interactions: Mechanisms and Applications,* Shingo Iwata and Toshio Tabuchi with Benno P. Warkentin
*Soil Analysis: Modern Instrumental Techniques, Second Edition,* edited by Keith A. Smith
*Soil Analysis: Physical Methods,* edited by Keith A. Smith and Chris E. Mullins
*Growth and Mineral Nutrition of Field Crops,* N. K. Fageria, V. C. Baligar, and Charles Allan Jones
*Semiarid Lands and Deserts: Soil Resource and Reclamation,* edited by J. Skujiņš
*Plant Roots: The Hidden Half,* edited by Yoav Waisel, Amram Eshel, and Uzi Kafkafi
*Plant Biochemical Regulators,* edited by Harold W. Gausman
*Maximizing Crop Yields,* N. K. Fageria
*Transgenic Plants: Fundamentals and Applications,* edited by Andrew Hiatt

*Environmental Chemistry of Selenium*, edited by William T. Frankenberger, Jr., and Richard A. Engberg

*Principles of Soil Chemistry: Third Edition, Revised and Expanded*, Kim H. Tan

*Sulfur in the Environment*, edited by Douglas G. Maynard

*Soil–Machine Interactions: A Finite Element Perspective*, edited by Jie Shen and Radhey Lal Kushwaha

*Mycotoxins in Agriculture and Food Safety*, edited by Kaushal K. Sinha and Deepak Bhatnagar

*Plant Amino Acids: Biochemistry and Biotechnology*, edited by Bijay K. Singh

*Handbook of Functional Plant Ecology*, edited by Francisco I. Pugnaire and Fernando Valladares

*Handbook of Plant and Crop Stress: Second Edition, Revised and Expanded*, edited by Mohammad Pessarakli

*Plant Responses to Environmental Stresses: From Phytohormones to Genome Reorganization*, edited by H. R. Lerner

*Handbook of Pest Management*, edited by John R. Ruberson

*Environmental Soil Science: Second Edition, Revised and Expanded*, Kim H. Tan

*Microbial Endophytes*, edited by Charles W. Bacon and James F. White, Jr.

*Plant–Environment Interactions: Second Edition*, edited by Robert E. Wilkinson

*Microbial Pest Control*, Sushil K. Khetan

*Soil and Environmental Analysis: Physical Methods, Second Edition, Revised and Expanded*, edited by Keith A. Smith and Chris E. Mullins

*The Rhizosphere: Biochemistry and Organic Substances at the Soil–Plant Interface*, Roberto Pinton, Zeno Varanini, and Paolo Nannipieri

*Woody Plants and Woody Plant Management: Ecology, Safety, and Environmental Impact*, Rodney W. Bovey

*Metals in the Environment*, M. N. V. Prasad

*Plant Pathogen Detection and Disease*, P. Narayanasamy

*Handbook of Plant and Crop Physiology, Second Edition, Revised and Expanded*, edited by Mohammad Pessarakli

*Additional Volumes in Preparation*

*Handbook of Postharvest Technology*, edited by A. Chakraverty, Arun S. Mujumdar, and G. S. V. Raghavan, and H. S. Ramaswamy

*Enzymes in the Environment: Activity, Ecology, and Applications*, Richard G. Burns and Richard Dick

*Plant Roots: The Hidden Half, Third Edition, Revised and Expanded*, Yoav Waisel, Amram Ehsel, and Uzi Kafkafi

*Environmental Chemistry of Arsenic*, edited by William T. Frankenberger, Jr.

*Biological Control of Major Crop Plant Diseases*, edited by Samuel S. Gnanamanickam

# Handbook of Plant and Crop Physiology

## Second Edition
## Revised and Expanded

edited by
## Mohammad Pessarakli
*The University of Arizona*
*Tucson, Arizona*

NOTTINGHAM
UNIVERSITY
AGRICULTURAL
AND FOOD
SCIENCES
LIBRARY

MARCEL DEKKER, INC.

NEW YORK · BASEL

**ISBN: 0-8247-0546-7**

This book is printed on acid-free paper.

**Headquarters**
Marcel Dekker, Inc.
270 Madison Avenue, New York, NY 10016
tel: 212-696-9000; fax: 212-685-4540

**Eastern Hemisphere Distribution**
Marcel Dekker AG
Hutgasse 4, Postfach 812, CH-4001 Basel, Switzerland
tel: 41-61-261-8482; fax: 41-61-261-8896

**World Wide Web**
http://www.dekker.com

The publisher offers discounts on this book when ordered in bulk quantities. For more information, write to Special Sales/Professional Marketing at the headquarters address above.

**Copyright © 2002 by Marcel Dekker, Inc. All Rights Reserved.**

Neither this book nor any part may be reproduced or transmitted in any form or by any means, electronic or mechanical, including photocopying, microfilming, and recording, or by any information storage and retrieval system, without permission in writing from the publisher.

Current printing (last digit):
10 9 8 7 6 5 4 3 2 1

**PRINTED IN THE UNITED STATES OF AMERICA**

In the memory of my beloved parents, Fatima and Vahab,
who regretfully did not live to see this work,
which in no small part resulted from their gift of many years of love.

# Preface

Like any other area in science, both the scope and depth of our knowledge of plant and crop physiology are rapidly expanding. Plant/crop physiologists are continuously making new discoveries. This phenomenon has resulted in the compilation of a large volume of information since the first edition of the *Handbook of Plant and Crop Physiology* was prepared and presented to scientists and professionals. The abundance of new data has necessitated that this unique, comprehensive source of information be revised to include all the new discoveries in the field. Like the first edition, the new edition of the *Handbook of Plant and Crop Physiology* is a unique, comprehensive, and complete collection of the topics in plant/crop physiology.

More than two-thirds of the material in the new edition is entirely new; these are included under new titles. The other one-third has been updated and substantially modified. This new edition consists of 12 parts while the first edition consisted of eight. Overall, about 80% of this book is new and a totally new volume has emerged.

The *Handbook of Plant and Crop Physiology* is needed to fill the gap in the available literature. In addition, it has long been recognized that physiological processes control plant growth and crop yields. Therefore, this handbook is prepared in a single volume to serve as a comprehensive resource and up-to-date reference to effectively cover the information relevant to plant/crop physiology that is scattered among plant/crop physiology books as well as plant physiology journals.

Several difficult decisions must be made when one plans to compile a handbook, such as the extent of content to include, the information to exclude, the depth to which the topics should be covered, and the organization of the selected content. I have chosen to include information that will be beneficial to students, instructors, researchers, field specialists, and any others interested in the areas of plant and crop physiology. In order to plan, implement, and evaluate comprehensive and specific strategies for dealing with plant and crop physiology problems, strategies must be based on a firm understanding of the facts and the principles.

The topics selected for discussion are those that I believe are relevant, and in which physiology plays the dominant role. The concepts have been presented to allow both beginning students and specialists of this discipline an opportunity to expand and refine their knowledge. Certain conclusions provided

*v*

throughout the text are related to the more significant and multifaceted problems of plant and crop physiology. They are presented to provide a concise guide to the most relevant goals of both the students and the specialists.

This practical and comprehensive guide has been prepared by 76 contributors from 17 countries, among the most competent and knowledgeable scientists, specialists, and researchers in agriculture. It is intended to serve as a resource for both lecture and independent purposes—for scientists, agriculture researchers, agriculture practitioners, and both educators and students in agricultural disciplines.

To facilitate the accessibility of desired information, the volume has been divided into 12 parts. Although the parts are interrelated, each serves independently to facilitate the understanding of the material presented therein. Each part also enables the reader to acquire confidence in his or her learning and use of the information offered.

Part I, Plants/Crops Growth Responses to Environmental Factors and Climatic Changes, consists of three chapters addressing these factors in detail. The seven chapters in Part II, Physiology of Plant/Crop Growth and Developmental Stages, cover plant physiological stages from seed germination to plant senescence and abscission. Part III, Cellular and Molecular Aspects of Plant/Crop Physiology, consists of five chapters that present in-depth information on cellular and molecular aspects of plant/crop organs. Part IV, Plant/Crop Physiology and Physiological Aspects of Plant/Crop Production Processes, contains eight chapters that link plant/crop physiology to production of food, feed, and medicinal compounds and discuss this relationship in detail. The four chapters in Part V, Plant Growth Regulators: The Natural Hormones (Growth Promoters and Inhibitors) and Plant Genes, address growth promoters, and growth inhibitor hormones as well as plant genes.

Since plants and crops, like other living things, at one time or another during their life cycle encounter biotic or abiotic stressful conditions, two parts [VI, Physiological Responses of Plants/Crops Under Stressful (Salt, Drought, and Other Environmental Stresses) Conditions and VII, Physiological Responses of Plants/Crops to Heavy Metal Concentration and Agrichemicals] are devoted to the physiological responses of plants and crops to stress. Several examples of empirical investigations of specific plants and crops grown under stressful conditions are presented.

The single—but thorough—chapter in Part VIII, Physiological Relationships Between Lower and Higher Plants, presents detailed information on this relationship.

The physiology of plant genetics is presented in two parts. Physiology of Lower-Plant Genetics and Development, Part IX, consists of one chapter that discusses developmental genetics in lower plants. Part X, Physiology of Higher-Plant/Crop Genetics and Development, contains four chapters that comprehensively review this subject. Part XI, Using Computer Modeling in Plant Physiology, consists of one chapter on computer simulation of plant and Crop allocation processes.

Finally, to extend the subject of plant/crop physiology beyond the earth, I included Part XII, Plant/Crop Physiology under Controlled Conditions, in Space, and on Other Planets. Its two chapters present the most recent available information on plant/crop physiology in controlled environment and perspectives for human life support on other planets.

Numerous figures, tables, and illustrations are included in this technical guide to facilitate comprehension of the presented materials. The index words further increase accessibility to the information.

It is hoped that an individual with a problem in the area of plant/crop physiology will turn to this practical and professional reference book and be able to promptly acquire the necessary assistance to solve that problem.

Like other fields, the area of plant/crop physiology has been growing so rapidly that all plant/crop physiologists are faced with the problem of constantly updating their knowledge. To grow with their profession, they will need to extend their interests and skills. In this regard, even a casual reading of the material in this handbook will help them to move ahead in the right direction.

*Mohammad Pessarakli*

# Acknowledgments

I would like to express my special appreciation for secretarial assistance from Mrs. Elenor R. Loya, Department of Soil, Water and Environmental Science, University of Arizona. Additional assistance by the secretarial staff of the Department of Plant Sciences, University of Arizona, is also greatly appreciated.

In addition, my sincere gratitude is extended to Mr. Russell Dekker (Chief Publishing Officer, Marcel Dekker, Inc.), who supported this project from its initiation to its completion. Certainly, this job would not have been completed as smoothly and rapidly without his most valuable support and sincere efforts. Also, Production Editor Ms. Dana Bigelow's patience and outstanding efforts in the careful and professional handling of this volume are greatly appreciated. The precision and accuracy of the copyeditor, Ms. Mary Prescott, are sincerely acknowledged.

The invaluable efforts of every one of the contributors are deeply appreciated. Their proficiency and knowledge in their areas of expertise made this significant task possible.

I thank my wife, Vinca, for her support in the completion of this work. Last, but not least, I would like to express my gratitude to my son, Mahdi, who had great patience and understanding and let me take time to complete this project that would have otherwise been spent with him.

# Contents

# Contributors

**Refat Abdel-Basset**   Faculty of Science, Assiut University, Assiut, Egypt

**Leon Hartwell Allen, Jr.**   U.S. Department of Agriculture–Agricultural Research Service, and Agronomy Department, University of Florida, Gainesville, Florida

**Timothy S. Artlip**   Appalachian Fruit Research Station, U.S. Department of Agriculture–Agricultural Research Service, Kearneysville, West Virginia

**Klaus Peter Bader**   Department of Cell Physiology, Faculty of Biology, University of Bielefeld, Bielefeld, Germany

**Fred E. Below**   Department of Crop Sciences, University of Illinois, Urbana, Illinois

**Wade L. Berry**   Department of Organismic Biology, Ecology, and Evolution, University of California, Los Angeles, California

**Martine Bertrand**   Marine Science and Technology Institute, Conservatoire National des Arts et Métiers, Cherbourg, France

**Bernard B. Bible**   Department of Plant Science, University of Connecticut, Storrs, Connecticut

**Elizabeth A. Bray**   Department of Botany and Plant Sciences, University of California, Riverside, California

**Donald P. Briskin**   Department of Natural Resources and Environmental Sciences, University of Illinois, Urbana, Illinois

**Robert Carpentier**   Groupe de Recherche en Énergie et Information Biomoléculaires, Université du Québec à Trois-Rivières, Trois-Rivières, Québec, Canada

**Calvin Chong**   Department of Plant Agriculture-Vineland, Ontario Agricultural College, University of Guelph, Vineland Station, Ontario, Canada

**Richard S. Criddle**   Department of Chemistry and Biochemistry, Brigham Young University, Provo, Utah

**Joel L. Cuello**   Department of Agricultural and Biosystems Engineering, The University of Arizona, Tucson, Arizona

**Irene S. Day**   Department of Biology, Colorado State University, Fort Collins, Colorado

**Ilia D. Denev**   Department of Plant Physiology and Molecular Biology, University of Plovdiv, Plovdiv, Bulgaria

**Frank G. Dennis, Jr.**   Department of Horticulture, Michigan State University, East Lansing, Michigan

**Donna M. Dubay**   Department of Botany and Plant Sciences, University of California, Riverside, California

**R. S. Dubey**   Department of Biochemistry, Faculty of Science, Banaras Hindu University, Varanasi, India

**Maria Gallo-Meagher**   Agronomy Department, University of Florida, Gainesville, Florida

**William Grierson**   Citrus Research and Education Center, University of Florida, Lake Alfred, Florida (retired)

**Bernard Grodzinski**   Division of Horticulture, Department of Plant Agriculture, University of Guelph, Guelph, Ontario, Canada

**Jean-Claude Guary**   Marine Science and Technology Institute, Conservatoire National des Arts et Métiers, Cherbourg, France

**Lee D. Hansen**   Department of Chemistry and Biochemistry, Brigham Young University, Provo, Utah

**Lyneen C. Harris**   Department of Chemistry and Biochemistry, Brigham Young University, Provo, Utah

**John E. Hendrix**   Department of Bioagricultural Sciences and Pest Management, Colorado State University, Fort Collins, Colorado

**Kwok Ki Ho**   Department of Biological Sciences, The National University of Singapore, Singapore, Republic of Singapore

**Benjamin Jacoby**   Department of Agricultural Botany, The Hebrew University of Jerusalem, Rehovot, Israel

**Chris Johansen***   International Crops Research Institute for the Semi-Arid Tropics (ICRISAT), Andhra Pradesh, India

**Hak-Yoon Ju**   Department of Plant and Animal Sciences, Nova Scotia Agricultural College, Truro, Nova Scotia, Canada

*Current affiliation:* Consultant in Agricultural Research and Development, Dhaka, Bangladesh

**Pawan K. Kasera**   Department of Botany, University of Jodhpur, Jodhpur, India

**Evangelos Demosthenes Leonardos**   Division of Horticulture, Department of Plant Agriculture, University of Guelph, Guelph, Ontario, Canada

**Lanfang He Levine**   Dynamac Corporation, Kennedy Space Center, Florida

**Antoaneta Ljubenova***   Department of Plant Physiology and Molecular Biology, University of Plovdiv, Plovdiv, Bulgaria

**Francesco Loreto**   Institute of Plant Biochemistry and Ecophysiology, National Research Council of Italy, Rome, Italy

**Monica A. Madore**   Department of Botany and Plant Sciences, University of California, Riverside, California

**Kenneth B. Marcum†**   Department of Plant Sciences, The University of Arizona, Tucson, Arizona

**Angelo Massacci**   Institute of Plant Biochemistry and Ecophysiology, National Research Council of Italy, Rome, Italy

**V. Wallace McCarlie**   Department of Botany and Range Science, Brigham Young University, Provo, Utah

**Adam Miller**   Department of Biology, Division of Science and Math, Lorain County Community College, Elyria, Ohio

**Ivan N. Minkov**   Department of Plant Physiology and Molecular Biology, University of Plovdiv, Plovdiv, Bulgaria

**Edmund R. Miranda**   Department of Botany and Plant Sciences, University of California, Riverside, California

**Sher Mohammed**   Department of Botany, Government Lohia (PG) College, Churu, India

**Nava Moran**   Department of Agricultural Botany, The Hebrew University of Jerusalem, Rehovot, Israel

**Joaquín Moreno**   Department of Biochemistry and Molecular Biology, University of Valencia, Burjassot, Valencia, Spain

**Syed Shamshad Mehdi Naqvi**   Nuclear Institute of Agriculture, Tando Jam, Pakistan

**Soek Ying Neo**   Institute of Molecular and Cell Biology, The National University of Singapore, Singapore, Republic of Singapore

**James W. O'Leary**   Department of Plant Sciences, The University of Arizona, Tucson, Arizona

**David J. Oliver**   Department of Botany, Iowa State University, Ames, Iowa

**Wattana Pattanagul**   Department of Botany and Plant Sciences, University of California, Riverside, California

---

*Current affiliation: University of the North, Sovenga, South Africa
†Current affiliation: Department of Tropical Plant and Soil Sciences, University of Hawaii at Mano'a, Honolulu, Hawaii

**Lola Peñarrubia** Department of Biochemistry and Molecular Biology, University of Valencia, Burjassot, Valencia, Spain

**Mohammad Pessarakli** Department of Plant Sciences, The University of Arizona, Tucson, Arizona

**Pramila Rajput** Department of Botany, University of Jodhpur, Jodhpur, India

**Idupulapati Madhusudana Rao** Soils and Plant Nutrition Unit, Centro Internacional de Agricultura Tropical (CIAT), Cali, Colombia

**A. S. N. Reddy** Department of Biology and Program in Cell and Molecular Biology, Colorado State University, Fort Collins, Colorado

**Vaka Subba Reddy** Department of Biology and Program in Cell and Molecular Biology, Colorado State University, Fort Collins, Colorado

**Steven Rodermel** Department of Botany, Iowa State University, Ames, Iowa

**Francisco F. de la Rosa** Department of Plant Biochemistry and Molecular Biology, University of Seville, Seville, Spain

**Benoît Schoefs\*** Department of Plant Physiology, University of South Bohemia, Budejovice, Czeck Republic

**David N. Sen** Department of Botany, University of Jodhpur, Jodhpur, India (retired)

**Roy Sexton** Department of Biological and Molecular Sciences, Stirling University, Stirling, Scotland

**Yun-Kang Shen** Laboratory of Photosynthesis, Shanghai Institute of Plant Physiology, Chinese Academy of Sciences, Shanghai, People's Republic of China

**Bruce N. Smith** Department of Botany and Range Science, Brigham Young University, Provo, Utah

**Martin Spalding** Department of Botany, Iowa State University, Ames, Iowa

**Dorothy L. Stradling** Department of Botany and Range Science, Brigham Young University, Provo, Utah

**Gary W. Stutte** Dynamac Corporation, Kennedy Space Center, Florida

**G. V. Subbarao**[†] Dynamac Corporation, Kennedy Space Center, Florida

**Tonya Thygerson** Department of Botany and Range Science, Brigham Young University, Provo, Utah

**Joseph C. V. Vu** U.S. Department of Agriculture–Agricultural Research Service, and Agronomy Department, University of Florida, Gainesville, Florida

**Jillian Walker** Department of Botany and Range Science, Brigham Young University, Provo, Utah

**John C. Wallace** Department of Plant Biology, University of New Hampshire, Durham, New Hampshire

---

\**Current affiliation:* Laboratoire Plasticité et Expression des Génomes Microbiens, Université Joseph Fourier, Grenoble, France

[†]*Current affiliation:* Japan International Research Center for Agricultural Sciences, Ibaraki, Japan

**Raymond M. Wheeler**   National Aeronautics and Space Administration, Kennedy Space Center, Florida

**Michael E. Wisniewski**   Appalachian Fruit Research Station, U.S. Department of Agriculture–Agricultural Research Service, Kearneysville, West Virginia

**Chengbin Xiang**   Department of Botany, Iowa State University, Ames, Iowa

**Da-Quan Xu**   Laboratory of Photosynthesis, Shanghai Institute of Plant Physiology, Chinese Academy of Sciences, Shanghai, People's Republic of China

**Neil C. Yorio**   Dynamac Corporation, Kennedy Space Center, Florida

# 1

# Time, Plant Growth, Respiration, and Temperature

**Bruce N. Smith, Lyneen C. Harris, V. Wallace McCarlie, Dorothy L. Stradling, Tonya Thygerson, Jillian Walker, Richard S. Criddle, and Lee D. Hansen**

*Brigham Young University, Provo, Utah*

## I. INTRODUCTION

The earth is very dynamic and has undergone dramatic changes in its history. All of the elements, including those common in living things, were synthesized from primordial hydrogen in the interior of stars [1]. Supernovas and other stellar instabilities dispersed many elements into space. Because hydrogen and the noble gases are greatly depleted on earth as compared with their cosmic abundances [2], it is likely that the chunks of matter giving rise to the protoplanet did not carry with them gaseous shells of their own. As a result of contraction and redistribution of materials in the developing planet, a hydrosphere and atmosphere developed that were highly reduced [3]. The surface of the earth today is much more oxidized, even to 21% $O_2$ in the atmosphere [4]. Was this oxidation linear or have there been fluctuations several times during the history of the earth resulting in major species extinctions [5]?

Plants have evolved to survive, thrive, and grow by adapting to ever-changing conditions. The sea was a stable, benign home for life during three fourths of the history of life on earth. Emergence on land exposed living things to a much greater range of environmental conditions [4]. Increasing biological diversity to exploit new environmental opportunities has given us the present distribution of life on earth [5]. Change continues today at an accelerated pace because of the impact of human activities. This chapter explores ways in which adaptations to environmental changes have occurred and how plant metabolism can be used to predict and better understand plant growth.

## II. PLANT GROWTH

Plants grow by the process of cell division or mitosis followed by cell enlargement and maturation. Cells then differentiate into tissues that make up the organs of the plant. Mitosis includes replication of organelles, synthesis of nuclear material, enzymes, etc. Cell enlargement consists largely of water uptake to form a large vacuole. Growth may be measured as change in mass, volume, or length of shoot or root. Crop productivity is often expressed not in biomass but in yield of the desired product: flower, fruit, seed, root, oil, protein, or specific chemical.

Water, sunlight, carbon dioxide, oxygen, and mineral elements from the soil are well known to be essential for sustained plant growth. If any of these things are deficient in the environment or present in excess (toxic amounts), plants may become stressed and even die. But plants have adapted to life in a variety of conditions [6]. Plants may play a role in modifying their environment and climate [7]. In addition, plants have complex relationships with other organisms in their communities including herbivores, pathogens, parasites, symbiotic or free-living nitrogen-fixing bacteria, and mycorrhizae. All of these factors can affect the rate of plant growth.

Photosynthesis supports all life on earth and in eukaryotes occurs exclusively in chloroplasts. All green tissues contain chloroplasts, but most photosynthesis, by far, occurs in leaves. $C_4$ plants have a photosynthetic rate that is two to three times faster than that of $C_3$ plants and 100-fold faster than that of Crassulacean acid metabolism (CAM) plants [8], but within each of these groups there is much variability in photosynthetic rate. Despite much effort to correlate this variation with growth rates, no consistent results have been obtained [9]. Thus, although photosynthesis is absolutely necessary for plant growth, the rate of photosynthesis does not predict the rate of plant growth [9]. Insufficient carbon assimilation does *not* explain why alpine plants are so small and why biomass accumulation per unit land area is so low [10]. Several investigators have suggested that respiration is a better predictor for plant growth [11].

## III.  RESPIRATION

McCree [12] reported that specific respiration rate and specific plant growth rate are linearly correlated with a positive slope and positive intercept. Thornley [13] then borrowed a model from microbiology [14] that equates the slope of such a plot to a growth coefficient and the intercept to a maintenance rate. This is represented in Eq. (1).

$$R = R_M + R_G \tag{1}$$

where $R$ is total respiration, $R_M$ is maintenance respiration or that necessary to maintain life processes, and $R_G$ is the respiration responsible for growth. This model has been widely used for 30 years but provides only an empirical fit to the data [15–17]. This model cannot predict plant growth rates from metabolic rates.

This chapter discusses another model linking plant respiratory metabolism with growth [18] that is testable, based on first principles, allows predictions of growth rates from metabolic rate measurements, and defines responses to subtle changes in environmental stress. The theory will be presented followed by several examples of applications.

## IV.  GROWTH AND RESPIRATION

Consider the overall growth reaction (2).

$$C_{substrate} + x(\text{compounds and ions of N, P, K, etc.}) + yO_2 \rightarrow$$
$$\varepsilon C_{str.biomass} + (1 - \varepsilon)CO_2 + yH_2O + \text{heat} \tag{2}$$

Reaction (2) is the sum of two reactions, that is, the catabolic reaction (3)

$$C_{substrate} + zO_2 \rightarrow CO_2 + \text{heat} \tag{3}$$

and the anabolic reaction (4)

$$\text{heat} + C_{substrate} + x(\text{compounds and ions of N, P, K, etc.}) \rightarrow C_{str.biomass} \tag{4}$$

that occur in the condition-dependent ratio $(1 - \varepsilon)/\varepsilon$, where $\varepsilon$ is the substrate carbon conversion efficiency. Reactions (3) and (4) are energy coupled through cyclic production and hydrolysis of ATP and redox cycling of NADH. Because the ratio of the rates of reactions (3) and (4) varies with conditions, reaction (3) must always produce an excess of ATP and NADH, as clearly explained in the book *Introduction to the Thermodynamics of Biological Processes* [19]. This necessitates both an ability to change the efficiency of production of ATP through such pathways as the alternative oxidase and a third reaction, the futile hydrolysis of ATP and oxidation of NADH as in reaction (5).

$$aATP + bNADH \rightarrow aADP + aP_i + bNAD \tag{5}$$

Note that $a$ and $b$ must always be greater than zero and that the rate of reaction (5) varies with conditions because catabolism and anabolism are not stoichiometrically coupled [19].

For the anabolic reaction (4) the initial system is lower in energy and higher in entropy than the final system. In symbolic notation,

$$\Delta E_{anab.} > 0 \tag{6}$$

and

$$\Delta S_{anab.} < 0 \tag{7}$$

Thus, the anabolic reaction must extract energy from the catabolic reaction and the catabolic reaction must increase the entropy of the surroundings more than the anabolic reaction decreases the entropy of the system. The system is defined by reaction (2). These conditions can be expressed in equation form as

$$\Delta E_{system} < 0 \tag{8}$$

and

$$\Delta S_{surr.} + \Delta S_{system} > 0 \tag{9}$$

Note that Eq. (9) is simply a statement of the second law of thermodynamics [19]. The value of $\Delta S_{surr.}$ is related to the heat ($Q$) exchanged between the system and surroundings and the absolute temperature ($T$) by Eq. (10).

$$\Delta S_{surr.} = Q/T \tag{10}$$

Neglecting pressure-volume work, which is negligible for most terrestrial biological systems [20], allows equating $Q$ to $-\Delta H$, the enthalpy change, where the minus sign indicates that heat goes from the system to the surroundings, and equating $\Delta E$ to $\Delta G$, the Gibbs free energy change. Substituting and rearranging in Eqs. (8), (9), and (10) provides the result

$$\Delta G_{system} = \Delta H_{system} - T \Delta S_{system} \tag{11}$$

where $\Delta G_{system}$ is the total energy change for the energy-coupled anabolic and catabolic reactions and must be less than zero for growth to occur.

Because the entropies of the products and reactants are nearly equal, the value of $T \Delta S_{system}$ for reaction (2) is small and can be either negative or positive. Thus, $\Delta G_{system}$ is negative as required for a spontaneous growth process only because $\Delta H_{system}$ is negative; i.e., metabolic heat must always be exothermic. This requires that growing organisms with aerobic metabolism must produce heat energy that is lost to the surroundings. This metabolic heat, which is absolutely required for growth, is not "wasteful," is path (condition) dependent, and should not be confused with the "maintenance rate" that appears as an energy compartment in the model used in the reviews [15–17].

## V.  CALORIMETRY

Respiration has usually been measured as the rate of oxygen uptake or carbon dioxide evolution. However, this is insufficient information [Eq. (11)] to predict growth and/or ability to handle stress from abiotic or biotic factors. In addition to gas exchange rate, the energy lost as heat must be measured. In some instances, where there is little or no change in substrate carbon conversion efficiency ($\varepsilon$), it is possible to predict plant growth from gas exchange measurements alone [Eq. (2)]. But if the efficiency of conversion of photosynthate to biomass changes, gas exchange measurements by themselves will be of limited utility. Measurements of both gas exchange and heat rates are necessary to determine both rate and efficiency of growth.

Using modern calorimeters, it is possible to make rapid, isothermal measurements of metabolic heat rate ($q$) and respiration rate ($R_{CO2}$) at several temperatures for small samples (~100 mg fresh weight) of plant tissues. Much can be learned from these two simple measurements.

Plant tissue (80–100 mg fresh weight) is placed in each of three ampules of the calorimeter (Hart Scientific model 7707 or Calorimetry Sciences Corporation MCDSC model 4100). After 15–20 min of thermal equilibration at the desired temperature, the metabolic heat rate ($q$) is measured for another 15–20 min. The ampules are removed from the calorimeter and a small vial filled with 40 μL of 0.40 M NaOH is placed in the calorimeter ampule with the tissue. Again, a 15–20 min thermal equibration is necessary, followed by measurement of the respiration rate ($R_{CO2}$) for 15–20 min. As the $CO_2$ and NaOH react in solution, additional heat is produced ($-108.5$ kJ mol$^{-1}$ is the heat of reaction for carbonate formation), giving the rate of $CO_2$ evolution ($R_{CO2}$) by the plant tissue. Next, the NaOH is removed and the heat rate ($q$) is measured as before. The tissue may then be run at another temperature. The difference in $q$ and $455R_{CO2}$ [21] can then be used to predict growth rate changes with temperature [see Eqs. (2) and (4)] under the assumption that carbohydrate is the substrate for reaction (2).

Substrate carbon conversion efficiency $\varepsilon$, described in Eq. (2), is related to the ratio $q/R_{CO2}$ as in Eq. (12) [21].

$$(\varepsilon/1 - \varepsilon)\Delta H_B = -q/R_{CO2} - (1 - \gamma_p/4)\Delta H_{O2} \tag{12}$$

where $\Delta H_B$ is the enthalpy change for the formation of biomass from photosynthate [Eq. (4)], $\gamma_p$ is the mean chemical oxidation state of the substrate carbon oxidized to $CO_2$, and $\Delta H_{O2}$ is Thornton's constant, with a value of $-455 \pm 15$ kJ mol$^{-1}$ of $O_2$.

Incorporating Thornton's constant and assuming carbohydrate substrate with $\gamma_p = 0$, the specific growth rate of structural biomass ($R_{SG}$) is related to the two measured variables as in Eq. (13).

$$R_{SG}\Delta H_B = 455R_{CO2} - q \tag{13}$$

## VI.  CATABOLISM AND ANABOLISM

Photosynthesis transforms energy from sunlight into energy-rich organic matter, i.e., carbohydrates. This organic matter then serves as the energy source for all life on earth. The energy is partially liberated in glycolysis (fermentation) or in the oxidative pentose phosphate cycle, both in the cytoplasm. Substrate-level ATP and reduced pyridine nucleotides are produced. This may have been the extent of energy conservation in anoxic early earth [3]. Once oxygen began to increase, mitochondrial activity provided a much higher rate of energy turnover, resulting in explosive adaptive radiation [5]. The key to rapid expansion of life on earth as well as growth of a single plant is rapid turnover of ATP/ADP—perhaps as much as 50% of the dry biomass of active tissues every 24 hr [22]. If an inhibitor blocks the cytochrome oxidase pathway or an uncoupler destroys the proton gradient across the inner mitochondrial membrane, there is a rapid increase in oxygen uptake and $CO_2$ production in response to the drop in ATP production.

Louis Pasteur showed that yeast cells would produce more $CO_2$ in nitrogen than in air. Plant biochemists showed that tissues committed to rapid growth (e.g., germinating seeds, meristematic tissue) would show the Pasteur effect whereas mature or senescing tissue would not. The control mechanism for respiration proposed was the ATP/ADP ratio [23]. In growing tissues, oxidative phosphorylation rapidly produces ATP, which is utilized just as rapidly in anabolic activities. Plants store energy not as ATP but rather as sucrose, starch, protein, or lipid. Of interest is that chloroplasts do not export ATP but mitochondria do. For growth, both ADP and ATP must be present.

## VII.  STRESS

Plants are subject to many forms of environmental stress. Some are abiotic, physicochemical, or density independent, such as temperature, drought, fire, and air pollution. Other sources of stress are biotic or density dependent, such as competition, herbivory, disease, and parasitism [24]. For each of these environmental factors there is a range or life zone that the plant can tolerate. If the tolerance range for a given stress factor is exceeded, the plant will suffer stress, and if the stress is severe enough, the plant may die. Short-term acclimation may be possible, and given enough time, natural selection may result in adaptation to the stress.

At the cellular and molecular level, the common theme of stress is the formation of free radicals—strong oxidants that can do significant damage to membranes and DNA. Free radicals include superoxide, hydrogen peroxide, and superhydroxide [25,26]. Air pollutants may themselves be strong oxidants, such as ozone, peroxyacetyl nitrate (PAN), and oxides of sulfur and nitrogen [27]. Heavy metals with more than one possible valence state can also serve as strong electron donors.

Chemical defences against free radicals include compounds that are strong reductants such as glutathione, phenols, flavonoids, and polyamines [25]. Enzymatic defenses against free radicals include superoxide dismutases, catalase, peroxidases, phenol oxidase, and ascorbic acid oxidase [26]. Excess light energy trapped by chlorophyll in a high-oxygen environment can do significant damage. The violaxanthin-zeaxanthin cycle plays a major role in helping to dissipate that energy [28]. A diminished capacity to defend against free radicals is thought to play a major role in tissue senescence [29]. Is it possible to learn from respiratory metabolism the influence of environmental stress before the plant shows visible symptoms? The answer is yes, as shown in the following sections.

## A.  Temperature Stress

Cheatgrass (*Bromus tectorum* L.) is a weedy annual first introduced into the Great Basin of the western United States in the late 19th century. It germinates in the fall, overwinters as seedlings, grows very rapidly in the early spring when moisture is abundant, flowers in May, sets seed, and drops them, completing the life cycle by early June. The dead grass then serves as fuel for wildfires. Cheatgrass seed survives fire well whereas competing native perennials do not, creating conditions for further spread of the weed.

Cheatgrass is a highly autogamous species with minimal levels of genetic variation. Nonetheless, genetic differentiation may arise in response to general and predictable differences among habitats that make a population-level response appropriate [30].

Characteristics of respiratory metabolism were examined in 11 subpopulations from different habitats [31]. Seeds from each subpopulation were germinated and metabolic heat rates and respiration rates determined calorimetrically at 5°C intervals from 5 to 45°C. From the experimental data, growth rates and efficiency of carbon conversion were calculated. Results are summarized in Table 1. One might suppose that the temperature response would follow the large range of altitudes of the 11 populations studied. That was not the case as the lowest elevation and warmest site, St. George at 850 m, had the lowest optimal growth temperature (10°C) and the lowest upper limit for growth (16°C). On the other hand, higher elevation sites had higher optimal growth temperatures and higher upper limit temperatures (Table 1). The explanation is that plants must be adapted to the microclimate in which they must survive. In St. George at 850 m, cheatgrass can grow only in the winter and very early spring, when temperatures are cool but water is available. In the dry, hot summer, survival is impossible. By contrast, mountain sites (2000 to 3000 m) have a shorter frost-free period, but water is available in summer when temperatures are often very warm. Cheatgrass has thus adapted to grow in warmer temperatures at high elevations.

Corn (*Zea mays* L.) varieties are grown worldwide. Growth rates of some of the cultivars are predicted to increase at low temperature, go through a maximum in the "normal" growth range, and then de-

**TABLE 1**  Metabolic Heart Rate ($q$) and Respiration Rate ($R_{CO2}$) Measured Every 5°C from 0 to 45°C for Germinated Cheatgrass (*Bromus tectorum* L.) Seed from 11 Populations at Different Elevations[a]

| Population (altitude, m) | Temperature response (°C) | | |
| --- | --- | --- | --- |
| | Low stress | Optimal | High stress |
| St. George (850) | 3–4 | 10 | 16 |
| Green River (1280) | 6 | 10 and 30 | 35 |
| White Rocks (1450) | 3–4 | 15 | >30 |
| Ephraim (1740) | 0–3 | 5 and 20 | 25 |
| Hobble Creek (1800) | 2–3 | 15 and 25 | >30 |
| Potosi (1850) | <0 | 15 | 27 |
| Castle Rock (1980) | 6 | 15 | 44 |
| Salina (2040) | 7 | 25 | 32 |
| Strawberry (2400) | 5 | 10 | 40 |
| Fairview (2770) | 15 | 20 | 26 |
| Nebo Summit (2850) | 5 | 15 | 20 |

[a] Populations are listed in order of altitude with the low-stress and high-stress temperatures indicated as well as the temperature for optimal growth.

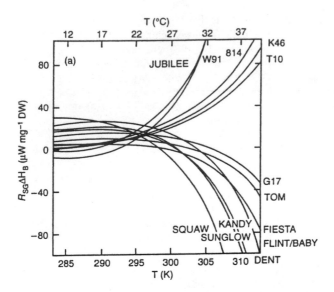

**Figure 1**   $R_{SG}\,\Delta H_B$ plotted against temperature based on metabolic measurements at 15 and 25°C for 12 cultivars of corn (*Zea mays* L.). Negative values of $R_{SG}\,\Delta H_B$ indicate temperatures at which growth does not occur and tissues are stressed. (From Ref. 32.)

crease with warmer temperatures (Figure 1) [32]. Growth rates of other cultivars are predicted to be nonexistent or very low at low temperatures, continuing to increase with temperature until tissue damage occurs (Table 2). Changes in predicted growth rate ($R_{SG}$) and efficiency ($q/R_{CO_2}$) with temperature are similar for a given cultivar [32]. The model, together with measurements of $q$ and $R_{CO_2}$ at two or more temperatures, may aid in selection of cultivars and in understanding adaptation of plants to climatic changes.

Soybean [*Glycine max* (L.) Merr.] is grown in many countries and climates. Twenty-two North American soybean cultivars from six different maturity groups were grown from seed under the same conditions. Measurement of metabolic heat rate of leaf tissue with a scanning calorimeter revealed that the slope of heat rate versus temperature showed abrupt changes reflecting shifts in metabolism [33]. The chilling response temperature for all cultivars was near 17.5°C. The maximum tolerable temperature for all cultivars was near 43.5°C. Differences in response to temperatures between the extremes relate to maturity group, follow latitudinal trends, and represent adaptation to different climates. Selection of cultivars of soybean for best growth in different climates has resulted in relatively rapid adaptation to local temperatures [33].

**TABLE 2**   Metabolic Heat Rate and Respiration Rate Measured Every 5°C from 5 to 40°C for Corn (*Zea mays* L.) Seedlings Grown from Seven "Older" Cultivars[a]

|  | Temperature response (°C) | | |
| --- | --- | --- | --- |
| Cultivar | Low stress | Optimal | High stress |
| Pula Janku | 5 | 20 | 32 |
| Santo Domingo | <5 | 20 | 33 |
| Black Popcorn | 5 | 20 | 30 |
| Loncho | <5 | 25 | 38 |
| Black Mexican Sw. | <5 | 25 | 30 |
| Santa Ana Blue | <5 | 20 | 40 |
| Minipopcorn | <5 | 20 | 27 |

[a] Low and high stress temperatures are indicated as well as the temperature for optimal growth.

*Artemisia tridentata* Nutt. or big sagebrush is one of the most widespread and economically important shrubs in western North America. Subspecies *vaseyana* grows at slightly higher, cooler, and drier sites than does *A. t. sp. tridentata*. Natural hybrids between the two subspecies are commonly found, for example, on a single hillside, where the parent populations are separated by 85 m in elevation and 1.1 km along the transect. In 1993, three gardens were established with seedlings from five populations along the transect planted in each garden [34]. Measurement of water potential and dark respiration by gas exchange did show differences [35]. Tissue was collected from plants in each garden at several different times of the year and analyzed using calorimetry, which proved to be more instructive. The results are shown in Table 3. All populations seem best adapted to their native environment and most stressed in environments different from their origin. Acclimation, showing phenotypic plasticity, occurred with change of season. Thus, metabolic distinctions can be made among closely related populations of plants grown on a single hillside in environments with only slight differences.

*Eurotia lanata* (Pursh) Moq. (Winterfat) is a small boreal cold-desert shrub that thrives in dry climates. Seeds were collected in populations from three different elevations and germinated. Metabolic rates were determined using calorimetry at temperatures from −10 to +20°C. Optimum temperature for germination, metabolism, and early seedling growth is about 10°C. Stress was noted near +20 and −5°C (Figure 2). Acclimation during germination had no effect, Differences between the three populations correlated with altitude rather than latitude.

## B. Drought

Metabolic response to temperature may also be measured during or following treatment with another environmental stress such as drought. The relative degree of drought tolerance was studied for six populations of small burnet (*Sanguisorba minor* Scop.) and six cultivars of alfalfa (*Medicago sativa* L.) grown

**TABLE 3**  Summary of Data Collected on Different Dates in 1998 on Plant Tissue from Gardens in Salt Creek Canyon (near Nephi, Utah) of Basin Big Sagebrush (*Artemisia tridentata* Nutt. ssp. *tridentata*), Mountain Big Sagebrush (*A. tridentata* ssp. *vaseyana*), and Hybrids Between Them[a]

| Garden-seed source (date) | Temperature response (°C) | | |
| | Low stress | Optimal | High stress |
| --- | --- | --- | --- |
| Basin-basin (Feb. 23, 1998) | 10 | 15 | 30 |
| Basin-basin (April 22, 1998) | <5 | 15–20 | 30 |
| Basin-basin (July 21, 1998) | <10 | 30–40 | >40 |
| Basin-hybrid (Feb. 23, 1998) | 10 | 15 | 20 |
| Basin-hybrid (April 22, 1998) | <5 | 20–25 | 30 |
| Basin-hybrid (July 21, 1998) | <5 | 30 | 35 |
| Basin-mountain (Feb. 23, 1998) | 5 | 15–25 | 30 |
| Basin-mountain (April 22, 1998) | 5 | 25–30 | 35 |
| Basin-mountain (July 21, 1998) | 10 | 20–35 | >35 |
| Hybrid-basin (Feb. 23, 1998) | <5 | 15–25 | 30 |
| Hybrid-basin (July 21, 1998) | 10 | 15 | 30 |
| Hybrid-hybrid (Feb. 23, 1998) | 5 | 10, 25 | 30 |
| Hybrid-hybrid (July 21, 1998) | 10 | 15–25 | >35 |
| Hybrid-mountain (Feb. 23, 1998) | <5 | 10, 25 | 35 |
| Hybrid-mountain (July 21, 1998) | <5 | 10, 20 | >25 |
| Mountain-basin (April 22, 1998) | <5 | 5–25 | 30 |
| Mountain-basin (July 21, 1998) | <10 | 10, 15 | 20 |
| Mountain-hybrid (March 11, 1998) | 5 | 10, 25 | 30 |
| Mountain-hybrid (April 22, 1998) | 15 | 20, 25 | 30 |
| Mountain-hybrid (July 21, 1998) | <5 | 10, 30 | >40 |
| Mountain-mountain (March 11, 1998) | 5 | 10, 25 | 30 |
| Mountain-mountain (April 22, 1998) | 15 | 25–35 | 40 |
| Mountain-mountain (July 21, 1998) | <5 | 10, 25 | 35 |

[a] Calorimetric measurements were made every 5 degrees from 5 to 45°C.

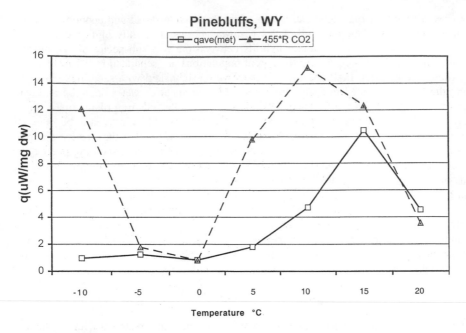

**Figure 2**   Metabolic heat rate ($q$, □) and respiration rate ($455R_{CO2}$, Δ) for winterfat (*Eurotia lanata* {Pursh} Moq.) seedlings from Pinebluffs, Wyoming measured at different temperatures. The data indicate optimal growth between 0 and 15°C and stress at temperatures below 0°C and above 18°C.

in common gardens under natural conditions and in the laboratory with different levels of moisture. Metabolic heat rate and respiratory rate were measured weekly. Both species grew best in early spring but remained green and metabolically active throughout the summer. Small burnet was much more drought tolerant than alfalfa. Differences among populations and among cultivers were detected in both common garden and laboratory conditions [36].

Cryptogamic crusts in deserts all over the world are communities composed of lichens, cyanobacteria, algae, mosses, and fungi found on or near the soil surface. Crusts are very susceptible to physical disturbance but if intact appear to play a role in providing nutrients, especially nitrogen, to higher plants [37]. Crusts, if they are present, also appear to increase the water holding capacity of the soil following infrequent precipitation events. Using calorimetric measurements of metabolism, we have learned that exposure of crusts to various levels of relative humidity had no effect, but liquid water caused immediate elongation of algal filaments. The temperature optimum for metabolism is about 15°C, indicating growth of cryptogamic crusts under cool, moist conditions.

## C.  Salt

Many desert playas are covered with water in the early spring. As the weather becomes warmer and drier, water evaporates, increasing the salt content of the soil from 7000 to almost 16,000 mM NaCl. Changes in respiratory metabolism during the growing season of four halophytes characteristic of cold desert playas were followed using calorimetry. In order of decreasing salt tolerance and metabolic activity, the species examined were the forbs *Salicornia rubra* and *S. utahensis*, the grass *Distichlis spicata*, and the shrub *Allenrolfea occidentalis*. These species are all well adapted to the environment in which they are found. Highest metabolic activity was found during May and June with lowest activity during the hot, dry month of August (Figure 3) [38].

*Salicornia utahensis* was grown in growth chambers in concentrations of NaCl ranging from 0 to 1.8 M. Metabolic rates were measured at temperatures from 5 to 45°C. Predicted growth was best at salt concentrations greater than 400 mM NaCl. The best growth was at 1000 mM NaCl but only at temperatures above 40°C (Figure 4).

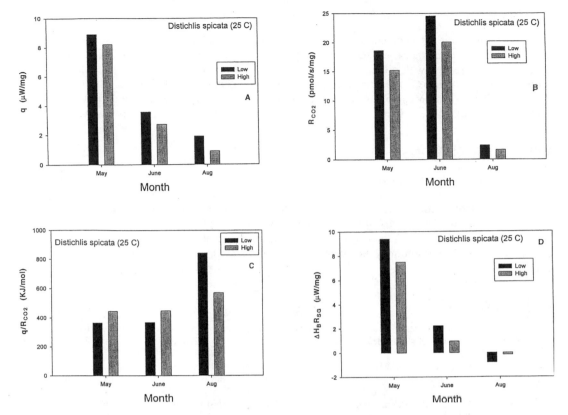

**Figure 3**   Leaf tissue was cut from *Distichlis spicata* (L.) Greene growing on a salt playa near Goshen, Utah. Plant collections were made during May, June, and August of 1997 from soil relatively high and low in NaCl. Isothermal calorimetric measurements were made at 25°C. (A) Metabolic heat rates ($q$). (B) Respiration rates ($R_{CO2}$). (C) Ratio of metabolic heat rate to respiration rate ($q/R_{CO2}$) or energy efficiency. Smaller numbers indicate greater efficiency. (D) Predicted specific growth rate ($\Delta H_B R_{SG}$). (From Ref. 38.)

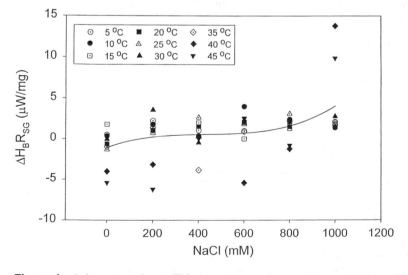

**Figure 4**   *Salicornia utahensis* Tidestr. was grown in several concentrations of NaCl at temperatures ranging from 5 to 45°C. Specific growth rate, $\Delta H_B R_{SG}$, calculated from metabolic heat and $CO_2$ rates is plotted as a function of the salt concentration in the growth medium.

## D.  Manganese

Methylcyclopentadienyl manganese tricarbonyl (MMT) is currently added to gasoline to replace tetraethyl lead as an antiknock fuel additive. Manganese concentrations in roadside soil and plants are increasing and correlated with distance from the roadway, traffic volume, plant type, and microhabitat. Radish (*Raphanus sativus* L.) seedlings were treated for 5 to 35 days with different levels of manganous chloride (0–1000 ppm). Metabolic heat rates and respiration rates, measured calorimetrically, indicated severe stress at Mn concentrations between 10 and 100 ppm and at temperatures above 20°C [39].

## VIII.  CONCLUSIONS

Plants acclimate or adapt to survive and grow in the presence of environmental stresses. The degree of adaptation to a particular stress can be monitored by measuring the rates of metabolic heat loss ($q$) and catabolism of photosynthate ($R_{CO2}$). Because growth and defense against environmental stresses rely on energy release from metabolic substrates, subtle degrees of adaptation can be determined using calorimetry. It is now possible to select rapidly populations or cultivars for growth in a particular microenvironment. The result may be increased food production and more effective environmental conservation.

## REFERENCES

1.   WA Fowler. Nuclear Astrophysics, Philadelphia: American Philosophical Society, 1967, pp 1–109.
2.   HE Suess. Chemical evidence bearing the origin of the solar system. Annu Rev Astron Astrophys 3:217–234, 1965.
3.   P Cloud. Cosmos, Earth, and Man. New Haven, CT: Yale University Press, 1978, pp 1–201.
4.   TJ Algeo, RA Berner, JB Maynard, SE Scheckler. Late Devonian oceanic anoxic events and biotic crises: "Rooted" in the evolution of vascular land plants?. GSA Today 5:45, 64–66, 1995.
5.   A Hallam. An Outline of Phanerozoic Biogeography. Oxford: Oxford University Press, 1994, pp 7–18.
6.   HG Jones. Plants and Microclimate. 2nd ed. Cambridge: Cambridge University Press, 1992, pp 163–263.
7.   T Volk. Gaia's Body. New York: Springer-Verlag, 1998, pp 1–187.
8.   JR Ehleringer. Photosynthesis and photorespiration: Biochemistry, physiology, and ecological implications. Hortscience 14:217–222, 1979.
9.   CJ Nelson. Genetic associations between photosynthetic characteristics and yield: Review of the evidence. Plant Physiol Biochem 26:543–556, 1988.
10.   C Korner, W Larcher. Plant life in cold climates. Symp Soc Exp Biol 42:25–57, 1988.
11.   RKM Hay, AJ Walker. An Introduction to the Physiology of Crop Yield. Essex, England: Longman Scientific and Technical, 1989, pp 87–106.
12.   KJ McCree. An equation for the respiration of white clover plants grown under controlled conditions. In: I Setlik, ed. Prediction and Measurement of Photosynthetic Productivity. Wageningen: Pudoc, 1970, pp 221–229.
13.   JHM Thornley. Respiration, growth and maintenance in plants. Nature 227:304–305, 1970.
14.   SJ Pirt. The maintenance erergy of bacteria in growing cultures. Proc R Soc LondSer B 163:224–231, 1965.
15.   RS Loomis, JS Amthor. Yield potential, plant assimilatory capacity, and metabolic efficiencies. Crop Sci 39:1584–1596, 1999.
16.   MGR Cannell, JHM Thornley. Modelling the components of plant respiration: Some guiding principles. Ann Bot 85:45–54, 2000.
17.   JHM Thornley, MGR Cannell. Modelling the components of plant respiration: Representation and realism. Ann Bot 85:55–67, 2000.
18.   RS Criddle, LD Hansen. Calorimetric methods for analysis of plant metabolism. In: RB Kemp, ed. Handbook of Thermal Analysis and Calorimetry. Vol 4. Amsterdam: Elsevier, 1999, pp 711–763.
19.   D Jou, JE Llebot. Introduction to the Thermodynamics of Biological Processes. Englewood Cliffs, NJ: Prentice-Hall, 1990, pp 90–95, 108–118.
20.   J Wrigglesworth. Energy and Life. London: Taylor & Francis, 1997, pp 11–12.
21.   LD Hansen, MS Hopkin, DR Rank, TS Anekonda, RW Breidenbach, RS Criddle. The relation between plant growth and respiration: A thermodynamic model. Planta 194:77–85, 1994.
22.   L Stryer. Biochemistry. 4th ed. New York: WH Freeman, 1995, pp 444–446, 559–565.
23.   H Beevers. Respiratory Metabolism in Plants. Evanston, IL: Row, Peterson, 1961, pp 141, 147–160.
24.   J Leavitt. Responses of Plants to Environmental Stresses. 2nd ed. Vol II. New York: Academic Press, 1980, pp 11–20.
25.   JG Scandalios. Oxygen stress and superoxide dismutases. Plant Physiol 101:7–12, 1993.
26.   CH Foyer, M Lelandais, KJ Kunert. Photooxidative stress in plants. Physiol Plant 92:696–717, 1994.

27. BN Smith, CM Lytle. Air pollutants. In: MNV Prasad, ed. Plant Ecophysiology. New York: John Wiley & Sons, 1997, pp 375–392.

28. WW Adams, B Demmig-Adams, AS Verhoeven, DH Barker. 'Photoinhibition' during winter stress: Involvement of sustained xanthophyll cycle–dependent energy dissipation. Aust J Plant Physiol 22:261–276, 1995.

29. EF Elstner. Mechanisms of oxygen activation in different compartments of plant cells. In: EJ Pell, KL Steffen, eds. Active Oxygen/Oxidative Stress and Plant Metabolism. Rockville, MD: American Society of Plant Physiologists, 1991, pp 13–25.

30. SE Meyer, PS Allen, J Beckstead. Seed germination regulation in *Bromus tectorum* (Poaceae) and its ecological significance. Oikos 78:475–485, 1997.

31. DJB Hemming, SE Meyer, BN Smith, LD Hansen. Respiration characteristics differ among cheatgrass (*Bromus tectorum* L.) populations. Great Basin Naturalist 59:355–360, 1999.

32. DK Taylor, DR Rank, DR Keiser, BN Smith, RS Criddle, LD Hansen. Modelling temperature effects on growth-respiration relations of maize. Plant Cell Environ 21:1143–1151, 1998.

33. DJB Hemming, TA Monaco, LD Hansen, BN Smith. Respiration as measured by scanning calorimetry reflects the temperature dependence of different soybean cultivars. Thermochim Acta 349:131–134, 2000.

34. BN Smith, S Eldredge, DL Moulton, TA Monaco, AR Jones, LD Hansen, ED McArthur, DC Freeman. Differences in temperature dependence of respiration distinguish subspecies and hybrid populations of big sagebrush: Nature versus nurture. Proceedings: Shrubland Ecotones, RMRS-P-11. Ogden, UT: USDA-Forest Service Rocky Mountain Research Station, 1999, pp 25–28.

35. ED McArthur, DC Freeman, JH Graham, H Wang, SC Sanderson, TA Monaco, BN Smith. Narrow hybrid zone between two subspecies of big sagebrush (*Artemisia tridentata:* Asteraceae). VI. Respiration and water potential. Can J Bot 76:567–574, 1998.

36. AR Jones, BN Smith, LD Hansen, SB Monsen, RB Stevens. Calorimetric study of the effects of water and temperature on the respiration and growth of small burnet and alfalfa. Proceedings: Ecology and Management of Pinyon-Juniper Communities Within the Interior West, RMRS-P-9. Ogden, UT: USDA Forest Service, Rocky Mountain Research Station, 1999, pp 134–137.

37. RD Evans, J Belnap. Long-term consequences of disturbance on nitrogen dynamics in an arid ecosystem. Ecology 80:150–160, 1999.

38. LC Harris, B Gul, MA Khan, LD Hansen, BN Smith. Seasonal changes in respiration of halophytes in salt playas in the Great Basin, USA. Wetlands Ecol Manage, in press.

39. AR Jones, CM Lytle, RL Stone, LD Hansen, BN Smith. Methylcyclopentadienyl manganese tricarbonyl (MMT), plant uptake and effects on metabolism. Thermochim Acta 349:141–146, 2000.

# 2

# Role of Temperature in the Physiology of Crop Plants: Pre- and Postharvest

**William Grierson**

*University of Florida, Lake Alfred, Florida (retired)*

## I. INTRODUCTION

### A. Importance of Temperature

Temperature, like the poor, is always with us but, like the poor, it is only too often overlooked. This is unfortunate, as temperature is a major factor in all things biological.

To a physicist, temperature is simply a manifestation of the kinetic energy of component atoms, ions, and molecules.

To a chemist, the role of temperature is epitomized by the "Q 10 rule," whereby, over some reasonable range, the rate of a chemical reaction approximately doubles with every 10°C increase in temperature.

But to a biologist, temperature is the supreme conductor of the orchestra of life, initiating specific reactions and modulating, integrating, or suppressing them just as the conductor of a great orchestra calls upon, modulates, or dismisses the diverse instruments, whose discrete voices are thereby integrated into one harmonious whole. Regardless of the crop, or of the physiological response being monitored, consideration of the role of temperature can often be the sine qua non in interpreting the phenomena being investigated. It has been said that "The scientist shows his intelligence . . . by his ability to discriminate between the important and the negligible" [1]. Only too often, temperature may appear to be a negligible factor when, unnoticed, it plays some critical role.

### B. Scope of This Chapter

Because every physiological and biochemical system of every crop plant is affected by temperature, it would be impossible to cover all its manifestations in a whole textbook, much less in a single chapter. Most aspects are, therefore, dealt with superficially. Specific examples are cited to indicate the types of relationships that invite further study and, when such study does not suffice, may inspire further research.

Citrus fruits, and most particularly the chilling injury (CI) syndrome, are represented in greater depth because the writer and his colleagues devoted many years to research on citrus, particularly the study of the basic mechanisms of chilling injury.

## C.  Definitions

### 1.  *Temperature*

Temperature per se does not need definition. Not all the work cited, however, deals with the temperature of the actual plant tissues involved. Often "temperature" refers to that recorded for the immediate vicinity of the plant or organ.

### 2.  *Crop Plants*

Crop plants are taken to be any grown for profit or pleasure, thus including ornamental plants grown for either indoor decoration or outdoor landscaping. No attempt has been made to include nondomesticated species.

## II.  ECOLOGICAL ROLE OF TEMPERATURE

Temperature obviously limits the geographical areas in which various crops can be grown. However, temperature per se is often not the only determinant: the effects of temperature extremes are usually associated with other factors such as availability of water, prevalence of high winds, and the duration and intensity of sunlight (insolation). An important aspect, as discussed in the following, is that limitations imposed by extremes of temperature differ sharply for annual versus perennial crops.

## A.  Extremes of Temperature

### 1.  *High-Temperature Limitations*

The limiting effect of high temperatures on crop production takes two principal forms: limitation of vegetative growth such as for cereal grains [2] and peanuts [3] and adverse effects on fruit settings [4]. Vegetable crops subject to very high transpiration losses, such as asparagus, lettuce, and all the *Brassica* species (cabbage, cauliflower, broccoli, brussels sprouts, etc.) are obviously limited by the excessive transpiration concurrent with exposure to extremely high temperatures. Tomato (*Lycopersicon esculentum* Mill.) is the quintessential example of a crop for which very high temperatures limit fruit setting. (In this regard, the small-fruited "cherry tomatoes" are more tolerant than the usual commercial varieties.) Plant breeders are having limited success in developing more heat-tolerant tomato varieties because heat tolerance and cold tolerance in fruit setting have only moderate heritability and such inheritance is complex [5].

A further complication is that the upper limit for fruit set can be correlated with humidity levels [6]. Successful breeding of truly heat-resistant tomatoes may well turn out to depend on the physiologists and biochemists more exactly defining the influence of temperature and humidity on the hormonal systems controlling anthesis, pollen tube activity, ovule receptivity, and, in some instances [5], parthenocarpy. A press account [7] reported that a major U.S. seed company has developed both a tomato and a zucchini that set fruit in temperatures as high as 35.6°C (96°F). For commercial purposes, assuming that the report is correct, this "high temperature fruit set" will have to be incorporated into varieties having commercially acceptable yield and eating quality.

Very high temperatures can also limit fruit setting of citrus fruits. In this case, intensity of insolation appears to be another limiting factor, because flowers within the leafy canopy, protected from direct exposure to sunlight, will usually set some fruit [8]. A less subtle effect of extremely high temperatures on fruit set of citrus is the "burning" or "scorching" of blossoms, particularly on young trees, that is occasionally reported from desert areas such as southern California, Arizona, and the Negev of Israel. Even without such drastic effects, fruit set of navel oranges is reported to be sharply affected by temperatures during the bloom period [9].

A high-temperature effect causing no visible symptoms is a cessation of growth even though nutrients and soil moisture are adequate, as reported for citrus trees during very hot weather in Arizona [10].

### 2.  *Low-Temperature Limitations*

The obvious limitation imposed by low temperature is killing of plant tissues by freezing. Most plant tissues can be destroyed by freezing temperatures suddenly imposed during a period of rapid growth. Some plants, given sufficient time under suitable conditions, can adapt themselves to freezing temperatures, and some cannot. This dichotomy is discussed in Sec. II.B.

### 3. Freezing of Plant Tissues

A more specific effect is the response to brief periods of freezing, or near-freezing, temperatures. The classical example, feared by fruit growers almost everywhere except in the tropics, is a freeze while the trees are in full bloom. This is much more drastic for deciduous fruit trees than for evergreen trees such as citrus. If the blossom-bearing wood is not damaged, such tropical or subtropical trees have a chance to replace fruit buds within the same bearing season, although yield and fruit quality may be impaired. As discussed later, this cannot happen with deciduous fruit trees.

A more subtle effect, to which green (English, garden) peas (*Pisum sativum*) are particularly susceptible, is low-temperature stunting of young plants. When such peas and snap (wax) beans (*Phaseolus vulgaris*) are growing side by side, immature pea plants may be permanently stunted by a brief chilly period from which the beans usually recover.

### 4. Microclimates

It is apparent to even the most casual observer that on a frosty night, cold air can drain into hollows, thereby sometimes limiting damage to such small "microclimate" areas. In addition, vegetation can be markedly different on the north and south sides of a steep valley because the exposures to sunlight are very different. Foehn winds provide striking examples of rather larger microclimates utilized for the growing of specialized crops. A classic example is the chinook of the Rocky Mountains of Washington State and British Columbia. Strong winds off the Pacific Ocean are forced to rise on encountering the coastal range. As the air rises rapidly, moisture condenses, releasing great amounts of latent heat and forming a bank of clouds (the "foehn wall") that drenches the western slopes. This sequence of events provides a mild, moist area ideal for such crops as cane fruits, crucifers, and many ornamentals. By the time the air mass has crossed the coastal range, it is very dry, and on its leeward descent adiabatic compression warms it rapidly, providing a sudden spring. The resultant microclimate is (provided irrigation water is available) ideal for the growing of stone fruits. Apricots are particularly well served by this microclimate because they have a very short rest period, with consequent susceptibility to spring frosts, which are virtually unknown in inland chinook areas. The chinook occurs on such a grandiose scale as to almost exceed definition as "microclimate." But the eponymous foehn winds in the Austrian Alps, the ghibli in the Tripolitanian Mountains of Libya, and the zonda in the Argentine Andes produce the same effects on a much more local scale.

The writer's master's thesis [11], dated 1940, includes a map of a microclimate area once known as the "fruit bowl of Canada." Thirty-five miles (56 km) long at its maximum and varying in width from 5 to 14 miles (8–22 km), the fruit-growing area of the Niagara peninsula once produced most of Canada's peaches, plums, cherries, pears, and small fruits and virtually all the wine grapes of eastern Canada. A high cliff (the Niagara escarpment) shelters this area on the south side. On the north, Lake Ontario moderates the temperature of the north winds in midwinter. In spring, the escarpment protects the orchards from unseasonable warm south winds that might induce too early a bloom, with consequent risk of a blossom freeze. Now, more than 50 years later, it is sad to return to the once overflowing "fruit bowl": this precious miracle of microclimate has been largely paved over with factories, shopping centers, and housing developments that could just as well have been located a few miles to the south, above the escarpment. Such squandering of invaluable microclimates is all too common everywhere.

What might be termed "mini-microclimates" occur within any local microclimate, as indicated by the surprising range of temperatures recorded within a single lemon orchard [12]. When studying such fine details as individual leaf temperatures, even heat conduction along thermocouple wires must be considered [13].

But microclimate effects can also manifest themselves in far more subtle ways, often involving vertical as well as horizontal temperature differences. When air temperatures are favorable for growth, it is easy to forget that soil temperatures can also be limiting. Soil temperatures, both above and below optimum range, have been shown to limit uptake of soil water by citrus trees to the extent that visible wilting occurs even when soil moisture is adequate [14]. When water uptake is limited, obviously the uptake of water-soluble ions can also be affected. Iron deficiency chlorosis of citrus trees has been reported to be exacerbated by soil temperatures below 12.8°C [15]. Such ion uptake limitation can also be critical in nutrition experiments in which air temperature is ignored. This relationship was confirmed in a controlled environment experiment with six varieties of spinach (*Spinacia oleracea*). The nitrate content of six cultivars of spinach grown at temperatures from 5 to 25°C varied significantly, not with whether nitrogenous fertilizer was applied but with the growing temperature [16].

Hazards from soil pathogens can depend directly on soil temperatures. All Florida citrus seemed to be doomed by a mysterious "spreading decline" until it was found that the cause was a nematode (*Radopholus similis*) that could be cultured only at subsoil temperatures. Because Florida laboratory temperatures normally exceed those of the soil below about 30 cm, cultures from diseased roots processed at ambient temperatures never indicated that *R. similis* was the causal agent [17,18].

## 5. *Annual Versus Perennial Crops*

Temperature limitations differ sharply for perennial and annual crops. For perennials (largely tree, vine, and bush crops, various grasses, and other pasture crops), ecological limits are usually set by *winter* temperatures. Few species are hardy enough to survive subarctic extremes of winter cold. In the tropics, the need for a cool winter rest period limits the cultivation of pome (e.g., apple and pear) and most drupe (e.g., peach, plum, cherry, apricot, almond, walnut, pecan olive) fruits. Coconut (which botanically is a drupe with a desiccated mesocarp and liquid endosperm) is a conspicuous exception. Conversely, the lack of winter freeze hardiness limits the potential growing areas for purely tropical fruits (banana, mango, avocado, durian, mangosteen, etc.), tropical ornamentals, and purely tropical grasses, including sugarcane.

This set of limits is in sharp contrast to those applicable to purely annual crops such as almost all vegetables and grains, and annual flowers, for which *summer* temperatures are critical. All these annual crops require is about 3–5 months of suitable growing weather. Vegetables grow luxuriantly in the warm, long summer days in Alaska; the subarctic winters are of no consequence for them.

## B.  Various Interactions with Temperature

In the years immediately prior to World War II, the writer was a young graduate student in Canada working on storage and ripening of pears. At that time, it was customary for Canadian housewives to put fruits on sunny windowsills to ripen them. Because it seemed illogical that light should hasten ripening, I decided to put a row of unripe pears on the laboratory windowsill and cover half of them with a black cloth. Fortunately, I checked pulp temperatures: those under the black cloth were several degrees warmer. Then I tried shading with a white-painted board. Better, but still quite a difference. By the time the next year's pear crop came in, I was in uniform on the other side of the Atlantic. I never did return to the sunlight-pear-ripening problem but have ever since been acutely aware that one way or another, temperature can be an interactant, wanted or not, in a great deal of plant research.

## 1.  *The "Day/Degrees" Concept*

A very useful concept for expressing heat units is "total day degrees": that is, the accumulated number of days (or sometimes hours) above a certain base temperature. Another version is the accumulated sum of diurnal maximum temperatures times the number of days. For the reverse (cold units), the usual figure is the total number of hours below a given temperature, such as 40°F or 5°C. The usefulness of such methods is not helped by overreliance on statistical analysis of findings based on an initial arbitrary decision. In the United States, for example, 40°F (4.4°C) and 45°F (7.2°C) have been common baseline temperatures for determining chilling hours. As the Fahrenheit scale is abandoned in favor of Celsius, 5 and 7.5°C are more likely to be used. With such baseline variations, apparent fine statistical differences can be deceptive.

Peaches afford an excellent example of the use of such methods. Florida peach breeders have very successfully extended the southern limits for commercial production of peaches by breeding "200 hour" peaches and nectarines, in contrast to the 400, even 600, hour peaches grown in districts with cold winters [19]. In more northern states, versions of this day/degrees concept are used to forecast blossom freeze risks for varieties in a given area [20] and date of bloom in others [21]. Readers interested in a highly sophisticated discussion of the mathematics involved are referred to correspondence in a 1991 issue of *HortScience* [22].

## 2. Freezing of Plant Tissues

As depicted in older texts, freeze injury and freeze resistance were simply explained: in freeze-suscepti-ble tissues, free water froze, forming crystals that disrupted cell membranes, whereas in freeze-resistant tissues the water was bound in the form of hydrophilic colloids. When this model was subjected to mod-ern research, however, little if any of it turned out be so simple. Interested readers are referred to two ex-cellent reviews [23,24]. Freeze-hardy plants have hormonally controlled mechanisms enabling them to respond to gradual changes in temperature and day length in preparation for winter. Such changes are ob-vious with deciduous trees, vines, and shrubs, which shed their leaves, often after having displayed dra-matic changes in leaf color. No such highly visible evidence is afforded by conifers, which, nevertheless, also need gradual autumnal climatic changes to induce similar hormone-controlled internal adaptation to prepare for winter [25]. But what of plants that survive a freeze without a prior hardening period? Ex-pressed very briefly, water in certain woody plants can supercool to a surprising extent, although this pro-tective mechanism is often negated by the presence of ice-nucleating bacteria [24]. Such bacteria are by no means ubiquitous, but they are very common and a real factor in freeze injury.

Exposure to freezing but nonlethal temperatures can cause various chemical changes in plant tissue. Only one is mentioned here. It is very common for oranges that survive a freeze to develop white crystals clearly visible between the segment membranes. These are hesperidin, the principal flavone in citrus fruits, and although their presence sometimes causes alarm, they are completely nontoxic. Up to the 1950s, growers placed much credence on estimations of fruit damage as judged by the amount of hes-peridin crystals. This mindset proved quite fallacious [26].

The once apparently simple field of tissue freezing is further complicated by work with detached plant parts. Celery pollen has been stored in viable condition at $-10°C$ for as long as 9 months [27]. The use of "cryoprotectants" has made possible prolonged, very low temperature storage of living tissue for in vitro tissue culture and propagation. Using such cryoprotectants as polyethylene glycol + glucose and dimethyl sulfoxide, such living material as apices of brussels sprouts [28] and *Rubus* [29] have been rapidly cooled, then held at $-196°C$ until needed for tissue culture propagation.

## 3. Dormancy, Bud Initiation, and Fruit Setting

Obviously, it is well that autumnal climatic changes prepare perennials of the temperate zone for the rigors of winter. It might seem that if no winter was to be expected, such plants could grow happily in eternal summer. Or so thought the planners of the huge (>1 hectare under glass) Devonian Gardens, located over a large shopping mall in Calgary, Alberta, Canada. Their concept had been to surround the clientele with familiar summer vegetation in the depths of Calgary's cold, snowy winter. It was a costly error. Deprived of their climate-induced cycle, the familiar native plants became spindly and unthrifty and soon began to die. The thousands of years of evolution that had fitted those plants for the rugged winter of the Rocky Mountain foothills had produced plants that could not do without it. Instead, the native plants had to be replaced with (as nearly as possible) "look-alikes" imported from Florida and California [30].

The dormancy of winter-hardened plants is deceptive. Essential physiological and morphological changes are progressing and will do so only at the low temperature to which evolution has adapted such plants. Spring bulbs (tulips, daffodils, narcissi, Easter lilies, etc.), brought indoors and kept in warm tem-peratures after flowering, will not bloom again. Such bulbs left in the winter ground (or held in correctly regulated cold storage) undergo histological changes clearly discernible under a dissecting microscope or even a powerful hand lens. By the time the bulbs are ready to start growing again in the spring, each one contains all the necessary floral parts, minute but discernible. It is by use of a series of very exact storage temperatures that today's scientific flower producers are able to have spring bulbs in bloom timed for such occasions as Mothers' Day and Easter. Such imposed temperature regimes are very precise: there are sharp differences in temperature requirements, not only among genera, but even between individual cul-tivars [31].

The same thing happens (on a truly microscopic scale) within the fruit buds of deciduous fruit trees and shrubs. This is why, as horticultural students, we could cut apple boughs in late spring, place them in water in a warm building, and, apparently miraculously, decorate our Easter dance with apple blossoms. The same phenomenon explains why a blossom freeze wipes out a deciduous tree fruit crop for a whole year. Those blossoms came from fruit buds initiated 10 or 11 months before, which had developed while dormant and apparently inactive during the winter months.

Even normally hardy plants, such as oak trees, need time to adapt to winter temperatures. A Florida neighbor of this author grew oak seedlings and shipped them to Michigan in late winter. They were immediately killed by freezing.

In addition to cold hardiness, cold hardening can induce disease resistance [32] in addition to the customary physiological effects.

It is very different with citrus fruits. For one thing, fruit buds on deciduous fruit trees are clearly recognizable to anyone cognizant in such matters. Fruit and leaf buds are indistinguishable on citrus trees [33], however, and the initiation of fruit bud development takes place only a few weeks before bloom. The citrus industry and the literature usually speak of "dormant" citrus trees, but such dormancy is in no way comparable to that of deciduous fruit trees. "Quiescent" is a far better term. Blooming of quiescent citrus trees is usually initiated by the termination of a long cool spell or drought [34]. The best and most uniform blooms come when mild stresses from cool weather and drought are relieved simultaneously. A mild winter, followed by a warm, moist spring, tends to give a straggly bloom, spread over many weeks or even months, with consequent poor yield, low fruit quality, and difficult harvesting.

When hormonal control of chilling injury was still a very new theory, a colleague and I sprayed a number of grapefruit trees with various combinations of growth regulators in November. We definitely affected susceptibility to chilling injury of the fruit harvested in the following fall, although not in any clearly discernible pattern [35]. What was tantalizing about the test was that with one treatment we got a highly significant increase in yield, which we felt we could not publish. Temperatures were so mild that winter that bloom straggled on and on for many weeks—except on one of our growth regulator treatments, for which the bloom was a "snow bloom," on schedule in mid-March. The treatment would become useful only if long-range weather forecasts were so precise that each November they could forecast whether temperatures between November and March would be uniformly, and atypically, mild.

Obviously, the occasional chilly spells so resented by winter tourists initiate the hormonal activity necessary for a desirably brief, early full bloom.

Even when fruit trees have bloomed satisfactorily, temperature can be a determinant of whether a good crop will be harvested. Most deciduous fruits need pollination, which is normally done by honeybees. It can be very difficult to get the attention of apple or pear growers whose trees are in full bloom if the temperature suddenly drops below that favored by the bees. If the temperature is not right, the bees just quit flying, and that can mean a very poor crop indeed. Even if the bees fly and pollen is spread, the pollen must germinate and the pollen tube grow down to the ovule, a process that can be severely restricted by unseasonably low temperatures [36]. And even when pollination has been successful, growth of individual grape berries (botanically, grapes are berries) can be restricted by both too high and too low temperatures [37]. Too high temperatures are more likely to affect fruit set of citrus than of deciduous fruits. In California, extremely high temperatures after fruit set can cause excessive fruit shedding of navel oranges [38]. In Florida, trouble is more apt to come from a combination of high temperature and high humidity, resulting in fungal invasion of the fruitlets [39].

Such problems are not limited to dessert fruits. The buying public having developed an unreasoning prejudice against seeds in fruits and vegetables of many types, parthenocarpy has become highly desirable. For some cucumber varieties, parthenocarpy can be induced with sprayes of chlorfuorenol—unless the night temperatures are too high. Night temperatures between 16 and 21°C have been reported as favorable, with parthenocarpy very much reduced when the thermometer reaches 21°C [40].

## 4. *Seed Dormancy and Germination*

A very helpful specialist in seed science whom I consulted on the preparation of this chapter sent me, in addition to various published papers, a page-long list (which he considers incomplete) of textbooks, symposia, and so on dealing with the handling and storage of seeds. With temperature so often a critical factor in storage and germination of seeds, this account can be only the briefest of introductions for the nonspecialist.

An important temperature-related difference should be noted between seed-bearing plants of the temperate zone and those originating in the tropics or subtropics. In areas that experience killing winter freezes, seeds *must not* germinate until the following spring. Exceptions to this principle are seeds of plants that bloom early enough in the spring to be able to establish mature plants before the onset of winter. The dandelion (*Taraxacum officinale*) is a familiar, and usually unwelcome, example. Seeds of plants that evolved in tropical areas need no such protective device and so usually (but not always) can be germinated immediately on separation from the plant [41]. The lack of true seed dormancy severely limited the spread of many tropical species when they were first discovered by early European explorers. Even in modern times, dispersal of such crops as cocoa (*Theobroma cacao* L.) has been difficult because the seeds not only are adapted to immediate germination but also are highly susceptible to chilling injury (see later) if held in cold storage.

A word on terminology: "dormancy" for seeds is used much as it is found in discussions of buds, bulbs, and so on. Seeds that will not respond to usually effective treatments are said (most appropriately) to be "recalcitrant." Some authorities designate as recalcitrant only seeds that do not survive desiccation [41]. Such distinctions are, however, beyond the terms of reference of this chapter, which is limited to the effects, direct or indirect, of temperature. "Stratification" is used (not very logically) for chilling treatments to break dormancy. Perhaps this comes from the old custom of filling a box with alternate layers of sand and seeds from peaches (or other stone fruit) and setting it outside, exposed to the coldest possible weather.

For some seeds it has been demonstrated that dormancy is purely mechanical, being enforced as long as the tough impermeable testa is intact [42]. In this regard, it used to be argued that hard freezing only splits the peach pits, thus mechanically releasing the seed to germinate. Our pomology lecturer at the Ontario Agricultural College, Guelph, settled this for us more than 50 years ago. At his direction, we compared germination of "stratified" peach pits from the preceding year with that of fresh peach pits we had carefully cracked. The result was quite fascinating. The seedlings from the stratified seeds were normal. Those from the fresh, but mechanically cracked, seeds resembled tiny pineapple plants, producing leaves with no internodes. Prolonged cold temperatures (most effectively between 2 and 6°C) are definitely essential in such "stratification."

Various treatments (such as presoaking) to encourage emergence of seeds used to be called "vernalization," presumably because it hastened the effects of spring. The term was brought into disrepute by claims of permanent genetic changes by the Soviet charlatan Trofim Lysenko [43]. Today, "priming" is appropriately used for seed treatments (involving temperature, solutes, etc.) in wet or dry media to accelerate germination. But if seeds have been primed, subsequent permissible holding temperatures may be affected. Primed tomato seeds have been reported to retain viability at 4°C, but at 30°C they deteriorated within 6 months [44]. Similarly, primed tomato seed was reported to retain viability at storage temperatures as high as 20°C for 18 months. However, the seed degenerated at 30°C, particularly when primed with potassium nitrate rather than with polyethylene glycol (PEG) [45].

Priming does not necessarily overcome adverse weather conditions, as shown by 3 years of unsuccessful trials with primed sugar beet seed in cold Idaho spring weather [46]. Current research developments, however, promise to overcome these ill effects of too early sowing when they are due to a combination of moisture imbibition and too low temperature. A review article [47] reported success in such circumstances when seeds of table beet (*Beta vulgaris* L.) were primed with PEG.

Imbibitional chilling injury is of particular concern for seeds of plants of tropical origin, such as cotton, corn (maize, *Zea mays*), tomato, and many legumes, which are susceptible to chilling injury. For their seeds, the onset of CI is related to rate of water uptake [47]. Treatment with materials (such as PEG) that delay imbibition can be helpful but is not temperature specific. This problem appears to be surmountable by use of temperature-sensitive polymeric seed coatings that become permeable to water at specifically selected temperatures [47,48].

Too hot temperatures can also impede germination. Florida celery growers have been able to surmount this problem by using high-temperature (30°C) priming in a solid matrix of calcined clay [49].

Recalcitrant seeds occur in all climates, and temperature can be a factor in achieving successful germination. Wild rice (*Zizania palustris*) is an excellent example. Deeply dormant at harvest, it will not germinate without prolonged cold treatment [50]. It is thus perfectly adapted to self-propagation in the Minnesota wetlands and as a food staple for Native Americans, who have depended on it over the centuries. Some of wild rice's reputation as a "recalcitrant seed" involves a supposed desiccation intolerance, but this misjudgment has been related to failure to understand the "novel relationship between seed viability, temperature, and moisture content" [51].

An interesting form of recalcitrance in tropical seeds is that some, such as kola (*Cola nitida*), must be aged for as long as 7–11 months, for which ambient temperatures are satisfactory [52]. This requirement accounts for how, for many centuries, the highly valued, but frail, caffeine-rich kola "nuts" (caffeine being a stimulant not prohibited to Muslims) were traded all over West Africa, wrapped in damp leaves and transported for weeks on the heads of slaves [53]. Such aging of tropical seeds is not necessarily completely temperature independent. Seeds of *Plantago ovata* (an annual herb grown in India), although completely recalcitrant at harvest, germinated freely after a single day at 15°C plus treatment with gibberellic acid (GA3) [54].

A record for temperature-related recalcitrance is held by American ginseng (*Panax quinquefolius*). It is no wonder that this wild herb has been hard to domesticate: it is reported [55] to need cool-warm-cool stratification over a period as long as 540 days to 18 months!

Temperature may or may not prove to be important in the storage and germination of a particular type of seed, but it can never be ignored as a possibly critical factor.

## 5.  Temperature-Induced Ethylene Effects

Ethylene ($C_2H_4$) is the universal growth regulator. Until the advent of the gas chromatograph, it was believed that biosynthesis of ethylene was confined to certain plant tissues (e.g., apples) and was not present in others (e.g., oranges). As analytical equipment improved, it became apparent that under various forms of stress any plant tissue can produce ethylene, and the extent of this effect is temperature dependent [56]. Among the more striking temperature-induced effects of endogenous ethylene are the "fall colors" in deciduous woodlands, which result from the reaction of ethylene with plant pigments.

Bright colors are not only attractive but, as long as consumers insist on relying on their eyes rather than their taste buds, they can be very valuable. Thus, temperature-modulated ethylene effects become essential tools in the marketing of certain fruits and vegetables. Citrus fruits afford an excellent example. Citrus fruits grown at sea level in the humid tropics, where the species originated, are all green: no brilliant oranges or yellows gleaming amid the jungle foliage.

But for centuries, citrus fruits have been grown in cooler, usually more arid, areas, principally around the Mediterranean Sea. There, the considerable stress of cool nights on a tropical fruit forces production of minute amounts of ethylene, with consequent loss of chlorophyll and development of carotenoids. Thus, we have the obvious "fact" that oranges should be orange and lemons should be yellow. This consumer prejudice presents citrus growers in milder climates such as Florida and Brazil with a very real, temperature-induced problem. In such districts, early varieties may mature and pass their optimum maturity without ever developing "typical varietal color."

It has long been axiomatic among Florida citrus growers that their fruit would not change color without "a week of cool nights" (which in many years comes after the early varieties are over). A 1942 study confirmed this [57]. No significant color break was observed as long as night temperatures were above 55°F (12.8°C), and a week of nights below 50°F (10°C) resulted in good orange color on early varieties of oranges. Grapefruit, however, responded to the stress of low night temperatures much less predictably.

In California, an ingenious experiment studied the effect of temperature on the coloring of Valencia (late) oranges under controlled conditions. Fruit-bearing branches were grafted onto young potted rootstocks, and air and soil temperatures were controlled separately [58]. Both variables were found to affect fruit color, the best orange color being achieved with 7°C soil temperature and 20°C air temperature. Internal analyses found no correlation between fruit color and fruit maturity.

Various attempts by this author to reproduce such temperature-induced color changes with detached fruit have been unsuccessful. Once the fruit has been detached from the tree, exogenous ethylene must be supplied and the effect is, again, sharply temperature dependent, but with a relationship quite different from that observed for attached (nonpicked) fruit. In an early Florida study [59], we found a very sharply defined optimum for chlorophyll destruction in oranges at 85°F (29.4°C) and a very ill-defined optimum for grapefruit at approximately the same temperature. Such ethylene "degreening" had no apparent effect on carotenoids; the degreened oranges were pale yellow. California packinghouses that commonly degreened at 75°F (23.9°C) reported development of a deep orange color, but the process took 8–10 days, a prohibitive period in Florida because of endemic stem-end rot (caused by *Diplodia natalensis*), which is strongly stimulated by ethylene.

Nearly 20 years after the Florida work just described, the carotenoid development/chlorophyll destruction effect was studied in detail with very much more sophisticated equipment [60]. This time ethylene-induced carotenoid accumulation was shown to be (1) temperature sensitive and (2) inhibited at 30°C and above. The work was continued and showed that very high levels of specifically identified carotenoids could be achieved with concentrations of ethylene as low as 0.1 ppm. However, induced carotenoid development took weeks, rather than days, hence was commercially unacceptable in a stem-end rot district.

I still do not know why, prior to picking, cool (below ca. 12°C) temperatures are necessary to destroy chlorophyll in the peel of citrus fruits, but warm (ca. 30°C) temperatures maximize the rate of ethylene-mediated chlorophyll disappearance after picking. This paradox does, however, emphasize something that is too often ignored or forgotten: prior to picking, a fruit is an integral part of the physiology of the plant as a whole.

## 6.   Temperature and Fruit Quality: Preharvest

There is no point in producing fruits commercially unless they are palatable, and in some instances palatability is strongly related to growing temperatures. Again, a citrus fruit, grapefruit (*Citrus paradisi*), will serve as a prime example, not so much because of its place in this writer's past research but because the internal and external qualities of grapefruit have been extensively studied. All growing districts base their quality standards on what they do best [61], and since Florida's climate is so unsuited to the production of grapefruit with a bright, colorful exterior, standards have been developed largely based on internal quality. These are expressed in terms of sugar (as degrees Brix), acid (as ratio of citric acid to Brix), and juice volume (as cubic centimeters per fruit) [62,63]. Internal quality obviously varies widely among growing districts, leading to some totally unprofitable studies in day/degree relationships. But even the most casual observations make it apparent that districts famed for the high quality of their grapefruit (such as the Rio Grande Valley of Texas and the Indian River district of Florida) are areas with warm winter nights, during which growth of the tree and of the fruit can continue uninterrupted. A controlled climate experiment with Redblush grapefruit in Florida confirmed this. Maximum internal quality was found in fruit from little trees, which were grown where night temperatures were not allowed to drop below 21°C [64].

The shape of grapefruit is very sharply associated with internal quality: the flatter the fruit, the higher the internal quality. The influences of day and night temperatures, and of day length, were studied under controlled conditions [65]. A 32/7°C (day/night) temperature regime produced severely "sheep-nosed" fruit of very low internal quality. A 32/24°C temperature regime produced flat fruit (axis length < diameter) of high internal quality. No correlation between fruit quality and day length was found.

## 7.   Wound Healing: Temperature × Humidity × Time

Some plant products have considerable ability to heal mechanical lesions after harvest. The ability depends on certain ranges of temperature and humidity, however, and the healing takes several days to complete. It has long been known that both sweet potatoes (*Ipomoea batatas*) and so-called Irish potatoes (*Solanum tuberosum*) can heal damage to their own tissue [66]; for this reason, it is advised that potatoes be harvested, then held for several days at ambient (or higher) temperature and very high humidity before being placed in cold storage, because such healing occurs only at high temperatures and humidities [67]. Similarly, when seed potatoes are cut into "planting pieces," they should be "cured" for several days prior to planting under the warmest conditions available. During this period of comparatively high temperature, a layer of suberized cells forms over the wounds.

A much more recent finding is that citrus fruits can heal shallow wounds into the flavedo (colored part of the peel), but only at very high relative humidity (ca. 95% RH) and temperatures as high as 28–29°C (which, fortunately, are the conditions recommended within Florida citrus degreening rooms). An unusual aspect of this healing of citrus fruits is that it involves lignification, not suberization, and it is associated with sharp increases in phenolic compounds and of the enzyme phenylalanine ammonialyase (PAL) [68].

In both these types of healing, the role of comparatively high temperatures is critical. Such wound healing should not, however, be confused with drying treatments, which are essentially catabolic rather than anabolic. The "curing" of onions prior to storage is an example of drying. The curing process aims at killing the outer layers of cells by heat and desiccation, a form of localized necrosis that would be disastrous with living products of most other types.

Attention is again drawn to the different physiological responses of plant organs on and off the mother plant. After a Florida hurricane, attached citrus fruits will heal severe wounds and continue to grow to maturity at normal ambient temperatures although badly scared. Fruits with similar injuries that become detached from the tree promptly rot. Various forms of squash (*Curcubita* spp.) carved with a gardener's initials when immature will grow to maturity with the initials as prominent scars. Any such wounds inflicted on detached fruits would cause decay.

## C. Temperature × Light Interactions

A factor that is easily overlooked in determining optimum temperature for a given response is light, which may play either a positive or a negative role.

Modern apple orchards are often based on clonal rootstocks rather than on seedling roots. The rootstocks must be rooted from cuttings, which is not always easy, and light can be a complicating factor. Rooting of M-26 clonal rootstocks has been reported to be maximum at 25°C, but only in the absence of light, which may inhibit rooting [69].

The prospect of establishing life support systems in space has led to the prospect of crop production under controlled conditions not necessarily corresponding to those in terrestrial horticulture. One such study with lettuce (*Lactuca sativa*) found that maintaining a constant day/night temperature at 25°C maximized growth, but only with intensified light during the "day" period [70].

Many plants are known to respond sharply to photoperiod (a misnomer: it is the period of unbroken darkness, not of light, that is controlling). A study of the effect of photoperiod on the growth of West Indian mahogany (*Swietenia mahagoni*), grown in southern Florida as an ornamental, found that its typical response to photoperiod was inhibited by low temperatures atypical of its native tropics [71]. Flowering of the annual ornamental *Rudbeckia fulgida* involves a cold treatment × photoperiod interaction [72], as is also reported for six herbaceous perennials [73].

The relationship between temperature and photoperiod and flowering of traditional ornamentals such as *Chrysanthemum* is now well understood by both professional and amateur growers. (But salable flower quality also depends on growing temperature [74].) With the increasingly common introduction of exotic ornamentals, specific responses (to temperature, light, watering, etc.) must be established for the new arrivals. One such exotic is "kangaroo paw" (*Anigozanthos manglesii*), for which very sharp interactions between day and night temperatures and between temperature and day length control flowering and even mortality [75].

Individual species within a genus may respond quite differently to interactions of temperature and light. A *Peperomia* species imported to Indiana from the Andean highlands was unable to adapt to the double change, in summer, of temperature and photoperiod. Another *Peperomia* species from the lowlands of Ecuador made the transition successfully [76].

Temperature-light interactions are not limited to higher plants. For example, sporulation of some fungi, such as the citrus pathogen *Diplodia natalensis* (*Physalospora rhodina*), needs not only optimum temperature but also exposure to light of high intensity (GE Brown, personal communication).

A complicating role for light is always a possibility in the investigation of temperature relationships.

## D. Temperature Control in Crop Production

### 1. Microclimate

Greenhouse (British "glasshouse," often a misnomer in this plastic age) production is the obvious example of microclimate temperature control. But greenhouse production has its own considerable expertise and literature. Thus, the examples of greenhouse research cited here are included only to illustrate specific situations in which individual control of air and soil temperatures is important.

Even outdoors, although climate (including temperature) is usually regarded as beyond the control of man, localized temperature control is sometimes effective on a microclimate scale. Vancouver, Canada, is a few miles north of the 49th parallel, about 60 miles farther north than Minot, North Dakota, with its legendary harsh winter temperatures. But constant foehn winds off the Pacific Ocean make Vancouver winters mild and wet, although sunshine is scant. When I had a garden there in the late 1940s, a neighbor used to say that I "cheated God" to bring in my lettuce and tomatoes earlier than anyone else. The bed in which the vegetables grew was banked toward the south at approximately 50 degrees, and the area between the plants was covered with flat stones, gathered from the nearby beach, to maximize soil heating from the weak late winter–early spring sun.

This management was, of course, an extreme example of microclimate modification for crop production. Nevertheless, it was no more than ingenious growers have done to survive inhospitable climates throughout the ages, as with pre-Columbian Andean potato growers. In recent years, the native peoples of the Andean Altiplano have learned to revive the methods of their ancestors, growing potatoes on high narrow beds at the foot of mountain slopes. On freezing nights, the cold air settles between the raised beds without damaging the aerial parts of the plants, whose subterranean portions are protected by the latent heat of the water accumulated in the troughs between the beds, an ancient example of sophisticated microclimate control.

Poinsettia is typical of an ornamental grown for a specific date; unless the plants are marketable at Christmas, their value drops dramatically. Growth of the plants can be sharply reduced by too cool air temperatures. Maintaining temperatures in a greenhouse in very cold weather is very expensive. However, it has been found that raising soil temperature to 23°C (which is much cheaper to do) could counteract the adverse effects of air temperature as low as 11.5°C [77].

Sometimes the reverse modification is needed. Flowering of *Alstroemeria* (lily-of-the-Incas) was stimulated by cooling the root zone with 10°C circulating water. There was also an interaction with light, supplementary lighting being essential in winter but harmful in spring and summer [78]. A beneficial lowering of root zone temperature explained an anomalous result with azaleas pot-grown outside on either clamshell mulch or black polyethylene. Placing the pots close together increased growth of azaleas in black pots but not in white pots. The beneficial effect was traced to a decrease in root zone temperatures by shading when the plants in black pots were placed close together [79].

Another unexpected root zone temperature effect was traced to the chilling effect of cold greenhouse irrigation water in winter. The effect was noted with roses and chrysanthemums and was sufficient to affect turgidity, stomate opening, and flowering. Such unforeseen deleterious temperature effects are particularly easy to overlook when they involve the temperatures of soil rather than air [80].

Root zone heating usually involves use of expensive fuel. This potential cost was halved in an ingenious system of pumping comparatively warm water from a well 100 m deep and circulating it through buried pipes [81].

Temperature, of course, affects more than plant growth. It is sometimes necessary to tread a fine line between temperatures optimum for growth and those that initiate or increase fungal attack. This can be a problem for Florida foliage growers in warm weather, as shown in a study of aerial blight (*Rhizoctonia solani*) infection of Boston fern (*Nephrolepis exalta*). Some plant quality had to be sacrificed if potting medium and air temperatures were to be regulated to restrict development of the pathogen [82].

Given sufficient irrigation water, many deserts will blossom as the rose. But sometimes the desert sun is too hot, with consequent potential for crop damage. An obvious remedy is to spray the crop with an overhead irrigation system. The cooling effect of such sprinkling is sharply dependent on initial air temperature. A California study [83] reported the following (the results have been converted from Fahrenheit to Celsius):

| Macroclimate temperature | Lowered by |
|---|---|
| 32°C | 2–3°C |
| 38°C | ≤5°C |
| 39°C | ≤7.5°C |

In addition to other benefits, the water spray at 39°C was reported as being successful in reducing excessive "June drop" of small fruitlets. But such spraying of water in extremely hot weather can cause localized injury due to the "lens effect" of standing drops of water on the leaves [84]. Lens effect injury can be avoided, and better temperature reduction obtained, by using nozzles that emit a fine mist instead of streams of water [85].*

Microclimate is being modified on a very large scale. Whole hectares are commonly covered with plastic sheeting, which may be black, white, or transparent. Plastic covering may be spread over raised beds, with the plants inserted through holes in the plastic; it may lie over individual rows secured along the sides, with or without some form of framing [86]; or it may be used as "floating row covers," sup-

---

* A recent report indicates use of such "hot weather misting" to improve color of apples in Washington State.

ported by the crop itself and rising as it grows [87]. Sometimes such plastic covering serves essentially for weed and soil moisture control. Often, however, some degree of temperature elevation is sought, and air and soil temperatures are commonly included in research reports. The elevation of temperature under plastic film will depend on both the climate and the type of plastic [88]. In sunny climates, temperature rise may be sufficient to provide effective disinfestation of pathogenic fungi [89].

## 2. Sunshading

Another form of large-scale microclimate control is by shading. A practice that started as "slat houses" for orchids and "cloth houses" for high-quality tobacco has developed into very considerable industries, usually growing ornamentals. A high proportion are foliage plants, grown under coarse-woven plastic material developed to give certain "percentages" of shade. Obviously, any modification of insolation (irradiance) also modifies temperature. It is remarkable that although research reports commonly pay considerable attention to the expression of the exact degree of shade [90], temperature differences are often not mentioned. It can be very helpful to include temperature as a variable, as demonstrated in a study of disease intensity under different levels of shade [91]. Research workers in this field are urged to routinely measure and report the temperature variations that inevitably accompany any modification of irradiance.

Shade conditions can be expected not only to lower daytime temperatures but also to raise night temperatures, particularly under cold night–clear sky conditions, in which ground-to-sky radiation can cause a very rapid, possibly harmful, drop in temperature near the ground. Similarly, the use of spectral filters can be expected to modify not only light quality but also temperature.

## 3. Freeze Protection

The first, most obvious, and least expensive protection against freeze injury is to select a planting site where injurious freezing is unlikely to occur. Because this is often not possible, freeze protection measures may be necessary. Burning fossil fuels should be regarded as a last resort—the fuels themselves are very expensive, and their use is often environmentally questionable. Only too often, freeze protection methods are ineffective because of ignorance of the following basic thermodynamic and meteorological principles.

1. Cold air will roll down a slope until arrested by some physical barrier, which then forms a "frost pocket."

2. Hot air rises vertically. It cannot be made to move up a slope.

3. Radiated heat travels in all directions uniformly but only in line-of-sight (straight) lines. Thus, to be warmed by irradiation from a heat source (such as an orchard heater), a plant must be able to "see" the heat source. Because radiated heat, like all forms of radiation, is subject to the inverse square law (i.e., intensity decreases proportionately to the square of the distance traveled), radiation warming decreases sharply with distance from the heat source.

4. The total heat content of a mass of air depends not only on its temperature (sensible heat) but also on its latent heat, the two together approximating its total energy content or enthalpy. Thus, total heat content can be very much greater for moist air than for dry air at the same temperature. Putting this in a different way: air masses at the same atmospheric pressure and conditions of 15°F ($-9.4$°C) and 100% RH, 20°F ($-6.7$°C) and 40% RH, and 25°F ($-3.9$°C) or 0% RH all have the same heat content of approximately 5.5 Btu per pound of dry air (ca. 3 kg cal kg$^{-1}$) [92].

5. The latent heat of evaporation is approximately 7.5 times as great as the latent heat of freezing. Thus, when spraying irrigation water for freeze protection (a common practice for Florida strawberries and various other crops), it is essential to freeze at least 7.5 times as much water as is evaporated [93]. In calm or near-calm weather, this is no problem. Continuing to spray after the onset of a brisk breeze, however, can be disastrous. An ingenious application of this principle is to use such evaporative cooling to delay the blooming of fruit trees until the danger of a blossom freeze is over. The blooming of apple trees was delayed by as long as 17 days by use of thermostatically controlled sprinkling whenever prebloom temperatures exceeded 7°C (44.6°F) [94].

6. Smoke from burning oil or other fuel does *not* form a protective shield. It used to be believed (particularly in California) that "smudge pots" could create a low cloud that reflected heat back

to the crop below. It is now known that the smoke particles are not in a size range suited to re-flect infrared emissions. It is, however, possible to generate very fine water fog with droplets of the appropriate size. An added benefit is that any fog droplets that freeze give off latent heat to the surrounding atmosphere.

7.   Freezes are classified as "convection freezes" or "advection freezes." Convection freezes occur with calm air and cloudless skies, conditions in which the earth is radiating heat to the sky, with consequent rapid cooling of the air near the ground. For orchard crops, it is beneficial to have bare ground to radiate ground heat to the trees. Weeds or cover crops trap such radiated heat at the expense of the trees. Convective conditions commonly result in atmospheric inversions, in which the lower air is colder than that at 10–30 m above the ground. In such conditions, "wind machines" mounted on tall towers or pylons can be beneficial. Helicopters have sometimes been used to achieve the same effect, particularly to prevent dangerously cold air from accumulating in the "frost pocket" hollows.

In an advective freeze, a wind strong enough to disrupt normal convection patterns freezes crops on the exposed higher ground, with much less freeze injury in the valleys and lowlands. Wind machines are worse than useless in an advective freeze, but rows of heaters placed at right angles to the wind direction can benefit crops for a considerable distance downwind.

A deadly interaction among temperature, humidity, and wind speed can occur in an advective freeze. Tender leaves and shoots can be killed, not by freezing, but by desiccation, if wind speed is high enough when the temperature approaches the freezing point of plant tissues under conditions of very low humidity (which frequently occur).

For further information on methods and principles of freeze protection, readers are referred to an ex-tensive chapter on freeze protection [95].

## E.   Incidental Effects of Temperature

Old Ecclesiastes said, "Of the making of many books there is no end," and a number of them probably could be written on the incidental effects of temperature. However, only a very few examples can be cited here to indicate how often temperature is an unforeseen or unplanned-for variable.

Temperature can move in mysterious ways, its wonders to perform, through its subtle influence on the activity of growth regulators. As noted earlier, fruit setting in tomato plants is inhibited by too high temperatures. A role for growth regulators in this high-temperature inhibition is indicated by a report [96] that relative levels of gibberellin and auxinlike growth regulators were sharply affected at high tempera-tures.

On a purely physical basis, temperature can be expected to affect gas diffusion rates, hence rates of photosynthesis and leaf respiration. However, not only can the physical effects of temperature be com-plicated by the metabolic effects of temperature on rates of photosynthesis and respiration, but such gas exchange is reported to be affected by an interaction between temperature and humidity [97]. Exact con-trol of temperature is routine, but equivalent accuracy in control of humidity can be difficult, and exact simultaneous control of temperature and humidity can be very challenging indeed.

Vegetable transplants usually benefit from hardening by controlled temperature and/or moisture stress before being planted out in the field [98]. This does not appear to be the case for sweet potato trans-plants, which are vine cuttings rather than seedlings. Transplants held at 13–18°C were reported to have greatly increased vitality and ultimately higher yields compared with transplants held at an ambient tem-perature of 26.7°C [99]. (That "26.7°C ambient" temperature is curiously exact and is possibly a transla-tion from "ca. 80°F ambient.")

A particularly intriguing example of an unexpected temperature effect is reported in a study of male sterility in the common bean (*Phaseolus vulgaris* L.) [100]. When, in the course of an atypically cool sum-mer, unexpected fertility was noted in supposedly male sterile plants, research was transferred to growth chambers. A day/night temperature regime of 30/18°C for an average of 12 days was sufficient to cause most unstable steriles to produce sterile buds. Day/night conditions of 18/7°C for an average of 14 days were effective in converting sterile to partially sterile phenotypes. Both temperature-stable and tempera-ture-unstable genotypes were identified; this is an excellent example of valuable research findings achieved by following up on a temperature-related anomaly revealed in a field study.

The literature abounds in such examples. Many mysteries would be elucidated if research workers routinely reported temperatures (whether controlled or not) and included such data in their research reports. Subsequent research workers, if alert to the multitudinous roles of temperature, will then be in a position to carry the research further, perhaps with the advantages of better funding or instrumentation.

## III. POSTHARVEST ROLE OF TEMPERATURE

## A. Handling, Storage, and Shipping Temperatures

It is all too often forgotten that crops are still alive after harvest. No matter how meticulously grown, most horticultural crops will not realize their full economic or nutritional potential unless handled at suitable temperatures after harvest. How important this is depends on both the frailty of the crop and time between harvest and consumption or processing. During this period, the importance of temperature and humidity depends very largely on the biological maturity of the plant part being harvested [101]. Temperature control is obviously of more consequence for asparagus than for coconuts. Only a very brief account of the principles involved can be given here. Attention is drawn to the U.S. Department of Agriculture handbook dealing with storage conditions for a very wide range of produce [102]. Most agronomic crops are far less sensitive to postharvest temperatures, but there are exceptions, such as potatoes (see Sec. III.D).

### 1. Fruits

Chapter 7 deals with the development and physiology of fruits, which, botanically, can mean any matured plant ovary from a grain of wheat to a watermelon. Thus, the comments here are very brief and are largely confined to temperature relationships of dessert fruits that are sometimes processed but more traditionally are eaten fresh. Bear in mind, however, that many products considered to be vegetables are botanically fruits: tomatoes, green (snap) beans, squash, bell peppers, and cucumbers are all botanically fruits.

Fruits can be classified according to their respiration pattern as climacteric or nonclimacteric [103]. Soon after harvest, climacteric fruits (e.g., apples, pears, bananas) produce ethylene in quantities sufficient to overcome the antidoting effect of internal carbon dioxide [104]. The result is a rapid rise in respiration rate, at the conclusion of which the fruit is senescent, overripe, and unpalatable. The useful life of a climacteric-type fruit is typically ended by senescence rather than by decay. Prompt refrigeration is thus critical for climacteric-type fruits. The more the climacteric rise in respiration can be suppressed, the longer the postharvest life of the fruit.

Nonclimacteric fruits (e.g., citrus and grapes) have no climacteric rise in postharvest respiration. At any constant temperature, their respiration rate remains constant. For such fruits, refrigeration functions more to prevent or delay the onset of decay than to lower respiration rate. For any type of fruit, one of the major functions of temperature regulation is to maintain fruit quality. This involves control of desiccation, minimization of flavor and texture loss, and prevention of off-flavors.

Selection of optimum storage temperatures for some fruits can be conditioned by susceptibility to chilling injury (see Sec. III.C). Particularly for long-term storage, avoidance of chilling injury can override considerations of respiration rate or decay.

Prevalence of storage disorders such as water core [105] and superficial scald [106] of apples can be affected not only by storage temperature but also by preharvest growing temperatures. Chemical composition of Satsuma mandarins varies with growing temperature [107]. Production of high-quality, low-acid grapefruit depends on uninterrupted warm winter night temperatures [108].

### 2. Seeds

Storage temperature and thus potential storage life are sharply conditioned by the tolerance of seeds to desiccation. "Orthodox" seeds that will survive desiccation (and often will desiccate on the plant) can be stored at very low (subfreezing) temperatures. "Recalcitrant" seeds that cannot survive desiccation are very difficult to store because they cannot survive low temperatures. These brief remarks oversimplify a complex situation. Readers needing to know more are referred to a very detailed review article by Ellis [41].

### 3. *Other Plant Organs*

The urgency of immediate postharvest temperature and humidity control is related to the maturity of the plant part involved [101]. Grain crops, mature root crops, and cabbage are typical of storage organs that enter a resting stage preparatory to winter. Their respiration rate is very low, and thus prompt postharvest refrigeration is of little consequence. Young actively growing tissues, such as asparagus, green peas, and sweet corn, have very high respiration rates that need to be reduced by refrigeration as soon as possible. The same is true of cut flowers, an intrinsically ephemeral product.

There is a tendency to forget the economic consequence of unrestricted respiration rate in crops for processing. Nevertheless, particularly when crops are paid for on the basis of sugar content, excessive respiration rates due to prolonged exposure to high temperature (as with truckloads of oranges waiting in the sun outside a Florida cannery) deplete sugar content, hence the cash value of the product. Even sugarcane stacked in the sun by the roadside after harvest is losing sugar for which the grower would otherwise be paid [109].

## B.  Prestorage "Curing": Temperature × Humidity × Time

Traditionally, those who handled horticultural crops for shipment or storage were advised to refrigerate as soon as possible after harvest. It is now known that there are marked exceptions to this general rule. One such exception is the group of products that need to be "cured" prior to storage to heal mechanical wounds (see Sec. II.B.7). The outstanding example is sweet potato, for which *Rhizopus* decay in cold storage was often calamitous until is was demonstrated that prior "curing" at ambient (or higher) temperature and very high humidity for several days healed wounds that otherwise would have been invasion sites for *Rhizopus* [110]. The same benefit can occur, although usually to a less marked extent, with other root and tuber crops.

## C.  The Chilling Injury Syndrome

Perhaps the most intriguing response of plants to temperature is the chilling injury syndrome exhibited by many plants of tropical origin (which include such familiar crops as cotton, soybeans, tomatoes, citrus, and cucumbers, commonly grown in the temperate zone). Morphological and biochemical responses of corn (*Zea mays* L.) to field chilling conditions have been reported in considerable detail [111]. CI-susceptible plants (and their detached plant organs) are severely injured by temperatures well above freezing. Critical temperatures vary, but typically injury occurs at temperatures below 10°C. Preharvest chilling injury can occasionally be troublesome, particularly with cotton seedlings [112] and mature, but unripe, tomatoes [113]. But CI is particularly important after harvest, not only because of the products lost due to incorrect storage or transit temperatures but also (perhaps more significantly) because of severe limitations on marketing. If Florida grapefruit could be stored and shipped at the same temperatures as Florida oranges, markets for grapefruit growers would be enormously expanded.

The symptoms of CI can be either superficial or metabolic. Superficial effects are typically various forms of peel injury, which may be uniform (e.g., the darkening of the peel of a banana held in a household refrigerator) or highly irregular (e.g., discrete, necrotic sunken areas of grapefruit or cucumbers, surrounded by healthy tissue).

The metabolic origin of CI is so profound that a remarkably precipient study demonstrated a parallel between behavior of mitochondria in CI-susceptible versus nonsusceptible plants and of mitochondria from poikilothermic (cold-blooded) versus homeothermic (warm-blooded) animals [114].

The tomato is an example of a climacteric-type fruit that is metabolically sensitive to CI. A mature green tomato that has been chilled will never ripen, even when treated with exogenous ethylene.

The literature on CI is dispersed among many types of plants and journals; moreover, research reports often deal solely with individual reactions or systems isolated from ecological considerations. Much of this literature up to 1986 has been reviewed [115].

Nevertheless, this account reviews the 25-year-long series of reports on grapefruit (and occasionally bananas, limes, and avocados, when grapefruit were out of season) at the University of Florida's Citrus Research and Education Center in Lake Alfred. There are several reasons for this duplication.

1. Grapefruit is uniquely suited for CI research in that fruit can be harvested from a single bloom on an individual tree for as long as 8 or 9 months (typically from September to May). Moreover,

the same plant (tree) can be harvested year after year. In eight seasons (1974–75 to 1981–82), the same 28 trees were randomly picked (north, south, east and west sides; upper and lower, inner and outer fruit) at 14-day intervals for a total of more than 100 pickings. We know of no comparable testbed material for CI research.

2.  A reporting method was developed whereby the results of each individual picking were reduced to a single value, thus greatly facilitating statistical analysis of multiple experiments [116–118].

3.  The program both sought immediate commercial results for the Florida citrus industry and provided training in basic research methods for a series of graduate students. Such training involved rigid adherence to the classical scientific method (i.e., constant testing and evaluation of hypotheses), evidence of which approach is singularly missing in many published reports on CI.

The initial hypothesis was that CI involved a breakdown of the respiratory system, resulting in toxic products of incomplete oxidation (typically acetaldehyde), which in turn caused the distinctive peel lesions. (Acetaldehyde was always detectable in the atmosphere around chilled fruit, and application of exogenous acetaldehyde caused superficially similar lesions). A report that hypobaric (vacuum) storage greatly prolonged the useful lifetime of various products (at their usual recommended storage temperatures) attributed this effect to the continual removal of endogenous ethylene [119]. So we tried hypobaric storage of bananas at chilling temperatures. CI was completely controlled, which we attributed to continual removal of toxic acetaldehyde [120]. The same effect was soon confirmed for limes and mitochondrial respiration of CI-susceptible citrus fruits (limes and grapefruit) versus CI-resistant Florida-grown Valencia oranges [121].

The hypothesis of the breakdown of the respiratory mechanism appeared to be true. (It still does, but it is now regarded as a secondary effect). In "micro" respiratory studies with 5-mm peel disks, the banana disks always chilled. In tissue culture, less than half the grapefruit peel disks chilled, which corresponded well to the curious pattern of CI-induced peel lesions [122]. An unsolved mystery is why, in fruits such as grapefruit and cucumber, the cells at the periphery of a necrotic lesion collapse and die while the immediately adjacent cells surrounding the lesion remain healthy. Carbon dioxide (a standard respiratory depressant) was found to minimize adenosine 5′-triphosphate (ATP) accumulation (apparent evidence for CI-induced impairment of the ATP/ADP energy transfer system). There was no correlation with CI and levels of three enzymes (pectinmethylesterase, polygalacturonase, and cellulase), which had been suspected of involvement in lesion formation [123,124].

Because "controlled atmosphere storage" has long been commercially used for other products, the effect of $CO_2$ in suppressing CI was investigated. Two treatments were tested: a prestorage treatment with very high levels (e.g., 25%) of $CO_2$ and also storage atmospheres developed under differentially permeable plastic films [125–127]. Success in suppressing CI was sometimes notable, but with three disconcerting caveats.

1.  The early-season sensitivity to CI, which traditionally had been considered to decrease with increasing fruit maturity, was reappearing in late-season, very mature grapefruit. An alert graduate student, Kazuhide Kawada, found that such late-season susceptibility to CI had been reported in some detail for California grapefruit as long as ago as 1936, but researchers had missed the paper because it had been given an inappropriate title [128].

2.  Although extremely effective in early and midseason, $CO_2$ had absolutely no protective effect on grapefruit picked after the new bloom (ca. mid-March).

3.  The length of delay between picking and postharvest treatments sometimes had more protective effect than the treatments being compared.

A new hypothesis was clearly called for, and the one produced was twofold: the tree and the fruit had to be considered as a whole (fruit off tress in full "growth flush" obviously behaved very differently from fruit from dormant trees), and the controlling mechanism between tree and fruit had to be growth regulators (GRs). A working hypothesis that CI was promoted by gibberellins and prevented by abscissic acid (ABA) was largely confirmed [129]. ABA, the protective "stress hormone," apparently can be developed either pre- or postharvest. Much of this material has been summarized elsewhere [130]. With this knowledge, it is easy to understand the protective effect of various prestorage treatments, not only for grapefruit

[131] but for a wide variety of other products such as CI-sensitive Australian oranges [132] and zucchini squash [133].

## D. Anomalous Chilling Injuries

Although the basic principles described above apply to a very wide range of CI-sensitive crops, there are other forms of low-temperature injury. Apples are susceptible to a wide range of temperature-related storage diseases that constitute a field of study outside this discussion, with one exception. Apples grown in North America generally tolerate storage temperatures close to freezing point (1–2°C). Apples, even of the same variety, grown in Britain or Northern Europe cannot tolerate such low temperatures, however, and formerly this disadvantage sharply limited their marketing season. Thus "controlled atmosphere" (CA) storage (then called "gas storage") was developed in England in the early 1930s. Initially, CA relied on raising carbon dioxide levels to suppress the respiratory climacteric. Later practice favors lowering oxygen to just above a level that would induce anaerobiosis [134]. Such CA storage has made possible the year-round marketing of apples. I have seen no explanation of why apples from the two sides of the Atlantic should respond so differently to storage temperatures, but the effect is real. Similar differences in response to temperature exist for other products from widely dispersed growing areas. For example, Valencia oranges grown in California and Australia are susceptible to chilling injury during long-term storage and shipment, whereas those from Florida and Brazil are not.

Potatoes are subject to an important temperature-related storage disorder that can be very costly for manufacturers of such products as potato chips and frozen ready-to-cook french fries. At temperatures below about 5°C, potatoes undergo reversible starch-sugar hydrolysis, which causes potato products to darken when the sugar caramelizes upon exposure to high cooking temperatures. Such discolored products are discounted or are unsalable. If chilled potatoes are held at room temperature for several days, however, the reverse (condensation) reaction will convert the sugar back to starch.

Another anomalous postharvest "chilling" hazard is physical and pathological rather than physiological. Some products, such as leafy vegetables, celery, and peaches, benefit from "hydro cooling" in refrigerated water. A marked exception is the tomato, which should never be immersed in water cooler than product temperature. The skin of the tomato is virtually impervious; gas exchange is through the porous stem scar. (A drop of molten wax on the stem scar of a green tomato will turn it into a self-contained "controlled atmosphere storage unit," thereby greatly delaying ripening.) When a warm tomato is immersed in cool water, contraction of its internal atmosphere draws nonsterile water in through the porous stem scar, with consequent greatly increased decay hazard [135]. The same problem obviously is possible with other products.

## IV. GLOBAL TEMPERATURE CHANGES

Since this chapter was first written, consideration of global temperature changes has become an international concern of quite extraordinary magnitude. Despite objections from many reputable scientists, both individually [136] and collectively [137], this has generated a popular media-driven controversy with consequent proposals for economically disastrous measures to reduce emissions of $CO_2$ in order to maintain the *status quo ante* for worldwide temperatures [138,139].

Apart from the notable disregard for scientific findings in many fields of endeavor, this is hubris in the classical Greek sense of arrogance that would challenge the gods.

Nearly a century ago, Svante Arrhenius showed that $CO_2$ is a "greenhouse gas" that transmits short-wave radiation but impedes long-wave (heat) radiation. However, any possibly deleterious effects on global temperatures from mankind's generation of $CO_2$ are very minor in comparison with the sun's dominant effects, short term through sunspots [140–143] and longer term due to irregularities in its axis [144,145]. Added to which is a gradual, but inexorable, change in the tilt of earth's own axis [146] and the precession of the equinoxes that so puzzled ancient astronomers [147].

Moreover, the climatic influence of the sun involves other variables, some as obvious as solar flares [148,149], others as arcane as very minor irregularities in its orbit that mathematical astronomers are only now beginning to explain.

Long before modern instrumentation, sunspots could be studied with no more equipment than a piece of smoked glass, isinglass, or other animal membrane—and the ancient Chinese left written records. Sunspots come and go, but they persist for long periods. (Galileo used them to time the rotation of the sun.) Mean earth temperatures vary directly with the number of sunspots. In 1922, an English lady, Annie Maunder, correlated sunspot frequency with climatic records [150]. When sunspots almost disappeared, a period known as the Maunder Minimum, the Northern Hemisphere suffered the "Little Ice Age." From about 1500 to 1900 AD sunspots were few, with intermittent minima during one of which England's Thames River froze and another when George Washington's army had the misfortune to be encamped at Valley Forge [151]. Evidence of the Little Ice Age, and also of the "Little Climatic Optimum" 500 years before, still lingers in deep rock temperatures [152].

Within the larger sunspot cycle is a minor, rather consistent, approximately 11 year, cycle. Curious evidence of this is afforded by the trading records of Canada's Hudson Bay fur company. Rythmic fluctuations in the populations of prey animals, largely arctic hares and lemmings, are echoed 1 year later in increases in pelts taken from carnivores, particularly the valuable white fox.

Geological evidence indicates wide variations in mean temperatures and $CO_2$ levels in past interglacial [153] and even postglacial, Holocene [154] periods. Some have been correlated with volcanism or meteor showers [155]. Archeology now indicates that collapse of some major Bronze Age civilizations was due to droughts associated with volcanic eruptions [156]. When Mount Krakatoa blew up in 1883, it lowered mean global temperature 0.27°C (0.5°F). The amounts of industrially released $CO_2$ are minor compared with those from such natural forces.

Moreover, global warming is not necessarily harmful [157,158]. During the 11th century sunspot maximum (the Little Climatic Optimum) Greenland supported a thriving farming community, as did the Orkney Islands. During the Little Ice Age the Greenlanders died and the Orkney Islanders struggled to survive. With today's sunspot plenitude, the Orkneys have become Scotland's major beef-producing county [159], although green pastures have yet to return to Greenland.

Supposed scientific calculations and much popular alarmism predict that a few degrees of global warming will cause disastrous flooding of many coastal areas and complete disappearance of low-lying Pacific Islands due to melting of the polar icecaps [160]. History shows otherwise. During the 1000-year cycle that included the Little Climatic Optimum and the Little Ice Age, sea levels did not change materially [161]. Some ice-freed coasts rose, some coastlines eroded and others accreted, and occasionally coastal subsidence became threatening. London is an example of the latter phenomenon. The considerable engineering feat of the Thames Barrier has been necessitated by slight, but inexorable, land subsidence and occasional coincidence of an abnormally high spring tide with a very strong northeast wind.

Apparently minor temperature changes can have drastic effects due to their influence on the winds. The El Niño phenomenon has had much publicity of late, although it is nothing new, as indicated by coral growth records going back over 100,000 years [143,162] and by ocean and lake sediments [163] for shorter periods. The apparent warming of hundreds of cubic miles of Pacific Ocean water is not due to enormous amounts of added heat but to failure of the trade winds that normally push the sun-warmed water toward the Philippines and Indonesia, without which they suffer devastating droughts.

Ground-penetrating radar shows that great mountain-fed rivers once transversed the Sahara Desert. Cave paintings and rock carvings [164] prove that 8000 years ago the Sahara was verdant and teeming with tropical wildlife. Such a scenario is now impossible with today's wind patterns.

Obviously, any practices that are deleterious to the environment should be curtailed wherever it is possible to do so without incurring unacceptable human and economic consequences. However, any climatic effects from emissions of $CO_2$ from consumption of fossil fuels are trivial by comparison with nature's inexorable forces.

Conclusion: There is no foreseeable reason why producers of crops need to modify where or how they grow them despite grossly exaggerated accounts of hazards from worldwide global warming.

## V.  CONCLUSION

With virtually any crop, from seed germination, bud sprouting, or anthesis to harvest, and after harvest to final consumption, temperature plays important, and sometimes unsuspected, roles.

# REFERENCES

1. H Zinsser. As I Remember Him. Cited from Bartlett's Familiar Quotations. 16th ed. Boston: Little, Brown, 1992.
2. K Al-Khatib, GM Paulsen. Crop Sci 39:119–125, 1999.
3. PO Craufurd, TR Wheeler, RH Ellis, RJ Summerfield, JH Williams. Crop Sci 39:136–142, 1999.
4. M Bonhomme, R Rageau, JP Richard, A Erez, M Gendraud. Sci Hort 80:157–171, 1999.
5. WL George Jr, JW Scott, WE Splitstoesser. Hortic Rev 6:65–84, 1984.
6. DW Kretchman. Ohio Agric Res Center Summ 26:5–6, 1968.
7. Associated Press. The Tampa Tribune, Dec. 7, 1992.
8. M Samedi, LC Cochran. HortScience 10:593, 1976.
9. FS Davies. Hortic Rev 8:129–120, 1986.
10. WC Cooper, RH Hilgeman, GE Rasmussen. Proc Fla State Hortic Soc 77:101–106, 1964.
11. WRF Grierson-Jackson. The storage and ripening of Bartlett pears, M.Sc. Agriculture thesis, University of Toronto, Aug. 26, 1940.
12. CC Daamen, WA Dugas, PT Prendergast, MJ Judd, KG McNaughton. Agr For Met 91:171–183, 1999.
13. M Tarnopolsky, I Seginer. Agr For Met 91:185–194, 1999.
14. AW Marsh. The Citrus Industry. Vol 3. Berkeley: University of California Press, 1973, pp 235–236.
15. HD Chapman. The Citrus Industry. Vol 2. Berkeley: University of California Press, 1968, pp 168–169.
16. DJ Cantliffe. J Am Soc Hortic Sci 97:674–676, 1972.
17. RF Suit, EP DuCharme. Plant Dis Rep 37:379–383, 1953.
18. RC Baines, SD Van Gundy, EP DuCharme. The Citrus Industry. Vol 4. Berkeley: University of California Press, 1978, pp 321–345.
19. WB Sherman, J Rodriguez, EP Miller. Proc Fla State Hortic Soc 97:320–322, 1984.
20. J Logan, DE Deyton, DW Lockwood. HortScience 25:1382–1384, 1990.
21. MCB Raseira, JN Moore. HortScience 22:216–218, 1987.
22. RK Seagel, S Wiest, D Linvil. HortScience 26:99–100, 1991.
23. MJ Burke, LV Gusta, HA Quamme, CJ Weiser, PH Li. Annu Rev Plant Physiol 27:507–528, 1976.
24. EN Ashworth. HortScience 21:1325–1328, 1986.
25. R Timmis. Pacific Forestry Research Center, Canadian Forestry Service Internal Report BC-35, 1972.
26. W Grierson, FW Hayward. Proc Am Soc Hortic Sci 73:278–288, 1959.
27. V D'Antonio, CF Quiros. HortScience 22:479–481, 1987.
28. T Harada, A Inaba, T Yakuwa, T Tamura. HortScience 20:678–680, 1985.
29. BM Reed, EB Lagerstedt. HortScience 22:302–303, 1987.
30. RS Benjamin. The Tampa Tribune, Jan. 10, 1993.
31. J van Tuyl. HortScience 18:754–756, 1983.
32. Å Ergon, SS Klemsdal, AM Tronsmo. Physiol Mol Pl Phys 53:301–310, 1998.
33. CE Abbott. Am J Bot 22:476–485, 1933.
34. CW Coggins Jr, HZ Hield. The Citrus Industry. Vol 2. Berkeley: University of California Press, 1968, p 383.
35. MA Ismail, W Grierson. HortScience 12:118–120, 1977.
36. M Vasilakakis, IC Porlingis. HortScience 20:733–735, 1985.
37. CR Hale, MS Buttrose. J Am Soc Hortic Sci 99:390–394, 1974.
38. HD Frost, RK Soost. The Citrus Industry Vol 2. Berkeley: University of California Press, 1968, p 299.
39. SM Southwick, FS Davies, NE El-Gholl, CI Schoulties. J Am Soc Hortic Sci 107:800–804, 1982.
40. BB Dean, LR Baker. Hort Science 18:349–351, 1983.
41. RH Ellis. HortScience 26:1119–1125, 1991.
42. CB Heiser Jr. Of Plants and People. Norman: University of Oklahoma Press, 1985, pp 185–186.
43. T Lysenko. The Science of Biology Today. New York: International Publishers, 1948.
44. CA Argerich, KJ Bradford, AM Tarquis. J Exp Bot 40:593–598, 1989.
45. AD Alvarado, KJ Bradford. Seed Sci Technol 16:601–612, 1988.
46. G Murray, JB Swensen, JJ Gallian. HortScience 28:31–32, 1993.
47. AG Taylor, J Prusinski, EJ Hill, MD Dickson. HortTechnology 3:336–344, 1992.
48. RF Stewart. U.S. Patent 5,129,180, assigned to Landec Laboratories, Inc., Menlo Park, CA, July 14, 1992.
49. CA Parera, P Qiao, DJ Cantliffe. Hort Science 28:20–22, 1993.
50. DA Kovach, KJ Bradford. Ann Bot 69:297–301, 1992.
51. DA Kovach, KJ Bradford. J Exp Bot 43:747–757, 1992.
52. GA Ashiru. J Am Soc Hortic Sci 94:429–432, 1969.
53. R Oliver. The African Experience. New York: Icon Editions, Harper Collins, 1991, pp 137–138.
54. DL McNeil, RS Duran. Trop Agric (Trinidad) 229–234, 1992.
55. LF Stoltz, JC Snyder. HortScience 20:261–262, 1985.
56. ME Saltveit Jr, DR Dilley. Plant Physiol 61:675–679, 1978.
57. CR Stearns Jr, GT Young. Proc Fla State Hortic Soc 55:59–61, 1942.
58. LB Young, LC Erickson. Proc Am Soc Hortic Sci 78:197–200, 1961.
59. W Grierson, WF Newhall. Proc Fla State Hortic Soc 66:42–46, 1953.

60.  I Stewart, TA Wheaton. J Agric Food Chem 20:448–449, 1972.
61.  W Grierson, SV Ting. Proc Int Soc Citric 21–27, 1978.
62.  WF Wardowski, J Soule, J Whigham, W Grierson. Florida Agricultural Extension Service Special Publication 99, 1991.
63.  W Grierson. Citrus Ind (Bartow, FL) 72(1):47–50, 1991.
64.  R Young, F Meredith, A Purcell. J Am Soc Hortic Sci 94:672–674, 1969.
65.  HK Wutscher. J Am Soc Hortic Sci 101:573–575, 1976.
66.  JW Eckert. In: Postharvest Biology and Handling of Fruits and Vegetables. New York: AVI Publishing (Van Nostrand Reinhold), 1975, pp 81–117.
67.  RE Hardenburg, AE Watada, CY Wang. U.S. Department of Agriculture Handbook 66, rev. 1985.
68.  MA Ismail, GE Brown. J Am Soc Hortic Sci 104:126–129, 1979.
69.  CL Le. HortScience 20:451–452, 1985.
70.  SL Knight, CA Miller. HortScience 18:462–463, 1983.
71.  TK Broschat, HM Donselman. HortScience 18:206–207, 1983.
72.  EJ Runkle, RD Heins, AC Cameron, WH Carlson. HortScience 34:55–58, 1999.
73.  AM Armitage, JM Garner. J Hortic Sci Biotech 74:170–174, 1999.
74.  RU Larsen, L Persson. Sci Hortic 80:73–89, 1999.
75.  A Hagiladi. HortScience 18:369–371, 1983.
76.  CB Heiser Jr. Of Plants and People. Norman: University of Oklahoma Press, 1985, pp 158–159.
77.  HW James, R McAvoy. HortScience 18:363–364, 1983.
78.  WC Lin. HortScience 19:515–516, 1984.
79.  GJ Keever, CS Cobb. HortScience 19:439–441, 1984.
80.  WJ Carpenter, HP Rasmussen. J Am Soc Hortic Sci 95:578–582, 1970.
81.  FJ Regulski. HortScience 18:476–478, 1983.
82.  AR Chase, CA Conover. HortScience 22:65–67, 1987.
83.  AW Marsh. The Citrus Industry. Vol 3. Berkeley: University of California Press, 1973, pp 272–274.
84.  JW Gerber, J Janick, D Martsolf, C Sacamano, EJ Stang, S Wiest. HortScience 18:402–404, 1983.
85.  AD Mathias, WE Coates. HortScience 21:1453–1455, 1986.
86.  CS Wells, JB Loy. HortTechnology 3:92–94, 1993.
87.  CS Wells, JB Loy. HortScience 20:800, 1985.
88.  WJ Lamont Jr. HortTechnology 3:35–39, 1993.
89.  TE Hartz, JE DeVay, CL Elmore. HortScience 28:104–106, 1993.
90.  JE Barrett. HortScience 20:812, 1985.
91.  RL Sealey, CM Kenerley, EL McWilliams. HortScience 25:293–294, 1990.
92.  W Grierson, WF Wardowski. HortScience 10:356–360, 1975.
93.  W Grierson. Proc Fla State Hortic Soc 77:87–93, 1964.
94.  JL Anderson, GL Ashcroft, EA Richardson, JF Alfaro, RE Griffin, GR Hanson, J Keller. J Am Soc Hortic Sci 100:229–231, 1975.
95.  FM Turrell. The Citrus Industry. Vol 2. Berkeley: University of California Press, 1973, p 338.
96.  CG Kuo, CT Tsai. HortScience 19:870–872, 1984.
97.  F Lorenzo-Minguez, R Ceulemans, R Gabriels, I Impens, O Verdonck. HortScience 20:1060–1062, 1985.
98.  W Grierson. Beneficial aspects of stress on plants. In: M Pessarakli, ed. Handbook of Plant and Crop Stress. New York: Marcel Dekker, 1993, pp 645–657.
99.  LE Hammett. HortScience 20:198–200, 1985.
100. S Estrada, MA Mutschler, FA Bliss. HortScience 19:401–402, 1984.
101. W Grierson, WF Wardowski. HortScience 13:570–574, 1978.
102. RE Hardenburg, AE Watada, CY Wang. U.S. Department of Agriculture Handbook 66, rev 1986.
103. JB Biale. CSIRO Food Preserv Q (Aust) 22(3):57–62, 1962.
104. SP Burg, EA Burg. Plant Physiol 42:144–152, 1967.
105. H Yamada, S Kobayashi. Sci Hortic 80:189–202, 1999.
106. WJ Bramlage, SA Weis. HortScience 32:808–811, 1997.
107. KB Marsh, AC Richardson, EA Macrae. J Hortic Sci Biotech 74:443–451, 1999.
108. W Grierson, SV Ting. Proc Int Soc Citric (Aust mtg), 21–27, 1978.
109. JI Lauritzen, RT Balch. U.S. Department of Agriculture Technical Bulletin 449, 1934.
110. JI Lauritzen, LL Harter. J Agric Res 33:527–539, 1926.
111. MJ Verheul, C Picatto, P Stamp. Eur J Agron 5:31–43, 1996.
112. A Rikin, D Atsmon, C Gitler. Plant Cell Physiol 20:1537–1546, 1979.
113. LL Morris. Proceedings of the Association of American Railroads, Freight Loss Conference, Feb. 5–7, 1953, 1953, pp 141–146.
114. JM Lyons, JK Raison. Comp Biochem Physiol 24(1):1–7, 1970.
115. AA Markhart III. HortScience 21:1329–1333, 1986.
116. W Grierson. Proc Trop Reg Am Soc Hortic Sci 18:66–73, 1974. Reprinted in Citrus Industry (Bartow, FL), 56(12):15–17, 19, 21–22, 1975.

117. W Grierson. Proc Trop Reg Am Soc Hortic Sci 23:290–294, 1979.
118. K Kawada, W Grierson, J Soule. Proc Fla State Hortic Soc 91:128–130, 1978.
119. SP Burg, EA Burg. Science 148:1190–1196, 1965; 153:314–315, 1966.
120. EB Pantastico, W Grierson, J Soule. Proc Trop Reg Am Soc Hortic Sci 11:82–91, 1967.
121. EB Pantastico, J Soule, W Grierson. Proc Trop Reg Am Soc Hortic Sci 12:171–183, 1968.
122. N Vakis, W Grierson, J Soule, LG Albrigo. HortScience 5:472–473, 1971.
123. N Vakis, W Grierson, J Soule. Proc Trop Reg Am Soc Hortic Sci 14:89–100, 1970.
124. N Vakis, J Soule, RH Biggs, W Grierson. Proc Fla State Hortic Soc 83:304–310, 1970.
125. W Grierson. Proc Trop Reg Am Soc Hortic Sci 15:76–88, 1971.
126. WF Wardowski, W Grierson, GJ Edwards. HortScience 8:173–175, 1973.
127. WF Wardowski, LG Albrigo, W Grierson, CR Barmore, TA Wheaton. HortScience 10:381–383, 1975.
128. EM Harvey, GL Rygg. J Agric Res 52:747–787, 1936.
129. K Kawada. Some physiological and biochemical aspects of chilling injury of grapefruit (*Citrus paradisi* Macf.) with emphasis on growth regulators. Ph.D. thesis, University of Florida, 1980.
130. W Grierson. In: WF Wardowski, S Nagy, W Grierson, eds. Fresh Citrus Fruits. New York: AVI Publishing (Van Nostrand Reinhold), 1986, pp 371–373.
131. TT Hatton, RH Cubbedge. HortScience 18:721–722, 1983.
132. BL Wild, CW Hood. HortScience 24:109–110, 1989.
133. GF Kramer, CY Wang. HortScience 24:995–996, 1989.
134. RM Smock, AM Neubert. Apples and Apple Products. New York: Interscience, 1950.
135. RK Showalter. HortTechnology 3:97–98, 1993.
136. JM Grove. http://www.cyf-kr.edu.pl/~ziniedzw/paper042.html, 1995.
137. EA Roberts Jr. Tampa Tribune, Dec. 7, 1997.
138. Anonymous. Tampa Tribune, Dec. 2, 1997.
139. J Jacoby. Int Herald Tribune, Nov. 7, 1998.
140. J Boryczka. http://www.cyf-kr.educ.pl/~ziniedzw/paper016.html, 1995.
141. J Carlisle. Environment 40. Nat Ctr Pub Pol Res, 1999.
142. BI Sazonov. http://www.cyf-kr.edu.pl/~ziniedzw/paper087.html, 1995.
143. KW Butzer. Encyc Britannica 4:730–741, 1974.
144. I Charvátová. http://www.cyf-kr.edu.pl/~ziniedzw/paper023.html, 1995.
145. J Strestik. http://www.cyf-kr.cdu.pl/~ziniedzw/paper023.html, 1995.
146. Anonymous. Geographica Natl Geogr Sept. 1999.
147. Anonymous. Encyc Britannica VIII:183–184, 1974.
148. A McClymont. http://www.edu.educ/gatherscatter/gswinter97/mcclymont.html, 1999.
149. G Withbroe. Cited by Boston Globe in Tampa Tribune, March 10, 1999.
150. Anonymous. Encyc Britannica IX:678, 1974.
151. RA Kerr. Science 284:2069, 1999.
152. V Cermák. http://www.cyf-kr.edu.pl/~ziniedzw/paper025.html, 1995.
153. MR Frogley, PC Tzedakis, TME Heaton. Science 285:1886–1889, 1999.
154. F Wagner, SJP Bohncke, DL Dilcher, WM Kürschner, B van Geel, H Vischer. Science 284:1971–1973, 1999.
155. RS Boyd. Knight Ridder Nwsps in Tampa Tribune, Aug. 17, 1999.
156. H Weiss, M-A Courty, W Wetterstrom, F Guichard, L Senior, R Meadow, A Curnow. Science 261:995–1004, 1993.
157. DT Avery. Reader's Digest 45–56, Aug. 1999.
158. SA Changnon. Cited by Assoc. Press in Tampa Tribune, Sept. 2, 1999.
159. Anonymous. Encyl. Britannica VII:584–585, 1974.
160. R Monastersky. Sci News 155:188–189, 1999.
161. F Mowat. West Viking, 319–321, McClelland and Steward, Canada, 1965.
162. F Pearce, New Scientist 164:37–38, 1999.
163. D Rodbell. Science 283:576, 1999.
164. D Carlson. Natl Geogr Sept: 82–89, 1999.

Internet citations courtesy of the late George Edwards.

# 3

# Crop Plant Responses to Rising CO$_2$ and Climate Change

**Joseph C. V. Vu and Leon Hartwell Allen, Jr.**

*U.S. Department of Agriculture–Agricultural Research Service, and University of Florida, Gainesville, Florida*

**Maria Gallo-Meagher**

*University of Florida, Gainesville, Florida*

## I. INTRODUCTION

The earth's atmospheric carbon dioxide concentration ([CO$_2$]) has fluctuated between 170 and 300 ppm over the past 160,000 years. However, since the start of the industrial revolution in Western Europe (1750–1800), atmospheric [CO$_2$] has increased from 280 to approximately 365 ppm at present [1,2]. The future [CO$_2$] depends on the degree to which CO$_2$ emissions are controlled. However, with the rapid increase in world population and economic activity, a doubling of the present atmospheric [CO$_2$], assuming a mean annual increase rate of 1.5 ppm, which was observed over the decade 1984–1993 [2], could be expected before the end of the 21st century [1,3,4]. A rise in atmospheric [CO$_2$] may have important effects on global climate. As CO$_2$ is responsible for 61% of global warming [5], a doubling of the atmospheric [CO$_2$] and a rise in other so-called greenhouse gases (methane, nitrous oxide, chlorofluorocarbons) would increase the mean global temperature, possibly as much as 4.5 to 6°C [6,7]. In addition, shifts in regional precipitation patterns as a result of rising atmospheric [CO$_2$] will probably result in decreased soil water availability in many areas of the world [3,8–11].

Atmospheric CO$_2$ is an essential compound for life on earth. Through photosynthesis plants obtain carbon for their growth and provide sustenance for other living things, ourselves included. In photosynthesis, solar energy is absorbed by a system of pigments, and inorganic atmospheric CO$_2$ is fixed and reduced into organic compounds. Reduction of carbon is a major function of photosynthesis and is quantified by realizing that total plant organic matter is about 45% carbon on a dry weight basis. The biochemistry of carbon reduction has attracted much research attention since the early 1950s, leading to recognition of different biochemical pathways for net carbon flow during plant photosynthesis. Humankind, however, has not devised ways to manipulate or control this process because many foundations of photosynthesis and knowledge of its regulatory mechanisms under environmental change are still not fully understood [12,13]. Rising atmospheric [CO$_2$] could benefit many economically important crops, especially the C$_3$; however, gains may or may not be realized in long-term growth because of the interaction of various environmental factors that complicate the issue [11,14].

This chapter focuses on the photosynthetic responses of crop plants to long-term elevated growth [CO$_2$]. The physiological, biochemical, and molecular aspects of photosynthetic acclimation to rising atmospheric [CO$_2$] and interactive effects of elevated [CO$_2$] with anticipated simultaneous increases in air temperature and/or decreases in soil water availability on leaf photosynthesis will be discussed. As the photosynthetic mechanism of a plant species is the major determinant of how it will respond to rising atmospheric [CO$_2$] [14], understanding the mechanisms of photosynthesis acclimation to rising [CO$_2$] and other environmental stresses could potentially be translated into a basic framework for improving the efficiency of crop production in a future climate-changed world.

## II.  THE MAJOR PATHWAYS OF PHOTOSYNTHESIS

Present understanding of photosynthetic carbon metabolism classifies terrestrial plants into three major photosynthetic categories: C$_3$, C$_4$, and Crassulacean acid metabolism (CAM). Each category possesses a unique set of anatomical, physiological, and biochemical features that allows them to adapt to a specific ecological niche [15]. It is estimated that approximately 95% of terrestrial plant species fix atmospheric CO$_2$ by the C$_3$ (i.e., photosynthetic carbon reduction, or PCR) pathway, while 1% fix CO$_2$ by the C$_4$ pathway and 4% by CAM [14].

### A.  The C$_3$ (Calvin) Cycle

In mesophyll cells of C$_3$ plants, CO$_2$ binding to its primary acceptor, ribulose-1,5-bisphosphate (RuBP), is catalyzed by RuBP carboxylase/oxygenase (Rubisco), and the product of this carboxylation process, 3-phosphoglycerate (PGA), is converted to other carbohydrates. In addition to the usual carboxylation reaction, Rubisco catalyzes an oxygenase reaction in which O$_2$ reacts with RuBP to give PGA and phosphoglycolate, a process known as photorespiration [16]. The oxygenase reaction and associated metabolism have an adverse effect on the efficiency of photosynthesis in C$_3$ plants, which results in a loss of CO$_2$, energy, and reducing potential [17]. The balance between carboxylation and oxygenation of RuBP depends on the relative concentrations of CO$_2$ and O$_2$ at the site of Rubisco in the mesophyll chloroplasts. A higher atmospheric [CO$_2$] will reduce photorespiration and therefore increase the leaf CO$_2$ exchange rate (CER) of C$_3$ plants (Figure 1).

### B.  The C$_4$ Pathway of CO$_2$ Fixation

C$_4$ plants have developed a biochemical mechanism to overcome the limitations of low atmospheric [CO$_2$] and photorespiration [15,18–20]. In C$_4$ plants, atmospheric CO$_2$ is first hydrated to bicarbonate by carbonic anhydrase in the cytosol of mesophyll cells; subsequently, it reacts with the three-carbon phosphoenolpyruvate (PEP) to give the C$_4$ acid oxaloacetate (OAA) in a reaction catalyzed by PEP carboxylase (PEPC). OAA is rapidly converted to malate in the mesophyll chloroplasts by NADP–malate dehydrogenase (NADP-MDH) or transaminated to aspartate in the mesophyll cytosol by aspartate

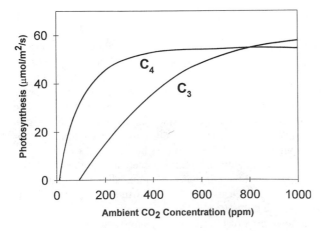

**Figure 1**  Photosynthesis of typical C$_3$ and C$_4$ plants versus ambient CO$_2$ concentration. Relative to C$_3$ plants, C$_4$ plants have a low CO$_2$ compensation point (the intercept on the abscissa), a high carboxylation efficiency (the initial slope of CO$_2$-response curve), and a near-saturation photosynthetic rate at current atmospheric [CO$_2$]. (Adapted from Refs. 15 and 31.)

aminotransferase, depending on the C$_4$ acid-decarboxylating mechanism of the C$_4$ plant [21]. These C$_4$ compounds are then transported to the bundle sheath cells, where they are decarboxylated to release CO$_2$ by one of the three C$_4$ acid-decarboxylating enzymes: NADP-malic enzyme (NADP-ME), NAD–malic enzyme (NAD-ME), or PEP carboxykinase [18,19,22].

In the NADP-ME species, which contain crops of global importance including sugarcane, maize, and sorghum, OAA is reduced in the mesophyll chloroplasts via NADP-MDH to malate, which is then transferred to the adjacent bundle sheath cells. In the bundle sheath chloroplasts, malate undergoes decarboxylation catalyzed via NADP-ME to produce CO$_2$ which is reassimilated by Rubisco in the conventional Calvin C$_3$ (PCR) cycle. In C$_4$ species in which NAD-ME is the major C$_4$ acid-decarboxylating enzyme (e.g., *Atriplex spongiosa, Portulaca oleracea, Amaranthus edulis*), aspartate from the mesophyll cells enters the bundle sheath mitochondria, where it is converted to OAA. OAA is then reduced to malate, which, in turn, is decarboxylated via NAD-ME, generating CO$_2$ to be assimilated by the PCR cycle. In species in which PEP carboxykinase is the primary decarboxylating enzyme (e.g., *Panicum maximum, Chloris gayana, Sporobolus fimbriatus*), aspartate from the mesophyll cells is converted to OAA in the bundle sheath cytosol, and OAA is subsequently decarboxylated producing CO$_2$, which is then assimilated by the PCR cycle.

Thus, the reactions that are unique to C$_4$ photosynthesis can be considered as an additional step to the conventional C$_3$ pathway. They operate to transfer CO$_2$ from mesophyll to bundle sheath cells through the intermediary of a dicarboxylic acid and consequently increase levels of CO$_2$ in bundle sheath cells specifically for refixation via Rubisco in the C$_3$ cycle [19]. Through this additional metabolic pathway, C$_4$ plants are able to concentrate CO$_2$ in the Rubisco-containing bundle sheath cells to levels up to 3 to 20 times higher than atmospheric [CO$_2$] [19,23–25]. Photosynthesis by C$_4$ plants is therefore near saturation at current atmospheric [CO$_2$], and a rise in atmospheric [CO$_2$] presumably may have little or no effect on C$_4$ photosynthesis (Figure 1).

## C. The CAM Pathway

CAM (Crassulacean acid metabolism) is a photosynthetic process, named after the family Crassulaceae, in which the accumulation of malic acid in the dark, a distinctive property of the process, was first observed [26]. CAM plants are widely distributed in arid and semiarid regions, where their contribution to community biomass production is significant [26–28]. Although many plants that exhibit CAM belong to the dicotyledonous Crassulaceae family (*Kalanchoe* spp., *Sedum* spp.), this photosynthetic process is also widespread in plants of other dicotyledonous families (Aizoaceae, Asclepiadaceae, Bataceae, Cactaceae, Caryophyllaceae, Chenopodiaceae, Compositae, Convolvulaceae, Euphorbiaceae, Plantaginaceae, Portulacaceae, Vitaceae) as well as the monocotyledonous families (Agavaceae, Bromeliaceae, Liliaceae, Orchidaceae) and even the Pteridophyte family (Polypodiaceae) [18].

CAM plants normally close their stomata during the day to prevent water loss. At night, the stomata are open, and atmospheric CO$_2$ enters the cytoplasm of chloroplast-containing cells of photosynthetic leaf or stem tissues and combines with PEP, a product of glucan metabolism, via PEPC to form OAA [18,29]. OAA is subsequently reduced by NAD–malate dehydrogenase to malate, which then accumulates in large vacuoles that are characteristic of the cells of CAM plants. During the daylight hours, stomata become closed, and malate is transported back into the cytoplasm, where it is decarboxylated by an NADP–malic enzyme. The CO$_2$ just released enters the chloroplasts, where it is fixed by Rubisco of the conventional C$_3$ cycle. Although CAM plants and C$_4$ plants share the two major carboxylating enzymes PEPC and Rubisco, the carbon reduction catalyzed by these enzymes differs temporally and spatially, respectively, for these two photosynthetic categories [26,28]. Furthermore, the $K_m$ (PEP) value of PEPC from CAM plants is less than one third that of C$_4$ plants [26]. Thus, the effects of elevated atmospheric [CO$_2$] on the uptake of CO$_2$ by CAM plants can be different than for C$_4$ plants [30].

## III. RISING ATMOSPHERIC CO$_2$ AND ITS INTERACTIONS WITH OTHER ENVIRONMENTAL VARIABLES

## A. Plant Responses to Rising CO$_2$

Research during the past 20 years on growth, as well as mechanisms and acclimation (down-regulation or up-regulation) in photosynthetic processes, as a result of long-term exposure to elevated [CO$_2$], has focused mainly on $C_3$ species. For $C_4$ and CAM plants, the mechanisms and the nature of interactive effects of elevated $[CO_2]$ and other adverse environmental conditions on growth and yield, and their fundamental physiology, biochemistry, and/or molecular biology, are still not well understood.

## 1. *$C_3$ Species*

The present atmospheric $[CO_2]$ limits the photosynthetic capability, growth, and yield of many agricultural crop plants, among which the $C_3$ species show the greatest potential for response to rising $[CO_2]$ [11,14,31–33]. Current atmospheric $CO_2$ and $O_2$ levels and $C_3$ Rubisco specificity factors translate into photorespiratory losses of 25% or more for $C_3$ species [14,34]. The projection that a rise in atmospheric $[CO_2]$ will reduce the deleterious effect of $O_2$ on $C_3$ photosynthesis but that it has a negligible effect on $C_4$ photosynthesis is indeed supported by experimental growth data. Exposure of $C_3$ plants to elevated $[CO_2]$ generally results in stimulated photosynthesis (Figure 1) and enhanced growth and yield [31–33,35]. A compilation of the existing data available from the literature for $C_3$ agricultural crops, including agronomic, horticultural, and forest tree species, shows an average enhancement in net $CO_2$ exchange rates up to 63% and growth up to 58% with a doubling of the present atmospheric $[CO_2]$ [31,32,36–38].

Long-term exposure to elevated $[CO_2]$ leads to a variety of acclimation effects, which include changes in the photosynthetic biochemistry and stomatal physiology and alterations in the morphology, anatomy, branching, tillering, biomass, and timing of developmental events as well as life cycle completion [14,33,39,40]. A greater number of mesophyll cells and chloroplasts have been reported for plants grown under elevated $[CO_2]$ [41,42]. With respect to leaf photosynthetic physiology and biochemistry, acclimation occurs, ranging from species-specific changes in the $A/C_i$ (assimilation rate versus intercellular $CO_2$) curves [43–45] to alterations in dark respiration [33] and biochemical components with Rubisco playing the leading role [46]. In terms of dark respiration, exposure of plants to elevated $[CO_2]$ usually results in lowering the dark respiration rate, which can be explained by both indirect and direct effects [33]. Whereas the mechanism for the indirect (acclimation) effect of elevated $[CO_2]$ on dark respiration may be related to changes in tissue composition, the direct effect appears to be an inhibition of the enzymes in the mitochondrial electron transport system [47,48].

Many $C_3$ species grown for long periods at elevated $[CO_2]$ show a down-regulation of leaf photosynthesis [45,49,50], and carbohydrate source-sink balance is believed to have a major role in the regulation of photosynthesis through feedback inhibition [51,52]. Source-sink imbalances may occur during exposure to elevated $[CO_2]$ when photosynthetic rate exceeds the export capacity or the capacity of sinks to use photosynthates for growth, resulting in an accumulation of carbohydrates in photosynthetically active source leaves [52–54]. Under elevated growth $CO_2$, although the extent to which starch and soluble sugars accumulate largely depends on the species, the increase of starch seems to be greater than that of soluble sugars in many plants, and a correlation between starch accumulation and inhibition of leaf photosynthesis has been more frequently observed [54]. Also, for many plant species, longer exposure to elevated $[CO_2]$ results in a down-regulation of Rubisco [33,44–46,55–66]. Both "coarse" control, through lowering of the enzyme protein content, and "fine" control, through decreasing the enzyme activation state, play a role in the down-regulation of Rubisco mediated by elevated $[CO_2]$. Coarse control suggests a reallocation of nitrogen resources away from Rubisco [14] as well as an optimization of $CO_2$ acquisition with utilization of the fixed carbon [67]. Down-regulation of Rubisco at elevated $[CO_2]$, however, is not a universal phenomenon, and claims of altering the enzyme activity need careful evaluation, as the basis on which Rubisco activity is expressed may vary or nullify the observation [14].

In addition to Rubisco, there are reports that elevated $[CO_2]$ affects the regulation of sucrose phosphate synthase (SPS) and acid invertase. In rice, leaf SPS activity, expressed on a leaf total soluble protein basis, is up-regulated in $CO_2$-enriched plants, suggesting an acclimation response to optimize the capacity for carbon utilization and export for this crop species [68]. On the other hand, activities of SPS, expressed on a leaf fresh weight basis, are down-regulated by high $[CO_2]$ in bean, cotton, cucumber, plantain, and wheat but up-regulated in pea, soybean, spinach, sunflower, and tomato [69]. Under elevated growth $[CO_2]$, leaf acid invertase activities are down-regulated in cotton, cucumber, parsley, pea, radish, soybean, spinach, tobacco, and wheat but up-regulated in bean, plantain, and sunflower [69].

Levels of soluble sugars in plant cells have been shown to influence the regulation of expression of several genes coding for key photosynthetic enzymes [70–75]. The buildup in carbohydrates may signal the repression, but does not directly inhibit the expression, of Rubisco and other proteins that are required

**TABLE 1** Effect of Long-Term Growth of Plants at Elevated [$CO_2$] on *rbcS* Transcript Abundance

| Species | *rbcS* mRNA (% of ambient $CO_2$)[a] | Reference |
|---|---|---|
| *Arabidopsis* | 40 | [69] |
| Bean | 85 | [69] |
| Cotton | 54 | [69] |
| Maize | 152 | [69] |
| Parsley | 60 | [69] |
| Pea | 45–110 | [69, 88] |
| Plantain | 125 | [69] |
| Radish | 83 | [69] |
| Rice | 83–94 | [90, 91] |
| Soybean | 73–87 | [69, 92] |
| Spinach | 135 | [69] |
| Sunflower | 69 | [69] |
| Tobacco | 92 | [69] |
| Tomato | 40–81 | [69, 86] |
| Wheat | 50–61 | [61, 89] |

[a] The percentage is expressed relative to the corresponding value for plants grown at ambient [$CO_2$].

for photosynthesis [52,54,70,71,76]. Although the signal transduction pathway for regulation of the sugar-sensing genes may involve phosphorylation of hexoses, derived from sucrose hydrolysis by acid invertase, via hexokinase [73–75,77–84], unknown gaps still exist between hexose metabolism and repression of gene expression at elevated growth [$CO_2$] [54,83]. However, future molecular genetic studies of *Arabidopsis* mutants with altered sensitivity to sugars may aid in elucidating steps along this signaling pathway [75].

Transcription of the Rubisco small subunit (*rbcS*), and to a lesser extent the large subunit (*rbcL*), appears to be strongly repressed by sucrose and glucose [85,86]. Reduced expression of Rubisco genes and differential response of other photosynthetic genes have been reported for a variety of crops grown at elevated [$CO_2$] [61,69,81,83,87–92]. Table 1 shows the influence of long-term growth at elevated [$CO_2$] on *rbcS* transcript levels for various crop plants. For many species, the expression level of *rbcS* transcripts, however, does not always correlate with the Rubisco protein content at elevated growth [$CO_2$] [69].

In tomato, transcript levels for Rubisco subunits, chlorophyll *a/b* binding protein (*Cab*), and Rubisco activase (*Rca*) decline with $CO_2$ enrichment, whereas those for core proteins in photosystems I and II remain unchanged [87,93]. In wheat, transcripts for Rubisco subunits and phosphoglycerate kinase of the flag leaves are sensitive to elevated $CO_2$, whereas those for sedoheptulose-1,7-bishosphatase and phosphoribulokinase are insensitive [89]. For tomato, despite a large accumulation of starch occurring in leaves of elevated $CO_2$–grown plants, transcript levels for ADP glucose pyrophosphorylase show little change [87]. Furthermore, although photorespiration decreases under elevated [$CO_2$] [52], responses of the enzymes and/or transcripts associated with the photorespiratory pathway have not been well investigated [83]. Elevated $CO_2$ has little effect on the transcript level of glycolate oxidase in tomato [87] but inhibits the accumulation of hydroxypyruvate reductase messenger RNA (mRNA) in cucumber [94].

## 2. C₄ Species

Although C₄ plants represent only 1% of the total plant species [14], their economic and ecological significance is substantial [95]. Over 100 genera contain plants that utilize the C₄ pathway, and about 21% of gross primary productivity (i.e., annual net $CO_2$ assimilation per unit ground area) is provided by C₄ plants on a global basis [96,97]. In many tropical regions, the food supply is primarily based on C₄ plants, including grasses providing grains for many tropical diets and pastures and rangelands supplying forage for livestock [98]. Sugarcane, maize, sorghum, millet, and amaranth are the most widely grown C₄ crops. On a land area basis, maize, millet, and sorghum account for 70, 55, and 46% of the cereals grown in Africa, South America, and North America, respectively [98]. C₄ plants dominate all tropical and subtropical grasslands, many temperate grasslands, and most disturbed landscapes in warm regions, and 8 of the world's 10 most invasive weeds are $C_4$ species [95,99,100].

In $C_4$ species, the presence of a $CO_2$-concentrating mechanism has led to a general assumption that there would be little or no increase in photosynthesis and growth with rising atmospheric $[CO_2]$. However, examination of the literature reveals a positive growth response to a doubling of $[CO_2]$ for a number of $C_4$ plants, although to a smaller extent (~14%) than for $C_3$ plants (40–58%) [31,32,36,37,101–105]. Therefore, it cannot be assumed that $C_4$ species will not respond to rising atmospheric $CO_2$. In spite of the growth stimulation, these $C_4$ plants often show little or no enhancement in leaf CER at elevated $[CO_2]$, which is in contrast to $C_3$ species [31,37,62,105–107]. Chamber-grown $C_4$ maize and sugarcane showed no differences in leaf photosynthetic rates between the ambient $CO_2$– and double-ambient $CO_2$–grown plants, although leaf area and total plant biomass of the $CO_2$-enriched plants increased 14% [104]. Also, maize plants grown in controlled chambers under a triple-ambient $CO_2$ atmosphere (1100 ppm) were only 10% higher in light-saturated rates of photosynthesis for mature leaves but 20 to 23% higher in total biomass and leaf area [108]. In a study conducted in naturally sunlit temperature-gradient greenhouses to investigate the effects of elevated $CO_2$ and high temperatures on growth and photosynthesis of sugarcane (cv. CP 73-1547), $[CO_2]$ at 700 ppm increased leaf area by 31% (Figure 2A), total aboveground dry weight by 21% (Figure 2B), and main stem juice volume by 83% (Figure 2C) when compared with plants grown at 360 ppm $[CO_2]$ [109,110]. Furthermore, growth of sugarcane plants under both elevated $CO_2$ (700 ppm) and temperature (4.5°C above baseline temperature control, which was 2°C above outside ambient) increased leaf area by 56%, total aboveground dry weight by 74%, and juice volume by 164% (Figure 2A, B, C). These increases occurred without an enhancement of leaf CER, measured at the growth $[CO_2]$ for the most expanded sections of the uppermost, fully expanded leaves (Figure 2D).

Causes of the observed growth stimulation by elevated $CO_2$ on $C_4$ plants remain uncertain, but factors that indirectly impinge on Rubisco may be involved in this enhanced growth [32,46]. First, a reduction in stomatal aperture and conductance is a common response to a doubling of atmospheric growth $[CO_2]$. This decrease occurs across a variety of $C_3$ and $C_4$ species, although there are cases of insensitive stomatal responses [14,33,46]. The reduction in stomatal aperture and conductance explains the reduction in transpiration observed in plants grown under elevated $[CO_2]$. This results in an improved water use efficiency (WUE) and tissue water status and a potentially increased growth and/or yield with no additional penalty in water consumption [33,111]. Under water-shortage conditions, an improvement in WUE induced by elevated $CO_2$ could delay soil drying and reduce drought inhibition of $C_4$ vegetation and thus enhance growth, and this has also been suggested as a factor in the improved photosynthesis and increased biomass of some $C_4$ species [112–115]. Second, adverse growth conditions such as low nitrogen, high salinity, or limited soil water availability may undermine the effectiveness of the $CO_2$-concentrating mechanism by increasing $CO_2$ leakage from the bundle sheath, thus making $C_4$ species more responsive to elevated atmospheric $CO_2$ [116,117]. Even a small, but consistent, percent stimulation in the $CO_2$ assimilation rate throughout the growth season could account for the growth enhancement seen in the $C_4$ species [37,62]. Third, elevated growth $[CO_2]$ can enhance tillering and leaf area, so that total plant photosynthesis is greater, even without an increase in $CO_2$ uptake rate per unit leaf area [14,39,46,118]. In addition, changes in dark respiration and photosynthate partitioning, which are still poorly understood for $CO_2$-enriched $C_4$ species, may explain part of the enhanced growth [37]. In maize, the increased capacity to synthesize and utilize sucrose and starch to produce extra energy by respiration could contribute to plant biomass enhancement under elevated growth $[CO_2]$ [108].

### 3. CAM Species

The response of CAM plants to elevated atmospheric $[CO_2]$ is less clear because studies examining the $CO_2$-enrichment responses of CAM plants are limited, with varying results being reported. Presumably, minimal response may be expected for plants that are capable of raising their daytime internal $CO_2$ levels as high as 10,000 ppm through decarboxylation of the $C_4$ malic acid accumulated during the previous evening period; however such a presumption is only partially corroborated [14,26]. Under a doubling of atmospheric $[CO_2]$, there was no enhancement in leaf CER and leaf area or total plant biomass for pineapple, an economically important CAM species, but these parameters were 20 to 44% higher for *Aechmea magdalanae* [119]. Plants of *Agave vilmoriniana* responded positively to $CO_2$ enrichment only when water supply during growth was limited [120,121]. Elevated $CO_2$ did not enhance either

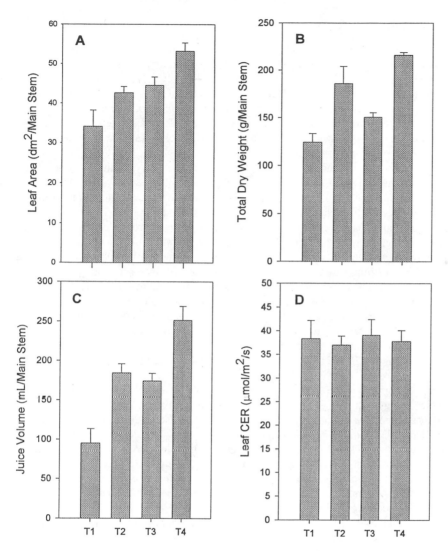

**Figure 2**  Leaf area (A), aboveground dry weight (B), juice volume (C), and leaf photosynthesis (D) of sugarcane (cv. CP 73-1547) grown in sunlit temperature-gradient greenhouses under [CO₂] of 360 and 700 ppm and temperatures at 2 and 6.5°C above outdoor ambient temperature (Ta). The four treatments were T1 = 360 ppm CO₂/Ta + 2°C; T2 = 360 ppm CO₂/Ta + 6.5°C; T3 = 700 ppm CO₂/Ta + 2°C; T4 = 700 ppm CO₂/Ta + 6.5°C.

plant biomass accumulation over several months of treatment [120] or leaf $CO_2$ assimilation rates monitored over the entire 24-hr diel period when plants were well irrigated [121]. In contrast, phase IV $CO_2$ assimilation that commences in the late daylight period and overall plant growth rates were consistently higher in the $CO_2$-enriched, water-limited plants [120,121]. Plants of *Kalanchoe* also did not show enhanced rates of $CO_2$ uptake under supranormal growth [CO₂] [122,123]. For *Agave deserti* and *Ferocactus acanthodes*, when growth [CO₂] was raised 300 ppm above ambient, short-term net $CO_2$ uptake over 24 hr and long-term dry biomass gain over 1 year were enhanced about 30% for both plants [124]. In *Opuntia ficus-indica*, long-term $CO_2$ enrichment increased net $CO_2$ uptake, water use efficiency, root growth, stem thickness, and biomass production but decreased activities of PEPC and Rubisco [30,125]. Doubling the ambient [CO₂] also increased levels of glucose, starch, and nocturnal malate production and enhanced activities of sucrose synthase and soluble starch synthase for this perennial CAM species [126].

## B. Rising $CO_2$ and Climate Warming

There have been a number of reviews regarding the effects of temperature on leaf photosynthesis [127–129] and the effects of interactions of rising atmospheric [$CO_2$] and temperature on growth, function, and development in $C_3$ plants [4,46,130,131]. Photosynthesis of $C_3$ plants, in addition to $CO_2$, is influenced by high growth temperature regimes, and Rubisco plays a central role in these responses [130]. Unfortunately, there is little experimental information on possible mechanisms of Rubisco regulation under interacting $CO_2$-temperature growth conditions [4]. Temperature and $CO_2$ have interactive effects because a rise in temperature reduces the activation state of Rubisco [64,132,133] (also see Table 2) and decreases both the specificity for $CO_2$ and the solubility of $CO_2$, relative to $O_2$ [130,134,135], resulting in increased photorespiratory $CO_2$ losses as the temperature rises. Consequently, a doubling of atmospheric [$CO_2$] and the concomitant inhibition of the Rubisco oxygenase reaction could partially offset the adverse effects of increased global temperature on $C_3$ photosynthesis [130]. However, the data in this regard are equivocal [53], and species-specific differences may be partially accounted for the differing results. In addition, these photosynthetic gains may or may not be realized in long-term growth and yield because growth and reproduction reflect the integrated temperature response of metabolism and developmental processes, not just photosynthesis [46]. In soybean, the enhancement effect on leaf photosynthetic rate due to doubling the growth [$CO_2$] increased linearly from 32 to 95% with increasing day temperatures from 28 to 40°C, whereas with rice it was relatively constant at 60% from 32 to 38°C [64]. In addition, although both elevated [$CO_2$] and temperature reduced Rubisco protein and activity, the reduction by either factor was greater for rice than for soybean [64]. Even within the same species, however, plant biomass and grain yield respond differently to increasing growth temperature. In the case of rice, plants grown at 34°C accumulated biomass and leaf area faster than plants at 28°C, but grain yield declined by about 10% for each 1°C rise above 26°C [136–138]. Similar scenarios have been reported for soybean [139] and wheat [140].

In citrus, the net CER measured at the [$CO_2$] used for growth is substantially enhanced by elevated [$CO_2$] [141–144]. At elevated growth [$CO_2$], the inhibitory effects of high leaf-to-air vapor pressure difference and decreased available soil water on citrus CER are lessened, and the $CO_2$ assimilation rate does not exhibit the midday depression commonly observed in trees grown under ambient [$CO_2$] [144,145]. In addition, elevated [$CO_2$] can compensate for the adverse effects of high temperature relative to the net photosynthetic rate [142,146], as seen in other crops [64]. In sour orange grown in Phoenix, Arizona, the mean daily leaf CER under summer conditions was about twofold greater for the elevated (700 ppm) $CO_2$ treatment in comparison with the control at 400 ppm $CO_2$ [142]. $CO_2$ enrichment enhanced sour orange leaf CER by 75% at a leaf temperature of 31°C, 100% at 35°C, and 200% at 42°C [146]. These degrees of enhancement are in the range of the predictions for an idealized $C_3$ plant, showing that a rise in temperature from 28 to 40°C increases enhancements in CER from 66 to 190% when atmospheric [$CO_2$] is raised from 350 to 650 ppm [130]. This is substantially greater than the 32–95% enhancement found with

**TABLE 2** Activation of Rubisco Extracted from Leaves of Soybean Plants Grown at 350 and 700 ppm $CO_2$ and Under Varying Day/Night Maximum/Minimum Air Temperature Regimes[a]

| Temperature regime (°C) | [$CO_2$] (ppm) | Degree of activation (%) | |
|---|---|---|---|
| | | Midday | Predawn |
| 28/18 | 350 | 66.3 ± 4.7 | 31.4 ± 2.2 |
| | 700 | 66.9 ± 4.2 | 39.0 ± 3.1 |
| 32/22 | 700 | 67.3 ± 2.8 | 31.9 ± 1.3 |
| 36/26 | 700 | 64.7 ± 3.9 | 21.9 ± 1.3 |
| 40/30 | 350 | 63.3 ± 4.1 | 19.3 ± 1.7 |
| | 700 | 65.2 ± 1.1 | 20.1 ± 0.7 |
| 44/34 | 700 | 52.7 ± 3.4 | 20.4 ± 1.3 |
| 48/38 | 700 | 41.5 ± 2.9 | 17.2 ± 1.2 |

[a] Uppermost, fully expanded leaves were sampled at predawn and midday, 48 days after planting. Activation is computed as the ratio of the initial to the corresponding total activity of midday-sampled leaves. Values are the mean ± standard error.

soybean when the growth $[CO_2]$ was raised to 700 ppm over the same temperature range [64]. The difference may be partially attributed to the fact that the temperature optimum of 32°C for soybean under ambient $[CO_2]$ is 7°C higher than that of the model (idealized) $C_3$ plant [64]. At an afternoon leaf temperature of 46°C, leaf CER of sour orange trees grown at ambient $[CO_2]$ declined to near zero, whereas the $CO_2$-enriched trees still maintained their CER at ~4 $\mu$mol/m²/sec [146]. Theoretically, a 300-ppm increase in atmospheric $[CO_2]$ could raise the temperature optimum of light-saturated CER of $C_3$ plants by 5°C [130].

The interactive effects of elevated $[CO_2]$ and temperature for $C_4$ species are not well understood. As discussed earlier, because of their $CO_2$-concentrating capability, it has been generally considered that $C_4$ plants would show little $CO_2$ stimulation irrespective of temperature [46,147]. However, with reports showing stimulation of biomass [32,62,104,107,109,110], the response of $C_4$ plants to both $CO_2$ and temperature deserves more attention. For the $C_4$ sugarcane, the degrees of enhancement in plant growth parameters are much greater under long-term exposure to both elevated $CO_2$ and temperature than to elevated $CO_2$ alone (Figure 2A–C).

For $C_3$ and $C_4$ plants adapted to similar climates, leaves of $C_4$ plants generally have a higher temperature optimum for photosynthesis as well as a higher overall photosynthetic rate at the temperature optimum [11,148–150]. At the current atmospheric $[CO_2]$, CERs of $C_4$ plants tend to increase with temperature to a greater extent than those of $C_3$ plants. Elevated $[CO_2]$ increases the temperature optimum of $C_3$ plants, bringing it closer to that of $C_4$ photosynthesis [130]. Besides, factors such as light regime, soil moisture, nutrient status, and plant developmental stage all modify the interactive responses to elevated $CO_2$ and temperature [4,46,62,151–155].

## C. Rising CO₂ and Limited Soil Water Availability

As atmospheric $[CO_2]$ rises, potential shifts in regional scale precipitation patterns could result in increased drought conditions in many areas of the world. Responses of plants to rising $[CO_2]$ in water deficit situations have been reviewed [156]. Despite our understanding of the responses of leaf photosynthesis to elevated $[CO_2]$ as well as to soil water deficit, the interactions of $CO_2$ enrichment and drought stress are still uncertain [11]. In particular, much less is known about the effects of rising $[CO_2]$ on the fundamental regulatory aspects of leaf photosynthesis in major agricultural crop plants subjected to drought [14,46,157,158]. A reduction in stomatal conductance is a common response of plants to elevated growth $[CO_2]$. Observations of a variety of $C_3$ and $C_4$ species indicate that a doubling of atmospheric $[CO_2]$ can also double the instantaneous WUE [11,156,159]. As the $[CO_2]$ is increased, the improvements in WUE are the results of increased assimilation rate and decreased water loss, with the latter being more important under water deficit situations [46]. The increase in WUE as a result of elevated $[CO_2]$ is likely to be more important than the increase in net photosynthesis per se, and the same may be true for drought-stressed plants grown in a $CO_2$-enriched atmosphere [157].

As soil water becomes less available, the relative enhancement of photosynthesis and growth by elevated $[CO_2]$ tends to be greater, which can alleviate drought stress and delay its onset [39,40,156,160]. A delay in the adverse effects of water deficit on leaf and canopy photosynthesis by elevated $[CO_2]$ has been reported for a number of $C_3$ plants, including soybean [161,162], sweet potato [163], groundnut [164], and rice [165–167]. Studies conducted on a variety of plant species indicate that elevated $[CO_2]$ may actually prevent plants from succumbing to the rigors of environmental stresses and enable them to maintain essential growth processes [168]. Soybean plants grown under high $[CO_2]$ transpire less and conserve more soil moisture than plants grown at ambient $[CO_2]$ [161]. $CO_2$ enrichment also enhances water conservation and midday xylem water potentials in drought-stressed sweet potato plants [163] and leaf water potentials of soybean [169]. For groundnut, elevated growth $[CO_2]$ has a similar beneficial effect on plants subjected to severe drought stress [164]. In rice, elevated $[CO_2]$ delays the adverse effects of severe drought on *rbcS* transcript abundance and activities of Rubisco and permits photosynthesis to continue for an extra day during the drought-stress cycle [91,166,167].

There is also evidence indicating that, under water deficit conditions, $C_4$ growth can respond as strongly to elevated $CO_2$ as does that of $C_3$ species. In the tallgrass prairie ecosystems, $C_4$ species show increased productivity under elevated $CO_2$ in dry years but not in wet years [113]. In drying soil, growth of maize also responds strongly to $CO_2$ enrichment [170].

## D.  Rising $CO_2$ and Light Intensity

Measurements of $CO_2$ enrichment effects on photosynthesis have usually been carried out with relatively high irradiance. In nature, photosynthesis occurs in both high and low light environments, and light-limited photosynthesis can account for half of the total carbon gain [33]. Several studies show that $C_3$ photosynthesis is enhanced by elevated [$CO_2$] even under light-limited conditions [62,171–174], and the enhancement rises with temperature [33]. Photosynthesis versus solar irradiance response curves show that soybean leaves grown and measured at 660 ppm $CO_2$ have lower light compensation points, steeper initial slopes, higher apparent quantum yields, and greater CER at light saturation than those adapted to and measured at 330 ppm $CO_2$ [172]. Canopy photosynthesis of soybean increases linearly with increases in growth [$CO_2$] from 160 up to 990 ppm as a result of improvements in leaf area index, leaf photosynthesis, and quantum yield [173].

Most studies of leaf photosynthetic acclimation to elevated $CO_2$ have focused on the most recently fully expanded, sunlit leaves. This may not reliably predict acclimation of the whole canopy at one specific crop developmental stage, as a difference in acclimation could occur between the uppermost, fully developed sunlit leaves and the older, shaded leaves located deeper within the canopy [65,66,89]. Studies of photosynthetic acclimation in a vertical profile of leaves through canopies of wheat [65,89] and sunflower [66] show that even at stages of development at which elevated $CO_2$ did not affect the carboxylation capacity in the uppermost fully expanded leaves, there was a decrease in the lower shaded leaves.

In a crop canopy, photosynthesis is light limited for all of the day for the interior or lower canopy leaves [33]. For a citrus canopy, although the absolute benefits of elevated $CO_2$ are greatest at high light intensity, the relative benefits are more significant at low light levels [175]. The positive direct effect of elevated growth $CO_2$ on citrus photosynthesis more than compensates for the negative self-shading effect produced by the high $CO_2$-induced proliferation of leaf area [175].

It is expected that the interaction between different growth $CO_2$ concentrations and light intensities will be different for $C_3$ and $C_4$ plants, as the $C_4$ photosynthetic pathway requires more energy than the $C_3$ pathway [62]. This extra energy is associated with the regeneration of PEP by the $C_4$ pathway in the mesophyll cells [176]. Studies with various species of $C_4$ weed grasses at elevated [$CO_2$] showed that growth at low light did not increase the growth-enhancing effects of $CO_2$ enrichment, whereas elevated [$CO_2$] and high growth irradiance significantly enhanced their net photosynthesis and early growth [177]. Assimilation-irradiance relationships for plants grown at ambient and elevated [$CO_2$] indicate that $CO_2$-enriched plants had higher light saturation values and greater rates at high irradiance levels [177]. Studies of *Panicum* species with different photosynthetic pathways showed that twice-ambient growth [$CO_2$] enhanced biomass at both low and high irradiance regimes for the $C_3$ (*P. laxum*) grass but only at high light for the $C_4$ (*P. antidotale*) species [62]. The elevated $CO_2$–grown $C_3$ plants had more leaves, greater total leaf area, longer main stems, more nodes, and more tillers than the ambient $CO_2$–grown plants under both low and high light treatments. These enhancements in biomass were not seen for the elevated $CO_2$–grown $C_4$ species under low light regimes. Only under high light did elevated $CO_2$ enhance stem elongation and shoot biomass accumulation in the $C_4$ plants [62]. In addition, there was no significant difference in leaf photosynthetic rates, measured at respective growth [$CO_2$], between the ambient and elevated $CO_2$–grown plants for both *P. laxum* and *P. antidotale*, although small but nonsignificant enhancements by elevated growth $CO_2$ were observed for the low light–treated $C_3$ and the high light–treated $C_4$ plants [62].

## E.  Rising $CO_2$ and Nitrogen Nutrition

As $CO_2$ is just one of many inorganic substrates required by plants, long-term response of plant photosynthesis and growth to elevated [$CO_2$] also depends on the availability of mineral nutrients and the way in which plants utilize them [154]. Most studies of elevated $CO_2$ and nutrient interactive effects have focused on nitrogen (N), because it is a common limitation in many natural and agroecosystems [178]. There have been many reports on the interaction between N nutrition and the response of photosynthesis, metabolism, and growth to elevated $CO_2$ [154,155,179–187]. In plants grown under elevated [$CO_2$], the overall N concentration usually decreases [55,60,188–190]. This overall N decrease under elevated growth [$CO_2$] might reflect either a higher N use efficiency due to reallocation of proteins, an ontogenetic drift leading to accelerated senescence as a result of faster growth, or inadequate N fertilization, uptake, and/or assimilation [154].

Acclimation of photosynthesis to long-term elevated growth [$CO_2$], which includes shifts to a decrease in the carboxylation capacity; a decline in Rubisco activity, content, and transcript level; an accumulation of nonstructural carbohydrates, especially starch; and a decrease of the N content in the plant, is usually more marked when the supply of N to plants during growth is limited [51,55,60,61,154,155, 179,185,191,192]. With respect to plant growth, whereas elevated [$CO_2$] typically leads to a marked increase in biomass in well-fertilized plants [46], this response changes with inadequate N fertilization [154]. However, species-specific differences will be encountered under N-limiting growth conditions. Elevated [$CO_2$] does not significantly enhance biomass of tobacco [155], rice [185], soybean [193], and several woody species [194–196] when the N supply is limited. In other N-limited grown species, elevated [$CO_2$] still increases plant biomass to some extent, but the stimulation is much less than in well-fertilized plants [51,55,180,188,192,197–199]. Occasionally, large stimulation of growth by elevated [$CO_2$] under N-limited conditions has also been observed [117,200].

## IV.  RISING CO₂ AND PLANT/LEAF DEVELOPMENTAL STAGE

The effect of elevated [$CO_2$] on plant growth also depends on plant age [154]. Most studies of the acclimation response under a $CO_2$ enrichment growth regime have focused on mature, fully expanded leaves. However, there is strong evidence from the literature that there may be interactions between leaf ontogeny and the degree of the acclimation response to elevated $CO_2$ exposure [50,65,93,154,201–205]. Leaves of dicots, during their ontogeny, undergo two distinct photosynthetic phases: a phase of increasing assimilation rates, which is correlated with import of nutrients and leaf expansion, and a prolonged senescence phase of declining assimilation rates, with a transient peak of maximal assimilation rates between the two phases [206]. In tobacco, both ambient (at 350 ppm) and high (at 950 ppm) $CO_2$–grown plants exhibit this photosynthetic pattern during leaf ontogeny; however, high $CO_2$–grown plants have a temporal shift to an earlier transition from the first phase of increasing photosynthesis to the senescence phase of declining photosynthesis [204]. These changes in photosynthetic rates are controlled largely by Rubisco activity, and the high $CO_2$–grown leaves also enter the stage of photosynthetic decline several days before their ambient $CO_2$–grown counterparts [204]. Studies of the effects of elevated $CO_2$ on photosynthesis and Rubisco in tomato during leaf ontogeny also reveal similar observations [50]. In addition, studies of other $C_3$ annual species also show that long-term exposure to elevated [$CO_2$] leads to an enhancement of the growth rate in young plants but not in older plants [207–209]. Similarly, for trees, increases in biomass are mostly due to increased growth rates during the first year of elevated $CO_2$ exposure, and growth is enhanced less or not at all in the subsequent years [196,210,211]. Therefore, any consideration of elevated [$CO_2$] effect on plant growth and physiology must also address time-dependent changes in the growth rate of plants [154].

The expression of $C_4$ photosynthetic characteristics is controlled by factors such as leaf age and leaf position. In some $C_4$ species, the first leaves show the normal $C_3$ type of photosynthesis, and this may cause such species to be responsive to high $CO_2$, at least in the short term [212]. In *Portulaca oleracea*, an NADP-ME $C_4$ dicot, there is a shift in the route of $CO_2$ assimilation toward a limited, direct entry of $CO_2$ into the PCR cycle in senescent leaves [213]. In *Flaveria trinervia*, also a $C_4$ dicot of the NADP-ME subgroup, an estimated 10 to 12% of the $CO_2$ entered the PCR pathway directly in young expanding leaves. However, $CO_2$ is apparently fixed entirely through the $C_4$ pathway in mature expanded leaves, and this partitioning pattern is attributed to the bundle sheath compartment in young leaves, which have a relatively high conductance to $CO_2$ [214].

In maize, an NADP-ME type monocot, pulse-chase experiments with mature and senescent leaf tissues show that the predominant $C_4$ acids malate and aspartate differ between the two leaf ages [215]. After a 10-sec chase, aspartate is the predominant $C_4$ acid in the mature leaves and malate is the major $C_4$ acid in the senescent leaves. In addition, the activity of Rubisco during leaf ontogeny in maize parallels the development in activity of this enzyme in $C_3$ plants [215]. Furthermore, a high $CO_2$ compensation point (22–24 ppm) is found in senescent leaves of maize, in contrast to values of 0 to 10 ppm for most $C_4$ plants [216]. Also in maize, the [14]C-labeling patterns of photosynthetic products in different sections of a developing leaf suggest that there may be some direct entry of $CO_2$ into the PCR pathway in the young tissues of the basal section, whereas the $C_4$ pathway functions in the more differentiated tissues of the center and top sections [217]. In addition, the activities of Rubisco and PEPC in maize leaves are found to vary according to leaf position, with activity of PEPC less than that of Rubisco in the lower leaves, whereas the upper leaves exhibit high levels of PEPC [218]. Moreover, in maize, bundle sheath cell walls of young and senescent leaves have a relatively high conductance, leading to a low capacity for $CO_2$ concentration in these bundle sheath cells during photosynthesis [219]. In the uppermost fully expanded leaves of sugarcane, CER, stomatal conductance, and activities of both PEPC and Rubisco increase from the base to the tip of the leaf [220]. Analyses of a range of leaf developmental stages in maize also indicate that when leaf chlorophyll and Rubisco protein contents are below a critical level, i.e., 50% or less compared with those found in mature leaves, the degree of photorespiration could approach that of $C_3$ plants [219].

In $C_4$ monocots and dicots, the vascular system features a radial pattern structure (Kranz type) around which photosynthetic bundle sheath and mesophyll cells are arranged [23,221,222]. Such compartmentalization for metabolic cooperation between mesophyll and bundle sheath cells is essential for the $C_4$ pathway. In $C_4$ dicots, bundle sheath cells generally have centripetally arranged chloroplasts, whereas in monocots the arrangement of the chloroplasts varies with the $C_4$ acid-decarboxylating enzyme subtype: centrifugal for NADP-ME species, centripetal for NAD-ME species, and random for PEP carboxykinase species [23,34]. In maize and sugarcane, chloroplasts of the bundle sheath and mesophyll cells are morphologically similar early in development; i.e., both contain granal stacks [34,223,224]. However, subsequent dedifferentiation of bundle sheath cell chloroplasts results in the agranal bundle sheath chloroplasts as seen in the mature leaves [223,224]. With respect to leaf ontogeny, leaf shape results from distinct patterns of cell division and expansion in both shoot apical meristem and leaf primordium [34]. Leaves of monocots are derived from the outer two layers of the shoot apical meristem, whereas those of dicots are derived from the outer three layers of the shoot apical meristem. The shape of monocot leaves is generated through polarized patterns of cell division and expansion that maintain cells in files. In maize, cell divisions occur throughout the leaf and become restricted to the leaf base only after initiation of the ligule at the boundary of leaf blade and leaf sheath. Dicot leaves, which are generally less uniform in shape than monocot leaves, are generated through less polarized divisions [34].

The expression of $C_4$ genes does not occur until Kranz anatomy has been established, and exclusive use of the $C_3$ photosynthetic pathway may occur prior to the full differentiation of Kranz anatomy [34]. Therefore, one of the proposed explanations for the biomass enhancement observed in $C_4$ plants grown at elevated $CO_2$ is that the "immature" $C_4$ pathway in young $C_4$ leaves has $C_3$-like characteristics, and thus photosynthesis of these young leaves is responsive to increasing $CO_2$ above current ambient levels [32,62,177,212]. This hypothetical explanation, however, may be species specific, as one study argues against this possibility by showing that the gas exchange parameters in young leaves of *Panicum antidotale* ($C_4$, NADP-ME) and *Panicum coloratum* ($C_4$, NAD-ME) do not have $C_3$-like characteristics [225].

## V.  CONCLUSION

In the 21st century, world agriculture is confronted with unprecedented environmental challenges. Erosion of the protective ozone layer, increased ultraviolet B (UV-B) irradiation, desertification, damage to long-established ecologies, greenhouse effects of rising atmospheric [$CO_2$] and temperature, and shifts in regional scale rainfall patterns [12,226] are all environmental concerns that will affect global agriculture on a scale never before encountered. A change in global climate and a rapidly expanding world population accelerate the demand for food, energy, and fresh water and threaten the ability of the world to feed itself [13,227–229]. As a consequence, the need to enhance the production efficiency of agricultural crops and their tolerance to warmer, more arid environmental conditions will escalate as competition for arable land and fresh water increases. However, we do not know what the net consequences of plant responses to these environmental changes will be, simply because we do not understand enough about how plants grow and their interactions with the environment to predict the effects of such changes [12,69]. Therefore, producing crops under climate change conditions will be a growing challenge in world agriculture.

It has been well recognized that increasing crop yields require an increase in photosynthesis, and genetic manipulation of photosynthetic processes has been the primary focus for crop improvement [13,229–233]. Thorough knowledge of crop growth and development and plant interactions with the environment, as well as new approaches and ambitious strategies, such as "reengineering" photosynthesis [13,229,230,233,234], "remodeling" Rubisco for more effectiveness [231], or "supercharging" photosynthesis of $C_3$ crop plants with $C_4$ genes [235–237], may all be required to improve crop efficiency at turning atmospheric carbon into food and maintaining world food supplies and nutrition.

## ACKNOWLEDGMENTS

We acknowledge the partial support provided by the U.S. Department of Agriculture (USDA-CRS NRICGP grants 91-37100-6594 and 94-01541), the U.S. Department of Energy (Interagency Agreement DE-AI02-93ER61720), and the U.S. Environmental Protection Agency Carbon Dioxide Research Division (Interagency Agreement DW12934099). Florida Agricultural Experiment Station Journal Series No. R–07652.

## REFERENCES

1. RT Watson, H Rodhe, H Oeschger, U Siegenthaler. Greenhouse gases and aerosols. In: JT Houghton, GJ Jenkins, JJ Ephraums, eds. Climatic Change. The IPCC Scientific Assessment. Cambridge: Cambridge University Press, 1990, pp 1–40.
2. D Schimel, D Ives, I Enting, M Heimann, F Joos, D Raynaud, T Wigley. CO$_2$ and the carbon cycle. In: JT Houghton, LG Meira Filho, BA Callendar, N Harris, A Kattenberg, K Maskell, eds. Climate Change 1995. Cambridge: IPCC Cambridge University Press, 1996, pp 65–131.
3. CD Keeling, TP Whorf, M Wahlen, J van der Plicht. Interannual extremes in the rate of rise of atmospheric carbon dioxide since 1980. Nature 375:660–670, 1995.
4. JIL Morison, DW Lawlor. Interactions between increasing CO$_2$ concentration and temperature on plant growth. Plant Cell Environ 22:659–682, 1999.
5. KP Shine, RG Derwent, DJ Wuebbles, JJ Morcrette. Radiative forcing of climate. In: JT Houghton, GJ Jenkins, JJ Ephraums, eds. Climatic Change. The IPCC Scientific Assessment. Cambridge: Cambridge University Press, 1990, pp 41–68.
6. JFB Mitchell, S Manabe, V Meleshko, T Tokioka. Equilibrium climate change and its implications for the future. In: JT Houghton, GJ Jenkins, JJ Ephraums, eds. Climate Change. The IPCC Scientific Assessment. Cambridge: Cambridge University Press, 1990, pp 131–172.
7. A Kattenberg, F Giorgi, H Grassl, GA Meehl, JFB Mitchell, RJ Stouffer, T Tokioka, AJ Weaver, TML Wigley. Climate models—Projections of future climate. In: JT Houghton, LG Meira Filho, BA Callendar, N Harris, A Kattenberg, K Maskell, eds. Climate Change 1995. Cambridge: IPCC Cambridge University Press, 1996, pp 285–357.
8. CA Wilson, JFB Mitchell. A doubled CO$_2$ climate sensitivity experiment with a global climate model, including a simple ocean. J Geophys Res 92:13315–13343, 1987.
9. J Hansen, I Fung, A Lacis, S Lebedeff, D Rind, R Ruedy, G Russell, P Stone. Global climate changes as forecast by the GISS 3-D model. J Geophys Res 98:9341–9364, 1988.
10. TML Wigley, SCB Raper. Implications for climate and sea level of revised IPCC emissions scenarios. Nature 357:293–300, 1992.
11. LH Allen Jr. Carbon dioxide increase: Direct impacts on crops and indirect effect mediated through anticipated climatic changes. In: KJ Boote, JM Bennett, TR Sinclair, GM Paulsen, eds. Physiology and Determination of Crop Yield. Madison, WI: American Society of Agronomy, 1994, pp 425–459.
12. JE Varner. Foreword: 101 reasons to learn more plant biochemistry. Plant Cell 7:795–796, 1995.
13. P Horton. Prospects for crop improvement through the genetic manipulation of photosynthesis: Morphological and biochemical aspects of light capture. J Exp Bot 51:475–485, 2000.
14. G Bowes. Facing the inevitable: Plants and increasing atmospheric CO$_2$. Annu Rev Plant Physiol Plant Mol Biol 44:309–332, 1993.
15. MSB Ku, Y Kano-Murakami, M Matsuoka. Evolution and expression of C$_4$ photosynthesis genes. Plant Physiol 111:949–957, 1996.
16. WL Ogren, G Bowes. Ribulose diphosphate carboxylase regulates soybean photorespiration. Nature (New Biol) 230:159–160, 1971.
17. WL Ogren. Photorespiration pathways, regulation and modification. Annu Rev Plant Physiol 35:415–442, 1984.
18. TW Goodwin, EI Mercer. Introduction to Plant Biochemistry. 2nd ed. Oxford: Pergamon Press, 1990.
19. MD Hatch. Resolving C$_4$ photosynthesis: Trials, tribulations and other unpublished stories. Aust J Plant Physiol 24:413–422, 1997.
20. MD Hatch. C$_4$ photosynthesis: A historical overview. In: RF Sage, RK Monson, eds. C$_4$ Plant Biology. San Diego: Academic Press, 1999, pp 17–46.
21. MD Hatch. C$_4$ photosynthesis: A unique blend of modified biochemistry, anatomy and ultrastructure. Biochim Biophys Acta 895:81–106, 1987.
22. MD Hatch. C$_4$ photosynthesis: An unlikely process full of surprises. Plant Cell Physiol 33:333–342, 1992.
23. R Kanai, G Edwards. Biochemistry of C$_4$ photosynthesis. In: RF Sage, RK Monson, eds. The Biology of C$_4$ Photosynthesis. New York: Academic Press, 1999, pp 49–87.
24. S von Caemmerer, RT Furbank. Modeling C$_4$ photosynthesis. In: RF Sage, RK Monson, eds. The Biology of C$_4$ Photosynthesis. New York: Academic Press, 1999, pp 173–211.

25.  JP Maroco, MSB Ku, GE Edwards. Utilization of $O_2$ in the metabolic optimization of $C_4$ photosynthesis. Plant Cell Environ 23:115–121, 2000.
26.  CB Osmond, JAM Holtum. Crassulacean acid metabolism. In: MD Hatch, NK Boardman, eds. The Biochemistry of Plants. A Comprehensive Treatise. Vol 8. Photosynthesis. New York: Academic Press, 1981, pp 283–328.
27.  CB Osmond. Crassulacean acid metabolism: A curiosity in context. Annu Rev Plant Physiol 29:379–414, 1978.
28.  PS Nobel. Environmental Biology of Agaves and Cacti. New York: Cambridge University Press, 1988.
29.  M Kluge. The flow of carbon in Crassulacean acid metabolism (CAM). In: M Gibbs, E Latzo, eds. Photosynthesis II. Photosynthetic Carbon Metabolism and Related Processes. Berlin: Springer-Verlag, 1979, pp 113–125.
30.  M Cui, PM Miller, PS Nobel. $CO_2$ exchange and growth of the Crassulacean acid metabolism plant *Opuntia ficus-indica* under elevated $CO_2$ in open-top chambers. Plant Physiol 103:519–524, 1993.
31.  BA Kimball, JR Mauney, FS Nakayama, SB Idso. Effects of elevated $CO_2$ and climate variables on plants. J Soil Water Conserv 48:9–14, 1993.
32.  H Poorter, C Roumet, BD Campbell. Interspecific variation in the growth response of plants to elevated $CO_2$: A search for functional types. In: C Korner, FA Bazzaz, eds. Carbon Dioxide, Populations, and Communities. New York: Academic Press, 1996, pp 375–412.
33.  BG Drake, MA Gonzalez-Meler, SP Long. More efficient plants: A consequence of rising atmospheric $CO_2$? Annu Rev Plant Physiol Plant Mol Biol 48:609–639, 1997.
34.  T Nelson, JA Langdale. Developmental genetics of $C_4$ photosynthesis. Annu Rev Plant Physiol Plant Mol Biol 43:25–47, 1992.
35.  LH Allen Jr, JT Baker, KJ Boote. The $CO_2$ fertilization effect: Higher carbohydrate production and retention as biomass and seed yield. In: F Bazzaz, W Sombroek, eds. Global Climate Change and Agricultural Production: Direct and Indirect Effects of Changing Hydrological, Pedological and Plant Physiological Processes. Rome: FAO, and Chichester: John Wiley & Sons, 1996, pp 65–100.
36.  BA Kimball. Carbon dioxide and agricultural yield: An assemblage and analysis of 430 prior observations. Agron J 75:779–788, 1983.
37.  H Poorter. Interspecific variation in the growth response of plants to an elevated ambient $CO_2$ concentration. Vegetatio 104/105:77–97, 1993.
38.  RJ Norby, SD Wullschleger, CA Gunderson, DW Johnson, R Ceulemans. Tree responses to rising $CO_2$ in field experiments: Implications for the future forest. Plant Cell Environ 22:683–714, 1999.
39.  LH Allen Jr. Plant responses to rising carbon dioxide and potential interactions with air pollutants. J Environ Qual 19:15–34, 1990.
40.  FA Bazzaz. The response of natural ecosystems to the rising global $CO_2$ levels. Annu Rev Ecol Syst 21:167–196, 1990.
41.  JF Thomas, CN Harvey. Leaf anatomy of four species grown under continuous $CO_2$ enrichment. Bot Gaz 144:303–309, 1983.
42.  JCV Vu, LH Allen Jr, G Bowes. Leaf ultrastructure, carbohydrates and protein of soybeans grown under $CO_2$ enrichment. Environ Exp Bot 29:141–147, 1989.
43.  JW Radin, BA Kimball, DL Hendrix, JR Mauney. Photosynthesis of cotton plants exposed to elevated levels of carbon dioxide in the field. Photosynth Res 12:191–203, 1987.
44.  WJ Campbell, LH Allen Jr, G Bowes. Effects of $CO_2$ concentration on rubisco activity, amount, and photosynthesis in soybean leaves. Plant Physiol 88:1310–1316, 1988.
45.  RF Sage, TD Sharkey, JR Seemann. Acclimation of photosynthesis to elevated $CO_2$ in five $C_3$ species. Plant Physiol 89:590–596, 1989.
46.  G Bowes. Photosynthetic responses to changing atmospheric carbon dioxide concentration. In: NR Baker, ed. Photosynthesis and the Environment. Dordrecht: Kluwer Academic Publishers, 1996, pp 387–407.
47.  MA Gonzalez-Meler, M Ribas-Carbo, JN Siedow, BG Drake. The direct inhibition of plant mitochondrial respiration by elevated $CO_2$. Plant Physiol 112:1349–1355, 1997.
48.  BG Drake, J Azcon-Bieto, J Berry, J Bunce, P Dijkstra, J Farrar, RM Gifford, MA Gonzalez-Meler, G Koch, H Lambers, J Siedow, S Wullschleger. Does elevated atmospheric $CO_2$ concentration inhibit mitochondrial respiration in green plants? Plant Cell Environ 22:649–657, 1999.
49.  EH Delucia, TW Sasek, BR Strain. Photosynthetic inhibition after long-term exposure to elevated levels of atmospheric carbon dioxide. Photosynth Res 7:175–184, 1985.
50.  RT Besford, LJ Ludwig, AC Withers. The greenhouse effect: Acclimation of tomato plants growing in high $CO_2$, photosynthesis and ribulose-1,5-bisphosphate carboxylase protein. J Exp Bot 41:925–931, 1990.
51.  WJ Arp. Effects of source-sink relations on photosynthetic acclimation to elevated $CO_2$. Plant Cell Environ 14:869–875, 1991.
52.  M Stitt. Rising $CO_2$ levels and their potential significance for carbon flow in photosynthetic cells. Plant Cell Environ 14:741–762, 1991.
53.  JF Farrar, ML Williams. The effects of increased atmospheric carbon dioxide and temperature on carbon partitioning, source-sink relations and respiration. Plant Cell Environ 14:819–830, 1991.

54. A Makino, T Mae. Photosynthesis and plant growth at elevated levels of $CO_2$. Plant Cell Physiol 40:999–1006, 1999.
55. SC Wong. Elevated atmospheric partial pressure of $CO_2$ and plant growth. I. Interactions of nitrogen nutrition and photosynthetic capacity in $C_3$ and $C_4$ plants. Oecologia 44:68–74, 1979.
56. JCV Vu, LH Allen Jr, G Bowes. Effects of light and elevated atmospheric $CO_2$ on the ribulose bisphosphate carboxylase activity and ribulose bisphosphate level of soybean leaves. Plant Physiol 73:729–734, 1983.
57. MA Porter, B Grodzinski. Acclimation to high $CO_2$ in bean. Carbonic anhydrase and ribulose bisphosphate carboxylase. Plant Physiol 74:413–416, 1984.
58. AJ Rowland-Bamford, JT Baker, LH Allen Jr, G Bowes. Acclimation of rice to changing atmospheric carbon dioxide concentration. Plant Cell Environ 14:577–583, 1991.
59. FX Socias, H Medrano, TD Sharkey. Feedback limitation of photosynthesis of *Phaseolus vulgaris* L. grown in elevated $CO_2$. Plant Cell Environ 16:81–86, 1993.
60. DT Tissue, RB Thomas, BR Strain. Long-term effects of elevated $CO_2$ and nutrients on photosynthesis and rubisco in loblolly pine seedlings. Plant Cell Environ 16:859–865, 1993.
61. AN Webber, GY Nie, SP Long. Acclimation of photosynthetic proteins to rising atmospheric $CO_2$. Photosynth Res 39:413–425, 1994.
62. O Ghannoum, S von Caemmerer, EWR Barlow, JP Conroy. The effect of $CO_2$ enrichment and irradiance on the growth, morphology and gas exchange of a $C_3$ (*Panicum laxum*) and a $C_4$ (*Panicum antidotale*) grass. Aust J Plant Physiol 24:227–237, 1997.
63. H Nakano, A Makino, T Mae. The effect of elevated partial pressures of $CO_2$ on the relationship between photosynthetic capacity and N content in rice leaves. Plant Physiol 115:191–198, 1997.
64. JCV Vu, LH Allen Jr, KJ Boote, G Bowes. Effects of elevated $CO_2$ and temperature on photosynthesis and Rubisco in rice and soybean. Plant Cell Environ 20:68–76, 1997.
65. CP Osborne, J LaRoche, RL Garcia, BA Kimball, GW Wall, PJ Pinter Jr, RL LaMorte, GR Hendrey, SP Long. Does leaf position within a canopy affect acclimation of photosynthesis to elevated $CO_2$? Analysis of a wheat crop under free-air $CO_2$ enrichment. Plant Physiol 117:1037–1045, 1998.
66. DA Sims, W Cheng, Y Luo, JR Seemann. Photosynthetic acclimation to elevated $CO_2$ in a sunflower canopy. J Exp Bot 50:645–653, 1999.
67. IE Woodrow. Optimal acclimation of the $C_3$ photosynthetic system under enhanced $CO_2$. Photosynth Res 39:401–412, 1994.
68. MW Hussain, LH Allen Jr, G Bowes. Up-regulation of sucrose phosphate synthase in rice grown under elevated $CO_2$ and temperature. Photosynth Res 60:199–208, 1999.
69. BD Moore, SH Cheng, J Rice, J Seemann. Sucrose cycling, Rubisco expression and prediction of photosynthetic acclimation to elevated atmospheric $CO_2$. Plant Cell Environ 21:905–915, 1998.
70. J Sheen. Feedback control of gene expression. Photosynth Res 39:427–438, 1994.
71. JC Jang, J Sheen. Sugar sensing in higher plants. Plant Cell 6:1665–1679, 1994.
72. KE Koch. Carbohydrate-modulated gene expression in plants. Annu Rev Plant Physiol Plant Mol Biol 47:509–540, 1996.
73. JC Jang, J Sheen. Sugar sensing in higher plants. Trends Plant Sci 2:208–214, 1997.
74. S Smeekens. Sugar regulation of gene expression in plants. Plant Biol 1:230–234, 1998.
75. JV Pego, AJ Kortsee, C Huijser, SCM Smeekens. Photosynthesis, sugars and the regulation of gene expression. J Exp Bot 51:407–416, 2000.
76. JJ Van Oosten, RT Besford. Acclimation of photosynthesis to elevated carbon dioxide through feedback regulation of gene expression: Climate of opinion. Photosynth Res 48:353–365, 1996.
77. EE Goldschmidt, SC Huber. Regulation of photosynthesis by end-product accumulation in leaves of plants storing starch, sucrose and hexose sugars. Plant Physiol 99:1443–1448, 1992.
78. A Krapp, B Hofmann, C Schafer, M Stitt. Regulation of the expression of *rbcS* and other photosynthetic genes by carbohydrates: A mechanism for the 'sink' regulation of photosynthesis? Plant J 3:817–828, 1993.
79. S Smeekens, F Rook. Sugar sensing and sugar-mediated signal transduction in plants. Plant Physiol 115:7–13, 1997.
80. JC Jang, P Leon, L Zhou, J Sheen. Hexokinase as a sugar sensor in higher plants. Plant Cell 9:5–19, 1997.
81. SH Cheng, BD Moore, JR Seemann. Effects of short- and long-term elevated $CO_2$ on the expression of ribulose-1,5-bisphosphate carboxylase/oxygenase genes and carbohydrate accumulation in leaves of *Arabidopsis thaliana* (L.) Heynh. Plant Physiol 116:715–723, 1998.
82. NG Halford, PC Purcell, DG Hardie. Is hexokinase really a sugar sensor in plants? Trends Plant Sci 4:117–124, 1999.
83. BD Moore, SH Cheng, D Sims, JR Seemann. The biochemical and molecular basis for photosynthetic acclimation to elevated atmospheric $CO_2$. Plant Cell Environ 22:567–582, 1999.
84. KE Koch, Z Ying, Y Wu, WT Avigne. Multiple paths of sugar-sensing and a sugar/oxygen overlap for genes of sucrose and ethanol metabolism. J Exp Bot 51:417–427, 2000.
85. J Sheen. Metabolic repression of transcription in higher plants. Plant Cell 2:1027–1038, 1990.
86. JJ Van Oosten, RT Besford. Sugar feeding mimics effect of acclimation to high $CO_2$—rapid down regulation of Rubisco small subunit transcripts but not of the large subunit transcripts. J Plant Physiol 143:306–312, 1994.

87. JJ Van Oosten, S Wilkins, RT Besford. Regulation of the expression of photosynthetic nuclear genes by high $CO_2$ is mimicked by carbohydrates: A mechanism for the acclimation of photosynthesis to high $CO_2$. Plant Cell Environ 17:913–923, 1994.

88. N Majeau, JR Coleman. Effect of $CO_2$ concentration on carbonic anhydrase and ribulose-1,5-bisphosphate carboxylase/oxygenase expression in pea. Plant Physiol 112:569–574, 1996.

89. GY Nie, DL Hendrix, AN Weber, BA Kimball, SP Long. Increased accumulation of carbohydrates and decreased photosynthetic gene transcript levels in wheat grown at an elevated $CO_2$ concentration in the field. Plant Physiol 108:975–983, 1995.

90. RW Gesch, KJ Boote, JCV Vu, LH Allen Jr, G Bowes. Changes in growth $CO_2$ result in rapid adjustments of ribulose-1,5-bisphosphate carboxylase/oxygenase small subunit gene expression in expanding and mature leaves of rice. Plant Physiol 118:521–529, 1998.

91. JCV Vu, RW Gesch, LH Allen Jr, KJ Boote, G Bowes. $CO_2$ enrichment delays a rapid, drought-induced decrease in Rubisco small subunit transcript abundance. J Plant Physiol 155:139–142, 1999.

92. JCV Vu, RW Gesch, LH Allen Jr, G Bowes, KJ Boote. Soybean grows well as high as 40°C daytime temperature under a doubling of atmospheric $CO_2$. Plant Biology 99: Abstract No. 518. Final Program of the Annual Meeting of the American Society of Plant Physiologists, Baltimore, July 24–28, 1999.

93. JJ Van Oosten, RT Besford. Some relationships between the gas exchange, biochemistry and molecular biology of photosynthesis during leaf development of tomato plants after transfer to different carbon dioxide concentrations. Plant Cell Environ 18:1253–1266, 1995.

94. GP Bertoni, WM Becker. Expression of the cucumber hydroxypyruvate reductase gene is down-regulated by elevated $CO_2$. Plant Physiol 112:599–605, 1996.

95. RF Sage, RK Monson. $C_4$ Plant Biology. San Diego: Academic Press, 1999, pp XIII–XV.

96. J Lloyd, GD Farquhar. $^{13}C$ discrimination during $CO_2$ assimilation by the terrestrial biosphere. Oecologia 99:201–215, 1994.

97. TE Cerling, JM Harris, BJ MacFadden, MG Leakey, J Quade, V Eisenmann, JR Ehleringer. Global vegetation change through the Miocene/Pliocene boundary. Nature 389:153–158, 1997.

98. RH Brown. Agronomic implications of $C_4$ photosynthesis. In: RF Sage, RK Monson, eds. $C_4$ Plant Biology. San Diego: Academic Press, 1999, pp 473–507.

99. AK Knapp, E Medina. Success of $C_4$ photosynthesis in the field: Lessons from communities dominated by $C_4$ plants. In: RF Sage, RK Monson, eds. $C_4$ Plant Biology. San Diego: Academic Press, 1999, pp 251–283.

100. RF Sage, DA Wedin, M Li. The biogeography of $C_4$ photosynthesis: Patterns and controlling factors. In: RF Sage, RK Monson, eds. $C_4$ Plant Biology. San Diego: Academic Press, 1999, pp 313–373.

101. JP Simon, C Potvin, BR Strain. Effects of temperature and $CO_2$ enrichment on kinetic properties of phosphoenol-pyruvate carboxylase in two ecotypes of *Echinochloa crusgalli* (L.) Beauv., a $C_4$ weed grass species. Oecologia 63:145–152, 1984.

102. C Potvin, BR Strain. Photosynthetic response to growth temperature and $CO_2$ enrichment in two species of $C_4$ grasses. Can J Bot 63:483–487, 1985.

103. C Potvin, BR Strain. Effects of $CO_2$ enrichment and temperature on growth in two $C_4$ weeds, *Echinochloa crus-galli* and *Eleusine indica*. Can J Bot 63:1495–1499, 1985.

104. LH Ziska, JA Bunce. Influence of increasing carbon dioxide concentration on the photosynthetic and growth stimulation of selected $C_4$ crops and weeds. Photosynth Res 54:199–208, 1997.

105. FB Fritschi, KJ Boote, LE Sollenberger, LH Allen Jr, TR Sinclair. Carbon dioxide and temperature effects on forage establishment: Photosynthesis and biomass production. Global Change Biol 5:441–453, 1999.

106. DH Greer, WA Laing, BD Campbell. Photosynthetic responses of thirteen pasture species to elevated $CO_2$ and temperature. Aust J Plant Physiol 22:713–722, 1995.

107. DR LeCain, JA Morgan. Growth, gas exchange, leaf nitrogen and carbohydrate concentrations in NAD-ME and NADP-ME $C_4$ grasses grown in elevated $CO_2$. Physiol Plant 102:297–306, 1998.

108. JP Maroco, GE Edwards, MSB Ku. Photosynthetic acclimation of maize to growth under elevated levels of carbon dioxide. Planta 210:115–125, 1999.

109. LH Allen Jr, JCV Vu, JD Ray. Sugarcane responses to carbon dioxide, temperature, and water table. 1998 Annual Meeting Abstracts. American Society of Agronomy, Baltimore, Oct. 18–22, 1998. Abstract page 20, Division A-3.

110. JCV Vu, LH Allen Jr, G Bowes. Growth and photosynthesis responses of sugarcane to high $CO_2$ and temperature. Plant Biology 98: Abstract No 448. Final Program of the Annual Meeting of the American Society of Plant Physiologists, Madison, WI, June 27–July 1, 1998.

111. AJ Jarvis, TA Mansfield, WJ Davies. Stomatal behaviour, photosynthesis and transpiration under rising $CO_2$. Plant Cell Environ 22:639–648, 1999.

112. BG Drake, PW Leadly. Canopy photosynthesis of crops and native plant communities exposed to long-term elevated $CO_2$. Plant Cell Environ 14:853–860, 1991.

113. CE Owensby, PI Coyne, JM Ham, LM Auen, AK Knapp. Biomass production in a tallgrass prairie ecosystem exposed to ambient and elevated $CO_2$. Ecol Appl 3:644–653, 1993.

114. AB Samarakoon, RM Gifford. Soil water content under plants at high $CO_2$ concentration and interactions with the direct $CO_2$ effects: A species comparison. J Biogeogr 22:193–202, 1995.

115. HW Polley, HB Johnson, HS Mayeux, CR Tischler. Are some of the recent changes in grassland communities a response to rising $CO_2$ concentrations? In: GW Koch, HA Mooney, eds. Carbon Dioxide and Terrestrial Ecosystems. New York: Academic Press, 1996, pp 177–195.

116. WD Bowman, KT Hubick, S von Caemmerer, GD Farquhar. Short-term changes in leaf carbon isotope discrimination in salt- and water-stress $C_4$ grasses. Plant Physiol 90:162–166, 1989.

117. SC Wong, CB Osmond. Elevated atmospheric partial pressure of $CO_2$ and plant growth. III. Interactions between *Triticum aestivum* ($C_3$) and *Echinochloa frumentacea* ($C_4$) during growth in mixed culture under different $CO_2$, N nutrition and irradiance treatments, with emphasis on below-ground responses estimated using the $\Delta^{13}C$ value of root biomass. Aust J Plant Physiol 18:137–152, 1991.

118. SD Smith, BR Strain, TD Sharkey. Effects of $CO_2$ enrichment on four Great Basin grasses. Func Ecol 1:139–143, 1987.

119. KP Hogan, AP Smith, LH Ziska. Potential effects of elevated $CO_2$ and changes in temperature on tropical plants. Plant Cell Environ 14:763–778, 1991.

120. SB Idso, BA Kimball, MG Anderson, SR Szarek. Growth response of a succulent plant, *Agave vilmoriniana*, to elevated $CO_2$. Plant Physiol 80:796–797, 1986.

121. SR Szarek, PA Holthe, IP Ting. Minor physiological response to elevated $CO_2$ by the CAM plant *Agave vilmoriniana*. Plant Physiol 83:938–940, 1987.

122. CB Osmond, O Bjorkman. Pathways of $CO_2$ fixation in the CAM plant *Kalanchoe daigremontiana*. II. Effects of $O_2$ and $CO_2$ concentration on light and dark $CO_2$ fixation. Aust J Plant Physiol 2:155–162, 1975.

123. JAM Holtum, MH O'Leary, CB Osmond. Effect of varying $CO_2$ partial pressure on photosynthesis and on carbon isotope composition of carbon-4 of malate from the Crassulacean acid metabolism plant *Kalanchoe daigremontiana* Hamet et Perr. Plant Physiol 71:602–609, 1983.

124. PS Nobel, TL Hartsock. Short-term and long-term responses of Crassulacean acid metabolism plants to elevated $CO_2$. Plant Physiol 82:604–606, 1986.

125. AA Israel, PS Nobel. Activities of carboxylating enzymes in the CAM species *Opuntia ficus-indica* grown under current and elevated $CO_2$ concentrations. Photosynth Res 40:223–229, 1994.

126. N Wang, PS Nobel. Doubling the $CO_2$ concentration enhanced the activity of carbohydrate-metabolism enzymes, source carbohydrate production, photoassimilate transport, and sink strength for *Opuntia ficus-indica*. Plant Physiol 110:893–902, 1996.

127. J Berry, O Bjorkman. Photosynthetic response and adaptation to temperature in higher plants. Annu Rev Plant Physiol 31:491–543, 1980.

128. S Falk, DP Maxwell, DE Laudenbach, NPA Huner. Photosynthetic adjustment to temperature. In: NR Baker, ed. Photosynthesis and the Environment. Dordrecht: Kluwer Academic Publishers, 1996, pp 367–385.

129. RC Leegood, GE Edwards. Carbon metabolism and photorespiration: Temperature dependence in relation to other environmental factors. In: NR Baker, ed. Photosynthesis and the Environment. Dordrecht: Kluwer Academic Publishers, 1996, pp 191–221.

130. SP Long. Modification of the response of photosynthetic productivity to rising temperature by atmospheric $CO_2$ concentrations: Has its importance been underestimated? Plant Cell Environ 14:729–739, 1991.

131. MUF Kirschbaum. The sensitivity of $C_3$ photosynthesis to increasing $CO_2$ concentration: A theoretical analysis of its dependence on temperature and background $CO_2$ concentration. Plant Cell Environ 17:747–754, 1994.

132. J Kobza, GE Edwards. Influences of leaf temperature on photosynthetic carbon metabolism in wheat. Plant Physiol 83:69–74, 1987.

133. AS Holaday, W Martindale, R Alred, AL Brooks, RC Leegood. Changes in activities of enzymes of carbon metabolism in leaves during exposure of plants to low temperature. Plant Physiol 98:1105–1114, 1992.

134. DB Jordan, WL Ogren. The $CO_2/O_2$ specificity of ribulose 1,5-bisphosphate carboxylase/oxygenase. Planta 161:308–313, 1984.

135. A Brooks, GD Farquhar. Effect of temperature on the $CO_2/O_2$ specificity of ribulose-1,5-bisphosphate carboxylase/oxygenase and the rate of respiration in the light. Planta 165:397–406, 1985.

136. JT Baker, LH Allen Jr, KJ Boote. Temperature effects on rice at elevated $CO_2$ concentration. J Exp Bot 43:959–964, 1992.

137. JT Baker, LH Allen Jr. Contrasting crop species responses to $CO_2$ and temperature: Rice, soybean and citrus. Vegetatio 104/105:239–260, 1993.

138. LH Allen Jr, JT Baker, SL Albrecht, KJ Boote, D Pan, JCV Vu. Carbon dioxide and temperature effects on rice. In: S Peng, KT Ingram, HU Neue, LH Ziska, eds. Climate Change and Rice. Berlin: Springer-Verlag, 1995, pp 258–277.

139. JT Baker, LH Allen Jr, KJ Boote, P Jones, JW Jones. Response of soybean to air temperature and carbon dioxide concentration. Crop Sci 29:98–105, 1989.

140. RAC Mitchell, VJ Mitchell, SP Driscoll, J Franklin, DW Lawlor. Effects of increased $CO_2$ concentration and temperature on growth and yield of winter wheat at two levels of nitrogen application. Plant Cell Environ 16:521–529, 1993.

141. WJS Downton, WJR Grant, BR Loveys. Carbon dioxide enrichment increases yield of Valencia orange. Aust J Plant Physiol 14:493–501, 1987.

142. SB Idso, BA Kimball, SG Allen. Net photosynthesis of sour orange trees maintained in atmospheres of ambient and elevated $CO_2$ concentration. Agric For Meteorol 54:95–101, 1991.

143. SB Idso, BA Kimball. Effects of atmospheric $CO_2$ enrichment on photosynthesis, respiration, and growth of sour orange trees. Plant Physiol 99:341–343, 1992.

144. M Brakke, LH Allen Jr. Gas exchange of citrus seedlings at different temperatures, vapor-pressure deficits, and soil water contents. J Am Soc Hortic Sci 120:497–504, 1995.

145. JCV Vu, G Yelenosky. Water deficit and associated changes in some photosynthetic parameters in leaves of 'Valencia' orange (*Citrus sinensis* [L.] Osbeck). Plant Physiol 88:375–378, 1988.

146. SB Idso, KE Idso, RL Garcia, BA Kimball, JK Hoober. Effects of atmospheric $CO_2$ enrichment and foliar methanol application on net photosynthesis of sour orange tree (*Citrus aurantium*; Rutaceae) leaves. Am J Bot 82:26–30, 1995.

147. RM Gifford. Interaction of carbon dioxide with growth-limiting environmental factors in vegetation productivity: Implications for the global carbon cycle. Adv Bioclimatol 1:24–58, 1992.

148. JR Ehleringer, RK Monson. Evolutionary and ecological aspects of photosynthetic pathway variation. Annu Rev Ecol Syst 24:411–439, 1993.

149. MKH Ebrahim, G Vogg, MNEH Osman, E Komor. Photosynthetic performance and adaptation of sugarcane at suboptimal temperatures. J Plant Physiol 153:587–592, 1998.

150. SP Long. Environmental responses. In: RF Sage, RK Monson, eds. $C_4$ Plant Biology. San Diego: Academic Press, 1999, pp 215–249.

151. HM Rawson. Plant responses to temperature under conditions of elevated $CO_2$. Aust J Bot 40:473–490, 1992.

152. JS Coleman, FA Bazzaz. Effects of $CO_2$ and temperature on growth and resource use of co-occurring $C_3$ and $C_4$ annuals. Ecology 73:1244–1259, 1992.

153. MJ Robertson, GD Bonnett, RM Hughes, RC Muchow, JA Campbell. Temperature and leaf area expansion of sugarcane: Integration of controlled-environment, field and model studies. Aust J Plant Physiol 25:819–828, 1998.

154. M Stitt, A Krapp. The interaction between elevated carbon dioxide and nitrogen nutrition: The physiological and molecular background. Plant Cell Environ 22:583–621, 1999.

155. M Geiger, V Haake, F Ludewig, U Sonnewald, M Stitt. The nitrate and ammonium nitrate supply have a major influence on the response of photosynthesis, carbon metabolism, nitrogen metabolism and growth to elevated carbon dioxide in tobacco. Plant Cell Environ 22:1177–1199, 1999.

156. JIL Morison. Response of plants to $CO_2$ under water limited conditions. Vegetatio 104/105:193–209, 1993.

157. MM Chaves, JS Pereira. Water stress, $CO_2$ and climate change. J Exp Bot 43:1131–1139, 1992.

158. RF Sage, CD Reid. Photosynthetic response mechanisms to environmental change in $C_3$ plants. In: RE Wilkinson, ed. Plant-Environment Interactions. New York: Marcel Dekker, 1994, pp 413–499.

159. LH Allen Jr, P Jones, JW Jones. Rising atmospheric $CO_2$ and evapotranspiration. In: Advances in Evapotranspiration. ASAE Publication 14–85. St. Joseph, MI: American Society of Agricultural Engineers, 1985, pp 13–27.

160. WJ Arp, BG Drake, WT Pockman, PS Curtis, DF Whighman. Interactions between $C_3$ and $C_4$ marsh plant species during four years of exposure to elevated atmospheric $CO_2$. Vegatatio 104/105:133–143, 1993.

161. HH Rogers, N Sionit, JD Cure, HM Smith, GE Bingham. Influence of elevated $CO_2$ on water relations of soybeans. Plant Physiol 74:233–238, 1984.

162. P Jones, JW Jones, LH Allen Jr. Seasonal carbon and water balances of soybeans grown under stress treatments in sunlit chambers. Trans Am Soc Agric Eng 28:2021–2028, 1985.

163. NC Bhattacharya, DR Hileman, PP Ghosh, RL Musser, S Bhattacharya, PK Biswas. Interaction of enriched $CO_2$ and water stress on the physiology and biomass production in sweet potato grown in open-top chambers. Plant Cell Environ 13:933–940, 1990.

164. SC Clifford, IM Stronach, AD Mohamed, SN Azam-Ali, NMJ Crout. The effects of elevated atmospheric carbon dioxide and water stress on light interception, dry matter production and yield in stands of groundnut (*Arachis hypogaea* L.). J Exp Bot 44:1763–1770, 1993.

165. JT Baker, LH Allen Jr, KJ Boote, NB Pickering. Rice responses to drought under carbon dioxide enrichment. 1. Growth and yield. Global Change Biol 3:119–128, 1997.

166. JT Baker, LH Allen Jr, KJ Boote, NB Pickering. Rice responses to drought under carbon dioxide enrichment. 2. Photosynthesis and evaporation. Global Change Biol 3:129–138, 1997.

167. JCV Vu, JT Baker, AH Pennanen, LH Allen Jr, G Bowes, KJ Boote. Elevated $CO_2$ and water deficit effects on photosynthesis, ribulose bisphosphate carboxylase-oxygenase, and carbohydrate metabolism in rice. Physiol Plant 103:327–339, 1998.

168. SB Idso. Three phases of plant response to atmospheric $CO_2$ enrichment. Plant Physiol 87:5–7, 1988.

169. LH Allen Jr, RR Valle, JW Jones, PH Jones. Soybean leaf water potential responses to carbon dioxide and drought. Agron J 90:375–383, 1998.

170. AB Samarakoon, RM Gifford. Elevated $CO_2$ effects on water use and growth of maize in wet and drying soil. Aust J Plant Physiol 23:53–62, 1996.

171. N Sionit, H Hellmers, BR Strain. Interaction of atmospheric $CO_2$ enrichment and irradiance on plant growth. Agron J 74:721–725, 1982.

172. R Valle, JW Mishoe, WJ Campbell, JW Jones, LH Allen Jr. Photosynthetic responses of "Bragg" soybean leaves adapted to different $CO_2$ environments. Crop Sci 25:333–339, 1985.

173. WJ Campbell, LH Allen Jr, G Bowes. Response of soybean canopy photosynthesis to $CO_2$ concentration, light, and temperature. J Exp Bot 41:427–433, 1990.

174. I Nijs, I Impens, P Van Hecke. Diurnal changes in the response of canopy photosynthetic rate to elevated $CO_2$ in a coupled temperature-light environment. Photosynth Res 32:121–130, 1992.

175. SB Idso, GW Wall, BA Kimball. Interactive effects of atmospheric $CO_2$ enrichment and light intensity reductions on net photosynthesis of sour orange tree leaves. Environ Exp Bot 33:367–375, 1993.

176. GE Edwards, JP Krall. Metabolic interactions between organelles in $C_4$ plants. In: AK Tobin, ed. Plant Organelles. Cambridge: Cambridge University Press, 1992, pp 97–112.

177. N Sionit, DT Patterson. Responses of $C_4$ grasses to atmospheric $CO_2$ enrichment. I. Effects of irradiance. Oecologia 65:30–34, 1984.

178. CB Field, FS Chapin III, PA Matson, HA Mooney. Responses of terrestrial ecosystems to the changing atmosphere: A resource-based approach. Annu Rev Ecol Syst 23:201–235, 1992.

179. E Delgrado, RAC Mitchell, MAJ Parry, SP Driscoll, VJ Mitchell, DW Lawlor. Interacting effects of $CO_2$ concentration, temperature and nitrogen supply to leaves on photosynthesis and composition of winter wheat leaves. Plant Cell Environ 17:1205–1213, 1994.

180. IF McKee, FI Woodward. $CO_2$ enrichment responses of wheat: Interaction with temperature, nitrate and phosphate. New Phytol 127:447–453, 1994.

181. R Pettersson, JS MacDonald. Effects of nitrogen supply on the acclimation of photosynthesis to elevated $CO_2$. Photosynth Res 39:389–400, 1994.

182. H Bassiridad, RB Thomas, JF Reynolds, BR Strain. Differential responses of root uptake kinetics of ammonium and nitrate to enriched atmospheric carbon dioxide concentration in field grown loblolly pine. Plant Cell Environ 19:367–371, 1996.

183. JM Bowler, MC Press. Effects of elevated carbon dioxide, nitrogen form and concentration on growth and photosynthesis of a fast- and slow-growing grass. New Phytol 132:391–401, 1996.

184. Tingey DT, MG Johnson, DL Phillips, DW Johnson, JT Ball. Effects of elevated $CO_2$ and nitrogen on the synchrony of shoot and root growth in ponderosa pine. Tree Physiol 16:905–914, 1996.

185. LH Ziska, W Weerakoon, OS Namuco, R Pamplona. The influence of nitrogen on the elevated carbon dioxide response in field grown rice. Aust J Plant Physiol 23:45–52, 1996.

186. S Ferrario-Mery, MC Thibaud, T Betsche, MH Valadier, CH Foyer. Modulation of carbon and nitrogen metabolism, and of nitrate reductase, in untransformed *Nicotiana plumbaginifolia* during $CO_2$ enrichment of plants grown in pots and hydroponic culture. Planta 202:510–521, 1997.

187. WJ Arp, JEM Mierlo, F Berendse, W Snijders. Interactions between elevated carbon dioxide concentration, nitrogen and water: Effects on growth and water use of six perennial plant species. Plant Cell Environ 21:1–11, 1998.

188. R Pettersson, JS MacDonald, I Stadenburg. Response of small birch plants (*Betula pendula* Roth.) to elevated $CO_2$ and nitrogen supply. Plant Cell Environ 16:1115–1121, 1993.

189. H Poorter, V Berkel, R Baxter, J den Hertog, P Dijkstra, RM Gifford, KL Griffin, C Rounet, J Roy, SC Wong. The effect of elevated carbon dioxide on the chemical composition and construction costs of leaves of 27, $C_3$ species. Plant Cell Environ 20:472–482, 1997.

190. MF Cotrufo, P Ineson, A Scott. Elevated carbon dioxide reduces the nitrogen concentration of plant tissues. Global Change Biol 4:43–54, 1998.

191. H Riviere-Rolland, P Contard, T Betsche. Adaptation of pea to elevated atmospheric $CO_2$: Rubisco, PEP carboxylase and chloroplast phosphate translocator at different levels of nitrogen and phosphate nutrition. Plant Cell Environ 19:109–117, 1996.

192. GS Rogers, PJ Milham, M Gillings, JP Conroy. Sink strength may be the key to growth and nitrogen response in N-deficient wheat at elevated carbon dioxide. Aust J Plant Physiol 23:253–264, 1996.

193. N Sionit. Response of soybean to two levels of mineral nutrition in $CO_2$ enriched atmosphere. Crop Sci 23:329–333, 1983.

194. D Eamus, PG Jarvis. The direct effects of increase in the global atmospheric $CO_2$ concentration on natural and commercial temperate trees and forests. Adv Ecol Res 19:1–55, 1989.

195. RB Thomas, JD Lewis, BR Strain. Effects of leaf nutrient status on photosynthetic capacity in loblolly pine seedlings grown in elevated atmospheric $CO_2$. Tree Physiol 14:947–950, 1994.

196. DT Tissue, RB Thomas, BR Strain. Atmospheric $CO_2$ enrichment increases growth and photosynthesis of *Pinus taeda*, a 4 year experiment in the field. Plant Cell Environ 20:1123–1134, 1997.

197. JP Conroy, PJ Milham, EWR Barlow. Effects of nitrogen and phosphorus availability on the growth response of *Eucalyptus grandis* to high $CO_2$. Plant Cell Environ 15:843–847, 1992.

198. RLE Gebauer, JF Reynolds, BR Strain. Allometric relations and growth in *Pinus taeda*: The effect of elevated $CO_2$ and changing N availability. New Phytol 134:85–93, 1996.

199. A Fangmeier, U Grueters, B Hoegy Vermehren, HJ Jaeger. Effects of elevated carbon dioxide, nitrogen supply and tropospheric ozone on spring wheat. II. Nutrients. Environ Pollut 96:43–59, 1997.

200. SJ Whitehead, SJM Caporn, MC Press. Effects of elevated carbon dioxide, nitrogen and phosphorus on the growth and photosynthesis of two upland perennial. New Phytol 135:201–211, 1997.

201. DQ Xu, RM Gifford, WS Chow. Photosynthetic acclimation in pea and soybean to high atmospheric $CO_2$ partial pressure. Plant Physiol 106:661–671, 1994.

202. GY Nie, SP Long, RL Garcia, BA Kimball, RL LaMorte, PJ Pinter Jr, GW Wall, AN Webber. Effects of free air $CO_2$ enrichment on the development of the photosynthetic apparatus in wheat, as indicated by changes in leaf proteins. Plant Cell Environ 18:855–864, 1995.

203. M Pearson, G Brooks. The influence of elevated $CO_2$ on growth and age-related changes in leaf gas exchange. J Exp Bot 46:1651–1659, 1995.

204. A Miller, CH Tsai, D Hemphill, M Endres, S Rodermel, M Spalding. Elevated $CO_2$ effects during leaf ontogeny. A new perspective on acclimation. Plant Physiol 115:1195–1200, 1997.

205. F Kauder, F Ludewig, D Heineke. Ontogenic changes of potato plants during acclimation to elevated carbon dioxide. J Exp Bot 51:429–437, 2000.

206. S Gepstein. Photosynthesis. In: LD Nooden, AC Leopold, eds. Senescence and Aging in Plants. San Diego: Academic Press, 1988, pp 85–109.

207. K Garbutt, WE Williams, FA Bazzaz. Analysis of the differential response of five annuals to elevated carbon dioxide during growth. Ecology 7:1185–1194, 1990.

208. R Baxter, M Grantley, TW Ashenden, J Farrar. Effects of elevated carbon dioxide on three grass species from montane pasture. II. Nutrient allocation and efficiency of nutrient use. J Exp Bot 45:1267–1278, 1994.

209. M Geiger, L Walch-Piu, J Harnecker, ED Schulze, F Ludewig, U Sonnewald, WR Scheible, M Stitt. Enhanced carbon dioxide leads to a modified diurnal rhythm of nitrate reductase activity and higher levels of amino acids in higher plants. Plant Cell Environ 21:253–268, 1998.

210. R Cuelemans, XN Jiang, BY Shao. Growth and physiology of one year old poplar under elevated atmospheric carbon dioxide levels. Ann Bot 75:609–617, 1994.

211. HSJ Lee, PG Jarvis. Trees differ from crops and from each other in their response to increases in $CO_2$ concentration. J Biogeogr 22:323–330, 1995.

212. DC Tremmel, DT Patterson. Responses of soybean and 5 weeds to $CO_2$ enrichment under 2 temperature regimes. Can J Plant Sci 73:1249–1260, 1993.

213. RA Kennedy, WM Laetsch. Relationship between leaf development, carboxylase enzyme activities and photorespiration in the $C_4$-plant *Portulaca oleracea* L. Planta 115:113–124, 1973.

214. BD Moore, SH Cheng, GE Edwards. The influence of leaf development on the expression of $C_4$ metabolism in *Flaveria trinervia*, a $C_4$ dicot. Plant Cell Physiol 27:1159–1167, 1986.

215. LE Williams, RA Kennedy. Photosynthetic carbon metabolism during leaf ontogeny in *Zea mays* L.: Enzyme studies. Planta 142:269–274, 1978.

216. LE Williams, RA Kennedy. Relationship between early photosynthetic products, photorespiration and stage of leaf development in *Zea mays*. Z Pflanzenphysiol 81:314–322, 1977.

217. JT Perchorowicz, M Gibbs. Carbon dioxide fixation and related properties in sections of the developing green maize leaf. Plant Physiol 65:802–809, 1980.

218. HM Crespo, M Frean, CF Cresswell, J Tew. The occurrence of both $C_3$ and $C_4$ photosynthetic characteristics in a single *Zea mays* plant. Planta 147:257–263, 1979.

219. Z Dai, MSB Ku, GE Edwards. $C_4$ photosynthesis. The effects of leaf development on the $CO_2$-concentrating mechanism and photorespiration in maize. Plant Physiol 107:815–825, 1995.

220. FC Meinzer, NZ Saliendra. Spatial patterns of carbon isotope discrimination and allocation of photosynthetic activity in sugarcane leaves. Aust J Plant Physiol 24:769–775, 1997.

221. WM Laetsch. The $C_4$ syndrome: A structural analysis. Annu Rev Plant Physiol 25:27–52, 1974.

222. WV Brown. Variations in anatomy, associations, and origins of Kranz tissue. Am J Bot 62:395–402, 1975.

223. WM Laetsch, I Price. Development of the dimorphic chloroplasts of sugar cane. Am J Bot 56:77–87, 1969.

224. SJ Kirchanski. The ultrastructural development of the dimorphic plastids of *Zea mays* L. Am J Bot 62:695–705, 1975.

225. O Ghannoum, K Siebke, S von Caemmerer, JP Conroy. The photosynthesis of young *Panicum* $C_4$ leaves is not $C_3$-like. Plant Cell Environ 21:1123–1131, 1998.

226. LH Allen Jr, JS Amthor. Plant physiological responses to elevated $CO_2$, temperature, air pollution, and UV-B radiation. In: GM Woodwell, FT Mackenzie, eds. Biotic Feedbacks in the Global Climatic System. Will the Warming Feed the Warming? New York: Oxford University Press, 1995, pp 51–84.

227. DO Hall, JI House. Biomass as a modern fuel. In: CP Mitchell, AV Bridgewater, eds. Environmental Impacts of Bioenergy. Newbury, UK: CPL Press, 1994, pp 81–114.

228. C Rosenzweig, D Hillel. Climate Change and the Global Harvest—Potential Impacts of the Greenhouse Effect on Agriculture. New York: Oxford University Press, 1998.

229. CC Mann. Crop scientists seek a new revolution. Science 283:310–314, 1999.

230. CA Raines, CJ Lloyd. Molecular biological approaches to environmental effects on photosynthesis. In: NR Baker, ed. Photosynthesis and the Environment. Dordrecht: Kluwer, 1996, pp 305–319.

231. CC Mann. Genetic engineers aim to soup up crop photosynthesis. Science 283:314–316, 1999.

232. JM Dunwell. Transgenic approaches to crop improvement. J Exp Bot 51:487–496, 2000.

233. MP Reynolds, M van Ginkel, JM Ribaut. Avenues for genetic modification of radiation use efficiency in wheat. J Exp Bot 51:459–473, 2000.

234. RA Richards. Selectable traits to increase crop photosynthesis and yield of grain crops. J Exp Bot 51:447–458, 2000.

235. K Ishimaru, H Ichikawa, M Matsuoka, R Ohsugi. Analysis of a $C_4$ pyruvate, orthophosphate dikinase expressed in $C_3$ transgenic *Arabidopsis* plants. Plant Sci 129:57–64, 1997.

236. K Ishimaru, Y Ohkawa, T Ishige, DJ Tobias, R Ohsugi. Elevated pyruvate orthophosphate dikinase (PPDK) activity alters carbon metabolism in $C_3$ transgenic potatoes with a $C_4$ maize *PPDK* gene. Physiol Plant 103:340–346, 1998.

237. MSB Ku, S Agarie, M Nomura, H Fukayama, H Tsuchida, K Ono, S Hirose, S Toki, M Miyao, M Matsuoka. High-level expression of maize phosphoenolpyruvate carboxylase in transgenic rice plants. Nat Biotech 17:76–80, 1999.

# 4

# Germination and Emergence

**Calvin Chong**

*University of Guelph, Vineland Station, Ontario, Canada*

**Bernard B. Bible**

*University of Connecticut, Storrs, Connecticut*

**Hak-Yoon Ju**

*Nova Scotia Agricultural College, Truro, Nova Scotia, Canada*

## I. INTRODUCTION

A seed (zygote) results from the fertilization or union of male and female gametes and is the reproductive structure of a plant. Thus, the regeneration or multiplication of plants from seed is termed sexual. Plants are also reproduced by asexual (vegetative) means from bulbs or pieces of stem, root, or other plant part [1].

A seed is essentially an embryo or young plant in the quiescent or dormant stage. In this state, the embryo has an extremely low metabolic rate. Most seeds can survive on their stored reserves for pro-longed periods. The seed is the primary means by which a plant reproduces itself at a later time when conditions are suitable.

During fertilization, the genes controlling plant characteristics regenerate and recombine in many different ways, resulting in seeds that may or may not mimic their parents. Seeds resulting from self-pollination may produce true-to-type specimens, whereas those resulting from cross-pollination usually do not. Cross-pollinated seeds provide genetic diversity for breeding and selection of new cultivars (cultivated varieties) and are often sources of new or novel plant material.

Some seedling plants are commonly grown as rootstocks for budding or grafting of cultivar fruit trees, nut trees, and woody landscape plants or to produce superior landscape specimens that are difficult to propagate asexually or for which asexual methods of propagation are unknown [2]. Regeneration from seed is the most economical way to grow large numbers of plants. Most forest species, vegetables, and flowering and other cultivated plants are grown from seeds.

A few species of plants produce seeds without undergoing fertilization. This form of reproduction is called apomixis and is characteristic of species such as *Poa pratensis* (Kentucky bluegrass), which rarely or never produces true seed. A plant grown from an apomictic seed is genetically identical to its parent.

## II. SEED MORPHOLOGY

There are two major classes of seed-bearing plants: angiosperms (flower-bearing), whose seeds are borne in ovules enclosed within the ovary or fruit, and gymnosperms (cone-bearing), whose seeds are borne in pairs at the base of scales of the cones. An example of a seed from each of these classes of seed-bearing plants is illustrated in Figure 1A and B.

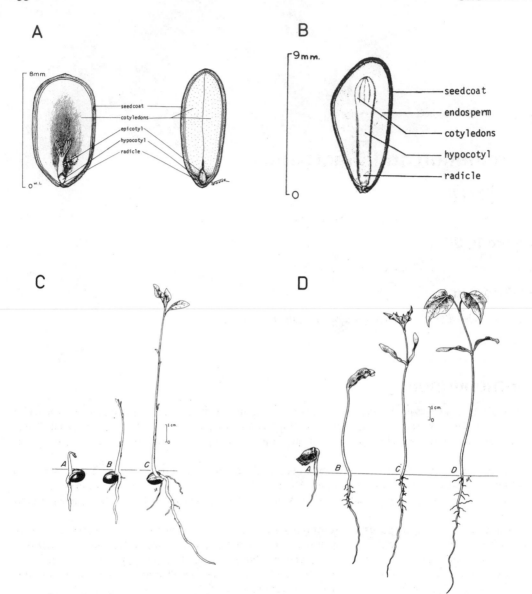

**Figure 1**   (A) Longitudinal section through seed of an angiosperm, *Albizia julibrissin* (silktree), 5×. (From Ref. 3.) (B) Longitudinal section through seed of a gymnosperm, *Pinus ponderosa* (ponderosa pine), 6×. (From Ref. 4.) (C) Hypogeous germination: development of *Lindera benzoin* (spicebush) 2, 3, and 10 days after germination. (From Ref. 5.) (D) Epigeous germination: development of *Acer platanoides* (Norway maple) 1, 3, 7, and 19 days after germination. (From Ref. 6.)

Most seeds consist of three parts: *embryo*, a miniature plant inside the seed; *endosperm*, stored food reserves for the growing embryo; and *seed coat* (testa), which encompasses and protects the embryo and endosperm from damage, excess water loss, and other unfavorable conditions.

The embryo has one or more miniature seed leaves (*cotyledons*), an embryonic stem (*plumule* or *epicotyl*), an embryonic root (radicle), and a *hypocotyl*, the transition zone between the embryonic stem and root. Among angiosperms, plants that have two cotyledons are classified as *dicotyledons* and those with a single cotyledon as *monocotyledons*. Gymnosperms may have as many as 15 cotyledons.

There are about 250,000 different seed-bearing plants in the world. Each species has its own morphologically unique form of seed, which can be identified by its size, shape, color, and other external features [7]. The endosperm contains stored food reserves composed of carbohydrates, proteins, oils, and other biochemical substances. All seeds contain stored food reserves. In some, the amount can be quite small. Generally, the larger the food reserve, the greater the vigor of the seedling. Plump seeds usually have more food reserves than small, shriveled seeds. Also, food reserves are found in the cotyledons of some species.

The seed coat may appear dull, highly glossy, smooth, wrinkled or pitted, hard or soft, thick or thin, or any combination of these characteristics. Many seeds also have attached wings or other appendages. In seeds having two seed coats, the inner membranous one is usually thin, transparent and physiologically active; that is, it restricts gaseous exchanges and movement of biochemical substances [8]. The outer seed coat is hard and thick. A non viable seed may contain an empty seed coat without an embryo or one that is reduced and shrunken. Seed coverings play an important role in protecting the seed and in influencing germination.

## III. SEED GERMINATION

Most seeds begin to germinate (resume activity) soon after being exposed to or planted in a moist, warm soil or germinating medium. The germination process begins with a swelling of the seed as it takes up or imbibes moisture. Usually, the radicle emerges first from the softened or ruptured seed coat, grows downward, and develops into the primary root system. The plumule grows upward to form the stem. During early growth, the young seedling derives its nourishment from the seed's cotyledons and/or endosperm. Cytokinins—members of the group of plant hormones, including kinetin, that act synergistically with auxins to promote cell division but, unlike auxins, promote lateral growth—promote the mobilization of the food reserves toward the developing shoot and to the root, which begins to function and absorb nutrients from the soil or medium.

In some instances, the cotyledon or cotyledons remain beneath the surface of the ground, *hypogeous* germination (Figure 1C), although in most species they push above the surface, *epigeous* germination (Figure 1D), turn green, and perform the functions of leaves, but are not true leaves. The food reserves continue to nourish the seedling until photosynthesis occurs at a rate capable of supporting the plant, usually when the first true leaves are formed. At this stage, germination is completed and, in most cases, seedlings are capable of independent existence and germination is completed.

## IV. PHYSIOLOGICAL AND ENVIRONMENTAL FACTORS

Each species of plant has its own unique requirements for moisture, temperature, oxygen, light, and other factors.

## A. Moisture

The need for moisture is the most important prerequisite for triggering germination. Whereas some seeds require little moisture for germination, others, such as those from *Nymphaea* spp. (water lilies) and other aquatic plants, must be completely submerged in water. Seeds with hard or impermeable seed coats require special treatment (softening or scarification) to allow efficient uptake of water. Generally, water reaches the seed through contact with the soil or germinating medium. Once the germination process begins, an adequate moisture level must be maintained as temporary drying can result in death of the seed or seedling. Too much moisture can cause the soil or germinating medium to become saturated and deprive the seed of oxygen, leading to death. Water uptake by seeds during germination has been described using a three-stage model [9]:

### 1. Stage I (Imbibition)

Dry seeds have a high negative water potential ($-100$ to $-200$ mPa) because of the colloidal properties of the seed coat. The surfaces of proteins, cellulose, starch, and other substances must first become hydrated. The uptake or imbibition of water in stage I is physical, resulting in softening or rupturing of the seed coat and an increase in the volume of the seed. As the seed imbibes water, its internal water potential rises.

## 2. Stage II (Active Metabolism and Hydrolysis)

Continuing water uptake activates stored enzymes and stimulates synthesis of new ones. These enzymes hydrolyze and transform some of the stored reserves into energy and lower molecular weight, soluble compounds, used for the production of more cells and tissues. These metabolic processes lower the water potential of the embryo and surrounding tissues.

Stage II appears to be a lag phase between uptake and growth. During this stage, the rate of water absorption is governed by the internal osmotic potential. The duration of the lag phase may represent the time required for weakening the restraints on embryo enlargement imposed by surrounding tissue to a degree necessary to allow further water uptake by the embryo. During stage II, exogenously applied plant hormones such as gibberellins or abscisic acid may assist with weakening or strengthening, respectively, of the tissues surrounding the embryo [10]. The generally slow response to gibberellin treatment is probably related to the time lag for enzyme synthesis.

## 3. Stage III (Visible Germination)

Rapid growth of the radicle and shoot defines the third stage. For this process to occur, the water potential of the external solution should not be lower than $-0.2$ to $-0.3$ mPa. Germinating solutions with water potentials of $-0.45$ to $-0.80$ mPa noticeably slow radicle emergence, and solutions of $-1.0$ mPa or lower severely restrict the expansion of radicle cells necessary for radicle protusion.

## B. Temperature

When moisture is adequate, the next most important requirement for germination is suitable temperature. Temperature affects the rate at which water is imbibed as well as the rate of metabolic processes such as the translocation of nutrients and hormones, cell division and elongation, and other physiological and biochemical processes.

According to Hartmann et al. [8], temperature is the single most important factor in the regulation of the timing of germination, because of its role in dormancy control and/or release, or climate adaptation. Generally, high temperatures induce or reinforce dormancy; low temperatures overcome dormancy.

Most seeds can tolerate prolonged hot weather if they are kept dry, and some can withstand even greater extremes of hot or cold [8]. Seeds of some species, such as forest pines with a very hard seed coat, germinate only after exposure to intense heat, such as that from a brush fire [8,11]. The heat shock from the dry heat fractures the seed coat, allowing penetration of water or exchange of gases or freeing the embryo from the physical constraint of the hard seed coat. Seeds are often placed in boiling water to control disease or to soften the seed coat without affecting seed viability.

Seeds of different species have been categorized into suitable temperature groups: cool-temperature tolerant, cool-temperature requiring, warm-temperature requiring, and alternating temperature [8,12].

The optimal temperature requirement for germination may be different from that for early seedling growth. In the greenhouse, propagating nursery, or seed-germinating laboratory, the usual practice is to shift the seedlings to a lower temperature regime, which makes them sturdy and more hardy for transplanting and growing [2,8].

## C. Oxygen

Most seeds require an adequate supply of oxygen during germination. Oxygen is required for respiration to oxidize starches, fats, and other food reserves, and its utilization is proportional to the amount of metabolic activity [13]. Thus, a germinating medium or seedbed should be loose, friable, and well aerated. Seeds sown in heavy soils may germinate poorly, especially during wet seasons, when the soil becomes saturated and often lacks sufficient oxygen. Deep planting is unfavorable to germination because the oxygen supply may be restricted or seedlings may be unable to reach the surface, especially if the soil or medium is hard or compacted.

## D. Light

Provided moisture and temperature are adequate, most seeds germinate equally well in darkness or light, particularly seeds of most agricultural food plants, which have been rigorously selected for ease of germination. Others are partially or completely inhibited by light or require it to germinate. Some species, such as *Betulla* spp. (birch), respond to long photoperiods and are categorized as "long-day" seeds. Small-seeded species, including numerous weeds, that are favored by light for germination need only low-intensity light. Photosensitive seeds should be sown upon or near the surface and shaded to prevent excessive moisture loss from the germinating surface. Reaction to light (photosensitivity) is mediated through phytochrome, a pigment that absorbs either red or far-red light [13].

**Phytochrome.** The phytochrome pigment system plays a key role in the photosensitivity of seeds and is of particular significance to many small-seeded species. Phytochrome is converted to Pfr (far-red light–absorbing form) by 660-nm (red) light and, in turn, can be reconverted to Pr (red light–absorbing form) by 730-nm (infrared) light.

Much of the phytochrome in quiescent seeds is in the Pfr form. However, within several hours after seeds are fully hydrated, conversion of Pfr to Pr can occur in the absence of light [14]. Because Pfr actively promotes germination and lack of Pfr inhibits germination, photosensitive seeds germinate in response to exposure to the Pfr-forming 660-nm light.

The ratio of 655–665 nm light to 725–735 nm light varies significantly in nature. Sunlight filtered through foliage has a low ratio of 655–665 nm light to 725–735 nm light because chlorophyll selectively absorbs 655–665 nm light while transmitting 725–735 nm light [15].

The ecological significance of the seed phytochrome system involves allowing shaded, light-sensitive seeds on or near the soil surface to remain dormant until the leaf canopy above the seed disappears [16]. Of course, even though fully hydrated, buried light-sensitive seeds of weedy species, for example, will also remain dormant until returned to the soil surface by tillage or other soil disturbances.

Interactions between light and temperature are known for some kinds of seed. For example, photosensitivity may be overcome by alternating high and low temperatures. Externally applied chemicals can also interact with light and temperature. Many nitrogenous compounds, including cyanide, nitric acid, ammonium salts, urea, thiourea, and particularly potassium nitrate (10–20 mM solutions), have been found to stimulate the germination of photosensitive seeds [17].

## V. ADAPTIVE FACTORS

### A. Life Cycle

In the life cycle of every sexually reproduced plant, the seed germinates and the plant makes its vegetative growth, flowers and bears seeds (physiological maturity), and sooner or later dies. The duration of the cycle determines the three broad categories of plants: annuals, biennials, and perennials.

In its natural habitat, an annual plant usually lives for only 1 year or one season, a biennial for 2 years, and a perennial for more than 2 years. A perennial will continue to grow more or less indefinitely and, once physiologically mature, will annually produce flowers, fruits, and seeds under suitable condition.

The distinction between annual and the other categories is not absolute. A biennial that seed prematurely within the first growing season is considered to be an annual by this outcome. An annual or biennial grown year-round in a greenhouse, or outdoors in a warmer climate, becomes a perennial. A perennial that would normally grow indefinitely in a warmer climate may be killed by frost in a colder climate.

Perennials are either *herbaceous* (having annual tops but perennial roots, crowns, or related underground structures) or *woody* (having biennial or perennial tops and perennial roots). Woody perennials consist mostly of trees, shrubs, and vines. These are readily distinguishable from herbaceous perennials, biennials, and annuals having nonwoody tops and/or roots, which are typically killed by frost in colder climates. Because of their longer juvenile periods, most woody species grown from seed do not start producing flowers and seeds until many years later [18]. These categories describe the pattern of adaptation, cultural requirements of plants, and, to a great extent also, their seed germination requirements. In general, but not always, seeds of woody perennials are more difficult to germinate or may possess more complex germination constraints than those of herbaceous types. Seeds of species within the same plant family or genera tend to have similar requirements for germination [19].

### B. Dormancy

Germination may be immediate or delayed. Seeds of many flowering garden annuals and trees, such as *Acer saccharinum* (silver maple), will germinate within a month after maturing and some with almost no delay when removed from their protective fruits or seed coverings. Occasionally, germination in some species occurs while the seed is still on the parent plant, apparently because of lack of inhibiting chemicals [20].

The germination of seeds, especially those of many woody trees and shrubs, is complex and erratic. In nature, every species has one or more mechanisms for preventing germination until the seed has been dispersed, and failure to germinate seeds in cultivation is often due to our lack of understanding of these mechanisms [8,21,22].

Dormancy is an all-inclusive term used to describe a "resting" state with reduced metabolic rate [23]. In this condition, a seed will not germinate because of constraints associated with the seed itself (physical or physiological) or with the external environment. Until a seed dies, it remains metabolically active even during prolonged storage under dry conditions or extreme temperatures. Dormancy is either primary or secondary (Table 1). Primary dormancy is an adaptation of the plant species to control the time and conditions for germination. Secondary or consequential dormancy is a further adaptation that prevents germination of an imbided or nondormant seed if other environmental conditions are unfavorable [8,24]. Geneve [24] categorized commonly grown small-seeded vegetable and flower species according to their dormancy type.

## VI.  TYPES OF DORMANCY

### A.  Primary

#### 1.  *Exogenous*

Viable seeds, especially of many woody species, may not germinate after considerable lengths of time even when the germination environment seems to be ideal. In most seeds with delayed germination, this condition is due to a hard seed coat restricting water absorption and gaseous exchange and/or due to the actual mechanical constraint by the seed coat or covering to the developing embryo [8]. Seeds from plants of the Cornaceae, Geraniaceae, Fabaceae, Malvaceae, and Convolvulaceae families are characterized by this condition [23]. Usually, this condition can be overcome by any method that softens, scarifies, or removes the seed coat or covering, including chemical or mechanical degradation (*scarification*).

#### 2.  *Endogenous*

In many species, delayed germination results from internal conditions of the embryo and food storage tissues, or a portion of these tissues, which must undergo certain developmental (rudimentary or immature embryos) or physiological changes before seeds will germinate [8,24].

Whereas seeds of many species possess germination inhibitors that are deactivated before germination occurs, seeds of other species may require "after-ripening," defined as a short period of dry storage after seeds are harvested, usually several weeks in duration (nondeep or transitory physiological dormancy) but possibly a few months or even several years for some species [8,24]. During this process, physiological and mechanical changes occur: growth-promoting hormones and enzymes are synthesized or activated, inhibiting hormones or related chemicals are deactivated, water is absorbed, and respiration increases.

The after-ripening requirement is quite common for seeds of herbaceous garden plants and vegetables but less common for trees and shrubs [21]. Other species may require a period of moist chilling, which may be moderate (intermediate physiological dormancy) or longer (deep physiological dormancy) in duration (Table 1). Tables 2 and 3 indicate the chilling requirements for selected flowering herbaceous and selected woody trees and shrubs, respectively.

#### 3.  *Double or Combinational*

Immature or rudimentary embryos are characteristic of many species of seeds with double dormancy, such as *Ginkgo biloba* (Maidenhair tree), *Ilex* spp. (holly), *Taxus* spp. (yew), *Viburnum* spp. (viburnum), and *Fraxinus nigra* (black ash). Seeds in this category do not germinate until dormancy related to external and internal physiological factors is overcome sequentially. In many other cases, the embryo may be fully developed and appears to be mature but may be dormant because of various physiological and chemical constraints [8,24]. Specific regions of the embryos may be responsible for this dormancy, such as the seed coat or a dormant radicle, hypocotyl, or epicotyl (Table 1).

## B.  Secondary

Viable seeds of many woody species may not germinate for considerable lengths of time, even when conditions are ideal. This dormancy condition prevents seeds in their natural habitat from germinating in unseasonal times. Readily germinable seeds should be stored promptly and properly until sown. This usually requires keeping the seeds dry and storing them in a refrigerator in sealed plastic bags or other sealed containers. In unseasonal times, seeds should be sown in a greenhouse or other favorable environment. When the weather warms sufficiently, seedlings grown inside may be moved to a cold frame for further growth [8].

## VII.  SEED TREATMENTS

The most simple and practical approach to overcoming seed dormancy is to sow in outdoor seed beds, allowing nature and its seasonal cycles to provide the appropriate conditions. Cultural practices, including good seed bed preparation, appropriate seeding rate and depth, and protection from pests, are all-important factors for success [8]. Seedlings are usually allowed to grow from one to two seasons in beds or containers and then transplanted to more permanent field locations or to larger containers [8]. However, good cultural practices may not always be reliable or successful.

Because of numerous interactions of the seed's heredity expression and the effect of environmental and other internal factors of the seed, the germination of different kinds of seeds, or even of different seeds of one kind, can be extremely variable [20].

Treatment of seeds may hasten or induce more uniform and greater germination. It is more often required with seeds of woody trees and shrubs than with those of herbaceous species. Rudolf [26] found that out of 400 species of woody plants, 33% had seeds that are commonly nondormant, 43% had seeds with internal dormancy, 7% had seeds with an impermeable seed coat, and 17% had more than one kind of seed dormancy.

Without treatment, dormant seeds of many woody species may not germinate or may do so sporadically over a prolonged period lasting 2 or 3 years, resulting in plants of irregular size and age in seed beds or flats. Treatment procedures usually involve one of two types: *scarification* or *stratification*. A combination of both procedures may be required for seeds with more complex dormancy (Tables 2 and 3).

## A.  Scarification

Scarification is any treatment applied to seed to enhance germination [41].

### 1.  Abrasion

Hard seed coats can be scratched or abraded with a file, sandpaper, or abrasive wheel or cracked with a hammer or vise grip. A small mechanical tumbler lined with sandpaper or filled with sand or gravel may be more practical for larger amounts of seed [42]. The quantity of seeds in the tumbler should be sufficient to allow all the seeds to be abraded. For very large quantities of seeds, a concrete mixer containing coarse sand or gravel may be quite effective. The sand or gravel must be a size that can be easily screened from the seeds [42].

Commercially designed machines are also available for scarifying large quantities of seed. These scarifiers usually abrade or scar the seeds between two rubber-faced surfaces or impel seeds against roughened surfaces such as sandpaper. The severity of the abrasion or impact must be controlled to prevent damage to the seed [34].

### 2.  Acid

Concentrated sulfuric acid ($H_2SO_4$ commercial grade, specific gravity 1.84) is often used because it is very effective. Caution is advised when using acid. NEVER ADD WATER TO ACID! Goggles and protective clothing should be worn. Care must be taken to avoid spilling. Skin that comes in contact with acid must be washed immediately under running cold water.

Using about twice the volume of acid over the seed in a glass container, stir gently with a glass rod during treatment. Duration of acid exposure will depend upon seed coat thickness. Fifteen minutes to 3 hr or more of exposure is required, depending on the species. Carefully pour off the acid and rinse seeds

**TABLE 1** Types of Seed Dormancy, Causal Factors, and Treatment Strategies

| Types of dormancy | Cause | Treatment strategies | Comments |
|---|---|---|---|
| A. Primary | Induced by seed-related factors although germination environment may be favorable. | | |
| 1. Exogenous | Regulated by factors outside the embryo, usually seed coat and other external covering. | | Seed coverings include capsules, hulls, pits or shells, and mucilaginous or fleshy parts of fruits that may become hardened. In nature or outdoor sowing, seed is "scarified" by alternate freezing and thawing, action of soil microbes; also by passage through intestines of animals or by fire. |
| Physical | Seed coat impervious to water or restricts aeration. | Soften, rupture, or remove; mechanical abrasion; alternate freezing and thawing. | |
| Mechanical | Hard seed covering constrains the embryo. | Rupture or remove; warm or cold treatment. | |
| Chemical | Seed coverings release chemical inhibitors or prevent leaching of inhibitors. | Remove seed coverings and/or leach with water. | Inhibitors include abscissic acid, phenols, and coumarins. In nature leaching access due to sufficient soil moisture and rainfall. |
| 2. Endogenous | Regulated by internal and physiological factors within the embryo. | | |
| Morphological | Rudimentary or immature embryo. | A period of warm temperature usually favors embryo development; sometimes cold treatment, alternating warm and cold or gibberellic acid; excise embryo and germinate aseptically. | |
| Physiological | Semipermeable inner seed coat or active membranes; restricted gaseous exchange and inhibitor movement; possible excess of abscissic acid. | | |
| Nondeep (transitory) | Lack of "after-ripening." | A short period of dry storage; sometimes short periods of chilling; alternating temperatures, potassium nitrate, or gibberellic acid. | "After-ripening" is defined as a period of dry storage required by some seeds to lose dormancy. This type of dormancy occurs in freshly harvested seeds of many herbaceous and some trees species. It often disappears with normal harvesting and handling of the seeds. |

| Type | Characteristic | Treatment | Remarks |
|---|---|---|---|
| Photodormant | | Expose to light (positive photodormancy) or darkness (negative photodormancy). | Some seeds may lose their requirement for light after dry storage. |
| Intermediate | Embryo germinates if seed coat removed. | Moderate periods of moist chilling < 2 months. | Most common in temperate herbaceous plant. |
| Deep | Embryo usually will not germinate if seed coat removed or will develop abnormally. | Long periods of moist chilling > 2 months; gibberellic acid. | Most common in temperate trees and shrubs. Abnormal development of unchilled deep dormant seeds includes enlargement and greening of the cotyledons, short thick radicles, no epicotyl growth, abnormal roots, physiological dwarfing. |
| 3. Double or combinational | Combinations of two or more types of primary dormancy; separate dormancy conditions for different parts of the embryo. | Eliminate dormancy conditions in proper sequence. | |
| Morphological Simple | Underdeveloped embryos plus physiological dormancy. | Warm to induce embryo development, then cold to eliminate physiological dormancy. | |
| Epicotyl | Epicotyl dormant; radicle and hypocotyl nondormant. | Warm to induce initial germination and growth of radicle and hypocotyl, then cold to release epicotyl dormancy. | |
| | Both epicotyl and radicle dormant. | Cold followed by warm for radicle growth, then another cold period to release epicotyl dormancy. | In nature, or when sown outdoors, such seeds require at least two full growing seasons to complete germination. |
| Exo-endodormancy | Example: seed coat dormancy plus intermediate/deep physiological dormancy. | Scarify then moist chill. | |
| B. Secondary | Induced by germination environment although seed itself not dormant. | | |
| 1. Thermodormancy | High temperature | Moist chilling; growth regulators. | Thermal dormant seeds will not germinate without treatment when temperatures become lower. |
| 2. Conditional | Unfavorable seasonal conditions. | Chilling or warm treatment. | Conditions include unfavorably high or low temperatures, water stress, prolonged light or darkness. Important survival adaptation for seeds in nature but less so in cultivation |

*Source:* Adapted from Refs. 8 and 24.

**TABLE 2** List of Selected Flowering Herbaceous Plants Showing Desired Medium Temperature Regimes for the Germinating and Posttransplant Stages and the Approximate Time for Germination under Greenhouse Conditions as well as the Need for Light or Seed Pretreatment

| Species | Plant type | Medium temperature (°C) | | Days to germinate | Light or dark | Pretreatment | Remarks |
|---|---|---|---|---|---|---|---|
| | | Germination | Posttransplant (night) | | | | |
| *Acanthus spinosus*<br>Bear's breeches<br>Acanthaceae | P | 10–12 | | 21 | | | Use fresh seed. |
| *Achillea filipendulina*<br>Yarrow<br>Asteraceae | P | 20–24 | 15 | 7–10 | L | | Lower germinating medium night temp. by 3°C.<br>See note 6. |
| *A. millefolium* | P-W | 18 | 10 | 5–8 | L | | Seeds may produce inferior plants; vegetative propagation may be more reliable. |
| *Aconitum* spp.<br>Monkshood<br>Ranunculaceae | P | 12–15 | | 30 | D | Sow, then freeze 3 wk | Fresh seed will germinate much more quickly than older, ripened seed. |
| *Actaea alba*<br>Baneberry<br>Ranunculaceae | P-W | 5–10 | 10 | 30 | D | Stratify 21°C 3 wk, freeze 5 wk, then sow | Drying of seed will delay germination. |
| *Adenophora confusa*<br>Lady bells<br>Campanulaceae | P | 21–23 | | 14–21 | L | | |
| *Adonis* spp.<br>Adonis<br>Ranunculaceae | P | 12–15 | | 30 | D | Sow, then freeze 3 wk | |
| *Agastache foeniculum* | P | 21–24 | 10–15 | 10–14 | D or L | None | |

**Definitions of plant type**

A = Annual.   A plant that normally complete its life cycle within one calendar year. Some may be considered short-lived perennials in areas of mild winters or with adequate protection (i.e., snow cover, mulch, sheltered location).

B = Biennial.   A plant that require 2 years to complete its life cycle, usually producing vegetative growth only in the first, then flowering and dying in the second.

P = Hardy herbaceous perennial.   A plant that lives longer than 2 years, normally producing flowers and seeds each year.

HHP = Half-hardy perennial.   A plant that may survive mild winters or harsher winters with adequate protection, i.e., snow cover, mulch, sheltered location.

P-W = Perennial wildflower.   A plant that is indigenous to North America or has become naturalized to the point of being considered so.

| Species | Type | °C | °C | Days | Light | Notes |
|---|---|---|---|---|---|---|
| *A. cana* Giant hyssop Labiaate | | | | | | |
| *Ageratum houstonianum* Floss flower Asteraceae | A | 21–24 | 12–16 | 7–10 | L | Lower germinating medium night temp. by 3°C. After transplanting, gradually reduce night temp. to 10°C. See note 6. |
| *Alcea rosea* Hollyhock Malvaceae | B | 18–21 | 10 | 12–18 | | |
| *Alchemilla mollis* Lady's mantle Rosaceae | P | 10–15 | 10 | 7–21 | D | Fresh seed germinates readily at 21°C. Old or purchased seed needs pretreatment. Self-sows easily. Stratify 21°C 3 wk, freeze 5 wk, then sow |
| *Amaranthus* spp. Joseph's coat Amaranthaceae | A | 21–24 | 15–18 | 8–10 | D or L | Good seed at 80–85% germination. |
| *Anacyclus depressus* Rocky mountain daisy Asteraceae | P | 15–20 | 10–15 | 14–21 | | Fresh seed germinates best. |
| *Anagalis monelli* | P | >10 | | 0–30 | D | |
| *A. linifolia* | A | >10 | | 0–30 | D | Can be directly sown to garden or container. |
| *A. arvensis* Pimpernel Primulaceae | P-W | 21 | 16 | 21–28 | D or L | |
| *Anaphalis margaritaceae* Pearly everlasting Asteraceae | P-W | 18–21 | 10–13 | 14–21 | L | |
| *Anchusa capensis* Bugloss, Alkanet Boraginaceae | A | 20–25 | 15–18 | 14–21 | D | Reduce night temp. of medium by 3°C for best germination. Can be directly sown to garden when soil has reached 20°C. |
| *A. azurea* | A | 20–25 | 15–18 | 14–21 | D | |

*continues*

**TABLE 2** *Continued*

| Species | Plant type | Medium temperature (°C) | | Days to germinate | Light or dark | Pretreatment | Remarks |
|---|---|---|---|---|---|---|---|
| | | Germination | Posttransplant (night) | | | | |
| *Anemone* spp.<br>Anemone<br>Ranunculaceae | P | 18–20 | 10–15 | 21 | | | Rub seed with dry sand to remove cottony covering. |
| *Angelica archangelica*<br>Wild parsnip<br>Apiaceae | B | 21–25 | 10–15 | 20–40 | L | None if fresh. | Use fresh seed. Viability decreases rapidly with time, i.e., seeds may not germinate after 6 months of storage. |
| *Antennaria* spp.<br>Pussytoes<br>Asteraceae | P | 22–24 | 15 | 14–21 | D or L | | Seed is very fine, cover lightly. |
| *Anthemis* spp.<br>Golden marguerite,<br>Chamomile<br>Asteraceae | P | 20 | 8–14 | 7–21 | | | |
| *Antirrhinum majus*<br>Snapdragon<br>Scrophulariaceae | HHP | 18–21 | 8–12 | 10–14 | L | Chill 0°C 7 days | Sow 6–8 wk before scheduled outdoor planting. Suscept-ible to damping-off disease. Use a presow fungicide drench. Keep medium moist at all times during germination. Lower temperature after germination. Plants self-sow easily. |
| *Aquilegia* spp.<br>Columbine<br>Ranunculaceae | P | 20–24 | 10 | 14–30 | L | Sow then refrigerate 3 wk | Reduce night temp of medium by 7–8°C during germination. |
| *Arabis* spp.<br>Rock cress<br>Brassicaceae | P | 21 | 10 | 15 | L | | |
| *Arctotis stoechadifolia*<br>African daisy<br>Asteraceae | A | 15–20 | 10–15 | 21–35 | D or L | | Sow 6–8 wk before scheduled outdoor planting. Use fresh seed. Viability decreases quickly. |

| Species / Common name / Family | | | | | | Treatment | Notes |
|---|---|---|---|---|---|---|---|
| *Arenaria montana* Sandwort Caryophyllaceae | P | 12–17 | | 15–20 | L | | Germination is sometimes low or erratic |
| *Arneria spp.* Thrift Plumbaginaceae | P | 15–18 | 10 | 14–21 | D or L | Soak in warm water 6–8 hr | |
| *Arnica spp.* Arnica Asteraceae | P | 12 | | 25–30 | | | |
| *Aruncus dioicus* Goat's beard Rosaceae | P | 22–24 | 10–15 | 14–21 | | If not sown fresh, stratify 5°C 4 wk | Fresh seed can be sown immediately. |
| *Asclepias tuberosa* Butterfly weed Asclepiadaceae | P-W | 20–24 | | 21–28 | D | If not sown fresh, stratify 5°C 3 wk | Germination is erratic. Fresh seed gives best results if sown immediately. |
| *Aster spp.* Aster Asteraceae | P | 21 | 10 | 15 | | | |
| *Astilbe chinensis* 'Pumila' False Spiraea Asteraceae | P | 15–21 | | 21–28 | | Sow, then stratify 21°C 2 wk, then 5°C 4 wk | |
| *A.* × *arendsii* Saxifragceae | P | 15–21 | | 21–28 | | | |
| *Astrantia major* Masterwort Apiaceae | P | 10–20 | 10 | 10–14 | | Stratify 2–5°C 4 wk | |
| *Aubrieta* sp. Rock cress Brassicaceae | P | 18 | 10 | 20 | D or L | | |
| *Aurinia saxitilis* Basket-of-gold Brassicaceae | P | 15 | 10 | 7–10 | L | | |
| *Baptisia australis* Wild indigo Fabaceae | P | 21–24 | | 5–10 | | Nick or file seed coat | When seeds turn black, collect and sow. |

*continues*

**TABLE 2** *Continued*

| Species | Plant type | Medium temperature (°C) | | Days to germinate | Light or dark | Pretreatment | Remarks |
|---|---|---|---|---|---|---|---|
| | | Germination | Posttransplant (night) | | | | |
| *Begonia semperflorens* | A | 21–24 | 16–18 | 15–20 | L | | Reduce night temp. of medium by 3°C for best germination. Fertilize with 100 ppm N immediately after germination Never allow seed surface to dry. |
| *B. × tuberhybrida* | A | 21–24 | 16–18 | 15–20 | L | | |
| Begonia | | | | | | | |
| Begoniaceae | | | | | | | |
| *Belamcanda chinensis* | P | 20–30 | | 14–21 | | | |
| Blackberry lily | | | | | | | |
| Iridaceae | | | | | | | |
| *Bellis perennis* | B | 20–24 | 10 | 6 | L | | Reduce night temp. of medium by 3°C for best germination. |
| English daisy | | | | | | | |
| Asteraceae | | | | | | | |
| *Bergenia cordifolia* | P | 13 | | 15–20 | | | |
| Bergenia | | | | | | | |
| Saxifragaceae | | | | | | | |
| *Boltonia asteroides* | P | 21 | 10 | 15 | D or L | | |
| Boltonia | | | | | | | |
| Asteraceae | | | | | | | |
| *Brachycome iberidifolia* | A | 21 | 10 | 10–18 | | | |
| Swan river daisy | | | | | | | |
| Asteraceae | | | | | | | |
| *Brassica oleraceae* var. *acephala* | A | 21–24 | 13–16 | 10 | L | | |
| Flowering cabbage, Kale | | | | | | | |
| Brassicaceae | | | | | | | |
| *Browallia speciosa* | A | 24 | 18–21 | 7–14 | L | | Reduce night temp. of medium by 3°C for best germination. Predrench medium with fungicide to prevent damping off; keep seedlings on the dry side. |
| Browallia | | | | | | | |
| Solanaceae | | | | | | | |

| Plant | Type | Temp. (°C) | | Days | Light | Notes |
|---|---|---|---|---|---|---|
| *Brunnera macrophylla* Siberian bugloss Boraginaceae | P | 21–24 | | 14–21 | D | |
| *Buphthalmum speciosum* Oxeye Asteraceae | A | 21–24 | | | D or L | |
| *Calendula officinalis* Pot marigold Asteraceae | A | 21 | 10 | 7–10 | D | |
| *Callistephus chinensis* Aster Asteraceae | A | 21 | 16 | 6–14 | D or L | Reduce night temp. of medium by 3°C for best germination. |
| *Campanula medium* | B | 20–24 | 10 | 6–16 | L | |
| *C. carpatica* | P | 20–24 | 10 | 6–16 | L | |
| *C. rotundifolia* Bell flower Campanulaceae | P-W | 20–24 | 10 | 6–16 | L | |
| *Catananche caerula* Cupid's dart Asteraceae | P | 21 | | 14–21 | L | |
| *Catharanthus roseus* Annual periwinkle Vinca Apocynoceae | A | 24–26 | 18–24 | 14–21 | D | Reduce night temp. of medium by 3°C for best germination. Be careful not to overwater. Plants are sensitive to high moisture levels. |
| *Celosia* spp. Cockscomb Amaranthaceae | A | 24–27 | 21 | 7–10 | L | Before sowing, drench medium with fungicide to prevent damping-off disease. Reduce night temp. of medium by 3°C for best germination. Plant growth will be stunted if temp. falls below 18°C. |

*continues*

**TABLE 2** *Continued*

| Species | Plant type | Medium temperature (°C) | | Days to germinate | Light or dark | Pretreatment | Remarks |
|---|---|---|---|---|---|---|---|
| | | Germination | Posttransplant (night) | | | | |
| *Centaurea cinerarea*<br>Dusty miller | HHP | 18–21 | 10–13 | 5–10 | D | | Use fresh seed for best results. |
| *C. cyanus*<br>Bachelor button<br>Asteraceae | A | 18–21 | 10–13 | 5–10 | D | | |
| *C. montana* | P | 18–21 | 10–13 | 5–10 | D | | |
| *Centranthus ruber*<br>Red valerian<br>Valerianaceae | P | 15 | 15 | 10–20 | | | |
| *Cephalaria gigantea*<br>Cephalaria<br>Dipsacaceae | P | 21–22 | 10–14 | 7–12 | D or L | If not sown fresh, stratify 5°C 6 wk | |
| *Cerastium tomentosum*<br>Snow-in-summer<br>Caryophyllaceae | P | 15 | 18 | 8–18 | D or L | | |
| *Ceratostigma plumbaginoides*<br>Plumbago<br>Plumbaginaceae | P | 10–20 | | 10–21 | | Stratify 5°C 4–6 wk | |
| *Chelone glabra* | P | 15 | | 10–14 | | Sow, then stratify 5°C 6 wk | |
| *C. lyonii*<br>Turtlehead<br>Scrophulariaceae | P | 15 | | 10–14 | | | |
| *Chiastophyllum oppositifolium*<br>Cotyledon<br>Crassulaceae | P | 13 | | 30 | L | Stratify 15–20°C 2–3 wk, then 5°C 4–6 wk | Seeds are tiny. Do not cover with medium. |
| *Chieranthus chieri*<br>Wallflower<br>Brassicaceae | B | 21 | 10 | 7–10 | D or L | | Can be directly sown to garden in summer. |
| *Chrysanthemum parthenium*<br>Feverfew | A | 18–21 | 10–13 | 5–10 | L | | Reduce night temp. of medium by 3°C for best germination. |
| *C. coccineum*<br>Painted daisy | P | 15–18 | 10 | 7–14 | L | | |

| | | | | | | |
|---|---|---|---|---|---|---|
| C. ptarmiciflorum<br>Dusty miller | P | 15–18 | 10 | 7–14 | D or L | |
| C. superbum<br>Shasta daisy<br>Chrysanthemum<br>Asteraceae | P | 15–18 | 10 | 7–14 | L | Plants self-sow. |
| Chrysogonum virginianum<br>Golden aster<br>Asteraceae | P | 22–24 | | 14–21 | D or L | |
| Chrysopsis sp.<br>Golden aster<br>Asteraceae | P | 22–24 | | 14–21 | D or L | Can be directly sown to garden in summer. |
| Cimicifuga simplex<br>Bugbane | P | 10 | | 7–30 | | Stratify 20–25°C 6–10 wk, then 1–5°C 6–8 wk |
| C. racemosa<br>Snakeroot<br>Ranunculaceae | P | 10 | | 7–30 | | Use fresh seed. Germination is erratic. |
| Clarkia spp.<br>Clarkia, Godetia<br>Onagaceae | A | 20–23 | 12–15 | 5–10 | D or L | Difficult to transplant. Best sown directly to garden or container. |
| Cleome hasslerana<br>Spiderflower<br>Capparaceae | A | Day 28<br>Night 21 | 21–24 | 5–14 | L | Alternate day and night temp. as noted |
| Codonopsis clematidea<br>Codonopsis<br>Campanulaceae | P | 20–25 | 10–15 | 14–21 | L | |
| Coleus × hybridus<br>Coleus<br>Lamiaceae | A | 21–24 | 16–18 | 10–21 | L | Presow drench with fungicide to prevent damping off. |
| Collinsia grandiflora<br>Blue lips<br>Scrophulariaceae | A | 15–20 | 10–15 | 10–15 | | |
| Consolida ambigua<br>Larkspur<br>Ranunculaceae | A | 15–20 | 13–16 | 15–21 | D | Difficult to transplant. Best shown directly to garden or container. |
| Convolvulus tricolor<br>Dwarf morning glory<br>Convolvulaceae | A | 21–24 | 21 | 10–16 | D or L | |

*continues*

**TABLE 2** Continued

| Species | Plant type | Medium temperature (°C) | | Days to germinate | Light or dark | Pretreatment | Remarks |
|---|---|---|---|---|---|---|---|
| | | Germination | Posttransplant (night) | | | | |
| *Coreopsis tinctoria* Tickseed Asteraceae | A | 20–24 | 10 | 7–21 | L | | Reduce night temp. of medium by 3°C for best germination. |
| *C. lanceolata* | P-W | 20–24 | 10 | 7–21 | L | | |
| *C. verticillata* | P | 20–24 | 10 | 7–21 | L | | |
| *Coronilla varia* Crown vetch Fabaceae | P | 21–23 | | 30 | D or L | Nick hard seed coat or soak in water 6–9 hr | |
| *Corydalis lutea* Yellow corydalis Fumariaceae | P | 22–25 | | 7–14 | | Stratify 22°C 6–8 wk, then 0–2°C 6–8 wk | Plants self-sow easily. Seed must be sown fresh. Vitality decreases rapidly with age. |
| *Cosmos sulphureus* Cosmos Asteraceae | A | 24 | 18 | 5–7 | D or L | | |
| *C. bipinnatus* | A | 24 | 18 | 5–7 | D or L | | |
| *Cuphea ignea* Cigar flower Lythraceae | A | 21 | | 8–10 | L | | |
| *Cynaglossum amabile* Chinese hound's tongue Boraginaceae | B | 18–23 | | 5–10 | D | | |
| *Dahlia* hybrids Dahlia Asteraceae | A | 21–24 | 13–16 | 7–10 | D or L | | |
| *Datura meteloides* Angel's trumpet Solanaceae | A | 21–26 | 15–20 | 30–50 | L | | Germination may be slow and erratic. |
| *Delphinium* spp. Delphinium Ranunculaceae | P | Day 27 Night 21 | 7–10 | 7–10 | D | Sow, then freeze 4–8 hr or more | Use fresh seed. Viability decreases rapidly with age. Alternate day and night temp. as noted. |
| *Dianthus chinensis* | P | 21 | 10–13 | 8–10 | D or L | | |
| *D. plumarius* | P | 21 | 10–13 | 8–10 | D or L | | |

| | Type | | | Days | Light | Pretreatment | Comments |
|---|---|---|---|---|---|---|---|
| *D. caryophyllus* <br> Pinks <br> Caryophyllaceae | P | 21 | 10–13 | 8–10 | D or L | | Self-sows easily. Fresh seed may germinate immediately. Older seed will need to be stratified. |
| *Dicentra* <br> Bleeding heart <br> Fumariaceae | P | 10–12 | | 21–28 | D or L | Freeze or stratify 15–18°C 4 wk, then 5°C 4 wk | |
| *Diascia barberae* <br> Twinspur <br> Scrophulariaceae | A | 21–25 | 10–12 | 5–10 | L | | |
| *Dictamnus albus* <br> Gas plant <br> Rutaceae | P | 12–15 | | 30–40 | | Stratify 3–5°C 6 wk | If fall-sown outdoor, germination will occur over a 2-year period and may be more productive than "artificial" stratification. |
| *Digitalis* <br> Foxglove <br> Scrophulariaceae | B | 21 | 10 | 7–10 | L | | Reduce night temp. of medium by 3°C for best germination. |
| *Dimorphotheca aurantiaca* <br> Cape marigold <br> Asteraceae | A | 16–18 | 16 | 7–10 | D or L | | Seeds are short lived and should not be stored. |
| *Disporum* <br> Fairy bells <br> Liliaceae | P | 15–20 | 10 | 10–14 | D or L | | |
| *Dodecatheon media* <br> Shooting star <br> Primulaceae | P | 15–22 | | 30+ | | Freeze or stratify 3–5°C 4–6 wk | Germination may be erratic. |
| *Doronicum* spp. <br> Leopard's bane <br> Asteraceae | P | 21 | 10 | 10–20 | L | | Reduce night temp. of medium by 3°C for best germination. |
| *Dorotheanthus bellidiformis* <br> (*Mesembryanthemum criniflorum*) <br> Livingston daisy <br> Aizoaceae | A | 18–22 | | 15–20 | D | | Place in dark location but do not cover because seed is fine. |
| *Droba aizoides* <br> Whitlow grass <br> Cruciferae | P | 20–22 | 10–12 | 10–14 | L | | |

*continues*

**TABLE 2** *Continued*

| Species | Plant type | Medium temperature (°C) | | Days to germinate | Light or dark | Pretreatment | Remarks |
|---|---|---|---|---|---|---|---|
| | | Germination | Posttransplant (night) | | | | |
| *Dryas octopetala* Mountain arens Rosaceae | P | 15–20 | | 50+ | | | |
| *Dyssodia tenuiloba* Dahlberg daisy Asteraceae | A | 18–21 | 15–18 | 10–15 | D or L | | |
| *Echinacea purpurea* Purple coneflower Asteraceae | P-W | 21–24 | 13–16 | 10–12 | L | | Reduce night temp. of germinating medium by 3°C for best germination. Seeds may not come true to type and may take 2 years to flower. |
| *Echinops ritro* Glove thistle Asteraceae | P | 20–24 | 10 | 8–20 | L | | |
| *Epimedium* spp. Barrenwort Berberidaceae | P | 5–10 | 7–15 | Variable | D | Sow, then stratify 20–30°C 6–8 wk then 4–7°C 6–8 wk | When sown outdoors in late summer, seeds are naturally stratified and germinate in early spring. |
| *Erigeron* spp. Fleabane Asteraceae | P | 13 | 10 | 15–20 | D or L | | |
| *Erinus alpinus* Liverbalsam Scrophulariaceae | P | 18–24 | 10–12 | 20–25 | | | |
| *Eriophyllum lanatum* Woolly sunflower Asteraceae | P | 20–24 | 10–15 | 4–8 | L | None | |
| *Eryngium* Sea holly Apiaceae | P | 18–23 | 15 | 5–10 | | Stratify 20–24°C 2–4 wk, then 3–5°C 4–6wk | Use fresh seed. After 2 wk, dormancy sets in and germination time may be extended over 1–2 years. |

| Name | | Temp | | Days | D/L | Treatment | Comments |
|---|---|---|---|---|---|---|---|
| *Erysimum* spp.<br>Alpine erysimum<br>Brassicaceae | B | 13–18 | 10 | 7–10 | | | |
| *Eschscholzia californica*<br>California poppy<br>Papaveraceae | A | 16–18 | 10–12 | 14 | D | | Plants self-sow easily. Difficult to transplant. Best sown directly to garden or container. |
| *Eupatorium* spp.<br>Boneset<br>Asteraceae | P | 20–24 | 15 | 2–3 wk | D or L | Stratify 5°C 4 wk | |
| *Euphorbia epithymoides* | P | 21–24 | 18–21 | 8–12 | D or L | Chill 7 days, then | |
| *E. marginata* | A | 21–24 | 18–21 | 8–12 | D or L | soak 4 hr in water | |
| *E. myrsinites*<br>Spurge<br>Euphorbiaceae | P | Day 27<br>Night 27 | 10 | 12–18 | L | | |
| *Filipendula* spp.<br>Meadowsweet<br>Rosaceae | P | 18–23 | 12–15 | | | | |
| *Gaillardia pulchella* | A | 20–24 | 10 | 8–12 | L | | Reduce night temp. of medium by 3°C for best germination. |
| *G. aristata*<br>Blanket flower<br>Asteraceae | P-W | 20–24 | 10 | 8–12 | L | | |
| *Gaura lindheimeri*<br>Gaura<br>Onagraceae | P | 18–23 | | 14–21 | | | Difficult to transplant because of a long tap root. |
| *Gazania* spp.<br>Gazania<br>Asteraceae | A | 15–17 | 13–16 | 10–14 | D | | |
| *Gentiana acaulis*<br>Gentian<br>Gentianaceae | P | 20–25 | 10 | 14–28 | | Sow, then freeze 3 wk | Likes cool growing condition. |
| *Geranium* spp.<br>Cranesbill<br>Geraniaceae | P | 20–22 | 10–12 | 7–28 | | Ripen fresh seeds 22–24°C 2–4 wk | Seeds are short lived and should not be stored. |
| *Gerbera jamesonii*<br>Transvaal daisy<br>Asteraceae | A | 20–22 | 16–18 | 7–14 | D | | Use fresh seed. Stored seed loses viability. |

*continues*

**TABLE 2** *Continued*

| Species | Plant type | Medium temperature (°C) Germination | Posttransplant (night) | Days to germinate | Light or dark | Pretreatment | Remarks |
|---|---|---|---|---|---|---|---|
| *Geum* spp. Avens Rosaceae | P | 20–24 | 10 | 10–20 | L | | Reduce night temp. of medium by 3°C for best germination. |
| *Gillenia trifoliata* Bowman's root Rosaceae | P | 20–24 | 10–12 | Variable | | Stratify −4 to +4°C 3 months | Stored seed (1 year) germinates best. |
| *Globularia* | P | 13 | 10 | 10–12 | | Stratify 5°C 21 days | |
| *G. cordifolia* Globe daisy Globulariaceae | P | 13 | 10 | 10–12 | | | |
| *Godetia* Farewell-to-spring Onagraceae | A | 21 | 10 | 10 | D | | |
| *Gomphrena globosa* Globe amaranth Amaranthaceae | A | 21–24 | 21 | 10–14 | D | Soak in water 24 hr | Germination may be erratic or slow. Predrench medium with fungicide to prevent damp-off. Keep seedlings relatively dry. |
| *Gypsophila repens* | P | 21–25 | 13 | 10–14 | L | | |
| *G. elegans* Baby's breath Caryophyllaceae | A | 21–25 | 13 | 10–14 | L | | |
| *Helenium hoopesii* | P-W | 21 | 13–16 | 12–16 | L | | Reduce night temp. of medium by 3°C for best germination. |
| *H. autumnale* Sneezeweed Asteraceae | P-W | 21 | 13–16 | 12–16 | L | | |
| *Helianthenum* spp. Rock rose Cistaceae | P | 20–24 | 10–20 | 15–25 | D or L | Chill 4°C 2 wk | Use fresh seed for best results. Seed stored longer than 1 year has much reduced viability. |

| Name | Type | | | | Light | Treatment | Comments |
|---|---|---|---|---|---|---|---|
| *Helianthus annuus* Sunflower Asteraceae | A | 20–30 | 15–20 | 14–21 | D or L | | Seeds best sown directly in garden because they grow large and rapidly. |
| *Helichrysum bracteatum* Strawflower Asteraceae | A | 24 | 16 | 7–10 | L | | |
| *Heliopsis* spp. Heliopsis Asteraceae | P | 20 | 10–12 | 7–14 | | | |
| *Heliotropium arborescens* Garden heliotrope Boraginaceae | A | 21–24 | 16–18 | 14–21 | L | | Reduce night temp. of medium by 3°C for best germination. |
| *Helleborus* spp. Hellebore Ranunculaceae | P | 20–23 | 5–10 | 30–60 | | Stratify 20°C 8–10 wk, then 2–5°C 8–10 wk, may need to repeat once | |
| *Hemerocallis* spp. Daylily Liliaceae | P | 15–22 | 10–12 | 21–50 | | Stratify 3–5°C 6 wk | |
| *Herniaria glabra* Rupture wort Illecebraceae | P | 21 | 10–15 | 10–12 | | | |
| *Hesperis matronalis* Dame's rocket Brassicaceae | P-W | 21–24 | 10 | 7–10 | L | | Reduce night temp. of medium by 3°C for best germination. |
| *Heuchera sanguinea* Coral bells, Alumroot Saxifragaceae | P | 8–21 | 10 | 10–18 | L | | |
| *Hibiscus trionum* Mallow Malvaceae | A | 21 | 10 | 14–20 | D | Scarify in $H_2SO_4$ 30 min or soak in water 48 hr | Hard seed coat may require scarification. |
| *Hosta* spp. Plantain lily Liliaceae | P | 21 | 5–15 | 15–20 | | | |

*continues*

**TABLE 2** *Continued*

| Species | Plant type | Medium temperature (°C) | | Days to germinate | Light or dark | Pretreatment | Remarks |
|---|---|---|---|---|---|---|---|
| | | Germination | Posttransplant (night) | | | | |
| *Hunnemannia fumariifolia* Mexican tulip poppy Papaveraceae | P | 21–24 | 15 | 15–20 | D | | |
| *Hypericum* spp. St. John's wort Hypericaceae | P | 21–23 | 10–15 | 25–30 | | | |
| *Hypoestes phyllostachya* Polka dot plant Acanthaceae | A | 21–24 | 15 | 10–12 | | | |
| *Hyssopus officinalis* Hyssop Labiatae | P | 20–24 | 10–15 | 14–21 | D or L | | |
| *Iberis umbellata* Candytuft Cruciferae | A | 21 | 10 | 7–14 | | | |
| *I. sempervirens* | P | 16–18 | 7–10 | 16–20 | L | | |
| *Impatiens* spp. Patience plant Balsaminaceae | A | 21–24 | 16–18 | 7–18 | L | | Keep surface of medium moist at all times during germination. |
| *Incarvillea delavayi* Hardy gloxinia Bignoniaceae | P | 13–18 | 13–15 | 25–30 | | | |
| *Inula* spp. Elecampane Asteraceae | P | 20–24 | 15–20 | 14–28 | L | | Germination may be erratic. |
| *Ipomoea tricolor* Morning glory vine Convolvulaceae | A | 21–28 | 15–20 | 7–10 | | Nick seed coat with a file or soak in water 24 hr | |
| *Iris* spp. Iris Iridaceae | P | 22 | 10–20 | 21–40 | | | Use fresh seed. Germination may be slow and erratic. |

| | Type | | | | Light | Treatment | Comments |
|---|---|---|---|---|---|---|---|
| *Jasione perennis*<br>Sheep's bit<br>Campanulaceae | P | 21 | 15 | 10–15 | | | |
| *Knautia macedonica*<br>Knautia<br>Dipsacaceae | P | 20–24 | 10–20 | 14–50 | L | | Expect erratic germination. |
| *Kniphofia* spp.<br>Torch lily<br>Liliaceae | P | 20–24 | 10 | 10–18 | L | Sow, then stratify 5°C 6 wk | Reduce night temp. of medium by 3°C for best germination. |
| *Kochia scoparia*<br>Burning bush<br>Chenopodiaceae | A | 21–24 | 18 | 6–9 | L | Soak in water 24 hr | Reduce night temp. of medium by 3°C for best germination. Water sparingly after germination. Directly sow to garden or transplant container. |
| *Lathyrus* spp.<br>Sweet pea<br>Fabaceae | A or P | 12–18 | 12–18 | 10–25 | D | Nick seed coat with a file or soak in water for 24 hr | |
| *Lavandula* spp.<br>Lavender<br>Lamiaceae | P | 20 | 18 | 20–30 | D of L | Stratify 5°C 30 days | |
| *Lavatera trimestris*<br>Tree mallow<br>Malvaceae | A | 21 | 18–20 | 15–20 | | | Difficult to transplant. Best sown directly to garden or container. |
| *Layia platyglossa*<br>Tidy tips<br>Asteraceae | A | 21–24 | 20 | 10–12 | | | |
| *Leontopodium alpinum*<br>Edelweiss<br>Asteraceae | P | 15–20 | 10–15 | 14–21 | L | | |
| *Lewisia cotyledon*<br>Lewisia<br>Portulacaceae | P | 21 | | 30 | | Stratify 3–5°C 3 wks | |
| *Liatris*<br>Blazing star, Gray feather<br>Asteraceae | P | 15–24 | 18–21 | 15–22 | | | Reduce night temp. of medium by 3°C for best germination. |

*continues*

**TABLE 2** *Continued*

| Species | Plant type | Medium temperature (°C) | | Days to germinate | Light or dark | Pretreatment | Remarks |
|---|---|---|---|---|---|---|---|
| | | Germination | Posttransplant (night) | | | | |
| *Ligularia dentata*<br>Golden ray<br>Asteraceae | P | 20–24 | 15–18 | 5–10 | D | None if fresh. Stratify stored seed 20–24°C 2–4 wk then 3–5°C 4–6 wk | Fresh seed recommended. Dormancy sets in quickly (2 wk after seed harvest). When sown outdoors, germination may occur 1–2 years later. |
| *Limonium sinuatum*<br>Statice<br>Plumbaginaceae | A | 21 | 18 | 14–18 | D | | |
| *Linaria maroccana*<br>Toad flax<br>Scrophulariaceae | A | 18–21 | 13 | 8–10 | | | Reduce night temp. of medium by 3°C for best germination. |
| *L. vulgaris* | P-W | 18–21 | 13 | 8–10 | | | |
| *Linum grandiflorum*<br>Flax<br>Linaceae | A | 18–21 | 10 | 8–18 | L | | |
| *L. perenne* | P | 15–18 | 10 | 8–18 | L | | |
| *L. flavum* | P | 18–21 | 10 | 8–18 | L | | |
| *Liriope* spp.<br>Lily-turf<br>Liliaceae | P | 18–21 | | 30 | | Soak in water 24 hr | |
| *Lisianthus russellianus (Eustoma grandiflora)*<br>Prairie gentian<br>Gentianaceae | A | 21–24 | 18 | 8–12 | L | | Seedlings grow slowly. |
| *Lithodora diffusa*<br>Lithodora<br>Boraginaceae | P | 20 | | 14–21 | | Scarify with sandpaper or file, or soak in warm water for 2 days | |
| *Lobelia erinus*<br>Lobelia<br>Lobeliaceae | A | 24 | 16 | 14–21 | L | | |
| *L. cardinalis* | P-W | 21 | 10–13 | 10–14 | L | Stratify 3–5°C 3 months | |
| *L. siphilitica* | P-W | 21 | 10–13 | 10–14 | L | | |

| Name | Type | Temp | | Days | Light | Treatment | Comments |
|---|---|---|---|---|---|---|---|
| *Lobularia maritima* <br> Sweet alyssum <br> Brassicaceae | A | 20–24 | 8–10 | 4–10 | L | | Can be directly sown to garden or container. Predrench medium with fungicide to prevent dampingoff disease. |
| *Lunaria annua* (*L. biennis*) <br> Honesty <br> Brassicaceae | B | 15–18 | 10 | 16–20 | D | | |
| *Lupinus* spp. <br> Lupine <br> Fabaceae | P | 21–24 | 10 | 16–20 | D | Soak in warm water 18 hr | Reduce night temp. of medium by 3°C for best germination. |
| *Lychnis chalcedonica* <br> Campion <br> Caryophyllaceae | P | 21 | 18 | 14–21 | L | | |
| *Machaeranthera tanacetifolia* <br> Tahoka daisy <br> Asteraceae | A | 21 | 10–12 | 25–30 | | Stratify 3–5°C 2 wk | |
| *Malva alcea* <br> Mallow <br> Malvaceae | B | 18–21 | 10 | 12–18 | | | |
| *Mathiola incana* <br> Stocks <br> Brassicaceae | A | 20 | 10–13 | 7–10 | L | | Before sowing, drench medium with fungicide to prevent damping-off disease. |
| *Meconopsis* <br> Welsh poppy <br> Papaveraceae | P | 18–22 | 10–12 | 20–25 | | | |
| *Melampodium paludosum* <br> African zinnia <br> Asteraceae | A | 21–25 | 15–18 | 7–10 | D | | |
| *Mimulus* × *hybridus* <br> Monkey flower <br> Scrophulariaceae | A | 18–21 | 16 | 14–21 | L | | Can grow on at low temp. (3–6°C) once established. Sow when days are at least 13 hr long to initiate flowering. |
| *Mirabilis jalapa* <br> Four o'clock <br> Nyctaginaceae | A | 21 | 10–12 | 7–10 | | | |

*continues*

**TABLE 2** *Continued*

| Species | Plant type | Medium temperature (°C) | | Days to germinate | Light or dark | Pretreatment | Remarks |
|---|---|---|---|---|---|---|---|
| | | Germination | Posttransplant (night) | | | | |
| *Moluccella laevis* Bells of Ireland Lamiaceae | A | 25 day 10 night | 16 | 25–30 | L | Chill 7°C 5 days | Difficult to transplant due to long tap root. Alternate day and night temp. as noted. |
| *Monarda fistulosa* Beebalm Lamiaceae | P-W | 21 | 18–21 | 14–21 | L | | Reduce night temp. of medium by 3°C for best germination. |
| *Myosotis* spp. Forget-me-not Boraginaceae | P | 18–21 | 10 | 8–12 | D | | |
| *Nemesia strumosa* Pouch nemesia Scrophulariaceae | A | 16 | 13–16 | 5–10 | D | | Sensitive to medium temp. above 18°C. |
| *Nemophila menziesii* Baby blue eyes Hydrophyllaceae | A | 12 | 13–15 | 7–12 | | | |
| *Nepeta × faassenii* Catmint Lamiaceae | P | 20–24 | 10 | 5–7 | | | |
| *Nicotiana alata* Flowering tobacco Solanaceae | A | 21 | 13 | 7–12 | L | | Reduce night temp. of medium by 3°C for best germination. Transplants can be grown very cool (5–7°C). |
| *Nierembergia hippomanica* Cupflower Solanaceae | A | 25 | 18–20 | 14 | | | |
| *Nigella damascena* Love-in-a-mist Ranunculaceae | A | 20–22 | 15 | 10–15 | | | Difficult to transplant. Best sown directly to garden or container. |
| *Ocimum basilicum* Basil Lamiaceae | A | 22 | 15 | 10–14 | | | |

| Name | Type | | | | D/L | Treatment | Remarks |
|---|---|---|---|---|---|---|---|
| *Oenothera argillicola* | P-W | 18–21 | 13–16 | 10–14 | D or L | | Reduce night temp. of medium by 3°C for best germination. |
| *O. missouriensis* | P-W | 20–24 | 18–21 | 10–21 | L | | |
| Evening primrose | | | | | | | |
| Onagraceae | | | | | | | |
| *Osteospermum* spp. | A | 20–24 | 7–10 | 7–10 | D or L | | |
| Star of the Veldt | | | | | | | |
| Asteraceae | | | | | | | |
| *Paeonia* spp. | P | 22 | 20 | 30–60 | | Sow at 21°, allow roots to emerge then refrigerate 3–5°C 8–10 wk | Vegetative propagation is preferred because it takes 5–7 years to produce a flowering plant from seed. |
| Peony | | | | | | | |
| Paeoniaceae | | | | | | | |
| *Papaver rhoeas* | A | 20–24 | 7–10 | 12–16 | D | | Reduce night temp. of medium by 3°C for best germination. |
| *P. orientalis* | A | 20–24 | 7–10 | 12–16 | L | | |
| Poppy | | | | | | | |
| Papaveraceae | | | | | | | |
| *Pelargonium* × *hortorum* | A | 21–24 | 16–18 | 5–18 | D or L | Soak in $H_2SO_4$ 4 min, may induce uniform germination | Presow drench medium with fungicide to prevent damping off. |
| Geranium | | | | | | | |
| Geraniaceae | | | | | | | |
| *Penstemon gloxinioides* | A | 16 | 10 | 7–14 | L | | |
| *P. strictus* | P-W | 16 | 10 | 7–14 | L | | |
| Penstemon, beard tongue | | | | | | | |
| Scrophulariceae | | | | | | | |
| *Petunia* × *hybrida* | A | 21–24 | 13–16 | 7–12 | L | | Keep surface of medium constantly moist until germination is complete. Seedling vigor varies widely between cultivars. |
| Petunia | | | | | | | |
| Solanaceae | | | | | | | |
| *Phacelia campanularia* | A | 12–15 | 12–15 | 4–10 | | | |
| California bluebell | | | | | | | |
| Hydrophyllaceae | | | | | | | |
| *Phlox drummondi* | A | 15–16 | 10–13 | 10 | D | Stratify 5°C 2 wk | Presow drench medium with fungicide to prevent damping off. |
| *P. paniculata* | P | 21 | | 25–30 | D | | |
| Phlox | | | | | | | |
| Polemoniaceae | | | | | | | |

*continues*

**TABLE 2** *Continued*

| Species | Plant type | Medium temperature (°C) | | Days to germinate | Light or dark | Pretreatment | Remarks |
|---|---|---|---|---|---|---|---|
| | | Germination | Posttransplant (night) | | | | |
| *Phuopsis stylosa* Crosswort Rubiaceae | P | 20–24 | 15–18 | 10–20 | D or L | | |
| *Physalis alkekengi* Chinese lantern Solanaceae | P | 21–24 | 18 | 12–24 | L | Stratify 5°C 4–6 wk | Germination may be slow—up to 6 months. |
| *Physostegia virginica* Obedient plant Lamiaceae | P-W | Day 24 Night 13 | 10 | 12–24 | D or L | | Alternate day and nigh temp. as noted. |
| *Platycodon grandiflorus* Balloon flower Campanulaceae | P | 18–22 | 10–15 | 14–21 | L | | |
| *Plumbago auriculata* Leadwort Plumbaginaceae | P | 20–24 | 15–18 | 7–14 | D or L | | |
| *Polemonium* spp. Jacob's ladder Polemoniaceae | P | 20–24 | 13 | 10–20 | D or L | | Reduce night temp. of medium by 3°C for best germination. |
| *Polygonatum* spp. Solomon's seal Liliaceae | P-W | 20–23 | 15–18 | 20–25 | | Stratify 5°C 6 wk | |
| *Polygonum* spp. Smart knotweed Polygonaceae | P | 21 | 18 | 21–28 | D or L | | Germination may be erratic. |
| *Portulaca grandiflora* Rose moss Portulacaceae | A | 21–24 | 16–18 | 7–10 | L | | Reduce night temp. of medium by 3°C for best germination. Seed can be directly sown to garden or container. |
| *Potentilla* spp. Cinquefoil Rosaceae | P | 21 | 10–12 | 15–20 | D or L | | Seeds lose viability quickly and should not be stored. |

| Plant | Life cycle | Temp | Second temp/time | Days | Light | Treatment | Comments |
|---|---|---|---|---|---|---|---|
| *Primula* spp. Primrose Primulaceae | P | 16–18 | 3 wk at 13, then 5–7 | 16–24 | L | Stratify 2–5°C 3–4 wk | Reduce posttransplant temp gradually. Sow seeds as soon as they ripen to enhance germination. |
| *Pulsatilla vulgaris* Pasque flower Ranunculaceae | P | 20–23 | 10–12 | 30–40 | | | Seed must be fresh or dormancy occurs. |
| *Ranunculus* spp. Buttercup Ranunculaceae | P | 15–16 | 10–12 | 6–8 wk | | | Fresh seed can be sown immediately. |
| *Ratibida columnifera* Coneflower Asteraceae | P-W | 20–24 | 13 | 17–20 | D or L | | Reduce night temp. of medium by 3°C for best germination. |
| *Reseda odorata* Mignonette Resedaceae | A | 21 | 15 | 5–10 | L | | Difficult to transplant. Best sown directly to garden or container. |
| *Racinis communis* Castor bean plant Euphorbiaceae | A | 21–24 | 15–18 | 15–20 | D | Nick seed coat or soak in warm water 24 hr | |
| *Rodgersia* spp. Rodgers flower Saxifragaceae | P | 21–23 | 10–15 | | D or L | | Seeds are tiny. Do not cover. |
| *Rudbeckia hirta* Cone flower, Gloriosa daisy Asteraceae | P-W | 21–24 | 16 | 7–10 | L | | Reduce night temp. of medium by 3°C for best germination. |
| *Salpiglossis sinuata* Painted tongue Solanaceae | A | 21–24 | 10–13 | 12–15 | D | | Seeds are very small. Do not cover. Place in dark to germinate. |
| *Salvia* spp. Sage Lamiaceae | A | 21–24 | 16 | 10–14 | L | | Seeds need light for only 48 hr, then can be covered. Reduce night temp. of medium by 3°C for best germination. Harden off seedlings at 13°C for 3–4 days before transplanting. Keep salt levels low in transplant medium. |

*continues*

**TABLE 2**  *Continued*

| Species | Plant type | Medium temperature (°C) | | Days to germinate | Light or dark | Pretreatment | Remarks |
|---|---|---|---|---|---|---|---|
| | | Germination | Posttransplant (night) | | | | |
| *Sanguinaria canadensis* Bloodroot Papaveraceae | P | 22–25 | 10–14 | 3 mo–2 yr | D | None if fresh. Stratify stored seed 22°C 2–4 wk, then 0–2°C 6 wk | Fresh seed recommended. Stored seed requires prolonged time for germination. Cover lightly if sown outdoors. |
| *Sanvitalia procumbens* Creeping zinnia Asteraceae | A | 21 | 16–18 | 10 | L | | Difficult to transplant. Best sown directly to garden or container. Reduce night temp. of medium by 3°C for best germination. Grow on the dry side after germination. |
| *Saponaria ocymoides* Soapwort Caryophyllaceae | P | 15–20 | 8–12 | 14 | D | Stratify 5°C 2–4 wk | |
| *Saxifraga* spp. Saxifrage Saxifragaceae | P | 15 | 15 | 14–21 | D or L | Stratify 0–3°C 6 | |
| *Scabiosa* spp. Pincushion flower Dipsacaceae | P | 21 | 13 | 10–12 | D or L | | Reduce night temp of medium by 3°C for best germination. Germination may be slow and erratic. |
| *Sedum* spp. Stonecrop Crossulaceae | P | 15–25 | 10 | 10–14 | L | | |
| *Sempervivum tectorum* Hens and chickens Crassulaceae | P | 21 | 10–13 | 10–14 | L | | Normally propagated vegetatively. |
| *Senecio cineraria* Dusty miller Asteraceae | HHP | 21–24 | 16–18 | 10 | L | | Predrench medium with fungicide to prevent damping off. |

| | | | | | | | |
|---|---|---|---|---|---|---|---|
| *Sidalcea malviflora* Prairie mallow Malvaceae | P | 20–24 | 10–14 | 5–9 | D or L | None if fresh | If seeds is more than 6 months old, sow and chill 4–10°C for 1–2 wk |
| *Silene armeria* Catchfly Caryophyllaceae | P | 22 | 12–15 | 15–20 | | | Seed is small. Cover lightly or not at all. |
| *Sisyrinchium* spp. Blue-eyed grass Iridaceae | P-W | 21–23 | 15 | 21–28 | D or L | | |
| *Stachys* spp. Betony Lamiaceae | P | 18–21 | 10 | 10–16 | L | | Reduce night temp. of medium by 3°C for best germination. Do not overwater. Germination may be erratic. |
| *Stokesia laevis* Stoke's aster Asteraceae | P | 22–25 | 15–18 | 28–40 | | Stratify 5°C 6 wk | |
| *Tagetes erecta* Asteraceae | A | 21–24 | 16–18 | 5–8 | D or L | | Reduce night temp. of medium by 3°C for best germination. *T. erecta* cultivars require 14 hr or more of darkness per day at the germination and early seedling stages for proper flowering. |
| *T. patula* Marigold Asteraceae | A | 21–24 | 16–18 | 5–8 | D or L | | |
| *Teucrium chamaedrys* Germander Lamiaceae | P | 21 | 15–18 | 25–30 | | | |
| *Thalictrum rochebrunianum* Meadow rue Ranunculaceae | P | 22 | 15–18 | 28–40 | | | |
| *Thermopsis* spp. False lupine Fabaceae | P | 22 | 15–18 | 15–20 | | Nick with a file or soak in H$_2$SO$_4$ 30 min | Seed vitality is quickly reduced as seed ages. Fresh seed may germinate without scarification. |

*continues*

**TABLE 2** *Continued*

| Species | Plant type | Medium temperature (°C) | | Days to germinate | Light or dark | Pretreatment | Remarks |
|---|---|---|---|---|---|---|---|
| | | Germination | Posttransplant (night) | | | | |
| *Thunbergia alata* Black-eyed Susan vine Acanthaceae | A | 21–24 | 16 | 12 | D or L | | Can be directly sown to garden or container. |
| *Thymus* spp. Thyme Lamaceae | P | 15–17 | 8–10 | 7–14 | L | | |
| *Tiarella cordifolia* Foamflower Saxifragaceae | P-W | 10–15 | 8–10 | 21–28 | | | |
| *Tithonia rotundifolia* Mexican sunflower Asteraceae | A | 22 | 18–22 | 14–21 | L | | |
| *Torenia fournieri* Wishbone flower Scrophulariaceae | A | 21–24 | 13–16 | 8–12 | L | | |
| *Tradescantia* spp. Spider wort Commelinaceae | P | 15–22 | 13 | 16–21 | D or L | | Reduce night temp. of medium by 3°C for best germination. |
| *Tropaeolum majus* Nasturtium Tropaealaceae | A | 18–21 | 16 | 10–14 | D or L | | Difficult to transplant. Best sown directly to garden or container. |
| *Ursinia anethoides* Ursinia Asteraceae | A | 20 | 10 | 10 | | | |
| *Valeriana officinalis* Valerian Valerianaceae | P | 15–20 | 10–20 | 14–21 | | | |
| *Venidium fastuosum* Monarch of the veldt Asteraceae | A | 21–24 | | 15–25 | L | | |
| *Verbascum chaixii* Mullein Scrophulariaceae | P | 18–23 | 15–18 | 15–20 | | | |

| | | | | | | |
|---|---|---|---|---|---|---|
| *Verbena* × *hybrida*<br>Garden verbena<br>Verbenaceae | A | 18 | 10–13 | 12–18 | D | Chill 5°C 7 days | Germination can be slow and erratic. Keep germinating medium on the dry side. Total darkness necessary until germination completed. Apply fungicide before sowing to prevent damping-off disease. |
| *Veronica* spp.<br>Speedwell<br>Scrophulariaceae | P | 18–21 | 7–10 | 10–16 | L | | Reduce night temp. of medium by 3°C for best germination. |
| *Viola* spp.<br>Pansy, Violets<br>Violaceae | P | 18–21 | 10 | 8–15 | D | | Reduce night temp. of medium immediately after germination. Apply fungicide before sowing to prevent damping off disease. |
| *Waldsteinia* spp.<br>Barren-strawberry<br>Rosaceae | P | 21–24 | 15–16 | | | | Germination is erratic. |
| *Xeranthemum anuum*<br>Immortelle<br>Asteraceae | A | 22 | 18 | 10–15 | | | Difficult to transplant. Best sown directly to garden or container. |
| *Zinnia* spp.<br>Zinnia<br>Asteracea | A | 24–27 | 21 | 5–10 | D or L | | Difficult to transplant. Best sown directly to garden or container. |

1. The list is limited to genera and species grown primarily in North America from seeds for their ornamental value. References used for the compilation of the table: 27–34.

2. If seeds require light for germination, maintain adequate moisture around the seeds during the germination period by a mist system, by careful watering with a fine spray, by use of a plastic cover over seed trays, or by watering from the bottom. Otherwise, cover the seed to a depth at least twice its size with vermiculite or sowing medium to ensure adequate moisture is provided. *Caution:* plastic covers cause heat buildup which, if allowed to reach high levels, could be detrimental to seeds.

3. Low temperature, such as that in a refrigerator, increases the longevity of most species of seed. Unless otherwise indicated, store dry in the original sealed packet or in a tightly closed plastic bag or sealed container. Silica gel or other hygroscopic substances may be added to resealed packages or containers to keep humidity low.

4. Freshly harvested (fresh) seeds of many herbaceous species will germinate if sown right away. The same seeds produced from a seed house or garden center have been stored for a period of time and may require chilling (stratification). Chilling temperature (near freezing to 5°C) in a refrigerator or cold room will usually suffice. However, freezing temperature (−10° to −20°C) such as in a household freezer is often recommended. Seeds may be frozen before or after sowing.

5. Fall sowing of herbaceous perennial seeds in outdoor beds or in flats or containers kept outdoors during the winter will satisfy their chilling (cold stratification) requirement to germinate the following spring. This practice may also replace a need for warm stratification or for scarification. Alternatively, germination can be scheduled to occur at any time after chilling seeds in a refrigerator.

6. Reducing the temperature of the germinating media at night by about 3°C may enhance germination of many herbaceous species. Gradually reducing the temperature during the post-transplant stage allows the development of short, sturdy plants, which are usually more desirable.

**TABLE 3** List of Selected Woody Perennial Trees and Shrubs Showing the Recommended Dates for Seed Collection, Seed Longevity and Storage Requirements, and Pretreatments to Overcome Seed Dormancy

| Species | Plant type | Collection date | Storage | Pretreatment | | | Remarks |
|---|---|---|---|---|---|---|---|
| | | | | Scarification | Stratification (months) | | |
| | | | | | Warm (20–30°C) | Cold (1–5°C) | |
| Abbies balsamea | ET | Aug.–Sept. | 5 years at 1–5°C and 9–12% moisture content | | | 2–3 | Collect cones before they shatter. Fresh seeds |
| A. concolor | ET | Sept.–Oct. | | | | 1–2 | germinate best. |
| A. fraseria | ET | Sept.–Oct. | | | | 1–2 | Light required for |
| A. homolepis | ET | Sept. | | | | 1–2 | germination. |
| A. nordmanniana | ET | Sept.–Oct. | | | | 1–2 | See notes 2, 3, and 4. |
| A. procera | ET | Sept. | | | | 1–2 | |
| A. veitchii | ET | Sept.–Oct. | | | | 1–2 | |
| Pinaceae | | | | | | | |
| Acer campestre | DT | Oct.–Nov. | 1–2 years at 1–5°C and 10–15% moisture content | | | 3–6 | |
| A. ginnala | DT | Oct.–Nov. | | | 1 | 3–4 | |
| A. negundo | DT | Sept.–Mar. | | | 1–2 | 2–3 | Unless otherwise indicated, stratify or fall-sow outdoors before seeds turn completely brown; otherwise, germination may be delayed by up to 1 year. |
| A. pensylvanicum | DT | Sept.–Oct. | | | | 3–4 | |
| A. platanoides | DT | Sept.–Oct. | | | | 3–4 | |
| A. pseudoplatanus | DT | Aug.–Oct. | | | | 3 | |
| A. rubrum | DT | May | | None—sow immediately after collection. | None | None | See notes 4 and 5. |
| A. saccharinum | DT | May | | None—see A. rubrum. | None | None | |
| A. saccharum | DT | Aug.–Oct. | | | | 2–3 | |
| Aceraceae | | | | | | | |

**Definitions of plant type**

DT = Deciduous tree. A woody plant that at maturity is more than 6 m in height, usually has a single stem more than 6 cm in diameter, is unbranched for 2 m in height, and has annual diameter growth. Leaves are shed at the end of each growing season.

DS = Deciduous shrub. A multistemmed woody plant that normally grows to less than 4 m and sheds its leaves at the end of each growing season.

ET = Evergreen tree. A woody tree that retains its needle-shaped leaves throughout the year. Older needles may drop after the second or third year.

ES = Evergreen shrub. A multistemmed woody plant that retains its needles throughout the year and attains a height usually less than 4 m.

BES = Broadleaf evergreen shrub. A multistemmed woody plant that retains its broad-leaved foliage throughout the year and attains a height usually less than 4 m.

| Species / Common name / Family | | Collection | Storage | Pretreatment | | | Comments |
|---|---|---|---|---|---|---|---|
| *Aesculus glabra* Ohio buckeye *A. hippocastanum* Horse chestnut Hippocastanaceae | DT | Sept.–Oct. | 6–12 months at 5°C | | | 4 | Stratify or fall-sow outdoors immediately after harvest to prevent moisture loss from seed. |
| *Ailanthus altissima* Tree of heaven Simaroubaceae | DT | Oct.–Apr. | Store at 2–5°C in sealed containers | 10-day water soak may aid germination. | | 2 | Fresh seeds may germinate without pretreatment. |
| *Albizia julibrissin* Silktree Fabaceae | DT | Fall | 18+ months at 4–8°C | $H_2SO_4$ 30 min or nick seed coat with a file. | | | Scarification not necessary if sown before seed coat hardens. |
| *Alnus glutinosa* European alder Betulaceae | DT | Fall–winter | 1–2 years at 1–5°C | | | 6 (dried seed only) | Fresh seeds will germinate without pretreatment. |
| *Amelanchier alnifolia* Serviceberry Rosaceae | DT | July–Aug. | | $H_2SO_4$ 15–30 min. | Fresh none Dried 3–6 | Fresh 3 Dried 3–6 | Collect fruits immediately upon ripening to avoid bird removal; macerate to extract the seeds. Fall-sow fresh seeds outdoors to germinate the following spring. See note 5. |
| *Aronia arbutifolia* *A. melanocarpa* *A. prunifolia* Rosaceae | DS DS DS | Aug.–Nov. Aug.–Nov. Aug.–Nov. | 2 years dry at 1–5°C | | | 2–3 2–3 2–3 | Macerate to extract seeds before sowing. |
| *Asimina triloba* Pawpaw Annonaceae | DT | Aug.–Sept. | | | | 2 | Collect fruits when flesh is soft. Entire fruit can be sown or seeds can be extracted by maceration. |
| *Berberis thunbergi* Japanese barberry Berberidaceae | DS | May–Sept. | 4–5 years at 1–5°C | | | 1–2 | Collect fruits when red. |

*continues*

**TABLE 3** *Continued*

| Species | Plant type | Collection date | Storage | Scarification | Pretreatment Stratification (months) Warm (20–30°C) | Cold (1–5°C) | Remarks |
|---|---|---|---|---|---|---|---|
| *Betula nigra* | DT | May–June | 1.5–2 years at room temp. if dried to 1–3% moisture content | | | None | Collect seeds while still green to avoid scattering by wind. Seeds are small and thin-coated, tend to lose viability quickly. Long-day photoperiod during germination eliminates need for stratification. |
| *B. papyrifera* | DT | Aug.–Sept. | | | | 1–2 | |
| *B. lenta* | DT | Aug.–Sept. | | | | 1–2 | |
| *B. pendula* | DT | July–Aug. | | | | 1 | |
| Birch Betulaceae | | | | | | | |
| *Carpinus betulus* | DT | Sept. | 1+ years at 1–5°C and 10% moisture content | | Fresh none Dried 2 | Fresh 3–4 Dried 2–4 | Fall-sow fresh outdoors to germinate the following spring. Viability may be low; test by soaking 6 hr in water and sowing only those which sink. |
| *C. caroliniana* | DT | | | | | | |
| Hornbean Betulaceae | | | | | | | |
| *Carya glabra* | DT | Sept.–Dec. | 3–5 years at 5°C and 3–5% moisture content | 1–2 days soak in H₂O at 20–25° | | 2–4 | Sow freshly harvested nuts immediately, or after 1–2 days of soaking at room temp.; germination may be erratic. Alternating day/night temps. of 30/20°C may aid germination. |
| Pignut | | | | | | | |
| *C. illinoensis* | DT | | | | | 2–4 | |
| Pecan | | | | | | | |
| *C. ovata* | DT | | | | | 2–4 | |
| Hickory | | | | | | | |
| *C. tomentosa* | DT | | | | | 2–4 | |
| Mockernut Juglandaceae | | | | | | | |

| Species / Common name / Family | | Collection | | Pretreatment | | | Remarks |
|---|---|---|---|---|---|---|---|
| *Castanea dentata* | DT | Aug.–Oct. | 1 year at 1–3°C and 40–45% moisture content | | | 1–3 | Collect nuts as soon as burrs open; viability is lost if allowed to dry. Cure nuts 1–7 days at 15–22°C out of sunlight. |
| *C. mollisima* Chestnut Fagaceae | DT | Aug.–Oct. | | | | 1–3 | No pretreatment necessary. |
| *Catalpa* spp. Catalpa Bignoniaceae | DT | Oct.–Mar. | 2 + years dry at 5°C | | | | |
| *Cedrus atlantica* | ET | Fall–winter | 3–6 years at 0–5°C and 10% moisture content | | | 1 | Collect seeds when comes break open. Pretreatment not usually necessary but germination may be erratic and slow. Spring sowing recommended. Store immediately upon collection; seeds are oily and may deteriorate quickly upon drying. |
| *C. deodara* Cedar Pinaceae | ET | Fall–winter | | | | 1 | |
| *Celtis occidentalis* Hackberry Ulmaceae | DT | Oct.–winter | 5–6 years at 0–5°C | | | 2–3 | |
| *Cercis canadensis* Eastern redbud Fabaceae | DT | July–Aug. | | $H_2SO_4$ for 30 min before cold treatment | | 2–3 | |
| *Chaenomeles japonica* Flowering quince Rosaceae | DS | Oct. | | | | 2–3 | Fall planting without scarification will result in good germination if seeds are fresh and not dried. |
| *Chamaecyparis nootkatensis* | ET | Sept.–winter | 2–5 years at 0°C and 10% moisture content | | 1 | 1 | Seedlings are extremely variable; vegetative propagation is recommended for uniform plants. |
| *C. lawsoniana* | ET | Sept.–winter | | | None | 1 | |
| *C. obtusa* Cypress Cupressaceae | ET | Sept.–winter | | | None | 1 | See notes 4 and 5. |

*continues*

**TABLE 3** *Continued*

| Species | Plant type | Collection date | Storage | Pretreatment | | | Remarks |
|---|---|---|---|---|---|---|---|
| | | | | Scarification | Stratification (months) | | |
| | | | | | Warm (20–30°C) | Cold (1–5°C) | |
| *Chionanthus virginicus* Fringetree Oleaceae | DT or DS | Oct. | 1–2 years at 0–5°C | | 1–3 | 2–3 | Collect seeds when fruits turn purple. Fall sowing will produce seedlings the second spring. See note 5. |
| *Cladrastis lutea* Yellowwood Fabaceae | DT | Oct.–Dec. | | $H_2SO_4$ for 30 min before or instead of cold treatment | | 3 | |
| *Cornus alternifolia* | DT | Aug.–Sept. | 3 years at 1–5°C | | 2–5 | 2–3 | Collect fruits as soon as ripened to avoid bird removal. Fleshy coverings may enhance dormancy and should be removed. See note 5. |
| *C. canadensis* | DS | Aug.–Sept. | | | 3 | 3 | |
| *C. florida* | DT | Oct. | | | | 3 | |
| *C. kousa* | DT | Sept.–Oct. | | | | 3–4 | |
| *C. mas* | DS | June–July | | | 4–5 | 3 | |
| *C. racemosa* | DS | Aug.–Sept. | | | | 3–4 | |
| *C. sericea* Dogwood Cornaceae | DS | Aug.–Sept. | | | | 3 | |
| *Cotoneaster apiculata* | BES | Oct. | 2 years dry at 5°C | $H_2SO_4$ 2 hr | | 3 | Warm stratification for 5–6 months before cold treatment may be used instead of acid scarification. Acid-treated seeds may be sown outdoors in fall. |
| *C. horizontalis* Cotoneaster Rosaceae | BES | Oct. | | $H_2SO_4$ 2 hr | | 3–4 | |
| *Crataegus arnoldiana* | DT | Oct.–winter | 3 years at 5°C | 4–5 hr $H_2SO_4$ | 3 | 6 | Macerate fruits in water to remove pulp from seeds. Freshly harvested seeds may need less pretreatment time. Seed coat may become impermeable after drying. See notes 4, 5, and 6. |
| *C. crus-galli* | DT | Oct.–winter | 3 years at 5°C | 2–3 hr $H_2SO_4$ | | 3 | |
| *C. mollis* | DT | Oct.–winter | 3 years at 5°C | | | 3 | |
| *C. phaenopyrum* Hawthorn Rosaceae | DT | Oct.–winter | 3 years at 5°C | | | 3 | |

| Species / Common name / Family | | Collection | Storage | Scarification | | | Remarks |
|---|---|---|---|---|---|---|---|
| *Cytisus scoparius* Scotch broom Fabaceae | BES | Sept. | 10+ years at 5°C | Repeated hot water soaks or 30 min in $H_2SO_4$. | | | Outdoor fall sowing may replace scarification. |
| *Diospyros virginiana* Persimmon Ebenaceae | DT | Oct.–Nov. | | | 2–3 | | |
| *Elaeagnus angustifolia* Russian olive Elaeagnaceae | DS or DT | Aug.–winter | 1–3 years at 5°C and 6–14% moisture content | | 2–3 | | |
| *Euonymus alatus* Winged euonymus Celastraceae | DS | Aug.–Sept. | Moist, cold storage | | 3–4 | | Collect fruits just as capsules start to split. Store moist. Drying reduces germination. Do not allow seeds to dry; sow outdoors immediately to germinate in spring. |
| *Fagus grandifolia* *F. sylvatica* Beech Fagaceae | DT DT | Oct.–Nov. Oct.–Nov. | 3+ years at −15°C and 8–10% moisture content | | 3 3 | | Separate sound seeds by flotation method; seed should not dry out. Fall sow outside or treat as soon as possible after collecting. See note 6. |
| *Fraxinus americana* *F. excelsior* *F. nigra* *F. ornus* *F. pennsylvanica* Ash Oleaceae | DT DT DT DT DT | Oct.–Nov. Oct.–Nov. Sept. Oct.–Nov. Oct.–Nov. | 7+ years at 5°C and 7–10% moisture content | | 2 3 3 3 2–6 | 1 3 2–3 2 2 | Collect and treat or sow as soon after ripening as possible. Older seed becomes more dormant. See note 5. Double stratification needed (warm moist and cool moist) for *F. nigra* and *F. pennsylvanica*. Seeds sown outdoor require 2–8 yr to germinate. |

*continues*

**TABLE 3** *Continued*

| Species | Plant type | Collection date | Storage | Pretreatment | | | Remarks |
|---|---|---|---|---|---|---|---|
| | | | | Scarification | Stratification (months) | | |
| | | | | | Warm (20–30°C) | Cold (1–5°C) | |
| *Ginkgo biloba* Maidenhair tree Ginkgoaceae | DT | Late fall | 1 year if stored moist | | 1–2 | 1–2 | Collect fruits when dropped. Remove seeds by maceration in water before storing or sowing. Fresh seeds will germinate but at lower percentage. See note 5. |
| *Gleditsia triacanthos* Honeylocust Fabaceae | DT | Sept.–winter | 2+ years at 0–5°C | H₂SO₄ 1–2 hr or hot water soak, leave overnight | | None | |
| *Gymnocladus dioicus* Kentucky coffeetree Fabaceae | DT | Winter | 5+ years dry | H₂SO₄ 2–4 hr or nick seed coat with a file | | None | Seeds are large. |
| *Halesia carolina* Carolina silverbell Styracaceae | DS | Oct.–Nov. | 1–2 years dry at 0–5°C | | (2) 6 following first prechilling | (1) 3 initially (3) 4–5 after warm | Very complex dormancy; require (1) cold–(2) warm–(3) cold in this sequence, but outcome not dependable. Most reliable germination is obtained by fall sowing outdoors. Germination will occur the second spring. See note 5. |
| *Hamamelis virginiana* Witch-hazel Hamamelidaceae | DS | Aug.–Oct. | 1+ years at 0–5°C | | 3 | 3 | Collect capsules while still yellowish and before open to avoid natural seed dispersal. Also, if fall sown early, germination may occur the first spring. If seed is ripe and dry, a second season will be needed. See note 5. |

| Species / Common name / Family | | Collection | Storage | Pretreatment | | | Remarks |
|---|---|---|---|---|---|---|---|
| *Ilex aquifolium* | BES | Oct.–winter | Dry, in sealed container | H$_2$SO$_4$ 30 min | None fall sow | | If fall sown outdoors, *Ilex* seed may require up to 3 years in nature to complete germination. |
| *I. glabra* | BES | Oct.–winter | Dry, in sealed container | H$_2$SO$_4$ 30 min | None fall sow | | Fresh seeds sown immediately after collection may hasten germination by up to one year for *I. vomitoria*. |
| *I. opaca* | BES | Oct.–winter | Dry, in sealed container | H$_2$SO$_4$ 30 min | 2–3 | 3 | |
| *I. verticillata* | BES | Oct.–winter | Dry, in sealed container | H$_2$SO$_4$ 30 min | 2–3 | 2 | Choose either acid treatment or fall sowing. |
| *I. vomitoria* Holly Aquifoliaceae | BES | Oct.–winter | Dry, in sealed container | H$_2$SO$_4$ 30 min | None–fall sow | | See note 5. |
| *Juglans nigra* Black walnut Juglandaceae | DT | Oct.–Nov. | 1+ years at 1–5°C and 20–50% moisture content | | | 3–4 | Protect against rodents if fall sown outdoors. |
| *Juniperus communis* | ES | Sept.–winter | 2+ years at 1–5°C and 10–12% moisture content | | | 3 | See note 5. |
| *J. virginiana* Juniper Cupressaceae | ES | Sept.–winter | | | 2–3 | 2–3 | |
| *Kalmia latifolia* Mountain laurel Ericaceae | BES | Oct.–Nov. | 10+ years dry at room temp. | | | | No special pretreatment required. Light required for germination. |
| *Koelreuteria paniculata* Golden rain tree Fabaceae | DT | Sept.–Oct. | 1–2 years dry at 1–5°C | 1–2 hr H$_2$SO$_4$ for fresh seed or soak in hot water and let cool for 24 hr | | 1 | Collect before capsules open. Seeds may germinate after acid treatment only, depending on the state of dormancy. Fall-sown fresh seed gives best results. |
| *Larix decidua* | DT (conifer) | Sept.–winter | 3+ years at 1–5°C | | | 1–2 | Fresh seeds may germinate without pretreatment if exposed to light after sowing. |
| *L. laricina* Larch Pinaceae | | Sept.–winter | | | | 1–2 | |

*continues*

**TABLE 3** *Continued*

| Species | Plant type | Collection date | Storage | Scarification | Pretreatment Stratification (months) Warm (20–30°C) | Cold (1–5°C) | Remarks |
|---|---|---|---|---|---|---|---|
| *Ligustrum amurense* | DS | Sept.–Nov. | Low temp., sealed container | | | 2–3 | Fresh seeds may germinate without cold treatment. Stratification times refer to stored seed. |
| *L. japonicum* | DS | Sept.–Nov. | | | | 2–3 | |
| *L. lucidum* | DS | Sept.–Nov. | | | | 2–3 | |
| *L. vulgare* Privet Oleaceae | DS | Sept.–Nov. | | | | 2–3 | |
| *Lindera benzoin* Spicebush Lauraceae | DS | Sept.–Oct. | <1 year–keep cool and sealed | | 1 | 3 | Extract seeds by maceration. Sow outdoors in fall for best germination in spring. |
| *Liquidambar styraciflua* Sweet gum Hamamelidaceae | DT | Sept.–Nov. | 4+ years sealed at 1–5°C and 10–15% moisture content | | | 1–3 | Pick fruit when green color fades to avoid natural seed dispersal. |
| *Liriodendron tulipifera* Tulip tree Magnoliaceae | DT | Aug.–Oct. | Several years if stored dry at 1–5°C | | | 2–3 | Low percentage of viable seeds per cone (10%). Sow fresh, untreated seed in fall to germinate following spring. |
| *Lonicera maackii* | DS | Sept.–Nov. | 15+ years at 1–5°C | | | 1–3 | Collect fruits soon after ripening to avoid bird removal. Extract seed by maceration of fresh, ripened fruits. |
| *L. tatarica* Honeysuckle Caprifoliaceae | DS | July–Aug. | | | | 1–3 | |
| *Maclura pomifera* Osage-orange Moraceae | DT | Sept.–Oct. | 3+ years at 1–5°C | Soak fresh seed in water for 2 days–replaces stratification. | | 1 | Collect fruits when dropped. Extract seeds by macerating fruits in water. Fruits may be allowed to ferment over winter in a heap. This makes seed separation easier and eliminates need for stratification. |

| | | | | | | Notes |
|---|---|---|---|---|---|---|
| *Magnolia acuminata* | DT | Aug.–Sept. | 2+ years at 1–5°C | 3–6 | | Collect when follicle opens exposing the seeds. |
| *M. grandiflora* | DT | Aug.–Sept. | | 3–6 | | Fresh seeds with fleshy pulp provide best results. |
| *M. virginiana* | DT | Aug.–Sept. | | 3–6 | | Macerate to remove pulp. Dried seeds lose viability. |
| Magnolia Magnoliaceae | | | | | | |
| *Malus baccata* | DT | Sept.–Oct. | 2+ years at 1–5°C and <11% moisture content | 1 | | Extract seeds from fruit by maceration before treating or sowing. |
| *M. coronaria* | DT | Sept.–Oct. | | 4 | | |
| *M. floribunda* | DT | Sept.–Oct. | | 2–4 | | |
| *M. ioensis* | DT | Sept.–Oct. | | 2 | | |
| Crabapple Rosaceae | | | | | | |
| *Metasequoia glyptostroboides* Dawn redwood Taxodiaceae | DT (conifer) | Oct.–Nov. | | 1 | | Collect seeds when cones begin to open. Fresh seeds need no treatment. |
| *Morus alba* Mulberry Moraceae | DT | June–Aug. | 1+ years at −10 to −20°C | 1–3 | | Harvest fruits as soon as ripened to avoid bird removal. Macerate to separate pulp from seeds. Fresh seeds can be sown immediately for reasonable germination. |
| *Myrica pensylvanica* Bayberry Myricaceae | BES | Oct.–winter | 10–15 years if dewaxed, dried and stored at 1–5°C in sealed containers | 1–3 | | Remove wax by rubbing over a screen before stratifying or sowing. |
| *Nyssa sylvatica* Tupelo Nyssaceae | DT | Sept.–Nov. | 1+ year dry at 1–5°C | 1–3 | | Pulp removal by maceration is often recommended but not necessary. |
| *Ostrya virginiana* Ironwood Betulaceae | DT | Aug.–Oct. | | 3–5 | 3 | Collect fruits when pale greenish brown and before seeds are naturally dispersed. |

*continues*

**TABLE 3**  *Continued*

| Species | Plant type | Collection date | Storage | Scarification | Pretreatment Stratification (months) Warm (20–30°C) | Cold (1–5°C) | Remarks |
|---|---|---|---|---|---|---|---|
| *Parthenocissus quinquefolia* Virginia creeper Vitaceae | DV | Sept.–winter | 2+ years at 1–5°C | | | 2 | Collect fruits when bluish black in color. Macerate to remove pulp. Collect fruits before they open and seeds are naturally dispersed. No pretreatment required. Light required for germination. |
| *Paulownia tomentosa* Empress tree Bignoniaceae | DT | Sept.–Oct. | | | | | |
| *Phellodendron amurense* Amur corktree Rutaceae | DT | Sept.–Nov. | Dry until sowing | | | 2 (only for stored seed) | Macerate fruit to remove pulp. No pretreatment necessary for fresh seeds. |
| *Picea abies* | ET | Sept.–Nov. | 5–20 years at 1–5°C and 4–8% moisture content | | | None | Collect cones before they shatter. Generally no pretreatment required so fall sowing is not recommended. Cold stratification of 1 month may hasten and unify germination, which takes place following spring sowing. See note 4. |
| *P. engelmanni* | ET | Sept.–Oct. | | | | None | |
| *P. glauca* | ET | Sept. | | | | None | |
| *P. omorika* | ET | Oct. | | | | None | |
| *P. pungens* | ET | Fall | | | | None | |
| *P. sitchensis* Spruce Pinaceae | ET | Oct.–spring | | | | None | |
| *Pinus cembra* | ET | Aug.–Oct. | 1+ years | For those requiring stratification first soak in water for 1–2 days | | 3–4 | Seed provenance is an important consideration to future survival and growth characteristics. Collect cones of most species as soon as ripened and starting to crack open; otherwise seeds will be dispersed. |
| *P. densiflora* | ET | Aug.–Oct. | 2–5 | | | 0–1 | |
| *P. mugo* | ES | Oct.–Dec. | 5 | | | None | |
| *P. nigra* | ET | Sept.–Nov. | 10+ | | | 0–2 | |
| *P. parviflora* | ET | Sept.–Nov. | | | | 3 | |
| *P. resinosa* | ET | Aug.–Nov. | 30 | | | 2 | |
| *P. strobus* | ET | Aug.–Sept. | 10 | | | 2 | |

| Species | Type | Sowing months | Storage | | | Remarks |
|---|---|---|---|---|---|---|
| P. sylvestris | ET | Sept.–Mar. | 15 | 1–3 | | Fresh seeds of some species requiring stratification may be sown without pretreatment with some success. |
| P. thunbergii | ET | Oct.–Dec. | 11 | 1–2 | | |
| P. virginiana | ET | Sept.–Nov. | 5+ at 1–5°C and 5–10% moisture content | 0–1 | | |
| Pine Pinaceae | | | | | | |
| Platanus occidentalis | DT | Nov.–Feb. | 1+ years at 1–5°C and 10–15% moisture content | None | | Cold stratification for 2 months may improve germination. |
| P. orientalis | DT | Nov.–Feb. | | None | | |
| Sycamore, Planetree Platanaceae | | | | | | |
| Populus deltoides | DT | May–June | 2–3 years if air dried 4 days and stored sealed at 5°C and 5–8% moisture content | None | | Clean seeds before sowing. Fresh seeds germinate rapidly. |
| Poplar | | | | | | |
| P. tremuloides | DT | May–June | | None | | |
| Aspen Salicaceae | | | | | | |
| Prunus armeniaca | DT | May–June | Many conditions tried. Air dried several years 0–5°C in sealed containers | 1–5 | | Extract seeds by maceration and flotation. Fall sowing eliminates need for prechilling but seeds should be sown early enough to satisfy this requirement before greezing occurs. Mulching the seed beds delays freezing. |
| Apricot | | | | | | |
| P. avium | DT | June–July | | 3–4 | 1 | |
| Mazzard cherry | | | | | | |
| P. besseyi | DT | July–Sept. | | 3 | | |
| Western sand cherry | | | | | | |
| P. cerasifera | DT | July–Aug. | | 3 | | |
| Myrobolan plum | | | | | | |
| P. domesica | DT | July–Oct. | | | | |
| Plum | | | | | | |
| P. persica | DT | July–Oct. | | 2 | | See notes 4, 5, and 6. |
| Peach | | | | | | |
| P. pumila | DT | July–Sept. | | 3 | | |
| Sand cherry | | | | | | |
| P. serotina | DT | Aug.–Sept. | | 4 | | |
| Black cherry | | | | | | |
| P. virginiana | DT | Aug.–Sept. | | 3 | | |
| Choke cherry Rosaceae | | | | | | |

*continues*

**TABLE 3** *Continued*

| Species | Plant type | Collection date | Storage | Scarification | Pretreatment Stratification (months) Warm (20–30°C) | Pretreatment Stratification (months) Cold (1–5°C) | Remarks |
|---|---|---|---|---|---|---|---|
| *Pseudotsuga menziesii* Douglas fir Pinaceae | ET | Aug.–Sept. | 3–4 months stored as cones dry and warm; 10–20 years at 0°C and 6–9% moisture content | | | 1 | Collect seeds when golden brown in color. Fall sowing not recommended as germination may begin too early. No pretreatment required, but if given, may enhance germination percentage and rate. |
| *Pyrus communis* Pear Rosaceae | DT | July–Oct. | 2–3 years at 1–5°C and 10% moisture content | | | 3–4 | Macerate fruit to extract seed. Presoaking seed in water for 24 hr may improve stratification results. |
| *Quercus* spp. | DT | Aug.–Dec. | 6 months or less for Black Oak group. White Oak group has almost no storage capabilities. | | | | With few exceptions, acorns of the White Oak (W) group have little or no dormancy and germinate immedaitely after dropping from the tree. Acorns of the Black Oak (B) group exhibit embryo dormancy. Outdoor fall sow for germination the following spring. |
| *Q. alba* (W) White oak | DT | Aug.–Dec. | | | | None | |
| *Q. bicolor* Swamp white oak (W) | DT | Aug.–Dec. | | | | None | |
| *Q. coccinea* (B) Scarlet oak | DT | Aug.–Dec. | | | | 1–2 | |
| *Q. imbricaria* (B) Shingle oak | DT | Aug.–Dec. | | | | 1–2 | |
| *Q. macrocarpa* (W) Bur oak | DT | Aug.–Dec. | | | | 0–2 | |
| *Q. muhlenbergii* (W) Chinkapin oak | DT | Aug.–Dec. | | | | None | See note 4. |
| *Q. nigra* (B) Water oak | DT | Aug.–Dec. | | | | 1–2 | |
| *Q. palustris* (B) Pin oak | DT | Aug.–Dec. | | | | 1–2 | |

| Species / Family | Type | Collection | Storage | Treatment | Strat. | Strat. | Notes |
|---|---|---|---|---|---|---|---|
| *Q. phellos* (B) Willow oak | DT | Aug.–Dec. | | | 1–2 | | |
| *Q. robur* (W) English oak | DT | Aug.–Dec. | | | None | | |
| *Q. rubra* (B) Red oak | DT | Aug.–Dec. | | | 1–2 | | |
| *Q. shumardii* (B) Shumard oak | DT | Aug.–Dec. | | | 1–2 | | |
| *Q. velutina* (B) Black oak | DT | Aug.–Dec. | | | 1–2 | | |
| *Q. virginiana* (W) Live oak Fagaceae | DT | Aug.–Dec. | | | None | | Collect seeds as soon as ripe before removal by birds. |
| *Rhamnus frangula* Buckthorn Rhamnaceae | DS | July–Oct. | 2+ years at 1–5°C | $H_2SO_4$ 20 min | 2–3 | | |
| *Rhododendron maximum* | ES | Sept.–Oct. | 2 years at room temp. and 4–9% moisture content | | None | | Collect seeds as soon as capsules begin to turn brown and before they open. Light is required for germination. |
| *R. catawbiense* Rhododendron Ericaceae | ES | July–Oct. | | | None | | |
| *Ribes alpinum* Alpine currant Saxifragaceae | ES | June–July | | | 3–6 | | Macerate fruit in water to remove fleshy pulp. |
| *Robinia pseudoacacia* Black locust Fabaceae | DT | Aug.–winter | Up to 10 years at 1–5°C | Nick with a file or 10–120 min in $H_2SO_4$ | | | Collect fruits before pods split open and seeds are dispersed. Seeds are large. |
| *Rosa canina* | DS | Sept.–Oct. | 2–4 years at 1–5°C | 45 min $H_2SO_4$ instead of warm stratification. | 3 | 3–5 | Collect seeds fresh for best results. Hips should be just turning red. Macerate in water to remove pulp. Fresh seeds can be fall sown immediately after cleaning. Alternative to warm stratification is scarification. See note 5. |
| *R. multiflora* | DS | Sept.–Oct. | | | 3 | 3–5 | |
| *R. rugosa* Rose Rosaceae | DS | Sept.–Oct. | | | 3 | 3–5 | |

*continues*

**TABLE 3** *Continued*

| Species | Plant type | Collection date | Storage | Scarification | Pretreatment Stratification (months) Warm (20–30°C) | Cold (1–5°C) | Remarks |
|---|---|---|---|---|---|---|---|
| *Salix discolor* *S. nigra* Pussy willow Salicaceae | DT DT | Apr.–May Apr.–May | 4–6 weeks | | | None None | Very short lived. Viability 10 days at room temp. Sow immediately after collection—no dormancy. |
| *Sambucus canadensis* Elder Caprifoliaceae | DS | June–Aug. | 2 years dry at 5°C | 10–20 min $H_2SO_4$ instead of warm stratification. | 2 | 3–5 | Collect fruits as soon as ripened to avoid bird removal. Macerate in water to remove pulp. Alternative to warm stratification is scarification. See note 5. |
| *Sassafras albidum* Sassafras Lauraceae | DT | Sept. | 2+ years at 5°C | | | 4 | Collect seeds when color changes to dark blue. Remove pulp by maceration. Fall sow as late as possible to prevent fall germination. |
| *Sciadopitys verticillata* Umbrella pine Taxodiaceae | ET | Fall. | 2 years at 5°C and 10% moisture content | | 3 | 3 | See note 5. |
| *Sorbus americana* American mountain ash *S. aucuparia* European mountain ash Rosaceae | DT DT | Aug.–Oct. | 2–8 years at 5°C and 6–8% moisture content | | | 3 3 | Collect fruit when just ripened to avoid bird removal and to reduce pretreatment time. |
| *Symphoricarpos orbiculatus* Coralberry Caprifoliaceae | DS | Sept.–Oct. | 2 years at 5°C | 20–60 min $H_2SO_4$ in place of or in addition to warm stratification | 3–4 | 4–6 | Soak fruit several days, then macerate to remove pulp. See note 4. |

| | | | | | | |
|---|---|---|---|---|---|---|
| *Syringa amurensis* | DS | Aug.–Oct. | | | 1–3 | Seeds may germinate without pretreatment. |
| *S. persica* | DS | Aug.–Oct. | | | 1–3 | |
| *S. vulgaris* | DS | Aug.–Oct. | 2 years dry at 1–5°C | | 1–3 | |
| Lilac | | | | | | |
| Oleaceae | | | | | | |
| *Taxodium distichum* | DT | Oct.–Dec. | 1+ years dry at 5°C | | 2–3 | Collect fruits as soon as ripened to avoid bird removal. Macerate to remove pulp and do not allow to dry. |
| Bald cypress | | | | | | |
| Taxodiaceae | | | | | | |
| *Taxus baccata* | ES | Aug.–Oct. | 5–6 years dry at 5°C | 5–7 | 2–4 | Seeds sown after collecting in July will germinate naturally the second spring. Seeds that are allowed to dry after harvest will take longer to germinate or need longer pretreatment times than fresh seed. See note 5. |
| *T. canadensis* | ES | July–Sept. | | 5–7 | 2–4 | |
| *T. cuspidata* | ES | Oct.–Nov. | | 5–7 | 2–4 | |
| Yew | | | | | | |
| Taxaceae | | | | | | |
| *Thuja occidentalis* | ES | Aug.–Sept. | 51 years at 1–5°C and 6–8% moisture content | | 2 | Collect seeds immediately as cones turn light brown and before seeds are naturally dispersed. Some seed lots require no pretreatment. Seeds may require light to germinate. |
| *T. orientalis* | | | | | | |
| Arborvitae | | | | | | |
| Cuppressaceae | | | | | | |

*continues*

**TABLE 3** *Continued*

| Species | Plant type | Collection date | Storage | Scarification | Pretreatment Stratification (months) Warm (20–30°C) | Cold (1–5°C) | Remarks |
|---------|-----------|-----------------|---------|---------------|------------------------------------------------------|--------------|---------|
| *Tilia american* Basswood | DT | Sept.–winter | 2–3 years at 1–5°C with 10–12% moisture content | HNO₃ 30 min then H₂SO₄ 30–60 mi | | 3 | For best results, harvest fruits as soon as ripened. Treat seeds immediately. Dried seed has deeper dormancy. Both fall and spring sowing result in germination over a 2–3 year period. Seeds have both hard seed coat and impermeable outer pericarp, hence the need for both acid scarification treatments. See note 5. |
| *T. cordata* Linden Tiliaceae | DT | Sept.–winter | | | | 3 | |
| *Tsuga canadensis* | ET | Sept.–winter | 4 years at 1–5°C and 6–9% moisture content | | | 1–4 | Spring sowing is recommended. Fall sowing eliminates need for chilling but results in poor seedling survival. See note 4. |
| *T. caroliniana* Hemlock Pinaceae | ET | Sept.–winter | | | | 1–4 | |

| Species | | Collection | Storage | | | Notes |
|---|---|---|---|---|---|---|
| Ulmus americana | DT | Mar.–June | 15 years at −3°C and 3–4% moisture content | 2–3 | | Sow seeds with wings attached. |
| U. parvifolia | DT | Sept.–Oct. | | 2–3 | | |
| Elm | | | | | | |
| Ulmaceae | | | | | | |
| Viburnum acerifolium | DS | July–Oct. | 3+ years dry at 1–5°C | 3–9 | 3 | Macerate fruits in water to remove pulp. |
| V. dentatum | DS | July–Oct. | | 10–12 | 1 | See note 5. |
| V. lantana | DS | July–Oct. | | None | 2 | |
| V. lentago | DS | July–Oct. | | 5–9 | 2–4 | |
| V. opulus | DS | July–Oct. | | 5 | 3 | |
| Viburnum | | | | | | |
| Caprifoliaceae | | | | | | |

1. The list is limited to genera and species in temperate North America grown primarily for ornamental purposes. The following references were used for the compilation of the table: 35–40.

2. If seeds require light for germination, maintain adequate moisture around the seeds during the germination period by a mist system, by careful watering with a fine spray, by use of a plastic cover over seed trays, or by watering from the bottom. Otherwise, cover the seed to a depth at least twice its size with vermiculite or sowing medium to ensure that adequate moisture is provided. Very small seeds are usually sown on top of the germinating medium and are not covered. *Caution*: plastic covers cause heat buildup, which if allowed to reach high levels, could be detrimental to seeds.

3. Low temperature, such as that in a refrigerator, increases the longevity of most species of seed. Unless otherwise indicated, store dry in the original sealed packet or in a tightly closed plastic bag or sealed container. Silica gel or other hygroscopic substances may be added to resealed packages or container to keep humidity low.

4. Fall sowing of woody perennial seeds in outdoor beds, or in flats or containers kept outdoors during the winter, will satisfy their chilling (cold stratification) requirement to germinate the following spring. This practice may also replace a need for warm stratification or for scarification. Alternatively, germination can be scheduled to occur at any time after chilling seeds in a refrigerator.

5. Seeds with complex dormancy may require both warm and cold stratification in sequence. If sufficient heat is gained by unstratified seeds after a late summer or fall seeding to satisfy the warm requirement, germination will occur the following spring. Otherwise, germination will occur the second spring.

6. Seeds must usually be removed from fleshy fruits. Mechanical devices such as a fruit press can be used to macerate the fruit. Seeds can then be separated by flotation, screening, fermentation, or other means. Nonviable seeds are light and usually float in water; viable seeds are heavier and sink.

several times with cold water to remove the acid. Stir the seeds carefully during rinsing. Decant the water, spread the seeds uniformly on old newspaper, and allow to dry at room temperature before sowing.

Properly treated seeds are firm as little water is absorbed. The length of time for acid treatment, if unknown, must be determined empirically to prevent seed injury. Although suitable for a small amount of seed, acid treatment may not be practical for large quantities because of the hazards of working with concentrated acid. Treatment with nitric acid or with other chemicals, including potassium or sodium hydroxide, sodium hypochlorite, hydrogen peroxide, alcohol, acetone, and various growth-regulating substances, may be effective for some seeds [2].

### 3. Hot Water

Soaking in hot water is a treatment commonly used with hard-seeded species. Soaking softens, and sometimes ruptures the seed coat and leaches naturally occurring substances that may inhibit germination.

Pour about five times the volume of hot water (75–100°C) over the seeds and allow them to soak in the gradually cooling water for 6 to 24 hr. The amount of swelling of the seeds will indicate the degree of water uptake. Occasionally, seeds are boiled in water for 2–5 min; however, this procedure is apparently injurious to seeds of most species. Also, seeds may be soaked in running water or by exposing them to frequent changes of water to leach inhibitors.

Soaking in water may not yield as consistently good results as acid treatment [8,39]. However, soaking is easier to do, is not hazardous, and needs no special equipment. Although damp or wet seeds are more difficult to sow, they should normally be sown immediately because drying may make the treatment ineffective.

### 4. Other Considerations

In temperate climates, abrasion of seeds may occur from soil particles as the soil is alternately frozen and thawed. In warm climates, seeds are ruptured by swelling. Also, organic acids in the soil or substances and enzymes excreted by soil microorganisms soften or degrade the seed coat to some degree.

Removal of the fleshy seed coating or passage through the intestine of animals is required to overcome dormancy of some species [8,11]. Dry heat may cause increased germination of some hard-seeded species by rupturing the seed coat or by denaturing seed coat inhibitors [11].

Commercial seed companies routinely treat seeds with chemical disinfectants and/or hot water to prevent infection by surface-borne fungi and bacteria. Hot water appears to be a good disinfectant.

## B. Stratification

### 1. Cold

The term *stratification* formerly applied to storing alternate layers of seeds with moist sand and subjecting them to the cold or, more generally, freezing temperatures [43]. Nowadays, seeds are sown or mixed in the substrate rather than in layers, although the term is still used. The major requirements for cold stratification (often referred to as moist chilling or cold treatment) are adequate moisture, aeration, low temperature, and time of exposure. During stratification, the levels of growth-promoting substances in the seeds increase and those of growth-inhibiting substance decrease [44].

In temperate climates, seeds of many species of plants sown directly in seed beds, or in flats kept outdoors, undergo natural cold stratification during the winter and are ready to germinate in the spring. Seeds sown in flats or containers, or simply mixed with moist medium, may at any time be "stratified artificially" in refrigerators or coolers. For small lots, a plastic bag may be used. A moist medium such as sand, peat moss, vermiculate, or a combination of these ingredients is mixed with seed and placed in the bag, which is then sealed and placed in a refrigerator. The plastic allows gaseous exchange but keeps in moisture. Once stratification has begun, seeds should not be allowed to dry because drying may reverse the process. Therefore, the medium should be inspected periodically during the process. Seeds may also be chilled or frozen without being mixed in a substrate.

Freezing is unessential but is sometimes recommended for certain herbaceous seeds. In general, best results occur with temperatures from just above freezing to 5°C. Although seeds may require higher stratification temperatures in milder climates, from a practical viewpoint, lower temperatures will help to prevent germination while seeds are being stratified. During stratification, root radicles may emerge from the seeds, indicating that they are ready to sprout. Exposure to freezing temperatures at this time may be injurious to the germinating seedlings.

Stratification, if required, varies with each species. The closer the required temperature and duration are to optimum, the better will be the outcome (Tables 2 and 3). Thus, knowing the proper temperatures and stratification duration can result in more effective and efficient seedling production. Refrigeration (artificial stratification) is more predictable than outdoor sowing (natural stratification) and refrigeration usually provides better and more consistent results.

### 2. Warm

Some seeds require moist, warm stratification. Seeds with double dormancy, such as those of *Taxus* spp. (yew) and *Viburnum* spp. (viburnum), require both warm and cold stratification, and others with even more complex dormancy such as *Halesia carolina* (Carolina silverbell) require cold-warm-cold exposures in this sequence (Tables 2 and 3).

During exposure to warm temperatures, usually between 20 and 30°C but possibly more or less depending on the species, immature or rudimentary embryos develop. If sufficient heat is gained after outdoor summer or fall seeding to satisfy the warm temperature requirement, germination will occur the first spring after seeding. Otherwise, germination will occur the second spring after seeding. Dormancy of some seeds is also broken by storage in hot, dry conditions [11].

### 3. Embryo Culture

The technique of in vitro embryo culture or embryo rescue is used by plant breeders and seed laboratories to obtain seedlings from otherwise nongerminable seeds that are not sufficiently mature when the fruit is ripe or from seeds with very complex dormancies [2]. The procedure involves aseptically excising the embryo from the seed and culturing it in a suitable, sterilized nutrient medium. When an immature or rudimentary embryo is cultured in this way, it may bypass the need for warm and/or cold stratification.

## C.  Growth Regulators

Seed dormancy and germination are believed to be controlled by the balance and interaction of growth-promoting and growth-inhibiting substances. These regulatory hormones accumulate in seeds during embryo development, although not necessarily in the embryo itself [8]. Gibberellic acid, in particular, appears to be essential for seed germination. It mobilizes food sources and stimulates growth of embryonic tissue. Differences in endogenous gibberellin concentrations of some cold-requiring seeds have been related to the amount of chilling exposure of the seeds [45]. Abscisic acid appears to be a specific antagonist of gibberellin action to promote germination. It is present in the seed coat, endosperm, or embryo.

Exogenously applied growth regulators sometimes influence seed germination. Some dormant seeds, particularly of wild plants that require light or cold for germination, may be induced to germinate by applying gibberellins. Seeds of these plants typically germinate in pockets of leaf mold where fungal activity releases gibberellins [21]. Ethylene also breaks dormancy and initiates germination, but its effect is not as well documented as that of other growth regulators.

According to Bell et al. [11], water-soluble chemical factors from charred wood or smoke may stimulate germination of some seeds. Interestingly, ethylene is a component of wood smoke. Ethylene released from seeds may also stimulate their germination [12]. Also, applied cytokinins overcome dormancy in many species. Conversely, abscisic acid often inhibits germination when applied to nondormant seeds. In some instances, the abscisic acid–induced germination inhibition can be reversed by cytokinin. Gibberellins do not usually reverse abscisic acid–induced germination inhibition. However, the combined application of gibberellin and cytokinin can induce germination of dormant seeds in a wider range of species than either chemical administered separately.

## D.  Priming

Postharvest treatments that improve germination and seedling vigor are termed seed enhancements [46]. Seed priming (osmoconditioning) is a seed enhancement technique that has proved effective for improving germination, seedling emergence, and yield of many early-planted, small-seeded vegetable and flower crops. Priming or controlled hydration refers to conditioning seeds in an aerated solution with a high solute content, which keeps the seed in a partially hydrated state [47]. Polyethylene glycol (PEG), an inert compound, is often used, although some systems use salt solutions of various compositions (Table 4). Primed seeds may be sown moist, dried, or even stored for later use.

Seeds treated with osmotic solutions ranging from $-1.0$ to about $-2.0$ mPa water potentials may germinate more rapidly and uniformly under a wider range of temperatures than untreated seeds. The water potential of the priming solution, priming temperature, and priming duration are all important [54] if radicle elongation is to be prevented while at the same time allowing most other germination processes to proceed.

Priming sometimes improves germination of aged seeds [55]. Reinvigoration of aged seeds during the priming process is associated with partial reversal of some lipid peroxidation [56]. However, priming of nonaged seeds ages them faster than untreated counterparts [57].

Osmoconditioning does not affect stage I water uptake (imbibition) because the priming solutions have much higher water potentials than the water potential of the colloid-like seed tissue. However, the stage II processes (active metabolism and hydrolysis) occur during the priming treatment. Thus, osmoconditioning enables the seed to absorb enough water to become metabolically active and accumulate reserves of sugars, amino acids, proteins, and other substances required for germination. The water potential of the priming solution is insufficient for visible germination to occur.

The primed seeds germinate uniformly and rapidly once the osmotic stress is relieved and the final phase of seed hydration occurs. The water potential of the osmoconditioning solution varies among osmotica and species (Table 4). Osmotic priming can substitute for the chilling requirement in certain species [58].

A more recent improvement of the seed priming technique, referred to as matriconditioning, involves the use of a protective gel or colloidal agent with a high water absorptive property instead of an osmotic solution [59,60]. Matriconditioning may be better suited than osmoconditioning for treating large amounts of seeds.

Seed priming, either osmoconditioning or matriconditioning, may be integrated with use of growth regulators and with fluid drilling (pregerminated seeds suspended in a protective gel) to improve plant emergence and performance under field conditions. Different gels are used and growth regulators, fertilizers, and pesticides are incorporated into the gels in attempts to increase the effectiveness of the technique.

**TABLE 4** Examples of Successful Seed Osmoconditioning Treatments for Selected Species

| | Osmoconditioning | | Osmotica | | | |
|---|---|---|---|---|---|---|
| Species | Temp. (°C) | Duration (days) | Chemical | Amount (g per per kg of $H_2O$) | Estimated water potential (mPa)[a] | Reference |
| *Beta vulgaris* | | | | | | |
| Sugar beet | 15 | 7 | PEG[b] 8000 | 302 | $-1.22$ | 48 |
| *Daucus carota* | | | | | | |
| Carrot | 15 | 28 | $K_3PO_4 + KNO_3$ | 21.65 20.6 | $-1.5$ | 49 |
| *Daucus carota* | | | | | | |
| Carrot | 15 | 14 | $K_2HPO_4 + KNO_3$ | 18.28 21.1 | $-1.69$ | 49 |
| *Allium cepa* | | | | | | |
| Onion | 15 | 14 | PEG 8000 | 342 | $-1.55$ | 50 |
| *Apium graveolens* | | | | | | |
| Celery | 15 | 14 | PEG 8000 | 273 | $-1.0$ | 51 |
| *Lycopersicum esculentum* | | | | | | |
| Tomato | 15 | 14 | $K_2HPO_4 + KNO_3$ | 15.67 11.92 | $-1.0$ | 49 |
| *Petroselinum crispum* | | | | | | |
| Parsley | 15 | 21 | PEG 8000 | 296 | $-1.17$ | 52 |

[a] Water potential estimates for PEG 8000 from Ref. 53.
[b] PEG, polyethylene glycol.

## E.  Other Treatments

The pelleting of seeds has been used for a long time with varied success. Although there are different ways of pelletizing seeds, in the simplest procedure, seeds are placed in a rotating drum and coated with a liquid binder and dust. The procedure results in uniform-sized, spherical pellets that facilitate more precise planting and often increased and more uniform germination [2,61].

Grass, vegetable, and flower seeds and sometimes seeds of woody species have been "seeded" in plastic rolls, tapes, or in water-absorbent, fibrous mats. The seeds are held in position by water-soluble adhesives [61]. These procedures simplify planting and may result in more uniform germination and seedling establishment. The roll, tape, or mat serves as a mulch and provides a more uniform germinating environment. Often fertilizers, inoculants, insecticides, fungicides, and other chemicals are added to improve the effectiveness of these products [2].

## VIII.  COLLECTION AND STORAGE

Seed companies and seed banks regularly conduct germination tests in controlled laboratory conditions to determine the relative viability of seed lots and to maintain quality control. These tests also determine the temperature and moisture limits for successful storage of each type of seed [8,24].

Each type of seed must be collected, handled, and stored differently. In theory, seeds are ready to harvest when there is no further increase in weight. Seeds from different species mature at different times of the year [39]. Some seeds that appear ripe may in fact contain undeveloped embryos. Fruits have many different shapes and sizes and may be fleshy or dry, dehiscent or indehiscent. The fleshy coverings of some fruit may contain substances that inhibit germination and must be removed. Removing such coverings lessens the chance for bacterial or fungal growth, which may effect seed viability. Freshly harvested seeds of some species may require no treatment, or less stratification time, compared with those that have been dried and/or stored. The seed coat of *Crataegus* spp. (hawthorns), although not impermeable when freshly collected, becomes so after drying (Table 3).

Under normal conditions, many seeds are relatively short lived or lose viability with time. Hellum [62] reported a 12% reduction in the rate of germination of *Pinus balsamifera* seeds stored at 7°C for 4 months. Seeds of *Acer saccharinum* (silver maple) remain viable for only a few days if they are not kept moist and cool (Table 3). *Salix* (willow) and *Populus* (poplar or aspen) seeds are viable for only 4 weeks, but many other seeds remain viable for several to 15 years and some longer. Because many woody species do not produce seed abundantly each year, commercial seed companies must collect and store seeds of these species for many years. Therefore, many different methods of collecting, handling, and storage are required. These methods have been described by other authors [2,8,34,39,63].

Under proper storage conditions, seeds of most species can be kept viable for 5-year periods. Keeping them dry, usually 5–12% moisture content, and keeping them cool are the most important factors affecting longevity and viability. A temperature range of 0–5°C is usually adequate for most species, although lower temperatures may be acceptable for some. Freeze-drying at temperatures below 1°C with moisture control appears to offer the best storage conditions [12] but is not an economical way to store most seeds.

## IX.  SUMMARY AND CONCLUSION

The geographic location or provenance of a seed can substantially influence its germinability. Seeds collected from different geographic sources may not germinate or perform uniformly under the same conditions. Those from a more southerly location may require a shorter stratification period to overcome dormancy and may result in plants that are less winter hardy in a more northerly location. Because of preharvest environmental conditions that affect seed maturation or seed-handling procedures and humidity and temperature of storage, which affect the permeability of the seed coat, seed treatments may yield different results between seed lots of the same species or from year to year. Therefore, treatments and other requirements (Tables 2 and 3) should be considered as guides and may need to be modified to compensate for variations in seed condition. Differences in germination requirements have evolved in response to species adaptation to changing environments or to selection pressure by cultivation and breeding.

Whereas seeds of most domesticated plants are selected for ease and predictability of germination, the germination requirements of wild species, or those closer to their wild ancestry, appear to be more clued to ecological and environmental influences. Because of the complex interactions of the preharvest and postharvest history of seeds and of the large number of seeds for which germination requirements are unknown or not fully characterized, germination studies will continue to challenge plant breeders and propagators.

## ACKNOWLEDGMENT

We appreciate the review of this chapter by Dr. Thomas D. Smith, Department of Plant Science, Nova Scotia Agricultural College, Truro, NS, Canada.

## REFERENCES

1. C Chong. In: BR Christie, ed. Plant propagation. Vol I. Boca Raton, FL: CRC Press, 1987, p 91.
2. B Macdonald. Practical Woody Plant Propagation for Nursery Growers. Vol I. Portland, OR: Timber Press, 1986.
3. HL Wick, GA Walters. Albizia. Agriculture Handbook No 450. Washington, DC: USDA Forest Service, 1974, p 203.
4. SL Krugman, JL Jenkinson. *Pinus*. Agriculture Handbook No 450. Washington, DC: USDA Forest Service, 1974, p 598.
5. KA Brinkman, HM Phipps. *Lindera*. Agriculture Handbook No 450. Washington, DC: USDA Forest Service, 1974, p 503.
6. DF Olson, WJ Gabriel. *Acer*. Agriculture Handbook No 450. Washington, DC: USDA Forest Service, 1974, p 187.
7. What You Should Know About Seeds. Ontario Factsheet AGDEX 500/010. Ontario: Ontario Ministry of Agriculture and Food, 1987.
8. HT Hartmann, DE Kester, FT Davies, RL Geneva. Plant Propagation: Principles and Practices. 6th ed. Englewood Cliffs, NJ: Prentice-Hall, 1997.
9. J Bewley, M Black. Seeds: Physiology of Development and Germination. New York: Plenum, 1985.
10. Y Liu. Hormones and Tomato Seed Germination. PhD thesis, Wageningen Agricultural University, Netherlands, 1996.
11. DT Bell, JA Plummer, SK Taylor. Bot Rev 59: 24, 1993.
12. RP Poincelot. Horticulture: Principles and Practical Applications. Englewood Cliffs, NJ: Prentice-Hall, 1980.
13. WG Hopkins. Introduction to Plant Physiology. 2nd ed. New York: John Wiley & Sons, 1999.
14. H Borthwick, S Hendricks, M Parker, E Toole, V Toole. Proc Natl Acad Sci U S A 38:662, 1952.
15. A Taylorson, H Borthwick. Weed Sci 17:48, 1969.
16. B Cumming. Can J Bot 41:1211, 1963.
17. S Hendricks, R Taylorson. Nature 237:167, 1972.
18. JL Farrar. Trees in Canada. Markham, Ontario: Fitzhenry and Whiteside Limited, 1995.
19. GP Steinbauer. Plant Physiol 12:813, 1937.
20. BM Pollock, VK Toole. After Ripening, Rest Period, and Dormancy. USDA Yearbook of Agriculture. Washington, DC: Government Printing Office, 1961, p 106.
21. NC Deno. Am Nurs 179(12):42, 1994.
22. NC Deno. Seed Germination Theory and Practice. 2nd ed. 139 Lenor Drive, State College, PA, 1993.
23. JP Mahlstede, ES Haber. Plant Propagation. New York: John Wiley & Sons, 1957.
24. RL Geneve. Seed Technol 20:236, 1998.
25. MA Dirr, CW Heuser Jr. The Reference Manual of Woody Plant Propagation. Athens, GA: Varsity Press, 1987.
26. PO Rudolf. Collecting and Handling Seeds of Forest Trees. USDA Yearbook of Agriculture. Washington, DC: Government Printing Office, 1961, p 221.
27. Seed and Plant Cultural Guide. St. Catharines, Ontario: Jack Van Klaveren Ltd, 1989.
28. AM Armitage. Herbaceous Perennial Plants. Athens, GA: Varsity Press, 1989.
29. V Ball, ed. Ball Red Book. 14th ed. Reston, VA: Reston Publishing Co, 1986.
30. J Bennett, T Forsyth. The Harrowsmith Annual Garden. Ontario: Camden House Publishing, 1990.
31. B Ferguson, ed. Color with Annuals. San Ramon, CA: Ortho Books, Chevron Chemical Co, 1987.
32. L Jelitto, W Schacht. Hardy Herbaceous Perennials. Vols I and II. Portland, OR: Timber Press, 1990.
33. HR Phillips. Growing and Propagating Wildflowers. Chapel Hill: University of North Carolina Press, 1985.
34. HK Tayama, TJ Roll, eds. Tips on Growing Bedding Plants. Bulletin FP-763, Ohio Cooperative Extension Service 1989.
35. MA Dirr. Manual of Woody Landscape Plants, rev. Champaign, IL: Stepes Publishing, 1977.
36. JA Young, CG Young. Seeds of Woody Plants in North America. Portland, OR: Dioscorides Press, 1992.

37. TA Fretz, PE Read, MC Peale. Plant Propagation Lab Manual. 3rd ed. St. Paul: Burgess Publishing Co, 1979.
38. RC Hosie. Native Trees of Canada. Ottawa: Queen's Printer, 1980.
39. PDA McMillan-Browse. Hardy Woody Plants from Seed. London: Grower Books, 1979.
40. JS Wells. Plant Propagation Practices. Chicago: American Nurseryman Publishing Co, 1985.
41. F Sauvageau. Silvicultural Terms in Canada. Cat. No. F042-170. Hull, Quebec: National Resources Canada, Canadian Forest Service, Policy, Economics and International Affairs Directorate, 1995.
42. S Hendricks, R Taylorson. Nature 237:167, 1972.
43. MC Kains, LM McQueston. Propagation of Plants. New York: Orange Judd Publishing Co, 1944.
44. HT Hartmann, WJ Flocker, AM Kofranek. Plant Science: Growth, Development and Utilization of Cultivated Plants. Englewood Dliffs, NJ: Prentice-Hall, 1981.
45. J Ross, J Bradbeer. Planta 100:288, 1971.
46. AG Taylor, PS Allen, MA Bennett, KJ Bradford, JS Burris, MK Misra. Seed Sci Res 8:245, 1998.
47. A Khan. Hortie Rev 13:131, 1992.
48. G Murray, J Swensen. HortScience 28:31, 1993.
49. R Haigh, E Barlow, F Milthorpe, P Sinclair. J Am Soc Hortic Sci 111:660, 1986.
50. PA Brocklehurst, J Dearman. Ann Appl Biol 105:391, 1984.
51. PA Brocklehurst, J Dearman. Ann Appl Biol 102:577, 1983.
52. W Heydecker, P Coolbear. Seed Sci Technol 5:353, 1977.
53. B Michel, O Wiggins, W Outlaw. Plant Physiol 72:60, 1983.
54. Z Cheng, KJ Bradford. J Exp Bot 50:89, 1999.
55. JG Van Pijlen, HL Kraak, RJ Bino, CH DeVos. Seed Sci Technol 29:823, 1995.
56. C Bailly, A Benamar, F Corbineau, D Côme. Physiol Plant 104:646, 1998.
57. G Hacisalihoglu, AG Taylor, DH Paine, MB Hildebrand, AA Khan. HortScience 34:1240, 1999.
58. TL Finnerty, JM Zajicek, MA Hussey. HortScience 27:310, 1992.
59. A Taylor, G Harman. Annu Rev Phytopathol 28:321, 1990.
60. R Madakadze, E Chirco, A Khan. J Am Soc Hortic Sci 118:330, 1993.
61. LH Purdy, JE Hamond, GB Welch. Special Processing and Treatment of Seeds. USDA Yearbook of Agriculture. Washington, DC: Government Printing Office, 1961, p 322.
62. AK Hellum. Lodgpole Pine Seed Extraction. Proceedings of Workshop on High-Quality Collection and Production of Conifer Seed, Northern Forest Research Centre, Can. For. Serv., Information Rpt. NOR-X235, 1981, p 38.
63. Seeds of Woody Plants of the United States. Agriculture Handbook No 450. Washington, DC: USDA Forest Service, 1974.

# 5

# Influence of Source Strength on Leaf Developmental Programming

**Steven Rodermel and Martin Spalding**

*Iowa State University, Ames, Iowa*

**Adam Miller**

*Lorain County Community College, Elyria, Ohio*

## I. INTRODUCTION

Dicot leaf ontogeny is a complex process that is regulated by a variety of exogenous factors (e.g., light) and endogenous factors (e.g., hormones, developmental signals) [1,2]. Striking alterations occur in photosynthetic rates during leaf development, and these changes have been used to monitor the progression of this process. In general, increasing photosynthetic rates are coincident with leaf expansion; a phase of maximal rates occurs at full expansion; and finally, a prolonged senescence phase of declining rates takes place in the fully expanded leaf [3]. The senescence phase is marked by a progressive yellowing of the leaf, loss of protein (most notably of Rubisco), and the translocation of resources to growing parts of the plant [4–8].

The changes that occur in photosynthetic rates during leaf ontogeny are reflected in marked changes in plastid form and function. During leaf expansion, chloroplasts develop from undifferentiated proplastids in the apical meristem and undergo a series of rapid divisions to form mature organelles [9]. Once full expansion is attained, chloroplasts differentiate into "gerontoplasts." This differentiation process comprises a progressive loss of pigments and organized lamellar structures and an accumulation of lipid-containing plastoglobuli [8]. Because most multimeric protein complexes in plastids are composed of subunits encoded by genes in the nucleus and the organelle [10], a central element of leaf developmental programming involves the integration and coordination of gene expression in the nuclear-cytosolic and chloroplast genetic compartments. The mechanisms are poorly understood.

### A. Carbohydrates and Leaf Development

It has been suggested that carbohydrates play a central role in regulating leaf development [11–15]. According to the "sink regulation of photosynthesis" hypothesis, a decrease in sink demand leads to a buildup of carbohydrates and an inhibition of photosynthesis in source leaves [16–18]. In some cases this inhibition occurs as a consequence of decreases in photosynthetic gene expression at the level of transcription [14,16,18–24]. We and others have extended this hypothesis to leaf development and have suggested that feedback inhibition of photosynthetic gene expression by carbohydrates is an important factor that regulates the initiation of the senescent decline in photosynthesis [12,15,25–28]. General support for this notion comes from studies showing that plants grown in elevated light intensities have enhanced rates of leaf senescence [29,30]. Additional support is provided by studies with transgenic plants that have increased internal sugar levels [11,26,31–35]. These plants have chlorotic or yellow leaves with reduced rates of photosynthesis.

## II. INCREASED SOURCE STRENGTH: ELEVATED $CO_2$ AND THE "TEMPORAL SHIFT" MODEL

Despite the attractiveness of the "feedback inhibition" hypothesis to explain the patterns of change that occur in photosynthetic rates during leaf development, very few studies have directly investigated the impact of carbohydrates on leaf developmental programming. We have previously examined various photosynthetic parameters during tobacco leaf development under conditions of increased source strength (carbohydrate production) [28]. In these experiments, individual leaves were examined under ambient $CO_2$ levels (approximately 350 μL/L) or enriched $CO_2$ concentrations (950 μL/L). Leaf 10 (counting up from the base) was chosen for analysis because of its large final size. The elevated $CO_2$ regime was initiated at the time of visible leaf emergence, and measurements were made at various time points until the leaf abscised. As illustrated in Figure 1, ambient $CO_2$–grown leaves exhibited increasing $CO_2$ exchange rates (CERs) up to day 12 (coincident with leaf expansion), a transient maximum (at full expansion), then a steady decline from day 14 onward. The high $CO_2$–grown leaves, on the other hand, attained a similar photosynthetic maximum, but they reached this maximum significantly earlier. The patterns of senescent decline in photosynthetic rates were comparable in both sets of leaves inasmuch as the duration of the senescence phase appeared to be unchanged. The major difference was that

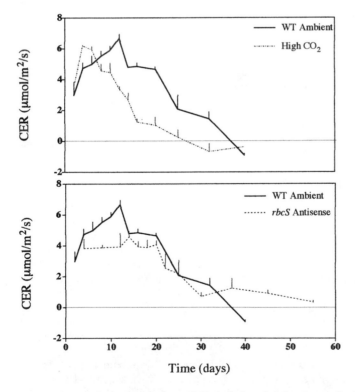

**Figure 1** Photosynthetic rates (CERs) during development of tobacco leaves grown under ambient or elevated $CO_2$ conditions (top panel) or compared with developmentally similar leaves from Rubisco antisense plants (bottom panel). The wild-type and antisense plants were maintained under identical growth conditions. For the increased source strength studies, wild-type plants were moved into high $CO_2$ when leaf 10 reached 1 cm in length. In all cases, day "1" status was assigned when the leaf reached 3 × 5 cm in length. Each point represents the average (± SD) of multiple measurements on leaves from at least four different plants. (Adapted from Refs. 27 and 28.)

CERs reached zero at day 25 in the high $CO_2$–grown leaves versus day 35 in the ambient-grown leaves. These results suggest that the decline in photosynthetic rates characteristic of senescence is initiated at an earlier time point in leaves that have an increased source strength; this is consistent with the feedback inhibition hypothesis. Interestingly, the onset of senescence in the high $CO_2$–grown leaves occurred while the leaves were still expanding.

An examination of other photosynthetic parameters provided further support for the notion that photosynthetic rates attain an earlier photosynthetic maximum in the elevated $CO_2$–grown leaves [28]. These included measurements of chlorophyll concentrations, Rubisco contents, and Rubisco activities. On the basis of these results, we proposed a "temporal shift model" to explain the phenomenon of "acclimation" (down-regulation of photosynthesis) that is frequently observed during growth of plants in elevated $CO_2$ [17]. In this model, the lower photosynthetic rates are the result of a shift in timing of the normal photosynthetic stages of leaf ontogeny to an earlier onset of senescence. Hence, when fully expanded leaves from ambient- versus high $CO_2$–grown plants are compared at a given day after leaf initiation (as in a typical "acclimation" experiment), lower photosynthetic rates are observed in the high $CO_2$–grown leaves because they are further along the progression of the senescence phase of development. Although there appear to be species-specific differences, the findings of Miller et al. [28] are in general agreement with other studies that have examined the impact of elevated $CO_2$ on leaf development [36–39].

If source strength has a regulatory role during leaf development, as suggested by the preceding studies, then it might be anticipated that a decreased source strength condition would have the opposite effect and delay the initiation of the senescence decline in photosynthesis. To address this question, we examined leaf development in Rubisco antisense mutants of tobacco [13,27,40,41]. These plants have a decreased source strength because of a specific reduction in Rubisco content.

## III. RUBISCO ANTISENSE MUTANTS

The Rubisco holoenzyme is composed of eight large subunit (LS) proteins coded for by single genes (*rbc*L) on each of the polyploid chloroplast DNAs and eight small subunit (SS) proteins coded for by a small multigene (*rbc*S) family in the nuclear DNA. To determine whether *rbc*L expression is responsive to SS protein concentrations, as suggested by the "cytoplasmic control principle" [42], we used antisense *rbc*S RNA to down-regulate the expression of *rbc*S messenger RNAs (mRNAs) and proteins in tobacco [43]. For these experiments, tobacco plants were transformed with a highly expressed member of the tobacco *rbc*S gene family cloned in reverse (antisense) orientation behind the cauliflower mosaic virus (CaMV) 35S promoter. The resulting transgenic plants had reduced *rbc*S mRNA and SS protein levels. The reductions in SS protein in these plants were matched by corresponding reductions in the accumulation of LS protein and Rubisco holoenzyme. This lack of overproduction of the LS indicated that there are stoichiometric alterations in the accumulation of the SS and LS in the mutant plants. In contrast to the decreases in LS protein, *rbc*L mRNA levels were unperturbed in the mutants. This indicates that LS protein amounts are regulated posttranscriptionally in these plants. The various transgenic plants had a range of Rubisco concentrations from 10 to 90% of normal, and the antisense *rbc*S RNA gene dosage correlated inversely with Rubisco content.

To examine the nature of the posttranscriptional defect in LS accumulation, mutant plants were pulse labeled with $^{35}$S-Met [43,44]. LS synthesis was markedly decreased during the pulse, suggesting that the antisense plants have a defect in *rbc*L mRNA translation. To pinpoint this defect, we examined polysome profiles of *rbc*L mRNAs [44]. We found that *rbc*L mRNAs are associated with fewer than normal polysomes in the antisense plants, suggesting that less LS accumulates because there is an impairment in the initiation step of *rbc*L mRNA translation. This impairment appears to be specific for *rbc*L mRNAs and not a general consequence of decreased plastid protein synthesis, inasmuch as the polysome distributions (and abundances) of other plastid mRNAs are not affected in the mutants.

Our current working hypothesis is illustrated in Figure 2. In this figure, the SS (directly or indirectly) affects the recruitment of ribosomes to *rbc*L mRNAs. For example, the SS could act as a translational activator: increased SS would increase *rbc*L mRNA translation initiation and thereby increase LS protein production (positive regulation). Alternatively, the LS (or its degradation products) could repress *rbc*L mRNA translation initiation when the LS is produced in excess of the SS (negative regulation). This mechanism would be similar to end-product inhibition at the translational level, as observed for some bac-

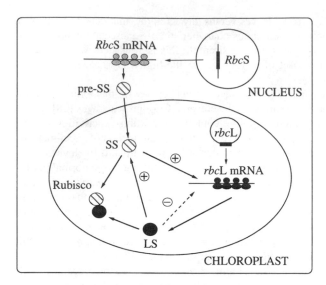

**Figure 2**   Models of mechanisms of control of Rubisco subunit accumulation by SS and LS abundance. Enhanced SS protein levels, either directly or indirectly, positively influence the recruitment of ribosomes onto *rbc*L mRNAs (positive regulation). Alternatively, reduced SS levels may inhibit ribosome recruitment, perhaps via negative feedback by excess LS or LS breakdown products (negative regulation, dashed line extending from LS protein). As demonstrated in conditions of LS limitation, as in *rbc*L mutants [40], LS concentrations influence SS protein stability (positive regulation). (Adapted from Ref. 44.)

terial genes. As such, it would represent a relic of the prokaryotic nature of the chloroplast and its endosymbiont origins.

## IV.   LEAF DEVELOPMENT IN THE RUBISCO ANTISENSE MUTANTS

The preceding studies on the Rubisco antisense mutants were conducted on plants growing in tissue culture medium supplemented with sucrose. Under these conditions, the antisense plants grew at a similar rate and were morphologically similar to wild-type plants. However, exogenously supplied sugars can result in altered patterns of growth and development [14]. Therefore, to study leaf development in the antisense mutants, we grew the plants on soil in the greenhouse [13,27,41]. Under these conditions, the antisense plants are impaired in their ability to fix carbon and to produce carbohydrates suitable for export [45,46].

### A.   Whole Plant Development

As a background for the leaf development experiments, we examined whole plant development in antisense plants with up to 80% reductions in Rubisco holoenzyme content [41]. We found that an early, slow-growth phase of shoot morphogenesis is markedly prolonged in the mutant plants (Figure 3). Leaf emergence is retarded during this phase, and a higher than normal number of (very small) leaves are produced. Following this phase, the wild-type and mutant plants have similar fast-growth phases in terms of leaf emergence rates and numbers, internode distances, leaf sizes, and leaf dry weights. Plant height, total leaf areas, and shoot dry weights are similar at flowering. Collectively, these data suggest that source strength regulates the duration of an early phase of tobacco shoot development and the transition to a later phase. This phase change may occur in response to the attainment of a threshold source strength, which is delayed in the mutant plants.

### B.   Canopy Leaf Development

Jiang and Rodermel [13] examined photosynthesis and photosynthetic gene expression in the antisense plants as a function of leaf nodal position on the plant. All of the leaves on antisense and wild-type plants

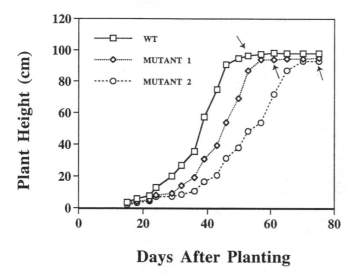

**Figure 3**  Growth of Rubisco antisense and wild-type (WT) tobacco. Plant height (cm) was plotted as a function of days after planting. The samples included WT and antisense plants with either 40% (Mutant 1) or 20% (Mutant 2) of WT Rubisco amounts. Arrows signify the beginning of flowering. (Adapted from Ref. 13.)

were sampled just prior to flowering. The leaves were fully expanded (and thus senescing) with the exception of the youngest ones at the top of the plant. These analyses showed that photosynthetic rates (CERs) are depressed in the antisense leaves but that the overall patterns of change are similar in the mutant and wild-type plants: after attaining a maximum in young fully expanded leaves, photosynthetic capacities decline progressively in the older leaves. Alterations in chlorophyll content and in intercellular $CO_2$ concentrations (Ci) did not closely parallel the changes in CER in either the wild type or mutant, suggesting that light harvesting and stomatal conductance do not strongly limit photosynthesis during leaf development in these plants. By contrast, the patterns of change in CER correlated well with changes in Rubisco initial and total activities as well as with changes in Rubisco content (Figure 4A). "Initial" Rubisco activities provide an estimate of the amount of activated enzyme in the leaf sample at the time of harvest, and "total" activities provide a measure of the amount of Rubisco that is capable of being activated in the sample.

The correlation between initial activities and Rubisco contents suggests that Rubisco activity is primarily a function of holoenzyme concentration in leaves from the antisense and wild-type plants, regardless of leaf nodal position. Consistent with this notion, the activation state of the enzyme (the ratio of initial to total activities) was similar in all of the leaves from both sets of plants. Collectively, these data suggest that Rubisco is a primary determinant regulating photosynthetic rates during leaf development, regardless of holoenzyme concentration. This is consistent with flux-control measurements on first fully expanded leaves of the antisense plants showing that Rubisco activity can explain ~70% of the control on photosynthetic rates under moderate to high light intensities [47,48].

Jiang and Rodermel [13] also examined the mechanism of Rubisco accumulation in the antisense plants: is it similar to that in plants growing in tissue culture on sucrose-containing medium? For these analyses, Rubisco subunit mRNA levels were measured by RNA gel blot analysis as a function of leaf nodal position (Figure 4B and C). As mentioned earlier, Rubisco concentrations in both sets of plants are highest in the youngest fully expanded leaves at the top of the plant and decrease progressively to the oldest leaves at the bottom of the plant (Figure 4A); LS and SS proteins are not present in excess in either set of plants. The RNA gel blot analyses showed that, in the wild type, the alterations in Rubisco abundance are due primarily to coordinate changes in *rbc*S and *rbc*L transcript accumulation. In the antisense plants, however, Rubisco concentrations appear to be controlled by the abundance of *rbc*S, but not *rbc*L, mRNAs; the levels and patterns of change in *rbc*L mRNA were normal in the mutants even though they accumulated less LS protein. This suggests that LS accumulation is regulated posttranscriptionally during antisense leaf development, mirroring the situation in the tissue

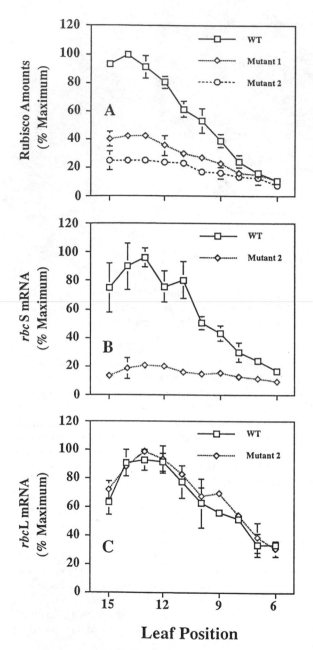

**Figure 4** Rubisco, *Rbc*S, and *rbc*L mRNA abundances during development in the wild-type and antisense plants. Measurements of relative amounts of Rubisco (A), *Rbc*S mRNA (B), and *rbc*L mRNA (C) were performed on all the leaves from wild-type and antisense plants with either 40% (Mutant 1) or 20% (Mutant 2) of wild-type Rubisco amounts. Leaf 15 is the youngest leaf at the top of plant just prior to flowering, and leaf 6 is the oldest leaf still remaining at the bottom of the plant. Each data point represents the means ($\pm$ SD) of three measurements from three different plants; abundances were calculated relative to the maximum value in the wild type. For the protein assays, soluble proteins were isolated from frozen leaf disks and electrophoresed through discontinuous 12.5% sodium dodecyl sulfate (SDS) polyacrylamide gels. Proteins were loaded on the gel on a leaf area basis. The gels were immunoblotted with tobacco SS and LS antibodies, and band intensities on the filters were quantified by phosphorimage analysis. LS and SS gave similar results; LS data are shown. To determine transcript abundances, RNAs were isolated from frozen leaf disks and equal amounts were applied to slot blot filters. The filters were probed with *Rbc*S and *rbc*L gene-specific probes. The filters were quantified by phosphorimage analysis. (Adapted from Ref. 13.)

culture–grown plants. We do not know whether this regulation occurs specifically at the level of *rbc*L mRNA translation initiation.

To gain insight into whether *rbc*L mRNA expression is limited by *rbc*L DNA template availability, Jiang and Rodermel [13] performed genomic Southern blot analyses. In these experiments, equal amounts of total cell DNA from the wild-type and mutant leaves were blotted onto filters and the filters were probed with *rbc*L DNA sequences. These analyses showed that *rbc*L DNA levels fall in concert with *rbc*L mRNA as one moves down the canopy. The patterns of change (and the magnitude of the change) were similar in both sets of plants. Although we do not know whether each of the polyploid chloroplast DNAs is equally capable of being transcribed in leaves from these plants, these data suggest that developmental controls on *rbc*L mRNA abundance may be exerted, at least in part, by *rbc*L template availability in both wild-type and Rubisco antisense tobacco plants.

The decreases in Rubisco appear to have little impact on the accumulation of proteins other than Rubisco [13]. This indicates that leaf developmental programming is generally insensitive to Rubisco concentrations. This conclusion is consistent with the observations of others who have examined protein accumulation in first fully expanded leaves of antisense plants with up to ~80% reductions in Rubisco content [46,49,50]. The protein accumulation profiles to date have relied on Western immunoblot analyses of relatively few "representative" proteins; it is now possible to conduct detailed proteomics analyses to confirm this generalization.

## C. Development of an Individual Leaf

Miller et al. [27] have recently studied the development of individual antisense and wild-type leaves. In these experiments, leaves were sampled throughout their ontogeny (similar to the elevated $CO_2$ studies). "Developmentally similar" leaves were used. These were leaves from node 13 in the antisense plants and from node 10 in the wild-type plants. Both of these leaves emerge during the fast-growth phase of shoot morphogenesis and have similar characteristics (final size, canopy position, photosynthetic rate) [41]. Because many of the analyses required destructive sampling, developmentally similar leaves were isolated from many plants. "Day 1" status was accorded to the leaves when they first attained a size sufficient for analysis (~3 cm in width and ~5 cm in length).

We first examined various photosynthetic parameters in the mutant and wild-type leaves. As illustrated in Figure 1, photosynthetic rates increased to a maximum on day 12 in the wild-type plants, then declined steadily until they fell below zero on day 35; shortly thereafter they abscised from the plant. Although maximal rates were somewhat lower in the antisense leaves, they were relatively constant until about day 20, after which they declined steadily until day 30. Thereafter, they remained fairly constant and did not fall below zero, even at day 55. Antisense leaves did not abscise until around day 60. Similar patterns of change were observed for chlorophyll concentrations, Rubisco contents, and Rubisco activities. Taken together, these data indicate that the antisense leaves are longer lived than wild-type leaves and that this increase in longevity is due to a prolongation of the senescence phase of development.

Much of this prolongation appears to be due to alterations in the expression of genes for photosynthetic proteins. During wild-type leaf development, we found that the senescence phase is marked by a progressive decline in the content of total cell protein, chloroplast rRNA, and chloroplast DNA. These parameters followed similar patterns of change in the antisense leaves, with the exception that the senescent declines were markedly prolonged. mRNAs for specific photosynthetic proteins also decreased during the senescence of wild-type and antisense leaves. For instance, *rbc*S and *rbc*L mRNAs decreased in parallel in the wild type, consistent with the hypothesis that coordinate changes in *rbc*S and *rbc*L mRNA abundance play a central role in determining Rubisco content during wild-type leaf ontogeny. As discussed earlier, this coordination has been observed in all of our developmental studies on wild-type tobacco [13,28,40]. By contrast, *rbc*S and *rbc*L mRNAs undergo a longer senescent decline in the antisense leaves. Also as observed in our earlier studies, the content of the holoenzyme in the antisense leaves appears to be regulated primarily at the level of *rbc*S transcript accumulation; i.e., LS protein accumulation is regulated posttranscriptionally.

## V. CONCLUSIONS

Examination of the Rubisco antisense mutants has revealed that decreased source strength regulates the duration and progression of tobacco leaf senescence. Increased source strength (elevated $CO_2$), on the other hand, results in an earlier onset of the senescence phase of leaf development. These findings suggest that leaf developmental programming is broadly responsive to a range of source strength conditions. This programming includes alterations in the patterns of gene expression that extend beyond photosynthetic gene expression, inasmuch as the abundances of chloroplast rRNA, chloroplast DNA, and total cell protein are all affected by source strength. On the other hand, source strength does not affect all elements of tobacco leaf development, as is evident from the similarity in the expansion rates of leaves from wild-type and antisense plants and leaves exposed to elevated $CO_2$ [28].

The mechanism by which source strength is sensed is likely to be complex. One hypothesis is that hexokinase acts as a sugar sensor, initiating a signal transduction pathway that modulates the expression of various nuclear genes (reviewed in Refs. 4, 6, 14, and 20). Chloroplast genes for subunits of chloroplast multimeric protein complexes (e.g., LS and SS of Rubisco) are also expressed coordinately in response to alterations in source strength during leaf development (very likely at the transcriptional level), and thus sugar sensing must involve regulatory communication between the nucleus and the plastid. These regulatory circuits are poorly defined (reviewed in Ref. 40). One further complication is that many other factors, e.g., hormones and light, influence the progression and duration of leaf development. Components of signal transduction pathways for all of these factors probably interact and share elements in common [15,51].

Our studies have shown that carbohydrates are able to regulate leaf developmental programming in a predictable manner, consistent with the idea of feedback inhibition of photosynthesis ("sink regulation" hypothesis). We suggest that in some cases a threshold source strength is sensed and that this regulates a developmental switch, for instance, a phase transition in shoot morphogenesis [41] or the onset of the senescence phase of leaf development [28]. In other cases, source strength is able to modulate the duration of development responses once they have commenced, e.g., the senescence phase of leaf development [27].

## REFERENCES

1. TP Brutnell, JA Langdale. Signals in leaf development. Adv Bot Res 28:161–195, 1998.
2. M Van Lijsebettens, J Clarke. Leaf development in *Arabidopsis*. Plant Physiol Biochem 36:47–60, 1998.
3. S Gepstein. Photosynthesis. In: LD Noodén, AC Leopold, eds. Senescence and Aging in Plants. San Diego: Academic Press, 1988, pp 85–109.
4. AB Bleecker, SE Patterson. Last exit: Senescence, abscission, and meristem arrest in *Arabidopsis*. Plant Cell 9:1169–1179, 1997.
5. V Buchanan-Wollaston. The molecular biology of leaf senescence. J Exp Bot 307:181–199, 1997.
6. S Gan, RM Amasino. Making sense of senescence: Molecular genetic regulation and manipulation of leaf senescence. Plant Physiol 113:313–319, 1997.
7. KN Lohman, S Gan, CJ Manoram, RM Amasino. Molecular analysis of natural leaf senescence in *Arabidopsis thaliana*. Physiol Plant 92:322–328, 1994.
8. P Matile. Chloroplast senescence. In: NR Baker, H Thomas, eds. Crop Photosynthesis: Spatial and Temporal Determinants. Amsterdam: Elsevier Science, 1992, pp 413–441.
9. JE Mullet. Chloroplast development and gene expression. Annu Rev Plant Physiol Plant Mol Biol 39:475–502, 1988.
10. L Bogorad. Possibilities for intergenomic integration: Regulatory crosscurrents between the plastid and nuclear-cytoplasmic compartments. Cell Cult Somatic Cell Genet Plants 7B:447–466, 1991.
11. JC Jang, P León, L Zhou, J Sheen. Hexokinase as a sugar sensor in higher plants. Plant Cell 9:5–19, 1997.
12. CZ Jiang, SR Rodermel, RM Shibles. Photosynthesis, Rubisco activity and amount, and their regulation by transcription in senescing soybean leaves. Plant Physiol 101:105–112, 1993.
13. CZ Jiang, SR Rodermel. Regulation of photosynthesis during leaf development in *RbcS* antisense DNA mutants of tobacco. Plant Physiol 107:215–224, 1995.
14. KE Koch. Carbohydrate-modulated gene expression in plants. Annu Rev Plant Physiol Plant Mol Biol 47:509–540, 1996.
15. A Wingler, A von Schaewen, RC Leegood, PJ Lea, WP Quick. Regulation of leaf senescence by cytokinin, sugars, and light. Plant Physiol 116:329–335, 1998.
16. J Sheen. Feedback control of gene expression. Photosynth Res 39:427–438, 1994.
17. M Stitt. Rising $CO_2$ levels and their potential significance for carbon flow in photosynthetic cells. Plant Cell Environ 14:741–762, 1991.
18. JJ VanOosten, RT Besford. Acclimation of photosynthesis to elevated $CO_2$ through feedback regulation of gene expression: Climate of opinion. Photosynth Res 48:353–365, 1996.
19. MC Criqui, A Durr, J Parmentier, J Marbach, J Fleck, E Jamet. How are photosynthetic genes repressed in freshly-isolated mesophyll protoplasts of *Nicotiana sylvestris*? Plant Physiol Biochem 30:597–601, 1992.

20. JC Jang, J Sheen. Sugar sensing in higher plants. Plant Cell 6:1665–1679, 1994.
21. A Krapp, B Hofmann, C Schäfer, M Stitt. Regulation of the expression of *rbcS* and other photosynthetic genes by carbohydrates: A mechanism for the "sink regulation" of photosynthesis? Plant J 3:817–828, 1993.
22. A Krapp, WP Quick, M Stitt. Ribulose-1,5-bisphosphate carboxylase-oxygenase, other photosynthetic enzymes and chlorophyll decrease when glucose is supplied to mature spinach leaves via transpiration stream. Planta 186:58–69, 1991.
23. A Krapp, M Stitt. Influence of high-carbohydrate content on the activity of plastidic and cytosolic isoenzyme pairs in photosynthetic tissues. Plant Cell Environ 17:861–866, 1994.
24. J Sheen. Metabolic repression of transcription in higher plants. Plant Cell 2:1027–1038, 1989.
25. LL Hensel, V Grbic, DA Baumgarten, AB Bleecker. Developmental and age-related processes that influence the longevity and senescence of photosynthetic tissues in *Arabidopsis*. Plant Cell 5:553–564, 1993.
26. M Stitt, U Sonnewald. Regulation of metabolism in transgenic plants. Annu Rev Plant Physiol Plant Mol Biol 46:341–368, 1995.
27. AC Miller, C Schlagnhaufer, M Spalding, S Rodermel. Carbohydrate regulation of leaf development: Prolongation of leaf senescence in Rubisco antisense mutants of tobacco. Photosynth Res 63:1–8, 2000.
28. A Miller, CH Tsai, D Hemphill, M Endres, S Rodermel, M Spalding. Elevated $CO_2$ effects during leaf ontogeny—a new perspective on acclimation. Plant Physiol 115:1195–1200, 1997.
29. LD Noodén, JW Hillsberg, MJ Schneider. Induction of leaf senescence in *Arabidopsis thaliana* by long days through a light-dosage effect. Physiol Plant 96:491–495, 1996.
30. B Veierskov. Irradiance-dependent senescence of isolated leaves. Physiol Plant 71:316–320, 1987.
31. N Dai, A Schaffer, M Petreikov, Y Shahak, Y Giller, K Ratner, A Levine, D Granot. Overexpression of *Arabidopsis* hexokinase in tomato plants inhibits growth, reduces photosynthesis, and induces rapid senescence. Plant Cell 11:1253–1266, 1999.
32. CD Dickinson, T Altabella, MJ Chrispeels. Slow-growth phenotype of transgenic tomato expressing apoplastic invertase. Plant Physiol 95:420–425, 1991.
33. EE Goldschmidt, SC Huber. Regulation of photosynthesis by end-product accumulation in leaves of plants storing starch, sucrose, and hexose sugars. Plant Physiol 99:1443–1448, 1992.
34. PG Jones, JC Lloyd, CA Raines. Glucose feeding of intact wheat plants represses the expression of a number of Calvin cycle genes. Plant Cell Environ 19:231–236, 1996.
35. A von Schaewen, M Stitt, R Schmidt, U Sonnewald, L Willmitzer. Expression of a yeast-derived invertase in the cell wall of tobacco and *Arabidopsis* plants leads to accumulation of carbohydrate and inhibition of photosynthesis and strongly influences growth and phenotype of transgenic tobacco plants. EMBO J 9:3033–3044, 1990.
36. RT Besford, LJ Ludwig, AC Withers. The greenhouse effect: Acclimation of tomato plants growing in high $CO_2$, photosynthesis and ribulose-1,5-bisphosphate carboxylase protein. J Exp Bot 41:925–931, 1990.
37. GY Nie, SP Long, RL Garcia, BA Kimball, RL Lamorte, PJ Pinter Jr, GW Wall, AN Webber. Effects of free-air $CO_2$ enrichment on the development of the photosynthetic apparatus in wheat, as indicated by changes in leaf proteins. Plant Cell Environ 18:855–864, 1995.
38. M Pearson, GL Brooks. The influence of elevated $CO_2$ on growth and age-related changes in leaf gas exchange. J Exp Bot 46:1651–1659, 1995.
39. JJ VanOosten, RT Besford. Some relationships between the gas exchange, biochemistry and molecular biology of photosynthesis during leaf development of tomato plants after transfer to different carbon dioxide concentrations. Plant Cell Environ 18:1253–1266, 1995.
40. S Rodermel. Subunit control of Rubisco biosynthesis—a relic of an endosymbiotic past? Photosynth Res 59:105–123, 1999.
41. CH Tsai, A Miller, M Spalding, SR Rodermel. Source strength regulates an early phase transition of tobacco shoot morphogenesis. Plant Physiol 115:907–914, 1997.
42. RJ Ellis. Protein synthesis by isolated chloroplasts. Biochim Biophys Acta 463:185–215, 1977.
43. SR Rodermel, MS Abbott, L Bogorad. Nuclear-organelle interactions: Nuclear antisense gene inhibits ribulose bisphosphate carboxylase enzyme levels in transformed tobacco plants. Cell 55:673–681, 1988.
44. S Rodermel, J Haley, CZ Jiang, CH Tsai, L Bogorad. A mechanism for intergenomic integration: Abundance of ribulose bisphosphate carboxylase small subunit protein influences the translation of the large subunit mRNA. Proc Natl Acad Sci U S A 93:3881–3885, 1996.
45. WP Quick, U Schurr, R Scheibe, ED Schulze, SR Rodermel, L Bogorad, M Stitt. Decreased ribulose-1,5-bisphosphate carboxylase-oxygenase in transgenic tobacco transformed with "antisense" *rbcS*. I. Impact on photosynthesis in ambient growth conditions. Planta 183:542–554, 1991.
46. WP Quick, U Schurr, K Fichtner, ED Schulze, SR Rodermel, L Bogorad, M Stitt. The impact of decreased Rubisco on photosynthesis, growth, allocation and storage in tobacco plants which have been transformed with antisense *rbcS*. Plant J 1:51–58, 1991.
47. M Lauerer, D Saftic, WP Quick, C Labate, K Fichtner, ED Schulze, SR Rodermel, L Bogorad, M Stitt. Decreased ribulose-1,5-bisphosphate carboxylase-oxygenase in transgenic tobacco transformed with "antisense" *rbcS*. VI. Effect on photosynthesis in plants grown at different irradiance. Planta 190:332–345, 1993.

48. M Stitt, WP Quick, U Schurr, ED Schulze, SR Rodermel, L Bogorad. Decreased ribulose-1,5-bisphosphate carboxylase-oxygenase in transgenic tobacco transformed with "antisense" *rbcS*. II. Flux-control coefficients for photosynthesis in varying light, $CO_2$, and air humidity. Planta 183:555–566, 1991.

49. GS Hudson, JR Evans, S von Caemmerer, YBC Arvidsson, TJ Andrews. Reduction of ribulose-1,5-bisphosphate carboxylase/oxygenase content by antisense RNA reduces photosynthesis in transgenic tobacco plants. Plant Physiol 98:294–302, 1992.

50. J Masle, GS Hudson, MR Badger. Effects of ambient $CO_2$ concentration on growth and nitrogen use in tobacco (*Nicotiana tabacum*) plants transformed with an antisense gene to the small subunit of ribulose-1,5-bisphosphate carboxylase/oxygenase. Plant Physiol 103:1075–1088, 1993.

51. L Zhou, JC Jang, TL Jones, J Sheen. Glucose and ethylene signal transduction crosstalk revealed by an *Arabidopsis* glucose-insensitive mutant. Proc Natl Acad Sci U S A 95:10294–10299, 1998.

# 6

# Ecophysiological Aspects of the Vegetative Propagation of Saltbush (*Atriplex* spp.) and Mulberry (*Morus* spp.)

**David N. Sen\* and Pramila Rajput**

*University of Jodhpur, Jodhpur, India*

## I.  INTRODUCTION

Nature has provided the phenomenon of reproduction to the living world in order to perpetuate species. The way in which reproduction in the plant kingdom is carried out may be broadly divided into two categories.

Vegetative propagation
Sexual reproduction

Some plants at times fail to complete their life cycle by means of seeds, yet they survive and perpetuate themselves. This is because nature has provided an alternative to sexual reproduction, that is, vegetative propagation. The latter is the most common method of propagation because of various advantages, such as maintenance of particular characteristics of the plants, relative seed in raising samplings in large numbers for plantation, adaptability to a particular habitat, development of resistance to pests and diseases, and drought tolerance to modify the growth of plant.

Propagation through seeds is mainly done to bring about a varied population for the purpose of selection and hybridization. Rooting of cut pieces of stem is a prerequisite for multiplication and survival. Trees in many cases fail to produce roots from cuttings and thus present difficulties. Among the factors affecting rooting of cuttings, the position of the shoot plays an important role.

The multiplication of species by vegetative means is practiced in forestry and horticulture to obtain plants of a desired genetic constitution for crossing in a breeding program for many reasons (to improve growth and yield, stem quality, wood quantity, resistance to pests and diseases, or other desirable characteristics and also to maintain the purity of types so evolved for commercial exploitation). This process has been used for quick multiplication for a number of plant species, which is important for afforestation purposes in arid zones, where quick growth and development of plants are very much needed.

In easily rooting species, the ability of stem cuttings to root varies considerably with the season. In many cases, profuse rooting occurs when cuttings are taken from trees in an active season. The seasonal rooting response of stem cuttings is related to the disappearance of starch. The hydrolytic activity is high when rooting occurs, but is not detached when cuttings fail to root [1]. Nanda [2] showed that the effectiveness of exogenously applied auxins varies with the season and that these differences may be ascribed to changes in a plant's nutritional and hormonal status during its annual cycle of growth.

---

\* retired

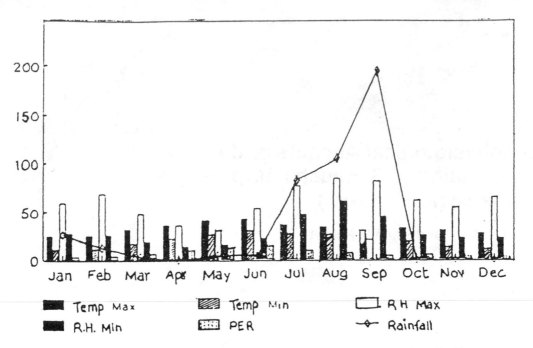

**Figure 1**  Climatic parameters at Jodhpur during 1992 RH = relative humidity; PER = potential evapotranspiration.

Propagation through cuttings is one of the most common methods of perpetuating vegetable species. It is restricted to varieties that are fully acclimatized to local conditions. Growers select plants that display the properties chosen for multiplication, such as nutritious leaf, higher yield, quick growth, and resistance to diseases, insect pests, and drought. Resistance to drought is an important property to be associated with other desirable characteristics in tropical and semiarid regions. Drought occurs frequently in semiarid tropical regions, hence crop loss due to scarcity of water also occurs frequently. In addition to the development of more suitable farming technology, evaluation of drought-resistant varieties is desirable.

The arid ecosystem environment offers an adaptive challange to the survival of plants: the only species that can survive possess adaptive mechanisms that enable them to adjust under strong climatic fluctuations [3]. According to Sen [3], ecophysiological studies are thus important for judging the ability of a particular species to adjust under prevailing climatic (Figure 1) and edaphic conditions.

Available soil moisture is used by the roots of annual and perennial plants from the end of the rainy season until early summer, by which time such moisture has been depleted. Later, a partial or total status quo is maintained in soil moisture, mainly in the open, with the result that water loss is eliminated by shedding or reduction of leaves by plants [3]. Whatever rainwater is retained by the soil is used by the roots of annual and perennial species from June–July to November–December [4]. Physiological studies are helpful in determining the individual and collective influence of different factors on vegetative propagation.

## A.  Significance of Vegetative Propagation

Rooting in stem cuttings can be important means of vegetative propagation for afforestation purposes. In arid zones, quick establishment of plants with ample root systems is a necessity.

In arid regions, water in the form of precipitation is available only in the rainy season, and the plants must be established in suitable conditions of soil moisture. Therefore, rooted stem cuttings are more useful than seed sowing because rooted cuttings are far better able to survive in the stressful environment of the desert than delicate seedlings.

There are various methods of multiplication of mulberry plants. In mulberry species, the stem cuttings readily form roots. Both grafting and layering need time for establishment. Hence, propagation of mulberry through stem cuttings is preferred.

Thimann and Behnke-Rogers [5] showed that the rooting of cuttings of many tree species is stimulated by synthetic growth substances. Bose [6] has developed easier and better methods of vegetative propagation by the use of growth substances for ornamental and fruit plants. Bose and Mukherjee [7] used some growth substances to improve rooting in cuttings of *Legerstroemia indica*. Prasad and Dikshit [8] obtained maximum success in rooting with cuttings of essential oil–producing plants treated with growth regulators. Teaotia and Pandey [9] obtained better results in rooting guava stem cuttings with the assistance of growth substances.

## B.  Factors Affecting Vegetative Propagation

More than 50% of the land surface of the developing countries is located in the arid and semiarid zones. In many of these countries, in which more than 80% of the population lives with agricultural and animal husbandry, a tragic and dangerous imbalance is developing between requirements for and available supply of food, fodder, and fuel [10–12]. Dwindling vegetation cover will adversely affect all facets of rural life in which trees and shrubs generally serve not only as fuel but also as shade and shelter for man, animal, and crops. In the long term, depletion of the natural vegetation will increase ecological fragility and contribute to gradual degradation of the resource base as well as the natural resources themselves [13–15].

A practice common among peasants is migration of cattle to neighboring states or within the state wherever fodder is available. This large-scale migration does immense harm to the delicate ecosystem. Animals usually strip all of the plants from the area; this causes poor regeneration and increased soil erosion, and more areas become barren. This necessitates the utilization of saline wastelands for fodder production as crop cultivation is impossible because of the high salt content of the soil. Enumeration of indigenous saline species showed that very few plants are palatable and their growth pattern is not at an acceptable level for fodder production.

Many taxa of the family Chenopodiaceae are indigenous to arid and saline regions of the world. Their ecological amplitude is very high, and various adaptive features at different levels of the plant life cycle are observed. Many are shrubs, and they offer a tremendous potential for human benefit in making the arid and semiarid lands of the world more productive and useful [16].

To revegetate the salt-affected soils and secondary salinized soils, plants that can survive in arid and saline conditions are needed. Shrubby halophytes of the genus *Atriplex* are particularly adapted to such conditions. The genus *Atriplex* includes several haloxeric fodder species very useful in arid zones. The primary driving force of all animals is the need to finding the right kind of food and enough of it. Food is the burning question in animal society, and the whole structure and activities of the community are dependent upon questions of food supply.

Saline and sodic soils are problems of individual localities, and their formation and causes of development must be considered before these soils are put to any economic use. Salt-tolerant plants have been used as forage in arid saline areas for millennia. The recognition of the value of certain salt-tolerant shrub grass species is reflected in their incorporation in pasture improvement programs in many salt-affected regions throughout the world. However, reproduction, survival, and multiplication under the inhospitable conditions of arid saline areas are basic needs for any halophytic or glycophytic species. In many halophytes, germination of seeds is usually retarded by high concentrations of salt in the soil [17–19]. Germination is the most important stage in the life cycle of any species growing in an arid saline environment. Seed germination in saline environments occurs mostly with high precipitation, when soil salinity levels are usually reduced [20–23]. It is also known that when seeds are sown in a saline environment, there is a decrease in the rate of germination, delaying completion of germination; moreover, there is a water potential below which germination does not occur [24–27]. In general, it is agreed that salinity affects germination by creating sufficiently low to inhibit water uptake (osmotic effect) and/or by providing conditions for the entry of ions that may be toxic to the embryo [28,29]. These constraints affect the different stages of seed germination and seed establishment to varying degrees.

Reduction of germination occurs when halophytes are subjected to salinities above 1% NaCl; increasing salt concentrations also delay germination [30]. Salinity or sodicity and water stress are the most important factors responsible for limiting seed germination and plant growth. To overcome the present environmental stress of saline areas, plants produce a variety of ecological adaptations. Propagation through vegetative means has been used as a method of multiplication for a number of plant species under arid saline conditions.

Among factors affecting rooting of cuttings, the position of the shoot plays an important role [31]. It is reported that without auxin treatment and without leaves, no roots were obtained in cutting of red *Hibiscus* and *Allamenda cathartica* [32].

Vegetative reproduction substitutes for or at least contributes to the reproductive potential of many plants. This statement is more applicable to various halophytic species that are restricted to narrow ecological limits, either in the production of disseminules or by their germination [33]. Self-layering species of *Atriplex* are at an advantage in establishing themselves in salt-affected soil, which they accomplish faster than other species: the growth of developing roots results in rapid penetration through the upper salty soil layers. Furthermore, roots developing at different nodes are not dependent on a direct supply of water from the soil [34,35]. Being well supplied with water by the parent plant, roots can penetrate layers of extreme salinity.

## C.  Vegetative Propagation in Saline Plants

The distribution of salinity varies spatially, temporally, qualitatively, and quantitatively. In addition, the responses of plants to salt stress vary during the life cycle of the individual [36,37]. Phenotype plasticity involving both morphological and physiological changes in response to episodic events is an important characteristic associated with the survival of long-lived plants under highly stressful environmental conditions. Transient reductions in yield in response to salinity may be the result of the adaptive reconstruction of growth habits of a plant. The heterogeneity of saline habitats leads to considerable genetic differentiation among populations as a result of natural selection: an all-purpose genotype capable of growing in a wide range of saline habitats probably does not exist [38].

The growth and productivity of *Atriplex* under conditions of low and erratic rainfall are exceptional, and the adaptation of this species to high salinity makes its introduction very suitable [39]. Agronomic testing, feeding trials, and development of the best agronomic practices are necessary in the evaluation of suitable species for introduction and mass propagation [40].

Normal vegetation, except for some halophytes, cannot survive on saline and sodic soils. Thus, areas having soils of these types are of limited agricultural use unless the salinity is quite mild. Increased salinity has rendered many lands unfit for cultivation. Plant species that are capable of accumulating large quantities of sodium in their tissues are the least sensitive to the presence of salt in the soil. The tolerance of a species to high amounts of absorbed or exchangeable sodium is modified by the pH of the soil and by the accumulation of $CO_2$.

With increasing human and animal populations and the need for greater crop and fodder production, nonproductive salt-affected lands may be used to grow nonconventional crops of economic value and also such food crops as pearl millet. It is desirable to choose species well suited to saline habitats and to calculate the most economical means of reclamation to make the salt-affected soils productive. The essential ingredients of technology for meeting these problems consist of the use of tolerant species, special planting techniques, and aftercare.

Cultivation of salt-affected areas with palatable halophytes is one of the most promising and ecologically safe approaches in the reclamation process. It also helps cattle breeders and farmers to improve a chronically stagnant economy. Selection of the most suitable halophytic species for introduction into saline land needs extensive research. Malcolm [41,42] and Sen et al. [43] have produced a guide to the selection of salt-tolerant shrubs for forage production from saline lands in southwestern Australia and India, respectively. Important selection parameters include:

1.  Growth and survival for a sufficient period in a representative environment
2.  Reproduction by seed or vegetative means
3.  Acceptable growth form for management use
4.  Production of biomass of sufficient quantity, quality, and acceptability to livestock
5.  Ease of establishment
6.  Persistence under a profitable management system

7. Effectiveness for erosion control, lowering ground water, and improvement of habitat for wildlife

In addition, the plant must be evaluated by another set of criteria before attempting its development on a crop scale.

1. Establishment:
   a. Seed germination percentage
   b. Vegetative propagation
   c. Seedling vigor and root establishment
   d. Need for supplemental water and nutrition
2. Hardiness under crop production densities:
   a. Insect and disease resistance
   b. Intra- and interspecific competition
3. Ecological traits:
   a. Ecotype variability from which to select stock for introduction
   b. Total genetic plasticity to different ecosystems

Many halophytic species appear to have significant economic potential for desert agriculture. In addition, the productivity of cultivated halophytes is high. Haloxeric species of the genus *Atriplex* are widely used as fodder crops in otherwise unusable saline wastelands in many parts of the world. Many *Atriplex* species are promising in the reclamation of the salt-affected lands. Use of salt-affected soils for uncontrolled grazing, subsistence cropping, or intensive fuel gathering results in degradation of the natural vegetation cover. This process may take decades to reverse, and the land may never be returned to its original condition. To slow such deterioration, new economically useful exotic species can be introduced in these areas. Forage-yielding xerohalophytes such as *Atriplex* can be suitable candidates for the management of saline wastelands because these plants can also be irrigated with brackish water. Land reclamation and rehabilitation in arid zones can be achieved by using salt-tolerant plant species for a number of different purposes suited to the local conditions.

Many halophytic species (e.g., *Arthrocnemum* spp., *Nitraria retusa, Salicornia* spp.) are capable of forming adventitious roots on their twigs. This ability varies among species and according to the season of the year [44]. Vegetative propagation is of great advantage in revegetating salt-affected soils. It favors more assured establishment in the field than direct seeding or seedling transplantation. Rooted stem cuttings of *Atriplex* are also helpful in raising a large number of plants with such desired properties as favorable growth habits, regeneration capacity, leafiness, and palatability.

Vegetative propagation of desert shrubs is a means of producing genetically identical individuals in species whose sexually produced offspring normally exhibit higher variability. Reduced variability of plant materials can increase experimental precision, and many genetically identical individuals are necessary for varietal testing. Reproduction of desirable parental characteristics such as high seed yield would be valuable in the establishment of seed nurseries. Vegetative propagation is also a method of producing transplants of species whose seeds do not germinate readily.

## II. VEGETATIVE PROPAGATION OF SALTBUSH (*Atriplex* spp.)

*A. amnicola* Paul G. Wilson (river saltbush or swamp saltbush) shows a remarkable high growth rate under desert conditions. The seedlings can be transplanted in the first week of October and can be irrigated with poor quality water. For the first 2–3 months the growth rate is slow, after which fast growth occurs. Enormous production of side branches during the winter season is a very distinctive feature, and these newly formed branches (stems) are soft, fleshy, and purplish pink in color. By mid-December plants attain a height of about 60–70 cm and lateral branches measure about 50–60 cm. Plants may show two types of growth patterns: (1) an erect type and (2) a prostrate spreading type. Two-year-old plants may cover an area of more than 2–5 m$^2$. These plants grow sideways and cover the ground very rapidly.

Rooted cuttings of *Atriplex* species are needed to establish a rapid plantation. Some *Atriplex* species are subdioecious, with at least three genders [45]. Moreover, rooted cuttings can be used to propagate superior individual plants for a variety of purposes, including breeding programs and provision of superior or uniform outplanting stock [46]. Observations made in the field have revealed that *A. amnicola* plants have a natural ability to produce rooted cuttings. During the monsoon season, *A. amnicola* was found to produce nodal roots from the lateral branches wherever they touched the ground. This ability is of great importance in binding the loose topsoil. It also helps the plant to recover speedily from grazing pressure and enables the plant to spread rapidly and multiply. Vegetative propagation is much easier in *A. amnicola* because its nodal root formation helps in the production of a large number of rooted cuttings for field planting.

The effects of different growth regulators used on stem cuttings for root regulation and axillary shoot growth in different seasons of the year (Figures 2–5) are described in Secs. A–D.

## A. Indole Acetic Acid (IAA)

Observations regarding the effect of indole acetic acid on root and shoot growth are presented in Figure 2. Indole acetic acid did not produce much beneficial effect on root and shoot growth; it promoted roots only when administered in lower concentrations. In higher concentrations (40 and 50 ppm) [47] during winter and at all concentrations in rainy seasons, root and shoot growth were affected severely: there was no root formation. IAA favored root growth only in lower concentrations (10 and 20 ppm) during the winter and summer seasons, respectively. Slight yellowing and drying effects on leaves were seen at higher concentrations.

## B. Naphthalene Acetic Acid (NAA)

Figure 3 shows that compared with other auxins, NAA caused the maximum initiation of roots in cuttings. Root growth was affected more favorably only at lower concentrations (10 ppm) during winter; at higher concentrations the roots produced were thinner and had a minimum number of secondary roots. A distinct effect of NAA on root growth was seen on comparing results from winter and summer. In the rainy season, the length of the root was less than during the rest of the year. Drastic inhibition of root and axillary

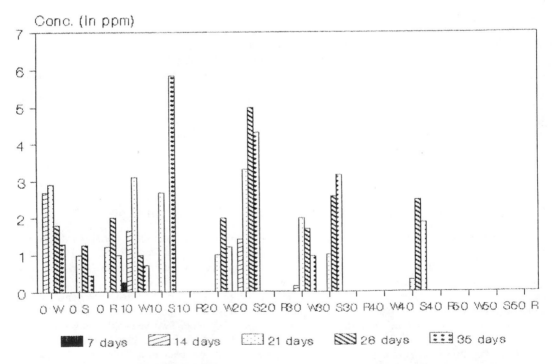

**Figure 2**   Effect of IAA on rooting of *Atriplex amnicola* during different seasons.

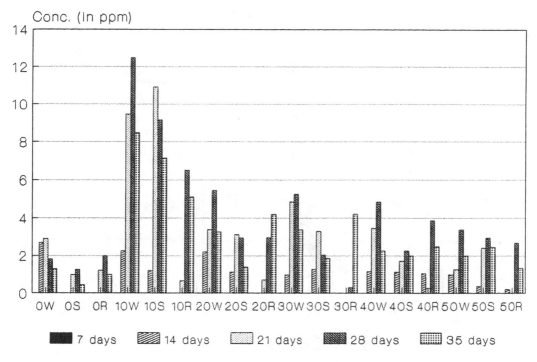

**Figure 3** Effect of NAA on rooting of *Atriplex amnicola* during different seasons.

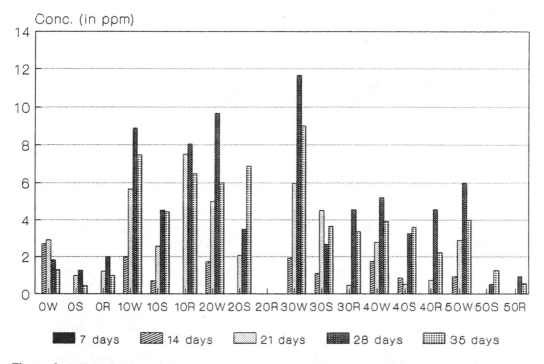

**Figure 4** Effect of IBA on rooting of *Atriplex amnicola* during different seasons.

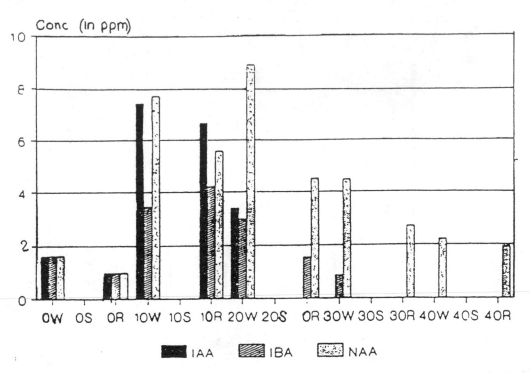

**Figure 5**  Effect of IAA, IBA, and NAA (10–50 ppm) on rooting of *Atriplex amnicola* in the field during winter (W), summer (S), and rainy season (R).

branch initiation and growth in the rainy season was observed. Interestingly, at almost all concentrations, very large numbers of roots were also produced on the internodal region.

## C.  Indole Butyric Acid (IBA)

It is evident from Figure 4 that the effect of IBA on rooting is next to that of NAA; that is, IBA promotes root growth better at lower concentration (30 ppm) than at higher ones and no distinct difference in the growth of axillary branches was observed. Root growth was maximum in winter at 30 ppm and with well-developed secondary roots. Very poor growth of roots and no initiation of axillary branches were observed in plants treated with IBA in summer.

## D.  Field Transfer and Establishment of Rooted Cuttings

The effect of growth regulators on root and shoot growth was observed by growing the cuttings in polyethylene bags for 35 days after treatment. It is clear from the results (Figure 5) that root growth was maximum at the higher concentration (20 ppm) of NAA, followed by IBA (10 ppm), and the least growth was obtained with IAA (10 ppm) after 35 days. In the control set, the roots were very much shorter than in the treated cuttings. The maximum development of roots with profuse secondary roots was observed with NAA and IBA. Whereas IAA suppressed the growth of roots and axillary branches during summer. NAA and IBA enhanced the growth of axillary branches to a maximum, but the number and the length of the roots were diminished in comparison with NAA. The maximum number of axillary branches was observed in winter and rainy seasons, the least in summer.

The propagation of stem cuttings of several saltbush species and a few species from other salt desert shrub genera was studied by Nord and Goodin [48], Wieland et al. [49], Ellern [50], and Wiesner and Johnson [51]. Although Nord and Goodin [48] and Ellern [50] observed a general trend for better rooting of saltbush (*Atriplex*) species in spring than in fall, no data were available for summer and winter. Nord and Goodin [48] noted better rooting of green stem tips than ripe wood cuttings, but Ellern [50] failed to find any difference in rooting of soft, green cuttings and young woody stem cuttings.

Nanda et al. [52] used IAA, IBA, and NAA to enhance the rooting response of stem cuttings of forest trees and investigated the possibility that even seasonal changes in the effectiveness of different auxins are governed by morphophysiological factors. Auxins enhanced the rooting of stem cuttings of *Populus nigra* and *Hibiscus rosa-sinensis* even during December–February, but these hormones failed to cause rooting in *Ficus infectoria* cuttings during the same period. It was observed that auxins enhanced the rooting more in winter, followed by the rainy season, and least in summer.

Indole acetic acid has been one of the most commonly used auxins, but different workers have obtained varying results [8,9,53,54]. Chatterjee [55] found that *Pogostemon potehouli*, an essential oil–yielding plant, responded more favorably to IAA than other auxins. Shanmugavelu [56] also obtained the maximum percentage of rooting in cuttings of certain shrubby plants with IAA. On the other hand, NAA gave favorable results in the induction of roots in cuttings of *Levendula, Ficus infectoria*, and *Hibiscus rosa-sinensis* [57]. The experimental results of our study showed that a large number of roots were produced at lower concentrations of NAA, IAA, and IBA.

A number of saltbush species may be established from cuttings, including *A, amnicola, A. nummularia, A. canescens, A. halimus, A. lentiformis, A. paludosa*, and *A. polycarpa* [58]. The cuttings should be taken at the peak of spring growth or in the autumn in a Mediterranean climate. The wood should be about 6 mm thick and 250 mm long, taken from young stems between two leaf axils. A rooting hormone (e.g., IBA) may be applied to encourage root growth before approximately half the stem is covered with a moist, sandy soil. The cuttings should root within 6 weeks and should be ready for transplanting in 10 weeks [59]. In our study, IBA also enhanced the rooting in *A. amnicola*.

According to Richardson et al. [46], fourwing saltbush cuttings could be rooted best in the summer, but *A. amnicola* rooted best in winter, followed by the rainy season and summer. According to Sharma and Sen [60] and Rajput and Sen [61], respectively, winter is most suitable for the vegetative propagation of *Tamarix* and *Atriplex*. The present results also support these views.

The results of field experiments showed that NAA is more effective than IBA and IAA. The increased appearance of new leaves with an increase in the percentage of rooting also points to better rooting possibilities, with the emergence of more new leaves on the cuttings. The greater number of roots per cutting and the greater number of leaves may also help the cuttings to survive when sown in natural conditions.

## III. VEGETATIVE PROPAGATION OF MULBERRY (*Morus* spp.)

Since the dawn of agriculture, one of the principal aims of human beings has been the control and promotion of plant growth to satisfy human needs. These two important aspects of people's work with plants in the struggle to increase production are by no means synonymous. Humans soon realized that lush green growth does not always produce the best crop in the form of fruit and seeds, and hence they were forced to evolve such well-known cultural methods as pruning, balanced manuring, and use of mineral fertilizers to regulate the nature and luxuriance of plant growth.

The naturally occurring (endogenous) growth substances are commonly known as plant hormones, while the synthetic ones are called growth regulators. A plant hormone (synonym: phytochrome) is an organic compound synthesized in one part of a plant and translocated to another part, where at very low concentrations it causes a physiological response. Plant hormones are identified as promoters (auxin, gibberellin, and cytokinin), inhibitors (abscisic acid, xanthoxin, and violaxanthin), and ethylene and other hypothetical growth substances (florigen, death hormone, etc.). They usually exist in plants and crops at a concentration lower than 1 μM; above this, they are generally considered supraoptimal [62].

Mulberry is propagated either through seeds or vegetatively. The latter is the more common method of propagation because of such advantages as maintenance of particular properties of the plant, relative speed in raising saplings in large numbers for plantation, adaptability to a particular habitat, and abilities to develop resistance to pests and diseases and to modify the growth of plants. Propagation through seeds has reached certain limitations. For example, triploid plants, which do not produce viable seeds, cannot be propagated. It is not possible to reproduce true to the type from a seed of biparental origin.

Mulberry is a highly heterozygous plant that is open for cross-fertilization. Therefore, the seeds that are formed through open pollination are natural hybrids. Seedling populations from such seeds provide wider chances for selection of superior types whose characteristics are perpetuated through vegetative propagation. Generally, the population thus obtained is a mixture of several clones. Each clone is heterozygous although homogeneous, and the same genotype is maintained because propagation is vegetative, Interclonal variations are due to heredity. Depending on climatic and soil conditions, different countries follow different modes of vegetative propagation. Hamada [63] described the methods used in Japan, which include (1) bark grafting (Fukurotsugi), (2) veneer grafting (Kiritsugi), (3) simple layers (Magedori), (4) continuous layers (Shumokudori), and (5) division (Shirodasmi), hardwood cuttings (Kojyosashiki), and softwood cuttings (Shinshosashiki), Generally, grafting is used in places where the temperature is 6°C in March and more than 25°C in July, with rainfall of 175 mm. Shirodasmi cottage is popular in places having temperatures less than 4°C in March and less than 25°C in July with rainfall lower than 175 mm. Propagation through hardwood and softwood cuttings is common in the northern districts and the southern region, respectively, of Japan [64]. In Italy [65], rooted grafting is a popular method of multiplying Japanese mulberry varieties.

In India, the most common method of propagating mulberry is through cuttings in multivoltine regions (e.g., Karnataka and West Bengal). Exotic varieties that are not established by cuttings are propagated through root grafts. Many of the indigenous varieties and well-acclimatized exotic varieties are propagated through cuttings. Bud grafting (budding) is used only when scion material is scarce. Whenever a large number of mulberry plants must be obtained in a shorter time than would be possible if they were started as a cutting, the method of layering is used. Layering allows the grower to fill in the gaps formed as a result of the failure to sprout of certain cuttings planted in pits of established plantations.

In univoltine areas (e.g., Kashmir), the mulberry is propagated through seedlings and the exotic varieties through root grafts. In India, the field-scale propagation through cuttings of Japanese varieties of mulberry is still a problem.

Propagation through seeds is used mainly to bring about a varied population for the purpose of selection and hybridization. Because mulberry flowers are open for cross-pollination, the seeds thus collected serve mainly as sources of stock material for grafting.

In general, a deficiency of hormone must be created experimentally (as by removing young leaves or using a hormone-deficient mutant) to show that adding a hormone has an effect. In this respect, the Mitscherlich law of diminishing return can be modified as follows: the increase in plant response produced by a unit increment of a deficient (limiting) hormone is proportional to the decrement of that hormone from the maximum.

Mulberry varieties that do not ordinarily produce roots from a cutting are induced to root with application of the requisite quantity of root hormones. The following chemicals are generally used, but their efficiency varies from species to species and from variety to variety: (1) IAA, (2) IBA, and (3) NAA.

The objective of using growth regulators is to increase the percentage of cuttings that form roots, hasten root initiation, and increase the number of roots per cutting. IBA and NAA have proved to be better in producing roots than other growth regulators.

The water requirement of mulberry does not differ greatly from species to species or from variety to variety. The plant must be capable of absorbing water from soils of low moisture regimes. Generally resistant plants should have well-developed root systems, hydrophilic colloids to absorb and hold water by imbibition, and adaptations to facilitate the lowering of transpiration. In this regard, certain Japanese varieties have a thick cuticle, sometimes a two-layered epidermis, a palisade parenchyma, and other beneficial characteristics.

Although many tropical species root profusely through cuttings, certain temperate varieties do not ordinarily produce roots. Root induction has been successfully achieved in the latter varieties by the (artificial) application of the requisite quantity of root hormones. However, the efficacy of the substances varies from species to species and from variety to variety.

Development of the root primordium depends on the relative amount of natural auxin present in the plant. Varieties that do not root apparently contain less auxin. The growth regulators act like auxins when applied in small quantities and move upward in mass translocation through the xylem when the bases of the cuttings are soaked in their solutions. The objective of treatment is to increase the percentage of cuttings that form roots, hasten root initiation, and increase the number of roots per cutting. Indole butyric acid and naphthalene acetic acid appear to be better at producing roots than other agents [66,67]. The chemicals may be applied by various methods, including direct application of a powder, soaking the cuttings in dilute solutions, dipping the cuttings in concentrated solutions, and application as a paste in lanolin.

**TABLE 1**  Effect of Different Concentrations of IAA and NAA on Bud Sprouting (BS), Initiation of Leaves (L), and Inflorescences (I) on Stem Cuttings of *M. alba* (Cultivated Variety 92)

| Concentration (ppm) | Total buds on cuttings | Jan. 25 | | | Feb. 01 | | | Feb. 03 | | | Feb. 11 | | |
|---|---|---|---|---|---|---|---|---|---|---|---|---|---|
| | | BS | L | I | BS | L | I | BS | L | I | BS | L | I |
| Control | 5 | 1 | | | 2 | 1 | | 3 | 1 | 3 | 3 | 4 | 6 |
| IAA | | | | | | | | | | | | | |
| 10 | 5 | 3 | | | 4 | 1 | | 4 | 2 | | 5 | 5 | 7 |
| 20 | 8 | 4 | | | 5 | 1 | | 5 | 1 | | 5 | 3 | 4 |
| NAA | | | | | | | | | | | | | |
| 10 | 7 | 2 | | | 5 | | | 5 | | | 5 | 3 | 2 |
| 20 | 9 | | | | 3 | | | 3 | | | 3 | 4 | 3 |

The action of many gibberellic acids (GAs) is similar to that of IAA, including elongation, promotion of cambial activity, induction of parthenocarpy, and stimulation of nucleic acid and protein synthesis. The GA3s vary greatly in their biological activity, and GA5 and GA7 are considered to have the widest range. In ferns, algae, and fungi, GA3s have also been shown to influence growth and development [68].

For the vegetative propagation experiments, mulberry cuttings were collected from both cultivated and wild varieties at Jodhpur (site Chopasni). Growth regulators used for root initiation in cuttings were NAA, IAA, IBA, and GA. Shoots of thick branches with well-developed buds were used for rooting experiments. Cuttings taken from parts with a high carbohydrate content have been reported to root more readily and profusely than cuttings selected from parts rich in nitrogen. Portions of the shoot that were too tender at the top and overmature at the base were rejected. Cuttings taken from young branches sprouted rapidly and profusely as compared with those taken from old parts. Cuttings of 7 to 10 cm usually of pencil thickness with three to four well-developed buds were prepared from the central portion of the clone with a slanting cut.

Table 1, which gives the results of experiments on rooting behavior in mulberry (cultivated variety), shows that the addition of IAA and NAA at lower concentrations almost always caused more bud sprouting. Also, the total leaves generally increased, together with the number of inflorescences. Increasing concentrations of hormones tended to decrease the values. Slightly higher values of these parameters were observed with 10 ppm than with 20 ppm IAA. Of the two auxins, IAA was more effective than NAA. It is also clear from Table 1 that in the case of NAA, a lower concentration is more effective than a higher one.

From the observations of the rooting behavior in a wild variety of mulberry (Table 2), we see that the lower concentration of IAA is more effective than the higher one. The maximum number of sprouting buds was 11; afterward the values remained constant. However, in the case of NAA, 20 ppm was more effective than 10 ppm. Comparatively, IAA was more effective than NAA and higher values were observed in the wild than in the cultivated variety.

Cuttings were immersed in different concentrations of growth regulators for 24 hr. During treatment, cuttings were kept inside the growth room. After this treatment, cuttings were washed in distilled water

**TABLE 2**  Effect of Different Concentrations of IAA and NAA on Bud Sprouting (BS), Initiation of Leaves (L), and Inflorescences (I) on Stem Cuttings of *M. indica* (Wild Variety 92)

| Concentration (ppm) | Total buds on cuttings | Jan. 25 | | | Feb. 01 | | | Feb. 03 | | | Feb. 11 | | |
|---|---|---|---|---|---|---|---|---|---|---|---|---|---|
| | | BS | L | I | BS | L | I | BS | L | I | BS | L | I |
| Control | 8 | 4 | | | 5 | | | 5 | 6 | 4 | 7 | 6 | 10 |
| IAA | | | | | | | | | | | | | |
| 10 | 16 | 9 | | | 11 | | | 11 | 9 | 26 | 11 | 8 | 28 |
| 20 | 16 | 7 | | | 11 | | | 11 | 3 | 4 | 11 | 4 | 6 |
| NAA | | | | | | | | | | | | | |
| 10 | 30 | 6 | | | 8 | | | 9 | 3 | 4 | 9 | 4 | 9 |
| 20 | 18 | 9 | | | 9 | | | 9 | 4 | 4 | 9 | 4 | 11 |

and individually transferred to test tubes filled with water. Three cuttings were used for each set, and each set was repeated three times for confirmation. The observations were recorded after a definite interval of time. Measurements of bud sprouting, number of leaves, inflorescence, and root initiation were observed in the cultivated variety (Table 3).

The lower concentration (10 ppm) of IAA led to a better response than the higher one (20 ppm). At 10 ppm the bud sprouting was 100%, whereas with 20 ppm it diminished to 4. Leaf initiation and inflorescence followed the same trend as the bud sprouting, being 8 and 11, respectively, in 10 ppm, and 8 and 6, respectively, in 20 ppm.

Similar results were obtained in NAA, with 100% bud sprouting in 10 ppm, slightly more than in IAA, being 5. Leaf initiation and inflorescence were also higher in 10 ppm compared with 20 ppm, being 9, 9, 8, and 5, respectively.

IAA again showed a beneficial effect at the lower concentration (10 ppm) as compared with the higher (20 ppm), producing 7 and 4 buds, respectively. Leaf initiation increased from 7 to 10 with increasing concentration, but the inflorescence did not show any change (Table 4).

The results with NAA showed effects similar to those with IAA. At 10 ppm, 8 buds sprouted out of 8 buds, whereas at 20 ppm the figures were 4 out of 7 buds. Leaf initiation showed a better response at the lower concentration than at the higher one, and a similar trend was also shown for inflorescence (Table 4).

Growth means an irreversible increase in the weight, area, or length of a plant or a particular tissue or organ of a plant, while development denotes the changing pattern of organization as growth progresses. Control over plant growth by the regulated exogenous supply of chemical substances may occur in different ways. It has become clear that total control of plants is vested not in a single hormonal type; rather, control is shared by a group of several specifically defined auxins, gibberellins, ethylene, and certain naturally occurring inhibitors such as phenols and abscisic acid. Thus, the plant growth regulators provide a very helpful tool for controlling physiological processes in plants.

NAA was found to be better than IAA in rooting by Jauhari and Rehman [53] in cuttings of sweet lime. It responded favorably on induction of roots in stem cuttings of many plants [69]. In the present study IAA was found to be more effective than NAA.

Stem cuttings of *Ipomoea pes-caprae* and species of *Morus* showed a large number of roots and buds in the higher concentration but with maximum suppression of growth, whereas lower concentrations resulted in only improvement in the growth of roots [70]. In our investigation also, the higher values were observed with lower concentrations of the growth regulators.

Under favorable environmental conditions, during the period of root development, a callus tissue develops at the basal end of a cutting: an irregular mass of parenchyma cells in various stages of lignification. Callus growth arises from cells and adjacent phloem, although various cortical and medullar cells

**TABLE 3**  Effect of Different Concentrations of IAA and NAA on Bud Sprouting (BS), Initiation of Leaves (L), and Number of Inflorescences (I) on Stem Cuttings of *Morus* spp. (Cultivated Var.) from Chopasni in Growth Room (1993)

| Concentration (ppm) | Total buds on cuttings | Dec. 12 | | | Dec. 21 | | | Dec. 30 | | | Jan. 10 | | |
|---|---|---|---|---|---|---|---|---|---|---|---|---|---|
| | | BS | L | I | BS | L | I | BS | L | I | BS | L | I |
| IAA | | | | | | | | | | | | | |
| 10 | 5 | 2 | | | 3 | 2 | | 4 | 6 | 9 | 5 | 8 | 11 |
| | 1 | 1 | | | 1 | 1 | | 1 | 2 | 1 | 1 | 2 | 1 |
| 20 | 7 | 2 | | | 3 | 3 | | 4 | 6 | 3 | 4 | 8 | 6 |
| | 1 | 1 | | | 1 | | | 1 | 2 | 1 | 1 | 1 | 2 |
| NAA | | | | | | | | | | | | | |
| 10 | 5 | 2 | | | 3 | 5 | 3 | 5 | 8 | 7 | 6 | 9 | 9 |
| | 1 | 1 | | | 1 | 2 | 1 | | 2 | 2 | 1 | 2 | 3 |
| 20 | 7 | 2 | | | 3 | 3 | 1 | 4 | 6 | 4 | 5 | 8 | 5 |
| | 2 | 2 | | | 1 | 1 | 1 | 1 | 2 | 1 | | 1 | 1 |
| Control | 4 | | | | 1 | | | 2 | | | 2 | 1 | 3 |
| | 1 | | | | 1 | | | 1 | | | | 1 | 1 |

**TABLE 4**  Effect of Different Concentrations of IAA and NAA on Bud Sprouting (BS), Initiation of Leaves (L), and Number of Inflorescences (I) on Stem Cuttings of *Morus* spp. (Wild Var.) from Chopasni in Growth Room (1993)

| Concentration (ppm) | Total buds on cuttings | Dec. 12 | | | Dec. 21 | | | Dec. 30 | | | Jan. 10 | | |
|---|---|---|---|---|---|---|---|---|---|---|---|---|---|
| | | BS | L | I | BS | L | I | BS | L | I | BS | L | I |
| IAA | | | | | | | | | | | | | |
| 10 | 7 | 2 | | | 3 | 1 | 1 | 5 | 4 | 4 | 7 | 7 | 8 |
| | 2 | 1 | | | 1 | 1 | | 1 | | | 1 | 1 | 2 |
| 20 | 8 | 1 | | 1 | 2 | 5 | 4 | 4 | 7 | 6 | 4 | 10 | 8 |
| | 1 | 1 | | | 1 | 1 | 2 | 1 | 2 | 1 | | 1 | 3 |
| NAA | | | | | | | | | | | | | |
| 10 | 8 | 3 | 3 | 1 | 5 | 8 | 9 | 6 | 9 | 6 | 8 | 10 | 8 |
| | 1 | 1 | | | 1 | 1 | | 1 | 2 | 2 | 1 | 1 | 2 |
| 20 | 7 | 2 | | | 3 | 1 | 1 | 4 | 3 | 2 | 4 | 4 | 3 |
| | 1 | 1 | | | 1 | 1 | | 1 | 1 | 1 | | 2 | 2 |
| Control | 5 | 1 | | | 2 | 2 | 2 | 3 | 3 | 3 | 3 | 4 | 4 |
| | 1 | 1 | | | 1 | 1 | 1 | 1 | 1 | 1 | 1 | 1 | 1 |

also contribute. Because root development and callus formation occur simultaneously, it is believed that the formation of callus is essential for root development. In reality, these two are entirely different phenomena. Sometimes roots develop even without callus from the nodes. Callus formation is sometimes beneficial in varieties that are slow to root because it provides a protective layer, preventing the cutting from becoming desiccated and decayed. Sometimes the callus interferes with the absorption of water by the cutting. In our investigations, rooting did not start, instead, callus formation was observed after 1 week of treatment. The callus was creamy white and had a granular texture.

The rate of sprouting of vegetative buds is of primary consideration in introducing a variety or species in an area. Mulberry varieties grown in Mysore and West Bengal sprout throughout the year, facilitating the attempts of sericulturists to rear the silkworms year-round. The axillary buds vary in size, shape, and position from variety to variety.

Thus the rooted stem cuttings are more useful than seed sowing because the survival of a rooted cutting is far better than that of the delicate seedlings in the stressful environment of the desert.

## IV.  SUMMARY

In India, propagation through stem cuttings is the most common method. It is restricted to varieties that are fully acclimatized to local environmental conditions. Plants that have qualities such as a nutritious leaf for silkworms, higher yield, quick growth for establishment, and resistance to diseases and insect pests and drought-resistant varieties such as *Atriplex* are selected.

Plant hormones are identified as growth promoters (auxins, gibberellins, and cytokinins). They usually exist in plants at a concentration lower then 1 μM, above which they are generally considered supraoptimal. Mulberry varieties that do not ordinarily produce roots from a cutting are induced to root with application of root hormones. The chemicals used for this purpose are IAA, NAA, and IBA.

The objective of using growth regulators in mulberry and *Atriplex* was to increase rooting in cuttings, hastening root initiation and increasing the number of roots per cutting. IAA and NAA were found better for producing roots. *Atriplex* produces the maximum number of roots with NAA, followed by IBA and least in IAA. The results of field experiments showed that NAA is more effective than IBA and IAA. The appearance of more new leaves with an increased percentage of rooting also points to better possibilities, with the emergence of more new leaves on the cuttings. The greater number of roots and leaves per cutting may also help the cuttings to survive when sown under natural field conditions.

These cuttings collected from wild and cultivated varieties dried out in all treatments with growth regulators in the summer season in both growth room and garden pot experiments. In the rainy season the experiments in the growth room failed as the cuttings dried without any sprouting, but in the garden pots lower concentrations produced better results even during the rainy season. In winter, the growth room experiments showed better responses in the treatments with both IAA and NAA.

Vegetative propagation with various treatments with different growth regulators is based on bud sprouting, leaf initiation, and inflorescence, as no root initiation was seen in mulberry, which showed only callus formation. It could be concluded that summer and rainy seasons are not suitable at all, and so vegetative propagation of mulberry should be done only in the winter season.

## ACKNOWLEDGMENTS

Financial assistance received from DoEn (Department of Environment), Central Silk Board and seeds from Texas Tech University and the University of Arizona are gratefully acknowledged.

## REFERENCES

1. Bala A, Nanda VK, Nanda KK. Seasonal changes in the rooting response of stem cuttings of *Dalbergia sisoo* and their relationship with biochemical changes. Indian J Plant Physiol 12:154–165, 1969.
2. Nanda KK. Use of auxins in rooting stem cuttings of some forest tree species. Final Report PL-480 Project A-7-F5-11, 1971.
3. Sen DN. Ecology of saline areas of Rajasthan and exploitation of saline ecosystem for increased productivity. Final Project Report, Department of Environment, Government of India, New Delhi, 1990.
4. Sen DN. Water relations of psammophyte-*Convolvulus microphyllus* Sieb. ex. Spreng. Ecophysiological Foundation of Ecosystem Productivity in Arid Zone, International Symposium, USSR, 1972, pp 79–83.
5. Thimann KV, Behnke-Rogers J. The use of auxins in the rooting of wood cutting. Maria Moors Cabot Foundation Publication 1, 1950.
6. Bose TK. Improvement in the method of vegetative propagation of *Amherstia nobilis* and *Petrea arborea*, Sci Cult 30:198–199, 1964.
7. Bose TK, Mukherjee D. Use of root promoting chemicals on cuttings of *Legerstroemia indica*. Sci Cult 34:217–218, 1968.
8. Prasad A, Dikshit AP. Vegetative propagation of essential oil–yielding plants. Sci Cult 29:460–461, 1963.
9. Teaotia SS, Pandey IC. Effect of growth substance on rooting of guava stem cuttings. Sci Cult 27:442–444, 1961.
10. FAO. Forestry and rural development. FAO Forestry Paper No. 26, FAO, Rome, 1981.
11. FAO. Tropical Forest Resources. FAO Forestry Paper No. 30, FAO, Rome, 1982.
12. FAO. Forestry beyond 200. Prospect and Problem in arid and semi-arid zones. Secretarial Note Committee on Forestry, COFO 84/5, May 1984, FAO, Rome, 1984.
13. Eckholm EP. The Other Energy Crisis: Firewood. Washington, DC: Worldwatch Institute, 1975.
14. Palmberg C. Genetic resources of fuelwood tree species for the improvement of rural living. Paper presented at the FAO/UNEP/IBPGR International conference on crop genetic Resources, Rome, April 6–10, 1981, FAO, Rome, 1981.
15. Palmberg C. A vital fuelwood genepool is in danger. Unasylva 33:22–30, 1981.
16. McKell CM, Blaisdal JP, Goodin JR. Wildland Shrubs—Their Biology and Utilization. Gen. Tech. Report INT-I. U.S. Forest Service, Washington, DC, 1972.
17. Rajpurohit KS, Sen DN. Soil salinity and seed germination under water stress. Trans Isdt Ucds 2:106–110, 1977.
18. Mohan Ram HY, Gupta P. Plant life under extreme environment. Curr Sci 72:306–315, 1997.
19. Sen DN, Mohammed S, Kasera PK. Biology of plants in saline environment. In: IS Grover, AK Thukral, eds. Environment and Development. Jodhpur: Scientific Publishers, 1997, pp 117–126.
20. Abd-El Samad HM, Shaddad MAK. Salt tolerance of soybean cultivars. Biol Plant 39:263–269, 1997.
21. McMahon A, Ungar IA. Phenology, distribution and survival of *Atriplex triangularis* Willd. in an Ohio salt pan, Am Midl Nat 100:1–14, 1978.
22. Sen DN, Mohammed S. Water relations of halophytes in Indian arid zone. In: M Dainel, ed. The Changing Landscape of Plant Sciences. Prof SD Sabnis Felicitation Volume. DehraDun: International Book Distributors, 1997, pp 283–298.
23. Sen DN, Mohammed S, Kasera PK. Biology of saline ecosystem and reclaiming unproductive land in Indian desert. In: AS Mehta, B Saran, eds. Environment and Plant Productivity. Proceedings of 4th APPSC Conference and Ist Neo Botanica Convention, Patna, 1998, pp 1–7.
24. Uhvits R. Effect of osmotic pressure in water absorption and germination of alfalfa seeds. Am J Bot 33:278–285, 1946.
25. Prisco JT, O'Leary JW. Osmotic and "toxic" effects of salinity on germination of *Phaseolus vulgaris* L. seeds. Turrialba 20:177–184, 1970.
26. Ungar IA. Halophyte seed germination. Bot Rev 44:233–264, 1978.

27. Sen DN, Mohammed S. General aspects of salinity and the biology of saline plants. In: M Pessarakli, ed. Handbook of Plant and Crop Stress. New York: Marcel Dekker, 1994, pp 125–145.

28. Bewley JD, Black M. Physiology and Biochemistry of Seeds in Relation to Germination, Viability. Dormancy and Environmental Control. New York: Springer-Verlag, 1982.

29. Shukla SP. Physiological Studies in the Halophyte *Salvadora persica* L. PhD thesis, Poona University, Pune, 1999.

30. Chapman VJ. Salt Marshes and Salt Deserts of the World. Bremerhaven, Germany: J Cramer, 1974.

31. Yamdagni N, Sen DN. Role of leaves present on the stem cuttings for vegetative propagation in *Portulaca grandiflora* L. Biochem Physiol Pflanz 164:447–449, 1973.

32. Van Overbeek J, Gordon SA, Gregory LE. Analysis of the function of the leaf in the process of root formation in cuttings. Am J Bot 33:100–107, 1946.

33. Waisel Y. Biology of Halophytes. New York: Academic Press, 1972.

34. Waisel Y, Pollack G. Estimation of water stresses in the active root zone of some native halophytes in Israel. J Ecol 57:789–794, 1969.

35. Sen DN, Kasera PK, Mohammed S. Perennation of weeds in Indian desert. In: RK Tandon, P Singh, eds. Biodiversity, Taxonomy and Ecology. Prof KMM Dakashini Commemoration Volume. Jodhpur: Scientific Publishers, 1999, pp 355–364.

36. Sen DN, Rajpurohit KS. Contributions to the Ecology of Halophytes. The Hague: W Junk, 1982.

37. Epstain E, Rains DW. Advances in salt tolerance. Plant Soil 99:17–29, 1987.

38. Houreou L. Ecoclimatic and biogeographic comparison between the rangelands of the isoclimatic Mediterranean arid zone of Northern Africa and the near East. Proceedings of the 2nd International Conference on Range Management in the Arabian Gulf, Kuwait, March 3–6, 1990, p 20.

39. Gallagher JL. Halophytic crops for cultivation at seawater salinity, Plant Soil 89:323–336, 1985.

40. Raheja PC. Salinity and aridity, new approaches to old problems. In: H Boyko, ed. Aridity and Salinity—A Survey of Soils and Land Use. The Hague: W Junk, 1966, pp 43–127.

41. Malcolm CV. Wheat belt Salinity—A Review of the Saltland Problem in South Western Australia. Technical Bulletin 52, Department of Agriculture of Western Australia, Perth, 1982, p 65.

42. Malcolm CV. Rehabilitation agronomy—Guidelines for revegetating degraded land. Proc Ecol Soc Aust 16:551–556, 1990.

43. Sen DN, Prakash BSV, Thomas TP. Wasteland management with special reference in introduction of *Atriplex* spp. In: SS Negi, ed. Advances in Forestry Research in India. Dehradun, India: International Book Distributors, Booksellers and Publishers, 1987, pp 221–240.

44. Waisel Y, Pollack G. Estimation of water stresses in the active root zone of some native halophytes in Israel. J Ecol 57:789–794, 1969.

45. McArthur ED, Freeman DC. Sex expression in *Atriplex canescens*: Genetics and environment. Bot Gaz 143:476–482, 1982.

46. Richardson SG, Barker JR, Crofts KA, Van Epps GA. Factors affecting root of stem cuttings of salt desert shrubs. J Range Manage 32:280–283, 1979.

47. Solanki DK. Ecological Studies on Mulberry under Semi-arid Zone of Rajasthan. PhD thesis, J.N.V. University, Jodhpur, 1996.

48. Nord EC, Goodin JR. Rooted Cuttings of Shrub species for Planting in California Wildlands. Research Note PSW 213. U.S. Department of Agriculture. Forest Service, Pacific Southwest Forest and Range Experiment Station, Berkeley, CA, 1970.

49. Wieland PAT, Frolich EF, Wallance A. Vegetative propagation of woody shrub species from the northern Mojave and southern Great Basin deserts. Madrono 21:149–152, 1971.

50. Ellern SJ. Rooted cuttings of saltbush (*Atriplex halimus* L.). J Range Manage 25:154–155, 1972.

51. Wiesner LE, Johnson WJ. Fourwing saltbush (*Atriplex canescens*) propagation techniques. J Range Manage 30:154–156, 1977.

52. Nanda KK, Anand VK, Kumar P. Some investigations of auxin effects on rooting of stem cuttings of forest plants. Indian For 96:171–187, 1970.

53. Jauhari OS, Rehman SF. Further investigation on rooting in cuttings of sweet lime (*Citrus limettoides* Tanaka), Sci Cult 24:432–434, 1959.

54. Nanda KK, Kochhar VK, Gupta S. Effect of auxins, sucrose and morphacting in the rooting of hypocotyl cuttings of *Impatiens balsamina* during different seasons. Biol Land Plants 1:181–187, 1972.

55. Chatterjee SK. A note on the vegetative propagation of some of the essential oil yielding plants newly introduced at Mungpoo, Darjeeling. Sci Cult 25:687–688, 1960.

56. Shanmugavelu KG. A note on the response of rooting of cuttings of the *Hibiscus rosa-sinensis* Linn. and *Allamanda cathartica* Linn. to the application of plant growth regulators. Sci Cult 26:136–137, 1960.

57. Everett RL, Meeuwing RO, Robertson JH. Propagation of Nevada shrubs by stem cuttings. J Range Manage 31:426–429, 1978.

58. McArthur ED, Plummer AP, Van Epps GA, Freeman DC, Jorgensen KR. Producing fourwing saltbush seed in seed orchards. In: DN Hyder, ed. Proceedings of the First International Rangeland Congress, Denver, August 14–18, Society for Range Management, 1978, p 406.

59.  de Kock GC. Drought resistant fodder shrub crops in South Africa. In: H Houeron, ed. Browse in Africa: The Current State of Knowledge. Addis Ababa: International Livestock Centre of Africa, 1980, pp 399–408.

60.  Sharma TP, Sen DN. Seasonal variation in the rooting capacity of *Tamarix* spp in inland salines of Indian arid zone. 76th Indian Science Congress, Madurai, Abstract, 1989, p 151.

61.  Rajput P, Sen DN. Role of certain growth regulators on the vegetative propagation of an exotic species *Atriplex amnicola* Paul G. Wilson. 4th All India Botanical Conference, Lucknow, 1991.

62.  Naqvi SSM. Plant hormones and stress phenomena. In: M Pessarakli, ed. Handbook of Plant and Crop Stress. New York: Marcel Dekker 1994, pp 383–400.

63.  Hamada S. Propagation of mulberry tree in Japan. J Silkworm 10:273–278, 1958.

64.  Taguchi S. Distribution of the propagating methods of mulberry trees from Japan and the relation to the various climates. J Sericult Sci Jpn 40:399–403, 1971.

65.  Lombardi PL. Attuale stato della gelsicolture italiana sistemazione per eventualli allevamenti successivi con rifermento anche ai tipi giapponesi impartati. J Silkworm 12:41–50, 1960.

66.  Sharma SS, Sen DN. Role of certain growth regulators on the vegetative propagation of a desert species—*Ipomoea biloba* Forsk. India For 101:625–633, 1975.

67.  Sen DN, Rajput P. Ecophysiology aspects of the vegetative propagation of saltbush (*Atriplex* spp.) and mulberry (*Morus* spp.). In: M Pessarakli, ed. Handbook of Plant and Crop Physiology. New York: Marcel Dekker, 1994, pp 177–191.

68.  Sehdev P. Ecology and biology of mulberry in Indian desert. MPhil dissertation, Jai Narain Vyas University, Jodhpur, 1992.

69.  Gardner FP, Pearce RB, Mitchell RL. Physiology of Crops Plants. Ames: Iowa State University Press, 1985.

70.  Kempanna C, Lingaraj DS, Chandrsekhariah SRP. Propagation of *Glricidia maculata*, H.B. and K. by air layering with the aid of growth regulators. Sci Cult 27:85–86, 1983.

# 7

# Fruit Development, Maturation, and Ripening

## William Grierson

*University of Florida, Lake Alfred, Florida (retired)*

## I. INTRODUCTION

### A. What Is a Fruit?

The biblical phrase "the precious fruits of the earth" can be taken far more literally than the epistle writer probably imagined. There is very little in agriculture that does not depend on the development of fruits. By definition, a fruit is the end product of a matured ovary. This end product can vary from being a single seed such as a grain of any cereal (e.g., wheat, rice, rye, oats, or barley) to being a fleshy, succulent structure (e.g., peach, pear, or watermelon). All nut crops, including peanuts (or "ground nuts"), are technically fruits, as are the products of oil palm, coconuts, rape (canola), flax (linseed), and other plants grown for extraction of edible or industrial oils. Even many root and pasture crops are dependent on fruit setting to provide seed for sowing the next crop. Root and tuber crops grown from vegetative propagules are an obvious exception, but their genetic improvement by plant breeders is dependent on flowering, pollination, and fruit setting to provide seed with which to start improved varieties. It should also be noted that many "vegetables," including tomatoes, peas, beans, cucumbers, squash, peppers (capsicums), eggplant (aubergine), and okra (lady's fingers), are botanically fruits.

### B. Scope of This Chapter

For the purposes of this chapter, only the products classified horticulturally as fruits are considered for detailed discussion. In general, these are fleshy products, characteristically high in sugars (the avocado being a notable exception) and although sometimes processed on a very large scale, traditionally eaten raw as dessert. Unlike vegetables, most are perennials grown on trees, vines, or shrubs (strawberries are the fruit of a perennial herbaceous plant). Melons are an exception, being annuals.

Whether annual or perennial, whether classified commercially as a fruit, vegetable, or cereal, it should always be remembered that until the instant of harvesting, a fruit is an integral part of the parent plant, participating in a common physiology and subject to the same ecological influences. As pointed out in Chapter 2, a fruit cannot be considered independent of the growth status of the parent plant or of the environment in which it was grown. A simpleminded quest for a single recommendation as to optimum postharvest conditions for a given type of fruit, regardless of growing district and preharvest climatic conditions, is doomed to failure.

## C.  Definitions

1. Fruit is the product of a matured ovary.

2. *Maturation* is the completion of the development of a fruit to the point at which it is physiologically mature enough to be separated from the parent plant. Typically, this is the point at which its seeds are viable. There is no necessary relationship with market maturity, for which immature fruits may be required (e.g., cucumber, okra) or for which arbitrary legal standards may be set for external color and/or sugar or acid content (e.g., citrus, grapes).

3. *Ripening* and maturation can be synonymous for nonclimacteric fruits (e.g., grapes, strawberries, and citrus) that are edible at the time of picking and have no postharvest ripening cycle. However, they are quite different for climacteric-type fruits: those that are considered unripe until they have entered on a distinctive postharvest respiratory rise in which ethylene is evolved, $CO_2$ output increases (sometimes as much a tenfold), tissues soften, starch/sugar or acid/sugar changes occur, and typical external color changes may be involved. Tomatoes, apples, pears, avocados, and bananas are typical climacteric-type fruits with distinctive postharvest ripening cycles.

4. *Berry* is used quite differently by plant scientists and by the general public. Botanically, a berry is the product of a single pistil, fleshy throughout, usually indehiscent, and homogeneous in texture [1]. Thus a grape is technically a berry, but a strawberry is not.

5. *Anthesis* is the stage of flowering at which pollination can take place, usually considered to be the initiation of fruit development.

6. *Parthenocarpy* in its narrowest sense is defined as the ability of a plant to develop fruit without sexual fertilization. More broadly, it is the ability to produce fruit without seeds [2].

Readers interested in further details of terminology are referred to two publications: Watada et al. for general terminology relating to developing horticultural crops [3] and Gortner et al. for the biochemical basis for terminology used in maturation and ripening of fruits [4].

## II.  PREREQUISITES FOR FRUIT FORMATION

As long ago as several hundred years BC, it was recognized that all fruit came from flowers. The ancient Greeks named one exception, the fig, "the only fruit not preceded by a flower." This was because they did not realize that the fig is an aggregate fruit with many minuscule flowers *inside* the enlarged, fleshy receptacle.

Flowers must be preceded by buds specifically differentiated for flower formation. In deciduous fruits, this starts some 10 or 11 months prior to bloom (i.e., initiation of fruit bud formation for the next year's crop starts almost as soon as the new crop is set). A study of the rate of flower bud development in deciduous fruits indicates that each species follows a sigmoidal growth pattern within a temperature range specific to that species [5]. In citrus fruits, fruit bud differentiation is initiated only a few weeks prior to bloom [6]. For both deciduous and citrus fruits, blossom formation occurs on wood at least 1 year old. ("Fruiting spurs" on apple trees may bear fruit almost every other year for a dozen years or more.) Grapes are in sharp contrast to this pattern. Skilled grape pruners remove almost all woody growth (canes) from the previous year, leaving only a few buds (how many depends on the variety, district, and vigor of the plant). From these few buds grow long canes on which leaf and fruit bud differentiation has to take place rapidly enough to provide for the current crop. Grape flower development has been described in detail [7]. Bud formation in tropical fruits is controlled mainly by water availability and temperature and thus can be less predictable than for deciduous fruits. An extreme example is papaya (*Carica papaya*). Although basically dioecious, under various temperature, moisture, and nutrient stresses, carpels can metamorphosize into stamens, and vice versa [8].

Thus flower bud initiation is a necessary precursor to fruit formation. Particularly after the landmark 1918 paper by Kraus and Kraybill on fruiting in the tomato [9,10], it was believed that flower bud initiation was dependent on the balance between carbohydrates and nitrogenous compounds in developing tissues (the C/N hypothesis). Within the last 50 years, it has been realized that in any plant, flower bud initiation and hence fruit formation are controlled by growth regulators (GRs). Development of GRs and the balance between them is, in turn, controlled by environmental forces, notably temperature and light.

Gibberellins were among the first GRs to become available in commercial quantities, thus greatly facilitating research showing that for a very wide range of plants, gibberellin could inhibit flower bud formation and sometimes induce parthenocarpy if applied after flower bud initiation [11–17]. For details of the histology of flower induction in apples, see Buban and Faust [18].

Research on the role of GRs in bud initiation has been facilitated by the finding that tracheal sap is a convenient source of naturally occurring GRs [19]. Abscisic acid (ABA) is now known to be very much involved, not only in flower bud formation but also in fruit development [20,21]. Growth regulators control messenger RNA (mRNA), which generates necessary enzymes de novo for fruit development following anthesis [15,22,23]. Much research has involved manipulation of bud differentiation and fruit development using exogenous application of both natural and synthetic GRs [24–26]. But flower bud initiation, and hence the entire cycle of flowering and fruiting, can be controlled solely by intelligent manipulation of temperature and light. A 39-week cycle (repeated at 4-week intervals) has been developed using dark and lighted cold rooms and greenhouse or nursery facilities to provide a continuous supply of three varieties of container-grown apples for year-round harvesting, a remarkable feat [27].

Normally, pollination is necessary for fruit set; however, there are notable exceptions. With the buying public increasingly demanding seedless fruits of various kinds, parthenocarpy has become highly prized for many types of fruits. This is certainly so for citrus fruits, for which pollination had long been deemed unnecessary, even undesirable, as it increases the number of seeds in supposedly "seedless" varieties. That was before the introduction of a number of human-made crosses such as the tangelos (tangerine × grapefruit). Tangelo varieties that are apparently fruitful when grown in small trial plots were almost completely barren when planted in large multihectare blocks. Thus it was found that for some hybrids, such as Orlando tangelo, pollination by some other variety was as necessary as it is for apples and pears [2]. A remarkable example of parthenocarpy is the navel orange, which has a small secondary fruitlet at the stylar end and which is always seedless. Fruit set of navel oranges, which is often uneconomically light in Florida, is sharply affected by ambient temperatures prior to and during fruit set [28]. Because seedless table grapes may fetch more than twice the price of seedy grapes, parthenocarpy is highly valued [29]. In the popular Thompson Seedless variety, fruit set is dependent on GRs involved in pollen tube development, even though the pollen tube does not reach, and hence does not fertilize, the ovule [30].

## III.  MORPHOLOGICAL CATEGORIES OF FRUITS

Fruits have evolved so many diverse forms that Soule lists 46 different morphological fruit types [1]. Although anatomical and taxonomic considerations cannot be ignored completely, only a few general categories can be considered within this context of fruit physiology. Nondessert fruits are discussed only insofar as is necessary to establish their place in the wide general category of fruits. For a detailed histological treatment of the various tissues that can be involved in fruit development, see Esau [31]. All fruits are the products of matured ovaries. Some, in addition, incorporate other floral parts. This is particularly true for fruits derived from inferior ovaries (epigyny), that is, fruits such as apple and pear, in which the other floral parts (stamens, petals, and sepals) are above the ovary.

## A.  Achene

An achene is a hard, dry, fully matured simple ovary. Achenes are usually thought of as "seeds" (although some may contain two seeds). A grain of wheat is an achene, each flower within a head of wheat (inflorescence) having matured individually to form an achene. A grain of corn (maize, *Zea mays*) is an achene, corn on the cob being an unusual example of an intact, nondehiscent inflorescence. Achenes are, in general, nutritious and have been utilized as foods since antiquity, not only in the form of our well-known cereal grains but also as such lesser known species as the sumpweed (*Iva annua*), gathered by native North Americans, and amaranth (*Amaranthus caudatus* and *A. quitensis*), a staple of the pre-Columbian Aztecs, the cultivation of which has persisted in remote Andean valleys and which is currently an interest of "health food" devotees [32].

## B. Typical Fruits from Superior Ovaries (Hypogyny)

### 1. *Grape*

The grape is the simplest of hypogynous fruits and one that conforms exactly to the botanical definition of a berry. Remnants of floral parts other than the ovary are absent or vestigial and the developed ovary tissue is fleshy, succulent, and homogeneous.

### 2. *Hesperidium*

The hesperidium is the highly specialized form of berry specific to citrus fruits. (Etymologically, the term *hesperidium* is a misnomer based on the assumption that the "Golden Apples of the Hesperides" in Greek mythology were oranges. However, citrus, as the etrog. *Citrus medica*, did not reach the Mediterranean area until historical times [33].)

The hesperidium (which is derived entirely from the ovary) has several sharply defined tissues (Figure 1). The usually five-lobed calyx remains attached unless the fruit naturally abscises; then it remains attached to the bearing branch.

The outer layer, or peel, includes the pigmented *flavedo* and the white or colorless *albedo*. The flavedo (Figure 1A, top left) consists of the epicarp proper, hypodermis, and the outer mesocarp. Embedded in it are the so-called oil glands, containing "essential oils" specific for each citrus species or hybrid. These are principally terpenes (mainly *d*-limonene) and are highly toxic to surrounding tissue if extruded due to rough handling of the fruit. The cells of the single-layered epicarp contain green chloroplasts that metamorphose into chromoplasts as the fruit degreens. Over the epicarp is the intact cuticle (Figure 1A, lower left), composed largely of cutin, and over it an outer layer of epicuticular wax deposited as eas-

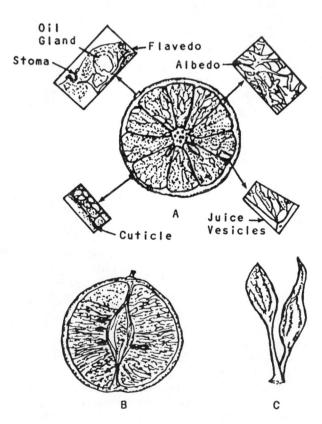

**Figure 1** Citrus fruit: (A) transverse section with enlarged views of the flavedo and cuticle on the left and of the albedo and juice vesicles attached to the outer tangential and radial locule walls on the right; (B) longitudinal section showing the lunate locules with seeds attached to the inner tangential wall next to the central axis; (C) separate juice vesicles. (From Ref. 34.)

ily dislodged platelets. (No citrus fruit is naturally shiny; the shine demanded by retail customers has to be applied as some form of approved wax or resin after washing, an operation that dislodges much of the natural nonshiny wax.) The cuticle is penetrated by numerous stomata, except in a narrow (ca. 3 mm) area around the calyx.

The albedo, or inner mesocarp (Figure 1A, top right), consists of a loose network of parenchymatous cells with large airspaces formed when small, originally spherical albedo cells retained their original points of contact as the fruit expanded. Thickness of the albedo can range from as little as 1 to 2 mm in some limes and tangerine hybrids to 2 cm or more in large shaddocks (pummelos, *Citrus grandis*).

The edible flesh of a mature citrus fruit is divided into segments, each derived from an ovary locule. The number of segments varies widely but is typically between 10 and 15. Each segment is surrounded by a tough endocarp membrane and filled with tightly packed juice sacs or vesicles (Figure 1A, lower right, and C). Each of these thin-walled juice sacs has a minute oil gland in its center and is attached by a fine stalk to vascular bundles in the radial segment walls. Except in parthenocarpic fruit, seeds are within the segments and attached to axial vascular bundles. Despite various varieties being sold as "seedless," few except navel oranges and Persian (Tahiti) limes (*Citrus latifolia*) are truly seedless. Purists prefer the term "sparsely seeded," for which citrus dealers show no enthusiasm at all.

The vascular system is a highly ramified network whereby every cell is connected to, or adjacent to, a cell in contact with a particular sector of the vascular system. In many types of citrus fruits, particularly seedless grapefruit and tangerines, the central "core" bundles separate as the fruit matures, leaving a considerable cavity in the center of the fruit (a complication in specific gravity separation of freeze-damaged fruit). For a more detailed discussion of citrus fruit anatomy, see Soule and Grierson [34].

### 3.  *Drupe*

Drupes start out as though they were going to be berries but then develop their typical hardened "pit." The resultant fruit is technically described as a "simple fruit with soft exterior, fleshy, usually indehiscent, with heterogeneous texture and the center with a hard, bony, or cartilaginous endocarp enclosing the seed proper" [1].

The most familiar drupe fruits are peach (and its genetic recessive, nectarine), plum, cherry, and apricot, and in the tropics, the mango. In all of these fruits, the edible portion is the fleshy mesocarp. Other, less obvious drupe fruits are coffee, in which the fleshy mesocarp (though edible) is discarded. It is an anomalous drupe, having two seeds enclosed in a parchment-like endocarp, the seeds being the "coffee beans" of commerce. Other drupes grown for their seeds are almond and pistachio. The most atypical of all drupes is the coconut, in which the dry, fibrous epicarp and mesocarp become the husk (the source of coir fiber used in brushes, matting, and rope). The large seed has edible white oily flesh and a liquid endosperm (the "coconut milk").

The epidermal characteristics among drupe species vary widely from the smooth epicuticular wax of the cherry, mango, or date to the "fuzzy" epidermis of the peach, whose soft "hairs" are extensions of epidermal cells. Particularly striking is the white "bloom" on the surface of some drupes, particularly plums. Electron microscopy shows this to be formed from an amorphous wax layer adjacent to the cuticle proper, together with crystalline granules of wax protruding from the surface. It is easily brushed off in routine handling. This is considered undesirable [35].

Drupe crops can be of purely temperate-zone origin with specific winter-chilling requirements (peach, plum, cherry, apricot) or purely tropical (mango, date). Intermediate is the pistachio, which has a brief winter chilling requirement but very limited freeze hardiness [36].

## C.  Typical Fruits from Inferior Ovaries (Epigyny)

In flowers of epigynous fruits, the other major floral parts, sepals, petals, and stamens, are fused at their bases and located above the ovary. As such fruits develop, nonovarian tissues become intrinsic parts of the fruit. It is often very difficult to discern ovarian from nonovarian tissue.

### 1.  *Pome Fruits*

All the pome fruits are members of the Rosaceae family, for example, apple, pear, quince, medlar, hawthorn, and the tropical loquat. A pome is defined as a fruit in which the papery or cartilaginous en-

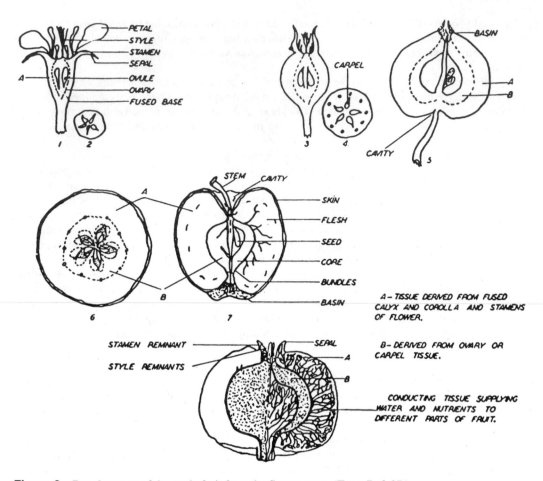

**Figure 2**    Development of the apple fruit from the flower stage. (From Ref. 37.)

docarp is embedded in the mesocarp, fused with and completely enveloped by the enlarged fleshy receptacle or the fused base of the sepals; the ripened ovary is only a small part of the total structure [1].

By far, the best known and most widely grown pome fruit is the apple (*Malus sylvestris*). Its flower parts are in fives: five sepals, five petals, five stamens, and five carpels making up the deeply embedded ovary (Figure 2). The parenchyma of the fused bases of the calyx, corolla, and stamens constitutes the major portion of the edible tissue of the mature fruit. The nonedible core is largely ovary tissue.

Although parthenocarpy is not unknown in some obscure varieties, fruit development normally starts at pollination. Because most apple varieties are self-infertile, pollen usually has to come from some other variety (cultivar). Fruit development is almost invariably dependent on fertilization and resultant seed formation. (Fortunately, the buying public's prejudice against seedy fruits does not include apples.) The hormonal control of fruit development was first indicated by the common observation that when seeds fail to develop in one or more of the five carpels, the fruit tends to grow lopsided. Most flowers never survive to form fruits. Only about 2 to 4% of the flowers in a normal bloom need to develop to provide as heavy a crop as the trees can bear.

The epidermis of the very young fruit is constantly growing, initially with very active cell division. After 4 or 5 weeks of development, cell division slows down and then ceases. As the fruit continues to expand, the epidermal cells flatten and elongate. As the fruit matures, these epidermal cells become surrounded by cuticle. The cuticle is covered by a layer of wax that is continuous in varieties with a natural shine but is deposited as irregular platelets in "nonshiny" varieties such as Golden Delicious and Grimes Golden. Today, most apples are artificially waxed, partly to retard shrinkage, but more because of the buying public's fascination with shine (even for fruits that are not naturally shiny). The edible parenchymatous tissue consists of large, thin-walled cells with a surprising volume, estimated at ca. 25%, taken up by airspaces [37].

**Comment on the Relation Between Fruit Structure and Handling Damage.** Because most apples have a continuous cuticle, they are very resistant to water damage and have even been stored successfully under water. If fungal spores enter through a break in the cuticle, fungal hyphae tend to spread so slowly that it is common to cut out an infected area and consume the rest of the apple. However, apples and pears have very little resistance to pressure, which can rupture parenchymatous cells. These are rich in oxidases and surrounded by air in the intercellular interstices. The results is an ugly dark brown bruise.

Because of its discontinuous waxy coat and occasional still-dividing epidermal cells [38], a citrus fruit has poor resistance to prolonged submersion in water. The structure of a citrus fruit, with its spongy albedo and radially oriented juice sacs, is very resistant to pressure from smooth surfaces. It does not bruise but is very susceptible to damage from sharp objects. Any rupture of epidermal oil cells releases "peel oil," toxic to adjacent tissue, with resultant ugly lesions ("oleocellosis"). Any fungal spores introduced into the albedo find a perfect culture medium. The spread of fungal mycelium is almost explosive.

## 2. Banana (*Musa sapientum*)

An interesting tropical fruit, the banana, is sometimes classified as a berry, which is clearly erroneous because nonovary tissue is involved (be it only as the nonedible skin of the fully mature fruit). Banana flowers are dioecious, the male flowers being borne within conspicuous purple bracts at the end of the long, hanging inflorescence. The female flowers are clustered in groups along the stem of the inflorescence. These groups of flowers develop into clusters of fruits called "hands," in which each individual fruit is referred to as a "finger." The general tendency is that the more hands there are on a bunch, the more fingers there are in each hand.

The female flower is inconspicuous and described as a "tepal," in which the components of the perianth are so similar in size, form, and coloration that sepals cannot be distinguished from petals [1]. The inconspicuous perianth is abscised immediately after the flower opens. Thus only ovary and receptacle remain.

Pollination is necessary for fruit set, but all commercial varieties are, nevertheless, parthenocarpic. Banana breeders thus have a double problem. When, for example, Panama disease was wiping out Gros Michel, the major commercial variety of Central America, they had to cross-fertilize with seedy resistant varieties. With that accomplished, backcrossing was necessary to eliminate the seeds while retaining necessary disease resistance. This was accomplished in a surprisingly short time.

Initially, the peel (which is receptacle tissue) weighs five times as much as the interior pulp. As the fruit grows, the endocarp develops fleshy protrusions into the locules forming the edible pulp. At full maturity, the edible pulp typically weighs twice as much as the inedible peel. However, few commercial bananas are allowed to reach full maturity. For long-distance shipment, bananas are picked at stages of development known by such terms as "$\frac{3}{4}$-full" and "$\frac{2}{3}$-full," terms describing a somewhat angular cross section. Fortunately, the highly climacteric banana will ripen to good eating quality even when harvested well short of physiological maturity. It is usual commercial practice for bananas to be shipped green and ripened in "ethylene degreening rooms" at destination. In addition to accelerating the natural climacteric ripening process, this ensures uniformity in ripening, a convenience in marketing.

## D. Aggregate Fruits

Aggregate fruits are compound fruits produced from many pistils in an inflorescence rather than from a single pistil. Temperate-zone aggregate fruits include strawberry, mulberry, and the various cane and bramble fruits. As mentioned previously, the fig is an aggregate fruit, with its minuscule flowers inside the vaselike receptacle and so indiscernible without dissection of the fruit. Among tropical fruits, breadfruit, pineapple, and cherimoya are aggregate fruits. Three familiar examples of aggregate fruit are discussed here.

## 1. Raspberry (*Rubus idaeus*)

The raspberry is an intrinsically frail fruit in that, unlike its near relative the blackberry, at picking the receptacle remains on the plant. The harvested fruit is thus hollow, formed only from adhering drupelets (miniature drupes), each from a separate floret within an inflorescence. Improving the inherent structural weakness of the raspberry has become a challenge for research workers [39,40].

### 2.  Strawberry (*Fragaria virginiana* × *F. chiloensis*)

The strawberry is an accessory fruit, one in which the conspicuous fleshy part is composed of tissues external to the pistil. (The *Annonas*, soursop, sweetsop, and cherimoya are tropical examples of accessory fruits.) The succulent flesh of the strawberry is receptacle tissue. The "seeds" embedded in its exterior surface are achenes and thus true fruits.

### 3.  Pineapple (*Ananas comosus*)

The pineapple is a multiple fruit, one formed from many pistils of an inflorescence. The pineapple fruit develops from separate lavender-colored flowers distributed around the length of the central axis of the inflorescence. The entire flowers become incorporated into the fruit, much of the flesh being formed from the fleshy bracts subtending each flower. Individual varieties are self-infertile; hence pineapples grown in monocultures of a single variety are always seedless. However, in areas such as the Caribbean, where small plots of various varieties are common, it is usual to have pineapples with occasional small black seeds.

## IV.  PHYSIOLOGICAL DEVELOPMENT

As a general principle, fruit development in terms of weight and volume tends to be sigmoidal. A period of very rapid cell division, but very little increase in fruit size (stage I), is followed by a period of rapid increase in size as small, newly differentiated, dense cells develop vacuoles and assume their roles as specific tissues (stage II). In the final stage, as the fruit reaches physiological maturity, increase in size slows and may even stop, although biochemical changes may continue (stage III). There are about as many variations on this pattern as there are different types of fruit, but the sigmoidal mode is usually discernible. The orange, apple, and apricot are discussed next as typical examples of the development of citrus, pome, and drupe fruits.

## A.  Hesperidium, e.g., Orange (*Citrus sinensis*)

The duration of growth and maturation varies sharply with variety. For early varieties such as Hamlin and navels, harvesting commonly starts 6 to 7 months after bloom. For the late Valencia variety, harvesting starts about 12 months after bloom. Harvesting can continue for a "tree storage" period lasting several months, during which late oranges have two crops on the tree at the same time. Herein lies a critical difference between citrus and deciduous fruits. The latter *must* be picked soon after maturation is complete or they will fall from the tree. Citrus fruits have no such sharply defined abscission period, something that is frustrating to would-be developers of mechanical harvesting equipment, but an enormous advantage in marketing the crop over a period of weeks or months in which the crop is "stored on the tree." Stages of development are shown in Figure 3.

Stage I lasts a month or less, during which cell division is extremely rapid but fruit enlargement is trivial. At this stage the cuticle has not yet developed, making the little fruitlets extremely vulnerable to superficial damage. In growing areas such as Florida, where stage I coincides with the strongest winds of the year, just brushing against an adjacent leaf causes major "windscars" on the mature fruit. This problem is exacerbated in areas such as Brazil and Florida, where rains in the postbloom period facilitate superficial infection of such windscars by waterborne spores of the melanose fungus (*Diaporthe citri*). Although most cell division takes place in this period, some cell division can continue in the peel until maturation, particularly with navel oranges, making such fruit very vulnerable to water damage [38].

Stage II is the period of cell (and hence fruit) enlargement. The fruit expands rapidly, as does $CO_2$ output per fruit, although $CO_2$ evolution per unit weight (the usual way of expressing respiration) declines sharply (Figure 3). During this period, the juice sacs are enlarging and developing their distinctive solutes. Increases in whole fruit and pulp radii and whole fruit, pulp, rind, and albedo volume during fruit development follow single sigmoidal patterns (four-parameter logistic function, $R^2 \geq 0.99$) [41]. Such solutes are initially high in organic acids and low in sugars. As the orange matures, sugars increase steadily while acids decline. Legal maturity standards for citrus fruits are usual in major producing areas. In this, every

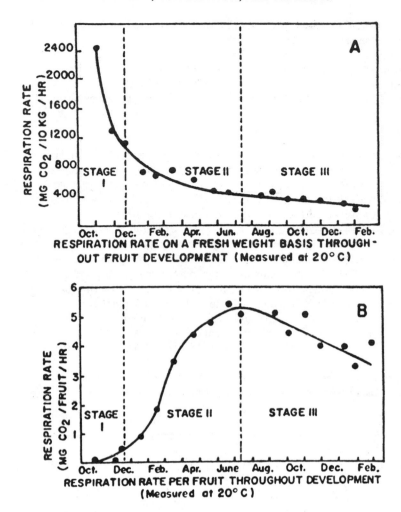

**Figure 3**   Respiration during fruit development of Valencia orange. (A) Expressed as $CO_2$ evolution per unit fresh weight; (B) expressed as $CO_2$ evolution per fruit, a form that more clearly defines the stages of fruit development. (From Ref. 34.)

district sets its standards according to what they do best [42]. European citrus districts, South Australia, California, and other districts with Mediterranean-type climates (cool winter nights, bright days, and low rainfall) can rely almost entirely on external standards to sell their oranges. Florida, with its blossom-period winds and humid, subtropical climate, cannot compete on appearance and so relies principally on standards based on the high sugar content of its oranges. These maturity standards are based not only on sugar content but also on the ratio of total soluble solids (TSS, mainly sugars) to acids (titratable as citric acid), with a sliding scale throughout the season [43,44] (Figure 3). At the beginning of the season, Florida oranges must have 8.0% TSS with a TSS/acid ratio of 10.5:1 (Figure 3). By the end of the season, this ratio may exceed 20:1, but with the proviso that (for fresh fruit sale) acid cannot be below 0.4% lest the oranges taste too insipid.

Regardless of growing district, consistent gradients occur within a citrus fruit, particularly in terms of sugar content. The vascular system extends down the central axis of the fruit, reaching the blossom (stylar, distal) end first, them ramifies back up the carpels to the stem (calyx, proximal) end of the fruit. Apparently as a consequence of this distribution of photosynthates, sugars are higher in the blossom end. A very thorough study reported that the proximal halves of mature California Valencia oranges averaged 7.2 g of sugar per liter of juice as compared with 9.5 g/L for the distal (blossom, stylar) halves, a difference clearly discernible by taste [45]. When sharing a grapefruit, canny citrus people give the stem-end half to their companion, retaining the blossom end half for themselves.

In the jungles of southeast Asia where citrus first evolved, all are still green when mature. The extent to which the expected orange or yellow colors develop depends on the growing area having cold enough nights to stress the fruits [46]. In subtropical areas such as Florida and Brazil, early varieties may mature while still green, necessitating postharvest removal of the green chlorophyll with ethylene [47].

## B. Pome, e.g., Apple

The typical growth curve of any main crop apple variety is only slightly sigmoidal. Very early varieties, such as Early Harvest, Yellow Transparent, and Melba, mature to acceptable eating quality before any deceleration of growth (Figure 4). Apples that mature this early are very frail and suitable only for local consumption. The longer it takes an apple variety to reach maturation, the more sigmoidal its growth curve. In general, the later an apple variety matures, the longer its potential marketing life.

Initially, all cells of the apple are alive. Cell division in the epidermis ceases at the end of stage I. Marked elongation and flattening of the epidermal cells occur throughout stage II, during which period the epidermal cells extrude waxy, cutinous material. In fully mature late-season apples, the epidermal cells are separated, dead or dying, embedded in the continuous cuticle (a heterogeneous polymer of fatty acids overlaid with a layer of wax). The cuticle can continue to develop after harvest. During the stage II growth period, the epidermis is penetrated by stomata that tend to cork over at full maturity. Under the epidermis in some varieties is the periderm, a thin layer of cork cambium. If the epidermis is injured early in stage II growth, as by mechanical abrasion or frost, the periderm develops a protective layer of corky cells: biologically an excellent protection for the fruit but a "grade-lowering defect" for the packer and the consumer.

Parenchyma tissue from the fused bases of the calyx, corolla, stamens, and receptacle constitutes the major part of the edible tissue of the mature fruit. Cell division having ceased at the end of stage I (usu-

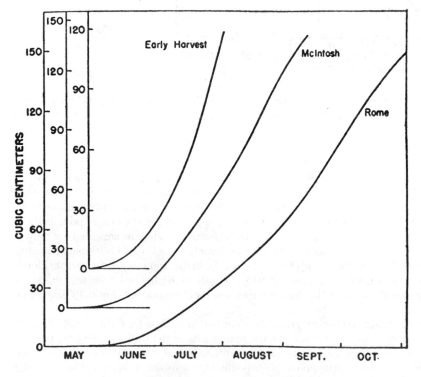

**Figure 4** Increase in the volume of Early Harvest, McIntosh, and Rome Beauty apples from full bloom to maturity. (From Ref. 37.)

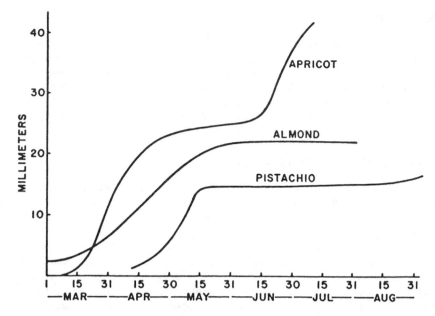

**Figure 5** Growth in diameter of fruits of Ne Plus almond, Royal apricot, and Kerman pistachio. (From Ref. 36.)

ally 3 to 5 weeks after anthesis), the considerable enlargement of the fruit comes from cell enlargement and their partial separation to form a considerable volume of air-filled intercellular spaces. Except for the petals (which abscise and fall after fruit set), all the original parts of the flower persist in the fully developed apple.

## C.  Drupe, e.g., Apricot (*Prunus armeniaea*)

The growth curve of the apricot, indeed of all fleshy, succulent drupes, is exaggeratedly sigmoidal (Figure 5). Stage II growth is interrupted by "pit hardening," in which the endocarp thickens and lignifies to form the hard, stony "pit" enclosing the seed. During this period, the fruit ceases to increase in size. Biochemical changes continue, but without cell enlargement. Morphological development in the peach (which is essentially similar to apricot) has been reported in considerable detail [48].

The apricot pit is smooth and, at maturity, quite free from the edible mesocarp tissue, being attached only at its proximal end by the persistent vascular system. In other drupes, the pit is seldom so separate, although in "freestone" peaches the deeply incised pit is nearly free from the edible mesocarp. In "clingstone" peaches, the endocarp and mesocarp interfaces adhere.

Such characteristics are of commercial significance. "Freestone" varieties (cultivars) are preferred for the fresh fruit market. Because of their considerably firmer flesh (mesocarp), clingstone varieties are preferred by the canneries. (A machine neatly removes the clingstone pits).

In the mango (*Mangifera indica*), the ultimate example of a "clingstone drupe," the pit is covered profusely with tough fibrous "hairs" that usually extend into the edible flesh. The date (*Phoenix dactylifera*), the ultimate "oasis crop," is a specialized drupe that develops so much sugar that its cells plasmolyze and ultimately die. Initially, most of the sugar is sucrose, but during maturation, all the sucrose converts to glucose and fructose. When fully mature, all that is still living is the embryo within the stony seed. After harvest, the date is therefore handled as a confection rather than as a fruit.

Very occasionally, a drupe may have multiple seeds within the boney endocarp. One such is the highly poisonous, but attractive seeming, fruit of the manchineel (*Hippomane mancinella*), the so-called poisonous guava.

Two familiar dessert nuts are the seeds of drupes. The almond (*Prunus communis*) and the pistachio (*Pistacia vera*) are drupes in which the mesocarp fails to develop any further after pit hardening, thus resulting in a growth curve that is definitely not sigmoidal (Figure 5).

## V.  POSTHARVEST LIFE

It should not be necessary to emphasize that fruits are still alive after harvest. However, a surprising number of people who make their livelihoods growing, packing, shipping, and selling fruit do not realize that they are handling living, breathing creatures, subject to specific diseases and the ravages of senescence. ("Gee, Doc, don't tell me these things are alive. They've been picked!") Moreover, effective postharvest handling is not merely a matter of maintaining the state of fruit quality at the time of picking. Properly handled, many fruits improve in eating quality after harvest. Others degenerate rapidly or slowly, depending on their innate physiology and the postharvest conditions to which they are subjected.

### A.  Climacteric Versus Nonclimacteric Fruits

The first step in proper postharvest handling of a given type of fruit lies in understanding its type of life cycle [49]. The climacteric rise in respiration of fruits such as apple, pear, avocado, mango, and banana represents a rapid depletion of potential postharvest life (Figure 6). For fruits such as pear, banana, and avocado, experiencing the climacteric is essential to the ripening that makes them truly edible. But it should be delayed as much as possible until the consumer is ready to eat that piece of fruit. Very prompt refrigeration is essential for orderly marketing of climacteric-type fruits, to delay or suppress the evolution of endogenous ethylene that initiates the climacteric rise. As the height of the climacteric is reduced, its duration is extended proportionately. Immediate temperature and humidity control is the first line of defense against expensive wastage. Humidity control is important if for no other reason than that a shriveled fruit ceases to be marketable. However, there are other physiological benefits also [50]. Even within a specific variety, response to such storage techniques as controlled atmosphere storage can be sharply influenced by cultural and climatic factors [51]. When the peak of the climacteric rise is past, the fruit becomes senescent. Although adequate reserves of respiratory substrate may be available, cellular organization breaks down, the cell membranes lose their integrity, and the fruit dies of old age [52,53]. Thus the challenge with climacteric-type fruits is to suppress and extend the respiratory rise.

Apples and pears are examples of climacteric-type fruits that have to be harvested within a very brief period but marketed for as long a period as correct storage procedures permit. Long-storing varieties have ample reserves of respiratory substrate and resilient respiratory systems. Under near-optimum conditions, late varieties such as Winesap can be kept year-round. Some, such as Northern Spy and Winter Banana, improve in eating quality during the first few months of storage.

The avocado (*Persea americana*) is an interesting climacteric-type fruit. Although strongly climacteric, the characteristic respiratory rise will not start until the avocado is picked. For many years research workers were convinced that when their instrumentation improved sufficiently, they would be able to identify a preharvest "climacteric inhibitor." Even with modern equipment, it has been impossible to identify any such inhibitor [54].

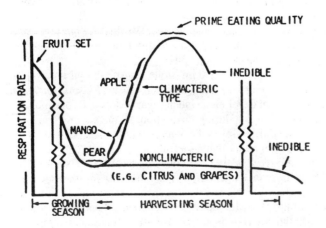

**Figure 6**   Climacteric and nonclimacteric life cycles for typical fruits. (From Ref. 34.)

Most varieties of pears (*Pyrus communis*) do not ripen to acceptable eating quality on the tree. Once picked, pears have to be *either* ripened for immediate use (preferably at 20 to 25°C) *or* held in cold storage at only a degree or two above their freezing point. Pears, particularly the popular Bartlett variety, will neither ripen nor store at intermediate temperatures, particularly in the range 8 to 12°C. Instead, they become rubbery in texture and virtually inedible.

This is necessarily an abbreviated and simplified account of the complex physiology of climacteric-type fruits. The extraordinary development of nonchemical analytical equipment has stimulated much postharvest research. Some surprising results are being encountered, such as a newly developed thornless blackberry being strongly climacteric [55]. Those interested in further reading are referred to a 1985 symposium [56], particularly the paper by McGlasson [57].

Handling of nonclimacteric fruits is very much simpler. There are no significant physiological changes involved in separation from the tree and no postharvest ripening cycle. Those signs along the Florida highways saying "TREE-RIPENED CITRUS," although misleading, are legally defensible, *all* citrus fruits being "tree ripened." With no climacteric rise to suppress, nonclimacteric fruits such as citrus of various types, grapes, and certain vegetables that are botanically fruits do not benefit nearly as much from prompt refrigeration as do climacteric-type fruits. Indeed, for fruits susceptible to chilling injury, delayed storage may be beneficial by enabling the fruit to adapt to lower storage temperatures [46]. Sooner or later, of course, any fruit can be expected to abscise if left on the tree long enough. Modern research shows this to be a surprisingly complicated biochemical and histological process [58]. Such abscission is always due to trace amounts of ethylene at the abscission zone. Typically, this is induced by ABA (abscisic acid), the growth regulator produced in response to such environmental stresses as low temperature or drought. Deciduous fruit trees have deciduous fruits that fall when fully mature. Such natural abscission can be delayed with "stop drop" sprays, but at a loss of some postharvest shelf life.

Citrus fruits, typical fruits of evergreen trees, have no such programmed abscission, making harvesting much more onerous than for deciduous fruits. [Typically, a Valencia orange must be removed with a pull force of 18 to 22 pounds (8 to 10 kg) as compared to ca. 4 to 5 pounds (1.8 to 2.5 kg) for a McIntosh apple.] Research [59] has shown that the abscission-causing ethylene in citrus fruits can also be triggered by endogenous IAA (indole acetic acid).

## B.  When to Harvest

This discussion is out of chronological order in terms of the fruit because it is necessary to understand something of postharvest fruit physiology before dealing with optimum picking dates for various types of fruits.

### 1.  Citrus Fruits

It is fairly simple to set legal maturity criteria for nonclimacteric fruits such as citrus and grapes. These undergo no considerable physiological change at harvest, nor do they abruptly abscise and fall. Maturity standards, either legal or voluntary, can be set in terms of sugar content, sugar/acid ratio, and juice yield. Moreover, citrus fruits can be "tree stored." Early tangerine varieties can be picked over a period of several weeks, at the end of which period they start to dry out rapidly. Orange varieties, particularly the late, main crop Valencia variety, can be picked over a period of 2 to 3 months, sometimes more. Grapefruit from a single bloom can be harvested over a period of 6 months or more. (As this is being written, the same Florida grapefruit that might have been picked in October 1992 are still being harvested during the first week of June 1993). This is a great convenience in marketing, provided that the shipper does not try to extend marketing by storing grapefruit that has already used up its storage potential during prolonged tree storage [60].

### 2.  Apples (Malus sylvestris)

Deciding on a harvesting date is very much more difficult for climacteric-type, temperate-zone deciduous fruits for which only a narrow window of opportunity is available. "It is exceedingly important that apples be harvested at the right time. The exact degree of maturity at which a given variety should be picked depends in large part on what disposition is to be made of the fruit. . . . If apples are picked too soon and then stored for any length of time they are subject to storage troubles such as bitter pit and scald. . . . Almost every measure or index of maturity has to be defined for not only a given variety but for a given location, season, and soil type".[37] That advice was published 50 years ago, and despite considerable research, not much has really changed since then. In the search for a reliable criterion (or combination of criteria) as a guide to optimum picking date for apples, research workers have investigated days from full bloom, ground color, pull test (ease of separation), pressure test (with various modifications of the original 1925 Magness-Taylor pressure tester [61]), soluble solids, iodine-starch pattern, seed color, and corking of lenticels. No criterion has proved consistent across varieties, growing districts, cultural methods, and seasonal variations in climate. This is in sharp contrast to maturity standards for nonclimacteric fruits such as grapes and citrus. Such variability makes legal maturity standards for climacteric fruits difficult to enforce legally, e.g. [62]. Rootstocks can have a significant effect on maturity criteria [63], as can use of spur-type scion selections. But it is nice to note that organic cultural methods are reported as not significantly affecting maturity criteria [64]. Harvesting criteria for each particular apple variety in each district still have be based largely on local experience and judgment. A user-friendly computer program has been developed to help growers and packers select optimum harvest dates [65].

A further complication is that "stop-drop sprays" have been used for many years to extend the possible harvesting season for apples [66]. Such prolonging of the harvesting period can be expected to reduce potential storage life. This is particularly true for the highly colored strains that usually have been selected from chance sports regardless of other fruit quality criteria. (This writer has grown weary of attending meetings at which nursery owners and produce merchandisers proclaim that their aim is to "Give the lady what she wants," a policy that all too often sacrifices eating quality for appearance.) Now it appears that the selection of the culturally profitable spur-type strains may also sometimes be at the expense of keeping quality [67].

### 3.  Pears (Pyrus communis)

The situation for pear harvesting is no more promising. Over 50 years ago, this writer was a graduate student participating in a massive 5-year project involving five pear orchards throughout Canada's Niagara Peninsula. A major objective was to establish a reliable maturity standard for harvesting Bartlett pears, particularly for research in a then very new and experimental controlled atmosphere storage. (This method was then called "gas storage," later renamed "controlled atmosphere" by Bob Smock of Cornell University.) As well as pressure test, starch-iodine pattern, and so on, this program included measuring respiration immediately after picking. Although variation among seasons and orchards excluded all other criteria, one remained consistent. The best quality and longest storage life were always from the picking at the nadir of fruit respiration on the tree [68]. Because that can only be determined retroactively, it cannot be used as an indication of when to pick for maximum quality. Developments such as growing pears with apple interstocks and on clonal apple roots [70] further complicate the prospect of finding generally applicable criteria to determine optimum picking time for pears. Localized growing areas, particularly in irrigated districts, may use some standard (pressure test is most common), but it seems unlikely that statewide legal maturity standards will ever be established such as have long been enforced for citrus fruits [43,44].

## VI.  SOME ECONOMIC CONSIDERATIONS

A few fruits, such as Tung Nut (Aleurites fordii), are grown for industrial use. Most, however, are grown for food or drink. When fruits are to be processed into food products or beverages, external appearance is of no consequence. But for fresh market sale "eye appeal" can be critical to profitability. Unfortunately, most customers "taste with their eyes" and usually will not purchase unattractive looking fruit. [Kiwi fruit (Actinidia deliciosa) is a conspicuous exception.]

## A.  Color

The public preference for highly colored fruits has led to considerable varietal selection for high color, regardless of internal quality. (The Delicious apple is a conspicuous example.) However, for the discerning buyer, background color can be a useful indicator of maturity of many fruits such as the Bartlett (Williams, Bon Chretien) pear. The change in background color from dark green to pale green or yellow indicates incipient ripening to edible quality.

**TABLE 1**  Thermodynamic Data for Citrus Fruits[a]

| Parameter | | Oranges | Grapefruit | Lemons |
|---|---|---|---|---|
| Specific gravity | | 0.98 | 0.88 | 0.95 |
| Specific heat (BTU/lb/°F) | | 0.86 | 0.88 | 0.89 |
| Thermal diffusivity (sq ft/hr) | | 0.0049 | 0.0047 | 0.0049 |
| Thermal conductivity: | | | | |
| $\quad$ BTU/hr/ft$^2$/°F/in. | | 2.95 | 3.00 | 2.85 |
| $\quad$ kcal/sec/cm$^2$/°C/cm | | 1.1 | 0.78 | 1.05 |
| Heat of respiration (BTU/ton/day) at: | | | | |
| $\quad$ 32°F | 0°C | 900 | 500 | 580 |
| $\quad$ 40 | 4.5 | 1400 | 1100 | 800 |
| $\quad$ 50 | 10.0 | 1300 | 1500 | 2300 |
| $\quad$ 60 | 15.5 | 5000 | 2800 | 3000 |
| $\quad$ 70 | 21.0 | 6200 | 3500 | 4100 |
| $\quad$ 80 | 27.5 | 8000 | 4200 | 6200 |
| $\quad$ 90 | 32 | 9900 | 6000 | 8000 |

[a] Values listed as means of data from various sources. Values vary with horticultural variety (cultivar), district, maturity, size of fruit, etc.
*Source*: Ref. 69.

## B.  Shape

Regardless of edibility, the buying public rejects misshapen fruits. But in some cases fruit shape can be a useful indicator of eating quality.

Mango, a strongly climacteric fruit, develops an irregular shape as it matures on the tree, one "shoulder" becoming considerably higher than the other. The more marked this irregularity, the more mature the mango and so the better chance that, once picked, it will ripen to good eating quality.

Grapefruit typically bloom irregularly, the first major bloom being followed by later blooms at irregular intervals of days, weeks, or even months. Grapefruit from the first bloom tend to be oblate with the axis often considerably less than the diameter. Such fruit are of superior eating quality. Successive blooms result in increasingly spherical fruit of decreasing internal quality. "Sheep-nosed" grapefruit (axis considerably longer than diameter) are avoided by discerning buyers.

More usually, market grades based on fruit shape are quite unrelated to organoleptic quality. That a banana should be curved and a cucumber should not epitomizes the illogic of many market grades.

## C.  Thermodynamic Properties

Increasingly, modern fruit distribution involves the use of refrigeration. (Physiological responses of fruit to low temperature are dealt with in Chapter 2.) Refrigeration can be inefficient or unnecessarily expensive when the refrigeration system used does not take into account the thermodynamic properties of the product, in this case fruits. Such data are curiously hard to find, being scattered among horticultural and engineering publications. Such data have been compiled for citrus fruits; see Table 1. The values for heat of respiration at various temperatures of such highly climacteric fruits as apples, pears, mangos, and bananas can be several times as high as for citrus fruits.

## VII.  CONCLUSIONS

There is very little in agriculture that one way or another is not dependent on successful fruit development. Among those who make their livelihoods growing and marketing dessert fruits, there are many who could profit from improved understanding of the complex biology of these gracious additions to our diet.

## REFERENCES

1.  Soule J. Glossary for Horticultural Crops. New York: Wiley, 1985.
2.  Krezdorn AH, Robinson FA. Proc Fla State Hortic Soc 71:86, 1958.

3. Watada AE, Herner RC, Kader AA, Romani RJ, Staby GL. HortScience 19:20, 1984.
4. Gortner WA, Dull GG, Krauss BH. HortScience 2:141, 1967.
5. Seeley SD. HortScience 27:1263, 1992.
6. Davenport TL. Hortic Rev 12:349, 1990.
7. Gerrath JM. Hortic Rev 13:315, 1991.
8. Arkle TD Jr, Nakasone NY. HortScience 19:832, 1984.
9. Kraus EJ, Kraybill HR. Oreg Agric Exp Stn Bull 149, 1918.
10. Cameron JS, Dennis FG. HortScience 21:1099, 1986.
11. Monselise SP, Halevy AM. Proc Am Soc Hortic Sci 84:141, 1964.
12. Crane JC, Primer PE, Campbell RC. Proc Am Soc Hortic Sci 75:129, 1960.
13. Hull J Jr, Lewis LN. Proc Am Soc Hortic Sci 74:93, 1959.
14. Griggs WH, Iwakiri IT. Proc Am Soc Hortic Sci 77:73, 1961.
15. Van Overbeek J. Science 152:721, 1966.
16. Marcelle R, Sironval C. Nature 197:405, 1963.
17. Crane JC. Proc Am Soc Hortic Sci 83:240, 1963.
18. Buban T, Faust M. Hortic Rev 4:174, 1982.
19. Saidha T, Goldschmidt EE, Monselise SP. HortScience 18:231, 1983.
20. Wen-Shaw Chen. HortScience 25:314, 1990.
21. Rudnicki R, Bukovac MJ. HortScience 19:655, 1984.
22. Grierson D, Slater A, Speirs J, Tucker GA. Planta 163:263, 1985.
23. Callahan A, Morgens P, Walton E. HortScience 24:356, 1989.
24. Erez A. HortScience 22:1240, 1987.
25. Gianfagna TJ, Marini R, Rachmiel S. HortScience 21:69, 1986.
26. Southwick SM, Davies FS. J Am Soc Hortic Sci 107:395, 1982.
27. Saltveit ME Jr. HortScience 18:919, 1983.
28. Davies FS. Hortic Rev 8:129, 1986.
29. Acuff G. Fruit Grow 113:34, 1993.
30. Ledbetter CA, Ramming DW. Hortic Rev 11:159, 1989.
31. Esau K. Plant Anatomy. New York: Wiley, 1965, chap 19.
32. Heiser CB Jr. Of Plants and People. Norman: University of Oklahoma Press, 1985, pp 107–109 and 163–172.
33. Grierson W. HortScience 5:1, 1970.
34. Soule J, Grierson W. Anatomy and physiology. In: WF Wardowski, S Nagy, W Grierson, eds. Fresh Citrus Fruits. Westport, CT: AVI, 1986.
35. Storey R, Price WE. Sci Hortic 81:279–286, 1999.
36. Crane JC, Iwakiri BT. Hortic Rev 3:376, 1981.
37. Smock RM, Neubert AM. Apples and Apple Products. New York: Interscience, 1950.
38. Scott FM, Baker KC. Bot Gaz 108:459, 1947.
39. Robbins J, Sjulin TM. HortScience 24:776, 1989.
40. Robbins J, Moore PP. HortScience 25:679, 1990.
41. Storey R, Treeby MT. J Hortic Sci Biotechnol 74:464–471, 1999.
42. Grierson W, Ting SV. Proc Int Soc Citric 21–27, 1978.
43. Wardowski W, Soule J, Whigham J, Grierson W. Fla Ext Spec Publ 99, 1991.
44. State of Florida. The Florida Citrus Code of 1949 (as amended). Florida Statutes, Chap. 601, 1949 et seq.
45. Haas ARC, Klotz LJ. Hilgardia 9:181, 1935.
46. Grierson W. Beneficial Aspects of Stress on Plants. In: M Pessarakli, ed. Handbook of Plant and Crop Stress. New York: Marcel Dekker, 1993, pp 645–657.
47. Grierson W, Cohen E, Kitagawa H. Degreening. In WF Wardowski, S Nagy, W Grierson, eds. Fresh Citrus Fruits. Westport, CT: AVI, 1986.
48. Gage J, Stutte G. HortScience 26:459, 1991.
49. Biale JB, Young RE. In: J Friend, RE Young, eds. Recent Advances in the Biochemistry of Fruits and Vegetables. London: Academic Press, 1981, pp 1–39.
50. Grierson W, Wardowski WF. HortScience 13:570, 1978.
51. Sharples RC, Johnson DS. HortScience 22:763, 1987.
52. Harker FR, Hallett IC. HortScience 27:1291, 1992.
53. Grierson D. HortScience 22:859, 1987.
54. Zauberman G, Fuchs Y, Ackerman M. HortScience 23:588, 1988.
55. Walsh CS, Popenoe J, Solomos T. HortScience 18:482, 1983.
56. Blanpied D, Yang SV, Reid M, McGlasson WB, Kader AA, Sherman M. HortScience 20:39, 1985.
57. McGlasson WB. HortScience 20:51, 1985.
58. Morrison LA, Webster BB. Hortic Rev 1:172, 1979.
59. Okuda H. J Hortic Sci Biotechnol 74:422–425, 1999.
60. Grierson W, Hatton TT. Proc Int Soc Citric 1:207, 1977.
61. Magness JR, Taylor GF. US Dept Agric Circ 350, 1925.

62. Picard A. The Globe and Mail (Canada), Sept. 13, 1999.
63. Brown GR, Wolfe D. HortScience 27:76, 1992.
64. DeEll DR, Prange RK. HortScience 27:1096, 1992.
65. Embree CG, MacLean BW, O'Regan RJ. HortScience 26:1560, 1991.
66. Marini RP, Byers RE, Sowers DL. HortScience 24:957, 1989.
67. Meheriuk M. HortScience 24:978, 1989.
68. Grierson-Jackson WRF. The Storage and Ripening of Bartlett Pears. MSc Agric thesis, University of Toronto, 1940.
69. Grierson W, Ben-Yehoshua S. In: WF Wardowski, S Nagy, W Grierson, eds. Fresh Citrus Fruits. New York: Van Nostrand Reinhold, 1986.
70. Westwood MN, Lombard PB, Bjornstad HO. HortScience 24:765, 1989.

# 8

# Dormancy: Manifestations and Causes

## Frank G. Dennis, Jr.

*Michigan State University, East Lansing, Michigan*

## I. IMPORTANCE OF DORMANCY

During their life cycles, plants are exposed to periods of stress caused by low or high temperatures, drought, or other environmental factors. In the course of evolution, complex defense mechanisms have developed for protection against such stresses. One such mechanism is dormancy. Simply defined, *dormancy* is the inability of an otherwise viable seed, whole plant, or meristem (a bud, apex, etc.) to grow.

Many plants adapted to the tropics do not become dormant; shoot growth occurs whenever environmental conditions permit. However, growth often occurs in flushes, and certain branches may be growing while others are not. In the dry topics, rainy seasons alternate with dry ones; here plants are adapted to growing when water is available, but growth slows or ceases during the dry season. Where cold and warm seasons alternate, as in the temperate zones, continuous growth is similarly impossible. Plants stop growing in the late summer or autumn, then resume growth again in the spring. In both the temperate and the polar regions another adaptation has occurred—plants develop resistance to low temperatures, or "cold hardiness," to permit survival at temperatures as low as −40°C or below. Perennial plants may be *deciduous* or *evergreen*; in the former the leaves abscise before winter begins, in the latter the leaves are functional throughout the year.

Tropical annuals will grow in any climatic zone where the growing season is long enough to allow them to mature. Thus green beans and marigolds can be cultivated from the equator to the arctic circle. In contrast, woody perennials will not survive outdoors if grown in an area where winters are too cold. Peach trees adapted to the temperate zone will grow poorly, or not at all, in the tropics for lack of "chilling" (see later), whereas mangos will not survive the low winter temperatures characteristic of the temperate zone.

Seed physiology may reflect the environmental conditions in the area of origin of the species. The seeds of plants native to the humid tropics need no dormancy provided that conditions are favorable for germination year-round. In contrast, seeds of plants adapted to the temperate zone often exhibit some degree of dormancy. If seeds shed at the end of the growing season were to germinate immediately, they would not survive the winter. Some species have circumvented this problem by having an abbreviated period of fruit development, permitting the shedding of seeds in early summer (silver maple, dandelion). In others, termed *winter annuals*, seeds germinate in late summer/early fall, and the seedlings develop sufficient cold hardiness to survive the winter and produce seed early the following year. Such seeds are dormant when shed but become capable of germination in the fall (see later).

Even when climatic factors do not dictate a need for seed dormancy, the characteristic provides a safeguard for survival. If all seeds germinated immediately, cataclysmic events such as fires and late freezes could destroy entire species, at least in local areas. Differing levels of dormancy in a seed population permit germination over a period of several years or even longer, depending on seed longevity.

## II. TYPES OF DORMANCY

Numerous types of dormancy exist. The many types of dormancy exhibited by plant organs have created problems in terminology and definition. This problem was summarized for seeds by Simpson [1]: "A precise definition of dormancy cannot be used in the general sense to apply to all seeds, but can only be given for each individual seed considered in the context of a precisely defined set of environmental conditions." Nevertheless, Lang et al. [2,3] and Lang [4] have attempted to classify the many types of dormancy into three main categories, based on the controlling factor(s): *ecodormancy*, when growth is prevented by environmental conditions, such as low or high temperature; *paradormancy*, when growth is prevented by conditions outside the meristem but within the plant; and *endodormancy*, when growth is prevented by conditions within the meristem itself. Examples of these types of dormancy are the failure of buds of trees to expand in the late winter, when low temperatures prevent growth (ecodormancy); their failure to grow in early winter, even when held in a warm greenhouse, because they have not been exposed to sufficient "chilling hours" (see later) to permit growth (endodormancy); and the failure of lateral buds to develop in an herbaceous or woody plant when the terminal bud is growing rapidly (paradormancy). In the buds of perennials, dormancy progresses gradually from paradormancy, also called apical dominance, through endodormancy to ecodormancy as the seasons progress from summer to fall to winter and spring.

These definitions are more applicable to whole plants or shoots than they are to seeds, and seed scientists have been less receptive to their use [5]. Is a dry bean seed, which exhibits no dormancy, ecodormant just because it will not grow without water? Does paradormancy exist in a seed? Does a single type of dormancy prevent growth, or are control mechanisms more complex? As we will see, dormancy is indeed a complex phenomenon in many systems.

I have spoken of dormancy in seeds and whole plants, but dormancy can occur in other structures as well. Bulbs, tubers, and corms—all organs that permit plants to survive unfavorable environmental conditions—also exhibit dormancy. This dormancy can be likened to bud dormancy, for all three structures contain buds, and bud development is the primary indication of the ending of their dormant period. In some respects the structures represent intermediates between whole plants and seeds in that they are more compact than the former but less compact than the seed, which has in addition a seed coat surrounding the embryo and closely associated parts. Most of the remainder of this chapter deals with seed and bud dormancy. Given the many aspects of dormancy, I will not address apical dominance in detail. Several reviews [6,7] provide information on this topic. Khan [8,9], Bewley and Black [10,11], and Bradbeer [12] provide thorough coverage of seed dormancy; Saure [13], Powell [14,15], and Martin [16] have reviewed many aspects of bud dormancy; and Dennis [17] and Lang [18] offer additional information on dormancy in general.

## III. SEED DORMANCY

### A. Induction of Dormancy

Some seeds do not become dormant until fully mature. The percentage germination of barley seeds increases with maturation up to a certain point, then declines (Table 1). Germination is further reduced when mature seeds are held at room temperature for 1 week, but it is stimulated by a brief exposure of moist seeds to low temperature [19]. Breeders sometimes take advantage of this by harvesting fruits before they reach maturity, when seeds or embryos can germinate without special treatment. Considerable research has focused on the physiological basis for the inability of immature seeds to germinate. Kermode [20] provides an analysis of the problem.

### B. Types of Seed Dormancy

Early investigators recognized that many factors could be responsible for the failure of seeds to germinate. One obvious cause of such failure is a nonviable embryo. Death of the embryo can occur during seed

**TABLE 1**  Effect of Stage of Development and Cold
Treatment on Germination of 'Cape $\times$ Coast' Barley

| Stage | Germination (%) |
|---|---|
| Milk stage | 5 |
| Yellow-ripe | 60 |
| Mature | 36 |
| Mature + stored 1 week | 1 |
| Mature + stratified for 2 days | 64 |

*Source*: Ref. 19.

development (abortion) or after shedding of the mature seed. Some seeds (silver maple, citrus) are very short-lived; if germination does not occur within a few weeks, the seed does not survive. Other seeds, including many nuts, as well as avocado and cacao, lose viability rapidly when dried; if stored, a high relative humidity should be maintained. Information on methods of evaluating and prolonging seed viability are available elsewhere (e.g., Bewley and Black [10]) and will not be discussed here.

By definition, a seed that is dormant has the potential to germinate (is *viable*) but requires exposure to certain treatments or environmental conditions before germination can occur (Table 2). Some fruits contain inhibitors that prevent seed germination. Seeds of tomato and cucumber, for example, will not germinate within the fruit; the pulp must be removed and the seeds washed before germination can occur. In other species (e.g., peach, cherry), the presence of a hard pit (endocarp or inner ovary wall) may limit germination. Although such seeds can germinate following the breaking of endodormancy by chilling (see later), germination is improved by endocarp removal. Neither of these conditions represents true seed dormancy, as control is external to the seed, but they are often discussed in relation to seed dormancy. Some of the conditions that break seed dormancy are given in Table 3.

Like the endocarp, the seed coat itself can prevent germination in some species, especially legumes such as alfalfa, locust, and redbud. The structure of the seed coat (testa) prevents the entry of water and its absorption by the embryo (imbibition); thus the embryo cannot germinate. The seed coat must be weakened, either naturally by abrasion or by exposure to fire or to HCl during passage through the gut of an animal, or artificially by *scarification* before imbibition can occur. Ground fires damage hard seed coats, thereby permitting germination of seeds that might otherwise remain dormant [31]. Scarification can be either mechanical, by rotating seeds with gravel or filing the seed, or chemical, by brief exposure to concentrated $H_2SO_4$. "Heat shock" by immersing seeds briefly in boiling water can be more effective than mechanical scarification in some species. For example, Bell et al. [31] reported that germination of seeds of

**TABLE 2**  Types of Seed Dormancy, Conditions That Break Dormancy, and Specific Examples

| Cause of dormancy | | Conditions that break dormancy | Species |
|---|---|---|---|
| A. | Control outside the seed | | |
| | 1. Inhibitors in the fruit | Seed removal, washing | Tomato, cucumber |
| | 2. Hard endocarp | Acid treatment, endocarp removal | Stone fruits |
| B. | Control by seed coat | | |
| | 1. Coat impermeable to $H_2O$ | Acid or mechanical scarification, fire | Some legumes |
| | 2. Coat impermeable to $O_2$ | Seed coat removal? | ? |
| C. | Morphologically immature embryo | Warm-moist storage<br>Cool-moist storage | Ginkgo, coconut<br>Cowparsnip |
| D. | Physiologically immature embryo | | |
| | 1. "Shallow" dormancy | Light, alternating temperature, dry storage | Lettuce, celery, oats |
| | 2. "Deep" dormancy | Cool, moist storage | Apple, peach |
| |    a. Epicotyl dormancy | Cool, moist storage | Tree peony |
| |    b. Double dormancy | Cool, moist storage | Trillium |
| C + D. | Hard seed coat plus deep dormancy[a] | Scarification, followed by cool, moist stratification | Redbud |

[a] Some authors use the term *double dormancy* for this phenomenon.

**TABLE 3** Optimum Conditions for Seed Germination in Selected Species

| Pretreatment | Light | Temperature Low | Temperature High | Temperature Alternating | Species | Ref. |
|---|---|---|---|---|---|---|
| None | | | + | | Bean, tomato | — |
| | + | | + | | Birch (*B. pubescens*) | 10 |
| | | + | | | Lettuce | 10 |
| | | | | + | Broadleafed dock | 21 |
| | + | | | + | *Lythrum salicaria*, tobacco | 22 |
| | | | + (25°C) | | *Pinus lambertiana* | 23 |
| Dry storage | | | | | Wild oats | 24 |
| | | | | | Rice | 25 |
| Scarification | | | | | Black locust | 26 |
| Chilling | | | | | Apple, *Pinus lambertiana* | 27, 23 |
| | + | | | | *Pinus strobus* | 28 |
| | + | + | | | *Delphinium ambiguum* | 29 |
| Scarification + chilling | | | | | Redbud | 30 |

*Acacia divergens* averaged 11, 28, and 90% for no treatment, mechanical scarification, and boiling in water for 30 sec, respectively. The coats of some seeds are impermeable to oxygen. In this case, scarification allows oxygen to penetrate to the embryo. Tran and Cavanagh [32] reviewed the structural aspects of seed dormancy, emphasizing seed coat impermeability and methods of increasing it. Microscopic examination of seeds indicated [33] that treatment with boiling water or fire did not soften the seed coat but affected the structure of the "lens" (strophiole) near the hilum, thereby allowing entry of water. In some species the seed coat, although permitting entry of water and oxygen, is a mechanical barrier to germination; on its removal the embryo germinates readily. The seed coat may also contain chemicals that inhibit germination.

Many factors can affect germination. Because of the many interactions possible, Karssen [34] cautioned that "an absolute requirement for any stimulatory factor hardly occurs." Therefore, one must be cautious in discussing any one factor in isolation. Nevertheless, several factors, light and temperature in particular, have pronounced effects.

## C. Temperature and Seed Dormancy

Dormancy is often temperature dependent. In some cultivars of lettuce and celery, for example, germination occurs readily between 10 and 20°C but declines to nil as temperature increases to 30°C (Figure 1A).

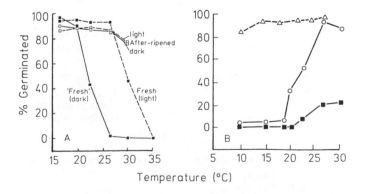

**Figure 1** Effect of temperature and light on germination of seeds of (A) lettuce and (B) *Betula pubescens*. 'Grand Rapids' lettuce seeds were tested immediately after harvest ("fresh") or after storage at about 18°C for 18 months ("after-ripened"). Birch seeds were tested in darkness (■), under a 20-hr photoperiod (△), or were exposed to red light for 15 min each day (○). (From Ref. 10.)

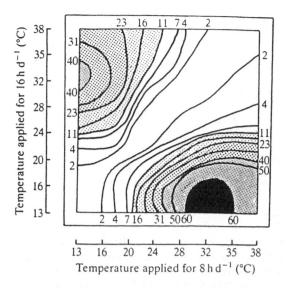

**Figure 2** Effects of constant and alternating temperatures on germination of seeds of signal grass (*Brachiaria humidicola*). Seeds were held at indicated temperatures for 40 days. Percentage germination is indicated at points where lines intersect perimeter of square, and is proportional to density of stippling. (From Ref. 35.)

In contrast, seeds of birch germinate better in darkness at high than at low temperature (Figure 1B), but exposure to light can markedly affect response. Other seeds germinate best when temperature is alternated on a daily cycle (Table 3). For example, when seeds of signal grass [*Brachiaria humidicola* (Rendle) Schweickerdt] are germinated at constant temperatures ranging from 13 to 38°C, germination does not exceed 2%, whereas daily alternation between 13 and 32°C results in 60% germination (Figure 2). Baskin and Baskin [36] reported that freshly harvested seeds of curled dock (*Rumex crispus* L.) remained "nondormant" for 2 years when buried 7 cm deep in moist soil. However, the seeds germinated in the light at alternating temperatures. Few seeds (<1%) germinated while buried. Therefore, the seeds would probably have remained dormant had they been held in darkness at constant temperature.

Seeds of certain species require prolonged exposure to relatively high temperatures before germination can occur. Chickweed (*Stellaria media* L.) and other "winter annuals" remain vegetative in the winter, then flower and produce seeds in the early summer. Such seeds remain dormant until fall, then germinate and repeat the cycle. Experiments have demonstrated that the periods at warm temperatures break dormancy, provided that the seeds are subsequently exposed to appropriate conditions, especially alternating temperatures and light [37]; temperatures below 20°C are ineffective in breaking dormancy regardless of subsequent treatment.

Exposure of such seeds to low soil temperatures in the autumn reintroduces dormancy (see Sec. III.H), so that they once again become incapable of germination. A seasonal pattern thus develops, with periods of high germinability in autumn alternating with periods of low germinability in the summer. The behavior of such seeds contrasts with that of seeds of summer annuals, such as *Polygonum persicaria* [38], in which chilling is essential for breaking secondary dormancy (see later) and which germinate readily in the late winter and spring but poorly in the summer and fall (Figure 3). Chilling temperatures are required for breaking dormancy in other seeds (see Sec. III.E).

## D.   Light and Seed Dormancy

Seed response to light has been studied intensively in 'Grand Rapids' lettuce. Seeds of this and a number of other cultivars of lettuce and celery germinate readily in the light at 25°C but fail to germinate in the dark (Figure 1A). A brief exposure of moist seeds to white or red light (660 nm) induces subsequent germination in darkness. The time of exposure required varies with species (Table 4). A brief exposure is effective only at high temperatures in birch, whereas a long exposure time is effective at all temperatures from 10 to 25°C (Figure 1B). However, if the brief red light treatment is followed by a

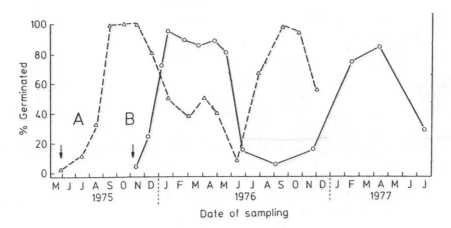

**Figure 3** Germination of *Veronica hederofolia* (A) and *Polygonum persicaria* (B) seeds at alternating temperatures following burial in the field for varying periods of time. Arrows indicate dates of burial. Veronica seeds were held at 4/10°C for 16/8 hr per day in darkness; *Polygonum* seeds were held at 12/22°C for 12/12 hr per day and were exposed to light during 12 hr at 22°C. (Adapted from Ref. 10, based on data of Karssen [38] and Roberts and Lockett [39].)

similar brief exposure to far-red light (730 nm), the effect of the red light treatment is nullified (Table 5). Alternating red with far-red light leads to germination or dormancy, depending on the wavelength of last exposure. This is a classic case of a phytochrome-controlled response. Cone and Kendrick [41] provide a thorough review of the role of phytochrome in seed germination. Certain chemicals, especially gibberellic acid, can substitute for red light treatment (see later). In seeds of some species, shade from a plant canopy can reduce germination, relative to seeds held in darkness, by reducing the ratio of red to far-red light [42].

## E.  Shallow Versus Deep Dormancy

Seeds that will germinate in response to environmental cues (light, alternating temperatures) are considered to have a *shallow* dormancy; those that require prolonged exposure to certain conditions (e.g., moist chilling) are considered to have a *deep* dormancy. Certain seeds will not germinate immediately after harvest but do so after several weeks or months of dry storage ("after-ripening") at room temperature (Table 6). This characteristic provides a safeguard against premature germination. In genotypes that do not possess this characteristic, germination can even occur on the plant, provided that moisture is abundant or rain occurs. This is an example of *viviparity* (Latin *vivus* = alive, plus *parere* = to give birth). The length of the dormant period in rice seeds is shortened as storage temperature is raised from 27 to 57°C [43]. Plotting the log of mean dormancy period ($y$) versus storage temperature ($x$) gives a straight line with negative slope (Figure 4). The depth of dormancy declines even at very low temperatures ($-75$°C) in seeds of orchard grass (*Dactylis glomerata*), although the rate of change is extremely slow [44].

**TABLE 4**  Time of Illumination Required to Break Dormancy in Seeds of Selected Species

| Time required | Species |
| --- | --- |
| Seconds or minutes | 'Grand Rapids' lettuce (*Lactuca sativa*) |
| Several hours | *Lythrum salicaria* |
| Days | *Kalenchoë blossfeldiana* |
| Long photoperiods | Begonia (*Begonia evansiana*) |
| Short photoperiods | Hemlock (*Tsuga canadensis*) |

*Source*: Adapted from Ref. 10.

**TABLE 5** Reversible Effects of Brief Exposures
to Red (R = 580 to 680 nm) and Far-Red
(FR > 700 nm) Radiation on Germination of
Lettuce Seed, cv. 'Grand Rapids', in Darkness

| Sequence | Germination (%) |
|---|---|
| Darkness | 8.5 |
| R (640–680 nm) | 98 |
| R-FR | 54 |
| R-FR-R | 100 |
| R-FR-R-FR | 43 |
| R-FR-R-FR-R | 99 |
| R-FR-R-FR-R-FR | 54 |
| R-FR-R-FR-R-FR-R | 98 |

*Source*: Ref. 40.

**TABLE 6** Time at Room Temperature for Dry After-Ripening of Seeds of Selected Species

| Time required (months) | Species | Alternative method |
|---|---|---|
| 1 | Brome grass (*Bromus secalinus*) | Chilling |
| 2–3 | Rice (*Oryza sativa*) | — |
| 12–18 | Lettuce (*Lactuca sativa*) | Light, chilling |
| 60 | Curled dock (*Rumex crispus* L.) | Light, chilling, alternating tempeature |

*Source*: Adapted from Ref. 10.

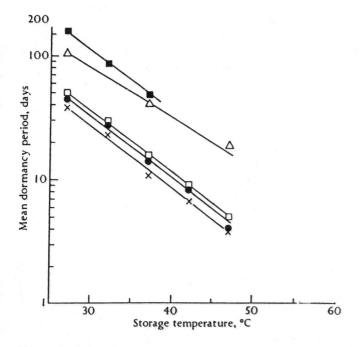

**Figure 4** Effect of storage temperature on mean dormancy period in rice. Each line represents a different cultivar. (Adapted from Ref. 43.)

If seed coat removal does not allow germination under favorable conditions, control obviously lies within the embryo (exalbuminous seeds) and/or endosperm (albuminous seeds). Albuminous seeds are composed primarily of endosperm; the embryo is relatively small. Warm, moist conditions for a period of 2 to 4 months following seed dispersal are usually required for coconut and ginkgo embryos to enlarge to the point where they are capable of germination. Some species [e.g., cowparsnip (*Heracleum sphondylium* L.)] require chilling for embryo development [45]; embryos develop very slowly at 15°C.

In exalbuminous seeds, the embryo is fully developed at maturity. However, many such embryos will not germinate, or germinate only sluggishly, when the seed coat is removed. Exposure to moisture and low temperatures (0 to 10°C) for periods of 1 to 20 weeks (cool, moist stratification) is often required to permit germination. Little or no growth of the embryo occurs during this time; the treatment alters the embryo's metabolism without affecting its morphology.

## F.  Epicotyl Dormancy

Some seeds [e.g., tree peony (*Paeonia suffruticosa* Haw.)] germinate readily without special treatment, but the epicotyl (shoot) will not elongate unless chilled [46]. Chilling prior to germination is ineffective.

## G.  Double Dormancy

More than one mechanism may prevent the germination of a seed. Certain legumes [e.g., redbud (*Cercis canadensis*)] not only have hard seed coats but their embryos must be chilled before germination can occur (Table 3). Scarification, followed by moist chilling, breaks their dormancy. In other seeds (e.g., *Trillium erectum*) the radicle and the epicotyl both require chilling, but the periods at low temperature must be sequential. The first period permits radicle protrusion, the second shoot emergence [47].

## H.  Thermodormancy and Secondary Dormancy

All of the types of dormancy just described are examples of *primary dormancy*, in which germination is prevented by conditions within the seed at the time it matures on the plant. *Thermodormancy* can be induced by exposure of seeds that are capable of germination at low temperatures (10 to 15°C) to high temperatures (25 to 30°C). This can occur in lettuce, for example, when soil temperatures are very high. *Secondary dormancy* is induced when a seed that is not dormant when shed, or whose dormancy has been partially broken, is exposed to unfavorable conditions, such as high temperature or drying. In seeds that are chilled for less than the required time, for example, premature exposure to high temperature can eliminate the effects of prior chilling.

## IV.  BUD DORMANCY

Following bud break in the spring, shoot growth is relatively slow at the beginning of the season, accelerates with time, then slows and eventually stops. This pattern tends to occur even at constant temperature. As noted before, growth tends to be cyclical. Even in the humid tropics flushes of growth occur in a more or less random fashion; one shoot on a tree may be growing rapidly while growth of another is negligible or nil. In contrast, growth of perennials in the temperate zone is synchronized. Growth ceases in mid- to late summer and the plants pass through a dormant period lasting for several months.

Fuchigami et al. [48] have described this pattern of growth as a sine wave (degree growth stage model), with 0° representing the end of ecodormancy/beginning of active growth; 90°, the end of active growth (maturity induction point = beginning of paradormancy); 180°, "vegetative maturity" (beginning of endodormancy); 270°, the time of deepest endodormancy; and 315°, the end of endodormancy/beginning of ecodormancy (Figure 5). Note that phase transition is gradual rather than abrupt; endodormancy does not end one day and ecodormancy begin the next; rather, there is a gradual transition from one phase to the next. During the early part of the summer, removal of the shoot apex and/or defoliation relieves apical dominance and permits growth of the lateral buds. This is true not only in woody plants but in many herbaceous ones as well. Horticulturists remove the apical portion ("pinch") chrysanthemums and petunias to force branching and thereby create more attractive plants. Arboriculturists use the same practice to stimulate the formation of lateral branches. At this time, the axillary buds are *paradormant* (see ear-

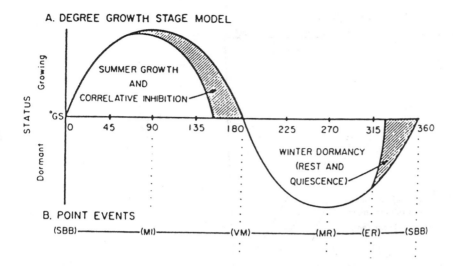

**Figure 5** Degree growth state model representing stages in the annual cycle of growth in woody plants. Five sequential growth stages [spring budbreak (SBB), maturity induction point (MI), vegetative maturity (VM) (= onset of endodormancy), maximum endodormancy (MR), and end of endodormancy (ER)] occur at 0, 90, 180, 270, and 315°C, respectively. (From Ref. 48.)

lier)—they are prevented from growing by the presence of the apex rather than by conditions within the buds themselves. As the summer progresses, the ability of the buds to grow following apex removal declines; paradormancy is gradually becoming endodormancy as control shifts from the apex to the buds themselves. By the end of the season, the buds no longer respond to apex removal; endodormancy is now fully established.

Many woody perennials (e.g., birch) exhibit a marked response to photoperiod, growing rapidly under long photoperiods, slowly or not at all under short photoperiods. This response is truly photoperiodic rather than being a function of total time of exposure to light per se and is an example of ecodormancy. When plants are grown under short days but the long night is interrupted by a brief period of light, they continue their growth. Under natural conditions, the effects of long days are often masked by other environmental limitations, such as water supply or competition among growing points. Thus mature trees of birch stop growth in midsummer, even though daylength is near its maximum.

Chilling temperatures appear to be required for buds to become fully endodormant. In some areas of the tropics and subtropics where temperatures never fall below 20°C, the buds of peaches, grapes, and apples can be forced to grow by defoliation soon after harvest. This permits production of two or more crops per year. The longer the interval between harvest and defoliation, the poorer the response. Trees that are not defoliated may eventually become endodormant; in the absence of chilling, they cease growth entirely and eventually die.

Endodormancy is normally broken by exposure to chilling temperatures. Optimum temperatures vary with species but generally range from 0 to 10°C; temperatures below 0°C have little or no effect. Considerable research has been done to determine the chilling requirements of fruit tree species and cultivars, and several models have been developed to predict when these requirements have been satisfied. For example, according to the Utah model [49], the number of chill units required for 'Elberta' peach and 'Delicious' apple are 800 and 1234, respectively [50]. A *chill unit* is defined as 1 hr of exposure to a temperature of 6°C; higher and lower temperatures between 0 and 13°C are less efficient, and temperatures above 13°C are inhibitory; thus adjustments must be made in calculation (Figure 6). This model, developed in the north temperate zone, may not apply in regions where diurnal temperature fluctuations are greater. Israeli scientists have therefore developed a "dynamic" model in which temperatures alternating between about 6 and 13–14°C are considered to have a greater effect than continuous cold in breaking dormancy [52]. Temperatures above 15°C are inhibitory unless the exposure time is less than a critical length. This model was more effective than the Utah model in predicting end of rest when used in Israel

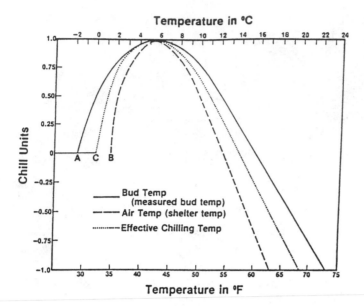

**Figure 6**   Curve used in estimating chill unit accumulation, based on the Utah model, for the breaking of bud dormancy in deciduous tree fruits. Effective chilling temperature is the mean of the two temperatures measured. Positive values are assigned to temperatures between −2 and 13°C, negative values to higher temperatures. (From Ref. 51.)

[52]. Species and cultivars vary greatly in the number of chilling hours required. For example, the chilling requirements of peach cultivars grown in Florida should not exceed 300 hr, whereas those grown in the northernmost parts of the United States may require 800 hr or more.

Bud dormancy is not confined to woody plants. Many herbaceous perennials must be chilled before growth can resume in the spring. Ornamental bulbs such as tulips and daffodils are planted in the fall. Cold soil temperatures provide the chilling required to allow normal stem elongation the following spring. If such bulbs are planted indoors, the flower stalks are much shorter and the flowers themselves may abort. Florists meet the demand for these flowers out of season by artificially chilling the bulbs, then forcing them in a warm greenhouse. Note that this period of cold temperature stimulates elongation of preexisting inflorescences and therefore differs from *vernalization*, in which chilling stimulates the *initiation* of flowers. In some species, however, including Dutch iris (*Iris* spp.) and Easter lily (*Lilium longiflorum*), vernalization indeed occurs. Although the rate of sprout development in onion bulbs is greater at 15°C than at higher or lower temperatures [53], Abdalla and Mann [54] established that the time required for sprouting was independent of storage temperature prior to transfer to 15°C. Thus onion differs from tulip in not requiring chilling for floral stalk elongation.

Similarly, some tubers (e.g., Jerusalem artichoke) must be chilled before buds can grow normally. This, of course, is not the case with crops, such as the potato, that originated in the tropics. Although potato has no chilling requirement, the tubers are dormant at harvest. Dry storage at room temperature for several weeks permits bud development; this parallels the response of seeds of several grains to "after-ripening" (see earlier).

## V.   METHODS FOR BREAKING OR PROLONGING DORMANCY

Dormancy or lack thereof can be troublesome to the plant grower. Waiting 6 to 10 weeks or more for seeds to be after-ripened or buds to be chilled may not be inconvenient in areas where cold temperatures prevent winter production but can reduce profitability in areas where crops can be grown year-round. In the latter areas, multiple cropping is practiced, with two or more crops being harvested each year. Thus yields will be maximized if no dormant periods intervene.

As noted earlier, in some areas of the tropics or subtropics, peach, apple, and grape can be multiple cropped, although a brief dormant period intervenes between foliations. The leaves must be removed to stimulate bud break, and chemicals, such as sodium chlorate, copper sulfate, or urea, are often applied to injure the leaves and induce premature abscission. In areas where multiple cropping is impossible, but chilling inadequate to completely relieve dormancy, other chemicals, such as combinations of dinitro-*O*-cresol and oils, are used to hasten bud break and concentrate the bloom period. Hydrogen cyanamide ($H_2NCN$), which releases HCN within the tissues, is a relatively new compound that has similar effects and has been extensively tested for this purpose [55,56].

In arid regions bud dormancy of some species can be broken by withholding water for several weeks, then irrigating. Asparagus growers in California and Peru can produce crops year-round using this method. Irrigation is also used in combination with rest-breaking chemicals and/or defoliation of deciduous fruit trees in tropical regions [57].

In areas where chilling is adequate but spring freezes often damage flowers and/or fruits, delaying bloom could provide protection. Evaporative cooling by misting with water can delay bloom; delays of 3 weeks or more are possible in arid climates [58,59]. However, side effects, such as poor fruit set, have limited commercial application. This method has also been tested in warm climates for cooling buds during the winter [60], thereby hastening the breaking of dormancy; again, commercial application has been limited.

Methods of weakening the integuments of seeds with hard seed coats to allow water to penetrate have already been discussed, as well as the effects of light and temperature on seeds with "shallow" dormancy. Several growth regulators, including both gibberellins (GAs) and cytokinins, promote germination in dormant or partially dormant seeds. GA is effective in stimulating germination in seeds with a shallow dormancy. Light-sensitive lettuce seeds, for example, will germinate in darkness when supplied with GA. Cytokinin, although generally effective in stimulating dark germination, can overcome the inhibitory effects of high temperatures. Abscisic acid (ABA) blocks germination in many seeds, regardless of environmental conditions. Khan [61] tested the effects of all three hormones and their combinations on the germination of light-sensitive lettuce seeds. The action of GA was blocked by ABA, but cytokinin counteracted the effect of ABA, thereby permitting germination when all three hormones were applied. From these data, Khan [62] proposed that the roles of GA, ABA, and cytokinin were primary, preventive, and permissive, respectively; GA is the primary stimulus, with cytokinin being essential only when ABA is present. Khan and others [63,64] have confirmed and extended these observations by using inhibitors of GA synthesis to block germination and demonstrating that in some cases, cytokinin and/or ethylene is required, in addition to GA, to overcome the inhibitory effects of stress caused by water deficit, salinity, and other conditions.

GA will also stimulate germination in some cold-requiring seeds, although some chilling is usually required before maximum response is obtained. Cytokinins are usually less effective. Both GA and cytokinins can hasten release from dormancy in buds of woody plants, as well as overcoming apical dominance during the early growing season. A combination of $GA_{4/7}$ and benzyladenine, for example, is currently available commercially to stimulate growth of lateral buds of conifers used for Christmas trees, thereby providing a more pleasing form.

Ethylene promotes germination in some weed species [e.g., redroot (*Amaranthus retroflexus*) and lamb's quarters (*Chenopodium album*)], but many species are not responsive [65]. Gibberellins and cytokinins have more general effects. A few cases are known in which ethylene breaks bud dormancy, but again, response is species dependent.

Several chemicals are effective in *prolonging* bud or seed dormancy. Potato tubers are regularly fumigated with 1-methyl-3-chlorophenylcarbamate (CIPC) to delay their sprouting during storage. Scientists are testing naturally occurring compounds as potential substitutes. Andean natives store potatoes in pits together with leaves of muña (plants of the genera *Minthostachys* and *Satureja*) to delay sprouting and reduce both weight loss and insect injury [66]. Trials with volatile components of readily available essential oils demonstrated that 1,8-cineole, found in eucalyptus oil, has promise in inhibiting both sprouting and fungal growth [67]. Application of maleic hydrazide to the foliage of onion plants several weeks before harvest inhibits sprouting of the stored bulbs [68]. The naturally occurring plant growth inhibitor ABA inhibits seed germination in many species [10], although its cost prohibits commercial use. It is less effective on buds, perhaps because of limited penetration and/or rapid metabolism.

Inhibitors of respiration, or more specifically, of cytochrome oxidase, can break dormancy in some seeds, including rice [69,70], barley [71], and lettuce [72], as well as in isolated apple embryos [73]. Apple embryos also respond to anaerobiosis; holding them in nitrogen for 2 weeks or longer permits subsequent germination in air [74]. Other reports indicate that high oxygen tension relieves dormancy in several grains [71,75–77]. The similar effects of these conditions that restrict versus promote respiration suggest that different mechanisms control dormancy at different times and dictate that caution be used in assigning causal effects to various external factors that influence dormancy. (See Roberts and Smith [78] for a hypothesis to explain these effects.)

## VI.   PHYSIOLOGICAL BASIS OF DORMANCY

Despite much effort by scientists, the mechanisms that control dormancy in plants remain a mystery. However, numerous theories have been proposed to account for the phenomenon. All physiological processes are ultimately controlled by genes, and progress is being made in identifying genes associated with dormancy. Seeds of *Arabidopsis thaliana* require dry storage to break dormancy, but mutants have been isolated that produce nondormant seeds [79,80]. The ability of such seeds to germinate has been associated with single-gene differences in their ability to synthesize ABA or GA (see later). In maize, genes have been identified that are responsible for preventing premature germination (viviparity) [81,82]. Again, these genes appear to regulate the synthesis of, or sensitivity to, ABA [83–86]. Skriver and Mundy [87] and Thomas [88] have reviewed the effects of these and related genes during embryo development. Single-gene control of dormancy has also been demonstrated in hazel (*Corylus avellana*) [89] and in peach (*Prunus persica*) [90], although no data are yet available on the mechanisms involved. More comprehensive information on genetic and molecular approaches to dormancy may be found in Lang [18] and King [91].

Although control of dormancy ultimately lies within the genome, such control must be exerted via physiological mechanisms. The many theories advanced to explain dormancy can be grouped into three general categories: nutritional/metabolic deficiencies, blocks to membrane permeability, and excesses or deficiencies of hormones. Briefly stated, these theories propose that the failure of a seed or bud to grow results from (1) deficiency of a nutrient(s) or of an enzyme(s) able to metabolize such a nutrient, (2) the inability of nutrients to reach shoot and/or root apices within the dormant organ, or (3) an excess of a growth inhibitor(s), a deficiency of a growth promoter(s), or an improper balance between the two within the meristem and/or adjacent tissues. In general, more attention has been devoted to hormone studies than to the other two areas of research. Seeds are more convenient for studying dormancy than are buds, for they are small, self-contained, and thus more easily manipulated.

## A.   Metabolic Aspects of Dormancy

As Bewley and Black [10] emphasized, "Dormancy cannot be equated with overall metabolic inactivity. . . ." Respiration rates of hydrated, dormant seeds of lettuce and cocklebur differ little from those of nondormant seeds prior to germination, and activity of hydrolytic enzymes is unlikely to be crucial, for little mobilization of reserves occurs prior to radicle emergence [10]. Nevertheless, many studies have compared the metabolism of dormant versus nondormant seeds and several investigators have proposed that dormant tissues are deficient in specific enzymes required for metabolism of carbohydrates, fats, and/or proteins.

### 1.   Nutrient Supply

Stokes [92] differentiated between two types of seed dormancy, with embryo dormancy ("true dormancy") being responsible for the first and lack of nutrients for the second (nonresting embryo). In the former, interruption of chilling by exposure to high temperature can negate the effect of previous chilling by inducing secondary dormancy, and the effects of two or more periods of chilling are less than additive. In the latter, the effects of chilling are additive and irreversible; interruption by high temperature does not negate the effects of prior exposure to low temperature.

The response of seeds of the second type is easier to explain, superficially, at least. The embryo is very small and grows at the expense of the surrounding seed tissues (endosperm and/or nucellus). Chilling stimulates the activity of enzymes that hydrolyze stored reserves, which the embryo cannot otherwise utilize, to compounds that can be used for growth. Thus in seeds of cowparsnip (*Heracleum spho-*

*ndylium*), embryos of seeds held at 15°C elongate for approximately 6 weeks, then stop growing when approximately half their full size [45]. Although the initial rate of growth is slower in seeds held at 2°C, elongation of the embryo continues logarithmically for 9 weeks. Parallel changes occur in the endosperm, but in reverse (i.e., the endosperm of seeds held at 2°C is consumed by the embryo, while that in seeds held at 15°C is not). If lack of suitable nutrients were responsible for the failure of embryos to develop at 15°C, one would expect that growth of excised embryos in vitro at 15°C could be stimulated by supplying appropriate nutrients. Stokes [93] observed that arginine and glycine concentrations in the endosperm were higher in seeds held at 2°C than in those held at 20°C. When embryos cultured in vitro at 20°C were supplied with glucose plus various sources of nitrogen, arginine and glycine were the most effective amino acids in supporting growth, although $KNO_3$ was the best source of nitrogen. From these and other data, Stokes [93] concluded that exposure to 2°C stimulated embryo growth by increasing the quantities of arginine and glycine available to the embryo.

A similar situation occurs in both black ash (*Fraxinus nigra*) [94] and European ash (*F. excelsior*) [95,96], except that chilling is not essential for embryo enlargement but is required for germination once embryos have reached full size. Stokes [92] provides other examples of seeds with similar requirements. Axes from dormant hazel embryos will grow in vitro when supplied with inorganic salts and sucrose [97], suggesting that failure of the intact embryo to germinate is due to inability to mobilize nutrients from the cotyledons [98,99]. Application of $GA_3$ both breaks embryo dormancy and permits mobilization of reserves, suggesting that gibberellin biosynthesis following chilling has a similar effect (see later).

## 2. Protein Metabolism

A group of proteins termed "late-embryogenesis-abundant" (*Lea*) proteins accumulates as seeds mature and become dehydrated (see Ref. 87). These appear to bind water, thereby protecting macromolecules such as nucleic acids (?) from dehydration and resultant denaturation. *Lea* proteins disappear during germination.

Several facts, summarized by Quatrano [100], suggest that such proteins play a role in dormancy: (1) embryos of viviparous mutants do not synthesize these proteins if cultured on a medium containing ABA; (2) dehydration of immature embryos induces the production of the proteins, possibly by stimulating the synthesis of ABA; and (3) treating mature seeds with ABA prevents both germination and the loss of *Lea* proteins.

Most studies of *Lea* proteins have involved species whose seeds either are nondormant or have a shallow dormancy, and no studies are known involving species with deeply dormant seeds. Therefore, the connection between such proteins and dormancy remains tenuous. ABA blocks germination while inducing or maintaining the synthesis of *Lea* proteins, but these two responses may be unrelated.

Protein metabolism has also been implicated as a factor in the breaking of dormancy. As already noted, holding *Heracleum sphondylium* seeds at 2°C permits the hydrolysis of reserve proteins and their transfer to the embryo, whereas holding them at 20°C does not [45]. In apple embryos, however, hydrolysis of reserve proteins occurs at both 5 and 20°C [101]. Furthermore, no proteolysis is observed in seeds held in the fruit at 0°C, although this treatment also breaks embryo dormancy. Similarly, Chen and Varner [102] reported that dormant and nondormant seeds of wild oats (*Avena fatua* L.) synthesize protein at similar rates.

Lewak et al. [103] suggested that an insufficient supply of amino acids may prevent germination in dormant apple seeds. Protease activity increases with chilling, reaching a maximum after 7 weeks, then declines to the level observed in nonchilled seeds. The authors suggested that germination is dependent on a supply of amino acids released by hydrolysis of proteins. However, they presented no data on the effects of amino acids on germination of dormant embryos.

Subsequent work (see later) emphasized the effects of dormancy-breaking treatments on the concentrations of specific proteins or polypeptides. The rationale for much of this work is that regardless of what substances control induction or breaking of dormancy, enzymes (proteins) must be synthesized before such compounds can be produced. Therefore, changes in protein content should precede changes in other compounds, be they carbohydrates or hormones. Protein analysis involves electrophoretic separation of extracted proteins, together with the use of radiolabeled amino acids as markers for newly synthesized polypeptides. Although no significant changes were observed in total soluble protein content of pear [104] or apple embryos [105] during chilling, Eichholtz et al. [105] observed an increase in the concentrations of four peptides in the embryonic axes of apple embryos held at 5°C. No changes were evident in the cotyledons at 5°C or in either axes or cotyledons at 20°C. The authors suggested that protein reserves might be mobilized to the axis during the breaking of dormancy.

Despite much research in this field, the picture remains confusing. Some workers have suggested that proteins found in dormant, but not in nondormant, seeds inhibit germination [106–108]. Mahhou and Dennis [109] reported reduced levels of large proteins (36 and 41 kDa) in the cotyledons of peach seeds stratified at 5°C, even when the embryonic axis was excised. These changes did not occur at 20°C. In some tissues, chilling increases the content of certain proteins (e.g., Ref. 110); in others, some proteins increase during chilling while others decrease [111–113]. Ried and Walker-Simmons [114] have presented evidence for heat-stable proteins in embryonic axes of dormant wheat seeds that are induced by treatment with ABA. Much higher concentrations of ABA are required to produce similar levels of proteins in nondormant embryos, suggesting that sensitivity to ABA may play a role in dormancy.

### 3. Synthesis of Nucleotides

The limited ability of dormant tissues to convert adenosine to nonadenylic nucleotides (NTP = sum of triphosphates of guanidine, cytosine, and uridine) has been suggested as a possible cause of dormancy. Correlations between the ability to convert adenosine to NTP and the dormant state have been reported in Jerusalem artichoke (*Helianthus tuberosum* L.) tubers [115], in apple embryos [116], and in buds or subapical tissues of ash [117], willow, and hazel [118].

## B. Permeability Changes

Several investigators have proposed that changes in membrane permeability are responsible for dormancy. To test this hypothesis, tissues are incubated with a weak acid [5,5-dimethyl-2,4-oxazolidinedione (DMO)]; only the undissociated form can pass through the cell membrane. Use of radioactive DMO permits determination of the ratio of the concentration of DMO within the cell ($C_i$) to the concentration in the intercellular spaces ($C_e$). Relative membrane permeability parallels the $C_i/C_e$ ratio. Using this method, Gendraud and Lefleuriel [119] observed a higher $C_i/C_e$ ratio in dormant than in nondormant tubers of Jerusalem artichoke. This implies less movement of nutrients to the meristematic tissues of dormant tubers. In similar studies, Ben Ismail [120] compared $C_i/C_e$ ratios in bud versus shoot tissues of apple during the dormant period. Higher ratios occurred in shoots than in buds during the fall and early winter, suggesting limited movement of solutes from shoots to buds. Thereafter, the ratio in the buds rose to levels higher than those observed in the shoots. Although the results parallel the expected response of intact trees or isolated shoots, bud development in single-node cuttings exposed to laboratory conditions was reduced only in samples collected in November.

## C. Role of Hormones

The role of hormones in seed dormancy is supported primarily by the effects of applied hormones in both inhibiting the germination of nondormant seeds (ABA) and stimulating the germination of dormant seeds (cytokinins, GAs). However, effective concentrations are often much higher than those found in the seeds themselves, and the response is seldom as great as one might expect. Although treatment with GA is effective in breaking dormancy in lettuce seeds, germination of peach seeds can be maximized only after some chilling has occurred [121]. Even then, the symptoms of insufficient chilling (abnormal leaves, etc.) are not eliminated. Furthermore, despite early reports to the contrary, few good correlations have been established between content of endogenous hormones and dormancy status.

Several hypotheses have been proposed regarding the role of hormones in seed dormancy. Germination is prevented by:

1.  High concentrations of growth inhibitors, (e.g., ABA)
2.  Inhibitory concentrations of auxin [indole-3-acetic acid (IAA)]
3.  Insufficient concentrations of growth promoters (GA, cytokinins)
4.  Both (1) and (3)

Modifications of these hypotheses propose that:

5.  Promoters are synthesized in seeds requiring chilling only following their return to warm temperatures [122].

6.  High levels of promoters are required only temporarily at the beginning of the "trigger" phase that ends dormancy [123].
7.  ABA blocks the action of GAs; if both are present, cytokinin must also be present to permit GA to act [61,62].

## 1.  Auxin

Nikolaeva [124] determined the content of presumed IAA (wheat coleoptile segment and mustard seed germination assays) in seeds and/or embryos of several tree species before, during, and after cold stratification. Activity (promotion of coleoptile section growth, inhibition of germination) declined as stratification was prolonged. Inhibitor activity in nondormant seeds was approximately half of that observed in dormant seeds. Nondormant seeds treated with the naturally occurring auxin IAA produced seedlings with symptoms similar to those of seedlings from insufficiently chilled seeds. From these and other data she concluded that high levels of IAA prevented germination of nonchilled seeds, and that chilling reduced the IAA concentration to the levels found in seeds that did not require chilling. Subsequent investigators have found little support for the role of auxin in dormancy. Most later research on hormones has focused on GAs and ABA.

## 2.  Gibberellins

Amen [123] proposed that seed dormancy could be divided into four phases. During the *induction* phase, levels of growth promoters decline and/or the seed coat becomes impermeable to oxygen; therefore, the seed becomes dormant. During the ensuing *maintenance* phase, germination is prevented by endogenous inhibitors. In the *trigger* phase, a factor that elicits germination but whose continued presence is not essential (the trigger, e.g., light) induces the production of a *germination agent*, whose continued presence is required for germination. In the final phase (*germination*), the germination agent [growth promoter(s)] provides the stimulus for radicle protrusion.

Much of the evidence for this scheme is based on the effects of exogenous growth regulators on germination; only a few studies have supported the hypothesis in terms of actual increases in seed hormone content following action by "triggers," including chilling, and light. In one such study, Williams et al. [122] could detect little change in GA content of hazel seeds during moist chilling at 5°C. However, levels rose rapidly once dormancy had been broken, provided that the seeds were returned to 20°C.

The gibberellin ($GA_4$) content of apple seeds rises during chilling but is no higher in fully chilled seed than in nonchilled seed [125]. This could, of course, be interpreted as supporting a "trigger" role for GA. Similar roles for both GA and cytokinin have been suggested in maple seeds [126].

## 3.  Abscisic Acid

Considerable effort has been directed toward elucidating the role of ABA in controlling dormancy in seeds. The ABA content of immature seeds of several species, including wheat [127] and rapeseed [128], rises to a maximum, then falls as the seeds mature and dry out. Although the concentration of ABA in the mature seed is low, desiccation reduces water content, thereby preventing germination.

The effects of ABA in preventing the germination of immature embryos in vitro plus the evidence for the role of ABA in viviparity, noted above, strongly imply that ABA is one of the factors preventing embryo germination. Seeds of the species investigated in these studies (e.g., maize, rapeseed) are nondormant or have only a shallow dormancy at harvest; similar relationships may not apply in seeds that exhibit deep dormancy.

In ash (*Fraxinus*) seeds, ABA content is low in *F. americana* relative to that in *F. ornus* [129]. Seeds of the former are nondormant, whereas the latter require moist chilling to break their dormancy. This dormancy again is correlated with ABA content. While the ABA content of seeds of three species of *Rosa* is negatively correlated with their germinability [130], the ABA content of seeds of several species of pear bears no relationship to depth of dormancy [131], nor does the ABA content of immature or mature seeds of *Avena fatua* (dormant) differ from that of seeds of *A. sativa* (nondormant) [132]. Differences in sensitivity to ABA could, of course, explain some of these discrepancies but have seldom been tested experimentally. Early results indicated that the levels of ABA or ABA-like inhibitors fell during moist chilling of ash [129] and several other species, including apple [133]. However, subsequent investigations indicated that ABA content either did not decline during low-temperature stratification [134] or that the decline was not temperature-dependent [135]. The concentration of ABA declines during soaking of lettuce seed, regardless of their germination capacity [136].

In many of these studies the entire seed was extracted. Karssen et al. [79] proposed that the GA and/or ABA content of the embryo may be more important than that of the whole seed. Using selected mutants of *Arabidopsis thaliana*, they demonstrated that embryos recessive for ABA production were nondormant even when the seed coat contained high levels of ABA. Later work with GA-deficient mutants led Karssen et al. [80] to propose that GA content is the critical factor in germination. The seeds of genotypes that cannot synthesize GAs remain dormant regardless of their ABA content.

Inhibitor content of buds has also been quantified in relation to dormancy. Again initial results were promising; the inhibitor content of buds of several species, as measured by bioassay, appeared to rise when plants were transferred from long to short photoperiods [137,138]. This work led to the identification of ABA by Ohkuma et al. [139] and Cornforth et al. [140]. As analytical instruments have become more sensitive and experiments more critical, however, the negative correlation between ABA content and growth response has not been confirmed [141,142]. In fact, one laboratory reported [143,144] that rapidly growing apices contained more ABA than did subapical tissues.

Coleman and King [145] reported a positive correlation between ABA content of tubers of 10 potato cultivars following 2 months of storage at 10°C and the time to 50% sprouting at 20°C. However, ABA content of three other cultivars actually *increased* during storage at three temperatures (2, 10, and 20°C), yet dormancy was broken in all cases, often when ABA content was near maximal.

## D.  New Approaches to the Understanding of Dormancy

Relatively little is known about how genes control seed and bud dormancy, but research in molecular biology is beginning to open the "black box." Studies of apical dominance, for example, are under way using transgenic plants that differ in the relative amounts of IAA and cytokinin synthesized. Plants with high IAA/cytokinin ratios exhibit strong apical dominance, and vice versa [146,147], suggesting that these hormones may indeed be responsible for this phenomenon. Genes for hormone synthesis in plants that exhibit seed and/or bud dormancy have been identified and can now be cloned. Once these can be inserted in the same or other species, rapid progress may be expected in elucidating the roles of such compounds in controlling dormancy.

## VII.  SUMMARY

Dormancy serves a protective function in permitting plant survival under extremes of temperature, water deficit, and other environmental stresses, and species differ in their manifestations of dormancy. Several types of dormancy are known, with control sometimes residing within the dormant organ, sometimes outside the organ. As would be expected, the conditions required to break dormancy differ with the type of dormancy exhibited and vary from changes in light intensity or photoperiod to exposure to low or alternating temperatures. Many theories have been proposed to explain the physiological basis of dormancy, but none has proved valid in accounting for all the facts known. New approaches, especially molecular biology, should provide new information in this important field.

## REFERENCES

1.  Simpson GM. In: ME Clutter, ed. Dormancy and Developmental Arrest. New York: Academic Press, 1978, p 167.
2.  Lang GA, Early JD, Aroyave NG, Darnell RL, Martin GC, Stutte GW. HortScience 20:809, 1985.
3.  Lang GA, Early JD, Martin GC, Darnell RL. HortScience 22:371, 1987.
4.  Lang GA. HortScience 22:817, 1987.
5.  Juntilla O. HortScience 23:805, 1988.
6.  Martin GC. HortScience 22:824, 1987.
7.  Cline MG. Bot Rev 57:318, 1991.
8.  Khan AA, ed. The Physiology and Biochemistry of Seed Dormancy and Germination. New York: North-Holland, 1977.
9.  Khan AA, ed. The Physiology and Biochemistry of Seed Development, Dormancy, and Germination. New York: Elsevier Biomedical Press, 1982.
10. Bewley JD, Black M. Physiology and Biochemistry of Seeds in Relation to Germination. Vol 2. Viability, Dormancy and Environmental Control. New York: Springer-Verlag, 1982.

11. Bewley JD, Black M. Seeds: Physiology of Development and Germination. New York: Plenum, 1985.
12. Bradbeer JW. Seed Dormancy and Germination. London: Blackie & Son, 1988.
13. Saure MC. Hortic Rev 7:239, 1985.
14. Powell LE. In: PJ Davies, ed. Plant Hormones and Their Role in Plant Growth and Development. Boston: Martinus Nijhoff, 1987, p 539.
15. Powell, LE. HortScience 22:845, 1989.
16. Martin GC. In: FC Steward, RGS Bidwell, eds. Plant Physiology: A Treatise. Vol X. Growth and Development. New York: Academic Press, 1991, p 183.
17. Dennis FG Jr. HortScience 29:1249, 1994.
18. Lang GA. HortScience 29:1255, 1994.
19. Moormann B. Kuehn-Arch 56:41. 1942 [cited by Stokes (92, p 783)].
20. Kermode AR. Crit Rev Plant Sci 9:155, 1990.
21. Totterdell S, Roberts EH. Plant Cell Environ 3:3, 1980.
22. Toole EH, Toole VK, Borthwick HA, Hendricks SB. Plant Physiol 30:473, 1955.
23. Stone EC. Forest Sci 3:357, 1957.
24. Naylor JM, Simpson GM. Can J Bot 39:281, 1961.
25. Roberts EH. J Exp Bot 13:75, 1962.
26. Chapman AG. J Forestry 34:66, 1936.
27. Luckwill LC. J Hortic Sci 27:53, 1952.
28. Toole VK, Toole EH, Borthwick HA, Snow AG. Plant Physiol 37:228, 1962.
29. Ezumah BS. PhD thesis, University of London, 1980 (cited by Bewley and Black [10]).
30. Dirr MA, Heuser CW Jr. The Reference Manual of Woody Plant Propagation. Athens, GA: Varsity Press, 1987.
31. Bell DT, Plummer JA, Taylor SK. Bot Rev 59:24, 1993.
32. Tran VN, Cavanagh AK. In: DR Murray, ed. Seed Physiology. Vol 2 Germination and Reserve Mobilization. New York: Academic Press, 1984 p 1.
33. Tran VN, Cavanagh AK. Aust J Bot 28:39, 1980.
34. Karssen CM. In: AA The Physiology and Biochemistry of Seed Development, Dormancy and Germination. New York: Elsevier Biomedical Press, 1982, p 243.
35. Goedert CO, Roberts EH. Plant Cell Environ 9:521, 1986.
36. Baskin JM, Baskin CC. Weed Sci 33:340, 1985.
37. Baskin JM, Baskin CC. New Phytol 77:619, 1976.
38. Karssen CM. Isr J Bot 29:65, 1980/81.
39. Roberts A, Lockett PM. Weed Res 18:41, 1978.
40. Borthwick HA, Hendricks SB, Parker MW, Toole EH, Toole VK. Proc Natl Acad Sci U S A 38:662, 1952.
41. Cone JW, Kendrick RE. In: RE Kendrick, GHM Kronenberg, eds. Photomorphogenesis in Plants. Dordrecht: Martinus Nijhoff/Dr W Junk, 1986, p 443.
42. Silvertown J. New Phytol 85:109, 1980.
43. Roberts EH. J Exp Bot 16:341, 1965.
44. Proberts RJ, Smith RD, Birch P. New Phytol 101:521, 1985.
45. Stokes P. Ann Bot 16:441, 1952.
46. Barton LV. Contrib Boyce Thompson Inst 5:451, 1933.
47. Barton LV. Contrib Boyce Thompson Inst 13:259, 1944.
48. Fuchigami LH, Weiser CJ, Kobayashi K, Timmis R, Gusta LV. In: PH Li, A Sakai, eds. Plant Cold Hardiness and Freezing Stress. Vol 2. New York: Academic Press, 1982, p 93.
49. Richardson EA, Seeley SD, Walker DR. HortScience 9:331, 1974.
50. Ashcroft GL, Richardson EA, Seeley SD. HortScience 12:347, 1977.
51. Anderson JL, Richardson EA, Kesner CD. Acta Hortic 184:71, 1986.
52. Erez A, Fishman S, Gat Z, Couvillon GA. Acta Hortic 232:76, 1988.
53. Komochi S. In: HD Rabinowitch, JL Brewster, eds. Onions and Allied Crops. Vol 1. Boca Raton, FL: CRC Press, 1990, p 89.
54. Abdallah AA, Mann LK. Hilgardia 35:85, 1963.
55. Shulman Y, Nir G, Fanberstein L, Lavee S. Sci Hortic 19:97, 1983.
56. George AP, Nissen RJ. Acta Hortic 279:427, 1990.
57. Bederski K. Acta Hortic 199:33, 1987.
58. Alfaro JF, Griffin RE, Hanson GR, Keller J, Anderson JL, Ashcroft GI, Richardson EA. Trans Am Soc Agric Eng 17:1025, 1974.
59. Anderson JL, Ashcroft GL, Richardson EA, Alfaro JF, Griffin RE, Hanson GR, Keller J. J Am Soc Hortic Sci 100:229, 1975.
60. Gilreath PR, Buchanan DW. J Am Soc Hortic Sci 104:536, 1981.
61. Khan AA. Science 171:853, 1971.
62. Khan AA. Bot Rev 41:391, 1975.
63. Khan AA, Andreoli C. In: D Côme, F Corbineau, eds. Vol 2. 4th International Workshop on Seeds. Paris: ASFIS, 1993, p 625.

64.  Khan AA, Xuelin H, Guangwen Z, Prusinski J. In: F Jiarui, AA Khan, eds. Advances in the Science and Technology of Seeds. New York: Science Press, 1992, p 313.
65.  Taylorson RB. Weed Sci 27:7, 1979.
66.  Aliaga TJ, Feldheim W. Ernahrung 9:254, 1985.
67.  Vaughn SF, Spencer FG. Am Potato J 68:821, 1991.
68.  Wittwer SH, Sharma RC. Science 112:597, 1950.
69.  Roberts EH. Physiol Plant 16:732, 1964.
70.  Roberts EH. Physiol Plant 16:745, 1964.
71.  Major W, Roberts EH. J Exp Bot 19:77, 1968.
72.  Taylorson RB, Hendricks SB. Plant Physiol 52:23, 1973.
73.  Perino C, Simond-Cote E, Côme D. C R Acad Sci 299(Ser 3):249, 1984.
74.  Tissaoui T, Côme D. Planta 111:315, 1973.
75.  Roberts EH. J Exp Bot 12:430, 1961.
76.  Durham V, Wellington PS. Ann Bot 25:197, 1961.
77.  Black M. Can J Bot 37:393, 1959.
78.  Roberts EH, Smith RD. In: AA Khan, ed. The Physiology and Biochemistry of Seed Dormancy and Germination. New York: North-Holland, 1977, p 385.
79.  Karssen CM, Brinkhorst-van der Swan DLC, Breekland AE, Koornneef M. Planta 157:158, 1983.
80.  Karssen CM, Zagorski S, Kepczynski J, Groot SPC. Ann Bot 63:71, 1989.
81.  Robertson DS. J Hered 66:67, 1975.
82.  McCarty DR, Carson CB, Stinard PS, Robertson DS. Plant Cell 1:523, 1989.
83.  Neill SJ, Horgan R, Parry AD. Planta 169:87, 1986.
84.  Neill SJ, Horgan R, Rees AF. Planta 171:358, 1987.
85.  Robichaud CS, Wong J, Sussex IM. Dev Genet 1:325, 1980.
86.  Robichaud C, Sussex IM. J Plant Physiol 126:235, 1986.
87.  Skriver K, Mundy J. Plant Cell 2:503, 1990.
88.  Thomas TL. Plant Cell 5:1401, 1993.
89.  Thompson MM, Smith DC, Burgess JE. Theor Appl Genet 70:687, 1985.
90.  Rodriguez-A J, Sherman WB, Scorza R, Okie WR, Wisniewski M. J Am Soc Hortic Sci, 1994.
91.  King J. The Genetic Basis of Plant Physiological Processes. New York: Oxford University Press, 1991.
92.  Stokes P. Encyc Plant Physiol 15:746, 1965.
93.  Stokes P. J Exp Bot 11:222, 1953.
94.  Steinbauer GP. Plant Physiol 12:813, 1937.
95.  Lakon G. Naturwiss Z Land Fortswirtsch 9:285, 1911.
96.  Villiers TA, Wareing PF. J Exp Bot 15:359, 1964.
97.  Jarvis BC, Frankland B, Cherry JH. Planta 83:257, 1968.
98.  Bradbeer JW, Pinfield NJ. New Phytol 66:5, 1967.
99.  Bradbeer JW, Pinfield NJ. New Phytol 66:515, 1967.
100. Quatrano RS. In: PJ Davies, ed. Plant Hormones and Their Role in Plant Growth and Development. Boston: Martinus Nijhoff, 1987, p 494.
101. Bouvier-Durand M, Dawidowicz-Grzegorsewska A, Thevenot C, Côme D. Can J Bot 62:2308, 1983.
102. Chen SSC, Varner JE. Plant Physiol 46:108, 1970.
103. Lewak S, Rychter A, Zarska-Maciejewska. B. Physiol Veg 13:13, 1975.
104. Alscher-Herman R, Khan AA. Physiol Plant 48:285, 1980.
105. Eichholtz DA, Robitaille HA, Herrmann KM. Plant Physiol 72:750, 1983.
106. Noland TL, Murphy JB. J Plant Physiol 124:1, 1986.
107. Hance BA, Bevington JM. Plant Physiol 96(Suppl):63, Abstract 414, 1991.
108. Li B, Foley ME. Plant Physiol 96(Suppl):63, Abstract 413, 1991.
109. Mahhou A, Dennis FG Jr. J Am Soc Hortic Sci 119:131, 1994.
110. Lang GA, Tao J. HortScience 26:733, Abstract 376, 1991.
111. Callaway AS, PhD thesis, University of Georgia, Athens, 1991.
112. Lin CH, Lee LY, Tseng M-J. Plant Physiol 96(Suppl):404, Abstract 405, 1991.
113. Di Nola L, Mischke CF, Taylorson RB. Plant Physiol 92:427, 1990.
114. Ried JL, Walker-Simmons MK. Plant Physiol 93:663, 1990.
115. Gendraud M. Physiol Veg 15:121, 1977.
116. Thomas F, Thévenot C, Gendraud M. CR Acad Sci 300(Ser 3):409, 1985.
117. Lavarenne S, Champciaux M, Barnola P, Gendraud M. Physiol Veg 20:371, 1982.
118. Champciaux M. Diplôme d'études approfondies, Clermont-Ferrand, France, 1979.
119. Gendraud M, Lafleuriel J. Physiol Veg 21:1125, 1983.
120. Ben Ismail MC. PhD thesis, Faculté des Sciences Agronomiques de Gembloux, Belgium, 1989.
121. Donoho CW Jr, Walker DR. Science 126:1178, 1957.
122. Williams PM, Bradbeer JW, Gaskin P, MacMillan J. Planta 117:101, 1974.
123. Amen RD. Bot Rev 34:1, 1968.

124.  Nikolaeva MG. Physiology of Deep Dormancy in Seeds (translated from the Russian by Z Shapiro). Jerusalem: Israel Program for Scientific Translations, 1969.
125.  Sinska I, Lewak S. Physiol Veg 8:661, 1970.
126.  Webb DP, van Staden J, Wareing PF. J Exp Bot 24:741, 1973.
127.  King RW, Salminen SO, Hill RD, Higgins TJV. Planta 146:249, 1979.
128.  Finkelstein RR, Tenbarge KM, Shumway JE, Crouch ML. Plant Physiol 78:630, 1985.
129.  Sondheimer E, Tzou DS, Galson EC. Plant Physiol 43:1443, 1968.
130.  Jackson GAD. Soc Chem Ind (London) Monogr 31:127, 1968.
131.  Dennis FG Jr, Martin GC, Gaskin P, Macmillan J. J Am Soc Hortic Sci 103:314, 1978.
132.  Berrie AMM, Buller D, Don R, Parker W. Plant Physiol 63:758, 1979.
133.  Rudnicki R. Planta 86:63, 1969.
134.  Ozga JA, Dennis FG Jr. HortScience 26:175, 1991.
135.  Balboa-Zavala O, Dennis FG Jr. J Am Soc Hortic Sci 102:633, 1977.
136.  Braun JW, Khan AA. Plant Physiol 56:731, 1975.
137.  Phillips IDJ, Wareing PF. J Exp Bot 10:504, 1959.
138.  Kawase M. Proc Am Soc Hortic Sci 78:532, 1961.
139.  Ohkuma K, Addicott FT, Smith DE, Thiessen WE. Tetrahedron Lett 29:2529, 1965.
140.  Cornforth JW, Milborrow BV, Ryback G, Wareing PF. Nature 205:1269, 1965.
141.  Lenton JR, Perry VM, Saunders PF. Planta 106:13, 1972.
142.  Powell LE. HortScience 11:498, 1976.
143.  Powell LE, Seeley SD. HortScience 5:327, 1970.
144.  Seeley SD. PhD thesis, Cornell University, Ithaca, NY, 1971.
145.  Coleman WK, King RR. Am Potato J 61:437, 1984.
146.  Medford JI, Horgan R, El-Sawi Z, Klee HJ. Plant Cell 1:403, 1989.
147.  Smigocki AC. Plant Mol Biol 16:105, 1991.

## ADDENDUM

The review of literature for this chapter was completed in the early 1990s, and the information has not been updated. In the intervening years, major advances have been made toward a better understanding of the mechanisms controlling dormancy, particularly in the area of molecular biology. At least two symposia have been devoted entirely to dormancy (Lang, 1996; Viemont and Crabe, 2000), and the many papers in the published proceedings are valuable contributions to the literature. In addition, several book chapters and reviews have appeared (e.g., Crabbe, 1994; Faust, et al., 1997; Fuchigami and Wisniewski, 1997; Li and Foley, 1997).

Lang's [18] review (see above) of molecular approaches to studying dormancy is a source of references to early work in this area. More recent work has dealt with gene insertion into woody plants (Rhode, et al., 1996), as well as herbaceous ones. Genetic studies have also provided new insights, including those of Foley and Fennimore (1998), of Chen, et al. (2000) with *Populus*, and of Koornneef, et al. (2000) with *Arabidopsis*. Numerous papers on methods of modeling the effects of environmental factors that regulate the breaking of dormancy have also appeared, e.g., Seeley (1996), and Haakinen (1999).

Some of these publications are listed below to supplement those in the original list of references.

1.  Chen THH, Davis J, Frewen BE, Howe GT, Bradshaw HD Jr. In J-D Viemont, J Crabbe, eds. Dormancy in Plants: From Whole Plant Behaviour to Cellular Control, Wallingford, Oxon, UK: CABI Publishing, 2000, p 319.
2.  Crabbe J. Encyclop Agr Sci 1:597, 1994.
3.  Faust M, Erez A, Rowland LJ, Wang SY, Norman HA. HortScience 32:623, 1997.
4.  Foley ME, Fennimore SA. Seed Sci Res 8:173, 1998.
5.  Fuchigami LH, Wisniewski M. HortScience 32:618, 1997.
6.  Geneve RL. Seed Technol 20:236, 1998.
7.  Haakinen R. Tree Physiol 19:613, 1999.
8.  Koornneef M, Alonso-Blanco A, Bentsink L, Blankestijn-de Vries H, Debeaujon I, Hanhart CJ, Leon-Kloosterziel KM, Peeters AJM, Raz V. In Viemont J-D and Crabbe J, eds. Dormancy in Plants: From Whole Plant Behaviour to Cellular Control. Wallingford, Oxon, UK: CABI Publishing, 2000, p 365.
9.  Lang GA, ed. Plant Dormancy: Physiology, Biochemistry and Molecular Biology. Wallingford, Oxon, UK: CABI Publishing, 1996.
10.  Rhode A, van Montagu M, Boerjan W. For Sci, 49:183, 1996.
11.  Li B, Foley M. Trends in Plant Sci 2:384, 1997.
12.  Seeley SD. In GA Lang, ed. Plant dormancy: Physiology, Biochemistry, and Molecular Biology. Wallingford, Oxon, UK: CABI Publishing, 1996, p 361.
13.  Viemont J-D, Crabbe J, eds. Dormancy in Plants: From Whole Plant Behaviour to Cellular Control. Wallingford, Oxon, UK: CABI Publishing, 2000.

# 9
# Senescence in Plants and Crops

## Lola Peñarrubia and Joaquín Moreno

*University of Valencia, Burjassot, Valencia, Spain*

## I. INTRODUCTION

*Senescence* has been defined as the endogenously controlled deteriorative changes that are natural causes of death in cells, tissues, organs, or organisms [1]. The differences from the term *aging* are well established, aging being all the degenerative changes that occur in time without reference to death as a consequence. Aging takes place during the entire life span of an organism, whereas senescence is considered the final developmental phase that culminates in death [1–3]. On the other hand, *apoptosis* or *programmed cell death* refers to an intracellular process by which the cell promotes a set of genetically determined self-destructive activities (including specific proteolysis and nuclear chromatin fragmentation) leading to its own death (i.c., dcath results from internal activity of the cell and not from ambient injuries). It is now accepted that many features of stress response and senescence at the cellular and molecular level are achieved through the operation of programmed cell death [4].

Senescence is a natural developmental process that may be considered as terminal differentiation because it usually takes place at the end of the life cycle of an organ or organism. However, different kinds of environmental stress, as well as pathogenesis (i.e., biotic stress), can induce senescence at any stage of the plant life cycle [5]. In general, the main biochemical changes associated with stress-induced senescence are almost identical to those of natural senescence. Accordingly, gene expression patterns are frequently coincident [6–8], or differ at the relative levels of isoenzymatic activities [9,10], but some particular genes may display specific expression in senescence processes induced by different factors (reviewed in Ref. 11). In any case, this differential expression does not usually result in significant variation at the physiological level. This may be due to the fact that all senescence responses result from triggering the same adaptive mechanisms (most of them aimed at the mobilization and transport of nutrients out of the decaying tissues) that are constitutively present in plants. These induced adaptive responses (collectively known as the *senescence syndrome*) are the hallmark of senescence, whatever the circumstances (aging, stress, or pathogen attack) that originated them.

Even if senescence is essentially a degenerative process, it is far from being a chaotic breakdown. On the contrary, senescence occurs as an orderly loss of functions and structures, comprising an array of biochemical and physiological processes whose ultimate goal is the efficient removal of nutrients from the decaying tissues. The sequence of events constitutes the senescence syndrome and includes the turnover of macromolecules and lipids and the transport of mobilized nutrients out of the senescing structures toward other parts of the plant, be these either growing organs, such as fruits or young leaves, or specialized storage tissues, such as the bark of deciduous trees. In this regard, it is not paradoxical that senescence promotes the rise of both degradative and protective enzymatic activities because the ordered dismantling needed for optimal exploitation of nutrients requires both specific degradation and protection against uncontrolled agents, an unavoidable by-product of breakdown.

Fruit ripening is another physiological process that is usually associated with senescence because it shares with truly senescent processes several metabolic features, especially the dismantling of chloroplastic components and structures. However, fruit ripening also has many particular metabolic characteristics, and its final goal as a physiological process is different from that of senescence, being the development of physicochemical and organoleptic properties that facilitate the spreading of seeds. This teleological difference is reflected in the fact that fruits usually continue to act as a sink of nutrients during ripening, whereas other senescing organs behave as a source. We will not consider the particular features of fruit ripening in this chapter.

Despite the inherent diversity of senescence, three broad phases or stages may be distinguished in a typical senescent process. First, there is a phase of selective degradation of certain molecules whose lysis does not cause a major impairment of the physiological function of the senescent structure. Therefore, the mobilized molecules may be thought as nutrient storage materials, and this stage may be termed *storage mobilization*. In some cases, senescence may be reversed during this phase by suitable changes in the environmental conditions. The second stage is characterized by the extension and generalization of breakdown to components that are central in maintaining the physiological function, which is consequently lost. Somewhere along this phase, which might be called *generalized breakdown*, the senescent process becomes irreversible and the cells are definitively targeted to death. Finally, once the senescent structure has been emptied of profitable nutrients, there is a third stage of *abscission* (i.e., shedding of the senescent part from the rest of the plant) and death. Abscission, a biochemically and physiologically complex process, is studied in another chapter and will not be discussed further here. Although exceptions or overlapping of stages may be found in many particular senescent processes, the preceding three-phase scheme may serve as a developmental outline that emphasizes the strategy of senescence.

Senescence of crop plants is of special interest because it encompasses phenomena of economic importance that occur both in the field and during storage and handling of plant products of commercial value. Moreover, on the basis of the current knowledge of the genetic control of senescent processes, it is already possible to manipulate several features of senescence, using recombinant DNA technology, for improving the quality of the crops. Tomatoes that are bruise resistant [12,13] or that do not overripen [14,15], soybean plants with increased seed yield [16], and tobacco with higher biomass [17] due to delayed senescence are among the first offspring of this approach that will surely bring great benefits to agriculture in the near future.

The aim of this chapter is to provide an outline of the natural patterns and features of senescence, with special emphasis on the control and development of the senescence process at the molecular level and the experimental approaches to its study. It is intended not to be exhaustive but rather representative of the current trends in the field, and it is specially devoted to crop plants. The reader interested in more detailed physiological information is referred to the excellent book by Noodén and Leopold [18].

## II.  PATTERNS OF SENESCENCE IN THE LIFE CYCLE OF PLANTS

Senescence is a very pervasive phenomenon that may be encountered in all plants and at all stages of the life cycle, related to developmental as well as adaptive functions. It shows a variety of patterns ranging from death of specific cells to the decline of the entire plant. Patterns of plant senescence may be illustrated with two extreme behaviors. There are plants, such as trees, in which survival of the individual over a long period (including several reproductive phases) is the most important commitment. These plants (termed *polycarpic*) usually undergo a periodic senescence, which is limited to older organs and combined with the growth of young ones. Other plants (e.g., the annuals) sustain only one reproductive phase and die with the development of fruits. The latter species (called *monocarpic*), in which individuals are subordinated to survival of the population, develop whole plant senescence directed to mobilization of nutrients to the growing fruits. Between these extremes, there is a broad spectrum of life cycles displaying intermediate senescence strategies. Nevertheless, the major food crops are typically monocarpic, storing a high proportion of the plant biomass in the harvestable fraction (fruits or seeds).

Leopold [1] distinguishes four possible patterns of senescence in different plants:

1.  Senescence of the whole plant at the end of the reproductive phase, which is typical of monocarpic plants.
2.  Senescence of the aerial parts of the plant maintaining the underground structures. Bulbs belong to this group.
3.  Senescence of the leaves only, the stems and roots remaining alive. This is the case with deciduous plants.
4.  Progressive senescence of the leaves and other organs along the stem beginning at the base. Annual plants usually senesce in this way.

Another way to categorize senescence patterns is on the basis of the structural level (i.e., cellular, tissue, organ, or organism level) at which they act, as Noodén [3] has proposed. In this regard, senescence is sometimes restricted to a single cell or a small number of cells within a tissue as, for example, in the formation of the root aerenchyma or the xylem tracheids [3]. Cellular senescence could serve either to clear away the cell in order to create a hollow structure (as in the former cases) or to nurture the neighboring cells with the breakdown products. The most spectacular examples of the latter strategy take place in pollinated ovaries during the first stages of embryo development, involving senescence of specific cells in an orderly sequence of events [3,19]. In all these cases, cellular death is clearly a result of an internal developmental program and is properly termed apoptosis.

Frequently, senescence affects whole organs of a plant. Although the global nutritional balance of the plant usually determines the onset of senescence of mature organs, these behave afterward as independent structures establishing internal (spatial and temporal) patterns of nutrient mobilization before abscission. Leaves, flowers, and fruits are the most studied and better known instances.

Leaves may follow different patterns of senescence. For example, in trees the oldest leaves may decline when the new leaves are growing as part of a progressive senescence; or all the leaves may senesce together seasonally. In the first case, the breakdown products serve directly as a nitrogen source for the new organs [20]; in the second case, the nutrients are stored in the branches waiting for the next growing period [21]. In any case, yellowing (due to destruction of photosynthetic pigments) is a conspicuous feature of leaf senescence, evidencing that chloroplasts are early and sensitive targets of the dismantling activities that take place during the senescence of green (i.e., photosynthetic) tissues.

Flowers usually have the shortest longevity among plant organs [22]. Flower parts such as calyx, perianth, androecium, gynoecium, and peduncle are interrelated but differ from each other in both structure and physiology. Senescence of flowers is a representative example of a kind where several components decay tightly enchained to growth and development of other structures. Pollination is a central event in flower development. Some parts of flowers, such as perianth, senesce after pollination, whereas others, such as ovaries, develop. Changes originated by pollination, collectively termed the *pollination syndrome*, include a number of developmental processes, such as perianth pigmentation changes, ovary maturation, and ovule differentiation, which are crucial to ensure fertilization and embryogenesis [23]. The signal brought forth by pollination may be a requisite for perianth senescence. The most spectacular cases are orchids, whose unpollinated flowers can stay fresh for 6 months waiting for a specific insect to be fecundated. However, factors other than pollination could cause flower senescence in other species, possibly as a result of regulation by an endogenous clock [24].

Ovaries senesce naturally if not stimulated either by pollination or by hormonal induction of parthenocarpic fruits. The most spectacular modifications during fruit ripening probably occur at the plastids and cell wall, where important changes in structure and composition take place [25]. Fruit softening is a consequence of the induction of specific cell wall hydrolases [26]. Besides, the ripening-associated color changes of fruits are a result of the transition from chloroplast to chromoplasts that are rich in red or yellow carotenoid pigments [27]. In most fruits, there is also a decrease in acidity during ripening, as well as an increase in sweetness in addition to changes in aroma produced by volatile odorant compounds [28].

Whole plant senescence, a typical feature of monocarpic plants, is characterized by a general mobilization of nutrients that are transported to the fruits, and especially to the seeds, which act as a strong sink. The molecular nature of the hormonal effectors that are responsible for whole plant senescence is controversial. Even if senescence and death are internally programmed, elimination of flowers and fruits may delay senescence of the whole plant. This fact has been interpreted in diverse ways, the most accepted idea being that hormonal factors produced by flowers and fruits affect the levels of signals responsible for senescence in the rest of the plant [2]. There is also evidence of roots playing a role as a source of growth substances during senescence of fruiting plants. Thus, simultaneous senescence of all plant parts is probably a result of interorgan signaling.

## III.  EXPERIMENTAL SYSTEMS TO STUDY SENESCENCE

The importance of having appropriate experimental systems and controls deserves special attention. Systems to study senescence have to be well defined and as close as possible to the natural environmental conditions of the plants under study. In this sense, intact plants are the best experimental systems and should be used whenever possible. Moreover, environmental conditions need to be under careful control. Usually, plant senescence experiments are carried out under optimal greenhouse or growth chamber conditions. However, stresses due to field suboptimal supplies of water and mineral nutrients, extreme temperatures, salt, ultraviolet (UV) light, and wounding may heavily modify the natural senescence behavior and should be taken into account. Light conditions are also crucial. For example, chlorophyll breakdown is strongly retarded by continuous illumination when compared with leaves kept in the dark. Even light quality (in relation to red or far-red components) is relevant because of the participation of phytochrome in light-mediated responses during senescence [29,30].

### A.  Nonintact Plants

Senescence studies of bulky plants, such as trees, under controlled laboratory conditions may require the use of simpler systems. The use of excised plant parts often produces important physiological and biochemical changes, especially in the detached organ but also in the rest of the plant. If the use of a nonintact plant system is unavoidable, precautions should be taken in analyzing the results because of the correlative nature of many plant processes. As information grows, there is more and more evidence of interrelation between all plant parts during senescence processes, probably through hormonal crosstalk [31]. Thus, it is always necessary to contrast the observations with the changes undergone by intact organs.

On the other hand, in testing the influence of some plant parts on the senescence of the remaining ones, a common procedure consists of the surgical elimination of these parts. Usually, the replacement of the excised organ with an agar block containing the putative hormonal effector or another type of exogenous application completes the experiment [32]. The conclusions that can be extracted from this class of work are limited. Exogenous application of any biological material is subjected to strong limitations such as uptake, sequestration, transport, and metabolism of the active material and the difficulty of quantitating the amount of it within the target tissue [33].

### B.  Mutants and Transgenic Plants

Currently, the use of intact (but genetically modified) plants is making it possible to dissect senescence without interferences from other processes. Therefore, mutants and transgenic plants are becoming widely accepted as the best tools for this purpose.

Once a natural or induced mutation affecting senescence has been spotted, the next goal is usually to identify the mutated gene and to characterize the phenotype at the molecular level in order to evaluate its functional role in the senescence process. In this regard, current protocols that use map-based cloning, large-scale complementary DNA (cDNA) sequencing, and polymerase chain reaction (PCR)-based methods for screening DNA insertion tagged populations of plants genetically engineered with transposons [34] or T-DNA from *Agrobacterium tumefaciens* [35,36] facilitate the rescue of the mutated gene (for a review see Ref. 37). Besides, in *Arabidopsis thaliana*, DNA microarrays (*chips*) to detect changes in gene expression [38] and seed collections of plants carrying deletions of single genes (*knockouts*) [39] are commercially available for identifying relevant genes and characterizing their roles in plant senescence. For convenience, most work on characterization of senescence-related genes by mutant analysis has been done in the small-genome weed *A. thaliana*, but this first step usually leads (or will lead in due time) to the characterization and eventual manipulation of homologous genes from crop plants.

As exhaustive mutational analysis in *Arabidopsis* has failed to provide any mutants that globally and completely block the senescence process [40,41], it appears that no single nonlethal gene regulates senescence as a whole. However, several mutations partially altering senescence or some aspects of the process have been described in *Arabidopsis* and other plants (reviewed in Ref. 42). Most of the mutations analyzed to date are varieties from natural cultivars of the "stay-green" type [42,43]. A mutation in a nuclear gene of *Festuca pratensis* results in a drastic reduction of chlorophyll loss without other senescence characteristics (such as decrease of protein and RNA content, rise of proteolytic activities, or degradation of plastid structure) being altered [44–46]. The mutated gene controls the thylakoid membrane disassembly in senescent leaves, impairing the degradation of thylakoid pigments, protein, and lipids [47]. It has been shown that this mutant is unable to carry out oxygenolytic cleavage of the porphyrin macrocycle [48]. Mutations in nuclear and organular genes altering chlorophyll loss or gas exchange during monocarpic senescence have also been described in soybean [49]. Plants mutated in two loci (*d*1 and *d*2) experience a significant delay in degradation of soluble proteins [50], plasma membrane, and chloroplasts [51]. Besides, a mutation that affects the light regulation of seedling development has been shown to interfere with the onset of leaf senescence [52].

Mutants that seem unable to control the rate and extent of cell death when exposed to different senescence-inducing agents have been described [53]. These mutants (*acd*1) exhibit accelerated cell death with rapid spreading of necrotic lesions in response to virulent and avirulent pathogens but also during aging of aseptic plants. Because these lesions are characteristic of the so-called *hypersensitive defense response* to pathogens, analysis of the mutants may provide an understanding at the molecular level of this response and its relationship to natural senescence.

The progress achieved in plant transformation techniques, and especially the easy *Agrobacterium tumefaciens*–mediated protocols for transforming *Arabidopsis* [54], has made transgenic plants a common tool. Once a senescence-related gene has been cloned, the function and physiological relevance of the gene product may be tested in transgenic plants. The most common strategy consists of the modification of a target endogenous protein level through either overexpression (introducing new copies of the gene under strong promoters) or decreasing its transcription by antisense technology. Similar procedures are employed for introducing heterologous genes under the control of endogenous promoters (or other suspected regulatory sequences) in host plants in order to test the effect of subtle manipulations at the promoter without an endogenous background of the reporter gene or with the aim of introducing novel and agriculturally desirable properties in transgenic crops. As an example (discussed in detail in the following), the use of transgenic plants whose hormone metabolism has been altered (thereby modifying hormonal levels endogenously) has shed light on the role of certain phytohormones, such as ethylene and cytokinin, thereby suggesting successful strategies for manipulating senescence.

## IV.   ULTRASTRUCTURAL, PHYSIOLOGICAL, AND BIOCHEMICAL CHANGES DURING SENESCENCE

### A.   Ultrastructural Changes

Characteristic changes ocurring during senescence in different plants share common features at the ultrastructural level. In green organs, chloroplasts are the organelles in which the first symptoms of senescence are observable. Following an ordered sequence of events, the chloroplast dismantling begins with swelling, unstacking, and degradation of thylakoids (first those of the lamellae, then the grana), appearance of lipid droplets and plastoglobuli, and finally fragmentation of the envelope [55]. In some cases, chloroplasts have been observed to fuse with vacuoles at the late stages of senescence [56]. The number and size of chloroplasts are reduced during senescence, and the rate of oxygen evolution decreases approximately in parallel with the chloroplast content [57]. Loss of starch is also characteristic of senescence and may result in deformation of the cells. This may explain the distortion of endocarp and mesocarp cells observed in senescent ovaries of pea [56].

Some extraplastidic membranes, such as those of the endoplasmic reticulum, also undergo early degradation, the smooth and rough fractions being degraded nonsimultaneously depending on the species [58]. Changes in the properties of lipid phase have been observed in senescing membranes using wide-angle x-ray diffraction and freeze-fracture electron microscopy [59]. Regions of the lipid bilayer switch from liquid crystalline to gel phase, rendering leaky membranes. Freeze-fracture electron microscopy shows the gel-phase domains as intramembranous particle-free regions that increase in size and number as senescence progresses [60].

At the nucleus, senescence causes a progressive condensation of the chromatin that is detectable through the fluorescence decrease of propidium iodide–stained nuclei. Condensation begins at early stages and proceeds until it becomes irreversible after the endonucleolytic fragmentation that is typical of late apoptotic processes [61].

In contrast, there are other structures, such as mitochondria, that remain intact until later phases, when some swelling or distortion of cristae becomes apparent. The plasmalemma integrity is also maintained until the final stages. Cells become progressively more vacuolate with age, and changes in the permeability of the tonoplast membrane, surrounding the vacuole, could allow the transfer of cytoplasmic material into the vacuole, favoring its degradation. Autophagic processes in which organelles become engulfed in vacuole-like structures have been observed [56]. In some cases, tonoplast rupture may cause the lysis of cells at very late phases [58,62]. With differences depending on cell type, these changes proceed sequentially until the whole cell is dismantled. For example, a study of the senescence of mesophyll cells of rice coleoptiles [63] has shown that all cells follow precisely the same temporal sequence of events, consisting of (1) degradation of chloroplast DNA, (2) condensation of the nucleus and decrease in the size of the chloroplast with degradation of ribulose-1,5-bisphosphate carboxylase/oxygenase and chloroplast inner membranes, (3) disorganization of the nucleus, and (4) complete loss of cellular components and distortion of the cell wall.

## B.  Physiological Changes

### 1.  Leaf Conductance and CO₂ Assimilation

Senescence produces closure of stomata leading to a decline in transpiration. It has been suggested that the stomata aperture may control the rate of leaf senescence [64]. Because stomata are the main sites of entrance for $CO_2$, it might be speculated that insufficient $CO_2$ supply could be the cause of the decreased photosynthetic assimilation observed during senescence. However, experimental measurement of $CO_2$ concentration in the substomatal cavity [65] suggested that $CO_2$ does not limit photosynthetic assimilation. Hence, stomatal closure may be more a consequence than a cause of lowered photosynthetic activity according to the optimal variation hypothesis, which proposes that stomatal conductance adapts to the photosynthetic capacity of the leaf [66]. It is remarkable that stomatal guard cells remain functional until very late stages of senescence, far beyond other leaf cells. This may be a result of their lack of symplastic connection with the surrounding cells [67].

### 2.  Respiration

Changes in respiratory rate of senescing fruits have been known for a long time. In detached apple fruits, the respiration rate decreases gradually until a sudden burst, termed *climacteric*, followed by a further decline in respiratory activity is observed. Fruits are divided into two categories, climacteric and nonclimacteric, depending on whether or not their respiration shows a sudden peak. In detached leaves and cut flowers, a climacteric-like rise in respiration has also been observed during senescence in some species but not in others [68]. The common metabolic feature in climacteric fruits is their ability to produce and respond to ethylene. It appears that the rise in respiration is a consequence of ethylene action and not of senescence as such. The main reason for this conclusion is that inhibition of both the biosynthesis and action of ethylene eliminates the rise in respiration without preventing eventual senescence. Besides, ethylene treatment enhances respiration of nonripening tomato mutants but does not promote the typical changes associated with ripening. It appears that ethylene enhances plant respiration by activating a preexistent enzymatic potential [68].

Respiratory pathways of senescent plants include glycolysis, the pentose pathway, tricarboxylic acid (TCA) cycle, and the electron transport pathway, in which some changes have been described [68], but also an alternative oxidase pathway that is enhanced during senescence [69]. In aging potato tuber slices the alternative oxidase has been characterized as an integral membrane protein synthetized de novo [70]. It has been suggested that the alternative pathway is activated when the cytochrome pathway is saturated or limited, allowing the TCA cycle to function using up excess carbohydrates [71]. In addition, a plant uncoupling mitochondrial protein (PUMP) has been shown to be induced by low-temperature stress and aging in potato tubers [72]. PUMP is homologous to the mammalian uncoupling protein of brown adipose tissue mitochondria and apparently uses the same mechanism of proton translocation via fatty acid recycling [73], thereby regulating respiratory uncoupling and thermogenesis during senescence and fruit ripening.

## C.  Biochemical Changes

### 1.  *Photosynthetic Pigments*

Color changes are important criteria for the visual evaluation of the advance of senescence, especially in fruits [27]. Breakdown of chlorophylls may be one of the earliest symptoms of senescence. However, chlorophyll decline is strongly retarded by continuous illumination in a process regulated by phytochrome [30]. The chlorophyll *a*/*b* ratio has been shown to decline with the advance of senescence [74,75], probably as a result of the nonsynchronous dismantling of lamellae and grana thylakoids and the asymmetrical distribution of photosystems between them. Carotenoids are lost at a much lower rate than chlorophylls [76]. This difference in degradation rate accounts for most of the color changes associated with leaf senescence and may reflect the persistence of the photoprotective role of carotenoids until later phases of the process. In ripening fruits, senescence is sometimes associated with de novo synthesis of both carotenoids and anthocyanins [27].

The pathway of chlorophyll degradation may be distinct for different species or even organs. In senescing barley leaves, chlorophyll *b* reduction seems to be the first and obligatory step of chlorophyll *b* breakdown [77]. This step is carried out by chlorophyll(ide) *b* reductase, a thylakoidal enzymatic activity that peaks earlier (day 2) than chlorophyllase (day 4) during dark-induced leaf senescence [77]. The activity of chlorophyllase, a thylakoidal enzyme that hydrolyzes the phytyl ester group, has been shown to correlate with chlorophyll degradation in maturing citrus fruits [78]. In *Arabidopsis thaliana*, two genes for chlorophyllase (*AtCLH1* and *AtCLH2*) have been characterized [79]. *AtCLH2* encodes a protein with a typical chloroplast targeting sequence, while *ATCLH1* has no defined location signal. The expression of *AtCLH1* (but curiously not of *AtCLH2*) is strongly induced by methyljasmonate, which is known to induce plant senescence and chlorophyll degradation [79]. Chlorophyll oxidase, a complex enzymatic system that renders chlorophyll *a*1 as a first step, may also be involved in chlorophyll turnover. The fact that chloroplast chlorophyllase is hindered by its membrane localization and that chlorophyll oxidase activity is dependent on free fatty acids, liberated by lipid hydrolysis, may provide a link between thylakoidal membrane dismantling and chlorophyll degradation inside the chloroplast [78,80]. Alternatively, the peroxidase–hydrogen peroxide pathway, which opens the chlorophyll porphyrin ring, can also be involved and has been shown to be the main catabolic pathway in detached spinach leaves [81]. In addition, direct photodamage represents a contribution, even if minor, to chlorophyll breakdown [55].

The existence of a *Festuca pratensis* mutant that does not exhibit chlorophyll loss during senescence supports the conclusion that the associated decline in photosynthetic capacity is not a result of chlorophyll breakdown. Studies of the proteins that are abnormally retained in the mutant indicate that all of them possess an associated tetrapyrrole prosthetic group (heme or chlorophyll) [46]. As it has been proposed that the degradation of porphyrins and their associated apoproteins is correlated [82], a lesion in the heme/chlorophyll catabolic pathway may be responsible for the phenotype of the mutant [46].

### 2.  *Nucleic Acids*

Nucleic acids are a rich source of nutrients (especially phosphate), which ought to be mobilized and exported from senescing organs before abscission. On the other hand, the genetic information stored in DNA should be preserved throughout the decay process because the completion of the senescence program requires the uninterrupted synthesis of specific messenger RNAs (mRNAs) until very late stages. Thus, the amount and integrity of DNA are usually maintained in senescent cells until the late phase of chromatin fragmentation. A decrease in nuclear DNA (about 20%) has been described at the final stages of senescence in soybean cotyledons [83] as well as tobacco and peanut leaves [84]. It has been shown that repeated sequences are selectively degraded while coding regions of nuclear DNA remain largely intact [85,86].

In general, senescence is a process of overall decline in RNA and protein synthesis, especially in the chloroplast [87]. Accordingly, total RNA has been found to be around 10-fold lower in a yellow leaf compared with a green one [88,89]. However, a nonspecific decline in RNA synthesis does not cause senescence and, furthermore, selective synthesis of specific mRNAs seems necessary for the progress of senescence (see later). The quantitative decline in RNA is explained mainly by the decrease in ribosomal RNA (rRNA), which is the most abundant cellular RNA in both the chloroplast and cytosol, and a major fraction of mRNAs that decay in parallel with total RNA maintaining their relative abundance [89]. This decrease correlates with the decline in protein synthesis. Variations in relative amounts of two phenylalanyl transfer RNAs (tRNAs) have been detected during senescence [90], and tRNA synthase activities are greatly reduced, probably limiting the translational capacity of senescing chloroplasts [91]. Ribosome-inactivating proteins have also been shown to increase in naturally senescent and stressed leaves [92].

A general increase in ribonuclease (RNase) activities, described in senescing tissues [93,94], could account for the generalized loss of RNA. Moreover, some of these activities are known to be selectively induced during senescence processes [95–97]. Free nucleotides produced as a result of RNase activity may be further metabolized to release phosphate. The fate of nitrogen bases resulting from nucleic acid catabolism is still unclear, although their degradation could take place in peroxisomes [98,99].

## 3. Intracellular Proteins

In terms of total protein content, leaf senescence is characterized by a progressive loss of proteins [85]. This loss may be attenuated if additional nitrogen is supplied to the plant [100] or sink organs are removed. In general, the demand for mineral nutrients by growing structures has been described as a regulatory factor in leaf senescence, except for phosphorus nutrition, which does not show any regulatory control on the process [101].

The patterns of protein loss are characteristic and independent of the cause of senescence. A wide range of specific proteins are degraded while others remain intact. In green organs, chloroplast proteins are principal targets of degradation during early phases of senescence. The loss of chlorophyll correlates with degradation of chlorophyll-carrying thylakoidal proteins, whose lysis is strongly retarded by continuous illumination. However, stromal proteins rapidly disappear under the same conditions, indicating that breakdown of membrane and soluble proteins is differently regulated by light [30].

The most abundant soluble protein in chloroplasts, ribulose-1,5-bisphosphate carboxylase/oxygenase (Rubisco), represents more than 50% of the chloroplast nitrogen and about 25% of that of the whole cell [102]. Rubisco is known to be extensively and selectively degraded at early stages of senescence in many plants [20,21,103–106]. The specific proteolysis of this enzyme accounts for up to 85% of the soluble protein lost in senescing barley leaves [107] and more than 90% of the nitrogen mobilized from leaves before abscission in apple trees [21]. Experiments with transgenic plants in which the level of Rubisco was decreased using antisense technology have firmly established the natural excess of Rubisco over the amount needed for performing its catalytic function and the correlation of the amount of enzyme with the nitrogen status of the plants [108]. Evidence for this "luxury" excess, together with the spectacular contribution of this enzyme to nutrient mobilization during senescence, supports the concept of Rubisco as a nitrogen storage protein [108,109].

Protein turnover is a common feature in every living organism; therefore, the global decrease observed during senescence has to be considered as an imbalance between the rates of protein synthesis and degradation [110]. Probably, both reduced synthesis and enhanced proteolysis are responsible for the protein loss associated with senescence. In this regard, synthesis of all thylakoidal proteins is known to be severely curtailed in senescing bean leaves, except for the D-1 protein of photosystem II [111]. On the other hand, increased protein breakdown may result from different mechanisms: de novo synthesis of proteolytic enzymes, activation of preexisting proteases, decompartmentalization of proteases and their substrates, or making the protein substrates susceptible to degradation.

References on increased proteolytic activities during senescence are abundant (reviewed in Refs. 2, 112–116); however, they more frequently report enhanced levels of preexistent proteases than the appearance of new activities specific to senescence [113]. Characteristically, increased expression of cysteine (or thiol) proteinases (which are typical apoptotic agents) has been associated with senescence of different flower parts such as tepals [117], petals [118], and ovaries [119,120] as well as with other developmental events that include programmed cell death, such as xylogenesis [121,122]. Moreover, levels of cysteine protease mRNAs have been shown to increase during leaf senescence of tomato [123], and cysteine proteases are known to be induced in suspension-cultured soybean cells by oxidative treatments that produce programmed cell death [124]. Although cysteine proteases involved in animal apoptosis hydrolyze peptidic bonds at aspartic acid residues (thus called *caspases*), it appears that plant enzymes do not follow that rule [124]. Increased levels of proteases have been reported to occur in pea leaf peroxisomes [125] and lytic vacuoles of vegetative tissues in *Arabidopsis* [126] during senescence. However, the implication of vacuolar or peroxisomal activities in hydrolysis of proteins from other organelles seems doubtful, at least throughout the first stages of senescence, when the compartmental integrity of the cell has been firmly established. Nevertheless, these compartmental barriers may be progressively overridden throughout senescence. For example, it has been shown in apoptotic animal cells that the activity of 15-lipoxygenase renders the membranes of organelles leaky to proteins, thus allowing a certain degree of protease exchange between cellular compartments [127]. In any case, the fusion of organelles with vacuoles during the final phases of cell death [56] surely results in those activities gaining access to the remaining substrates and producing the ultimate breakdown of the organellar structure. Similar compartmental restrictions may limit the activity of the proteasome, which is present only in nucleus and cytoplasm of intact plant cells. Although some evidence of proteasome processing of ubiquitin-labeled proteins during plant senescence has been gathered (reviewed in Ref. 128), expression of the proteasome $\alpha$-type subunit has been shown to decrease during flower and leaf senescence in tobacco, suggesting that the plant proteasome is likely to play a regulatory role in developing tissues rather than be involved in massive senescent degradation [129].

In instances in which no correlation between senescence and increased proteolytic activities has been found [112], the loss of protein during senescence may be due to natural turnover after arrest of synthesis or to modification of the proteins that may label them for proteolysis. In this regard, the case of the chloroplastic $CO_2$-fixing enzyme, Rubisco, appears to be paradigmatic. In some species, such as corn and wheat, the enzyme seems to be degraded as a result of natural turnover [2,130]. However, in most species the turnover rate of Rubisco is negligible under nonsenescent conditions [131,132] but changes dramatically with the onset of senescence. It has been shown that the susceptibility of Rubisco to proteases increases markedly through oxidation of sulfhydryl groups belonging to critical cysteine residues of the enzyme [133–135]. This suggests that Rubisco degradation may be induced by the oxidative conditions developed inside the functionally impaired chloroplast during senescence [136]. Evidence for in vivo oxidation of Rubisco has been found in different organisms under stress-induced senescence [137–139]. Moreover, a chloroplastic proteolytic activity that is activated by oxidative conditions has been described [140]. This suggests that alteration of the redox state of the chloroplast may provide a general mechanism for triggering a selective protein degradation during senescence or other processes that arrest chloroplast function.

Certain enzymatic activities other than hydrolases are also known to be enhanced during senescence. For example, transition of leaf peroxisomes to glyoxisomes is a well-characterized phenomenon associated with senescence. Accordingly, enzymatic activities of markers of the glyoxylate cycle have been shown to increase dramatically during darkness-induced senescence of spinach leaves [141] and pumpkin cotyledons [142]. Cytosolic glutamine synthetase also increases about fourfold during senescence of rice leaves [143]. This rise surely facilitates the mobilization of nitrogen by enhancing the synthesis of the major transported amino acid (glutamine) [144]. Interestingly, transgenic overexpression of glutamine synthetase leads to accelerated development and early senescence of plants grown in ammonium-rich medium [145]. In addition, different isozymes of threonine dehydratase, an enzyme that probably plays a role in nitrogen remobilization, are specifically synthesized in senescing tomato leaves [146].

## 4. Free Radicals and Antioxidant Enzymes

The relationship between the internal production of free radicals and senescence is well established [147]. Free radicals derived from oxygen such as superoxide, hydroxyl, peroxyl, and alkoxy radicals, as well as other molecular forms of incompletely reduced oxygen, are known as reactive oxygen intermediates (ROIs). ROIs are considered both as primary mediators of oxidative damage during senescence and as signals that trigger cellular defensive responses. In particular, superoxide has been shown to mediate the spreading of the hypersensitive response (leading to cell death) in mutants of *Arabidopsis thaliana* that exhibit spontaneous lesions [148]. ROIs may be generated as by-products of some enzymatic reactions, but plants specifically produce these radicals as a consequence of photosynthesis [149]. Production of ROIs is enhanced during late stages of senescence because of the impairment of the electron flow between the two photosystems, which limits the availability of photosynthetic power [150]. Free radicals are known to induce the breakdown of nucleic acids, polysaccharides, and proteins [151] and enhance the ethylene production pathway [152]. In addition, ROIs and other free radicals may initiate a chain reaction in membranes leading to extensive lipid peroxidation and subsequent alterations of fluidity and permeability [147]. Under normal conditions, oxygen-detoxifying enzymes, such as superoxide dismutases, catalases, peroxidases, and glutathione reductases, are present in plant cells to prevent the damage by these toxic species [153–156]. Superoxide is eliminated by the superoxide dismutase, which produces hydrogen peroxide. Catalases convert hydrogen peroxide into water and oxygen, whereas peroxidases reduce hydrogen peroxide to water and oxidize a variety of substrates. Ascorbate peroxidase is a hydrogen peroxide–scavenging enzyme that is specific to higher plants and algae [157]. This enzyme protects chloroplasts and other cell constituents from damage by hydrogen peroxide and derived hydroxyl radicals. Ascorbate peroxidase, glutathione reductase, and dehydroascorbate reductase remove hydrogen peroxide through a pathway termed *photoscavenging* [154]. Nevertheless, in certain cases some of these activities decrease with the progression of senescence, thereby favoring a parallel rise of ROIs [158,159]. In other instances, the soluble activity of protective enzymes decreases but there is an increase of wall-bound activity. This is the case for peroxidase during senescence of stigmas and styles in *Citrus* [160]. It has been shown that the reduction of ascorbate contents in transgenic potato plants by antisense inhibition of the GDP-mannose pyrophosphorylase accelerates senescence [161].

As a consequence of proteolysis, metal cofactors present in metalloproteins are released to the cell internal environment. Some of these free metal ions, such as iron and copper, may be pernicious to the plant because they catalyze the production of ROIs [151]. On the other hand, these metals have to be mobilized to growing parts, being needed as nutrients by the developing organs. Metallothioneins, metal-binding proteins that are known to be induced during senescence [88,162] and are localized in vascular tissues, may fulfill the task, chelating metal ions (thereby protecting against ROIs) and favoring their directed transport.

The localization of the metabolism of ROIs in specific compartments, such as peroxisomes, could also serve to protect the cell under normal conditions [149]. It has been stated that the number of peroxisomes increases with oxidative stress [163] and superoxide radicals have been localized in glyoxisomes, a special kind of peroxisomes [164]. Nevertheless, membrane deterioration during senescene and the subsequent loss of compartmentation may contribute to an extension of the effects of ROIs inside the cell.

## 5. *Membranes*

One of the most characteristic changes during senescence is the progressive loss of membrane integrity. The major classes of lipids in plasma membrane and tonoplast are phospholipids, sterols, and ceramide monohexosides [165]. Common changes during senescence include a decrease in the total phospholipid and protein content, an increase in neutral lipids, and generalized oxidation [165]. Sterols also decline with physiological aging [166]. As a consequence of all of these, the physicochemical properties of the membranes, such as lipid fluidity, phase transition temperature, and nonbilayer lipid structure, are progressively altered during senescence [59,167–169]. The bilayer destabilization leads to a generalized failure of membrane functions, including loss of selective permeability and intracellular compartmentation, as well as membrane-associated enzymes and receptors [59,170]. Unfortunately, all these alterations complicate the experimental isolation of senescent membranes due to the changes in density and surface charge and the loss of marker enzymes [171].

The main enzymes implicated in lipid degradation are phospholipase D, phosphatidate phosphatase, and lipolytic acyl hydrolase. Most of the enzymes implicated in this process possess both membranous and cytosolic forms, which are differently regulated. The sequential action of these enzymes produces polyunsaturated fatty acids (PUFAs), which are substrates for lipoxygenase [172]. Lipoxygenase is a dioxygenase that catalyzes the oxidation of PUFA to fatty acid hydroperoxide, which is a precursor of volatile compounds that provide the typical flavor that characterizes wounded tissues [165]. Lipoxygenase activity has been described to increase during senescence in different plant organs [173,174]. The degree of partitioning of lipoxygenase between cytosol and membranes seems to be an important factor in the peroxidative damage because this enzyme seems to favor the oxidative injury of membranes by superoxide radicals. Delta 9 desaturase, an enzyme that has been shown to increase in senescing petals [175], may also play a role in the degradation of saturated fatty acids of membrane lipids.

The fact that some products of lipid peroxidation could serve as $Ca^{2+}$ ionophores, together with structural changes in the lipid phase, renders senescent membranes leaky to $Ca^{2+}$. $Ca^{2+}$ is stored in compartments such as apoplast and vacuole by the action of adenosinetriphosphatases (ATPases) that maintain its cytoplasmic concentration below micromolar levels under steady-state conditions. However, cytosolic $Ca^{2+}$ increases during senescence as a consequence of the decrease in the efficiency of ATPases, along with $Ca^{2+}$ leakage from the storage compartments. In turn, the increase in cytosolic $Ca^{2+}$ triggers a profusion of secondary effects. Among them, $Ca^{2+}$ may influence directly phospholipase D and phosphatidic acid phosphatase and indirectly affect lipoxygenase [176]. In addition, calcium and spermine have been shown to cause a decrease in membrane fluidity of tomato microsomes and increase phospholipase D activity by a mechanism attributable to the biophysical effect of the cations on the membranes [177].

## V. CHANGES IN GENE EXPRESSION

Senescence is a controlled process of disorganization that must be regulated by a set of genes acting in concert. Even if senescence is characterized by a global decrease in total RNA, specific mRNAs have been described to decrease or increase their levels on both a leaf area basis and an RNA mass basis [88,178]. Thus, senescence seems to begin with turning off and on of specific genes. The expression of genes can be regulated at several levels. Among them, transcriptional activation and repression are better known because they can be easily detected through the differential screening of cDNA libraries (for reviews, see Refs. 11, 40, 178). However, posttranscriptional regulation at the level of poly(A) tail shortening [179] or ribosome inactivating proteins [92] has been also reported during senescence. It is not our aim in this section to include an exhaustive list of genes that are transcriptionally regulated during senescence; therefore, only general trends and some illustrative examples are discussed.

Among the wealth of down-regulated genes, those coding for ATP sulfurylase [180], a photosystem II polypeptide [181], and a few stromal enzymes [182] have been characterized. On the other hand, transcriptionally activated genes are expected to be more relevant in defining the typical features of senescence and are, therefore, more intensely studied. Genes up-regulated during senescence are often termed senescence-associated genes (SAGs). SAGs may be classified according to their pattern of expression during leaf development or according to their putative functions based on sequence homology with other already cloned genes [40,178]. Regarding temporal patterns of expression, 10 different classes of senescence-related genes have been described in *Brassica napus* [178]. Some of these genes are also expressed in other phases of leaf development, especially in young tissues with elevated rates of metabolism. Although the pattern of expression can provide some hints about their function, sequence similarity to other characterized genes is usually more revealing. According to this latter (admittedly presumptive) evidence, SAGs can be divided into two main functional categories: those related to nutrient mobilization and those involved in cell protection. Exceptionally, some SAGs could play both roles, and the function of some others remains unknown.

### A. SAGs Involved in Nutrient Mobilization

According to the central role played by nutrient mobilization during senescence, SAGs related to this process are the most numerous. Moreover, these messages are relatively abundant (as corresponds to their extensive function) and, thus, they are easier to detect. Indeed, most of the clones that have been described to be up-regulated during senescence have been shown to be related to macromolecular breakdown and/or nutrient transport from senescent tissues to growing organs.

Among degradative enzymes, enhanced expression of cysteine proteases has been described in different plant species and senescence systems [117,118,120,122,123,180,183–185]. Certain clones show senescence-specific expression (typically *SAG12* from *Arabidopsis thaliana*) [17,88], and others respond selectively to specific senescence-inducing factors [10].

Other types of proteases, such as peroxisomal serine proteases [125], and vacuolar processing enzymes [126] (probably involved in regulatory activation of hydrolases in lytic vacuoles) are also up-regulated during senescence processes. Expression of proteins involved in the ubiquitin-mediated proteolytic pathway (which targets proteins for specific proteasome degradation) is also enhanced [186–188]. However, the expression of subunits of the 26S proteasome has been described to decline during senescence of tobacco leaves and flowers [129] but to increase during spinach cotyledon senescence [189], leading to controversial assumptions about the implication of proteasomal degradation in these processes.

In protein degradation in senescent chloroplasts, expression of some subunits of the protease system denominated Clp has been reported to be increased during natural senescence [190,191]. These protease complexes are assembled from a combination of subunits that are encoded in both the nuclear and chloroplast genomes. In contrast to its mRNA, protein levels of ERD1, one of these Clp-like protease subunits, decline during senescence, suggesting that it is unlikely to play a major role in chloroplast senescence [192]. Moreover, other Clp protease subunits are constitutively expressed throughout *Phaseolus* development [182], thus questioning their specific function in senescence.

Several RNase genes from different species have been shown to be up-regulated during senescence [40]. In *Arabidopsis thaliana*, RNS2 is a message coding for an S-like RNase that has similarity to self-incompatibility RNases (S-RNases). Unlike these, RNS2 is expressed in leaf and petals during senescence [95]. Because RNS2 and other RNase clones from *Arabidopsis* and tomato are also expressed under phosphate starvation [96,97], it has been proposed that they could play a role in the mobilization of this compound during senescent processes.

The genes encoding the 3-ketoacyl-CoA thiolase and other glyoxysomal proteins, which are probably implicated in fatty acid mobilization, have been reported to have increased expression during leaf senescence [193,194]. Besides, messages for a cytosolic glutamine synthetase and a β-glucosidase are also known to be induced [195,196].

## B. SAGs Involved in Cell Protection

In order to optimize the recycling of materials from the decaying tissues, senescing cells must stay alive during the general mobilization phase and have to be protected against multiple senescence-induced stresses. In this regard, the expression of many genes related to stress, or to stress signaling through the ethylene pathway, is known to be induced during senescence. Among them are genes implicated in the antioxidative response, such as ascorbate oxidase [196], anionic peroxidase [184], and glutathione S-transferase [180,197]; several genes encoding pathogenesis-related proteins [7,181]; and dark-inducible genes [198,199]. Besides, metallothioneins have been reported to be up-regulated in senescent leaves [88,162], where they may act in detoxifying metals (released from degraded metalloenzymes) and protecting against oxidative stress [178,200]. In addition to metallothioneins, other metal-binding proteins, such as a blue copper-binding protein (BCB) and a copper chaperone (CCH), are distinctly expressed during senescence, although their particular roles remain unclear [201,202].

## C. SAGs of Unknown Function

Among genes of unidentified function, *SAG13* is an excellent senescence marker because of its timely expression pattern. Sequence comparison studies indicate a certain similarity of this clone to short-chain alcohol dehydrogenase genes. Curiously, a member of the same highly diverged family is *TASSELSEED2*, a gene involved in programmed cell death in maize [203].

## VI. SENESCENCE REGULATION AND SIGNALING

Senescence, in a broad sense, encompasses a number of processes with a common final outcome: the breakdown of cells through an endogenously organized program that optimizes nutrient economy and ensures functional takeover. Temporal and spatial coordination of senescence events requires a complex signaling network, whose analysis is further complicated by the intrinsic diversity of senescence processes. Generally speaking, three types of factors are known to affect the onset and progression of plant senescence: (1) internal factors, related to development and aging; (2) environmental conditions, especially those causing plant stress; and (3) hormonal factors related to intercommunication with the rest of the plant.

Developmental processes involving cell death related to differentiation of tissues and directed to morphogenetic goals (e.g., xylogenesis or postpollination changes in ovaries) are typical cases of apoptosis, most likely governed by morphogens. In contrast, natural senescence related to aging may be triggered by nutritional and metabolic imbalances. The senescence of old leaves is a paradigmatic case. It has been suggested that, regarding the onset of senescence, plant leaves may be considered as autonomous, their life span being evolutively fixed from a compromise between recurse allocation and payoff expectations [41,204]. Leaves usually produce photosynthetic assimilates in excess of their own needs, exporting them to the rest of the plant. However, because their productivity declines with aging, old leaves may barely contribute to their own nutrition, no longer benefiting the plant as a whole. In that instance, they are said to be at the *compensation point*, and those leaves are hypothesized to be targeted to senesce [204]. In this regard, the level of photosynthates (or other nutrients), evaluated at the leaf phloem, could act directly or indirectly as a signal to trigger the senescence-associated dismantling of the leaf. Metabolic imbalances caused by restricted photosynthetic activity [6,205], but also by the strong nitrogen demand from growing organs [206], have been proposed to act as signals triggering the senescence program in old leaves. Nevertheless, the signaling mechanisms by which these imbalances are supposed to be perceived are currently indeterminate. The metabolic signal hypothesis is supported by several indirect experimental observations. Victorin, a fungal toxin whose only demonstrated effect at the molecular level is the inhibition of a photorespiratory enzyme, is known to induce a decay process in oat leaves displaying all typical symptoms of senescence (chlorophyll loss, Rubisco degradation, and chromatin fragmentation, as well as ethylene and $Ca^{2+}$ signaling) [207]. Moreover, in agreement with the hypothesis of a photosynthate sensor, overexpression of *Arabidopsis* hexokinase, a key regulatory enzyme in sugar metabolism, induces rapid senescence in transgenic tomato plants [208].

Adverse environmental conditions caused by both biotic and abiotic factors (such as viral infections [209], ozone exposure [210] or continuous darkness [199]) have been shown to induce senescence. Besides, stress can advance or accelerate natural senescence to different degrees depending on the type and intensity of the stress and the developmental stage of the plants [211]. Stress processes may have a direct influence on the onset of leaf senescence through the putative photosynthate sensor signaling because they usually cause a decline in photosynthetic efficiency. However, overproduction of ROIs (a common consequence of stress processes) may also work as a signal triggering secondary responses inside the cell, especially at the membrane level. This is particularly true in case of oxidative processes such as the stress caused by ozone pollution (reviewed in Ref. 212).

Most plant growth regulators (phytohormones) are known to influence senescence, either promoting (ethylene, abscisic acid, and jasmonates) or inhibiting (cytokinins, auxins, and gibberellins) it. However, cytokinins and ethylene have been demonstrated to exert the greatest influence on plant senescence processes. A decline of cytokinins is observed during natural leaf senescence, and the exogenous application of cytokinins to excised leaves prevents senescence [213]. Moreover, delayed leaf senescence in tobacco plants expressing the maize homeobox gene *knotted1* under control of the promoter of a senescence-regulated gene (*SAG12*) is accompanied by increased cytokinin content [214]. Perhaps the most conclusive experiment on the effect of cytokinins in senescence has been the transformation of tobacco plants with an enzyme of cytokinin synthesis (the *Agrobacterium tumefaciens* isopentenyl transferase gene) under the SAG 12 promoter. These plants show an autoregulatory production of cytokinins (thereby avoiding the deleterious effects of overproduction) and a concomitant delay of plant leaf senescence without any other phenotypic alterations [17].

Ethylene accelerates the onset of senescence in some plant species. The *Arabidopsis* mutant etr1-1, which is insensitive to ethylene because of an inactive receptor, shows a delay of several days in chlorophyll loss [215]. Moreover, leaf senescence is transiently delayed in transgenic tomato plants with blocked ethylene biosynthesis due to antisense expression of the 1-amino cyclopropane-1-carboxylic acid (ACC) oxidase gene, but once the senescence process is started, the expression pattern of SAGs does not differ from that of wild-type plants [216]. These results indicate that ethylene influences senescence timing and, because the ethylene concentration increases during stress conditions, this could be a mechanism for adjusting the speed of the plant response to the environment.

In the last instance, control of senescence has to be exerted through regulation of gene expression acting at both the nuclear and the chloroplast genome. Some of the most abundant proteins in the chloroplast, such as Rubisco and chlorophyll *a/b* binding proteins, are composed of different subunits encoded in both the nuclear and the chloroplastic genome. A certain degree of coordination in the expression of the genomes must exist during senescence. Indeed, nuclear control of cell senescence has been postulated according to several facts [2]. First, there are mutations in the nuclear genome that alter the senescence syndrome. Second, chloroplast senescence is prevented by enucleation. Third, selective inhibitors of nuclear RNA synthesis inhibit senescence-related processes. In contrast, specific inhibitors of organelle RNA polymerases do not inhibit senescence. Something similar happens with the inhibitors of protein synthesis: cycloheximide (an inhibitor of the 80S cytoplasmic ribosomes) blocks a variety of senescence-related changes, whereas chloramphenicol (an inhibitor of the 70S organella ribosomes) does not delay senescence. It has been suggested that the control of the nuclear-encoded chloroplast RNA polymerase could be the key step in this coordination [2].

Information on the secondary signals that propagate and diversify senescence responses inside the cell acting at the gene expression (transcriptional) level may be gathered, in principle, from the analysis of the SAG promoters. However, the nonexistence of mutants lacking the whole senescence syndrome and the diversity of expression patterns among SAGs indicate an intrinsic redundancy and/or complexity of this process. Indeed, SAG promoters have proved difficult to analyze. Some of them display several regulatory boxes responding to developmental stage, stress, or/and hormones. Among those involved in natural senescence, SAG12 from *Arabidopsis thaliana* is one of the best senescence markers found to date [40]. The dissection of this promoter has allowed the identification of a senescence-specific region responsible for its expression [217]. This sequence remains functional when expressed in heterologous systems such as tobacco [17], and it is conserved in orthologue genes from *Brassica* [218], suggesting that some senescence regulatory mechanism may be conserved among higher plant species. This region has been shown to bind nuclear proteins, perhaps senescence-specific transcription factors, which remain to be characterized.

On the other hand, some habitual components of cellular signal transduction pathways have also been reported to participate in plant senescence. For example, active MAP kinases have been identified in senescent maize leaves [219], GTP binding and protein phosphorylation are present in senescing *Arabidopsis thaliana* [220], $Ca^{2+}$ chelators have been shown to prevent the senescence syndrome in victorin-treated oat leaves [207], and overexpression of the transcription factor AmMYB308, which inhibits phenolic acid metabolism, is known to induce premature senescence in tobacco [221]. Moreover, homologues of the prohibitin family, which are mitochondrial proteins that regulate the replicative life span, have also been found in plants [222], and the plant homologue of the defender against apoptotic death gene is known to be down-regulated during senescence of flower petals [223].

In summary, current knowledge indicates that senescence regulation is a highly complex process involving a multitude of signals that propagate and diversify the cellular responses in a dense network, where even the main pathways remain poorly understood.

## VII.  MONITORING SENESCENCE

To follow senescence, it is necessary to find the appropriate parameters to measure the evolution of the process. No single measurement is definitive, although in some cases certain parameters may be adequate for particular tissues. However, single measurements should be checked against other parameters whenever possible.

The loss of chlorophyll is one of the most obvious changes during senescence of green organs, although the existence of mutants in which senescence proceeds without chlorophyll loss indicates that this change is not crucial to the process [5,46]. Precautions should be taken during the extraction of chlorophyll because some protocols (e.g., involving acetone) may lead to its degradation [224]. Radiolabeling of chlorophyll has allowed a more sensitive measurement of its disappearance and the identification of the degradation products [225].

Biochemical changes that precede chlorophyll loss include a decline in photosynthetic capacity [55] and lowering of protein content. The progressive impairment of photosystem II may be evaluated by monitoring its photochemical efficiency from chlorophyll fluorescence quenching curves [226,227]. Besides, laser-induced fluorescence imaging is a sensitive and noninvasive technique that can be used to assess the in vivo photosynthetic activity of green tissues [228], although it requires sophisticated instrumentation. In contrast, the degradation of the $CO_2$-fixing enzyme (Rubisco), a preferent target of proteases, is an easy and widely used parameter to follow senescence in photosynthetically active organs [229]. The decrease in total protein levels may also be measured. To avoid the interferences inherent in some methods, they may be assayed through dye binding to protein adsorbed on washed paper disks [230] or by nitrogen determination in digests [231].

Senescence may also be followed through measurements related to oxidative damage, such as extent of lipid peroxidation [232] or levels of protective enzymatic activities (typically superoxide dismutase and/or ascorbate peroxidase) [157,233]. Furthermore, specific assays have been developed to measure total antioxidant power (e.g., the FRAP assay) [234]. Leakage of the cell membranes may be also a significant parameter but occurs late in the senescence process.

Besides, in cases in which a hormone is directly implicated in senescence, measurement of the hormonal levels can be a good approach to detect early symptoms of decay. This is the case for some fruits, such as tomato or avocado, where a burst of ethylene precedes the onset of ripening [235].

If it is necessary to use exclusively biochemical assays, a combination of different parameters may be the best solution for precise monitoring of senescence. For example, Pastori and Trippi [236] have utilized chlorophyll loss, lipid peroxidation, and cell electrolyte leakage to study senescence in maize, and Oh et al. [237] used amount of Rubisco large subunit, RNase, and peroxidase activities, together with photosystem II efficiency and chlorophyll content, to monitor senescence in *Arabidopsis thaliana*.

Whenever possible, gene expression analysis of selected senescence markers will provide more sensitive and accurate monitoring of the senescence process. Because expression patterns display some variability between species and between senescence-inducing treatments, it might be advisable to check the particular case by screening an array of selected markers in order to choose the most appropriate one. In principle, one may use any gene product (at the level of mRNA or protein) that shows a significant variation throughout the relevant senescence process. Among markers of declining mRNA levels, the most widely utilized are the messengers corresponding to the chlorophyll a/b binding protein (CAB) and the small subunit of the Rubisco enzyme (rbcS) [88,201,215] because homologous oligonucleotide probes for these genes are readily available in a wide variety of plants (where at least one of these genes has been sequenced). Among the messages whose levels are increased during senescence, those corresponding to the genes *LSC54* from *Brassica napus* and *SAG12* and *SAG13* from *Arabidopsis thaliana* are claimed to be the most senescence specific [4]. *LSC54* (encoding a metallothionein) is expressed in leaves and flowers of *B. napus* exclusively during senescence [89]. In contrast, the homologous gene in *A. thaliana* (*SAG17*) is constitutively expressed at a moderate level that rises with senescence [88]. Another useful gene, *SEN1* from *A. thaliana*, exhibits senescence-dependent expression but with a different intensity in natural or hormone (abscisic acid or ethylene) induced senescence [199]. The practical utility of these markers in other plants is somewhat hindered by the need for identification and characterization of the homologous genes or the use of less specific heterologous probes.

## VIII.  SUMMARY AND CONCLUDING REMARKS

Senescence appears as an ordered dismantling of structures and components from plant parts whose functional contribution has become unnecessary and which are therefore directed to abscission and death. Aside from functional advantages that may be derived in special cases from senescence of certain structures, the principal goal of senescence is to recover nutrients from the decaying tissues, withdrawing them to the surviving parts before abscission. Thus, senescence is essentially a physiological strategy of nutritional economy.

Natural senescence of plant organs is probably triggered by a nutritional imbalance leading to certain metabolic alterations (sensed locally by an unknown mechanism), which begin a transduction cascade involving multiple intermediate signals (hormones, ROIs, $Ca^{2+}$, transcription factors, etc.). Primary signals activate a set of endogenous adaptive responses (mostly directed to nutrient salvage before programmed death or to protection of nonsenescing nearby tissues), which are executed through secondary signals switching off and on specific genes in a functionally coordinated temporal and spatial pattern. Stress-induced senescence appears to elicit the same adaptative responses through interference by adverse environmental conditions or pathogenesis somewhere along the signal transduction pathway of natural senescence. In any case, these endogenous responses lead to the typical alterations that are characteristic of all senescence processes, including breakdown of photosynthetic pigments and selected macromolecules, progressive deterioration and loss of functions of membranes, and, in the final stage, degeneration of cell internal structure.

Considering the wealth of information currently being gathered through molecular analysis at the gene expression level, our understanding of senescence is expected to improve in the coming years. This will probably uncover the signaling mechanisms, clarify the bounds between senescence and related processes such as stress responses and fruit ripening, and extend the possibilities of genetic engineering of the senescence features of crops for nutritional and commercial benefit.

## ACKNOWLEDGMENTS

This work was supported by grants BIO99-1201-C02-02 and PB98-1445 from DGICYT.

# REFERENCES

1.  AC Leopold. Aging and senescence in plant development. In: KV Thimann, ed. Senescence in Plants. Boca Raton, FL: CRC Press, 1980, pp 1–12.
2.  R Sexton, HW Woolhouse. Senescence and abscission. In: MB Wilkins, ed. Advanced Plant Physiology. New York: John Wiley & Sons, 1984, pp 469–497.
3.  LD Noodén. The phenomena of senescence and aging. In: L D Noodén, AC Leopold, eds. Senescence and Aging in Plants. London: Academic Press, 1988, pp 2–51.
4.  S Gan, RM Amasino. Making sense of senescence. Plant Physiol 113:313–319, 1997.
5.  H Thomas, JL Stoddart. Leaf senescence. Annu Rev Plant Physiol 31:83–111, 1980.
6.  GA King, KM Davies, RJ Stewart, WM Borst. Similarities in gene expression during the post-harvest induced senescence of spears and natural foliar senescence of asparagus. Plant Physiol 108:125–128, 1995.
7.  C Hanfrey, M Fife, V Buchanan-Wollaston. Leaf senescence in *Brassica napus*: expression of genes encoding pathogenesis-related proteins. Plant Mol Biol 30:597–609, 1996.
8.  BF Quirino, J Normanly, RM Amasino. Diverse range of gene activity during *Arabidopsis thaliana* leaf senescence includes pathogen-independent induction of defense-related gene. Plant Mol Biol 40:267–278, 1999.
9.  BJ Pogson, CG Downs, KM Davies. Differential expression of two 1-aminocyclopropane-1-carboxylic acid oxidase gene in broccoli after harvest. Plant Physiol 108:651–657, 1995.
10. R Khanna-Chopra, B Srivalli, YS Ahlawat. Drought induces many forms of cysteine proteases not observed during natural senescence. Biochem Biophys Res Commun 255:324–327, 1999.
11. HG Nam. The molecular genetic analysis of senescence. Curr Opjn Biotechnol 8:200–207, 1997.
12. RE Sheehy, MK Kramer, WR Hiatt. Reduction of polygalacturonase activity in tomato fruit by antisense RNA. Proc Natl Acad Sci USA 85:8805–8809, 1988.
13. CJS Smith, CF Watson, J Ray, CR Bird, PC Morris, W Schuch, D Grierson. Antisense RNA inhibition of polygalacturonase gene expression in ripening tomatoes. Nature 33:724–726, 1988.
14. AJ Hamilton, GW Lycett, D Grierson. Antisense gene that inhibits synthesis of the hormone ethylene in transgenic plants. Nature 346:284–287, 1990.
15. PW Oeller, L Min-Wong, LP Taylor, DA Pike, A Theologis. Reversible inhibition of tomato fruit senescence by antisense RNA. Science 254:437–439, 1991.
16. JJ Guiamet, JA Terri, LD Noodén. Effect of nuclear and cytoplasmic genes altering chlorophyll loss on gas exchange during monocarpic senescence in soybean. Plant Cell Physiol 31:1123–1130, 1990.
17. S Gan, RM Amasino. Inhibition of leaf senescence by autoregulated production of cytokinin. Science 270:1986–1988, 1995.
18. LD Noodén, AC Leopold, eds. Senescence and Aging in Plants. London: Academic Press, 1988.
19. TA Steeves, I M Sussex. Patterns in Plant Development. 2nd ed. Cambridge: Cambridge University Press, 1989.
20. J Moreno, JL García-Martinez. Nitrogen accumulation and mobilization in *Citrus* leaves throughout the annual cycle. Physiol Plant 61:429–434, 1984.
21. S Kang, JS Titus. Qualitative and quantitative changes in nitrogenous compounds in senescing leaf and bark tissues of apple. Physiol Plant 50:285–290, 1980.
22. S Mayak, AH Halevy. Flower senescence. In: L D Noodén, AC Leopold, eds. Senescence and Aging in Plants. London: Academic Press, 1980, pp 131–156.
23. GN Drews, RB Goldberg. Genetic control of flower development. Trends Genet 5:256–261, 1989.
24. H Kende, B Baumgartner. Regulation of aging in flowers of *Ipomea tricolor* by ethylene. Planta 116:279–289, 1974.
25. CJ Brady. Fruit ripening. Annu Rev Plant Physiol 38:155–178, 1987.
26. RL Fischer, AB Bennett. Role of cell wall hydrolases in fruit ripening. Annu Rev Plant Physiol Plant Mol Biol 42:675–703, 1991.
27. EE Goldschmidt. Pigment changes associated with fruit maturation and their control. In: KV Thimann, ed. Senescence in Plants. Boca Raton, FL: CRC Press, 1980, pp 207–218.
28. MJC Rhodes. The maturation and ripening of fruits. In: KV Thimann, ed. Senescence in Plants. Boca Raton, FL: CRC Press, 1980, pp 157–205.
29. YN Behera, B Biswal. Leaf senescence in fern: effect of duration, intensity and quality of light. Environ Exp Bot 30:181–186, 1990.
30. K Okada, Y Inoue, K Satoh, S Katoh. Effect of light on degradation of chlorophyll and proteins during senescence of detached rice leaves. Plant Cell Physiol 33:1183–1191, 1992.
31. LD Noodén. Whole plant senescence. In: LD Noodén, AC Leopold, eds. Senescence and Aging in Plants. London: Academic Press, 1988, pp 391–439.
32. LM Behera, NK Choudhury. Effect of organ excision and kinetin treatment on chlorophyll content and DCPIP photoreduction activity of chloroplasts of pumpkin cotyledons. J Plant Physiol 137:53–57, 1990.
33. H Klee, M Estelle. Molecular genetic approaches to plant hormone biology. Annu Rev Plant Physiol Plant Mol Biol 42:529–551, 1991.

34. H-P Döring, P Starlinger. Molecular genetics of transposable elements in plants. Annu Rev Genet 20:175–200, 1986.

35. KA Feldmann. T-DNA insertion mutagenesis in *Arabidopsis*: mutational spectrum. Plant J 1:71–82, 1991.

36. C Koncz, K Németh, GP Rédei, J Schell. T-DNA insertional mutagenesis in *Arabidopsis*. Plant Mol Biol 20:963–976, 1992.

37. S Gibson, C Somerville. Isolating plant genes. Trends Biotechnol 11:306–313, 1993.

38. DM Kehoe, P Villand, S Somerville. DNA microarrays for studies of higher plants and other photosynthetic organisms. Trends Plant Sci 4:38–41, 1999.

39. PJ Krysan, JC Young, F Tax, MR Sussman. Identification of transferred DNA insertions within *Arabidopsis* genes involved in signal transduction and ion transport. Proc Natl Acad Sci U S A 93:8145–8150, 1996.

40. LM Weaver, E Himelblau, RM Amasino. Leaf senescence: gene expression and regulation. In: JK Setlow, ed. Genetic Engineering. Vol 19. New York: Academic Press, 1997, pp 215–234.

41. AB Bleecker, SE Patterson. Last exit: senescence, abscission, and meristem arrest in *Arabidopsis*. Plant Cell 9:1169–1179, 1997.

42. H Thomas, CM Smart. Crops that stay green. Ann Appl Biol 123:193–219, 1993.

43. LD Noodén, JJ Guiamet. Genetic control of senescence and aging in plants. In: EL Schneider, JW Rowe, eds. Handbook of the Biology of Aging. 4th ed. San Diego: Academic Press, 1996, pp 94–118.

44. H Thomas. Leaf senescence in a non-yellowing mutant of *Festuca pratensis*. I. Chloroplast membrane polypeptides. Planta 154:212–218, 1982.

45. H Thomas. Leaf senescence in a non-yellowing mutant of *Festuca pratensis*. II. Proteolytic degradation of thylakoid and stroma polypeptides. Planta 154:219–223, 1982.

46. TGE Davies, H Thomas, BJ Thomas, LJ Rogers. Leaf senescence in a nonyellowing mutant of *Festuca pratensis*. Plant Physiol 93:588–595, 1990.

47. H Thomas. Sid: a mendelian locus controlling thylakoid membrane disassembly in senescent leaves of *Festuca pratensis*. Theor Appl Genet 73:551–555, 1987.

48. F Vicentini, S Hörstensteiner, M Schellenberg, H Thomas, P Matile. Chlorophyll breakdown in senescent leaves: identification of the biochemical lesion in a stay-green genotype of *Festuca pratensis* Huds. New Phytol 129:247–252, 1995.

49. JJ Guiamet, E Schwartz, E Pichersky, LD Noodén. Characterization of cytoplasmic and nuclear mutations affecting chlorophyll and chlorophyll-binding proteins during senescence in soybean. Plant Physiol 96:227–231, 1991.

50. JJ Guiamet, MC Giannibelli. Nuclear and cytoplasmic stay-green mutations of soybean alter the loss of leaf soluble proteins during senescence. Physiol Plant 96:655–661, 1996.

51. JJ Guiamet, MC Giannibelli. Inhibition of the degradation of chloroplast membranes during senescence in nuclear stay-green mutants of soybean. Physiol Plant 91:395–402, 1994.

52. J Chory, P Nagpal, CA Peto. Phenotypic and genetic analysis of det2, a new mutant that affects light-regulated seedling development in *Arabidopsis*. Plant Cell 3:445–460, 1991.

53. JT Greenberg, FM Ausubel. *Arabidopsis* mutants compromised for the control of cellular damage during pathogenesis and aging. Plant J 4:327–341, 1993.

54. SJ Clough, AF Bent. Floral dip: a simplified method for *Agrobacterium*-mediated transformation of *Arabidopsis thaliana*. Plant J 16:735–743, 1998.

55. S Gepstein. Photosynthesis. In: LD Noodén, AC Leopold, eds. Senescence and Aging in Plants. London: Academic Press, 1988, pp 85–109.

56. Y Vercher, J Carbonell. Changes in the structure of ovary tissue and in the ultrastructure of mesocarp cells during ovary senescence or fruit development induced by plant growth substances in *Pisum sativum*. Physiol Plant 81:518–526, 1991.

57. M Kura-Hotta, H Hashimoto, K Satoh, S Katoh. Quantitative determination of changes in the number and size of chloroplasts in naturally senescing leaves of rice seedlings. Plant Cell Physiol 31:33–38, 1990.

58. JL Stoddart, H Thomas. Leaf senescence. In: D Boulter, B Parthier, eds. Nucleic Acids and Proteins in Plants I (Encyclopedia of Plant Physiology New Series, Vol 14A). Berlin: Springer Verlag, 1982, pp 592–636.

59. JE Thompson, CD Froese, Y Hong, KA Hudak, MD Smith. Membrane deterioration during senescence. Can J. Bot 75:867–879, 1997.

60. KA Platt-Aloia, WW Thomson. Freeze-fracture evidence of gel phase lipid in membranes of senescing cowpea cotyledons. Planta 163:360–369, 1985.

61. IE O'Brien, BG Murray, BC Maguley, BA Morris, IB Ferguson. Major changes in chromatin condensation suggest the presence of an apoptotic pathway in plant cells. Exp Cell Res 241:46–54, 1998.

62. Y Vercher, A Molowny, J Carbonell. Gibberellic acid effects on the ultrastructure of endocarp cells of unpollinated ovaries. Physiol Plant 71:302–308, 1987.

63. N Inada, A Sakai, H Kuroiwa, T Kuroiwa. Three-dimensional analysis of the senescence program in rice (*Oryza sativa* L.) coleoptiles. Investigations of tissues and cells by fluorescence microscopy. Planta 205:153–164, 1998.

64. KV Thimann, SO Satler. Relation between senescence and stomatal closure: senescence in light. Proc Natl Acad Sci U S A 76:2295–2298, 1979.

65.  VA Wittenbach. Effect of pod removal on leaf photosynthesis and soluble protein composition of field-grown soybeans. Plant Physiol 73:121–124, 1983.
66.  IR Cowan, GD Farquhar. Stomatal function in relation to leaf metabolism and environment. In: DH Jenning, ed. Integration of Activity in the Higher Plant. London: Cambridge University Press, 1977, pp 471–505.
67.  E Zeiger, A Schwartz. Longevity of guard cell chloroplasts in falling leaves: implication for stomatal function and cellular aging. Science 218:680–682, 1982.
68.  T Solomos. Respiration in senescing plant organs: its nature, regulation and physiological significance. In: LD Noodén, AC Leopold, eds. Senescence and Aging in Plants. London: Academic Press, 1988, pp 112–146.
69.  AC Liden, HE Akerlund. Induction and activation of the alternative oxidase of potato tuber mitochondria. Physiol Plant 87:134–141, 1993.
70.  C Hiser, L McIntosh. Alternative oxidase of potato is an integral membrane protein synthesized de novo during aging of tuber slices. Plant Physiol 93:312–318, 1990.
71.  GG Laties. The cyanide-resistant alternative path in higher plant respiration. Annu Rev Plant Physiol 33:519–555, 1982.
72.  IL Nantes, MM Fagian, R Catisti, P Arruda, IG Maia, AE Vercesi. Low temperature and aging-promoted expression of PUMP in potato tuber mitochondria. FEBS Lett 457:103–106, 1999.
73.  P Jezek, AD Costa, AE Vercesi. Reconstituted plant uncoupling mitochondrial protein allows for proton translocation via fatty acid cycling mechanism. J Biol Chem 272:24272–24278, 1997.
74.  P Siffel, J Kutik, NN Lebedev. Spectroscopically analyzed degradation of chlorophyll-protein complexes and chloroplast ultrastructure during yellowing of leaves. Photosynthetica 25:395–407, 1991.
75.  J Hidema, A Makino, Y Kurita, T Mae, K Ojima. Changes in the levels of chlorophyll and light-harvesting chlorophyll *a/b* protein in rice leaves aged under different irradiances from full expansion through senescence. Plant Cell Physiol 33:1209–1214, 1992.
76.  AJ Young, R Wellings, G Britton. The fate of chloroplast pigments during senescence of primary leaves of *Hordeum vulgare* and *Avena sativa*. J Plant Physiol 137:701–705, 1990.
77.  V Scheumann, S Schoch, W Rudiger. Chlorophyll *b* reduction during senescence of barley seedlings. Planta 209:364–370, 1999.
78.  D Amir-Shapira, EE Goldschmidt, A Altman. Chlorophyll catabolism in senescing plant tissues: in vivo breakdown intermediates suggest different degradative pathways for *Citrus* fruit and parsley leaves. Proc Natl Acad Sci U S A 84:1901–1905, 1987.
79.  T Tsuchiya, H Ohta, K Okawa, A Iwamatsu, H Shimada, T Masuda, K Takamiya. Cloning of chlorophyllase, the key enzyme in chlorophyll degradation: finding of a lipase motif and the induction by methyl jasmonate. Proc Natl Acad Sci U S A 96:15362–15367, 1999.
80.  B Luthy, E Martinoia, P Matile, H Thomas. Thylakoid-associated chlorophyll oxidase: distinction from lipoxygenase. Z Pflanzenphysiol 113:423–434, 1984.
81.  N Yamauchi, AE Watada. Regulated chlorophyll degradation in spinach leaves during storage. J Am Soc Hortic Sci 116:58–62, 1991.
82.  VJ Dwarki, VNK Francis, GJ Bhat, G Padmanaban. Regulation of cytochrome P-450 messengers RNA and apoprotein levels by heme. J Biol Chem 262:16958–16962, 1987.
83.  DY Chang, JP Miksche, SS Dhillon. DNA changes involving repeated sequences in senescing soybean (*Glycine max*) cotyledon nuclei. Physiol Plant 64:409–417, 1985.
84.  JB Harris, VG Schaefer, SS Dhillon, JP Miksche. Differential decline in DNA in aging leaf tissues. Plant Cell Physiol 23:1267–1273, 1982.
85.  CJ Brady. Nucleic acid and protein synthesis. In: L D Noodén, AC Leopold, eds. Senescence and Aging in Plants. London: Academic Press, 1988, pp 147–181.
86.  FB Abeles, LJ Dunn. Restriction fragment length polymorphism analysis of DNA from senescing cotyledon tissue. Plant Sci 72:13–18, 1990.
87.  R Tomas, A Vera, M Martin, B Sabater. Changes in protein synthesis without evidence of DNA methylation in barley chloroplasts during leaf growth and development. Plant Sci 85:71–77, 1992.
88.  KN Lohman, S Gan, MC John, RM Amasino. Molecular analysis of natural leaf senescence in *Arabidopsis thaliana*. Physiol Plant 92:322–328, 1994.
89.  V Buchanan-Wollaston. Isolation of cDNA clones for genes that are expressed during leaf senescence in *Brassica napus*. Plant Physiol 105:839–846, 1994.
90.  H Pfitzinger, L Marechal-Drouard, DTN Pillay, JH Weil, P Guillemaut. Variations during leaf development of the relative amounts of two bean (*Phaseolus vulgaris*) chloroplast phenylalanyl transfer RNA which differ in their minor nucleotide content. Plant Mol Biol 14:969–976, 1990.
91.  C Jayabaskaran, M Kuntz, P Guillemaut, JH Weil. Variations in the levels of chloroplast transfer RNA and aminoacyl transfer RNA synthetases in senescing leaves of *Phaseolus vulgaris*. Plant Physiol 92:136–140, 1990.
92.  F Stirpe, L Barbieri, P Gorini, P Valbonesi, A Bolognesi, L Polito. Activities associated with the presence of ribosome-inactivating proteins increase in senescent and stressed leaves. FEBS Lett 382:309–312, 1996.
93.  A Blank, TA McKeon. Expression of three RNase activities during natural and dark-induced senescence of wheat leaves. Plant Physiol 97:1409–1413, 1991.

94.   PJ Green. The ribonucleases of higher plants. Annu Rev Plant Physiol Plant Mol Biol 45:421–445, 1994.
95.   CB Taylor, PA Bariola, SB del Cardayre, RT Raines, PJ Green. RNS2—a senescence-associated RNAase of *Arabidopsis* that diverged from the sRNAases before speciation. Proc Natl Acad Sci U S A 90:5118–5122, 1993.
96.   PA Bariola, CJ Howard, CB Taylor, MT Verburg, VD Jaglan, PJ Green. The limitation. *Arabidopsis ribonuclease* gene *RNS1* is tightly controlled in response to phosphate. Plant J 6:673–685, 1994.
97.   A Lers, A Khalchitski, E Lomaniec, S Burd, PJ Green. Senescence-induced RNases in tomato. Plant Mol Biol 36:439–449, 1998.
98.   F Vicentini, P Matile. Gerontosomes, a multi-functional type of peroxisome in senescent leaves. J Plant Physiol 142:50–56, 1993.
99.   GM Pastori, LA del Rio. An activated-oxygen-mediated role for peroxisomes in the mechanism of senescence of *Pisum sativum* L. leaves. Planta 193:385–391, 1994.
100.  P Millard, CM Thomson. The effect of the autumn senescence of leaves on the internal cycling of nitrogen for the spring growth of apple trees. J Exp Bot 40:1285–1289, 1989.
101.  SJ Crafts-Brandner. Phosphorus nutrition influence on leaf senescence in soybean. Plant Physiol 98:1128–1132, 1992.
102.  JR Evans, JR Seemann. The allocation of protein nitrogen in the photosynthetic apparatus: cost, consequences and control. In: WS Briggs, ed. Photosynthesis. New York: Alan Liss, 1989, pp 183–205.
103.  LW Peterson, RC Huffaker. Loss of RDP carboxylase and increase of proteolytic activity in senescence of detached primary barley leaves. Plant Physiol 55:1009–1015, 1975.
104.  VA Wittenbach. Breakdown of ribulose bisphosphate carboxylase and change in proteolytic activity during dark-induced senescence of wheat seedlings. Plant Physiol 62:604–608, 1978.
105.  R Shurtz-Swirski, S Gepstein. Proteolysis of endogenous substrates in senescing oat leaves. Plant Physiol 78:121–125, 1985.
106.  SJ Crafts-Brandner, ME Salvucci, DE Egli. Changes in ribulosebisphosphate carboxylase oxygenase and ribulose 5-phosphate kinase abundances and photosynthetic capacity during leaf senescence. Photosynth Res 23:223–230, 1990.
107.  JW Friedrich, RC Huffaker. Photosynthesis, leaf resistance, and ribulose-1,5-bisphosphate carboxylase degradation in senescing barley leaves. Plant Physiol 65:1103–1107, 1980.
108.  WP Quick, K Fichtner, R Wendler, E-D Schulze, SR Rodermel, L Bogorad, M Stitt. Decreased ribulose-1,5-bisphosphate carboxylase-oxygenase in transgenic tobacco transformed with antisense rbcS. IV. Impact on photosynthesis in conditions of altered nitrogen supply. Planta 188:522–531, 1992.
109.  RC Huffaker, BL Miller. Reutilization of ribulose bisphosphate carboxylase. In: HW Siegelman, ed. Photosynthetic Carbon Assimilation. New York: Plenum Press, 1978, pp 139–152.
110.  HW Woolhouse. Biochemical and molecular aspects of senescence. In: H Smith, D Grierson, eds. Molecular Biology of Plant Development. Oxford: Blackwells, 1982, pp 256–287.
111.  MJ Droillard, NJ Bate, SJ Rothstein, JE Thompson. Active translation of the D-1 protein of photosystem II in senescing leaves. Plant Physiol 99:589–594, 1992.
112.  RC Huffaker. Proteolytic activity during senescence of plants. New Phytol 116:199–231, 1990.
113.  U Feller, A Fischer. Nitrogen metabolism in senescing leaves. Crit Rev Plant Sci 13:241–273, 1994.
114.  J Callis. Regulation of protein degradation. Plant Cell 7:845–857, 1995.
115.  Z Adam. Protein stability and degradation in the chloroplast. Plant Mol Biol 32:773–783, 1996.
116.  RD Vierstra. Proteolysis in plants: mechanisms and functions. Plant Mol Biol 32:275–302, 1996.
117.  V Valpuesta, NE Lange, C Guerrero, MS Reid. Up-regulation of a cysteine protease accompanies the ethylene-insensitive senescence of daylily (*Hemerocallis*) flowers. Plant Mol Biol 28:575–582, 1995.
118.  ML Jones, PB Larsen, WR Woodson. Ethylene-regulated expression of a carnation cysteine proteinase during flower petal senescence. Plant Mol Biol 28:505–512, 1995.
119.  M Cercós, J Carbonell. Purification and characterization of a thiol-protease induced during the senescence of unpollinated ovaries of *Pisum sativum*. Physiol Plant 88:275–280, 1993.
120.  M Cercós, S Santamaria, J Carbonell. Cloning and characterization of TPE4A, a thiol-protease gene induced during ovary senescence and seed germination in pea. Plant Physiol 119:1341–1348, 1999.
121.  ZH Ye, JE Varner. Induction of cysteine and serine proteases during xylogenesis in *Zinnia elegans*. Plant Mol Biol 30:1233–1246, 1996.
122.  FX Xu, ML Chye. Expression of cysteine proteinase during developmental events associated with programmed cell death in brinjal. Plant J 17:321–327, 1999.
123.  R Drake, I John, A Farrell, W Cooper, W Schuch, D Grierson. Isolation and analysis of cDNAs encoding tomato cysteine proteases expressed during leaf senescence. Plant Mol Biol 30:755–767, 1996.
124.  M Solomon, B Belenghi, M Delledone, E Menachem, A Levine. The involvement of cysteine proteases and protease inhibitor genes in the regulation of programmed cell death in plants. Plant Cell 11:431–443, 1999.
125.  S Distefano, JM Palma, M Gomez, LA del Rio. Characterization of endoproteases from plant peroxisomes. Biochem J 327:399–405, 1997.
126.  T Kinoshita, K Yamada, N Hiraiwa, M Kondo, M Nishimura, I Hara-Nishimura. Vacuolar processing enzyme is up-regulated in the lytic vacuoles of vegetative tissues during senescence and under various stressed conditions. Plant J 19:43–53, 1999.

127. K van Leyen, RM Duvoisin, H Engelhardt, M Wiedmann. A function for lipoxygenase in programmed organelle degradation. Nature 395:392–395, 1998.

128. WR Belknap, JR Garbarino. The role of ubiquitin in plant senescence and stress responses. Trends Plant Sci 1:331–335, 1996.

129. AR Bahrami, JE Gray. Expression of a proteasome alpha-type subunit gene during tobacco development and senescence. Plant Mol Biol 39:325–333, 1999.

130. C-Z Jiang, SR Rodermel, RM Shibles. Photosynthesis, Rubisco activity and amount, and their regulation by transcription in senescing soybean leaves. Plant Physiol 101:105–112, 1993.

131. LW Peterson, GE Kleinkopf, RC Huffaker. Evidence for lack of turnover of RuDP carboxylase in barley leaves. Plant Physiol 51:1042–1045, 1973.

132. T Mae, A Makino, K Ohira. Changes in the amount of ribulose bisphosphate carboxylase synthesized and degraded during the life span of rice leaf (*Oryza sativa* L.). Plant Cell Physiol 24:1079–1086, 1983.

133. L Peñarrubia, J Moreno. Increased susceptibility of ribulose-1,5-bisphosphate carboxylase/oxygenase to proteolytic degradation caused by oxidative treatments. Arch Biochem Biophys 281:319–323, 1990.

134. C García-Ferris, J Moreno. Redox regulation of enzymatic activity and proteolytic susceptibility of ribulose 1,5-bisphosphate carboxylase/oxygenase from *Euglena gracilis*. Photosynth Res 35:55–66, 1993.

135. J Moreno, RJ Spreitzer. C172S substitution in the chloroplast-ebcoded large subunit affects stability and stress-induced turnover of ribulose-1,5-bisphosphate carboxylase/oxygenase. J Biol Chem 274:26789–26793, 1999.

136. J Moreno, L Peñarrubia, C García-Ferris. The machanism of redox regulation of ribulose-1.5-bisphosphate carboxylase/oxygenase turnover. A hypothesis. Plant Physiol Biochem 33:121–127, 1995.

137. RB Ferreira, DD Davies. Conversion of ribulose-1,5-bisphosphate carboxylase to an acidic and catalitically inactive form by extracts of osmotically stressed *Lemna minor* fronds. Planta 179:448–455, 1989.

138. RA Mehta, TW Fawcett, D Porath, AK Mattoo. Oxidative stress causes rapid membrane translocation and in vivo degradation of ribulose-1,5-bisphosphate carboxylase/oxygenase. J Biol Chem 267:2810–2816, 1992.

139. C García-Ferris, J Moreno. Oxidative modification and breakdown of ribulose 1,5-bisphosphate carboxylase/oxygenase induced in *Euglena gracilis* by nitrogen starvation. Planta 193:208–215, 1994.

140. T Kubawara, Y Hashimoto. Purification of a dithiothreitol-sensitive tetrameric protease from spinach PS II membranes. Plant Cell Physiol 31:581–589, 1990.

141. R Landolt, P Matile. Glyoxisome-like microbodies in senescent spinach leaves. Plant Sci 72:159–164, 1990.

142. L De Bellis, M Nishimura. Development of enzymes of the glyoxilate cycle during senescence of pumpkin cotyledons. Plant Cell Physiol 32:555–562, 1991.

143. K Kamachi, T Yamaya, T Hayakawa, T Mae, K Ojima. Changes in cytosolic glutamine synthetase polypeptide and its mRNA in a leaf blade of rice plants during natural senescence. Plant Physiol 98:1323–1329, 1992.

144. K Kamachi, T Yamaya, T Mae, K Ojima. A role for glutamine synthetase in the remobilization of leaf nitrogen during natural senescence in rice leaves. Plant Physiol 96:411–417, 1991.

145. R Vincent, V Fraisier, S Chaillou, MA Limami, E Deleens, B Phillipson, C Douat, JP Boutin, B Hirel. Over-expression of a soybean gene encoding cytosolic glutamine synthetase in shoots of transgenic *Lotus corniculatus* L. plants triggers changes in ammonium assimilation and plant development. Planta 201:424–433, 1997.

146. I Szamosi, DL Shaner, BK Singh. Identification and characterization of a biodegradative form of threonine dehydratase in senescing tomato (*Lycopersicon esculentum*) leaf. Plant Physiol 101:999–1004, 1993.

147. JE Thompson, RL Legge, RF Barber. The role of free radicals in senescence and wounding. New Phytol 105:317–344, 1987.

148. T Jabs, RA Dietrich, JL Dang. Initiation of runaway cell death in an *Arabidopsis* mutant by extracellular superoxide. Science 273:1853–1856, 1996.

149. EF Elstner. Oxygen activation and oxygen toxicity. Annu Rev Plant Physiol 33:73–96, 1982.

150. GI Jenkins, HW Woolhouse. Photosynthetic electron transport during senescence of the primary leaves of *Phaseolus vulgaris* L. II. The reactivity of photosystems one and two and a note on the site of reduction of ferricyanide. J Exp Bot 32:989–997, 1981.

151. B Halliwell, JMC Gutteridge. Free Radicals in Biology and Medicine. 3rd ed. Oxford: Clarendon Press, 1999.

152. A Kacperska, M Kubacka-Zebalska. Formation of stress ethylene depends both on ACC synthesis and on the activity of free radical–generating systems. Physiol Plant 77:231–237, 1989.

153. H Greppin, C Penel, T Gaspar. Molecular and Physiological Aspects of Plant Peroxidases. Geneva: University of Geneva, 1986.

154. IK Smith, TL Vierheller, CA Thorne. Properties and functions of glutathione reductase in plants. Physiol Plant 77:449–456, 1989.

155. C Bowler, M Van Montagu, D Inzé. Superoxide dismutase and stress tolerance. Annu Rev Plant Physiol Plant Mol Biol 43:83–116, 1992.

156. L Guan, J G Scandalios. Characterization of the catalase antioxidant defense gene *cat1* of maize and its developmentally regulated expression in transgenic tobacco. Plant J 3:527–536, 1993.

157. K Asada. Ascorbate peroxidase—a hydrogen peroxide–scavenging enzyme in plants. Physiol Plant 85:235–241, 1992.

158.  GNM Kumar, NR Knowles. Changes in lipid peroxidation and lipolytic and free-radical scavenging enzyme activities during aging and sprouting of potato (*Solanum tuberosum*) seed-tubers. Plant Physiol 102:115–124, 1993.

159.  M Kar, P Streb, B Hertwig, J Feierabend. Sensitivity to photodamage increases during senescence in excised leaves. J Plant Physiol 141:538–544, 1993.

160.  FR Tadeo, E Primo-Millo. Peroxidase activity changes and lignin deposition during the senescence process in *Citrus* stigmas and styles. Plant Sci 68:47–56, 1990.

161.  R Keller, FS Renz, J Kossmann. Antisense inhibition of the GDP-mannose pyrophosphorylase reduces the ascorbate content in transgenic plants leading to developmental changes during senescence. Plant J 19:131–141, 1999.

162.  J Zhou, PB Goldsbrough. Functional homologs of fungal metallothionein genes from *Arabidopsis*. Plant Cell 6:875–884, 1994.

163.  JM Palma, M Garrido, MI Rodriguez-García, LA Del Rio. Peroxisome proliferation and oxidative stress mediated by activated oxygen species in plant peroxisomes. Arch Biochem Biophys 287:68–74, 1991.

164.  LM Sandalio, VH Fernandez, FL Ruperez, LA Del Rio. Superoxide free radicals are produced in glyoxisomes. Plant Physiol 87:1–4, 1988.

165.  G Paliyath, MJ Droillard. The mechanisms of membrane deterioration and disassembly during senescence. Plant Physiol Biochem 30:789–812, 1992.

166.  MM Olsson, C Liljenberg. Effects of physiological and ontogenetical ageing on sterol levels and composition in pea leaves. Phytochemistry 29:765–768, 1990.

167.  G Paliyath, JE Thompson. Evidence for early changes in membrane structure during postharvest development of cut carnation flowers. New Phytol 114:555–562, 1990.

168.  K Lohner. Effects of small organic molecules on phospholipid phase transitions. Chem Phys Lipids 57:341–362, 1991.

169.  CL Duxbury, RL Legge, G Paliyath, RF Barber, JE Thompson. Alterations in membrane protein conformation in response to senescence-related changes in membrane fluidity and sterol concentration. Phytochemistry 30:63–68, 1991.

170.  A Carruthers, DL Melchior. How bilayer lipids affect membrane protein activity. Trends Biol Sci 11:331–335, 1986.

171.  S Yoshida, M Uemura, T Niki, A Sakai, LV Gusta. Partition of membrane particles in aqueous two-polymer phase system and its practical use for purification of plasma membranes from plants. Plant Physiol 72:105–114, 1983.

172.  EJ Pell, KL Steffen. Active oxygen/oxidative stress and plant metabolism. Current Topics in Plant Physiology. An American Society of Plant Physiologists Series. Vol 6. Rockville, MD, 1991.

173.  KP Pauls, JE Thompson. Evidence for the accumulation of peroxidized lipids in membranes of senescing cotyledons. Plant Physiol 75:1152–1157, 1984.

174.  MA Rouet-Mayer, JM Bureau, C Laurière. Identification and characterization of lipoxygenase isoforms in senescing carnations. Plant Physiol 98:971–978, 1992.

175.  M Fukuchi-Mizutani, K Savin, E Cornish, Y Tanaka, T Ashikari, T Kusumi, N Murata. Senescence-induced expression of a homologue of delta 9 desaturase in rose petals. Plant Mol Biol 29:627–635, 1995.

176.  F Cheour, J Arul, J Makhlouf, C Willemot. Delay of membrane lipid degradation by calcium treatment during cabbage leaf senescence. Plant Physiol 100:1656–1660, 1992.

177.  DJ McCormac, JF Todd, G Paliyath, JE Thompson. Modulation of bilayer fluidity affects lipid catabolism in microsomal membranes of tomato fruit. Plant Physiol Biochem 31:1–8, 1993.

178.  V Buchanan-Wollaston. The molecular biology of leaf senescence. J Exp Bot 48:181–199, 1997.

179.  H Wang, HM Wu, AY Cheung. Pollination induces mRNA poly(A) tail-shortening and cell deterioration in flower transmitting tissue. Plant J 9:715–727, 1996.

180.  CM Smart, SE Hosken, H Thomas, JA Greaves, BG Blair, W Schuch. The timing of maize leaf senescence and characterization of senescence-related cDNAs. Physiol Plant 93:673–682, 1995.

181.  I John, R Hackett, W Cooper, R Drake, A Farrell, D Grierson. Cloning and characterization of tomato leaf senescence-related cDNAs. Plant Mol Biol 33:641–651, 1997.

182.  SJ Crafts-Brandner, RR Klein, P Klein, R Holzer, U Feller. Coordination of protein and mRNA abundances of stromal enzymes and mRNA abundances of the Clp protease subunits during senescence of *Phaseolus vulgaris* (L.) leaves. Planta 200:312–318, 1996.

183.  M Schaffer, R Fischer. Analysis of mRNAs that accumulate in response to low temperature identifies a thiol protease gene in tomato. Plant Physiol 87:431–417, 1988.

184.  C Tournaire, S Kushnir, G Bauw, D Inze, B Teyssendier de la Serve, JP Renaudin. A thiol protease and anionic peroxidase are induced by lowering cytokinins during callus growth in *Petunia*. Plant Physiol 111:159–168, 1996.

185.  C Guerrero, M de la Calle, MS Reid, V Valpuesta. Analysis of the expression of two thiolprotease genes from daylily (*Hemerocallis* spp.) during flower senescence. Plant Mol Biol 36:565–571, 1998.

186.  P Genschik, A Durr, J Fleck. Differential expression of several E2-type ubiquitin carrier protein genes at different developmental stages in *Arabidopsis thaliana* and *Nicotiana sylvestris*. Mol Gen Genet 244:548–556, 1994.

187.   JE Garbarino, WR Belknap. Isolation of a ubiquitin-ribosomal protein gene (*ubi3*) from potato and expression of its promoter in transgenic plants. Plant Mol Biol 24:119–127, 1994.

188.   JE Garbarino, T Oosumi, WR Belknap. Isolation of a polyubiquitin promoter and its expression in transgenic potato plants. Plant Physiol 109:1371–1378, 1995.

189.   N Ito, K Tomizawa, K Tanaka, M Matsui, RE Kendrick, T Sato, H Nakagawa. Characterization of 26S proteasome alpha- and beta-type and ATPase subunits from spinach and their expression during early stages of seedling development. Plant Mol Biol 34:307–316, 1997.

190.   K Nakashima, T Kiyosue, K Yamaguchi-Shinozaki, K Shinozaki. A nuclear gene, *erd1*, encoding a chloroplast-targeted Clp protease regulatory subunit homolog is not only induced by water stress but also developmentally up-regulated during senescence in *Arabidopsis thaliana*. Plant J 12:851–861, 1997.

191.   K Nakabayashi, M Ito, T Kiyosue, K Shinozaki, A Watanabe. Identification of *clp* genes expressed in senescing *Arabidopsis* leaves. Plant Cell Physiol 40:504–514, 1999.

192.   LM Weaver, JE Froehlich, RM Amasino. Chloroplast-targeted ERD1 protein declines but its mRNA increases during senescence in *Arabidopsis*. Plant Physiol 119:1209–1216, 1999.

193.   A Kato, M Hayashi, Y Takeuchi, M Nishimura. CDNA cloning and expression of a gene for 3-ketoacyl-CoA thiolase in pumpkin cotyledons. Plant Mol Biol 31:843–852, 1996.

194.   IA Graham, CJ Leaver, SM Smith. Induction of malate synthase gene expression in senescent and detached organs in cucumber. Plant Cell 4:349–357, 1992.

195.   N Sakurai, T Hayakawa, T Nakamura, T Yamaya. Changes in the cellular localization of cytosolic glutamine synthetase protein in vascular bundles of rice leaves at various stages of development. Planta 200:306–311, 1996.

196.   C Callard, M Axelos, L Mazzolini. Novel molecular markers for late phases of the growth cycle of *Arabidopsis thaliana* cell-suspension cultures are expressed during organ senescence. Plant Physiol 112:705–715, 1996.

197.   MH Walter, J-W Liu, J Wunn, D Hess. Bean ribonuclease-like pathogenesis-related protein gene (*Ypr10*) displays complex patterns of developmental, dark-induced and exogenous-stimulus-dependent expression. Eur J Biochem 239:281–293, 1996.

198.   Y Azumi, A Watanabe. Evidence for a senescence-associated gene induced by darkness. Plant Physiol 95:577–583, 1991.

199.   SA OH, SY Lee, IK Chung, C-H Lee, HG Nam. A senescence-associated gene of *Arabidopsis thaliana* is distinctly regulated during natural and artificially induced leaf senescence. Plant Mol Biol 30:739–754, 1996.

200.   M García-Hernández, A Murphy, L Taiz. Metallothioneins 1 and 2 have distinct but overlapping expression patterns in *Arabidopsis*. Plant Physiol 118:387–397, 1998.

201.   A Van-Gysel, M Van-Montagu, DA Inzé. Negatively light-regulated gene from *Arabidopsis thaliana* encodes a protein showing high similarity to blue copper-binding proteins. Gene 136:79–85, 1993.

202.   E Himelblau, H Mira, S-J Lin, VC Culotta, L Peñarrubia, RM Amasino. Identification of a functional homolog of the yeast copper homeostasis gene *ATX1* from *Arabidopsis*. Plant Physiol 117:1227–1234, 1998.

203.   A DeLong, A Calderonurrea, SL Dellaporta. Sex determination gene *TASSELSEED2* of maize encodes a short-chain alcohol dehydrogenase required for stage-specific floral organ abortion. Cell 74:757–768, 1993.

204.   AB Bleecker. The evolutionary basis of leaf senescence: method to the madness? Curr Opin Plant Biol 1:73–78, 1998.

205.   LL Hensel, V Grbic, DA Baumgarten, AB Bleecker. Developmental and age-related processes that influence the longevity and senescence of photosynthetic tissues in *Arabidopsis*. Plant Cell 5:553–564, 1993.

206.   R Hayati, DB Egli, SJ Crafts-Brandner. Carbon and nitrogen supply during seed filling and leaf senescence in soybean. Crop Sci 35:1063–1069, 1995.

207.   DA Navarre, TJ Wolpert. Victorin induction of an apoptotic/senescence-like response in oats. Plant Cell 11:237–249, 1999.

208.   N Dai, A Schaffer, M Petreikov, Y Shahak, Y Giller, K Ratneer, A Levine, D Granot. Overexpression of *Arabidopsis* hexokinase in tomato plants inhibits growth, reduces photosynthesis, and induces rapid senescence. Plant Cell 11:1253–1266, 1999.

209.   A Smart. Gene expression during leaf senescence. New Phytol 126:419–448, 1994.

210.   JD Miller, RN Arteca, EJ Pell. Senescence-associated gene expression during ozone-induced leaf senescence in *Arabidopsis*. Plant Physiol 120:1015–1024, 1999.

211.   EJ Pell, MS Dann. Multiple stress–induced foliar senescence and implications for whole-plant longevity. In: HA Mooney, WE Winner, EJ Pell, eds. Response of Plants to Multiple Stresses. San Diego: Academic Press, 1991, pp 189–204.

212.   L Peñarrubia, J Moreno. Molecular mechanisms of plant responses to elevated levels of tropospheric ozone. In: M Pessarakli, ed. Handbook of Plant and Crop Stress. 2nd ed. New York: Marcel Dekker, 1999, pp 769–793.

213.   S Gan, RM Amasino. Cytokinins in plant senescence: from spray and pray to clone and play. Bioessays 18:557–656, 1996.

214.   N Ori, MT Juarez, D Jackson, J Yamaguchi, GM Banowetz, S Hake. Leaf senescence is delayed in tobacco plants expressing the maize homeobox gene *knotted1* under the control of a senescence-activated promoter. Plant Cell 11:1073–1080, 1999.

215. V Grbic, AB Bleecker. Ethylene regulates the timing of leaf senescence in *Arabidopsis*. Plant J 8:595–602, 1995.
216. I John, R Drake, A Farrell, W Cooper, P Lee, P Horton, D Grierson. Delayed leaf senescence in ethylene deficient ACC-oxidase antisense tomato plants—molecular and physiological analysis. Plant J 7:483–490, 1995.
217. Y-S Noh, RM Amasino. Identification of a promoter region responsible for the senescence-specific expression of *SAG12*. Plant Mol Biol 41:181–194, 1999.
218. Y-S Noh, RM Amasino. Regulation of developmental senescence is conserved between *Arabidopsis* and *Brassica napus*. Plant Mol Biol 41:195–206, 1999.
219. T Berberich, H Sano, T Kusano. Involvement of a MAP kinase, ZmMPK5, in senescence and recovery from low-temperature in maize. Mol Gen Genet 262:534–542, 1999.
220. GV Novikova, IE Moshkov, AR Smith, ON Kulaeva, MA Hall. The effect of ethylene and cytokinin on guanosine 5′-triphosphate binding and protein phosphorylation in leaves of *Arabidopsis thaliana*. Planta 208:239–246, 1999.
221. L Tamagnone, A Merida, N Stacey, K Plaskitt, A Parr, CF Chang, D Lynn, JM Dow, K Roberts, C Martin. Inhibition of phenolic acid metabolism results in precocious cell death and altered cell morphology in leaves of transgenic tobacco plants. Plant Cell 10:1801–1816, 1998.
222. PJ Coates, DJ Jamienson, K Smart, AR Presott, PA Hall. The prohibitin family of mitochondrial proteins regulate replicative lifespan. Curr Biol 7:607–610, 1997.
223. D Orzaez, A Granell. Programme of senescence in petals and carpels of *Pisum sativum* L. flowers and its control by ethylene. FEBS Lett 404:275–278, 1997.
224. Y Okatan, GM Kahanak, LD Noodén. Characterization and kinetics of soybean maturation and monocarpic senescence. Physiol Plant 52:330–338, 1981.
225. C Peisker, H Thomas, F Keller, P Matile. Radiolabelling of chlorophyll for studies on catabolism. J Plant Physiol 136:544–549, 1990.
226. U Schreiber, U Schliwa, W Bilger. Continuous recording of photochemical and non-photochemical chlorophyll fluorescence quenching with a new type of modulation fluorometer. Photosynth Res 10:51–62, 1986.
227. V Raggi. $CO_2$ accumulation, respiration and chlorophyll fluorescence in peach leaves infected by *Taphrina deformans*. Physiol Plant 93:540–544, 1995.
228. HK Lichtenthaler, JA Miehé. Fluorescence imaging as a diagnostic tool for plant stress. Trends Plant Sci 2:316–320, 1997.
229. L Peñarrubia, J Moreno, JL García-Martínez. Proteolysis of ribulose-1,5-biphosphate carboxylase/oxygenase in leaves of *Citrus* explants throughout the annual cycle and its regulation by ethylene. Physiol Plant 73:1–6, 1988.
230. HWJ Van den Broek, LD Noodén, JS Sevall, J Bonner. Isolation, purification and fractionation of nonhistone chromosomal proteins. Biochemistry 12:229–236, 1973.
231. BD Derman, DC Rupp, LD Noodén. Mineral distribution in relation to fruit development and monocarpic senescence in *Anoka* soybeans. Am J Bot 65:205–213, 1978.
232. JMC Gutteridge, B Halliwell. The measurement and mechanism of lipid peroxidation in biological systems. Trends Biochem Sci 15:129–135, 1990.
233. RA Allen. Dissection of oxidative stress tolerance using transgenic plants. Plant Physiol 107:1049–1054, 1995.
234. IFF Benzie, JJ Strain. The ferric reducing ability of plasma (FRAP) as a measure of antioxidant power: the FRAP assay. Anal Biochem 239:70–76, 1996.
235. KS Rowan, HK Pratt, RN Robertson. Relationship of high-energy phosphate content, protein synthesis, and the climacteric rise in the respiration of ripening avocado and tomato fruits. Aust J Biol Sci 2:329–335, 1958.
236. GM Pastori, VS Trippi. Antioxidative protection in a drought-resistant maize strain during leaf senescence. Physiol Plant 87:227–231, 1993.
237. SA Oh, J Park, GI Lee, KH Paek, SK Park, HG Nam. Identification of three genetic loci controlling leaf senescence in *Arabidopsis thaliana*. Plant J 12:527–535, 1997.

# 10
# Abscission

**Roy Sexton**

*Stirling University, Stirling, Scotland*

## I.  GENERAL FEATURES OF ABSCISSION

## A.  Definitions

During its life a plant will shed many of its organs, such as leaves, fruit, petals, buds, bud scales, and bark. Two distinct processes contribute to this loss:

1.  *General attrition* is responsible for the detachment of dead or dying tissues such as bark, old branches, and roots. In these cases, large mechanical forces such as the wind or differential growth of the stem rupture an inherently weak region of tissue, usually producing an irregular tear.

2.  *Abscission* is involved in the loss of leaves, fruit, flowers, and floral parts. It is an active metabolic process resulting in the weakening of anything from 1 to 20 rows of cells in genetically determined abscission zones. Cell wall breakdown is an important element in the loss of structural integrity. Unlike attrition, abscission is under very precise internal hormonal control.

This chapter is concerned with the mechanism, regulation, and agricultural importance of abscission.

## B.  Abscission Zones

The shedding of leaves, flowers, fruit, and so on, occurs as a result of the weakening of abscission zones (AZs). These bisect the base of leaf petioles, leaflets, petals, styles, flower buds, axillary buds, young fruit, and the nodes of very young stems. Abscission zones are not inherently weak and a 0.5-kg weight can be hung on a bean leaf AZ without rupturing it. After abscission has been induced, the same AZ will weaken so that it will break at the slightest touch. This progressive loss of structural integrity occurs over 72 hr after an initial 18-hr lag [1]. Although leaf and fruit abscission usually takes up to 3 days, the process can be extremely rapid and some petals [2] and flower buds [3,4] are shed 1 to 4 hr after the inductive treatment.

## C.  Weakening Process

The scanning electron micrograph in Figure 1 shows a fracturing abscission zone at the base of a bean leaflet. The discrete nature of the fracture line reveals that loss of structural integrity is restricted to just

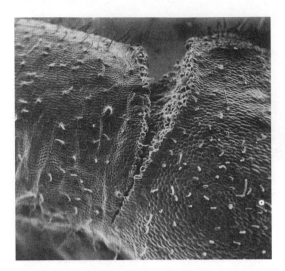

**Figure 1** Scanning electron micrograph of a fracturing leaf abscission zone from bean. Note that the fracture is confined to only one or two rows of cells. The tissue on the left is senescing and its diameter is contracting (note folds), while that on the right is still turgid and is enlarging. This differential growth causes stresses at the abscission zone interface which help separate the cells in the separation layer and rupture the stele. (From Ref. 1.)

one or two rows of cells [1]. If the scar faces are examined after fracture, separation seems to have occurred along the line of the middle lamella, leaving the intact rounded cells covering the surface (Figure 2).

Light microscope and transmission electron microscope (TEM) observations implicate cell wall breakdown as a major factor in the separation of these intact cells [3] (Figure 3). Degradation is particularly prominent in the central region of the wall and involves not only the middle lamella but also the adjacent areas of the primary wall [5]. The fracture bisects all the tissues in a stem or petiole, and studies have shown that wall degradation occurs in all the different living cell classes along the fracture line, including the epidermis and phloem [6].

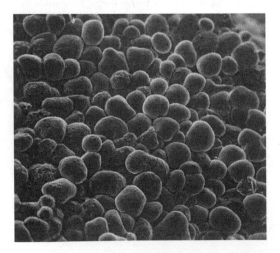

**Figure 2** Scanning electron micrograph of part of a bean abscission zone fracture surface. Note that the cells on the exposed scar are round and turgid, having separated as a result of breakdown of the central areas of the wall. (From Ref. 1.)

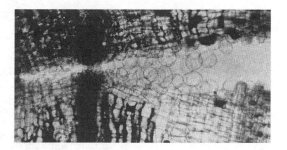

**Figure 3**  Longitudinal fresh section through an abscising leaf abscission zone. The separation layer that runs through the center of the micrograph is full of round, separated, turgid cells. The expansion of these cells in the petiole cortex results in stretching and subsequent rupture of the xylem vessels in the dark vascular trace. (From Ref. 9.)

## D.   Mechanical Forces and Separation

Although the walls of living cells in the abscission zone (AZ) are enzymically degraded, mechanical forces are necessary both to facilitate cell separation and to rupture the xylem [3]. External forces such as the wind and gravity may be involved, although they are usually not sufficient by themselves. Weisner in 1871 [7] showed that if all the living tissues in a petiole are severed, leaves will often remain attached by the xylem for long periods, despite these external agencies.

Many mechanisms have evolved to generate the forces required to cause complete separation [3]. In bean, the growth of cells on the stem side of the abscission zone, coupled with shrinkage on the distal side, has been implicated in producing stresses at the AZ interface that facilitate rupture [8] (Figure 1). Another common system involves the rounding up and osmotic expansion of the separating cells stretching and breaking the xylem [9] (Figure 3). The squirting cucumber provides another rather bizarre example where internally hydrostatic pressures rupture the abscission zone at the base of the fruit, allowing it to shoot away like a water-propelled rocket [10].

## E.   Why Is Weakening Restricted to the Separation Layer?

The positions of abscission zones are a genetically determined characteristic of a given species. For instance, the blackberry (*Rubus fruticosus*) is shed by an abscission zone across the base of the fruit, leaving the white receptacle or plug within the berry. In the closely related raspberry (*R. idaeus*), each of the 70 or so drupelets in the berry has an AZ at its base, so when the berry is detached it leaves the receptacle attached to the plant. In hybrids like the loganberry, the blackberry position is dominant [11]. Plant breeders have also produced varieties that lack the normal AZs, such as the lupin cultivar, which cannot abscise its leaves [12].

## F.   Are There Specialized Abscission Cells?

Sections through AZs show that their anatomy is not very different from adjacent regions of the petiole or pedicel which they bisect [3]. They often have subtle characteristics that allow the general region to be distinguished [3]. The cells are frequently smaller than those in adjacent tissues, and this close packing can make AZs rather darkly pigmented when viewed externally. The stele usually divides into separate bundles before it enters the zone. Abscission zones lack lignification; sclerenchymatous fibers are often replaced by collenchyma.

These features are all thought to have evolved to facilitate rupture. It is assumed that lignin is reduced because it makes walls less susceptible to enzymic attack [3]. The close-packed angular AZ cells expand as the walls are degraded and they round up (Figure 3) [9]. Their enlargement has been implicated in producing the forces that rupture the xylem [9]. The branching of the xylem can be explained because under asymmetrical loading a number of thin, separated strands are more readily ruptured than is one large central bundle.

The cell separation process does not usually involve the entire AZ but only a narrow layer of cells across it. The cells actually involved are known as the *separation layer*. One might expect that the separation layer would contain a distinct specialized class of *abscission cells* that degrade their walls as a result of the abscission signal. However, the cells that will be involved in separation cannot be picked out from their neighbors by simple microscopic examination [3].

One of the enigmas of abscission is why these separation layer cells degrade their walls when apparently identical cells on either side do not. It has been suggested that separation layer cells are biochemically distinct without there being any structural manifestations of the difference [3,12]. An alternative hypothesis envisages that potentially all cells can degrade their walls but that the abscission triggering signal is restricted to just a few rows [3]. McManus and Osborne [13] have used immunological methods to demonstrate specific proteins in the AZ prior to separation. This observation gives credibility to the hypothesis that there are discrete abscission cells.

Abscission zones can be formed very early in the development of some organs; for instance, the minute leaves and flowers inside dormant buds can already have responsive AZs. Why AZs develop at specific positions is not understood. The smaller cells found in them are created by more persistent cell division in the region [14]. Interspecific chimeras between two species of tomato with different AZ positions have been used to study the differentiation of the AZ. It seems that the position of the AZ is dictated by the genetic status of the inner cell lineages, the outer cells responding to signals produced by them [15].

## G. Loss of Abscission Zone Responsiveness

Abscission zones are not necessarily active throughout the entire life of the organ. An unfertilized orange flower can be shed by activating the AZ in the pedicel, but after fruit development starts, this abscission zone becomes inactive and will not respond to identical inductive stimuli [16] (also Figure 8). In peach flowers there are three AZs [17], each of which is active at different phases of flower and fruit development.

Two explanations have been put forward to account for the loss of AZ responsiveness. The first envisages that there is a loss of some vital component of the response machinery, such as a hormone receptor. The second proposes that the response still occurs but that the cell walls of the separation layer are modified by substances such as lignin or suberin that make them resistant to attack by wall hydrolases. In orange pedicels, evidence seems to support the latter hypothesis [18].

## H. Adventitious Abscission Zones

Although virtually all abscission takes place at precisely predictable sites, there are a few interesting cases where it occurs rather randomly. An example of this type of *adventitious* abscission is the *shot hole effect*, where diseased or damaged areas of the leaf blades of *Prunus* species are abscised, leaving holes [19] (Figure 4). The leaves appear as though a shotgun has been fired through them—hence the name. Adventitious AZs also form in internodes of stems of *Impatiens* [20] and mulberry [21], where fracture can occur at variable positions along the internode.

At first sight the ability to induce abscission at random positions seems to contradict the hypothesis that there are discrete preprogrammed abscission cells at genetically defined positions. Since cell division precedes fracture in adventitious zones, it is possible to argue that these new cells differentiate into an AZ, which is then induced to abscise. This need for differentiation of an AZ before abscission can take place might explain why adventitious stem abscission in *Impatiens* takes so much longer (5 to 14 days) [20] than normal leaf abscission (22 to 36 hr) at preformed zones [5]. In a very interesting series of experiments, Warren Wilson et al. [20] have shown that the position of adventitious AZs in *Impatiens* can be modified in a predictable manner by manipulating auxin gradients in the tissue.

## I. Protection of the Fracture Surface

After fracture has occurred, cells in what remains of the AZ on the plant divide and suberize to form the scar that protects the wound [3,22]. The broken xylem vessels become blocked with gums or tyloses and the phloem sieve plates are callosed over to prevent pathogen entry [3]. The antimicrobial enzymes chitinase and $\beta$1–3 glucanase are produced in bean abscission zones [23] and are likely to be one part of an

**Figure 4**  Adventitious abscission zones on the leaf of ornamental cherry. Three areas of the leaf have been wounded by heating, and abscission zones have formed in the living tissue around them. The cell separation, which will result in excision of the damaged area, is evident as a white line. (From R. Sexton, unpublished.)

arsenal of compounds produced in the fracture surface to prevent infection. These protective mechanisms must be very efficient, as relatively few diseases originate at abscission scars.

## II.  REGULATION OF ABSCISSION

## A.  Inductive Stimuli

Normally, the induction of abscission appears to be an integral part of the senescence program accompanying the yellowing of leaves and ripening of fruit. In most abscission systems the process can be accelerated and will take place prematurely in the absence of senescence. For instance, pollination can dramatically accelerate petal abscission [24,25]. In cyclamen, all pollinated flowers shed their corollas in 5 days, whereas unpollinated flowers retained theirs even after 23 days [24]. Accelerated floral abscission is thought to have evolved to prevent wasted visits of scarce pollinators to fertilized flowers.

Leaf loss in temperate species accompanies senescence, which in turn is induced by environmental factors such as photoperiod changes, low temperatures, and drought. Factors that affect the leaf blade adversely can cause premature shedding. These include frost damage, drought [26], bacterial or fungal attack [27,28], damage by herbivores [29], mineral deficiencies, toxins, excessive shading, darkness [30], and competition with younger leaves. Leaf fall is not invariably linked to lamina senescence, and water-stressed ivy plants will shed leaves with the same chlorophyll content as those still attached to normal healthy plants.

Fruit appear to be abscised at several distinct stages of development [31]. Immature fruit can be shed in large numbers in what appears to be a natural thinning process. Sometimes, this seems immensely wasteful, and in species as diverse as oak and avocado less than 10% of potential fruit mature [32,33]. Some young fruit are shed because seed development is defective, although in avocado many embryos in abscised fruit remain viable [34]. In fruit, water deficits [35], mineral deficiencies, pathogen [36] and herbivore attack, and frost damage can also be factors that precipitate premature abscission.

In many cases the reasons why such huge numbers of young fruit are lost are not clear, although competition between developing fruit is certainly an important factor. The chances of a fruit being shed can be reduced dramatically by removing other fruit from the same plant. In soybean, the fate of fruit that will be shed (50 to 80%) is probably determined before fertilization takes place, on the day that the flower reaches anthesis. In the flowers that will be lost, there is a failure for *sink intensity* to increase, so they do not accumulate photoassimilate from source leaves [37].

During the growth phase of a fruit, abscission seems to be inhibited by the presence of the developing seeds. Mature fruit are usually induced to abscise as one of the terminal events in the ripening program. In some species, abscission is not complete but serves just to loosen the fruit so that birds and small animals can detach them. The manipulation of fruit abscission is very important in developing mechanical harvesting.

Flower bud loss is a serious problem in some crops and decorative plants. In lupins, the development of young fruit at the base of the flower spike seems to induce the loss of buds at the apex [38] (Figure 5). Removing some flower buds usually decreases the likelihood of abscission in those that remain. Disease, water stress, waterlogging, mechanical shaking, and frost damage are also reported to enhance bud abscission.

It is not widely recognized that the cessation of stem growth in several tree species involves abscission of the growing apex. In *Tilia* [39] and *Salix* [40] photoperiod seems to be a primary determinant of when abscission occurs, but position of the branch in the canopy, its orientation, conditions for root growth, competition from other apices, and other climatic factors provide modifying influences [21]. There is very little literature concerned with the inductive conditions that lead to the loss of bud scales, stigmas, anthers, and sepals.

## B.   Experimental Induction of Abscission

Experiments early in the 20th century showed that removal of the leaf blade resulted in rapid abscission of the subtending petiole (Figure 6). This was a conveniently reproducible system to study abscission, and seedlings of bean, cotton, and *Coleus* were commonly used. The need for faster synchronized abscission led to the *explant* technique. Here, the abscission region was removed by cuts 1 to 2 cm on either side of the zone. The isolated piece of tissue was kept in a sealed container often over 2% agar until it abscised 2 to 3 days later [41]. This explant system has become extremely popular because it provides a lot of ma-

**Figure 5**   Influence of maturing pods on floral abscission. Lupin flower spike buds open from the base upward (right). If the basal flowers are fertilized, the apical buds abscise, leaving bare stem (center). If the basal flowers are removed, the apical buds remain and develop into pods (left). (From R. Sexton, unpublished.)

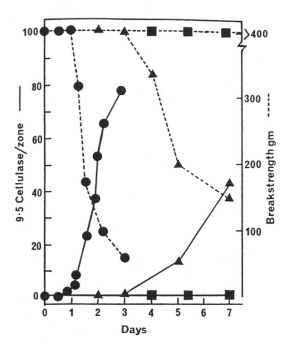

**Figure 6**  Effects of auxin and ethylene on abscission. The structural integrity (break strength) of bean leaf abscission zones was measured at various times after deblading (triangles). After a 72-hr lag, the force necessary to rupture the zone slowly decreases. Adding ethylene speeds up this process by reducing the length of the lag and increasing the rate of weakening (circles). Adding IAA to the end of the petiole inhibits abscission completely (squares). The corresponding levels of 9.5 cellulase in the absission zones are also shown. (From Ref. 1.)

terial for biochemical studies and regulators can readily be applied directly to the abscising region (Figure 6).

## C.  Early Experiments with Regulation

After the demonstration that removal of the leaf blade would cause the abscission of the remaining leaf stalk, it was proposed that reduction in photosynthate supply from the leaf caused the AZ cells to collapse and fracture to occur. Kuster in 1916 [42] discovered that a tiny fragment of blade left attached to the petiole was enough to prevent abscission. As it seemed unlikely that this small area of leaf was providing sufficient nutrients, it was suggested that the healthy blade produced a hormonal factor that prevented abscission [42].

The discovery that auxin was produced by young leaf blades led Laibach and his student Mai [43] to investigate whether Kuster's inhibitor was auxin. They demonstrated that pollen rich in auxin applied to the cut end of a debladed petiole delayed abscission. La Rue [44] repeated the experiment with synthetic indoleacetic acid (IAA) (Figure 6). A few years later, workers using IAA to induce parthenocarpic development of fruit noticed that the treatment also delayed fruit abscission [45]. As a result, a general hypothesis was put forward that abscission resulted from a reduction of the amounts of auxin in the abscission zone caused by a reduced auxin supply from the senescing distal organ (reviewed in Ref. 45).

At the beginning of the 20th century it had been found that traces of the gas used for illumination would cause the abscission of leaves, petals, and fruits. The active component, identified as ethylene (ethene), promoted abscission at very low concentrations of 1 to 8 ppm [46] (Figure 6). Some 30 years later it was shown that ethylene was synthesized by most plants [47,48], and Milbrath et al. [49] demonstrated that ethylene produced by apples would defoliate roses. However, auxin was in vogue at the time and it was to be 20 years before the role of ethylene as a natural regulator of abscission was taken seriously.

In 1955, Osborne [50] showed that diffusates from senescent petioles contained a soluble factor that accelerated explant abscission. Soon afterward, Van Stevenick [51] demonstrated that developing pods at the base of a lupin inflorescence stimulated abscission of the flowers above them (Figure 5). He succeeded in extracting an abscission stimulator from the young pods [51]. A third group headed by Addicott [52] identified a growth inhibitor that was present in young cotton fruits approaching abscission. The substance was purified [52], shown to accelerate abscission, and characterized as abscisic acid (ABA). Subsequently, lupin pod extracts were also shown to contain ABA [53]. After a number of correlations were reported between increasing ABA levels and abscission [54,55], ABA became accepted as a third potential regulator of the process.

## D.   IAA and the Control of Abscission

The demonstration that auxin would inhibit leaf abscission (Figure 6) was followed by several attempts to measure its levels in naturally abscising systems. It was found that both the extractable and diffusible auxin levels dropped rapidly as the leaves yellowed and abscission approached [45,56,57]. Similar correlations were observed between low auxin levels and fruit abscission [58].

A simple model emerged which suggested that if the auxin levels in the AZ remained above a critical level, abscission was inhibited [58]. Factors that promoted abscission, such as aging, frost damage, and water stress, were thought to lower the levels of free auxin in the zone. It emerged that the rate of auxin transport from the distal organ was a major influence on IAA levels in the AZ [59], although rates of synthesis and degradation are also implicated.

Modifications of the simple auxin concentration theory were necessary when it became clear that the levels of auxin on the stem or proximal side of the AZ also influenced abscission. Jacobs [60] demonstrated that the presence of young auxin-producing leaves on the stem seemed to accelerate loss of debladed petioles. Removal of the apical bud and young leaves delayed abscission below it, the influence of the apical bud being restored if replaced by a supply of IAA [60,61].

As a result of Jacobs' observations, a number of groups showed that if auxin was applied to the proximal side of an AZ, it accelerated abscission [62], while if applied to the distal side, it delayed weakening. This led Addicott et al. [63] to propose the *gradient theory*, where the direction of the auxin gradient across the AZ was important, not the absolute concentration. Auxin approaching the zone from the distal direction inhibited abscission, whereas that moving from the stem accelerated the process [45].

The gradient theory was subsequently challenged because very high levels of auxin applied to the stem side would often inhibit abscission [64]. In 1964, Abeles and Rubinstein [65] reported that auxin applications would promote the synthesis of ethylene, a potent accelerator of abscission. As a consequence, it is possible to attribute the accelerating effect of proximal auxin to increased ethylene production coupled to a failure of proximal IAA to reach the AZ before abscission was under way. Abeles [66] argued that proximally applied auxin would move to the zone much more slowly than would distal applications, as its movement by diffusion would be opposed by basipetal auxin transport. Higher auxin concentrations applied proximally would diffuse to the zone more rapidly, accounting for the inhibition sometimes observed. Morris [67] has shown that the ethylene synthesis inhibitor aminooxyacetic acid (AOA) will inhibit the accelerating effect of proximal auxin additions, adding weight to Abeles' explanation. Morris [67] also proposes that auxin applied proximally induces the synthesis of the ethylene precursor aminocyclopropane carboxylic acid (ACC), which in turn diffuses to the zone and promotes ethylene synthesis.

Although ethylene production by proximal auxin applications offers an explanation of the acceleration of abscission, it does not account entirely for the speeding effect of the apex in Jacobs' [60] experiments, and as a result, the gradient theory still has its advocates.

## E.   Stage 1 and Stage 2 Responses

In 1963 Rubinstein and Leopold [68] discovered that if auxin was added distally more than 12 hr after deblading leaves, it accelerated abscission rather than preventing it. They put forth the view that explants went through two stages after excision. In stage 1, auxin additions would inhibit abscission and prolong the stage. If auxin levels fell, stage 2 was entered, when auxin accelerated weakening.

The finding that auxin induced ethylene formation was to provide an explanation of the opposing effects of auxin in stages 1 and 2 [65]. It was proposed that in stage 1 the AZ cells were sensitive to auxin but insensitive to ethylene and that in stage 2 the reverse applied. If auxin was added in stage 1, it prolonged the length of the ethylene-insensitive condition. If the auxin levels fell below a critical threshold, stage 2 was entered, when auxin no longer had any effect and the additional ethylene accelerated weakening. Some support for this hypothesis has come from experiments that reduced the accelerating effect of auxin in stage 2 by removing ethylene or inhibiting its production [66].

It is widely assumed that a loss of responsiveness to auxin in stage 2 is due to a loss of auxin receptors. Jaffe and Goren [69] have reported that during stage 2 there is not only a reduction in IAA's ability to retard abscission but also in its power to evoke an $H^+$ ion efflux. They speculate that these two diverse processes may become ineffective because a common component of the response machinery (such as a receptor) is lost. Another possible explanation is that auxin is simply not reaching the separation zone cells if additions are delayed because of a decline in auxin transport. When, after abscission, auxin is added to the cells of the fracture surface, it still inhibits production of the wall-degrading enzymes, suggesting that at least one system is still auxin sensitive [70,71].

The role of auxin in the regulation of abscission has been eclipsed by work on ethylene. However, it should be remembered that IAA additions will completely prevent any effect of ethylene for extended periods of time and that IAA should therefore be included in any model of abscission control.

## F.  Ethylene Accelerates Abscission

The ethylene-induced acceleration of abscission is probably the most consistently demonstrated of all plant growth regulator responses (Figure 6). It has been shown to induce shedding of a wide variety of organs and in a huge range of plant species (see the lists in Ref. 72). The threshold concentrations necessary to induce the response are between 0.1 and 5 µL/L [72]. Other unsaturated hydrocarbons, such as propene, acetylene, and butene, will act as ethylene substitutes, but they are much less effective [73]. Some analogues, such as 2,5-norbornadiene, are competitive inhibitors [74,75].

There can be dramatic changes in the responsiveness of AZs to ethylene. We have already seen how distal auxin additions to fruit and leaf AZs makes them insensitive to ethylene. There are also well-documented changes in natural sensitivity. For instance, Halevy et al. [24] showed that unfertilized cyclamen flowers will not shed their corollas in ethylene, whereas fertilized flowers will. Similarly, styles of orange would not abscise in the presence of ethylene until fertilization had occurred [76].

It is assumed that the presence of both the gas and its receptors are required for ethylene action to occur.

$$C_2H_4 + \text{receptor} \rightarrow \text{ethylene-receptor complex} \rightarrow \text{abscission}$$

Factors that increase sensitivity, such as water deficit, aging, and ethylene itself, are thought to increase the levels of the receptor, while auxin reduces it [72].

## G.  Is Ethylene a Natural Regulator of Abscission?

The early observations that ethylene accelerated abscission were treated as a curious artifact, and even after it was shown that plants naturally evolve ethylene, this gas was not widely envisaged as a natural regulator. Part of the problem was that young leaves seemed to produce more ethylene than older ones [78]. After the demonstration that auxin inhibited the accelerating effects of ethylene, Barlow [79] proposed that it was the auxin/ethylene balance in the tissue that was important. In young leaves there was sufficient auxin to inhibit ethylene's abscission accelerating effect, while in old leaves there was not.

In 1962, a review by Burg [80] argued that the concentrations of ethylene in plants were such that they could easily control abscission. The first claims that the gas was a natural regulator of abscission were based on demonstrations that increased ethylene production rates were correlated with abscission. Such parallels were shown in a wide variety of leaves, fruit, and flowers, although a few authors found no simple correlation (reviewed in Ref. 72).

Of course, correlations do not constitute proof of involvement. To implicate ethylene firmly in abscission, it is necessary to show that endogenous ethylene concentrations increase above the threshold concentrations of 0.5 to 1.0 μL/L that are necessary to cause accelerated abscission if added exogenously [72]. Because AZs are very tiny, it is difficult to extract enough gas to make these measurements. A strong correlation exists between ethylene production rates and internal concentrations. Using this relationship, it was estimated that ethylene production rates of 3 to 5 μL of ethylene per kilogram per hour were necessary to trigger abscission [83], and these were subsequently shown to be exceeded in many abscising systems [72,82]. There have been some direct measurements of the gas concentration in AZs which showed levels above the threshold [83,84]. Raspberry fruit are unusually well suited to these measurements, having 70 to 100 abscission zones enclosed within the fruit. Ethylene levels around these zones showed that concentrations were less than the threshold level of 0.5 μL/L in green fruit but exceeded it in ripening, abscising fruit [85] (Figure 7).

A clever alternative approach was adopted by Jackson et al. [86]. They measured the rate of ethylene production in senescent bean leaves just prior to abscission and then applied (2-chlorethyl)phosphonic acid (CEPA) to younger petioles to generate similar amounts. This treatment caused abscission.

Reducing the levels of internal ethylene in AZs has also been used to establish ethylene's role. Early experiments employed potassium permanganate or mercuric perchlorate to absorb the gas, and there are several reports of delayed abscission as a result [87]. A more effective approach has been to use hypobaric or low pressures, which increase diffusive loss [72]. Aminoethoxyvinyl glycine (AVG), an inhibitor of ethylene biosynthesis, has been shown to slow natural abscission [89–91]. Transgenic tomato plants have been produced that synthesize very little ethylene, but unfortunately, their abscission behavior was not recorded [92].

Inhibitors of ethylene action such as silver ions, which inhibit ethylene responses such as fruit ripening and floral senescence, are also very effective at preventing abscission [93–96]. The mechanism of the $Ag^+$ effect is not understood, although interaction with the ethylene receptor is assumed. Norbornadiene

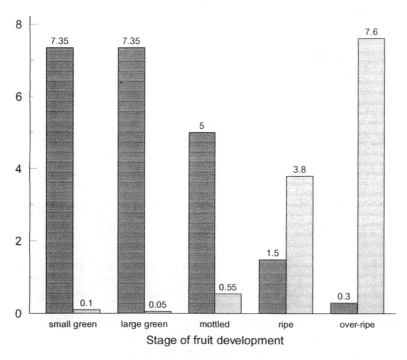

**Figure 7**  Correlation between the internal ethylene concentrations in raspberry fruit and the onset of abscission. The concentration of ethylene (in μL/L) (pale columns) around the abscission zones within fruit at various stages of ripening are shown. When green fruit progress to the mottled and ripe stages, the concentrations exceed the 0.25 μL/L threshold necessary to induce abscission experimentally in green fruit. The fruit removal force (Newtons) required to break the abscission zones is also plotted (dark columns). Note that it starts to decline in fruit that are mottled or riper, where ethylene levels exceed 0.25 μL/L. (From Ref. 85.)

**Figure 8**  Inhibition of sweetpea floral abscission by the ethylene antagonist DACP [77]. Sweetpea buds exposed to DACP for 16 h (far left) are just beginning to abscise after 144 hr in a vase; the untreated controls (second left) shed their buds completely 48 hr earlier. Opened flowers (far right) had abscised all their petals by 144 hr while those exposed to DACP (middle right) were still firmly attached. Note that the unopened buds are shed at an abscission zone at the base of the pedicel, in contrast to older flowers, where this zone is inactive and the petals alone are lost, leaving the remainder of the flower attached. (From R. Sexton, unpublished.)

is a competitive inhibitor and also interferes with natural abscission [74,75]. Sisler [77] has shown that diazocyclopentadiene inhibits ethylene responses, probably by binding irreversibly to the receptor in the light. This substance inhibits sweet pea abscission (Figure 8).

As a result of this wealth of data, there is widespread acceptance that ethylene is involved in natural abscission. Whether it acts directly or indirectly is a more contentious question. There is evidence that the movement of the ethylene precursor ACC in the xylem from water-stressed roots to the leaves may be an important mechanism for inducing abscission [97].

## H.   Is Ethylene Having a Direct or an Indirect Effect?

It has been shown that ethylene lowers auxin concentrations in the abscission zone by reducing synthesis [98] and transport [99] of the hormone and increasing loss by conjugation and breakdown [100]. As a result, it was proposed that ethylene speeds abscission indirectly by lowering auxin levels [72].

In a series of simple but elegant experiments, Beyer [101] showed that ethylene was actually involved in abscission both directly and indirectly. A system was developed whereby the leaf blade and abscission zone could be exposed to ethylene independently. If the zone or leaf blade was treated alone, abscission did not occur; however, if both were treated, the lamina was shed. Beyer showed that treatment of the leaf blade with ethylene reduced auxin transport down the petiole by over 80%. The effect could be mimicked by auxin transport inhibitors and could be reversed if the supply of auxin from the blade was augmented. This suggested that ethylene served indirectly to reduce the levels of auxin in the AZ by interfering with auxin movement from the blade. Abscission would not occur, however, if auxin transport alone was impeded: ethylene also had to be present at the zone, suggesting a second direct role in abscission induction.

## I.   Does Ethylene Induce Abscission or Merely Accelerate It?

The question of whether ethylene induces abscission or just acts as an accelerator has not been resolved. If ethylene is the inducer, removing it or interfering with ethylene action should not just slow the process down but should stop it entirely. In the majority of experiments of this type, either abscission does eventually occur or observations are not continued long enough to distinguish between stopping and slowing.

## J.   Abscisic Acid

After Addicott's group had isolated ABA from young cotton fruit, a number of correlations were reported in which increases in ABA seemed to be associated with abscission [54,55,57,102–104]. In contrast, there is a series of papers in which no simple relationship was shown to exist between endogenous ABA and leaf [105,106], fruit, or flower bud [107–109] abscission. However, correlations cannot prove or disprove involvement, and it could be argued that increases in tissue sensitivity to the hormone could induce abscission without a change in ABA concentration. Another general problem encountered when measuring hormone levels is that the value obtained represents an overall mean concentration for the piece of material and ignores important local variations. For instance, a very thin (2 mm) slice made to remove the AZ from a petiole will at best contain only 10% abscission zone cells, and the hormone concentrations in these could be very different from those in the adjacent, contaminating senescing tissue.

In a review of ABA action, Milborrow [110] expressed surprise that among the hundreds of plants sprayed with ABA, more did not show an abscission response. By way of a reply, Addicott [55] listed a considerable number of cases where a response does occur. Abscisic acid is much more effective when applied to explants, and some scientists believe that it acts by promoting ethylene formation. For instance, Sagee et al. [111] have reported that ABA is ineffective in the presence of the ethylene synthesis inhibitor AVG. The current consensus seems to be that ABA is not directly involved in abscission. However, there is some evidence that it may be important in cereal seed shedding, which is one of the few ethylene-insensitive systems [112].

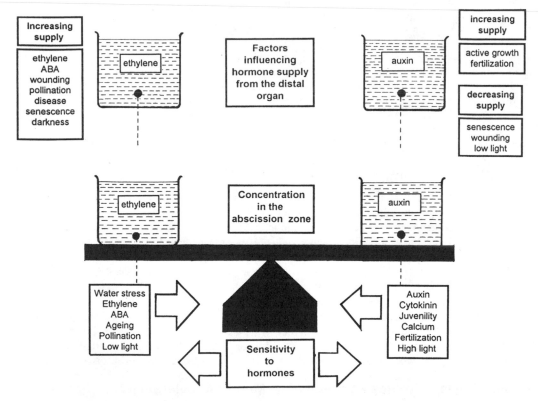

**Figure 9**  Modern representation of the ethylene auxin balance theory. The relative concentrations of auxin and ethylene in the abscission zone are represented by the weight of beakers on either side of a balance. If the left-hand side goes down, the induction of cell separation occurs and abscission takes place. The weight of the containers is influenced predominately by supply from the distal organ, such as a leaf or fruit. Some factors that influence this supply are shown. The balance can also be tipped in either direction by moving the fulcrum from side to side. This represents changes in the tissue's sensitivity to either hormone. Moving the fulcrum to the right increases sensitivity to ethylene, and some factors that do this, such as water stress and aging, are shown. Moving the fulcrum to the left increases sensitivity to auxin and decreases ethylene responsiveness. Adding auxin itself or cytokinin will cause this to occur. (From R. Sexton, unpublished.)

## K.  Other Potential Regulators

Both gibberellic acid and cytokinins will influence abscission, although they are thought to be less important than the other plant hormones [55,113]. Cytokinins can delay abscission, probably by indirectly delaying senescence [114]. Gibberellic acid will accelerate abscission [115,116], there being some debate as to whether or not the effect is mediated by ethylene [66,117]. Long-chain unsaturated fatty acids such as linolenic acid also enhance abscission [118]. Experiments on bean abscission zones showed that the accelerating effect of the $C_{18}$ unsaturated fatty acids was mediated by the production of fatty acid hydroperoxides and that ethylene was not involved. It is not clear whether these compounds are involved in the regulation of natural abscission, but they do accumulate in some senescent tissues.

## L.  Regulation of Abscission: Summary

As far as one can judge from the literature, the consensus is that the regulation of abscission directly involves the concentrations of IAA and ethylene and the sensitivity of AZ tissue to them. Other factors that influence the process do so through these agencies.

Figure 9 illustrates a *balance model* based on the relative AZ concentrations of auxin on the right and ethylene on the left. If the left-hand side of the balance goes down, weakening starts, whereas if the right end is down, the process is inhibited. The concentrations of ethylene and auxin in the AZ are influenced by a variety of factors, some of which are shown at the top of the diagram. The balance can also be affected by the position of the fulcrum, which can move from the center toward either end. Moving this to the right represents a decrease in the sensitivity to auxin and an increase in sensitivity to ethylene, and vice versa. Changing sensitivity probably involves the amount of receptors, and some of the factors that change it are illustrated.

## III.  CELL BIOLOGY OF ABSCISSION

## A.  Mechanism of Abscission: Early Theories

Early botanists believed that abscission was due to the formation of a corky layer on the stem side of the AZ which cut the supply of sap to the separation layer and caused the cells in it to collapse. The anatomist Inman in 1848 [119] opposed this idea, suggesting that the process was a *vital* one in which the cells of the separation layer remained *plump, fresh, and apparently living*. A few years later, Von Mohl [120] demonstrated that abscission would take place without the formation of a layer of periderm, and as a result, it became widely accepted that the process involved living cells [120].

Two theories emerged to account for the phenomenon. The *turgor theory* proposed that the solute concentration in the separation zone cells increased as a result of starch degradation. The increased turgor pressure generated in the cells caused them to round up, tearing the wall along the line of the middle lamella. Kendall in 1918 [121] claimed to disprove this hypothesis when he showed that cell separation did not always begin at the cell corners and that some separating cells did not round up at all.

One of the first anatomical changes observed to occur after the induction of abscission was increased rates of cell division in the region of the separation layer [3]. It was assumed that this was an important part of the weakening process until Gawadi and Avery [122] showed that abscission would occur in the absence of cell division, as is frequently observed if ethylene is used to accelerate the process.

Beginning in the 1920s, scientists assumed that the newly discovered wall-degrading enzymes were involved in cell separation, although some researchers believed that wall acidification was also implicated. Separation zone cells were reported to have very active respiration [117] and protein synthesis [3,123]. These observations fueled speculation that the synthesis and secretion of wall-degrading enzymes was all-important. The turgor mechanism retained some advocates, as it seemed to be the only way to account for the abscission of some petals. This took place so rapidly (<1 hr) that it was difficult to believe that protein synthesis, secretion, and wall breakdown could all occur in such a short interval [124].

The finding by Horton and Osborne [125] that cellulase (endo-β-1,4-glucan 4-glucan hydrolase) increased in weakening AZs, coupled with the demonstration that both protein and RNA synthesis inhibitors prevented abscission [3,126], led to the current widespread belief that abscission involve the induction of wall-degrading enzymes. It has also been shown that protein synthesis inhibitors will stop rapid abscission of petals, removing one of the last objections to the involvement of wall hydrolases [2].

## B.  Nature of Cell Wall Breakdown

The evidence that wall breakdown is involved in abscission is almost entirely anatomical [3,5]. There are very few biochemical analyses of abscission zone cell walls, although Morre [127] reported an 11% loss of the wall material during weakening and Taylor et al. [128] reported a depolymerization of pectins.

Low-power observations of the fracture surfaces usually show that they are covered in intact rounded cells [6] (Figures 2 and 3). When washed from the surface, the cells still have their permeability barriers intact and can be plamolyzed [5]. They are not protoplasts, but retain part of the cell wall, which is still resilient enough to prevent them bursting when turgid (Figure 10). The burst cells sometimes seen over limited areas of the fracture surface are probably ruptured by the mechanical forces that facilitate separation in many abscission systems.

Electron microscope observations show that breakdown of the wall is not restricted to the middle lamella but involves adjacent areas of the primary wall [5] (Figure 10). Both of these swell during AZ weakening, leaving a layer of undigested wall around the protoplast. The swollen areas of wall still contain intact cellulose microfibrils, suggesting that the wall matrix and middle lamella are attacked [3,5,129]. Both x-ray microprobe analysis and autoradiography have shown that $Ca^{2+}$ is lost from the wall during cell separation [130]. It is not clear if $Ca^{2+}$ is lost as a consequence of wall hydrolysis or whether its active removal contributes to wall weakening [3,5].

The separation layer bisects the petiole or pedicel and therefore crosses many different tissues. In a study of *Impatiens* leaf abscission, cell wall breakdown was recorded around cells of the epidermis, collenchyma, cortex, xylem parenchyma, phloem seive tubes, and transfer cells [6]. In situ hybridization studies [131] indicate that cellulase mRNA is induced in a variety of different cell classes in the separation layer. Comparisons of AZs in leaves and fruit suggest that there may be differences in the nature and extent of wall breakdown [129].

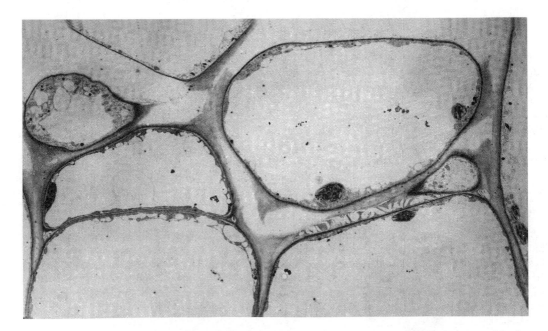

**Figure 10**   Electron micrograph of the separation layer cells from a weakened leaf abscission zone of *Impatiens*. Note that the cell walls have degraded, allowing the cells to separate. The inner layer of the wall remains intact and the cytoplasm is apparently normal. (From Ref. 5.)

## C.  Protein Synthesis Is a Prerequisite for Abscission

Anatomists often observed that protein accumulated in separating abscission zone cells [3] and EM observations of the cytoplasm showed it to be rich in organelles, particularly Golgi and rough endoplasmic reticulum (RER) [70,132]. Studies of the incorporation of labeled amino acids and nucleotides demonstrated very active synthesis of proteins and RNA in the AZ [123,126,133]. Some of the proteins synthesized during abscission are thought to play an essential role in the process, as it has been widely demonstrated that both transcriptional and translational protein synthesis inhibitors block abscission [3].

Both qualitative and quantitative changes in the protein profiles of separating AZ cells have been reported [2,134,135]. Complementary abscission-related changes in mRNA populations have also been observed [137,138]. In addition, Poovaiah et al. [136] demonstrated that the pattern of protein phosphorylation is altered in abscising zones.

## D.  Cell Wall Hydrolases and Their Control

After the initial observations that cellulase increased in separation layers, there was some confusion because it was not appreciated that more than one form of cellulase was found in AZs. Lewis and Varner [139] concluded that a cellulase isoenzyme with an alkaline isoelectric point (9.5 cellulase) was specifically involved in bean leaf abscission. It was formed de novo in AZs and its activity seemed correlated with weakening [139] (Figure 6). Antibodies raised against 9.5 cellulase were used to discriminate between it and the other isoenzymes involved in normal growth [140]. The 9.5 cellulase was localized in the separation layer and adjacent stele of bean AZs [140,141], and its production was accelerated by ethylene and inhibited by IAA [142] (Figure 6). This increase in specific cellulase isoforms has now been reported in many different abscission systems, including leaves [140], flowers [4], and fruit [16,143].

Despite its familiar name, 9.5 cellulase will not attack crystalline cellulose. It is assayed by its ability to break down soluble carboxymethylcellulose and is really a Cx-cellulase [144]. Its natural substrate is not known, but is likely to be a β-1,4-glucan in the wall matrix [145]. By itself, bean 9.5 cellulase will not cause cells to separate, but it acts synergistically with pectinase [146].

On the basis of the anatomical observations, one might expect polygalacturonases (PGs) to be involved in breakdown of the middle lamella. PG increases have been reported in *Impatiens*, *Sambucus*, tomato, orange, and peach [5,143,147]. As with cellulase, there are abscission-specific isoforms of PG. Transgenic plants have been used to show that the PG associated with ripening in tomato is not the same as that involved in abscission [148]. Both exo- and endo-cleaving PGs increase, although the endocleaving enzyme is probably more important.

The cellulase from bean and *Sambucus* abscission zones have been cloned [138,149,150]. The bean cDNA has been sequenced and has 64% identically matched nucleotides to the cellulase from avocado fruit [150]. The partial sequence of the *Sambucus* cDNA shows that it is very similar [138]. They share consensus sequences with a series of E2-type cellulase from microbial and other plant sources [144]. Bonghi et al. [143] have made use of this similarity by employing avocado fruit cellulase cDNA as a heterologous probe. It hybridized to 1.8- and a 2.2-kb mRNAs, which accumulated in ethylene-treated peach abscission zones.

Tucker et al. [149] have studied the expression of bean 9.5 cellulase. In situ hybridization showed that the cellulase mRNA was confined to the separation layer and the adjacent stele [151] (Figure 11). Northern blot analysis indicated that cellulase mRNA was virtually absent from uninduced abscission zones but increased as they weakened in ethylene [149] (Figure 12). This increase was dependent, at least in the short term, on the presence of ethylene. Indoleacetic acid suppressed the increase even in an ethylene atmosphere. Removal of ethylene after cellulase mRNA had started to accumulate, in conjunction with inhibition of any endogenous ethylene with norbornadiene, caused the cellulase mRNA levels to decline to very low levels [149]. Indoleacetic acid administered to the fracture surfaces after abscission had occurred inhibited further accumulation of cellulase mRNA in the fracture surface cells even in the presence of ethylene [71]. This suggests that expression of 9.5 cellulase is under the joint control of both IAA and ethylene. The cellulase gene complete with its upstream sequences has now been cloned [71].

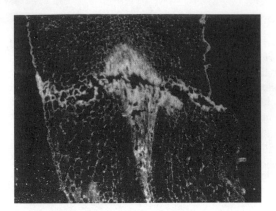

**Figure 11**  Dark-field micrograph of a thin longitudinal section through an abscising abscission zone of bean. The section was hybridized to a ³⁵S-labeled cellulase cDNA probe. The hybridization signal is seen as bright light reflecting silver grains in the separation layer and in the central vascular traces. (From Ref. 131.)

## IV.  AGRICULTURAL MANIPULATION OF ABSCISSION

### A.  Methods of Manipulating Abscission

There are a great number of crops for which the control of abscission is desirable. For example, too little natural thinning of young fruit can result in large numbers of small unmarketable fruits, while too much abscission results in uneconomical yields.

Techniques are slowly emerging which allow the manipulation of abscission. They can be categorized as follows:

1.  *Understanding the physiological basis of abscission.* Perhaps the best way of controlling abscission is to understand the physiological basis of its induction. For instance, many ornamental plants

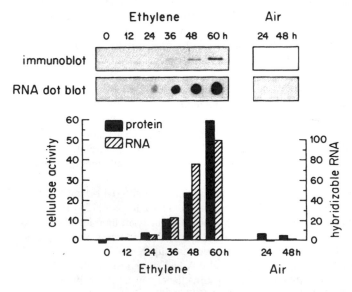

**Figure 12**  Time course of accumulation of cellulase protein and mRNA. The cellulase activity in bean abscission zones kept in ethylene and air has been plotted against time. The amount of cellulase cDNA probe binding to RNA from the same preparations is also shown. The photographs are of a Western immunoblot of abscission zone proteins probed with cellulase antibodies and RNA dot blots probed with ³²P-labeled cellulase cDNA. (From Ref. 149.)

suffer from bud drop and flower shattering during transit and marketing [152]. This can be reduced by avoiding the inductive conditions that lead to abscission: high temperatures [153], low light intensities [154], ethylene pollution [152], fertilization of flowers [24,25], and mechanical perturbation [155]. In many cases the physiological basis of agriculturally important abscission is not yet fully understood, the extent of young fruit thinning being an important example.

2. *Genetics of abscission behavior.* Geneticists have been able to breed varieties with different abscission characteristics. For instance, among the many raspberry cultivars, there is wide variation in the extent to which fruit abscission has progressed at the time the fruit are ready for harvest [156]. Cultivars that do not drop ripe berries when the bushes are shaken are suitable for hand picking, while more easily detached varieties are selected for mechanical harvesting.

3. *Recombinant DNA technologies and the manipulation of abscission.* Being able to manipulate abscission using transgenic plant technologies is a very real prospect. Oeller et al. [92] have already produced nonripening transgenic tomato plants in which the synthesis of ethylene is blocked. This was achieved by expressing antisense RNA for ACC synthase. One would predict that this strategy should produce slow or nonabscising plants that would abscise to order if treated with ethylene-generating sprays. Similarly, ethylene production has been reduced in plants producing antisense ethylene-forming enzyme RNA [157]. An alternative approach is to overexpress in plants the ACC deaminase gene from bacteria, destroying the ethylene precursor ACC as it is formed [158,159].

It might prove possible to manipulate abscission by producing more IAA in the distal tissues of transgenic plants. The Ti plasmid IAA synthesis genes have already been cloned and there are leaf- and fruit-specific promoters to drive them. An alternative strategy for preventing abscission could be to produce transgenic plants expressing antisense RNA for the wall-degrading enzymes cellulase and polygalacturonase. This approach has been used to reduce softening in fruit [160,161].

4. *Accelerating abscission by increasing ethylene production.* Ethylene-releasing sprays such as Ethephon [(2-chloroethyl)phosphonic acid] [162,163] and Etacelasil [164], which release ethylene spontaneously, are very effective at causing abscission. Their use is not always straightforward, as they can have unfortunate side effects. For instance, ethephon used to promote fruit abscission often induces undesirable leaf fall. Attempts have been made to increase ethylene production using the plant's natural substrate ACC [164], although the method is not widely used.

Ethylene production can be increased by wounding the plant. The protein synthesis inhibitor cycloheximide has been employed to damage and induce consequent abscission of oranges prior to mechanical harvesting. Initial trials with ethylene-releasing sprays were unsuccessful because they damaged the trees by defoliating them, so a method of localizing ethylene production in the fruit was sought. It was found that cycloheximide damaged the peel of the fruit, causing wound ethylene formation, which in turn induced abscission [165]. The abscission of cotton leaves prior to boll harvest is also achieved by damaging the leaves and inducing ethylene formation [166,167]. The mechanism of action of some thinning agents, such as insecticides carbaryl and oxamyl, which are used to thin apples [168,169], is not fully understood.

5. *Inhibiting abscission by reducing ethylene production and sensitivity.* Abscission can be inhibited by reducing natural ethylene production or interfering with ethylene action. A reduction of natural fruit thinning has been reported using AVG and AOA to inhibit ethylene formation [89,90,170,171]. Silver thiosulfate [152] has been widely employed to prevent abscission of ornamental flowers to such an extent that it is a pollution hazard in some horticultural areas. The ethylene antagonist DACP [77] also seems extremely effective (Figure 8), but unfortunately, it is an explosive gas, which may limit its field use!

6. *Auxin and control of abscission in the field.* Auxin sprays have been employed to prevent abscission. Indoleacetic acid is not used because it is rapidly degraded in the plant and synthetic auxin analogues such as 1-naphthaleneacetic acid (NAA), picloram, and 2,4D are preferred. Preharvest drop of apples has been treated in this way [172]. Rather perversely, NAA sprays are used to thin young apple fruit [169]. In this case, the effect seems to be caused indirectly by interfering with sugar translocation from sprayed leaves to the developing fruit [168,173].

## B.  Flower Shattering

The abscission or shattering of whole buds or floral parts is a major problem in a number of ornamental plants [152,174]. These include *Zygocactus* [175], *Fuschia, Calceolaria,* snap-dragon, sweet peas [176], lilies [154,177], geraniums [155], *Pelargonium* [153], cyclamen [24], *Impatiens, Bougainvillea*, delphiniums, and foxgloves [25]. There are many other delightful flowers that are not marketed commercially because of these difficulties. The problems usually occur during transit and retailing, when low light intensities, water stress, high temperatures, and ethylene buildup contribute to the problem. The concentration of ethylene in mixed cool stores may reach high levels, particularly if ethylene-generating fruit are enclosed with the flowers. Motor exhausts (0.25% ethylene) have been shown to raise ethylene levels to inductive concentrations in auction halls [152] and the trucks used for transport [179,180].

Silver thiosulfate (0.5 to mM) has been extremely effective at reducing this loss, either when sprayed directly on the plants or *pulsed* through the transpiration stream of cut flowers [152,174]. There are rumors that its use may be banned because of pollution problems. Auxin analogues have been used with flowers [167,178], but petals seem rather unresponsive.

## C.  Fruit Thinning and Harvesting

Some plants produce vastly more flowers than will mature into fruit. This is particularly true of many fruit trees, such as orange [31], apple, apricot, mango [33], avocado [34,181], and cherry, where a natural thinning process occurs. Only 0.2% of Washington Navel orange flowers develop into fruit [31]. Floral and young fruit abscission is also a particular problem in leguminous crops such as soybean [37,182], field bean (*Vicia*) [166], lupins (38), French bean (*Phaseolus*) [183], and cowpeas [184]. In other crops, such as pistachio, alternate bearing is a difficulty, where a heavy crop of fruit appears to cause excessive abscission of the subsequent year's buds [108]. Flowers and fruit are lost in a succession of abscission episodes [31]. These are classified as follows:

1.  *Bud drop* that occurs before the flowers reach anthesis
2.  *Flower or young fruit drop* immediately after anthesis
3.  *Enlarging fruit or June drop*
4.  *Mature fruit or preharvest drop*

The loss of flower buds can be considerable. In oranges, it can be up to 33% and is attributed to nutritional causes such as zinc deficiency [31]. In apricots, a failure to fulfill chilling requirements in areas such as South Africa, Turkey, and Israel results in almost complete loss of flowers. This can be overcome by spraying the trees with GA to break dormancy or by growing varieties with a short chilling requirement [31].

Loss of opened flowers can usually be attributed to a failure of flower development or failure of fertilization. In Shamouti orange, 77% of the flowers that were shed had abnormalities, often with aborted pistils [31]. Benzyladenine has been used to prevent flower and young fruit drop [182,185].

The June drop of enlarging fruit can be very significant in citrus crops, being as high as 21% in lemons, 45% in Shamouti oranges, and 60% in clementines [31]. The main causes [31] of this drop are:

1.  *Abnormalities* in or lack of fertilization and zygote abortion or degeneration.
2.  *Competition* for photosynthates and mineral nutrients between fruit and vegetative apices; fruit with fewer seeds on weaker branches tend to be shed first.
3.  *Water stress* can be a major cause of abscission [102] in arid areas because fruit desiccate first, having a higher (less negative) water potential than leaves. Hail and wind damage are also climatic factors.
4.  *Invasion of fruit by fungal pathogens* or by herbivorous larvae can cause abscission. *Runoff* of blackcurrants, which results in the premature abscission of apparently healthy fruitlets, can reduce yield by 50% [90]. Evidence suggests that symptomless infection with *Botrytis* causes elevated ethylene production, which, in turn, induces abscission [75].

Aminoethoxyvinyl glycine (AVG), the ethylene synthesis inhibitor, has been used to prevent fruit thinning in apples [170,171]. When natural thinning is not vigorous enough and too many fruit are set, sprays are used to reduce the crop and get fewer bigger fruit. Apples have been thinned with ethephon [167,186], carbaryl [167], oxamyl [169], and NAA [173]. Ethephon has also been used with pecans [187], peaches [188], prunes [189], and pears [190].

During maturation of the crop, abscission of the ripe fruit occurs. This preharvest drop is undesirable as far as hand picking is concerned. It is common practice to prevent it by using NAA sprays in pears and apples and with 2,4D [167,172] and NAA in grapefruit and oranges [31]. Trials with NAA have also been conducted to try to reduce the loss of grapes from bunches before and after harvest [191]. Spraying with AVG 1 month before harvest delayed the preharvest drop of apples [89].

Harvesting can be carried out using machines that beat, shake, or blow fruit from plants. The greater the force needed to separate the fruit, the greater the damage to the fruit and the plant. Ethylene-releasing sprays have been used in trials to accelerate and synchronize abscission prior to harvest of grapes [192], oranges [165], apples [172], olives [193], raspberries [194], and blackberries. Ethephon promotes both abscission and reddening in peppers and its feasibility is being assessed as an aid to once-over harvesting [195]. Although these methods have been successful as an aid to mechanical harvesting [196], they can produce unfortunate side effects, such as the shedding of leaves [196], shoot dieback, gummosis [165,196], and excessive fruit drop prior to harvest [192]. The problem of leaf loss has been overcome successfully by the use of calcium acetate sprays [197,198]. Ethephon can give inconsistent results [164] because ethylene production is very dependent on ambient temperature [196] and the pH of the cell sap [163,168]. 2-Chloroethylmethylbisphenylmethoxysilane (CGA) [196] and Etacelasil (164) [2-chlorotris(2-methoxyethoxy)silane] may prove more reliable ethylene-releasing agents.

It is important to realize that using ethylene-generating sprays will accelerate abscission only if natural ethylene production is subsaturating. For this reason, adding more ethylene to ripening raspberries is counterproductive as it has no effect on abscission of ripe fruit but instead causes immature green fruit to redden and abscise [85].

## D.  Leaf Loss

Leaf fall is induced in several crops where the foliage interferes with the mechanical harvesting of fruit. In cotton both desiccants and ethylene-releasing sprays are employed prior to boll harvest [166,167]. Defoliation of nursery trees is also practiced prior to shipping [152,164,174].

Leaf fall is a problem in some display plants, such as *Ficus benjamina* [199]. *Radermachera* [200], *Philodendron* [201], and potted roses. It is induced by low light intensities, water stress [199], and ethylene pollution during retailing. These difficulties are usually solved using silver thiosulfate [200], but the loss of holly leaves is treated using the auxin analogue NAA [202].

## E.  Other Uses

The ability to control abscission has been put to a number of rather unusual uses. During the storage of lemons the short stem (button) left attached to the fruit abscises and allows entry of the fungus *Alternaria* via the scar. This problem is overcome by adding the isopropyl ester of 2,4D [203]. The reverse problem is encountered with bananas, where the failure of perianth abscission makes some varieties of fruit less attractive [204]. Ethephon sprays have been used to reduce mistletoe infestations of Black Spruce by causing the abscission of 90 to 100% of the mistletoe shoots [205]. Twig abscission has been induced in white oak in an attempt to improve timber quality [206].

## V.  SUMMARY

Abscission occurs at genetically determined abscission zones, where the induction of wall-degrading enzymes such as cellulase and polygalacturonase weakens a restricted band of cells called the separation layer. It seems likely that the separation layer cells are specifically preprogrammed to respond to the inductive stimuli. The generation of mechanical forces to facilitate separation of the loosened separation zone cells and rupture the xylem is an important component of the process. After abscission, the wound is protected from microbial attack by the formation of chitinase and β-1,3-glucanase, division and suberization of the surface cells, and blockage of the vascular traces.

The control of abscission seems to depend on the relative concentrations of IAA and ethylene. Auxin is an inhibitor of abscission, and ethylene accelerates and synchronizes the process. It is not clear if one or another of these regulators or the relative concentrations of both are responsible for the induction of abscission. Other hormones, such as ABA, probably have indirect effects via ethylene and IAA.

The manipulation of abscission is important in agriculture. Methods are being developed to control the thinning of fruit, the shattering of flowers, and mature fruit drop as an aid to harvesting.

## REFERENCES

1. Sexton R, Tucker ML, del Campillo E, Lewis LN. In: DJ Osborne, MB Jackson, eds. Cell Separation in Plants. NATO ASI Series H35. Berlin: Springer-Verlag, 1989, p 69.
2. Eversen KB, Clark DG, Singh A. In: JC Peche et al., eds. Cellular and Molecular Aspects of the Plant Hormone Ethylene. Boston: Kluwer, 1993, p 278.
3. Sexton R, Roberts JA. Annu Rev Plant Physiol 33:133, 1982.
4. Roberts JA, Schindler CB, Tucker GA. Planta 160:164, 1983.
5. Sexton R, Burdon JN, Reid JSG, Durbin ML, Lewis LN. In: WM Dugger, S Bartnicki-Garcia, eds. Structure, Function and Biosynthesis of Plant Cell Walls. Proceedings of the 7th Annual Symposium on Botany, University of California at Riverside. Baltimore: Waverly Press, 1984, p 195.
6. Sexton R. Planta 128:49, 1976.
7. Weisner J., 1871, quoted by Facey V, New Phytol 49:103, 1950.
8. Wright M, Osborne DJ. Planta 120:163, 1974.
9. Sexton R, Redshaw AJ. Ann Bot 48:745, 1981.
10. Jackson MB, Osborne DJ. Can J Bot 50:1465, 1972.
11. Mackenzie KAD. Ann Bot 62:249, 1988.
12. Osborne DJ. In: AK Matoo, JC Suttle, eds. Boca Raton, FL: The Plant Hormone Ethylene. CRC Press, 1991, p 193.
13. McManus MT, Osborne DJ. Physiol Plant 79:471, 1990.
14. Halliday MB, Wangermann E. New Phytol 71:649, 1972.
15. Szymkowiak EJ, Sussex IM. In: DJ Osborne, MB Jackson, eds. "'1 Separation in Plants. NATO ASI Series H35. Berlin: Springer-Verlag, 1989, p 363.
16. Greenberg J, Goren R, Riov J. Physiol Plant 34:1, 1975.
17. Rascio N, Casadoro G, Ramina A, Masia A. Planta 164:1, 1985.
18. Huberman M, Goren R, Zamski E. Physiol Plant 59:445, 1983.
19. Samuel G. Ann Bot 41:375, 1927.
20. Warren Wilson J, Warren Wilson PM, Walker ES. Ann Bot 62:235, 1988.
21. Suzuki T. Physiol Plant 82:483, 1991.
22. Biggs AR, Northover J. Can J Bot 63:1547, 1985.
23. del Campillo E, Lewis LN. Plant Physiol 98:955, 1992.
24. Haley AH, Whitehead CS, Kofranek AM. Plant Physiol 75:1090, 1984.
25. Stead AD, Moore KG. Planta 157:15, 1983.
26. Jordan WR, Morgan PW, Davenport TL. Plant Physiol 50:756, 1972.
27. Ketring DL, Melouk HA. Plant Physiol 69:789, 1982.
28. Jankiewicz LS. Bull Soc Bot Fr 127 Actual Bot 1:165, 1980.
29. Hendrickson RM, Dysart RJ. J Econ Entomol 76:1075, 1983.
30. Cacker LE, Zhao SY, Decoteau DR. J Exp Bot 38:883, 1987.
31. Kaska N. In: DJ Osborne, MB Jackson, eds. Cell Separation in Plants. NATO ASI Series H35. Berlin: Springer-Verlag, 1989, p 309.
32. Feret PP, Kreh RE, Merkle SA, Oderwald RG. Bot Gaz 143:216, 1982.
33. Nunez-Elisea R, Davenport TL. Plant Physiol 82:991, 1986.
34. Skene KGM, Barlass M. Ann Bot 52:667, 1983.
35. Guinn G. Crop Sci 22:580, 1982.
36. Williamson B, McNicol RJ, Young K. Grower Oct:21, 1989.
37. Brun WA, Betts KJ. Plant Physiol 75:187, 1984.
38. Van Stevenick RFN. J Exp Bot 8:373, 1957.
39. Pigott CD. New Phytol 97:575, 1984.
40. Junttila O. Physiol Plant 38:278, 1976.
41. Addicott FT, Lynch RS, Livingston GA, Hunter JK. Plant Physiol 24:537, 1949.
42. Kuster E. Dtsch Bot Ges 34:184, 1916.
43. Mai G. Jahr Wiss Bot 79:681, 1934.
44. La Rue CD. Proc Natl Acad Sci U S A 22:254, 1936.
45. Addicott FT. Biol Rev 45:485, 1970.
46. Doubt S. Bot Gaz 63:209, 1917.
47. Zimmerman PW, Crocker W, Hitchcock AE. Proc Natl Soc Hortic Sci 27:53, 1930.
48. Crocker W. Growth of Plants. New York: Reinhold, 1948.
49. Milbrath JA, Hansen E, Hartman H. Science 91:100, 1940.
50. Osborne DJ. Nature 176:1161, 1955.

51.  Van Stevenick RFM. Nature 183:1246, 1959.
52.  Ohkuma K, Lyon JL, Addicott FT, Smith OE. Science 142:1592, 1963.
53.  Cornforth JW, Milborrow BV, Ryback G, Rothwell K, Wain RL. Nature 211:742, 1966.
54.  Davis LA, Addicott FT. Plant Physiol 49:644, 1972.
55.  Addicott FT. Abscission. Berkeley, CA: University of California Press, 1982.
56.  Roberts JA, Osborne DJ. J Exp Bot 32:875, 1981.
57.  Elkinawy M. Physiol Plant 62:593, 1984.
58.  Carns HR. Annu Rev Plant Physiol 17:295, 1966.
59.  Davenport TL, Morgan PW, Jordan WR. Plant Physiol 65:1023, 1980.
60.  Jacobs WP. Annu Rev Plant Physiol 13:403, 1962.
61.  Jacobs WP. Plant Hormones and Plant Development. Cambridge: Cambridge University Press, 1979.
62.  Addicott FT, Lynch RS. Science 114:688, 1951.
63.  Addicott FT, Lynch RS, Carns HR. Science 121:644, 1955.
64.  Guar BK, Leopold AC. Plant Physiol 30:487, 1955.
65.  Abeles FB, Rubinstein B. Plant Physiol 39:963, 1964.
66.  Abeles FB. Physiol Plant 20:442, 1967.
67.  Morris DA. J Exp Bot 44:261:807, 1993.
68.  Rubinstein B, Leopold AC. Plant Physiol 38:262, 1963.
69.  Jaffe MT, Goren R. Bot Gaz 140:378, 1979.
70.  Osborne DJ, McManus MT, Webb J. In: JA Roberts, GA Tucker, eds. Ethylene and Plant Development. London: Butterworth, 1985, p 197.
71.  Tucker MJ, Matters GL, Koehler SM, Kemmerer EC, Baird SL, Sexton R. In: JC Peche et al., eds. Cellular and Molecular Aspects of the Plant Hormone Ethylene. Boston: Kluwer, 1993, p 265.
72.  Sexton R, Lewis LN, Trewavas AJ, Kelly P. In: A Roberts, GA Tucker, eds. Ethylene and Plant Development. London: Butterworth, 1985, p 173.
73.  Abeles FB, Gahagan HE. Plant Physiol 43:1255, 1968.
74.  Sisler EC, Goren R, Huberman M. Physiol. Plant 63:114, 1985.
75.  McNicol RJ, Williamson B, Young K. Acta Hortic 262:209, 1989.
76.  Goldschmidt EE, Leshem B. Am J Bot 58:14, 1971.
77.  Sisler EC, Blankenship SM, Fearn JC, Haynes R. In: JC Peche et al., eds. Cellular and Molecular Aspects of the Plant Hormone Ethylene. Boston: Kluwer, 1993, p 182.
78.  Hall WC. Bot Gaz 113:310, 1952.
79.  Barlow HWB. Rep 13th Int Hortic Congr 1:145, 1952.
80.  Burg SP. Annu Rev Plant Physiol 13:265, 1962.
81.  Burg SP. Plant Physiol 43:1503, 1968.
82.  Morgan PW, He C-J, Drew MC. Plant Physiol 100:1587, 1992.
83.  Ben-Yehoshua S, Aloni B. Bot Gaz 135:41, 1974.
84.  Morgan PW, Durham JI. Plant Physiol 66:88, 1980.
85.  Burdon JN, Sexton R. Ann Bot 66:111, 1990.
86.  Jackson MB, Hartley C, Osborne DJ. New Phytol 72:1251, 1973.
87.  Jackson MB, Osborne DJ. Nature 225:1019, 1970.
88.  Sexton R, Burdon JN, Bowmer JM. In: JC Peche et al., eds. Cellular and Molecular Aspects of the Plant Hormone Ethylene. Boston: Kluwer, 1993, p 317.
89.  Bangerth F. J Am Soc Hortic Sci 103:401, 1978.
90.  Williams MW. HortScience 15:76, 1980.
91.  Davenport TL, Manners MM. J Exp Bot 33:815, 1982.
92.  Oeller PW, Wong LM, Taylor LP, Pike DA, Theologis A. Science 254:437, 1991.
93.  Beyer EM. Plant Physiol 58:268, 1976.
94.  Cameron AC, Reid MS. Sci Hortic 19:373, 1983.
95.  Reid MS. Acta Hortic 167:57, 1985.
96.  Ketring DL, Melouk HA. Plant Physiol 69:789, 1982.
97.  Tudela D, Primo-Millo E. Plant Physiol 100:131, 1992.
98.  Ernest LC, Valdovinos JG. Plant Physiol 48:402, 1971.
99.  Beyer EM. Plant Physiol 52:1, 1973.
100.  Riov J, Goren R. Plant Cell Environ 2:83, 1979.
101.  Beyer EM. Plant Physiol 55:322, 1975.
102.  Guinn G. Crop Sci 22:580, 1982.
103.  Guinn G, Brummett DL. Plant Physiol 86:28, 1988.
104.  Guinn G. Plant Physiol 69:349, 1982.
105.  Davenport TL, Jordan WR, Morgan PW. Plant Physiol 59:1165, 1977.
106.  Peterson JC, Sacalis JN, Durkin DJ. J Am Soc Hortic Sci 105:793, 1980.
107.  Ramina A, Masia A. J Am Soc Hortic Sci 105:465, 1980.
108.  Takeda F, Crane JC. J Am Soc Hortic Sci 105:573, 1980.

109.   Porter NG. Physiol Plant 40:50, 1977.
110.   Milborrow BV. Annu Rev Plant Physiol 25:259, 1974.
111.   Sagee O, Goren R, Riov J. Plant Physiol 66:750, 1980.
112.   Sargent JA, Osborne DJ, Dunford SM. J Exp Bot 35:1663, 1984.
113.   Sexton R, Woolhouse HW. In: MB Wilkins, ed. Advanced Plant Physiology. Essex, UK: Longman, 1984, p 469.
114.   Kuang A, Peterson CM, Dute RR. J Exp Bot 43:1611, 1992.
115.   Chatterjee SK, Leopold AC. Plant Physiol 39:334, 1964.
116.   Morgan PW. Planta 129:275, 1976.
117.   Marynick MC. Plant Physiol 59:484, 1977.
118.   Ueda J, Morita Y, Kato J. Plant Cell Physiol 32:983, 1991.
119.   Inman T. Proc Lit Philos Soc Liverpool 36:89, 1848.
120.   von Mohl H. Bot Z 18:273, 1860.
121.   Kendall JN. Univ Calif Publ Bot 5:347, 1918.
122.   Gawadi AG, Avery GS. Am J Bot 37:172, 1950.
123.   Abeles FB, Holm RE. Plant Physiol 41:1337, 1966.
124.   Sexton R, Struthers WA, Lewis LN. Protoplasma 116:179, 1983.
125.   Horton RF, Osborne DJ. Nature 214:1086, 1967.
126.   Abeles FB, Leather GR, Forrence LE, Craker LE. HortScience 6:371, 1971.
127.   Morre DJ. Plant Physiol 43:1545, 1968.
128.   Taylor JE, Webb STJ, Coupe SA, Tucker GA, Roberts JA. J Exp Bot 44:93, 1993.
129.   Zanchin C, Bonghi G, Casadoro G, Ramina A, Rascio N. New Phytol 123:555, 1993.
130.   Stosser R, Rasmussen HP, Bukovac MJ. Planta 86:151, 1969.
131.   Tucker ML, Baird SL, Sexton R. Planta 186:52, 1991.
132.   Sexton R, Jamieson GGC, Allan MHIL. Protoplasma 99:55, 1977.
133.   Abeles FB, Holm RE. Ann NY Acad Sci 144:367, 1967.
134.   Reddy ASN, Friedmann M, Poovaiah BW. Plant Cell Physiol 29:179, 1988.
135.   del Campillo E, Lewis LN. Plant Physiol 98:955, 1992.
136.   Poovaiah BW, Friedmann M, Reddy ASN, Rhee JK. Physiol Plant 73:354, 1988.
137.   Kelly P, Trewavas AJ, Lewis LN, Durbin ML, Sexton R. Plant Cell Environ 10:11, 1987.
138.   Roberts JA, Taylor JE, Coupe SA, Harris N, Webb STJ. In: JC Peche et al., eds. Cellular and Molecular Aspects of the Plant Hormone Ethylene. Boston: Kluwer, 1993, p 272.
139.   Lewis LN, Varner JE. Plant Physiol 46:194, 1970.
140.   Sexton R, Durbin ML, Lewis LN, Thompson WW. Nature 283:873, 1980.
141.   del Campillo E, Reid PD, Sexton R, Lewis LN. Plant Cell 2:245, 1990.
142.   Durbin ML, Sexton R, Lewis LN. Plant Cell Environ 4:67, 1981.
143.   Bonghi C, Rascio N, Ramina A, Casadoro G. Plant Mol Biol 20:839, 1992.
144.   Lashbrook CC, Bennett AB. In: JC Peche et al., eds. Cellular and Molecular Aspects of the Plant Hormone Ethylene, Boston: Kluwer, 1993, p 123.
145.   Hatfield R, Nevins DJ. Plant Cell Physiol 27:541, 1986.
146.   Lewis LN, Linkins AE, O'Sullivan S, Reid PD. Proceedings of the 8th International Conference on Plant Growth Substances. Tokyo: Hirokawa Publishing Co, 1974, p 708.
147.   Taylor JE, Webb STJ, Coupe SA, Tucker GA, Roberts JA. J Exp Bot 44:93, 253, 1993.
148.   Taylor JE, Tucker GA, Lasslett Y, Smith CJS, Arnold CM, Watson CF, Schuch W, Grierson D, Roberts JA. Planta 183:133, 1990.
149.   Tucker ML, Sexton R, del Campillo E, Lewis LN. Plant Physiol 88:1257, 1988.
150.   Tucker ML, Milligan SB. Plant Physiol 95:928, 1991.
151.   Tucker ML, Baird SL, Sexton R. Planta 186:52, 1991.
152.   Reid MS. Acta Hortic 167:57, 1985.
153.   Cameron AS, Reid MS. Sci Hortic 19:373, 1983.
154.   Durieux AJB, Kamerbeek GA, Van Meeteren U. Sci Hortic 18:287, 1983.
155.   Armitage AM, Heins R, Dean S, Carlson W. J Am Soc Hortic Sci 105:562, 1980.
156.   Burdon JN, Sexton R. Sci Hortic 43:95, 1990.
157.   Gray JE, Picton S, Fray R, Hamilton AJ, Smith H, Barton S, Grierson D. In: JC Peche et al., eds. Cellular and Molecular Aspects of the Plant Hormone Ethylene. Boston: Kluwer, 1993, p 82.
158.   Sheehy RE, Ursin V, Vanderpan S, Hiatt WR. In: JC Peche et al., eds. Cellular and Molecular Aspects of the Plant Hormone Ethylene. Boston: Kluwer, 1993, p 106.
159.   Klee HJ, Hayford MB, Kretzmer KA, Barry GF, Kishore GM. Plant Cell 3:1187, 1991.
160.   Smith CJS, Watson CF, Ray J, Bird CR, Morris PC, Schuch W, Grierson D. Nature 334:724, 1988.
161.   Sheehy RE, Kramer M, Hiatt WR. Proc Natl Acad Sci U S A 85:8805, 1988.
162.   Edgerton LJ, Greenhalgh WJ. J Am Soc Hortic Sci 94:11, 1969.
163.   Lavee S, Martin GC. J Am Soc Hortic Sci 106:14, 1981.
164.   Lurssen K, Konze J. In: JA Roberts, GA Tucker, eds. Ethylene and Plant Development. London: Butterworth, 1985, p 363.

165. Cooper WC, Henry WH. J Agric Food Chem 19:559, 1971.
166. Osborne DJ. Outlook Agric 13:97, 1984.
167. Addicott FT. In: LJ Audus, ed. Herbicides: Physiology, Biochemistry and Ecology. Vol 1. London: Academic Press, 1976, p 191.
168. Knight JN. J Hortic Sci 58:371, 1983.
169. Meyer RH. HortScience 17:658, 1982.
170. Greene DW. J Am Soc Hortic Sci 108:415, 1983.
171. Child RD, Williams RR. J Hortic Sci 58:365, 1983.
172. Edgerton LJ. HortScience 6:378, 1971.
173. Schneider GW. J Am Soc Hortic Sci 103:455, 1978.
174. Reid MS. HortScience 20:45, 1985.
175. Cameron AC, Reid MS. HortScience 16:761, 1981.
176. Mor Y, Reid MS, Kofranek AM. J Am Soc Hortic Sci 109:866, 1984.
177. van Meeteren U, de Proft M. Physiol Plant 56:236, 1982.
178. Mor Y, Halevy AH, Kofranek AM, Ried MS. J Am Soc Hortic Sci 109:494, 1984.
179. Abeles FB. In: JA Roberts, GA Tucker, eds. Ethylene and Plant Development. London: Butterworth, 1985, p 287.
180. Schouten SP. In: JA Roberts, GA Tucker, eds. Ethylene and Plant Development. London: Butterworth, 1985, p 353.
181. Adato I, Gazit S. J Exp Bot 28:636, 1977.
182. Peterson CM, Williams JC, Kuang A. Bot Gaz 151:322, 1990.
183. Zehni MS, Morgan DG, Saad FA. Nature 227:628, 1970.
184. Ojehomon OO. J Exp Bot 23:751, 1972.
185. Kuang A, Peterson CM, Dute RR. J Exp Bot 43:1611, 1992.
186. Jones KM, Koen TB, Meredith RJ. J Hortic Sci 58:381, 1983.
187. Wood BW. HortScience 18:53, 1983.
188. Martin GC, Nishijima C. J Am Soc Hortic Sci 97:561, 1972.
189. Sibbett GS, Martin GC. HortScience 17:665, 1982.
190. Knight JN. J Hortic Sci 57:61, 1982.
191. Ergenoglu F. In: DJ Osborne, MB Jackson, eds. Cel Separation in Plants. NATO ASI Series H35. Berlin: Springer-Verlag, 1989, p 323.
192. Hedberg PR, Goodwin PB. Am J Enol Vitic 31:109, 1980.
193. Martin GC. In: DJ Osborne, MB Jackson, eds. Cell Separation in Plants. NATO ASI Series H35. Berlin: Springer-Verlag, 1989, p 331.
194. Knight JN. Acta Hortic 60:99, 1976.
195. Batal KM, Granberry DM. HortScience 17:944, 1982.
196. Olien WC, Bukovac MJ. J Am Soc Hortic Sci 107:1085, 1982.
197. Martin GC, Cambell RC, Carlson RM. J Am Soc Hortic Sci 105:34, 1980.
198. Iwahori S, Oohata JT. Sci Hortic 12:265, 1980.
199. Peterson JC, Sacalis JN, Durkin DJ. J Am Soc Hortic Sci 105:788, 1980.
200. Wang Y-T, Dunlap JR. HortScience 25:233, 1990.
201. Marousky FJ, Harbaugh BK. J Am Soc Hortic Sci 104:876, 1979.
202. Milbrath JA, Hartman H. Oreg Agric Stn Bull 413, 1942.
203. Einset JW, Lyon JL, Johnson P. J Am Soc Hortic Sci 106:531, 1981.
204. Israeli Y, Blumenfeld A. HortScience 15:187, 1980.
205. Livingston WH, Brenner ML. Plant Dis 67:909, 1983.
206. Chaney WR, Leopold AC. Can J For Res 2:492, 1972.

# 11

# Cell Cycle Regulation in Plants

## A. S. N. Reddy and Irene S. Day

*Colorado State University, Fort Collins, Colorado*

## I. INTRODUCTION

Cell division is one of the fundamental processes of growth and development of plants and animals. The time and place of cell division in an organism play a critical role in many developmental processes. The development of a complex organism with a defined form and structure requires tightly regulated cell growth and proliferation as well as transitions from cycling state to quiescent state and vice versa. In order to duplicate the genetic material and produce two daughter cells, the cell goes through a set of orderly events generally referred to as the cell cycle. The cell cycle consists of four distinct phases called gap1 ($G_1$), synthetic phase (S), gap2 ($G_2$), and mitosis (M). In the $G_1$ phase cells prepare for S phase, during which DNA synthesis takes place and the cell replicates its chromosomes [1,2]. The completion of S phase leads into another gap phase ($G_2$). Upon completion of $G_2$, cells enter mitosis (M phase), where duplicated chromosomes segregate into two daughter cells [3]. However, it should be pointed out that in some rare instances cycling cells have only two phases (M and S) without intervening gap phases ($G_1$ and $G_2$). For example, the first 13 nuclear division cycles during *Drosophila* embryo development do not have any gap phases [4]. Similarly, nuclear division cycles during early endosperm development in plants seem to lack gap phases [5].

Normal proliferating cells in $G_1$ can continue to cycle or revert to quiescent ($G_0$) state. The decision to undergo another round of DNA synthesis and continue to cycle or to exit cell cycle to enter into a quiescent state ($G_0$) is made during $G_1$ phase [1]. Cells in $G_0$ state either terminally differentiate or can be activated to reenter the cell cycle. These switches in and out of $G_1$ are primarily controlled by extracellular factors such as hormones and other mitogens [1]. However, once the cells enter into S phase, the cell cycle events become independent of extracellular factors, leading to mitosis and production of two daughter cells. These events are mostly regulated by internal controls. Stringent control of decision points in the cell cycle is vital for normal growth and development of organisms [1,4,6,7]. Deregulation of the regulatory mechanisms that control decision points in the cell cycle results in uncontrolled cell division leading to abnormal growth. The biochemical and molecular mechanisms that regulate the cell cycle are of great interest not only to help us understand how cells divide during normal growth and development of organisms but also to get insights into abnormal growth processes such as cancer. Knowledge derived from cell cycle regulation in plants should enhance the ability to manipulate growth and developmental processes in plants and could have practical implications. For instance, regeneration of plants is very critical for crop improvement through genetic engineering [8,9]. However, the ability to regenerate a whole plant from differentiated somatic tissues varies considerably from species to species [5]. The induction of cell division in differentiated cells ($G_0$ to $G_1$ transition) is the first critical step in the regeneration process. Hence, studies on cell cycle regulation are likely to provide some clues to mechanisms that regulate plant regeneration [10,11].

In yeast and animal systems considerable advances have been made in our understanding of the control of different phases of the cell cycle using yeast and animal systems [12–15]. The combination of genetic, biochemical, and molecular approaches has resulted in identification of decision points in the cell cycle and key regulatory proteins that control progression through the decision points. A number of excellent reviews describing the cell cycle regulation in fungi [13,16,17], insects [4], and mammalian cells [12,14,15,18] are available. Cell cycle research in plants is in its early stages. However, research during the last several years shows that at least some of the key cell cycle regulatory proteins are structurally and functionally conserved between plants and other unicellular and multicellular eukaryotes. Our goal here is to summarize what is known about cell cycle regulation in plants and some of the unique aspects of cell cycle in plants. Because of limited information with plant systems and considerable similarity in cell cycle regulation across phylogenetically divergent species, it is necessary that we present an overview of cell cycle regulation in fungi and animal systems.

Largely based on genetic analysis in yeast, the eukaryotic cell cycle is believed to be regulated at two major decision points—a point late in $G_1$ called START, which is where a cell commits itself to DNA replication, and $G_2$/M phase transition [16]. Studies with fungi and animal systems indicate that both these transitions as well as progression of cells through S phase are controlled by protein kinases whose activity is regulated in a very complex manner [13,14,16].

## II. KEY PROTEINS INVOLVED IN CELL CYCLE REGULATION

### A. Cyclin-Dependent Kinases

In multicellular organisms there is a family of closely related protein kinases that function at different cell cycle transitions. This family of protein kinases is called cyclin-dependent kinases (Cdks) as the activity of these enzymes is dependent on interaction with a member of the cyclin family of proteins (see later). Cdks catalyze the transfer of phosphate from ATP to specific serine or threonine residues on regulatory and structural proteins, the aggregate modification of which drives cells through cell cycle checkpoints [19]. The first vertebrate Cdk, p34$^{CDC2}$ (Cdk1), was identified as the catalytic subunit of maturation-promoting factor (MPF) [20]. In vertebrates, nine Cdks (including Cdk1) have been identified based on their sequence and ability to complement yeast mutants or to interact with cyclins (Table 1) [21–23]. Cdk2 closely resembles Cdk1 [24–26]. Cdk3, Cdk5, and Cdk6 were identified in humans on the basis of their sequence similarity to a conserved stretch of residues (PSTAIRE motif) in Cdk1 [27]. Cdk5 is the only Cdk that is active exclusively in nondividing cells [28]. Cdk4 was first identified as a member of the protein-serine kinase family and designated as p34$^{PSK-J3}$ [29]. The same gene was later isolated from mouse macrophage cells in early $G_1$ and classified as Cdk4 as it was found to act in a cyclin-dependent manner [30]. A *Xenopus* p$^{34cdc2}$-related protein was shown to be a subunit of CAK (cdc2 activating kinase), an enzyme necessary for the activation of p34$^{cdc2}$ by phosphorylation [31]. When this protein was shown to associate with a novel cyclin (cyclin H), it was classified as Cdk7 [32]. Cdk8 is a 53-kDa protein containing sequence motifs and subdomains of serine/threonine-specific kinases [23]. Cdk9 has a modified PSTAIRE motif (PITALRE) and has been shown to be involved in transcription regulation rather than cell cycle control as have other Cdks such Cdk8 [33,34]. Cdk7 has also been shown to be involved in transcription regulation as well as cell cycle control [35–37]. Table 1 lists the known Cdks, conserved motifs, and, where known, the cyclin or other regulating protein they associate with. The interactions and functions will be explained in more detail in the following sections.

### B. Cyclins

Cyclins, a family of proteins named for their cyclical expression and degradation, play an important role in the cell division cycle. Cdks by themselves are inactive and are activated by their association with cyclins [38]. Cyclins were first discovered in clams and sea urchins as a class of proteins that accumulate to high levels in interphase and are abruptly destroyed at the end of M phase [38–40]. Proper timing of cyclin expression is controlled at the transcriptional level [41] and by ubiquitin-mediated degradation of cy-

**TABLE 1**  Cyclin-Dependent Kinases and Their Regulatory Subunits Identified in Vertebrates

| Name | PSTAIRE motif | Amino acid identity to cdc2 kinase domain (%) | Regulatory subunit | Major phosphorylation sites TY[a] | T[b] |
|---|---|---|---|---|---|
| cdc (Cdk1) | PSTAIRE | 100 | Cyclin A, B types | GEGTYGV | RVYTHEV |
| Cdk2 | PSTAIRE | 65 | Cyclin A, D types | GEGTYGV | RTYTHEV |
| Cdk3 | PSTAIRE | 66 | Unknown | GEGTYGV | RTYTHEV |
| Cdk4 | PV/ISTVRE | 44 | Cyclin D types | GVGAYGT | MALTPVV |
| Cdk5 | PSSALRE | 57 | p35,[c] Cyclin D types | GEGTYGT | RCYSAEV |
| Cdk6 | PLSTIRE | 47 | Cyclin D types | GEGAYGK | MALTSVV |
| Cdk7 | NRTALRE | 40 | Cyclin H, p36[d] | GEGQFAT | RAYTHQV |
| Cdk8 | SACRE | 36[e] | Cyclin C | GRGTYGH | Missing |
| Cdk9 | PITALRE | 42 | Cyclin K, T types | GQGTFGEV | NRYTNRV |

[a] T-14, Y-15 in p34$^{cdc2}$.

[b] T-161 in p34$^{cdc2}$.

[c] p35 is a brain-specific activator of Cdk5 and is not similar to cyclins.

[d] p36 is not structurally similar to cyclins.

[e] Overall sequence identity.

clins [42]. Cyclins are involved in activation of Cdks and are probably involved in substrate specificity of Cdks [43–45].

Fourteen different types of cyclins have been identified in vertebrates (Table 2). The defining feature of all cyclins is a conserved 100-amino-acid domain called the cyclin box, which contains the Cdk-binding site [46]. Cyclins have been generally divided into two groups—mitotic cyclins, which are involved in the $G_2/M$ transition, and $G_1$ cyclins, involved at Start and the $G_1/S$ transition. Mitotic cyclins have the motif $RXXL(X)_{2-4}N$, called a destruction box, that has been implicated in their destruction through the ubiquitin pathway [47]. $G_1$ cyclins do not have this motif but alternatively have PEST (proline, glutamic acid, serine, and threonine-rich) motifs that are associated with protein instability and are thought to allow protein levels to closely parallel messenger RNA (mRNA) abundance [48]. The mitotic and $G_1$ cyclins also differ in their overall structure. Mitotic cyclins have approximately 200 amino acids (including the destruction box) N-terminal to the cyclin box, whereas $G_1$ cyclins mostly extend C-terminal (including PEST sequences) [49]. Cyclins A and B are considered mitotic cyclins. Cyclin A is involved in S

**TABLE 2**  Cyclins in Vertebrates

| Name | Cdk partner | Function[a] | Expression[b] |
|---|---|---|---|
| A | Cdk1/Cdk2 | S and $G_2/M$ | Peak at $G_2/M$ |
| B1 | Cdk1 | $G_2/M$ | Peak at $G_2/M$ |
| B2 | Cdk1 | $G_2/M$ | Peak at $G_2/M$ |
| B3 | Cdk1/Cdk2 | $G_2/M$ (and S?) | Peak at $G_2/M$ |
| C | Cdk8 | $G_1$ | Peak early $G_1$ |
| D1 | Cdk4/6 and Cdk2/5[c] | $G_1$ | Predominantly $G_1$ |
| D2 | Cdk4 | $G_1$ | Constant |
| D3 | Cdk4/6 | $G_1$ | Peak at $G_1/S$ |
| E | Cdk2 | $G_1/S$ | Peak at $G_1/S$ |
| F | ?[d] | ? | Peaks at $G_2$ |
| G | ? | ? | Responds to growth stimuli |
| H | Cdk7 | Multiple phases | Constant |
| I | ? | ? | Constant |
| K | Cdk9 | Transcription | |

[a] Phase of cell cycle functions in.

[b] Refers to protein levels.

[c] Cdk4/6 appear to be the primary partners of cyclin D1.

[d] ?, not known.

phase and $G_2/M$ phase transition events, and cyclin B is implicated at the $G_2/M$ transition [13,50–52]. Cyclin F, which is most closely related to A and B, also fluctuates with the cell cycle peaking at $G_2$ but it lacks the destruction box motif of mitotic cyclins and contains PEST sequences as do $G_1$ cyclins [48]. Cyclins C, D, and E were identified in humans based on their ability to complement a yeast mutant lacking $G_1$-type (CLN) cyclins [52–54]. D cyclins do not fluctuate with the cell cycle but are responsive to growth factors and nutrient supply, and cyclins C and E accumulate periodically, peaking at different times in $G_1$. Cyclin C levels rise about twofold early in $G_1$ and decrease slowly through S, $G_2$, and M phases [54]. Studies indicate that cyclin C may be associated with the transcription apparatus and may be involved in relaying growth-regulatory signals [23]. Cyclin E peaks in late $G_1$ and is involved at the $G_1/S$ transition [53]. Cyclin G has neither a destruction box nor a PEST sequence and does not fluctuate with the cell cycle. Rather, it responds to growth stimuli and is a transcriptional target of the p53 tumor suppressor protein [55,56]. Cyclin H is most closely homologous to cyclin C and is implicated in the control of multiple cell cycle transitions [32]. Cyclin I was isolated from human brain but also expressed in skeletal and heart muscle [57]. Its expression is not cell cycle dependent. Cyclin K was isolated in a yeast two-hybrid screen with Cdk9 [58]. Table 2 lists the cyclins in vertebrates, their known Cdk partners, and possible function(s) of the cyclin-cdk complex.

## III. ONSET OF M PHASE

Research in the past several years with yeast and mammalian systems using various approaches indicates that the onset of M phase is regulated by a mechanism that is common to all eukaryotic cells [14,59]. The regulation of the $G_2/M$ transition was the first to be elucidated and is the best understood in yeast and vertebrates. Regulation is coupled to mechanisms monitoring time, cell mass, growth rate, and the completion of chromosome replication [19]. Entry into mitosis is characterized by a structural reorganization of the cell including chromosome condensation, disassembly of the nuclear lamina and other intermediate filament systems, arrest of membrane traffic and nuclear envelope breakdown, reorganization of microtubules to form a mitotic spindle apparatus, and rearrangements of the actomyosin cytoskeleton for cell rounding and cytokineses [19].

### A. Key Proteins Involved in $G_2/M$ Phase Transition

#### 1. *p34 Protein Kinase*

p34 protein kinase was identified genetically in the fission yeast (*Saccharomyces pombe*) as the product of a cell division cycle gene (*cdc2*) that encodes a 34-kDa protein [60–62]. Homologues of this gene have been found in budding yeast (p34$^{CDC28}$) [63–65], several vertebrates [15,22], invertebrates [4], and plants [66,67] and are shown to be highly conserved both structurally and functionally among all eukaryotes. p34 protein kinase genes from evolutionarily distant multicellular organisms including vertebrates and plants have been shown to complement yeast mutants in this gene [22,66]. Hence, it is considered to be a universal regulator of mitosis in eukaryotic cells [13,14,59]. In vertebrates, Cdk1/mitotic-cyclin complex (MPF) is the center of regulation but other proteins are also involved (Figure 1). The molecular mechanism of M-phase induction involves activation of Cdk1. When activated, kinase activity is directed against serine and threonine residues in substrates. The consensus phosphorylation target is S/T-P-X-Z (X = polar amino acid, Z = basic amino acid) [68]. Table 3 lists some substrates of Cdk/cyclin complexes. Several of these are M-phase substrates, some of which have been shown to be in vivo as well as in vitro substrates. Lamins, histone H1, nucleolin, caldesmon, and the regulatory light chain of myosin II are examples of in vivo substrates involved in the M-phase transition [21,68].

#### 2. *Cyclins and Regulation of p34 Protein Kinase Activity*

Figure 1 is a model of MPF activation during the $G_2/M$-phase transition. The level of p34 protein kinase is fairly constant during the cell cycle of dividing cells in yeast [69] and vertebrates [38]. However, the activity of this kinase increases significantly prior to the onset of M phase [13,14]. In animals, cyclin B accumulates during $G_2$, associates with Cdk1, and is abruptly destroyed at mitosis [70,71]. Cyclin B accumulation and association with Cdk1 have been shown to be required for MPF activation but are not sufficient to activate MPF [70]. The activity of MPF is also regulated by phosphorylation and dephosphory-

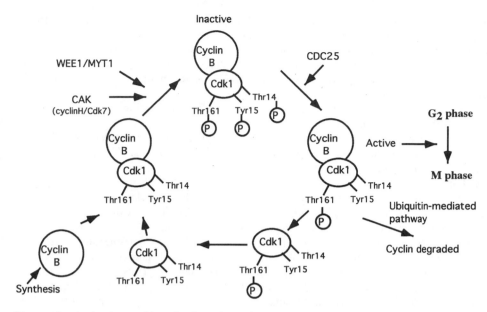

**Figure 1**   Activation and inactivation of vertebrate MPF (cyclinB/Cdk1). Cyclin B and Cdk1 association leads to phosphorylation of Thr 161 by CAK (Cdk activating kinase) and Tyr 15 and Thr 14 and WEE1/MYT1 kinases, respectively. Phosphorylation of Tyr 15 and Thr 14 is inactivating. Dephosphorylation of Tyr15 and Thr14 by the dual-action phosphatase CDC25 activates the complex. Cyclin is degraded through the ubiquitin-mediated pathway. Cdk1 is inactivated and presumably dephosphorylated.

**TABLE 3**   Some Known Substrates of Cdk/Cyclin Complexes

| G$_2$/M and M phase | | G$_1$/S and S phase | |
|---|---|---|---|
| Substrate | Protein type | Substrate | Protein type |
| Histone H1 | Chromatin-associated protein | p53 | Transcription factor |
| HMG I, Y, P1 | Chromatin-associated protein | E2F | Transcription factor |
| NO38, nucleolin | Chromatin-associated protein | PH04 | Transcription factor |
| Nuclear lamins factors | Cytoskeletal protein | p110Rb | Regulator of transcription |
| Caldesmon factors | Cytoskeletal protein | p107 | Regulator of transcription |
| Vimentin | Cytoskeletal protein | RNA polymerase II | Transcription enzyme |
| Neurofilament H | Cytoskeletal protein | Simian virus 40 T antigen | Protein implicated in replication |
| Myosin regulatory light chain | Cytoskeletal protein | DNA polymerases | Protein implicated in replication |
| SW1, SW15 | Transcription factor | Ribonuclotide reductase | Protein implicated in replication |
| c-myb | Transcription factor | Replication protein A | Protein implicated in replication |
| EF-1 beta | Translation factor | Subunits of yeast origin recognition complex | Protein implicated in replication |
| EF-1 gamma | Translation factor | | |
| CENP-E | Motor protein | | |
| MAP4/MAP1B | Microtubule-associated protein | | |
| p60$^{c\text{-}src}$ | Protein kinase | | |
| CKII (alpha/beta) | Protein kinase | | |
| p150abl | Protein kinase | | |
| Cyclin B | Cyclin | | |
| Cdc 25 | Protein phosphatase | | |
| PtP1b | Protein phosphatase | | |
| Rab1Ap/Rab4Ap | GTP-binding proteins | | |
| Rap1GAP | G protein | | |

lation events. In yeast and vertebrates, phosphorylation of a threonine, Thr167 in yeast and Thr 161 in *Xenopus*, is necessary for activation of MPF [72]. In budding yeast a monomeric kinase, CAK1p, has been identified as responsible for this activating phosphorylation [35]. In vertebrates, the activating phosphorylation of this threonine has been shown to be achieved by a kinase originally called CAK (cyclin-dependent kinase activating kinase), recognized now as Cdk7, which together with its cyclin regulating subunit, cyclin H, phosphorylates conserved threonines in other Cdks [32,35]. Cdks 2, 3, 4, and 6 and Cdk7 itself have this conserved threonine, whereas Cdk5 has a serine (Cdk7 is a serine/threonine kinase) residue and Cdk8 is missing the motif completely (Table 1).

In yeast, negative regulation of MPF is achieved by phosphorylation of Tyr15 on cdc2 by *MIK1* and *WEE1* gene products [73], whereas in animals phosphorylation of both Tyr15 and Thr14 is required to maintain MPF in an inactive state [72]. In 1993 a human WEE1 kinase was isolated that phosphorylates Cdk1 on Tyr15 but not Thr14 [74,75], and in 1997 a Wee1-type kinase, MYT1, was shown to phosphorylate Thr14 [76]. Cyclin association with Cdk1 is necessary for phosphorylation of these residues [73,77]. These residues are in the ATP-binding subdomain element GEGTYGV of Cdk1 and it is assumed that phosphorylation of these sites interferes with ATP binding [21]. As can be seen in Table 1, Cdks 2, 3, 5, and 8 have both sites conserved, Cdks 4 and 6 have only the Tyr15 site, and neither site is present in Cdk7. Dephosphorylation of Tyr15 in yeast and Tyr15 and Thr14 in animals is necessary to activate inactive MPF [14]. A tyrosine phosphatase coded by cdc25 in yeast dephosphorylates Tyr15 [78], while its homologue in animals dephosphorylates both Tyr15 and Thr14 [15,79,80]. Activation of CDC25 is dependent on phosphorylation by cyclin B/Cdk1 causing a positive feedback loop [81]. Another type of kinase, polo-like kinase, has also been shown to be involved in activation of CDC25. Inactivation of CDC25 and thus maintenance of Tyr phosphorylation can cause $G_2$ delay in response to DNA damage [59]. Inactivation occurs by phosphorylation of CDC25 on a conserved serine residue by the protein kinases CHK1 and CHK2 in vertebrates [82,83]. Cyclin A also associates with Cdk1 to promote entry into mitosis. It is destroyed earlier in mitosis than cyclin B [84]. Inhibitors of cylin/Cdk complexes have been found (see later). The inhibitor p21 inhibits cyclin A/cdc2 in early $G_2$ [85].

Besides regulation by cyclin synthesis and destruction and specific phosphorylation-dephosphorylation of Cdks, studies have shown that controlling the subcellular localization of Cdk-cyclins is also essential for proper cell cycle coordination [86]. Cyclin A is constitutively nuclear but cyclins B1 and B2 accumulate in the cytoplasm and as cells enter prophase B1 is transported to the nucleus [87]. Cyclin D1, on the other hand, increases in the nucleus during $G_1$ but is transported as a Cdk/cyclin D1 complex to the cytoplasm in S phase [88].

### 3. Exit from M Phase

The activation of MPF induces the cell to divide and also sets the stage for its inactivation by activating the cyclin degradation system [84]. Destruction of cyclins in M phase inactivates p34 protein kinase and is required for transition from mitosis to interphase [12–14]. Sudden destruction of cyclins just prior to anaphase is mediated by the ubiquitin pathway of protein degradation [47,89,90]. In addition to inactivation of p34 protein kinase, reentry into the interphase requires dephosphorylation of proteins involving protein phosphatase action. Protein phosphatases that are required in late mitosis have been identified in yeast ("defective in sister chromatid disjoining"—dis; "bypass of wee suppression"—bws1) and *Aspergillus* ("blocked in mitosis"—bimG) [91]. Inactivation of MPF is necessary for the cell to complete cytokinesis and return to a new interphase but not sufficient to inactivate cyclin degradation [92]. Studies suggest that $G_1$ cyclin/Cdk activity is required to inactivate mitotic-cyclin destruction [92,93]. Although it was thought that cyclin degradation was necessary for movement from metaphase to anaphase, experiments showed that cyclin degradation (MPF inactivation) was not required for sister chromatid separation but separation was linked to ubiquitin-mediated proteasome degradation [94,95]. A chromosome-tether protein was proposed as a candidate for the necessary degradation.

## B.  M Phase Regulatory Proteins in Plants

Cell cycle research in plants at the biochemical and molecular level started relatively recently and is greatly benefiting from the tools and information obtained with fungal and animal systems. The obvious first step was to find out which of the known cell cycle regulatory components are conserved in plants. Research during the last several years has yielded some information indicating that at least some of the key cell cycle regulatory proteins (e.g., p34 protein kinase and cyclins, mitogen-activated protein kinase) are present and highly conserved whereas the presence of various other proteins is yet to be explored. The availability of these genes will help in studying the detailed regulation of various components involved in the cell cycle.

## 1. Plant Cdks

The first evidence of a $p34^{cdc2}$ (cdc2) homologue in plants came from studies with antibodies to an internal peptide containing the highly conserved EGVPSTAIREISLLKE motif found only in Cdks, the carboxy-terminal 127 amino acids of the human cdc2 homologue or the whole *Saccharomyces pombe* cdc2 protein [96]. A 34-kDa protein was detected by the antibodies in cell extracts from oats, *Arabidopsis*, and algae. Phosphorylation of the 34-kDa protein identified in the green algae *Chlamydomonas* was correlated to the time of commitment to divide and dephosphorylation to the end of mitosis. In another study, monoclonal antibodies to cdc2 were also used to identify the presence of a 34-kDa protein in 10 different plant species [97]. In some species there was a polymorphism around 34 kDa. In protein extracts from mitotically nonsynchronous cell populations, this polymorphism has been attributed to distinct phosphorylation states of the protein [97].

Since 1990, molecular cloning techniques have been utilized to isolate cdc2 homologues. Table 4 lists the isolated genes and some of their characteristics. Homologues have been identified in alfalfa

**TABLE 4** Plant Cyclin-Dependent Kinases

| Plant | Gene name | No. of aa[a] | MW[b] | PSTAIRE[c] | Phosphorylation sites | | Rescue[f] | Reference |
|---|---|---|---|---|---|---|---|---|
| | | | | | TY[d] | T[e] | | |
| Alfalfa | *cdc2MsA* | 294 | 33,886 | PSTAIRE | GEGTYGV | RTFTHEV | cdc2 | 98 |
| | *cdc2MsB* | 294 | 33,864 | PSTAIRE | GEGTYGV | RTFTJEV | cdc28 | 98 |
| | *cdc2MsC* | 509 | 57,000 | PITALRE | GEGTYGM | ANLTNRV | NR | 99 |
| | *cdc2MsD* | 311 | 35,000 | PPTALRE | GEGTYGK | KSYTHEI | NR | 99 |
| | *cdc2MsE* | 414 | g | SPTAIRE | g | LSENGVV | NR | 99 |
| | *cdc2MsF* | 316 | 36,000 | PPTTLRE | GEGTYGK | KKYTHEI | NR | 99 |
| *Antirrhinum* | *cdc2a* | NR | NR | PSTAIRE | GEGTYGV | RTFTHEV | cdc2 | 100 |
| | *cdc2b* | NR | NR | PSTAIRE | GegAYGV | RTFTHEY | cdc2 | 100 |
| | *cdc2c* | NR | NR | PPTALRE | GEGTYGK | KSYTHEI | no | 100 |
| | *cdc2d* | NR | NR | PPTTLRE | GEGTYGV | RTFTHEV | no | 100 |
| *Arabidopsis* | *cdc2a* | 294 | 34,008 | PSTAIRE | GEGTYGV | RTFTHEV | cdc28 | 101 |
| | | | | | | | cdc2 | 103 |
| | *cdc2b* | 309 | 35,295 | PPTALRE | GEGTYGK | KSYTHEI | NR | 102 |
| | *cdc2c* | 644 | 72,300 | PSTAIRE | GQGTYSS | NQLTSRV | NR | 104 |
| Maize | *cdc2ZmA* | 294 | 33,812 | PSTAIRE | GEGTYGV | RTFTHEV | cdc28 | 105 |
| | *cdc2ZmB* | g | g | PSTAIRE | g | | g | 105 |
| Mothbean | *cdc2* | 294 | 33,961 | PSTAIRE | GEGTYGV | RTFTHEI | NR | 106 |
| N. spruce | *cdc2Pa* | 294 | 33,702 | PSTAIRE | GEGTYGV | RTFTHEV | NR | 107 |
| Pea | *cdkPs1* | 294 | 33,864 | PSTAIRE | GEGTYGV | RTFTHEV | cdc28 | 97 |
| | *cdkPs2* | 294 | 33,900 | PSTAIRE | NR | | cdc28 | 21 |
| | *cdkPs3* | 18 | 57,524 | PITAIRE | NR | | NR | 21 |
| Petunia | *cdc2Pet* | g | g | PSTAIRE | g | | g | 108 |
| Rice | *cdc20s-1* | 294 | 34,049 | PSTAIRE | GEGTYGV | RTFTHEV | cdc28 | 109 |
| | *cdc20s-2* | 292 | 33,671 | PSTAIRE | GEGTYGV | RTFTHEV | h | 109 |
| | *R2* | 424 | 47,582 | NFTALRE | GEGTYGV | RNFTHQV | h | 110 |
| Soybean | *cdc2-S5* | 294 | 33,940 | PSTAIRE | GEGTYGV | RTFTHEV | cdc28 | 111 |
| | *cdc2-S6* | 294 | 33,950 | PSTAIRE | GEGTYGV | RTFTHEV | cdc28 | 111 |
| Tobacco | *cdc2Nt1* | 294 | 33,900 | PSTAIRE | GEGTYGV | RTFTHEV | cdc28 | 112 |
| Tomato | *cdkA1* | 294 | 33,700 | PSTAIRE | GEGTYGV | RTFTHEV | NR | 113 |
| | *cdkA2* | 294 | 33,700 | PSTAIRE | GEGTYGV | RTFTHEV | NR | 113 |

[a] Predicted number of amino acids in derived polypeptide.
[b] Predicted molecular weight of derived polypeptide.
[c] Composition of "PSTAIRE" domain.
[d] T-14, Y-15 in $p34^{cdc2}$.
[e] T-161 in $p34^{cdc2}$.
[f] Rescue of temperature-sensitive (ts) *cdc28* mutant of *S. cerevisiae* or ts *cdc2* mutant of *S. pombe*.
[g] Data unavailable due to truncated clone.
[h] Failed to rescue *S. cerevisiae cdc28* mutants.
NR, not reported.

[98,99], *Antirrhinum* [100], *Arabidopsis* [101–104], maize [105], mothbean [106], Norway spruce [107], pea [21,97], *Petunia* [108], rice [109,110], soybean [111] tobacco [112], and tomato [113]. These species represent plants as diverse as both monocots and dicots in angiosperms and a gymnosperm. The PSTAIRE motif is conserved in many of the plant homologues but is modified in others. Five types of Cdks can be identified on the basis of their sequences [66]. A-type Cdks have the conserved PSTAIRE motif and are most closely related to Cdk1 and Cdk2. B type have a modified PSTAIRE motif (PP-TALRE). The other three types are represented by only one or two members and are not well character-ized. An analysis done by Huntley and Murray [114] suggests that B-type Cdks in plants form two sub-groups, one with a PPTALRE motif and one with a PPTTLRE motif. The protein products of most plant cdc2 homologues are predicted to be near the expected 34 kDa except alfalfa cdc2MsC, *Arabidopsis* cdc2c/At, pea cdkPs3, and rice R2 (Table 4). In yeast, only one Cdk is involved in both $G_1$/S and $G_2$/M transitions, but in animals multiple Cdks are involved [115]. At least two different genes have been iso-lated for many of the plant species (Table 4). Southern analysis in *Petunia* suggests the possibility of more than one homologue [108]. Other tobacco complementary DNA (cDNA) fragments were isolated that had modified PSTAIRE motifs [112], and Southern analysis in Norway spruce suggests a family of 10 genes, some of which were identified as psuedogenes [107]. Olomoucine, a chemical inhibitor of Cdk1/Cdk2 ki-nases, reversibly arrested *Petunia* protoplasts at $G_1$ and *Arabidopsis* cell suspension cells at $G_1$ and $G_2$, suggesting that Cdk1/Cdk2 kinases are involved in both $G_1$/S and $G_2$/M transitions [116]. Using the same inhibitor, similar results were obtained in *Vicia faba* [117]. An alfalfa Cdk, CDC2Ms, was activated at the $G_1$/S transition when phosphate-starved cells reentered the cell cycle and remained active through S, $G_2$, and M phases, again suggesting that one Cdk is involved in both checkpoints [118]. However, a study of the cell cycle phase specificity of putative Cdk variants in alfalfa showed a fluctuation of transcript lev-els and amounts and activities of kinases in different cell cycle phases, which suggests the involvement of more than one Cdk [99]. Some Cdks may be involved in non–cell cycle functions as has been shown in yeast [119] and vertebrates [120]. The *Arabidopsis* Cdk, CDC2b, was shown to be involved in seedling growth via regulation of hypocotyl cell elongation and cotyledon cell development [121].

Expression of plant Cdks has been correlated with proliferative tissues [101,105,107,108,112,122] and the competence to divide [123,124]. In situ hybridization studies using an *Arabidopsis* cdc2 cDNA sequence as a probe showed that transcripts accumulated in leaf primordia, vegetative shoot apical meristem, flower meristem, root meristematic regions, and pericycle [123]. In root tips, where expres-sion is high in rapidly proliferating cell files and low in the quiescent center, specific distribution of ex-pression in the meristems parallels the pattern of mitotic activity. Hemerly et al. [125] introduced a fu-sion gene consisting of the promoter of the *Arabidopsis* cdc2 homologue, *CDC*2a, fused with the β-glucuronidase gene (*gus, uidA* gene from *E. coli*) into *Arabidopsis* plants. Histochemical GUS anal-ysis showed a positive correlation between *CDC*2a mRNA levels and the proliferative state of cells. However, *CDC*2a expression was not restricted to dividing cells. *CDC*2a expression was observed in some nondividing, differentiated tissue. In contrast, in animal cells Cdk1 is expressed in proliferating cells but not in differentiated, nonproliferating cells [126,127]. It is suggested that the ability of plant cells to dedifferentiate and reenter the cell cycle may be linked to the low-level expression of cdc2. The expression pattern of the four rice Cdks showed *CDC2Os*-1, *CDC2Os*-2, and *R2* uniformly in the di-viding regions of the root apex with *CDC2Os*-1 and *CDC2Os*-2 also expressed in differentiated cells [128]. *CDC2Os*-3 was detected only in patches in the dividing region. In *Petunia* the level of the cdc2 homologue *CDC2Pet* was higher in 4C nuclei than in 2C nuclei even in nonproliferating cells, and so a high level of Cdk may not indicate the proliferative state in tissues that have a high number of 4C nu-clei in mature cells [108]. Studies of the expression of plant cdc2 homologues during the cell cycle have given different results. *Arabidopsis CDC2aAt* and tobacco *CDC2Nt*1 mRNA levels were constant throughout the cell cycle whereas *Arabidopsis CDC2bAt* was preferentially expressed during S and $G_2$ phases of the cell cycle [112,124,129]. *Petunia CDC2Pet* mRNA levels increased during $G_2$ compared with $G_0$-$G_1$, rice *CDC2Os*-3 levels are highest from $G_2$ to M phase, and *Antirrhinum CDC2c* and *Cdc2d* also fluctuate with the cell cycle, being expressed at the highest levels in $G_2$/M [100,108,128,130]. Of the six alfalfa Cdks identified, four showed expression throughout the cell cycle (cdc2MsA–C and *CDC2MsE*), while *CDC2MsD* and *F* fluctuated with the cell cycle with the highest amount in $G_2$ to M phases [99].

Histone H1 kinase (H1K) activity has been associated with fractions from plant cell extracts. Differ-ent molecular weight fractions of protein from pea cell extracts were analyzed for the presence of a 34-

kDa protein that cross-reacted with a cdc2 antibody to see if the p34 was complexed with other proteins. Fractions containing the p34 were used for in vitro H1K studies. Kinase activity was highest in the higher molecular mass fractions and was correlated with the presence of dephosphorylated p34 [97]. The protein p13$^{suc1}$ is a protein that binds Cdks tightly enough to be used as a nonspecific Cdk affinity purification reagent [21]. It was first identified in yeast and later found to be a component of the p34$^{cdc2}$ protein kinase [131]. A human homologue of this protein forms a hexamer. It was suggested that this hexamer acts as a hub for Cdk multimerization in vivo [132]. Protein fractions precipitated using p13$^{suc1}$-Sepharose beads also displayed H1K activity in maize, wheat, and pea [133–135]. H1K assays defined a minor peak at $G_2$/M and a much stronger peak at $G_1$/S in extracts from synchronous pea root tips and alfalfa cell cultures [21,136]. More recently, histone H1 kinase specifically associated with A-type Cdks has been analyzed in alfalfa [99,118], *Arabidopsis*, and tobacco [66]. Most results showed high kinase activity in S, $G_2$, and M phases and largely reduced activity in $G_1$. On the other hand, the B-type Cdks showed a peak of H1 kinase activity in M phase [66,99]. A protein kinase that phosphorylated the heat shock protein AtHSF1 in *Arabidopsis* was identified as CDC2a, suggesting a possible regulatory interaction between heat shock response and cell cycle control in plants [137].

The WEE1-type phosphorylation site sequence in the species for which it was reported is generally conserved (GEGTYGV), the notable exception being *Antirrhinum* CDC2b, which has a nonconserved substitution of Thr-14 by Ala [130]. The CAK phosphorylation site sequence has more variation, with RTFTHEV being the most common sequence motif (Table 4). Evidence of the phosphorylation of these sites is increasing. Multiple signals have been found around 34 kDa in immunoblots of plant proteins probed with cdc2 antibody, which could be due to different phosphorylation states [97,98,122,138] but could also be due to cross-reactivity with other members of the Cdk family. Yamaguchi et al. [36] found that rice R2 could complement CAK deficiency in budding yeast and could phosphorylate rice CDC20s-1 in vitro and the C-terminal domain of the large subunit of RNA polymerase II. Phosphorylation of RNA polymerase II as well as Cdks by CAK has also been demonstrated in animals [139,140]. An *Arabidopsis* Cdk (CAKAt) was isolated by complementation of a CAK-deficient yeast mutant [141]. It was able to phosphorylate human Cdk2 but not RNA polymerase II. As phosphorylation by CAK of both Cdks and RNA polymerase II has been shown in animals, CAKAt is unique in not phosphorylating RNA polymerase II. However, a CAK identified in *Saccharomyces cerevisiae* has the same capability to phosphorylate the yeast Cdk, Cdc28p but not RNA polymerase II [35]. Evidence that plant Cdks are phosphorylated can also be seen from studies of cells arrested in $G_2$ by the absence of cytokinin [142]. The Cdk complexes had reduced kinase activity and high phosphotyrosine content. A p34$^{cdc2}$-like protein isolated from cytokinin-depleted cells was treated in vitro with yeast CDC25 phosphatase, which led to its dephosphorylation and activation. The cytokinin requirement of *N. plumbaginifolia* cells could be alleviated by expression of the cdc25 gene from yeast [143]. Mutations in the *Arabidopsis* Cdks, CDC2aAt and CDC2bAt, in which the Thr-14 and Tyr-15 were substituted for nonphosphorylatable residues, showed an increase in histone H1 kinase activity, which supports the negative regulation of p34 kinase activity by phosphorylation [144].

In plant cell division, plant cells form a preprophase band (PPB), a dense array of microtubules that aggregate at the periphery of the location where the new cell wall will form at cytokinesis [145]. Two reports have linked cdc2 homologues with this band [105,138]. In onion root tip cells, immunofluorescence microscopy using an antibody raised against the PSTAIRE motif of cdc2 revealed general staining in the cytoplasm throughout the cell cycle with more intense staining during prophase in an area reminiscent of the PPB. Double staining in prophase cells with PSTAIR and tubulin antibodies showed that the cdc2 homologue band was always located with the microtubule band but the cdc2 homologue band was narrower than the microtubule band [138]. Immunofluorescence microscopy studies in maize using similar antibodies showed localization of maize cdc2-like protein to the nucleus during interphase and early prophase [105]. Colocalization with the PPB was also found for some early prophase cells in both the root tip and subsidiary cell mother cells that give rise to the subsidiary cells of the stomatal complex in leaves. Association studies indicate that the PPB is necessary for localization of the kinase rather than vice versa. These two studies suggest a role for a cdc2-like kinase in establishing the division site of plant cells. Another microtubule-based structure, the spindle, was also shown to be affected by Cdk/cyclin function. Blocking of Cdks in *Vicia faba* root tips by olomoucine resulted in cells with abnormally short and dense kinetochore microtuble fibers that were randomly arranged in the vicinity of the kinetochores and chromosome arms [117].

## 2. Plant Cylins

The first cyclins identified in plants were from carrot and soybean [146]. One partial length carrot clone and two soybean clones (one full length, one partial length) were reported. Their sequences were mitotic-like but, as has been shown to be the case with other plant cyclins isolated to date, they could not be classified as either A- or B-type cyclins [146]. Since that report, cyclins have been identified in *Adiantum capillus-veneris* L. [147], alfalfa [148–150], *Antirrhinum* [100], *Arabidopsis* [21,124,151–156], *Brassica* [157], *Catharanthus roseus* [158], *Lupinus luteus* [159,160], maize [161–163], pea [21], rice [164,165], *Sesbania rostrata* [166], soybean [167], and tobacco [168,169]. See Table 5.

The majority of the plant cyclins were reported to have mitotic-like sequences based on the presence of the destruction box motif in their amino terminal end and homology to animal mitotic-cyclin sequences. Comparison of these mitotic-like cyclins with animal A- and B-type cyclins showed that they had sequence homologies to both types and so they could not be definitively classified as A- or B-type cyclins [101,161,167]. Comparative analysis of a large number of plant cyclins that have been isolated has revealed that the A/B family of cyclins possesses two distinct B-type groups and three distinct A-type

**TABLE 5** Plant Cyclins

| Plant | Gene name | Aa[a] | Dest motif[b] RXXLX$_{2-4}$N/PEST | Reference |
|---|---|---|---|---|
| *Adiantum* | Adica;CycA1;1 | 532 | RAALANLTN | 147 |
| Alfalfa | Medsa;CycB2;1 | 428 | RRALGVIN | 148 |
| | Medsa;CycB2;2 | 434 | RALGGINQ | 148, 149 |
| | Medsa;CycB2;3 | 428 | RRALGVIN | 150 |
| | Medsa;CycA2;1 | 452 | RAILQDVTN | 149 |
| | Medsa;CycD3;1 | 386 | PEST | 219 |
| *Antirrhinum* | Antma;CycB1;1 | 473 | RRALGDIGN | 100 |
| | Antma;CycB1;2 | 441 | RRALGDIGN | 100 |
| *Arabidopsis* | Arath;CycB1;1 | 428 | RQVLGDIGN | 124 |
| | Arath;CycB1;2 | 445 | RRALGDIGN | 152, 153 |
| | Arath;CycB1;3 | 414 | RRALGDIGN | 153, 154 |
| | Arath;CycB2;1 | 428 | RRVLRVIN | 151 |
| | Arath;CycB2;2 | 429 | RRALGVINH | 151 |
| | Arath;CycA2;1 | 445 | AKALGVSN | 151 |
| | Arath;CycA2;2 | 437 | RAVLKDVSN | 151 |
| | Arath;CycD1;1 | 334 | PEST | 155 |
| | Arath;CycD2;1 | 383 | PEST | 155 |
| | Arath;CycD3;1 | 376 | PEST | 155 |
| *Brassica*[c] | Brana;CycA2;1 | 434 | RAVLGDISN | 157 |
| | Brana;CycA1;1 | 425[d] | RAPLGNITN | 157 |
| *Catharanthus* | Catro;CycA3;1 | 372 | RVVLGELKN | 158 |
| | Catro;CycB1;1 | 436 | RRALGDIGN | 158 |
| Carrot | Dauca;CycA3;1 | 341[e] | RVVLGEISN | 146 |
| *Lupinus* | Luplu;CycB1;1 | 429 | RRVLKDIGN | 159 |
| | Luplu;CycB1;2 | 454 | RVVLGDIGN | 159, 160 |
| | Luplu;CycB1;3 | 460 | RRALGDIGN | 159, 160 |
| | Luplu;CycB1;4 | 452 | RKALGDIGN | 159, 160 |
| Maize | Zeama;CycA1;2 | 503 | Not reported | 162 |
| | Zeama;CycB1;1 | 420[e] | RAPLGDIGN | 161 |
| | Zeama;CycA1;1 | 456[e] | RASVGSLGN | 161 |
| | Zeama;CycB2;1 | 424 | RRALSDIKN | 161 |
| | Zeama;CycB1;2 | 445 | RRALGDIGN | 161 |
| Pea | Pissa;CycA2;1 | 472 | RAALHDIGN | 21 |
| | Pissa;CycA2;2 | 449 | RAGLTDVTN | 21 |
| | Pissa;CycB1;1 | 566 | RAILHDVTN | 21 |
| Rice | Orysa;CycA1;1 | 508 | RVALSNISN | 164 |
| | Orysa;CycB2;1 | 420 | RRPLRDINN | 164 |
| | Orysa;CycB2;2 | 419 | RRALRDIKN | 165 |
| *Sesbania* | Sesro;CycB1,1 | 445 | RKALGDIGN | 166 |

**TABLE 5** *Continued*

| Plant | Gene name | Aa[a] | Dest motif[b] RXXLX$_{2-4}$N/PEST | Reference |
|-------|-----------|-------|----------------------------------|-----------|
| Soybean | Glyma;CycA3;1 | 348 | RVVLGELPN | 167 |
| | Glyma;CycA2;1 | 469 | RAVLSDISN | 167 |
| | Glyma;CycA1;1 | 484 | RPPLSNLTN | 167 |
| | Glyma;CycB1;3 | 440 | RRVLQDIGN | 167 |
| | Glyma;CycB1;2 | e | RRALGDIGN | 146 |
| | Glyma;CycB1;1 | 454 | RKALGDIGN | 146 |
| Tobacco | Nicta;CycA1;1 | 483 | RPALTNISN | 169 |
| | Nicta;CycA1;2 | 482 | RPALTNISN | 174 |
| | Nicta;CycA2;1 | 493 | RAVLKDMKN | 169 |
| | Nicta;CycA3;1 | 371 | RVVLGELIN | 174 |
| | Nicta;CycA3;2 | 383 | RVVLGEIQN | 174 |
| | Nicta;CycA3;3 | 314 | RVVLGEIRN | 174 |
| | Nicta;CycB1;1 | 447 | RRALDIGN | 168 |
| | Nicta;CycB1;2 | 473 | RKALGDIGN | 169 |
| | Nicta;CycD2;1 | 354 | PEST | 223 |
| | Nicta;CycD3;1 | 373 | PEST | 223 |
| | Nicta;CycD3;2 | 367 | PEST | 223 |

[a] Predicted number of amino acids in derived polypeptide.
[b] Type of destruction motif present and sequence if RXXLX$_{2-4}$ N.
[c] Six other cyclin-box sequences were obtained by a PCR analysis of putative positives from a screening of a genomic library.
[d] Two other possible start sites following the first stop codon would give polypeptides of different aa number and MW.
[e] Uncertain due to truncated clone.

groups [170]. A third group of cyclins (discussed later), the D-like cyclins, also can be classified into three groups. In 1996 it was proposed by Renaudin et al. [170] that a uniform naming system be used for plant cyclins. Cyclin names used in this chapter conform to this system.

　　Expression of cyclins found in plants has been studied in many different ways and it is hard to compare between systems. However, some overall conclusions can be drawn from the different studies. Analysis of expression of mitotic-like cyclin mRNA in different plant tissues, cell suspension cultures, and calli using Northern or RNA dot blots has led to the conclusion that this mRNA is more highly expressed in tissues that contain meristematic regions (roots, young leaves, flower buds, callus, cell suspension) than in tissues that do not (stem, old leaves) [151,157,161]. These studies also showed that within a species, different cyclin mRNAs can be expressed in different amounts in the same tissues. In *Brassica*, Brana;CycA2;1 and Brana;CycA1;1 are both expressed in young leaves and apical meristem but only Brana;CycA1;1 is expressed in roots. Similarly, in *Arabidopsis*, Arath;CycB2;2 is expressed only in roots but Arath;CycB2;1 is expressed equally in all tissue [151]. Arath;CycA2;1 and Arath;CycA2;2 are also expressed largely in roots but Arath;CycA2;2 is also expressed at lower levels in several tissues. In soybean, Glyma;CycB1;1 was expressed at the highest levels in all tissues and Glyma;CycA3;1 and Glyma;CycA2;1 at the lowest levels [167]. The transcripts of Glyma;CycA3;1 and Glyma;CycB1;1 appeared to be most abundant in root tips and nodules, whereas Glyma;CycA1;1 was more abundant in shoot apices but the evidence was not conclusive [167].

　　In situ hybridization has also been used to study expression of plant cyclins. Expression of cyclins in *Antirrhinum* showed that Antma;cycB1;1 and Antma;cycB1;2 were expressed in only some cells in meristematic regions, suggesting that only cells in specific stages of the cell cycle expressed the genes [100]. Cells in mitosis expressed both cyclin genes in prophase and metaphase. Some cells in interphase also expressed cyclin genes but the exact stage ($G_1$, S, or $G_2$) could not be determined. In roots of 4-day-old *Arabidopsis* seedlings, Arath;CycB2;1&2 and Arath;CycA2;1&2 expression was restricted to the root apical meristem and strong signals were detected during the formation of lateral roots [151]. In-depth in situ studies of Arath;CycB1;1 correlate its expression with meristematic tissues such as the root tip, shoot apices, axillary buds of the inflorescence, and pericycle [171]. In situ hybridization studies of rice A-type cyclin Orysa;CycA1;1 showed expression from $G_2$ to early M phase, whereas expression of the B-type cyclins Orysa;CycB2;1 and Orysa;CycB2;2 lasted until the end of mitosis [164]. In soybean, in situ hybridization has shown that Glyma;CycB1;1 is expressed in $G_2$ to M phases and Glyma;CycA1;1 is expressed from late S to $G_2$ phase, suggesting a B- and A-type function for these cyclins, respectively [167]. A third soybean cyclin gene's (Glyma;CycA3;1) expression was limited to S phase, suggesting that it is a novel class of plant cyclin which correlates to its A3 group classification [167]. Human mitotic cyclin B antibodies recognized two proteins in synchronized *Allium cepa* L. root meristem cells that were expressed during $G_2$ with a maximum at late $G_2$ to early M phase, and degraded in the late hours of mitosis, suggesting a cyclin B–like pattern of expression [172].

Expression of the cyclins has also been studied as a function of the cell cycle using cell cycle inhibitors to halt the cells at a particular stage and then either testing their mRNA expression or letting them divide synchronously following inhibition. Various plant cyclins have been correlated with specific cell cycle stages. In alfalfa, Medsa;CycB2;1 and Medsa;CycB2;2 mRNA showed maximal expression during $G_2$ and M phase with Medsa;CycB2;1 appearing earlier than Medsa;CycB2;2 in $G_2$ [148]. Expression of Medsa;CycB2;1 and Medsa;CycB2;2 was also correlated with the growth phase of cell suspension cultures, being expressed during the logarithmic stage but not the stationary phase [148]. The *Arabidopsis* cyclin Arath;CycB1;1 was expressed in higher amounts in $G_2$ nuclei than in $G_0$-$G_1$ nuclei separated by flow cytometry [124]. Arath;CycB1;1 expression was decreased in roots treated with the cell cycle inhibitor hydroxyurea, which holds cells at $G_1$/S, indicating that its expression is during $G_2$/M as found in nuclei from $G_2$ cells. The Arath;CycB2;1 and Arath;CycB2;2 messages were identified in S to metaphase, and the Arath;CycA2;1 and Arath;CycA2;2 messages were identified in late $G_1$ to metaphase [151]. Promoter analysis of Arath;CycB1;1 showed an increase in the rate of transcription upon exit of the S phase, a peak at the $G_2$-to-M transition and during mitosis, and a decrease upon exit from the M phase; similar analysis of Arath;A;2;1 showed low transcription during $G_1$ with a slow increase in S, a peak at the $G_2$ and $G_2$-to-M transition, and down-regulation before early metaphase [173]. Tobacco cyclin mRNA for Nicta;CycA1;1 and Nicta;CycA2;1 was detectable through S, $G_2$, and M phases (A type–like expression), and Nicta;CycB;2 was detectable from $G_2$ to M (B type–like expression) [169]. Reichheld et al. [174] isolated five cDNA clones for A-type cyclins in tobacco that have multiple expression patterns through the cell cycle, suggesting different roles for different cyclins. Promoter analysis of Nicsy;CycB1;1, which is homologous to Nicta:B1;1, showed that the 1149-bp 5'-flanking region is sufficient to regulate expression in a cell cycle regulated manner [175]. The reporter gene exhibited the same pattern of a peak just before mitosis and disappearance immediately after anaphase as did endogenous Nicsy;CycB1;1 and its homologue Nicta;CycB1;1 [175,176]. These studies substantiate the classification of A- and B-type cyclins by sequence and indicate that cyclins in plants have roles similar to those of their counterparts in animals. A few exceptions were also noted. Adica;CYCA1;1, while phylogenetically falling in the A-type cyclins, was not expressed at the onset of S phase of the first cycle of germinating spores and became detectable after S phase and accumulated during the second $G_1$ phase [147]. Also, Medsa;CycA2;1, although mitotic-like in sequence, was found to be present in all cell cycle stages [149]. This suggests that plants may have some different roles than their animal counterparts.

Promoter analysis of the *Catharanthus roseus* B-type cyclin Catro;CycB1;1 (CYM) showed that the promoter could direct M phase–specific transcription of a β-glucuronidase reporter gene [158]. Mutational analysis of the promoter showed that a 9-bp element is essential for M phase-specific promoter activity [177]. The promoter contained three similar elements, and when these elements were fused to a heterologous promoter, they were sufficient for M phase–specific expression. Similar elements (called M-specific activators, MSAs) were found in other B-type cyclin promoters including Glyma;CycB1;3, Nicta;CycB1;3 (Nt-CYM), Arath;CycB1;1, and Arath;CycB2;1. MSA-like sequences were also found in two other M phase–specific tobacco kinesin-like proteins, suggesting that MSA may be a common cisacting promoter element that controls M phase–specific expression of cell cycle–related genes in plants [177].

Cell cycle–dependent proteolysis of the mitotic cyclins has been demonstrated in yeast and animals [42,47,89,90,178]. Plant ubiquitin/proteasome-mediated degradation of mitotic-like A- and B-type cyclins in tobacco was studied using the N-terminal domains of the cyclins containing the destruction box motif [179]. Fusions of the domains to chloramphenicol acetyltransferase (CAT) reporter gene caused an oscillation of the fusion proteins in a cell cycle–specific manner. Mutations in the destruction box abolished cell cycle–specific proteolysis. Cyclin A-CAT proteolysis was turned off during S phase, whereas cyclin B-CAT proteolysis was turned off during late $G_2$ phase. As further evidence, a known proteasome inhibitor, MG132, blocked the cells during metaphase and the cyclin-CAT fusion proteins remained stable [179].

Functional studies have been done on a few of the cloned cyclins. The established functional test for cyclins is injection of mRNA or protein into *Xenopus* oocytes [21,39]. The $G_2$-arrested oocytes undergo germinal vesicle breakdown in a dose-dependent manner when injected with cyclin mRNA. This test was used for Glyma;CycB1;1, Arath;CycB1;1, and the maize cyclins [124,146,161]. Using a relatively new test for function, the *Arabidopsis* cyclin Arath;CycB1;2; the tobacco cyclins Nicta;CycA1;1, Nicta; CycA2;1, and Nicta;CycB1;1; the maize cyclin Zeama;CycA1;2; the alfalfa cyclin Medsa;CycA2;1; and the rice cyclins Orysa;CycB2;1, Orysa;CycB2;2, and Orysa;CycA1;1 were able to complement a yeast CLN ($G_1$) cyclin-minus mutant [149,153,162,164,169]. Although the cyclins complement $G_1$ cyclins, mitotic cyclins in animals can also replace the function of the $G_1$ cyclins in yeast [54]. A study using a p34cdc2/cyclinB-like kinase from *Chlamydomonas* injected into *Tradescantia virginiana* stamen hairs demonstrated the effects of this complex on plant cell division [180]. Microinjection caused rapid disassembly of the preprophase band of microtubules and chromatin condensation and nuclear envelope breakdown were accelerated, similar to the initiation of nuclear division by the maturation- or mitosis-promoting factor of animal cells.

### 3. Other Proteins Involved in the $G_2$/M Transition

A plant homologue of a SUC1/CKS1-type protein (CKS1At) was isolated in *Arabidopsis* using *CDC2aAt* (a Cdk kinase) as a bait in a yeast two-hybrid screen [181]. Studies of human *SUC1/CKS1* gene suggest that the SUC1/CKS1 protein may function as a docking factor for both positive and negative regulators of Cdk complexes [182]. CKS1At could bind both CDC2aAt and CDC2bAt and could rescue yeast mutant in the cks1 gene. Mutants of CDC 2aAt and CDC2bAt in the region involved in the interaction of human Cdk2 with the CKS1Hs protein [182] abolished the binding of CKS1At with CDC2aAt [144].

Plant homologues of the kinetochore protein SKP1, which is required for cell cycle progression in mammals, have been identified in *Arabidopsis* [183]. Human SKP1 associates with the cyclin A/Cdk2 complex, and in yeast it is shown to be an essential part of the ubiquitin complex that marks proteins for destruction. An SKP1 homologue was identified in orchids and *Arabidopsis* and its expression was highly correlated with meristem activity [183].

A WEE1 homologue (ZmWee1) was isolated from maize [184]. WEE1 is a kinase whose phosphorylation provides negative regulation of Cdks (see earlier). Overexpression of ZmWEE1 in yeast inhibited cell division, and recombinant ZmWEE1 could inhibit activity of a cyclin-dependent kinase from maize [184].

## IV. PROGRESSION THROUGH $G_1$ AND S PHASES

In yeast, the same protein kinase (p34cdc2/CDC28) is responsible for regulating START, a point in $G_1$ where a cell commits itself to DNA synthesis and the $G_2$/M transition. $G_1$ cyclins (CLN genes) interact with p34cdc2/CDC28 to drive the cell through this $G_1$ restriction point (START) to enter into S phase [16]. However, in multicellular organisms progression through $G_1$ and S phase seems to be much more complex and appears to be controlled by a family of Cdks that are structurally related to p34 kinase [24,27,115,185–187]. Each of the Cdks seems to associate with a specific type of cyclin(s) to be activated and appears to be involved in a specific phase of the cell cycle (Figure 2 and Table 2).

## A. Proteins Involved in Animal $G_1$ and S Phases

### 1. Cyclin/Cdk Interactions

Two major players in $G_1$ are the D cyclins and cyclin E. The D-type cyclins (D1, 2, 3) interact with Cdk4 or Cdk6 and act in middle to late $G_1$ [88,188,189]. The D-type cyclins are good candidates for activating the Cdk required to pass through the R point (Figure 2). In mammalian cells, the D-type cyclins are induced in a cell lineage–specific manner and are synthesized as long as growth factor stimulation persists and are degraded rapidly when mitogens are withdrawn [190]. Cyclin D-bound Cdk4 undergoes activating phosphorylation by CAK (cyclin H/Cdk7) on its conserved threonine (Table 1) [191].

The D-type cyclins can bind directly to retinoblastoma protein (Rb) in complexes with either Cdk4 or Cdk6 [191]. The expression of D1 is also dependent on the presence of pRb, suggesting the existence of a regulatory loop between pRb and cyclin D1 [192]. pRb binds to transcription factors such as E2F,

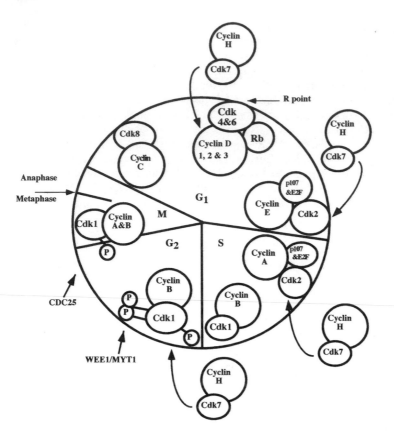

**Figure 2** Cyclin and cyclin-dependent kinase complexes in the vertebrate cell cycle. Rb (retinoblastoma protein) is known to be phosphorylated by a cyclin D/Cdk4. p107 is a homologue of Rb, and E2F is a transcription elongation factor. Cyclin H/Cdk7 complex is the Cdk activating kinase (Cak). WEE1/MYT1 are kinases and CDC25 is a phosphatase involved in inactivation/activation of cyclinB/Cdk1.

negatively regulating their activities. Phosphorylation of pRb by D cyclins reverses the growth-suppressive effect of pRb by releasing E2F. Active E2F can then trigger transcription of E2F-driven genes, which probably include cyclin E and cyclin A [193]. D-type cyclins contain a motif LXCXE (X = any amino acid) through which they bind pRb [155]. This domain is also present in viral proteins that bind pRb. D-type cyclins have also been shown to associate with proliferating cell nuclear antigen (PCNA), an acidic, nonhistone nuclear protein and an auxiliary protein of DNA polymerase-$\delta$ shown to be present only in proliferating mammalian cells and not in nondividing cells [194].

Cdk2 is considered to be the Cdk most directly involved in DNA replication [195]. Cyclins E and A sequentially activate Cdk2. The Cdk activity required for DNA replication has been defined as S phase–promoting factor (SPF) [196]. In the early *Xenopus* embryo, cyclin E/Cdk2 is sufficient to support entry into S phase, but later in development, cyclin A/Cdk2 provides a significant additional quantity of SPF [196]. Activity of the cyclin E/Cdk2 complex is dependent on phosphorylation of Thr 160 (in human cyclin E) by cyclin H/Cdk7. Mammalian cells that fail to proliferate because of loss of anchorage show a decrease in phosphorylation of this threonine [197]. Overexpression of cyclin E causes premature entry into S phase [198], and microinjection of antibodies against cyclin E prevents S phase initiation [199]. Cdk2 complexes are negatively regulated by T14/Y15 phosphorylation [72], and a mammalian CDC25 homologue (CDC25A) regulates $G_1$ progression [200].

Cyclin E is degraded once cells enter S phase and Cdk2 forms complexes with cyclin A (Figure 2) [199]. Cyclin A activates Cdk2 shortly after cyclin E and is essential for progression through S phase [22]. Both cyclin E and A complexes associate with the retinoblastoma (Rb)-related protein p107 and the transcription factor E2F [49]. Cyclin E/Cdk2 interaction with E2F activate transcription, whereas cyclinA/Cdk2 phosphorylation results in the loss of E2F binding activity [193].

The way components of SPF are regulated is being elucidated. An Rb-like protein has been tied to control of cyclin A expression [201]. Protein kinase C has been implicated in the regulation of CAK activity toward cyclin/Cdk2 complexes [202]. In human fibroblasts, the activation of protein kinase C in late $G_1$ causes cell cycle arrest at least in part through down-regulation of CAK-mediated Cdk2 phosphorylation. The suppression of CAK activity was accompanied by a decrease in the message levels of both cyclin H and Cdk7 (the components of CAK).

Cyclin C is a third cyclin type isolated by rescue of $G_1$ cyclin minus yeast [54]. Its activity had been assigned to $G_1$. An eighth Cdk (Cdk8) was isolated in a search for human protein kinases with a possible role in cell cycle control [23]. This kinase was shown to associate with cyclin C in vitro and probably in vivo. The cyclin C/Cdk8 pair is structurally related to SRB10-SRB11, a Cdk/cyclin pair shown to be a part of the RNA polymerase II holoenzyme of *S. cerevisiae* [139]. It is proposed that cyclin C/Cdk8 might be functionally associated with the mammalian transcription apparatus and perhaps be involved in relaying growth-regulatory signals [23].

## 2.  *Inhibitors of Cyclin/Cdk Complexes*

As discussed before, the sequential formation, activation, and subsequent inactivation of a series of cyclin/Cdk complexes (Figure 2) govern progression of cells through the cell cycle. Studies have identified regulatory proteins that bind to cyclin/Cdk complexes and inhibit their activity. These proteins are termed Cdk inhibitory proteins (CKIs) [203]. The first CKIs were identified in yeast [204–206]. CKIs with different roles have also been identified in vertebrates. One of these CKIs, p21, was identified in three separate studies. It was isolated as a protein that interacted with cyclin/Cdk2 complexes and inhibited their activity [207], a protein whose expression was increased in senescent cells [208], and a protein that could be induced by the tumor repressor p53 [209]. p21 inhibits cyclin A/Cdk2, cyclin A/Cdk1, cyclin E/Cdk2, cyclin D1/Cdk4, and cyclin D2/Cdk4 complexes [85,210]. The tumor suppressor p53 is involved in regulation of p21. The p21 promoter has a p53 binding site and p21 transcription is activated by wild-type p53, which appears to be essential to the p53-mediated arrest of the cell cycle in $G_1$ in response to DNA damage [200]. The tumor suppressor pRb might also be transcriptionally activated by p53 and there may be a direct protein-protein interaction between p53 and pRB [85]. In light of the phosphorylation of pRb by cyclin/Cdk complexes causing the release of transcription factor E2F at the $G_1$/S transition, p53 may be involved in arrest by increasing the p21 inhibitor of cyclin/Cdk complexes, increasing the amount of pRb and directly interacting with it and in some way limiting its ability to be phosphorylated. Various viral oncogene products can promote cell growth by abrogating activity of either p53 or pRb. p21 is also regulated in other ways. Three studies in muscle indicate that p21 is up-regulated in a non–p53-dependent manner during skeletal muscle differentiation [211–213] or by signal transducers and activator of transcription (STAT) proteins.

A second CKI, p27, also interacts with multiple Cdks including cyclin E/Cdk2, cyclin D/Cdk4, and cyclinA/Cdk2 complexes [22,214]. The antimitogenic factor transforming growth factor β (TGFβ) induces reversible arrest of target cells in late $G_1$. This arrest has been correlated with inhibition of the cyclin E/Cdk2 complex [215], and in cells arrested by TGFβ, cyclin E/Cdk2 is associated with p27 [216]. Cyclin D/Cdk4 complexes also interact with p27 and may sequester it [217]. As reviewed by Sherr [191], p27 is titrated by cyclin D/Cdk complexes, inhibiting them until a threshold is exceeded. TGFβ inhibits Cdk4 synthesis, which would raise the effective level of "free" p27 allowing inhibition of cyclin E/Cdk2 activity.

Both p21 and p27 preferentially associate with the cyclin/Cdk complex rather than with the individual kinases. Two closely related proteins, p16 and p15, that are structurally and functionally distinct from p21 and p27 target Cdk4 and Cdk6 subunits and prevent their binding to cyclins [22]. p16 and p15 are representatives of a family of 15 to 20-kDa proteins with loosely conserved ankyrin motifs, some members of which have been isolated and are differentially expressed in response to a number of antiproliferative signals [185].

## B.  $G_1$/S Phase Plant Proteins

Studies indicate that plants have several distinct p34-like protein kinases [218,219]. Two cdc2 homologues from alfalfa (CDC2A and CDC2B) appear to regulate different phases of the cell cycle. CDC2A could complement only $G_2$/M transition, whereas CDC2B complemented $G_1$/S function [219]. A study in maize endosperm gives evidence that two types of Cdks are involved in cell cycle regulation [220]. H1K activity was associated with a protein fraction recovered from a p13$^{suc1}$-agarose column known to bind cdc2 kinases. Activity of this kinase correlated with early endosperm development when mitosis is still taking place. Addition of extract from endosperm with low cdc2 H1K activity caused the inactivation of kinase from early stage endosperm, indicating the presence of an inhibitor in the extract [220]. A second H1K activity in maize endosperm was associated with a maize cdc2-related protein precipitated with GST (glutathione-*S*-transferase) fusions to E2F-1 and E1A. This suggests that different cyclin-dependent kinase complexes are involved in cell cycle regulation in maize: an M phase kinase that can interact with p13$^{suc1}$ and an S phase kinase that can interact with the human transcription factor E2F-1 and the adenovirus E1A proteins. In alfalfa cell extracts, two temporally separable S-phase H1K peaks could be resolved by precipitation with anti–human cyclin antibodies (early S peak) or p13$^{suc1}$ (later S peak) [136].

A new group of plant cyclins have been identified in *Arabidopsis* [155], alfalfa [221], pea [222], and tobacco [223]. In *Arabidopsis*, three cDNA clones (Arath;CycD1;1, Arath;CycD2;1, Arath;CycD3;1) were identified that could complement $G_1$ cyclin function in the yeast mutants [155]. Sequence analysis revealed that they did have cyclin characteristics but not the mitotic destruction box motif. Rather, like animal $G_1$ cyclins, they have PEST sequences. When compared with databases, they showed the greatest relatedness to mammalian D-type cyclins. Cell cycle inhibitor studies indicated that Arath;CycD3;1 is expressed prior to DNA synthesis and Arath;CycD1;1 and Arath;CycD2;1 are expressed in $G_1$. Using cDNAs in DNA hybridization studies, hybridizing bands were found in tobacco, Jerusalem artichoke, cauliflower, and *Antirrhinum* [155]. In situ hybridization in alfalfa detected a D-type cyclin (Medsa;CycD3;1) in meristem tissue in a subset of interphase cells [221]. Of particular note was the presence of a conserved motif, LXCXE, for Rb binding [224]. This domain, which is present in animal $G_1$ cyclins, was found in all three *Arabidopsis* clones and in the alfalfa Medsa;CycD3;1 clone but not in plant mitotic-like cyclin sequences [155,221]. Three tobacco D-type cyclins isolated using cyclin D cDNA probes also contain the LXCXE motif [223]. One of these, Nicta;CycD3;2, was induced in $G_1$ following a stationary phase and then remained at a constant level in synchronous cells. The other two (Nicta;CycD2;1 and Nicta;CycD3;1) accumulated during mitosis, which is not typical of D-type cyclins and suggests a mitotic role for these cyclins. A fourth cyclin D (Arath;CycD4;1) was isolated from *Arabidopsis* in a two-hybrid interaction screen using a Cdk, CDC2aAt, as a bait [156]. This was significant as cyclin/Cdk pairs have not previously been identified in plants. In situ hybridization studies suggested a role for this D-type cyclin in developmental processes [156]. Based on sequence similarity, a C-type cyclin was isolated from rice, but expression and functional analyses have not been performed [225].

Human Rb protein can bind in vitro to translated *Arabidopsis* cyclin D(δ) [226] and Rb-related proteins have been isolated in maize [227–230], tobacco [231], and *Arabidopsis* [232]. The maize Rb proteins have been shown to bind to plant cyclin D and to be phosphorylatable by mammalian Rb-kinases and seem to be involved in developmental processes [227–230]. The tobacco Rb-like protein was shown to be phosphorylated in insect cells by a tobacco Cdc2-kinase/cyclin D complex [231]. Another player in the cyclin D/Cdk, Rb regulation of the cell cycle has been isolated from a plant. Using the tobacco Rb protein ZmRb1 in a yeast two-hybrid screen of a wheat cDNA library, an interacting clone was identified that showed homology to E2F family members [233]. The plant E2F was shown to be expressed in proliferating cultured cells and in differentiated tissues and was up-regulated early in S phase. These data suggest that control of the $G_1$/S transition in plants is similar to animal $G_1$/S cell cycle control.

Proliferating cell nuclear antigen (PCNA), which associates with cyclin D/Cdk complexes in animals, has been isolated in rice and *Catharanthus* and peas [234,235]. Like its animal counterpart, plant PCNA is also preferentially expressed in proliferating cells and was not detectable in quiescent cells [234]. In a synchronized population of cells, PCNA expression was highly expressed in S phase. Shimizu and Mori [235] found that PCNA associates with a pea cyclin D (Pissa;CycD3;1) during dormancy but not in growing buds, suggesting a possible mechanism for arrest at $G_1$.

Inhibitors of the cell cycle are also being isolated in plants. A yeast two-hybrid screen with *Arabidopsis* Cdc2a as a bait identified a cyclin-dependent kinase inhibitor called ICK1 [236,237]. It contains an important consensus sequence found in the mammalian Cdk inhibitor p27 (Kip1), but the rest of the sequence shows little similarity to any known Cdk inhibitor. ICK1 was also identified in a two-hybrid screen with the *Arabidopsis* CycD3 and was shown to be induced by abscisic acid, and along with its induction there was a decrease in cdc2-like histone H1 kinase activity [236].

## C.  Endoreduplication Cycle

The endoreduplication cycle, during which consecutive doublings of the genomic DNA occur in the absence of chromatin segregation and cytokinesis, is common in plants. Both local (endopolyploidy restricted to specialized cell types such as endosperm) and systemic somatic polyploidy have been reported [238–242]. Jacqmard et al. [243] investigated the presence of cell cycle–related genes in mitotically dividing cells and endoreduplicating tissues of *Arabidopsis*. They found that Cdks CDC2aAt and CDC2bAt and cyclin Arath;CycB1;1 were present only in mitotically dividing cells while CKS1At (see Sec. III.B.3) was present in both mitotic cells and endoreduplicative cells, suggesting that CKS1At may play a role in both the mitotic and endoreduplication cycle. An H1K activity in maize endosperm was associated with a maize cdc2-related protein precipitated with GST (glutathione-*S*-transferase) fusions to E2F-1 and E1A. The H1K activity was higher during the period of endoreduplication [220].

A plant Rb protein, ZmRb, undergoes changes in level and phosphorylation state concomitant with endoreduplication and it is phosphorylated in vitro by an S-phase kinase from endoreduplicating endosperm cells [228]. Another cell cycle–related gene from maize, *ZmWee1*, is highly expressed during endoreduplication, suggesting a possible role in this process [184].

Endoreduplication requires exit from the mitotic cycle and transformation of the cell cycle to the endocycle. A homologue of CCS52, a protein involved in mitotic cyclin degradation, was isolated from *Medicago sativa* root nodules [244]. Overexpression of CCS52 in yeast triggered mitotic cyclin degradation, cell division arrest, endoreduplication, and cell enlargement. Expression of CCS52 in *Medicago* was enhanced in differentiating cells undergoing endoreduplication [244].

## V.  ROLE OF CALCIUM AND CALMODULIN IN CELL CYCLE REGULATION

Calcium, a key intracellular messenger in both plants and animals, has been shown to regulate many different processes in plants [245–247]. Calmodulin, a calcium-binding protein found in all eukaryotes, is one of the primary mediators of calcium action (see Chapter 35 for more information on calmodulin). For over a decade, calcium and calmodulin have been implicated in controlling cell proliferation in eukaryotic cells including plants [246,248–252]. Calcium is essential for the growth of all eukaryotic cells. It has been shown that cells require the presence of millimolar levels of extracellular calcium to proliferate [253,254].

Progression of normal cells through the cell cycle is found to be associated with transient changes in intracellular calcium concentration [248,249,251,255]. Neoplastic cells, which can proliferate in the absence of external calcium, contain a higher level of intracellular calcium than normal cells [256]. Manipulation of cytosolic calcium concentration has been shown to affect cell cycle events [257–259]. By determining the level of intracellular calcium during different stages of the cell cycle, it has been demonstrated that rapid and transient increases in intracellular calcium occur at specific stages of the cell cycle in plant and animal cells [260–263]. Calcium transients are observed at the awakening from quiescence, $G_2/M$ transition, as the cells completed mitosis, and both sides of $G_1/S$ boundary [249,252]. Mitotic events such as breakdown of nuclear envelope, chromatin condensation, and onset of anaphase have been correlated with a transient increase in intracellular calcium [251,259,260]. Furthermore, these mitotic events could be induced prematurely by artificially elevating cytosolic calcium, whereas chelation of intracellular calcium by calcium chelating agents blocked the nuclear envelope breakdown and the metaphase/anaphase transition, suggesting that an increase in cytosolic calcium is required for these mitotic events to take place [257–259]. Blocking of intracellular calcium prior to the $G_1/S$ boundary results in inhibition of DNA synthesis [249]. This suggests that calcium transients are critical for the progression of cells from $G_1$ to S phase of the cell cycle.

Studies with both plant and animal tissues have revealed a higher level of calmodulin in dividing cells as compared with nondividing cells [245,249,264]. An increased level of calmodulin mRNA, protein, and activity is observed in meristematic tissues of the plants [245,265,266]. In vertebrates and lower eukaryotic cells, a twofold increase in the intracellular calmodulin concentration is observed at the $G_1/S$ boundary [267–269]. Stimulation of quiescent cells to reenter the proliferative state elevated the amount of calmodulin. Furthermore, transformed mammalian cell lines have been shown to contain elevated levels of calmodulin [270,271]. To study the effect of altered levels of calmodulin on the cell cycle, Rasmussen and Means [272,273] manipulated the levels of calmodulin by stably transforming mouse cell lines with vectors that constitutively or inducibly express either calmodulin sense or antisense RNA. A transient increase in calmodulin resulted in acceleration of proliferation, whereas a decrease in calmodulin caused a transient cell cycle arrest. Constitutive elevation of intracellular calmodulin levels in these cells shortened the cell cycle due to the reduction in the length of $G_1$. Calcium and calmodulin level determinations during different stages of the cell cycle and the data on the effect of an elevated or reduced level of calmodulin on the cell cycle indicate that three specific points in the cell cycle ($G_1$/S, $G_2$/M, and metaphase/anaphase) are sensitive to calcium and calmodulin (Figure 3). Overexpression of calmodulin in *Aspergillus nidulans* increased growth rate by decreasing cell cycle time, whereas a reduced level of calmodulin prevented entry into mitosis [249].

Calcium and calmodulin have multiple functions and regulate a variety of processes including some housekeeping functions [245,266,274]. Hence, it has been argued that the observed effects of calcium and calmodulin manipulations on the cell cycle may not affect specific control points but could be due to the requirement of calcium and calmodulin for many housekeeping functions. Studies with unicellular fungi (yeast and *A. nidulans*), which are amenable to genetic manipulations, indicate that calcium and calmodulin regulate specific decision points during the cell cycle [249]. However, the mechanisms by which calcium and calmodulin control of cell cycle are beginning to be elucidated.

## A.   Mode of Calcium and Calmodulin Action in Regulating $G_2$/M Transition

Repression of calmodulin synthesis, thereby calmodulin levels, or reduced extracellular calcium in *Aspergillus* cells blocked entry into mitosis [275,276]. Under these conditions tyrosine dephosphorylation of p34 protein kinase that is needed for its activation is blocked and the activity of NIMA (never in mitosis mutant) protein kinase, a protein kinase required for the $G_2$/M transition in *Aspergillus*, is also reduced. Effects of reduced calmodulin and calcium could be reversed by elevating their levels. These studies with *Aspergillus* indicate that calcium and calmodulin are required for activation of p34 kinase and another protein kinase called NIMA that are associated with the $G_2$/M transition [249]. The activation of p34 kinase and NIMA protein kinase by calcium and calmodulin could be due to direct interaction of NIMA protein kinase and the enzyme responsible for tyrosine dephosphorylation of p34 kinase with calcium/calmodulin complex or indirect interaction through proteins that bind to the calcium/calmodulin complex. The NIMA protein kinase and tyrosine phosphatase involved in p34 activation did not bind to calcium/calmodulin and the activity of immunoprecipitated NIMA kinase was not affected by calcium and calmodulin. These results indicate that the activation of p34 and NIMA kinases could be mediated by the proteins that bind to the calcium/calmodulin complex. Over two dozen calmodulin-binding proteins

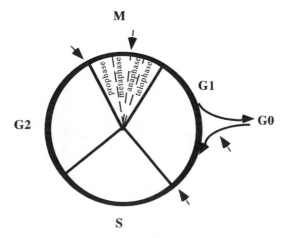

**Figure 3**   The phases of the cell cycle that require calcium and calmodulin. Arrows indicate the control points that are regulated by calcium/calmodulin.

have been identified in animal systems [277–279]. Some preliminary results suggest that a calmodulin-dependent protein kinase, a multifunctional enzyme that requires calcium and calmodulin for its activation, could be a likely candidate in mediating the calcium/calmodulin effect on NIMA protein kinase and NIMT (a *cdc25* homologue) of *Aspergillus* [122]. The purified calmodulin-dependent protein kinase has been shown to phosphorylate NIMA kinase and NIMT in vitro in a calcium/calmodulin-dependent manner. Furthermore, B-type cyclins that are known to associate with CDC25 proteins and regulate their activity [280] have been found to act as substrates for calcium/calmodulin-dependent protein kinase in vitro [122]. However, the effect of this phosphorylation on the activity of these enzymes is not known.

Human p54(cdc25-c) dephosphorylates cyclinB/Cdk1 and triggers mitosis. A study of the activation of p54(cdc25-c) by phosphorylation indicates that a calcium/calmodulin-dependent step may be involved in its initial activation [281]. The calcium/calmodulin-dependent protein kinase (CaM kinase) II could phosphorylate p54(cdc25-c) in vitro and increase its phosphatase activity. An inhibitor of the CaM kinase II resulted in a cell cycle block at $G_2$ phase. The Cdk1 remained tyrosine phosphorylated in the blocked cells.

Studies with plants indicate that there are a number of calmodulin-binding proteins in plants [246,282]. The identity and function of some of these proteins are being elucidated [279,283]. A cDNA that encodes a calcium/calmodulin-dependent protein kinase has been isolated from plants [284]. In addition to calcium/calmodulin-dependent protein kinase, plants contain a unique calcium-regulated protein kinase that requires calcium but not calmodulin [calcium-dependent and calmodulin-independent protein kinase, also called calcium-dependent protein kinase (CDPK)] [285,286] and appears to be present in all plants. A kinesin-like calmodulin-binding protein (KCBP) was isolated from *Arabidopsis* and other flowering plants [282,287,288]. KCBP has two unique domains that are not present in known kinesin-like proteins (molecular motors that move along microtubules): a calmodulin-binding domain at the C-terminus following the motor domain and a myosin tail homology domain in the tail [282,289,290]. KCBP binds calmodulin in a calcium-dependent manner at physiological calcium concentration [282] and the binding of calmodulin inhibits KCBP from binding microtubules or dissociates the preformed KCBP/MT complex [290–292]. KCBP has been immunolocalized in association with the preprophase band, spindle apparatus, and phragmoplast [293]. A non–calcium/calmodulin-regulated kinesin-like protein in humans (HsEg5) is phosphorylated by a cyclin/Cdk complex [294]. Whether any of the calcium, calcium/calmodulin-regulated protein kinases and calmodulin-binding proteins other than KCBP are involved in plant cell cycle regulation is not known.

## B.  Calcium/Calmodulin in Metaphase/Anaphase Transition

Several lines of evidence indicate that calcium and calmodulin are required for the metaphase-anaphase transition [260–263]. A transient increase in cytosolic free calcium at the onset of anaphase has been demonstrated. As indicated earlier, one of the critical events that take place during the metaphase-anaphase transition is inactivation of p34 kinase due to degradation of cyclins. Studies indicate that calcium and calmodulin could be involved in degradation of cyclins [122]. It has been demonstrated that micromolar concentrations of calcium induce cyclin B degradation in metaphase-arrested *Xenopus* egg extracts [295]. The addition of a synthetic peptide that binds to the calcium/calmodulin complex, prior to raising the calcium level in the extract, blocked cyclin degradation and inactivation of p34 kinase [295]. The inhibition of cyclin degradation by micromolar concentration of calcium with calcium/calmodulin-binding peptide could be reversed by adding calmodulin, suggesting that the calcium action is mediated by calmodulin. Furthermore, by using appropriate inhibitors the involvement of calpain, a calcium-dependent protease, and protein kinase C was eliminated. These results indicate that calcium and calmodulin are involved in cyclin degradation in *Xenopus* eggs. It is known that cyclins are degraded by ubiquitin-dependent proteolysis [47]. Proteasome activity was shown to be influenced by calcium specifically during the metaphase-anaphase transition in ascidian meiotic cycle [296] and a subunit of the proteasome was shown to bind calcium [297].

## C.  Calcium in the $G_1$/S Transition

D-type cyclins do not have the ubiquitin/proteasome destruction box motif of the mitotic cyclins. Instead they have PEST sequences that are typical of short-lived proteins. Loss of cyclin D1 induced by serum starvation was reversed by inhibitors of the calcium-dependent protease calpain [298]. Conflicting data from another study indicated that cyclin D1 was more likely degraded by the proteasome than calpain although both could degrade it [299]. It was concluded that in human fibroblast cells the protease that mediates the progression from late $G_1$ to S phase is calpain, not the proteasome [300].

## VI. PHYTOHORMONES AND CELL DIVISION

Phytohormones, especially auxins or cytokinins, have been shown to be intimately involved in cell division control in plants [5]. In many plants these hormones, singly or in combination, induce cell division in dedifferentiated noncycling cells. It has been well established from plant tissue culture studies that auxin and cytokinins are necessary for inducing cell division. Also, apical meristems that contain the cycling cells contain high levels of auxin. Addition of these hormones to differentiated cells that have ceased to divide results in dedifferentiation and reentry of these cells into the cell cycle [301,302].

Using auxin-dependent tobacco suspension cultures, seven different auxin-inducible cDNA clones have been isolated and characterized [303,304]. mRNA corresponding to these clones is rapidly induced when quiescent cells are triggered to undergo cell division by an exogenous supply of auxin. Takahashi et al. [301,305] isolated two auxin-induced cDNAs named *parA* and *parB* (protoplast auxin regulated) from tobacco mesophyll protoplasts. Addition of auxins and cytokinins can induce cell division in tobacco mesophyll protoplasts, which are differentiated cells that have ceased to divide. Expression of *par* genes was not detected in differentiated cells, whereas they are expressed in protoplasts that are cultured in the presence of auxin. Both *parA* and *parB* genes are expressed during the transition from $G_0$ to S phase of in vitro cultured protoplasts [301,305]. Furthermore, the expression of *par* genes was observed prior to initiation of DNA synthesis. *parB* has been identified as glutathione *S*-transferase [305]. Although this enzyme is mostly known to be involved in detoxification of xenobiotics, studies have indicated its involvement in cell proliferation [269,306,307]. The role of the *parB*-coded enzyme in tobacco mesophyll protoplasts is not yet known. It is somewhat intriguing that none of the auxin-regulated genes are similar to genes implicated in cell division, nor do cell cycle phase specific genes show any homology to known key cell cycle regulatory genes [301,303,305,308].

Two genes responsible for auxin mutants have been identified as genes that are involved in auxin signal transduction and are homologous to enzymes known to be involved in regulating the stability of key cell cycle regulatory proteins such as the cyclin-dependent kinase inhibitor SIC1p [309]. The gene products AXR1 and TIR1 are homologous to proteins involved in the ubiquitination of SIC1p, targeting it for destruction and thereby the release of Cdk inhibition. Auxin has also been shown to induce a mitogen-activated protein kinase in *Petunia*, PMEK1 [310].

In studies using p34 protein kinase cDNAs and antibodies, it has been shown that auxin induces p34 protein kinase mRNA and protein [111,122,123,134,311]. However, it should be noted that the auxin effect on p34 protein kinase mRNA and protein was studied after a long time following the auxin treatment (the earliest time point is 1 day), whereas most of the auxin-regulated cDNAs that are implicated in cell division [301,304,305] have been isolated from the libraries that are made after several hours of auxin treatment. This and other factors such as posttranscriptional regulation and abundance of mRNA corresponding to known key cell regulatory proteins in relation to other auxin-regulated genes could account for the absence of known key cell cycle regulatory genes in the pool of auxin-induced cDNAs. In a few plant systems that have been tested, addition of auxin has been shown to induce the expression of p34 protein kinase at both the mRNA and protein level [98,111,122,123]. A severalfold increase, in p34 protein kinase was observed during auxin-induced cell division in carrot cotyledons [122]. In addition, a soybean p34 kinase highly expressed in roots (cdc2-S6, Table 4) is up-regulated by *Rhizobium* infection, which leads to cell division and nodulation in alfalfa [111].

The expression of mitotic-like cyclins Arath;CycA;2;1&2 and Arath;CycB2;1&2 was greatly decreased if auxin was withdrawn from cell cultures, but no effect was shown on the expression of Arath;CycB1;1 [151]. In situ hybridization studies of Arath;CycB1;1 showed that the increase in expression was due to stimulation of cell division by the addition of hormone but not directly by the hormone itself [171].

D-type cyclins in plants would be expected to respond to growth factors such as hormones. To investigate this, auxin, cytokinin, and sucrose were withdrawn from *Arabidopsis* cell suspension cultures

[155]. The levels of Arath;CycD3;1 did not change with withdrawal or readdition of all three substrates. However, when cytokinin alone was added, the expression of Arath;CycD3;1 increased fourfold and this was somewhat enhanced with addition of sucrose. Auxin was antagonistic to the increase induced by cytokinin. The levels of Arath;CycD2;1 decreased on removal and increased with addition of sucrose but were independent of hormone. These results suggest that these cyclins respond to growth stimulators and carbon source much like animal D-type cyclins. D-cyclins may in turn form active kinase complexes targeting Rb homologues, causing inactivation and dissociation from E2F, which could then up regulate expression of S phase–specific genes. A recent study confirmed that Arath;CycD3;1 is induced by cytokinin and showed that constitutive expression of Arath;CycD3;1 allowed induction and maintenance of cell division in the absence of exogenous cytokinin [312]. Riou-Khamlichi et al. [312] suggest that cytokinin activates *Arabidopsis* cell division through induction of Arath;CycD3;1 at the $G_1$/S transition. In alfalfa, expression of the D-type cyclin Medsa;CycD3;1 was induced 12 hr following addition of auxin and cytokinin to pieces of fully differentiated leaves [221]. Expression of another alfalfa cyclin, Medsa;CycA2;1, was induced only 4 hr after addition [149]. Medsa;CycA2;1, although mitotic-like in sequence, is expressed in $G_1$ and responds to growth regulator and so may have a $G_1$ function.

Cytokinin has also been implicated in the stimulation of the tyrosine dephosphorylation and activation of Cdc2-like H1 histone kinase [142]. Addition of auxin and cytokinin to pith parenchyma cells resulted in a greater than 40-fold increase in a Cdc2-like protein with high H1 histone kinase activity. Without cytokinin the amount of protein increased, but it was inactive and contained a high amount of phosphotyrosine. This inactive protein could be activated by addition of bacterially produced yeast CDC25 phosphatase. Zeatin has been shown to be necessary for the $G_2$/M transition in tobacco cells [313]. An inhibitor of cytokinin biosynthesis inhibited mitosis at the $G_2$/M transition and this block could be overcome only by addition of zeatin. On the other hand, zeatin was not restrictive for the occurrence of the $G_1$/S transition in tobacco cells [314]. Furthermore, addition of cytokinin at early $G_1$ blocked the cycle at $G_1$/S, suggesting that down-regulation of the zeatin type of cytokinins is important for the $G_1$/S transition [314]. A somewhat conflicting report finds that both auxin (dichlorophenoxyacetic acid) and cytokinin ($N^6$-benzyladenine) are necessary for release from a block at $G_1$/S in *Petunia* [310]. Auxin alone could not stimulate *CDC2Pet* transcript accumulation but together with cytokinin there was an increase in transcript. Different phytohormones may have different effects on different tissues. In legume lateral root formation, auxin but not cytokinins causes cells in the $G_2$ phase to reenter the cell cycle giving rise to a lateral root primordium, while cytokinin inhibits lateral root formation and mimics Nod factors by activating inner root cortical cells to form a nodule primordia [315].

Other phytohormones such as abscisic acid (ABA) and gibberellic acid (GA) have been implicated in cell division control in certain plant systems [316–320]. In deepwater rice, GA induces growth and part of this growth is found to be due to stimulation of cell division [316]. GA induces cell division in the intercalary meristem of rice internodes in cells that are arrested at $G_2$ [165]. ABA is implicated in inhibiting cell division in endosperm of cultured maize kernels, maize root tips, pea buds, and in pollen mother cells [317–320]. ABA was shown to induce ICK1 (a putative Cdk inhibitor) in *Arabidopsis*, which resulted in a reduction of Cdc2-like H1 kinase activity [237].

## VII. SYNCHRONIZATION OF PLANT CELLS

Synchronized cell populations are essential to study biochemical and molecular events that take place during different phases of the cell cycle. Much of the information about cell cycle regulatory proteins in animals was obtained by studying the level or activity of a given protein during different phases of the cell cycle. Cells in meristems of plants have different cell cycle times and are highly asynchronous [5]. However, at a certain stage during the life cycle of a plant, cells divide synchronously for several cycles. For instance, microspore mother cells in anthers progress through meiosis synchronously. The first few divisions in the embryo and free nuclear divisions in endosperm are also synchronous. Natural synchrony, which occurs rarely, was found to be not appropriate for biochemical studies for various reasons [321]. Hence, several methods have been developed to obtain synchronized populations of cells in plant tissues and cultured cells. These methods include growing cultured cells after treating the cell with DNA synthesis inhibitors (e.g., aphidicolin, hydroxyurea, 5-aminouracil and fluorodeoxyuridine) or growing cells in some nutrient-limiting medium [321,322]. However, only a few methods have been found to be effective in inducing synchronization in plant cells; either the majority of the methods were found to be only partially effective or the agents that cause synchrony were found to have toxic effects on cell metabolism. Among the DNA synthesis inhibitors, aphidicolin is found to be the most effective in inducing synchronous growth in suspension cultures as well as in differentiated tissues. However, because of endogenous aphidicolin-inactivating activity in plant cells, which varies between cell types and plants, the concentration of aphidicolin and length of the incubation should be determined empirically in each case. Treatment of cells with aphidicolin, a mycotoxin that specifically blocks nuclear DNA replication by inhibiting DNA polymerase $\alpha$, causes accumulation of cells at the $G_1/S$ boundary of the cell cycle [323]. The effect of this inhibitor is reversible, hence removal of aphidicolin from the medium results in synchronous resumption of DNA synthesis. In several plant cells aphidicolin was shown to arrest about 80–95% of cells in $G_1$, which were found to move synchronously through the first round of mitosis after $G_1/S$ arrest [148,322]. The tobacco Bright Yellow (BY-2) cell line is one of the most well-characterized cell culture systems and can be synchronized efficiently using these inhibitors [324].

In suspension cultures of *Catharanthus roseus*, double phosphate starvation effectively induces synchrony [321]. This system is already helping to identify some of the phase-specific changes in mRNA and proteins [234,325]. In suspension cultures of *Datura*, hydroxyurea, another inhibitor of DNA synthesis, reversibly arrested the cells at the $G_1/S$ boundary [321].

Other cell cycle inhibitors have been found that block specific stages of the cell cycle by inhibiting cell cycle proteins. Olomoucine, a purine analogue that inhibits Cdks at micomolar concentrations while having little effect on other protein kinases, inhibits both the $G_1/S$ and $G_2/M$ transitions in plants [116]. Two structurally modified olomoucine-like molecules, bohemine and roscovitine, inhibit Cdks 10 to 100-fold better than olomoucine [117]. Roscovitine was found to block the cell cycle prior to entry into S and M phases [310,326].

Synchronization of plant cells with the preceding methods coupled with flow cytometry should greatly expedite the progress in cell cycle research in plants [327]. During the last 10 years flow cytometry has been increasingly used in analyzing plant cells. Protoplasts and isolated nuclei are amenable to flow cytometry. However, when protoplasts are used some modifications in methods and instrumentation are necessary because of their large size (20–75 $\mu$m) [327,328]. Developments in the use of flow cytometry for plant protoplasts have opened new avenues to analyze cell cycle regulatory proteins. Using multiparameter analysis, one could monitor the levels of two or more desired proteins during different phases of the cell cycle [329].

## VIII. CELL CYCLE IN PLANT DEVELOPMENT

Cell division is one of the primary determinants of various aspects of development in multicellular eukaryotic organisms. The regulatory mechanisms that determine various aspects of the cell cycle (e.g., which of the cells in an organism should undergo cell division, the timing and the plane of cell division in these cells, and which cells should remain quiescent and reenter the cell cycle) play a critical role in plant developmental processes such as embryogenesis, seed germination, and flowering. Hence, investigating these regulatory mechanisms will not only help us understand cell cycle regulation but also enable us to elucidate developmental programming in plants. Various developmental processes that involve the cell cycle are unique to plants. Unlike that in animals, cell division in higher plants is restricted to meristematic regions (shoot apical meristem, root apical meristem, and lateral meristem). The primary meristems such as shoot and root apical meristems continuously divide and contribute to the production of new organs and growth of the plants. Furthermore, shoot apical meristem can lose its indeterminate vegetative growth to become determinate floral meristem. The transition from vegetative meristem to floral meristem involves shortening of the cell cycle time as well as synchronization of the cell cycle [5]. In plants, during the course of normal development, quiescent cells become proliferative. For instance, lateral meristems (pericycle and cambium), auxillary buds, and cambium retain their ability to undergo cell division and enter into the cell cycle in response to some developmental cues. The root apex in plants contains, in addition to dividing cells, a group of cells called the quiescent center, which do not normally undergo cell division. However, if the root meristem is damaged, cells in the quiescent center reenter the cell cycle and form new meristem. In addition, if the cells from the quiescent center are cultured in vitro in the presence of hormones, they can undergo cell division and regenerate into whole plants. Pericycle cells retain the ability to divide and are responsible for the formation of lateral roots at vascular poles. Unlike animal cells, plant cells are unique in that they are totipotent. In several plant systems terminally differentiated nondividing somatic cells can dedifferentiate, divide, and regenerate into a whole plant. Reinitiation of cell division in differentiated and nondividing cells is a central feature in plant regeneration. Cytokinesis, a process by which cytoplasm is divided, is considerably different in plants as compared with other organisms. In plant cells, cytokinesis is initiated by forming a phragmoplast (made of mircrotubules and actin) between daughter cells, which is followed by deposition of cell wall material.

Some of the proteins involved in cell cycle control have appeared to be involved in development. A D-type cyclin, Arath;CycD4;1, was expressed along with CDC2aAt upon mitogenic stimulation following starvation [156]. In situ hybridization with Arath;CycD4;1 showed expression during vascular tissue development, embryogenesis, and formation of lateral roots. In pea auxilary buds, Pissa;CycD3;1 was found in dormant buds while Pissa;CycB1;2 and Cdc2 proteins were not [235]. Pissa;CycD3;1 interacted with PCNA during dormancy but not in growing buds, suggesting a means of regulation of this D-type cyclin. An *Arabidopsis* cdc2 homologue (CDC2cAt) was isolated that is divergent from the other cdc2 homologues in *Arabidopsis*, having unusual N- and C-terminal ends [104]. Its expression is restricted to flowers, weakly in buds and strongly in mature flowers. Its promoter has high homologies with a transcription factor that was previously immunolocalized in the epidermal cell layer of petals, suggesting that CDC2cAt is involved in flower development [104]. Two tomato Cdks were found to be expressed between anthesis and 5 days after anthesis (DPA) but their maximum kinase activity was obtained between 5 and 20 DPA, suggesting a posttranslational regulation of Cdk at the temporal and spatial levels during early tomato fruit development [113].

The retinoblastoma protein (Rb) is a Cdk substrate that when phosphorylated is involved in regulation of the cell cycle [224,330,331]. Rb has been shown to be involved in differentiation in animals by repressing transcription via E2F, and it also seems to act as a transcriptional coactivator in differentiating cells, possibly through its interaction with other proteins involved in transcriptional activation [332]. The Rb pathway may be involved in differentiation and development in plants [229,333]. ZmRb showed a gradient of accumulation that correlated with the gradient of cell proliferation in maize leaves, being abundant in the more differentiated cells whereas it is almost undetectable in the basal proliferative zone [229]. Plant Rb protein can interact with RbAp48, a protein that binds Rb and is present in chromatin assembly and histone deacetylation complexes and thus may also negatively regulate the expression of E2F-regulated genes by directing chromatin alterations [334].

Analysis of p34 protein kinase mRNA in roots has shown high levels of p34 protein kinase mRNA in meristem and all pericycle cells but not in the quiescent center. In pericycle, p34 protein kinase mRNA is expressed uniformly in all the prericycle cells, although lateral roots are initiated only at the vascular poles [123]. These results suggest that lateral root initiation opposite to vascular poles is controlled by a mechanism other than p34 protein kinase transcription. In situ hybridization studies during lateral root formation showed that induction of Arath;CycB1;1 was a very early event and that its accumulation might be one of the limiting factors for activation of cell division [171]. Before stem nodule development in *Sesbania rostrata*, Sesro;CycB1;1 transcripts were absent in cortical cells whereas Cdk gene *Cdc2-1Sr* transcripts were found in all cells [166]. After infection with *Azorhizobium caulinodans*, Sesro;CycB1;1 transcripts were expressed in a patchy pattern in the cortex of the root primordium. As discussed in Sec. III.B, cyclin transcripts appeared in only some cells of meristematic tissues [100,128,171], suggesting that cyclin accumulation could influence cell cycle timing. Doerner et al. [335] studied the affect of the expression of an *Arabidopsis* cyclin (Arath;CycB1;1) under the control of the cdc2a promoter. In roots, this ectopic expression accelerated root growth without altering the pattern of lateral root development. Normal and mutant forms of an *Arabidopsis* cdc2 homologue were expressed in *Arabidopsis* and tobacco [336]. Overexpression of normal Cdc2a did not affect cell division. Thus, it appears that cyclin expression may be the limiting factor for growth [335]. Expressed mutant Cdc2aAt completely abolished cell division in *Arabidopsis*. When expressed in tobacco, a few mutants were able to survive. They had reduced H1K activity and a lower number of cells than normal plants. The morphogenesis, histogenesis, and developmental timing were not affected, indicating that the developmental controls defining shape can act independently of cell division rates [336].

On the other hand, evidence suggests that cell division can be regulated by cell fate specification

[337]. In the *Arabidopsis* root epidermis, hairless cells are longer than hair cells at maturity. The cell division rate in the hairless cells slows down, allowing the cells to reach their normal larger size [338]. Furthermore, mutation of a gene controlling hair cell fate (*TTG*) causes ectopic root hair formation and all cell sizes are similar. Coupled with the finding that growth and morphogenesis can be uncoupled from cell division, this indicates that developmental processes control cell division rather than cell division controlling development. Organogenesis in flowering plants results from patterned control of the number, place, and plane of cell divisions. Of the cloned mutants that affect the various modes of cell division in mersistems, none appear to be homologues of the major cell cycle control genes, and so pattern control genes may be acting at some distance to regulate the cell cycle machinery [339].

A

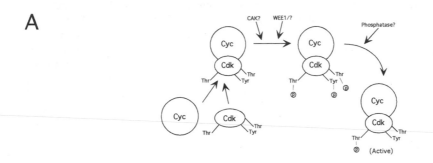

B

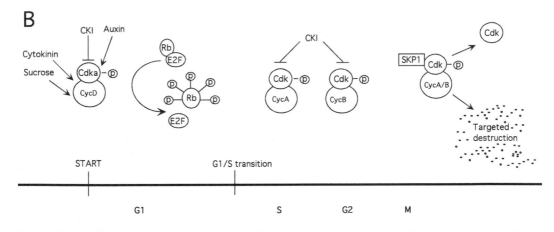

**Figure 4** Model of plant cell cycle control. (A) General proteins involved in cyclin/cyclin-dependent kinase regulation. A cyclin and its Cdk partner form a complex. Negative regulation is by phosphorylation of Thr14 and Tyr15 by a Weel-type kinase as isolated in maize [184]. Activating phosphorylation is on a Thr in the RTFTHEV (Table 4) motif by either a CakAt-like kinase [141] or an R2-like CAK [36]. The negative phosphates on Thr14 and Tyr 15 are removed by a phosphatase that has yet to be identified but whose existence is expected [143,144]. Active cyclin/Cdk phosphorylates appropriate substrates. (B) Cell cycle progression in plants. During G$_1$, D-type cyclins may be induced by hormones and other mitogens [149,221]. Some plant Cdks have also been shown to be induced by auxins [111,122,123]. Formation of active CyclinD/Cdk (Cdc2a in *Arabidopsis* [156]) complexes results in phosphorylation of Rb protein [227–230] causing the release of E2F transcription factors [233], which, in turn, activate the genes necessary for S-phase progression. Inhibitors of cyclin/Cdk (CKI) such as the ABA-inducible protein ICK1 isolated in *Arabidopsis* [236] may inhibit various cyclin/Cdk complexes in the cell cycle. A-type cyclins appear at S phase and B-type toward G$_2$ [151,164,167,169]. No specific Cdk partners have been established; however, different timing of expression of genes encoding differend Cdks has been reported [99,100]. Activated cyclin/Cdk complexes phosphorylate substrates involved in cell cycle progression. At the end of mitosis cyclin is marked for destruction by the ubiquitin/proteasome pathway possibly involving SKP1 protein [183,340].

# IX. CONCLUSIONS

Several major themes are emerging from the investigations of cell cycle regulatory mechanisms using different model systems ranging from simple eukaryotes (yeast and *Aspergillus*) to complex metazoans including vertebrates, invertebrates, and plants. First, it is increasingly evident that a few key proteins are critical in controlling the decision points in the cell cycle and these key proteins are highly conserved in all eukaryotes, indicating the universality of these key components. Second, the activity of certain protein kinases appears to play a key role in regulating the transition points between different phases of the cell cycle. Third, the mode of regulation of these key proteins may vary across phylogenetically divergent species. Finally, the regulatory mechanisms that control the cell cycle are far more complex in multicellular organisms than in unicellular organisms.

Cell cycle research in plants is in its early stages but because of the developments and tremendous progress made with fungi, vertebrates, and invertebrate systems and the highly conserved nature of some of the key cell cycle regulatory proteins, progress has been expedited in finding similarities and differences in regulatory mechanisms between plants and other eukaryotic organisms. Some of the key proteins known to be involved in yeast and mammalian systems have been identified in plants. Figure 4 is a model of what has been discovered in plants and how they might be involved in cell cycle control. Because no in vivo cyclin/Cdk partnerships have been shown and no in vivo substrates for Cdk phosphorylation have been identified, much is left to be discovered about cell cycle control in plants.

Although the cell cycle is common to all eukaryotes, it is controlled by different hormones or growth factors in plants and animals. Hence, although some key proteins are highly conserved across phylogenetically divergent species, it is likely that different regulatory mechanisms exist in plants and animals. As cell division is so fundamental to growth and development, it is bound to be an exciting area of research in plant biology. Advances in molecular and cell biology offer new approaches to investigate this very complex process. Manipulation of cell cycle regulatory proteins in cultured cells and transgenic plants should provide more insights into cell cycle regulation in plants as well as in plant development.

# ACKNOWLEDGMENTS

Research in our laboratory is supported by grants from the National Science Foundation, Agricultural Experiment Station, USDA, NASA, Plant Biotechnology Laboratory, Colorado Biotechnology Research Institute, and Colorado RNA Center.

# REFERENCES

1. AB Pardee. $G_1$ events and regulation of cell proliferation. Science 246:603–608, 1989.
2. RA Laskey, MP Fairman, JJ Blow. S phase of the cell cycle. Science 246:609–614, 1989.
3. JR McIntosh, MP Koonce. Mitosis. Science 246:622–628, 1989.
4. PH O'Farrell, BA Edgar, D Lakich, CF Lehner. Directing cell division during development. Science 246:635–640, 1989.
5. D Francis. The cell cycle in plant development. New Phytol 122:1–20, 1992.
6. LH Hartwell, TA Weinert. Checkpoints: controls that ensure the order of cell cycle events. Science 246:629–634, 1989.
7. L Hartwell. Defects in a cell cycle checkpoint may be responsible for the genomic instability of cancer cells. Cell 71:543–546, 1992.
8. R Fraley. Sustaining the food supply. Biotechnology 10:40–43, 1992.
9. CS Gasser, RT Fraley. Genetically engineering plants for crop improvement. Science 244:1293–1299, 1989.
10. C Bergounioux, C Perennes, SC Brown, C Sarda, P Gadal. Relation between protoplast division, cell cycle stage and nuclear chromatin structure. Protoplasma 142:127–136, 1988.
11. S Kartzke, H Saedler, P Meyer. Molecular analysis of transgenic plants derived from transformations of protoplasts at various stages of the cell cycle. Plant Sci 67:63–72, 1990.
12. AW Murray, MW Kirschner. Dominoes and clocks: the union of two views of the cell cycle. Science 246:614–621, 1989.
13. P Nurse. Universal control mechanism regulating onset of M-phase. Nature 344:503–508, 1990.
14. C Norbury, P Nurse. Animal cell cycles and their control. Annu Rev Biochem 61:441–470, 1992.
15. KW Kohn. Molecular interaction map of the mammalian cell cycle control and DNA repair systems. Mol Biol Cell 10:2703–2734, 1999.
16. SI Reed. G1-specific cyclins: in search of an S-phase-promoting factor. Trends Genet 7:95–99, 1991.

17. NR Morris. Lower eukaryotic cell cycle: perspective on mitosis from the fungi. Curr Opin Cell Biol 2:252–257, 1990.

18. G Draetta. Cell cycle control in eukaryotes: molecular mechanisms of cdc2 activation. Trends Plant Sci 15:378–383, 1990.

19. EA Nigg. Targets of cyclin-dependent protein kinases. Curr Opin Cell Biol 5:187–193, 1993.

20. J Gautier, C Norbury, M Lohka, P Nurse, J Maller. Purified maturation-promoting factor contains the product of a *Xenopus* homolog of the fission yeast cell cycle control gene *cdc2 +*. Cell 54:433–439, 1988.

21. TW Jacobs. Cell cycle control. Annu Rev Plant Physiol Plant Mol Biol 46:317–339, 1995.

22. EA Nigg. Cyclin-dependent protein kinases: key regulators of the eukaryotic cell cycle. Bioessays 17:471–480, 1995.

23. J-P Tassan, M Jaquenoud, P Leopold, SJ Schultz, S Schultz, EA Nigg. Identification of human cyclin-dependent kinase 8, a putative protein kinase partner for cyclin C. Proc Natl Acad Sci USA 92:8871–8875, 1995.

24. L-H Tsai, E Harlow, M Meyerson. Isolation of the human *cdk2* gene that encodes the cyclin A– and adenovirus E1A–associated p33 kinase. Nature 353:174–177, 1991.

25. J Rosenblatt, H Debondt, J Jancarik, DO Morgan, SH Kim. Purification and crystallization of human cyclin-dependent kinase-2. J Mol Biol 230:1317–1319, 1993.

26. J Paris, R LeGuellee, A Couturier, K LeGuellee, F Omili, J Camonis, S MacNeil, M Philippe. Cloning by differential screening of a *Xenopus* cDNA coding for a protein highly homologus to cdc2. Proc Natl Acad Sci U S A 88:1039–1043, 1991.

27. M Meyerson, GH Enders, C-L Wu, L-K Su, C Gorka, C Nelson, E Harlow, L-H Tsai. A family of human cdc2-related protein kinases. EMBO J 11:2909–2917, 1992.

28. LH Tsai, T Takahashi, VS Caviness, E Harlow. Activity and expression pattern of cyclin-dependent kinase-5 in the embryonic mouse nervous system. Development 119:1029–1040, 1993.

29. SK Hanks. Homology probing: identification of cDNA clones encoding members of the protein-serine kinase family. Proc Natl Acad Sci U S A 84:388–392, 1987.

30. H Matsushime, ME Ewen, DK Strom, J-Y Kato, SK Hanks, MF Roussel, CJ Sherr. Identification and properties of an atypical catalytic subunit (p34$^{PSK-J3}$/cdk4) for mammalian D type G1 cyclins. Cell 71:323–334, 1992.

31. J Shuttleworth, R Godrey, A Colman. p40$^{MO15}$ a cdc2-related protein kinase involved in negative regulation of meiotic maturation of *Xenopus* oocytes. EMBO J 9:3233–3240, 1990.

32. RP Fisher, DO Morgan. A novel cyclin associates with MO15/CDK7 to form the CDK-activating kinase. Cell 78:713–724, 1994.

33. G De Falco, A Giordan. CDK9 (PITALRE): a multifunctional cdc2-related kinase. J Cell Physiol 177:501–506, 1998.

34. V Leclerc, P Leopold. The cyclin C/Cdk8 kinase. Prog Cell Cycle Res 2:197–204, 1996.

35. P Kaldis. The cdk-activating kinase (CAK): from yeast to mammals. Cell Mol Life Sci 55:284–296, 1999.

36. M Yamaguchi, M Umeda, H Uchimiya. A rice homolog of Cdk7/MO15 phosphorylates both cyclin-dependent protein kinases and the carboxy-terminal domain of RNA polymerase II. Plant J 16:613–619, 1998.

37. M Hirst, MS Kobor, N Kuriakose, J Greenblatt, I Sadowski. GAL4 is regulated by the RNA polymerase II holoenzyme-associated cyclin-dependent protein kinase SRB10/CDK8. Mol Cell 3:673–678, 1999.

38. AW Murray, MW Kirschner. Cyclin synthesis drives the early embryonic cell cycle. Nature 339:275–280, 1989.

39. KI Swenson, KM Farrel, JV Ruderman. The clam embryo protein cyclin A induces entry into M phase and the resumption of meiosis in *Xenopus* oocytes. Cell 47:861–870, 1986.

40. C Norbury, P Nurse. Cyclins and cell cycle control. Curr Biol 1:23–24, 1991.

41. C Koch, K Nasmyth. Cell cycle regulated transcription in yeast. Curr Opin Cell Biol 6:451–459, 1994.

42. RW King, RJ Deshaies, JM Peters, MW Kirschner. How proteolysis drives the cell cycle. Science 274:1652–1659, 1996.

43. BA Schulman, DL Linstron, E Harlow. Substrate recruitment to cyclin-dependent kinase 2 by a multipurpose docking site on cyclin A. Proc Natl Acad Sci U S A 95:10453–10458, 1998.

44. DS Peeper. A- and B-type cyclins differentially modulate substrate specificity of cyclin-cdk complexes. EMBO J 12:1947–1954, 1993.

45. J Roberts. Evolving ideas about cyclins. Cell 98:129–132, 1999.

46. JHA Nugent, CE Alfa, T Young, JS Hyams. Conserved structural motifs in cyclins identified by sequence analysis. J Cell Sci 99:669–674, 1991.

47. M Glotzer, AW Murray, MW Kirschner. Cyclin is degraded by the ubiquitin pathway. Nature 349:132–138, 1991.

48. C Bai, R Richman, SJ Elledge. Human cyclin F. EMBO J 13:6087–6098, 1994.

49. J Pines. Cyclins and their associated cyclin-dependent kinases in the human cell cycle. Biochem Soc Trans 21:921–925, 1993.

50. J Marx. The cell cycle: spinning farther afield. Science 252:1490–1492, 1991.

51. J Pines, T Hunter. Human cyclin A is adenovirus E1A–associated protein p60 and behaves differently from cyclin B. Nature 346:760–763, 1990.

52. Y Xiong, T Connolly, B Futcher, D Beach. Human D-type cyclin. Cell 65:691–699, 1991.
53. A Koff, F Cross, A Fisher, J Schumacher, K Leguellec, M Philippe, JM Roberts. Human cyclin E, a new cyclin that interacts with two members of the *CDC2* gene family. Cell 66:1217–1228, 1991.
54. DJ Lew, V Dulic, SI Reed. Isolation of three novel human cyclins by rescue of G1 cyclin (Cln) function in yeast. Cell 66:1197–1206, 1991.
55. K Okamoto, D Beach. Cyclin G is a transcriptional target of the p53 tumor suppressor protein. EMBO J 13:4816–4822, 1994.
56. K Tamura, Y Kanaoka, S Jinno, A Nagata, Y Ogiso, K Shimizu, T Hayakawa, H Nojima, H Okayama. Cyclin G: a new mammalian cyclin with homology to fission yeast Cigl. Oncogene 8:2113–1228, 1993.
57. T Nakamura, R Sanokawa, YF Sasaki, D Ayusawa, M Oishi, N Mori. Cyclin I: a new cyclin encoded by a gene isolated from human brain. Exp Cell Res 221:534–542, 1995.
58. TJ Fu, J Peng, G Lee, DH Price, O Flores. Cyclin K functions as a CDK9 regulatory subunit and participates in RNA polymerase II transcription. J Biol Chem 274:34527–34530, 1999.
59. R Ohi, KL Gould. Regulating the onset of mitosis. Curr Opin Cell Biol 11:267–273, 1999.
60. P Nurse, P Thuriaux. Regulatory genes controlling mitosis in the fission yeast *Schizosaccharomyces pombe*. Genetics 96:627–637, 1980.
61. V Simanis, P Nurse. The cell cycle control gene *cdc2 +* of fission yeast encodes a protein kinase potentially regulated by phosphorylation. Cell 45:261–268, 1986.
62. J Hindley, G Phear. Sequence of the cell division gene *CDC2* from *Schizosaccharomyces pombe*; patterns of splicing and homology to protein kinases. Gene 31:129–134, 1984.
63. MG Lee, P Nurse. Complementation used to clone a human homologue of the fission yeast cell cycle control gene *cdc2*. Nature 327:31–35, 1987.
64. A Lorincz, S Reed. Primary structure homology between the product of yeast cell division control gene *CDC28* and vertebrate oncogenes. Nature 307:183–185, 1984.
65. D Beach, B Durkacz, P Nurse. Functionally homologous cell cycle control genes in budding and fission yeast. Nature 300:706–709, 1982.
66. V Mironov, L De Veylder, M Van Montagu, D Inze. Cyclin-dependent kinases and cell division in plants—the nexus. Plant Cell 11:509–522, 1999.
67. RR Fowler, S Eyre, W Scott, A Slater, MC Elliott. The plant cell cycle in context. Mol Biotechnol 10:123–153, 1998.
68. S Moreno, P Nurse. Substrates for p34$^{cdc2}$: in vivo veritas? Cell 61:549–551, 1990.
69. B Durkacz, A Carr, P Nurse. Transcription of the *CDC2* cell cycle control gene of the fission yeast *Schizosaccharomyces pombe*. EMBO J 5:369–373, 1986.
70. J Pines, T Hunter. Isolation of a human cyclin cDNA: evidence for cyclin mRNA and protein regulation in the cell cycle and for interaction with p34$^{cdc2}$. Cell 58:833–846, 1989.
71. J Gautier, J Minshull, M Lokha, M Glotzer, T Hunt, J Maller. Cyclin is a component of maturation-promoting factor from *Xenopus*. Cell 60:487–494, 1990.
72. Y Gu, J Rosenblatt, DO Morgan. Cell cycle regulation of CDK2 activity by phosphorylation of Thr160 and Tyr15. EMBO J 11:3995–4005, 1992.
73. MJ Solomon, M Glotzer, TH Lee, M Philippe, MW Kirschner. Cyclin activation of p34$^{cdc2}$. Cell 63:1013–1024, 1990.
74. CH McGowan, P Russell. Human Wee1 kinase inhibits cell division by phosphorylating p34$^{cdc2}$ exclusively on Tyr 15. EMBO J 12:75–85, 1993.
75. LL Parker, H Piwnica-Worms. Inactivation of the p34$^{cdc2}$-cyclin B complex by the human WEE1 tyrosine kinase. Science 257:1955–1957, 1992.
76. A Fattacy, RN Booher. Myt1: a Wee1-type kinase that phosphorylates Cdc2 on residue Thr14. Prog Cell Cycle Res 3:233–240, 1997.
77. LL Parker, S Atherton-Fessler, MS Lee, S Ogg, JL Falk, KI Swenson, H Piwnica-Worms. Cyclin promotes the tyrosine phsophorylation of p34$^{cdc2}$ in a wee 1+ dependent manner. EMBO J 10:1255–1263, 1991.
78. P Russell, P Nurse. *cdc25$^+$* functions as an inducer in the mitotic control of fission yeast. Cell 45:145–153, 1986.
79. B Sebastian, A Kakizuka, T Hunter. Cdc25M2 activation of cyclin-dependent kinases by dephosphorylation of threonine-14 and tyrosine-15. Proc Natl Acad Sci U S A 90:3521–3524, 1993.
80. U Strausfeld, JC Labbé, D Fesquet, JC Cavadore, A Picard, K Sadhu, P Russell, M Dorée. Dephosphorylation and activation of a p34$^{cdc2}$/cyclin B complex in vitro by human CDC25 protein. Nature 351:242–248, 1991.
81. I Hoffmann, PR Clarke, MJ Marcote, E Karsenti, G Draetta. Phosphorylation and activation of human cdc25-C by cd2-cyclin B and its involvement in the self-amplification of MPF at mitosis. EMBO J 12:53–63, 1993.
82. A Blasina, IV De Weyer, MC Laur, AE Parker. A human homologue of the checkpoint kinase Cds1 directly inhibits Cdc25 phosphatase. Curr Biol 9:1–10, 1999.
83. S Matsuoka, M Huang, SJ Elledge. Linkage of ATM to cell cycle regulation by the Chk2 protein kinase. Science 282:1893–1897, 1998.
84. RW King, PK Jackson, MW Kirschner. Mitosis in transistion. Cell 79:563–571, 1994.
85. LS Cox, DP Lane. Tumour suppressors, kinases and clamps: how p53 regulates the cell cycle in response to DNA damage. Bioessays 17:501–508, 1995.

86.  J Yang, S Kornbluth. All aboard the cyclin train: subcellular trafficking of cyclins and their partners. Trends Cell Biol 4:207–210, 1999.
87.  J Pines, T Hunter. Human cyclins A and B1 are differentially located in the cell and undergo cell cycle–dependent nuclear transport. J Cell Biol 115:1–17, 1991.
88.  V Baldin, J Lukas, MJ Marcote, M Pagano, G Draetta. Cyclin D1 is a nuclear protein required for cell cycle progression in G1. Genes Dev 7:812–821, 1993.
89.  T Hunt. Destruction's our delight. Nature 349:100–101, 1991.
90.  AW Murray, MJ Solomon, MW Krischner. The role of cyclin synthesis and degradation in the control of maturation promoting factor activity. Nature 339:280–286, 1989.
91.  MS Cyert, J Thorner. Putting it on and taking it off: phosphoprotein phosphatase involvement in cell cycle regulation. Cell 57:891–893, 1989.
92.  A Amon, S Irniger, K Nasmyth. Closing the cell cycle circle in yeast: G2 cyclin proteolysis initiated at mitosis persists until the activation of G1 cyclins in the next cycle. Cell 77:1037–1050, 1994.
93.  JA Knoblich, K Sauer, L Jones, H Richardson, R Saint, CF Lehner. Cyclin E controls S phase progression and its down-regulation during *Drosophila* embryogenesis is required for the arrest of cell proliferation. Cell 77:107–120, 1994.
94.  SL Holloway, M Glotzer, RW King, AW Murray. Anaphase is initiated by proteolysis rather than by the inactivation of maturation-promoting factor. Cell 73:1393–1402, 1993.
95.  W Surana, A Amon, C Dowzer, J McGrew, B Byers, K Nasmyth. Destruction of the CDC28/CLB mitotic kinase is not required for the metaphase to anaphase transition in budding yeast. EMBO J 12:1969–1978, 1993.
96.  PCL John, FJ Sek, MG Lee. A homolog of the cell cycle control protein p34$^{cdc2}$ participates in the division cycle of *Chlamydomonas,* and a similar protein is detectable in higher plants and remote taxa. Plant Cell 1:1185–1193, 1989.
97.  HS Feiler, TW Jacobs. Cell division in higher plants: A *cdc2* gene, its 34-kDa product, and histone H1 kinase activity in pea. Proc Natl Acad Sci U S A 87:5397–5401, 1990.
98.  H Hirt, A Páy, J Györgyey, L Bakó, K Németh, L Bögre, RJ Schweyen, E Heberle-Bors, D Dudits. Complementation of a yeast cell cycle mutant by an alfalfa cDNA encoding a protein kinase homologous to p34$^{cdc2}$. Proc Natl Acad Sci U S A 88:1636–1640, 1991.
99.  Z Magyar, T Meszaros, P Miskolczi, M Deak, A Feher, S Brown, E Kondorosi, A Athanasiadis, S Pongor, M Bilgin, L Bako, C Koncz, D Dudits. Cell cycle phase specificity of putative cyclin-dependent kinase variants in synchronized alfalfa cells. Plant Cell 9:223–235, 1997.
100. PR Fobert, ES Coen, GJP Murphy, JH Doonan. Patterns of cell division revealed by transcriptional regulation of genes during the cell cycle in plants. EMBO J 13:616–624., 1994.
101. PCG Ferreira, AS Hemerly, R Villarroel, M Van Montagu, D Inzé. The *Arabidopsis* functional homolog of the p34$^{cdc2}$ protein kinase. Plant Cell 3:531–540, 1991.
102. Y Hirayama, Y Imajuku, T Anai, M Matsui, A Oka. Identification of two cell-cycle-controlling *cdc2* gene homologs in *Arabidopsis thaliana*. Gene 105:159–165, 1991.
103. Y Imajuku, T Hirayama, H Endoh, A Oka. Exon-intron organization of the *Arabidopsis thaliana* protein kinase genes CDC2a and CDC2b. FEBS Lett 304:73–77, 1992.
104. P Lessard, JP Bouly, S Jouannic, M Kreis, M Thomas. Identification of cdc2cAt: a new cyclin-dependent kinase expressed in *Arabidopsis thaliana* flowers. Biochim Biophys Acta 1445:351–358, 1999.
105. J Colasanti, S Cho, S Wick, V Sundaresan. Localization of the functional p34$^{cdc2}$ homolog of maize in root tip and stomatal complex cells: association with predicted division sites. Plant Cell 5:1101–1111, 1993.
106. Z Hong, G-H Miao, DPS Verma. p34$^{cdc2}$ protein kinase homolog from mothbean (*Vigna aconitifolia*). Plant Physiol 101:1399–1400, 1993.
107. A Kvarnheden, K Tandre, P Engstrom. A *cdc2* homologue and closely related processed retropseudogenes from Norway spruce. Plant Mol Biol 27:391–403, 1995.
108. C Bergounioux, C Perennes, AS Hemerly, LX Qin, C Sarda, D Inze, P Gada. A *cdc2* gene of *Petunia hybrida* is differentially expressed in leaves, protoplasts and during various cell cycle phases. Plant Mol Biol 20:1121–1130, 1992.
109. J Hashimoto, T Hirabayashi, Y Hayano, S Hata, Y Ohashi, I Suzuka, T Utsugi, A Toh-e, Y Kikuchi. Isolation and characterization of cDNA clones encoding cdc2 homologues from *Oryza sativa*: a functional homolog and cognate variants. Mol Gen Genet 233:10–16, 1992.
110. S Hata. cDNA cloning of a novel cdc2/CDC28-related protein kinases from rice. FEBS Lett 279:149–152, 1991.
111. G-H Miato, Z Hong, DPS Verma. Two functional soybean genes encoding p34*cdc2* protein kinases are regulated by different plant developmental pathways. Proc Natl Acad Sci U S A 90:943–947, 1993.
112. YY Setiady, M Sekine, N Hariguchi, H Kouchi, A Shinmyo. Molecular cloning and characterization of a cDNA clone that encodes a cdc2 homolog from *Nicotiana tobacum*. Plant Cell Physiol 37:369–376, 1996.
113. J Joubes, T-H Phan, D Just, C Rothan, C Bergounioux, P Raymond, C Chevalier. Molecular and biochemical characterization of the involvement of cyclin-dependent kinase A during the early development of tomato fruit. Plant Physiol 121:857–869, 1999.
114. RP Huntley, JAH Murray. The plant cell cycle. Curr Opin Plant Biol 2:440–446, 1999.

115. J Pines. Cyclins and cylin-dependent kinases: take your partners. Trends Biochem Sci 18:195–197, 1993.
116. N Glab, B Labidi, L Qin, C Trehin, C Bergounioux, L Meijer. Olomoucine, an inhibitor of the cdc2/cdk2 kinases activity, blocks plant cells at the G1 to S and G2 to M cell cycle transitions. FEBS Lett 353:207–211, 1994.
117. P Binarova, J Dolezel, P Draber, E Heberle-Bors, M Strnad, L Bogre. Treatment of *Vicia faba* root tip cells with specific inhibitors to cyclin-dependent kinases leads to abnormal spindle formation. Plant J 16:697–707, 1998.
118. L Bogre, K Zwerger, I Meskiene, P Binarova, V Csizmadia, C Planck, E Wagner, H Hirt, E Heberle-Bors. The cdc2Ms kinase is differentially regulated in the cytoplasm and in the nucleus. Plant Physiol 113:841–852, 1997.
119. A Kaffman, I Herskowitz, R Tjian, EK O'Shea. Phosphorylation of the transcription factor PH04 by a cyclin-CDK complex, PH080-PH085. Science 263:1153–1156, 1994.
120. J Lew, QQ Huang, Z Qi, RJ Winkefein, R Aebersold, T Hunt, J Wang. A brain-specific activator of cyclin-dependent kinase 5. Nature 371:423–426, 1994.
121. T Yoshizumi, N Nagata, H Shimada, M Matsui. An *Arabidopsis* cell cycle–dependent kinase-related gene, *CDC2b*, plays a role in regulating seeding growth in darkness. Plant Cell 11:1883–1896, 1999.
122. JR Gorst, PCL John, FJ Sek. Levels of p34$^{cdc2}$-like protein in dividing, differentiating and dedifferentiating cells of carrot. Planta 185:304–310, 1991.
123. MC Martinez, J Jorgensen, MA Lawton, C Lamb, PW Doerner. Spatial pattern of *cdc2* expression in relation to meristem activity and cell proliferation during plant development. Proc Natl Acad Sci U S A 89:7360–7364, 1992.
124. A Hemerly, C Bergounioux, M Van Montagu, D Inzé, P Ferreira. Genes regulating the plant cell cycle: isolation of mitotic-like cyclin from *Arabidopsis thaliana*. Proc Natl Acad Sci U S A 89:3295–3299, 1992.
125. AS Hemerly, P Ferreira, J Engler, M Van Montagu, G Engler, K Inze. *cdc2a* expression in *Arabidopsis* is linked with competence for cell division. Plant Cell 5:1711–1723, 1993.
126. W Krek, EA Nigg. Mutations of p34$^{cdc2}$ phosphorylation sites induce premature mitotic events in HeLa cells: evidence for a double block to p34$^{cdc2}$ kinase activation in vertebrates. EMBO J 10:3331–3341, 1991.
127. CF Lehner, PH O'Farrell. Expression and function of *Drosophila* cyclin A during embryogenic cell cycle progression. Cell 56:957–968, 1989.
128. M Umeda, C Umeda-Hara, M Yamaguchi, J Hashimoto, H Uchimiya. Differential expression of genes for cyclin-dependent protein kinases in rice plants. Plant Physiol 119:31–40, 1999.
129. G Segers, I Gadisseur, C Bergounious, J De Almeida Engler, A Jacqmard, M Van Montagu, D Inze. The *Arabidopsis* cyclin-dependent kinase gene *cdc2bAt* is preferentially expressed during S and G2 phases of the cell cycle. Plant J 10:601–612, 1996.
130. PR Fobert, V Gaudin, P Lunness, ES Coen, JH Doonan. Distinct classes of *cdc2*-related genes are differentially expressed during the cell division cycle in plants. Plant Cell 8:1465–1476, 1996.
131. L Brizuela, G Draetta, D Beach. p13$^{suc1}$ acts in the fission yeast cell division cycle as a component of the p34$^{cdc2}$ protein kinase. EMBO J 6:3507–3514, 1987.
132. HE Parge, AS Arvai, DJ Murtari, SI Reed, JA Tainer. Human CKsHs2 atomic structure: a role for its hexameric assembly in cell cycle control. Science 262:387–395, 1993.
133. TJ Jacobs. Control of the cell cycle. Dev Biol 153:1–15, 1992.
134. PCL John, FJ Sek, J Hayles. Association of the plant p34$^{cdc2}$-like protein with p13$^{suc1}$—implications for control of cell division cycles in plants. Protoplasma 161:70–74., 1991.
135. J Colasanti, M Tyers, V Sundaresan. Isolation and characterization of cDNA clones encoding a functional p34$^{cdc2}$ homologue from *Zea mays*. Proc Natl Acad Sci U S A 88:3377–3381, 1991.
136. Z Magyar, L Bako, L Bogre, D Dedeoglu, T Kapros. Active *cdc2* genes and cell cycle phase–specific cdc2-related kinase complexes in hormone-stimulated alfalfa cells. Plant J 4:151–161, 1993.
137. A Reindl, F Schoffl, J Schell, C Koncz, L Bako. Phosphorylation by a cyclin-dependent kinase modulates DNA binding of the *Arabidopsis* heat-shock transcription factor HSF1 in vitro. Plant Physiol 115:93–100, 1997.
138. Y Mineyuki, M Yamashita, Y Nagaham. p34$^{cdc2}$ kinase homologue in the preprophase band. Protoplasma 162:182–186, 1991.
139. SM Liao, J Zhang, DA Jeffery, AJ Koleske, CM Thompson, DM Chao, M Viljoen, HJ van Vuuren, RA Young. A kinase-cyclin pair in the RNA polymerase II holoenzyme. Nature 374:193–196, 1995.
140. LJ Cisek, JL Corden. Phosphorylatin of RNA polymerase by the murine homologue of the cell-cycle control protein cdc2. Nature 339:679–684, 1989.
141. M Umeda, RP Bhalerao, J Schell, H Uchimiya, C Koncz. A distinct cyclin-dependent kinase-activating kinase of *Arabidopsis thaliana*. Proc Natl Acad Sci U S A 95:5021–5026, 1998.
142. K Zhang, DS Letham, PC John. Cytokinin controls the cell cycle at mitosis by stimulating the tyrosine dephosphorylation and activation of p34cdc2-like H1 histone kinase. Planta 200:2–12, 1996.
143. PCL John. Cytokinin stimulation of cell division: essential signal transduction is via Cdc25 phosphatase. J Exp Bot 49(Suppl):91, 1998.
144. A Porceddu, L De Veyldera, J Hayles, M Van Montagu, D Inze, V Mironov. Mutational analysis of two *Arabidopsis thaliana* cyclin-dependent kinases in fission yeast. FEBS Lett 446:182–188, 1999.
145. BES Gunning, SM Wick. Preprophase bands, phragmoplasts, and spatial control of cytokinesis. J Cell Sci 2(Suppl):157–179, 1985.

146. S Hata, H Kouchi, I Suzuka, T Ishii. Isolation and characterization of cDNA clones for plant cyclins. EMBO J 10:2681–2688, 1991.

147. K Uchida, T Muramatsu, K Tachibana, T Kishimoto, M Furuya. Isolation and characterization of the cDNA for an A-like cyclin in *Adiantum capillus-veneris* L. Plant Cell Physiol 37:825–832, 1996.

148. H Hirt, M Mink, M Pfosser, L Bogre, J Gyorgyey, C Jonak, A Gartner, D Dudits, E Heberle-Bors. Alfalfa cyclins: differential expression during the cell cycle and in plant organs. Plant Cell 4:1531–1538, 1992.

149. I Meskiene, L Bogre, M Dahl, M Pirck, DTC Cam Ha, I Swoboda, E Heberle-Bors, G Ammerer, H Hirt. *cycMs3*, a novel B-type alfalfa cyclin gene, is induced in the G$_0$-to-G$_1$ transition of the cell cycle. Plant Cell 7:759–771, 1995.

150. A Savoure, A Feher, P Kalo, G Petrovics, G Csanadi, J Szecsi, G Kiss, S Brown, A Kondorosi, E Kondorosi. Isolation of a full-length mitotic cyclin cDNA clone *CycIIIMs* from *Medicago sativa*: chromosomal mapping and expression. Plant Mol Biol 27:1059–1070, 1995.

151. PCG Ferreira, AS Hemerly, JA Engeler, C Bergounioux, S Burssens, M VanMontagu, G Engler, D Inze. Three discrete classes of *Arabidopsis* cyclins are expressed during different intervals of the cell cycle. Proc Natl Acad Sci U S A 91:11313–11317, 1994.

152. IS Day, ASN Reddy. Cloning of a family of cyclins from *Arabidopsis thaliana*. Biochim Biophys Acta 1218:115–118, 1994.

153. IS Day, ASN Reddy, M Golovkin. Isolation of a new mitotic-like cyclin from *Arabidopsis*: complementation of a yeast cyclin mutant with a plant cyclin. Plant Mol Biol 30:565–575, 1996.

154. IS Day, AS Reddy. Isolation and characterization of two cyclin-like cDNAs from *Arabidopsis*. Plant Mol Biol 36:451–461, 1998.

155. R Soni, JP Carmichael, ZH Shah, JAH Murray. A family of cyclin D homologs from plants differentially controlled by growth regulators and containing the conserved retinoblastoma protein interaction motif. Plant Cell 7:85–103, 1995.

156. L De Veylder, J de Almeida Engler, S Burssens, A Manevski, B Lescure, M Van Montagu, G Engler, D Inze. A new D-type cyclin of *Arabidopsis thaliana* expressed during lateral root primordia formation. Planta 208:453–462, 1999.

157. S Szarka, M Fitch, S Schaerer, M Moloney. Classification and expression of a family of cyclin gene homologues in *Brassica napus*. Plant Mol Biol 27:263–275, 1995.

158. M Ito, C Marie-Claire, M Sakabe, T Ohno, S Hata, H Kouchi, J Hahimoto, H Fukuda, A Komamine, A Watanabe. Cell-cycle-regulated transcription of A- and B-type plant cyclin genes in synchronous cultures. Plant J 11:983–992, 1997.

159. J Deckert, J Jelenska, EA Gwozda, AB Legocki. The isolation of lupine cDNA clone coding for putative cyclin protein. Biochimie 78:90–94, 1996.

160. J Deckert, J Jelenska, Z Zaborowska, AB Legocki. Isolation and classification of a family of cyclin gene homologues in *Lupinus luteus*. Acta Biochim Pol 44:37–42, 1997.

161. J Renaudin, J Colasanti, H Rime, Z Yuan, V Sundaresan. Cloning of four cyclins from maize indicates that higher plants have three structurally distinct groups of mitotic cyclins. Proc Natl Acad Sci U S A 91:7375–7379, 1994.

162. W Hsieh, SM Wolniak. Isolation and characterization of a functional A-type cyclin from maize. Plant Mol Biol 37:121–129, 1998.

163. Y Sun, BA Flannigan, JT Madison, TL Setter. Alternative splicing of cyclin transcripts in maize endosperm. Gene 195:167–175, 1997.

164. M Umeda, N Iwamoto, C Umeda-Hara, M Yamaguchi, J Hashimoto, H Uchimiya. Molecular characterization of mitotic cyclins in rice plants. Mol Gen Genet 262:230–238, 1999.

165. M Sauter. Differential expression of a CAK (cdc2-activating kinase)–like protein kinase, cyclins and *cdc2* genes from rice during the cell cycle and in response to gibberellin. Plant J 11:181–190, 1995.

166. S Goormachtig, M Alves-Ferreira, M Van Mantagu, G Engler, M Holsters. Expression of cell cycle genes during *Sebania rostrata* stem nodule development. Mol Plant Microbe Interact 10:316–325, 1997.

167. H Kouchi, M Sekine, S Hata. Distinct classes of mitotic cyclins are differentially expressed in the soybean shoot apex during the cell cycle. Plant Cell 7:1143–1155, 1995.

168. L Qin, L Richard, C Perennes, P Gadal, C Bergounioux. Identification of a cell cycle–related gene, cyclin, in *Nicotiana tabacum* (L.). Plant Physiol 108:425–426, 1995.

169. YY Setiady, M Sekine, N Hariguchi, T Yamamoto, H Kouchi, A Shinmyo. Tobacco mitotic cyclins: cloning, characterization, gene expression and functional assay. Plant J 8:949–957, 1995.

170. J-P Renaudin, JH Doonan, D Freeman, J Hashimoto, H Hirt, D Inze, T Jacobs, H Kouchi, P Rouze, M Sauter, A Savoure, DA Sorrell, V Sundaresan, JAH Murray. Plant cyclins: a unified nomenclature for plant A-, B- and D-type cyclins based on sequence organization. Plant Mol Biol 32:1003–1018, 1996.

171. PCG Ferreira, AS Hemerly, JA Engler, MV Montagu, G Engler, D Inze. Developmental expression of the *Arabidopsis* cyclin gene *cyc1At*. Plant Cell 6:1763–1774, 1994.

172. S Chadudhuri, S Ghosh. Monoclonal antibody raised against human mitotic cyclin B1 identifies cyclin B–like mitotic proteins in synchronized onion (*Allium cepa* L.) root meristem. Cell Biol Int 21:159–166, 1997.

173. O Shaul, V Mironov, S Burssnes, M Van Montagu, D Inze. Two *Arabidopsis* cyclin promoters mediate dis-

tinctive transcriptional oscillation in synchronized tobacco BY-2 cells. Proc Natl Acad Sci U S A 93:4868–4872, 1996.

174.  J-P Reichheld, N Chaubet, WH Shen, J-P Renaudin, C Gigot. Multiple A-type cyclins express sequentially during cell cycle in *Nicotiana tabacum* BY2 cells. Proc Natl Acad Sci U S A 93:13819–13824, 1996.

175.  C Trehin, I-O Ahn, C Perennes, F Couteau, E Lalanne, C Bergounioux. Cloning of upstream sequences responsible for cell cycle regulation of the *Nicotiana sylvetris CycB1;1* gene. Plant Mol Biol 35:667–672, 1997.

176.  L-X Qin, C Perennes, L Richard, M Bouvier-Durand, C Trehin, D Inze, C Bergounioux. G2- and early-M-specific expression of the *Ntcyc1* cyclin gene in *Nicotiana tabacum* cells. Plant Mol Biol 32:1093–1101, 1996.

177.  M Ito, M Iwase, H Kodama, P Lavisse, A Komamine, R Nishihama, Y Machida, A Watanabe. A novel cis-acting element in promoters of plant B-type cyclin genes activates M phase–specific transcription. Plant Cell 10:331–341, 1998.

178.  A Hershko, D Ganoth, V Sudakin, A Dahan, LH Cohen, FC Luca, JV Ruderman, E Eytan. Components of a system that ligates cyclin to ubiquitin and their regulation by the protein kinase cdc2. J Biol Chem 269:4940–4946, 1994.

179.  P Genschik, MC Criqui, Y Parmentier, A Derevier, J Fleck. Cell cycle–dependent proteolysis in plants: identification of the destruction box pathway and metaphase arrest produced by the proteasome inhibitor MG132. Plant Cell 10:2063–2075, 1998.

180.  J Hush, L Wu, PC John, LH Hepler, PK Hepler. Plant mitosis promoting factor disassembles the microtubule preprophase band and accelerates prophase progression in *Tradescantia*. Cell Biol Int 20:275–287, 1996.

181.  L De Veylder, G Segers, N Glab, G Casteels, M Montagu, D Inze. The *Arabidopsis* Cks1At protein binds the cyclin-dependent kinases Cdc2aAt and Cdc2bAt. FEBS Lett 412:446–452, 1997.

182.  Y Bourne, MH Watson, MJ Hickey, W Holmes, W Rocque, SI Reed, JA Tainer. Crystal structure and mutational analysis of the human CDK2 kinase complex with cell cycle–regulatory protein CksHs1. Cell 84:863–874, 1996.

183.  R Porat, P Lu, SD O'Neill. *Arabidopsis* SKP1, a homologue of a cell cycle regulator gene, is predominantly expressed in meristematic cells. Planta 204:345–351, 1998.

184.  Y Sun, BP Dilkes, C Zhang, RA Dante, NP Carneiro, KS Lowe, R Jung, WJ Gordon-Kamm, B Larkins. Characterization of maize (*Zea mays* L.) Wee1 and its activity in developing endosperm. Proc Natl Acad Sci U S A 96:4189–4185, 1999.

185.  CJ Sherr. Mammalian G1 cyclins. Cell 73:1059–1065, 1993.

186.  SJ Elledge, MR Spottswood. A new human p34 protein kinase, CDK2, identified by complementation of a *cdd28* mutation in *Saccharomyces cerevisiae*, is a homolog of *Xenopus* EG1. EMBO J 10:2653–2659, 1991.

187.  SJ Elledge, R Richman, FL Hall, RT Williams, N Lodgson, JW Harper. *CDK2* encodes a 33-kDa cyclin A–associated protein kinase and is expressed before *CDC2* in the cell cycle. Proc Natl Acad Sci U S A 89:2907–2911, 1992.

188.  H Matsushime, DE Quelle, SA Shurtleff, M Shibuya, CJ Sherr, JY Kato. D-type cyclin–dependent kinase activity in mammalian cells. Mol Cell Biol 14:2066–2076, 1994.

189.  M Meyerson, E Harlow. Identification of G1 kinase activity for cdk6, a novel cyclin D partner. Mol Cell Biol 14:2077–2086, 1994.

190.  CJ Sherr. The ins and outs of RB: coupling gene expression to the cell cycle clock. Trends Cell Biol 4:15–18, 1994.

191.  CJ Sherr. G1 phase progression: Cycling on cue. Cell 79:551–555, 1994.

192.  H Muller, J Lukas, A Schneider, P Warthoe, J Bartek, M Eilers. Cyclin D1 expression is regulated by the retinoblastoma protein. Proc Natl Acad Sci U S A 91:2945–2949, 1994.

193.  P Lavia, P Jansen-Durr. E2F target genes and cell-cycle checkpoint control. Bioessays 21:221–230, 1999.

194.  G Prelich, C Tan, M Kostura, MB Mathews, AG So, KM Downey, B Stillman. Functional identity of proliferating cell nuclear antigen and a DNA polymerase-delta auxiliary protein. Nature 326:517–520., 1987.

195.  F Fang, J Newport. Evidence that the G1-S and G2-M transitions are controlled by different Cdc2 proteins in higher eukaryotes. Cell 66:731–742, 1991.

196.  UP Strausfeld, M Howell, P Descombes, S Chevalier, RE Rempel, J Adamczewski, JL Maller, T Hunt, JJ Blow. Both cyclin A and cyclin E have S-phase promoting (SPF) activity in *Xenopus* egg extracts. J Cell Sci 109:1555–1563, 1996.

197.  F Fang, G Orend, N Watanabe, T Hunter, E Ruoslahti. Dependence of cyclin E-CDK2 kinase activity on cell anchorage. Science 271:499–502, 1996.

198.  M Ohtsubo, JM Roberts. Cyclin-dependent regulation of $G_1$ in mammalian fibrobalsts. Science 259:1908–1912, 1993.

199.  KA Heichman, Roberts JM. Rules to replicate by. Cell 79:557–562, 1994.

200.  T Hunter, J Pines. Cyclins and cancer II: cyclin D and Cdk inhibitors come of age. Cell 79:573–582, 1994.

201.  A Kramer, C-P Carstens, WE Fahl. A novel CCAAT-binding protein necessary for adhesion-dependent cyclin A transcription at the G1/S boundary is sequestered by a retinoblastoma-like protein in G0. J Biol Chem 271:6579–6585, 1996.

202.  K Hamada, N Takuwa, W Zhou, M Kumada, Y Takuwa. Protein kinase C inhibits the CAK-CDK2 cyclin-de-

pendent kinase casade and G1/S cell cycle progression in human diploid fibroblast. Biochim Biophys Acta 1310:1149–1156, 1996.

203.  M Peter, I Herskowitz. Joining the complex: cyclin-dependent kinase inhibitory proteins and the cell cycle. Cell 79:181–184, 1994.

204.  M Peter, I Herskowitz. Direct inhibition of the yeast cyclin-dependent kinase Cdc28-Cln by Far1. Science 265:1228–1231, 1994.

205.  MD Mendenhall. An inhibitor of p34$^{CDC28}$ protein kinase activity from *Saccharomyces cerevisiae*. Science 259:216–219, 1993.

206.  E Schwob, T Bohn, MD Mendenhall, K Nasmyth. The B-type cyclin kinase inhibitor p40$^{SIC1}$ controls the G1 to S transition in *S. cerevisiae*. Cell 79:233–244, 1994.

207.  JW Harper, GR Adami, N Wei, K Keyomarsi, SJ Elledge. The p21 cdk-interacting protein Cip 1 is a potent inhibitor of G1 cyclin-dependent kinases. Cell 75:805–816, 1993.

208.  A Noda, Y Ning, SF Venable, OM Pereira-Smith, JR Smith. Cloning of senescent cell–derived inhibitors of DNA synthesis using an expression screen. Exp Cell Res 211:90–98, 1994.

209.  V Dulic, WK Kaufmann, SJ Wilson, TD Tisty, E Lees, JW Harper, SJ Elledge, SI Reed. p53-dependent inhibition of cyclin-dependent kinase activities in human fibroblasts during radiation-induced G1 arrest. Cell 76:1013–1023, 1994.

210.  Y Xiong, GJ Hannon, H Zhang, D Casso, R Kobayashi, D Beach. p21 is a universal inhibitor of cyclin kinases. Nature 366:701–704, 1993.

211.  O Halevy, BG Novitch, DB Spicer, SX Skapek, J Rhee, GJ Hannon, D Beach, AB Lassar. Correlation of terminal cell cycle arrest of skeletal muscle with induction of p21 by MyoD. Science 267:1018–1021, 1995.

212.  SB Parker, G Eichele, P Zhang, A Rawis, AT Sands, A Bradley, EN Olson, JW Harper, SJ Elledge. p53-independent expression of p21$^{Cip1}$ in muscle and other terminally differentiating cells. Science 267:1024–1027, 1995.

213.  SX Skapek, J Rhee, DB Spicer, AB Lassar. Inhibition of myogenic differentiation in proliferating myoblasts by cyclin D1–dependent kinase. Science 267:1022–1024, 1995.

214.  AA Russo, PD Jeffrey, AK Patten, J Massague, NP Pavletich. Crystal structure of the p27$^{Kip1}$ cyclin-dependent kinase-2. Nature 382:325–331, 1996.

215.  A Koff, M Ohtsuki, K Polyak, JM Roberts, J Massague. Negative regulation of G1 in mammalian cells: inhibition of cyclin E–dependent kinase by TGF-beta. Science 260:536–539, 1993.

216.  K Polyak, J Kato, MJ Solomon, CJ Sherr, J Massague, JM Roberts, A Koff. p27$^{Kip1}$, a cyclin-Cdk inhibitor, links transforming growth factor-beta and contact inhibition to cell cycle arrest. Genes Dev 8:9–22, 1994.

217.  H Toyshima, T Hunter. p27, a novel inhibitor of G1 cyclin-Cdk protein kinase activity, is related to p21. Cell 78:67–74, 1994.

218.  TW Jacobs, ME Prewett, BK Buerr, HS Feiler, J Dunphy, H Chen, J Poole. Protein kinases in cell division control. 12th Annual Missouri Plant Biochemistry, Molecular Biology and Physiology Symposium, University of Missouri, 1993, pp 67–68.

219.  L Bogre, C Jonak, DT Cam Ha, T Murbacher, S Kiegerl, L Bako, C Planck, M Pfosser, A Pay, I Meskiene, M Mink, J Gyorgyey, K Plame, E Wagner, D Dudits, E Heberle-Bors, H Hirt. Alfalfa cell cycle control elements: cdc2, cyclins, MAP kinases. 12th Annual Missouri Plant Biochemistry, Molecular Biology and Physiology Symposium, University of Missouri, 1993, pp 71–71.

220.  G Grafi, BA Larkins. Endoreduplication in maize endosperm: involvement of M phase–promoting factor inhibition and induction of S phase–related kinases. Science 269:1262–1264, 1995.

221.  M Dahl, I Meskiene, L Bogre, DT Cam Ha, I Swoboda, R Hubmann, H Hirt, E Heberle-Bors. The D-type alfalfa cyclin gene *cycMs4* complements G1 cyclin-deficient yeast and is induced in the G1 phase of the cell cycle. Plant Cell 7:1847–1857, 1995.

222.  S Shimizu, H Mori. Analysis of cycles of dormancy and growth in pea axillary buds based on mRNA accumulation patterns of cell cycle–related genes. Plant Cell Physiol 39:255–262, 1998.

223.  DA Sorrell, B Combettes, N Chaubet-Gigot, C Gigot, JA Murray. Distinct cyclin D genes show mitotic accumulation or constant levels of transcripts in tobacco bright yellow-2 cells. Plant Physiol 119:343–352, 1999.

224.  J Kato, HH Matsushime, SW Hiebert, ME Ewen, CJ Sherr. Direct binding of cyclin D to the retinoblastoma gene product (pRb) and pRb phosphorylation by the cyclin D–dependent kinase CDK4. Genes Dev 7:331–342, 1993.

225.  J Hashimoto, H Tanaka, K Yamamoto, T Sasaki. A cDNA clone encoding a C-type cyclin from rice. Plant Physiol 112:862–863, 1996.

226.  J Greenwood, R Huntley, J Makkerh, C Riou-Khamlichi, C Cockcroft, D Freeman, G Davies, N Kilby, J Doonan, JAH Murray. Cell cycle control in plants: role of D cyclins. 7th Annual *Arabidopsis* Conference, Norwich, UK, 1996, p P222.

227.  RA Ach, T Durfee, AB Miller, P Taranto, L Hanley-Bowdoin, PC Zambryski, W Gruissem. RRB1 and RRB2 encode maize retinoblastoma-related proteins that interact with a plant D-type cyclin and geminivirus replication protein. Mol Cell Biol 17:5077–5086, 1997.

228.  G Grafi, RJ Burnett, T Helentjaris, BA Larkins, JA DeCaprio, WR Sellers, WG Kaelin Jr. A maize cDNA encoding a member of the retinoblastoma protein family: involvement in endoreduplication. Proc Natl Acad Sci U S A 93:8962–8967, 1996.

229. R Huntley, S Healy, D Freeman, P Lavender, S de Jager, J Greenwood, J Makker, E Walker, M Jackman, Q Xie, AJ Bannister, T Kouzarides, C Gutierrez, JH Doonan, JA Murray. The maize retinoblastoma protein homologue ZmRb-1 is regulated during leaf development and displays conserved interactions with G1/S regulators and plant cyclin D (CycD) proteins. Plant Mol Biol 37:155–169, 1998.

230. Q Xie, AP Sanz-Burgos, GJ Hannon, C Gutierrez. Plant cells contain a novel member of the retinoblastoma family of growth regulatory proteins. EMBO J 15:4900–4908, 1996.

231. H Nakagami, M Sekine, H Murakami, A Shinmyo. Tobacco retinoblastoma-related protein phosphorylated by a distinct cyclin-dependent kinase complex with Cdc2/cyclin D in vitro. Plant J 18:243–252, 1999.

232. D Inze, C Gutierrez, NH Chua. Trends in plant cell cycle research. Plant Cell 11:991–994, 1999.

233. E Ramirez-Parra, Q Xie, MB Boniotti, C Gutierrez. The cloning of plant E2F, a retinoblastoma-binding protein, reveals unique and conserved features with animal G(1)/S regulators. Nucleic Acids Res 27:3527–3533, 1999.

234. H Kodama, M Ito, N Ohnishi, I Suzuka, A Komamine. Molecular cloning of the gene for plant proliferating-cell nuclear antigen and expression of the gene during the cell cycle in synchronized cultures of *Catharanthus roseus* cells. Eur J Biochem 197:495–503, 1991.

235. S Shimizu, H Mori. Changes in protein interaction of cell cycle–related genes during the dormancy-to-growth transition in pea axillary buds. Plant Cell Physiol 39:1073–1079, 1998.

236. H Wang, Q Qi, P Schorr, AJ Cutler, WL Crosby, LC Fowke. ICK1, a cyclin-dependent protein kinase inhibitor from *Arabidopsis thaliana* interacts with both Cdc2a and CycD3, and its expression is induced by abscisic acid. Plant J 15:501–510, 1998.

237. H Wang, LC Fowke, WL Crosby. A plant cyclin-dependent kinase inhibitor gene. Nature 386:451–452, 1997.

238. DW Galbraith, KR Harkins, S Knapp. Systemic endopolyploidy in *Arabidopsis thaliana*. Plant Physiol 96:985–989, 1991.

239. EJ De Rocher, KR Harkins, DW Galbraith, HJ Bohnert. Developmentally regulated systemic endopolyploidy in succulents with small genomes. Science 250:99–101, 1990.

240. W Nagl. DNA endoreduplication and polyteny understood as evolutionary strategies. Science 261:614–615, 1976.

241. W Nagl. Nuclear organization. Annu Rev Plant Physiol 27:39–69, 1976.

242. F D'Amato. Endopolyploidy as a factor in plant development. Caryologia 17:41–52, 1964.

243. A Jacqmard, L De Veylder, G Segers, J de Almeida Engler, G Bernier, M Van Montagu, D Inze. Expression of *CKS1At* in *Arabidopsis thaliana* indicates a role for the protein in both the mitotic and the endoreduplication cycle. Planta 207:496–504, 1999.

244. A Cebolla, JM Vinardell, E Kiss, B Olah, F Roudier, A Kondorosi, E Kondorosi. The mitotic inhibitor ccs52 is required for endoreduplication and ploidy-dependent cell enlargement in plants. EMBO J 18:4475–4484, 1999.

245. BW Poovaiah, ASN Reddy. Calcium messenger system in plants. CRC Cri Rev Plant Sci 6:47–103, 1987.

246. BW Poovaiah, ASN Reddy. Calcium and signal transduction in plants. CRC Cri Rev Plant Sci 12:185–211, 1993.

247. DM Roberts, A Harmon. Calcium modulated proteins: targets of intracellular calcium signals in higher plants. Annu Rev Plant Physiol Plant Mol Biol 43:375–414, 1992.

248. PK Hepler. Calcium transcients during mitosis: observations in flux. J Cell Biol 109:2567–2573, 1989.

249. KP Lu, AR Means. Regulation of the cell cycle by calcium and calmodulin. Endocr Rev 14:40–58, 1993.

250. M Whitaker, R Patel. Calcium and cell cycle control. Development 108:525–542, 1990.

251. M Whitaker. Calcium and mitosis. Prog Cell Cycle Res 3:261–269, 1997.

252. L Santella. The role of calcium in the cell cycle: facts and hypotheses. Biochem Biophys Res Commun 244:317–324, 1998.

253. JF Whitfield, JP MacManus, RH Rixon, AL Boynton, T Yoydale, S Swierenga. The roles of calcium and cyclic AMP in cell proliferation. In Vitro 12:1–18, 1976.

254. JF Whitfield, AL Boynton, JP MacManus, RH Rixon, M Sikorska, B Tsang, PR Walker. The roles of calcium and cyclic AMP in cell proliferation. Ann N Y Acad Sci 339:216–240, 1980.

255. RB Silver. Calcium and cellular clocks orchestrate cell division. Ann N Y Acad Sci 582:207–221, 1990.

256. ML Veigl, TC Vanaman, WD Sedwick. Calcium and calmodulin in cell growth and transformation. Biochim Biophys Acta 738:21–48, 1984.

257. JG Izant. The role of calcium ions during mitosis. Calcium participates in the anaphase trigger. Chromosoma 88:1–10, 1983.

258. JPY Kao, JM Alderton, RY Tsien, RA Steinhardt. Active involvement of $Ca^{2+}$ in mitotic progression of Swiss 3T3 fibroblasts. J Cell Biol 111:183–196, 1990.

259. RA Steinhardt, J Alderton. Intracellular free calcium rise triggers nuclear envelope breakdown in the sea urchin embryo. Nature 332:364–366, 1988.

260. M Poenie, J Alderton, RY Tsien, RA Steinhardt. Changes of free calcium levels with stages of the cell division cycle. Nature 315:147–149, 1985.

261. M Poenie, J Alderton, RA Steinhardt, RY Tsien. Calcium rises abruptly and briefly throughout the cell at the onset of anaphase. Science 233:886–899, 1986.

262. CH Keith, R Ratan, FR Maxfield, A Bajer, ML Shelanski. Local cytoplasmic calcium gradients in living mitotic cells. Nature 316:848–850, 1985.

263. PK Hepler, DA Callaham. Free calcium increases during anaphase in stamen hair cells of *Tradescantia*. J Cell Biol 105:2137–2143, 1987.

264. DM Roberts, TJ Lukas, DM Watterson. Structure, function and mechanism of action of calmodulin. CRC Cri Rev Plant Sci 4:311–339, 1986.

265. S Muto, S Miyachi. Production of antibody against spinach calmodulin and its application to radioimmunoassay for plant calmodulin. Z Pflanzenphysiol 114:421–431, 1984.

266. PK Jena, ASN Reddy, BW Poovaiah. Molecular cloning and sequencing of a cDNA for plant calmodulin: signal-induced changes in the expression of calmodulin. Proc Natl Acad Sci U S A 86:3644, 1989.

267. JG Chafouleas, WE Bolton, H Hidaka, AE Boyd 3d, AR Means. Calmodulin and the cell cycle: involvement in regulation of cell-cycle progression. Cell 28:41–50, 1982.

268. JG Chafouleas, L Legace, W Bolton, AE Boyd 3d, AR Means. Changes in calmodulin and its mRNA accompany reentry of quiescent (G0) cells into the cell cycle. Cell 36:73–81, 1984.

269. Y Sasaki, H Hidaka. Calmodulin and cell proliferation. Biochem Biophys Res Commun 104:451–456, 1982.

270. JG Chafouleas, RL Pardue, BR Brinkely, JR Dedman, AR Means. Regulation of intracellular levels of calmodulin and tubulin in normal and transformed cells. Proc Natl Acad Sci U S A 78:996–1000, 1981.

271. DC LaPorte, S Gidwitz, MJ Weber, DR Storm. Relationship between changes in the calcium dependent regulatory protein and adenylate cyclase during viral transformation. Biochem Biophys Res Commun 86:1169–1173, 1979.

272. CD Rasmussen, AR Means. Calmodulin is involved in regulation of cell proliferation. EMBO J 6:3961–3968, 1987.

273. CD Rasmussen, AR Means. Calmodulin is required for cell-cycle progression during G1 and mitosis. EMBO J 8:73–82, 1989.

274. AR Means, MFA VanBerkum, IC Bagchi, KP Lu, CD Rasmussen. Regulatory functions of calmodulin. Pharmacol Ther 50:255–270, 1991.

275. KP Lu, CD Rasmussen, GS May, AR Means. Cooperative regulation of cell proliferation by calcium and calmodulin in *Aspergillus nidulans*. Mol Endocrinol 6:365–374, 1992.

276. KP Lu, SA Osmani, AR Means. Activation of the cell cycle regulated NIMA protein kinase at G2 required calcium/calmodulin. J Cell Biol 115:426, 1991.

277. CB Klee. Concerted regulation of protein phosphorylation and dephosphorylation by calmodulin. Neurochem Res 16:1059–1065, 1991.

278. AR Rhoads, F Friedberg. Sequence motifs for calmodulin recognition. FASEB J 10:331–340, 1997.

279. WA Snedden, H Fromm. Calmodulin, calmodulin-related proteins and plant responses to the environment. Trends Plant Sci 3:299–304, 1998.

280. K Galaktionov, D Beach. Specific activation of cdc25 tyrosine phosphatases by B-type cyclins: evidence for multiple roles of mitotic cyclins. Cell 67:1181–1194, 1991.

281. R Patel, M Holt, R Philipova, S Moss, H Schulman, H Hidaka, M Whitaker. Calcium/calmodulin-dependent phosphorylation and activation of human Cdc25-C at the G2/M phase transition in HeLa cells. J Biol Chem 19:7958–7968, 1999.

282. ASN Reddy, F Safadi, SB Narasimhulu, M Golovkin, X Hu. A novel plant calmodulin-binding protein with a kinesin heavy chain motor domain. J Biol Chem 271:7052–7060, 1996.

283. RE Zielinski. Calmodulin and calmodulin-binding proteins in plants. Annu Rev Plant Physiol Plant Mol Biol 49:697–725, 1998.

284. B Watillion, R Kettmann, P Boxus, A Burny. A calcium/calmodulin-binding serine/threonine protein kinase homologous to the mammalian type II calcium/calmodulin-dependent protein kinase is expressed in plant cells. Plant Physiol 101:1381–1384, 1993.

285. JF Harper, MR Sussman, GE Schaller, C Putnam-Evans, H Charbonneau, AC Harmon. A calcium-dependent protein kinase with a regulatory domain similar to calmodulin. Science 252:951–954, 1991.

286. JH Choi, K-L Suen. Isolation and sequence analysis of a cDNA clone for a carrot calcium-dependent protein kinase: homology to calcium/calmodulin-dependent protein kinases and to calmodulin. Plant Mol Biol 17:581–590, 1991.

287. ASN Reddy, SB Narasimhulu, F Safadi, M Golovkin. A plant kinesin heavy chain–like protein is a calmodulin-binding protein. Plant J 10:9–21, 1996.

288. W Wang, D Takezawa, SB Narasimhulu, ASN Reddy, BW Poovaiah. A novel kinesin-like protein with a calmodulin-binding domain. Plant Mol Biol 31:87–100, 1996.

289. VS Reddy, ASN Reddy. A plant calmodulin-binding motor is part kinesin and part myosin. Bioinformatics 15:1055–1057, 2000.

290. BE Deavours, ASN Reddy, RA Walker. $Ca^{2+}$ calmodulin regulation of the *Arabidopsis* kinesin-like calmodulin-binding protein. Cell Motil Cytoskeleton 40:408–416, 1998.

291. SB Narasimhulu, YL Kao, ASN Reddy. Interaction of *Arabidopsis* kinesin-like calmodulin-binding protein with tubulin subunits: modulation by $Ca^{2+}$-calmodulin. Plant J 12:1139–1149, 1997.

292. SB Narasimhulu, ASN Reddy. Characterization of microtubule binding domains in the *Arabidopsis* kinesin-like calmodulin binding protein. Plant Cell 10:957–965, 1998.

293. J Bowser, ASN Reddy. Localization of a kinesin-like calmodulin-binding protein in dividing cells of *Arabidopsis* and tobacco. Plant J 12:1429–1437, 1997.

294. A Blangy, L Arnaud, EA Nigg. Phosphorylation by p34cdc2 protein kinase regulates binding of the kinesin-related motor HsEg5 to the dynactin subunit p150. J Biol Chem 272:19418–19424, 1997.

295. T Lorca, S Galas, D Fesquet, A Devault, J-C Cavadore, M Dorée. Degradation of the proto-oncogene product p39$^{mos}$ is not necessary for cyclin proteolysis and exit from meiotic metaphase: requirement for a Ca$^{2+}$-calmodulin dependent event. EMBO J 10:2087–2093, 1991.

296. H Kawahara, H Yokosawa. Intracellular calcium mobilization regulates the activity of 26 S proteasome during the metaphase-anaphase transition in the ascidian meiotic cell cycle. Dev Biol 166:623–633, 1994.

297. C Realini, M Rechstgeiner. A proteasome activator subunit binds calcium. J Biol Chem 270:29664–29667, 1995.

298. YH Choi, SJ Lee, P Nguyen. Regulation of cyclin D1 by calpain protease. J Biol Chem 270:28479–28484, 1997.

299. J Langenfeld, H Kiyokawa, D Sekula, J Boyle, E Dimitrovsky. Posttranslational regulation of cyclin D1 by retinoic acid: a chemoprevention mechanism. Proc Natl Acad Sci U S A 94:12070–12074, 1997.

300. L Santella, K Lyozuka, L De Riso, E Carafoli. Calcium, protease action, and the regulation of the cell cycle. Cell Calcium 23:123–130, 1998.

301. Y Takahashi, H Kuroda, T Tanaka, Y Machida, I Takebe, T Nagata. Isolation of an auxin-regulated gene cDNA expressed during the transition from G0 to S phase in tobacco mesophyll protoplasts. Proc Natl Acad Sci U S A 86:9279–9283, 1989.

302. Y Takahashi, Y Niwa, Y Machida, T Nagata. Location of the cis-acting auxin-responsive region in the promoter of the *par* gene from tobacco mesophyll protoplasts. Proc Natl Acad Sci U S A 87:8013–8016, 1990.

303. EJ van der Zaal, J Memelink, AM Mennes, A Quinn, KR Libbenga. Auxin-induced mRNA species in tobacco cell cultures. Plant Mol Biol 10:145–157, 1987.

304. EJ van der Zaal, FNJ Droog, CJM Boot, LAM Hensgens, JHC Hoge, RA Schilperoort, KR Libbenga. Promoters of auxin-induced genes from tobacco can lead to auxin-inducible and root specific expression. Plant Mol Biol 16:983–998, 1991.

305. Y Takahashi, T Nagata. *parB:* an auxin-regulated gene encodidng glutathione *S*-transferase. Proc Natl Acad Sci U S A 89:56–59, 1992.

306. K Satoh, A Kitahara, Y Soma, Y Inaba, I Hatayama, K Sato. Purification, induction, and distribution of placental glutathione transferase: a new marker enzyme for preneoplastic cells in the rat chemical hepatocarcinogenesis. Proc Natl Acad Sci U S A 82:3964–3968, 1985.

307. Y Li, T Seyama, AK Godwin, TS Winokur, RM Lebovitz, MW Lieberman. *MTrasT24*, a metallothionein-ras fusion gene, modulates expression in cultured rat liver cells of two genes associated with in vivo liver cancer. Proc Natl Acad Sci U S A 85:344–348, 1988.

308. H Kodama, N Kawakami, A Watanabe, A Komamine. Phase-specific polypeptides and poly (A)$^+$ RNAs during the cell cycle in synchronous cultures of *Catharanthus roseus* cells. Plant Physiol 89:910–917, 1989.

309. O Leyser. Auxin signaling: protein stability as a versatile control target. Curr Biol 8:R305–R307, 1998.

310. C Trehin, S Planchais, N Glab, C Perennes, J Tregear. Cell cycle regulation by plant growth regulators: involvement of auxin and cytokinin in the re-entry of *Petunia* protoplasts into the cell cycle. Planta 206:215–224, 1998.

311. PCL John, JP Carmichael, DW McCurdy. p34$^{cdc2}$ homologue level, cell division, phytohormone responsiveness and cell differentiation in wheat leaves. J Cell Sci 97:627–630, 1990.

312. C Riou-Khamlichi, R Huntley, A Jacqmard, JAH Murray. Cytokinin activation of *Arabidopsis* cell division through a D-type cyclin. Science 283:1541–1544, 1999.

313. F Laureys, W Dewitte, E Witters, M Van Montagu, D Inze, H Van Onckelen. Zeatin is indispensable for the G2-M transition in tobacco BY-2 cells. FEBS Lett 426:29–32, 1998.

314. F Laureys, R Smet, M Lenjou, D Van Bockstaele, D Inze, H Van Onckelen. A low content in zeatin type cytokinins is not restrictive for the occurrence of G1/S transition in tobacco BY-cells. FEBS Lett 460:123–128, 1999.

315. H Hirt. In and out of the plant cell cycle. Plant Mol Biol 31:459–464, 1996.

316. M Sauter, H Kende. Gibberellin-induced growth and regulation of the cell division cycle in deepwater rice. Planta 188:362–368, 1992.

317. PN Myers, TL Setter, JT Madison, JF Thompson. Abscisic acid inhibition of endosperm cell division in cultured maize kernels. Plant Physiol 94:1330–1336, 1990.

318. ES Ober, TL Setter, JT Madison, JF Thompson, PS Shapiro. Influence of water deficit on maize endosperm development. Plant Physiol 97:154–164, 1991.

319. PW Barlow, PE Pilet. The effect of abscisic acid on cell growth, cell division and DNA synthesis in the maize root meristems. Physiol Plant 62:125–132, 1984.

320. HS Saini, D Aspinall. Sterility in wheat (*Triticum aestivum* L.) induced by water deficit or high temperature: possible mediation by abscisic acid. Aust J Plant Physiol 9:529–537, 1982.

321. S Amino, R Fujimura, A Komamine. Synchrony induced by double phosphate starvation in a suspension culture of *Catharanthus roseus*. Physiol Plant 59:393–396, 1983.

322. F Sala, MG Galli, G Pedrali-Noy, S Spadari. Synchronization of plant cells in culture and in meristems by aphidicolin. Methods Enzymol 118:87–96, 1986.

323. S Spadari, F Sala, G Pedrali-Noy. Aphidicolin, a specific inhibitor of nuclear DNA replication in eukaryotes. Trends Biochem Sci 7:29–32, 1982.

324. T Nagata, Y Nemoto, S Hasezawa. Tobacco BY-2 cell line as the "Hela" cell in the cell biology of higher plants. Int Rev Cytol 132:1–30, 1992.

325. H Kodama, M Ito, T Hattori, K Nakamura, A Komamine. Isolation of genes that are preferentially expressed at the $G_1$/S boundary during the cell cycle in synchronized cultures of *Catharanthus roseus* cells. Plant Physiol 95:406–411, 1991.

326. S Planchais, N Glab, C Trehin, C Prennes, J-M Bureau, L Meijer, C Bergounioux. Roscovitine, a novel cyclin-dependent kinase inhibitor, characterizes restriction point and G2/M transition in tobacco By-2 cell suspension. Plant J 12:191–202, 1997.

327. MH Fox, DW Galbraith. Application of flow cytometry and sorting to higher plant systems. In: MR Melamed, T Lindmo, ML Mendelsohn, eds. Flow Cytometry and Sorting. New York: Wiley-Liss, 1990, pp 633–650.

328. DW Galbraith. Isolation and flow cytometric characterization of plant protoplasts. Methods Cell Biol 33:527–547, 1990.

329. S Bruno, HA Crissman, KD Bauer, Z Darzynkiewicz. Changes in cell nuclei during S phase: progressive chromatin condensation and altered expression of the proliferation-associated nuclear proteins Ki-67, cyclin (PCNA), p105, and p34. Exp Cell Res 196:99–106, 1991.

330. LR Bandara, JP Adamczewski, T Hunt, NB La Thangue. Cyclin A and the retinoblastoma gene product complex with a common transcription factor. Nature 352:249–254, 1991.

331. T Durfee, K Becherer, P-L Chen, S-H Yeh, Y Yang, AE Kilburn, W-H Lee, SJ Elledge. The retinoblastoma protein associates with the protein phosphatase type 1 catalytic subunit. Genes Dev 7:555–569, 1993.

332. P Zhang. The cell cycle and development: redundant roles of cell cycle regulators. Curr Opin Cell Biol 11:655–662, 1999.

333. C Gutierrez. The retinoblastoma pathway in plant cell cycle and development. Curr Opin Plant Biol 1:492–497, 1998.

334. SM de Jager, JAH Murray. Retinoblastoma proteins in plants. Plant Mol Biol 41:295–299, 1999.

335. P Doerner, J-E Jorgensen, R You, J Stepuhn, C Lamb. Control of root growth and development by cyclin. Nature 380:520–523, 1996.

336. AS Hemerly, J de Almeida Engler, M van Montagu, G Engler, D Inze, PCG Ferreira. Dominant negative mutants of Cdc2 kinase uncouple cell division from iterative plant development. EMBO J 14:3925–3936, 1995.

337. B Sheres, R Heidstra. Digging out roots: pattern formation, cell division, and morphogenesis in plants. Curr Top Dev Biol 45:207–247, 1999.

338. F Berger, CY Hung, L Dolan, J Schiefelbein. Control of cell division in the root epidermis of *Arabidopsis thaliana*. Dev Biol 194:235–245, 1998.

339. EM Meyerowitz. Genetic control of cell division patterns in developing plants. Cell 88:299–308, 1997.

340. M Yang, Y Hu, M Lodhi, WR McCombie, H Ma. The *Arabidopsis SKP1-LIKE1* gene is essential for male meiosis and may control homologue separation. Proc Natl Acad Sci U S A 96:11416–11421, 1999.

# 12

# Chlorophyll Biosynthesis During Plant Greening

## Benoît Schoefs

*University of South Bohemia, Budejovice, Czech Republic\**

## I. INTRODUCTION

The development and the maintenance of life on earth are predominantly dependent on photosynthesis, which transforms the radiant energy, coming from the sun, into the chemical energy stored in various molecules.

In photosynthetic eukaryotic organisms, this process takes place in the chloroplast. At the heart of the photosynthetic process are chlorophyll (Chl) and carotenoid (Car) pigments, which are principally, if not completely, associated with proteins (reviewed in Ref. 1).

In angiosperms, chloroplast formation is a light-dependent process, which starts from the proplastid stage. This stage is characterized by the presence of few internal membranes and of starch. In further development of the proplastid, two pathways are possible: in the light, proplastids directly differentiate into chloroplasts, whereas in the dark, they develop into etioplasts [2,3]. The light dependence of the chloroplast formation lies in the absolute requirement of light for enzymatic transformation of protochlorophyllide *a* (Pchlide) to chlorophyllide *a* (Chlide)[†]. In contrast, green algae and most of the other eukaryotic groups of land plants are able to form chloroplasts in the absence of light (reviewed in Ref. 3). This ability lies in the additional presence of a light-independent enzyme that transforms Pchlide to Chlide (reviewed in Refs. 7 and 8).

In the first part of this chapter, the chlorophyll biosynthetic pathway is briefly described. Then the transformation of Pchlide under the impact of the first photons in angiosperms, cultivated in conditions similar to those found in the natural environment and in fields, is discussed. The third part concerns how Chl is synthesized during greening and in green plants. The fourth part of this chapter describes light-dependent and light-independent Chl formation in gymnosperms, and the last part gives information about the regulation of Chl synthesis.

## II. OUTLINE OF THE CHLOROPHYLL BIOSYNTHETIC PATHWAY

Chl biosynthesis starts with the formation of δ-aminolevulinic acid (δ-ALA), the universal precursor of tetrapyrroles. All photosynthetic organisms, except those of the α-proteobacterial group, synthesize δ-

---

\* *Current affiliation:* Université Joseph Fourier, Grenoble, France.
[†] Dark-grown tissues can contain some traces of chl. It seems that they are deposited in the embryo during its formation rather than synthesized in vivo during dark growth [4]. For a full discussion of the possibility of Chl synthesis in the dark in angiosperms tissues, see Adamson et al. [5]. There is no protochlorophyll b in nonilluminated tissues [6].

ALA using the C-5 pathway (reviewed in Refs. 7 and 9). This pathway starts with the activation of glutamate by a transfer RNA (tRNA Glu) molecule. This complex is then reduced to glutamate 1-semialdehyde, which is transaminated to δ-ALA (Figure 1) (reviewed in Refs. 7 and 9). In photosynthetic eukaryotes, all the enzymes catalyzing the formation of protoporphyrin IX (Proto-IX) appear to be localized only in chloroplasts (reviewed in Ref. 9). From the finding of an $Fe^{2+}$-chelatase in chloroplasts, it was deduced that chloroplastic hemes are synthesized there [10], whereas mitochondrial hemes are synthesized in mitochondria. Therefore, the $Fe^{2+}$ or $Mg^{2+}$ insertion inside Proto-IX constitutes the reaction at which the pathways yielding hemes and open tetrapyrroles and those yielding Chl and bacteriochlorophylls diverge (reviewed in Refs. 7 and 9). After insertion of $Mg^{2+}$, Mg-Proto-IX undergoes several specific chemical modifications ending with Chl or Bchl formation. The most important steps are (1) the formation of the isocyclic ring, typical of Bchl and Chl; (2) the reduction of Pchlide to Chlide; and (3) the esterification of Chlide to Chl.

All the reactions of the Chl biosynthetic pathway are catalyzed by enzymes that can be called "normal" in the sense that they transform their substrate to a product when they are in contact. There is, how-

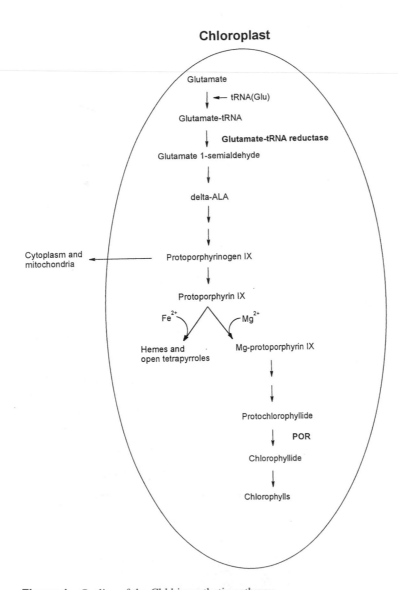

**Figure 1**  Outline of the Chl biosynthetic pathway.

ever, one exception: the light-dependent NADPH:Pchlide oxidoreductase (LPOR, EC 1.3.1.33). This enzyme has two amazing properties: (1) it requires light for activity and (2) in the absence of light, the enzyme forms stable ternary complexes with its cofactor (NADPH) and its substrate (Pchlide) without reacting with them [11]. It is therefore obvious that the transformation of Pchlide to Chlide constitutes a strong regulating point of Chl synthesis, especially in plants, which are unable to synthesize Chl in the dark (see Sec. III.D). The reactions catalyzed by PORs are the main topic of this chapter.

## III. TRANSFORMATION OF PCHLIDE TO CHLIDE UNDER A FIRST ILLUMINATION IN ANGIOSPERMS: INFLUENCE OF THE GROWTH CONDITION

In angiosperms, chloroplast biogenesis invariably begins with the photoreduction of photoactive Pchlide to Chlide because the formation of the first Chlide molecules initiates the synthesis of the chloroplast-encoded proteins, which will be used for the assembly of the photosynthetic apparatus [12].

Most of the data about the development of the photosynthetic apparatus including pigment biogenesis have been obtained using etiolated plants (reviewed in Refs. 13 and 14). Although the etiolated plants—Dubrunfaut [15] considered them ill—cannot be taken as a model for plants that develop in nature, they can probably be used to study chloroplast development in plants cultivated in the field. In fact, modern agricultural methods bury the seeds deep in the soil and, therefore, the leaves start to grow almost in the absence of light. In situ measurements demonstrate that in these conditions, the leaves perceive light when they reach a level approximately 2 mm below the soil surface [16] (reviewed in Ref. 17). It is likely that at this moment the proplastids are already developed into etioplasts. Even if the seeds fall on the ground, in the natural environment, the embryonic leaves can hardly see the light before germination. When the appropriate conditions exist, the seed germinates, i.e., the radicle emerges from the seed [18] (Figure 2). This event modifies the light environment of the embryonic leaves because the radicle can conduct light to them as an optic guide would do [19]. In the literature, the terms designating the material used for greening experiments are often confusing. Therefore, throughout this chapter, the terms old and young leaves were used to designate etiolated leaves with etioplasts and embryonic leaves with proplastids, respectively.

It is important to note that plant species can be classified in different groups on the basis of the Pchlide chemical form (either monovinyl or divinyl) accumulated during the night and Chlide chemical form produced at daybreak and later [20]. Spectroscopic measurements using isolated LPOR have indicated that the mechanism of Pchlide reduction is not significantly affected by the group to which a plant belongs [21]. In addition, the presence of an 8-ethyl or an 8-vinyl at the Pchlide ring B does not significantly influence the spectral properties of the different Pchlide forms in the red region (600–800 nm). In contrast, in the blue region (400–500 nm), significant differences can be observed [22–24].

**Figure 2**  Germination of seeds of *Acer* sp. in nature. Only the radicle is out of the seeds. Pictures taken on March 14, 1999 in the park of the Hluboka castle (Hluboka nad Valtavou, Czech Republic).

## A. Chl Formation in Plants Growing in Conditions Similar to the Natural Ones

Using radiolabeled precursors, it was shown that the first Pchlide molecules are synthesized 12 hr after the imbibition of cucumber cotyledons [25]. In situ spectroscopic investigations, especially in vivo absorbance measurements, at these early developmental stages are very difficult because the leaves are very small [26,27] and contain only traces of pigments [28]. Using 77 K fluorescence detection, nonphotoactive and photoactive Pchlide were detected in situ during the second photoperiod of greening in bean leaves, i.e., when the radicle emerges from the seed [14,27]. At this developmental stage, the ratio of photoactive to nonphotoactive Pchlide is in favor of the nonphotoactive form [26,29]. Similar results have been obtained with other angiosperm tissues (*Pisum sativum* [30], *Triticum aestivum* [31]). Photoactive Pchlide is a ternary complex containing Pchlide, NADPH, and a photoenzyme, the so-called LPOR. In vivo, individual ternary complexes form aggregates (reviewed in Ref. 32). Reconstitution experiments suggest that LPOR activity requires at least LPOR dimers [33], which could correspond to the photoactive Pchlide $P_{638-645}$* observed in vivo. Aggregation of these dimers yields the formation of large aggregates also observed in vivo, i.e., $P_{650-657}$. Both $P_{638-645}$ and $P_{650-657}$ have been isolated [34]. The behaviors of the two photoactive Pchlides are very difficult to analyze separately. Consequently, in this chapter they have been considered as a single entity that is denoted by $P_{638,650-657}$. Nonphotoactive Pchlide is denoted by $P_{628-633}$. The biochemical state of nonphotoactive Pchlide is less clear because it is spectrally and chemically heterogeneous [28,35–38]. Actually, several different nonphotoactive Pchlide forms have been characterized more: (1) free Pchlide (emission at approximately 625 nm [39], (2) a monomeric Pchlide-LPOR complex (emission at approximately 634 nm [39], and (3) an aggregate similar to the photoactive Pchlide but with NADP$^+$ instead of NADPH ($P_{642-649}$ [40]).

When a young dark-grown leaf (e.g., 2 days old) is illuminated by a short and saturating flash, $P_{638,650-657}$ is transformed to the Chlide. The reaction consists of the reduction of ring D of Pchlide (Figure 3). The Chlide formed has absorbance and fluorescence emission maxima at 676 and 688 nm, respectively ($C_{676-688}$) [29,41] (Figure 4A). $C_{678-688}$ is an aggregate similar to $P_{638,650-657}$ but containing Chlide and NADP$^+$ instead of Pchlide and NADPH [42]. Then the major part of the Chlide is liberated from the active site of the enzyme, leaves the aggregate, and a new Chlide spectral form, $C_{670-675}$, appears. The liberation of Chlide from the active site of the enzyme is indicated by the simultaneous regen-

---

* $P_{X-Y}$ and $C_{X-Y}$ mean Pchlide and Chlide absorbing at $X$ nm and emitting fluorescence at $Y$ nm at 77 K, respectively.

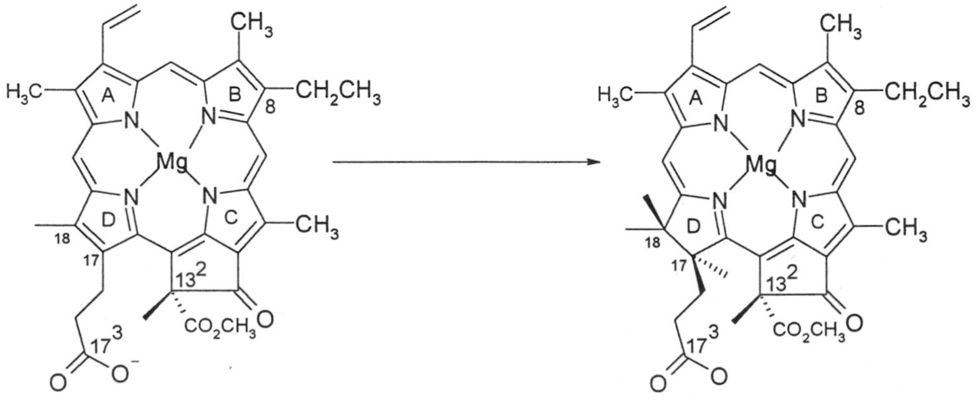

**Protochlorophyllide *a***          **Chlorophyllide *a***

**Figure 3**  Scheme of the photoreduction of Pchlide to Chlide.

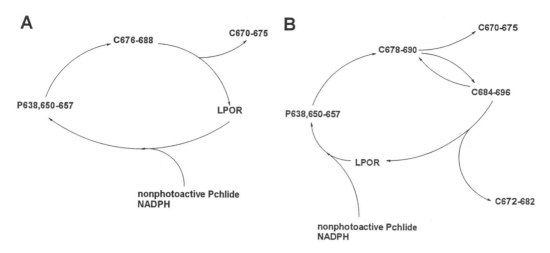

**Figure 4**   The Pchlide-Chlide cycle in leaves developing (A) under natural conditions and (B) under conditions similar to those found in the field.

cration of photoactive Pchlide occurring during Chlide liberation [29]. If the spectroscopic properties of $C_{670-675}$ are well defined, its biochemical state remains to be determined precisely and it is not clear whether $C_{670-675}$ corresponds to a free pigment or to a pigment-protein complex. In this case, the protein moiety cannot be LPOR because it has been reused for the regeneration of photoactive Pchlide. Almost nothing is known about the regeneration process. Analyses of excitation spectra have indicated that $P_{642-649}$ is an intermediate [43].

The remaining part of $C_{676-688}$ is transformed to another spectral form of Chlide ($C_{684-696}$). From the biochemical point of view, $C_{684-696}$ is similar to $C_{676-688}$ but contains NADPH instead of NADP$^+$ [42]. These events are summarized in Figure 4A, which displays the so-called Pchlide-Chlide cycle. Similar results have been obtained with *Spirodela oligorrhiza*, a plant that does not develop etioplasts when cultivated in darkness [44].

## B.   Chl Formation in Plants Cultivated for a Long Time in the Dark (i.e., Etiolated Leaves)

During dark growth, proplastids develop to etioplasts, which are characterized by the presence of a prolamellar body (PLB) and some single perforated membranes called prothylakoids (reviewed in Ref. 3). Simultaneously with the differentiation of proplastids to etioplasts, photoactive Pchlide is accumulated [26,45] into the PLB, where LPOR is by far the most abundant protein [46,47]. Etiolated leaves contain the same spectral forms of Pchlide as the young leaves, i.e., $P_{638,650-657}$ and $P_{628-633}$, but the ratio of photoactive to nonphotoactive Pchlide is in favor of the photoactive form [29,48].

The first product of the photoreduction of photoactive Pchlide in etiolated leaves, $C_{678-690}$, has slightly different spectral properties than found in young leaves (see Sec. III.A). This minor difference in the position of the absorbance and fluorescence maxima (77 K) may reflect a slightly different environment of the pigment.

The absorbance and fluorescence kinetics of the Pchlide photoreduction are monoexponential when the process is studied on the second time scale. The rate constants of the kinetics are identical in young and old leaves, indicating that the photoreduction mechanism is identical [45]. The formation of $C_{678-690}$ is preceded by the formation of several nonfluorescent intermediates (reviewed in Ref. 49), whose chemical structure remains unknown.

In etiolated leaves, only a minor part of $C_{678-690}$ is transformed to $C_{670-675}$ [29,50]. The major part is transformed to $C_{684-696}$, which is an efficient fluorescence quencher at room temperature [51]. $C_{684-696}$ formation, which occurs readily in the dark after the initial phototransformation step, can revert to $C_{678-690}$ under illumination [52,53] (Figure 4B). $C_{684-696}$ is the photoreceptor for this transformation [53]. Once the reaction reverted, $C_{678-690}$ can be transformed back to $C_{684-696}$ in the dark, which can be transformed again to $C_{678-690}$ by new illumination and so on. Therefore $C_{678-690}$ and $C_{684-696}$ form a cycle. In vitro experiments have shown that the $C_{678-690}$ to $C_{684-696}$ transformation required NADPH [42]. Consequently, when the cycle is turning, NADPH is consumed. It is probably oxidized at each light-triggered $C_{684-696} \rightarrow C_{678-690}$ conversion [54]. We showed that this cycle is involved in the photoprotection of newly formed Chlide against photooxidation [55,56]: when Chlide is in the $C_{678-688}$ conformation, it is readily photodestroyed, whereas in the $C_{684-696}$ form, this is not the case. This is in line with the action spectrum of the oxygen uptake by Chlide in shortly illuminated plastids established by Redlinger and McDaniel [57]. It should be emphasized that a photoprotection mechanism is needed at this stage because carotenoids, although present in the etiolated leaves [28,34,58], do not protect the newly formed Chlide against photodestruction [59]. It should be noted here that unprotected Chl(ide) is very reactive with oxygen when illuminated and generates activated oxygen species, which are able to destroy cellular and subcellular structures (reviewed in Ref. 60). The photoprotection mechanism is specifically NADPH dependent [55]. Because in the two spectral forms of Chlide-LPOR aggregates involved in the cycle, i.e., $C_{678-688}$ and $C_{684-696}$, the Chlide is still bound to the enzyme, we can conclude that LPOR is involved in the transformation of $C_{684-696}$ to $C_{676-688}$. The involvement of LPOR in this process is further supported by the increase in Chlide photoprotection in *Arabidopsis* overexpressing LPOR [61]. When the aggregates are dissociated, Chl(ide) is partially released from LPOR and is esterified. Both events occur during the Shibata shift (reviewed in Ref. 32). These events are summarized in Figure 4B.

During the Shibata shift, photoactive Pchlide is regenerated. This process has not yet been extensively studied. It was shown that an aggregate similar to photoactive Pchlide, but containing $NADP^+$ (i.e., P642-649) instead of NADPH, is formed very rapidly after the photoreduction [62]. Experiments using inhibitors of protein synthesis have shown that LPOR is partly reused to regenerate photoactive Pchlide [63,64]. On the other hand, full regeneration requires protein synthesis [65]. Although it is established that at least enzymes involved in the δ-ALA synthesis are involved, the exact number of proteins synthesized de novo remains undetermined.

## C. Chlorophyll Formation in Partially Green Leaves and in Fully Green Leaves

Nonilluminated leaves contain far less Pchlide than fully mature green leaves contain Chl (reviewed in Ref. 66). Therefore, Chl should be produced during greening. In Secs. III.A and III.B the arguments in favor of the involvement of aggregates of LPOR-Pchlide *a*-NADPH complexes in Chlide *a* formation during the first illumination were presented. Although some evidence suggests that the same types of aggregates are involved in Chlide *a* formation during leaf greening and also in green leaves [27,67,68], no firm proof was given in these works. Therefore it was crucial to determine whether Chl is formed during greening and in green leaves according to the set of reactions illustrated in Figure 4. If so—i.e., similar spectral forms of photoactive Pchlide are used to synthesize Chl during greening—a steady-state amount of photoactive complexes should be detected when the plants are illuminated with nonsaturating light. This was demonstrated by in situ fluorescence [69,70] and by absorbance measurements [7,71]. The amount of photoactive Pchlide detected during greening is directly related to the light intensity used for cultivation [71]. The photoactive Pchlide involved in greening has a 77 K emission maximum slightly shifted to the blue (653 nm) [69,70]. Under a saturating flash, the pool of photoactive Pchlide, not photoreduced by the light used to drive greening, is transformed to $C_{678-690}$, which is in turn transformed to $C_{684-696}$ and subsequently to $C_{672-682}$. The duration of these shifts, similar in nature to those observed in etiolated leaves, is dramatically accelerated when compared with the etiolated material [69]. Therefore, it can be concluded that the cycle presented in Figure 4B also describes the reactions leading to Chl production in greening and green leaves.

This conclusion is in sharp contradiction to the view expressed by Lebedev and Timko [72], who proposed the existence of a cycle similar to Figure 4A. It also apparently runs against the measurements of the variations of LPOR messenger RNA (mRNA) and LPOR amounts during greening. In fact, both dramatically decrease during the first hours of greening (Figure 5) (reviewed in Ref. 73). This last contradiction vanished when it was found that most of the angiosperms contain two LPORs, denoted LPORA and LPORB [74–76] (reviewed in Refs. 9 and 14). Exceptions have been found in cyanobacteria [77],

A

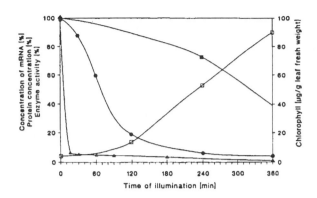

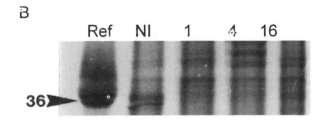

B

**Figure 5** (A) The effect of light on (•) the relative amount of LPOR mRNA, (■) the relative LPOR concentration, (▲) the specific LPOR activity, and (□) the relative Chlide content. (From Ref. 73.) (B) Sodium dodecyl sulfate–polyacrylamide gel electrophoresis of the proteins contained in (NI) nonilluminated 2-day-old bean leaves and leaves after (1) 1 hr, (4) 4 hr, and (16) 16 hr of greening. The band corresponding to LPOR is indicated by the arrow. Ref indicates the protein used as a standard.

*Chlamydomonas reinhardtii* [78], and pea [79], which contain only one LPOR gene. In dark-grown leaves, LPORA is usually more abundant than LPORB [79]. In contrast to LPORA, whose expression appears to be correlated with Pchlide synthesis [80], LPORB is constitutively expressed [76]. In plants cultivated under a light/dark regime, LPORA concentration shows diurnal variations [81], which might be correlated with the variations in the amount of Chl in leaves from plants cultivated in the field [82]. The accumulation of LPORA and Pchlide at the end of the dark phase of each light/dark cycle [28,81] correlates with the observation of small PLB during this period [83].

All POR proteins have a high degree of homology [77]. LPORA and LPORB amino acid sequences, as deduced from complementary DNA (cDNA) clones, are very related proteins presenting more than 75% homologies. The homology increases to 82% when the deduced amino acid sequences of the mature proteins are compared. The two sequences are much more divergent within the signal sequences [76]. Electron microscopy and spectroscopic measurements have demonstrated that LPORB and LPORA are able to induce the formation of a regular PLB and also to drive the Pchlide-Chlide cycle [84].

## D.  Chlorophyll Formation in Plants That Contain Both the Light-Dependent and the Light-Independent Protochlorophyllide Oxidoreductase

The fact that most of the eukaryotic nonangiosperms and green algae are able to green in the dark has been recognized for a long time (e.g., Ref. 85). The ability to synthesize Chl in the dark correlates with the additional presence of a light-independent Pchlide oxidoreductase (DPOR). Biochemical and molecular genetic data indicate that LPOR and DPOR are not related. In fact, DPOR is probably formed by three different subunits (reviewed in Refs. 8, 9, and 14). Most of the data on Chl formation in plants containing LPOR and DPOR have been obtained from gymnosperms.

Plastids from dark-grown pine cotyledons are differentiated into grana and thylakoids and also contain a PLB (reviewed in Ref. 3). They are called etiochloroplasts. LPOR has been found in PLB as well as thylakoids and grana [86]. Forreiter and Apel [87] demonstrated that etiochloroplasts, as the etioplasts, contain two LPORs: the first one (36 kDa) is associated with the PLB, and the second one (38 kDa) is found in the thylakoids. Although direct evidence for the analogy of these two LPORs to LPORA and LPORB is lacking, it can be assumed that LPORA is located in the PLB and LPORB is located in thylakoids. Analyses of fluorescence spectra of dark-grown pine tissues indicate the same spectral and chemical heterogeneity of nonphotoactive and photoactive Pchlides as in dark-grown higher plants (cotyledons [36,88], primary needles [24]). Spectroscopic investigations of the Pchlide-Chlide cycle in angiosperms are difficult because of the presence of emission bands corresponding to photosystems I and II (PSI and PSII) [88]. In order to determine the fate of the Chlide resulting from the photoreduction in primary needles, pine seeds were cultivated in the dark and in the presence of norflurazon, an inhibitor of $\beta$-carotene synthesis (reviewed in Ref. 89). In the absence of carotenoid, neither PSI nor PSII assembled [24]. When Pchlide photoreduction was triggered in such plants, the first product of the photoreduction, $C_{676-688}$, was rapidly transformed to $C_{670-675}$ [24]. Therefore a Pchlide-Chlide cycle similar to the one observed in embryonic angiosperm leaves (Figure 2) seems to operate in these conditions.

## IV.  REGULATION (AN ASSAY)

### A.  Amount of Pchlide in Nonilluminated Plastids

The total Pchlide and photoactive Pchlide accumulation curves during the development in the dark are sigmoidal. They reach a stationary level after approximately 7 to 10 days of growth depending on the species and on the growth conditions (reviewed in Ref. 13). These levels correspond to the maximum amount of Pchlide that a definite species is able to accumulate naturally for given growth conditions. The arrest of Pchlide accumulation cannot be explained by feedback inhibition of $\delta$-ALA synthesizing enzymes by Pchlide because they are not very sensitive to Pchlide [90,91]. In contrast, the preferential accumulation of photoactive Pchlide during dark growth can be explained by the fact that the import and processing of LPORA precursor (pLPORA) from the cytoplasm into the plastids are dependent on the availability of the nonphotoactive Pchlide [80].

According to this model, when Pchlide synthesis stops, the import of pLPORA is blocked and photoactive Pchlide is no longer actively accumulated. In contrast, this model cannot explain why under application of exogenous $\delta$-ALA nonphotoactive Pchlide is accumulated [92]. In fact, the saturation level observed when plants are developing in the dark does not correspond to the maximum capacity of Pchlide accumulation inside the plastids. Upon addition of exogenous $\delta$-ALA, dark-grown leaves are able to accumulate much more Pchlide than untreated leaves. Consequently, their yellow color, due to carotenoids, is masked and the leaves appear green! This Pchlide, however, is nonphotoactive [92]. It is relevant to add here that when leaves fed with $\delta$-ALA are illuminated, the accumulated nonphotoactive Pchlide produced so much activated oxygen species that the leaf can be bleached. This is especially obvious with *tigrina* mutants of barley, which "naturally" overproduce nonphotoactive Pchlide (for pictures, see Ref. 93). Because carotenoids do not protect Pchlide from photo-oxidation (reviewed in Ref. 59), the simultaneous arrest of Pchlide and LPOR accumulation during dark growth can be understood as a mechanism to avoid production of activated oxygen species.

### B.  Regulation of the Chlorophyll Accumulation During Greening

Usually, plastids from dark-grown leaves do not contain polypeptides belonging to the photosynthetic apparatus but contain their corresponding transcripts [94]. A very elegant study demonstrated that mRNAs start to be translated by polysomes into the plastids during dark growth but the translation cannot be completed because some cofactor(s) is (are) missing [95]. Eichacker et al. [12] demonstrated that the missing cofactor is not light itself but Chlide (plus phytol). In fact, these authors incubated lysed etioplasts in the dark with exogenous Chlide plus phytol and observed the synthesis of several polypeptides encoded by the chloroplastic genome in complete darkness!

On the other hand, Franck et al. [96] demonstrated that the appearance of variable fluorescence after one single millisecond flash is detected only when the extent of Pchlide phototransformation is higher than 40%. In these conditions, $C_{684-696}$ is preferentially formed. This suggests that $C_{684-696}$ formation (see Secs. III.B and III.C) is essential for further assembly of PSII. This is confirmed by the observation that in young leaves $C_{684-696}$ is formed only in low quantities (Figure 4A) even when the percentage of photoreduction is 100% and consequently the development of the photosynthetic apparatus is very slow [27,29,67] (see also later).

The results of these experiments emphasize the central role of the Pchlide-Chlide cycle in the biogenesis of the photosynthetic apparatus. The cycle not only is used to produce Chl but also acts as the primary regulator of the synthesis of polypeptides. It also explains why the reaction centers of PSI and PSII are synthesized before the antennae, which slowly accumulate thereafter [97].

In gymnosperms, the Chlide produced in the dark continuously activates the transcription of the polypeptides required for the assembly of the photosynthetic apparatus. Therefore, the 77 K fluorescence spectra of dark-grown gymnosperm cotyledons [36,88] presented the typical bands of PSI and PSII. A similar observation was made with dark-grown primary needles [24], a tissue noted for its inability to synthesize Chl in the dark (e.g., Ref. 93).

It can be deduced from several experiments that the expression of several nuclear genes involves the presence of functional plastids [98,99]. Barbato et al. [100] proposed that the presence of Chl also regulates the light-harvesting chlorophyll *a/b* binding protein (*Cab*) CP29 maturation during greening. The idea that a signal originating from the chloroplast activates nuclear gene synthesis emerges from these studies (reviewed in Ref. 101). However, the nature of the signal remains unknown. Pchlide precursors can correspond to such a signal. This can be deduced from experiments with *Chlamydomonas* incubated with a metal chelator [108]. In this condition, the Chl biosynthetic pathway is impaired and Mg-protoporphyrin monomethyl ester is accumulated [102] with the consequence that the light-dependent accumulation of *Cab* proteins [103,104] and of the small subunits of ribulose-1,5-bisphosphate carboxylase (Rubisco) are inhibited [105]. The mechanism(s) of action remains uncertain. The presence of Pchlide precursors could decrease the amount of mRNA [103] or interfere with the light-dependent transcription [104,106]. Another possibility for regulation is the very different affinity of Mg-chelatase for Proto-IX [107]. This implies that when Mg-chelatase is active, Proto-IX is preferentially used to synthesize Chl. Chl hemes are turning over [109]. When heme degradation is higher than heme formation, the δ-ALA synthesis is stimulated because heme inhibits δ-ALA formation stoichiometrically [110–112]. Consequently, the heme concentration will rise again and the δ-ALA formation will be partially inhibited. Such a mechanism has been proposed for bacteria by Lascelles and Hatch [113] but also seems to occur in plastids [9] regardless of their developmental stage.

After the initial Pchlide reduction, Chl synthesis in angiosperms shows three phases [114]: the lag phase, the phase of rapid accumulation, and the stationary phase.

The length of the lag phase is dependent on the developmental stage. Precise measurements of the length of the lag phase as a function of the bean leaf age [28,115,116] have indicated that there exists a developmental stage, i.e., 3 days old, for which the lag phase is very short. Leaves below this stage, i.e., younger leaves, can accumulate Chl only after a very long lag phase. Above this stage, the older the leaves, the longer the lag phase. A similar conclusion was reached with wheat leaves [117]. Using etiolated material, it was shown that the factor limiting Chl accumulation during greening is the synthesis of δ-ALA. In fact incubation of seedlings with this compound abolished the lag phase [118]. However, this is not the case in young seedlings [41,115]. Therefore, the long lag phase observed in young seedlings is partially due to another factor(s). It is important to mention that during this period, a minimal but functional photosynthetic apparatus is assembled very rapidly after the onset of the illumination. However, the $F_0$ level of induction kinetics remains very high during all this period, suggesting that most of the Chl remains not integrated with the photosynthetic units [27,67]. There are several lines of evidence that during greening Chl synthesis is coordinated with those of Car and polypeptides composing the photosynthetic apparatus [119–122]. Therefore, the long lag phase can be a consequence of either a deficiency of the Chl biosynthesis itself or of other pathways (carotenoids, synthesis of *Cab* proteins; $CO_2$ fixation, chemical energy production, etc). Interestingly, Chl, carotenoids, and leaf dry weight, which reflect the actual $CO_2$ fixation, present the same lag phase whatever the leaf developmental stage [122]. This experimental fact can be explained as follows: Chl phytol and carotenoids are both synthesized from geranylgeraniol, which itself is synthesized from the simultaneously fixed $CO_2$ [123,124]. Therefore, if the $CO_2$ fixation activity is low, the pigment synthesis is low. Interestingly, it has been shown that phytochrome controls the length of the lag phase [125] at the level of δ-ALA synthesis (reviewed in Refs. 126 and 127),

Time of illumination [h]

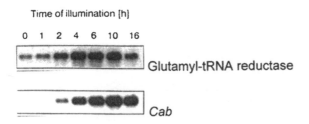

**Figure 6** Northern blot analysis of glutamyl-tRNA reductase and *cab* mRNAs during the first 16 hr of greening of 6-day-old dark-grown barley seedlings. (From Ref. 132.)

the transcription of small subunit Rubisco [128], and *cab* gene transcription (reviewed in Ref. 129) as well as the expression of phytoene synthase [130]. Horwitz et al. [131] demonstrated that the abundance of *Cab* mRNA is not the factor limiting Chl accumulation.

## C. In Fully Matured Leaves

It has been observed that in plants many cellular activities occur with a daily rhythmicity. These rhythms are called circadian. The input factors such as light and temperature are connected to the central oscillator(s), which generates output rhythms via a range of signaling pathways. Interestingly, it has been shown that the glutamyl tRNA reductase (see Figure 1) and *cab* genes present circadian variations in barley. The phase at which these genes are expressed is slightly earlier than that of the *cab* genes (Figure 6) [132,133]. It is important to note that among all the genes expressed according to the circadian rhythms, only promoters of *Arabidosis cab2* and wheat *cab1* have been shown to confer clock regulation to a reporter gene (reviewed in Ref. 134).

In plants cultivated under a light/dark regime, the amount of Chl slightly decreases during the dark period [28]. Therefore, Chl should be resynthesized at the beginning of each day [82]. This can be achieved by accumulation during the dark period of aggregates of LPORA-Pchlide-NADPH complexes according to the mechanism identified by Reinbothe et al. [80]. This regeneration also triggers the formation of small prolamellar bodies [83].

## ACKNOWLEDGMENTS

The author thanks the Ministry of Education, Youth and Sports of the Czech Republic (grant VS96085) and the Grant Agency of the Czech Republic (grant 204/98/P120) for their financial support.

## REFERENCES

1. BR Green, DG Dunford. The chlorophyll-carotenoid proteins of oxygenic photosynthesis. Annu Rev Plant Physiol Plant Mol Biol 47:685–694, 1996.
2. JM Whatley. Variations in the basic pathway of chloroplast development. New Phytol 78:407–420, 1977.
3. J Hudak. Photosynthetic apparatus. In: M Pessarakli, ed. Handbook of Photosynthesis. New York: Marcel Dekker, 1997, pp 27–49.
4. B Böddi, A Lindsten, C Sundvist. Chlorophylls in dark-grown epicotyls and stipula of pea. J Photochem Photobiol B Biol 48:11–16, 1999.
5. H Adamson, RG Hiller, J Walmsley. Protochlorophyllide reduction and greening in angiosperms: an evolutionary perspective. J Photochem Photobiol B Biol 41:201–221, 1997.
6. V Scheumann, H Klement, M Helfrich, U Oster, S Schoch, W Rüdiger. Protochlorophyllide b does not occur in barley etioplasts. FEBS Lett 445:445–448, 1999.
7. B Schoefs, M Bertrand. Chlorophyll biosynthesis. In: M Pessarakli, ed. Handbook of Photosynthesis. New York: Marcel Dekker, 1997, pp 49–69.
8. G Armstrong. Greening in the dark: light-independent chlorophyll synthesis from oxygenic photosynthetic bacteria to gymnosperms. J Photochem Photobiol B Biol 43:87–100, 1998.
9. SI Beale. Enzymes of the chlorophyll biosynthesis. Photosynth Res 60:43–73, 1999.
10. KS Chow, DP Singh, AR Walker, AG Smith. Two different genes encode ferrochelatase in *Arabidopsis*: mapping, expression and subcellular targeting of the precursor proteins. Plant J 15:531–541, 1998.

11. TW Griffiths. Reconstitution of chlorophyllide formation by isolated membranes. Biochem J 174:681–692, 1978.

12. LA Eichacker, J Soll, P Lauterbach, W Rüdiger, RR Klein, JE Mullet. In vitro synthesis of chlorophyll A in the dark triggers accumulation of chlorophyll A apoproteins in barley etioplasts. J Biol Chem 265:13566–13571, 1990.

13. Z Sestak. Effects of age on protochlorophyllide and chlorophyllide formation (a review). In C Sironval, M Brouers, eds. Protochlorophyllide Reduction and Greening, The Hague: Martinus Nijhoff/Dr W Junk Publishers, 1984, pp 365–375.

14. B Schoefs. Light-dependent and light-independent protochlorophyllide reduction. Photosynthetica 36:481–496, 1999.

15. M Dubrunfaut. Statique de la lumière dans les phénomènes de la vie des végétaux et des animaux. C R Hebd Sceances Acad Sci Paris 66:425–427, 1868.

16. JT Wolley, EW Stoller. Light penetration and light-induced seed germination in soil. Plant Physiol 61:597–600, 1978.

17. H Smith. Sensing the light environment: the functions of the phytochrome family. In RE Kendrick, GHM Kronenberg, eds. Photomorphogenesis in Plants. 2nd ed. Dordrecht: Kluwer Academic Publishers, 1994, pp 377–416.

18. D Côme. Problèmes posés par la germination et ses obstacles. Bull Soc Fr Physiol Vég 14:3–9, 1968.

19. S Mandoli, W Briggs. Fiber optics in plants. Sci Am 251:90–99, 1984.

20. CA Rebeiz, R Parham, DA Fasoula DA, IM Ionnides. Chlorophyll *a* biosynthetic heterogeneity. In DJ Chadwick, K Ackrill, eds. Biosynthesis of Tetrapyrroles Pigments. Chichester: Wiley, 1994, pp 177–193.

21. M Bertrand, B Seyfried, H Senger. In vitro photoreduction of monovinyl- and divinyl-protochlorophyllide. Physiol Plant 79:A65, 1990.

22. FC Belanger, CA Rebeiz. Chloroplast biogenesis 30. Chlorophyll(ide) (E459F675) and chlorophyllide (E449F675) the first detectable product of divinyl and monovinyl protochlorophyllide photoreduction. Plant Sci Lett 18:343–350, 1980.

23. B Schoefs, C Funk, B Andersson. Spectral changes during photoactive protochlorophyllide regeneration in spinach cotyledons. Arch Physiol Biochem 105:15, 1997.

24. B Schoefs, F Franck. Chlorophyll synthesis in dark-grown pine primary needles. Plant Physiol 118:1159–1168, 1998.

25. CA Rebeiz, M Yaghi, M Abou-Had'dar, PA Castelfranco. Protochlorophyll biosynthesis in cucumber (*Cucumis sativus* L.) cotyledons. Plant Physiol 46:57–63, 1970.

26. S Klein, JA Schiff. The correlated appearance of prolamellar bodies, protochlorophyll(ide) species and the Shibata shift during developmental bean etioplast in the dark. Plant Physiol 49:619–626, 1972.

27. B Schoefs, M Bertrand, F Franck. Plant greening: biogenesis of the photosynthesis apparatus in bean leaves irradiated shortly after the germination. Photosynthetica 27:497–504, 1992.

28. B Schoefs, M Bertrand, Y Lemoine. Changes of the photosynthetic pigment content in bean leaves during the first photoperiod of greening. Comparison between etiolated (10-d-old) and young (2-d-old) leaves. Photosynth Res 57:203–213, 1998.

29. B Schoefs, F Franck. Photoreduction of protochlorophyllide to chlorophyllide in 2-d old dark-grown bean (*Phaseolus vulgaris* cv. Commodore) leaves. Comparison with 10-d old dark-grown (etiolated) leaves. J Exp Bot 44:1053–1057, 1993.

30. Z-H He, J Li, C Sundqvist, P Timko. Leaf developmental age controls expression of genes encoding enzymes of chlorophyll and heme biosynthesis in pea (*Pisum sativum*). Plant Physiol 106:537–546, 1994.

31. S Younis, M Ryberg, C Sundqvist. Plastid development in germinating wheat (*Triticum aestivum*) is enhanced by gibberellic acid and delayed by gabaculine. Physiol Plant 95:336–346, 1996.

32. M Ryberg, N Artus, B Böddi, A Lindsten, B Wiktorsson, C Sundqvist. Pigment-protein complexes of chlorophyll precursors. In: JH Argyroudi-Akoyunoglou, ed. Regulation of Chloroplast Biogenesis. New York: Plenum Press, 1992, pp 217–225.

33. GEM Martin, MP Timko, HM Wilks. Purification and kinetic analysis of pea (*Pisum sativum* L.) NADPH:protochlorophyllide oxidoreductase expressed as fusion with maltose-binding protein in *Escherichia coli*. Biochem J 325:139–145, 1997.

34. MA Ouazzani Chahdi, B Schoefs, F Franck. Purification and characterization of photoactive complexes of NADPH:protochlorophyllide reductase from wheat. Planta 206:673–680, 1998.

35. CE Cohen, CA Rebeiz. Chloroplast biogenesis. 34. Spectrofluorometric characterization in situ of protochlorophyllide species in etiolated tissues of higher plants. Plant Physiol 67:98–103, 1981.

36. B Schoefs, M Bertrand, F Franck. Spectral heterogeneity of photoinactive protochlorophyllide in dark-grown bean leaf and pine cotyledons. In: P Mathis, ed. Photosynthesis: from Light to Biosphere. Vol 3. Dordrecht: Kluwer Academic Publishers, 1995, pp 1013–1016.

37. F Bovey, T Ogawa, K Shibata. Photoconvertible and nonphotoconvertible forms of protochlorophyll(ide) in etiolated leaves. Plant Cell Physiol 15:1133–1137, 1974.

38. B Böddi, F Franck. Room temperature fluorescence spectra of protochlorophyllide and chlorophyllide forms in etiolated leaves. J Photochem Photobiol B Biol 41:73–82, 1997.

39. B Böddi, K Kipetik, AD Kapori, J Fidy, C Sundqvist. The two spectroscopically different short-wavelength protochlorophyllide forms in pea epicotyls are both monomeric. Biochim Biophys Acta 1365:531–540, 1998.

40. B El Hamouri, C Sironval. A new non-photoreducible protochlorophyll(ide)-protein P649-642: NADPH mediation of its transformation to photoreducible P657-650. FEBS Lett 103:343–347, 1979.
41. B Schoefs. Photoreduction of Protochlorophyllide *a* to Chlorophyllide *a* During the Biogenesis of the Photosynthetic Apparatus in Higher Plants. Ann Arbor, MI: Bell & Howell, 2000.
42. B El Hamouri, M Brouers, C Sironval. Pathway from photoinactive P633-628 protochlorophyllide to the P696-682 chlorophyllide in cucumber etioplast suspension. Plant Sci Lett 21:375–379, 1981.
43. B Schoefs, F Franck, M Bertrand. Spectroscopic properties of regenerated photoactive protochlorophyllide after a flash at two different leaf developmental stages. In: JC Merlin, S Turrell, JP Huvenne, eds. Proceedings of the 6th International Conference on the Spectroscopy of Biological Molecules, Dordrecht: Kluwer Academic Publishers, 1995, pp 611–612.
44. DJ McCormac, CA Marwood, D Bruce, BM Greenberg. Assembly of photosystem I and II during the early phase of light-induced development of chloroplasts from proplastids in *Spirodela oligorrhiza*. Photochem Photobiol 63:837–845, 1996.
45. B Schoefs, H-P Garnir, M Bertrand. Comparison of the photoreduction of protochlorophyllide to chlorophyllide in leaves and cotyledons from dark grown bean as a function of age. Photosynth Res 41:405–417, 1994.
46. E Selstam, A Widell, LB Johansson. A comparison of prolamellar bodies from wheat, Scots pine and Jeffrey pine. Pigment spectra and properties of protochlorophyllide oxidoreductase. Physiol Plant 70:209–214, 1987.
47. A Lindsten, M Ryberg, C Sundqvist. The polypeptide composition of highly purified prolamellar bodies and prothylakoids from wheat (*Triticum aestivum*) as revealed by silver staining. Physiol Plant 72:167–176, 1988.
48. C Sironval, Y Kuyper, J-M Michel, M Brouers. On the primary photoact in the conversion of protochlorophyllide into chlorophyllide. Studia Biophys 5:43–50, 1967.
49. TW Griffiths. Protochlorophyllide photoreduction. In: H Scheer, ed. Chlorophylls. Boca Raton, FL: CRC Press, 1991, pp 433–450.
50. C Sironval. The protochlorophyllide-chlorophyllide cycle as a source of photosynthetically active chlorophylls. In: G Akoyunoglou, ed. Photosynthesis. Vol 5: Chloroplast Development. Philadelphia: Balaban Inetrnational Science Service, 1981, pp 3–14.
51. M Jouy, C Sironval. Quenching of the fluorescence emitted by P695-684 at room temperature in etiolated illuminated leaves. Planta 147:123–133, 1979.
52. S Bauer, W Siegelman. Photoconversion of chlorophyllide 684 to chlorophyllide 678. FEBS Lett 20:352–354, 1972.
53. F Franck, Y Inoue. Light-driven reversible transformation of chlorophyllide P696-684 into chlorophyllide P688-676 in illuminated etiolated leaves. Photobiochem Photobiophys 8:85–96, 1984.
54. F Franck, GH Schmid. On the correlation between oxygen uptake in plastids of greening etiolated oat leaves and pigment photooxidation. Z Naturforsch 40c:699–704, 1985.
55. B Schoefs, F Franck. Role of NADPH-protochlorophyllide-reductase in photoprotection of newly formed chlorophyllide (Chlide). In: P Mathis, ed. Photosynthesis: from Light to Biosphere. Vol 3. Dordrecht: Kluwer Academic Publishers, 1995, pp 1009–1012.
56. F Franck, B Schoefs, X Barthélemy, B Mysliwa-Kurdziel, K Strzalka, R Popovic. Protection of native chlorophyll(ide) forms of photosystem II against photodamage during early stages of chloroplast differentiation. Acta Physiol Plant 17:123–132, 1995.
57. TE Redlinger, MC McDaniel. Light-mediated oxygen uptake measured in wheat etioplast. Plant Physiol 60:452–456, 1977.
58. C Lütz. Distribution of carotenoids and lipids in separated prolamellar bodies and prothylakoids of etioplasts from *Avena sativa* L. Z Pflanzenphysiol 104S:43–52, 1981.
59. WL Butler. Chloroplast development: energy transfer and structure. Arch Biochem Biophys 92:287–299, 1961.
60. M Bertrand, B Schoefs. Photosynthetic pigment metabolism in plants during stress. In: M Pessarakli, ed. Handbook of Plant and Crop Stress. 2nd ed. New York: Marcel Dekker, 1999, pp 527–543.
61. U Sperling, B van Cleve, G Frick, K Apel, G Armstrong. Overexpression of light-dependent PORA and PORB in plants depleted of endogenous POR by far-red light enhances seedlings survival in white-light and protects against photooxidative damage. Plant J 12:649–658, 1997.
62. F Franck, B Bereza, B Böddi. Protochlorophyllide-NADP$^+$ and protochlorophyllide-NADPH complexes and their regeneration after flash illumination in leaves and etioplast membranes of dark-grown wheat. Photosynth Res 59:53–61, 1999.
63. S Süzer, K Sauer. The site of photoconversion of protochlorophyllide to chlorophyllide in barley seedlings. Plant Physiol 48:60–64, 1971.
64. JK Hoober, WJ Stegemann. Control of the synthesis of a major polypeptide of chloroplast membranes in *Chlamydomonas reinhardtii*. J Cell Biol 56:1–12, 1973.
65. RG Alscher, SP Hawkes, K Sauer. The association of protein synthesis with protochlorophyllide holochrome regeneration in dark-grown barley leaves. Biochem Biophys Res Commun 73:240–247, 1976.
66. J Catsky, Z Sestak. Photosynthesis during leaf ageing. In: M Pessarakli, ed. Handbook of Photosynthesis. New York: Marcel Dekker, 1997, pp 633–661.
67. AA Shlyk, GY Savchenko, VG Averina. Investigation of the kinetics of photoreduction of protochlorophyllide in green leaves by the spectrofluorographic method. Biofizika 14:119–129, 1969.

68. B Schoefs, F Franck. Photosystem II assembly in 2-day-old bean leaves during the first 16 hrs of greening. C R Acad Sci Paris 313:441–445, 1991.

69. F Franck, X Barthélemy, K Strzalka. Spectroscopic characterization of protochlorophyllide photoreduction in the greening leaf. Photosynthetica 29:185–194, 1993.

70. F Franck, B Schoefs. Chlorophyll synthesis in relation to the assembly of photosystems. In: B Schoefs, F Franck, J Aghion, eds. Biology, Biochemistry and Molecular Biology of Photosynthesis. Bull Soc R Sci Liege 65:269–278, 1996.

71. F Franck, K Strzalka. Detection of photoactive protochlorophyllide-protein complex in the light during the greening of barley. FEBS Lett 309:73–77, 1992.

72. N Lebedev, MP Timko. Protochlorophyllide photoreduction. Photosynth Res 58:5–23, 1998.

73. R Schulz, H Senger. Protochlorophyllide reductase: a key enzyme in the greening process. In: M Ryberg, C Sundqvist, eds. Pigment-Protein Complexes in Plastids: Synthesis and Assembly. New York: Academic Press, 1993, pp 179–218.

74. G Mayer, H Bliedung, K Kloppstech. NADPH-protochlorophyllide oxidoreductase: reciprocal regulation in mono- and dicotyledonean plants. Plant Cell Rep 2:26–29, 1983.

75. GA Armstrong, S Runge, G Frick, U Sperling, K Apel. Identification of NADPH:protochlorophyllide oxidoreductases A and B: a branched pathway for light-dependent chlorophyll biosynthesis in *Arabidopsis thaliana*. Plant Physiol 108:1505–1517, 1995.

76. H Holtorf, S Reinbothe, C Reinbothe, B Bereza, K Apel. Two routes of chlorophyllide synthesis that are differentially regulated by light in barley. Proc Natl Acad Sci U S A 92:3254–3258, 1995.

77. JY Suzuki, CE Bauer. A prokaryotic origin for light-dependent chlorophyll biosynthesis of plant. Proc Natl Acad Sci U S A 92:3749–3753, 1995.

78. J Li, MP Timko. The *pc-1* phenotype of *Chlamydomonas reinhardtii* results from a deletion mutation in the nuclear gene for NADPH:protochlorophyllide oxidoreductase. Plant Mol Biol 30:15–37, 1996.

79. AJ Spano, ZII Ile, H Michel, DF Hunt, MP Timko. Molecular cloning, nuclear gene structure and developmental expression of NADPH:protochlorophyllide oxidoreductase in pea (*Pisum sativum* L.). Plant Mol Biol 18:967–972, 1992.

80. S Reinbothe, S Runge, C Reinbothe, B van Cleve, K Apel. Substrate-dependent transport of the NADPH:protochlorophyllide oxidoreductase into isolated plastids. Plant Cell 7:161–172, 1995.

81. H Holtorf, K Apel. The regulation of NADPH-protochlorophyllide oxidoreductases A and B in green barley plants kept under a diurnal light/dark cycle. Planta 199:289–295, 1996.

82. C Sironval, MR Michel-Wolwertz. Quelques particularités du métabolisme des chlorophylles. In: La Photosynthèse. Paris: Edition du CNRS, 1963, pp 317–342.

83. CC Rebeiz, CA Rebeiz. Ultrastructural study of chloroplast development during photoperiodic greening. In: G Akoyunoglou, H Senger, eds. Regulation of Chloroplast Differentiation. New York: Alan R Liss, 1986, pp 389–397.

84. U Sperling, F Franck, B van Cleve, G Frick, K Apel, G Armstrong. Etioplast differentiation in *Arabidopsis*: both *PORA* and *PORB* genes restore the prolamellar body and photoactive protochlorophyllide F655 to the *cop*1 photomorphogenic mutant. Plant Cell 10:283–296, 1998.

85. E Morren. Dissertation sur les feuilles vertes et colorés envisagées spécialement au point de vue de la chlorophylle et de l'érythophylle. Gand, Belgium, 1858.

86. E Selstam, A Widell. Characterization of prolamellar bodies, from dark-grown seedlings of Scots pine, containing light- and NADPH-dependent protochlorophyllide oxidoreductase. Physiol Plant 67:345–352, 1986.

87. C Forreiter, K Apel. Light-independent and light-dependent protochlorophyllide-reducing activities and two distinct NADPH-protochlorophyllide oxidoreductase polypeptides in mountain pine (*Pinus mungo*). Planta 190:536–543, 1993.

88. MR Michel-Wolwertz. Chlorophyll formation in cotyledons of *Pinus jeffreyi* during germination in the dark. Occasional accumulation of protochlorophyll(ide) forms. Plant Sci Lett 8:125–134, 1977.

89. PM Bramley. Carotenoid biosynthesis: a target site for bleaching herbicides. Biochem Soc Trans 22:625–629, 1994.

90. JD Weinstein, SI Beale. Enzymatic conversion of glutamate to δ-aminolevulinic acid in soluble extracts of the unicellular green alga, *Chlorella vulgaris*. Arch Biochem Biophys 237:454–464, 1985.

91. D-D Huang, W-Y Wang. Genetic control of chlorophyll biosynthesis: regulation of δ-aminolevulinic acid synthesis in *Chlamydomonas*. Mol Gen Genet 205:217–220, 1986.

92. C Sundqvist. Transformation of protochlorophyllide formed from exogenous δ-aminolevulinic acid in continuous light and in flashlight. Physiol Plant 22:147–156, 1969.

93. D. von Wettstein, S Gough, CG Kannangara. Chlorophyll biosynthesis. Plant Cell 7:1039–1057, 1995.

94. RR Klein, JE Mullet. Regulation of chloroplast-encoded chlorophyll-binding protein translation during higher plant chloroplast biogenesis. J Biol Chem 261:11138–11145, 1986.

95. RR Klein, HS Mason, JE Mullet. Light-regulated translation of chloroplast proteins. I. Transcripts of PsaA-PsaB, PsbA, and RbcL are associated with polysomes in dark-grown and illuminated barley seedlings. J Cell Biol 106:289–301, 1988.

96. F Franck, P Eulaffroy, R Popovic. Formation of long-wavelength chlorophyllde (Chlide695) is required for the assembly of photosystem II in etiolated barley leaves. Photosynth Res 51:107–118, 1997.

97. DJ Kyle, S Zalik. Development of the photochemical activity in relation to pigment and membrane protein accumulation in chloroplasts of barley and its virescens mutant. Plant Physiol 69:1392–1400, 1988.

98. SP Mayfield, WC Taylor. Carotenoid-deficient maize seedlings fail to accumulate light-harvesting chlorophyll *a/b* binding protein (LHCP) mRNA. Eur J Biochem 144:79–84, 1984.

99. JA Sullivan, JC Gray. Plastid translocation is required for the expression of nuclear photosynthesis genes in the dark and in roots of the pea *lip*1 mutant. Plant Cell 11:901–910, 1999.

100. R Barbato, ML Di Paolo, F Rigoni, GM Giacometti. Identification of a stable precursor of the apoprotein of the PSII antenna complex CP29. J Plant Physiol 136:468–471, 1990.

101. WC Taylor. Regulatory interactions between nuclear and plastic genomes. Annu Rev Plant Physiol Plant Mol Biol 40:211–233, 1989.

102. J Duggan, M Gassman. Induction of porphyrin synthesis in etiolated bean leaves by chelators of iron. Plant Physiol 53:206–215, 1974.

103. U Kittsteiner, H Brunner, W Rüdiger. The greening process in cress seedlings. II. Complexing agents and 5-aminolevulinate inhibit accumulation of *cab*-mRNA coding for the light-harvesting chlorophyll *a/b* protein. Physiol Plant 81:190–196.

104. U Oster, H Brunner, W Rüdiger. The greening process in cress seedlings. V. Possible interference of chlorophyll precursors, accumulated after thujaplicin treatment, with light-regulated expression of Lhc genes. J Photochem Photobiol 36:255–261, 1996.

105. U Johanningmeier. Possible control of transcript levels by chlorophyll precursors in *Chlamydomonas*. Eur J Biochem 177:417–424, 1988.

106. U Kittsteiner, H Brunner, W Rüdiger. The greening process in cress seedlings. I. Pigment accumulation and ultrastructure after application of 5-aminolevulinate and complexing agents. Physiol Plant 81:190–196, 1991.

107. R Guo, M Luo, JD Weinstein. Magnesium-chelatase from developing pea leaves. Characterization of a soluble extract from chloroplast and resolution into three required protein fractions. Plant Physiol 116:605–615, 1998.

108. U Johanningmeier, SH Howell. Regulation of light-harvesting chlorophyll-binding protein mRNA accumulation in *Chlamydomonas reinhardtii*. Possible involvement of chlorophyll precursors. J Biol Chem 259:13541–13549, 1984.

109. PA Castelfranco, JTO Jones. Protoheme turnover and chlorophyll synthesis in greening barley tissue. Plant Physiol 55:485–490, 1975.

110. D Jahn. Complex formation between glutamyl-tRNA synthase and glutamyl-tRNA reductase during tRNA-dependent synthesis of 5-aminolevulinic acid in *Chlamydomonas*. FEBS Lett 314:77–80, 1992.

111. PA Castelfranco, X Zheng. Regulation of 5-aminolevulinic acid biosynthesis in developing chloroplasts. IV. An endogenous inhibitor from the thylakoid membranes. Plant Physiol 97:1–6, 1991.

112. J Thomas, JD Weinstein. Measurements of heme efflux and heme content in isolated chloroplasts. Plant Physiol 94:1414–1423, 1990.

113. J Lascelles, TP Hatch. Bacteriochlorophyll and heme synthesis in *Rhodopseudomonas spheroides*: possible role of heme in regulation of the branched biosynthetic pathways. J Bacteriol 98:712–720, 1969.

114. JI Liro. Die Photochemische Chlorophyllbindung bei den Phanerogamen. Ann Acad Sci Fenn Ser A1:1–147, 1908.

115. EC Sisler, WH Klein. The effect of age and various chemicals on the lag phase of chlorophyll synthesis in dark-grown bean seedlings. Physiol Plant 16:315–322, 1963.

116. G Akoyunoglou, JH Argyroudi-Akoyunoglou. Effect of intermittent and continuous light on chlorophyll formation in etiolated plants at various age. Physiol Plant 22:288–295, 1966.

117. AN Misra, SM Sahu, F Dilnawaz, P Mohapatra, M Misra, NK Ramaswamy, TS Desai. Photosynthetic pigment-protein content, electron transport activity and thermoluminescence properties of chloroplasts along the developmental gradient in greening wheat (*Triticum aestivum* L.) leaves. In: G Garab, ed. Photosynthesis: Mechanisms and Effects. Vol 4. Dordrecht: Kluwer Academic Publishers, 1998, pp 3179–3182.

118. PA Castelfranco, PM Rich, SI Beale. The abolition of the lag phase in greening cucumber cotyledons by exogenous δ-ALA aminolevulinic acid. Plant Physiol 53:615–618, 1974.

119. K Apel, K Kloppstech. The effect of the light on the biogenesis of the light-scattering chlorophyll *a/b* protein. Evidence for the requirement of chlorophyll *a* for the stabilization of the apoprotein. Planta 150:426–430, 1980.

120. J Bennett. Biosynthesis of the light-harvesting chlorophyll *a/b* protein. Polypeptide turnover in darkness. Eur J Biochem 118:61–70, 1981.

121. FG Plumley, GW Schmidt. Light-harvesting chlorophyll *a/b* complexes: interdependent pigment synthesis and protein assembly. Plant Cell 7:689–704, 1995.

122. B Schoefs, M Bertrand, Y Lemoine. Pigment biosynthesis during the first photoperiod of greening (16 h) of 2-d-old and 10-d-old bean leaves. In: G. Garab, ed. Photosynthesis: Mechanisms and Effects. Vol III. Dordrecht: Kluwer Academic Publishers, 1998, pp 3277–3280.

123. LJ Rodgers, SPJ Shah, TW Goodwin. Compartmentation of terpenoid biosynthesis in green plants. Biochem J 114:395–405, 1966.

124. A Heintze, J Görlach, C Leuschner, P Hoppe, P Hagelstein, D Schulze-Siebert, G Schulze. Plastidic isoprenoid synthesis during chloroplast development. Plant Physiol 93:1121–1127, 1990.

125. HI Virgin. Action spectrum for the elimination of the lag phase in chlorophyll formation in previously dark-grown leaves of wheat. Physiol Plant 14:439–452, 1961.
126. H Kasemir. Light control and chlorophyll accumulation in higher plants. In: W Shropshire, H Mohr, eds. Encyclopedia of Plant Phsyiology. 2nd ed. Vol 16B. Berlin: Springer Verlag, 1983, pp 662–667.
127. HI Virgin, HS Egneus. Control of plastid development in higher plants. In: W Shropshire, H Mohr, eds. Encyclopedia of Plant Phsyiology. 2nd ed. Vol 16B. Berlin: Springer Verlag, 1973, pp 289–311.
128. SL Berry-Lowe, RB Meagher. Transcriptional regulation of a gene encoding the small subunit of ribulose-1,5-bisphosphate carboxylase in soybean tissue is linked to the phytochrome response. Mol Cell Biol 5:1910–1917, 1985.
129. WB Terzaghi, AR Cashmore. Light-regulated transcription. Annu Rev Plant Physiol Plant Mol Biol 46:445–474, 1995.
130. J von Lintig, R Welsch, M Bonk, G Guiliano, A Batschauer, H Kleining. Light-dependent regulation of carotenoid biosynthesis occurs at the level of phytoene synthase expression and is mediated by phytochrome in *Sinapis alba* and *Arabidopsis*. Plant J 12:625–634, 1997.
131. BA Horwitz, WF Thompson, WR Briggs. Phytochrome regulation of greening in *Pisum*. Chlorophyll accumulation and abundance of mRNA for the light-harvesting chlorophyll *a/b* binding proteins. Plant Physiol 86:299–305, 1988.
132. O Bougry, B Grimm. Members of a low-copy number gene family encoding glutamyl-tRNA reductase are differentially expressed in barley. Plant J 9:867–878, 1996.
133. PE Jensen, RD Willows, BL Petersen, UC Votknecht, BM Stummann, CG Kannangara, D von Wettstein, KW Henningsen. Structural genes for Mg-chelatase subunits in barley: *xantha-f*, *-g* and *-h*. Mol Gen Genet 250:383–394, 1996.
134. JA Kreps, SA Kay. Coordination of plant metabolism and development by the circadian clock. Plant Cell 9:1235–1244, 1997.

# 13

# Structure and Function of Photosynthetic Membranes in Higher Plants

## Ilia D. Denev and Ivan N. Minkov

*University of Plovdiv, Plovdiv, Bulgaria*

## I. INTRODUCTION

The biological conversion of light quanta energy into chemical energy is known as photosynthesis. The photosynthesis takes place in the chloroplasts. Electron micrographs (Figure 1) of chloroplasts of higher plants revealed that they consist of a double membrane envelope enclosing a complex of inner membranes known as thylakoids. The thylakoid membrane system of one chloroplast is believed to be formed from one continuous membrane, which divides the inner chloroplast volume into two separate spaces: extrathylakoid (stroma) and intrathylakoid (thylakoid lumen) [1–3]. The biochemical part of photosynthesis takes place in the stroma, which contains all the enzymes of the $CO_2$ fixation pathway.

The thylakoids of higher plant chloroplasts are probably the most complexly organized of all biological membranes. Their main function is to capture light quanta and to drive series of redox reactions that produce ATP and oxygen and reduce ferredoxin. The thylakoid membranes are 5–7 nm thick and consist of a lipid bilayer where complexes of proteins, pigments, and some other minor components are situated. The predominant parts of thylakoid proteins and pigments are organized into intrinsic membrane-spaning supramolecular complexes. Their transverse and lateral distribution is highly asymmetric, and this is one of the basic features of photosynthetic membranes [4]. The main molecular supracomplexes in thylakoid membranes are the complex photosystem I, the complex photosystem II, the cytochrome $b_6f$ complex, and ATP synthase [5].

## II. COMPLEX PHOTOSYSTEM II

### A. Supramolecular Organization of Photosystem II

Photosystem II (PSII) is a multisubunit membrane protein complex that catalyzes the light-driven oxidation of water and reduction of plastoquinone. PSII contains about 25 polypeptides (Figure 2) [6]. The minimal PSII supramolecular complex, capable of carrying out stable charge separation, is the PSII reaction center (RC) complex. The RC complex contains at least six proteins: the D1, D2, cytochrome $b_{559}$ (two polypeptides), PsbI, and probably PsbL polypeptides [7]. These proteins bind several cofactors: tetra-Mn cluster, nonheme Fe, two pheophytins and two quinones ($Q_A$ and $Q_B$) per RC complex, six chlorophylls, and two pheophytins [8]. Two of the RC chlorophylls make up the chlorophyll special pair (P680) and two accessory chlorophylls, whose monomers presumably participate in electron transfer between the

**Figure 1** Transmission electron micrograph of a *Haberlea rhodopensis* (Friv.) chloroplast (magnification ×
20,000).

P680 and pheophytin. The second pair of chlorophylls is coordinated by symmetry-related histidine
residues at the periphery of the RC complex. These peripheral accessory chlorophylls are excitonically
coupled to the antennae chlorophylls and to the P680 [8–10]. In addition to their light-harvesting function, it was proposed that the peripheral accessory chlorophyll(s) ($Chl_Z$) function in an electron transfer
cycle around PSII (including $Q_B$ and cytochrome $b_{559}$), which protects PSII from accumulating long-lived
$P680^+$ states and consequently from D1 protein degradation [10].

## B.  Subunits of PSII Reaction Center Complex

### 1.  *PsbA-D1 Protein*

D1 protein is a highly conserved RC protein with a molecular mass of about 38 kDa depending on the
species [6]. From hydropathy plots and comparison with the L subunit of the reaction center of purple
bacteria, it is assumed that D1 contains five transmembrane helices (I to V) and two surface helices between II and IV (luminal) and between IV and V (stromal). The N-terminal threonine can be reversibly
phosphorylated [11]. D1 protein binds the majority of cofactors involved in PSII-mediated electron

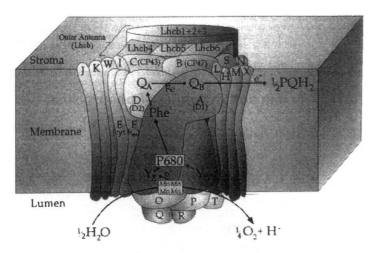

**Figure 2**  Scheme for PSII of higher plants emphasizing subunit composition and primary and secondary
electron transfer steps that occur in the reaction center. (From Ref. 6.)

transport:

Tyr161 is Yz.
His198 binds P680.
Tyr126, Tyr147, Ala150, and Glu130—probably pheophytin.
Tyr254, Phe255, and Gly256 probably interact with $Q_B$.
Asp170, Glu189, Gls165, Ala344, His109, His332, and His377 bind tetramanganese cluster.
His215 and His272 probably bind nonheme iron [6,12,13].

The tight involvement of D1 proteins in primary photochemistry makes them a major target of photoinduced damage [14,15]. This damage leads to photoinhibition and reduction of photosynthetic efficiency. The damaged D1 is a subject of proteolytic degradation [16]. The major cleavage site on D1 is on the stromal side of the thylakoid membrane [17]. However, the D1 protein is the protein in PSII with the highest turnover rate [18]. This phenomenon might be linked to the requirement to repair PSII after it has been damaged by photoinhibitory light [19].

## 2.  PsbD-D2 Protein

The D2 protein is homologous to the D1 protein, although it has a higher molecular mass of about 39.5 kDa. Like D1, the D2 protein consists of five transmembrane helices and has similar surface helices [6,13]. N-terminal threonine can also undergo reversible phosphorylation [6,11].

The D2 protein binds lesser cofactors involved in primary electron transport, although it does contain inactive cofactors [6]. The D2 protein binds via His198 the P680 and the $Q_A$, probably by Thr218, Phe253, and Trp254 [13]. The His215 and His269 probably form ligands for the nonheme iron and Glu69 forms a ligand for the Mn cluster [7].

## 3.  PsbE and PSbF-Cytochrome $b_{559}$ Proteins

The PsbE and PsbF proteins are the α and β subunits of cytochrome $b_{559}$. The proteins have molecular masses of 9.3 and 4.4 kDa, respectively [20]. Hydropathy plots revealed single transmembrane helices that each of the proteins forms [21]. The two proteins are closely associated with D1 and D2 proteins and probably form a heterodimer that binds a heme via the single histidine residue contained in their sequences [6,22].

There have been many speculations about the function of Cyt $b_{559}$. It was found that Cyt $b_{559}$ is more closely associated with D1 than with the D2 protein, and it was proposed that Cyt $b_{559}$ may have a function in an electron transfer cycle around PSII (including $Q_B$, and $Chl_Z$) that protects PSII from accumulation of long-lived $P680^+$ states and consequently from D1 protein degradation [10].

## 4.  PsbI and PsbL Proteins

Both PsbI and PsbL are small proteins with molecular masses of 4.2 and 4.4 kDa, respectively. Their amino acid sequences are highly conservative and reveal single transmembrane helices. They are located very close to the D1, D2 proteins but their functions are not clear [23,24]. Perhaps PsbI protein binds chlorophyll [20], and PsbL seems to be required for the normal function of the $Q_A$ site [25].

# C.   Subunits of the PSII Light-Harvesting Apparatus

Each reaction center is connected to a large set of chlorophyll proteins, which together form the light-harvesting apparatus. It consists of the inner antenna and light-harvesting complex (LHCII). The inner antenna is built up from CP47 (CPa-1, psbB) and CP43 (CPa-2, psbC) proteins, about 10–15 chlorophyll molecules, and carotenoids (β-carotene and lutein) [26].

## 1.  PsbB-CP47 and PsbC-CP43 Proteins

CP47 and CP43 proteins are important constituents of the RC core complex. Their absence can have a serious impact on both PSII assembly and water oxidation function [27]. CP47 and CP43 proteins have about 500 and 470 amino acid residues and their molecular masses are 56 and 50 kDa, respectively [6]. Their structures are similar in many ways. They both possess six transmembrane helices and large (200 and 150 amino acids, respectively) luminal loops between helices V and VI [26]. Both proteins contain a considerable number of histidine residues, which bind chlorophyll and β-carotenes. For example, CP47 binds 15 chlorophylls *a* and 3 β-carotenes [14,26]. It is thought that these pigments form a core light-harvesting system for the reaction center. However, the luminal loops probably play some role in water oxidation reactions [6].

Despite the similarities, CP43 differs from CP47 in two aspects:

1. The CP43 N-terminal threonine can be reversibly phosphorylated [11].
2. The CP43 association with RC is weaker and it can be removed from the isolated core to yield a CP47-RC complex [28].

## 2. LHCII Proteins

At least six LHCs are energetically connected with PSII—LHCII a, b, c, d. Their proteins are encoded by the nuclear gene family lhcIIb [29]. The major LHCII, LCHIIb, is a trimeric complex, binding approximately 65% of Chl *a*, and 40% of Chl *b* is in LHCIIb [30,31]. It plays a crucial role in capturing light energy for photosynthesis and in regulation of energy flow within the photosynthetic apparatus. Two populations of LHCIIb were identified, containing a mixture of polypeptides originally designed as "LHCII-27" and LHCII-25" and subsequently identified as the *lhcb1* and *lhcb2* gene products [32]. The proximal antenna consists almost exclusively of Lhcb1 and its associated pigments, while the peripheral antenna contains both Lhcb1 and Lhcb2 [33]. Changes in LHCIIb content in response to the environment reflect varying quantities of the peripheral antenna per PSII. There are no changes in the size of the proximal antenna [34]. Both Lhcb1 and Lhcb2 proteins bind chlorophyll and xanthophyll chromophores. It is commonly believed that in pigment binding and the light-harvesting function the various Lhcb1 and Lhcb2 protein constituents of LHCIIb are structurally and functionally equivalent [35]. However, Walter and Horton [36] found that Lhcb2 protein, isolated from low light–grown plants, specifically binds at least one additional chlorophyll *a* compared with Lhcb1 and alters energy transfer characteristics. Therefore, the differences in the functioning of LHCIIb from high and low light–grown plants are a direct consequence of the changes in polypeptide composition.

The other LHC proteins, a, c, and d (also known as CP29, CP26, and CP24), bind about 5% of PSII chlorophylls and probably link LHCIIb with the inner antenna [31].

## D. Subunits of Water Oxidation Complex

Another important group of proteins are those involved in water oxidation processes. Despite the fact that ligands for Mn binding have been identified with the D1 and D2 proteins, other proteins are also likely to be involved. It was already mentioned that CP47 and CP43 proteins have very large hydrophilic loops exposed on the luminal surface [26]. There is every reason to believe that one or both of these loops maintain the structural integrity of the Mn cluster. In addition, PSII contains a 33-kDa protein that, together with a few smaller proteins, acts to stabilize the Mn cluster.

## 1. PsbO Protein

The product of the *psbO* gene, often referred to as the 33-kDa manganese-stabilizing protein, actually has a molecular mass of about 26.5 kDa [37]. It is a hydrophilic protein with a high content of β-sheets [38]. Cross-linking studies indicate that it is closely localized to the luminal loop of CP47 [39] and to the PsbE and PsbI proteins [40]. The 33-kDa protein does not bind Mn directly but the tetramanganese cluster itself is a nonequilibrium system. Therefore, it requires a suitable protein matrix for its stabilization and the PsbO protein plays a crucial role in stabilization and maintenance of an optimal environment for water oxidation [41].

## 2. Other Proteins

Three other proteins were found to be in close contact with PsbO; these are the PsbP, PsbQ, and PsbR extrinsic proteins, localized in the thylakoid lumen. The PsbP and PsbQ proteins have molecular masses of about 23 and 17 kDa, respectively. Their functions seem to be to optimize the $Ca^{2+}$ and $Cl^-$ levels needed for water oxidation. Both proteins were located with the PsbO protein [6,12]. The PsbR protein was also found in the vicinity of the water splitting site. Its molecular mass is 10.2 kDa [42] but its function is unclear.

In addition to the proteins already mentioned, several others were found. They have a low molecular mass between 3 and 8 kDa, one transmembrane helix, and unknown function.

## E.  Electron and Proton Transport Within Photosystem II

Photosystem II functions as a water-plastoquinone oxidoreductase. The primary electron donor of PSII is P680. Following excitation, P680 transfers an electron to pheophytin and is subsequently reduced by $Tyr_Z$. The $P680^+/P680$ midpoint potential is unusually oxidizing (1.2 Ev) and can drive the oxidation of water (0.8 Ev). In contrast to the primary electron donor Chls of all other reaction center types, only the midpoint potential of the PSII primary donor Chl(s) is shifted in a positive direction relative to free Chl. It is evident that the structural organization of P680 and its interactions with the D1 and D2 proteins determine the unusual redox properties of P680. The structural organization of P680 is not known; however, P680 has been reported to have properties of a Chl monomer, a Chl dimer, as well as a chlorin multimer (reviewed in Ref. 10).

Most of the data suggest that P680 is a Chl dimer [10,43]. However, the absence of an appreciable red shift in the $Q_y$ absorbance band as expected for a chlorophyll dimer and the Stark effect and hole burning measurements do not show features predicted for a chlorophyll dimer (reviewed in Ref. 10). To account for these conflicting observations, Schelvis et al. [43] proposed that P680 is a chlorophyll dimer with monomeric properties. The monomeric properties are attributed to an antiparallel or asymmetric orientation of the chlorophyll $Q_Y$ transition moments.

Alternatively, P680 has been proposed to be a multimer of excitonically interacting chlorins (including the chlorophyll spectral pair, chlorophyll monomers, and pheophytin) [44]. Following optical excitation, the P680 excited state rapidly equilibrates with most of the pigments in the reaction center and charge separation occurs from the equilibrated state. Bleaching of the pheophytin $Q_X$ transition also occurs within 300 fsec after excitation of PSII reaction centers [45]. These observations, in conjunction with theoretical predictions of the dipole-dipole coupling strengths between the accessory chlorophyll monomers, pheophytins, and the $Chl_{SP}$, suggest that all three chlorin groups interact excitonically [45]. Significantly, the multimer model is not particularly dependent on the spatial organization of the pigments as long as the chlorins have overlapping $Q_Y$ transitions and are sufficiently close to allow rapid (<100 fsec) excited state equilibration. This results in a spatially heterogeneous excited state.

When P680 absorbs a photon, it donates $e^-$ to the first stable electron acceptor, pheophytin (Phe). The state $P680^+Phe^-$ is referred to as the primary radical pair. It has an electrochemical potential of 1.7 Ev [46]. For efficient charge separation the back reaction with $P680^+$ must be limited, which is achieved mostly by rapid transfer of $e^-$ to the second electron acceptor $Q_A$. This is a plastoquinone molecule, which is tightly bound to the D2 protein and functions as a one-electron carrier and does not normally undergo protonation. In contrast, the second plastoquinone, D1-bound $Q_B$, can accept two electrons and two protons. Therefore, for the complete reduction of $Q_B$ two primary charge separations take place at P680. In its fully reduced state $Q_B$, this (already plastoquinol) molecule is released from the binding site on the D1 protein into the lipid matrix of the membrane [6].

The described two-electron acceptor site processes contrast with a four-electron gate on the oxidizing (donor) site of PSII. Here four turnovers of primary charge separation are needed to create the four oxidizing equivalents required for the conversion of two molecules of water to a molecule of oxygen [41]. The model of water splitting (Figure 3) is known as the S-state model [47]. However, the exact location and organization of the catalytic center of the complex are still not known [6]. The preferred model, based on data from x-ray and resonance techniques, is that the cluster is composed of two di-$\mu$-oxobridged dinuclear Mn units linked via an oxybridge and a pair of amino acids providing carboxalato bridges [48]. The reaction scheme of S-state transitions suggests that D-Tyr161 ($Y_Z$) in the oxidized state is capable of extracting protons as well as electrons from water molecules bound to the Mn cluster [41,49]. The abstraction of $4H^+$ and $4e^-$ by four turnovers of $Y_Z$ therefore maintains electroneutrality while at the same time accumulates the oxidizing potential to create dioxygen on the fourth turnover as required by the S-state model [6,47]. The extracted protons are released in the thylakoid lumen when the electrons are transferred via $Y_Z$ to P680 to compensate for electron deficiency, which is a result of $Q_B$ reduction [6].

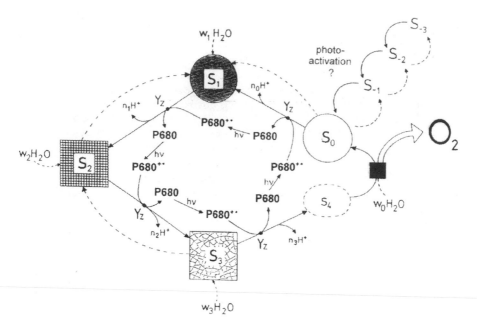

**Figure 3**  Functional scheme of photosynthetic water oxidation. (From Ref. 41.)

## III.  COMPLEX CYTOCHROME $b_6f$

### A.  Supramolecular Organization

The cytochrome $b_6f$ complex acts as plastoquinol:plastocyanin oxidoreductase. The complex contains as many as seven polypeptide subunits. The four "large" subunits of 18–32 kDa, products of the *petA–D* gene, are cytochrome $f$ (Cyt $f$), cytochrome $b_6$ (Cyt $b_6$), the Rieske iron-sulfur protein, and subunit IV. They bind a $c$-type heme, two $b$-type hemes, and a high-potential iron-sulfur center, respectively. Subunit IV does not contain a prosthetic group [50]. The latter three subunits are involved in the binding of the electron donor plastoquinol. There are three smaller subunits, each with one transmembrane helix [50,51]. Furthermore, a single chlorophyll $a$ molecule with so far unknown function is part of the complex [52,53].

The anticipated approximate cross-section view of the cytochrome $b_6f$ complex is presented in Figure 4. It is based on the solved structure of Cyt $f$ [54], dimensions of the extrinsic domain of the Rieske

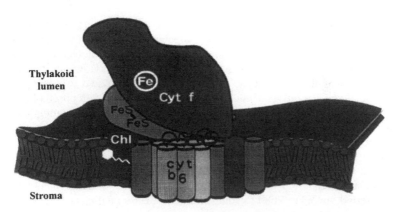

**Figure 4**  Anticipated cross section of the cytochrome $b_6f$ complex orthogonal to the plane of the membrane. (According to Ref. 51.)

iron-sulfur protein [51,55], and the consensus membrane folding pattern of Cyt $b_6$ and subunit IV [51,56–58]. The monomeric $b_6f$ complex is inferred to contain 11 transmembrane helices: 1 (Cyt $f$) helix, 4 (Cyt $b_6$) helices, 3 (subunit IV) helices, and 3 helices of three smaller subunits. Several types of data support the idea that in vivo cytochrome $b_6f$ is associated in a dimer [51].

## B.   Subunits of Cytochrome $b_6f$ Complex

### 1.   *PetA Subunit (Cyt f)*

The largest cytochrome $b_6f$ subunit is the organelle *petA* gene product, Cyt $f$, with a molecular mass of about 32 kDa [59]. It covalently binds $c$-type cytochrome and contains a docking site of the electron acceptor plastocyanin. Martinez et al. [54,60] reveled that Cyt $f$ has an elongated ($2.5 \times 3.5 \times 75$ nm) structure with a large and a small domain with a predominant β-strand motif of the large domain. The extended structure may be necessary for Cyt $f$ to make contact with the plastocyanin. The $c$-heme is ligated by the N-terminal α-amino group, which is unique to all heme proteins. Another unusual structure is an internal extended (1.1 nm) linear water chain, which may function as a luminal exit port for protons translocated by the Cyt $b_6f$ complex. In addition, it was found that the interface between large and small domains contains five basic residues: Lys58, 65, and 66 (large domain) and Lys187 and Arg209 (small domain). Because the complementary plastocyanin has two corresponding regions with negative surface-situated residues, it was inferred that in the beginning successful docking of Cyt $f$ and plastocyanin involves a long-range electrostatic attraction. This is followed by rearrangement of the protein structure during which metal centers become close enough for rapid intracomplex electron transfer [61].

### 2.   *PetC Subunit*

The Rieske protein is a product of the nuclear *petC* gene. The product PetC has a molecular mass of about 19 kDa [62]. It binds a high-potential iron-sulfur center. Information about its secondary structure was derived from the known structure of the Cyt $bc_1$ mitochondrial protein. However, despite the similarities in the Rieske protein family, the existing differences make the complete Rieske–Cyt $b_6f$ protein secondary structure still unclear [51]. The existing data based on the circular dichroism spectrum suggest that Rieske protein has only one α-helix, while β-sheets are 52–60%, β-turns 7–25%, and random coils up to 40% [51,63]. The crystal structure of the luminal part of Rieske protein has been determined [64]. It shows two domains: a small "cluster-binding" subdomain that comprises the 2Fe-2S cluster and a large subdomain. Two cysteines and two histidines coordinate the Fe-S cluster. The two histidines are exposed to the solvent, whereas the rest of the cluster is shielded by two loops covalently linked by a disulfide bridge and by a third proline loop [64]. In the arrangement of the Rieske protein along the luminal surface of the membranes, the Fe-S cluster faces the plastoquinol binding site, which would be formed by cytochrome $b_6$, subunit IV, and the Rieske protein motifs. It was proposed that a flexible hinge at the luminal side of its helix could allow the Rieske protein to orient in two configurations with respect to the membranes: a "relaxed" configuration in which it is close enough to the cytochrome $f$ in the lumen to allow electron transfer between the two and a "tight" configuration in which it is close enough to cytochrome $b_6$ at the thylakoid membrane surface to allow electron transfers from the plastoquinol binding site [65].

### 3.   *PetB Subunit*

The second larger Cyt $b_6f$ complex protein is cytochrome $b_6$. It is a 24-kDa product of the organelle *petB* gene. Cyt $b_6$ binds two $b$-type hemes. It is capable of interacting with plastoquinol as well as with products of its oxidation—semiquinone and plastoquinone [51]. Some authors suggest that it could also bind the chlorophyll $a$ molecule [50]. The consensus membrane-folding pattern of Cyt $b_6$ reveals four transmembrane helices.

### 4.   *Other Subunits*

The structure and function of other proteins (subunit IV, PetG, PetL, and PetM) are unclear. It was found that subunit IV is a product of the *petD* organelle gene [51] with a molecular mass of 17.5 kDa and three transmembrane helices. The three other proteins have molecular masses between 3.4 and 4 kDa and one transmembrane helix. They appear to be necessary for stable assembly of the complex [51].

## C.   Electron and Proton Transport Within the Cytochrome $b_6f$ Complex

The cytochrome $b_6f$ complex mediates the electron transport between photosystem II and photosystem I via plastoquinol to plastocyanin and cyclic electron flow around PSI via ferredoxin to plastocyanin. The binding of plastoquinol is a function of Cyt $b_6$, subunit IV, and Rieske proteins [51]. The oxidation of plastoquinol is described by the concept of the Q-cycle. According to this concept, cytochrome $b_6f$ acts as a plastoquinol oxidizing and plastoquinone reducing enzyme. Although the details of this dual function are still debated [66], in general it means that cytochrome $b_6f$ binds plastoquinol and plastoquinone. For every electron transported along the linear path from plastoquinol to plastocyanin, another one is used for reduction of the plastoquinone. This process is coupled with injection of two protons in the thylakoid lumen. The oxidation of a second molecule of plastoquinol leads to complete reduction of bound plastoquinone to plastoquinol via acceptance of two protons from the stromal site of thylakoid. In summary, the linear transport of two electrons would lead to release of four protons in the thylakoid lumen, oxidation of two molecules of plastoquinol, and coupled reduction of one molecule of plastoquinone to plastoquinol. Under natural conditions, depending on light intensities, the ratio can vary and a strong decrease appears under high light [66]. Mechanistically, this could be explained by proton channels connecting the plastoquinol binding site alternatively to the luminal or stromal side of the cytochrome $b_6f$ complex, giving rise to a proton slip reaction at high transmembrane $\Delta$pH [66]. This scheme is supported by several discoveries:

> The appearance of an internal five-water chain, which has the properties of a proton wire and serves as a long-distance proton thanslocation line, was suggested by Martinez et al. [60] and Cramer et al. [51].
> The existence of two binding places at the Qo-plastoquinol/plastoquinone binding site was proved by use of inhibitors [67].
> A flexible hinge was found in the Rieske protein that allows it to orient in two configurations: the first close enough to cytochrome $b_6$ to allow electron transfers from the plastoquinol binding site and the second close enough to the cytochrome $f$ in the lumen to allow electron transfer to its heme was also found [65].

## IV.   COMPLEX PHOTOSYSTEM I

## A.   Supramolecular Organization of Photosystem I Complex

Photosystem I (PSI) is a pigment-protein complex that functions as a plastocyanin:ferredoxin oxidoreductase [68]. The holo-PSI complex is composed of light-harvesting complex I (LHCI) and a core complex. The holo-PSI contains up to 17 subunits, 100–200 molecules chlorophyll per P700 (chlorophyll $a/b$ ratio is greater than 5), 10–15 molecules $\beta$-carotene, 2 phylloquinones, and 3 (4Fe-4S) clusters [68,69]. Some lipids may also be integral components of the PSI complex [68]. Limited information about LHCI is available. It is accepted that two sets of light-harvesting pigment-protein complexes deliver energy to PSI. LHCI is specific for PSI and probably bound to PSI in a fixed stoichiometry [68]. LHCII serves as a light-harvesting complex for PSII as well as PSI and its association with PSI and PSII is variable [70]. Figure 5 summarizes the current knowledge of the PSI subunit structure.

The core complex drives the electron transfer from plastocyanin, which is located in the luminal space, to ferredoxin, which is situated in the chloroplast stroma. The core complex combines both integral and peripheral subunits. It consists of 13 subunits—from A to N (gene products of *psaA* to *psaN*) without M, which was found only in cyanobacteria [69].

## B.   Subunit Organization of the PSI Reaction Center (Core) Complex

### 1.   PSI-A and PSI-B Subunits

The backbone of PSI is a heterodimer consisting of the PSI-A and PSI-B subunits. It is known as P700–chlorophyll *a*–protein1 and, together with PSI-C, builds all the core complex pigments and electron transfer subunits [68,69].

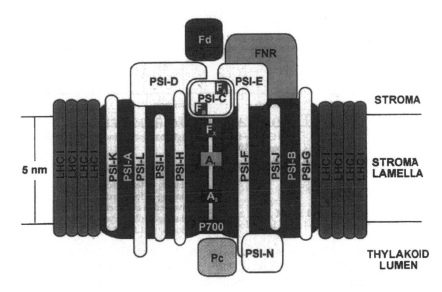

**Figure 5**  Schematic model of the PSI reaction center complex in higher plants. (From Ref. 68.)

PSI-A and PSI-B are products of chloroplast genes *psaA* and *psaB*. They are the largest PSI proteins with molecular masses of about 84 and 83 kDa, respectively. They are integral membrane proteins and have 11 transmembrane helices each. They bind P700, a chlorophyll *a* dimer, which is the primary electron donor in PSI. They also bind the primary acceptor $A_0$, which is a chlorophyll *a* monomer; the $A_1$ acceptor, vitamin $K_1$ (phylloquinone); and the Fx electron acceptor, which is an iron-sulfur cluster (4Fe-4S) [68]. The PSI-A, PSI-B heterodimer also binds about 100 molecules of chlorophyll *a* and 10–15 molecules of β-carotene, the internal antenna [69].

## 2.  Other Integral Subunits

The other, smaller integral subunits are PSI-G, PSI-I, PSI-J, PSI-K, and PSI-L. PSI-I and PSI-J are chloroplast-encoded proteins with molecular masses of 4 and 5 kDa, respectively. They have a single transmembrane helix. The rest of the proteins are nuclear encoded and have masses of 9, 11, and 18 kDa (PSI-K, PSI-G, and PSI-L) and two transmembrane helices [69]. Functions of all these proteins in higher plants are unknown.

There is one more protein, which is thought to be integral—PSI-F [71]. It is a nuclear-encoded protein with a molecular mass of about 17 kDa. Its probable function is as a part of the docking for plastocyanin [72,73]. However, the PSI-F protein has a large luminal domain [74].

## 3.  Lumenal-Site Subunit

The only subunit entirely located on the luminal side is the nuclear-encoded PSI-N. This is a relatively small protein with a molecular mass of 10 kDa. Its function is not clear [69].

## 4.  Stromal-Site Subunits

Subunits PSI-C, PSI-D, and PSI-E are localized on the stromal side of the PSI complex.

PSI-C is a product of the *psaC* chloroplast gene with a molecular mass of about 9 kDa. It binds two iron-sulfur clusters (4Fe-4S), $F_A$ and $F_B$, and is clearly involved in the transport of electrons [68].

The PSI-D protein with a molecular mass of 18 kDa is nuclear encoded. It is a docking site of ferredoxin, flavodoxin, and it is also responsible for the binding of PSI-C to the core complex [75–78].

PSI-E is a 10-kDa nuclear-encoded protein. According to some authors [76,77], it is involved in binding of ferredoxin and flavodoxin. Andersen et al. [75,79] found that the PSI-E subunit in barley interacts with ferredoxin:NADP$^+$ oxidoreductase. The cyclic electron transport is affected in PSI-E PSI [80].

The PSI-H subunit is more hydrophobic than the other three stromal exposed PSI subunits and may in fact be an integral membrane protein. It is a nuclear-encoded, 11-kDa protein with unknown function. There is a speculation that PSI-H could be responsible for the stronger binding of PSI-D [76].

## C.  Subunits of the LHCI Complex

Light-harvesting PSI complex has four distinct polypeptide components present in equimolar ratios [68]. All the proteins are products of nuclear genes (*lhca1*, *lhca2*, *lhca3*, and *lhca4*) with molecular masses of 22 kDa for Lcha1 and Lhca4, 23 kDa for Lhca2, and 25 kDa for Lhca3. The consensus membrane-folding pattern of all LHCI proteins revealed three transmembrane helices [69]. The polypeptides bind Chl *a*, Chl *b*, and xanthophyll. In contrast to LHCII, proteins of LHCI are tightly bound to the PSI core complex, forming dimers with two different emission maxima at 680 nm (Lhca2 and 3) and at 730 nm (Lhca1 and 4) [68]. The exact localization of LHCI on PSI is still not completely elucidated. More investigations are also needed to determine the subunit composition and stoichiometry of the LCHI dimer as well as the points of attachment of these dimers [68].

## D.  Electron Transport Within Photosystem I

Photosystem I functions as a plastocyanin:ferredoxin oxidoreductase. The primary electron donor of PSI is P700, which is a chlorophyll *a* dimer. Following excitation, P700 transfers an electron to the $A_0$ acceptor (chlorophyll *a* monomer) and is subsequently reduced by plastocyanin. The electron transport within PSI is so fast that there is a little fluorescence from antenna chlorophyll at physiological temperatures [5]. The next acceptor is $A_1$, vitamin $K_1$ (phylloquinone); then the electron passes through the $F_X$ electron acceptor, which is an iron-sulfur cluster (4Fe-4S), and by alternative working two iron-sulfur clusters (4Fe-4S), $F_A$ and $F_B$. The last step is reduction of ferredoxin docked on the stromal site of the membrane [69]. Electrons from plastocyanin compensate the electron deficiency of P700. The plastocyanin is a water-soluble 10.5-kDa protein and it is a mobile carrier of electrons between the two membrane-embedded supramolecular complexes—cytochrome $b_6f$ and PSI [4,81].

According to the available structural data, the existence of two probable electron transfer sites in plastocyanin has been inferred. One site is the so-called hydrophobic patch around His87 (a copper-coordinating residue on the "north pole" of the protein), which acts as the active site for redox interactions with PSI [82]. The other one is referred to as the acidic patch around Tyr83 on the "east face" of the protein, which could act as the entry port of electrons coming from cytochrome *f* [61]. PSI participates in two types of electron transport, cyclic and noncyclic. It is generally thought that grana-localized PSI participates in noncyclic electron transport from water via PSII, plastoquinone pull, cytochrome $b_6f$, and plastocyanin to ferredoxin and consequently to NADPH. The stroma-lamellae–localized PSIs participate in cyclic electron flow via reduced ferredoxin, cytochrome $b_6f$, plastocyanin, and back to P700 [4].

## V.  THE CHLOROPLAST ATP SYNTHASE

The chloroplast ATP synthase belongs to the family of $F_1$-type adenosinetriphosphatases (ATPases), which are also present in bacteria and mitochondria [83]. It generates ATP from ADP and inorganic phosphate (Pi) using energy delivered from a trans-thylakoid electrochemical proton gradient $\Delta\mu H^+$ [84].

Our view of the supramolecular structure of ATP synthase is based mainly on investigations of bacterial and mitochondrial ATP synthases. Because both the primary sequences of their subunits and their functional characteristics are very conservative, it is reasonable to infer the supramolecular organization of the chloroplast ATP synthase from data obtained with the enzyme from other sources [69].

The ATP synthase has long been described as the association of two distinct sectors: a membrane-embedded $CF_0$ and a catalytic sector $CF_1$ located on the stromal surface of the thylakoid membranes. The whole enzyme comprises nine subunits. The $CF_0$ is built up by four subunits (I, II, III, and IV) in an assumed stoichiometry of 1:1(2):9–12:1. The $CF_1$ is built of five subunits: $\alpha$, $\beta$, $\gamma$, $\delta$, and $\epsilon$ in the stoichiometry 3:3:1:1:1.

## A.  The $CF_0$ Supramolecular Organization

The most abundant subunit of $CF_0$ is subunit III. This is a small, hydrophobic (8 kDa), chloroplast-encoded (*atpH*) protein [69]. Based on the resolved structure of the bacterial homologue of subunit III, subunit c, it was proposed that it has two membrane-traversing helices and a more polar loop region exposed to the outer (stromal) site of the membrane [85]. At subunit c, Asp61 situated in the second transmembrane helix is known to protonate and deprotonate during $H^+$ transport. Extensive cross-linking studies with native bacterial $F_0$ indicated that subunits c (respectively subunits III) are arranged in a ring with a diameter of about 5.5 nm (7 nm for $CF_0$) [69]. For $CF_0$ this is supported by electron and atomic force spectroscopy observations [86,87]. Helix 1 at all subunits is situated inside the ring, while helix 2 is outside [88]. This packing supports the suggestion that the proton binding site is formed at the packed interface of two units, with Asp61 at the front face of one subunit interacting with Ala24, Ile28, and Ala61 at the back face of a second subunit [84].

The binding of the loop region to subunits $\gamma$ and $\epsilon$ of $CF_1$ is proposed to force rotation of subunit $\gamma$ as proton transport drives rotation of the subunit III oligomer ring.

Subunit I is a chloroplast-encoded (*atpF*) protein with a molecular mass of about 21 kDa and one transmembrane helix. Subunit II is similar to it but is a nuclear-encoded protein. It has a molecular mass of 16 kDa and a single transmembrane helix. The organization of subunits I and II and their function are also proposed on the basis of the resolved structure of the bacterial $F_0$ complex. Two subunits of b are present of $F_0$ forming a dimer placed outside the c-oligomeric ring [84,89]. They play the role of a stator holding $\alpha_3\beta_3$ subunits of $F_1$ ($CF_1$) fixed to the stationary $F_0$ ($CF_0$) subunits as $c_{12}$-$\gamma$,$\epsilon$ subunits rotate as a unit [84]. The cytoplasmic domain of b subunits (subunits I and II at $CF_0$, respectively) binds to the $\delta$ subunit of $F_1$ ($CF_1$) [90]. The interactions between subunit b and subunits $\delta$ and $\alpha$ at the top of $F_1$ were demonstrated by Rodger and Capaldi [91]. To reach the top of $F_1$, subunit b is estimated to extend 11 nm from the surface of the membrane [91]. Subunit b, like subunits I and II from $CF_0$, is anchored in the membrane via a single transmembrane helix at the N-terminal end [92].

Subunits I and II are bound to subunit IV probably without close contact with the subunit III oligomer ring. As with bacterial ATP synthase, the major stator component tightly anchored in the membrane is subunit IV (the bacterial homologue is subunit a). Subunit IV is a chloroplast-encoded protein (*atpI*) with a molecular mass of 25 kDa and four transmembrane helices. It was proposed that subunit IV plays a central role together with the subunit III ring of a functional conductance to protons of the assembled enzyme. Nothing is known about the arrangement of the subunit IV transmembrane domains and the mechanisms of its participation in proton transport in chloroplasts [69].

## B.  Supramolecular Organization of CF₁

The $\alpha_3$:$\beta_3$:$\gamma$ complex is the active site of ATP synthesis. The $\alpha$ subunit is a chloroplast-encoded (*atpA*) protein with molecular mass 55 kDa. It is entirely situated outside the membrane. It is accepted that the $\alpha$ subunit participates in nucleotide binding and has a regulatory role [69].

The $\beta$ subunit is also a chloroplast-encoded protein (*atpB*) with a mass of about 54 kDa. It is localized outside the membrane and is a catalytic site of ATP synthase [69].

The subunit is a nuclear-encoded (*atpC*) protein with a molecular mass of 35 kDa. Its C- and N-terminal regions are situated in the core of the $\alpha_3$:$\beta_3$ ring. The other part of the $\gamma$ subunit protrudes below the C-terminal domain of the $\alpha$ and $\beta$ subunits. Together with the $\epsilon$ subunit, the protruding portion of the $\gamma$ subunit is tightly bound to the surface-exposed loops of the subunit III oligomer. The role of the $\gamma$ subunit is as a transducer of rotational energy from the subunit III oligomer rotor [69,84].

The present arrangement of the enzyme stems from the study of a mitochondrial $\alpha_3$:$\beta_3$ subcomplex [93]. The complex consists of a hexameric ring of alternating $\alpha$ and $\beta$ subunits that surrounds the $\alpha$-helical domain formed of both C- and N- terminal regions of the $\gamma$ subunit. The top of the $\alpha\beta$ assembly, distal to the membrane, consists of a $\beta$-barrel composed of the N-terminal portion, covering a central nucleotide-binding domain, followed by the C-terminal, $\alpha$-helical domain. Three groups of $\alpha\beta$-heterodimers are thus formed, which can adopt three distinct nucleotide binding conformations corresponding to empty sites, ADP/Pi binding site, and ATP tight binding sites. These structural features support a rotatory mechanism wherein the central $\gamma$ subunit rotates within the $\alpha\beta$-hexamer, driving the enzyme through three successive configurations that are required for ATP synthesis/hydrolysis [69,84].

## C.  Function of ATP Synthase

The chloroplast ATP synthase generates ATP from ADP and inorganic phosphate using energy delivered from a trans-thylakoid electrochemical proton gradient $\Delta\mu H^+$ [84]. At the membrane-embedded $CF_0$ portion an energy-releasing proton transport takes place when the energy-consuming ATP synthesis occurs at the extrinsic $CF_1$ portion [94]. The hypothetical mechanisms of ATP synthesis are proposed on the base of data derived from investigations of ATP synthases from different sources.

Proton transport in $CF_0$ is supposedly mediated by the 12 copies of the proton-binding subunit III (subunit c), which are arranged as a ring [69,84]. The association and dissociation of protons take place at the strictly conserved carboxyl residues on subunit III (its localization was discussed before). Rotation of the ring is driven by proton binding to the residue via a thylakoid lumen inlet channel supposedly formed by subunit a (IV) [69,84]. The protonated binding site then moves from the stator interface to the lipid phase of the membrane, where after 12 steps it reaches an outlet channel with access to the cytoplasmic, $CF_1$ binding side of the membrane. Agr210 on transmembrane helix 4 of subunit a (bacterial homologue of $CF_0$ subunit IV) is proposed to promote proton release to the outlet channel [84].

The $\gamma$ and $\epsilon$ subunits are proposed to remain fixed to the top of the subunit c (III) ring so that rotation of the ring also drives rotation of the $\gamma$ subunit within the $\alpha_3\beta_3$ subunits of $CF_1$.

Although the detailed structures of the ring and stator portion of $CF_0$ ($F_0$) are under debate [95], there is agreement that the translocation of one proton $\gamma$ drives the ring and therefore the $\gamma$ subunit around by 30° [89,96]. After four of these steps, i.e., after a turn of 120°, a newly synthesized ATP molecule is released from a binding site on $CF_1$ [94,95]. The subunit $\gamma$ acts as an elastic element like a cylindrical torsional bar that rotates eccentrically within the $\alpha_3\beta_3$ subunits, causing conformational changes of the active centers.

At any time all three $\alpha\beta$ pairs display different conformational stages, representing open (O), loosely closed (L), and tightly closed (T) binding sites [94]. Substrate exchange with the medium is practically restricted to the open site with competitive binding between ATP and ADP or Pi and random binding of ADP and Pi. The rotation of the $\gamma$ unit 120° within the $\alpha_3\beta_3$ hexameric ring drives the concerted binding change $O \rightarrow L \rightarrow T \rightarrow O$. Conversion of ADP and Pi into ATP and $H_2O$ is a couplet with the $L \rightarrow T$ transition [89,94].

## VI. THE LIPID MATRIX OF THE THYLAKOID MEMBRANE

Thylakoid lipids form about 50% of the mass of membranes and act as a fluid matrix for the functional supramolecular complexes. The fatty acid tails form the hydrophobic, central core of the membrane and hydrophilic heads of lipids are situated at the surface. The lipids are not equally distributed between the two monolayers as well as in the lateral direction [97]. Because the thylakoid lipids are highly unsaturated, the membranes are very fluid at physiological temperatures. Fluidity allows high lateral mobility of pigment-protein complexes through the membranes. However, because of the high contents of proteins (50% of the mass) the diffusion coefficient of individual molecules is limited to $10^{-10}$–$10^{-9}$ $m^2$ $sec^{-1}$ [4,98].

Thylakoid lipids are a complex mixture containing about 80% galactolipids such as monogalactosyldiacyl glycerol (50 mol% total lipids) and digalactosyldiacyl glycerol (25 mol%), which are electrically neutral. The remainder is mainly phosphatidylglycerol (10–15 mol%) and sulfoquinovosyl giacilglycerol (5–10 mol%), charged under physiological pH [97,99]. The chloroplast lipids are highly unsaturated; the predominant fatty acid is linolenic (C18:3). However C16:3 fatty acids are also present in some groups of plants [100]. A fatty acid specific to the thylakoid membranes is *trans*-3-hexadecanoic acid (C16:1). It is a component of phosphatidylglycerol [97,99,100].

## VII. TOPOLOGY OF THYLAKOID MEMBRANE

All photosynthetic organisms house an elaborately folded network of thylakoid membranes that convert solar energy into biochemically useful forms [5]. In higher plants the continuous thylakoid membrane network is differentiated into regular domains of closely appressed membranes in granal stacks that are interconnected by nonappressed single membranes—the stroma thylakoids (Figure 6) [4,5]. The grana itself consists of an appressed central core domain, a curved domain such as the margins of grana lamellae, and two end membranes, the outer membranes on terminal grana lamellae [4]. The stroma thylakoids are planar membranes, although some authors distinguish the necklike connections between grana and stroma lamellae as a separate domain [101]. All these domains differ in their enrichment of supramolecular complexes and functions [4]. According to earlier models, PSII complexes are concentrated in grana when

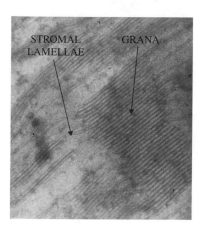

**Figure 6**   Transmission electron micrograph of a *Haberlea rhodopensis* (Friv.) chloroplast with granal stack and stromal lamellae (magnification $\times$ 50,000).

PSI complexes are mainly situated in stromal lamellae [102]. Later it was found that two types of both photosystems occur: grana-situated PSII (PSIIα) and PSI (PSIα) and stroma lamellae–situated PSII (PSIIβ) and PSI (PSIβ) [103]. The two types differ not only in localization but also in structure and function. The α type contains more chlorophyll, larger light-harvesting complexes, respectively, and is involved in linear electron transport [104]. The stromal (β) photosystems have smaller (or lack of?) light-harvesting complexes and carry out cyclic electron transport [104].

It has also been confirmed that cytochrome $b_6 f$ and ATP synthase are situated in both granal and stromal lamellae [4,105]. Therefore, granal photosystems carry out the noncyclic photophosphorylation and the stromal photosystems the cyclic type [5].

According to the latest models of the thylakoid membranes, 80% of the membrane is in the form of grana and 20% consists of stroma lamellae [3,4]. Linear electron transport occurs in the grana, where PSIIα, localized in the core domain of the grana, cooperates with PSIα in the peripheral curved domain. There is no sharp border between photosystems. This intermixing allows faster electron exchange via cytochrome $b_6 f$ complexes localized in the peripheral domain. The ATP synthase is also situated within the peripheral domains [4].

In each grana there is more chlorophyll associated (via LHCII) with PSII than PSI (ratio about 60:40) [4]. However, LHCII is very functionally flexible. This flexibility is provided by a relatively complex assembly of protein subunits and interactions between them, controlled by protonation, xanthophyll de-epoxidation, and phosphorylation. On the other hand, LCHII and its assembly with other protein complexes are vital for grana packing and grana stability [31].

The stroma lamellae contain PSIβ, cytochrome $b_6 f$ complexes, and ATP synthase and carry out cyclic electron transport and photophosphorylation. The role of PSIIβ is unclear. Several alternatives have been suggested: that PSIIβ poises the cyclic electron flow around PSIβ, that PSIIβ is a precursor of PSIIα, and that it is a stage in the repair cycle of PSIIα [4].

## REFERENCES

1.  J. Heslop-Harrison. Structure and morphogenesis of lamellar system in grana-containing chloroplasts. Membrane structure and lamellar architecture. Planta 60:243–260, 1963.
2.  SW Thorne, JT Duniec. The physical principles of energy transduction in chloroplast thylakoid membranes. Q Rev Biophys 16:197–278, 1983.
3.  P-O Arvidsson, C Sundby. A model for the topology of the chloroplast thylakoid membrane. Aust J Plant Physiol 26:687–694, 1999.
4.  P-A Albertsson. The structure and function of the chloroplast photosynthetic membrane—a model for the domain organization. Photosynth Res 4:141–149, 1995.
5.  JM Anderson. Insights into the consequences of grana staring of thylakoid membranes in vascular plants: a personal perspective. Aust J Plant Physiol 26:625–639, 1999.

6. J Barber, J Nield, EP Morris, D Zheleva, B Hankamer. The structure, function and dynamics of photosystem two. Physiol Plant 100:817–827, 1997.

7. WFJ Vermaas, S Styring, WP Schröder, B Andersson. Photosynthetic water oxidation: the protein framework. Photosynth Res 38:249–263, 1993.

8. C Eijckelhoff, JP Dekker. A routine method to determine the chlorophyll alpha, pheophytin alpha and beta-carotene contents of isolated photosystem II reaction center complexes. Photosynth Res 52:69–73, 1997.

9. P Manna, R LoBrutto, C Eijckelhof, JP Dekker, W Vermaas. Role of Arg180 of the D2 protein in photosystem II structure and function. Eur J Biochem 251:142–154, 1998.

10. HJ van Gorkom, J Schelvis. Kok's oxygen clock: what makes it tick? The structure of P680 and consequences of its oxidizing power. Photosynth Res 38:297–301, 1993.

11. HP Michael, DF Hunt, J Shabonowitz, J Bennett. Tandem mass spectrometry reveals that three photosystem II proteins from spinach chloroplasts contain $N$-acetyl-$O$-phosphothreonine at their $NH_2$ termini. J Biol Chem 263:1123–1140, 1988.

12. RJ Debus. The manganese and cadmium ions of photosynthetic oxygen evolution. Biochim Biophys Acta 1102:269–352, 1992.

13. H Michel, J Deisenhofer. Relevance of the photosynthetic reaction centre from purple bacteria to the structure of PSII. Biochemistry 27:1–7, 1988.

14. J Barber. Molecular basis of the vulnerability of photosystem II to damage by light. Aust J Plant Physiol 22:201–208, 1995.

15. J Barber, B Andersson. Too much of a good thing: light can be bad for photosynthesis. Trends Biochim Sci 17:61–66, 1992.

16. BM Greenberg, V Gaba, AK Mattoo, M Edelman. Identification of primary in vivo degradation product of the rapidly-turning over 32 kDa protein of photosystem II EMBO J 6:2865–2869, 1987.

17. J de Las Rivas, B Anderson, J Barber. Two sites of primary degradation of the D1-protein induced by acceptor or donor side photoinhibition in PSII core complex. FEBS Lett 301:246–252, 1992.

18. AK Mattoo, H Hoffman-Falk, J Marder, M Edelman. Regulation of protein metabolism: degradation of the rapidly metabolized 32 kDa protein of the chloroplast membrane. Proc Natl Acad Sci U S A 81:1380–1384, 1984.

19. D Kyle, I Ohad, CJ Arntzen. Membrane proteins damage and repair. Selective loss of quinone protein function in chloroplast membranes Proc Natl Acad Sci U S A 81:4070–4074, 1984.

20. J Sharma, M Panico, J Barber, HR Morris. Characterization of the low molecular weight PSII reaction center subunits and their light-induced modifications by mass spectrometry. J Biol Chem 272:3935–3943, 1997.

21. S Alizadeh, PJ Nixon, A Telfer, J Barber. Isolation and characterization of the PSII reaction centre complex from a double mutant of *Chlamydomonas reinhardtii*. Photosynth Res 43:165–171, 1995.

22. GT Babcock, WR Widger, WA Cramer, WA Oertlind, JG Metz. Axial ligands of chloroplast cytochrome $b_{559}$—identification and requirement for a heme-crosslinked polypeptide structure. Biochemistry 24:3638–3645, 1985.

23. M Ikeuchi, Y Inoue. A new photosystem II reaction centre component (4.8 kDa protein) encoded by the chloroplast genome. FEBS Lett 241:99–104, 1988.

24. M Ikeuchi, K Takio, Y Inoue. N-terminal sequence of PSII low molecular mass proteins: 5 and 4.1 kDa components of the oxygen evolving core complex from higher plants. FEBS Lett 242:263–269, 1989.

25. K Kitamura, S Ozawa, T Shiina, Y Toyoshima. L protein, encoded by *psbL*, restores normal functioning of the primary quinone acceptor, $Q_A$, in isolated D1/D2/CP47/Cyt$b_{559}$/I PSII reaction centre core complex. FEBS Lett 354:113–116, 1994.

26. TM Bricker. The structure and function of Cpa-1 and Cpa-2 in photosystem II. Photosynth Res 24:1–13, 1990.

27. WFJ Vermaas, M Ikeuchi, Y Inoue. Protein-composition of the PSII core complex in genetically engineering mutants of the cyanobacteria *Synechocystis* sp. PCC 6803. Photosynth Res 17:97–113, 1988.

28. JP Decker, SD Betts, FC Yocum, EJ Boekema. Characterization by electron microscopy of isolated particles and crystals of the CP47-D1-D2-cytochrome $b_{559}$ complex of photosystem II. Biochemistry 29:3220–3225, 1990.

29. BR Green, E Pichersky, K Kloppstech. The chlorophyll $a/b$–binding light-harvesting antennas of green plants: the story of an extended gene family. Trends Biochem Sci 16:181–186, 1991.

30. GF Peter, JP Thornber. Biochemical composition and organization of higher plant photosystem II light-harvesting pigment proteins. J Biol Chem 266:16745–16754, 1991.

31. P Horton. Are grana necessary for regulation of light harvesting? Aust J Plant Physiol 26:659–669, 1999.

32. S Jansson, E Selstam, P Gustavsson. The rapidiy phosphorylated 25 kDa polypeptide of the light-harvesting complex of photosystem II is encoded by type 2 *cab-II* genes. Biochim Biophys Acta 1019:110–114, 1990.

33. M Spangfort, B Anderson. Subpopulations of the main chlorophyll $a/b$ light-harvesting complex of photosystem II—isolation and biochemical characterization. Biochim Biophys Acta 977:163–170, 1989.

34. P Mäepää, B Andersson. Photosystem II heterogeneity and long-term acclimation of light-harvesting. Z Naturforsch 44C:403–406, 1989.

35. B Hankamer, J Barber, EJ Boekema. Structure and membrane organization of photosystem II in green plants Annu Rev Plant Physiol Plant Mol Biol 48:641–671, 1997.

36. RG Walter, P Horton. Structural and functional heterogeneity in the major light-harvesting complex of higher plants. Photosynth Res 61:77–89, 1999.
37. PJ Nixon, DA Chisholm, BA Diner. Isolation and functional analysis of random and site directed mutants of photosystem II. In: PR Shewry, S Gutteridge, eds. Plant Protein Engineering. Vol 2. Cambridge: Cambridge University Press, 1992, pp 93–141.
38. Xu, J Nelson, TM Bricker. Secondary structure of the 33 kDa extrinsic protein of PSII: a far UV-CD study. Biochim Biophys Acta 1188:427–431, 1992.
39. WR Odom, TM Bricker. Interaction of Cpa-1 with the manganese-stabilizing protein of photosystem II: identification of domains cross-linked by 1-ethyl-3-[3-(dimethyl-amino)propyl]carbodiimid. Biochemistry 31:5616–5620, 1992.
40. IS Enami, S Mitsuhashi, S Takahashi, MB Ikeuchi, S Katoh. Evidence from crosslinking for a close association of the extrinsic 33 kDa protein with the 9.4 kDa subunit of cytochrome $b_{559}$ and the 4.8 kDa product of the *psbI* gene in oxygen-evolving photosystem II complex from spinach. Plant Cell Physiol 33:291–297, 1992.
41. G Renger. Mechanistic and structural aspects of photosynthetic water oxidation. Physiol Plant 100:828–841, 1997.
42. A Lautner, R Klein, U Ljungberg, H Reilander, D Bartling, B Andersson, H Reinke, K Beyreuther, RG Herrmann. Nucleotide sequence of cDNA clones encoding the complete precursor for the 10 kDa polypeptide of PSII from spinach. J Biol Chem 263:10077–10081, 1988.
43. J Schelvis, P van Noort, T Aartsma, H van Gorkom. Energy transfer, charge separation and pigment arrangement in the reaction center of photosystem II. Biochim Biophys Acta 1184:242–250, 1994.
44. TA Roelofs, SLS Kwa, R van Grondelle, JP Dekker, AR Holzwarth. Primary processes and structure of the photosystem II reaction center: II. Low-temperature picosecond fluorescence kinetics of a $D_1$-$D_2$-cyt-$b_{559}$ reaction center complex isolated by short Triton exposure. Biochim Biophys Acta. 1143:147–157, 1993.
45. LB Giorgi, JR Durrant, S Alizadeh, PJ Nixon, DM Joseph, T Rech, J Barber, G Porter, DR Klug. Comparison of primary electron transfer in photosystem II reaction centers isolated from the higher plant *Pisum sativum* and the green alga *Chlamydomonas reinhardtii*. Biochim Biophys Acta. 1186:247–251, 1994.
46. M Haumann, W Junge. Photosynthetic water oxidation: a simplex-scheme of its partial reactions. Biochim Biophys Acta 1411:86–91, 1999.
47. B Kok, B Forbush, M McGloin. Cooperation of changes in photosynthetic evolution. I. A linear four step mechanism. Photochem. Photobiol 11:457–475, 1970
48. RD Britt. Oxygen evolution. In: DR Ort, CF Yocum, eds. Oxygenic Photosynthesis: The Light Reactions. Dordrecht: Kluwer Academic Publishers, 1996, pp 137–164.
49. C Tommos, XS Tang, K Wamcke, CW Hoganson, S Styring, J McCraken, BA Diner, GT Babcock. Spin-density distribution, conformation and hydrogen bonding of the redox-active tyrosine $Y_z$ in PSII from multiple electron magnetic-resonance spectroscopies: implications for photosynthetic oxygen evolution. J Am Chem Soc 177:10325–10335, 1995.
50. J Dietrich, W Kühlbrandt. Purification and two-dimensional crystallization of highly active cytochrome $b_6f$ complex from spinach. FEBS Lett 463:97–102, 1999.
51. WA Cramer, GM Soriano, H Zhang, MV Ponamarev, JL Smith. The cytochrome $b_6f$ complex. Novel aspects. Physiol Plant 100:852–862, 1997.
52. D Huang, RM Everly, RH Cheng, JB Heymann, H Schägger TS Baker, WA Cramer. Characterization of the Cyt $b_6f$ complex as a structural and functional dimer. Biochemistry 33:4401–4409, 1994.
53. Y Pierre, C Breyton, D Kramer, JL Popot. Purification and characterization of the Cyt $b_6f$ complex from *Chlamydomonas reinhardtii*. J Biol Chem 270:29342–29349, 1995.
54. SE Martinez, D Huang, A Szczepaniak, WA Cramer, JL Smith. Crystal structure of chloroplast Cyt *f* reveals a novel Cyt fold and unexpected heme ligation. Structure 2:95–105, 1994.
55. S Iwata, M Saynovits, TA Link, H Michel. Structure of a water-soluble fragment of the "Rieske" iron-sulfur protein of the bovine heart mitochondria Cyt $bc_1$ complex determined by MAD phasing at 1.5 Å resolution. Structure 4:567–579, 1996.
56. AR Crofts, H Robinson, K Andrews, S van Doren,. E Berry. Catalytic sites for reduction and oxidation of quinones. In: S Papa, B Chance, L Ernster, eds. Cytochrome Systems: Molecular Biology and Bioenergetics. New York: Plenum, 1987, pp 617–624.
57. A Szczepaniak, WA Cramer. Thylakoid membrane protein topography: location of the termini of the chloroplast Cyt $b_6$ on the stromal side of the membrane. J Biol Chem 265:17720–17726, 1990.
58. W Widger, WA Cramer, R Herrmann, A Trebst. Sequence homology and structural similarity between the Cyt *b* of mitochondrial complex III and the chloroplast $b_6f$ complex: position of the Cyt *b* hemes in the membrane. Proc Natl Acad Sci U S A 81:674–678, 1984.
59. T Matsumoto, M Matsuo, Y Matsuda. Structural analysis and expression during dark-light transitions of a gene for Cyt *f* in *Chlamydomonas reinhardtii*. Plant Cell Physiol 32:863–872, 1991.
60. SE Martinez, M Ponamarev, WA Cramer, JL Smith, The heme redox center of chloroplast Cyt *f* is linked to a buried five-water chain. Protein Sci 5:1081–1092, 1996.
61. AB Hope. Electron transfers amongst cytochrome *f*, plastocyanin and photosystem I: kinetics and mechanisms. Biochim Biophys Acta 1456:5–26, 2000.

62.  C de Vitry. Characterization of the gene of the chloroplast Rieske iron-sulfur protein in *Chlamydomonas rein-hardtii*. Indications for an uncleaved targeting sequence. J Biol Chem 269:7603, 1994.
63.  P Manavalan, WC Johnson Jr. Variable selection method improves the prediction of protein secondary structure from circular dichroism. Anal Biochem 167:76–85, 1987.
64.  H Zhang, CJ Carrell, D Huang V Sled, T Ohnishi, JL Smith, WA Cramer. Characterization and crystallization of the lumen-side domain of the chloroplast Rieske iron-sulfur protein. J Biol Chem 271:31360–31366, 1996.
65.  Z Zhang, L Huang, V Shulmeister, Y Chi, K Kyeong, L Hung, A Crofts, E Berry, S Kim. Electron transfer by domain movement in cytochrome $bc_1$. Nature. 392:677–684, 1998.
66.  S Berry, B Rumberg. Proton to electron stoichiometry in electron transport of spinach thylakoids. Biochim Biophys Acta 1410:248–261, 1999.
67.  RP Barbagallo, G Finazzi, G Forti. Effects of inhibitors on the activity of the cytochrome $b(6)f$ complex: evidence for the existence of two binding pockets in the lumenal site. Biochemistry 38:12814–12821, 1999.
68.  HV Scheller, H Navel, BL Møller, Molecular aspects of photosystem I. Physiol Plant 100:842–851, 1997.
69.  FA Wollman, L Minai, R Nechushtai. The biogenesis and assembly of photosynthetic proteins in thylakoid membranes. Biochim Biophys Acta 1411:21–85, 1999.
70.  BR Green, DG Dunford. The chlorophyll-carotenoid-proteins of oxygenic photosynthesis. Annu Rev Plant Physiol Plant Mol Biol 47:685–714, 1996.
71.  S Jansson, B Andersen, HV Scheller. Nearest neighbor analysis of higher plant photosystem I holocomplex. Plant Physiol 112:409–420, 1996.
72.  M Hippler, R Ratajczak, W Haehnel. Identification of the plastocyanin binding subunit of photosystem I. FEBS Lett 250:280–284, 1989.
73.  J Farah, F Rappaport, Y Choquet, P Juliot, ID Rochaix. Isolation of a psaF-deficient mutant of *Chlamydomonas reinhardtii*: efficient interaction of plastocyanin with the photosystem I reaction center is mediated by the PsaF subunit. EMBO J 14:4976–4984, 1995.
74.  J Kruip, D Bald, B Hankamer, J Nield, AF Boostra, J Barber, E Boekema, M Rögner. Localization of subunits in PSI, PS2 and in a PS2/light-harvesting-supercomplex. In: P Mathis, ed. Photosynthesis: From Light to Biosphere. Vol II. Dordrecht: Kluwer Academic Publishers, 1995, pp 405–408.
75.  B Andersen, B Koch, HV Scheller. Structural and functional analysis of the reducing site of photosystem I. Physiol Plant 84:154–161, 1992.
76.  H Naver, MP Scott, B Andersen, BL Møller, HV Scheller. Reconstruction of barley photosystem I reveals that the N-terminus of the PSI-D subunit is essential for tight binding of PSI-C. Physiol Plant 95:19–26, 1995
77.  U Mühlenhoff, JD Zhao, DA Bryant. Interaction between photosystem I and flavodoxin from the cyanobacterium *Synechococcus* sp. PCC 7002 as revealed by chemical cross-linking. Eur J Biochem 235:324–331, 1996.
78.  VP Chitnis, Y Jung, L Albee, JH Golbeck, PR Chitnis. Mutational analysis of photosystem I polypeptides. J Biol Chem 271:11772–11780, 1996.
79.  B Andersen, HV Scheller, BL Møller. The PSI-E subunit of photosystem I binds ferredoxin:$NADP^+$ oxidoreductase. FEBS Lett 311:169–173, 1992.
80.  L Yu, J Zhao, U Mühlenhoff, DA Bryant, JH Golbeck. PsaE is required for in vivo cyclic flow around photosystem I in the cyanobacterium *Synechococcus* sp. PCC 7002. Plant Physiol 103:171–180, 1993.
81.  Hervás, JA Navarro, B De la Cerda, A Diaz, MA de la Rosa. Reduction of photosystem I by cytochrome $c_6$ and plastocyanin: molecular recognition and reaction mechanism. Bioelectrochem Bioenerg 42:249–254, 1997.
82.  K Sigfridsson, S Young, O Hansson. Structural dynamics in the plastocyanin–photosystem 1 electron-transfer complex as revealed by mutant studies. Biochemistry 35:1249–1257, 1996.
83.  N Nelson. Structural conservation and functional diversity of V-ATPases. J Bioenerg Biomembr 24:407–414, 1992.
84.  RH Fillingame, W Jiang, OY Dmitriev. Coupling $H^+$ transport to rotary catalysis in F-type ATP synthases: structure and organization of the transmembrane rotary motor. J Exp Biol 203:9–17, 2000.
85.  RH Fillingame. Coupling $H^+$ transport and ATP synthesis in F0F1 ATP synthases: glimpses of interacting parts in a dynamic molecular machine. J Exp Biol 200:217–224, 1997.
86.  R Birkenhäger, M Hoppert, G Deckers-Hebestreint, F Mayer, K Altendorf. The F0 complex of the *Escherichia coli* ATP synthase. Investigation by electron spectroscopic imaging and immunoelectron microscopy. Eur J Biochem 230:58–67, 1995.
87.  S Singh, P Turina, CJ Bustamante, JD Keller, RA Capaldi. Topographical structure of membrane-bound *Escherichia coli* F0F1 ATP synthase in aqueous buffer. FEBS Lett 397:30–34, 1996.
88.  PC Jones, W Jiang, RH Fillingame. Arrangement of the multicopy $H^+$-translocating subunit c in the membrane sector of the *Escherichia coli* F0F1 ATP synthase. J Biol Chem 273:17178–17185, 1998.
89.  RH Fillingame, PC Jones, W Jiang, FI Valiyaveetil, OY Dmitriev. Subunit organization and structure in the F0 sector of *Escherichia coli* F0F1 ATP synthase. Biochim Biophys Acta 1365:135–142, 1998.
90.  SD Dunn, J Chandler. Characterization of a $b_2\delta$ complex from *Escherichia coli* ATP synthase. J Biol Chem 273:8646–8651, 1998.
91.  AJW Rodgers, RA Capaldi. The second stalk composed of the b- and δ- subunit connects F0 to F1 via an α-subunit in *Escherichia coli* ATP synthase. J Biol Chem 273:29406–29410, 1998.
92.  O Dmitriev, PC Jones, W Jiang, RH Fillingame. Structure of the membrane domain of subunit b of the *Escherichia coli* F0F1 ATP synthase. J Biol Chem 274:15598–15604, 1999.

93. N Hu, DA Mills, B Huchzermeyer, J Richter. Inhibition by tentoxin of cooperativity among nucleotide binding sites on chloroplast coupling factor 1. J Biol Chem 268:8536–8540, 1993.

94. O Panke, B Runmberg. Kinetic modeling of rotary CF0F1-ATP synthase: storage of elastic energy during energy transduction. Biochem Biophys Acta 1412:118–128, 1999.

95. DA Cherepanov, AY Mulkidjanian, W Junge. Transient accumulation of elastic energy in proton translocating ATP synthase. FEBS Lett 449:1–6, 1999.

96. W Junge, H Lill, S Engelbrecht. ATP synthase: an electrochemical transducer with rotatory mechanics. Trends Biochem Sci 22:420–423, 1997.

97. JM Anderson. Photoregulation of the composition, function and structure of thylakoid membranes. Annu Rev Plant Physiol 37:93–136, 1986.

98. M Blackwell, C Gibas, S Gyqax, D Roman, B Wagner. The plastoquinone diffusion coefficient in chloroplasts and its mechanistic implications. Biochim Biophys Acta 1183:533–543, 1994.

99. A Benson. Plant membrane lipids. Annu Rev Plant Physiol 15:1–16, 1974.

100. E Selstam, A Widell-Wigge. Chloroplast lipids and the assembly of membranes. In: C Sundqvist, M Rybery, eds. Pigment-Protein Complexes in Plastid: Synthesis and Assembly. New York: Academic Press, 1993, pp 241–277.

101. LA Staehelin. Chloroplast structure and supramolecular organization of photosynthetic membranes. In: LA Staehelin, CJ Arnzen, eds. Encyclopedia of Plant Physiology. New Series. Vol 19. Berlin: Springer-Verlag, 1986, pp 1–84.

102. PV Sane, JD Googchild, RB Park. Characterization of chloroplast photosystems 1 and 2 separated by a nondetergent method. Biochim Biophys Acta 216:162–178, 1970.

103. B Andersson, JM Anderson. Lateral heterogeneity in the distribution of chlorophyll-protein complexes of the thylakoid membranes of spinach chloroplasts. Biochim Biophys Acta 593:427–440, 1980.

104. JM Anderson. Distribution of the cytochromes of spinach chloroplast between the appressed membranes of grana stacks and stroma-exposed thylakoids. FEBS Lett 138:62–66, 1982.

105. AN Webber, KA Platt-Aloia, RL Heath, WW Thomson. The marginal regions of thylakoid membranes: a partial characterization by Tween 20 solubilization of spinach thylakoids. Physiol Plant 72:288–297, 1988.

# 14

# Bioenergetic Aspects of Photosynthetic Gas Exchange and Respiratory Processes in Algae and Plants

**Klaus Peter Bader**

*University of Bielefeld, Bielefeld, Germany*

**Refat Abdel-Basset**

*Assiut University, Assiut, Egypt*

## I. INTRODUCTION

### A. General Scientific Features

The nuclear reaction that takes place deep inside the sun consists of the fusion of hydrogen ($H_2$) to helium (He). According to up-to-date estimates about $5 \times 10^9$ kg of molecular hydrogen is converted every second with the difference in the masses of the involved molecules being emitted as radiation equivalent to approximately 6000 K photons. The resulting energy can be calculated to a value of 1400 kW m$^{-2}$. Per year, this process supplies earth with the immense energy amount of $56 \times 10^{23}$ J. Even under the assumption that about 50% of the radiation might be reflected by clouds and gases of the atmosphere (thus not reaching the earth's surface) and that 50% of the passing radiation is in the long-wavelength region with a low energy content (e.g., infrared), energy of about $14 \times 10^{23}$ J is available. This gigantic amount, however, is "mere" radiation energy, which cannot be readily used by (nearly) any biological organism. Only one type of organism is capable of converting this abiotic energy into a biologically useful form, and consequently these organisms are called photoautotrophs. Among these are cyanobacteria, green algae, photosynthetic bacteria, and higher plants.

Similarly, rough estimates give a value of $3 \times 10^{21}$ J for the overall biomass produced by autotrophic organisms and $13 \times 10^{18}$ J for the energy uptake by mankind. In any case and without overestimating the correctness of the given values, it is clear that less than 1% of the available radiation energy is converted into biomass! Of this value, less than 1% is actually taken up as nutrition (of any type) by mankind! Thus, the significance and the importance of photosynthesis (and the organisms involved) can hardly be overvalued.

In evolutionary terms, blue-green algae (cyanobacteria) must be mentioned in the first place because they were the organisms that "invented" an "improved" and most important form of photosynthesis, namely oxygenic photosynthesis, about 3–4 billion years ago. It must be emphasized that this process took place in a reducing atmosphere that consisted of nitrogen and carbon dioxide (possibly some hydrogen) with virtually no oxygen present. This generally accepted view, however, does not completely hold, as small but substantial amounts of oxygen must have been present at the time. Detailed analyses of the biosynthesis of essential pigments such as chlorophyll have shown that at least one step in the biosynthesis sequence requires molecular oxygen! The reaction from coproporphyrinogen III to protoporphyrinogen IX, i.e., the formation of the vinyl groups from the propionic acid side chains, is catalyzed by the coprogen oxidative decarboxylase only in the presence of oxygen [1]! Moreover, it has been stated that the oxidation of water in principle requires catalytic amounts of oxygen with a cooperative mode of binding for its functioning. In an absolutely anaerobic atmosphere, the oxygen-evolving complex (OEC) does not operate [2]. Photolytic reactions involving, e.g., ultraviolet (UV) light might have played a role in generating the necessary low oxygen partial pressure at least in ecological niches such as lakes or puddles [3].

With the modern discussion of an increasing carbon dioxide concentration of the atmosphere in mind, it should be noted that the $CO_2$ partial pressure of the early atmosphere amounted to some percent (instead of ppm!). At the time, the photosynthetic organisms were restricted to inorganic salts as electron donors for light-induced electron transport, and it is assumed that iron and sulfide compounds played a substantial role. However, on a long-term scale this situation was problematic and unfavorable for the evolution of higher photosynthetic organisms because (1) the energy required for the oxidation of such compounds was relatively high and (2) the availability of sufficiently high amounts of these electron sources was limited. It appeared necessary to evolve an improved system with less energy required and—most important—to find a ubiquitous electron supply. Such an electron source was finally found with the simple and almost ubiquitously available and disposable molecule of water ($H_2O$). In this context, the evolutionary significance of hydrogen peroxide in some ecological niches functioning (transitorily) as an intermediate electron donor between inorganic salts and molecular water has been proposed and discussed [e.g., 3–5]. However, another problem came up based on some unique properties of the water molecule. In this context, only the extremely high stability of this molecule will be mentioned. Even in the modern world with all the technical facilities available today, drastic reaction conditions such as electric current or high temperatures of about 2000°C or more are needed technically to oxidize the molecule. Photosynthetic organisms perform oxygenic photosynthesis under physiological conditions and at room temperature! Although many scientific details of the mechanisms of photosynthetic reactions have been worked out, many questions remain to be elucidated. Some of the relevant parameters will be referred to in the respective sections of this chapter. From the multitude of relevant investigations and reviews, only a few will be mentioned here [3–10] (G Renger, submitted).

Although plants are capable of utilizing abiotic radiation energy (see earlier) and transforming it to biologically useful forms, they also operate (at the same time or under specific conditions) many oxidative (respiratory) processes. Among these are (completely) different and independent reactions—dark respiration, alternative respiration, photorespiration, chlororespiration, and concerted reactions that have been termed maintenance respiration. It is clear that these reactions require a substantial but "reasonable" oxygen partial pressure. It must be emphasized that oxygen as such is by far not the "positive" molecule for plants that it is for animals and human beings; in many cases it is a problematic gas whose partial pressure has to be strictly regulated in order to avoid detrimental effects. Thus, the evolution of oxygenic photosynthesis—finally reaching an ambient partial pressure of 21% $O_2$—made the situation more and more complex and was in principle a type of ecological catastrophy for the early anaerobic organisms. It must be kept in mind that the process of water "splitting" oxidizes $H_2O$ with $O_2$ being released as a waste product, thus substantially increasing the partial pressure of molecular oxygen at endogenous physiological sites where (unregulated) oxidative processes appeared highly problematic and might result in the undesired oxidation of sensitive vicinal components. (The reaction center pigment of photosystem II, $P_{680}^+$ has a high positive redox potential of about 1.2 V.) It should be added that in the absence of molecular oxygen the decay of the $P_{680}$ triplet state turned out to be much slower; i.e., the excitation state was more stable than under oxygenic conditions [11]. Moreover, it is well known that reactive oxygen species are formed inside the oxygen-evolving complex; $P_{680}$ in the triplet state forms singlet oxygen, and this singlet oxygen is not least formed by direct recombination of the radical pair $P_{680}^+$ $Pheo^-$ [12]. (The exceptionally high turnover rate of the D1 protein can at least in part be explained by the need for protection against such detrimental oxidative processes.)

It has been shown that photosystem II of photosynthetic species can interact with atmospheric oxygen, forming a peroxidic component such as hydrogen peroxide (see earlier). This reaction might be suited to lower the internal partial pressure of oxygen and at the same time to supply an additional electron donor for the OEC as hydrogen peroxide has been shown to be effectively oxidized by photosystem II [13–17]. Thus, the interaction of the OEC with molecular oxygen might help to keep the oxygen partial pressure in the immediate vicinity of the enzymatic process at a low level. Taking the arguments together, it is clear that oxygenic photosynthesis requires small amounts of oxygen but at the same time has to limit the resulting increasing partial pressure via some regulatory mechanism. The binding of atmospheric oxygen to the OEC with the subsequent formation of a peroxidic component might well represent such a regulatory mechanism, and this peroxide may have played an essential role in evolution as a "transitory" electron donor [3,5]. In this context, it should be mentioned that hemoglobin was described to exist in various plants and crops such as barley [18] which might also be seen in context with the fine regulation of the oxygen content in plants. This hemoglobin has an oxygen dissociation constant of 3 nmol $L^{-1}$ and seems to act as a direct oxygenase and/or to regulate the energy status of the plant under conditions of low oxygen [18].

In this chapter we have tried to summarize substantial features of the complex mechanisms of gas exchange reactions that have to be (coarse- and fine-) regulated in plant physiology. Although we did not restrict the presentation to oxygen exchange reactions, it appears clear that this gas represents the most complicated task of regulation for a photosynthetic organism as evolution and uptake reactions take place in the immediate vicinity of each other and, in some cases, even concomitantly. In some organisms such as cyanobacteria a similar problem exists for hydrogen gas exchange reactions; e.g., hydrogen oxidation (i.e., hydrogen uptake) and proton reduction (forming molecular hydrogen) occur virtually in parallel or in highly regulated transitions [19,20].

## B. Technical Aspects of Plant Physiological Gas Exchange

In photosynthesis, molecular water is oxidized and oxygen is produced as a waste product of the reaction. However, physiological processes in photosynthetic organisms include (various) reactions that require or include the oxidation or oxygenation of compounds so that technically an oxygen uptake from the surrounding atmosphere takes place. On the basis of the overall oxygen gas exchange, the immanent problem for scientific investigation is obvious. Normally, oxygen gas analyses are carried out using oxygen electrodes, which, however, cannot discriminate between evolution and concomitant uptake processes. Consequently, such systems quantify the overall balance or difference following an "event" (e.g., illumination) by simply adding up positive and negative changes. Thus, an important goal of plant physiological investigations has been to develop techniques allowing the simultaneous recording of gas evolution and uptake reactions, e.g., in a liquid reaction assay and from an artificial gas atmosphere over the aqueous phase without interference. (For reasons of clarity, the interference of an evolution signal increasing the atmospheric partial pressure and thus affecting the composition of the gas phase is neglected here.) With respect to this and other requirements of plant physiological investigations, mass spectrometry has been shown to be specifically well suited and the possible applications of this technique in biology in general and in plant physiology in particular have been described [e.g., 21,22]. Early instruments, however, suffered from both limited sensitivity and insufficient time resolution of the signals. Moreover, an improved response of the mass spectrometer to dynamic changes of gas partial pressure was required, in particular following the first application of short (in the region of $\mu$sec) flash illumination techniques. Now, quite a few instruments with setups specifically adapted to the needs and requirements of plant physiological investigations exist and important studies concerning gas exchange reactions in plants and algae have been performed in photosynthesis research laboratories.

## II. PHOTOSYNTHESIS

The terminus photosynthesis defines and summarizes the complex process(es) by which radiation energy of light is used to form carbohydrates according to the simple formula

$$CO_2 + D^{2-}/2H^+ \rightarrow CH_2O + DO$$

where D(onor) means any reduced photo-oxidizable compound. Thus, in the case of anoxygenic photosynthesis D might be a sulfide or a ferric salt, whereas $D^{2-}/2H^+ = H_2O$ for oxygenic photosynthesis. In the course of the light-induced electron transport from an ultimate donor through two consecutively operating photosystems, energy is conserved in the form of ATP and reducing equivalents ($NADPH_2$) are built. By now, essentially two gas exchange reactions can be investigated that reflect the photosynthetic capacity of a plant, namely the evolution of molecular oxygen as a waste product of the water oxidation and the decrease of the carbon dioxide partial pressure of the surrounding atmosphere due to carbon dioxide assimilation ($CO_2$ uptake).

Taking the trivial formula from before, one essential question still has to be clarified, namely the origin of the evolved oxygen in the case of oxygenic photosynthesis. In the early phases of photosynthesis research, it was not obvious that the light-evolved oxygen originated from water and not from carbon dioxide. By means of mass spectrometry and the application of stable oxygen isotopes containing water ($H_2{}^{18}O$), it became clear that the transformation of carbon dioxide to carbohydrates did not entail any liberation of molecular oxygen because the isotopic oxygen showed up exclusively in the gas phase so that apparently the water from the aqueous part of the reaction assays had been oxidized. (Consequently, in the case of $C^{18}O_2$ no liberation of isotopic oxygen was detectable.) Again surprisingly, the simplest carbohydrate, formaldehyde—the molecule that is structurally identical to the chemical formula known from all photosynthetic schemes, $CH_2O$—was never observed in the course of $CO_2$ assimilation. Instead, the rather complex molecule ribulose-1,5-bisphosphate was found to serve as carbon dioxide acceptor catalyzed by the enzyme known as ribulose-1,5-bisphosphate carboxylase/oxygenase (Rubisco). This enzyme is relatively unique with respect to its bifunctionality: here it catalyzes the assimilation of carbon dioxide (carboxylase function), but it can, depending on the reaction conditions, also react with molecular oxygen (oxygenase function)—at first glance a useless and even lavish and wasteful reaction.

Relevant details of this process and the significance of the phenomenon, e.g., for crop yield, are discussed in following sections of this chapter. The photosynthetic electron transport includes specific carriers (redox components) that operate sequentially via one or two photosystems. These components have been relatively well investigated by now and are described in textbooks and reviews. Therefore, we concentrate in this chapter on relevant mechanistic details and aspects of photosynthetic water cleavage as an extraordinary (without disregarding others) evolutionary achievement in plant physiology.

## A.   Mechanism of Water Oxidation

In the introduction we listed quite a few bioenergetic estimates to describe the importance of photosynthesis for life in general. Figure 1 illustrates the significance of oxygenic photosynthesis for the evolution of higher life forms. The picture is based on the classical experiment by Joseph Priestley, who designed this setup in 1780 to demonstrate that the mouse survived better in a closed system under an artificial gas atmosphere when a green plant was added to the system provided that the plant was illuminated. The plant, in this case mint, had "restored" the air (as cited in Ref. 23). Under this condition, water was oxidized, oxygen was evolved, and this oxygen served for the respiration of the mouse. (The plant also profits from this system as the mouse breathes out carbon dioxide, which increases the amount of substrate $CO_2$ for the plant.)

In recent years, evolutionary aspects of photosynthesis and many details of the water oxidation mecha-

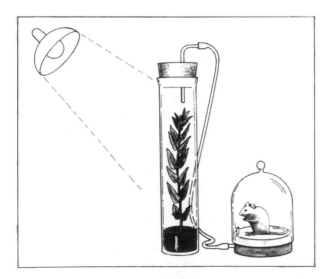

**Figure 1**   Cartoon based on the experiments by Joseph Priestley demonstrating the mutual interrelationship between an animal and an illuminated plant with respect to the oxygen and carbon dioxide gas exchange.

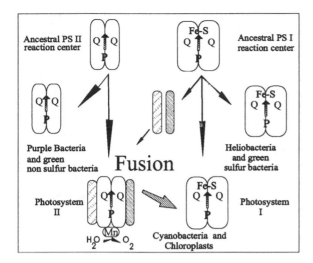

**Figure 2** A heterologous fusion model for the evolution of oxygenic photosynthesis based on phylogenetic analysis. (Modified from Ref. 8.)

nism have been worked out and published. Figure 2 depicts a "heterologous" fusion model of ancestral forms of anoxygenic types of both photosystems (PSI and PS II) in early photosynthetic organisms leading to the evolution of the "modern" oxygen-evolving photosystem II. This model derives the core polypeptides D1 and D2 from homodimeric forms similar to the L and M polypeptides in purple bacteria. Gene fragmentation from common ancestors with subsequent duplication is thought to account for the development of the CP43 and CP47 core polypeptides within photosystem II (see Ref. 8 and references therein).

The present idea about the arrangement of the involved electron transport complexes and redox components within the thylakoid membrane of plants and other eukaryotes is summarized in Ref. 24 and Figure 3. The major complexes PSII(OEC), cytochrome $b_6/f$ (Cyt $b_6/f$), and PSI are linked by the mobile

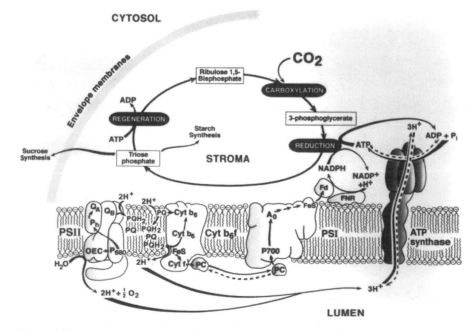

**Figure 3** Schematic diagram of the electron transport system of oxygenic photosynthesis in eukaryotic organisms. (Courtesy of Donald R. Ort.)

plastoquinone pool ($PQ/PQH_2$) and plastocyanin (PC), respectively. The proton gradient that is built up during a light phase "drives" the formation of ATP via another membrane-spanning complex, the ATP synthase. The ultimate electron donor of the redox chain is molecular water, whose oxidation takes place at the luminal side of the thylakoid membrane. Upon illumination, the reaction supplies electrons, which are fed into the photosynthetic electron transport chain; protons (which in some organisms are subsequently reduced again to give molecular hydrogen); and molecular oxygen. A detailed model of photosystem II is shown in Figure 4. The photosynthetic electron transport through photosystem II can, in parts, be effectively followed by absorption spectroscopy. By means of this technique the dependence of redox reactions in the region of photosystem II on the temperature has been analyzed by Renger and his coworkers by recording absorption changes at 830 nm (Figure 5). The results clearly showed that the direct light-dependent electron flow from $Y_z$ to $P680^+$ is virtually independent of temperature (at least in the range between 0 and 33°C). Interestingly, however, the electron abstraction from the OEC up to the formation of $S_3$ (see later) is not invariant or steady but showed a significant change of $E_A$ at a discrete temperature (inset in Figure 5). From these and other experiments it could be concluded that the reaction coordinates of the OEC remained essentially constant and unmodified during evolution (from cyanobacteria to higher plants); this interpretation implies that the basic functions of photosynthesis were optimized in the early stages after invention [9,10].

It can easily be imagined that the oxidation of molecular water requires more than one oxidation step (absorption of one photon), but in the early phases of photosynthesis research illumination was always done with continuous light lasting for at least seconds. Thus, single oxidation steps could not be followed. As early as in 1955, technically remarkable experiments by Allen and Franck [25] showed that photosynthetic preparations lost their capacity to photoevolve molecular oxygen as the consequence of one short light pulse if a sufficiently long dark adaptation period preceded the light phase. Following the improvement of highly sensitive electrode systems and the development of illumination regimes with short light flashes triggered at 1 Hz or more by suitable pulse generators, the phenomenological studies by Joliot and coworkers [26] became possible (Figure 6). The observations have been described and unequivocally explained by the so-called Kok model [9,10,27] (Figure 7). The model says that four photons have to be absorbed and five (four tangible) redox states (S states $S_i$ with i = 0–4) have to be consecutively attained before molecular oxygen is evolved from the dark reaction out of $S_4$ and the reaction center "falls" back to the ground state $S_0$. The oscillation that can be observed in the course of a so-called oxygen evo-

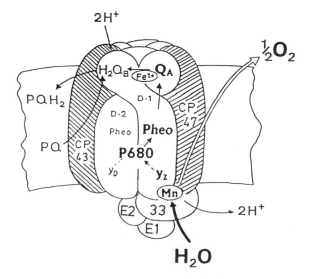

**Figure 4**   Detailed cartoon of the photosystem II complex integrated in the thylakoid membrane: P680, reaction center; Pheo, pheophytin; PQ, plastoquinone; D1/D2, intrinsic membrane-spanning polypeptides; E1, E2, E3, extrinsic polypeptides with molecular masses of 16, 23, and 33 kDa, respectively; $Y_Z/Y_D$, redox active tyrosines of polypeptides D1 and D2, respectively, with $Y_Z$ directly participating in the electron transport; CP, core protein. (From Ref. 10.)

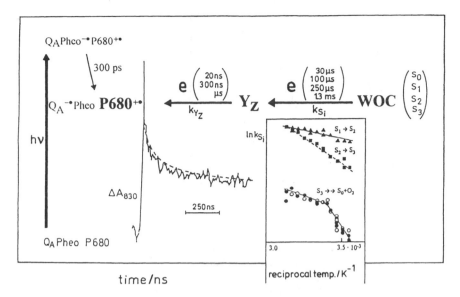

time/ns

**Figure 5** Absorption changes at 830 nm (left side) as a function of time and reciprocal half lifetimes as a function of reciprocal temperature (right side) in PSII fragments from *Synechococcus vulcanus* Copeland. The signal on the left side represents a trace monitored at 33°C, and the dashed curve symbolizes the data at 0°C. The fast decay of the 830 nm absorption change due to $Pheo^{-\bullet}$ oxidation is not resolved and is symbolized by a spike. (From Ref. 10.)

lution pattern is rapidly damped out because of three transition parameters, $\alpha$, $\beta$, and $\gamma$. $\alpha$ means that a single reaction center does not change its redox state upon the flash ($S_i \rightarrow S_i$ = "miss"), whereas in the case of $\gamma$ two oxidation steps occur within the lifetime of the flash ($S_i \rightarrow S_{i+2}$ = "double hit"). It is obvious that the transition parameter $\gamma$ is heavily dependent on the duration of the flash. Xenon flashes with a lifetime of about 5 μsec have been shown to be sufficiently short to result in little double hit contribution (~1–4%). In order to exclude this parameter properly, dye (oxazine, rhodamine, etc.) laser flashes with a lifetime of < 10 nsec have proved useful. The transition parameter $\beta$ finally reflects the successful (intended) transition from $S_i$ to $S_{i+1}$.

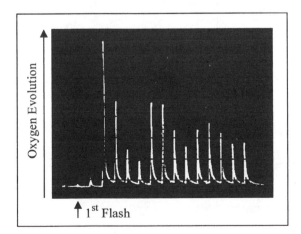

**Figure 6** Typical oxygen evolution pattern induced by a train of short (5 μsec) saturating light flashes following an extensive dark adaptation as observed with higher plant chloroplasts or green algae. (Such phenomenological studies have been explained by the so-called Kok model depicted in Figure 7.) Flash frequency, 3.3 Hz; dark adaptation time, 15 min.

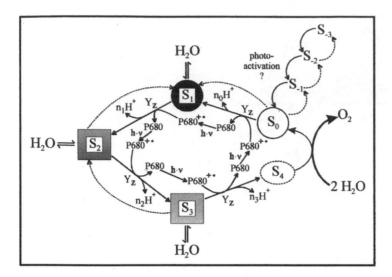

**Figure 7**   Functional scheme of the Kok model of photosynthesis. By successive absorption of light quanta, the reaction centers involved cycle through about five redox states ($S_0$–$S_4$) before molecular oxygen is liberated. $S_1$ is (together with $S_0$) stable in the dark, and this explains why the maximum amplitude is observed with the third flash of a sequence. Under specific conditions and depending on the organism investigated, "overreduced" states $S_{-1}/S_{-2}/S_{-3}$ have to be inserted before the "ground state" $S_0$. (Modified from G Renger, submitted, 2000.)

It is clear that photosynthetic water oxidation also includes the liberation of protons. These protons, however, do not necessarily originate directly from the substrate water, at least not before a water molecule is definitively oxidized as the consequence of a turn of the Kok cycle. In any case, the theoretical function of water as the exclusive source of protons would have been difficult to bring into line with the isotope experiment illustrated in Figure 8. Rather, most of the protons are initially liberated as a consequence of protolytic reactions of specific redox cofactors (Mn ligands) or via the deprotonation

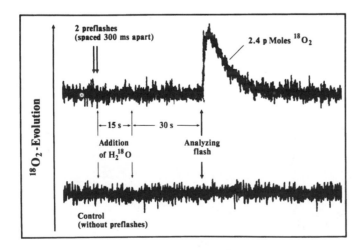

**Figure 8**   Photosynthetic water oxidation from $H_2$ $^{18}O$ in the blue-green alga *Oscillatoria chalybea*. This mass spectrometric experiment showed that flash-induced water oxidation can be observed as a consequence of a single analyzing flash to to a prefabricated $S_3$ redox state (two preflashes). Oxygenic photosynthesis does not necessarily require a water/oxygen intermediate or partially oxidized water molecules. (From Ref. 30.)

of specific amino acids. The reaction rate of the proton release strongly depended on the redox conditions as well as on the pH in specific cases [28,29]. Manganese as an essential cofactor in photosystem II has some specific properties that made it extraordinarily well suited for an important role in the redox system of photosynthesis. The most significant one might be that manganese occurs in four different valence states in covalent binding situations with oxygen; in complexes six and maximally seven different valence states have been described. In fact, it has been found, e.g., by analyses of manganese $K$-edge X-ray absorption near edge structure (XANES) measurements, that during the light-induced cycling through the "Kok clock" valence states from the essential manganese cluster within photosystem II appear to change systematically. (Unfortunately, the scientific results are not unequivocal and there is still a strong debate about the details and even about whether there is really a clear correlation between a given valence state and a corresponding S state of the OEC.)

One of the essential questions in this context was whether substrate water is oxidized with or without an obligatory intermediate and in which state of the cycle a specific water molecule has to be irreversibly bound in order to be oxidized upon a following flash. For a long time this question could not be answered because substrate water molecules could not be discriminated depending on their binding in different redox states. Only mass spectrometric analyses with the application of stable isotopes containing water molecules ($H_2$ $^{18}O$) resolved this problem (Figure 8) [30,31]. The trick was to add the $H_2$ $^{18}O$ only after preflashing the photosynthetic assay with (in the initial experiment) two preflashes. In this way the reaction centers were transferred from the dark stable $S_1$ to $S_3$. Onto this prefabricated $S_3$, substrate water in the form of $H_2$ $^{18}O$ was provided and one analyzing flash was fired. As can be seen from Figure 8, this analyzing flash yielded a significant amount of isotopic molecular oxygen ($^{18}O_2$). This result meant that substrate water had been exchanged in the highly oxidized $S_3$ state! The important conclusion for the mechanism of photosynthetic water oxidation was that there is not necessarily an oxygen precursor or partially oxidized water, which had generally been assumed earlier. Molecular water can be oxidized by the appropriate redox conditions and a *single* flash [30,31].

Mass spectrometry proved important in plant physiological research because of further advantages and specificities. With the choice of a suitable isotope distribution and composition of both liquid and gas phases in a given reaction assay, it became possible to directly record oxygen evolution and oxygen uptake reactions independently, concomitantly, and nearly without interference. Experiments have shown for the first time the blue light–enhanced respiration of algae under the conditions of running photosynthesis, i.e., during light-induced oxygen evolution [32]. This result was achieved by recording photosynthetic oxygen evolution from $H_2$ $^{16}O$ as $^{16}O_2$ ($m/e = 32$) and the respiratory oxygen uptake from an artificially installed $^{18}O_2$ atmosphere as $^{18}O_2$ ($m/e = 36$).

Appropriate mass spectrometric assays have been developed and applied for measurements of different light-induced gas exchange reactions in algae and plants. The carbon dioxide metabolism has been investigated, e.g., by means of the mass spectrometric setup of Badger [33]. In our laboratory, nitrogen fixation by blue-green algae (cyanobacteria) could be directly followed and quantified as $^{15}N_2$ uptake at $m/e = 30$ depending on the presence (or rather the absence) of a combined nitrogen source in the medium. Trivially, light-induced nitrogen fixation was observed only with cultures grown without an N source [34].

The oxygen partial pressure of the atmosphere (natural or artificial) seems to be much more relevant for optimal functioning of photosynthesis than was originally expected. Normally, the oxygen concentration was considered essential "only" for the reaction rates of respiratory processes (see later). It appeared, however, that oxygenic photosynthesis did not function in the complete absence of molecular oxygen (Figure 9). Small but distinct amounts appeared necessary to catalyze a normal water oxidation reaction [2]. It was shown in our laboratory that about four molecules of $O_2$ are required per reaction center and that oxygen is bound in a cooperative manner (Figure 10). Our observation immediately calls to mind the binding properties of hemoglobin in human physiology and zoophysiology. (It was mentioned in the introduction that hemoglobin even appears to play a direct role in plant physiology [18].) The requirement of catalytic amounts of oxygen for the functioning of the OEC has been described in detail for cyanobacteria, but it can be observed with cell suspensions from higher plants and also with green algae. Thus, it might be taken as a general feature of oxygenic photosynthesis. One of the recent ideas about how oxygen, molecular water, and hydrogen peroxide might interact in the immediate vicinity of the oxygen-evolving complex has been extensively discussed [2].

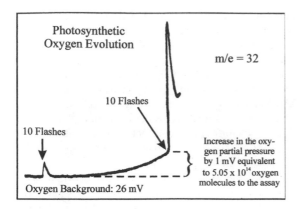

**Figure 9** Mass spectrometric analysis of the photosynthetic oxygen evolution in *Oscillatoria chalybea* as a function of the oxygen partial pressure of the surrounding gas atmosphere. The cumulative signals were induced by a train of 10 short (5 μsec) flashes fired at a frequency of 3.3 Hz. The signal at the left was obtained at an oxygen background signal of 26 mV equivalent to about $10.1 \times 10^{15}$ oxygen molecules. Under completely anaerobic conditions, no photolytic activity can be observed at all. Addition of $5 \times 10^{14}$ oxygen molecules resulted in the enhanced oxygen evolution signal shown at the right. (From Ref. 2.)

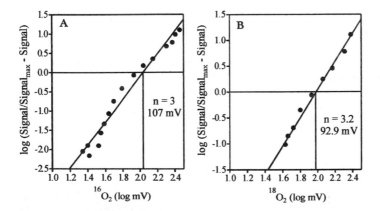

**Figure 10** Hill plots of the dependence of the photosynthetic oxygen evolution of $^{16}O_2$ (mass 32) and $^{18}O_2$ (mass 36) on increasing oxygen partial pressure of the atmosphere. [The mixed isotope molecule $^{16}O^{18}O$ (mass 34) gave a value of $n = 3.1$ and is not shown in the figure.] The value $n$ corresponds to the Hill coefficient calculated for the respective isotope. The mV values in the graphs represent the oxygen partial pressure at half-maximal oxygen evolution ($S_{0.5}$). (From Ref. 2.)

## III. RESPIRATORY PROCESSES

### A. Dark Respiration

In general terms, the expression "dark respiration" might mean any plant physiological reaction linked to the consumption of molecular oxygen without(?) the participation of light. A correct and detailed definition cannot easily be given because the process as a whole comprises quite a few independent reactions that (even separately) participate in the oxidation of carbohydrates: glycolysis, oxidative decarboxylation, tricarboxylic acid (TCA) cycle, NADH oxidation. In this chapter we will not specifically describe and elaborate on this process; rather, the reader is referred to one of the many chapters in modern physiological textbooks and reviews [e.g., 35–37]. However, we want to stress some of the specific points that play a role in modern plant physiology research, in particular with respect to environmental problems and to interference with photosynthesis.

Because photosynthesis and respiration collaborate to fulfill energy needs of plant cells, it can be expected that these processes interfere—one affecting the other. In photosynthetic prokaryotes, the electron transport systems of both photosynthesis and respiration even take place in the same membranes, which might represent the highest challenge for coordinate regulation. Thus, photosystem II–generated electrons together with those produced from substrate oxidation form a common electron pool for the cellular energy-consuming pathways. In higher plants and algae, despite the separation of photosynthesis and respiration in different cell organelles (chloroplasts and mitochondria), they have been shown to interact quite efficiently.

Apart from the interference of electron transport reactions, other photosynthetically relevant factors such as $CO_2$ and light are reported to have an impact on respiration. It was found that $CO_2$ enrichment increased the root respiration of wheat by 24% [38]. The dark respiration rate of gametophytes of the tropical epiphytic fern *Pyrrosia piloselloides* increased substantially with increasing $CO_2$ concentrations during growth [39]. The general stimulation of the rate of shoot respiration in plants by $CO_2$ enrichment was clearly time dependent [40–42]. Unfortunately, stimulation of respiratory activity by an elevated carbon dioxide partial pressure was not consistently observed and described in the literature. Thus, in the case of strawberry (*Fragaria xananassa*) leaves, high $CO_2$ concentrations (up to 900 ppm) did not significantly affect the dark respiration rate [43]. The rate of both shoot and root respiration in *Plantago major* increased with increasing internal nitrogen concentration but was not affected by $CO_2$ [44]. In accordance with these finding, it was reported [45] that shoot and root respiration per unit dry weight was positively correlated with the nitrogen content as a common phenomenon but was again not altered by the atmospheric $CO_2$ concentration. Elevated $CO_2$ levels slightly but significantly increased dark respiration in *Abutilon* but had no significant effect on dark respiration in *Ambrosia* [46]. In shoot and root respiration, it was found that the use of carbon compounds often decreased when the atmospheric $CO_2$ concentration increased [41,47] and thus may have contributed to an increased relative growth rate at elevated $CO_2$. The effect of carbohydrates on the expression of respiratory genes at the transcriptional level has been described [48].

Light as an environmentally relevant factor affects respiration (in)directly via its impact on photosynthesis. (In photosynthetic systems, the mechanism might involve the export of energy out of the chloroplasts under excess high light conditions.) Respiration (nonphotorespiratory mitochondrial $CO_2$ release) of tobacco leaves can be inhibited by light with a following stimulation in the dark. The inhibition of respiration in the light took about 50 sec and was even evident at 3 $\mu$mol photons $m^{-2}$ $sec^{-1}$ regardless of the light quality (red, blue or white) in tobacco leaves. Accordingly, two peaks of $CO_2$ release were exhibited by tobacco leaves after switching off the light [49]. The initial $CO_2$ liberation was observed at 15–20 sec (the photorespiratory postillumination burst) and the second at 180–250 sec (light-enhanced dark respiration, LEDR) following the offset of light. The increases of both LEDR and the light-induced inhibition were positively correlated with each other and also positively correlated with the increasing irradiance during the predark period, suggesting a dependence on the preceding photosynthesis. It has been proposed that the cytochrome $b_6/f$ complex should not be involved in respiratory electron transport because respiratory oxygen uptake was not suppressed by far-red illumination in *Synechocystis* PCC 6803 cells grown photoautotrophically [50].

Mitochondrial oxidation of respiratory substrates is usually catalyzed by the cytochrome oxidase or by the so-called alternative oxidase. The alternative path, which is classically known to be cyanide resistant but salicylhydroxamic acid (SHAM) sensitive, oxidizes respiratory substrates and produces more heat energy than is produced by the cytochrome pathway, which generates more ATP. Generally, changes in heat development are increased or decreased in close correspondence with the rate of respiration (i.e., the oxygen consumption rate). Thus, in soybean, $dQ/dO_2$ (heat production per oxygen consumed) ratios were higher when the alternative pathway activity was higher [51]. Both the capacity and the activity of the alternative pathway were found to be much higher in cotyledon-purified mitochondria (CPM) than in hypocotyl-purified mitochondria (HPM) and, accordingly, $dQ/dO_2$ ratios were again higher in CPM. In 4-day-old roots, respiration of soybean proceeded almost entirely via cytochrome $c$ oxidase (COX). By day 17, however, more than 50% of the flux occurred via alternative oxidase (AOX), which resulted in a substantial decrease in the theoretical yield of ATP synthesis and concomitantly root relative growth rate [52]. Decreases in whole-root respiration during growth of soybean seedlings can be largely explained by decreases in maximal rates of electron transport via COX. In the case of increased AOX, the ubiquinone pool can be maintained in a moderately reduced state.

In wheat (*Triticum aestivum*) the initial growth during the first 21–24 hr showed no sensitivity to KCN. Salicylhydroxamic acid and disulfiram as inhibitors of the alternative path were, however, almost completely inhibitory if added at any time until at least day 4 or for 3 days after inhibition, respectively [53]. The alternative path was dominant and decreased with the concomitant development of the cytochrome path, indicating that the initial growth of germinating wheat seedlings depends essentially on the alternative path. In sunflower plants, the highest respiration rates were observed in young leaves followed by old and mature leaves [54]. Cyanide had no effect on young leaves but it enhanced respiration in mature and old leaves. SHAM reduced respiration in young leaves, indicating that the major portion of respiration at this stage is based on alternative respiration, which coincides with the results obtained in the case of wheat. In extracts from whole roots of different ages, the ubiquinone pool was maintained at 50 to 60% reduction, whereas the pyruvate content fluctuated without a consistent trend. The amount of mitochondrial protein on a dry-mass basis, however, did not vary significantly with root age.

In rice plants, switching from the cytochrome pathway to the alternative cyanide-resistant respiratory pathway can be exogenously induced, e.g., by application of the rice blasticide SSF126. This chemical (like others) catalyzes the transformation of the high-molecular-weight form of the oxidase to the low-molecular-weight form in which the alternative pathway is preferentially operational. Thus, application of this chemical is suited to artificially switching between the two pathways, affecting the recovery from the rice blast symptoms [55].

## B. Photorespiration

Photorespiration is still a somewhat enigmatic process whose significance for plant physiology is not really understood. It is based on the bifunctionality of the enzyme responsible for carbon dioxide assimilation, namely ribulose-1,5-bisphosphate carboxylase/oxygenase (Rubisco). The reaction of the enzyme with atmospheric oxygen did not appear to be of any obvious advantage for the plant as it did not result in a net catabolic gain with respect to the carbohydrate content of the cells. On the contrary, it looked like a mere waste of energy and a waste of carbon compounds as a consequence of the oxygenase reaction—carbon dioxide is evolved in the light (instead of being taken up) and oxygen is taken up (instead of being evolved). However, detailed bioenergetic investigations of the process have shown that in fact photorespiration is essential for plants and that inhibiting it or switching to nonphotorespiratory conditions is detrimental for plants (compare Sec. V of this chapter). The counteracting partial reactions of photosynthesis and photorespiration are illustrated schematically in Figure 11.

Following the oxygenation of the C5 compound ribulose bisphosphate, photorespiration comprises a cyclic series of reactions with the participation of three different organelles—chloroplasts, peroxisomes, and mitochondria—in a reaction sequence called the C2 cycle from the initial compound phosphoglycolate, the smaller product of the oxygenation reaction. (The splitting of the oxygenated five-carbon compound produces a C2 structure, phosphoglycolate, together with a C3 compound, "normal" 3-phosphoglycerate (3-PGA), the "same" as that is formed after carboxylation.) Specific conditions such as high temperatures or low $CO_2/O_2$ ratios steer the system toward higher rates of oxygenation and lower carboxylation. Details of the overall cycle can be found in every modern textbook on plant physiology. Principally, the photorespiratory C2 cycle can be understood as recovery of three fourths of the carbon from the formed phosphoglycolate because two glycine molecules are converted to one serine inside the mitochondria so that finally "only" one carbon of four is lost as carbon dioxide. The same reaction step yields

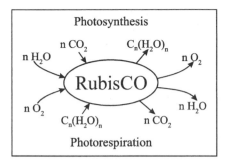

**Figure 11**   Schematic diagram summarizing the counteracting effects of photosynthesis and photorespiration with respect to the oxygen and carbon dioxide gas exchange. (The photorespiratory $NH_3$ liberation is omitted.)

ammonia, which cannot easily be taken as a measure for photorespiratory activity as $NH_3$ (for reasons of both bioenergetic economy and toxicology) is rapidly refixed via the GOGAT system (GOGAT = glutamine-2-oxoglutarate aminotransferase).

Many crops have been investigated with respect to their photorespiration rates—not least with the background idea of the search for increased photosynthetic yields under conditions of low photorespiration. This theory, however, could not be substantiated and in many cases even the contrary turned out to be correct. Nonphotorespiratory conditions did not result in higher yields of plants and crops; they were in general detrimental for the plants and sometimes even lethal. In *Dunaliella tertiolecta*, photorespiratory metabolism was quantified by determining the concentrations of extracellular dissolved glycolate or intracellular free pools of serine and glycine as the parameter in field studies [56]. In this case, the amount of glycolate was light dependent and reached 100 nmol ($10^6$ cells)$^{-1}$ for a cell concentration of around 1.5 $\times$ $10^8$ cells $L^{-1}$ which "disappeared" from the dissolved phase in the dark. Under photorespiratory conditions, i.e., elevated oxygen partial pressure, the activities of glycolate oxidase, hydroxypyruvate reductase, and catalase were decreased 10–25% by elevated $CO_2$ in late reproductive growth of soybean. Serine concentrations were concomitantly decreased at elevated $CO_2$ concentrations [57]. In spinach leaves, the required reducing equivalents for serine reduction to glycerate in the peroxisomes were provided by mitochondria via the malate-oxaloacetate (OAA) shuttle, in which OAA was reduced in the mitochondrial matrix by NADH generated during glycine oxidation [58]. Redox equivalents can be transferred from the mitochondria to peroxisomes for glycerate formation in the photorespiratory cycle because a very low reductive state of the NADH/NAD system prevails in the cytosol of mesophyll cells during photosynthesis. The rate of peroxisomal glycerate formation and the malate/OAA ratio were similar in both a reconstituted system of spinach and the cytosol of mesophyll cells of intact illuminated spinach leaves. The malate/OAA ratio was in equilibrium with an NADH/NAD ratio equivalent to 8.8 $\times$ $10^{-3}$ [58].

In C3 plants, the competition between $CO_2$ and $O_2$ on the active site of Rubisco is limiting for the carbon dioxide assimilation rates in the sense that elevated $CO_2/O_2$ ratios enhance photosynthesis with simultaneously inhibited photorespiration and vice versa. Lowering the $O_2$ partial pressure or elevating that of $CO_2$ (2% $O_2$ or 1000 ppm $CO_2$, respectively) is conventionally used to inhibit photorespiration significantly. When $CO_2$ fixation by Rubisco is limited in C4 plants, an increase in the $CO_2$ concentration in bundle sheath cells via the C4 pathway may further reduce the oxygenase activity of Rubisco. Decreased oxygenase activity of Rubisco decreases the inhibition of photosynthesis under high partial pressures of $O_2$ while it increases $CO_2$ leakage and overcycling of the C4 pathway [59].

Generally, an increasing external $CO_2$ concentration leads to an immediate increase in the internal $CO_2$ concentration in the leaf, accelerated photosynthetic activity, and—at least transiently—repressed photorespiration. Under these conditions, net photosynthesis of soybean increased 56% on average and photorespiration decreased 36% in terminal mainstem leaves [57]. Under permanently elevated concentrations, however, the down-regulation of photosynthesis counteracts this effect and photosynthesis may come down to normal levels. A decrease in the activity and quantity of Rubisco and a decrease of messenger RNA (mRNA) encoding Rubisco activase and chlorophyll-binding proteins contribute to acclimation to elevated $CO_2$ values [60–62]. However, in some cases the rate of photosynthesis per unit leaf area was increased by 20–30% in plants exposed to doubling of the ambient $CO_2$ concentration [60,63]. Moreover, elevated $CO_2$ treatment prevented the ozone-induced suppression of net photosynthesis and photorespiration of soybean, which amounted to 30 and 41%, respectively, in the ambient $CO_2$ level [57]. Such photosynthetic increase is frequently attributed to inhibition of the oxygenase activity of Rubisco, i.e., inhibition of photorespiration.

Electron transport rates of photorespiratory systems are expected to be different from nonphotorespiratory ones because of the exclusion of the consumption of reducing equivalents in the carbon reduction pathway. Thus, suppression of the photorespiratory pathway will in turn affect electron transport; e.g., the resulting net oxygen evolution under nonphotorespiratory conditions is expressed as percent stimulation of the oxygen gas exchange under photorespiratory conditions and taken as a measure of the photorespiration rate of the respective organism (see Warburg effect in plant physiology textbooks). The light-dependent linear electron transport was decreased more than 90% at a photon flux density of 800 $\mu$mol m$^{-2}$ sec$^{-1}$ under conditions of inhibited photosynthesis and photorespiration (by either HCN or glycolaldehyde) in intact leaves of spinach (*Spinacia oleracea* L.) and sunflower (*Helianthus annuus* L.) [64]. Concomitantly, nonphotochemical quenching of chlorophyll fluorescence was increased after inhibition of $CO_2$ assimilation and photorespiration to dissipate excess excitation energy. Despite the effective nonphotochemical energy quenching, appreciable oxygen-dependent photoinactivation was observed not only of photosystem II but also of photosystem I; it was significantly reduced or even completely absent when the oxygen concentration of the atmosphere was reduced from 21% to 1%. This observation illustrates the importance of Mehler reactions in trapping excess electrons under these conditions [64].

The photosynthetic electron transport rates usually exceed the capacity of carbon reduction and usually there is an "excess" of electrons that might be used, e.g., for $NO^{3-}$ and $NO^{2-}$ reduction or even for a reduction of the quinone pool. Laisk and Edwards [65] evaluated the photosynthetic linear electron transport rate in excess of that used for $CO_2$ reduction in *Sorghum bicolor* Moench. [NADP–malic enzyme (ME)–type C4 plant], *Amaranthus cruentus* L. (NAD-ME–type C4 plant), and *Helianthus annuus* L (C3 plant) leaves at different $CO_2$ and $O_2$ concentrations. Under high light intensities there was a large excess of electron transport at 10–100% $O_2$ in the C3 plant because of photorespiration but very little in *Sorghum* and somewhat more in *Amaranthus*, showing that photorespiration is suppressed more in the NADP-ME– and less in the NAD-ME–type species. In C4 plants, such excess was very sensitive to the presence of $O_2$ in the gas phase, rapidly increasing between 0.01 and 0.1% $O_2$; at 2% $O_2$ it was about two thirds of that at 21% $O_2$. This shows the importance of the Mehler-type $O_2$ reduction as an electron sink compared with photorespiration in C4 plants [65]. However, the rate of the Mehler reaction is still too low to account fully for the extra ATP that is needed in C4 photosynthesis. In a mutant of *Festuca pratensis*, the calculated electron flux through the photosystem was substantially higher than in the wild type and more electrons were directed into the photorespiratory chain [66]. Treatment of the plants with the photorespiratory inhibitors phosphinothricin (PPT) and aminooxyacetic acid (AOA) for more than 1 hr induced a depletion in the ratios of $F_v/F_m$, $F_v/F_o$, and $F_m/F_o$—in spite of the existence of a good linear correlation between the photochemical efficiency of PSII and the quantum yield of $CO_2$ assimilation [67].

C4 photosynthesis has long been known to be virtually $O_2$ insensitive. However, a dual inhibitory effect of $O_2$, below or above the optimum partial pressure (5 kPa), on the net rate of $CO_2$ assimilation among species representing all three C4 subtypes from both monocots and dicots was found and described [68]. Apparently, inhibition of net $CO_2$ assimilation with increasing $O_2$ partial pressure above the optimum has to be associated with photorespiration, whereas inhibition at suboptimal $O_2$ concentrations may be caused by a reduced supply of ATP to the C4 mechanism. In C4 plants, inhibition of photochemical reactions such as PSII quantum yield, increased state of reduction of $Q_A$, and decreased efficiency of open PSII centers could account for photosynthesis inhibition under low $O_2$ partial pressure [68]. Photorespiration appears to buffer the quantum efficiency of $CO_2$ assimilation from changes associated with decreases in the rate of $CO_2$ fixation resulting from imbalances in photosynthetic photon Feux density (PPFD) absorption by PSI and PSII [69]. A photorespiratory response to oxygen has also been reported for the leaves of maize plants [69]. However, in this case the authors argued that the possible occurrence of photorespiration in maize leaves, which could result from an inhibition of the $CO_2$ concentrating mechanism, cannot account for the decrease in the quantum efficiency of $CO_2$ assimilation. Atmospheric levels of $O_2$ (20 kPa) caused increased inhibition of photosynthesis as a result of higher levels of photorespiration in the C4 cycle–limited mutant of *Amaranthus edulis* (a phosphoenolpyruvate carboxylase–deficient mutant). Thus, the optimal $O_2$ partial pressure for photosynthesis was reduced from approximately 5 to 1–2 kPa

$O_2$, becoming similar to that of C3 plants [59]. Therefore, the higher $O_2$ requirement for optimal C4 photosynthesis must be specifically associated with the C4 function. With the Rubisco-limited *Flaveria bidentis* (an antisense transformation of the small subunit of Rubisco as a C3 cycle–limited transformant), there was less inhibition of photosynthesis by supraoptimal levels of $O_2$ than in the wild type. The optimum $O_2$ partial pressure for C4 photosynthesis at 30°C, atmospheric $CO_2$ levels, and half-full sunlight (1000 μmol quanta $m^{-2}$ $sec^{-1}$) was about 5–10 kPa [68]. Photosystem II activity, measured as chlorophyll *a* fluorescence, however, was not inhibited by $O_2$ levels above the optimum for $CO_2$ assimilation but was inhibited by suboptimal ones [68].

Photorespiration, by definition, is a light-dependent evolution of $CO_2$ and thus it can be traced by a $CO_2$ evolution signal instantly following a light phase. This signal usually lasts for up to 1–2 min. The rate and time of photorespiratory $CO_2$ postirradiation burst in wheat leaves are suppressed by the PSII inhibitor 3-(3,4-dichlorophenyl)-1,1-dimethylurea (DCMU) [7]. This postirradiation burst is dependent on the electron transport system and on the PSII activity. However, the relationship between PSII electron transport and $CO_2$ assimilation remained similar throughout state transitions in maize leaves [69]. In wheat leaves, the carbon needed for long-term $CO_2$ evolution in the $CO_2$-free air might be derived not only directly from the pool of intermediates in the Calvin cycle but also indirectly from a remotely fixed reserve of photosynthates in the leaf via a photorespiratory carbon oxidation–mediated mobilization process [7]. Such a mobilization process of photosynthates probably played an important role in the coordination of photochemical reactions and carbon assimilation during photosynthesis in C3 plants under photoinhibitory conditions. In addition, photorespiratory losses of $CO_2$ in transgenic tobacco plants or subunit 1 of cotton seed (*Gossypium hirsutum*) were significantly reduced with increasing catalase activities at 38°C, indicating that the stoichiometry of photorespiratory $CO_2$ formation per glycolate oxidized normally increases at higher temperatures because of enhanced peroxidation [70]. The Calvin cycle metabolites, and especially those requiring ATP and/or NADPH for their metabolism such as 3-PGA or triose-P, would control the photosynthetic electron transport capacity when photorespiration is blocked. Under low-temperature conditions (18°C), there was nearly complete loss of $O_2$ sensitivity of photosynthesis at normal ambient levels of $CO_2$ in the flag leaf of rice (*Oryza sativa*), in contrast to the large enhancement of photosynthesis by supra-atmospheric levels of $CO_2$ and subatmospheric levels of $O_2$. These conditions induce a suppression of photorespiration; i.e., there is no limitation in utilizing the initial product of $CO_2$ assimilation (triose-P) as predicted from the kinetic properties of Rubisco [71].

## C. Maintenance Respiration

In general, respiration can be defined as a mechanism to gain energy equivalents from the oxidation of an appropriate substrate. The energy is then used for various physiological demands with respect to the overall energy budget of a growing cell. The utilization of assimilates for the synthesis and maintenance of plant materials can be described by two respiratory components: growth respiration and maintenance respiration [72]. A third component of respiration can be related to energy costs for ion uptake against a concentration gradient, and this is termed ion respiration. Growth respiration represents the cost of converting assimilates into new structural plant constituents [73,74], while the maintenance coefficient represents the energy required to maintain biomass. It is likely that maintenance respiration is dependent upon the tissue composition, the growth environment, and the temperature in particular [72]. The most important processes utilizing energy of maintenance respiration may be protein turnover, compartmentation, and secretion and repair of membranes.

Three methods for determining the maintenance respiration coefficient are described in the literature, each of which is based on a different rationale. These methods are the dark decay method [72], the dynamic method [72], and the zero-growth-rate method [75]. In the dark decay method, the plants are kept in the dark and respiration rates are followed. During the dark period, respiration rates decrease with time until a minimal steady state is attained. Such decline in respiration is ascribed to the fact that under dark conditions the available substrate pools (sugars, organic acids, fatty acids, etc.) are gradually consumed. Following the quantitative depletion of these pools, respiratory rates become minimal just to keep the cells alive. In other words, no respiratory energy is diverted for growth or yield. Under specific conditions and in comparison with earlier work in the literature [e.g., 72], 60 hr of darkness was sufficient for the experimental plants to reach such a minimal rate of respiration. For guidance, these values were about 27, 26, and 18 (mg $CO_2$ (g dry wt)$^{-1}$ $d^{-1}$) for sunflower, maize, and broad bean plants, respectively [76]. The dynamic method assumes that maintenance respiration is represented by the dark $CO_2$ efflux when the net daytime uptake is zero. With net $CO_2$ uptake being zero, there are few readily available assimilates for growth during the following night. With the zero growth rate (or alternatively gross uptake), it is assumed that maintenance occurs when the growth rate or the daily carbon accumulation rate is zero; thus, the total daily assimilate production is utilized in maintenance. In the latter two methods, the maintenance coefficient is calculated by extrapolation because practically zero uptake in the light or zero growth rates, respectively, would not occur.

Schwarz and Gale [77] pointed out that the consumption of assimilates for maintenance processes and possibly their diversion from growth requirements may increase under the demand of environmental stress leading to higher respiration rates. Under conditions of environmental stress, plants may shift energy expenditure from growth to maintenance respiration and repair in order to accommodate the energetic costs of stress [78,79]. This increase in maintenance respiration may be a characteristic feature of salt tolerance insofar as it suggests an ability to divert assimilates and respiratory energy to maintain the biomass [77,80]. For instance, a native salt-tolerant species of *Lycopersicon pennellii* exhibited an increase in maintenance expenditure and a domesticated salt-tolerant species (*L. esculentum*) showed a decrease in root maintenance respiration under exposure to saline conditions [81]. However, Stavarek and Rains [82] described reduced values of maintenance respiration in *Medicago sativa* under conditions of salinity. Wild barley is more tolerant to sulfate salinity than is cultivated barley [83]. Thus, the increase in maintenance respiration may serve as a criterion for evaluation of the ability of a plant to cope with stress. Moreover, it may help to evaluate the cost that the plant must pay for adaptation in terms of allocation of resources [77].

Stress factors other than salinity were also found to affect maintenance and growth respiration. In this respect, Amthor and Cumming [84] found that leaves of *Phaseolus vulgaris* exposed to ozone exhibited a 15% increase in maintenance respiration. Similar patterns have been reported with *Cucumis sativus* under conditions of chilling stress [85]. However, low dark respiration rates and low specific leaf area of the tropical *Pandanus* species have been regarded as important characteristics for growth and survival in environments where resource levels are low and the likelihood of tissue damage is high [86]. Ahmed [87], working with *Chlorella fusca*, reported that maintenance respiration increased when the algae had been exposed to salinity. Huang and Redmann [88] reported similar results based on experiments with wild and cultivated barley plants, depending on the $Ca^{2+}$ availability. On the basis of maintenance respiration coefficient values, the sensitivity of three plants can be arranged as follows: broad bean > sunflower > maize. Broad bean was the most sensitive one and exhibited the highest value of maintenance respiration [76]. Maize, however, exhibited no response of maintenance respiration to water stress but reacted to salinity, whatever the level used [77]. $Ca^{2+}$ lowered the values of maintenance respiration in sunflower, broad bean, and maize plants. Accordingly, Ahmed [87] found that salinized *Chlorella fusca* decreased its maintenance respiration when exposed to $Ca^{2+}$. In barley, maintenance respiration was significantly reduced by low $Ca^{2+}$ treatment but was slightly increased by high $Ca^{2+}$ treatment [88], which might indicate that maintenance respiration can be minimized by appropriate concentrations of $Ca^{2+}$. The explanation might be that $Ca^{2+}$ is a structural component of cell walls and membranes and an indispensable cofactor of photosystem II in addition to its physiological role in the regulation of enzyme activities. It was found that $Ca^{2+}$ reduces respiration in general and maintenance respiration in particular, not least because of its importance for maintaining membrane integrity.

In addition to environmental factors, a variety of internal plant factors affect the magnitude of the portion of respiratory energy that is used for maintenance. For instance, in field populations of *Phragmites australis*, respiratory activity was inversely related to the age of the rhizomes. In the case of 3-year-old rhizomes, values went down to about $2.66 \pm 0.40$, $2.28 \pm 0.40$, and $2.72 \pm 0.40$ µmol $CO_2$ (g dry wt)$^{-1}$ hr$^{-1}$. The specific location played only a minor role in this context [89]. Maintenance respiration rates varied with the tissue size of stems and branches of 9-year-old loblolly pine (*Pinus taeda* L.) but were constant with respect to the nitrogen content of the tissue [90]. In this context, root respiration may account for as much as 60% of total soil respiration [91]. Small lateral roots at the distal end of the root system have much greater tissue nitrogen concentrations than larger roots, and this led to the hypothesis that the smallest roots have significantly higher rates of respiration than larger roots. Nitrogen content in the roots might explain 70% of the observed variation in respiration in sugar maple (*Acer saccharum* Marsh.). The nitrogen values in any case appeared to be a better indicator of root function than, e.g., morphological parameters such as the root diameter. The carbon budget of the lowest Scots pine (*Pinus sylvestris* L.) branches subject to abscission was analyzed for a cost-benefit approach [92]. All but one of the branches studied were found to be nonproductive (the budget was negative) over the growing season. Following a decrease in photosynthetic capacity in July, the cumulative budget became negative and the branches died, indicating that a negative carbon budget corresponds to the onset of abscission of the lowest branches. It has been suggested that maintenance respiration per unit biomass is frequently not constant. Rather, it decreases as biomass increases, so that the ratio of photosynthesis versus respiration is often approximately constant [93]. In these investigations, the photosynthesis/respiration ratio was measured as a quantitative measure of the capability of frost hardening in wintering cereal plants. It was found that the average level of the photosynthesis/respiration ratio in the hardened *Triticum aestivum* plants at $0°C$ was two times higher than that in the control plants at $18°C$ [94].

Experiments suggested that a clear separation of respiration into growth and maintenance components might not be a useful concept because they cannot be unambiguously measured or defined in terms of biochemical processes [95]. Accordingly, growth yield calculations from biochemical pathway analysis, from biomass molecular composition, from biomass heat of combustion, and from biomass elemental composition do not include all of the energy costs for biosynthesis; thus, they are not accurate measures of the carbon cost for plant growth. Improper definitions of growth respiration relations are impeding the use of physiological properties for prediction of plant growth as a function of environmental variables. In accordance with the preceding argument, it was concluded that the failure to account for tissue nitrogen effects on respiration rates results in serious errors when estimating annual maintenance costs [90].

## D. Chlororespiration

The phenomenon of chlororespiration is based on experiments in the 1960s in the laboratories of Bessel Kok, who demonstrated a significant effect of light not only on photosynthesis but also on the respiratory activity of algae that might be interpreted in the sense of a link between the photosynthetic and the respiratory electron transport chains [96]. Later, this observation was substantiated by Pierre Bennoun, who demonstrated respiratory activity in the chloroplast membranes in a green alga (*Chlamydomonas*). The idea was that there should be an electron transfer from NAD(P)H via the plastoquinone pool to oxygen so that $PQH_2$ acts as the component common to both the photosynthetic and the respiratory electron pathways [97]. Achim Trebst's group partially purified the respiratory enzyme NAD(P)H-plastoquinone oxidoreductase (NDH complex) and demonstrated that NAD(P)H could feed electrons into the photosynthetic transport chain [98,99]). For a long time the question was whether the phenomenon was restricted to lower photosynthetic organisms or might also exist in higher plants. Some but not too much evidence from physiological and biochemical data was given to support the latter possibility, e.g., from fluorescence studies in tobacco (*Nicotiana tabacum*), peas (*Pisum sativum*), and maize (*Zea mays*). Interestingly, "chlororespiratory activity" was stronger under conditions of nitrogen deficiency [100–103].

Comparative analyses with cyanobacteria, green algae, and higher plants were carried out to investigate the effect of cyanide on light-induced redox reactions of the Cyt $b_6/f$ complex. The authors described a clear enhancement of the reduction rate of oxidized cytochrome $f$ by appropriate concentration of cyanide, and no significant differences were obtained for pro- and eukaryotes [103]. Moreover, it was observed that coding genes in the liverwort chloroplast exactly corresponded to the genes of mitochondrial NADH dehydrogenase components [104]. This type of activity might also explain the nonphotochemical reduction of plastoquinone in pea leaves [105]. In investigations with sunflower (*Helianthus annuus* L.), leaves exhibited an increased minimal fluorescence following a light-to-dark transition. The increase could be reversed by far-red illumination and inhibited by rotenone or methyl viologen. It was interpreted in the sense that the capacity of the plant for plastoquinone reduction might directly affect the chloroplast ATP synthase [106]. Antimycin A [an inhibitor of the ferredoxin quinone reductase (FQR)] inhibited the cyclic electron flow around photosystem I via NDH in tobacco measured as an enhanced dark fluorescence after a light phase. Interestingly, these reactions were also inhibited by Amytal (an inhibitor of mitochondrial NADH dehydrogenase [NDH]) and by nigericin. The inhibition of plastoquinone reduction showed a biphasic behavior that was taken as evidence for two different mechanisms; the inhibition at low Antimycin A concentrations might be correlated with FQR and the one at higher concentrations with NDH [107].

Mass spectrometric analyses in our laboratory using a mixed isotope composition of the reaction assays have been performed with chloroplasts from *Nicotiana tabacum* to investigate the oxygen gas exchange in the presence of (normal) $H_2\ ^{16}O$ substrate water and $^{18}O_2$ in an artificial gas atmosphere [108]. Upon illumination with 10 short (5 $\mu$sec) saturating xenon flashes and under identical conditions with respect to the background signals of both oxygen isotopes, the tobacco chloroplasts evolved oxygen (as $^{16}O_2$) equivalent to 334 $\mu$mol and (simultaneously) took up 233 $\mu$mol oxygen (as $^{18}O_2$). The photosystem II acceptor silicomolybdate quantitatively inhibited the oxygen uptake signal and so did DBMIB (2,5-dibromomethyl-6-isopropyl-$p$-benzoquinone), a chemical that efficiently oxidizes the plastoquinone pool. The results were interpreted in the sense that a chlororespiratory process in which the plastoquinone pool appears to be involved must be the source for the light-induced oxygen uptake. Theses results were extended by Kowallik's group using chloroplasts from peas [109]. Figure 12 shows parallel recordings of both oxygen evolution (at $m/e = 32$) and oxygen uptake (at $m/e = 36$) in continuous light. (The constant decline in the $^{18}O_2$ signal is technically explained as due to the unidirectional gas flow from the measuring cell to the ion source of the mass spectrometer.) Both the photoevolution of oxygen and the light-induced oxygen uptake are dependent on the intensity of light but independent of the light quality (Figure 13). Also, the light-dependent $^{18}O_2$ uptake increased with increasing light intensity. The effect was identical in blue ($\lambda = 679$ nm) and in red ($\lambda = 679$ nm) light. This effect closely resembled the increase in photosynthetic oxygen evolution in both spectral regions and no matter what the tested fluence rate was. Opposing arguments that such light-induced oxygen uptake might be linked to photorespiration or Mehler-type reactions rather than to chlororespiration may be ruled out by the observation that oxygen consumption occurred even at very low light intensities (Figure 13).

### 1. Physiological Significance

The function of chlororespiration might be related to the idea that under specific conditions algae and plants require a type of valve or overflow mechanism to remove excess reduction equivalents within the chloroplast [110] and supply oxidative pyridine nucleotides for chloroplastic metabolism in the dark [111]. Biochemical and molecular biological studies have described details of the plastid-specific NAD(P)H-plastoquinone oxidoreductase (*ndh* genes) and the homology to the mitochondrial NADH-ubiquinone oxidoreductase in higher plants. Mutagenesis experiments showed that uninjured *ndh* genes are essential for the viability of tobacco plants. Transformants with deleted genes lacked a rapid fluorescence rise in the dark following illumination—the signal that is supposed to indicate the transient reduc-

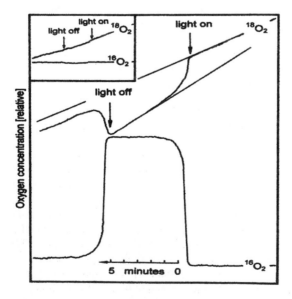

**Figure 12**  Mass spectrometric recording of the oxygen gas exchange of isolated chloroplasts from peas (*Pisum sativum*) in white light of 4 $\mu$mol $- $ m$^{-2}$ $-$ sec$^{-1}$. Oxygen evolution has been recorded as $H_2\ ^{16}O$ oxidation at $m/e = 32$, whereas the concomitant oxygen consumption was measured at $m/e = 36$ from an artificial $^{18}O_2$ gas atmosphere over the aqueous phase of the reaction assay. The inset shows an identical experiment with heat -denatured chloroplasts (10 min, 100°C) as a technical control. (From Ref. 109.)

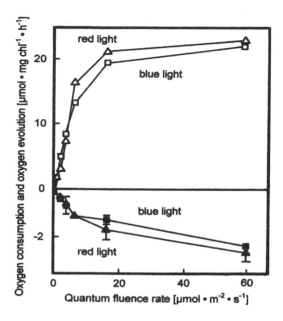

**Figure 13** Dependence of the quantum fluence rate of blue light ($\lambda = 448$ nm) and of red light ($\lambda = 679$ nm) of $^{16}O_2$ evolution and $^{18}O_2$ consumption of isolated chloroplasts from *Pisum sativum*. (From Ref. 109.)

tion of PQ by reduction equivalents from the stroma. High amounts of accumulated starch were observed in transformants with deletions within the *ndh*C-K-J region, and this result was explained by suboptimal oxidation of glucose in both glycolysis and the oxidative pentose phosphate pathway [110]. In largely identical experiments with tobacco plants defective in NAD(P)H dehydrogenase, altered chlorophyll emission behavior ($F_v/F_m$) was taken as evidence for enhanced sensitivity to photoinhibition in the case of the transformants. Repetitive illumination of the *ndh*-defective plants at high light intensity even led to severe responses with respect to the pigments; i.e., the plants showed strong chlorosis and were much less able to recover from the treatment than the wild type, which also showed a smaller effect [107,112]. In the case of the diatom *Phaeodactylum tricornutum*, it was shown that chlororespiration appeared to contribute to the proton gradient needed for the formation of diatoxanthin and that this proton gradient was as effective as is that in the case of a light-driven $\Delta$pH [113]. A type of control of photosynthesis by varying reaction rates of chlororespiration appeared to be much more significant (and important for the plants) under heat stress, i.e., under conditions of elevated temperatures [103].

Interesting investigations of the distribution of chlororespiratory activities within a plant came from molecular biological and immunological analyses. Fragments of NAD(P)H-plastoquinone oxidoreductase from barley were expressed in *Escherichia coli* and antisera against a protein of approximately 70 kDa were prepared. From these experiments, enhanced *ndh*F levels were calculated for etiolated tissue in relation to greening leaves. The values were higher in roots than in leaves, and on a timely basis *ndh*F values decreased during senescence. Photo-oxidative treatments generally increased the levels [114]. Chlororespiration appears to control and regulate the activity of photosystem II as it "mediates" the overall rate of electron flow through the transport chain between the two photosystems and affects the redox condition within the sequence. At least in green algae it was observed that, e.g., acetate enhances chlororespiration rates. Under these conditions the photosystem II activity is down-regulated, probably in order to avoid overreduction at specific sites (where, e.g., destructive reactive oxygen species might otherwise be produced). In principle, this corresponds to any condition of heterotrophic growth where the water-splitting capacity is decreased because of the presence of reduced carbon sources [115]. Generally, it was suggested that chlororespiration from the onset of illumination serves to prevent any overreduction of the transport carriers involved in the electron transport system of higher plants (maize) correlated with lower Calvin cycle rates [102]. Structural evidence for the necessity for a chlororespiratory mechanism in plants might as well be derived from the complete nucleotide sequencing of the genome of *Epifagus* *virginiana* (a nonphotosynthetic parasitic plant) showing that all photosynthetic genes are missing and that the plant does not contain any of the genes that have been described in the context of chlororespiration up to now [116].

## IV.  ENVIRONMENTAL ASPECTS

Plant physiological investigations play an essential role in the modern detection and quantification of the detrimental effects of air pollutants, plant protective chemicals, and other (possibly problematic) substances. Fluorescence emission parameters have been shown to reflect directly the "fitness" of plants, and in many cases injuries or damage by pollutants or toxic substances, early phases of diseases, etc. have been inferred from effects on fluorescence long before macroscopic parameters (e.g., lesions) on leaf surfaces could be observed. The techniques are based principally on the classical work by Duysens and Sweers [117], who observed and described the specificity of room temperature fluorescence for photosystem II and many details of the process. Since then, fluorescence emission measurements have been used to investigate the effects of chemicals of any type on plants. In other cases, the technique has been ("inversely") adapted to examine the load, e.g., of local water (in lakes, brooks, or rivers) by simply adding the respective probe to plants or algae and recording the resulting fluorescence behavior of the test organism. One of the main advantages is that the industry now offers simple, compact, and handy instruments and the result is obtained almost immediately.

In our laboratory, we have shown that simple spray application of plant protective chemicals such as triforine fungicides and pyrethroid insecticides (in any case, nonherbicides) to intact leaves of tobacco (*Nicotiana tabacum*) results in an increase of the maximal fluorescence emission ($F_{max}$) and a much less pronounced Kautsky effect [118,119]. This was a completely unexpected result because fungicides and insecticides were not supposed to have any negative effect on plants. In the case of the pyrethroid insecticides, the site of inhibition of fenvalerate could even be localized (by means of oxygen gas exchange analyses) and shown to be identical to that of the standard herbicide(!) diuron (DCMU), namely $Q_B$, on the acceptor side of photosystem II. The ferricyanide Hill reaction was clearly inhibited, whereas the silicomolybdate Hill reaction and the DPiP Mehler reaction were not affected [118,119]. Mass spectrometric techniques have been applied not only for gas analyses in plant physiology but also for the detection and enrichment of compounds, e.g., in whole plants. Thus, polycyclic aromatic carbohydrates have been detected and analyzed at $m/e = 128$–$202$ in probes from plant tree bark. These measurements appeared suited for estimation of pollution (e.g., from traffic) in both urban and rural environments [120].

Recent experiments have dealt with the effects of inorganic compounds such as metal ions and salts on plant physiological processes. Such parameters can be evaluated in terms of both optimal nutrient supply and pollution of the environment, depending on the concentration of the respective molecule. Algae and cyanobacteria are very sensitive to modifications of their aqueous environment and are in many cases well suited for such studies with respect to physiological gas exchange reactions. The oxygen evolution flash pattern (see Sec. II) of *Oscillatoria chalybea* was substantially modified by addition of various inorganic salts (Figure 14). Although we carried out the experiments with rather intact protoplasts, the photosynthetic apparatus was apparently salt limited in the sense that higher chloride concentrations were needed for optimal water oxidation. In this context, manganese (although an indispensable cofactor of the OEC) played only a minor role as manganese sulfate yielded a smaller stimulation of the oxygen amplitudes. Manganese chloride gave higher yields equivalent to those observed with sodium chloride (result not shown). Calcium also stimulated the oxygen evolution rates to a substantial extent. These experiments are under investigation at present and will be discussed in terms of mechanistic implications for the OEC (S Spiegel, KP Bader, submitted). In these cases, however, the required concentrations were rather high and ranged up to 400 mM in the in vitro assays.

Interesting results have come from the laboratory of Yoshihiro Shiraiwa, who has investigated the effects of the trace element selenium on physiological and biochemical parameters in marine coccolithophorids. Addition of selenium to the culture medium of *Emiliana huxleyi*, *Gephyrocapsa oceanica*, and *Helladosphaera* sp. had significant effects on growth, oxygen gas exchange reactions, and chlorophyll content [121]. Figure 15 shows the complex gas exchange analysis in the presence or in the absence of 10 nM selenite, respectively, for two consecutively (one out of the other) inoculated cultures. Interestingly, the Se-deficient ("first") culture did not grow at all with respect to an increase in cell number fol-

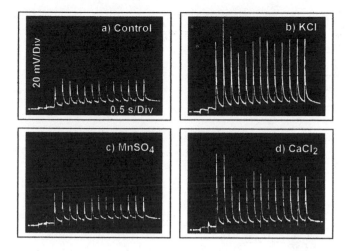

**Figure 14** Effect of inorganic salts (KCl, MnCl$_2$, MnSO$_4$, and CaCl$_2$) on photosynthetic oxygen evolution amplitudes as a consequence of short (5 μsec) saturating flashes in thylakoid preparation from the blue-green alga *Oscillatoria chalybea*. Flash frequency, 3.3 Hz; dark adaptation, 15 min. (From S Spiegel, KP Bader, submitted, 2000.)

lowing transfer to a fresh medium without selenium but showed normal rates in the presence of the metal (Figure 15F). Clear effects were also observed on the oxygen gas exchange, whereas the pigmentation was not significantly affected. Selenium is a specific component of so-called selenoenzymes such as glutathione peroxidase and is covalently bound to this enzyme in the form of selenocysteine. Functionally, it appears to be relevant and even indispensable for the redox buffering of the respective organisms; the effects are still under investigation (see Ref. 121 and references therein).

## V. REFLECTIONS ON CROP YIELD

One of the most prominent goals of modern agriculture and applied plant physiology is to increase the yield of agricultural crops. In principle, this goal can be approached by techniques of plant breeding and/or exogenous applications of compounds such as fertilizers. One of the most obvious and seemingly even trivial ideas was that there should be a direct link between an increase in crop yield and a decrease in photorespiration. Surprisingly, this inverse correlation could not be substantiated. Even direct breeding experiments selecting exclusively for high yields of crops did not decrease the rate of photorespiration! This was one of the most convincing pieces of evidence that the process of photorespiration is by no means the wasteful and useless phenomenon it was long taken for. Moreover, crop mutants that were defective with respect to their photorespiratory activity grew well under nonphotorespiratory conditions (e.g., artificial low oxygen partial pressure in a laboratory setup). As soon as the plants were transferred to the ambient atmosphere, however, the mutation immediately proved lethal. Thus, the process of photorespiration appears to have a beneficial effect for plants at least under appropriate conditions (justifying the high energy demand of the reaction), and this has been discussed and investigated in recent years.

It must be kept in mind, however, that on an evolutionary level Rubisco developed its bifunctionality at a time when there was no substantial oxygen in the atmosphere. Thus, although current investigations might suggest it, a protective function of the oxygenation of ribulose bisphosphate against detrimental concentrations of oxygen cannot have been the original "idea" behind this mechanism. Surprisingly, the oxygenase function of Rubisco also requires preceding activation of the enzyme by a carbon dioxide molecule that is bound to Lys201 of the large subunit of Rubisco, thus forming a carbamate together with the binding of magnesium ions [122,123]. In some cases, elevated productivity was reported, e.g., for mutant genotypes of tobacco selected for under conditions of low carbon dioxide concentrations: The leaf area per plant was larger and photosynthetic rates were higher with similar rates of respiration. However, none of the described positive modifications could be correlated with a decrease in

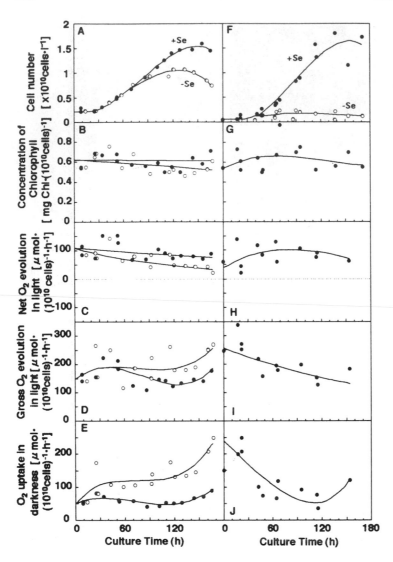

**Figure 15**  Effect of selenium on physiological parameters of *Emiliania huxleyi*: changes in cell number (A and F), the concentration of chlorophyll (B and G), net photosynthesis (C and H), gross photosynthesis (D and I), and dark respiration (E and J) from the first (A–E) and second (F–J) cultures. Aliquots from cells grown for 170 hr in selenium-deficient medium (first culture) were taken and inoculated into fresh deficient medium (second culture). Filled and open symbols represent the presence and the absence of 10 nM selenite, respectively. Lines were drawn according to curve fit functions of Deltagraph 4.0 (Deltapoint Inc. Monterey, CA). (From Ref. 121.)

photorespiration; instead, higher amounts of assimilates, larger leaves, and a better leaf carbon balance leading to improved light absorption and carbon accumulation with a resulting increase in dry matter production have been discussed [124]. Thus, the interpretation still appears valid that high reaction rates of photorespiration seem indispensable for an agriculturally relevant C3 plant and that this process should not be the principal (negative) target for plant breeding and/or physiological experiments aimed at increasing the average yield of crops.

Interesting and promising models have been presented by Marcelis et al. [125] for horticultural crops. Most of the models presented refer to photosynthesis with respect to leaf area development, light interception, and interrelationship with growth and maintenance respiration. Following reports on increases in vegetative growth and the net yield of crops together with a decrease in water use with the spray application of methanol, systematic field experiments were performed with winter wheat (*Triticum aestivum* L.), spring barley (*Hordeum vulgare* L.), and others and the results were unfortunately not unequivocally confirmed [126]. More experiments may be required to substantiate the observations, in particular with respect to a possible inhibitory effect on the water requirement and uptake. In any case, water supply is of course an important parameter in many parts of the world. Of specific interest are experiments on cassava suggesting that selection of appropriate plants and/or breeding experiments in this direction might be promising. *Manihot esculenta* showed effective growth and high yields under conditions of especially irregular rainfall. Under conditions of severe drought, the plants largely maintained their leaf area and were capable of effective and faster growth following rain in comparison with other crops. The plants had modified growth periods with photosynthesis essentially taking place in the morning so that the stomata (which are specifically sensitive to drought in this plant) could be closed during the day, avoiding leaf dehydration [127].

## VI. CONCLUDING REMARK

We feel that even the few examples and results presented in this chapter make clear that improved knowledge is needed not only of many details but also of basic parameters of the interaction between plant physiology, the environment, and specific chemicals that play a role, e.g., in applied science or in agriculture. A reintensification of plant physiological investigations might be more than helpful in the understanding of interrelationships leading to new developments, e.g., in ecological plant protection linked to modern biochemistry.

## ACKNOWLEDGMENTS

The authors wish to express their gratitude to the Alexander von Humboldt-Stiftung (R. A.-B.) and to the Deutsche Forschungsgemeinschaft (K.P.B.) for generous financial support.

## REFERENCES

1. L Bogorad. The biosynthesis of chlorophylls. In: LP Vernon, GR Seely, eds. The Chlorophylls. New York: Academic Press, 1966, pp 481–510.
2. KP Bader, GH Schmid. Cooperative binding of oxygen to the water-splitting enzyme in the filamentous cyanobacterium *Oscillatoria chalybea*. Biochim Biophys Acta 1456:108–120, 2000.
3. CP McKay, H Hartmann. Hydrogen peroxide and the evolution of oxygenic photosynthesis. Origins Life Evol Biosphere 21:157–163, 1991.
4. JF Kasting, HD Holland, JP Pinto. Oxidant abundances in rainwater and the evolution of atmospheric oxygen. J Geophys Res 90:10497–10510, 1985.
5. KP Bader. Physiological and evolutionary aspects of the $O_2/H_2O_2$-cycle in cyanobacteria. Biochim Biophys Acta 1188:213–219, 1994.
6. JHA Nugent. Oxygenic photosynthesis—electron transfer in photosystem I and photosystem II. Eur J Biochem 237:519–531, 1996.
7. J Xiong, S Subramaniam, Govindjee. A knowledge-based three dimensional model of the photosystem II reaction center of *Chlamydomonas reinhardtii*. Photosynth Res 56:229–254, 1998.
8. J Xiong, K Inoue, CE Bauer. Tracking molecular evolution of photosynthesis by characterization of a major photosynthesis gene cluster from *Heliobacillus mobilis*. Proc Natl Acad Sci USA 95:14851–14856, 1998.
9. RE Blankenship, H Hartmann. The origin and evolution of oxygenic photosynthesis. Trends Biochem Sci 23:94–97, 1998.
10. G Renger. Studies on structure and mechanism of photosynthetic water oxidation. In: GA Peschek, W Löffelhardt, G Schmetterer, eds. The Phototrophic Prokaryotes. New York: Kluwer Academic/Plenum Publishers, 1999, pp 35–50.
11. AW Rutherford. How close is the analogy between the reaction centre of photosystem II and that of purple bacteria? Biochem Soc Trans 14:15–17, 1986.
12. PJ Booth, B Crystall, LB Giorgi, J Barber, DR Klug, G Porter. Thermodynamic properties of D1/D2 cytochrome *b*-559 reaction centres investigated by time-resolved fluorescence measurements. Biochim Biophys Acta 1016:141–152, 1990.
13. H-E Akerlund. Oxygen flash-yield sequence of inside-out thylakoids—effects of salt-washing and readdition of a 23 kDa protein. In: C Sybesma, ed. Advances in Photosynthesis Research. Dordrecht: Martinus Nijhoff/Dr W Junk, 1984, pp I.3.391–I.1.394.

14. J Mano, M Takahashi, K Asada. Oxygen evolution from hydrogen peroxide in photosystem II. Flash-induced catalatic activity of water-oxidizing photosystem II membranes. Biochemistry 26:2495–2501, 1987.
15. SP Berg, M Seibert. Is functional manganese involved in hydrogen-peroxide-stimulated anomalous oxygen evolution in $CaCl_2$-washed photosystem II membranes? Photosynth Res 13:3–17, 1987.
16. WP Schröder, H-E Akerlund. $H_2O_2$ accessibility to the photosystem II donor side in protein-depleted inside-out thylakoids measured as flash-induced oxygen production. Biochim Biophys Acta 848:359–363, 1986.
17. J Johansen. A possible role for hydrogen peroxide as a naturally occurring electron donor in photosynthetic oxygen evolution. Biochim Biophys Acta 933:406–412, 1988.
18. RD Hill. What are hemoglobins doing in plants? Can J Bot 76:707–712, 1998.
19. R Abdel-Basset, KP Bader. Physiological analyses of the hydrogen gas exchange in cyanobacteria. J Photochem Photobiol B Biol 43:146–151, 1998.
20. R Abdel-Basset, S Spiegel, KP Bader. Saturation of cyanobacterial photoevolution of molecular hydrogen by photosynthetic redox components. J Photochem Photobiol B Biol 47:31–38, 1999.
21. H Budzikiewicz, H Eckau, HH Inhoffen. Zur photosynthese gruener pflanzen-1: versuche mit $H_2^{18}O$ und $K_2C^{18}O_3$ an *Chlorella pyrenoidosa* CHICK. Z Naturforsch, 24b:1147–1152, 1969.
22. K Biehler, H Fock. Evidence for the contribution of the Mehler-peroxidase reaction in dissipating excess electrons in drought-stressed wheat. Plant Physiol 112:265–272, 1996.
23. EI Rabinowitch. Photosynthesis. Sci Am 178:25–35, 1948.
24. DR Ort, CF Yocum. Electron transfer and energy transduction in photosynthesis: an overview. In: DR Ort, CF Yocum, eds. Advances in Photosynthesis. Vol 4, Oxygenic Photosynthesis: The Light Reactions. Dordrecht: Kluwer Academic Publishers, 1996, pp 1–9.
25. FL Allen, J Franck. Photosynthetic evolution of oxygen by flashes of light. Arch Biochim Biophys 58:124, 1955.
26. P Joliot, G Barbiéri, R Chabaud. Un nouveau modèle des centres photochimique due système II. Photochem Photobiol 10:309–329, 1969.
27. B Forbush, B Kok, M McGloin. Cooperation of charges in photosynthetic $O_2$-evolution—II. Damping of flash yield, oscillation, deactivation. Photochem Photobiol 14:307–321, 1971.
28. J Lavergne, W Junge. Proton release during the redox cycle of the water oxidase. Photosynth Res 38:279–296, 1993.
29. M Haumann, A Mulkidjanian, W Junge. Tyrosine-Z in oxygen-evolving photosystem II: a hydrogen-bonded tyrosinate. Biochemistry 38:1258–1267, 1999.
30. KP Bader, P Thibault, GH Schmid. Study on the properties of the $S_3$-state by mass spectrometry in the filamentous cyanobacterium *Oscillatoria chalybea*. Biochim Biophys Acta 893:564–571, 1987.
31. KP Bader, G Renger, GH Schmid. A mass spectrometric analysis of the water splitting reaction. Photosynth Res 38:355–361, 1993.
32. KP Bader, GH Schmid, G Ruyters, W. Kowallik. Blue light enhanced respiratory activity under photosynthetic conditions in *Chlorella*. Z Naturforsch 47c:881–888, 1992.
33. MR Badger, K Palmqvist, J-W Yu. Measurement of $CO_2$ and $HCO_3^-$-fluxes in cyanobacteria and microalgae during steady state photosynthesis. Physiol Plant 90:529–536, 1994.
34. KP Bader, A Röben. Mass spectrometric detection and analysis of nitrogen fixation in *Oscillatoria chalybea*. Z Naturforsch 50c:199–204, 1995.
35. D Walker. Excited leaves. Tansley Review No 36. New Phytol 121:325–345, 1992.
36. SA Robinson, M Ribas-Carbo, D Yakir, L Giles, Y Reuveni, JA Berry. Beyond SHAM and cyanide: opportunities for studying the alternative oxidase in plant respiration using oxygen isotope discrimination. Aust J Plant Physiol 22:487–496, 1995.
37. SB Idso, BA Kimball. Tree growth in carbon dioxide enriched air and its implications for global carbon cycling and maximum levels of atmospheric carbon dioxide. Global Biogeochem Cycles 7:537–555, 1993.
38. O Monje, B Bugbee. Adaptation to high $CO_2$ concentration in an optimal environment: radiation capture, canopy quantum yield and carbon use efficiency. Plant Cell Environ 21:315–324, 1998.
39. BL Ong, CKK Koh, CY Wee. Effects of $CO_2$ on growth and photosynthesis of *Pyrrosia piloselloides* (L) Price gametophytes. Photosynthetica 35(1):21–27, 1998.
40. H Poorter, S Pot, H Lambers. The effect of an elevated atmospheric $CO_2$ concentration on growth, photosynthesis and respiration of *Plantago major*. Physiol Plant 73:553–559, 1988.
41. H Poorter, RM Gifford, PE Kriedemann, SC Wong. A quantitative analysis of dark respiration and carbon content as factors in the growth response of plants to elevated $CO_2$. Aust J Bot 40:501–513, 1992.
42. J den Hertog, I Stulen, H Lambers. Assimilation, respiration and allocation of carbon in *Plantago major* as affected by atmospheric $CO_2$ levels. Vegetatio 104/105:369–378, 1993.
43. K Chen, G Hu, N Keutgen, M Blanke, F Lenz. Effects of $CO_2$ concentration on strawberry. II. Leaf photosynthetic function. J Appl Bot 71(5–6):173–178, 1997.
44. J den Hertog, I Stulen, F Posthumus, P Hendrik. Interactive effects of growth-limiting N supply and elevated atmospheric $CO_2$ concentration on growth and carbon balance of *Plantago major*. Physiol Plant 103:451–460, 1998.
45. H Lambers, AHJ Frejsen, H Poorter, T Hirose, A van der Werf. Analysis of growth based on net assimilation rate and nitrogen productivity. Their physiological background. In: H Lambers, ML Cambridge, H Konings,

TL Pons, eds. Causes and Consequences of Variation in Growth Rate and Productivity of Higher Plants. The Hague: SPB Academic Publishing, 1990, pp 1–17.

46. T Hirose, DD Ackerly, MB Traw, D Ramseier, FA Bazzaz. $CO_2$ elevation, canopy, photosynthesis, and optimal leaf area index. Ecology 78:2339–2350, 1997.

47. JA Bunce. Responses of respiration to increasing atmospheric carbon dioxide concentrations. Physiol Plant 90:427–430, 1994.

48. SA Felitti, DH Gonzalez. Carbohydrates modulate the expression of the sunflower cytochrome $c$ gene at the mRNA level. Planta 206:410–415, 1998.

49. OK Atkin, JR Evans, K Siebke. Relationship between the inhibition of leaf respiration by light and enhancement of leaf dark respiration following light treatment. Aust J Plant Physiol 25:437–443, 1998.

50. T Endo. Cytochrome $b/f$ complex is not involved in respiration in the cyanobacteria *Synechocystis* PCC 6803 grown photoautotrophically. Bioscience Biotechnol Biochem 61:1770–1771, 1997.

51. LS Geraldes, JD Arrabaca. Respiration and heat production by soybean hypocotyl and cotyledon mitochondria. Plant Physiol Biochem 35:897–903, 1997.

52. AH Millar, OK Atkin, RI Menz, B Henry, G Farquhar, DA Day. Analysis of respiratory chain regulation in roots of soybean seedlings. Plant Physiol 117:1083–1093, 1998.

53. K Kasai, N Mori, C Nakamura. Changes in the respiratory pathways during germination and early seedling growth of common wheat under normal and NaCl-stressed conditions. Cereal Res Commun 26:217–224, 1998.

54. M Pandey, GC Srivasta. Effect of BA and ABA treatment on CN-sensitive and CN-resistant respiration during leaf ontogeny in sunflower. Indian J Plant Physiol 2:297–299, 1997.

55. A Mitzutani, N Miki, K Nabana. Defense mechanism of rice plant to respiratory inhibition by a promising candidate blasticide, SSF126. Pestic Biochem Physiol 60(3):187–194, 1998.

56. C Leboulanger, L Serve, L Comellas, H Jupin. Determination of glycolic acid released from marine phytoplankton by post-derivatization gas chromatography mass spectrometry. Phytochem Anal 9(1):5–9, 1998.

57. FL Booker, CD Reid, HS Brunschoen, EL Fiscus, JE Miller. Photosynthesis and photorespiration in soybean (*Glycine max* (L.) Merr.) chronically exposed to elevated carbon dioxide and ozone. J Exp Bot 48:1843–1852, 1997.

58. AS Raghavendra, S Reumann, HW Heldt. Participation of mitochondrial metabolism in photorespiration: reconstituted system of peroxisomes and mitochondria from spinach leaves. Plant Physiol 116:1333–1337, 1998.

59. JP Maroco, MSB Ku, PJ Lea, LV Dever, RC Leegood, RT Furbank, GE Edwards. Oxygen requirement and inhibition of C4 photosynthesis: an analysis of C4 plants deficient in the C3 and C4 cycles. Plant Physiol 116:823–832, 1998.

60. G Bowes. Growth at elevated $CO_2$: photosynthetic responses mediated through Rubisco. Plant Cell Environ 14:795–806, 1991.

61. T Besford. Photosynthetic acclimation in tomato plants grown in high $CO_2$. Vegetatio 104/105:441–448, 1993.

62. J-J Van Oosten, D Wilkins, RT Besford. Regulation of the expression photosynthetic nuclear genes by $CO_2$ is mimicked by carbohydrates: a mechanism for the acclimation of photosynthesis to high $CO_2$. Plant Cell Environ 17:913–923, 1994.

63. JD Cure, B Acock. Crop responses to carbon dioxide doubling: a literature survey. Agric for Meteorol 38:127–145, 1986.

64. C Wiese, LB Shi, U Heber. Oxygen reduction in the Mehler reaction is insufficient to protect photosystems I and II of leaves against photoinactivation. Physiol Plant 102:437–446, 1998.

65. A Laisk, GE Edwards. Oxygen and electron flow in photosynthesis: Mehler reaction, photorespiration and $CO_2$ concentration in the bundle sheath. Planta 205:632–645, 1998.

66. SAH Kingston, H Thomas, CH Foyer. Chlorophyll *a* fluorescence, enzyme and antioxidant analyses provide evidence for the operation of alternative electron sinks during leaf senescence in a stay-green mutant of *Festuca pratensis*. Plant Cell Environ 20:1323–1337, 1997.

67. MMA Gonzalez, R Matamala, J Penuelas. Effects of prolonged drought stress and nitrogen deficiency on the respiratory $O_2$ uptake of bean and pepper leaves. Photosynthetica 34:505–512, 1997.

68. JP Maroco, MSB Ku, PJ Lea, GE Edwards. Oxygen sensitivity of C4 photosynthesis: evidence from gas exchange and chlorophyll fluorescence analyses with different C4 subtypes. Plant Cell Environ 20:1525–1533, 1997.

69. JR Andrews, NR Baker. Oxygen-sensitive differences in the relationship between photosynthetic electron transport and $CO_2$ assimilation in C3 and C4 plants during state transitions. Aust J Plant Physiol 24:495–503, 1997.

70. LF Brisson, I Zelitch, EA Havir. Manipulation of catalase levels produces altered photosynthesis in transgenic tobacco plants. Plant Physiol 116:259–269, 1998.

71. TL Winder, J Sun, TW Okita, GE Edwards. Evidence of the occurrence of feedback inhibition of photosynthesis in rice. Plant Cell Physiol 39:813–820, 1998.

72. KJ McCree, JH Silsbury. Growth and maintenance requirements of subterranean clover. Crop Sci 18:13–18, 1978.

73. JHM Thornley. Respiration, growth and maintenance in plants. Nature 227:304–305, 1970.

74. FWT Penning De Vries. The cost of maintenance respiration in plant cells. Ann Bot 39:77–92, 1975.

75. JHM Thornley. Mathematical models in plant physiology. London: Academic Press, 1976, p 318.

76.  AMA El-Hakimi. Calcium-drought interactions on carbon strategy, membrane disorders and some related metabolic processes of *Helianthus annus* L., *Vicia faba* L., *Zea mays* L. PhD thesis, Assiut University, Assiut, Egypt, 1999, p 233.

77.  M Schwarz, J Gale. Maintenance respiration and carbon balance of plants at low levels of sodium chloride salinity. J Exp Bot 130:933–941, 1981.

78.  JS Amthor. The role of maintenance respiration in plant growth. Plant Cell Environ 7:561–569, 1984.

79.  GJ Taylor. Maximum potential growth rate and allocation of respiratory energy as related to stress tolerance in plants. Plant Physiol Biochem 27:605–611, 1989.

80.  SM Shone, J Gale. Effect of sodium chloride stress and nitrogen source on respiration, growth and photosynthesis in lucerne (*Medicago sativa* L.). J Exp Bot 34:1117–1125, 1983.

81.  EL Taleisnik. Salinity effects on growth and carbon balance in *Lycopersicon esculentum* and *pennellii* L. Physiol Plant 73:213–218, 1987.

82.  SJ Stavarek, DW Rains. Effect of salinity on growth and maintenance costs of plant cells. UCLA Symp Mol Cell Biol 22:129–143, 1985.

83.  CG Suhayda, RE Redmann, BL Harvey, AL Cipywnyk. Comparative responses of cultivated and wild barley species to salinity stress and calcium supply. Crop Sci 32:154–163, 1992.

84.  JS Amthor, JR Cumming. Low level of ozone decrease bean leaf maintenance respiration. Can J Bot 66:724–726, 1988.

85.  RK Szaniawski. Homeostasis in cucumber plants during low temperature stress. Physiol Plant 64:212–216, 1985.

86.  CE Lovelock. Observation of the photosynthetic physiology of tree species within the C3 monocotyledon genus *Pandanus* and comparison with dicotyledon C3 tree species. Aust J Bot 46:103–110, 1998.

87.  HA Ahmed. Salinity-calcium interactions on growth and some metabolic activities of *Chlorella fusca*. MSc thesis, Assiut University, Assiut, Egypt, 1994, p 130.

88.  J Huang, RE Redmann. Carbon balance of cultivated and wild barley under salt stress and calcium. Photosynthetica 32:23–35, 1996.

89.  H Cizkova, V Bauer. Rhizome respiration of *Phragmites australis*: effect of rhizome age, temperature and nutrient status od the habitat. Aquat Bot 61:239–253, 1998.

90.  CA Maier, SJ Zarnoch, PM Dougherty. Effects of temperature and tissue nitrogen on dormant season stem and branch maintenance respiration in a young loblolly pine (*Pinus takeda*) plantation. Tree Physiol 18(1):11–20, 1998.

91.  KS Pregitzer, MJ Laskowski, AJ Burton. Variation in sugar maple root respiration with root diameter and soil depth. Tree Physiol 18:665–670, 1998.

92.  J Witowski. Gas exchange of the lowest branches of young Scots pine: a cost-benefit analysis of seasonal branch carbon budget. Tree Physiol 17:757–765, 1997.

93.  JA Bunce. Growth rate, photosynthesis and respiration in relation to leaf area index. Ann Bot 63:459–463, 1989.

94.  VV Klimov. Increased photosynthesis: respiration ratio at low temperatures as an important prerequisite for cold acclimation of winter wheat. Fiziol Rast 45:419–424, 1998.

95.  LD Hansen, RW Breidenbach, BN Smith, JR Hansen, RS Criddle. Misconceptions about the relation between plant growth and respiration. Bot Acta 111:255–260, 1998.

96.  G Hoch, OvH Owens, B Kok. Photosynthesis and respiration. Arch Biochem Biophys 101:171–180, 1963.

97.  P Bennoun. Evidence for a respiratory chain in the chloroplast. Proc Natl Acad Sci USA 79:4352–4356, 1982.

98.  D Godde. Evidence for a membrane bound NADH-plastoquinone-oxidoreductase in *Chlamydomonas reinhardtii* CW 15. Arch Microbiol 131:197–202, 1982.

99.  D Godde, A Trebst. NADH as electron donor for the photosynthetic membrane of *Chlamydomonas reinhardtii*. Arch Microbiol 127:245–252, 1980.

100. G Garab, F Lajkó, L Mustárdy, L Márton. Respiratory control over phyotosynthetic electron transport in chloroplasts of higher plant cells: evidence for chlororespiration. Planta 179:349–358, 1989.

101. S Damdinsuren, M Osaki, T Tadano. Quenching of chlorophyll alpha fluorescence by oxygen in normal air in higher plant leaves. Soil Sci Plant Nutr 41:529–537, 1995.

102. S Damdinsuren, M Osaki, T Tadano. Quenching of chlorophyll alpha fluorescence by oxygen in normal air in maize leaves grown under nitrogen deficiency conditions. Soil Sci Plant Nutr 41:539–546, 1995

103. F Lajko, A Kadioglu, G Borbely, G Garab. Competition between the photosynthetic and the (chloro)respiratory electron transport chains in cyanobacteria, green algae and higher plants. Effect of heat stress. Photosynthetica 33:217–226, 1997.

104. K Ohyama, T Kohchi, T Sano, Y Yamada. Newly identified groups of genes in chloroplasts. Trends Biochem Sci 13:19–22, 1988.

105. QJ Groom, DM Kramer, AR Crofts, DR Ort. The non-photochemical reduction of plastoquinone in leaves. Photosynth Res 36:205–215, 1993.

106. TS Feild, L Nedbal, DR Ort. Nonphotochemical reduction of the plastoquinone pool in sunflower leaves originates from chlororespiration. Plant Physiol 116:1209–1218, 1998.

107. T Endo, T Shikanai, F Sato, K Asada. NAD(P)H dehydrogenase-dependent, antimycin A–sensitive electron donation to plastoquinone in tobacco chloroplasts. Plant Cell Physiol 39:1226–1231, 1998.

108. WI Gruszecki, KP Bader, GH Schmid. Light-induced oxygen uptake in tobacco chloroplasts explained in terms of chlororespiratory activity. Biochim Biophys Acta 1188:335–338, 1994.
109. N Grotjohann, D Messdaghi, W Kowallik. Oxygen uptake during photosynthesis of isolated pea chloroplasts. Z Naturforsch 54c:209–219, 1999.
110. W Kofer, H-U Koop, G Wanner, K Steinmüller. Mutagenesis of the genes encoding subunits, A, C, H, I, J and K of the plastid NAD(P)H-plastoquinone-oxidoreductase in tobacco by polyethylene glycol–mediated plastome transformation. Mol Gen Genet 258:166–173, 1998.
111. JW Poskuta. Relationship between photosynthesis and respiration in plants. Wiadomosci Bot 36:47–52, 1992.
112. T Endo, T Shikanai, A Takabayashi, K Asada. The role of chloroplastic NAD(P)H dehydrogenase in photoprotection. FEBS Lett 457:5–8, 1999.
113. T Jakob, R Goss, C Wilhelm. Activation of diadinoxanthin de-epoxidase due to a chlororespiratory proton gradient in the dark in the diatom *Phaeodactylum tricornutum*. Plant Biol 1:76–82, 1999.
114. R Catala, B Sabater, A Guera. Expression of the plastid *ndh*F gene product in photosynthetic and non-photosynthetic tissues of developing barley seedlings. Plant Cell Physiol 38:1382–1388, 1997.
115. T Endo, K Asada. Dark induction of the non-photochemical quenching of chlorophyll fluorescence by acetate in *Chlamydomonas reinhardtii*. Plant Cell Physiol 37:551–555, 1996.
116. KH Wolfe, CW Morden, JD Palmer. Function and evolution of a minimal plastid genome from a non-photosynthetic parasitic plant. Proc Natl Acad Sci USA 89:10648–10652, 1992.
117. LNM Duysens, HE Sweers. Mechanism of two photochemical reactions in algae as studied by means of fluorescence. In: Microalgae and Photosynthetic Bacteria. Japanese Society of Plant Physiology. Tokyo: University of Tokyo Press, 1963, pp 353–372.
118. KP Bader, J Schüler. Inhibition of the photosynthetic electron transport by pyrethroid insecticides in cell cultures and thylakoid suspensions from higher plants. Z Naturforsch 51c:721–728, 1996.
119. KP Bader, R Abdel-Basset. Adaptation of plants to anthropogenic and environmental stress: the effects of air constituents and plant-protective chemicals. In: M Pessarakli, ed. Handbook of Plant and Crop Stress. 2nd ed. New York: Marcel Dekker, 1999, pp 973–1010.
120. A Sturaro, G Parvoli, L Doretti. Plant bark tree as a passive sampler of polycyclic aromatic hydrocarbons in an urban environment. J Chromatogr 643:435–438, 1993.
121. A Danbara, Y Shiraiwa. The requirement of selenium for the growth of marine coccolithophorids, *Emiliania huxleyi*, *Gephyrocapsa oceanica* and *Helladosphaera* sp. (Prymnesiophyceae). Plant Cell Physiol 40:762–766, 1999
122. GH Lorimer, MR Badger, TJ Andrews. The activation of ribulose-1,5-bisphosphate carboxylase by carbon dioxide and magnesium ions. Equilibria kinetics, a suggested mechanism and physiological implications. Biochemistry 15:529–536, 1976.
123. GH Lorimer, MR Badger, TJ Andrews. [cf5]D-Ribulose-1,5-bisphosphate carboxylase-oxygenase. Improved methods for the activation and assay for catalytic activities. Anal Biochem 78:66–75, 1977.
124. H Medrano, AJ Keys, DW Lawlor, MAJ Parry, J Azcon-Bieto, E Delgado. Improving plant production by selection for survival at low $CO_2$ concentrations. J Exp Bot 46:1389–1396, 1995.
125. LFM Marcelis, E Heuvelink, J Goudriaan. Modelling biomass production and yield of horticultural crops: a review. Sci Hortic (Amst) 74:83–111, 1998.
126. SL Albrecht, CL Douglas Jr, EL Klepper, PE Rasmussen, RW Rickman, RW Smiley, DE Wilkins, DJ Wysocki. Effects of foliar methanol applications on crop. Crop Sci 35:1642–1646, 1995.
127. J Itani, T Oda, T Numao. Studies on mechanisms of dehydration postponement in cassava leaves under short-term soil water deficits. Plant Prod Sci 2:184–189, 1999.

# 15

# Diffusive Resistances to $CO_2$ Entry in the Leaves and Their Limitations to Photosynthesis

**Angelo Massacci and Francesco Loreto**

*National Research Council of Italy, Rome, Italy*

## I. LEAF RESISTANCES TO $CO_2$ DIFFUSION

### A. Definitions

Driven by a gradient between $CO_2$ molar fraction in the air and at the active sites of the ribulose-1,5-bisphosphate carboxylase (Rubisco) in the chloroplasts, $CO_2$ diffuses through leaf stomata in the intercellular air spaces present in the mesophyll, then crosses the cell wall and plasmalemma and the chloroplast membrane, and finally reaches the chloroplasts. Each of these steps (Figure 1) constitutes a physical resistance to $CO_2$ diffusion that progressively reduces the $CO_2$ molar fraction [1]. The drop of $CO_2$ caused by the stomatal resistance has been calculated from gas exchange parameters exploiting the fact that $CO_2$ and water vapor share the same diffusion pathway to and from the substomatal cavities (for a review see Ref. 2). The molar fraction of $CO_2$ in the intercellular spaces ($c_i$) can, therefore, be estimated as follows:

$$c_i = c_a - Ar_s \tag{1}$$

where $c_a$ is the ambient $CO_2$ molar fraction, $A$ is the photosynthetic rate, and $r_s$ is the stomatal resistance.

The drop in $CO_2$ between the intercellular air spaces and the Rubisco active sites is not paralleled by the flux of water and is more difficult to calculate. However, the balance between photosynthesis and photorespiration and the correct estimation of the catalytic properties of Rubisco through gas exchange are based upon as reliable as possible calculations of the chloroplastic $CO_2$ molar fraction ($c_c$).

Photosynthesis has been modeled assuming that $c_i$ is the same as $c_c$ [3]. Alternatively, it has been assumed that mesophyll resistance reduces $c_c$ to the compensation point between photosynthesis and photorespiration [4].

Three methods have been used to estimate in vivo the further resistances between the intercellular spaces and the chloroplasts. These studies have indicated that a further drop of $CO_2$ molar fraction into the leaf may be significant and is attributable to what has been called wall resistance [5,6], liquid phase resistance [7], mesophyll resistance [8,9], internal resistance [10,11], $CO_2$ transfer resistance [12], or mesophyll diffusional resistance [13].

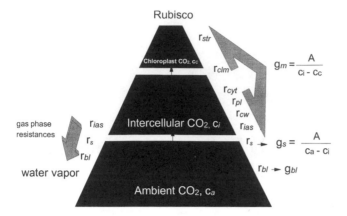

**Figure 1**   Representation of the resistances to $CO_2$ diffusion within a leaf. Resistances caused by diffusion through stomata lower the $CO_2$ molar fraction from ambient ($c_a$) to intercellular ($c_i$). Resistances encountered in the mesophyll lower the $CO_2$ molar fraction from $c_i$ to chloroplastic $CO_2$ ($c_c$). Stomatal and mesophyll conductances can be calculated when $CO_2$ fluxes and molar fraction gradients are known. The $CO_2$ fraction that reaches Rubisco in the chloroplasts drives the photosynthetic process.

## B.   Location of Mesophyll Resistance

Stomatal resistance is attributable only to stomatal movements; it is commonly believed that the resistances encountered by diffusion in the mesophyll are shared between gas phase and liquid phase, but there is no consensus about the relevance of each component. On the one hand, it has been found that decreasing the gas phase resistances by partially substituting helium for the air sometimes stimulates photosynthesis. This would indicate that the gas phase resistances to diffusion are substantial [14]. On the other hand, Loreto et al. [8] found only a limited association between mesophyll resistances and porosity. Moreover, it has been pointed out that $CO_2$ diffusion in liquid is 10,000 times less than in air [7]. Although it is likely that the path length in the liquid phase is short [15], liquid phase resistances may also be considerable. Parkhurst [7] concluded on the basis of porosity and path length that the two resistances may be similar, but gas phase resistances may be prevalent when the mesophyll cells are tightly packed and $CO_2$ entry in the leaf is structurally limited by stomatal distribution (i.e., more in hypostomatous than in amphistomatous leaves). However, this viewpoint has been challenged by Syvertsen et al. [11]. Their diffusional model predicts that diffusion resistances are higher in the liquid phase than in the gas phase in hypostomatous tree species.

In summary, all methods used to estimate the resistances to diffusion within the leaf do not distinguish between resistances encountered at the cell wall, plasmalemma, cytoplasm, or chloroplast membrane. They are also ineffective in partitioning between gas and liquid phase resistances. Because of this, we will denote the series of resistances between intercellular air spaces and Rubisco sites by the generic term mesophyll resistance ($r_m$), which includes both gas and liquid phase resistances. These resistances are directly proportional to the gradient of carbon molar fraction and inversely proportional to the flux density of carbon (the photosynthetic rate $A$):

$$r_m = (c_i - c_c)/A \tag{2}$$

The total diffusion resistance ($r_{tot}$) is the sum of the stomatal and mesophyll resistances. Because the diffusion resistances are expressed in term of flux, it is frequently convenient to use their inverses, i.e., conductances. By knowing the stomatal ($g_s$) and mesophyll ($g_m$) conductances, the total conductance to $CO_2$ can be calculated:

$$g_{tot} = 1/g_s + 1/g_m \tag{3}$$

We will not address the biophysics of the diffusion in the leaf, which has already been extensively described by Parkhurst [7]. Comprehensive reviews of the methods used and of their accuracy have also appeared [16]. This chapter will focus on the ecological significance of diffusion resistances, including the species-specific capacity for diffusion of gases in trees and herbaceous plants, and the effect of environmental stresses, leaf ontogeny, and increasing $CO_2$ concentration on diffusion resistances. In conclusion, the photosynthesis limitations caused by diffusion resistances are investigated.

## II.   SPECIES-SPECIFIC LEAF RESISTANCES

## A.   Species-Specific Capacity for Stomatal Conductance

The capacity for leaf $g_s$ often varies significantly between different species, within leaves of genotypes of the same species, and between leaves of the same plant depending on leaf age and insertion [17]. This large variation may be attributed to the leaf morphological characteristics, which, in turn, are mainly controlled by the leaf water status. Stomatal conductances shown in Figure 2 have been selected from measurements on plants under nonlimiting conditions (see figure legend) for each species and have been primarily grouped according to a general classification as hydrophytes, mesophytes, and xerophytes.

Among hydrophytes, we take as an example *Phragmites australis* and *Carex* spp., Graminaceae species from wetland habitats. Leaves of these plants are broad and flat and have many invaginations of the upper epidermis, facilitating gas exchanges. Stomata are mostly in the lower epidermis and bulliform cells are present in the upper epidermis. The mesophyll tissue is highly lacunose [18,19]. The $g_s$ of hydrophytes on average is 0.20–0.30 mol m$^{-2}$ sec$^{-1}$.

The mesophyte group includes many species typical of land habitats with no severe moisture and temperature stresses. It includes woody trees and grass species. The latter are further distinct as C$_3$ and C$_4$ species. Despite the differences in biochemistry and anatomy, there is no significant difference in $g_s$ between these subgroups. Stomatal conductance is, for instance, around 0.2 mol m$^{-2}$ sec$^{-1}$ in the woody mesophyte *Prunus avium* [20], in C$_3$ herbaceous species such as wheat [21], and in C$_4$ leaves such as those of sorghum [22]. Leaves of both these C$_3$ and C$_4$ herbaceous species are amphistomatous, but C$_4$ plants present a special bundle sheath surrounding the veins containing large, conspicuous chloroplasts and from which the mesophyll radiates [23].

The last group, the xerophytes, includes many plant species (particularly trees) typical of arid or semiarid environments. These species have a very thick cuticle, epidermis, and palisade. The lower epidermis layer often creates an invagination that contains many trichomes as well as all of the stomata [24]. Leaves of these plants have low $g_s$ as in the case of the Mediterranean sclerophyllous tree *Quercus ilex* (0.1–0.15 mol m$^{-2}$ sec$^{-1}$) [8] and of the boreal tree *Larix x eurolepis* (0.1 mol m$^{-2}$ sec$^{-1}$) [25]. *Larix* leaves are characteristic because they show xerophytic features despite vegetating in a cold habitat. In

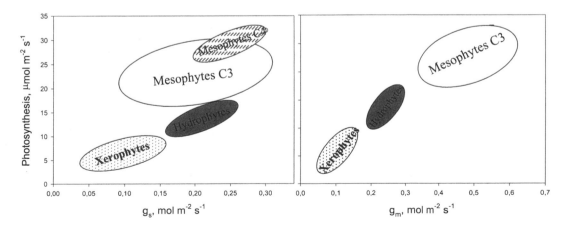

**Figure 2**   Relationship between photosynthesis and stomatal ($g_s$) and mesophyll ($g_m$) conductances in different groups of plants. Data for hydrophytes (*Carex* spp. and *Phragmites* spp.) are unpublished. Data for C3 and C4 mesophytes are from Refs. 9, 12, 13, 21, 48 and 4, 22, 39, respectively. Data for xerophytes are from Refs. 5, 6, 10, 19, 25, 50.

fact, the leaves have sunken stomata in channels of the lower epidermis filled with wax and attached to subsidiary cells resembling the crypt invagination (M. Centritto et al., unpublished).

The data set of Figure 2 shows that a clear specificity exists for the capacity of $g_s$ that reflects substantially the adaptation of stomata and leaves to the water status of the different habitats. This is in agreement with the interpretation that the dominant stimulus for adaptation is the prevention of water loss impairment of growth and the relevant biophysical and biochemical processes [26].

## B.   Species-Specific Capacity for $g_m$

Mesophyll conductance has the same order of magnitude of stomatal conductance, and reported values range between 0.02 and 0.7 mol m$^{-2}$ sec$^{-1}$. Loreto et al. [8] showed an empirical relationship between $g_s$ and $g_m$ with $g_m = 1.4g_s$. However, mesophyllous plants (included in the hydrophyte and mesophyte groups of Figure 2) apparently have a higher $g_m$ than sclerophyllous trees when $g_s$ is similar. The reason for this difference is unknown. Plants with hypostomatous leaves, thick mesophyll, or several layers of palisade cells, such as those showing xerophytic adaptations (see Figure 2), may have high internal resistances to $CO_2$ diffusion. Tree leaves, in fact, possess all of the described features, and the low $g_m$ observed in these plants may be simply caused by the longer distance between substomatal sites and Rubisco active sites compared with that of herbaceous leaves. Trees may also have a high density of cells and, consequently, a low porosity in the mesophyll, which would make the $CO_2$ path toward the chloroplasts more tortuous and difficult. Nobel [1] observed that the morphological parameter that correlates better with mesophyll resistances is the ratio between mesophyll cell wall area and leaf area. Following an early suggestion of Laisk et al. (27), Evans et al. [12] pointed out that diffusion resistances increase when the chloroplast surface exposed to the intercellular spaces decreases while the total mesophyll surface exposed to intercellular spaces does not change. Sharkey et al. [28] noticed that $g_m$ was low in mutants of *Nicotiana* characterized by cupped chloroplasts, a feature that made inhomogeneous chloroplasts adhere to cell walls and may have created further resistance to gas diffusion while crossing the cytoplasm. Although results do not conclusively identify the factors involved in the relationship between leaf anatomy and diffusion resistances, they suggest that a reduction of both photosynthesis and $g_m$ may in fact be caused by anatomical changes related to chloroplast exposure to air spaces and shape.

## III.   EFFECT OF LEAF RESISTANCES ON PHOTOSYNTHESIS

## A.   Effect of $g_s$ on Photosynthesis

Mostly hydrophytes and xerophytes show a good proportionality between the capacity for $g_s$ and photosynthesis. In mesophytes, however, the relationship is not so clear as shown by the large variation of $g_s$ in plants having similar photosynthesis (Figure 2). In addition, $C_4$ plants may have much higher photosynthesis than $C_3$ plants, but a comparable $g_s$, because of the biochemical mechanism that concentrates $CO_2$ to very high levels inside the mesophyll.

Thus, there are indications that $g_s$ may limit photosynthesis. The limitation is clear at low $g_s$ and less evident at high $g_s$ despite the outlined differences in the leaf morphological characteristics. Wong et al. [29] showed that the relationship between conductance and photosynthesis is due to the tendency of plants to maintain a proportionality between the calculated internal and the measured external $CO_2$ molar fractions. This observation, made on plants of several species grown under different light environments and subjected to different nitrogen nutrition, was further investigated. Farquhar and Wong [30] explained that this proportionality depends on the response of stomata to the pool size of a photosynthetic substrate. Subsequently, Jarvis and Davies [31] held this concept in their model of the stomata response to photosynthesis and developed the idea that stomata respond to "a signal in proportion to the degree to which the photosynthetic capacity is realized."

In Figure 3 the relationship between $A$ and $g_s$ passing through the origin is shown for three different genotypes of sorghum grown in the field under irrigated and nonirrigated conditions. Similar results have been reported for three genotypes of wheat under the same water conditions of sorghum [21]. In both cases, it appears that the proportionality between $A$ and $g_s$ holds. Therefore, even when the leaf water deficit reduces stomatal conductance, the stomatal limitations probably contribute to the overall reduction

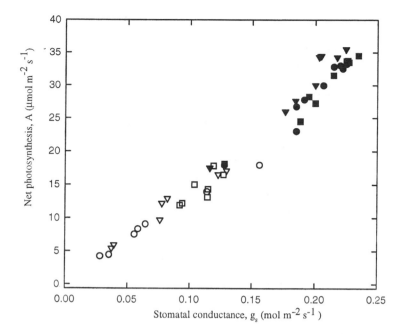

**Figure 3**   Relationship between photosynthesis (*A*) and stomatal conductance (*g*$_s$) in leaves of sorghum geno-types (□, SV1; ○, Red Swazi; ▽, 8504H) grown in the field near Rome under irrigated (filled) and nonirrigated (open) conditions and measured before anthesis at different times of the day.

of photosynthesis independently of drought sensitivity of the different species and genotypes. On the other hand, changes in light exposure, inorganic nutrition, leaf age, and general acclimation or long-term stress effects have been suggested to produce differences in the slope of the correlation between *A* and *g*$_s$ and, consequently, may modify the stomatal limitations of photosynthesis [17]. Finally, several examples of a lack of correlation between *A* and *g*$_s$ can be found in the literature, probably reflecting a lack of control of stomatal conductance on photosynthesis and the strong involvement of nonstomatal limitations [32,33]. As an example, it has been shown [33] that the inhibition of adenosinetriphosphatase (ATPase) is responsible for nonstomatal effects on photosynthesis in plants subjected to water stress.

## B.   Effect of *g*$_m$ on Photosynthesis

Because sclerophyllous plants also have low stomatal conductance (Figure 2), it has been suggested that photosynthesis in these plants may be limited by the inherently high resistances to gas diffusion and by the consequently low molar fraction of $CO_2$ in the chloroplasts. However, similarly to what has been observed for *g*$_s$ (see Sec. III.A), this implies that photosynthesis should drop proportionally less than *g*$_m$. Although *g*$_m$ correlates fairly well with photosynthesis irrespective of leaf sclerophycity [8,34], the slope of this relationship is not as steep in mesophyllous plants as in sclerophyllous plants. This is particularly evident when *g*$_m$ is higher than 0.4 mol m$^{-2}$ sec$^{-1}$, when increments of photosynthesis are poorly coupled with increments of *g*$_m$ (e.g., Figure 2) [16]. We also noticed that the correlation between *g*$_m$ and photosynthesis is loose when results obtained with herbaceous species are pooled together (data from Refs. 8 and 9 and S. Delfine et al. unpublished results). In mesophytic plants there seems to be a poor association between photosynthesis and stomatal conductance as well. For these plants the cumulative resistance to $CO_2$ diffusion offered by stomata and the mesophyll is low and probably unable to effectively limit photosynthesis. It should be noticed, however, that the relationship between *g*$_m$ and photosynthesis mirrors that between *g*$_s$ and photosynthesis in xerophytic plants and even in aquatic species such as *Carex* and *Phragmites*. When both conductances are reduced concurrently, restrictions on $CO_2$ diffusion inside the leaves become particularly high. Thus, we conclude that *g*$_m$ may contribute to limit photosynthesis only when it decreases *c*$_c$ significantly. Under these conditions, photosynthesis may be progressively limited by the competition between the high concentration of $O_2$ and the low concentration of $CO_2$, which, in turn, favor photorespiration [see Eq. (2) for further comments].

To make the chloroplast environment favorable to $CO_2$ fixation, the gas composition can be changed by either increasing the $CO_2$ or decreasing the $O_2$ molar fraction. It is conceivable that species characterized by low $g_s$ and low $g_m$, such as trees and sclerophyllous plants, will greatly benefit from exposure to $CO_2$ molar fractions higher than ambient. It has been demonstrated that although the $g_m$ of *Quercus ilex* and *Citrus aurantium* does not change with increasing $CO_2$ molar fraction, photosynthesis of these sclerophyllous plants is by far more sensitive to $CO_2$ than photosynthesis of herbaceous plants [8]. Thus, the photosynthetic capacity of trees may exceed that of herbaceous plants, but this does not result in higher photosynthesis at ambient $CO_2$ because of the low amount of substratum reaching the chloroplasts. If this is true, trees may have an evolutionary advantage over other plants during the current trend of atmospheric $CO_2$ increase. However, it should be pointed out that such an advantage may be lost whether leaf photosynthesis is limited by end-product removal in plants grown at high $CO_2$ [35] or whether other factors begin to be limiting for plant growth. For instance, fast growth could lead to depletion of the soil content of N and to early competition for light caused by canopy closure.

## IV. EFFECT OF ENVIRONMENTAL STRESSES ON LEAF RESISTANCES AND PHOTOSYNTHESIS

### A. Effect of Environmental Stresses on $g_s$

Under high evaporative atmospheric demand, stomata closure minimizes leaf water loss to avoid immediate restrictive effects on the rates of biochemical and biophysical processes. At the same time, an adequate supply of $CO_2$ through the stomata to the carboxylation sites is required for optimal performance of photosynthesis when energetic resources and nutrients are not limiting. These two contrasting stimuli affect the resulting degree of stomatal opening when the leaf is not water stressed. Indeed, Jones [36] suggested that the actual dominant stimulus that drives stomatal movements is that to prevent irreversible damage to leaves. It has been shown that this stimulus is amplified when leaves are just mildly water stressed and thus photosynthesis becomes limited by the supply of $CO_2$ to the carboxylation sites if other factors are not concurrently limiting [26,37]. Cornic and Massacci [38] have reviewed results showing that when leaf water deficit is induced slowly, the photosynthetic apparatus becomes very resistant to drought and limitation to photosynthesis might be completely attributed to resistances to $CO_2$ diffusion inside the leaves, primarily stomata closure but also mesophyll resistances (see Sec. IV.B and Figure 4). Results of Figure 3 for sorghum and of Di Marco et al. [21] for wheat represent slow field development of water stress that does not alter the proportionality between changes in $A$ and $g_s$. Recently, Meyer and Genty [39] showed that also under severe water stress rapidly developed the reduction of stomatal conductance is the main limitation of photosynthesis and the main cause of heterogeneous stomatal closure in *rosa rubiginosa* L. Slowly developing salt stress apparently mimics water stress because their effect is a coordinate reduction of photosynthesis and both stomatal and mesophyll conductances to $CO_2$ diffusion (Figure 4).

The interaction of temperature stress with $g_s$ and $A$ is much more complex than that attributable to water deficit and salt stress. In fact, changes of a few degrees Celsius have a great effect on evapotranspiration and thus on $g_s$ through a hydraulic feedback [40]. Besides, if such changes are fast, they may unevenly alter the water status of some stomata areoles and induce heterogeneous variations in their degree of closure. This may lead to underestimation of the actual rate of photosynthesis and evapotranspiration and to artifactual changes in the relationship between $A$ and $g_s$ [41,42]. The occurrence of heterogeneous stomatal closure may also impair the calculation of mesophyll conductance (see Sec. IV.B).

In $C_3$ plants the effects of temperature changes on $A$ and $g_s$ are even more complicated by the more competitive response of photorespiration with respect to photosynthesis for their substrata [43] and by the frequent occurrence of feedback limitation of photosynthesis by accumulation of phosphorylated substrata in the cytosol [44]. These temperature effects on photosynthesis as well as those typical of elevated temperatures (alteration of photosynthetic protein and membrane conformation) have, however, only an indirect effect on $g_s$.

In $C_4$ plants, on the other hand, a decrease of temperature may also increase $g_s$ and lead, in the case of these plants, to leaf wilting because of the partial loss of stomatal control [45].

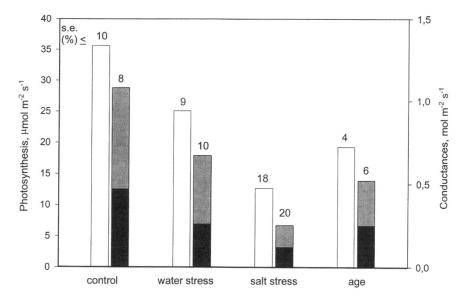

**Figure 4**   Reduction of photosynthesis (white bars) and total conductance as partitioned in mesophyll (dark bars) and stomatal (gray bars) components following environmental stresses or in aging leaves. Data from Refs. 9, 48, and 49 are reported.

An increased ambient $CO_2$ molar fraction might be considered a particular stress factor for stomata. Besides, the interaction of a $CO_2$ increase in the atmosphere with increased air temperature is well documented. The effect per se of an increase of $CO_2$ is a reduction of stomata opening. The mechanism underlying this response is not fully understood, and we invite readers to consult specialized papers dealing with this basic topic [46]. However, this reduction of $g_s$ is very unlikely to be solely responsible for the limitation of photosynthesis that certain species show under high $CO_2$, particularly because the feedback limitation mechanism for sugar accumulation also seems to make a significant contribution under these conditions [47].

## B.   Effect of Environmental Stresses on $g_m$

Almost all of the currently published studies assessed $g_m$ in leaves grown and exposed to nonstressful conditions. Growth under environmental constraints that cause leaf anatomy changes probably cause changes in mesophyll resistances as well, as a few studies seem to indicate (Figure 4).

The effect of a saline environment on the growth of olive leaves is an increase of leaf thickness and doubling of the palisade layers, associated with a decrease of $g_m$ [5]. A reduction of $g_m$ was also noticed after growing spinach with saline water [48,49]. In this case, $g_m$ decreased from a value typical of mesophyllous leaves (about 0.50 mol m$^{-2}$ sec$^{-1}$). However, no increase in leaf thickness was noticed in spinach leaves, but the spongy cells became more appressed and the intercellular spaces were significantly reduced with respect to control leaves. The authors, therefore, suggested that $g_m$ is dependent on leaf porosity and that a low porosity increases the tortuosity of the path leading to chloroplasts. Delfine et al. [49] also demonstrated that the reduction of $g_m$ under salinity could be partially reversed if the internal salt concentration was lowered by irrigation with salt-free water. This is the only report indicating that reduction of mesophyll conductance under stress conditions is not a permanent feature.

An association between photosynthesis and $g_m$ was also found in leaves of spinach and chestnut exposed to increasing water stress [50,51]. On the contrary, we did not notice any change in the $g_m$ of *Rosa* leaves exposed to rapid dehydration (B. Genty, S. Meyer, E. Brugnoli, F. Loreto, unpublished). The last experiment was done by using video images of chlorophyll fluorescence. This technique can map and quantify heterogeneities in stomatal closure and photosynthetic activity of leaves [52] that frequently occur in stressed leaves and cause errors in $c_i$, $c_c$, and $g_m$ calculations [53]. These results suggest that if the stress is rapid it cannot change the $g_m$ of leaves. On the contrary, slowly developing stresses, which are more common in nature, cause a wide range of changes in leaf anatomy that, in turn, seem to affect gas diffusion resistances. However, no conclusive indication of the stress effect on $g_m$ can be inferred until heterogeneities of stomatal opening are quantitatively assessed because methods used to estimate $g_m$ are sensitive to this error.

Aging is another factor likely to affect leaf anatomy; consequently, $g_m$ was expected to change in ageing leaves (Figure 4). Mesophyll conductance of wheat leaves also decreased dramatically during leaf senescence and reached a value (0.15 mol m$^{-2}$ sec$^{-1}$) typical of trees and sclerophyllous plants [9], but the morphological parameter associated with this reduction was unclear.

Reduction of photosynthetic capacity in aging leaves may be related to low $g_m$. However, age also causes a reduction of leaf nitrogen [12] and of the amount of Rubisco [9]. Therefore, it is difficult to partition between age-related metabolic and diffusive limitation of photosynthesis. Intuitively, if the reduction of $g_m$ were as strong and immediate as that shown by aging wheat leaves, it might significantly contribute to the age-related reduction of photosynthesis. But we have also found that the reduction of photosynthesis in aging leaves of *Quercus ilex* L., a tree species characterized by a constitutively low $g_m$ [8], is caused by reduced activity of Rubisco rather than by further reduction of $g_m$, as previously speculated by Di Marco et al. [53, and unpublished results]. Therefore, the available information indicates that leaf age is likely to affect $g_m$ only in mesophyllous leaves characterized by constitutively low diffusive resistances.

## V. CONCLUDING REMARKS

Knowledge of the variation of the cumulative diffusive resistances to $CO_2$ entry in leaves of different species enables us to understand how the plasticity of leaf characteristics to the environment contributes to the photosynthetic performances in contrasting habitats. This knowledge is essential to accomplish the goals of modern ecophysiology applied to agriculture: to widen the geographical distribution of valuable plants, to optimize plant phenotypes in response to predictable climate changes, and to achieve stable crop yields in areas exposed to environmental constraints. We have presented results indicating the likely effects of intrinsic (plant species, age) and environmental (water availability, salt stress, suboptimal temperatures, $CO_2$ changes) factors on the two main components of diffusive resistances to $CO_2$: stomatal and mesophyllous. We have shown a generally clear association between these two resistances and have highlighted the possible combined effect of both resistances in limiting carbon uptake and photosynthesis, particularly under stress conditions. These conclusions may be of use for modeling photosynthesis and predicting plant growth in a changing environment.

## REFERENCES

1. PS Nobel. Leaves and fluxes. In: PS Nobel, ed. Physicochemical and Environmental Plant Physiology. San Diego: Academic Press, 1999, pp 293–349.
2. GD Farquhar, TD Sharkey. Stomatal conductance and photosynthesis. Annu Rev Plant Physiol 33:317–345, 1982.
3. S von Caemmerer, GD Farquhar. Some relationships between the biochemistry of photosynthesis and the gas exchange of leaves. Planta 153:376–387, 1981.
4. HG Jones, DO Hall, JE Corlett, A Massacci. Drought enhances stomatal closure in response to shading in sorghum (*Sorghum bicolor*) and in millet (*Pennisetum americanum*). Aust J Plant Physiol. 22:1–6, 1995.
5. G Bongi, F Loreto. Gas exchange properties of salt-stressed olive (*Olea europea* L.) leaves. Plant Physiol 90:1408–1416, 1989.
6. J Lloyd, JP Syvertsen, PE Kriedemann, GD Farquhar. Low conductances for $CO_2$ diffusion from stomata to the sites of carboxylation in leaves of woody species. Plant Cell Environ 15:873–899, 1992.
7. DF Parkhurst. Diffusion of $CO_2$ and other gases inside leaves. New Phytol 126:449–479, 1994.
8. F Loreto, PC Harley, G Di Marco, TD Sharkey. Estimation of mesophyll conductance to $CO_2$ flux by three different methods. Plant Physiol 98:1437–1443, 1992.
9. F Loreto, G Di Marco, D Tricoli, TD Sharkey. Measurements of mesophyll conductance, photosynthetic electron transport and alternative electron sinks of field grown wheat leaves. Photosynth Res 41:397–403, 1994.
10. D Epron, D Godard, G Cornic, B Genty. Limitation of net $CO_2$ assimilation rate by internal resistances to $CO_2$ transfer in the leaves of two tree species (*Fagus sylvatica* L. and *Castanea sativa* Mill.). Plant Cell Environ 18:43–51, 1995.

11. JP Syvertsen, J Lloyd, C McConchie, PE Kriedemann, GD Farqhuar. On the relationship between leaf anatomy and $CO_2$ diffusion through the mesophyll of hypostomatous leaves. Plant Cell Environ 18:149–157, 1995.

12. JR Evans, S von Caemmerer, BA Setchell, GS Hudson. The relationship between $CO_2$ transfer conductance and leaf anatomy in transgenic tobacco with a reduced content of Rubisco. Aust J Plant Physiol 21:475–495, 1994.

13. A Laisk, F Loreto. Determining photosynthetic parameters from leaf $CO_2$ exchange and chlorophyll fluorescence: Rubisco specificity factor, dark respiration in the light, excitation distribution between photosystems, alternative electron transport rate and mesophyll diffusion resistance. Plant Physiol 110:903–912, 1996.

14. DF Parkhurst, KA Mott. Intercellular diffusion limits to $CO_2$ uptake in leaves. Plant Physiol 94:1024–1032, 1990.

15. I Terashima, M Ishibashi, K Ono, K Hikosaka. Three resistances to $CO_2$ diffusion: leaf-surface water, intercellular spaces and mesophyll cells. In: P Mathis, ed. Photosynthesis: From Light to Biosphere. Dordrecht: Kluwer Academic, 1995, pp 537–542.

16. JR Evans, S von Caemmerer. Carbon dioxide diffusion inside leaves. Plant Physiol 110:339–346, 1996.

17. ED Schulze, AE Hall. Stomatal responses, water loss and $CO_2$ assimilation rates of plants in contrasting environments. In: OL Lange, PS Nobel, CB Osmond, H Zieger, eds. Physiological Plant Ecology II. Water Relations and Carbon Assimilation. New York: Springer-Verlag, 1982, pp 181–230.

18. SW Young, PJ Milligan, DH Davies. Correlation of concentration of water nutrients with anatomical features of three species of aquatic grass. Arch Hydrobiol 140:419–430, 1997.

19. A Fahn. The leaf. In: A Fahn, ed. Plant anatomy. Oxford: Pergamon Press, 1974, pp 235–282.

20. M Centritto, F Magnani, H Lee, P Jarvis. Interactive effects of elevated [$CO_2$] and water stress on cherry (*Prunus avium*) seedlings. II. Photosynthetic capacity and water relations. New Phytol 141:141–152, 1999.

21. G Di Marco, A Massacci, R Gabrielli. Drought effects on photosynthesis and fluorescence in hard wheat cultivars in the field. Physiol Plant 74:385–390, 1988.

22. A Massacci, A Battistelli, F Loreto. Effects of drought stress on photosynthetic characteristics, growth and sugar accumulation of field-grown sweet sorghum. Aust J Plant Physiol 23:331–340, 1996.

23. M Freeling, B Lane. The maize leaf. In: M Freeling, V Walbot, eds. The Maize Handbook. New York: Springer-Verlag, 1994, p 1728.

24. F Manes, G Astorino, M Vitale, F Loreto. Morphofunctional characteristics of *Quercus ilex* L. leaves of different ages and their ecophysiological behaviour during different seasons. Plant Biosyst 131:149–158, 1997.

25. AP Sandford, PG Jarvis. Stomatal responses to humidity in selected conifers. Tree Physiol 2.89–103, 1986.

26. HG Jones, RA Sutherland. Stomatal control of xylem embolism. Plant Cell Environ 6:607–6112, 1991.

27. A Laisk, V Oja, M Rahi. Diffusion resistance of leaves in connection with their anatomy. Fiziol Rast 47:40–48, 1970.

28. TD Sharkey, TL Vassey, PJ Vanderveer, RD Vierstra. Carbon metabolism enzymes and photosynthesis in transgenic tobacco (*Nicotiana tabacum* L.) having excess phytochrome. Planta 185:287–296, 1991.

29. SC Wong, IR Cowan, GD Farquhar. Leaf conductance in relation to rate of $CO_2$ assimilation. I Influence of nitrogen nutrition, phosphorus nutrition, photon flux density, and ambient partial pressure of $CO_2$ during ontogeny. Plant Physiol 78:821–825, 1985.

30. GD Farquhar, SC Wong. An empirical model of stomatal conductance. Aust J Plant Physiol 11:191–210, 1984.

31. AJ Jarvis, WJ Davies. The coupled response of stomatal conductance to photosynthesis and transpiration. J Exp Bot 49:399–406, 1998.

32. WM Kaiser. Effects of water deficit on photosynthetic capacity. Physiol Plant 71:142–149, 1987.

33. W Tezara, VJ Mitchell, SD Driscoll, DW Lawlor. Water stress inhibits plant photosynthesis by decreasing coupling factor and ATP. Nature 401:914–917, 1999.

34. S von Caemmerer, JR Evans. Determination of the average partial pressure of $CO_2$ in chloroplasts from leaves of several C3 plants. Aust J Plant Physiol 18:287–305, 1991.

35. FX Socias, H Medrano, TD Sharkey. Feedback limitation of photosynthesis of *Phaseolus vulgaris* L. grown in elevated $CO_2$. Plant Cell Environ 16:81–86, 1993.

36. HG Jones. Stomatal control of photosynthesis and transpiration. J Exp Bot 49:387–398, 1998.

37. HG Jones. Partitioning stomatal and non-stomatal limitations to photosynthesis. Plant Cell Environ 8:95–104, 1985.

38. G Cornic, A Massacci. Leaf photosynthesis under drought stress. In: NR Baker, ed. Photosynthesis and Environment. Dordrecht: Kluwer Academic, 1996, pp 347–366.

39. S Meyer, B Genty. Heterogeneous inhibition of photosynthesis over the leaf surface of Rosa rubiginosa L during water stress and abscissic acid treatment: induction of a metabolic component by limitation of $CO_2$ diffusion. Planta 219:126–131, 1999.

40. AD Friend. Use of a model of photosynthesis and leaf microenvironment to predict optimal stomatal conductance and leaf nitrogen partitioning. Plant Cell Environ 14:895–905, 1991.

41. KA Mott, TN Buckley. Stomatal heterogeneity. J Exp Bot 49:407–417, 1998.

42. DWG Van Kraalingen. Implications of non-uniform stomatal closure on gas exchange calculations. Plant Cell Environ 13:1001–1004, 1990.

43. PJ Lea, RC Leegood. Plant Biochemistry and Molecular Biology. Chichester: John Wiley & Sons, 1994, pp 27–46.

44.  RC Leegood, GE Edwards. Carbon metabolism and photorespiration: temperature dependence in relation to other environmental factors. In: NR Baker, ed. Photosynthesis and Environment. Dordrecht: Kluwer Academic, 1996, pp 191–221.
45.  LA Mustardy, TT Vu, A Faludi-Daniel. Stomatal response and photosynthetic capacity of maize leaves at low temperature. A study on varietal differences in chilling sensitivity. Physiol Plant 55:31–34, 1982.
46.  JIL Morison. Stomatal response to increased $CO_2$ concentration. J Exp Bot 49:443–452, 1998.
47.  TD Sharkey. Feedback limitation of photosynthesis and the physiological role of ribulose bisphosphate carboxylase carbamylation. Bot Mag Tokyo 2:87–105, 1990.
48.  S Delfine, A Alvino, M Zacchini, F Loreto. Consequences of salt stress on diffusive conductances, Rubisco characteristics and anatomy of spinach leaves. Aust J Plant Physiol 25:395–402, 1998.
49.  S Delfine, A Alvino, MC Villani, F Loreto. Restrictions to $CO_2$ conductance and photosynthesis in spinach leaves recovering from salt stress. Plant Physiol 119:1101–1106, 1999.
50.  F Loreto, S Delfine, A Alvino. On the contribution of mesophyll resistance to $CO_2$ diffusion to photosynthesis limitation during water and salt stress. Acta Hortic 449:417–422, 1997.
51.  M Lauteri, A Scartazza, MC Guido, E Brugnoli. Genetic variation in photosynthetic capacity, carbon isotope discrimination and mesophyll conductance in provenances of *Castanea sativa* adapted to different environments. Funct Ecol 11:675–683, 1997.
52.  B Genty, S Meyer. Quantitative mapping of leaf photosynthesis using chlorophyll fluorescence imaging. Aust J Plant Physiol 22:277–284, 1994.
53.  PC Harley, F Loreto, G Di Marco, TD Sharkey. Theoretical considerations when estimating the mesophyll conductance to $CO_2$ flux by analysis of the response of photosynthesis to $CO_2$. Plant Physiol 98:1429–1436, 1992.
54.  G Di Marco, F Manes, D Tricoli, E Vitale. Fluorescence parameters measured concurrently with net photosynthesis to investigate chloroplastic $CO_2$ concentration in leaves of *Quercus ilex* L. J Plant Physiol 136:538–543, 1990.

# 16

# Mineral Nutrient Transport in Plants

**Benjamin Jacoby and Nava Moran**

*The Hebrew University of Jerusalem, Rehovot, Israel*

## I. INTRODUCTION

This chapter deals with the absorption and accumulation of mineral nutrient ions by plant cells and with their primary translocation in the plant. The permeability of the phospholipid bilayer of biological membranes to mineral ions is very low. Transport proteins—carriers and channels—embedded in the phospholipid bilayer facilitate the transport of mineral ions across the membranes.

Plant cells accumulate all essential mineral ions to higher concentrations than those present in their environment (Table 1). This accumulation is selective, as evidenced by the different accumulation ratios of the ions shown in Table 1. Some questions that arise are:

How is passage through the impermeable lipid bilayer accomplished?
How is accumulation against the concentration gradient accomplished?
How is metabolic energy coupled to such transport?
What is the mechanism of selectivity?
How is vectorial transport accomplished?

These questions are dealt with in the sections that follow.

## II. DEFINITIONS

At the outset, let us define some basic terms used in this chapter.

*Electrochemical potential of solute j—$\bar{\mu}_j$ (J mol$^{-1}$):* This is the Gibbs free energy [1] of the solute $j$:

$$\bar{\mu}_j = \bar{\mu}_j^* + 2.3RT \log a_j + z_j F\psi + P\overline{V} \tag{1}$$

where $R$ is the gas constant (8.314 J mol$^{-1}$), $T$ is temperature in kelvin, $a_j$ is the chemical activity of $j$ ($a_j = \gamma_j c_j$, where $\gamma_j$ is the activity coefficient and $c_j$ the chemical concentration of $j$), $z_j$ is the electrical charge (ionic charge) of $j$, $F$ is the Faraday constant (9.649 $10^{-4}$ J mol$^{-1}$ V$^{-1}$), $\psi$ is the electrical potential (V), $P$ is the pressure in excess of atmospheric pressure (MPa), $\overline{V}$ is the partial molal volume of $j$ (m$^3$ mol$^{-1}$), and $\bar{\mu}_j^*$ is the electrochemical potential of $j$ in the standard state ($a = 1.0$, $\psi = 0$, $P = 0$).

**TABLE 1**  Composition of Pond Water and of the Sap of the Alga *Nitella clavata* Growing in the Pond

| Ion | Pond (mol m$^{-3}$) | Sap (mol m$^{-3}$) | Ratio: sap/pond |
|---|---|---|---|
| Mg$^{2+}$ | 1.5 | 5.5 | 3.6 |
| Ca$^{2+}$ | 0.7 | 7.0 | 10 |
| Na$^+$ | 1.2 | 49 | 41 |
| K$^+$ | 0.5 | 49 | 97 |
| H$_2$PO$_4^-$ | 0.008 | 1.7 | 212 |
| Cl$^-$ | 1.0 | 101 | 101 |
| SO$_4^{2-}$ | 0.34 | 6.5 | 20 |

*Electrochemical potential difference:* The driving force for the transport of solutes across plant cell membranes. It is the electrochemical potential difference across the membrane ($\Delta\bar{\mu}_j = \bar{\mu}_j^i - \mu_j^o$, where i is inside and o is outside). Contributions of the pressure term ($\Delta P\bar{V}$) to the electrochemical potential difference of ions across biological membranes are usually negligible compared with the electrical contribution and hence can generally be ignored [2]. The electrochemical potential difference across a membrane then is

$$\Delta\bar{\mu}_j^{(i-o)} = 2.3RT \log \frac{a_j^i}{a_j^o} + z_jF(\psi^i - \psi^o) \qquad (2)$$

*Flux of solute $J_j$* (mol sec$^{-1}$ m$^{-2}$): This is the unidirectional rate of solute movement across a unit membrane area. The net flux $J^{net}$, or uptake, is the difference between the influx ($J^{in}$) and efflux ($J^{out}$): $J^{net} = J^{in} - J^{out}$

*Accumulation of a solute:* Specifies a higher concentration (not necessarily higher electrochemical potential) of the solute inside.

*Active transport:* Transport of a solute against its electrochemical potential gradient [3]. Such transport always needs energy input.

*Passive transport:* Transport of a solute along its electrochemical potential gradient.

*Metabolic transport:* Any transport that depends on metabolic energy supply; it is inhibited by inhibitors of energy metabolism [4].

*Electrogenic transport:* Transport of an ion unaccompanied by equal opposite charge, thus creating an electrical potential difference or changing it.

*Electrophoretic transport:* Transport of an ion in response to a preexisting electrical potential difference. Such transport is also electrogenic and results in a change of the electrical potential difference.

## III. FREE SPACE AND OSMOTIC VOLUME

In experiments on the time course of salt uptake by plant tissues, two phases are revealed: (1) a rapid initial phase that is completed within a few minutes and (2) a slower phase that may proceed for several hours at a constant rate. The initial rapid uptake is into the free space [5], namely the extramembranal space of the plant tissue. The free space consists of the cell walls and the intercellular spaces. Uptake into the free space is reversible and nonmetabolic. All the anions and part of the cations that are absorbed in the first uptake phase can be washed out with water, and the remaining cations can be exchanged with another cation. Uptake in the second phase is into the osmotic volume [5], namely the space that is surrounded by plasma membranes. The latter is usually a metabolic process.

Cation exchange in the free space results from the presence of immobile negative charges in the cell walls. Dissociated carboxylic groups, in particular those of polygalacturonic acid, are responsible for these charges [5]. The presence of immobile negative charges in the cell wall, adjacent to the external aqueous phase, results in an electrical potential difference, the Donnan potential [6].

After equilibration of the external solution with the free space, the electrochemical potential differences of cations and anions in the free space (FS) and external solution (sol) is zero. Hence, from Eq. (2):

$$0 = 2.3RT \log \frac{a_j^{FS}}{a_j^{sol}} + z_j F(\psi^{FS} - \psi^{sol})$$

The Donnan potential ($E_D$) is the electrical potential difference ($E_D = \psi^{FS} - \psi^{sol}$), and:

$$E_D = \frac{2.3RT}{z_j} \log \frac{a_j^{sol}}{a_j^{FS}} \tag{3}$$

The Donnan potential in cell walls is negative. Equation (3) then shows that cations ($z$, positive) will accumulate in the negatively charged cell walls (Donnan phase) and that the anion concentration in the latter phase will be lower than in the (adjacent) aqueous phase. Donnan potentials from $-7$ to $-289$ mV have been calculated for various cell wall–solution systems [5]. The Donnan potential changes with the dissociation of the charged sites: it decreases with salt concentration and increases with the dissociation constants of the various cations.

# IV.  ELECTRICAL POTENTIALS AT PLANT CELL MEMBRANES

## A.  Proton Gradients: Uniport and Cotransport

Metabolic solute transport in plant cells is usually energized by an electrochemical potential gradient of protons ($\Delta\bar{\mu}_{H^+}$) across the membranes and is facilitated by channels and carriers. The proton electrochemical potential difference is formed by active proton transport, from the cytoplasm to the free space and to the vacuoles. This proton transport is catalyzed by membrane-embedded electrogenic proton pumps that catalyze the transformation of chemical energy in adenosine-5′-triphosphate (ATP) and pyrophosphate (PPi), to an electrochemical proton gradient.

Metabolic transport in plant cells that is driven by the electrochemical proton gradient is termed *uniport* [7] (also see Sec. V.A) or *cotransport* [7] (also see Sec. V.B.3). Uniport is passive and it occurs via channels in the direction of the electrical potential gradient of the solute. Cotransport of solutes is active and derives its energy from concomitant passive transport of another ion. In plants the cotransported ion is, in most instances, a proton.

## B.  The Membrane Potential

The membrane potential ($E_M$) is the electrical potential difference across a membrane: it consists of a diffusion potential and a potential difference resulting from the action of electrogenic pumps. Diffusion potentials result from different diffusion velocities of anions and cations across a membrane. Membrane potentials of plant cells are measured with reference to the cytoplasm ($E_M = \Psi^i - \Psi^o$), where inside (i) is always the cytoplasm and outside (o) is the free space (with reference to the plasma membrane) and the vacuole (with reference to the tonoplast). These conventions will be maintained throughout this chapter. Accordingly, under physiological conditions, the membrane potential is negative at both membranes (positive charges in the free space as well as the vacuole). An increase in the electrical potential difference, or hyperpolarization, is synonymous with a decrease of $E_M$ (to more negative values), and depolarization is synonymous with an increase of $E_M$.

## C.  The Diffusion Potential

The unidirectional flux ($J$) of a solute $j$ across a membrane depends on the driving force ($\Delta\bar{\mu}_j$) and the membrane permeability ($P_j$) of the solute: $J = P_j \Delta\bar{\mu}_j$ (mol sec$^{-1}$ m$^{-2}$). Thus, if a salt with different anion and cation permeabilities diffuses across the membrane, an excess charge of the more permeable ion is transported and a diffusion potential ($E_M^D$) results. This potential will then retard the diffusion of the more permeable ion. A small anion-cation concentration difference creates a rather large diffusion potential (see Sec. VI); therefore, practically equivalent amounts of ions of both kinds will pass through the membrane.

The diffusion potential depends on the relative permeabilities of all the cations and anions in the system. Equation (4), the Goldman-Hodgkin-Katz equation [8,9], calculates the diffusion potential for $Na^+$, $K^+$, and $Cl^-$. These are often the quantitatively most important ions in biological systems, and they determine the diffusion potential:

$$E_M^D = 2.3 \frac{RT}{F} \log \frac{P_K[K^+]^o + P_{Na}[Na^+]^o + P_{Cl}[Cl^-]^i}{P_K[K^+]^i + P_{Na}[Na^+]^i + P_{Cl}[Cl^-]^o} \qquad (4)$$

where $P_K$, $P_{Na}$, and $P_{Cl}$ are the membrane permeabilities of $K^+$, $Na^+$, and $Cl^-$, respectively, and brackets [ ] designate concentration (mol m$^{-3}$). The diffusion potential across a membrane is the membrane potential that can be measured when metabolic transport is inhibited.

Let us calculate the diffusion potential across the plasma membrane for the following situation: $[KCl]^o = 10$ mM, $[NaCl]^o = 10$ mM, $[K^+]^i = 100$ mM, $[Na^+]^i = 10$ mM, and $[Cl^-]^i = 110$ mM; the relative permeabilities of $K^+$, $Na^+$, and $Cl^-$ are 1, 0.2, and 0.01, respectively. The temperature is 30°C (303 K) and $2.3RT/F = 60$ mV. Then:

$$E_M^D = 60 \log \frac{10 + (0.2 \times 10) + (0.01 \times 110)}{100 + (0.2 \times 10) + (0.01 \times 20)} = -53.5 \text{ mV}$$

## D. Proton Motive Force

A proton gradient across a membrane consists of an electrical component and a chemical proton concentration gradient, the pH difference across the membrane. The relations of these two components can be defined by replacing $j$ in Eq. (2) with $H^+$ and $a_j$ with $[H^+]$:

$$\Delta \bar{\mu}_{H^+} = 2.3RT \log \frac{[H^+]^i}{[H^+]^o} + zF(\psi^i - \psi^o) \qquad (5)$$

or when $(\psi^i - \psi^o)$ is replaced by $E_M$, $\log [H^+]^i/[H^+]^o$ by $-\Delta pH^{(i-o)}$ (or $\Delta pH^{o-i}$), and 1 is substituted for $z$ (the charge of protons):

$$\Delta \bar{\mu}_{H^+} = FE_M + 2.3RT \, \Delta pH^{(o-i)} \text{ (J mol}^{-1}) \qquad (6)$$

The electrochemical proton gradient can be expressed in electrical units (V) instead of energy units (J mol$^{-1}$) by division of Eq. (6) with $F$; $\Delta \bar{\mu}_{H^+}/F$ is then the proton motive force (pmf) [2,10] defined as follows:

$$\text{pmf} = E_M + \frac{2.3RT}{F} \Delta pH^{(o-i)} \qquad (7)$$

When the values for $R$ and $F$ ($R = 8.3$ J mol$^{-1}$ K$^{-1}$, $F = 96.49$ J mol$^{-1}$ mV$^{-1}$) are substituted and the protonmotive force is calculated at 30°C ($T = 303$ K), Eq. (7) becomes

$$\text{pmf} = E_M + 60\Delta pH \text{ (mV)}$$

Thus, when $E_M = -120$ mV, $pH^i = 7$, and $pH^o = 6$, a pmf of $-180$ mV is obtained.

## E. Differentiation of Active and Passive Transport

Nonelectrolytes, such as sugars, are actively transported whenever they are accumulated in the cell to a higher concentration than outside. For nonelectrolytes, the electrical component of the electrochemical potential [Eq. (2)] nullifies and the equation becomes a function of the concentration ratio only. The electrical component of Eq. (2) is important for ions; they can passively accumulate in response to an electrical potential difference. Suitable transport proteins (channels) may facilitate such passive accumulation. Cations may passively accumulate in the negatively charged cytoplasm and anions in the positively charged vacuole. Passive ion accumulation is metabolic because energy metabolism is needed to maintain the necessary membrane potential. The possible passive accumulation ratio of ion $j$ ($a_j^i/a_j^o$) depends on the ionic charge of the accumulated ion and on the membrane potential. This ratio can be derived from Eq. (2). On assuming equilibrium, namely $\Delta\bar{\mu}_j = 0$, and substituting $E_M$ for $(\psi^i - \psi^o)$, Eq. (2) becomes

$$0 = 2.3RT \log \frac{a_j^i}{a_j^o} + z_j F E_M$$

or:

$$\log \frac{a_j^o}{a_j^i} = \frac{z_j F}{2.3RT} E_M \tag{8}$$

at 30°C, $F/2.3RT = 1/60$ (mV) and

$$\log \frac{a_j^o}{a_j^i} = \frac{z_j}{60} E_M$$

This is one form of the Nernst equation [2]; it gives the relation between the membrane potential and the expected ion accumulation ratio at equilibrium. Another form of this equation gives the Nernst potential $(E^N)$ at 30°C:

$$E_N = \frac{60}{z_j} \log \frac{a_j^o}{a_j^i} \tag{9}$$

This is the membrane potential needed to sustain equilibrium at a certain ion accumulation ratio. Thus at $-120$ mV, $a^o/a^i$ would be $10^{-2}$ for $K^+$, $10^{-4}$ for $Ca^{2+}$, and $10^2$ for $Cl^-$.

The Nernst equation can be employed to determine whether specific ions have been transported passively. Such an analysis must be performed for tissues in the steady state, namely when net transport has ceased $(J^{in} = J^{out})$. For this analysis, the membrane potential and the concentration ratio of the ion at the steady state must be known. Active transport is assumed when the measured $E_M$ differs from the expected $E_N$.

Whether transport is active can also be determined for the non–steady state, but the flux ratio must be known. Using [3] and Theorell [11] have shown that the ratio of passive fluxes $(J^{in}/J^{out})$ of a solute is proportional to the electrochemical potential difference. The Ussing-Theorell equation is

$$\Delta\bar{\mu} = 2.3RT \log \frac{J^{in}}{J^{out}} \tag{10}$$

Substituting Eq. (2) for $\Delta\bar{\mu}_j$ and $E_M$ for $(\psi^i - \psi^o)$ in Eq. (2) gives

$$2.3RT \log \frac{a^i}{a^o} + zF E_M = 2.3RT \log \frac{J^{in}}{J^{out}}$$

Dividing by $2.3RT$ and writing the equation for 30°C results in

$$\log \frac{J^{in}}{J^{out}} = \log \frac{a^i}{a^o} + \frac{z}{60E_M} \tag{11}$$

If a flux ratio is larger than expected from this relation, transport is assumed to be active.

# V.  TRANSPORT PROTEINS

Ion transport across the membranes of plant cells is usually energized by an electrochemical potential gradient of protons $(\Delta\bar{\mu}_{H^+})$ and is facilitated by channels and carriers. The proton electrochemical potential gradient is formed by primary active [7] proton transport from the cytoplasm to the free space and to the vacuoles. This proton transport is catalyzed by membrane-embedded electrogenic proton pumps that catalyze the transformation of chemical energy in ATP and PPi to an electrochemical proton gradient.

Ion transport in plant cells that is driven by the electrochemical proton gradient is termed uniport or cotransport [7] (also see Secs. V.A and V.B.3). Uniport is passive and it occurs via channels in the direction of the electrical potential gradient of the solute. Cotransport of solutes is secondary active [12] and derives its energy from concomitant passive transport of another ion. In plants the cotransported ion is, in most instances, a proton.

## A.  Channels

### 1.  Structure and Basic Properties

Channels are multisubunit proteins that span the membrane. They allow passive fluxes of solutes, usually ions, or water, and in some cases also small solutes such as glycerol and urea [13]. Water channels (aquaporins) have been shown to conduct ions in some rare cases, but they are not the subject of this discussion. The tertiary conformation of the main (alpha) subunit of the ion channel creates a hydrophilic pore across the membrane. A narrow region within the pore forms a "filter" that determines the selectivity of the channel toward different ions. The selectivity is based on the ion charge and the size, usually of its dehydrated form, and its ability to bind (not too tightly) to polar or charged groups in the filter [13,14]. Thus, there are $K^+$ channels, $Ca^{2+}$ channels, cation-nonselective channels, anion channels, etc. Gating also controls transport across channels, meaning that various external and internal conditions regulate the conformation of channel proteins, thus affecting the probability that they are in the open state. Some channels can be gated electrically, by membrane potential [15,16]; other channels chemically, by reversibly binding specific molecules such as the second messengers, inositol 1,4,5-trisphosphate ($IP_3$) [17]; and yet other channels mechanically, by membrane stretching [18,19]. Finally, channels differ in their conductivity. Transport rates through an open channel are relatively high, about $107$–$108$ $sec^{-1}$ [20], and $100$–$1000$ times higher, per transport unit, than those of carriers [21]. Net fluxes of ions through open channels can be recorded as electrical current. The "patch-clamp" method permits the recording of such currents through a single open ion channel or through the sum of the open channels in the membrane of the whole cell (Figure 1) [22]. The whole-cell current thus depends on the probability of the channels be-

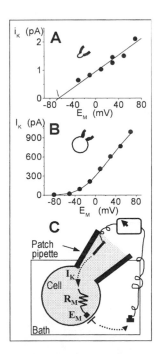

**Figure 1**   Voltage dependence of $K^+$ channel gating. (A) A linear current-voltage relationship (an *I-V* plot) for a single open ion channel in a plasma membrane patch excised from a protoplast of the mimosa *Samanea*. The line represents an "Ohm's law" for the current through an open ion channel: $i_K = \gamma (E_M - E_M^D)$ [22]. The slope is $\gamma$, the conductance of the single channel. The intersection (marked by an arrow) is $E_M^D$, the diffusion potential of the ions to which the channel is permeable [Eq. 4]. Inset: the excised-patch recording configuration. (B) An *I-V* plot for a whole cell (see the recording configuration in the inset). The whole-cell current ($I_K$) is the sum of the single-channel currents, and the curvature of the *I-V* plot represents the increasing number of channels opening with increased depolarization. (C) An electrical circuit representing the patch-clamp recording configuration. $E_M$ is the applied potential difference across the membrane. $R_M$ is the resistance (1/conductance) of the membrane. $I_K$ is the current through open $K^+$ channels in the membrane.

ing in the open conformation. The relation between the amount of current flowing through the membrane at different membrane potentials and the values of these potentials is then analyzed to learn about the conductivity, the gating, and the selectivity of the channels [23–25].

The direction of the net ion flux that is facilitated by opening of voltage-gated channels depends on a number of factors. These include the charge of the transported ion and the prevailing Nernst potential ($E_N$) for the specific ion.

Let us consider the effect of depolarization and hyperpolarization on $K^+$ and $Cl^-$ fluxes via voltage-gated channels in the plasma membrane of a plant cell. We assume that $K^+$ has been accumulated passively in the cytoplasm and is at equilibrium across the plasma membrane ($E_M = E_N$ for $K^+$, for example, $-100$ mV) and that $Cl^-$ has been accumulated actively in the cytoplasm and would be at equilibrium only at a very positive membrane potential (for example, 70 mV). Activation of $K^+$ channels by depolarization (for example, to 0 mV) would then induce $K^+$ efflux from the cytoplasm to the free space. Net $K^+$ efflux occurs because depolarization increases $E_M$ above $E_N$ for $K^+$. Depolarization-activated opening of a $Cl^-$ channel would also induce $Cl^-$ efflux. This is because even at $E_M = 0$, the electrochemical potential of $Cl^-$ inside is larger than outside.

Let us now consider hyperpolarization-activated $K^+$ and $Cl^-$ channels. Hyperpolarization (changing $E_M$ to more negative than the $E_N$ for $K^+$) creates an inward-directed electrochemical potential gradient for $K^+$ and results in net $K^+$ influx. Activation of a $Cl^-$ channel by hyperpolarization would, however, again result in net Cl efflux through the open channel. This is because the outward-directed electrochemical potential gradient of $Cl^-$ would increase with the decrease in membrane potential.

Using patch clamp, a large variety of ion channels has been detected in the plasma membranes, tonoplasts, and even membranes of chloroplasts isolated from plant tissues, ranging from mesophyll to root hairs and pollen tubes. Among these channels, the best characterized are, of course, the most easily accessible channels, those in the plasma membranes.

## 2. Plasma Membrane Channels

$K^+$ CHANNELS    There are two major kinds of voltage-activated $K^+$ channels in the plasma membrane of plant cells. One kind is activated by hyperpolarization and conducts $K^+$ into the cells (termed also "inward rectifying" channels). The second type is activated by depolarization and facilitates $K^+$ efflux ("outward rectifying" channels). The gating of both types of channels is affected strongly by protons. Hyperpolarization-activated $K^+$-influx channels in the stomatal guard cell membrane are activated by lowering external pH (external acidification) [26,27] and those in phloem cells by increasing external pH [28]. Depolarization-activated $K^+$ channels of stomatal guard cells are inhibited by protons [26,29]. $Ca^{2+}$ inhibits the hyperpolarization-activated $K^+$ channels in stomatal guard cells but does not affect the depolarization-activated $K^+$ channels in these cells.

The hyperpolarization-activated $K^+$ channels are not ideally selective for $K^+$ and admit other cations, such as $Rb^+$, $Na^+$ and $Ca^{2+}$ [30,31]. The reported $Na^+$ permeability relative to $K^+$ permeability in $K^+$-influx channels ranged between 1/100 and 1/10 [30,32]. The depolarization-activated $K^+$ channels were usually less $K^+$ selective [33–36]. Since the electrochemical potential gradient of $Ca^{2+}$ is always inward (the usual concentration ratio of $Ca^{2+}$ across the plasma membrane is at least $10^4$), even a tiny permeability of the $K^+$ channels to $Ca^{2+}$ may allow a nonnegligible influx of $Ca^{2+}$ into the cytosol.

A number of genes of channels have been cloned from plants (*Arabidopsis thaliana*, maize, potato). They were expressed in yeast cells, in oocytes of the frog *Xenopus*, and in insect cell lines. Patch-clamp and related electrophysiological methods were used to characterize the channels in these foreign (heterologous) systems (reviewed in Ref. 37). Molecular identification of a native (in situ) plant channel with a channel clone has not been yet established unequivocally for any one of these plant channels. However, the *Arabidopsis* KAT1 channel clone [38] tends to be indentified with the guard-cell $K^+$-influx channels [39]. Similarly, the AKT2/3 type of channels [40,41] are probably identical to the $K^+$-influx channels of the phloem elements [28,42] and the *Arabidopsis* SKOR1 channel clone [43] with $K^+$-efflux channels in xylem parenchyma (reviewed in Ref. 37). The ZMK1 channel clone from corn (of the AKT1 family) was identified with the hyperpolarization-activated $K^+$ channels in corn coleoptiles and implicated in gravitropism [42].

$Ca^{2+}$ CHANNELS   Depolarization-activated, $Ca^{2+}$-selective channels were described [44]. $Ca^{2+}$ channels have been implicated in signaling evoked by various environmental stimuli, including mechanical stimuli [18] (see reviews in Refs. 45 and 46).

NONSELECTIVE CATION CHANNELS   A depolarization-activated nonselective cation channel that conducts $Ca^{2+}$ was demonstrated in guard cells [47], in seed endosperm cells [48], and in roots of cereals [49]. These channels are probably the major pathways for entry of $Na^+$ and $Ca^{2+}$ into cells. When the external $Na^+$ concentration is high (saline conditions) and the electrochemical potential gradient of $Na^+$ is inward, depolarization-activated cation channels may conduct considerable fluxes of $Na^+$ into the cytosol [49]. A tobacco plant overexpressing a putative nonselective cation channel became sensitive to $Pb^{2+}$ [50].

ANION CHANNELS   Two types of $Cl^-$ channels have been described in the guard cell plasma membrane: rapidly gated anion channels (R type) and slowly gated ones (S type) [51]. It has not been resolved whether they are indeed separate molecular entities or two interconverting facets of the same channel [52]. They are permeable to various additional anions, including nitrate, malate, and sulfate [53–55]. The gating of the S-type anion channel of guard cells is further regulated by these anions [56], as well as by the plant hormones auxin and abscisic acid [57,58] (see also review in Ref. 59).

In addition to their obvious function in nutrient uptake, anion channels are implicated in signaling in photomorphogenesis, in osmoregulation of stomatal guard cells, and in interactions with pathogens and symbionts [57–62]. A gene homologous to the animal $Cl^-$ channel has been cloned from *Arabidopsis* [63,64], but its identification with a native (in situ) anion channel remains uncertain [37].

### 3. Tonoplast Channels

Vacuoles make up over 90% of the cell volume in most plant cells and serve as a major storing compartment for minerals. Nevertheless, our knowledge of the tonoplast channels is much less complete. A total of two channels, both $Ca^{2+}$ activated, have been identified that are capable of conducting $K^+$; these are the voltage-independent vacuolar channels specific for $K^+$ (VK) and the slow vacuolar (SV) channels [65]. The SV channels are slowly activated by depolarization (of the cytosolic side relative to the vacuolar side) and are permeable to $K^+$ and $Ca^{2+}$. They may be involved in the mechanism of "$Ca^{2+}$-activated $Ca^{2+}$-release" [66]. The fast-activating (FV) channels are cation-conducting channels and dominate the ion conductance of the vacuolar membrane at physiological $Ca^{2+}$ concentrations. They may serve for the uptake of $NH_4^+$ and $K^+$ into the vacuole [67] (see also reviews in Refs. 68–70).

Two distinct types of vacuolar calcium channels facilitating $Ca^{2+}$ release from the vacuole to the cytosol are activated by two types of second messengers: $IP_3$ [17] and cADP-ribose [71]. Both have been implicated in various abscissic acid responses [72–74] (see also reviews in 75–78). $IP_3$ is produced by the phosphatidylinositol cascade [79]. cADP-ribose is produced from $NAD^+$ by ADP-ribosyl-cyclase (see references in Ref. 77).

Much less is known about anion channels. Malate and chloride channels have been described in the vacuole [80,81].

## B. Carriers

### 1. Transport Kinetics

Carriers catalyze the transport of specific solutes across the membrane. This transport is often vectorial (unidirectional). The relation between solute concentration and unidirectional flux in plant tissues was described by Epstein and Hagen [82]. They used terms introduced by Michaelis and Menten [83] for enzyme kinetics (the relation between the chemical activity of the substrate and the velocity of enzyme-catalyzed reactions). Carrier-catalyzed transport (or enzyme-catalyzed reaction) is modeled as follows:

$$[E] + [S^0] \underset{k_2}{\overset{k_1}{\longleftrightarrow}} [ES] \overset{k_3}{\longleftrightarrow} [E] + [S^i] \qquad (12)$$

where [S], [E], and [ES] are the concentrations of the transported solute (or substrate), the carrier (or enzyme), and the carrier-solute (or enzyme-substrate) complex, respectively, and $k_1$, $k_2$, and $k_3$ are the rate constants of the reactions. Assuming that $k_3$ is much smaller than $k_1$, the transport velocity ($v$), or influx, is given by

$$v = k_3[ES] \tag{13}$$

Maximal transport velocity $V_{max}$ should be attained at saturating solute concentration, when all carrier molecules are occupied by the solute:

$$V_{max} = k_3[E_T] \tag{14}$$

where $[E_T]$ is the sum of occupied and unoccupied carriers or the total carrier concentration ($[E_T] = [E] + [ES]$). The affinity of the carrier for the solute is the reciprocal of the dissociation constant of the carrier-solute complex or the reciprocal of the Michaelis-Menten constant ($K_m$):

$$K_m = \frac{k_2}{k_1} = \frac{[E][S^\circ]}{[ES]} \tag{15}$$

$K_M$ and $V_{max}$ customarily define transport kinetics. To calculate the transport velocity, $[E]$ in Eq. (15) is replaced by ($[ET] - [ES]$), and the equation is rearranged:

$$K_m = \frac{([E_T] - [ES])[S^\circ]}{[ES]} = \frac{[E_T][S^\circ]}{[ES]} - [S^\circ] \quad \text{and} \quad [ES] = \frac{[E_T][S^\circ]}{K_m + [S^\circ]}$$

The latter equation can be substituted for $[ES]$ in Eq. (13), that is, $v = k_3[E_T][S]/(K_M + [S])$, and by replacing ($k_3[E_T]$) with $V_{max}$ [Eq. (14)] the Michaelis-Menten equation is obtained:

$$v = \frac{V_{max}[S]}{K_m + [S]} \tag{16}$$

This equation can be transformed into a linear function of the reciprocals of $v$ (same as $J^{o \to i}$) and $S$ (same as $a_j^o$):

$$\frac{1}{v} = \frac{K_M}{V_{max}[S]} + \frac{1}{V_{max}} \tag{17}$$

where $K_M/V_{max}$ is the slope of the line and $1/V_{max}$ is the Y-axis intercept. Substitution of $V_{max}/2$ for $v$ shows that $K_M$ is the solute activity when the velocity is half-maximal. Various graphic analyses [84] show that competitive inhibition (inhibition by competition for carrier sites) of transport increases the apparent $K_M$ (decreases apparent affinity) but does not affect $V_{max}$.

Strict compliance with Michaelis-Menten kinetics is usually obtained only in a narrow solute concentration range and, in particular, in the low concentration ranges up to about 0.7 mM. In broader concentration ranges, multiphasic uptake kinetics [85] are ordinarily encountered. The reason for multiphasic kinetics has not been resolved; various possible explanations have been discussed [85–90].

## 2. Primary Active Transport

The term primary active transport [7] is reserved for active transport that is directly driven by energy-rich metabolites, such as ATP, pyrophosphate, or electron donors. Cotransport is classified as secondary active because it derives its energy from the electrochemical potential difference that is produced by primary active transport.

Two types of adenosinetriphosphatase (ATPase), at the plasma membrane and the tonoplast, respectively, and an inorganic pyrophosphatase (PPiase) at the tonoplast are known to generate an electrochemical proton potential difference, namely a proton motive force, at these membranes [32]. In addition, a $Ca^{2+}$-ATPase performs primary active $Ca^{2+}$ transport. Proton motive force is also generated by vectorial electron transport across the inner membranes of mitochondria and chloroplasts. This proton motive force is primarily used for ATP synthesis, catalyzed by a third type of ATPase (an ATP synthase) [91].

PROTON MOTIVE FORCE GENERATING ATPases   The proton motive force-generating $H^+$-ATPase at the plasma membrane transports protons actively from the cytoplasm to the free space. At the tonoplast, the V-ATPase and the PPiase pump protons actively from the cytosol into the vacuole. Both ATPases need $Mg^{2+}$ to function; their substrate, indeed, is Mg-ATP. These ATPases differ in their evolution, in their homology, and in some of their characteristics from bacterial and animal ATPases [92].

The plasma membrane $H^+$-ATPase forms a phosphorylated intermediate during ATP hydrolysis and belongs to the P-ATPase family [92]. Similar intermediates are formed by other ion-transporting P-ATPases. These ATPases are specifically inhibited by orthovanadate ions. The plasma membrane $H^+$-ATPase has a functional molecular mass of 200 kDa and is apparently composed of two 100-kDa subunits, each forming at least 8, and possibly 10, transmembrane helices [93]. Optimal pH for this enzyme is 6.5; its activity is enhanced (maximally doubled) by $K^+$, but the enzyme is not directly involved in $K^+$ transport [94]. The $H^+$-ATPase is a major component of the plasma membrane. In root cells with high transport activity, there are about $10^6$ molecules of the enzyme per cell, with a turnover number of 20–100 $sec^{-1}$. This results in proton fluxes of 10–100 pmol $sec^{-1}$ $cm^{-2}$ [95].

The tonoplast $H^+$-ATPase of plant cells does not form a phosphorylated intermediate and is not inhibited by vanadate. It belongs to the family of V-ATPases, operating as proton pumps at endomembranes of eukaryotic cells [91]. The vacuolar $H^+$-ATPase is inhibited by the antibiotic bafilomycin [96] and by $NO_3^-$. It is stimulated by $Cl^-$ and not by monovalent cations; the pH optimum is 7.9 [92]. A functional molecular mass of 750 kDa was assigned to the enzyme, which is composed of at least 10 different subunits [97] and exhibits four potential transmembrane helices [92].

$Ca^{2+}$-ATPase    Another primary active carrier is the enzyme $Ca^{2+}$-ATPase. This enzyme is located in the plasma membrane [98], in the tonoplast [99], and in the endoplasmic reticulum [100]. The $Ca^{2+}$-ATPase forms a phosphorylated intermediate, is inhibited by vanadate [101,102], and belongs to the P-ATPase family [103]. $Ca^{2+}$-ATPases transport $Ca^{2+}$ out of the cytoplasm to the free space, into the vacuole, and into endoplasmic reticulum vesicles. $Ca^{2+}$-ATPases, together with $Ca^{2+}/H^+$ antiporters (see Mineral Cations in Sec. V.B.3), are involved in $Ca^{2+}$ homeostasis of the cytosol and regulate the $Ca^{2+}$ activity there at about 0.1 $\mu M$ [104]. Cytosolic $Ca^{2+}$ [103] activity rises only transiently in response to certain stimuli [105]. Alignment of eukaryotic $Ca^{2+}$-ATPases shows that the plant and animal genes are related [99] and can be divided into two types of pumps, those that are stimulated by calmodulin and those that are not [106]. Calmodulin is a calcium-modulated protein involved in many $Ca^{2+}$-regulated processes [107].

A spinach plasma membrane $Ca^{2+}$-ATPase was identified as a 120-kDa polypeptide. Calmodulin increased the $V_{max}$ for $Ca^{2+}$ pumping into inside-out spinach plasma membrane vesicles (homologous to pumping from the cytosol to the free space) threefold and decreased the $K_m$ from 1.6 to 0.9 mM. During trypsin treatment (limited proteolysis) the amount of the 120-kDa polypeptide diminished and a 109-kDa polypeptide appeared. The appearance of the 109-kDa polypeptide correlated with increased enzyme activity and loss of calmodulin regulation. Limited proteolysis increased the $V_{max}$ for $Ca^{2+}$ pumping more than calmodulin [108]. The contribution of the $Ca^{2+}$-ATPase to the electrical potential difference across the plasma membrane is insignificant because its activity is two orders of magnitude lower than that of the $H^+$-ATPase [95].

Three genes encoding calmodulin-stimulated $Ca^{2+}$-ATPases have been cloned from plants, *ACA1* and *ACA2* from *Arabidopsis thaliana* and *BCA1* from *Brassica oleracea*. None localized to the plasma membrane; they are expressed in endomembranes and the tonoplast [109]. A further three genes encoding calmodulin-insensitive $Ca^{2+}$-ATPases have been cloned from plants, *LCA1* from *Lycopersicum esculentum*, *pH27* from *Nicotiana tabacum*, and *ACA3* (*ECA1*) from *Arabidopsis thaliana*. *LCA1* was localized to the tonoplast and the plasma membrane and both *pH27* and *ACA3* to endomembranes [109].

ABC TRANSPORTERS    An additional group of primary active solute transporters have been demonstrated. They belong to the ABC (ATP-binding cassette) superfamily [110]. These transporters were originally identified in bacterial and animal cells. The ABC transporters use $Mg^{2+}$-ATP as a direct energy source for transport; they form a phosphorylated intermediate during hydrolysis and belong to the P-type ATPases. All ABC transporters consist of an integral membrane sector and a cytosol-orientated ATP binding domain. The membrane sector of the ABC transporters provides the pathway for transport and determines the molecular specificity of the transporter.

Two major subclasses of ABCs have been molecularly identified in plants: MRPs (multiple drug resistance–associated proteins) and MDRs (multidrug resistance proteins, so named according to their animal prototypes). Only MRPs have been defined functionally. MRPs are localized to the vacuole and function in detoxification. A mineral ion–related function of MRPs is detoxification of $Cd^{2+}$ and perhaps other heavy metals. MRP functions in detoxification by transport of metal phytochelatins to the vacuole

[111]. Phytochelatins are glutathione-derived peptides (poly γ-glutamylcystein followed by a C-terminal glycine) [112].

PYROPHOSPHATASE   Another primary proton pump, an $H^+$-PPiase ($H^+$-pyrophosphatase), functions in parallel with the V-ATPase to create a proton gradient across the tonoplast [113]. The reaction is

$$H_4P_2O_7 + 2H_2O \xrightarrow{\text{PPiase}} 2H_3PO_4 + (H^+)^o + (OH^-)^i$$

This enzyme has not been identified in endomembranes of animal cells. The molecular mass of the $H^+$-PPiase is 81 kDa, and it appears to comprise a single polypeptide [32]. The substrate of this enzyme is MgHPPi or MgPPi, and it is stimulated by additional free $Mg^{2+}$ [114]. The optimal pH for the $H^+$-PPiase is 8.5–9.0, depending on the $Mg^{2+}$ concentration [115]. The enzyme is not inhibited by vanadate or $NO_3^-$ [116]. The proton transport by both the PPiase and the V-ATPase is regulated by the $\Delta\bar{\mu}_{H^+}$ across the tonoplast. The PPiase exhibits an almost absolute requirement for $K^+$ on its cytoplasmic face [117]. On the basis of this and additional data [117], it was proposed that the PPiase serves to catalyze the coordinated translocation of both $H^+$ and $K^+$ from the cytosol into the vacuole [113]. Under chill or hypoxic stress, the transcript level and activity of the PPiase can increase to counter the impaired activity of the V-ATPase as ATP levels drop and the latter enzyme dissociates [118].

## 3. Cotransport

Cotransport is the secondary active transport [12] of a specific solute. It is coupled to transmembrane ion gradients formed by primary active transport. Plant cells mostly employ the proton motive force as the energy source for active solute cotransport. The general tenet of this transport is similar to that of a pulley: protons are carried "downhill" (passively) across the membrane, while the other solute is carried "uphill" (actively). The direction of passive proton transport is from the free space or vacuole to the cytoplasm. The cotransport of protons and the other solute in the same direction is called symport; when protons and the other solute are cotransported in opposite directions, the phenomenon is termed antiport [7].

Cotransporters are perceived as membrane-embedded transport proteins [119]. Conformation changes are supposed to expose the solute binding sites alternatively to the inside or outside. Two principal modes of cotransport were proposed [120]. According to the simultaneous model of Jauch and Läger, a proton binds to a proton binding site when the site is exposed to the medium, where the chemical potential of protons is high. This supposedly induces a conformation change, resulting in increased affinity of the binding site for the cotransported solute. The increased affinity facilitates binding of the cotransported solute, even when its chemical potential is low. A symported solute binds on the same side of the membrane as the proton, and an antiported one binds on the opposite side. Binding of the cotransported solute is supposed to induce another conformation change that exposes the proton to the inside. The proton is then released inside, where the proton electrochemical potential is lower. A cotransported solute is now also exposed to the inside and an antiported solute to the outside. The release of the proton decreases the affinity of the binding site for the cotransported solute and it is also released, at the side of its higher electrochemical potential (inside and outside for symport and antiport, respectively).

The stoichiometry of protons and cotransported solute differs in the various cases and is not always known. In some cotransport systems, the number of cotransported protons equals the negative charges of symported anions or positive charges of antiported cations. This results in electroneutral transport, exclusively driven by the $\Delta pH$ component of the proton motive force. In other cotransport systems, an excess of protons is cotransported and electrogenic transport results. The latter kind of transport is driven by both components of the proton motive force, namely the $\Delta pH$ and the $E_M$.

For a known accumulation ratio of a specific cotransported ion, the minimal required number of cotransported protons per ion can be estimated when both $E_M$ and $\Delta pH$ are known [10]. At the steady state, the following relation between the electrochemical potential difference of a solute $j$ and the number ($n$) of symported protons should apply [10]:

$$\Delta\bar{\mu}_j = -n\Delta\bar{\mu}_{H^+} \quad \text{or} \quad \Delta\bar{\mu}_j = -nF\,\text{pmf} \tag{18}$$

Substitution of Eq. (2) for $\Delta\bar{\mu}_j$, and substitution of Eq. (7) for pmf at 30°C gives

$$2.3RT \log \frac{a_j^i}{a_j^o} + z_jF(\psi^i - \psi^o) = -nF(E_M - 60\Delta pH)$$

Substitution of $E_M$ for $(\psi^i - \psi^o)$ and division by $F$ results in

$$\frac{2.3RT}{F} \log \frac{a_j^i}{a_j^o} + z_j E_M = -n(E_M - 60\Delta pH)$$

When we define $\Delta pj = -\log \dfrac{a_j^i}{a_j^o}$ and substitute 60 mV for $2.3RT/F$ (30°C), we have $-60\Delta pj + z_j E_M = -n(E_M - 60\Delta pH)$ and Eq. (19) is obtained:

$$\Delta pj = \frac{n+z}{60} E_M - n\Delta pH \tag{19}$$

For antiport, the steady-state situation is $\Delta\overline{\mu}_j = nF$ pmf, and

$$\Delta pj = \frac{z-n}{60} E_M + n\Delta pH \tag{20}$$

Equation (19) can be employed to analyze $Cl^-$-$H^+$ symport. If electroneutral transport is assumed ($n = 1$ and $z$-$n = 0$) $\Delta pCl$ will equal $\Delta pH$, and such transport would not depend on the membrane potential. When a 100-fold $Cl^-$ accumulation is found ($\Delta pCl = -2$), a relatively large $\Delta pH$ of 2 would be needed for electroneutral transport. However, if $n = 2$ is assumed and $E_M = -120$ mV, no $\Delta pH$ would be needed for a $\Delta pCl$ of $-2$.

## MINERAL CATIONS

**Sodium.** The first evidence in higher plants for cotransport energized by the proton motive force was that for $Na^+/H^+$ antiport in barley roots [121]. Much of the information about $Na^+/H^+$ antiport at the plasma membrane of higher plants emanated from experiments with intact plant tissues (see Ref. 122). The $Na^+/H^+$ antiporter was investigated at the level of plasma membrane vesicles in the halotolerant unicellular alga *Dunaliella salina* [123,124], in *Atriplex nummularia* [125,126], and in *Gossypium hirsutum* [126]. The $Na^+/H^+$ antiporter in *Dunaliella* plasma membrane vesicles had a $K_m$ for $Na^+$ of about 16 mM and was inhibited by amiloride, an inhibitor of $Na^+/H^+$ antiport in animal cells [123]. The $V_{max}$ of the antiport increased when the cells had been adapted to a high NaCl concentration or to ammonia at a high pH; it decreased in LiCl-adapted cells [127]. The increase of $V_{max}$ was interpreted as overproduction of the $Na^+/H^+$ antiporter and was correlated with overproduction of a 20-kDa and a 50-kDa polypeptide. The $Na^+/H^+$ antiport in *Dunaliella* was specific for $Na^+$ in comparison with $K^+$, $Cs^+$, and $Li^+$ [123]. However, in plasma membrane vesicles of *Atriplex* [125,126] and *Gossypium* [126], similar dissipation of pH gradients was found with $Na^+$ and $K^+$ ions, and a high concentration of one of these ions was needed for activity. An additive effect was obtained when $Na^+$ was added to saturating $K^+$ concentrations and vice versa. It was suggested that separate antiporters for $K^+$ and $Na^+$ may operate.

Evidence for $Na^+/H^+$ antiport at the tonoplast emanated first from experiments of Blumwald and Poole [128] with tonoplast vesicles. They demonstrated amiloride-sensitive $Na^+/H^+$ amiport in tonoplast vesicles from *Beta vulgaris* storage tissue. At a constant $\Delta pH$ the apparent $K_M$ for $Na^+$ increased from 7.5 to 26.6 mM when the internal pH decreased from about 7.5 to 6.5. In tonoplast vesicles from suspension cultured cells of *B. vulgaris* [129] and from *Plantago maritima* roots [130], $Na^+/H^+$ antiport activity increased in response to NaCl in the growth medium. In *B. vulgaris* [129] such increased activity was associated with an increase in $V_{max}$ without a change of $K_m$. Sodium antiport activity was also increased by cultivation of the *B. vulgaris* cells in the presence of amiloride [131]. In both cases the increase of $Na^+/H^+$ antiport activity was accompanied by synthesis of a 170-kDa polypeptide. Polyclonal antibodies against this polypeptide almost completely inhibited the $Na^+/H^+$ antiport activity [131]. Garbarino and DuPont [132] induced $Na^+/H^+$ exchange in barley roots by pretreatment with sodium salts. The half-time of induction was 15 min and it was attributed to the activation of a preexisting protein. Unlike its effect on the relatively salt-tolerant crops beet and barley, NaCl treatment did not activate the antiporter in rice roots [133].

**Potassium.** Potassium concentrations in soil solutions range from 1 μM to 10 mM, with many soils falling in the range of 0.3 to 5.0 mM, while intracellular $K^+$ levels are maintained at 100 to 200 mM [109]. Hence, assuming an $E_M$ of $-120$ mV, plants may obtain $K^+$ by uniport from most soils, but a high-affinity transporter may be needed when the $K^+$ concentration in the soil solution is less than 1.0 mM [134]. At much more negative potentials $K^+$ channels may suffice for a high-affinity $K^+$ influx [135].

In the marine alga *Chara australis* $Na^+$-$K^+$ symport was demonstrated. Such symport employs the naturally existing and inward-directed electrochemical $Na^+$ gradient for high-affinity $K^+$ transport against its electrochemical potential gradient. Utilization of the natural $Na^+$ gradient for symport of various solutes was suggested for other halophytes and ions, namely $Na^+$-$NO_3^-$ symport in the marine higher plant *Zoospora maritima* [136] and $Na^+$-urea, $Na^+$-sugar, and $Na^+$-lysine symport in charophyte algae [137,138]. However, all investigated terrestrial plants were able to sustain growth and $K^+$ uptake in the absence of $Na^+$ [139]. The latter authors suggested that in terrestrial species $Na^+$-coupled $K^+$ transport has no or limited physiological significance, whereas in certain aquatic angiosperms and algae it plays a significant role. Two complementary DNAs (cDNAs) encoding high-affinity $K^+$ antiporters, HKT1 [140] and KUP [141], with $K_m$ values of 29 and 22 $\mu$M were isolated from wheat and *Arabidopsis*, respectively. Both transporters were up-regulated by $K^+$ deficiency [141,142]. The wheat HKT1 functioned as an $H^+$-$K^+$ symporter with an $H^+/K^+$ ratio of 1.0. Potassium transport by the wheat HKT1 was stimulated by micromolar $Na^+$ concentrations and $K^+$-activated high-affinity $Na^+$ uptake. However, $K^+$ uptake was increasingly inhibited by $Na^+$ concentrations above 1.0 mM [143].

**Calcium.** Kasai and Muto [144] suggested that $Ca^{2+}/H^+$ antiport operates at the plasma membrane, in addition to the $Ca^{2+}$-ATPase. Genes encoding a high-affinity $Ca^{2+}/H^+$ antiporter (*CAX1*) and a low-affinity $Ca^{2+}/H^+$ antiporter (*CAX2*) have now been cloned from *Arabidopsis* [145]. When expressed in a hypersensitive yeast strain, *CAX1* catalyzed $\Delta$pH-dependent $Ca^{2+}$ transport with a $K_m$ of 13 $\mu$M. Hence, *CAX1* along with $Ca^{2+}$-ATPase is proposed to keep cytosolic $Ca^{2+}$ below 1 $\mu$M in resting plant cells.

A $Ca^{2+}$-ATPase and $Ca^{2+}/H^+$ antiport are apparently also involved in the transport of $Ca^{2+}$ from the cytosol to the vacuole. The $Ca^{2+}$ concentration in plant cell vacuoles is in the millimolar range. Accordingly, the apparent $K_M$ values for $Ca^{2+}$ of the $Ca^{2+}/H^+$ antiporter in tonoplasts from oat roots [146,147] and *B. vulgaris* [148] were 10–14 mM and 42–200 $\mu$M, respectively. In the latter case, the $K_M$ varied with the pH in the tonoplast vesicles. The $Ca^{2+}/H^+$ antiport created an 800 to 2000-fold $Ca^{2+}$ gradient in tonoplast vesicles [149]; an $H^+/Ca^{2+}$ stoichiometry of 3 was indicated [150].

MINERAL ANIONS The accumulation of anions in the negatively charged cytosol should be an active process whether they are transported to the cytosol across the plasma membrane from the free space or across the tonoplast from the vacuole. Anion-proton symport is expected to expedite such transport. The early evidence for symport of protons with $Cl^-$ [151], $NO_3^-$ [152,153], $H_2PO_4^-$ [154,155], and $SO_4^{2-}$ [156] emanates from experiments that showed a transient depolarization of the plasma membranes, or alkalization of the medium [155], upon addition of these anions to intact roots. In all these investigations the degree of transient depolarization was related to the transport rate of the anions. Transient depolarization indicates electrogenic transport with a stoichiometry exceeding 1 for $H^+$/anionic charge transported. Depolarization is transient because the membrane potential is restored by increased proton pump activity. Sakano [155] measured the stoichiometry of $H_2PO_4^-$ and $H^+$ uptake (transient external alkalinization) in *Caranthus roseus* cell cultures and calculated an $H^+/H_2PO_4^-$ stoichiometry of 4.

**Chloride.** Proton-chloride symport was shown [157] for barley roots that were de-energized by anaerobic pretreatment. In such roots, $Cl^-$ influx could be induced by application of an artificial pH gradient (acid outside). However, $Cl^-$ influx was not induced by a low pH outside in the absence of a $\Delta$pH, obtained with a weak acid such as acetic acid. In the latter case the high membrane permeability of acetic acid molecules resulted in the dissipation of the initial pH gradient. Sanders [151] reviewed $Cl^-$ transport in plants and concluded that the $H^+/Cl^-$ stoichiometry of symport should be at least 2. Such stoichiometry is consistent with electrogenic $Cl^-$ transport and with transient plasma membrane depolarization upon application of $Cl^-$. The symport of protons with $Cl^-$ into isolated barley root plasma membrane vesicles was also characterized [158]. The ATP-mediated acidification of the vesicles was strongly dependent on the presence of permeant anions. Also, $^{36}Cl^-$ transport into the vesicles depended on the electrochemical potential generated by the $H^+$-ATPase.

**Phosphate.** The symport of phosphate and protons across the plasma membrane was demonstrated in outside-out *B. vulgaris* plasma membrane vesicles (SR Stutz, B Jacoby, unpublished). Phosphate uptake and dissipation of the pH gradient occurred concomitantly when $H_2PO_4^-$ was added to these vesicles in the presence of an artificial pH gradient. The initial rate of Pi transport depended on the magnitude of the $\Delta pH$. Phosphate transport into the vesicles was also mediated by a negative (inside) electrical gradient (negative $E_M$) in the absence of a $\Delta pH$. This indicated electrogenic $H^+$-phosphate symport and an $H^+/H_2PO_4^-$ stoichiometry $> 1.0$.

Two cDNAs (*AtPT1* and *AtPT2*) encoding plant phosphate transporters have been isolated from phosphate-starved *Arabidopsis thaliana* roots [159]. The corresponding protein belongs to a family of transporter proteins with 12 putative transmembrane proteins. It is highly homologous to Pi transporters isolated from yeasts and fungi. When expressed in a Pi uptake–deficient yeast mutant, it exhibited high-affinity phosphate transport activity. The transcripts of both genes were expressed in roots but were not detectable in leaves. Similar genes, *LePT1* and *LePT2*, were highly expressed in tomato roots [160]; only *LePT1* was expressed in the leaves as well. The expression was induced by phosphate starvation but not by nitrate, potassium, and iron starvation. The transcripts were localized to the root epidermis cells and *LePT1* also to leaf palisade cells. When *LePT1* was expressed in a Pi uptake–deficient yeast mutant [161] it exhibited a $K_m$ of 31 mM, but the expressed enzyme was still active at submicromolar Pi concentration and mediated highest uptake at pH 5.0. The transport activity of *LePT1* depended on an electrochemical proton gradient.

**Nitrate.** Nitrate-$H^+$ symport was characterized in plasma membrane vesicles isolated from maize [162] and cucumber [163] roots (nitrate was labeled with $^{36}ClO_3^-$, an $NO_3^-$ analogue). The transport was driven by an artificially imposed $\Delta pH$. The initial rate of nitrate transport depended on the magnitude of the $\Delta pH$. The imposed $\Delta pH$ affected the $K_m$ but not the $V_{max}$. Nitrate transport was higher into vesicles isolated from both nitrate- and starvation-induced cucumber plants than that observed in vesicles obtained from uninduced plants [163].

An additional kind of nitrate transport, namely $Na^+$-$NO_3^-$ symport, was demonstrated in cells of the marine plant *Zostera marina* L. [136]. In this marine environment the innate $Na^+$ gradient is utilized for the active transport of $NO_3^-$. The plasma membranes of *Z. marina* cells were depolarized in the presence of both $Na^+$ and $NO_3^-$ ions but not in the presence of one of these ions alone. The depolarization was inhibited by monensin, a sodium ionophore. The depolarization indicated a transport stoichiometry of at least $2Na^+/NO_3^-$.

The *CHL1* (*NRT1*) gene of *Arabidpsis* encodes a nitrate-inducible nitrate transporter. This transporter is thought to be a component of the low-affinity nitrate uptake system in plants [164]. It functions in the low-affinity (10 mM) as well as in the high-affinity (0.1 $\mu$M) concentration range. It may be a dual-affinity nitrate transporter. An additional gene encoding a putative high-affinity nitrate transporter (*GmNRT2*) was isolated from a soybean root cDNA library and sequenced [165]. It is related to high-affinity nitrate transporters in *Chlamydomonas reinhardtii* and *Aspergillus nidulans* and putative high-affinity nitrate transporters in barley and tobacco. Expression of the gene was selectively regulated by different N sources. The expression was barely detected in $NH_4^+$ grown plants; it was higher in N-deprived plants and highest in $NO_3^-$ grown plants. Induction resulted in a fourfold increase of $NO_3^-$ uptake from 0.1 $\mu$M external $NO_3^-$ and occurred within 1 hr.

**Sulfate.** Sulfate uptake by *Brassica napus* root plasma membrane vesicles was characterized [166]. It was driven by an artificially imposed $\Delta pH$. The $K_m$ for sulfate was strongly pH dependent (at constant $\Delta pH$). It decreased from 1.0 $\mu$M at pH 5.0 outside to 64 $\mu$M at pH 6.4. The initial rate of sulfate uptake and the equilibrium concentration in vesicles isolated from sulfate-starved roots were approximately twofold greater than observed in those isolated from sulfate-fed plants.

A cDNA encoding a high-affinity sulfate transporter was isolated from barley [167]. This cDNA, designated *HVST1*, encodes a polypeptide that has a high sequence homology with other identified eukaryotic sulfate transporters. The $K_m$ for $SO_4^{2-}$ was 6.9 nM when the *HVST1* cDNA was expressed in a yeast mutant.

**Mineral Anion Transport to Vacuoles.** The electrochemical potential gradients of the major anions and of $H^+$ across the tonoplast suggest anion-proton symport from the vacuole to the cytosol. However, evidence for such symport is limited. Blumwald and Poole [168] formed an ATPase-dependent proton motive force in *B. vulgaris* tonoplast vesicles and subsequently added $Cl^-$ or $NO_3^-$. Both ions dissipated the electrical component of the proton motive force (the $E_M$), but only $NO_3^-$ dissipated the $\Delta pH$ as well. They concluded that both anions entered the vesicles by uniport, dissipating the $E_M$, and that only $NO_3^-$ was excreted again by symport with protons, thus dissipating the $\Delta pH$. In similar experiments of Schumaker and Sze [169] with oat root tonoplast vesicles, both $Cl^-$ and $NO_3^-$ dissipated the $E_M$ as well as the $\Delta pH$.

## 4.  Uptake of Heavy Metals

Genes encoding proteins involved in the transport of copper, manganese, zinc, and iron have been identified [109]. In most cases it is not known whether these proteins have a direct role in the transport across the plasma membrane or whether they function as carriers or channels. More detailed information is available for iron uptake.

IRON   The low solubility of Fe-bearing minerals restricts the available Fe pools in most soils. The free Fe concentration in soil solutions is usually less than $10^{-15}$ M [170]. Plants employ two distinct and in all known cases mutually exclusive strategies for solubilization and absorption of Fe [171].

All plants except grasses employ a procedure termed strategy I to acquire Fe. These plants reduce and solubilize Fe(III) prior to transport of Fe(II) across the plasma membrane of root cells. The initial reduction and solubilization are carried out by two plasma membrane–bound enzymes, an $H^+$-ATPase and an Fe(III) chelate reductase. Proton release, Fe(III) chelate reductase, and Fe(II) transport activities are all enhanced under Fe deficiency [109]. Proton release lowers the rhizosphere pH and increases Fe(III) solubility. In *Arabidopsis* one of the genes encoding $H^+$-ATPase, *AHA2*, is up-regulated in response to Fe deficiency and may be involved in this acidification. An FRO2-encoding gene was isolated from *Arabidopsis* roots. The FRO2 protein belongs to a superfamily of flavocytochromes that transport electrons across membranes [17?]. The authors showed that FRO2 is allelic to the *frd1* mutations that impair Fe(III) chelate reductase activity. Introduction of functional FRO2 complemented the frd1-1 phenotype in transgenic plants. Iron deficiency induces the expression of a further *Arabidopsis* gene; it encodes IRT1 (iron-regulated transporter). Expression of IRT1 in yeast restored iron-limited growth to a yeast mutant defective in Fe uptake [173]. Yeasts expressing IRT1 possess an iron uptake system that is specific for Fe(II) over Fe(III) and other potential substrates such as Cu(I), Cu(II), Mn(II), and Zn(II). However, $Cd^{2+}$ inhibited $Fe^{2+}$ uptake by IRT1. It is proposed that IRT1 is an $Fe^{2+}$ transporter and that it may transport $Cd^{2+}$ as well.

Grasses employ another procedure, termed strategy II, for Fe acquisition [171]. These plants synthesize and secrete phytosiderophores. These are low-molecular-weight, Fe(III)-specific ligands. The phytosiderophores involved in strategy II are mucigenic acids, namely nonproteinous amino acids synthesized from methionine [174]. They posses a high chelation affinity for Fe(III) but not for other polyvalent cations. Iron transport is regulated by a specific uptake system that transports the phytosiderophore-Fe(III) complex across the plasma membrane [175]. Nicotine amine (NA) and the enzyme nicotine amine aminotransferase (NAAT) are implicated in the synthesis of mucigenic acids. NAAT-encoding cDNAs were identified in barley and NAAT was strongly induced under Fe deficiency [109]. In order to identify the Fe(III)–mucigenic acid transporter, a yeast mutant *ctr1* that is unable to grow on Fe-deficient media was transformed with a barley cDNA expression library. A clone designated SFD1 (suppressor of ferrous uptake defect) that could use Fe(III)–mucigenic acid as an Fe source was isolated.

# VI.  CHARGE BALANCE

Uniport, cotransport, and primary active transport may all result in the transport of unbalanced charges; that is, they may all perform electrogenic transport. Primary active transport generates a membrane potential, and electrogenic cotransport and uniport dissipate it. Continued electrogenic contransport and uniport are sustained by persistent proton motive force turnover (dissipation, followed by regeneration by primary active proton transport). An extremely small imbalance of cation and anion fluxes results in a considerable $E_M$. The ion concentration difference that sustains the membrane potential of plant cells (up to about $-240$ mV at the plasma membrane) can be calculated from its relation to the capacitance $C$ and the electrical charge $Q_M$ of the membrane: $Q_M = CE_M$ [2]. For a spherical cell with a radius of 50 $\mu$m, such a calculation shows that an uncompensated ion concentration difference of about 1.5 $\mu$M is needed to sustain a membrane potential of $-240$ mV. The total cytosolic salt concentration may be assumed to be about 50 mM. An anion concentration difference of about 1.5 $\mu$M then constitutes an anion excess of only 0.003% and is tantamount to equivalent anion and cation concentrations.

Considerably different amounts of the anion and cation of a salt are absorbed by plant cells, this in spite of near anion and cation balance in the cells. The difference is compensated by proton fluxes, resulting in transient pH shifts in the cells. Plant cells regulate the cytosolic pH at about 7.0 [176]. Synthesis or decomposition of organic acids accomplishes such regulation in response to nonstoichiometric anion and cation absorption.

When plants are transferred to a salt solution with $K_2SO_4$, they usually absorb an excess of $K^+$ over $SO_4^{2-}$ equivalents. The excess of positive charge in the cytosol is compensated by proton excretion, which results in a transient elevation of cytosolic pH. An elevated cytoplasmic pH induces respiratory $CO_2$ to form bicarbonate; this then enhances phosphoenolpyruvate carboxylase activity and consequent malic acid synthesis. The consumption of phosphoenolpyruvate enhances its replenishment, via glycolytic formation of phosphoglyceric acid from triose phosphate. Thus, replacing KOH with $K_2$-malate [177] (Figure 2A) adjusts the elevated cytosolic pH.

Excess uptake of anionic equivalents may occur when plants are presented with $Ca(NO_3)_2$. The transient acidification of the cytoplasm by cotransported protons is regulated by activation of malic enzyme and malic acid decomposition [176] (Figure 2B).

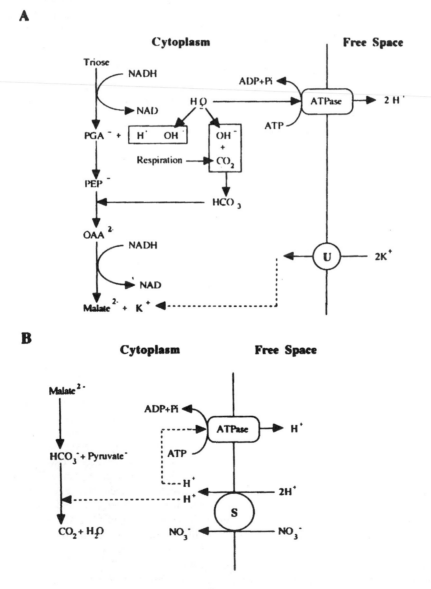

**Figure 2**   Charge compensation and pH regulation during cation (A) and anion (B) uptake by plant tissues when the uptake is not balanced by a counterion. U, uniport; S, symport; NAD, nicotinamide adenine dinucleotide; PGA, phosphoglyceric acid; PEP, phosphoenolpyruvate; OAA, oxaloacetic acid. Details in text.

## VII.  TRANSPORT TO THE SHOOT

Since the classical experiments of Stout and Hoagland [178], it has been accepted that mineral ions are transported from roots to shoots in the xylem. This pathway also applies to organic nitrogen compounds that are products of ammonium fixation and of symbiotic $N_2$ fixation [179]. The mechanism of this transport is mass flow in the aqueous xylem solution. The control of mineral ion transport to the shoots occurs during their passage across the roots to the xylem.

### A.  Transport Across the Root

Water moves across the root in the symplast, in the apoplast, or in both. The root symplast [180] comprises the plasmodesmata-connected cytoplasmic continuum of the root cells; it does not include the vacuoles. Symplastic mass flow across the root traverses two membranes: one upon entering the symplast and another upon exiting to the xylem. Biological membranes do not constitute a serious barrier to the diffusion of water, but they impede the free passage of the mineral ions dissolved in the water. Thus, symplastic movement of solutes across roots can be regulated by metabolically controlled transport across these membranes [181].

Apoplastic water flow across the root occurs in the root cell-wall continuum. The apoplast [180] is the free space continuum throughout the plant. This continuum is interrupted by the Casparian bands, which are hydrophobic incrustations in the radial and transverse walls of the endodermis [182]. Water bypasses the Casparian bands by movement across the membrane into the endodermal symplast and out again. Mineral ions that are dissolved in the water also bypass the Casparian bands and migrate across the membranes of endodermal cells. However, membrane transport of the mineral ions is involved. Whatever course mineral ions choose, they traverse two membranes, where control may occur. Solutes moving to the xylem must be absorbed into the symplast, from the medium at the root surface, or from the apoplast, by either cortical cells or the cortical face of the endodermis. The solutes move out again into the stelar apoplast or the xylem, at the stelar face of the endodermis or from the stelar parenchyma, respectively. These membrane transport processes are qualitatively and quantitatively controlled.

Vacuoles of root cells are another site for selective regulation of solute transport across the root. Vacuoles accumulate various solutes, thus removing them from the symplastic stream. In particular, most of the cellular $Ca^{2+}$ is sequestered in the vacuoles, while the cytosolic $Ca^{2+}$ concentration is maintained very low [104]. Under saline conditions, much of the $Cl^-$ is also sequestered in the vacuoles, where it serves for turgor regulation. The same applies to $Na^+$ in plants that have conserved the $Na^+/H^+$ antiporter that is needed for $Na^+$ transport to the vacuole [122]. Other solutes are stored in the vacuoles when available in excess of requirement. They may then be released again upon demand [183].

### B.  Effects of Transpiration

Haberlandt [184] concluded more than a century ago that transpiration is not of major importance for plant mineral nutrition. This has since been reconfirmed repeatedly (see, e.g., references in Refs. 185 and 90). Indeed, the amount of solutes transported to the tops of plants should not be affected by the rate of water flow if membrane transport-dependent delivery of solutes to the xylem is the rate-limiting process. A low solute concentration would be expected in the xylem sap at high transpiration rates and a high concentration when transpiration is low. This seems essentially to be the situation under conditions of low external salt concentration and low salt status of the roots [185] and for solutes that are recognized by the transport proteins (not xenobiotics).

Broyer and Hoagland [185] stated in 1943 that delivery of ions to the xylem is metabolically controlled, whereas upward movement in the xylem is passive mass flow. They found that in plants of high salt status the delivery of the salt to the xylem is rapid and the velocity of mass flow in the xylem may be the rate-limiting process. Consequently, variation in the rate of transpiration may affect salt transport to the shoots under saline conditions. A good correlation between transpiration and transport to the shoot was also found for some nonessential elements such as cadmium [186] and silicon [187] and for xenobiotic organic compounds [188]. The transport proteins of the root may not recognize these solutes. They are apparently transported in a fraction of the apoplastic mass flow that bypasses the Casparian strips [189,190]. It was suggested that such bypass occurs when solutes enter at sites of secondary-root emergence [189,190] or through the apical region of the root [191]. Calcium ions that reach the xylem sap may also utilize these latter pathways. Most of the $Ca^{2+}$ that enters the symplast via $Ca^{2+}$ channels is sequestered in the vacuoles and thus prevented from moving to the xylem. The relative amount of water bypassing the Casparian strips varies between a few percent [192] and more than 10% [186]. It seems to increase under conditions of stress damage [192]. For metabolically transported solutes at low external concentration, the amount transported in the bypass flow is an insignificant fraction of the total transport to the xylem. It may, however, constitute the whole amount of a xenobiotic solute that is transported to the tops [186].

## C.  Resorption and Exchange Binding

Solute transport in xylem vessels is driven strictly by mass flow. Nevertheless, the composition of xylem sap varies along its path. Two processes, absorption from the xylem sap by adjacent living cells and binding by xylem walls, are responsible for these variations: The walls of xylem vessels, like other cell walls, contain immobile negative charges. These charges constitute a Donnan phase that retains cations moving in the vessels. In particular, polyvalent cations such as free $Ca^{2+}$ [193], $Zn^{2+}$ [194], and $Fe^{2+}$ and $Fe^{3+}$ [195] are retained. Polyvalent cation retention can be diminished, or prevented, by chelating agents [195,196] and by displacement with similar or other cations [196].

Stout and Hoagland [178] demonstrated absorption of solutes from the xylem by surrounding tissues. Such absorption of $Na^+$ is rather pronounced and selective in some plants [197] and has been shown to depend on energy metabolism [198,199]. Xylem-parenchyma transfer cells are apparently involved in $Na^+$ absorption from the xylem sap and its transfer to the phloem [200,201].

## VIII.  SUMMARY

Mineral nutrients in the root cortex-apoplast equilibrate with those in the root medium. These mineral ions are then transported into the root symplast, moving across the plasma membranes of epidermis, cortex, or endodermis cells. Depending on the specific ion, this transport is facilitated by passive uniport through channels, by carrier-mediated cotransport with protons, or by primary active transport.

Once in the root symplast, the ions may be transported into the vacuole, across the tonoplast, or excreted to the xylem, across the plasma membranes of xylem parenchyma cells. Some mineral ions, such as $Na^+$ and $Ca^{2+}$, may also be reexcreted from the symplast to the apoplast. In the xylem, mineral ions and other solutes move by mass flow, primarily to the leaf apoplast. There, the ions are again absorbed into the symplast and vacuoles of leaf cells. Secondary transport from the leaves to sinks occurs in the phloem, together with the products of photosynthesis. Metabolic transport across the roots, to the xylem, regulates the amount of mineral ions conveyed to the tops. Normally, this amount is very little affected by the velocity of xylem sap flow.

All the membrane transport processes mentioned depend on energy metabolism. A part of the $Ca^{2+}$ transport is driven by a primary active $Ca^{2+}$-ATPase. Most of the mineral ion-transport processes are driven by the proton motive force, which is composed of an electrical potential gradient and a pH gradient across the membrane. At the plasma membrane, a P-type ATPase generates the proton motive force. At the tonoplast, two pumps that function in parallel, a V-type ATPase and a pyrophosphatase, generate it. The latter may also pump potassium into the vacuole. Membrane-bound channels or carriers utilize the proton motive force and facilitate the membrane transport of mineral ions. Genes encoding some of the transport proteins have now been sequenced.

## ACKNOWLEDGMENT

During the composition of this chapter, BJ's laboratory was supported by the Endowment Fund for Basic Sciences: Charles H. Revson Foundation, and NM's laboratory was supported by grant 108/97 from the Israel Science Fundation, both administered by the Israel Academy of Sciences and Humanities.

## REFERENCES

1.  DG Nicholls. Bioenergetics. London: Academic Press, 1982, p 189.
2.  PS Nobel. Physicochemical and Environmental Plant Physiology. San Diego: Academic Press, 1991.

3. HH Ussing. The distinction by means of tracers between active transport and diffusion. Acta Physiol Scand 19:43–56, 1949.
4. U Lüttge. Stofftransport der Pflanzen. Berlin: Springer Verlag, 1973, p 280.
5. GE Briggs, RN Robertson. Apparent free space. Annu Rev Plant Physiol 8:11–29, 1957.
6. FG Donnan. Theorie der Membrangleichgewichte und Membranpotetiale bei vorhandensein von nicht dialysieren Elektrolyten. Ein Beitrag zur physikalisch-chemischen Physiology. Z Electrochem 17:572, 1911.
7. P Mitchell. Chemiosmotic coupling in oxidative and photosynthetic phosphorylation. Biol Rev 41:445–502, 1966.
8. DE Goldman. Potential impedence and rectification in membranes. J Gen Physiol 27:37–60, 1943.
9. AL Hodgkin, B Katz. The effect of sodium ions on the electrical activity of the giant axon of the squid. J Physiol (Lond) 108:37–77, 1949.
10. R Serrano. The Plasma Membrane ATPase of Plants and Fungi. Boca Raton, FL: CRC Press, 1985, p 174.
11. T Theorell. Membrane electrophoresis in relation to bio-electrical polarization effects. Arch Sci Physiol 3:205–219, 1949.
12. FM Harold. Ion currents and physiological functions in microorganisms. Annu Rev Microbiol 31:181–203, 1977.
13. B Hille. Ionic Channels of Excitable Membranes. Sunderland, MA: Sinauer Associates, 1992, p 607.
14. DA Doyle, JM Cabral, RA Pfuetzner, A Kuo, JM Gulbis, SL Cohen, BT Chait, R Mackinnon. The structure of the potassium channel: molecular basis of $K^+$ conduction and selectivity. Science 280:69–77, 1998.
15. N Moran, G Ehrenstein, K Iwasa, C Bare, C Mischke. Ion channels in plasmalemma of wheat protoplasts. Science 226:835–838, 1984.
16. RJ Hedrich, JI Schroeder. The physiology of ion channels and electrogenic pumps in higher plant cells. Annu Rev Plant Physiol 40:539–569, 1989.
17. J Alexandre, JP Lassalles, RT Kado. Opening of $Ca^{2+}$ channels in isolated red beet vacuoles by inisitol 1,4,5-trisphosphate. Nature 342:567–570, 1990.
18. DJ Cosgrove, R Hedrich. Stretch-activated chloride, potassium, and calcium channels coexisting in plasma membranes of guard cells of *Vicia faba* L. Planta 186:143–153, 1991.
19. LC Falke, KL Edwards, BG Pickard, S Misler. A stretch-activated anion channel in tobacco protoplasts. FEBS Lett 237:141–144, 1988.
20. M Tester. Plant ion channels: whole-cell and single-channbel studies. New Phytol 114:305–340, 1990.
21. SD Tyerman. Anion channels in plants. Annu Rev Plant Physiol Plant Mol Biol 43:351–375, 1992.
22. AL Hodgkin, AF Huxley. A quantitative description of membrane current and its application to conduction and excitation in nerve. J Physiol (Lond) 117:500–544, 1952.
23. OP Hamill, A Marty, E Neher, B Sakman, FJ Sigworth. Improved patch-clamp techniques for high-resolution current recording from cells and cell-free membrane patches. Pfluegers Arch 391:85–100, 1981.
24. RL Satter, N Moran. Ionic channels in plant cell membranes. Physiol Plant 72:816–820, 1988.
25. JM Ward. Patch-clamping and other molecular approaches for the study of plasma membrane transporters demystified. Plant Physiol 114:1151–1159, 1997.
26. MR Blatt. $K^+$ channels of stomatal guard cells. Characteristics of the inward rectifier and its control by pH. J Gen Physiol 99:615–644, 1992.
27. N Ilan, A Schwartz, N Moran. External protons enhance the activity of the hyperpolarization-activated K channels in guard cell protoplast of *Vicia faba*. J Membr Biol 154:169–181, 1996.
28. I Marten, S Hoth, R Deeken, P Ache, KA Ketchum, T Hoshi, R Hedrich. AKT3, a phloem-localized $K^+$ channel, is blocked by protons. Proc Natl Acad Sci U S A 96:7581–7586, 1999.
29. N Ilan, A Schwartz, N Moran. External pH effects on the depolarization-activated K channels in guard cell protoplasts of *Vicia faba*. Plant Physiol 89:1184–1192, 1994.
30. JI Schroeder. $K^+$ transport properties of $K^+$ channels in the plasma membrane of *Vicia faba* guard cells. J Gen Physiol 92:667–683, 1988.
31. KA Fairley-Grenot, SM Assmann. Permeation of $Ca^{2+}$ through $K^+$ channels in the plasma membrane of *Vicia faba* guard cells. J Membr Biol 128:103–113, 1992.
32. FJM Maathuis, D Sanders. Plant membrane transport. Curr Opin Cell Biol 4:661–669, 1992.
33. N Moran, D Fox, RL Satter. Interaction of the depolarization-activated K channel of *Samanea saman* with inorganic ions: A patch-clamp study. Plant Physiol 94:424–431, 1990.
34. N Moran. Membrane-delimited phosphorylation enables the activation of the outward-rectifying K channels in a plant cell. Plant Physiol 111:1281–1292, 1996.
35. LA Romano, H Miedema, SM Assmann. $Ca^{2+}$-permeable, outwardly-rectifying $K^+$ channels in mesophyll cells of *Arabidopsis thaliana*. Plant Cell Physiol 39:1133–1144, 1998.
36. H Stoeckel, K Takeda. Calcium-sensitivity of the plasmalemmal delayed rectifier potassium current suggests that calcium influx in pulvinar protoplasts from *Mimosa pudica* L. can be revealed by hyperpolarization. J Membr Biol 146:201–209, 1995.
37. K Czempinski, N Gaedeke, S Zimmermann, B Muller Rober. Molecular mechanisms and regulation of plant ion channels. J Exp Bot 50(Special Issue SI):955–966, 1999.

38. JA Anderson, SS Huprikar, LV Kochian, WJ Lucas, RF Gaber. Functional expression of a probable *Arabidopsis thaliana* potassium channel in *Saccharomyces cerevisiae*. Proc Natl Acad Sci U S A 89:3736–3740, 1992.

39. RL Nakamura, WL-J Mckendree, RE Hirsch, JC Sedbrook, RF Gaber, MR Sussman. Expression of an *Arabidopsis* potassium channel gene in guard cells. Plant Physiol 109:371–374, 1995.

40. Y Cao, JM Ward, WB Kelly, AM Ichida, RF Gaber, JA Anderson, N Uozumi, JI Schroeder, NM Crawford. Multiple genes, tissue specificity, and expression-dependent modulation contribute to the functional diversity of potassium channels in *Arabidopsis thaliana*. Plant Physiol 109:1093–1106, 1995.

41. KA Ketchum, CW Slayman. Isolation of an ion channel gene from *Arabidopsis thaliana* using the H5 signature sequence from voltage-dependent $K^+$ channels. FEBS Lett 378:19–26, 1996.

42. K Philippar, I Fuchs, S Hoth, H Luethen, M Boettger, D Becker, R Hedrich. Auxin-induced upregulation of inwardly-rectifying $K^+$ channels represents a key-step in maize coleoptile growth and gravitropic bending. Nature 96:12186–12191, 1999.

43. F Gaymard, B Lacombe, D Bouchez, D Bruneau, J Boucherez, N Michaux-Ferriere, JB Thibaud, H Sentenac. Identification and disruption of a plant Shaker-like outward channel involved in $K^+$ release into the xylem sap. Cell 94:647–655, 1998.

44. P Thuleau, JM Ward, R Ranjeva, JI Schroeder. Voltage-dependent calcium-permeable channels in the plasma membrane of a higher plant cell. EMBO J 13:2970–2975, 1994.

45. M Pineros, M Tester. Calcium channels in higher plant cells: selectivity, regulation and pharmacology. J Exp Bot 48:551–577, 1997.

46. PJ White. Calcium channels in the plasma membrane of root cells. Ann Bot 81:173–183, 1998.

47. JI Schroeder, S Hagiwara. Repetitive increases in cytosolic $Ca^{2+}$ of guard cells by abscisic acid activation of nonselective $Ca^{2+}$ permeable channels. Proc Natl Acad Sci U S A 87:9305–9309, 1990.

48. H Stoeckel, K Takeda. Calcium-activated, voltage-dependent, non-selective cation currents in endosperm plasma-membrane from higher-plants. Proc R Soc 237:213–231, 1989.

49. SD Tyerman, IM Skerrett. Root ion channels and salinity. Sci Hortic 78:175–235, 1999.

50. T Arazi, R Sunkar, B Kaplan, H Fromm. A tobacco plasma membrane calmodulin-binding transporter confers $Ni^{2+}$ tolerance and $Pb^{2+}$ hypersensitivity in transgenic plants. Plant J 20:171–182, 1999.

51. JI Schroeder, BU Keller. Two types of anion channel currents in guard cells with distinct voltage regulation. Proc Natl Acad Sci U S A 89:5025–5029, 1992.

52. P Dietrich, R Hedrich. Interconversion of fast and slow gating modes of GCAC1, a guard cell anion channel. Planta 195:301–304, 1994.

53. C Schmidt, I Schroeder Julian. Anion selectivity of slow anion channels in the plasma membrane of guard cells: large nitrate permeability. Plant Physiol 106:383–391, 1994.

54. M Skerrett, SD Tyerman. A channel that allows inwardly directed fluxes of anions in protoplasts derived from wheat roots. Planta 192:295–305, 1994.

55. R Hedrich. Voltage-dependent chloride channels in plant cells: identification, characterization, and regulation of a guard cell anion channel. In: WB Guggino, ed. Current Topics in Membranes. San Diego: Academic Press, 1994, pp 1–33.

56. P Dietrich, R Hedrich. Anions permeate and gate GCAC1, a voltage-dependent guard cell anion channel. Plant J 15:479–487, 1998.

57. I Marten, G Lohse, R Hedrich. Plant growth hormones control voltage-dependent activity of anion channels in plasma membrane of guard cells. Nature 353:758–762, 1991.

58. ZM Pei, K Kuchitsu, JM Ward, M Schwarz, JI Schroeder. Differential abscisic acid regulation of guard cell slow anion channels in *Arabidopsis* wild-type and *abi1* and *abi2* mutants. Plant Cell 9:409–423, 1997.

59. JM Ward, ZM Pei, JI Schroeder. Roles of ion channels in initiation of signal transduction in higher plants. Plant Cell 7:833–844, 1995.

60. MH Cho, EP Spalding. An anion channel in *Arabidopsis* hypocotyls activated by blue light. Proc Natl Acad Sci U S A 93:8134–8138, 1996.

61. H Barbier Brygoo, JM Frachisse, J Colcombet, S Thomine. Anion channels and hormone signalling in plant cells. Plant Physiol Biochem 37:381–392, 1999.

62. A Gelli, VJ Higgins, E Blumwald. Activation of plant plasma membrane $Ca^{2+}$-permeable channels by race-specific fungal elicitors. Plant Physiol 113:269–279, 1997.

63. M Hechenberger, B Schwappach, WN Fischer, WB Frommer, TJ Jentsch, K Steinmeyer. A family of putative chloride channels from *Arabidopsis* and functional complementation of a yeast strain with a *CLC* gene disruption. J Biol Chem 271:33632–33638, 1996.

64. C Lurin, D Geelen, H barbierbrygoo, J Guern, C Maurel. Cloning and functional expression of a plant voltage-dependent chloride channel. Plant Cell 8:701–711, 1996.

65. R Hedrich, A Kurkdjian. Characterization of an anion-permeable channel from sugar beet vacuoles: effect of inhibitors. EMBO J 7:3661–3666, 1988.

66. JM Ward, JI Schroeder. Calcium-activated $K^+$ channels and calcium-induced calcium release by slow vacuolar ion channels in guard cell vacuoles implicated in the control of stomatal closure. Plant Cell 6:669–683, 1994.

67. LI Bruggemann, II Pottosin, G Schonknecht. Selectivity of the fast activating vacuolar cation channel. J Exp Bot 50:873–876, 1999.
68. EAC Macrobbie. Signal transduction and ion channels in guard cells. Philos Trans R Soc Lond 353:1475–1488, 1998.
69. BJ Barkla, O Pantoja. Physiology of ion transport across the tonoplast of higher plants. Annu Rev Plant Physiol Plant Mol Biol 47:159–184, 1996.
70. GJ Allen, D Sanders. Control of ionic currents in guard cell vacuoles by cytosolic and luminal calcium. Plant J 10:1055–1069, 1996.
71. GJ Allen, SR Muir, D Sanders. Release of $Ca^{2+}$ from individual plant vacuoles by both insp3 and cyclic ADP-ribose. Science 268:735–737, 1995.
72. MR Mcainsh, C Brownlee, AM Hetherington. Abscisic acid–induced elevation of guard cell cytosolic $Ca^{2+}$ precedes stomatal closure. Nature 343:186–188, 1990.
73. Y Wu, J Kuzma, E Marechal, R Graeff, HC Lee, R Foster, NH Chua. Abscisic acid signaling through cyclic ADP-ribose in plants. Science 278:2126–2130, 1997.
74. CP Leckie, MR Mcainsh, GJ Allen, D Sanders, AM Hetherington. Abscisic acid–induced stomatal closure mediated by cyclic ADP-ribose. Proc Natl Acad Sci U S A 95:15837–15842, 1998.
75. GG Cote, YG Yueh, RC Crain. Phosphoinositide turnover and its role in plant signal transduction. In: BB Biswas and B Biswas, eds. Myoinositol-Phosphates, Phosphoinositides and Signal Transduction. London: Plenum, 1996, pp 317–343.
76. N Moran, YG Yueh, RC Crain. Signal transduction and cell volume regulation in plant leaflet movements. News Physiol Sci 11:108–114, 1996.
77. CP Leckie, MR Mcainsh, L Montgomery, AJ Priestley, I Staxen, AAR Webb, AM Hetherington. Second messengers in guard cells. J Exp Bot 49:339–349, 1998.
78. P Thuleau, JI Schroeder, R Ranjeva. Recent advances in the regulation of plant calcium channels: evidence for regulation by G-proteins, the cytoskeleton and second messengers. Curr Opin Plant Biol 1:424–427, 1998.
79. MJ Berridge. Inositol triphosphate and calcium signalling. Nature 315:315–325, 1993.
80. ZM Pei, JM Ward, JF Harper, JI Schroeder. A novel chloride channel in *Vicia faba* guard cell vacuoles activated by the serine/threonine kinase, CDPK. EMBO J 15:6564–6574, 1996.
81. CM Cheffings, O Pantoja, FM Ashcroft, JAC Smith. Malate transport and vacuolar ion channels in CAM plants. J Exp Bot 48:623–631, 1997.
82. E Epstein, CE Hagen. A kinetic study of the absorption of alkali cations by barley roots. Plant Phsiol 27:457–474, 1952.
83. L Michaelis, ML Menten. Die Kinetik der Invertase Wirkung. Biochem Z 49:333–369, 1913.
84. DA Baker, JL Hall. Introduction and general principles. In: DA Baker, JL Hall, eds. Solute Transport in Plant Cells and Tissues. Harlow: Longman Scientific & Technical, 1988, pp 1–27.
85. P Nissen. Uptake mechanisms inorganic and organic. Annu Rev Plant Physiol 25:53–79, 1974.
86. ADM Glass, J Dunlop. The influence of potassium content on the kinetics of potassium influx into excised ryegrass [*Lolium multiflorum*] and barley roots. Planta 141:117–119, 1978.
87. GGJ Bange. Multiphasic kinetics in solute absorption: an intrinsic property of the transport system. Z Pflanzenphysiol 91:75–78, 1979.
88. FN Dalton. Dual pattern of potassium transport in plant cells: a physical artifact of a single uptake mechanism. J Exp Bot 35:1723–1732, 1984.
89. AC Borslapp. The use of model-fitting in the interpretation of 'dual' uptake isotherms. Plant Cell Environ 6:407–416, 1983.
90. W Tanner, H Beevers. Does transpiration have an essential function in long-distance ion transport in plants? Plant Cell Environ 13:745–750, 1990.
91. N Nelson. Organellar proton ATPases. Curr Opin Cell Biol 4:654–660, 1992.
92. N Nelson, L Taiz. The evolution of $H^+$-ATPases. Trends Biochem 14:113–116, 1989.
93. DP Briskin, JB Hanson. How does the plant plasma membrane $H^+$-ATPase pump protons? J Exp Bot 43:269–289, 1992.
94. DP Briskin. The plasma membrane ATPase of higher plant cells: biochemistry and transport function. Biochim Biophys Acta 1019:95–109, 1990.
95. R Serrano. Structure and function of proton translocating ATPases in plama membranes of plants and fungi. Biochim Biophys Acta 947:1–28, 1988.
96. EJ Bowman. Bafilomycins: a class of inhibitors of membrane ATPases from microorganisms and animal cells. Proc Natl Acad Sci U S A 85:7972–7976, 1988.
97. U Lüttge, R Ratajczak. The physiology, biochemistry and molecular biology of the plant vacuole ATPase. Adv Bot Res 25:253–296, 1997.
98. JL Giannini, J Ruiz-Christin, DP Briskin. Calcium transport in sealed vesicles from red beet (*Beta vulgaris* L.) storage tissue. Characterization of a $Ca^{2+}$ uptake into plasma membrane vesicles. Plant Physiol 85:1137–1142, 1987.
99. S Malmström, P Askerlund, MG Palmgren. A calmodulin-stimulated $Ca^{2+}$-ATPase from plant vacuolar membranes with a putative regulatory domain at its N-terminus. FEBS Lett 400:324–328, 1997.

100. TJ Buckhout. ATP-dependent $Ca^{2+}$ transport in endoplasmic reticulum isolated from roots of *Lepidium sativum* L. [gardencress]. Planta 159:84–90, 1983.

101. JL Giannini, LH Gildensoph, I Reynolds Niesman, DP Briskin. Calcium transport in sealed vesicles from red beet (*Beta vulgaris* L.) storage tissue. Characterization of a $Ca^{2+}$-pumping ATPase associated with the endoplasmic reticulum. Plant Physiol 85:1129–1136, 1987.

102. LE Williams, SB Schueler, DP Briskin. Further characterization of the red beet plasma membrane $Ca^{2+}$-ATPase using GTP as an alternative substrate. Plant Phsiol 92:747–754, 1990.

103. LA Wimmers, NN Ewing, AB Bennett. Higher plant calcium ATPase: primary structure and regulation of mRNA abundance by salt. Proc Natl Acad Sci U S A 89:9205–9206, 1992.

104. S Gilroy, A Trewavas. Signal sensing and signal transduction across the plasma membrane. In: C Larsson, IM Møller, eds. The Plasma Membrane: Structure, Function and Molecular Biology. Berlin: Springer Verlag, 1990, pp 203–232.

105. BW Poovaiah, ASN Reddy. Calcium and signal transduction in plants. Crit Rev Plant Sci 12:185–211, 1993.

106. JV Møller, B Juul, M Le Maire. Structural organization, ion transport, and energy transduction of P-type ATPase. Biochim Biophysica Acta 1:1–51, 1996.

107. DM Roberts, TJ Lucas, DM Watterson. Structure, function and mechanism of action of calmodulin. CRC Crit Rev Plant Sci 4:311–393, 1986.

108. M Olbe, M Sommarin. The spinach plasma membrane $Ca^{2+}$ pump is a 120-kDa polypeptide regulated by calmodulin-binding to a terminal region. Physiol Plant 103:35–44, 1998.

109. TC Fox, ML Guerinot. Molecular biology of cation transport in plants. Annu Rev Plant Physiol Plant Mol Biol 49:669–696, 1988.

110. PA Rea, Z-S Li, Y-P Lu, YM Drozdowicz, E Martinoia. From vacuolar GS-X pumps to multispecific ABC transporters. Annu Rev Plant Physiol Plant Mol Biol 49:727–760, 1998.

111. D Salt, E Rauser. MgATP-dependent transport of phytochelatins across the tonoplast of oat roots. Plant Physiol 107:1293–1301, 1995.

112. E Rauser, P Meuwly. Retention of cadmium in roots of maize seedlings. Role of complexation by phytochelatins and related thiol peptides. Plant Physiol 109:195–202, 1995.

113. PA Rea, RJ Poole. Vacuolar $H^{+}$-translocating pyrophosphatase. Annu Rev Plant Physiol Plant Mol Biol 44:157–180, 1993.

114. E Johannes, HH Felle. The role of $Mg^{2+}$ in proton transport by the tonoplst pyrophosphatase in *Riccia fluitans* vacuoles. Physiol Plant 77:326–331, 1989.

115. P Malsowski, H Maslowska. Purification and some properties of proton-translocating pyrophosphatase from microsomal vesicles of corn seedlings. Biochem Physiol Pflanz 182:73–84, 1987.

116. PA Rea, CJ Griffith, D Sanders. Differential susceptibilities of tonoplast ATPase and PPiase to irreversible inhibition by chaotropic anions. In: B Marin, ed. Plant Vacuole. New York: Plenum Press, 1987.

117. JM Davies, PA Rea, D Sanders. Vacuolar proton-pumping pyrophosphatase in *Beta vulgaris* shows vectorial activation by potassium. FEBS Lett 278:66–68, 1991.

118. JM Davies. Vacuolar energization: pumps, shunts and stress. J Exp Bot 48:633–641, 1997.

119. W Tanner, T Caspari. Membrane transport carriers. Annu Rev Plant Physiol Plant Mol Biol 47:595–626, 1996.

120. P Jauch, P Läuger. Electronic properties of the sodium-alanine cotransporter in pancreatic acinar cells. II. Comparison with transport models. J Membr Biol 94:117–127, 1986.

121. A Ratner, B Jacoby. Effect of $K^{+}$, its counter anion, and pH on sodium efflux from barley root tips. J Exp Bot 24:231–238, 1976.

122. H Mennen, B Jacoby, H Marschner. Is sodium proton antiport ubiquitous in plant cells? J Plant Physiol 137:180–183, 1990.

123. A Katz, HR Kaback, M Avron. $Na^{+}H^{+}$ antiport in isolated plasma membrane vesicles from the halotolerant alga *Dunaliella salina*. FEBS Lett 202:141–144, 1986.

124. A Katz, U Pick, M Avron. Characterization and reconstitution of the $Na^{+}H^{+}$ antiporter from the plasma membrane of the halotolerant alga *Dunaliella*. Biochim Biophys Acta 983:9–14, 1989.

125. Y Braun, M Hassidim, HR Lerner, L Reinhold. Evidence for a $Na^{+}H^{+}$ antiporter in membrane vesicles isolated from roots of the halophyte *Atriplex nummularia*. Plant Physiol 87:104–108, 1988.

126. M Hassidim, HR Lerner, L Reinhold. $Na^{+}H^{+}$ and $K^{+}H^{+}$ antiport in root membrane vesicles isolated from the halophyte *Atriplex* and the glycophyte cotton. Plant Physiol 94:1975–1801, 1990.

127. A Katz, U Pick, M Avron. Modulation of $Na^{+}H^{+}$ antiporter activity by extreme pH and salt in the halotolerant alga *Dunaliella salina*. Plant Physiol 100:1224–1229, 1992.

128. E Blumwald, RJ Poole. $Na^{+}H^{+}$ antiport in isolated tonoplast vesicles from storage tissue of *Beta vulgaris*. Plant Physiol 78:163–167, 1985.

129. E Blumwald, RJ Poole. Salt tolerance in suspension cultures of sugar beet. Induction of $Na^{+}H^{+}$ antiport activity at the tonoplast by growth in salt. Plant Physiol 83:884–887, 1987.

130. M Staal, FJM Maathuis, TM Elzenga, JHM Overbeek, HBA Prins. $Na^{+}H^{+}$ antiport activity in tonoplast vesicles of the salt-tolerant *Plantago maritima* and the salt-sensitive *Plantago media*. Physol Plant 82:179–184, 1991

131. BJ Barkla, E Blumwald. Identification of a 170-kDa protein associated with the vacuolar $Na^+H^+$ antiport of *Beta vulgaris*. Proc Natl Acad Sci U S A 88:11177–11181, 1991.

132. J Garbarino, FM DuPont. Rapid induction of $Na^+H^+$ exchange activity in barley root tonoplasts. Plant Physiol 89:1–4, 1989.

133. A Fukuda, Y Yazaki, T Ishikawa, S Koike, Y Tanaka. $Na^+H^+$ antiporter in tonoplast vesicles from rice roots. Plant Cell Physiol 39:196–201, 1998.

134. FJM Maathuis, D Sanders. Energization of potassium uptake in *Arabidopsis thaliana*. Planta 191:302–307, 1993.

135. L Bruggemann, P Dietrich, D Becker, I Dreyer, K Palme, R Hedrich. Channel-mediated high-affinity $K^+$ uptake into guard cells from *Arabidopsis*. Proc Natl Acad Sci U S A 96:3298–3302, 1999.

136. MJ Gracia-Sanches, MP Jaime, A Ramus, D Sanders. $Na^+$-dependent $NO_3^-$ transport at the plasma membrane of leaf cells of the marine higher plant *Zoestera marina* L. Plant Physiol, in press.

137. NA Walker, D Sanders. Sodium-coupled solute transport in *Charophyte* algae: a general mechanism for transport energization in plant cells? Planta 185:443–445, 1991.

138. NA Walker. Sodium coupled symports in the plasma membrane of plant cells. In: MR Blatt, RA Leigh, D Sanders, eds. Membrane Transport in Plants and Fungi. Soc. Exp. Biol. Symp. XLVII. Cambridge: Company of Biologists, 1994, pp 179–192.

139. FJM Maathuis, D Verlin, FA Smith, D Sanders, JA Fernandez, NA Walker. The physiological relevance of $Na^+$-coupled $K^+$-transport. Plant Physiol 112:1609–1616, 1996.

140. DP Schachtman, JI Schroeder. Structure and transport mechanism of a high-affinity potassium uptake transporter from higher plants. Nature 370:655–658, 1994.

141. EJ Kim, JM Kwak, N Uozumi, JI Schroeder. *Atkup1:* an *Arabidopsis* gene encoding high-affinity potassium transport activity. Plant Cell 10:51–62, 1998.

142. TB Wang, W Gassman, F Rubio, JI Schroeder, ADM Glass. Rapid up-regulation of *HKT1*, a high affinity potassium transporter gene in roots of barley and wheat following withdrawal of potassium. Plant Physiol 118:651–659, 1998.

143. F Rubio, W Gassmann, JI Schroeder. Sodium-driven potassium uptake by the plant potassium transporter HKT1 and mutations conferring salt tolerance. Science 270:1660–1663, 1995.

144. M Kasai, S Muto. Calcium ion pump and calcium ion/hydrogen ion antiporter in plasma membrane vesicles isolated by aqueous two-phase partitioning from corn leaves. J Membr Biol 114:133–142, 1990.

145. KD Hirschi, RG Zhen, KW Cunningham, PA Rea, GR Fink. CAX1, an $H^+/Ca^{2+}$ antiporter from *Arabidopsis*. Proc Natl Acad Sci U S A 93:8782–8786, 1996.

146. KS Schumaker, H Sze. A $Ca^{2+}/H^+$ antiport system driven by the proton electrochemical gradient of a tonoplast $H^+$-ATPase from oat roots. Plant Physiol 79:1111–1117, 1985.

147. KS Schumaker, H Sze. Calcium transport into the vacuole of oat roots: characterization of $H^+/Ca^{2+}$ exchange activity. J Biol Chem 261:12172–12178, 1986.

148. E Blumwald, RJ Poole. Kinetics of $Ca^{2+}/H^+$ antiport in isolated tonoplast vesicles from storage tissue of *Beta vulgaris* L. Plant Physiol 80:727–731, 1986.

149. FM Dupont, DS Bush, JJ Windle, RL Jones. Calcium and proton transport in membrane vesicles from barley roots. Plant Physiol 94:179–188, 1990.

150. S Blackford, PA Rea, D Sanders. Voltage sensitivity of $H^+/Ca^{2+}$ antiport in higher plant tonoplasts suggests a role in vacuolar calcium accumulation. J Biol Chem 265:9617–9620, 1990.

151. D Sanders. Gradient-coupled chloride transport in plant cells. In: GA Gerencser, ed. Chloride Transport Coupling in Biological Membranes. Amsterdam: Elsevier, 1984, pp 64–120.

152. WR Ullrich, A Novacky. Nitrate-dependent membrane potentials and their induction in *Lemna gibba* GI. Plant Sci Lett 22:211–217, 1981.

153. ADM Glass, JE Shaff, LV Kochian. Studies of the uptake of nitrate in barley. IV. Electrophysiology. Plant Physiol 99:99–463, 1992.

154. ECI Ullrich, A Novacky, AJE van Bel. Phosphate uptake in *Lemna gibba* G1: energetics and kinetics. Planta 161:46–52, 1984.

155. K Sakano. Proton/phosphate stoichiometry in uptake of inorganic phosphate by cultured cells of *Catharanthus roseus* (L.) G. Don. Plant Physiol 93:479–482, 1990.

156. B Lass, CI Ullrich Eberius. Evidence for proton/sulfate cotransport and its kinetics in *Lemna gibba* G1. Planta 161:53–60, 1984.

157. B Jacoby, B Rudich. Proton-chloride symport in barley roots. Ann Bot 46:493–498, 1980.

158. K Yamashita, Y Yamamoto, H Matsumoto. Characterization of an anion transporter in the plasma membrane of barley roots. Plant Cell Physiol 37:949–956, 1996.

159. US Muchal, JM Pardo, KG Raghothama. Phosphate transporters from the higher plant *Arabidopsis thaliana*. Proc Natl Acad Sci U S A 93:10519–10523, 1996.

160. C Liu, US Muchal, M Uthappa, AK Kononowicz, KG Raghotama. Tomato phosphate transporter genes are differentially regulated in plant tissues by phosphorus. Plant Physiol 116:91–99, 1998.

161. P Daram, S Brunner, BL Persson, N Amrhein, M Bucher. Functional analysis and cell-specific expression of a phosphate transporter from tomato. Planta 206:255–233, 1998.

162. J Ruiz-Cristin, DP Briskin. Characterization of a $H^+/NO_3^-$ symport associated with plasma membrane vesicles of maize roots using 36clo3- as a radiotracer analog. Arch Biophys Biochem 285:74–82, 1991.

163. R Wang, D Liu, M Crawford N. The *Arabidopsis* CHL1 protein plays a major role in high-affinity nitrate uptake. Proc Natl Acad Sci U S A 95:15134–15139, 1998.

164. G Klobus, J Jerzykiewcz, J Buczek, AK Kononovicz, KG Raghothamaq. Characterization of the the nitrate transporter in root plasma membrance of *Cucumis sativus* L. Physiol Plant 20:323–328, 1998.

165. G Amarasinghe, Y De Bruxielle, M Braddon, I Onyeocha, BG Forde, MK Uvardi. Regulation of *GMNRT2* expression and nitrate transport activity in roots of soybean (*Glycine max*). Planta 206:44–52, 1998.

166. MJ Hawkesford, JC Davidian, C Grignon. Sulphate/proton cotransport in plasma-membrane vesicles isolated from roots of *Brassica napus* L.: increased transport in membranes isolated from sulphur-starved plants. Planta 190:297–304, 1993.

167. FW Smith, MJ Hawkesford, PM Ealing, DT Clarkson, BPJ Vanden, AR Belcher, AGS Warrilow. Regulation of expression of a cDNA from barley roots encoding a high affinity sulphate transporter. Plant J 12:875–884, 1997.

168. E Blumwald, RJ Poole. Nitrate storage and retrieval in *Beta vulgaris:* effects of nitrate and chloride on proton gradients in tonoplast vesicles. Proc Natl Acad Sci U S A 82:3683–3687, 1985.

169. KS Schumaker, H Sze. Decrease of pH gradients in tonoplast vesicles by $NO_3^-$ and $Cl^-$: evidence for $H^+$-coupled anion transport. Plant Physiol 83:490–496, 1987.

170. H Marschner. Mineral Nutrition of Higher Plants. London: Academic Press, 1955.

171. V Römheld, H Marschner. Evidence for a specific uptake system for iron siderophores in roots of grasses. Plant Physiol 80:175–180, 1986.

172. J Robinson, CM Procter, EL Connolly, ML Guerinot. A ferric-chelate reductase for iron uptake from soils. Nature 397:694–697, 1999.

173. D Eide, M Broderius, J Fett, L Guerinot Mary, G Klobus, J Jerzykiewicz, J Buczek. A novel iron-regulated metal transporter from plants identified by functional expression in yeast. Proc Natl Acad Sci U S A 93:5624–5628, 1996.

174. S Mori, N Nishizawaw. Methionine as a dominant precursor of phytosiderophores. Plant Cell Physiol 28:1082–1092, 1996.

175. JF Ma. K Nomoto. Effective regulation of iron acquisition in graminaceous plants. The role of muginetic acids as phytosiderophores. Physiol Plant 97:609–617, 1996.

176. FA Smith, JA Raven. Intracellular pH and its regulation. Annu Rev Plant Physiol 30:89–311, 1979.

177. B Jacoby, GG Laties. Bicarbonate fixation and malate compartmentation in relation to salt-induced stoichiometric synthesis of organic acid. Plant Physiol 47:525–531, 1971.

178. PR Stout, DR Hoagland. Upward and lateral movement of salt in certain plants as indicated by radioactive isotopes of potassium, sodium and phosphorus absorbed by roots. Am J Bot 26:320–324, 1939.

179. JS Pate. Transport and partioning of nitrogenous solutes. Annu Rev Plant Physiol 31:313340, 1980.

180. E Münch. Versuche über den Saftkreislauf. Ber Deutsch Bot Ges 45:340–356, 1927.

181. MG Pitman. Ion transport into the xylem. Annu Rev Plant Physiol 28:71–88, 1977.

182. DT Clarkson, AW Robards. The endodermis, its structural development and physiological role. In: JG Torrey, D Clarkson, eds. The Development and Function of Roots. London: Academic Press, 1975, pp 415–446.

183. DT Clarkson. Movement of ions across roots. In: DA Baker, JL Hall, eds. Solute Transport in Plant Cells and Tissues. Harlow: Longman Scientific and Technical, 1988, pp 251–304.

184. G Haberlandt. Anatomisch-pysiologische Untersuchungen über das tropische Laubblatt. Sitzungsber Akad Wiss Wien I 101:785–816, 1892.

185. JL Broyer, DR Hoagland. Metabolic activities of roots and their bearing on the relation of upward movement of salts and water in plants. Am J Bot 30:261–273, 1943.

186. RT Hardiman, B Jacoby. Absorption and translocation of Cd in bush beans (*Phaseolus vulgaris*). Physiol Plant 61:670–674, 1984.

187. HP Jones, KA Handreck. Studies of silica in the oat plant. III. Uptake of silica from the soil by the plant. Plant Soil 23:79–96, 1965.

188. MGT Shone, DT Clarkson, J Sanderson, AV Wood. A comparison of the uptake and translocation of some organic molecules and ions in higher plants. In: WP Anderson, ed. Ion Transport in Plants. New York: Academic Press, 1973, pp 771–582.

189. EB Dumbroff, DR Peirson. Probable sites for passive movement of ions across the endodermis. Can J Bot 49:35–38, 1971.

190. CA Peterson, ME Emanuel, GB Humphreys. Pathway of movement of apoplastic fluorescent dye tracers through the endodermis at the site of secondary root formation in corn (*Zea mays*) and broad bean (*Vicia faba*). Can J Bot 59:618–625, 1981.

191. DT Clarkson, JB Hanson. Mineral nutrition of higher plants. Annu Rev Plant Physiol 31:239–298, 1980.

192. AR Yeo, MA Yeo, TJ Flowers. The contribution of an apoplastic pathway to sodium uptake by rice roots in saline conditions. J Exp Bot 38:1141–1153, 1987.

193. CW Bell, O Biddlph. Translocation of calcium. Exchange versus mass flow. Plant Physiol 38:610–614, 1963.

194. JF Mcgrath, AD Robson. The movement of zinc through excised stems of seedlings of *Pinus radiata* D. Don. Ann Bot 54:231–242, 1984.
195. JC Brown, LC Tiffin. Iron stress as related to the iron and citrate occurring in stem exudate. Plant Physiol 40:315–325, 1965.
196. B Jacoby. The effect of the roots on calcium ascent in bean stems. Ann Bot 31:725–730, 1967.
197. B Jacoby. Mechanisms involved in salt tolerance of plants. In: M Pessarakli, ed. Handbook of Plant and Crop Stress. New York: Marcel Dekker, 1999, pp 97–123.
198. B Jacoby. Function of bean roots and stems in sodium retention. Plant Physiol 39:445–449, 1964.
199. B Jacoby. Sodium retention in exicised bean stems. Physiol Plant 18:730–739, 1965.
200. D Kramer, A Lauchli, AR Yeo, J Gullasch. Transfer cells in roots of *Phaseolus coccineus:* ultrastructure and possible function in exclusion of sodium from the shoot. Ann Bot 41:1031–1040, 1977.
201. B Jacoby. Sodium recirculation and loss from *Phaseolus vulgaris* L. Ann Bot 43:741–744, 1979.

# 17

# Sodium: A Functional Nutrient in Plants

**G. V. Subbarao\* and Gary W. Stutte**

*Dynamac Corporation, Kennedy Space Center, Florida*

**Raymond M. Wheeler**

*National Aeronautics and Space Administration, Kennedy Space Center, Florida*

**Wade L. Berry**

*University of California, Los Angeles, California*

## I.  INTRODUCTION

Sodium nutrition of plants has remained a fascinating and elusive topic despite several decades of intensive research efforts, particularly during the 1960s and 1970s. Using Arnon and Stout's [1] definition of "essential nutrient" as modified by Epstein [2] as the standard to evaluate essentiality, Na has still not been shown to meet their criteria to be an essential nutrient for all higher plants (certain types of $C_4$ plants may be an exception). This is despite the fact that in some plants, internal Na tissue levels can become extremely high, nearly reaching K in tissue concentrations [3]. Sodium and K are chemically and structurally similar monovalent cations. The hydrated Na ion has a radius of 0.358 nm, the K ion 0.331 nm; thus, physically, there appears to be no size limitation for them to be taken up through the same ion channels [4]. The Na concentration in the earth's crust is similar to that of K (2.8% vs. 2.6%) [5,6]. Sodium levels are very high in many irrigation waters and in some cases approach 10 times that of K (Table 1). Many halophytic plants have taken advantage of this close similarity between Na and K and have adapted to grow in high-salt (NaCl) areas (see review by Glenn et al. [7]) where other less well adapted plants (i.e., glycophytes) are limited in growth because of the high salinity stress [8]. Many nonhalophytic plants can utilize Na under conditions of limited K availability for a number of non–K-specific metabolic functions [3,9,10]. Glycophytic plants such as beets, celery, turnips, and spinach can utilize Na to such a degree that it is possible for farmers to substitute the relatively inexpensive Na as a fertilizer for K fertilizer [9].

One of the most noteworthy features of Na metabolism in plants is the remarkable difference among species in accumulating or excluding Na from their tissues. Despite the the physical and chemical similarity between Na and K, many higher plants have developed a very high degree of selectivity for the uptake of K even in the presence of large amounts of Na. The same high degree of discrimination is also often found in the transport of Na from the roots to the shoot; thus, seeds and fruits of most plants are very low in Na (see Chapter 44 on mechanisms of salinity tolerance by Subbarao and Johansen). This trait conserves K, very often a limiting resource for plant growth, but limits the transfer of Na from the soil through plants on to animals, thus promoting soil salinization and Na deficiency in herbivores.

Sodium is an essential element for animals (including humans) and must be present in relatively large amounts in the diet. Sodium, the principal electrolyte in animal systems, plays an important role in maintaining the ionic balance of body tissues; its osmotic characteristics are utilized in the blood

---

\* *Current affiliation:* Japan International Research Center for Agricultural Sciences, Ibaraki, Japan.

**TABLE 1**  Chemical Characteristics and Comparison of Sodium and
Potassium Concentrations in Soils, Natural Waters, and Plants

|  | Sodium | Potassium |
|---|---|---|
| Atomic number | 11 | 19 |
| Atomic weight | 22.9 | 39.5 |
| Concentration in lithosphere (ppm) | 28.3 | 25.9 |
| Mineral soils: (% as the oxides) |  |  |
|    Tropical | 0.01–0.5 | 0.1–2 |
|    Temperate | 0.01–1.0 | 0.1–4 |
| Soil solution (mM) | 0.4–150 | 0.2–10 |
| Soil solution in field soils (mM) | — | 0.08–1.6 |
| Seawater (mM) | 480 | 10 |
| Rivers of North America (mM) | 0.4 | 0.04 |
| Rivers of Australasia (mM) | 0.13 | 0.04 |
| Plant foliage |  |  |
| Glycophytes[a] | 0.2–2.0 | 15–50 |
| Halophytes[b] | 25–154 | 10–33 |

[a] Grown in 5 mM K + 1 mM Na (g kg$^{-1}$ dwt).
[b] Grown in 5–8 mM K + 295–340 mM Na (g kg$^{-1}$ dwt).
*Source*: Ref. 6.

stream for regulating osmotic pressure within the cells and body fluids, where it protects against excessive loss of water [11]. In contrast, the principal electrolyte for plants is K, and even in ecosystems where there is a predominance of Na, plants exhibit a strong preference for K. Because of this contrast in electrolyte requirements, insufficient Na is available in the edible portions of most plants for herbivores. Therefore the dietary requirements of herbivores for Na must be met from external supplements, such as salt licks.

It is possible under the appropriate management conditions (such as low K and high Na in the rooting media) for many crop plants to accumulate significant concentrations of Na in their edible tissues. Under such conditions, a number of vegetative crops can supply a significant portion of an animal's dietary requirements for Na. Such an increase in plant tissue Na would be especially desirable in meeting the metabolic Na requirements of grazing animals where it is difficult to supply Na through external supplements [12]. For example, in intensively farmed areas of New Zealand, the Na levels in pastures are insufficient to meet the metabolic requirements of grazing animals. Thus, improving Na levels in pasture crops could play an important role in meeting the dietary Na needs, which in turn could substantially enhance their appetite, daily food intake, and weight gain [12]. In our present industrial society, there are relatively few instances in which human Na requirements are not met from diet. This is mostly because of Na additions (both for taste and as a preservative) to most processed foods.

Problems of secondary salinization associated with irrigated agriculture can partially be related to the long-term affect of continued discrimination against Na during nutrient uptake by plants. This secondary salinization could be limited if the appropriate management procedures were applied, such as leaching, restrictive K fertilization, and biomass removal, along with using genetic strategies to secure efficient plants for Na mobilization. This chapter summarizes the current level of understanding of Na metabolism in plants. Although Na is not a major nutrient in most plants, there is some degree of utilization of Na in many if not all plants. The high degree of similarity between K and Na (physical and chemical properties) and the extensive use of Na by a number of salt-tolerant plants suggest that there is a potential for Na supplementing or replacing K in many, if not all non–K-specific monovalent plant functions. However, to increase significantly Na utilization by plants requires the implementation of suitable nutrient management practices and/or appropriate genetic strategies. Such a utilization could be useful in the management of certain saline agricultural systems and the management of low-Na pastures. Sodium cycling is also a problem in other biosystems including the bioregenerative life support systems being developed by the National Aeronautics and Space Administration (NASA) for space. In such a closed system all waste products must be recycled, including Na-containing wastes, to grow plants that provide food, $O_2$, and clean water for the system (see Chapter 48 by Wheeler et al.).

## II.  SODIUM AND ESSENTIALITY

According to Arnon and Stout [1], the following conditions should be met for an element to be considered as an essential nutrient:

The organism cannot complete its life cycle without it.
Its action must be specific and cannot be replaced by any other element.
Its effect on the organism must be direct.

This set of criteria was expanded by Epstein [2] to include

The element is part of an essential compound, or metabolite.

These criteria are based exclusively on the ecological considerations of survival and reproduction, where high yield and biomass production may or may not be an important aspect or even associated with nutrient essentiality. It is possible that some nutrients such as Na and Si may promote maximal biomass production without meeting the preceding requirement for essentiality. In addition, not all metabolic functions require a unique nutrient to function. Many essential metabolic processes can function equally well with a number of different chemically and physically similar elements. It appears that it is possible for similar elements such as Na and K to replace each other fully in certain nonspecific metabolic functions. Thus, even though an element may function completely in an essential function (may be even more effectual than any other element), it would not be considered an essential nutrient unless it has a unique function that it alone can meet. It could be argued, at least from agronomic considerations, that additional levels of essentiality should be differentiated to denote nutrients that may be required for maximal yield or are able to replace other nutrients in certain essential metabolic functions, reducing the critical level of an essential nutrient.

Based on the expanded criteria of Arnon and Stout [1], Na has been shown to meet the criteria for essentiality only for certain $C_4$ plant species, such as *Atriplex vesicaria, A. tricolor, Kochia childsii,* and *Panicum miliaceum* [4,13–19]. In the absence of Na, these $C_4$ plant species grew poorly, showed visual deficiency symptoms such as chlorosis and necrosis, or failed to form flowers. Supplying 100 μM Na enhanced growth and alleviated visual symptoms. Even for extreme halophytes, in which Na is beneficial if not essential, it is required only at the micronutrient level [20]. Even in $C_4$ families, Amaranthaceae, Chenopodiaceae, and Cyperaceae, in which Na has been shown to be essential, the amount of Na required is at a level typical of a micronutrient [21]. For $C_4$ species such as maize, sorghum, and sugarcane, Na has not been shown to be essential [22].

Despite the fact that Na is not essential for many species, application of Na to the growth medium has been shown to stimulate the growth of asparagus, barley, broccoli, brussels sprouts, caraway, carrot, chicory, cotton, flax, millet, oat, pea, rutabaga, tomato, vetch, wheat, cabbage, celeriac, horseradish, kale, kohlrabi, mustard, radish, rape, celery, marigold, sugarbeet, red beet, Swiss chard, and turnip [13,23–25]. Visual leaf symptoms of low Na on sugarbeet, marigold, and red beet appear as a dull dark green color, rapid wilting in drought, and a tendency for leaves to grow out horizontally from the crown. In some cases marginal intervenal scorch may develop, similar to that of K deficiency [24]. The presence of Na tends to reduce K content in the leaves of sugarbeet and red beet. Several researchers believe that Na promotes growth and vigor of sugarbeet, which results in increased yield [26,27]. However, the specific function of Na in the metabolic processes of these crops is still unknown. There is not sufficient evidence to indicate that Na is essential for these crops. But this is open to further investigations because it appears that if Na is required, it is required at micronutrient levels, and that the older analytical methods (before 1955 and the use of flame photometry or atomic absorption) were not sufficiently sensitive to evaluate adequately very low concentrations of Na. It is unknown therefore whether the deficient solutions employed in these early physiological experiments were completely free of Na. In addition, there is such a large potential for Na contamination that every step must be checked for Na. Thus, the question of essentiality is still open and needs to be readdressed with modern analytical techniques.

## III.  CONCEPT OF "FUNCTIONAL NUTRIENT"

To overcome the difficulty associated with this limited definition of "essentiality," Nicholas [28] suggested the term "functional or metabolism nutrient," which is defined as "any mineral element that functions in plant metabolism irrespective of whether or not its action is specific." The term functional nutrient seems appropriate, but the definition needs to be much more specific. It seems that it would be desirable to define a functional nutrient as one that is essential for maximal biomass production or can function in an essential metabolic process, reducing the critical level of an essential nutrient, without having a unique function itself, as defined by Arnon and Stout [1]. The remaining portion of the chapter deals with this issue, and we will try to present evidence to support the notion that Na should be considered as a functional nutrient.

## IV.  UPTAKE AND TOLERANCE OF SODIUM

### A.  Uptake Mechanisms

The concept of dual mechanisms has been widely recognized in the absorption of alkali cations by plant roots [29–32]. Because of the chemical similarity between K and Na, it is generally assumed that K and Na compete for common absorption sites in the root. Selective ion transport or mechanism 1 K transport is effective at very low external K concentrations [33], with a maximum rate at an external K concentration of 1 mM [29]. Sodium even in 20-fold excess fails to compete significantly with K under mechanism 1. Mechanism 1 depends on metabolic energy derived from adenosine triphosphate (ATP). At higher concentrations of K (up to 50 mM), mechanism 2 becomes important [29,30]. Mechanism 2 does not discriminate K from Na, and thus Na can competitively inhibit the absorption of K [33]. Also, mechanism 2 does not directly require metabolic energy to function and is thought to operate through diffusive forces, which involve ion channels. Sodium uptake in plants is believed to be primarily through mechanism 2 [34].

Inward-rectifying K channels ($K_{in}$ channels) have been reported in different root cells, including cortical, root hair, stelar, and xylem parenchyma cells, that can sense K concentrations [35–41]. These ion channels transport at rates between $10^6$ and $10^8$ ions per second per channel protein. Transport is "passive," in which diffusion of ions through the channel is a function of both the membrane voltage and the concentration difference across the membrane; thus, uptake is not directly coupled to the input of other forms of free energy [42]. Also, selectivity is not absolute and many channels can conduct a range of ions, although not all at the same efficiency [42]. This property is reflected in the so-called ionic selectivity sequence for the channel, which can have physiological significance. Thus, some K channels conduct Na to a finite extent and can affect the degree to which plants withstand high Na (i.e., salinity) in the root zone [43]. Another property of ion channels is their ability to reside in "open" or "closed" conformational states, which either permit or prevent ion permeation. This conformational switching can occur in response to ligands or to a change in membrane voltage after which channels activate or deactivate [44]. The control of activation by membrane voltage or by ligands such as Ca may be the key to understanding the role of these ion channels in cell biology [42].

Although, it is widely believed that mechanism 1 does not have much affinity to transport Na in the presence of adequate K, for some crops such as beets this mechanism may be transporting Na independent of the external concentration [45]. Several *Atriplex* species take up Na in preference to K. In these species, Na competes with K during uptake, but K does not compete with Na [46–48]. Thus, mechanisms of Na transport at low concentrations and in the presence of K are open to further investigation.

### B.  Regulation, Translocation, and Partitioning

Plant species vary widely in their ability to absorb and translocate Na to the shoot [49]. Generally, species that absorb Na and translocate it freely to the shoot are termed *natrophiles* [50]. Most plant species do not readily absorb Na but prefer K instead and are termed *natrophobes* [49–51]. If natrophobes take up Na, it is usually retained in the root and with relatively little translocation to the shoot [49]. Natrophobes translocate Na, but only when subjected to high concentrations ($\geq 100$ mM) in the root zone, which usually results in a major growth reduction and/or death of the plants as a result of Na toxicity [8,52] (also see Chapter 44 by Subbarao and Johansen). Physiological mechanisms that regulate Na levels of these natrophobes are addressed elsewhere (see Chapter 44 by Subbarao and Johansen).

The capability to translocate significant amounts of Na to the shoot even among natrophiles vary widely among plant species, with halophytes representing the extreme on the high end. The general assumption about Na tolerance in plants is that they compartmentalize the absorbed Na in vacuoles and use it as an inorganic osmoticum in place of or along with K. It is widely believed that the cytoplasm does not tolerate high levels of Na as it interferes with normal metabolic functioning [53]. This seems to be true for both natrophiles and natrophobes, with the only difference between these groups being the ability of natrophiles to compartmentalize the absorbed Na effectively. Natrophobes that have limited or no ability to compartmentalize Na spend substantial amounts of energy in preventing Na from entering the plant (see Chapter 44 by Subbarao and Johansen). These Na-excluding mechanisms are a drain on the plant's carbon and energy resources and thus natrophobes can suffer substantial growth reductions or death when grown in high-Na environments [8] (see chapter by Subbarao and Johansen).

Plant cell cytosol typically contains about 100 mM K and rarely tolerates Na levels above 20 mM [54–56]. Enzymes isolated from salt-sensitive *Phaseolus* and salt-tolerant *Atriplex* and *Salicornia* are equally sensitive to NaCl when bioassayed [53]. This was established for four different enzymes, which included the rather salt-sensitive aspartate transaminase as well as salt-tolerant glucose-6-phosphate dehydrogenase [53]. Furthermore, growth of *Phaseolus* and *Atriplex* in saline cultures failed to alter the specific activity or NaCl sensitivity of the enzymes [53]. This indicates that even natrophilic species, such as *Atriplex* and *Salicornia*, cannot tolerate high Na levels in their cytoplasm. They maintain this relatively constant cytoplasmic level of Na by compartmentalizing high levels of Na in their vacuoles [53]. This is an important survival feature of halophytic plants under saline conditions [57] (see Chapter 44 by Subbarao and Johansen).

Most crop plants belong to the category of *glycophytes*. Some crop plants such as sugar beet, red beet, Swiss chard, celery, and turnip have a substantial ability to translocate Na to the shoot. For red beet, a considerable buildup of Na in the tops can occur whenever Na is present in the nutrient solution [3]. Our studies indicate that Na is absorbed from the nutrient medium and translocated to the shoot relatively freely in red beet but not in either spinach or lettuce (G.V. Subbarao and R.M. Wheeler, unpublished).

The extent to which Na is taken up by plants is influenced by other nutrients, particularly K and N [58]; however, this does vary with species [59,60]. The rate of transpiration can influence uptake and movement of some ions in plants [61]. For example, Pitman [62,63] showed that higher rates of transpiration increased the ratio of K to Na reaching the leaves of barley and white mustard (*Sinapis alba* L.). In tomato, the ability of roots to exclude Na from the rest of the plant decreased rapidly as the level of K in the nutrient solutions fell [64].

## V.  SODIUM FUNCTION IN METABOLISM

### A.  C₄ Metabolism

In the Calvin cycle of $C_4$ plants, $CO_2$ is concentrated in the bundle sheath cells. An extensive flow of metabolites between mesophyll and bundle sheath cells is required to operate this $CO_2$ concentration mechanism [4]. Sodium deficiency has been reported to impair this conversion of pyruvate to phosphoenolpyravate (PEP), which takes place in the mesophyll chloroplasts [4]. In certain $C_4$ species, e.g., *Amaranthus tricolor*, the $C_3$ metabolites alanine and pyruvate accumulate, whereas $C_4$ metabolites PEP, malate, and aspartate decrease under Na deficiency [21]. Sodium deficiency leads to a reduction in photosystem II (PSII) activity and ultrastructure changes in mesophyll chloroplasts but not in bundle sheath chloroplasts of *A. tricolor* and *Kochia childsii* [65,66]. Resupplying Na to these species restored PSII activity [22] in thechloroplast. In $C_4$ species, nitrate assimilation also appears to be confined to the mesophyll cells. For *A. tricolor*, nitrate reductase activity is substantially decreased in leaves of Na-deficient plants but is restored after resupplying Na [67]. Sodium reportedly enhances nitrate uptake by roots and nitrate assimilation in leaves of *A. tricolor* [68].

Using isolated chloroplasts of *Panicum miliaceum* (a $C_4$ species), it was shown that Na enhanced pyruvate uptake, indicating that $Na^+$/pyruvate was cotransported through the envelope into the chloroplast by light-stimulated Na efflux pumps [4]. In contrast, no Na effect on pyruvate uptake rates was noted in mesophyll chloroplasts of *Zea mays*. Additional evidence suggests that in $C_4$ species of the $NADP^+$-

ME (malic enzyme) type (such as *Zea mays* and *Sorghum bicolor*), the $H^+$/pyruvate cotransport system operates instead of the $Na^+$/pyruvate cotransport system in mesophyll chloroplasts isolated from *P. miliaceum* [22]. This stresses the necessity of differentiating between metabolic types of $C_4$ species when studying the role of Na.

## B. Replacing Potassium Functions

Next to N, K is the mineral nutrient required in largest amounts by plants for metabolic functions and growth [69]. Plant nutritionists have been intrigued for decades by this high requirement for K. Potassium plays a vital role in a wide range of biochemical and biophysical processes in plants. It is a highly mobile charge carrier, it neutralizes the effects of anions, and it plays an important role in enzyme activation and membrane transport. The process that is generally regarded as being the most sensitive to K is the maintenance of turgor pressure and, as a consequence, cell expansion. Biochemically, most K-requiring enzymes need only 10–50 mM K for maximum activity, and other monovalent cations can sometimes substitute for K [70,71]. It is reported that protein synthesis in vitro in plant systems is maximal at 100 mM K or higher [56,72,73]. Potassium exists as a monovalent cation in biological systems and does not participate in covalent bonding. Moreover, it forms only weak coordination complexes because of its small ionic charge, low electronegativity, and its completed $3p$ electron shell [69]. However, K is the major cytoplasmic cation in cells and plays a number of interrelated and integrated roles [56,74]. These include:

Cofactor in enzyme activation, especially protein synthesis (translation)
Stabilization of the active conformation of enzymes and possibly membranes
Cytoplasmic volume regulation
Energy conservation across membranes
Cytoplasmic pH regulation

In general, K functions in plants can be summarized as both biophysical (non–K-specific role as an osmoticum in the vacuole) and biochemical (specific and nonspecific roles in the cytoplasm). According to prevalent concepts, the need of monovalent cations can be completely filled by K, but some functions can also be exercised by Na and other monovalent cations, thus reducing the total amount of K required by the plant.

### 1. Internal Osmoticum

The large central vacuole of the plant cell (which occupies nearly 90% of the cell's volume) provides a large buffer volume, primarily of inorganic ions for satisfying the osmotic requirements of terrestrial plants without maintaining a large volume of cytosol filled with energy-expensive organic solutes [75]. The peripheral cytosol layer facilitates the distribution of chloroplasts close to the cell surface, thus maximizing the penetration of light to the photosynthetic apparatus [75]. Potassium makes a major contribution to the solute potential ($\Psi s$) of the cell [76,77]. Investigations of 200 plant species by Iljin [78] have shown that the contribution of K to the total $\Psi s$ varied from 66 to 90%. The accompanying anions are primarily $P_i$, Cl, $NO_3$, and $SO_4$ [69,75]. Although K salts are the most common inorganic osmotica in the vacuole, there is no known absolute requirement for K in this compartment as vacuoles contain high concentrations of many other solutes such as Na salts [79], sugars [75,80], and amino acids [81]. If K salts are the only vacuolar solutes, the $\Psi s$ of the vacuole containing 200 mM K (and associated anions) would be about $-0.9$ MPa [69]. Reported sap $\Psi s$ varied from $-0.7$ to 1.2 MPa for plants not subjected to salt or water stress [77,82], suggesting that turgor maintenance in the vacuole is generated primarily by K salts under K-sufficient conditions. Examples of the contribution of the K to the total leaf sap $\Psi s$ for red beet, spinach, and lettuce under K-sufficient conditions are given in Table 2.

The vacuole is considered to be a storage organelle, and the nutrients accumulated within it are largely removed from active metabolism. Nevertheless, they are important to generate and maintain cell turgor [75,83,84]. It has been shown that the vacuolar K levels can be highly variable (10–200 mM), determined mostly by the external K concentrations in the root zone. In contrast, cytoplasmic levels are relatively stable, near 200 mM for most plant species [56,76,77,84–90]. In well-fertilized crops, the vacuolar K concentration can be high, on the order of 200 mM. Because the vacuole occupies nearly 80–90% of a mature plant cell volume, it has most of the cell K in K-sufficient plants, providing a K concentration

**TABLE 2** Leaf Sap K Levels, Osmotic Potential ($\Psi$s), and Contribution of K to the Leaf Sap $\Psi$s for Red Beet, Spinach, and Lettuce Grown Under 5.0 mM K Using Nutrient Film Technique

| Plant species | Leaf sap K (mM) | Leaf sap $\Psi$s (MPa) | % contribution of K to the lamina $\Psi$sap |
|---|---|---|---|
| Red beet (*Beta vulgaris*) | 435 | 0.97 | 96 |
| Spinach (*Spinacea oleracea*) | 242 | 0.91 | 57 |
| Lettuce (*Lactuca sativa*) | 244 | 0.98 | 53 |

*Source*: Subbarao et al., unpublished data.

similar to that of the averaged whole cell [88]. Conceptually, at least 90% of the plant's K (when grown under K-sufficient conditions) can be replaced by Na (if Na is equivalent in function to K) without affecting the specific functions of K in the cytoplasm or the turgor of the cell.

Growth responses of halophytes to Na reflect the high salt requirement for osmotic adjustment [6,48]. For the halophyte *Salicornia herbacea* [91], both Na and K are effective in promoting hypocotyl elongation. Potassium has only 87% of the effect of Na in *S. herbacea*, although both ions are effective in glycophytes but at much lower concentrations (about 10 mM as opposed to 100 to 200 mM in halophytes). In halophytes, Na accumulation and its contribution to $\Psi$s reach a maximum [48]. There is evidence that this reflects the ability of the tonoplast in the leaf cells to restrict Na efflux from vacuoles [20,92,93]. In *Ricinus communis*, K initially contributes to vacuole and cell expansion but can be replaced by Na following maturation of the root tissues [90].

Many halophytes osmotically compensate for high external osmotic potential by accumulating Na salts, often NaCl from the environment [74,94]. The sap $\Psi$s of $-2$ to $-5$ MPa required to maintain turgor in halophytes under seawater salinity ($\Psi$s $-2.3$ MPa) can be accounted for by the 400 to 700 mM Na and Cl concentrations in the sap [20,95]. Many members of Chenopodiaccac and halophytes show marked selectivity for Na over K at low concentrations of both ions [96].

Biophysical functions of K in cells are nonspecific [89], and Na may at times be more suitable than K [6]. With a limited K supply, Na, Mg, and Ca can replace K in the vacuole as an alternative inorganic osmoticum [6]. Our studies with red beet have shown that Na can replace K for vacuolar function; nearly 95% of the plant's K was replaced by Na (Table 3). For crops such as spinach and lettuce, the ability to substitute Na for K is low (G.V. Subbarao et al., unpublished results). Thus, it appears that Na can fill the "biophysical" function of K provided the plants have the ability to take up this element, translocate it to the shoot, and compartmentalize it in their vacuoles.

In contrast, the location of major metabolic processes such as protein synthesis, photosynthesis, and glycolysis within the cytoplasmic compartment places restrictions on the type and concentrations of solute in this compartment [56,74]. Solutes that accumulate in cytoplasm must not disrupt metabolism and must be maintained at concentrations that permit the various processes to proceed at favorable rates. There is evidence that mechanisms exist for regulating cytoplasmic concentrations of a range of ions including K, Na, H, and $P_i$ [56,77,89,97–99]. For all eukaryotic organisms, the composition of cytoplasm appears to be highly conservative during evolution. Despite wide variations in the concentrations of K, Na, and Cl in the vacuolar compartment, the cytoplasm is characterized by 100 to 200 mM K, with little potential for Na replacing K in cytoplasm [56,74,90].

**TABLE 3** Effect of Replacing K with Na in Hydroponic Solution on Leaf Na Levels of Red Beet

| % Na substituted for K in nutrient solution (from a total of 5 mM) | K in sap (mM) | Na + K in sap (mM) | % Na to total Na + K in sap |
|---|---|---|---|
| 0 | 414 | 415 | 0.2 |
| 75 | 83 | 222 | 63 |
| 95 | 25 | 237 | 89 |
| 98 | 13 | 294 | 96 |

*Source*: Subbarao and Wheeler, unpublished data.

### 2. Stomatal Function

In most plant species, $K^+$ is the dominant cation responsible for turgor changes in guard cells during ion-induced stomatal movement [4]. An increase in the K concentration of guard cells increases $\Psi s$ and results in uptake of water from the adjacent cells. The corresponding increase in turgor of the guard cells results in stomatal opening. Closure of stomata in the dark is correlated with K efflux and a corresponding decrease in the $\Psi s$ of the guard cells. Selectivity of transport systems for K over Na provides a fundamental limitation on the degree to which Na can substitute for this stomatal function in plants [6]. Thus, although Na can substitute for K in vacuolar osmotic adjustment for a number of plant species, it does not appear to be able to carry out this role for stomatal turgor [100,101]. However, for *Commelina communis*, Na was able to replace K and was even more effective than K in stomatal function [102]. Thus, Na can have a role in stomatal physiology of such plants even if it is not an obligate role [70].

However, from the differences between plant species with respect to the membrane permeability for K and Na, one can suppose that in plant species with high permeability to Na (e.g., *Beta vulgaris*), K is at least partially replaceable for this stomatal function [9]. Thus, Na may act as an alternative cation to K for stomatal opening [103–109]. Our results with red beet indicate that stomatal conductance is nearly normal even when nearly 95% of the plant's K was replaced by Na and Na levels in the leaf sap approached 200 mM (G.V. Subbarao et al., unpublished).

### 3. Photosynthesis

Potassium is the dominant counterion for the light-induced $H^+$ influx across the thylakoid membranes [110] and for the establishment of the transmembrane pH gradient necessary for the synthesis of ATP [111–113]. Also, in the formation of chloroplast structure, the translocation and storage of assimilates (sucrose) in the sink tissue seem to depend on adequate K concentrations in the tissue [6].

In sugar beet, chloroplasts often contain high concentrations of Na [9], and Na and K are incorporated in the chloroplasts to a similar extent [114]. Because much of the leaf Na in sugar beet is concentrated in the chloroplasts, it is hypothesized that Na may have been involved in photosynthesis [115]. In chloroplasts of *Limonium vulgare*, the Na content is even higher than K. Considering the beneficial effect of Na in *Beta vulgaris*, it is possible that Na participates in photophosphorylation [9]. A prerequisite for this function is high membrane permeability for Na. The high mobility of Na in crops such as sugar beet suggests that this prerequisite is fulfilled. However, Na is unable to replace K in chlorophyll synthesis in spinach, lettuce, and sugar beet [116,117]. For sugar beet, Na was able to replace K for chloroplast multiplication [9]. Nevertheless, photosynthetic rates of sugar beet declined substantially during K deficiency even when Na was present [108,109]. In red beet, however, leaf photosynthetic rates were nearly normal despite high levels of Na in leaf lamina (up to 100 g kg$^{-1}$ dwt = dry weight) [3]. Chlorophyll fluorescence and leaf Na levels are presented for red beet, spinach, and lettuce (Table 4). For red beet, leaf chlorophyll levels and chlorophyll fluorescence ($F_v/F_m$) are not affected at tissue Na concentrations of 76 g kg$^{-1}$ dwt (Table 4). In contrast, chlorophyll fluorescence is decreased substantially at leaf Na levels of 39.8 g kg$^{-1}$ dwt in lettuce (Table 4).

### 4. Counterion in Long-Distance Transport

Potassium is often the dominant counterion in long-distance transport as well as being the counterion during storage of $NO_3^-$ in vacuoles [4]. As a consequence of $NO_3$ reduction in leaves, the remaining counterion, K, requires the stoichiometric synthesis of organic acids (e.g., malate) for charge balance and pH

**TABLE 4**  Leaf Na Levels and Chlorophyll Fluorescence in Red Beet, Spinach, and Lettuce[a]

| Plant species | Leaf Na concentration (g kg$^{-1}$ dwt) | Chlorophyll fluorescence ($F_v/F_m$) ratio |
|---|---|---|
| Red beet (*Beta vulgaris*) | 76.0 | 0.74 |
| Spinach (*Spinacea oleracea*) | 28.9 | 0.75 |
| Lettuce (*Lactusa sativa*) | 39.8 | 0.69 |

[a] Red beet is grown for 42 days after planting (DAP); spinach and lettuce are grown for 30 DAP using nutrient-film technique where K and Na levels in the nutrient solutions were 0.25 and 4.75 mM, respectively.
*Source*: Subbarao and Wheeler unpublished data.

homeostasis. Part of this newly formed K-malate may be retranslocated to the roots for subsequent reutilization of K as a counterion for $NO_3$ transport. The high mobility of K in the phloem and its continuous circulation within the plant are indications of a function for K in the long-distance transport processes of higher plants [9,118]. This function of K as a counterion in long-distance transport may not be specific and other cations should be able to replace K in this function provided they are phloem mobile. For several crops, e.g., maize, Na is not phloem mobile [9]. But in crops such as sugar beet, Na is reported to be phloem mobile and thus could be as effective as K for the long-distance transport functions [9]. Our studies with red beet indicate that $NO_3$ levels in shoot were not affected by large drops in the plant's tissue K concentration when Na was available as an alternative ion (G.V. Subbarao et al., unpublished data). This indicates that in red beet, Na may be able to replace the K in this function. Replacing the plant K with Na, particularly at higher degrees of substitution ($>90\%$), has resulted in higher levels of $NO_3$ in leaves of red beets, spinach, and lettuce (G.V. Subbarao et al., unpublished). Nitrate reductase was reported to require K for its activation [119,120].

## 5. *Enzyme Activation*

Potassium has a direct metabolic role within the cytoplasm [4,121]. Several enzymes are activated by K ions, and that activation was generally maximal at a concentration of about 100 mM, the same as the normal K range in cytoplasm [70]. Potassium ions have an important role in protein and starch synthesis [77] as well as in respiratory and photosynthetic metabolism [76]. The precise mechanisms of activation are not known, but it is important to note that Na is frequently (but not always) less effective as an activating cation for these enzymes [77]. Regardless of the fluctuations of K levels in the vacuole compartment, it seems that plant cells maintain cytoplasmic K levels in the range 100 to 200 mM [69]. Cytoplasmic K levels are thought to be affected only under severe K deficiency afer the vacuolar K pools have been exhausted. Thus, the cytoplasm is very conservative in its K requirements. Because most of the metabolic processes and enzyme action are located in the cytoplasm, it is thought that most of the cytoplasmic enzymes require K for their functions.

Protein synthesis [77,122,123] and oxidative phosphorylation [124] are equally inhibited by high Na in vitro whether the organelles are isolated from glycophytes or halophytes [53]. Starch synthetase has a requirement of about 50 mM K for normal functioning. Other monovalent cations such as rubidium, cesium, and ammonium are about 80% as effective as K, while Na is only about 20% as effective at maintaining starch synthetase activity [125]. Several reports indicate that K is needed for the normal functioning of starch synthetase in sweet potato, taro, white potato, wheat, bush beans, field corn, soybeans, peas, and rice [126–129]. Sodium seems unable to replace this K function even in sugar beet. Potassium deficiency results in the accumulation of solute carbohydrates and reducing sugars due to the inhibition of starch synthesis [70]. Glucose transport across the plasmalemma of sugar beet storage cell protoplasts is faster in the presence of KCl than in the presence of NaCl [130]. Sodium, however, is more effective in catalyzing the transport of sucrose across the tonoplast into the vacuole and in stimulating sucrose accumulation in the storage tissue [130]. This effect of Na on sucrose storage seems to be related to stimulation of adenosinetriphosphatase (ATPase) activity at the tonoplast of beet storage cells [4,130].

Vacuolar ATPase is substrate specific and Mg dependent and is distinguished from nonspecific phosphatase [75]. Beet root ATPase is stimulated about 100% by both Na and K ions. The highest ATPase activity is obtained in sugar beet and *Avicennia* roots with combinations of K and Na but not with either K or Na alone [131]. This agrees with observations that growth of sugar beet is highest when both Na and K are present in the growing medium. Green and Taylor [132] proposed that ATPase activation requires two binding sites, one for K and one for Na, and that maximal activation is obtained when both sites are occupied. During in vitro studies of the effect of high concentrations of KCl and NaCl on the $K^+,Na^+$-ATPases, genotypic differences were found suggesting differences in cellular localization of $K^+$ and $Na^+$ [133,134].

Using isolated mitochondria from *Brassica rapa*, it was shown that the esterification of phosphate and $PO_4/O_2$ ratio are increased more by Na than K [135]. But it is unclear whether this increased rate of phosphate metabolism has some connection with the formation of complexes with ATP. It is also unclear whether an increased rate is desirable in terms of overall plant growth. Sodium is able to form stable complexes with polyphosphates [136]. The effect of Na on the enzymatic activity of pyruvic kinase is small in comparison with that of K [9]. Also, Na is not effective in activating acetic thiokinase from spinach leaves [137].

The existence of isozymes has not been considered when evaluating whether Na could replace K dif-

**Total dry weight (% of K control)**

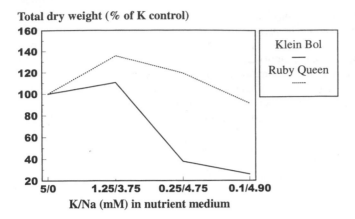

**Figure 1**   Growth response of two red beet varieties to substitution of Na for K in the nutrient medium. Plants are harvested at 42 days after sowing. (From Ref. 3.)

ferentially. In rice, four isozymes of malic dehydrogenase have been reported [138] and several isozymes of α-amylase have been isolated from seeds [139]. Because these isozymes have different properties, they may have different ionic requirements for full activity and they may differ in their ability to function under high levels of Na instead of K [9].

## VI.  GROWTH STIMULATION BY SODIUM

Growth stimulation by Na has both practical and scientific interest as it raises the possibility of applying inexpensive, low-grade Na fertilizers [4]. The presence of Na in the growing medium has been reported to have beneficial effects on the growth of numerous plants [25,140–143]. This stimulation is particularly large for members of Chenopodiaceae. In the case of sugar beet, red beet, and spinach, Na stimulated growth even when there appeared to be adequate K in the nutrient medium [23,142,144–150]. Many field experiments showed that Na fertilization has improved the growth and yield of sugar beet [4,6,149]. However, the positive effects varied with cultivar, soil type, and climatic conditions [143,151–154]. Our studies with red beet showed that maximum growth is observed when both Na and K are present rather than either of them alone [3]. Also, varietal differences are observed for the optimal ratio of K and Na in the nutrient medium, with some varieties appearing to prefer higher levels of Na (Figure 1). However, no such growth stimulation is observed for spinach (Figure 2).

**Total dry weight (% of K control)**

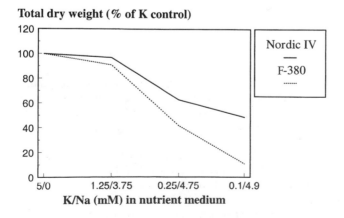

**Figure 2**   Growth response of two spinach cultivars to substitution of Na for K in the nutrient medium. Plants are harvested at 42 days after sowing. (From G.V. Subbarao et al., unpublished data.)

Among nonchenopods, tomato has been reported to respond positively to additional Na in the nutrient medium [155]. For example, there was a 12% increase in the dry weight of tomato after the addition of 1 mM NaCl in the nutrient medium [155]. Sodium alleviated symptoms of K deficiency and decreased the critical foliar K concentration at the symptoms appeared [64]. Also, there is some evidence that potato responds positively to Na. On the basis of a series of field trials on sandy soils, Na application improved potato yields up to 6% for plots where adequate K was given and nearly 10% in plots where K fertilization was not given [24]. Barley, oats, and carrots showed positive responses to supplemental Na at low K levels [13,25,26], and alfalfa, flax, and celery showed a moderate response to Na at adequate K levels [25,26].

## A.  Sodium Improves the Quality

For a few crops, Na has been reported to improve the quality of the product. For celery, Na improved resistance to blight (*Septoria petroselini appli*), crispness, and thus market value [156]. Also, taste tests demonstrated that addition of Na tended to reduce a strong celery flavor (improved flavor) [13,24]. Sodium can also improve the taste of carrots by increasing their sweetness [24].

## VII.  SODIUM AND POTASSIUM INTERACTIONS

Crops vary widely in their ability to substitute Na for K in their growth requirements. For instance, crops that have halophytic ancestors or crops that evolved near seashores typically have a high potential to substitute Na for K. The following list of plants and their origins illustrates this phenomenon:

Beets—evolved in sandy soils near the sea in the Canary Isles, Persia, Babylon, and western India (cited in Harmer and Benne [13])

Cabbage—found in rocky coastlines (cited in Harmer and Benne [13]) on the Isle of Lolland in Denmark, the island of Heligoland, Germany, the south of England and Ireland, the Channel Isles, islands off the coast of Charente, France, and on the north coast of the Mediterranean near Nice, Genoa, and Lucca [96]

Horseradish—known in Holland as sea radish, grows wild in the salty soils in the east of Russia

Turnip—common in the sand on the seacoast in Sweden, Holland, and England

Celery—in damp places from Sweden to Algeria, Egypt, Abyssinia, and in Asia (cited in Harmer and Benne [13])

Harmer and Benne [13] and Harmer et al. [157] presented good surveys of the potential of Na substitution for K in plants. In addition to variation between species, benefits of Na uptake can vary between genotypes of the same species such as sugar beet [158,159], celery [160], red beet [3], and tomato [161].

## A.  Sodium Replacement of Potassium

On the basis of their growth response to or tolerance for Na, crop plants have been classified into four response groups (Table 5) or three levels of tolerance [162]. Plants that discriminate less against Na are likely to have a higher ability to utilize Na for their monovalent cation requirements. Accordingly, group 1 plants, which do not respond favorably to Na even under K deficiency, have little potential for utilizing Na (Table 5). For group 2 plants, there is slight potential (about 10%) for replacing some K with Na in these functions and thus tissue K with Na (Table 5). The largest potential for replacing K with Na in these functions lies in group 3 and group 4 plants (Table 5). From their known levels of tolerance to external Na and the reported translocation of Na to the shoot and edible portion of the plant, we have estimated the amount of tissue K in edible plant parts that is potentially replaced by Na with minimum effects on growth (Table 6).

## B.  Influence of Sodium on Critical Potassium Levels

Sodium has a major role in determining critical K levels (the tissue K level at which 95% of the maximum yield can be achieved) [164]. For crops that have a capacity to substitute Na for K in metabolic func-

**TABLE 5**   Effect of Sodium Applied as a Nutrient on Several Crops

| Degree of benefit in deficiency of K | Degree of benefit in deficiency of K | Degree of benefit in sufficiency of K | Degree of benefit in sufficiency of K |
|---|---|---|---|
| None to slight | Slight to medium | Slight to medium | Large |
| **Group I** | **Group II** | **Group III** | **Group IV** |
| Buckwheat | Asparagus | Cabbage | Celery |
| Corn | Barley | Celeriac | Mangel |
| Lettuce | Broccoli | Horseradish | Sugar beet |
| Onion | Brussels sprouts | Kale | Swiss chard |
| Parsley | Caraway | Kohlrabi | Red beet |
| Parsnip | Carrot | Mustard | Turnip |
| Peppermint | Chicory | Radish | |
| Potato | Cotton | Rape | |
| Rye | Flax | | |
| Soybean | Millet | | |
| Spinach | Oat | | |
| Squash | Pea | | |
| Strawberry | Rutabaga | | |
| Sunflowers | Tomato | | |
| White bean | Vetch | | |
| | Wheat | | |

*Source*: Ref. 13.

tions, minimum tissue K levels can be determined only when Na is supplied adequately. For red beet, it was shown that tissue K levels of certain tissues can drop from $100 \text{ g kg}^{-1}$ dwt (normal tissue K levels for these tissues with unlimited availability of K and Na) to $4 \text{ g kg}^{-1}$ dwt when low K and high Na are provided [3]. This decrease in tissue K occurred without any short-term effect on growth, suggesting that $4 \text{ g kg}^{-1}$ dwt in the tissue is near the critical K level for these tissues. For spinach, these low K levels are about $30 \text{ g kg}^{-1}$ dwt, whereas for lettuce the low K levels are near $65 \text{ g kg}^{-1}$ dwt (Figure 3). Similarly, for red beet, the leaf Na levels reach $100 \text{ g kg}^{-1}$ dwt without negative effects on growth rates, whereas for spinach the level was $17 \text{ g kg}^{-1}$ dwt and for lettuce it was $4.9 \text{ g kg}^{-1}$ dwt (Figure 4). For Rhodes grass,

**TABLE 6**   Concentrations of K in the Edible Portions of Some Crops and Estimates of Na Replacement [Source: Ref. 10]

| Crop | K (%) in edible plant part | % of K replaced with Na (estimate) |
|---|---|---|
| Red beet | 10[b] | 90 |
| Chard | 10[a] | 90 |
| Celery | 10[a] | 75 |
| Lettuce | 10[b] | 25 |
| Spinach | 9[b] | 50 |
| Radish | 9 | 25 |
| Tomato | 5 | 25 |
| Potato | 2.5 | 40 |
| Sweet potato | 2.5 | 40 |
| Wheat | 0.5 | 1 |
| Rice | 0.2[c] | 1 |
| Soybean | 1.7[c] | 1 |
| Peanut | 0.7 | 1 |

[a] Estimations based on our experience with some of the above crops and other published sources.
[b] Based on studies at Kennedy Space Center.
[c] From Duke and Atchley [163].
*Source*: Ref. 10.

**Dry matter production (% of K control)**

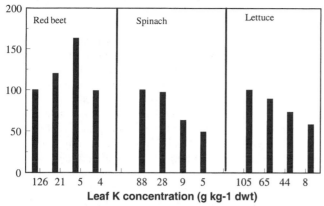

**Figure 3**   Leaf K levels of red beet, spinach, and lettuce in presence of adequate levels of Na in the nutrient medium. (From G.V. Subbarao et al., unpublished data.)

the critical leaf K levels were progressively reduced from 27 g kg$^{-1}$ dwt in plants not fertilized with Na to 5 g kg$^{-1}$ dwt in plants receiving 400 mg Na per pot [165].

It has been suggested that as the tissue K concentration declines, the concentration in the cytoplasm is preferentially maintained for operation of K-dependent processes in the cytoplasm [56,74]. Therefore, initially all changes in tissue K concentration are likely to be at the expense of vacuolar K, with other solutes being diverted to the vacuole to maintain Ψs [56]. Leigh and Wyn Jones [89] argued that the cytoplasmic K concentration would be expected to decline to 15 g kg$^{-1}$ dwt or less, which agrees with the values of critical K concentrations of 5 to 20 g kg$^{-1}$ dwt reported for various tissues in a number of crops [89]. This hypothesis may also explain the effects of other cations, such as Na and Mg, on tissue K concentrations. When these other cations are abundant in tissue, critical K concentrations range from 10 to 20 g kg$^{-1}$ dwt, but when they are low, K values can increase up to 40 to 70 g kg$^{-1}$ dwt depending on the species [141,166]. For Italian ryegrass, the leaf K optimum decreased from 35 dwt to 8 g kg$^{-1}$ dwt when Na was provided as an alternative ion [141]. For fodder beet, sugar beet, red beet, oats, barley, ryegrass, English ryegrass, turnips, lupins, red and white clover, potatoes, kale, and rapeseed, the optimal K levels were lower when Na was supplied [3,9,24].

**Dry matter production (% of control)**

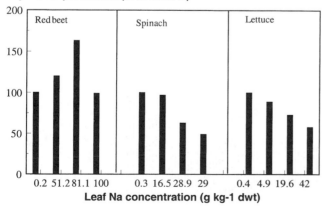

**Figure 4**   Leaf Na levels and total dry matter production (expressed as percent of control) for red beet, spinach, and lettuce for the same tissues as in Figure 3.

## VIII.   SODIUM, OTHER PERSPECTIVES

### A.   Sodium Tissue Levels

Regardless of Na supply, most crops translocate very little Na to reproductive structures such as seeds, fruits, or storage roots, which are the edible portions of many staple crops [167]. This is the case for cereals such as wheat and rice, fruit crops such as tomatoes, tubers such as potatoes, and roots such as sweet potatoes. The main reason for this low Na concentration in the reproductive or storage organs is that they are fed mostly through phloem, where there is heavy discrimination against Na translocation.

In contrast, the edible portions of leafy salad crops are vegetative structures that maintain a heavy xylem flow, thus allowing Na to accumulate in their tissues. If plant nutrient supplies are modified to favor Na uptake (e.g., by reducing K supply in the presence of Na), then it is possible for some leafy crops to accumulate Na in edible structures at reasonably high levels without adversely affecting productivity or quality. This approach could be adapted for greenhouse crops such as lettuce or spinach, where a significant portion of the K salts could be replaced with relatively inexpensive Na salts. Also, for field crops such as beets, celery, and radish, there is a large potential to replace K with Na. Sodium levels of the edible portions could also be increased by substituting some Na fertilizers for K fertilizers [3,10,168].

### B.   Genetic Variation in Tissue Sodium

As noted previously, there are genotypic differences within species for Na uptake and partitioning to edible structures, namely in sugar beet, red beet, and celery. Some wild relatives of the cultivated tomato translocate substantially higher levels of Na to fruits (M. Shannon, personal communication). Some of the tomato lines derived from crosses of these wild relatives accumulate nearly 10 times higher Na levels in the fruits than the cultivated tomatos growing under identical conditions (Figure 5).

There are considerable differences among forage species in the contents of Na in their shoots, even when the supplies to their roots are similar [50,60]. Such differences could be due to selectivity during absorption [50] or to differences in transport to shoots [49]. Identification and testing of germplasm adapted to low-K conditions could be useful in understanding the physiological roles of K as well as in providing breeding materials for forage production on soils low in K and where K fertilizers are costly or unavailable [169]. The Na content of forage and pasture crops is also important in animal nutrition [4].

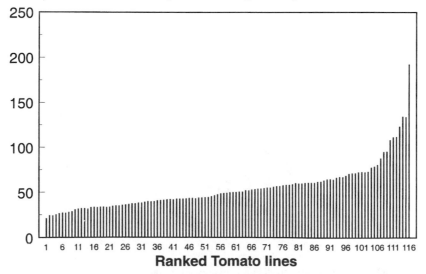

**Figure 5**   Sodium levels of tomato fruits derived from the cross between L. esculutum X L. cheesmani. (From M. Shannon, unpublished data from USDA Salinity Laboratory, Riverside.)

The Na requirement for lactating dairy cows is approximately 2 g kg$^{-1}$ dwt in forage [12], which is higher than the average Na content of natrophobic pasture species [50]. In contrast, the K content (20 to 25 g kg$^{-1}$ dwt) in natrophobic species is usually adequate or in excess of animal needs. An adequate Na content can increase the acceptability of forage to animals and enhances daily feed intake [170].

## C. Agricultural Production Systems—Secondary Salinization

Secondary salinization associated with irrigated agriculture is a serious problem, threatening the long-term sustainability of many production areas [171]. Because of the inherent limitations associated with most engineering approaches, large tracts of cropland are becoming saline despite the enormous resources committed to these projects. Nearly 2.12 million ha of irrigated cropland out of total of 14 million ha under irrigation in Pakistan became saline after only a few years of irrigation [172]. Similarly, 40% of the irrigated cropland in Iraq and Iran has been affected by secondary salinization [173], as has nearly a third of the irrigated cropland in India [174]. Secondary salinization is a constant threat to the agriculture-based economy of California, where irrigated agriculture plays a critical role in the production of fruits and vegetables [175]. According to estimates of the Food and Agriculture Organization (FAO) and the United Nations Educational, Scientific, and Cultural Organization (UNESCO), as much as half of all the existing irrigation systems of the world are under the influence of secondary salinization, alkalization, and waterlogging. This phenomenon is common not only in established irrigation systems but also in areas where irrigation has only recently began. Worldwide, nearly 10 million ha of irrigated lands are abandoned yearly because of secondary salinization resulting from irrigated agriculture [176].

The long-term survival of irrigated agricultural production systems depends on tackling salinity problems in an integrated manner. There needs to be a proper balance of nutrient management and irrigation coupled with the biological option of genetic improvement of the salinity tolerance of crops. Most crops discriminate against Na uptake, favoring K instead (see Chapter 44 by Subbarao and Johansen), which leads to buildup of Na salts in soils as irrigation continuously delivers some Na salts, although at low concentrations. Integrating crops that have the ability to take up significant amounts of Na into the cropping systems could help bring about a more balanced flow of Na through these irrigated systems. This could be particularly relevant to high-intensity irrigated systems, where vegetable crops are grown under intensive fertilizer and irrigation management. Use of crops such as red beet, celery, pac-choi, Swiss chard, and horseradish coupled with judicious limitation of K application to facilitate Na uptake could help to bring a positive shift in the salt balance of these regions. As noted earlier, a major portion of K could be replaced by Na without negative effects on crop growth rates. Our preliminary estimations indicate that red beet could remove as much as 900 kg of NaCl ha$^{-1}$ (about 60- to 80-day growing period), assuming that Na levels in the plant (both tops and roots) reach a moderately high level of 50 g kg$^{-1}$ dwt and a productivity of about 70 Mt fresh wt ha$^{-1}$ (of both tops and tubers) (I. Goldman, personal communication).

## D. Closed Life Support Systems for Space

Advanced life support systems (ALSS) being studied for space travel must ultimately strive for self-sufficiency in providing food, potable water, and a breathable atmosphere for humans. Such closed life support systems could form the basis for human colonies on the lunar or Martian surface [177]. Bioregenerative components of such systems would use plants to regenerate oxygen, food, and clean water through photosynthesis and transpiration. However, to minimize resupply costs, waste materials need to be completely recycled to provide nutrients for sustained plant production. For example, inedible plant biomass could be processed in bioreactors with the effluent nutrients used in the food production systems [178–181]. Human wastes (both solid and liquid) would also need to be processed as a source of nutrient and water inputs. If the requirements of plants and humans were the same, then nutrient cycling from one component to another would not be a problem. However, some elements such as Na are needed in relatively high levels for human metabolism but are absorbed and utilized by plants in only limited amounts. This discrepancy in Na metabolism between humans and plants could pose a threat to the system's long-term equilibrium if cycling is incomplete and external Na is used to meet the metabolic requirements of humans (Figure 6).

In a functional bioregenerative system, human urine would be one of the waste products recycled back to the plant production systems as a source of water and nutrients, especially N. Nearly 900 mmol

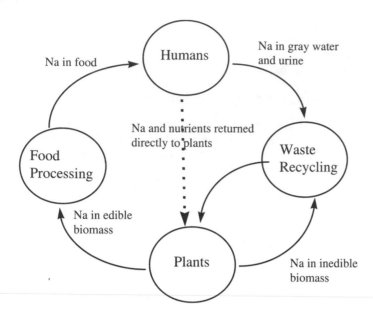

**Figure 6** Some considerations for Na flows in closed systems in space where crops might be used for life support.

of N per person per day is excreted in urine by humans who eat a typical Western diet [182]. Along with this, however, nearly 180 mmol of Na is excreted in urine [182]. Thus, Na inputs from human liquid wastes (which include urine, gray water, etc.) could amount to nearly 200 mmol per person per day. A nutrient solution volume of 17 L m$^{-2}$ for a hydroponic crop production system and 40 m$^2$ growing are per person are assumed (A. Drysdale, personal communication), Na inputs from waste recycling could lead to 0.28 mM increase per day [10,168]. If plants fail to remove this Na and recycle it to the crews through the food chain, Na concentrations could reach 100 mM in the nutrient solutions within 1 year from the urine input alone; this would kill most crops, resulting in a collapse of the food production system [10,168]. Thus, appropriate nutrient management approaches, coupled with crop selection strategies, are required to manage the cycling of Na between humans and plants in tightly closed systems. One solution to this would involve getting Na into the edible portions of the crops and returning Na to the humans through the food loop so that there would be no reason for an external supply of Na (Figure 6).

## IX. CONCLUDING REMARKS

In spite of many decades of research, the full role of Na in plant metabolism remains unresolved. The determination that Na is an essential nutrient under Arnon and Stout's [1] definition has been convincingly established only for certain $C_4$ plants but not for all higher plants. Numerous reports indicate a significant stimulation of the growth of selected crops in response to Na even under conditions of apparent K sufficiency. There is convincing evidence that Na can substitute for K in nonspecific functions such as an osmoticum during cell enlargement and a counterion in long-distance transport. The functional role of Na in plant nutrition is large, multifaceted, and sufficiently important in many biosystems and plant production that there needs to be a category of essentiality to designate such a role. The term functional nutrient seems to be appropriate, when defined as a nutrient that is essential for maximal yield and functional in metabolic functions without being unique for any one function.

## ACKNOWLEDGMENTS

We would like to acknowledge the extensive support and discussions we had with Dr. Mike Shannon (USDA Salinity Laboratory, Riverside, CA) during the course of preparing this chapter, which has helped in our interpretation of various issues related to sodium nutrition. Also, we would like to thank Dr.

William Knott, Biomedical Office, Kennedy Space Center for his support and encouragement in this project. This research was supported through the National Research Council and NASA Life Support Contract (NAS 10-20701).

# REFERENCES

1. DI Arnon, PR Stout. The essentiality of certain elements in minute quantity for plants with special reference to copper. Plant Physiol 14:371–375, 1939.
2. E Epstein. Mineral metabolism. In: J Bonner, JE Varner, eds. Plant Biochemistry. New York: Academic Press, 1965, pp 438–466.
3. GV Subbarao, RM Wheeler, GW Stutte, LH Levine. How far can sodium substitute for potassium in redbeet? J Plant Nutr 22:1745–1761, 1999.
4. H Marschner. Mineral Nutrition of Higher Plants. London: Academic Press, 1995.
5. VM Goldschmidt. Geochemistry. Oxford: Clarendon Press, 1954, p 730.
6. TJ Flowers, A Lauchli. Sodium versus potassium: substitution and compartmentation. In: A Pirson, MH Zimmermann, eds, Encyclopedia of Plant Physiology, New Series, Vol 15B, Berlin: Springer-Verlag, 1983, pp 651–681.
7. EP Glenn, JJ Brown, E Blumwald. Salt tolerance and crop potential of halophytes. Crit Rev Plant Sci 18:227–255, 1999.
8. H Greenway, R Munns. Mechanisms of salt tolerance in nonhalophytes. Annu Rev Plant Physiol Plant Mol Biol 31:149–190, 1980.
9. H Marschner. Why can sodium replace potassium in plants? Proceedings of 8th Colloquim of International Institute of Potash Institute, Berlin, 1971, pp 50–63.
10. GV Subbarao, RM Wheeler, GW Stutte. Feasibility of substituting sodium for potassium in crop plants for advanced life support systems. Life Support Biosphere Sci, in press.
11. TR Harrison. Harrison's Principles of Internal Medicine. 12th ed. New York: McGraw-Hill, 1991.
12. GS Smith, KR Middleton. Sodium and potassium content of top-dressed pastures in New Zealand in relation to plant and animal nutrition. N Z J Exp Agric 6:217–225, 1978.
13. PM Harmer, EJ Benne. Sodium as a crop nutrient. Soil Sci Soc Am J 60:137–148, 1945.
14. G Barbier, J Chabannes. Accumulation du sodium dans les racines des plantes. R Acad Sci 232:1372–1374, 1951.
15. PF Brownell, JG Wood. Sodium as an essential micronutrient element for *Atriplex vesicaria* Heward. Nature 179:365–366, 1957.
16. PF Brownell. Sodium as an essential micronutrient element for a higher plant (*Atriplex vesicaria*). Plant Physiol 40:460–468, 1965.
17. AM Alekseev, AA Abdurakhamanov. On the influence of ions of mineral nutrient salts on the state of the protoplasm of plant cells. Fiziol Rast 13:414–419, 1966.
18. PF Brownell, ME Jackman. Changes during recovery from sodium deficiency in *Atriplex*. Plant Physiol 41:617–622, 1966.
19. PF Brownell, CJ Crossland. The requirements of sodium as a micronutrient by species having the $C_4$ dicarboxylic photosynthetic pathway. Plant Physiol 49:794–797, 1972.
20. TJ Flowers, PF Troke, AR Yeo. The mechanism of salt tolerance in halophytes. Annu Rev Plant Physiol 28:89–121, 1977.
21. M Johnston, CP Grof, PF Brownell. The effect of sodium nutrition on the pool sizes of intermediates of the $C_4$ photosynthetic pathway. Aust J Plant Physiol 15:749–760, 1988.
22. J Ohnishi, U Flugge, HW Heldt, R Kanai. Involvement of $Na^+$ in active uptake of pyruvate in mesophyll chloroplasts of some $C_4$ plants. Plant Physiol 94:950–959, 1990.
23. WE Larson and WH Pierre. Interaction of sodium and potassium on yield and cation composition of selected crops. Soil Sci Soc Am J 76:51–64, 1953.
24. JJ Lehr. Sodium as a plant nutrient. J Sci Food Agric 4:460–471, 1953.
25. AH Montasir, HH Sharoubeem, GH Sidrak. Partial substitution of sodium for potassium in water cultures. Plant Soil 25:181–194, 1966.
26. E Troug, KC Berger, OJ Attoe. Response of nine economic plants to fertilization with sodium. Soil Sci Soc Am J 76:41–50, 1953.
27. JJ Lehr. The importance of sodium for plant nutrition. Soil Sci 63:479, 1947.
28. DJD Nicholas. Minor mineral nutrients. Annu Rev Plant Physiol 12:63–90, 1961.
29. E Epstein. The essential role of calcium in selective cation transport by plant cells. Plant Physiol 36:437–444, 1961.
30. E Epstein, DW Rains, OE Elzam. Resolution of dual mechanisms of potassium absorption by barley roots. Proc Natl Acad Sci U S A 49:684–692, 1963.
31. FJM Maathuis, D Sanders. Energization of potassium uptake in *Arabidopsis thaliana*. Planta 191:302–307, 1993.

32. FJM Maathuis, D Sanders. Mechanisms of potassium absorption by higher plant roots. Physiol Plant 96:158–168, 1996.
33. DW Rains, E Epstein. Transport of sodium in plant tissue. Science 148:1611, 1965.
34. DW Rains. Salt transport by plants in relation to salinity. Annu Rev Plant Physiol 23:367–388, 1972.
35. GP Findlay, SD Tyerman, A Garrill, M Skerrett. Pump and $K^+$ inward rectifiers in the plasmalemma of wheat root protoplasts. J Membr Biol 139:103–116, 1994.
36. W Gassmann, JI Schroeder. Inward-rectifying $K^+$ channel currents in root hairs of wheat: a mechanism for aluminum-sensitive low-affinity $K^+$ uptake and membrane potential control. Plant Physiol 105:1399–1408, 1994.
37. SA Vogelzang, HBA Prins. Patch clamp analysis of the dominant plasma membrane $K^+$ channel in root cell protoplasts of *Plantago media* L: its significance for the P and K state. J Membr Biol 141:113–122, 1994.
38. LH Wegner, K Raschke. Ion channels in the xylem parenchyma of barley roots. Plant Physiol 105:799–813, 1994.
39. FJM Maathuis, D Sanders. Contrasting roles in ion transport of two $K^+$ channel types in root cells of *Arabidopsis thaliana*. Planta 197:456–464, 1995.
40. SK Roberts, M Tester. Inward and outward $K^+$-selective currents in the plasma membrane of protoplasts from maize root cortex and stele. Plant J 8:811–825, 1995.
41. LH Wegner, A DeBoer. Two inward $K^+$ channels in the xylem parenchyma cells of barley roots are regulated by G-protein modulators through a membrane-delimited pathway. Planta 203:506–516, 1997.
42. FJM Maathuis, AM Ichida, D Sanders, JI Schroeder. Roles of higher plant $K^+$ channels. Plant Physiol 114:1141–1149, 1997.
43. D Schachtman, S tyerman, B Terry. The $K^+/Na^+$ selectivity of a cation channel in the plasma membrane of root cells do not differ in salt-tolerant and salt-sensitive wheat species. Plant Physiol 97:598–605, 1991.
44. B Hille. Ionic Channels of Excitable Membranes. Sunderland: Sinauer Associates, 1992.
45. V Tullin. Response of the sugar beet to common salt. Physiol Plant 7:810–834, 1954.
46. DW Rains E Epstein. Preferential absorption of potassium by leaf tissue of the mangrove, *Avicennia marina*: an aspect of halophylic competence in coping with salt. Aust J Biol Sci 20:847–857, 1967.
47. A Mozafar, JR Goddin, JJ Oertli. Na- and K-interactions in increasing the salt tolerance of *Atriplex halimus* L. I. Yield characteristic and osmotic potential. Agron J 62:478–481, 1970.
48. EP Glenn, RPfister, JJ Brown, TL Thompson, J O'Leary. Na and K accumulation and salt tolerance of *Atriplex canescens* (Chenopodiaceae) genotypes. Am J Bot 83:997–1005, 1996.
49. MGT Shone, DT Clarkson, J Sanderson. The absorption and translocation of sodium by maize seedlings. Planta 86:301–314, 1969.
50. GS Smith, KR Middleton, AS Edmonds. Sodium nutrition of pasture plants. I. Translocation of sodium and potassium in relation to transpiration rates. New Phytol 84:603–612, 1980.
51. DC Whitehead, LHP Jones. The effect of replacing potassium by sodium on cation uptake and transport to the shoots in four legumes and Italian ryegrass. Ann Appl Biol 71:81–89, 1972.
52. JM Cheeseman. Mechanisms of salinity tolerance in plants. Plant Physiol 87:547–550, 1988.
53. H Greenway, CB Osmond. Salt responses of enzymes from species differing in salt tolerance. Plant Physiol 49:256–259, 1972.
54. DJ Walker, RA Leigh, AJ Miller. Potassium homeostasis in vacuolate plant cells. Proc Natl Acad Sci U S A 93:10510–10514, 1996.
55. A Amtmann, D Sanders. Mechanisms of $Na^+$ uptake by plant cells. Adv Bot Res 29:75–112, 1999.
56. RG Wyn Jones. Cytoplasmic potassium homeostasis: review of the evidence and its implications. In: D Oosterhuis, G Berkowitz, eds. Frontiers in Potassium Nutrition: New Perspectives on the Effects of Potassium on Physiology of Plants. Saskatoon: Potash and Phosphate Institute of Canada, 1999, pp 13–22.
57. TJ Flowers, AR Yeo. Ion relations of salt tolerance. In: DA Baker, JL Hall, eds. Solute Transport in Plant Cells and Tissues. New York: Longman, 1988, pp 392–416.
58. JWS Reith, RHE Inkson, W Holmes, DS Maclusky, D Reid, RG Heddle, GJF Copeman. The effect of fertiliser on herbase production. II. The effect of nitrogen, phosphorus, and potassium on botanical and chemical composition. J Agric Sci (Camb) 63:209, 1964.
59. JJ Lehr. The sodium content of meadow grass in relation to species and fertilization. Proceedings of the 8th International Grassland Congress, 1960, pp 101–103.
60. G Griffith, RJK Walters. The sodium and potassium content of some grass genera, species and varieties. J Agric Sci (Camb) 67:81, 1966.
61. PE Weatherley. Ion movement within the plant. In: IH Rorison, ed. Ecological Aspects of the Mineral Nutrition of Plants. Oxford: Blackwell, 1969, pp 323–340.
62. MG Pitman. Transpiration and selective uptake of potassium by barley seedlings (*Hordeum vulgare* cv. Boliva). Aust J Biol Sci 18:987, 1965.
63. MG Pitman. Uptake of potassium and sodium by seedlings of *Sinapsis alba*. Aust J Biol Sci 19:257, 1966.
64. RT Besford. Effect of replacing nutrient potassium by sodium on uptake and distribution of sodium in tomato plants. Plant Soil 50:399–409, 1978.
65. M Johnston, CPL Grof, PF Brownell. Chlorophyll *a/b* ratios and photosystem activity of mesophyll and bundle sheath fractions from sodium-deficient $C_4$ plants. Aust J Plant Physiol 16:449–457, 1989.

66. CPL Grof, M Johnston, PF Brownell. Effect of sodium nutrition on the ultrastructure of chloroplasts of $C_4$ plants. Plant Physiol 89:539–543, 1989.
67. D Ohta, T Matoh, E Takahashi. Early responses of sodium-deficient Amaranthus tricolor L. plants to sodium application. Plant Physiol 84:112–117, 1987.
68. D Ohta, S Yasuoka, T Matoh, E Takahashi. Sodium stimulates growth of Amaranthus tricolor L. plants through enhanced nitrate assimilation. Plant Physiol 89:1102–1105, 1989.
69. TC Hsiao, A Lauchli. Role of potassium in plant-water relations. Adv Plant Nutr 2:281–312, 1986.
70. HJ Evans, GJ Sorger. Role of mineral elements with emphasis on the univalent cations. Annu Rev Plant Physiol 17:47–76, 1966.
71. RG Wyn Jones, A Pollard. Proteins, enzymes and inorganic ions. In: DL Laidman, RG Wyn Jones eds. Encyclopedia of Plant Physiology, New Series, Vol 15B, Berlin: Springer-Verlag, 1983, pp 528–562.
72. WS Pierce, N Higinbotham. Compartments and fluxes of $K^+$, $Na^+$ and $Cl^-$ in Avena coleoptile cells. Plant Physiol 46:660–673, 1970.
73. TS Gibson, J Spiers, CJ Brady. Salt-tolerance in plants. II. In vitro translation of mRNAs from salt-tolerant and salt-sensitive plants on wheat germ ribosomes. Responses to ions and compatible organic solutes. Plant Cell Environ 7:579–587, 1984.
74. RA Leigh, RG Wyn Jones. Cellular compartmentation in plant nutrition: the selective cytoplasm and the promiscuous vacuole. Adv Plant Nutr 2:249–279, 1986.
75. GJ Wagner. Compartmentation in plant cells: the role of the vacuole. In: LL Creasy, G Hrazdina, eds. Cellular and Subcellular Localization in Plant Metabolism. New York: Plenum, 1982, pp 1–45.
76. A Lauchli, R Pfluger. Potassium transport through plant cell membranes and metabolic role of potassium in plants. In: Potassium Research—Review and Trends, Berlin: International Potash Institute, 1979, pp 111–163.
77. RG Wyn Jones. CJ Brady J Speirs. Ionic and osmotic relations in plant cells. In: DC Laidman, RG Wyn Jones, eds. Recent Advances in the Biochemistry of Cereals. New York: Academic Press, 1979, pp 63–103.
78. WS Iljin. Zusammensetzung der Salze in der Pflanze auf verschiedenen Standorten. Kalkpflanzen. Beih Bot Zbl 50:95–137, 1932.
79. DMR Harvey, JL Hall, TJ Flowers, B Kent. Quantitative ion localization within Suaeda maritima leaf mesophyll cells. Planta 151:555–560, 1981.
80. RA Leigh, T ApRees, WA Fuller, J Banfield. The location of acid invertase activity and sucrose in vacuoles of storage root of beetroot (Beta vulgaris). Biochem J 178:539–547, 1979.
81. M Thom, A Maretzki, E Komor. Vacuoles from sugarcane suspension cultures. I. Isolation and partial characterization. Plant Physiol 69:1315–1319, 1982.
82. WJ Cram. Negative feedback regulation of transport in cells. The maintenance of turgor, volume and nutrient supply. In: U Luttge, MG Pitman, eds. Encyclopedia of Plant Physiology, New Edition, Vol 2A, Transport in Plant Cells, Berlin: Springer Verlag, 1976, pp 284–316.
83. F Marty, D Branton, RA Leigh. Plant vacuoles. In: NE Tolbert, ed. The Biochemistry of Plants: A Comprehensive Treatise, Vol 1. New York: Academic Press, 1980, pp 625–658.
84. RA Leigh, A Deri Tomos. An attempt to use isolated vacuoles to determine the distribution of sodium and potassium in cells of storage roots of redbeet (Beta vulgaris L.). Planta 159:469–475, 1983.
85. MG Pitman. The determination of the salt relations of the cytoplasmic phase in beet root tissue. Aust J Biol Sci 16:647–668, 1963.
86. RJ Poole. Effect of sodium on potassium fluxes at the cell membrane and vacuole membrane of red beet. Plant Physiol 47:731–734, 1971.
87. AES Macklon. An examination, by compartmental flux analysis, of the development of sodium and chloride absorption capacities in beet root disks. J Exp Bot 27:651–657, 1976.
88. MG Pitman, A Lauchli, R Stelzer. Ion distribution in roots of barley seedlings measured by electron probe X-ray microanalysis. Plant Physiol 68:673–679, 1981.
89. RA Leigh, RG Wyn Jones. A hypothesis relating critical potassium concentrations for growth to the distribution and functions of this ion in the plant cell. New Phytol 97:1–13, 1984.
90. WD Jeschke, O Wolf. External potassium supply is not required for root growth in saline conditions: experiments with Ricinus communis L. Grown in a reciprocal split-root system. J Exp Bot 39:1149–1167, 1988.
91. H Kawasaki, H Takada, S Kamisaka. Requirement of sodium chloride for the action of gibberellic acid in stimulating hypocotyl elongation of a halophyte Salicornia herbacea L. Plant Cell Physiol 19:1415–1425, 1978.
92. AR Yeo. Salt tolerance in the halophyte Suaeda maritima L. Dum.: intracellular compartmentation of ions. J Exp Bot 32:487–497, 1981.
93. WD Jeschke. $K^+$-$Na^+$ exchange at cellular membranes, intracellular compartmentation of cations, and salt tolerance. In: Salinity Tolerance in Plants: RC Staples, GH Toennissen, eds. Strategies for Crop Improvement. New York: Wiley, 1984, pp 37–66.
94. RG Wyn Jones, J Gorham. Osmoregulation. In: OL Lange, PS Nobel, CB Osmond, H Zeigler, eds. Encyclopedia of Plant Physiology, Vol 12c. Berlin: Springer-Verlag, 1983, pp 35–58.
95. TJ Flowers. Halophytes. In: DA Baker, JL Hall, eds. Ion Transport in Plant Cells and Tissues. Amsterdam: Elsevier, 1975.
96. R Collander. Selective absorption of cations by higher plants. Plant Physiol 16:691–720, 1941.

97. FA Smith, JA Raven. Intracellular pH and its regulation. Annu Rev Plant Physiol 30:289–311, 1979.

98. RB Lee, RG Ratcliffe. Phosphorus nutrition and the intracellular distribution of inorganic phosphate I pea root tips: a quantitative study using $^{31}$P-NMR. J Exp Bot 34:1222–1244, 1983.

99. BC Loughman, RG Ratcliffe. Nuclear magnetic resonance and the study of plants. In: PB Tinker, A Lauchli. Advances in Plant Nutrition, Vol 1. New York: Praeger, 1984, pp 241–283.

100. GD Humble, TC Hsiao. Specific requirement of potassium for light-activated opening of stomata in epidermal strips. Plant Physiol 44:230–234, 1969.

101. TC Hsiao. Stomatal ion transport. In: U Luttge, MG Pitman, eds. Encyclopedia of Plant Physiology, New Series, Vol 2B. Berlin: Springer-Verlag, 1976, pp 193–221.

102. AS Raghavendra, JM Rao, VSR Das. Replaceability of potassium by sodium for stomatal opening in epidermal strips of stomatal opening in epidermal strips of *Commelina benghalensis*. Z Pflanzenphysiol 80:36–42, 1976.

103. M Fujino. Role of adenosine triphosphate and adenosine triphosphatase in stomatal movement. Sci Bull Fac Educ Nagasaki Univ 18:1–47, 1967.

104. GD Humble, TC Hsiao. Light-dependent influx and efflux of potassium of guard cells during stomatal opening and closing. Plant Physiol 46:483–487, 1970.

105. DA Thomas. The regulation of stomatal aperture in tobacco leaf epidermal strips. I. The effect of ions. Aust J Biol Sci 23:961–979, 1970.

106. RA Fischer. Stomatal opening: role of potassium uptake by guard cells. Science 160:784–785, 1968.

107. RA Fischer. Role of potassium in stomatal opening in the leaf of *Vicia faba*. Plant Physiol 47:555–558, 1971.

108. N Terry, A Ulrich. Effects of potassium deficiency on the photosynthesis and respiration of leaves of sugar beet. Plant Physiol 51:783–786, 1973.

109. N Terry, A Ulrich. Effects of potassium deficiency on the photosynthesis and respiration of leaves of sugar beet under conditions of low sodium supply. Plant Physiol 51:1099–1101, 1973.

110. M Tester, MR Blatt. Direct measurement of $K^+$ channels in thylakoid membranes by incorporation of vesicles into planar lipid bilayers. Plant Physiol 91:249–252, 1989.

111. RA Dilley, LP Vernon. Ion and water transport processes related to light-dependent shrinkage of spinach chloroplasts. Arch Biochem Biophys 111:365–375, 1965.

112. B Rumburg, E Reinwald, H Schroder, U Sippel. Correlation between electron flow, proton translocation and phosphorylation in chloroplasts. Naturwissenschatten 55:77–79, 1968.

113. W Wu, J Peters, GA Berkowitz. Surface charge–mediated effects of $Mg^{2+}$ on $K^+$ flux across the chloroplast envelope are associated with regulation of stromal pH and photosynthesis. Plant Physiol 97:580–587, 1991.

114. G Mix, H Marschner. Mineralstoffverteilung zwischen chloroplasten und ubrigem blattgewebe. Z Pflanzenphysiol 73:307–312, 1974.

115. JV Moscolov, VA Aleksandrovskaya. Physiological role of sodium in plants. Chem Abstr 56:1784, 1962.

116. JS Knypl, KM Chylinska. Comparison of the stimulatory effect of potassium on growth, chlorophyll and protein synthesis in the lettuce cotyledons with the effects produced by other univalent ions. Biochem Physiol Pflanz 163:52–63, 1972.

117. H Marschner, JV Possingham. Effects of $K^+$ and $Na^+$ on growth of leaf discs of sugar beet and spinach. Z Pflanzenphysiol 75:6–16, 1975.

118. EA Kirkby, MJ Armstrong, JE Leggett. Potassium recirculation in tomato plants in relation to potassium supply. J Plant Nutr 3:955–966, 1981.

119. CH Suelter. Enzymes activated by monovalent cations. Science 168:289–295, 1970.

120. W Pfluger, W Wiedemann. Der Einflus monovalent kationen auf die nitratreduktion von *Spinacia oleracea* L. Z Pflanzephysiol 85:125–133, 1977.

121. DT Clarkson, JB Hanson. The mineral nutrition of higher plants. Annu Rev Plant Physiol 31:239–298, 1980.

122. JL Hall, TJ Flowers. The effect of salt on protein synthesis in the halophyte *Suaeda maritima*. Planta 110:361–368, 1973.

123. HJ Evans, RA Wildes. Potassium and its role in enzyme action. In: Potassium in Biochemistry and Physiology. Berlin: International Potash Institute, 1972.

124. TJ Flowers. Salt tolerance in *Suaeda maritima* (L) Dum. A comparison of mitochondria isolated from green tissues of *Suaeda* and *Pisum*. J Exp Bot 101:101–110, 1974.

125. RE Nitsos, HJ Evans. Effects of univalent cations on the activity of particular starch synthetase. Plant Physiol 44:1260–1266, 1969.

126. T Akatsuka, OE Nelson. Granule bound adenosine diphosphate glucose-starch glucosyl transferase of maize seeds. J Biol Chem 241:2280–2286, 1966.

127. VN Nigam, A Fridland. Studies on glycogen synthesis in pigeon liver homogenates. Biochem J 105:505–513, 1967.

128. J Preiss, E Greenberg. Biosynthesis of starch in *Chlorella pyrenoidosa*. I. Purification and properties of the adenosine diphosphoglucose:α-4-glucosyl transferase from *Chlorella*. Arch Biochem Biophys 118:702–708, 1967.

129. T Murata, T Akazawa. Enzymatic mechanism of starch synthesis in sweet potato root. I. Rquirement of potassium ions for starch synthetase. Arch Biochem Biophys 126:873–879, 1968.

130. J Willenbrink, S Doll, HP Getz, S Meyer. Zuckeraufnahme in isolierten vakuolen und protoplasten aus dem speichergewebe von beta-ruben. Ber Dtsch Bot Ges 97:27–39, 1984.

131. G Hanson, A Kylin. ATP-ase activities in homogenates from sugar-beet roots, relation to $Mg^{++}$ and $(Na^+ + K^+)$-stimulation. Z Pflanzenphysiol 60:270–275, 1969.

132. AL Green, CB Taylor. Kinetics of (Na+K) stimulated adenosine triphosphatase (ATPase) of rabbit kidney microsome. Biochem Biophys Res Commun 14:118–123, 1964.

133. A Kylin, G Hansson. Transport of sodium and potassium and properties of (sodium + potassium) activated adenosine triphosphatase: possible connection with salt tolerance in plants. Eighth Coll. Int. Potash Institute, Berlin, 1971, pp 64–68.

134. G Hanson. Patterns of ionic influences on sugar beet ATPases. PhD dissertation, University of Stockholm, Stockholm, 1975.

135. BH Shah, RT Wedding. Sodium ion influence on phosphorylations associated with oxidation of succinate by turnip root mitochondria. Science 160:304–306, 1968.

136. O Lamm, H Malmgren. Measurements of the degree of dispersion of a highly polymerized metaphosphate prepared according to Tammann. Z Anorg Allg Chem 245:103–120, 1940.

137. AJ Hiatt, HJ Evans. Influence of certain cations on activity of acetic thiokinase from spinach leaves. Plant Physiol 35:673–677, 1960.

138. V Rocha, IP Ting. Tissue distribution of microbody, mitochondrial, and soluble malate dehydrogenase isoenzymes. Plant Physiol 46:754–756, 1970.

139. Y Tanaka, T Ito, T Akazawa. Enzymatic mechanism of starch breakdown in germinating rice seed. III. α-Amylase isoenzymes. Plant Physiol 46:650–654, 1970.

140. MC Williams. Effect of sodium and potassium salts on growth and oxalate content of halogeton. Plant Physiol 35:500–509, 1960.

141. LO Hylton, A Ulrich, DR Cornelius. Potassium and sodium interrelations in growth and mineral content of Italian ryegrass. Agron J 59:311–314, 1967.

142. AM El-Sheikh, A Ulrich, TC Broyer. Sodium and rubidium as possible nutrients for sugar beet plants. Plant Physiol 42:1202–1208, 1967.

143. AP Draycott, MJ Durrant. Response by sugar beet to potassium and sodium fertilizers, particularly in relation to soils containing little exchangeable potassium. J Agric Sci (Camb) 87:105–112, 1976.

144. JJ Lehr, JM Wybenga. Exploratory pot experiments on sensitiveness of different crops to sodium. Plant Soil 3:251–261, 1955.

145. PBH Tinker. The effects of nitrogen, potassium and sodium fertilizers on sugar beet. J Agric Sci (Camb) 65: 207–212, 1965.

146. WD Jeschke. $K^+$-$Na^+$ exchange and selectivity in barley root cells: effect of $Na^+$ on the $Na^+$ fluxes. J Exp Bot 28:1289–1305, 1977.

147. WD Jeschke. $K^+$-$Na^+$ selectivity in roots, localisation of selective fluxes and their regulation. In: E Marre, O Ciferri, eds. Regulation of Cell Membrane Activities in Plants. Amsterdam: Elsevier, 1977, pp 63–78.

148. MA Nunes, MA Dias, M Correia, MM Oliveira. Further studies on growth and osmoregulation of sugarbeet leaves under low salinity conditions. J Exp Bot 35:322–331, 1984.

149. GFJ Milford, WF Cormack, MJ Durrant. Effects of sodium chloride on water status and growth of sugar beet. J Exp Bot 28:1380–1388, 1977.

150. NH Peck, JP VanBuren, GE MacDonald, M Hemmat, RF Becker. Table beet plant and canned root responses to Na, K, and Cl from soils and from applications of NaCl and KCl. J Am Soc Horiet Sci 112:188–194, 1987.

151. AP Draycott, JAP Marsh, PBH Tinker. Sodium and potassium relationships in sugar beet. J Agric Sci (Camb) 74:568–573, 1970.

152. G Judel, H Kuhn. Uber die wirkung einer natriumdungung zu zuckerruben bei guter versogung mit kalium in gefabversuchen. Zucker 28:68–71, 1975.

153. MJ Durrant, AP Draycott, GFJ Milford. Effect of sodium fertilizer on water status and yield of sugar beet. Ann Appl Biol 88:321–328, 1978.

154. AP Draycott, SM Bugg. Response by sugarbeet to various amounts and times of application of sodium chloride fertilizer in relation to soil type. J Agric Sci 98:579–592, 1982.

155. JT Wooley. Sodium and silicon as nutrients for the tomato plant. Plant Physiol 32:317–321, 1957.

156. A Pardossi, G Bagnoli, F Malorgio, CA Campiotti, F Tognoni. NaCl effects on celery (*Apium graveolens* L.) grown in NFT. Sci Hortic 81:229–242, 1999.

157. PM Harmer, EJ Benne, WM Laughlin, C Key. Factors affecting crop response to sodium applied as a common salt on Michigan muck soil. Soil Sci 76:1–17, 1953.

158. H Marschner, A Kylin, PJC Kuiper. Differences in salt tolerance of three sugar beet genotypes. Physiol Plant 51:234–238, 1981.

159. H Marschner, PJC Kuiper, A Kylin. Genotypic differences in the response of sugar beet plants to replacement of potassium by sodium. Physiol Plant 51:239–244, 1981.

160. PM Harmer. Muck soil management for sugar beet production. Mich Agric Exp Stn Cir Bull 187, 1943.

161.  DW Rush, E Epstein. Comparative studies on the sodium, potassium, and chloride relations of a wild halo-phytic and a domestic salt-sensitive tomato species. Plant Physiol 68:1308–1313, 1981.
162.  LA Richards. Diagnosis and Improvement of Saline and Alkali Soils. USDA Handbook No 60, 1954.
163.  JA Duke, AA Atchley. CRC Handbook of Proximate Analysis Tables of Higher Plants. Boca Raton, FL: CRC Press, 1986.
164.  DJ Greenwood, DA Stone. Prediction and measurement of the decline in the critical-K, the maximum-K and total cation plant concentrations during the growth of field vegetable crops. Ann Bot 82:871–881, 1998.
165.  FW Smith. The effect of sodium on potassium nutrition and ionic relations in Rhodes grass. Aust J Agric Res 25:407–414, 1974.
166.  GS Smith, DR Lauren, IS Cornforth, MP Agnew. Evaluation of putrescine as a biochemical indicator of potassium requirements of lucerne. New Phytol 91:419–428, 1982.
167.  BK Watt, AL Merrill. Composition of Foods. USDA Handbook No 8, 1975.
168.  GV Subbarao, C Mackowiak, RM Wheeler. Recycling of Na in adanced life support: strategies based on crop production systems. Life Support Biosphere Sci 6:153–160, 1999.
169.  SC Jarvis. Sodium absorption and distribution in forage grasses of different potassium status. Ann Bot 49:199–206, 1982.
170.  H Ziegler. Nature of transported substances. In: MH Zimmermann, JA Milburn, eds. Encyclopedia of Plant Physiology, New Series, Vol 1. Berlin: Springerg-Verlag, 1975, pp 59–100.
171.  F Ghassmi, A Jakeman, H Nix. Salinization of Land and Water Resources: Human Causes, Extent, Management and Case Studies. Sydney, Australia: University of South Wales Press, 1995.
172.  N Ahmad. A review of salinity-alkalinity status of irrigated soils of West Pakistan. Agrokem Talajtan 14(suppl):117–154, 1965.
173.  VA Kovda. Problems of combating salinization of irrigated soils. UNEP, 1980.
174.  IP Abrol. Salinity management: favorale water balance vital. In: Hindu Survey of Indian Agriculture, 1991, pp 49–51.
175.  E Epstein, JD Norlyn. Seawater based crop production: a feasibility study. Science 197:249–251, 1977.
176.  I Szabolics. Salt-Affected Soils. Boca Raton, FL: CRC Press, 1989.
177.  RW MacElroy, J Bredt. Controlled ecological life support systems. Life support systems in space travel: current concepts and future directions of CELSS. NASA Conf Publ 2378 XXV COSPAR Meeting, Graz, Austria, 1985.
178.  JL Garland, CL Mackowiak, JC Sager. Hydroponic crop production using recycled nutrients from inedible crop residues. SAE Tech Paper 932173, 1993.
179.  CL Mackowiak, JL Garland, JC Sager. Recycling crop residues for use in recirculating hydroponic crop production. Acta Hortic 440:19–24, 1996.
180.  CL Mackowiak, JL Garland, RG Strayer, BW Finger, RM Wheeler. Comparison of aerobically treated and untreated crop residue as a source of recycled nutrients in a recirculating hydroponic system. Adv Space Res 18:281–287, 1996.
181.  CL Mackowiak, RM Wheeler, GW Stutte, NC Yorio, JC Sager. Use of biologically reclaimed minerals for continuous hydroponic potato production in a CELSS. Adv Space Res 20:1815–1820, 1997.
182.  DF Putnam. Composition and concentrative properties of human urine. NASA CR-1802, 1972.

# 18
# Nitrogen Metabolism and Crop Productivity

**Fred E. Below**

*University of Illinois, Urbana, Illinois*

## I. INTRODUCTION

Among the mineral nutrient elements, nitrogen (N) most often limits the growth and yield of nonleguminous crop plants, which require relatively large quantities of N (from 1.5 to 5% of the plant dry weight) for incorporation into numerous organic compounds. These compounds include proteins, nucleic acids, chlorophyll, and growth regulators, all of which have crucial roles in plant growth and development. The N composition of plant tissues also has important nutritional consequences, since plants are a major source of proteins in the diet of humans and animals. Because N deficiency can seriously decrease yield and crop quality, elaborate steps are often taken to assure that adequate N levels are available to plants.

Although plants can absorb small amounts of N from the atmosphere through their foliage, by far the greater part of it is acquired from specific forms in the soil such as nitrate ($NO_3^-$) or ammonium ($NH_4^+$). Most soils, however, do not have sufficient N in available form to support desired production levels. Therefore, addition of N from fertilizer is typically needed to maximize crop yields; this requirement has resulted in the development of a large N fertilizer industry. Some estimates suggest that N fertilizer accounts for 80% of all fertilization costs and 30% of all energy costs associated with crop production [1].

Although it is well accepted that sufficient N is needed to obtain high yields, growers each year must determine how much fertilizer N to apply. This problem results from the complex cycle of N in the environment, which can allow loss from the rooting zone. It is further complicated by mechanistic inconveniences associated with fertilizer N application and by uncertainty related to weather conditions, especially water availability. Unused fertilizer N is economically wasteful and can become an environmental hazard if it is lost from the soil. Excessive use of fertilizer N has been implicated in the contamination of ground water by $NO_3$ [2–5], which represents a potential health hazard to humans and animals [6,7]. As public awareness focuses on environmental quality, there are increasing pressures on growers to improve N management.

Additional knowledge regarding N use by crop plants is clearly one way to help improve N fertilizer management. Although complex, factors such as N use that limit or enhance crop productivity do so by affecting specific physiological processes within the plant. A better understanding of how N governs crop growth and yield will add to information required to improve N management and will help to minimize the adverse environmental impact of N fertilizer use.

## II.  NITROGEN ACQUISITION BY CROP PLANTS

## A.  Nitrogen Availability

Under natural conditions, N enters the soil environment as the result of biological fixation and/or decomposition of animal or plant residues. Most (>90%) of the N in soils is contained in organic matter, which is relatively stable and not directly available to plants. Although a portion of the N in organic matter can be made available through mineralization by soil microorganisms, the amount released is variable depending on management practices and environmental conditions. In addition, the release is normally too slow to meet the needs of a growing crop, with only 2–3% of the N converted to available forms per year. As a result, addition of N from chemical fertilizers is usually required to optimize crop growth and yield.

Nitrogen is unique among the mineral nutrients in that it can be absorbed by plants in two distinct forms, as either the anion $NO_3^-$ or the cation $NH_4^+$. Although numerous N fertilizer formulations are available that contain varying proportions of $NO_3$-N to $NH_4$-N, ammoniacal fertilizers are used more extensively because they are lower in cost [8]. However, $NO_3$ is the predominant form of N absorbed by plants, regardless of the source of applied N [9,10]. This preference is due to two groups of chemoautotropic soil bacteria, which rapidly oxidize $NH_4$ to $NO_3$ (nitrification) in warm, well-aerated soils that are favorable to crop growth.

The form of N ($NH_4$ or $NO_3$) can affect the availability of N to the plant as a result of differences in mobility of each form in the soil solution. In soil, the positively charged $NH_4^+$ ion is bound to negatively charged soil particles and is relatively immobile. In contrast, the negatively charged $NO_3^-$ ion is repelled by soil particles, which aids in its movement to plant roots. Even though $NO_3$ is the N form most available to plants, however, it can be more readily lost from the rooting zone because it is susceptible to leaching and denitrification [11]. Both these economically and environmentally undesirable processes (i.e., leaching and denitrification) perpetuate a large amount of the uncertainty associated with N fertilizer management.

In the United States, N fertilizer recommendations are usually based on the past crop history of the field and expected yield goal and, to a lesser extent, on formulas calculated to estimate the soil's capacity for N mineralization [12,13]. Other factors (e.g., fertilizer cost and value of the crop) must also be considered [14]. While generally sound, problems with fertilizer recommendations can arise if the yield goal is unrealistic or if growers fail to assess accurately the capacity of the soil to supply the crop with N. As a result, two main types of test have been developed to measure soil N: tests to determine the soil's potential to mineralize N from organic matter and direct measurement of residual inorganic N.

Several techniques have been developed to measure mineralization of soil N, which are collectively known as N availability indices [15,16]. These methods estimate the potential for organic N to be mineralized and involve either incubations [16–18] or some type of chemical extraction [19,20]. Some studies have shown that these tests can provide reasonable estimates of potentially mineralizable N [21,22]. They have not been widely used for making N fertilizer recommendations, however, because of difficulty in conducting the measurements and lack of supporting data to help interpret the results.

The other approach to assessing the soil N supply involves measuring the level of inorganic N in the soil profile and then adjusting the fertilizer N recommendation to account for N that is already present [15,16,19]. One such test for maize, known as the "late spring nitrate test," takes some of the uncertainty associated with N cycling into account by not removing soil samples until after the crop has been established (plants are in the early vegetative stage), when the potential for N loss is lessened. Based on soil analysis and yield response to applied N, a soil $NO_3$-N concentration in excess of 20–25 mg kg$^{-1}$ (ppm) is considered adequate for maximum yield of maize, whereas lower values indicate the need for additional fertilizer N [23–26]. Although good at identifying situations in which no fertilizer N is required, the test does not work as well when the degree of responsiveness to fertilizer N application must be predicted or when a high percentage of the N is available as $NH_4$ [27]. In addition, this technique cannot be used if all the N is applied preplant or if the N is knifed in as anhydrous ammonium. As a result, tests based on plant characters have also been developed as a way of assessing the soil N supply.

An advantage of plant measurements is that they integrate the effects of soil N availability and plant N uptake, regardless of the N source or application method. Additionally, because they are based on the plant, rather than the soil, plant measurements are more likely to reflect the direct impact of N availability on growth and yield. Tissue testing of plants to compare N concentrations with critical levels is a well-established procedure to document crop N status. These tests involve measuring organic N (also called reduced N) in the leaves [28,29] or inorganic N ($NO_3$) in the stems [30,31] and can be used to determine deficiencies as well as excessive applications of N. This technology, however, has typically been used in diagnostic work rather than as a management tool because the measurements are usually made too late to permit corrective N applications.

Leaf chlorophyll measurements have been advocated as a means of taking advantage of the close association between chlorophyll and leaf N concentration to assess soil N availability and plant N status [32,33]. The development of a handheld leaf chlorophyll meter (SPAD-502, Minolta Camera Co.) allows rapid and nondestructive measurement of leaf greenness [34,35], and some evidence suggests that this technique can be used as a management tool for making fertilizer N recommendations [36,37]. Other work, however, has shown that widespread calibration of chlorophyll meters to determine crop N status may not be practical, given differences in leaf greenness among cultivars and/or effects on the readings of growth stage, N form, and management practices [32,38]. As a result, normalization procedures may be necessary to standardize chlorophyll meter readings across cultivars, locations, and growth stages by comparing readings from well-fertilized rows with those from the test area [32].

## B. Nitrogen Accumulation

### 1. Nitrogen Uptake

Plants acquire the vast bulk of their N from the soil via the root system. This process involves the movement of inorganic N ($NO_3$ and $NH_4$) across membranes, transport or storage within the plant, and ultimately assimilation into organic compounds. The uptake of both N forms is generally considered to require metabolic energy mediated by enzyme permeases located in or on the plasmalemma of external root cells. Absorption of both forms is affected by the ion's concentration in the external solution, with the uptake rate exhibiting diminishing returns in response to increasing internal concentrations. Absorption is also affected by external factors such as temperature and pH (see Section II.B.2).

The consequences to plant metabolism from the uptake of $NO_3$ and $NH_4$ are vastly different because of differences in the charge of $NO_3$ and $NH_4$. With $NH_4$ nutrition, plants absorb cations in excess of anions, resulting in a net efflux of $H^+$ from the root and an acidification of the external medium [9,10]. Conversely, with $NO_3$ nutrition, plants absorb an excess of anions, which causes the medium to become more alkaline [9,10]. Also because of these differences in charge, the mechanisms for uptake by plant roots differ for $NO_3$ and $NH_4$.

In evaluations of N uptake, plants that have depleted their N supply (both in solution and in storage) are typically used to observe all phases of uptake and the influence of N in inducing the uptake system. For $NO_3$-depleted plants, the pattern of $NO_3$ uptake generally exhibits a two-phase pattern, with an initial lag period followed by an exponential increase in uptake [39–41]. The initial lag in $NO_3$ uptake is in contrast to that observed with many other ions [40,41] and suggests the induction of a specific $NO_3$ transporter by $NO_3$. The accelerated phase of $NO_3$ uptake is also indicative of induction because it is dependent on a critical $NO_3$ concentration in the root, in a manner similar to enzyme induction by its substrate [41,42]. In addition, the accelerated phase is restricted by inhibitors of protein or RNA synthesis or by conditions that limit or inhibit respiration [40,43]. Collectively, these studies show that the $NO_3$ uptake system is dynamic and capable of adjusting to changes in the level of $NO_3$ in the root environment.

The uptake of $NO_3$ is an active process, which must overcome an unfavorable electrochemical gradient between the soil and the root. However, because of this gradient, $NO_3$ can also efflux (or leak) back out of the root. Efflux has been described as a passive diffusion process [44] or as a carrier-mediated process [45] but in either case dependent on the internal concentration of $NO_3$ in the root. As a result, the net accumulation of $NO_3$ is a function of the difference between influx and efflux. As might be expected, efflux is greatest when high concentrations of $NO_3$ have been accumulated by root tissues [46,47].

Unlike $NO_3$ uptake, the absorption of $NH_4$ does not exhibit a prolonged lag under N-depleted conditions [48], although uptake can also be characterized by two main phases [40]. The initial phase of $NH_4$ is insensitive to low temperatures or metabolic inhibitors, hence is thought to occur passively [40,49]. In contrast, the second phase of $NH_4$ uptake involves metabolic energy and is sensitive to low temperatures and inhibitors [49]. In some plant species, the active phase of $NH_4$ uptake is also multiphasic, exhibiting uptake and growth rates associated with deficiency, luxury consumption, and toxicity [40,50,51].

Although the process of $NH_4$ uptake is not completely understood, it is clear that passive and active uptake must occur by different mechanisms. For passive uptake, the positively charged $NH_4$ ion may be absorbed by a uniport following the electrochemical gradient across the plasmalemma [52,53]. Conversely, since membrane permeability to $NH_3$ is greater than that of $NH_4$, passive uptake could also occur by nonspecific diffusion of $NH_3$ gas [54,55]. In the soil solution, the distribution of $NH_4^+$ and $NH_3$ is a function of an equilibrium relationship driven by pH. At neutral (or lower) soil pH values, more than 99% of the total ammoniacal N is in the protonated ($NH_4$) form, which would result in limited absorption of gaseous $NH_3$ by the root [54]. While aboveground plant parts can also absorb gaseous $NH_3$ through stomata, the amounts acquired are limited in unpolluted air [56,57]. In addition, because high concentrations of $NH_3$ are toxic to plant growth, especially roots [58], it seems unlikely that passive $NH_3$ absorption is a major source of N for plant growth. Therefore, the bulk of ammoniacal N absorbed by plants is likely the result of active uptake of $NH_4$. The mechanism of active $NH_4$ uptake, which has not been clearly established, appears to be carrier regulated, as indicated by saturation kinetics and the depression of uptake by factors that limit energy metabolism [39,52].

## 2.  Factors Affecting Nitrogen Uptake

The uptake of $NO_3$ and $NH_4$ can be affected by internal factors, such as N and carbohydrate status, and by external factors, such as temperature, $O_2$ level, and rhizosphere pH. Plant species and stage of plant development can also influence N uptake. When the uptake of a specific N form is affected differentially by these factors, contrasting patterns of N uptake and growth can result, depending on the form of N available to the plant.

Although $NH_4$ uptake does not appear to be affected by the presence of $NO_3$ [59,60], there are many reports of $NH_4$-induced inhibition of $NO_3$ uptake [40,43,61–63]. However, there are also cases in which $NH_4$ appeared to have little or no effect on $NO_3$ uptake [64,65] or even resulted in a stimulation in uptake [66]. Although the precise manner by which $NH_4$ inhibits $NO_3$ uptake is not clear, possibilities include (1) a decrease in $NO_3$ reduction, resulting in feedback inhibition of $NO_3$ uptake, (2) an alteration in the rate of activation or synthesis of the $NO_3$ uptake system, thereby restricting influx, and/or (3) an acceleration in $NO_3$ efflux. For a description of $NO_3$ reduction and nitrate reductase, see Section III.B.3. Various lines of evidence support each of these mechanisms.

Although some researchers have shown a decrease in the level of extractable nitrate reductase by $NH_4$ treatment [67–70], others have shown that $NH_4$ or products of $NH_4$ assimilation do not interfere with nitrate reductase [71–73]. In addition, the ability of $NH_4$ to inhibit $NO_3$ uptake in plants without detectable nitrate reductase activity [74] and the lack of proportional changes in activity and $NO_3$ uptake in response to $NH_4$ [63,75,76] further indicate that a change in nitrate reductase is not the main mechanism responsible for $NH_4$-induced inhibition in $NO_3$ uptake.

Alternatively, $NH_4$ or one of its assimilation products may interact with $NO_3$ transporters at either the external or internal surfaces of the plasmalemma and inhibit the activation or synthesis of the $NO_3$ absorption system [43,53,62]. One possibility is that $NH_4$ or the high acidity adjacent to the plasmalemma resulting from $NH_4$ uptake in excess of $NO_3$ uptake causes an alteration in membrane permeability, thereby restricting the capacity for $NO_3$ absorption [77]. Another possibility is that $NH_4$ may inhibit net $NO_3$ uptake by increasing $NO_3$ efflux [78]; yet others suggest that $NO_3$ influx, not efflux, is inhibited by $NH_4$ [79–81]. Although additional research is needed to elucidate the exact mechanism(s) involved in $NH_4$-induced inhibition of $NO_3$ uptake, the identification of genotypic variation for the extent of this inhibition [82–84] indicates that the process is under genetic control.

Many studies have shown that $NO_3$ uptake is more sensitive to low temperatures than is the uptake of $NH_4$ [85–90]. For example, at temperatures below 9°C, perennial ryegrass plants absorbed more than 85% of their total N as $NH_4$, while the proportion decreased to only 60% absorption as $NH_4$ at temperatures of 17°C or above [86]. Although the reason for the preferential uptake of $NH_4$ over $NO_3$ at low temperatures is unclear, physical changes in the membrane may be responsible, rather than differences in temperature sensitivity of the two transport systems [85]. Alternatively, because temperature has a strong influence on the rate of nitrification, it is reasonable to assume that the largest amounts of $NH_4$ will occur in cool soils. Thus, the greater uptake of $NH_4$ at low soil temperatures may be partly the result of more $NH_4$-induced inhibition of $NO_3$ uptake.

Another important difference between $NO_3$ and $NH_4$ uptake lies in the sensitivity to pH of these two N forms. The maximal uptake of $NH_4$ occurs at neutral pH values, and uptake is depressed as the pH falls

[74,91,92]. The limitation in $NH_4$ uptake at low pH can lead to N stress and a decrease in growth when $NH_4$ is the only form of N supplied to the plant [92,93]. This problem is further exacerbated by the decrease in rhizosphere pH associated with the uptake of $NH_4$. Compared with $NH_4$ uptake, the opposite pH optimum occurs for $NO_3$, where more rapid uptake occurs at pH values of around 4–5 and uptake is depressed at higher pH values [9,74]. The reduction in $NO_3$ uptake at high pH may be related to a competitive effect of $OH^-$ ions on the $NO_3$ uptake system [74]. Similar to $NH_4$, the alkalinity generated from $NO_3$ uptake could further restrict $NO_3$ uptake. Thus, the consequences of absorbing $NO_3$ or $NH_4$ can have rather detrimental effects on the subsequent uptake as a result of differences in the optimum pH for uptake of the ion absorbed.

In addition to environmental and soil factors, the stage of plant development may influence the relative proportions of uptake between $NO_3$ and $NH_4$. Some evidence suggests that plants absorb $NH_4$ more rapidly than $NO_3$ during early vegetative growth, and the reverse situation occurs and more $NO_3$ is absorbed than $NH_4$ as growth progresses [9,94,95]. Possibly, young plants may lack a completely functional systems for $NO_3$ uptake and assimilation [96]. Alternatively, changes in the carbohydrate status of the root during plant development could alter the N form that is preferentially absorbed [97,98].

## 3. Nitrogen Assimilation

Regardless of the form absorbed, the inorganic N must be assimilated into organic forms, typically amino acids, to be of use to the plant. Because $NH_4$ is toxic to plant tissues at relatively low levels, it is rapidly assimilated in the roots and the N translocated as organic compounds. In contrast, $NO_3$ can be assimilated in the root, stored in the vacuoles of root cells, or transported to the shoot, where it can also be stored or assimilated. Nitrate storage and translocation play important roles in N metabolism inasmuch as $NO_3$ in the vacuole can be made available for assimilation when external sources of N are depleted. However, relatively little is known about factors that regulate the entry and exit of $NO_3$ in the vacuole.

Whereas $NH_4$ can be used directly for amino acid synthesis, $NO_3$ must first be reduced to $NH_4$. The reduction of $NO_3$ to $NH_4$ is an energy-requiring process occurring by two main partial reactions. The first step involves a two-electron reduction of $NO_3^-$ to $NO_2^-$ and is catalyzed by the enzyme nitrate reductase, while the second step involves a six-electron reduction of $NO_2^-$ to $NH_4^+$ catalyzed by nitrite reductase. Of these two enzymes, nitrate reductase is considered to be the rate-limiting step in the assimilation of $NO_3$ because it initiates the reaction and is the logical point of control when $NO_3$ is available. Nitrate reductase is also induced by its substrate $NO_3$; it has a short half-life, and its activity varies diurnally and with environmental factors that affect the flux of $NO_3$ to the sites of induction and assimilation [61,99,100].

The reduction of $NO_3$ by nitrate reductase can occur in either the root or the shoot, and in both cases, the energy is derived from the oxidation of carbohydrates [61]. The extent to which $NO_3$ is reduced in roots and shoots varies widely with plant species and environmental conditions [101,102]. Based on the contribution of total $NO_3$ reduction by the roots, plants can be classified into three main groups:

Species in which the root is the major site for reduction
Species exhibiting $NO_3$ reduction in both the root and the shoot
Species in which the shoot is the primary site for reduction

These three classifications are roughly typified by woody plants, perennial herbs, and fast-growing annuals, respectively [101,102]. Although many studies have indicated a cytosolic location for nitrate reductase [61,100,103], others have suggested that nitrate reductase is associated with chloroplasts, microbodies, or the plasmalemma [100,104–106].

Two main types of nitrate reductase, which differ in the electron donor, have been identified in higher plants [61,99,100]. One nitrate reductase uses NADH (reduced nicotinamide dinucleotide), while another nitrate reductase uses NADH or NADPH (reduced nicotinamide dinucleotide phosphate). Essentially, all higher plants contain the NADH-specific nitrate reductase, and it is the only form of nitrate reductase in some species [99]. In contrast, other plant species contain both an NADH-specific and an NADPH-bispecific nitrate reductase [107]. In some plant species, the NADH-specific nitrate reductase is found in both leaves and root and constitutes the majority of the total nitrate reductase activity, while the NADPH-bispecific form is found only in the roots [99].

Like $NO_3$ reduction, the reduction of $NO_2^-$ to $NH_4^+$ can occur in either the root or the shoot; the cellular location and the electron donor, however, vary depending on the site of reduction. In the shoot, $NO_2^-$ reduction occurs in the chloroplast and is coupled to the light reaction of photosynthesis by the use of reduced ferredoxin as the electron donor [61]. Nitrite reduction in the root occurs in a plastid, which is analogous to a chloroplast of the leaf, but the reaction differs from shoot reduction in the following respects: (1) the nitrite reductase of the root is similar but not identical to the leaf enzyme, (2) the electron donor is a ferredoxin-like protein that is not identical to the leaf protein, and (3) the root ferredoxin is reduced by NADPH and a corresponding enzyme, with the energy supplied from the oxidation of carbohydrates [108].

The $NH_4^+$ that results from both $NO_3$ assimilation and the $NH_4^+$ absorbed directly by the roots is assimilated by the glutamate synthase cycle, which involves two reactions operating in succession and catalyzed by the enzymes glutamine synthetase and glutamate synthase [109]. A characteristic of this pathway is the cyclic manner in which the amino acid glutamate acts as both acceptor and product of ammonia assimilation. In this cycle, $NH_4^+$ is incorporated into glutamine by glutamine synthetase, which attaches $NH_3$ to the carboxy group of glutamate, using energy supplied by ATP. In leaf cells, this reaction occurs in chloroplasts, and in roots it most likely occurs in plastids [110]. In the chloroplast, the light-trapping system provides the energy to regenerate ATP, while in root cells, other enzyme systems oxidize carbohydrates to provide the energy for ATP regeneration.

Another isoform of glutamine synthetase is found in the cytoplasm of both leaf and root cells and is not identical to the plastid enzyme [111,112]. The cytoplasmic enzyme can assimilate any free $NH_3$ or $NH_4^+$ regardless of its origin (from either deamination of amino acids or absorption from the soil). Thus, in addition to producing the key intermediate, glutamine, the glutamine synthetase reaction is a detoxification process that avoids injury from the accumulation of $NH_4^+$ or $NH_3$.

Following the formation of glutamine, the amino group ($—NH_2$) is transferred to $\alpha$-oxo glutarate via glutamate synthase to form two molecules of glutamate. This reaction can occur in shoots or roots, and in both cases the enzyme is located in plastids [109,111,112]. There are three isoforms of glutamate synthase in plant cells, which utilize different electron donors [108]. In leaf chloroplasts, the electron donor is reduced ferredoxin derived directly from the trapping of light energy. Conversely, the electron donor in root cells is NADH or NADPH, where the energy to reduce the oxidized form of the pyridine-linked nucleotides is derived from oxidation of carbohydrates [109,111,112].

A further series of related reactions mediated by specific transaminases transfers the amino group from glutamate to the 2-oxy group ($=O$) of a 2-oxoacid. Biochemical modification of glutamate, glutamine, and the array of amino acids produced by the transaminase reactions generates the 20 amino acids required for protein synthesis. These amino acids can also be anabolized into a variety of complex nitrogenous compounds (e.g., chlorophyll, growth regulators, alkaloids, nucleic acids) that are involved in plant growth and metabolism.

## C.  Timing of Nitrogen Accumulation

Like dry matter production, the seasonal accumulation of N by crop plants can be divided into three main phases:

> An initially slow accumulation due to limited crop biomass
> A period of rapid, nearly linear accumulation that coincides with the onset of rapid plant growth
> A cessation of N accumulation with advancing maturity

Examination on a daily rate basis generally reveals two periods of rapid N accumulation, corresponding to late vegetative growth and the onset of linear seed fill [113,114]. Although the maximum accumulation rate usually occurs during linear vegetative growth, it can be delayed by a delay in the availability of N [115]. The period of maximum N accumulation can also be affected by such other factors as planting date, irrigation, and climate [114,115].

Under most conditions, the majority of plant N accumulated by cereal plants is acquired during vegetative growth. Numerous reports in the literature for maize and wheat show cases of 75% or more of the total plant N accumulation having occured by anthesis [116–120]. However, there is some indication that continued accumulation of N during grain fill can be a beneficial trait, especially for high-yielding genotypes in good growing environments [121,122]. For example, the hybrid FS854, which holds the world record yield for maize, 23.2 Mg ha$^{-1}$ [123], has been shown to accumulate a substantial proportion of its N during grain fill [119,124,125]. The proportion of plant N accumulated after anthesis, however, is highly influenced by growing season [117], soil N level [126], and cultivar [120,126,127].

More extensive early-season N uptake may also be the result of a larger supply of N in the soil, because available soil N often exhibits a marked decline coincident with rapid vegetative growth, with the lowest N levels occurring around anthesis [128]. However, whereas N applications made during the early stages of reproductive development can increase protein percentage [129,130], there is often little or no response in terms of grain yield [131,132]. Similarly, foliar N sprays have the least impact on increasing plant N levels when applied around anthesis because the additional N interferes with the metabolism of indigenous N [124]. Collectively, these findings suggest that there is a level of genetic control over N accumulation and distribution that is independent of the availability of N. Complicating the understanding of how N availability and genetics determine plant growth, however, is the inability to control stringently the supply of N in the soil.

In the few cases of the use of hydroponic culture to deprive maize plants of N at anthesis, yield either has been unaffected [133] or has decreased only modestly [134,135]. Similarly, grain yield in soils is generally affected more by the N supply before anthesis than after [136,137]. This evidence, and the lack of ability to increase yields of most maize cultivars with postanthesis foliar N sprays [138,139], suggests that N has its main impact on yield before anthesis. However, redistribution of previously accumulated N from vegetative to reproductive plant parts could minimize the need for postanthesis N uptake. For both maize and wheat, the grain typically contains about 70% of the total N in the plant at maturity, with more than half of it coming from remobilization from other plant parts [119,120,140,141]. Thus, because of extensive changes in the distribution of N among plant parts, it is difficult to separate the effects of the timing of N accumulation from the contribution of N to grain development and yield.

## III. PHYSIOLOGICAL ROLES FOR NITROGEN IN CROP PRODUCTIVITY

### A. Importance of Nitrogen to Plant Growth

Crop growth and productivity involve the integrated effect of a large number of components and metabolic processes that act, with variable intensity, throughout the life cycle of the crop. The interdependence of N and C metabolism creates additional problems in describing an independent role for N in achieving maximum crop productivity. Nevertheless, four major roles for N have been proposed for attaining high yields of rice [142] and maize [122], and these roles appear to be valid for many crops:

    Establishment of photosynthetic capacity
    Maintenance of photosynthetic capacity
    Establishment of sink capacity (the number and potential size of seeds)
    Maintenance of functional sinks throughout seed development

Each of these roles is discussed briefly with reference to the potential impact on crop productivity.

The objective in establishing photosynthetic capacity is to ensure that the supply of N does not limit development of the photosynthetic apparatus (enzymes, pigments, and other compounds needed for photosynthesis). Within limits, and if no other restrictive factors are present, an increase in N supply increases the growth, the composition of N and chlorophyll, and the photosynthetic capacity of leaves [143–145]. Nitrogen supply has also been shown to regulate the synthesis of photosynthetic carboxylating enzymes by affecting transcription and/or the stability of messenger RNA [146,147]. Collectively, these effects result in greater light interception, higher canopy photosynthesis, and higher yield. However, because little N is accumulated by the leaf after it has reached full expansion [148], a sufficient supply of N must be available throughout the development of each leaf if the individual leaves are to attain their full genetic potential for photosynthetic capacity.

To achieve high yields, plants must not only establish photosynthetic capacity but also continue photosynthesis throughout the grain-filling period. Thus, once established, sufficient N must be available to maintain the photosynthetic apparatus. This role is particularly important because dry matter accumulation in cereal grains is dependent on current photosynthesis [119,149,150]. Most of the N in the leaf is associated with proteins in the chloroplast—60% in $C_4$ plants and up to 75% in $C_3$ plants [122,151,152]—and these proteins are subject to breakdown and remobilization of the resultant amino acids [153,154].

Thus, as leaves age and senesce, their capacity for photosynthesis declines, with a correspondingly negative effect on assimilate supply and yield.

Many studies have shown a concurrent loss of photosynthetic activity and organic N from the leaves, especially during seed development [155–158]. An example of this relationship for maize leaves is shown in Figure 1: the losses in leaf N and photosynthesis were initiated at or near pollination and declined nearly linearly during the grain-filling period. Although it is clear that the loss of N from the leaf impairs photosynthetic activity, management practices that increase the N supply (such as supplementary side dressing or foliar sprays of N) do not automatically increase leaf N status and photosynthetic activity [159–162]. The absence of these effects is probably attributable to several key photosynthetic enzymes (the large subunit of RuBPCase) that are encoded for and synthesized by the chloroplast [163]. After full leaf expansion, the chloroplast loses much of its ability to synthesize these proteins, regardless of the availability of N [148,163]. This phenomenon indicates that the application of supplementary N to maintain photosynthetic activity may be of limited value until a technique is found that will reactivate protein synthesis in the chloroplast.

Another important role for N in assuring high productivity of crop plants is establishment of reproductive sink capacity. Sink capacity of a cereal plant is a function of the number and the potential size of grains. Grain number is dependent on the number of ears per unit area, the number of florets per ear, and the proportion of florets that develop into grain [149,164,165], and the potential size of individual grains depends on the number of endosperm cells and starch granules [166–169]. In either case, reproductive initials, like all growing tissues, are characterized by high concentrations of N and high metabolic activities. This need could indicate that sufficient amounts of both C and N assimilates are required for full expression of the genetic potential for initiation and early development of grains.

For cereal crops, grain number is usually more closely related to yield than other yield components [149,164,165]. Consequently, many studies have shown that N-induced yield increases are the result of more grains per plant [170–173]. For wheat, this enhancement is related to an increase in tiller production and survival [174,175] and to a lesser extent to a decrease in floret abortion [176]. In contrast, for maize, N supply affects kernel number primarily by decreasing kernel abortion [172,177]. An example of the effect of N supply on kernel number and kernel abortion of maize is shown in Figure 2: kernel number increases as the N supply is increased from a deficient to a sufficient level, which is associated with a decrease in kernel abortion. Other studies, however, have indicated that N supply can also affect individual grain weights [178,179], perhaps by means of a change in endosperm cell number [180].

Although the number of ears and grains is usually the yield component most affected by N supply, increases in kernel weight can also affect yield [149,164]. Because vegetative development in cereal crops is negligible after flowering, the N subsequently acquired, or remobilized from the vegetation, is used exclusively for grain development. This need for N is demonstrated by the fact that adequately fertilized cereal crops typically contain from 9 to 13% protein in the grain. Indeed, some

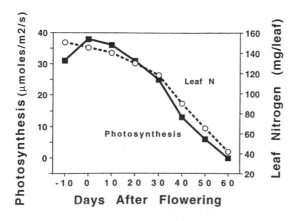

**Figure 1**  Changes in photosynthesis and N content of a selected leaf (the first leaf above the ear) of maize during the grain filling period. Values presented are for adequately fertilized plants (200 kg N ha$^{-1}$) averaged over two hybrids at the University of Illinois research farm in 1985 and 1986.

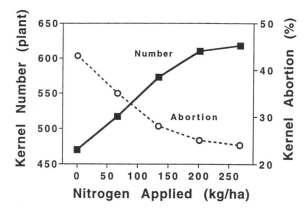

**Figure 2**   The effect of N fertilizer rate on kernel number and kernel abortion of maize. Values are averaged over two hybrids at the University of Illinois research farm in 1990.

workers have suggested that the deposition and/or accumulation of storage protein by the kernel is a factor regulating grain development [181–183]. This suggestion is based on the positive correlation between storage protein, kernel weight, and grain yield [182] and on genetic studies showing reduced levels of storage protein (zein) and starch in zein-deficient mutants of maize [184]. An alternative explanation, however, is that the availability of N within the plant and to the grain is positively associated with kernel development, and as such the amount of storage protein deposited is only an accurate reflection of the N supply [122].

Other needs for N by developing kernels could include embryo growth and the initial and continued synthesis of enzymes needed for energy generation and the deposition of storage products in the kernel. Embryo development could affect the kernel's hormonal balance because a large portion of kernel phytohormones are produced by the embryo [185,186]. Because several of the key classes of phytohormones either contain N (auxins, cytokinins, polyamines) or are synthesized from amino acids (auxins, ethylene, polyamines), an adequate supply of N may be needed for their production. With regard to storage product formation, provision of N to developing maize kernels has been shown to increase their capacity to synthesize proteins and to utilize sugars for the biosynthesis of starch [187]. Nitrogen supply also exerts a marked effect on endosperm enzymology and on the deposition of storage proteins in the endosperm [187,188]. Thus, it appears that at least a portion of the yield increase produced by N fertilization results from a modification of kernel metabolism in response to N supply.

## B.   Interactions of Carbon and Nitrogen

Grain yield of crops is primarily a function of the plant's ability to acquire, metabolize, and utilize C and N assimilates and its genetic potential for maximum grain production. For cereal crops, the relative abundance of C versus N in the plant (approximately 44% C vs. 1.5% N) dictates a predominant role for photosynthesis in achieving maximum yields. However, as discussed in Sec. III.A, the metabolism of N plays a major role in the production of C assimilates and in their utilization for reproductive development. In addition, as evidenced by the use of reduced ferredoxin in $NO_2^-$ reduction and $NH_4^+$ assimilation (see Sec. II.B.3), C and N interact at numerous points in plant metabolism [189]. This interdependence in C and N metabolism creates problems when one is attempting to describe an independent role for either C or N in achieving maximum productivity.

Grain composition offers a prime example of the complexities involved in understanding how C and N interact to affect productivity. A negative relationship between grain yield and protein percentage is widely noted in cereals, especially in cultivars selected for abnormally high or low percentages of grain protein [190,191]. The higher metabolic cost associated with the synthesis of protein than with carbohydrate has been proposed to explain this relationship [192,193]. However, evidence showing that carbohydrate supply does not normally limit kernel development [194–196] and progress toward identification and breeding of high-protein, high-yielding cereals [197,198] make this explanation seem unlikely. In addition, source-sink alteration experiments have indicated that C- and N-storage processes in cereal kernels seem to operate autonomously [199–201].

Further complicating attempts to understand the relation between yield and protein concentration is the tendency for individual grain weight to vary with grain number [173]. A negative relationship between grain number and grain weight is often observed [202], which may be due in part to the relative supplies of C and N from the vegetative plant [203]. There is some indication that the composition of assimilates (C and N) channeled to the kernel by the mother plant controls the amounts of starch and protein accumulated in maize grain [204,205]. In other work, however, it is shown that the genotype of the kernel primarily dictates the range of grain composition, with external factors modulating the phenotype within this range either to a large or small degree [206]. Alternatively, zygotic factors and the source supply may interact to control grain composition [207]. Compensation phenomena also complicate our understanding of how C and N relationships control grain composition because plants can make up for a lack of current assimilate (both C and N) with enhanced remobilization from the vegetation [172].

On a whole-plant basis, the supply of N often appears more limiting for grain development than does the supply of carbohydrate. For maize, the capacity of the plant to supply N to the ear was more limiting than the capacity to provide photosynthate, inasmuch as the net remobilization of vegetative N occurred earlier and was much more extensive than the remobilization of vegetative carbohydrate [119,140,208]. Although shading plants during grain fill decreased yield and enhanced the remobilization of both dry matter and N, the availability of newly reduced N was still more limiting to grain fill than current photosynthate [209]. Similarly, although supplemental illumination to the lower two-thirds of the canopy increased carbohydrate status and yield, these effects could not be separated from an enhancement in the total accumulation, and the tissue concentration, of N [210]. Collectively, these data suggest that the availability of N to and within the plant is more variable than the availability of photosynthate and at least as limiting to grain development.

## IV.  CROP RESPONSE TO APPLIED NITROGEN

## A.  Growth and Yield Response

Increases in crop productivity due to fertilizer N additions may be realized as dry matter yield, protein yield, or an improvement in quality factors. For cereal crops, grain yield and protein quality exhibit a typical pattern in response to N supply that can be divided into three main components [211]:

1.  Grain yield and protein content (total amount present) and concentration (protein per unit weight) increase in unison with increasing N supply.
2.  Grain yield reaches a plateau, but protein continues to increase with additional increments of N.
3.  Grain protein content peaks, grain yield begins to decline, and protein concentration continues to rise with further increases in N supply.

Responses to applied N are affected by many environmental [212,213], cultural [214,215], and soil factors [14], with the result that the response curves can vary considerably at different locations. For example, in a fertile soil with a high residual N supply, applications of N may have no effect or may even decrease crop yields. Alternatively, if some factor other than N, such as soil moisture or another nutrient, is limiting, then applications of N fertilizer will not increase growth and yield even if the supply of soil N is low. The optimal economic N rate also depends on the soil type and the ratio of fertilizer N costs to the value of the crop. In general, the fertilizer N rate required for maximum yield and the economic optimal N rate are lower for soils with higher organic matter such as silt loams than for sandy soils [14].

Despite the variation associated with crop response to fertilizer N, an example of a general pattern for maize is presented in Table 1. In the absence of other limiting factors, addition of fertilizer N will increase maize yields in a curvilinear fashion. As with other growth inputs, response to N fertilizer decreases as more and more fertilizer is added. As a result, plants are always the most efficient at utilizing fertilizer N when it is available at low levels (see Sec. IV.D for a discussion of N use efficiency). In this example, 202 kg of N per hectare increased yield by 4.8 Mg, compared with no N application, with nearly half (2.1 Mg) of this increase coming from the first 67-kg increment of N (Table 1). Although the second and third increments also increased yield, the size of these increases diminished successively (1.7 and 0.9

**TABLE 1** General Effect of N Fertilizer Rate on Grain Yield and N Recovery of Maize Grown on Highly Fertile Silt Loam Soil in Illinois[a]

| Nitrogen rate (kg ha$^{-1}$) | Grain yield (Mg ha$^{-1}$) | N in crop (kg ha$^{-1}$) | N removed with grain (kg ha$^{-1}$) |
|---|---|---|---|
| 0 | 8.4 | 124 | 83 |
| 67 | 10.5 | 170 | 119 |
| 134 | 12.2 | 212 | 150 |
| 202 | 13.1 | 241 | 170 |
| 267 | 13.2 | 262 | 180 |

[a] Values are averaged over two hybrids grown at the University of Illinois research farm in 1990.

Mg, respectively). In contrast, the fourth 67-kg increment increased yield by only 0.1 Mg, which would not be considered economically (or environmentally) sound. Therefore, although it is obvious that fertilizer N was needed to maximize yield, the optimal rate required is much less clear.

In addition to demonstrating the need for fertilizer N, the foregoing data show that a reasonable yield (8.4 Mg ha$^{-1}$) was produced without adding any N to the soil (Table 1). The crop accumulated 124 kg ha$^{-1}$ from the soil, of which 83 kg was removed with the grain. For heavily fertilized plants (267 kg ha$^{-1}$ N), these values increased to 262 kg of plant N and 180 kg of grain N. When the level of fertilizer N limited grain yield (at N rates of 134 kg ha$^{-1}$ or less), more N was removed with the grain than was provided by the fertilizer. Conversely, N levels in excess of those needed for maximum yields (202 kg ha$^{-1}$ and up) resulted in the removal of less N than had been applied. This situation greatly increases the potential for accumulation of residual N (usually as $NO_3$) in the soil. Based on this example, and other published reports [216,217], it is suggested that the N level that just maximizes grain yield also results in the best balance between fertilizer N added and the amount removed with the grain.

Similarly to grain yield, the largest increase in total plant N accumulation, and the greatest plant recovery, occurred with the first increment of N. For example, the first 67 kg of fertilizer N increased plant N accumulation by 46 kg, representing a plant recovery of 68% (calculated from data of Table 1). Of this 46 kg of plant N, 36 kg was removed with the grain for an N removal recovery of just over 50%. In contrast, plant N accumulation was increased by just 21 kg (32% recovery) for the fourth N increment, of which only 10 kg (15% recovery) was removed with the grain. These data demonstrate the inherent inefficiency with which fertilizer N is recovered by the maize plant, emphasizing the potential for environmental damage at excessive rates.

## B. Genotypic Variation

Different cultivars grown at the same location can exhibit different response patterns to N fertilization, and such variation has been observed for wheat [218,219], rice [220], sorghum [221], and maize [126,127]. However, as might be expected, this variation is highly affected by the environment and growing conditions and is most apparent under controlled conditions (e.g., in hydroponics) [135,222]. Interest in identifying genetic differences in responsiveness to N fertilizer is intensifying, as producers and agricultural consultants see genotypic variation as one way to fine-tune N fertilizer management. There is also a desire to develop or identify genotypes that perform well under a low N supply or, conversely, to find genotypes that will respond to high fertility conditions.

From a botanical standpoint, plants can vary in their use of N in two major ways: in how much N the plant uses to produce maximum yield or in when (i.e., at what stage during the growing season) the plant acquires its N. An example of this type of variation is depicted for maize in Figure 3. In this example, the low N response type produced its maximum yield at 120 kg N ha$^{-1}$ compared with an N requirement of 200 kg ha$^{-1}$ for the high N type (Figure 3, left). Although high N types are usually capable of producing the highest yields, the low N types may outyield the high N types at low levels of soil N. Cultivars can also differ in their timing of N acquisition: some accumulate the majority of their N before flowering, whereas others may have a substantial requirement for N accumulation after flowering (Figure 3, right).

A cultivar's N acquisition pattern can affect N accumulation and productivity because plants acquiring most of their N by flowering should be less subject to fluctuations in the N supply during grain fill. These types may be more consistent from year to year because adverse growing conditions usually occur

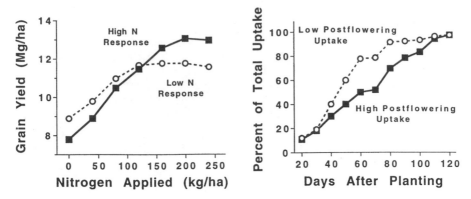

**Figure 3** Representation of the two major ways in which maize cultivars can vary in their use of N. Cultivars can differ in the amount of fertilizer N required for maximum yield (high or low N response types; left) or in their timing of N acquisition (high or low postflowering uptake types; right).

after flowering. However, these types might also be more susceptible to N deficiency early in the growing season and less likely to recover from such setbacks. Further complicating the ability to characterize a cultivar's response to fertilizer N is the fact that these two strategies can interact to determine the final N use. In other words, the amount of N required for maximum yield may or may not be related to when the N is accumulated by the plant.

Because of its economic value and high requirement for fertilizer N, much of the recent effort in identifying genotypic variation for N use has been directed toward maize. Although differences in N use among maize genotypes have been reported for inbreds [223] and open-pollinated populations [121], there is considerable controversy regarding whether these differences can be used to improve N fertilizer management of hybrids. Whereas some studies have reported large differences among maize hybrids in their response to fertilizer N [127,224], others have shown no or limited differences [117,225]. Similarly, a separate large-scale study did not observe hybrid × N rate interactions, although it was noted that hybrids in an individual location responded differently to the level of soil-applied N [226]. In an attempt to reconcile these differences, hybrids were divided into breeding groups based on their inbred parents, and differences in the N rate required for maximum yield were observed between, but not within, the groups [126]. Collectively, these studies suggest that maize hybrids do respond differently to the level of fertilizer-applied N but that the growing environment and the hybrid's genetic makeup can markedly influence the response.

## C.  Form of Nitrogen

As mentioned earlier, N can be utilized by plants as either $NO_3$ or $NH_4$, although under production conditions the greater amount is thought to be absorbed as $NO_3$. Thus, enhancing the supply of soil N as $NH_4$ is one way to improve N fertilizer management (i.e., by minimizing the potential for N losses). Increasing the supply of $NH_4$ in soils could also enhance plant performance: a survey of the literature shows numerous examples of improved vegetative growth and N accumulation when growing cereal plants are provided with mixtures of $NO_3$ and $NH_4$ compared with $NO_3$ alone [64,84,227–229]. Similar results have been reported for tomato [230], soybean [231], and sunflower [232]. These data imply that seedlings of many plant species cannot acquire sufficient N for maximum vegetative growth when N is supplied solely as $NO_3$.

Although much of the earlier work involved vegetative growth, later studies show that cereal crops supplied with both $NO_3$ and $NH_4$ (mixed-N nutrition) also produce higher yields than those supplied with only $NO_3$ [233–238]. For a variety of maize hybrids grown in field hydroponics, an equal mixture of $NO_3$ and $NH_4$ increased the yield by 11–14% [217,235,239] compared with plants grown with only $NO_3$. Even greater mixed-N–induced yield increases (average of 21–43%) have been reported for hydroponically or pot-grown spring wheat [236,238,240,241].

Although these responses were obtained with hydroponics or pot culture, where a finer degree of control over the $NO_3/NH_4$ ratio is possible, there is evidence that enhancing the supply of $NH_4$ to cereal crops is beneficial under production conditions [242–247]. Several reports [243,246,247] have shown moderate yield increases of 6–11% when maize plants were grown under field conditions designed to provide mixed-N nutrition. However, not all environments [243] or hybrids [247] are responsive to mixed N, indicating that factors other than the availability of $NH_4$ can affect the N use and productivity of maize. Other work shows that cultivars differ in their physiological strategy for achieving mixed N–induced yield increases and in the magnitude of response [235,238,247].

In most cases, mixed N–induced yield increases are the result of more grains per plant [235,237,238,248], although increases in individual grain weight have also been reported [217,233]. For wheat, the additional kernels are primarily achieved by increasing the number of grain-bearing tillers [238,240,248] and to a lesser extent the number of grains per tiller [248]. Alteration of the N form at anthesis showed that mixed N supplied continuously, or during vegetative growth only, increased yield and tillering over all $NO_3$ plants, but mixed N during reproductive growth only did not [240]. Thus, for wheat, it appears that mixed N–induced increases in yield potential occur during the early stages of plant development, when tillers are being formed.

In contrast to wheat, the main effect of enhanced $NH_4$ on maize is an increase in the number of grains per plant through more kernels per ear [235,239], although there is also a tendency for increased prolificacy [247,249]. Additional kernels per ear result primarily from a decrease in kernel abortion [246,250] and sometimes an increase in ovules per ear [247]. These findings suggest a direct physiological effect of N form on kernel development, inasmuch as all studies presumably supplied a more than adequate level of N (either as $NO_3$ or mixed N) to the plant. These results also suggest that mixed N–induced yield increases are associated with events that occur during ovule initiation and pollination rather than with processes occurring during the grain-filling period.

Support for a pre-grain-fill effect of mixed N on productivity has been obtained by transfer experiments in which N was supplied either as all $NO_3$ or as an equal mixture of $NO_3$ and $NH_4$ until anthesis, whereupon half the plants in each group were switched to the other N form [233,239]. In both sets of experiments, yield was increased over all $NO_3$ plants when mixed N was available continuously or only before anthesis but not when it was available only after anthesis. Similarly, Reddy et al. [134] reported that the N form supplied before, but not after, anthesis affected growth and nutrient status of maize. These studies and other data [251] suggest that late vegetative and early reproductive development are the most crucial times to supply mixed N to the plant.

Although the physiological basis for improved productivity with mixed N is not understood, additional plant N accumulation has been implicated. As in reports for vegetative growth [64,228], cereal plants grown to maturity with mixed N typically contain more plant N (both content and concentration) than those grown with $NO_3$ alone. These results have been observed for plants grown hydroponically [233,235,238–240] and in soils [241,244,247,252]. Like the results of earlier work with seedlings, these data suggest that when N is supplied primarily as $NO_3$, cereal crops are unable to acquire sufficient N for maximal productivity.

Although it is unclear exactly how this additional N (from mixed nutrition) enhances productivity, it is well known that N supply and plant N status affect tillering in wheat [174,175], and kernel abortion and prolificacy in maize [172,209,253,254]. Alternatively, a certain level of $NH_4$ may exert a direct effect on reproductive development, with a corresponding change in plant metabolism. For example, $NH_4$ nutrition has been reported to stimulate sucrose uptake by maize kernels, which in turn increased the production and translocation of assimilates from the leaves [182]. Similarly, $NO_3$-N was uniformly assimilated throughout the plant, whereas $NH_4$-N was assimilated in the root and preferentially exported as organic N to meristematic regions like the ear [255,256]. Indicative of an enhanced supply of N to the ear is an increase in grain protein concentration under mixed N, compared with plants grown on all, or predominantly, $NO_3$ [134,239,247,257].

In addition to enhanced N accumulation, mixed N–induced increases in reproductive development and yield may be related to energy status. Because assimilation of $NH_4$ requires a third as many ATP equivalents as does $NO_3$ [258], plants acquiring a large percentage of their N as $NH_4$ may expend less total energy, especially if $NO_3$ is assimilated in the root [259]. Although the physiological impact of this energy saving is unclear, it seems possible that any effect would be largest for crops such as maize, which require high levels of N. However, based on cost estimates for $NO_3$ assimilation [173], Alexander et al. [233] concluded that mixed N–induced increases could not be explained solely on the basis of energetics.

Alternatively, partitioning effects may play a role in altered growth: $NH_4$ must be assimilated immediately by the root, resulting in greater amounts of nitrogenous compounds in the roots and an altered partitioning of carbon between the root and the shoot [260]. Enhanced movement of sugars out of the leaves has been shown to relieve feedback inhibition of photosynthesis due to carbohydrate accumulation [261], which has been theorized to result in higher photosynthesis under mixed-N conditions [182]. However, whereas some studies have reported higher photosynthetic rates for $NH_4$-grown plants [262,263], others have shown greater photosynthesis for $NO_3$-grown plants [264,265]. In addition, field studies have shown equivalent (or greater) rates, and a similar duration, of canopy photosynthesis, but lower grain yields for maize plants supplied with predominantly $NO_3$ than with mixed N [247]. These findings, and the observation that mixed-N nutrition alters dry matter partitioning between shoots and roots [228,246] and between vegetative and reproductive fractions [235,239,246,247], suggest that altered partitioning may be more important than photosynthesis in the enhanced productivity observed with mixed-N nutrition.

Other studies have shown that additional physiological processes are beneficially altered by mixed-N nutrition. For example, increasing the proportion of N used by the plant as $NH_4$ usually results in an increase in anion uptake, especially for P [258,266,267]. Because of its acidifying effect on the rhizosphere [268], enhanced uptake of $NH_4$ may also make trace elements like iron and zinc more available [170,269]. In addition, mixed-N nutrition has been shown to increase root branching [84,270] and the supply of cytokinins to the shoot [246] compared with $NO_3$-grown plants. It is also possible that by utilizing both N forms, plant cells are able to control their intracellular pH more tightly [271,272].

The experiments discussed in this section show that mixed-N nutrition can increase crop productivity as the result of alterations in several important physiological processes such as reproductive development, N acquisition, dry matter production, and assimilate partitioning. Indirect effects on other mineral nutrients and on endogenous phytohormone balance may also be important. For maximum yield enhancement, mixed N needs to be available during the period when reproductive potential is determined and set. Thus, although additional work is needed to further elucidate the physiological basis for mixed N–induced increases in crop growth and yield, the prospect of using mixed N to improve fertilizer use efficiency is encouraging.

## Ɔ. Nitrogen Use Efficiency

The efficient use of N is an important goal in maximizing yield in ways that have a minimal impact on the environment. Various methods have been used to define and characterize nitrogen use efficiency (NUE), so care must be taken to specify the method or definition that is used [13]. These methods can reflect agronomic, economic, or environmental perspectives, and they can be characterized on an incremental basis, on a cumulative basis, or as a yield efficiency index [13].

From an agronomic perspective, NUE refers to three main functions detailing the relationships between:

N availability and yield
N availability and N recovered
Yield and N recovered

To calculate these values requires measurements of grain yield, the total nitrogen in the plant, and the total available soil N. However, because the soil N availability and the total N recovered by the plant are difficult to determine in field experiments, the N content in the aboveground plant parts and the N rate supplied as fertilizer are typically used. In all cases, the most accurate estimates subtract the yield or plant N accumulated in unfertilized plots from the values obtained in fertilized plots.

The relationship between yield and N rate is most often referred to as "yield efficiency" or "agronomic efficiency" and is defined as the yield increase per unit of applied N for a specific portion of the yield response curve. The yield efficiency is a function of the efficiencies of N recovery and N utilization, which are known as "recovery efficiency" and "physiological efficiency," respectively. The recovery efficiency represents the N accumulated by the plant per unit of applied N, while the physiological efficiency is the grain produced per unit of N accumulated by the plant. Physiological efficiency integrates the effect of plant factors on N use and yield, while recovery efficiency is a measure of how much fertilizer N is absorbed by the plant. The yield efficiency for N use can be improved by increasing the recovery efficiency, or the physiological efficiency, or both.

From a soil standpoint, the overall NUE depends on the interaction of factors responsible for N loss (leaching, denitrification, volatilization, and immobilization) with such N management variables as N rate, N source, N placement, and timing of N application. In conjunction with soil factors, NUE from a plant standpoint depends on the processes associated with the absorption, translocation, assimilation, and redistribution of N. The NUE is greatest at low levels of N and is highly influenced by soil type, which determines the mineralization and N loss characteristics [13,173,214]. The NUE can also be influenced by plant characters such as tissue N concentration and the size and number of reproductive sinks [173,253].

Another measure of NUE uses data on yield and plant N content, without correcting for dry matter or N accumulation by unfertilized plants [273]. This procedure was developed to assess genotypic variation in response to N supply, where evaluation of a large number of genotypes by traditional methods is constrained by the size of the necessary field experiments. The procedure denotes dry weight and N values as a series of ratios, all expressed in the same unit—often grams per plant [273]. As with the traditional measures of NUE, this method defines NUE as grain production per unit of fertilizer N; there are two main components: the efficiency of N absorption (uptake efficiency), and the efficiency with which the N absorbed is utilized to produce grain (utilization efficiency). The uptake efficiency is denoted as the N in the plant divided by the fertilizer N applied, while the utilization efficiency is the grain produced divided by the N in the plant. Thus, the overall NUE can be expressed as a product function of uptake and utilization efficiencies.

Further subdivision of uptake and utilization efficiencies can be made to reflect more specific aspects of plant N use (e.g., translocation, remobilization, distribution, timing of N acquisition) [273]. In addition, converting the appropriate dry matter and N ratios to logarithms provides a means of partitioning variation in NUE into the proportion attributable to each of its components. Such data have shown that genotypic differences in NUE of eight maize hybrids were primarily the result of differences in N utilization efficiency when the crops were grown with a low supply of N and differences in uptake efficiency at a high N supply [273]. The data also showed that either high or low values of NUE could be attained by different combinations of uptake and utilization efficiency. Similar cases of such variation have been noted for wheat [219]. The NUE and its components have also been shown to vary as a function of N fertilizer rate and the timing of N availability [126,173,274]. Collectively, these data emphasize that each of the plant traits involved in the acquisition and utilization of N is subject to genetic diversity, which may contribute to N use and crop productivity in different degrees under different environmental conditions.

# V. ENVIRONMENTAL ASPECTS OF NITROGEN FERTILIZER USE

In addition to being removed with the crop, N can be lost from agricultural ecosystems in large amounts as the result of several processes. These include leaching, denitrification, volatilization, surface runoff, and soil erosion. Nitrogen can also be temporarily removed from the available soil pool because of adsorption, fixation, and microbial immobilization. The economic implications of these losses are self-evident, especially when they are large enough to limit crop productivity. These losses can also have environmental consequences with regard to water and air quality.

Losses of N are highly affected by which ionic N form ($NO_3^-$ or $NH_4^+$) predominates in the soil [275,276]. Both forms are lost by soil erosion, but only $NH_4$ is lost directly to the atmosphere from volatilization. Ammonium can also be temporarily removed from the plant-available N pool by cation exchange with soil particles, fixation by clay lattices of the soil, fixation by organic matter, immobilization into microbial biomass, and conversion to $NO_3$. Conversely, $NO_3$-N is not readily used by soil microbes, nor does it bind to soil particles or organic matter. It is, however, subject to losses from leaching and denitrification.

The least controllable of these N losses, which are determined by soil type and rainfall, are the leaching and denitrification of $NO_3$. Leaching is a physical process that occurs because $NO_3^-$ is repelled by negatively charged soil colloids and readily moves with soil water. However, if too much downward movement of water occurs, $NO_3$ can be leached below the plant's rooting zone, ultimately to accumulate in ground water [2]. In denitrification, a separate microbial process that occurs under waterlogged or anaerobic conditions, $NO_3$ is converted to gaseous compounds, which are lost to the atmosphere. Both leaching and denitrification are economically and environmentally undesirable and add a large degree of uncertainty to N fertilizer management.

The problems of leaching and denitrification have stimulated the identification and development of nitrification inhibitors that block microbial conversion of $NH_4$ to $NO_3$ [277,278]. Research has shown that the use of these inhibitors can reduce losses of fertilizer N, especially under soil and weather conditions that favor N loss [276,279,280]. For several reasons, however, a consistent yield increase from the use of nitrification inhibitors is not always observed [276,280]. These reasons include:

A lack of opportunity for the nitrification inhibitor to express its potential for reducing N loss
Inadequate duration of the inhibitory effect
Inadequate experimental sensitivity to permit statistical detection of small benefits that may occur
Adverse effects on other soil microorganisms
Genetic differences among cultivars to N level or to $NH_4$ nutrition

Although nitrification inhibitors were originally developed to minimize N losses, they have also been proposed as a means of altering the predominant form of N in the soil [245]. The use of ammoniacal fertilizers along with nitrification inhibitors may alter plant nutrition by supplying a greater proportion of the N to the plant as $NH_4$. Enhancing the supply and utilization of $NH_4$-N may also be beneficial to plant growth, as several crop species have been shown to absorb more N and to grow more rapidly when supplied with mixtures of $NO_3$ and $NH_4$ (see Sec. IV.C).

Urease inhibitors represent another approach to preventing fertilizer N loss [281,282]. When applied to the soil surface, urea [$(NH_2)_2CO$] is rapidly cleaved to $NH_4^+$ and $CO_2$ by the action of urease enzymes present in the soil and plant residue. This conversion gives rise both to high $NH_4$ levels and to elevated soil pH, two properties that are conducive to volatilization of N as $NH_3$. Urease inhibitors temporarily decrease the activity of urease enzymes, maintaining urea-applied N as urea for several days. Because the uncharged urea molecule is quite mobile in soil, rainfall can move surface-applied urea into the soil profile, where it can hydrolyze with less opportunity for N losses via volatilization. As with nitrification inhibitors, use of a urease inhibitor will generally be effective only when the crop can respond to the N conserved by the inhibitor and when environmental conditions are conducive to large losses of surface-applied urea (such as warm soils with abundant plant residue). Conversely, urease inhibitors are of limited value when urea-based fertilizers can be easily and inexpensively incorporated into the soil during or immediately following their application [281,282]. They also require sufficient rainfall within a few days to facilitate urea movement into the soil.

In summary, although the use of fertilizer N has the potential for negative environmental consequences, several cultural practices can be used to minimize this possibility. These practices include:

Use of N rates appropriate for the historical productivity of the land and the yield of the crop being grown
Timing of N applications to better fit plant N needs
Specific placement of N-containing fertilizers
Use of appropriate N sources
Use of nitrification inhibitors to slow the breakdown of $NH_4$ to $NO_3$
Use of urease inhibitors to minimize volatilization of surface-applied urea
Taking into account the soil's capacity to supply the crop with N
Adequate fertilization with other mineral nutrients to maximize the plant's use of N.

## VI.  CONCLUSIONS

Because of the high requirement of crop plants for elemental N and its numerous important roles in growth and development, N is the mineral nutrient element that most often limits crop productivity. Because N mineralization from the soil is normally too low to support desired production levels, soil N levels are typically increased through fertilization. However, the complex cycle of N in the environment causes uncertainty in N fertilizer management, increasing the chances for economic loss and environmental damage. Nitrogen use and productivity of crop plants is also complex, resulting from an interaction of biochemical, physiological, and morphological processes in the plant.

Nitrogen is unique among the mineral nutrients in that it can be absorbed by plants in two distinct forms, as either the anion $NO_3^-$ or the cation $NH_4^+$. The form of N absorbed has a pronounced effect on the mechanisms for uptake, transport, assimilation, and storage and, in some cases, on the physiology and morphology of crop productivity. The use of a specific N form also can be affected differentially by environmental or culture factors, such as plant N status, temperature, and pH. While N is usually applied as $NH_4$-fertilizer, the nitrification process renders $NO_3$-N the soil form most available to the crop. In addition, $NO_3$-N is the N form most susceptible to losses from the crop's rooting zone. Several relatively new techniques have been developed in an attempt to better assess the soil N supply; however, their usefulness is still being evaluated. Plant-based estimates of soil N supply are also receiving attention.

Although the maximum rate of N accumulation usually occurs during vegetative growth, the timing of N acquisition can be altered by cultural and environmental factors. Extensive redistribution of N among plant parts further confuses our understanding of when N has the greatest impact on crop productivity. Nevertheless, the major roles for N in crop productivity can be divided into four general areas:

Establishment of photosynthetic capacity
Maintenance of photosynthetic capacity
Establishment of sink capacity
Maintenance of sink capacity

Although the relative abundance of C and N in the plant dictates a predominant role for photosynthesis in the productivity of cereal crops, some evidence suggests that the availability of N to and within the plant is more variable than the availability of photosynthate and at least as limiting to grain development.

Improved crop productivity from N fertilization can result from increases in dry matter yield and/or improvements in quality factors. In either case, increases with N supply follow the law of diminishing returns; thus N is used most efficiently when available at low levels. Cultivars grown at the same location may exhibit different responses to N supply that result from differences in how much N they need for maximum yield or when in the life cycle they mainly acquire their N. Supplying N as mixtures of $NO_3^-$ and $NH_4^+$ can also increase productivity as the result of alterations in important plant processes (e.g., reproductive development, N acquisition, dry matter production, assimilate partitioning). The efficient use of N is an important goal in strategies to maximize yield potential while minimizing negative effects of fertilizer N on the environment. Several methods have been used to assess N use efficiency, and its components, in crop plants.

The use of nitrogen by crop plants is dictated by a complex interaction of plant metabolism with cultural and environmental factors that alter the availability of N. Each of the plant processes involved in the acquisition and utilization of N is under genetic control, and each may contribute to varying degrees depending on the environmental conditions. A better understanding of these processes will undoubtedly help in developing strategies to improve the management of fertilizer nitrogen.

## ACKNOWLEDGMENT

The author expresses his sincere gratitude to P. S. Brandau for providing critical comments on the text and for help with the graphics.

## REFERENCES

1. PJ Stangel. In: RD Hauck, ed. Nitrogen in Crop Production. Madison, WI: ASA, CSSA, SSSA, 1984, p 23.
2. SR Aldrich. In: RD Hauck, ed. Nitrogen in Crop Production (R. D. Hauck, ed.), Madison, WI: ASA, CSSA, SSSA, 1984, p 663.
3. P Newbould. Plant Soil 115:297, 1989.
4. JS Schepers, KD Frank, C Bourg. J Fertil Issues 3:133, 1986.
5. RF Spalding, ME Exner, CW Lindau, DW Eaton. Hydrology 58:307, 1982.
6. CA Black. Reducing American exposure to nitrate, nitrite, and nitroso compounds. Comments from CAST, Ames, IA, 1989.
7. HI Shuval, N Gruener. Am J Public Health 62:1045, 1972.
8. DA Russel. In: RD Hauck, ed. Nitrogen in Crop Production. Madison, WI: ASA, CSSA, SSA, 1984, p 183.
9. RH Hageman. In: RD Hauck, ed. Nitrogen in Crop Production. Madison, WI: ASA, CSSA, SSA, 1984, p 67.

10. RJ Haynes, KM Goh. Biol Rev 53:465, 1978.
11. FE Allison. Adv Agron 18:219, 1966.
12. SA Barber. In: FL Patterson, ed. Agronomic Research for Food. Madison, WI: ASA, 1976, p 13.
13. BR Bock. In: RD Hauck, ed. Nitrogen in Crop Production. Madison, WI: ASA, CSSA, SSA, 1984, p 273.
14. SL Oberle, DR Keeney. J Prod Agric 3:522, 1990.
15. DR Keeney. In: AL Page, ed. Methods of Soil Analysis. Madison, WI: ASA, 1982, p 711.
16. G Stanford. Nitrogen in Agricultural Soils. Madison, WI: ASA, 1982, p 651.
17. KL Sahrawat. Adv Agron 36:415, 1983.
18. G Stanford, JN Carter, SJ Smith. Soil Sci Soc Am Proc 38:99, 1974.
19. RH Fox, WP Piekielek. Soil Sci Soc Am Proc 42:751, 1978.
20. K Németh, IQ Makhdum, K Koch, H Beringer. Plant Soil 53:445, 1979.
21. R Saint-Fort, KD Frank, JS Schepers. Commun Soil Sci Plant Anal 21:1945, 1990.
22. A Van Diest. Plant Soil 64:115, 1985.
23. GD Binford, AM Blackmer, ME Cerrato. Agron J 84:53, 1992.
24. AM Blackmer, D Pottker, ME Cerrato, J Webb. J Prod Agric 2:103, 1989.
25. RH Fox, GW Roth, KV Iverson, WP Piekielek. Agron J 81:971, 1989.
26. FR Magdoff, WE Jokela, RH Fox, GF Griffin. Commun Soil Sci Plant Anal 21:1103, 1990.
27. HM Brown, RG Hoeft, ED Nafziger. Evaluation of soil $NO_3$-N profile for prediction of N fertilizer requirement. Proceedings of Illinois Fertilizer Conference, 1992, Urbana, IL, pp 21–30.
28. SW Melsted, HL Motto, TR Peck. Agron J 61:17, 1969.
29. RE Voss, JJ Hanway, LC Dumenil. Agron J 62:726, 1970.
30. GD Binford, AM Blackmer, NM El-Hout. Agron J 82:124, 1990.
31. TC Knowles, TA Doerge, MJ Ottman. Agron J 83:353, 1991.
32. JS Schepers, DD Francis, M Vigil, FE Below. Commun Soil Sci Plant Anal 23:2173, 1992.
33. M Takebe, T Yoneyama, K Inada, T Murakami. Plant Soil 122:295, 1990.
34. LM Dwyer, M Tollenaar, L Houwing. Can J Plant Sci 71:505, 1991.
35. UL Yadava. HortScience 21:1449, 1986.
36. M Takebe, T Yoneyma. Plant Soil 122:295, 1990.
37. CW Wood, DW Reeves, RR Duffield, KL Edmisten. J Plant Nutr 15:487, 1992.
38. WP Piekielek, RH Fox. Agron J 84:59, 1992.
39. SS Goyal, RC Huffaker. Plant Physiol 82:1051, 1986.
40. RJ Haynes. Mineral Nitrogen in the Plant-Soil System. Orlando, FL: Academic Press, 1986, p 303.
41. WA Jackson. In: DR Neilsen, JG MacDonald, eds. Nitrogen in the Environment. New York: Academic Press, 1978, p 45.
42. WA Jackson, RJ Volk, TC Tucker. Agron J 64:518, 1972.
43. DT Clarkson. Fundamental, Ecological, and Agricultural Aspects of Nitrogen Metabolism in Higher Plants. Dordrecht: Martinus Nijhoff, 1986, p 3.
44. MA Morgan, RJ Volk, WA Jackson. Plant Physiol 51:267, 1973.
45. CE Deane-Drummond, ADM Glass. Plant Physiol 73:100, 1983.
46. WA Jackson, KD Kwik, RJ Volk, RG Butz. Planta 132:149, 1976.
47. RH Teyker, WA Jackson, RJ Volk, RH Moll. Plant Physiol 86:778, 1988.
48. MA Morgan, WA Jackson. Physiol Plant 73:38, 1988.
49. H Sasakawa, Y Yamamoto. Plant Physiol 62:665, 1978.
50. RA Joseph, T van Hai, J Lambert. Physiol Plant 34:321, 1975.
51. P Nissen, NK Fageria, AJ Rayar, MM Hassan, T van Hai. Physiol Plant 49:222, 1980.
52. WA Ullrich, M Larsson, CM Larsson, S Lesch, A Novacky. Physiol Plant 61:369, 1984.
53. WA Ullrich, In: WR Ullrich, PJ Aparicio, PJ Syrtee, F Castillo, eds. Inorganic Nitrogen Metabolism. Berlin: Springer-Verlag, 1987, p 32.
54. D Kleiner. Biochim Biophys Acta 639:41, 1981.
55. HW Scherer, CT MacKown, JE Leggett. J Exp Bot 35:1060, 1984.
56. VN Faller. J Plant Nutr Soil Sci 131:120, 1972.
57. GD Farquhar, PM Firth, R Wetselaar, B Weir. Plant Physiol 66:710, 1980.
58. AC Bennett. In: EW Carson, ed. The Plant Root and Its Environment. Charlottesville: University of Virgina, 1974, p 669.
59. K Mengel, M Viro. Soil Sci Plant Nutr 24:407, 1978.
60. LJ Youngdahl, R Pacheco, JJ Street, GLP Viek. Plant Soil 69:225, 1982.
61. L Beevers, RH Hageman. In: A Läuchi, RL Bieleski, eds. Inorganic Plant Nutrition. Berlin: Springer-Verlag, 1983, p 351.
62. WA Jackson, WL Pan, RH Moll, EJ Kamprath. In: CA Neyra, ed. Biochemical Basis of Plant Breeding, Vol II. Boca Raton, FL: CRC Press, 1986, p 73.
63. CT MacKown, WA Jackson, RJ Volk. Plant Physiol 69:353, 1982.
64. LE Schrader, D Domska, PE Jung, LA Peterson. Agron J 64:690, 1972.
65. DD Warncke, SA Barber. Agron J 65:950, 1973.

66. LM Bernardo, RB Clarck, JW Maranville. J Plant Nutr 7:1401, 1984.
67. JC Lycklama. Acta Bot Neerl 12:361, 1963.
68. CT MacKown, RJ Volk, WA Jackson. Plant Sci Lett 24:295, 1982.
69. TO Orebamjo, GR Stewart. Planta 122:37, 1975.
70. JW Radin. Plant Physiol 55:178, 1975.
71. PL Minotti, DC Williams, WA Jackson. Planta 86:267, 1969.
72. A Oaks, M Aslam, IL Boesel. Plant Physiol 59:391, 1977.
73. HS Srivastava. Phytochemistry 31:2941, 1992.
74. KP Rao, DW Rains. Plant Physiol 57:55, 1976.
75. NLV Datta, M Rao, S Guha-Mukherjee, SK Sorory. Plant Sci Lett 20:305, 1981.
76. P Mehta, HS Srivastava. Phytochemistry 19:2527, 1980.
77. PL Minotti, DC Williams, WA Jackson. Crop Sci 9:9, 1969.
78. CE Deane-Drummond, ADM Glass. Plant Physiol 73:105, 1983.
79. P Oscarson, B Ingemarson, M Ugglas, CM Larson. Planta 170:550, 1987.
80. RB Lee, DT Clarkson. J Exp Bot 37:1753, 1986.
81. RB Lee, MC Drew. J Exp Bot 37:1768, 1986.
82. AJ Bloom, J Finazzo. Plant Physiol 81:67, 1986.
83. WL Pan, WA Jackson, RH Moll. Plant Physiol 77:560, 1985.
84. XT Wang, FE Below. Crop Sci 32:997, 1992.
85. DT Clarkson, AJ Warner. Plant Physiol 64:557, 1979.
86. DT Clarkson, MJ Hooper, LJP Jones. Plant Cell Environ 6:535, 1986.
87. JC Lycklama. Acta Bot Neerl 12:361, 1963.
88. JH MacDuff, MJ Hooper. Plant Soil 91:303, 1986.
89. H Sasakawa, Y Yamamoto. Plant Physiol 62:665, 1978.
90. T Yoneyama, Y Akiyama, K Kumazawa. Soil Sci Plant Nutr 23:85, 1977.
91. JR Magalhaer, DM Huber. J Plant Nutr 12:985, 1989.
92. JK Vessey, LT Henry, S Chaillou, CD Raper. J Plant Nutr 13:95, 1990.
93. L Tolley-Henry, CD Raper. Plant Physiol 82:54, 1986.
94. EJ Hewitt. In: EA Kirkby, ed. Nitrogen Nutrition of Plants. Leeds, England: Waverly Press, 1970, p 68.
95. ED Spratt. Agron J 66:57, 1974.
96. ED Spratt, JKR Gasser. Can J Soil Sci 50:263, 1970.
97. G Michael, P Martin, I Owissia. In: EA Kirkby, ed. Nitrogen Nutrition of Plants. Leeds, England: Waverly Press, 1970, p 22.
98. HM Reisenauer. In: DR Neilsen, JG MacDonald, eds. Nitrogen in the Environment. New York: Academic Press, 1978, p 157.
99. T Hoff, BM Stummann, KW Henningsen. Physiol Plant 84:616, 1992.
100. LP Solomonson, MJ Barber. Annu Rev Plant Physiol Plant Mol Biol 41:225, 1990.
101. M Andrews. Plant Cell Environ 9:511, 1986.
102. GR Stewart, N Sumear, M Patel. In: WR Ullrich, PJ Aparicio, PJ Syrtee, F Castillo, eds. Inorganic Nitrogen Metabolism. Berlin: Springer-Verlag, 1987, p 39.
103. K Vaughn, WH Campbell. Plant Physiol 88:1354, 1988.
104. K Kamachi, Y Amemiya, N Ogura, H Nakagawa. Plant Cell Physiol 28:333, 1987.
105. SH Lips, Y Avissar. Eur J Biochem 29:20, 1972.
106. MR Ward, HD Grimes, RC Huffaker. Planta 177:470, 1989.
107. MG Redinbaugh, WH Campbell. Plant Physiol 68:115, 1981.
108. A Oaks, B Hirel. Annu Rev Plant Physiol 36:345, 1985.
109. BJ Miflin, PJ Lea. In: BJ Miflin, ed. The Biochemistry of Plants. New York: Academic Press, 1980, p 169.
110. RB Lee. Plant Cell Environ 3:65, 1980.
111. BJ Miflin, PJ Lea. In: ID Boulter, B Parthiev, eds. Nucleic Acids and Proteins in Plants. New York: Springer-Verlag, 1982, p 5.
112. GR Stewart, AF Mann, PA Fenten. In: BJ Miflin, ed. The Biochemistry of Plants. New York: Academic Press, 1980, p 271.
113. DL Karlen, RL Flannery, EJ Sadler. Agron J 80:232, 1988.
114. A Olness, GR Benoit, K Van Sickle, J Rinke. J Agron Crop Sci 164:42, 1990.
115. MP Russelle, RD Hauck, RA Olson. Agron J 75:293, 1983.
116. RB Austin, MA Ford, JA Edrich, RD Blackwell. J Agric Sci 88:159, 1977.
117. LG Bundy, PR Carter. J Prod Agric 1:99, 1988.
118. JJ Heitholt, LI Croy, NO Maness, HT Nguyen. Field Crops Res 23:133, 1990.
119. JC Swank, FE Below, RJ Lambert, RH Hageman. Plant Physiol 70:1185, 1982.
120. DA Van Sanford, CT MacKown. Crop Sci 27:295, 1987.
121. BI Muruli, GM Paulsen. Maydica 26:63, 1981.
122. RH Hageman, FE Below. YP Abrol, ed. Nitrogen in Higher Plants. Taunton, England: Research Studies, 1990, p 313.

123. WL Nelson, HF Reetz. Crops Soils 38(8):5, 1986.
124. FE Below, RJ Lambert, RH Hageman. Agron J 76:777, 1984.
125. SJ Crafts-Brandner, FE Below, JE Harper, RH Hageman. Plant Physiol 74:360, 1984.
126. KD Smiciklas, FE Below. Maydica 35:209, 1990.
127. CY Tsai, DM Huber, DV Glover, HL Warren. Crop Sci 24:277, 1984.
128. DT Walters, GL Malzer. Soil Sci Soc Am J 54:115, 1990.
129. DW Altman, WL McCuistion, WE Kronstad. Agron J 75:87, 1983.
130. EL Deckard, RJ Lambert, RH Hageman. Crop Sci 13:343, 1973.
131. LD Maddux, PL Barnes. J Fertil Issues 2:124, 1985.
132. FM Rhoads, A Manning. Soil Crop Sci Soc Fla Proc 45:50, 1986.
133. JW Friedrich, LE Schrader, EV Nordheim. Agron J 71:461, 1979.
134. KS Reddy, HA Mills, JB Jones. Agron J 83:201, 1991.
135. LE Gentry, FE Below. Crop Sci 33:491, 1993.
136. VV Rendig, TW Crawford. J Sci Food Agric 36:645, 1985.
137. CJ Pearson, BC Jacobs. Aust J Agric Res 38:1001, 1987.
138. FE Below, RJ Lambert, RH Hageman. Agron J 76:773, 1984.
139. JS Tomar, AF MacKenzie, RG Mehuys, I Alli. Agron J 80:802, 1988.
140. FE Below, LE Christensen, AJ Reed, RH Hageman. Plant Physiol 68:1186, 1981.
141. RT Weiland, TC Ta. Aust J Plant Physiol 19:77, 1992.
142. Y Murata, S Matsushima. In: LT Evans, ed. Crop Physiology: Some Case Histories. Cambridge: Cambridge University Press, 1975, p 73.
143. JR Evans. Plant Physiol 72:297, 1983.
144. SC Huber, T Sugiyama, RS Alberte. Plant Cell Physiol 30:1063, 1989.
145. SC Wong, IR Cowan, GD Farquhar. Plant Physiol 78:821, 1985.
146. B Sugiharto, K Miyata, H Nakomoto, H Sasakawa, T Sugiyama. Plant Physiol 92:963, 1990.
147. B Sugiharto, T Sugiyama. Plant Physiol 98:1403, 1992.
148. T Mae, A Makino, K Ohira. Plant Cell Physiol 24:1079, 1983.
149. LT Evans, IF Wardlaw, RA Fisher. In: LT Evans, ed. Crop Physiology: Some Case Histories. Cambridge: Cambridge University Press, 1975, p 101.
150. SR Simmons, RJ Jones. Crop Sci 25:1004, 1985.
151. MJ Dalling. In: JE Harper, LE Schrader, RW Howell, eds. Exploitation of Physiological and Genetic Variability to Enhance Crop Productivity. Rockville, MD: American Society of Plant Physiologists, 1985, p 55.
152. K Morita, M Kono. Soil Sci Plant Nutr 20:79, 1974.
153. JA Friedrich, RC Huffaker. Plant Physiol 65:1103, 1980.
154. K Morita. Ann Bot 46:297, 1980.
155. P Boon-Long, DB Egli, JE Leggett. Crop Sci 23:617, 1983.
156. SJ Crafts-Brandner, CG Poneleit. Plant Physiol 84:255, 1987.
157. JR Evans. Plant Physiol 72:297, 1983.
158. A Makino, T Mae, K Ohira. Plant Cell Physiol 25:429, 1984.
159. FE Below, SJ Crafts-Brandner, RH Hageman. Plant Physiol 79:1077, 1985.
160. KJ Boote, RN Gallaher, WK Robertson, K Hinson, LC Hammond. Agron J 70:787, 1978.
161. HJ Harder, RE Carlson, RH Shaw. Agron J 74:759, 1982.
162. R Killorn, D Zourarakis. J Prod Agric 5:142, 1992.
163. HJ Thomas, L Stoddart. Annu Rev Plant Physiol 31:83, 1980.
164. WG Duncan. LT Evans, ed Crop Physiology: Some Case Histories. Cambridge: Cambridge University Press, 1975, p 23.
165. M Tollenaar. Maydica 22:49, 1977.
166. R Capitanio, E Gentinetta, M Motto. Maydica 28:365, 1983.
167. VM Reedy, TB Daynard. Maydica 28:339, 1983.
168. BK Singh, CF Jenner. Aust J Plant Physiol 9:83, 1982.
169. BK Singh, CF Jenner. Aust J Plant Physiol 11:151, 1984.
170. KM Goh, RJ Haynes. Mineral Nitrogen in the Plant-Soil System. Orlando, FL: Academic Press, 1986, p 379.
171. BD Jacobs, CJ Pearson. J Exp Bot 43:557, 1992.
172. JH Lemcoff, RS Loomis. Crop Sci 26:1017, 1986.
173. R Novoa, RS Loomis. Plant Soil 58:177, 1981.
174. JF Power, J Alessi. J Agric Sci 90:97, 1978.
175. JH Spiertz, NM De Vos. Plant Soil 75:162, 1983.
176. SM Thomas, GN Thorne, I Pearman. Ann Bot 42:827, 1978.
177. A Mozafar. Agron J 82:511, 1990.
178. HV Eck. Agron J 76:421, 1984.
179. I Pearman, SM Thomas, GN Thorne. Ann Bot 41:93, 1977.
180. JR Czyzewicz, FE Below. Agron Abstr 1993, p 110.

181. H Doll, M Kreis. In: JHJ Spiertz, T Kramen eds. Crop Physiology and Cereal Breeding. Wageningen, Netherlands: Centre for Agricultural Publishing and Documentation, 1979, p 173.
182. CY Tsai, DM Huber, HL Warren. Crop Sci 18:399, 1978.
183. CY Tsai, DM Huber, HL Warren. Plant Physiol 66:330, 1980.
184. CY Tsai, BA Larkins, DV Glover. Biochem Gen 16:883, 1978.
185. AA Khan, R Verbeck, EC Waters, HA van Onckelen. Plant Physiol 51:641, 1973.
186. G Michael, H Beringer. Physiological aspects of crop productivity. 15th Colloquium of the International Potash Institute, Berne, Switzerland, 1980, p 85.
187. GW Singletary, FE Below. Plant Physiol 92:160, 1990.
188. GW Singletary, DC Doehlert, CM Wilson, MJ Muhitch, FE Below. Plant Physiol 94:858, 1990.
189. I Stulen. In: H Lambers, JJ Neeteson, I Stulen, eds. Fundamental, Ecological, and Agricultural Aspects of Nitrogen Metabolism in Higher Plants. Dordrech: Martinus Nijhoff, 1986, p 261.
190. JW Dudley, RJ Lambert, IA de la Roche. Crop Sci 17:111, 1977.
191. J Mesdag. In: JHJ Spiertz, T Kramer, eds. Crop Physiology and Cereal Breeding. Wageningen, Netherlands: Centre for Agricultural Publishing and Documentation, 1979, p 166.
192. FWT Penning de Vries, AHM Brunsting, HH van Laar. J Theor Biol 45:399, 1974.
193. CR Bhatia, R Rabson. Science 194:1418, 1976.
194. P Chevalier, SE Lingle. Crop Sci 23:272, 1983.
195. AJ Reed, GW Singletary. Plant Physiol 91:986, 1989.
196. BK Singh, CF Jenner. Aust J Plant Physiol 11:151, 1984.
197. MC Cox, CO Qualset, DW Rains. Crop Sci 25:430, 1985.
198. PI Payne. In: J Daussant, I Mosse, J Vaughan, eds. Seed Proteins. New York: Academic Press, 1983, p 223.
199. RJ Jones, SR Simmons. Crop Sci 23:129, 1983.
200. YZ Ma, CT MacKown, DA Van Sanford. Crop Sci 30:1099, 1990.
201. F Zink, G Michael. Z Acker Pflanzenbau 154:203, 1986.
202. LT Evans, IF Wardlaw. Adv Agron 28:301, 1976.
203. DT Canvin. Genetic Improvement of Seed Proteins. Washington, DC: US National Academy of Sciences, 1976, p 172.
204. R Reggiani, C Soave, N Di Fonzo, E Gentinetta, F Salamini. Genet Agric 39:221, 1985.
205. CI Tsai, I Dweikat, CY Tsai. Maydica 35:391, 1990.
206. CS Wyss, JR Czyzcwicz, FE Below. Crop Sci 31:761, 1991.
207. DC Doehlert, RJ Lambert. Crop Sci 31:151, 1991.
208. JB Cliquet, E Deleens, A Mariotti. Plant Physiol 94:1547, 1990.
209. AJ Reed, GW Singletary, JR Schussler, DE Williamson, AL Christy. Crop Sci 28:819, 1988.
210. MJ Ottman, LF Welch. Agron J 80:619, 1988.
211. DH Sander, WH Allaway, RA Olsen. In: RA Olsen, KJ Krey, eds. Nutritional Quality of Cereal Grains: Genetic and Agronomic Improvement. Madison, WI: ASA, 1987, p 45.
212. M Asghari, RG Hanson. Agron J 76:911, 1984.
213. SE Hollinger, RG Hoeft. Agron J 78:818, 1986.
214. SL Oberle, DR Keeney. J Prod Agric 3:527, 1990.
215. ED Nafziger, RL Mulvaney, DL Mulvaney, LE Paul. J Fertil Issues 1:136, 1984.
216. FE Broadbent, AB Carlton. In: DR Nielsen, JG MacDonald, eds. Nitrogen in the Environment. New York: Academic Press, 1978, p 1.
217. JS Schepers, FE Below. Influence of corn hybrids on nitrogen uptake and utilization efficiency. Proceedings of the 42nd Annual Corn and Sorghum Industry Research Conference, Chicago, 1987, pp 172–186.
218. DT Gehl, LD Bailey, CA Grant, JM Sadler. Can J Plant Sci 70:51, 1990.
219. DA Van Sanford, CT MacKown. Theor Appl Genet 72:158, 1986.
220. FE Broadbent, SK De Datta, EV Laureles. Agron J 79:786, 1987.
221. VB Ogunlela, PN Okoh. Fertil Res 21:67, 1989.
222. P Chevalier, LE Schrader. Crop Sci 17:897, 1977.
223. LG Balko, WA Russell. Agron J 72:723, 1980.
224. RJ Sabata, SC Mason. J Prod Agric 5:137, 1992.
225. MB Hatilitligil, WA Compton. Fertil Res 5:321, 1984.
226. CAC Gardner, PL Bax, DJ Bailey, AJ Cavalieri, CR Clausen, GA Luce, JM Meece, PA Murphy, TE Piper, RL Segebart, OS Smith, CW Tiffany, MW Trimble, BN Wilson. J Prod Agric 3:39, 1990.
227. L Gashaw, LM Mugwira. Agron J 73:47, 1981.
228. LE Gentry, XT Wang, FE Below. J Plant Nutr 12:363, 1989.
229. WJ Cox, HM Reisenauer. Plant Soil 38:363, 1973.
230. R Ganmore-Neumann, U Kafkafi. Agron J 72:758, 1980.
231. TW Rufty, CD Raper, WA Jackson. Bot Gaz 144:466, 1983.
232. GS Weissman. Plant Physiol 39:947, 1964.
233. KG Alexander, HM Miller, EG Beauchamp. J Plant Nutr 14:31, 1991.
234. RS Antil, DS Yadav, V Kumar, M Singh. Trop Plant Sci Res 1:353, 1983.

235.  FE Below, LE Gentry. J Fertil Issues 4:79, 1987.
236.  BR Bock. J Fertil Issues 4:68, 1987.
237.  JJ Camberto, BR Bock. Plant Soil 113:79, 1989.
238.  JA Heberer, FE Below. Ann Bot 63:643, 1989.
239.  FE Below, LE Gentry. Crop Sci 23:163, 1992.
240.  FE Below, JA Heberer. J Plant Nutr 13:667, 1990.
241.  JJ Camberto, BR Bock. Agron J 82:463, 1990.
242.  FG Adriaanse, JJ Human. S Afr J Plant Soil 3:57, 1986.
243.  KL Barber, LD Maddux, DE Kessel, GM Pierzynski, BR Bock. Soil Sci Soc Am J 56:1166, 1992.
244.  J Hagin, SR Olsen, A Shaviv. J Plant Nutr 13:1211, 1990.
245.  JR Huffman. J Agron Educ 18:93, 1989.
246.  KD Smiciklas, FE Below. Plant Soil 142:307, 1992.
247.  KD Smiciklas, FE Below. Crop Sci 32:1220, 1992.
248.  JJ Camberto, BR Bock. Agron J 82:467, 1990.
249.  WL Pan, EJ Kamprath, RH Moll, WA Jackson. Soil Sci Soc Am J 48:1101, 1984.
250.  FE Below, LE Gentry, KD Smiciklas. Role of mixed N in enhancing productivity of maize. Proceedings of
      Symposium, Division S-8, Soil Science Society of America, Denver, CO, 1991, pp 1–11.
251.  LE Gentry, FE Below. Agron Abstr 1992, p 278.
252.  PE Jung, LA Peterson, LE Schrader. Agron J 64:668, 1972.
253.  EL Anderson, EJ Kamprath, RH Moll. Agron J 76:397, 1984.
254.  M Motto, RH Moll. Maydica 28:53, 1983.
255.  GS McNaughton, MR Presland. J Exp Bot 34:880, 1983.
256.  T Yoneyama, Y Akiyama, K Kumazawa. Soil Sci Plant Nutr 23:85, 1977.
257.  WL Pan, JJ Camberato, WA Jackson, RH Moll. Plant Physiol 82:247, 1986.
258.  L Salsac, S Chillou, JF Morot-Gaudry, C Lesaint, E Jolivet. Plant Physiol Biochem 25:805, 1987.
259.  JA Raven. New Phytol 101:25, 1985.
260.  OAM Lewis, B Fulton, AAA van Zelewski. In: WR Ulrich, PJ Aparicio, PJ Syrett, F. Casrillo, eds. Inorganic
      Nitrogen Metabolism. Berlin: Springer-Verlag, 1987, p 240.
261.  TF Neales, LD Incoll. Bot Rev 34:107, 1968.
262.  NP Hall, R Reggiani, J Franklin, AJ Keys, PJ Lea. Photosynth Res 5:361, 1984.
263.  GF Morot-Gaundry, F Thuillier, C Lesaint, S Chaillou, E Jolivet. Physiol Veg 23:257, 1985.
264.  AM Amory, CF Cresswell. Ann Bot 54:719, 1984.
265.  OAM Lewis, MIM Soares, SH Lips. In: H Lambers, JJ Neeteson, I Stulen, eds. Fundamental, Ecological and
      Agricultural Aspects of Nitrogen Metabolism in Higher Plants. Dordrecht: Martinus Nijhoff, 1986, p 295.
266.  GJ Blair, HM Miller, WA Mitchell. Agron J 62:530, 1970.
267.  EP Papanicolaou, VD Skarlou, C Nobeli, NS Katranis. J Agric Sci 101:687, 1983.
268.  RW Smiley. Soil Sci Soc Am Proc 38:795, 1974.
269.  E Schnug, A Fink. In A Scaife, ed. Plant Nutrition 1982. Slough, England: Commonwealth Agriculture Bu-
      reau, 1982, p 583.
270.  RH Teyker, DC Hobbs. Agron J 84:694, 1992.
271.  J Gerendas, RG Ratcliffe, B Sattelmacher. J Plant Physiol 137:125, 1990.
272.  JA Raven, FA Smith. New Phytol 76:415, 1976.
273.  RH Moll, EJ Kamprath, WA Jackson. Agron J 74:562, 1982.
274.  EL Anderson, EJ Kamprath, RH Moll. Crop Sci 25:598, 1985.
275.  KC Cameron, RJ Haynes. Mineral Nitrogen in the Plant-Soil System. Orlando, FL: Academic Press, 1986, p
      166.
276.  PC Scharf, MM Alley. J Fertil Issues 5:109, 1988.
277.  RD Hauck. In: JJ Meisinger, GW Randall, ML Vitosh, eds. Nitrification Inhibitors—Potentials and Limita-
      tions. Madison, WI: ASA and SSSA, 1980, p 19.
278.  DM Huber, HL Warren, CY Tsai. Bioscience 27:523, 1977.
279.  HF Chancy, EJ Kamprath. Agron J 74:656, 1982.
280.  RG Hoeft. In: RD Hauck, ed. Nitrogen in Crop Production. Madison, WI: ASA, CSSA, SSA, 1984, p 561.
281.  LL Hendrickson. J Prod Agric 5:131, 1992.
282.  RD Voss. In: RD Hauck, ed. Nitrogen in Crop Production. Madison, WI: ASA, CSSA, SSA, 1984, p 571.

# 19

# Quantifying Immediate Carbon Export from Source Leaves

**Evangelos Demosthenes Leonardos and Bernard Grodzinski**

*University of Guelph, Guelph, Ontario, Canada*

## I. INTRODUCTION

Carbon, hydrogen, and oxygen constitute 96% of a plant's dry weight. Source leaves are the primary sites of C reduction and the main organs exporting reduced C to growing sinks. It is well known that in almost all species, sugars, starch, and amino acids accumulate in leaves during the daytime and export of assimilates derived from these reserves occurs both concurrently with photosynthesis and subsequently during night periods [1–5]. Our overall knowledge of translocation processes has been derived from diverse experimental approaches [6–11]. For example, imaging techniques, which include light, electron, and fluorescence microscopy using dyes or proteins, provide valuable qualitative data on intercellular connections and export [12–16]. Generally, these imaging techniques are destructive. However, procedures using isotopes of carbon (e.g., mass isotopes, $^{13}C$, and radioisotopes, $^{11}C$ and $^{14}C$) to study export can be both quantitative and noninvasive [7,17–19]. Phloem sap exudation from cut sieve tubes or from aphid stylectomy has provided a practical means of sampling mobile assimilates [20]. Collection of apoplastic fluids [21–22] and measurements of pH and membrane potential [23] further demonstrate the physiological and biochemical interactions that operate intercellularly as sugars are loaded or unloaded from the phloem. More recently, molecular techniques have led to characterization of sugar transporters [24] and the engineering of transgenic plants that can be designed to provide important information regarding the role of specific export processes in the leaves [11,25].

Most researchers who have studied translocation [6,5,26–29] acknowledge that it is very difficult to quantify simultaneously (1) C assimilation by the leaf, (2) C recycling within the leaf, (3) temporary C storage within the leaf, and (4) immediate C efflux rate via the phloem. It is even more complex to relate any of these leaf processes to daily export patterns and relative growth rates (RGRs) at the whole plant level [30].

One of our interests has been to examine the importance of product removal from the leaf in processes such as C recycling in that organ specifically during periods of active photosynthesis and photorespiration [18,31–34]. By knowing what is happening quantitatively to export concurrently with $CO_2$ fixation, we are better able to evaluate concepts such as feedback inhibition of photosynthesis and quantify intercellular movements of metabolites that link long-distance export to sinks and the primary metabolism of the leaf (i.e., photosynthesis).

In leaves of $C_3$ plants, the stroma of the chloroplast is the site of the reductive pentose pathway (i.e., the Calvin cycle) and $CO_2$ fixation [35]. The primary *inorganic* substrate of ribulose-1,5-bisphosphate carboxylase-oxygenase (Rubisco) is $CO_2$. The primary *organic* substrate is ribulose 1,5-bisphosphate (RuBP). Although the Calvin cycle fixes *inorganic* substrate $CO_2$, it also regenerates the acceptor RuBP. The regeneration of RuBP provides a feedback mechanism for control of the rate of photosynthesis. However, metabolism and recycling of reduced C and N occur in subcellular compartments other than the chloroplast and even outside the cell that may have been the primary site of C reduction. During photorespiration, for example, the C in the phosphoglycolate molecule that is generated by the oxygenase reaction is not recycled conservatively within the chloroplast [36,37] or in the cell [32]. Similarly, the sucrose that is the major phloem mobile leaf product is not synthesized in the chloroplast but in the cytosol [11,27,38]. In fact, phloem mobile sugars need not be synthesized in the same cell that initially fixed the $CO_2$. It suffices here to note that the auxiliary phloem sugars such as raffinoses are made in phloem cells distant from the site of $CO_2$ fixation [39]. It also suffices here to note that intercellular exchanges of assimilates preceding export from the leaf are more complex in $C_4$ and $C_3$-$C_4$ intermediate types than in the leaves of $C_3$ species and that sucrose synthesis and $CO_2$ fixation occur in separate cells and tissues [38,40,41]. Clearly, we are dealing with a complex set of processes when we try to understand how net photosynthesis, which is measured as substrate utilization (e.g., µmol of $CO_2$ fixed per second per $m^2$), is affected by sink demand and product removal (µmol of $CO_2$ exported per second per $m^2$). Subcellular, cellular, and tissue level interactions are all involved. Is there any unifying mechanism that truly explains how photosynthesis in the chloroplast of a chlorenchyma cell and movement in the phloem sieve cell are linked?

The literature suggests that source photosynthesis and sink demand for assimilates are linked mechanistically, in part, by the operation of specific site exchanges in the light such as that mediated by operation of the phosphate translocator at the chloroplast membrane [11,25,27,38]. It is known that reserves of sugars and starch buffer sink demand and affect growth and development [4,5]. Can we hope to correlate primary leaf parameters such as photosynthetic efficiency and leaf export capacity with sink demand? We have developed a working hypothesis which states that in most plants the immediate export rates from leaves during photosynthesis best predict RGR. If this hypothesis is valid, then in many cases the RGR of a plant should be tightly correlated with the capacity of the leaves to export C in the light as the C is being fixed (i.e., immediately). How can one measure or estimate immediate C flux through a complex organ such as the leaf?

We have defined immediate export as the direct or instantaneous flux of C from $CO_2$ to assimilates to the phloem [18]. This operational definition of immediate export excludes for practical purposes export arising from storage reserves either during the same photoperiod or during subsequent night periods [5]. Again, we emphasize that it is important to export total leaf assimilate supply at different times in the day. The specific question that is being addressed is, how important to the maximum operation of the C and N reduction pathways in the leaf tissue is the immediate rate of assimilate removal via the translocation stream in the light? The $V_{max}$ of a single enzyme step in a complex reaction pathway can provide useful information about the kinetics and the importance of that enzyme step. To obtain $V_{max}$ the enzyme rate needs to be measured quantitatively. Similarly, the importance of immediate export rate from the leaf during photosynthesis needs to be quantified experimentally.

## II.   METHODOLOGIES USED TO ESTIMATE IMMEDIATE C EXPORT

In this chapter we describe how steady-state $^{14}CO_2$ labeling of leaf tissue is achieved and why data obtained when $^{14}C$ isotopic equilibrium exists between photosynthesis and export provide useful estimates of the immediate mass efflux of C. To appreciate the advantages and disadvantages of steady-state $^{14}CO_2$ labeling, one needs to consider other methodologies that have been used to quantify C export [6–10].

### A.   Gas Exchange and Differential Dry Weight Analysis

Perhaps the least costly method to estimate mass rates of export from source leaves is to measure the net $CO_2$ exchange rate with an infrared gas analyzer (IRGA) and the changes in the dry weight of the leaf over time [3,42–46]. The IRGA quantifies the net amount of C fixed and thus the total dry weight that should have been retained in the leaf in the absence of export. Photorespiratory $CO_2$ losses are accounted for because net $CO_2$ exchange is measured. The difference in dry weight between samples in the light and the theoretical amount of C that should have been retained can only be due to export. Whereas the IRGA measurements are nondestructive, the differential dry weight estimations of leaf tissue are destructive, and many small samples need to be taken from several leaves. Subsampling reduces disturbance of a single leaf and eliminates the large variability in dry weight within the same leaf and heterogeneity among different leaves. A disadvantage of this protocol is that it is difficult to evaluate short-term changes in export. Also, measurable changes in leaf dry weight do not occur rapidly, usually taking several hours. Nevertheless, destructive sampling does provide the tissue necessary for biochemical characterization of the leaf intermediates and, coupled with labeling procedures, can provide important information regarding the specific pools of assimilates contributing to leaf metabolism and export [18].

## B. Isotopes of C

In most instances 95% of the photoassimilates being transported via the phloem are carbohydrates such as sucrose. The mass (e.g., $^{13}C$) and radioactive (e.g., $^{11}C$, $^{14}C$) isotopes of C introduced as labeled $CO_2$ are excellent tools for directly tracing the movement of photoassimilates in plants. Two of the advantages of using radioactive and mass isotopes are that information about metabolism can be obtained and a less invasive method of measuring C export can be achieved. However, sufficient time is required for proper labeling of the pools of the primary phloem mobile assimilates, and care must be exercised in calculating for instrument sensitivity and isotope discrimination.

### 1. The Mass Isotope $^{13}C$

The mass isotope $^{13}C$ has been used extensively in studies of isotopic discrimination as a tool to distinguish photosynthetic pathways in $C_4$, $C_3$, and $C_3$-$C_4$ intermediate plants [47–49]. Mass isotopes such as $^{13}C$ and $^{15}N$ are currently not used as extensively as they should be to study export patterns (Ref. 50 and references therein). The problem at the moment is the lack of an inexpensive detection method for the mass isotopes that does not require destructive sampling [51,52]. In most studies, a destructive sampling step is required to determine the enrichment level of the assimilates in samples that are often prepared for analysis by mass spectrophotometry coupled to gas or liquid chromatography. However, it is quite clear that the use of mass isotopes to probe assimilate translocation can and should be coupled with improving technologies such as nuclear magnetic resonance (NMR) imaging [19,53,54]. NMR imaging in medicine has revolutionized remote sensing of tissues [55]. With steady-state labeling of leaves with mass isotopes such as $^{13}CO_2$ we should be able to replace other methods of quantifying export as well as monitor partitioning and allocation patterns. It is theoretically possible to analyze export of labeled assimilates from leaves, movement within individual bundles of phloem cells, and sink metabolism noninvasively.

### 2. The Radioisotope $^{11}C$

The short-lived radioisotope $^{11}C$ has been used extensively as a noninvasive probe of translocation processes [17,56–63]. However, because of the short half-life of $^{11}C$ (i.e., 20.4 min), experiments are restricted in time and must be performed in proximity to a particle accelerator. The main advantage of $^{11}C$ is that the isotope emits $\beta^+$ particles of much higher energy than the $\beta^-$ particles that are emitted from $^{14}C$, and thus, translocation of the $^{11}C$-labeled intermediates is easier to monitor remotely using Geiger-Müller (GM) detectors. However, heavy shielding of the detectors is required [64], which limits an analysis of partitioning of label within different tissues in the same organ [54]. Nevertheless, $^{11}C$ has been used to study directional movements and translocation speed in stems [56,63,65]. In other studies, transfer function and compartmental analysis have been used to quantify $^{11}C$-photoassimilate export from source leaves [62,66,67]. Given the cost, time, and effort of setting up labeling experiments, the overall value of using short-lived isotopes must be considered carefully. Although the procedure is noninvasive and the short half-life of $^{11}C$ and $^{13}N$ permits repetition of tests using the same leaf [58,60,64], a significant disadvantage of these isotopes is their short half-life, which makes analysis of labeled assimilates by current biochemical techniques very difficult. Sap samples and tissue extracts must be purified and analyzed immediately. We have been able to use $^{13}N$ (half-life 10 min) fed as $^{13}NH_3$ to probe leaf photorespiration directly [58]. In the first hour, $^{13}N$-labeled glutamate, alanine, serine, and glycine were detected (Grodzinski and Lapointe, unpublished) and export of these photorespiratory intermediates (using $^{14}C$)

was observed [31,68]. It is impossible to follow the fate of the intermediates using short-lived isotopes during pulse-chase experiments that extend into a normal day-night cycle.

### 3.   The Radioisotope $^{14}C$

The most widely used radioisotope in the study of leaf photosynthesis and phloem translocation is $^{14}C$. The main advantage of $^{14}C$ is its commercial availability in different forms (e.g., $^{14}CO_2$, $^{14}C$-sucrose) and its long half-life (5730 years). It can be used in both short- and long-term studies of export and partitioning. Although $^{14}C$ emits $\beta^-$ (negatron) of low energy, translocation of $^{14}C$-labeled assimilates can be monitored in a noninvasive manner with GM detectors and products can be analyzed following sampling [18,69]. Unfortunately, GM detectors cannot measure $^{14}C$ in tissues farther than 1 mm from the plant surface and a correction must be made for each leaf [18,69]. The manner in which $^{14}CO_2$ is introduced and export of $^{14}C$ assimilates is monitored varies. Both pulse-chase and steady-state labeling protocols have been used to obtain useful data regarding early export of photoassimilates.

Pulse-chase labeling with $^{14}CO_2$ is the most common procedure used to study translocation of $^{14}C$-photoassimilates [31,70–75]. Typically, the label is fed to a source leaf as $^{14}CO_2$ and either its disappearance from the leaf or appearance in sink tissue is analyzed. Sizes and rates of turnover of sugar pools (e.g., transport and vacuolar sucrose pools) in the light have been estimated by monitoring the translocatory efflux of $^{14}C$ from leaves pulse labeled with $^{14}CO_2$ and employing compartmental models [72,73,76,77]. However, pulse-chase experiments may not provide precise measurements of the mass transfer rate of C during photosynthesis because the specific activity of $^{14}C$ in the pools changes dramatically during chase periods, especially in leaves subjected to different environmental conditions [69,78]. We have used pulse-chase experiments primarily to follow the export and respiration of reserves during a dark period, when no label can be incorporated into the transport products directly via photosynthesis [79,80]. Steady-state labeling has been the method used to quantify the mass transfer rate of immediate export during photosynthesis.

During steady-state labeling, transport pools of sugars achieve isotopic equilibrium with the $^{14}CO_2$, which is supplied continuously at a constant specific activity [18,69]. There are different protocols for establishing steady-state labeling with $^{14}CO_2$ [7,18,34,69,81]. By calculating export fluxes only when isotopic equilibrium has been achieved, errors associated with determining immediate export rates using non-steady-state labeling and pulse-chase experiments are significantly reduced [18,78].

## III.   STEADY-STATE $^{14}CO_2$ LABELING AND MEASUREMENT OF IMMEDIATE C EXPORT RATES DURING PHOTOSYNTHESIS

### A.   Open-Flow Gas Analysis System

We have used open-flow gas analysis systems similar to that in Figure 1 to establish steady-state $^{14}CO_2$ labeling conditions [33,34,63,65]. During the leaf gas exchange analysis and labeling experiments, plants were held in a growth chamber in which irradiance, temperature, and humidity were controlled. The middle portion of a leaf was enclosed in a brass leaf chamber that had been chrome plated to reduce problems associated with water exchange [82]. The leaf chamber consisted of a top part (16 cm$^2$ exposing area through a glass window) and a bottom part in which was mounted a GM detector (model EWGM, window area 6.8 cm$^2$, Bicron Corp., Newbury, OH). Both upper and lower sections of the leaf chamber were designed as water circulating jackets for leaf temperature control. Leaf and gas stream temperatures were measured with two thermistors (YSI 44003 A, YSI Inc., Yellow Spring, OH) inside the leaf chamber. Photosynthetic photon flux density (PPFD) (400–700 nm) was provided by three 1,000 W metal halide lamps (Sylvania, GTE, Toronto, ON, Canada) and measured with a Li-Cor quantum sensor (model LI-189, Li-Cor Inc., Lincoln, NE) positioned at the surface level of the leaf. The desired $CO_2$ concentration (35 or 90 Pa) was obtained by mixing $CO_2$–free air that had been passed through soda lime with pure $CO_2$ using two mass flow controllers (Side-Trak, Sierra Instruments, Inc., Monterey, CA). The $CO_2$ concentration in the gas stream entering and exiting the leaf chamber was measured with an IRGA (model 6262, Li-Cor Inc.). Humidity in the gas entering the leaf chamber was controlled by first passing the gas stream through a gas bubbler placed in a temperature-controlled water bath (model RTE-9, Neslab Instruments Inc., Portmouth, NH). The dew point in the gas stream entering and exiting the leaf chamber was monitored with a digital humidity analyzer (Dew point meter;

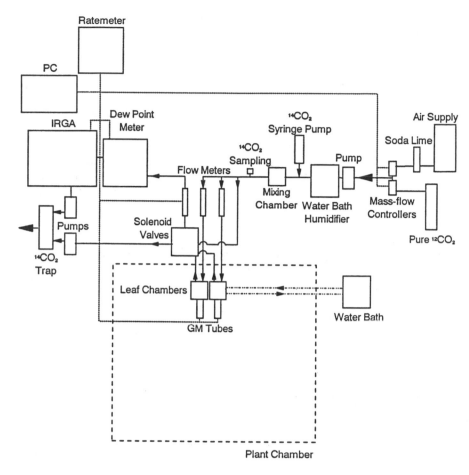

**Figure 1**    A simplified schematic of the open-flow gas analysis and $^{14}CO_2$ labeling system used to measure photosynthesis and C export rates in source leaves. Gas, water, and electrical lines are represented by solid, dashed-dotted, and dotted lines, respectively.

model 991, Dev-All, EG&G, Waltham, MA). The flow rate in each chamber was maintained constant between 0.25 and 0.70 L min$^{-1}$ by flow meters (Cole-Parmer Instrument Comp., Niles, IL) and precisely measured by an electronic flowmeter (Top-Trak, Sierra Instruments, Inc., Monterey, CA). A personal computer was used to control, monitor all devices, and log data.

## B.    $^{14}CO_2$ Feeding and Monitoring of $^{14}C$ Retention

An acclimation period of 15–30 min was usually required before a steady rate of photosynthesis was monitored after inserting the leaf in the chamber. Only after this period was the $^{14}CO_2$ supplied. During steady-state $^{14}CO_2$ labeling, $^{14}CO_2$ generated from NaH$^{14}CO_3$ was injected into the gas stream by using a precision syringe pump (model 341A, Sage Instruments, Cambridge, MA). The specific activity of $^{14}CO_2$ in the gas stream was measured at regular intervals (20 min) by trapping 3 mL of the inlet gas in 2.5 mL of ethanolamine/ethylene glycol monomethyl ether (1:2 v:v). This specific activity varied among experiments from 0.3 to 3.0 kBq μmol$^{-1}$ C depending on the leaf photosynthetic rate or the $CO_2$ concentration. However, during each feeding the specific activity of the $^{14}CO_2$ was constant.

The GM detector positioned under the leaf surface was used to monitor the radioactivity accumulated in the source leaf during the feeding period. The GM output through a rate meter (model 8731-32, Nuclear Chicago Co., Des Plaines, IL) was recorded and the counts were corrected for the total radioactivity recovered in the leaf at the end of the experiment. The total radioactivity recovered in the leaf extract was determined by liquid scintillation counting (model LS-6800, Beckman Instruments Inc, San Ramon, CA). The C export rate was calculated as the difference between the C fixation rate measured continuously by the IRGA and the C retention rate estimated by the GM trace and corrected for the efficiency of the GM tube.

## C.  GM Detector Counting Efficiency

Because of the difference in morphological, anatomical, and biochemical leaf characteristics (e.g., leaf thickness, venation pattern, partitioning) among species that we have examined, the counting efficiency of the GM detectors varied between 0.1 and 1.0%. The counting efficiencies of the GM detectors for a representative number of different photosynthetic types of *Panicum* species [30,34] are shown in Figure 2. In spite of the low counting efficiency (i.e., 0.5–1.0%), for each species there was a high linear correlation between the radioactivity determined by destructive analysis and that counted by the GM detector. The coefficient of determination ($r^2$) varied from 0.79 to 0.98. These data support the view that the GM detectors can be used to monitor $^{14}C$ in the leaf nondestructively.

## D.  Calculation of Concurrent Export During Steady-State $^{14}CO_2$ Feeding

The theories behind determining mass fluxes of C during photosynthesis using either the method of differential weight analysis or $^{14}CO_2$ steady-state labeling coupled with net gas exchange are very similar. Figure 3 shows the net $CO_2$ assimilation calculated from the photosynthetic rate obtained from the IRGA for representative $C_3$, $C_3$-$C_4$ intermediate, and $C_4$ species that transport sucrose (Figure 3A, B, and C, respectively) and a $C_3$ species that translocates auxiliary sugars as well as sucrose (Figure 3D). The IRGA was used to estimate the rate of assimilation (dashed line) throughout the experiment. In each case, the photosynthetic rate was constant before $^{14}CO_2$ was supplied at a constant specific activity. The retention of $^{14}C$ was measured nondestructively with the GM detector and was corrected with measurements of radioactivity made by destructive sampling at the end of the feeding period (solid line in Figure 3A–D). Immediate export of $^{14}C$-assimilates (dotted line) was calculated as the difference between fixation and retention rate given by the data derived from the IRGA and the GM detector during an appropriate period (shaded area in Figure 3A–D).

In order to determine the appropriate period, a series of destructive experiments were designed to approximate the time required for transport pools to reach isotopic equilibrium. When the leaf was sampled during a typical 2-hr feeding period, the pattern of $^{14}C$ partitioning in the transport sugars indicated that isotopic equilibrium between the $^{14}CO_2$ in the air stream and the major $^{14}C$-translocates was generally not achieved in the first hour (Figure 3M–P). A period of 60 to 90 min was usually required before the specific activity of the major sugar (sucrose) reached a steady level. The sugar pools in the $C_4$ species generally reached isotopic equilibrium earlier than in the $C_3$ and $C_3$-$C_4$ intermediates species. Normally, the data between 90 and 120 min were used to calculate values for photosynthesis and the corresponding concurrent export rate (shaded area in Figure 3A–D). During this period the $^{14}C$-sucrose pool was in isotopic equilibrium with the $^{14}CO_2$ being assimilated. Similarly, in species such as *C. sativus* the $^{14}C$-stachyose pool (auxiliary phloem mobile sugar that was used as a marker of transport) was in isotopic equilibrium. Figure 3I, J, and K show the accumulation of sucrose, which is the main form of assimilates being exported in the *Panicum* species. In *C. sativus* auxiliary sugars accumulated as well as sucrose (Figure 3L). However, in some species there was a large pool of labeled hexoses [e.g., in the $C_4$ species *P. miliaceum* and in the $C_3$ species *C. sativus* (data not shown)]. In all species $^{14}C$ accumulated in sugar and starch (Figure 3E–H) that sustain metabolic requirements within the leaf and export during subsequent periods of light or darkness [30,63,79,83]. The fate of these pools and their contribution to export and respiration could be determined in pulse-chase experiments.

The rates of photosynthesis and immediate export obtained when isotopic equilibrium was first established (e.g., 90–120 min) were the data sets we used to evaluate changes in immediate export rates in leaves challenged with environmental stresses [18,33,34,63], or diseases [78,80]. These data also provide comparisons of immediate export capacity among leaves with naturally different $CO_2$ fixation pathways (i.e., $C_3$, $C_3$-$C_4$ intermediate, and $C_4$) [30,33,34] or transgenics with specifically altered C metabolism [83]

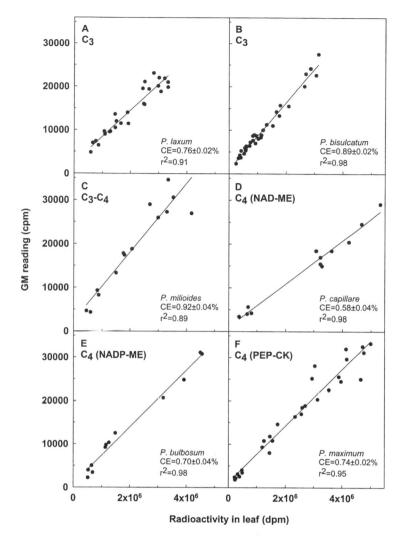

**Figure 2** Counting efficiency of the GM detectors in experiments with *Panicum* species. Radioactivity was measured nondestructively by monitoring $^{14}C$ accumulation in the leaf with a GM detector and by scintillation counting after destructive sampling of the leaf. Data shown are those for two $C_3$ species, *P. laxum* (A) and *P. bisulcatum* (B); a $C_3$-$C_4$ intermediate species, *P. milioides* (C); an NAD-ME $C_4$ species, *P. capillare* (D); an NADP-ME $C_4$ species, *P. bulbosum* (E); and a PEP-CK $C_4$ species, *P. maximum* (F). Each point is the measurement of one leaf. The data are from measurements made at different $CO_2$ levels (35 and 90 Pa) and at the end of different experimental periods (30, 60, 90, 120 min and 17 hr). Feeding periods were usually 120 min, but in pulse-chase experiments we frequently extended the period of noninvasive monitoring of $^{14}C$ retention to 17 hr. Respiration data in these pulse-chase experiments were used to correct for total export (see Refs. 63, 79, and 80). Counting efficiency (CE) is the value of radioactivity obtained by the GM detector divided by the radioactivity determined after destructive analysis of the leaf tissue times 100. Values for CE are means $\pm$ SE. The variation explained by a linear model fitted to the data is indicated by the coefficient of determination ($r^2$).

## IV.  CASE STUDIES

## A.  Photosynthesis and Export Under Stress

As already pointed out, the primary functions of a "source" leaf are to fix light energy and provide that energy in the form of photoassimilates for plant growth. The leaf is not a homogeneous structure, and in

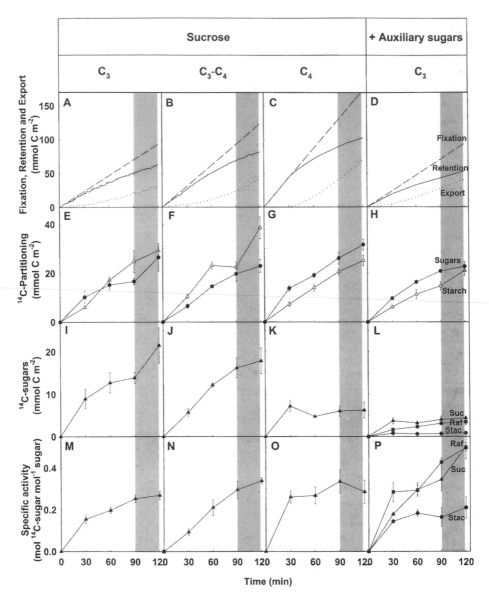

**Figure 3** Total $^{14}C$ fixation, $^{14}C$ retention, $^{14}C$ export, and $^{14}C$ partitioning in major intermediates during a 2-hr $^{14}CO_2$ feeding of source leaves. Data are those for species that transport sucrose [a $C_3$ species, *P. bisulcatum* (A, E, I, and M); a $C_3$-$C_4$ intermediate species, *P. decipiens* (B, F, J, and N); and a $C_4$ species, *P. miliaceum* (C, G, K, and O] and for a species that transports auxiliary sugars as well as sucrose [a $C_3$ species, *C. sativus* (D, H, L, and P)]. Measurements were made under saturating irradiance (1500, 1600, and 1750 $\mu$mol m$^{-2}$ sec$^{-1}$ for the $C_3$, $C_3$-$C_4$ intermediate, and $C_4$ species, respectively), 35 Pa $CO_2$, 21 kPa $O_2$, and 30°C. Cumulative net C fixation (dashed line in A, B, C, and D) was calculated from IRGA data, whereas $^{14}C$ retention in the leaf was measured both nondestructively by monitoring $^{14}C$ with a GM detector continuously (solid line) and destructively at the end of the feeding. Export (dotted line) was estimated as the difference between total fixation (dashed line) and $^{14}C$ retention in the leaf (solid line). Panels E, F, G, and H show partitioning of total $^{14}C$ in the total sugar fraction (●) and in ethanol insolubles (starch, △). Panels I, J, K, and L show partitioning of total $^{14}C$ into sugars (sucrose and auxiliary sugars), and panels M, N, O, and P show the specific activity of the transport sugars, sucrose (Suc, ▲), raffinose (Raf, ■), and stachyose (Sta, ●). Each point is the average of at least four leaves on four different plants and each error bar represents the SE of the mean. The shaded areas represent the 90–120 min period during which export was estimated.

the light photosynthesis and photorespiration occur simultaneously while C is being exported from the tissue. Our analyses showed that in healthy source leaves when photorespiration was suppressed by low $O_2$ and high $CO_2$ levels, both photosynthesis and immediate export increased [18,63]. Significantly, using our steady-state $^{14}CO_2$ labeling procedure, we were able to show that leaf warming resulted in a reduction in immediate export prior to any inhibition of the photosystems that would have altered the C-fixation processes per se [18,63]. These data challenge conventional literature which focuses on the concept of disruption of thylakoid membranes in chloroplasts and loss of the photosystem activity being the primary limiting event during heat stress of leaf tissue. One applied outcome of the studies of immediate export at elevated leaf temperatures [18] has been a reexamination of the effect of high temperature on the maintenance of valuable greenhouse crops, such as alstroemeria [63] and roses [79]. In the case of roses, for example, our data provide an explanation for reduced production of flowers at high temperatures. Photosynthesis is maintained but export during photosynthesis is inhibited, leading to poorer flower development. Carbohydrate limitation does not appear to be compensated for by nighttime export of stored reserves because most of the C in roses is exported in the light even when the plants have acclimated to high $CO_2$ and more C is stored in the light [79].

In another study of bean leaves infected with bacterial blight (i.e., *Xanthomonas*), the immediate export rate was reduced before a drop in photosynthesis was observed [78]. This conclusion was not readily evident from other analyses of export. In a related study with geranium infected with bacterial blight, photosynthesis and immediate and subsequent nighttime export rates were inhibited more at high $CO_2$ than at ambient levels even though bacterial numbers in the leaves were lower in plants grown at high $CO_2$ [80]. The data show that classical definitions of disease severity (virulence) based on bacteria numbers [e.g., colony-forming units (CFU)] are not accurate. In plants that were acclimated to high $CO_2$ but not infected, the immediate and daily export rates were greater than those in control plants grown at ambient $CO_2$. Collectively, these results are interesting because they provide some insight into the acclimation processes that might operate in plants growing in a "$CO_2$-enriched world" as well as the manner in which plants compete with pathogens or symbiotes for assimilates. Using $^{14}C$ pulse-chase analysis, it appears in geranium that the healthy uninfected control plants [80], like the healthy rose plants [79], export the bulk of the $^{14}C$ fed during a light period (i.e., during photosynthesis). Long-term $CO_2$ enrichment increased the storage of sugars and starch in the leaves but proportionally both ambient-grown plants and those that were acclimated to $CO_2$ enrichment had similar patterns of immediate (daytime) and nighttime export. Nighttime respiration was generally higher in the $CO_2$-enriched plants. Using destructive weight analyses to study diurnal export patterns in castor bean that had been acclimated to either ambient or elevated (700 $\mu$l L$^{-}$1) $CO_2$, Grimmer and Komor [46] also showed that in both ambient- and high $CO_2$–grown plants more export occurred during the light period than during the dark period. The rates of export at night and respiration at night were also greater in the high $CO_2$–acclimated leaves, but during $CO_2$ enrichment the C content (as a percentage) was not altered by $CO_2$ enrichment.

Growth rate and plant development might be markedly altered as more or less of the total transport proceeds during the day versus the night [5,46,80,83]. A number of studies with both $C_3$ and $C_4$ species have shown the magnitude of export during daytime rather than during the night and therefore the importance of export that occurs concurrently with photosynthesis [1–3,46,79,80].

## B.  Immediate Export in Natural Photosynthetic Variants

Differences in photosynthetic capacity are recognized to occur naturally among genetically similar species and genera. Anatomical, biochemical, and physiological differences among $C_3$, $C_3$-$C_4$ intermediate, and $C_4$ photosynthetic pathways in genera such as *Panicum* and *Flaveria* are well documented [40,84–86]. One advantage of examining the leaf photosynthesis and export patterns in *Panicum* and *Flaveria* more closely is that these genera represent a range of naturally occurring variants of photosynthetic types. Both genera contain $C_3$-$C_4$ intermediate types in addition to species with well-defined $C_3$ and $C_4$ traits. Although the impact of these different photosynthetic pathways on photosynthetic and photorespiration rates has been examined, studies that provide measurements of export capacity during photosynthesis in leaves with different photosynthetic types are limited and have often employed different techniques to quantify translocation [2,33,34,70].

On the basis of the rate of disappearance of $^{14}C$ during a 6-hr chase period, Hofstra and Nelson [70] showed that the $C_4$ species corn and sorghum had higher export rates than $C_3$ species such as soybean and

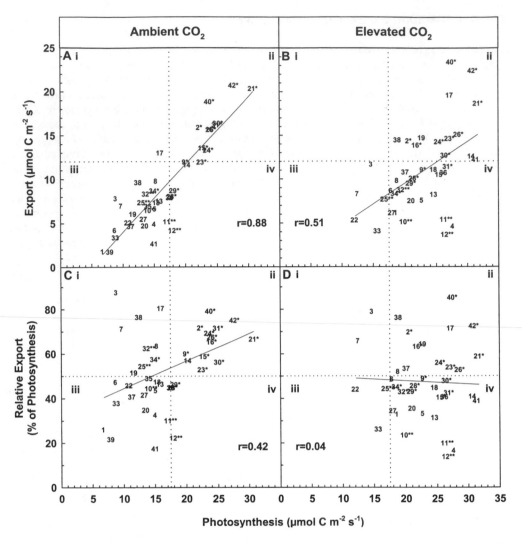

**Figure 4** Photosynthesis and immediate export rates of source leaves of 42 species measured at ambient $CO_2$ (35 Pa, A and C) and at short-term exposure to elevated (90 Pa, B and D) $CO_2$ levels. Export rates were expressed in absolute values (A and B) and as relative C efflux rate calculated as a percentage of the photosynthesis rate (C and D). Each panel was divided into four areas (quadrants) so that the data among panels could be compared easily. The species included $C_3$ types (numbers alone) as well as $C_4$ and $C_3$-$C_4$ intermediate types, which are indicated by * and ** (next to the numbers), respectively. The 42 species were 1, *Alstroemeria* sp. cv. Jacqueline; 2*, *Amaranthus retroflexus*; 3, *Apium graveolens*; 4, *Avena sativa* L. cv. Elgin; 5, *Capsicum annuum* cv. Cubico; 6, *Chrysanthemum morifolium*; 7, *Coleus blumei*; 8, *Cucumis sativus* cv. Revenue; 9*, *Flaveria bidentis*; 10**, *Flaveria chloraefolia*; 11**, *Flaveria floridana*; 12**, *Flaveria linearis*; 13, *Flaveria pringlei*; 14, *Flaveria robusta*; 15*, *Flaveria trinervia*; 16*, *Gomphrena globosa*; 17, *Helianthus annuus*; 18, *Hordeum vulgare*; 19, *Nepeta faassenii*; 20, *Nicotiana tabacum*; 21*, *Panicum antidotale*; 22, *Panicum bisulcutum*; 23*, *Panicum bulbosum*; 24*, *Panicum capillare*; 25**, *Panicum decipiens*; 26*, *Panicum dichotomiflorum*; 27, *Panicum laxum*; 28*, *Panicum leavifolium*; 29*, *Panicum makarikariense*; 30*, *Panicum maximum*; 31*, *Panicum miliaceum*; 32**, *Panicum milioides*; 33, *Panicum trichanthum*; 34*, *Panicum virgatum*; 35, *Phaseolus vulgaris*; 36, *Pisum sativum* cv. Improved Laxton's Progress; 37, *Rosa hybrida* cv. Samantha; 38, *Salvia splendens* cv. Bonfire; 39, *Sandersonia aurantiaca*; 40*, *Sorghum bicolor* cv. Sudan; 41, *Triticum aestivum* cv. Karat; 42*, *Zea mays*. Measurements were made under light saturating conditions and at the growing temperature (25°C except for the *Panicum* species, which was 30°C). Each point is an average of at least four leaves on four different plants. The SEs of the means are not shown for graphical clarity. Starting from top to bottom and from left to right, species' numbers that overlap within each panel were: A (30*, 31), (16*, 26*), (15*, 24*), (9*, 14), (34*, 32**), (28*, 30), (13, 18), (35, 10**, 5), and (22, 37); B (36, 15*), (32**, 6, 34*), and (27, 1); C (24*, 26*, 16*), (32**, 8), (18, 13), (29*, 28*, 30), and (10**, 5); and D (19, 16*), (34*, 25**), (32**, 29*), and (31*, 15*, 36).

tomato. Leaves of six $C_3$ dicots and four $C_4$ monocots from different families were compared. Their conclusion was substantiated by Gordon [2], who employed differential weight analysis and infrared gas analysis to estimate immediate export. Using a steady-state $^{14}CO_2$ labeling protocol to measure photosynthesis and immediate export rates, we have examined over 42 monocot and dicot species from different families and genera including a number from the genera *Panicum* and *Flaveria* [33,34]. We examined immediate export both as an absolute rate and as a rate relative to the fixation rate (i.e., as a percentage of photosynthesis). Collectively, the data for all 24 species showed that the faster photosynthesis was, the faster immediate export was (Figure 4). At ambient $CO_2$, there was a high correlation coefficient ($r = 0.88$) between the rate of photosynthesis and the absolute rate of immediate export (Figure 4A). Among all species and within each of the *Panicum* (#21–34) and *Flaveria* (#9–15) genera, photosynthesis and export rates of the $C_4$ species were higher than those of $C_3$ species (Figure 4Aii) [34]. Previous studies also showed that at ambient $CO_2$, $C_4$ species have higher translocation rates than $C_3$ species [2,70,87–89]. However, the concept that leaves with a functional $C_4$ metabolic pathway inherently export newly fixed C more readily than those with $C_3$ metabolism was challenged [33]. At ambient $CO_2$, the percentage of C exported immediately relative to photosynthesis was high in a number of $C_3$ dicot species (#3, 7, 8, 19, 38; Figure 4Ci) [33] that produce auxiliary transport sugars [39]. The notable exception was sunflower (#17; Figure 4Ci), which not only translocated sucrose but also had a relatively high immediate export capacity [33,70]. Among the *Flaveria* species, *F. robusta* (#14; Figure 4Cii), a $C_3$ that also translocated only sucrose, seemed to have a relatively high export flux.

An interesting finding was that the $C_3$-$C_4$ intermediate species can be very different in their ability to export C immediately [34]. When immediate $^{14}C$ efflux was examined relative to the rate of $^{14}C$ assimilation, "type I" $C_3$-$C_4$ intermediate *Panicum* species (#25\*\*, 32\*\*) exported newly acquired $^{14}C$ as quickly as the $C_4$ species (#21\*, 23\*, 24\*, 26\*, 28\*, 29\*, 30\*, 31\*, 34\*) did (Figure 4C). In contrast to this pattern, among the *Flaveria* species, the "type II" $C_3$-$C_4$ intermediates (#10\*\*, 11\*\*, 12\*\*) had the lowest export rates of the three photosynthetic types [34]. Collectively, the data in Figure 4A and C show that the $C_3$-$C_4$ intermediate type I and type II species of the two genera behave differently with respect to immediate export. The reason for this difference in immediate export capacity remains unclear.

In both type I and type II $C_3$-$C_4$ intermediate species, special anatomy and biochemistry lead to reduced rates of photorespiration compared with those of $C_3$ species [40,90–92]. In leaves of type I $C_3$-$C_4$ intermediates, the mitochondrial enzyme glycine decarboxylase is localized in the bundle sheath cells [90,91]. Photorespired $CO_2$ that is released in the bundle sheath may be refixed by Rubisco before escaping from the leaf and result in reduced rates of apparent photorespiration at ambient $CO_2$ [92]. Anatomical features such as partially developed Kranz anatomy and localization of a higher number of organelles (e.g., mitochondria) in the bundle sheath cells would further facilitate the refixation of $CO_2$ [40,91,93–96]. In addition to compartmentation of glycine decarboxylase in the bundle sheath cells [90,91], some elements of $C_4$ metabolism are found in the type II $C_3$-$C_4$ intermediate species [40,97,98]. Although not as well developed as in $C_4$ *Flaveria* species, aspects of Kranz-type anatomy are also evident [40,93]. Clearly, both anatomical and biochemical characteristics need to be considered to explain why the type I $C_3$-$C_4$ intermediate *Panicum* species export newly fixed $^{14}C$ as quickly as their $C_4$ cousins whereas the type II $C_3$-$C_4$ intermediate *Flaveria* species exported less $^{14}C$ (Figure 4C).

Consistent with the expected suppression of photorespiration and the increased availability of $CO_2$ for fixation [84,92,99,100], short-term $CO_2$ enrichment increased photosynthesis in all $C_3$, and type I and type II $C_3$-$C_4$ intermediate species (Figure 4B). In most species except for a few $C_4$ species, the absolute export rate increased at high $CO_2$ but not proportionally with photosynthesis (Figure 4B). In all species the relative export rates decreased under $CO_2$ enrichment (Figure 4D). Collectively, these data indicate that during $CO_2$ enrichment all species tended to accumulate excess C in their leaves in the light. Plant productivity of $C_3$ and $C_3$-$C_4$ intermediate and $C_4$ species depends on many factors including the ability of the leaves to export C immediately [4,5,30,40]. More data are required to determine whether these extra reserves of C support export and new growth under sustained $CO_2$ enrichment [46,79].

## V. SUMMARY

Over the last half of the 20th century the availability of radioisotopes of C (e.g., $^{14}C$) led to the elucidation of major photosynthetic processes in algae and higher plants. For example, the discovery of the Calvin cycle [35] helped to define the manner in which net $CO_2$ assimilation occurs in all plants [11,86] and predicted the involvement of secondary regeneration cycles such as photorespiration [32,36,37]. Although the first function of the leaf as the major site of C fixation is fairly well understood today, the second function of the leaf as a source of reduced C for developing sinks is not well understood at the whole plant level [4,5,101]. Terms such as source strength and sink demand define concepts affecting C partitioning and allocation in plant tissue [101]. However, a limited number of methodologies provide quantitative data regarding fundamental C fluxes and exchanges such as the immediate export rate from source leaves. The mass ($^{13}$C) and radioactive ($^{11}$C, $^{14}$C) isotopes of C are valuable probes for quantifying assimilate movements within the plant. The potential exists to use the mass isotopes such as $^{13}$C and $^{15}$N more extensively; however, to date, user-friendly noninvasive techniques for mature plants have not been devised. Although the radioisotope $^{11}$C emits particles of sufficient energy to be used to study phloem transport in a noninvasive manner, its use has been restricted. The more stable form of C, $^{14}$C, has limitations as a noninvasive probe. However, it remains a powerful tool in studying export during photosynthesis. The steady-state labeling methodology outlined in this chapter, which depends on measurements of immediate export being made when the transport sugar pools are in isotopic equilibrium with the $^{14}CO_2$ being assimilated, provides estimates of mass transfer rates of C through export during photosynthesis.

As pointed out in this chapter, plant productivity in natural photosynthetic variants ($C_3$, $C_3$-$C_4$ intermediate, and $C_4$ types) depends on many factors including the ability of the leaves to export C [4,5,30,40]. During the last 25 years, researchers have attempted to increase productivity of plants through genetic engineering by altering the primary metabolic steps involved in the reduction of $CO_2$. It has been proposed that modifications of properties of key photosynthetic enzymes such as Rubisco [25,102,103] or the transfer of $C_4$ genes into $C_3$ species will alter leaf photorespiratory and photosynthetic capacity [104]. Furbank and Taylor [102] noted that there are many challenges in the area of photosynthesis to use the large bulk of data on the enzymes of the pathway and their regulation. Both fixation and export are functions of the leaf. Much more data are needed at the whole plant level. Integration of photosynthesis and export processes and not merely enzymes of C metabolism are required to understand how specific site mutations affect diurnal patterns of C partitioning. With the integration of techniques such as the use of metabolic engineering coupled with traditional biochemical and physiological approaches, we may develop the means to improve photosynthetic performance, assimilate partitioning, and growth in higher plants.

## REFERENCES

1. BR Fondy, DR Geiger. Plant Physiol 70:671, 1982.
2. AC Gordon. In J Cronshaw, JW Lucas, RT Giaquinta, eds. Plant Biology, Vol 1, Phloem Transport. New York: Alan R Liss, 1986, p 499.
3. W Kalt-Torres, PS Kerr, H Usuda, SC Huber. Plant Physiol 83:283, 1987.
4. IF Wardlaw. New Phytol 116:341, 1990.
5. DR Geiger, JC Servaites. Annu Rev Plant Physiol Plant Mol Biol 45:235, 1994.
6. MJ Canny. Phloem Translocation. Cambridge, UK: Cambridge University Press, 1973.
7. DR Geiger. Methods Enzymol 69:561, 1980.
8. JA Milburn, J Kallarackal. In (DA Baker, JA Milburn, eds. Transport of Photoassimilates. New York: John Wiley & Sons, 1989, p 264.
9. G Olrich, E Komor. Methods Enzymol 174:288, 1989.
10. JF Farrar. In DO Hall, JMO Scrulock, HR Bolhar-Nordenkampf, RC Leegood, SP Long, eds. Photosynthesis and Production in a Changing Environment. London: Chapman & Hall, 1993, p 232.
11. HW Heldt. Plant Biochemistry and Molecular Biology. Oxford, UK: Oxford University Press, 1997.
12. CA Peterson, HB Currier. Physiol Plant 22:1238, 1969.
13. YV Gamalei. Fiziol Rast 32:866, 1985.
14. M Knoblauch, AJE van Bel. Plant Cell 10:35, 1998.
15. R Côté, JM Gerrath, CA Peterson, B Grodzinski. Plant Physiol 100:1640, 1992.
16. MD Fricker, KJ Oparka. J Exp Bot 50:1089, 1999.
17. PEH Minchin. Short-Lived Isotopes in Biology. Proceeding of an International Workshop on Biological Research with Short-Lived Isotopes. DSIR, Lower Hutt, New Zealand, 1986.
18. J Jiao, B Grodzinski. Plant Physiol 111:169, 1996.
19. J Verscht, B Kalusche, J Köhler, W Köckenberger, A Metzler, A Haase, E Komor. Planta 205:132, 1998.
20. J Weibull, F Ronquist, S Brishammar. Plant Physiol 92:222, 1990.
21. G Ntsika, S Delrot. Physiol. Plant 68:45, 1986.
22. IJ Tetlow, JF Farrar. J Exp Bot 44:929, 1993.
23. S Delrot. Plant Physiol 68:706, 1981.

24.  C Kühn, L Barker, L Bürkle, WB Frommer. J Exp Bot 50:935, 1999.
25.  WB Frommer, U Sonnewald. J Exp Bot 46:587, 1995.
26.  MH Zimmermann, H Ziegler. In: MH Zimmermann, JA Milburn, eds. Transport in Plants, Phloem Transport, Encyclopedia of Plant Physiology, New Series, Vol 1. New York: Springer-Verlag, 1975, p 480.
27.  M Stitt, S Huber, P Kerr. In: PK Stumpf, EE Conn, eds. The Biochemistry of Plants: A Comprehensive Treatise. Vol 10, Photosynthesis (MD Hatch, NK Boardman, eds.) London: Academic Press, 1987, p 328.
28.  AJE van Bel. Annu Rev Plant Physiol Plant Mol Biol 44:253, 1993.
29.  R Turgeon. Trends Plant Sci 1:418, 1996.
30.  ED Leonardos. Implications of Photosynthetic Pathways on C Export in Source Leaves of $C_3$, $C_3$-$C_4$ Intermediate and $C_4$ Panicum and Flaveria Species. PhD thesis, University of Guelph, Guelph, 1999.
31.  M Madore, B Grodzinski. Plant Physiol 76:782, 1984.
32.  B Grodzinski, Bioscience 42:517, 1992.
33.  B Grodzinski, J Jiao, ED Leonardos. Plant Physiol 117:207, 1998.
34.  ED Leonardos, B Grodzinski. Plant Cell Environ 23:839–851, 2000.
35.  JA Bassham, M Calvin. The Path of Carbon in Photosynthesis. Englewood Cliffs, NJ: Prentice-Hall, 1957.
36.  DW Husic, HD Husic, NE Tolbert. CRC Crit Rev Plant Sci 5:45, 1987.
37.  NN Artus, SC Somerville, CR Somerville. CRC Crit Rev Plant Sci 4:121, 1986.
38.  SC Huber, JL Huber. Annu Rev Plant Physiol Plant Mol Biol 47:431, 1996.
39.  R Turgeon. In: MA Madore, WJ Lucas, eds. Carbon Partitioning and Source Sink Interactions in Plants, Current Topics in Plant Physiology. Vol 13. Rockville, MD: American Society of Plant Physiologists, 1995, p 195.
40.  GE Edwards, MSB Ku. In: MD Hatch, NK Boardman, eds. The Biochemistry of Plants. Vol 10, Photosynthesis. London: Academic Press, 1987 p. 275.
41.  R Ohsugi, SC Huber. Plant Physiol 84:1096, 1987.
42.  N Terry, DC Mortimer. Can J Bot 50:1049, 1972.
43.  LC Ho. J Exp Bot 27:87, 1976.
44.  JE Silvious, DF Kremer, DR Lee. Plant Physiol 62:54, 1978.
45.  DL Hendrix, SC Huber. Plant Physiol 81:584, 1986.
46.  C Grimmer, E Komor. Planta 209:275, 1999.
47.  RK Monson, JA Teeri, MSB Ku, J Gurevitch, LJ Mets, S Dudley. Planta 174:45, 1988.
48.  GD Farquhar, JR Ehleringer, KT Hubick. Annu Rev Plant Physiol Plant Mol Biol 40:503, 1989.
49.  S von Caemmerer, KT Hubick. Planta 178:475, 1989.
50.  H Schnyder, R de Visser. Plant Physiol 119:1423, 1999.
51.  T Hirano, N Uchida, T Azuma, T Yasuda. Jpn J Crop Sci 66:675, 1997.
52.  JW Warringa, MJ Marnissen. Neth J Agric Sci 45:505, 1997.
53.  PC Lauterbur. Nature 242:190, 1973.
54.  B Kalusche, J Verscht, G Gebauer, E Komor, A Haase. Planta 208:358, 1999.
55.  LW Jelinski. In: FA Bovey, ed. Nuclear Magnetic Spectroscopy. New York: Academic Press, 1988, p 489.
56.  RG Thompson, DS Fenson, RR Anderson, W Leiper. Can J Bot 57:845, 1979.
57.  PEH Minchin. J Exp Bot 30:1171, 1979.
58.  B Grodzinski, S Jahnke, R Thompson. J Exp Bot 35:678, 1984.
59.  WF Pickard, PEH Minchin. J Exp Bot 41:1361, 1990.
60.  R Côté, RG Thompson, B Grodzinski. J Exp Bot 23:819, 1992.
61.  WF Pickard, PEH Minchin, MR Thorpe. J Exp Bot 44:1491, 1993.
62.  N Keutgen, GW Roeb, PEH Minchin, F Führ. J Exp Bot 46:489, 1995.
63.  ED Leonardos, MJ Tsujita, B Grodzinski. Physiol Plant 97:563, 1996.
64.  DS Fensom, EJ Williams, DP Aikman, JE Dale, J Scobie, KWD Ledingham, A Drinkwater, J Moorby. Can J Bot 55:1787, 1977.
65.  L Woodrow, RG Thompson, B Grodzinski. J Exp Bot 39:667, 1988.
66.  PEH Minchin, JH Troughton. Annu Rev Plant Physiol 31:191, 1980.
67.  P Young. Recursive Estimation and Time-Series Analysis. An Introduction. Berlin: Springer-Verlag, 1984.
68.  M Madore, B Grodzinski. J Plant Physiol 121:79, 1985.
69.  DR Geiger, BR Fondy. Plant Physiol 64:361, 1979.
70.  G Hofstra, CD Nelson. Planta 88:103, 1969.
71.  Y Mor, AH Halevy. Plant Physiol 45:177, 1979.
72.  SC Farrar, JF Farrar. New Phytol 100:271, 1985.
73.  SC Farrar, JF Farrar. New Phytol 103:645, 1986.
74.  A Moing, F Carbonne, MH Rashad, J-P Gaudillere. Plant Physiol 100:1878, 1992.
75.  E Jeannette, J-P Rocher, J-L Prioul. Physiol Plant 94:319, 1995.
76.  J Moorby, PD Jarman. Planta 122:155, 1975.
77.  CJ Bell, LD Incoll. J Exp Bot 33:896, 1982.
78.  J Jiao, P Goodwin, B Grodzinski. Can J Bot 74:1, 1996.
79.  J Jiao, B Grodzinski. J Am Soc Hortic Sci 123:1081, 1998.
80.  J Jiao, P Goodwin, B Grodzinski. Plant Cell Environ 22:15, 1999.

81. Y Shishido, H Challa, J Krupa. J Exp Bot 38:1044, 1987.
82. M Dixon, J Grace. Plant Cell Environ 5:323, 1982.
83. B Grodzinski, J Jiao, VL Knowles, WC Plaxton. Plant Physiol 120:887, 1999.
84. GE Edwards, MSB Ku, MD Hatch. Plant Cell Physiol 23:1185, 1982.
85. GE Edwards, DA Walker. $C_3$ and $C_4$: Mechanisms and Cellular and Environmental Regulation of Photosynthesis. Oxford: Blackwell Scientific, 1983.
86. MD Hatch. Biochim Biophys Acta 895:81, 1987.
87. WM Lush, LT Evans. Aust J Plant Physiol 1:417, 1974.
88. RN Gallaher, DA Ashley, RH Brown. Crop Sci 15:55, 1975.
89. WM Lush. Planta 130:235, 1976.
90. S Rawsthorne, CM Hylton, AM Smith, HW Woohouse. Planta 176:527, 1988.
91. CM Hylton, S Rawsthorne, AM Smith, DA Jones, HW Woolhouse. Planta 175:452, 1988.
92. Z Dai, MSB Ku, GE Edwards. Planta 198:563, 1996.
93. RH Brown, PW Hattersley. Plant Physiol 91:1543, 1989.
94. RH Brown, JH Bouton, LL Ridgsby, M Rigler. Plant Physiol 71:425, 1983.
95. RH Brown, LL Ridgsby, DE Akin. Plant Physiol 71:437, 1983.
96. S Rawsthorne. Plant J 2:267, 1992.
97. SBM Ku, RK Monson, RO Littlejohn, H Nakamoto, DB Fisher, GE Edwards. Plant Physiol 71:944, 1983.
98. RK Monson, BD Moore, MSB Ku, GE Edwards. Planta 168:493, 1986.
99. SBM Ku, GE Edwards. Plant Cell Physiol 19:665, 1978.
100. SBM Ku, J Wu, Z Dai, RA Scott, C Chun, GE Edwards. Plant Physiol 96:518, 1991.
101. JF Farrar. Plant Cell Environ 16:1013, 1993.
102. RT Furbank, WC Taylor. The Plant Cell 7:797, 1995.
103. SM Whitney, S von Caemmerer, GS Hudson, TJ Andrews. Plant Physiol 121:579, 1999.
104. CC Mann. Science 283:314, 1999.

# 20

# Production-Related Assimilate Transport and Partitioning

**John E. Hendrix**

*Colorado State University, Fort Collins, Colorado*

## I.  INTRODUCTION

The prime assimilatory process is photosynthesis. Assimilation products must be transported to a plant part (sink) if it is to grow. It is important to maximize the proportion of total assimilate pool that is partitioned into plant parts that are harvested for their economic value. Therefore, any procedure one might use to enhance partitioning of assimilate into these sinks will serve to increase economic gain. This involves production of assimilates, loading of assimilates into phloem, their transport through phloem (translocation), and their unloading from phloem into appropriate sinks and incorporation into sink materials. Also, if pesticides are to be used in the most effective and benign manner, their transport must be understood.

If phloem function and assimilate partitioning are to be related to crop productivity, several questions must be addressed:

How are materials partitioned among plant parts?
What materials are translocated and why these and not others?
What are the rate and velocity of assimilate translocation?
What is the mechanism of translocation through sieve tubes?
How are materials loaded into and unloaded from sieve tubes?
What is the mechanism of xenobiotic transport?
How are these processes controlled?

Then we must learn how genetics and environment interact to control partitioning. It is apparent that the study of gene expression and activity as well as control of proteins coded for by these genes is becoming critical in our efforts to advance understanding of crop productivity.

These topics are not fully understood, but they have been studied extensively and discussed in several reviews and symposium publications (see, e.g., Refs. 1–10). The effort here is directed toward relating these topics to crop productivity.

## II.  CROP PRODUCTIVITY

## A.  Assimilate Partitioning

### 1.  General Considerations

Total biomass production of a plant is dependent on the balance between photosynthesis and respiration. Therefore, it might seem appropriate to develop genetic or cultural strategies to control these two processes. Cultural practices have been directed primarily toward increasing total biomass by increasing plant density and use of fertilizer and/or water, with the assumption that increased biomass would result in increased economic productivity. Yet it has long been known that excessive use of nitrogen or excessively high plant densities often lower production of economically important plant parts.

At critical stages in the life of a plant, environment strongly influences the development of economically important plant parts. Therefore, cultural practices have the potential of influencing harvest index (HI). Understanding the timing of development becomes increasingly important as certain inputs such as water, pesticides, and fertilizer become more expensive, difficult to obtain, or use restricted. For the most part, genetic selection of crop plants over the past century has altered biomass partitioning (changing HI) but has not resulted in an increase in biomass [11,12].

Assimilate partitioning (see reviews by Wardlaw [9] and Pollock et al. [13]) includes the partitioning of all assimilated materials among plant parts. One of the most comprehensive models for carbon and nitrogen partitioning was developed by Pate et al. [14]. This topic is important for both theoretical and applied plant physiology. Control of the incorporation of assimilates into economically important components of a crop determines economic reward.

Processes that control assimilate partitioning are cell-to-cell transport, including transfer of materials between xylem and phloem, loading and unloading of vascular tissues; long-distance translocation through vascular tissues; and metabolic sequestration of materials such that they are either temporarily or permanently eliminated from transport processes. Several of these topics are discussed in detail in other sections but are included here to emphasize their impact on carbon partitioning.

### 2.  Developmental Considerations

As plants develop, assimilates are partitioned differently at different times [15]. During seed germination, the radical elongates first. A day or two later, the plumule growth rate increases. Metabolic control of the development of source-sink relations in cereal seedlings was discussed by Thomas and Rodriguez [16]. As each leaf matures in sequence, it converts from sink to source [17] and enzyme activity is modified to accommodate these changes [18]. In plants of determinate growth habit (e.g., corn, wheat, barley, sunflower), essentially all vegetative growth is completed at flowering. For a short time there are few growing sinks, so assimilate is stored in vegetative parts. As fruit and seeds enlarge, not only is current photosynthate partitioned into reproductive parts but also, depending on circumstances, stored materials are remobilized. This remobilization of assimilate from stems of wheat has been well illustrated (see, e.g., Refs. 19 and 20). However, tomato plants remobilized proportionately less assimilate from vegetative parts to fruit; most of what they remobilized was from leaves [21]. During these changes, direction of translocation through the phloem often changes from primarily downward to primarily upward.

Patterns for indeterminate annuals appear to be similar, although more complex, for most of their reproductive growth occurs with the first group of flower, yet they do have a capacity for continued vegetative growth and to form later fruit if early fruit is lost. In perennials, there are several variations of partitioning patterns. In many plants, however, flowering is followed by early rapid vegetative growth using assimilates from the previous season. The major portion of assimilate is then used in reproductive growth. Then assimilate is stored [22]. In some plants such as elm, reproductive growth is completed in early spring before appreciable vegetative growth occurs. Many other variations could be cited for perennial plants. For example, *Agave* develops much like a determinate annual but with the pattern extending over several years.

With development, the chemical mix of materials translocated is modified. Pate [23] reported increasing concentrations of nitrogenous compounds in phloem sap during seed growth of lupine. Glad et al. [24] reported a similar changing pattern of sieve tube sap composition for grape.

## 3. Harvest Index

The mechanism by which increase in economic productivity has been attained is by increasing total dry matter production and by increasing the proportion of assimilate partitioned into the economically important plant part (i.e., increasing Hl). It should be pointed out that Hl is somewhat misleading when total assimilate partitioning is considered, for Hl usually accounts for only aboveground parts of a crop. Dunan and Zimdahl [25] reported that roots accounted for 13.6% and 16.4% of the dry mass of barley and oats, respectively. Pate et al. [14] did the most complete carbon balance study known. Unfortunately, their study plant was *Lupinus albus*, which has a rather large tap root. Also, in considering their data, one must realize that gas exchange studies account for net photosynthesis of the shoot, yielding low values for actual photosynthesis and shoot respiration, while root gas exchange represents 24-hr root respiration. Considering those limitations, Pate and coworkers reported that about 44% of the carbon fixed by net photosynthesis was used by roots, with about one fourth of that going into growth. In addition, the root nodules used about 12% of the photosynthate in growth and respiration. Nodules used additional carbon in the formation of nitrogenous compounds that were supplied to the rest of the plant. These computations did not account for carbon that may be lost as exudate, in sloughed cells, or use by other organism. Buwalda [26], in his review of perennial crops, stated that mycorrhizal fungi account for 5 to 10% of total carbon acquired by photosynthesis. He also stated, "For mature plants, root growth is . . . a relatively small sink for carbon." Reports of high root/shoot ratios for plants that have perennial roots and annual shoots [27] do not negate Buwalda's statement. Increasing partitioning to roots may enhance water and nutrient absorption, thereby increasing production efficiency.

The genetic component of increased crop productivity has not been assessed with regard to the proportion of assimilate partitioned to roots; rather an increase in Hl based on analysis of aboveground parts only has been demonstrated. These changes in Hl have been associated with shorter plants for both small grains and soybeans [11,12,28] but not with increased rates of photosynthesis (carbon fixed per unit time and leaf area).

Economic yield is also increased by increasing total dry matter yield without altering Hl. Increase in total yield, as well as Hl, is influenced by cultural practice, environment, and genotype. Gifford [12] compiled data for several crops to determine the basis of increased crop production over the years. Those data indicate that total shoot yield and Hl have increased for all of the crops he studied. Hl was increased by genetic selection for high yield.

## 4. Source–Sink Ratios

It is also clear that an alteration in the pattern of assimilate partitioning occurs in response to crop thinning or removal of plant parts by pruning, herbivory, or violent weather. Various manipulative experiments have been performed to develop an understanding of the control of assimilate partitioning. One involved bean seedlings with fully expanded primary leaves as the source and a small, rapidly expanding first trifollate as a sink. The experimenters removed the terminal leaflet of the trifoliate, leaving the two lateral leaflets as sinks for each primary leaf, respectively [29]. Using $^{14}CO_2$, they demonstrated that each leaflet received ~80% of its carbon from the nearest primary leaf. When one primary leaf was removed, the amount of assimilate translocated into the two leaflets did not decrease. The remaining primary leaf became the source for both, doubling its export to the leaflets without changing its rate of photosynthesis.

Loss of leaf area from insect, hall, or experimental desiccation [30] during grain or seed filling does not decrease production in proportion to the loss of leaf area because there is an increase in the utilization of stored carbohydrates. In addition, over longer time periods than in the experiments cited, remaining leaf tissue increases its rate of photosynthesis. Water stress can also influence partitioning [31].

How does this redirection of partitioning occur? The short-term response is as indicated by the work already discussed [29]. We should think of the vascular system of plants as a pipeline distribution system that runs vertically in the stem with interconnections at nodes. Pressure is greatest where loading is greatest and least where unloading is greatest. Flow follows pressure gradients. In addition, greatest resistance in the system is across the stem at the nodal interconnections. Therefore, with no perturbations within the system, most translocation is vertical with little movement across nodal interconnections [32]. However, if the system is altered by removal of sources or sinks, cross-movement becomes significant. Longer term adjustments are made in photosynthetic rates and in utilization of stored assimilates.

## 5.  Carbohydrate Metabolism

Of course, a plant's development and response to the environment are mediated by gene expression. Therefore, the demonstration by Xu et al. [33] that the expression of different isozymes of carbohydrate metabolism is either enhanced or suppressed by the environment is of great importance but not surprising. For the establishment of economic sinks in many plants, starch synthesis is essential (see review by Smith et al. [34]). Also, the understanding of sugar-metabolizing enzymes is illustrated for sugarcane in that differences in expression of acid invertase and sucrose phosphate synthase determine the accumulation of sucrose in different cultivars [35]. In addition, Geiger et al. [36] reported that invertase activity in developing bean pods and corn grains increased before the rate of assimilate import increased, indicating that utilization of sucrose enhanced its import (also see Sturm [37]). It is also interesting to note that these events occurred a few days before the time of abortion, suggesting that assimilate supply is critical in inhibiting abortion. Zeng et al. [38] demonstrated that activity of invertase in maize roots was markedly lowered by anoxia (0% $O_2$) or hypoxia (3% $O_2$). Because low $O_2$ would limit the root's ability to metabolize sugars, this seems reasonable, for lowered use of sucrose would lower the sink capacity; however, the activity of sucrose synthase was not lowered by these treatments. This supports Sturm's view [37] that invertase activity is important in controlling sink strength.

In her review, Koch [39] categorized "famine" and "feast" genes. "Famine" genes enhance supply and suppress utilization of assimilates, whereas "feast" genes operate in the opposite way. An ability to understand and control these genes may enhance productivity and HI. Koch et al. [40] also reported that there are two genes for sucrose synthase in maize roots. One is maximally expressed in a sugar-depleted environment, the other in a high-sugar environment. Furthermore, they are expressed in different cells, potentially having an impact on assimilate partitioning.

## 6.  Floral Evocation

For plants from which sexual parts are harvested, maximum potential harvest is established at flower formation. Clearly, plants are variable in the stage of development at which reproduction is controlled. Some plants continue to flower after fruit form, but most of those later flowers abscise if the earlier fruit has not been removed. Ornamentalists have known for centuries that removal of old flowers and young fruit (carbohydrate sinks) markedly increases subsequent flowering. It is reasonable to conclude that assimilate supply is critical in determining degree of flowering, fruit, and seed set.

Hendrix et al. [41] demonstrated that the content of fructan in wheat inflorescences 7 days before anthesis was highly correlated with grain number. Bodson [42] and Bodson and Outlaw [43] reported that accumulation of carbohydrates in buds of *Sinapsis* was associated with floral evocation. In their review, Bodson and Bernier [44] stated:

> Available evidence suggests that an early change in carbohydrate concentration in the apical bud is critical to floral initiation, but that this modification is not sufficient. alone to trigger initiation. It is not possible to conclude whether assimilate accumulation in the reproductive structures is responsible for inflorescence development since the timing of events that are integral parts of reproductive development is generally very poorly known.

Several workers [45–47] have demonstrated that floral induction of the long-day plants *Sinapsis alba* and *Arabidopsis thallana* by one photoinductive cycle resulted in an increase in sucrose concentration in sieve tube exudate. Ishiora et al. [48] were able to induce floral development on cultured *Pharbitis* apices by raising sucrose and/or lowering ammonium concentrations. Nitrate did not inhibit floral induction. One might conclude that floral induction induces greater sucrose translocation. However, inductive photoperiods altered the diurnal timing of carbohydrate supplied from photosynthesis. That alteration, alone, may account for altered concentrations of sucrose in exudate. One could just as easily conclude that a floral inductive signal increases sink capacity of buds, thereby enhancing flowering, or that once floral development has been initiated, the degree of reproductive development is controlled by assimilate supply. It also appears that the importance of carbohydrate supply extends to shortly after anthesis, a period of potential abortion. Once the abortion period has passed, studies cited subsequently indicate that the capacity of sinks rather than sources is limiting.

It appears that in all crops, assimilate supply and partitioning patterns determine the degree of flowering and/or fruit set. The amount of fruit, seed, or grain that develops determines subsequent patterns of assimilate partitioning and may even feed back on supply by affecting photosynthesis. From an evolutionary standpoint, these patterns seem reasonable, for a plant will be more successful in producing progeny if it produces only the number of seeds that assuredly will be viable. So, at some stage, a plant must sense the number of seeds that it can successfully supply, even if a disaster, such as drought or loss of leaves to insects, should occur. Once the critical developmental stages have passed, a plant is committed to seeds that are set, and seed number is determined. One possible mechanism of the abortion process is suggested by the demonstration [49,50] that more basal sites on inflorescences are more adequately vascularized than more distal sites. Thus, a more distal location would receive adequate assimilate supply only if that supply was larger or if the more proximal sites were unoccupied. From the economic standpoint, the floral development period is critical, for it is during this period that maximum potential sink capacity is established. Subsequent events may diminish that capacity, but capacity cannot be increased when that critical time has passed.

## 7.  *Productivity of Specific Crops*

SMALL GRAIN    Shanahan et al. [51] demonstrated that grain number was a stronger indicator of winter wheat yields than grain size, even in the plains of eastern Colorado, where low moisture and high temperatures during grain filling often result in large variability in grain size.

Bremner and Rawson [52] reported that position within the spike and within each spikelet had a large influence on grain mass. In control plants, the largest grains within the spike were located about one third the distance from the base. Within each spikelet, the most basal grain was largest. Removal of some grains 9 days after anthesis resulted in some increase in size, primarily at the ends of the spikes. In no case did the most distal grain on the thinned spikelet attain the mass of those at the base of the unthinned spikelet. These investigators interpreted the results as indicating that the basal position of each spikelet had a more adequate vascular system supplying assimilates rather than suggesting a limiting supply of assimilates within the plants. That conclusion is supported by others [49,50].

Further evidence that sink capacity is limiting to production was presented by several others. Blade and Baker [53] demonstrated that changing the source/sink ratio by lowering plant density, removing developing grain, or removing the flag leaf had little influence on the mass of individual grains. Other work supports the conclusion that wheat [54] and oats [55] are sink limited during grain filling.

To illustrate the impact of lowered supply of photoassimilate on wheat grain production, Fischer [56] conducted several shading experiments. In his more severe treatments, shaded plants received only 35% of the natural light during a single 21-day treatment period. When the shading period was centered during vegetative development or early in floral development, there was little impact on grain production when compared with unshaded controls. The shading period centered near the midpoint of floral development lowered grain production to just over 40% of controls, shading centered at anthesis lowered yield to 80% of controls, whereas subsequent shading periods had progressively less impact, with the final period, which extended to maturity, having little impact. Work by Wardlaw [57] and Caldiz and Sarandon [58] supports these results. Supporting data were also provided by Kiniry [59] in the demonstration that shading sorghum plants during inflorescence development resulted in production of far fewer grains than in controls. Removal of the shade at anthesis resulted in larger grains than in controls, but not nearly enough larger to compensate for loss in number.

High temperature alone [60] or in combination with low light prior to anthesis resulted in lower kernel number [57,61] than high temperature at any other developmental stage. Either of these environmental impacts would lower assimilate accumulation during inflorescence development.

Willenbrink et al. [62] demonstrated that shading (50%) starting at 22 days after anthesis had little impact on grain mass or number; loss of assimilate from photosynthesis was made up by added mobilization of stored fructan from vegetative parts. They further demonstrated that removing about two thirds of the grains increased individual grain mass only slightly. These data added support to the idea that grain production is sink limited during grain filling.

High temperature during grain filling inhibited starch synthesis [63] and shortened the grain-filling period [64,65], thereby lowering production of wheat by lowering grain mass [66]. Tashiro and Wardlaw [67] obtained similar results for rice and in addition demonstrated that starch content was more negatively influenced at elevated temperatures than protein content. Jenner and coworkers [68–70], studying the impact of high temperatures on starch-synthesizing enzymes of wheat and barley, found lower activity whether high temperature was applied in vivo or in vitro. They also reported that plants treated with high temperatures were able to recover some lost enzyme activity after being returned to a cool environment.

It appears that environmental impacts during grain filling do not lower yield as much as comparable environments during inflorescence development. A further indication that assimilate supply during vegetative growth does not limit grain production was the observation that grazing of wheat prior to floral initiation did not lower grain yield [71].

MAIZE   Maize appears to be more closely balanced than small grain on the basis of source-sink capacity during grain filling. Jones and Simmons [72] demonstrated that removing a portion of the developing ear had little if any impact on grain size, whereas defoliation of the plants 12 days after midsilking resulted in lower grain number and mass per grain and defoliation 24 days after midsilking resulted in only lowered mass per grain. Defoliation at either date resulted in rapid depletion of carbohydrates from stems, while control plants had repartitioned about 40% of their maximum stem mobilizable carbohydrates at maturity. Maize is more sensitive than wheat to loss of photosynthetic capacity during grain filling, for stem-stored carbohydrates are mostly sugars [72] rather than larger polymers, resulting in a lower storage capacity. Shading corn plants during reproductive development lowered the quantity of harvested grain more that shading at any other time [73]. Unlike the results for wheat, shading during the vegetative stage caused significantly lower grain production. None of the shading treatments caused lower stover yield. Setter and Flannigan [74] demonstrated that decreasing source by shading early in grain development resulted in a correlation between lowered number of endosperm nuclei and lowered grain dry mass. Work by Hueros et al. [75] indicated that development of transfer cells at the base of developing grains is critical to grain filling and prevention of abortion, probably because these cells are involved in supplying assimilate to developing grain. Another interesting phenomenon in maize is the relationship of carbohydrate metabolism in "sweet corn." A lowered ability to accumulate starch [76,77] resulted in lower grain mass, which was then reflected in lower HI.

LEGUMES   It is more difficult to associate the pattern of carbohydrate partitioning and accumulation with reproductive activity in indeterminate plants. Aufhammer and coworkers [78–80] demonstrated that unlike the effect in wheat, lowering the number of seed-bearing sites by removal of flower buds from *Vicia faba* did not lower production, for these plants were able to compensate by increasing seed number and/or size at other sites. In addition, they demonstrated that removal of basal flower buds increased the fruit set at more distal positions. It is likely that some minimum carbohydrate concentration must be met if buds are to develop into flowers and then into fruit. Sage and Webster [81] demonstrated that more distal buds, flowers, or fruit of a raceme of *Phaseolus* are more likely to abort than those at the more proximal positions. These studies imply that development at more distal positions was limited by insufficient assimilate supply. Mauk and Breen [82] were able to support that hypothesis with data of $^{14}C$ assimilation studies. This appears to be parallel to the conclusion of Fischer and HilleRisLambers [50] that more distal reproductive sites are limited by vascularization. White et al. [83] reported that genetically determined seed size and yield (including HI) are negatively correlated in *Phaseolus vulgaris*. This suggests that the growth of small seeds in a basal fruit on a raceme or on a basal raceme of a stem results in less depletion of the carbohydrate supply, thus allowing more fruit and/or seeds to develop. However, Stockman and Shibles [84] showed that neither increasing light intensity nor removing leaves altered flower and pod abscission. They concluded that carbohydrate supply was not immediately involved in abscission.

Kelly and Spanswick [85] demonstrated that competition for assimilates between vegetative and reproductive growth can affect seed size. Using nearly isogenic lines of pea, they found that the lines that stopped vegetative growth early and even lost their leaves produced larger seeds, even though their total seasonal photosynthesis was less than that of lines that maintained active vegetative growth and photosynthesis longer.

Wiles and Wilkerson [86] studied soybean production in competition with cocklebur. They reported that little production loss occurred if the cocklebur plants were removed before the fifth week of plant growth, about the time the soybean plants started to flower. Losses increased as time of competition continued through the 16th week, when flowering was complete. No additional loss of seed production occurred when cocklebur was allowed to compete to harvest. These results are supported by result for shading of indeterminate *Vicia faba* [79,80] beginning at the start of flowering. Those treatments resulted in lowering of seed number but not seed size. As with the shading studies cited earlier for wheat, these data support the hypothesis that the period of flower development is the most critical portion of the life of crop plants if sexual parts are economically important.

A positive relationship between seed number and yield, as indicated earlier for grain, has been demonstrated for soybeans [87]. Guffy et al. [88] also demonstrated a stronger correlation with nitrogen supply (nodulated vs. nonnodulated) than total assimilate. It should be pointed out that nitrogen fixation is an energy- (assimilate-) expensive process [89]; therefore, competition by weeds, cited before, could result in lowered $N_2$ fixation rates. Salado-Navarro et al. [90] observed that 80 to 90% of the plant's protein is in seeds at harvest, further indicating the importance of nitrogen partitioning to soybean production. It should also be noted that much of this nitrogen that accumulates in soybean seeds is stored temporarily in specialized cells within leaves [91].

TREE CROPS    It is common for some fruit trees, especially certain apple varieties, to have a pattern of alternate years of heavy and light crops [92]. Floral initiation occurs in late summer, a time when a large crop would compete for carbohydrates. This would result in low flower bud formation and few flowers the following spring. During the subsequent summer of a small crop, there would be little competition for carbohydrates by the crop, so large numbers of flower buds would form. A common practice has been to break this alternate year cycle by thinning the heavy crop. Unfortunately, a frost during flowering one year can restart this cycle. This scenario is supported by Ryugo [93]; however, Westwood [92] seems to favor a concept of hormonal control. It is likely that there is an interaction between carbohydrate supply and hormones that provides control of floral development.

Treating apple or peach trees by shading (10% natural light) or with a photosynthetic inhibitor [94,95] induced abscission of fruit comparable to a "June drop" but markedly increased the proportion of fruit lost. Minchin et al. [96] demonstrated that lowering the supply of photosynthate available to two similar young apple fruit on a single spur resulted in one being favored in the receipt of the limited supply. If this limitation of assimilate had continued, it seems likely that the less favored fruit would abscise. These observations indicate that a similar mechanism based on assimilate supply controls sexual reproduction in trees as in annual plants.

Miller and Walsh [97] compared partitioning of assimilate in peach trees where fruit had been either thinned or not. This is a matter of economic importance, for thinning is a common cultural practice used to obtain larger fruit, which have greater economic value. The HI of unthinned trees was 0.50 but that of thinned trees only 0.37. It is interesting that the economically valuable part of fruit, the fleshy mesocarp, is markedly increased by thinning, but the size of the energy-expensive seed is not increased nearly as much. In a closely related tree crop, almond, from which the seeds are marketed, thinning is not practiced for it would result in economic loss. Girdling of table grapes has been a common practice for many years for it is effective in trapping carbohydrate above the girdle to increase fruit size.

Little work has been done on evergreen trees except for citrus. In their review, Goldschmidt and Koch [98] reported that citrus trees differentiate flower buds in the winter, for, unlike deciduous trees, they have a continuing supply of photosynthate. Even so, a supply of stored assimilate seems to be important to floral development, for some varieties bear crops in alternate years. In "off" years 'Wilking' mandarin trees accumulated more starch and produced more flowers than in "on" years. The impact of thinning on fruit size and total production is as reported above for other fruit.

VEGETATIVE CROPS    Unlike growth of seeds and fruit, growth of vegetative sinks appears to be limited by source capacity rather than sink capacity. Working in a cool climate, Engels and Marschner [99–101] reported on a series of experiments with potato. They demonstrated that current photosynthate was rapidly used in tuber growth. In addition, tubers that initiated only 2–4 days after the first initiated tubers were at a large disadvantage in accumulating assimilate. Fourteen days after first tuber initiation, tubers that initiated only 4 days later were about one tenth the size of earlier tubers and were growing more slowly.

When Engels and Marschner altered source/sink ratios by removing over half of the tuber mass, total tuber growth rate ($cm^3$ $plant^{-1}$) returned to the previous rate within 4 days. When 50% of the leaves were removed with no reduction in tuber mass, tuber growth rate was halved almost at once. Both of these experiments support their conclusions about immediate use of photosynthate in tuber growth and that growth of potato tubers was source limited. However, Midmore et al. [102,103] reported that shading potato plants, especially early in development, enhanced tuber production. This would seem to contradict the preceding study, but Midmore's work was done in Peru, where soil temperatures were reportedly as high as 32°C for the unshaded plots. It seems likely that lowering soil temperature was more important than loss of light in a region of high temperatures and high irradiance. They used this study to demonstrate the desirability of intercropping potatoes and corn. Although differences between vegetative and sexual crops have been emphasized, the demonstration that high sucrose concentrations enhanced potato tuber formation by cultured potato stems [104] is reminiscent of the impact of sugar supply on flower bud differentiation [42,43].

In a study of sweet potatoes, Nakatani et al. [105] made several reciprocal grafts between various clones to obtain differences in source and sink capacities. They reported that tuber dry mass was correlated with leaf area. Because there was no difference in rate of photosynthesis (mg $CO_2$ $dm^{-2}$ $hr^{-1}$), they concluded that rate of tuber growth was limited by the source. This principle of source limitation probably applies to forage crops as well.

## B. Metabolic Consequences of Source/Sink Manipulation

During the day, as much as half of the carbon fixed in current photosynthesis is incorporated into starch within chloroplasts [106]. This is critical for the well-being of a plant, for starch degradation at night supplies carbohydrate for respiration of that leaf and for export to the rest of the plant to maintain respiration and growth. The importance of this reserve starch was illustrated by transgenic *Arabidopsis* with limited ability to synthesize starch [107]. As a result of that metabolic deficiency, soluble carbohydrates accumulated in leaves, inhibited photosynthesis, and resulted in little carbohydrate being available for night export, thereby limiting growth to daytime.

When sinks are limiting, accumulation of assimilate can feed back and inhibit photosynthesis [106]. Claussen and Lenz [108], working with eggplant, demonstrated that removal of fruit resulted in accumulation of sucrose and starch in source leaves. They also reported that this treatment lowered the activity of sucrose phosphate synthase and the rate of photosynthesis. These changes were observed over periods of days to weeks. Azcon-Bieto [109] demonstrated inhibition of photosynthesis within hours in wheat leaves when export of assimilate was inhibited. The feedback mechanisms of assimilates on the enzymes of photosynthesis are quite complex and not completely understood. For reviews on this subject, see Refs. 106 and 110–114. In addition to direct feedback on photosynthesis, Mandahar and Garg [115] reported that removal of sinks from barley plants resulted in loss of chlorophyll from flag leaves.

## C. Environmental Control

This section is intended not to cover in full detail the many factors of the environment that influence plant productivity but rather to relate environmental control to factors discussed earlier. Adjustments in environment, genetics, or cropping cycle to enhance floral development and seed or fruit set are based on the realization that sink development will increase HI and very likely yield.

Plants have an ability to adjust partitioning when exposed to different environments. Logendra et al. [116] exposed tomato plants to photoperiods of 8, 16, or 20 hr each day. Those exposed to the shortest light periods retained a higher proportion of their photosynthate for later export during the long dark period, thereby maintaining some supply for growth and maintenance during that dark period. This occurred even though the total amount of carbohydrate accumulated in the light period by leaves of plants with the shortest photoperiod was less than that of those exposed to longer photoperiods. Grange [117] obtained similar results for pepper.

Increasing $CO_2$ concentration in the atmosphere can increase crop yields only if plants are able to increase sinks in response to increased photosynthetic opportunity. If they are not able to increase their sinks, assimilate accumulation will inhibit photosynthesis and no production increase will occur. Nie et al. [118] found that wheat exposed to an elevated $CO_2$ concentration (550 vs. 360 $\mu$mol $mol^{-1}$) had higher leaf concentrations of carbohydrate and lower expression of messenger RNA (mRNA) of photosynthetic enzymes. However, Wang and Nobel [119] demonstrated that an elevated $CO_2$ concentration induces *Opuntia* to form larger sinks, increases tissue concentrations of carbohydrate, and enhances activity of nonphotosynthetic enzymes of carbohydrate metabolism. Jitla et al. [120] demonstrated that rice grain yield was increased if an elevated $CO_2$ concentration was supplied from the time of planting but not if that supply was delayed until only 15 days later. The importance of the early growth period is that it aids the establishment of sinks. These data should be used in the interpretation of other experiments in which elevated concentrations of $CO_2$ were used for specific time periods. Also, this appears to be in conflict with shading experiment reported previously for wheat.

If projected warming occurs, temperatures may move out of the desired range for sink establishment, and therefore lower productivity would result. Such an environmental change may require changing crops or time of planting. Impacts of increasing $CO_2$ concentration are extensively discussed by Bowes [121].

For factors that can be controlled but have a cost, such as water, pesticides, and fertilizer, cost/benefit ratios, both economic and environmental, can be enhanced by an awareness of the most critical time for development of economic sinks. Therefore, if some input is limited, emphasis should be placed on the critical phase of plant development. One must also be aware that the critical phase might be at very different calendar dates for different crops. For example, floral initiation in annual crops occurs in late spring or early summer, whereas initiation occurs in late summer for the next season's crop on woody perennials. No matter how favorable the conditions are during seed and fruit growth, partitioning of assimilates into those economically important parts can occur only if those parts form earlier in development.

## D.  Plant Disease and Partitioning

Recently, the emphasis of crop protection has changed. Using either traditional genetic techniques or recombinant DNA, genes for the production of materials toxic to pests are being incorporated into crop plants. However, this approach is not always successful, for these materials can be toxic to humans or farm animals or even to the plant itself [122,123]. Therefore, we will probably need to continue to use chemical pest control, but the timing will be much more important so that the most economic benefit can be obtained with the smallest amount of material, lowest possible rate of resistance development by pests, and lowest possible environmental impact. Timing of pesticide application has emphasized the vulnerable part of pest life cycles. Timing of crop vulnerability to pests must also be considered, for economically important plant parts are not equally vulnerable at all portions of a plant's life cycle. In addition, we will be able to lower total pesticide use by avoiding many prophylactic applications if we learn how pesticides are transported through plants, how plant parts are targeted, and use appropriate timing of application. It seems reasonable to apply this principle to transgenic plants that produce their own pesticide such that this production is confined to vulnerable plant parts at appropriate times.

Control of assimilate partitioning and/or availability of assimilates may be a mechanism by which some pathogens alter a plant in the manifestation of disease symptoms. Blunt ear syndrome of corn, manifested by stunted development of ears with little evidence of disease in the vegetative parts of the plant [124,125], may result from disruption of normal partitioning. Failure of plants to accumulate carbohydrates in stems or supply it to developing ears would lead to suppression of flower development, discussed earlier, and result in this syndrome.

Dickinson et al. [126] created a transgenic tomato plant that had the appearance of one infected with *Clavibacter michiganensis* subsp. *sepedonicus*, the causative agent of potato ring rot (CA Ishimura, personal communication). The transgenic tomato plants expressed an apoplastic yeast invertase. This resulted in the hydrolysis of most of the apoplastic sucrose of source leaves. Because tomato uses the apoplastic pathway of phloem loading (see later), little sucrose was available to be loaded and the resultant hexoses were reabsorbed by mesophyll cells. These cells accumulated large amounts of starch and became chlorotic, a typical symptom of this disease and of excess starch accumulation in chloroplasts. Von Schaewen et al. [127], using the same principle, had similar results with tobacco but much less severe symptoms on *Arabidopsis*.

More recently, Balachandran et al. [128,129] showed that partitioning was altered in transgenic tobacco by the expression of the movement protein of tobacco mosaic virus. The transgenic plants had lower dry mass of roots and stems compared with leaves, again indicating interference with assimilate export. Furthermore, unloading of the phloem is influenced by parasitism. It has been shown that higher plant parasites such as *Cuscuta* [130,131] and a parasitic root nematode [132] induced phloem unloading.

These results indicate that at least some disease conditions are caused by the disruption of normal partitioning. Use of transgenic plants to mimic disease physiology may make it possible to learn the mechanisms of pathogenesis and to develop procedures to interfere with such mechanisms.

## III.  WHAT IS TRANSLOCATED AND WHY

### A.  Sugars and Sugar Alcohols

Sugars and sugar alcohols carry 60 to 95% of the translocated carbon and have a total concentration ranging up to 180 mg mL$^{-1}$ [133]. Sucrose is the primary translocated carbohydrate in most plants [134–137]. In many plants, sucrose represents essentially all of the carbohydrate translocated. Other nonreducing oligosaccharides, such as raffinose, stachyose [134,138–141], gentianose, umbelliferose [136], and fructan [138], are translocated in some plants.

Sugar alcohols, along with sucrose, are translocated in a few groups of plants. Specifically, sorbitol is translocated by several members of the Rosaceae and Oleaceae [134,142], and mannitol is translocated by celery [143]. Even the giant alga, *Macrocystis* [144] and *Fucus* [145] translocated mannitol through "sieve cells" [146].

Occasionally, there are reports of glucose and fructose being isolated from phloem exudate [133]. However, these may be artifacts that resulted from hydrolysis of sucrose by enzymes released from damaged cells or a portion of the sap being supplied by damaged cells other than sieve tubes [147]. Sucrose hydrolysis should result in a 1:1 glucose-to-fructose ratio; however, Glad et al. [24] reported a ratio of about 2:1 from sieve tube exudate of grape. Nevertheless, Swanson and El Shishiny [148] obtained data from grape labeled with $^{14}CO_2$ and concluded that grape did not translocate hexoses. Wang and Nobel [137] reported significant amounts of hexoses, especially fructose, in phloem exudate of *Agave*. However, the existence of fructan hydrolases in the sap suggests that these hexoses may be an artifact (N Wang, personal communication).

In general, most if not all of the sugars and sugar alcohols translocated by plants are nonreducing. Why should that be? Arnold [149] proposed that these are "protected" molecules. Molecules acted upon by only a few enzymes would be favored for translocation because activity of only these few enzymes would need to be suppressed within sieve tubes to maintain the protected state.

Why do some plants translocate only sucrose while others translocate a mixture of sucrose, larger oligosaccharides, and/or sugar alcohols? Handley et al. [150] suggested that plants translocate sugars different from those accumulated in storage tissue, thereby maintaining a concentration gradient for translocated sugar from sieve tubes to sink cells. Examples of such a system are represented for cucurbits [150] and legumes [1,135]. Even some parasites and symbionts operate similarly, accumulating carbohydrates not found in appreciable concentrations in their hosts [130,151,152]. Obviously, this does not apply to all plants, for sugar beets and sugarcane translocate and store sucrose, but they store it within the vacuole that would keep it away from the transport system. It has been proposed that selection of the carbohydrate that is translocated is based on phloem loading mechanisms (discussed in the following).

In spite of the variability observed in translocated carbohydrates, all plants appear to translocate some carbon as sucrose. There is no known study that accounts for this observation. A possible explanation involves the observation that callose synthesis at sieve plates occurs rapidly [153]. This implies that the required enzymes must be present at all times but are usually inactive. If injury activated the sucrose synthase that Nolte and Koch [154] located in companion cells, the UDP-glucose substrate of callose synthesis would be provided and an inhibitor of callose synthesis, UDP, would be removed [155]. Neither larger oligosaccharides nor sugar alcohols could provide appropriate conditions so easily.

### B.  Other Organic Compounds

Although most of the carbon is translocated through phloem as carbohydrate, other organic compounds are translocated in significant quantities. Ziegler [156] summarized the literature then available for nitrogenous compounds in sieve tube sap. Reported concentrations varied from 0.8 to 137 μmol mL$^{-1}$. Pate [157] observed sieve tube concentrations of amino acids and amides in the phloem of lupine up to 21 mg mL$^{-1}$ and reported that asparagine accounted for over half of that content and glutamine was second most concentrated. In sieve tube exudate of yucca, glutamine and glutamic acid predominated [133], and in grape exudate, glutamine predominated [24]. Interestingly, the same amino acids predominate in the translocation steam of giant algae [144]. However, no single amino acid or small group of amino acids were dominant in the phloem exudate of *Opuntia* [158]. Glad et al. [24] also reported that molar quantities of amino acids approximated those of sugar. It should be noted, however, that on average each translocated amino acid molecule is much smaller than sucrose; therefore, mass-based quantities strongly favored sucrose. As with sugars, the concentrations and ratios of concentrations of various nitrogenous compounds in sieve tubes vary markedly from the values for the same compounds in mesophyll cells [159]. Other nitrogenous compounds such as ureides and alkaloids are translocated through the phloem in certain species [23].

During development, the ratio of amino acids to sugar in the translocation stream increased [23,24,157]. This would be expected, for as seeds developed, more nitrogenous compounds would be required by those sinks. A significant portion of the nitrogenous compounds in phloem was derived from xylem, presumably transferred in stems and minor veins of leaves. This system appears to provide control for partitioning of nitrogen compounds required for seed development. Amino acid concentration of sieve tube exudate increased from 14 $\mu g \ mL^{-1}$ 1 week after anthesis to 21 $\mu g \ mL^{-1}$ 6 weeks later, while the sucrose concentration fell from 112 to 80 $\mu g \ mL^{-1}$ [23,157]. Environment also influences what is transported; for example, water-deficient alfalfa markedly increased the concentration of proline in phloem sap [160].

Ziegler [156] reviewed several studies that found large amounts of protein in sieve tube exudate of members of the Cucurbitaceae. In the same volume, Eschrich and Heyser [161] concluded that protein in the exudate resulted from surging induced by a sudden release of turgor when cuts were made. That conclusion is supported by $^{14}C$ labeling studies of selected members of this family indicating that most of the carbon is carried as sucrose, raffinose, and stachyose [139,140]. More recently, a study using $^{35}S$ demonstrated that a small amount of protein was translocated through sieve tubes [162]. Even though translocation of protein is not quantitatively important, it appears to be involved in maintaining the functional integrity of the sieve tubes [163] and may be involved in control of sink development, thereby affecting partitioning indirectly [163,164]. Furthermore, nucleic acids have been isolated from phloem sap. Some are viral and represent the mechanism by which viruses become systemic. There is evidence that viral RNA enters sieve tubes via plasmodesmata by a mechanism normally used by plant mRNA [165]. This plant mRNA may be important in integrating growth, development, and assimilate partitioning [164,166]. Movement of macromolecules through plasmodesmata into and through phloem has been reviewed [167].

Reduction of nitrate and sulfate in mesophyll cells produces $OH^-$ [168,169], which is neutralized by the $H^+$ supplied as organic acids are synthesized. Acid anions are then loaded into the phloem. Although the concentration of organic anions in sieve tube sap of yucca was only 7 me/mL [133], these anions are exceedingly important in maintaining the pH of sieve tube sap [168], for, together with amino acids, they produce salts of weak acids and strong bases with inorganic cations in sieve tubes. This yields a pH near 8.0, which facilitates phloem loading and may be critical in the control of enzymes that might otherwise degrade translocated saccharides.

Many other organic compounds are undoubtedly translocated through phloem. Data indicate that auxins [170], cytokinins [171], and gibberellins [172] as well as abscisic acid [173], salicylic acid [174], and jasmonic acid [175] are phloem mobile. These substances are in such low concentrations within sieve tubes that their osmotic influence on transport process is minimal; therefore, they move passively with the flow through this system. However, they undoubtedly influence translocation of assimilates indirectly by modifying the metabolic activity of phloem sinks [176] and, possibly, even of the membrane processes of sieve tubes and companion cells [177–180].

## C.  Inorganics

Most plant scientists consider phloem as the conduit for organic materials, yet are also aware of the concept of phloem mobility of inorganic nutrients. As already noted, the combination of inorganic cations and organic anions contributes to the control of sieve tube pH.

The concentration of inorganic materials of *Yucca* [133] and lupine [181] sieve tube sap was about 2 mg/mL. This low concentration is, nevertheless, an extremely important component of the phloem transport system, for nutrients that enter mature leaves via the transpiration stream could not be repartitioned within plants were they not transported through phloem. Organs such as expanding buds, young leaves, and developing flowers and fruit that get most of their nutrients through phloem would receive inadequate supplies of inorganic nutrients were they not translocated through phloem. Therefore, mobility of inorganic nutrients through phloem is critical to plant growth and development.

Standard classroom deficiency experiments, often used to indicate mobility of nutrients in phloem, are based on the assumption that the initial appearance of deficiency symptoms in older leaves indicates phloem mobility of the nutrient being tested, whereas the initial appearance of symptoms in younger leaves indicates phloem immobility. Analysis of phloem exudates generally supports that conclusion. However, this is not always so. In addition, one must not use absolute concentrations of phloem exudate to determine mobility because requirements for different nutrients vary by orders of magnitude.

A more reasonable approach is to compare phloem to xylem concentrations in plants that were adequately supplied with nutrients. Table 1 [181–185] offers such a comparison for studies in which concentrations of nutrients of both xylem and phloem exudate were obtained from the same plant within any one experiment. The following assumptions were used to construct Table 1:

1. Over short periods of time, the same amount of potassium is translocated into a leaf through the xylem and out through the phloem.

2. Therefore, mobility of any nutrient is made relative to potassium by computing the ratio of phloem (P) to xylem (X) concentration $(P_N/X_N)$ of any nutrient (N), dividing by the same ratio for potassium (K), and multiplying by 100:

$$\text{mobility} = \frac{(P_N/X_N)}{(P_K/X_K)} \times 100 \tag{1}$$

Equation (1) sets potassium equal to 100, whereupon phloem mobility of any other nutrient becomes a percentage of potassium. When that was applied to data from several studies (Table 1), phloem mobilities of sodium, calcium, and manganese were low, iron and zinc were intermediate, and magnesium was variable. A low value should reflect the accumulation of that nutrient in a leaf as it ages, as does calcium [186,187].

One must wonder why this approach to analysis of phloem mobility of iron disagrees with the standard nutrient deficiency experiment. It may be that under deficiency conditions Iron is rapidly incorporated into cellular components as it enters a leaf and, therefore, is unavailable to phloem, whereas with sufficient supply, some iron is available to be translocated through phloem. Loneragan et al. [188] reported that copper supplied to leaves of copper-deficient plants was retained in the leaves, but when supplied to copper-sufficient plants, it was exported rapidly. That principle may also apply to iron. In addition, variability in sufficiency of magnesium in various experiments may explain the variable results for that nutrient (Table 1). Certainly, the availability of organic anions, including amino acids, is critical to maintaining solubility of iron, copper, and zinc [189].

Data of Gorham et al. [190], reproduced in Table 2, support the foregoing conclusions regarding phloem mobility of inorganic ions, for ions with high phloem mobility were at higher concentrations in the phloem sink, while ions of lower mobility were at higher concentrations in xylem sinks. It seems likely that this pattern has survival significance by protecting the next generation (seeds) from enzyme-inhibit-

**TABLE 1** Relative Phloem Mobility[a]

| Nutrient | Lupinus angustifola[b] | Lupinus alba[b] | Quercus rubra[c] | Ricinus cammus[d] | Lupinus alba[e] |
|---|---|---|---|---|---|
| K | 100 | 100 | 100 | 100 | 100 |
| Na | 20 | 12 | 7 | 9 | 23 |
| Mg | 173 | 18 | 5 | NA[f] | 49 |
| Ca | 9 | 7 | 2 | 4 | 0.4 |
| Fe | 69 | 32 | NA | NA | NA |
| Mn | 15 | 14 | NA | NA | NA |
| Zn | 78 | 85 | NA | NA | NA |

[a] Kylem/phloem concentrations of several ions divided by the same ratio of concentrations for potassium, thus setting all values for potassium equal to 100; values for other nutrients as a percentage of phloem mobility of potassium. See Eq. (1) in the text.
[b] From Ref. 181.
[c] From Ref. 182, compiled by Pate [23].
[d] From Refs. 183 and 184, compiled by Pate [23].
[e] From Ref. 185.
[f] NA, data not available.

**TABLE 2**   Chemical Composition of Leaves and Florets[a]

| Nutrient | Leaves (mol m$^{-3}$)[b] | Florets (mol m$^{-3}$)[b] | Florets/leaves ($K = 100$)[c] |
|---|---|---|---|
| Potassium | $72 \pm 6$ | $133 \pm 21$ | 100 |
| Sodium | $360 \pm 18$ | $56 \pm 3$ | 8 |
| Calcium | $35 \pm 9$ | $25 \pm 2$ | 39 |
| Magnesium | $37 \pm 5$ | $45 \pm 2$ | 66 |
| Chloride | $320 \pm 15$ | $51 \pm 9$ | 9 |

[a] Data for *Aster tipolium*.
[b] Concentrations on plant water basis ($\pm$ standard errors, $n = 3$) from Gorham et al. [190].
[c] Florets/leaves adjusted such that potassium equals 100, other elements as a percentage of potassium.

ing ions [191]. It should be noted that the cells of embryos have essentially no vacuole in which to sequester ions. Other studies comparing accumulation of nutrients in phloem sinks (fruit) with accumulation in xylem sinks (leaves) lead to similar conclusions about relative mobility of various ions [192–195]. One must take care in the interpretation of phloem exudate data, for the location of sampling can be critical. Pearson et al. [196] have shown that Mn and Zn enter wheat grains through phloem but much of the transfer from xylem to phloem occurs within the inflorescence. However, Grusak [197] reported that iron, supplied as Fe(III) citrate to leave of pea plants, was loaded into phloem and exported to developing seeds.

It should also be noted that an understanding of relative phloem-xylem mobility of nutrients leads one to an understanding of which plant parts best supply various materials for human nutrition. In addition, toxic metal ions are generally more mobile in xylem than in phloem [198]; therefore leafy materials grown on sites contaminated with such materials are more likely to be toxic to humans and livestock than are fruit and seeds [195].

Inorganic anions are not as concentrated in the phloem as the cations, for much of the negative charge is accounted for by organic ions. Van Die and Tammes [133] reported inorganic chloride, phosphate, and sulfate in sieve tube exudate of yucca. About 75% of the phosphorus was combined into organic ions, and, undoubtedly, a significant portion of the sulfur was in amino acids. Nitrate is seldom reported to be a component of the phloem sap, but it is occasionally reported in low concentrations [146,199,200].

Wolterbeek and Van Die [201], using neutron activation analysis, were able to identify small amounts of several other inorganics in phloem exudate, including rubidium, copper, bromine, vanadium, and even gold. These data provide no information regarding relative phloem-xylem mobility of these materials.

Until recently, there has been conflicting evidence concerning the phloem mobility of boron. However, Blevins and Lukaszewski [202] have concluded that boron is phloem mobile, probably as a cyclic diester with mannitol, sorbitol, or other di- and polyols. Furthermore, Brown et al. [203] demonstrated that wild-type tobacco plants transported very little boron from mature leaves whereas plants transgenically modified to synthesize sorbitol transported large amounts of boron from leaves. This observation probably explains earlier conflicts in that plants that normally transport sugar alcohols would exhibit much higher phloem mobility of boron than plants that do not transport polyols.

## IV.  MECHANISM OF LONG-DISTANCE TRANSPORT

As early as 1900, one of the predominant hypotheses of assimilate translocation was pressure flow through sieve tubes [1]. In 1927, Munch [204] proposed an osmotic model for the generation of pressure in phloem. This came to be known as the Munch pressure-flow hypotheses. Even so, textbooks published in the early 1930s [205,206] made no mention of a pressure-flow mechanism.

Over the subsequent decades several different mechanisms of long-distance transport through phloem were proposed [207–210]. Arguments were put forth that the microanatomy of sieve tubes could not support a flow mechanism because the holes in sieve plates are blocked with "slime plugs" and, even if they are not blocked, sieve plates would create far too much resistance for flow to occur at observed rates. During the 1950s and 1960s, a substantial amount of physiological data was interpreted as refuting a flow mechanism. When plants were supplied with $^{14}CO_2$, the profile of $^{14}C$ in the transport system decreased from source to sink in a pattern that would be expected for diffusion, although rates and velocities of movement were too great to be accounted for by diffusion. Furthermore, when plants were double labeled with $^{14}$C-sugar, $^{3}$H$_2$O, and/or $^{32}$p, it was found that no two nuclides moved together as was predicted for a flow system. Horwitz [211] and Biddulph [212] attempted to explain these data with a flow model through differential exchange between sieve tubes and adjacent, nontransporting cells. Nonetheless, several alternative proposals were put forth during this period. Included was a proposal for electro-osmotic pumping by removal of $K^+$ from the downstream side of sieve plates and cycling it to the upstream side [213–215], which would require metabolic energy. The resultant potential gradient across a sieve plate would drive $K^+$ through small pores in slime plugs, causing a solution flow. Thaine and coworkers [216,217] reported observing "transcellular strands" of cytoplasm moving in both directions in each sieve tube. Canny [218] developed a model of assimilate transport based on this system that would result in bidirectional movement within one sieve tube. Trip and Gorham [219] supplied different leaves of a squash plants with either $^{14}$CO$_2$ or $^{3}$H-sugars and found both nuclides in the same sieve tube. They interpreted their data as supporting bidirectional movement by transcellular strands.

Both the electo-osmotic and transcellular strand models would require metabolic energy along the translocation pathway. For many plants, such as bean and squash, cooling of only the translocation stream, but not source or sink, inhibited translocation [220,221] and respiration. However, Swanson and Geiger [222] demonstrated that chilled (1°C) petioles of sugar beets translocated sugars at a nonchilled rate after only a few minutes of acclimation. In addition, Sij and Swanson ([223] and personal observations), demonstrated that squash, a chilling-sensitive plant, translocated carbon through petioles exposed to an N$_2$ atmosphere even though such environment eventually caused tissue death. Furthermore, Peterson and Currier [224], using a fluorescent dye, demonstrated that bidirectional movement within one sieve tube was unlikely.

Concurrently, data were accumulating regarding the concentration of solutes and the pressure in sieve tubes [225–227]. Many of the proposed mechanisms could be eliminated by considering specific mass transport (g dry wt cm$^{-2}$ hr$^{-1}$) and velocity of translocation (cm hr$^{-1}$). Crafts and Crisp [1] compiled values for specific mass transport per unit cross section of phloem that varied from 0.14 to 4.8 (average 3.6) g dry wt cm$^{-2}$ hr$^{-1}$. Phloem is composed of many cells in addition to functional sieve tubes; therefore, rates must be two to four times those computed values. Assuming that the total solute in the sieve tubes is 18% w/v [133], probably a low estimate; using the minimum and maximum (and average) specific mass transfer rates; and assuming that the area of the sieve tubes is one half to one fourth the total phloem area [228], velocities of 1.6 to 106 cm hr$^{-1}$ (average 40 to 80) are obtained. Similar velocities have been obtained using labeled materials. At average velocity, this would result in material moving from one end of a sieve tube element to the other in 2 sec (sieve tube element length of 0.03 cm measured from micrograph in Ref. 229).

Cataldo et al. [230,231] explained the differential movement of $^{14}$C-sucrose and $^{3}$H$_2$O by differential lateral exchange. Then Christy and Ferrier [232] presented a mathematical model of phloem translocation by flow that is in agreement with empirical observation, confirming, in principle, the model developed by Horwitz [211]. Knoblauch and Van Bel [233] were able to observe sieve tubes transporting a dye in a flow and then observe that damage induced blockage of transport by accumulation of protein on sieve plates. All of these observations taken together should elevate the pressure flow "hypothesis" to "theory" status.

Briefly, as understood today, assimilates are loaded into a sieve tube–companion cell complex [ST-CC] against a free energy gradient using metabolic energy. This causes the sieve tubes to have an osmotic potential more negative than other cells (except companion cells) in the source. Water follows osmotically, causing pressure to develop. In sinks, assimilates are unloaded and water follows. These processes generate a pressure drop from source to sink and flow results.

## V.  PHLOEM LOADING

### A.  General Considerations

Phloem loading refers to the transfer of assimilate into the ST-CC from photosynthetic cells or cells involved in temporary storage. It has been a difficult subject to study, for cells and sites involved in transport cannot easily be isolated from portions of the system supplying assimilates. Therefore, it is a relatively new field of study and has been reviewed [234,235].

In discussing phloem loading, several characteristics of the system must be considered: (1) loading is selective, (2) concentrations of oligosaccharides and certain amino acids are higher in the ST-CC than in other cells, (3) as a result, the ST-CC has an osmotic potential more negative than that of adjacent cells, and (4) these conditions (2 and 3) result in the ST-CC having a higher pressure than other source cells.

Pathways available for phloem loading are apoplastic (cell wall) and symplastic (plasmodesmata). The conditions listed above indicate that metabolic energy must be involved in phloem loading. This energy requirement is most easily explained by invoking the apoplastic pathway. However, the mechanism of transport through plasmodesmata is not sufficiently understood to specify a mechanism of energy input for the symplastic pathway. In 1987, Delrot [236] and Van Bel [237] separately published papers in which they discussed the merits of each proposed pathway.

## B.  Apoplastic Pathway

Serious early steps in developing understanding of phloem loading were taken by Gunning and Pate [238] and Gunning [239] in their study of transfer cells associated with the phloem of minor veins of several families. Transfer cells have a large number of cell wall intrusions toward the interior of cells resulting in a large surface between the symplast and apoplast. Other plant structures that transfer materials between apoplast and symplast contain transfer cells [240]; therefore, it seemed reasonable to assume that phloem transfer cells are involved in phloem loading.

Sucrose supplied to the apoplast of sugar beet leaves was readily translocated through the phloem [241]. Glaquinta [242] using sugar beet and Robinson and Hendrix [243] using wheat demonstrated that asymmetrically labeled sucrose supplied to mature leaves did not have its label randomized as would be expected if supplied sucrose was hydrolyzed and then resynthesized before phloem loading. Taken together, these data clearly indicate that sucrose can be absorbed from the apoplast directly into the translocation system.

Giaquinta [244] demonstrated that [$^{14}$C]sucrose supplied to sugar beet leaf disks was accumulated in minor veins, and he also demonstrated that the nonpenetrating sulfhydryl reagent $p$-chloromercuribenzenesulfonic acid (PCMBS) inhibited absorption of sucrose without altering absorption of glucose, fructose, or 3-$O$-methylglucose. In addition PCMBS did not alter the rate of photosynthesis [244,245]. At about the same time, Gunning [239] cited several papers which indicated that companion cells that differentiated as transfer cells were involved in phloem loading.

Giaquinta also demonstrated that the optimum apoplastic pH for phloem loading was between 5.0 and 6.0 [245,246]. Furthermore, he demonstrated [245] that changing the apoplastic pH from 5.0 to 8.0 more than doubled $K_m$ for sucrose absorption but did not change $V_{max}$. Bush [247], using plasma membrane vesicles, demonstrated a 1:1 relationship between $H^+$ and sucrose transport. More recently, Lemoine et al. [248] demonstrated that plasma membrane vesicles from mature sugar beet leaves accumulated four times as much sucrose as vesicles from immature leave when a proton motive force was applied. Even though these vesicles were from entire leaves, there was a clear indication of differences between sink and source leaf plasma membranes. On that basis, a model for phloem loading was proposed in which a sieve tube plasmalemma adenosinetriphosphatase (ATPase) pumps protons out of the ST-CC into the apoplast [249]. Sucrose secreted into the apoplast by mesophyll cells would then be loaded into the ST-CC complex by a sucrose-proton cotransporter against the sucrose concentration gradient using the free energy gradient of protons that had been established by the ATPase. The charge gradient established by that proton-ATPase would also account for the high concentration of potassium in sieve tubes (discussed earlier).

Expression of an $H^+$-sucrose symport has been demonstrated in *Arabidopsis* and tobacco plants [250,251]; also, suppression of this protein inhibited sucrose transport [251]. Shakya and Sturm [252] demonstrated the existence of two sucrose-$H^+$ symports, one expressed in source leaves and the other expressed in storage cells. Furthermore, Botha and Cross [253] used dye infusion and plasmodesmata frequency along with electrical potential differences to demonstrate the physiological isolation of the ST-CC from other leaf cells.

All of these data support the mechanistic model proposed in the early 1980s for apoplastic phloem loading as reviewed by Giaquinta [2,3]. Furthermore, work on the isolation of sucrose binding and $H^+$-ATPase proteins from membranes of source leaves together with characterization of the mode of control of these proteins [254–257] will aid us in understanding phloem loading. Efforts are also being made to characterize proteins that may be involved in phloem loading of noncarbohydrate materials [258].

## C.  Symplastic Pathway

The apoplastic pathway of phloem loading was favored by most phloem physiologists in the early 1980s. A few, however, held a different view. Lucas and coworkers repeated much of Giaquinta's work and extended it to include additional tests and additional species. Their studies [259–263] led to the conclusion that absorption of exogenously supplied sugars was accomplished by a sugar retrieval system that is normally used to absorb sugars that leak from cells. Some of this sugar would be loaded into phloem. These investigators further concluded that the symplast is the prime pathway of phloem loading for many plants. They also suggested that partitioning of sucrose between cytosol and vacuole of source cells may control availability of assimilate to the phloem loading system [263].

Contrary to Giaquinta's [244] findings that PCMBS inhibited vein loading in sugar beet leaves whether they were supplied with $^{14}CO_2$ or [$^{14}C$]sucrose, Madore and Lucas [264] found that PCMBS inhibited vein loading of *Ipomoea tricolor* leaf disks at pH 5.0 only when [$^{14}C$]sucrose was supplied but not when the system was labeled via photosynthesis by $^{14}CO_2$. From all of these data, Madore and Lucas concluded that apoplastic phloem loading occurs in some species whereas symplastic phloem loading occurs in others. Turgeon and coworkers [265–267] have since presented data that support this conclusion.

Van Bel, Gamalel, and coworkers pointed out that the frequency of plasmodesmata connecting ST-CC complex with other leaf cells varies by close to two orders of magnitude [268–270]. Therefore, they categorized vein types into those that have frequent plasmodesmata connections between ST-CC complex and other cells as type 1 veins and those with few such connections as type 2 veins. They also indicated that there are intermediate types, so they added subcategories (types 1/2, 2a, and 2b). Type 1 vein companion cells have an intermediary cell structure, companion cells of 2b vein have a transfer cell structure, and 2a companion cells retain the type 2 characteristic of containing few plasmodesmata between companion cells and phloem parenchyma, yet have smooth walls. Various vein types are represented in Figure 1.

It was then demonstrated that plants in which PCMBS inhibited phloem loading of carbon from photosynthesis contain type 2 minor veins and plants in which PCMBS did not inhibit phloem loading have type 1 minor veins [268,271]. These workers concluded that plants with type 1 vein anatomy use the symplastic pathway as the prime pathway of phloem loading, those with type 2 anatomy use the apoplastic

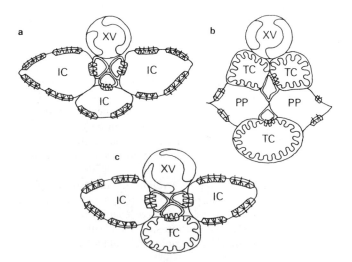

**Figure 1**   Typical minor vein structures: (a) open type (type 1) with intermediary cells (IC) as companion cell (*Hydrangea petiolaris*), (b) closed type (type 2b) with transfer cells (TC) as companion cells (*Pisum satiuum*), and (c) composite type (1/2b) with intermediary and transfer cell in a single minor vein (*Acanthus mollis*). XV, xylem vessel; PP, phloem parenchyma; the central unmarked elements are sieve elements. (From Ref. 269.)

pathway, and plants with intermediate anatomy probably use either or both pathways. Other studies support this hypothesis of a dual nature of phloem loading [249,272].

Furthermore, Van Bel [273] proposed that the type 1 (symplastic loading) is evolutionarily primitive and type 2 is evolutionarily advanced. He also pointed out that type 1 is more prevalent in tropical rain forests, type 2a predominates in steppe and deciduous forest communities, and type 2b (containing transfer cells) predominates in cold deserts and arctic-alpine communities.

For plants using the symplastic pathway, one must wonder how plasmodesmata can provide the selective control of phloem loading and use metabolic energy to generated osmotic and pressure gradients between ST-CC and other leaf cells. To be able to understand the functioning of plasmodesmata, their structure must be understood. Plasmodesmata have been studied extensively and the literature on that subject has been reviewed periodically (see, e.g., Refs. 274–276). The earlier papers were devoted to structure and distribution of plasmodesmata. Later papers also discussed function.

Briefly, plasmodesmata are tubes of cytoplasm that connect the symplast of adjacent cells through the cell wall with the plasmalemma surrounding these tubes. On average, the diameters of plasmodesmata are about an order of magnitude narrower than connections between sieve tube members through sieve plates. Each plasmodesma varies in diameter across the cell wall, with the narrowest portion being near the ends where they connect with the main body of each cell (neck region). A desmotubule, which is an extension of (or attached to) endoplasmic reticulum of each cell, extends through each plasmodesma. (Despite the name, desmotubules do not appear to be open tubes.) Between the desmotubule and plasmalemma, within the "cytoplasmic sleeve," is an extension of the cytosol. Surrounding the desmotubule, and apparently adhering to it, are spherical cytoplasmic sleeve subunits that almost completely occlude the space between the desmotubule and plasmalemma in the neck region.

Plasmodesmata vary in their degree of branching; many appear unbranched whereas others are highly branched [277]. Plasmodesmatal connections between sieve tubes and companion cells have a single pore into a sieve tube but are branched such that multiple pores enter a companion cell [261,278]. No suggestion has been put forth to explain a functional basis for branching.

The most likely pathway for materials moving through plasmodesmata appears to be through the cytoplasmic sleeve. Electron micrographic measurements of the most restricted region, the neck, and experimental studies of movement of various-sized dye molecules support that hypothesis. Madore and coworkers [261,279] and Van Kesteren et al. [280] demonstrated movement of fluorescent dyes from mesophyll into vascular bundles via plasmodesmata. Madore et al. [261] also showed that continuous plasmodesmatal connections exist from mesophyll into sieve tubes. However, Erwee and Goodwin [281] demonstrated that $Ca^{2+}$ could induce blockage such that cell-to-cell movement of dye did not occur. Robards and Lucas [276] pointed out that the exclusion limit for diffusion of dye through plasmodesmata varies from 376 Da for roots and stems to 870 Da between mesophyll and bundle sheath cells of C plants. However, they stated that "dye-coupling results do not establish that . . . phloem loading occur(s) via this symplastic pathway. However, . . . modeling of the phloem system must incorporate the finding that plasmodesmata within the vascular bundle are not vestigial."

Another approach to studying phloem loading involved the introduction of yeast invertase genes into tomato [126], tobacco, and *Arabidopsis* [127]. This enzyme was secreted into the apoplast and there hydrolyzed any sucrose present. In tobacco and tomato, this genetic alteration resulted in accumulation of carbohydrate in mature leaves, major inhibition of carbohydrate translocation, stunted growth, and other malformations that are characteristic of some plant diseases. *Arabidopsis* was only slightly affected. These data indicate that the prime (exclusive?) pathway of phloem loading for tobacco and tomato is apoplastic, whereas *Arabidopsis* may have a symplastic alternative.

We are still faced with the problem of proposing a mechanism of symplastic phloem loading that meets the criteria for phloem loading discussed earlier. However, some data suggest that not all those criteria need be met. Turgeon and Medville [282] suggested that, for some plants, mesophyll cells are the osmotic/pressure origin of the mass flow process, as originally proposed by Munch. This is based on their observation of no osmotic gradient between mesophyll and the ST-CC in willow. In addition, Richardson et al. [283] found that sieve tube exudate of cucurbits had a rather low concentration of osmotically active substances. If these exudation values are close to in vivo values, one could eliminate the criteria specifying that large osmotic and pressure differences must exist between ST-CC complex and mesophyll cells. But they [147] did question the assumption that phloem exudate represents the in vivo contents of sieve tubes. Others [226,227], using a plasmolytic method, reported that *Cucurbita* and *Coleus,* plants with type 1 anatomy, had quite negative osmotic potentials in their ST-CC complex when compared with mesophyll cells, thereby demonstrating that all of the previously listed criteria for phloem loading must be retained.

Van Bel [284] and Turgeon [285] proposed a mechanism by which plasmodesmata could act as an osmotic trap. In that scheme, sucrose and galactinol, the precursors of raffinose and stachyose, would diffuse from mesophyll cells into intermediary or companion cells, where the larger oligosaccharides would be synthesized. Resultant oligosaccharides would be too large to return to the mesophyll cells through plasmodesmata but could penetrate the larger diameter plasmodesmata into sieve tubes. This would provide the osmotic gradient reported for cucurbits and *Coleus*. Such a system is supported by the distribution of raffinose and stachyose in various leaf cells of melon [286]. However, as pointed out by Haritatos et al. [286], failure of melon to translocate galactinol is not explained by this model, nor is the fate of *myo*-inositol, the product of raffinose and stachyose synthesis. It may be that there is an efficient retrieval (apoplastic?) system for these materials. One must also wonder how a higher concentration of hexose and a lower concentration of sucrose are maintained in mesophyll cells than in the ST-CC [286] if sucrose can diffuse through connecting plasmodesmata unless the tonoplast is the diffusion-limiting structure.

The mechanism proposed by Van Bel [284] and Turgeon [285] would require that much of the carbon translocated by plants with type 1 anatomy would be as large oligosaccharides. Correlation between species that transport various sugars (Ref. 34 and others) and the vascular bundle types [269] is quite strong, although there are striking exceptions. For example, *Fraxinus ornus*, the most extreme of the type 1 group, transports not only the higher oligosaccharides but also significant amounts of mannitol. In addition, both grape and willow have type 1 anatomy yet translocate sucrose as the primary carbohydrate [134,148]. This mechanism of phloem loading suggested by Van Bel [287] also does not explain the observation that the ratio of various amino acids in sieve tube exudate does not match that of mesophyll cells [159]. This lack of correspondence, however, could be accounted for by transfer of amino acids from xylem to phloem [257] using an apoplastic pathway.

## D.   Summary, Phloem Loading

Briefly, the apoplastic model of phloem loading requires that assimilates within mesophyll cells move symplastically from cell to cell via plasmodesmata until they reach minor vein parenchyma cells. Assimilates are then transferred to the apoplast, where they are absorbed by the ST-CC complex using the $H^+$-sucrose symport and driven by the proton gradient maintained by the $H^+$-ATPase. It is seen as essential that movement be symplastic up to the vein, for water moving from the vein to the point of transpiration in the apoplast would move assimilate away from the phloem. This system requires type 2 vein anatomy.

The mechanism for the symplastic pathway is not so clear. This system requires type 1 vein anatomy. The pathway is through plasmodesmata from mesophyll cells into sieve tubes. A mechanism of size trapping for selection of raffinose family oligosacchrides has been proposed. However, this does not explain the mechanism for selection against other materials. It would be attractive to suggest that there are proteins that transport selected materials through plasmodesmata as reported for nucleic acids [164,166]; however, there is no datum suggesting such a mechanism. In addition, data on the diffusion of dyes must be explained, although these may be an artifact [276].

Much work is need before we understand the processes of the plasmodesmata. Current studies in various laboratories using molecular procedures offer the greatest promise for solving problems of symplastic phloem loading as well as other processes involving cell-to-cell communication. This new information will be useful in understanding both mechanisms of phloem loading, for even the apoplastic pathway appears to involve transport through plasmodesmata up to the ST-CC.

We are left with the conclusion that both symplastic and apoplastic phloem loading do occur but that the mechanism of only the apoplastic pathway is well understood (see Ref. 287), for McLean et al. [288] in their review stated that "our understanding of plasmodesmatal structure and function is still naive."

## VI.   PHLOEM UNLOADING

Assimilates are unloaded from phloem of the plant producing the assimilate either into vegetative cells via the apoplast or plasmodesmata or into a developing embryo and/or endosperm (i.e., a separate plant) [289–293]. Patrick [293] has classified these as apoplastic, symplastic, and maternal/filial. In special situations, unloading includes transfer of assimilates to a parasite or symbiont. Clearly, the pathway must be apoplastic if the sink is a separate organism, for no plasmodesmata exist between parent and seed or between a host and parasite.

Thorne [290] presented several mechanisms of phloem unloading:

1.  Sucrose is passed into sink cells via plasmodesmata. Once in sink cells, sucrose is degraded into hexose in either the cytoplasm or vacuole of those cells. This type of unloading is reported to occur in sugar beet sink leaves, corn root tips, and bean endocarp.

2.  Sucrose is unloaded into the apoplast, where it is degraded into hexoses. Hexoses are transferred into sink cell cytoplasm, where sucrose is resynthesized. Sucrose is then accumulated in the vacuole. This mechanism has been reported for assimilate movement into corn and sorghum kernels, sugarcane stalks, and into some parasites [130].

3.  Sucrose is unloaded from sieve tubes into the apoplast, transferred into sink cell cytoplasm, and then accumulated in vacuoles. This has been reported for sugar beet tap root, legume seeds, and wheat grain.

Eschrich [294] has added a fourth mechanism: carbohydrates follow any of these pathways, then are incorporated into starch. It could be argued that any metabolic removal of sugar would be analogous. One could also visualize various combinations of steps of listed processes to obtain even more categories.

From a thermodynamic standpoint, no energy should be required for transfer of sugar across sieve tube or sink cell plasmalemma or through the plasmodesmata [e.g., 295], for other processes maintain appropriate gradients along the pathways such that diffusion can account for movement. Wang and Fisher [296] presented evidence that metabolic energy is not used in movement of sucrose into the endosperm cavity of developing wheat grains; however, metabolic energy may be a mechanistic requirement in some circumstances. A possible exception may exist for citrus fruit, for Koch and Avigne [297] reported that the rate of sucrose imported into juice sacs peaked early in fruit development while it was moving against its concentration gradient; then the rate fell as equillibrium was approached. It then rose to a second higher peak after the concentration in vascular bundles had exceeded that in juice sacs. This would imply a requirement for metabolic energy early in developmental, though the energy-requiring step may involve accumulation of sucrose in vacuoles. This bimodal accumulation pattern suggests that two separate mechanisms of sucrose transfer were involved. Furthermore, metabolic requirements of corn root tips beyond the terminus of mature sieve tubes cannot be satisfied by diffusion through plasmodesmata [298]. Therefore, it was proposed that solution flow through plasmodesmata might be the mechanism of supply, although current ideas on structure of plasmodesmata would seem to make that unlikely.

As stated earlier, unloading of assimilates into a developing embryo must involve transfer of assimilates into the apoplast. Thorne and Rainbird [299] developed a technique for studying this process. While leaving the seed coat attached to the parent plant, they removed the embryo. They demonstrated that unloading of assimilate into agar in the seed coat proceeded at a rate comparable to that of unloading into an intact embryo. They also demonstrated that the nonpenetrating sulfhydryl reagent PCMBS inhibited unloading into agar-filled seed coat cavity but not transfer of assimilates into the seed coat, thus indicating that transfer of assimilates into seed coat cells was symplastic. However, NaF and NaAsO$_2$ inhibited both processes, indicating an energy dependence of a portion of the unloading process. Wang and Fisher [300] developed an analogous process for studying transfer of sucrose to wheat endosperm cavities. The technique of embryo removal has been used extensively to study the unloading process, and data indicate that assimilates are transferred from sieve tubes symplastically through several cell layers before being deposited into the apoplast, either at the surface of the embryo/endosperm or several cell diameters from that surface. Once at the embryo/endosperm surface, assimilates are absorbed into developing cells [299,301]. Oparka [302] has referred to this as "apoplastic unloading sink" rather than "apoplastic unloading." He restricts the latter to unloading of assimilates into the apoplast directly from sieve tubes.

A compilation from the literature by Fisher and Oparka [303] indicates that symplastic unloading of the ST-CC is more common than apoplastic; however, eventual transfer to the apoplast must occur during seed growth.

## VII.  XENOBIOTIC TRANSPORT

Understanding of the transport of xenobiotics is essential to the judicious development and use of systemic pesticides. An example of failure to use such information was the program in which solutions of benomyl (Benlate) were injected into trunks of elm trees in an attempt to control Dutch elm disease. Benlate is soluble at the pH of xylem (5.0–6.0) but essentially insoluble at sieve tube pH (~8.0). As a result, the Benlate was translocated through xylem to leaves, providing some temporary protection to those structures. However, no protection was provided to any part of the plant supplied by the phloem, i.e., roots and even newly developed xylem.

The first extensive study of xenobiotic translocation was done by Crafts [304]. He was able to demonstrate whether a material was transported through the xylem, phloem, neither, or both. It has been generally assumed that once a xenobiotic has entered the translocation system, it moves passively with the flow, which is driven by processes independent of the xenobiotic. In addition, the xenobiotic must not leak from the translocation stream too easily or it will not travel far.

An important mechanism for phloem translocation of xenobiotics is ion trapping of weak acids (Ref. 305, cited in Ref. 306). The basis of ion trapping is as follows: a weak acid is protonated in the lower pH of the apoplast and ionized in the higher pH of sieve tubes. The protonated form readily penetrates the lipid portion of membranes. Once in the sieve tube, however, the ionized form would not easily penetrate the plasmalemma of sieve tubes (i.e., is ion trapped) and, therefore, is carried to the sink with the flow.

Tyree et al. [306] proposed a model for phloem mobility of substances that does not ionize. This model is based on the idea that there is an optimal membrane permeability for translocation of a xenobi-

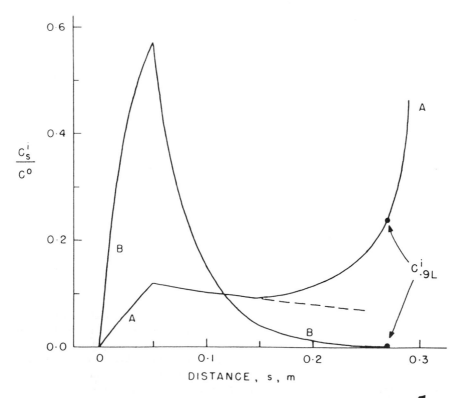

**Figure 2**   Theoretical distribution of xenobiotic in sieve tube of a "linearized" plant 0.3 m. Concentration of xenobiotic in the sieve tube is plotted against distance as a fraction of the concentration in the source leaf apoplast. Source is assumed to be 0.05 m long. In this calculation, the sieve tube radius times sap velocity, $rV$, is $1.5 \times 10^{-9}$ m sec$^{-1}$. Curve R is for the optimum permeability, $P$, of about $2 \times 10^{-9}$ m sec$^{-1}$. In curve B, P is 10 times larger, $2 \times 10^{-8}$ m sec$^{-1}$. Dashed line extending from curve A shows how concentration would decline if $V$ in the root remained constant instead of decreasing. (From Ref. 306.)

otic. If permeability is too low, very small amounts of the material will enter the sieve tubes. If permeability is too high, much xenobiotic would enter the sieve tubes but would leak out rapidly once it moved away from the supply region. If the supply region consists of leaves, the material would be carried back into the leaves in xylem; therefore, the xenobiotic would never reach the sink (target). Figure 2 provides models of movement of two materials of different membrane permeabilities. Kleier [307] combined ion trapping and permeability models to predict the mobility of xenobiotics through the phloem.

For all of these models, the required low membrane permeability of xenobiotics (for movement through phloem) results in an accumulation of the xenobiotic in the sink region within sieve tubes (see Figure 2). This buildup results from continuous flow of solution into the sink that is driven by unloading of osmotically active materials and water following the movement of osmoticum. Because some leakage of xenobiotic does occur, its accumulation in sink phloem results in significant delivery of xenobiotic into sink cells.

It seems reasonable to develop pesticides that attach to sites used by materials normally loaded into and unloaded from phloem. These xenobiotics would then use the loading and unloading mechanisms of metabolites. Other than auxin herbicides, to date, only glyphosate has been reported to use a phosphate transporter [308], and this use occurred not by design but by serendipity.

## VIII. EXTRATERRESTRIAL AGRICULTURE

If humans are to inhabit a lunar base or travel to Mars, their needs must be met, at least in part, as they are on Earth, by plants. Interest in this topic is illustrated by support that the National Aeronautics and Space Administration (NASA) has provided for research (see, e.g., Refs. 309–313 and references therein).

If food is to be provided most effectively for space travelers, a high HI is needed. A NASA-funded project on wheat is illustrative. In an attempt to achieve the highest possible food and $O_2$ production by wheat, temperature at or near the optimum for photosynthesis was used, together with high irradiance. An HI of 0.25 was obtained, but when the temperature was lowered, an HI of nearly 0.50 was obtained (FB Salisbury, personal communication). It is also important that both volume and energy be used efficiently. To get maximum production per unit volume, a high irradiance level must be provided. However, as irradiance level was increased, efficiency of light utilization in photosynthesis fell [309]. Therefore, a satisfactory irradiance-volume compromise must be developed. This need for efficiency of volume utilization drove the development of a "super dwarf" (25 cm) variety of wheat [313]. Under controlled conditions an HI of 0.52, greater than obtained with semidwarf varieties, was obtained.

If NASA's goals of human exploration of deep space are to be attained, methods of providing a crew with plant-produced food, $O_2$, and pure water from transpiration, as well as recycling nutrients through plants and removal of $CO_2$, must be developed in a way that most effectively uses both volume and energy.

## IX. SUMMARY

Understanding the mechanisms of assimilate transport is useful or even essential if crop inputs, such as pesticides, fertilizer, and water, are to be used economically. Understanding these processes is also helpful in developing the most effective use of both traditional and molecular genetics in enhancing crop production.

Assimilate partitioning is the most critical component of translocation processes that determine economic productivity. When the economic product is the result of sexual reproduction, establishment of sinks is the critical step in productivity, for it is during floral development that maximum yield potential is established. Once established, that potential might be lowered but it cannot be increased. Genetic and environmental factors interact to limit yield. It appears that sink formation is the limiting component of crop production in many situations, for experimental evidence supports the proposition that during grain/seed filling, the systems are sink limited, even to the extent that there is a feedback of assimilate on the rate of photosynthesis. For crops in which the economic product is vegetative, it appears that genetic and environmental determinants affect the establishment of sinks; once sinks are established, however, their growth is more likely to be source limited.

Over the last several decades, the mechanism of long-distance translocation of assimilates through sieve tubes has been established as an osmotically driven pressure flow system. Several different nonreducing oligosaccharides and sugar alcohols are translocated by this system. For many plants, sucrose appears to be the only carbohydrate translocated; others translocate a mixture of sugars and sugar alcohols, but sucrose seems to be a component of all. Many different nitrogenous compounds are also translocated through the phloem. These include most, if not all, of the protein amino acids and amides as well as several ureides. Most common are aspartic and glutamic acid and their amides. Nitrate is of little, if any, significance in nitrogen translocation through phloem.

Various inorganic materials are translocated through the phloem. By far the most concentrated, other than water, is potassium. Together with weak organic anions and amino acids, inorganic cations form a salt that buffers the pH of sieve tube sap at about 8.0.

It now appears that there are two pathways of phloem loading. In the apoplastic route, assimilates pass into the cell wall before entering sieve tubes. The symplastic pathway follows the plasmodesmata from cell to cell and finally into sieve tubes. Based on the anatomy of the vascular bundles, it appears that many plants use primarily one or the other pathway, whereas other plants may to be able to use either or both. Both mechanisms result in a higher concentration of sugars in the sieve tube–companion cell complex than in other cells. The mechanism of concentrating the carbohydrates is understood for the apoplastic pathway but not for the symplastic pathway.

Phloem unloading also follows either pathway. Because there is no symplastic connection between a parent plant and the seed it is forming, transfer of assimilates into a developing seed must be apoplastic, although the assimilates may not enter the apoplast directly from sieve tubes. When assimilates are being transferred to cells within the same plant, they may follow either the symplastic or the apoplastic pathway. In most cases, it appears that chemical modifications maintain free energy gradients for carbohydrates along the unloading pathway. Judicious use of xenobiotics requires an understanding of transport processes, including the special problems involved in loading and unloading of these substances.

Exploration of space requires the recycling of human wastes into materials used by humans. Efficient use of plants is required in this system. Getting a high proportion of the photoassimilates partitioned into edible parts is an important component of that efficiency.

## ACKNOWLEDGMENTS

I wish to express my appreciation to all who contributed research cited in this chapter, as well as the many others not cited whose work contributed to my background understanding. I wish, also, to thank those who sent reprints for consideration and especially those with whom I discussed their research and who aided me in interpretation of their work. Special thanks go to Dr. Cecil Stushmoff for reading the manuscript and offering helpful comments.

## REFERENCES

1. RS Crafts, CE Crisp. Phloem Transport in Plants. San Francisco: Freeman, 1971.
2. RT Giaquinta. Ber Dtsch Bot Ges 93:187, 1980.
3. RT Giaquinta. Annu Rev Plant Physiol 34:347, 1983.
4. RL Heath, J Preiss, eds. Regulation of Carbon Partitioning in Photosynthetic Tissue. Rockville, MD: American Society of Plant Physiologists, 1985.
5. J Cronshaw. Annu Rev Plant Physiol 32:465, 1981.
6. J Cronshaw, WJ Lucas, RT Giaquinta, eds. Plant Biology, Vol I, Phloem Transport. New York: Liss, 1986.
7. L Ho. Annu Rev Plant Physiol Plant Mol Biol 39:355, 1988.
8. DA Baker, JA Milburn eds. Transport of Assimilates. New York: Longman, 1989.
9. IF Wardlaw. New Phytol 116:341, 1990.
10. MA Thorpe, PEH Minchin. In: E Zanski, AA Schaffer, eds. Photoassimilate Distribution in Plants and Crops, Source-Sink Relationships. New York: Marcel Dekker, 1996, p 2261.
11. RB Rustin, J Bingham, RD Blackwell, LT Evans, MA Ford, CL Morgan, M Taylor. J Agric Sci 94:675, 1980.
12. RM Gifford. In: J Cronshaw, WJ Lucas, RT Giaquinta, eds. Plant Biology, Vol I, Phloem Transport. New York: Liss, 1986, p 435.
13. CJ Pollock, JF Farrar, AJ Gordon, eds. Carbon Partitioning Within and Between Organisms. Oxford: BIOS Science Publishers, 1992.
14. JS Pate, DB Layzell, DL McNell. Plant Physiol 63:730, 1979.
15. JF Farrar. In: CJ Pollock, JF Farrar, AJ Gordon, eds. Carbon Partitioning Within and Between Organisms. Oxford BIOS Science Publishers, 1992, p 163.

16. BR Thomas, RL Rodriguez. Physiol Plant 106:1235, 1994.
17. R Turgeon. Annu Rev Plant Physiol Plant Biochem 40:119, 1989.
18. W-H Cheng, DH Im, PS Chourey. Plant Physiol 111:1021, 1996.
19. CJ Bell, LD Incoll. J Exp Bot 41:949, 1990.
20. DJ Davidson, PM Chevaller. Crop Sci 32:186, 1992
21. JD Hewett, M Murrush. J Am Soc Hortic Sci 111:142, 1986.
22. JD Keller, WH Loescher. J Am Soc Hortic Sci 114:969, 1989.
23. JS Pate. In: IF Wardlaw, JB Passioura eds. Transport and Transfer Processes in Plants. New York: Academic Press, 1976, p 253.
24. C Glad, J-I Regnard, Y Querou, D Brun, J-F Morol-Gaudry. Vitis 31:131, 1992.
25. CM Dunan, RL Zimdanl. Weed Sci 39:558, 1991.
26. JG Buwalda. Environ Exp Bot 33:131, 1993.
27. MM Caldwell. In: PJ Gregory, JV Lake, DR Rose, eds. Root Development and Function—Effects on the Physical Environment. Cambridge: Cambridge University Press, 1987, p 167.
28. MS Lin, RL Nelson. Crop Sci 28:218, 1988.
29. C Borchers-Zampini, AB Giamm, J Hoddinott, CA Swanson. Plant Physiol 65:1116, 1980.
30. RH Blum, H Poiarkova, G Golan, J Mayer. Field Crop Res 6:51, 1983.
31. ED Schultz, K Schilling, S Nagarajah. Oecologia 58:169, 1983.
32. Y Shishido, Y Hori. Tohoku J Agric Res 28:82, 1977.
33. J Xu, WT Avigne, DR McCarty, DE Koch. Plant Cell 8:1209, 1996.
34. AM Smith, K Denyer, CR Martin. Plant Physiol 107:673, 1995.
35. YJ Zhu, E Komor, PH Moore. Physiol. Plant 115:609, 1997.
36. DR Geiger, KE Koch, W-J Shieh. J Exp Bot 47:1229, 1996.
37. A Sturm. Plant Physiol 121:1, 1999.
38. Y Zeng, Y Wu, T Avigne, KD Koch. Plant Physiol 121:599, 1999.
39. KE Koch. Annu Rev Plant Physiol Plant Mol Biol 47:509, 1996.
40. KE Koch, KD Nolte, ER Duke, DR McCarty, WT Avigne. Plant Cell 4:59, 1992.
41. JE Hendrix, JC Linden, DH Smith, CW Ross, IK Park. Aust J Plant Physiol 13:391, 1986.
42. M Bodson. Planta 135:19, 1977.
43. M Bodson, WH Outlaw Jr. Plant Physiol 79:420, 1985.
44. M Bodson, G Bernier. Physiol Veg 23:491, 1985.
45. P Lejeune, G Bernier, J-M Kinet. Plant Physiol Biochem 29:153, 1991.
46. L Corbesier, P Lejeune, G Bernier. Planta 206:131, 1998.
47. C Perilleux, G Bernier. New Phylol 135:59, 1997.
48. N Ishioda, S Tanimoto, H Harada. J Plant Physiol 138:573, 1991.
49. Z Natrova, L Natr. Field Crop Res 31:121, 1993.
50. RA Fischer, D HilleRisLambers. Aust J Agric Res 29:443, 1978.
51. JF Shanahan, DH Smith, JA Welsh. Agon J 76:611, 1984.
52. PM Brcmncr, HM Rawson. Aust J Plant Physiol 5:61, 1978.
53. SF Blade, RJ Baker. Crop Sci 31:1117, 1991.
54. B Borghi, M Corbellini, M Cattaneo, ME Fornasari, L Zucchelli. J Argon Crop Sci 157:245, 1986.
55. DM Peterson. Field Crop Res 7:41, 1983.
56. RA Fischer. Crop Sci 15:607, 1975.
57. IF Wardlaw. Aust J Plant Physiol 21:731, 1994.
58. DO Caldiz, SJ Sarandon. Agronomie 8:327, 1988.
59. JR Kiniry. Agron J 80:221, 1988.
60. IJ Warrington, RL Dunstone, LM Green. Aust J Agric Res 28:11, 1977.
61. RA Fischer. J Agric Sci 105:447, 1985.
62. J Willenbrink, GD Bonnett, S Willenbrink, IF Wardlaw. New Phytol 139:471, 1998.
63. SS Bhuller, CF Jenner. Aust J Plant Physiol 13:605, 1986.
64. I Sofield, LT Evans, MG Cook, IF Wardlaw. Aust J Plant Physiol 4:785, 1977.
65. IF Wardlaw, L Moncur. Aust J Plant Physiol 22:391, 1995.
66. IF Wardlaw, CW Wrigley. Aust J Plant Physiol 21:695, 1994.
67. T Tashiro, EF Wardlaw. Aust J Plant Physiol 18:259, 1991.
68. JS Hawker, CF Jenner. Aust J Plant Physiol 20:197, 1993.
69. MAB Wallwork, SJ Logue, LC MacLeod, CF Jenner. Aust J Plant Physiol 25:173, 1998.
70. CF Jenner, K Denyer, J Guerin. Aust J Plant Physiol 22:703, 1995.
71. SC Torbit, RB Gill, AW Aldredge, JC Liever. J Wildl Manage 57:173, 1993.
72. RJ Jones, SR Simmons. Crop Sci 23:129, 1983.
73. DMN Mbewe, RB Hunter. Can J Plant Sci 66:53, 1986.
74. TL Setter, BA Flannigan. Ann Bot 64:481, 1989.
75. G Hueros, E Gomez, N Cheikh, J Edwards, M Weldon, F Salamini, R Thompson. Plant Physiol 121:1143, 1999.

76. JW Gonzales, AM Rhodes, D Dickinson. Plant Physiol 58:28, 1976.
77. DB Dickinson, CD Boyer, JG Velu. Phytochemistry 22:1371, 1983.
78. W Aufhammer, E Nalborezyk, B Geyer, I Gotz, C Mack, S Paluck. J Agric Sci 112:419, 1989.
79. W Aufhammer, I Gotz-Lee. J Agric Sci 113:317, 1989.
80. W Aufhammer, I Gotz-Lee. J Agric Sci 116:409, 1991.
81. TL Sage, BD Webster. Bot Gaz 148:35, 1987.
82. CS Mauk, PJ Breen. J Am Soc Hortic Sci 111:416, 1986.
83. JW White, SP Sing, C Pino, MJ Rios B, I Buddenhagen. Field Crop Res 28:295, 1992.
84. YM Stockman, R Shibies. Iowa State J Res 61:35, 1986.
85. MO Kelly, RM Spanswick. Plant Physiol 114:1055, 1997.
86. LJ Wiles, GG Wilkerson. Agric Syst 35:37, 1991.
87. HD Guffy, JD Hesketh, RL Nelson, RL Bernard. Biotronics 20:19, 1991.
88. RD Guffy, BL Vasilas, JD Hesdeth. Biotronics 21:1, 1992.
89. AJ Gordon, In: CJ Pollock, JF Farrar, AJ Gordon, eds. Carbon Partitioning Within and Between Organisms. Oxford: BIOS Science Publishers, 1992, p 133.
90. LR Salado-Navarro, K Hinson, TS Sinclair. Crop Sci 25:451, 1985.
91. VR Franceschi, RT Giaquinta. Planta 159:515, 1983.
92. MN Westwood. Temperate-Zone Pomology, Phynology and Culture. 3rd ed. Portland, OR: Timber Press, 1993.
93. K Ryugo. Fruit Culture: Its Science and Art. New York: Wiley, 1988.
94. RE Byers, JA Barden, RE Polomoki, RW Young, DH Carbaugh. J Am Soc Hortic Sci 115:14, 1990.
95. RE Byers, CG Lyons Jr, KS Yoder, JA Barden, RW Young, J Hortic Sci 60:465, 1985.
96. PEH Minchin, MR Thorpe, JN Wunsche, JW Palmer, RF Picton. J Exp Bot 48:1401, 1997.
97. AN Miller, CS Walsh. J Am Soc Hortic Sci 113:309, 1988.
98. ED Goldschmidt, KE Koch. In: E Zamski, AA Schaffer, eds. Photoassimilate Distribution in Plants and Crops, Source-Sink Relationships. New York: Marcel Dekker, 1996, p 797.
99. CH Engels, H Marschner. J Exp Bot 37:1804, 1986.
100. CH Engels, H Marschner. J Exp Bot 37:1813, 1986.
101. CH Engles, H Marschner. Potato Res 30:177, 1987.
102. DJ Midmore, J Roca, D Berrios. Field Crop Res 18:141, 1988.
103. JD Midmore, D Berrios, J Roca. Field Crop Res 18:159, 1988.
104. X Xu, AAM Van Lammeren, E Vermeer, D Vreugdenhll. Plant Physiol 117:575, 1998.
105. MA Nakatani, A Oyanagi, Y Watanabe. Jpn J Crop Sci 57:535, 1988.
106. DR Geiger, JC Servaites. Annu Rev Plant Physiol Plant Mol Biol 45:235, 1994.
107. J Sun, TW Okita, GE Edwards. Plant Physiol 119:267, 1999.
108. W Claussen, F Lenz. Z Pflanzenphysiol 109:459, 1993.
109. J Azcon-Bieto. Plant Physiol 73:681, 1983.
110. M Stitt. In: J Cronshaw, WJ Lucas, RT Giaquinta, eds. Plant Biology, Vol I, Phloem Transport. New York: Liss, 1986, p 331.
111. M Stitt. Annu Rev Plant Physiol Plant Mol Biol 41:153, 1990.
112. CH Foyer. Plant Physiol Biochem 26:483, 1988.
113. M Stitt, WP Quick. Physiol Plant 77:633, 1987.
114. U Sonnewald, L Willmitzer. Plant Physiol 99:1267, 1992.
115. CL Mandahar, ID Garg. Photosynthetica 9:407, 1975.
116. S Logendra, JD Putman, HW Janes. Sci Hortic 42:75, 1990.
117. RI Grange. J Exp Bot 36:1749, 1985.
118. G Nie, DL Hendrix, AN Webber, BA Kimball, SP Long. Plant Physiol 108:975, 1995.
119. N Wang, PS Nobel. Plant Physiol 110:893, 1996.
120. DS Jitla, GS Rogers, SP Seneweera, AS Basdra, RJ Oldfield, JP Conroy. Plant Physiol 115:15, 1997.
121. G Bowes. Annu Rev Plant Physiol Plant Mol Biol 44:309, 1993.
122. DA Smith. In: JA Bailey, JW Mansfield, eds. Phytoalexins. New York: Halsted, 1982, p 218.
123. M Friedman, GM McDonald. Crit Rev Plant Sci 16:55, 1997.
124. CH Pearson, HM Golus. Tech Rep TR90-7, Agric Exp Stn, Colorado State University, Ft. Collins, 1990.
125. WA Fithian. Etiology and Management of Blunt Ear Syndrome of Corn, M.S. thesis, Colorado State University, Ft. Collins, 1999.
126. CD Dickinson, T Altabella, MJ Chrispeels. Plant Physiol 95:420, 1991.
127. R Von Schaewen, M Stitt, R Schmidt, U Sonnewald, L Willmitzer. EMBO J 9:3033, 1990.
128. S Balachandran, RJ Hull, Y Vaadia, S Wolf, WJ Lucas. Plant Cell Environ 18:1301, 1995.
129. S Balachandran, RJ Hull, RR Martins, Y Vaadia, WJ Lucas. Plant Physiol 114:475, 1997.
130. WE Seel, I Cechin, GA Vincent, MC Press. In: CJ Pollock, JF Farrar, AJ Gordon eds. Carbon Partitioning Within and Between Organisms. Oxford: BIOS Science Publishers, 1992, p 199.
131. P Wolswinkel. Ann Bot 50:863, 1982.
132. A Bockenhoff, DRM Prior, FMW Grundler, KJ Oparka. Plant Physiol 112:1421, 1996.

133. J Van Die, PML Tammes. In: MH Zimmermann, JA Milburn, eds. Encyclopedia of Plant Physiology, Vol I, Transport in Plants, Part I. New York: Springer-Verlag, 1975, p 196.
134. MH Zimmermann, H Ziegler. In: MH Zimmermann, JA Milburn, eds. Encyclopedia of Plant Physiology, Vol I, Transport in Plants, Part I. New York: Springer-Verlag, 1975, p 480.
135. O Kandler, H Hoff. In: FA Loewus, W Tanner, eds. Encyclopedia of Plant Physiology, Vol 13(A). New York: Springer-Verlag, 1982, p 348.
136. JA Milburn, DA Baker. In: DA Baker, JA Milburn, eds. Transport of Photoassimulates. New York: Longman, 1989, p 345.
137. N Wang, PS Nobel. Plant Physiol 116:707, 1998.
138. LM Cruz-Perez, D Durkin. Proc Am Soc Hortic Sci 85:414, 1964.
139. JA Webb, PR Gorham. Plant Physiol 39:663, 1964.
140. JE Hendrix. Plant Physiol 43:1631, 1968.
141. JE Hendrix. Plant Sci Lett 25:1, 1982.
142. KL Webb, JWA Burley. Science 137:766, 1962.
143. JM Davis, WH Loescher. Physiol Plant 79:656, 1990.
144. K Schmitz, LM Srivastava. Plant Physiol 63:995, 1979.
145. M Diouris. Phycologia 28:504, 1989.
146. BC Parker. J Phycol 2:38, 1966.
147. PT Richardson, DA Baker, LC Ho. J Exp Bot 33:1239, 1982.
148. CA Swanson, EDH El-Shishiny. Plant Physiol 33:33, 1958.
149. WN Arnold. J Theor Biol 21:13, 1968.
150. LW Handley, DM Pharr, RF McFeeters. Plant Physiol 72:489, 1983.
151. JM Daly. In: R Heilefuss, PH Williams, eds. Encyclopedia of Plant Physiology, Vol 4. New York: Springer-Verlag, 1976, p 450.
152. Y Shachar-Hill, PE Pfeffer, D Douds, SF Osman, LW Doner, RG Ratcliffe. Plant Physiol 108:7, 1995.
153. DH Webster, HB Currier. Can J Bot 66:1215, 1968.
154. KD Nolte, KE Koch. Plant Physiol 101:899, 1993.
155. DL Morrow, WJ Lucas. Plant Physiol 84:565, 1987.
156. H Ziegler. In: MH Zimmermann, JA Milburn, eds. Encyclopedia of Plant Physiology, Vol I, Transport in Plants, Part I. New York: Springer-Verlag, 1975, p 59.
157. JS Pate. In: IF Wardlaw, JB Passiouru eds. Transport and Transfer Processes in Plants. New York: Academic Press, 1976, p 447.
158. N Wang, PS Nobel. Ann Bot 75:525, 1995.
159. D Schobert, E Komer. Planta 177:342, 1989.
160. C Girousse, R Bournoville, J-L Bonnemain. Plant Physiol 111:109, 1996.
161. W Eschrich, W Heyser. In: MH Zimmermann, JA Milburn, eds. Encyclopedia of Plant Physiology, Vol I, Transport in Plants, Part I. New York: Springer-Verlag, 1975, p 181.
162. DB Fisher, Y Wu, MSB Ku. Plant Physiol 100:1433, 1992.
163. Y Ishiwatari, T Fujiwara, KC McFarland, D Memoto, M Chino, WJ Lucas. Planta 205:12, 1998.
164. S Balachardran, Y Xiang, C Schobert, GA Thompson, WJ Lucas. Proc Natl Acad Sci USA 94:14150, 1997.
165. V Citovsky. Plant Physiol 102:1071, 1993.
166. B Xoconostle-Cazares, Y Xiang, R Ruiz-Medrano, H-L Wang, J Monzer, B-C Yoo, KC McFarland, VR Franceschi, WJ Lucas. Science 283:94, 1999.
167. S Ghoshroy, R Lartey, J Sheng, V Citovsky. Annu Rev Plant Physiol Plant Mol Biol 48:27, 1997.
168. JA Raven, FA Smith. New Phytol 76:415, 1976.
169. EA Kirkby, RH Knight. Plant Physiol 60:349, 1977.
170. MH Goldsmith, DA Cataldo, J Karn, T Brenneman, P Trip. Planta 116:301, 1974.
171. JS Taylor, B Thompson, JS Pate, CA Atkins, RP Pharis. Plant Physiol 94:1714, 1990.
172. M Katsumi, DE Foard, BO Phinney. Plant Cell Physiol 24:379, 1983.
173. ES Ober, TL Setter. Plant Physiol 98:352, 1992.
174. I Raskin. Plant Physiol 99:799, 1992.
175. PE Stanswick. Plant Physiol 99:804, 1992.
176. U Petzold, S Peschel, I Dahse, G Adam. Acta Bot Neerl 41:469, 1992.
177. JW Patrick. In: F Wardlaw, JB Passioura, eds. Transport and Transfer Processes in Plants. New York: Academic Press, 1976, p 433.
178. I Sakeena, MA Salam. Indian J Plant Physiol 31:428, 1988.
179. S Jahnke, D Bler, JJ Estruch, JP Beltran. Planta 180:53, 1989.
180. V Borkovec, S Prochazka. J Agron Crop Sci 169:229, 1992.
181. JS Pate. In: MH Zimmermann, JA Milburn, eds. Encyclopedia of Plant Physiology, Vol I, Transport in Plants, Part I. New York: Springer-Verlag, 1975, p 441.
182. J Van Die, PCM Willense. Acta Bot Neerl 24:237, 1975.
183. SM Hall, DA Baker. Planta 106:131, 1972.
184. SM Hall, DA Baker, JA Milburn. Planta 100:200, 1971.

185. WD Jeschke, CA Atkens, JS Pate. J Plant Physiol 117:319, 1985.
186. D Lamb. Plant Soil 45:477, 1976.
187. GJ Waughman, DJ Bellamy. Ann Bot 47:144, 1981.
188. JF Longeragan, K Snowball, AD Robson. In: IF Wardlaw, JB Passioura, eds. Transport and Transfer Processes in Plants. New York: Academic Press, 1976, p 463.
189. GL Mullins, LE Sommers, TL Housley. Plant Soil 96:377, 1986.
190. J Gorham, LL Hughes, RG WynJones. Plant Cell Environ 3:309, 1980.
191. A Pollard, RG WynJones. Planta 144:291, 1979.
192. GA Mitchell, FT Bingham, AL Page. J Envion Qual 7:166, 1978.
193. DM Maynard, OA Lorenz, V Magnifico. J Am Soc Hortic Sci 105:79, 1980.
194. LJ Sikora, RL Chaney, NH Frankos, CM Murry. J Agric 6:1281, 1980.
195. EM Romney, A Wallace, AK Schulz, J Kinnear, RA Wood. Soil Sci 132:40, 1981.
196. JN Pearson, Z Rengel, CF Jenner, RD Graham. Physiol Plant 95:449, 1995.
197. MA Grusak. Plant Physiol 104:649, 1994.
198. GS Dollard, NW Lepp. Z Pflanzenphysiol 97:409, 1980.
199. H Hayashi, M Chino. Plant Cell Physiol 26:325, 1985.
200. H Hahashi, M Chino. Plant Cell Physiol 27:1387, 1986.
201. B Wolterbeek, J Van Die. Acta Bot Neerl 29:307, 1980.
202. DG Blevins, KM Lukaszewski. Annu Rev Plant Physiol Plant Mol Biol 49:481, 1998.
203. PH Brown, N Bellaloui, H Hu, A Dandekar. Physiol Plant 119:17, 1999.
204. E Munch. Ber Beut Bot Ges 44:68, 1927.
205. EC Miller. Plant Physiology. New York: McGraw-Hill, 1931.
206. O Raber. Principles of Plant Physiology. New York: Macmillan, 1933.
207. CA Swanson. In: FC Steward, ed. Plant Physiology. A Treatise, Vol II, Plants in Relation to Water and Solutes. New York: Academic Press, 1959, p 481.
208. FL Milthorpe, J Moorby. Annu Rev Plant Physiol 20:117, 1969.
209. AJ Peel. Transport of Nutrients In Plants. New York: Wiley, 1974.
210. S Aronoff, J Dainty, PA Gorham, LM Srivastava, CA Swanson, eds. Phloem Transport. New York: Plenum, 1975.
211. L Horwitz. Plant Physiol 33:81, 1958.
212. O Biddulph. In: FC Steward, ed. Plant Physiology: A Treatise, Vol II, Plants in Relation to Water and Solutes. New York: Academic Press, 1959, p 553.
213. DS Fensom. Can J Bot 35:573, 1957.
214. DC Spanner. J Exp Bot 9:332, 1958.
215. DJF Bowling. Planta 80:21, 1968.
216. R Thaine. Nature 192:772, 1961.
217. R Thaine, ME de Maria, HIM Sarisalo. J Exp Bot 26:91, 1975.
218. MJ Canny. Ann Bot 25:152, 1962.
219. P Trip, PR Gorham. Plant Physiol 43:877, 1968.
220. JA Webb. Can J Bot 49:717, 1971.
221. RT Giaquinta, DR Gelger. Plant Physiol 51:372, 1973.
222. CA Swanson, DR Gelger. Plant Physiol 42:751, 1967.
223. JW Sij, CA Swanson. Plant Physiol 51:368, 1973.
224. CA Peterson, HB Currier. Physiol Plant 22:1238, 1969.
225. JP Wright, DB Fisher. Plant Physiol 65:1133, 1980.
226. DG Fisher. Planta 169:141, 1986.
227. R Turgeon, PK Hepler. Planta 179:24, 1989.
228. RI Grange, AJ Peel. Planta 124:191, 1975.
229. K Esau. Plant Anatomy. New York: Wiley, 1953.
230. DA Cataldo, AL Christy, CL Coulson, JM Ferrier. Plant Physiol 49:685, 1972.
231. DA Cataldo, AL Christy, CL Coulson. Plant Physiol 49:690, 1972.
232. AL Christy, JM Ferrier. Plant Physiol 52:531, 1973.
233. M Knoblauch, AJE Van Bel. Plant Cell 10:35, 1998.
234. MA Grusak, DO Beebe, R Turgeon. In: E Zamski, AA Schaffer, eds. Photoassimilate Distribution in Plants and Crops, Source-Sink Relationships. New York: Marcel Dekker, 1996, p 209.
235. JW Ward, C Kuhn, M Tegeder, WB Frommer. Int Rev Cytol 178:41, 1998.
236. S Delrot. Plant Physiol Biochem 25:667, 1987.
237. AJE Van Bel. Plant Physiol Biochem 25:677, 1987.
238. BES Gunning, JS Pate. Protoplasma 68:107, 1969.
239. BES Gunning. Sci Prog Oxf 64:539, 1977.
240. E Schnepf, E Proso. Protoplasma 89:105, 1976.
241. DR Geiger, SA Sovonick, TL Shock, RJ Fellows. Plant Physiol 54:892, 1974.
242. RT Giaquinta. Plant Physiol 60:339, 1977.

243. NL Robinson, JE Hendrix. Plant Physiol 71:701, 1983.
244. RT Giaquinta. Plant Physiol 57:872, 1976.
245. RT Giaquinta. Plant Physiol 63:744, 1979.
246. S Deirot, J-L Bonnemain. Plant Physiol 67:560, 1981.
247. DR Bush. Plant Physiol 93:1590, 1990.
248. R Lemoine, O Gallet, C Gaillard, W Frommer, S Delrot. Plant Physiol 100:1150, 1992.
249. RT Giaquinta. Nature 267:369, 1977.
250. E Truernit, N Sauer. Planta 196:564, 1995.
251. L Burkie, JM Hibberd, WP Quick, C Kuhn, B Hirner, WB Frommer. Plant Physiol 118:59, 1998.
252. R Shakya, A Sturm. Plant Physiol 118:1473, 1998.
253. CEJ Botha, RAM Cross. Physiol Plant 99:355, 1997.
254. Z-S Li, O Gallet, C Gaillard, A Lemoine, S Delrot. FEBS 286:117, 1991.
255. Z-S Le, AM Noubhani, A Bourbouloux, S Delrot. Biochim Biophys Acta 1219:389, 1994.
256. AM Noubhani, S Sakr, M-H Demis, S Delrot. Biochim Biophys Acta 1281:213, 1996.
257. G Roblin, S Sakr, J Bonmort, S Delrot. FEBS Lett 424:165, 1998.
258. D Rentsch, M Laloi, I Rouhara, E Schmelzer, S Delrot, WB Frommer. FEBS Lett 370:264, 1995.
259. JW Maynard, WJ Lucas. Plant Physiol 69:734, 1982.
260. JW Maynard, WJ Lucas. Plant Physiol 70:1436, 1982.
261. MA Madore, JW Dross, WJ Lucas. Plant Physiol 82:432, 1986.
262. C Wilson, WJ Lucas. Plant Physiol 84:1088, 1987.
263. MR Madore, WJ Lucas. In: DA Baker, JA Milburn, eds. Transport of Photoassimilates, New York: Wiley, 1989, p 49.
264. MA Madore, WJ Lucas. Planta 171:197, 1987.
265. R Turgeon, DU Beebe. Plant Physiol 96:349, 1991.
266. R Turgeon, E Gowan. Plant Physiol 84:1244, 1994.
267. LE Wimmers, R Turgeon. Planta 186:2, 1991.
268. CEJ Botha, AJE Van Bel. Planta 187:359, 1992.
269. AJE Van Bel, YV Gamalei, A Ammerlaan, LPM Bik. Planta 186:518, 1992.
270. YV Gamalei. Trees 5:50, 1991.
271. LL Flora, MA Madore. Planta 198:171, 1996.
272. D Schmitz, B Cuypers, M Moll. Planta 171:19, 1987.
273. AJE Van Bel. Acta Bot Neeri 41:121, 1992.
274. AW Robards. Annu Rev Plant Physiol 26:13, 1975.
275. AW Robards. In: BES Gunnings, AW Robards, eds. Intercellular Communication in Plants: Studies on Plasmodesmata. New York: Springer-Verlag, 1976, p 2.
276. AW Robards, WJ Lucas. Annu Rev Plant Physiol Plant Mol Biol 41:369, 1990.
277. YD Gamalel. Soviet Plant Physiol (Eng trans) 28:649, 1981.
278. JW Oross, MA Grusak, WJ Lucas. In: J Cronshaw, WJ Lucas, RT Giaquinta, eds. Plant Biology Vol I, New York: Liss, 1986, p 477.
279. MA Madore, WJ Lucas. In: J Cronshaw, WJ Lucas, RT Giaquinta, eds. Plant Biology Vol I, Phloem Transport. New York: Liss, 1986, p 129.
280. WJP Van Kesteren, C Van der Schoot, AJE Van Bel. Plant Physiol 88:667, 1988.
281. MG Erwee, PB Goodwin. Protoplasma 122:162, 1984.
282. R Turgeon, R Medville. Proc Natl Acad Sci USA 95:12055, 1998.
283. PT Richardson, DA Baker, LC Ho. J Exp Bot 35:1575, 1984.
284. AJE Van Bel. In: CJ Pollock, JF Farrar, AJ Gordon, eds. Partitioning Within and Between Organisms. Oxford: Bios Science Publishers, 1992, p 53.
285. R Turgeon. Trends Plant Sci 1:418, 1996.
286. E Haritatos, F Keller, R Turgeon. Planta 198:614, 1996.
287. AJE Van Bel. Annu Rev Plant Physiol Plant Mol Biol 44:253, 1993.
288. BG McLean, FD Hempel, PC Zambryski. Plant Cell 9:1043, 1997.
289. JH Thorne. Annu Rev Plant Physiol 36:317, 1985.
290. JH Thorne. In: J Cronshaw, WJ Lucas, RT Giaquinta, eds. Plant Biology, Vol I, Phloem Transport. New York: Liss, 1986, p 211.
291. W Eschrich. In: DR Baker, JA Milburn, eds. Transport of Photoassimilates. New York: Wiley, 1989, p 206.
292. JW Patrick. Physiol Plant 78:298, 1990.
293. JW Patrick. Annu Rev Plant Physiol Plant Mol Biol 48:191, 1997.
294. W Eschrich. In: J Cronshaw, WJ Lucas, RT Giaquinta, eds. Plant Biology, Vol I, Phloem Transport. New York: Liss, 1986, p 225.
295. DB Fisher, N Wang. Plant Physiol 109:587, 1995.
296. N Wang, DB Fisher. Plant Physiol 109:579, 1995.
297. KE Koch, WT Avigne. Plant Physiol 93:1405, 1990.
298. MS Bret-Harte, WK Silk. Plant Physiol 105:19, 1994.

299.   JH Thorne, RM Rainbird. Plant Physiol 72:268, 1983.
300.   N Wang, DB Fisher. Plant Physiol 104:7, 1994.
301.   N Wang, DB Fisher. Plant Physiol 104:17, 1994.
302.   KJ Oparka. Plant Physiol 94:393, 1990.
303.   DB Fisher, KJ Oparka. J Exp Bot 47:1141, 1996.
304.   AS Crafts. Ann N Y Acad Sci 144:357, 1967.
305.   CE Crisp. Insecticides. Proceedings of the 2nd IUPAC Congress on Pesticidal Chemicals, Vol 1, Tel Aviv Israel, p 211, 1972.
306.   MT Tyree, CA Peterson, LD Edgington. Plant Physiol 63:367, 1979.
307.   DA Kleier. Plant Physiol 86:803, 1985.
308.   M-H Denis, S Delrot. Physiol Plant 87:569, 1993.
309.   BG Bugbee, FB Salisbury. Plant Physiol 88:869, 1988.
310.   DW Ming, DL Henninger, eds. Lunar Base Agriculture: Soils for Plant Growth. Madison, WI: ASA, SSA, 1989.
311.   BG Bugbee, O Monje. BioScience 42:494, 1992.
312.   KA Corey, RM Wheeler. Bioscience 42:503, 1992.
313.   DL Bishop, BG Bugbee. J Plant Physiol 153:558, 1998.

# 21

# Phloem Transport of Solutes in Crop Plants

**Edmund R. Miranda, Wattana Pattanagul, and Monica A. Madore**

*University of California, Riverside, California*

## I. INTRODUCTION

Many plant parts, including many flowers, fruits, and seeds, do not contain chlorophyll and are therefore not photosynthetically competent. Other plant parts, particularly underground roots, rhizomes, and tubers, are located on the plant in areas where light reception is insufficient to drive photosynthesis. In plant parts such as meristems, stems, and developing leaves, modification, incomplete development, or insufficient number of plastids also limits photosynthetic competence. Photosynthetic activity is therefore found to be largely confined to organs located in areas of maximal light interception and containing fully functional chloroplasts. In higher plants, these organs are represented by mature, fully expanded leaves.

The consequence of this separation of the plant body into photosynthetically competent and non-photosynthetic organs is that photosynthesizing leaves become the sole "source" of photosynthetically produced biomolecules (photoassimilates) for the rest of the plant. Thus, to supply the demands of non-photosynthetic plant parts, which act as competing "sinks" for photosynthetic products, leaves must produce photoassimilates in amounts far in excess of what is required simply for maintenance of leaf metabolism. In higher plants, the delivery of photoassimilates from source to sink regions within the plant body is accomplished by translocation in the phloem tissues.

In crop plants, phloem transport is a particularly important physiological process, for with very few exceptions, the agronomically important plant parts that are harvested from our major agricultural crop plants are sink tissues. From a physiological standpoint, what this ultimately means is that the ability of a particular crop plant to carry out photosynthesis during a growth season will only partly determine the final harvestable yield of that crop. The phloem transport process will be of equal importance, for it is this process that determines just how efficiently photosynthetically produced nutrients are made available to the plant part to be harvested. A complete understanding of phloem transport and its regulation is therefore basic to our understanding of crop physiology.

## II. PHLOEM STRUCTURE

It is beyond the scope of this chapter to provide more than a general description of the anatomy of the phloem transport system. Readers should consult a general plant anatomy textbook (e.g., Ref. 1) or reviews of phloem structure [2–4] for more details regarding the anatomy, morphology, and differentiation of vascular tissues. This chapter emphasizes the phloem structure of the minor veins of leaves, the key interface of the phloem transport system with the photosynthetic tissues.

## A.   General Feature of Phloem Tissues

### 1.   Sieve Elements

Phloem tissues in general consist of several structurally distinct cell types: sieve elements, companion cells, parenchyma cells, and fibers [1]. The most characteristic cells are the sieve elements, which are linked end to end to form the conduit for the long-distance movement of solutes (Figure 1). Unlike xylem tracheids, which are dead at maturity, functional sieve elements are living cells. During maturation of the sieve element, the tonoplast and nucleus degenerate and all ribosomes disappear. Mitochondria and plastids assume a parietal position next to the plasma membrane. Plastids accumulate either starch or protein inclusions. Proteinaceous strands (P protein) may also be present in the cell lumen [1–4].

   The end walls of the sieve element are modified to form the sieve plate (Figure 1). Contiguous sieve elements are interconnected to form a sieve tube through strands of protoplasm, which pass through the plasma membrane–lined sieve plate pores. The side walls of adjacent sieve tubes may also contain sieve areas connecting the protoplasts of the neighboring sieve elements [1–4]. The cytoplasmic compartments of the sieve elements, therefore, form a continuum through which solutes can be moved.

### 2.   Companion Cells

Companion cells are associated with sieve elements and arise concurrently with the sieve elements by division of a common mother cell. Unlike the sieve elements, companion cells retain their nuclei and vacuoles and are characterized by densely staining cytoplasm containing numerous free ribosomes and many highly differentiated mitochondria and plastids [1–4]. These structural features are indicative of high metabolic activity, and it is thought that the companion cells act to maintain the structural integrity of the sieve elements, which lose metabolic capability as a result of the structural changes that occur during maturation. The protoplasts of the companion cells are connected to sieve elements by numerous branched (on the companion cell side) plasmodesmata, providing cytoplasmic connection for metabolite exchange

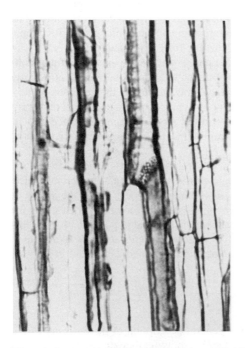

**Figure 1**   Longitudinal section of a squash (*Cucurbita pepo* L.) stem, showing the sieve elements (SE). Arrow indicates a sieve plate. (Paraffin section slide courtesy of D. A. DeMason.)

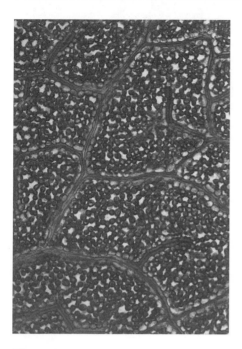

**Figure 2**   Paradermal section of a lilac (*Syringa vulgaris* L.) source leaf, showing the arrangement of the minor venation within the photosynthetic tissues. (Paraffin section slide courtesy of D. A. DeMason.)

between the two cell types. Because of the high degree of symplastic continuity between the companion cell and sieve element, these are often referred to as sieve element–companion cell (SE-CC) complexes [2–4].

## B.  Minor Vein Structure

The venation of source leaves is designed such that individual photosynthetic mesophyll cells are never more than a few cells away from a minor vein (Figure 2). This arrangement drastically reduces the distance that assimilates must travel from the sites of photosynthesis to the phloem transport system [5]. It is in the SE-CC complexes of the leaf minor veins that loading of the phloem transport system with photosynthetic products is initiated. Not surprisingly, the companion cells within the minor veins are very much larger than the sieve elements with which they are associated (Figure 3), which more than likely reflects the added metabolic activity imparted by the phloem loading process.

Minor vein companion cells form a key interface between the photosynthetic tissues of the leaf and the conduits of the phloem system. Based on ultrastructural differences, Gamalei [6], van Bel and Garnalei [7], and van Bel [8] have categorized three classes of minor vein companion cells within source leaves (Figure 4).

Type 1 companion cells (Figure 4A), referred to as "intermediary cells" [9–11], are characterized by large numbers of plasmodesmata, which link the cytoplasm of these cells to that of the adjacent photosynthetic cells. These cells are specialized for the symplastic transfer of assimilates from the photosynthetic cells to the sieve elements. Companion cells of this type are common in many species of horticultural importance, including tree species such as olive [12], most woody ornamental vines and shrubs [6–8], culinary herbs and ornamentals of the mint family such as coleus [13], and the cucurbit vine crops [14,15]. Crop species that have this companion cell type tend to be of tropical or subtropical origin [16].

Type 2a companion cells (Figure 4B) lack the extensive plasmodesmatal connections to the photosynthetic tissues that are typical of type 1 cells. Assimilates produced in leaves with this type of companion cell, therefore, do not have an elaborate symplastic pathway through which to travel into the SE-CC complex and must be released from the cytoplasm of the photosynthetic cells [6]. Assimilates are

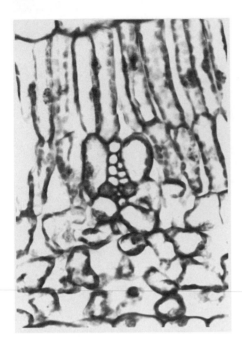

**Figure 3**  Cross section of a lilac (*Syringa vulgaris* L.) source leaf showing a minor vein. C, companion cell; S, sieve element; X, xylem; MC mesophyll (photosynthetic) cell. The companion cell in this species is a type 1 (intermediary) cell. (Paraffin section slide courtesy of D. A. DeMason.)

transferred into the cell wall space (the apoplast) of the SE-CC complex and are taken up across the plasma membrane of the SE-CC complex for export in the phloem [17,18]. Leaves with this type of minor vein configuration, therefore, use a membrane transport mechanism for phloem loading. Plant species with type 2a companion cells are almost exclusively herbaceous annuals of temperate origin [16]. Most of the major crop species of agronomic importance fall into type 2a.

Type 2b companion cells (Figure 4C) are similar to type 2a in that they lack extensive symplastic connections to the photosynthetic tissues [6–8]. However, the cell walls of type 2b companion cells are characterized by extensive wall ingrowths, which greatly amplify the plasma membrane surface exposed to the apoplastic space. This type of companion cells is referred to as a "transfer cell" [18], for it has apparently been modified to facilitate the transfer of assimilates from the apoplast into the SE-CC complex. Type 2b companion cells are again typical of temperate herbaceous crops [16] and are a particularly common characteristic of legume species [19].

## III.  LONG-DISTANCE TRANSPORT

The vascular system of higher plants can be regarded as a series of parallel conduits of xylem and phloem tissue, which run the length of the plant body from root to shoot and permeate all major plant organs. Despite the co-occurrence of xylem and phloem tissues in the vascular strands, these tissues are functionally quite distinct.

The main function of the xylem is the transport of water and dissolved mineral nutrients to the shoot following uptake from the soil by the roots. Xylem transport is unidirectional (upward from roots to shoots) and is driven by the water potential gradient created by evaporation of water from leaves (transpiration) [20]. Rates of transport can be of the order of 10 cm min$^{-1}$ [20]. The transport pathway is formed by the cell walls of the xylem vessels and tracheids, which are dead at maturity [20]. Xylem transport therefore does not require cells with a living protoplast. Xylem sap consists primarily of water, with low levels of dissolved solutes (Table 1). Most noticeably, sugars are absent from xylem sap (Table 1).

In constrast, phloem transport occurs through living cells, the sieve elements [1]. The rates of transport are much lower than one sees in the xylem (on the order of 1 cm min$^{-1}$) and can occur in either an upward or downward direction [21]. Rate and direction of transport are dictated by differences in solute concentrations between sites of solute synthesis (sources) and solute consumption (sinks) within the plant body. It is these solute concentration differences that provide the driving force for phloem transport [22]. In contrast to xylem saps, phloem saps contain very high solute levels (Table 1) and particularly high levels of sugars, amino acids, and potassium.

## A. The Munch Pressure Flow Mechanism

The pressure flow mechanism first postulated by Munch [22] provides the best explanation for the driving force for phloem transport presently available based on our knowledge of rates of transport and

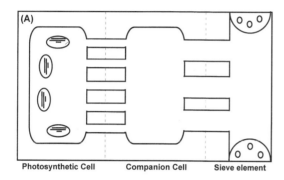

Photosynthetic Cell     Companion Cell     Sieve element

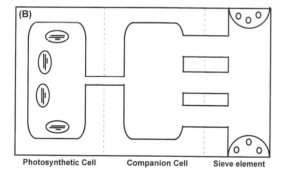

Photosynthetic Cell     Companion Cell     Sieve element

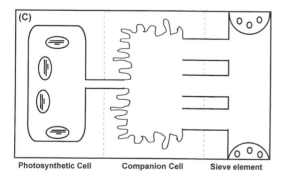

Photosynthetic Cell     Companion Cell     Sieve element

**Figure 4** Diagrammatical representations of the classes of minor vein companion cells found in source leaves: (A) Type 1 companion cell (intermediary cell), (B) type 2a companion cell, and (C) type 2b companion cell (transfer cell).

**TABLE 1** Typical Ranges for Components of Xylem and Phloem Saps in Higher Plants

| Substance | Concentrations ($\mu g\ mL^{-1}$) | |
| | Xylem | Phloem |
| --- | --- | --- |
| Sugars | Absent | 140,000–210,000 |
| Amino acids | 200–1000 | 900–10,000 |
| P | 70–80 | 300–550 |
| K | 200–800 | 2800–4400 |
| Ca | 150–200 | 80–150 |
| Mg | 30–200 | 100–400 |
| Mn | 0.2–6.0 | 0.9–3.4 |
| Zn | 1.5–7.0 | 8–23 |
| Cu | 0.1–2.5 | 1.0–5.0 |
| B | 3.0–6.0 | 9–11 |
| $NO_3^-$ | 1500–2000 | Absent |
| $NH_4^+$ | 7–60 | 45–846 |

*Source*: Data from Refs. 31 and 36.

phloem structure. Phloem loading in source tissues leads to the very high solute concentrations characteristic of the phloem. The high solute levels create a water potential gradient within the sieve element, and water moves into the sieve element from the adjacent xylem tissues. Because the sieve element is a living cell and has a functional plasma membrane, this influx of water creates a very high hydrostatic pressure within the sieve element. At the sink end, the solutes are removed from the sieve element for use by the sink cells, and the hydrostatic pressure is reduced. This combination of solute loading at the source end and solute unloading at the sink end of the phloem system creates a strong hydrostatic pressure gradient. Because the sieve elements are linked end to end by open sieve plates, water containing the dissolved solutes passes through the pores of the sieve plates in response to the pressure gradient and solutes are moved by this bulk flow from source to sink.

## B.  Solutes Translocated in the Phloem

A mature, fully expanded leaf not only is the primary site of photosynthesis but also has the highest rate of transpiration. As a result, a significant percentage of the dissolved mineral nutrients present in the xylem sap will end up in leaves, not in the agronomically important plant parts. The phloem of the minor veins of leaves is therefore very important, not only for the transport of photosynthate produced in the leaves but also for the redistribution of mineral elements delivered by the xylem. In addition, phloem transport plays a major role in the transduction of developmental and environmental stimuli via the transport of growth regulators and systemic signal molecules.

### 1.  Carbohydrates

Carbohydrates translocated in the phloem are all low-molecular-weight nonreducing sugars (Figure 5A) or sugar alcohols (Figure 5B). The disaccharide sucrose (Figure 5A) is ubiquitous in the phloem of crop plants. However, many important crop species transport sugars in addition to sucrose [23]. Plant species that possess type 1 companion cells (Figure 4A) all translocate the raffinose family oligosaccharides such as raffinose and stachyose [6–8,13–15,24], which are galactoside derivatives of sucrose (Figure 5A). Some members of the Rosaceae, including tree crops such as apples, cherries, plums, and apricots, also translocate significant quantities of the sugar alcohol sorbitol [23,25,26]. Members of the Apiaceae, such as celery, transport the sugar alcohol mannitol in addition to sucrose [25,27–29]. Still other plant species, such as olive [12] and euonymus [23], translocate both raffinose family sugars and a sugar alcohol (mannitol and dulcitol, respectively).

### 2.  Nitrogen-Containing Compounds

Most protein amino acids are found in phloem saps (Table 2) [30–33]. The predominant amino acids tend to be those having a high ratio of nitrogen to carbon (Figure 5C), particularly the amides asparagine and

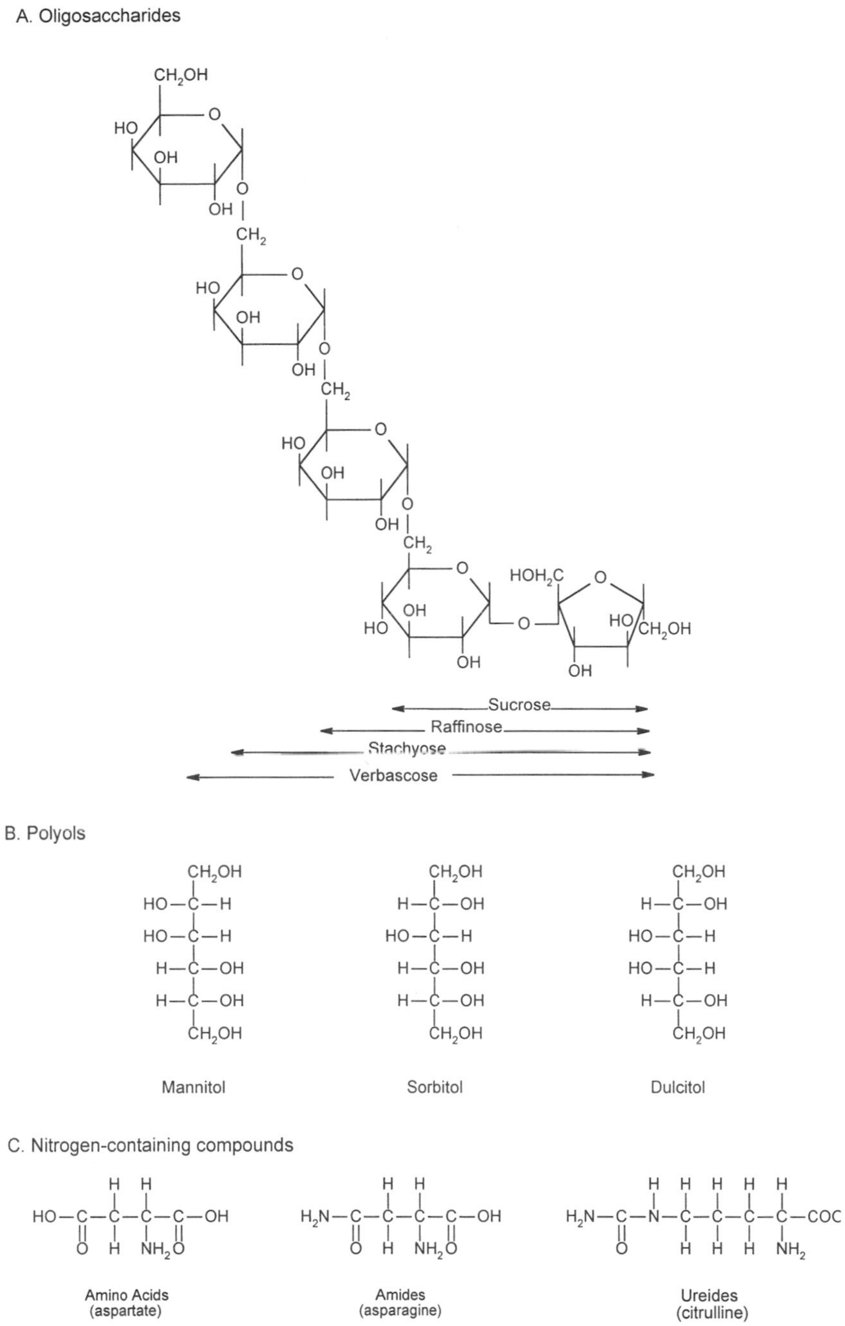

A. Oligosaccharides

B. Polyols

C. Nitrogen-containing compounds

**Figure 5**   Chemical structures of (A) phloem-mobile sugars, (B) sugar alcohols, and (C) nitrogenous compounds found in phloem saps.

glutamine [2]. The amino acid composition of the phloem sap can vary greatly depending on the species (Table 2) or environmental conditions [34]. In addition to amides, some species, particularly the nitrogen-fixing legumes, transport small quantities of the ureides allantoin and allatonic acid in the phloem [30]. Nonprotein amino acids, such as canavanine and ornithine, other ureides such as citrulline, and polyamines such as putrescine may also be found in limited quantities in phloem saps of many plant

**TABLE 2**   Typical Range of Amino Acid Composition of
Phloem Sap

| Amino acid | Phloem sap concentration (mM) |
|---|---|
| Aspartate | 2–20 |
| Glutamate | 7–25 |
| Asparagine | 2–275 |
| Glutamine | 10–25 |
| Serine | 5–15 |
| Glycine | Trace–6 |
| Homoserine | 0–trace |
| Citrulline | 0–20 |
| Histidine | 0–trace |
| Arginine | Trace–5 |
| Threonine | 1–10 |
| Alanine | 1–8 |
| Proline | 5–15 |
| Tyrosine | 0.5–2.0 |
| Valine | 0–9 |
| Methionine | 0–trace |
| Cysteine | 0–1 |
| Isoleucine | 2–6 |
| Leucine | 0–6 |
| Phenylalanine | 3–5 |
| Tryptophan | 0–trace |
| Ornithine | 0–trace |
| Lysine | 1–3 |

*Source*: Data from Refs. 31–33.

species [30]. Feeding experiments with cadaverine and spermidine have exhibited reduced phloem sap levels of putrescine, which demonstrates a competitive effect between various polyamines [35]. Among the purported roles of polyamines one function may be in stabilization of biomembranes, which may be important for the regulation of growth in plants.

## 3.  *Mineral Nutrients*

Many of the same mineral ions found in xylem saps are also found in phloem saps (Table 2) [30,31,36,37], indicating that these nutrients can be removed from the xylem and loaded into the phloem transport system. Many sink tissues, being only poorly supplied with these nutrients by the xylem because of low transpiration rates, must depend on phloem transport for much of their mineral requirements [30,37].

In general, the relative mobility of mineral ions in the phloem can be determined by the site at which deficiency symptoms first appear. Some ions (e.g., boron and calcium) are only poorly loaded into the phloem [30,31,36,37]. In these cases, deficiency symptoms appear predominantly in sink tissues such as fruits and young leaves, which must depend on transpiration and xylem movement for a supply of these minerals [37]. In contrast, in the cases of minerals that are highly phloem mobile (e.g., magnesium, potassium), deficiency symptoms appear first in the mature leaves [37]. This indicates a remobilization of minerals from the mature leaves and delivery of these elements to the sink leaves via phloem transport.

Mineral ions may be translocated as free elemental ionic forms (e.g., $K^+$, $Cl^-$) but frequently may exist in other chemical forms (e.g., phosphate, sulfate, ammonium). Notably, although free nitrate is a common constituent of xylem saps, it is never found in phloem saps (Table 2) [30,31,36]. Mineral elements may also be combined into organic complexes (e.g., ferric chelates, zinc peptides, phosphate esters, sulfur-containing amino acids) for transport in the phloem [30,36]. For example, *S*-methylmethionine (SMM) in phloem has been reported to provide more than half of the sulfur needed for grain protein synthesis in wheat [38]. The enzymes involved in synthesis of SMM have shown strong amino acid sequence homology with those of *Arabidopsis* and maize, and one can speculate that a transgenic approach to increase the copy number of SMM genes might reduce the fertilizer levels required on croplands.

## 4. Growth Regulators

All classes of naturally occurring plant growth regulators (auxins, gibberellins, cytokinins, abscisic acid) can be recovered in phloem saps, indicating that these compounds are normally translocated in the phloem [30,39].

Cytokinins have been reported to regulate shoot development, possibly through regulation of sink activity and changes in resource partitioning patterns. Zeatin and zeatin riboside have been reported to be the dominant transportable forms of cytokinin in the plant [40,41]. Although the root is believed to be the primary site of synthesis of cytokinins, Kamboj et al. [41] showed that zeatin riboside was the predominant form in the roots while zeatin predominated in the phloem sap. This indicated that zeatin riboside was the predominant form translocated from roots via the xylem and that zeatin itself was the major form transported via the phloem. The site of synthesis of zeatin in the phloem sap is still unknown. It has been suggested that origin is through synthesis within the mature leaf followed by subsequent loading into the phloem, through direct exchange with xylem cytokinin metabolites or through recirculation from the roots.

## 5. Systemic Signals

Grafting experiments using source leaves have indicated that other growth factors apart from the known growth regulators are also translocated in the phloem. These include floral initiation signals, cold hardiness–inducing signals, and pathogen resistance factors [30]. The chemical nature of these systemic signals is only beginning to be deciphered. Salicylic acid, which appears to be one prime candidate as a signaling molecule for these responses in some plant species, is thought to be phloem mobile [42,43]. In addition, a phloem-mobile peptide, systemin, has been shown to induce pathogen resistance [44]. The plant growth regulator abscisic acid is a likely candidate as a phloem-mobile cold hardiness–inducing factor [45].

Phloem tissues may also be capable of limited synthesis of phloem-specific proteins, whose function is unknown, but they may also be involved in signaling [46,47]. The localization of sucrose synthase within phloem tissue [48,49] suggests that this enzyme may be involved in the signaling pathway that leads to callose synthesis [48] in response to wounding or pathogen invasion. It is likely that many more signaling mechanisms will be discovered in the phloem.

## 6. Xenobiotics

A number of man-made chemicals of agronomic importance, including many herbicides and pesticides [30], are also translocated in the phloem. Combinations of lipid permeability and acid dissociation constants (pKa) are predictors of phloem mobility that have been validated for many compounds in various plant systems. One particularly good systemic herbicide is glyphosate (N-phosphonomethylglycine), which is highly mobile in the phloem (Table 3).

The limitations of phloem mobility appear to be due mostly to failure of the applied chemical to cross cuticular barriers, retention along the phloem path (limitation of lateral efflux along translocation path)

**TABLE 3** Symplastic (Phloem) Transported Herbicides

| Herbicide class | Typical representative | Chemical structure |
|---|---|---|
| Phenoxy herbicides | 2,4-D | 2,4-Dichlorophenoxyacetic acid |
| | 2,4,5-T | 2,4,5-Trichlorophenoxyacetic acid |
| Benzoic acids | Dicamba[a] | 3,6-Dichloro-2-methoxybenzoic acid |
| | 2,3,6-TBA | 2,3,6-Trichlorobenzoic acid |
| Picolinic acids | Picloram[a] | 4-Amino-3,5,6-trichloropicolinic acid |
| | Triclopyr | [(3,5,6-Trichloro-2-pyridyl)oxy]acetic acid |
| Chlorinated aliphatics | Dalapon | 2,2-Dichloropropionic acid |
| Triazoles | Amitrole[a] | 3-Amino-s-triazole |
| Organic arsenicals | DSMA | Disodium methanoarsenate |
| Glyphosate | | N-Phosphonomethylglycine |
| Sulfonylureas | Chlorsulfuron | 2-Chloro-N-[(4-methoxy-6-methyl-1,3,5-triazin-2-yl) aminocarbonyl]benzenesulfonamide |

[a] Also transported apoplastically (xylem).

[50], and failure to cross biological membranes. It is proposed that ionized groups or low lipophilicity prevents entry into the SE-CC complex [51,52]. It is important to note that not all herbicides need be mobile but can be herbicidal simply if they disrupt transport of photoassimilates from source to sink tissues. Chlorsulfuron is one such herbicide that has been shown to disrupt transport of photoassimilates, but the mode of its action is still unknown [53].

### 7. *Protein and Viral Movement in the Phloem*

Researchers have investigated the cell-to-cell trafficking of macromolecules from the companion cells into the sieve tube via plasmodesmata [54]. It has long been known that cell-to-cell movement of small molecules can occur via the symplasm, but more recently it has been demonstrated that large macromolecules (viral-encoded movement proteins [55,56], messenger RNA [57] and plant macromolecules [58]) can apparently enter the phloem via plasmodesmata.

Viral coat proteins are believed to interact with endogenous plasmodesmal proteins to increase the size exclusion limit (SEL) and allow movement of virus particles from infected to uninfected neighboring cells. It appears that these proteins found in the phloem are exclusively synthesized in the companion cells [59]. Phloem sap proteins of *Cucurbita maxima* have the ability to induce an increase in the SEL of plasmodesmata by greater than 20 kDa [54]. It is hypothesized that macromolecules greater than 20 kDa may partially unfold to facilitate transport through the plasmodesmata. The ability to transport large proteins or nucleic acids via the phloem translocation pathway may help explain how many pathogenic and developmental processes are controlled.

## IV. PHLOEM LOADING

As indicated by structural differences, there appear to be two pathways by which assimilates can be transferred from the photosynthetic cells to the minor vein SE-CC complexes in the source leaf. In species possessing the type 1 minor vein configuration, this transport can occur by a symplastic route through the numerous plasmodesmata that interconnect the photosynthetic cells and the phloem transport system. In plants with type 2 configurations, which lack a high degree of symplastic interconnection, transport can occur by a transmembrane route via the apoplast.

## A. Apoplastic Phloem Loading

The textbook model of phloem loading in source leaves consists of a sequence of events starting with cell-to-cell transport of assimilates, primarily sucrose, through mesophyll cell plasmodesmata to a site close to the SE-CC complex. At this point, sucrose is unloaded into the apoplast, where it is actively accumulated into the SE-CC complex by a proton-sucrose symport mechanism (Figure 6A). The apoplastic proton symport model of phloem loading [17,18] affords a very satisfactory mechanism for establishing the high concentration gradient required within the phloem to drive phloem transport by Munch pressure flow.

One of the key demonstrations of the apoplastic loading pathway is the inhibition of sucrose-proton cotransport by inhibitors such as *p*-chloromercuriphenylsulfonic acid (PCMBS). In isolated leaf plasma membrane systems [60,61] and leaf tissues [62], this compound has been shown to bind to the sucrose carrier [63] and to prevent transfer of sucrose. In many plants, this compound will also inhibit the delivery of photosynthetically produced sucrose from the photosynthetic cells to the minor veins of leaf tissues [62]. This finding lends significant support to the apoplastic loading theory.

Interestingly, though, only species possessing the type 2 minor vein configurations show this sensitivity to PCMBS [29,64]. In these species, which include most of the important agronomic crops, sucrose is the only sugar transported in the phloem. Therefore, apoplastic phloem loading probably best explains delivery of sucrose to the phloem in most agronomic species. However, although there is appreciable experimental evidence in support of the apoplastic pathway, it is now becoming apparent that this model may not hold for all crop plants.

## B. Symplastic Phloem Loading

In plants with the type 1 minor vein configuration, where a symplastic route through plasmodesmata is available for the delivery of endogenously produced photoassimilates to the minor veins, phloem loading

**A. Apoplastic**

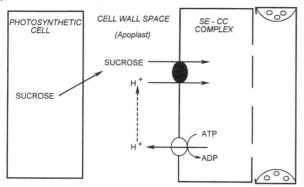

**B. Symplastic**

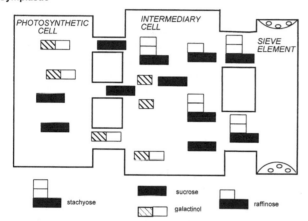

**Figure 6**   Diagrammatic representations of the processes involved in loading of the phloem via the apoplast or symplast. (A) In apoplastic loading, sucrose leaves the photosynthetic cell and enters the cell wall space. It is then taken up across the plasma membrane of the sieve element–companion cell (SE-CC) complex by a sucrose carrier (solid circle), which transports sucrose in conjunction with a proton ($H^+$). The proton gradient is established by proton extrusion via a plasma membrane ATPase (open circle). (B) In symplastic loading, disaccharides (sucrose, galactinol) are passed through the plasmodesmata from the photosynthetic cells to the intermediary cells, where the galactose residues (□) of galactinol are transferred to sucrose to form the trisaccharide (raffinose) and tetrasaccharide (stachyose) with the release of *myo*-inositol (■). The tri- and tetrasaccharides then pass into the sieve element (SE) but are prevented from passing back into the photosynthetic cell by the smaller diameter of the plasmodesmata connecting this cell to the intermediary cell.

is not affected by PCMBS [12,64,65]. This is true despite the obvious sensitivity to inhibition by PCMBS of the uptake of exogenously supplied sugars into leaf tissues [12,66]. This observation suggests that although a PCMBS-sensitive proton symport mechanism may exist in type 1 plants, it is not utilized for loading of assimilates into the phloem.

   If an apoplastic step is not involved in phloem loading in type 1 plants, some alternative mechanism must be invoked to create the high sieve element solute levels necessary for phloem transport. Plasmolysis studies clearly show that high solute levels do exist in the phloem of type 1 plants [67]. The answer to this dilemma may come from the observations that all type 1 plants export raffinose family oligosaccharides such as stachyose in the phloem [6–8] and that synthesis of the oligosaccharides destined for export most likely occurs within the intermediary cells characteristic of the type 1 morphology [68–71].

   The biochemistry of raffinose sugar biosynthesis is somewhat anomalous compared with that of other sugars in that the galactose donor is not a sugar nucleotide but a simple disaccharide, galactinol [72]. (For more details of this biochemistry, readers are referred to the chapter on carbohydrate synthesis in this volume.) A hypothetical model reconciles both the odd ultrastructural features (i.e., the ubiquitous presence of the symplastic links between the photosynthetic cells and the minor vein intermediary cells) and the rather unusual carbohydrate biochemistry of type 1 plants [73].

The crux of this model (Figure 6B) is the hypothesis that the pore size of the plasmodesmata connecting the intermediary cells within the photosynthetic cells is wide enough to allow passage of disaccharides such as sucrose and galactinol only, not their oligosaccharide products, the tri- and tetrasaccharides raffinose and stachyose (Figure 6B). Thus, when stachyose and raffinose are synthesized within the intermediary cells, they cannot move anywhere, except into the adjacent sieve tubes. This "polymerization trap" model remains to be proved [74], and the details of compartmentation of the stachyose reactions within the intermediary cell need to be elucidated, but the model does give a feasible explanation of how symplastically linked cells might operate in establishing a solute gradient. In *Cucumis melo* L., the sugar levels in individual cells are consistent with the operation of a polymer trap [75].

However, many plant species can be classified as having type 1 minor vein anatomy and yet do not transport raffinose oligosaccharides. One example is parsley, which translocates only sucrose and an even smaller molecular weight polyol, mannitol [29], and yet appears to have an "open" minor vein structure. Similarly, willow (*Salix babylonica*) possesses a symplastically linked minor vein structure and yet translocates only sucrose [76]. Plasmolysis and sugar distribution studies on willow leaves showed no positive concentration gradient between the mesophyll and the minor veins of this species, and so it was concluded that short-distance transport of photoassimilates must be entirely diffusional into the phloem. In this case, long-distance transport would be reliant upon the capacity of sink tissues to remove solute from phloem and maintain a positive pressure potential gradient. In contrast, Moing et al. [77] have suggested, on the basis of PCMBS sensitivity, that peach leaves, which transport sucrose and sorbitol and which have a symplastically open minor vein system, still apparently use an apoplastic pathway for phloem loading.

## C.   Loading of Other Solutes

Because sugars are the predominant solutes translocated in the phloem, most of what is known about phloem loading concerns the movement of sugars into the phloem. Relatively little is known of either the pathways taken or the mechanisms used to load the other component solutes characteristically found in phloem saps. There is evidence for the operation of proton–amino acid transporters in plant tissues [61], but whether these are phloem tissue specific is not known. It is likely that active accumulation of the potassium ion takes place in exchange for protons, but the carrier(s) involved has not yet been characterized. How other ions enter the phloem is not clear. It is quite likely that further studies will reveal that phloem sap composition in crop plants is determined by means of a combination of apoplastic and symplastic transport.

## V.   REGULATION OF PHLOEM TRANSPORT

Despite the great numbers of different sink tissues and organs that constitute a typical plant, most plants tend to maintain a balanced ratio of shoot tissue to root tissue. This indicates that the plant has some means of regulating the amount of photoassimilate that is delivered to developing roots and shoots and that some metabolic control exists to control the direction of phloem transport. Because rate and direction of phloem transport are dictated by the solute gradients between sources and sinks, the regulatory mechanisms can exist either at the source end (where assimilates are loaded) or at the sink end (where assimilates are removed from the phloem).

## A.   Regulation by Sources

Source leaves are the primary sites of photoassimilate production, but the plant faces a dilemma with respect to allocation choices of photoassimilates. Because the photosynthetic period does not encompass the entire diurnal period but the demand for assimilates does, source tissues must conserve part of the carbon fixed during photosynthesis for use during nonphotosynthetic periods. The role of the source leaf in controlling phloem transport is therefore one of allocation, assigning fixed carbon to export or storage pools in such a manner that export can be maintained at some "set point" level throughout the diurnal period [78,79]. In addition, the source must be able to accumulate enough reserve carbon to allow for environmental conditions (clouds, water stress, temperature fluctuations) that may interfere with photosynthetic processes even in the light [80]. How the plant decides which carbon is destined for export and which is to be stored is not fully understood.

Many tightly regulated metabolic steps control the accessibility of photosynthetically fixed carbon to the phloem transport system. Control at the source end is governed largely by rates of photosynthetic incorporation of $CO_2$, but for photosynthetic rate to have any direct effect on the rate of phloem transport, the carbon must be fixed into phloem-mobile intermediates (predominantly sucrose in most agronomically important crops). The flow of carbon into soluble sugars, which are synthesized in the cytoplasm of the photosynthetic cell, is regulated by complex biochemical interactions, which direct the export of fixed carbon out of the chloroplast [81]. Carbon not released from the chloroplast is retained as insoluble starch and will not be immediately available for phloem transport [82]. In addition, once synthesized, soluble sugars can be siphoned off into the vacuole for storage, and this carbon also would not be available for phloem transport [78].

The phloem loading process, which establishes the high solute level in the phloem, must therefore compete with storage processes also occurring in the chloroplast and vacuole, which can divert substantial amounts of photosynthetically fixed carbon from the phloem loading site. Control of phloem transport by source tissues is therefore exerted largely by control of the availability of phloem-mobile solutes and not directly by the rate of photosynthesis per se. Indeed, environmental factors that reduce rates of photosynthesis do not necessarily result directly in lowered rates of phloem transport. This is because stored carbon, either in the source tissues or in storage tissues along the pathway, can be mobilized to maintain the high solute levels in the phloem [78].

One key contribution of source leaf metabolism, arising from the combination of photosynthetic activity and membrane transport between cellular compartments, is therefore the control of amounts and probably the types (sugars or amino acids) of assimilates that have access to the loading sites. Source leaf metabolism, therefore, directly regulates the overall composition of the phloem sap [34], including both organic and inorganic constituents. In general, though, the set point for rates of phloem transport is established in the sink tissues, where these nutrients are utilized.

## B.   Regulation by Sinks

A typical higher plant has a myriad of sink tissues that depend on the source leaves for photoassimilates. Reproductive sinks (flowers, seeds, fruits) are of prime agronomic importance, and as a result most studies of sink regulation of phloem transport have tended to focus on carbon partitioning to these sinks. However, reproductive sinks represent only a small proportion of potential sinks on a plant, and we are now beginning to realize that during the growth period, carbon partitioning to other sinks, particularly temporary vegetative sinks, can be important in determining final crop yield.

### 1.   Vegetative ("Buffering") Sinks

During the translocation process, carbon is continuously diverted from the phloem to surrounding parenchyma cells for temporary storage. Parenchyma tissues of leaves, petioles, stems, and roots can all act as sinks for assimilates, which are usually stored in the form of starch. These stored reserves can be drawn on and reloaded into the phloem under conditions of reduced photosynthesis [78–80,82] (e.g., during adverse environmental conditions) or when sink demand increases (e.g., during the reproductive phase of plant growth) [83]. An amplified version of this type of sink activity is seen in perenniating organs such as tubers and taproots and also in ray cells of woody species, in which large amounts of carbon are diverted to storage to allow for regrowth of vegetative tissues in the next growing season. The phenomenon of alternate bearing in perennial tree crops may also reflect this type of sink activity: that is, carbon diverted to vegetative storage sinks in nonbearing years may be utilized for crop production in the subsequent bearing year.

The vegetative "buffering" sinks, therefore, have the unique property of being able to act both as sinks for assimilates and as sources of assimilates for phloem transport, depending on the carbon needs of the plant at a particular growth phase or under the prevailing environmental conditions. Sinks of these types can, therefore, regulate phloem transport by coarse control of the assimilates available to the sieve elements along the phloem transport path.

Like minor vein phloem loading, sieve tube unloading into sinks can occur by either symplastic or apoplastic routes (Figure 7B). Unloading into buffering sinks such as the sugar beet taproot [84] and sugarcane stem [85] occurs via an apoplastic route. Unloading of sugar into the apoplast from the sieve tubes lowers the hydrostatic pressure and also promotes the flow of water out of the sieve element, thus allowing bulk flow to occur from source to sink. Unloaded solutes are then taken up into the sink cell, where compartmentation into the vacuole or conversion to insoluble starch (Figure 7A) can further dissipate the hydrostatic pressure between the phloem and the sink organ. In some cases, sucrose is hydrolyzed prior to uptake into the sink cell, and in other cases it may be taken up intact, then hydrolyzed in the vacuole (Figure 7A).

## 2. Terminal Sinks

Unlike the buffering sinks, terminal sinks, as the name implies, act as sinks only for assimilates. Carbon partitioned to terminal sinks is unavailable for remobilization out of those sinks, usually because it is incorporated into structural, as opposed to storage, components. Prime examples of terminal sinks are reproductive tissues, such as fruits and seeds, and rapidly growing meristems. Carbon partitioned to these sinks cannot be reaccessed by the plant, even if the carbon is stored in a conventional storage form such as starch. Sinks of these types, therefore, can exert a strong regulatory influence on phloem transport by controlling the low end of the hydrostatic pressure gradient created in the sieve tubes.

Phloem unloading in terminal sinks occurs by either apoplastic or symplastic routes (Figure 7). In rapidly growing meristematic organs such as developing roots [86] and leaves [87,88], unloading occurs via the symplast. The conversion of imported assimilates to insoluble structural components, principally cellulose, and other polymers (protein, nucleic acid, etc.) and their rapid utilization as respiratory sub-

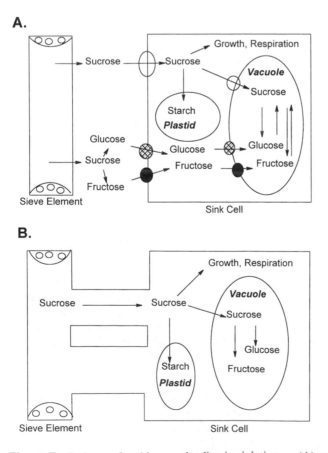

**Figure 7**  Pathways for phloem unloading in sink tissues: (A) apoplastic and (B) symplastic.

strates reduce the hydrostatic pressure in the phloem and allow continued phloem transport by bulk flow. In addition, in expanding cells, some of the water for expansion may come from the phloem, allowing further dissipation of the phloem turgor pressure.

## VI. MOLECULAR APPROACHES TO PHLOEM TRANSPORT: SUCROSE TRANSPORTERS

Developments in molecular techniques have allowed characterization of sucrose transporters and construction of transgenic plants to evaluate the impact of sucrose transporters on phloem loading and unloading [89–91]. Surprisingly, there is as yet no information on species with a type 1 minor vein configuration. The availability of these genes should greatly facilitate localization of the gene products within source leaves with either type 1 or type 2 companion cells and should allow confirmation of the role of sucrose-proton contransport in phloem loading in different plant species.

So far, studies of the sucrose carrier genes have shown that antisense mutants of the *SUT*1 gene in tobacco and potato accumulate soluble sugars and starch in source leaves [92,93], which clearly demonstrates the importance of this class of transporter for facilitating transport of sucrose into the phloem cells. Other studies have demonstrated impaired flowering and tuber yield [94], and it is possible that other developmental processes will be found to be affected by sucrose transport capabilities of the plant.

There is no question of the strength of molecular techniques for potentially increasing crop biomass or yield, but it is important to note that manipulation of genes involved in biochemical pathways may be difficult to achieve. Self-regulation, biofeedback, and alternative pathways for precursors, intermediates, and metabolites of sucrose production and degradation products may be altered in response to changes in sucrose export, delivery, and utilization patterns in source tissues such that the "theoretical" trait desired is buried within emergent properties of engineered crop species.

## VII. FUTURE PERSPECTIVES

Our understanding of the physiology of carbon allocation, partitioning, and phloem transport in crop plants is still evolving. With the advent of and continued progress in molecular biology techniques, we may be able to answer once-difficult phloem transport questions. The ability to modify plants genetically in highly specific ways through molecular approaches should continue to revolutionize the study of crop physiology. In combination with conventional physiology studies, these techniques should allow significant progress to be made in our understanding of assimilate transport processes in crop plants, and indeed there is still much to be learned.

## REFERENCES

1. K. Esau. Anatomy of Seed Plants. New York: Wiley, 1977, pp 157–182.
2. MV Parthasarathy. In: MH Zimmermann, JA Milburn, eds. Transport in Plants, Vol 1. Phloem Transport. Berlin: Springer-Verlag, 1975, pp 3–38.
3. RF Evert. Bioscience 32:789, 1982.
4. HD Behnke. In: DA Baker, JA Milburn, eds. Transport of Photoassimilates. Harlow, Essex: Loughman Scientific and Technical, 1989, pp 79–137.
5. K Esau. Anatomy of Seed Plants. New York: Wiley, 1977, pp 321–322.
6. YV Gamalei. Trees 5:50, 1991.
7. AJE van Bel, YV Gamalei. Plant Cell Environ 15:265, 1992.
8. AJE van Bel. Annu Rev Plant Physiol Plant Mol Biol 44:253, 1993.
9. YV Gamalei. Trees 3:96, 1989.
10. DG Fisher. Planta 169:141, 1986.
11. R Turgeon, DU Beebe, E Gowan. Planta 191:446, 1993.
12. LL Flora, MA Madore. Planta 189:484, 1993.
13. DG Fisher. Plant Cell Environ 11:639, 1988.
14. R Turgeon, JA Webb, RF Evert. Protoplasma 83:217, 1975.
15. K Schmitz, B Cuypers, M Moll. Planta 171:19, 1987.
16. AJE van Bel. Acta Bot Neerl 41:121, 1992.
17. RT Giaquinta. Annu Rev Plant Physiol 34:347, 1983.

18. S Delrot. In: DA Baker, JA Milburn, eds. Transport of Photoassimilates. Harlow, Essex: Loughman Scientific and Technical, 1989, pp 167–205.
19. BES Gunning, JS Pate. In: AW Robards, ed. Dynamic Aspects of Plant Ultrastructure. New York: McGraw-Hill, 1974, pp 441–480.
20. MH Zimmermann. Xylem Structure and the Ascent of Sap. Berlin: Springer-Verlag, 1983, pp 4–62.
21. TG Mason, CJ Lewin. Sci Proc R Dublin. Soc 18:203, 1926.
22. E Munch. Die Stoffbewugungen in der Pflanze. Jena: Gustav Fischer, 1930.
23. MH Zimmermann, H Ziegler. In: MH Zimmermann, JA Milburn, eds. Transport in Plants, Vol 1, Phloem Transport. Berlin: Springer-Verlag, 1975, pp 480–503.
24. MA Madore. Plant Physiol 93:617, 1990.
25. DH Lewis, DC Smith. New Phytol 66:143, 1967.
26. RJ Redgwell, RL Bieleski. Phytochemistry 17:407, 1978.
27. ME Rumpho, GE Edwards, WH Loescher. Plant Physiol 73:869, 1983.
28. WH Loescher, RH Tyson, JD Everard, RJ Redgwell, RL Bieleski. Plant Physiol 98:1396, 1992.
29. LL Flora, MA Madore. Planta 198:171, 1996.
30. H Ziegler. In: MH Zimmermann, JA Milburn, eds. Transport in Plants, Vol I, Phloem Transport. Berlin: Springer-Verlag, 1975, pp 59–100.
31. PJ Hocking. Ann Bot 45:633, 1980.
32. C Girousse, JL Bonnemain, S Delrot, R Bournville. Plant Physiol Biochem 29:41, 1991.
33. DE Mitchell, MV Gadus, MA Madore. Plant Physiol 99:959, 1992.
34. DE Mitchell, MA Madore. Plant Physiol 99:966, 1992.
35. F Antognoni, S Fornale, C Grimmer, E Komor, N Bagni. Plant 204:520, 1998.
36. BJ Shelp. Ann Bot 61:83, 1988.
37. K Mengel, EA Kirby. Principles of Plant Nutrition. Bern, Switzerland: International Potash Institute, 1982, pp 207–219.
38. F Bourgis, S Roje, ML Nuccio, DB Fisher, MC Tarczynski, C Li, C Herschbach, H Rennenberg, MJ Pimenta, T Shen, DA Gage, D Hanson. Plant Cell 11:1485, 1999.
39. E Komor, L Liegl, C Schobert. Planta 191:252, 1993.
40. SM Hall, DA Baker. Planta 106:131, 1972.
41. JS Kamboj, PS Blake, DA Baker. Plant Growth Regul 25:123, 1998.
42. I Raskin. Annu Rev Plant Physiol Plant Mol Biol 43:439, 1992.
43. J Malemy, DF Klessig. Plant J 2:643, 1992.
44. G Pearce, D Strydom, S Johnson, CA Ryan. Science 253:895, 1991.
45. TL Setter, WA Brun, ML Brenner. Plant Physiol 67:774, 1981.
46. DB Fisher, Y Wu, MSB Ku. Plant Physiol 100:1433, 1992.
47. T Sakuth, C Schobert, A Pecsvaradi, A Eichholtz, E Komor, G Orlich. Planta 191:207, 1993.
48. KD Nolte, KE Koch. Plant Physiol 101:899, 1993.
49. T Martin, W Frommer, M Salanoubat, L Willmitzer. Plant J 4:367, 1993.
50. E Grimm, A Grube, S Jahnke, S Neumann. Planta 197:11, 1995.
51. MA Ross, CA Lambi. Applied Weed Science, Burgess, Minneapolis MN, 1985, pp 157–198.
52. FC Hsu, DA Kleier. J Exp Bot 47:1265, 1996.
53. S Kim, S Han, WH Vanden Born. Weed Sci 45:470, 1997.
54. S Balachandran, Y Xiang, C Schobert, GA Thompson, WJ Lucas. Proc Natl Acad Sci U S A 94:14150, 1997.
55. WJ Lucas, S Balachandran, J Park, S Wolf. J Exp Bot 47:1119, 1996.
56. S Santa Cruz. Trends Microbiol 7:237, 1999.
57. B Xoconostle-Cazares, Y Xiang, R Ruiz-Medrano, H Wang, J Mozner, B Yoo, KC McFarland, VR Francheschi, WJ Lucas. Science 283:94, 1999.
58. LA Mezitt, WJ Lucas. Plant Mol Biol 32:251, 1996.
59. C Kuhn, VR Franceschi, A Schulz, R Lemoine, WB Frommer. Science 275:1298, 1997.
60. DR Bush. Photosynth Res 32:155, 1992.
61. DR Bush. Ann Rev Plant Physiol Plant Mol Biol 44:513, 1993.
62. S Bourquin, JL Bonnemain, S Delrot. Plant Physiol 92:97, 1990.
63. JW Reisemeier, L Willmitzer, WB Frommer. EMBO J 11:4705, 1992.
64. AJE van Bel, YV Gamalei, A Ammerlaan, LPM Bik. Planta 186:518, 1992.
65. LA Weisberg, LE Wimmers, R Turgeon. Planta 175:1, 1988.
66. R Turgeon, E Gowan. Plant Physiol 94:1244, 1990.
67. R Turgeon, PK Heples. Planta 179:24, 1989.
68. MA Madore. Planta 187:537, 1992.
69. K Schmitz, U Holthaus. Planta 169:529, 1986.
70. U Holthaus, K Schmitz. Planta 185:479, 1991.
71. DU Beebe, R Turgeon. Planta 188:354, 1992.
72. O Kandler. In: A San Pietro, FA Greer, TJ Army, eds. Harvesting the Sun. New York: Academic Press, 1967, pp 131–152.

73. R Turgeon. In: JL Bonnemain, S Delrot, WJ Lucas, J Dainty, eds. Phloem Transport and Assimilate Compartmentation. Nantes, France: Ouest Editions Presses Academique, 1991, pp 18–22.
74. R Turgeon, DU Beebe. Plant Physiol 96:349, 1991.
75. E Haritatos, F Keller, R Turgeon. Planta 198:614, 1996.
76. R Turgeon, R Medville. Proc Natl Acad Sci U S A 95:12055, 1998.
77. A Moing, F Carbonne, B Zipperlin, L Svanella, J Gaudillere. Phys Plant 101:489, 1997.
78. DR Geiger, BR Fondy. In: JL Bonnemain, S Delrot, WJ Lucas, J Dainty, eds. Phloem Transport and Assimilate Compartmentation. Nantes, France: Ouest Editions Presses Academique, 1991, pp 1–9.
79. DL Hendrix, RI Grange. Plant Physiol 95:228, 1991.
80. DR Geiger, JC Servaites. In: HA Mooney, WE Winner, EJ Pell, eds. Responses of Plants to Multiple Stresses. New York: Academic Press, 1991, pp 103–127.
81. M Stitt, Plant Physiol 84:201, 1987.
82. JC Servaites, BR Fondy, B Li, DR Geiger. Plant Physiol 90:1168, 1989.
83. VR Franceschi. In: J Cronshaw, WJ Lucas, RT Giaquinta, eds. Plant Biology, Vol I, Phloem Transport. New York: Liss, 1986, pp 399–409.
84. RE Wyse. Plant Physiol 63:828, 1978.
85. KT Glaziou, KR Gayler. Plant Physiol 49:912, 1972.
86. RT Giaquinta, W Lin, NL Sadler, VR Franceschi. Plant Physiol 72:362, 1983.
87. J Gougler-Schmalstig, DR Geiger. Plant Physiol 79:237, 1985.
88. B Ding, MV Parthasarathy, K Niklas, R Turgeon. Planta 176:307, 1988.
89. E Truernit, N Sauer. Planta 196:564, 1995.
90. R Lemoine, C Kuhn, N Thiele, S Delrot, WB Frommer. Plant Cell Environ 19:1124, 1996.
91. A Schulz, C Kuhn, JW Riesmeier, WB Frommer. Planta 206:533, 1998.
92. C Kuhn, WP Quick, A Schulz, JW Riesmeier, U Sonnewald, WB Frommer. Plant Cell Environ 19:1115, 1996.
93. L Burkle, JM Hibberd, WP Quick, C Kuhn, B Hirner, WB Frommer. Plant Physiol 118:59, 1998.
94. B Kuhn, L Barker, L Burkle, WB Frommer. J Exp Bot 50:935, 1999.

# 22

# Carbohydrate Synthesis and Crop Metabolism

**Wattana Pattanagul, Edmund R. Miranda, and Monica A. Madore**

*University of California, Riverside, California*

## I. INTRODUCTION

Plants are capable of producing all organic materials required for growth, metabolism, and reproduction from very simple inorganic molecules obtained from the atmosphere and the soil. Using light energy trapped by chlorophyll in the process of photosynthesis, these inorganic molecules (principally $CO_2$, phosphate, and nitrate or ammonia) are incorporated within the chloroplasts of mature leaves in a number of relatively simple biomolecules (e.g., triose phosphate, amino acids), which are then used elsewhere in the cell for respiration or for the construction of the more complex biomolecules (e.g., complex saccharides, proteins, nucleic acids) required for growth and metabolism. In agronomic crop species, the incorporation of fixed carbon into carbohydrates is particularly important, for carbohydrate production largely determines the yield of crop plants.

Plant carbohydrates can be classified into two forms: structural and nonstructural. The form of carbohydrate in a particular plant part will also to a large extent determine its agronomic usage. Structural carbohydrates, as the name implies, are polymers that help to form the rigid plant cell wall and give support to the plant body. These carbohydrates are in the form of permanent, usually extracellular structures, and carbon incorporated into structural elements is in general not available for further metabolism by the plant. An exception may occur in some seeds, in which cell wall polysaccharides can be metabolized by the germinating seedling [1,2].

Structural carbohydrates, of which cellulose is a prime example, have agronomic importance as components of livestock feed and as sources of fiber for industrial purposes (e.g., cotton). In terms of human nutrition, although they represent an important source of dietary fiber, they are of no direct nutritional value, as the enzymes for their metabolism are lacking in humans. The same is also true of many of the nonstructural carbohydrates. Indeed, of the many nonstructural carbohydrates characteristic of plant organs, only two forms, sucrose and starch, are directly metabolizable by humans.

Both structural and nonstructural carbohydrates form a large proportion of the dry weight of plant organs. The way in which carbon is partitioned among different carbohydrate types in a particular crop, therefore, becomes an important determinant of the agronomic value of that crop, particularly from a nutritional standpoint. In light of this, it is evident that a basic understanding of carbohydrate synthesis and its control in crop plants is central to our understanding of crop physiology.

## II.  CARBOHYDRATE FORMATION IN SOURCE LEAVES

Triose phosphates, the first products of the photosynthetic process, represent key metabolic intermediates, for they are the immediate precursors of all carbohydrates synthesized in the source leaves. The type of carbohydrates synthesized is, in turn, regulated by compartmentation of triose phosphates within chloroplastic and cytoplasmic pools within the source leaf. Because the major photosynthetic tissues are fully matured leaves, cell wall synthesis (i.e., the synthesis of structural carbohydrate) is of little importance in these tissues. Photosynthetic carbon is instead partitioned to nonstructural carbohydrates, which, in leaves of higher plants, may take the form of insoluble polymers (starch), soluble polymers (fructans), and soluble low-molecular-weight carbohydrates (sucrose, raffinose family oligosaccharides, simple monosaccharides, polyols, and cyclitols).

## A.   Starch

Starch is an insoluble glucan polymer that exists as granules within the chloroplast where it is formed. Starch consists of two molecular species: amylose, an essentially linear ($\alpha$-1,4)-glucose polymer, and amylopectin, in which linear $\alpha$-1,4-glucans are linked via $\alpha$-1,6 linkages to form a highly branched structure. The proportion of amylose to amylopectin in starch grains varies in different plant species and in different cultivars of the same species [3–5].

In source leaves, starch is often called "assimilatory" starch because it is a major reserve of photosynthetically fixed carbon [4]. During nonphotosynthetic periods (i.e., at night), this starch is mobilized and utilized to support growth and maintenance of the plant. Starch granules form by apposition of newly formed polymers onto existing grains; they are degraded by the reverse process. Therefore, increases and decreases in the size of starch grains are seen during photosynthetic and nonphotosynthetic periods, respectively [3,6].

As indicated in Figure 1, starch formation in source leaves begins by the assimilation of $CO_2$ by photosynthesis, with the subsequent formation of triose phosphates (triose-P). Further operation of the photosynthetic carbon reduction (PCR) cycle results in the formation of fructose-6-phosphate (Fru-6-P). If

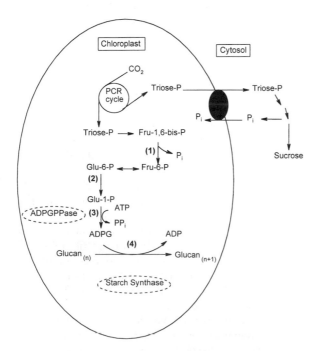

**Figure 1**   Pathway of starch synthesis in the chloroplasts of source leaves. Reaction 1, fructose-1,6-bisphosphatase; reaction 2, phosphoglucomutase; reaction 3, ADPG pyrophosphorylase (ADPGPPase); reaction 4, starch synthase.

conditions are right (i.e., if enough carbon exists in the PCR cycle to allow sufficient regeneration of ribulose-1,5-bisphosphate to maintain the photosynthetic $CO_2$ fixation rate), some carbon may be diverted out of the PCR cycle for starch synthesis. Fru-6-P is converted to glucose-6-phosphate (Glu-6-P) via a chloroplastic form of the enzyme hexose phosphate isomerase (reaction 1 in Figure 1), and Glu-6-P is converted to glucose 1-phosphate (Glu-1-P) by chloroplastic phosphoglucomutase (reaction 2). Glu-1-P is then converted to a sugar nucleotide, adenosine diphosphoglucose (ADPG), via the enzyme ADPG pyrophosphatase (ADPGPPase, reaction 3). The sugar nucleotide ADPG then acts as the glucose donor for the reaction catalyzed by starch synthase (reaction 4), which lengthens the glucan chain by one $\alpha$-1,4-linkage. A further enzyme, the branching enzyme (not shown), is responsible for creation of the $\alpha$-1,6 linkages of amylopectin. There appear to be multiple enzyme forms of both starch synthase and branching enzyme, which may be related to the structural asymmetries associated with the starch molecule [3–5].

## 1. Regulation of Starch Synthesis in Leaves

Regulation of starch synthesis in leaves is at the level of the enzyme ADPGPPase [4,5,7]. This enzyme is allosterically controlled by levels of 3-phosphoglyceric acid (3-PGA), the initial product of $CO_2$ fixation, by the PCR cycle, which activates the enzyme, and inorganic phosphate ($P_i$), which inactivates it. The ratio of 3-PGA to $P_i$ in the chloroplast thus determines the activity of the ADPGPPase enzyme. Consequently, starch synthesis is promoted during periods of high photosynthetic rate, during which high levels of 3-PGA are formed and $P_i$ is rapidly incorporated into ATP and other phosphorylated intermediates of the PCR cycle. Details of the allosteric control mechanism of this enzyme are not fully understood. There is evidence that at least two lysine sites, Lys404 and Lys441, on the small subunit of the protein might be involved in the regulation of the enzyme. Site-directed mutagenesis of these lysines results in a decrease in affinity for both its activator, 3-PGA, and its inhibitor, $P_i$, and thus results in decreased enzyme activity [8].

In addition, the starch synthesis rate is coupled to the sucrose synthesis rate through the export of triose-P out of the chloroplast. As indicated in Figure 1, this export occurs in strict exchange with the import of $P_i$ via operation of the phosphate translocator of the chloroplast membrane [9]. Thus, conditions that favor triose-P export out of the chloroplast (i.e., high rates of cytosolic sucrose synthesis) result in a low PGA/$P_i$ ratio inside the chloroplast and inhibit the formation of starch through inhibition of ADPGPPase activity [4,5,7]. Conversely, under conditions of reduced sucrose synthesis, cytosolic levels of $P_i$, a product of the sucrose synthetic pathway (Figure 1), are low, preventing the export of triose-P from the chloroplast. The resulting reduction in import of $P_i$ coupled with reduced export of triose-P raise the PGA/$P_i$ ratio and activates the ADPGPPase [4,5,7].

# B. Sucrose

As already indicated, the synthesis of sucrose and starch in photosynthesizing leaves is coupled with the operation of the phosphate translocator of the chloroplast membrane. Unlike starch synthesis, which occurs in the chloroplast, synthesis of the disaccharide sucrose ($\alpha$-D-glucose-1,2-$\beta$-D-fructofuranoside, Figure 2) occurs in the cytosol of the photosynthetic cell from triose-P that is exported to this compartment via the phosphate translocator [9–11].

As indicated in Figure 3, once in the cytosol, triose-P is converted to fructose 1,6-bisphosphate (Fru-1,6-bis-P), which is dephosphorylated to fructose-6-phosphate (Fru-6-P) via a specific fructose 1,6-bisphosphatase (FBPase, reaction 1 in Figure 3). In a series of reactions paralleling that seen in the chloroplast for starch synthesis, Fru-6-P can then be converted to glucose 1-phosphate (Glu-1-P) and then to a sugar nucleotide, in this case uridine diphosphoglucose (UDPG), via the enzyme uridine diphosphoglucose pyrophosphorylase (UDPGPPase, reaction 2). This glucose residue of this sugar nucleotide is then transferred to Fru-6-P in the reaction catalyzed by sucrose-phosphate synthase (SPS; reaction 3). The sucrose phosphate produced in this reaction is finally converted to sucrose by sucrose phosphate phosphatase (reaction 4), resulting in the release of $P_i$ to the cytosol.

## 1. Regulation of Sucrose Formation in Leaves

To prevent inhibition of photosynthetic carbon fixation during sucrose synthesis, the export of triose-P, which is also required to run the PCR cycle, must be controlled [10]. Specifically, to maintain optimal $CO_2$ fixation rates, no more than one triose-P molecule out of six produced photosynthetically can leave

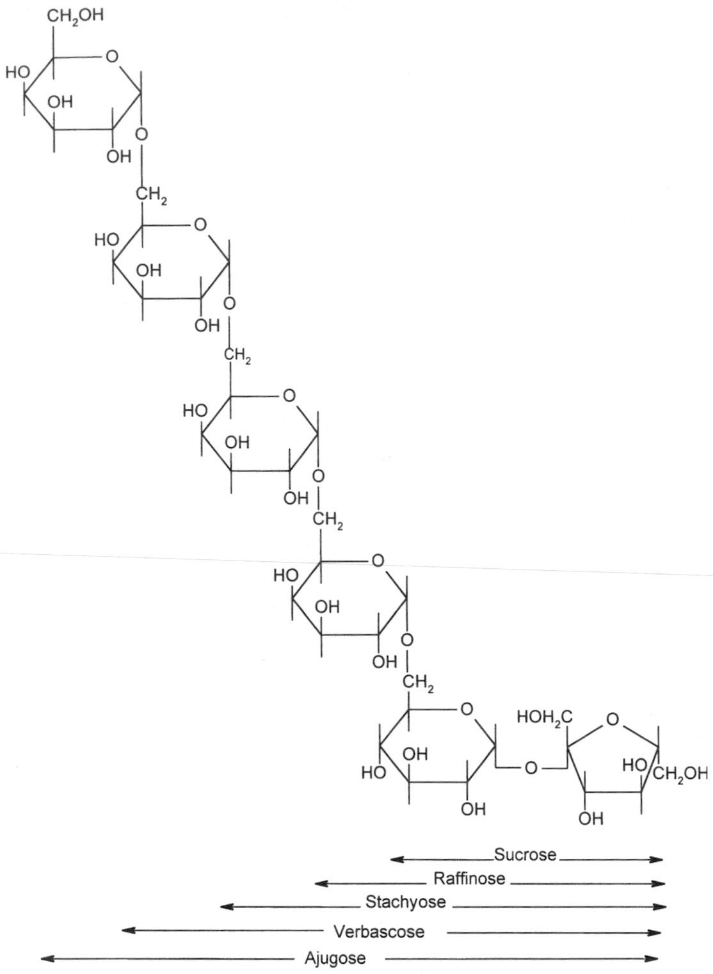

**Figure 2**   Chemical structure of sucrose and related raffinose family oligosaccharides.

the PCR cycle. As indicated in Figure 3, the synthesis of sucrose results in the formation of $P_i$ in the cytosol. [Although not indicated in the diagram, the inorganic pyrophosphate ($PP_i$) released by reaction 1 can also be converted to $P_i$ by the action of an inorganic pyrophosphate [10].] Because this production of $P_i$ could result in a large drain of triose-P from the chloroplast, the synthesis of sucrose in the cytosol (or perhaps more correctly the synthesis of cytosolic $P_i$) must be coordinated with ongoing photosynthetic rates in the chloroplast. Sucrose synthesis is, therefore, a tightly regulated metabolic reaction in the photosynthetic cell [10].

At present, there appear to be at least two different strategies for regulation of sucrose production in green plant cells: one mechanism involving regulation of cytosolic FBPase [10,11], which provides hexose phosphates for UDPG and sucrose formation, and the other involving regulation of the SPS enzyme itself [12,13].

REGULATION OF CYTOSOLIC FBPASE ACTIVITY    The cytosolic FBPase reaction (reaction 1, Figure 3) is the first irreversible step in carbon flow to sucrose and is subject to strong inhibition by a specific metabolite, fructose 1,2-bisphosphate (F-2,6-bis-P). Formation of F-2,6-bis-P in the cytosol is controlled by a specific Fru-6-P,2-kinase, and degradation is controlled by a Fru-2,6-bis-P phosphatase. The total concentration of F-2,6-bis-P is, therefore, a net result of the combined activities of these two enzymes. The amount of F-2,6-bis-P can, therefore, control the flow of carbon to sucrose by modulating the activity of the FBPase [10,11].

The kinase enzyme, which forms the inhibitor, is activated by Fru-6-P and $P_i$, the two products of the FBPase reaction. The same two products also inactive the phosphatase enzyme, which degrades the inhibitor. Thus, the FBPase can indirectly inhibit its own activity, inasmuch as increased activity of the FBPase will lead eventually to increased levels of Fru-1,2-bis-P through increased synthesis and a slower rate of degradation. Conversely, the kinase, which forms the inhibitor, is inactivated by PGA and triose-P (in the form of dihydroxyacetone phosphate), so that when high levels of triose-P are being exported to the cytosol, the inhibition of FBPase by F-2,6-bis-P is relieved and carbon flow to sucrose can continue [10,11].

REGULATION OF SPS ACTIVITY    Regulation of sucrose phosphate synthase (SPS, reaction 3, Figure 3) occurs by a number of different mechanisms. One mechanism is via coarse control through enhanced or decreased synthesis of SPS protein. A second mechanism is the fine control by metabolites acting as allosteric effectors. It is now generally accepted that SPS is activated by a high Glu-1-P/$P_i$ ratio, a form of allosteric feed-forward control. The SPS enzyme may also be inactivated by high levels of sucrose, a form of feedback control [14].

Finally, light-dark modulation of SPS activity can be regulated by covalent modification through reversible protein phosphorylation. The major regulatory phosphorylation site of SPS has been identified as Ser158. Phosphorylation of Ser158 is both necessary and sufficient for the inactivation of SPS activity, and dephosphorylation of this site causes its activation. SPS kinase, which deactivates the enzyme, is strictly $Ca^{2+}$ dependent. Thus, when cytosolic $Ca^{2+}$ levels are reduced in the light, SPS kinase is deactivated. At the same time, phospho-SPS is dephosphorylated/activated by SPS protein phosphatase, which is inhibited by $P_i$ [14].

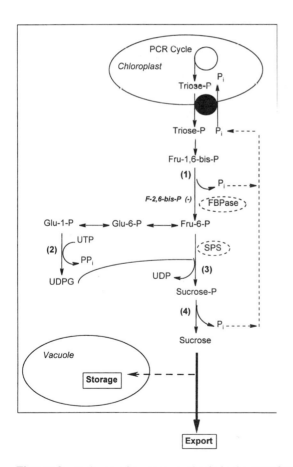

**Figure 3**    Pathway of sucrose synthesis in the cytoplasm of photosynthetic leaf cells. Reaction 1, cytoplasmic fructose-1,6-bisphosphatase (FBPase); reaction 2, UDPG pyrophosphorylase; reaction 3, sucrose phosphate synthase (SPS); reaction 4, sucrose phosphate phosphatase.

Species differences in SPS regulation are now being reported. Interestingly, the type of SPS regulation seen in a given plant appears to be correlated with the type of carbohydrate stored diurnally in its leaves. For example, in maize and spinach, which accumulate both sucrose and starch as temporary storage reserves, SPS is subject to allosteric control via Glu-1-P and $P_i$ and also appears to be under phosphorylation control. In contrast, in soybean, in which starch alone is accumulated as a storage carbohydrate, there appears to be no such regulation of SPS [12,13].

## C. Fructans

In some plant species, water-soluble polymers known as fructans accumulate as carbohydrate storage products. Fructans, as the name implies, are linear and branched polymers of fructose; in leaves, they are derived from photosynthetically produced sucrose. Fructans have as their core starting component a single molecule of sucrose, to which chains of fructose residues are attached. The type of linkage between adjacent fructose residues, as well as the point of attachment of the fructose chains to the sucrose molecule, determines the type of fructan accumulated in a given plant [15–17].

Fructans have some physiological significance and a variety of applications. Small fructans have a sweet taste, whereas longer fructan chains form emulsions with a fat-like texture and a neutral taste. The human digestive system does not contain enzymes capable of hydrolyzing fructans; therefore, there is strong interest in the food industry in developing fructans for use as low-calorie food ingredients [18]. In plants, other than being a major reserve carbohydrate, fructans have been implicated as protective agents imparting tolerance to water deficit and low temperatures [19].

### 1. Fructan Structure

Three major classes of fructans are found in agronomically important crop plants—the isokestose or inulin series, the kestose or phlein series, and the neokestose series, each of which is named for its characteristic trisaccharide sucrosyl-fructose [15–17]. In the isokestose series, which is synthesized in members of the Asteraceae such as Jerusalem artichoke (*Helianthus tuberosus* L.), fructose residues are attached to the fructosyl residue of sucrose in nonreducing β-2,1-linkages. Fructans of the isokestose series, therefore, have the general form

$$Glu\text{-}1,2\text{-}Fru\text{-}1, (2\text{-}Fru\text{-}1)_n, 2\text{-}Fru$$
$$\uparrow$$
$$Sucrose$$

where $n_{max}$ is approximately 35.

Fructans of the kestose series, which are common in many temperate grass species including wheat and barley, consist of fructose residues joined by β-2,6 linkages and have the general form

$$Glu\text{-}1,2\text{-}Fru\text{-}6, (2\text{-}Fru\text{-}6)_n, 2\text{-}Fru$$
$$\uparrow$$
$$Sucrose$$

where $n_{max}$ is approximately 250.

Fructans of the neokestose series, which have been isolated from asparagus (*Asparagus officinalis* L.), have fructose residues joined to both the glucose and the fructose residues of sucrose and have the general form

$$Fru\text{-}2, (1\text{-}Fru\text{-}2)_m\text{-}1\text{-}Fru\text{-}2, 6\text{-}Glu\text{-}1,2\text{-}Fru\text{-}1, (2\text{-}Fru\text{-}1)_n, 2\text{-}Fru$$
$$\uparrow$$
$$Sucrose$$

where $m_{max}$ and $n_{max}$ are each approximately 10 residues.

Branched fructans also occur in nature, and frequently there is more than one fructan series in the same plant.

### 2. Fructan Synthesis in Leaves

The biosynthetic pathway leading to fructan synthesis differs substantially from that leading to sucrose or starch synthesis in that the fructosyl donor is not a sugar nucleotide. Instead, sucrose itself acts as the fructosyl donor to create the fructosylsucrose isokestose by the reaction catalyzed by the enzyme sucrose:sucrosylfructose transferase (SST, reaction 1, Figure 4). The glucose released by the SST reaction is thought to reenter the general cytoplasmic hexose phosphate pool following phosphorylation (Figure 4). Chain elongation then proceeds by the reaction catalyzed by another enzyme, fructan:fructosyltransferase (FFT, reaction 2, Figure 4), which utilizes the fructosylsucrose as a fructose donor to another fructosylsucrose. Distinct FFT enzymes can be isolated from plant tissues, which can form β-1,2 or β-2,6 linkages to fructose or glucose residues of fructosylsucrose formed by the SST reactions [13,14]. Both SST and FFT were purified to homogeneity and incubation of these enzymes with sucrose resulted in the formation of a fructan polymer length of up to 20 fructosyl residues.

Both SST and FFT enzymes appear to be localized exclusively in plant vacuoles, where fructan accumulation occurs [15,16]. Vacuolar fructan synthesis lowers the sucrose concentration in the cell and prevents sugar-induced feedback inhibition of photosynthesis. Continuous illumination or feeding sucrose to excised leaves of fructan-accumulating species induces fructan synthesis, suggesting a correlation between high sucrose levels and the induction of fructan synthesis. In leaves, fructan levels are usually low, but fructans can accumulate in response to environmental conditions that serve to elevate carbohydrate levels—for example, in response to low temperatures [20]. Experimentally, cereal and grass leaves can be induced to form large quantities of fructan following excision, which eliminates phloem transport of sucrose, and continuous illumination, which promotes sucrose synthesis [15,21]. Fructan pools can also form important reserve sources for use during grain filling or other periods of high sink demand [22]. The genetic machinery for fructan synthesis, therefore, is present in leaves, although the growth conditions or developmental cues a plant experiences may not always result in activation of this machinery.

Fructans appear to be a form of readily accessible carbon and are degraded by the action of a fructan hydrolase, β-fructofuranosidase. This enzyme degrades the fructan polymers by removing the terminal fructose residue, resulting in the release of free fructose. It is thought that fructan metabolism in the vac-

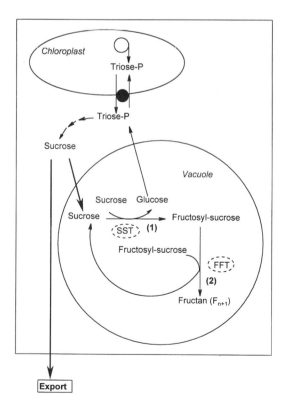

**Figure 4** Pathway of fructan synthesis in the vacuole of photosynthetic leaf cells. Reaction 1, sucrose sucrosyl transferase (SST); reaction 2, fructan:fructan fructosyltransferase (FFT).

uole of photosynthetic cells may serve to buffer chloroplast from the adverse changes in cytosolic metabolites that occur when phloem transport is limited. In addition, utilization of the vacuole provides a larger compartment for short-term carbohydrate storage than either the chloroplast or the cytoplasm, while polymerization avoids the osmotic problems that would occur if the large amounts of carbon partitioned into fructan were stored in the form of sucrose [15–17]. The mechanisms that control carbon partitioning into fructans are not yet established. Also, the question of how fructan chain length is determined has not been elucidated.

## D. Polyols

Polyhydroxy alcohols, or polyols, are probably ubiquitous in all plant species, but only in relatively few plant families are these compounds found to be synthesized from photosynthetically fixed carbon in source leaves [23,24]. The most commonly occurring polyols are derivatives of hexose sugars in which the aldose or ketose group has been reduced to a hydroxyl group. Thus, mannitol, sorbitol, and dulcitol (Figure 5) are the polyol equivalents of the hexoses glucose, fructose, and galactose, respectively.

Formation of a polyol from a hexose sugar requires reduction of the aldehyde or ketone group. In higher plants, this reduction takes place through a hexose phosphate intermediate, as indicated in Figure 6. In source leaves of celery [25,26] and privet [25], reduction of mannose-6-P to mannitol-1-P is catalyzed by the enzyme mannonse-6-P reductase (M6PR), which utilizes NADPH as reductant. Similarly, in leaves of apple, peach, pear, apricot [27], and loquat [28], an aldose 6-phosphate reductase catalyzes the reduction of glucose-6-P to sorbitol-6-P, again using NADPH. A similar NADPH-dependent enzyme is also present in *Euonymus* leaves, producing dulcitol [29].

As indicated in Figure 6, synthesis of polyols always occurs in addition to sucrose synthesis, not in substitution for it. The regulatory mechanisms that control the allocation of carbon between sucrose and polyols are not yet known. Immunological evidence clearly indicates that polyol synthesis is a cytoplasmic event [30], but the regulation of this biosynthetic pathway has not been deciphered. Both sucrose and polyols are exported in the phloem and/or may be stored in the vacuole for later export, but again, the regulation of compartmentation between storage and export pools is not fully understood.

Because polyols are not rapidly utilized by source leaf tissues, which lack enzymes to reconvert them to hexose or hexose phosphate, they are particularly useful as storage and transport forms of carbon in source leaves. Polyols may also play a role as compatible solutes in source leaves, allowing continuation of photosynthetic activity and carbon metabolism under adverse environmental conditions such as water stress. Also, the intriguing hypothesis has been put forward that the utilization of reductant in polyol synthesis allows recycling of NADPH between the chloroplast and cytosol, preventing photoinhibition under stress conditions. This possibility could also explain the unusually high photosynthetic rates commonly seen in polyol-synthesizing plants [23,24]. The underlying mechanisms that regulate the synthesis of polyols in source leaves have not yet been established.

Because of the apparent properties of polyols in promoting stress tolerance in plants, there have been attempts to transform genetically plants that normally do not make polyols with polyol synthesis genes. To enhance polyol production and accumulation, a bacterial gene for mannitol synthesis has been successfully transformed into tobacco [31,32]. Transgenic plants that synthesized mannitol appeared to grow better under salt stress, supporting the conclusion that mannitol might be involved in stress tolerance. The actual mechanism is unknown, but it is believed that mannitol may either act as an osmoticum or have

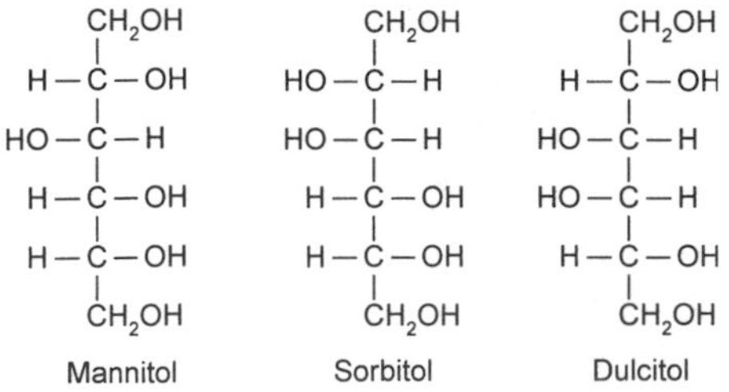

**Figure 5**   Structure of commonly occurring plant polyols.

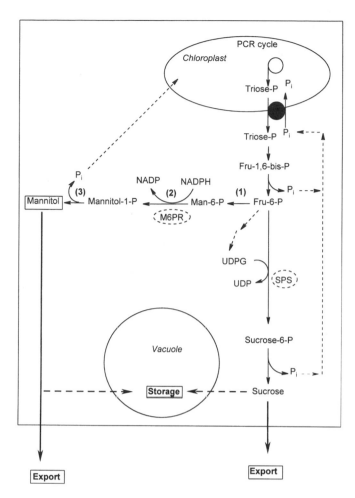

**Figure 6** Pathway of polyol (mannitol) synthesis in the cytoplasmic of photosynthetic leaf cells. Reaction 1, mannose-6-phosphate isomerase; reaction 2, mannose-6-phosphate reductase (M6PR); reaction 3, mannitol phosphate phosphatase.

some other physiological significance, for example, as a free radical scavenger imparting protection against hydroxyl radicals generated during stress [33,34].

## E. Raffinose Family Oligosaccharides

Like the polyols, raffinose family oligosaccharides are probably ubiquitous in the plant kingdom. There are a large number of plant families in which these oligosaccharides are synthesized in leaves and used as translocatable forms of carbon [35]. However, of the many plant species in which translocation of these sugars may occur, only a few, such as the cucurbit vine crops, are of major agronomic importance. As a result, this biochemical pathway of carbohydrate formation has been relatively neglected by crop physiologists. This is truly unfortunate, for evidence now clearly indicates that the synthesis of raffinose family oligosaccharides is quite different from that of other known soluble carbohydrates in a number of ways.

The raffinose family oligosaccharides, of which raffinose and stachyose (Figure 2) are the most common examples, are all simple galactosides of sucrose. The addition to the sucrose molecule of the galactose residues (which are linked by $\alpha$-1,6 linkages to the glucose moiety of sucrose) occurs, as was seen with fructans, without the direct use of a sugar nucleotide. Instead, galactinol, a novel galactoside of *myo*-inositol, is used as the galactose donor [36]. What is particularly unique about the raffinose family oligosaccharides is that the transfer of galactose residues to sucrose probably does not occur in the pho-tosynthetic cell where sucrose is synthesized [37–39].

As indicated in Figure 7, the synthesis of raffinose family oligosaccharides is now believed to take place in two separate leaf cell types: the photosynthetic mesophyll cell and the modified phloem com-panion cell, or intermediary cell, which is characteristically found in leaves in which these oligosaccha-rides are synthesized [40,41]. As far as is known, production of sucrose in leaves in these plants occurs in much the same fashion as in other plant species. In the cucurbit vine crops, sucrose synthesis does not appear to be light regulated [42], which suggests that SPS is not controlled by protein phosphorylation. The actual mechanisms controlling sucrose production in raffinose oligosaccharide–synthesizing plants have not been elucidated. There is evidence, however, that sucrose synthesis occurs within the cytoplasm of the photosynthetic cells [37,43].

In plants that synthesize the raffinose family oligosaccharides, sucrose is used as a phloem-mobile and a storage carbohydrate. However, it must also be used as the sucrose backbone for the synthesis of the raffinose oligosaccharides. The way in which mobile or storage sucrose pools are kept separated from metabolizable sucrose pools is not clear, but compartmentation within the two different cell types in-volved in raffinose oligosaccharide biosynthesis may be occurring.

To further complicate partitioning in these leaves, carbon must also be diverted *away* from the su-crose biosynthetic pathway to allow formation of the galactose donor, galactinol. In some plants, this may

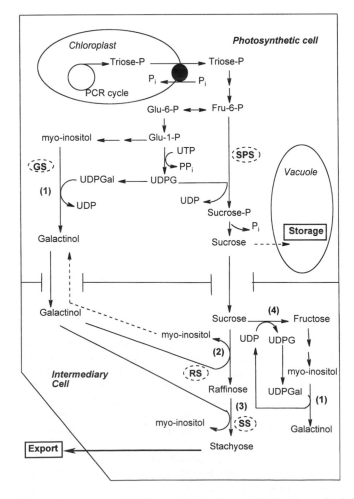

**Figure 7**  Pathway of raffinose family oligosaccharide biosynthesis in leaf tissues. Reaction 1, galactinol syn-thase (GS); reaction 2, raffinose synthase (RS); reaction 3, stachyose synthase (SS).

occur within the photosynthetic cell, probably through the conversion of UDP glucose to uridine diphos-phogalactose (UDPGal), the galactose donor used by galactinol synthase (GS, reaction 1, Figure 7). Galactinol and sucrose then cross into the intermediary cell, via the abundant plasmodesmata that inter-connect these cells with the photosynthetic cells, where raffinose oligosaccharide synthesis takes place via the operation of raffinose synthase (RS, reaction 2) and stachyose synthase (SS, reaction 3).

In other plant species, there is evidence that galactinol synthesis may also take place within the intermediary cell [38]. In this case, as indicated in Figure 7, sucrose alone may leave the photosynthetic cell, to be used both as the sucrose moiety of the raffinose sugars and for the synthesis of galactinol. Metabolism of sucrose may take place via sucrose synthase (reaction 4), which yields UDPG, from which UDPGal could be synthesized. In squash leaves, immunological data indicate the presence in the intermediary cells of both stachyose synthase (SS, reaction 3) and galactinol synthase (GS, reaction 1), but the complete details concerning the location of the biosynthetic enzymes in this pathway remain to be established. In fact, because these oligosaccharides can also serve a storage function in plant tissues, it is likely to prove that the entire pathway leading to raffinose oligosaccharide synthesis occurs both in the photosynthetic cells, where they are used for storage, and in the intermediary cell, where they are used for transport [38,44–47].

Although raffinose and stachyose are synthesized via the preceding reactions, there is evidence indicating that verbascose and higher degree of polymerization (DP) raffinose family oligosaccharides (RFO) do not use galactinol as a galactosyl donor. Cold-induced RFO accumulation in *Ajuga reptans* L. is associated with an increase in the activity of a novel vacuolar enzyme, galactan:galactan galactosyl-transferase (GGT) [45,46]. This enzyme catalyzes galactosyl transfer from one raffinose family oligosac-charide to another, resulting in the formation of galactosides one higher and one lower in degree of polymerization than the two starting substrates [47], as shown in Figure 8.

The regulation of the raffinose pathway in source leaves is not fully understood. From preliminary reports, the key regulating enzyme in the pathway would appear to be galactinol synthase (GS, reaction 1, Figure 7). This enzyme catalyzes the first committed step in the biosynthesis of RFOs and is therefore potentially a good metabolic control point. GS activity has been shown to increase in response to illumination and to changes in photoassimilate export rate [48], and levels of GS messenger RNA (mRNA) also increased when plants were exposed to cold and desiccation, a condition that also induces RFO accumulation [49].

Stachyose synthase (SS, reaction 3, Figure 7), which has higher activity in fruiting than in vegetative plants [50], may also be under some form of metabolic control. During the dark period, export of stachyose in muskmelon declines and the plant becomes predominantly a sucrose transporter [51], an observation suggesting that some form of light regulation of these enzymes is a possibility. In this context,

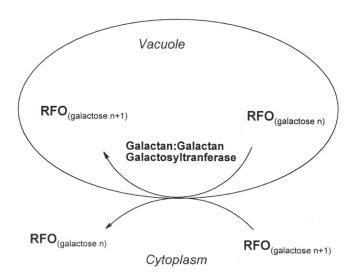

**Figure 8** Pathway for production of high-molecular-weight (high DP) raffinose family oligosaccharides.

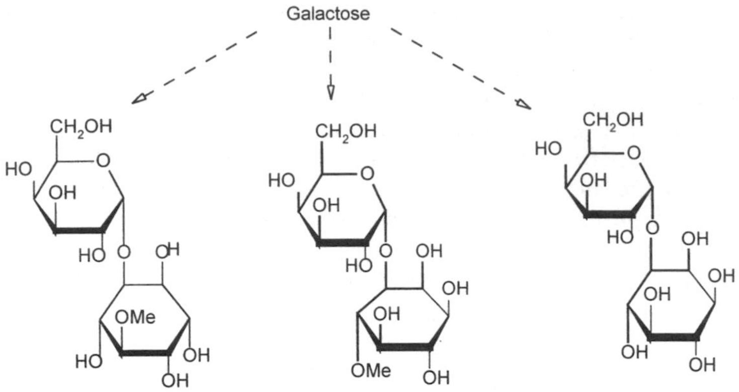

Galactose

| Galactopinitol | Galactosylononitol | Galactinol |

**Figure 9** Structures of some common galactosylcyclitols.

the common observation that GS requires a reducing environment for full activity suggests that this enzyme may be under some form of redox control. Certainly, further study is warranted.

## F.  Cyclitols and Galactosylcyclitols

Cyclitols are similar to the linear polyols already discussed except that their carbon skeletons are cyclized to form a ring structure (Figure 9). The most common cyclitols are inositols, of which *myo*- and *chiro*-inositol are the most prevalent. In addition to these simple forms, these cyclitols are often encountered as their methylated derivatives, pinitol and ononitol, and may be further complexed with galactose to form galactosylcyclitols [52]. The most prevalent galactosylcyclitol in nature is galactinol, which is probably a reflection of its important role in synthesis of raffinose oligosaccharides.

The methylated derivatives often accumulate in source leaves during periods of water stress [53–56], suggesting that like the linear polyols, cyclitols may play some as yet unknown role in stress tolerance.

## G.  Seven-Carbon Sugars

Whereas most plants synthesize carbohydrates based on a hexose framework, certain plant species are able to synthesize and translocate significant amount of novel seven-carbon (C7) sugars. Despite their infrequent occurrence in the plant kingdom, these sugars appear to have important physiological functions in the species in which they are found. The best studied of the C7 sugars are the ketoheptuloses, sedoheptulose and mannoheptulose, and their polyol forms, volemitol and perseitol (Figure 10). Sedoheptulose, in its mono- and bisphosphorylated form, is an important intermediate in the Calvin cycle. Its polyol

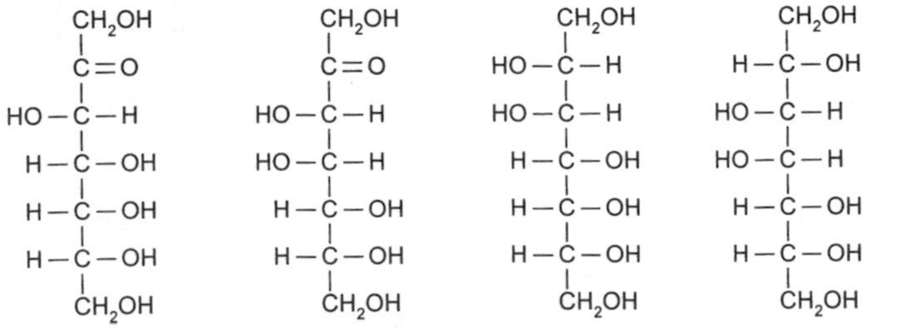

| Sedoheptulose | Mannoheptulose | Volemitol | Perseitol |
| (D-altro-2-Heptulose) | (D-manno-2-Heptulose) | (D-glycero-D-manno-Heptitol) | (D-glycero-D-galacto-Heptitol) |

**Figure 10** Structures of common seven-carbon heptuloses and their polyhydroxy derivatives.

equivalent, volemitol, is less common and appears to be restricted to certain section of the genus *Primula* [57]. In this species, the polyol is a product of photosynthesis and appears to be phloem mobile. An enzyme that catalyzes the reduction of sedoheptulose to volemitol has been characterized [57].

Mannoheptulose and its polyol form, perseitol, are found in all parts of the avocado (*Persea americana* Mill.) tree [58,59]. In avocado, these C7 sugars are apparently important in metabolic processes associated with fruit development as well as respiratory processes associated with postharvest physiology and fruit ripening [59]. Because of their localized distribution in the plant kingdom, the biosynthesis and metabolic pathways of these sugars have not yet received much attention. Mannoheptulose, a very potent inhibitor of hexokinase reactions in respiration, does hold some promise as an antitumor agent, so further study of these sugars is again warranted and needed.

## III. CARBOHYDRATE FORMATION IN NONPHOTOSYNTHETIC (SINK) TISSUES

The soluble and insoluble forms of carbohydrate that have been listed are all used as temporary storage reserves in the leaf. However, certain of the soluble forms, such as sucrose, the raffinose family oligosaccharides, and the polyols, are also phloem mobile and can be delivered to nonphotosynthetic tissues to support growth and development of these plant parts [35]. It is commonly found, therefore, that even nonphotosynthetic tissues will contain some or all of the same carbohydrates that commonly occur in phloem sap. However, because carbohydrates are also required for growth processes such as respiration or cell wall synthesis, sink tissues are also equipped with enzymes for breakdown, interconversion, and metabolism of whatever phloem-mobile carbohydrates are supplied to them. As a result, it is also not uncommon to find carbohydrates that are in fact quite different from those supplied to the sink by phloem transport.

Carbohydrates formed in sink tissues may act as storage reserves and, as occurs in source leaves, they are found to be compartmentalized in specialized cells or cellular compartments such as the plastids or vacuoles. In most agronomic crops, it is these storage reserves that are of economic importance—for example, the yield of seeds, grains, and storage roots or tubers is dictated principally by the size of their carbohydrate reserves at harvest. The enzymes for synthesis of common storage carbohydrates, including soluble and polymeric forms, are therefore found in a range of plant tissues, not just the mature leaves. Research into carbohydrate metabolism in nonphotosynthetic tissues is showing that the controlling factors in the regulation of carbohydrate synthesis are often surprisingly similar to those in photosynthetic cells.

### A. Starch

Starch synthesis by sink tissue is probably one of the most important plant biochemical reactions in terms of human nutrition because starch, particularly from grain crops (where it can make up 70% of the dry weight), is a major provider of nutritional calories in the human diet everywhere [15]. Despite the importance of starch biosynthesis in crop plants, we actually know very little concerning the biochemical details of starch deposition in sink tissues.

Starch deposition in sink tissues occurs at the expense of imported assimilates and appears to require the conversion of phloem-delivered solutes into a usable hexose phosphate form [60]. For phloem-derived sucrose, there are at least two pathways by which this conversion occurs: by invertase hydrolysis to hexose followed by phosphorylation to hexose phosphate:

*Invertase:* Sucrose → glucose + fructose

and by reversal reaction of sucrose synthase to provide fructose and UDPG:

*Sucrose synthase:* Sucrose + UDP → UDPG + fructose

Depending on the sink tissues, there is evidence for operation of both these pathways in sinks. Formation of hexose-P takes places by direct phosphorylation by hexokinase reactions or, in the case of UDPG, by reversal of the UDPG pyrophosphorylase reaction:

*UDPPase:* UDPG + PP$_i$ → Glu-1-P + UTP

It is now believed that synthesis of starch in the amyloplasts of sink tissues takes place from hexose-P imported from the cytoplasm [60]. The subsequent enzymatic steps appear to be similar to the series found in source leaves. Somewhat surprisingly, the regulation of ADPGPPase by 3-PGA and $P_i$ appears to be a constitutive property of this enzyme from all sources studied so far [4,5,7], even though for most sinks 3-PGA is not a major amyloplast metabolite.

## B.  Sucrose

Synthesis of sucrose is an important physiological function in some sink tissues—for example, in ripening fruits, where stored starch is metabolized to sugar during the ripening process. Similar metabolism of starch or fructans also occurs upon sprouting of perennating organs such as tubers and bulbs. During germination of seeds, conversion to sucrose of stored carbohydrates, and also other reserves such as wall materials and oils, occurs in endosperm tissues and cotyledons and is an essential process for growth and emergence of the embryo. The regulation of sucrose synthesis in sink tissues is under active investigation in many research laboratories, but comprehensive details of control of sucrose synthesis in sinks are still lacking.

## C.  Fructans

There are many reports of fructan accumulation in vegetative sinks of fructan-accumulating plants. Indeed, fructan accumulation is far more common in vegetative storage sinks than it is in source leaves, where experimental manipulation must often be used to induce fructan accumulation. Fructans accumulate naturally to particularly high levels in overwintering organs such as bulbs and tubers and also in stems of grasses, possibly in response to environmental cues. In wheat plants, stem fructans form a pool of carbon reserves that can be drawn on during grain filling [22]. The exact regulatory mechanisms controlling fructan accumulation from imported solutes in sinks are not yet known.

## D.  Polyols

Although sink tissues such as seeds may contain polyols, these are usually only trace components, not major carbohydrates. In polyol-translocating species, vegetative tissues such as petioles, stems, and roots may accumulate polyols, but there is no evidence that this occurs by directly synthesis rather than by simple import from the phloem. Polyol metabolism in sink tissues is, therefore, poorly understood and needs to be addressed in further research [61].

## E.  Raffinose Family Oligosaccharides

Raffinose oligosaccharides are prevalent in seeds of numerous plant species, even those that do not use these sugars as phloem-mobile compounds. Thus, although the biosynthetic pathway of raffinose oligosaccharide synthesis is not operative in the source leaves of many plants, it is encoded in the genome and is expressed in the developing seeds. It is thought that these oligosaccharides, which accumulate during seed drying, may allow maintenance of seed cellular membranes during desiccation. Despite their prevalence and the possibility that they perform an essential metabolic role in the dry seed, the regulatory mechanisms controlling synthesis of these sugars from imported assimilates in seeds have received relatively little attention. Indeed, current research appears to be focusing on genetics and molecular protocols for removal of this biochemical pathway from seeds because the raffinose oligosaccharides are major antinutritional factors in many seed crops used for human consumption. It will be interesting to see what success these approaches achieve if indeed the raffinose pathway turns out to be necessary for seed viability.

Raffinose oligosaccharides also accumulate in some vegetative storage tissues, where they may simply reflect accumulation of imported stachyose [62]. However, there are numerous cases in which these sugars appear to be synthesized de novo in vegetative tissues including, interestingly, the sugar beet taproot. Although the first isolation of galactinol was from this tissue [63] and although raffinose produced in sugar beet is a major "contaminant," which interferes with sucrose crystallization, there has been little study of this de novo biochemical pathway in vegetative tissues. Certain environmental stresses, such as low temperature, can also result in accumulation of these oligosaccharides in vegetative tissues, including source leaves in which they would not normally be synthesized. The metabolic function and cellular location of these induced oligosaccharides are not known. Clearly, more research is also needed in the area of raffinose family oligosaccharide synthesis in sink tissues.

## F. Cyclitols and Galactosylcyclitols

Like the RFOs [64], these metabolites, and especially the galactosylcyclitols, often occur in high amounts in dry seeds, with certain legume seeds being particularly good sources. Seeds of *Vigna angularis* L. contain galactosylononitol [65] and seeds of *Lens culinaris* L. contain a range of these compounds including galactopinitol A, galactopinitol B, digalactopinitiol, ciceritol, and fagopyritol [66]. In addition to these cyclitols, seeds of both species contain high levels of RFOs, particularly stachyose. There is now strong evidence that stachyose synthase, which catalyzes the synthesis of stachyose in the RFO pathway, is also responsible for the galactosyltransferase reactions producing the galactosylcyclitols [67].

## G. Structural Carbohydrates

A key structural component of plant cells that differentiates them from animal cells is the presence of a rigid, highly structured cell wall. The primary plant cell wall consists largely of cellulose and related heteroglycan polymers [68] and can therefore be regarded as a very specialized carbohydrate. The formation of cell wall material requires a tremendous input of carbon and is therefore one of the most important uses of photosynthetically fixed carbon in a developing plant tissue. From an agronomic standpoint, the cell wall gives rise to many important fibers (e.g., cotton) with many industrial applications and also is an important fiber component of animal feeds.

The principal cell wall component is the carbohydrate cellulose, which is a linear glucose polymer containing β-1,4 linkages. Synthesis of cellulose occurs at the plasma membrane by terminal complexes consisting of cellulose synthase associated with sucrose synthase subunits. Sucrose delivered to the plasma membrane is degraded by sucrose synthase to liberate UDPG and fructose. UDPG is then used for cellulose microfibril chain elongation catalyzed by the enzyme cellulose synthase (UDPG: glucan synthase) [69,70]. The pectins and hemicelluloses of the wall matrix are heterogeneous polysaccharides and, unlike cellulose, are synthesized mainly in the Golgi apparatus by a series of glycosyltransferase reactions. Because of the complexity and heterogeneity of these carbohydrates, understanding of their pathways of biosynthesis is still fragmented.

## IV. FUTURE PERSPECTIVES

The advent of molecular biology techniques now allows us to study carbohydrate biochemistry in ways that before were impossible. With these techniques, we are now able to create transgenic plants with enhanced or novel carbohydrate production or to knock out completely genes responsible for carbohydrate pathway enzymes. With these types of experiments has come the realization that there is a bigger picture to view in terms of carbohydrates in plants, for it is now becoming quite evident that carbohydrates play many surprising roles in plant metabolism. We are only just beginning to discover and appreciate the new and exciting roles that carbohydrates play as sensors of environmental cues, as signal transduction intermediates, and as regulators of gene expression.

## REFERENCES

1. P Hamer. Physiol Veg 23:107, 1985.
2. DA DeMason, MA Madore, KN Chandra Sekhar, MJ Harris. Protoplasma 166:177, 1992.
3. J Preiss, C Levi In: J Preiss, ed. The Biochemistry of Plants, Vol 3, Carbohydrates: Structure and Function. New York: Academic Press, 1980, pp 371–423.
4. J Preiss. In: J Preiss, ed. The Biochemistry of Plants, Vol 14, Carbohydrates. New York: Academic Press, 1988, pp 182–249.
5. P John. Biosynthesis of the Major Crop Products. Chichester: Wiley, 1992, pp 32–53.
6. M Steup. In: J Preiss, ed. The Biochemistry of Plants, Vol 14, Carbohydrates. New York: Academic Press, 1988, pp 255–295.
7. DM Stark, KP Timmerman, GG Barry, J Preiss, GM Kishore. Science 258:287–292, 1992.

8.    MA Ballicora, Y Fu, NM Nesbitt, J Preiss. Plant Physiol 118:265, 1998.
9.    U Heber, HW Heldt. Annu Rev Plant Physiol 32:139–167, 1981.
10.   M Stitt. Plant Physiol 84:201–201, 1987.
11.   M Stitt, SC Huber, P Kerr, MD Hatch, NK Boardman, eds. The Biochemistry of Plants, Vol 10, Photosynthesis. New York: Academic Press, 1987, pp 327–408.
12.   SC Huber, JLA Huber, RW McMichael. In: CJ Pollock, JF Farrar, AJ Godon, eds. Carbon Partitioning Within and Between Organisms. Oxford: BIOS Scientific Publishers, 1992, pp 1–26.
13.   SC Huber. Plant Physiol 99:1275–1278, 1992.
14.   SC Huber, JL Huber. Ann Rev Plant Physiol Plant Mol Biol 47:431–444, 1996.
15.   CJ Pollock, AJ Cairns. Annu Rev Plant Physiol Plant Mol Biol 42:77–101, 1991.
16.   CJ Pollock, NJ Chatterton. In: J Preiss, ed. The Biochemistry of Plants, Vol 14, Carbohydrates. New York: Academic Press, 1988, pp 109–140.
17.   P John. Biosynthesis of the Major Crop Products. Chichester: Wiley, 1992, pp 55–69.
18.   I Vijn, S Smeekens. Plant Physiol 120:351–359, 1999.
19.   A Vágújfalvi, I Kerepesi, G Galiba, T Tischner, J Sutka. Plant Sci 144:85–92, 1999.
20.   P Bancal, JP Gaudillere. New Phytol 124:375–379, 1993.
21.   U Simmen, D Obenland, T Boller, A Wiemken. Plant Physiol 101:459–468, 1993.
22.   MR Shakiba, B Edhaie, MA Madore, JG Waines. J Genet Breed 50:91, 1996.
23.   WH Loescher, JK Fellman, TC Fox, JM Davis, RJ Redgwell, RA Kennedy. In: RL Heath, J Preiss, eds. Regulation of Carbon Partitioning in Photosynthetic Tissue. Rockville, MD: American Society of Plant Physiologists, 1985, pp 309–332.
24.   WH Loescher. Physiol Plant. 70:553–557, 1987.
25.   WH Loescher, RH Tyson, JD Everard, RJ Redgwell, RL Bieleski. Plant Physiol 98:1396–1402, 1992.
26.   ME Rumpho, GE Edwards, WH Loescher. Plant Physiol 73:869–873, 1983.
27.   FB Negm, WH Loescher. Plant Physiol 67:139–142, 1981.
28.   FB Negm. Plant Physiol 80:972–977, 1986.
29.   M Hirai. Plant Physiol 67:221–224, 1981.
30.   JD Everard, VR Franceschi, WH Loescher. Plant Physiol 102:345–356, 1993.
31.   M Tarczynski, RG Jensen, HJ Bohnert. Science 259:508–510, 1993.
32.   B Karakas, P Ozias-Akins, C Stushnoff, M Suefferheld, M Reiger. Plant Cell Environ 20:609–616, 1997.
33.   B Shen, RG Jensen, HJ Bohnert. Plant Physiol 115:527–532, 1997.
34.   DM Pharr, JMH Stoop, JD Williamson, ME Studer Feusi, MO Massel, MA Conkling. Hortscience 30:1182–1188, 1995.
35.   MH Zimmerman, H Ziegler. In: MH Zimmerman, JA Milburn, eds. Transport in Plants, Vol I, Phloem Transport. Berlin: Springer-Verlag, 1975, pp 480–503.
36.   O Kandler. In: A San Pietro, FA Greer, TJ Army, eds. Harvesting the Sun. New York: Academic Press, 1967, pp 131–152.
37.   K Schmitz, U Holthaus. Planta 169:529–535, 1985.
38.   DU Beebe, R Turgeon. Planta 188:354–361, 1992.
39.   LL Flora, MA Madore. Planta 189:484–490, 1993.
40.   R Turgeon. In: JL Bonnemain, S Delrot, WJ Lucas, J Dainty, eds. Phloem Transport and Assimilate Compartmentation. Nantes, France: Quest Edutions Presses Academique, 1991, pp 18–22.
41.   AJE van Bel. Annu Rev Plant Physiol Plant Mol Biol 44:253–281, 1993.
42.   SC Huber, TH Neilsen, JLA Huber, DM Pharr. Plant Cell Physiol 30:277–285, 1989.
43.   MA Madore, JA Webb. Can J Bot 60:126–130, 1982.
44.   MA Madore. Planta 187:537–541, 1992.
45.   M Bachmann, P Matile, F Keller. Plant Physiol 105:1335–1345, 1994.
46.   M Bachmann, F Keller. Plant Physiol 109:991–998, 1995.
47.   N Sprenger, F Keller. Plant J 21:249, 2000.
48.   NS Robbins, DM Pharr. Plant Physiol 85:592–597, 1985.
49.   JJ Liu, DC Krenz, AF Galvez, BO de Lumen. Plant Sci 134:11–20, 1998.
50.   U Holthaus, K Schmitz. Planta 184:525–531, 1991.
51.   DE Mitchell, MV Gadus, MA Madore. Plant Physiol 99:959–965, 1992.
52.   DV Phillips, DE Dougherty, AE Smith. J Agr Food Chem 30:456, 1982.
53.   CW Ford. Phytochem 21:1149, 1982.
54.   F Keller, MM Ludlow. J Exp Bot 44:1351, 1993.
55.   GA Gilbert, C Wilson, MA Madore. Plant Physiol 115:1267, 1997.
56.   W Pattanagul, MA Madore. Plant Physiol 121:987, 1999.
57.   B Häfliger, E Kindhause, F Keller. Plant Physiol 119:191–197, 1999.
58.   X Liu, PW Robinson, MA Madore, GW Witney, ML Arpaia. J Am Soc Hortic Sci 124:671–675, 1999.
59.   X Liu, PW Robinson, MA Madore, GW Witney, ML Arpaia. J Am Soc Hortic Sci 124:676–681, 1999.
60.   T Ap Rees. In: CJ Pollock, JF Farrar, AJ Godon, eds. Carbon Partitioning Within and Between Organisms. Oxford: BIOS Scientific Publishers, 1992, pp 115–132.

61. JH Stoop, DM Pharr. Arch Biochem Biophys 298:612–619, 1992.
62. F Keller. Plant Physiol 98:442–445, 1992.
63. RJ Brown, RF Serro. J Am Chem Soc 75:1040–1044, 1953.
64. PM Dey. In: PM Dey, JB Harborne, eds. Methods in Plant Biochemistry, Vol 2, Carbohydrate. New York: Academic Press, 1990, pp 189–217.
65. T Peterbauer, A Richter. Plant Physiol 117:165, 1998.
66. G Hoch, T Peterbauer, A Richter. Arch Biochem Biophys 366:75, 1999.
67. J Frias, A Bakhsh, DA Jones, AE Arthur, C Vidal-Valverde, MJC Rhodes, CL Hedley, J Exp Bot 50:469, 1999.
68. A Bacic, PJ Harris, BA Stone. In: J Preiss, ed. The Biochemistry of Plants, Vol 14, Carbohydrates. New York: Academic Press, 1988, pp 297–371.
69. DP Delmer, BA Stone. In: J Preiss, ed. The Biochemistry of Plants, Vol 14, Carbohydrates. New York: Academic Press, 1988, pp 373–419.
70. DM Gibeaut, NC Carpita. FASEB J 8:904–913, 1994.

# 23

# Production of Phytomedicinal Chemicals by Plants

## Donald P. Briskin

*University of Illinois, Urbana, Illinois*

## I. INTRODUCTION

Throughout history, plants have provided a rich source for the development of human medicines. Through empirical discovery, humans have continually identified plants yielding beneficial health effects. This practice appears universal, for no human cultures have been found that lack the use of plants in their health practices [1]. Indeed, plant-based medicines remain the most widely used pharmaceutical materials in most third world countries, where they are considered to be as efficacious for many ailments and more affordable than those of "Western medicine" [2]. Up through the early 20th century, plant medicinal products represented a significant component in conventional medicine, but their use declined with the development of modern pharmaceuticals containing pure chemical compounds. Many of these modern pharmaceuticals may have been based upon active chemicals isolated from plants, and the development of synthetic or semisynthetic derivatives led to drugs with even higher levels of potency. Unlike the earlier traditional plant remedies, these modern pharmaceuticals could also be patented, which represented a clear advantage to the pharmaceutical industry. However, the enhanced potency of one or a few active chemicals in these modern pharmaceuticals frequently resulted in problematic side effects. Moreover, these drugs were often expensive [3].

Over the past decade there has been a strong resurgence in the interest in and use of medicinal plants and phytomedicines, especially in North America. Surveys of phytomedicinal use by the American public have shown an increase from about 3% of the population in 1991 to over 37% in 1998 [4,5]. At present, the North American market for plant medicinal products has reached about $3 billion a year [5]. Once the domain of health food and specialty stores, phytomedicinals have reemerged into the mainstream as evidenced by their availability for sale at a wide range of retail outlets, the extent of their advertisement in the popular media, and the entrance of several major pharmaceutical companies into the business of producing phytomedicinal products [4,5]. No doubt a major contributing factor to this great increase in phytomedicinal use in the United States has been the passing of federal legislation in 1994 (Dietary Supplement Health and Education Act or DSHEA) that facilitated the production and marketing of phytomedicinal products [4].

Given the current and future demand for phytomedicines, production of medicinal plants as "alternative crops" could provide important new opportunities in agriculture. However, in comparison with studies of most food crops, very little fundamental research has been conducted on the physiology and biochemistry of medicinal plants. Hence, from the perspective of plant physiology, extensive opportunities exist for basic research on medicinal plants and the study of their phytomedicinal chemical production. This chapter presents a discussion of some fundamental aspects of phytomedicinal chemical production by plant cells with an overview of several medicinal plants that have received considerable use and attention over the past decade.

## II.  PLANT SECONDARY METABOLISM AND PHYTOMEDICINES

The beneficial pharmacological actions of plant materials typically result from the combinations of secondary products that are present in the plant. That the medicinal actions of plants are unique to particular plant species or groups would be consistent with this concept, as the combinations of secondary products in a particular plant species are often taxonomically distinct [6,7]. This is in contrast to primary products, such as the carbohydrates, lipids, proteins, heme chlorophyll, and nucleic acids, that are common to all plant species and are involved in the primary metabolic processes of building and maintaining plant cells [7,8]. Although plant secondary products have historically been defined as chemicals that do not appear to have a vital biochemical role in the process of building and maintaining plant cells, more current research has shown a pivotal role of these chemicals in the ecophysiology of plants. In this respect, secondary products can have a defensive role against herbivory, pathogen attack, and interplant competition or an attractant role toward beneficial organisms such as pollinators or symbionts [6,8,9]. Plant secondary products can also have protective actions in relation to abiotic stresses such as those associated with changes in temperature, water status, light levels, ultraviolet (UV) exposure, and mineral nutrients [8]. Furthermore, recent work has indicated potential roles of secondary products at the cellular level as plant growth regulators, as modulators of gene expression, and in signal transduction [8].

Although secondary products can have a variety of functions in plants, it is likely that their ecological function may have some bearing on potential medicinal effects for humans. For example, secondary products involved in plant defense through cytotoxicity toward microbial pathogens could prove useful in antimicrobial phytomedicine in humans, if not too toxic. Likewise, secondary products involved in defense against herbivores through neurotoxin activity could have beneficial effects in humans (as antidepressants, sedatives, muscle relaxants, or anesthetics) through their action on the central nervous system. In order to fulfill functions in promoting the ecological survival of plants, structures of secondary products have evolved to interact with molecular targets affecting the cells, tissues, and physiological functions in other competing microorganisms, plants, and animals (see Ref. 9 for discussion). In this respect, some plant secondary products may exert their action by resembling endogenous metabolites, ligands, hormones, signal transduction molecules, or neurotransmitters and thus have beneficial medicinal effects on humans because of similarities in their potential target sites (e.g., central nervous system, endocrine system) for action [6,8]. As noted by Wink [7], this development of structural similarity between plant secondary products and the endogenous substances of other organisms could be termed "evolutionary molecular modeling."

## III.  CHEMICAL "SYNERGISMS" IN THE ECOLOGICAL FUNCTION OF SECONDARY PRODUCTS AND THE BENEFITS OF PHYTOMEDICINES

In contrast to synthetic pharmaceuticals based on single chemicals, many phytomedicinals exert their beneficial effects through several chemical compounds acting additively or synergistically at single or multiple target sites associated with a physiological process. As pointed out by Tyler [3], this synergistic or additive pharmacologic effect can promote pharmacological effectiveness without the problematic side effects associated with the predominance of a single xenobiotic compound in the body. In this respect, Kaufman et al. [8] extensively document how synergistic interactions underlie the effectiveness of a number of phytomedicines. This theme of multiple chemicals acting in an additive or synergistic manner probably has its origin in the functional role of secondary products in promoting plant survival [9]. For example, in the role of secondary products as defense chemicals, a mixture of chemicals having additive or synergistic effects at multiple target sites would not only ensure effectiveness against a wide range of herbivores or pathogens but also decrease the chances of these organisms developing resistance or adaptive responses [7–9].

## IV. PROPERTIES, USES, AND PHYTOMEDICINAL CHEMICALS OF SEVERAL FREQUENTLY USED MEDICINAL PLANTS

Although thousands of medicinal plants are utilized in Western and non-Western medical approaches, a relatively small number have received considerable use and interest in the United States over the past decade (see Ref. 4). As shown in Table 1, nine medicinal plants represented about 80% of the total market for U.S. sales of medicinal plant products in 1999 [10]. The sales for each of these plants increased over the previous year, although with a wide range in the extent of increase. In 1999, sales of kava, saw palmetto, St. John's wort, and ginseng demonstrated the largest increase over the previous year [10], and this probably reflects both increased visibility for these medicinal plants in the popular media (and commercial advertisement) and recognition of their effectiveness. The data shown in Table 1 are for 1999 sales, but it should be noted that interest in these nine particular medicinal plants remained constant over at least the previous 5 years. The plants shown in Table 1 have remained within the top 10–15 plants dominating medicinal plant sales, although the percentage of the market and relative increase have varied from year to year [4]. What follows is an overview of these nine plants that have dominated medicinal plant use and interest in the United States, focusing on their biochemical characteristics and the pharmacological actions of their plant secondary product chemicals.

## A. Ginkgo

*Ginkgo biloba* is the last living relative of a primative family of gymnosperms (Ginkgoaseae); all other species exist only as fossils [11,12]. Ginkgo trees are widely used as ornamentals worldwide because of their hardiness and appearance. The therapeutic use of *G. biloba* dates back about 2000 years in traditional Chinese medicine [11,13]. On the other hand, the phytomedicine utilized today is based on acetone extraction of the fan-shaped leaves and further purification of active constituents [14–16]. Clinical studies have supported the effectiveness of *G. biloba* in improving peripheral and cerebrovascular circulation [11,13,16–19]. A main use of *G. biloba* is in the management of cognitive decline associated with disturbances in brain blood circulation (i.e., vascular insufficiency dementia) that can occur in the elderly (Refs. 15, 17, and 18 and references therein). In addition, *G. biloba* extracts have been useful in the treatment of tinnitus and vertigo and for improving circulation in the legs [11,15,16]. It should be noted that the effectiveness of *G. biloba* extracts in improving the cognitive performance of young healthy individuals is less certain.

The active constituents present in extracts of *G. biloba* leaves have been shown to be a mixture of terpene lactones and flavonoids [11,12,16,19,20]. Most commercial preparations of *G. biloba* are leaf extracts standardized to about 5 to 7% terpene lactones and 22 to 27% flavonoids [18–21]. The prominent terpene lactones in *G. biloba* extracts are ginkgolides A, B, C, J, and M and bilobalide (Figure 1). While the ginkgolides are considered to be diterpenes and bilobalide is considered to be a sesquiterpene, the latter compound most likely represents a product of ginkgolide metabolism [11,19]. Studies by Cartayrade et al. [22] have shown that although *G. biloba* leaves represent sites of ginkgolide (and bilobalide) accu-

**TABLE 1** 1999 Sales of Medicinal Plant Products in the United States

| Plant | % of total sales | Change from 1998 (%) |
|---|---|---|
| *Ginkgo biloba* | 21.0 | 6.0 |
| St. John's wort | 15.9 | 23.2 |
| Ginseng | 12.3 | 15.6 |
| Garlic | 11.2 | 11.4 |
| *Echinacea*/goldenseal | 9.1 | 0.2 |
| Saw palmetto | 6.1 | 37.8 |
| Kava | 2.4 | 456.0 |
| Valerian | 1.2 | 3.2 |

*Source*: Data from Ref. 10.

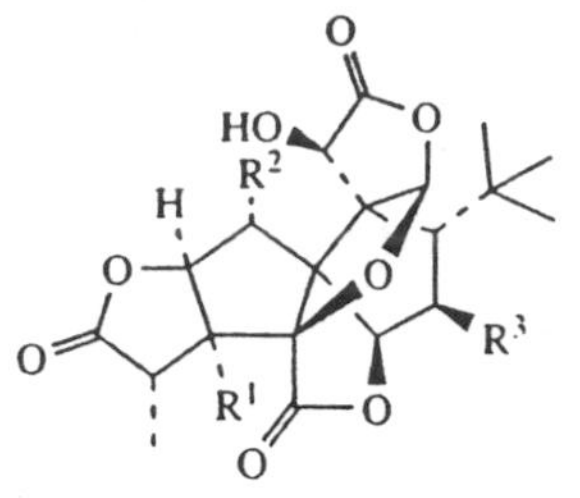

bilobalide

|  | R¹ | R² | R³ |
|---|---|---|---|
| ginkgolide A | OH | H | H |
| ginkgolide B | OH | OH | H |
| ginkgolide C | OH | OH | OH |
| ginkgolide J | OH | H | OH |
| ginkgolide M | H | OH | OH |

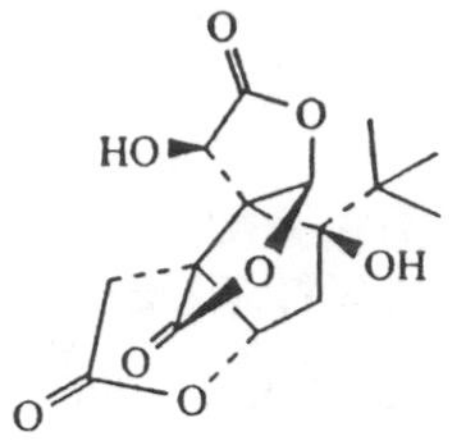

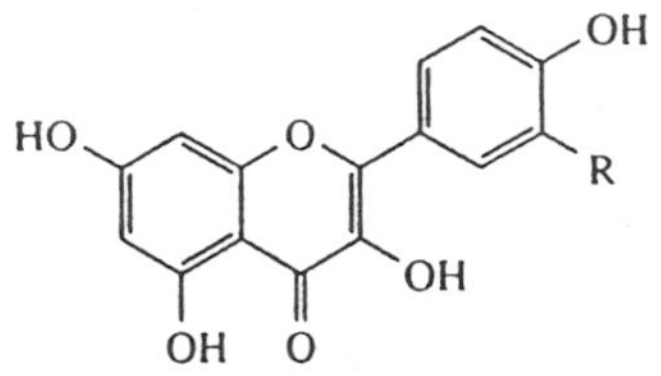

R = H, kaempferol
R = OH, quercetin

**Figure 1** Ginkgolides and flavonoids present in *Ginkgo biloba*. (Adapted from Ref. 19.)

mulation, biosynthesis of these compounds takes place in the roots. Moreover, these authors demonstrated that for at least ginkgolides A to C, biosynthesis occurs in a sequential manner (ginkgolide A → ginkgolide B → ginkgolide C) through successive addition of hydroxyl groups [22].

Pharmacological studies have demonstrated that the ginkgolides (especially ginkgolide B) are potent antagonists of platelet-activating factor (PAF), a bioregulatory molecule involved in blood platelet activation and inflammatory processes [13,15,16,19]. The flavonoids present in *Ginkgo* extracts exist primarily as glycosylated derivatives of kaempferol and quercetin [11,12,16,18,19,21] (Figure 1). These flavonoid glycosides have been shown to be extremely effective free radical scavengers [11,13,15,19]. It is believed that the collective action of these components leads to a reduction in damage and improved functioning of the blood vessels [13,15,16,21].

## B. St. John's Wort

The use of St. John's wort (*Hypericum perforatum* L.) for depression has its origins in the medical traditions of Europe dating back to well before the 1600s [23]. In Germany, St. John's wort is currently one of the most widely used prescription medications for depression. Moreover, St. John's wort is extensively used in the United States as a nonprescription botanical supplement [15,23,24]. For production of the botanical medicine, the aerial portion of the plant is harvested and dried just after flowering, and then an alcohol-water extract is produced (Refs. 15, 18, 23, and 24 and references therein).

Naphthodianthrones such as hypericin and pseudohypericin (Figure 2) are predominant components in St. John's wort extracts, and most St. John's wort phytomedicinals are currently standardized according to their hypericin content [15,23,24]. These chemicals are localized in dark glandular structures mainly located on the margins of St. John's wort leaves and flower petals and appear to serve in the defense against insect herbivory [25]. Although there is some evidence that biosynthesis of St. John's wort naphthodianthrones involves the polyketide pathway, few details are currently known (Refs. 19 and 26 and references therein). The production of napthodianthrones in St. John's wort can be influenced by environmental factors such as light and soil mineral nutrients [23]. Although there is strong evidence that hypericin and pseudohypericin contribute to the antidepressant action of St. John's wort, it is unclear whether this is associated with its activity as a monoamine oxidase inhibitor [23,26]. Inhibition of monoamine oxidase is one mechanism by which some antidepressants operate to increase levels of neurotransmitters such as serotonin, norepinephrine, or dopamine [15].

The prenylated phloroglucinol derivative hyperforin (Figure 2) can also be a predominant component in extracts of the flowers and leaves of St. John's wort, and there is evidence that this phytochemical contributes to the antidepressant action of this plant [15,24]. In human clinical studies, the hyperforin content in St. John's wort extracts correlated with the level of antidepressant action [27]. This chemical appears to block synaptic reuptake of serotonin, dopamine, and norepinephrine [28]. Blocking reuptake of neurotransmitters elevates their synaptic concentration. This is another mechanism by which synthetic antidepressants may operate [15,24]. Although hyperforin also has antibacterial activity [23], it is uncertain whether or not this compound is effective in defense functions for the plant.

## C. Ginseng

The name "ginseng" can often lead to some confusion because of its use for different plants with different phytochemical constituents. True ginsengs are plants in the genus *Panax*, of which Asian ginseng (*Panax ginseng*) and American ginseng (*Panax quinquefolium*) have received the most interest for phytomedicinal use [12,15,29,30]. These plants are low-growing perennial shade plants that generate a bulky storage root that is used medicinally. On the other hand, *Eleuthrococcus senticosis*, a completely different plant not even in the genus *Panax*, is sometimes referred to as Russian or Siberian "ginseng." Roots from this shrubby tree, native to regions of Siberia and northern China, were studied in the former Soviet Union as a substitute for Asian ginseng [12,15].

Interest in the use of ginseng and *Eleuthrococcus* is due to their purported "adaptogen" or "tonic" activities. Such activities are thought to increase the body's capacity to tolerate external stresses, leading to increased physical or mental performance [15]. For Asian ginseng, this adaptogenic activity was noted over 3500 years ago in traditional Chinese medicine, and for American ginseng, use as an adaptogen-type tonic was familiar to many Native American tribes [31,32]. Whereas an extensive literature documenting adaptogenic effects in laboratory animal systems exists, results of human clinical studies have tended to be more conflicting and variable [13,15,21,29,30]. However, there is evidence that extracts of ginseng and *Eleuthrococcus* can have an immunostimulatory effect in humans, and this may contribute to the adaptogen or tonic effects of these plants [15,18,21].

The major secondary products present in ginseng roots are an array of triterpene saponins, collectively called ginsenosides [13,19,29,33]. The ginsenosides are glycosylated derivatives of two major aglycones, panaxadiol and panaxatriol [12,19,33]. At present, 30 ginsenosides have been identified, of which the ginsenosides Rb1, Rb2, Rc, Rd, Re, Rf, Rg1, and Rg2 shown in Figure 3 are considered to be the most relevant for pharmacological activity [12,19,21,29,33]. Different ginseng species have different proportions of ginsenosides in root tissue, and this may be related to reported differences in the pharmacological properties of these plant materials [20,21,30,33]. Moreover, within a particular ginseng species, levels of particular ginsenosides can be affected by environmental factors such as soil mineral nutrient supply [34]. From laboratory studies, it has been suggested that the pharmacological target sites for these compounds may involve the hypothalmic-pituitary-adrenal axis because of the observed effects on serum levels of adrenocorticotropic hormone (ACTH) and corticosterone [13]. However, it should also be noted that the overall effects of the ginsenosides can be quite complex because of their potential for multiple actions even within a single tissue [20,30].

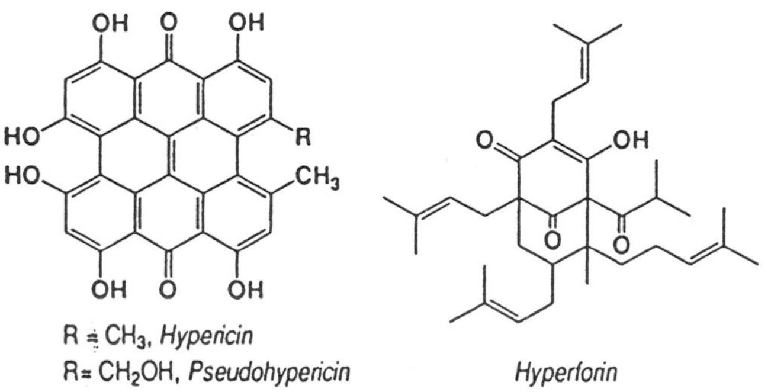

R = CH₃, *Hypericin*
R = CH₂OH, *Pseudohypericin*

*Hyperforin*

**Figure 2** Hypericin, pseudohypericin, and hyperforin from St. John's wort (*Hypericum perforatum*). (Adapted from Ref. 12.)

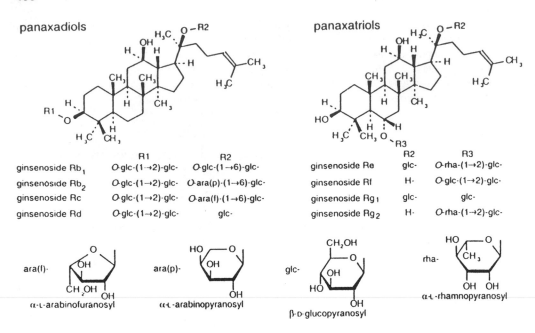

**Figure 3** The ginsenosides from ginseng (*Panax* sp.) as panaxadiol and panaxatriol derivatives. (Adapted from Ref. 21.)

Although extracts of *Eleuthrococcus* roots have been reported to have effects similar to those of the ginsenosides in animal systems, the active constituents may be quite different [19]. In *Eleuthrococcus*, the active constituents are thought to be lignanglycosides (eleutheroside E) or phenylpropane glycosides (eleutheroside B) [12,19]. However, the pharmacological action of these compounds remains unclear and little information is available on their biosynthetic pathway (Ref. 19 and references therein).

## D.  Garlic

Garlic is truly an ancient medicine, being used by more cultures and over a longer time period than any other medicinal plant. Garlic has been used by many cultures as an herbal medicine for infections, digestive problems, expelling parasites, and promoting health. Many of these effects of garlic are the basis for its use today. The earliest recorded use of garlic was in about 3000 BC by the Sumerians of Mesopotamia and by people of ancient India [35,36]. Garlic was an important herbal medicine of the Egyptians, and its use was recorded in a medicinal text known as the Ebers Papyrus in about 1550 BC [36]. The use of garlic was also described in ancient texts of Indian Ayurvedic medicine and traditional Chinese medicine [36]. The Greek physician Hippocrates was a strong proponent of the use of garlic for treatment of infections, parasites, pneumonia, and cancer [36]. Interestingly, Dioscorides, a Roman often considered the founder of the science of pharmacy, noted in about AD 1 that in addition to its antiseptic action, garlic was useful for "clearing the arteries." At present, a major interest in the use of garlic stems from its effects on lowering cholesterol levels and decreasing blood pressure [15,35,36]. Prior to the development of modern antibiotics, garlic was extensively used for its antiseptic properties in reducing infection [35,36]. These beneficial effects of garlic appear to be due to sulfur-containing compounds present in the bulb that are also responsible for the culinary flavoring uses of this plant.

Garlic (*Allium sativum*) is a member of the lily family and originated in central Asia [15,36]. However, today most garlic is grown commercially in many parts of the world for culinary and medicinal purposes, with the world production being about 2 million tons per year [35]. On a worldwide basis, 60% of garlic is grown in Asia, 20% is grown in Europe, 10% is grown in Africa, and 10% is grown in North America [35].

The portion of the garlic plant utilized for culinary and medicinal purposes is the multisegmented bulb, and different modes of postharvest processing are utilized. Following harvest, most garlic is dried for the production of garlic powders that can be used for culinary spices or production of medicines such as garlic tablets or capsules [35]. Other postharvest preparation methods used to generate medicinal forms of garlic include oil extraction of the bulb to generate garlic oils and steam distillation to generate an essential oil preparation from volatile compounds present in the bulb [35,37]. As oil extraction selects for compounds soluble in oil and steam distillation selects for more volatile compounds, the array of active chemicals may differ in these preparations [37].

The pharmacological activity of garlic appears to be associated with the sulfur-containing chemicals that are accumulated in the bulb. Of the sulfur-containing compounds present in the bulb, about 95% can be found in two classes of compounds: the γ-glutamyl-S-alkyloysteines and the S-alkylcysteine sulfoxides (Figure 4A). Of these compounds alliin is most prominent, amounting to about 6–14 mg/g fresh weight [36,37]. This is followed by γ-glutamyl-S-t-1-propenylcysteine (3–9 mg/g fresh weight), γ-glutamyl-S-allylcysteine (2–6 mg/g fresh weight), methiin (0.5–2 mg/g fresh weight), and cycloalliin (0.5–1.5 mg/g fresh weight) [36]. Although alliin and methiin may be prominent compounds in the intact bulb, these chemicals are not present in any garlic preparations for which the garlic cloves have been crushed or powdered. This is due to the activity of the enzyme alliinase, which is released upon crushing of the tissue. A major result of this alliinase activity is the production of allicin (Figure 4B) and this is sig-

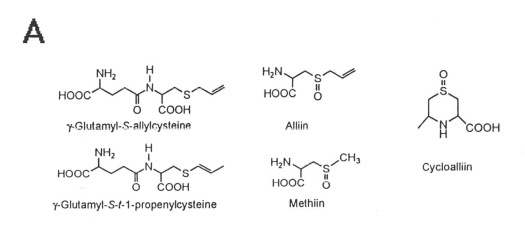

**Figure 4**  The γ-glutamyl-S-alkyulcysteines and S-alkylcysteines present in garlic (*Allium sativa*) (A) and the reaction conducted by alliinase with the crushing of garlic bulb tissue (B). (Adapted from Refs. 15 and 36.)

nificant because it appears that allicin (generated by alliinase activity on alliin) has a significant role in the pharmacological activity of this medicinal plant [15,36]. In contrast, the γ-glutamyl-$S$-alkylcysteines and cycloallin are not altered during crushing of the tissue (Ref. 36 and references therein).

Allicin appears to play a major role in the beneficial antimicrobial activity of this medicinal plant. In this respect, crushed garlic or juice expressed from garlic cloves has been shown to have potent antimicrobial activity against a variety of both gram-positive and gram-negative bacteria including *Escherichia*, *Salmonella*, *Staphylococcus*, *Streptococcus*, *Klebsiella*, *Proteus*, *Bacillus*, *Mycobacterium*, and *Clostridium* [15,36]. There is also substantial evidence that allicin forms the basis for garlic's effects in lowering blood serum cholesterol and triglyceride levels [36]. Allicin has been shown to be an inhibitor of hydroxymethylglutaryl coenzyme A (HMG CoA) reductase, which catalyzes a key step in cholesterol biosynthesis [15,36]. Clinical studies have suggested that allicin may lower blood serum triglyceride levels by increasing rates of lipid catabolism (Ref. 36 and references therein). Although it is clear that garlic can decrease blood pressure in both humans and animals, this is not due to allicin and the active chemical is currently unknown [21,36].

## E. Echinacea

Although there are 11 species in the genus *Echinacea*, this term is typically used to describe a phytomedicine produced from the aerial portion of *Echinacea purpurea* ("purple coneflower"), roots of *Echinacea pallida* ("pale-purple coneflower"), roots of *Echinacea angustifolia* ("narrow-leaf coneflower"), or a combination of these materials [18,21]. These plants are herbaceous perennials native to North America and were originally used in Native American herbal traditions for wound healing, infections, and rattlesnake bite [12,31]. Use of this phytomedicine was subsequently introduced to Europe in the early 1900s, and current interest lies in its use for colds, flu-like infections and upper respiratory infections [15,21]. The best studied and effective versions of this phytomedicine involve the expressed juice of the aerial portion of *E. purpurea* and an alcohol extract of *E. angustifolia* roots [21]. A number of studies suggest that *Echinacea*-based phytomedicines may be beneficial in reducing the symptoms and perhaps duration of upper respiratory infections (Refs. 15 and 21 and references therein). Pharmacological studies (>350 to date) have provided strong evidence for effects of *Echinacea* extracts in modulating immune system capacity including stimulation of the phagocytic activity of human lymphocytes, stimulation of fibroblasts for new tissue production, increased respiration, and elevated mobility of leukocytes [15,21,38]. Extracts of *Echinacea* also appear to inhibit both tissue and bacterial hyaluronidase, and this action is thought to aid in localization of infection, preventing its spread to other regions of the body (Refs. 21 and 38 and references therein).

The immunostimulatory activity of *Echinacea* preparations appears to result from the combined effects of a complex array of constituents (Figure 5) including a series of alkylamides as isobutylamides, caffeic acid derivatives (chicoric acid, cynarin, echinacoside—not present in *E. purpurea*), a series of polyalkynes (polyacetylenes), a series of polyalkenes, and high-molecular-weight (high-MW) polysaccharides including heteroxylans (approximate MW 35,000) and arabinorhamnogalactans (approximate MW 45,000) [12,15,21,38]. Pharmacological studies have shown effects of the alkamides, polyalkynes, and caffeic acid derivatives in stimulating white blood cell phagocytosis (Refs. 18 and 21 and references therein). The high-molecular-weight polysaccharide components also appear active in stimulating phagocytosis as well as promoting production of interferon [21]. Furthermore, through their inhibitory effect on 5-lipoxygenase, the alkylamide constituents may provide anti-inflammatory activity. Although details of the biosynthetic pathways of these active constituents are unknown, the hydrocarbon portions of alkylamides and polyalkynes most likely represent desaturation products of long-chain fatty acids (e.g., oleic acid) shortened through β-oxidation (see Ref. 19 for a discussion of these processes). Chicoric acid, cynarin, and echinacoside most likely arise as conjugated products of caffeic acid generated through the shikimic acid pathway. Although factors that influence the biosynthetic pathways of these chemicals are unknown, it has been shown that the phytomedicinal content of *Echinacea* can be affected by environmental conditions (soil nitrogen and potassium) and the developmental state of the plant [39].

Currently, most *Echinacea* phytomedicines containing *E. angustifolia* are standardized according to their echinacoside content because the caffeic acid derivative is a marker chemical unique to *Echinacea* root. However, such standardization does not consider the complex chemical interactions and possible synergistic effects necessary for the beneficial affects of this phytomedicinal on humans.

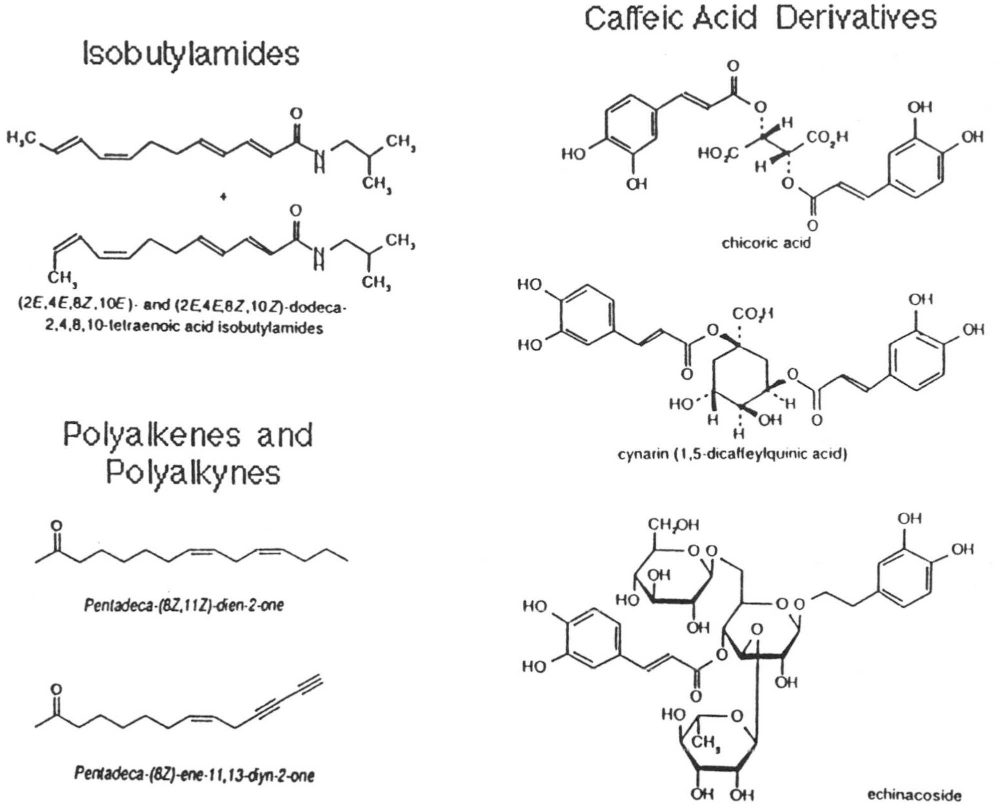

## Isobutylamides

(2E,4E,8Z,10E)- and (2E,4E,8Z,10Z)-dodeca-
2,4,8,10-tetraenoic acid isobutylamides

## Polyalkenes and Polyalkynes

Pentadeca-(8Z,11Z)-dien-2-one

Pentadeca-(8Z)-ene-11,13-dyn-2-one

## Caffeic Acid Derivatives

chicoric acid

cynarin (1,5-dicaffeylquinic acid)

echinacoside

**Figure 5** Examples of immunostimulating phytochemicals present in *Echinacea* extracts. (Adapted from Refs. 12 and 21.)

## F. Goldenseal

Goldenseal (*Hydrastis canadensis*) is a member of the Ranunculus family that is native to the North American continent [31,38]. This plant is a herbaceous perennial that grows primarily in damp areas that are shaded, such as low areas in forested regions [31,40,41]. Goldenseal has a short horizontal rhizome with multiple slender roots, and it is the rhizome that serves as the source for production of the phytomedicinal [41]. The use of goldenseal as a medicinal plant originated with Native American tribes, primarily the Cherokee peoples [12,31]. In addition to its use as a medicine for infection and respiratory problems, goldenseal rhizome was used for production of a bright gold dye used for clothing [31,40,41].

Current interest in goldenseal is related to its use for infections such as upper respiratory infections [40,42]. Often, goldenseal is used in conjunction with *Echinacea* for treatment of symptoms related to colds and flu [42]. Despite wide interest in its use as an antimicrobial agent, relatively little work has been conducted on this plant. As yet, the pharmacological activity of this medicinal plant has not been demonstrated in any well-constructed clinical trial [42]. Hence, most evidence for the beneficial effects of this medicinal plant for infections such as respiratory illness is anecdotal. However, from the limited in vitro studies that have been conducted, the isoquinoline alkaloids, berberine and hydrastine, would appear to play a major role in the pharmacological activity of this medicinal plant (Figure 6). These compounds have strong antibacterial action against many organisms including *Staphylococcus, Streptococcus, Salmonella*, and *Proteus* [12]. Although it has also been claimed that goldenseal has immunostimulatory activity similar to that of *Echinacea* (Ref. 42 and references therein), this has never been confirmed.

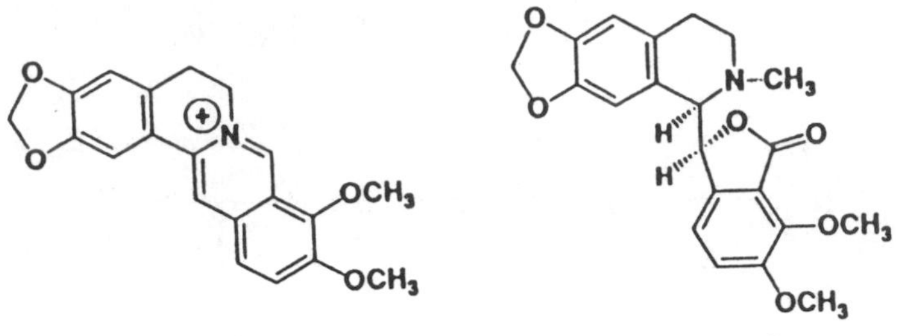

Berberine                                    (−)-β-Hydrastine

**Figure 6**   Berberine and hydrastine present in the rhizome and roots of goldenseal (*Hydrastis canadensis*). (Adapted from Ref. 12.)

## G.   Saw Palmetto

Saw palmetto (*Serenoa repens*), also known as "Sabal," is a small fan palm tree that is native to the southeastern coastal regions of the United States. The use of this plant by humans dates back to Native American tribes who populated this region, although their use for this this plant was primarily as a food source [43]. For medicinal purposes, the berries are utilized to produce a phytomedicinal extract that is useful for the treatment of symptoms associated with noncancerous swelling of the prostate (benign prostatic hyperplasia or BPH) [15,42–44]. This phytomedicinal extract is generally produced by extracting the ripe berries with either hexane or liquid carbon dioxide [12,15,43]. Although most research supporting the usefulness of saw palmetto as an herbal treatment for benign prostatic hyperplasia was conducted in Europe starting in the 1960s [12,15,42,44], this use for extracts of the berries was suggested by Felter and Lloyd in the United States in 1898 [45]. The use of saw palmetto as an alternative herbal treatment for benign prostatic hyperplasia has become popular in the United States because of its low cost and the absence of many undesirable side effects (e.g., impotence) associated with the types of synthetic pharmaceuticals utilized in the treatment of this condition [42].

Saw palmetto extracts that are effective for symptomatic treatment of benign prostatic hyperplasia appear to be a complex mixture of primarily hydrophobic components [12]. However, extensive research on the chemical composition of saw palmetto extracts has not yet been conducted. Nevertheless, it is known that the major components present in these extracts are saturated and unsaturated fatty acids, free and esterified plant sterols (e.g., β-sitosterol, campesterol, cycloartenol), polyprenols, and anthranilic acid [12,15,44]. Nearly 50% of the fatty acids contain 14 or fewer carbons [12].

Although the cause of benign prostatic hyperplasia is not well understood, the most widely favored hypothesis involves an alteration in steroid hormone metabolism in prostate cells resulting in increased levels of dihydrotestosterone and an increase in the estrogen/androgen ratio [15,44]. Key enzymes in prostate cells related to this process include 5α-reductase, which converts testosterone to dihydrotestosterone, and aromatase, which converts testosterone to estrogens [15,44]. In this respect, several laboratory studies have shown that extracts of saw palmetto berries are inhibitors of these enzymes [12,44]. In addition to potential effects in decreasing alterations in prostatic steroid hormone metabolism, saw palmetto extracts appear to have anti-inflammatory activity that would also be beneficial for symptoms of benign prostatic hyperplasia [12,44].

## H.   Kava-Kava

The use of kava originated in the "Oceania" island communities encompassed by Polynesia, Melanesia, and Micronesia [46]. In these cultures "kava" (also known as "kava-kava," "ava," and "awa") refers to an intoxicating beverage used in rituals and ceremonies that is produced from the mashed rhizome and roots of the woody shrub *Piper methysticum*. However, the term kava is now used as the common name for *P. methysticum* as well as the phytomedicinal produced from root-rhizome extracts. Worldwide, kava is receiving considerable attention as a phytomedicinal treatment for anxiety, nervous tension, agitation, and/or insomnia. Clinical studies have shown that the effectiveness of kava is comparable to that of sedatives such as benzodiazapines but without the development of either physical or psychological dependence [15,46].

From intensive chemical and pharmacological studies of kava root-rhizome extracts over the past century, several key active constituents have been identified [12,46]. The pharmacological activity of this plant appears to be associated with a family of styrylpyrones called kavapyrones (or kavalactones) that have effects on several neurotransmitter systems including those involving glutamine, γ-aminobutyric acid (GABA), dopamine, and serotonin [15]. Although the kavapyrones shown in Figure 7 represent the predominant pharmacologically active components in kava root extracts, a total of 18 have been identified to date [47]. These remaining kavapyrones appear to be derivatives of either kawain, yangonin, or dihydromethysticin [12,47]. Although details of kavapyrone biosynthesis are still lacking, evidence from other systems such as *Equisetum arvense* gametophytes suggests that styrylpyrones may arise from a triketide produced by successive condensation of two malonyl CoA molecules with a phenylpropanoid CoA ester (Ref. 48 and references therein). This is similar to reactions catalyzed by chalcone synthase except that two rather than three successive condensations involving malonyl CoA are involved. Studies have shown that kavapyrone levels in kava roots are influenced by environmental factors. In cultivated kava plants, kavapyrone levels appear to increase with irrigation and mineral nutrient supplementation and decrease with shading [49]. Moreover, varietal differences in kava also appear to have a role in determining the overall level of kavapyrone production [49].

Interestingly, kava plants are sterile and plantation production involves propagation from stem cuttings [46]. From genetic studies, it has been suggested that kava represents a sterile relative of *Piper wichmannii* (native to New Guinea), which was distributed across the South Pacific islands with human migration and through somatic mutation became sterile (Ref. 49 and references therein). This sterilty of kava, its limited growth habitat range (South Pacific tropics), the time required for growth before root harvest (about 8 years), and the high world demand have raised concerns about potential overharvesting. Although attempts have been made to grow kava in tissue culture for propagation and possible in vitro phytochemical production, little success has been achieved [50].

## I. Valerian

Valerian (*Valeriana officinalis*) has had a long history of use as a relaxing sedative and as treatment for insomnia [51]. For example, this use of valerian was noted in Dioscorides' *De Materia Medica* written in Roman times. During World War I, valerian was used as a treatment for shell shock [19]. Valerian is a tall perennial growing in damp swampy areas that is native to Europe and temperate regions of Asia [15,21,51]. The plant used traditionally for medicinal purposes is only one of 250 species in the genus *Valeriana* [15]. For use as a phytomedicinal, hot-water infusions (teas) or alcohol-water extracts (tinctures) are generally produced from the roots of this plant [51].

The major compounds present in active valerian extracts are a series of sesquiterpenes and iridoids (Figure 8). Major sesquiterpenes include valerenic acid and valeranone. The iridoids are present as a series of closely related epoxy-iridoid esters called the valeopotriates, the major compounds being valtrate, isovaltrate, acevaltrate, and didrovaltrate [12,19,21]. Overall, the valeopotriate level in dried valerian roots ranges from 0.4 to 2% [15,19,21]. The valeopotriates are labile compounds that break down under conditions of moisture, heat, or acidity to liberate unsaturated aldehydes such as baldrinal and isovaleric acid [12,51]. Production of isovaleric acid during the drying of valerian roots contributes to the unpleasant odor of this plant material [15,19].

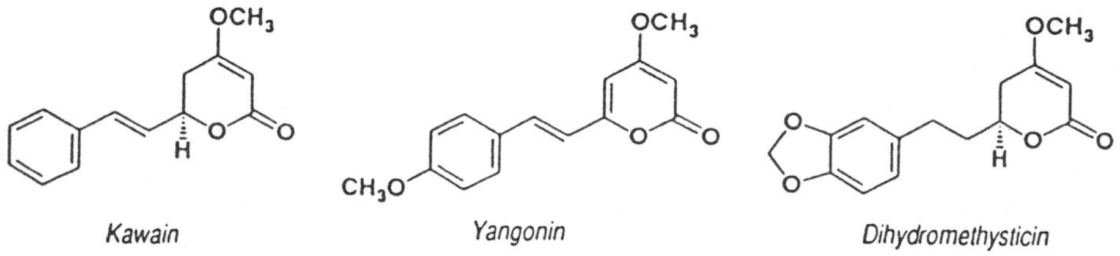

**Figure 7** Examples of kavapyrones from *Piper methysticum* (kava). (Adapted from Ref. 12.)

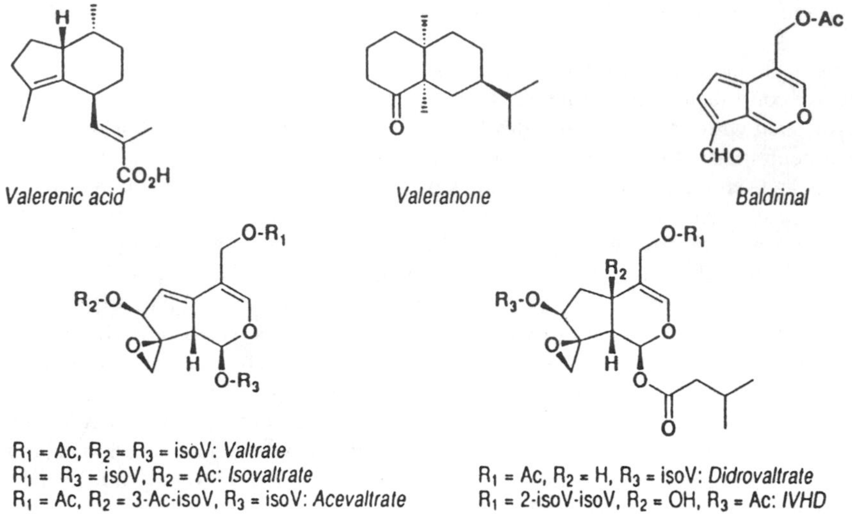

Valerenic acid                          Valeranone                          Baldrinal

R$_1$ = Ac, R$_2$ = R$_3$ = isoV: *Valtrate*
R$_1$ = R$_3$ = isoV, R$_2$ = Ac: *Isovaltrate*
R$_1$ = Ac, R$_2$ = 3-Ac-isoV, R$_3$ = isoV: *Acevaltrate*

R$_1$ = Ac, R$_2$ = H, R$_3$ = isoV: *Didrovaltrate*
R$_1$ = 2-isoV-isoV, R$_2$ = OH, R$_3$ = Ac: *IVHD*

**Figure 8**  The sesquiterpenes and iridoids present in root extracts of valerian (*Valeriana officinalis*). (Adapted from Ref. 19.)

An array of clinical studies have demonstrated the effectiveness of valerian as a sedative [15,51,52]. However, despite extensive study and its use for about 2000 years, the basis for this pharmacological activity is less clear [42]. Although it had been thought that the sedative activity of valerian was associated with the valeopotriates, more recent studies have suggested that the sesquiterpenes may play an important role [15,19,51]. Valerone and valerenic acid have been shown to block degradation and synaptic reuptake of the neurotransmitter GABA, thus increasing its level in the synapse [51,53]. In addition, there is some evidence that sesquiterpenes may bind to benzodiazepine receptors and potentially exert effects similar to those of these drugs [e.g., diazepam (Valium)] [12,51]. Nevertheless, the role of sesquiterpenes in the action of valerian has been questioned because of the relative level of these components in typical extracts [42].

## V.  PHYTOMEDICINAL CHEMICAL PRODUCTION IN THE CONTEXT OF PLANT CHEMICAL ECOLOGY

Given that many phytochemicals that have pharmaceutical effects can play a defensive role in the plants from which they are produced, it is worthwhile to consider briefly plant secondary product metabolism in the context of chemical ecology. This has been an area of strong interest over the past 35 years, especially as it relates to plant-herbivore interactions (Ref. 54 and references therein). Of primary interest would be identifying the factors that could potentially determine the levels and patterns of phytomedicinal chemical production in plants. In this respect, several models have been developed to rationalize relationships between environmental factors (biotic and abiotic) and plant secondary product production.

In models related to "optimal defense theory," a central concept is that plant defense chemistry (secondary product metabolism) is expensive from a biochemical standpoint and diverts resources that would otherwise be utilized for growth and development [55–59]. Hence, plants will utilize chemical defenses to protect the regions of the plant that will provide the greatest return from an ecological standpoint [55,58]. This would certainly include reproductive structures as well as storage root structures in perennial plants that are indispensible for plant survival. On the other hand, less critical regions of the plant that could be more easily replaced would receive less defensive chemistry. In addition, some models associated with optimal defense theory also consider plant chemical defense to be dependent on the vulnerability of particular plants to herbivory over ecological time frames [56]. That is, plants that are more susceptible to herbivore attack will have a greater extent of defensive chemistry.

In models related to the "growth rate hypothesis," plant defense chemistry is viewed as occurring in a compensatory manner with plant growth. Here, the metabolic cost of replacing plant tissues consumed by herbivores is considered in the context of the costs of plant secondary product metabolism [60]. Under conditions of high resource availability where plant growth rates are high, levels of defensive secondary products would be expected to be low as the higher rate of growth would assure adequate survival from herbivory. On the hand, in resource-limited environments where suboptimal growth occurs, production of defensive secondary products would be elevated. Several studies have provided data in support of this model [60–63]. For example, in studies conducted by McKey [55], tree species grown in nutrient-poor soils that limited growth had higher levels of defensive phenolic compounds than the same trees growing in nutrient-rich soils.

The "carbon-nutrient balance" model was developed by Bryant et al. [61] to explain the effects of soil nutrient supply and light levels on secondary product metabolism. A central theme of this model is that the carbon/nitrogen (C/N) ratio of the plant under a given set of environmental conditions will have a strong bearing on the types and levels of secondary products generated by the plant. In this model, production of "carbon-based" secondary products (e.g., phenolics, terpenes, and other chemicals having only C, O, and H as part of their structure) would be directly proportional to the C/N ratio, whereas production of nitrogen-based secondary products (e.g., alkaloids, cyanogenic glycosides, nonprotein amino acids) would be inversely proportional to the C/N ratio. For example, under conditions of adequate light and low nitrogen supply, the C/N ratio of the plant would increase and this would lead to increased production of carbon-based secondary products. On the other hand, with conditions such as low light and adequate nitrogen that lead to a decrease in the plant C/N ratio, increased production of nitrogen-containing secondary products would be expected. As simple as this concept is, a substantial number of studies appear to support the C/N balance model for regulation of secondary product metabolism. Several studies have shown that soil fertilization tends to increase levels of nitrogen-based secondary products [64–66] and decrease levels of carbon-based secondary products [67–74]. Likewise, a decrease in light levels has been shown to increase levels of nitrogen-based secondary products [75] and decrease levels of carbon-based secondary products [76–80].

Although these models point to different relationships with factors governing production and distribution of secondary metabolites, it is apparent that two of the three general models place strong emphasis on the growth environment in determining levels and patterns of plant secondary metabolite production. These two models (growth rate hypothesis and carbon-nitrogen balance hypothesis) are also amenable to experimental testing, and data in support of each have been presented. Although no single model may be able to account for levels and patterns of secondary product production (see Ref. 54 for discussion), these results point to the importance of considering how the growth environment of a particular medicinal plant could have an impact on phytomedicinal chemical production and hence the quality of a botanical medicine.

# VI.  PERSPECTIVE AND OUTLOOK

Although medicinal plants have had long-standing use throughout human history and are of considerable interest as alternatives to synthetic pharmaceuticals, there is a paucity of basic knowledge of the physiology and biochemistry of these plants. With only a few exceptions, many widely used medicinal plants have not received the extensive physiological, biochemical, and genetic characterization received by food crops or model plant systems such as *Arabidopsis*. Although some active chemicals may have been identified in these plants, the pathways for their biosynthesis and the factors (biotic and abiotic) regulating their biochemical production are, in many cases, unclear.

At present, a major concern about the use of phytomedicinals regards the maintenance of consistent medicinal quality in botanical medicines [81]. Whereas the focus has tended to be on quality control in herbal manufacturing practices (good manufacturing practices or GMPs), variation in phytomedicinal content because of environmental effects on secondary plant metabolism in the plant material can also be a significant factor in determining the quality of the plant material entering the botanical medicine production process. In this respect, a set of guidelines for good agricultural practices" (GAPs) has been suggested that takes into consideration the importance of standardizing growth conditions for optimal phytomedicinal chemical production by a medicinal plant (see Ref. 82 for discussion). These guidelines would call for development of comprehensive protocols for medicinal plant production to ensure quality and reproducibility in the raw material utilized in the production of botanical medicines. A major objective of these guidelines is harmonizing phytopharmaceutical quality requirements within the European Union countries [82].

The use of molecular and biotechnology approaches to medicinal plants would also have wide application and promise, especially with regard to such topics as the modification of phytomedicinal chemical pathways [6,83], growth and propagation of medicinal plants in vitro, and the in vitro production of phytomedicinals in large-scale tissue culture systems such as bioreactors [84].

## REFERENCES

1.  B Griggs. Green Pharmacy. The History and Evolution of Western Herbal Medicine. Rochester, VT: Healing Arts Press, 1997.
2.  WHO. Guidelines for the Appropriate Use of Herbal Medicines. Manila: World Health Organization, 1998.
3.  VE Tyler. Phytomedicines: back to the future. J Nat Prod 62:1589–1592, 1999.
4.  P Brevoort. The blooming U.S. botanical market. A new overview. Herbalgram 44:33–46, 1998.
5.  V Glaser. Billion-dollar market blossoms as botanicals take root. Nat Biotechnol 17:17–18, 1999.
6.  RA Dixon. Plant natural products: the molecular genetic basis of biosynthetic diversity. Curr Opin Biotechnol 10:192–197, 1999.
7.  M Wink. Introduction: biochemistry, role and biotechnology of secondary products. In: M Wink, ed. Biochemistry of Secondary Product Metabolism. Boca Raton, FL: CRC Press, 1999, pp 1–16.
8.  PB Kaufman, LJ Cseke, S Warber, JA Duke, HL Brielmann. Natural Products from Plants. Boca Raton, FL: CRC Press, 1999.
9.  M Wink, O Schimmer. Modes of action of defensive secondary metabolites. In: M Wink, ed. Functions of Plant Secondary Metabolites and Their Exploitation in Biotechnology. Boca Raton, FL: CRC Press, 1999, pp 17–112.
10. M Blumenthal. Herb market levels after five years of boom: 1999 sales in mainstream market up only 11% in first half of 1999 after 55% increase in 1998. Herbalgram 47:64–65, 1999.
11. H Huh, EJ Staba. The botany and chemistry of *Ginkgo biloba* L. J Herb Spices Med Plant 1:91–124, 1992.
12. J Bruneton. Pharmacognosy, Phytochemistry, Medicinal Plants. Paris: Lavisior, 1995.
13. KC Huang. Pharmacology of Chinese Herbs. Boca Raton, FL: CRC Press, 1999.
14. J Kleijinen, P Knipschild. *Ginkgo biloga*. Lancet 340:1136–1139, 1992.
15. V Schultz, R Hänsel, V Tyler. Rational Phytotherapy. A Physician's Guide to Herbal Medicine, Berlin: Springer-Verlag, 1998.
16. B Ahlemeyer, J Krieglstein. Neuroprotective effects of *Ginkgo biloba* extract. In: LD Lawson, R Bauer, eds. Phytomedicines of Europe. Chemistry and Biological Activity. Washington, DC: American Chemical Society, 1998, pp 210–220.
17. J Kleijinen, P Knipschild. *Ginkgo biloba* for cerebral insufficiency. Br J Clin Pharmacol 34:352–358, 1992.
18. M Blumenthal, A Goldberg, A Binckmann. Herbal Medicine. Expanded Commission E Monographs. Boston: Integrative Medicine Communications, 2000.
19. PM Dewick. Medicinal Natural Products. West Sussex, UK: John Wiley & Sons, 1997.
20. O Sticher. Quality of *Gingko* preparations. Planta Med 59:2–11, 1993.
21. World Health Organization. WHO Monographs on Selected Medicinal Plants, Vol 1. Geneva: WHO, 1999.
22. A Cartayrade, E Neau, C Sohier, JP Balz, JP Carde, J Walter. Ginkgolide and bilobalide biosynthesis in *Ginkgo biloba*. I. Studies of synthesis, translocation and accumulation of ginkgolides and bilobalide. Plant Physiol Biochem 35:859–868, 1997.
23. R Upton. St. John's Wort. Quality Control, Analytical and Therapeutic Monograph. Santa Cruz, CA: American Herbal Pharmacopoeia Comendium, 1998.
24. HD Reuter. Chemistry and biology of *Hypericum perforatum* (St. John's wort). In: LD Lawson, R Bauer, eds. Phytomedicines of Europe. Chemistry and Biological Activity. Washington, DC: American Chemical Society, 1998, pp 287–298.
25. RB Fornasiero, A Bianchi, A Pinetti. Anatomical and ultrastructural observations in *Hypericum perforatum* L. leaves. J Herb Spice Med Plant 5:21–33, 1998.
26. A Nahrstedt, V Butterwick. Biologically active and other chemical constituents of the herb of *Hypericum perforatum* L. Pharmacopsychiatry 30:129–134, 1997.
27. G Laakman, C Schule, T Baghai, M Kieser. St. John's wort in mild to moderate depression—the relevance of hyperforin for the clinical efficacy. Pharmacopsychiatry 31:54–59, 1998.
28. SS Chatterjee, SK Bhattacharya, M Wonneman, A Singer, WE Muller. Hyperforin as a possible antidepressant component of *Hypericum* extracts. Life Sci 63:499–510, 1998.
29. O Sticher. Biochemical, pharmaceutical and medical perspectives of ginseng. In: LD Lawson, R Bauer, eds. Phytomedicines of Europe. Chemistry and Biological Activity. Washington, DC: American Chemical Society, 1998, pp 221–240.

30. AS Attele, JA WU, CS Yuan. Ginseng pharmacology—multiple constituents and multiple actions. Biochem Pharmacol 58:1685–1693, 1999.
31. AR Hutchens. Indian Herbology of North America. Boston: Shambala, 1991, pp 143–145.
32. S Fulder. The Book of Ginseng and Other Chinese Herbs for Vitality. Rochester, VT: Healing Arts Press, 1993.
33. JF Cui. Identification and quantification of ginsenosides in various commercial ginseng preparations. Eur J Pharm Sci 3:77–85, 1995.
34. TSC Li, G Mazza. Correlations between leaf and soil mineral concentrations and ginsenoside contents in American ginseng. Hortscience 34:85–87, 1999.
35. GR Fenwick, AB Hanley. The genus *Allium*—Part 1. Crit Rev Food Sci Nutr 22:199–235, 1985.
36. LD Lawson. Garlic: a review of its medicinal effects and indicated active compounds. In: LD Lawson, R Bauer, eds. Phytomedicines of Europe. Chemistry and Biological Activity. Washington, DC: American Chemical Society, 1998, pp 176–209.
37. HD Reuter, A Sendl. *Allium sativum* and *Allium ursinum*. Chemistry, pharmacology and medicinal applications. In: H Wagner, NR Farnsworth, eds. Economic and Medicinal Plants Research, Vol 6. London: Academic Press, 1994, pp 55–113.
38. R Bauer. *Echinacea*: Biological effects and active principles. In: LD Lawson, R Bauer, eds. Phytomedicines of Europe. Chemistry and Biological Activity. Washington, DC: American Chemical Society, 1998, pp 140–157.
39. SE El-Gengaihi, AS Shalaby, EA Agina, SF Hendawy. Alkylamides of *Echinacea purpurea* L. as influenced by plant ontogony and fertilization. J Herb Spices Med Plant 5:35–41, 1998.
40. JK Crellin, J Philpott. A Reference Guide to Medicinal Plants. Durham, NC: Duke University Press, 1990, pp 232–234.
41. M Barlow. From the Shepard's Purse. HongKong: Everbest Printing Co, 1990, pp 72–73.
42. V Tyler. Importance of European phytomedicinals in the American market: an overview. In: LD Lawson, R Bauer, eds. Phytomedicines of Europe. Chemistry and Biological Activity. Washington, DC: American Chemical Society, 1998, pp. 2–12.
43. BC Bennett, JR Hicklin. Uses of saw palmetto (*Serenoa repens*, Arecaceae) in Florida. Econ Bot 52:381–393, 1998.
44. H Schilcher. Herbal drugs in the treatment of benign prostatic hyperplasia. In: LD Lawson, R Bauer, eds. Phytomedicines of Europe. Chemistry and Biological Activity. Washington, DC: American Chemical Society, 1998, pp. 62–73.
45. IIW Felter, JU Lloyd. King's American Dispensatory. 18th ed. 3rd rev. Vol II. Portland, OR: Eclectic Medical Publications, 1983 (reprint of original 1898), pp 1750–1752.
46. YN Singh, M Blumenthal. Kava. An overview. Herbalgram 39:33–55, 1997.
47. X-G He, L Longze, L-Z Lian. Electrospray high performance liquid chromatography–mass spectrometry in phytochemical analysis of kava (*Piper methysticum*) extract. Planta Med 63:70–74, 1997.
48. J Schröder. A family or plant-specific polyketide synthases: facts and predictions. Trends Plant Sci 2:373–378, 1997.
49. V Lebot, E Johnson, QY Zheng, D McKern, DJ McKenna. Morphological, phytochemical and genetic variation in Hawaiian cultivars of 'awa (kava, *Piper methysticum*, Piperaceae). Econ Bot 53:407–418, 1999.
50. M Taylor, L Taufa. Decontamination of kava (*Piper methysticum*) for in vitro propagation. Acta Hortic 461:267–274, 1998.
51. P Morazzoni, E Bombardelli. *Valeriana officinalis:* traditional use and recent evaluation of activity. Fitoterapia 66:99–112, 1995.
52. H Schultz, C Stolz, J Muller. The effect of valerian extract on sleep polygraphy in poor sleepers: a pilot study. Pharmacopsychiatry 27:147–151, 1994.
53. MS Santos, F Ferreira, AP Cunha. An aqueous extract of valerian influences the transport of GABA in synaptosomes. Planta Med 60:278–279, 1994.
54. MR Berenbaum. The chemistry of defense: theory and practice. Proc Natl Acad Sci USA 92:2–8, 1995.
55. D Mckey. Adaptive patterns in alkaloid physiology. Am Nat 108:305–320, 1974.
56. P Feeney. Plant apparency and chemical defense. Recent Adv Phytochem 10:1–40, 1976.
57. DF Rhoades, RG Cates. Toward a general theory of plant antiherbivore chemistry. Recent Adv Phytochem 10:168–213, 1976.
58. DF Rhoades. Evolution of plant chemical defense against herbivores. In: GA Rosenthal, DH Jansen, eds. Herbivores: Their Interaction with Secondary Plant Metabolites. Orlando, FL: Academic Press, 1979, pp 3–45.
59. T Fagerstrom, S Larsson, O Tenow. On optimal defense in plants. Funct Ecol 1:73–81, 1987.
60. DH Jansen. Tropical backwater rivers, animals and mast fruiting by the Dipterocarpaceae. Biotropica 6:69–103, 1974.
61. JP Bryant, FS Chapin, DR Klein. Carbon/nutrient balance of boreal plants in relation to vertebrate herbivory. Oikos 40:357–368, 1983.
62. PD Coley, JP Bryant, FS Chapin. Resource availability and plant antiherbivore defense. Science 230:895–899, 1985.
63. PD Coley. Interspecific variation in plant anti-herbivore properties: the role of habitat quality and rate of disturbance. New Phytol 106:251–263, 1987.

64. E Nowacki, M Jurzysta, P Gorski, D Nowacka, GR Waller. Effect of nitrogen nutrition on alkaloid metabolism in plants. Biochem Physiol Pflanz 169S:231–240, 1976.
65. WJ Mattson. Herbivory in relation to plant nitrogen content. Annu Rev Ecol Syst 11:119–161, 1980.
66. J Gershenzon. Changes in levels of plant secondary metabolites under water and nutrient stress. In: BN Timmermann, C Steelink, FA Leows, eds. Phytochemical Adaptations to Stress. New York: Plenum, 1984, pp 51–64.
67. CA Mihaliak, DE Lincoln. Growth pattern and carbon allocation to volatile leaf terpenes under nitrogen-limiting conditions in *Heterotheca subaxillaris* (Asteraceae). Oecologia 66:423–426, 1985.
68. RH Waring, AJS McDonald, S Larsson, T Ericsson, A Wiren, E Arwidsson, A Ericson, T Lohammar. Differences in chemical composition of plants grown at constant relative growth rates with stable mineral nutrition. Oecologia 66:157–160, 1985.
69. JP Bryant. Feltleaf willow–snowshoe hare interactions: plant carbon/nutrient balance and floodplain succession Ecology 68:1319–1327, 1987.
70. JP Bryant, FS Chapin, P Reichardt, T Clausen. Response of winter chemical defense in Alaska paper birch and green alder to manipulation of plant carbon/nutrient balance. Oecologia 72:510–514, 1987.
71. JP Bryant, TP Clausen, PB Reichardt, MC McCarthy, RA Werner. Effect of nitrogen fertilization upon the secondary chemistry and nutritional value of quaking aspen (*Populus tremuloides* Michx) leaves for the large aspen tortrix (*Choritoneura conflictana* (Walker)). Oecologia 73:513–517, 1987.
72. DC Lightfoot, WG Whitford. Variation in insect densities on desert creosotebush: is nitrogen a factor? Ecology 68:547–557, 1987.
73. PW Price, GL Waring, R Julkunen-Tiitto, J Tahvanainen, HA Mooney, TP Craig. Carbon-nutrient balance hypothesis in within-species phytochemical variation of *Salix lasiolepis*. J Chem Ecol 15:1117–1131, 1989.
74. PG Waterman S Mole. Extrinsic factors in influencing production of secondary metabolites in plants. In: EA Bernays, ed. Insect-Plant Interactions, Vol 1. Boca Raton, FL: CRC Press, 1989, pp 107–134.
75. B Van Horne, TA Hanley, RG Cates, JD McKendrick, JD Horner. Influences of seral stage and season on leaf chemistry of southeastern Alaska deer forage. Can J For Res 18:90–99, 1988.
76. S Woodhead. Environmental and biotic factors affecting the phenolic content of different cultivars of *Sorqhgum bicolor*. J Chem Ecol 7:1035–1047, 1981.
77. PG Waterman, JAM Ross, DB McKey. Factors affecting levels of some phenolic compounds, digestibility, and nitrogen content of the mature leaves of *Barteria fistulosa* (Passifloraceae). J Chem Ecol 10:387–401, 1984.
78. SA Larsson, A Wiren, L Lundgren, T Ericsson. Effects of light and nutrient stress susceptibility to *Galerucella lineola* (Coleoptera). Oikos 47:205–210, 1986.
79. SJ Mole, AM Ross, PG Waterman. Light-induced variation in phenolic levels in foliage of rain-forest plants I. chemical changes. J Chem Ecol 14:1–21, 1988.
80. CM Nichols-Orians. The effect of light on foliar chemistry, growth and susceptibility of seedlings of a canopy tree to an attine ant. Oecologia 86:552–560, 1991.
81. HB Matthews, GW Lucier, KD Fisher. Medicinal herbs in the United States: research needs. Environ Health Perspect 107:773–778, 1999.
82. A Mathe, C Franz. Good agricultural practices and the quality of herbal medicines. J Herb Spices Med Plant 6:101–113, 1999.
83. CL Nessler. Metabolic engineering of plant secondary products. Transgenic Res 3:109–115, 1994.
84. NJ Walton, AW Alfermann, JC Rhoades. Production of secondary metabolites in cell and differentiated organ cultures. In: M Wink, ed. Functions of Plant Secondary Metabolites and Their Exploitation in Biotechnology. Boca Raton, FL: CRC Press, 1999, pp 17–112.

# 24

# Plant Growth Hormones: Growth Promotors and Inhibitors

**Syed Shamshad Mehdi Naqvi**

*Nuclear Institute of Agriculture, Tando Jam, Pakistan*

## I. INTRODUCTION

Since the dawn of agriculture, one of principal aims of human beings has been the control and promotion of plant growth to satisfy human needs. These two important aspects of people's work with plants in the struggle to increase production are by no means synonymous. Humans soon realized that lush green growth does not always produce the best crop in the form of fruit and seeds and hence were forced to evolve such well-known cultural methods as pruning, balance manuring, and mineral fertilizers to regulate the nature and luxuriance of plant growth. This control of the plant development pattern—this adjustment of balance between root and shoot and between leaf and flowering—was until about seven decades ago mediated by appropriate and very large empirical combinations of what may be called dietary and surgery. More than 60 years ago proof was given of what Julius Sachs [1] in 1880 postulated: that endogenous substances regulate the growth of various plant organs. In 1926, Went [2], in Holland, provided convincing evidence of a diffusible substance (auxin) from oat (*Avena sativa*) seedlings that promoted growth of these seedlings. At the same time, Kurosawa [3], in Japan, discovered another substance (gibberellin) from cell-free fungus (*Gibberella fujikuroi*) filterate that promoted growth of rice (*Oryza sativa*) seedlings. But it was not until 1955 that Skoog and his associates [4] discovered kinetin in an autoclaved sample of herring sperm DNA, which was active in what Wiesner in 1892 called cell division factor [5].

In both scientific and popular literature, these chemicals in plant and crop physiology have come to be known as plant growth regulators. Starting, as do many great scientific advances, in a few unobtrusive laboratories as purely academic observations, the volume of research on these substances by both plant physiologists and chemists has swelled to enormous proportions.

The naturally occurring (endogenous) growth substances are commonly known as plant hormones, while the synthetic ones are called growth regulators. A *plant hormone* (synonym: phytohormone) is an organic compound synthesized in one part of a plant and translocated to another part, where in very low concentrations it causes a physiological response. Plant hormones are identified as promotors (auxin, gibberellin, and cytokinin), inhibitors (abscisic acid, xanthoxin, and violaxanthin), ethylene, and other hypothetical growth substances (florigen, death hormone, etc.). They usually exist in plants and crops at a concentration lower than 1 µM; above this they are generally considered supraoptimal [6].

The mechanism(s) by which hormones trigger a response is still far from well understood. Specific receptors have been proposed, but no proof for their function in mediating hormone action has been given.

There is, however, considerable evidence that gene expression is controlled by these hormones, but how it is done biochemically is largely unknown [7]. These hormones are found in all actively growing plant parts; young leaves and apical buds are particularly high in auxin, whereas young roots are high in gibberellins and cytokinins. Fruits and seeds are generally rich in all growth hormones. Therefore, these hormones are ubiquitous in plants and crops and are generally not species specific.

In general, a deficiency of hormone must be created experimentally (as by removing young leaves or using a hormone-deficient mutant) to show that adding a hormone has an effect. In this respect, the Mitscherlich law of diminishing returns [8] can be modified as follows: the increase in plant response produced by a unit increment of a deficient (limiting) hormone is proportional to the decrement of that hormone from the maximum.

To deal in depth and breadth with the entire field within our space limitations here is not possible. Therefore, we aim in this chapter to discuss advances in our understanding that are relevant to plant and crop management.

## II. AUXINS

*Auxin* is a Greek word derived from *auxein*, which means "to increase." It is a generic term for chemicals that typically stimulate cell elongation, but auxins also influence a wide range of growth and development response. The existence of growth-regulating chemicals that control plant growth, and the interrelations between their parts, was the outcome of experiments on root and shoot responses to external stimuli. Ciesielski, working with roots, and Charles and Francis Darwin, working with shoots, observed that in both organs the tip controlled the growth rate of the immediate growing axes as well as the regions located some distance away [9]. The Darwins performed simple experiments on the photoresponse of canary grass (*Phalaris canariensis*) and oat coleoptiles. When the tip was unilaterally illuminated, a strong positive curvature (growth toward light) along the growing axes resulted. If the tip was shielded by an opaque cap and only the lower part was exposed unilaterally, curvature generally did not result. From these and other empirical experiments, the Darwins concluded that "when seedlings are freely exposed to a lateral light, some influence is transmitted from the upper to the lower part, causing the latter to bend." Boysen-Jensen [10], working with *Avena* coleoptiles, concluded that "the transmission of the irritation is of a material nature produced by concentration changes in the coleoptile tip." Paál [11] corroborated his findings by demonstrating that "the stem tip is the seat of the growth regulating center. In it, a substance (or mixture) is formed and internally secreted, and this substance equally distributes on all sides, moving downwards through the living tissue." But it remained for Went [2] to make the definitive discovery of auxin and to determine it quantitatively by *Avena* curvature bioassay (Figure 1). The chemical isolation and characterization were, however, done by Kogl et al. [12]. The details of the development of the concept and discovery of indoleacetic acid (IAA), as outlined above, are described in two classical works [10,13].

It was not until 1946 that a good chemical identification of IAA was made in a higher plant [14]. IAA has come to be recognized as perhaps the only true auxin of plants and crops. This auxin has also been isolated from culture filterates of bacteria, fungi, and yeast plasmolysate, but its role in these organisms is less clear.

Besides IAA, plants contain three other compounds that are structurally similar and elicit many of the same responses as that of IAA: 4-chloroindoleacetic acid (CIIAA), phenylacetic acid (PAA), and indolebutyric acid (IBA). However, their physiological significance and transport properties remain obscure at present. Engvild [15] has advanced the idea of death hormones and suggested CIIAA as one of them. Four additional compounds—indoleacetaldehyde (IAALD), indoleacetamide (IAM), indoleacetonitrile (IAN), and indole ethanol—are also found in a range of plants, but they are readily converted to IAA in vivo. The enzymes aldehyde dehydrogenase and indoleacetamide hydrolase, which catalyze the conversion of IAALD and IAM, respectively, to IAA, are active in plant tissues in which workers have detected IAALD or IAM. Similarly, IAN, found in the Cruciferae and Gramineae families, is also accompanied by the enzyme nitrilase, which is involved in the conversion of IAN to IAA. These similar circumstances, in a range of plants, indicate that IAA is the true active free auxin in plants. Furthermore, free auxin forms are probably the most immediately utilizable by plants in their growth processes.

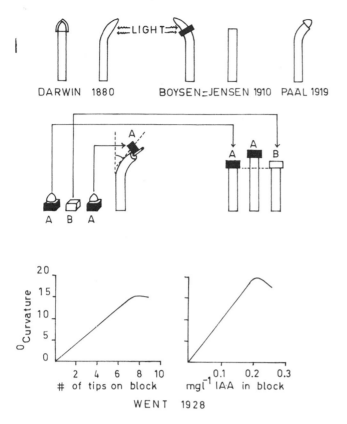

**Figure 1**  Diagrammatic representation of the major experiments leading to the discovery and quantification, by bioassay, of the auxin.

## A.  Chemical Nature

In 1882, Nencki and Sieber discovered that indole-3-acetic acid was a constituent of human urine, which in 1934 was confirmed by Kögl and his coworkers, with the additional information that it was active in promoting the growth of some plant tissues or organs [16]. Within a year, it was also isolated from yeast plasmolysate and from the culture filterates of *Rhizopus suinus* [10,13]. However, its first isolation from a crop plant [i.e., from immature maize (*Zea mays*) kernels] was made by Haagen-Smit et al. [14]. It is now commonly accepted that IAA is perhaps the only endogenous auxin in plants and crops. Interestingly, chemists were aware of IAA long before plant scientists became aware of it. IAA was first synthesized by a German chemist in 1904 [17], but it was not suspected to have biological activity.

There are many purely synthetic compounds that mimic physiological actions similar to that of IAA. They are chemically diverse but can be classified in five major categories: indole acids, naphthalene acids, chlorophenoxy acids, benzoic acid, and picolinic acid derivatives. Two compounds belonging to the first group, indolebutyric acid and indolepropionic acid, are not exclusively synthetic; they have also been reported to be present in some plant species. The well-known naphthaleneacetic acid and β-naphthoxyacetic acid belong to the second group. The best known among the chlorophenoxy acid group are 2,4-dichlorophenoxyacetic acid (2,4-D), 2,4,5-trichlorophenoxyacetic acid (2,4,5-T), and 2-methyl-4-chlorophenoxyacetic acid (MPCA), which are known to be very powerful defoliants and herbicides when used in higher concentrations. In the benzoic group, the common synthetic auxins are the 2,3,6- and 2,4,6-trichlorobenzoic acids and dicamba, which is a powerful herbicide and is effective in some species of deep-rooted perennials, which are not readily killed by 2,4-D. Among the picolinic acid series, the best known is picloram (4-amino-3,5,6-trichloropicolinic acid), which is known to be the most powerful selective herbicide. However, the trend in recent years is to use natural plant hormones to regulate crop growth for greater production. This is because they act in low concentration and are fully degraded, therefore do not pose environmental and/or ecological threats.

## B. Metabolism

The hormone IAA has been studied for more than six decades, yet it remains unclear how it is synthesized or degraded in plants. Because of the structural similarities, the amino acid tryptophan is commonly considered to be a precursor to IAA. To date, biosynthetic pathways from L-tryptophan by way of tryptamine [18], indole-3-pyruvate and indole-3-acetaldehyde [19], indole-3-aldoxine and indole-3-acetonitrile [20], and indole-3-acetamide [21] have been proposed. Nevertheless, some workers have reported that D-tryptophan may also be an effective precursor for IAA biosynthesis [7,21]. However, in *Lemna gibba*, D-tryptophan was not converted to IAA and also the rate of conversion from L-tryptophan was far lower than expected for a direct precursor [22]. However, it remains unclear which pathway does function in plants. Wright et al. [23], using the tryptophan auxotroph maize mutant *orange pericarp*, have questioned the idea that tryptophan is a precursor of auxin and suggest a nontryptophan pathway as a primary route of IAA biosynthesis. It is therefore possible that plants and crops use more than one route for in vivo IAA biosynthesis.

It is reasonable to assume that plants have mechanisms to regulate the levels of auxin to maintain balanced growth. This is done by controlling the rate of synthesis as well as by degradation or by forming conjugates (bound). The enzyme IAA oxidase with its several isoenzymes, which usually have the characteristics of peroxidases, is known to catalyze the reaction. Two pathways of degradation are known in many plants. The first involves oxidation by $O_2$, leading to loss of the carboxyl group as $CO_2$ and usually 3-methyleneoxyindole as a principal product. In the second pathway the carboxyl group of IAA remains intact, but carbon at the second position of the heterocyclic ring is oxidized to oxindole-3-acetic acid. In some species, however, carbons 2 and 3 are oxidized to form dioxindole-3-acetic acid [24]. Lee and Starratt [25] have shown that soybean (*Glycine max*) callus and hypocotyl tissues were capable of oxidizing [$^{14}$C]IAA via the carboxylative pathway to indole-3-methanol glucoside as a major product. However, details of these degradative pathways are still unclear. Synthetic auxins and IAA conjugates are also not destroyed by these enzymes.

In auxin conjugates the carboxyl group is covalently combined with other molecules in the cell to form derivatives that do not allow easy extraction. Many IAA conjugates are known, including the indole-3-acetylaspartic acid (IAAsp), indole-3-acetylglutamic acid (IAAGlu), and the esters IAA-*myo*-inositol (IAIns) and indole-3-acetylglucose (IAGlu). These conjugates, along with the free IAA, have also been found in IAA-overproducing transgenic and wild-type tobacco (*Nicotiana tabacum*) plants [26]. These conjugates are not active per se but on hydrolysis release free IAA.

## C. Transport

The integrity of the complicated structure of plants depends, to a great extent, on regulations that coordinate the various parts of the whole plant. Because production centers and action sites are often located at different places in the plant body, auxin transport takes place. Ever since Went first demonstrated the basipolar movement of auxin and its quantitative description by van der Weij [13], physiologists have been interested in studying various parameters of its transport as a way to understand the general phenomena of polarity in plant development. With the availability of high-specific-activity $^{14}$C-labeled auxin coupled with liquid scintillation spectrometry (>90% counting efficiency), it became possible to perform more complex experiments and to obtain more reliable data from a single segment than was feasible previously [27,28]. In dicots and monocots alike, auxin moves predominantly in the basipolar direction [28]. However, in the young vegetative *Coleus blumei* internode, the auxin applied moves with a 3:1 ratio in the basipetal to acropetal direction, but this changed to 1.3:1.0 when the plants flowered [27,29]. Using paper chromatography, evidence was obtained, for the first time, that the auxin collected apically was the auxin applied at the basal end [27–29]. In excised root segments the polar transport was in the acropetal direction, but it appears that in the apical segments of intact roots (with caps), auxin moves basipetally [30]. However, in both organs, polarity was maintained regardless of the tissue orientation.

The other characteristics of auxin transport are that it is metabolically dependent and moves basipolarly with a velocity generally ranging from 10 to 20 mm/hr, depending on the plant species tested. In short-term (4 hr or so) experiments, it moved unchanged with similar velocity in intact and isolated segments. It could also move laterally following tropic stimulation [27,31,32].

In intact plants, auxin moves in two distinctly different systems. From the apex it moves toward the roots, and movement from the young leaves and meristematic regions of shoots resembles polar transport. This transport requires living cells and is interferred with by inhibitors such as triiodobenzoic acid (TIBA) and abscisic acid (ABA), anoxia, and low temperature. Auxin was not usually translocated through the epidermis, cortex, pith, or vascular bundles but instead through parenchyma cells in contact with vascular bundles [33]. Cell-to-cell auxin transport across the tissues is now considered by some investigators to be chemiosmotically coupled to the electrochemical potential of auxin and proton [7]. The second mode of transport is along with the assimilates exported from the leaves. This movement lacks polarity and the auxin can move in any direction with a velocity of 100 to 240 mm/hr, depending on the location of the metabolic sink and the water status of plants. This indicates that auxin supplied to or from the mature leaves enters the sieve tubes and is transported rapidly with assimilates. The physiological importance of this system for delivering endogenous auxin over long distances has not been investigated.

## D. Biological Activity

Biological activities of the applied auxin are so diverse that compiling a complete list is quite difficult. A number of responses at the molecular, cellular, organ, and whole plant levels have been described which are known to be influenced by the exogenous application of IAA. But to what extent these are under the control of endogenous IAA has not been established unequivocally. However, there are a few examples, such as control of the elongation of stamen filament of *Gaillardia grandiflora* and the photoinhibition of mesocotyl growth, that correlate well with the endogenous IAA levels [32].

Auxin response is related to concentration, which is normally extremely low. In plants, free IAA is on the order of $10^{-8}$ g/kg fresh weight. The endogenous level of auxin is important in determining the course of development [34] (Figure 2). Changing concentration can convert root meristem to shoot meristem and vice versa [35]. A high concentration is inhibitory, while a low concentration is stimulatory, and both are important. Commonly, the highest concentration of auxin is found in the meristematic regions.

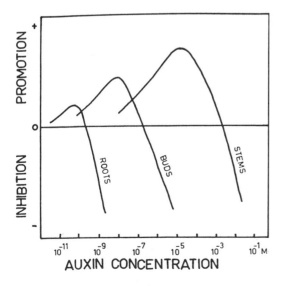

**Figure 2** Growth responses of plant organs to various concentrations of auxin. (From Ref. 34.)

## 1. Bioassay

Only a bioassay can detect the physiological activity of a substance at hormonal concentrations, and thus development of a quantitative bioassay provided the beginning for all hormonal work [13]. A number of bioassays have been developed to measure the activity of auxins. Well known among them are the (1) *Avena* coleoptile curvature test, (2) *Avena* coleoptile section test, (3) split pea (*Pisum sativum*) stem curvature test, and (4) cress root inhibition test. Modern instruments of separation and quantitation, such as high-performance liquid chromatography (HPLC) and gas chromatography coupled with mass spectrometry (GC/MS), are commonly used. Another extremely sensitive detection method is immunoassay (a type of bioassay), and commercial kits are available for determining picogram quantities of plant hormones [36].

## 2. Tropisms

In nature, the orientation of shoot and root is of crucial importance to seedlings developing from seeds oriented at all angles in the soil. For survival, shoots therefore need to be oriented toward the light so that photosynthesis can begin before the stored food reserves are depleted, and roots must be oriented toward the gravitational vector, to obtain water and ions and to secure anchorage and mechanical support. Shoots are thus considered to be positively phototropic, and roots, negatively phototropic. On the other hand, shoots are negatively gravitropic and roots positively gravitropic. These two responses are of great ecological importance and also have relevance to plant and crop productivity.

Tropisms (from the Greek word *trope*, "turn") have been divided into three phases: perception, transduction, and response. To explain the transduction and response phases, caused by photo or gravity stimulation, the Cholodny-Went theory states: "Growth curvatures, whether induced by internal or by external factors, are due to an unequal distribution of auxin between the two sides of the curving organ. In the tropisms induced by light or gravity the unequal auxin distribution is brought about by a transverse polarization of the cells, which results in lateral transport of the auxin."

Although the validity of the Cholodny-Went theory has been questioned without an alternative explanation [37], others have come to its defense [38,39]. Li et al. [40], using auxin-responsive messenger RNAs (mRNAs) called small auxin up RNAs (SAURs) as a molecular probe, have supported the idea of asymmetric distribution of auxin at the sites of action during tropistic response. However, Jaffe et al. [41], using the pea mutant 'Ageotropum', found that roots appear to be neither negatively phototropic nor positively gravitropic but grow in the direction of increasing soil moisture.

PHOTOTROPISM   When the exposure of light falling on plant organs becomes differential, a curvature develops, so that reorientation takes place in such a way that the organ is evenly illuminated. This response of the plant or its organ(s), where the plane of curvature is determined by the spatial relationship of the organ and the light stimulus, is known as *phototropism*.

The Cholodny-Went theory suggests that unilateral illumination causes auxin to move laterally to the darkened side, causing the organ to curve toward light. Bioassay and [14]C-labeled auxin studies using etiolated maize coleoptile tips have shown that there was no difference in auxin yield between evenly illuminated segments (first or second positive range) and their dark control segments. But unilateral illumination caused asymmetry in auxin yield between the two halves of the tip segments [42,43]. These observations demonstrated the consequence rather than the cause of the lateral asymmetry in auxin transport. In such a cause-and-effect relationship, it is important to differentiate between the two. Using [14C]IAA and maize coleoptiles, Naqvi and Gordon [44], studying the [14C]IAA transport kinetics of etiolated maize coleoptiles, provided the first evidence that bilateral illumination, in the first positive range, caused retardation of basipetal auxin transport intensity (capacity) without affecting the velocity. The lateral asymmetry observed was thus a consequence of the resultant concentration gradient. Based on their observations and other evidence, Naqvi and Engvild [31] proposed that "photolysis of a carotenoid (violaxanthin) produces compounds (similar to or identical with ABA) that inhibit the basipetal auxin transport. Unilateral stimulation produces an asymmetry of inhibition and, hence, the curvature." It is now known that in maize coleoptiles, ABA is the most dominant hormone after IAA [45]. Therefore, any change in its concentration would influence auxin transport. Further support for the lateral transport theory came from the observation that in bean (*Phaseolus vulgaris*), auxin transport inhibitor DPX-1840 inhibited basipetal as well as lateral transport, resulting in growth retardation and the loss of phototropic response [46]. These results suggest that basipetal and lateral transports are essential for photo-stimulated differential growth.

GRAVITROPISM   Frank [47] was perhaps the first to establish that curvatures due to gravitational stimulus were directly connected with growth and described this phenomenon by the term *geotropism*. The precise gravity-perceiving mechanism is still unknown. Since the root cap (terminal 500 μm) appears to be the seat of perception, it was thought for a long time that amyloplasts (organelles filled with starch grains) were, in fact, the gravity-perceiving mechanism. However, the use of a mutant lacking amyloplasts in the cap, yet exhibiting nearly normal gravitropic response, has ruled out this mechanism [48]. Whatever may be the mechanism of perception, the ultimate response is asymmetrical growth, resulting in a positive (root) or negative (shoot) curvature.

Strong evidence suggests that auxin controls the gravitropic response and the Cholodny-Went theory has been invoked to explain the growth asymmetry that results in the shoot bending away from, and the root toward, the gravity vector. It is well established that when shoot segments are placed horizontally, more auxin is recovered from the lower than from the upper half. Using bioassay and $^{14}$C-labeled auxin, differences between the upper and lower halves of the maize coleoptiles were demonstrated, but the activity on the upper halves did not differ from that of the vertical halves [38]. However, Naqvi and Gordon [49], working on $^{14}$C-labeled auxin transport kinetics, demonstrated that horizontal reorientation of maize coleoptiles reduced the transport intensity (capacity) of the upper half without affecting the velocity. Thus a lateral auxin concentration gradient enhanced movement from the upper toward the lower half. This lateral movement has also been demonstrated in the maize mutant amylomaiz [50]. McClure and Guilfoyle [51], using a molecular biology approach, have shown a clear correlation between auxin-controlled gene expression and the gravitropic response of soybean hypocotyls. Molecular genetic studies on the phenotype of auxin-resistant mutants have further substantiated that auxin played an important role in root gravitropism [52].

## 3.   *Apical Dominance*

The integrity of the complicated form of higher plants depends to a great extent on regulations that integrate the various component parts. This gives a characteristic form or shape that is repeatable in time and space. The stems assume characteristic geometry due largely to the extent of biochemical influence exerted by the apex on the development of lateral (axillary) bud meristems. This phenomenon of apical dominance (growth correlation or compensatory growth, i.e., preventing or slowing of lateral bud growth by the apex) is of major importance in integration of the plant body, and parallel examples are found in mosses and ferns [53]. Awareness of this role of the apex has undoubtedly influenced pruning practices in horticulture and crop production. Interested readers are referred to excellent reviews on the subject [54,55].

In a pioneering work using *Vicia faba*, Thimann and Skoog [56] demonstrated that auxin diffused in agar blocks from the excised shoot apices, *Rhizopus* filterate, or human urine can partially inhibit lateral bud growth. Later, Leopold [57] showed that when the shoot apex of 'Wintex' barley (*Hordeum vulgare*) was destroyed, tillering was profuse unless the apex destroyed was replaced by an auxin. These observations indicated that auxin from the apex exerts an influence on *Vicia* lateral bud growth as well as on barley tillering. Studies on inhibiting auxin emanating from an apical bud, using inhibitors such as TIBA or morphactin, have shown that the lateral bud growth was effectively enhanced. These studies thus provide evidence in support of a direct role, but others question it and assign an indirect role to auxin in controlling this phenomenon [54]. In intact tobacco, petunia, and *Arabidopsis thaliana* plants, use of transgene-mediated auxin and/or ethylene deficiencies, along with mutants insensitive to auxin or ethylene, supports the idea that apical dominance is the result of the auxin/cytokinin ratio rather than auxin-induced ethylene production [58].

## 4.   *Root Formation*

The most apparent auxin control of cell division is the formation of roots. Early evidence indicating the presence of active buds on cuttings to promote root development below led to the identification of auxin as the root-forming hormone [13]. The ease with which roots can form on cuttings varies enormously; shoot cuttings from some plants produce roots simply if their basal cut end is left in water, whereas other species do so only rarely. Root formation shows polarity and always occurs at the morphological basal end, even if the cuttings are inverted upside down. Because auxin moves basipolarly, it was logical to believe that root formation at the basal end is a consequence of the movement of auxin to the lower tissues. Removal of rich sources of auxin (i.e., buds and young leaves) reduces the number of lateral roots formed. This capacity is restored, however, if auxin is substituted for these organs. Tissue culture studies have provided further support by showing that a higher auxin/cytokinin ratio induces root formation and that changing it converts shoot meristem to root meristem and vice versa [35].

To preserve genetic homogeneity, vegetative propagation (cloning) is important in horticulture, floriculture, forestry, and in the conventional breeding and/or biotechnology of higher plants. Therefore, basal treatment with indolebutyric acid (IBA) and synthetic auxin naphthaleneacetic acid (NAA) is commonly used to induce adventitious root formation in hard-to-root species.

## III. GIBBERELLINS

The unique property of gibberellins (GAs)—that of increasing the growth of plants by greatly elongating the cells—was discovered by Kurosawa [3]. Studying the symptoms of the rice disease "bakanaebyo" ("foolish seedling disease"), Takahashi et al. [59] observed that the causal pathogen was a soilborne fungus, *G. fujikuroi*, the sexual or perfect stage of *Fusarium moniliforme*, which caused infected seedlings to grow abnormally taller and to fall over due to their spindly stem structure. They observed further that when a pure culture filterate was sprayed onto rice seedlings, it produced the same abnormal growth. This suggested that the abnormal growth of the infected seedlings was caused by a soluble substance(s) produced by the fungus. Other Japanese biologists showed that the excessive growth was not confined to rice but that the filterate could induce it in many other species. According to Takahashi et al. [59,60], in 1938 Yabuta and Surniki isolated two crystalline active substances from the culture filterates and called them gibberellin A and B.

Western scientists became interested in gibberellin research in early 1950 and succeeded in isolating an active principle from *G. fujikuroi*. The growth-promoting activity of this compound was similar to that of the GAs isolated by Japanese investigators, but the chemical nature was clearly different. Therefore, it was named gibberellic acid ($GA_3$) [59]. The concentration of GAs is usually highest in immature seeds, reaching up to 18 mg/kg fresh weight in *Phaseolus* species [61]. However, it decreases rapidly as the seeds mature. In general, roots contain higher amounts of GAs than the shoots, and vegetative tissues contain a comparatively low level of GAs, depending on the types of tissues and their stages of development.

### A. Chemical Nature

As of 1993, 84 gibberellins were listed, of which 25 are from fungi, 73 from higher plants, and 14 common to both [7,59]. Among these, 68 are free and 16 are known to occur in conjugated form [60]. All gibberellins are acidic diterpenoids having an *ent*-gibberellane carbon skeleton and are designated $GA_1$, $GA_2$, $GA_3$, . . ., $GA_{84}$. They differ from one another mainly in the numbers and positions of substituent groups on the ring system and in the degree of saturation in the A ring. Free GAs are divided into two groups: those possessing an *ent*-gibberellane skeleton (20 carbons) or *ent*-20 nongibberellane (19 carbons) mono-, di-, or tricarboxylic acids. The terms C-20 and C-19 denote compounds that have retained and lost, respectively, carbon atom 20, and generally, C-19 GAs are more active than C-20 GAs. They are grouped in either four- or five-ring systems. The fifth ring is the lactone ring attached to ring A, which is not present in the *ent*-gibberellane. The carboxyl group at C-7 seems to be essential for biological activity. They also seem to be rather stable in plants and are readily interconverted to form glycosides by conjugating with sugars.

### B. Metabolism

Gibberellins are diterpene, belonging to a large group of naturally occurring compounds in plants known as terpenoids. All terpenoids are basically built up from isoprene units, which are five-carbon (5C) compounds.

The linking of two units yields a monoterpene (C-10), of three a sesquiterpene (C-15), of four a diterpene (C-20).

Our knowledge regarding the biosynthesis of GA in plants and crops stems from feeding $^{14}$C-labeled acetic acid and mevalonate to *G. fujikuroi* through the culture medium. It was observed that $^{14}$C from these two compounds was incorporated into gibberellic acid ($GA_3$) [59]. Graebe et al. [62], using a cell-free system [the endosperm nucellus of wild cucumber (*Marah macrocarpus*; formerly *Echinocystis macrocarpa* Greene)], reported incorporation of [$^{14}$C]mevalonic acid into kaurene, kaurenol, and geranylgeraniol. The pathway commonly accepted is 3-acetyl-CoA → mevalonic acid → isopentenyl pyrophosphate (a five-carbon terpenoid) → geranylgeranyl pyrophosphate (a 20-carbon compound) → copalyl pyrophosphate → kaurene → kaurenol → kaurenal → kaurenoic acid → $GA_{12}$-aldehyde. The $GA_{12}$-aldehyde is a branch point to the formation of various GAs. Pathways to various GAs differ mainly in the position and sequence of hydroxylation, and more than one pathway can operate in the same plant. The details of the pathways are covered comprehensively in several excellent works [59,60].

The metabolism of GAs in plant tissue is not well understood, and very meager information exists regarding its eventual fate. There is evidence that considerable interconversion of gibberellins (i.e., one GA can be converted to another GA) takes place in the plant. Immature seeds from "summer"-grown *Pisum sativum* were fed with $GA_9$, which was metabolized to $GA_{51}$ and dihydro-$GA_{31}$ and its conjugate. But in "winter"-grown seeds, the metabolites were $GA_{20}$ and $GA_{51}$. Another metabolite, gibberellethione, was isolated from immature seeds of *Pharbitis nil* [60]. Degradation of commonly used $GA_3$ appears to be slow. However, during the active growth phase, most of the gibberellins are metabolized to inactive forms by hydroxylation or by conjugation with glucose to form glucosides.

## C.  Transport

Gibberellins are known to be synthesized in all young, actively growing organs, vegetative or reproductive, including immature and mature seeds. Understanding their transport within the plant pertains primarily to work with excised coleoptile, stem, or petiole segments in a donor-tissue-receiver system. Transport has generally been observed to be nonpolar, but occasionally, basipolar movement has been reported [28] with a velocity up to 1 mm/hr. However, information regarding endogenous movement is rather indirect. It has been noted to occur in the phloem by the same mechanism and in a pattern similar to that with which other assimilates move. Gibberellins have been isolated from phloem sieve tube saps as well as from the xylem stream. Experimental evidence using $^{14}$C-labeled GA shows an interchange between phloem and xylem [17]. This suggests that GA is transported both symplastically and apoplastically. Its phloem transport rate was similar to that of other assimilates. In analogy with the source-to-sink movement of assimilates in phloem, perhaps the polar movement observed was to a growth center rather than to the morphological base.

## D.  Biological Activity

The physiological properties of these highly active compounds are wide ranging, but extensive growth and de novo enzyme synthesis are the most significant. The GAs act synergistically with other hormones in what might be called a system approach. The best known response is the stimulation of internode growth of dwarf maize, pea, and bush bean, which after treatment with GA attains the normal height. In some cases, but not all, dwarfism does in fact seem to be correlated with endogenous GA deficiency. The most detailed analysis, at the molecular level, has been done with dwarf mutant of maize, known as dwarf-5 ($d_5$) [59,60]. The height of the dwarf-5 mutant is about one fifth that of its parent, due to a single gene mutation causing a deficiency of GA. When treated with GA, the mutant attains the height of normal maize. The action of many GAs is similar to that of IAA, including cell elongation, promotion of cambial activity, induction of parthenocarpy, and stimulation of nucleic acid and protein synthesis. The GAs vary greatly in their biological activity, and $GA_3$ and $GA_7$ are considered to have the widest range. In ferns, algae, and fungi, GAs have also been shown to influence growth and development.

The content of GAs varies depending on the types of tissues and their stages of growth. Tissues other than seeds usually contain very low amounts (e.g., 0.3 μg/kg in young bamboo shoots) [60]. Roots are considered to be the richest source of GAs, and most GAs, as such or in bound form, are supplied to the shoot.

### 1.  Bioassay

A number of bioassays, such as elongation of dwarf maize, pea, and rice seedlings and of lettuce (*Lactuca sativa*) hypocotyl, *Avena* leaf segments, and chlorophyll retention in *Rumex* leaf disks [63,64], have been used to determine gibberellin activity in plant extracts. These basic responses have generally been used in various modifications with different plant materials. These assays are convenient, specific, easy to use, and detect nanogram levels of gibberellins. Nishijima and Katsura [65] have improved the dwarf rice bioassay to detect picogram quantities of gibberellins. Among the commonly used bioassays, the barley endosperm assay [66] has generally been preferred. In this bioassay, sterile de-embryoed barley (cv. Himalaya) half-seed (endosperm) is incubated in $GA_3$ solutions for 24 to 48 hr. The presence of $GA_3$ stimulates $\alpha$-amylase synthesis in the aleurone layers (two to four layers of live but undividing cells), which breaks up the starch and builds up the reducing sugars. The amount of reducing sugar, analyzed colorimetrically, is dependent on the $GA_3$ concentration in the test solution.

## 2. Growth Promotion

The isolation of gibberellins from plants, combined with their physiological responses to applied GAs, suggests that they do play a role in the regulation of various phases of their development. At the same time, if a process fails to respond to a certain GA, it could not be used as evidence that GA is not required. It may be that a different GA is required to elicit the response. In *Silene*, $GA_3$ fails while $GA_7$ induces flowering under noninductive conditions, suggesting the involvement of gibberellin in this process. Several species of conifers show little or no elongation to $GA_3$ treatment, but they do respond to a mixture of $GA_4$ and $GA_7$ [67].

During the vegetative growth phase, mitotic activity in subapical meristem is regulated by gibberellins. A reduction in its level causes a severe imbalance between internode and leaf growth, resulting in a form of growth called a *rosette*, first noted in *Hyoscymus niger* and later in many other plants. In plants such as cabbage (*Brassica oleracea capitata*), leaf development is profuse and internode growth is retarded during the vegetative phase. But before the start of reproductive growth, a marked elongation of the internode, called *bolting*, takes place. When treated with $GA_3$ during their rosette phase, such plants bolt and flower, whereas nontreated plants remain rosetted. There is evidence that endogenous gibberellin levels are higher in the bolted plants than in the rosetted plants. In addition, higher concentrations have been found in the bolted long-day *Rudbeckia speciosa* and cold-requiring *Chrysanthimum morifolium* cv. Shuokan than in their nonbolted forms [63]. Thus it appears that the influence of gibberellin in such a response includes the stimulation of cell division as well as cell elongation.

For many crop species there are genetic dwarf mutants that are deficient in gibberellin. Dwarfs of rice, maize, and peas phenotypically attained the height of normal varieties when treated with gibberellin. These mutants have been used successfully for gibberellin bioassay and in breeding programs for increasing crop productivity. Dwarf rice responded to as little as 4 pg of $GA_3$ per plant [64]. Five different gibberellin-synthesis mutants are known which are underproducing dwarf mutants. Each mutant has a mutation on a different gene, and each gene controls a different enzyme needed for gibberellin synthesis. The work of MacMillan and Phinney [68] suggests that only $GA_1$ controls elongation in maize, and all five dwarf mutants lack the enzyme(s) that can convert other gibberellins to $GA_1$. Other evidence also indicates that $GA_1$ is the main gibberellin needed by dwarf rice, rape (*Brassica napus*), peas, sweetpeas, tomato (*Lycopersicon esculentum*), and some wheat (*Triticum aestivum*) cultivars for stem elongation. Mutants are not only lacking $GA_1$, but $GA_1$-overproducing mutants with abnormally long internodes have been reported in *Brassica rapa* (syn. *campestris*) [69].

## 3. Dormancy of Buds and Seeds

Buds and seeds of many plants show the ability to retain viability while having limited metabolic activity and no observable growth during an unfavorable season. This physiological condition is commonly known as dormancy, and plants that grow in regions with a pronounced season usually adopt this strategy in late summer or early fall. Buds in dormant conditions are relatively more cold and drought tolerant than are actively growing buds. Similarly, seeds of many noncrop plant species remain dormant when they mature and will not germinate even if favorable conditions are provided. Dormancy of buds and seeds must be broken at a time when conditions are suitable for their growth and germination, respectively, during the spring. Long-day or brief red-light exposures have been found to break seasonal dormancy in many species. Gibberellins have also been found effective in overcoming both kinds of dormancy in buds as well as in seeds. Treatment with gibberellins has been observed to substitute effectively for long-day, low-temperature, or red-light exposure requirements. Due to ease in handling, much more is known about seed dormancy, but it is likely that much of this information may also be applicable to buds.

### 4. Mobilization of Nutrients

Endosperm is the nutrient storage organ of the seed for the developing embryo. Soon after the axis becomes active and root and shoot develop, the nutrient reserve (minerals, fats, starch, and proteins) is mobilized to support the juvenile seedling growth. This supply continues until the root develops the ability to absorb nutrient ions from the soil and the shoot system begins the photosynthetic process. Large molecules of proteins, fats, and starch have problems being translocated; therefore, they need to be metabolized to smaller molecules such as amino acids or amides and sugars, which can readily be translocated from source to sink. Gibberellins are known to play a key role in the hydrolysis of starchy endosperm of cereal seeds. In 1960, Yomo in Japan and Paleg in Australia independently observed that $GA_3$ stimulated the degradation of de-embryoed barley endosperm [60]. A few years later it was demonstrated that barley embryo axis produces GAs, and it has further been shown by GC/MS as well as by immunochemical methods that grains also contain a large number of GAs [70]. The type of gibberellin is species specific, but $GA_1$ and $GA_3$ are important in barley. The involvement of a living aleurone layer in the degradation of starchy endosperm in grasses, including barley, was recognized more than a century ago, but the experimental evidence was not provided until much later.

When isolated aleurone layers are incubated with $GA_3$, a large number of enzymes (e.g., $\alpha$-amylase, protease, ribonuclease, esterase, $\beta$-1,3-glucanase, acid phosphatase, glucosidase, peroxidase) were secreted in the incubation media [70]. Later studies confirmed that $GA_3$ was required for the de novo synthesis of $\alpha$-amylase, $\beta$-1,3-glucanase, protease, and ribonuclease. Aleurone layers of barley, wheat, and wild oat respond to $GA_3$ by synthesizing hydrolytic enzymes, but most maize cultivars and some cultivated oat cultivars do not respond identically. Thus there is considerable genetic variability regarding the gibberellin response in cereal seeds. However, the role of gibberellins in the mobilization of food is not as clear in dicots and gymnosperms. The food reserves in these two classes of plants could be starch or fat, where added GAs may or may not influence degradation.

## IV. CYTOKININS

The discovery of cytokinins was an outgrowth of tissue culture research by Skoog and associates. The isolation and identification of kinetin (6-furfurylaminopurine) from aged or autoclaved herring sperm DNA, and its promotion of cytokinesis (cell division) at concentrations as low as 1 $\mu$g/L, greatly stimulated research in the field of plant growth and development. Although kinetin does not occur naturally, its discovery greatly supported the concept of the existence of a cell division factor, postulated by Wiesner in 1892.

Haberlandt is generally credited as the pioneer in providing experimental evidence for the hypothetical cell division factor. In 1913 he demonstrated that phloem diffusates could cause cell division in potato (*Solanum tuberosum*) parenchyma cells. Later, in 1921, he reported that cell division induced by wounding was prevented if the cut surface was washed and that leaf juice spread over the washed cut surface would restore it [71]. In the early 1940s, Van Overbeek [71a] observed that coconut milk could sustain the growth of isolated *Datura* embryo. According to Koshizima and Iwamura [72], subsequent work by Stewart and his associates established that coconut milk markedly stimulated the growth of carrot (*Daucus carota*) root explants by cell division. They did succeed in isolating the growth-inducing factor, but it was a mixture rather than a single compound that was effective in their carrot bioassay system. According to Koshizima and Iwamura [72], Skoog and associates, in the late 1940s, using aseptically isolated slabs of nondividing mature stem piths of tobacco plants (var. Wisconsin No. 38), observed cell division in the presence of vascular tissues. However, the first endogenous cytokinin was isolated from maize kernels and was named zeatin (Z) [6-(4-hydroxy-3-methyl-2-*trans*-butanylamino) purine]. Germinating seeds, roots, sap streams, developing fruits, and tumor tissues are rich in cytokinins [73]. There are 25 free cytokinins reported from higher plants [72] and some are active in causing maximum tobacco callus growth at concentrations as low as 0.004 $\mu$M.

## A. Chemical Nature

All the known endogenous cytokinins are substituted purines attached to the $N^6$ position of the adenine ring. These $N^6$-substituted adenines can be classified into two groups according to their carbon skeleton of $N^6$ substituents: as $N^6$-isoprenoid and $N^6$-benzyladenine analogues.

The major group of cytokinins are $N^6$-isoprenoid adenine analogues and can be divided into three groups. The first subgroup consists of zeatin and its derivatives—ribosides, glucosides, and nucleotides—whose $N^6$-isoprenoid side chain is either 4-hydroxy-3-methyl-2-*trans*-butenylaminopurine or its *cis* isomer. The second subgroup of $N^6$-isoprenoid analogues consists of dihydrozeatin (diH)Z and its ribosides and glucosides. The third subgroup includes 6-(3-methyl-2-butenylamino)purine and $N^6$-($\Delta^2$-isopentenyl)adenine (2iP) and its ring substitution products.

The second and minor group of cytokinins (i.e., $N^6$-benzyladenine analogues) was first synthesized as 6-benzylaminopurine (BAP) with high biological activity. Later it was found to exist in a number of plant species.

Certain nonpurine compounds, such as 8-azakinetin, benzimidazole, $N,N'$-diphenylurea, and 2-benzthiozolyloxyacetic acid, have also been reported to have cytokininlike activity. Of these, three are synthetic and only $N,N'$-diphenylurea occurs naturally in plants. It has been suggested by some workers that these so-called urea cytokinins may also be considered true cytokinins, but most hormone physiologists do not agree [72,73]. These compounds may not be active as such, but they may serve as precursors or inducers for the commonly accepted cytokinins.

## B. Metabolism

Miura and Miller [74] have suggested that all plant cells are capable of synthesizing cytokinins provided that the mechanisms to do so are "switched on." However, this does not mean that cytokinins are biosynthesized in the entire plant. Evidence suggests that actively dividing regions of plants are the sites of cytokinin biosynthesis. Because the root system possesses the most actively dividing regions, these regions are considered to be the major sites of cytokinin production.

Compared with other aspects of cytokinin physiology, little is known about their biosynthesis, which is comparatively quite complicated. The circumstances of its discovery and its effect on cell division and protein synthesis have somehow closely associated free cytokinins with RNA and DNA. The production in plants can be accounted for either by the turnover of cytokinin-containing transfer RNA (tRNA), by de novo biosynthesis, or by both mechanisms. It has also been reported to be present in ribosomal RNA (rRNA) [75].

The major cytokinin-active base in tRNA, [9R]iP, is formed by the condensation of adenine with an appropriate donor of the $N^6$ substituent during posttranscriptional processing. The $\Delta^2$-isopentenyl pyrophosphate (IPP) is the immediate precursor (donor) of the $\Delta^2$-isopentenyl side chain of $N^6$-($\Delta^2$-isopentenyl)adenosine in tRNA. A cell-free enzyme system, isopentenyl AMP synthase, has been isolated from cultured autotrophic tobacco tissue which forms cytokinin from adenosine monophosphate (AMP) and IPP as substrate [76]. The $\Delta^2$-isopentenyl pyrophosphate is a product of mevalonic acid (MVA) (an important precursor of carotenoids, abscisic acid, gibberellins, sterols, and other isoprenoid compounds) via $\Delta^3$-IPP.

Besides the formation of free cytokinins from tRNA, there is strong evidence that they are also formed by de novo biosynthesis. Beutelmann [77] supplied labeled adenine to moss callus cells and obtained labeled cytokinin that cochromatographed with 2iP, but no labeled cytokinin was detectable from tRNA. Similar results were obtained from the cytokinin-autotroph tobacco callus tissues, *Vinca rosea* crown gall tissues, and synchronously dividing tobacco callus cells [78].

The amount of cytokinin present in the tissue is regulated by conversion to a diversity of metabolites by the following reactions: (1) trans-hydroxylation of the terminal methyl group on the side chain, (2) side-chain reduction, (3) isoprenoid side-chain cleavage, (4) O-glucosylation, (5) N-glucosylation, (6) ring substitution by alanine moiety, and (7) base-ribonucleoside-ribonucleotide interconversion. These types of reactions have been observed in a number of plant species as well as in crown gall tissues. These reactions have also been obtained from tissues exogenously supplied with the hormone. Three enzymes, cytokinin oxidase [molecular weight (MW) 88,000], cytokinin 7-glucosyltransferase (MW 46,500), and (9-cytokinin)alanine synthase (MW 64,500), have been purified and characterized [78]. The free base, nucleotide, and nucleoside forms of cytokinins appear to be easily interconverted in plant tissues. Incorporation of labeled cytokinin bases into ribosides (ribonucleoside) and ribotides (riboside 5'-phosphates) have been observed in a number of plant species. Five enzyme systems, purified from wheat germ, may be responsible for this interconversion. These are (1) adenosine phosphorylase, (2) adenosine kinase, (3) adenine phosphoribosyltransferase, (4) (5'-ribonucleotide phosphohydrolase) 5'-nucleotidase, and (5) adenosine ribohydrolase (adenosine nucleosidase) [72]. The ribosides are considered to be the transloca-tion form, while ribotides are associated with uptake and transport across the cell membrane forms of cy-tokinins. Thus, enzymatic regulation of bases, ribosides, and ribotides plays an important role in keeping adequate levels of free and active forms of cytokinins in plants and crops.

## C.  Transport

It is paradoxical that plant parts that are meristematic or that otherwise have growth potential (young leaves, buds and internodes, and developing fruits and seeds) are known to be the primary center of pro-duction as well as the main sink for the endogenous cytokinins or its metabolites.

The detection of cytokinin, from the xylem exudate and phloem sap of a large number of plants and crops, clearly indicates that nonliving as well as living tissues are involved in translocation. The polarity of cytokinin movement is acropetal in the xylem, whereas it moves bidirectionally in the phloem. Thus, in phloem, cytokinins move not only from organ to organ in the aerial portion but also from shoot to root and vice versa [79]. Thus, the velocity of its movement is the same as that of other assimilates. However, under in vitro conditions (donor-tissue-receiver system), 6-benzylaminopurine (BAP) is very poorly translocated [80].

## D.  Biological Response

Cytokinins are known to evoke a diversity of responses when applied exogenously to whole plants, plant tissues, or plant organs. Like other hormones, cytokinins have been used as a tool to investigate their role as endogenous controllers of plant growth and development. Like other hormones, cytokinins influence a multitude of morphological and physiological processes, among them seed germination, cell division and cell elongation, promotion of cotyledonary and leaf growth, control of apical dominance, delayed senescence, and morphology of cultured tissues.

As pointed out earlier, deficiency of a hormone must exist either experimentally or genetically to show that adding hormone has an effect. Because cytokinins occur in all meristematic as well as in po-tential growing tissues and organs, it is not possible to create experimental deficiency. Genetically engi-neered cytokinin-overproducing tobacco and *Arabidopsis* plants have been used to study the phenomena of apical dominance [81]. But in the absence of comparison with cytokinin-deficient mutants, the con-clusion is equivocal. Thus, clear evidence is yet lacking which shows that specific physiological processes in plants and crops are under the control of endogenous cytokinins.

### 1.  *Bioassay*

The physicochemical methods for isolation, purification, and identification of endogenous cytokinins have improved rapidly, but it appears that bioassays will always be an integral part of the identification process. Therefore, a number of bioassays have been developed and used by various investigators to test the bioactivities of endogenous as well as synthetic cytokinins. They include (1) lettuce seed germination, (2) radish (*Raphanus sativus*) leaf disk expansion, (3) *Xanthium* leaf disk (chlorophyll preservation), (4) soybean callus, (5) carrot phloem, (6) tobacco pith callus, (7) cucumber (*Cucumis sativus*) cotyledon greening, (8) radish cotyledon expansion, (9) *Amaranthus* betacyanin, (10) barley leaf senescence, and (11) oat leaf senescence [82,83].

### 2.  *Germination*

Seeds that require preexposure to light for germination are called photodormant. Red-light exposure stim-ulates, and far-red exposure inhibits, germination of a number of species, including lettuce (cv. Grand Rapids) seeds. But the seeds of most crops do not require light because of natural selection against such a requirement. Cytokinin-imbibed seeds germinate better in dark than do unimbibed lettuce seeds. Simi-larly, cytokinin together with gibberellin effectively breaks the photodormancy of celery (*Apium grave-olens*) seeds, but it was not as effective alone [84]. This indicates that red-light exposure may cause en-hancement in the hormone level either by biosynthesis or by release from a bound form. However, such information is lacking because the level of hormones in the radicle or hypocotyl cells, responsible for germination, has not yet been determined.

### 3.  Organ Development

The totipotency of plant cells was demonstrated in a classical paper by Skoog and Miller [85] showing that a balance between cytokinin and auxin controlled bud and root formation in tobacco pith explant. At high concentrations of both hormones, cells often grow amorphously without differentiation. But a high cytokinin/auxin ratio causes induction of shoots, whereas a high auxin/cytokinin ratio enhances root formation [35].

Lateral development of axillary buds is inhibited by the presence of apical bud. This was shown by excision of the apical bud to remove correlative inhibition. But in the presence of apical bud, soaking the entire shoot in cytokinin enhanced lateral bud growth to a large extent [86]. Treatment with hadacidin, an inhibitor of purine synthesis, inhibited axillary bud growth following decapitation [87]. Because adenine treatment could not reverse this inhibition, the inhibitor may not be specific. In the absence of cytokinin-deficient mutants in higher plants, there is only indirect evidence that an endogenous cytokinin level regulated the development of axillary buds. Medford et al. [81], using genetically engineered cytokinin-over-producing tobacco and *Arabidopsis* plants, observed that the most significant morphological change of high cytokinin levels was that it caused extensive growth of the axillary buds (Figure 3). Thus, there is strong evidence that cytokinin and auxin balance is important in organ differentiation and its further development.

### 4.  Delayed Senescence

In plants and crops, the process of senescence is encountered at all stages of their life cycle. When a functional mature leaf is excised from the main body of a plant, it switches on to its death program. Progressive degradation of RNA, proteins, lipids, and chloroplast leading to the loss of chlorophylls starts when the leaf dies. Once started, the degradation of cell constituents continues even if the cut end is dipped in mineral salts solution. This process of senescence (i.e., breakdown of cell constituents and yellowing of leaf), leading to ultimate death, is accelerated further if the leaf is kept in darkness. In many plant species, adventitious roots are formed at the cut end of the petiole, which decelerates the degradative process of the metabolites in the leaf blades. Because the supply of mineral salts did not influence the degradative process, roots being the major source of cytokinin supply [79], this hormone may have been responsible for delaying the process. But different species show a diversity of response to cytokinins, auxins, or gibberellins, in terms of loss of chlorophyll and protein, in experimental systems using detached leaves or leaf disks [88]. However, two lines of evidence suggest that cytokinins may play an important role in de-

A                                                                              B

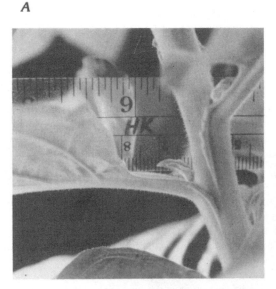

**Figure 3**   Axillary bud growth (twelfth node from apex) in (A) wild-type and (B) transgenic cytokinin-over-producing tobacco plants. (From Ref. 81; with permission from copyright owner.)

laying senescence. The cytokinin contents of rooted leaf blades rise substantially and cytokinins can partially replace these roots [7]. Thimann [89] observed that when cut leaves of many species, including oats, were floated in cytokinin solution in the dark, the light requirement for delaying senescence was effectively replaced. He suggested that treatment with cytokinins maintained the integrity of the cell membrane. Studies have also shown that cytokinin, auxin, and/or ABA influence stomatal movement [6] and that its closure accelerates senescence [89].

Cytokinins are also known to act as a sink for the transport of solutes from older to younger part(s) of a plant. Leopold and Kawase [90] demonstrated this very clearly. They painted the primary (oldest) leaves of a bean plant with benzyladenine at 4-day intervals. These leaves started senescing as soon as the trifoliate leaves above expanded and thus died off first. However, the treated leaves in their experiment lived longer than the untreated first trifoliate leaves because cytokinins do not readily move except in the xylem stream. Many variations in the experimental design and test species have indicated that they act as a sink for solutes in the potential or actively growing parts of the plant [7].

## V. ETHYLENE

The effect of ethylene on plant growth was noted as early as 1858 by the behavior of plants exposed to illuminating gas [91]. Nevertheless, the Russian scientist Neljubow [92] is credited with having identified the active growth-regulating component of the illuminating gas as ethylene. In the presence of ethylene, etiolated pea plants exhibit inhibition of elongation, an increase in diameter, and horizontal growth of shoots. In the literature, these three responses are known as the *triple response* and are still sometimes used to identify and measure ethylene response. However, Cousins [93] was the first to observe that gases released from oranges caused premature ripening of banana. But it was not until 1934 that Gane [94] provided evidence that ethylene was produced autocatalytically by ripening fruits.

### A. Chemical Nature

Ethylene is the simplest organic compound.

Its structural simplicity and the fact that it is gaseous in nature make it a unique plant hormone. It is a symmetric molecule having one double bond; the biological activity seems to be related to its unsaturated bond, which is attached to a terminal carbon atom.

### B. Metabolism

The task of unraveling the biosynthesis of ethylene was not an easy one. Feeding experiments using various radioactive materials with ethylene-producing plant tissues were unsuccessful in identification of the pathway. However, based on model nonenzymatic system, Lieberman and Mapson [95] proposed methionine as the precursor. Subsequently, Lieberman et al. [96] demonstrated the in vivo conversion of [$^{14}$C]methionine to [$^{14}$C]ethylene in apple (*Malus domestica*) tissues. Studies with [$^{14}$C]methionine have shown that the C-1 atom is converted to $CO_2$, C-2 to formic acid, and C-3 and C-4 to ethylene and the sulfur atom is retained in the tissue. Because the ethylene production system is extremely labile and is completely lost by tissue disruption, the characterization has been made at the living tissue level (Figure 4) [97]. In these studies, climactaric fruit slices or plugs and auxin-treated stem segments of etiolated pea and mungbean (*Phaseolus aureus*) seedlings have been used extensively [98].

Earlier studies on the metabolism of ethylene, conducted with improper precautions, had led to the conclusion that the compound was metabolically inert. However, Beyer [99], employing proper precautionary measures, convincingly demonstrated that ethylene was metabolized by plants, and the metabolic products of the dark-grown aseptic pea seedlings were identified as $CO_2$ and ethylene oxide. In addition to these two gaseous metabolites, ethylene was metabolized to a number of nonvolatile soluble products,

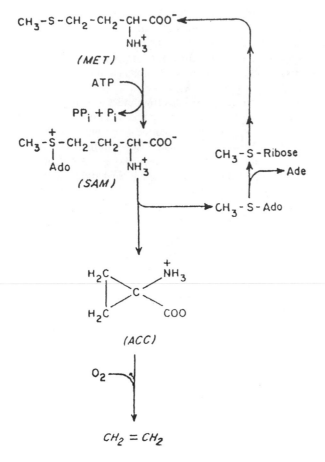

**Figure 4**   Pathway of ethylene biosynthesis. (Modified from Ref. 97.)

including free ethylene glycol and its glucose conjugate, plus oxalate and a number of unidentified products [98]. However, metabolism to volatile products and metabolism of nonvolatile products are independent of each other. Pea, tomato, cotton, carnation, and morning glory tissues have also been observed to metabolize ethylene [98]. Studies based on nonbiological systems suggest that copper ($Cu^+$) is involved in ethylene oxidation.

## C.  Biological Response

Ethylene elicits biologically spectacular responses at very low concentrations. As a plant hormone, it is unique in its structural simplicity and in being gaseous in nature. Whether the term *hormone* should be applied to ethylene, in that its *translocation* in the gas phase seems nonspecific, is a subject of debate. But being moderately water soluble, it moves rapidly between tissues, with minimum hindrance, in either the gaseous or the liquid phase. Therefore, there can be no doubt that it is a natural mobile growth regulator. Unlike other hormones, it is not transported directionally but accomplishes its integrative function by diffusing rapidly through the tissues.

At physiological concentrations, ethylene inhibits stem and root extension growth, but there are instances where it increases the growth rate in *Callitriche platycarpa* stem, in *Helianthus* petiole, and in rice stems and roots [100]. The myriad of plant responses and functions, including seed germination, cell division, epicotyl curvature, seedling growth, flowering, fruit ripening, response to stress, and senescence, are known to be influenced by ethylene. The diversity of the processes in a wide variety of plants makes it difficult to assign the hormone a definitive role.

Its production is regulated by a number of developmental and environmental factors. Ethylene production is induced at germination, ripening of fruits, and senescence (abscission) by auxins and by wounding and other chemical stress. It is also produced autocatalytically in the climactaric fruits. Many of the effects of IAA, such as apical dominance and stomatal movement, are attributed to IAA-induced ethylene production [58,101]. However, evidence with transgene-mediated auxin/or ethylene deficiencies and mutants insensitive to either of the hormones has ruled out the notion that auxin-induced ethylene was involved in the inhibitory influence that the apical bud exerts on the growth of lateral buds [58].

## 1. Emergence and Seedling Growth

A seed is considered to have germinated when its radicle emerges through the outer covering. Cell division and elongation start at about this time, and to most seed scientists it is the completion of germination. But from the crop physiologists' point of view it is extended to include the processes that ensure seedling establishment. Several fungal species and some bacteria are known to produce ethylene, including those growing in the soil [98]. The ethylene released by these soilborne microorganisms is suspected to influence seed germination, retard soilborne diseases, and regulate seedling growth. Thus, these stages can be separated in two phases: radicle protrusion (phase I) and the subsequent events related to seedling growth, dependent on seed reserves (phase II).

Two types of germination have been recorded in plants and crops: *epigean*, whereby the cotyledons emerge above ground (*Phaseolus vulgaris*; garden bean), and *hypogean* (*Pisum sativum*; garden pea), where the cotyledon remains below the soil surface. In the former case, the cotyledons emerge above ground with the growing tip, due to the elongation of hypocotyl forming a hook or arch. With the emergence of hypocotyl hook and exposure to light, symmetrical growth takes place and the hook straightens up. In the second case, the plumule is arched or recurved near the apex, to protect the shoot tip, and when it is pushed through the soil, exposure to light causes the epicotyl to straighten up. As it turned out, localized production of ethylene, at a rate of about 6 μL/kg per hour (etiolated pea seedlings), was found responsible for the formation and maintenance of the hook [17]. As the etiolated seedling emerges from the soil or is exposed to white light, a transient decrease in ethylene is observed with concomitant straightening of the shoot [17]. A similar explanation applies to seedlings that show *epigean* germination (garden bean) and develop a hypocotyl hook. It has further been observed that green tissues of seedlings are not as sensitive to ethylene as are etiolated tissues. Ethylene responses of emerging dicot seedlings have a survival value for the crop. Shortly after germination the hook is formed, in response to endogenous ethylene, which helps the cotyledons or young leaves to emerge safely out of the normal soil. Under compacted soils the hook and the primary root become unusually thick (i.e., they grow in diameter). This response is elicited by the organs, perhaps due to enhanced ethylene synthesis by imposed mechanical stress.

However, plants growing with their roots and stems submerged respond to ethylene by eliciting enhanced growth. This occurs due to accumulation of ethylene because of slower diffusion of the gas out of the tissue and through the water. Among the species are *Regnellidium diphyllum* (water fern), *Nymphoides peltata*, *Rananculus sceleratus*, and *Callitriche platycarpa* (star wort), which experience submergence at least part of the time during their growth. During submergence the stems elongate rapidly to keep leaves and upper stem parts buoyant. Submergence causes ethylene accumulation, which causes stems or petioles to grow rapidly [102,103]. Deepwater rice exhibits similar phenomena, and internode lengths of up to 0.6 m have been recorded and the plant completes its life cycle in several meters of water [102]. These contrasting responses support the notion that depending on the condition, similar cells respond differently to the same hormone [104].

## 2. Stress Ethylene

A low level of ethylene is produced by plant species, but when tissues are injured by a variety of stresses (wounding, pathogens, mechanical, chemical, temperature extremes, etc.), ethylene production increases severalfold. This enhanced production, frequently referred to as wound- or stress-induced ethylene, injures not only the tissues but also the site of ethylene production [105]. The intensity of enhancement in stress ethylene has been related to gamma radiation dosage, and it has also been suggested as a rapid assay method for bacterial toxins [106]. The enhancement in its production has also been observed to correlate with the number and size of foliar lesions induced by tobacco mosaic virus [106]. Visual injury was suggested as the most sensitive and rapid technique to evaluate the plant response to acute pollution abuses [107]. Although visual evaluation is rapid, it is also subjective. Measurement of pollution-induced ethylene surge before any visual symptom appears points to its sensitivity and superiority for physico-chemical determinations. However, the ethylene production surge is not long-lasting; rather, it is a short-lived phenomenon. Therefore, it has been suggested by Craker [108] that it acts as a trigger mechanism that initiates the biochemical change(s) expressing the response.

Auxin at supraoptimal concentrations ($10^{-5}$ to $10^{-3}$ M) acts as a natural factor in enhancing ethylene production, and depending on species and severity of stress, its biosynthesis starts after a lag of 30 to 60 min, can continue up to 48 hr, and then declines to normal level. The mechanism of IAA-induced ethylene production has been studied in pea and mungbean seedlings. The detailed studies have shown that IAA stimulates ethylene production by enhancing conversion of SAM (*S*-adenosylmethionine) to ACC (1-aminocyclopropane-1-carboxylic acid) through its effect on the enzyme ACC synthase [109]. Thus at supraoptimal concentrations, auxin per se does not cause growth inhibition, but it is the induction of enhanced ethylene production that inhibits growth [91].

A number of developmental processes have been listed by Abeles [110], where auxin-induced ethylene synthesis is considered to mediate auxin action. Therefore, it seems reasonable to assume that like other stresses, auxin acts similarly in accelerating ethylene production.

## 3. Senescence

It can be considered that programmed changes in the metabolic processes may ultimately lead to the death of a tissue, organ, or the whole plant. In nature, we experience three categories of senescence: sequential, where the oldest leaves senesce first; synchronous, where all the leaves senesce simultaneously (as in deciduous trees); and senescence of the whole plant after the completion of seed production (as in monocarpic crops). Simons [111] believes it is likely that the various types of senescence may result from different control mechanisms in the leaves. Studies have commonly been conducted with detached leaves, where experimental conditions may influence the result. Thus, it is very difficult to relate results from a single experimental system to the system operating in situ. Two major biochemical events, extensive proteolysis and chlorophyll loss, have been observed consistently at the beginning of the process. Leaf senescence may be induced or accelerated by a number of environmental factors, including competition for space, light, and nutrients; pollution; biotic or abiotic stresses; or it may be genetically programmed. At the cellular level it does seem to be controlled by endogenous growth regulators. It is now known that in most cases, auxin, cytokinin, ethylene, and ABA play a role in the regulation of senescence in plants and cytokinins, and auxin can delay senescence in a number of plant species. Thus, according to their actions, they have been classified as senescence promoters and retardants [112]. Ethylene plays an important role in accelerating leaf, petals, and fruit senescence, and auxin and cytokinins act as retardants. It is now well accepted that a balance between auxin and ethylene is a crucial factor in the retention or nonretention of leaves and/or fruits. Premature fruit drop is common in a number of important fruit trees, such as apple and mango (*Mangifera indica*). In mango, fruit drop occurs at all stages of its development but is extensive (>90%) during the first 2 to 4 weeks after fertilization [113]. This stage coincides with the maximum ethylene production by the fruitlet pericarp [114]. Therefore, treatments with $Ag^+$, $Co^{2+}$, or synthetic auxin NAA (naphthaleneacetic acid), which regulates ethylene balance, have been observed to enhance the number of harvested fruits significantly [113,115]. In some studies, salicylic acid (an inhibitor of ethylene biosynthesis) has also been observed to reduce mango fruit drop (S. S. M. Naqvi, unpublished results).

## VI. ABSCISIC ACID

The pioneering work pointing to the possibility that plant growth and development are regulated by levels of both promoter (auxin) and inhibitor is generally credited to Hemberg. Using *Avena* bioassay, he observed that potato peels contained high levels of growth inhibitors [116]. In the same year, he demonstrated further the presence of a similar inhibitor that could be correlated with the levels and degree of ash (*Fraxinus excelsior*) bud dormancy [117].

Employing paper chromatography to analyze plant growth substances from plant extracts, Bennet-Clark et al. [118] observed growth inhibitory activity at $R_f$ 0.6 to 0.7. This was later shown to be present in a number of plant species and the levels responded to the changes in environmental conditions. As pointed out by Hemberg [116,117], these results supplemented the physiological importance of the growth inhibitor, which was named "inhibitor β" [119,120].

Independent investigations of two different physiological phenomena in two different laboratories across the Atlantic Ocean led to the identification and discovery of abscisic acid (ABA) as the causal agent. Wareing and his associates in Wales worked for two decades or more on the seasonal changes in bud dormancy in woody plants, particularly of sycamore (*Acer pseudoplatanus*), and identified a compound that was named *dormin* [121]. Concurrently, a team led by Carns and Addicott in California, working on natural control of abscission in cotton, identified two compounds, which they named abscisin I and abscisin II [122,123]. However, by 1965 these two independent but diverse paths converged on the discovery that ABA was the hormone involved in both phenomena [124]. Like other hormones, abscisic acid is also ubiquitous among vascular plants and has been found to occur in some mosses, algae, and fungi.

## A. Chemical Nature

The naturally occurring enantiomorph is (S)-ABA, which is a sesquiterpenoid (a 15-carbon compound) and by its biogenesis is related to monoterpenes, diterpenes (gibberellins), carotenoids, and triterpenes. Endogenous (S)-ABA is optically active, having one center of asymmetry at C-1′, while synthetic ABA is racemic and composed of equal amounts of (S)- and (R)-enantiomers. The synthetic (R)-ABA accounts for 50% of the racemic mixtures of ABA and has biological activity equal to that of the natural (S)-ABA (Figure 5) in most cases, except in stomatal closure, where it is inactive. Because the catabolism of the (S)- and (R)-enantiomers is different, it is necessary to identify which compound is being used (R, S, or RS). In such a situation, care must be taken to use only natural (S)-ABA for metabolic studies.

## B. Metabolism

The typical sesquiterpene nature of ABA indicates that its endogenous synthesis is through mevalonic acid (MVA) as a precursor. Two pathways for its biosynthesis have been suggested. First is via farnesyl pyrophosphate, from which GAs are also derived. Through this pathway, MVA is converted to mevalonate 5-phosphate $\rightarrow$ mevalonate-5-pyrophosphate $\rightarrow$ $\Delta^3$-isopentenyl pyrophosphate (IPP). This compound is converted either directly to geranyl pyrophosphate or through 3,3-dimethylallyl pyrophosphate (DMAPP) to geranyl pyrophosphate $\rightarrow$ farnesyl pyrophosphate and finally, to ABA. However, use of radioactive MVA has yielded low amounts of ABA in only a few systems [125]. The second pathway is known to occur through the degradation of certain (40-carbon) carotenoids. Although this pathway is indirect, it seems to produce major amounts of ABA via ABA-aldehyde in perhaps all plants [126,127]. Zeevaart et al. [128], using various tissues incubated in an atmosphere containing $^{18}O_2$, have demonstrated that xanthophylls rather than farnesyl pyrophosphate are the precursors of ABA. In the xanthophyll cycle, 9′-*cis*-neoxanthin is converted to xanthoxin $\rightarrow$ ABA-aldehyde, which is finally oxidized to ABA.

Abscisic acid is catabolized to more polar compounds by conjugation, oxidation, hydroxylation, or isomerization. However, it seems that each plant species has its own system to regulate its free ABA level. This regulation is further dependent on the kind of organ tested as well as the physiological state. This regulation of ABA may operate through conjugation with sugar(s) to form glucoside or glycosyl ester or an acylated form. It can also be inactivated by oxidation to more polar free acids such as phaseic and dihydrophaseic acids. Both of these metabolites possess low or no growth-regulating activities and are derived via 6′-hydroxymethyl ABA.

## C. Transport

Abscisic acid is known to be translocated through the xylem and phloem to the actively growing regions (apical buds, root tips) and also in the parenchyma outside the vascular tissues. Thus, ABA is not translocated polarly; instead, it moves bidirectionally short as well as long distances [129].

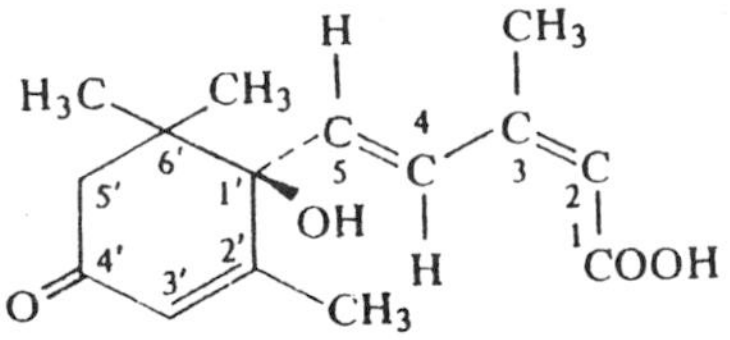

**Figure 5** Structure of natural abscisic acid.

## D.   Biological Activity

It is reasonable to assume that if growth and developmental processes proceeded uncontrolled, the result could be a distinct disadvantage. Plants have localized areas of cell division, the meristem, which is usually located at the tips of the growing organ(s). There are obvious structural limitations to an unlimited elongation of stem internodes, leaves, or in fact nearly any plant organ. Thus, the programmed plasticity exhibited by growth and developmental processes is certainly advantageous for plant survival, as shown by their ability to become dormant or otherwise restrain their growth or reproductive activities to match the alterations in their external environment. Plants do this by producing ABA to adjust their shoot and root growth accordingly. Like other hormones, abscisic acid also has multiple physiological effects in influencing plant growth and development. The concentrations of ABA varies widely, from 3 to 5 μg/kg in aquatic plants to 10 μg/kg in avocado (*Persea americana*) fruit mesocarp, and in the leaves of temperate crop plants it is usually between 50 and 500 μg/kg [130].

### 1.   Bioassay

The main impact of bioassays has been to study the growth-regulating properties of endogenous substances and aid in their isolation and identification in a pure form by existing physicochemical methods. Despite the superiority of the physicochemical methodologies, the sheer simplicity of the bioassay is likely to continue indefinitely to be used as an analytical tool. An enzyme-amplified immunoassay, with a sensitivity of 0.05 to 2.5 pg of ABA, has been developed to measure the ABA content of mesophyll and/or guard cells [131]. The majority of the bioassays have exploited the growth-inhibiting properties of the ABA, and such diverse materials as *Lemna*, oat first internode or wheat coleoptile, lettuce and cucumber hypocotyls, and rice seedlings were used. Stomatal closure response of *Commelina communis* and barley and inhibition of hydrolases in barley aleurone layers have also been used effectively to detect the presence or absence of ABA [132,133]. Depending on the test employed, these bioassays have been able to detect ABA levels ranging from $10^{-6}$ to $10^{-11}$ M.

### 2.   Growth

Abscisic acid was the first inhibitory hormone known to be involved in the regulation of growth along with growth promotors. At a concentration of $10^{-7}$ to $10^{-5}$ M it inhibits the growth of wheat coleoptiles, barley shoots, bean axes, and the second leaf sheaths of rice seedlings [134]. It is generally accepted that abscisic acid inhibits shoot growth, but its effect on root growth is contradictory and needs satisfactory resolution. Exogenous ABA has been observed to promote as well as inhibit root growth [135,136], and endogenous ABA has been shown to elicit similar responses. Under water stress conditions, the endogenous ABA level increases manyfold, which in turn has been implicated to affect shoot and root growth differentially (i.e., reducing shoot while maintaining root growth) [137].

Mulkey et al. [135] observed a triphasic response to ABA treatment: a period of promotion lasting 12 hr, followed by a similar period of inhibition (12 hr) and gradual recovery to about 80% of the normal growth rate after 24 hr Robertson et al. [138] dried sunflower roots to enhance the endogenous ABA level or treated them with exogenous ABA and observed a similar triphasic response. The initial transitory increase in length was related to the initial rise in water potential in the root apices, followed by an inhibition and then a partial recovery in root elongation. On the other hand, Saab et al. [137], using fluridone (an inhibitor of carotenoid biosynthesis) and a mutant deficient in carotenoid biosynthesis (*vp*5) to reduce the endogenous ABA level in maize seedlings, concluded that ABA played a direct role in the inhibition of shoot growth and in the maintainance of root elongation. However, Creelman et al. [139] concluded that under water-deficit conditions, at all the internal ABA concentrations tested, root growth was inhibited less than hypocotyl growth. But Plaut and Carmi [140] attributed the root response to the hydrotropic nature of the organ, which induces it to reach to the wet soil rather than to any other factor(s).

Abscisic acid was also projected as causing differential growth in gravitropic responses of roots. But later work using norflurazon or fluridone (inhibitors of carotenoid biosynthesis) and viviparous maize mutant *vp*-9 (lacking ABA biosynthesis) showed that a drastic reduction in endogenous ABA level did not alter root gravitropic response [141,142].

### 3.   Dormancy

Dormancy can be considered as the ability to retain viability while having minimal metabolic activity and no visible growth. Plants and crops have evolved this strategy as a mechanism of survival to cope with the pronounced seasonal changes unfavorable for their normal growth and development. Walton [143] concluded that "a role for ABA on the induction of bud and seed dormancy has been neither unequivocally demonstrated nor disproven." This is still valid with regard to bud dormancy in woody species. A seasonal change in the ABA content of leaves, stem apices, and xylem sap of *Salix viminalis* [144] and in the buds and stems of *Acer saccharum* [145] was observed. But these workers concluded that ABA did not play a role in the photoperiodic control of bud dormancy. These works have received further support from the evidence that cessation of seedling growth in *Salix* spp. is not regulated through the effect of day length on ABA levels [146,147]. It is, however, possible that short-day conditions may have altered tissue sensitivity to ABA [146]. These studies indicate further that in the control of bud dormancy, factors other than ABA are possibly involved.

The results in the case of seeds are, however, different. Several studies indicate that ABA treatment prevents vivipary (precocious germination of the developing embryo) in immature seeds. In vitro studies have shown that a high percentage of germination was obtained when the ABA content of immature soybean embryo was less than 4 μg/g fresh weight [148]. Similar studies with cultured immature embryos of wheat [149], soybean [148,150,151], cotton [152], rapeseed [153], and maize [154] have shown that exogenous ABA not only prevented precocious germination but often caused embryo growth and storage protein accumulation. However, with the maturity of embryos, the endogenous level of ABA and the sensitivity to exogenous hormone declined. Convincing evidence for the control of seed dormancy by ABA has been provided by a reciprocal cross between wild-type and ABA-deficient *Arabidopsis* mutants as well as with the treatments of wild-type young maize kernels with fluridone. The reciprocal crosses indicated that maternal ABA had a minor role [155], while treatment with fluridone induced precocious germination [156].

## 4. Stomatal Control

The discovery that ABA plays a leading role in the regulation of stomatal movement generated the interest of many workers [157,158]. Abscisic acid–deficient mutants are known in tomato, potato, pea, and *Arabidopsis* [125], which reverts phenotypically to the wild types when treated with ABA. The response is quite rapid, and after exogenous ABA application to the cut leaf bases, it takes 3 to 9 min to close stomata in maize, sugar beet (*Beta vulgaris*), and *Rumex obtusifolia* [17]. The magnitude of stomatal response to ABA is, however, dependent on the concentration of $K^+$ in the incubation media [159]. It has been estimated that when the stomates are closed, $K^+$ concentration of the epidermal cells ranges from 250 to 450 mol/m$^3$, but when $K^+$ concentration falls to about 100 mol/m$^3$, it opens. Harris and Outlaw [160] have measured ABA levels in isolated guard cells using an enzyme-amplified immunoassay and observed that water stress caused at least a 20-fold increase (up to 8 fg per cell). This may suggest that ABA causes stomates to close by inhibiting an energy-dependent (ATP/cAMP) proton pump in the guard cell plasma membrane. Thus ABA exerts two major biochemical effects. One is its effect on altering plasma membranes, which by shutting off the proton pump stops influx of $K^+$, causing $K^+$ and water to leak out. This reduces guard cell turgor, causing the stomates to close. However, the evidence of ABA role in stomatal regulation discussed above is not unequivocal and there is evidence that suggests the involvement of other factor(s), including hormone(s) and/or modification in tissue sensitivity [6,161].

## VII. CONCLUSIONS

Currently, five classes of hormone—auxin, gibberellin, cytokinin, ethylene, and abscisic acid—are known to be ubiquitous in higher plants and crops. Some of them have also been found to be produced by bacteria, fungi, bryophytes, and pteridophytes. They influence a myriad of plant functions and responses, and presumably any one process is influenced by the balance of the existing complement of hormones. Hormone physiologists generally classify auxin, gibberellin, and cytokinin as growth promotors and ethylene and abscisic acid as growth inhibitors. Although plant and crop hormones regulate a wide range of growth and developmental processes, their diversity makes it difficult to assign a definitive role to them from observations on plant responses. They influence each other's level and thus play important roles in a network of feedback control mechanisms modulating normal growth and development and thus preventing odd overgrowths. At times, each of them can act as a promote or inhibitor, or vice versa, in this network of feedback control mechanisms. Therefore, the categorization seems rather conjectural.

New research in molecular biology and biotechnology/genetic engineering has opened the door to exciting approaches. Mutants are available that are either synthesis or response mutants, and genetically engineered plants, which overproduce auxin, cytokinin, and/or ethylene, have also been developed. Using a molecular biological approach, investigators have provided support to the classical Cholodny—Went theory of differential growth elicited by tropistic responses. Similarly, transgenic plants have been used to support the theory that the auxin/cytokinin ratio, not the auxin-stimulated ethylene, controls the process of correlative growth (apical dominance, compensatory growth). Thus, studies using advanced technologies have supported theories advanced many years ago with the response techniques developed earlier. It is therefore encouraging that use of such mutants and/or transgenic plants is generating information on the biochemical and cellular processes that modulate plant growth and development [162].

## REFERENCES

1. J Sachs. Arb Bot Inst (Wurzburg) 2:452, 1880.
2. FW Went. Recl Trav Bot Neerl 25:1, 1928.
3. E Kurosawa. J Nat Hist Soc (Formosa) 16:213, 1926.
4. CO Miller, F Skoog, MH von Saltza, FM Strong. J Am Chem Soc 77:1392, 1955.
5. J Wiesner. Die Elementarstrucktur und das Waschstum der labenden Substanz. Vienna: Holder, 1892.
6. SSM Naqvi. In: M Pessarakli, ed. Handbook of Plant and Crop Stress. New York: Marcel Dekker, 1994, p 383.
7. FB Salisbury, CW Ross. Plant Physiology. Belmont, CA: Wadsworth, 1992.
8. FP Gardner, RB Pearce, RL Mitchell. Physiology of Crop Plants. Ames: Iowa State University Press, 1985.
9. J Heslop-Harrison. In: F Skoog, ed. Plant Growth Substances 1979. Berlin: Springer-Verlag, 1980 p 3.
10. P Boysen-Jensen. In: Growth Hormones in Plants (translated and revised by GS Avery Jr, PR Burkholder, eds). New York: McGraw-Hill, 1936.
11. A Paal. Jahrb Wiss Bot 58:406, 1919.
12. F Kögl, AJ Haagen-Smit, H Erxleben. Z Physiol Chem 288:90, 1934.
13. FW Went, KV Thimann. Phytohormones. New York: Macmillan, 1937.
14. AJ Haagen-Smit, WB Dandliker, SH Wittwer, AE Murneek. Am J Bot 33:118, 1946.
15. KC Engvild. Physiol Plant 77:282, 1989.
16. RL Wain, CH Fawcett. In: FC Steward, ed. Plant Physiology: A Treatise, Vol 5A. New York: Academic Press, 1969, p 231.
17. TC Moore. Biochemistry and Physiology of Plant Hormones. New York: Springer-Verlag, 1979.
18. RH Phelps, L Sequeira. Plant Physiol 42:1161, 1967.
19. RA Gibson, EA Schneider, F Wightman. J Exp Bot 23:381, 1972.
20. J Ludwig-Muller, W Hilgenberg. Physiol. Plant 74:240, 1988.
21. M Kawaguchi, S Fujioka, A Sakurai, YT Yamaki, K Syono. Plant Cell Physiol 34:121, 1993.
22. BG Baldi, BR Maher, JP Slovin, JD Cohen. Plant Physiol 95:1203, 1991.
23. AD Wright, MB Sampson, MG Neuffer, L Michalczuk, JP Slovin, JD Cohen. Science 254:988, 1991.
24. DM Reinecke, RS Bandurski. In: PJ Davies, ed. Plant Hormones and Their Role in Plant Growth and Development. Boston: Martinus Nijhoff, 1987, p 24.
25. TT Lee, AN Starratt. Physiol Plant 84:209, 1992.
26. F Sitbon, A Ostin, S Sundberg, O Olsson, G Sandberg. Plant Physiol 101:313, 1993.
27. SM Naqvi, PhD thesis, Princeton University, Princeton, NJ, 1963.
28. WP Jacobs. Plant Hormones and Plant Development. Cambridge: Cambridge University Press, 1979.
29. SM Naqvi, SA Gordon. Plant Physiol 40:116, 1965.
30. ML Evans. Plant Physiol 95:1, 1991.
31. SM Naqvi, KC Engvild. Physiol Plant 30:283, 1974.
32. RS Bandurski, HM Nonhebel. In: MB Wilkins, ed. Advanced Plant Physiology. London: Pitman, 1984, p 1.
33. R Aloni. In: PJ Davies, ed. Plant Hormones and Their Role in Plant Growth and Development. Boston: Martinus Nijhoff, 1987, p 363.
34. AC Leopold. Auxins and Plant Growth. Berkeley: University of California Press, 1960.
35. VJ Philip, J Padikkala. J Plant Physiol 135:233, 1989.
36. VC Pence, L Caruso. In: PJ Davies, ed. Plant Hormones and Their Role in Plant Growth and Development. Boston: Martinus Nijhoff, 1987, p 240.
37. RD Firn, J Digby. Annu Rev Plant Physiol 31:131, 1980.
38. BG Pickard. Annu Rev Plant Physiol 36:55, 1985.
39. IR MacDonald, W Hart. Plant Physiol 84:568, 1987.
40. Y Li, G Hagen, TJ Guilfoyle. Plant Cell 3:1167, 1991.
41. MJ Jaffe, H Takahashi, RL Biro. Science 230:445, 1985.
42. WR Briggs. Plant Physiol 38:237, 1963.
43. BG Pickard, KV Thimann. Plant Physiol 39:341, 1964.
44. SM Naqvi, SA Gordon. Plant Physiol 42:138, 1967.
45. EW Weiler, RS Jourdan, W Conrad. Planta 153:561, 1981.

46. S Marumo. In: N Takahashi, ed. Chemistry of Plant Hormones. Boca Raton, FL: CRC Press, 1986, p 9.
47. AB Frank. Beitrage zur Pflanzenphysiologie. I. Ueber die durch Schwerkeft Verusachte Bewegung von Pflanzentheilen. Leipzig: W Engelmann, 1868.
48. ML Evans, R Moore, K-H Hasenstein. Sci Am 255(6):100, 1986.
49. SM Naqvi, SA Gordon. Plant Physiol 41:1113, 1966.
50. R Hertel, RK de la Fuente, AC Leopold. Planta 88:204, 1969.
51. BA McClure, T Guilfoyle. Science 243:91, 1989.
52. H Klee, M Estelle. Annu Rev Plant Physiol Plant Mol Biol 42:529, 1991.
53. KV Thimann. Annu Rev Plant Physiol 14:1, 1963.
54. GC Martin, HortScience 22:824, 1987.
55. IA Tamas. In: PJ Davies, ed. Plant Hormones and Their Role in Plant Growth and Development. Boston: Martinus Nijhoff, 1987, p 393.
56. KV Thimann, F Skoog. Proc R Soc (Lond) B114:317, 1934.
57. AC Leopold. Am J Bot 36:437, 1949.
58. CP Romano, ML Cooper, HJ Klee. Plant Cell 5:181, 1993.
59. N Takahashi, BO Phinney, I MacMillan. Gibberellins. Berlin: Springer-Verlag, 1990.
60. N Takahashi, I Yamaguchi, H Yamane. In: N Takahashi, ed. Chemistry of Plant Hormones. Boca Raton, FL: CRC Press, 1986, p 57.
61. RC Durley, J MacMillan, DM Reid, BH Most. Phytochemistry 10:1891, 1971.
62. JE Graebe, DJ Dennis, CD Upper, CA West. J Biol Chem 240:1847, 1965.
63. RM Devlin, FH Witham. Plant Physiology. Boston: PWS Publishers, 1983.
64. Y Murakami. Jpn Agric Res Q 5(2):5, 1970.
65. T Nishijima, N Katsura. Plant Cell Physiol 30:623, 1989.
66. RL Jones, JE Varner. Planta 72:155, 1967.
67. RP Pharis, LT Evans, RW King, LN Mander. In: E Lord, G Bernier, eds. Plant Production: From Floral Induction to Pollination. American Society of Plant Physiologists Symposium Series, Vol 1. Rockville, MD: ASPP, 1989, p 29.
68. J MacMillan, BO Phinney. In: DJ Cosgrove, DP Knievel, eds. Physiology of Cell Expansion During Plant Growth. Rockville, MD: American Society of Plant Physiologists, 1987, p 156.
69. SB Rood, PH Williams, D Pearce, N Murofushi, LN Mander, RP Pharis. Plant Physiol 93.1168, 1990.
70. RL Jones, J MacMillan. In: MB Wilkins, ed. Advanced Plant Physiology. London: Pitman, 1984, p 21.
71. G Haberlandt. Beitr Allg Bot 2:1, 1921.
71a. J Van Overbeek. Science 152:721, 1966.
72. K Koshizima, H Iwamura. In: N Takahashi, ed. Chemistry of Plant Hormones. Boca Raton, FL: CRC Press, 1986, p 153.
73. DS Letham. Phytochemistry 5:269, 1966.
74. GA Miura, CO Miller. Plant Physiol 44:1035, 1969.
75. BJ Taller, N Murai, F Skoog. Plant Physiol 83:755, 1987.
76. CM Chen, DK Melitz. FEBS Lett 107:15, 1979.
77. P Beutelmann. Planta 112:181, 1973.
78. DS Letham, LMS Palni. Annu Rev Plant Physiol 34:163, 1983.
79. J Van Staden, JE Davey. Plant Cell Environ 2:93, 1979.
80. PE Pilet. In: F Wightman, G Setterfield, eds. Biochemistry and Physiology of Plant Growth Substances. Ottawa: Runge Press, 1968, p 993.
81. JI Medford, R Horgan, Z El-Sawi, HJ Klee. Plant Cell 1:403, 1989.
82. DS Letham. Annu Rev Plant Physiol 18:343, 1967.
83. R Horgan. In: MB Wilkins, ed. Advanced Plant Physiology. London: Pitman, 1984, p 53.
84. TH Thomas. J Plant Growth Regul 8:255, 1989.
85. F Skoog, CO Miller. Symp Soc Exp Biol 11:118, 1957.
86. M Wickson, KV Thimann. Physiol Plant 11:62, 1958.
87. PKW Lee, B Kessler, KV Thimann. Physiol Plant 31:11, 1974.
88. R Sexton, HW Woolhouse. In: MB Wilkins, ed. Advanced Plant Physiology. London: Pitman, 1984, p 469.
89. KV Thimann. In: WW Thomson, EA Nothnagel, RC Huffaker, eds. Plant Senescence: Its Biochemistry and Physiology. Rockville, MD: American Society of Plant Physiologists, 1987, p 1.
90. AC Leopold, M Kawase. Am J Bot 51:294, 1964.
91. SP Burg. Proc Natl Acad Sci USA 70:591, 1973.
92. DN Neljubow. Beth Bot Centralbl 10:128, 1901.
93. HH Cousins. Agricultural Experiments: Citrus. Jamaica Department of Agriculture Annual Report, 1910, p 7.
94. R Gane. Nature 134:1008, 1934.
95. M Lieberman, LW Mapson. Nature 204:343, 1964.
96. M Lieberman, A Kunishi, LW Mapson, DA Wardale. Plant Physiol 41:376, 1966.
97. SF Yang. Hortscience 120:41, 1985.
98. H Imaseki. In: N Takahashi, ed. Chemistry of Plant Hormones. Boca Raton, FL: CRC Press, 1986, p 249.

99.   EM Beyer. In: A Roberts, GA Tucker, eds. Ethylene and Plant Development. London: Butterworth, 1985, p 125.
100.  M Zeroni, MA Hall. In: J MacMillan, ed. Encyclopedia of Plant Physiology (NS): Hormonal Regulation of Development I, Vol. 9. Berlin: Springer-Verlag, 1980, p 511.
101.  LK Levitt, DB Stein, B Rubinstein. Plant Physiol 85:318, 1987.
102.  MB Jackson. In: JA Roberts, GA Tucker, eds. Ethylene and Plant Development. London: Butterworth, 1985, p 241.
103.  I Ridge. In: JA Roberts, GA Tucker, eds. Ethylene and Plant Development. London: Butterworth, 1985, p 229.
104.  DJ Osborne, MT McManus, J Webb. In: JA Roberts, GA Tucker, eds. Ethylene and Plant Development. London: Butterworth, 1985, p 197.
105.  FB Abeles. Ethylene in Plant Biology. New York: Academic Press, 1973.
106.  DT Tingey, C Standley, RW Field. Atmos Environ 10:969, 1976.
107.  JA Dunning, WW Heck. Environ Sci Technol 7:824, 1973.
108.  LE Craker. Environ Pollut 1:299, 1971.
109.  H Yoshii, H Imaseki. Plant Cell Physiol 22:369, 1981.
110.  FB Abeles. In: JA Roberts, GA Tucker, eds. Ethylene and Plant Development. London: Butterworth, 1985, p 1.
111.  EW Simons. Symp Soc Exp Biol 21:215, 1967.
112.  LD Noodén. In: KV Thimann, ed. Senescence in Plants. Boca Raton, FL: CRC Press, 1980, p 219.
113.  SSM Naqvi, SM Alam, S Mumtaz. Aust J Exp Agric 30:433, 1990.
114.  R Nunez-Elisea, TL Davenport. Plant Physiol 82:991, 1986.
115.  SSM Naqvi, SM Alam, S Mumtaz. Pak J Bot 24:197, 1992.
116.  T Hemberg. Physiol Plant 2:24, 1949.
117.  T Hemberg. Physiol Plant 2:37, 1949.
118.  TA Bennet-Clark, MS Tambiah, NP Kefford. Nature 169:452, 1952.
119.  TA Bennet-Clark, NP Kefford. Nature 171:645, 1953.
120.  NP Kefford. J Exp Bot 6:245, 1955.
121.  PF Wareing, CF Eagles, PM Robinson. In: JP Nitsch, ed. Régulateurs naturels de la croissance végétale. Vol 9. Paris: CNRS, 1964, p 377.
122.  WC Liu, HR Carns. Science 134:384, 1961.
123.  K Ohkuma, JL Lyon, FT Addicott, OE Smith. Science 142:1592, 1963.
124.  FT Addicott, JL Lyon, K Ohkuma, WE Thiessen, HR Carns, OE Smith, JW Cornforth, BV Milborrow, G Ryback, PF Wareing. Science 159:1493, 1968.
125.  JAD Zeevaart, RA Creelman. Annu Rev Plant Physiol Plant Mol Biol 39:439, 1988.
126.  CD Rock, JAD Zeevaart. Plant Physiol 93:915, 1990.
127.  RK Sindhu, DH Griffin, DC Walton. Plant Physiol 93:689, 1990.
128.  JAD Zeevaart, CD Rock, F Fantauzzo, TG Heath, DA Gage. In: WJ Davies, HG Jones, eds. Abscisic Acid: Physiology and Biochemistry. Oxford: BIOS Scientific Publishers, 1991, p 39.
129.  O Wolf, WD Jeschke, W Hartung. J Exp Bot 41:593, 1990.
130.  BV Milborrow. In: MB Wilkins, ed. Advanced Plant Physiology. London: Pitman, 1984, p 76.
131.  MJ Harris, WH Outlaw, R Mertens, EW Weiler. Proc Natl Acad Sci USA 85:2584, 1988.
132.  K Dörffling, D Tietz. In: FT Addicott, ed. Abscisic Acid. New York: Praeger, 1983 p 23.
133.  CH Lin, YL Lin, YJ Chow. J Plant Growth Regul 7:161, 1988.
134.  N Hirai. In: N Takahashi, ed. Chemistry of Plant Hormones. Boca Raton, FL: CRC Press, 1986, p 201.
135.  TJ Mulkey, ML Evans, KM Kuzmanoff. Planta 157:150, 1983.
136.  PE Pilet, PW Barlow. Plant Growth Regul 6:217, 1987.
137.  IN Saab, RE Sharp, J Pritchard, GS Voetberg. Plant Physiol 93:1329, 1990.
138.  JM Robertson, KT Hubick, EC Yeung, DM Reid. J Exp Bot 41:325, 1990.
139.  RA Creelman, HS Mason, RJ Benson, JS Boyer, JE Mullet. Plant Physiol 92:205, 1990.
140.  Z Plaut, A Carmi. Plant Physiol (Abstr), 99:160, 1992.
141.  R Moore, K Dickey. J Exp Bot 36:1793, 1985.
142.  LJ Feldman, PS Sun. Physiol Plant 67:472, 1986.
143.  DC Walton. Annu Rev Plant Physiol 31:453, 1980.
144.  R Alvin, S Thomas, PF Saunders. Plant Physiol 62:779, 1978.
145.  EB Dumbroff, DB Cohen, DP Webb. Physiol Plant 45:211, 1979.
146.  RS Barros, SJ Neill. Planta 168:530, 1986.
147.  LG Johansen, P-C Oden, O Juntilla. Physiol Plant 66:409, 1986.
148.  RC Ackerson. J Exp Bot 35:414, 1984.
149.  BA Triplett, RS Quatrano. Dev Biol 91:491, 1982.
150.  RC Ackerson. J Exp Bot 35:403, 1984.
151.  AJ Eisenberg, JP Mascarenhas. Planta 166:505, 1985.
152.  DL Hendrix, JW Radin. J Plant Physiol 117:211, 1984.
153.  RR Finkelstein, KM Tenbarge, JE Shumway, ML Crouch. Plant Physiol 78:630, 1985.

154. SJ Neill, R Horgan, AF Rees. Planta 171:358, 1987.
155. CM Karseen, DLC Brinkhorst-van der Swan, AE Breekland, M Koornneef. Planta 157:158, 1983.
156. F Fong, JD Smith, DE Koehler. Plant Physiol 73:899, 1983.
157. STC Wright, RWP Hiron. In: DJ Carr, ed. Plant Growth Substances 1970. Berlin: Springer-Verlag, 1972, p 291.
158. K Raschke. In: G Zeiger, D Farquhar, IR Cowan, eds. Stomatal Function. Stanford, CA: Stanford University Press, 1987, p 253.
159. PJ Snaith, TA Mansfield. Plant Cell Environ 5:309, 1982.
160. MJ Harris, WH Outlaw. Physiol Plant 78:495, 1990.
161. EV Kearns, SM Assmann. Plant Physiol 102:711, 1993.
162. JD Hamill. Aust J Plant Physiol 20:405, 1993.

# 25

# The Activation Sequence-1 Cognate Promoter Elements Play Critical Roles in the Activation of Defense-Related Genes in Higher Plants

**Chengbin Xiang**

*Iowa State University, Ames, Iowa*

## I. INTRODUCTION

Sessile plants rely on environmental cues to program their development and growth. Meanwhile, they are unable to escape from unfavorable, sometimes hostile, environmental conditions. In order to survive, higher plants have evolved and deployed sophisticated systems that sense the changes in their living environment and activate corresponding defense mechanisms. These adaptive responses are crucial for plant survival and crop productivity. The genetic basis for the plasticity and capacity of these defense mechanisms has shaped the current geographic distribution of higher plants on the earth. One central process of adaptive responses is the activation of defense-related genes in response to biotic and abiotic stresses in their living environment. Increasing evidence has shown that the activation sequence-1 (*as-1*) cognate promoter elements were recruited by the promoters of plant defense-related genes as well as by the gene promoters of plant pathogens. A number of studies indicated that elements of this type play a critical role in the activation of defense genes. This review chapter focuses on the general features of this class of *cis*-acting promoter elements and discusses how diverse stress signals converge on this type of element to activate defense genes against biotic and abiotic stresses.

## II. THE *as-1*–TYPE *CIS*-ACTING ELEMENTS ARE EXPLOITED BY PLANT PATHOGENS

The *as-1* DNA element was first identified in the $-75$ region of the 35S promoter of cauliflower mosaic virus [1]. The *as-1* element is composed of two tandem repeats of a TGACG motif that resembles the mammalian activator protein-1 (AP-1) binding site (Figure 1). Similar TGACG motif–containing promoter elements were also found in the gene promoters of plant pathogens. These include the *nos-1* element ($-131$ to $-111$) of the nopaline synthase gene promoter [2], the *mas-1* element of the mannopine synthase [9], and the *ocs* element ($-193$ to $-173$) of the octopine synthase gene promoter of *Agrobacterium tumefaciens* T-DNA [3].

A tobacco nuclear protein factor, activation sequence factor-1 (ASF-1) was found to interact specifically with the *as-1* element [1]. This factor can also bind to a number of *cis*-acting elements containing the TGACG motif such as *nos-1* and *ocs-1*. A single *as-1* element is sufficient to confer root expression when fused to a leaf-specific promoter [1] and is responsive to multiple stress-related stimuli.

| Elements | DNA sequence | Reference |
|----------|-------------|-----------|
| as-1 | TGACGTAAGGGATGACGCAC | 1 |
| nos-1 | TGAGCTAAGCACATACGTCA | 2 |
| ocs-1 | AAACGTAAGCGCTTACGTAC | 3 |
| GNT35 | TTAGCTAAGTGCTTACGTAT | 4 |
| as103-1 | ATAGCTAAGTGCTTACGTAT | 5 |
| as114-4 | TTACGCAAGCAATGACATCT | 5 |
| as107-2 | TGACGAATGCGATGACCTCT | 5 |
| parB | TGAGGTCATTACTTACCTAA | 6 |
| GH2/4 | TGATGTAAGAGATTACGTAA | 7 |
| Wheat GstA1 | ATCCGTACCAACGCACGTGT | 8 |
| PR-1a | ACGTCATCGAGATGACGGCC | 9 |
| PR-1 | TTACGTCATAGATGTGGCGG | 10 |
| lox1 | TGACTTCATTCATGACGATT | 11 |
| GSH1 | TGACTTCATCTTTGACGGCT | This study |
| | | |
| AP-1 site | TGACATTGCTAATGGTGACAAAGC | 12 |

**Figure 1**  The *as-1* cognate promoter elements in plant pathogen promoters and plant defense-related gene promoters. The sequences are compiled from the references cited. The half site core sequences are underlined. The two repeats can be in any orientation. The AP-1 binding site of mouse GST Ya gene is also shown for comparison.

Upon pathogen attack, plants respond by activating their defense mechanisms. The activated transcription machinery of plants is exploited by pathogens to establish colonization on their host. It is apparent that plant viral and bacterial pathogens have coevolved with higher plant defense systems and recruited *as-1*–type *cis* elements for their gene promoters in order to counteract plant defense systems. This indicates that *as-1*–type elements are critical in regulating defense-related genes under biotic stress conditions.

## III.   THE *as-1*–TYPE ELEMENTS ARE ONE CLASS OF STRESS-RESPONSIVE ELEMENTS WIDELY USED BY DEFENSE-RELATED GENES IN HIGHER PLANTS

Although *as-1*–type *cis* elements were initially identified and characterized in plant pathogens *Agrobacterium tumefaciens* and cauliflower mosaic virus, it is conceivable and expected that this type of *cis* element is authentic to the promoters of plant genes that play important roles in defense against both biotic and abiotic stresses. Indeed, more and more plant genes have been found to possess *as-1*–type *cis* elements in their promoters.

## A.   The *as-1*–Type Elements Are Found in the Promoters of Pathogenesis-Related Genes of Higher Plants

Pathogenesis-related (PR) genes are coordinately induced in higher plants during the onset of systemic acquired resistance (SAR). SAR is a well-characterized plant defense mechanism [13] that is triggered upon pathogen infection or by exposure to salicylic acid (SA). The promoter of the *PR-1* gene, one of the genes encoding PR proteins, has been subjected to extensive functional analyses [10,14]. Strompen et al. [14] identified a region of 139 bp (from −691 to −553) in the *PR-1a* gene promoter responsive to SA and revealed an *as-1*–like element with two TGACG motifs within this region. TGA1a, a bZIP transcription factor known to bind the *as-1* element specifically, bound this element in vitro with specificity and affinity similar to those of *as-1* in gel shift assays. Mutations in this element in the context of the *PR-1a* promoter caused significant reduction of reporter gene activity. Lebel et al. [10] delineated the SA-responsive region from −640 to −610 within the *PR-1* promoter using linker-scanning mutagenesis. A bZIP transcription factor binding site was identified in the −640 region and a consensus binding site for the transcription factor nuclear factor κB (NF-κB) was identified in the −610 region. In vivo footprinting results showed tight correlation with the functionality of these promoter regions. It has been demonstrated that TGA2, another bZIP transcription factor in *Arabidopsis* (see later), binds to the −640 region of the *PR-1* promoter [15]. A similar element was identified in another PR gene (*PRB-1b*) promoter [16]. These studies have firmly established the functional importance of the *as-1*–type elements in activating defense genes in response to the signals that arise upon pathogen attack. It is conceivable that this class of *cis* elements may reside in the promoters of other PR genes whose expression is regulated similarly to *PR-1*.

## B. The *as-1*–Type Elements Are Found in the Stress-Related Glutathione *S*-Transferase Gene Promoters

Glutathione *S*-transferases (GSTs) are enzymes catalyzing the conjugation of glutathione to a broad range of hydrophobic, electrophilic metabolites and xenobiotics such as herbicides [17,18]. This class of enzymes plays important roles in protecting plants from oxidative stress, xenobiotic toxicity, and heavy metal toxicity. In addition, GSTs serve as ligandins for nonenzymatic binding and intracellular transport of auxins and antimicrobial compounds (for review, see Ref. 19). GSTs are encoded by a large gene family in higher plants and their expression is strongly activated in response to stress. Most, if not all, GST genes whose promoters have been analyzed possess one or more *as-1*–type elements in their promoter. These include tobacco *GNT1* and *GNT35* [4,20]; *parA*, *parB*, and *parC* [6,21–24]; soybean *GH2/4*(7); wheat *GstA1* (8); and *Arabidopsis GST6* [25,26]. The inducibility of the GST gene promoters by various stress conditions and stress-related signal molecules is tightly correlated with the *as-1*–type elements within their promoters. All the evidence accumulated so far for GSTs demonstrates that *as-1*–type elements also play very important roles in activating defense genes against abiotic stresses. It is predicted that other abiotic stress-related genes possess *as-1*–type elements in their promoters.

## C. An *as-1* Element Is Found in the Promoter of *Arabidopsis GSH1* Encoding the Rate-Limiting Enzyme, γ-Glutamylcysteine Synthetase, for Glutathione Biosynthesis

The tripeptide glutathione (GSH) is enzymatically synthesized by two ATP-dependent reactions catalyzed by γ-glutamylcysteine synthetase encoded by *GSH1* [27] and GSH synthetase encoded by *GSH2* [28], respectively. GSH is a major antioxidant in higher plants and plays a pivotal role in protecting plants from oxidative stress through the ascorbate-GSH cycle [29], xenobiotics and cytotoxic metabolites through GST-catalyzed conjugation [19], and heavy metals through phytochelatins [30]. We have demonstrated that all the genes involved in GSH synthesis and recycling (GSH disulfide reduction by GSH reductase encoded by *GR*) are coordinately up-regulated in response to heavy metals and jasmonic acid (JA) treatments [31]. Analysis of the *GSH1* promoter revealed that the transcription up-regulation by heavy metals and JA is mediated by an *as-1* element. Removal of this element rendered the promoter nonresponsive to heavy metals and JA (C. Xiang and D.J. Oliver, unpublished results). The *as-1* element in the *GSH1* promoter provides further evidence that *as-1*–type elements are directly involved in the activation of the genes against abiotic stress. It is conceivable that *as-1*–type elements may reside in the promoters of other genes involved in GSH synthesis and recycling because of their coordinated expression.

## D. The *as-1*–Type Elements Are Found in the Promoter of TGA bZIP Transcription Factors

TGA factors are a subclass of bZIP transcription factors that specifically interact with and activate *as-1*–type elements (see later). The *as-1*–type elements are found in the promoter of the genes encoding TGA bZIP transcription factors. Multiple *as-1* elements were found in the promoter of TGA3 [32] and TGA6 (C. Xiang and E. Lam, unpublished results). The promoter of tobacco TGA1a also contains several such elements [33]. This suggests that TGA bZIP transcription factors may autoregulate their own expression and play critical roles in regulating the expression of stress-related genes in higher plants.

## E. The *as-1*–Type Elements Are Found in Other Defense-Related Plant Gene Promoters

Lipoxygenases are a group of enzymes involved in the synthesis of stress signal molecules such as abscisic acid, traumatic acid, and JA [34]. A short region responsive to JA was identified in the promoter of barley lipoxygenase 1 gene [11]. Within this region, a 36-bp fragment contains inverted repeats of the TGACG motif of the *as-1* element. Mutations within the motif abolished JA-responsive expression, therefore identifying this element as a JA-responsive element. The *as-1*–type elements are strongly responsive to JA (see later). JA induces the expression of defense-related genes encoding the proteinase inhibitors [35] phenylalanine ammonia lyase [36], lipoxygenases [37], a thionin with potent antifugal activity [38], and a ribosomal-inactivating protein [39]. It is predictable that the *as-1*–type elements may reside in these JA-responsive gene promoters.

The presence of *as-1*–type elements in the promoters of various defense-related genes indicates the importance of this class of *cis* elements in plant defense mechanisms. It is now clear that *as-1*–type cognate elements are one class of multiple stress-responsive elements that play critical roles in regulating the expression of defense-related genes in response to biotic and abiotic stresses.

As the *Arabidopsis* genome sequencing nears completion, it will soon be possible to locate all *as-1*–type elements in the whole genome of *Arabidopsis*. It is expected that most, if not all, defense-related genes contain one or more such elements in their promoters. On the other hand, if a gene with unknown functions contains *as-1* elements in its promoter, one may predict that the function of the gene is defense related.

## IV.   THE *as-1*–TYPE ELEMENTS ARE STRONGLY RESPONSIVE TO DIVERSE STRESS-RELATED STIMULI

The *as-1*–type elements are responsive to a wide spectrum of stress-related stimuli. The *as-1*–type elements, when fused to a minimal CaMV 35S promoter, are responsive to stress stimuli in activating reporter gene expression [26,40–43]. Figure 2 shows that the *nos-1* element–driven *uidA* reporter gene expression in transgenic *Arabidopsis* behaves similarly to the endogenous *GST6* gene and is strongly induced by JA, $H_2O_2$, the synthetic auxin 2,4-D, the heavy metal Cu, and SA. Interestingly, the *PR-1* gene is also responsive to the synthetic auxin 2,4-D. Figure 3 shows that tobacco genes coding for GSTs (*parA*,

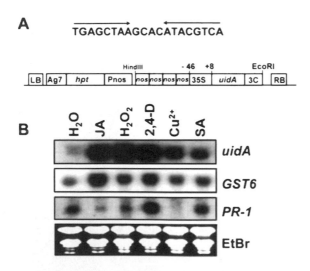

**Figure 2**   Induction of the expression of *nos-1* element–driven reporter gene *uidA* and defense-related genes with *as-1*–type elements in their promoter. (A) Illustration (not drawn to scale) of the GUS reporter gene (*uidA*) under the control of the synthetic tetrameric *nos-1* elements (*nos*) and the minimal 35S promoter of CaMV. Only the T-DNA region of this binary vector is shown. RB and LB stand for the right and left border sequences of the T-DNA, Pnos for the promoter of nopaline synthase gene, *hpt* for hygromycin resistance gene, and 3C and Ag7 for terminators. (B) Northern blot analysis of the steady-state transcript levels in response to JA, SA, 2,4-D, Cu, and $H_2O_2$ treatments. Two-week-old *Arabidopsis* plants grown in liquid culture were treated with 5 mM $H_2O_2$ or 100 $\mu$M JA, SA, Cu, or 2,4-D for 6 hrs. Total RNA was isolated and analyzed as previously described (31) using cDNAs for *GST6* and *PR-1* or *uidA* DNA as probes. The ethidium bromide- (EtBr) stained RNA gel is shown for equal loading.

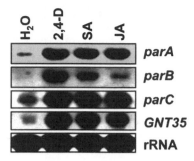

parA
parB
parC
GNT35
rRNA

**Figure 3**  Northern blot analysis of steady-state transcript levels of GST genes in response to JA, SA, and 2,4-D treatments in tobacco seedlings. One-week-old tobacco plants grown on MS agar plates were treated with 100 μM JA, SA, or 2,4-D for 6 hr. Total RNA was isolated and analyzed as previously described [31] using cDNAs for *parA, ParB, parC,* and *GNT35* and 18S rRNA for equal loading.

*parB, parC,* and *GNT35*) are also highly responsive to JA, SA, and 2,4-D. The induction of gene expression by auxins was also reported for a soybean GST [44] and tobacco GSTs [5,22,24,43,45]. Activated transcription is responsible for the induction of the expression of these genes as demonstrated using actinomycin D, an inhibitor for RNA polymerase II, as shown in Figure 4. Actinomycin D treatment completely abolishes the accumulation of the transcripts.

Overwhelming evidence has demonstrated that *as-1*–type elements are responsive to the signals originated from diverse stress-related stimuli. These stress-related stimuli include auxin and xenobiotics, JA, SA, heavy metals, $H_2O_2$ and oxidative stress, and biotic and abiotic environmental stresses that may elicit the production of stress signal molecules such as JA, SA, and $H_2O_2$.

It has been well documented that xenobiotics such as herbicides and the synthetic auxins 2,4-D and α-naphthaleneacetic acid (NAA) induce the expression of GST genes whose products are directly involved in the detoxification of these hydrophobic and electrophilic organic compounds (19). Auxins act as phytohormones at low physiological concentrations, and normally their concentration is under tight control. When present above physiological concentrations, plant cells may sense them as cytotoxins that have to be detoxified. These compounds are structurally diverse. The only chemical similarity shared among these compounds is that they all contain or are able to form, through metabolism, a "Michael acceptor" (carbon-carbon double bonds adjacent to an electrophilic group) [46]. This feature may provide the common ground by which these structurally diverse chemicals activate the same suite of genes. It can be speculated that a component(s) in the signal transduction pathway leading to the expression of these

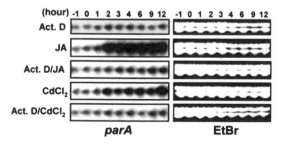

parA                    EtBr

**Figure 4**  GST genes are transcriptionally activated. Two-week-old *Arabidopsis* plants grown in liquid culture were pretreated with 0.5 mM actinomycin D for 1 hr before 100 μM JA or Cd was added. The coincubation of actinomycin D with either JA or Cd was continued for indicated time periods (Act. D/JA and Act. D/CdCl₂). Treatments with actinomycin D, JA, and CdCl₂, respectively, were controls. Total RNA was isolated and analyzed as previously described [31] using cDNA *parA*. The EtBr-stained RNA gel is shown for equal loading.

genes may have one or more sulfhydryl groups exposed on the surface. The nucleophilic SH groups would be apparent targets for the Michael acceptor of these compounds in a way similar to the conjugation to GSH. The conjugation of these compounds to the signaling components could somehow activate the signaling pathways leading to strong induction of GST gene expression.

The activation of *as-1*–type elements in response to stress signal molecules, $H_2O_2$, jasmonic acid, and salicylic acid, is also well demonstrated [25,26,31,40,43,47]. A wealth of information has been accumulated for the responses of defense genes to these stress signal molecules that are the subjects of a number of reviews [13,48,49]. $H_2O_2$ and salicylic acid are endogenous signal molecules that are elicited upon pathogen infection [13], and jasmonic acid production is triggered upon wounding and insect feeding [35,50]. In addition, the production of these signal molecules may be elicited and accelerated by abiotic stress conditions. Cellular $H_2O_2$ concentration can be increased on exposure to heavy metals and oxidative stress conditions.

However, variations of the responsiveness to the stress signals do exist from promoter to promoter of defense genes. The different responsiveness may result from the variation in the cognate *as-1*–type elements in different defense gene promoters. Of course, other *cis* elements in the promoter context also contribute to the expression. It is the combinatorial effect of all the *cis* elements that determines a promoter activity [51].

## V. THE *as-1*–TYPE ELEMENTS SPECIFICALLY INTERACT WITH A SUBCLASS OF BZIP TRANSCRIPTION FACTORS CALLED TGA FACTORS

The bZIP proteins are transcription factors that contain a basic region for specific DNA contact and a leucine zipper domain for dimerization. All bZIP factors bind to specific DNA sequences as homo- or heterodimers [52]. These transcription factors are believed to contribute to the efficiency with which RNA polymerase II binds and initiates transcription at the promoter of a gene. They are generally activators of transcription in response to external stimuli either constitutively or in a regulated manner, usually through posttranslational modifications such as phosphorylation. However, bZIP proteins can also be repressors in some cases [53,54]. Many bZIP factors are also expressed in a cell type–specific or developmentally regulated fashion.

In vitro studies using gel retardation assays have firmly established the specific interaction between *as-1*–type elements and a subclass of bZIP transcription factors called TGAs in plants [1,2,55,56]. Lam and Lam [57], with supershift assays using specific antisera raised against specific TGA factors, demonstrated that TGA factors are integral components of the in vivo *as-1* element–binding activity called ASF-1.

The TGA factors are highly conserved in higher plants. A small gene family exists for the TGA bZIP transcription factors in *Arabidopsis*. Six members of the *Arabidopsis* TGA family of bZIP transcription factors have been cloned and characterized. These genes were designated as *TGA1* [58], *TGA2* [59] *TGA3* [32], *OBF4* and *OBF5* [60] and *TGA6* [61]. Based on sequence homology, TGA family members may be divided into three subgroups: (1) *TGA2, OBF5*, and *TGA6*; (2) *TGA1* and *OBF4*; and (3) *TGA3*, which shows less than 66.2% homology in DNA sequence to the other known TGA family members [61]. The seventh member of this gene family, *TGA7*, was recently isolated [62].

The multiple gene family of TGA factors provides functional redundancy and subtle variations that may be required for fine-tuning the regulation of target genes. Theoretically, these seven bZIP factors are able to produce 28 different types of dimers. These dimer combinations could generate further diversity in the types of transcriptional regulation mediated by the TGA factors. Distinct DNA binding properties have been shown in vitro for different TGA factors, although the basic region for DNA contact is well conserved among different members of the TGA family [57,61]. If the same is true in vivo, preferential binding could introduce additional subtlety in regulation, which may be required in selectively activating or repressing target genes through interaction with a particular *cis* element in the context of the target gene promoters in response to different environmental stress conditions. Besides, variant TGA factor binding sites are indeed present in the promoters of defense-related genes [20,45].

## VI. SIGNALS ORIGINATED FROM DIVERSE STRESS-RELATED STIMULI CONVERGE ON A SINGLE *CIS* ELEMENT

How signals originated from diverse stress-related stimuli converge on a single *cis* element is a very important and interesting biological question. Answering this question should contribute tremendously to our understanding of the fundamental process of transcriptional activation and signal transduction pathways that biotic and abiotic stresses use to regulate the genes of defense systems.

A simplified scheme of signal transduction pathways leading to the activation of *as-1*–type elements is illustrated in Figure 5. In this scheme, the production of endogenous stress signal molecules (JA, SA, and $H_2O_2$) is elicited upon pathogen attack, wounding by insect feeding, or adverse abiotic stress conditions. These signals are amplified and transduced through corresponding signal transduction pathways leading to the TGA factors bound to the *as-1*–type elements and finally activate their transcription in the nucleus. Auxins and xenobiotics may directly activate a signaling component through the conjugation by its Michael acceptor as discussed before.

Unfortunately, little is known about the signal transduction pathways. The genetic approach has been playing a significant role in elucidating these cellular processes. Several JA-responsive mutants were isolated in *Arabidopsis* [63–65]. Only one signaling component in the JA signaling pathway has been isolated by position cloning of the *coi 1* locus. The predicted amino acid sequence of the COI1 has similarity to the F-box proteins that appear to function by targeting repressor proteins for removal by ubiquitination [66].

Significant progress has been made in the field of plant-microbe interaction concerning the SA signaling pathways. SAR induction requires the signal molecule SA. The isolation and characterization of *Arabidopsis* mutants nonresponsive to the inducers of SAR [67] in *NPR1* (also known as *NIM1*) started off a series of exciting new discoveries in this field. *NPR1* has been cloned and shown to contain ankyrin repeat domains known for protein-protein interaction [68,69]. To understand the biological function of NPR1, Zhang et al. [15] performed a yeast two-hybrid screen using NPR1 as bait and isolated NPR-1–interacting proteins that turned out to be TGA bZIP transcription factors. Interestingly, all three TGA factors AHBP-1B (TGA2), OBF5 (TGA5), and TGA6 with strong interaction with NPR1 fall into the same subgroup as mentioned earlier. These results suggest that NPR1 regulates *PR-1* gene expression by inter-

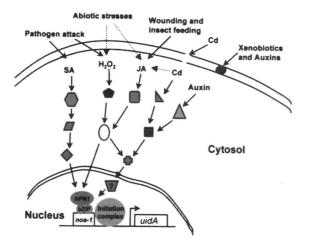

**Figure 5** A simplified model for the signal transduction pathways that originate from diverse stress-related stimuli and eventually converge on a single *cis* element. To illustrate the complexity of the signal transduction pathways that originate from diverse external stress-related stimuli (e.g., pathogen attack, wounding and insect feeding, abiotic stresses, and exposure to heavy metals and organic xenobiotics), a single *nos-1* element is shown to drive the expression of the reporter gene *uidA* and to be the ultimate signal recipient. The *nos-1* element may be bound by two TGA dimers with which NPR1 and/or OBP-1 can interact. As a result, the complex assembled on the *nos-1* element is able to interact positively with the transcription initiation complex and activate transcription. However, how signals are transduced to NPR1 and OBP-1 remains largely unknown.

acting with TGA factors that bind to the *as-1* cognate element in the *PR-1* gene promoter. Indeed, NPR1 stimulates the binding of TGA factors to their binding sites in vitro [62]. It appears that NPR1 preferentially interacts with TGA factors [70]. These exciting results were further confirmed and extended by the work of Despres et al. [62]. It was found that NPR1 can also interact with TGA3 but not TGA1 or TGA4 (OBF4). This result coincides with the subgroups within this gene family, suggesting that NPR1 has preference for TGA factors. It should be noted that NPR1 is not the only protein factor that interacts with TGA factors. An OBF-binding factor, OBP1, has been cloned by Zhang et al. [71] OBP-1 itself is a sequence-specific DNA-binding protein that enhances the binding of OBF to *ocs* elements. Li et al. [72] identified and cloned a negative regulator of SAR, SN1, in a screen for *npr1-1* suppressor, indicating that the induction of SAR involves activation and derepression. A new component of the SA signal transduction pathway was also characterized [73].

It is interesting to speculate that the biological significance of the preferential interaction may play a role in selecting target genes in response to different stress signals. It is even more attractive if considering the variant TGA factor binding sites in the promoters of target genes and the distinct DNA binding properties of different TGA factors together with the preferential interaction of NPR1 with TGA factors. It has been observed that the expression of abiotic stress-related genes such as GSTs was also altered in *npr1* mutants in response to stress signal molecules other than SA. More interestingly, there are clear differences in the expression of these genes in different *npr1* alleles (unpublished results). These observations indicate that NPR1 may be a converging point of different signaling pathways and/or a channel for cross talks between signaling pathways.

## VII. PERSPECTIVES

The signal perception and transduction processes in plant defense systems are very complex and still remain largely unknown. The interconnectedness among different pathways makes it even more complicated (for review, see Refs. 74, 75). There is no doubt that more mutants are required for dissecting these pathways and functional genomic tools should also prove extremely useful in analyzing the global regulation by stress signals as well as identifying target genes of a particular signaling pathway.

To dissect stress signaling pathways completely, new strategies for screening for more mutants have to be designed. A single *as-1*–type elements–driven reporter gene system has greater advantages in that it responds to various stress signals and avoids the complications of using a full-length promoter. A large-scale mutagenesis in the genetic background shown in Figure 2 has been carried out, and many putative $H_2O_2$-responsive mutants have been isolated. Screens for JA-, SA-, and 2,4-D–responsive mutants are being conducted. Characterization of these mutants should reveal more signaling components and shed new light on this exciting research field.

## REFERENCES

1.  E Lam, PN Benfey, PM Gilmartin, R-X Fang, N-H Chua. Site-specific mutations alter in vitro factor binding and change promoter expression pattern in transgenic plants. Proc Natl Acad Sci U S A 86:7890–7894, 1989.
2.  E Lam, F Katagiri, N-H Chua. Plant nuclear factor ASF-1 binds to an essential region of the nopaline synthase promoter. J Biol Chem 265:9909–9913, 1990.
3.  H Fromm, F Katagiri, N-H Chua. An octopine synthase enhancer element directs tissue-specific expression and binds ASF-1, a factor from tobacco nuclear extracts. Plant Cell 1:977–984, 1989.
4.  BJ van der Zaal, FNJ Droog, CJM Boot, LAM Hensgens, JHC Hoge, RA Schilperoort, KR Libbenga. Promoters of auxin-induced genes from tobacco can lead to auxin-inducible and root tip–specific expression. Plant Mol Biol 16:983–998, 1991.
5.  F Droog, A Spek, A van der Kooy, A de Ruyter, H Hoge, K Libbenga, P Hooykaas, B van der Zaal. Promoter analysis of the auxin-regulated tobacco glutathione *S*-transferase gene *Nt103-1* and *Nt103-35*. Plant Mol Biol 29:413–429, 1995.
6.  Y Takahashi, T Sakai, S Ishida, T Nagata. Identification of auxin-responsive elements of parB and their expression in apices of shoot and root. Proc Natl Acad Sci U S A 92:6359–6363, 1995.
7.  T Ulmasov, G Hagen, T Guilfoyle. The soybean *GH2/4* gene contains an *ocs*-like promoter element responsive to diverse agents and encodes a glutathione *S*-transferase. Plant Mol Biol 26:1055–1064, 1994.
8.  R Dudler, C Hertig, G Rebman, J Bull, F Mauch. A pathogen-induced wheat gene encodes a protein homologous to glutathione *S*-transferases. Mol Plant-Microbe Interact 4:14–18, 1991.

9.  D Feltkamp, R Masterson, J Starke, S Rosahl. Analysis of the involvement of ocs-like bZip-binding elements in the differential strength of the bidirectional *mas1'2'* promoter. Plant Physiol 105:259–268, 1994.

10. E Lebel, P Heifetz, L Thorne, S Uknes, J Ryals, E Ward. Functional analysis of regulatory sequences controlling *PR-1* gene expression in *Arabidopsis*. Plant J 16:223–233, 1998.

11. J Rouster, R Leah, J Mundy, V Cameron-Mills. Identification of a methyl jasmonate–responsive region in the promoter of a lipoxygenase 1 gene in barley. Plant J 11:513–523, 1997.

12. RS Friling, S Bergelson, V Daniel. Two adjacent AP-!–like binding sites form the electrophile-responsive element of the murine glutathione *S*-transferase Ya subunit gene. Proc Natl Acad Sci. U S A 89:668–672, 1992.

13. J Ryals, U Neuenschwander, M Willitis, A Molina, H-Y Steiner, M Hunt. Systemic acquired resistance. Plant Cell 8:1809–1819, 1996.

14. G Strompen, R Gruner, UM Pfitzner. An as-1–like motif controls the level of expression of the gene for the pathogenesis-related protein 1a from tobacco. Plant Mol Biol 37:871–883, 1998.

15. Y Zhang, W Fan, M Kinkema, X Li, X Dong. Interaction of NPR1 with basic leucine zipper protein transcription factors that bind sequences required for salicylic acid induction of the *PR-1* gene. Proc Natl Acad Sci U S A 96:6522–6528, 1999.

16. G Sessa, Y Meller, R Fluhr. A GCC element and G-box motif participate in ethylene-induced expression of the *PRB-1b* gene. Plant Mol Biol 28:145–153, 1995.

17. T Neuefeind, P Reinemer, B Bieseler. Plant glutathione *S*-transferases and herbicide detoxification. Biol Chem 378:199–205, 1997.

18. DP Dixon, I Cummins, DJ Cole, R Edwards. Glutathione-mediated detoxification systems in plants. Curr Opin Plant Biol 1:258–266, 1998.

19. K Marrs. The functions and regulation of glutathione *S*-transferases in plants. Annu Rev Plant Physiol Plant Mol Biol 47:127–158, 1996.

20. BJ van der Zaal, FNJ Droog, FJ Pieterse, PJJ Hooykaas. Auxin-sensitive elements from promoters of tobacco GST genes and a consensus *as-1*–like element differ only in relative strength. Plant Physiol 110:79–88, 1996.

21. Y Takahashi, T Nagata. parB: an auxin-regulated gene encoding glutathione *S*-transferase. Proc Natl Acad Sci U S A 89:56–59, 1992.

22. Y Takahashi, H Kuroda, T Tanaka, I Takebe, T Nagata. Isolation of an auxin-regulated gene cDNA expressed during the transition from $G_0$ to S phase in tobacco mesophyll protoplasts. Proc Natl Acad Sci U S A 86:9279–9283, 1989.

23. Y Takahashi, M Kusaba, Y Hiraoka, T Nagata. Characterization of the auxin-regulated *par* gene from tobacco mesophyll protoplasts. Plant J 1:327–332, 1991.

24. Y Takahashi, Y Niwa, Y Machida, T Nagata. Location of the *cis*-acting auxin-responsive region in the promoter of the *par* gene from tobacco mesophyll protoplasts. Proc Natl Acad Sci U S A 87:8013–8016, 1990.

25. W Chen, G Chao, KB Singh. The promoter of a $H_2O_2$-inducible, *Arabidopsis* glutathione *S*-transferase gene contains closely linked OBF- and OBP1-binding sites. Plant J 10:955–966, 1996.

26. W Chen, KB Singh. The auxin, hydrogen peroxide and salicylic acid induced expression of the arabidopsis GST6 promoter is mediated in part by an ocs element. Plant J 19:667–677, 1999.

27. MJ May, CJ Leaver. *Arabidopsis thaliana* γ-glutamylcysteine synthetase is structurally unrelated to mammalian, yeast, and *Escherichia coli* homologues. Proc Natl Acad Sci U S A 91:10059–10063, 1995.

28. C Wang, DJ Oliver. Cloning of the cDNA and genomic clones for glutathione synthetase from *Arabidopsis thaliana* and complementation of a gsh2 mutant in fission yeast. Plant Mol Biol 31:1093–1104, 1996.

29. G Noctor, CH Foyer. Ascorbate and glutathione: keeping active oxygen under control. Annu Rev Plant Physiol Plant Mol Biol 49:249–279, 1998.

30. E Grill, EL Winnacker, MH Zenk. Phytochelatins, a class of heavy-metal-binding peptides from plants, are functionally analogous to metallothioneins. Proc Natl Acad Sci U S A 84:439–443, 1987.

31. C Xiang, DJ Oliver. Glutathione metabolic genes coordinately respond to heavy metals and jasmonic acid in *Arabidopsis*. Plant Cell 10:1539–1550, 1998.

32. ZH Miao, X Liu, E Lam. *TGA3* is a distinct member of the TGA family of bZIP transcription factor in *Arabidopsis thaliana*. Plant Mol Biol 29:1–11, 1994.

33. H Fromm, F Katagiri, N-H Chua. The tobacco transcription activator TGA1a binds to a sequence in the 5′ upstream region of a gene encoding a TGA1a-related protein. Mol Gen Genet 229:181–188, 1991.

34. RA Creelman, ML Tierney, JE Mullet. Jasmonic acid/methyl jasmonate accumulate in wounded soybean hypocotyls and modulate wound gene expression. Proc Natl Acad Sci U S A 89:4938–4941, 1992.

35. EE Farmer, CA Ryan. Interplant communication: airborne methyl jasmonate induces synthesis of proteinase inhibitors in plant leaves. Proc Natl Acad Sci U S A 87:7713–7716, 1990.

36. H Gundlach, MJ Muller, TM Kutchan, MH Zenk. Jasmonic acid is a signal transducer in elicitor-induced plant cell cultures. Proc Natl Acad Sci U S A 89:2389–2393, 1992.

37. E Bell, JE Mullet. Lipoxygenase gene expression is modulated in plants by water deficit, wounding, and methyl jasmonate. Mol Gen Genet 230:456–462, 1991.

38. I Anderson, W Becker, K Schluter, J Burges, B Parthier, K Apel. The identification of leaf thionin as one of the main jasmonate-induced proteins of barley. Plant Mol Biol 19:193–204, 1992.

39. B Chaudhry, F Muller-Uri, V Cameron-Mills, S Gough, D Simpson, K Skriver, J Mundy. The barley 60 kDa jasmonate-induced protein (JIP60) is a novel ribosome-inactivating protein. Plant J 6:815–824, 1994.

40. B Zhang, KB Singh. *Ocs* element promoter sequences are activated by auxin and salicylic acid in *Arabidopsis*. Proc Natl Acad Sci U S A 91:2507–2511, 1994.

41. Y Kim, K Buckley, MA Costa, G An. A 20 nucleotide upstream element is essential for the nopaline synthase (*nos*) promoter activity. Plant Mol Biol 24:105–117, 1994.

42. X Liu, E Lam. Two binding sites for the plant transcription factor ASF-1 can respond to auxin treatments in transgenic tobacco. J Biol Chem 269:668–675, 1994.

43. C Xiang, ZH Miao, E Lam. Coordinated activation of as-1–type elements and a tobacco glutathione *S*-transferase gene by auxins, salicylic acid, methyl-jasmonate and hydrogen peroxide. Plant Mol Biol 32:415–426, 1996.

44. G Hagen, N Uhrhammer, TJ Guilfoyle. Regulation of expression of an auxin-induced soybean sequence by cadmium. J Biol Chem 263:6442–6446, 1988.

45. FNJ Droog, PJJ Hooykaas, KR Libbenga, EJ van der Zaal. Proteins encoded by an auxin-regulated gene family of tobacco share limited but significant homology with glutathione *S*-transferase and one member indeed shows in vitro GST activity. Plant Mol Biol 21:965–972, 1993.

46. P Talalay, MJ De Long, HJ Prochaska. Identification of a common chemical signal regulating the induction of enzymes that protect against chemical carcinogenesis. Proc Natl Acad Sci U S A 85:8261–8265, 1988.

47. XF Qin, L Holuigue, DM Horvath, NH Chua. Immediate early transcription activation by salicylic acid via the cauliflower mosiac virus *as-1* element. Plant Cell 6:863–874, 1994.

48. C Lamb, RA Dixon. The oxidative burst in plant disease resistance. Annu Rev Plant Physiol Plant Mol Biol 48:251–275, 1997.

49. RA Creelman, JE Mullet. Biosynthesis and action of jasmonates in plants. Annu Rev Plant Physiol Plant Mol Biol 48:355–381, 1997.

50. RA Creelman, JE Mullet. Oligosaccharins, brassinolides, and jasmonates: nontraditional regulators of plant growth, development, and gene expression. Plant Cell 9:1211–1223, 1997.

51. KB Singh. Transcriptional regulation in plants: the importance of combinatorial control. Plant Physiol 118:1111–1120, 1998.

52. WH Landschulz, PE Johnson, SL McKnight. The leucine zipper: a hypothetical structure common to a new class of DNA binding protein. Science 240:1759–1764, 1988.

53. AC Vincent, K Struhl. ACR1, a yeast ATF/CREB repressor. Mol Cell Biol 12:5394–5405, 1992.

54. IG Cowell, A Skinner, HC Hurst. Transcriptional repression by a novel member of the bZIP family of transcription factors. Mol Cell Biol 12:3070–3077, 1992.

55. F Katagiri, E Lam, NH Chua. Two tobacco DNA-binding proteins with homology to the nuclear factor CREB. Nature 340:727–730, 1989.

56. J Tokuhisa, K Singh, ES Dennis, WJ Peacock. A DNA-binding protein factor recognises two binding domains within the octopine synthase enhancer element. Plant Cell 2:215–224, 1990.

57. E Lam, YKP Lam. Binding site requirements and differential representation of TGA factors in nuclear ASF-1 activity. Nucleic Acids Res 23:3778–3785, 1995.

58. U Schindler, H Beckmann, AR Cashmore. TGA1 and G-box binding factors: two distinct classes of *Arabidopsis* leucine zipper proteins compete for the G-box-like element TGACGTGG. Plant Cell 4:1309–1319, 1992.

59. T Kawata, T Imada, H Shiraishi, K Okada, Y Shimura, M Iwabuchi. A cDNA clone encoding HPB-1b homologue in *Arabidopsis thaliana*. Nucleic Acids Res 20:1141, 1992.

60. B Zhang, RC Foley, KB Singh. Isolation and characterization of two related *Arabidopsis ocs*-element bZIP binding proteins. Plant J 4:711–716, 1993.

61. C Xiang, Z Miao, E Lam. DNA-binding properties, genomic organization and expression pattern of TGA6, a new member of the TGA family of bZIP transcription factors in *Arabidopsis thaliana*. Plant Mol Biol 34:403–415, 1997.

62. C. Despres, C DeLong, S Glaze, E Liu, PR Fobert. The arabidopsis NPR1/NIM1 protein enhances the DNA binding activity of a subgroup of the TGA family of bZIP transcription factors. Plant Cell 12:279–290, 2000.

63. PE Staswick, W Su, SH Howell. Methyl jasmonate inhibition of root growth and induction of a leaf protein are decreased in an *Arabidopsis thaliana* mutant. Proc Natl Acad Sci U S A 89:6837–6840, 1992.

64. BJF Feys, CE Benedetti, CN Penfold, JG Turner. *Arabidopsis* mutants selected for resistance to the phytotoxin coronatine are male sterile, insensititive to methyl jasmonate, and resistant to a bacterial pathogen. Plant Cell 6:751–759, 1994.

65. S Berger, E Bell, JE Mullet. Two methyl jasmonate–insensititive mutants show altered expression of AtVsp in response to methyl jasmonate and wounding. Plant Physiol 111:525–531, 1996.

66. DX Xie, BF Feys, S James, M Nieto-Rostro, JG Turner. *COI1*: an *Arabidopsis* gene required for jasmonate-regulated defense and fertility. Science 280:1091–1094, 1998.

67. H Cao, SA Bowling, S Gordon, X Dong. Characterization of an *Arabidopsis* mutant that is nonresponsive to inducers of systemic acquired resistance. Plant Cell 6:1583–1592, 1994.

68. H Cao, J Glazebrook, JD Clarke, S Volko, X Dong. The *Arabidopsis NPR1* gene that controls systemic acquired resistance encodes a novel protein containing ankyrin repeats. Cell 88:57–63, 1997.

69. J Ryals, K Weymann, K Lawton. The *Arabidopsis* NIM1 protein shows homology to the mammalian transcription factor I kappa B. Plant Cell 9:425–439, 1997.

70. JM Zhou, Y Trifa, H Silva, D Pontier, E Lam, J Shah, DF Klessig. NPR1 differentially interacts with members of the TGA/OBF family of transcription factors that bind an element of the *PR-1* gene required for induction by salicylic acid. Mol Plant-Microbe Interact 13:191–202, 2000.
71. B Zhang, W Chen, RC Foley, M Büttner, KB Singh. Interactions between distinct types of DNA binding proteins enhance binding to *ocs* element promoter sequence. Plant Cell 7:2241–2252, 1995.
72. X Li, Y Zhang, JD Clarke, Y Li, X Dong. Identification and cloning of a negative regulator of systemic acquired resistance, SNI1, through a screen for suppressors of *npr1-1*. Cell 98:329–339, 1999.
73. J Shah, P Kachroo, DF Klessig. The *Arabidopsis* ssi1 mutation restores pathogenesis-related gene expression in npr1 plants and renders defensin gene expression salicylic acid dependent. Plant Cell 11:191–206, 1999.
74. X Dong. SA, JA, ethylene and disease resistance in plants. Curr Opin Plant Biol 1:316–323, 1998.
75. K Maleck, RA Dietrich. Defense on multiple fronts: how do plants cope with diverse enemies? Trends Plant Sci 4:215–219, 1999.

# 26

# Multilevel Regulation of Glutathione Homeostasis in Higher Plants

**Chengbin Xiang and David J. Oliver**

*Iowa State University, Ames, Iowa*

## I. INTRODUCTION

Glutathione (GSH), the tripeptide γ-glutamylcysteinyl glycine, is ubiquitous and the most abundant non-protein thiol in plant cells [1]. The unique structure of GSH gives this molecule the exceptional stability, high water solubility, and redox properties that make GSH well suited to perform diverse functions. GSH is proposed to play important roles in defense. These include scavenging reactive oxygen species (ROS) through the ascorbate-GSH cycle [2–5], detoxifying a variety of organic electrophilic contaminants through glutathione S-transferases [6] and heavy metals through the phytochelatins (PCs) synthesized by PC synthase [7–9] upon heavy metal exposure [10,11]. In addition, GSH is important in organic sulfur storage, transport, and metabolism [12].

GSH is enzymatically synthesized from glutamate, cysteine, and glycine by two ATP-dependent reaction [13]. The first reaction is rate limiting and forms γ-glutamylcysteine (γ-EC) from glutamate and cysteine by the enzyme γ-EC synthetase (γ-ECS) [14], which is encoded by *GSH1* [15]. GSH is then synthesized by the ligation of γ-EC and glycine in the reaction catalyzed by the enzyme GSH synthetase encoded by *GSH2* [16]. The end product, GSH, is believed to feedback inhibit the rate-limiting enzyme, γ-ECS, so that a steady-state cellular GSH level is metabolically controlled. When GSH is oxidized as part of its antioxidant activity, it forms GSH disulfide (GSSG) and constitutes a major cellular redox buffer. GSSG, the oxidized form of GSH, is reduced back to GSH with reducing equivalents from reduced nicotinamide-adenine dinucleotide phosphate (NADPH) by GSH reductases that are encoded by *GR1* and *GR2* [17].

Because of the diverse functions of GSH in plants, it is not surprising that its synthesis is controlled by a network of complex regulatory mechanisms. This multilevel regulation provides the ability for GSH synthesis to respond to all the different environmental stress factors that GSH must mitigate. These complex regulatory mechanisms are starting to be uncovered. This review chapter focuses on the regulation of GSH synthesis in response to environmental stress. A working model is presented to depict the complexity of GSH homeostasis and autoregulation of GSH synthesis. GSH transport and turnover, which contribute to GSH homeostasis in the whole plant body, are also discussed.

## II. REGULATION OF GSH HOMEOSTASIS

GSH homeostasis in the whole plant body is dynamically established through a concerted interplay of synthesis, transport, utilization, and degradation. Little is known about the regulation of these processes.

Multilevel regulation of GSH synthesis and homeostasis has been implicated in a number of studies [18–20]. To date, at least five levels of control of steady-state GSH concentrations have been identified or implicated. These are (1) control of the transcription of the genes for GSH synthesis, (2) posttranscritional regulations including translational controls, (3) feedback inhibition of GSH formation at $\gamma$-EC synthetase, (4) rate limitation of GSH synthesis by $\gamma$-EC synthetase activity, and (5) substrate availability. These temporal regulatory mechanisms are further complicated by spatial regulations at the whole plant level and GSH transport and turnover.

## A.  Transcriptional Regulation

### 1.  The Genes for GSH Synthesis and Recycling Are Coordinately Up-Regulated in Response to Heavy Metals and Jasmonic Acid

It was previously demonstrated that *Arabidopsis* plants treated with cadmium or copper responded by increasing transcription of the genes for GSH synthesis, $\gamma$-ECS and GSH synthetase, as well as GSH reductase [20]. The response was specific for the metals whose toxicity is mitigated through phytochelatins. Other toxic and nontoxic metals did not alter messenger RNA (mRNA) levels. Feeding experiments suggested that neither oxidative stress, as results from exposure to $H_2O_2$, nor oxidized or reduced glutathione levels were responsible for activating transcription of these genes.

Jasmonic acid (JA), a naturally occurring growth regulator that has important roles in plant development and in insect and disease resistance (for review see Refs. 21 and 22), also activated the same suite of genes. This suggests that JA may be a general stress signal molecule and may be involved in the signal transduction pathway for copper and cadmium.

The up-regulation of the transcripts by JA and heavy metals was dependent on de novo protein synthesis. The elevated transcript accumulation was transcriptionally controlled. Interestingly, these genes respond to heavy metals and JA in a coordinated manner [20].

### 2.  The Responsiveness to Heavy Metals and JA Appears to Be Mediated by an as-1–Like Element in the GSH1 Promoter

Part of the complexity in understanding the regulation of $\gamma$-ECS activity is that the transcription of *GSH1* is regulated by both heavy metals and JA. An integrated approach was taken to investigate the molecular mechanisms by which JA and heavy metals control the expression of *GSH1*. Promoter deletion analysis was conducted to locate the heavy metal– and JA-responsive *cis* elements in the *GSH1* promoter. *Arabidopsis* genomic clones for *GSH1* were isolated. The *GSH1* promoter and its 5' coding region were sequenced. Promoter deletion analysis for *GSH1* identified a region responsive to JA and heavy metals. An apparent *as-1*–type element is located within this region. Removal of this element renders the promoter nonresponsive to JA or heavy metals (C. Xiang and D. J. Oliver, unpublished results). The *as-1*–type elements are known to respond to multiple external stimuli, including JA and heavy metals [23,24]. This *as-1*–type element in the *GSH1* promoter may be responsible for the up-regulation by JA and heavy metals. It is speculated that other genes regulated coordinately with *GSH1* may also possess *as-1*–like elements in their promoters.

### 3.  Heavy Metal and JA Signaling Pathways Are Parallel

How does the *GSH1* promoter respond to both JA and heavy metals? One interpretation of the data is that JA is an intermediate in the heavy metal signal transduction pathway. To ascertain whether JA and heavy metals share a common signal transduction pathway in the activation of *GSH1* or whether there are two different pathways that converge on a common *cis* element, two *Arabidopsis* mutants, *jar1* and *fad3-2fad7-2fad8* (18:3), were analyzed. The mutant *jar1* is JA unresponsive [25] and the triple mutant *fad3-2fad7-2fad8* is deficient in the jasmonate precursor, linolenic acid [26]. RNA gel blot analysis demonstrates that heavy metal signaling and JA signaling pathways are independent of each other. In both the JA-deficient triple mutant and *jar1*, the *GSH1* gene is still able to respond to heavy metal treatment in the same way as wild-type plants (Xiang and Oliver, unpublished results). Clearly, JA is not a mandatory intermediate in the heavy metal response pathway. However, these results cannot rule out the possibility that the exposure to heavy metals triggers the production of JA in vivo.

In order to dissect the JA and heavy metal signaling pathways, a major effort is currently under way to isolate *Arabidopsis* mutants deficient in JA or heavy metal responsiveness by using the reporter line described earlier.

## B. Translational Control

### 1. The Translation of GSH1 mRNA Is Regulated in Response to JA and $H_2O_2$

The existence of posttranscriptional control for GSH synthesis was demonstrated previously [20]. First, jasmonic acid treatment increased mRNA levels and the capacity for GSH synthesis but did not alter the GSH content in unstressed plants, suggesting that posttranscriptional regulation is involved. Second, oxidative stress in the form of $H_2O_2$ did not up-regulate the transcript level of GSH metabolic genes but did raise the GSH concentration in plants [27,28] (Xiang and Oliver, unpublished data). These observations strongly suggest that some additional control mechanisms must be functioning, possibly at the translational or posttranslational levels. Posttranscriptional regulation of *GSH1* expression was also implicated from a study using suspension cell culture [19].

To investigate further the control mechanisms for the expression of *GSH1* in response to JA and $H_2O_2$ [20], the steady-state transcript and protein levels of the endogenous γ-ECS were analyzed. It was found that JA increased the steady-state level of *GSH1* transcript at least 10-fold but did not significantly alter the γ-ECS protein level. In contrast, $H_2O_2$ treatment had no appreciable effect on *GSH1* transcript level but did elevate the endogenous γ-ECS protein level after 3 hr or longer exposure (Xiang and Oliver, unpublished results). This disconnection between the levels of *GSH1* mRNA and γ-ECS protein clearly demonstrates the dramatic change in the translational efficiency of *GSH1* mRNA or in γ-ECS protein turnover in response to $H_2O_2$ and JA.

The regulation observed for the endogenous *GSH1* expression was confirmed for the reporter gene *uidA* under the control of the *GSH1* gene promoter and its entire 5′ untranslated region (5′UTR) in the transgenic *Arabidopsis* plants. The RNA gel blot analysis of the reporter lines revealed that the transgene behaved like the endogenous *GSH1* and the level of GUS mRNA increased following treatment with JA but not with $H_2O_2$. Although the level of GUS mRNA increased with JA treatment, GUS activity as well as GUS protein level did not significantly change. When the transgenic reporter plants were exposed to $H_2O_2$, the transcript level remained unchanged, but the GUS protein level increased substantially (Xiang and Oliver, unpublished results).

Given that these are two unrelated proteins, a change in protein stability for both γ-ECS and GUS is unlikely under these conditions. Therefore, the control mechanism uncovered is most likely at the translational level but not at a posttranslational level. Taken together, these results strongly favor a translational control mechanism modulating the de novo synthesis of the rate-limiting enzyme for GSH synthesis from the existing *GSH1* mRNA.

### 2. The Translation of GSH1 mRNA May Be Modulated by a Redox-Sensitive Binding Activity Specific for the 5′UTR of GSH1 mRNA

The previous observations on the translational control of *GSH1* mRNA suggest a mechanism that might be similar to that of the iron-responsive element/iron regulatory proteins (IRE/IRPs)-regulated iron homeostasis in animal cells, where cellular iron availability controls the translation of ferritin mRNA [29]. The IRP binds to the 5′UTR of ferritin mRNA and blocks its translation when the cellular iron concentration is low. When excess iron is present, IRP dissociates from ferritin mRNA, allowing the synthesis of ferritin, the iron storage protein. JA increased the mRNA level for both γ-ECS and GUS, but this elevated transcript level was not paralleled by an increased protein level, suggesting that its translation was blocked. Exposure to $H_2O_2$ increased de novo synthesis of γ-ECS and GUS from the existing mRNA, probably by releasing this block on translation. It is speculated that a protein(s) might bind the *GSH1* mRNA and the binding of this protein might be altered by $H_2O_2$.

Because the *GSH1:uidA* construct contains the 5′UTR but lacks the *GSH1* coding region and 3′UTR, this putative *cis* element should be within the 5′UTR of the *GSH1* mRNA. Using RNA gel retardation assays and the 5′UTR of *GSH1* mRNA as a probe, a specific binding activity was identified in protein extracts from *Arabidopsis*. This binding activity is specific for the 5′UTR of *GSH1* mRNA because only RNA containing the 5′UTR of *GSH1* mRNA competes for binding with the factor. The RNA for luciferase and the 5'UTRs of plant mRNA for the ribosomal protein S15 and *phyA* do not compete in the binding assays (Xiang and Oliver, unpublished results).

More important, the in vitro binding of this factor(s) is sensitive to the GSH/GSSG ratio, a parameter that, in turn, is dependent on the $H_2O_2$ concentration. Inclusion of 5 mM GSSG in the assay reaction completely abolished the binding, and dithiothreitol (DTT) and GSH enhanced the binding. However, $H_2O_2$ did not significantly alter the binding, indicating that the effect of $H_2O_2$ might be indirect through changing the GSH/GSSG ratio (Xiang and Oliver, unpublished results). These results suggest that at high GSH/GSSG ratios, as would be found in unstressed plant cells, the factor binds to the 5'UTR of *GSH1* mRNA and restricts the de novo synthesis of γ-ECS, therefore preventing the synthesis of excessive GSH that is not needed. In contrast, at low GSH/GSSG ratios, as would be found following $H_2O_2$ or $Cu^{2+}$ treatments or under oxidative stress, the binding of the factor decreases, derepressing the translation of *GSH1* mRNA. Subsequently, de novo synthesis of γ-ECS would increase, and as a result, more GSH would be synthesized to replenish the reduced GSH pool within cells.

This is consistent with the observation of the oxidative stress–stimulated accumulation of the steady-state γ-ECS protein level in $CuCl_2$-treated *Arabidopsis* tissues. $Cu^{2-}$ is known to decrease the GSH/GSSG ratio and to cause oxidative stress as well as to increase the *GSH1* transcript level in our system [20].

### 3. Significance of Translational Control by Oxidative Stress

Translational control is the regulation of the efficiency of translation of mRNAs, either globally on general protein synthesis or selectively on a specific mRNA or a subset of mRNAs. Both activation and repression of mRNA translation occur very quickly as compared with transcriptional regulation, thus affording cells directness, rapidity, and reversibility in regulating gene expression in response to constantly changing living conditions [30]. Numerous studies have indicated translational control of individual or suites of genes in plants (for review, see Ref. 31, 32, and 33). But there are very few cases in which the molecular mechanism is understood. In higher plants, adaptive responses to oxidative stress are widespread and crucial for plant survival and crop productivity. However, translational control has not been well explored in these responses.

The translational control of *GSH1* mRNA by oxidative stress is significant in that the specific RNA-binding complex acts as both an oxidative stress sensor and a modulator of translation. The GSH/GSSG ratio would serve as a signal of oxidative stress and a switch to regulate GSH synthesis by modulating the translation of *GSH1* mRNA. It is interesting to speculate that translational control by GSH/GSSG may regulate the expression of a suite of enzymes whose activity is modulated by oxidative stress. GSH reductase (GR) expression appears to be translationally regulated in maize bundle sheath cells [34]. Although there is GR mRNA in this tissue, there is little or no GR protein. These authors propose that the cellular redox potential possibly mediated through the $NADP^+$/NADPH ratio controls the translation of GR mRNA. Thus it is possible that cellular redox status may control the translation of a suite of proteins involved in protecting plants from oxidative stress.

## C.   Metabolic Regulation

### 1. Feedback Inhibition of γ-EC Synthetase by GSH in Vivo

The role of feedback inhibition by GSH on γ-EC synthetase has been extrapolated from in vitro results [14]. Whether this feedback inhibition operates in vivo in higher plants has not been verified. A molecular genetic approach was taken to assess whether the feedback inhibition of γ-ECS by GSH functions in vivo.

The transgenic *Arabidopsis* plants overexpressing *Arabidopsis* γ-ECS indicate that large increases in γ-ECS protein level do not necessarily result in parallel increases in GSH level (Xiang and Oliver, unpublished results). Similar results were reported for poplar plants overexpressing *Escherichia coli* γ-ECS [35]. These observations are indicative of metabolic regulation of γ-ECS activity in vivo, most likely the feedback inhibition by GSH. Another way to demonstrate whether such regulation operates in vivo is to examine the metabolic flux through the γ-ECS–catalyzed reaction in the plant tissues with lowered GSH levels

Antisense RNA techniques were used to repress the expression of *GSH2* encoding GSH synthetase [16]. Transgenic *Arabidopsis* plants expressing the *GSH2* complementary DNA (cDNA) in the antisense orientation under the regulation of CaMV 35S promoter were generated; confirmed by genomic DNA, RNA, and protein gel blot analyses; and analyzed for thiol content. Among the herbicide-resistant T2 segregates containing the anti-*GSH2* and *bar* selectable marker gene, the GSH level dropped to about 60% of the wild-type level. These plants had an average 90-fold increase in γ-EC levels and a 25% increase in total thiols compared with the wild type. The herbicide-sensitive segregates, on the other hand, showed a thiol composition similar to that of wild-type plants [36]. This observation is consistent with GSH controlling its own synthesis by feedback control on γ-EC synthetase. The lowered GSH levels would relax the feedback control, resulting in the observed elevation in γ-EC concentrations as well as total thiols. These results demonstrate that the same feedback mechanism observed in vitro also works in vivo in higher plants.

## 2. Substrate Availability

The control by substrate availability and γ-EC synthetase activity was demonstrated in poplars by Foyer and colleagues [4,35,37–42]. They showed that the amount of γ-EC synthetase but not of GSH synthetase controls the GSH level in these transgenic plants. Glycine, the substrate for GSH synthetase, was shown to be limiting in the dark. The change of glycine pool size diurnally correlates well with the change of the GSH pool size. In plants overexpressing γ-EC synthetase, the amount of GSH can be increased substantially by feeding cysteine, suggesting that under these conditions the availability of this amino acid limits synthesis of GSH.

As a key substrate, cysteine can be an important limiting factor of GSH synthesis. The amino acid cysteine occupies a central position in sulfur metabolism in higher plants. It serves as a hub linking sulfur assimilation and utilization. Accumulating evidence suggests that there is coordination between these pathways (see later).

# D. Temporal and Spatial Regulation

## 1. Expression Pattern of Arabidopsis GSH1 Gene and GSH Synthesis

To appreciate fully the complexity of multilevel regulation, the spatial control of *GSH1* expression must be investigated. The expression pattern of *GSH1* should provide clues to where GSH is synthesized. To address this problem, the expression pattern of the *GSH1* promoter was examined in the same *Arabidopsis* transgenic lines used for translational control analysis. The expression of *GSH1* is developmentally regulated and tissue specific. In germinating seeds and young seedlings, the *GSH1* promoter activity is primarily localized in roots, indicating early onset of GSH synthesis in this organ. As seedlings develop into the rosette stage, the activity appears in leaves and remains high in roots. In mature plants, the activity is localized in rosette leaves, flower buds, flowers, and young siliques. Strong GUS staining is always observed in the root system of young seedlings and in mature roots and is consistent with the γ-ECS protein level. In flower buds, only stigma and immature anthers show GUS activity. In flowers, GUS activity is exclusively localized in anthers, stigma, and receptacle. GUS activity is localized in the tip and the base of young siliques. The spatial pattern of GUS activity in flowering plants is rather intriguing and may reflect the complexity of GSH synthesis at the whole plant level.

This expression pattern is also similar to that of the gene coding for the bZIP transcriptional factor TGA6 that specifically binds *as-1*–type elements in vitro [43]. Because to the nature of the translational reporter fusion, the GUS activity should report the localization of γ-ECS. The spatial expression pattern of γ-ECS during development indicates that the expression of γ-ECS is subjected to mutilevel control and GSH homeostasis in a whole plant body is complex. These data also indicate that GSH transport between cells and among organs must be functioning in order to maintain an effective GSH level in each cell.

RNA gel blot analysis of steady-state transcript levels for *GSH1* and *GSH2* in different tissues primarily agrees with the GUS expression pattern. The transcripts for *GSH1* as well as *GSH2* were detected in all tissues examined with relatively higher levels in siliques, roots, and inflorescence stems. γ-ECS protein is present in all tissues analyzed, although the protein levels vary from tissue to tissue. The disjunction between transcription and translation is obvious in these data; notably, the protein level for γ-ECS is highest in roots while the *GSH1* transcript level is lower there than in siliques and inflorescence stems, where less γ-ECS protein is present. These data again indicate the multilevel regulation of GSH synthesis.

Intriguingly, GSH levels on a fresh weight basis vary significantly from tissue to tissue. It appears that reproductive organs (flower buds, flowers, and siliques) have higher GSH levels than other organs, although the transcript levels for *GSH1* are low in flower buds and flowers. Leaves, the source tissue, are generally thought to be the major site of GSH synthesis; however, they have the lowest GSH level among all tissues. The differences in GSH levels in tissues may indicate the intrinsic differences in GSH requirements of different organs and active GSH transport among these organs.

## 2.  *Transport and Turnover of GSH*

Transport of GSH in higher plants, both intracellular and long-distance interorgan, is another important level of regulation of GSH homeostasis. Up to now, the existence of intracellular GSH transporters has not been demonstrated. Although significant progress has been made on the regulation of sulfur uptake and assimilation by GSH [44,45] and GSH transport between organs has been studied [46] the regulation of long-distance interorgan GSH transport deserves further investigation in greater detail.

One of the biological functions assigned to GSH is the storage and transport of reduced sulfur (cysteine). Therefore, GSH has to be degraded in order to make cysteine available for protein synthesis and other utilization. Unfortunately, very little is known about the regulation of GSH degradation in higher plants, although the pathway(s) is known and similar to that in animals [47–50]. The DNA sequence information for γ-glutamyl transpeptidase and 5-oxo-L-prolinase is available from GenBank. The 5-oxo-L-prolinase is extremely conserved through evolution in that its more than 3000 bp coding sequence is intronless. This single open reading frame–encoded polypeptide shares high amino acid sequence homology with that in other organisms from prokaryotes to human. Cloning of the genes encoding the key enzymes of the GSH degradation pathway will provide necessary tools to probe the regulation mechanisms for GSH breakdown.

## III.  COORDINATION BETWEEN GSH AND CYSTEINE SYNTHESIS

Cysteine is a key substrate for GSH synthesis. It has been expected that there should be coordination between sulfur uptake/assimilation and GSH synthesis [5]. Indeed, the transcription of genes for GSH synthesis (γ-ECS) and for sulfur assimilation (5′-adenylylsulfate [APS] reductases) is coordinated in response to heavy metals in plants [20,51]. The expression of mRNA and enzyme activity for the sulfur assimilation enzyme APS reductase was analyzed in *Brassica juncea* exposed to $CdSO_4$. In plants exposed to varying Cd doses up to 200 μM the level of APS reductase mRNA increased linearly with the dose but the activity lagged behind the mRNA and was inhibited above 50 μM. When the plants were treated with Zn, Pb, Cu, and Hg, APS reductase mRNA and enzyme activity increased synchronously [51]. Figure 1 shows that the expression of *APR* genes encoding the rate-limiting enzyme, APS reductase, for cysteine synthesis coordinates with that of *GSH1* for GSH synthesis in response to JA treatment. *GSH1* and two *APR* genes analyzed in this experiment were induced in 1 hr by JA treatment. The induction kinetics of the *APR* genes was similar to that of *GSH1*. The induction by JA appears to be dose dependent. Unlike *GSH1*, the *APR* genes are strongly responsive to $H_2O_2$.

These results suggest that the genes encoding the rate-limiting enzymes for both pathways are coordinately regulated in response to environmental stresses. The cross talk between sulfur uptake/assimilation and GSH synthesis is currently under investigation.

## IV.  MULTILEVEL REGULATION OF GSH HOMEOSTASIS IN RESPONSE TO ENVIRONMENTAL STRESS

A working model for the multilevel regulation of *GSH1* expression and GSH homeostasis based on the experimental results is illustrated in Figure 2.

At the transcriptional level, heavy metals and JA up-regulate the transcription of the genes for GSH synthesis. The transcriptional control of *GSH1* expression in response to heavy metals and JA is mediated by the *cis* elements that have been recruited by *GSH1* promoter. These include the *as-1*–type element re-

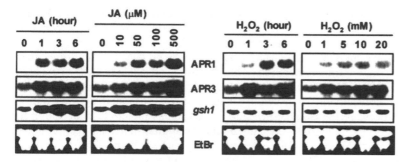

**Figure 1** Evidence for the coordination between cysteine and GSH synthesis pathways. *Arabidopsis* plants grown in liquid culture were treated with 100 μM JA for the indicated time or for 3 hr with the indicated concentrations of JA and with 5 mM $H_2O_2$ for the indicated time or for 3 hr with the indicated concentrations of $H_2O_2$. Total RNA isolation and gel blot analysis were performed as described [20]. The filter was sequentially probed with [32]P-labeled cDNA for *APR1, APR2*, and *gsh1*. Ethidium bromide–stained gel is shown for RNA equal loading.

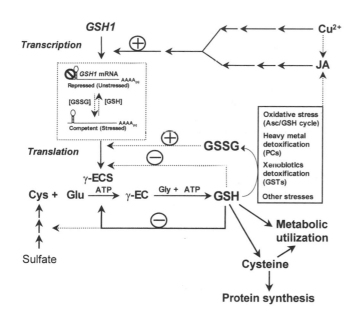

**Figure 2** Multilevel regulation of GSH homeostasis. GSH homeostasis is dynamically established through a concerted interplay of synthesis, transport, utilization, and degradation in plant cells. GSH is synthesized by γ-EC synthetase and GSH synthetase using Glu, Cys, Gly, and ATP as substrates. GSH is utilized to combat oxidative stress by the ascorbate (Asc)/GSH cycle, to synthesize phytochelatins (PCs) for detoxification of heavy metals and to detoxify xenobiotics by GSH *S*-transferases (GSTs). GSH homeostasis is regulated at multiple levels. At the transcriptional level, heavy metal Cu and JA use parallel signaling pathways to up-regulate *GSH1*. At the translational level, a redox-sensitive RNA-binding complex specific for the 5′ UTR of *GSH1* mRNA modulates the translation of *GSH1* mRNA. The redox-sensitive nature gives this RNA-binding complex the ability to sense the change in GSH/GSSG ratio, an indicator of oxidative stress, and to modulate the translation accordingly. At the metabolic level, GSH feedback inhibits γ-EC synthetase. In addition, transport and turnover of GSH contribute to GSH homeostasis. As the transport and storage form of organic sulfur, GSH is exported from source tissue to sink tissue, where it is degraded to release cysteine for protein synthesis and other utilization. Sulfate uptake and assimilation pathways that make cysteine the key substrate for GSH synthesis are likely to be coordinately regulated with the pathway for GSH synthesis. The circled plus sign indicates positive effects, and the minus sign indicates negative regulation. Dashed lines indicate uncertain but possible regulatory mechanisms. The stop sign indicates the RNA-binding factor bound to the 5′ UTR of *GSH1* mRNA.

sponsive to heavy metals and JA and possibly other *cis* elements to be identified in the promoter (Xiang and Oliver, unpublished data). JA and heavy metals appear to use parallel signaling pathways that eventually converge on a small region in the *GSH1* promoter.

At the translational level, the translation efficiency of the mRNAs produced by increased transcription may be altered in response to the cellular redox state. In unstressed cells, the mRNA could be repressed by a protein factor(s) present in the cytosol that binds to the 5'UTR of these mRNAs (at least to the mRNA for *GSH1*) and prevents their translation. Oxidative stress–generated ROS as well as heavy metals and xenobiotics decrease the GSH/GSSG ratio and cause this protein to dissociate from the mRNA molecules. Once this protein has dissociated, the mRNAs are translated and the rate-limiting enzyme for GSH synthesis is produced. This binding activity may serve as both a redox sensor and a modulator for the translation of *GSH1* mRNA in accordance with oxidative stress that the plants are experiencing. This translation control model is consistent with the observation that neither GSH, GSSG, nor $H_2O_2$ affected the transcript level of GSH metabolic genes [20]. Using this model, it can be better interpreted that translational control is functioning in a more direct and more rapid response to the fluctuating GSH/GSSG ratio in plant cells when compared with the much slower response through transcriptional control.

Given the fine metabolic control mechanisms (feedback inhibition and substrate availability) and possible posttranslational modification of GSH metabolic enzymes [18], plant cells have evolved multilevel regulatory mechanisms for GSH homeostasis. The significance of the multilevel regulation model is severalfold. First, it enables both long-term (transcriptional control) and short-term (translational) responses to the fluctuating oxidative stress status. Second, the translational control affords cells reversibility and flexibility in response to the GSH/GSSG ratio, an indicator of oxidative stress status. Third, the redox-sensitive nature of the RNA-binding complex is particularly important in that it may act as an oxidative stress sensor modulating the rate-limiting enzyme level for GSH synthesis and ensuring that an optimal GSH/GSSG ratio is maintained. It is likely that other genes in this pathway are also regulated by this regulatory mechanism. Sessile higher plants have evolved sophisticated oxidative stress sensing and modulating mechanisms that give plant cells the plasticity required for their adaptive responses in a constantly changing environment.

In addition, GSH synthesis coordinates with cysteine synthesis and utilization of GSH. The coordination between GSH synthesis and cysteine formation is also reflected metabolically. Sulfate uptake and assimilation are thought to be controlled by GSH [44,45]. As the storage form of organic sulfur, GSH has to be transported from source tissues to sink tissues and degraded in a controlled manner to release cysteine for protein synthesis and other metabolic utilization. These processes further increase the complexity of GSH homeostasis at the whole plant level.

## V. PERSPECTIVES

Taken together, these results demonstrate that the expression of $\gamma$-ECS is subject to multilevel regulation and indicate that GSH homeostasis at the whole plant level is very complex. Temporal multilevel regulation is further complicated with spatial control mechanisms and transport within the plant body. It is interesting and challenging to pursue how GSH synthesis and transport are regulated and coordinated among different organs and tissues at the whole plant level.

Preliminary results have shown that the genes encoding the rate-limiting enzymes for both pathways are coordinated at least at the transcript level. It is expected that these two pathways are simultaneously activated in response to environmental stresses where large consumption of GSH takes place. It has been demonstrated that in response to heavy metal exposure, both cysteine synthesis and GSH synthesis are elevated in *Arabidopsis* plants (Xiang and Oliver, unpublished results). Future investigation is also required to demonstrate at the molecular level the coordination between sulfate uptake and GSH synthesis.

Elucidation of the RNA-binding complex of the 5'UTR of *GSH1* mRNA should shed light on the molecular mechanisms by which oxidative stress controls the synthesis of GSH. An integrated approach combining molecular, biochemical, and genetic means has been taken to identify the components of the RNA-binding complex. Once the identity of the binding complex is resolved, other interesting questions, such as whether this binding complex regulates other mRNA species, especially oxidative stress-related transcripts, can be addressed.

## ACKNOWLEDGMENT

This research was funded by grants from the Competitive Grants Program of the USDA/NRI (99-35100-7545). The authors thank T. Leustek for providing *APR* cDNA clones and the *Arabidopsis* Biological Resource Center for the *Arabidopsis* cDNA library, EST clones, and seeds.

## REFERENCES

1. L Bergmann, H Rennenberg. Glutathione metabolism in plants. In: LJ De Kok, ed. Sulfur Nutrition and Assimilation in Higher Plants. The Hogue, the ••: SPB Academic Publishing bv, 1993, pp 109–123.
2. RA Larson. The antioxidants of higher plants. Phytochemistry 27:969–978, 1988.
3. RG Alscher. Biosynthesis and antioxidant function of glutathione in plants. Physiol Plant 77:457–464, 1989.
4. CH Foyer, P Descourvieres, KJ Kunert. Protection against oxygen radicals: an important defense mechanism studied in transgenic plants. Plant Cell Environ 17:507–523, 1994.
5. G Noctor, CH Foyer. Ascorbate and glutathione: keeping active oxygen under control. Annu Rev Plant Physiol Plant Mol Biol 49:249–279, 1998.
6. KA Marrs. The functions and regulation of glutathione *S*-transferases in plants. Annu Rev Plant Physiol Plant Mol Biol 47:127–158, 1996.
7. S Clemens, EJ Kim, D Neumann, JI Schroeder. Tolerance to toxic metals by a gene family of phytochelatin synthases from plants and yeast. EMBO J 15:3325–3333, 1999.
8. SB Ha, AP Smith, R Howden, WM Dietrich, S Bugg, MJ O'Connell, PB Goldsbrough, CS Cobbett. Phytochelatin synthase genes from *Arabidopsis* and the yeast *Schizosaccharomyces pombe*. Plant Cell 11:1153–1164, 1999.
9. OK Vatamaniu, S Mari, YP Lu, PA Rea. AtPCS1, a phytochelatin synthase from *Arabidopsis*: isolation and in vitro reconstitution. Proc Natl Acad Sci U S A 96:7110–7115, 1999.
10. E Grill, EL Winnacker, MH Zenk. Phytochelatins, a class of heavy-metal-binding peptides from plants, are functionally analogous to metallothioneins. Proc Natl Acad Sci U S A 84:439–443, 1987.
11. MH Zenk. Heavy metal detoxification in higher plants— a review. Gene 179:21–30, 1996.
12. T Leustek, K Saito. Sulfate transport and assimilation in plants. Plant Physiol 120:637–644, 1999.
13. A Meister, ME Anderson. Glutathione. Annu Rev Biochem 52:711–760, 1983.
14. R Hell, L Bergmann. γ-Glutamylcysteine synthetase in higher plants: catalytic properties and subcellular localization. Planta 180:603–612, 1990.
15. MJ May, CJ Leaver. *Arabidopsis thaliana* γ-glutamylcysteine synthetase is structurally unrelated to mammalian, yeast, and *Escherichia coli* homologues. Proc Natl Acad Sci U S A 91:10059–10063, 1995.
16. C Wang, DJ Oliver. Cloning of the cDNA and genomic clones for glutathione synthetase from *Arabidopsis thaliana* and complementation of a *gsh2* mutant in fission yeast. Plant Mol Biol 31:1093–1104, 1996.
17. A Kubo, T Sano, H Saji, K Tanaka, N Kondo, K Tanak. Primary structure and properties of glutathione reductase from *Arabidopsis thaliana*. Plant Cell Physiol 34:1259–1266, 1993.
18. MJ May, T Vernoux, CJ Leaver, MV Montagu, D Inzé. Glutathione homeostasis in plants: implications for environmental sensing and plant development. J Exp Bot 49:649–667, 1998.
19. MJ May, T Vernoux, R Sánchez-Fernández, MV Montagu, D Inzé. Evidence for posttranscriptional activation of γ-glutamylcysteine synthetase during plant stress responses. Proc Natl Acad Sci U S A 95:12049–12054, 1998.
20. C Xiang, DJ Oliver. Glutathione metabolic genes coordinately respond to heavy metals and jasmonic acid in *Arabidopsis*. Plant Cell 10:1539–1550, 1998.
21. RA Creelman, JE Mullet. Oligosaccharins, brassinolides and jasmonates: nontraditional regulators of plant growth, development, and gene expression. Plant Cell 9:1211–1223, 1997.
22. RA Creelman, JE Mullet. Biosynthesis and action of jasmonates in plants. Annu Rev Plant Physiol Plant Mol Biol 48:355–381, 1997.
23. T Ulmasov, G Hagen, T Guilfoyle. The soybean *GH2/4* gene contains as *ocs*-like promoter element responsive to diverse agents and encodes a glutathione *S*-transferase. Plant Mol Biol 26:1055–1064, 1994.
24. C Xiang, Z Miao, E Lam. Coordinated activation of *as-1*–type elements and a tobacco glutathione *S*-transferase gene by auxin, salicylic acid, methyl-jasmonate and hydrogen peroxide. Plant Mol Biol 32:415–426, 1996.
25. PE Staswick, W Su, SH Howell. Methyl jasmonate inhibition of root growth and induction of a leaf protein are decreased in an *Arabidopsis thaliana* mutant. Proc Natl Acad Sci U S A 89:6837–6840, 1992.
26. M McConn, RA Creelman, E Bell, JE Mullet, J Browse. Jasmonate is essential for insect defense in *Arabidopsis*. Proc Natl Acad Sci U S A 94:5473–5477, 1997.
27. IK Smith. Stimulation of glutathione synthesis in photorespiring plants by catalase inhibitors. Plant Physiol 79:1044–1047, 1985.
28. MJ May, CJ Leaver. Oxidative stimulation of glutathione synthesis in *Arabidopsis thaliana* suspension cultures. Plant Physiol 103:621–627, 1993.

29. MW Hentze, LC Kuhn. Molecular control of vertebrate iron metabolism: mRNA-based regulatory circuits operated by iron, nitric oxide, and oxidative stress. Proc Natl Acad Sci U S A 93:8175–8182, 1996.
30. MB Mathews, N Sonenberg, JWB Hershey. Origin and targets of translational control. In: JWB Hershey, MB Mathews, N Sonenberg, eds. Translational Control. Cold Spring Harbor, NY: Cold Spring Harbor Laboratory 1996, pp 1–29.
31. DR Gallie. Translational control of cellular and viral mRNAs. Plant Mol Biol 32:145–158, 1996.
32. DR Gallie. J Bailey-Serres. Eyes off transcription! The wonderful world of post-transcriptional regulation. Plant Cell 9:667–673, 1997.
33. J Bailey-Serres, DR Gallie eds. A Look Beyond Transcription—Mechanisms Determining mRNA Stability and Translation in Plants. Rocksville MD: American Society of Plant Physiologists. 1998.
34. GM Pastori, PM Mullineaux, CH. Foyer. Post-transcriptional regulation prevents accumulation of glutathione reductase protein and activity in the bundle sheath cells of maize. Plant Physiol 122:667–676, 2000.
35. G Noctor, M Strohm, L Jouanin, KJ Kunert, CH Foyer, H Rennenberg. Synthesis of glutathione in leaves of transgenic poplar overexpressing γ-glutamylcysteine synthetase. Plant Physiol 112:1071–1078, 1996.
36. C Xiang, D Bertrand. Glutathione synthesis in *Arabidopsis*: multilevel controls coordinate responses to stress. In: Sulfur nutrition and sulfur assimilation in higher plants: molecular biochemical and physiological aspects. C Brunold, H Rennenberger, LJ De Kok, I Stulen, JC Davidian, eds. Bern: Paul Haupt. 2000, pp. 409–412.
37. CH Foyer, N Souriau, S Perret, M Lelandais, KJ Kunert, C Pruvost, L Jouanin. Overexpression of the glutathione reductase but not glutathione synthetase leads to increases in antioxidant capacity and resistance to photoinhibition in poplar trees. Plant Phyiol 109:1047–1057, 1995.
38. CH Foyer, H Lopez-Delgado, JF Dat, IM Scott. Hydrogen peroxide– and glutathione-associated mechanisms of acclimatory stress tolerance and signalling. Physiol Plant 100:241–254, 1997.
39. M Strohm, L Jouanin, KJ Kunert, C Pruvost, A Polle, CH Foyer, H Rennenberg. Regulation of glutathione synthesis in leaves of transgenic polar (*Populus tremula* × *P. alba*) overexpressing glutathione synthetase. Plant J 7:141–145, 1995.
40. ACM Arisi, G Noctor, CH Foyer, L Jouanin. Modulation of the thiol contents in poplars (*Populus tremula* × *P. alba*) over-expressing enzymes involved in glutathione synthesis. Planta 202:357–369, 1997.
41. G Noctor, ACM Arisi, L Jouanin, MH Valadier, Y Roux, CH Foyer. The role of glycine in determining the rate of glutathione synthesis in poplar. Possible implications for glutathione production during stress. Physiol Plant 100:255–263, 1997.
42. G Noctor, ACM Arisi, L Jouanin, KJ Kunert, H Rennenberg, CH Foyer. Glutathione: biosynthesis, metabolism and relationship to stress tolerance explored in transformed plants. J Exp Bot 49:623–647, 1998.
43. C Xiang, Z Miao, E Lam. DNA binding properties, genomic organization and expression pattern of TGA6, a new member of the TGA family of bZIP transcription factors in *Arabidopsis thaliana*. Plant Mol Biol 34:403–415, 1997.
44. AG Lappartient, B Touraine. Demand-driven control of root ATP sulfurylase activity and $SO_4^{2-}$ uptake in intact canola. Plant Physiol 111:147–157, 1996.
45. AG Lappartient, JJ Vidmar, T Leustek, ADM Glass, B Touraine. Inter-organ signaling in plants: regulation of ATP sulfurylase and sulfate transporter genes expression in roots mediated by phloem-translocated compound. Plant J 18:89–95, 1999.
46. H Rennenberg, K Schmitz, L Bergmann. Long-distance transport of sulfur in *Nicotiana tabacum*. Planta 147:57–62, 1979.
47. H Rennenberg, R Steinkamp, J Kesselmeier. 5-Oxo-prolinase in *Nicotiana tabacum*: catalytic properties and subcellular localization. Physiol Plant 62:211–216, 1981.
48. R Steinkamp, H Rennenberg. γ-Glutamyltranspeptidase in tobacco suspension cultures: catalytic properties and subcellular localization. Physiol Plant 61:251–256, 1984.
49. R Steinkamp, B Schweihofen, H Rennenberg. γ-Glutamyltranspeptidase in tobacco suspension cultures: catalytic properties and subcellular localization. Physiol Plant 69:499–503, 1987.
50. A Meister. Glutathione metabolism and its selective modification. J Biol Chem 263:17205–17208, 1988.
51. S Lee, T Leustek. The affect of cadmium on sulfate assimilation enzymes in *Brassica juncea*. Plant Sci 141:201–207, 1999.

# 27
# Genes Associated with Orchid Flower

**Soek Ying Neo and Kwok Ki Ho**

*The National University of Singapore, Singapore, Republic of Singapore*

## I. INTRODUCTION

Orchids are a globally important flower crop. It has been estimated that in the year 2000 there was a worldwide demand of 1600 million units of planting material, and the majority of this was produced by the tissue culture method and used for mass cultivation in Southeast Asian countries [1]. To date, clonal propagation of orchids has been extended to more than 43 genera [2]. Moreover, the physiology of some orchid species in *Cymbidium, Dendrobium, Phalaenopsis,* and *Oncidium* has been studied intensively over the past decades, and this has led to the establishment of recommendations for flower induction and storage [2]. Despite the advances made in these areas, numerous questions related to flower production remain to be investigated. Examples are how to increase harvestable flower yield, how to control flowering to meet market demand, how to prolong the vase life of cut flowers, and how to enhance their aesthetic properties such as color and architecture. Such questions can be addressed with a better understanding of the genetic factors affecting such physiological processes including flower production and development, flowering time, senescence, and pigmentation. Careful examination of the orchid literature published over the past decade, particularly from a survey of the molecular database, indicates that researchers are beginning to reveal genes closely associated with these processes (Table 1) and are poised to make further progress in the next few years. In this chapter, we describe what is known to date about these orchid genes in relation to our current knowledge of their counterparts in other plant systems. It is hoped that this chapter will stimulate interest in further investigation of these and other genes associated with the biology of the orchid flower.

## II. FLOWER DEVELOPMENT

Over the past decade, there has been considerable progress in our understanding of floral development at the molecular level. From mutational studies of flower development in some plant species, a large number of genes associated with the transition of the apical vegetative meristem to the floral meristem have been identified [24]. For some, their roles in controlling the transition and organogenesis have been clearly demonstrated, whereas for others, their roles remain to be defined. It is now appreciated that the process of flower development involves two classes of consecutively acting regulatory genes that encode transcription factors. First, meristem-identity genes, expressed throughout the incipient floral primordia,

**TABLE 1**   Genomic and cDNA Clones from Orchids

| Identity | Accession number | Reference |
|---|---|---|
| ACC oxidase | *Doritaenopsis* sp. L07912 | 3 |
| | *Doritaenopsis* sp. L37103 | 4 |
| | *Dendrobium crumenatum* AF038840 | 5 |
| ACC synthase | *Phalaenopsis* Z77854 | 6 |
| | *Doritaenopsis* sp. L07882; L07883 | 3 |
| | *Dendrobium crumenatum* U64031 | 7 |
| Acyl-CoA oxidase | *Phalaenopsis* U66299 | 8 |
| Bibenzyl synthase | *Bromheadia finlaysoniana* AJ131830 | 9 |
| | *Phalaenopsis* sp. X79903; X79904 | 10 |
| Chalcone synthase | *Phalaenopsis* sp. `True Lady' U88077 | 11 |
| | *Bromheadia finlaysoniana* AF007097; AF007098, AF007099 | 12 |
| 4-Coumarate:CoA ligase | *Vanilla plamifolia* X75542 | 13 |
| | *Cymbidium hybrida* AF017451 | 14 |
| Cysteine proteinase | *Phalaenopsis* sp. `hybrid SM 9108' U34747 | 15 |
| Cytochrome P-450 monooxygenase | *Phalaenopsis* sp. `hybrid SM 9108' U34744 | 15 |
| Ethylene receptor | *Phalaenopsis* sp. `KCbutterfly' AF113541 | 16 |
| | *Phalaenopsis* sp. `True Lady' AF055894 | 17 |
| Flavanone 3-hydroxylase | *Bromheadia finlaysoniana* X89199 | 18 |
| Methyltranferase | *Vanilla planifolia* X78703; X73587; | 19 |
| | X87099; X69192 | 20 |
| Phenylalanine ammonia-lyase | *Bromheadia finlaysoniana* X99997 | 21 |
| Putative transcription factor | *Dendrobium grex* Madame Phong-In AF198174; AF198175; AF198176; AF107588 | 22 |
| | *Aranda* Deborah X69107 | 23 |
| | *Phalaenopsis* sp. `hybrid SM 9108' U34743 | 15 |
| *S*-Adenosyl homocysteine hydrolase | *Phalaenopsis* sp. X79905 | 10 |
| 0108; 0126 (function unknown) | *Phalaenopsis* sp. `hybrid SM 9108' U34745; U34746 | 15 |

are thought to be involved in flower initiation. Second, homeotic genes, expressed later, are responsible for the identity of individual floral organs. On the basis of studies of homeotic mutants that develop floral organs at positions normally occupied by other organs, floral homeotic genes are divided into three different groups that cooperatively determine floral organ formation: the AGAMOUS group, the APETALA3/PISTILLATA group, and the APETALA1/AGL1 group [25]. The majority of the homeotic genes have a highly conserved DNA binding domain known as the MADS box motif (MCM1, AGAMOUS, DEFICIENS, SRF) within the N-terminus region [26–31]. In vitro binding studies have shown that the MADS box proteins bind to a consensus sequence of $CC(A/T)_6GG$, although each protein possesses a distinct binding specificity. The DNA target sequence of the MADS box proteins is located in the promoters of numerous genes, including MADS box genes themselves and the genes controlled by them [32]. Another conserved region within plant MADS box proteins is the K box domain, an approximately 70-amino-acid region that has the potential to form an amphipathic α-helical structure that is thought to facilitate the formation of a functional MADS transcription factor dimer [33,34].

Studies of MADS box proteins have principally been performed in dicotyledonous species such as *Arabidopsis, Antirrhinum*, and *Petunia* [35]. It is now clear that flower development in monocotyledons such as orchids and gymnosperms is also based on a conserved organization of developmental factors in which MADS box proteins play a crucial role. The first orchid MADS box gene was isolated from *Aranda deborah* by screening the mature flower complementary DNA (cDNA) library with an *agamous* cDNA probe from *Arabidopsis* [23]. The gene, designated *om1*, is expressed in the petals and sepals of mature flowers but not in young developing inflorescences or young floral buds. The 250-amino-acid O-MADS protein encoded by the *om1* gene possesses a MADS box domain that is highly similar (>95%) to those of tomato and *Arabidopsis*. Four other MADS box cDNAs have now been identified in *Dendrobium grex* Madame Thong-IN during the floral transition [22]. Of these, two cDNAS, *otg7* and *DOMADS1*, both encode the same protein. They also possess very similar nucleotide sequences (97% similar) except that *DOMADS1* is a longer transcript with more 5′ and 3′ untranslated regions. DOMADS1, DOMADS2, and DOMADS3 are of different sizes, comprising 174 amino acids, 247 amino acids, and 220 amino acids, respectively, and are thought to belong to the APETALA1/AGL9 subfamily of the MADS box gene family. Sequence comparison reveals that DOMADS1 shares 54% and 67% similarity with DOMADS2 and DOMADS3, respectively, whereas DOMADS2 and DOMADS3 share only 49% similarity (Figure 1). Interestingly, O-MADS is more similar to DOMADS1 (78% similarity) and DOMADS3 (61% similarity) than DOMADS2 (40% similarity), suggesting that O-MADS, DOMADS1, and DOMADS3 may share some functional similarity and that DOMADS2 may play a different role than the others.

Substantial gaps remain in our knowledge of how these orchid floral homeotic genes are controlled. Although meristem-identity genes such as *FLOWERING LOCUS T* (*FT*) and *LEAFY* (*LFY*) have been identified in *Arabidopsis*, the orchid homologues have yet to be isolated. Furthermore, the link between meristem-identity and homeotic genes is unclear. Recent studies in *Arabidopsis* suggest that *FT* and *LFY* act in a parallel fashion to transduce floral initiating signals such as photoperiod [37,38]. The FT protein is similar to the sequence of TERMINAL FLOWER 1 (TFL1), an inhibitor of flowering that also shares sequence similarity with membrane-associated mammalian proteins including phosphatidylethanolamine-binding protein (PEBP). As PEBP is a precursor of hippocampal cholinergic neurostimulating peptide, this raises the intriguing possibility that peptide molecules may be generated as the transmissible signals during flower development. Isolation of the orchid homologues of *Arabidopsis FT* and *TFL1* using heterologous probes may provide some insights.

```
           1                                                            60
DOMADS1    MGRGRVEMKRIENKINRQVTFAKRRTGLLKKAYELSVLCDVEVALIIFSNRGKLYEFCSS
O-MADS     MGRGRVELKMIENKINRQVTFAKRRKRLLKKAYELSVLCDAEVALIIFSNRGKLYEFCSS
DOMADS3    MGRGRVELKRIENKINRQVTFAKRRNGLLKKAYELSVLCDAEVALIVFSNRGRLFEFCSS
DOMADS2    MGRGRVQLKRIENKINRQVTFSKRRSGLLKKAHEISVLCDAEVALIVFSNKGKLYEYSTD
           ******::* ***********:***. *****:*:***** *****:***:*:*:*:..:.

           120
DOMADS1    RSMLKTLEKYQKCSDGAPEMTMTSRETQSSQ-VEYLKLKSQVEALQRSQRNLLGEDLNPL
O-MADS     TSMLKTLEKYQKCNFGSPESTIISRETQSSQ-QEYLKLKNRVEALQRSQRNLLGEDLGPL
DOMADS3    TSMTKTLERYQKCSYNASESAVPSKDAQNSY-HEYLTLKAKVEYLQRSQGNLLGEDLIEL
DOMADS2    SSMEKILERYERYSYAERALFSNEANPQADWRLEYNKLKARVESLQKSQRHLMGEQLDSL
           **  * **:*:: .          . :.*  .  ** .** :** **:** :*:**:*  *

           180
DOMADS1    GGKDLDQLERQLEASLKQIISTRMQYMLDQLGDLQQRELLLFETNKSLGTRVS--AL---
O-MADS     GSKELEQLERQLDSSLRQIRSTRTQFMLDQLADLQRREQMLCEANKTLKRRFE--ESSQA
DOMADS3    SSKELDQLELQLEMSLKQIRSTKTQLMLDQLCDIKRKEQMLHEANRALSMKLK--EDGPE
DOMADS2    SIKELQRLEQQLESSLKFIRSRKTQLILHSISELQKMEKILLEQNKTLEKEIIAKEKAKA
           . *:*::** **: **: * *  : * :*..: :::: * :* * *::*  ..

           240
DOMADS1    ------------------------------------------------------------
O-MADS     NQQQV-WDPSNTHAVGYGRQPAQHHGEAFYHPLEC-EPTLQIGYHSDITMATATASTVNN
DOMADS3    IPLELSWPGGETNGSSERQQP---QSDKFFQPLPCSNPSLQIGYSP--------------
DOMADS2    LVQHAPWEKQNQSQYSSALPP--VISDSVPTPTSR---TFQARANEEESPQPQ-LRVSNT

           255
DOMADS1    ---------------
O-MADS     YMPPGWLGQISGSYE
DOMADS3    ---------------
DOMADS2    LLPPWMLSHMNGQ--
```

**Figure 1** Comparison of amino acid sequences among O-MADS (EMBL/GenBank Accession X69107) from *Aranda deborah* [23] and DOMADS1 (EMBL/GenBank Accession AF198174), DOMADS2 (EMBL/GenBank Accession AF198175), and DOMADS3 (EMBL/GenBank Accession AF198176) from *Dendrobium grex* Madame Thong-IN [22]. Alignment was performed by the CLUSTAL W program [36]. The stars indicate perfectly conserved sequences among the genes, colons represent conservation of strong groups, and dots represent conservation of weak groups.

## III.  POLLINATION OF FLOWERS

The study of flower pollination has long attracted interest because of the developmental changes that take place following the pollination event and the fact that these changes contribute to successful reproduction. The major physiological and morphological changes in orchid flowers following pollination have been described in great detail [39]. Of these changes, ovule differentiation, perianth senescence, and ethylene biosynthesis have been the central thrust of molecular studies. The model orchid system *Phalaenopsis* has been the source of many of the genes involved in these changes to date.

### A.  Ovule Development

Orchid is unusual in the sense that in many of its species, ovule development is induced by pollination. Following induction, the development of the megasporocyte and megagametophyte is not much different from that in many other plant species, and the subsequent maturation of the embryo sac conforms to the *Polygonum* type [40]. Characterization of gene expression associated with ovule differentiation in *Phalaenopsis* sp. SM9108 has been aided by defining the timing and hormonal stimuli of each developmental transition stage [41]. The isolation of genes *O39, O40, O108, O126, O141* [15], and *O138* [42] from *Phalaenopsis* is further facilitated by establishing stage-specific cDNA libraries derived from the archesporial cell, megaspore mother cell, and mature embryo sac and screening for cDNA clones differentially expressed at each of these stages.

*O39* encodes a 84-kDa putative homeobox transcription factor and is expressed in the ovule from formation of the ovule primordium at early stages through to various stages of ovule tissue differentiation. On the basis of its pattern of expression and its strong similarity at the N-terminal to the homeobox DNA binding motif of transcription factors, *O39* is thought to be involved in the initiation of ovule development. Alternatively, *O39* may be induced in response to the commitment to ovule differentiation, and it may in turn regulate a subset of genes involved in the developmental pathway to ensure appropriate differentiation. Using *O39* as a probe to screen an *Arabidopsis* floral bud cDNA library resulted in identification of the *Arabidopsis* homologue *ATML1* [43]. *ATML1* is expressed specifically in the L1 layer of the meristem from the earliest stages of meristem patterning and throughout shoot development, suggesting that *ATML1* may be involved in setting up morphogenetic boundaries of positional information necessary for controlling cell specification and pattern formation. *ATML1* is also thought to provide an early molecular marker for the establishment of both apical-based radical patterns during plant embryogenesis. As with other homeobox protein families, *Phalaenopsis* O39 has little sequence similarity to other homeobox protein families outside the homeobox region. A notable exception is *Arabidopsis* GLABRA2 (GL2), which regulates trichome differentiation, and ATML1. O39, together with the *Arabidopsis* GLABRA2 (GL2) and ATML1, is proposed to define a new class of plant homeodomain-containing proteins designated HD-GL2.

*O40* encodes a cytochrome P450 monooxygenase of predicted molecular mass 48 kDa. The large superfamily of cytochrome P450 monooxygenases are membrane-bound enzymes that catalyze the oxidation of diverse and often overlapping substrates of both endogenous and xenobiotic origin in bacteria, fungi, plants, and animals [44,45]. O40 is the first orchid cytochrome P450 identified, and by convention it represents the second of the CYP78 gene family and consequently is named CYP78A2 by the Cytochrome P450 Gene Nomenclature Committee. In situ hybridization has shown that O40 messenger RNA (mRNA) is not specific to ovules but is found exclusively in the pollen tubes intertwined with the ovules during early ovule development. The function of O40 is as yet unclear, although it is thought to play a role in the biosynthesis of a hormone involved in intercellular communication, similar to its homologue in the animal systems.

In contrast to *O39* and *O40*, *O108, O126*, and *O141* are selectively expressed in maturing ovules. The 15-kDa putative protein encoded by *O108* contains a consensus ATP/GTP binding site at the C-terminal. *O108* is expressed in actively dividing cells of the ovule, exclusively in the outer layer of the outer integument and in the female gametophyte of mature ovules. When *O108* was initially cloned, it showed no obvious similarity to any other gene with a known function in the database. *O108* was then thought to be important for successful seed production, perhaps by facilitating signal transduction between the ovule and pollen tube. Based on the sequence of O108, the *Arabidopsis* homologue *Atskp1* has been identified [46]. *Atskp1* is expressed in all actively dividing cells of the plant body, particularly in plant meristem.

Apart from *Arabidopsis Atskp1, Phalaenopsis O108* also shares significant homology with yeast *SKP1* that encodes a novel kinetochore protein required for cell cycle progression [47,48]. Because ovule development is initiated from a quiescent meristem in the orchid ovary that becomes activated by pollination, it is likely that some of the ovule-associated genes may be involved in cell cycle regulation, and *O108* appears to be a potential candidate. Further experiments, such as complementation assays using yeast skp1 mutants with O108, may establish the functional relationship between yeast skp1 and O108.

*O126* encodes an 18-kDa glycine-rich putative protein that contains a signal peptide sequence similar to those of other glycine-rich proteins that are thought to be structural components of the cell wall. This suggests that the *O126* protein is a component of a specialized cell wall in the ovule. Wang et al. [42] reported an *O138* gene regulated in a stage- and tissue-specific manner, but the sequence is not available in the EMBL/GenBank database. *O141* encodes a putatative 40-kDa cysteine proteinase that is most similar to endopeptidases of the papain family found in seed or fruit. Cysteine proteinases occur widely in plants and are induced by various stress conditions, including cold, heat, salt, and drought, and by wounding [49]. The expression of cysteine proteinase genes is also associated with ripening in a number of fruits such as tomato and citrus, where they are predicted to play a role in fruit development. In situ hybridization indicates that *O141* is specifically expressed in the outer integument of ovules during seed formation. Because the formation of seed coat requires the degeneration of integument cells, it is conceivable that *O141* may be involved in the developmentally regulated programmed cell death in a manner akin to the mammalian cysteine proteases called caspases [50]. In agreement with this hypothesis, recent findings in soybean cells show that cysteine proteases are involved in the regulation of programmed cell death in plants [51].

## B. Perianth Senescence

Pollination-induced senescence is a well-documented phenomenon in many flowers [52,53]. For *Phalaenopsis*, the longevity of intact unpollinated flowers can reach up to 3 months, but once pollinated, the petals start to show signs of visible senescence sometimes within 1 day. Such an early display of senescence symptoms has been taken as an indication that some pollination signals move through the flower, eventually reaching the petals well before pollen germination and fertilization, both of which begin much later after pollination. There is substantial evidence supporting a role for an increase in sensitivity to ethylene following pollination in pollination-induced petal senescence [54–57]. This "sensitivity signal" is thought to be the first signal moving into the petals, much earlier than the ethylene biosynthesis signal. In *Phalaenopsis*, direct involvement of GTP-binding proteins, calcium, and protein phosphorylation has been implicated in the regulation of ethylene sensitivity [58]. Studies suggest that short-chain saturated fatty acids (SCSFAs), particularly octanoic acid, may be the ethylene "sensitivity factors" produced following pollination, and they are likely to act by altering the properties of the lipid bilayer membrane [59]. Other candidate molecules for ethylene sensitivity factors include auxin, but not ethylene, 1-aminocyclopropane-1-carboxylate (ACC), systemin, lipoxygenase, and jasmonates [41,60–62]. More recently, a small molecular substance with molecular weight (MW) below 3000 that is distinct from auxin has been extracted from the pollinia of *Phalaenopsis* [63]. Initial characterization of this substance indicates that it is water soluble, unlikely to be proteinaceous in nature, and may be separated into at least five different fractions by high-performance liquid chromatography (HPLC). The exact nature of the primary pollen signals awaits further purification of these peaks and more detailed chromatographic analysis such as gas chromatography coupled with mass spectrometry.

Clearly, the ethylene sensitivity pathway is closely associated with loss of membrane integrity, indicating membrane changes. In connection with this, β-oxidation of fatty acids plays an essential role in the lipid metabolism of the cell membrane and is carried out exclusively by glyoxysomes in plants [64]. Catalysis of the dehydrogenation of fatty acyl-coenzyme (CoA) into hydrogen peroxide is achieved by acyl-CoA oxidase, the rate-limiting enzyme of peroxisomal β-oxidation [65]. Although acyl-CoA oxidase has been identified in various plant species, little is understood about its expression and regulation at the molecular level. A *Phalaenopsis* cDNA *pOACO31* has been isolated by screening a library of poly(A)$^+$ RNA extracted from petals 1 day after pollination [8]. It encodes a 699-amino-acid putative peroxisomal acyl-CoA oxidase protein named PACO1 and is expressed specifically in petals after induction by pollination. PACO1 shares significant sequence similarity with the peroxisomal acyl-CoA oxidase from human, rat, and yeast acyl-CoA, particularly within 13 conserved regions and a putative flavin mononucleotide (FMN) binding site [66]. The genomic structure of PACO1 encompasses approximately 14 kb and is divided into seven exons by six introns [67]. Southern blot analysis indicates that the gene occurs in one copy or a low number of copies per haploid genome, and Northern analysis shows two species of alternatively spliced mRNA. Understanding the complexity of the PACO1 gene structure and its regulation should further elucidate the role of β-oxidation in flower senescence.

A major event occurring after pollination is an increase in ethylene production, which is generally accepted to be a primary signal for pollination-induced senescence [40,53]. After the physical event of pollination, ethylene evolution occurs in the stigma, possibly due to triggering by pollen-borne auxin and other factors, followed by endogenous synthesis of ACC. The details of ethylene biosynthesis and the interorgan regulation are discussed in the next section.

## C.  Ethylene Biosynthesis in Pollinated Flowers

The ethylene biosynthetic pathway involves the conversion of S-adenosylmethionine to ACC by ACC synthase. The ethylene precursor ACC is then converted to ethylene by ACC oxidase [68]. ACC synthase and ACC oxidase are therefore key enzymes in ethylene biosynthesis, although ACC synthase is generally considered to be rate limiting. Both ACC synthase and ACC oxidase genes are encoded by multigene families and have been well characterized in many plant species [69,70]. ACC synthase genes exhibit differential tissue specificity and are regulated by environmental or hormonal stimuli [68,70]. ACC oxidase genes are considered in most cases to be mostly constitutive [71]. However, there is compelling evidence that ACC oxidase genes are differentially expressed and are highly regulated in floral organs [69,72,73]. In addition, both ACC synthase and ACC oxidase contribute to a positive feedback loop wherein ethylene treatment leads to increased ethylene production [74].

In orchids, three full-length ACC synthase genes, Ds-ACS1, Ds-ACS2 [3] and Pt-ACS1 [6], have been identified in Phalaenopsis. Their predicted protein sequences comprise 425, 444, and 445 amino acids, respectively. In addition, an ACC synthase gene DC-ACS, which is predicted to encode a 435-amino-acid protein, has been isolated from Dendrobium crumenatum [7]. Ds-ACS1 and Ds-ACS2 appear to be different transcripts of the same gene, and their predicted protein sequences show 97% similarity (Figure 2). In addition, Ds-ACS1 is reported to be homologous to Phal-ACS1 [75]. Pt-ACS1 shares 68%, 67%, and 74% similarity with Ds-ACS1, Ds-ACS2, and DC-ACS, respectively (Figure 3). DC-ACS is highly similar to Ds-ACS1 (84% similarity) and Ds-ACS2 (85% similarity). The partial sequences of two other distinct Phalenopsis ACC synthase genes, Phal-ACS2 and Phal-ACS3, have been cloned by reverse transcription–polymerase chain reaction (RT-PCR) [75], indicating that Phalaenopsis ACC synthase consists of at least three gene members. These ACC synthase genes show differential expression in floral organs in response to pollination and various chemical stimuli. Phal-ACS2 and Phal-ACS3 genes are expressed within 1–2 hr after pollination in the stigma and ovary, respectively, whereas the expression of Phal-ACS1 is detected only in the stigma 6 hr after pollination. Furthermore, Phal-ACS2 and Phal-ACS3 are induced in response to primary pollination signals such as auxin, whereas Ds-ACS1 is induced in response to secondary pollination signals such as ACC. This suggests that orchid flowers have at least two different types of ACC synthase genes, one responding to the primary pollination signal and another amplifying the primary signal by triggering and/or sustaining autocatalytic ethylene production. The auxin-induced Phal-ACS3 mRNA accumulation in the ovary is severalfold less than that induced by pollination, indicating that an unknown pollination factor may have a synergistic effect with auxin. The differential expression of Phal-ACS1 and Phal-ACS2 in the stigma of pollinated orchids suggests that their combined expression may be responsible for the de novo synthesis of ACC required for sustained ethylene production in the stigma. Similar regulation of ACC synthase in carnation has been shown [76].

For ACC oxidase, two genes, OAO1 [3,73] and D-ACO2 [4], have been isolated from Phalaenopsis and one gene, DCACO, has been identified in Dendrobium [5]. Their predicted protein sequences consist of 317, 318, and 325 amino acids, respectively. Sequence comparison indicates that these ACC oxidase proteins are highly similar to each other. D-ACO2 shares 94% similarity with OAO1, and DCACO shares 85% similarity with both OAO1 and D-ACO2 (Figure 3). The regulation of ACC oxidase activity and mRNA expression within the floral organs in Phalaenopsis orchid after pollination have been examined. Rapid induction of ACC oxidase activity and high accumulation of the OAO1 mRNA is observed following pollination. Furthermore, ACC oxidase gene expression is regulated by ethylene.

```
          1                                                           60
Ds-ACS1   MSKMFGKEVPLSKMAVSKAHGEGSPYFAGWKAYEENPYDVVGNPDGVIQMGLAENQLSFD
Ds-ACS2   MSKMFGKEVPLSKMAVSKAHGEGSPYFAGWKAYEENRYDAVGNPDGVIQMGLAENQLSFD
DC-ACS    MSKEFGIEAPLSKIAVSKAHGEDSPYFAGWKADSPYDAVDNPNGVIQMGLAENQLSFD
Pt-ACS1   MARGIGLGAPLSKIAVSKAHGEDSPYFAGWKAYDEDPYDPVANPTGVIQMGLAENQVSFD
          *::  :*  .****:******** .**********:*: ** * ** ***********:***
                                                                    120
Ds-ACS1   LLEEYLELHPEAFSWASDSSSFRENALFQDYHGLQTLRQALASFMEKIRGGRSKFDANRI
Ds-ACS2   LLEEYLELHPEAFSWASDSSSFRENALFQDYHGLQTLRQALASFMEKIRGGRSKFDANRI
DC-ACS    LLEEYLEQHPEASSWGSGMSSFRENALFQDYHGLQTFRKAMASFMEKIRGGRSKFDPDRI
Pt-ACS1   LLEDYLEQHPETASWSSGISGFKENALFQDYHGLQSFRKAMASLMEQIRGKRVKFNPDRM
          ***:*** ***: **.*. *.*:*********** :*.*:**:**:*** * **:.:*:
                                                                    180
Ds-ACS1   VLTAGATAANEILTFILADRGDALLVPTPYYP-------------------GSNGFQLT
Ds-ACS2   VLTAGATAANEILTFILADRGDALLVPTPYYPGFLRDLQWRTGVTIFPVHCYSSNGFQLT
DC-ACS    VLTAGATAANELLTFILADRGDALLVPTPYYPGFLRDLQWRTGVTIFPVHCHSSNGFQLT
Pt-ACS1   VLTAGATAANELLTFILADPGDAVLVPTPYYPGFDRDLQWRTGVKIFPVHCHSSNGFQLT
          ***********:******* ***:********                   *******
                                                                    240
Ds-ACS1   LSSLEKAYAEAKASNFNVRGLLMTNPCNPLGTSASLSLLQDIIHFISDKNIHLISDEIYS
Ds-ACS2   LSSLEKAYAEAKASNFNVRGLLMTNPCNPLGTSASLSLLQDIIHFISDKTIHLISDEIYS
DC-ACS    LASLESAYADAKASNFNVKGLLITNPCNPLGTVASLSLLQDIILFISDKNIHLISDEIYF
Pt-ACS1   LPSLQSAYSIAEASNLRVKALLITNPSNPLGTTMPRSLLEDILNFISQKNIHLISDEIYS
          *.**:.**: *:***.*:.**:**:***.*****  ***.:**: ***:*.*********
                                                                    300
Ds-ACS1   GSVFSSTNLFSISDLITDAIS---EQVHIVYSLSKDLGLPGFRVGALYSYNDRVVKTARR
Ds-ACS2   GSVFSSTNLFSISDLITDAIS---EQVHIVYSLSKDLGLPGFRVGALYSYNDRVVKTARR
DC-ACS    GCVFSSTNLFSISDLITNAVS---YQIHIVYSLSKDLGLPGFRVGALFSYNDRVVKTARR
Pt-ACS1   GSVFSSPEFISVAEVVEASQHKNCDGVHIVYSLSKDLGLPGFRVGTIYSYNDRVVTTARR
          *.****.:;:*::::: :  :****************:::*******.****
                                                                    360
Ds-ACS1   MSSFSLVSSQTQKLLSFMLSDEEFTVRYIEKNRERLRERYELVVNGLKEAGIECLKGEAG
Ds-ACS2   MSSFSLVSSQTQRLLSFMLSDEEFTVRYIEKNRERLRERYELVVNGLKEAGIECLKGEAG
DC-ACS    MSSFSLVSSQTQKLLAFMLSDEEFTVNYIKKNRERLRERYELVVGGLKEAGIECLKGEAG
Pt-ACS1   MSSFSLVSSQTQKMLASMLSDEEFTVKYIKTNRERLRRHGYIVEGLKDAGIECLEGNAG
          ************:;*: *********.**:. ******.*:   :* ***:******:*:**
                                                                    420
Ds-ACS1   LFCWVNMEKLMEEETKEGRARLWKVIIDDLKLNISPCEECCCABPGWFRLCFANMSRETL
Ds-ACS2   LFCWVNMEKLMEEETKEGEAEPWKVIIDDLKLNISPGSSCCCAEPGWFRLCFANMSRETL
DC-ACS    LFCWVNMEELMEDKTEEGELRLWKVMVDELKLNISPGSSCCCSEPGWFRVCFANTSRETL
Pt-ACS1   LFCWMNLTQMLEEKSMEGELRLWKMILSEVKLNVSPGSSCYCAEPGWFRVCFANMTKETL
          ****:*:  :::*::: *** . **::::.:;****.****** *;******:**** ::***
                                                                    428
Ds-ACS1   EVALKRLKDFAQKKVAAKKKKMNNVYFC
Ds-ACS2   EG-VKRLKDFAQKKVAAKKKKMNNVYFC
DC-ACS    EVALRRLKDFAR-------KKTKNI---
Pt-ACS1   EVALRRTKDFVEK---IKGRKSNVIFRS
          *  ::* ***..     :*  :  :
```

**Figure 2** Comparison of amino acid sequences among Ds-ACS1 (EMBL/GenBank Accession L07882), Ds-ACS2 (EMBL/GenBank Accession L07883) from *Doritaenopsis* sp. [3] DC-ACS (EMBL/GenBank Accession U64031) from *Dendrobium crumenatum* [7] and Pt-ACS1 (EMBL/GenBank Accession Z77854) from *Phalaenopsis* [6]. Alignment was performed by the CLUSTAL W program [36]. The stars indicate perfectly conserved sequences among the genes, colons represent conservation of strong groups, and dots represent conservation of weak groups.

Despite the cloning of these ACC synthase and ACC oxidase genes in orchid, the nature of pollen-pistil interactions leading to the expression of these genes in the floral organs is not well clarified. Based on the current data, the latest model [3,75,77] for the interorgan regulation of ACC synthase and ACC oxidase gene expression in the pollinated orchid flower has been modified. This model, which is similar to the one proposed in carnation [78] suggests that auxin and other pollen-borne factors are the primary inducers of ACC synthase gene expression in the stigma and ovary, and the precursor ACC is the main signal translocated among the floral organs. However, the molecular events that follow ethylene perception and gene expression are not clearly understood.

The important role of ethylene in agriculture has led to intensive investigation of signal transduction of this compound. The first putative ethylene receptor gene *ETR1*, thought to be the first component of the ethylene signal transduction cascade, was isolated from *Arabidopsis* [79]. It encodes a protein resembling the two-component signal transducer in prokaryotes [80]. Subsequently, more than 30 putative ethylene receptors have been isolated from different plant species [69,81,82]. Sequence comparison of the *Arabidopsis* and tomato ethylene receptors reveals the existence of a second type of re-

```
         1                                                             60
OAO1     ----------MELLQGSQRPAAMALLRDACENWGLYELLNHGISHELMNRVETVNKEHY-
D-ACO2   MESGSFPVINMELLQGSQRPAAMALLRDACENWGFFELLNHGISHELMNRVEAVNKEHY-
DCACO    ----------MELLEGSRRSDAMAVLRDACENWGFFELLNHGISHEPNEQSETVNKEHYR
                   ****:**:*. ***:**********:;********** ;: *:******
                                                                     120
OAO1     ---RRFREQRFKEFASKTLDTVENVEPENLDWESTFFLRHLPTSNISQIPDLDDDCRSTM
D-ACO2   ---RRFREQRFKEFASKTLDSVENVDPDNLDWESTFFLRHLPTSNISQIPDLDDDCRATM
DCACO    EHYRRFREQRFKEFAAKTLDSGENVGDNLDWESTFFLRHLPTSNISQVPDLDEDCGSTM
            ************:****:.***: :************************:****:**:**
                                                               180
OAO1     KEFALELENLAERLLDLLCEDLGLEKGYLKKVFCGGSDGLPTFGTKVSNYPPCPKPELIK
D-ACO2   KEFARELEKLAERLLDLLCEDLGLEKGYLKRVFCGGSDGLPTFGTKVSNYPPCPKPDLIK
DCACO    KEFALGLEKLAERLLDLLCENLGLEKGYLKRVFCGGSDGLPTFGTKVSNYPPCPKPELIK
         ****  **:***********:**********:***************************:***
                                                                 240
OAO1     GLRAHTDAGGIILLFQDDKVSGLQLLKDGEWIDVPPVRHSIVVNIGDQLEVITNGKYKSV
D-ACO2   GLRAHTDAGGIILLFQDDKVSGLQLLKDREWIEVPPLRYSIVVNIGDQLEVITNGKYKSV
DCACO    GLRAHTDAGGIILLFQDDTVSGLQLLKDEEWIDVPPMRHSIVVNIGDQLEVITNGKYKSV
         *****************.********* ***:***:*:*****************
                                                                   300
OAO1     LHRVVAQTDGNRMSIASFYNPGSDAVIFPAPALVEKEAEEKEEKKKEIYPKFVFQDYMNL
D-ACO2   LHRVVAQTDGNRMSIASFYNPGSDAVIFPAPALVEKEAE--EEEKKEIYPKFVFQDYMNL
DCACO    MHRVVAQTNGNRMSIASFYNPGSDAVIFPAPELVEKEAA---EKKKETYPKFVFEDYMKL
         :*******:**************************:*****   *:*** ******:***:*
                                       331
OAO1     YIRKKFEAKEPRFEAMKSMEIVMSSQPIPTA
D-ACO2   YIRKKFEAKEPRFEAMKSMEIVMSSQPIPTA
DCACO    YVRQKFEAKEPRFEAMKTMDAVISSQPIPTA
         *:*:.**************:**: *:********
```

**Figure 3** Comparison of amino acid sequences among OAO1 (EMBL/GenBank Accession L07912) from *Doritaenopsis* sp. [3], D-ACO2 (EMBL/GenBank Accession L37103) from *Doritaenopsis* sp. [4] and DC-ACO (EMBL/GenBank Accession AF038840) from *Dendrobium crumenatum* [5]. Alignment was performed by the CLUSTAL W program [36]. The stars indicate perfectly conserved sequences among the genes, colons represent conservation of strong groups, and dots represent conservation of weak groups.

ceptor ERS, which shares a high degree of identity with the N-terminal domain and putative histidine protein kinase domain of ETR1 but lacks the C-terminal response regulator domain present in ETR1. Using the N-terminal domain of the *ETR1* gene as a probe, an orchid ethylene receptor homologue PER1 [16] and another distinct putative orchid ethylene receptor [17] have been isolated in *Phalaenopsis* sp. 'True Lady' and *Phalaenopsis* sp. 'KCbutterfly', respectively. Sequence comparison indicates that PER1 represents a novel ERS type of ethylene receptor in that it lacks the C-terminal domain, but its N-terminus and histidine kinase domain are more closely related to *Arabidopsis* ETR1 than to ERS. These putative orchid ethylene receptors are also highly similar to ETR1 from bacteria such as *E. coli* and *Pseudomonas* that is responsible for environmental sensing for the presence of nutrients or changes in osmolarity [83]. Whether these orchid ethylene receptors are responsible for the perception of ethylene signal remains to be clarified.

## IV.  FLORAL COLORATION

Flavonoids are secondary metabolites, and the importance of some of these in plant pigmentation has been well documented [84]. There are several types of flavonoid molecules, the chalcones, flavanones, flavones, flavonols, and anthocyanins. Among the different flavonoids, anthocyanins are the most common and important. Some of these are highly colored and tend to have marked effects in colors or patterns. The biosynthesis of flavonoids has been reviewed several times [85–87] and will not be described in detail here. Briefly, it starts with the condensation of three malonyl-CoA molecules with one molecule of hydroxycinnamic acid. The resulting products, chalcones, are converted to flavanones, which give rise to flavones, flavonols, and anthocyanins.

Our understanding of the molecular and biochemical determinants affecting floral coloration of orchid is not as advanced as that of some other ornamental plants. Nevertheless, studies are beginning to shed light on some of the genes that encode enzymes involved in the flavonoid pathway of orchid petal tissue. To date, the enzymes for which cDNA clones have been isolated are phenylalanine ammonia-lyase

(PAL), chalcone synthetase (CHS), the flavanone 3-hydroxylase (F3H), and dihydroflavonol 4-reductase (DFR).

The first enzyme, PAL, is the only one not directly associated with the flavonoid pathway. It is a key enzyme catalyzing the first step from phenylalanine to hydroxycinnamic acid, the deamination of phenylalanine to cinnamic acid. Hydroxycinnamic acid is required for flavonoid biosynthesis. Two cDNA clones were isolated for *Bromheadia finlaysoniana* PAL but only one was sequenced to completion [21]. In several other species, PAL is encoded by a small family of genes, although loblolly pine contains only a single PAL gene [88,89].

The second enzyme, CHS, catalyzes the formation of chalcones as indicated earlier. Three cDNA clones were isolated for *B. finlaysoniana* CHS [12], and evidence was obtained for multiple CHS genes. Further evidence indicated the presence of transcripts in all floral organs and plant parts. However, definitive experiments to differentiate the transcripts corresponding to individual cDNA clones have not been reported. The analysis of CHS genes in different plant species has shown variability in copy number ranging from 1 to 10 copies per genome [90–94]. Individual members of CHS multigene families can be differently regulated. In the case of petunia, it is known that only one of the CHS genes is strongly expressed in petal tissue [95]. A cDNA clone was also obtained for *Phalaenopsis* sp. 'True lady' CHS [11]. The deduced amino acid sequence of the latter CHS is 60–64% identical to that of *B. finlaysoniana* CHS. We did the sequence comparison using an ALIGN program [36]. Based on results of Southern hybridization, the *Phalaenopsis* CHS gene was reported to exist in multiple copies.

The third enzyme, F3H, is responsible for the conversion of flavanones to dihydroflavonols, the relative expression of its gene affecting the production of different anthocyanins. A cDNA clone was reported for *B. finlaysoniana* F3H [18].

The last enzyme, DFR, catalyzes the first committed step to anthocyanin biosynthesis by converting dihydroflavonols into leucoanthocyanidins, the immediate precursors for anthocyanins. One cDNA clone each was obtained for *B. finlaysoniana* DFR [96] and *Cymbidium hybrida* DFR [14], and the deduced amino acid sequence of the former DFR is 85% homologous to that of the latter DFR. Evidence was also obtained for a single copy of the DFR gene in the petal tissue of either orchid.

Although the number of molecular studies of anthocyanin biosynthesis in orchid is limited, the cloned gene for *C. hybrida* DFR has already been used in a transgenic approach to understand the molecular factors affecting the color range of some species of *Cymbidium* flowers that conspicuously lack orange-colored flowers [97]. Pelargonidin is the anthocyanin that normally leads to orange pigmentation. When the *C. hybrida* DFR gene was transformed into a DFR⁻ petunia line, the *C. hybrida* DFR was found not to efficiently reduce dihydrokaempferol to leucopelargonidin, a substrate required for pelargonidin production. These results suggested that the lack of orange-colored *Cymbidium* flowers is due to the loss of preference for dihydrokaempferol as a substrate, a phenomenon also observed in the DFR preparations from *Petunia, Lycoperscion*, and *Nicotiana* plants [98].

In addition to the preceding cloned genes, there are orchid genes isolated from nonflower tissue, and some of these are relevant to the understanding of floral coloration. The availability of these genes will help in the isolation of their counterparts in petal tissue. The enzymes for which cDNA clones have been isolated from nonflower tissue include 4-coumarate:coenzyme A ligase [13], caffeic acid *O*-methyltransferases [19,20], *S*-adenosylhomocysteine hydrolase [10], bibenzyl synthase [9,10], and vacuolar $H^+$-ATP synthase 16-kDa proteolipid subunit [99]. Except for the last enzyme, the rest are involved in pathways that interact with that of the flavonoid. By virtue of their catalytic activities, these enzymes affect either directly or indirectly the biosynthesis of flavonoids. The role of vacuolar $H^+$-ATP synthase in floral coloration has not been investigated before. This enzyme pumps protons across vacuolar membrane at the expense of ATP and brings about a pH change in the vacuole or cytosol. It is also known that the anthocyanins are stored in the vacuole, whose pH has a marked effect on the anthocyanin secondary structures and thus on the resultant floral coloration. Taken together, it is hard not to imagine that the vacuolar $H^+$-ATP synthase can influence floral coloration.

## V.  CONCLUSION

This chapter focuses on the orchid flower genes and cites many more recent references than an earlier survey concluded in 1994 [100]. Although our knowledge of each of the research areas reviewed is still rudimentary, the cloning of these genes will pave the way to address some of the questions raised in the introduction. Moreover, the results provided by the study of *Cymbidium hybrida* DFR have suggested a transgenic approach to produce orange-colored *Cymbidium* flowers, i.e., the introduction of a DHK-catalyzing DFR into a DHK accumulating *Cymbidium* and subsequent crossing with elite lines [97]. The development of transgenic plants with altered flower color, however, will depend on a more efficient transformation system than those presently available. Orchid cells are recalcitrant to tissue culture manipulations in that plant regeneration from dedifferentiated cells has yet to be achieved. A report [101] on particle bombardment of *Cymbidium* orchid has attempted to address these drawbacks by adopting a liquid culture system to stimulate active proliferation of the meristematic tissues before and after particle bombardment. This system may provide a step toward genetic engineering of orchids with improved characteristics.

# REFERENCES

1. CS Hew, Yong JWH. The Physiology of Tropical Orchids in Relation to the Industry. Singapore: World Scientific, 1997, pp 1–330.
2. J Arditti, R Ernst. Micropropagation of Orchids. New York: John Wiley & Sons, 1993, pp 1–640.
3. SD O'Neill, JA Nadeau, XS Zhang, AQ Bui, AH Halevy. Interorgan regulation of ethylene biosynthetic genes by pollination. Plant Cell 5:419–432, 1993.
4. JA Nadeau, SD O'Neill. Nucleotide sequence of a cDNA encoding 1-aminocyclopropane-1-carboxylate oxidase from senescing orchid petals. Plant Physiol 108:833–834, 1995.
5. XH Yang, EC Pua, CJ Goh. Isolation of a cDNA (EMBL/GenBank Accession AF038840) clone encoding 1-aminocyclopropane-1-carboxylate oxidase from *Dendrobium crumenatum*. Plant Gene Register PGR99-015. Plant Physiol 119:805, 1999.
6. Y-Y Do, PL Huang. Sequence of a cDNA coding for a 1-aminocyclopropane-1-carboxylate synthase homolog from *Phalaenopsis*. DNA Seq 8:263–266, 1998.
7. XH Yang, EC Pua, CJ Goh. Isolation of a cDNA (EMBL/GenBank Accession U64031) clone encoding 1-aminocyclopropane-1-carboxylate synthase from *Dendrobium crumenatum*. Plant Gene Register PGR96-088. Plant Physiol 112:863, 1996.
8. Y-Y Do, PL Huang. Isolation of a cDNA (EMBL/GenBank Accession U66299) involved in pollination-induced flower senescence in *Phalaenopsis* which encodes for a protein homologous to human peroxisomal acyl-CoA oxidase. Plant Gene Register PGR96-089. Plant Physiol 112:863, 1996.
9. SH Lim, CY Lee, CF Liew, CE Ong. EMBL/GenBank. Accession AJ131830. 1999.
10. R Preisig-Muller, P Gnau, H Kindl. The inducible 9,10-dihydrophenanthrene pathway: characterization and expression of bibenzyl synthase and *S*-adenosylhomocysteine hydrolase. Arch Biochem Biophys 317:201–207, 1995.
11. WS Hsu, YY Do, P-L Huang. Isolation and characterization of a cDNA (EMBL/GenBank Accession U88077) encoding chalcone synthase homolog from *Phalaenopsis*. Plant Gene Register PGR97-068. Plant Physiol 113:1465, 1997.
12. CF Liew, CJ Goh, CS Loh, SH Lim. Cloning and characterization of full-length cDNA clones encoding chalcone synthase from the orchid *Bromheadia finlaysoniana*. Plant Physiol Biochem 36:647–656, 1998.
13. P Brodelius, ZT Xue. Isolation and characterization of a cDNA from cell suspension cultures of *Vanilla planifolia* encoding 4-coumarate:coenzyme A ligase. Plant Physiol Biochem 35:497–506, 1997.
14. H Yi, E Johnson, BJ Oh, G Choi. EMBL/GenBank Accession AF017451, 1998.
15. JA Nadeau, XS Zhang, J Li, SD O'Neill. Ovule development: identification of stage-specific and tissue-specific cDNAs. Plant Cell 8:213–239, 1996.
16. YY Do, YC Chen, PL Huang. Molecular analysis of a cDNA (EMBL/GenBank Accession AF055894) encoding ethylene receptor that expresses in *Phalaenopsis* petals. Plant Gene Register PGR99-047. Plant Physiol 119:1567, 1999.
17. IJ Chai, BH Lee, WK Wang, CC Liang, CY Lin. EMBL/GenBank. Accession AF113541. 1999.
18. CF Liew, CJ Goh, CS Loh, SH Lim. Cloning and nucleotide sequence of a cDNA (EMBL/GenBank Accession X89199) encoding flavanone 3-hydroxylase from *Bromheadia finlaysoniana* (Lindl.). Plant Gene Register PGR95-062. Plant Physiol 109:339, 1995.
19. ZT Xue, PE Brodelius. Kinetin-induced caffeic acid *O*-methyltransferases in cell suspension cultures of *Vanilla planifolia Andr.* and isolation of caffeic acid *O*-methyltransferase cDNAs. Plant Physiol Biochem 36:779–788, 1998.
20. ZT Xue, P Kjellbom, P Brodelius. EMBL/GenBank Accession X69192. 1999.
21. CF Liew, CJ Goh, CS Loh, SH Lim. Cloning and nucleotide sequence of a cDNA (EMBL/GenBank Accession X99997) encoding phenylalanine ammonia-lyase from *Bromheadia finlaysoniana*. Plant Gene Register PGR96-087. Plant Physiol 112:863, 1996.
22. H Yu, CJ Goh. EMBL/GenBank. Accession AF107588, AF198174, AF198175, AF198176. 1999.

23. ZX Lu, M Wu, CS Loh, CY Yeong, CJ Goh. Nucleotide sequence of a flower-specific MADS box cDNA clone from orchid. Plant Mol Biol 23:901–904, 1993.
24. D Weigel. The genetics of flower development: from floral induction to ovule morphogenesis. Annu Rev Genet 29:19–39, 1995.
25. ES Coen, EM Meyerowitz. The war of the whorls: genetic interactions controlling flower development. Nature 353:31–37, 1991.
26. C Norman, M Runswick, R Pollock, R Treisman. Isolation and properties of cDNA clones encoding SRF, a transcriptional factor that bind to the *c-fos* serum response element. Cell 55:989–1003, 1988.
27. S Passmore, GT Maine, R Elble, C Christ, BK Tye. A *Saccharomyces cerevisiae* protein involved in plasmid maintenance is necessary to mating of MAT cell. J Mol Biol 204:593–606, 1988.
28. JL Riechmann, EM Meyerowitz. MADS domain proteins in plant development. Biol Chem 378:1079–1101, 1997.
29. P Shore, AD Sharrocks. The MADS box family of transcription factors. Eur J Biochem 229:1–13, 1995.
30. H Sommer, JP Beltran, P Huijser, H Pape, WE Lonning, H Saedler, Z Schwarz-Sommer. *Deficiens*, a homeotic gene involved in the control of flower morphogenesis in *Antirrhinum majus*: the protein shows homology to transcription factors. EMBO J 9:605–613, 1990.
31. MF Yanofsky, H Ma, JL Bowman, GN Drews, KA Feldmann, EM Meyerowitz. The protein encoded by the *Arabidopsis* homeotic gene *AGAMOUS* resembles transcriptional factors. Nature 346:35–39, 1990.
32. G Theissen, H Saedler. MADS box genes in plant ontogeny and phylogeny: Haeckel's "biogenetic law" revisited. Curr Opin Genet Dev 5:628–639, 1995.
33. H Ma, MF Yanofsky, EM Meyerowitz. *AGL1-AGL6*, an *Arabidopsis* gene family with similarity to floral homeotic and transcription factor genes. Genes Dev 5:484–495, 1991.
34. Z Schwarz-Sommer, I Hue, P Huijser, PJ Flor, R Hansen, F Tetens, WE Lonning, H Saedler, H Sommer. Characterization of the *Antirrhinum* floral homeotic MADS box gene *DEFICIENS*: evidence for DNA binding and autoregulation of its persistent expression throughout flower development. EMBO J 11:251–263, 1992.
35. L Colombo, AF van Tunen, HJM Dons, GC Angenent. Molecular control of flower development in *Petunia hybrida*. Adv Bot Res 26:229–250, 1997.
36. JD Thompson, DG Higgins, TJ Gibson. CLUSTAL W: improving the sensitivity of progressive multiple sequence alignment through sequence weighting, position-specific gap penalties and weight matrix choice. Nucleic Acids Res 22:4673–4680, 1994.
37. I Kardailsky, VK Shukla, JH Ahn, N Dagenais, SK Christensen, JT Nguyen, J Chory, MJ Harrison, D Weigel. Activation tagging of the floral inducer *FT*. Science 286:1962–1965, 1999.
38. Y Kobayashi, H Kaya, K Goto, M Twabuchi, T Araki. A pair of related genes with antagonistic roles in mediating flowering signals. Science 286:1960–1962, 1999.
39. J Arditti. Fundamentals of Orchid Biology. New York: John Wiley & Sons, 1992, pp 1–691.
40. SD O'Neill, JA Nadeau. Postpollination flower development. Hortic Rev 19:1–58, 1997.
41. XS Zhang, SD O'Neill. Ovary and gametophyte development are coordinately regulated following pollination by auxin and ethylene. Plant Cell 5:403–418, 1993.
42. L Wang, XS Zhang, HW Zhong, QZ Li. Expression and sequence analysis of a cDNA relative to orchid ovule development. Acta Bot Sin 41:276–279, 1999.
43. P Lu, R Porat, JA Nadeau, SD O'Neill. Identification of a meristem L1 layer–specific gene in *Arabidopsis* that is expressed during embryonic pattern formation and defines a new class of homeobox genes. Plant Cell 8:2155–2168, 1996.
44. RP Donaldson, DG Luster. Multiple forms of plant cytochrome P-450. Plant Physiol 96:669–674, 1991.
45. DW Nebert, FJ Gonzalez. P450 genes: structure, evolution and regulation. Annu Rev Biochem 56:945–993, 1987.
46. R Porat, P Lu, S O'Neill. *Arabidopsis* SKP1, a homologue of a cell cycle regulator gene, is predominantly expressed in meristematic cells. Planta 204:345–351, 1998.
47. C Bai, P Sen, K Hofmann, L Ma, M Goebl, W Harper, SJ Elledge. SKP1 connects cell cycle regulators to the ubiquitin proteolysis machinery through a novel motif, the F-box. Cell 86:263–274, 1996.
48. C Connelly, P Heiter. Budding yeast SKP1 encodes an evolutionary conserved kinetochore protein required for cell cycle progression. Cell 86:275–285, 1996.
49. B Turk, V Turk, D Turk. Structural and functional aspects of papain-like cysteine proteinases and their protein inhibitors. Biol Chem 378:141–150, 1997.
50. V Cryns, J Yuan. Proteases to die for. Genes Dev 12:1551–1570, 1998.
51. M Solomon, B Belenghi, M Delledonne, E Menachem, A Levine. The involvement of cysteine proteases and protease inhibitor genes in the regulation of programmed cell death in plants. Plant Cell 11:431–444, 1999.
52. LD Nooden. The phenomena of senescence and aging. In: LD Nooden, AC Leopold, eds. Senescence and Aging in Plants. San Diego: Academic Press, 1988, pp 2–51.
53. AD Stead. Pollination-induced flower senescence: a review. Plant Growth Regul 11:13–20, 1992.
54. AH Halevy. Recent advances in postharvest physiology of flowers. J Kor Soc Hortic Sci 39:652–655, 1998.

55. AH Halevy, CC Whitehead, AN Kfranek. Does pollination induce corolla abscission of cyclamen flowers by promoting ethylene production? Plant Physiol 75:1090–1093, 1984.

56. R Porat, A Borochov, AH Halevy, SD O'Neill. Pollination-induced senescence in *Phalaenopsis* petals. The wilting process, ethylene production and sensitivity to ethylene. Plant Growth Regul 15:129–136, 1994.

57. R Porat, AH Halevy, M Serek, A Borochov. An increase in ethylene sensitivity following pollination is the initial event triggering an increase in ethylene production and enhanced senescence of *Phalaenopsis* orchid flowers. Physiol Plant 93:778–784, 1995.

58. R Porat, A Borochov, AH Halevy. Pollination-induced senescence in *Phalaenopsis* petals. Relationship of ethylene sensitivity to activity of GTP-binding proteins and protein phosphorylation. Physiol Plant 90:679–684, 1994.

59. AH Halevy, R Porat, H Spiegelstein, A Borochov, L Botha, CS Whitehead. Short-chain saturated fatty acids in the regulation of pollination-induced ethylene sensitivity of *Phalaenopsis* flowers. Physiol Plant 97:469–474, 1996.

60. AH Halevy, CC Whitehead. Pollination-induced corolla abscission and senescence and the role of short-chain saturated fatty acids in the process. In: DJ Osborne, MB Jackson, eds. Cell Separation in Plants. NATO ASI Series, Vol 35. Berlin: Springer-Verlag, 1989, pp 221–331.

61. G Pearce, D Strydom, S Johnson, CA Ryan. A polypeptide from tomato leaves induces wound-inducible proteinase inhibitor proteins. Science 253:895–898, 1991.

62. R Porat, N Reiss, R Atzorn, AH Halevy, A Borochov. Examination of the possible involvement of lipoxygenase and jasmonates in pollination-induced senescence of *Phalaenopsis* and *Dendrobium* orchid flowers. Physiol Plant 94:205–210, 1995b.

63. R Porat, JA Nadeau, JA Kirby, EG Sutter, SD O'Neill. Characterization of the primary pollen signal in the postpollination syndrome of *Phalaenopsis* flowers. Plant Growth Regul 24:109–117, 1998.

64. CL Escher, F Widmer. Lipid mobilization and gluconeogenesis in plants: do glyoxylate cycle enzyme activities constitute a real cycle? A hypothesis. Biol Chem 378:803–813, 1997.

65. H Kindl. Fatty acid degradation in plant peroxisomes: function and biosynthesis of the enzymes involved. Biochimie 75:225–230, 1993.

66. Y-Y Do, PL Huang. Characterization of a pollination-related cDNA from *Phalaenopsis* encoding a protein which is homologous to human peroxisomal acyl-CoA oxidase. Arch Biochem Biophys 344:295–300, 1997.

67. Y-Y Do, PL Huang. Gene structure of PACO1, a petal senescence–related gene from *Phalaenopsis* encoding peroxisomal acyl-CoA oxidase homolog. Biochem Mol Biol Int 41:609–618, 1997.

68. S Yang, N Hoffman. Ethylene biosynthesis and its regulation in higher plants. Annu Rev Plant Physiol 35:155–189, 1984.

69. PR Johnson, JR Ecker. The ethylene gas signal transduction pathway: a molecular perspective. Annu Rev Genet 32:227–254, 1998.

70. TI Zarembinski, A Theologis. Ethylene biosynthesis and action: a case in conservation. Plant Mol Biol 26:1579–1597, 1994.

71. H Kende. Ethylene biosynthesis. Annu Rev Plant Physiol Plant Mol Biol 44:283–307, 1993.

72. K Manning. The ethylene forming enzyme system in carnation flowers. In: JA Roberts, GA Tucker, eds. Ethylene and Plant Development. London: Butterworth, 1985, pp 83–92.

73. JA Nadeau, XS Zhang, H Nair, SD O'Neill. Temporal and spatial regulation of 1-aminocyclopropane-1-carboxylate oxidase in the pollination-induced senescence of orchid flowers. Plant Physiol 103:31–39, 1993.

74. H Kende, JAD Zeevaart. The five "classical" plant hormones. Plant Cell 9:1197–1120, 1997.

75. AQ Bui, SD O'Neill. Three 1-aminocyclopropane-1-carboxylate synthase genes regulated by primary and secondary pollination signals in orchid flowers. Plant Physiol 116:419–428, 1998.

76. ML Jones, WR Woodson. Differential expression of three members of the 1-aminocyclopropane-1-carboxylate synthase gene family in carnation. Plant Physiol 119:755–764, 1999.

77. SD O'Neill. Pollen signaling and interorgan regulation of the postpollination syndrome of flowers. In: AG Stephenson, TH Kao, eds. Pollen-pistil Interactions and Pollen Tube Growth. Rockville, MD: American Society of Plant Physiologists, 1994, pp 161–177.

78. ML Jones, WR Woodson. Interorgan signaling following pollination in carnations. J Am Soc Hortic Sci 124:598–604, 1999.

79. C Chang, SF Kwok, AB Bleecker, EM Meyerowitz. *Arabidopsis* ethylene-response gene *etR1*: similarity of product to two-component regulators. Science 262:539–544, 1993.

80. SM Wurgler-Murphy, H Saito. Two-component signal transducers and MAPK cascades. Trends Biochem Sci 22:172–176, 1997.

81. J Hua, E Meyerowitz. Ethylene responses are negatively regulated by a receptor gene family in *Arabidopsis thaliana*. Cell 94:261–271, 1998.

82. CC Lashbrook, DM Tieman, HK Klee. Differential regulation of the tomato ETR gene family throughout plant development. Plant J 15:243–252, 1998.

83. J Parkinson. Signal transduction schemes of bacteria. Cell 73:857–871, 1993.

84. KM Davis, KE Schwinn. Flower color. In: RL Geneve, JE Preece, SA Merkle eds. Biotechnology of Ornamental Plants. Wallingford, UK: CAB International, 1997, pp 259–294.

85. G Forkmann. Flavonoids as flower pigments: the formation of the natural spectrum and its extension by genetic engineering. Plant Breeding 106:1–26, 1991.
86. C Martin, T Gerats. The control of flower colouration. In: BR Jordon, ed. The Molecular Biology of Flowering. Wallingford, UK: CAB International, 1993, pp 219–255.
87. AJ van Tunen, JNM Mol. Control of flavonoid synthesis and manipulation of flower colour. In: D Grierson, ed. Developmental Regulation of Plant Gene Expression. Glasgow: Blackie, 1991, pp 94–130.
88. HJ Joos, K Hahlbrock. Phenylalanine ammonia-lyase in potato (*Solanum tuberosum*): genomic complexity, structural comparison of two selected genes and modes of expression. Eur J Biochem 203:621–629, 1992.
89. RW Whetten, RR Sederoff. Phenylalanine ammonia-lyase from loblolly pine. Purification of the enzyme and isolation of complementary DNA clones. Plant Physiol 98:380–386, 1992.
90. P Franken, U Niesbach-Klosgen, U Weydemann, L Marechal-Drouard, H Saedler, U Wienand. The duplicated chalcone synthase genes *C2* and *Whp* (white pollen) of *Zea mays* are independently regulated; evidence for translational control of *Whp* expression by the anthocyanin intensifying gene *in*. EMBO J 10:2605–2612, 1991.
91. CL Harker, THN Ellis, ES Coen. Identification and genetic regulation of the chalcone synthase multigene family in pea. Plant Cell 2:185–194, 1990.
92. RE Koes, CE Spelt, JNM Mol, AGM Gerats. The chalcone synthase multigene family of *Petunia hybrida* (V30): sequence homology, chromosomal localization and evolutionary aspects. Plant Mol Biol 10:159–169, 1987.
93. RE Koes, CE Spelt, JNM Mol. The chalcone synthase multigene family of *Petunia hybrida* (V30): differential light-regulated expression during flower development and UV light induction. Plant Mol Biol 12:213–225, 1989a.
94. TB Ryder, SA Hedrick, JN Bell, X Laing, SD Clouse, CJ Lamb. Organization and differential activation of a gene family encoding the plant defense enzyme chalcone synthase in *Phaseolus vulgaris*. Mol Gen Genet 210:219–233, 1987.
95. RE Koes, CE Spelt, JNM Mol. Cloning and molecular characterization of chalcone synthase multigene family of *Petunia hybrida*. Gene 81:245–257, 1989b.
96. CF Liew, CS Loh, CJ Goh, SH Lim. The isolation, molecular characterization and expression of dihydroflavonol 4-reductase cDNA in the orchid *Bromheadia finlaysoniana*. Plant Sci 135:161–169, 1998.
97. ET Johnson, H Yi, B Shin, BJ Oh, H Cheong, G Choi. *Cymbidium hybrida* dihydroflavonol 4 reductase does not efficiently reduce dihydrokaempferol to produce orange pelargonidin-type anthocyanins. Plant J 19:81–85, 1999.
98. W Heller, G Forkmann. Biosynthesis of flavonoids. In: JB Harborne, ed. The Flavonoids: Advances in Research Since 1986. London: Chapman & Hall, 1993, pp 499–536.
99. CF Liew, AL Lim, MY Loo, S Swarup. A cDNA (EMBL/GenBank Accession AF193814) encoding the 16-kilodalton proteolipid subunit of the vacuolar $H^+$-ATP synthase from the orchid *Dendrobium crumenatum*. Plant Gene Register PGR99-180. Plant Physiol 121:1383, 1999.
100. AR Kuehnlc. Molecular biology of orchids. In: J Arditti, AM Pridgeon, eds. Orchid Biology: Reviews and Perspectives VII. Dordrecht, The Netherlands: Kluwer Academic Publishers, 1997, pp 75–115.
101. J Yang, HJ Lee, DH Shin, SK Oh, JH Seon, KY Paek, KH Han. Genetic transformation of *Cymbidium* orchid by particle bombardment. Plant Cell Rep 18:978–984, 1999.

# 28
# Biology and Physiology of Saline Plants

## David N. Sen* and Pawan K. Kasera

*University of Jodhpur, Jodhpur, India*

## Sher Mohammed

*Government Lohia (PG) College, Churu, India*

## I.  INTRODUCTION

Worldwide agricultural production is greatly affected by a number of environmental hazards, major among which is salinity associated with aridity. Saline lands are not only distributed in desert and semidesert regions but also frequently occur in fertile alluvial plains, river, valleys, and coastal regions, close to densely populated areas and irrigation systems. These soils cover an area of about 1 billion ha on our planet earth, out of which 75 million ha lic in Southwest Asia. Iran tops the latter group with nearly 27 million ha, followed by India (23.8 Mha), Pakistan (10.5 Mha), Iraq (6.7 Mha), Afghanistan (3 Mha), and Turkey (2.5 Mha) [1]. It is therefore important to examine saline ecophysiology in terms of the environment and adaptability of plant species found therein.

The interactions between salinity and soil water and climatic conditions change the plant's ability to tolerate salinity. A basic understanding of the interaction between salinity and environment is necessary for an accurate assessment of salt tolerance. In addition to precipitation, temperature and atmospheric humidity can markedly influence salt tolerance [2].

An excess of salts in the soil inhibits plant growth in various ways. When rain failure is prolonged, it affects vegetation by reducing the level of the water table and increasing salinity by capillary movement. Temperature is one of the most critical factors of the environment that exerts a pronounced effect on all the physiological activities by controlling the rate of chemical reactions.

Salinity and aridity are the two oldest enemies of agriculture. The salinity stress problem arises when semiarid or arid lands are subjected to cultivation because saline soils, excessive use of chemical fertilizers, and excessive irrigation have turned hundreds of hectares of cultivated fertile lands into saline lands [3]. It is believed that about 10% of the total surface area of the world is salt affected [4]. About 15% of the arid and semiarid lands are affected by salts, and one third of all agricultural lands are also becoming saline [5]. The rapidly growing demand for increased food, fiber, and fuel in the presence of rapidly declining availability of agricultural land due to increased soil salinity makes it imperative that crop production under saline conditions be significantly increased. It is believed that halophytes have potential value for agriculture and could be grown in these degraded lands.

The most serious problem of the arid/inland saline areas is the scarcity of water. In hot deserts, most of the rainfall occurs during the summer, which leads to a high rate of evapotranspiration. During that time, the effective utilization of water by plants is also high. In fact, only a small group of higher plants,

---

* retired

the halophytes, can grow under saline conditions. These halophytic species that live under conditions of high salinity exhibit succulence, which might resort to other physiological adaptations to overcome the adverse saline environment in the soil. Their extreme tolerance to salinity is related to their ability to maintain a high salt concentration within their cells. In desert areas salinity is often very prominent, caused by the input of sodium chloride and other salts for a long period of lack of drainage. On such saline soils, typical associations dominated by Chenopodiaceae often develop along the salt gradients [6]. Plant species growing in an area may provide useful information regarding the degree of salinization and consequent soil deterioration. Such information may be helpful in more effective planning for practical uses of wastelands.

## A.  Studies of Saline Areas and Their Vegetation Cover

In the early 19th century, an enormous amount of work was done on this special group of plants, and this topic has been discussed periodically in many reviews and books covering manifold aspects [7–35]. The major findings reveal that the salt-affected soils have multiple effects on plants. Seed germination, mortality, and growth of halophytes are controlled in nature by the interaction of soil salinity and moisture. The ionic and toxic effects of various salts, especially of NaCl, play a major role in halophytism. Increased osmotic stress due to drought and the high rate of evaporation during the summer months may cause rapid changes in the density and diversity of species in halophyte communities [15,16]. Seasonal precipitation is often a major factor determining soil water potentials. This factor in turn affects the establishment of seedlings, often increasing the rate of mortality during drought periods [17].

Although the salt accumulation nature of halophytes has been recognized for many years, only during the last three decades was it shown that sodium is essential for the growth of some Chenopodiaceae [18]. Strogonov [19] assumed that the survival of plants in saline environments depends upon altered biochemical reactions and on a quantitative ratio between the toxic and protective (i.e., proline) compounds. Stewart and Lee [20] supported the view that proline functions as a source of solute for intracellular osmotic adjustment.

## II.  BIOLOGY OF NATIVE HALOPHYTES

Despite the wide distribution of halophytes in various climatic regions, their taxonomic, structural, and behavioral uniformity is striking. Salinity is known to affect many aspects of plants, which induces numerous changes in their morphology. These changes would be adaptations that increase the chance of plants to endure stress imposed by salinity or damage and disrupt the normal equilibrium of life processes [11]. In saline soils, the most common adverse features are delayed germination, high mortality of seedlings, and poor growth of crops. Plants are stunted, less vigorous, and give poor yield.

The precise effects of salinity and the sites where salinity may affect plants are not easily assessed. Because both salt combination and salt concentration differ from one habitat to another, the term "salinity" usually has a loose meaning. In certain cases, it is not the absolute amount of a certain ion that may affect plants but rather the composition and total concentration of salts. Certain species of plants may be found in sites where the sodium chloride concentration is beyond their theoretical tolerance but where high concentrations of calcium, potassium, or sulfate are found as supplementary ions. These ions moderate the toxic effects of sodium and chloride, thus enabling plants to exist.

The ecological limits for distribution of plant communities depend upon the presence of soluble salts in the water or soil. The water of the habitat is the dominant ecological factor that determines the distribution of species. Thus, the distribution of a halophytic community appears to be limited by salinity and the depth of the water table as well as by the competitive ability of members of the next community in the development of halophytic vegetation [31,32]. It is suggested that because sharp boundaries are observed between halophyte communities even when there is only a gradual change in the physiochemical environment, biotic interactions may play a significant role in determining the distribution pattern of species and the composition of zonal communities [33].

## III.  ADAPTATIONS OF SALINE PLANTS

Not only are halophytes in their saline environments exposed to salt stress, but the root may also be exposed to osmotic water stress and low oxygen pressure stress [36]. They must adjust their tissue water potentials to a level that is lower than that of the soil water potential in the habitat where they are growing; enabling the plants to absorb water. Without sufficient moisture, halophytes can be stunted, and reproduction becomes very limited [37]. Halophytes take up ions to increase the osmotic levels in their tissues, which permits moisture to move from the soil into the tissues. On the other hand, excess salt ions can produce a toxic effect to the plant cells. Some of the mechanisms used by halophytes to counter the potential toxic effect of high concentration of ions involve exclusion of salts by the roots, dilution of the ions through succulence, synthesis of organic osmotic compounds that can reduce the need for salt ions, and compartmentalization of the excess salt ions into tissues, organs, or cell vacuoles.

Plants have evolved two very different strategies in adapting to high levels of sodium salts in their environment. One strategy is to exclude the salts from the interior of the leaf cells, and the other is to include the salts within the leaf cells but sequester most of them in the cell vacuoles of those cells. In both cases, the end result is to maintain a relatively low cytoplasmic sodium concentration [38]. These two broad categories of plants are referred to as salt secreting and salt accumulating. Those that exclude salts from the leaf cells are able to tolerate high levels of the salts in the root environment but at the expense of reduced growth. Most of them avoid salinity, some evade it, and a few others tolerate it. Most plants avoid salinity by limiting reproduction, growth, and germination during specific parts of the year, by limiting the uptake of salt, and by allowing roots to penetrate into nonsaline soils. Evasion of salt has been achieved through the accumulation of salts into certain specific cells and trichomes or secretion of excess salts through especially mechanized salt-secreting glands [12,32,39]. Secretion of ions by special salt glands or bladder hairs, release through the cuticle or in the guttation fluid, and retransportation via the phloem are examples of these mechanisms.

The exclusion of ions by roots can be a factor in salt tolerance, but some type of osmotic compound needs to be produced in the plant for it continue to absorb water from the saline soil. In dicot halophytes, root exclusion is not an effective mechanism, although there may be some ion regulation at the root level [40].

Salt resistance and salt tolerance on the cellular level as well as the formative effects of salinity producing halosucculent leaves and/or stems also need to be taken into consideration. The genetic background regulating compartmentalization of solutes and formation of compatible solutes has to be regarded in connection with the adaptation on the higher levels of complexity. There are still several questions open concerning the growth and development of halophytes, e.g., root architecture in saline habitats and formation of mycorrhizae, hormonal balance and growth regulation, mineral uptake, and selectivity [41].

Xerosucculents are characterized by a thick cuticle and a cover of waxy layers, such as displayed in *Suaeda fruticosa*, *Salsola baryosma*, and *Haloxylon recurvum*. Cuticle and waxy layers have also been reported on the leaf surfaces of *Cressa cretica*, *Aeluropus lagopoides*, *Sporobolus helvolus*, and *Chloris virgata*. Some halophytes, such as *C. cretica*, *C. virgata*, *S. helvolus*, and *A. lagopoides*, show an additional mode of adaptation to their habitat. The leaves and stems of these plants remain covered with hairs (trichomes), giving the plant a grayish appearance. Their effectiveness in reducing water loss is small, but they are able to protect the leaf surface against dust [12].

Another approach for accumulating ions and at the same time preventing them from becoming toxic to the photosynthetic cells is through the accumulation of excess ions in tissues that eventually die. The dead tissues containing the excess salts act as a storage region for the salts. In the case of *Heliotropium curassavicum*, during the dry conditions, more and more salts are accumulated in their fleshy leaves and these salt-saturated leaves dry up to keep the osmotic level of the plant balanced. New leaves sprouts, and this process continues during the whole life span of the plant [24].

Rajput [26] reported that in *Atriplex* species with increased duration of leaching, the amount of ions leached from the leaves also increased and the maximum value was observed after 72 hr. Maximum values of leacheable $Na^+$ and $Cl^-$ were observed in young leaves of *A. halimus* and *A. nummularia*, respectively, during the summer, and a minimum value of $Na^+$ was observed in young leaves of *A. argentina* during the rainy season, probably because of leaching of the salts from the leaves by rainwater.

Young leaves and stems of *Atriplex* spp. contained quite high values of sodium and chloride in comparison with mature parts because of fleshiness and the presence of more salt bladders in the epidermis of tender plant parts. Young leaves of *A. triangularis* also contained more $Na^+$ and $Cl^-$ than mature leaves with intact bladders [26].

Halophytic cells need to have high osmotic pressure and at the same time prevent the excess ions from inhibiting the enzymatic processes. If the excess ions are stored in the vacuole, the metabolic activity can be carried on in the cytoplasm, where the ion content is lower. The lower salt concentration prevents organelles such as chloroplasts from being damaged by excess ions [5]. When a change in metabolism results in a change in the ability to resist stress conditions, anthocyanin may develop in the leaves or stems of plants. Some halophytes of the Indian region such as *S. fruticosa*, *S. baryosma*, *Trianthema triquetra*, and *Zygophyllum simplex* exhibit these characteristics under osmotic stress conditions.

The effect of salinity as a specific and dominant factor in a saline environment determines to a great extent the ability of halophytes to reproduce and perpetuate. Information regarding the germination behavior of Indian halophytes is still scant. Rajpurohit and Sen [42] concluded that under field conditions the highest germination percentage in *C. cretica*, *S. fruticosa*, *S. baryosma*, *Sesuvium sesuvioides*, and *T. triquetra* can be achieved after rain that is heavy enough to leach out the salt from the closed environment of the seeds. Several authors found that the increase in salinity leads to dormancy of seeds in halophytes and glycophytes [42,43].

In spite of the preceding discussion, how halophytes handle salts is still not fully understood. Breckle [41] and Weber [37] stated that we know many mechanisms, and it has been estimated that over 1000 genes are turned on or off as a response to salinity. This does not count the genes that are "hard wired" into the enzyme system.

## IV.  MECHANISMS OF SALT TOLERANCE

Whether it is drought, cold, heat, salt, metal, or any other (pollution stress) stress or a combination of some or all of these, the end result is a dehydration stress. Levitt [44] speculated that plants may have a general mechanism of resistance to every stress. At the morphological level, wilting, leaf rolling, and decrease in stomatal aperture, succulence, leaflessness, etc. constitute a general mechanism for conserving water. At the physiological level, reduction in evapotranspiration and decrease in water potential are major manifestations.

Biochemical studies have revealed similarities in processes induced by various abiological stresses, leading to accumulation of compounds such as ascorbate, glutathione, α-tocopherol, betaine, proline and other amino acids, quaternary ammonium compounds, polyamines, sucrose, polyols (mannitol, sorbitol, pinitol), and oligosaccharides in plant tissue. In addition, changes in the activity of certain key enzymes, gene expression, and biosynthesis of abscisic acid (ABA) have been noted [45].

For many plants there is a correlation between increases of metabolites and osmotic stress tolerance, but the mechanisms that cause this protection are not clear. During salt stress, cells retained more of the six-carbon polyols than glycerol. To understand the role of glycerol in salt tolerance, salt-tolerant suppressor mutants were isolated from the glycerol-deficient strains, and results compare with the "osmotic adjustment" concept typically applied to accumulating metabolites in plants. The accumulation of polyols may have dual functions: facilitating osmotic adjustment and supporting redox control [46].

High selectivity during nutrient uptake is one mechanism used by halophytes to avoid a deficiency of essential ions such as K, Mg, and Ca. The results of Koyro et al. [47] suggest that *Laguncularia racemosa* developed an intercellular and intracellular K and Ca buffer that enabled it to grow and to exclude Na. However, higher salinities led to high energy metabolism and decreased growth, starch content, and nectar production.

The most characteristics associated with salt tolerance are under polygenic control, as observed by Jefferies and Rudmik [48]. However, halophytes exhibit a high level of physiological plasticity in this respect, and there is evidence in some that morphogenetic changes occur in response to salinity [48].

Wu and Seliskar [49] stated that the response of the plasma membrane $H^+$ adenosinetriphosphatase (ATPase) in *Spartina patens* suggests that this species has evolved mechanisms that can regulate this important enzyme when cells are exposed to NaCl.

## V.  SALT STRESS AND ROLE OF PROLINE

Water stress produces numerous metabolic irregularities in plants [44]. An increased proline concentration in water-stressed plants is due either to the inhibition of protein oxidation or to the breakdown of protein from its precursors [50]. In the Indian desert, 65 plant species were examined for proline content [51]; of these, 54 showed the presence of proline. These studies further revealed that some of the well-adapted desert plants do not accumulate proline at all [51]. Proline accumulation in plants is not governed by the environment but rather by the plants' internal factors [52].

The accumulation of proline in plants is correlated with the extent of the water stress in the plant. Mechanisms of regulation of proline accumulation during normal plant development are quite different from those operating during the abiotic stress response [53]. Treichel [54] determined that the activity level of delta-1-pyrroline-5-carboxylate reductase increases with progressive adaptation to NaCl stress. The increase in proline concentration is associated with the activity of delta-1-pyrroline-5-carboxylate reductase. It has been suggested that proline at a high concentration acts as a source of solute for intracellular osmotic adjustment [20] and a storage compound for both nitrogen and carbon for utilization in growth after stress [50]. The accumulation of proline upon dehydration related to water deficit or increasing osmotic pressure is the most frequent and extensive response of saline plants [55]. In halophytes, a positive correlation was found between the proline content and the $Na^+ + Cl^-$ levels in the cell sap [54].

Venkatesalu et al. [56] grew *Sesuvium portulacastrum*, a salt marsh halophyte, at different salinity levels. The total amino acids decreased with increased salinity. Proline and glycinebetaine levels increased as the salinity level increased. Proline concentration in *Cressa cretica* increased with an increase in salinity levels [57]. The proline concentration increases greatly in the growing regions of maize (*Zea mays*) primary roots at low water potentials, largely as a result of an increased net rate of proline deposition [58]. Naidoo and Naidoo [59] reported that concentrations of proline in roots and shoots increased significantly with salinity increase in *Sporobolus virginicus*. The proline content increased under saline conditions over control in *Vigna radiata* and *Cicer arietinum* as reported by Muthukumaraswamy and Panneerselvam [60] and Muthukumaraswamy et al. [61], respectively. Joshi and Khairatkar [62] observed that asparagine, aspartic acid, glutamic acid, phenylalanine, proline, glutamine, glycine, serine, and threonine constituted major fractions of amino acids in 40-day-old seedlings of *Juncus* spp. The concentrations of asparagine, aspartic acid, glutamic acid, and proline increased while the others decreased in response to salinity stress. However, the total amount of amino acids increased under saline conditions.

Garcia et al. [63] reported that in rice (*Oryza sativa*) proline either has no effect or in some cases promotes the effect of NaCl on growth inhibition, chlorophyll loss, and induction of a highly sensitive marker for plant stress, the osmotically regulated Sa/T gene. However, a high concentration (10 mM) of proline prevents NaCl-induced chlorophyll loss in blades, preserves its integrity, and enhances growth. Proline does not play an important role in salt tolerance in rice.

In *Carthamus tinctorius*, an increased proline content under NaCl saline conditions did not help to maintain growth because productivity at flowering was less than the control value at all levels of NaCl salinity [64]. Thus, it appears that an increased proline content in safflower under NaCl salinity helps in survival and not in maintaining growth. However, increased proline at Electrical Conductivity (ECe) 5.0 mS cm$^{-1}$ of NaCl helps in increasing productivity at maturity of this variety [64].

The data on proline accumulation in the leaves of halophytes in the Indian desert revealed that plants that grow in saline areas exhibited higher proline during winter followed by summer and least in rainy seasons (Table 1) [65]. All plant species at site I (Pachpadra salt basin) accumulated more proline as compared with sites II (Didwana salt lake) and III (Jodhpur nonsaline), which may be due to the high salinity of this habitat. Because site I is more saline than the latter two, it can be concluded that salt stress caused more proline. Perhaps free proline contents play an essential role in survival of these plants.

Sangwan et al. [66] reported that accumulation of free proline in the calli derived from seedlings of *Cicer arietinum* under the influence of chloride salinity was more than the sulfate salinity and was increased with increasing concentration of salts. This indicates that the proline production depends upon the ions and the degree of stress and the plant species on which the stress is imposed, as reported by Yang et al. [67]. Free proline accumulation under chloride salinity may also be attributed to the fact that oxidation of proline was inhibited more by chloride ions than sulfate ions as reported by Stewart et al. [68]. One of

**TABLE 1** Seasonal Variations in Proline (μg/g Fresh Weight) Content in Halophytes Growing at Different Sites (I–III)

| Species | Pachpadra (site I) | | | Didwana (site II) | | | Jodhpur (site III) | | |
|---|---|---|---|---|---|---|---|---|---|
| | Rainy | Winter | Summer | Rainy | Winter | Summer | Rainy | Winter | Summer |
| *Aeluropus lagopoides* | 6.2 | 26.6 | —[a] | 24.6 | — | — | — | — | — |
| *Cressa cretica* | 0.5 | 108.0 | 3.8 | 4.8 | — | — | — | — | — |
| *Salsola baryosma* | 0.1 | 7.1 | 5.7 | — | — | — | 0.1 | 7.0 | 2.7 |
| *Sesuvium sesuvioides* | 2.0 | — | — | 4.3 | — | — | 1.8 | — | — |
| *Sporobolus helvolus* | 5.3 | 100.0 | 7.9 | 6.1 | 8.7 | 6.2 | — | — | — |
| *Suaeda fruticosa* | 1.4 | 11.9 | 7.0 | 2.7 | 9.9 | 6.6 | 1.9 | 5.8 | 5.0 |
| *Trianthema triquetra* | 3.4 | 74.7 | — | 3.7 | 14.3 | — | 5.2 | 15.6 | 8.2 |
| *Zygophyllum simplex* (G)[b] | 6.4 | 64.2 | — | — | — | — | — | — | — |
| *Z. simplex* (R)[c] | 5.7 | 109.9 | — | — | — | — | — | — | — |

[a] Plant not seen.
[b] Green strain.
[c] Red strain.
*Source:* Ref. 65.

the critical functions of proline is to act as an osmoprotectant against stress in plants [69–71]. At higher levels of salinity, the proline accumulation was greater, perhaps as a last resort for plants to survive and to avoid osmotic death.

## VI. SOIL-PLANT RELATIONSHIP

All desert plants and most saline plants are totally dependent upon the availability of water in the rainy season. This water controls their seed germination, seedling growth, and plant survival. Rainfall leaches salts down the soil profile, as far down as the ground water, with a compensating upward movement as a result of capillary action. Decreases in soil moisture and the intensity of evaporation lead to an increase in soil salinity [72,73].

Halophytes studied by the authors were found to absorb salts continuously from their surrounding environment. *S. baryosma, S. sesuvioides, S. fruticosa, T. triquetra,* and *Z. simplex* continue to accumulate salt in their tissues. *A. lagopoides, C. cretica,* and *S. helvolus* secrete excess salt through the entire shoot. Fine streaks of white salt are seen on the stem and leaves throughout. On the basis of the ion analyses, $Na^+$ and $K^+$ were among the major cations and $Cl^-$ among the anions absorbed by these halophytes in large quantities (Table 2) [74]. Considering the habit as well as the $Cl^-$ content of individual halophytes, it is concluded that (1) the $Cl^-$ amounts absorbed by the leaves of *Z. simplex* (11–18%) and *S. fruticosa* (13–19%) were nearly equal but comparatively much higher than in other species, (2) the internal $Cl^-$ content of ion-accumulating species was higher than that of the two ion-secreting grass species (*A. lagopoides* and *S. helvolus*), and (3) among the ion-accumulating species, *S. baryosma* (6–11%), *S. sesuvioides* (3–7%), and *T. triquetra* (3–8%) accumulated much less $Cl^-$. Because $Cl^-$ is the dominant ion present in the medium at both saline sites, it can be concluded that *S. fruticosa, T. triquetra, S. baryosma, C. cretica,* and *Z. simplex* are well suited to these habitats, and thus they are the most salt tolerant species [75,76].

The water and salt stress on halophytes usually changes during the season. In order to survive under these changing conditions, halophytes must also make changes in their tissue osmotic potential during the growing season. Harward and McNulty [77] found that $Na^+$ and $Cl^-$ concentrations in the *Salicornia rubra* plant changed from 700 mM at the start of the growing season to 2 M at the end of the growing season. The $Na^+$ and $Cl^-$ accounted for almost 93% of the total osmotic pressure of *S. rubra* cell sap.

**TABLE 2**  Range of Ionic Content (mg/g) Accumulated by Leaves of Halophytes Growing at Different Sites (I–III)

| Species | I | | | | II | | | | III | | | |
|---|---|---|---|---|---|---|---|---|---|---|---|---|
| | $Na^+$ | $K^+$ | $Ca^{2+}$ | $Cl^-(\%)$ | $Na^+$ | $K^+$ | $Ca^{2+}$ | $Cl^-(\%)$ | $Na^+$ | $K^+$ | $Ca^{2+}$ | $Cl^-(\%)$ |
| *Aeluropus lagopoides* | 55–100 | 5–14 | 1–22 | 5.8–6.3 | 10 | 7 | 10 | 4 | —[a] | — | — | — |
| *Cressa cretica* | 20–110 | 1–31 | 3–30 | 4–10 | 42–49 | 9–14 | 19–21 | 4–8 | — | — | — | — |
| *Salsola baryosma* | 44–291 | 27–78 | 4–43 | 6–11 | — | — | — | — | 37–241 | 8–49 | 2–50 | 1.4–2.9 |
| *Sesuvium sesuvioides* | 16–100 | 14–36 | 5–37 | 3–7 | — | — | — | — | 72–106 | 9–21 | 2–20 | 2–3 |
| *Sporobolus helvolus* | 7–70 | 4–37 | 1–89 | 3–5 | 22–28 | 4–12 | 5–7 | 1–4 | — | — | — | — |
| *Suaeda fruticosa* | 43–313 | 10–45 | 3–55 | 13–19 | 45–119 | 7–63 | 5–26 | 7–21 | 48–315 | 8–20 | 1–37 | 7–15 |
| *Trianthema triquetra* | 29–115 | 3–33 | 2–28 | 3–8 | 27–112 | 7–34 | 3–87 | 1–11 | 22–225 | 9–24 | 1–30 | 1.9–2.2 |
| *Z. simplex* (G)[b] | 41–91 | 8–35 | 4–65 | 2–8 | — | — | — | — | — | — | — | — |
| *Z. simplex* (R)[c] | 42–105 | 8–15 | 5–52 | 11–18 | — | — | — | — | — | — | — | — |

[a] Plant not seen.
[b] Green strain.
[c] Red strain.
*Source:* Ref. 74.

One of the characteristics distinguishing halophytes from glycophytes is their capacity to accumulate selectively large quantities of ions in their cells without disrupting metabolic processes [78]. Maliwal [79] found that absorption of Na increased with increasing chloride salt concentration rather than sulfate salinity (0.78 to 15.40 dS/m) in wheat (*Triticum* spp.). A number of halophytes have been analyzed for their ionic contents [80], and the most strongly accumulated ions were $Na^+$ and $Cl^-$ with a few species having relatively high sulfate and $K^+$ concentrations. Albert and Popp [81] concluded that the $K^+$ content was generally higher in monocotyledons, whereas the $Na^+$ content was higher in dicotyledons. The $K^+/Na^+$ ratio in plant tissues of *Sesbania rostrata* was always higher than in the substrate, indicating selectivity for $K^+$ uptake, a characteristic generally considered unique to halophytes [82]. At certain concentrations, potassium is reported to inhibit the growth of halophytes such as *Suaeda* and *Atriplex*, while isomotic sodium promotes growth [83]. Abbas [84], using monthly samples of *Zygophyllum quatarense* populations, found that plants from a saline habitat had a higher chloride content than plants from a nonsaline habitat. Chellappan [85] determined the mineral distribution in *Sesuvium portulacastrum* and found that the $Na^+$ content increased significantly with increasing NaCl concentration. Hamada [86] and Saha and Gupta [87] also observed an increase in Na concentration with increasing salinity in wheat and sunflower, respectively.

The tolerances to water and salt stress of *Atriplex canescens* are linked through a common mechanism of Na uptake for osmotic adjustment in this species [88]. Egan and Ungar [89] observed that in *Atriplex prostrata*, plant growth parameters decreased with a lowering of the medium osmotic potential, and $K^+$ salts were more inhibitory than $Na^+$ salts. The ion content of plant tissue generally increased with a lowering of osmotic potential. It is suggested that halophytes such as *A. triangularis* may use $Na^+$ as an osmoticum to adjust the vacuolar water potential but were unable to use $K^+$ for this function because of a specific ion toxicity.

Ion uptake by plants was largely dependent upon their availability in the soil. When these ions fluctuated in the soil by upward or downward movements, their uptake by plants was also affected. As with soil salinity, a higher quantity of elements in plants was observed during dry periods in the Indian arid zone [23].

## VII.  GERMINATION ECOLOGY OF HALOPHYTES

The presence of excess salt in the soil is one of the critical factors that adversely affects seed germination under such conditions, thereby preventing plant species from inhabiting the saline environments successfully [10,12,15]. Halophytes show a reduction in germination when subjected to salinities above 1% NaCl, and increasing salt concentrations also delay germination [8]. Keiffer and Ungar [90] observed that prolonged exposure to saline solutions can inhibit or stimulate germination in certain species, and the resulting germination and recovery responses are related to the duration and intensity of their exposure to salt in their natural habitats. Rajpurohit [21], Jhamb [22], and later Mohammed and Sen [91] carried out a detailed study of this important aspect in the Indian desert. These investigators collected seeds from four different sites (Pachpadra, site I; Didwana, site II; Jodhpur, site III; and Luni, site IV) and studied the effect of various salts on seed germination. Various salts that are common in saline soils were selected, including NaCl, $Na_2SO_4$, $MgSO_4$, KCl, and $CaCl_2$. A single salt solution at concentrations of 100, 1000, 5000, and 10,000 ppm was used. The aim of this investigation was to understand how a plant species from different localities behaves with a particular salt and whether the inhibition of germination is due to the osmotic or toxic effect of different ions. To make a distinction between the osmotic and ionic or the combined effect of these two factors on seed germination inhibition, the seeds that remained ungerminated in the saline-medium at 10,000 ppm were transferred to distilled water individually to determine the additional germination per treatment.

The results revealed that the germination percentage varied with different salt solutions [92]. Higher concentrations of all the salts directly affected germination, and the germination percentage was reduced. The seeds of C. cretica collected only from site I showed dormancy, and no germination was observed in the control. No salt solution could improve the germination percentage. The maximum (40%) germination was recorded with 100 ppm $Na_2SO_4$. The seeds of Z. simplex had severe dormancy because no germination was observed either in the control or in any salt solution [93]. However, Khan and Ungar [94] reported that growth regulator treatments increased germination to over 80% in nonsaline conditions in Zygophyllum simplex.

After 10 days of salt treatments, ungerminated seeds from the 10,000 ppm concentration of five salt solutions were individually transferred to distilled water. It was discovered [91] that the germination inhibition in saline media was due to osmotic stress or specific ion toxicity because the germination percentage increased when seeds were transferred from salt solution to distilled water (Table 3) [91,95]. Variation in temperature appears to play an important role in recovery of germination of halophytes from salt stress when seeds are transferred to distilled water [96].

Mohammed and Sen [91] proved that higher concentrations of salts retarded germination because of osmotic effects, as the process of seed germination speeded up after transfer to nonsaline medium. There was up to 80% recovery of germination for seeds of Suaeda fruticosa that initially did not germinate in 500 mM NaCl [97]. This may be of significance under natural conditions, especially for inland desert salines, because seeds that could not germinate under extreme salinity stress may have evolved a mechanism to germinate rapidly when the salt stress is relieved [10,15,98–100]. Although NaCl is the major salt in most salt-affected soils, other salts also present in the soil play a combined role in the salt tolerance of a species at the time of germination.

Gibberellic acid and kinetin significantly alleviated the inhibitory effects of salinity on germination of seeds in Arthrocnemum indicum but over different salinity ranges and to different degrees. Both growth regulators significantly increased the rate of germination over most salinities, but the effect of gibberellic acid was more pronounced than that of kinetin [101].

Haloxylon recurvum and H. salicornicum are two characteristic halophytes of the Indian desert. Extremely fast germination in the seeds of these two species, commonly occurring within an hour, has been

**TABLE 3** Additional Mean Germination Percentage of Some Halophytic Species Observed After Transfer of Ungerminated Seeds from 10,000 ppm Concentration of Each Salt Solution to Distilled Water After 10 Days

| Species | Site | Salt solutions | | | | |
|---------|------|------|---------|-------|-------|-----|
| | | NaCl | $Na_2SO_4$ | $MgSO_4$ | $CaCl_2$ | KCl |
| Salsola baryosma | I | 36 | 16 | 23 | 10 | 40 |
| | III | 53 | 13 | 23 | 20 | 40 |
| Sesuvium sesuvioides | I | 20 | 10 | 20 | 70 | 40 |
| | II | 10 | 13 | 16 | 20 | 50 |
| | III | 13 | 16 | 10 | 30 | 10 |
| | IV | 16 | 20 | 13 | 60 | 10 |
| Trianthema triquetra | I | 16 | 30 | 20 | 13 | 30 |
| | II | 10 | 10 | 13 | 10 | 10 |
| | III | 13 | 13 | 10 | 10 | 10 |
| Suaeda fruticosa | I | 36 | 26 | 16 | 20 | 20 |
| | II | 40 | 26 | 30 | 16 | 26 |
| | III | 40 | 30 | 10 | 30 | 26 |

Source: Ref. 91 and 95.

**TABLE 4**   Effect of Time on Seed Germinability in Two *Haloxylon* spp.

| Species | Time | Initiation of germination (min.) | Germination (%) |
|---|---|---|---|
| *H. salicornicum* | 09 AM | 60 | 26.6 |
| (wt. 62.6 mg/100 seeds) | 10 AM | 105 | 20.0 |
| | 11 AM | 90 | 13.3 |
| | 12 Noon | 180 | 26.6 |
| | 01 PM | 75 | 13.3 |
| | 02 PM | 60 | 26.6 |
| | 03 PM | 45 | 33.3 |
| *H. recurvum* | 09 AM | 120 | 06.6 |
| (wt. 93 mg/100 seeds) | 10 AM | —[a] | — |
| | 11 AM | — | — |
| | 12 Noon | — | — |
| | 01 PM | 75 | 20.0 |
| | 02 PM | — | — |
| | 03 PM | — | — |

[a] Nil.
*Source:* Ref. 29.

reported for the first time in any Indian plant species from our laboratory (Table 4) [29]. The fast germination indicates that these halophytes show an adaptive strategy, as the availability of water with reduced NaCl content in soil during the rainy season exists for a short period. This is because evaporation of moisture under bright sunlight and heat increases the salt content by capillary movement [29].

Ungar [102] studied the ecology of halophyte seed banks and found that in unpredictable environments not all of the seeds germinate. In inland areas seed banks may be large, but some coastal seed banks may be very large and others small. Because annuals have only one chance to reproduce, the seed bank for annuals may be more significant than for perennials. Several studies have indicated that seeds of glycophytes and halophytes respond in a similar manner to increased salinity stress in relation to both a reduction in the total number of seeds germinating and a delay in the initiation of the germination process and that seeds of many halophytes remain dormant because of low water potentials [8,103]. The success of annual halophyte populations is greatly dependent on the germination responses of their seeds. Seed germination usually occurs early in the growing season or during a period when soil salinity levels are reduced, allowing the establishment of seedlings prior to the period of highest salt stress [80].

A significant characteristic of halophyte seeds, which distinguishes them from the glycophytes, is their ability to maintain seed viability for extended periods of time during exposure to hypersaline conditions and then to initiate germination when the salinity stress is reduced [8,104,105]. The enforced dormancy response of halophyte seeds to saline conditions is of selective advantage to plants growing in highly saline habitats. These seeds could withstand high salinity stress and provide a viable seed bank for recruitment of new individuals. However, the seed germination would be limited to periods when the soil salinity levels were within the species tolerance limits [104].

Storage of seeds in the soil is a significant factor in recruitment of the plant population in some saline habitats because their establishment is often difficult as their edaphic conditions are continuously changing [80,106]. One-year-old seeds of *Atriplex nummularia* germinated better than freshly harvested seeds [107]. Although several researchers have determined that a large seed bank of viable seeds does exist in many saline habitats, it is not clear how seeds from the individual species respond to high salinity over time [10,108].

Various reports [109,110] indicate that the source from which seeds were obtained may be very critical in determining their germination response when exposed to saline conditions. A very distinct feature of the present study was that the seeds of different sites exhibited different behavior toward a particular salt solution. It was also observed that the levels of salt tolerance vary within the different populations of seeds. It is most probable that genetic selection has taken place for increased salt tolerance in the evolution of at least some taxa found growing in both saline and nonsaline environments [111]. It was also found from the present data that seed germination is controlled by both osmotic and ionic factors [92,112].

## VIII. PHYSIOLOGY OF METABOLIC PRODUCTS

Plants regulate various aspects of their growth in a synchronized form with a high degree of organization involving coordination of many components. Regulation of various metabolic processes has direct control with respect to regulation of catalysis, action, and transport. To understand the metabolic activities, it becomes necessary to study multienzyme systems because metabolic processes in plant systems occur due to specific enzyme activity [64].

In living cells, the intense chemical activity is due to the activity of numerous specific enzymes, which leads to consideration of the interdependence of physiological processes [113]. Under saline conditions, growth is related not only to osmotic and nutritional effects but also to the disturbances in their normal physiological and metabolic processes [114]. Also, under salt stress, the salt induces a decrease or increase in enzyme activity [115], which in turn reflects several metabolic processes.

The biochemical processes inside the leaf cells generally regulate the growth and development of plants. But toxicity influences early metabolic changes, such as enzyme synthesis, to a greater extent [116]. The adaptation of glycophytes to saline soil in adverse conditions is possible mainly because of changeability of their metabolism and chemical properties of the protoplasm [114]. Oxidative, photosynthetic, and photorespiratory enzymes are important because of their various interrelationships in the process of growth and development. Physiologists and biochemists have tried to correlate the possible role of these enzymes with relative metabolic processes of the plants under saline conditions.

Ribulose 1,5-bisphosphate Carboxylase (RuBP-Case) is the main enzyme of $CO_2$ fixation in $C_3$ plants, and Phosphoenolpyruvate Carboxylase (PEP-Case) and RuBP-Case are important enzymes of $C_4$ plants. Popova et al. [117] reported that NaCl stress imposed through the root medium for 8 days decreased the activity of RuBP-Case in *Hordeum vulgare*. Sudhakar et al. [118] studied the response of a few Calvin cycle enzymes to salinity shock in vitro and observed a decline in RuBP carboxylase activity in 10-day-old seedlings of *Dolichos uniflorus* subjected to NaCl or $Na_2SO_4$ treatment, indicating that RuBP-Case was more sensitive to salt shock than other enzymes and NaCl was more toxic than $Na_2SO_4$.

Bankar [64] reported that when plants of *Carthamus tinctorius* were grown at ECe 5.0 to 15 mS cm$^{-1}$ of NaCl, the activity of RuBP-Case decreased with increasing concentrations of NaCl in the growth medium. Saha and Gupta [87] reported an increase in peroxidase activity with increasing NaCl salinity in sunflower seedlings. Sankhla and Huber [119] studied the effect of NaCl on the activities of photosynthetic enzymes in wheat, *Lemna minor* and *Pennisetum typhoides*, and reported that the salt tolerance of RuBP-Case and PEP-Case varies with species.

According to Poljakoff-Mayber [120], the enzymes do not behave identically under saline conditions, and enzymes located in certain places or on certain membranes in the cell may be salt tolerant, whereas others may be salt sensitive.

Peroxidase is an oxidative enzyme and it is also essential for the conversion of $H_2O_2$ to $H_2O$ and O in photorespiration. According to Strogonov [9], peroxidase plays an important role in adaptation of plants to saline conditions by regulating toxic accumulation of $H_2O_2$.

It was reported by Seemann and Sharkey [121] that salinization lowered the RuBP pool size in *Phaseolus vulgaris*. The biochemical basis for this reduction under salt stress is unknown. One reason may be inhibition of ATP synthesis under saline conditions. In addition, the rate of photosynthesis at any given pool size was lower for leaflets from the salinized plants than the control leaves. Thus, it can be concluded that salinity reduces the photosynthetic capacity of leaves by reducing the pool of RuBP as an effect on the RuBP regeneration capacity and secondly by reducing the activity of RuBP-Case by an unknown mechanism when RuBP is in limited supply.

Salinity is known to affect almost all the aspects of plant metabolism. The leaves of plants subjected to water stress often showed a decrease in starch, which is usually followed by an increase in sugar contents [45,122,123]. A similar trend was also observed by these investigators in halophytes of the Indian desert [122,124]. These investigators observed that plant species from site II, which is less saline, showed a maximum sugar content during the summer, when plant water stress was higher than in winter or rainy seasons. The plant species at site I, which is extremely saline, had a higher sugar content during the rainy season, followed by the winter, and the sugar content was least in summer (Table 5) [124]. These varying observations of sugar content may be due to the higher salinities at a particular site. The level of soluble sugars decreased with increased salinity levels, as observed by Gill and Singh [125] in different varieties of paddy (*Oryza sativa*). Naidoo and Naidoo [59] found that in *Sporobolus virginicus*, $CO_2$

**TABLE 5** Seasonal Variations in Total Sugar (mg/g Dry Weight) and Crude Protein (% Dry Weight) Contents in Halophytes Growing at Sites I and III

| Species | Total sugar | | | | | | Crude protein | | | | | |
|---|---|---|---|---|---|---|---|---|---|---|---|---|
| | Site I | | | Site III | | | Site I | | | Site III | | |
| | R[a] | W[a] | S[a] | R | W | S | R | W | S | R | W | S |
| *Aeluropus lagopoides* | 34 | —[b] | — | — | — | — | 12 | — | — | — | — | — |
| *Cressa cretica* | 19 | 11 | 17 | — | — | — | 21 | 19 | 17 | — | — | — |
| *Salsola baryosma* | 33 | 16 | 18 | 5 | 10 | 18 | 20 | 16 | 15 | 36 | 16 | 22 |
| *Sesuvium sesuvioides* | 22 | 9 | — | 16 | 17 | — | 20 | 16 | — | 19 | 9 | — |
| *Sporobolus helvolus* | 29 | 8 | 14 | — | — | — | 18 | 12 | 8 | — | — | — |
| *Suaeda fruticosa* | 39 | 35 | 18 | 5 | 14 | 25 | 29 | 25 | 19 | 28 | 25 | 17 |
| *Trianthema triquetra* | 10 | 4 | — | 5 | 15 | 19 | 19 | 17 | — | 21 | 15 | 12 |
| *Zygophyllum simplex* (red strain) | 18 | 25 | — | — | — | — | 18 | 17 | — | — | — | — |

[a] R, rainy; W, winter; S, summer.
[b] Plant not available.
*Source*: Ref. 124.

exchange, conductance, transpiration, and internal $CO_2$ increased in salinities up to 100 mol/m$^3$ and thereafter decreased. Chandrashekar and Sandhyarani [126] found a decrease in sugar and starch content at different increasing salinity levels in *Crotalaria striata*. Abd-El Samad and Shaddad [127] reported that the sensitivity of a soybean cultivar was due to decreased saccharide content under NaCl salinity. Muthukumaraswamy and Panneerselvam [60] found that NaCl salinity decreased the accumulation of starch, sugar content, and activity of α-amylase in *Raphanus sativus* seedlings.

Bankar [64] reported that the carbohydrate content was more than the control value at all levels of NaCl salinization and maximum content was recorded at ECe 7.5 mS cm$^{-1}$ in *Carthamus tinctorius* stem. Under $Na_2SO_4$ salinization, the content was less than control at all levels. In leaves, the carbohydrate content increased up to ECe 10.0 mS cm$^{-1}$ and decreased further under NaCl salinity. With $Na_2SO_4$ treatment, the carbohydrate content increased up to ECe 12.5 mS cm$^{-1}$ and decreased at higher salinity levels. The results of Zidan and Elewa [128] revealed that soluble carbohydrate remained unchanged at low and moderate levels of NaCl in four plant species of Umbelliferae. Thus, these investigators [122,123] concluded that the maximum sugar content at site I during the rainy season may be due to low soil salinity because of leaching of the salts during rainfall compared with summer and winter.

From the preceding account it is clear that in some plants soluble sugar increases, whereas in others it decreases and in some others its content remains unaffected. In general, plants use soluble sugars as an osmoticum under saline conditions. Hence, the plants that can tolerate low or medium levels of salt stress synthesise more soluble sugars and tolerate salt stress. The plants that fail to increase soluble sugar biosynthesis could not tolerate salts.

The nutritive pattern of plants is very important when fodder values and productivity are taken into consideration. Root zone salinization presents a challenge to plant productivity that is effectively countered by salt-tolerant halophytic plants but, unfortunately, much less successfully by major crop plants. The way in which salt affects plant metabolism was reviewed by Volkmar et al. [129]. Protein synthesis and turnover in growing plants is a basic component of metabolic regulation that provides a way to vary the enzymatic complement during the response to environmental conditions [130]. Protein is the most important constituent of cells from both structural and functional points of view. Changes in the ion content of plant cells induced changes in the activity of certain metabolic systems. Such changes may have serious consequences for membrane proteins.

Vera-Estrella et al. [131] observed that increasing concentrations of NaCl stimulated the activities of tonoplast and plasma membrane H$^+$-ATPases in *Mesembryanthemum crystallinum*. Immunodetection of the ATPases showed that the increased activity was not due to changes in protein amount that could be attributed to treatment conditions. A specific role for these mechanisms in salt adaptation is supported by the inability of mannitol-induced water stress to elicit the same responses and the absence of enzyme activity and protein expression associated with Crassulacean acid metabolism in the cells. Under conditions of extreme salinity, proteins are precipitated. The protein content of various plant tissues generally declined under drought or saline conditions because of increased proteolysis and decreased protein synthesis [12]. Muthukumaraswamy et al. [61] observed that the protein content decreased in root, shoot, and leaf in *Cicer arietinum* seedlings under NaCl salinity. Venkatesalu et al. [132] and Chandrashekar and Sandhyarani [133] reported an increase in protein content with increasing NaCl salinity in *Sesuvium portulacastrum* and *Crotalaria striata*, respectively. Muthuchelian et al. [134] reported that salt stress increased the protein content in *Erythrina variegata* seedlings. Thus, in many plants protein synthesis is stimulated under saline conditions. Strogonov [9] reported higher protein content in maize under sulfate salinity. According to him in general, salt-tolerant plants maintain protein synthesis under saline conditions but salt-susceptible plants do not. Our findings revealed that the protein content of plants at both saline and nonsaline sites was maximum during the rainy season, when plant water status was higher than in winter or summer [23,122,123].

## IX.  STOMATAL BEHAVIOR

Water relations and stomatal behavior are important indices that reflect the ability of plants to economize essential requirements under prevailing climatic and edaphic conditions [135–138]. In plants adapted to dry environments, anatomical and morphological changes at the leaf and whole plant levels prevent metabolic imblance and help to improve water relations [139]. Stomata show amazing versatility in their reaction to the environment and respond to all factors that are of physiological importance. Stomata are known to play a pivotal role in productivity of plants. Thus, it is necessary to study stomatal characteristics. The plants require adaptive mechanisms to help their survival under saline stress. Leaves are the major sites of transpiration and photosynthesis in higher plants. In relation to salinity-induced water stress, one might expect the principal structural and metabolic modifications in leaves to be associated with a tendency to minimize transpiration rate and the occurrence of photosynthetic pathways with high water use efficiency.

According to Perera et al. [140], stomatal opening was suppressed by increasing NaCl concentration in *Aster tripolium*, which has no glands or cannot excrete salt, indicating that salt accumulation in cell vacuoles increases. Na ions in apoplast around guard cells, causing partial closure, reducing transpiration and increasing water use efficiency, reducing flow of salt to leaves, and not affecting new photosynthate synthesis and growth. The increasing supplies of $Ca^{2+}$ ions reduced the effect of salinity on stomatal conductance in the whole plant (*Aster tripolium*) as well as in the isolated epidermis. This finding is consistent with the well-established role of calcium in increasing resistance to salinity. In the presence of high calcium, the plants can tolerate a greater salt intake, and hence there is a reduced need for transpiration to be restricted by partial stomatal closure. Ayala and O'Leary [141] observed decreased stomatal conductance with increasing salinity that increased the transpiration rate at a low salinity level in *Salicornia bigelovii*. Lakshmi et al. [142] reported a decrease in stomatal conductance in *Morus alba* under saline conditions.

Robinson et al. [143] noted that when saline plants were subjected to 200 mM NaCl, stomatal conductance was reduced by 70%, which decreased actual photosynthesis. However, in white mangrove (*Laguncularia racemosa*) the number of stomata and salt glands per leaf area was increased in high NaCl [144]. The study of Gulzar and Khan [145] revealed that the water relations of perennial halophytes showed similar patterns of variation in all parameters, and plants at the coastal locations appeared to be more stressed than plants at inland locations in Pakistan. Plant physiological responses to high NaCl included an increased chlorophyll *a/b* ratio (to enable the plants to cover the high energy demands for adaptation to salt stress), an increase in soluble protein contents with rising salt stress, a tendency for carbohydrates to accumulate in foliage, and a negative influence of high soil salinity on secondary compound (phenol, hydrolyzed or condensed tannins) metabolism [144].

Villiers et al. [146] reported that the net leaf photosynthetic rate and leaf stomatal conductance decreased with increasing salinity, while the intercellular $CO_2$ concentration increased. Both stomatal closure and inhibition of biochemical processes probably caused the reduced leaf photosynthetic rates. The stomatal indices suggested that the trend toward an increase in number of stomata per unit leaf area with an increase in salinity was not due to decreased epidermal cell size.

The study of Bankar [64] revealed that the pattern of stomatal opening in *C. tinctorius* is similar to that of $C_3$ plants. The stomata remained closed when the heat was maximum in this plant. Thus, plants adapt to saline conditions by conserving water. Plants also adapt to saline conditions by decreasing the number of stomata, which leads to conservation of water.

According to Perera et al. [147], x-ray microanalysis revealed that the sodium content of the stomatal guard cells of *Aster tripolium* remained much lower than that of other leaf cells when the plants were grown at high salinity levels. In contrast, large amounts of sodium accumulated in epidermal and subsidiary cells and particularly in the mesophyll tissue, suggesting that a mechanism exists to limit the extent of its entry into guard cells. Even in plants grown at high salinity, the content of potassium was much higher than that of sodium in the guard cells, consistent with the view that this is a major ion involved in determining stomatal movements in this halophyte. It is suggested that the acquisition by the guard cells of some ability to restrict the intake of sodium ions may be an important component of sodium-driven regulation of transpiration and hence of salinity tolerance in *A. tripolium* [147]. A detailed study of stomatal behavior to determine monthly and seasonal variations in the state of water balance in some saline plants of the Indian desert and their relationship with soil moisture conditions was carried out by Mohammed [23,137] and Sen and Mohammed [138].

Rainwater is the only source of available moisture in the desert of northwestern Rajasthan, India. Although the monsoon season starts here by mid-June and extends to October, the rains are very erratic and scant. The occurrence of a rather long intermission between successive rain showers, sometimes ranging from a few days to weeks, is not uncommon here. Whatever rainwater is retained by the soil is exploited by the roots of annual and perennial species from the months of June and July to November and December. After this, annual species start to disappear because of moisture scarcity in the upper soil layers and their inability to exploit moisture from the deeper soil layers because of their shallow root system. By this time, perennial species also start showing various symptoms of water shortage, which are reflected by a remarkable reduction in the transpiring surface. The maximum soil moisture was recorded in the rainy season. All halophytic annual species, such as *S. sesuvioides*, *T. triquetra*, and *Z. simplex*, completely disappear with the depletion of soil moisture. Winter showers and premonsoon rains, although scant in quantity, play a significant role in improving the water status of soil and plants [137,138].

In order to maintain lower water potentials within the cell, dicotyledon halophytes normally make the necessary osmotic adjustment by accumulating $Na^+$ and $Cl^-$ ions. The cellular basis of salt tolerance in halophytes depends upon the compartmentation of ions necessary for osmoregulation in vacuoles and upon osmotic adjustment of the cytoplasm by compatible solutes. The central role played by $Na^+$ and $Cl^-$ in osmotic adjustment suggests that the transport of these ions and its regulation must be of primary importance in the physiology of the plant as a whole. The decreases in transpiration rate per unit area of leaf help to lower the ion input into leaves. Any linked reductions in photosynthesis appear to be due to decreases in stomatal frequency [148]. In the grasses, potassium and sugars are used to make the osmotic adjustment. However, in succulent halophytes, the major use of organic compounds such as vacuolar solutes seems to be precluded on the basis of energetic grounds [37]. Kurban et al. [149] reported that with increasing salinity levels, the membrane permeability decreased in *Alhagi pseudoalhagi*, whereas in *Vigna radiata* it slightly increased at 9.1 dS m$^{-1}$. The leaf water potential and the osmotic potential decreased in both plants along with the seawater salinity levels. The contributions of organic and inorganic solutes to the osmotic adjustment differed in different species.

The physiological traits involved in leaf water relations were evaluated in *Avicennia germinans* seedlings by Suarez et al. [150]. They concluded that the leaves of seedlings adapt to hypersaline soils by increasing solute concentration and cell elasticity. It is suggested that both processes allow leaf water uptake and turgor maintenance over a large range of soil water potential.

All the halophytic species studied [23,24] could adjust themselves by changing their osmotic potentials rapidly with a greater range in osmotic potentials of surrounding soil. This is in agreement with Waisel (12), who stated that it is probably true that the great majority of the halophytic plants belong to the adjustable group and that their osmotic adjustment occurs rapidly. Recovery from osmotic stress occurs faster in salt-accumulating halophytes than in salt-enduring halophytes [151]. Osmotic adjustment or osmoregulation enables plants to maintain growth as the plant water potential decreases. Adjustment occurs through decreases in osmotic potential by solute accumulation in the cells as the leaf water potential decreases. In this condition the net result is that the cell turgor potential is kept relatively high, thus maintaining turgor-dependent processes, such as leaf growth and stomatal opening [152]. Further, it emerged from our study [23,24] that during the rainy season, the higher moisture in the soil and the leaching of salts resulted in an increase in the osmotic potential of the soil that led to

**TABLE 6**   Seasonal Variations in Stomatal Density (mm²) in Leaves of Different Plant Species from Two Sites[a]

| Species | Pachpadra (site I) | | | | | | Jodhpur (site III) | | | | | |
| | R | | W | | S | | R | | W | | S | |
| | U | L | U | L | U | L | U | L | U | L | U | L |
|---|---|---|---|---|---|---|---|---|---|---|---|---|
| *Cressa cretica* | 129 | 125 | 53 | 39 | 44 | 35 | —[b] | — | — | — | — | — |
| | (23) | (24) | (17) | (11) | (9) | (7) | | | | | | |
| *Salsola baryosma* | — | 205 | — | 53 | — | 35 | — | 188 | — | 99 | — | 103 |
| | | (13) | | (17) | | (12) | | (17) | | (14) | | (17) |
| *Sesuvium sesuvioides* | 56 | 52 | — | — | — | — | 43 | 20 | — | — | — | — |
| | (35) | (33) | | | | | (30) | (22) | | | | |
| *Suaeda fruticosa* | 46 | 43 | 30 | 29 | 25 | 35 | 100 | 82 | 59 | 69 | 52 | 39 |
| | (33) | (28) | (24) | (24) | (16) | (27) | (23) | (16) | (27) | (23) | (23) | (25) |
| *Trianthema triquetra* | 59 | — | 40 | — | — | — | 102 | — | 56 | — | 63 | — |
| | (31) | | (25) | | | | (28) | | (24) | | (26) | |
| *Zygophyllum simplex* (red strain) | 59 | 63 | 26 | 29 | — | — | — | — | — | — | — | — |
| | (33) | (33) | (15) | (16) | | | | | | | | |

[a] Values in parentheses are stomatal index (SI). R, rainy; W, winter; S, summer; U, upper; L, lower.
[b] Plant absent.
*Source*: Ref. 137.

the increased osmotic potential of the plants [65]. However, with the increase in salt concentration and the decrease in soil moisture, plants try to adjust themselves to drought by accumulation of salts. Thus, the accumulation of salts in plants decreases their osmotic potentials to the level of highest stress, so that plants are able to take up maximum water during the hot summer, resulting in a gradual increase in their osmotic potentials. Some plants exhibit lower values of osmotic potentials during winter periods associated with low temperature and/or high salinity. Thus, it is clear from the study of Mohammed [23] that the water relations of plants are directly related to the amount of moisture present in the soil [135,136]. The study [23] showed higher stomatal density and opening during the rainy season when soil water status remained higher (Table 6) [137]. More water was lost when water was abundant in the soil, that is, during the rainy season.

## X.  SUMMARY AND CONCLUSIONS

Salinity, water, and temperature are most critical factors that determine the vegetation pattern of a region. The excess of salts in the soil inhibits plant growth. Halophytes survive under conditions of high salinity and exhibit succulence, which might resort to other physiological adaptations to overcome the adverse saline environment in the soil. The ionic and toxic effects of various salts, especially of NaCl, play a major role in halophytism. Salinity is known to affect many aspects of plants and induce numerous changes in their morphology. Most of the halophytes avoid salinity, some evade it, and a few others tolerate it. Xerosucculents are characteristics of halophytes showing a thick cuticle and a cover of waxy layers, such as *Suaeda fruticosa*, *Salsola baryosma*, and *Haloxylon recurvum*. The accumulation of proline in plants is correlated with the extent of the water stress in the plant. In halophytes, a positive correlation is seen between the proline content and the amount of $Na^+$ and $Cl^-$ in the cell sap. Salt stress induces accumulation of more proline in halophytes and perhaps plays an essential role in their survival. Ion uptake by plants is largely dependent upon the availability of the ions in the soil. Higher level of minerals in halophytes has been observed during dry periods. Although NaCl is the major salt present in most salt-affected soils, other salts such as $MgCl_2$, $MgSO_4$, and $Na_2SO_4$ are also present and play a combined role in the salt tolerance of a species at the time of seed germination. Seed germination is controlled by both osmotic and ionic factors. Maximum germination of seeds in halophytes has been reported during the rainy season due to leaching of salts in deeper soil layers through rainfall. Because of the leaching action, there is a decrease in soil salinity. Salinity also affects almost all the aspects of plant metabolism. Maximum values of carbohydrate and crude protein in halophytes have been observed during the rainy season, when plant water status is higher than in winter or summer. The maximum stomatal opening and water loss are noted during the rainy season.

# REFERENCES

1. S Szablics. Salt Affected Soils. Boca Raton, FL: CRC Press, 1989.
2. GV Subbarao, C Johansen. Potential for genetic improvement in salinity tolerance in legumes: pigeon pea. In: M Pessarakli, ed. Handbook of Plant and Crop Stress. New York: Marcel Dekker, 1994, pp 581–595.
3. SM Alam. Nutrient uptake by plants under stress conditions. In: M Pessarakli, ed. Handbook of Plant and Crop Stress. New York: Marcel Dekker, 1994, pp 227–246.
4. VA Kovda, I Szabolcs. Modelling of soil salinization and alkalization. Agrokem Talajtan 28(suppl):1979.
5. MA Khan, IA Ungar. Biology of Salt Tolerant Plants. Chelsea, Michigan: Book Crafters, 1995.
6. SW Breckle. Studies on halophytes from Iran and Afghanistan. III. Ecology of halophytes along salt-gradients. Proc R Bot Soc (Edinb) 89B:203–215, 1986.
7. H Boyko. Salinity and Aridity: New Approaches to Old Problems. The Hague: Dr W Junk, 1966.
8. VJ Chapman. Salt Marshes and Salt Desert of World. Bremerhaven, Germany: L Cramer, 1974.
9. BP Strogonov. Physiological Basis of Salt Tolerance of Plants. Jerusalem: Israel Prog Sci Transl, 1964.
10. IA Ungar. Population ecology of halophyte seeds. Bot Rev 53:301–334, 1987.
11. A Poljakoff-Mayber, J Gale. Plants in Saline Environments. Berlin: Springer-Verlag, 1975.
12. Y Waisel. Biology of Halophytes. New York: Academic Press, 1972.
13. DN Sen, KS Rajpurohit. Contributions to the Ecology of Halophytes. The Hague: Dr W Junk, 1982.
14. DN Sen, RB Jhamb, DC Bhandari. Utilization of saline areas of western Rajasthan through suitable plant introduction. In: SD Mishra, DN Sen, I Ahmad, eds. Proc Natl Symp Eval Environs. Jodhpur: Geobios International, 1985, pp 348–360.
15. IA Ungar. Inland halophytes of United States. In: R Reimold, W Queen, eds. Ecology of Halophytes. New York: Academic Press, 1974, pp 235–305.
16. IA Ungar. The effect of salinity and temperature on seed germination and growth of *Hordeum jubatum* L. Can J Bot 52:1357–1362, 1974.
17. IA Ungar. Salinity tolerance of inland halophytic vegetation of North America. Bull Soc Bot 120:217–222, 1973.
18. PF Brownell. Sodium as an essential micronutrient element for a higher plant (*Atriplex vesicaria*). Plant Physiol 40:460–468, 1965.
19. BP Strogonov. Structure and Function of Plant Cells in Saline Habitats. New York: John Wiley & Sons, 1974.
20. GR Stewart, JA Lee. The role of proline accumulation in halophytes. Planta 12:279–289, 1974.
21. KS Rajpurohit. Soil salinity and its role on phytogeography of western Rajasthan. PhD dissertation, University of Jodhpur, Jodhpur, 1980.
22. RB Jhamb. Biology of halophytes. PhD dissertation, University of Jodhpur, Jodhpur, 1984.
23. S Mohammed. Comparative studies of saline and nonsaline vegetation in Indian arid zone. PhD dissertation, University of Jodhpur, Jodhpur, 1988.
24. DN Sen. Ecology of saline areas of Rajasthan and exploitation of saline ecosystem for increased productivity. DOEn Final Technical Report, University of Jodhpur, Jodhpur, 1990.
25. TP Sharma. Ecology and biology of saline ecosystem in Indian desert. PhD dissertation, University of Jodhpur, Jodhpur, 1991.
26. P Rajput. Ecological studies on introduced species of *Atriplex* in Indian desert. PhD dissertation, University of Jodhpur, Jodhpur, 1992.
27. TP Thomas. Ecology of certain halophytes in Indian desert with special reference to introduction of *Atriplex* spp. PhD dissertation, University of Jodhpur, Jodhpur, 1992.
28. AK Gehlot. Ecology of seed germination behaviour in Indian desert with special reference to saline habitats. PhD dissertation, Jai Narain Vyas University, Jodhpur, 1996.
29. TP Sharma, DN Sen. A new report on abnormally fast germinating seeds of *Haloxylon* spp.—an ecological adaptation to saline habitat. Curr Sci 58:382–385, 1989.
30. S Mohammed, DN Sen. Vegetation patterns in saline areas of Indian arid zone. Bull Life Sci Dibrugarh 4:1–8, 1994.
31. JF Reed. The relation of the *Spartinetum glabrae* near Beaufort, North Carolina, to certain edaphic factors. Am Midl Nat 38:605–614, 1947.
32. DN Sen, S Mohammed, PK Kasera. Biology of plants in saline environment. In: IS Grover, AK Thukral, eds. Environment and Development. Jodhpur: Scientific Publishers, 1998, pp 117–126.
33. IA Ungar. Are biotic factors significant in influencing the distribution of halophytes in saline habitats? Bot Rev 64:176–199, 1998.
34. S Aziz. Population biology of desert perennial halophyte *Cressa cretica*. M Phil dissertation, University of Karachi, Karachi, Pakistan, 1994.
35. AK Gehlot, DN Sen. Effect of different salts on seed germination of *Haloxylon recurvum* (Moq.) Bunge ex Boiss. A halophyte of Indian arid zone. Ann Arid Zone 35:61–64, 1996.
36. A Poljakoff-Mayber, HR Lerner. Plants in saline environments. In: M Pessarakli, ed. Handbook of Plant and Crop Stress. New York: Marcel Dekker, 1994, pp 65–96.
37. DJ Weber. 1995. Mechanisms and reactions of halophytes to water and salt stress. In: MA Khan, IA Ungar, eds. Biology of Salt Tolerant Plants. Chelsea, Michigan: Book Crafters, 1995, pp 170–180.

38.  DN Sen, S Mohammed, PK Kasera. Some ecological observations on plants of Indian desert. J Indian Bot Soc (Platinum Jubilee Vol) 74A:539–554, 1995.

39.  DN Sen, S Mohammed, PK Kasera. Biology of saline ecosystem and reclaiming unproductive land in Indian desert. In: AS Mehta, B Saran, eds. Environment and Productivity, Proceedings 4th APPSC Conference and New Botanica Convention, Patna, India, 1998, pp 1–7.

40.  TJ Flowers, MA Hajibagheri, NJW Clipson. Halophytes. Q Rev Biol 61:313–336, 1986.

41.  SW Breckle. How do halophytes overcome salinity? In: MA Khan, IA Ungar, eds. Biology of Salt Tolerant Plants. Chelsea, Michigan: Book Crafters, 1995, pp 199–213.

42.  KS Rajpurohit, DN Sen. Soil salinity and seed germination under water stress. Trans Isdt Ucds 2:106–110, 1977.

43.  RB Jhamb, DN Sen. Seed germination behaviour of halophytes in Indian desert. I. *Suaeda fruticosa* (Linn.) Forsk. Curr Sci 53:100–101, 1984.

44.  J Levitt. Responses of Plants to Environmental Stresses. Vol 2, Academic Press, New York: 1980.

45.  HY Mohan Ram, P Gupta. Plant life under extreme environments. Curr Sci 72:306–315, 1997.

46.  B Shen, S Hohmann, RG Jensen, HJ Bohnert. Role of sugar alcohols in osmotic stress adaptation. Replacement of glycerol by mannitol and sorbitol in yeast. Plant Physiol 121:45–52, 1999.

47.  HW Koyro, L Wegmann, H Lehmann, H Lieth, A Hamdy. Physiological mechanisms and morphological adaptation of *Laguncularia racemosa* to high NaCl salinity. In: H Lieth, A Hamdy, eds. CIHEAM International Conference Valenzano, Bari, Italy, 1997, pp 51–78.

48.  RL Jefferies, T Rudmik. The response of halophytes to salinity: An ecological perspective. In: RC Staples, GH Toenniessen, eds. Salinity Tolerance in Plant Strategies for Crop Improvement. New York: John Wiley & Sons, 1984, pp 213–227.

49.  J Wu, DM Seliskar. Salinity adaptation of plasma membrane $H^+$-ATPase in the salt marsh plant *Spartina patens*: ATP hydrolysis and enzyme kinetics. J Exp Bot 49:1005–1013, 1998.

50.  NM Barnetts, AW Naylor. Amino acids and protein metabolism in *Bermuda* grass during water stress. Plant Physiol 41:1222–1230, 1966.

51.  S Mohammed, DN Sen. Proline accumulation in arid zone plants. J Arid Environ 13:231–236, 1987.

52.  S Mohammed, DN Sen. Environmental changes and proline content in some desert plants. J Arid Environ 19:241–243, 1990.

53.  AP Stines, DJ Naylor, PB Hoj, RV Heeswijck. Proline accumulation in developing grapevine fruit occurs independently of changes in the levels of $\Delta$ 1-pyrroline-5-carboxylate synthetase mRNA or protein. Plant Physiol 120:923–931, 1999.

54.  S Treichel. The influence of NaCl on delta-1-pyrroline-5-carboxylate reductase in proline accumulating cell suspension cultures of *Mesembryanthemum nodiflorum* and other halophytes. Physiol Planta 67:173–181, 1986.

55.  B Heuer. Osmoregulatory role of proline in water- and salt-stressed plants. In: M Pessarakli, ed. Handbook of Plant and Crop Stress. New York: Marcel Dekker, 1994, pp 363–381.

56.  V Venkatesalu, RR Kumar, KP Chellappan. Sodium chloride stress on organic constituents of *Sesuvium portulacastrum* L., a salt marsh halophyte, J Plant Nutr 17:1635–1645, 1994.

57.  MA Khan, A Seemi, S Aziz. Some aspects of salinity, plant density and nutrient effects on *Cressa cretica* L. J Plant Nutr 21:769–784, 1998.

58.  PE Verslues, RE Sharp. Proline accumulation in maize (*Zea mays* L.) primary roots at low water potentials. II. Metabolic source of increased proline deposition in the elongation zone. Plant Physiol 119:1349–1360, 1999.

59.  G Naidoo, Y Naidoo. Salt tolerance in *Sporobolus virginicus*: the importance of ion relations and salt secretion. Flora 193:337–344, 1998.

60.  M Muthukumaraswamy, R Panneerselvam. Growth and biochemical changes of green gram under NaCl stress. Geobios 24:107–111, 1997.

61.  M Muthukumaraswamy, I Karikalan, SN Rajan, R Panneerselvam. Effect of NaCl stress on protein, proline and protease ATP activities in chikpea seedlings. Geobios 24:119–123, 1997.

62.  AP Joshi, PP Khairatkar. Seed germination, amino acids and sugars in seedlings of *Juncus maritimus* and *J. acutus* under salt stress. J Indian Bot Soc 74:15–17, 1995.

63.  AB Garcia, JD Engler, S Iyer, T Gerats, MV Montagu, AB Caplan. Effects of osmoprotectants upon NaCl stress in rice plants. Plant Physiol 115:159–169, 1997.

64.  NN Bankar. Physiological studies in safflower (*Carthamus tinctorius* L.) Cv. Bhima under saline conditions. PhD dissertation, University of Pune, Pune, 1999.

65.  DN Sen, S Mohammed. Proline accumulation in some halophytes in Indian desert. In: BN Prasad, GPS Ghirmire, VP Agrawal, eds. Role Biotechnology in Agriculture. New Delhi: Oxford & IBH Publishing Co, 1992, pp 129–137.

66.  V Sangwan, S Babber, TM Varghese. Correlation between proline accumulation and salt tolerance in chickpea (*Cicer arietinum*) calli. J Indian Bot Soc 75:259–262, 1996.

67.  YM Yang, RJ Newton, FR Miller. Salinity tolerance in *Sorghum*. III. Cell culture response to sodium chloride in *S. bicolor* and *S. halepense*. Crop Sci 30:781–785, 1990.

68.  CR Stewart, SF Boggers, D Aspinall, LG Paleg. Inhibition of proline accumulation by water stress. Plant Physiol 59:930–932, 1977.

69. D Aspinall, LG Paleg. Proline accumulation: physiological aspects. In: G Paleg, D Aspinall, eds. The Physiology and Biochemistry of Drought Resistance in Plants. New York: Academic Press, 1981, pp 205–241.

70. IS Sheoran, HS Nainawatee. Metabolic changes in relation to environmental stress. In: R Singh, ed. Plant Biochemistry Research in India. New Delhi: Society of Plant Physiology and Biochemistry, 1990, pp 157–178.

71. S Mohammed, PK Kasera, DD Chawan, DN Sen. Osmotic potential in the leaf sap of halophytes in Indian arid zone. J Indian Bot Soc 77:179–184, 1998.

72. EA Jackson, G Blackburn, ARP Clarke. Seasonal changes in soil salinity at Trintinara, South Australia. Aust J Agric Res 7:20–24, 1956.

73. S Mohammed, PK Kasera, DN Sen. Soil characteristics of inland salines in Indian desert. Geophytology 27:103–106, 1998.

74. DN Sen, S Mohammed. General aspects of salinity and the biology of saline plants. In: M Pessarakli, ed. Handbook of Plant and Crop Stress. New York: Marcel Dekker, 1994, pp 125–145.

75. S Mohammed, DN Sen. Ecophysiological studies on saltwort in Indian desert. Ann Arid Zone 31:115–118, 1992.

76. S Mohammed, DN Sen. Metabolic and mineral responses of *Salsola baryosma* (Roem. et Schult.) Dandy: a halophyte of Indian inland salines. J Indian Bot Soc 71:161–163, 1992.

77. MR Harward, I McNulty. Seasonal changes in ionic balance in Salicornia rubra. Utah Acad Proc 42:65–69, 1965.

78. RL Jefferies. Osmotic adjustment and the response of halophytic plants to salinity. Bioscience 31:42–46, 1981.

79. GL Maliwal. Response of wheat varieties to chloride and sulphate dominant salinity. Indian J Plant Physiol 2:225–228, 1997.

80. IA Ungar. Ecophysiology of Vascular Halophytes. Ann Arbor, MI: CRC Press, 1991.

81. R Albert, M Popp. Chemical composition of halophytes from the Neusiedler Lake region in Austria. Oecologia 27:157–170, 1986.

82. K Mahmood. Effects of salinity, external $K^+/Na^+$ ratio and soil moisture on growth and ion content of *Sesbania rostrata*. Biol Plant 41:297–302, 1998.

83. TJ Bhandal, CP Malik. Potassium estimation uptake and its role in the physiology and mechanism of flowering plant. Int Rev Cytol 110:205–254, 1988.

84 JA Abbas. Monthly variation in chloride accumulation by *Zygophyllum gatarense* from saline and non-saline habitats of Bahrain. In: MA Khan, IA Ungar, eds. Biology of Salt Tolerant Plants. Chelsea, Michigan: Book Crafters, 1995, pp 183–189.

85. KP Chellappan. Growth and mineral distribution of *Sesuvium portulacastrum* L., a salt marsh halophyte under sodium chloride stress. Compend Soil Sci Plant Anal 25:2797–2805, 1994.

86. AM Hamada. Effect of NaCl, water stress or both on gas exchange and growth of wheat. Bio Plant 38:405–412, 1996.

87. K Saha, K Gupta. Effect of NaCl salinity on ethylene production and metabolism in sunflower seedlings. Indian J Plant Physiol 2:127–130, 1997.

88. EP Glenn, JJ Brown. Effect of soil salt levels on the growth and water use efficiency of *Atriplex canescens* (Chenopodiaceae) varieties in drying soil. Am J Bot 85:10–16, 1998.

89. TP Egan, IA Ungar. Effect of different salts of sodium and potassium on the growth of *Atriplex prostrata* (Chenopodiaceae). J Plant Nutr 21:2193–2205, 1998.

90. CH Keiffer, IA Ungar. The effect of extended exposure to hypersaline conditions on the germination of five inland halophyte species. Am J Bot 84:104–111, 1997.

91. S Mohammed, DN Sen. Germination behaviour of some halophytes in Indian desert. Indian J Exp Biol 28:545–549, 1990.

92. S Mohammed, DN Sen. Polymorphism as an adaptation in seeds of halophytes in Indian desert. In: DN Sen, S Mohammed, eds. Marvels of Seed, Proceedings of International Seed Symposium, Jodhpur, India. Jodhpur: Jodhpur University Press, 1991, pp 207–212.

93. S Mohammed, DN Sen. Effect of $GA_3$ and different nitrates on seed germination of *Zygophyllum simplex* (Linn.), an inland halophyte of Indian desert. Proc Natl Acad Sci India 62:393–397, 1992.

94. MA Khan, IA Ungar. Alleviation of seed dormancy in the desert forb *Zygophyllum simplex* L. from Pakistan. Ann Bot 80:395–400, 1997.

95. S Mohammed, DN Sen. Effect of different salts on seed germination of *Suaeda fruticosa* (L.) Forsk. from different localities in Indian desert. J Indian Bot Soc 70:91–93, 1991.

96. MA Khan, IA Ungar. Effects of thermoperiod on recovery of seed germination of halophytes from saline conditions. Am J Bot 84:279–283, 1997.

97. MA Khan, IA Ungar. Germination of the salt tolerant shrub *Suaeda fruticosa* from Pakistan: salinity and temperature responses. Seed Sci Technol 26:657–667, 1998.

98. MA Khan, B Gul. High salt tolerance germinating dimorphic seeds of *Arthrocnemum indicum*. Int J Plant Sci 159:826–832, 1998.

99. MA Khan, IA Ungar. Effects of light, salinity and thermoperiod on the seed germination of halophytes. Can J Bot 75:835–841, 1997.

100. MA Khan, IA Ungar. Effect of salinity on seed germination of *Triglochin maritima* under various temperature regimes. Great Basin Nat 59:144–150, 1999.

101. MA Khan, IA Ungar, B Gul. Action of compatible osmotica and growth regulators in alleviating the effect of salinity on the germination of dimorphic seeds of *Arthrocnemum indicum* L. Int J Plant Sci 159:313–317, 1998.

102. IA Ungar. Seed bank ecology of halophytes. In: MA Khan, IA Ungar, eds. Biology of Salt Tolerant Plants. Chelsea, Michigan: Book Crafters, 1995, pp 65–79.

103. J Philipupillai, IA Ungar. The effect of seed dimorphism on the germination and survival of *Salicornia europaea* L. populations. Am J Bot 71:542–549, 1984.

104. IA Ungar. Germination ecology of halophytes. In: DN Sen, KS Rajpurohit, eds. Contributions to the Ecology of Halophytes. The Hague: Dr W Junk, 1982, pp 143–154.

105. SRJ Woodell. Salinity and seed germination patterns in coastal plants. Vegetatio 61:223–230, 1985.

106. S Mohammed, DN Sen. Storage and salt effects on seed germination of *Sesuvium sesuvioides* (Fencl.) Verdc., a halophyte of Indian desert. Proc Natl Acad Sci India 62:597–601, 1992.

107. MA Turk. Oldman saltbush seed treatment for germination improvement. Agric Trop Subtrop 31:53–59, 1998.

108. MA Leck. Wetland seed banks. In: MA Leck, VT Parker, RL Simpson, eds. Ecology of Soil Seed Banks. New York: Academic Press, 1989.

109. JP Workman, NE West. Germination of *Eurotia lanata* in relation to temperature and salinity. Ecology 48:659–661, 1969.

110. RW Kingsbury, A Radlow, PJ Mudie, J Rutherford, R Radlow. Salt stress in *Lasthenia glabrata*, a winter annual composite endemic to saline soils. Can J Bot 54:1377–1385, 1976.

111. LD Clarke, NE West. Germination of *Kochia americana* in ralation to salinity. J Range Manage 22:286–287, 1969.

112. S Mohammed, DN Sen. A report on polymorphic seeds in halophytes. 1. *Trianthema triquetra* L. in Indian desert. Curr Sci 57:616–617, 1988.

113. HE Street, W Cockburn, eds. Plant Metabolism. Oxford: English Language Book Society and Program Press, 1972, p 247.

114. BP Strogonov, PA Henckel. Physiology of Plants Consuming Saline Water. Proceedings of Symposium on Salinity Problems in Arid Zones, Tehran, 1961, pp 145–151.

115. E Hasson-Porath, A Poljakoff-Mayber. The effect of salinity in the growth medium on carbohydrate metabolism in pea root stripe. Plant Cell Physiol 9:195–203, 1968.

116. KA Malik, SS Sahukat. Effect of NaCl salinity on growth and peroxidase activity on *Triticum aestivum* L. var. Chanab 70. Pak J Bot 18:29–35, 1986.

117. LP Popova, ZG Stoinova, LT Maslenkova. Involvement of abscisic acid in photosynthetic process in *Hordeum vulgare* L. during salinity stress. J Plant Growth Regul 14:211–218, 1995.

118. C Sudhakar, BR Ramanjulu, PR Veeranjaneyulu. Responses of some calvin cycle enzymes subjected to salinity shock in vitro. Indian J Exp Biol 35:665–671, 1997.

119. N Sankhla, W Hüber. Ecophysiological studies on Indian arid zone plants: IV. Effect of salinity and gibberellin on the activities of photosynthetic enzymes and $^{14}CO_2$ fixation products in the leaves of *Pennisetum typhoides* seedlings. Biochem Physiol Pflanzen 166:181–187, 1974.

120. A Poljakoff-Mayber. Biochemical and physiological responses of higher plants to salinity stress. Environ Sci Res 23:245–269, 1982.

121. JR Seemann, TD Sharkey. Salinity and nitrogen effects on photosynthesis, RuBP-C and metabolite pool sizes in *Phaseolus vulgaris*. Plant Physiol 82:555–560, 1986.

122. DN Sen, S Mohammed. Eco-physiological studies of *Fagonia cretica* L. in Indian desert. In: SK Agarwal, RK Garg, eds. Environmental Issues and Research in India, Prof LN Vyas Commemoration Volume. Udaipur, India: Himanshu Publications, 1987, pp 61–83.

123. S Mohammed, PK Kasera, DD Chawan, DN Sen. Eco-physiology of *Cassia italica* (Mill) Lamk. ex Anders. in Indian desert. Sci Cult 64:233–234, 1998.

124. S Mohammed, DN Sen. Seasonal variations in sugar and protein contents of halophytes in Indian desert. Ann Arid Zone, 33:249–251, 1994.

125. KS Gill, OS Singh. Effect of salinity on carbohydrate metabolism during paddy (*Oryza sativa* L.) seed germination under salt stress condition. Indian J Exp Biol 23:384–386, 1985.

126. KR Chandrashekar, S Sandhyarani. Effect of salinity stress on germination, carbohydrate, protein and proline contents of *Crotalaria striata* DC. seeds. Acta Bot Indica 23:59–62, 1995.

127. HM Abd-El Samad, MAK Shaddad. Salt tolerance of soybean cultivars. Biol Plant 39:263–269, 1997.

128. MA Zidan, MA Elewa. Effect of salinity on germination, seedling growth and some metabolic changes in four plant species (Umbelliferae). Indian J Plant Physiol 38:57–61, 1995.

129. KM Volkmar, Y Hu, H Steppuhn. Physiological responses of plants to salinity: a review. Can J Plant Sci 78:19–27, 1998.

130. RC Huffakar, L Paterson. Protein turnover in plants and possible means of its regulation. Annu Rev Plant Physiol 25:363–392, 1974.

131. R Vera-Estrella, BJ Barkla, HJ Bohnert, O Patoja. Salt stress in *Mesembryanthemum crystallinum* L. cell suspensions activates adaptive mechanisms similar to those observed in the whole plant. Planta 207:426–435, 1999.

132. V Venkatesalu, RR Kumar, KP Chellappan. Growth and mineral distribution of *Sesuvium portulacastrum* L., a salt marsh halophyte under sodium chloride stress. Indian J Plant Physiol 25:2797–2805, 1994.

133. KR Chandrashekar, S Sandhyarani. Salinity induced chemical changes in *Crotalaria striata* DC. plants. J Plant Physiol 1:44–48, 1996.

134. K Muthuchelian, C Murugan, R Harigovindan, N Nedunchezhian, G Kulandaivelu. Ameliorating effect of triacontanol on salt stressed *Erythrina variegata* seedlings: changes in growth, biomass, pigments and solute accumulation. Biol Plant 38:133–136, 1996.

135. PG Jarvis. Comparative plant water relation. Ann Arid Zone 6:75–91, 1967.

136. DN Sen, DD Chawan, KD Sharma. Ecology of Indian desert. V. On the water relations of *Salvadora* species. Flora 161:463–471, 1972.

137. S Mohammed. Stomatal behaviour and water loss in halophytes of Indian arid zone. In: DD Chawan, ed. Environment and Adaptive Biology of Plants, Prof. D.N. Sen Commemoration Volume. Jodhpur: Scientific Publishers, 1995, pp 151–165.

138. DN Sen, S Mohammed. Water relations of halophytes in Indian arid zone. In: M Dainel, ed. The Changing Landscape of Plant Sciences, Prof. S. D. Sabnis Felicitation Volume. Dehra Dun: International Book Distributors, 1997, pp 283–298.

139. FI Pugnaire, LZ Endolz, J Pardos. Constraints by water stress on plant growth. In: M Pessarakli, ed. Handbook of Plant and Crop Stress. New York: Marcel Dekker, 1994, pp 247–259.

140. LKRR Perera, TA Mansfield, AJC Malloch. Stomatal responses to sodium ions in *Aster tripolium*: a new hypothesis to explain regulation in aboveground tissues. Plant Cell Environ 17:335–340, 1994.

141. F Ayala, JW O'Leary. Growth and physiology of *Salicornia bigelovii* Torr. at suboptimal salinity. Int J Plant Sci 156:197–205, 1995.

142. A Lakshmi, SK Ramanjulu, K Veeranjaneyulu, C Sudhakar. Effect of NaCl on photosynthesis parameters in 2 cultivars of mulberry (*Morus alba* L.) cultivars (S 30 and K 2, salt tolerant and salt sensitive, respectively). Photosynthetica 32:285–289, 1996.

143. MF Robinson, AA Very, D Sanders, TA Mansfield. How can stomata contribute to salt tolerance? Ann Bot 80:387–393, 1997.

144. L Wegmann. The influence of salt stress on the morphology, physiology and economic use of the white mangrove *Laguncularia racemosa*. PhD dissertation, Tierarztliche Hochschule Hannover, Germany, 1998.

145. S Gulzar, MA Khan. Diurnal water relations of inland and coastal halophytic populations from Pakistan. J Arid Environ 40:295–305, 1998.

146. AJ De-Villiers, IV Teichman, MW Van-Rooyen, GK Theron. Salinity induced changes in anatomy, stomatal counts and photosynthetic rate of *Atriplex semibaccata* R. Br. S Afr J Bot 62:270–276, 1996.

147. LKRR Perera, DLR De-Silva, TA Mansfield. Avoidance of sodium accumulation by the stomatal guard cells of the halophyte, *Aster tripolium*. J Exp Bot 48:707–711, 1997.

148. TJ Flowers. Physiology of halophytes. Plant Soil 89:41–56, 1985.

149. H Kurban, H Saneoka, K Nehira, R Adilla, K Fujita. Effect of salinity on growth and accumulation of organic and inorganic solutes in the leguminous plants: *Alhagi pseudoalhagi* and *Vigna radiata*. Soil Sci Plant Nutr 44:589–597, 1998.

150. N Suarez, MA Sabrado, E Medina. Salinity effect on the leaf water relations components and ion accumulation patterns in *Avicennia germinans* (L.) seedlings. Oecologia 114:299–304, 1998.

151. H Greenway. Growth stimulation by high chloride concentration in halophytes. Isr J Bot 17:169–177, 1968.

152. EY Sambo, MJ Ashton. Evidence for osmotic adjustment in *Phalaris tuberosa* L. cvv. Australian and Sirosa. Aust J Plant Physiol 12:481–486, 1985.

# 29

# Role of Physiology in Improving Crop Adaptation to Abiotic Stresses in the Tropics: The Case of Common Bean and Tropical Forages

**Idupulapati Madhusudana Rao**

*Centro Internacional de Agricultura Tropical (CIAT), Cali, Colombia*

## I. INTRODUCTION

In the course of evolution, plants have developed so that their life cycle, growth habit, and other expressions of survival are adapted to specific environmental conditions. The term "adaptation" can be used to define processes conferred by genetic attributes that serve to "fit" the plant to the ambient conditions of temperature, light, and mineral and water availability. Understanding the genetic and physiological mechanisms by which plants cope with changes in environmental conditions is critical for creating efficient strategies to develop stress-resistant cultivars for sustainable production systems.

Improved crop yields achieved by plant breeders are mainly attributable to changes that fall into two categories [1,2]: (1) agronomic change through improved genetic adaptation to overcome major biotic (e.g., pests and diseases) and abiotic (e.g., temperature, drought, mineral deficiency and toxicity, and salinity) constraints to crop production and (2) raising the genetic yield potential per se above that of standard cultivars in the same environment. Most advances have occurred for the first category, and these in the 20th century. They include resistance to various pests and diseases, selective resistance to herbicides and pesticides, elimination of seed dormancy and susceptibility to lodging and shattering, and modification of timing of crop life cycles so that cultivars are better adapted to their environments. Although raising the genetic yield potential is equally important, research progress has been less than dramatic for most crops.

Plant physiological research contributes to both agronomic change and cultivar improvement [3], as illustrated by the following examples: advances in plant nutrition have led to more effective fertilizer practices, understanding of plant-water relations to better irrigation management, and the discovery of plant growth substances to selective herbicides and regulators. However, with cultivar improvement, empirical selection for improved crop adaptation and yield potential has continued effectively without heavy reliance on selection criteria or screening techniques from plant physiology. Crop yield is the result of a series of genetically controlled physiological and biochemical processes. Plant and crop physiologists face the challenge of understanding the mechanisms underlying these processes and of discovering how they interrelate and interact with diverse edaphic and climatic (abiotic) environments.

Complex plant processes such as photosynthesis, respiration, water transport, nutrient uptake, assimilate partitioning, or morphogenesis can be broken down into a number of elementary processes [4]. However, these processes are highly interrelated and each has complex and versatile regulatory systems, allowing plants to adapt to and withstand wide environmental variations. Research on physiological aspects of crop adaptation to abiotic environments should aid plant breeders, geneticists, and molecular biologists to manipulate crop genotypes for improved yield potential, stress resistance, and nutritional quality.

Improved adaptation of a crop to its environment can be achieved by two general approaches: the growth environment may be altered, or the plant genotype may be improved. Often a combined approach is the most effective.

Plant growth is closely related to the assimilation of carbon, the element's partitioning into different plant structures, and its loss through respiration, all of which must be accompanied by water and nutrient uptake. Assimilated carbon enters a pool of carbohydrates, and from there it is used either in respiration or in the growth of assimilatory and supportive structures. Partitioning of dry matter into leaves has a positive feedback on plant productivity because of its effects on total leaf area, but it inevitably increases demand for nutrients and water under conditions in which too few carbohydrates are available for root growth. These simultaneous parallel requirements need to be balanced by the plants.

Several authors have reviewed research efforts on improving plant adaptation to different climatic and edaphic stress factors [5–18]. In general, these reviews discuss physiological processes in detail relative to whole plant stress tolerance. The most successful approaches to improving crop and forage adaptation to abiotic stresses have historically used field-based evaluations to identify tolerant cultivars, followed by breeding and selection of genotypes that combine performance in stressful environments with other desirable plant attributes.

In this chapter, I have attempted to evaluate the role of physiological research in improving crop adaptation to abiotic stresses in the tropics, using case studies of common bean (*Phaseolus vulgaris* L.) and tropical forages.

## II.   ROLE OF PHYSIOLOGICAL RESEARCH IN CROP IMPROVEMENT

Plant physiology explores the full range of plant behavior, whereas crop physiology concentrates on how cultivars and related genotypes differ and how one may excel others under particular environmental or stress conditions [3]. A plant's genetic characteristics determine its potential maximum size, rate of photosynthesis, rate of dry matter production, and the form and nature of its storage organs, including those that are usually harvested for food or feed. Environmental factors such as water availability, temperature, photoperiod, light intensity, and availability of nutrients determine to what extent this potential can be reached. The main challenge is to recognize improved genotypes and to determine where energy-dependent inputs (fertilizer, irrigation, pesticides, etc.) can be used with greatest effect and efficiency. Attempts to answer such problems form the basis of crop physiology.

An effective crop improvement program for genetically enhancing crop adaptation to abiotic stress factors would involve (1) identifying germplasm tolerant of the abiotic stress factors of interest, (2) characterizing plant traits and mechanisms responsible for superior genetic adaptation, (3) determining mechanisms of inheritance for key plant traits, (4) identifying quantitative trait loci (QTLs) associated with key traits involved in stress tolerance for which marker-assisted selection in populations is feasible, and (5) developing an integrated genetic enhancement scheme.

Physiological research can make substantial contributions to crop improvement through characterizing germplasm for yield potential, making physiological analyses of yield potential, identifying key physiological traits, integrating physiological tools, and developing resource use–efficient genotypes for sustainable cropping systems.

## A.   Characterizing Germplasm for Yield Potential

In the 20th century, the plant breeder has at hand several new techniques that both speed up breeding and increase the range of genetic variation [19]. These include "classical" genetic modification methods such as inducing mutations by treatment with chemicals or x-rays; anther and ovule culture, which allows the production of completely homozygous plants, thereby cutting out some of the requirements for selfing; and embryo rescue techniques, which permit previously incompatible species to produce viable offspring. The second type includes "cellular" modification, which generates somatic variation through tissue culture by producing novel hybrids through cell fusion. These increase the gene pool available to breeders and can produce results much more rapidly than hybridization in plants with long generation times. The third type (embryo rescue) includes "molecular" techniques that involve the insertion and integration of a short segment of alien DNA into the plant genome. The process of inserting and integrating DNA is known as genetic engineering, genetic manipulation, genetic modification, transformation, or transgenesis.

Yield potential is defined as the yield of a cultivar when grown in an environment to which it is adapted, with unlimited nutrients and water and with pests, diseases, weeds, lodging, and other stresses effectively controlled [20]. Evans and Fischer [20] distinguished yield potential from potential yield, which they defined as the maximum yield that can be reached by a crop in a given environment, as determined, for example, by simulation models with plausible physiological and agronomic assumptions. Evans [1] assessed the progress in yield potential for many crops in many environments by growing historical series of leading cultivars side by side. Several studies indicated that both the rate of progress and the extent of increase in yield potential have differed greatly among crops [20–25]. A major outcome of these efforts was the realization that improvement in yield potential was greater for cereals and cotton than for grain legumes and root and tuber crops.

Increases in yield in recent years, across all crops, owe as much to innovation and improvements in agronomy as to plant breeding, more with some crops and less with others, more at some stages and less at others [26]. Crop yields could continue rising because of agronomic innovation and improvement on the one hand and breeding for improved stress resistance on the other, especially as the global environment changes. Breeders of a range of crops in most agricultural environments have devised technologies for crossing and testing that have successfully improved yields. Miflin [2] pointed out that, in the excitement of the tremendous advances in genetics across all organisms, it is important not to forget the role of the environment in crop performance and that food comes from successful phenotypes.

## B. Physiology of Yield Potential

Conceptually, high yield can be achieved by (1) maximizing the extent and duration of solar radiation interception, (2) using the captured energy efficiently in photosynthesis, (3) partitioning assimilates in ways that provide optimal proportions of economic product to other plant structures, and (4) maintaining those plant organs at a minimum cost of energy [27]. A key requirement for achieving high and stable yields is flexibility in morphogenesis and acclimation of physiological systems to overcome biotic and abiotic constraints. A retrospective analysis of the physiological basis of genetic yield improvement in temperate-climate maize indicated that a large proportion of yield improvement may be attributable to the capacity of newer hybrids to better tolerate stress conditions [25]. Increased stress tolerance in maize was associated with lower plant-to-plant variability. Studies of the physiological basis of yield improvement in soybean suggested that recently released cultivars not only supply more photoassimilates during the seed-filling period than old cultivars but also display improved $N_2$ fixation and better tolerance of the stress of high plant populations [28].

## C. Identifying Important Physiological Traits

The effectiveness of selection for physiological traits depends on factors such as heritability, genetic correlation between traits, inputs required for measuring a trait, intensity of selection, and the manner in which the selection is integrated into the breeding program [29]. Studies of plant response to different climatic and edaphic stress factors indicate that genetic variation is available for a number of important physiological traits [6,11,16–18]. Plant breeders have tried to incorporate this genetic variation into cultivars that exhibit whole plant stress tolerance. Most breeders are not convinced that selection based on physiological traits will give better results, believing that improvements in field experimentation and computerization will ensure continued success of empirical selection for stress tolerance.

Among the several reasons why breeders have seldom adopted physiological traits as selection criteria [11,16] are that (1) the genetic control of stress tolerance is poorly understood; (2) if understood, stress tolerance is often controlled by multiple genes; and (3) variation for stress tolerance usually exhibits a large environmental component or large genotype-by-environment interaction, making direct selection for a physiological trait in a single environment difficult. Yield increases associated with a particular trait are small, and breeders have not been convinced that selecting for the trait is more efficient than selecting for yield. Furthermore, stress tolerance at one developmental stage does not always confer tolerance at another stage. In addition, many methods proposed by physiologists to monitor stress tolerance are based on the performance of individual cells, tissues, organs, or individual plants and do not provide a good indication of the whole plant response to stress when grown in a spaced-plant nursery or in a competitive environment in the field. Ceccarelli et al. [30] argued that selection for a single trait is often unsuccessful, particularly in unpredictable environments where the frequency, timing, and severity of stresses are unknown.

Simulation modeling can make an important contribution to improving plant adaptation to stressful environments. Our ability to assess accurately the interaction of numerous processes over a crop's life cycle is limited, and the development of models can remove much of the "hunch taking" in selecting relevant physiological traits for genetic manipulation [31,32]. Seed yield can be described as the rate of photosynthate accumulation, the intensity or fraction of current assimilate allocated to seed, the duration of photoassimilate partitioning to seed, and the extent of remobilization of previously assimilated materials to the seed. Boote and Tollenaar [33] used crop growth simulation to evaluate hypothetical yield response to many genetic traits. Using a modeling approach, they made a systematic evaluation of the importance of plant traits as they affect the five "P's" of yield potential: prior events (vegetative canopy with sufficient tillering and fruiting sites), photosynthesis, partitioning, pod- or grain-filling period, and prior accumulation and remobilization of photosynthates and minerals. They found that of the five P's listed, duration of the pod-filling period is the most likely to account for past, present, and future yield increase. They suggested that yield improvement could also come from increased stress tolerance to the extent that photosynthesis is maintained, seed fill is longer, and mobilization is slower.

## D.   Integrating Physiological Tools and Molecular Genetics for Crop Improvement

The use of genetics in plant biology aims at the physiological and molecular genetic characterization of the phenotypic variation for the trait under study [34]. Testing possible associations between physiological and biochemical traits by comparing plant phenotypes and looking for correlations between them is not highly reliable [35]. Advances in molecular marker technologies offer powerful alternative methods to examine the relationships between traits. Using these techniques, it became clear that even for highly complex traits such as crop yield, a small number of QTLs explained a large part of the genetic variability [36,37]. Information from various genetic linkage maps will have to be integrated to facilitate comparison between detected QTLs and known major genes on the conventional genetic map [38].

The combination of genetics and plant physiology allows genetic markers to be associated with specific responses to stress [39]. Abiotic stress work on gene pools of small-grain cereals such as barley frequently shows that adaptive and developmental genes are strongly associated with the stress response [40]. Using barley as a model plant for application of molecular markers, Forster et al. [40] expressed concern that much of the genetic variation for improving abiotic stress tolerance has been lost during domestication, selection, and modern breeding, leaving pleiotropic effects of the selected genes for crop development and adaptation. Their work indicated that transfer of such genes from primitive landraces and related wild species is critical in matching improved cultivars to their targeted agronomic environments. The application of marker technologies to the redomestication of crops by exploiting the potential gold mine of favorable alleles existing in the crop's wild relatives provides the best relatively short-term opportunity for achieving the necessary advances in crop performance [2,41].

The ability to map DNA sequences physically to specific locations on a chromosome has advantages over more widely used genetic mapping procedures [42]. Stuber et al. [43] indicated that new investigations, using DNA-based marker technology as a tool for plant geneticists and plant breeders, will continue to add evidence on the projected role of markers, not only for identifying useful genes (or chromosomal segments) in various germplasm sources but also for transferring these genes into desired cultivars or lines. They also pointed out that the synergy of empirical breeding, marker-assisted selection, and genomics will "produce a greater effect than the sum of the various individual actions."

The need for integrating the knowledge available for different crops has never been greater. Improvements in crop simulation techniques and in the understanding of crop genetics suggest the possibility of integrating genetic information on physiological traits into crop simulation models [44]. In view of the increasing demand for food by the world's growing population, the development and improvement of crop yield will play a crucial role in the future. The advances in agricultural biotechnology have revolutionized the genetic analysis and improvement of crop plants and provided not only geneticists but also physiologists, agronomists, and plant breeders with valuable new tools to identify traits of economic, environmental, and nutritional importance. The integration of knowledge and biotechniques into the plant breeder's set of tools for cultivar development makes plant breeding more precise and shortens the time needed for cultivar development.

## E. Need for Physiologically Superior Genotypes for Sustainable Cropping Systems

The plant genetic approach to improving adaptation to major abiotic constraints is ecologically clean, energy conserving, and much more economical for resource-poor farmers in the tropics than modifying the soil and crop environment. Hence, it is compatible with national and international goals of economical food production; conservation of soils, water, and energy; and pollution control.

The shallow rooting ability of less adapted crop and forage cultivars is generally believed not only to reduce nutrient acquisition from low-fertility acid soils but also to increase susceptibility to seasonal drought. Developing genotypes that can root more deeply under adverse conditions is an important research objective for improving genetic adaptation to low-fertility soils.

Some of the benefits that can be obtained by integrating stress-resistant cultivars into cropping systems include fewer input requirements, reduced production costs, and reduced environmental pollution and soil degradation. Improved genetic adaptation to low-fertility soils will reduce nutrient requirements of crop and forage cultivars and minimize maintenance fertilizer applications through one of two pathways [45]: (1) deeper root growth → more efficient uptake of nutrients from subsoil → less leaching of nutrients, and (2) more biomass production → less seepage, less leaching → more intensive nutrient cycling → maintenance of higher soil organic matter content → less erosion owing to better soil protection by vegetation and mulch. Lynch [46] argued that the degree and extent of nutritional limitations to crop productivity and the economic and ecological liabilities of intensive fertilization are such that, eventually, nutrient-efficient crops will be an important part of integrated nutrient management of cropping systems.

One important objective of modern agriculture is to maximize crop productivity, preferably within a sustainable cropping system. As the cropping systems vary from high-input to low-input ones, crop improvement strategies have to be modified accordingly. A careful analysis of major factors limiting production in each system should dictate whether improvement can be based on breeding strategies, crop management, or soil management. International agricultural research centers are focusing more on "sustainable" yields rather than maximum yields [47], that is, on achieving high, sustainable crop yields within a societal framework that imposes significant managerial constraints on the farmer [48]. Some farmers may well have to abandon the goal of maximum crop yields as a result of new economic and environmental realities. Physiologically superior genotypes are needed to achieve resource use efficiency and profitability while minimizing environmental degradation.

## III. CASE STUDIES

## A. Common Bean

The common bean (*Phaseolus vulgaris* L.) was originally a crop of the New World, but it is now grown extensively in all major continental areas. The genus *Phaseolus* was domesticated in the upland regions of Latin America more than 7000 years ago [49–52]. It is the world's most important food legume, with an annual production value of over U.S.$10 billion. Latin America produces nearly half (5.1 million tons) of the world's supply (11.6 million tons from 14.3 million ha) of dry beans [53]. Beans are grown in a wide range of environments [54] from sea level to elevations of more than 3000 m [55]. Bean production is often relegated to marginal environments, such as those characterized by steep, erosion-prone slopes or by low soil fertility with seasonal droughts. Nearly 80% of dry bean production occurs on small-scale farms in the developing countries of tropical Latin America and Africa. Women are the primary bean growers on small farms in Africa. Widely known as the "poor man's meat," the crop provides an inexpensive source of protein for low-income consumers. Bean consumption is highest in eastern and southern Africa, where beans are the second most important protein source after maize and the third most important caloric source after cassava and maize. Beans are also nutritionally important in Central America, Mexico, and Brazil.

The cropping system used ranges from the highly mechanized, irrigated, and intensive production of monocropped bush beans to complex associations of indeterminate or climbing beans with maize, other cereals, sugarcane, coffee, or plantain [56]. Soil and crop management inputs in such multiple-cropping systems are often limited, with the result that seed yield can range from less than 500 kg ha$^{-1}$ in parts of Latin America and Africa to as much as 5000 kg ha$^{-1}$ under experimental conditions.

Research to enhance common bean genetically is complicated by the diversity of edaphic and climatic conditions under which the crop is grown, compounded by highly specific local preferences for particular grain types or colors. However, great progress has been achieved in developing genotypes resistant to several biotic constraints [56]. Whereas the success in improving genetic adaptation to major abiotic constraints has been substantial, progress in improving yield potential has been limited indeed. Different perspectives on the reasons why are given in previous reviews [53–59].

A world collection of beans, comprising more than 40,000 accessions, is held at the Centro Internacional de Agricultura Tropical (CIAT) in Cali, Colombia. This collection includes indigenous wild and weedy specimens, unimproved landraces, pure lines of *Phaseolus vulgaris*, and numerous related species. *Phaseolus vulgaris* was domesticated several times in the pre-Colombian era in both Mesoamerica and the Andean region, resulting in at least two major gene pools of cultivated bean, one Mesoamerican and one Andean [60]. These two gene pools are distinguished by yield potential, morphology [61,62], isozymes [63], DNA molecular markers [64–66], and physiological traits related to photosynthesis [67,68].

Singh [69] developed a key for identifying different growth habits in common bean. Bush types fall into three groups: type I plants form a determinate inflorescence at the end of stems and branches. Typically, they have a low number of nodes and a short flowering period and are early maturing. Type II and type III plants have indeterminate growth, the stems and branches ending in a vegetative guide. Type II plants are erect, have little guide development, and are usually intermediately maturing. Type III plants have a more prostrate growth, are usually strongly branched, and show moderate ability to climb if given support. Type IV plants have indeterminate growth and very weak and excessively long stems and branches that possess strong climbing ability. Type I is considered to have the lowest yield potential [54]. In general, indeterminate cultivars provide greater yield stability than determinate cultivars [70,71].

The process of matching growth habit to changing environment, economy, and technology is a major challenge in genetically enhancing the common bean. Vandenberg and Nleya [72] indicated that common bean germplasm suitable for direct harvest systems could be developed by introducing parents that can contribute to the genetic enhancement of pod distribution in the overall plant canopy. They have identified the following plant traits that may optimize canopy structure at harvest: (1) long internodes in the lower stem, (2) consistent internode elongation under a wide range of environmental conditions, (3) reduced stem stunting during early season growth, (4) increased stem length, (5) increased stem strength, particularly in the more basal internodes, (6) reduced pod length without decreasing seed size, (7) increased pod curvature so that tips do not extend below the cutter bar, (8) long upright peduncles, (9) flowering beginning on upper nodes, (10) high fertility at the upper nodes, and (11) a sufficient number of main stem nodes to maximize productivity in the available growing season.

White and Izquierdo [73] discussed physiological processes that determine bean yield and applied that information to analyze limitations to yield potential and stress tolerance. They identified several characteristics that may possibly confer general stress adaptation: an ability for recuperative growth, presumably by remobilizing carbohydrate or nitrogen (N) reserves and having an indeterminate growth habit; good competitive ability; high tissue concentrations of phenolic compounds with inhibitory effects on a broad range of pathogens or pest organisms; greater partitioning of photoassimilates to root growth; and buffer ability for adequate pod retention and seed filling.

Large-seeded bush bean cultivars usually give lower yields than small-seeded ones, especially in warm, tropical environments. Andean genotypes are predominantly large-seeded, whereas small seededness is associated with the Mesoamerican region of domestication [54,61]. Large-seeded genotypes tend to have a lower relative growth rate (RGR) than small-seeded types [54,74]. This poorer performance of large-seeded lines is not limited to RGR because seed yield is also often negatively associated with seed size among bush bean cultivars [75].

Research was conducted to examine the physiological basis for the lower productivity of the large-seeded Andean genotypes [76,77]. Andean lines were found to have less yield and slower seed growth rate (land area basis) than lines of Mesoamerican background [76]. Studies showed that values for large-seeded Andean lines were smaller than for small-seeded Mesoamerican lines on such attributes as vegetative growth, single-leaf carbon exchange rate (CER), internal leaf anatomy, RGR, net assimilation rate (NAR), specific leaf weight (SLW), specific leaf N (SLN), leaf thickness, and mesophyll-cell surface area per unit leaf area exposed to air. These studies also found a positive correlation between RGR and CER and postulated that the smaller RGR of the large-seeded Andean lines is a function of their slower CER, which results from thinner leaves with less photosynthetic apparatus per unit leaf area.

White et al. [78] tested the effect of growth habit on the yield of large-seeded bush cultivars and concluded that the mere change in stem type from determinate to indeterminate growth habit did not increase the yield potential or stability of the large-seeded indeterminate near-isogenic lines. The single gene change affecting growth habit was therefore in itself not sufficient to improve yield when the rest of the plant's genetic composition remained unaltered. White et al. [78] suggested the following yield-increasing traits, which could be genetically manipulated in the indeterminate growth habits: increased number of nodes and branches, delayed and extended flowering periods, and the ability to recover from stress at flowering through regrowth. They also pointed out the need to develop an optimal ideotype for the large-seeded indeterminate beans, based on architectural components and physiological mechanisms that contribute to greater seed yields and stress adaptation.

## 1. Adaptation to Temperature and Photoperiod

Both temperature and photoperiod have strong effects on growth and development in the common bean [79,80]. In the tropics, high air temperature is normally accompanied by high soil temperature in the rooting zone (top 20 cm of soil). Poor root formation due to high temperatures can lead to drought stress. In semiarid regions, high temperatures and drought often act together to reduce bean yields significantly. The effects of high temperatures include flower fall, abortion, reduced pollen grain viability, impaired pollen tube formation in the styles, and reduced seed size [81,82].

Two components contribute to plant adaptation to high temperature [8]: (1) heat avoidance, in which plant tissues subjected to high solar radiation or hot air have lower temperatures than control plants, and (2) heat tolerance, whereby essential plant functions are maintained when tissues become hot.

Beans are grown in a very wide range of latitudes and the mean air temperature varies between 14 and 35°C. Temperatures of air and rooting zone can determine seed germination, root growth, taproot formation, and flowering. High temperatures negatively affect pollen-stigma interaction, pollen germination, pollen tube growth, and fertilization. Consequently, if plants are exposed to high temperatures for 1 to 6 days before flowering, pod set is very low [83].

Extreme temperatures, that is, lower than 10°C and higher than 40°C, can result in a poor germination rate [84]. White and Montes-R. [85] characterized the germination response of 20 genotypes of common bean by fitting cumulative counts, using a maximum-likelihood analysis. They found that the germination rate increased from a base temperature typically near 8°C to an optimal development temperature ($T_O$) of 29 to 34°C. Base temperature did not differ among common bean genotypes. Mesoamerican germplasm showed slightly higher $T_O$ than Andean germplasm, but $T_O$ varied widely within each of the two gene pools. The only accession of tepary bean (*P. acutifolius*) evaluated, 'Sonora 32', was the most tolerant of high temperatures at germination.

Bean cultivars tolerant of cold conditions are needed for mountainous regions and bean-growing regions in the higher latitudes. Beans suffer cold stress either during seed germination or, later, at the pod-filling stage. Beans possess significant genetic variability for cold tolerance [86]. Cold-tolerant cultivars can be selected, using a laboratory test, at 12°C in the $F_2$ generation when appropriate parents and/or sources are identified [87]. Selecting bean lines for cold tolerance at seedling stage is possible [88]. The Universidade Federal de Lavras, Brazil, is attempting to breed cultivars that are cold tolerant during early growth stages and at maturity.

Selecting for improved adaptation to high temperatures in beans is possible [89]. Masaya and White [79] suggest that the most difficult part of developing bean cultivars adapted to extreme temperatures is not so much the search for adequate physiological response to temperatures as for resistance to associated biotic constraints such as fungal diseases (e.g., web blight) under hot, humid conditions and various root rots under cool conditions. Researchers at the Instituto Pernambucano de Pesquisa Agropecuária (IPA,

Brazil) have identified several heat-tolerant lines (e.g., HF 465-63-2). Their leading line, IPA 7, is not only heat tolerant but also resistant to a major fungal disease (*Macrophomina* spp.) that affects bean performance under drought [90].

Various studies [54,79,91] of the effects of photoperiod on common bean indicate that adaptation of common bean is strongly affected by photoperiod, and the species shows considerable genetic variation in photoperiod response. White and Laing [92] characterized this variation among 4000 bean genotypes and found that about 60% of genotypes were photoperiod sensitive and that small-seeded and bush-indeterminate materials had the highest proportion of day-neutral genotypes. The authors also constructed a frequency distribution of photoperiod response in 3060 genotypes, which showed three distinct peaks, suggesting simple genetic control.

Interaction between air temperature and photoperiod plays a great role in bean production [79]. The small-seeded day-neutral genotypes are physiologically the most efficient, especially at warmer sites and higher latitudes [74]. Many highland cultivars are poorly adapted to lowland areas at the same latitude because of temperature differences between altitudes. Acosta-Gallegos et al. [93] determined the effect of sowing date on the growth and seed yield in highland environments. They found that variation in maturity had a more consistent effect on growth and seed yield than on days to flowering, indicating that growth and seed yield are more affected by the duration of the reproductive phase than by the duration of the vegetative phase. The authors suggested that part of the variation in growth and seed yield may be due to genotypic differences in photoperiod or temperature response. Highland bean cultivars can produce flowers and seeds when exposed to a 6-hr daylight regime [73]. Temperatures of the rooting zone and subsoil may be key factors in determining days to flowering, seed germination, and tap or lateral root formation.

## 2. Physiological Response to Water Deficits

About 60% of common beans produced worldwide are grown in regions subjected to water stress, making drought the second largest contributor after disease to yield reduction [94,95]. Increased adaptation of common bean genotypes to soil water deficits would contribute to both stability and expansion of production in drought-prone environments such as northeast Brazil and the central highlands of Mexico. Bean cultivars adapted to drought would require less water for irrigation and would therefore contribute to the conservation of an important natural resource. The short growing season reduces the common bean's water requirements to levels below those of other species generally considered as more drought adapted [96].

"Drought resistance" is a general term encompassing a diversity of mechanisms that enable plants to survive and produce in periods of dry weather. "Drought tolerance" involves the maintenance of a positive turgor pressure at low tissue water potential. Drought tolerance mechanisms include osmotic adjustment and dehydration tolerance achieved via protoplasm resistance. "Drought avoidance" is the maintenance of a high tissue water potential (i.e., maintenance of green, turgid tissue) during a period of a high evaporative demand or a period of increasing soil water deficit.

Under rain-fed conditions, water deficit can occur more than once during a crop's growth cycle, caused by erratic patterns of rainfall distribution, or may kill the crop [95]. The intensity and duration of stress determine the degree of yield reduction relative to its yield potential. Research approaches that have most successfully improved drought performance (1) used realistic soil conditions, (2) tested with adequate water and with limited water, (3) understood the sources of crop failure in the proposed growing area, and (4) targeted a limited number of traits for genetic improvement [14].

Although common bean is not a drought-tolerant species [54], it is grown over a wide range of habitats where it is exposed to seasonal droughts and wide fluctuations in soil moisture availability between years. Research efforts on common bean adaptation to drought involve studying the effects of water stress on plant growth, development, and seed yield [73,97,98]; developing field screening methods [99–101]; evaluating and identifying sources of drought tolerance in germplasm [94,102–104]; and evaluating physiological traits related to underlying mechanisms of adaptation to drought [105–110].

Performance under drought can be evaluated in terms of three discrete groups of characteristics: morphological, physiological, and phenological [111]. Loss of leaf area is the most important morphological adaptation and results from a reduced number of leaves, reduced size of younger leaves, inhibited expansion of developing foliage, or leaf loss accentuated by senescence, all of which result in decreased seed yield [105]. Through field screening, some relatively drought-tolerant lines of bean germplasm were identified, such as BAT 477, A 195, and BAT 1289 [112]. The superior adaptation of BAT 477 to water deficits was attributed to dehydration postponement through greater root length density and deeper soil moisture extraction [100]. Castonguay and Markhart [113], on measuring saturated rates of photosynthesis in water-stressed leaves of common and tepary beans, found that genotypic variability in drought tolerance between common and tepary beans is not related to differences in mesophyll tolerance of dehydration. Tepary bean relies more on dehydration postponement than on drought tolerance. Severe drought impaired N mobilization, harvest index, and water use efficiency in common bean [110].

Grafting diverse shoot genotypes on selected root genotypes of common bean and evaluating yield under soil water deficits showed variation in shoot genotype. However, the effect of shoot genotype on growth and yield under water deficits was found to be small compared with that of root genotype [114]. Field research under rain-fed conditions indicated that water use efficiency (based on carbon isotope discrimination) is not a promising indicator of adaptation to water deficit in the common bean [96]. Other physiological traits such as shoot dry weight and leaf N concentration appeared the most promising, being based on heritability, strong general combining ability effects, and correlations with seed yield across trials [108]. Phenotypic plasticity is considered another mechanism contributing to increased performance under drought [109]. This particular attribute allows genotypes to shorten their growing cycle dramatically at later planting dates to avoid drought conditions or low temperatures later in the growing season.

Identification of a shoot trait or traits that reflect rooting ability and adaptation to drought will minimize labor-intensive root measurements in a breeding program. Studies of other grain legumes such as peanut and soybean have indicated that water use efficiency is negatively associated with certain shoot traits such as specific leaf area (leaf area per unit leaf dry weight) and leaf ash content [12,115]. The decrease in specific leaf area in drought-adapted genotypes may also be related to the accumulation of nonstructural carbohydrates in leaves. Understanding the relationships between grain yield and shoot traits such as specific leaf area, leaf ash, and leaf nonstructural carbohydrates, using contrasting genotypes, may help identify the specific shoot traits related to adaptation to drought in common bean.

## 3.  Breeding for Improved Adaptation to Drought

Progress in breeding for adaptation to drought in common bean has been slow, although several selection criteria for resistance to drought have been identified [94,116,117]. Because seed yield is the most important economic trait, the most practical method for improving performance is through the direct measurement of yield-related characteristics [118]. Studies on inheritance of seed yield of the common bean under rain-fed conditions in contrasting environments by White et al. [108] indicated that an efficient system for breeding for increased seed yield under drought can be developed by using early generation yield testing of population bulks. They suggested that potential parents adapted to drought should first be tested for combining ability in environments of the region before using them extensively in hybridization and selection programs. Singh [119] reported an increase in yield under drought through hybridization between races and gene pools, involving high-yielding and water stress–tolerant progenitors derived from different origins, such as those found in the Mexican highlands. New sources for drought tolerance were found in cultivars grown in Jalisco and Durango, Mexico (S. Singh, personal communication).

Schneider et al. [111] evaluated the performance of two common bean populations consisting of 78 and 95 recombinant inbred lines (RILs) under conditions with and without drought. They examined seed yield under drought, yield potential, drought susceptibility index, harvest index, and geometric mean as potential indicators of drought-resistant genotypes. Among the plant traits measured, they found that the 100-seed weight was the most highly heritable trait in both populations. They also found that the geometric mean of the two drought treatments (with and without) was the single strongest indicator of performance under both drought and no-drought treatments. On the basis of this study, they suggested that the most effective breeding strategy would involve selection based, first, on the geometric mean, followed by selection based on yield under drought stress. Using the RILs of the same two populations, Schneider et al. [120] studied the possibility of molecular marker–assisted selection (MAS) to improve drought resistance in common bean. Using one-way analysis of variance and multiple regression, they identified four RAPD (random amplified polymorphic DNA) markers in one population and five in another that were consistently and significantly associated with yield under drought, yield without drought, and/or geometric mean yield across a broad range of environments. From this study, they concluded that the relative value of MAS is inversely proportional to the heritability of the trait under examination.

As mentioned before, bean cultivars show great diversity in phenology, and this is a major factor determining yield potential and adaptation to rain-fed environments [73]. Many landraces and cultivars of the Mexican highlands showed a plastic response to planting date such that, with increasingly later plantings, they became earlier flowering [109]. This phenological plasticity appears to be an adaptive strategy by common bean to rain-fed environments. In drought-stricken northeast Brazil, BAT 477 can tolerate drought very well. This line combined drought tolerance with resistance to *Macrophomina* disease that prevails in drought areas. A more recent bred line, SEA 5, was found to be better adapted to drought than BAT 477. This was mainly attributed to its ability to partition a greater proportion of assimilates to grain production [121].

## 4.  *Physiological Responses to Low Phosphorus Supply*

Phosphorus (P) deficiency is widespread, covering an area estimated at over 2 billion ha [122]. Cochrane et al. [123] estimated that 86% of the tropical soils of Latin America have levels of P less than 7 ppm (Bray II) available in the topsoil. These soils have a high capacity [124] to fix P in forms that are mostly unavailable to plants, thus imposing agronomic and economic constraints. Application of P fertilizer is common practice and is necessary if agricultural productivity is not to be seriously limited. Improved cultivars with genetic adaptation to low-P soils may be a viable alternative or complement to P fertilization, particularly for crop-livestock systems in the tropics [18,125].

The two major components of P efficiency are P acquisition efficiency and P use efficiency [18]. Phosphorus acquisition efficiency refers to the plant's ability to acquire greater amounts of P per unit root length, whereas P use efficiency refers to the plant's ability to produce yield per unit of acquired P from soil. With a given P supply in soil, P acquisition per plant might be improved in at least three ways: (1) with a root system that provides greater contact with P, (2) with greater uptake per unit of root due to enhanced uptake mechanisms, and (3) with an ability to use insoluble organic or inorganic P forms that are relatively unavailable or poorly available to plants [7,18,126,127]. Association with arbuscular mycorrhizae (AM) significantly affects each of these attributes [128]. Uptake of P by upland rice, pigeonpea, and groundnut, which all associate with AM, was found to be higher than that by buckwheat, castor, cotton, maize, sorghum, and soybean in soils with low P availability [129].

Among the edaphic stresses, P deficiency is the primary constraint to common bean production in the tropics and subtropics, limiting seed yield on at least 60% of the bean-producing areas of Latin America and Africa [130,131]. The symbiotic nitrogen fixation (SNF) of common bean is more affected by P deficiency [132] than that of other crops such as soybean [133]. High SNF in common bean was reported to be related to nodule number, nodule mass, late nodule senescence, early nodulation, and secondary nodulation [134,135]. Screening 220 lines (Andean and Mesoamerican origin) under greenhouse conditions resulted in the identification of contrasting lines that may be useful for further improvement of SNF potential and adaptation to P-deficient soils [132]. Beans with SNF and tolerance of P deficiency were mostly found among late-flowering, type IV lines but included three early-flowering, type III lines. High P concentration in seeds produces seedlings less dependent on soil P supply and therefore could enhance nodulation and SNF of common bean [136].

Substantial genetic variation in P efficiency in common bean has been demonstrated under both field and greenhouse conditions [58,121,137–156] (Table 1). Beebe et al. [154] studied the relationship between geographic origin and response to low P supply in soil in a selection of 364 genotypes drawn from the gene bank held at CIAT. They found highly significant variation in P efficiency among genotypes in all growth habits. Wild beans usually performed relatively poorly, indicating that P efficiency traits in common bean have been acquired during or after domestication.

Attempts by Singh et al. [145] to improve P efficiency in common bean were not successful because of the confounding effects of other edaphic and climatic factors rather than because of P deficiency per se. Thus, the interactions of genotype × season, genotype × P levels, and genotype × season × P levels underscore the difficulty of relying on yield performance as a sole criterion for selection in a breeding program [148]. Identifying specific mechanisms of P efficiency would be far more reliable and preferable. If a multigenic character such as P efficiency could be resolved into physiological mechanisms governed by discrete traits, these traits could be tagged with molecular markers more reliably than P efficiency could be measured as a quantitative trait by seed yield trials [46].

The genetic control of P efficiency is well known to be complex because of the involvement of this important nutrient in several aspects of plant metabolism [157,158]. The most pronounced effect of P de-

**TABLE 1**  List of Genotypes and Advanced Breeding Lines of Beans (*Phaseolus vulgaris* L.) with Superior Adaptation to Some Abiotic Stresses

| Abiotic stress factor | Germplasm accessions and advanced lines |
|---|---|
| Low-N tolerance | AFR 44; AFR 403; BAT 25; BAT 85; Carioca; Diacol Calima; MUS 97; PAI 112; PEF 4; PEF 16; Porrillo Sintetico; RAO 55 |
| Low-P tolerance | A 321; ACC 433; AFR 300; AFR 475; AFR 544; BAT 85; BAT 477; Carioca; DOR 375; G 1937; G 3153; G 5053; G 7300; G 11702; G 12105; G 16106; G 19842; G 21212; MMS 224; MUS 18; PAI 112; PEF 14; RWR 382, VAX 1; XAN 76 |
| Low-K tolerance | DOR 375; EMP 84; ICA Pijao; MUS 97; Porrillo Sintetico; RAO 52; XAN 76 |
| Al tolerance | 7/4 ACC; AFR 300; AFR 344; AFR 476; AND 740; AND 773; CAL 98; F-15; FEB 190; FEB 192; Muhinga; MUS 18; Ntekerabasilumu; RAB 94; RAB 475; RAO 55; Superba |
| Mn tolerance | A 120; A 197; AFR 13; AFR 298; AFR 378; AFR 476; AFR 531; AFR 544; AND 829; AND 871; Argentino; BAT 271; Calima; EMP-84; H6 Mulatinho; CAL 96; Carioca; DOR 404; MCM 5001; NEPA 29; NEPA 38; PAD 126; Pintado; PVA 774; SUG 69; XAN 76 |
| Low-soil-fertility tolerance | Carioca; G 12871; G 21212; XAN 76; RAO 55; OBA 1 |
| Drought | A 54; A 170; A 195; Apetito; BAT 336; BAT 477; BAT 1289; Bayo Criollo del Llano; Bayo Rio Grande; Durango 5; Durango 222; Favinha; Gordo; Guanajuato 31; Mulatinho Vagem Roxa; Rim do Porco; San Cristobal 83; SEA 5; V 8025 |

*Source*: Adapted from Refs. 95, 108, 119, 121, 131, and 148.

ficiency on plant growth is reduced leaf expansion so that relatively more dry matter is apportioned to the roots than to the shoots [159]. Phosphorus deficiency in common bean causes substantial reductions in the shoot-to-root dry weight ratio [148,160] and leaf growth rate, whereas the rate of photosynthesis per unit leaf area decreased only slightly [161,162]. The slightness of the decrease was attributed to the enhancement of the inorganic phosphate ($P_1$) recirculation during glycolic and phosphoenolpyruvate metabolism. In bean plant tissues, as $P_1$ concentration decreases, root carbohydrate content [160] and reduced pyradine nucleotide concentrations increase [163,164]. These changes are accompanied by a decline in total respiration rate and increased cyanide resistance [165], resulting in a lower concentration of ATP in the roots [166]. The decrease in $P_1$ concentration of bean leaves and roots leads to decreased rates of nitrate uptake and increased nitrate accumulation in roots, accompanied by alterations in nitrate distribution between shoots and roots [167]. Thus, P deficiency not only affects the common bean's SNF potential [132] but also alters its ability to assimilate and translocate nitrate.

Studies on plant nutrition indicated that plant adaptation to low-P soils is not specific to soil type in common bean, and that results derived for one soil type may be extrapolated to other soils [150,151]. Mesoamerican and Andean germplasm responded differently to P availability in soil. Mesoamerican types were more responsive to added P in terms of seed yield. The ability of crop plants to remobilize P from vegetative to reproductive organs may form an important mechanism that allows plants to improve the use of P acquired from soil [18,126]. Common bean lines with a low P concentration in shoot tissues retain more P in roots and older leaves under P-deficient conditions than do lines with a high P concentration [168]. The greater remobilization of P in bean lines with a high P concentration could be attributed to higher P requirements to maintain normal metabolic activity in growing tissues. Using a split-root system and a $^{32}P$ tracer, Snapp and Lynch [169] measured patterns of P remobilization from roots and leaves of common bean and suggested that P retention may allow roots to sustain nutrient and water uptake to late in the ontogeny.

Lynch and Beebe [148] hypothesized that the existing genetic variation for P efficiency in bean germplasm, especially variation that is agronomically useful, is largely due to variation in P acquisition efficiency rather than P use efficiency. Substantial genetic variation in the growth and architecture of bean root systems was observed, with some evidence that P-efficient genotypes have a vigorous, highly branched, root system with many growing points [170]. The P status of bean plants greatly influenced local root growth patterns and P uptake from localized P patches [171].

Several theoretical and empirical lines of evidence indicate that root architecture (the three-dimensional shape of the root system over time) may be the basis for genetic differences in P efficiency in beans [148,172–174]. Phosphorus availability regulates many features of root architecture, including adventitious rooting, aerenchyma formation, basal root elongation, basal root growth angle, lateral rooting, root hair density, and root hair length [174].

SimRoot, an explicit geometric model of bean root growth, confirmed that root architectural traits can influence the relationship between root C costs and P acquisition [175]. Root growth responded dynamically to P stress through changes in the proliferation of lateral roots and the geotropic response of basal roots. After further research, Lynch and associates defined root gravitropism as a potentially beneficial trait for P efficiency in common bean [176,177].

## 5.  Breeding for Improved Adaptation to Low Phosphorus Supply

Selection criteria to improve P efficiency in common bean can be based on physiological traits and mechanisms. These include the abilities to (1) mobilize P within the plant, (2) set pods and mobilize photoassimilates to seeds, (3) minimize storage of P in seed (phytic acid), and (4) modify root architecture to exploit greater soil volume [95]. Genetic studies of P efficiency in bean showed that the P efficiency ratio (dry matter produced per unit of tissue P content) differs among specific crosses according to parents [137]. These studies also showed that maternal inheritance was of minor importance. Narrow-sense heritability estimates derived from parent offspring regression in bean families of efficient $\times$ inefficient lines were estimated to be about 40% [138], and in other studies they were high for total dry matter yield in all families tested [178]. Studies of broad-sense heritability estimates for total dry matter yield showed that efficiency in P use was a highly heritable trait (range 0.68 to 0.86) in bean [178,179]. Epistasis (primarily, additive $\times$ additive and dominance $\times$ dominance gene effects) made significant contributions to the efficiency of P use in bean [178]. Quantitative inheritance patterns and transgressive segregation for root dry matter yields were also observed [179]. Dominance variance was more important than additive variance for P efficiency in four out of six families used in the experiments. Urrea and Singh [180] found heritability for seed yield under low P supply in soil to be 0.61, based on regression of $F_3$ populations on the corresponding $F_2$ populations.

Phosphorus use efficiency has been transferred from an exotic germplasm to an adapted variety by Schettini et al. [181] using the inbred backcross line method. They derived several tolerant lines from the P-efficient donor (PI 206002) combined with the desirable recurrent parent 'Sanilac'. They showed that lines that performed well in nutrient solution culture could also perform well in a field test with soil having a moderately deficient P supply. Saborío and Beebe [182] made efforts to breed for tolerance of low P in soils of Costa Rica at two locations. They started with 10 segregating populations derived from crosses with varying structures and incorporating parents from the highlands of Mexico and Peru. They selected 14 lines and coded them as TLP (tolerant of low P) lines. Among these TLP lines, TLP 28 and TLP 29 were superior in their adaptation to low-P soils.

Posada et al. [183] studied the heritability and mechanisms of tolerance of low-P soils in Mesoamerican and Andean cultivars of common bean. They evaluated 12 parents, 6 each of Andean and Mesoamerican types, and 27 of their $F_2$ populations for P uptake and biomass (shoot and root) production under conditions of low and high P. In both parents and progeny, they observed significant differences in traits associated with low-P tolerance, including high P uptake and efficient internal use through efficient P partitioning. These characteristics can therefore be used in a breeding program to improve low-P tolerance of agronomically desirable bean cultivars.

At CIAT headquarters, more than 7000 bean germplasm accessions were evaluated for soil constraints, especially low P [148] (S. Beebe, personal communication). Wide differences have been found in both P uptake and P use efficiency. Physiological studies have been combined with QTL analysis to elucidate mechanisms of P uptake [174,184]. Recombinant inbred lines (RILs) of a mapping population were evaluated in a greenhouse hydroponic test for a suite of traits for which the parental genotypes were found to be contrasting. These traits were basal root number, length, and dry weight; root hair density and coverage on the root surface; and hydrogen ion ($H^+$) exudation from roots [184] (J. Lynch, S. Beebe, and X. Yan, unpublished results). QTLs were identified for all traits, and these were often associated with QTLs that contributed to root length and P uptake in the field, as measured on the same RILs. Thus, P acquisition reflects the interaction of several plant mechanisms. In this experiment, QTL analysis was a critically important tool for dissecting different P acquisition mechanisms.

Another population of RILs was developed from the cross BAT 881 × G 21212 to study yield potential under low-P conditions [184] (S. Beebe, personal communication). Linkage groups were established and QTL analysis carried out. One linkage group was particularly important for yield at low P, and a long segment of more than 80 cM appeared to carry several QTLs for yield. The entire segment accounted for more than 300 kg/ha at low P—a remarkable effect under very difficult production conditions. Field studies indicated that G 21212 is particularly efficient in mobilizing photosynthates to grain when grown in P-deficient soil [121]. Thus, it is possible that the QTLs identified are unlike those that were identified for P acquisition.

The AFLP (amplified fragment length polymorphism) technique, combined with selective genotyping, was used to map QTLs associated with tolerance of low P in rice [185]. Molecular markers for QTLs may serve to select the desired traits, but whether a breeder bases selection on a molecular marker or on a trait will depend on the relative ease and cost of each approach. In any case, physiological analysis and QTL analysis are highly complementary approaches. The use of QTL analysis strengthens physiological analysis of traits. On the other hand, QTLs per se often express variably across environments, and physiological analysis can offer the breeder more understanding of the biological significance of a given QTL and its potential value. The possibilities of using other species as sources for improving P efficiency in common bean, by wide crossing or genetic transformation, are still to be exploited.

## 6. Selecting for Improved Adaptation to Other Soil Constraints

Knowledge of specific nutritional requirements of common bean can help determine more precisely the amount of fertilizer needed to overcome soil constraints and maintain productivity over time. Substantial progress was made in defining the nutritional requirements of common bean [58,143]. Table 2 shows critical nutrient values for soil and plant analysis compared with normal range in plant tissue to detect edaphic constraints to bean production in the tropics.

Studies of genotypic variation of common bean for tolerance of various edaphic constraints have been reviewed [95,148,149,155]. Results demonstrate the feasibility of selecting and breeding for tolerance of certain edaphic constraints [95,186] (Table 1). In most regions, N deficiency limits common bean and associated crop production. Tolerance of low N supply in soil has several components [187]. These include rate and duration of N acquisition, efficiency of N use in vegetative growth, timing of transition to reproductive growth, rate and duration of N accumulation in seeds, and efficiency of N use in seed formation. Studies on photosynthetic N use efficiency in relation to leaf longevity in common bean indicate

**TABLE 2** Critical Nutrient Values for Soil and Plant Analyses to Detect Edaphic Constraints to Common Bean (*Phaseolus vulgaris* L.) Production in the Tropics

| Element | Critical Soil | Critical Plant | Normal range for plants |
|---|---|---|---|
| N | | 25 (g kg$^{-1}$) | 52–54 (g kg$^{-1}$) |
| P | 11–15 (mg kg$^{-1}$) | 2.0 (g kg$^{-1}$) | 4.0–6.0 (g kg$^{-1}$) |
| K | 0.15 (cmol kg$^{-1}$) | 15 (g kg$^{-1}$) | 15–35 (g kg$^{-1}$) |
| Ca | 4.5 (cmol kg$^{-1}$) | 5.0 (g kg$^{-1}$) | 15–25 (g kg$^{-1}$) |
| Mg | 2.0 (cmol kg$^{-1}$) | 2.0 (g kg$^{-1}$) | 3.5–13 (g kg$^{-1}$) |
| S | 4–15 (mg kg$^{-1}$) | | 1.6–6.4 (g kg$^{-1}$) |
| B (mg kg$^{-1}$) | 0.4–0.6 | 20 | 10–50 |
| Zn (mg kg$^{-1}$) | 0.8 | 15 | 35–100 |
| Mn (mg kg$^{-1}$) | 5.0 | 20 | 50–400 |
| Cu (mg kg$^{-1}$) | 0.6 | 5.0 | 5–15 |
| Fe (mg kg$^{-1}$) | 2.0 | 5.0 | 100–800 |
| Exch. Al (cmol kg$^{-1}$) | 1.0 | 50 | 100–800 |
| Al saturation (%) | 50 (organic soil) | | |
| | 10 (mineral soil) | | |
| pH | 5.0–7.8 | | |
| | 5.5–6.5 (optimal) | | |
| Mn toxicity (mg kg$^{-1}$) | 20–80 | | |

*Source*: Adapted from Refs. 58 and 143.

that leaf longevity is an important factor in leaf N use efficiency but that genetic variation for shoot architecture is not important in determining the N use efficiency of individual leaves [188].

Common bean is widely regarded as weak in nodulation and $N_2$ fixation [55,134]. This is partly the result of the marginal soil conditions under which it is commonly grown, partly a result of competition from indigenous but often ineffective soil rhizobia, and partly a result of selection for early flowering and short growth season in many areas. Bush types fix less $N_2$ than indeterminate and climbing types [189,190]. There is significant genetic variability within growth types for $N_2$-fixing ability [191]. Several research centers have breeding programs under way to improve $N_2$ fixation in common bean [192]. Because seed yield gains from $N_2$ fixation have proved limited, even monocropped beans are often fertilized with N fertilizers [193].

Genetic analyses of $F_1$, $F_2$, and backcross progenies from diverse germplasm of snap bean grown in nutrient solutions with a low K supply indicated a single gene control of K efficiency (dry matter yield per unit K absorbed) [194]. Reciprocal $F_1$ progenies from crosses between efficient and inefficient strains showed no maternal effects. The gene for K efficiency was homozygous recessive in the efficient genotypes. Differential responses among strains grown at low K appeared to be associated with K use rather than K uptake or high accumulation of K and did not appear to be associated with Na substitution for K in the plant [194,195].

In a pan-African effort, Wortmann et al. [149] evaluated 280 entries from African bean-breeding programs for tolerance of low availabilities of soil N, P, and K, and toxicities of Al and Mn. Several entries were identified as tolerant of each of the stresses (Table 1). Especially promising were RWR 382, RAO 55, ACC 433, XAN 76, and MMS 224 for low-P tolerance; ICA Pijao and EMP 84 for low-K tolerance; Muhinga, Ntekerabasilumu, and 7/4 ACC for tolerance of Al toxicity; and MCM 5001 and XAN 76 for tolerance of Mn toxicity. Several varieties, including XAN 76, RAO 55, and OBA 1, performed well under several edaphic stresses.

In Latin America, Thung et al. [142] proposed a field screening method to evaluate Al tolerance in beans, using seed yield as a selection parameter. Attempts were made to screen bean germplasm for Al resistance, using nutrient solutions [196,197]. The response of seven cultivars of beans exposed to toxic levels of Al was assessed, using root elongation rate and callose accumulation in 5-mm root tips as early markers of Al injury [197]. Based on root elongation rate, which is very sensitive to Al toxicity, Massot et al. [197] identified 'F-15' and 'Superba' as the most Al-tolerant cultivars. Callose synthesis correlated positively with internal Al concentration and negatively with root elongation rate. Results indicated that while both callose accumulation and root elongation rate could be useful in classifying the bean cultivars for Al tolerance, root elongation rate is the more sensitive parameter.

Field screening of 5000 accessions of germplasm collection and bred lines over the past few years in an Al-toxic soil at Quilichao, Colombia, has resulted in the identification of 77 genotypes for further testing and analysis [121]. Among the 77 genotypes tested, four bred lines, A 774, VAX 1 (interspecific), FEB 190, and FEB 192, were found to be outstanding in their adaptation to Al-toxic soil conditions. Grain yield of a bred line, A 774, was 60% greater than that of a widely adapted cultivar, Carioca, with no lime treatment. A 774, VAX 1, and FEB 190 were also responsive to lime and P application. To date, the genetic control of Al resistance has not been elucidated, much less the association of mechanisms studied, as has been possible with P.

Sources of genetic tolerance of Mn toxicity were identified in common bean using three growing conditions: nutrient solution culture, silica sand culture, and Mn-amended soil [198]. Six genotypes (Argentino, BAT 271, Calima, EMP 84, H6 Mulatinho, and Pintado) out of 25 screened were tolerant of a toxic level of Mn in solution culture. The tolerance of Mn observed in solution culture correlated with tolerance observed in the silica sand system. Some genotypes that performed very well in solution culture and silica sand suffered severe yield reduction in Mn-amended soil. This study indicated that screening of genotypes in solution culture is useful to identify sources of tolerance of Mn toxicity, but performance of those genotypes in soil might be confounded by other edaphic stress factors common to low-fertility tropical soils. In another study, González and Lynch [199] characterized the mechanisms of Mn tolerance in common bean using two contrasting (tolerant and sensitive) bean genotypes. They demonstrated that Mn compartmentation occurs at both the tissue and organelle level and that Mn accumulation in the epidermis-enriched fraction could contribute to Mn tolerance in common bean.

Common bean shows genetic variability for tolerance of soil salinity [95,200]. Pessarakli [200] discussed the effects of salt stress on dry matter production, total N, $^{15}N$, crude protein, and water uptake by three green bean cultivars. Among the three cultivars, the Tender Improved variety was the least and the Slim Green variety the most severely affected by salinity in all aspects of the stress. Zaiter and Mahfouz [201] showed that cultivar Badrieh is tolerant of salinity, whereas T No. 1 is susceptible. These authors indicated that foliar injury symptoms could be used as criteria for salinity tolerance. In the sand culture system, cultivar Badrieh did not show symptoms at 4 dS m$^{-1}$, whereas T No. 1 was found to be sensitive.

## B.  Tropical Forages

Beef and milk provide 10% of total caloric consumption and 25% of protein consumption in Latin America [202] and are important components in the diet of all economic strata in the tropics [203]. The livestock industry in tropical America is far more important relative to other areas of the developing world (East Asia and Africa). The vast grasslands (240 million ha) that have developed on low-fertility acid soils (mainly Oxisols and Ultisols) in tropical America offer considerable potential for increasing livestock productivity and hence lowering the cost of ruminant animal products [204,205]. Undernutrition is the key constraint to increased livestock productivity on these low-fertility acid soils. Cattle are dependent on native pastures and, as a result of acid-soil stress, good quality forage is scarce. Using lime to improve soil chemical properties is not economically feasible because of the low unit value of forage. Therefore, the selection or breeding of tropical forages (legumes and grasses) adapted to low-fertility acid soils is considered as the most viable approach to increasing pasture (and cattle) productivity. In Brazil alone, 50 million ha of introduced pastures have been sown in the tropical savannas in the last 30 years [206].

The low-fertility, acid soils of tropical grasslands have low levels of available nutrients and high levels of soluble Al and Mn, and soil pH is lower than 5.0 [205]. The native pastures in low-fertility acid soils have poor productivity and nutritive value, and animal performance is correspondingly low. Growing improved tropical forage grasses (with the $C_4$ photosynthetic pathway) and legumes (with $C_3$) in associations can make introduced pastures productive [207].

The main role of the legume in a grass-legume association is to improve forage quality [207]. Pasture legumes are rich in N and provide an extra source of protein for grazing animals, particularly in the dry season when grasses supply little nutrition. The legume directly contributes to animal production by providing protein-rich forage. It can also improve productivity of low-fertility acid soils by increasing the amount of N available in the soil for associated grasses [208,209]. But legumes in tropical pastures do not readily persist under grazing and are more difficult for farmers to manage. To give sound management advice and help farmers benefit from forage legumes, researchers must understand how nutrient supply influences the physiology of interactions between grasses and legumes grown in association [210].

With good management, grass-legume pastures can increase nutrient cycling, greatly improve animal production, markedly increase soil biological activity, and also sequester significant amounts of organic carbon deep in the soil [211–213]. Pasture productivity should then increase with time as long as the grass-legume balance is maintained. Furthermore, in the Colombian savannas, improved grass-legume pastures have clearly boosted yields of subsequent crops of upland rice [209,214]. The success of crop-pasture rotation has established the need for a different group of tropical forages to take advantage of the higher soil fertility resulting from crop fertilization.

### 1.  Identifying Grasses and Legumes Adapted to Low-Fertility Acid Soils

Exploiting the natural variability of forage germplasm has been an important research strategy for CIAT and its collaborators to identify tropical grass and legume species adapted to the various ecosystems in acid-soil regions [205,215]. Germplasm is gathered from a wide range of conditions throughout the low-fertility acid-soil regions of tropical America, Southeast Asia, and Africa.

The germplasm bank held at CIAT contains over 20,000 accessions from more than 700 species in 150 genera. In cooperation with national research institutions, CIAT screens this germplasm for tolerance of high soil Al and acidity, low P availability, and tolerance of diseases and insects. Ecotypes that pass this first screening are characterized in terms of tolerance of grazing, minimum nutrient requirements, nutritive value, dry-season performance, and compatibility in grass-legume mixtures. Subsequently, highly promising ecotypes are assembled into pastures, relevant establishment technology is developed, and cattle liveweight gains measured. The most promising pasture combinations undergo long-term productivity and economic evaluation, and the respective technological packages are further adapted to the requirements of the predominant farm system in the area. The result is a new generation of forage grasses and legumes for these regions [205].

**TABLE 3**  Tropical Forage Grasses and Legumes Formally Released as Commercial Cultivars for
Livestock Production on Low-Fertility Acid Soils of Tropical Latin America

| Species | CIAT accession number | Countries where released |
|---|---|---|
| **Grasses** | | |
| *Andropogon gayanus* | 00621 | Colombia, Brazil, Venezuela, Panama, Costa Rica, Perú, México, Cuba, Honduras, Nicaragua, Guatamela |
| *Brachiaria brizantha* | 06780 | Brazil, Cuba, Venezuela, México, Costa Rica |
| *B. Brizantha* | 26646 | Colombia |
| *B. decumbens* | 00606 | Cuba, México, Panamá, Costa Rica |
| *B. dictyoneura* | 06133 | Colombia, Venezuela, Panamá, Costa Rica |
| *B. humidicola* | 00679 | Ecuador, Venezuela, Colombia, Panamá, México |
| *Panicum maximum* | 26900 | Brazil |
| *P. maximum* | 06962 | Brazil |
| **Legumes** | | |
| *Arachis pintoi* | 17434 | Colombia, Brazil, Honduras, Costa Rica |
| *A. pintoi* | Multiline | Panamá |
| *A. pintoi* | 18744 | Costa Rica |
| *Centrosema acutifolium* | 05277 | Colombia |
| *C. pubescens* | 00438 | Honduras |
| *Desmodium ovalifolium* | 00350 | Brazil |
| *Pueraria phaseoloides* | 09900 | México |
| *Stylosanthes capitata* | 10280 | Colombia |
| *S. guianensis* | 00184 | Perú |
| *S. guianensis* | 02950 | Brazil |
| *S. guianensis* | 02243 | Brazil |
| *S. macrocephala* | 01281 | Brazil |
| *Cliforia ternatea* | 20692 | México, Honduras |

*Source*: Adapted from Refs. 205, 215, and 216.

Selections are made at four major field screening sites, which represent the major ecosystems where Oxisols or Ultisols predominate. Promising grass and legume accessions are further tested in collaboration with the International Tropical Pastures Evaluation Network (RIEPT, its Spanish acronym) throughout the continent at more than 200 sites. Final selections are made for evaluation under grazing, on-farm validation, and for eventual commercial release [205,215,216] (Table 3). The released grass and legume cultivars require low fertilizer inputs, typically between 10 and 30 kg P ha$^{-1}$ at planting, and fungicides and insecticides are not applied because the forage cultivars are either resistant to or tolerant of major pests and diseases.

In the Colombian savannas, a clear advantage has been demonstrated, both at the Carimagua research station and on farm, in the performance of animals grazing on grass-legume pastures [207]. On-farm trials in the Colombian Eastern Plains documented the excellent performance of forage cultivars recently released to supplement native grasslands under conditions of farmer management [217]. An economic analysis of the investment showed that the marginal rate of return of the grass-legume associations was 31% [203]. Similar trends are evident in results obtained throughout the humid tropics of Peru, Ecuador, and Brazil. Some of the same species and accessions have also shown remarkably good adaptation to the moderately acid soils of the Central American hillsides and humid Caribbean coasts. Furthermore, many of these species have shown that they possess the ability to respond to increases in soil fertility.

## 2. Physiological Aspects of Grass-Legume Associations

Several studies were conducted by CIAT researchers to determine differences in growth and development of tropical forage grasses such as *Andropogon gayanus*, *Panicum maximum*, and *Brachiaria* species and tropical forage legumes such as *Stylosanthes* species, *Centrosema* species, *Desmodium ovalifolium*, and *Arachis pintoi* grown as either monocrops or grass-legume associations under grazing [210,218–222].

Because most selection among accessions of newly introduced species is carried out on the basis of high dry matter yields, other things being equal, the outcome of the process is grasses with high potential growth rates due to $C_4$ photosynthesis [221]. The legumes not only have the less efficient $C_3$ photosynthesis but also have the added burden of having to provide energy to the symbiont organisms in the nodules. The outcome of these differences is that, unless the legume has some other advantage, it will inevitably be dominated by the grass component [223].

It has been a major challenge to find tropical forage legumes that could persist in association with aggressive grasses in tropical pastures. Among the several tropical forage legumes tested over the past 20 years. *Arachis pintoi* has been the only tropical legume that has persisted with such aggressive grasses as *Brachiaria* species over longer periods [224]. Because of its growth habit, environmental adaptation, and grazing tolerance, *A. pintoi* has been considered a pasture legume ideotype for grass-legume associations in the tropics [221].

## 3. Adaptive Responses to Low Nutrient Supply

An essential part of germplasm selection and improvement is to identify morphological, physiological, and biochemical mechanisms by which forage plants adapt to acid-soil conditions. Although most tropical grass and legume cultivars show wide adaptation to a range of edaphic and climatic conditions, their environmental adaptation has not been well studied. Only recently have attempts been made to understand the physiological and biochemical bases of their adaptation to abiotic constraints [210,219, 225–229].

CIAT researchers conducted a series of investigations to identify plant attributes of tropical forage ecotypes that help plants acquire and efficiently use nutrients from low-fertility acid soils [205,228,230–237]. The outcome of this research has been an improved understanding of the physiological basis of forage grass and legume adaptation to acid soils. This research is essential for improving selection and breeding; identifying plant-soil, plant-plant, and soil-plant-animal nutrient interactions in forage-based production systems; and assisting the identification of ecological niches for forage germplasm.

Low supply of nutrients, particularly P, N, and Ca, greatly limits forage adaptation and production in acid soils. Widespread adoption of forage cultivars depends on their efficiently acquiring nutrients from the soil and using them for growth. Identifying plant attributes that confer adaptation to low-fertility acid soils is needed to develop tropical forages rapidly through agronomic evaluation and genetic improvement. Plant attributes appear to be linked to different strategies to acquire and use nutrients [7,18,127]. Understanding these linkages is fundamental in integrating plant attributes into a selection index. Plant attributes (indices) conferring adaptation to low-fertility acid soils must be identified to develop rapid and reliable screening procedures.

Adaptation of forage plants to acid soils involves changes in the partitioning of biomass between shoots and roots in response to growth conditions [205,231]. Greenhouse studies were conducted to determine the effects of acid-soil stress and nutrient supply on biomass production, dry matter partitioning between shoots and roots, and nutrient uptake, transport, and use efficiency in several forage grasses and legumes adapted to acid soils [231]. Soil texture and fertility (nutrient supply) affected biomass production and dry matter partitioning between plant parts. Forage grasses had higher biomass production in a clay loam soil, especially at lower fertility levels. This higher production was attributed to the higher organic matter content and N availability in the clay loam soil.

In contrast, legumes, because of their nitrogen-fixing capacity, showed similar biomass production in both types of soil at low fertility levels. The effect of soil fertility on the allocation of fixed carbon by grasses and legumes is manifested in their root production. At the higher fertility level, root production in grasses was higher in the sandy loam soil. In contrast, legumes showed little increase in root production in either soil type. The change in their allocation of fixed carbon toward shoot growth probably helped improve the nitrogen-fixing ability of legume roots.

In addition to changes in allocation of dry matter, grasses and legumes showed marked differences in uptake and use efficiency of nutrients [231]. Efficiency of P uptake in legume roots was twice that of grasses. However, the N and Ca use efficiency of grasses was about four times that of legumes. The superior efficiency of legumes for P uptake was probably a result, in part, of the higher activity of the enzyme acid phosphatase in their roots, thus favoring mobilization of P from organic sources in acid soils [234].

Larger root systems with greater surface area, typical of tropical grasses, are generally believed to be better for acquiring P per unit soil surface area than are the smaller roots typical of tropical legumes. Larger root systems are often associated with plants that are better competitors for nutrients, water, light, etc., such as tropical grasses [213], and grasses may competitively exclude companion species of legumes. Legumes with smaller root systems, however, may have a greater capacity to absorb P rapidly than grasses with larger root systems [234]. The higher P uptake efficiency in legumes could be attributed to the ability of legume root systems to modify the chemistry of the rhizosphere by exuding organic acids, Al and Fe chelators, reducing agents, or enzymes such as phosphatases. The superior compatibility of certain forage legumes (e.g., *Arachis pintoi*) with the aggressive grasses may be due to their ability to acquire P from less available forms of P (aluminum-bound P and organic P) in the low-fertility acid soils.

## 4. Adaptive Responses to Low Phosphorus Supply

In highly weathered acid soils, such as the Oxisols and Ultisols of tropical America, P is often the most limiting nutrient for pasture establishment and production [238,239]. For pasture establishment in tropical soils, yield increases following P applications are common [238].

Rao et al. [232] showed that the grass *B. dictyoneura* and legume *A. pintoi* differ significantly in their responsiveness to soluble-P fertilizer application in acid soils. When grown either as monocrop or in association, the grass responded more than the legume to an increase in P supply on either sandy loam or clay loam soil. The most striking effects of low P supply on shoot growth and development are reductions in leaf expansion and leaf surface area [232]. Leaf expansion is strongly related to the extension of epidermal cells, and this process may be particularly impaired when the P content of epidermal cells is low [240]. However, the P concentration in legume epidermal cells may be greater than that in grass leaves, which would contribute to the legume's greater leaf expansion.

Root attributes such as length, surface area, fineness (radius), and density of root hairs are considered to influence strongly plant adaptation to low-P soils [18]. This is because soil P is supplied to plants mainly by diffusion and the P diffusion coefficient is very low [241,242]. The most prominent root characteristic of the grass is high root length that results in a large root surface area and a high ratio of root surface to shoot dry weight. These two parameters greatly enhance P acquisition and P supply to the shoots. This is because a finely divided and rapidly developing root system provides better access to less mobile soil nutrients, such as P [243].

The response of *B. dictyoneura* to applied P was greater than that of *A. pintoi* in terms of both shoot and root biomass production [232]. This increased response to P supply in the grass was associated with higher P use efficiency (grams of forage produced per gram of total P uptake). However, P uptake efficiency (milligrams of P uptake in shoot biomass per unit root length) was several times higher in the legume than in the grass [234]. These differences in P acquisition between the grass and legume could result from their differences in ability to use sources of less available P from low-fertility acid soils.

Rao et al. [236] tested this hypothesis and found marked differences between the legume and grass in their ability to use sources of relatively less available inorganic and organic P. The legume produced greater amounts of leaf area when grown on sources of sparingly soluble P than did the grass. Increased availability of P from calcium-bound P (Ca-P) was of greater advantage to the grass than to the legume, enabling the grass to dominate when grown in association [232,244]. Grass and legume growth, as measured by shoot and root biomass per unit soil surface area, responded very differently to different P sources. The response of the two plants to sources of relatively less available aluminum-bound P (Al-P) and organic P was similar.

Increasing P supply to an Oxisol improves fine root production of the grass to a greater extent than in the legume [232]. Under monoculture, although source of P and soil type affected the grass's total root length but not that of the legume [236,237], root length values were several times higher for the grass than for the legume. Specific root length (SRL) values were also greater for the grass than for the legume, regardless of soil type and P source. Under association, the proportion of legume roots was greater under no P supply and lower under Ca-P supply. The ratio of root length to shoot biomass was also markedly greater for the grass than for the legume. These results indicate that the grass was more efficient than the legume in producing root length and root biomass, regardless of P source and soil type. When grown in association with legumes, the grass can effectively compete with the legume by exploring a greater volume of soil to acquire nutrients and water.

Otani and Ae's [245] field and greenhouse studies indicated that P uptake by crops in soils where P availability is high is strongly related to root length, but this relationship is lost in soils with low P availability or where soil volume is limited. Their results also suggested that peanut (*Arachis hypogaea* L.) uses additional mechanisms beyond root length to increase P uptake. The forage legume *A. pintoi*, a perennial peanut and wild relative of the cultivated peanut, shows remarkable adaptation to less available P forms, such as Al-P and organic P, by producing greater leaf area with less root length [236,237].

Rao et al. [237] showed that *A. pintoi* is more efficient in acquiring Al-P and organic P from acid soils than is *B. dictyoneura*. They also showed that the association of the two forages could increase the total P acquisition from low-fertility acid soils. The legume has at least three attributes that are important for its efficiency in acquiring P from acid soils: (1) high storage capacity for inorganic P, (2) a favorable ratio of P uptake per unit root length, and (3) high activity of acid phosphatase in the root and a capability for using P from organic P sources. These attributes could form a self-controlling system for acquiring P from low-P acid soils. The authors speculated, furthermore, that the superior compatibility of *A. pintoi* with aggressive grasses such as *Brachiaria* species may be due to its ability to acquire P from less available forms. The mechanism by which *A. pintoi* accesses the sparingly soluble inorganic P remains unknown.

Rao et al. [237] tested the relationships between root and shoot attributes and showed a significant positive correlation between the level of inorganic P in the legume's roots and key shoot attributes such as leaf area production, shoot biomass, and shoot P uptake. This observation indicates that measuring the level of inorganic P in roots may serve as a selection method to evaluate differences in adaptation of tropical forage legumes to P-deficient acid soils.

## 5. Selecting for Improved Adaptation to Other Soil Constraints

Defining nutritional requirements of different grass and legume ecotypes can help to reduce the amount of fertilizer needed to establish pastures rapidly and to maintain productivity over time. Research conducted at CIAT and elsewhere generated valuable information on both internal (plant) and external (soil) critical nutrient requirements for several tropical forage species [225–227,246–248]. Internal P and Ca requirements for plant growth of *Brachiaria* species are much lower than those of *P. maximum* (Table 4). Among the *Brachiaria* species, *B. humidicola* appears to require lower internal concentrations of P, Ca, and K. The external P requirements of *B. brizantha* were much higher than those of *B. decumbens* when grown in low-fertility Latosol in Brazil [228]. Although nutrient requirements of *A. pintoi*, compared with other tropical forage legumes, are relatively low to moderate [227], it has higher critical levels for P and Ca (Table 4).

Diagnosis of mineral nutrient disorders using visual symptoms has been developed for *B. decumbens* and *A. pintoi* [227,228]. Several greenhouse and field experiments on *Brachiaria* species demonstrated striking responses in terms of forage yield to P applications but not to lime applications [228]. Responses

**TABLE 4** Values or Ranges of Critical Concentrations of Various Nutrients in Shoot Dry Matter of Different Tropical Forage Grasses and Legumes, Compared with Animal Requirements

| Forage species | N | P | K | Ca | Mg | S | Zn | Cu |
|---|---|---|---|---|---|---|---|---|
| | | | (g kg$^{-1}$) | | | | (mg kg$^{-1}$) | |
| **Grasses** | | | | | | | | |
| *Andropogon gayanus* | 13.0 | 1.0 | 9.0 | 2.5 | 2.0 | 1.4 | 18 | 6.0 |
| *Brachiaria brizantha* | | 0.9 | 8.2 | 3.7 | | | | |
| *B. decumbens* | | 1.0 | 8.3 | 3.7 | | 1.6 | | |
| *B. humidicola* | | 0.8 | 7.4 | 2.1 | | 1.4 | | |
| *Panicum maximum* | | 1.7 | — | 6.0 | | 1.5 | | |
| **Legumes** | | | | | | | | |
| *Stylosanthes* spp. | 21–31 | 1.7–2.5 | 7.8–9.2 | 8–14 | 2.3–3.1 | 1.1–1.5 | 16–24 | 4–6 |
| *Centrosema pubescens* | | 1.6–2.0 | 7.5–14 | 6–13 | 2.4–4.6 | 1.5–1.9 | 20–25 | 4–6 |
| *Arachis pintoi* | | 2.3 | 5.0 | 18 | | 1.1 | 16 | 5 |
| **Animal requirements** | | **>1.2** | **>6.0** | **>1.8** | **>1.0** | **>1.0** | **>20** | **>7** |

*Source*: Adapted from Refs. 225–228 and 248.

to N and Ca were very much site specific. The lack of clear response to micronutrients may be attributed to their efficiency in acquisition from low amounts of soil reserves and/or their efficiency in using acquired amounts.

Considerable progress has been made in selecting germplasm accessions of tropical forage grasses and legumes adapted to low-fertility acid soils [205,216,225,226,246,247,249–252]. Such accessions will make the most efficient use of scarce fertilizer inputs, permitting a decrease in the required rates of fertilizer application until minimal but adequate quality and yield are attained. The relative importance of different soil nutrients in influencing growth and productivity of adapted plants may depend on the plants' physiological adaptation to low-fertility acid soils.

At CIAT, efforts to screen forage germplasm for tolerance of toxic levels of Al and Mn in nutrient solution [225,253] led to the identification of several promising grass and legume accessions. Brazilian workers successfully developed solution culture techniques to identify acid-soil tolerance in *Leucaena leucocephala*, a multipurpose forage tree species [254]. They also found that cotyledon retention on seedlings was the most reliable, nondestructive, and easily assessed indicator of Al tolerance [255,256]. Improving acid-soil tolerance was also a breeding objective for two forage legumes, *Centrosema pubescens* and *L. leucocephala* [257,258], although commercial cultivars were not developed.

An ongoing *Brachiaria* breeding program at CIAT aims to combine superior acid-soil adaptation, found in *B. decumbens* cv. Basilisk, with resistance to spittlebugs, found in *B. brizantha* cv. Marandú. Both species are natural tetraploid apomicts but produce fertile pollen, which can be used to pollinate the closely related sexual species *B. ruziziensis* once its normal diploid chromosome number is doubled [259]. Among the resulting hybrids, both apomictic and sexual plants are found. Hence, combining genes of the two apomictic species becomes possible. Field evaluation was conducted for 43 genetic recombinants from a breeding population, four parents, and eight germplasm accessions for their tolerance of low-fertility acid soils. The study identified two genetic recombinants that combined several desirable attributes, such as superior leaf area and leaf biomass, greater N content in leaves, and greater partitioning of N and P to leaves, that would contribute to adaptation and persistence in low-fertility acid soils [235].

Continued progress in the selection and improvement of *Brachiaria* genotypes will depend on identifying plant attributes that contribute to tolerance of low-fertility acid soils and on developing rapid and reliable screening methods [231,235]. Adapted and persistent genotypes are able to acquire key nutrients (e.g., N, P, and Ca) in a soil environment characterized by low pH and high Al [231–237]. Adaptation may be through several root and shoot attributes [228], including (1) maintenance of root growth at the expense of shoot growth; (2) acquisition and use of N (both forms), nitrate, and ammonium (e.g., *B. humidicola*); (3) ability to acquire N through associative biological fixation (*B. decumbens*); (4) ability to acquire P through an extensive root system and association with vesicular-arbuscular mycorrhizae; and (5) development of an extensively branched root system (more root tips), which facilitates greater acquisition of Ca (*B. ruziziensis*).

CIAT [260] conducted research to elucidate the physiological basis of acid-soil adaptation in *Brachiaria* and to develop a high-throughput screening procedure to evaluate genetic recombinants of *Brachiaria* for this trait. Results from these studies indicated that the high level of adaptation to acid soils of *B. decumbens* cv. Basilisk is due to its superior resistance to toxic levels of Al, combined with excellent adaptation to P and N deficiencies. A particularly important finding was that *B. decumbens* cv. Basilisk exhibits a level of Al resistance markedly superior to that of Al-resistant varieties of crops such as maize, upland rice, or wheat. Unexpectedly, secretion of organic acids—an apparently widespread mechanism for Al resistance [261]—is unlikely to be the principal mechanism conferring such outstanding resistance [260]. Patterns of accumulation of Al and callose in root apices suggested that mechanisms excluding Al from apices contribute to the superior resistance of *B. decumbens*. These physiological studies led to the development of a rapid and reliable screening procedure to evaluate Al resistance of those genetic recombinants, making it possible to improve the efficiency of genetic improvement of *Brachiaria* germplasm [260]. This procedure is currently being incorporated into the breeding program.

## 6. Contribution of Adapted Grasses and Legumes to Sustainable Production Systems

Tropical forage species contribute to the sustainability of land management by helping regenerate degraded soils and replenish the N supply of the production system. Forage plants as cover crops can control weeds and reduce soil erosion. Grass and legume cultivars adapted to low-fertility acid soils support stable and productive farming systems via their contributions to soil enhancement [208,209,212,262,263] and nutrient cycling [264,265]. This is mainly attributed to their generally efficient use of external nutrients and inherently higher and more stable production potential under low fertilizer inputs [205,231]. Introduced tropical grass and legume cultivars develop deep and abundant rooting systems that penetrate well into A1-toxic subsoils, giving them greater potential to survive seasonal droughts and reduce nutrient leaching [211,213,266].

The rooting ability of acid soil–adapted grass and legume cultivars has several consequences. Soil physical conditions are improved, as shown by higher rates of water infiltration and increased stability and size of soil aggregates [267,268]. Both roots and aboveground litter contribute to the quantity and quality of soil organic matter, which, in turn, improves soil biological activity. These improvements in soil quality attributes lead to significant increases in grain yield of acid soil–adapted upland rice [205,209].

The ability of acid soil–adapted grasses and legumes to root profusely would also help these species to become established in degraded and compacted soils with minimum tillage. This ability to reclaim degraded lands may be particularly valuable for small farmers in the humid tropics, who often do not have farm machinery. The vigorous rooting ability of grasses and legumes help prevent soil runoff in regions where rains are intense and abundant. Some tropical forage legume species can also be used as improved and accelerated fallow to complement or substitute the fallow of native species, which grow more slowly [269].

## IV. FUTURE PERSPECTIVES

This chapter showed that considerable progress is being made in improving genetic adaptation of common bean and tropical forages to major abiotic constraints in the tropics. It also highlighted the role of physiological studies for improving genetic adaptation to major abiotic constraints. Crop physiology has been called the "retrospective science" by one plant breeder because physiologists elucidate what the breeders have already achieved [270]. This is because the links between physiology and genetics have not been established. This situation is likely to change in the future [2], when knowledge of plant physiological processes will become extremely important in screening for and measuring phenotypic traits.

Advances in agricultural biotechnology open a new and exciting perspective for dissecting and understanding the complex regulation of physiological traits and mechanisms controlling crop adaptation to abiotic stresses. As high-density molecular maps become more readily available for a range of food and feed crops, physiologists need only to screen the parents of the available mapping populations for variation in the expression of the trait(s) of interest and then to score the appropriate mapping population for the trait. As Prioul et al. [271] pointed out, it is time for plant and crop physiologists to study marker-characterized segregating populations and marker-specific-near-isogenic lines instead of improved cultivars.

To develop the new technologies, the disciplines of physiology, genetics, molecular biology, and breeding will need to be brought together. The integrated team efforts would contribute toward developing food and feed crops that would overcome major abiotic stresses, particularly in the tropics (Figure 1). We are likely to see continued significant progress in our understanding and ability to modify stress tolerance by molecular engineering, using both model and crop plants, based on understanding how stress affects plant biochemistry and physiology through gene expression [272]. The onset of genomics will provide massive amounts of information, but success will depend on using that information efficiently to improve crop phenotypes [273]. Screening for and measuring important phenotypic traits are crucial to the full exploitation of the opportunities offered by molecular marker technology.

A novel area that holds promise, once a physiological mechanism is identified, is the candidate gene approach [274,275]. This detects expression of a tolerance response to the stress in question. The key responses may be those that occur days or weeks after exposure to a "realistic" stress, not minutes after the imposition of a treatment, which is ecologically irrelevant [275]. The candidate gene approach would permit much more precision, both in mapping of important genes and in defining the physiological basis of yield improvement. One can envisage situations in which the improved adaptation to major abiotic constraints combined with adaptation to biotic constraints and improved nutritional quality could have a tremendous impact on food security and human nutrition in the tropics [2,273,276].

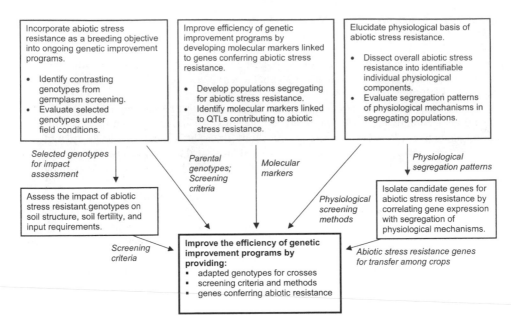

**Figure 1** Contribution of physiological studies to improve the efficiency of genetic enhancement of crops to abiotic stress factors.

## ACKNOWLEDGMENTS

The ideas presented here developed from discussions with many colleagues, and I wish to thank them for their inputs. Particular thanks go to Drs. Steve Beebe, John Miles, Mabrouk El-Sharkawy, and Peter Wenzl for constructive comments that helped improve the manuscript. I gratefully acknowledge the contributions of past and present colleagues of the Beans and Tropical Forages projects at CIAT. I also thank CIAT's collaborators from several national programs in Latin America and Africa for their valuable contributions to selection and breeding efforts over the years. I am very grateful to Elizabeth de Páez for editorial corrections of the manuscript.

## REFERENCES

1.  LT Evans. Crop Evolution, Adaptation and Yield. Cambridge: Cambridge University Press, 1993.
2.  B Miflin. Crop improvement in the 21st century. J Exp Bot 51:1–8, 2000.
3.  LT Evans. Physiological aspects of varietal improvement. In: JP Gustafson, ed. Gene Manipulation in Plant Improvement. New York: Plenum, 1984, pp 121–146.
4.  H Lambers, FS Chapin III, TL Pons. Plant Physiological Ecology. New York: Springer-Verlag, 1998: p 540.
5.  A Blum. Genetic and physiological relationship in plant breeding for drought tolerance. Agric Water Manage 7:195–205, 1983.
6.  MM Ludlow, RC Muchow. A critical evaluation of traits for improving crop yields in water-limited environments. Adv Agron 43:107–153, 1990.
7.  H Marschner. Mechanisms of adaptation of plants to acid soils. Plant Soil 134:1–20, 1991.
8.  AE Hall. Breeding for heat tolerance. Plant Breed Rev 10:129–168, 1992.
9.  CL Noble, N Rogers. Arguments for the use of physiological criteria for improving the salt tolerance in crops. Dev Plant Soil Sci 50:127–135, 1993.
10. M Ashraf. Breeding for salinity tolerance in plants. Crit Rev Plant Sci 13:17–42, 1994.
11. SP Loss, KHM Siddique. Morphological and physiological traits associated with wheat yield increases in mediterranean environments. Adv Agron 52:229–276, 1994.
12. GV Subbarao, C Johansen, AE Slinkard, RC Nageswara Rao, NP Saxena, YS Chauhan. Strategies for improving drought resistance in grain legumes. Crit Rev Plant Sci 14:469–523, 1995.
13. GN Amzallag, HR Lerner. Physiological adaptation of plants to environmental stresses. In: M Pessarakli, ed. Handbook of Plant and Crop Physiology. New York: Marcel Dekker, 1995, pp 557–576.

14. JS Boyer. Advances in drought tolerance in plants. Adv Agron 56:187–218, 1996.
15. NC Turner. Further progress in crop water relations. Adv Agron 58:293–338, 1997.
16. MA Sanderson, DW Stair, MA Hussey. Physiological and morphological responses of perennial forages to stress. Adv Agron 59:171–224, 1997.
17. AK Joshi. Genetic factors affecting abiotic stress tolerance in crop plants. In: M Pessarakli, ed. Handbook of Plant and Crop Stress. 2nd ed. New York: Marcel Dekker, 1999, pp 795–826.
18. IM Rao, DK Friesen, M Osaki. Plant adaptation to phosphorus-limited tropical soils. In: M Pessarakli, ed. Handbook of Plant and Crop Stress. 2nd ed. New York: Marcel Dekker, 1999, pp 61–96.
19. AF Raybold, AJ Gray. Genetically modified crops and hybridization with wild relatives: a UK perspective. J Appl Ecol 30:199–219, 1993.
20. LT Evans, RA Fischer. Yield potential: its definition, measurement, and significance. Crop Sci 39:1544–1551, 1999.
21. RB Austin. Yield of wheat in the United Kingdom: recent advances and prospects. Crop Sci 39:1604–1610, 1999.
22. DN Duvick, KG Cassman. Post–green revolution trends in yield potential of temperate maize in the North-Central United States. Crop Sci 39:1622–1630, 1999.
23. S Peng, KG Cassman, SS Virmani, J Sheehy, GS Khush. Yield potential trends of tropical rice since the release of IR8 and the challenge of increasing rice yield potential. Crop Sci 39:1552–1559, 1999.
24. MP Reynolds, S Rajaram, KD Sayre. Physiological and genetic changes of irrigated wheat in the post–green revolution period and approaches for meeting projected global demand. Crop Sci 39:1611–1621, 1999.
25. M Tollenaar, J Wu. Yield improvement in temperate maize is attributable to greater stress tolerance. Crop Sci 39:1597–1604, 1999.
26. LT Evans. Adapting and improving crops: the endless task. Philos Trans R Soc Lond Ser B 352:41–46, 1997.
27. RS Loomis, JS Amthor. Yield potential, plant assimilatory capacity, and metabolic efficiencies. Crop Sci 39:1584–1596, 1999.
28. JE Specht, DJ Hume, SV Kumudini. Soybean yield potential—a genetic and physiological perspective. Crop Sci 39:1560–1570, 1999.
29. BR Buttery, RI Buzzell, WI Findlay. Relationships among photosynthetic rate, bean yield and other characters in field-grown cultivars of soybean. Can J Plant Sci 61:191–198, 1981.
30. S Ceccarelli, E Acevedo, S Grando. Analytical breeding for stress environments: single traits, architecture of traits or architecture of genotypes. Euphytica 56:169–185, 1991.
31. J Moorby. Can models hope to guide change? Ann Bot 60:175–188, 1987.
32. R Shorter, RJ Lawn, GL Hammer. Improving genetic adaptation of crops—a role for breeders, physiologists and modellers. Exp Agric 27:155–175, 1991.
33. KJ Boote, M Tollenaar. Modeling genetic yield potential. In: KJ Boote, JM Bennett, TR Sinclair, GM Paulson, eds. Physiology and Determination of Crop Yield. Madison, WI: ASA-CSSA-SSSA, 1994, pp 533–565.
34. M Koornneef, C Alonso-Blanco, AJM Peters. Genetic approaches in plant physiology. New Phytol 137:1–8, 1997.
35. J-L Prioul, S Quarrie, M Causse, D Vienne. Dissecting complex physiological functions through the use of molecular quantitative genetics. J Exp Bot 48:1151–1163, 1997.
36. MD Edwards, CW Stuber, JF Wendel. Molecular-marker-facilitated investigations of quantitative-trait loci in maize. 1. Numbers, distribution and types of gene action. Genetics 116:113–125, 1987.
37. MD Edwards, T Helentjaris, S Wright, CW Stuber. Molecular-marker-facilitated investigations of quantitative-trait loci in maize. 4. Analysis based on genome saturation with isozyme and restriction fragment length polymorphism markers. Theor Appl Genet 83:765–774, 1992.
38. M Yano, T Sasaki. Genetic and molecular dissection of quantitative traits in rice. Plant Mol Biol 35:145–153, 1997.
39. BP Forster, JR Russell, RP Ellis, LL Handley, D Robinson, CA Hackett, E Nevo, R Waugh, DC Gordon, R Keith, W Powell. Locating genotypes and genes for abiotic stress tolerance in barley: a strategy using maps, markers and the wild species. New Phytol 137:141–147, 1997.
40. BP Forster, RP Ellis, WTB Thomas, AC Newton, R Tuberosa, D This, RA El-Enein, MH Bahri, M Ben Salem. The development and application of molecular markers for abiotic stress tolerance in barley. J Exp Bot 51:19–27, 2000.
41. SD Tanksley, SR McCouch. Seed banks and molecular maps: unlocking genetic potential from the wild. Science 277:1063–1066, 1997.
42. MW Humphreys, I Pasakinskiene, AR James, H Thomas. Physically mapping quantitative traits for stress-resistance in the forage grasses. J Exp Bot 49:1611–1618, 1998.
43. CW Stuber, M Polacco, ML Senior. Synergy of empirical breeding, marker-assisted selection, and genomics to increase crop yield potential. Crop Sci 39:1571–1583, 1999.
44. JW White, G Hoogenboom. Simulating effects of genes for physiological traits in a process-oriented crop model. Agron J 88:416–422, 1996.
45. WJ Horst. Fitting maize into sustainable cropping systems on acid soils of the tropics. Presented at the Consultants Meeting. Vienna, Austria: FAO/IAEA Division, 1999.

46.  J Lynch. The role of nutrient-efficient crops in modern agriculture. J Crop Prod 1:241–264, 1998.
47.  ET York Jr. Global perspectives on international agricultural research. In: KJ Boote, JM Bennett, TR Sinclair, GM Paulson, eds. Physiology and Determination of Crop Yield. Madison, WI: ASA, CSSA, and SSSA, 1994, pp 1–17.
48.  TR Sinclair. Limits to crop yields? In: KJ Boote, JM Bennett, TR Sinclair, GM Paulson, eds. Physiology and Determination of Crop Yield. Madison, WI: ASA, CSSA, and SSSA, 1994, pp 509–532.
49.  L Kaplan. Archeology and domestication in American *Phaseolus* beans. Econ Bot 19:358–368, 1965.
50.  L Kaplan. What is the origin of the common bean? Econ Bot 35:240–254, 1981.
51.  L Kaplan, LN Kaplan. *Phaseolus* in archeology. In: P Gepts, ed. Genetic Resources of *Phaseolus* Beans: Their Maintenance, Domestication, Evolution and Utilization. Dordrecht: Kluwer Academic Publishers, 1988, pp 25–142.
52.  P Gepts, D Debouck. In: A Schoonhoven, O Voysest, eds. Common Bean: Research for Crop Improvement. Wallingford, UK: CAB International and Cali, Colombia: CIAT, 1991, pp 7–53.
53.  SP Singh. Production and utilization. In: SP Singh, ed. Common Bean Improvement in the Twenty-First Century. Dordrecht: Kluwer Academic Publishers, 1999, pp 1–24.
54.  DR Laing, PG Jones, JHS Davis. Common bean (*Phaseolus vulgaris* L.). In: PR Goldsworthy, NM Fisher, eds. The Physiology of Tropical Field Crops. New York: John Wiley & Sons, 1984, pp 305–351.
55.  PH Graham, P Ranalli. Common bean (*Phaseolus vulgaris* L.). Field Crops Res 53:131–146, 1997.
56.  A van Schoonhoven, O Voysest, eds. Common Beans: Research for Crop Improvement. Wallingford, UK: CAB International and Cali, Colombia: CIAT, 1991.
57.  MW Adams, DP Coyne, JHC Davis, PH Graham, CA Francis. Common bean (*Phaseolus vulgaris* L.). In: RJ Summerfield, EH Roberts, eds. Grain Legume Crops. London: Collins, 1985, pp 433–476.
58.  NK Fageria, VC Baligar, CA Jones. Growth and mineral nutrition of field crops. New York: Marcel Dekker, 1991, pp 280–318.
59.  SP Singh. Common bean improvement in the tropics. Plant Breed Rev 10:199–269, 1992.
60.  AM Evans. Commentary upon plant architecture and physiological efficiency in the field bean. In: D Wall, ed. Potential of Field Beans and Other Food Legumes in Latin America. Cali, Colombia: CIAT, 1973, pp 279–286.
61.  SP Singh, P Gepts, DG Debouck. Races of common bean (*Phaseolus vulgaris*, Fabaceae). Econ Bot 45:379–396, 1991.
62.  E Garcia, CB Peña-Valdivia, JR Rogelio Aguirre, JM Muruaga. Morphological and agronomic traits of a wild population and improved cultivar of common bean (*Phaseolus vulgaris* L.). Ann Bot 79:207–213, 1997.
63.  SP Singh, R Nodari, P Gepts. Genetic diversity in cultivated common bean: I. Allozymes. Crop Sci 31:19–23, 1991.
64.  RO Nodari, SM Tsai, RL Gilbertson, P Gepts. Towards an integrated linkage map of common bean. 2. Development of an RFLP-based linkage map. Theor Appl Genet 85:513–520, 1993.
65.  J Tohme, DO González, S Beebe, MC Duque. AFLP analysis of gene pools of a wild bean core collection. Crop Sci 36:1375–1384, 1996.
66.  S Beebe, PW Skroch, J Tohme, MC Duque, F Pedraza, J Nienhuis. Structure of genetic diversity among common bean landraces of Middle American origin based on correspondence analysis of RAPD. Crop Sci 40:264–273, 2000.
67.  J Lynch, A Gonzalez, J Tohme, J Garcia. Variation for characters related to leaf photosynthesis in wild bean populations. Crop Sci 32:633–640, 1992.
68.  A Gonzalez, J Lynch, JM Tohme, SE Beebe, RE Macchiavelli. Characters related to leaf photosynthesis in wild populations and landraces of common bean. Crop Sci 35:1468–1476, 1995.
69.  SP Singh. A key for identification of different growth habits of *Phaseolus vulgaris* L. Annu Rep Bean Improv Coop 25:92–95, 1982.
70.  JS Beaver, CV Paniagua, DP Coyne, GF Freytag. Yield stability of dry bean genotypes in the Dominican Republic. Crop Sci 25:923–926, 1985.
71.  JD Kelly, MW Adams, GV Varner. Yield stability of determinate and indeterminate dry bean cultivars. Theor Appl Genet 74:516–521, 1987.
72.  A Vandenberg, T Nleya. Breeding to improve plant type. In: SP Singh, ed. Common Bean Improvement in the Twenty-First Century. Dordrecht: Kluwer Academic Publishers, 1999, pp 167–183.
73.  JW White, J Izquierdo. Physiology of yield potential and stress tolerance. In: A van Schoonhoven, O Voysest, eds. Common Beans: Research for Crop Improvement. Wallingford, UK: CAB International and Cali, Colombia: CIAT, 1991, pp 287–382.
74.  JW White, SP Singh, C Pino, MJ Rios B, I Buddenhagen. Effects of seed size and photoperiod response on crop growth and yield of common bean. Field Crops Res 28:295–307, 1992.
75.  JW White, A Gonzalez. Characterization of the negative association between seed yield and seed size among genotypes of common bean. Field Crops Res 23:159–175, 1990.
76.  PJ Sexton, JW White, KJ Boote, Yield-determining processes in relation to cultivar seed size of common bean. Crop Sci 34:84–91, 1994.
77.  PJ Sexton, CM Peterson, KJ Boote, JW White. Early-season growth in relation to region of domestication, seed size, and leaf traits in common bean. Field Crops Res 52:69–78, 1997.

78. JW White, J Kornegay, J Castillo, CH Molano, C Cajiao, G Tejada. Effect of growth habit on yield of large-seeded bush bean cultivars of common bean. Field Crops Res 29:151–161, 1992.
79. P Masaya, JW White. Adaptation to photoperiod and temperature. In: A van Schoonhoven, O Voysest, eds. Common Beans: Research for Crop Improvement. Wallingford, UK: CAB International and Cali, Colombia: CIAT, 1991, pp 445–500.
80. DH Wallace, KS Yourstone, JP Baudoin, J Beaver, DP Coyne, JW White, RW Zobel. Photoperiod × temperature interaction effects on the days to flowering of bean (*Phaseolus vulgaris* L.). In: M Pessarakli, ed. Handbook of Plant and Crop Physiology. New York: Marcel Dekker, 1995, pp 863–891.
81. J Haterlein, CD Clayberg, LD Teare. Influence of high temperature on pollen grain viability and pollen tube growth in the styles of *Phaseolus vulgaris* L. J Am Soc Hortic Sci 105:12–14, 1980.
82. ML Weaver, R Timm, MJ Silbernagel, DW Burki. Pollen staining and high temperature tolerance of bean. J Am Soc Hortic Sci 110:797–799, 1985.
83. Y Gross, J Kigel. Differential sensitivity to high temperature of stages in the reproductive development of common bean (*Phaseolus vulgaris* L.). Field Crops Res 36:201–212, 1994.
84. G Hoogenboom, DS McSmith. Temperature sensitivity of common bean during germination. Annu Rep Bean Improv Coop 35:72–73, 1992.
85. JW White, C Montes-R. The influence of temperature on seed germination in cultivars of common bean. J Exp Bot 44:1795–1800, 1993.
86. HZ Zaiter, E Baydoun, M Sayed-Hallak. Genotypic variation in common bean in response to cold temperature stress. Annu Rep Bean Improv Coop 36:66–67, 1993.
87. ST Otubo, MAP Ramalho, AFB Abreu, JB dos Santos, B Griffing, JL Jinks, HS Pooni. Genetic control of low temperature tolerance in germination of the common bean (*Phaseolus vulgaris* L.). Euphytica 89:313–317, 1996.
88. D Genchev. Screening for cold tolerance in dry bean (planting-emergence). Annu Rep Bean Improv Coop 31:210–212, 1988.
89. GC Shonnard, P Gepts. Genetics of heat tolerance during reproductive development in common bean. Crop Sci 34:1168–1175, 1994.
90. P Miranda, CJ da Anunciação Filho, DG da Cruz, VF dos Santos. Tolerância de cultivares de feijão comum (*Phaseolus vulgaris* L.) a alta temperatura. Pesq Agropec Pernambucana 8:73–80, 1997.
91. DH Wallace. Physiological genetics of plant maturity, adaptation, and yield. Plant Breed Rev 3:21–167, 1985.
92. JW White, DR Laing. Photoperiod response of flowering in diverse genotypes of common bean (*P. vulgaris*). Field Crops Res 22:113–128, 1989.
93. JA Acosta-Gallegos, P Vargas-Vazquez, JW White. Effect of sowing date on the growth and seed yield of common bean (*Phaseolus vulgaris* L.) in highland environments. Field Crops Res 49:1–10, 1996.
94. JW White, SP Singh. Breeding for adaptation to drought. In: A van Schoonhoven, O Voysest, eds. Common Beans: Research for Crop Improvement. Wallingford, UK: CAB International and Cali, Colombia: CIAT, 1991, pp 501–560.
95. M Thung, IM Rao. Integrated management of abiotic stresses. In: SP Singh, ed. Common Bean Improvement in the Twenty-First Century. Dordrecht: Kluwer Academic Publishers, 1999, pp 331–370.
96. JW White. Implications of carbon isotope discrimination studies for breeding common bean under water deficits. In: JR Ehleringer, AE Hall, GD Farquhar, eds. Stable Isotopes and Plant Carbon-Water Relations. San Diego: Academic Press, 1993, pp 387–398.
97. JS Robins, CE Domingo. Moisture deficits in relation to the growth and development of dry beans. Agron J 48:67–70, 1956.
98. JA Acosta-Gallegos, J Kohashi-Shibata. Effect of water stress on growth and yield of indeterminate dry-bean (*Phaseolus vulgaris*) cultivars. Field Crops Res 20:81–93, 1989.
99. G Bascur, MA Oliva, D Laing. Termometria infrarroja en selección de genotipos de frijol (*Phaseolus vulgaris*, L) resistentes a sequía: II. Crecimiento y productividad. Turrialba 35:49–53, 1985.
100. BN Sponchiado, JW White, JA Castillo, PG Jones. Root growth of four common bean cultivars in relation to drought tolerance in environments with contrasting soil types. Exp Agric 25:249–257, 1989.
101. JW White, JA Castillo. Relative effect of root and shoot genotypes on yield of common bean under drought stress. Crop Sci 29:360–362, 1989.
102. PM da Silveira, CM Guimaraes, LF Stone, J Kluthcouski. Aviliacao de cultivares de feijao para resistencia a seca baseada em dias de estresse de água no solo. Pesqui Agropecu Bras 16:693–699, 1981.
103. DE Miller, DW Burke. Response of dry beans to daily deficit sprinkler irrigation. Agron J 75:775–778, 1983.
104. J Jara-R. Respuesta a sequía de cinco variedades de frijol (*Phaseolus vulgaris* L.): estudio preliminar. Agrociencia (Chile) 6:95–101, 1990.
105. JA Acosta-Gallegos. Selection of common bean (*Phaseolus vulgaris*) genotypes with enhanced drought tolerance and biological nitrogen fixation. PhD dissertation (Diss Abstr 88-24816). Michigan State University, East Lansing, MI, 1988.
106. JR Ehleringer, S Klassen, C Clayton, D Sherrill, M Fuller-Holbrook, Q Fu, TA Cooper. Carbon isotope discrimination and transpiration efficiency in common bean. Crop Sci 31:1611–1615, 1991.
107. WY Kao, MP Comstock, JR Ehleringer. Variation in leaf movements among common bean cultivars. Crop Sci 34:1273–1278, 1994.

108. JW White, JA Castillo, JR Ehleringer, JA Garcia-C, SP Singh. Relations of carbon isotope descrimination and other physiological traits to yield in common bean (*Phaseolus vulgaris*) under rainfed conditions. J Agric Sci (Camb) 122:275–284, 1994.

109. JA Acosta-Gallegos, JW White. Phenological plasticity as an adaptation by common bean to rainfed environments. Crop Sci 35:199–204, 1995.

110. EF Foster, A Pajarito, JA Acosta-Gallego. Moisture stress impact on N partitioning, N remobilization and N-use efficiency in beans (*Phaseolus vulgaris*). J Agric Sci (Camb) 124:27–37, 1995.

111. KA Schneider, R Rosales-Serna, F Ibarra-Perez, B Cazares-Enriquez, JA Acosta-Gallego, P Ramirez-Vallejo, N Wassimi, JD Kelly. Improving common bean performance under drought stress. Crop Sci 37:43–50, 1997.

112. JW White. Preliminary results of the Bean International Drought Yield Trial (BIDYT). In: JW White, JWG Hoogenboom, F Ibarra, SP Singh, eds. Research on Drought Tolerance in Common Bean. Working Document No 41. Cali, Colombia: CIAT, 1988, pp 126–145.

113. Y Castonguay, AH Markhart III. Saturated rates of photosynthesis in water-stressed leaves of common bean and tepary bean. Crop Sci 31:1605–1611, 1991.

114. JW White, JA Castillo. Evaluation of diverse shoot genotypes on selected root genotypes of common bean under soil water deficits. Crop Sci 32:762–765, 1992.

115. RC Nageswara Rao, GC Wright. Stability of the relationship between specific leaf area and carbon isotope discrimination across environments in peanut. Crop Sci 34:98–103, 1995.

116. CM Guimarães. Efeitos fisiologicos do estresse hídrico. In: MJO Zimmermann, M Rocha, MR Yamada, eds. Cultura do feijoeiro. Piracicaba, Brazil: Associação Brasileira para Pesquisa de Potássio e do Fosfato, 1988, pp 157–174.

117. JW White, R Ochoa, MF Ibarra P, SP Singh. Inheritance of seed yield, maturity and seed weight of common bean (*Phaseolus vulgaris*) under semi-arid rainfed conditions. J Agric Sci (Camb) 122:265–273, 1994.

118. JA Acosta-Gallegos, MW Adams. Plant traits and yield stability of dry bean (*Phaseolus vulgaris*) cultivars under drought stress. J Agric Sci (Camb) 117:213–219, 1991.

119. SP Singh. Selection for water-stress tolerance in interracial populations of common bean. Crop Sci 35:118–124, 1995.

120. KA Schneider, ME Brothers, JD Kelly. Marker-assisted selection to improve drought resistance in common bean. Crop Sci 37:51–60, 1997.

121. CIAT. Bean Improvement for Sustainable Productivity, Input Use Efficiency, and Poverty Alleviation. Annual Report—Project IP-1. Cali, Colombia: CIAT, 1999, p 139.

122. T Fairhurst, R Lefroy, E Mutert, N Batjes. The importance, distribution and causes of phosporus deficiency as a constraint to crop production in the tropics. Agrofor Forum 9:2–8, 1999.

123. TT Cochrane, LG Sánchez, JA Porras, LG de Azevedo, CL Garver. Land in Tropical America. Cali, Colombia: CIAT, 1985.

124. PH LeMare. Sorption of isotopically exchangeable and non-exchangeable phosphate by some soils of Colombia and Brazil, and comparisons with soils of southern Nigeria. J Soil Sci 33:691–707, 1982.

125. IM Rao, DK Friesen, WJ Horst. Opportunities for germplasm selection to influence phosphorus acquisition from low-phosphorus soils. Agrofor Forum 9:13–17, 1999.

126. J Caradus. Mechanisms improving nutrient use by crop and herbage legumes. In: VC Baligar, RR Duncan, eds. Crops as Enhancers of Nutrient Use. San Diego: Academic Press, 1990, pp 253–311.

127. RB Clark, RR Duncan. Improvement of plant mineral nutrition through breeding. Field Crops Res 27:219–240, 1991.

128. RT Koide. Nutrient supply, nutrient demand and plant response to mycorrhizal infection. New Phytol 117:365–386, 1991.

129. T Otani, N Ae. Phosphorus (P) uptake mechanisms of crops grown in soils with low P status. I. Screening crops for efficient P uptake. Soil Sci Plant Nutr 42:155–163, 1996.

130. CIAT (Centro Internacional de Agricultura Tropical). Constraints to and Opportunities for Improving Bean Production. A Planning Document 1993–98 and an Achieving Document 1987–92. Cali, Colombia: CIAT, 1992, p 178.

131. CS Wortmann, RA Kirkby, CA Eledu, DJ Allen. Atlas of Common Bean (*Phaseolus vulgaris* L.) Production in Africa. Cali, Colombia: CIAT, 1998, p 133.

132. V Vadez, JH Lasso, DP Beck, JJ Drevon. Variability of $N_2$-fixation in common bean (*Phaseolus vulgaris* L.) under P deficiency is related to P use efficiency: $N_2$-fixation tolerance to P deficiency. Euphytica 106:231–242, 1999.

133. DW Israel. Investigation of the role of phosphorus in symbiotic dinitrogen fixation. Plant Physiol 84:835–840, 1987.

134. FA Bliss. Breeding common bean for improved biological nitrogen fixation. Plant Soil 152:71–79, 1993.

135. JA Kipe-Nolt, H Vargas, KE Giller. Nitrogen fixation in breeding lines of *Phaseolus vulgaris* L. Plant Soil 152:103–106, 1993.

136. MG Teixeira, JGM Guerra, DL de Almeida, AP Araújo, AA Franco. Effect of seed phosphorus concentration on nodulation and growth of three common bean cultivars. J Plant Nutr 22:1599–1611, 1999.

137. G Whiteaker, GC Gerloff, WH Gabelman, D Lindgren. Intraspecific differences in growth of beans at stress levels of phosphorus. J Am Soc Hortic Sci 101:472–475, 1976.

138. DT Lindgren, WH Gabelman, GC Gerloff. Variability of phosphorus uptake and translocation in *Phaseolus vulgaris* L. under phosphorus stress. J Am Soc Hortic Sci 102:674–677, 1977.

139. JG Salinas. Differential response of some cereal and bean cultivars to Al and P stress in an oxisol of central Brazil. PhD dissertation, North Carolina State University, Raleigh, 1978.

140. GC Gerloff, WH Gabelman. Genetic basis of inorganic plant nutrition. In: A Lauchli, RL Bieleski, eds. Inorganic Plant Nutrition. Encyclopedia of Plant Physiology, New Series, Vol 15B. New York: Springer-Verlag, 1983, pp 453–480.

141. IP Oliveira, MDT Thung, J Kluthcouski, H Aidar, JRP Carvalho. Avaliação de cultivares de feijão quanto à eficiência no uso de fósforo. Pesquisa Agropecuária Brasileira 22:39–45, 1987.

142. M Thung, J Ortega, O Erazo. Breeding methodology for phosphorus efficiency and tolerance to aluminum and manganese toxicities for beans (*Phaseolus vulgaris* L.). In: JG Salinas, LM Gourley, eds. Sorghum for Acid Soils. Proceeding of a Workshop on Evaluating Sorghum for Tolerance to Al-Toxic Tropical Soils in Latin America. Cali, Colombia: INTSORMIL/ICRISAT/CIAT, 1987, pp 207–222.

143. CA Flor, MT Thung. Nutritional disorders. In: HF Schwartz, MA Pastor Corrales, eds. Bean Production Problems in the Tropics. Cali, Colombia: CIAT, 1989, pp 653–691.

144. PAA Pereira, FA Bliss. Selection of common bean (*Phaseolus vulgaris* L.) for $N_2$ fixation at different levels of available phosphorus under field and environmentally-controlled conditions. Plant Soil 115:75–82, 1989.

145. SP Singh, CA Urea, JA Gutierrez, J Garcia. Selection for yield at two fertility levels in small seeded common bean. Can J Plant Sci 69:1011–1017, 1989.

146. M Thung. Phosphorus: a limiting nutrient in bean (*Phaseolus vulgaris* L.) production in Latin America and field screening for efficiency and response. In: N El-Basam, M Dambroth, BC Loughman, eds. Genetic Aspect of Plant Mineral Nutrition. Dordrecht: Kluwer Academic Publishers, 1990, pp 501–521.

147. L Youngdahl. Differences in phosphorus efficiency in bean genotypes. J Plant Nutr 13:1381–1392, 1990.

148. JP Lynch, SE Beebe. Adaptation of beans (*Phaseolus vulgaris* L.) to low phosphorus availability. HortSci 30:1165–1171, 1995.

149. CS Wortmann, L Lunze, VA Ochwoh, J Lynch. Bean improvement for low fertility soils in Africa. Afr Crop Sci J 3:469–477, 1995.

150. X Yan, JP Lynch, S Beebe. Genetic variation for phosphorus efficiency of common bean in contrasting soil types: I. Vegetative response. Crop Sci 35:1086–1093, 1995.

151. X Yan, JP Lynch, S Beebe. Genetic variation for phosphorus efficiency of common bean in contrasting soil types: I. Yield response. Crop Sci 35:1094–1099, 1995.

152. VD Aggarwal, SK Mughogho, RM Chirwa, S Snapp. Field-based screening methodology to improve tolerance of common bean to low-P soils. Commun Soil Sci Plant Anal 28:1623–1632, 1997.

153. AP Araujo, MG Texeira, DL De Almeida. Phosphorus efficiency of wild and cultivated genotypes of common bean (*Phaseolus vulgaris* L.) under biological nitrogen fixation. Soil Biol Biochem 29:951–957, 1997.

154. S Beebe, J Lynch, N Galwey, J Tohme, I Ochoa. A geographical approach to identify phosphorus-efficient genotypes among landraces and wild ancestors of common bean. Euphytica 95:325–336, 1997.

155. J Ribet, FM Arboleda, D Beck. Mejoramiento de frijol común para adaptación y fijación simbiótica de nitrógeno en suelos de baja P: una revisión. In: SP Singh, O Voysest, eds. Taller de mejoramiento de frijol para el siglo XXI. Cali, Colombia: CIAT, 1997, pp 23–56.

156. AP Araujo, MG Texeira, DL De Almeida. Variability of traits associated with phosphorus efficiency in wild and cultivated genotypes of common bean. Plant Soil 203:173–182, 1998.

157. IM Rao. Role of phosphorus in photosynthesis. In: M Pessarakli, ed. Handbook of Photosynthesis. New York: Marcel Dekker, 1996, pp 173–194.

158. C Cakmak, C Engels. Role of mineral nutrients in photosynthesis and yield formation. In: Z Rengl, ed. Mineral Nutrition of Crops: Fundamental Mechanisms and Implications. New York: Food Products Press, 1999, pp 141–168.

159. IM Rao, N Terry. Photosynthetic adaptation to nutrient stress. In: M Yunus, U Pathre, P Mohanty, eds. Probing Photosynthesis: Mechanism, Regulation and Adaptation. London, UK: Taylor & Francis, 2000, pp 378–396.

160. C Cakmak, C Hengeler, H Marschner. Partitioning of shoot and root dry matter and carbohydrates in bean plants suffering from phosphorus, potassium and magnesium deficiency. J Exp Bot 45:1245–1250, 1994.

161. A Kondracka, A Rychter. The role of Pi recycling process during photosynthesis in phosphate-deficient bean plants. J Exp Bot 48:1461–1468, 1997.

162. M Mikulska, J-L Bomsel, AM Rychter. The influence of phosphate deficiency on photosynthesis, respiration and adenine nucleotide pool in bean leaves. Photosynthetica 35:79–88, 1998.

163. AM Rychter, DD Randall. The effect of phosphate deficiency on carbohydrate metabolism in bean roots. Physiol Plant 91:383–388, 1994.

164. IM Juszczuk, AM Rychter. Changes in pyridine nucleotide levels in leaves and roots of bean plants (*Phaseolus vulgaris* L.) during phosphate deficiency. J Plant Physiol 151:399–404, 1997.

165. AM Rychter, M Mikulska. The relationship between phosphate status and cyanide-resistant respiration in bean roots. Physiol Plant 79:663–667, 1990.

166. AM Rychter, M Cheauveau, J-L Bomsel, C Lance. The effect of phosphate deficiency on mitochondrial activity and adenylate levels in bean roots. Physiol Plant 84:80–86, 1992.

167. A Gniazdowska, A Krawczak, M Mikulska, AM Rychter. Low phosphate nutrition alters bean plants' ability to assimilate and translocate nitrate. J Plant Nutr 22:551–563, 1999.

168. WH Gabelman, GC Gerloff. The search for an interpretation of genetic controls that enhance plant growth under deficiency levels of a macronutrient. Plant Soil 72:335–350, 1983.

169. S Snapp, JP Lynch. Phosphorus distribution and remobilization in bean plants as influenced by phosphorus nutrition. Crop Sci 36:929–935, 1996.

170. JP Lynch, J van Beem. Growth and architecture of seedling roots of common bean genotypes. Crop Sci 33:1253–1257, 1993.

171. S Snapp, R Koide, J Lynch. Exploitation of localized phosphorus-patches by common bean roots. Plant Soil 177:211–218, 1995.

172. KL Nielsen, JP Lynch, AG Jablokow, PS Curtis. Carbon cost of root systems: an architectural approach. Plant Soil 165:161–169, 1994.

173. J Lynch. Root architecture and plant productivity. Plant Physiol 109:7–13, 1995.

174. JP Lynch, KM Brown. Regulation of root architecture by phosphorus availability. In: JP Lynch, J Deikman, eds. Phosphorus in Plant Biology: Regulatory Roles in Molecular, Cellular, Organismic, and Ecosystem Processes. Rockville, MD: American Society of Plant Physiologists, 1999, pp 148–156.

175. JP Lynch, KL Nielsen, RD Davis, AG Jablokow. SimRoot: modelling and visualization of botanical root systems. Plant Soil 188:139–151, 1997.

176. AM Bonser, J Lynch, S Snapp. Effect of phosphorus deficiency on growth angle of basal roots in *Phaseolus vulgaris*. New Phytol 132:281–288, 1996.

177. H Liao, G Rubio, X Yan, JP Lynch. Gravitropic response of bean root system to phosphorus deficiency. In: JP Lynch, J Deikman, eds. Phosphorus in Plant Biology: Regulatory Roles in Molecular, Cellular, Organismic, and Ecosystem Processes. Rockville, MD: American Society of Plant Physiologists, 1999, pp 329–331.

178. I Fawole, WH Gabelman, GC Gerloff, EV Nordheim. Heritability of efficiency of phosphorus utilization in beans (*Phaseolus vulgaris* L.) grown under phosphorus stress. J Am Soc Hortic Sci 107:94–97, 1982.

179. I Fawole, WH Gabelman, GC Gerloff. Genetic control of root development in beans (*Phaseolus vulgaris* L.) grown under phosphorus stress. J Am Soc Hortic Sci 107:98–100, 1982.

180. C Urrea, SP Singh. Heritability of yield of common bean in soil with low phosphorus content (abstract). In: S Beebe, ed. Current Topics in Breeding of Common Bean. Cali, Colombia: CIAT, 1989, p 392.

181. TM Schettini, WH Gabelman, GC Gerloff. Incorporation of phosphorus efficiency from exotic germplasm into agriculturally adapted germplasm of common bean (*Phaseolus vulgaris* L.). Plant Soil 99:175–184, 1987.

182. A Saborío, S Beebe. Mejoramiento para tolerancia a bajo fosforo en Costa Rica. In: RA Villalobos, D Beck, eds. Mejoramiento y manejo agronómico del frijol común (*Phaseolus vulgaris*) para adaptación en suelos de bajo fósforo. San Jose, Costa Rica: University of Costa Rica, 1995, pp 9–18.

183. H Posada, J Kornegay, D Beck. Estudio de herencia y mecanismos de tolerancia en lineas de frijol (*Phaseolus vulgaris* L.) andinas y mesoamericanas a bajo fosforo en el suelo. In: RA Villalobos, D Beck, eds. Mejoramiento y manejo agronómico del frijol común (*Phaseolus vulgaris*) para adaptación en suelos de bajo fósforo. San Jose, Costa Rica: University of Costa Rica, 1995, pp 41–57.

184. CIAT. Bean Project Annual Report 1998. Working Document No 179. Cali, Colombia: CIAT, 1999, p 202.

185. JJ Ni, P Wu, D Senadhira, N Huang. Mapping QTLs for phosphorus deficiency tolerance in rice (*Oryza sativa* L.). Theor Appl Genet 97:1361–1369, 1998.

186. SP Singh. Bean genetics: nonbiotic stresses. In: A van Schoonhoven, O Voysest, eds. Common Beans: Research for Crop Improvement. Wallingford, UK: CAB International, and Cali, Colombia: CIAT, 1991, pp 232–233.

187. J Lynch, JW White. Shoot nitrogen dynamics in tropical common bean. Crop Sci 32:392–397, 1992.

188. J Lynch, NS Rodriguez H. Photosynthetic nitrogen-use efficiency in relation to leaf longevity in common bean. Crop Sci 34:1284–1290, 1994.

189. PH Graham. Some problems of nodulation and symbiotic nitrogen fixation in *Phaseolus vulgaris* L. Field Crops Res 4:93–112, 1981.

190. RJ Rennie, GA Kemp. $N_2$ fixation in field beans quantified by "N isotope dilution." II. Effect of cultivars of beans. Agron J 75:645–649, 1983.

191. JA Kipe-Nolt, H Vargas. Components of nitrogen fixation in *Phaseolus vulgaris* L (abstract). In: S Beebe, ed. Current Topics in Breeding of Common Bean. Cali, Colombia: CIAT, 1989, p 403.

192. FA Bliss, G Hardarson, eds. Enhancement of Biological Nitrogen Fixation of Common Bean in Latin America. Dordrecht: Kluwer Academic Publishers, 1993, p 160.

193. RA Henson, FA Bliss. Effect of fertilizer N application timing on common bean production. Fertil Res 29:133–138, 1991.

194. PF Shea, WH Gabelman, GC Gerloff. The inheritance of efficiency in potassium utilization in snap beans (*Phaseolus vulgaris* L.). Proc Am Soc Hortic Sci 91:286–293, 1967.

195. PF Shea, GC Gerloff, WH Gabelman. Differing efficiencies of potassium utilization in strains of snap beans, *Phaseolus vulgaris* L. Plant Soil 28:337–346, 1968.

196. M do Cl Braccini, A de Lucca e Braccini, HEP Martinez, PRG Pereira, PCR Fontes. Tecnicas de avaliação da toxidez do alumínio em plântulas de feijoeiro (*Phaseolus vulgaris* L.) cultivadas em solução nutritiva. Rev Ceres 42:3–16, 1996.

197. N Massot, M Llugany, C Poschenreider, J Barceló. Callose production as indicator of aluminum toxicity in bean cultivars. J Plant Nutr 22:1–10, 1999.

198. A González, J Lynch. Tolerance of tropical common bean genotypes to manganese toxicity: performance under different growing conditions. J Plant Nutr 22:511–525, 1999.

199. A González, J Lynch. Subcellular and tissue Mn compartmentation in bean leaves under Mn toxicity stress. Aust J Plant Physiol 26:811–822, 1999.

200. M Pessarakli. Response of green beans (*Phaseolus vulgaris* L.) to salt stress. In: M Pessarakli, ed. Handbook of Plant and Crop Stress. New York: Marcel Dekker, 1999, pp 827–842.

201. HZ Zaiter, B Mahfouz. Salinity effect on root and shoot characteristics of common and tepary beans evaluated under hydroponic solution and sand culture. Annu Rep Bean Improv Coop 36:80, 1993.

202. FAO. Food Balance Sheets. Rome, Italy: FAO, 1999.

203. JM Toledo, GA Nores. Tropical pasture technology for marginal lands of tropical America. Outlook Agric 15:2–9, 1986.

204. LR Humphreys. Tropical pasture and fodder crops. 2nd ed. New York: Longman Scientific and Technical, 1987, p 155.

205. IM Rao, RS Zeigler, R Vera, S Sarkarung. Selection and breeding for acid-soil tolerance in crops: upland rice and tropical forages as case studies. BioSci 43:454–465, 1993.

206. AH Zimmer, HCM Macedo, AN Kichel, VPB Euclides. Sistemas integrados de producción agropastoril. In: EP Guimaraes, JI Sanz, IM Rao, MC Amézquita, E Amézquita, eds. Sistemas agropastoriles en sabanas tropicales de America Latina. Cali, Colombia: CIAT and Brasilia, Brazil: EMBRAPA, 1999, pp 245–283.

207. CE Lascano. Managing the grazing resource for animal production in savannas of tropical America. Trop Grassl 25:66–72, 1991.

208. IM Rao, MA Ayarza, RJ Thomas. The use of carbon isotope ratios to evaluate legume contribution to soil enhancement in tropical pastures. Plant Soil 162:177–182, 1994.

209. RJ Thomas, MJ Fisher, MA Ayarza, JI Sanz. The role of forage grasses and legumes in maintaining the productivity of acid soils in Latin America. In: R Lal, JB Stewart, eds. Soil Management: Experimental Basis for Sustainability and Environmental Quality. Adv Soil Sci Series. Boca Raton, FL: Lewis Publishers, 1995, pp 61–83.

210. MJ Fisher, IM Rao, RJ Thomas, CE Lascano. Grasslands in the well-watered tropical lowlands. In: J Hodgson, AW Illius, eds. The Ecology and Management of Grazing Systems. Wallingford, UK: CAB International, 1996, pp 393–425.

211. MJ Fisher, IM Rao, MA Ayarza, CE Lascano, JI Sanz, RJ Thomas, RR Vera. Carbon storage by introduced deep-rooted grasses in the South American savannas. Nature 371:236–238, 1994.

212. MJ Fisher, RJ Thomas, IM Rao. Management of tropical pastures in acid-soil savannas of south America for carbon sequestration. In: R Lal, JM Kimble, RF Follett, BA Stewart, eds. Management of Carbon Sequestration in Soil. Boca Raton, FL: CRC Press, 1998, pp 405–420.

213. IM Rao. Root distribution and production in native and introduced pastures in the South American savannas. In: JE Box Jr, ed. Root Demographics and Their Efficiencies in Sustainable Agriculture, Grasslands, and Forest Ecosystems. Dordrecht: Kluwer Academic Publishers, 1998, pp 19–42.

214. JI Sanz, RS Zeigler, S Sarkarung, DL Molina, M Rivera. Sistemas mejoradas arrozpasturas para sabana nativa y pasturas degradadas en suelos acidos de América del Sur. In: EP Guimaraes, JI Sanz, IM Rao, MC Amézquita, E Amézquita, eds. Sistemas agropastoriles en sabanas tropicales de America Latina. Cali, Colombia: CIAT and Brasilia, Brazil: EMBRAPA, 1999, pp 232–244.

215. RS Zeigler, S Pandey, J Miles, LM Gourley, S Sarkarung. Advances in the selection and breeding of acid-tolerant plants: rice, maize, sorghum and tropical forages. In: RA Date, NJ Grundon, GE Rayment, ME Probert, eds. Plant-Soil Interactions at Low pH: Principles and Management. Dordrecht: Kluwer Academic Publishers, 1995, pp 391–406.

216. JW Miles, SL Lapointe. Regional germplasm evaluation: a portfolio of germplasm options for the major ecosystems of tropical America. In: W Hardy, ed. Pastures for the Tropical Lowlands. CIAT's Contribution. Cali, Colombia: CIAT, 1992, pp 9–28.

217. R Vera, C Seré. Onfarm results with *Andropogon gayanus*. In: JM Toledo, R Vera, C Lascano, JM Lenné, eds. *Andropogon gayanus* Kunth: A Grass for Tropical Acid Soils. Cali, Colombia: CIAT, 1990, pp 303–333.

218. CA Jones, D Peña, A Carabaly. Effects of plant water potential, leaf diffusive resistance, rooting density and water use on the dry matter production of several tropical grasses during short periods of drought stress. Trop Agric 57:211–219, 1980.

219. JM Toledo, MJ Fisher. Physiological aspects of *Andropogon gayanus* and its compatibility with legumes. In: JM Toledo, R Vera, C Lascano, JM Lenné, eds. *Andropogon gayanus* Kunth: A Grass for Tropical Acid Soils. Cali, Colombia: CIAT, 1990, pp 65–98.

220.  Z Baruch, MJ Fisher. Factores climáticos y de competencia que afectan el desarrollo de la planta en el establecimiento de una pastura. In: CE Lascano, JM Spain, eds. Establecimiento y renovación de pasturas. Cali, Colombia: CIAT, 1991, pp 103–142.

221.  MJ Fisher, P Cruz. Some ecophysiological aspects of *Arachis pintoi*. In: PC Kerridge, W Hardy, eds. The Biology and Agronomy of Forage *Arachis*. Cali, Colombia: CIAT, 1994, pp 53–70.

222.  MJ Fisher, PC Kerridge. The agronomy and physiology of *Brachiaria* species. In: JW Miles, BL Maass, CB do Valle, eds. *Brachiaria*: Biology, Agronomy and Improvement. Cali, Colombia: CIAT and Brasilia, Brazil: EMBRAPA, 1996, pp 43–52.

223.  MJ Fisher, PK Thornton. Growth and competition as factors in the persistence of legumes in pastures. In: GC Marten, AG Matches, RF Barnes, RW Brougham, RJ Clements, GW Sheath, eds. Persistence of Forage Legumes. Madison, WI: American Society of Agronomy, 1989, pp 293–309.

224.  CE Lascano. Nutritive value and animal production of forage *Arachis*. In: PC Kerridge, W Hardy, eds. The Biology and Agronomy of Forage *Arachis*. Cali, Colombia: CIAT, 1994, pp 109–121.

225.  JG Salinas, SR Saif. Nutritional requirements of *Andropogon gayanus*. In: JM Toledo, R Vera, C Lascano, JM Lenné, eds. *Andropogon gayanus* Kunth: A Grass for Tropical Acid Soils. Cali, Colombia: CIAT, 1990, pp 99–155.

226.  JG Salinas, PC Kerridge, RM Schunke. Mineral nutrition of *Centrosema*. In: R Schultze-Kraft, RJ Clements, eds. *Centrosema*: Biology, Agronomy and Utilization. Cali, Colombia: CIAT, 1990, pp 119–149.

227.  IM Rao, PC Kerridge. Mineral nutrition of forage *Arachis*. In: PC Kerridge, W Hardy, eds. The Biology and Agronomy of Forage *Arachis*. Cali, Colombia: CIAT, 1994, pp 71–83.

228.  IM Rao, PC Kerridge, M Macedo. Nutritional requirements of *Brachiaria* and adaptation to acid soils. In: JW Miles, BL Maass, CB do Valle, eds. *Brachiaria*: Biology, Agronomy and Improvement. Cali, Colombia: CIAT and Brasilia, Brazil: EMBRAPA, 1996, pp 53–71.

229.  T Kanno, MC Macedo, JA Bono. Growth responses of *Brachiaria decumbens* cv. Basilisk and *Brachiaria brizantha* cv. Marandu to phosphorus supply. Grassland Sci 45:1–8, 1999.

230.  IM Rao, WM Roca, MA Ayarza, E Tabares, R García. Somaclonal variation in plant adaptation to acid soil in the tropical forage legume, *Stylosanthes guianensis*. Plant Soil 146:21–30, 1992.

231.  IM Rao, MA Ayarza, R Garcia. Adaptive attributes of tropical forage species to acid soils I. Differences in plant growth, nutrient acquisition and nutrient utilization among $C_4$ grasses and $C_3$ legumes. J Plant Nutr 18:2135–2155, 1995.

232.  IM Rao, V Borrero, J Ricaurte, R Garcia, MA Ayarza. Adaptive attributes of tropical forage species to acid soils. II. Differences in shoot and root growth responses to varying phosphorus supply and soil type. J Plant Nutr 19:323–352, 1996.

233.  M Li, M Osaki, IM Rao, T Tadano. Secretion of phytase from the roots of several plant species under phosphorus-deficient conditions. Plant Soil 195:161–169, 1997.

234.  IM Rao, V Borrero, J Ricaurte, R Garcia, MA Ayarza. Adaptive attributes of tropical forage species to acid soils. III. Differences in phosphorus acquisition and utilization as influenced by varying phosphorus supply and soil type. J Plant Nutr 20:155–180, 1997.

235.  IM Rao, JW Miles, JC Granobles. Differences in tolerance to infertile acid soil stress among germplasm accessions and genetic recombinants of the tropical forage grass genus, *Brachiaria*. Field Crops Res 59:43–52, 1998.

236.  IM Rao, V Borrero, J Ricaurte, R Garcia. Adaptive attributes of tropical forage species to acid soils. IV. Differences in shoot and root growth responses to inorganic and organic phosphorus sources. J Plant Nutr 22:1153–1174, 1999.

237.  IM Rao, V Borrero, J Ricaurte, R Garcia. Adaptive attributes of tropical forage species to acid soils. V. Differences in phosphorus acquisition from inorganic and organic phosphorus sources. J Plant Nutr 22:1175–1196, 1999.

238.  WE Fenster, LA León. P fertilizer management for establishment and maintenance of improved pastures in acid and infertile soils of tropical America. In: PA Sánchez, LE Tergas, eds. Pasture Production in Acid Soils of the Tropics. Cali, Colombia: CIAT, 1979, pp 119–134.

239.  PA Sánchez, JG Salinas. Low input technology for managing Oxisols and Ultisols in tropical America. Adv Agron 34:280–406, 1981.

240.  MT Treeby, RFM van Steveninck, HM Devries. Quantitative estimates of phosphorus concentrations within *Lupinus luteus* leaflets by means of electron probe x-ray microanalysis. Plant Physiol 85:331–334, 1987.

241.  SA Barber. Soil Nutrient Bioavailability: A Mechanistic Approach. New York: John Wiley & Sons, 1984.

242.  A Jungk, N. Claassen. Ion diffusion in the soil-root system. Adv Agron 61:53–110, 1997.

243.  N Claassen, SA Barber. Simulation model for nutrient uptake from soil by a growing plant root system. Agron J 68:961–964, 1976.

244.  DF Coates, PC Kerridge, CP Miller, WH Winter. Phosphorus and beef production in northern Australia. 7. The effect of phosphorus on the composition, yield and quality of legume-based pasture and their relation to animal production. Trop Grassl 24:209–220, 1990.

245.  T Otani, N Ae. Sensitivity of phosphorus uptake to changes in root length and soil volume. Agron J 88:371–375, 1996.

246.  VT Paulino, DP Anton, MT Colozza. Problemas nutricionais do genero *Brachiaria* e algumas relacoes com o comportamento animal. Zootecnia 25:215–263, 1987.

247. E Malavolta, VT Paulino. Nutriciao mineral e adubacao do genero *Brachiaria*. In: VT Paulino, JVS Pedreira, DFV Camargo, NMF Meirelles, D Bianchini, PRP Oliveira, eds. 2° Encontro para Discussao sobre Capins de Genero *Brachiaria*. Nova Odessa, SP, Brazil: EMBRAPA, 1991, pp 45–135.

248. LR McDowell, J Velásquez-Pereira, G Valle. Minerales para ruminantes en pastoreo en regiones tropicales 1997. Tercera edición. Gainesville, FL: University of Florida, 1997, p 84.

249. D Thomas, B Grof. Some pasture species for the tropical savannas of South America. I. Species of *Stylosanthes*. Herb Abstr 56:445–454, 1986.

250. D Thomas, B Grof. Some pasture species for the tropical savannas of South America. II. Species of *Centrosema*, *Desmodium*, and *Zornia*. Herb Abstr 56:511–525, 1986.

251. D Thomas, B Grof. Some pasture species for the tropical savannas of South America. III. *Andropogon gayanus*, *Brachiaria* spp. and *Panicum maximum*. Herb Abstr 56:557–565, 1986.

252. CB do Valle, Avaliacao de germoplasma e melhoramento genetico de Braquiarias. In: VT Paulino, JVS Pedreira, DFV Camargo, NMF Meirelles, D Bianchini, PRP Oliveira, eds. 2° Encontro para Discussao sobre Capins de Genero *Brachiaria*. Nova Odessa, SP, Brazil: EMBRAPA, 1991, pp 301–342.

253. CIAT. Tropical Pastures Program Annual Report 1979. Cali, Colombia: CIAT, 1980, p 130.

254. AM Maluf, PS Martins, WR Maluf. Avaliacao de populacoes de *Leucaena* para tolerancia ao aluminio. Pesqui Agropecu Bras 19:859–866, 1984.

255. AM Maluf, PS Martins, WR Maluf. Avaliacao de populacoes de *Leucaena* para tolerancia ao aluminio. III. Criterios para avaliacao de tolerancia. Pesqui Agropecu Bras 19:1131–1134, 1984.

256. AM Maluf, PS Martins, WR Maluf. Persistencia de cotiledones na plantula como parametro para avaliacao de tolerancia ao aluminio em *Leucaena leucocephala*. Pesqui Agropecu Bras 20:355–360, 1985.

257. EM Hutton. Breeding and selecting *Leucaena* for acid tropical soils. Pesqui Agropecu Bras 19:263–274, 1984.

258. EM Hutton. *Centrosema* breeding for acid tropical soils, with emphasis on efficient Ca absorption. Trop Agric (Trinidad) 62:273–280, 1985.

259. I Gobbe, A Swenne, BP Louant. Diploides naturales et autotetraploides induits chez *Brachiaria ruziziensis* Germain ct Evrard criteres d'identification. Agron Trop 36:339–346, 1981.

260. CIAT. Tropical Grasses and Legumes: Optimizing Genetic Diversity for Multipurpose Use. IP-5 Project Annual Report. Cali, Colombia: CIAT, 1999, p 175.

261. LV Kochian. Cellular mechanisms of aluminum toxicity and resistance in plants. Annu Rev Plant Physiol Plant Mol Biol 46:237–260, 1995.

262. DK Friesen, IM Rao, RJ Thomas, A Oberson, JI Sanz. Phosphorus acquisition and cycling in crop and pasture systems in low fertility tropical soils. Plant Soil 196:289–294, 1997.

263. T Decaens, AF Rangel, N Asakawa, RJ Thomas. Carbon and nitrogen dynamics in ageing earthworm casts in grasslands of the eastern plains of Colombia. Biol Fertil Soils 30:20–28, 1999.

264. RM Boddey, IM Rao, RJ Thomas. Nutrient cycling and environmental impact of *Brachiaria* pastures. In: JW Miles, BL Maass, CB do Valle, eds. *Brachiaria*: Biology, Agronomy and Improvement. Cali, Colombia: CIAT and Brasilia, Brazil: EMBRAPA, 1996, pp 72–86.

265. MJ Fisher, IM Rao, RJ Thomas. Nutrient cycling in tropical pastures, with special reference to neotropical savannas. In: Proceedings of the XVIII International Grassland Congress (Vol 3) held during June 8–19 in Winnipeg and Saskatoon, Canada, 1999, pp 371–382.

266. T Kanno, MC Macedo, VPB Euclides, JA Bono, JDG Santos Jr, MC Rocha, GR Beretta. Root biomass of five tropical grass pastures grazing in Brazilian savannas. Grass Sci 45:9–14, 1999.

267. AJ Gijsman. Soil aggregate stability and soil organic matter fractions under agropastoral systems established in native savanna. Aust J Soil Res 34:891–907, 1996.

268. AJ Gijsman, RJ Thomas. Evaluation of some physical properties of an oxisol after conversion of native savanna into legume-based or pure grass pastures. Trop Grassl 30:237–248, 1996.

269. LT Szott, CA Palm, PA Sanchez. Agroforestry in acid soils of the humid tropics. Adv Agron 45:275–301, 1991.

270. LT Evans. Greater crop production: whence and whither? In: JC Waterlow, DG Armstrong, L Fowden, R Riley, eds. Feeding a World Population of More Than Eight Billion People: A Challenge to Science. New York: Oxford University Press, 1998, pp 89–97.

271. JL Prioul, S Quarrie, M Causse, D de Vienne. Dissecting complex physiological functions through the use of molecular quantitative genetics. J Exp Bot 48:1151–1163, 1997.

272. I Wincov. New molecular approaches to improving salt tolerance in crop plants. Ann Bot 82:703–710, 1998.

273. BJ Miflin. Crop biotechnology. Where now? Plant Physiol 123:17–28, 2000.

274. JL Prioul, S Pelleschi, M Séne, C Thévenot, M Causse, D de Vienne, A Leonardi. From QTLs for enzyme activity to candidate genes in maize. J Exp Bot 50:1281–1288, 1999.

275. TJ Flowers, ML Koyama, SA Flowers, C Sudhakar, KP Singh, AR Yeo. QTL: their place in engineering tolerance of rice to salinity. J Exp Bot 51:99–106, 2000.

276. MA El-Sharkawy. Drought-tolerant cassava for Africa, Asia, and Latin America. Breeding projects work to stabilize productivity without increasing pressures on limited natural resources. Bioscience 43:441–451, 1993.

# 30
# Adaptive Components of Salt Tolerance

**James W. O'Leary**

*The University of Arizona, Tucson, Arizona*

## I. INTRODUCTION

Plants have evolved two very different strategies in adapting to high levels of sodium salts in their environments. One strategy is to exclude the salts from the interior of the leaf cells, and the other includes the salts within the leaf cells but sequesters most of them in the vacuoles of those cells. In both cases, the end result is to maintain the cytoplasmic sodium concentration relatively low. This is accomplished in the former case by either preventing entry of the ions into the plant at the root surface or preventing them from being transported in the xylem from the roots to the leaves. In the latter case, entry and transport to the leaves are not prevented or severely restricted, and the problem is handled primarily at the tonoplast level of the leaf cells themselves. The latter strategy seems to have been more effective when adapting to the most extreme saline habitats, but the former seems to have been manipulated more successfully during directed selection by plant breeders. Even though the exclusion process is not perfect in most plants, and some might argue that there is no sharp line separating the two categories, for simplicity's sake in the following discussion, these two broad categories of plants will be referred to as *excluders* and *includers,* respectively.

Although both strategies are effective, there are some important differences between the two types of plants. Those that exclude salts from the leaf cells are able to tolerate high levels of those salts in the root environment but at the expense of reduced growth. That is, as cultivars or ecotypes within a species are developed with increasing ability to exclude the sodium salts and thereby survive at increasingly higher concentrations of those salts in their environment, growth is reduced to well below what it is in the absence of those salts, even within the range of the relatively low salt concentrations characteristic of irrigated agriculture. On the other hand, the plants that not only allow the salts to reach the leaves but contain them at relatively high concentrations typically show increased growth with increasing level of external salinity within this range of salinities. It seems to me that this difference provides a unique opportunity to address the question of what constitutes salt tolerance, a philosophy developed further later in this discussion.

The excluders include virtually all crop plants and most, if not all, monocotyledonous halophytes, plus many dicotyledonous halophytes. The includers are limited to a relatively small number of dicotyledonous halophytes. It is somewhat surprising that the domesticated plants whose salt tolerance has been increased most successfully by selection and breeding are the monocotyledonous species, all of which are excluders to some degree, but the plants that have been selected by nature to tolerate the most saline habitats are the includers, not the excluders. Why has increased ability to tolerate high salt concentration in leaf cells not been a target for plant breeders? The answer to that question and my feelings about the prospects of using that approach successfully to increase salt tolerance in crop plants will also be addressed in the following discussion.

The typical approach to studying salt tolerance is to compare plants (both sensitive and tolerant plants) subjected to excess salinity with plants not subjected to salinity, looking for responses to the added salt. Examples of such responses are production of unique proteins or large amounts of presumed compatible osmotic solutes, such as proline and glycinebetaine. The difficulty with such an approach is that it is difficult to distinguish the responses that are truly adaptive from those that are reflections of metabolic lesions.

For example, even though there has been a substantial amount of research over the years devoted to comparison of plant responses to growth-inhibiting salinity and nonsaline conditions, and the production and accumulation of putative compatible osmotic solutes such as proline and glycinebetaine have been investigated almost exhaustively in numerous plant species [1–4], the role of those solutes in salt tolerance has not yet been clearly demonstrated. The pathways, and control points therein, of synthesis and degradation of such solutes have been studied in great detail, yet there is now growing concern whether production of those solutes for osmotic adjustment is of any adaptive or other beneficial value [5,6]. That is, they may be produced in large quantities as a result of disruptions in metabolism (metabolic lesions) in response to stress, or they may simply accumulate as a result of a lower utilization of photosynthate in stressed plants.

It seems appropriate to suggest that it is time to take a fresh look at the salt tolerance question and consider some new approaches. For example, rather than continuing to focus on the plants that are not especially salt tolerant and trying to decide how to make them more tolerant, it may be more productive to devote more effort to trying to find out what makes the highly salt tolerant plants so tolerant. The plants to which I refer are halophytes, and one advantage they provide is the opportunity to compare growth at suboptimal salinity with optimal salinity, an approach that is not possible with present crop plants or other glycophytes. The optimum salinity for growth in crop plants and other glycophytes is zero, with decreased growth as salinity increases beyond a few mol/m$^3$. In contrast, in many (but not all) halophytes, the optimum salinity for growth has shifted to 50 to 200 mol/m$^3$, with decreased growth occurring at both higher and lower salinities. Thus, if one compares the responses of such plants to less than optimum salinity with responses in plants grown at optimum salinity, it might be possible to distinguish the responses that are truly adaptive from those that are the result of lesions or other types of damage. The hypothesis is that adaptation in these halophytes involves some processes having optimum performance at a salinity level well above zero, while at lower salinities these processes do not function as well. A process that does not function as well in a plant growing at 50 mol/m$^3$ as it does in a plant growing at 200 mol/m$^3$ certainly is not being altered by excess salinity and a priori would seem to be involved in the better growth of that plant at the higher salinity. The challenge, then, is to identify those processes.

Unfortunately, most research so far has focused on the effects of excess salinity rather than inadequate salinity, but there are a few examples in which those studies did include the suboptimal salinity levels as well. A quick survey of some of those studies may give us a hint about where the attention should be concentrated in studies involving comparison of suboptimal and optimal salinity levels.

## II. PLANT RESPONSES TO SALINITY

### A. Effect of Salinity Level on Growth

That some halophytes grow better at an appreciable salinity level than they do in fresh water was acknowledged by Chapman [7] and Waisel [8] in their comprehensive reviews of halophytes, although they implied that this represented a minority of halophytes. Barbour's [9] thorough review of the older literature specifically relating growth to salinity, however, revealed that such observations were fairly numerous, and the classic review by Flowers et al. [10] listed several species that showed greater growth at salinity levels equivalent to 50 to 200 mM NaCl than in nonsaline conditions. In fact, the general feeling now that this is such a common response among halophytes is reflected in a subsequent review by Flowers et al. [11], in which they emphasized a few species that do not show such a response rather than listing those that do.

In general, monocotyledonous halophytes do not have growth optima at substantial salinity levels (i.e., they show a steady decline in growth with any increase in salinity). There are a few reports in the literature (e.g., Refs 12 and 13) indicating that some monocots do have greater growth at salinity levels greater than 50 mM NaCl, but the overwhelming body of experimental evidence supports the generalization that monocotyledonous halophytes do not require substantial salinity levels for optimum growth [11,14].

Thus this review and discussion are concerned primarily with dicotyledonous halophytes. There are some problems with interpreting results of previous studies, however. In most cases, the intent was not to determine optimum salinity levels for growth, so the intervals between imposed salinity treatments were often large. Some halophytes have been reported to respond to extremely small amounts of Na, with growth increases two- or threefold in response to 1 mM NaCl [15,16]. So if the lowest treatment level is 50 or 100 mM NaCl, for example, the response to this relatively high level cannot be distinguished from the response due to satisfying the "need" for the trace amount of Na. However, unless specific measures are taken to exclude Na, it is usually present in most nutrient solutions at such low levels due to contamination. This should be verified by analysis of the base nutrient solution used, and if there is no Na present, NaCl should be added to give 1 to 2 mM Na in the control solution. Further complicating interpretation is the failure, in many cases, to correct for the weight of salt in the tissue. Halophytes characteristically accumulate substantial quantities of salt in their shoots, easily 30 to 50% of the total dry weight [17], so much of the difference in dry weight between plants grown in nonsaline conditions versus those grown in some substantial salinity can be due to the increased salt content in the latter. Nevertheless, when care is taken to account for the weight of salt in the tissue and other factors, it is clear that there are many halophytes for which growth is maximum at salinity levels on the order of 50 to 200 mM NaCl or equivalent (ca. 3000 to 12,000 ppm). As part of an intensive halophyte domestication program [18], we screened 150 diverse species, in several families, and 57 (38%) of them had greater growth at 170 mM NaCl (ca. 10,000 ppm) than they did on nonsaline nutrient solution. We termed those euhalophytes, and the others miohalophytes. A detailed report of 10 of each type, representing 19 genera and 10 families, is given in Glenn and O'Leary [19].

## B.  Cause of the Growth Reduction at Low Salinity

The emphasis in the reviews by Barbour [9] and Flowers et al. [10,11] was on growth *stimulation* between 1–2 mM and 50–200 mM NaCl. Even in the later review by Rozema [20], the difference is still viewed as a stimulation effect. This is not surprising because that focus is emphasized in virtually all graphical comparisons of growth at various salinities. Growth is usually plotted as a percentage, with growth at zero salinity equal to 100%. When this is done, the extreme halophytes always show growth greater than 100% over the salinity range between 1–2 or zero NaCl and 50–200 mM NaCl. However, when actual growth (as either rate or final biomass) is plotted, it is clear that what has happened is a shift of the response curve, similar to what occurs in plants adapted to extremes of other environmental parameters, such as light or temperature. It is difficult, if not impossible, to generate a pair of curves comparing adapted and nonadapted plants for salinity as has been done for light and temperature [21]. However, a comparison of relative growth rates (RGRs) for 10 euhalophytes and 10 miohalophytes (as defined above) [19] showed that the average of the maximum RGR for the miohalophytes was 0.43 g/g per week, and it occurred at zero salinity. The maximum RGR for the euhalophytes occurred at 180 mM salinity, and it was 0.42 g/g per week, almost exactly equal to the maximum RGR for the miohalophytes. The average RGR for the euhalophytes at zero salinity was 0.33 g/g per week. If these data were presented by comparing them on a percentage basis, setting growth at zero salinity equal to 100%, the miohalophytes would show a steady decline in RGR with increasing salinity, but the euhalophytes would show a "stimulation" effect, having growth at 180 mM salinity equal to 127% of their RGR at zero salinity. What actually happened is that the euhalophytes still had about the same maximum RGR as the miohalophytes, but they were able to achieve it at a substantial salinity level (180 mM). The "price" paid by the euhalophytes, however, is loss of the ability to maintain the same RGR at lower salinity. That is, the entire response curve has shifted.

Thus, it should not be surprising that there is not much information available about the cause for the growth *reduction* at low salinity. Munns et al. [14] did acknowledge that maybe we should think in terms of a growth reduction at zero NaCl rather than a growth stimulation at 50 to 200 mM. Their suggestion for the cause of the difference in growth is increased water deficit at the lower salinity. They feel that "the improvement in growth above 1 mM NaCl is most likely related to improved water relations of the leaves, due to accumulation of Cl and Na." This feeling that less growth of halophytes at suboptimal levels of salinity is due to lack of sufficient solutes for generating turgor or even possibly to reduced root hydraulic conductivity is widely shared [10,20,22].

## C.  Water Relations at Low Salinity

As already mentioned, there is a paucity of data on water relation parameters that allow one to compare the water status of plants at suboptimal versus optimal salinity. As one might expect, the osmotic potential in leaves typically declines with increasing salinity of the growth medium (e.g., Refs. 23 and 24), although Matoh et al. [25] found there to be no difference in osmotic potential of *Phragmites* at 0 and 100 mM NaCl. The calculated turgor pressure in *Atriplex* fell to almost 0 by 9:00 AM in control plants, but in the salinized plants it stayed positive all day long, albeit dropping slightly at midday [23]. However, when Clipson et al. [26] measured turgor directly with a pressure probe in *Suaeda,* they found it to be about the same at all salinities. Downton [27] measured osmotic and water potentials in *Avicennia* and calculated turgor pressure. In all salt treatments (10 to 100% seawater), turgor was about 0.8 MPa, but at zero salinity, it was only about 0.2 MPa. Growth was less at zero than at all salinities. In young seedlings of *Salicornia* grown at low light in the laboratory, Stumpf et al. [28] found the turgor to be almost zero (0.02 MPa) when grown on nonsaline conditions, but at salinities of 170 and 340 mM, the turgor pressures were 0.53 and 0.99 MPa, respectively. On the other hand, Weeks [29] found that in greenhouse-grown *Salicornia* the osmotic and water potentials both parallelled the decline in salinity, with the result being almost no difference in calculated turgor pressure across the entire range from 17 to 1020 mM salinity. The turgor pressure was close to 1.0 MPa at all treatment levels, but the growth at the lowest salinity level was very poor, and the growth was high and no different between 170 and 1020 mM salinity. Some of the discrepancies are undoubtedly due to methodology problems. It is difficult enough to obtain reasonably accurate measures for water and osmotic potentials in any plants, so that the calculated turgor pressures can be accepted with reasonable confidence, but with halophytes it is even more problematic, especially succulent halophytes. Nevertheless, based at least partly on the direct measure of turgor with a pressure probe in *Suaeda* by Clipson et al. [26] mentioned before, Munns [30] acknowledged that inadequate turgor is probably not a likely cause for the lower growth of halophytes at suboptimal salinity.

Hydraulic conductivity of roots typically falls with salinity in both halophytes [23,31] and nonhalophytes [32–34]. Thus it would not be expected to find that the hydraulic conductivity of halophytes at suboptimal salinity is less than at optimal salinity. However, there really have not been enough measurements of conductivity at the appropriate salinity levels to allow any conclusive statements at this point. Nevertheless, Munns et al. [14] speculated that the relatively low values for root hydraulic conductivity in halophytes, in general, coupled with the low root/shoot ratio in halophytes, in general, could account for the low turgor, if it in fact occurs, in the plants at suboptimal salinity. The difficulty with such a scenario is that at least in the few observations cited above, the root hydraulic conductivity in halophytes is usually higher in the plants at suboptimal salinity, and also the root/shoot ratio is usually higher in those plants [35,36]. Thus the plants at optimal salinity would be more likely to be at a disadvantage in this context. It is clear that there is a need for measurement of all of these parameters in the same plants at optimal and suboptimal salinity levels.

## D.  Growth Component Analysis

Most of the difference in dry weight production between optimal and suboptimal salinities is accounted for by difference in shoot growth (e.g., Refs. 35 and 36), but there are some reports of root growth being affected as well [27,37]. In addition to leaf size being reduced at suboptimal salinity, the leaf number can be less [38]. However, Longstreth and Strain [39] found no differences in leaf area or specific leaf weight in *Spartina* at 10 versus 0.5 ppt salinity. The difference in total leaf area between the two salinities is thought to be a major cause of the growth difference between the two by Munns et al. [14]. Osmond et al.

[21] feel the same way. They analyzed the data from the studies of Gale et al. [40] and Kaplan and Gale [23] and concluded that even though photosynthetic rate per unit leaf area was less at 72 mM NaCl than at zero NaCl, the leaf area was greater enough that the total photosynthetic capacity per plant was increased. Plotting of photosynthetic capacity per plant as a function of salinity over the range from 0 to 360 mM NaCl yielded a curve that closely paralleled the growth response curve over the same salinity range. It should be noted, however, that Clipson [41] found that leaf area in *Suaeda* decreased with all additions of salinity. Nevertheless, this may indicate the importance of photosynthate partitioning in determining the growth response. It may be that an important difference between the plants at the two salinity levels is how much photosynthate is reinvested in new photosynthetic surface. Unfortunately, there are not many data available that bear on that question. A thorough growth component analysis, similar to that done by Aslam et al. [42] in which they compared growth at optimal with growth at supraoptimal salinities, is needed.

## E.  Photosynthesis at Low Salinity

Because growth depends on substrate availability as well as sufficient turgor, it is reasonable to question the effect of the less optimum salinity on photosynthesis. The effect of salinity on photosynthesis in halophytes has been investigated, and despite the fact that the focus usually was on comparing optimum versus excess salinity, in some cases data were obtained for a range of salinity levels that enable one to compare photosynthesis rates at less than optimum salinities with those at optimum levels. In *Salicornia,* arguably the most salt-tolerant $C_3$ vascular plant, photosynthesis was higher at −3.2 MPa osmotic potential [43] or 342 mM salinity [35] when measurements were made at several salinity levels. However, Kuramoto and Brest [44] found that photosynthesis decreased at all levels of salinity, and Pearcy and Ustin [45] found no differences from 0 to 450 mM. Kuramoto and Brest [44] found the same response with *Batis maritima, Spartina foliosa,* and *Distichlis spicata.* Kemp and Cunningham [16] also found a steady decrease in photosynthesis with increasing salinity in *Distichlis spicata.* Longstreth and Strain [39] found no difference in photosynthesis at different salinity levels in *Spartina alterniflora.* In *Atriplex nummularia,* photosynthesis was higher at leaf water potentials of −1.5 to −2.0 MPa than at either higher or lower water potentials [47], and in *Sporobolus airoides* photosynthesis was higher at 1.0 MPa than at zero salinity [48]. In *Lepochloa fusca,* grown in the absence of NaCl or with NaCl at 250 mM, photosynthesis was higher in the presence of added NaCl than when it was absent at 32 or 39°C, but the reverse was true when the temperature was 19°C [49]. The data of Hajibagheri et al. [50] showed that photosynthesis in *Suaeda* was lower at 170 mM than at either 340 or 680 mM salinity.

In those few cases where photosynthesis was found to be lower at the lower salinity, there was insufficient information to determine whether the lower photosynthetic rates at the lower salinities (or higher water potentials) were due to stomatal or nonstomatal effects. In fact, even in the cases where investigators have demonstrated reduced photosynthetic rates at excessive salinity levels in halophytes, the picture is unclear. Some have attributed the reduced photosynthesis to reduced leaf conductance [51–54], while others have concluded that photosynthesis was reduced independently of changes in stomatal conductance [45,55,56]. The results of Schwartz and Gale [57] in which *Atriplex halimus* growing at 170 mM NaCl had a much greater increase in growth in response to increasing $CO_2$ than that of plants growing in nonsaline conditions are often cited as support for the view that the stomatal effect predominates at the higher salinity. On the other hand, Demming and Winter [58] found that even isolated chloroplasts from *Mesembryanthemum crystallinum* growing at different salinites showed reduced $CO_2$ fixation with increasing salinity. They found that electron transport was much less sensitive than $CO_2$ fixation, suggesting a direct effect of salinity on biochemical processes. Pearcy and Ustin [45] found that salinity did not affect the initial slope of the $CO_2$ response curve, but it did affect the $CO_2$-saturated photosynthetic capacity in *Spartina.* However, *Salicornia* photosynthetic capacity seemed to be relatively independent of salinity.

Furthermore, in some cases it has been concluded that growth was reduced by some factors other than photosynthesis, and the net effect was due to reduced photosynthetic surface rather than reduced photosynthetic rate per unit leaf surface [59–61]. Flowers [62] concluded that the general interpretation of the available evidence is that photosynthetic rates per unit leaf area are decreased or little affected by increases in salinity, and the rates per unit of chlorophyll appear to be either unaffected or to *increase.* He also indicated that the results of Kemp and Cunningham [46] with *Distichlis* showed that the reduced photosynthesis at increasing salinity is due to changes in stomatal frequency.

Some of the differences may be due to real differences among species, but some are also the result of differences in experimental conditions and differences in the manner in which the data are expressed [41,56]. It depends on whether the photosynthetic rate is expressed on a leaf area, leaf weight, or chlorophyll basis. Depending on which is used, the photosynthetic rate may be shown to be lower or higher [61].

## F.  Summary

It should be clear from this brief survey that it currently is no easier to explain the cause for the reduced growth at suboptimal salinity that it is to explain the reduced growth at supraoptimal salinity. In the former case, however, the reason is largely due to the fact that not nearly as much research has been directed at the question as in the latter case. Nevertheless, based on the limited database available, it does not seem likely that the reduced growth at suboptimal salinity is due to insufficient turgor. Neither does it seem that there is insufficient production of substrates for growth. In fact, with the few exceptions noted before, the evidence seems primarily to indicate that the rate of photosynthesis at suboptimal salinity is higher than at optimal salinity. If that is the case, and yet the plants are significantly smaller than those growing at optimal salinity, the obvious question that comes to mind is: Where is all the carbon going? Part of the problem may be that all of the photosynthetic rate measurements are instantaneous values, whereas the growth data are integrated values. The total carbon fixed per day, as well as the total carbon lost to respiration during the ensuing night period, needs to be determined under those salinity levels. A total carbon balance needs to be determined, in other words.

Partitioning of the photosynthate may be more important than the total amount produced. As described here, there is very little information available that bears on that point, particularly as concerns the plants growing at suboptimal salinity. There is almost no information available on hormone metabolism at the various salinity levels. Because the reduced growth at suboptimal salinity seems to be due to an apparent overall stunting of the plant, the problem may be one of growth regulation. There is a lack of information on this topic. Much needs to be done yet. That is clear.

Despite it being no easier to explain the cause of reduced growth at supoptimal salinity than it is to explain the reduced growth at supraoptimal salinity, it is clear that there are physiological differences between the plants growing at those salinities. It is likely that the causes for the growth reduction in each case are different. Thus, increased attention to the response of highly salt tolerant halophytes to suboptimal salinity is warranted.

## III.  CONCLUSIONS AND RECOMMENDATIONS

The decreasing availability of fresh water for agriculture, coupled with the increasing demand for plant-based agricultural commodities, makes the eventual use of increasingly saline water in agriculture a certainty. There are abundant reserves of saline water within reasonable pumping distance from the surface available in many areas of the world, especially in areas where successful crop production depends on irrigation. The limitation to eventual use of that water to irrigate crops is the availability of sufficiently salt tolerant crops. The long-term survival of agriculture in such areas is dependent on development of such crops. Even though progress has been made in increasing salt tolerance of some crops over the years, the pace of continued improvement and the ultimate maximum tolerance that can be achieved through conventional breeding approaches may not be sufficient to fulfill this need. Thus, other approaches that offer promise are required. The ability now to transfer genetic information between widely different types of plants makes the possibility of moving traits associated with increased tolerance of environmental stresses from alien genotypes into crop plants highly likely, *if* the required traits can be identified. Study of the physiology of highly salt tolerant "wild" plants (halophytes) is a necessary but not sufficient step. For the reasons described here, a completely different approach to studying salt tolerance in those plants is desirable as an *additional* approach to this important problem. This paradigm shift in our conceptual approach to analysis of salt tolerance has the strong potential for opening up a new line of research that could be highly productive. It could provide the means whereby the emerging molecular genetic techniques can be applied to what heretofore has been viewed as one of the most intractable problems at the whole plant level—stress resistance.

# REFERENCES

1. D Aspinall, LG Paleg. In: LG Paleg, D Aspinall, eds. The Physiology and Biochemistry of Drought Resistance in Plants. New York: Academic Press, 1981, pp 205–241.
2. CR Stewart. In: LG Paleg, D Aspinall, eds. The Physiology and Biochemistry of Drought Resistance in Plants. New York: Academic Press, 1981, pp 243–259.
3. RG Wyn Jones, R Storey. In: LG Paleg, D Aspinall, eds. The Physiology and Biochemistry of Drought Resistance in Plants. New York: Academic Press, 1981, pp 171–204.
4. NC Turner. Aust J Plant Physiol 13:175, 1986.
5. RG Wyn Jones, J Gorham. In: OL Lange, PS Nobel, CB Osmond, H Ziegler, eds. Physiological Plant Ecology, Vol III, Responses to the Chemical and Biological Environment. Berlin: Springer-Verlag, 1983, pp 35–58.
6. R Munns. Aust J Plant Physiol 15:717, 1988.
7. VJ Chapman. Salt Marshes and Salt Deserts of the World. New York: Interscience, 1960.
8. Y Waisel. Biology of Halophytes. New York: Academic Press, 1972.
9. M Barbour. Am Midl Nat 84:105 1970.
10. TJ Flowers, PF Troke, AR Yeo. Annu Rev Plant Physiol 28:89–121, 1977.
11. TJ Flowers, MA Hajibaheri, NJW Clipson. Q Rev Biol 61:313, 1986.
12. DA Adams. Ecology 44:445, 1963.
13. AJ Macke, I Ungar. Can J Bot 49:515, 1971.
14. R Munns, H Greenway, GO Kirst. In: OL Lange, PS Nobel, CB Osmond, H Ziegler, eds. Physiological Plant Ecology III: Responses to the Chemical and Biological Environment. Berlin: Springer-Verlag, 1983, pp 59–135.
15. RF Black. Aust J Biol Sci 13:249, 1960.
16. MC Williams. Plant Physiol 35:500, 1960.
17. JW O'Leary. In: E Whitehead, C Hutchinson, B Timmermann, R Varady, eds. Arid Lands Today and Tomorrow. Boulder, CO: Westview Press, 1988, pp 773–790.
18. JW O'Leary. In: RC Staples, GH Toenniessen, eds. Salinity Tolerance in Plants: Strategies for Crop Improvement. New York: Wiley, 1984, pp 285–300.
19. EP Glenn, JW O'Leary. Plant Cell Environ 7:253, 1984.
20. J Rozema. Aquat Bot 39:17, 1991.
21. CB Osmond, O Bjorkman, DJ Anderson. Physiological Processes in Plant Ecology: Toward a Synthesis with Atriplex. Berlin: Springer-Verlag, 1980.
22. DH Jennings. Biol Rev 51:453, 1976.
23. A Kaplan, J Gale. Aust J Biol Ser 25:895, 1972.
24. JA Bolanos, DJ Longstreth. Plant Physiol 75:281, 1984.
25. T Matoh, N Matsushita, E Takahashi. Physiol Plant 72:8, 1988.
26. NJW Clipson, AD Tomos, TJ Flowers, RG Wyn Jones. Planta 165:392, 1985.
27. WJS Downton. Aust J Plant Physiol 9:519, 1982.
28. DK Stumpf, JT Prisco, JR Weeks, VA Lindley, JW O'Leary. J Exp Bot 37:160, 1986.
29. JR Weeks. PhD dissertation, University of Arizona, Tucson, 1986.
30. R Munns. Plant Cell Environ 16:15, 1993.
31. RS Ownbey, BE Mahall. Physiol. Plant 57:189, 1983.
32. JW O'Leary. Isr J Bot 18:1  9, 1969.
33. JW O'Leary. In: J. Kolek, ed. Structure and Function of Primary Root Tissues. Bratislava, Czechoslovakia: Veda, 1974, pp 309–314.
34. RJ Joly. Plant Physiol 91:1261, 1989.
35. FS Abdulrahman, GJ Williams III. Oecologia 48:346, 1981.
36. G Naidoo, R Rughunanan. J Exp Bot 41:497, 1990.
37. KC Blits, JL Gallagher. Plant Cell Environ 13:419, 1990.
38. TF Neales, PJ Sharkey. Aust J Plant Physiol 8:165, 1981.
39. DJ Longstreth, BR Strain. Oecologia 31:191, 1977.
40. J Gale, R Naaman, A Poljakoff-Mayber. Aust J Biol Sci 23:947, 1970.
41. NJW Clipson, PhD. thesis, University of Sussex, 1984; cited in TJ Flowers. Plant Soil 89:41, 1985.
42. Z Aslam, WD Jeschke, EG Barrett-Lennard, TL Setter, E Watkin, H Greenway. Plant Cell Environ 9:571, 1986.
43. BL Tiku. Physiol Plant 37:23, 1976.
44. RT Kuramoto, DE Brest. Bot Gaz 140:295, 1979.
45. RW Pearcy, SL Ustin. Oecologia 62:68, 1984.
46. PR Kemp, GL Cunningham. Am J Bot 68:507, 1981.
47. AT Pham Thi, C Pimentel, J Vieira da Silva. Photosynthetica 16:334, 1982.
48. HM El-Sharkawi, BE Michel. Photosynthetica 9:277, 1975.
49. J Gorham. Plant Cell Environ 10:191, 1987.
50. MA Hajibagheri, DMR Harvey, TJ Flowers. Plant Sci Lett 34:353, 1984.
51. GD Farquhar, MC Ball, S von Caemmerer, Z Roksandic. Oecologia 52:121, 1982.
52. RD Guy, DM Reid, HR Krouse. Can J Bot 64:2693, 1986.
53. RD Guy, DM Reid, HR Krouse. Can J Bot 64:2700, 1986.

54. LB Flanagan, RL Jeffries. Plant Cell Environ 11:239, 1988.
55. MC Ball, GD Farquhar. Plant Physiol 74:1, 1984.
56. DJ Longstreth, JA Bolanos, JE Smith. Plant Physiol 75:1044, 1984.
57. M Schwarz, J Gale. J Exp Bot 35:193, 1984.
58. B Demming, K Winter. Planta 159:66, 1983.
59. TM DeJong. Oecologia 36:59, 1978.
60. J Gale, A Poljakoff-Mayber. Aust J Biol Sci 23:937, 1970.
61. K Winter. In: RL Jeffries, AJ Davy, eds. Ecological Processes in Coastal Environments. Oxford: Blackwell, 1979.
62. TJ Flowers. Plant Soil 89:41, 1985.

# 31

# Growth and Physiological Adaptations of Grasses to Salinity Stress

**Kenneth B. Marcum***

*The University of Arizona, Tucson, Arizona*

## I. INTRODUCTION

Degradation of arable lands is a major constraint to agriculture worldwide, with soil salinization, particularly in irrigated areas, playing a major role [1]. Nearly 10% of the earth's total land surface, or 954 Mha, is covered with salt-affected soil [2], of which from 60 to over 100 Mha is currently salt-affected as a result of human activity [3,4]. In addition, the extent of soil salinization is continually increasing. Ten to 20 Mha of irrigated agricultural lands deteriorate to zero productivity each year because of salt buildup [5,6]. Much of this land, although now too saline for conventional agriculture, is currently, or has the potential to be, utilized for growing salt-tolerant or halophytic forage species, of which the Poaceae (grasses) play a prominent role [7].

Critical water shortages are occuring in urban areas, resulting in restrictions on the use of potable water for irrigation of landscaped areas. For example, in a number of western U.S. states, laws have been passed that require the use of sewage effluent or other secondary, saline water sources for the irrigation of turfgrass landscapes [8,9].

The Poaceae, represented by over 7500 species, inhabit the earth in greater numbers and have greater range of climatic adaptation than any other plant family [10,11]. Therefore, it is not surprising that grasses show extreme range in salinity tolerance, from salt sensitive to extremely salt tolerant (halophytic). Grasses range in classification through salt sensitive, e.g., annual bluegrass (*Poa annua* L.); moderately salt sensitive, e.g., meadow foxtail (*Alopecurus pratensis* L.); moderately salt tolerant, e.g., dallisgrass (*Paspalum dilatatum* Poir.); and salt tolerant, e.g., bermudagrass [*Cynodon dactylon* (L.) Pers.] [12–14]. Other grasses are recognized as true halophytes, e.g., cordgrasses (*Spartina* sp.), coastal dropseed [*Sporobolus virginicus* (L.) Kunth], and saltgrasses (*Distichlis spicata* L.) [15,16].

The goal of this chapter is to investigate the range of salinity tolerance and physiological adaptations to salinity present in grasses utilized for forage, grazing, soil stabilization, or turf. This chapter will describe the responses to salinity of three grass species representing the full range of salt tolerance present in the Poaceae. Data from a study that compared three grass species will be presented: salt-sensitive buffalograss [*Buchloë dactyloides* (Nutt.) Engelm.], salt-tolerant bermudagrass, and halophytic desert saltgrass [*Distichlis spicata* var. *stricta* (Torr.) Beetle] [17]. Response differences highlighting important physiological mechanisms of salt tolerance will be discussed and results cross-referenced to other studies.

---

\* *Current affiliation:* University of Hawaii at Mano'a, Honolulu, Hawaii.

## II.  GROWTH RESPONSE AND RELATIVE SALINITY TOLERANCE

### A.  Shoot Growth Responses

Salt tolerance in plants depends not only on genotype but also on cultural and environmental conditions. Therefore, salt tolerance can be determined not with certainty but only on a relative basis [13,16]. Relative salinity tolerance is generally quantified as the salt level resulting in a 50% reduction in shoot growth (yield) or, alternatively, the threshold salinity, i.e., salinity level where yield begins to decline, followed by the rate, or slope, of yield reduction [16,18].

Fifty percent shoot growth reduction occurred at 150 mM NaCl salinity (approximately 12 dS m$^{-1}$ EC$_{soln}$) for buffalograss, 330 mM (26 dS m$^{-1}$) for bermudagrass, and >600 mM (>46 dS m$^{-1}$) for desert saltgrass (Figure 1). Reid et al. [19] also reported 50% shoot growth decline at 12 dS m$^{-1}$ for three buffalograss cultivars. In another study, six natural populations of buffalograss had an average 50% shoot growth reduction at 13 dS m$^{-1}$ [20]. Previous data for bermudagrass are more variable, perhaps because of the genetic diversity within this genus [21]. Fifty percent shoot growth reductions for bermudagrass cultivars and/or accessions have been reported as 24 and 33 dS m$^{-1}$ [22], 24 and 31 dS m$^{-1}$ [23], and 17 to 22 dS m$^{-1}$ [24]. The halophytic nature of saltgrass is apparent from other sources [15,16,25]. In several studies, shoot growth of desert saltgrass was not affected by salinities up to 40 dS m$^{-1}$ [26,27].

These results place buffalograss in the moderately salt-sensitive category [14], similar in tolerance to Kentucky bluegrass (*Poa pratensis* L.) [28], various grama grasses (*Bouteloua* spp. Lag.) [17], chewings fescue (*Festuca rubra* L.) [29], and bahiagrass (*Paspalum notatum* var. *saurae* Parodi) [30]. Bermudagrass is considered salt tolerant [12–14,16]. Other studies have shown it similar in tolerance to St. Augustinegrass [*Stenotaphrum secundatum* (Walt)) Kuntze] [31,32], tall wheatgrass [*Thinopyrum*

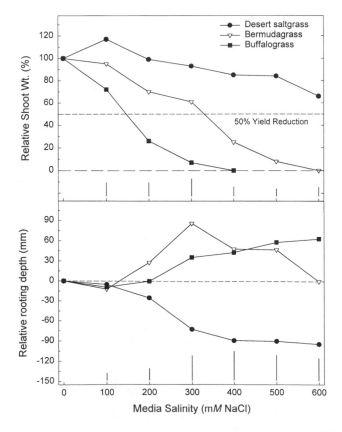

**Figure 1**   Relative shoot dry weight [(treatment/control) × 100] and relative rooting depth (treatment minus control) of three grasses exposed to increasing salinity levels in solution culture. Vertical bars represent LSD ($P < .05$) values for mean comparison at each salinity level.

*elongatum* (Host) D.R. Dewey] [33], seashore paspalum (*Paspalum vaginatum* Swartz) [22,34,35], and Kallar (brown beetle) grass [*Diplachne fusca* (L.) Beauv.] (syn. *Leptochloa fusca* L. Kunth.) [36,37]. In contrast, desert saltgrass is a halophyte [13,15]. Various studies have revealed other halophytic grasses, including *Spartina* spp. [38,39], *Sporobolus virginicus* (L.) Kunth [40,41], and *Sporobolus airoides* (Torr.) Torr. [17].

Salt-sensitive plants (glycophytes) and moderately salt-tolerant plants (mesophytes) generally have a flat yield response to salinity prior to a threshold salinity level beyond which shoot growth declines. In contrast, highly salt-tolerant plants often display stimulated shoot and root growth at moderate salinity levels, followed by yield decline [13,16,18]. Increased shoot growth (relative to control) at moderate salinity (100 mM NaCl, or 8 dS m$^{-1}$) was evident in desert saltgrass (Figure 1). However, bermudagrass and buffalograss displayed progressive shoot growth reductions at all salinity levels. Salt-stimulated shoot growth has been observed in other salt-tolerant or halophytic grasses. Shoot growth peaked at 90 mM NaCl (8 dS m$^{-1}$), then declined in *Halopyrum mucronatum* (L.) Stapf., a perennial grass found on coastal dunes of Pakistan [42]. Shoot growth was stimulated with increasing salinity up to 25 mM NaCl (2.5 dS m$^{-1}$), then declined, in two of six *Sporobolus species* studied (*S. stapfianus* and *S. pellucidus*) [43]. Shoot growth of *Sporobolus virginicus* increased up to 150 mM NaCl (12 dS m$^{-1}$), then declined [40]. However, Naidoo and Naidoo [44] reported no shoot increase with increasing salinity for this species. In addition, shoot stimulation at low to moderate salinity has sometimes been reported in certain salt-tolerant (although not halophytic) grasses, such as bermudagrass [45], seashore paspalum [18,22,34], and St. Augustinegrass [34].

## B.  Root Growth Responses

Root growth stimulation (increased root mass, rooting depth, or both) in salt-tolerant grasses is typically a more common, accentuated response to moderate salinity stress than shoot growth stimulation [16]. The net result is generally an increase in root/shoot ratios, which may be a salinity tolerance mechanism to counter low external water potential by increasing plant absorptive area [46,47]. Increased rooting depth, relative to control plants, was observed in bermudagrass and desert saltgrass under salinity stress (Figure 1). However, relative rooting depth declined at high salinity for bermudagrass but not desert saltgrass. Root weight (data not shown) also increased, being highly correlated with rooting depth ($r = 0.83$). In contrast, rooting of buffalograss progressively declined with increasing salinity stress. Rooting decline under salinity stress has been previously reported in buffalograss [48], and in other moderate to salt-sensitive grasses, such as Kentucky bluegrass [49], bahiagrass [22], chewings fescue [50], and sideoats grama [*Bouteloua curtipendula* (Michx.) Torr.] [17].

Root stimulation has been observed in a number of salt-tolerant as well as halophytic grasses. Root dry weights increased linearly with increasing salinity up to 450 mM NaCl (35 dS m$^{-1}$) in *Sporobolus virginicus*, resulting in a root/shoot ratio of 2.2, relative to 0.5 (control) [40]. Blits and Gallagher [41] reported a doubling in root mass of *S. virginicus* grown in seawater relative to fresh water. Although root growth (length) increased under moderate salinity stress, relative to control, shoot growth declined in rhodesgrass (*Chloris gayana* L.) [51], bermudagrass [52], and zoysiagrasses (*Zoysia japonica* Steud. and *Z. matrella* [L.] Merr.) [31].

## III.  PHYSIOLOGICAL ADAPTATIONS TO SALINITY

## A.  Ion Exclusion

It has long been accepted that the major causes of plant growth inhibition under salinity stress are osmotic stress (osmotic inhibition of plant water absorption) and specific ion effects, including toxicities and imbalances [53–55]. In comparison with salt-tolerant or halophytic dicotyledonous plants, monocots (including Poaceae) tend to exclude saline ions from shoots, thereby minimizing toxic effects [56–58]. Saline ion exclusion from shoots was strongly associated with salinity tolerance among three grasses representing the range of salinity tolerance present in the Poaceae (Figure 2). Chloride and Na$^+$ accumulated to high levels in buffalograss shoots but was maintained at concentrations similar to those in the growth media in bermudagrass and halophytic desert saltgrass shoots, particularly at high salinity. Salinity tolerance of other grasses has been related to saline ion exclusion. Salinity tolerance in *Sorghum halepense*

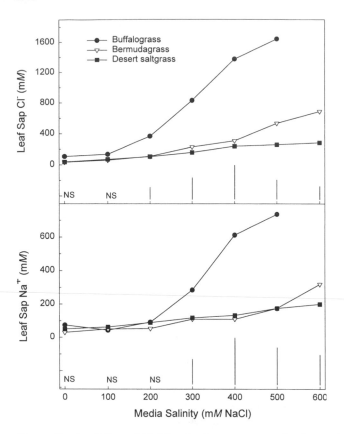

**Figure 2**    Leaf Cl⁻ and Na⁺ levels of grasses exposed to increasing salinity levels in solution culture. Vertical bars represent LSD ($P < .05$) values for mean comparison at each salinity level.

(L.) Pers. relative to *Sorghum bicolor* (L.) Moench was associated with shoot Cl⁻ concentration [59]. Similarly, salt-tolerant tall wheatgrass accessions excluded Na⁺ and Cl⁻ from shoots (while maintaining fairly high K⁺ contents) to a greater extent than salt-sensitive crested wheatgrass [*Agropyron desertorum* (Fisch. ex Link) Schult.] accessions [60]. In contrast, salt-tolerant weeping [*Puccinellia distans* (L.) Parl] and Lemmon [*P. lemmoni* (Vasey) Scribn.] alkaligrasses were found to accumulate more Na⁺ and Cl⁻ in shoots than did moderately salt-tolerant creeping bentgrass (*Agrostis stolonifera* L.) [61].

Saline ion exclusion also appears to be an important factor influencing intraspecies salinity tolerance, i.e., at the cultivar or accession level. For example, salt-sensitive populations were found having, at a given test salinity, higher shoot Na⁺ and Cl⁻ than coastal (or other saline site) salt-tolerant accessions in chewings fescue [50], red fescue (*Festuca rubra* L.) [62], bermudagrass [45], and creeping bentgrass [63]. However, this was not the case for eight natural populations of brown beetlegrass. There was no correlation between salinity tolerance and shoot saline ion concentrations or the soil salinity level of the original collection sites [36]. Relative salinity tolerance of zoysiagrass (*Zoysia* spp.) cultivars and accessions has been successfully predicted on the basis of shoot Na⁺ concentrations occurring under salt stress [64,65].

## B.    Osmotic Adjustment and Ion Regulation

Osmotic stress due to lack of osmotic adjustment, resulting in reduced water absorption and physiological drought, has long been considered a major cause of salinity injury in plants [47,61,66]. Maintenance of cell turgor and plant growth requires sufficient increase in sap osmolality to compensate for external osmotic stress, the process of osmoregulation, or osmotic adjustment [66,67]. In a saline environment, osmotic adjustment is needed to avoid osmotic stress, yet this may result in ion toxicity [56,68].

It has been noted that monocots (relative to salt-tolerant dicots), including Poaceae, tend to restrict saline ion uptake. This has been suggested to cause cell dehydration and reduced growth under saline conditions because of lack of osmotic adjustment [56,57,69]. Indeed, declining shoot water content is commonly observed in grasses under salinity stress [31,53,70,71], although a slight increase in shoot succulence at moderate salinity has been noted in some grass halophytes [40–42]. However, complete osmotic adjustment occurred in bermudagrass, buffalograss, and desert saltgrass, sap osmolalities being maintained below (more negative than) media osmolality (Figure 3). In fact, salt-sensitive buffalograss osmotically adjusted to a much greater degree than salt-tolerant desert saltgrass and bermudagrass. Among seven grasses, shoot sap osmolality was highly negatively correlated with salinity tolerance and root growth under salt stress ($r > -0.8$) [17]. Complete osmotic adjustment under salinity stress has been reported previously in a range of grasses [31,72–74]. In these studies, the shoot sap osmolality level was negatively correlated with salinity tolerance. In other words, in salt-tolerant grasses, osmotic adjustment, although complete, is nevertheless minimized; i.e., shoot sap osmolality is maintained close to saline media levels. Therefore, the importance of osmotic adjustment as a mechanism of salinity tolerance is currently being questioned [75].

Although salinity tolerance in grasses is clearly associated with saline ion exclusion, $Na^+$ and $Cl^-$ have been instrumental in shoot osmotic adjustment in a number of studies, constituting the majority of osmotically active solutes [17,31,36,40,42,76]. Among seven grasses, shoot $Na^+$ and $Cl^-$ concentrations were highly correlated with osmotic adjustment ($r = 0.9$) [17]. Therefore, although saline ion exclusion is clearly critical for salinity tolerance in grasses, saline ion *regulation*, rather than exclusion, may be a more apt description of the salinity tolerance mechanism operating in grasses.

Saline ion regulation in grasses may occur in several ways. Selectivity for $K^+$ over $Na^+$ may occur by selective $K^+$ absorption–vacuolar $Na^+$ compartmentation in root cortical cells or endodermis or by selective saline ion extrusion through specialized salt glands or bladders [66,77–79]. In glycophytic grasses, tissue $Na^+$ may be reabsorbed from the xylem via mature xylem parenchyma cells in roots or shoots and translocated back to soil [80–82]. Alternatively, ion partitioning may occur, whereby saline ions are redistributed to mature, senescing leaves or other organs [83–86].

## C.  Glandular Ion Excretion

Salt glands or bladders are present in a number of salt-adapted species, which eliminate excess saline ions from shoots by excretion [87–89]. Multicellular epidermal salt glands are present in several families of dicotyledons, e.g., Frankeniaceae, Plumbaginaceae, Aviceniaceae, and Tamaricaceae [89,90]. Within the Poaceae, bicellular epidermal salt glands have been reported to occur in over 30 species within the tribes Chlorideae, Eragrosteae, Aeluropodeae, and Pappophoreae [91–93], all members of the subfamily Chloridoideae, according to Gould and Shaw [10]. However, if the taxonomic system proposed by Clayton and Renvoize [94] is followed, grass species having functional salt glands occur only in two tribes, Eragrostideae and Cynodonteae, both also belonging to the subfamily Chloridoideae.

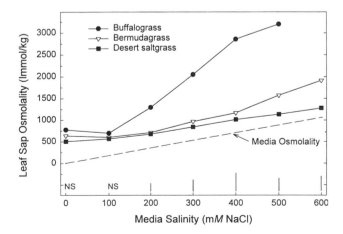

**Figure 3**   Leaf sap osmolality of grasses exposed to increasing salinity levels in solution culture. Vertical bars represent LSD ($P < .05$) values for mean comparison at each salinity level.

Salt glands of the Poaceae are, in outward appearance, similar to leaf epidermal bicellular microhairs. Although microhairs resembling salt glands have been observed in all grass subfamilies except Pooideae [88,95], functioning salt glands have been found only within the subfamily Chloridoideae [92,93]. This is probably due to an ultrastructural modification hypothesized to be responsible for salt excretion in the Poaceae, a series of parallel, invaginated plasma membrane channels within the gland's basal cell, [88,96,97], observed only in certain Chloridoid grasses (i.e., of the Chloridoideae) [93]. These membranes are actually infoldings of the plasmalemma that originate adjacent to the wall separating the cap and basal cells, forming open channels in the direction of ion flow. Ultracytochemical localization of adenosine triphosphatase (ATPase) activity within salt gland basal cells of *Sporobolus virginicus* supports the hypothesis of active ion loading at these sites [98]. In addition, there are numerous mitochondria associated with the parallel membranes, probably involved in providing an energy supply for channel ion loading [96,97,99].

Salt glands in the Poaceae are structurally distinct from the multicellular glands of dicots, consisting of a basal cell attached, or embedded, in the leaf epidermis and a cap cell [90,92] (Figure 4A). The glands are characterized by cutinized cell walls and are often surrounded by papillae. Although the basic, bicellular structure is the same in all Chloridoid species, their appearance varies [88] (Figure 5). In some species, glands are sunken into the epidermis with the basal cell totally embedded, e.g., desert saltgrass. In others, the basal cell is semi embedded, e.g., bermudagrass. Finally, the basal cell may extend out from the epidermis, with the gland lying recumbent to the leaf surface, e.g., buffalograss. Salt glands of Poaceae are quite small (usually 25–70 µm in length), although size may vary substantially, from imbedded to elongated, protruding types. Glands range in size from 15 µm in length in desert saltgrass [17], to 35 µm in Manilagrass [74], to 70 µm in buffalograss [17] (Figure 5). Salt glands have been found on both abaxial and adaxial leaf surfaces of excreting Chloridoid species [17,74,92]. Glands are longitudinally arranged in parallel rows atop intercostal regions of leaves, adjacent to rows of stomates (Figure 4B).

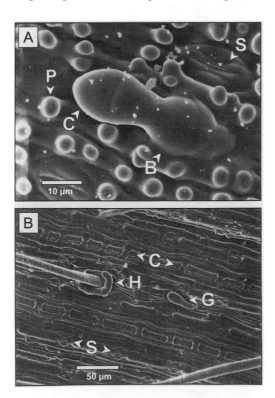

**Figure 4**  Scanning electron micrographs of adaxial leaf surfaces. (A) Salt gland of *Zoysia japonica*. C, cap cell; B, basal cell; P, papillae; S, stomate. (B) Overview of buffalograss (*Buchloë dactyloides*) leaf surface, showing location of salt gland relative to other structures. C, costal zone of leaf epidermis; G, salt gland; H, macrohair; S, stomata.

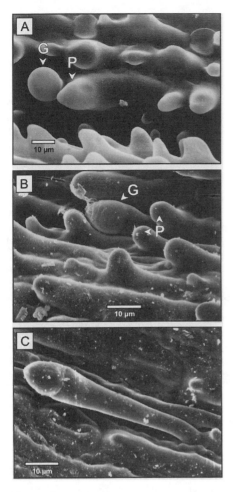

**Figure 5** Scanning electron micrographs of adaxial leaf surfaces. (A) Salt gland of desert saltgrass (*Distichlis spicata* var. *stricta*). Only cap cell is visible—basal cell is embedded in epidermis. G, salt gland; P, papillae. (B) Salt gland of bermudagrass (*Cynodon dactylon*). G, salt gland; P, papillae. (C) Salt gland of buffalograss (*Buchloë dactyloides*).

Evidence that salt gland ion excretion is an active, metabolically driven process is varied, including effects of temperature [100], light [101], oxygen pressure [88], and metabolic inhibitors [102] on excretion rate as well as selectivity of ion excretion. Excretion is typically highly selective for $Na^+$ and $Cl^-$ [102–104], although other ions may be excreted in minute amounts, such as $K^+$, $Ca^{2+}$, and $Mg^{2+}$ [34,44,74,88]. Comparison of salt gland excretion rates among studies is difficult because of the varying influence of environmental factors, such as light and temperature, cumulative days of exposure to salt stress, and plant factors such as leaf age [85]. Also, units of measurement differ, one fundamental difference being whether excretion rates are based on leaf area or leaf weight. Finally, excretion rate is not static but is influenced by saline ion concentrations in the growing media. Increasing media salinity generally stimulates excretion up to an optimal level above which excretion rate may decline [88]. Maximum excretion rate was reported to occur at 150 to 200 mM media NaCl (8–13 dS $m^{-1}$) in moderately tolerant Chloridoid species such as bermudagrass, Rhodesgrass, goosegrass [*Eleusine indica* (L.) Gaertn.], and Kallar grass [92,102,104]. However, excretion was maximal at 200 mM NaCl (17 dS $m^{-1}$) in desert saltgrass and *Spartina* spp. [88] and 300 mM NaCl (23 dS $m^{-1}$) in *Sporobolus virginicus* [40].

Among seven grasses belonging to the subfamily Chloridoideae, shoot $Na^+$ and $Cl^-$ concentrations were negatively correlated, whereas salt tolerance was positively correlated with salt gland $Na^+$ and $Cl^-$

**TABLE 1**   Leaf Salt Gland $Cl^-$ and $Na^+$
Excretion Rates[a] of Three Chloridoid Grasses: Ion
Excretion Measured in Plants Exposed to 200 mM
NaCl

| Grass | $Cl^-$ | $Na^+$ |
|---|---|---|
| Buffalograss | 39 | 36 |
| Bermudagrass | 191 | 163 |
| Desert saltgrass | 1267 | 1200 |
| $LSD^b_{0.05}$ | 56 | 72 |

[a] Excretion rates in $\mu$mol ion/g leaf dry wt/week.
[b] Fishers Protected Least Significant Difference

excretion rates [17]. Table 1 shows ion excretion rates for three of the Chloridoid grasses: buffalograss (salt sensitive), bermudagrass (moderately salt tolerant), and desert saltgrass (halophytic). Note that desert saltgrass had $Na^+$ and $Cl^-$ excretion rates 32 and 34 times higher, respectively, than buffalograss. Similar strong correlations between salt gland excretion rates, shoot $Na^+$ and $Cl^-$ concentrations, and salinity tolerance were observed among three Chloridoid grasses in another study [34]. Relative order of salinity tolerance again followed saline ion excretion rates, with *Zoysia matrella* (highly salt tolerant) having an $Na^+$ excretion rate of 730 compared with bermudagrass (salt tolerant) at 660 and *Zoysia japonica* (moderately salt sensitive) at 360 $\mu$mol/g leaf dry wt/week, respectively. Sodium and $Cl^-$ excretion rates were negatively correlated with shoot concentrations but positively correlated with leaf salt gland density and salinity tolerance among 57 zoysiagrass species accessions [64,74]. Excretions rates of various *Zoysia* spp. reported range from 130 $\mu$mol $Na^+$/g leaf dry wt/week in salt-sensitive *Zoysia japonica* to 730 $\mu$mol $Na^+$/g leaf dry wt/week in salt-tolerant *Zoysia matrella*, with gland densities ranging from 28/mm$^2$ leaf surface in salt-sensitive *Zoysia japonica* to 100/mm$^2$ in salt-tolerant *Zoysia macrostachya* Franch. & Sav.

## D.   Ion Compartmentation and Compatible Solutes

In vitro studies have shown that enzymes of both glycophytes and halophytes have similar sensitivities to salt, being inhibited at concentrations above 100–200 mM (approximately 8–17 dS m$^{-1}$) [71,105]. Therefore, salt-tolerant plants growing under saline conditions must restrict the level of ions in the cytoplasm. As preceding data have illustrated, salt-tolerant grasses utilize inorganic ions for a large part of their osmotic adjustment under saline growing conditions, as the ability to accumulate organic solutes on a whole cell basis is metabolically expensive and therefore limited [66,77]. Salt-tolerant plants that successfully accumulate saline ions for osmotic adjustment above concentrations of 100–200 mM do so by compartmentalizing them within the vacuole, which typically makes up 90 to 95% of a mature plant cell's volume [106]. Evidence exists for salinity inducing a $K^+/Na^+$ exchange across the tonoplast mediated by $Na^+/H^+$ antiport activity, resulting in saline ion compartmentation in vacuoles [78,79]. Under these conditions, the osmotic potential of the cytoplasm is maintained by the accumulation of organic solutes that are compatible with enzyme activity, termed "compatible solutes" [107,108]. Under highly saline conditions, relatively few organic solutes, including glycinebetaine, proline, and certain polyols and cyclitols, can be accumulated in sufficient concentrations to adjust the cytoplasm osmotically without inhibiting enzymes [75]. Evidence exists for the cytoplasmic localization of these compounds [107,109,110]. Of these, glycinebetaine and proline typically accumulate in grasses [111].

**TABLE 2**   Leaf Sap Glycinebetaine and Proline Levels (mM) of grasses exposed to 0 and 300 mM NaCl

| Grass | Glycinebetaine | | Proline | |
|---|---|---|---|---|
| | 0 mM | 300 mM | 0 mM | 300 mM |
| Buffalograss | 9.0 | 18.9 | 1.7 | 5.9 |
| Bermudagrass | 6.1 | 38.5 | 0.7 | 2.7 |
| Desert saltgrass | 11.0 | 62.2 | 0.6 | 1.8 |
| $LSD_{0.05}$ | 0.6 | 4.1 | 0.8 | 1.0 |

**TABLE 3** Estimated[a] Contribution to Cytoplasmic Osmotic Adjustment of Glycinebetaine and Proline, in mosmol kg$^{-1}$ (Osml) and as a Percentage (%) of Total Osmolality, of Plants Grown at 300 mM NaCl

| Grass | Glycinebetaine | | Proline | |
|---|---|---|---|---|
| | Osmol | % | Osmol | % |
| Buffalograss | 209 | 9.7 | 59 | 2.8 |
| Bermudagrass | 378 | 39.2 | 27 | 2.7 |
| Desert saltgrass | 625 | 73.7 | 17 | 2.0 |

[a] Estimate assumes glycinebetaine and proline are located in the cytoplasm, constituting 10% of total cell volume, with an osmotic coefficient of 1.0 for each compound.

Total leaf Na$^+$ + Cl$^-$ levels exceeded 200 mM in all three Chloridoid grasses grown at moderate to high salinity (Figure 2), necessitating vacuolar ion compartmentation for survival. Glycinebetaine levels increased under salinity in all grasses, reaching highest levels (62 mM) in desert saltgrass (Table 2). Although proline concentrations also increased under salinity, maximum levels occurred in salt-sensitive buffalograss, reaching only 6 mM. Assuming that glycinebetaine and proline are located in the cytoplasm (see earlier), which occupies 10% of the total cell volume, the contributions of glycinebetaine and proline to cytoplasmic osmotic adjustment can be calculated (Table 3). Glycinebetaine made substantial contributions to cytoplasmic osmotic adjustment in desert saltgrass (73%) and bermudagrass (39%) only. In contrast, proline contributions were insignificant in all grasses.

Shoot sap glycinebetaine concentrations in grasses grown at 200 mM NaCl (17 dS m$^{-1}$) ranged from 2 mM in centipedegrass [*Eremochloa ophiuroides* (Munro) Hack.] to 89 mM in bermudagrass [34]. St. Augustinegrass, seashore paspalum, *Zoysia japonica*, and *Zoysia matrella* were intermediate (in order from low to high) in shoot glycinebetaine concentrations. Given the same assumptions as before, glycinebetaine made substantial contributions to cytoplasmic osmotic adjustment in all grasses except centipedegrass. Centipedegrass stands alone among these grasses as being very salt sensitive. As before, proline contributions were too small to contribute to cytoplasmic osmotic adjustment, with the possible exception of bermudagrass, although contributions would still be minor. In the halophyte *Sporobolus virginicus*, glycinebetaine accumulated to 126 mM in shoots of plants grown at high salinity (450 mM NaCl or 35 dS m$^{-1}$), possibly contributing 93% of total cytoplasmic osmotic adjustment [40]. In contrast, proline levels were 11 times lower (11 mM) at this salinity, forming an insubstantial contribution to cytoplasmic osmotic adjustment. Other studies involving *Sporobolus virginicus* support the importance of glycinebetaine as a compatible solute relative to proline. Quaternary ammonium compounds (predominately glycinebetaine and possibly other related betaines) accumulated to 48 μmol g$^{-1}$ dry weight in shoots of *Sporobolus virginicus* grown in seawater, while proline levels reached only 1.6 μmol g$^{-1}$ dry weight [41]. Similarly, proline levels were insufficient to contribute significantly to cytoplasmic adjustment of *Sporobolus virginicus* grown in 80% seawater [44]. In lines of tall wheatgrass grown at 20 dS m$^{-1}$ total salinity, glycinebetaine accumulated to 45 μmol g$^{-1}$ fresh weight in shoots, compared with only 1 μmol for proline [70].

Whereas glycinebetaine concentrations under salinity were positively correlated with salinity tolerance among seven Chloridoid grasses, proline concentrations were negatively correlated, suggesting that glycinebetaine, but not proline, acts as a compatible solute [17]. Although both compounds have traditionally been considered compatible solutes, more recent evidence has favored the role of glycinebetaine. For example, (1) glycinebetaine is excluded from the hydration sphere of enzyme proteins and thus tends to stabilize their tertiary structure [112], (2) corn (*Zea mays* L.) mutants lacking a critical enzyme for glycinebetaine biosynthesis also lack salt tolerance [113], and (3) exogenously applied glycinebetaine has enhanced the salinity tolerance of glycophytes such as rice (*Oryza sativa* L.) [114]. In contrast, proline accumulation has been considered by some investigators merely a result of plant injury because of its universally rapid appearance following any type of stress [115,116].

## IV. SUMMARY

The Poaceae, represented by over 7500 species, show extreme range in salinity tolerance, from salt sensitive to extremely salt tolerant (halophytic). In this chapter, the range of salinity tolerance and physiological adaptations to salinity present in grasses were described, focusing on three grass species representing the range of salt tolerance present in the Poaceae: salt-sensitive buffalograss [*Buchloë dactyloides* (Nutt.) Engelm.], salt-tolerant bermudagrass, and halophytic desert saltgrass [*Distichlis spicata* var. *stricta* (Torr.) Beetle].

Salinity tolerance in the Poaceae, indicated by 50% growth reduction, ranges from 4 dS m$^{-1}$ (e.g., annual bluegrass) to 40+ dS m$^{-1}$, essentially seawater (e.g., desert saltgrass). Although shoot growth decline with increasing salinity is typical, shoot growth may be stimulated by moderate salinity in highly salt-tolerant or halophytic grasses. However, root growth stimulation under moderate salinity is much more common in salt-tolerant grasses, resulting in increased root/shoot ratios and therefore increased water absorption/transpiration area, which may be an adaptive mechanism to saline osmotic stress.

It has long been accepted that the major causes of plant growth inhibition under salinity stress are osmotic stress (osmotic inhibition of plant water absorption) and specific ion effects, including toxicities and imbalances. In a number of studies, salinity tolerance in the Poaceae has been related to shoot saline ion exclusion. However, studies have shown that complete osmotic adjustment does occur under salt stress, even in salt-sensitive grasses. Because the predominant osmotica utilized are typically saline ions, ion regulation, rather than ion exclusion, may be a more apt description of the mechanism of salt tolerance occurring in the Poaceae. Grasses regulate saline ion concentrations by vacuolar ion compartmentation at the root or shoot or by excretion via specialized salt glands, although ion reabsorption by xylem/phloem and redistribution to roots or senescing leaves may play a minor role.

Bicellular leaf epidermal salt glands occur in many Chloridoid grasses. Basal cells have specific ultrastructural modifications, including parallel partitioning membranes, allowing active, selective saline ion excretion. Excretion rates, which may be substantial, are dependent on media salinity level and are typically highly selective for Na$^+$ and Cl$^-$. More recently, salinity tolerance of Chloridoid grasses has been related to salt gland excretion rate and leaf salt gland density.

Enzymes of higher plants, salt sensitive and tolerant alike, are inhibited by saline ion concentrations above 100–200 mM. Under salt stress, grasses typically accumulate saline ions to well above these levels for shoot osmotic adjustment, necessitating Na$^+$ and Cl$^-$ compartmentation in vacuoles, which constitute 90–95% of mature cell volume. Remaining cytoplasmic osmotic adjustment is achieved by certain organic osmotica compatible with cell enzymes, termed compatible solutes. Glycinebetaine and proline typically accumulate in salt-stressed grasses and have been proposed as compatible solutes. However, more recent evidence has supported glycinebetaine, not proline, as a functional compatible solute.

## REFERENCES

1.  F Ghassemi, AJ Jakeman, HA Nix. Salinisation of Land and Water Resources. Wallingford Oxon, UK: CAB International, 1995, pp 1–16.
2.  M Pessarakli, I Szabolcs. Soil salinity and sodicity as particular plant/crop stress factors. In: M Pessarakli, ed. Handbook of Plant and Crop Stress. 2nd ed. New York: Marcel Dekker, 1999, pp 1–16.
3.  LR Oldeman, VWP van Engelen, JHM Pulles. The extent of human-induced soil degradation. In: LR Oldeman, RTA Hakkeling, WG Sombroek, eds. World Map of the Status of Human-Induced Soil Degradation: An Explanatory Note. Wageningen: International Soil Reference and Information Centre, 1991, pp 27–33.
4.  I Szabolcs. Salt-Affected Soils. Boca Raton, FL: CRC Press, 1989, pp 5–30.
5.  A Hamdy. Saline irrigation: assessment and management techniques. In: R Choukr-Allah, CV Malcolm, A Hamdy, eds. Halophytes and Biosaline Agriculture. New York: Marcel Dekker, 1996, pp 147–180.
6.  R Choukr-Allah. The potential of halophytes in the development and rehabilitation of arid and semi-arid zones. In: R Choukr-Allah, CV Malcolm, A Hamdy, eds. Halophytes and Biosaline Agriculture. New York: Marcel Dekker, 1996, pp 3–13.
7.  F Ghassemi, AJ Jakeman, HA Nix. Salinisation of Land and Water Resources. Wallingford Oxon, UK: CAB International, 1995, pp 291–335.
8.  Arizona Department of Water Resources. Modifications to the Second Management Plan: 1990–2000. Phoenix, AZ: 1995, pp 1–74.
9.  California State Water Resources Control Board. Porter-Cologne Act Provisions on Reasonableness and Reclamation Promotion. Sacramento: 1993, pp 1–20.
10. FW Gould, RB Shaw. Grass Systematics. 2nd ed. College Station, TX: Texas A&M University Press, 1983, pp 1–15.
11. AS Hitchcock. Manual of the Grasses of the United States. 2nd ed. New York: Dover, 1971, pp 1–14.
12. US Salinity Laboratory Staff. Diagnosis and improvement of saline and alkali soils. In: LA Richards, ed. USDA Handbook 60. Washington, DC: Government Printing Office, 1954, pp 100–130.

13. EV Maas. Salt tolerance of plants. Appl Agric Res 1(1):12–26, 1986.
14. KB Marcum. Salinity tolerance in turfgrasses. In: M Pessarakli, ed. Handbook of Plant and Crop Stress. 2nd ed. New York: Marcel Dekker, 1999, pp 891–906.
15. JA Aronson. Haloph: A Data Base of Salt Tolerant Plants of the World. Tucson, AZ: Office of Arid Land Studies, University of Arizona, 1989, pp 68–70.
16. EV Maas, GJ Hoffman. Crop salt tolerance—current assessment. J Irrig Drainage Div ASCE 103:115–132, 1977.
17. KB Marcum. Salinity tolerance mechanisms of grasses in the subfamily Chloridoideae. Crop Sci 39:1153–1160, 1999.
18. RN Carrow, RR Duncan. Salt-Affected Turfgrass Sites—Assessment and Management. Chelsea, MI: Ann Arbor Press, 1998, pp 83–99.
19. SD Reid, AJ Koski, HG Hughes. Buffalograss seedling screening in vitro for NaCl tolerance. HortSci 28:536, 1993.
20. L Wu, H Lin. Salt tolerance and salt uptake in diploid and polyploid buffalograss (*Buchloe dactyloides*). J Plant Nutr 17:1905–1928, 1994.
21. JMJ de Wet, JR Harlan. Biosystematics of *Cynodon* L. C. Rich. (Gramineae). Taxon 19:565–569, 1970.
22. AE Dudeck, CH Peacock. Salinity effects on growth and nutrient uptake of selected warm-season turf. Int Turfgrass Soc Res J 7:680–686, 1993.
23. LE Francois. Salinity effects on three turf bermudagrasses. HortScience 23:706–708, 1988.
24. AE Dudeck, S Singh, CE Giordano, TA Nell, DB McConnell. Effects of sodium chloride on *Cynodon* turfgrasses. Agron J 75:927–930, 1983.
25. JD Butler, JL Fults, GD Sanks. Review of grasses for saline and alkali areas. Int Turfgrass Res J 2:551–556, 1974.
26. RT Parrondo, JG Gosselink, CS Hopkinson. Effects of salinity and drainage on the growth of three salt marsh grasses. Bot Gaz 139:102–107, 1978.
27. PR Kemp, GL Cunningham. Light, temperature and salinity effects on growth, leaf anatomy and photosynthesis of *Distichlis spicata* (L.) Greene. Am J Bot 68:507–516, 1981.
28. GL Horst, RM Taylor. Germination and initial growth of Kentucky bluegrass. Agron J 75:679–681, 1983.
29. LJ Greub, PN Drolsom. Salt tolerance of selected grass species and cultivars as affected by soil type, soil phosphorus and level of salt application. Agron J Abstr 69.111, 1997.
30. AE Dudeck, CH Peacock. Salinity effects on warm-season turfgrasses. Proceedings of 33rd Annual Florida Turfgrass Conference 33:22–24, 1985.
31. KB Marcum, CL Murdoch. Growth responses, ion relations, and osmotic adaptations of eleven $C_4$ turfgrasses to salinity. Agron J 82:892–896, 1990.
32. MAL Smith, JE Meyer, SL Knight, GS Chen. Gauging turfgrass salinity responses in whole-plant microculture and solution culture. Crop Sci 33:566–572, 1993.
33. MC Shannon. Testing salt tolerance variability among tall wheatgrass lines. Agron J 70:719–722, 1978.
34. KB Marcum, CL Murdoch. Salinity tolerance mechanisms of six $C_4$ turfgrasses. J Am Soc Hortic Sci 119:779–784, 1994.
35. D Pasternak, A Nerd, Y de Malach. Irrigation with brackish water under desert conditions. IX. The salt tolerance of six forage crops. Agric Water Manage 24:321–334, 1993.
36. NWM Warwick, GM Halloran. Variation in salinity tolerance and ion uptake in accessions of brown beetle grass [*Diplachne fusca* (L.) Beauv.]. New Phytol 119:161–168, 1991.
37. M Ashraf, MI Naqvi. Responses of three arid zone grass species to varying Na/Ca rations in saline sand culture. New Phytol 119:285–290, 1991.
38. MW Hester, IA Mendelssohn, KL McKee. Intraspecific variation in salt tolerance and morphology in the coastal grass *Spartina patens* (Poaceae). Am J Bot 83:1521–1527, 1996.
39. PM Bradley, JT Morris. Relative importance of ion exclusion, secretion and accumulation in *Spartina alterniflora* Loisel. J Exp Bot 42:1525–1532, 1991.
40. KB Marcum, CL Murdoch. Salt tolerance of the coastal salt marsh grass, *Sporobolus virginicus* (L.) Kunth. New Phytol 120:281–288, 1992.
41. KC Blits, JL Gallagher. Morphological and physiological responses to increased salinity in marsh and dune ecotypes of *Sporoblus virginicus* (L.) Kunth. Oecologia 87:330–335, 1991.
42. MA Khan, IA Ungar, AM Showalter. Effects of salinity on growth, ion content, and osmotic relations in *Halopyrum mucronatum* (L.) Stapf. J Plant Nutr 22:191–204, 1999.
43. JN Wood, DF Gaff. Salinity studies with drought-resistant species of *Sporobolus*. Oecologia 78:559–564, 1989.
44. G Naidoo, Y Naidoo. Salt tolerance in *Sporobolus virginicus*: the importance of ion relations and salt secretion. Flora-Jena 193:337–344, 1998.
45. PS Ramakrishnan, R Nagpal. Adaptation to excess salts in an alkaline soil population of *Cynodon dactylon* (L.) Pers. J Ecol 61:369–381, 1973.
46. LA Donovan, JL Gallagher. Morphological responses of a marsh grass, *Sporobolus virginicus* (L.) Kunth., to saline and anaerobic stresses. Wetlands 5:1–13, 1985.

47. L Bernstein, HE Hayward. Physiology of salt tolerance. Annu Rev Plant Physiol 9:25–46, 1958.
48. L Wu, H Lin. Salt concentration effects on buffalograss germplasm seed germination and seedling establishment. Int Turfgrass Res J 7:883–828, 1993.
49. WA Torello, AG Symington. Screening of turfgrass species and cultivars for NaCl tolerance. Plant Soil 82:155–161, 1984.
50. AH Khan, C Marshall. Salt tolerance within populations of chewing fescue (*Festuca rubra* L.). Commun Soil Sci Plant Anal 12:1271–1281, 1981.
51. Y Waisel. The stimulating effects of NaCl on root growth of Rhodes grass (*Chloris gayana*). Physiol Plant 64:519–522, 1985.
52. RC Ackerson, VB Youngner. Responses of bermudagrass to salinity. Agron J 67:678–681, 1975.
53. H Greenway, A Gunn, DA Thomas. Plant response to saline substrates. VIII. Regulation of ion concentration in salt sensitive and halophytic species. Aust J Biol Sci 19:741–756, 1966.
54. JW O'Leary. Physiological basis for plant growth inhibition due to salinity. In: WG McGinnies, BJ Goldman, P Paylore, eds. Food, Fiber and the Arid Lands. Tucson, AZ: University of Arizona Press, 1971, pp 331–336.
55. L Bernstein, HE Hayward. Physiology of salt tolerance. Annu Rev Plant Physiol 9:25–46, 1958.
56. J Gorham, RG Wyn Jones, E McDonnell. Some mechanisms of salt tolerance in crop plants. Plant Soil 89:15–40, 1985.
57. R Albert, M Popp. Chemical composition of halophytes from the Neusiedler Lake region in Austria. Oecologia 27:157–170, 1977.
58. J Gorham, PJ Randall, E Delhaize, RA Richards, R Munns. Genetics and physiology of enhanced K/Na discrimination. Genetic Aspects of Plant Mineral Nutrition. Dev Plant Soil Sci 50:151–158, 1993.
59. YW Yang, RJ Newton, FR Miller. 1990. Salinity tolerance of *Sorghum*: I. Whole plant response to sodium chloride in *S. bicolor* and *S. halepense*. Crop Sci 30:775–781, 1990.
60. RC Johnson. Salinity resistance, water relations, and salt content of crested and tall wheatgrass accessions. Crop Sci 31:730–734, 1991.
61. MA Harivandi, JD Butler, L Wu. Salinity and turfgrass culture. In: DV Waddington, RN Carrow, RC Shearman, eds. Turfgrass. Madison, WI: ASA, CSSA, and SSSA, 1992, pp 207–229.
62. NJ Hannon, HN Barber. The mechanism of salt tolerance in naturally selected populations of grasses. Search 3:259–260, 1972.
63. L Wu. The potential for evolution of salinity tolerance in *Agrostis stolonifera* L. and *Agrostis tenuis* Sibth. New Phytol 89:471–486, 1981.
64. KB Marcum, SJ Anderson, MC Engelke. Salt gland ion secretion: a salinity tolerance mechanism among five zoysiagrass species. Crop Sci 38:806–810, 1998.
65. YL Qian, MC Engelke, MJV Foster. Salinity effects on zoysiagrass cultivars and experimental lines. Crop Sci 40:488–492, 2000.
66. J Levitt. Responses of plants to environmental stresses. Vol II. New York: Academic Press, 1980, pp 35–50.
67. JA Hellebust. Osmoregulation. Annu Rev Plant Physiol 27:485–505, 1976.
68. AR Yeo. Salinity resistance: physiologies and prices. Physiol Plant 58:214–222, 1983.
69. J Gorham, LL Hughes, RG Wyn Jones. Chemical composition of salt-marsh plants from Ynys-Mon (Anglesey): the concept of physiotypes. Plant Cell Environ 3:309–318, 1980.
70. R Weimberg, MC Shannon. Vigor and salt tolerance in 3 lines of tall wheatgrass. Physiol Plant 73:232–237, 1988.
71. H Greenway, R Munns. Mechanisms of salt tolerance in nonhalophytes. Annu Rev Plant Physiol 31:149–190, 1980.
72. CH Peacock, AE Dudeck. A comparative study of turfgrass physiological responses to salinity. Int Turfgrass Res J 5:821–829, 1985.
73. RG Wyn Jones, J Gorham. Use of physiological traits in breeding for salinity tolerance. In: FWG Barer, ed. Drought Resistance in Cereals. Wallingford, UK: CAB International, 1989, pp 95–106.
74. KB Marcum, CL Murdoch. Salt glands in the Zoysieae. Ann Bot 66:1–7, 1990.
75. J Gorham. Mechanisms of salt tolerance of halophytes. In: R Choukr-Allah, CV Malcolm, A Hamdy, eds. Halophytes and Biosaline Agriculture. New York: Marcel Dekker, 1996, pp 31–53.
76. RJ Daines, AR Gould. The cellular basis of salt tolerance studied with tissue cultures of the halophytic grass *Distichlis spicata*. J Plant Physiol 119:269–280, 1985.
77. D Kramer. Cytological aspects of salt tolerance in higher plants. In: RC Staples, GH Toenniessen, eds. Salinity Tolerance in Plants. New York: John Wiley & Sons, 1984, pp 3–15.
78. WD Jeschke. $K^+$–$Na^+$ exchange at cellular membranes, intracellular compartmentation of cations, and salt tolerance. In: RC Staples, GH Toenniessen, eds. Salinity Tolerance in Plants. New York: John Wiley & Sons, 1984, pp 37–66.
79. J Garbarino, FM Dupont. NaCl salinity induces $K^+/Na^+$ antiport in tonoplast vesicles from barley roots. Plant Physiol 86:231–236, 1988.
80. WD Jeschke. Univalent cation selectivity and compartmentation in cereals. In: DL Laidman, RG Wyn Jones, eds. Recent Advances in the Biochemistry of Cereals. New York: Academic Press, 1979, pp 37–61.

81.  AR Yeo, D Kramer, A Lauchli, B Gullasch. Ion distribution in salt-stressed mature *Zea mays* roots in relation to ultrastructure and retention of sodium. J Exp Bot 28:17–30, 1977.
82.  EL Taleisnik. Sodium accumulation in *Pappophorum*. I. Uptake, transport and recirculation. Ann Bot 63:221–228, 1989.
83.  AR Yeo, TJ Flowers. Mechanisms of salinity resistance in rice and their role as physiological criteria in plant breeding. In: RC Staples, GH Toenniessen, eds. Salinity Tolerance in Plants. New York: John Wiley & Sons, 1984, pp 151–171.
84.  H Lessani, H Marschner. Relation between salt tolerance and long-distance transport of sodium and chloride in various crop species. Aust J Plant Physiol 5:27–37, 1978.
85.  WD Jeschke, S Klagges, A Hilpert, AS Bhatti, G Sarwar. Partitioning and flows of ions and nutrients in salt-treated plants of *Leptochloa fusca* L. Kunth. I. Cations and chloride. New Phytol 130:23–35, 1995.
86.  AS Bhatti, S Steinert, G Sarwar, A Hilpert, WD Jeschke. Ion distribution in relation to leaf age in *Leptochloa fusca* (L.) Kunth. (Kallar grass). I. K, Na, Ca and Mg. New Phytol 123:539–545, 1993.
87.  TJ Flowers, PF Troke, AR Yeo. The mechanisms of salt tolerance in halophytes. Annu Rev Plant Physiol 28:89–121, 1977.
88.  N Liphschitz, Y Waisel. Adaptation of plants to saline environments: salt excretion and glandular structure. In: DN Sen, KS Rajpurohit, eds. Tasks for Vegetation Science. Vol 2: Contributions to the Ecology of Halophytes. The Hague: W Junk, 1982, pp 197–214.
89.  Y Waisel. Biology of Halophytes. New York: Academic Press, 1972, pp 141–165.
90.  A Fahn. Secretory tissues in vascular plants. New Phytol 108:229–257, 1988.
91.  EL Taleisnik, AM Anton. Salt glands in *Pappophorum* (Poaceae). Ann Bot 62:383–388, 1988.
92.  N Liphshchitz, Y Waisel. Existence of salt glands in various genera of the Gramineae. New Phytol 73:507–513, 1974.
93.  V Amarasinghe, L Watson. Variation in salt secretory activity of microhairs in grasses. Aust J Plant Physiol 16:219–229, 1989.
94.  WD Clayton, SA Renvoize. Genera Graminum, Grasses of the World. London: HMSO Books, 1986.
95.  V Amarasinghe, L Watson. Comparative ultrastructure of microhairs in grasses. Bot J Linn Soc 98:303–319, 1988.
96.  JW Oross, WW Thomson. The ultrastructure of *Cynodon* salt glands: the apoplast. Eur J Cell Biol 28:257–263, 1982.
97.  JW Oross, WW Thomson. The ultrastructure of the salt glands of *Cynodon* and *Distichlis* (Poaceae). Am J Bot 69:939–949, 1982.
98.  Y Naidoo, G Naidoo. Cytochemical localisation of adenosine triphosphatase activity in salt glands of *Sporobolus virginicus* (L.) Kunth. S Afr J Bot 65:370–373, 1999.
99.  CA Levering, WW Thomson. The ultrastructure of the salt gland of *Spartina foliosa*. Planta 97:183–196, 1971.
100.  G Pollak, Y Waisel. Ecophysiological aspects of salt excretion in *Aeluropus litoralis*. Physiol Plant 47:177–184, 1979.
101.  G Pollak, Y Waisel. Salt secretion in *Aeluropus litoralis* (Willd.) Parl. Ann Bot 34:879–888, 1970.
102.  J Wieneke, G Sarwar, M Roeb. Existence of salt glands on leaves of Kallar grass (*Leptochloa fusca* L. Kunth.). J Plant Nutr 10:805–820, 1987.
103.  MO Arriaga. Salt glands in flowering culms of *Eriochloa* species (Poaceae). Bothalia 22(1):111–117, 1992.
104.  W Worku, GP Chapman. The salt secretion physiology of a Chloridoid grass, *Cynodon dactylon* (L.) Pers., and its implications. Sinet 21:1–16, 1998.
105.  RG Wyn Jones, CJ Brady, J Speirs. Ionic and osmotic relations in plant cells. In: DL Laidman, RG Wyn Jones, eds. Recent Advances in the Biochemistry of Cereals. New York: Academic Press, 1979, pp 63–103.
106.  TJ Flowers. Physiology of halophytes. Plant Soil 89:41–56, 1985.
107.  RG Wyn Jones. Phytochemical aspects of osmotic adaptation. In: BN Timmerman, ed. Recent Advances in Phytochemistry. Vol 3. Phytochemical Adaptations to Stress. New York: Plenum, 1984, pp 55–78.
108.  RG Wyn Jones, J Gorham. Osmoregulation. In: OL Lange, PS Nobel, CB Osmond, H Ziegler, eds. Physiological Plant Ecology III. Responses to the Chemical and Biological Environment. Berlin: Springer-Verlag, 1983, pp 35–58.
109.  RA Leigh, N Ahmad, RG Wyn Jones. Assessment of glycinebetaine and proline compartmentation by analysis of isolated beet vacuoles. Planta 153:34–41, 1981.
110.  D Aspinall, LG Paleg. Proline accumulation: physiological aspects. In: LG Paleg, D. Aspinall, eds. Physiology and Biochemistry of Drought Resistance in Plants. Sydney: Academic Press, 1981, pp 205–241.
111.  D Rhodes, AD Hanson. Quaternary ammonium and tertiary sulfonium compounds in higher plants Annu Rev Plant Physiol Plant Mol Biol 44:357–384, 1993.
112.  PH Yancy. Compatible and counteracting solutes. In: K Strange, ed. Cellular and Molecular Physiology of Cell Volume Regulation. Boca Raton, FL: CRC Press, 1994, pp 81–109.
113.  HC Saneoka, C Nagasaka, DT Hahn, WJ Yang, GS Premachandra, RJ Joly, D Rhodes. Salt tolerance of glycinebetaine-deficient and -containing maize lines. Plant Physiol 107:631–638, 1995.

114. P Harinasut, K Tsutsui, T Takabe, M Nomura, S Kishitani, T Takabe. Glycinebetaine enhances rice salt tolerance. In: P Mathis, ed., Photosynthesis: From Light to Biosphere. Vol IV. Dordrecht: Kluwer Academic Press,
1995, pp 733–736.

115. TD Colmer, E Epstein, J Dvorak. Differential solute regulation in leaf blades of various ages in salt-sensitive
wheat and a salt-tolerant wheat X *Lophopyrum elongatum* (Host) A. Löve amphiploid. Plant Physiol
108:1715–1724, 1995.

116. S Mumtaz, SSM Maqvi, A Shereen, MA Khan. Proline accumulation in wheat seedlings subjected to various
stresses. Acta Physiol Plant 17:17–20, 1995.

# 32

# Physiological Mechanisms of Nitrogen Absorption and Assimilation in Plants Under Stressful Conditions

**R. S. Dubey**

*Banaras Hindu University, Varanasi, India*

**Mohammed Pessarakli**

*The University of Arizona, Tucson, Arizona*

## I. INTRODUCTION

Nitrogen is one of the most essential elements for plant growth and development. It is a constituent of many biomolecules such as proteins, nucleic acids, amino acids, coenzymes, vitamins, and pigments. Because of its high requirement by plants and its complete absence in the bedrock, it has a special place in plant nutrition. Nitrogen supply in the soil is often the most important factor limiting plant growth and yield. In the soil, N availability is due to the application of N fertilizers, biological action of $N_2$ fixing organisms, or natural fertilization. With the advent of modern agricultural practices, inorganic N fertilizers have become the major input to the soil. In our quest to achieve sustainable food production, to meet the increasing food requirements for global population, excessive uses of various forms of N fertilizers are still likely in the near future.

Ammonium ($NH_4^+$) and nitrate ($NO_3^-$) are available forms of N that can be absorbed by plants [1,2]. However, $NO_3^-$ is the predominant form of N available to the most cultivated plants grown under normal field conditions. Availability of nitrogenous nutrients, especially $NO_3^-$, is considered rate limiting for plant growth and crop production. Application of $NO_3^-$ in the soil medium induces $NO_3^-$ uptake and its assimilation to ammonium by assimilatory enzymes.

Plants are often exposed to various kinds of harsh environmental conditions that adversely affect their growth and metabolism. Adverse environmental conditions, such as soil salinity, drought, heat, cold, and excessive heavy metal content in the soil, create considerable stress in growing plants [3] and severely affect N absorption by the roots and its assimilation in the plant [1,2,4–15]. In order to adapt to changing environments, higher plants show well-defined metabolic alterations in response to nutrient availability in the environment [16]. For instance, $NO_3^-$ in the soil induces the system of its uptake, assimilation, transport, etc. [16]. The biochemical events leading to the uptake of $NO_3^-$ by plants are not well defined [17]. However, the process of $NO_3^-$ reduction involving the enzymes nitrate reductase (NR) and nitrite ($NO_2^-$) reductase (NIR) has been extensively studied in the diverse plant species, and these enzymes have been well characterized regarding their physicochemical properties and their subcellular localizations [18,19]. Environmental stresses influence N nutrition in plants by inhibiting N uptake [1–6,12–15,20–36] as well as its assimilation [18,27,31,33–35,38–62].

This review chapter focuses on various N sources and their mode of absorption and assimilation by plants and also presents information on the effects of different conditions on the N uptake and metabolism processes.

## II.  NITROGEN SOURCES, THEIR ABSORPTION AND ASSIMILATION

### A.  Sources of Nitrogen

Different forms of N that are absorbed by plants from the soil are $NO_3^-$, $NH_4^+$, and organic compounds such as amino acids and urea. The two major forms of soil N are $NO_3^-$ and $NH_4^+$. Of these two forms, $NO_3^-$ is the more abundant and under normal conditions most of the N absorbed by plant roots from the soil is in the form of $NO_3^-$, which is further reduced to $NH_4^+$ in the plant tissue. Frota and Tucker [30], Saad [31], and Pessarakli et al. [45] found that beans (*Phaseolus vulgaris* L.), $C_3$ plants, under either normal or stress conditions absorbed more $NO_3^-$ than $NH_4^+$. Sometimes, in the soil, the $NH_4^+$ form of N is abundant because of biological $N_2$ fixation by symbiotic association of $N_2$-fixing organisms, free-living soil bacteria, and blue-green algae. Although N fertilizers in ammoniacal forms are widely used in agricultural fields, ammonium in the soil is readily oxidized to $NO_3^-$ by nitrifying bacteria present in the soil. Certain plant species such as those inhabiting acidic soils and all $C_4$ plants (i.e., grasses) show a preference for the $NH_4^+$ form of N. Because of the deficiency of $NO_3^-$ in acid soils and the specific physiological, metabolic, and photosynthetic pathways of $C_4$ plants, such plants prefer $NH_4^+$ over $NO_3^-$. Plants that have low intrinsic $NO_3^-$ reductase activity also absorb $NH_4^+$ in preference to $NO_3^-$.

Because ammonia is toxic to the plants and interferes with various metabolic processes inside the cell, after absorption $NH_4^+$ ions are rapidly assimilated into amino acids, amides, etc. in the roots. Plant species that have an efficient ammonia-detoxifying system grow well on the $NH_4^+$ form of N [63]. These plants can detoxify $NH_3$ by forming $NH_4^+$ salts of organic acids. The majority of the plant species grow better when N is supplied as a mixture of both $NO_3^-$ and $NH_4^+$ forms in the soil [1]. Botella et al. [1] studied the uptake of $NO_3^-$ and $NH_4^+$ by wheat plants grown in nutrient solutions containing $NO_3^-$, $NH_4^+$ or $NO_3^- + NH_4^+$, with 1 mM (control) and 60 mM (saline) NaCl each. These investigators found that under saline conditions, the addition of both nitrogen forms was beneficial because higher nitrogen uptake rates resulted in better growth and development of the plant. Botella et al. [2] in another N uptake study, using $NO_3^-$, $NH_4^+$, and $NO_3^- + NH_4^+$, reported that the best N source for wheat growth was a mixture of $NO_3^-$ and $NH_4^+$, especially under saline conditions. Certain plants can absorb either $NH_4^+$ or $NO_3^-$ ions depending on the pH of the nutrient medium.

It has been suggested by certain groups of investigators that nutrient media containing both $NO_3^-$ and $NH_4^+$ forms of N in a proper combination are more suitable for the growth of the cells as well as the plants compared with the either form alone [1,2,64]. In many vegetable crops, $NH_4^+$ is taken up in preference to $NO_3^-$ when its concentration is above 10% of the total N in the nutrient solution [65]. Especially at a low root temperature, $NH_4^+$ is regarded as a safe source of N [65]. Genotypic diversity occurs in plants for N use efficiency. Kafkafi [65], when evaluating 12 genotypes of pearl millet (*Pennisetum glaucum*) plants, observed that genotypes varied greatly for uptake, translocation, and nitrogen use efficiency (NUE). Genotypes with lower NUE and translocation indices for N showed lower grain yields.

In plant roots, the initial product of $NH_4^+$ assimilation is glutamine whether $NH_4^+$ or $NO_3^-$ is absorbed by the roots. Other products of assimilation are asparagine, citrulline, amino acid, allantoin, and certain other soluble nitrogenous compounds [66]. The assimilation products are then translocated to various organs of the plants through xylem and phloem vessels. In some plants (i.e., tomatoes), $NO_3^-$ after absorption through the roots is reduced to $NH_4^+$ in the root itself, whereas in others (i.e., grasses), it may be transported as $NO_3^-$ to different organs.

### B.  Absorption and Assimilation of Nitrogen

Nitrate is the predominant form of N available in the soil, regardless of the $NO_3^-$ or $NH_4^+$ forms of fertilizers used. Availability of $NO_3^-$ in the soil is considered rate limiting for plant growth [16]. Systems related to the uptake, intracellular transport, and translocation of $NO_3^-$ are directly affected by the soil $NO_3^-$ level [17]. In response to environmental $NO_3^-$, root tissues proliferate and a general increase in root growth and metabolism occurs [16]. The uptake of $NO_3^-$ by plant roots is an active process involving a high-affinity $NO_3^-$ transport system where $NO_3^-$ acts as a signal for these events [67]. The uptake system is inducible with $NO_3^-$ and can be blocked with inhibitors of RNA and protein synthesis [17] as well as amino acid–modifying reagents [68]. This suggests that plasma membrane proteins are involved in the transport process.

According to Redinbaugh and Campbell [16], when plant roots sense exogenous $NO_3^-$, the primary response involves the transcription of genes encoding $NO_3^-$ transport proteins followed by the synthesis of these proteins. A model proposed by these investigators suggests that a constitutive $NO_3^-$ sensor protein system is the first component that detects and senses $NO_3^-$ in the environment. Binding of the $NO_3^-$ with sensor induces certain regulatory proteins, which in turn initiate the transcription of primary response genes by RNA polymerase II. The resulting transcripts are further translated into proteins such as $NO_3^-$ transporters, $NO_3^-$ translocators, and $NO_3^-$ assimilatory enzymes. Biochemical events leading to secondary responses such as root proliferation enhanced respiration in response to environmental $NO_3^-$ are not yet clearly understood [16]. Because environmental $NO_3^-$ is the only inducer for the synthesis of NR, NIR, and $NO_3^-$ transport proteins, it appears that $NO_3^-$ induces these processes at the plasma membrane level before entering the cell [17]. However, in plants, a specific receptor for $NO_3^-$ has not yet been identified.

Certain investigators suggest that the prime enzyme of $NO_3^-$ reduction, nitrate reductase, plays a significant role in the uptake of nitrate [69]. A plasma membrane–bound NR has been detected in the roots of barley (*Hordeum vulgare* L.) seedlings and it is observed that $NO_3^-$ transport is inhibited by anti-NR immunoglobulin G (IgG) fragments [70]. Uptake of $NO_3^-$ by plant roots is also dependent on the $NO_3^-$ reduction process or reduced $NO_3^-$ products in the shoot [71]. At a high intracellular concentration of $NO_3^-$ or in the presence of $NH_4^+$ in the growth medium, $NO_3^-$ uptake tends to decline. Production of malate in the shoot also influences $NO_3^-$ uptake. In order to neutralize the alkaline conditions due to $NO_3^-$ reduction in shoots, malate is produced. This is further transported to roots with $K^+$ as a counter ion, and in turn, as a result of oxidation, bicarbonate ions are formed in roots. These are exchanged for $NO_3^-$ in the external environment.

Nitrate uptake and its reduction activities in the plant tissues are coordinately regulated. In plant tissues, $NO_3^-$ induces increase in NR activity. The activity of NR increases in root cells in response to exogenous $NO_3^-$. Besides the $NO_3^-$ reduction process, which is the primary response to $NO_3^-$ uptake, plants have systems for translocation of $NO_3^-$ within and between the cells [69]. After uptake, $NO_3^-$ may be translocated to the vacuole in the cells, where it can be accumulated and serve as an $NO_3^-$ reserve [72]. The intracellular $NO_3^-$ translocation process possibly requires a tonoplast $NO_3^-$ translocator that is different from the membrane $NO_3^-$ transporter [16]. The distinct $NO_3^-$ translocators present at the symplasm-xylem interface control the translocation of $NO_3^-$ from root to xylem and then to different organs of the plant [16]. Although the translocator proteins appear to be different from transport proteins and are encoded by different genes, all three processes—$NO_3^-$ transport, translocation, and reduction—are coordinately regulated.

Although both forms of N ($NO_3^-$ and $NH_4^+$) are taken up by plants, only $NH_4^+$ is incorporated into organic molecules in the plant tissues by an enzymatic process. The primary step in the reduction process involves the reduction of $NO_3^-$ to $NO_2^-$ catalyzed by the enzyme NR. Ammonium, either directly absorbed by plant roots or as a result of $NO_3^-$ reduction, is further assimilated and incorporated into the amide amino group of glutamine by the action of glutamine synthetase and subsequently into glutamic acid by glutamate synthase. These two enzymes are responsible for the assimilation of most of the $NH_4^+$ derived from $NO_3^-$ reduction under normal growth conditions. An alternative route of $NH_4^+$ assimilation into glutamate involves the reductive amination of $\alpha$-ketoglutarate catalyzed by a mitochondrial enzyme glutamate dehydrogenase. Other amino acids, such as alanine and aspartic acid, are further synthesized from glutamic acid by transamination reactions.

## III. NITROGEN ABSORPTION AND ASSIMILATION UNDER DIFFERENT STRESSES

Crops growing in adverse environmental conditions of salinity, drought, high or low temperature, low light, and heavy metal–containing soils suffer severe losses in yield [3,5,6,8,12,73–76]. Harsh environmental conditions interfere with normal growth, metabolism [15], and protein synthesis of plants [77], and plants respond to these stresses by different types of physiological and biochemical adjustments. Like various physiological processes, N uptake, translocation, and assimilation are severely affected by different types of stresses [1,2,4–6,12–15,20–62,78–80]. Because the availability of nitrogenous nutrients in the soil, their uptake, and their assimilation are directly related to each other as well as to the growth and yield of the crops, considerable efforts have been made by various groups of investigators to study the possible implications of various stress conditions for N nutrition in plants [1,2,4–6,12–15,20–62,72]. In the following sections, the influence of diverse environmental stresses on the overall process of N uptake is summarized.

## A.   Salinity

Soil salinity is one of the major environmental stresses affecting crop productivity. Effect of salinity on plants may vary depending on the developmental stage of the plant [11,73] as well as the types and concentration of salts [1–6,12,74–76]. The responses to salinity on N uptake differ in different plant species and also depend on the type and extent of salinity. In the majority of the plant species studied, salinization in the soil affects N uptake, whereas in the halophytes and in many salt-tolerant crop species no significant effect of NaCl on $NO_3^-$ uptake is observed [81]. Barley (*Hordeum vulgare* L.) plants growing under saline conditions show reduced growth [81,82] as well as decreased N uptake [81]. In young barley seedlings, salinity severely inhibited $NO_3^-$ uptake, whereas little effect on $NO_3^-$ reduction was observed [78]. In wheat (*Triticum aestivum* L.) plants, reduction in growth was even more than that in barley at higher NaCl salinity levels [82], and uptake of N decreased with increasing salinity [81]. However, by increasing the N supply to the soil, the effect of salinity was alleviated [81]. Khalil et al. [83] found similar results for cotton (*Gossypium hirsutum* L.) and corn (*Zea mays* L.) plants. Soltani et al. [79] observed that when barley seedlings were grown in the presence of 200 mM NaCl, growth of the seedlings decreased with a concomitant reduction in the uptake and translocation of N compared with nonsalinized seedlings.

When seedlings of maize genotype differing in drought resistance were grown at $-0.84$ MPa NaCl salinity, the supply of reduced N for the synthesis of amino acids and proteins in the tissues was reduced [80]. The effect was more pronounced in drought-resistant genotypes, in which salinity reduced the activity of the metabolic pathway supplying reduced N accompanied with a corresponding reduction in the relative growth rate of the seedlings. Reduced growth in terms of dry-matter production and decreased absorption of N have been reported by several investigators for various plant species with different degrees of salt tolerance [2–6,12–14,22–26,29–32,35,36,41,42,44–47,51,73–75].

Under salinization, reduced uptake of N by crops appears to be due to more intake of $Na^+$ and $Cl^-$ by the roots. Increased levels of $Na^+$ in the plant tissues cause nutrient imbalance and displace $Ca^{2+}$ from the exchange sites on the membranes and cell walls [81]. Chloride present at more than 100 mM in the saline medium inhibits $NO_3^-$ uptake, possibly because of increased accumulation of Cl in the roots [84]. Smith [85] observed that $NO_3^-$ uptake in barley was dependent on the internal rather than the external concentrations of Cl. Reduced uptake of N could lead to N deficiency in plants and thus could become a limiting factor for growth of plants under saline conditions [84]. Because salinity leads to N deficiency, fertilization of plants growing in a saline environment with increasing doses of nitrogenous fertilizers has proved beneficial. It minimizes salt-induced damage and apparently provides salt tolerance [81]. However, in certain crop species such as corn, rice (*Oriza sativa* L.), wheat, and spinach (*Spinacia oleracea* L.) with excess application of nitrogenous fertilizers a decrease in salt tolerance is observed [81].

The presence of $Ca^{2+}$ in the medium increases $NO_3^-$ uptake under saline conditions. Ward et al. [84] observed that NaCl decreased $NO_3^-$ uptake in barley seedlings, whereas the uptake rate increased with increasing level of $Ca^{2+}$ between 1.0 and 3.0 mM in the saline medium. These investigators observed 31 to 35% more uptake of $NO_3^-$ by increasing $Ca^{2+}$ in the medium compared with a salt-free (control) medium. Manganese and $Mg^{2+}$ also enhanced $NO_3^-$ uptake under saline conditions, but $Ca^{2+}$ was more effective than these two ions [84]. The presence of $Ca^{2+}$ in the saline medium possibly decreased $Na^+$ as well as $Cl^-$ uptake and also reduced membrane disruption in saline solutions, leading to increased $NO_3^-$ uptake [84]. Calcium plays a significant role in maintaining the integrity of the root membranes; thus its deprivation, under salinization, decreases ion transport and $NO_3^-$ uptake by disrupting the $NO_3^-$ transporter that is located in the plasmalemma of roots [84]. Under saline conditions, $Ca^{2+}$ has been shown to increase the activity of the $NO_3^-$ transporter [84].

In halophytes, salinity either induces uptake and accumulation of $NO_3^-$ or has no effect on these processes [86]. Halophytes can accumulate inorganic ions such as $Na^+$, $K^+$, $Cl^-$, and $NO_3^-$ in excess compared with nonhalophytes under saline conditions. *Atriplex, Salicornia*, and *Suaeda maritima* plants show higher uptake of $Na^+$, $Cl^-$, $SO_4^{2-}$, and $NO_3^-$ in saline environments than under nonsaline conditions [86]. According to Flowers et al. [86], even in conditions of low salinity, the levels of $K^+$, $NO_3^-$, and $SO_4^{2-}$ are very high in halophytes compared with other plants. These investigators believe that high uptake of $NO_3^-$ by halophytes under salinization is related to the intrinsic properties of these plants as they are adapted to grow and show normal metabolic functions at high ion concentrations.

The prime enzyme of $NO_3^-$ assimilation, NR, which catalyzes the conversion of $NO_3^-$ to $NO_2^-$, was shown to play a major role in the uptake of $NO_3^-$ by serving as an $NO_3^-$ transporter [87]. This enzyme has been studied extensively by various groups of investigators for its behavior in different plant species under salinization. Evidence indicates that the uptake and utilization of applied $NO_3^-$ are largely dependent on its assimilation inside the plant tissues [88]. The effects of salinity on NR activity are varied and depend on the type of salinity as well as the plant species. Nitrate reductase is highly sensitive to various types of environmental stresses including salinity [81]. In many salt-sensitive plant species, NR activity decreased under NaCl salinity [89–91]. Plaut [89] observed decreased activity of NR in cell-free extracts as well as intact tissues of wheat (*Triticum aestivum* L.) seedlings when NaCl was applied to the nutrient medium and the enzyme was assayed after 24 hr of exposure to salinity stress. Similarly, while studying the effects of salinity on N metabolism on wheat plants, Abdul-Kadir and Paulsen [90] observed decreased NR activity under salinization. In pea (*Pisum sativum* L.) seedlings, an isosmotic concentration of NaCl suppressed NR activity and caused accumulation of $NO_3^-$ in the plant tissue, and in wheat seedlings an isosmotic salinity level decreased NR activity without significant accumulation of $NO_3^-$ in the tissues [81]. In young barley plants [78] and rice (*Oriza sativa* L.) seedlings [92], decreased NR activity has been observed under salinization. In pasture plants, NaCl salinity reduced growth with a concomitant decrease in NR activity [93–95].

Lal and Bhardwaj [91] observed that after 15 days of salinization of field pea (*Pisum sativum* L.) with NaCl and $CaCl_2$ (1:1), there was significant suppression of NR activity accompanied by decreases in total-N as well as protein-N and increases in $NO_3^-$-N and $NH_4^+$-N. According to these investigators, NaCl as well as $CaCl_2$ salinity impaired $NO_3^-$ assimilation in pea plants, leading to accumulation of $NO_3^-$ and $NH_4^+$ in the tissues. Tewari and Singh [94], while conducting stress studies in lentil (*Lens esculenta* Moench), observed that with increasing exchangeable sodium percentage (ESP) in the cell, there appeared a continuous decrease in NR as well as NIR activities in plants up to 60 days after sowing. Genotypes of rice plants differing in salt tolerance show varying behaviors of NR as well as NIR under salinity stress [39]. Katiyar and Dubey [39], while studying the mode of N assimilation under salinization in the seedlings of two sets of rice cultivars differing in salt tolerance, observed decreased NR activity in seedlings of salt-sensitive cultivars. When desalted enzyme extracts from nonsalinized rice seedlings were assayed for NR activity in the presence of 1 M NaCl in the assay medium, strong suppression of the enzyme activity was observed. Other investigators noticed similar suppression of NR activity in salt-sensitive genotypes of rice seedlings when grown in saline medium [38,90].

Several possible explanations have been suggested for the decreased NR activity in salt-sensitive plants under saline stress [89,96]. A plausible reason appears to be the inhibition of enzyme induction under salinization [94]. As NR is a substrate-inducible enzyme, under saline conditions $NO_3^-$ uptake by the plants is reduced. This causes limited $NO_3^-$ availability in the plant tissues so that NR induction is suppressed, which results in decreased NR activity [96].

Several investigators have emphasized that, under salinity stress, enhanced translocation of $NO_3^-$ and assimilates takes place from roots to shoots and from flag leaves to developing grains [78,97].

In certain plant species, an increase in NR activity has been observed due to salinity [38,39,81,98]. *Salicornia europeaca* plants and corn (*Zea mays* L.) seedlings showed increased NR activity when grown in a salinized medium [81]. Joshi [98], while conducting experiments on *Cajanus cajan* plants, observed that NaCl salinity stimulated NR activity in the leaves of plants, whereas $Na_2SO_4$ salinity inhibited the enzyme activity. In *Cajanus* plants a gradual increase in NR activity was observed with increase in NaCl salinity of the soil in the range 2.5 to 10.0 dS $m^{-1}$ [98]. In seedlings of *Phaseolus aconitifolius*, Sankhla and Huber [99] observed increased in vivo NR activity with salinization.

Rice plants differing in salt tolerance show varying behaviors of NR activity [38,39,95]. Salt-sensitive genotypes of rice plants showed decreased NR activity under salinization, whereas an increased NR level under salinity was observed in salt-tolerant genotypes [38,39]. Katiyar and Dubey [39] observed a marked increase in in vivo NR activity in roots as well as shoots of salt-tolerant rice cultivars CSR-1 and CSR-3 with a salinity level up to 14 dS m$^{-1}$ NaCl compared with nonsalinized plants. The higher NR level in seedlings of salt-tolerant cultivars suggests that salinity may promote synthesis or induction of the enzyme in seedlings of such cultivars. Salt-tolerant crop cultivars thus appear to have better adaptability to saline stress by exhibiting efficient NO$_3^-$ reduction under salinization.

The glutamine synthetase (GS)/glutamine-oxoglutarate amido transferase (GOGAT) pathway, which is the route of ammonia assimilation in plants under normal conditions, is adversely affected by salinization [100,101]. Miranda-Ham and Loyola-Vargas [100] observed that when *Canavalia ensiformis* plants were subjected to NaCl salinity stress, the activity of GS decreased markedly in roots as well as shoots. In the roots of salt-sensitive pea plants, decreased GS activity was observed under salinization [81]. Katiyar [101], while studying the behavior of GOGAT in situ in the two sets of rice cultivars differing in salt tolerance, observed that an NaCl salinity level of 14 dS m$^{-1}$ was inhibitory to the enzyme. In tolerant genotypes of crop plants, GOGAT is more tolerant to NaCl than in sensitive genotypes [101]. Decreased activities of GS and GOGAT under salinization suggest possible impairment of N assimilation and/or amino acid biosynthesis by this pathway as a result of salinity.

The glutamate dehydrogenase (GDH) enzyme plays an important role in ammonia assimilation under stress conditions by detoxifying the ammonia that tends to accumulate under such conditions. In many crop species examined, under salinization, the activity of GDH increased [81,101–103]. However, in certain cases, it remained comparable to that in the controls [100] or was decreased by NaCl salinity [81,91]. In the obligate halophyte *Suaeda maritima*, Boucaud and Billard [102] observed an increase in GDH activity with 25 mM NaCl. Similarly, in peanut leaves a salinity-induced increase in GDH activity was observed by Rao et al. [103]. Sharma and Garg [104], while studying the amination and transamination events in wheat plants, observed that plants grown at 8 and 16 dS m$^{-1}$ NaCl showed increased activity of GDH in leaves as well as roots. Seedlings of rice cultivars differing in salt tolerance, when raised under increasing levels of NaCl salinity, showed a marked increase in GDH activity both in vivo as well as in vitro compared with controls [101]. The effects were greater in the sensitive than in the tolerant cultivars. Increased activity of GDH with salinization suggests a possible role of this enzyme in ammonium assimilation under saline conditions [105]. It is suggested that saline conditions favor increased accumulation of ammonium and related compounds. Thus, the GS/GOGAT pathway of ammonium assimilation is impaired, and under such conditions, an increased level of GDH imparts adaptive value to plants by detoxifying and assimilating more ammonium [105]. Rice (*Oryza sativa* L.) crop cultivars appear to have better adaptability to saline stress by exhibiting efficient NO$_3^-$ reduction under salinization.

Effects of soil salinity on the N content of plant species are varied and depend on the species, the organs studied, as well as the type of salinity. Lal and Bhardwaj [91] observed decreases in the total-N and protein-N content of 15-day-old *Pisum sativum* seedlings salinized with a mixture of NaCl and CaCl$_2$ with 4 and 8 dS m$^{-1}$ salinity. However, an increase in soluble forms of N (NO$_3^-$ and NH$_4^+$) was observed with salinity. A similar decrease in the content of pea seedlings growing at isosmotic levels ($-0.1$ to $-0.5$ M Pa) of NaCl, CaCl$_2$, and Na$_2$SO$_4$ salts was observed by Singh et al. [106]. These investigators observed a decrease in N content with increasing salt stress and found that salt stress was more harmful to N content than water stress. While investigating the effects of salinization on nodulation and N fixation in pea plants, Siddiqui et al. [107] reported decreases in nodule N and total plant N and a significant reduction in the N$_2$-fixing efficiency of the nodules with increasing level of salinity. In *Vigna radiata* plants a salinity level of 8 dS m$^{-1}$ was lethal to plant growth and nodulation. Increasing the salinity level in such plants from 0 to 4 and 6 dS m$^{-1}$ decreased the nodulation and the N content of roots, stem, and leaves [108].

In certain cases, increased N contents have been observed in various plant species subjected to salinity by several investigators [3,6,22–35,42–47,49,51,83,90–98]. Sharma et al. [6] observed that N concentrations in grains and N uptake in grains and straw increased with an exchangeable sodium percentage (ESP) up to 25. Pessarakli and Tucker [26] found that the N contents of cotton shoots and roots increased with NaCl salinity up to $-0.8$ MPa osmotic potential of the nutrient solution. At the low level of salinity ($-0.4$ MPa osmotic potential), plants contained significantly higher total-N [26] as well as crude protein [47] compared with the controls (nonsalinized plants). Khalil et al. [83] reported similar increases in the total-N concentration of cotton and corn under salt stress conditions. In *Cajanus cajan*, a protein-rich leguminous crop, a salinity treatment of 10 dS m$^{-1}$ NaCl caused about a 43% increase in total-N and protein content in the leaves of 3-month-old plants over controls. A similar salinity treatment with $Na_2SO_4$ caused a decrease in N and protein content compared with controls [109]. It is suggested that the two salts NaCl and $Na_2SO_4$ show specific ion effects on the N metabolism of *Cajanus cajan* [109]. *Phaseolus vulgaris* plants, when grown in greenhouse conditions and irrigated with water containing 44, 88, and 132 mM NaCl, showed increases in the total N content of leaves with increasing salinity [110]. Addition of 4 or 8 mM $CaCl_2$ or $CaSO_4$ in the NaCl treatment medium further increased the leaf N content in such plants, indicating that $Ca^{2+}$ addition helps in maintaining the selective permeability of the membranes [110].

## B. Water Stress

Water availability is one of the most limiting environmental factors affecting crop productivity. In semi-arid tropics, the occurrence of drought or water deficit in the soil is common, whereas crop plants of temperate and tropical regions undergo seasonal periods of water stress, especially during the summer. The plant responses to water stress depend on the severity and the duration of stress and the growth stage of the plant [111]. Low water potential in the soil as well as inside the plant inhibits plant growth, reduces developmental activities of cells and tissues, decreases the uptake of essential nutrient elements, and causes a variety of morphological and biochemical modifications. Plants growing in water-stressed environments show reduced N uptake [3,4,7–11,13,20,30,31,35,40,48,51,76] from the culture medium or soil and decreased activities of N assimilatory enzymes [112–114].

When the water potential inside the plant declines below a threshold value, stomata closure takes place, which causes reductions in transpiration and water transport through the plant. This, in turn, affects the roots directly so that the roots are unable to accumulate or absorb $NO_3^-$ as effectively as when transpiration is normal [115]. At low water potential, the ability of roots to supply $NO_3^-$ to the transpiration stream decreases, leading to a decrease in $NO_3^-$ concentration of the xylem sap [115]. Under nonstressed conditions, in a freely transpiring plant, a continuous movement of $NO_3^-$ from the roots to the leaves ($NO_3^-$ flux) is maintained. This $NO_3^-$ flux decreases during water stress.

It was suggested by Viets [116] that, under water stress conditions, roots are unable to take up much nutrient from the soil because of lack of root activity and slow rates of ion diffusion and water movement. When examining the uptake of various nutrients by wheat varieties, Rao and Ramamoorthy [117] observed a 39% drop in N uptake of six improved varieties of wheat when moisture stress was imposed at different stages of plant growth. According to these investigators, the uptake of N was affected by applied stress mainly through restricted movement of water under such conditions.

Water stress causes a decrease in leaf $NO_3^-$ content as well as $NO_3^-$ flux from the roots to the leaves [115]. When the water-stressed plants were rewatered, $NO_3^-$ flux increased but not the leaf $NO_3^-$ content. When water-stressed plants were fertilized with more $NO_3^-$, the $NO_3^-$ flux increased and plant performance as well as grain yield improved [113]. Kathju et al. [113] observed that when wheat plants were grown under low ($N_0P_0$) and high ($N_{80}P_{80}$) fertility conditions and water stress was imposed at various stages of the plant's life cycle, increasing intensities of stress adversely affected leaf metabolism and plant performance. However, the performance of plants was better under high-fertility conditions at all stages with different intensities of water stress. Similar observations by other investigators also indicate that $NO_3^-$ application can partly alleviate water stress–associated damage in plants [118,119]. Lahiri [118] demonstrated that N application to the soil reduced the adverse effect of drought on dry matter and grain yield of pearl millet. Sorghum (*Sorghum halepense* L.) plants, when fertilized with N, recover faster after relief from water stress [119]. Although fertilized plants experienced water stress severely, they recovered from stress more quickly than unfertilized ones. Such observations have far-reaching consequences in the sense that in dry-land agriculture, where water is a limiting factor, fertilizer application can be considered for drought mitigation management [119].

Considerable studies have been performed by various groups of investigators to examine the behaviors of $NO_3^-$ assimilatory enzymes in plants under water stress conditions. In these studies, NR has received the most attention. The activity of NR is sensitive to the water potential of the plant and decreases with decreasing water potential [115]. Even under mild water stress conditions, NR activity declines rapidly compared with other N assimilatory enzymes [63].

In various crop species examined, NR activity has often been shown to decline with water stress [89,117,120,121]. In field-grown wheat plants, imposition of water stress caused a gradual decline in NR activity in leaves [113]. Kathju et al. [113] observed that in wheat plants, increasing the intensity of water stress progressively for 3 to 9 days reduced NR activity. These investigators also reported that under both low and high NP fertility conditions, water stress reduced NR activity at different growth stages of plants. However, activity was always greater in highly fertilized plants than those with low fertilizer treatments. A slow decline in NR activity with water stress may be attributed to a partially maintained $NO_3^-$ flux inside the plant despite increased stomatal resistance and decreased rate of transpiration under stress conditions.

In maize plants, desiccation leads to a steady decrease in NR activity with a concomitant decrease in leaf water potential, leaf $NO_3^-$ content, and $NO_3^-$ flux [120,121]. Water-stressed maize plants when re-watered recovered partially, showed increased NR activity and increased $NO_3^-$ flux [115]. In maize callus tissue, a decrease in relative humidity caused a gradual decrease in NR activity [119]. While examining NR activity in different organs of two chickpea (*Cicer arietinum* L.) varieties in relation to soil moisture stress, Wasnik et al. [120] observed a significant reduction in leaf NR activity due to moisture stress. In cacao (*Theobroma cacao* L.) plants, which experience a periodic drought during January to May in the coastal regions of India, water stress induced by withholding irrigation for 7 days caused a substantial decrease in NR activity in the seedlings [122].

Several explanations have been put forth for decreased NR activity in plant parts subjected to water stress [63,115,121]. The most plausible explanation, suggested by Morilla et al. [121], is that reduction in NR activity in *Zea mays* plants subjected to water stress is due to a decline in the rate of synthesis of NR protein rather than its increased rate of degradation or a direct effect of water potential on enzyme activity. According to these investigators, desiccation of plants leads to a decrease in leaf water potential. This, in turn, decreases $NO_3^-$ flux and causes slow delivery of $NO_3^-$ to the transpiration stream. Thus, movement of $NO_3^-$ to the induction site is prevented, resulting in decreased NR activity. These investigators believe that decreased NR activity in water-stressed plants is primarily due to a decrease in $NO_3^-$ flux and not a decrease in water potential or the $NO_3^-$ content of leaves [115]. Similarly, Singh and Sawhney (63) suggested that the decline in NR activity during water stress is due to a lowered capacity of tissues to synthesize NR protein because of degradation of polyribosomes to monoribosomes. More conclusive evidence is still required to ascertain whether decreased NR activity in water-stressed plants is due to a decreased rate of enzyme synthesis or an increased rate of enzyme degradation.

Certain measures have been suggested by different groups of investigators to overcome partially the effects of drought stress. By increasing soil fertility, especially with nitrogenous fertilizers, the adverse effects of drought can be substantially alleviated [118]. More N fertilizer application to water-stressed plants improved $NO_3^-$ uptake and increased NR activity. Such plants showed better performances and grain yield compared with low-fertilized plants [113]. Similarly, in plants such as sesame (*Sesamum indicum* L.), moderate water stress, when imparted at an early vegetative stage, partly helped to overcome the adverse effects of subsequent severe stress [123]. Such prestressed plants maintain high plant water status, show higher activities of NR, and have better plant performance. Foliar application of chemicals such as chlormequat, cycocel, or ABA also increased the relative water content in wheat [124] and cacao [122] plants, and such plants showed increased NR activity.

In water-stressed plants, activities of enzymes of ammonium assimilation remain high as evident from little or no accumulation of $NH_4^+$ in the leaves of such plants [125]. However, the pathway of ammonium assimilation under stress conditions depends on the plant species, growth stage, and the plant organs studied. It has been shown that water stress lowers the activity of GOGAT in the root nodules of alfalfa (*Medicago sativa* L.) and cicer plants [126,127]. In these plants, GOGAT is more sensitive than GS to water stress. Koundal and Chopra [127] reported a decline in NADH-GOGAT and GS activities in nodules of chickpea plants with water stress with a greater percentage decline in the activity of GOGAT than GS compared with the nodules of unstressed plants. Rewatering of such plants caused increases in GOGAT and GS activities, whereas the NR activity remained comparable to that in controls [127]. These observations indicate that in alfalfa and chickpea nodules, ammonium may be assimilated by the GDH pathway under water stress. At the flowering stage in *Brassica* and in the shoots of *Poterium*, increased GDH activity has been observed with water stress [105].

## C. Light

Light remarkably influences N uptake and its assimilation. Decreased light intensity reduces the uptake of $NO_3^-$, causes $NO_3^-$ accumulation in the tissues, and decreases the rate of its reduction by lowering the activities of $NO_3^-$ reducing enzymes [63,97]. Plants grown under low light intensity show decreased $NO_3^-$

uptake to the extent that the $NO_3^-$ uptake per gram of fresh weight production is also decreased [97]. During the day, roots show higher rates of $NO_3^-$ uptake than during the night [128]. Blom-Zandstra et al. [97] observed that lettuce genotypes (*Lactuca sativa* L.), differing in $NO_3^-$ accumulation, when grown under light with decreasing intensity, showed decreased $NO_3^-$ uptake with a concomitant decrease in growth. In such plants, $NO_3^-$ uptake per plant decreased proportionally even more than fresh weight production with decline in light intensity.

It has been shown that the uptake of $NO_3^-$ by roots is dependent on the continued flux of soluble carbohydrates from the shoot [129]. During the day period, because of the metabolic activity of roots as well as greater demand for carbohydrates from the shoot pool, translocation of carbohydrates from shoot to root is greater, which parallels the higher uptake of $NO_3^-$ by roots under daylight conditions [130]. Increased rates of $NO_3^-$ uptake by roots are observed due to diurnal variations associated with changes in day-night or seasonal conditions [131]. Interruption of the dark period for 3 hr using light of low intensity from an incandescent lamp, resulted in a two fold increase in $NO_3^-$ uptake in soybean (*Glycine max* L.) plants compared with the day period [130]. Raper et al. [130] suggested that the light-induced increase in $NO_3^-$ uptake by plant roots is phytochrome mediated. This, in turn, alters the permeability of plasma membranes and enhances starch degradation by increasing the activity of starch-degrading enzymes. This leads to an increase in the availability of soluble carbohydrates for translocation from shoots to roots.

In plant cells, the bulk of $NO_3^-$ is stored in vacuoles in the form of a storage pool [132]. This is a metabolically inactive pool of $NO_3^-$ and is not available for the induction of cytosolic NR; however, it plays a significant role as osmoticum along with organic acids and sugars that are located in the vacuoles [97]. The metabolically active pool of $NO_3^-$ is present in the cytosol [64]. It is believed that light affects the movement of $NO_3^-$ from the storage to the metabolic pool [106]. Nitrate taken up in the dark accumulates largely in vacuoles, and when such dark-kept plants are illuminated, the proportion of $NO_3^-$ in the metabolic pool increases [63]. In the light, $NO_3^-$ taken up by plants enters the metabolic pool, where it is available for NR induction. Thus, the processes of $NO_3^-$ uptake and NR induction are interrelated and both are dependent on light. It was suggested by Aslam et al. [133] that the transfer of $NO_3^-$ from the storage to the metabolic pool is mediated by phytochrome. Light thus regulates the availability of $NO_3^-$ in the metabolic pool.

Plants grown at low light intensities accumulate $NO_3^-$ largely in vacuoles, where it serves as an osmoticum [97]. Accumulation of $NO_3^-$ is inversely related to the accumulation of organic compounds, and in this way accumulating $NO_3^-$ may compensate for the shortage of photosynthates as a result of a decreased rate of photosynthesis under shade conditions [97]. Plants growing in insufficient light conditions thus show a twofold demand for N, one for the metabolic pool, which after reduction can be used for protein synthesis, and the other for the storage pool, which acts as an osmoticum [97]. The distribution of N between organic-N and nitrate-N changes in plants grown in the shade. Decreasing light intensity decreases the organic-N level in vacuoles and increases the nitrate-N level. Lettuce genotypes differing in the extent of $NO_3^-$ accumulation, when grown under shade conditions, show increased $NO_3^-$ concentration in the cell sap in both sets of cultivars accompanied by a decreased concentration of organic-N [97].

Light has a marked stimulatory effect on the reduction of $NO_3^-$ by regulating the synthesis as well as the functioning of NR. Leaves of shade-grown plants show a very low level of NR activity, but when such plants are transferred to light the NR activity increases severalfold [63]. As with NR, in photosynthetic tissues light plays a significant role in regulating the activity of NIR [63].

Several regulatory mechanisms for light-mediated enhancement of NR activity have been postulated. Based on the inhibitor studies and the labeling experiments, it has been suggested that light promotes de novo synthesis of both NR and NIR. Illumination of leaves leads to increased protein synthesis, indicating that light enhances the production of NR in leaves [64]. Certain investigators suggest that light-mediated enhancement of NR activity is due to enhanced uptake of $NO_3^-$ by plants in light [134]. Light enhances the movement of $NO_3^-$ from the storage pool to the metabolic pool [133], where $NO_3^-$ becomes available for the induction of NR activity. Sharma and Sopory [135] observed that in maize seedlings the NR activity increased by more than 300% on treatment with red light and kinetin. These investigators suggested that the light-induced increase in NR activity is mediated via phytochrome. Phytochrome action does not appear to be mediated by hormones; however, there appears to be an overlap in the signal transduction chains of phytochrome and plant hormones [135]. According to Sawhney and Nalik [136], some early events of photosynthesis, such as the Hill reaction, cause redox changes in green tissues and create favorable intracellular conditions for the synthesis of NR. These findings indicate that light influences the level of NR and NIR in plants, but a unified mechanism leading to the mode of action of light in regulating the activities of these enzymes is still awaited.

## D. Temperature

Nitrogen uptake, metabolism, and assimilation, like any other physiological and biochemical processes, are strongly related to temperature. Optimum absorption and assimilation occurs at the normal temperature. Any deviation from the normal temperature range adversely affects N absorption, metabolism, and assimilation.

### 1. High Temperature

High temperatures affect seed germination and plant growth, yield attributes and induce many metabolic alterations in crops. Different plant species have different optimum temperature ranges for growth and yield. Even a small increase in soil temperature affects the growth and nutrient uptake in plants. A rise in temperature beyond the optimum growth temperature impairs the rate of uptake as well as the assimilation of nutrients by exerting a profound influence on the activities of assimilatory enzymes. In many plant species, the effects of day-night temperatures on the uptake of various nutrients have been studied extensively [137,138].

$C_4$ plants grown under conditions of high temperature and high humidity show enhanced efficiency in N use compared with $C_3$ plants [138]. $C_4$ plants such as corn (*Zea mays* L.) and sorghum (*Sorghum bicolor* L.) and $C_3$ plants such as barley, rice, wheat, and oats (*Avena sativa* L.) were grown either for 7 days at 20 or 28°C or for 3 weeks at 26°C. Greater accumulation of $NO_3^-$ was observed in $C_3$ than in $C_4$ plants under any of the three conditions tested [138]. However, N supplied as $NO_3^-$ was more efficiently assimilated into protein in $C_4$ than in $C_3$ plants [138]. Lowering the temperature in both sets of plants from 28 to 20°C caused accumulation of $NO_3^-$ as well as a lower protein-to-nitrate ratio [138]. The greater efficiency of $C_4$ cereals toward $NO_3^-$ uptake and assimilation compared with $C_3$ plants at all tested temperature levels appears to be due to the highly organized cellular structure and spatial organization of N assimilatory enzymes in $C_4$ plants [138].

In young corn seedlings, day-night temperatures of 30/30°C are regarded as optimum for $NO_3^-$ uptake [137]. Polisetty and Hageman [137], while examining the effects of three temperature treatments, 30/20°C, 30/30°C, and 35/35°C day-night temperatures, on $NO_3^-$ uptake in corn seedlings, observed that the amount of $NO_3^-$ taken up during the night was about 4- and 3-fold greater for 30/30°C over 30/20°C and 35/35°C, respectively, whereas during the light period, $NO_3^-$ uptake increased by 1.5- and 1.3-fold, respectively. This suggests that optimum $NO_3^-$ uptake by corn seedlings occurs at 30/30°C and that either an increase or decrease in the temperature leads to a decrease in $NO_3^-$ uptake. The decreased $NO_3^-$ uptake above optimum growth temperatures appears to be due to the impairment of the $NO_3^-$ uptake process as well as the inhibition of root and shoot development due to the increase in temperature [137].

Partitioning of N in different parts of the plants is affected by increasing temperature. In rice, the absolute N content per kernel was comparatively stable in the temperature range from 24/19°C to 33/28°C, whereas beyond this temperature a decline in the N content of the kernels was observed [139]. The varieties of rice differ in sensitivity to higher temperature. Japonica varieties of rice were more sensitive than indica types during kernel development [140]. The highest concentration of N in terms of percentage of dry weight was recorded in rice kernels in the temperature range from 33/28°C to 39/34°C [139]. A decrease in the N content of shoots in soybean plants was reported by Hafeez et al. [141] due to an increase in temperature from 30 to 48°C compared with the control plants growing at 30°C.

The NR is sensitive to higher temperatures. Temperatures above a certain optimum affect the level of NR in plants as well as inhibit its activity. The magnitude of inactivation or lowering of the NR level by higher temperature varies according to the species [142]. Chandra and Pareck [142] observed that in sorghum plants an increase in temperature by 6 to 11°C caused more than a 60% reduction in NR activity at the vegetative stage and 30% at the anthesis stage. Corn seedlings maintained at 15–20°C showed six times more NR activity than at 25 to 30°C [63]. In barley seedlings, induction of NR did not take place when seedlings were maintained at 41°C. Seedlings growing at 24°C, when transferred to 43°C, lost 70% of their NR activity [63]. Whether high temperature causes inactivation, decreased synthesis, or increased degradation of NR still remains to be investigated.

Among the enzymes of ammonium assimilation, GS appears to be sensitive to higher temperatures, whereas GDH is comparatively heat stable (50 to 70°C) in many plant species [105]. Stability of GDH at higher temperatures appears to be of adaptational significance for plants growing at elevated temperatures as such plants may possibly assimilate $NH_4^+$ by the GDH pathway instead of the normal GOGAT/GS pathway.

## 2.  Low Temperature

Low-temperature treatment of plants below the optimum growth temperature reduced N uptake [72,142,143], decreased N partitioning in the young shoots [144], induced remobilization of N from older leaves to younger ones [144], and adversely affected the process of N assimilation [145].

In *Lolium multiflorum* and *Lolium perenne* grasses, a decrease in the rate of $NO_3^-$ uptake was observed with short-term exposure of roots to low-temperature treatment by decreasing the temperature from 25 to 15°C [72]. A similar decrease in $NO_3^-$ uptake by *Cicer arietinum* plants was observed at 16°C soil temperature compared with 22°C [142]. Macduff and Jackson [143] observed that when the root temperature of barley plants was lowered by 3°C, maintaining a common day-night air temperature of 25/15°C, $NO_3^-$ uptake by the roots decreased with a concomitant decrease in the total-N content of the plants. In barley plants, at all temperatures tested, $NH_4^+$ uptake was more than $NO_3^-$ uptake [143]. At low root temperatures, $NH_4^+$ is regarded as a safe source of N, whereas it appears to be harmful at higher temperatures [65]. Decreased uptake of $NO_3^-$ with decreasing temperature indicates that $NO_3^-$ uptake is sensitive to temperature.

Low root temperatures drastically affect the partitioning of N within the whole plant [144]. Walsh and Layzell [144] reported that when 35-day-old soybean plants were exposed to 15°C temperature for 4 days, N partitioning in the young shoots decreased 52 to 61% compared with that in control plants grown at 25°C. In treated plants, mature leaves maintained an N level similar to that in controls. In another experiment, Rufty et al. [128] observed a similar N partitioning pattern in soybean plant when roots were treated with low temperature. Besides reduced N uptake and disproportionate partitioning of N, low-temperature treatment of roots caused remobilization of N from older leaves to the young shoots. Walsh and Layzell [144] observed about 22% remobilization of N from mature leaves of soybean plants by 11 days of temperature treatment at 15°C compared with N present in leaves at 4 days of treatment. It appears that the remobilized N from older leaves supports growth of the new shoots under low-temperature stress conditions. Increased remobilization of N to the new shoots and proportionally less N partitioning indicates that cold-tolerant cultivars have increased partitioning of N in the shoots. This also suggests that tolerance to low temperature can be increased by increasing the N supply to young shoots [144].

Low-temperature treatment of roots decreased the rate of $NO_3^-$ flux to the leaves and, in turn, decreased NR activity [64]. Barley and maize seedlings, when grown at 20°C for 7 days, showed a drastic reduction in NR activity compared with seedlings grown at 28°C [138]. Because NR is a substrate-inducible enzyme, the level of $NO_3^-$ in the active pool has a major role in regulating leaf NR activity. Decreased $NO_3^-$ uptake by the roots as a result of the chilling treatment would ultimately lead to a decrease in NR activity. However, contrary to this in certain cases, an increased rate of ion uptake and root NR activity has been observed at a low temperature [145]. Vogel and Dawson [145] reported that when 2-week-old black alder (*Alnus glutinosa*) seedlings were exposed to chilling temperatures of $-1$ to 4°C for 2 hr during the night, immediately after chilling in vivo, NR activities of roots and shoots increased significantly compared with activities in prechilled plants. The apparent increase in NR activity following chilling appears to be due to the increased activity of a constitutive NR enzyme that is reported to be present in many $N_2$-fixing plants.

## E.  Metal Toxicity

Heavy metals such as $Cd^{2+}$, $Zn^{2+}$, $Cu^{2+}$, $Pb^{2+}$, and $Al^{3+}$ are major environmental pollutants that spread to the soil via sewage sludge, waste disposal practices, or airborne pollution. They cause plant growth to deteriorate, cause reduced $NO_3^-$ uptake by plants, and have direct inhibitory effects on enzymes of N assimilation. In corn seedlings, a direct adverse effect of cadmium on $NO_3^-$ uptake was reported by Volk and Jackson [146]. Industrial areas in many countries that are polluted with the heavy metals show reduced N concentration in leaves of plants. Pahlsson [147], while investigating the effects of pollutants in two industrialized belts of Sweden, observed reduced N content in the leaves of polluted trees compared with those growing in nonpolluted areas. It is suggested that the elevated level of heavy metals in the soil has a direct deteriorative effect on the growth of finer roots and root hairs of the plants contributing to reduced N uptake from the soil [147]. Nitrogen deficiency occurs in plants growing in soil with a high level of heavy metals, which also results in disturbed carbohydrate metabolism. Levels of total carbohydrate and especially starch and sucrose increase in the leaves of such plants [147].

Indiscriminate use of acid-forming nitrogenous fertilizers causes acidity in the soil. In acid soils below pH 5.0, $Al^{3+}$ toxicity is a major problem. In such soils, the $NH_4^+$ form of N predominates and $NO_3^-$ availability is limited. Uptake of many essential nutrients including $NO_3^-$ is reduced by $NH_4^+$ and $Al^{3+}$ toxicity in the soil. It has been suggested that plants differ in their sensitivity to $Al^{3+}$ and that $Al^{3+}$-tolerant plants are characterized by efficient use of $NO_3^-$ in the presence of $NH_4^+$. Such plants have the capacity to increase the pH of their growth medium [148]. When genotypes of sorghum plants differing in Al tolerance were grown with different $NO_3^-/NH_4^+$ ratios (39:1, 9:1, and 3:1) with 0 or 300 μM Al in the medium, Al-sensitive cultivar ICA-Natiama showed a greater reduction in $NO_3^-$ and $NH_4^+$ uptake than the Al-tolerant cultivar SC-283 when the plants were grown with $Al^{3+}$. When the plants were grown without $Al^{3+}$, the sensitive cultivar showed greater $NH_4^+$ uptake than the tolerant one [149]. This shows that uptake of $NO_3^-$ and $NH_4^+$ is reduced because of $Al^{3+}$ toxicity. It is suggested that differences in $NO_3^-$ and $NH_4^+$ uptake by plants are associated with changes in solution pH. As long as $NH_4^+$ is in solution, pH decreases, and it increases when $NH_4^+$ is depleted from the solution [149].

Among the enzymes of N assimilation, NR is the most sensitive to heavy metal toxicity. Cadmium ion ($Cd^{2+}$), $Cu^{2+}$, and $Pb^{2+}$ drastically inhibit NR activity. Elemental cadmium has a strong affinity for —SH groups and thus it inhibits the activity of many enzymes including NR. Muthuchlian et al. [150] reported that when etiolated leaf segments of *Vigna sinensis* were treated with $Cd^{2+}$ up to 10 μM, a stimulation of NR activity was observed. Beyond this level $Cd^{2+}$ strongly suppressed NR activity, and there was complete inhibition of activity with 1 mM $Cd^{2+}$. In similar experiments, $Cu^{2+}$ caused 91% inhibition of NR activity. A purified NR preparation from barley seedlings was inhibited up to 80 to 100% with 1 mM $Cu^{2+}$, $Zn^{2+}$, and $Co^{2+}$ [151]. In general, the inhibitory effects of $Cd^{2+}$ and $Cu^{2+}$ appear to be due to interference with the sulfhydryl sites of the enzymes [150].

In germinating pea (*Pisum sativum*) seeds $Pb^{2+}$ retarded the utilization of N reserves from cotyledons and decreased the activities of N assimilatory enzymes NR, GS, and GDH, whereas NIR remained relatively insensitive [152]. Mittal and Sawhney [152] reported about a 50% depression in NR activity of pea seeds 5 days after germination with a medium containing 1.0 mM $Pb^{2+}$. The activities of GS and GDH were suppressed by 70, 43, 45, and 30%, respectively, compared with controls. Decreased activities of NR, GDH, and aminotransferase in germinating pea seeds disturb the respiratory activity because of restricted generation of organic acids from amino acids. This would otherwise facilitate the operation of the tricarboxylic acid (TCA) cycle even under the partially anaerobic conditions existing during germination of seeds.

## IV. ACCUMULATION OF NITROGENOUS COMPOUNDS IN STRESSED PLANTS

Plants subjected to environmental stresses accumulate a number of soluble nitrogenous compounds. These compounds accumulate in high concentration and have specific roles in plants under stress conditions. Several investigators have observed accumulations of these compounds in a variety of plant species [1,3,6,12,15,22–33,41–47,49–51,53–59,61,62,77–80,83,84,91,106,109,11 0,113]. Several review articles have been published on the effects of various stresses on the accumulation and metabolism of these compounds [1,3,6,12,15,35,40,48,77,153–156].

Soluble nitrogenous compounds that accumulate most widely in stressed plants are the amino acids proline, arginine, glycine, serine, alanine, and leucine; the quaternary ammonium compounds glycine betaine, β-alanine betaine, stachydrine, trigonelline, and homostachydrine; the amides glutamine and asparagine; the imino acids pipecolic acid and 5-hydroxypipecolic acid; the diamines putrescine, N-carbamyl putrescine, and agmatine; and the polyamines spermine and spermidine. For the complete list of these compounds, readers are referred to the review article by Rabe [48]. When subjected to stress, plant species show accumulation of these compounds depending on the type of stress, extent of stress, and the types of plant species. With most stresses the amino acid proline and the quaternary ammonium compound glycine betaine accumulate and are regarded as components of the stress tolerance mechanism.

These compounds also contribute to osmotic balance in the cytoplasm when electrolytes are lower in cytoplasm than vacuoles and play a protective role for enzymes in the cytoplasm in the presence of high level of electrolytes [155].

## A.  Amino Acids

Plants subjected to most stressful environments show increased levels of total free amino acids [1,3,6,12,15,35,77,153,156,157]. Proline seems to accumulate in larger amounts than the other amino acids in response to salinity [153,157], water stress [158], temperature stress [159], mineral deficiency [154], pathogenesis, and anoxia [160]. In most of the plants studied, salinity and water stresses caused substantial increases in proline levels of the plant tissues. Proline along with the other soluble nitrogenous compounds serves as an osmoregulator in plants. A proline level up to 600 mM did not inhibit enzyme activities [155]. Higher plants differ markedly in their capacity to accumulate proline. Proline-accumulating species when grown in NaCl-free environments contain low levels of proline, but the level increases in the presence of salinity [153]. In salt-stressed plants, proline accumulation results from its increased synthesis and decreased utilization [153]. As a result of water stress, free proline accumulated appreciably in leaves and other tissues. The functional role of proline accumulation appears to be as a cytoplasmic osmoticum to lower cell water potential, provide hydration to biopolymers, and serve as an energy and N source under adverse environmental conditions [161].

In addition to proline, other amino acids that accumulate under salt and water stress are arginine, glycine, serine, alanine, leucine, and valine. Salt-stressed rice plants accumulated arginine, alanine, leucine, and valine in addition to proline [157]. Under salinity stress conditions, the level of these amino acids is higher in salt-tolerant plants than in salt-sensitive species [157]. Water-stressed plants accumulate proline, alanine, arginine, and phenylalanine, which have a distinct correlation with the stress tolerance mechanism [162]. Crop plants such as barley and radish, when grown under low-temperature conditions, accumulated a substantial amount of proline [153,159]. Other amino acids, such as serine, glycine, and alanine, also accumulated appreciably in several plant species grown under low-temperature conditions [153]. Barley and radish plants exposed to high temperature accumulated proline [159]. In lemon and orange leaves, infection by *Phytophthora* spp. or anaerobiotic conditions caused increased levels of proline, arginine, and total free amino acids [160]. Most of the mineral deficiencies cause increases in the level of free amino acids. The types of amino acids accumulated depend on the nature of the mineral deficiency [156]. Copper deficiency caused substantial accumulation of proline and serine in citrus plants [163], whereas iron deficiency resulted in accumulation of arginine, lysine, histidine, and serine in citrus and macadamia plants [164]. The basic amino acid arginine has been shown to accumulate under a variety of stress conditions such as Mg, K, S, Ca, Fe, Mn, and Zn deficiencies, osmotic stress, acid stress, excess ammonium in the growth medium, and infection by pathogens [156].

## B.  Quaternary Ammonium Compounds

Among quaternary ammonium compounds, glycine betaine (trimethylammonio-2-acetic acid) accumulates most widely in stressed plants. It is the predominant nitrogenous compound accumulating under salinity stress [165]. Together with proline, glycine betaine serves as a compatible cytoplasmic solute and has an important role as an osmoregulator in salinity stress. Betaine levels up to 1.0 M do not inhibit enzyme activity in vitro. Another compound, β-alanine betaine (trimethylammonio-3-propanoic acid), also accumulates in laboratory-grown salinized plants and in plants growing in saline habitats. Glycine betaine accumulated in many species, whereas β-alanine betaine was restricted to halophytes of *Plumbaginaceae* [163]. Plants differ in their capacity to accumulate betaine. Certain other quaternary ammonium compounds such as stachydrine, homostachydrine, and trigonelline accumulated in alfalfa plants in response to water stress, salinity, and abscisic acid treatments, respectively [166].

## C.  Amides and Imino Acids

Glutamine and asparagine together with amino acids and certain imino acids such as pipecolic acid and 5-hydroxypipecolic acid accumulate in plants subjected to saline stress; however, their level is much lower than that of proline or betaine. In plants such as *Agrostis stolonifera*, asparagine accumulation was greater than proline at a high salinity level [163]. Accumulation of amides may exceed proline in certain species under water stress [153]. Mineral deficiencies of K, P, Mg, S, Cl, and Zn cause accumulation of asparagine and glutamine along with the basic amino acid arginine [156]. Deficiencies of many micronutrients cause accumulation of asparagine. In certain cases, e.g., Zn deficiency, asparagine accumulation may occur up to 50-fold [153]. The imino acid pipecolic acid accumulated in Mg-, Cl-, K-, and Fe-deficient plants [153,156]. Saline conditions also favor the accumulation of pipecolic acid. Hydroxypipecolic acid accumulated in *Limonium* plants subjected to salt stress [153].

## D.  Nonprotein Amino Acids

Among nonprotein amino acids, citrulline, ornithine, and γ-aminobutyric acid accumulate in plants under certain stresses. Deficiency of mineral nutrient elements such as K in *Sesamum* and P in *Citrus* caused accumulation of citrulline and ornithine [156]. Barley plants subjected to water stress accumulated ornithine in the leaves [153]. In conditions of anaerobiosis, the most striking response was the accumulation of γ-aminobutyric acid [167]. Following anaerobiosis, γ-aminobutyric acid accumulated rapidly in leaves due to increased decarboxylation of glutamate and decreased transamination of 4-aminobutyrate [167]. Accumulation of γ-aminobutyric acid has also been observed in copper-deficient citrus plants [163] and in the leaves of tomato plants infected with tobacco mosaic virus [168].

## E.  Diamines and Polyamines

Accumulation of the diamine putrescine and stimulation of its biosynthetic enzyme arginine decarboxylase have been reported in several forms of environmental stresses, for example, deficiencies of nutrients such as K and Mg [169] and Ca [86], salinity [86,170], water stress [158], ammonium toxicity [171], $SO_2$ fumigation [172], and acid stress [173]. It has been suggested that putrescine accumulation under stress conditions has adaptive significance as it serves as an organic ion and can compensate partly for $K^+$ in $K^+$-deficient plants [153]. Many plants accumulate the polyamines spermine, spermidine, and agmatine under deficiency of K [169], P [174], Mg [86], S, Ca, and Mn [174]; salinity stress [170]; acid stress [165]; and ammonium toxicity [171]. Concentrations of these amines are very low in nonstressed plants, but stress conditions induce a severalfold increase in their level. The type of polyamines accumulated depends on the type of stress as well as the plant species. In detached oat leaves, osmotic treatment induces a rise in the level of putrescine and stimulation of arginine decarboxylase activity. Other species show increases in the levels of spermidine as well as spermine and decline of putrescine as well as its biosynthetic enzymes. It is suggested that changes in the level of putrescine under stress conditions might be important in regulating the ionic environment within the cell [173].

It appears from the preceding discussion that various environmental stresses induce accumulation of soluble nitrogenous compounds, the extent and nature of the compounds accumulated depending on the type of stress and the plant species. Levels of accumulation of some of these compounds, i.e., amino acids and betaine, are associated with the stress sensitivity or tolerance of the plant species. For instance, salt-tolerant plants possess inbuilt higher levels of the amino acid proline [153], glycine betaine [165], and β-alanine betaine [153]. Accumulation of these compounds is greater in tolerant plants than in sensitive ones. The sensitive species have low levels of these compounds in the nonstressed plants and show less accumulation under salinization [153]. Functions of these nitrogenous compounds are diverse. Amino acids and betaine accumulating under salt and water stresses serve as osmoregulators, protect biomolecules, decrease the water potential of the cytoplasm, and improve moisture uptake. Accumulation of amides and the amino acid arginine appears to have a role in detoxifying ammonia, which attains elevated levels during mineral deficiencies, water stress, low-temperature stress, etc. Rabe [156] has advocated that most of the nitrogenous compounds that accumulate during environmental stresses serve to detoxify the cell of ammonia.

## V.  CONCLUDING REMARKS

Nitrogen is one of the most essential elements in plant nutrition; however, its availability is limited under harsh environmental conditions of salinity, water deficit, extremes of temperature, metal toxicities, etc. These stresses considerably reduce $NO_3^-$ uptake, metabolism, and protein synthesis and drastically affect crop yields. The processes of $NO_3^-$ uptake, its translocation inside and between the cells, and its reduction are coordinately regulated. The pronounced effects of most of the stresses include decreased $NO_3^-$ uptake and inhibition of the activity of the key enzyme of nitrate assimilation, NR. The NR is $NO_3^-$ inducible and its activity is subject to regulation by a variety of environmental parameters that are readily influenced under stresses. Levels of $NO_3^-$ and NR inside the tissues are directly related to plant health and yield. Genotypes of plants differing in stress tolerance show different behaviors of NR and other N assimilatory enzymes. For instance, a salt-tolerant variety shows increased $NO_3$ uptake and a high level of NR that is further stimulated by salinization, whereas a salt-sensitive variety shows decreased $NO_3^-$ uptake and decreased activity of NR with salinization. This is suggestive of stress tolerance and sensitivity as complex phenomena, depending on the genetic and biochemical makeup of the species.

In spite of the extensive studies that have been performed, our knowledge of the biochemical mechanisms underlying the uptake of $NO_3^-$ by plants, the process of its assimilation, and the regulation of enzymes of $NO_3^-$ assimilation is still incomplete. Little information is available regarding molecular events of $NO_3^-$ uptake, the $NO_3^-$ sensor protein system, signal transduction of environmental $NO_3^-$, $NO_3^-$ induction regulatory proteins, primary responsive genes that are transcribed and translated as a result of $NO_3^-$ induction, etc. Besides this, the nature of the $NO_3^-$ transporters, $NO_3^-$ translocaters, events involving overall induction of NR by $NO_3^-$ and regulation of NR and other enzymes of $NO_3^-$ and $NH_4^+$ assimilation under various environmental stress conditions such as light, water, temperature, and salinity stresses needs to be examined in detail.

Although all the adverse environmental conditions discussed so far reduce $NO_3^-$ uptake and inhibit NR activity, the precise biochemical mechanisms involved in these events remain to be investigated. More extensive investigations are required to unveil the role of light and other factors in regulation of the NR level in plants. Environmental stresses adversely affect the behavior of enzymes of $NO_3^-$ and $NH_4^+$ assimilation. In certain cases, such as salinity and water stresses, suppression of the GS/GOGAT pathway and a sustained level of induction of the GDH pathway of ammonium assimilation are observed. Uptake and metabolism of N are triggered in stressed environments so that specific soluble nitrogenous compounds accumulate and provide adaptive value to the plants. Further information is required regarding the molecular structures and catalytic properties of enzymes involved in $NO_3^-$ and $NH_4^+$ assimilation. The precise effects of various environmental conditions in vivo as well as in vitro on behaviors of these enzymes, the sequences of events leading to accumulation of nitrogenous compounds, and the functional roles of these compounds in stressed plants need to be studied in greater detail.

## REFERENCES

1. MA Botella, A Cerda, SH Lips. Kinetics of $NO_3^-$ and $NH_4^+$ uptake by wheat seedlings, effect of salinity and nitrogen source. J Plant Physiol 144:53–57, 1994.
2. MA Botella, V Martinez, M Nieves, A Cerda. Effect of salinity on the growth and nitrogen uptake by wheat seedlings. J Plant Nutr 20:793–804, 1997.
3. M Pessarakli, ed. Handbook of Plant and Crop Stress. 2nd ed. New York: Marcel Dekker, 1999.
4. FE Broadbent, T Nakashima, DF Rolston. Effects of salinity and moisture gradients on nitrogen uptake by sorghum and wheat. Soil Sci 146:232–240, 1988.
5. MA Hamdia, HM El-Komy. Effect of salinity, gibberellic acid, and azospirillum on growth and nitrogen uptake of *Zea mays*. Biol Plant 40:109–120, 1998.
6. GK Sharma, M Singh, V Kumar. Effect of exchangeable sodium percentage (ESP) on yield, nitrogen uptake and ammonia volatilization losses in upland and waterlogged rice (*Oryza sativa* L.). Crop Res (Hisar) 8(1):45–51, 1994.
7. B Paseban-Islam, MR Shakiba, M Moghaddam-Vahed, J Jozan. Storage and remobilization of non-structural carbohydrates of stem in spring wheat isolines under drought/water deficit stress. Iran J Agric Sci 29:733–742, 1999.
8. SA Bonos, JA Murphy. Growth responses and performance of Kentucky bluegrass under summer stress. Crop Sci 39:770–774, 1999.
9. M Sari-Gorla, P Krajewski, N di Fonzo, M Villa, C Frova. Genetic analysis of drought conditions tolerance in maize by molecular markers. II. Plant height and flowering. Theor Appl Genet 99:289–295, 1999.
10. ER Carter, MK Theodorou, P Morris. Responses of *Lotus corniculatus* to environmental change. 2. Effect of elevated $CO_2$, temperature and drought on tissue digestion in relation to condensed tanoin and carbohydrate accumulation. J Sci Food Agric 79:1431–1440, 1999.
11. E Pawelzik, E Delgado. Effect of drought stress on the discoloration of potatoes. Kartoffelbau 50(9/10):358–360, 1999.

12. M Pessarakli. Response of green beans (*Phaseolus vulgaris* L.) to salt stress. In: M Pessarakli, ed. Handbook of Plant and Crop Stress. 2nd ed. New York: Marcel Dekker, 1999, pp 827–842.

13. SM Alam. Nutrient uptake by plants under stress conditions. In: M Pessarakli, ed. Handbook of Plant and Crop Stress. 2nd ed. New York: Marcel Dekker, 1999, pp 285–313.

14. SR Grattan, CM Grieve. Mineral nutrient aquisition and response by plants grown in saline environments. In: M. Pessarakli, ed. Handbook of Plant and Crop Stress. 2nd ed. New York: Marcel Dekker, 1999, pp 203–229.

15. E Rabe. Altered nitrogen metabolism under environmental stress conditions. In: M Pessarakli, ed. Handbook of Plant and Crop Stress. 2nd ed. New York: Marcel Dekker, 1999, pp 349–363.

16. MG Redinbaugh, WH Campbell. Physiol Plant 82:640, 1991.

17. CM Larsson, B Ingemarsson. In: JR Kinghorn, JL Wray, eds. Molecular and Genetic Aspects of Nitrate Assimilation. New York: Oxford Science Publishers, 1989, p 4.

18. WH Campbell. Physiol Plant 74:214, 1988.

19. JL Wray. In: JA Kinghorn, JL Wray, eds. Molecular and Genetic Aspects of Nitrate Assimilation. Oxford: Oxford Science Publications, 1989, p 244.

20. SM Alam, In: M. Pessarakli, ed. Handbook of Plant and Crop Stress. New York: Marcel Dekker, 1993, pp 227–246.

21. SR Grattan, CM Grieve. In: M Pessarakli ed. Handbook of Plant and Crop Stress. New York: Marcel Dekker, 1993, pp 203–226.

22. M Pessarakli. Dry matter yield, nitrogen-15 absorption, and water uptake by green beans under sodium chloride stress. Crop Sci 31:1633–1640, 1991.

23. M Pessarakli, JT Huber, TC Tucker. Dry matter yield, nitrogen uptake, and water absorption by sweet corn under salt stress. J Plant Nutr 12:279–290, 1989.

24. M Pessarakli, TC Tucker. Nitrogen-15 uptake by eggplant under sodium chloride stress. Soil Sci Soc Am J 52:1673–1676, 1988.

25. M Pessarakli, TC Tucker. Dry matter yield and nitrogen-15 uptake by tomatoes under sodium chloride stress. Soil Sci Soc Am J 52:698–700, 1988.

26. M Pessarakli, TC Tucker. Uptake of nitrogen-15 by cotton under salt stress. Soil Sci Soc Am J 49:149–152, 1985.

27. G Palfi. Plant Soil 22:127–135, 1965.

28. V Hernando, L Jimeno, C Cadahia. An Edafol Agrobiol 26:1147–1159, 1967.

29. TS Mahajan, KR Sonar. J Maharashtra Agric Univ 5:110–112, 1980.

30. JNE Frota, TC Tucker. Soil Sci Soc Am J 42:753–756, 1978.

31. R Saad. PhD dissertation, University of Arizona; University Microfilms, Ann Arbor, MI. Diss Abstr B 40:4057, 1979.

32. M Pessarakli. Uptake of Nitrogen by cotton (*Gossypium hirsutum* L.) under salt stress. PhD dissertation, University of Arizona; University Microfilms, Ann Arbor, MI. Diss Abstr B 42:286, 1981.

33. GW Langdale, JR Thomas. Agron J 63:708–711, 1971.

34. AA Luque, FT Bingham. Plant Soil 63:227–237, 1981.

35. M Pessarakli. Response of green beans (*Phaseolus vulgaris* L.) to salt stress. In: M Pessarakli, ed. Handbook of Plant and Crop Stress. New York: 1993, pp 415–430.

36. C Torres, FT Bingham. Proc Soil Sci Soc Am 37:711–715, 1973.

37. HS Srivastava. Phytochemistry 19:725, 1980.

38. UK Pandey, RDL Srivastava. Indian J Plant Physiol 32:175, 1989.

39. S Katiyar, RS Dubey. J Agron Crop Sci 169:289–297, 1992.

40. RS Dubey. In: M Pessarakli, ed. Handbook of Plant and Crop Stress. New York: Marcel Dekker, 1993, pp 277–299.

41. RS Dubey, M Rani. J Agron Crop Sci 162(2):97–106, 1989.

42. AS Gupta (A Sen-Gupta). Diss Abstr Int B Sci Eng 47(12):I 4728B, 1987.

43. R Krishnamurthy, KA Bhagwat. Indian J Exp Biol 27:1064–1066, 1989.

44. M Pessarakli, JT Huber, Biomass production and protein synthesis by alfalfa under salt stress. J Plant Nutr 14:283–294, 1991.

45. M Pessarakli, JT Huber, TC Tucker. Protein synthesis in green beans under salt stress with two nitrogen sources. J Plant Nutr 12:1361–1378, 1989.

46. M Pessarakli, JT Huber, TC Tucker. Protein synthesis in green beans under salt stress conditions. J Plant Nutr 12:1105–1121, 1989.

47. M Pessarakli, TC Tucker. J Plant Nutr 8:1025–1045, 1985.

48. E Rabe. In: M Pessarakli, ed. Handbook of Plant and Crop Stress. New York: Marcel Dekker, 1993, pp 261–276.

49. E Rabe. J Hortic Sci 65:231–243, 1990.

50. S Ramagopal. Plant Cell Rep 5:430–434, 1986.

51. JNE Frota, TC Tucker. Soil Sci Soc Am J 42:743–746, 1978.

52. A Golan-Goldhirsh, B Hankamer, SH Lips. Plant Sci (Limerick) 69(1):27–32, 1990.

53. SM Abdul-Kadir, GM Paulsen. J Plant Nutr 5:1141–1151, 1982.

54. GV Udovenko, VN Sinel'nikova, GV Khazova. Dokl Akad Nauk USSR 192:1395–1397, 1970.
55. GV Udovenko, VN Sinel'nikova, GV Khazova. Agrokhimiya 3:23–31, 1971.
56. HM Helal, K Mengel. Landwirtschftliche Forschungsanstalt Buntehof, Hannover, German Federal Republic. Plant Soil 51:457–462, 1979.
57. I Kahane, A Poljakoff-Mayber. Plant Physiol 43:1115–1119, 1968.
58. R Krishnamurthy, M Anbazhagan, KA Bhagwat. Indian J Plant Physiol 30:183–188, 1987.
59. J Wieneke, R Fritz. Acta Univ Agric Brno Fac Agron 33:653–657, 1985.
60. CA Morilla, JS Boyer, RH Hageman. Plant Physiol 51:8817–8824, 1973.
61. JT Prisco, JW O'Leary. Rev Braz Biol 30:317–321, 1970.
62. NM Iraki, RA Bressan, NC Carpita. Plant Physiol 91:54–61, 1989.
63. P Singh, SK Sawhney. In: SL Mehta, ML Lodha, PV Sane, eds. Recent Advances in Plant Biochemistry. New Delhi: ICAR Publications, 1989, p 141.
64. L Beevers, RH Hageman. In: BJ Miflin, ed. The Biochemistry of Plants. Vol 5. New York: Academic Press, 1980, p 115.
65. U Kafkafi. J Plant Nutr 13:1291, 1990.
66. CA Atkins. Proceedings of International Congress of Plant Physiology, New Delhi, India, 1990, pp 1022–1026.
67. ADM Glass, MY Siddiqui, TJ Ruth, TW Rufty. Plant Physiol 93:1585, 1990.
68. M Ni, L Beevers. J Exp Bot 41:987, 1990.
69. WA Jackson, WL Pan, RH Moll, EJ Kamprath. In: C Neyra, ed. Biochemical Basis of Plant Breeding. Vol 2. Boca Raton, FL: CRC Press, 1986, p 73.
70. MR Ward, M Aslam, RC Huffaker. Plant Physiol 80:520, 1986.
71. A Benzioni, Y Vaadia, SH Lips. Physiol Plant 24:288, 1971.
72. DT Clarkson. In: H Lambers, JJ Neeteson, I Stulen, eds. Fundamental, Ecological and Agricultural Aspects of Nitrogen Metabolism in Higher Plants. Dordrecht: Martinus Nijhoff, 1988, p 3.
73. AN Khan, RH Qureshi, N Ahmad, A Rashid. Response of cotton cultivars to salinity in various growth development stages. Sarhad J Agric 11:729–731, 1995.
74. FA de Oliveira, TGS da Campos, BC Oliveira. Effect of saline substrate on germination, vigor and growth of herbaceous cotton. Engenharia Agric 18(2):1–10, 1998.
75. WJ Li, HZ Dong, QZ Guo, JQ Pang, J Zhang. Physiological response of a good Upland hybrid and its parent to PEG and NaCl stresses. China Cottons 25(6):7–10, 1998.
76. R Vulkan-Levy, I Ravina, A Mantell, H Frenkel. Effect of water supply and salinity on Pima cotton. Agric Water Manage 37(2):121–132, 1998.
77. RS Dubey. Protein synthesis by plants under stressful conditions. In: M Pessarakli, ed. Handbook of Plant and Crop Stress. 2nd ed. New York: Marcel Dekker, 1999, pp 365–397.
78. M Aslam, RC Huffakar, DW Rains. Plant Physiol 76:321, 1984.
79. A Soltani, M Hajji, C Grignon. Agronomie 10:857, 1990.
80. YI Mladenova. Influence of salt stress on primary metabolism of *Zea mays* L. seedlings of model genotypes. Proceedings of the Third International Symposium on Genetic Aspects of Plant Mineral Nutrition, Braunschweig, Germany. Plant Soil 123:217, 1990.
81. SK Sharma, IC Gupta. Saline Environment and Plant Growth. New Delhi: Agro Botanical Publishers 1986, p 92.
82. M Pessarakli, TC Tucker, K Nakabayashi. Growth response of barley and wheat to salt stress. J Plant Nutr 14:331–340, 1991.
83. MA Khalil, A Fathi, MM Elgabaly. Soil Sci Soc Am Proc 31:683–686, 1967.
84. MR Ward, M Aslam, RC Huffaker. Plant Physiol 80:520, 1986.
85. TA Smith. Phytochemistry 12:2093, 1973.
86. TJ Flowers, PF Troke, AR Yeo. Annu Rev Plant Physiol 28:89, 1977.
87. RG Butz, WA Jackson. Phytochemistry 16:409, 1977.
88. TVR Nair, SR Chatterjee, YP Abrol. Plant Physiol Biochem 10(s):176, 1983.
89. Z Plaut. Physiol Plant 30:212, 1974.
90. SM Abdul-Kadir, GM Paulsen. J Plant Nutr 5:1141, 1982.
91. RK Lal, SN Bhardwaj. Indian J Plant Physiol 30:165, 1987.
92. RS Dubey, S Katiyar, R Nittal. Proceedings International Conference on Plant Physiology, Banaras Hindu University, India, 1991, pp 189–194.
93. GR Smith, KR Middleton. New Physiol 84:613, 1980.
94. TN Tewari, BE Singh. Plant Soil 136:225, 1991.
95. R Krishnamurthy, M Anbazhgan, KA Bhagwat. Curr Sci 56:489, 1987.
96. M Lacuesta, B Gonzalez-More, C Gonzalez-Marue, A Munoz-Rueda. J Plant Physiol 136:410, 1990.
97. Blom-Zandstra, JEM Lampe, FHM Ammerlaan. Physiol Plant 74:147, 1988.
98. S Joshi. Indian J Plant Physiol 30:223, 1987.
99. N Sankhla, W Huber. Z Pflanzenphysiol 76:467, 1975.
100. Ma de Lourdes Miranda-Ham, VM Loyola-Vargas. Plant Cell Physiol 29:747, 1988.
101. S Katiyar. PhD dissertation, Banaras Hindu University, India, 1990.

102. J Boucaud, JB Billard. Physiol Plant 44:31, 1978.
103. GG Rao, JK Ramiah, GR Rao. Indian J Exp Biol 19:771, 1981.
104. SK Sharma, OP Garg. Indian J Plant Physiol 38:407, 1985.
105. HS Srivastava, RP Singh. Phytochemistry 21:997, 1982.
106. M Singh, BB Singh, PC Ram. Biol Plant 32:232, 1990.
107. S Siddiqui, S Kumar, HR Sharma. Indian J Plant Physiol 28:369, 1985.
108. N Garg, IS Dua, SK Sharma, OP Garg. Res Bull Pubjab Univ (India) 39:187, 1988.
109. S Joshi. Indian J Plant Physiol 30:223, 1987.
110. M Khavan-Kharazian, WF Campbell, JJ Jurinak, LM Dudley. *Phaseolus vulgaris* L. Arid Soil Res Rehab 5:97, 1991.
111. J Levitt. Plant Responses to Environmental Stress. Vol 2. New York: Academic Press, 1980.
112. JA Morgan. Interaction of water supply and N in wheat. Plant Physiol 76:112, 1984.
113. S Kathju, SP Vyas, BK Garg, AN Lahiri. Fertility induced improvement in performance and metabolism of wheat under different intensities of water stress. Proceedings of International Congress of Plant Physiology 88, New Delhi, India, 1990, pp 854–858.
114. SK Sinha, DJ Nicholas. In: LG Paleg, D Aspinall, eds. Physiology and Biochemistry of Drought Resistance in Plants. New York: Academic Press, 1981, p 145.
115. DL Shaner, JS Boyer. Plant Physiol 58:505, 1976.
116. FGJ Viets. In: RM Hagan et al., eds. Irrigation of Agricultural Lands. Agronomy, Vol 11. Madison, WI: American Society of Agronomy, 1967, p 458.
117. ACS Rao, B Ramamoorthy. Indian J Plant Physiol 23:269, 1980.
118. AN Lahiri. In: NC Turner, PJ Kramer, eds. Adaptation of Plants to Water and Higher Temperature Stress. New York: John Wiley & Sons, 1980, p 341.
119. DG Rao, V Balasubramanian. Indian J Plant Physiol 29:61, 1986.
120. KG Wasnik, PB Varade, AK Bagga. Indian J Plant Physiol 31:324, 1988.
121. GA Morilla, JS Boyer, RN Hageman. Plant Physiol 51:817, 1973.
122. D Balasimha. Plant Physiol Biochem 10:69, 1983.
123. SP Vyas, BK Garg, S Kathju, AN Lahiri. Proceedings International Congress of Plant Physiology, New Delhi, India, 1990, pp 880–884.
124. R Sairam, PS Deshmukh, DS Shukia. Anal Plant Physiol 3:98, 1989.
125. AD Hanson, WD Hitz. Annu Rev Plant Physiol 33:163, 1982.
126. RG Groat, CP Vance. Plant Physiol 67:1198, 1981.
127. RK Koundal, RK Chopra. Biochem Physiol Pflanz 163:69, 1989.
128. TW Rufty, CD Raper, WA Jackson. New Phytol 88:607, 1981.
129. CD Raper Jr, DL Osmond, M Wann, WW Weeks. Bot Gaz 139:289, 1978.
130. CD Raper Jr, JK Vessey, LT Henry, S Chaillou. Plant Physiol Biochem 29:205, 1991.
131. CR Clement, MJ Hooper, LHP Jones, EL Leafe. J Exp Bot 29:1173, 1978.
132. RC Granstedt, RC Huffaker. Plant Physiol 70:410, 1982.
133. M Aslam, A Oaks, RC Huffaker. Plant Physiol 58:588, 1976.
134. L Beevers, RH Hageman. In: AG Giese, ed. Photophysiology. Vol VIII. New York: Academic Press, 1972, p 65.
135. AK Sharma, SK Sopory. Plant Physiol Biochem 15:107, 1988.
136. SK Sawhney, MS Nalik. Biochem J 130:475, 1972.
137. R Polisetty, RH Hageman. Indian J Plant Physiol 32:359, 1989.
138. A Oaks, X He, M Zoumadakis. Nitrogen use efficiency in $C_3$ and $C_4$ cereals. Proceedings of International congress of Plant Physiology '88, New Delhi, India, 1990, pp 1038–1045.
139. T Tashiro, IF Wardlaw. Aust J Plant Physiol 18:259, 1991.
140. S Yoshida, T Hara. Soil Sci Plant Nutr 23:93, 1977.
141. FY Hafeez, S Asad, KA Malik. Environ Exp Bot 31:289, 1991.
142. R Chandra, RP Pareck. Legume Res 13:95, 1990.
143. JH Macduff, SB Jackson. J Exp Bot 41:237, 1991.
144. B Walsh, DB Layzell. Plant Physiol 80:249, 1986.
145. CS Vogel, JS Dawson. Physiol Plant 85:551, 1991.
146. RJ Volk, WA Jackson. Mercury and cadmium interaction with nitrate absorption by illuminated corn seedlings. Environ Health Perspect 4:103, 1973.
147. AM Pahlsson. Effects of heavy-metal and $SO_2$ pollution on the concentrations of carbohydrates and nitrogen in tree leaves. Can J Bot 67:2106, 1989.
148. CD Foy, RL Chaney, MC White. The physiology of metal toxicity in plants. Annu Rev Plant Physiol 29:511, 1978.
149. L Galvez, RB Clark. Nitrate and ammonium uptake and solution pH changes for aluminum tolerant and aluminum sensitive sorghum genotypes. Plant Soil 134:179, 1991.
150. K Muthuchalian, SNV Rani, K Paliwal. Indian J Plant Physiol 31:169, 1988.
151. Y Oji, T Hamano, Y Ryema, Y Mike, N Wakichi, S Okamoto. J Plant Physiol 119:247, 1985.

152. S Mittal, SK Sawhney. Plant Physiol Biochem 17:73, 1990.
153. GR Stewart, F Larher. Accumulation of amino acids and related compounds in relation to environmental stress. In: BJ Miflin, ed. The Biochemistry of Plants. New York: Academic Press, 1980, p 609.
154. CR Stewart. Proline accumulation: biochemical aspects. In: LG Paleg, D Aspinall, eds. Physiology and Biochemistry of Drought Resistance in Plants. New York: Academic Press, 1981, p 243.
155. H Greenway, R Munns. Mechanisms of salt tolerance in nonhalophytes. Annu Rev Plant Physiol 31:149, 1980.
156. E Rabe. Stress physiology: the functional significance of the accumulation of nitrogen-containing compounds. J Hortic Sci 65:231, 1980.
157. RS Dubey, N Rani. J Agric Crop Sci 162:97, 1989.
158. NN Barnett, AW Naylor. Plant Physiol 41:1222, 1966.
159. TM Chu, D Aspinall, LG Paleg. Aust J Plant Phisiol 1:87, 1974.
160. CA Labaanuskas, LH Stolzy, MF Handy. J Soc Hortic Sci 99:497, 1974.
161. D Aspinall, LG Paleg. In: LG Paleg, D Aspinall, eds. Physiology and Biochemistry of Drought Resistance in Plants. New York: Academic Press, 1981, p 203.
162. VK Rai, J Singh, PS Thakur, S Banyal. Plant Physiol Biochem 10:161, 1983.
163. I Stewart. Proc Am Soc Hortic Sci 81:244, 1962.
164. IN Gilfillan, WW Jones. Proc Am Soc Hortic Sci 93:210, 1968.
165. R Storey, RG Wyn-Jones. Plant Sci Lett 4:161, 1975.
166. G Parameshwara, LG Paleg, D Aspinall, GP Jones. Proceedings of International Congress of Plant Physiology, New Delhi, India, 1990, pp 1014–1021.
167. JG Streeter, JF Thompson. Plant Physiol 49:572, 1972.
168. P Cooper, IW Selman. Ann Bot 38:625, 1974.
169. LC Basso, TC Smith. Phytochemistry 13:275, 1974.
170. S Katiyar, RS Dubey. Trop Sci 30, 1990.
171. H Klein, A Priebe, HJ Jager. Z Pflanzenkr Pflanzenschutz 83:555, 1979.
172. A Priebe, H Klein, HJ Jager. J Exp Bot 29:1043, 1978.
173. TA Smith, C Sinclair. Ann Bot 31:103, 1967.
174. C Sinclair. Nature 213:214, 1967.

# 33

# Induction of Proteins in Response to Biotic and Abiotic Stresses

## Timothy S. Artlip and Michael E. Wisniewski

*U.S. Department of Agriculture–Agriculture Research Service, Kearneysville, West Virginia*

## I. INTRODUCTION

The responses of plants and crops to environmental stresses generally involve some alteration in protein synthesis. Protein-based responses include qualitative changes in general protein synthesis or the up- or down-regulation of specific proteins. These changes depend on the nature, duration, and severity of the stress. The primary focus of this chapter will be on proteins that are induced or increase in abundance in response to an environmental stress. Although decreases of specific proteins can be physiologically significant, most research has centered on inducible proteins on the assumption that they confer some protection to the organism that leads to increased survival.

Two broad areas of stress will be considered: biotic (pathogen, herbivore) and abiotic (physical environment). Each confronts the plant with a particular set of challenges. In general, biotic stresses engender active structural and biochemical responses to herbivory or microbial attack. Abiotic stresses, conversely, induce the plant to adjust its metabolism (acclimation) or alter its pattern of growth in order to avoid sustained exposure to the stress. As will be seen, biotic and abiotic stresses sometimes result in similar plant responses and some degree of commonality is evident. Here, the biotic stresses are considered first.

It must be noted that because of the rapid, continuing progress in the field, a comprehensive treatise on all the relevant literature is beyond the scope of this chapter. Instead, this chapter represents a general overview, and the reader is encouraged to seek out the appropriate references for further information.

## II. BIOTIC STRESS

### A. General Concepts

Interactions between plants and most pathogens are compatible or incompatible, depending on gene-for-gene interactions between the plant and the pathogen. The gene-for-gene interactions are generally between plasma membrane–bound receptors (R or resistance-gene products) and products that may be expressed by the pathogen. A pathogen-expressed product that is recognized by a plant R-gene product is termed an avirulent one, and the interaction results in an incompatible interaction between the plant and the pathogen [1]. This in turn leads to both local and systemic acquired resistance in the plant (see later).

Conversely, lack of recognition (a compatible reaction) of such a product by the plant is characterized by pathogen virulence and results in infection of the plant by the pathogen and subsequent plant disease.

A plant has both constitutive and inducible defenses. A type of inducible biotic stress defense response is systemic acquired resistance (SAR), an increase in resistance throughout the entire plant to attack by a broad spectrum of pathogens, typically effected over a period of days to a week following initial pathogen attack. The region immediately under attack, however, achieves local acquired resistance (LAR), which may be separate from SAR. Among the first symptoms of an incompatible interaction between a plant and a pathogen is the hypersensitive response (HR), a component of both LAR and SAR. The HR is a localized, rapid necrosis of the tissue at the infection site, in part due to a phenomenon referred to as an oxidative burst [2]. The oxidative burst produces reactive oxygen species or intermediates (ROS, ROI) that are apparently necessary but not sufficient for cell death to occur. Such ROS also have antimicrobial effects, although their contribution to defense is still a matter of conjecture [2]. Rather than being a symptom of infection and resulting from trauma, sufficient data have been collected to suggest strongly that the HR is an example of programmed cell death (PCD), which is under coordinated, genetic control [3]. It should also be noted that colonization of roots by rhizogenic bacteria has also been shown in some species to give rise to a heightened resistance to foliar pathogens. This type of SAR has also been termed induced systemic resistance (ISR) [4].

The oxidative burst stimulates the synthesis of salicylic acid (SA), which is associated with SAR and appears to be largely responsible for transmitting the induction of defense responses throughout the plant [2]. SAR is manifested by changes in protein synthesis, de novo synthesis, and increased synthesis of specific defense proteins or proteins involved in metabolic biosynthetic pathways. These changes are generally similar for fungal, bacterial, and viral pathogens, with many proteins or biosynthetic pathways induced in common. Some of the proteins and enzymes that accumulate during LAR and SAR are involved in lignification of the cell wall. Lignified cell walls are highly resistant to cell wall–degrading enzymes produced by many invading pathogens and therefore prevent or limit infection. Another general response is the production of a class of nonproteinaceous compounds called phytoalexins, which appear to act as antimicrobial compounds. The compounds are structurally complex, with chemical derivation paralleling and specific to the species in which they are synthesized. The following sections are a brief summary of the better known biotic defense responses.

## B.  Cell Wall Modifications

Cell wall–modifying proteins encompass two groups, proteins that alter the cell matrix and enzymes involved in lignification. The cell wall matrix proteins are typically hydroxyproline-rich glycoproteins (extensins) or glycine rich. They appear to function by providing a framework for the cross-linking of carbohydrate (pectin, cellulose) or polyphenolic (lignin, suberin) moieties [5,6]. These proteins are inducible by ethylene (wounding), fungal elicitors, or viral infection [6]. The cell wall–modifying enzymes are mostly peroxidases, which catalyze the suberization and lignification of cell walls [6]. They are involved in the normal synthesis of cell walls but are also inducible by fungal elicitors [5] and may act in concert with enzymes involved in the biosynthesis of phenolic compounds [6]. The thickening of the cell walls then serves to wall off the pathogen and acts as a deterrent to further invasion.

Phenolic compounds utilized in the modification of cell walls have the same biosynthetic origin as the isoflavonoid phytoalexins: the phenylpropanoid pathway. This pathway is, in turn, a branch of the shikimic acid pathway, responsible for the synthesis of aromatic amino acids [7]. The precursor in the phenylpropanoid pathway, phenylalanine, is converted to 4-coumaryl-coenzyme A (CoA) by the involvement of phenylalanine-ammonia lyase (PAL), cinnamate-4-hydroxylase (C4H), and 4-coumaryl-CoA ligase (4CL) [7]. These enzymes, present during normal metabolism, increase dramatically and rapidly upon exposure to pathogens. This has been shown to occur at the level of both gene transcription and translation [6]. 4-Coumaryl-CoA serves as a branch point for the synthesis of lignin as well as of (iso)flavonoids. Hydroxylation of 4-coumaryl-CoA produces caffeic acid, which is successively modified to make ferulic and sinapic acids, as well as coniferyl alcohol [7]. Peroxidase-catalyzed polymerization of the alcohols corresponding to ferulic and sinapic acids, as well as coniferyl alcohol, gives rise to lignin [6,8]. Cinnamyl alcohol dehydrogenase is integral to this process and has been reported to be induced by fungal elicitors in several systems [9–12] and by ozone in spruce [13]. Inducibility by ozone indicates that the enzyme superoxide dismutase (SOD) may also be involved in this process. One of the products of SOD is $H_2O_2$, a substrate for peroxidases. Fungal attack or elicitor application also induces SOD in several plant species or cultivars that display an HR. Bowler et al. (Ref. 14 and references therein) hypothesize that SOD activation may aid in the strengthening of cell walls or possibly kill pathogens directly through $H_2O_2$, although the $H_2O_2$ involved in lignification may arise from other sources (reviewed in Ref. 15).

## C. Phytoalexins

Phytoalexins have been isolated from a number of plant species and appear to be specific to a particular plant family. For example, the Leguminosae produce (iso)flavonoids primarily, whereas the Solanaceae produce sesquiterpenes and the Umbelliferae mainly manufacture coumarin derivates [6]. Phytoalexins are toxic to both specific pathogens and the plants themselves and thus may contribute to the necrosis associated with the HR [16]. It should be noted, however, that questions remain as to whether phytoalexin synthesis is merely a response to infection as very few studies have conclusively demonstrated a role for phytoalexins in the defense response to pathogens [17]. In addition, several phytoalexins are also required for normal growth and development [18]. Although it is almost certain that some phytoalexins play a role in the defense response, they probably act in concert with other defense responses or proteins [17,18].

### 1. (Iso)flavonoid-Derived Phytoalexins

Chalcone synthase (CHS) is the start of the (iso)flavonoid branch of the phenylpropanoid pathway. This enzyme is active in normal growth, development, and metabolism of plants and is ultimately responsible for many plant pigments (e.g., anthocyanins) [7]. In the Leguminosae, where it has been studied extensively, CHS is highly inducible by pathogen attack [19]. This is accomplished by the differential regulation of several isozymes, some of which are constitutive while others are specific to pathogenic attack [20]. Numerous other enzymes associated with this class of phytoalexins are inducible at both the transcript and protein levels in response to biotic and abiotic stresses [18]. Although evidence for the specific mechanisms by which isoflavonoid phytoalexins achieve their toxicity is generally lacking, accumulated evidence suggests that they cause dysfunctions in the plasma membrane or tonoplast [21].

### 2. Coumarin-Derived Phytoalexins

Coumarins are also derivatives of the phenylpropanoid pathway but the branch mechanisms remain unresolved [12]. Hahlbrock and Scheel [12] suggested that glucosides or glucose esters are key intermediates, providing a means of safely sequestering potentially self-toxic compounds until they are needed. As with the (iso)flavonoid-synthesizing enzymes, coumarin pathway enzymes are stimulated by elicitor treatment.

### 3. Terpenoid-Derived Phytoalexins

The other major family of phytoalexins is derived from the terpenoid biosynthetic pathway. This pathway also operates during normal development and metabolism, producing such compounds as abscisic acid (ABA), giberellins, chlorophyll, carotenoids, and phytosterols [22]. Terpenoids arise via mevalonic acid, with hydroxymethyl glutarate reductase (HMGR) as the putative key regulatory enzyme for the entire pathway. The activity of several critical enzymes in the central pathway, including HMGR, increases as a result of pathogen attack, as evidenced by increases in gene transcription and translation. Data also exist concerning the de novo synthesis of several enzymes specific to terpenoid phytoalexin biosynthesis. For example, farnesyl pyrophosphate transferase and casbene synthase have been shown to be synthesized de novo in response to elicitors, leading to the formation of diterpenoid phytoalexins [23–25]. Research has uncovered evidence for an alternative terpenoid biosynthetic pathway in higher plants that does not involve mevalonic acid (reviewed in Ref. 26), but it is not known whether this pathway is important in biotic stress defense.

## D. Pathogenesis-Related (PR) Proteins

Much of the data regarding PR proteins is correlative rather than causal; however, there are reports that strongly suggest that PR proteins are an important component of the defense response in vivo. The PR proteins are compositionally quite diverse, falling into 10 or 11 families as defined by Kombrink and Somssich [27] or van Loon et al. [28], respectively. Although there is general agreement in their classification schemes, Kombrink and Sommsich combine the PR-4 and PR-5 families of Van Loon et al., insert another PR-4 family, and omit the PR-11 family. The classification scheme of Kombrink and Somssich will be followed, as their rationale seems better justified when considering the apparent functions and relatedness of the family members.

## 1. PR-1

This family currently has no function assigned to its members despite considerable effort and characterization at the gene, transcript, and protein levels [27]. The members are grouped together on the basis of sequence data.

## 2. PR-2 (β-1,3-glucanses)

This family is well characterized. Glucanases act by hydrolyzing β-1,3 glucan residues. This type of carbohydrate dominates fungal cell walls along with chitin [29]. Glucanase expression, as determined by transcript and protein analyses, increases greatly upon exposure of plant tissue to pathogens or fungal elicitors [5]. The family is subdivided based on pI values and cellular and extracellular location (i.e., isoforms) that reflect differences in enzymatic and antifungal properties [27].

It has been shown that the isoforms can be under differential regulation [30–32]. Non–pathogen-induced expression of glucanases has been reported in roots, stems, and flowers [5], with certain isoforms specific to the normal development of the plant [32]. Glucanases are apparently induced in concert with chitinases (PR-3 family) in a number of species [5]. The combination of the two enzymes has been shown to have direct antifungal properties in vitro [33].

## 3. PR-3 (Chitinases)

This family is also well characterized. Chitinases hydrolyze β-1,4 acetylglucosamine linkages of chitin polymers, which are a primary constituent of fungal walls [34]. Although plant secondary walls reportedly contain chitin, those linkages resist chitinase, possibly due to glycolipid modification of the linkages [35]. Chitinases are also induced by pathogens and microbial elicitors as well as by abiotic stresses such as heavy metals and salt [36]. Kombrink and Somssich (Ref. 27 and references therein) indicate that four classes of this family exist, based on primary structure, pI, enzymatic activity and antifungal properties.

## 4. PR-4 (Chitin-Binding; Win-like Proteins)

This family is poorly understood and the grouping is based on sequence homology and a chitin-binding motif. Although proteins have been isolated and their genes analyzed, no obvious function can be ascribed to them; however, antimicrobial activity has been demonstrated in vitro. The chitin-binding abilities of these proteins appear to be minimal [27].

## 5. PR-5 (Thaumatin-like Proteins; Osmotins)

These proteins are structurally related to the sweet-tasting protein thaumatin, originally isolated from *Thaumatococcus daniellii* Benth. [37]. Thaumatin and other members of this family have sequence homology with the maize trypsin/α-amylase inhibitor family of proteins [38]. These inhibitors have binding sites with varying specificity to both serine proteinases and α-amylases in insects, animals, and pathogens [39], and this family of proteins is also considered to be a part of the defense response of plants.

The thaumatin-like family has also been shown to have another defensive capability. A well-characterized member of this family of proteins is osmotin, a 26-kDa salinity-inducible protein initially isolated from cultured tobacco cells [40]. In whole tobacco plants, the expression of this protein is complex. Fungal infection and salt induce the accumulation of osmotin in a tissue-specific manner with some accumulation also induced by ethylene and tobacco mosaic virus [41]. In contrast, osmotin messenger RNA (mRNA), but not protein, is inducible by ABA, wounding, and several abiotic stresses [41,42]. Vigers et al. [43] showed that osmotin and the serologically related proteins zeamatin (from maize) and PR-5 (from tobacco) had antifungal activity, causing the rapid bursting of hyphal tips. This is probably due to a membrane-permeating ability that osmotin, along with a related class of proteins termed permatins, has been shown to have [44]

## 6. PR-6 (Proteinase Inhibitors)

Because of their small molecular mass, proteinase inhibitors are generally referred to as polypeptides rather than structurally complex proteins. They are also classified as inducible defense molecules. They act against the proteolytic activity of microbe-secreted proteases as well as the proteinases found in the digestive tract of animals, particularly insects. Consequently, animals are not deterred in the short term from consuming plant tissue. Prolonged exposure to the inhibitors, however, will contribute to starvation of the animal [39]. There are several classes of proteinases based on mechanism of action [45]. They are named after the active residue or cofactor responsible for the proteolytic cleavage (i.e., serine-, cysteine-, aspartic- and metalloproteinases). Not surprisingly, there are inhibitors for each of these proteinase classes and all apparently work by competitive inhibition. Pathogen attack, fungal elicitors, and wounding are all capable of inducing all of the types (classes) of proteinase inhibitors [39].

### 7.  PR-7 (Proteinases)

Despite the logical need for proteinases either as a component of the HR or to counter pathogens, little information exists about this family [27]. An alkaline endoproteinase has been shown to be inducible in response to viroid infection and *Phytophthora* infection in tomato leaves (summarized in Ref. 27).

### 8.  PR-8 (Lysozymes/Class III Chitinases)

This class consists of class III chitinases bearing no structural similarity to the chitinases from the PR-3 family. The PR-8 chitinases do bear sequence similarity to a bifunctional lysozyme/chitinase (Ref. 27 and references therein). Basic and acidic isoforms exist, with the acidic isoforms devoid of lysozyme activity. Curiously, the class III chitinases demonstrate little or no antifungal properties, leaving their contributions to pathogen defense debatable except for the lysozyme activity of the basic isoforms [27].

### 9.  PR-9 (Peroxidases)

This family of enzymes varies widely in function and structure but all catalyze the oxidation of a substrate in the presence of $H_2O_2$. As such, the activity of this family is diverse, ranging from lignification and suberization of cell walls (see earlier) to potentially participating in signal transduction (see later). Some of these processes occur in the absence of pathogen or related stimuli and are part of normal growth and development [27].

### 10.  PR-10 (Intracellular PR proteins)

Little is known of this class aside from sequence similarity and size range (16 to 18 kDa [27]). It has been speculated that they are intracellular in nature, which may be consistent with observed ribonuclease sequence similarities and activities (Ref. 46 and references therein). Lo et al. [46] presented data indicating a role for these proteins in degrading stress- or pathogen-specific RNAs as well as in developmental regulation of normal plant metabolism.

## E.  Non-PR Proteins

Numerous other peptides and proteins have been identified across a variety of plant species that accumulate in response to pathogens. For example, Segura et al. [47] reported that a peptide from potato, snakin-1, is active against both a bacterium and a fungus and has homology to hemotoxic snake venoms. These other peptides and proteins have not been yet been included in a formal classification scheme. This is due to very limited knowledge of the extent of their expression or the fact that most reports concern changes in a particular gene transcript with no attendant protein data. In addition, SA appears to be crucial for the expression of most, if not all, of the defense molecules already described, while the groups that follow do not necessarily require SA for induction.

### 1.  Thionins

Thionins are a family of small (approximately 5 kDa), sulfur-rich polypeptides shown to exist in several families of plants and have a putative role in plant defense [48]. Thionins were originally described as toxic factors in many cereal seeds [49] and classified as $\alpha$- or $\beta$-. More recently, they have been shown to be synthesized in leaves as well, with the highest abundance occurring in the epidermal cell walls [49]. Because the cell wall of the epidermis is often the primary site of pathogenic attack, thionins may act as a first line of defense. Thionin expression is regulated both developmentally and by pathogen attack, with pathogenic attack causing enhanced transcription and translation of thionin mRNA [49]. Non–pathogen-induced expression of thionins is limited to the period of growth preceding emergence from the soil. Thionin toxicity is hypothesized to arise from an amphipathic tertiary structure that may cause an increase in membrane permeability [48]. It is also believed that self-toxicity to the plant cell is minimized by the presence of a proteolytically cleaved precursor that may shield the cell's membrane system from the active domains of the thionins [49].

## 2. Plant Defensins

Plant defensins are another widespread family of small, sulfur-rich proteins, structurally unrelated to the thionins, although originally classified as γ-thionins [50]. They have antimicrobial activity in vitro (Ref. 50 and references therein). They are found primarily in seeds, although research indicates that they are also expressed in vegetative and fruit tissues [51,52] (Wisniewski et al., unpublished). The family has homologues found in mammals and insects, where they have also been shown to have antimicrobial activities. Plant defensins are all similar in size and have several conserved residues, including eight Cys, two Gly, an aromatic residue, and a Glu residue. Broekaert et al. [50] indicate that plant defensins can be broadly classified into two groups based on the in vitro morphogenic effects on treated fungal hyphae. One group causes reduced hyphal elongation with an increase in hyphal branching. The other group does not induce marked morphological distortions. The different classifications are based on the activity against particular fungi [50]. Plant defensins are generally not active against bacteria. *Arabidopsis thaliana* has been shown to have five plant defensin genes, some of which are constitutively expressed, while others are differentially regulated [52]. This is consistent with the observations that some plant defensins are constitutive while others are stimulated by pathogen attack [50]. It has been shown that jasmonic acid (see later) can induce the expression of plant defensins as well as thionins [53]. Terras et al. [54] indicate that the two radish (*Raphinus sativus* L.) defensins are inducible by methyl jasmonate but not by SA. This report and others suggest that plant defensins may form a portion of the biotic stress defense repertoire separate from SAR. In turn, jasmonic acid functions in a signaling capacity both in normal development and in response to wounding and pathogen attack independently of salicylic acid (see later).

## 3. Lectins

Lectins constitute a large family of proteins occurring in numerous plant taxa. Lectins are primarily confined to seeds, bark and vegetative organs and are typically viewed as nitrogen storage proteins. All lectins share the property of binding particular carbohydrates, and many have been shown to have activities that can be considered defensive in nature. The evidence for active synthesis in response to pathogen attack, however, is not strong. Van Damme et al. [55] summarized much of what is known about lectins in a comprehensive review. Briefly, the following lectin families appear to have some form of putative defensive capability. Legume lectins are suggested to bind to gut endothelial cells in animals, where they elicit noxious effects. Chitin-binding lectins are a broad subfamily, some of whose members fall into the class I chitinase family (see PR-3). Type 2 ribosome-inactivating proteins (RIPs) bind carbohydrates in addition to inactivating ribosomes. It is felt that this class of lectins confers defense against mammalian herbivores but members of this class also affect some insects. In addition, this class may act as suicide proteins upon a breach of their vacuolar containment, disrupting protein synthesis and inducing cell death. Monocot mannose-binding lectins are poorly understood but are presumed to have activity against insects, possibly by binding gut proteins with mannose moieties. The jaculin subfamily of lectins is also poorly understood, but feeding trials with insects have demonstrated an inhibitory effect. The Cucurbitaceae phloem lectins appear to interact with another phloem protein upon rupture of the phloem, forming a gel, thus blocking the phloem. The presumed function of this protein would be to prevent the transport of infectious or pathogenic agents via the phloem.

## 4. Alternative Oxidase

The alternative oxidase (AOX) is the terminal oxidase of the cyanide-resistant alternative pathway found in mitochondria. It has been shown that both gene transcript and protein levels increase upon infection in tobacco resistant to tobacco mosaic virus. AOX is inducible by SA and sufficient correlative evidence exists to implicate AOX in interrupting viral movement and replication, either directly or indirectly [56–58]. More research is required to determine the mechanisms involved and how widespread this activity is across different plant families and viral pathogens.

## 5. Glutathione S-Transferases

Glutathione S-transferases (GSTs) catalyze the conjugation of glutathione to a variety of substrates that are typically hydrophobic, electrophilic, and cytotoxic in nature, thus detoxifying them [59]. Some class

I and class III GSTs have been shown to be inducible at the transcriptional and translational levels by various environmental stresses that have oxidative stress in common, including pathogen attack [59]. The oxidative burst produces $H_2O_2$, which among other actions (see earlier) induces the synthesis of SA, several GSTs, and glutathione peroxidases [59]. SA inhibits the catalase that normally degrades $H_2O_2$ [57,59], thus these enzymes may help detoxify lipid peroxides formed via $H_2O_2$. The utilization of glutathione in these processes causes an up-regulation of its synthesis, and the elevated glutathione levels have been shown to induce a number of the biotic stress defense molecules already described [57].

## F.  Alkaloids

Alkaloids are a diverse group of nonproteinaceous compounds reported from many plant taxa. Whereas much is known about their chemistry, biosynthesis, and medicinal and toxicological properties, it is currently unknown to what extent these compounds play a defensive role in biotic stress. As noted by Kutchan [60], many alkaloids have cytotoxic effects, particularly toward insects. The biosynthesis of these proteins is probably induced by various factors, but this area clearly requires further study.

## G.  Signal Transduction

An aspect of alterations in protein synthesis that has been alluded to but not described is that of the signal transduction pathways responsible for the expression of the various pathogen-induced proteins. Indeed, a complex set of signal transduction pathways or networks exist which coordinate a plant's defensive response to biotic stress. It is often observed that several different types of proteins may be synthesized in response to a particular pathogen, whereas another pathogen engenders a much different protein response, and such responses are dependent on the tissue, tissue status, and numerous other conditions. Thus, it is important to consider the signal transduction pathways or networks involved in the induction of the proteins already discussed. Figure 1 has been compiled to help provide a context. Many of the pathways illustrated have been shown to exist only in a small number of plant species (e.g., systemin, see later), and some are still speculative (e.g., the roles of AOX and gentisic acid in viral resistance). It should be noted that some abiotic stresses also induce the expression of some of the defense-related genes. In particular, the enzymes responsible for phenylpropanoid-derived phytoalexins have been shown to be induced by ultraviolet light and heavy metals. As mentioned before osmotin is also induced by salt stress.

### 1.  Elicitors

Specific molecules (i.e., elicitors) are important in initiating the signal transduction pathways. Specific proteins (avirulence factors), fragments of fungal cell walls, and fragments of plant cell walls hydrolyzed by pathogens or produced by phytophagous insects serve as triggers of the defense system. Presumably, this is accomplished by the elicitors binding to specific receptors that in turn initiate the various signal transduction pathways (Figure 1). In some cases, the receptors have been identified and localized to the plasma membrane.

### 2.  Signal Molecules

NITRIC OXIDE.   Nitric oxide (NO) has been shown to be an integral part of the biotic stress response network as shown in Figure 1. Its mode of action is manifested in at least two ways, through transcriptional activation of various genes within the network and the hypersensitive response, with cyclic GMP (cGMP) and cyclic ADP-ribose (cADPR) as possible intermediates [61–63]. The hypersensitive response is not completely understood but may rely in part on direct synthesis of both NO and superoxide by NO synthase. These compounds can synergistically destroy many cellular structures [64]. NO synthase has also been shown to be induced in resistant plants [63]. Thus, NO synthase, although not specifically mentioned earlier, could also be considered a member of the cadre of defense proteins.

SALICYLIC ACID.   Salicylic acid (SA) is a derivative of the phenylpropanoid pathway and is widespread in plant species [65]. It is a key intermediary for numerous aspects of the plant defense response, including LAR and SAR [57] (see Figure 1). Evidence for their role is quite strong in dicots and less so in monocots. SA may also enable communication between infected and healthy plants via its volatile methyl ester, methyl salicylate (MeSA) [66]. Although such a communication system between sessile organisms has obvious benefits, its utility under physiological conditions has not yet been proved.

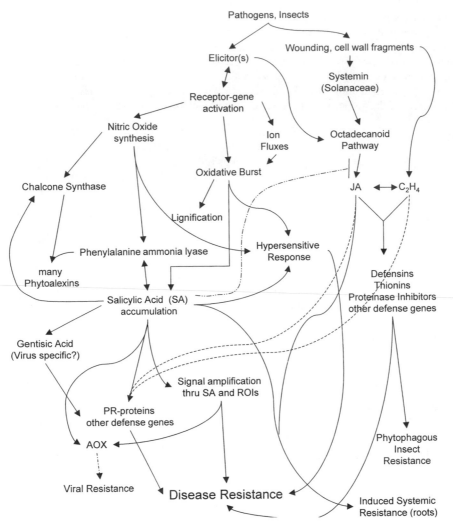

**Figure 1**    Biotic stress defense signal transduction model. SA, salicylic acid; JA, jasmonic acid; $C_2H_4$, ethylene; ROI, reactive oxygen species; AOX, alternative oxidase. Solid lines indicate strong evidence for the pathway, (-···-) indicates an inhibitory pathway of unknown mechanism, (—) indicates that some genes are expressed by combinations of JA, $C_2H_4$, and SA. (Adapted from Refs. 2, 53, 56, 61, 71.)

Some aspects of defense gene expression by SA have been elucidated. Genes for transcription factors, such as ethylene response binding element protein 1 (EREBP1) and NPR-1, are SA inducible [67–69]. Despres et al. [69] have shown that interactions of NPR1 with combinations of other transcription factors are necessary for PR-1 expression during SAR and ISR. In addition, some data suggest that combinations of SA with JA or ethylene are necessary for the induction of some PR proteins [70,71]. This suggests that the genes for those proteins have regulatory elements for transcription factors specific to SA, JA, and ethylene and require combinations of specific transcription factors to enact transcription.

SYSTEMIN.    Systemin is an 18-amino-acid polypeptide arising from the carboxy terminal region of a 200-amino-acid precursor, prosystemin [72,73]. Although the systemin transcript is constitutively produced at low levels, it is inducible by wounding [74]. Systemin has been shown to be released in tomato leaves in response to pathogenic attack, as a phloem-mobile, long-range signaling molecule, and is responsible for the synthesis of numerous biotic stress defense proteins. The signal transduction pathway by which it operates is the octadecanoid pathway, which is membrane lipid derived, generating jasmonic acid [74] (Figure 1 and see later). Systemin protein or prosystemin complementary DNAs (cDNAs) have been found in several species of solanaceous plants. A survey of recorded systemin sequences in Gen-Bank found no examples outside the Solanaceae, but it is possible that this polypeptide or its equivalent exists in other plant families.

JASMONIC ACID.   As noted before, jasmonic acid (JA) and its methyl ester (methyl jasmonate, MeJA) are products of the lipid-derived octadecanoid pathway [53]. It is thought that the interaction of elicitors, wounding, and/or cell wall fragments with a membrane receptor initiates the biosynthetic pathway resulting in production of JA with or without the action of systemin [53,74]. JA and MeJA participate in several aspects of normal growth and development as well as in several biotic and abiotic stress defense responses [53]. Numerous proteins appear to be regulated by JA and MeJA, including thionins, defensins, some proteinase inhibitors, genes involved in phytoalexin biosynthesis, and other defense genes, including osmotin. As noted, JA and MeJA may interact with SA and ethylene to induce some PR proteins. Interestingly, SA has been noted to inhibit JA synthesis [53], suggesting some degree of signal modulation. MeJa has also been proposed to function as a means of interplant communication, akin to MeSA [75], but its significance under physiological conditions again remains uncertain.

GENTISIC ACID.   Gentisic acid (GeA) is a derivative of salicylic acid. As noted by Belles et al. [76], GeA has been reported to have antifungal properties in vitro. Belles et al. [76] reported that GeA synthesis is induced in tomato in response to viral infection and can induce PR proteins that exogenously applied SA could not. The authors hypothesized that GeA may have a role complementary to that of SA (Figure 1). Further research is necessary to determine what specific role(s) GeA has in biotic stress defense responses and whether this compound is active in other species.

PROTEIN KINASES.   Protein kinases act either directly in the activation of various defense proteins or indirectly through transcription factor activation or synthesis. In some cases, MAP (mitogen-activated protein) kinase cascades are involved. Indeed, Kumar and Klessig [77] have reported differential induction of MAP kinases in tobacco by NO, SA, ethylene, and JA. Most protein kinases are also counteracted or modulated by specific phosphatases.

ION FLUXES.   Ion fluxes, particularly $Ca^{2+}$, have been implicated in many aspects of signal transduction. $Ca^{2+}$ has been shown to activate individual proteins, and transcription factors, usually in concert with protein kinases and MAP kinase cascades [78]. Changes in $Ca^{2+}$ are also important in normal growth and development as well as in signal transduction of several abiotic stress responses (Ref. 78 and references therein). A full accounting of such fluxes and subsequent signaling events, however, is beyond the scope of this chapter.

ETHYLENE.   Ethylene is produced upon wounding or infection by pathogens. Exogenous application of ethylene induces several PR proteins, indicating a role in biotic stress defense responses. Several lines of evidence suggest that induced ethylene may be a symptom rather than a cause of defense responses. It is likely that ethylene modulates such responses, acting together with JA or SA to induce several PR proteins [71].

OTHER SIGNALS.   It has been shown that action potentials akin to those observed in animal neural systems exist in plants as a consequence of wounding. Induction of several defense protein transcripts and proteins has been demonstrated to result from such action potentials. However, the mechanisms and the place of action potentials in the signaling network are not understood [79].

## III.  ABIOTIC STRESSES

Abiotic stresses also induce a diverse array of proteins and, in some cases, similar proteins are induced by different stresses. As noted before, biotic and abiotic stresses may induce similar or identical proteins as well. Frequently considered abiotic stresses include light, temperature, nutrients, salinity, and air and water pollutants. It is well known that an excess or deficit of any of these factors can greatly reduce growth and reproduction [80].

### A.  Heat Stress

Perhaps the most studied response to environmental stress in plants is the response to elevated temperature. An increase of five degrees or more above the optimal growth temperatures of a plant defines heat stress or, more commonly, "heat shock" [81]. For many temperate climate crops, temperatures in excess of 32 to 33°C constitute a heat shock [81]. Many of the protein synthetic responses to elevated temperature are found throughout the eukaryotic kingdom as well as in prokaryotes.

## 1. Decreased Translation

Quantitatively, a decline in overall protein synthesis occurs as a result of the translational repression of most mRNAs during severe heat shock [82]. Many explanations have been suggested, including general instability of polysomes at high temperature [83], loosely bound translational factors, changes in the cytoskeleton, and inhibition at the initiation or elongation step [82]. Some of the translationally repressed mRNAs are sequestered during the heat shock and are expressed after the stress is relieved [82].

## 2. Heat Shock Proteins/Molecular Chaperones

Qualitatively, several classes of proteins with different molecular masses are rapidly (20 min to 3 hr) and preferentially translated [82,84]. These proteins have been termed heat shock proteins (HSPs) and grouped by their molecular masses: low molecular weight (LMW; 15 to 30 kDa), which are seen only in plants, and the HSP60, HSP70, HSP90, and HSP110 classes. The appearance of these proteins has been correlated with enhanced thermotolerance as well as some measure of cross protection to other environmental stresses [82,85]. The HSPs are typically seen regardless of whether the temperature rise is rapid or slow [86] and have been shown to occur under field conditions [86,87]. In plants, 27- and 70-kDa proteins produced in response to elevated temperature also appear in response to variety of environmental stresses [88,89]. The 70-kDa HSPs are also seen to have seasonal expression patterns in some woody plant species, being more prevalent in autumn and winter months than in summer months in the species examined [90]. It has been suggested that these commonly produced proteins constitute a form of general-purpose stress tolerance.

The intensive study of HSPs and their constitutively synthesized heat shock cognates (HSCs) has led to the more general biological description of HSPs as molecular chaperones. Molecular chaperones comprise several classes (Ref. 91 and references therein) and act by assisting the self-assembly of nascent polypeptides into their correctly folded tertiary structures. HSPs are generally considered to be molecular chaperones, although in some cases the function of a particular HSP class remains in question. HSPs are thought to act by preventing the aggregation of nonfunctional proteins resulting from heat denaturation [92]. Homologues of the high MW HSPs have been reported in the cytoplasm, mitochondria, and chloroplasts. The HSP60 class appears to be restricted to mitochondria and chloroplasts, despite its nuclear origin. The function of the LMW HSPs is poorly understood even though they show a distribution similar to that of the higher MW HSPs.

Regulation of HSP synthesis occurs at the level of both transcription and translation. HSP70 provides a paradigm for such regulation [93,94]. A transcription factor, termed HSF, exists in a latent, monomeric form prior to heat shock. Upon heat shock, the monomer trimerizes and then binds to the appropriate promoter element (HSE) along with other transcription factors, resulting in transcription of HSP70 mRNA. Attenuation of transcription is apparently due to HSF being bound by HSP70 and possibly other factors. Phosphorylation may play a role in both the trimerization and the attenuation processes. In concert with this, HSP70 mRNA is stabilized at elevated temperatures and is efficiently translated, unlike most other mRNAs that are translationally repressed. Multiple HSFs have been reported from several plant species, suggesting functional differences and activities [94]. This is not surprising, considering that HSP70 is synthesized in response to multiple environmental stresses as well as normal growth and development [95].

## 3. Other Inducible Proteins

Other proteins or mRNAs also increase in abundance during elevated temperature but are not considered HSPs. They include several glycolytic enzymes [96], protein kinases [97], and ubiquitin [96,98,99]. Veirling [81] suggested that glycolytic enzymes and protein kinases are involved in metabolic readjustment. The activation or deactivation of regulatory proteins and enzymes by phosphorylation could be especially important. Ubiquitin is involved in protein degradation, and its enhanced expression is probably required to remove aberrant proteins resulting from damage to translational machinery or thermally denatured proteins.

## B.  Decreased Water Status

Several abiotic stresses have a common element between them, namely decreased cell water status. They include water deficit ("drought," "water stress"), salinity, and low temperature. Although each of these stresses imposes unique perturbations, numerous proteins are induced in common, and these stresses appear to share some signaling pathways. Hence, these stresses have been combined under one topic area, and each will be considered in turn.

### 1.  Water Deficit

PHYSIOLOGY.   Water deficit affects plants on several levels. Numerous studies have demonstrated that cell expansion and growth are among the first processes to decline under water deficit. With progressive water deficit, photosynthesis is adversely affected, and eventually assimilate partitioning. On the cellular level, membranes and proteins can be damaged by a reduction in hydration (see later) and an increase in reactive oxygen species or peroxidation.

Resistance to water deficit is manifested in four general ways: timing of growth to avoid water deficit, morphological adaptations, physiological adaptations, and metabolic alterations. The first three are complex processes and are incompletely understood, but significant progress has been made in understanding specific metabolic alterations.

Quantitative and qualitative changes in the synthesis of proteins have been reported to occur in plants in response to water deficit. Reductions in polyribosome stability have been reported [100–103], as well as changes in transcription [104]. Many, but not all, de novo synthesized proteins also appear in response to ABA application, supporting the role of ABA as a mediator in some water deficit–related responses (see later). Isolation of the genes responsible for these proteins has been accomplished in a number of species by cDNA cloning techniques. The proteins have been placed in several gene families on the basis of sequence homology and similar expression patterns. Numerous abbreviations have arisen in describing these genes, including rab (responsive to ABA), RD (responsive to desiccation), ERD (early dehydration inducible), lea (late embryogenesis abundant), and em (early maturation). The nomenclature used to describe water deficit stress is also diverse: drought, drought stress, water stress, and osmotic stress have all been used, but these may not be physiologically equivalent.

Seed maturation has been used as a model for metabolic alterations resulting from desiccation or dehydration. While this approach has identified many desiccation-related proteins, seed maturation represents a very specific type of water deficit and occurs simultaneously with other developmental events. As such, seed maturation will be considered only to a limited context.

WATER DEFICIT–INDUCIBLE PROTEINS

**LEA Proteins.**   LEA genes were first described in a survey of cDNAs from developing cotton seeds [105]. As analyzed by Dure [106], 18 different groups have been recognized on the basis of sequence homology both within cotton and between other species. Four groups, termed D-19, D-113, D-7, and D-11 (see later), have been characterized to varying extents. Little functional significance has been established for these groups, although Dure's analysis [106] suggests that their secondary and tertiary structures indicate that they could act as "reverse chaperones," facilitate counterion storage, or to maintain the hydration status of proteins and membranes.

**LEA D-11/RAB/Dehydrins.**   This group has been considerably studied at many levels [107]. Dehydrins are widespread, occurring in every higher plant species examined. Sizes range from 9 to 200 kDa, with a highly conserved 15-amino-acid consensus (EKKGIMDKIKEKLPG) located near the carboxy terminus. This consensus is repeated up to 11 times, depending on the specific dehydrin. An analysis of the sequences and induction stimuli suggests that several subclasses exist, some of which are principally low-temperature inducible rather than inducible strictly by water deficit, salinity, or ABA [107].

Secondary structure predictions indicate that dehydrins could form an amphipathic α-helix, which has been proposed to interact with and stabilize proteins or membranes [106,107]. Low water status reduces the hydration of biomolecules such as proteins, which can lead their denaturation [108] and to the disruption of membranes [109]. Dehydrins have been proposed to ameliorate these consequences by reducing hydrophobic aggregations or inappropriate interactions [107]. Although their specific function is yet to be demonstrated, several genetic studies have indicated a role in water deficit or cold tolerance. For example, Ismail et al. [110] reported cosegregation of dehydrin alleles with chilling tolerance in cowpea seedling emergence.

Constitutive expression of specific dehydrins has been reported in several species [90,107,111,112]. Thus, it is possible that some dehydrins fulfill a role(s) in normal growth and development. Alternatively, they may be synthesized as a constitutive defense against rapid changes in water status.

**Metabolic or Osmotic Adjustment Proteins.**   See section on salinity stress.

**Transport Proteins.**   Transport proteins include those involved in ion transport (see section on salinity) and those involved in water transport (aquaporins). Aquaporins have been characterized at both the transcriptional and translational levels, with differential expression reported [113]. Although water deficit and salinity have been shown to induce aquaporin transcripts, the overall function of aquaporins in regard to lowered water status as well as normal growth and development is still unresolved [113].

**Heat Shock Proteins.**   Water deficit has also been shown to induce several classes of HSPs, including HSP70 [88,89]. They probably play roles similar to those needed during elevated temperatures (see earlier).

**Protein Degradation.**   Ubiquitin and polyubiquitin are inducible by water deficit [114,115]. Given that proteins can be denatured by dehydration, these polypeptides probably play a role similar to that seen during elevated temperature (see before). It has also been shown that proteases can be drought inducible [114,115]. It is likely that the proteases either destroy denatured proteins or recycle amino acids for proteins needed in response to water deficit.

**Lipid Transfer Proteins.**   Lipid transfer proteins (LTPs) catalyze the transfer of several classes of phospholipids and/or glycolipids between membrane vesicles (in vitro) or their deposition in the cell wall [116]. LTPs have been shown to be induced by water deficit in the aerial portions of tomato and pea [117,118]. The data indicate a need to either increase membrane fluidity or decrease water loss via increased epidermal impermeability. These proteins are also inducible by low temperature and salinity stress [116].

## 2.  Salinity Stress

PHYSIOLOGY.   Although salinity stress is related to water deficit by a decrease in water status, the presence of excess ions also appears to be detrimental to many plant processes. Thus, plants subjected to salinity stress appear to face two stresses at the same time. Plants vary in their ability to survive salt stress, with tolerant plants generally either sequestering ions in the vacuole or synthesizing osmotically active compounds (i.e., osmoregulation). Nontolerant plants typically attempt to exclude excess ions via active transport. In the case of nontolerant *Arabidopsis*, a $Ca^{2+}$ sensor has been discovered that may be important for active transport [119]. Liu and Zhu [119] noted that $Ca^{2+}$ fluxes detected by this sensor may potentiate the regulation of $K^+$ and $Na^+$ transport systems, as occurs with animal homologues.

MULTIPLE PROTEIN RESPONSES.   Like water deficit, salt stress results in a general decrease in protein synthesis (e.g., Ref. 120), which is correlated with a loss of polysomes in vitro [121]. In turn, many proteins and transcripts have been reported to increase or be synthesized de novo in response to salt stress. Many of these proteins or transcripts are also inducible by water deficit or ABA.

OSMOTIC ADJUSTMENT.   Another aspect shared by salt stress and water deficit is osmotic adjustment, wherein organic or inorganic osmotically active solutes (osmolytes) are accumulated. This accumulation creates a lower solute potential, which allows a plant cell to maintain a higher water content than in the absence of these osmolytes. Many different organic molecules have been described as accumulating during water deficit or salt stress, including quaternary amines, polyols, and sugars [122], as well as inorganic $K^+$ and $Cl^-$ ions [123]. The ability of these organic molecules to balance ions sequestered in the vacuole and to stabilize enzymes incubated with salt solutions has resulted in describing these compounds as compatible solutes. Although osmotic adjustment does occur in response to water deficit or salinity stress, Hare et al. [124] contend that osmolyte accumulation is generally insufficient to lower solute potential significantly. In contrast, they suggest that the primary benefits of such osmolytes are metabolic in nature, either as compatible solutes, sensors of photosynthate partitioning, or buffering redox potentials. Further research is required to assess this hypothesis.

**Sugars.**   Two sugars have received attention, sucrose and trehalose. Sucrose synthesis is well characterized, and induction of sucrose synthase and sucrose-phosphate synthase transcription and translation has been observed in response to water deficit in several species [114]. In contrast, genes for trehalose synthesis have only recently been described in many species, with trehalose capacity probably present in all angiosperms [125]. Crowe et al. (Ref. 108 and references therein) have established that these sugars stabilize membranes and proteins in the presence of low water potentials and may play a crucial role in plant survival during decreased water status.

**Quaternary Amines.**   The most studied molecules of this class are proline and glycine betaine. Numerous studies have demonstrated that proline accumulates in response to water deficit or salinity stress. Arguments have been made that the generally inhibited metabolism resulting from these stresses reduces the demand for proline in protein synthesis. However, sufficient evidence indicates that proline accumulation is active rather than passive [126]. Hu et al. [127] showed that mRNA for the enzyme $\Delta^1$-pyrroline-5-carboxylate synthetase is increased in salt-stressed roots, and Delauney and Verma [126] suggested that it may be a rate-limiting step for proline synthesis. Delauney and Verma [126] also indicated that $\Delta^1$-pyrroline-5-carboxylate reductase activity and mRNA increase with salt stress, but there is doubt as to whether it is involved in the NaCl-dependent regulation of proline synthesis. Glycine betaine originates from choline rather than via amino acid biosynthesis and is found in 10 flowering plant families. McCue and Hanson [128] have shown that betaine dehydrogenase (BADH), the last step in the synthesis of glycine betaine, is salt inducible at both the protein and mRNA levels in sugar beets and spinach. ABA can also stimulate BADH protein and mRNA synthesis but at lower levels than via salt stress [129].

$Ca^{2+}$-ATPase.   $Na^+$ has been implicated in some of the difficulties faced by plants during salt stress. It reportedly displaces $Ca^{2+}$ from membranes, possibly reducing membrane stability [130]. It is more likely, however, that the primary injuries are from displaced $Ca^{2+}$ increasing the cytoplasmic $Ca^{2+}$, which could cause a disruption of signal transduction pathways requiring regulated levels of the ion [131]. Wimmers et al. [132] have shown that the mRNA for a $Ca^{2+}$-adenosinetriphosphatase (ATPase) is increased in abundance in response to elevated NaCl concentrations in tomato. They suggested that this ATPase may act to maintain proper levels of $Ca^{2+}$, thus mitigating the effects of $Na^+$.

### 3.   Cold Stress

PHYSIOLOGY.   Cold stress may be considered as a composite of two separate stresses: chilling (generally, temperatures from 4 to 15°C) and freezing. Chilling stress has in general been attributed to effects at the plasma membrane, manifested by electrolyte leakage from tissues (e.g., Ref 133). Williams [134] summarized data that indicated that leakage could be due to phase transitions caused by the presence of minor lipid components in the membrane or, alternatively, failure to seal critical intrinsic membrane proteins into the cell membrane by non–bilayer-forming lipids. Another explanation, that of lipid peroxidation (e.g., during photoinhibition), does not appear to be a cause of leakage. Hodgson and Raison [135] demonstrated that neither superoxide dismutase activity nor lipid peroxidation appears to increase at moderate photon flux levels.

Freezing stress appears to be the result of two components. The first is intracellular, in which ice crystals can pierce the plasma membrane (immediately lethal). The second is extracellular, in which the low water potential of ice in the intercellular spaces and cell wall can remove water from the cell (i.e., desiccation). Plants that achieve freezing tolerance apparently mitigate intracellular ice formation through biochemical means (i.e., changes in proteins and carbohydrates) [136]. Desiccation via freezing is ameliorated by both biochemical and biophysical changes, particularly in woody plant species [137].

Tolerance to chilling is apparently a prerequisite for tolerance to freezing. Chilling tolerance is an inducible response, dependent on day length and temperature [138,139], and is accompanied by an increase in the ABA content of cells (e.g., Ref. 140). Low temperatures also induce numerous proteins or their mRNAs, and evidence exists that some of these proteins are necessary for chilling tolerance. For example, Mohapatra et al. [141] compared alfalfa cultivars and cold-inducible gene products. They found a high correlation coefficient between the $LT_{50}$ (the temperature that is lethal to 50% of the treated plants) and the relative amounts of a particular low temperature–induced mRNA.

COMPARISONS WITH HEAT SHOCK.   Unlike heat shock, general protein synthesis does not appear to cease in response to chilling (Ref. 136 and references therein). In addition, there appears to be little conserved with heat shock in terms of the types of synthesized proteins [136]. However, some of the HSPs, or their transcripts, are also cold inducible [142–147]. Jaenicke [148] indicated that the stability of proteins is limited by both high *and* low temperatures. Hence, the presence of HSPs may not be that unusual. However, conclusive proof of a role for HSPs during low-temperature stress is currently lacking.

LOW TEMPERATURE–INDUCIBLE PROTEINS.   The low temperature–inducible protein literature is replete with various acronyms depending on the reporting laboratory's designation. For example,
in *Arabidopsis*, several gene families have been described, the members of which are variously referred
to as cold-responsive (COR), cold-inducible (KIN), responsive to desiccation (RD), low temperature inducible (LTI), early-dehydration inducible (ERD), and cold acclimation protein (CAP).

Low temperature–inducible proteins and mRNAs fall into several categories, including various signaling molecules or transcription factors (described in the following), metabolic enzymes, heat shock proteins (see earlier), and many hydrophobic or hydrophilic gene products. The proteins in the latter categories generally resemble or belong to the LEA or dehydrin protein classes (see earlier). Interestingly,
several of these proteins show sequence homology and activities resembling those of the antifreeze (thermal hysteresis) proteins of certain cold-water fish [149–152] (see also Ref. 153). Many of these proteins
have shown the ability to inhibit ice propagation or recrystallization either in vitro or in vivo.

The metabolic enzymes constitute a small portion of the inducible proteins, and those characterized
are primarily associated with glycolytic or fermentative pathways. The fermentative enzymes could serve
to compensate for a reduction in oxidative ATP production from mitochondrial membrane disruption.
Glycolytic enzymes could function by providing sugars to stabilize membranes or as compatible solutes
(see preceding sections). Synthesis of compatible solutes in the form of quaternary amines is also possible. Kishitani et al. [154] reported the accumulation of glycine betaine in barley plants during low-temperature acclimation and freezing tolerance. It is likely that many of the glycine betaine biosynthetic enzymes are induced, as Kishitani et al. [154] also noted the induction of betaine dehydrogenase (BADH).

Numerous studies [155] have demonstrated that alterations in membrane lipid composition occur
during cold acclimation. Several cold-inducible genes or gene products involved in this process have been
characterized. An ω-3 desaturase, whose activity could increase membrane fluidity, has been described
[156]. In addition, at least one nonspecific lipid transfer protein from barley is low temperature inducible
[116,157]. As noted previously, LTPs could also act to increase membrane fluidity.

### 4. Low Water Status Signal Transduction

The emerging view of signal transduction is one of a network of pathways with considerable cross talk.
Figure 2 has been compiled to provide context for the reader in understanding how the various stresses
sharing diminished water status interact. The initial perception step(s) for these stresses is still a matter of
conjecture. Some researchers have posited mechanosensors or stretch-activated channels, followed by
$Ca^{2+}$ fluxes coupled with various cation transporters and cation ATPases [158].

Regardless of the nature of initial perception, abscisic acid (ABA) has been shown to be a key intermediary in the expression of many, but not all, genes induced by decreased water status. Shinozaki and
Yamaguchi-Shinozaki [159] proposed a model consisting of two ABA-dependent and two ABA-independent signaling pathways (Figure 2). Pathway I results in the synthesis of a variety of transcription factors, which in turn bind to appropriate promotor elements, leading to gene expression. Pathway II appears
to be specific for a family of bZIP transcription factors that bind to abscisic acid response elements
(ABREs) in the promotors of specific genes. Wu et al. [160] uncovered portions of the ABA signaling
pathway, particularly the use of cADPR (cyclic ADP-ribose), two specific phosphatases, possibly IP$_3$ (inositol 1,4,5-triphosphate), as well as $Ca^{2+}$ fluxes. The $Ca^{2+}$ fluxes lead to activation of various protein
kinases and phosphatases, which directly or indirectly result in the induction or activation of the pathway
I and II transcription factors. It should be noted that low temperature alone also stimulates ABA but considerably less than water deficit or salinity stress [161]. This may account for the relatively weak induction of some otherwise highly ABA-inducible proteins by low temperature [157].

The ABA-independent pathway IV is primarily a function of low temperature [112,157,162]. Murata and Los [163] postulated that the initial perception may be via a change in membrane fluidity, possibly coupled with a change in the conformation of a membrane-bound protein. The low-temperature signal is then probably transduced via $Ca^{2+}$ fluxes, specific protein kinases, possibly an MAP kinase
cascade, and inactivation of specific phosphatases [153]. Gilmour et al. [164] proposed a model in which
these early events activate a protein termed ICE (inducer of CBF expression), where ICE and CBF are
proposed to be transcription factors. CBFs 1, 2, and 3 have been characterized and are also known as
DREB 1B, 1C, and 1A, respectively. These CBFs/DREBs bind to a promotor element described as the
CRT/DRE (C-repeat/ drought responsive element), hence CBF stands for CRT/DRE binding factor, and
DREB is an acronym for DRE binding [164]. Many low temperature–inducible genes must contain the

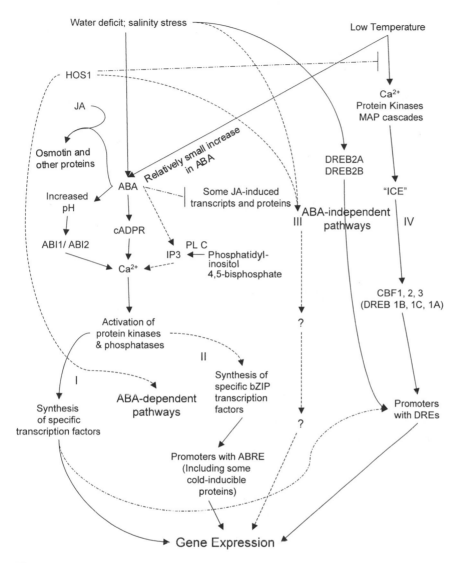

**Figure 2**  Decreased water status signal transduction model. ABA, abscisic acid; ABI, abscisic acid insensitive (a phosphatase); JA, jasmonic acid; cADPR, cyclic ADP-ribose; PL C, phospholipase C; IP3, inositol 1,4,5-triphosphate; bZIP, transcription factor with bZIP motif; ABRE, abscisic acid response element; DREB, drought-responsive element binding; ICE, inducer of CBF expression; CBF, C-repeat/DRE binding factor; MAP, mitogen-activating protein. Solid lines indicate strong evidence for the pathway, (- - - -) indicates an inhibitory pathway of unknown mechanism, (—-) indicates an incompletely understood pathway, (-··-) indicates a speculative pathway. (Adapted from Refs. 153, 159, 160, 164, 166, 167.)

DRE because a transgenic plant constitutively expressing CBF was shown to express many DRE-controlled proteins and to have a general elevation in cold tolerance [165].

DREs have been found in the promoters of genes generally associated with inducibility by low water status or pathway IV and negatively correlated with inducibility by ABA. In fact, several DREBs not corresponding to CBFs also bind to the DRE. They are dehydration inducible but are not regulated by low temperature. It is not known whether these DREBs constitute a portion of the poorly defined pathway III, but that would be a reasonable speculation.

An additional level of control has been demonstrated by Ishitani et al. [166], who indicated that pathway IV may not be totally independent of ABA, as shown by some intriguing transgenic plant analyses as well as by the observation of some ABA induction by low temperatures [140,161]. Ishitani et al. [167]

further showed that a gene product they termed HOS1 acts as a negative regulator of pathway IV at an early stage and appears to interact positively with both the ABA-dependent and ABA-independent signal transduction pathways. The nature of HOS1 remains elusive but is indicative of the extent of cross talk between these signal transduction pathways.

## C. Oxygen Deprivation

### 1. Physiology

The phrase *oxygen deprivation* is a general term for an area of study that has had a considerable variation in terminology. *Anaerobic* means $O_2$-free, *anoxia* refers to $O_2$ levels so low that ATP production by oxidative phosphorylation is essentially nil, and *hypoxia* defines $O_2$ levels that limit ATP production by mitochondria [168]. The actual levels of $O_2$ that correspond to these states are highly dependent on the tissue utilized and the physiological process under investigation. In general, $O_2$ levels from 2 to 10% (compared with the normal atmospheric concentration of 21%) result in a hypoxic state. For additional information, see Ref. 169. Low-$O_2$ environments are associated with excess water in the soil and relatively low diffusibility of $O_2$ in water compared with air. Proper soil aeration is prevented, which leads subsequently to consumption of available $O_2$ by aerobic organisms.

The effects of $O_2$ deprivation on protein synthesis are similar to those encountered during heat shock. In maize, there is a decrease in normal aerobic protein synthesis, associated with a loss of polysomes [170–172]. This is followed by the concomitant synthesis of approximately 20 proteins [170] under transcriptional and posttranscriptional regulation [172,173]. There may be one protein commonly expressed under both oxygen deprivation and heat shock [171].

### 2. Anaerobic Polypeptides

The preferentially synthesized proteins may be divided into two temporally regulated groups. Members of the first group are translated primarily during the first 5 hr of anoxia and are referred to as transition polypeptides (TPs) [170]. These proteins are stable, lasting long after their synthesis declines [171]. The second group, the anaerobic polypeptides (ANPs), begins to appear after approximately 90 min of anoxia, with synthesis continuing for several days, until cell death [170]. It has been noted that some plants differ in their tolerance to anoxia, e.g., maize (tolerant) and soybean (less tolerant). Both Sachs [171] and Hwang and Van Toai [174] speculated that a possible reason for the difference in anoxia tolerance is the number or types of ANPs synthesized.

Rather than acting to ameliorate protein denaturation, as in heat shock, most of these proteins are apparently involved in maintaining cellular ATP levels. In particular, several of the ANPs have been identified as glycolytic or fermentative enzymes. They include sucrose synthase, phosphoglucoisomerase, aldolase, alcohol dehydrogenase (ADH), and pyruvate decarboxylase (PDC) (summarized by Drew [169,175]).

Of these enzymes, ADH is the best characterized. Andrews et al. [176] examined *Adh* gene expression and enzyme activity in several tissues of maize under different $O_2$ concentrations. They showed that *Adh* gene expression is maximal with anoxia or extreme hypoxia (i.e., 0 to 4% $O_2$) in both root tips and axes. However, *Adh* transcripts did not always parallel ADH activity. The authors concluded that hypoxia is apparently crucial to increased ADH induction and activity. They suggested that a delay between *Adh* induction and enhanced activity provides a mechanism for survival during the anoxic state that would follow hypoxia. Indeed, Drew [175] indicated that in maize, tolerance to anoxia can be improved by exposure to hypoxic conditions. The regulatory pathway by which *Adh* is induced is not completely understood; however, common sequence elements have been found between *Adh*1 and aldolase gene promoter regions [173].

### 3. Ethylene Synthesis

Another important enzyme that increases during $O_2$ deprivation is 1-aminocarboxylate-1-cyclopropane synthase (ACC synthase). This enzyme catalyzes the rate-limiting step in the synthesis of ethylene, which increases dramatically in response to hypoxia [177]. One of ethylene's actions is to stimulate the formation of aerenchyma in the stem (and possibly roots), thus providing more $O_2$ to deprived tissues. Zarembinski and Theologis [178] reported that in rice, several ACC synthase genes are induced by anoxia, and that differential expression occurs under different hormonal and environmental signals. They suggested that the multiplicity of responses is consistent with the various aspects of ethylene in development and responses to other environmental stimuli (e.g., wounding).

### 4. Oxygen Radicals

A third enzyme worth noting is superoxide dismutase (SOD). The action of SOD in a low-$O_2$ environment may seem counterintuitive; however, the generation of superoxide radicals, leading to lipid peroxidation, has been implicated in postanoxic tissue damage [179]. In anoxia-intolerant *Iris germanica* rhizomes, lipid peroxidation was widespread compared with anoxia-tolerant *Iris pseudoacorus* rhizomes. Monk et al. [180] demonstrated that SOD activity rises after extended anoxia in *Iris pseudoacorus* rhizomes but not in *Iris germanica* or *Glyceria maxima*. This suggests that one of the strategies of tolerance to anoxia may be to synthesize proteins that anticipate the return of an aerobic environment [175].

### 5. Signal Transduction

This signal transduction pathway or network is incompletely understood. Drew [169] indicated that some evidence for an $O_2$ sensor exists, but not conclusively. In contrast, changes in $Ca^{2+}$ fluxes definitely appear to play a role, with changes in cytoplasmic pH or decreased energy metabolism possibly acting to induce these fluxes [169].

Regulation of aerenchyma formation, via ethylene, is apparently complex and an example of programmed cell death. Drew et al. [181] propose a model in which elevated levels of ethylene eventually activate phospholipase C, which then catalyzes a lipid-derived signal cascade culminating in the protein kinase–mediated phosphorylation of some target proteins involved in PCD. It is quite likely that this signal transduction pathway interacts with others to ensure its specificity to the roots.

## D. Air Pollution

Air pollution is still emerging as an area of research for alterations of protein synthesis. Ozone ($O_3$) and sulfur dioxide ($SO_2$) are usually considered to be the primary culprits in damage due to air pollution. However, these molecules eventually generate toxic oxygen species via the reaction $H_2O_2 + O_2^- \Rightarrow O_2 + OH^-$.

Changes in both protein [182] and mRNA [183,184] are known to occur, generally resulting in the action of antioxidants or detoxifying enzymes to minimize the damage to cellular membranes or macromolecules. Of the enzymes, superoxide dismutase (SOD) is the best characterized, catalyzing the reaction $2O_2^- + 2H^+ \Rightarrow H_2O_2 + O_2$. Bowler et al. [14] summarized information indicating that SOD protection in response to $O_3$ stress is often contradictory and extremely dependent on the conditions and plant species under consideration. Indeed, Badiani et al. [185] reported that with *Phaseolus vulgaris* L., fluctuations in the level of antioxidants and detoxifying enzymes occur during the day. Previously, SOD in *P. vulgaris* was reported to be unaffected by ozone treatment [186]. In contrast, there appears to be much better evidence for a role in $SO_2$ protection.

Of the antioxidant molecules, glutathione (GSH) has received the most attention. Glutathione is a polypeptide of the sequence γ-glutamyl-cysteinyl-glycine and acts to maintain the redox state of cysteine groups in proteins via its cysteinyl side chain. Reduction of protein disulfide bonds results in the formation of oxidized glutathione (GSSG). Glutathione, its synthetic enzymes, and glutathione reductase are known to be induced by $O_3$ [187]. As with SOD, however, the actual value of enhanced glutathione levels is equivocal, again being species and condition dependent.

## E. Metal Ion Stress

Heavy metal stress is frequently encountered by plants in areas of industrial pollution, as a result of mining activity, or, in the case of Al, acidic soils in the tropics and subtropics. In animals, the primary means of heavy metal sequestration is by the metallothionein family of proteins [188]. These proteins are Cys rich, relying on those residues to chelate metal ions. Although evidence of such proteins exist in plants, the primary defense against heavy metal toxicity relies on polypeptide equivalents, the phyochleatins. These polypeptides are inducible by heavy metals and are found throughout the plant kingdom [189].

Phytochelatins are related to glutathione, having the primary structure (γ-Glu-Cys)$_n$-Gly or (γ-Glu-Cys)$_n$-β-Ala, where $n = 2$ to 11 [188]. There is evidence for a metal-inducible phytochelatin synthase, catalyzing the transfer of γ-glutamylcysteine to glutathione, thus generating the (γ-Glu-Cys)$_n$ portion of the molecule [189].

In acidic soils, Al primarily affects root growth, and several hypotheses exist regarding the mechanism(s) of injury [190]. Plants apparently wield a variety of exclusion mechanisms, but the evidence for tolerance mechanisms is contradictory [190]. Some data exist for metallothionein-like proteins, as well as inducibility of PAL and proteinase inhibitors, but these are probably not the primary means of resistance [190]. Research indicates that Al-resistant *Arabidopsis* mutants utilize the exudation of Al-chelating organic acids or perhaps alkanization of the rhizosphere [191,192]. Undoubtedly, several enzyme biosynthetic and transport pathways must be activated or induced for such activity.

## F. UV Radiation

### 1. Ultraviolet-Absorbing Compounds

The responses of plants to UV radiation are of increasing concern because of the depletion of UV-absorbing ozone in the upper atmosphere. Research on the changes in protein synthesis related to UV has centered on the transcription and translation of enzymes involved in the flavonoid and anthocyanin biosynthetic pathways. Aside from their roles in plant defense against pathogens and as pigments, flavonoids also absorb UV. Chalcone synthase (CHS) and to some extent phenylammonia lyase (PAL) are the best studied. Wingender et al. [193] examined the promotor region of the *chs* gene from parsley and determined that two elements exist for the induction of CHS by UV in addition to an element for the elicitor induction. Subsequent work indicated that UV photoreceptors are responsible for the initial perception, and additional photoreceptors are required for anthocyanin or flavonoid biosynthesis in parsley [194].

Given the induction of CHS and PAL, it is perhaps unsurprising that other biotic stress genes are also induced by UV. Conconi et al. [195] noted the UV induction of several jasmonic acid–inducible genes and speculated that UV-induced lipid peroxidation may stimulate the octadecanoid pathway leading to JA (see earlier). Any function of these proteins against UV damage is unlikely, and Conconi et al. [195] suggested that diversion of C and N into these proteins may result in lowered fitness.

### 2. DNA Repair

A second area of interest is DNA repair mechanisms. As is well known, UV radiation induces various lesions in DNA. The best studied are cyclobutane-type pyrimidine dimers, which have been the only type of DNA lesions reported in plants [194]. The dimers can be repaired via photoreactivation (photolyase), excision repair, or recombinatorial repair [196,197]. The latter type of repair has not been reported in plants [194], and very little research has apparently been reported on excision repair in plants. Photoreactivation has been reported in several species (e.g., gingko [198], tobacco [199], pinto bean [200], and maize pollen [201]. Pang and Hays [202] reported on the presence of a photolyase activity in *Arabidopsis thaliana*. They indicated that the putative photolyase has a requirement for visible light, with an optimum of between 375 and 400 nm, which is similar to that of maize pollen photolyase. Pang and Hays [202] further suggested that *Arabidopsis* may actually have two photolyases, one similar to that found in *E. coli* and one similar to that reported in pinto bean.

## IV. CAVEATS

A final caution should be made: changes in expression at the transcriptional level do not necessarily equate to changes at the translational level. For example, LaRosa et al. [41] reported that in tobacco, osmotin mRNA is strongly induced by NaCl, water deficit, wounding, ABA, ethylene, and tobacco mosaic virus. In contrast, osmotin protein levels are weakly stimulated by all but NaCl and water deficit. In addition, Artlip et al. [203] reported on the appearance of a dehydrin transcript without the appearance of the corresponding dehydrin protein. Clearly, posttranscriptional regulation exists and needs to be considered before drawing conclusions based solely on transcript data.

An exception to this caution may be transcription factors, which frequently display apparent disparities between transcription and translation. Rather than posttranscriptional regulation, many transcription factors undergo rapid degradation or targeted proteolysis for activation. This is well known from cell cycle research, where mutants deficient in specific transcription factor degradation are seriously compromised. A report by Becker et al. [204] noted the necessity for a functional proteasome complex for induction of competence for elicitation of defense responses in cucumber hypocotyls, suggesting the need to regulate the amount of time a particular transcription factor can be active.

## V.  SUMMARY

The protein synthetic responses of plants to environmental stresses are diverse, in many cases yielding a specific set of proteins that presumably assist in ameliorating the stress. Ongoing research efforts will continue to uncover previously unknown environmental stress response proteins or provide a better understanding of the proteins and signal transduction networks described in this chapter.

It is clear that plants employ extensive signal transduction pathways and that components of these pathways may participate in both biotic and abiotic stress responses. For example, cyclic ADP-ribose appears to play a role not only in ABA signal transduction but also in nitric oxide signal transduction. Ethylene response element binding proteins are important not only for ethylene-mediated responses, such as aerenchyma formation, but also for some aspects of biotic stress responses. Phopholipase C appears to be a common point for both ABA signaling and aerenchyma formation as well. JA and ABA have both synergistic and antagonistic effects, with an "interface" between wounding and salt stress [205]. Zhou et al. [206] have demonstrated some degree of cross talk between the glucose and ethylene signal transduction pathways, and Kovtun et al. [207] have shown that an oxidative stress–activated MAP kinase cascade serves as a link between many abiotic stresses and auxin signal transduction. $Ca^{2+}$ fluxes are documented for most if not all of the signal transduction pathways [208]. An important question that will need to be resolved is how the various signal transduction pathways are coordinated and how specificity is established.

## REFERENCES

1.   KE Hammond-Kosack, JDG Jones. Annu Rev Plant Physiol Plant Mol Biol 48:575–607, 1997.
2.   C Lamb, RA Dixon. Annu Rev Plant Physiol Plant Mol Biol 48:251–275, 1997.
3.   JT Greenberg. Annu Rev Plant Physiol Plant Mol Biol 48:525–545, 1997.
4.   LC van Loon, PAHM Bakker, CMJ Pieterse. Annu Rev Phytopathol 36:453–484, 1998.
5.   JF Bol, HJM Linthorst, BJC Cornelissen. Annu Rev Phytopathol 28:113–138, 1990.
6.   RA Dixon, MJ Harrison. Adv Genet 28:165–234, 1990.
7.   TW Goodwin, EI Mercer. Introduction to Plant Biochemistry. Elmford, NY: Pergamon Press, 1983.
8.   RB Herbert. The Biosynthesis of Secondary Metabolites. New York: Chapman & Hall, 1981.
9.   C Grand, F Sarni, CJ Lamb. Eur Biochem 169:73–77, 1987.
10.   MH Walter, J Grima-Pettenati, C Grand, AM Boudet, CJ Lamb. Plant Mol Biol 15:525–526, 1990.
11.   MM Campbell, BE Ellis. Planta 186:409–417, 1992.
12.   K Hahlbrock, D Scheel. Annu Rev Plant Physiol Plant Mol Biol 40:347–369, 1989.
13.   K Smith, A Polle, H Rennenberg. Stress Responses in Plants: Adaptation and Acclimation Mechanisms. New York: Wiley-Liss, 1990, pp 201–225.
14.   C Bowler, M Van Montagu, D Inze. Annu Rev Plant Physiol Plant Mol Biol 43:83–116, 1992.
15.   R Cross, OTG Jones. Biochim Biophys Acta 1057:281–298, 1991.
16.   A Bell. Annu Rev Plant Physiol 32:21–81, 1981.
17.   R Hammerschmidt. Annu Rev Phytopathol 37:285–306, 1999.
18.   RA Dixon, NL Paiva. Plant Cell 7:1085–1097, 1995.
19.   TB Ryder, CL Cramer, JN Bell, MP Robbins, RA Dixon, CJ Lamb. Proc Natl Acad Sci U S A 81:5724–5728, 1984.
20.   TB Ryder, SA Hedrick, JN Bell, X Liang, SD Clouse, CJ Lamb. Mol Gen Genet 210:219–233, 1987.
21.   A Smith, SW Banks. Phytochemistry 25:979–995, 1986.
22.   R Threlfall, IM Whitehead. Ecological Chemistry and Biochemistry of Plant Terpenoids. Oxford: Clarendon Press, 1991, pp 159–163.
23.   MW Dudley, MT Dueber, CA West. Plant Physiol 81:335–342, 1986.
24.   MW Dudley, TR Green, CA West. Plant Physiol 81:343–348, 1986.
25.   J McGarvey, R Croteau. Plant Cell 7:1015–1026, 1995.
26.   M Rohmer. Nat Prod Rep 16:565–574, 1999.
27.   E Kombrink, IE Somssich. The Mycota V. Part A, Plant Relationships. Berlin: Springer-Verlag, 1997, pp 107–128.
28.   LC van Loon, WS Pierpoint, T Boller, V Conejero. Plant Mol Biol Reporter 12:245–264, 1994.
29.   JH Burnett. Fungal Walls and Hyphal Growth. Cambridge: Cambridge University Press, 1979, pp 1–10.
30.   A Ward, DW Lawlor. J Exp Bot 41:309–314, 1990.

31. MD van de Rhee, R Lemmers, JF Bol. Plant Mol Biol 21:451–461, 1993.
32. F Cote, JR Cutt, A Asselin, DF Klessig. Mol Plant Microbe Interact 4:173–181, 1991.
33. F Mauch, B Mauch-Mani, T Boller. Plant Physiol 88:936–942, 1988.
34. S Bartnicki-Garcia. Annu Rev Microbiol 22:87–108, 1968.
35. N Benhamou, A Asselin. Biol Cell 67:341–345, 1989.
36. B Collinge, KM Kragh, JD Mikkelsen, KK Neilsen, U Rasmussen, K Vad. Plant J 3:31–40, 1993.
37. H van Der Wel, K Loeve. Eur J Biochem 31:221–225, 1972.
38. M Richardson, S Valdes-Rodriguez, A Blanco-Labra. Nature 327:432–434, 1987.
39. CA Ryan. Annu Rev Phytopathol 28:425–449, 1990.
40. NK Singh, CA Bracker, PM Hasegawa, AK Handa, S Buckel, MA Hermodson, E Pfankoch, FE Regnier, RA Bressan. Plant Physiol 85:529–536, 1987.
41. PC LaRosa, Z Chen, DE Nelson, NK Singh, PM Hasegawa, RA Bressan. Plant Physiol 100:409–415, 1992.
42. AK Kononowicz, KG Raghothama, AM Casas, M Reuveni, A-EA Watad, D Liu, RA Bressan, PM Hasegawa. Plant Responses to Cellular Dehydration. Rockville, MD: American Society of Plant Physiologists, 1993, pp 144–158.
43. AJ Vigers, WK Roberts, CP Selitrennikoff. Mol Plant Microbe Interact 4:315–323, 1991.
44. AJ Vigers, S Wiedemeann, WK Roberts, M Legrand, CP Selitrennikoff, B Fritig. Plant Sci 83:155–159, 1992.
45. AJ Barrett. Proteinase Inhibitors. Amsterdam: Elsevier, 1986, pp 3–20.
46. S-CC Lo, JD Hispkind, RL Nicholson. Mol Plant Microbe Interact 12:479–489, 1999.
47. A Segura, M Moreno, F Madueno, A Molina, F Garcia-Olmedo. Mol Plant Microbe Interact 12:16–23, 1999.
48. H Bohlmann, K Apel. Annu Rev Plant Physiol Plant Mol Biol 42:227–240, 1991.
49. AG Darvill, P Albersheim. Annu Rev Plant Physiol 35:243–275, 1984.
50. WF Broekaert, FRG Terras, BPA Cammue, RW Osborn. Plant Physiol 108:1353–1358, 1995.
51. RG Terras, S Torrekens, F Van Leuven, RW Osborn, J Vanderleyden, BPA Cammue, WF Boekaert. FEBS Lett 316:233–240, 1993.
52. P Epple, K Apel, H Bohlmann. FEBS Lett 400:168–172, 1997.
53. RA Creelman, JE Mullet. Annu Rev Plant Physiol Plant Mol Biol 48:355–381, 1997.
54. FRG Terras, IAMA Penninckx, IJ Goderis, WF Boekaert. Planta 206:117–124, 1998.
55. EJM Van Damme, WJ Peumans, A Barre, P Rouge. Crit Rev Plant Sci 17:575–692, 1998.
56. AM Murphy, S Chivasa, DP Singh, JP Carr. Trends Plant Sci 4:155–160, 1999.
57. DA Dempsey, J Shah, DF Klessig. Crit Rev Plant Sci 18:547–575, 1999.
58. K Maleck, K Lawton. Curr Opin Biotechnol 9:208–213, 1998.
59. KA Marrs. Annu Rev Plant Physiol Plant Mol Biol 47:127–158, 1997.
60. TM Kutchan. Plant Cell 7:1059–1070, 1995.
61. J Dangl. Nature 394:525–526, 1998.
62. M Delledonne, Y Xia, RA Dixon, C Lamb. Nature 394:585–588, 1998.
63. J Durner, D Wendehenne, DF Klessig. Proc Natl Acad Sci U S A 95:10328–10333, 1998
64. Y Xia, JL Zweier. Proc Natl Acad Sci U S A 94:6954–6958, 1997.
65. I Raskin. Annu Rev Plant Physiol Plant Mol Biol 43:439–463, 1992.
66. V Shulaev, P Silverman, I Raskin. Nature 385:718–721, 1997.
67. D Lebel, P Heifitz, L Thorne, S Uknes, J Ryals, E Ward. Plant J 16:223–234, 1998.
68. DM Horvath, DJ Huang, N-H Chua. Mol Plant Microbe Interact 11:895–905, 1998.
69. C Despres, C DeLong, S Glaze, E Liu, PR Fobert. Plant Cell 12:279–290, 2000.
70. X Dong. Curr Opinion Plant Biol 1:316–323, 1998.
71. P Reymond, EE Farmer. Curr Opin Plant Biol 1:404–411, 1998.
72. G Pearce, D Strydom, S Johnson, CA Ryan. Science 253:895–898, 1991.
73. B McGurl, G Pearce, M Orozco-Cardenas, CA Ryan. Science 255:1570–1573, 1992.
74. CA Ryan, G Pearce. Annu Rev Cell Dev Biol 14:1–17 1998.
75. EE Farmer, CA Ryan. Proc Natl Acad Sci U S A 87:7713–7716, 1990.
76. JM Belles, R Garro, J Fayos, P Navarro, J Primo, V Conejero. Mol Plant Microbe Interact 12:227–235, 1999.
77. D Kumar, DF Klessig. Mol Plant Microbe Interact 13:347–351, 2000.
78. AJ Trewavas, R Malho. Curr Opin Plant Biol 1:428–433, 1998.
79. L Sticher, B Mauch-Mani, JP Metraux. Annu Rev Phytopathol 35:235–270, 1997.
80. JS Boyer. Science 218:443–448, 1982.
81. E Veirling. Annu Rev Plant Physiol Plant Mol Biol 42:579–620, 1991.
82. S Lindquist. Annu Rev Biochem 55:1151–1191, 1986.
83. VA Bernstam. Annu Rev Plant Physiol 29:25–42, 1978.
84. F Schoffl, G Baumann, E Raschke, M Bevan. Philos Trans R Soc Lond B Biol Sci 314:453–468, 1986.
85. MM Sachs, T-HD Ho. Plants Under Stress. New York: Cambridge University Press, 1989, pp 157–180.
86. JA Kimpel, JL Key. Plant Physiol 79:672–678, 1985.
87. JJ Burke, JL Hatfield, RR Klein, JE Mullet. Plant Physiol 78:394–398, 1985.
88. E Czarnecka, L Edelman, F Schoffl, JL Key. Plant Mol Biol 3:45–58, 1984.
89. JJ Heikkila, JET Papp, GA Schultz, JD Bewley. Plant Physiol 76:270–274, 1984.

90. M Wisniewski, TJ Close, T Artlip, R Arora. Physiol Plant 96:496–505, 1996.
91. JA Mienyk. Plant Physiol 121:695–703, 1999.
92. C Forreiter, M Kirschner, L Nover. Plant Cell 9:2171–2181, 1997.
93. RI Morimoto. Genes Dev 12:3788–3796, 1998.
94. F Schoffl, R Prandl, A Reindl. Plant Physiol 117:1135–1141, 1998.
95. CL Guy, Q-B Li. Plant Cell 10:539–556, 1998.
96. S Lindquist, EA Craig. Annu Rev Genet 22:631–677, 1988.
97. S Moisyadi, HM Harrington. Plant Physiol Suppl 93:88, 1990.
98. AH Christensen, PH Quail. Plant Mol Biol 12:619–632, 1989.
99. TJ Burke, J Callis, RD Vierstra. Mol Gen Genet 213:435–443, 1988.
100. RS Dhindsa, RE Cleland. Plant Physiol 55:778–781, 1975.
101. RS Dhindsa, RE Cleland. Plant Physiol 55:782–785, 1975.
102. PR Rhodes, K Matsuda. Plant Physiol 58:631–639, 1976.
103. HS Mason, K Matsuda. Physiol Plant 64:95–104, 1985.
104. K Skriver, J Mundy. Plant Cell 2:503–512, 1990.
105. GA Galau, DW Hughes, L Dure III. Plant Mol Biol 7:155–170, 1986.
106. L Dure III. Plant Responses to Cellular Dehydration. Rockville, MD: American Society of Plant Physiologists, 1993, pp 91–103.
107. TJ Close. Physiol Plant 100:291–296, 1997.
108. H Crowe, LM Crowe, SB Leslie, E Fisk. Plant Responses to Cellular Dehydration. Rockville, MD: American Society of Plant Physiologists, 1993, pp 11–20.
109. PL Steponkus, M Uemura, MS Webb. Plant Responses to Cellular Dehydration. Rockville, MD: American Society of Plant Physiologists, 1993, pp 37–47.
110. AM Ismail, AE Hall, TJ Close. Proc Natl Acad Sci U S A 96:13566–13570, 1999.
111. M Robertson, PM Chandler. Plant Mol Biol 26:805–816, 1994.
112. B Welin, A Olson, M Nylander, ET Palva. Plant Mol Biol 26:131–144, 1994.
113. C Maurel. Annu Rev Plant Physiol Plant Mol Biol 48:399–429, 1997.
114. J Ingram, D Bartels. Annu Rev Plant Physiol Plant Mol Biol 47:377–403, 1996.
115. A Campalans, R Messeguer, A Goday, M Pages. Plant Physiol Biochem 37:327–340, 1999.
116. J-C Kader. Annu Rev Plant Mol Biol 47:627–654, 1997.
117. JM Colmenero-Flores, F Campos, A Garciarrubio, AA Covarrubias. Plant Mol Biol 35:393–405, 1997.
118. MB Trevino, MA O'Connel. Plant Physiol 116:1461–1468, 1998.
119. J Liu, J-K Zhu. Science 280:1943–1945, 1998.
120. S Ramagopal, JB Carr. Plant Cell Environ 14:47–56, 1991.
121. CJ Brady, TS Gibson, EWR Barlow, J Speirs, RG Wyn Jones. Plant Cell Environ 7:571–578, 1984.
122. RG Wyn Jones. Chem Br 21:454–459, 1984.
123. JM Morgan. Annu Rev Plant Physiol 35:299–319, 1984.
124. PD Hare, WA Cress, J Van Staden. Plant Cell Environ 21:535–554, 1998.
125. O Goddijn, S Smeekens. Plant J 14:143–146, 1998.
126. AJ Delauney, DPS Verma. Plant J 4:215–223, 1993.
127. C-A Hu, AJ Delauney, DPS Verma. Proc Natl Acad Sci U S A 89:9354–9358, 1992.
128. KF McCue, AD Hanson. Plant Mol Biol 18:1–11, 1992.
129. KF McCue, AD Hanson. Aust J Plant Physiol 19:555–564, 1992.
130. AC Leopold, RP Willing. Salinity Tolerance in Plants: Strategies for Crop Improvement. New York: Wiley-Interscience, 1984, pp 67–76.
131. A Lauchli. Calcium in Plant Growth and Development. Rockville, MD: American Society of Plant Physiologists, 1990, pp 26–35.
132. LE Wimmers, NN Ewing, AB Bennett. Proc Natl Acad Sci U S A 89:9205–9209, 1992.
133. A Minchin, EW Simon. J Exp Bot 24:1231–1235, 1973.
134. WP Williams. Philos Trans R Soc Lond B Biol Sci 326:555–567, 1990.
135. RAJ Hodgson, JK Raison. Planta 185:215–219, 1991.
136. CL Guy. Annu Rev Plant Physiol Plant Mol Biol 41:187–223, 1990.
137. M Wisniewski, R Arora. Cytology, Histology, and Histochemistry of Fruit Tree Diseases. Boca Raton, FL: CRC Press, 1993, pp 299–320.
138. CJ Weiser. Science 169:1269–1278, 1970.
139. GR Gray, L-P Chauvin, F Sarhan, NPA Huner. Plant Physiol 114:467–474, 1997.
140. STC Wright. J Exp Bot 26:161–167, 1975.
141. SS Mohapatra, L Wolfraim. RJ Poole, RS Dhindsa. Plant Physiol 89:375–380, 1989.
142. RK Yacoob, WG Filion. Can J Genet Cytol 28:1125–1131, 1986.
143. RK Yacoob, WG Filion. Biochem Cell Biol 65:112–119, 1987.
144. VV Kuznetsov, JA Kimpel, G Goekjian, JL Key. Sov Plant Physiol 292:6–8, 1987.
145. M Cabane, P Calvet, P Vincens, AM Boudet. Planta 190:346–353, 1993.
146. JV Anderson, Q-B Li, DW Haskell, CL Guy. Plant Physiol 104:1359–1370, 1994.

147.  CL Guy, Q-B Li. Plant Cell 10:539–556, 1998.
148.  R Jaenicke. Philos Trans R Soc Lond B Biol Sci 326:535–551, 1990.
149.  S Kurkela, M Franck. Plant Mol Biol 15:137–144, 1990.
150.  S Kurkela, M Borg-Franck. Plant Mol Biol 19:689–692, 1992.
151.  W-C Hon, M Griffith, P Chong, DSC Yang. Plant Physiol 104:971–980, 1994.
152.  D Worrall, L Elias, D Ashford, M Smallwood, C Sidebottom, P Lillford, J Telford, C Holt, D Bowles. Science 282:115–117, 1998.
153.  MF Thomashow. Annu Rev Plant Physiol Plant Mol Biol 50:573–599, 1999.
154.  S Kishitani, K Watanabe, S Yasuda, K Arakawa, T Takabe. Plant Cell Environ 17:89–95, 1994.
155.  M Uemura, PL Steponkus. Plant Cold Hardiness: Molecular Biology, Biochemistry. New York: Plenum, 1997, pp 171–179.
156.  S Gibson, V Arondel, K Iba, C Somerville. Plant Physiol 106:1615–1621, 1994.
157.  AJ White, MA Dunn, K Brown, MA Hughes. J Exp Bot 45:1885–1892, 1994.
158.  AJ Netting. J Exp Bot 51:147–158, 2000.
159.  K Shinozaki, K Yamaguchi-Shinozaki. Plant Physiol 115:327–334, 1997.
160.  Y Wu, J Kuzma, E Marechal, R Graeff, HC Lee, R Foster, N-H Chua. Science 278:2126–2130, 1997.
161.  V Lang, E Mantyla, B Welin, B Sundberg, ET Palva. Plant Physiol 104:1341–1349, 1994.
162.  K Nordin, P Heino, ET Palva. Plant Mol Biol 16:1061–1071, 1991.
163.  N Murata, DA Los. Plant Physiol 115:875–879, 1997.
164.  SJ Gilmour, DG Zarka, EJ Stockinger, MP Salazar, JM Houghton, MF Thomashow. Plant J 16:433–442, 1998.
165.  KR Jaglo-Ottosen, SJ Gilmour, DG Zarka, O Schabenberger, MF Thomashow. Science 280:104–106, 1998.
166.  M Ishitani, L Xiong, B Stevenson, J-K Zhu. Plant Cell 9:1935–1949, 1997.
167.  M Ishitani, L Xiong, H Lee, B Stevenson, J-K Zhu. Plant Cell 10:1151–1161, 1998.
168.  A Pradet, JL Bomsel. Plant life in Anaerobic Environments. Ann Arbor, MI: Ann Arbor Science, 1978, pp 89–105.
169.  MC Drew. Annu Rev Plant Physiol Plant Mol Biol 48:223–250, 1997.
170.  MM Sachs, M Freeling, R Okimoto. Cell 20:761–767, 1980.
171.  MM Sachs. Plant Life Under Oxygen Deprivation: Ecology, Physiology and Biochemistry. The Hague: SPB Academic Publishing, 1990, pp 129–140.
172.  J Bailey-Seres, M Freeling. Plant Physiol 94:1237–1243, 1990.
173.  S Dennis, JC Walker, DJ Llewellyn, JG Ellis, K Singh, JG Tokuhisa, DR Wolstenhome, WJ Peacock. NATO Advanced Study Institute on Plant Molecular Biology (1987: Carlsberg Laboratory). New York: Plenum, 1987, pp 407–417.
174.  SY Hwang, TT Van Toai. Plant Soil 126:127–132, 1990.
175.  MC Drew. Soil Sci 154:259–268, 1992.
176.  DL Andrews, BG Cobb, JR Johnson, MC Drew. Plant Physiol 101:407–414, 1993.
177.  MC Drew, MB Jackson, S Giffard. Planta 147:83–88, 1979.
178.  TI Zarembinski, A Theologis. Mol Biol Cell 4:363–373, 1993.
179.  MIS Hunter, AM Hetherington, RMM Crawford. Phytochemistry 22:1145–1147, 1983.
180.  LS Monk, KV Fagerstedt, RMM Crawford. Plant Physiol 85:1016–1020, 1987.
181.  MC Drew, C-J He, PW Morgan. Programmed Cell Death in Animals and Plants. Oxford: BIOS Scientific Publishers, 2000, pp 183–192.
182.  P Guillemaut, F Weber-Lotfi, D Blache, M Prost, B Rether, A Dietrich. Physiol Plant 85:215–222, 1992.
183.  D Ernst, M Schraudner, C Langebartels, H Sandermann Jr. Plant Mol Biol 20:673–682, 1992.
184.  S Karpinski, G Wingsle, B Karpinska, J-E Hallgren. Physiol Plant 85:689–696, 1992.
185.  M Badiani, G Schenone, AR Paolacci, I Fumagalli. Plant Cell Physiol 34:271–279, 1993.
186.  CP Chanway, VC Runeckles. Can J Bot 62:236–240, 1984.
187.  WE Rauser. Annu Rev Biochem 59:61–86, 1990.
188.  JC Steffins. Annu Rev Plant Physiol Plant Mol Biol 41:553–575, 1990.
189.  E Grill, S Loffler, EL Winnaccer, MH Zenk. Proc Natl Acad Sci U S A 86:6838–6842, 1989.
190.  LV Kochian. Annu Rev Plant Physiol Plant Mol Biol 46:237–260, 1995.
191.  PB Larsen, J Degenhardt, C-Y Tai, LM Stenzler, SH Howell, LV Kochian. Plant Physiol 117:9–18, 1998.
192.  J Degenhardt, PB Larsen, SH Howell, LV Kochian. Plant Physiol 117:19–27, 1998.
193.  R Wingender, H Rohrig, C Horicke, J Schell. Plant Cell 2:1019–1026, 1990.
194.  AE Stapleton. Plant Cell 4:1353–1358, 1992.
195.  A Conconi, MJ Smerdon, GA Howe, CA Ryan. Nature 383:826–829, 1996.
196.  KC Smith. The Science of Photobiology. New York: Plenum, 1989, pp 111–118.
197.  A Kornberg, TA Baker. DNA Replication. New York: WH Freeman, 1989, p 771.
198.  GP Howland. Nature 254:160–161, 1975.
199.  JE Trosko, VH Mansour. Mutat Res 7:120–121, 1969.
200.  N Saito, H Werbin. Photochem Photobiol 9:389–393, 1969.
201.  M Ikenaga, S Kondo, T Fujii. Photochem Photobiol 19:109–115, 1974.
202.  Q Pang, JB Hays. Plant Physiol 95:536–543, 1991.

203. TS Artlip, AM Callahan, CL Bassett, ME Wisniewski. Plant Mol Biol 33:61–70, 1997.
204. J Becker, R Kempf, W Jeblick, H Kauss. Plant J 21:311–316, 2000.
205. A Moons, E Prinsen, G Bauw, M Van Montagu. Plant Cell 9:2243–2259, 1997.
206. L Zhou, J-C Jang, TL Jones, J Sheen. Proc Natl Acad Sci U S A 95:10294–10299, 1998.
207. Y Kovtun, W-L Chiu, G Tena, J Sheen. Proc Natl Acad Sci U S A 97:2940–2945, 2000.
208. A Trewavas. Plant Physiol 120:1–6, 1999.

# 34

# Physiological Responses of Cotton (*Gossypium hirsutum* L.) to Salt Stress

**Mohammad Pessarakli**

*The University of Arizona, Tucson, Arizona*

## I. INTRODUCTION

No plant species or animals are immune from stress. Any species at least once during its life cycle is subjected to stress. Nutrient uptake and utilization as well as water absorption by plants are adversely affected under stressful conditions. Plant growth and metabolism are usually impaired under such conditions, resulting in decreased crop yields.

Among the essential nutrient elements, nitrogen is one of the most widely limiting elements for crop production, and when plants are subjected to stress, N uptake and utilization are likely to be more severely affected than any other mineral nutrient. Regarding nutrient uptake and metabolism, plant species behave differently under stressful conditions. The adverse effects of stress are usually less severe on salt-tolerant plants such as cotton than on the salt-sensitive species such as beans.

Since the first publication of this chapter in 1994, numerous studies have been conducted on cotton and the findings have already been published [1–31]. Most of these studies were concerned with the cellular and molecular aspects of this plant [3,6,7,9,10,15–18,21,25–27,29,30]. Li et al. [3] studied the effect of salt stress on the activity of protective enzymes in cotton seedlings and concluded that the adaptation of cotton seedlings to salt stress was expressed by roots, but the difference in salt tolerance between cultivars was expressed by cotyledons. Studying stomatal density and size under salinity stress conditions, Jafri and Ahmad [4] reported that a decrease in stomatal density under salt stress was compensated by an increase in stomatal size and mesophyll surface area. Adaptation to a saline environment was adjusted by increasing mesophyll surface area to ensure normal exchange of gases and photosynthetic activities under the stress condition.

Evaluating several stages of cotton growth and development, including seedling, preflowering, flowering, and boll formation stages, Khan et al. [5] found that the seedling stage was the most sensitive one as compared with other growth development stages. In their study [5], the lowest seed cotton yield was found at this stage of growth. At all growth stages, the yield of the salt-tolerant cultivar was less affected than that of the salt-sensitive one by salinity stress. Renu and Goswami [6] studied the activities of several enzymes at various stages of growth in cotton treated with GA-3 (gibberellic acid) and NaCl. These investigators [6] observed that nitrate reductase activity in cotyledonary leaves decreased with salt stress and maximum activity was observed at the first stage. While salinity invariably resulted in an increase in the cellulase and protease activity at all stages, GA-3 alone as well as its interaction with NaCl increased the nitrate reductase activity. In another study, Renu and Goswami [7] found that NaCl decreased the total chlorophyll and carotenoid content in cotyledonary leaves of cotton. However, the carotenoid content decreased more slowly than the chlorophyll.

Lin et al. [8] observed that with increasing NaCl concentration, the protein content in cotton seedlings decreased while the enzyme activity and the soluble sugar content increased. These findings [8] indicated that changes in metabolism led to synthesis of large amounts of proline and soluble sugars to maintain the osmotic pressure. Examining a wide range of species of various genera in the family Malvaceae, Gorham [10] detected the zwitterionic quaternary ammonium compound glycine betaine in all but 3 of over 100 species. In a more limited range of the species, particularly of *Gossypium*, glycine betaine accumulated to concentrations sometimes in excess of 100 mM in response to water deficit or salinity stress. In Gorham's [10] study, glycine betaine concentrations were highest in young tissues and accumulated to about 10% of the total nitrogen.

To improve salt tolerance in cotton, Shen et al. [11] grew this plant after the seeds were soaked in paclobutrazol solution. They found that under salt stress the growth rate and chlorophyll, soluble sugar, and proline contents of cotton seedlings grown from seed soaked in paclobutrazol solution were higher than those of the controls. The results of these investigators [11] also showed a significant improvement in the water relations of these plants. From this study [11], it was concluded that seed treatment with paclobutrazol can mitigate the effects of salt stress and promote salt tolerance in cotton.

According to Qadir and Shams [12], in a pot culture study, imposed salinity stress had a deleterious effect on germination and vegetative growth with significant differences among the cotton genotypes. Leaf area, stem thickness, and shoot and root weights decreased with increasing substrate salinity level. Leidi and Saiz [14] studied physiological responses of two cotton cultivars previously selected on the basis of growth under salinity. They postulated that the higher tolerance was the result of several traits such as higher $Na^+$ uptake and water content. These investigators [14] also suggested that adaptation through adequate but tightly controlled ion uptake, typical of some halophytes, along with efficient ion compartmentation and redistribution would result in an improved water uptake capacity under salt stress conditions and lead to maintenance of higher growth rates.

Zhu and Zhang [15] studied antitranspiration and antigrowth activities of xylem sap of several plants including maize, sunflower, cotton, and castor bean subjected to various stress treatments, such as soil drying, flooding, and salinity. All xylem sap samples showed an increased concentration of proteins when plants were either soil dried, salt treated, or flooded. As a result, the protein transportation flux in xylem sap was also increased. In an experiment conducted on a salt-sensitive cultivar of cotton, Lin et al. [16] found greater relative reductions in root length and root fresh weight than in hypocotyl length of seedlings grown in 75 mM NaCl. This indicates that the root was more severely affected than the hypocotyl by the salt stress.

Banks et al. [17] studied the antioxidant response of several salt-sensitive and salt-tolerant cotton cultivars to salt stress during fiber development. The results of their study [17] indicated that salt treatment reduced fiber growth in all the cultivars, except the most salt-tolerant one. Glutathione-*S*-transferase activity significantly increased in all the cultivars when treated with NaCl. Regarding effects of salt stress on enzymes in cotton plant, experiments of Fowler et al. [18] revealed increased levels of the antioxidants peroxidase, catalase, ascorbate peroxidase, superoxide dismutase, glutathione reductase, and glutathione-*S*-transferase in NaCl-stressed cotton callus and plants.

In a greenhouse experiment, Oliveira et al. [19] studied the effects of different salinity levels of irrigation water (0, 2000, 4000, 6000, and 8000 mg/L 70% NaCl and 30% $CaCl_2$ solution) on germination and growth periods of different cotton cultivars. These investigators [19] found that salt concentrations above 2000 mg/L decreased germination, vigor, plant height, and salt concentrations above 4000 mg/L decreased cotton yield and dry weight. According to Kasumov et al. [21], salt stress stimulated root respiration by separating oxidation and phosphorylation. The antioxidant activity of roots decreased abruptly, resulting in uncontrolled acceleration of free radical processes.

Vulkan-Levy et al. [24] carried out an experiment on the effect of water supply and salinity on Pima cotton. These workers [24] found that an increase in water salinity caused a decrease in the seed cotton yield and the salinity threshold increased with an increasing amount of water. Delayed fluorescence and a decrease in intermittent amplitudes in the early stages of salt stress imposed on cotton plants were observed by Ganieva et al. [25]. This phenomenon is an indication of a decrease in photosystem II (PSII) activity. It may be related to damage to chlorophyll (Chl) in the PSII donor site and a decrease in Chl *b* molecules leading to an increase in the Chl *a/b* ratio.

When cotton callus tissues were exposed to salt stress, Banks et al. [27] observed an increase in the activity of antioxidant enzymes, including ascorbate peroxidase and glutathione reductase. According to these investigators [27], these responses suggest that the up-regulation of the activity of these enzymes in response to salt stress is due to de novo transcription of the genes encoding the two enzymes and not to translation of the existing transcripts or mobilization of existing enzyme pools.

The most recent findings on cotton plants were reported by Feng et al. [28], Murray et al. [29], Gossett et al. [30], and Rajguru et al. [31]. Feng et al. [28] studied the effects of salt stress on VA (vesicular arbuscular) mycorrhizal formation and of inoculation with VAM (vesicular-arbuscular mycorrhizal) fungi on saline tolerance of plants, including cotton, maize, soybean, and melon, grown on soils containing NaCl. They found that at a given NaCl level, cotton, maize, and soybean plants incubated with VAM had a higher biomass than noninoculated plants. These investigators [28] suggested that the VAM fungi–plant symbiosis might play an important role in survival of plants grown on saline soils. Also, inoculation with VAM fungi could enhance crop production in plants grown on saline soils and reduce the loss of plant yield caused by salt stress. Gossett et al. [30] reported that the total antioxidant enzyme response to NaCl stress in cotton callus tissue is somewhat specific to the combined effects of $Na^+$ and $Cl^-$ ions. Although Rajguru et al. [31] showed that salt treatment reduced ovule fresh weight in several cotton cultivars, superoxide desmolase activity increased in most of the cultivars under the salt stress condition. Glutathione-*S*-transferase activity significantly increased in all the cultivars treated with NaCl.

Several studies indicated that decreases in plant growth and crop yields under stress conditions have been associated with impairment of nutrient and water uptake, abnormal metabolism, and inhibition of plant protein synthesis [32–77]. In these studies, salt and/or water stress impaired growth and incorporation of nutrients (i.e., N) into the protein and increased accumulation of inorganic-N in plants. Reduction of nutrient uptake and utilization by plants was also reported by several investigators in earlier studies [78–84]. Uptake of N and P by plants was inhibited by high NaCl and $Na_2SO_4$ concentrations in the root medium, and the excess amount of absorbed $Na^+$ depressed $NH_4^+$ absorption in these studies. Absorption and metabolism of ammonium ($NH_4^+$) and nitrate ($NO_3^-$) in red kidney beans (*Phaseolus vulgaris* L.) was significantly reduced under salt or water stress [82–84]. In all of the preceding studies, reduction of root permeability and the consequent decrease in water and nutrient uptake under high electrolyte concentrations were stated as the cause of this abnormality in water and nutrient absorption and metabolism. Nevertheless, low levels of salts in the presence of N, P, and K stimulated growth and increased yield of cotton, *Gossypium hirsutum* L. [62,63,85–88]. With further increase in salinity, dry-matter yield decreased, but it increased with the addition of N at each salinity level. Moreover, plants continued to accumulate N under saline conditions in spite of the reduction in yield and dry-matter production.

Soil salinity did not inhibit N absorption by bermudagrass (*Cynodon dactylon* L.), a high-salt-tolerant plant [89], and stress had little or no effect on the rate of $NO_3^-$ uptake by barley (*Hordeum vulgare* L.), another high-salt-tolerant crop, except at the highest osmotic pressure, lowest osmotic potential ($-0.54$ MPa) of the rhizosphere [43,90]. Also, NaCl in the culture solution did not influence $NO_3^-$ uptake by tomato (*Lycopersicon esculentum* Mill.), a medium-salt-tolerant plant [80].

Abdul-Kadir and Paulsen [91] reported that the soluble protein and free amino acid content of wheat (*Triticum aestivum* L.) plants were not consistently affected by $MgSO_4$, $MgCl_2$, and NaCl. Udovenko et al. [92,93] found that under salt stress the non–protein-N fraction increased in beans, peas (*Lathyrus hirsutus* L.), barley, and wheat, whereas the protein-N fraction changed irregularly. These investigators [92,93] concluded that the response of N metabolism to salt stress is similar in plants with varying salt tolerance. An increased in the soluble-N fractions and free amino acid levels and decrease in protein-N content of cotton plants under medium ($-0.8$ MPa osmotic potential) and high ($-1.2$ MPa osmotic potential) levels of salinity were reported by Pessarakli and Tucker [63]. However, these investigators found that the low level of salinity ($-0.4$ MPa osmotic potential) slightly enhanced dry-matter production and protein content of the plants. On the other hand, this level of salinity ($-0.4$ MPa osmotic potential) and lower ($-0.25$ MPa osmotic potential) of the culture solution substantially decreased protein content of red kidney beans [83,84], green beans [57,58,94,95], and alfalfa, *Medicago sativa* L. [56]. Impaired N metabolism and decreased protein content of a number of plants under stress conditions have also been reported by several other investigators [96–102]. Rabe [64,65] and Dubey [38,39] reviewed altered N metabolism and protein synthesis, respectively, in plants under stressful conditions. These authors reported that N metabolism and protein synthesis in plant species were severely affected under stress.

Water stress induced by Carbowax also caused a marked reduction in protein synthesis by plants [83,84]. Although these studies were conducted on red kidney beans, a salt-sensitive plant, salt (NaCl) stress resulted in an appreciably greater reduction in $^{15}N$ incorporation in the protein fraction than water stress created by the Carbowax treatment. These results [83,84] indicated inhibition of N utilization caused by an ionic effect in addition to the osmotic effect of either NaCl or Carbowax.

Although investigations on the effects of salt and/or water stress on nutrient (i.e., N) absorption, utilization, metabolism, and protein synthesis mostly indicated a reduction in the absorption rate of N and decrease in the protein content of plants, a few controversial results make generalization difficult. These and other inconsistent results that demonstrated either an increase or no effect on the nutrient (i.e., N) absorption and metabolism can probably be explained as resulting from a dilution effect. This was suggested by Frota and Tucker [82,83], Saad [84], Pessarakli and Tucker [60–63], and Pessarakli [55,86,94,95], in which plant growth was affected more than the nutrient uptake and metabolism by salt stress and as a result the relative concentration of N was higher for the stressed plants.

Although the mechanisms by which salinity stress or drought adversely affect plant growth are still controversial, it is generally agreed that impairment of N absorption and metabolism is a critical factor. For a detailed review of the adverse effects of stress on plants and crops, readers are referred to the most comprehensive source, the new edition [103] and the original edition [104] of the *Handbook of Plant and Crop Stress*.

If it could be determined at what particular stage of growth high salinity most negatively affects plant growth and metabolism, the mechanisms by which these adverse effects occur might be identified and the detrimental effects prevented. In this regard, in addition to this work, several investigators have already attempted to study the effects of stress at different stages of plant growth [62,63,86,88,105–126].

The purpose of this investigation was to determine the physiological effects of salt stress on growth in terms of dry-matter production, nitrogen ($^{15}NH_4^+$) absorption and metabolism, protein synthesis, and water uptake by cotton plants at two stages of growth.

## II.  RESULTS AND DISCUSSION

### A.  Dry-Matter Production of Cotton Plants

At both the vegetative and the reproductive stages of growth, salt stress (particularly at medium and higher NaCl levels) drastically reduced dry-matter production (Table 1). The results of Khan et al. [5] showing a substantial decrease in cotton yield at different stages of growth under salinity stress confirm this finding. The report of Qadir and Shams [12] indicating that the imposed salinity stress had a deleterious effect on the germination and vegetative growth of cotton plants is also in agreement with the present work. According to these investigators [12], leaf area, stem thickness, and shoot and root weights decreased with increasing substrate salinity level. Kurth et al. [127] also observed adverse effects of both NaCl and $CaCl_2$ salinity on cotton growth in terms of cell enlargement and cell production. Several other

**TABLE 1**  Dry Matter Production of Cotton Plants Subjected to NaCl Salinity During Vegetative and Reproductive Stages of Growth

| Growth stage | Treatment, osmotic potential (MPa) | Plant dry weight/pot (two plants) (g) | | |
|---|---|---|---|---|
| | | Shoots | Roots | Total |
| Vegetative | Control | 5.42 | 0.93 | 6.35 |
| | −0.4 | 3.79 | 0.97 | 4.76 |
| | −0.8 | 2.71 | 0.77 | 3.48 |
| | −1.2 | 1.71 | 0.39 | 2.10 |
| | LSD (.05)[a] | 1.43 | 0.20 | 1.58 |
| Reproductive | Control | 20.13 | 3.90 | 24.03 |
| | −0.4 | 16.72 | 3.98 | 20.70 |
| | −0.8 | 12.11 | 3.52 | 15.63 |
| | −1.2 | 7.60 | 2.42 | 10.02 |
| | LSD (.05)[a] | 4.22 | 0.83 | 4.60 |

[a] The least significant difference among the means at the .05 probability level.
*Source*: Ref. 62.

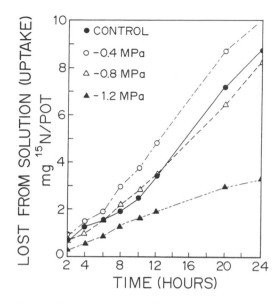

**Figure 1**   Solution loss of $^{15}$N (uptake) by cotton plants under various NaCl salinity conditions during the vegetative stage of growth. (From Ref. 62.)

investigators [2,8,10,15,17,19,24,31] reported similar reductions in different parameters of cotton growth and development under salinity stress that support the results of the present work.

In the present work, the dry-matter production of the stressed plants was highly negatively correlated with increasing levels of salinity at both stages of growth ($r$ of $-0.98$ to $-0.96$). Reduction of plant growth at higher levels of salinity has also been reported by other investigators for other salt-tolerant plants, such as barley [43,128,129], mangrove, *Avicennia marina* [51], and other halophytes including *Suaeda maritima* L. [130]. Several other investigators, in stress physiology, found that the growth of various plant species substantially decreased under stressful conditions [34,35,37,40,44,45,47,48, 50–61,69–76,81–86,88,95,104–126,131–153]. The present study showed that the shoot dry weight was reduced more by increasing salinity than the root dry weight. This is supported by the findings of several other investigators [55–61,82–84,86,94,95,103,104,128,129] and is consistent with the common knowledge in plant physiology that plant roots under stress conditions grow more and penetrate deeper in the soil or in the root medium in search of water and nutrients. Other studies also indicated a substantial reduction in shoot growth under stress conditions. For example, sodium chloride stress severely decreased shoot growth of rice, *Oryza sativa* L., cultivar GR-30 [154], and *Lactuca sativa* plants [155].

In the present study, the effect of salinity was more pronounced at the vegetative growth than at the reproductive growth stage. Other studies have also indicated that plants at earlier stages of growth were more sensitive to stress than those at later stages of growth [5,12,50,86,88,105–126,147,156–160]. Abnormal plant growth was also observed in experiments using sufficient amounts of salts other than sodium chloride [81,123,127,130,161,162] as well as under drought stress conditions [71,107,109,124, 125,163,164]. This is an indication of the adverse effects of stress on plant growth regardless of the source of the salt or the type of the stress.

## B.   Nitrogen Absorption by Cotton Plants

### *1.   Nitrogen ($^{15}$N) Absorption and Concentration in Plant Tissues*

The mean values for $^{15}$NH$_4^+$ absorption by cotton plants for 24 hr uptake time, under normal Hoagland solution (control) and salt (NaCl) stress conditions, obtained by analyzing solution samples indicated that low and medium levels of salinity did not significantly decrease the rate of $^{15}$N absorption (Figures 1 and 2). In fact, absorption was increased slightly at the vegetative stage at a low salinity level ($-0.4$ MPa osmotic potential). Similar amounts of NaCl drastically reduced the uptake rate of $^{15}$N in red kidney beans

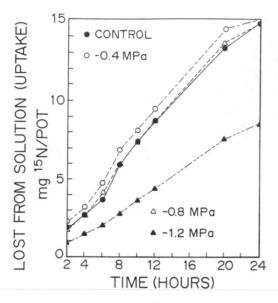

**Figure 2**    Solution loss of $^{15}$N (uptake) by cotton plants under various NaCl salinity conditions at the beginning of the reproductive stage of growth. (From Ref. 62.)

[82,84], green beans [55,57,58,94,95], alfalfa [56], and eggplant, *Solanum melongena* L. [60]. Such differences reflect variations in the salt tolerance of these different plant types. The high level of salinity ($-1.2$ MPa) appears to have caused a substantial reduction in the N absorption rate of cotton plants. The effect of the high salinity level on $^{15}$N uptake was more pronounced at the vegetative stage than at the reproductive stage of growth. The values for $^{15}$N uptake obtained by total N analysis of the plant materials (Tables 2 and 3) indicated essentially the same pattern as the solution loss data (Figures 1 and 2). Total amounts of $^{15}$N recovered in plants generally accounted for 95 to 99% of the apparent solution loss.

**TABLE 2**    Distribution of $^{15}$N Absorbed as Ammonium in Cotton Shoots and Roots Under Different NaCl Salinity Levels During Vegetative Stage of Growth

| Plant parts | Treatment, osmotic potential (MPa) | $^{15}$N uptake/pot (two plants) (mg) Uptake time, hr | | |
|---|---|---|---|---|
| | | 6 | 12 | 24 |
| Shoots | Control | 0.98 | 1.61 | 4.97 |
| | $-0.4$ | 0.82 | 1.71 | 4.78 |
| | $-0.8$ | 1.02 | 1.63 | 4.36 |
| | $-1.2$ | 0.36 | 0.88 | 1.81 |
| | LSD (.05)[a] | 0.43 | 0.49 | 0.45 |
| Roots | Control | 0.73 | 1.29 | 3.22 |
| | $-0.4$ | 1.39 | 1.84 | 4.20 |
| | $-0.8$ | 0.67 | 1.27 | 3.23 |
| | $-1.2$ | 0.34 | 0.57 | 1.24 |
| | LSD (.05)[a] | 0.48 | 0.93 | 0.77 |
| Total | Control | 1.71 | 2.90 | 8.19 |
| | $-0.4$ | 2.21 | 3.55 | 8.98 |
| | $-0.8$ | 1.69 | 2.90 | 7.59 |
| | $-1.2$ | 0.70 | 1.45 | 3.05 |
| | LSD (.05)[a] | 0.32 | 0.95 | 0.98 |

[a] The least significant difference among the means at the .05 probability level.
*Source:* Ref. 62.

**TABLE 3**   Distribution of $^{15}$N Absorbed as Ammonium in Cotton Shoots and Roots Under Different NaCl Salinity Levels at the Beginning of the Reproductive Stage of Growth

| Plant parts | Treatment, osmotic potential (MPa) | $^{15}$N uptake/pot (two plants) (mg) Uptake time, hr | | |
|---|---|---|---|---|
| | | 6 | 12 | 24 |
| Shoots | Control | 1.77 | 4.11 | 8.57 |
| | −0.4 | 1.91 | 3.78 | 8.64 |
| | −0.8 | 1.49 | 3.55 | 8.45 |
| | −1.2 | 1.15 | 2.18 | 4.67 |
| | LSD (.05)[a] | 0.43 | 0.64 | 1.67 |
| Roots | Control | 1.32 | 2.44 | 5.49 |
| | −0.4 | 1.62 | 3.07 | 5.68 |
| | −0.8 | 1.58 | 2.85 | 5.65 |
| | −1.2 | 1.01 | 1.81 | 3.39 |
| | LSD (.05)[a] | 0.39 | 0.65 | 1.80 |
| Total | Control | 3.09 | 6.54 | 13.99 |
| | −0.4 | 3.53 | 6.85 | 14.32 |
| | −0.8 | 3.07 | 6.40 | 14.10 |
| | −1.2 | 2.16 | 3.99 | 8.06 |
| | LSD (.05)[a] | 0.21 | 0.60 | 1.71 |

[a] The least significant difference among the means at the .05 probability level.
*Source*: Ref. 62.

The concentration of $^{15}$N in roots was higher than in shoots for all treatments at both stages of growth (Table 4). Adsorption of ammonium ions to the root surfaces or infusion of ions into apparent free space within roots, as suggested by Pessarakli and Tucker [60–63] and Pessarakli [55,86,94,95], could be a possible reason for the higher $^{15}$N concentrations in cotton roots. At both stages of growth, the $^{15}$N concentration of shoots significantly increased for salinized plants (−0.8 MPa) compared with the controls. This can be explained as a dilution effect as suggested by Frota and Tucker [82], Saad [84], Pessarakli and Tucker [60–63], and Pessarakli [55,94,95], as being due to a greater reduction in plant growth than $^{15}$N absorption under stress conditions. The relative translocation of $^{15}$N from roots to shoots was not appreciably affected by salt concentration.

## 2.   Total N Uptake by Plants

Total N uptake by plants decreased as the culture medium became more saline (Table 5). The reduction in total N uptake values at −0.8 MPa osmotic potential was to approximately 50 and 70% of the control

**TABLE 4**   Nitrogen ($^{15}$N) Concentration of Cotton Plants During Vegetative and Reproductive Stages of Growth as Influenced by NaCl Salinity

| Growth Stage | Treatment, osmotic potential (MPa) | $^{15}$N concentration (mg $^{15}$N/kg dry wt) | |
|---|---|---|---|
| | | Shoots | Roots |
| Vegetative | Control | 917 | 3462 |
| | −0.4 | 1261 | 4330 |
| | −0.8 | 1609 | 4195 |
| | −1.2 | 1059 | 3180 |
| | LSD (.05)[a] | 384 | 586 |
| Reproductive | Control | 422 | 1408 |
| | −0.4 | 517 | 1427 |
| | −0.8 | 698 | 1605 |
| | −1.2 | 615 | 1401 |
| | LSD (.05)[a] | 109 | 245 |

[a] The least significant difference among the means at the .05 probability level.
*Source*: Ref 62.

**TABLE 5**  Total N Uptake of Cotton Plants During Vegetative and Reproductive Stages of Growth as Influenced by NaCl Salinity

| Growth stage | Treatment, osmotic potential (MPa) | Total N/pot (two plants) (mg) | | |
|---|---|---|---|---|
| | | Shoots | Roots | Total |
| Vegetative | Control | 185.1 | 20.5 | 205.6 |
| | −0.4 | 114.8 | 23.0 | 137.8 |
| | −0.8 | 89.8 | 17.8 | 107.6 |
| | −1.2 | 43.1 | 7.2 | 50.3 |
| | LSD (.05)[a] | 15.9 | 5.3 | 17.7 |
| Reproductive | Control | 371.2 | 63.6 | 434.8 |
| | −0.4 | 333.8 | 66.7 | 400.6 |
| | −0.8 | 245.9 | 62.0 | 307.9 |
| | −1.2 | 156.5 | 41.6 | 198.1 |
| | LSD (.05)[a] | 96.1 | 15.0 | 105.6 |

[a] The least significant difference among the means at the .05 probability level.
*Source*: Ref. 62.

values at the vegetative and the reproductive stages of growth, respectively. At the higher salinity level (−1.2 MPa), the reduction in N uptake was proportionately greater at the vegetative stage than at the reproductive stage of growth. Reduced N uptake at the vegetative stage was largely due to the reduction in dry-matter production, except at the high salinity level, where the N concentration was reduced appreciably (Tables 1 and 5). At the reproductive stage, N concentration was essentially the same at all salinity levels, indicating that plants had adjusted somewhat to salinity and its effect on N uptake. The dry weights at the high salinity level were still less than 50% of the control levels, as were values for the total N uptake. Although generally these observations are similar to the previous discussion of $^{15}N$ data, some small deviations are apparent. Concentration data for $^{15}N$ (Table 4) indicate that short-term $^{15}N$ concentrations increased in the shoots at the −0.8 MPa salinity level.

## C.  Nitrogen Metabolism and Assimilation in Cotton Plants

### 1.  *Protein-N Content of Plants*

After a 24-hr exposure to $^{15}NH_4^+$, the protein-$^{15}N$ content of plants treated with a high level of NaCl (−1.2 MPa osmotic potential) was significantly less than in either the control, low, or medium NaCl treatments (Table 6). This is in agreement with the observations of Lin et al. [8] that with increasing NaCl concen-

**TABLE 6**  Concentration of $^{15}N$ Fractions and Protein-$^{15}N$ to Nonprotein-$^{15}N$ Ratio of Cotton Shoots Influenced by NaCl Stress for Two Stages of Growth After 24 hr of Uptake

| Osmotic potential (MPa) | Protein-N (mg) | Total soluble-N (mg) | Ammonium plus amide-N (mg) | Free amino acid-N (mg) | Protein-$^{15}N$ to nonprotein-$^{15}N$ ratio |
|---|---|---|---|---|---|
| Vegetative | | | | | |
| control | 489.6 | 318.6 | 6.97 | 129.0 | 1.54 |
| −0.4 | 560.9 | 463.9 | 9.37 | 189.1 | 1.21 |
| −0.8 | 544.3 | 741.3 | 21.96 | 287.1 | 0.73 |
| −1.2 | 230.4 | 523.2 | 13.67 | 273.1 | 0.44 |
| LSD (.05)[a] | 68.3 | 120.4 | 2.10 | 51.9 | 0.24 |
| Reproductive | | | | | |
| control | 229.9 | 145.5 | 4.87 | 60.6 | 1.58 |
| −0.4 | 232.8 | 204.8 | 5.31 | 65.9 | 1.14 |
| −0.8 | 217.3 | 363.6 | 19.71 | 96.4 | 0.60 |
| −1.2 | 98.8 | 393.6 | 9.64 | 108.1 | 0.25 |
| LSD (.05)[a] | 46.6 | 76.2 | 1.83 | 11.6 | 0.47 |

[a] The least significant difference among the means at the .05 probability level.
*Source*: Ref. 63.

tration, the protein content in cotton seedlings decreased. The depressing effects of salt on the protein content of cotton plants at high levels of NaCl could be attributed to decreased amino-N incorporation into protein as reported for red kidney beans [83,84], rice [40,79,101], and other plants [38,39,64–66, 92,93,165,166]. The decrease in polyribosome levels, as reported for corn, *Zea mays* L. [165], and for barley and pea shoots [166], is probably another reason for the decrease in protein synthesis in cotton shoots.

The low level of NaCl ($-0.4$ MPa osmotic potential of the nutrient solution) significantly increased the protein content of cotton shoots at the vegetative stage of growth. This is in agreement with the findings of Renu and Goswami [2] on uptake and accumulation of labeled $^{14}C$ photosynthates in cotyledonary leaf of cotton treated with gibberellic acid under salt stress. According to these investigators [2], the low levels of salinity stimulated $^{14}CO_2$ uptake and accumulation of carbohydrates in the cotyledonary test, whereas high salinity decreased it. The report of Zhu and Zhang [15] showing an increased concentration of proteins when maize, sunflower, cotton, and castor bean plants were either soil dried, salt treated, or flooded is also in support of the present study. However, the same level ($-0.4$ MPa) of salt stress substantially decreased protein synthesis in a number of other plants with lower degrees of salt tolerance [56–58,83,84,94,95,98]. Impaired N metabolism with the consequence of reduced protein content of several other plant species with various degrees of salt tolerance under stress conditions has been reported by several investigators [39,40,43,49,50,65,66,68,72,77,79,89,91–93,96,97,99–102,165,167].

Water stress is also known to impair N metabolism and reduce protein synthesis in plants [39,65,66,68,71,83,84,98,99,166]. In addition to the osmotic effect of salt, the specific ion effects of $Na^+$ and/or $Cl^-$ have certainly contributed appreciably to the inhibition of $^{15}N$ incorporation into protein. However, this study was not designed to distinguish specifically between osmotic and ionic effects.

The rates of $^{15}N$ incorporation into protein as measured by the concentration of $^{15}N$ in the protein fraction at 6, 12, and 24 hr of exposure to $^{15}NH_4^+$ appear to have been influenced only by the high level of NaCl at both growth stages (Table 7). The rate of incorporation was reduced at the $-1.2$ MPa salt level by factors of 2.5 and 2.8 at the vegetative and reproductive stages of growth, respectively. The rate of $^{15}N$ incorporation into protein at the vegetative stage was approximately 2.5 times greater than the rate at the reproductive stage, based on the $^{15}N$ concentration. The total $^{15}N$ in the protein fraction was greater at the reproductive than at the vegetative stage of growth, apparently because of the much greater amount of shoot dry weight (Table 1). Furthermore, the total protein-$^{15}N$ decreased with increased salt level at both stages of growth. This reflects the combined effect of salt on shoot growth and $^{15}N$ incorporation into protein.

**TABLE 7**  Slope ($b$) and Intercept ($a$) of the Regression Lines for Concentration of $^{15}N$ Fraction in Cotton Shoots Influenced by NaCl Salinity Versus 6, 12, and 24 hr Exposure Time for Two Stages of Growth[a]

| | | Growth stage | | | |
|---|---|---|---|---|---|
| | | Vegetative | | Reproductive | |
| 15N Fraction | Treatment, osmotic potential (MPa) | $b$ | $a$ | $b$ | $a$ |
| Protein | Control | 24.22 | $-99.08$ | 10.38 | 24.65 |
| | $-0.4$ | 26.44 | $-81.19$ | 10.43 | 23.18 |
| | $-0.8$ | 24.71 | $-61.10$ | 9.36 | $-11.58$ |
| | $-1.2$ | 9.89 | $-6.80$ | 3.61 | $+11.56$ |
| Soluble | Control | 13.20 | $-13.70$ | 5.14 | 17.53 |
| | $-0.4$ | 21.28 | $-58.50$ | 0.06 | $-1.79$ |
| | $-0.8$ | 30.64 | $-19.25$ | 16.36 | $-28.95$ |
| | $-1.2$ | 20.88 | $+30.25$ | 16.17 | $+2.36$ |
| Ammonium plus amide | Control | 0.32 | $-0.53$ | 0.19 | $-0.06$ |
| | $-0.4$ | 0.40 | $-0.16$ | 0.23 | $-0.17$ |
| | $-0.8$ | 0.93 | $-0.49$ | 0.80 | $+0.58$ |
| | $-1.2$ | 0.51 | $+1.71$ | 0.30 | $+2.52$ |
| Free amino | Control | 6.03 | $-22.79$ | 2.65 | $-1.05$ |
| | $-0.4$ | 9.63 | $-48.44$ | 2.73 | $+1.13$ |
| | $-0.8$ | 13.88 | $-58.63$ | 4.41 | $-9.63$ |
| | $-1.2$ | 14.24 | $-76.61$ | 5.12 | $-15.17$ |

[a] Correlation coefficient, $r$, values lie between 0.92 and 1.00.
*Source*: Ref. 63.

## 2. Total Soluble-N Content of Plants

Soluble N compounds that should be in an ethanol extract of plant tissues include $NO_3^-$, $NH_4^+$, amides, amino acids, amine, amino sugars, peptide, alkaloids, nucleotide, chlorophyll, and even some fats. Because only the $NH_4^+$ form of N was used in this investigation, $^{15}N$ from $^{15}NH_4^+$ exposure should not be found in the $NO_3^-$ form in the plant tissues in this study.

The total soluble-$^{15}N$ concentration of the plant tissues increased with NaCl concentration at $-0.8$ MPa osmotic potential after 24 hr exposure to $^{15}NH_4^+$ at the vegetative growth stage, then declined at $-1.2$ MPa (Table 6). At the reproductive growth stage, total soluble-$^{15}N$ increased in a similar manner but did not decline at the highest salinity level. Thus, the decrease in protein-$^{15}N$ at the $-1.2$ MPa salinity level did not result from a shortage of soluble-$^{15}N$ compounds. The rates of $^{15}N$ incorporation into the total soluble-$^{15}N$ fraction as indicated by the slope of the regression of the $^{15}N$ tissue concentration versus time (Table 7) followed the same pattern as described before for the 24-hr uptake time. Accumulation of soluble-N compounds in plants under stress conditions has also been reported by several other investigators for various plant species [43,49,53–61,65,66,69,70,72–76,79,81–86,89–96,101,102,153].

The amounts of soluble-$^{15}N$ reflect both concentrations in the tissue and dry-matter production (Table 1). Significantly less total soluble-$^{15}N$ was found after 24 hr with $-1.2$ MPa salinity at the vegetative growth stage. With the $-0.8$ MPa salinity, a larger amount of total soluble-$^{15}N$ was observed than with other treatments. At the reproductive stage, the amounts of total soluble-$^{15}N$ were equal at the high and low levels of salinity, with the intermediate salinity levels resulting in higher quantities of total soluble-$^{15}N$ in the plant parts.

Although the rate of $^{15}NH_4^+$ absorption was severely curtailed by high salinity at both growth stages [62], growth was not restricted by decreased total soluble-N concentration. However, impairment of soluble-N utilization at high salinity was reflected in a severe decrease in the protein concentration of plants. It is not clear whether this lower protein concentration was a cause of reduced growth. Growth was reduced at the lower salinity levels without a reduction in the protein concentration.

The ratio of protein-N to non–protein-N (soluble-N) is further evidence for the decrease in the protein-N content of the NaCl-treated plants. A substantial decrease in the ratio of protein-$^{15}N$ to non–protein-$^{15}N$ was observed for the plants subjected to a high level of NaCl ($-1.2$ MPa osmotic potential) compared with the controls at both stages of growth (Table 6). At both stages of growth, the values for the $-0.8$ MPa osmotic potential of the NaCl-treated plants were significantly lower than the controls.

## 3. Ammonium Plus Amide-N Content of Plants

At both stages of growth, significantly higher concentrations of ammonium plus amide-$^{15}N$ accumulated in the shoots of the plants subjected to NaCl stress compared with the controls (Table 6). The concentration of ammonium plus amide-$^{15}N$ increased with increasing salinity to a maximum at $-0.8$ MPa osmotic potential. Because the absorption rate of $^{15}NH_4^+$ did not change appreciably at these salinity levels, this increased accumulation of ammonium plus amide-N must have resulted from a reduced rate of utilization; however, reduced growth is another possible consideration. The concentrations of ammonium plus amide-N at the $-1.2$ MPa stress were lower than at the $-0.8$ MPa osmotic potential. These values reflect markedly reduced absorption rates at the $-1.2$ MPa stress [62]. The rate of $^{15}N$ utilization also decreased, allowing a higher $^{15}N$ concentration than commensurate with absorption rate. Slopes for regressions of ammonium plus amide-$^{15}N$ and time of uptake for each salinity level (Table 7) indicate a rate of accumulation pattern similar to the concentrations indicated for the 24 hr exposure time (Table 6).

## 4. Free Amino-N Content of Plants

Free amino acids would be expected to constitute the major portion of the total ethanol soluble-N compounds from plant tissues. In this study, the amino-N and ammonium plus amide-N accounted for 30 to 55% of the total soluble-N. The ninhydrin release method for free amino-N determination was used in this investigation. This method, however, can result in poor recoveries of a number of amino acids [168]. In Kennedy's [168] investigation, recoveries varied from 2 to 60% for 12 amino acids with complete recovery of 14 others. In the present study, the apparent low recovery of amino-$^{15}N$ from cotton tissues by the ninhydrin release method is consistent with the results of Kennedy [168] when all aspects of the methodology are considered. Even with the low recovery, however, the relative effects of NaCl salinity on amino acid formation and utilization should be valid.

**TABLE 8** Influence of NaCl Salinity on Water Absorption by Cotton Plants During the 24- hr [15]N Uptake Period for the Vegetative and Reproductive Stages of Growth

| Growth stage | Treatment, osmotic potential (MPa) | Water uptake/pot (two plants) (mL) Uptake time, hr | | |
|---|---|---|---|---|
| | | 6 | 12 | 24 |
| Vegetative | Control | 125.1 | 160.0 | 202.5 |
| | −0.4 | 87.5 | 122.5 | 172.5 |
| | −0.8 | 57.5 | 120.5 | 135.0 |
| | −1.2 | 40.0 | 70.0 | 87.5 |
| | LSD (.05)[a] | 46.0 | 23.4 | 28.2 |
| Reproductive | Control | 165.0 | 275.0 | 490.0 |
| | −0.4 | 147.5 | 245.0 | 430.0 |
| | −0.8 | 130.0 | 167.5 | 215.0 |
| | −1.2 | 75.0 | 107.5 | 145.0 |
| | LSD (.05)[a] | 78.7 | 102.8 | 205.5 |

[a] The least significant difference among the means at the .05 probability level.
*Source*: Ref. 62.

After a 24-hr exposure to [15]$NH_4^+$, a higher concentration of amino-[15]N was found in the NaCl-stressed plants, as compared with the controls, at both growth stages (Table 6). This increased concentration was sufficient to equal or exceed the reduction in dry weight, except at the highest NaCl level (−1.2 MPa osmotic potential), when dry weight was reduced most drastically. Yet, at the higher NaCl salinity level, the amino-[15]N concentration either remained constant or increased slightly as protein concentration values declined to less than half of the values for all other treatments. Slopes of the regression lines of the amino acid and exposure time (Table 7) reflect rate of amino acid accumulation. Thus, the incorporation of amino acids into protein was impaired by a high level of NaCl. This level of salinity that was required for interference with protein formation in cotton, a relatively high-salt-tolerant plant, was much higher than reported for green beans [57,58,94,95], red kidney beans [83,84,98], soybeans, *Glycine max* L. [102], peas [97], alfalfa [56], corn [68,165], rice [40,49,79,101], and wheat [91], which all have lower degrees of salt tolerance than cotton.

## D. Total Water Uptake by Plants

At both stages of growth (except for the −0.4 MPa osmotic potential during the reproductive stage), salt-stressed plants absorbed significantly less water than the controls (Table 8). Plants at −0.4 MPa stress did not exhibit a statistically significant difference in water uptake during the reproductive stage of growth compared with the controls. Reduction in water absorption by plants due to salinity stress has been reported by many investigators [15,22,24,55–62,74,82,84,86,94,95,129,170–172]. These investigators generally agreed that root permeability of plants (expressed as hydraulic conductivity of the root system) was decreased significantly under salt stress. This is an explanation for the reduction in water absorption rate and may contribute to a similar reduction in nutrient uptake and consequently reduction in crop yield under salinity conditions.

## III. SUMMARY AND CONCLUSIONS

Cotton plants grown in normal (control) and NaCl-treated Hoagland solutions were studied at two stages of growth (vegetative and reproductive). Plant growth in terms of dry-matter production was measured. Nitrogen absorption (total-N and [15]N) and water uptake were determined. Plant parts (shoots and roots) were analyzed separately for N content and distribution of [15]N in ammonium plus amide-N, free amino-N, total soluble-N, and protein-N after the plants were provided [15]$NH_4NO_3$ in nutrient solutions for 6, 12, and 24 hr.

Dry-matter production of the cotton plants was significantly reduced by decreasing the osmotic potential (increasing salinity) of the nutrient solution. The low and medium levels of salinity did not have a significant effect on the [15]N absorption rate, but the high salt levels caused a substantial reduction in the

[15]N uptake rate. The [15]N concentration of the roots was higher than that of the shoots, particularly under stress conditions. The [15]N concentration in plants increased with increasing salinity levels. The concentration of [15]N in plants in terms of the ratio of plant total [15]N content to dry matter produced (mg [15]N/kg dry matter) was significantly higher for moderately stressed than for control plants. This indicates that plants continued to accumulate [15]N under salt stress conditions in spite of the reduction in dry-matter production. Total water absorbed by plants decreased linearly with increasing salinity. This reduction was even more appreciable than the reduction in [15]N absorption rate. The effect of salinity was more pronounced at the vegetative than at the reproductive stage of growth.

The metabolism of [15]N in salinized cotton plants was adversely affected under medium and high levels of NaCl, at both vegetative and reproductive stages of growth. Significant accumulations of all soluble-[15]N fractions occurred when plants were subjected to medium and high levels of NaCl compared with the controls. The $-0.4$ MPa osmotic potential of the culture solution enhanced protein synthesis at the vegetative growth stage. Only the $-1.2$ MPa osmotic potential significantly decreased the protein-[15]N content of plants as compared with the controls and any other level of NaCl. Protein synthesis was impaired by a large excess of NaCl in the nutrient solution, which inhibited $NH_4^+$ metabolism.

Consequently, under salt stress conditions of sufficient magnitude, plant growth, N absorption and metabolism, protein synthesis, and water absorption will be altered. This will result in the failure of plants to fully utilize nutrients and water. Salinity levels in excess of those causing drastic interference with plant growth, nutrient (i.e., N) absorption and metabolism, and water uptake in salt-sensitive plants such as beans do not appreciably interfere with these factors in cotton, a relatively high-salt-tolerant plant. This indicates a link between salt tolerance, growth, nutrient (i.e., N) absorption and metabolism, and water uptake. Although the contribution of osmotic and specific ion effects cannot be distinguished from this study, it is likely that both were involved.

## REFERENCES

1. H Zhong, A Lauchli. Spatial distribution of solutes, K, Na, Ca and their deposition rates in the growth zone of primary cotton roots: effects of NaCl and CaCl$_2$. Planta Heidelberg 194(1):34–41, 1994.
2. M Renu, CL Goswami. Uptake and accumulation of labelled ([14]C) photosynthates in cotyledonary leaf of *Gossypium hirsutum* L. cv. H. 777 with gibberellic acid under salt stress. New Bot 21(1/4):115–119, 1994.
3. FG Li, FL Li, XL Li. Effect of salt stress on the activity of protective enzymes in cotton seedling. J Hebei Agric Univ 17(3):52–56, 1994.
4. AZ Jafri, R Ahmad. Effect of soil salinity on leaf development, stomatal size and distribution in cotton (*Gossypium hirsutum* L.) Pak J Bot 27:297–303, 1995.
5. AN Khan, RH Qureshi, N Ahmad. Response of cotton cultivars to salinity in various growth development stages. Sarhad J Agric 11:729–731, 1995.
6. M Renu, CL Goswami. Trends in activity of some enzymes in cotton cotyledonary leaves with GA-3 and NaCl. Crop Res Hisar 10:201–205, 1995.
7. M Renu, CL Goswami. Response of chloroplastic pigments to NaCl and GA-3 during cotton cotyledonary leaf growth and maturity. Agric Sci Dig 15(3):146–150, 1995.
8. JD Lin, ZY Zhu, BX Fan. Physiological reaction of cotton varieties under different levels of salt stress. China Cottons 22(9):16–17, 1995.
9. MC Lucas, T Fowler, DR Gossett. Glutathione *S*-transferase activity in cotton plants and callus subjected to salt stress. National Cotton Council of America Proceedings, Vol 2, Memphis, 1996, pp 1177–1178.
10. J Gorham. Glycinebetaine is a major nitrogen-containing solute in the Malvaceae. Phytochemistry (Oxf) 43:367–369, 1996.
11. FF Shen, CY Yin, FZ Gao, Y Yu. The promotion of salt tolerance of cotton seedlings from seed soaked in MET solution. China Cottons 23(5):9–10, 1996.
12. M Qadir, M Shams. Some agronomic and physiological aspects of salt tolerance in cotton (*Gossypium hirsutum* L.). J Agron Crop Sci 179(2):101–106, 1997.
13. A Nadler, B Heuer. Soil moisture levels and their relation to water potentials of cotton leaves. Aust J Agric Res 48:923–932, 1997.
14. ED Leidi, JF Saiz. Is salinity tolerance related to Na accumulation in Upland cotton (*Gossypium hirsutum* L.) seedlings? Plant Soil 190:67–75, 1997.
15. X Zhu, J Zhang. Anti-transpiration and anti-growth activities in the xylem sap from plants under different types of soil stress. New Physiol 137:657–664, 1997.
16. H Lin, SS Salus, KS Schumaker. Salt sensitivity and the activities of the H$^+$-ATPases in cotton seedlings. Crop Sci Soc Am 37(1):190–197, 1997.

17. SW Banks, SN Rajguru, DR Gossett, EP Millhollon. Antioxidant response to salt stress during fiber development. 1997 Proceedings Beltwide Cotton Conferences, New Orleans, January 6–10, 1997, Vol 2, 1997.
18. T Fowler, C Lucas, D Gossett. Glutathione *S*-transferase isozymes in control and salt-adapted cotton callus. 1997 Proceedings Beltwide Cotton Conferences, New Orleans, January 6–10, 1997, Vol 2, 1997, pp 1377–1379.
19. FA de Oliveira, TGS da Campos, BC Oliveira. Effect of saline substrate on germination, vigor and growth of herbaceous cotton. Engenharia Agric 18(2):1–10, 1998.
20. WJ Li, HZ Dong, QZ Guo, JQ Pang, J Zhang. Physiological response of a good Upland hybrid and its parent to PEG and NaCl stresses. China Cottons 25(6):7–10, 1998.
21. NA Kasumov, ZT Abbasova, G Gunduz. Effects of salt stress of the respiratory components of some plants. Turk J Bot 22:389–396, 1998.
22. F Moreno, E Fernandex-Boy, F Cabrera, JF Fernandez, MJ Palomo, IF Biron, B Belido. Irrigation with saline water in the reclaimed marsh soils of south-west Spain: impact on soil properties and cotton crop. In: R Ragah, G Pearce, eds. Proceedings of the International Workshop at the Tenth ICID Afro-Asian Regional Conference on Irrigation and Drainage, Denpasa, Bali, Indonesia, July 19–26, 1998. Jakarta, Indonesia: Indonesian National Committee on Irrigation and Drainage (INACID), 1998.
23. JD Liu, WW Ye, BX Fan. Research on stress resistance in cotton and its utilization in China. China Cottons 25(3):5–6, 1998.
24. R Vulkan-Levy, I Ravina, A Mantell, H Frenkel. Effect of water supply and salinity on Pima cotton. Agric Water Manage 37(2):121–132, 1998.
25. RA Ganieva, SR Allahverdiyev, NB Guseinova, NI Kavakli, S Nafisi. Effects of salt stress and synthetic hormone polystimuline K on the photosynthetic activity of cotton (*Gossypium hirsutum* L.). Turk J Bot 22(4):217–221, 1998.
26. DR Gossett, B Bellaire, SW Banks, MC Lucas, A Manchandia, EP Millhollon. The influence of abscisic acid on the induction of antioxidant enzymes during salt stress. 1998 Proceedings Beltwide Cotton Conferences, San Diego, January 5–9, 1998, Vol 2, 1998.
27. SW Banks, DR Gossett, A Manchandia, B Bellaire, MC Lucas, EP Millhollon. The influence of alpha-amanitin on the induction of antioxidant enzymes during slat stress. 1998 Proceedings Beltwide Cotton Conferences, San Diego, January 5–9, 1998, Vol 2, 1998.
28. G Feng, D Bai, M Yang, X Li, F Zhang. Effect of salinity on VA mycorrhiza formation and of inoculation with VAM fungi on saline-tolerance of plants. Chin J Appl Ecol 10(1):79–82, 1999.
29. AK Murray, DS Munk, J Wroble, GF Sassenrath-Cole. Myo-inositol, sucrosyl oligosaccharide metabolism and drought stress in developing cotton fibers, in-vivo, in-vitro, and in plants. In: P Dugger, D Richter, eds. Proceedings Beltwide Cotton Conference, Orlando, FL, January 3–7, 1999, Vol 1. Memphis: National Cotton Council, 1999, pp 518–520.
30. CDR Gossett, B Bellaire, SW Banks, MC Lucas, A Manchandia, EP Millholon. Specific ion effects on the induction of antioxidant enzymes in cotton callus tissue. In: P Dugger, D Richter, eds. Proceedings Beltwide Cotton Conference, Orlando, FL, January 3–7, 1999, Vol 1. Memphis: National Cotton Council, 1999, pp 540–542.
31. SN Rajguru, SW Banks, DR Gossett, MC Lucas, TE Fowler Jr, EP Millholon. Antioxidant response to salt stress during fiber development in cotton ovules. J Cotton Sci 3(1):11–18, 1999.
32. SM Alam. Nutrient uptake by plants under stress conditions. In: M Pessarakli, ed. Handbook of Plant and Crop Stress, 2nd ed. New York: Marcel Dekker, 1999, pp 285–313.
33. SM Alam. Nutrient uptake by plants under stress conditions. In: M Pessarakli, ed. Handbook of Plant and Crop Stress. New York: Marcel Dekker, 1993, pp 227–246.
34. SM Alam. Pak J Sci Ind Res 33:292–294, 1990.
35. SM Alam, SSM Naqvi, AR Azmi. Pak J Sci Ind Res 32:110–113, 1989.
36. M Aslam, RH Qureshi. Int Rice Res Newslett 14(3):25, 1989.
37. VN Bhivare, JD Nimbalkar. Plant Soil 80:91–98, 1984.
38. RS Dubey. Protein synthesis by plants under stressful conditions. In: M Pessarakli, ed. Handbook of Plant and Crop Stress, 2nd ed. New York: Marcel Dekker, 1999, pp 365–397.
39. RS Dubey. Protein synthesis by plants under stressful conditions. In: M Pessarakli, ed. Handbook of Plant and Crop Stress. New York: Marcel Dekker, 1993, pp 277–299.
40. RS Dubey, M Rani. J Agron Crop Sci 162(2):97–106, 1989.
41. SR Grattan, CM Grieve. Mineral nutrient aquisition and response by plants grown in saline environments. In: M Pessarakli, ed. Handbook of Plant and Crop Stress, 2nd ed. New York: Marcel Dekker, 1999, pp 203–229.
42. SR Grattan, CM Grieve. Mineral nutrient aquisition and response by plants grown in saline environments. In: M Pessarakli, ed. Handbook of Plant and Crop Stress. New York: Marcel Dekker, 1993, pp 203–226.
43. AS Gupta (A Sen-Gupta). Diss Abstr Int B Sci Eng 47(12):I 4728B, 1987.
44. FY Hafeez, Z Aslam, KA Malik. Plant Soil 106(1):3–8, 1988.
45. A Hamdy. Proceedings 15th ICIDEuropean Regional Conference, Dubrovnik, Yugoslavia, Int Comm Irrig Drain 1988(2):144–156, 1988.

46. PB Kavi-Kishor. Plant Cell Environ 12:629–634, 1989.
47. AH Khan, MY Ashraf. Acta Physiol Planta 10:257–264, 1988.
48. R Krishnamurthy, M Anbazhagan, KW Bhagwat. Oryza 24(1):66–69, 1987.
49. R Krishnamurthy, KA Bhagwat. Indian J Exp Biol 27:1064–1066, 1989.
50. N Mehta, S Bharti. Indian J Plant Physiol 26:322–325, 1983.
51. G Naidoo. New Phytol 107:317–325, 1987.
52. AS Nigwekar, PD Chavan. Acta Soc Bot Pol 56(1):93–99, 1987.
53. TA Omran. Alexandria J Agric Res 31:449–459, 1986.
54. R Pandey, PS Ganapathy. J Exp Bot 35:1194–1199, 1984.
55. M Pessarakli. Crop Sci 31:1633–1640, 1991.
56. M Pessarakli, JT Huber. J Plant Nutr 14:283–294, 1991.
57. M Pessarakli, JT Huber, TC Tucker. J Plant Nutr 12:1361–1378, 1989.
58. M Pessarakli, JT Huber, TC Tucker. J Plant Nutr 12:1105–1121, 1989.
59. M Pessarakli, JT Huber, TC Tucker. J Plant Nutr 12:279–290, 1989.
60. M Pessarakli, TC Tucker. Soil Sci Soc Am J 52:1673–1676, 1988.
61. M Pessarakli, TC Tucker. Soil Sci Soc Am J 52:698–700, 1988.
62. M Pessarakli, TC Tucker. Soil Sci Soc Am J 49:149–152, 1985.
63. M Pessarakli, TC Tucker. J Plant Nutr 8:1025–1045, 1985.
64. E Rabe. Altered nitrogen metabolism under environmental stress conditions. In: M Pessarakli, ed. Handbook of Plant and Crop Stress, 2nd ed. New York: Marcel Dekker, 1999, pp 349–363.
65. E Rabe. Altered nitrogen metabolism under environmental stress conditions. In: M Pessarakli, ed. Handbook of Plant and Crop Stress. New York: Marcel Dekker, 1993, pp 261–276.
66. E Rabe. J Hortic Sci 65:231–243, 1990.
67. RK Rabie, K Kumazawa. Soil Sci Plant Nutr 34:385–392, 1988.
68. S Ramagopal. Plant Cell Rep 5:430–434, 1986.
69. M Salim. J Agron Crop Sci 66:204–209, 1991.
70. M Salim. J Agron Crop Sci 162:35–42, 1989.
71. J Shalhevet, TC Hsiao. Irrig Sci 7:249–264, 1986.
72. MC Shannon, JW Gronwald, M Tal. J Am Soc Hortic Sci 112:416–423, 1987.
73. SK Sharma. Indian J Plant Physiol 32:200–205, 1989.
74. SM Shukr-Almashhadany, SMS Almashhadany. Diss Abstr Int B Sci Eng 47(2):442B, 1986.
75. G Singh, HS Gill, IP Abrol, SS Cheema. Field Crops Res 26(1):45–56, 1991.
76. A Sinha, SR Gupta, RS Rana. Plant Soil 95:411–418, 1986.
77. S Subbanaidu-Ramagopal (S Ramagopal). J Plant Physiol 132:245–249, 1988.
78. HE Dregne. New Mexico Agric Exp Stn Res Rep 94, 1964.
79. G Palfi. Plant Soil 22:127–135, 1965.
80. V Hernando, L Jimeno, C Cadahia. An Edafol Agrobiol 26:1147–1159, 1967.
81. TS Mahajan, KR Sonar. J Maharashtra Agric Univ 5:110–112, 1980.
82. JNE Frota, TC Tucker. Soil Sci Soc Am J 42:753–756, 1978.
83. JNE Frota, TC Tucker. Soil Sci Soc Am J 42:743–746, 1978.
84. R Saad, PhD dissertation, University of Arizona; University Microfilms, Ann Arbor, MI. Diss Abstr B 40:4057, 1979.
85. MA Khalil, A Fathi, MM Elgabaly. Soil Sci Soc Am Proc 31:683–686, 1967.
86. M Pessarakli. PhD dissertation, University of Arizona; University Microfilms, Ann arbor, MI. Diss Abstr B 42:286, 1981.
87. A Golan-Goldhirsh, B Hankamer, SH Lips. Plant Sci (Limerick) 69(1):27–32, 1990.
88. GL Maliwal, KV Paliwal. Agric Sci Dig India 4(3):147–149, 1984.
89. GW Langdale, JR Thomas. Agron J 63:708–711, 1971.
90. AA Luque, FT Bingham. Plant Soil 63:227–237, 1981.
91. SM Abdul-Kadir, GM Paulsen. J Plant Nutr 5:1141–1151, 1982.
92. GV Udovenko, VN Sinel 'nikova, GV Khazova. Dokl Akad Nauk USSR 192:1395–1397, 1970.
93. GV Udovenko, VN Sinel 'nikova, GV Khazova. Agrokhimiya 3:23–31, 1971.
94. M Pessarakli. Response of green beans (*Phaseolus vulgaris* L.) to salt stress. In: M Pessarakli, ed. Handbook of Plant and Crop Stress, 2nd ed. New York: Marcel Dekker, 1999, pp 827–842.
95. M Pessarakli. Response of green beans (*Phaseolus vulgaris* L.) to salt stress. In: M Pessarakli, ed. Handbook of Plant and Crop Stress. New York: Marcel Dekker, 1993, pp 415–430.
96. HM Helal, K Mengel. Landwirtschftliche Forschungsanstalt Buntehof, Hannover, German Federal Republic. Plant Soil 51:457–462, 1979.
97. I Kahane, A Poljakoff-Mayber. Plant Physiol 43:1115–1119, 1968.
98. JT Prisco, JW O'Leary. Rev Brazil Biol 30:317–321, 1970.
99. NM Iraki, RA Bressan, NC Carpita. Plant Physiol 91(1):54–61, 1989.
100. S Katiyar, RS Dubey. J Agron Crop Sci 165(1):19–27, 1990.
101. R Krishnamurthy, M Anbazhagan, KA Bhagwat. Indian J Plant Physiol 30(2):183–188, 1987.

102. J Wieneke, R Fritz. Acta Univ Agric Brno A Fac Agron 33:653–657, 1985.
103. M Pessarakli, ed. Handbook of Plant and Crop Stress, 2nd ed. New York: Marcel Dekker, 1999.
104. M Pessarakli, ed. Handbook of Plant and Crop Stress. New York: Marcel Dekker, 1993.
105. M Ashraf, E Rasul. Plant Soil 110(1):63–67, 1988.
106. AR Azmi, SM Alam. Acta Physiol Planta 12:215–224, 1990.
107. AR Bal, NC Chattopadhyay. Biol Planta 27(1):65–69, 1985.
108. AR Bal, YC Joshi, A Qadar. Curr Agric 10(1–2):65–69, 1986.
109. LK Chugh, MS Kuhad, IS Sheoran. Anal Biol 4(1–2):20–24, 1988.
110. PS Curtis, A Lauchli. Crop Sci 25:944–949, 1985.
111. KS Datta, J Dayal. Indian J Plant Physiol 31:357–363, 1988.
112. EB Dumbroff, AW Cooper. Bot Gaz 135(3):219–224, 1974.
113. LE Francois, EV Maas, TJ Donovan, VL Young. Agron J 78:1053–1058, 1986.
114. LE Francois, TJ Donovan, K Lorenz, EV Maas. Agron J 81:707–712, 1989.
115. DP Heenan, LG Lewin, DW McCaffery. Aust J Exp Agric 28:343–349, 1988.
116. EV Maas, JA Poss. Irrig Sci 10:313–320, 1989.
117. EV Maas, JA Poss. Irrig Sci 10:29–40, 1989.
118. GL Maliwal, KV Paliwal. Legume Res 5(1):23–30, 1982.
119. A Mozafar, JR Goodin. Plant Soil 96:303–316, 1986.
120. A Nukaya, M Masui, A Ishida. J Jpn Soc Hortic Sci 53(2):168–175, 1984.
121. OAA Osman, A Lauchli, A El-Beltagy. Proceedings of the first conference of the agric. develop. res., Cairo, December 19–21, 1987, Vol II. Agron Hortic Soil Sci Rural Sociol 126–145; Cairo, Ain Shams University, 1987.
122. L Prakash, G Prathapasenan. Aust J Plant Physiol 15:761–768, 1988.
123. HY Ryu, HC Choi, CH Cho, ST Lee. Res Rep Rural Dev Admin (SUWEON) 30(3 RICE):1–15, 1988.
124. U Schmidhalter, JJ Oertli. Plant Soil 132(2):243–252, 1991.
125. AR Sepaskhah. Can J Plant Sci 57:925–927, 1977.
126. MC Shannon, GW Bohn, JD McCreight. HortScience 19:828–830, 1984.
127. E Kurth, GR Cramer, A Lauchli, E Epstein. Plant Physiol 82:1102–1106, 1986.
128. S Al-Khafaf, A Adnan, NM Al-Asadi. Agric Water Manage 18(1):63–76, 1990.
129. M Pessarakli, TC Tucker, K Nakabayashi. J Plant Nutr 14:331–340, 1991.
130. NJW Clipson. J Exp Bot 38:1996–2004, 1987.
131. V Balasubramanian, SK Sinha. Physiol Planta 36:197–200, 1976.
132. VA Bastianpillai, C Stark, J Unger. Beitr Trop Landwirtsch Veterinaermed 20:359–363, 1982.
133. S Bouraima, D Lavergne, ML Champigny. Agronomie 6:675–682, 1986.
134. GS Cabuslay, LC Blanco, S Akita. Jpn J Crop Sci 60:271–277, 1991.
135. GF Craig, CA Atkins, DT Bell. Plant Soil 133:253–262, 1991.
136. PS Curtis, HL Zhong, A Lauchli, RW Pearcy. Am J Bot 75:1293–1297, 1988.
137. KS Datta, J Dayal, CL Goswami. Anal Biol 3(1):47–53, 1987.
138. WJS Downton. Aust J Plant Physiol 9:519–528, 1982.
139. J Gorham, E McDonnell, E Budrewicz, RG Wyn-Jones. J Exp Bot 36:1021–1031, 1985.
140. S Joshi, JD Nimbalkar. Plant Soil 74:291–294, 1983.
141. S Kannan, S Ramani. J Plant Nutr 11:435–448, 1988.
142. DK Kishore, RM Pandey, R Ranjit-Singh. Prog Hortic 17:289–297, 1985.
143. OAM Lewis, EO Leidi, SH Lips. New Phytol 111:155–160, 1989.
144. AF Radi, MM Heikal, AM Abdel-Rahman, BAA El-Deep. Rev Roum Biol 33(1):27–37, 1988.
145. R Rai, SV Prasad. Soil Biol Biochem 15:217–219, 1983.
146. P Reddell, RC Foster, GD Bowen. New Phytol 102:397–408, 1986.
147. BA Roundy, JA Young, RA Evans. Agric Ecosyst Environ 25(2–3):245–252, 1989.
148. AM Shaheen, MM El-Sayed. Minufiya Agric Res 8:363–383, 1984.
149. MGT Shone, J Gale. J Exp Bot 34:1117–1125, 1983.
150. C Stark. Beitr Trop Landwirtsch Veterinaermed 23(1):33–38, 1985.
151. EL Taleisnik. Physiol Planta 71:213–218, 1987.
152. C Torres, FT Bingham. Proc Soil Sci Soc Am 37:711–715, 1973.
153. E Zid, M Boukhris. Oecologia Planta 12:351–362, 1977.
154. L Prakash, G Prathapasenan. Biochem Physiol Pflanz 184(1–2):69–78, 1989.
155. D Lazof, N Bernstein, A Lauchli. Bot Gaz 152(1):72–76, 1991.
156. PS Curtis. Diss Abstr Int B Sci Eng 47(2):476B–477B, 1986.
157. LE Francois. J Am Soc Hortic Sci 112:432–436, 1987.
158. CR Hampson, GM Simpson. Can J Bot 68:529–532, 1990.
159. KS Gill. Plant Physiol Biochem India 14(1):82–86, 1987.
160. NH Karim, MZ Haque. Bangladesh J Agric 11(4):73–76, 1986.
161. E Aceves-N, LH Stolzy, GR Mehuys. Plant Soil 42:619–627, 1975.
162. EH Hansen, DN Munns. Plant Soil 107:95–99, 1988.

163. TJ Keck, RJ Wagenet, WF Campbell, RE Knighton. Soil Sci Soc Am J 48:1310–1316, 1984.
164. WJ Zimmerman. Anal Bot 56:689–699, 1985.
165. CA Morilla, JS Boyer, RH Hageman. Plant Physiol 51:8817–824, 1973.
166. PR Rhodes, K Matsuda. Plant Physiol 58:631–635, 1976.
167. TS Gibson. Plant Soil 111:25–35, 1988.
168. IR Kennedy. Anal Biochem 11:105–110, 1965.
169. JW O'Leary. In: J Kolek, ed. Structure and Function of Primary Root Tissues. Bratislava: Veda, Publishing House of the Slovak Academy of Science, 1974, pp 309–314.
170. VR Babu, V Ramesh-Babu. Seed Res 13(1):129–135, 1985.
171. M Salim. J Agron Crop Sci 166:285–287, 1991.
172. R Tipirdamaz, H Cakirlar. Doga Biyol Ser 14(2):124–148, 1990.

# 35

# Calcium as a Messenger in Stress Signal Transduction

**A. S. N. Reddy and Vaka Subba Reddy**

*Colorado State University, Fort Collins, Colorado*

## I. INTRODUCTION

Plant growth in the natural environment is often adversely affected by a number of factors. These include environmental factors such as low temperature, heat, drought, wind, ultraviolet light, anoxia, and high salinity and biological factors such as pathogens (bacteria, viruses, and fungi). Abiotic and biotic factors that limit growth and development of plants and eventually productivity are considered stress factors. Crop losses due to these various abiotic and biotic stresses are in the billions of dollars annually. It has been estimated that stress factors (abiotic and biotic) depress the yield of agronomically important crops in the United States by 78%, of which about 70% is due to unfavorable environmental conditions [1,2]. Plants possess built-in mechanisms to cope with the abiotic and biotic stress factors. Plant scientists have been studying the effects of various stresses on plants to better elucidate the mechanisms by which plants respond to stress signals. It is hoped that knowledge derived from the increased understanding of plant responses to biotic and abiotic stresses would eventually help in developing new plant varieties that are resistant to these stress factors. Advances in molecular and cellular biology are offering a variety of new approaches to investigate plant responses to stresses.

There has been increasing interest in understanding the biochemical and molecular basis of plant stress tolerance and in the identification of genes involved in stress tolerance. Several genes that are involved in biotic stress tolerance have been identified and used to obtain transgenic plants with enhanced resistance to these stresses [3–8]. Because of the complex nature of plant responses to abiotic stresses, the information about the biochemical and molecular mechanisms that contribute to resistance is limited. Studies suggest that the regulation of expression of specific genes as well as changes in the levels of certain osmolytes are an integral part of plant adaptation to stressful environmental conditions [2–7,9–11]. Effects of different stresses on various physiological processes and gene expression are reviewed in other chapters of this book, hence will not be covered here. The mechanisms by which stress signals induce changes in gene expression and affect biochemical pathways are beginning to be understood.

Plants, unlike other organisms, are sessile and, therefore, have developed mechanisms to sense and respond to the stress signals so that they can adapt to or tolerate adverse environmental conditions. However, different plant species differ in their ability to adapt to environmental variables. The mechanisms by which plant cells perceive and transduce stress signals are beginning to be elucidated. In animal cells, messengers such as cyclic nucleotides and calcium play vital roles in signal transduction pathways. The role of divalent calcium ($Ca^{2+}$) cations as one of the key signaling molecules has been known for a long time in animal systems. In plants also, $Ca^{2+}$ has been implicated for decades in regulating various physiological processes during growth and development. However, research during the last two decades strongly indicates that $Ca^{2+}$ plays an important messenger role in transducing a variety of hormonal and environmental signals. Several comprehensive reviews on various aspects of $Ca^{2+}$ messenger system in plants have appeared [3–7,12–19]. In this chapter, we will mainly focus on the role of $Ca^{2+}$ as a messenger in stress signal transduction. A number of criteria have been used to consider a chemical or an ion as a messenger in signal transduction in living cells. These include (1) quantitative changes in the concentration of putative messenger in response to a signal prior to a response, (2) presence of receptors to sense the changes in the level of messenger, (3) induction of a signal-induced response by changing the levels of putative messenger in the absence of a primary signal, and (4) blocking signal-induced responses in the presence of a primary signal by blocking the changes in the level of putative messenger. Evidence obtained in recent years indicates that $Ca^{2+}$ satisfies all these criteria as a messenger molecule in transducing various stress signals.

## II.  STRESS-INDUCED CHANGES IN CYTOSOLIC CALCIUM LEVELS

Despite initial technical problems in measuring cytoplasmic calcium ($[Ca^{2+}]_{cyt}$) in plant cells, tremendous progress has been made in this area [3,13,19–23]. Different methods using $Ca^{2+}$-binding fluorescent dyes (fura-2, indo-1); $Ca^{2+}$-selective electrodes; aequorin, a $Ca^{2+}$-binding photoprotein; and green fluorescent protein–based cameleon are employed to measure signal-induced changes in $[Ca^{2+}]_{cyt}$ [20,24–29]. Transgenic plants expressing targeted aequorin to different organelles combined with the use of different pharmacological agents that block or release $Ca^{2+}$ have greatly facilitated measurement of the $Ca^{2+}$ concentration in different plant cell compartments in response to different stress stimuli [30,31]. Calcium measurement studies indicate that the concentration of $Ca^{2+}$ in the cytoplasm of plant cells, as in animal cells, is maintained low in the nanomolar range (100 to 200 nM) [3]. However, the $Ca^{2+}$ concentration in the cell wall and in organelles is in the millimolar range [4,7,13,19].

Transgenic *Arabidopsis* plants expressing cameleon have been used to measure $Ca^{2+}$ transients in guard cells in response to abscisic acid (ABA) and plasma membrane polarization and the $Ca^{2+}$ concentration [20]. Cameleon, a chimeric protein, consists of an enhanced cyan fluorescent protein (CFP), a calmodulin (CaM), a CaM-binding domain (CBD) from myosin light chain kinase, and an enhanced yellow fluorescent protein (YFP). Increase in $Ca^{2+}$ level activates CaM, which in turn interacts with CBD and brings two GFPs closer by intramolecular arrangement and allows fluorescence resonance energy transfer (FRET) from CFP to YFP. Hence, in the presence of elevated $Ca^{2+}$ levels, excitation of CFP permits emission by YFP at 535 nm. Using cameleon, Allen et al. [20] have measured $[Ca^{2+}]_{cyt}$ spikes in guard cells of *Arabidopsis* in response to ABA and high levels of $Ca^{2+}$. Despite the existence of a large electrochemical gradient for $Ca^{2+}$ entry into the cytoplasm, plant cells maintain their $[Ca^{2+}]_{cyt}$ concentration between 0.1 to 1.5 $\mu$M [20]. Maintenance of low $[Ca^{2+}]_{cyt}$ levels requires active pumping of $Ca^{2+}$ to the apoplast or organelles. Using different approaches, a number of signals including stress signals have been shown to elevate $[Ca^{2+}]_{cyt}$. Table 1 shows various signals that have been shown to change the level of $[Ca^{2+}]_{cyt}$. Here, we describe only stress-induced changes in $[Ca^{2+}]_{cyt}$ levels. Elevation of $[Ca^{2+}]_{cyt}$ in response to other signals has been discussed elsewhere [3,4,7,19,24].

## A.  Abiotic Stress Signals

There are several reports indicating that cold and salt stress affects $Ca^{2+}$ homeostasis in plants [60,61,71]. Increasing evidence obtained during the last several years suggests that abiotic stress signals rapidly elevate the level of $[Ca^{2+}]_{cyt}$. Knight et al. [26] for the first time used transgenic plants expressing apoaequorin to reconstitute aequorin and measure changes in $[Ca^{2+}]_{cyt}$ in response to various signals in tobacco seedlings. These studies have shown that signals such as cold, touch, and wind that are known to influence markedly plant growth and development [10,72], elevate the levels of $[Ca^{2+}]_{cyt}$ [26,37,55]. In vivo imaging of cold-induced changes in $[Ca^{2+}]_{cyt}$ indicated that cotyledons and roots of a seedling are highly responsive whereas hypocotyls are relatively insensitive to cold shock [28]. Furthermore, it has been shown that cold shock causes wavelike $Ca^{2+}$ increases in the cells of cotyledons.

**TABLE 1**   Effect of Environmental, Chemical, Hormonal, and Biotic Stress Stimuli on Changes in Cytosolic Calcium Levels in Higher Plants

| Signal | Methods used to measure free cytosolic $Ca^{2+}$[a] | Effect on $[Ca^{2+}]_{cyt}$[b] | Response[c] | Reference |
|---|---|---|---|---|
| Cold | 1, 2 | ↑ | *COR* gene expression, proline synthesis, changes in membrane lipid profile and cold acclimatization | 32–34 |
| Drought | 1, 2 | ↑ | Gene expression, synthesis of osmoprotectants, and osmotolerance | 31,35,36 |
| Wind | 1, 2 | ↑ | Morphogenesis | 37,38,39 |
| Touch | 1, 2 | ↑ | Thigmomorphogenesis, cytoskeletal organization | 26,40–44 |
| Heat shock | 1, 2 | ↑ | Thermotolerance | 45,46 |
| Hypoosmotic stress | 2 | ↑ | Osmoadaptation | 47–50 |
| Red light | 1 | ↑ | Photomorphogenesis | 51 |
| Abscisic acid | 1, 2, 3 | ↑, ↓ | Stomatal closure | 20,52–55 |
| Gibberellic acid | 2 | ↑ | α-Amylase secretion | 25,56 |
| Auxin | 1 | ↑ | Cell elongation and cell division | 57 |
| Jasmonic acid | 1 | ↑ | Induction of wound-regulated genes expression | 58 |
| NaCl | 1, 2, 3 | ↑ | Gene expression and osmolyte synthesis, $K^+$ uptake | 26,31,59–63 |
| Ozone stress | 2 | ↑ | Production of AOSs | 64 |
| Oxidative stress ($H_2O_2$) | 1, 2 | ↑ | Production of AOSs and HR and cell death | 65–67 |
| Pathogens and elicitors | 1, 2 | ↑ | Phytoalexin biosynthesis and induction of HR | 26,66–69 |
| NOD factors | 1 | ↑ | Nodular formation and root hair curling | 70 |

[a] Cytosolic free calcium levels are measured using injection of fluorescent indicator dyes (1), transgenic plants expressing aequorin (2), or cameleon (3).
[b] Increase (↑) or decrease (↓) in cytosolic free calcium levels.
[c] *COR*, cold-regulated; AOSs, active oxygen species; HR, hypersensitive response; and $[Ca^2]_{cyt}$, cytosolic free calcium.

A similar approach has been used to measure elevated $[Ca^{2+}]_{cyt}$ levels in *Arabidopsis* [36,42,73] and in the moss plant *Physcomitrela patens* [40]. Furthermore, cold-induced $Ca^{2+}$ elevation is dependent on microtubule organization [74]. Although the precise $Ca^{2+}$ stores that release $Ca^{2+}$ in response to cold stress are not well defined, a plasma membrane–associated $Ca^{2+}$ channel was found to be stimulated by cold treatment [75,76] and vacuolar $Ca^{2+}$ store release was found to be via the inositol 1,4,5-triphosphate ($IP_3$) channel [36]. The contribution and sensitivity of different $Ca^{2+}$ stores in elevating levels of $[Ca^{2+}]_{cyt}$ in response to cold stress might explain the dynamic regulation of $Ca^{2+}$ levels by internal stores.

Monitoring of signal-induced changes in the presence of $Ca^{2+}$ channel blockers or channel openers of plasma membrane or organelles indicated that different signals use distinct $Ca^{2+}$ stores in elevating $[Ca^{2+}]_{cyt}$. For instance, cold-induced $Ca^{2+}$ increase is inhibited by plasma membrane channel blockers but is not affected by organellar channel blockers. However, wind-induced $Ca^{2+}$ increase is blocked by organellar $Ca^{2+}$ channel blockers whereas plasma membrane channel blockers did not have any effect [37,39]. These studies indicate that the extracellular $Ca^{2+}$ contributes to cold-induced elevation of $Ca^{2+}$ and internal $Ca^{2+}$ stores contribute to wind-induced increase in $[Ca^{2+}]_{cyt}$.

Using luminescence-imaging technology, the spatial and temporal pattern of elevated $[Ca^{2+}]_{cyt}$ in response to low temperature has been demonstrated in transgenic plants expressing aequorin [32]. These authors showed that a cold-induced signal was transmitted from root to aerial tissues with a lag period of 3 min. Proline, a small imino acid, accumulates in cold-stressed cells and acts as an osmoprotectant [77]. Using $Ca^{2+}$ chelators such as EGTA and channel blockers such as lanthanum, it has been shown that $Ca^{2+}$ is necessary for cold-induced accumulation of proline in *Amaranthus* [78] and tomato seedlings and cell cultures [79]. Furthermore, γ-aminobutyric acid (GABA) is also accumulated in response to several stresses. In asparagus cells, cold stress–induced $Ca^{2+}$ levels regulate the activity of L-glutamate decarboxylase activity synthesizing GABA [80].

Drought stress induces the activity of several $Ca^{2+}$-signaling components, suggesting a role of $Ca^{2+}$ in protecting plants from water loss [35,81]. In *Arabidopsis*, dehydration induces the accumulation of $IP_3$, which in turn causes the release of $Ca^{2+}$ from internal stores [82]. Further, transgenic *Arabidopsis* seedlings in which aequorin is targeted to the cytoplasmic face of the tonoplast membrane [36] showed release of vacuolar $Ca^{2+}$ in response to mannitol treatment [31]. These results indicate the involvement of $IP_3$ in increasing levels of $[Ca^{2+}]_{cyt}$ and that $Ca^{2+}$ is released from vacuolar source [31]. Indirect evidence has been obtained for the role of $Ca^{2+}$ in mannitol-treated rice cell cultures [83] and *Arabidopsis* seedlings [31]. These authors tested the mannitol-induced expression of *RAB* and *AtP5CS1* genes in the presence $Ca^{2+}$ channel blockers such as verapamil or lanthanum or the $Ca^{2+}$ chelator EGTA. The expression of these genes in treated cultures and plants is less than that of untreated counterparts, indicating a role of $Ca^{2+}$ in drought tolerance. With the help of pharmacological antagonists, elevated $[Ca^{2+}]_{cyt}$ has been observed in response to hypoosmotic stress in the alga *Nitella flexilis* [84,85], *Dunaliella salina* [86], *Lamprothamium* [87], *Fucus* zygote [47,48], and suspension cultures of *Nicotiana tabacum* [49,50].

Harmful high-intensity light and ultraviolet (UV) light cause severe damage in plants. The photoreceptors (primarily phytochrome, cryptochrome, and UV-B light photoreceptor) sense the light quantity and activate downstream signal transduction pathways. Using a pharmacological approach combined with the measurement of ultraviolet light–induced *CHS* gene expression, Christie and Jenkins [88] provided evidence for the involvement of elevated levels of $[Ca^{2+}]_{cyt}$ in response to UV-A/blue and UV-B light in *Arabidopsis* cell cultures. Although these two light-induced signal transduction pathways differ in their transducing downstream pathways, both signaling pathways induced elevated levels of $[Ca^{2+}]_{cyt}$. Further work from the same laboratory confirmed the presence of $Ca^{2+}$ involvement in UV-A/blue or UV-B light signal transduction pathways [89]. Involvement of $Ca^{2+}$ in phytochrome-controlled signal transduction mechanisms has also been reported [90,91].

Presoaking of maize seeds before germination or maize seedlings in $CaCl_2$ solution greatly enhanced thermotolerance of these seedlings after they were exposed to a higher temperature (50°C). Conversely, the seedlings treated with $Ca^{2+}$ chelator (EGTA) and channel blockers (lanthanum, verapamil) and CaM inhibitors (CPZ and W7) had significantly reduced thermotolerance. Further, seedlings treated with the same inhibitors showed enhanced thermotolerance when supplemented with $Ca^{2+}$. These results indicate the requirement of $Ca^{2+}$ in thermotolerance [46,92,93]. Using the fluorescent dye indo-1, Bisyaseheva et al. [45] showed a fourfold increase in the levels of $[Ca^{2+}]_{cyt}$ in response to heat shock in pea mesophyll protoplasts. Further evidence for the involvement of $Ca^{2+}$ in heat stress has been obtained from $Ca^{2+}$ measurement studies using transgenic seedlings expressing aequorin [46]. These measurements showed increased levels of $[Ca^{2+}]_{cyt}$ in tobacco seedlings in 5 to 35 min after treatment at elevated temperatures (39, 43, or 47°C) [46]. The expression of *TCH* genes in heat shock–treated cultured *Arabidopsis* cells revealed that their expression is $Ca^{2+}$ dependent.

Mechanical stimuli such as wind, touch, wound, and rain perturb plant growth and development. Plants respond to these stimuli and accordingly alter their morphogenesis (e.g., stunted growth, stem thickening) [94,95]. Studies on thigmotropism in plants revealed that the changes in $Ca^{2+}$ levels play a role in plant responses to touch. The calcium chelating agent EGTA and the $Ca^{2+}$ channel blocker lanthanum inhibited the rubbing- and touch-induced growth in soybean stems [96] and *Mimosa pudica* [97]. The direction of maize root growth changes upon mechanical stimulation. This thigmomorphogenic effect can be inhibited by treatment with gadolinium, an inhibitor of the stretch-activated $Ca^{2+}$ channel in maize roots [98]. Touch-stimulated coiling in *Bryonica diocia* was inhibited by treatment with the $Ca^{2+}$ adenosinetriphosphatase (ATPase) inhibitor erythrosine B, suggesting the involvement of $Ca^{2+}$ [99]. Mechanical stimulation induced the expression of several $Ca^{2+}$-binding proteins such as CaM or CaM-related proteins in *Arabidopsis* [100], *Brassica napus* [101], *Bryonia* [102], potato [103], and tomato [104].

In *Vigna radiata*, touch stimulus differentially affects the expression of CaM isoforms, suggesting the specificity of the signal transduction pathway [105]. Several CaM and CaM-like proteins (CLPs) were identified in *Arabidopsis* [100,106,107]. Of these, transcripts of *AtCaM1*, *-2*, *-3*, and touch (*TCH*) genes are inducible severalfold by mechanical stimuli [100,108]. The expression of *TCH* genes is enhanced by touch, wind, and darkness [41,42,109,110], suggesting a role for $Ca^{2+}$ in transducing these signals in *Arabidopsis*. The increased activity of CaM and CLPs may serve two important functions in thigmomorphogenesis: sequestration of increased $[Ca^{2+}]_{cyt}$ and regulation of the cytoskeletal network by associating with microtubule-associated proteins [43,111].

The strength of artificial wind application on transgenic aequorin seedlings correlated with the $[Ca^{2+}]_{cyt}$ levels [112]. Mechanical stimuli enhance the $[Ca^{2+}]_{cyt}$ in various tissues including meristematic, differentiated, and root cap zones [113]. Although the subcellular contribution of the $Ca^{2+}$ signal is not understandable at this time, research indicates that mitochondrial, endoplasmic reticulum (ER), cell wall, and vacuolar stores contribute to $[Ca^{2+}]_{cyt}$ [37,38,114]. In tobacco, expression of one specific CaM isoform is inducible by cold and wind. By targeting aequorin to cytoplasm (aequorin) and nuclear (nucleoplasmin-aequorin) organelles, Van der Luit et al. [39] showed that distinct cellular $Ca^{2+}$ pools respond to wind and cold stimuli in the expression of the *NpCaM1* gene. Wind and cold stimuli induce $[Ca^{2+}]_n$ and $[Ca^{2+}]_{cyt}$, respectively. These results suggested that different $Ca^{2+}$ transients employ distinct signal pathways in *NpCaM1* gene expression [39].

Soil containing high sodium chloride causes osmotic stress resulting in dehydration of the plant cell and interferes with the nutritional uptake of other components and water [115] essential for growth and development. Calcium has been implicated in drought and salinity stress. Calcineurin, a $Ca^{2+}$/CaM-dependent protein phosphatase, is involved in salinity tolerance in plants, suggesting a role for $Ca^{2+}$ in salt tolerance [116]. Evidence for the involvement of $Ca^{2+}$ in salt stress also comes from the Salt-Overlay-Sensitive3 (SOS3) mutant of *Arabidopsis*. It is sensitive to sodium and lithium ions but not to osmotic stress. But high levels of $Ca^{2+}$ in the medium support the growth and development of these mutants, suggesting the important role of $Ca^{2+}$ in salt stress [117]. The *SOS3* gene was cloned and found to encode a protein similar to calcineurin B, a regulatory subunit of $Ca^{2+}$-dependent protein phosphatase [63]. These results suggest that SOS3 acts as a $Ca^{2+}$ sensor in salt tolerance in *Arabidopsis*. Salt stress induces several $Ca^{2+}$-regulated signal components such as $Ca^{2+}$-ATPase in tobacco cells [71] and tomato [118] and expression of the *AtCP1* [119], *phosphatidylinositol-specific phospholipase C* [82], and certain *CDPKs* [120] in *Arabidopsis* and mung bean [121].

Treatment of corn root protoplasts with sodium chloride resulted in elevated $[Ca^{2+}]_{cyt}$ (1.1 to 1.8 μM) [61]. Using $^{45}Ca^{2+}$, it has been shown that NaCl-treated algae, such as *Chara* and *Dunaliella*, elevated their $[Ca^{2+}]_{cyt}$ and the magnitude of elevation was in direct proportion to the concentration of NaCl [86,122]. The increased $[Ca^{2+}]_{cyt}$ in sodium chloride–treated barley protoplasts and wheat aleurone cells was measured with the help of the $Ca^{2+}$-binding fluorescent dyes indo-1 and fluo-3 [62,123]. Further evidence for the elevated $[Ca^{2+}]_{cyt}$ in response to NaCl was obtained with tobacco [26] and *Arabidopsis* [31] seedlings expressing aequorin. The same group provided evidence for the vacuolar origin of elevated $[Ca^{2+}]_{cyt}$ in response to NaCl stress.

Plants produce active oxygen species (AOSs) ($O_2^{\bullet}$, OH•, and $H_2O_2$) when they are exposed to different stresses such as anoxia, UV-B radiation, or ozone [124]. The AOSs damage cellular macromolecules and cause severe irreversible damage in plants. To prevent such oxidative damage, plants elevate certain enzymatic activities including superoxide dismutase, catalase, peroxidase, glutathione synthase, and alcohol dehydrogenase as defense mechanisms to remove AOSs and to increase oxidative tolerance [65,124–126]. The pollutant gas ozone induces $[Ca^{2+}]_{cyt}$ in pinto bean leaves [127]. Using transgenic aequorin *Arabidopsis* plants, the effect of $O_3$ gas on $[Ca^{2+}]_{cyt}$ has been studied [64]. In these plants, biphasic $[Ca^{2+}]_{cyt}$ transients were reported. Further, with the help of lanthanum chloride and EGTA, the activity of AOS-induced glutathione synthase was found to require a second $[Ca^{2+}]_{cyt}$ transient peak [64]. Using fura-2, McAinsh et al. [128] showed that the levels of $[Ca^{2+}]_{cyt}$ were elevated in the guard cells of *Commelina communis* treated with $H_2O_2$ and methyl viologen, which produce AOSs in these plants. Transgenic tobacco seedlings expressing aequorin also revealed that $H_2O_2$ induced $[Ca^{2+}]_{cyt}$ levels [65].

Deprivation of $O_2$ in flooded soils causes the plants to activate anaerobic respiration and alcohol dehydrogenase [129,130]. It was demonstrated that anoxia induced $^{45}Ca^{2+}$ uptake in maize roots and $Ca^{2+}$ chelators inhibited this uptake. Therefore, survival of maize seedlings in anoxia stress is dependent on

$Ca^{2+}$. Further continuation of this work in maize suspension cultures led the authors to suggest that a mitochondrial $Ca^{2+}$ store contributes to anoxia-induced $[Ca^{2+}]_{cyt}$ [131]. Measurement of $[Ca^{2+}]_{cyt}$ in intact *Arabidopsis* seedlings expressing aequorin also provided direct evidence for the elevated $[Ca^{2+}]_{cyt}$ levels in response to anoxia treatment [132].

The most abundant metal in the earth's crust is the trivalent cation aluminum, which is highly toxic to plant metabolism and growth [133,134]. Aluminum induces transient elevations of $[Ca^{2+}]_{cyt}$ levels in wheat protoplasts [135] and in *Arabidopsis* seedlings [136]. In contrast to these results, Jones et al. [137] provided evidence for reduced levels of $[Ca^{2+}]_{cyt}$ in response to aluminum in tobacco BY2 cell cultures. However, a study using aequorin-expressing *Arabidopsis* plants has confirmed that roots treated with higher levels of aluminum showed decreased levels of $[Ca^{2+}]_{cyt}$. These results suggested that aluminum inhibits the $Ca^{2+}$-permeable channels, thereby creating aluminum toxicity [138]. However, the severity of aluminum toxicity depends on the concentration of $Ca^{2+}$, indicating that the higher levels of $Ca^{2+}$ in the soil could prevent the aluminum toxicity.

The levels of ethylene and ABA are known to change in response to stress factors [55,139]. Using $Ca^{2+}$-binding fluorescent dyes, it has been demonstrated that ABA, gibberellic acid, and auxin increase $[Ca^{2+}]_{cyt}$ [24,25,52–55,62,140,141]. However, transgenic plants with reconstituted aequorin did not show changes in $[Ca^{2+}]_{cyt}$ in response to hormones and heat shock [26]. This could be due to various factors such as sensitivity of the method, localized nature of the response because of restricted target cells, and stability of aequorin at high temperature. ABA-induced changes in $[Ca^{2+}]_{cyt}$ have been attributed to both $Ca^{2+}$ release from internal stores and $Ca^{2+}$ influx from external stores [53,141]. Wounding of plants by herbivores and other mechanical stress factors induce jasmonic acid (JA) [142]. To understand the involvement of $Ca^{2+}$ in JA-induced gene expression, Leon et al. [58] analyzed the JA- and wound-induced gene expression and $Ca^{2+}$ levels in wound- and jasmonic acid–treated *Arabidopsis*. The JA-responsive genes, *JR1* to *JR3*, are inducible by JA and wounding, whereas wound-responsive genes, *WR3* and *acyl CoA oxidase (ACO)*, are inducible by wound only [58], indicating the presence of JA-independent and -dependent wound-induced signal transduction pathways in *Arabidopsis*. These authors have shown that the increased levels of $[Ca^{2+}]_{cyt}$ exerted two opposing actions on the expression of *WR3* and *ACO* genes (up-regulated) and *JR* genes (down-regulated) in response to the wound signal. Therefore, $Ca^{2+}$ has opposing actions in these two-signal transduction pathways [58]. However, the mechanism(s) by which stress and mechanical signals such as cold, salt, and wind cause changes in $[Ca^{2+}]_{cyt}$ is not clear.

## B.  Biotic Stress Signals

Fungal elicitors that induce defense response in plants also elevate $[Ca^{2+}]_{cyt}$ [26,66,68,69]. However, the magnitude and kinetics of $Ca^{2+}$ transients induced by touch, cold, and fungal elicitors were found to be different [26,28,37]. In fact, artificial elevation of $[Ca^{2+}]_{cyt}$ by $Ca^{2+}$ ionophore A23187 influences the production of phytoalexins in soybean and carrot cell cultures. Fungal elicitor treatment of carrot protoplasts resulted in elevated levels of $[Ca^{2+}]_{cyt}$ as measured by the influx of $^{45}Ca^{2+}$ [143]. Using CaM antagonists, phenothiazines and naphthalenesulfonamides, Vogeli et al. [144] showed the inhibition of sesquiterpene phytoalexin biosynthesis in tobacco suspension cultures. These authors suggested a requirement for elevated $[Ca^{2+}]_{cyt}$ in the biosynthesis of defense compounds in response to pathogens.

Apoptosis is a process that occurs in plants and animals in which certain cells undergo programmed cell death. The characteristic features of apoptosis include cell shrinkage, plasma membrane blebbing and bleeding, and nuclear condensation [66,145]. Avirulent *Pseudomonas syringae* caused a hypersensitive reaction (HR) and programmed cell death in soybean cells and leaf tissue (pv. *glycinea*) and in *Arabidopsis* leaves (pv. *tomato*). Apoptosis is also observed in tobacco cells treated with the fungal peptide cryptogein and soybean cells treated with $H_2O_2$ [66]. At the early stages of HR, levels of AOSs are produced at the site of infection [68]. The AOSs stimulate rapid influx of $Ca^{2+}$ into the site of infection and initiate the HR in order to develop resistance against invading pathogens [66,68]. These findings provided evidence for the role of $Ca^{2+}$ in the apoptosis signal transduction pathway [66]. Parsley protoplasts treated with an oligopeptide elicitor, derived from the cell wall of *Phytophthora sojae*, also induced rapid increase in $[Ca^{2+}]_{cyt}$ levels [69]. These authors also identified a plasma membrane–located $Ca^{2+}$ channel named LEAC (large conductance elicitor-activated ion channel) using patch clamp analysis with parsley protoplasts [69]. A study by Xu and Heath [146] provided direct evidence for the elevation of $Ca^{2+}$ levels in epidermal cells of cowpea plant infected with cowpea rust fungus, *Uromyces vignae*. These authors observed the elevation of $[Ca^{2+}]_{cyt}$ by use of the green-1-dextran $Ca^{2+}$ reporter dye at the site of infection before fungus penetrates into the cell wall and the onset of HR [146]. During the interaction of *Arabidopsis* and *Pseudomonas syringae*, the resistance gene product, RPM1, functions immediately and elevates the $Ca^{2+}$ levels as monitored by the aequorin transgenic method [147,148]. Similar results were obtained for the *C. fulvum*–tomato interaction [149]. Identification and characterization of the *Arabidopsis* and rice *gp91^{phox}* homologues, *RbohA* (for respiratory burst oxidase homologue A), provided evidence for the downstream target for the elevated levels of $Ca^{2+}$ in oxidative burst [150]. The *RbohA* shows high similarity to human *gp91^{phox}* (*phox* for phagocyte oxidase). It has been found that the human *gp91^{phox}* is a plasma membrane–bound neutrophil phagocyte oxidase and is involved in the generation of superoxide radicles via its NADPH oxidase activity. In addition to the *gp91^{phox}* region, the plant *gp91^{phox}* (*RbohA*) also contains two EF hand motifs that are not present in human *gp91^{phox}*, suggesting that $Ca^{2+}$ modulates the formation of superoxide radicals through *gp91^{phox}* EF hands during the oxidative burst in plants [150].

Systemin has been shown to be an important mediator in wound-induced activation of defense genes in tomato [151,152]. Addition of systemin to the cell cultures of tomato causes rapid alkalinization of the medium [153]. However, the $Ca^{2+}$ channel inhibitor lanthanum and protein kinase inhibitors K252a and staurosporine inhibit the systemin-induced signal process, suggesting the involvement of a $Ca^{2+}$-dependent protein kinase in systemin-induced signal transduction in tomato [153]. Tomato plants expressing prosystemin showed higher levels of CaM transcripts, indicating the role of $Ca^{2+}$-dependent CaM in the defense response [151]. Further, Flego et al. [154] showed a correlation between increased $Ca^{2+}$ concentration in plants and increased resistance to the bacterial pathogen *Erwinia carotovora*. Using a dextran-linked $Ca^{2+}$ indicator dye, elevated $Ca^{2+}$ spikes were measured in developing nodules of alfalfa induced by *Rhizobium meliloti* nodulation factors, suggesting the participation of $Ca^{2+}$ in nodule formation [70]. Accumulating evidence indicates the involvement of a $Ca^{2+}$ signal in plant defense responses such as phytoalexin biosynthesis, induction of defense-related genes, and hypersensitive cell death. However, the exact mechanisms by which $Ca^{2+}$ regulates these processes are poorly understood [18,26,27,66, 69,144,155,164]. Various components of the $Ca^{2+}$-mediated signal transduction pathways are discussed later in this chapter. Influx of $Ca^{2+}$ ions from extracellular or intracellular $Ca^{2+}$ stores seems to contribute to signal-induced changes in $[Ca^{2+}]_{cyt}$. Based on the type of signal or cell type, both processes could be involved in raising $[Ca^{2+}]_{cyt}$. The elevated $Ca^{2+}$ interacts with other proteins and signal-transducing components located downstream of the signal cascades. The presence of the several of the downstream $Ca^{2+}$-based signal components has been reported in plants.

## III. CALCIUM-SENSING MECHANISMS

Changes in free $Ca^{2+}$ concentration in the cytoplasm are believed to regulate various cellular processes at the biochemical and molecular level, eventually leading to a physiological response. A transient $Ca^{2+}$ increase in the cytoplasm in response to abiotic and biotic stress factors is sensed by an array of $Ca^{2+}$-binding proteins. Once $Ca^{2+}$ sensors decode the stress signal, $[Ca^{2+}]_{cyt}$ levels are restored to the resting level by $Ca^{2+}$ efflux into cellular organelles such as vacuoles, ER, and mitochondria or the cell exterior. Decoding of the $Ca^{2+}$ signal to the metabolic machinery is accomplished through intracellular $Ca^{2+}$ receptors or $Ca^{2+}$-binding proteins. All $Ca^{2+}$-binding proteins, except annexins, contain a 29-residue helix-loop-helix structure called an EF hand that binds to $Ca^{2+}$ with high affinity [18,165,166]. However, different $Ca^{2+}$-binding proteins differ in the number of EF hand motifs and their affinity for $Ca^{2+}$ with dissociation constants ($K_d$s) ranging from $10^{-5}$ to $10^{-9}$ M. Binding of $Ca^{2+}$ to the receptor results in a conformational change in the receptor that enables it to interact with other proteins and modulate their function and/or activity. A number (over 150) of $Ca^{2+}$-binding proteins have been identified and characterized in animals [16,166]. Of these, only a few are present in all eukaryotic cells and are believed to be involved in mediating $Ca^{2+}$ action, whereas the majority of them (e.g., troponin C and parvalbumin) are found in specific tissues and play restricted roles. Studies in plants have identified several $Ca^{2+}$-binding proteins including CaM, CaM-related proteins [19,106,167], protein kinases, phosphatases, phospholipases, proteinases, and other proteins [3,6,13,16,18,159]. In the following sections we discuss $Ca^{2+}$-binding proteins involved in stress signal transduction pathways in plants.

## A. Protein Kinases and Phosphatases

Signals either directly or through messengers such as $Ca^{2+}$ regulate the activity of protein kinases and protein phosphatases that in turn regulate phosphorylation and dephosphorylation, respectively, of many proteins in the cell. The enzyme activity or biological property of many proteins is strongly influenced by (de)phosphorylation status. It has been well established that $Ca^{2+}$-regulated protein phosphorylation plays a pivotal role in signal transduction in animal cells. Several $Ca^{2+}$/CaM-dependent protein kinases [168,169], protein kinase C, and a $Ca^{2+}$- and phospholipid-dependent protein kinase [170] mediate the $Ca^{2+}$ effects on protein phosphorylation in animal cells. $Ca^{2+}$-regulated protein phosphorylation is believed to be involved in signal amplification, in obtaining sustained responses, as well as in producing diverse responses to transient rises in $[Ca^{2+}]_{cyt}$ [168,169]. Demonstration of a central role for $Ca^{2+}$-regulated protein kinases in $Ca^{2+}$ signaling in the animal system led plant scientists to investigate the presence of $Ca^{2+}$-regulated protein kinases. Current evidence indicates that there are different types of $Ca^{2+}$-regulated protein kinases: a $Ca^{2+}$-dependent and CaM-independent protein kinase (CDPK) and $Ca^{2+}$/CaM-dependent protein kinases CCaMK and CaM K II (Figure 1). The CDPK type protein kinase is unique to plants and has been well characterized as compared with CaM kinases [18,171].

CDPK is ubiquitous in plants. It has been purified to homogeneity from soybean [176] and partially purified from a number of other plant systems [18]. Calcium directly binds to the CDPK and stimulates the kinase activity by about 100-fold, whereas CaM did not have any significant effect on the kinase activity [177]. Using primers that correspond to amino acid sequences obtained from proteolytic fragments of a purified CDPK, complementary DNA (cDNA) encoding CDPK has been isolated from soybean [172]. The predicted amino acid sequence of soybean CDPK has revealed a unique structural organization different from that of all the known protein kinases, indicating that it represents a new family of protein kinases that are unique to plants. The deduced primary structure of the CDPK contains a protein kinase catalytic domain followed by a CaM-like region with four $Ca^{2+}$ binding motifs (Figure 1). The kinase domain of CDPK shows significant homology with the mammalian $Ca^{2+}$/CaM-dependent protein

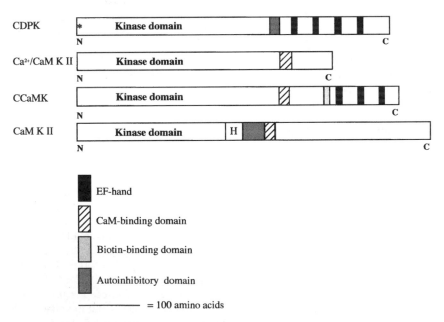

**Figure 1** Diagrammatic representation of the catalytic and regulatory domains of representative examples of calcium and calcium/calmodulin-dependent protein kinases. CDPK (508 amino acids), calcium-dependent and calmodulin-independent protein kinase [172]; $Ca^{2+}$/CaM K II (415 aa), calcium/calmodulin-dependent protein kinase II from plants [173]; CCaMK (520 aa), calcium and calcium/calmodulin-dependent protein kinase [174]; and CaM K II (565 aa), calcium/calmodulin-dependent protein kinase II from animals [175]. The hinge region that separates the kinase domain from the autoinhibitory domain in animal CaM K II is denoted by H. The myristoylation motif in CDPK is indicated by an asterisk (*).

kinase II (CaM K II) catalytic domain. The region that joins the kinase domain to the CaM-like region corresponds to the autoinhibitory/CaM-binding region of CaM K II and prevents kinase activity in the absence of $Ca^{2+}$ [172]. The cDNAs that encode CDPKs have also been isolated from other systems [178,179]. The activity of an *Arabidopsis* CDPK that is expressed in *Escherichia coli* is stimulated by $Ca^{2+}$ [178,179].

Immunological and cloning studies as well as Southern analyses of soybean and *Arabidopsis* genomic DNAs suggest that there are several isoforms of CDPK in plants [18]. Using a polymerase chain reaction (PCR) strategy, Urao et al. [120] cloned two *CDPK* cDNA sequences, *AtCDPK1* and *AtCDPK2*. The transcripts of these two genes are highly inducible by drought and high salt but not by low temperature or heat stress, suggesting the specificity of CDPK's induction in response to different stress factors. The *E. coli* expressed AtCDPK2 protein phosphorylates casein and myelin basic protein in a $Ca^{2+}$-dependent manner. Accumulating evidence suggests that there are more than 40 *CDPK*s in the *Arabidopsis* genome [180] and they are classified into seven groups on the basis of their sequence domain organization (myristoylation, PEST, and the number of EF hand motifs) [181]. Furthermore, these CDPKs differ in their affinity for $Ca^{2+}$. For example, AtCDPK1 differs from AtCDPK2 in its $Ca^{2+}$-stimulated activity, although both of them possess four EF hand motifs [181]. Studies indicate that, besides $Ca^{2+}$, lipids are involved in the regulation of CDPK activity [178,182,183]. A carrot calmodulin-like domain protein kinase, DcCPK1, resembles animal protein kinase C (PKC) in its activation by $Ca^{2+}$ and certain phospholipids, suggesting that lipids regulate the activity of some CDPKs and perform specific biological functions in plants [184]. The molecular weight of purified CDPKs from different plant systems ranges from 35,000 to 90,000 [18]. The wide range in the size and differences in their substrate specificity suggest that there could be multiple isoforms and functions. It is also possible that some of the small enzymes are derived from larger ones by proteolytic cleavages, which has been shown to be the case in oat [182]. A mutation in the *CDPK* gene did not reveal a phenotype, suggesting functional complementation among CDPKs [181].

In vitro and in vivo protein phosphorylation studies have demonstrated $Ca^{2+}$-regulated protein phosphorylation in a number of plant systems. *Arabidopsis* and soybean CDPKs phosphorylate α-TIP, a tonoplast intrinsic protein [185], and nodulin-26, respectively, that is involved in the formation of nodule [186]. Another CDPK has been shown to phosphorylate a guard cell vacuolar chloride channel [187]. It has been shown that alfalfa and *Arabidopsis* seedlings treated with W7 (a potent inhibitor of CaM and CDPKs) were unable to acclimatize and tolerate cold and freezing temperatures. These results suggest the involvement of $Ca^{2+}$- or CaM-dependent protein phosphorylation events in these process [188,189]. Using a PCR strategy, Botella et al. [121] isolated a cDNA clone encoding CDPK from *Vigna radiata*. The corresponding messenger RNAs (mRNAs) are highly inducible by wounding, $CaCl_2$, indoleacetic acid (IAA), and NaCl treatments [121]. These results provide evidence for the phosphorylation events in $Ca^{2+}$-mediated stress signal transduction cascades in plants. In an elegant experimental system, Sheen [190] has shown that *Arabidopsis* AtCDPK1 and AtCDPK1a are involved in regulating the expression of stress-inducible genes. Furthermore, phosphatases counteract these responses, suggesting that involvement of $Ca^{2+}$-regulated phosphorylation is necessary for stress-induced gene expression (also see Sec. V).

A cDNA that shows significant similarity to mammalian *CaM K II* has been isolated from plants by screening an expression library with radiolabeled CaM [173,191] (Figure 1). However, the biochemical properties of this plant CaM K II homologue are not known. A $Ca^{2+}$/CaM-dependent protein kinase (CCaMK) was cloned and characterized from lily [174] and tobacco [192]. A comparison of the sequence analysis revealed the presence of an N-terminal catalytic domain, a centrally located CaM-binding domain, and a C-terminus visinin-like domain containing three conserved EF hands (Figure 1). Biochemical studies of CCaMK established that $Ca^{2+}$/CaM stimulates CCaMK activity and in the absence of CaM, $Ca^{2+}$ promotes autophosphorylation of CCaMK. The phosphorylated form of CCaMK possesses more kinase activity than the nonphosphorylated form [193]. These authors suggested involvement of CCaMK in male gametophyte development. The same research group showed differential regulation of tobacco CCaMKs by CaM isoforms [192]. These studies indicate the presence of CaM-regulated protein kinases in plants, although how widely these kinases are distributed and their exact role are not clear.

Fungal elicitor–induced cytosolic $Ca^{2+}$ has been implicated in changes in the phosphorylation status of proteins in tomato suspension cultures [194]. The cell nuclei in cowpea plants infected with cowpea rust fungus are shown to migrate to the fungal penetration site. Calcium chelators as well as protein kinase inhibitors inhibit such nuclear movement, suggesting the involvement of a $Ca^{2+}$-dependent phosphorylation cascade in nuclear migration [146].

There is some evidence that $Ca^{2+}$-regulated protein phosphorylation is involved in cold stress and in host-pathogen interaction. In a freezing-tolerant cultivar of *Medicago sativa*, elevated cytosolic $[Ca^{2+}]_{cyt}$ levels in response to low temperature stimulated phosphorylation of several proteins including protein kinases as revealed by a phosphoprotein profile [195]. Treatment of parsley suspension cultures with fungal elicitors resulted in rapid and transient phosphorylation of specific proteins. These fungal elicitor–induced changes in phosphorylation have been shown to be dependent on the presence of $Ca^{2+}$ in the medium [196]. Fungal elicitor–induced protein phosphorylation, phytoalexin production, and mRNAs for phenylalanine ammonia lyase and 4-coumarate:CoA ligase were greatly reduced in $Ca^{2+}$-deprived cells [196]. Furthermore, addition of $Ca^{2+}$ to the cultures restored the inhibitory effects of $Ca^{2+}$ deprival, suggesting the participation of $Ca^{2+}$-dependent protein phosphorylation in fungal elicitor–induced responses.

The UV-B–regulated *CHS* gene expression in *Arabidopsis* cell cultures is inhibited in the presence of pharmacological agents such as W7 (inhibitor of CaM or CDPKs), K252a and staurosporine (inhibitors of protein kinases), and okadaic acid (inhibitor of protein phosphatases 1 and 2A). These results suggest a role for kinases and phosphatases in UV-B–mediated *CHS* gene expression in *Arabidopsis* cell cultures [88]. UV-B light also causes tremendous induction of anthocyanin pigment biosynthesis in rice [197]. It would be interesting to test the involvement of a $Ca^{2+}$-dependent signal cascade in monocots in response to UV-B radiation.

Calcium-dependent protein kinases and phosphatases and $Ca^{2+}$/CaM-dependent protein kinases have been shown to be involved in the response to thigmotropism in soybean cells. In this process, kinases and phosphatases play a crucial role in alterations of the actin and microtubule network that influence cell shape. These findings have been obtained using pharmacological inhibitors such as W7, calmidazolium, okadaic acid, or inhibitors specific to CaM-dependent phosphatase 2B [198]. Mastoparan- and hypoosmotic stress–induced cytosolic free $Ca^{2+}$ levels are elicited through the $IP_3$ system in tobacco cell cultures expressing apoaequorin [49,50]. The elevated $[Ca^{2+}]_{cyt}$ levels activate three different protein kinases (50, 75, and 80 kDa) whose activities are inhibited by neomycin and staurosporine treatments. In both processes, elevated levels of free $[Ca^{2+}]_{cyt}$ are essential to establish the phosphorylation and dephosphorylation process [49,50]. These studies together suggest the involvement of $Ca^{2+}$-dependent protein kinases in a variety of stress-responsive mechanisms.

Dephosphorylation of specific proteins involved in the signal transduction cascade is found to be important in $Ca^{2+}$ signaling. Wheat aleurone cells treated with gibberellic acid (GA) induce $[Ca^{2+}]_{cyt}$ levels as well as activation of cellular hydrolases. Okadaic acid (OA), a protein phosphatase inhibitor, partially inhibits ABA action and does not inhibit hypoxia-related stress responses. JA-dependent and -independent would-induced gene expression in *Arabidopsis* has also shown involvement of reversible phosphorylation events. Okadaic acid and staurosporine-sensitive protein phosphatase (type 2A) and kinases positively regulate the genes expressed through the JA-dependent pathway. However, JA-independent wound-induced gene expression relies on a phosphoprotein as this pathway is inhibited by staurosporine and activated by okadaic acid [58]. The existence of $Ca^{2+}$/CaM-dependent protein phosphatase (PP2B) and its involvement in stress signal transduction pathway have been reported in plants [116]. These findings indicate that the phosphorylation and dephosphorylation events are important in $Ca^{2+}$ mediated signaling in plants.

## B.  Calmodulin and Calmodulin-Related Proteins

Calmodulin is found in all eukaryotic organisms and is a well-characterized $Ca^{2+}$ receptor in both animal and plant cells. In plants, CaM, CaM-related proteins, $Ca^{2+}$-dependent protein kinases, and other $Ca^{2+}$-binding proteins are believed to sense the changes in $[Ca^{2+}]_{cyt}$. All these proteins are called $Ca^{2+}$ sensors or receptors. Calmodulin was first discovered in animals as an activator of cyclic nucleotide phosphodiesterase [199] and subsequently isolated and characterized from all eukaryotic organisms. It is, in fact, the discovery of CaM in plants that led plant scientists to propose a messenger role for $Ca^{2+}$ in plant cells. Calmodulin has been isolated and characterized from many different plants [13,16,19,106,167]. Gene structure and expression of CaMs from a number of plants have been analyzed [106].

Molecular cloning of CaM genes from plants indicates the presence of a small gene family that encodes different CaM isoforms. CaM is a low-molecular-weight protein of 148 amino acids that is highly conserved between plants and animals. CaMs from all eukaryotes, except from budding yeast [200], have four helix-loop-helix motifs (also known as EF hand domains) that bind to four Ca$^{2+}$ ions with high affinity (Figure 2). Plant CaM is structurally and functionally very similar to animal CaM [200]. Calmodulin has no enzyme activity, but when bound to Ca$^{2+}$ it can modulate the activity and function of numerous unrelated target proteins such as enzymes and certain structural proteins [19,106,167]. The crystal structure of CaM revealed that it has two globular domains, each with a pair of EF hands, connected by a central helix [207] and also provided the structural basis for its interaction with target proteins [208]. The

**A)**

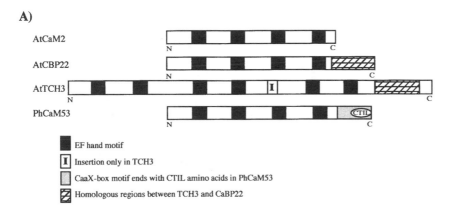

**B)**

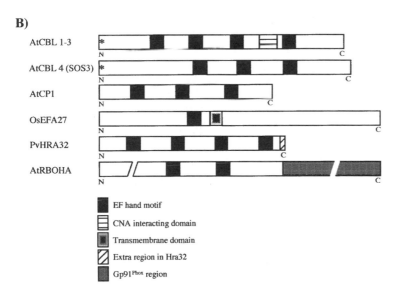

**Figure 2**    Schematic structural diagrams showing the features of calmodulin and other EF hand proteins from plants. (A) Calmodulin and calmodulin-related proteins. AtCaM2 (148 aa), *Arabidopsis* calmodulin 2 [201]; AtCBP22 (191 aa), *Arabidopsis* calcium-binding protein 22 [202]; AtTCH3 (324 aa), *Arabidopsis* touch 3 [41]; PhCaM53 (184 aa), *Petunia hybrida* calmodulin 53 [203]. (B) Other EF hand proteins involved in various stress responsive process in plants (see the text for details). AtCBL1 to 3 (213, 226, and 226 aa, respectively), *Arabidopsis* calcineurin B–like-1, -2, and -3 proteins [204]; AtCBL4 (191 aa), *Arabidopsis* CBL4 or salt-overly sensitive3 (SOS3) protein [63]; AtCP1 (155 aa), salt-induced *Arabidopsis* calcium-binding protein [119]; OsEFA27 (244 aa), rice EF hand protein responsive to abscisic acid [205]; PvHRA32 (161 aa), bean hypersensitive reaction associated [206]; and AtRBOHA (944 aa), *Arabidopsis* respiratory burst oxidase homologue A [150]. The myristoylation motif in CBLs is indicated by an asterisk (*). The AtRBOHA protein (944 aa) did not permit the depiction of its entire length. Therefore, interruptions in the protein are denoted by (//) in the N- and C-termini.

binding of $Ca^{2+}$ to CaM ($10^{-5}$ to $10^{-6}$ M) results in a conformational change in such a way that the hydrophobic pockets of CaM are exposed in each globular end, which can then interact with proteins (affinity in the nanomolar range) and regulate the activity of several unrelated target enzymes. The hydrophobic pockets that are exposed upon $Ca^{2+}$ binding are believed to interact with hydrophobic regions along the amphipathic helix of CaM target proteins [209,210]. In addition to CaM, studies indicate the presence of CaM-like proteins in plants [100,106,167,202,211–213]. However, the function of these proteins in the $Ca^{2+}$ signaling pathway(s) is not fully characterized as compared with that of CaM. These nonconserved CaM-like or CaM-related proteins differ from the conserved CaM in containing more than 148 amino acids and EF hand motifs (three to six) showing limited homology with CaM [213]. Hence, it is likely that these proteins are functionally distinct and are involved in controlling different $Ca^{2+}$-mediated cellular functions.

## 1. Effect of Stress Signals on the Expression of Calmodulin and Calmodulin-Related Proteins

In plant cells, CaM and CaM-like proteins are highly responsive to physical and hormonal signals such as touch, wounding, light, and auxin [100,214] that are known to raise free $[Ca^{2+}]_{cyt}$ levels. In *Arabidopsis*, the expression of CaM and CaM-like proteins is rapidly induced by several stimuli such as touch, wind, rain, and wounding [100,211]. Signal-induced changes in CaM and CaM-like proteins may play a significant role in cell growth and physiology. Small changes in CaM levels have been shown to affect drastically the progression of the cell cycle in animal cells [215] (see Chapter 11). Touch and wind signals have been shown to cause rapid and transient increases in free $[Ca^{2+}]_{cyt}$ that occur prior to observed changes in target gene expression [3,26,37,100,211,212]. Hence, it is likely that signal-induced changes in the $[Ca^{2+}]_{cyt}$ level could be involved in the expression of CaM and CaM-like proteins to activate downstream targets and to modify the cellular response. In several plant systems, there are multiple CaM genes that code for identical proteins or contain a few conservative changes [106,108,167,216,217]. These small changes in amino acid composition of CaM isoforms may contribute to differential interaction of each CaM isoform with target protein. For instance, a kinesin-like CaM-binding protein (KCBP) interacts differentially with different CaM isoforms from *Arabidopsis* [218]. Although the *Arabidopsis* CaM isoforms differ in few amino acids, our studies revealed that the *Arabidopsis* CaM2 isoform has a twofold higher affinity toward KCBP [218]. This study with CaM and target protein (KCBP) from the same system provides a clue to the differential affinity of CaM isoforms and CaM-related proteins with target proteins. However, these CaM isoforms may or may not differ much in their function or affinity toward $Ca^{2+}$ as compared with CaM-related proteins that may be involved in diverse cellular processes. Differential expression of CaM and CaM-related genes in response to different stimuli is also an important and convenient mechanism for cells for tuning $Ca^{2+}$-mediated stress signal transduction cascades.

Although the results are preliminary, studies of CaM and CaM-related gene expression in response to different stimuli indicate that different CaM isoforms are involved in mediating a specific signal [108,211,216]. Three of the six *Arabidopsis Cam* genes (*Cam1, -2, and -3*) are inducible by touch stimulation [108,216], indicating the presence of different *cis*-regulatory elements in their promoters. Another example is the expression of *TCH* genes (*Cam*-related genes from *Arabidopsis*) in response to various mechanical, chemical, and environmental stimuli [211]. These authors also provided direct evidence that increased $[Ca^{2+}]_{cyt}$ levels in response to various abiotic stimuli are a prerequisite for the up-regulation of the *TCH* genes. Braam and his coworkers have hypothesized two schemes for the *TCH* gene expression in response to various stimuli. Divergent stress signals induce *TCH* gene expression via different cascades involving various receptors in the signal pathway and interaction with different *cis*-regulatory elements on *TCH* genes. Alternatively, these different cascades may converge at one point before the onset of *TCH* gene expression and induce the *TCH* gene expression through a common *cis*-regulatory element of the *TCH* genes. Further studies should help understand the function of the *TCH* genes in $Ca^{2+}$-mediated stress signal transduction cascades in plants.

The presence of multiple CaM isoforms in plants adds further complexity to the $Ca^{2+}$-mediated network and points to their differential sensitivity to elevated $[Ca^{2+}]_{cyt}$ levels in response to different stress stimuli. In potato, only one of eight CaM isoforms (PCaM1) is inducible by touch [103]. A striking example of differential regulation of CaMs comes from studies with soybean CaM isoforms. In soybean, five CaM isoforms (SCaM1 to -5) have been identified. SCaM1, -2, and -3 are highly conserved compared with other plant CaM isoforms including *Arabidopsis* CaM isoforms, whereas SCaM4 and -5 are divergent and showed differences in 32 amino acids with the conserved group [219]. Surprisingly, these divergent CaM isoforms are specifically induced by fungal elicitors or pathogen [220]. These results provided evidence for the differential regulation of CaM isoforms in plants. The *Vigna radiata Cam* genes also differentially respond to a touch stimulus, *Cam1* being more responsive than *Cam2* [105]. In hexaploid bread wheat, 10 distinct genes encode three CaM isoforms, some of which are differentially expressed in response to different stimuli [221,222].

## 2.  *Calmodulin-Binding Proteins*

Calmodulin itself has no enzymatic activity. It controls various cellular activities by modulating the activity or function of a number of proteins. Hence, the role of CaM in a given cell or a tissue is determined by the presence of its target proteins. Calmodulin is multifunctional because of its ability to interact with and control the activity of a variety of target proteins (also called CaM-binding proteins). Therefore, characterization of CaM-binding proteins (CBPs) is the prerequisite in dissecting $Ca^{2+}$/CaM-mediated signaling pathways in plants. In animal systems, the activity of over 30 enzymes has been shown to be regulated by CaM in a $Ca^{2+}$-dependent manner [168,210,223,224]. These include several protein kinases, a protein phosphatase (also known as calcineurin), a plasma membrane $Ca^{2+}$-ATPase, adenyl cyclases, cyclic $3',5'$-nucleotide phosphodiesterase, motor proteins, inositol triphosphate kinase, transcriptional factors, nitric oxide synthase, and some structural proteins. Identification and characterization of CaM target proteins in animal cells have helped in elucidating the mechanisms by which $Ca^{2+}$/CaM regulate various biochemical and molecular events leading to a physiological response. The amino acid sequence of the CaM-binding domain in different CaM target proteins is not conserved [208]. However, CaM-binding motifs from different CaM-binding proteins form characteristic basic amphipathic α-helices [209,223]. The amino acids in the CaM-binding domain, when arranged in a helical wheel, form an amphipathic helix with several positive residues on one side and a number of hydrophobic residues on the other side. In most cases, binding of CaM to its target proteins requires $Ca^{2+}$. However, CaM binds to some proteins (e.g., neuromodulin, myosins) in the absence of $Ca^{2+}$ [225]. Furthermore, a CaM with no $Ca^{2+}$-binding activity has been shown to rescue a mutation in the only *Cam* gene in budding yeast [226], suggesting that CaM may perform some functions in the absence of $Ca^{2+}$.

Gel overlay studies with plant proteins indicate the presence of a number of CaM-binding proteins in plants [227,228]. Some enzymes and proteins that are activated by CaM have been identified in plants. These include NAD kinase [29], $Ca^{2+}$-ATPase [229,230], nuclear NTPases [231], glutamate decarboxylase [232,233], transporter-like proteins [234], $Ca^{2+}$/CaM kinases [235], kinesin-like protein [218,236–242], elongation factor-1α [111], transcription factor [243], glyoxalase I [244,245], and heat shock–inducible proteins TCB48 and TCB60 [246,247]. In fact, CaM was first discovered in plants as an activator of NAD kinase [248,249]. New approaches to isolating CaM-binding proteins by screening expression libraries with labeled CaM [$^{35}$S-labeled, biotinylated, or horseradish peroxidase (HRP)-conjugated CaM] as probes [250–253] have greatly aided in isolating and characterizing cDNAs encoding CaM-binding proteins. Several cDNAs have been isolated with this approach [106,254–256] and one of the isolated clones was found to have significant similarity to *E. coli* glutamate decarboxylase (GAD) [256]. Using a gel overlay assay with SCaM4 and -5, Lee et al [250] have shown that the two CaM isoforms compete for several CaM-binding proteins of total protein extracts prepared from various soybean tissues. However, the molecular identity and function of these CBPs from soybean are not known [250].

The expression of some of the CaM-binding proteins is regulated by heat and wind [246,254–257]. The plant GAD is unique in having a CaM-binding domain at its C-terminus [256] (Figure 3). Members of this protein from bacteria do not contain a CaM-binding domain (CBD). Although the catalytic core that catalyzes the conversion of glutamate in to γ-aminobutyric acid (GABA) and $CO_2$ is conserved across bacteria, plants, and animals, the $Ca^{2+}$/CaM regulation of GAD appears to be unique to plants [233]. At least three forms of GADs were identified in mammals [259], and the product GABA serves as an inhibitory neurotransmitter in animal systems. GAD has now been cloned and characterized from a number of plant species including petunia [256], *Arabidopsis* [260], rice [261], soybean [262], and asparagus [80]. Although the CBD is not conserved, all plant GADs isolated so far possess CBD and are regulated by $Ca^{2+}$/CaM [233]. Particularly interesting is the presence of two GAD isoforms (GAD1 and GAD2) in *Arabidopsis*. The CBDs of these two isoforms are different, raising the possibility of functional diversity and regulation by $Ca^{2+}$/CaM [260].

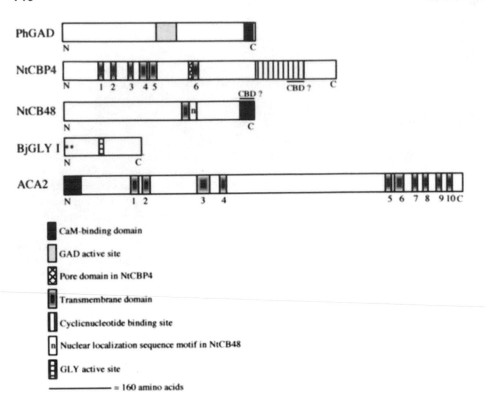

**Figure 3** Schematic representation of calmodulin-binding proteins involved in stress tolerance in plants. Numbers denote transmembrane domains. The location of calmodulin-binding domain (CBD) in NtCBP4, NtCB48, and BjGLY I has not been identified (CBD?). The myristoylation motifs in BjGLY I are indicated by asterisks (*). PhGAD (500 aa), *Petunia hybrida* glutamic acid decarboxylase [232]; NtCBP4 (708 aa), *Nicotiana tabacum* calmodulin-binding protein 4 [258]; NtCB48 (499 aa), *Nicotiana tabacum* calmodulin-binding protein 48 [246]; BjGLY I (185 aa), *Brassica juncea* glyoxalase I [244]; and ACA2 (1014 aa), *Arabidopsis* $Ca^{2+}$-ATPase 2 [229].

In plants, GABA is involved in normal plant growth and development. The detailed molecular analysis of GAD using the transgenic approach by Fromm and coworkers is discussed later in this chapter. Induction of GABA synthesis in response to various environmental stimuli [including cold shock, mechanical stress, anoxia, water stress, heat shock, hormonal treatment, and TMV (tobacco mosaic virus) infection] has been reviewed [263,264]. Soybean leaves produce 9-, 11-, and 18-fold increased levels of GABA in response to touch, rolling, and crushing, respectively, within 30 sec [265] and GABA levels are also inducible by cold treatment [266]. These environmental stresses also elevate the $[Ca^{2+}]_{cyt}$ level, and its participation in stress signal transduction is known. The concomitant increased $[Ca^{2+}]_{cyt}$ levels and GABA synthesis via GAD activity in response to environmental stresses raise the possibility of the involvement of GAD in $Ca^{2+}$-mediated stress signal cascades in plants.

Another $Ca^{2+}$/CaM-regulated enzyme is NAD kinase. Although the gene encoding NAD kinase has not been cloned it has been shown that its activity depends on the activated CaM [248,249]. The NAD kinase catalyses the conversion of NAD to NADP, an energy-generating molecule. The role of NAD kinase activity is vital to living organisms, especially when energy is in demand under stress conditions. It has been reported that NAD kinase plays a role in the oxidative burst and in the formation of active oxygen species (AOSs) [29]. Production of AOSs, which are known for their role against invading pathogens, is a costly event in plants and requires the input of NADPH molecules. A detailed analysis of the action of mutated CaM in the production of AOSs in transgenic tobacco plants is discussed later in this chapter. Quantitative retention of superoxide dismutase activity on a CaM-Sepharose affinity column has also been reported [267]. However, these authors did not show its regulation by $Ca^{2+}$/CaM. A unique microtubule motor protein, KCBP, containing a CaM-binding domain at its C-terminus and other unique domains in the N-terminus has been identified from *Arabidopsis* [218,236], tobacco [242], and potato [237,268–272]. Furthermore, the *Arabidopsis* CaM isoforms show differential affinity for this motor protein [218]. Disruption of KCBP revealed its involvement in trichome cell shape morphology in *Arabidopsis* [273,274]. Immunolocalization and microinjection studies indicate a role for KCBP in cell division [240,241].

Using differential screening of cDNA libraries prepared from NaCl-treated and-untreated tomato seedlings, a cDNA was isolated for glyoxylase I [245] and its transcripts are inducible by either NaCl, mannitol, or abscisic acid treatments in tomato [245]. Glyoxalase I, which catalyzes the conversion of toxic methylglyoxal to nontoxic metabolite in plants, has been purified using CaM-Sepharose affinity column chromatography from *Brassica juncea* [275]. Its activity is stimulated by $Ca^{2+}$ alone (2.6-fold) or $Ca^{2+}$/CaM (5 μM/145 nM) (7-fold). Further, the concentration of CaM required to stimulate the glyoxalase activity to the same extent in the presence of $Mg^{2+}$ (29 nM) and $Ca^{2+}$ (145 nM) is varied, suggesting that CaM has a cumulative effect on the metal-dependent activation of glyoxalase [275]. Using antibodies raised against purified *B. juncea* glyoxalase I as a probe, a cDNA sequence has been isolated from the cDNA expression library constructed for mRNA isolated from mannitol-treated *B. juncea* [244]. The full-length glyoxalase cDNA (*BjGly* I, Figure 3) is 784 bp in length and consists of an 558-bp ORF [185 amino acids (aa)]. Although $Ca^{2+}$/CaM stimulate its activity, the CaM-binding domain is not mapped. The binding sites of two cofactors ($Zn^{2+}$ and glutathione) along with the putative serine/threonine phosphorylation sites are shown to be conserved in glyoxalase I [244]. Further, its enzyme activity is stimulated by stress such as polyethylene glycol (PEG), light, and phytohormones in *B. juncea* [276].

Harrington and his coworkers [246] have successfully isolated several putative cDNA clones encoding CaMBPs by screening an expression library constructed from cell cultures of heat-shocked tobacco with [$^{35}$S]calmodulin. Of these, one cDNA sequence encodes a 499-amino-acid long protein, NtCBP48 (Figure 3). The sequence analysis predicts that it contains a centrally located transmembrane domain and nuclear localization sequence motif [246]. Using bacterially expressed deletion mutants, purification through CaM affinity chromatography, and their interaction with CaM on gel mobility shift assay, these authors have mapped the CaM-binding domain of CBP48 to a region of 40 amino acids in the C-terminus (Figure 3). The corresponding gene transcripts are rapidly inducible by heat shock (38°C) reaching maximum in 1.5 hr after treatment, indicating a role for NtCBP48 in heat stress [246].

## C.   Other Calcium-Binding Proteins Involved in Stress

A new family of $Ca^{2+}$-binding proteins, called calcineurin B–like (CBL) proteins, has been identified from *Arabidopsis* [63,204,277]. They are similar to the regulatory B subunit of calcineurin and the neuronal $Ca^{2+}$ sensor (NCS). Studies with *Arabidopsis* revealed the presence of at least six genes encoding highly similar but functionally distinct AtCBL proteins [204]. Although CaM and CBLs belong to two different groups, CBLs also contain four EF hand motifs and their activity depends on $Ca^{2+}$ binding (Figure 2). Drought, cold, and wound stress signals elevate *AtCBL1* gene transcripts, whereas *AtCBL2* and *AtCBL3* are constitutively expressed [204]. The *AtCBL4* (also called *SOS3*) is involved in the salt stress tolerance mechanism [63]. Further, CBL-interacting protein kinases (CIPK1 to 4) have also been identified using a yeast two-hybrid screen [277]. These protein kinases belong to the serine/threonine class of kinases and show high homology to SNF1 and AMPK from yeast and mammalian systems, respectively. These CIPKs interact with CBLs, but not with CaMs, in a $Ca^{2+}$-dependent manner [277]. Several isoforms of AtCBLs and CIPKs were also identified. However, CBL-CIPK isoform specificity is not determined and future studies in this direction should open up new avenues of research.

Using differential screening of a rice cDNA library constructed from ABA-treated rice seedlings, a cDNA sequence encoding a 27.4-kDa protein, EFA27, has been isolated [205]. It contains a single conserved EF hand motif, a characteristic $Ca^{2+}$-binding domain (Figure 2). Sequence analysis and a $Ca^{2+}$-binding assay with *E. coli*–expressed and–purified protein indicate that it is a new member of a $Ca^{2+}$-binding protein family that is induced by ABA and osmotic stress [205]. The *EFA27* gene transcripts are inducible in response to salt and dehydration stress and to an ABA signal. Sequence homology searches indicate that there are several *EFA27* gene homologues in *Arabidopsis* [205], suggesting the existence of similar proteins in phylogenetically distant species. Further, EF hand motif–containing protein phosphatases were also identified in the ABA signaling pathway [278]. Another $Ca^{2+}$-binding protein, AtCP1, has been identified in *Arabidopsis* [119]. It contains three EF hand motifs (Figure 2) and the bacterially expressed protein showed a characteristic $Ca^{2+}$-shifted electrophoretic mobility pattern. High levels of transcript are observed in flowers and roots compared with leaves and siliques in *Arabidopsis* [119]. The *AtCP1* gene transcripts are highly inducible by NaCl treatment but not by ABA treatment, indicating the specificity of this unique $Ca^{2+}$-binding protein in responding to various stress factors.

The *RbohA* (*respiratory burst oxidative homolog A*) gene from *Arabidopsis* and rice is also a $Ca^{2+}$-binding protein (Figure 2). The sequence features of this gene indicate its involvement in the oxidative burst [150]. Another cDNA, *PvHra32*, a transcript highly expressed during the hypersensitive reaction in bean cell cultures challenged with *Pseudomonas syringae*, has been shown to encode a small $Ca^{2+}$-binding protein (161 aa) [206]. The Hra32 protein contains four EF motifs and shows 51% sequence homology with a salt-inducible, three EF hands–containing $Ca^{2+}$-binding protein, AtCP1, from *Arabidopsis* [119]. Cytosolic $Ca^{2+}$ levels in both these processes are inducible and a correlation has been established between increased $Ca^{2+}$ levels and these $Ca^{2+}$-binding proteins in response to salt stress.

## D. Phospholipase C

A cDNA sequence, namely *AtPLC1*, was isolated from *Arabidopsis* using a PCR-based strategy with primers to conserved regions of animal phospholipases [82]. AtPLC1 encodes a 64-kDa protein with characteristic features of phosphatidylinositol-specific phospholipase C (PI-PLC) activity [82]. The *E. coli*–expressed PI-PLC hydrolyzes phosphatidylinositol-4,5-biphosphate to $IP_3$ and diacyl glycerol with an absolute requirement for $Ca^{2+}$ (1 μM) [82]. $IP_3$ has been shown to stimulate $Ca^{2+}$ release from the vacuolar store [279], and diacyl glycerol is an activator for protein kinase C activity. Hirayama et al. [82] showed that *AtPLC1* gene expression is induced by stresses including dehydration, salinity, and low temperature. Under these stresses, $Ca^{2+}$ and PLC may work in parallel in coordinating the environmental stress–inducible signal transduction pathway and cellular response. For example, coordinated action of $IP_3$-$Ca^{2+}$ in closure of guard cells under osmotic stress has been established. Furthermore, the presence of *AtPLC* gene activity in *Arabidopsis* raises the possibility of the presence of protein kinase C in plants. Another *AtPLC2* gene is shown to be constitutively expressed in *Arabidopsis* [280]. Three PI-PLC isoforms (*StPLC1* to -3) have been isolated from guard cell–enriched tissue of potato [281]. The expression pattern of the *StPLC1* and -2 genes suggests their involvement in drought stress in potato [281]. The soybean plasma membrane–associated PLC is unique in that the $Ca^{2+}$-binding domain spans across the so-called X and Y domains [282]. The existence of a small multigene PLC family has been identified in soybean [282]. The soybean PLC showed phospholipase activity and complemented the lethal mutant phenotype of yeast lacking PLC activity. Immunolocalization of PLC in the overexpressing transgenic PLC plants suggests that its distribution is associated with plasma membrane and cytosol [282].

## E. Protease

A 75-kDa cysteine class of $Ca^{2+}$-dependent protease (CDP) has been purified from *Arabidopsis* root cultures [283,284]. Its activity is specifically dependent on $Ca^{2+}$ but not on other divalent cations such as $Mg^{2+}$, $Sr^{2+}$, and $Zn^{2+}$. Calcium chelator EGTA inhibits the CDP activity. The concentration of $Ca^{2+}$ required to activate ACDP is more than the physiological cytosolic $Ca^{2+}$ levels. However, in other reported animal CDPs it has been shown that inositol phospholipids reduced the $Ca^{2+}$ required for CDP activity [285]. In animals, the CDPs or calpains are involved in many cellular functions by controlling proteolysis of other enzymes and structural proteins including protein kinases, myofibrillar proteins, cytoskeletal proteins, transcription factors, hormone receptors, and growth factors [285–287]. However, in plants, the role of CDPs in $Ca^{2+}$ signaling pathways is not known.

## IV. CALCIUM AND GENE EXPRESSION

It is becoming increasingly clear that the regulation of expression of specific genes is involved in a plant's ability to adapt or develop resistance to abiotic and biotic stress factors such as cold, heat, salinity, and pathogens [3,10,11]. The role of $Ca^{2+}$ in regulating gene expression, at both the transcriptional and the translational level, has been well documented in animal cells [288–295]. Much of the gene regulation by $Ca^{2+}$ is accomplished by $Ca^{2+}$-regulated protein kinases. Transcriptional regulation of expression of specific genes by $Ca^{2+}$ is due to regulation of phosphorylation of transacting factors by $Ca^{2+}$/CaM-dependent protein kinase II (CaM K II) [291] or $Ca^{2+}$/phospholipid-dependent protein kinase (protein kinase C) [295]. Calcium influences translation at the initiation as well as elongation steps. One of the elongation factors (eEF-2, eurkaryotic elongation factor-2) is a substrate for a $Ca^{2+}$/CaM-dependent protein kinase III (CaM K III) and phosphorylation of eEF-2 makes it inactive, resulting in the inhibition of translation [296–298]. Depletion of $Ca^{2+}$ results in a decrease in the rate of initiation of protein synthesis [299–301]. Inhibition of protein synthesis by $Ca^{2+}$ depletion is correlated with dephosphorylation of a ribosome-associated protein (26 kDa) [302].

Although there is a great deal of information on the involvement of $Ca^{2+}$ in regulating various physiological processes [12,13], very little is known about the role of $Ca^{2+}$ in regulating gene expression in plants. Manipulation of $[Ca^{2+}]_{cyt}$ by various means is shown to affect the expression of specific genes in plants. Lam et al. [303] presented some evidence indicating that $Ca^{2+}$ and CaM mediate the induction of expression of chlorophyll *a/b* binding (cab) genes. Ten percent of light-induced cab mRNA could be induced in the dark by increasing the intracellular level of $Ca^{2+}$ by ionomycin [303]. Microinjection studies with tomato phytochrome mutant (*aurea*) clearly show that $Ca^{2+}$ is involved in the expression of specific genes [91]. In the *aurea* mutant, light-regulated genes are not expressed. Injection of a plasmid construct containing cab promoter, a light-responsive promoter, fused to GUS reporter (cab-GUS) into cells of the *aurea* mutant showed no expression of GUS gene. However, coinjection of cab-GUS with $Ca^{2+}$ or $Ca^{2+}$-activated CaM resulted in the expression of the GUS gene. These results suggest the involvement of $Ca^{2+}$ and CaM in regulating the cab promoter activity [91]. Partial development of chloroplasts in the *aurea* mutant, which requires the expression of several genes, could be obtained by microinjection of $Ca^{2+}$ and CaM into these cells [304]. These results indicate that $Ca^{2+}$ regulates the expression of genes involved in chloroplast development.

In *Arabidopsis*, CaM and CaM-related genes (*TCH1*, *TCH2*, *TCH3*, and *TCH4*) are strongly induced by various mechanical stimuli such as touch and wounding [100]. It has been demonstrated that increased external $Ca^{2+}$ or heat shock rapidly induced the expression of touch-induced CaM-related genes (*TCH2*, *TCH3*, and *TCH4*), whereas the *TCH1* gene, which codes for CaM, is not significantly induced [212]. Heat shock, in the presence of EGTA, a $Ca^{2+}$ chelator, did not show induction of *TCH* genes. This EGTA effect is reversed by $Ca^{2+}$ replenishment. Based on these results, it was suggested that heat shock elevates $[Ca^{2+}]_{cyt}$ levels, which in turn regulates the expression of *TCH2*, *-3*, and *-4* genes. Heat shock is known to increase cytosolic $Ca^{2+}$ in animal cells [305–307]. Evidence for the involvement of $Ca^{2+}$ in heat shock stress in plants has been obtained from various studies [45,46,92,93]. The calcium effect on touch-induced CaM-related genes is specific because magnesium, another divalent ion, did not have any effect. Furthermore, increased $Ca^{2+}$ did not affect the expression of the heat shock–induced gene. In the case of *Arabidopsis*, the expression of one of the *Cam* genes (*TCH1*) is not significantly affected by an increase in external $Ca^{2+}$. There are multiple *Cam* genes in *Arabidopsis* and it is not known whether any of these genes are affected by changes in $[Ca^{2+}]_{cyt}$ levels [108,216,217].

Studies show that $Ca^{2+}$ is involved in stress-induced (both abiotic and biotic) gene expression. Plant-pathogen interaction, chemical elicitors, and a number of other stress factors have been shown to stimulate ethylene production in plants [308]. Ethylene is involved in the expression of some of the pathogenesis-related proteins including a chitinase. Depletion of $Ca^{2+}$ by $Ca^{2+}$ chelator blocked ethylene-induced chitinase synthesis, whereas artificial elevation of cytosolic $Ca^{2+}$ with $Ca^{2+}$ ionophore (ionomycin) or an inhibitor of microsomal $Ca^{2+}$-ATPase induced chitinase synthesis in the absence of ethylene [309]. These results indicate that $Ca^{2+}$ mediates the induction of the chitinase gene by ethylene. More recently, it has been shown that ethylene induced the phosphorylation of specific proteins although it is not known whether $Ca^{2+}$ is involved in this ethylene-regulated protein phosphorylation [310]. Using inhibitors of protein kinases and protein phosphatase, it was concluded that protein phosphorylation is one of the intermediate events involved in ethylene signal transduction. Fungal elicitors that are known to induce pathogenesis-related proteins have been shown to elevate cytosolic $Ca^{2+}$ [26].

In some plants, freezing tolerance can be developed by exposing them to nonfreezing low temperature [10]. This process, which is known as cold acclimation or cold hardening, is associated with changes in gene expression [10,195,311,312]. Expression of some of the cold-regulated genes is positively correlated with the ability of plants to develop freezing tolerance [195,312]. Earlier studies have shown elevation of cytosolic $Ca^{2+}$ in response to cold shock [26,27]. In alfalfa, low temperature–induced freezing tolerance is completely abolished by the $Ca^{2+}$ channel blocker lanthanum and verapamil and partially by

EGTA [195]. Furthermore, a CaM inhibitor (W7) but not its inactive analogue (W5) inhibited the development of cold acclimation–induced freezing tolerance. These results suggest that an increase in cytosolic $[Ca^{2+}]_{cyt}$ is necessary for developing cold-induced freezing tolerance and that CaM is involved in freezing tolerance. Accumulation of cold-induced mRNAs in alfalfa was partially blocked by lanthanum, a $Ca^{2+}$ channel blocker, and a CaM inhibitor (W7) completely blocked the expression of cold-regulated genes [195]. Lanthanum and W7 affect low temperature–induced changes in protein phosphorylation. However, the effects of these antagonists on phosphorylation are more severe and are not restricted to cold-induced changes.

As described earlier, there are several reports implicating $Ca^{2+}$ in regulating the expression of specific genes, including its own receptors, in plant cells. Calcium-regulated protein phosphorylation is likely to be involved in $Ca^{2+}$-regulated gene expression (see Sec. V) [106,190]. It is not yet known to what extent the changes in the $[Ca^{2+}]_{cyt}$ levels are reflected in changes in free $Ca^{2+}$ concentration in the nucleus. Studies show that there is a $Ca^{2+}$ gradient between the nucleus and cytoplasm indicating the presence of regulatory mechanisms that control $Ca^{2+}$ movement into and out of nucleus [313–315]. ATP stimulates $Ca^{2+}$ uptake into nuclei and studies implicate CaM involvement in this uptake process [314]. Currently, little is known about the participation of nuclear $Ca^{2+}$ stores in increasing cytosolic $Ca^{2+}$ and vice versa.

## V.  APPROACHES TO DECIPHER CALCIUM SIGNALING PATHWAYS IN STRESS SIGNAL TRANSDUCTION

The availability of well-characterized stress-induced genes and reporter genes such as GUS and GFP is facilitating plant biologists in the search for the intermediate components and mechanisms involved in stress signal transduction. Furthermore, availability of mutant plants that show increased sensitivity or resistance to stresses together with identification of these mutant genes is further advancing our quest to understand $Ca^{2+}$ signaling pathways. In this section we summarize three main approaches (cell biological, genetic, and transgenics) that have been used to elucidate stress signal transduction.

### A.  Cell Biological Approaches

Using isolated protoplast or cell culture systems, plant biologists have obtained important insights into the role of $Ca^{2+}$ in stress signal transduction. To study the effect of $Ca^{2+}$-activated CDPKs and phosphatases on the expression pattern of stress inducible genes, Sheen [190] used a cell biological approach. In this elegant experiment, maize protoplasts were utilized to monitor the transient expression of reporter and effector genes driven by stress-inducible promoter in the presence of elevated levels of CDPKs or phosphatases. Constitutive promoter (CaMV 35S) was used to direct the expression of effector genes (*CDPK*s and phosphatases), whereas the barley ABA (osmotic) stress–responsive gene promoter *HVA1* was used to express reporter genes (either *GFP* or *LUC*). A reporter gene (*GUS* or *GFP*) driven by the ubiquitin (*UBI*) promoter sequence, which is not inducible by any stress, was used as a control.

Initially, enhanced expression of GFP was observed in maize protoplasts harboring the *HVA1-GFP* reporter construct under cold, salt, dark, and ABA stresses. The same extent of GFP expression was also observed without stress treatments but in the presence of both $Ca^{2+}$ and $Ca^{2+}$ ionophore ($Ca^{2+}$-ionomycin or $Ca^{2+}$-A23187). However, in both experiments GFP expression was not observed in the maize protoplasts harboring the *UBI-GFP* reporter construct, indicating that $Ca^{2+}$ is involved in stress-induced gene expression [190]. Because $Ca^{2+}$ is able to induce the expression of a stress-inducible gene, Sheen has tested the effect of CDPKs on the expression of the *HVA1-LUC* gene. The maize protoplasts were cotransformed with reporter (*HVA1-LUC*) along with the truncated versions (containing the kinase domain) of one of the eight CDPK (*35S-PK*) constructs and the effect of eight effector protein kinases was monitored by quantifying luciferase activity. These results revealed that of the eight *Arabidopsis* CDPKs tested (ATCDPK1, ATCDPK1a, AK1/ATCDPK, ATCDPK2, ATPKa, ATPKb, ASK1, and ASK2), ATCDPK1 and ATCDPK1a activated the expression of the *LUC* gene driven by *HVA1* promoter. Furthermore, cotransformation of *HVA1-LUC* along with the constitutively expressed ATCDPK1a and PP2C, a protein phosphatase, or with the combination of CDPK and PP2C null mutants into maize protoplasts showed decreased or abolished LUC activity, indicating the involvement of phosphorylation and dephosphorylation events in the signal transduction leading to the activation of *HVA1-LUC* [190].

The oxidative burst in plants as a result of compatible plant-pathogen interaction leads to a hypersensitive response and cell death at the infection site in order to build up resistance in the neighboring cells against the invading microorganisms. Lamb and his colleagues [66] have provided evidence for the involvement of $[Ca^{2+}]_{cyt}$ in these events. *Pseudomonas syringae* pv. *glycinea* (*Psg*), carrying the *avrA* gene, causes an oxidative burst, hypersensitive response, and cell death in soybean cv. William 82, harboring its corresponding resistance gene *Rpg2* [67]. The early event in the *Psg* pathogen–mediated infection process in soybean is the production of AOSs including $H_2O_2$. Levine et al. [66] have taken advantage of soybean (*Psg* pv. *glycinea* and $H_2O_2$) and tobacco (cryptogein elicitor derived from *Phytophthora cryptogea*) cell culture systems as well as *Arabidopsis* seedlings (*Psg* pv. *tomato*) and pharmacological agents to show that $H_2O_2$-mediated influx of $Ca^{2+}$ is necessary and sufficient to induce cell death in response to their compatible avirulent pathogen strains or fungal-derived peptides. Such a pathogen-mediated cell death process can be initiated in the absence of either *Psg* (*avrA*) or $H_2O_2$ by $Ca^{2+}$ and ionophore A23187 but not ionophore alone and inhibited by $Ca^{2+}$-depleted medium as well as treatment with $Ca^{2+}$ channel blockers ($La^{3+}$), indicating the role of $Ca^{2+}$ in developing HR and cellular death in soybean cells [66]. Furthermore, these authors have shown that the $H_2O_2$-induced cell death process in soybean cell cultures is strongly inhibited by treatment with specific protein kinase inhibitors, staurosporine, K252A and AEBSF; partially inhibited by leupeptin; and not inhibited by inhibitors such as H7, H89, TPCK, TLCK, ANLM, and YVAD-CMK. These results indicate the role of specific CDPKs and phosphorylation events mediated by $Ca^{2+}$. However, $Ca^{2+}$ influx does not stimulate glutathione-*S*-transeferase induction as revealed by the fact that the treatment of soybean cells with A23187 did not induce GST transcripts and that $H_2O_2$-induced GST was not inhibited by $La^{3+}$, indicating that $Ca^{2+}$ elicits a specific signal pathway prior to the onset of the cellular death process in soybean cells.

Cho and his colleagues have used biochemical, cell biological, and transgenic approaches to address the functional differences and significance of the conserved (SCaM1, -2, and -3) and divergent (SCaM4 and -5) CaM isoforms from soybean [219,220,250,316]. They have analyzed these isoforms to address their interaction with the putative CaM-binding proteins isolated from plant protein extracts at different developmental stages on gel overlay assay, differences in in vivo distribution of target proteins using in situ hybridization with specific antibodies raised against conserved and divergent SCaM isoforms, ability to activate NAD kinase and phosphodiesterase, and finally their participation in the disease resistance mechanism using a transgenic approach (see later under Transgenic Approaches). Although SCaM isoforms show similar patterns, they differ in their relative affinity in interacting with CaM-binding proteins. Further, the isoforms show differences in their relative abundance in vivo. The conserved isoforms are relatively abundant in their expression compared with divergent forms. All CaM isoforms activate phosphodiesterase (PDE) but they differ in their activation of NAD kinase, calcineurin, and nopaline synthase, indicating $Ca^{2+}$/CaM specificity between CaM isoforms and target proteins. Recently, we have provided evidence for the differential interaction of CaM isoforms with a kinesin-like microtubule-associated protein from *Arabidopsis* [218]. These results also suggested that all CaM isoforms may bind to CaMBPs but with different affinity. Taken together, these findings suggest the existence of reciprocal regulatory mechanisms between $Ca^{2+}$/CaM and CBPs.

The studies just presented provide strong evidence for the role of $Ca^{2+}$, CDPKs, phosphatases, CaMs, and CaMBPs in stress-induced signal transduction cascades. However, further experiments are necessary to identify the intermediate components that induce the expression of target genes such as *HVA1* and cellular responses such as physiological cell death and to identify and functionally characterize CaMBPs.

## B. Molecular Genetic Approaches

Increases in free $[Ca^{2+}]_{cyt}$ levels in response to a variety of abiotic and biotic stresses initiate a complex signal network ranging from sensing to activation of a number of stress-responsive genes to an ultimately altered physiological process. To address the downstream signal components in the stress-induced $Ca^{2+}$-mediated signal cascades, a number of mutants with increased sensitivity or tolerance to stresses have been isolated. For example, cold acclimation is a process in which plant species show enhanced resistance to freezing after they are exposed to low-temperature conditions [33]. Molecular and genetic approaches led to identification of a common set of genes that respond to cold, drought, and ABA [33,317], which are well-known stress factors that elevate $[Ca^{2+}]_{cyt}$ [3].

The genes that are involved in the cold acclimatization process are called *COR* (cold responsive) genes and encode hydrophilic polypeptides with similar properties [33]. These include *COR6.6* (*KIN*), *COR15a*, *COR78* (*LT178*), *WCS19*, *CORa*, *CAS15a*, COR47, and *HVA1*, and these gene products potentially promote tolerance to freezing. The expression of some of these genes is regulated by CBF (CRT/DRE binding factor) transcriptional activators through binding of CRT/DRE motifs present in their promoter region [318,319]. The *HVA1* gene has been extensively used to test whether its expression is inducible by $Ca^{2+}$-mediated stress signals [190]. Thomashow and his colleagues have studied extensively the functions of the *COR15a* and *CBF1* genes in the cold acclimation process. Constitutive expression of *COR15a* [320] and *CBF1* [321] independently in nonacclimated *Arabidopsis* led to improved tolerance to low temperatures. Acclimation involves changes in the membrane lipid profile including phospholipids, sterols, cerebrosides, and accumulation of sucrose, other simple sugars, and proline, which stabilize membranes against freeze-induced damage [322]. Although the exact mechanism of action of the COR regulon in freeze tolerance is not known, the COR polypeptides are probably involved in the regulation of the enzymes that produce lipids, sugars, and other membrane-stabilizing components during acclimation process.

Mutants that show an altered response to a particular stress are invaluable resources in identifying components of the signal transduction pathway(s). *Arabidopsis* mutants sensitive to freezing (*sfr*) even after cold acclimation have been identified [323]. One of these mutants, *srf6*, showed reduced transcript levels for the *COR* genes containing a *CRT/DRE* sequence motif including *KIN1*, *COR15a*, and *LT178* [34]. However, the *sfr6* mutant is not defective in the expression of *CBF* genes (*CBF1* to *CBF3*) that activate the expression of *KIN1*, *COR15a*, and *LT178* (*CRT/DRE cis*-containing genes) and *AtP5CS1* (which does not contain the *CRT/DRE* element and is involved in proline biosynthesis). Further, the $Ca^{2+}$-sensing mechanism in *sfr6* is intact, which means that upon cold treatment the $[Ca^{2+}]_{cyt}$ levels are similar to those in the wild type. These authors considered SFR6 an essential component that promotes the interaction between CBF and *CRT/DRE* responsive genes, which are activated in response to cold, drought, and ABA stress events [34]. These studies indicate the involvement of other components in $Ca^{2+}$-mediated signal transduction pathways in plants.

Zhu and his colleagues [324] have screened ethyl methanesulfonate (EMS)-treated transgenic *Arabidopsis* lines (*RD29A-LUC*) expressing bioluminescence in response to cold and osmotic stress (NaCl and ABA). The LUC coding sequence is under the regulation of the *RD29A* (*COR78* or *LT178*) gene promoter, which harbors both *CRT/DRE* and *ABRE cis*-regulatory DNA elements, and thus responds to cold and osmotic stresses. Using this genetic approach, they isolated many genetic mutants that are defective in one or a combination of cold or osmotic stress (NaCl or ABA), which will be invaluable in dissecting the complex network of stress signaling in plants. Characterization of these mutants and cloning of mutant genes would provide valuable insights into the stress signal transduction.

The plant stress–responsive hormone ABA modulates several cellular activities such as guard cell closure and activation or repression of several genes in response to drought, cold, and high salt [325,326]. ABA is also involved in seed maturation, desiccation, dormancy, and germination. Unequivocally, $Ca^{2+}$ has been implicated in ABA-mediated guard cell closure in response to environmental stresses [53,141,327]. Genetic approaches have been employed to isolate mutants that are defective in ABA responses in several plant species [278,328–331]. Molecular and genetic approaches in *Arabidopsis* led to identification of many ABA-insensitive mutants including *abi1* and *ab12* mutants, which encode protein phophatases of type 2C (PP2Cs). Furthermore, a site-directed mutant of PP2C together with ABA-responsive gene expression in maize protoplast has also shown the role of PP2C in ABA signal transduction [332]. Accumulated research investigations indicate that elevated $[Ca^{2+}]_{cyt}$ levels act as a second messenger in ABA-mediated cellular responses [52,190,332,333]. Recently, Allen et al. [334] provided direct evidence for the role of $Ca^{2+}$ in ABA-mediated stomatal closure using ABA-insensitive mutants (*abi1* and *abi2*) of *Arabidopsis*. These mutants are unable to elevate sufficient $[Ca^{2+}]_{cyt}$ levels in response to ABA treatment as measured by fluorescent $Ca^{2+}$-binding dyes and stomata remain open. However, treatment of *abi* mutants with extracellular $Ca^{2+}$ restored ABA-induced stomatal closure, indicating the role of $Ca^{2+}$ in ABA-mediated stomatal closure. The downstream events following $[Ca^{2+}]_{cyt}$ elevation, such as kinase-activated S-type anion channels, membrane depolarization, and loss of turgor pressure, remain functional in *abi* mutants [334]. Further research would provide the identity of the missing components in this pathway.

About one third of arable lands contains high levels of NaCl in their soils [335]. Such high-salinity soils cause accumulation of $Na^+$ ions and less uptake of $K^+$. This situation in plant cells causes severe abnormalities in cellular activities including inhibition of photosynthesis, reduction in protein synthesis and potassium content, and increase in $Na^+$ and organic solutes such as proline, glycine betaine, and polyols. Zhu and his colleagues [63,117,336–338] have made significant contribution to our understanding of salt tolerance. They provided a molecular view of the relationship between $Ca^{2+}$, potassium, and sodium cations in salt tolerance mechanisms in *Arabidopsis* using a novel genetic approach [63,117,336, 337,338]. These authors screened a large number of EMS or fast neutron mutagenized M2 seed or T-DNA insertion *Arabidopsis* lines (~267,000) to isolate salt overly sensitive (*sos*) mutants using a root-bending assay on agar plates containing 50 mM NaCl [336]. Using this genetic screening approach, Zhu et al. discovered three genes, namely *SOS1*, *SOS2*, and *SOS3*, and numerous alleles for each gene.

Construction of combinations of double *sos* mutants revealed that they function in a linear pathway and exhibit similar phenotypes in that they are all hypersensitive to $Na^+$ and $Li^+$ and are unable to grow on a low-$K^+$ culture medium. They differ in their sensitivity to 100 mM NaCl (*sos1* being the most sensitive followed by *sos2* and *sos3*). The concentration of NaCl required to inhibit root growth 50% ($I_{50}$) was shown to be ~4 mM, ~10 mM, and ~40 mM for *sos1*, *sos2*, and *sos3*, respectively, as compared with the wild type, which requires 100 mM [117,336]. Interestingly, the normal growth pattern of *sos* mutants is restored by inclusion of $K^+$ in the growth medium along with 50 mM NaCl (50 mM $K^+$ for *sos1* and *sos2* and 1 mM for *sos3*). These abnormal growth patterns in the presence of NaCl could be mitigated by the addition of increased levels of $Ca^{2+}$ in the same medium. Calcium restores NaCl-inhibited growth partially in *sos1* (at 10 mM) and significantly in *sos3* (at 2 mM) mutant seedlings, indicating that $Ca^{2+}$ plays a crucial role in selective uptake of $K^+$ to $Na^+$. These results provided a direct relationship between three cations in a salt tolerance mechanism using the genetic approach.

On the basis of these results, Zhu and his colleagues predicted that these *SOS* genes encode regulatory proteins that control $Na^+/K^+$ uptake and a $Ca^{2+}$ sensor that controls $K^+$ nutrition [337]. Indeed, the *SOS3* gene encodes a $Ca^{2+}$ sensor protein that shows high affinity to $Ca^{2+}$ [63]. Its significant homology to calcineurin (CaN) of yeast [339,340] and neural $Ca^{2+}$ sensors (CNS) of animals [341] raises the possibility that the *SOS3* gene product might regulate the $K^+/Na^+$ transport system via $Ca^{2+}$-mediated activation of phosphatase and/or inhibition of a kinase signal cascade and confers salt tolerance. Using a transgenic approach, Pardo et al. [116] were able to express constitutively both catalytic and regulatory components of yeast CaN in tobacco with a 35S promoter and showed that the transgene enhances the salt tolerance of tobacco. Complementation of plant $Ca^{2+}$ pumps (ACA2 and ECA1) in yeast mutants also confirmed the similar functional $Ca^{2+}$ machinery between yeast and plants [229,342]. Together, these results provide direct molecular genetic evidence for the participation of $Ca^{2+}$-binding proteins in adapting salt tolerance in higher plants.

## C.  Transgenic Approaches

To address the functions of $Ca^{2+}$ and its interacting components at the whole plant level, researchers from a number of laboratories have developed transgenic plants with the genes involved in $Ca^{2+}$-mediated stress signaling cascades. The transgenes include the genes encoding stress-inducible transcriptional factors, CaM specific isoforms, $Ca^{2+}$ antiporters, and CaM-binding proteins (Table 2). In this section, we have presented some known examples of transgenic approaches that unravel the role of $Ca^{2+}$ in initiating stress-responsive mechanisms in plants.

Plants maintain an asymmetric distribution of $Ca^{2+}$ (millimolar versus nanomolar levels in organelles and cytosol, respectively) to avoid its toxic effects on cellular metabolism. This kind of $Ca^{2+}$ homeostasis is achieved by high-affinity $Ca^{2+}$ pumps ($Ca^{2+}$-ATPases) and low-affinity $Ca^{2+}/H^+$ antiporters. These $Ca^{2+}$ pumps and antiporters also play a crucial role in raising and restoring $[Ca^{2+}]_{cyt}$ levels in response to various stimuli. A vacuolar localized $Ca^{2+}/H^+$ antiporter (*calcium exchanger*) has been identified from *Arabidopsis* (*CAX1* and *CAX2*) [345] and mung bean [346]. The *Arabidopsis CAX1* gene functionally complements yeast mutant defective in its antiporter activity [345]. Its transcripts are inducible by extracellular $Ca^{2+}$ levels, $Na^+$, $K^+$, $Ni^{2+}$, PEG, and $Zn^{2+}$ but not by plant hormones. In order to gain more insight into the action of the *CAX1* gene product on $Ca^{2+}$ homeostasis, Hirschi [343] constitutively expressed the *Arabidopsis CAX1* coding region under the control of the 35S promoter in tobacco plants. The transgenic plants expressing the *CAX1* gene in sense orientation showed stunted growth with a poorly developed root system, necrosis of leaves and apical meristem, hypersensitivity to $K^+$ and $Mg^{2+}$ ions, and sensitivity to cold [343]. However, these abnormalities could be reversed by $Ca^{2+}$ sup-

**TABLE 2**   Summary of Genes Used in Developing Transgenic Plants to Decipher Their Role(s) in Calcium-Mediated Signaling Pathways in Stress Signal Transduction

| Gene | Gene product | Transgenic plant/promoter | Biological function | Reference |
|------|--------------|---------------------------|---------------------|-----------|
| *AtCax1* | $Ca^{2+}H^+$ antiporter | Tobacco/CaMV35S | Pumping of $Ca^{2+}$ to vacuoles | 343 |
| *AtCbf1* | CRT/DRE-binding factor | Tobacco/CaMV35S | Induction of *CRT/DRE*[a] *cis* element containing gene expression (*COR* genes) | 321 |
| *AtDreb1a* | DRE-binding factor 1A | Tobacco/RD29A | Induction of *CRT/DRE cis* element containing gene expression (*COR* genes) | 344 |
| *BjGly I* | Glyoxalase I | Tobacco/CaMV35S | Tolerance to NaCl, methylglyoxal | 244 |
| *PhGad* | Glutamate decarboxylase | Tobacco/CaMv35S | Synthesis of GABA and stress tolerance | 232 |
| *NtCbp4* | Membrane-bound cation channel | Tobacco/CaMV35S | $Ni^{2+}$ tolerance | 258 |
| *GmCam4* | Soybean CaM 4 isoform | Tobacco/CaMV35S | PR gene expression and biotic stress tolerance | 220 |
| *GmCam5* | Soybean CaM5 isoform | Tobacco/CaMV35S | PR gene expression and biotic stress tolerance | 220 |
| *VU-3* | Mutated form of synthetic CaM (K115R) | Tobacco/CaMV35S | Activation of NAD kinase and production of AOSs | 29 |
| *PhCam53* | *Petunia hybrida* CaM53 isoform | Transient and ectopic expression in tobacco and *Arabidopsis* | Importance of prenylation domain and cellular localization | 203 |

[a] *CRT/DRE*, C repeat/drought responsive element; *COR*, cold-regulated genes; GABA, γ-aminobutyric acid; PR, pathogen-related; and AOSs, active oxygen species.

plementation. Plants harboring the antisense *CAX1* or vector alone showed a wild-type pattern without any symptoms, but these symptoms could be induced by $Ca^{2+}$ depletion in the growth medium. Further, consistent with these findings, constitutive antiport activity was observed in sense transgenic tobacco plants compared with control plants. These results indicate that the hyperantiport activity of the *CAX1* gene causes severe depletion in free cytosolic $Ca^{2+}$ levels due to pumping of $Ca^{2+}$ from cytosol to vacuole. This nonavailability of cytosolic $Ca^{2+}$ impairs mitigating the stress-related signal transduction cascades and related cellular adjustments in the transgenic tobacco plants expressing the *CAX1* gene [343].

To investigate the role of distinct $Ca^{2+}$ pathways in mitigating cold and wind stress stimuli, transgenic tobacco plants expressing aequorin in cytoplasm (MAQ 2.4) and nucleus (MAQ 7.11) were constructed [39]. These transgenic plants allowed the authors to monitor $Ca^{2+}$ changes in different cellular (cytoplasm and nucleus) compartments in response to specific stimulus. Using these transgenic plants, these authors have measured the $Ca^{2+}$-mediated *NpCaM1* gene expression and corresponding increase in $Ca^{2+}$ changes in cytosol and nuclear compartments in response to cold and wind shocks [39]. Comparison of $Ca^{2+}$ dynamics and $Ca^{2+}$-induced *NpCaM1* gene expression revealed that two independent $Ca^{2+}$ signaling pathways are evolved in perceiving and transducing wind and cold stress stimuli. One is specific to wind and operates through nuclear $Ca^{2+}$ and the other is specific to cold and operates through cytoplasmic $Ca^{2+}$. The results obtained from this transgenic approach to measure raises in $Ca^{2+}$ levels established the occurrence of distinct $Ca^{2+}$ signal cascades and $Ca^{2+}$ respones to specific stress stimuli in plants.

As already discussed, the primary event in cellular activity in response to drought, cold, and salt stress is elevated $[Ca^{2+}]_{cyt}$ levels. Once the signal elevates $[Ca^{2+}]_{cyt}$, $Ca^{2+}$-sensing proteins are activated, leading to expression of several target genes. The target genes include *rd29A*, *kin1*, *cor.6.6*, *cor47*, *cor15a*, and *erd10* and their gene products are believed to be protective in response to various stresses and together referred to as regulon [33]. Interestingly, genes in this family possesses a common *cis*-regulatory DNA sequence motif, *CRT/DRE*, in their promoter region. Further, the *trans*-acting protein factors (CBF1, DREB1A) that bind to the *CRT/DRE* motif and up-regulate their gene expression have also been identified [63,321,347]. Because the expression of a battery of stress-inducible *cor* genes is regulated through a common *cis* element (*CRT/DRE*) by a single transcriptional factor (CBF1 or DREB1A), Thomashow and Yamaguchi-Shinozaki and their coworkers have used a transgenic approach to overexpress transcriptional factors (CBF1 or DREB1A) from either constitutive promoter, *CaMV35S-CBF1* [321] and *CaMV35S-DREB1A* [344], or stress-inducible promoter (*rd29A-DREB1A*) [344]. Although growth retardation has been a common phenomenon observed in the plants expressing transcriptional factors from constitutive promoters, the *cor* gene expression is hyperinducible in response to cold, drought, and salt stresses. However, the plants expressing transcriptional factor (DREB1A) from the *rd29A* promoter did not show negative effects on plant growth but the *cor* gene expression is super up-regulated only when stress conditions prevail. Both kinds of transgenics showed significant resistance to cold, drought, and salt stress responses [321,344].

To dissect the role of specific $Ca^{2+}$-activated CaM isoforms and their target enzymes and/or proteins in plant defense mechanisms [29,220,232] or in heavy metal tolerance [258], transgenic tobacco plants expressing genes encoding specific CaM isoforms or mutated forms of CaM or CaM-binding proteins were generated. Although preliminary experiments with a pharmacological approach indicated a role for CaM isoforms in plant defense responses, the specificity of CaM antagonists is questionable in unequivocally elucidating $Ca^{2+}$-mediated signal pathways. These pharmacological agents can also influence other cellular activities not related to $Ca^{2+}$/CaM signaling [16,19]. Therefore, the sense transgenic plants proved to be valuable tools in studying the function of CaM isoforms and CaM-target proteins in planta.

Transgenic plants (VU-3) harboring a mutated form of CaM (amino acid, K115R in the CaM sequence abolished posttranslational modification of trimethylation at 115R) showed increased levels of resistance to a variety of stimuli such as treatment with cellulase, bacteria-derived elicitor harpin, mechanical stress, and osmotic stress [29]. Transgenic VU-3 cell lines treated with cellulase or mechanical stress produced severalfold higher levels of AOSs compared with control cell lines. Diphenyleneiodonium, an inhibitor of AOSs production, abolishes AOS synthesis in VU-3 cell lines. Production of AOSs has been shown to initiate a battery of defense responses against invading pathogens [66,68,69,146]. The enhanced resistance activity of VU-3 plants is primarily attributed to their ability to show hyperactivation of NAD kinase (a $Ca^{2+}$/CaM-regulated enzyme) [29]. The only molecular difference between VU-3 and control NC lines is a mutation at a single amino acid residue (K115R) in the CaM protein sequence. This modification showed hyperactivation of its target NAD kinase. This elegant study provided evidence that NAD kinase may be one of the targets of CaM in plant defense responses. Certain CaM isoforms contain a 34-amino-acid Caax box motif (prenylation domain) at their C-termini with CTIL (PhCaM53) [203] or CVIL (OsCaM63) [348] amino acid residues. Although most CaM isoforms are soluble proteins (based on protein sequence) [349], transient expression of GFP fused with the full-length CaM53 or mutated CaM53 (without the CaaX box motif) in tobacco and petunia protoplasts showed functional significance of the prenylation domain (CaaX-box) [203,350]. The cells harboring GFP-CaM53 with an intact prenylation domain localized to plasma membrane, whereas the GFP mutant CaM53 (without a CaaX box) localized to the nucleus. Although CaM53 does not contain NLS motifs, its localization to nucleus suggested a dual role played by the Caax box in localizing CaM53. Further, the pattern of distribution of CaM53 may depend on the prenylation status of the cell or the requirement and/or availability of other NLS protein to interact and move into the nucleus to exert its different cellular activities [203]. Ectopically expressed prenylated CaM53 in tobacco showed stunted growth and a necrotic phenotype. However, ectopic expression of neither the nonprenylated form nor the Caax box motif alone did not result in such altered morphogenic alterations, indicating that high levels CaM could perturb specific cellular machinery involved in growth and development [203]. Together, these studies suggest functional diversity of CaM isoforms in coordinating $Ca^{2+}$ signaling pathways in higher plants.

Using a similar transgenic approach, another $Ca^{2+}$/CaM target enzyme, glutamate decarboxylase (GAD), was expressed in tobacco. The tobacco transgenic plants expressing GAD or GAD devoid of CBD (27-amino-acid region in the C-terminus) (GAD-CBD) (Figure 3) revealed different phenotypes and levels of GAD's enzymatic product and substrate, GABA and Glu, respectively. Transgenic tobacco plants expressing GAD showed a normal wild-type phenotype with moderately increased GABA and reduced Glu. However, the transgenic plants bearing GAD-CBD exhibit severe morphological disorders in plant growth and development. This phenotype is due to anatomical abnormalities in cell elongation in stem cortex and parenchyma tissues. This phenotype was shown to correlate with accumulation of extremely high levels of GABA and low levels of Glu [232]. This is a very important observation in light of the plants' ability to produce high levels of GABA in response to a variety of stresses [264,351].

Fromm and his colleagues [258] have isolated a cDNA sequence, *NtCBP4*, encoding a protein with a CaM-binding property from tobacco. Sequence comparison indicates its close resemblance to nonselective membrane-bound cation channels from animals, its plant homologues were isolated from *Arabidopsis* [352] and barley [234], and it has been shown that it is located on plasma membrane [234,258]. Constitutive expression of *NtCBP4* cDNA under the 35S promoter in sense and antisense orientation in tobacco exhibited a normal phenotype under normal growth conditions. However, when subjected to different metal toxicities, NtCBP4 antisense and GAD-expressing transgenics along with the wild-type plants showed symptoms associated with $Ni^{2+}$ toxicity such as retarded root and shoot growth and reduced chlorophyll content (chlorotic) although all contained the same amount of NtCBP4 protein. In contrast, the *NtCBP4* sense transgenic plants grew and developed normally even at 200 μM $NiCl_2$. However, sense plants showed hypersensitivity to $Pb^{2+}$ [258]. This is the first report that a protein (NtCBP4) showing CaM-binding capacity is involved in metal tolerance. Further, the CBD is speculated to be located at the C-terminus near the end of the cyclic nucleotide binding site (Figure 3).

Another example of involvement of $Ca^{2+}$/CaM regulation in the plant disease resistance response comes from plants expressing divergent CaMs [220]. Constitutive expression of divergent CaM isoforms SCaM4 and -5 under the 35S promoter in tobacco has yielded the important biological functional significance of the most divergent CaM isoforms identified so far. Conserved CaMs (SCaM1, -2, and -3) are not inducible by stress, whereas divergent CaM isoforms are normally expressed at a low level and are highly inducible in response to $Ca^{2+}$-mediated pathogen attack (*Fusarium solani*, *Phytophthora parasitica* var *nicotianae*) or elicitor contact (derived from *P. parasitica*). Transgenic plants expressing high levels of SCaM4 and -5 showed constitutive expression of SA-related gene expression independent of SA throughout their life cycle and showed enhanced disease resistance to a wide spectrum of pathogens (*Phytophthora parasitica* var *nicotiana*, *Pseudomonas syringae* pv. *tabaci*, and a virulent viral pathogen, TMV) [220]. These studies revealed the existence of CaM isoform specificity in their target activation in $Ca^{2+}$-mediated signal transduction processes in plants. Furthermore, studies revealed that the CaM isoforms differ in their affinity for the same target protein [106,167,174,192,193,201,218,316], highlighting the significance of the existence of multiple CaM isoforms in mediating specific $Ca^{2+}$-mediated signal transduction pathways. However, the CaM target proteins involved in various processes remain to be characterized.

In order to understand the biological significance of a small CaM-binding protein, BjGLY 1, Veena et al. [244] raised transgenic tobacco with the *BjGly I* gene under the CaMV35S promoter. The transgenic plants showed a normal phenotype as antisense *BjGly I* transgenic and wild-type plants. However, transgenic plants overexpressing the *BjGly 1* gene showed elevated levels of *BjGly I* gene transcripts when subjected to NaCl (50 to 800 mM), mannitol (100 to 800 mM), or $ZnCl_2$ [244]. Further, leaf disks of transgenic plants were assayed for their tolerance to abiotic stresses (NaCl, mannitol, methylglyoxal). The results revealed that the sense transgenic leaf disks showed tolerance with an intact chlorophyll content and without any deterioration from the normal phenotype. In contrast, the control antisense and wild-type leaf disks showed a bleaching effect (loss of chlorophyll) and did not survive these stresses. These results suggest that BjGly I plays a role in conferring tolerance to salt, methyl glyoxal, and water stresses (abiotic) in plants.

## VI.  CONCLUSIONS

The reports described in this chapter clearly indicate an important messenger role for $Ca^{2+}$ in transducing a variety of stress signals. Stress-induced rapid and transient changes in cytosolic $Ca^{2+}$ have been documented in several cases. At least two classes of proteins (CaM and CDPK) are found in all the plants that seem to play a broad role in mediating $Ca^{2+}$ action. Several other $Ca^{2+}$-sensing proteins (calcineurin B-like, channels, antiporters) that mediate $Ca^{2+}$ stress tolerance or regulate $Ca^{2+}$ changes in the cytoplasm have been functionally characterized. In addition, several proteins that are downstream targets of $Ca^{2+}$/CaM have been detected, and in some cases their involvement in stress-related processes has been demonstrated. Accumulating and ongoing research activities from different laboratories indicate the presence of several such proteins in plants. Identification of the biological function of $Ca^{2+}$ and $Ca^{2+}$/CaM-binding proteins is a necessary step in further elucidating the role $Ca^{2+}$. Calcium-mediated stress signals have been shown to modulate various components located downstream in the signal pathway in order to evoke specific cellular activity. In most cases, the identity of the full cascade (beginning from changes at the cytosolic level to the final physiological response) has not been investigated. The $Ca^{2+}$ signaling cascade in plants seems to work as a complex grid and interact with several components in many possible combinations.

The present understanding of the $Ca^{2+}$ signlaing cascade in stress signal transduction is depicted in Figure 4. The resting $[Ca^{2+}]_{cyt}$ level is about 100 nM. The calcium stores such as plasma membrane, vac-

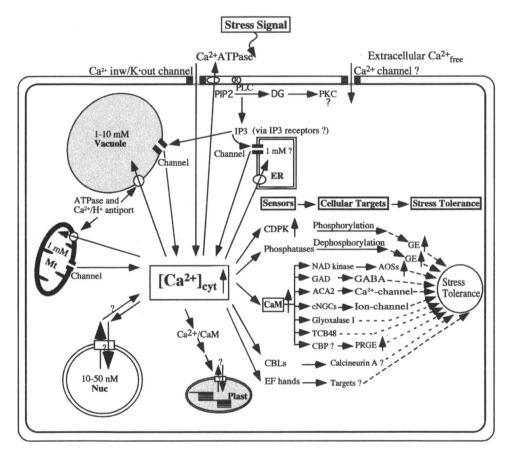

**Figure 4**   Schematic illustration of calcium-mediated signal pathways in stress signal transduction. The transport of calcium into and out of the cellular compartments (plasma membrane, vacuole, ER, and mitochondria) in response to a stress signal is indicated with arrows (the arrows point to the direction of calcium flow). The IP3 (inositol-1,4,5-triphosphate) system and its sensitive channels in vacuole and ER are shown. The estimated concentration of calcium in different organelles is indicated in mM [4,6,326,353]. The vertical arrows ($\uparrow$) indicate increased levels of respective components. The question mark (?) denotes lack of information. $[Ca^{2+}]_{cyt}$, free cytosolic calcium; ER, endoplasmic reticulum; Mt, mitochondria; Plast, plastids; Nuc, nucleus; R, an unidentified receptor on plasma membrane; PLC, phospholipase C; PIP2, phosphatidylinositol-4,5-bisphosphate; IP3, inositol-1,4,5-triphosphate; DG, diacyl glycerol; PKC, protein kinase C; GE, gene expression; AOSs, active oxygen species; GABA, γ-aminobutyric acid; PRGE, pathogen-related gene expression; CDPK, calcium-dependent protein kinase; CaM, calmodulin; CBLs, calcineurin B–like proteins; EF-hands, EF hand motifs containing proteins; GAD, glutamate decarboxylase; ACA2, *Arabidopsis* $Ca^{2+}$-ATPase; cNGCs, cyclic nucleotide gated channels; TCBP48, tobacco calmodulin-binding protein 48; and CBP, calmodulin-binding protein.

uole, ER, and mitochondria contain $Ca^{2+}$ in the mM range. To maintain such a steep electrochemical gradient between cytoplasm and organelles, the membranes are equipped with pumps and channels that open and pump $Ca^{2+}$ ($10^{-6}$ $Ca^{2+}$ ions/sec) to elevate $[Ca^{2+}]_{cyt}$ in response to the signal and maintain $Ca^{2+}$ homeostasis (Figure 4). Further, these channels and pumps are gated by other messengers such as $IP_3$, cADP-ribose, or voltage that regulate the $Ca^{2+}$ release in response to a specific stress stimulus [4,326]. Once the $[Ca^{2+}]_{cyt}$ levels are raised, an array of high-affinity $Ca^{2+}$-binding proteins sense and initiate a specific signal cascade via their cellular targets to generate appropriate responses to cope with the stresses (Figure 4) [4,7,19,167,353]. These sensors include CDPKs, CaMs, calcineurin B–like proteins, and other EF hand–containing proteins that modulate downstream target components. For example, CaMs regulate the activity of CaM-binding proteins, which in turn participate in stress tolerance mechanisms (Figure 4).

## ACKNOWLEDGMENTS

Research in our laboratory is supported by grants from the National Science Foundation, Agricultural Experiment Station, USDA, NASA, Plant Biotechnology Laboratory, Colorado Biotechnology Research Institute, and Colorado RNA Center. We thank Dr. Irene S. Day and Dr. Farida Safadi for comments on the manuscript.

## REFERENCES

1.  JS Boyer. Plant productivity and environment. Science 218:443–448, 1982.
2.  HJ Bohnert, E Sheveleva. Plant stress adaptations—making metabolism move. Curr Opin Plant Biol 1:267–274, 1998.
3.  H Knight. Calcium signaling during abiotic stress in plants. Int Rev Cytol 195:269–324, 2000.
4.  AJ Trewavas, R Malho. Signal perception and transduction: the origin of the phenotype. Plant Cell 9:1181–1195, 1997.
5.  N Smirnoff. Plant resistance to environmental stress. Curr Opin Biotechnol 9:214–219, 1998.
6.  D Sanders, C Brownlee, J Harper. Communicating with calcium. Plant Cell 11:691–706, 1999.
7.  A Trewavas. How plants learn? Proc Natl Acad Sci U S A 96:4216–4218, 1999.
8.  BJC Cornelissen, LS Melchers. Strategies for control of fungal diseases with transgenic plants. Plant Physiol 101:709–712, 1993.
9.  GL Matters, JG Scandalios. Changes in plant gene expression during stress. Dev Genet 7:167–175, 1986.
10. CL Guy. Cold acclimation and freezing stress tolerance: role or protein metabolism. Annu Rev Plant Physiol Plant Mol Biol 41:187–223, 1990.
11. C Lin, WW Guo, E Everson, MF Thomashow. Cold acclimation in *Arabidopsis* and wheat. Plant Physiol 94:1078–1093, 1990.
12. PK Hepler, RO Wayne. Calcium and plant development. Annu Rev Plant Physiol 36:397–439, 1985.
13. BW Poovaiah, ASN Reddy. Calcium messenger system in plants. CRC Cri Rev Plant Sci 6:47–103, 1987.
14. BW Poovaiah, JJ McFadden, ASN Reddy. The role of calcium ions in gravity signal perception and transduction. Physiol Plant 71:401, 1987.
15. BW Poovaiah, GM Glenn, ASN Reddy. Calcium and fruit softening: physiology and biochemistry. Hortic Rev 10:107, 1988.
16. DM Roberts, TJ Lukas, DM Watterson. Structure, function, and mechanism of action of calmodulin. CRC Cri Rev Plant Sci 4:311–339, 1986.
17. AJ Trewavas, S Gilroy. Signal transduction in plant cells. Trends Genet 7:356–361, 1991.
18. DM Roberts, AC Harmon. Calcium modulated proteins: targets of intracellular calcium signals in higher plants. Annu Rev Plant Physiol Plant Mol Biol 43:375–414, 1992.
19. BW Poovaiah, ASN Reddy. Calcium and signal transduction in plants. CRC Cri Rev Plant Sci 12:185–211, 1993.
20. GA Allen, JM Kwak, SP Chu, J Llopis, RY Tsien, JF Harper, JI Schroeder. Cameleon calcium indicator reports cytoplasmic calcium dynamics in *Arabidopsis* guard cells. Plant J 19:735–747, 1999.
21. A Miyawaki, O Griesbeck, R Heim, RY Tsien. Dynamic and quantitative $Ca^{2+}$ measurements using improved cameleons. Proc Natl Acad Sci U S A 96:2135–2140, 1999.
22. RY Tsien, A Miyawaki. Seeing the machinery of live cells. Science 280:1954–1955, 1998.
23. A Miyawaki, J Llopis, R Heim, JM McCaffery, JA Adams, M Ikura, RY Tsien. Fluorescent indicators for $Ca^{2+}$ based on green fluorescent proteins and calmodulin. Nature 388:882–887, 1997.
24. DS Bush. Calcium regulation in plant cells and its role in signaling. Annu Rev Plant Physiol Plant Mol Biol 46:95–122, 1995.
25. S Gilroy, RL Jones. Gibberellic acid and abscisic acid coordinately regulate cytoplasmic calcium and secretory activity in barley aleurone protoplasts. Proc Natl Acad Sci U S A 89:3591–3595, 1992.

26.  MR Knight, AK Campbell, SM Smith, AJ Trewavas. Transgenic plant aequorin reports the effects of touch and cold-shock and elicitors on cytoplasmic calcium. Nature 352:524–526, 1991.

27.  MR Knight, AK Campbell, SM Smith, AJ Trewavas. Recombinant aequorin as a probe for cytosolic free $Ca^{2+}$ in *Escherichia coli*. FEBS Lett 282:405–408, 1991.

28.  MR Knight, ND Read, AK Campbell, AJ Trewawas. Imaging calcium dynamics in living plants using semi-synthetic recombinant aequorins. J Cell Biol 121:83–90, 1993.

29.  SA Harding, SH Oh, DM Roberts. Transgenic tobacco expressing a foreign calmodulin gene shows an enhanced production of active oxygen species. EMBO J 16:1137–1144, 1997.

30.  CH Johnson, MR Knight, T Kondo, P Masson, J Sedbrook, A Haley, A Trewavas. Circadian oscillations of cytosolic and chloroplastic free calcium in plants. Science 269:1863–1865, 1995.

31.  H Knight, AJ Trewavas, MR Knight. Calcium signalling in *Arabidopsis thaliana* responding to drought and salinity. Plant J 12:1067–1078, 1997.

32.  AK Campbell, AJ Trewavas, MR Knight. Calcium imaging shows differential sensitivity to cooling and communication in luminous transgenic plants. Cell Calcium 19:211–218, 1996.

33.  MF Thomashow. Role of cold-responsive genes in plant freezing tolerance. Plant Physiol 118:1–8, 1998.

34.  H Knight, EL Veale, GJ Warren, MR Knight. The *sfr6* mutation in *Arabidopsis* suppresses low-temperature induction of genes dependent on the *CRT/DRE* sequence motif. Plant Cell 11:875–886, 1999.

35.  I Johansson, C Larsson, B Ek, P Kjellbom. The major integral proteins of spinach leaf plasma membranes are putative aquaporins and are phosphorylated in response to $Ca^{2+}$ and apoplastic water potential. Plant Cell 8:1181–1191, 1996.

36.  H Knight, AJ Trewavas, MR Knight. Cold calcium signaling in *Arabidopsis* involves two cellular pools and a change in calcium signature after acclimation. Plant Cell 8:489–503, 1996.

37.  MR Knight, SM Smith, AJ Trewavas. Wind-induced plant motion immediately increases cytosolic calcium. Proc Natl Acad Sci U S A 89:4967–4971, 1992.

38.  C Thonat, N Boyer, C Penel, JC Courduroux, T Gaspar. Effects of mechanical stimulation on localization of annexin-like proteins in *Bryonia dioica* internodes. Protoplasma 176:133–137, 1993.

39.  AH van der Luit, C Olivari, A Haley, MR Knight, AJ Trewavas. Distinct calcium signaling pathways regulate calmodulin gene expression in tobacco. Plant Physiol 121:705–714, 1999.

40.  AJ Russell, MR Knight, DJ Cove, CD Knight, AJ Trewavas, TL Wang. The moss, *Physcomitrella patens*, transformed with apoaequorin cDNA responds to cold shock, mechanical perturbation and pH with transient increases in cytoplasmic calcium. Transgenic Res 5:167–170, 1996.

41.  ML Sistrunk, DM Antosiewicz, MM Purugganan, J Braam. Arabidopsis *TCH3* encodes a novel calcium-binding protein and shows environmentally induced and tissue specific regulation. Plant Cell 1553–1565, 1994.

42.  DH Polisensky, J Braam. Cold-shock regulation of the *Arabidopsis TCH* genes and the effects of modulating intracellular calcium levels. Plant Physiol 111:1271–1279, 1996.

43.  RJ Cyr. Calcium/calmodulin affects microtubule stability in lysed protoplasts. J Cell Sci 100:311–317, 1991.

44.  RJ Cyr, BA Palevitz. Organization of cortical microtubules in plant cells. Curr Opin Cell Biol 7:65–71, 1995.

45.  AE Biyaseheva, YG Molotkovskii, LK Mamonov. Increase of free $Ca^{2+}$ in the cytosol of plant protoplast in response to heat stress as related to $Ca^{2+}$ homeostasis. Rus Plant Physiol 40:540–544, 1993.

46.  M Gong, AH van der Luit, MR Knight, AJ Trewavas. Heat-shock-induced changes in intracellular $Ca^{2+}$ level in tobacco seedlings in relation to thermotolerance. Plant Physiol 116:429–437, 1998.

47.  AR Taylor, NFH Manison, C Fernandez, J Wood, C Brownlee. Spatial organization of calcium signaling involved in cell volume control in the *Fucus* rhizoid. Plant Cell 8:2015–2031, 1996.

48.  AR Taylor, NFH Manison, C Brownlee. Regulation of channel activity underlying cell volume and polarity signals in *Fucus*. J Exp Bot 48:579–588, 1997.

49.  K Takahashi, M Isobe, S Muto. An increase in cytosolic calcium ion concentration precedes hypoosmotic shock–induced activation of protein kinases in tobacco suspension culture cells. FEBS Lett 401:202–206, 1997.

50.  K Takahashi, M Isobe, S Muto. Mastoparan induces an increase in cytosolic calcium ion concentration and subsequent activation of protein kinases in tobacco suspension culture cells. Biochim Biophys Acta 1401:339–346, 1998.

51.  PS Shacklock, ND Read, AJ Trewavas. Cytosolic free calcium mediates red-light induced photomorphogenesis. Nature 358:753–755, 1992.

52.  MR McAinsh, C Brownlee, AM Hetherington. Abscisic acid–induced elevation of guard cell cytosolic $Ca^{2+}$ precedes stomatal closure. Nature 343:186–188, 1990.

53.  MR McAinsh, C Brownlee, AM Hetherington. Visualizing changes in cytosolic $Ca^{2+}$ during the response of stomatal guard cells to abscisic acid. Plant Cell 4:1113–1122, 1992.

54.  S Gilroy, MD Fricker, ND Read, AJ Trewavas. Role of calcium in signal transduction of *Commelina* guard cells. Plant Cell 3:333, 1991.

55.  M Wang, B Van Duijn, AW Schram. Abscisic acid induces a cytosolic calcium decreases in barley aleurone protoplasts. FEBS Lett 278:69–74, 1991.

56.  DS Bush, RL Jones. Cytoplasmic calcium and α-amylase secretion from barley aleurone protoplasts. Eur J Cell Biol 46:466–469, 1988.

57.  CA Gehring, HR Irving, RW Parish. Effects of auxin and abscisic acid on cytosolic calcium pH in plant cells. Proc Natl Acad Sci U S A 87:9645, 1990.

58.  J Leon, E Rojo, E Titarenko, JJ Sanchez-Serrano. Jasmonic acid–dependent and –independent wound signal transduction pathways are differentially regulated by $Ca^{2+}$/calmodulin in *Arabidopsis thaliana*. Mol Gen Genet 258:412–419, 1998.

59.  J Lynch, GR Cramer, A Laüchli. Salinity reduces membrane-associated calcium in corn root protoplasts. Plant Physiol 83:390–394, 1987.

60.  J Lynch, A Laüchli. Salinity affects intracellular calcium in corn root protoplasts. Plant Physiol 87:351–356, 1988.

61.  J Lynch, VW Polito, A Läuchli. Salinity stress increases cytoplasmic $Ca^{2+}$ activity in maize root protoplasts. Plant Physiol 90:1271–1274, 1989.

62.  DS Bush. Effects of gibberellic acid and environmental factors on cytosolic calcium in wheat aleurone cells. Planta 199:566–574, 1996.

63.  J Liu, JK Zhu. A calcium sensor homolog required for plant salt tolerance. Science 280:1943–1945, 1998.

64.  H Clayton, MR Knight, H Knight, MR McAinsh, AM Hetherington. Dissection of the ozone-induced calcium signature. Plant J 17:575–579, 1999.

65.  AH Price, A Taylor, SJ Ripley, A Griffiths, AJ Trewavas, MR Knight. Oxidative signals in tobacco increase cytosolic calcium. Plant Cell 6:1301–1310, 1994.

66.  A Levine, RI Pennell, ME Alvarez, R Palmer, C Lamb. Calcium-mediated apoptosis in a plant hypersensitive disease resistance response. Curr Biol 6:427–437, 1996.

67.  A Levine, R Tenhaken, R Dixon, C Lamb. $H_2O_2$ from the oxidative burst orchestrates the plant hypersensitive disease resistance response. Cell 79:583–593, 1994.

68.  CJ Lamb, RA Dixon. The oxidative burst in plant disease resistance. Annu Rev Plant Physiol Plant Mol Biol 48:251–275, 1997.

69.  S Zimmermann, T Nurnberger, J-M Frachisse, W Wirtz, J Guren, R Hedrich, D Scheel. Receptor-mediated activation of a plant $Ca^{2+}$-permeable ion channel involved in pathogen defense. Proc Natl Acad Sci U S A 94:2751–2755, 1997.

70.  DW Ehrhardt, R Wais, SR Long. Calcium spiking in plant root hairs responding to *Rhizobium* nodulation signals. Cell 85:673–681, 1996.

71.  E Perez-Prat, ML Narasimhan, ML Binzel, MA Botella, Z Chen, V Valpuesta, RA Bressan, PM Hasegawa. Induction of a putative calcium-ATPase mRNA in sodium chloride adapted cells. Plant Physiol 100:1471–1478, 1992.

72.  MJ Jaffe. Thigmomorphogenesis: the response of plant growth and development to mechanical stimulation. Planta 114:143–157, 1973.

73.  BD Lewis, G Karlin-Neumann, RW Davis, EP Spalding. $Ca^{(2+)}$-activated anion channels and membrane depolarizations induced by blue light and cold in *Arabidopsis* seedlings. Plant Physiol 114:1327–1334, 1997.

74.  C Mazars, L Thion, P Thuleau, A Graziana, MR Knight, M Moreau, R Ranjeva. Organization of cytoskeleton controls the changes in cytosolic calcium of cold-shocked *Nicotiana plumbaginifolia* protoplasts. Cell Calcium 22:413–420, 1997.

75.  JP Ding, BG Pickard. Mechanosensory calcium-selective cation channels in epidermal cells. Plant J 3:83–110, 1993.

76.  JW Ding, R Andersson, B Hultberg, V Soltesz, S Bengmark. Modification of reticuloendothelial function by muramyl dipeptide–encapsulated liposomes in jaundiced rats treated with biliary decompression. Scand J Gastroenterol 28:53–62, 1993.

77.  N Verbruggen, R Villarroel, M Van Montagu. Osmoregulation of a pyrroline-5-carboxylate reductase gene in *Arabidopsis thaliana*. Plant Physiol 103:771–781, 1993.

78.  S Bhattacharjee, AK Mukherjee. Divalent calcium in heat and cold stress induced accumulation of proline in *Amaranthus lividus* Linn. Geobios 22:203–207, 1995.

79.  B De, S Bhattacharjee, AK Mukherjee. Short-term heat shock and cold shock induced proline accumulation in relation to calcium involvement in *Lycopersicon esculentum* (Mill.) cultured cells and seedlings. Indian J Plant Physiol 1:32–35, 1996.

80.  E Cholewa, AJ Cholewinski, BJ Shelp, WA Snedden, AW Bown. Cold-shock stimulated gamma-amminobutyric acid synthesis is mediated by an increase in cytosolic $Ca^{2+}$ not by an increase in cytosolic $H^+$. Can J Bot 75:375–382, 1997.

81.  A Pestenacz, L Erdei. Calcium-dependent protein kinases in maize and sorghum induced by polyethylene glycol. Physiol Plant 97:360–364, 1996.

82.  T Hirayama, C Ohto, T Mizoguchi, K Shinozaki. A gene encoding a phosphatidylinositol-specific phospholipase C is induced by dehydration and salt stress in *Arabidopsis thaliana*. Proc Natl Acad Sci U S A 92:3903–3907, 1995.

83.  A Leonardi, DS Heimovaara, M Wang. Differential involvement of abscisic acid in dehydration and osmotic stress in rice cell suspension. Physiol Plant 93:31–37, 1995.

84.  M Tazawa, OT Yoko, T Mimura, M Kikuyama. Intracellular mobilization of $Ca^{2+}$ and inhibition of cytoplasmic streaming induced by transcellular osmosis in internodal cells of *Nitella flexilis*. Plant Cell Physiol 35:63–72, 1994.

85. M Tazawa, K Shimada, M Kikuyama. Cytoplasmic hydration triggers a transient increase in cytoplasmic $Ca^{2+}$ concentration in *Nitella flexilis*. Plant Cell Physiol 36:335–340, 1995.

86. JH Ko, SH Lee. Role of calcium in the osmoregulation under salt stress in *Dunaliella salina*. J Plant Biol 38:243–250, 1995.

87. Y Osaki, N Iwasaki. Injection of a $Ca^{2+}$-chelating agent into the cytoplasm retards the process of turgor regulation upon hypotonic treatment in the alga, *Lamprothamnium*. Plant Cell Physiol 32:185–194, 1991.

88. JM Christie, GL Jenkins. Distinct UV-B and UV-A/blue light signal transduction pathways induce chalcone synthase gene expression in *Arabidopsis* cells. Plant Cell 8:1555–1567, 1996.

89. JC Long, GI Jenkins. Involvement of plasma membrane redox activity and calcium homeostasis in the UV-B and UV-A/blue light induction of gene expression in *Arabidopsis*. Plant Cell 10:2077–2086, 1998.

90. C Bowler, N-H Chua. Emerging themes of plant signal transduction. Plant Cell 6:1529–1541, 1994.

91. G Neuhaus, C Bowler, R Kern, N-H Chua. Calcium/calmodulin-dependent and -independent phytochrome signal transduction pathways. Cell 73:937–952, 1993.

92. M Gong, Y Li, X Dai, M Tian, ZG Li. Involvement of calcium and calmodulin in the acquisition of heat-shock induced thermotolerance in maize seedlings. J Plant Physiol 24:371–379, 1997.

93. M Gong, SN Chen, YQ Song, ZG Li. Effect of calcium and calmodulin on intrinsic heat tolerance in relation to antioxidant systems in maize seedlings. Aust J Plant Physiol 24:371–379, 1997.

94. Y Erner, MJ Jaffe. Thigmomorphogenesis: membrane lipid and protein changes in bean plants as affected by mechanical perturbation and Etherel. Plant Physiol 58:187–203, 1983.

95. AJ Trewavas, MR Knight. Mechanical signaling, calcium and plant form. Plant Mol Biol 26:1329–1341, 1994.

96. RS Jones, CA Mitchell. Calcium ion involvement in growth inhibition of mechanically stressed soybean (*Glycine max*) seedlings. Physiol Plant 76:598–602, 1989.

97. NA Campbell, WW Thomson. Effects of lanthanum and ethylenediaminetetraacetate on leaf movements of *Mimosa*. Plant Physiol 60:635–639, 1977.

98. B Millet, BG Pickard. Gadolinium ion is an inhibitor suitable for testing the putative role of stretch-activated ion channels in geotrophism and thigmotropism. Biophys J 53:115–121, 1988.

99. H Liss, C Bockelmann, N Werner, H Fromm, EW Weiler. Identification and purification of the calcium-regulated $Ca^{2+}$-ATPase from the endoplasmic reticulum of a higher plant mechanoreceptor organ. Physiol Plant 102:561–572, 1998.

100. J Braam, RW Davis. Rain-, wind-, and touch-induced expression of calmodulin and calmodulin-related genes in *Arabidopsis*. Cell 60:357–364, 1990.

101. SA Oh, JM Kwak, IC Kwun, HG Nam. Rapid and transient induction of calmodulin-encoding gene(s) of *Brassica napus* by a touch stimulus. Plant Cell Rep 15:586–590, 1996.

102. JP Galaud, JJ Lareyre, N Boyer. Isolation and sequence analysis of *Bryonica* calmodulin after mechanical perturbation. Plant Mol Biol 23:839–946, 1993.

103. D Takezawa, ZH Liu, G An, BW Poovaiah. Calmodulin gene family in potato: developmental and touch-induced expression of the mRNA encoding a novel isoform. Plant Mol Biol 27:693–703, 1995.

104. N Depege, C Thonat, C Coutand, JL Julien, N Boyer. Morphological responses and molecular modifications in tomato plants after mechanical stimulation. Plant Cell Physiol 38:1127–1134, 1997.

105. JR Botella, RN Arteca. Differential expression of two calmodulin genes in response to physical and chemical stimuli. Plant Mol Biol 24:757–766, 1994.

106. RE Zielinski. Calmodulin and calmodulin-binding proteins in plants. Annu Rev Plant Physiol Plant Mol Biol 49:697–725, 1998.

107. T Ito, M Hirano, K Akama, Y Shimura, K Okada. Touch-inducible genes for calmodulin and a calmodulin-related protein are located in tandem on a chromosome of *Arabidopsis thaliana*. Plant Cell Physiol 36:1369–1373, 1995.

108. IY Perera, RE Zielinski. Structure and expression of the *Arabidopsis* CaM-3 calmodulin gene. Plant Mol Biol 19:649–664, 1992.

109. W Xu, MM Purugganan, DH Polisensky, DM Antosiewicz, SC Fry, J Braam. *Arabidopsis* TCH4, regulated by hormones and the environment, encodes a xyloglucan endotransglycosylase. Plant Cell 7:1555–1567, 1995.

110. DM Antosiewicz, DH Polisensky, J Braam. Cellular localization of the $Ca^{2+}$ binding TCH3 protein of *Arabidopsis*. Plant J 8:623–636, 1995.

111. NA Durso, RJ Cyr. A calmodulin-sensitive interaction between microtubules and a higher plant homolog of elongation factor-1 alpha. Plant Cell 6:893–905, 1994.

112. A Haley, AJ Russell, N Wood, AC Allan, M Knight, AK Campbell, AJ Trewavas. Effects of mechanical signaling on plant cell cytosolic calcium. Proc Natl Acad Sci U S A 92:4124–4128, 1995.

113. V Legue, E Blancaflor, C Wymer, G Perbal, D Fantin, S Gilroy. Cytoplasmic free $Ca^{2+}$ in *Arabidopsis* roots changes in response to touch but not gravity. Plant Physiol 114:789–800, 1997.

114. B Klusener, G Boheim, H Liss, J Engelberth, EW Weiler. Gadolinium-sensitive, voltage-dependent calcium release channels in the endoplasmic reticulum of a higher plant mechanoreceptor organ. EMBO J 14:2708–2714, 1995.

115. JE Werner, RR Finkelstein. *Arabidopsis* mutants with reduced response to NaCl and osmotic stress. Physiol Plant 93:659–666, 1995.

116. JM Pardo, MP Reddy, S Yang, A Maggio, GH Huh, T Matsumoto, MA Coca, M Paino-D'Urzo, H Koiwa, DJ Yun, AA Watad, RA Bressan, PM Hasegawa. Stress signaling through $Ca^{2+}$/calmodulin-dependent protein phosphatase calcineurin mediates salt adaptation in plants. Proc Natl Acad Sci U S A 95:9681–9686, 1998.

117. J Liu, JK Zhu. An *Arabidopsis* mutant that requires increased calcium for potassium nutrition and salt tolerance. Proc Natl Acad Sci U S A 94:14960–14964, 1997.

118. LE Wimmers, NN Ewing, AB Bennett. Higher plant $Ca^{(2+)}$-ATPase: primary structure and regulation of mRNA abundance by salt. Proc Natl Acad Sci U S A 89:9205–9209, 1992.

119. HJ Jang, KT Pih, SG Kang, JH Lim, JB Jin, HL Piao, I Hwang. Molecular cloning of a novel $Ca^{2+}$-binding protein that is induced by NaCl stress. Plant Mol Biol 37:839–847, 1998.

120. T Urao, T Katagiri, T Mizoguchi, K Yamaguchi-Shinozaki, N Hayashida, K Shinozaki. Two genes that encode $Ca^{(2+)}$-dependent protein kinases are induced by drought and high-salt stresses in *Arabidopsis thaliana*. Mol Gen Genet 244:331–340, 1994.

121. JR Botella, JM Arteca, M Somodevilla, RN Arteca. Calcium-dependent protein kinase gene expression in response to physical and chemical stimuli in mungbean (*Vigna radiata*). Plant Mol Biol 30:1129–1137, 1996.

122. RJ Reid, M Tester, FA Smith. Effects of salinity and turgor on calcium influx in *Chara*. Plant Cell Environ 16:547–554, 1993.

123. D Bittisnish, D Robinson, M Whitecross. Membrane associated and intracellular free calcium levels in root cells under NaCl stress. In: Plant Membrane Transport: The Current Position. Proceedings of the Eighth International Workshop on Plant Membrane Transport, 1989, pp 681–682.

124. D Inze, M Van Montagu. Oxidative stress in plants. Curr Opin Biotechnol 6:153–158, 1995.

125. DJ Kliebenstein, RA Monde, RL Last. Superoxide dismutase in *Arabidopsis*: an eclectic enzyme family with disparate regulation and protein localization. Plant Physiol 118:637–650, 1998.

126. TK Prasad, MD Anderson, BA Martin, CR Stewart. Evidence for chilling-induced oxidative stress in maize seedlings and a regulatory role for hydrogen peroxide. Plant Cell 6:65–74, 1994.

127. FJ Castillo, RL Heath. Calcium ion transport in membrane vesicles from pinto bean leaves and its alteration after ozone exposure. Plant Physiol 93:1504–1510, 1990.

128. MR McAinsh, H Clayton, TA Mansfield, AM Hetherington. Changes in stomatal behavior and guard cell cytoplasmic free calcium in response to oxidative stress. Plant Physiol 111:1031–1042, 1996.

129. CC Subbaiah, DS Bush, MM Sachs. Elevation of cytosolic calcium precedes anoxic gene expression in maize suspension-cultured cells. Plant Cell 6:1747–1762, 1994.

130. CC Subbaiah, J Zhang, MM Sachs. Involvement of intracellular calcium in anaerobic gene expression and survival of maize seedlings. Plant Physiol 105:369–376, 1994.

131. CC Subbaiah, DS Bush, MM Sachs. Mitochondrial contribution to the anoxic $Ca^{2+}$ signal in maize suspension-cultured cells. Plant Physiol 118:759–771, 1998.

132. JC Seedbrook, PJ Kronebusch, GG Borisy, AJ Trewavas, PH Masson. Transgenic aequorin reveals organ-specific cytosolic $Ca^{2+}$ responses to anoxia in *Arabidopsis thaliana* seedlings. Plant Physiol 111:243–257, 1996.

133. PJ Jackson, PJ Unkefer, E Delhaize, LV Robinson. Mechanism of trace metal tolerance in plants. In: Environmental Injury to Plants, San Diego, 1990, pp 231–255.

134. LV Kochian. Cellular mechanisms of aluminium toxicity and resistance in plants. Annu Rev Plant Physiol Plant Mol Biol 46:303–311, 1995.

135. S Lindberg, H Strid. Aluminium induces rapid changes in cytosolic pH and free calcium and potassium concentrations in root protoplasts of wheat (*Triticum aestivum*). Physiol Plant 99:405–414, 1997.

136. DL Jones, S Gilroy, PB Larsen, SH Howell, LV Kochian. Effect of aluminum on cytoplasmic $Ca^{2+}$ homeostasis in root hairs of *Arabidopsis thaliana* (L.). Planta 206:378–387, 1998.

137. DL Jones, LV Kochian, S Gilory. Aluminium induced a decrease in cytosolic calcium concentration in BY-2 tobacco cell cultures. Plant Physiol 116:81–89, 1998.

138. C Plieth, B Sattelmacher, UP Hansen, MR Knight. Low-pH-mediated elevations in cytosolic calcium are inhibited by aluminium: a potential mechanism for aluminium toxicity. Plant J 18:643–650, 1999.

139. PW Morgan. Effects of abiotic stresses on plant hormone system. In: RG Alscher, ed. Stress Responses in Plants: Adaptation and Acclimation Mechanisms. New York: Wiley-Liss, 1990, pp 113–146.

140. A Kuo, S Cappelluti, M Cervantes-Cervantes, M Rodriguez, DS Bush. Okadaic acid, a protein phosphatase inhibitor, blocks calcium changes, gene expression, and cell death induced by gibberellin in wheat aleurone cells. Plant Cell 8:259–269, 1996.

141. JI Schroeder, S Hagiwara. Repetitive increases in cytosolic $Ca^{2+}$ of guard cells by abscisic acid activation of nonselective $Ca^{2+}$-permeable channels. Proc Natl Acad Sci U S A 87:9305–9309, 1990.

142. T Hildmann, M Ebneth, H Pena-Cortes, JJ Sanchez-Serrano, L Willmitzer, S Prat. General roles of abscisic and jasmonic acids in gene activation as a result of mechanical wounding. Plant Cell 4:1157–1170, 1992.

143. M Bach, J-P Schnitzler, HU Seitz. Elicitor-induced changes in $Ca^{2+}$ influx, $K^+$ efflux, and 4-hydroxybenzoic acid synthesis in protoplasts of *Dacus carota* L. Plant Physiol 103:407–412, 1993.

144. U Vogeli, R Vogeli-Lange, J Chappell. Inhibition of phtoalexin biosynthesis in elicitor-treated tobacco cell-suspension cultures by calcium/calmodulin antagonists. Plant Physiol 100:1369–1376, 1992.

145. SJ Martin, DR Green, TG Cotter. Dicing with death: dissecting the components of the apoptosis machinery. Trends Biochem Sci 19:26–30, 1994.

146.  H Xu, MC Heath. Role of calcium in signal transduction during the hypersensitive response caused by basid-iospore-derived infection of the cowpea rust fungus. Plant Cell 10:585–598, 1998.

147.  DC Boyes, J Nam, JL Dangl. The *Arabidopsis thaliana* RPM1 disease resistance gene product is a peripheral plasma membrane protein that is degraded coincident with the hypersensitive response. Proc Natl Acad Sci USA 95:15849–15854, 1998.

148.  M Grant, J Mansfield. Early events in host-pathogen interactions. Curr Opin Plant Biol 2:312–319, 1999.

149.  T Romeis, P Piedras, S Zhang, DF Klessig, H Hirt, JD Jones. Rapid *Avr9*- and *Cf-9*-dependent activation of MAP kinases in tobacco cell cultures and leaves: convergence of resistance gene, elicitor, wound, and salicy-late responses. Plant Cell 11:273–287, 1999.

150.  T Keller, HG Damude, D Werner, P Doerner, RA Dixon, C Lamb. A plant homolog of the neutrophil NADPH oxidase gp91$^{phox}$ subunit gene encodes a plasma membrane protein with $Ca^{2+}$ binding motifs. Plant Cell 10:255–266, 1998.

151.  DR Bergey, CA Ryan. Wound- and systemin-inducible calmodulin gene expression in tomato leaves. Plant Mol Biol 40:815–823, 1999.

152.  DR Bergey, M Orozco-Cardenas, DS de Moura, CA Ryan. A wound- and systemin-inducible polygalactur-onase in tomato leaves. Proc Natl Acad Sci U S A 96:1756–1760, 1999.

153.  A Schaller, C Oecking. Modulation of plasma membrane $H^+$-ATPase activity differentially activates wound and pathogen defense responses in tomato plants. Plant Cell 11:263–272, 1999.

154.  D Flego, M Pirhonen, H Saarilahti, TK Palva, ET Palva. Control of virulence gene expression by plant calcium in the phytopathogen *Erwinia carotovora*. Mol Microbiol 25:831–838, 1997.

155.  RA Dixon, MJ Harrison, CJ Lamb. Early events in the activation of plant defense responses. Annu Rev Phy-topathol 32:479–501, 1994.

156.  T Nurnberger, D Nennstiel, T Jabs, WR Sacks, K Hahlbrock, D Scheel. High affinity binding of a fungal oligopeptide elicitor to parsley plasma membranes triggers multiple defense responses. Cell 78:449–460, 1994.

157.  T Nurnberger, C Colling, K Hahlbrock, T Jabs, A Renelt, WR Sacks, D Scheel. Perception and transduction of an elicitor signal in cultured parsley cells. Biochem Soc Symp 60:173–182, 1994.

158.  D Scheel. Resistance response physiology and signal transduction. Curr Opin Plant Biol 1:305–310, 1998.

159.  IE Somssich, K Hahlbrock. Pathogen defense in plants—a paradigm of biological complexity. Trends Plant Sci 3:86–90, 1998.

160.  T Jabs, M Tschöpe, C Colling, K Hahlbrock, D Scheel. Elicitor-stimulated ion fluxes and $O_2^-$ from the oxida-tive burst are essential components in triggering defense gene activation and phytoalexin synthesis in parsley. Proc Natl Acad Sci U S A 94:4800–4805, 1997.

161.  P Piedras, KE Hammond-Kosack, K Harrison, JDG Jones. Rapid *Cf-9*- and Avr-9-dependent, production of ac-tive oxygen species in tobacco suspension cultures. Mol Plant Microbe Interact 11:1155–1166, 1998.

162.  MM Atikson, SL Midland, JJ Sims, NT Keen. Syringolide 1 triggers $Ca^{2+}$ influx, $K^+$ influx, and extracellular alkalization in soybean cells carrying the disease-resistance gene *RPg4*. Plant Physiol 112:297–302, 1996.

163.  A Gelli, VJ Higgins, E Blumwald. Activation of plant plasma membrane $Ca^{2+}$ permeable channels by race-specific fungal elicitors. Plant Physiol 113:269–279, 1997.

164.  E Blumwald, GS Aharon, BC Lam. Early signal transduction pathways in plant-pathogen interactions. Trends Plant Sci 3:342–346, 1998.

165.  TN Davis. What is new with calcium? Cell 71:557–564, 1992.

166.  ND Moncrief, RH Krestinger, M Goodman. Evolution of EF-hand calcium-modulated proteins. I. Relationship based on amino acid sequences. J Mol Evol 30:522–562, 1990.

167.  WA Snedden, H Fromm. Calmodulin, calmodulin-related proteins and plant responses to the environment. Trends Plant Sci 3:299–304, 1998.

168.  RJ Colbran, TR Soderling. Calcium/calmodulin-dependent protein kinase II. Curr Top Cell Regul 31:181, 1990.

169.  AC Nairn, HC Hemmings, P Greengard. Protein kinases in the brain. Annu Rev Biochem 54:931–976, 1985.

170.  Y Nishizuka. The molecular heterogeneity of a protein kinase C and its implication in cellular regulation. Na-ture 334:661, 1988.

171.  SK Sopory, M Munshi. Protein kinases and phosphatases and their role in cellular signaling in plants. CRC Cri Rev Plant Sci 17:245–318, 1998.

172.  JF Harper, MR Sussman, GE Schaller, C Putnam-Evans, H Charbonneau, AC Harmon. A calcium-dependent protein kinase with a regulatory domain similar to calmodulin. Science 252:951–954, 1991.

173.  B Watillon, R Kettmann, P Boxus, A Burny. A calcium/calmodulin-binding serine/threonine protein kinase ho-mologous to the mammalian type II calcium/calmodulin-dependent protein kinase is expressed in plant cells. Plant Physiol 101:1381–1384, 1993.

174.  S Patil, D Takezawa, BW Poovaiah. Chimeric plant calcium/calmodulin-dependent protein kinase gene with a neural visinin-like calcium-binding domain. Proc Natl Acad Sci U S A 92:4897–4901, 1995.

175.  CR Lin, MS Kapiloff, S Durgerian, K Tatemoto, AF Russo, P Hanson, H Schulman, MG Rosenfeld. Molecu-lar cloning of a brain-specific calcium/calmodulin-dependent protein kinase. Proc Natl Acad Sci U S A 84:5962–5966, 1987.

176. C Putnam-Evans, AC Harmon, MJ Cormier. Purification and characterization of a novel calcium-dependent protein kinase from soybean. Biochemistry 29:2488–2495, 1990.

177. AC Harmon, C Putnam-Evans, MJ Cormier. A calcium-dependent but calmodulin-independent protein kinase from soybean. Plant Physiol 83:830–837, 1987.

178. JF Harper, BM Binder, MR Sussman. Calcium and lipid regulation of an *Arabidopsis* protein kinase expressed in *Escherichia coli*. Biochemistry 32:3282–3290, 1993.

179. K-L Suen, JH Choi. Isolation and sequence analysis of a cDNA clone for a carrot calcium-dependent protein kinase: homology to calcium/calmodulin-dependent protein kinases and to calmodulin. Plant Mol Biol 17:581–590, 1991.

180. CWM Chan, MR Sussman. Gene disruptions of calcium-dependent protein kinases in *Arabidopsis thaliana*: towards understanding their in vivo functions. Proceedings 10th International *Arabidopsis* Meeting 6–7, 1999.

181. JS Satterlee, MR Sussman. Unusual membrane-associated protein kinases in higher plants. J Membr Biol 164:205–213, 1998.

182. GE Schaller, AC Harmon, MR Sussman. Characterization of a calcium- and lipid-dependent protein kinase associated with the plasma membrane of oat. Biochemistry 31:1721–1727, 1992.

183. BM Binder, JF Harper, MR Sussman. Characterization of an *Arabidopsis* calmodulin-like domain protein kinase purified from *Escherichia coli* using an affinity sandwich technique. Biochemistry 33:2033–2041, 1994.

184. PK Farmer, JH Choi. Calcium and phospholipid activation of a recombinant calcium-dependent protein kinase (DcCPK1) from carrot (*Daucus carota* L.). Biochim Biophys Acta 1434:6–17, 1999.

185. KD Johnson, MJ Chrispeels. Tonoplast-bound protein kinase phosphorylates tonoplast intrinsic protein. Plant Physiol 100:1787–1795, 1992.

186. CD Weaver, B Crombie, G Stacey, DM Roberts. Calcium-dependent phosphorylation of symbiosome membrane proteins from nitrogen-fixing soybean nodules. Plant Physiol 95:222–227, 1991.

187. Z-M Pie, JM Ward, JF Harper, JI Schroeder. A novel chloride channel in *Vicia faba* guard cell vacuoles activated by the serine/threonoine kinase, CDPK. EMBO J 15:6564–6574, 1996.

188. AF Monroy, RS Dhindsa. Low-temperature signal transduction: induction of cold acclimation-specific genes of alfalfa by calcium at 25°C. Plant Cell 7:321–331, 1995.

189. S Tahtiharju, V Sangwan, AF Monory, RS Dhindsa, M Borg. The induction of kin genes in cold-acclimating *Arabidopsis thaliana*: evidence of a role for calcium. Planta 203:442–447, 1997.

190. J Sheen. $Ca^{2+}$-dependent protein kinases and stress signal transduction in plants. Science 274:1900–1902, 1996.

191. B Watillon, R Kettmann, P Boxus, A Burny. Cloning and characterization of an apple (*Malus domestica* [L.] Borkh) calmodulin gene. Plant Sci 82:201–212, 1992.

192. Z Liu, M Xia, BW Poovaiah. Chimeric calcium/calmodulin-dependent protein kinase in tobacco: differential regulation by calmodulin isoforms. Plant Mol Biol 38:889–897, 1998.

193. D Takezawa, S Ramachandran, V Paranjape, BW Poovaiah. Dual regulation of a chimeric plant serine/threonine kinase by calcium and calcium/calmodulin. J Biol Chem 271:8126–8132, 1996.

194. T Xing, VJ Higgins, E Blumwald. Race-specific elicitors of *Cladosporium fulvum* promote translocation of cytosolic components of NADPH oxidase to the plasma membrane of tomato cells. Plant Cell 9:249–259, 1997.

195. AF Monroy, F Sarhan, RS Dhindsa. Cold-induced changes in freezing tolerance, protein phosphorylation, and gene expression. Plant Physiol 102:1227–1235, 1993.

196. A Dietrich, JE Mayers, K Hahlbrock. Fungal elicitor triggers rapid, transient and specific protein phosphorylation in parsley cell suspension cultures. J Biol Chem 265:6360–6368, 1990.

197. VS Reddy, KV Goud, RP Sharma, AR Reddy. Ultraviolet-B-responsive anthocyanin production in a rice cultivar is associated with a specific phase of phenylalanine ammonia lyase biosynthesis. Plant Physiol 105:1059–1066, 1994.

198. S Grabski, E Arnoys, B Bush, M Schindler. Regulation of actin tension in plant cells by kinases and phosphatases. Plant Physiol 116:279–290, 1998.

199. WY Cheung. Calmodulin plays a pivotal role in cellular regulation. Science 207:19–27, 1980.

200. TN Davis, MS Urdea, FR Masiarz, J Thorner. Isolation of the yeast calmodulin gene: calmodulin is an essential protein. Cell 47:423–431, 1986.

201. B Liao, MC Gawienowski, RE Zielinski. Differential stimulation of NAD kinase and binding of peptide substrates by wild-type and mutant plant calmodulin isoforms. Arch Biochem Biophys 327:53–60, 1996.

202. V Ling, RE Zielinski. Isolation of an *Arabidopsis* cDNA sequence encoding a 22 kDa calcium-binding protein (CaBP-22) related to calmodulin. Plant Mol Biol 22:207–214, 1993.

203. M Rodriguez-Concepcion, S Yalovsky, M Zik, H Fromm, W Gruissem. The prenylation status of a novel plant calmodulin directs plasma membrane or nuclear localization of the protein. EMBO J 18:1996–2007, 1999.

204. J Kudla, Q Xu, K Harter, W Gruissem, S Luan. Genes for calcineurin B–like proteins in *Arabidopsis* are differentially regulated by stress signals. Proc Natl Acad Sci U S A 96:4718–4123, 1999.

205. G Frandsen, F Muller-Uri, M Nielsen, J Mundy, K Skriver. Novel plant $Ca^{(2+)}$-binding protein expressed in response to abscisic acid and osmotic stress. J Biol Chem 271:343–348, 1996.

206. JL Jakobek, JA Smith-Becker, PB Lindgren. A bean cDNA expressed during a hypersensitive reaction encodes a putative calcium binding protein, Mol Plant Microbe Interact 12:712–719, 1999.

207. YS Babu, CE Bugg, WJ Cook. Structure of calmodulin refined at 2.2 Å resolution. J Mol Biol 204:191–204, 1988.
208. A Crivici, M Ikura. Molecular and structural basis of target recognition by calmodulin. Annu Rev Biophys Biomol Struct 24:85–116, 1995.
209. KT O'Neil, WF DeGrado. How calmodulin binds its targets: sequence-independent recognition of amphiphilic α-helices. Trends Biochem Sci 15:59–64, 1990.
210. AR Means, MFA VanBerkum, I Bagchi, KP Lu, CD Rasmussen. Regulatory functions of calmodulin. Pharmacol Ther 50:255–270, 1991.
211. J Braam, ML Sistrunk, DH Polisensky, W Xu, MM Purugganan, DM Antosiewicz, P Campbell, KA Johnson. Plant responses to environmental stress: regulation and functions of the *Arabidopsis TCH* genes. Planta 203:S35–S41, 1997.
212. J Braam. Regulated expression of the calmodulin-related *TCH* genes in cultured *Arabidopsis* cells: induction by calcium and heat shock. Proc Natl Acad Sci U S A 89:3213–3216, 1992.
213. D Bartling, H Butler, EW Weiler. *Arabidopsis thaliana* cDNA encoding a novel member of the EF-hand superfamily of calcium-binding proteins. Plant Physiol 102:1059, 1993.
214. PK Jena, ASN Reddy, BW Poovaiah. Molecular cloning and sequencing of a cDNA for plant calmodulin: signal-induced changes in the expression of calmodulin. Proc Natl Acad Sci U S A 86:3644–3649, 1989.
215. KP Lu, AR Means. Regulation of the cell cycle by calcium and calmodulin. Endocr Rev 14:40–58, 1993.
216. V Ling, I Perea, RE Zielinski. Primary structures of *Arabidopsis* calmodulin isoforms deduced from the sequences of cDNA clones. Plant Physiol 96:1196–1212, 1991.
217. MC Gawienowski, D Szymanski, IY Perera, RE Zielinski. Calmodulin isoforms in *Arabidopsis* encoded by multiple divergent mRNAs. Plant Mol Biol 22:215–225, 1993.
218. VS Reddy, F Safadi, RE Zielinski, ASN Reddy. Interaction of a kinesin-like protein with calmodulin isoforms from *Arabidopsis*. J Biol Chem 274:31727–31733, 1999.
219. SH Lee, JC Kim, MS Lee, WD Heo, HY Seo, HW Yoon, JC Hong, SY Lee, JD Bahk, I Hwang, J Cho. Identification of a novel divergent calmodulin isoform from soybean which has differential ability to activate calmodulin-dependent enzymes. J Biol Chem 270:21806–21812, 1995.
220. WD Heo, SH Lee, MC Kim, JC Kim, WS Chung, HJ Chun, KJ Lee, CY Park, HC Park, JY Choi, MJ Cho. Involvement of specific calmodulin isoforms in salicylic acid–independent activation of plant disease resistance responses. Proc Natl Acad Sci U S A 19:766–771, 1999.
221. Y Yang, J Dowling, QC Yu, P Kouklis, DW Cleveland, E Fuchs. An essential cytoskeletal linker protein connecting actin microfilaments to intermediate filaments. Cell 86:655–665, 1996.
222. T Yang, S Lev-Yadun, M Feldman, H Fromm. Developmentally regulated organ-, tissue-, and cell-specific expression of calmodulin genes in common wheat. Plant Mol Biol 37:109–120, 1998.
223. AR Rhoads, F Friedberg. Sequence motifs for calmodulin recognition. FASEB J 11:331–340, 1997.
224. CB Klee. Concerted regulation of protein phosphorylation and dephosphorylation by calmodulin. Neurochem Res 16:1059–1065, 1991.
225. ASN Reddy. Calcium: silver bullet in signaling. Plant Sci 160:381–404, 2001.
226. JR Geiser, D van Tuinen, SE Brockerhoff, MM Neff, TN Davis. Can calmodulin function without binding calcium? Cell 65:949–959, 1991.
227. V Ling, SM Assmann. Cellular distribution of calmodulin and calmodulin-binding proteins in *Vicia faba* L. Plant Physiol 100:970–978, 1992.
228. S-H Oh, H-Y Steiner, DK Dougall, DM Roberts. Modulation of calmodulin levels, calmodulin methylation, and calmodulin binding proteins during carrot cell growth and embryogenesis. Arch Biochem Biophy 297:28–34, 1992.
229. JF Harper, B Hong, I Hwang, HQ Guo, R Stoddard, JF Huang, MG Palmgren, H Sze. A novel calmodulin-regulated $Ca^{2+}$-ATPase (ACA2) from *Arabidopsis* with an N-terminal autoinhibitory domain. J Biol Chem 273:1099–1106, 1998.
230. S Malmstrom, P Askerlund, MG Palmgren. A calmodulin-stimulated $Ca^{2+}$-ATPase from plant vacuolar membranes with a putative regulatory domain at its N-terminus. FEBS Lett 400:324–328, 1997.
231. HL Hsieh, CG Tong, C Thomas, SJ Roux. Light-modulated abundance of an mRNA encoding a calmodulin-regulated, chromatin-associated NTPase in pea. Plant Mol Biol 30:135–147, 1996.
232. G Baum, S Lev-Yadun, Y Fridmann, T Arazi, H Katsnelson, M Zik, H Fromm. Calmodulin binding to glutamate decarboxylase is required for regulation of glutamate and GABA metabolism and normal development in plants. EMBO J 15:2988–2996, 1996.
233. WA Snedden, N Koutsia, G Baum, H Fromm. Activation of a recombinant petunia glutamate decarboxylase by calcium/calmodulin or by a monoclonal antibody which recognizes the calmodulin binding domain. J Biol Chem 271:4148–4153, 1996.
234. RC Schuurink, SF Shartzer, A Fath, RL Jones. Characterization of a calmodulin-binding transporter from the plasma membrane of barley aleurone. Proc Natl Acad Sci U S A 95:1944–1949, 1998.
235. S Ramachandiran, D Takezawa, W Wang, BW Poovaiah. Functional domains of plant chimeric calcium/calmodulin-dependent protein kinase: regulation by autoinhibitory and visinin-like domains. J Biochem 121:984–990, 1997.

236.  ASN Reddy, F Safadi, SB Narasimhulu, M Golovkin, X Hu. A novel plant calmodulin-binding protein with a kinesin heavy chain motor domain. J Biol Chem 271:7052–7060, 1996.

237.  ASN Reddy, SB Narasimhulu, F Safadi, M Golovkin. A plant kinesin heavy chain–like protein is a calmodulin-binding protein. Plant J 10:9–21, 1996.

238.  ASN Reddy, SB Narasimhulu, IS Day. Structural organization of a gene encoding a novel calmodulin-binding kinesin-like protein from *Arabidopsis*. Gene 204:195–200, 1998.

239.  VS Reddy, ASN Reddy. A plant calmodulin-binding motor is part kinesin and part myosin. Bioinformatics 15:1055–1057, 1999.

240.  ASN Reddy. Molecular motors and their functions plants. Intl Rev Cytol Cell Biol, in press.

241.  ASN Reddy. A novel calcium/calmodulin-regulated microtubule motor protein from plants: role in trichome morphogenesis and cell division. In: SK Sopory, R Oelmuller, SC Maheswari, eds. Signal Transduction in Plants: Current Advances. New York: Kluwer Academic/Plenum, in press.

242.  W Wang, D Takezawa, SB Narasimhulu, ASN Reddy, BW Poovaiah. A novel kinesin-like protein with a calmodulin-binding domain. Plant Mol Biol 31:87–100, 1996.

243.  MC Gawienowski, RE Zielinski. Calmodulin isoforms differentially enhance the binding of cauliflower nuclear proteins and recombinant TGA3 to a region derived from the *Arabidopsis* CaM-3 promoter. Plant Cell 8:1069–1077, 1996.

244.  Veena, VS Reddy, SK Sopory. Glyoxalase I from *Brassica juncea*: molecular cloning, regulation and its overexpression confer tolerance in transgenic tobacco under stress. Plant J 17:385–395, 1999.

245.  J Espartero, I Sanchez-Aguayo, JM Pardo. Molecular characterization of glyoxalase-I from a higher plant; upregulation by stress. Plant Mol Biol 29:1223–1233, 1995.

246.  Y-T Lu, MAN Dharmasiri, HM Harrington. Characterization of a cDNA encoding a novel heat-shock protein that binds to calmodulin. Plant Physiol 108:1197–1202, 1995.

247.  S Dash, W Niemaczura, HM Harrington. Characterization of the basic amphiphilic alpha-helix calmodulin-binding domain of a 61.5 kDa tobacco calmodulin-binding protein. Biochemistry 36:2025–2029, 1997.

248.  S Muto, S Miyachi. Properties of a protein activator of NAD kinase from plants. Plant Physiol 59:55–60, 1977.

249.  JM Anderson, H Charbonneau, HP Jones, RO McCann, MJ Cormier. Characterization of the plant nicotinamide adenine dinucleotide kinase activator protein and its identification as calmodulin. Biochemistry 19:3113–3120, 1980.

250.  SH Lee, MC Kim, WD Heo, JC Kim, WS Chung, CY Park, HC Park, YH Cheong, CY Kim, SH Lee, KJ Lee, JD Bahk, SY Lee, MJ Cho. Competitive binding of calmodulin isoforms to calmodulin-binding proteins: implications for the function of calmodulin isoforms in plants. Biochim Biophys Acta 1433:56–67, 1999.

251.  H Fromm, N-H Chua. Cloning of plant cDNAs encoding calmodulin-binding proteins using [35]S-labeled recombinant calmodulin as a probe. Plant Mol Biol Rep 10:199–206, 1992.

252.  AP Fordham-Skelton, F Safadi, M Golovkin, ASN Reddy. A non-radioactive method for isolating complementary DNAs encoding calmodulin-binding proteins. Plant Mol Biol Rep 12:355–363, 1994.

253.  B Liao, RE Zielinski. Production of recombinant plant calmodulin and its use to detect calmodulin-binding proteins. Methods Cell Biol 49:487–500, 1995.

254.  H Lu, O Flores, R Weinmann, D Reinberg. The nonphosphorylated form of RNA polymerase II preferentially associates with the preinitiation complex. Proc Natl Acad Sci U S A 88:10004–10008, 1991.

255.  ASN Reddy, D Takezawa, H Fromm, BW Poovaiah. Isolation and characterization of two cDNAs that encode for calmodulin-binding proteins from corn root tips. Plant Sci 94:109–117, 1993.

256.  G Baum, Y Chen, T Arazi, H Takatsuji, H Fromm. A plant glutamate decarboxylase containing a calmodulin binding domain: cloning, sequence and functional analysis. J Biol Chem 268:19610–19617, 1993.

257.  Y Lu, HM Harrington. Isolation of tobacco cDNA clones encoding calmodulin-binding proteins and characterization of a known calmodulin-binding protein. Plant Physiol Biochem 32:413–422, 1994.

258.  T Arazi, R Sunkar, B Kaplan, H Fromm. A tobacco plasma membrane calmodulin-binding transporter confers $Ni^{2+}$ tolerance and $Pb^{2+}$ hypersensitivity in transgenic plants. Plant J 20:171–182, 1999.

259.  MG Erlander, AJ Tobin. The structural and functional heterogeneity of glutamic acid decarboxylase: a review. Neurochem Res 16:215–226, 1991.

260.  M Zik, T Arazi, WA Snedden, H Fromm. Two isoforms of glutamate decarboxylase in *Arabidopsis* are regulated by calcium/calmodulin and differ in organ distribution. Plant Mol Biol 37:967–975, 1998.

261.  N Aurisano, A Betani, R Reggiani. Involvement of calcium and calmodulin in protein and amino-acid-metabolism in rice roots under anoxia. Plant Cell Physiol 36:1525–1529, 1995.

262.  WA Snedden, T Arazi, H Fromm. Calcium/calmodulin activation of soybean glutamate decarboxylase. Plant Physiol 108:543–549, 1995.

263.  A Bown, B Shelp. The metabolism and function of γ-aminobutyric acid. Plant Physiol 115:1–5, 1997.

264.  V Satya Narayana, PM Nair. Metabolism enzymology and possible roles of 4-aminobutyric acid in higher plants. Phytochemistry 29:367–375, 1990.

265.  AI Ramputh, AW Bown. Rapid γ-aminobutyric acid synthesis and the inhibition of the growth and development of oblique banded leaf-roller larvae. Plant Physiol 111:1349–1352, 1996.

266.  W Wallace, J Secor, L Schrader. Rapid accumulation of gamma-aminobutyric acid and alanine in soybean leaves in response to an abrupt transfer to lower temperature, darkness, or mechanical manipulation. Plant Physiol 75:60–66, 1984.

267.  M Gong, Z-G Li. Calmodulin-binding proteins from *Zea mays* germs. Phytochemistry 40:1335–1339, 1995.
268.  SB Narasimhulu, Y-L Kao, ASN Reddy. Interaction of *Arabidopsis* kinesin-like calmodulin-binding protein with tubulin subunits: modulation by $Ca^{2+}$-calmodulin. Plant J 12:1139–1149, 1997.
269.  SB Narasimhulu, ASN Reddy. Characterization of microtubule binding domains in the *Arabidopsis* kinesin-like calmodulin-binding protein. Plant Cell 10:957–965, 1998.
270.  H Song, M Golovkin, ASN Reddy, SA Endow. In vitro motility of AtKCBP, a calmodulin-binding kinesin-like protein of *Arabidopsis*. Proc Natl Acad Sci U S A 94:322–327, 1997.
271.  Y-L Kao, BE Deavours, KK Phelps, R Walker, ASN Reddy. Bundling of microtubules by motor and tail domains of a kinesin-like calmodulin-binding protein from *Arabidopsis*: regulation by $Ca^{2+}$/calmodulin. Biochem Biophys Res Commun 267:201–207, 2000.
272.  J Bowser, ASN Reddy. Localization of a kinesin-like calmodulin-binding protein in dividing cells of *Arabidopsis* and tobacco. Plant J 12:1429–1438, 1997.
273.  DG Oppenheimer, MA Pollock, J Vacik, DB Szymanski, B Ericson, K Feldmann, D Marks. Essential role of a kinesin-like protein in *Arabidopsis* trichome morphogenesis. Proc Natl Acad Sci U S A 94:6261–6266, 1997.
274.  DG Oppenheimer. Genetics of plant cell shape. Curr Opin Plant Biol 1:520–524, 1998.
275.  R Deswal, SK Sopory. Glyoxalase I from *Brassica juncea* is a calmodulin stimulated protein. Biochim Biophys Acta 1450:460–467, 1999.
276.  R Deswal, SK Sopory. Purification and partial characterization of glyoxalase I from a higher plant, *B. juncea*. FEBS Lett 282:277–280, 1991.
277.  J Shi, KN Kim, O Ritz, V Albrecht, R Gupta, K Harter, S Luan, J Kudla. Novel protein kinases associated with calcineurin B–like calcium sensors in *Arabidopsis*. Plant Cell 11:2393–2406, 1999.
278.  K Meyer, MP Leube, E Grill. A protein phosphatase 2C involved in ABA signal transduction in *Arabidopsis thaliana*. Science 264:1452–1455, 1994.
279.  GG Cote, RC Crain. Biochemistry of phosphoinositides. Annu Rev Plant Physiol 44:333–356, 1993.
280.  T Hirayama, N Mitsukawa, D Shibata, K Shinozaki. *AtPLC2*, a gene encoding phosphoinositide-specific phospholipase C, is constitutively expressed in vegetative and floral tissues in *Arabidopsis thaliana*. Plant Mol Biol 34:175–180, 1997.
281.  J Kopka, C Pical, JE Gray, B Muller-Rober. Molecular and enzymatic characterization of three phosphoinositide-specific phospholipase C isoforms from potato. Plant Physiol 116:239–250, 1998.
282.  J Shi, RA Gonzales, MK Bhattacharyya. Characterization of a plasma membrane–associated phosphoinositide-specific phospholipase C from soybean. Plant J 8:381–390, 1995.
283.  F Safadi, DL Mykles, ASN Reddy. Partial purification and characterization of a $Ca^{2+}$-dependent proteinase from *Arabidopsis* roots. Arch Biochem Biophys 348:143–151, 1997.
284.  ASN Reddy, F Safadi, J Beyette, DL Mykles. Calcium-dependent proteinase activity in root cultures of *Arabidopsis*. Biochem Biophys Res Commun 199:1089–1095, 1994.
285.  T Saido, H Sorimachi, K Suzuki. Calpain: new perspectives in molecular diversity and physiological-pathological involvement. FASEB J 8:814–822, 1994.
286.  DL Mykles, DM Skinner. Calcium-dependent proteinases in crustaceans. In: RL Mellgren, T Murachi, eds. Intracellular calcium-dependent proteolysis. Boca Raton, FL: CRC Press, 1990, pp 139–154.
287.  H Sorimachi, S Kimura, K Kinbara, J Kazama, M Takahashi, H Yajima, S Ishiura, N Sasagawa, I Nonaka, H Sugita, K Maruyama, K Suzuki. Structure and physiological functions of ubiquitous and tissue-specific calpains. Adv Biophys 33:101–122, 1996.
288.  EJ Resendez, JW Attenello, A Grafsky, CS Chang, AS Lee. Calcium ionophore A23187 induces expression of glucose-regulated genes and their heterologous fusion genes. Mol Cell Biol 5:1212–1219, 1985.
289.  C Stratowa, WJ Rutter. Selective regulation of trypsin gene expression by calcium and by glucose starvation in a rat exocrine pancreas cell line. Proc Natl Acad Sci U S A 83:4292–4296, 1986.
290.  BA White, C Bancroft. Regulation of gene expression by calcium. In: Calcium and Cell Function. New York: Academic Press, 1987.
291.  PK Dash, KA Karl, MA Colicos, R Prywes, ER Kandel. cAMP response element–binding protein is activated by $Ca^{2+}$/calmodulin- as well as cAMP-dependent protein kinase. Proc Natl Acad Sci U S A 88:5061–5065, 1991.
292.  MS Kapiloff, JM Mathis, CA Nelson, CR Lin, MG Rosenfeld. Calcium/calmodulin-dependent protein kinase mediates a pathway for transcriptional regulation. Proc Natl Acad Sci U S A 88:3710–3714, 1991.
293.  BM Rayson. $[Ca^{2+}]_i$ regulates transcription rate of the $Na^+/K^+$-ATPase 1 subunit. J Biol Chem 266:21335–21338, 1991.
294.  M Sheng, MA Thompson, ME Greenberg. CREB: a $Ca^{2+}$-regulated transcription factor phosphorylated by calmodulin-dependent kinases. Science 252:1427–1430, 1991.
295.  M Wegner, Z Cao, MG Rosenfeld. Calcium-regulated phosphorylation within the leucine zipper of C/EBPβ. Science 256:370–373, 1992.
296.  AG Ryazanov, EA Shestakova, PG Natapov. Phosphorylation of elongation factor 2 by EF-2 kinase affects rate of translation. Nature 334:170–173, 1988.
297.  RZ Ryazanov, AS Spirin. Phosphorylation of elongation factor 2: a key mechanism regulating gene expression in vertebrates. New Biol 2:843–850, 1990.

298. O Nygard, A Nilsson, U Carlberg, L Nilsson, R Amons. Phosphorylation regulates the activity of the eEF-2-specific Ca$^{2+}$- and calmodulin-dependent protein kinase III. J Biol Chem 266:16425–16530, 1991.

299. MA Brostrom, X Lin, C Cade, D Gmitter, CO Brostrom. Loss of a calcium requirement for protein synthesis in pituitary cells following thermal or chemical stress. J Biol Chem 264:1638–1643, 1989.

300. CO Brostrom, KV Chin, WL Wong, C Cade, MA Brostrom. Inhibition of translational initiation in eukaryotic cells by calcium ionophore. J Biol Chem 264:1644–1649, 1989.

301. K-V Chin, C Cade, CO Brostrom, EM Galuska, MA Brostrom. Calcium-dependent regulation of protein synthesis at translational initiation in eukaryotic cells. J Biol Chem 262:16509–16514, 1987.

302. EH Fawell, IJ Boyer, MA Brostrom, CO Brostrom. A novel calcium-dependent phosphorylation of a ribosome-associated protein. J Biol Chem 264:1650–1655, 1989.

303. E Lam, M Benedyk, N-H Chua. Characterization of phytochrome-regulated gene expression in a photoautotrophic cell suspension: possible role for calmodulin. Mol Cell Biol 9:4819–4823, 1989.

304. C Bowler, G Neuhaus, H Yamagata, N-H Chua. Cyclic GMP and calcium mediate phytochrome phototransduction. Cell 77:73–81, 1994.

305. MA Stevenson, SK Calderwood, GM Hahn. Rapid increases in inositol trisphosphate and intracellular Ca$^{++}$ after heat shock. Biochem Biophys Res Commun 137:826–833, 1986.

306. IAS Drummond, SA McClure, M Poenie, RY Tsein, RA Steinhardt. Large changes in intracellular pH and calcium observed during heat shock are not responsible for the induction of heat shock proteins in *Drosophila melanogaster*. Mol Cell Biol 6:1767–1775, 1986.

307. SK Calderwood, MA Stevenson, GM Hahn. Effects of heat on cell calcium and inositol lipid metabolism. Radiat Res 113:414–425, 1988.

308. SF Yang, NE Hoffman. Ethylene biosynthesis and its regulation in higher plants. Annu Rev Plant Physiol 35:155–189, 1984.

309. V Raz, R Fluhr. Calcium requirement for ethylene-dependent responses. Plant Cell 4:1123–1130, 1992.

310. V Raz, R Fluhr. Ethylene signal is transduced via protein phosphorylation events in plants. Plant Cell 5:523–530, 1993.

311. C Lin, MF Thomashow. DNA sequence analysis of a complementary DNA for cold-regulated *Arabidopsis* gene *cor*15 and characterization of the COR15 polypeptide. Plant Physiol 100:546–547, 1992.

312. SS Mohapatra, L Wolfraim, RJ Poole, RS Dhindsa. Molecular cloning and relationship to freezing tolerance of cold-acclimation-specific genes of alfalfa. Plant Physiol 89:375–380, 1989.

313. DA Williams, PL Becker, FS Fay. Regional changes in calcium underlying contraction of single smooth muscle cells. Science 235:1644, 1988.

314. P Nicotera, DJ McConkey, DP Jones, S Orrenius. ATP stimulates Ca$^{2+}$ uptake and increases the free Ca$^{2+}$ concentration in isolated rate liver nuclei. Proc Natl Acad Sci U S A 86:453–457, 1989.

315. BD Birch, DL Eng, JD Kocsis. Intranuclear Ca$^{2+}$ transients during neurite regeneration of an adult mammalian neuron. Proc Natl Acad Sci U S A 89:7978–7982, 1992.

316. SH Lee, HY Seo, JC Kim, WD Heo, WS Chung, KJ Lee, MC Kim, YH Cheong, JY Choi, CO Lim, MJ Cho. Differential activation of NAD kinase by plant calmodulin isoforms. The critical role of domain I. J Biol Chem 272:9252–9259, 1997.

317. MA Hughes, MA Dunn. The molecular biology of plant acclimation to low temperature. J Exp Bot 47:291–305, 1996.

318. K Yamaguchi-Shinozaki, K Shinozaki. A novel *cis*-acting element in an *Arabidopsis* gene is involved in responsiveness to drought, low-temperature, or high-salt stress. Plant Cell 6:251–264, 1994.

319. SS Baker, KS Wilhelm, MF Thomashow. The 5′-region of *Arabidopsis thaliana COR15a* has *cis*-acting elements that confer cold-, drought-, and ABA-regulated gene expression. Plant Mol Biol 24:701–713, 1994.

320. PL Steponkus, M Uemura, RA Joseph, SJ Gilmour, MF Thomashow. Mode of action of the *COR15a* gene on the freezing tolerance of *Arabidopsis thaliana*. Proc Natl Acad Sci U S A 95:14570–14575, 1998.

321. KR Jaglo-Ottosen, SJ Gilmour, DG Zarka, O Schabenberger, MF Thomashow. *Arabidopsis* CBF1 overexpression induces *COR* genes and enhances freezing tolerance. Science 280:104–106, 1998.

322. TJ Anchordoguy, AS Rudolph, JE Carpenter, JH Crowe. Modes of interaction of cryoprotectants with membrane phospholipids during freezing. Cryobiology 24:324–331, 1987.

323. G Warren, R McKown, AL Marin, R Teutonico. Isolation of mutations affecting the development of freezing tolerance in *Arabidopsis thaliana* (L.) Heynh. Plant Physiol 111:1011–1019, 1996.

324. M Ishitani, L Xiong, B Stevenson, JK Zhu. Genetic analysis of osmotic and cold stress signal transduction in *Arabidopsis*: interactions and convergence of abscisic acid–dependent and abscisic acid–independent pathways. Plant Cell 9:1935–1949, 1997.

325. J Ingram, D Bartels. The molecular basis of dehydration tolerance in plants. Annu Rev Plant Physiol Plant Mol Biol 47:377–403, 1996.

326. MJ Chrispeels, L Holuigue, R Latorre, S Luan, A Orellana, H Pena-Cortes, NV Raikhel, PC Ronald, A Trewavas. Signal transduction networks and the biology of plant cells. Biol Res 32:35–60, 1999.

327. AC Allan, MD Fricker, JL Ward, MH Beale, AJ Trewavas. Two transduction pathways mediate rapid effects of abscisic acid in *Commelina* guard cells. Plant Cell 6:319–328, 1994.

328. M Koornneef, G Reuling, CM Karssen. The isolation and characterization of abscisic acid–insensitive mutants of *Arabidopsis thaliana*. Physiol Plant 61:377–383, 1984.

329. S Cutler, M Ghassemian, D Bonetta, S Cooney, P McCourt. A protein farnesyl transferase involved in abscisic acid signal transduction in *Arabidopsis*. Science 273:1239–1241, 1996.

330. J Leung, M Bouvier-Durand, PC Morris, D Guerrier, F Chefdor, J Giraudat. *Arabidopsis* ABA response gene *ABI1*: features of a calcium-modulated protein phosphatase. Science 264:1448–1452, 1994.

331. J Leung, S Merlot, J Giraudat. The *Arabidopsis ABSCISIC ACID-INSENSITIVE2 (ABI2)* and *ABI1* genes encode homologous protein phosphatases 2C involved in abscisic acid signal transduction. Plant Cell 9:759–771, 1997.

332. J Sheen. Mutational analysis of protein phosphatase 2C involved in abscisic acid signal transduction in higher plants. Proc Natl Acad Sci U S A 95:975–980, 1998.

333. Y Wu, J Kuzma, E Marechal, R Graeff, HC Lee, R Foster, NH Chua. Abscisic acid signaling through cyclic ADP-ribose in plants. Science 278:2126–2130, 1997.

334. GJ Allen, K Kuchitsu, SP Chu, Y Murata, JI Schroeder. *Arabidopsis abi1-1* and *abi2-1* phosphatase mutations reduce abscisic acid–induced cytoplasmic calcium rises in guard cells. Plant Cell 11:1785–1798, 1999.

335. E Epstein, JD Norlyn, DW Rush, RW Kingsbury, DB Kelly, GA Cunningham, AF Wrona. Saline culture of crops: a genetic approach. Science 210:399–404, 1980.

336. SJ Wu, L Ding, J-K Zhu. *SOS1*, a genetic locus essential for salt tolerance and potassium acquisition. Plant Cell 8:617–627, 1996.

337. JK Zhu, J Liu, L Xiong. Genetic analysis of salt tolerance in *Arabidopsis*. Evidence for a critical role of potassium nutrition. Plant Cell 10:1181–1191, 1998.

338. E Epstein. How calcium enhances plant salt tolerance? Science 280:1906–1907, 1998.

339. T Nakamura, Y Liu, D Hirata, H Namba, S Harada, T Hirokawa, T Miyakawa. Protein phosphatase type 2B (calcineurin)-mediated, FK506-sensitive regulation of intracellular ions in yeast is an important determinant for adaptation to high salt stress conditions. EMBO J 12:4063–4071, 1993.

340. I Mendoza, F Rubio, A Rodriguez-Navarro, JM Pardo. The protein phosphatase calcineurin is essential for NaCl tolerance of *Saccharomyces cerevisiae*. J Biol Chem 269:8792–8296, 1994.

341. NC Schaad, E De Castro, S Nef, S Hegi, R Hinrichsen, ME Martone, MH Ellisman, R Sikkink, F Rusnak, J Sygush, P Nef. Direct modulation of calmodulin targets by the neuronal calcium sensor NCS-1. Proc Natl Acad Sci U S A 93:9253–9258, 1996.

342. I Hwang, IF Harper, F Liang, H Sze. Calmodulin activation of an endoplasmic reticulum–located calcium pump involves an interaction with the N-terminal autoinhibitory domain. Plant Physiol 122:157–168, 2000.

343. KD Hirschi. Expression of *Arabidopsis CAX1* in tobacco: altered calcium homeostasis and increased stress sensitivity. Plant Cell 11:2113–2122, 1999.

344. M Kasuga, Q Liu, S Miura, K Yamaguchi-Shinozaki, K Shinozaki. Improving plant drought, salt, and freezing tolerance by gene transfer of a single stress-inducible transcription factor. Nat Biotechnol 17:287–291, 1999.

345. KD Hirschi, RG Zhen, KW Cunningham, PA Rea, GR Fink. CAX1, an $H^+/Ca^{2+}$ antiporter from *Arabidopsis*. Proc Natl Acad Sci U S A 93:8782–8786, 1996.

346. H Ueoka-Nakanishi, Y Nakanishi, Y Tanaka, M Maeshima. Properties and molecular cloning of $Ca^{2+}/H^+$ antiporter in the vacuolar membrane of mung bean. Eur J Biochem 262:417–425, 1999.

347. EJ Stockinger, SJ Gilmour, MF Thomashow. *Arabidopsis thaliana* CBF1 encodes an AP2 domain–containing transcriptional activator that binds to the C-repeat/DRE, a *cis*-acting DNA regulatory element that stimulates transcription in response to low temperature and water deficit. Proc Natl Acad Sci U S A 94:1035–1040, 1997.

348. C Xiao, H Xin, A Dong, C Sun, K Cao. A novel calmodulin-like protein gene in rice which has an unusual prolonged C-terminal sequence carrying a putative prenylation site. DNA Res 6:179–181, 1999.

349. A Fraichard, E Perotti, O Gavin, A Chanson. Subcellular localization, distribution and expression of calmodulin in *Zea mays* roots. Plant Sci 118:157–165, 1996.

350. M Rodriguez-Concepcion, S Yalovsky, W Gruissem. Protein prenylation in plants: old friends and new targets. Plant Mol Biol 39:865–870, 1999.

351. A Bown, B Shelp. The metabolism and function of γ-amino butyric acid. Plant Physiol 115:1–5, 1997.

352. C Kohler, T Merkle, G Neuhaus. Characterization of a novel gene family of putative cyclic nucleotide– and calmodulin-regulated ion channels in *Arabidopsis thaliana*. Plant J 18:97–104, 1999.

353. AJ Trewavas, R Malho. $Ca^{2+}$ signaling in plant cells: the big network! Curr Opin Plant Biol 1:428–433, 1998.

# 36

# Regulation of Gene Expression During Abiotic Stresses and the Role of the Plant Hormone Abscisic Acid

**Elizabeth A. Bray**

*University of California, Riverside, California*

## I. INTRODUCTION

The plant response to the environment is a complex set of processes. Changes in the environment result in plant responses at many different levels: morphological, physiological, cellular, and metabolic. The type of response depends on the source of the stress, the duration and severity of the stress, the genotype of the stressed plant, the stage of development, and the organ and cell type in question. To fully understand plant stress, the mechanism by which these responses are regulated and the function of the responses must be characterized and understood. To further understand plant stress, researchers have turned to the study of molecular responses. In this chapter, stresses discussed are water deficit, salt stress, and low-temperature stress. A common link between these stresses is loss of cellular water.

As a result of the complex nature of the responses, it is difficult to determine the role of the various responses with respect to resistance of the plant to the loss of water. Responses may be adaptive, contributing to the ability of plant to withstand stress, may not be involved in adaptation, or may be a result of injury. Among the many responses to changes in the environment are alterations in the pattern of gene expression. Changes in gene expression are an important part of the plant response to the environment. Although some responses, possibly short-term metabolic and physiological responses, may not require changes in gene expression, the majority of responses to the environment are predicted to require alterations in gene expression.

As there is a complex of responses to the environment, the mechanisms that control the plant response to the environment are also expected to be complex. It is important to understand the cues from the environment that are detected as stress and to understand the signaling mechanism(s) within the plant at the whole plant and cellular levels. One plant signal that is prominent in plant stress studies is the plant hormone abscisic acid (ABA). The concentration of ABA increases in the plant during stress [1]. The best studied of the changes in ABA concentration is in response to drought stress. ABA levels increase in response to lowered water potential, and it is postulated that the loss of turgor is the trigger that induces ABA biosynthesis [2]. ABA levels have also been observed to rise in response to salt stress and low temperature. The signal ABA is common to all of the stresses discussed in this chapter, although there are differences in the pattern and magnitude of ABA accumulation in response to the different stresses. These similarities have led to the suggestion that some of the responses to the various stresses are similar and play similar roles in the ability of the plant to withstand periods of water deficit imposed by different environmental stresses.

A number of genes have been identified that are induced when plants or plant parts are subjected to stresses resulting in cellular water deficit. Yet the function or role that most of these gene products play is still elusive. Some of the changes in gene expression may be adaptive, having functions that promote plant survival during water deficit, but this cannot be assumed. The function of many genes can be predicted based on the amino acid sequence of the gene product deduced. Clues to function may also be obtained from expression characteristics such as timing of expression and organ, tissue, cellular, and subcellular location. However, the significance of gene expression with respect to stress tolerance cannot always be predicted.

At this time, most researchers have concentrated on identifying genes that respond to stress, beginning to determine the function, and studying their regulation. In this chapter, genes that are regulated by water deficit will be categorized based on functional predictions. Utilization of the techniques of plant molecular biology is providing new insights into plant responses to the environment with respect to the function of these changes and the regulation of these responses and the commonalty of the response among different stresses and among different species.

## II.  PREDICTED FUNCTIONS OF STRESS- AND ABA-INDUCED GENE PRODUCTS

Many different genes, encompassing many classes of gene products, are induced by abiotic stresses. To fully understand the significance of these changes in gene expression, the function of the stress-induced gene products and the mechanism by which these genes are regulated must be understood. Many of these stress-induced genes are responsive to ABA application. In terms of function, these genes can be divided into different classes based on their DNA sequence, expression characteristics, and/or predicted functions. Unfortunately, at this writing, there are few examples in which the in vivo function of the gene product has been demonstrated. In many cases, functions have been predicted based on deduced amino acid sequence, but no biochemical or physiological data have been obtained to prove the function in vivo. Therefore, several of the gene classifications are based on predicted functions, and as the true functions are determined these categories may need to be corrected. The classifications used for abiotic stress–induced genes are hydrophilic gene products (Figure 1), enzymes, those with other predicted functions, and those for which a function has not been predicted.

### A.  Hydrophilic Proteins Predicted to Have a Protective Function

A number of genes induced during periods of water deficit have been identified that encode proteins which are overwhelmingly hydrophilic, are soluble upon boiling, and are therefore expected to be located in the cytosol. These characteristics have led to the prediction that these gene products are involved in protecting cellular structures and components from dehydration associated with water deficit, salinity, and low-temperature stress. Many of these genes were first shown to be expressed during seed desiccation, the period of seed development following maturation, and are referred to as late embryogenesis abundant (*lea*) genes [3,4]. Dure et al. [4] established three groups of *lea* genes based on the publication of homologous genes found in seeds of other species. This classification is useful because of the preponderance of different names for genes within these classes. However, it must be acknowledged that the name *lea* for genes expressed during stress in vegetative organs may not be appropriate because a number of these stress-induced genes are not expressed during seed development. It was suggested by Dure et al. [4] that the name WSP (water stress protein) be applied to proteins of this class for that reason. However, this naming system has not been adopted, possibly because the name WSP also has problems; this name does not acknowledge that these proteins may be induced by other stresses, such as low temperature or salt stress, and it does nothing to add to our understanding of the function of these genes. The genes discussed in this section on hydrophilic gene products include genes that have been identified after periods of stress in vegetative organs and share significant DNA sequence homology with *lea* genes from cotton. In addition, several new genes have been identified that are stress induced, overwhelmingly hydrophilic, yet have not been found in desiccating cotton seeds.

### 1.  Em Family (Group 1)

This family of proteins is hydrophilic with no notable structural domains predicted by the amino acid sequence [5]. Cysteine and tryptophan are not found in the proteins of this group. The entire deduced pro-

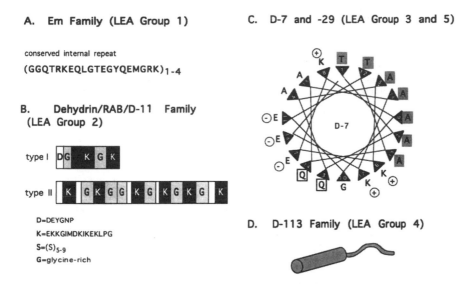

A. Em Family (LEA Group 1)

conserved internal repeat

(GGQTRKEQLGTEGYQEMGRK)<sub>1-4</sub>

B. Dehydrin/RAB/D-11 Family (LEA Group 2)

type I

type II

D=DEYGNP
K=EKKGIMDKIKEKLPG
S=(S)<sub>5-9</sub>
G=glycine-rich

C. D-7 and -29 (LEA Group 3 and 5)

D. D-113 Family (LEA Group 4)

**Figure 1**　Unique aspects of several of the overwhelmingly hydrophilic proteins that accumulate during periods of water deficit. (A) Conserved element that may be repeated in the Em family of proteins [3–7]. (B) Conserved amino acid motifs found in the dehydrin/RAB/D-11 family of hydrophilic proteins. Two types have been noted, those with the polyserine region (S) and those without [3,8–10,19,24]. (C) Conserved 11-amino-acid repeat found in D-7 which may form an amphipathic α-helix. At position 12 in the wheel, the amino acid sequence is repeated. Positive and negative amino acids are noted with plus and minus signs. Amino acids with an amide group are boxed. Apolar amino acids are shown in the shaded box. Note that one face is apolar and the other has a pattern on negatively charged, amide-containing, positively charged amino acids [3,25]. A similar pattern of amino acids with these characteristics is repeated in D-29, although the exact amino acids are not conserved [3,4,25,26]. (D) The D-113 family of proteins has a conserved structure with an α-helix in the amino-terminal domain followed by a random coil region at the carboxy terminus [19,26].

tein sequence is well conserved among all the genes that have been identified in monocots and dicots. An internal repeat of a hydrophilic amino acid motif GGQTRKEQLGEEGYREMGHK was found in the barley genes and may be repeated up to four times [6] (Figure 1). A similar 20-amino-acid motif is duplicated in the cotton gene *leaA2*, corresponding to the cDNA D-32 [7], although most of the genes identified thus far contain only one of these motifs, including the gene first published, D-19 [3]. It has been suggested that group 1 proteins function in a water-binding capacity, creating a protective aqueous environment [6].

## 2. *Dehydrin/RAB/D-11 Family (Group 2)*

Group 2 proteins have been called dehydrin [8], RAB (responsive to ABA) [9], and D-11 [3]. These proteins are also overwhelmingly hydrophilic. There is a characteristic lysine-rich region with the consensus amino acid sequence EKKGIMDKIKEKLPG, which is repeated at least two times, once at the carboxy terminus and once internally [10]. These genes are expressed in seeds and in response to water deficit, salt stress, and low-temperature treatments in vegetative tissues [11,12]. These genes are also ABA regulated. The mRNA has been found in all stressed organs that have been investigated. In tomato, maize, and *Arabidopsis*, family members have been shown to be regulated by elevated levels of endogenous ABA during periods of dehydration in leaves [13–15]. In *Craterostigma plantagineum*, the protein DSP16 is localized in the cytoplasm, as determined by immunocytolocalization [16]. Close et al. [10] have observed that dehydrins are localized in the cytoplasm and the nucleus of aleurone layers, but primarily in the cytoplasm of root and shoot cells.

This class of genes has been identified more frequently than any other stress-induced gene in plants. Members of this family of proteins have been found in grasses and in dicots, and there is also immunological evidence that these proteins accumulate in *Anabaena* [17,18]. Tables of these genes are presented in Close et al. [10] and Dure [19]. It has been recognized that this class of proteins can be divided into at least two types. In type I, the internal lysine-rich signature motif is adjacent to a polyserine region. The polyserine region is a site for phosphorylation [20,21]. The lysine-rich signature motif is also found at the carboxy terminus. There is another conserved motif, DEYGNP, that is found one or two times near the amino terminus. However, gene products with the consensus lysine-rich repeat have been identified that do not have the polyserine region, referred to as type II in Figure 1. These genes are made up of the consensus lysine-rich repeat alternating with glycine-rich repeats. The glycine-rich repeats do not have a consensus sequence; they are characterized only by the abundance of glycine residues. These genes have been identified in wheat [22,23] and in alfalfa [24]. Interestingly, the alfalfa genes are induced by low-temperature treatments but not by drought stress or by ABA application [24]. In wheat, the gene, which encodes a 39-kDa protein, is induced preferentially by low-temperature treatments.

### 3. Groups 3 and 5

Group 3 proteins, represented by D-7 from cotton, and group 5 proteins, represented by D-29 from cotton, contain repeated tracts of 11 amino acids [4]. Although the 11-mer repeat has diverged among different species, a functional consensus was derived based on the polarity, charge, or methylation of the amino acid (Figure 1). The periodicity of 11 indicates that there may be an amphiphilic α-helix formed by polar and apolar amino acids aligned on different faces of the α-helix [3,4,25]. The D-7 protein is found uniformly in the mature cotton embryo, calculated to be from 200 to 300 μM in the cytoplasm immediately prior to desiccation [26]. Interestingly, in *Craterostigma plantagineum*, a protein with a similar 11-mer repeat was found to be localized in the chloroplast [16]. The abundance of these proteins has been used to rule out several possible functions—these proteins are not expected to function as enzymes, structural (architectural) proteins, regulatory proteins, or as ion or water transport proteins [26]. The prediction has been made that these proteins function in the sequestration of ions during cellular dehydration [25].

### 4. Group 4

It has been discovered that another gene that is expressed in drying cotton seeds accumulates to high levels [26]. This protein, D-113, has a homologue in tomato, LE25 [11,27], whose mRNA is expressed in vegetative tissues in response to drought stress, and elevated levels of endogenous ABA are required for its expression [14]. Members of the D-113 family are biased toward alanine residues and contain the random coil–promoting residues glycine and threonine. It is predicted that the amino-terminal portion of the molecule is made up of an α-helix (Figure 1). The remainder of the molecule has no predicted structure [19]. It has been proposed that D-113 behaves as a surrogate water film in the desiccated state, stabilizing the intracellular surface of seeds [26]. It is uncertain if this would also be a role for D-113–like proteins in vegetative tissues of plants such as tomato because these tissues are not capable of surviving desiccation.

### 5. LT178, 65/RD29 Family

An additional family of genes, which have thus far been identified only in *Arabidopsis*, are also induced by water deficit–based stresses and ABA [28–30]. There are two genes in *Arabidopsis* that are adjacent to each other in the genome. The deduced amino acid sequences result in 77.9- and 64.5-kDa proteins that are overwhelmingly hydrophilic [30]. These genes were identified by three different laboratories and have been named *lti78* and *lti65* [28], *rd29A* and *rd29B* [29,30], and *cor78* [31]. Thus far, no specific predictions about the function of these proteins have been made.

### 6. Glycine-Rich Proteins

A family of genes in alfalfa has been characterized which are regulated by ABA, low temperature, and drought. The gene products are predicted to be hydrophilic and are characterized by glycine-rich repeats [31–33]. There are no specific predictions for their role other than a general role in protection from cellular dehydration.

### 7. KIN1 and KIN2/COR6.6

Another protein family was identified during low-temperature stress in *Arabidopsis* and is encoded by at least two genes in *Arabidopsis* [34–36]. These proteins are of low molecular weight (6.6 kDa), are rich in alanine, glycine, and lysine, and are largely hydrophilic. These genes are also induced by ABA treatments [35]. The *kin2* gene is expressed strongly in response to drought stress and salinity [36]. A low degree of similarity with fish antifreeze protein was noted [35]; however, there is no functional evidence for the role of these proteins in a stress response.

A variety of genes that encode hydrophilic proteins are expressed in response to abiotic stresses. Many of the protein families have different structures, leading to predictions that these proteins are involved in ion sequestration, water binding, and other protective roles in the cytoplasm. At this time, biochemical studies are needed to confirm these predictions.

## B. Genes Encoding Enzymes

The activity or the amount of many enzymes has been shown to be altered during drought stress [37]. In many cases the activity of an enzyme has been studied during stress but the gene has not been characterized. In this chapter, only enzymes in which the gene has been cloned and the mRNA corresponding to that gene has been shown to be elevated in response to drought are discussed. In the coming years, the number of enzymes that are cloned and shown to be regulated during stresses that impose water deficits is sure to grow. Although an enzymatic function can be predicted from the deduced amino acid sequence of a cloned gene, the role during stress may not be obvious or ensured by the presence of a transcript during stress. Until additional experiments are done, it cannot be determined if the activity of these enzymes promotes stress adaptation. Enzymes induced by water stress and involved in osmolyte accumulation, protein degradation, CAM, ion transport, and signal transduction are described.

### 1. Osmolyte Accumulation

During periods of water deficit, compatible solutes accumulate in the cell, resulting in a lower cellular osmotic potential. If the cellular water potential is more negative than that of the cells environment, water will be taken up by the cell. This process, called osmotic adjustment, may occur in the field after a long-term drought stress [38]. Compatible solutes, or osmolytes, include inorganic ions; organic ions; soluble carbohydrates, including polyols; amino acids, particularly proline; and quaternary ammonium compounds such as betaines [37]. Several genes have been identified which code for enzymes that may be involved in the accumulation of osmolytes during osmotic stress; these include enzymes in the biosynthetic pathway of proline [39–41], glycine betaine [42], and polyols [43,44]. The plant biosynthetic pathway for proline contains two genes; one is a bifunctional enzyme, $\Delta^1$-pyrroline-5-carboxylate synthetase, which has both $\gamma$-glutamyl kinase and glutamic-$\gamma$-semialdehyde dehydrogenase activities [40], and the other is $\Delta^1$-pyrroline-5-carboxylate reductase (P5CR) [39–41]. Both of these genes are induced by salt stress, indicating that they may play a role in osmotic adjustment. In pea, P5CR is induced in roots, but not in shoots, of pea seedlings in response to salt stress [41]. The gene coding for betaine aldehyde dehydrogenase, the last step in glycine betaine synthesis, has been isolated and is induced by salt stress [42]. A gene encoding *myo*-inositol O-methyl transferase (*imt1*) was isolated from the facultative CAM plant *Mesembryanthemum crystallinum* and is induced by salt stress [44], drought, and low temperature [45]. This enzyme is involved in synthesis of the polyol, pinitol. Tobacco plants transformed with *imt1* driven by the 35S promoter accumulated ononitol, an osmolyte not synthesized in wild-type tobacco. Although stress studies were not reported, these results indicate that the enzyme encoded by *imt1* is a step in the biosynthetic pathway of pinitol [46]. Aldose reductase, an enzyme involved in sorbitol synthesis, has been identified in barley seeds, although this gene is not expressed in dehydrated barley leaves [43]. An NADPH-dependent aldose reductase is induced by ABA in bromegrass suspension cells with increased freezing tolerance [47]. In response to a number of stresses, enzymes involved in osmolyte accumulation are induced. These enzymes are probably involved in the accumulation of osmolytes, which promotes uptake of water into the cells through osmotic adjustment.

Vegetative storage protein genes, *vsp*, are induced by environmental stresses such as water deficit and wounding. The VSPs have been shown to accumulate preferentially in the vacuoles of paraveinal mesophyll cells of soybean. It has been shown that two of the proteins, VSP$\alpha$ and VSP$\beta$, are acid phosphatases with the highest substrate specificity for tetrapolyphosphates [48]. Staswick et al. [49] reported that although the VSPs have acid phosphatase activity, they are not the major acid phosphatase in leaves. DeWald et al. [48] propose that VSPs are involved in amino acid uptake and temporary sequestration of amino acids in paraveinal mesophyll cells. Therefore, these proteins may play a role in osmotic adjustment during stress.

## 2.  Proteases

Proteases have also been shown to be induced by abiotic stresses. A thiol protease is induced in pea by water deficit [50]. Two additional proteinases, cysteine proteinases, were identified in drought-stressed *Arabidopsis* [51]. These genes have the catalytic sites typical of cysteine proteinases and have amino-terminal signal peptides. They are not induced by ABA or temperature stress but are strongly induced by salt and drought stress [51]. These genes are similar to a cysteine protease induced by low temperature in tomato [52]. Their function during drought stress is not confirmed, but they may be involved in the degradation of polypeptides denatured during stress, processing of precursor proteins to the mature form, or degradation of vacuolar proteins, after which the amino acids may be used in synthesis of stress-induced proteins or osmotic adjustment [50,51].

## 3.  Induction of CAM

In plants that have the capacity for Crassulacean acid metabolism (CAM), a switch from C3 metabolism to CAM may occur during periods of stress or in response to development. In *Mesembryanthemum crystallinum*, there is a 10- to 20-fold increase in phosphoenolpyruvate carboxylase (PEPCase) activity in response to salt stress [53]. PEPCase is the enzyme responsible for the primary fixation of $CO_2$ into oxaloacetate, which is subsequently converted to malate during CAM. In *M. crystallinum*, PEPCase is represented by two genes, one of which is salt stress induced. However, induction does not occur in salt-stressed suspension cultures of *M. crystallinum* [54]. CAM coupled with stomatal conductance provides a means for the improved stress tolerance that may occur in specialized plants.

## 4.  Plasma-Membrane $H^+$-ATPase

During salt stress or water deficit, the concentration of ions in the cytoplasm must be controlled. The import of ions into the cytoplasm must be limited to the capacity to compartmentalize $Na^+$ and $Cl^-$ into the vacuole. Active transport of ions is driven by an $H^+$ electrochemical gradient that is generated by the plasma-membrane $H^+$-ATPase [55]. Carriers or channels facilitate the active or passive transport of these ions, with active transport driven by the $H^+$-ATPase. During development of water deficit in soybean seedlings, there is an increase in $H^+$-ATPase mRNA levels in the roots only. This correlates with the ability of growth maintenance in this organ; in shoots in which growth is not maintained during water stress, there is not an increase in $H^+$-ATPase mRNA levels [56]. This may indicate that growth during stress is dependent on increased activity of ATPase and is associated with ion transport. In the halophyte *Atriplex nummularia*, it has been demonstrated that the plasma-membrane $H^+$-ATPase is regulated by NaCl [57]. The ATPase activity may have an important role in stress tolerance and should be studied further.

## 5.  Protein Kinase

For plants to respond to the environment, mechanisms must have evolved that signal changes from the environment at the cellular level. The pathway of information transfer from the environment to the cell, resulting in alterations in gene expression, is the signal transduction pathway. Many different component pathways may be required to achieve alterations in gene expression. In one type of signal transduction pathway, a protein kinase controls activation or deactivation of proteins by phosphorylation. A cDNA, PKABA1, corresponding to a protein kinase that is induced by ABA, has been isolated [58]. The deduced amino acid sequence has 12 catalytic subdomains found in serine/threonine protein kinases that are thought to be critical for phosphorylation. Two mRNAs hybridize to PKABA1 in dehydrated seedlings, and the accumulation of these mRNAs corresponds to an increase in ABA concentration. The rate of accumulation is similar to the accumulation of other drought- and ABA-induced genes; therefore, it is proposed that this kinase is involved in the phosphorylation of other ABA-induced gene products [58]. Although there are many other possibilities for the action of this gene product, the function cannot be determined until its substrate specificity is characterized.

## C.  Genes That Have Other Predicted Cellular Functions

For a plant to survive periods of stress, many developmental, physiological, and metabolic functions may need to be altered. A number of unique changes in gene expression have been identified, using gene cloning techniques. Many of these would not have been identified using other types of studies. Genes have been identified that are involved in antifungal activity, protection from freeze-thaw inactivation, transport of water, RNA binding, and gene regulation.

## 1.  Antifungal Proteins

In addition to the direct effects of water stress, environmental stress may contribute to the susceptibility of the plant to pathogens. Genes encoding osmotin and nonspecific lipid transfer proteins have antifungal activity and are induced in response to water deficit–based stresses.

A protein was first discovered that was prominent in tobacco cells adapting to high concentrations of salt. The accumulation of the protein is correlated with osmotic adaptation and it was therefore named osmotin [59,60]. This protein has a signal sequence and is localized in vacuole inclusion bodies. It is a basic homologue of family 5 pathogenesis-related proteins. This protein family, including osmotin, has been shown to have antifungal activity [60–62]. Members of this family have been shown to permeabilize the fungal plasma membrane. Transgenic tobacco plants overproducing osmotin have been shown to be more tolerant of fungal attack than are control plants [60]. It is proposed that there is a specific interaction between osmotin and the membrane, possibly with a portion of the molecule interacting with the membrane, with the rest forming an ion or water channel that permeabilizes the fungal membrane [60]. It is not certain if this is the only role of osmotin or if it also plays a direct role in salt tolerance.

Proteins that have the capacity to transfer lipids from liposomes to mitochondria in vitro have been studied in plants. A class of these proteins transfers a number of different classes of lipids and has been called nonspecific lipid transfer proteins (nsLTPs) [63]. Three genes that are homologous to nsLTP genes have been found to be induced by stress in aerial plant parts: two from tomato [64,65] and one from barley [66]. However, the role of these genes during stress has not been determined. Since it was determined that nsLTPs are expressed in the epidermis of the shoot, it has been suggested that they play a role in cuticle formation [67]. It has now been demonstrated that an nsLTP-like protein isolated from radish seeds has antifungal activity; it inhibited fungal hyphae growth but did not affect spore germination [68]. Therefore, the role of nsLTPs during stress may be in the protection of the shoot from fungal attack. However, further studies are required to elucidate the function of nsLTP-like proteins during stress.

## 2.  Protection from Freeze-Thaw Inactivation

Another type of protection is the protection of enzymes from freeze-thaw inactivation. The *cor15* gene is induced by cold acclimation in *Arabidopsis* [69]. The polypeptide that is encoded by this gene is targeted to the chloroplast [70]. It is 100 times more effective than BSA at protecting lactate dehydrogenase from freeze-thaw inactivation [71]. This preliminary evidence indicates that this gene encodes a protein with a potential protective role in freeze-induced dehydration.

## 3.  Protein Degradation

Cellular function may also be protected during stress by preventing protein degradation or degrading proteins that are no longer functional. Heat shock proteins that are induced by water deficit [72,73] may be involved in the refolding of proteins to regain their function, or the prevention of protein aggregation [74] during stress. Protease inhibitors induced by water deficit may protect against proteases released after cellular disruption and membrane disorganization as a result of stress. The gene *Bnd22*, induced after prolonged dehydration stress in *Brassica napus*, has some characteristics of Künitz trypsin inhibitor, although it does not have all the signature amino acids. It is also expressed in a manner that is different from other Künitz trypsin inhibitors; it is not expressed in seeds, it is expressed only in the aerial parts of the plant after a prolonged drought stress [75]. Further studies are needed to determine if this protein is a protease inhibitor that plays a role during water deficit. Ubiquitin, which was also shown to be induced by water deficit [72], may target proteins for degradation that cannot regain function after water deficit.

## 4.  Major Intrinsic Proteins

A class of proteins, major intrinsic proteins, have been identified which may form transmembrane channels and be involved in the transport of ions, other metabolites, or water across membranes. These proteins have six membrane-spanning domains that are postulated to form a channel. Drought-induced examples of this class of proteins have been isolated from pea [52] and *Arabidopsis* [76–79]. These genes are most closely related to each other, but are also similar to *nod26*, tonoplast intrinsic protein (TIP) from bean, bovine major intrinsic proteins (MIP), and glycerol facilitator protein [77]. *Arabidopsis* has at least four genes in this family, each with different expression characteristics. Two members of this family are induced by decreased turgor, TMPα and TMPβ [78,79]. It is predicted that the expression pattern reflects functional specialization of the various family members [77]. During drought stress, these channels may facilitate transport of ions, metabolites, or even water. Maurel et al. [80] demonstrated that γ-TIP is a water-channel protein when expressed in *Xenopus* oocytes. Although γ-TIP is not expressed during water stress, these results present some interesting possibilities for the drought-induced members of this class of proteins. It must be determined if the water stress–induced proteins are also water-channel proteins and in which membranes these proteins are located.

### 5.   RNA-Binding Protein

A gene from maize, MA16, with many characteristics of the *leas*, contains a glycine-rich repeat and is developmentally regulated in seeds and induced in dehydrated vegetative tissues [81]. Unlike the LEAs, this protein contains a consensus sequence, RGFGFVTF, which is conserved in RNA-binding proteins. Ludevid et al. [82] used in vitro ribohomopolymer binding assays to confirm that MA16 has RNA-binding activity. In these studies, MA16 preferentially binds poly(G). There are many different functions for ribonucleoproteins, including regulation of alternative splicing, pre-mRNA processing, and mRNA translation and stability. At this time the stress response cannot be predicted because the RNAs that are associated with MA16 in vivo have not been determined. But it is possible that this protein has an important regulatory or protective function during plant stress.

### 6.   Gene Regulation

In addition to signaling changes in the environment, the signal transduction pathway must include the induction or activation of transcription factors and activators that are required for transcription induction. A transcription factor that is involved in the induction of some of the ABA-induced genes has been identified and isolated [83]. This protein, EmBP-1, is a bZIP-type transcription factor and recognizes DNA sequences containing the DNA sequence CACGTGGC. A gene, *Alfin-1*, has been isolated from salt-tolerant alfalfa cells that has zinc finger motifs and may be a sequence-specific DNA-binding protein [84]. An H1 histone-like protein [85] and a nonhistone chromosomal protein [86] have also been shown to be induced by drought stress. The roles of these proteins have not been elucidated.

## D.   Genes for Which a Function Has Not Been Predicted

In addition to the genes that have a function that can be predicted from the DNA sequence and amino acid sequence, there are also many stress-induced genes for which no function has been predicted. A few of these types of genes are discussed here.

### 1.   LEA Group 6

With the identification of *rab28* [87], a new *lea* group can be formed. Two representatives of group 6 have been identified, one from cotton, D-34 [3], and the other from maize, *rab28* [87]. The predicted gene product is different from the other LEA proteins, which are overwhelmingly hydrophilic; it has a balanced hydrophobicity plot. RAB28 is predicted to have four to six α-helices and a globular structure. Functional predictions have not been made.

### 2.   RD22

A gene, *rd22*, induced by water stress, salinity, and ABA in *Arabidopsis* [88], was determined to be similar to an unidentified seed protein (USP) from *Vicia faba* [89]. RD22 is induced early during seed development but not during the late stages of embryogenesis, like USP. The homology between RD22 and USP is found in the carboxy-terminal portion of the protein [88]. RD22 has an amino-terminal hydrophobic region with five repeated sequences in the amino terminus. The consensus is TnVnVGnG-GVnnnnnnKGK, which is predicted to contain a β sheet–turn–β sheet structure [88]. The significance of this structure or the function of this gene product is not known.

### 3.   Germinlike Proteins

A root-specific protein that is induced in barley by salt stress was found to be similar to germin [90]. Germin is a protein of unidentified function that accumulates during the onset of growth following germination in wheat [91]. Germinlike proteins of barley accumulated only in roots after salt stress. The proteins are soluble upon boiling and are found mainly in the soluble fraction using cellular fractionation studies [90]. A similar protein was found in *Mesembryanthemum crystallinum* roots; however, it was found to decrease in response to salt stress. It is proposed that germinlike proteins are sensors of water status, and the expression of these genes may be involved in the control of growth, depending on plant water status [92].

### 4. Early Light-Induced Proteins

In desiccated *Craterostigma plantagineum* leaves, a gene, *dsp*-22, that is targeted to the chloroplast is induced [93]. The deduced amino acid sequence is similar to a protein called early light-induced protein (ELIP) from pea and barley [94,95]. The protein consists of alternating regions of hydrophobic and hydrophilic domains. There is a putative transit peptide at the amino terminus. Light is essential for the accumulation of this protein. ABA induction does not occur in the dark. DSP-22 may be involved in photoprotection of the photosystems or in maintaining assembled photosynthetic structures essential for resuming active photosynthesis following rehydration [93].

This description of genes that are induced by water deficit elicited by different types of abiotic stresses serves to indicate that there are indeed many changes in genes expression in response to changes in the environment. There is a complex of molecular processes that are altered by the environment. The functions of the majority of the genes that have been characterized fall in the broad category of protection of cellular function. Protection is predicted to come from hydrophilic proteins in the cytoplasm, osmolyte accumulation, degradation of denatured proteins, and protection from pathogens. Genes involved in the regulation of other genes that are induced by stress have also been identified. These responses occur in many different plants, species that are tolerant and those that are not, and in response to different stresses that cause water deficit.

## III. METHODS TO EVALUATE THE ADAPTIVE ROLES OF STRESS-INDUCED GENES

The expression of specific genes during stress implies that the genes are involved in stress tolerance. However, this may not be a valid assumption. It is possible that gene induction, in addition to promoting stress tolerance, is a result of an injury or a coincidence because of similar signal transduction pathways initiated by different stresses. Therefore, techniques must be developed to evaluate the adaptive significance of the expression of these genes. In many cases, to begin this evaluation it must first be determined if the gene products in fact accumulate during stress. In many of the studies that have been completed thus far, only the accumulation of transcripts has been studied. This does not always ensure that the protein will accumulate. Using osmotin as an example, it was shown that osmotin mRNA is not always translated even though it accumulates [96]. Once it has been established that the protein accumulates, further studies can be completed at the biochemical, genetic, and molecular levels. Several molecular techniques may be exploited to study the role of specific genes.

### A. Under- and Overexpression of Specific Genes

One method to investigate the function of drought-induced gene is to over- and underexpress the genes in transgenic plants. The rationale behind this strategy is that the altered expression of these genes will alter stress tolerance if these genes play an essential role in tolerance. Three *lea*-like genes from *Craterostigma plantagineum* were overexpressed in tobacco driven by the 35S promoter from cauliflower mosaic virus (CaMV) [97]. The plants were tested for physiological traits that might indicate a difference in stress tolerance. An ion leakage test after PEG stress of leaflets was used to test for drought tolerance. However, no differences were observed between the transgenic plants and the wild type. It is possible that the parameters measured were not useful for detecting differences in drought tolerance, that these proteins alone are not sufficient by themselves and need other proteins and/or osmolytes [97], or that these proteins do not function in drought tolerance.

The antisense strategy, in which antisense RNAs accumulate in transgenic plants in order to eliminate a specific protein, might also be used to evaluate gene function. This technique can be used to construct specific single gene mutants. At this time, antisense plants for stress-induced genes have not been evaluated and reported. However, the same problems as encountered above may also occur. There are multiple responses to stress, and the elimination of a single aspect of the stress response may not have a significant effect on plant stress tolerance.

## B.  Introduction of Foreign Genes

When a function is understood that promotes adaptation to stress, genes that have been identified previously may be exploited to improve plant stress tolerance. Thus far, there is one example of this. A gene isolated from *E. coli, mtlD*, which encodes mannitol-1-phosphate dehydrogenase, is involved in mannitol catabolism and leads to the production of fructose-6-phosphate. It was hypothesized that if this gene were expressed in plants, the enzyme may catalyze the reverse reaction, resulting in the synthesis of mannitol-1-phosphate, which would be a substrate for general phosphatases, resulting in the synthesis and accumulation of mannitol [98]. Transgenic tobacco plants, with *mtlD* driven by the 35S CaMV or NOS promoter, resulted in the accumulation of mannitol in young leaves and roots [98]. After 30 days of exposure to salinity, transgenic plants producing mannitol had a greater shoot height and root length than those of control plants [99]. Therefore, this is the first example of a transgenic plant, altered with a microbial gene, which has a greater resistance to plant osmotic stress. Further trials are required to determine the agricultural applicability of these plants. Other genes might also be exploited, especially for other strategies of osmotic adjustment. The use of these strategies in combination may improve stress tolerance of transgenic plants as well as our understanding of the importance of water-deficit responses to plant adaptation to the environment.

## IV.  ABA INDUCES SPECIFIC GENES DURING WATER DEFICIT

The concentration of ABA is altered by changes in the environment. Large increases in ABA concentration have been documented in response to drought stress, with lesser increases in ABA concentration occurring in response to salt and low-temperature stress. These changes occur at the cellular level, but the ABA that is transported throughout the plant at the whole plant level is also changed.

The ABA biosynthetic pathway occurs through the carotenoid biosynthetic pathway. The compound 9'-*cis*-neoxanthin is cleaved to result in the postcleavage intermediate to ABA, xanthoxin. Xanthoxin is oxidized to ABA-aldehyde, which is converted to ABA by ABA-aldehyde oxidase [100]. The step in the ABA biosynthetic pathway that is regulated by stress is likely to be the cleavage step, although this has not been proved because this gene has not been isolated and/or an assay has not been developed for that enzymatic step. The rate of ABA biosynthesis during stress is limited by the production of xanthoxin, not the conversion of xanthoxin to ABA [100].

The trigger that is recognized by the cell to induce ABA biosynthesis is not understood. ABA accumulation is correlated with a reduction in turgor to near zero [2]. Therefore, it is thought that the cellular mechanism for turgor perception is linked to the ABA biosynthetic pathway through a signal transduction pathway. Inhibition of transcription and translation prevented ABA accumulation in response to stress [10], indicating that these processes are required for the cell to recognize stress, or the ABA biosynthetic enzymes must be synthesized for ABA to accumulate. More effort is needed to understand the mechanism of stress-induced ABA accumulation.

Although it is not certain how ABA biosynthesis is controlled, it has been demonstrated that ABA is part of the signaling mechanism during stress that induces specific genes. This has been demonstrated using mutants that are deficient in ABA biosynthesis [14,15,87]. For most studies it is convenient to apply ABA and determine if the application of ABA causes an accumulation of specific transcripts. In the cases where it does, those genes are found to be ABA responsive, indicating that gene induction may occur in response to ABA. However, these studies do not prove that ABA is an endogenous signal used during specific stresses to induce particular genes. Application studies indicate only that ABA is one of the signals that the gene is capable of responding to. Inhibitors of carotenoid biosynthesis result in a decreased level of ABA [100], and responses of the plant that are reduced by inhibitor application have been used to analyze the role of ABA. But as with the use of other inhibitors, effects that are not caused directly by the reduction in ABA concentration may also occur, because carotenoid concentration is also reduced after application of these inhibitors. Mutants that are deficient in ABA biosynthesis and cannot accumulate ABA during stress can be used to identify genes that require elevated levels of endogenous ABA for expression. Mutants in tomato, maize, and *Arabidopsis*, which have specific blocks in the ABA biosynthetic pathway, have been used for this purpose. In each of these cases, several genes have been identified which require ABA for expression [14,15,87].

## A. Identification of Genes That Require ABA for Expression

Mutants that are blocked in the ABA biosynthetic pathway have proved to be useful in the identification of genes that require elevated levels of ABA for expression. mRNAs for specific water deficit–induced genes do not accumulate in response to a water deficit in ABA-deficient mutants as they do in the wild type. For example, the ABA-deficient mutant of tomato, *flacca*, is blocked in the last step of the ABA biosynthetic pathway [102]. This mutant does not accumulate as much ABA during periods of water deficit as does the wild type. After stress, the mutant accumulates 6% of the ABA that the wild type accumulates [103]. There are fewer proteins accumulating in *flacca* leaves than in the wild type during drought stress. Application of ABA to *flacca* restores the accumulation of this set of ABA-induced proteins [103]. The accumulation of three mRNAs was shown to be dependent on the accumulation of ABA during stress [14]. These mRNAs accumulated only in the drought-stressed wild type and were not detected in the drought-stressed ABA-deficient mutant. Mutants in maize and *Arabidopsis* have been used similarly to identify additional ABA-requiring genes (Table 1).

## B. Genes That Are Responsive to ABA but Do Not Require ABA

The ABA-deficient mutants have been used to define an additional set of genes, those that are responsive to ABA but do not require ABA for expression [30,104,105]. These genes are induced by ABA application, but unlike the ABA-requiring genes, they are induced by low-temperature and water-deficit treatments in the ABA-deficint mutants of *Arabidopsis*. Therefore, it has been concluded that these genes do not require elevated levels of endogenous ABA for expression but are capable of responding to ABA and may be called ABA-responsive genes. These results indicate that there are two pathways that can be followed to induce these genes, but it is unknown if the pathways converge or if there are two entirely separate pathways. ABA applications have also been used to show that there are a number of water deficit–induced genes that do not respond to ABA application [52,76]. These genes may be induced directly by the drought stress, or they may be controlled by other signaling mechanisms operating during water deficit.

## C. DNA Elements That Confer ABA Responsiveness

The conditions under which a gene is induced is controlled by the DNA elements acting within each gene. Therefore, to understand the mechanism of regulation of a specific gene, the DNA elements that confer responsiveness and the factors that recognize those elements must be identified and characterized. Studies have been initiated on genes that are regulated by ABA to understand the basis of ABA-regulated gene expression during stress and seed development. A region of *Em*, a member of the group 1 *lea* family from wheat, was identified which confers ABA inducibility upon a minimal 35S CaMV promoter [106,107]. A chimeric gene was constructed with a 646-bp segment of *Em* and the reporter gene β-glucuronidase (GUS). When this gene was introduced into rice protoplasts, GUS activity was increased 15- to 30-fold after ABA

**TABLE 1**  Genes That Have Been Demonstrated to Be Regulated by Elevated Levels of Endogenous ABA Resulting from Environmental Stress Using ABA-Deficient Mutants of Maize, *Arabidopsis,* and Tomato

| Genotype | Gene designation | Gene family | Stress induction[a] | Refs. |
|---|---|---|---|---|
| Maize | *rab*17 | *dhn/rab*/group 2 | D | 13 |
|  | *rab*28 | D-34/group 6 | D | 87 |
| *Arabidopsis* | *rab*18 | *dhn/rab*/group 2 | D, L | 15 |
|  | *lti*65 | *rd29/cor78* | D, L | 28,29,104 |
| Tomato | *le*4 | *dhn/rab*/group 2 | D, S, L | 11,14 |
|  | *le*16 | nsLTP | D, S, L | 14,64 |
|  | *le*20 | H1-histone | D, S, L | 85 |
|  | *le*25 | D-113/group 4 | D, S, L | 11,14 |

[a] D, water deficit; L, low-temperature stress; S, salinity.

application [106]. Further delineation of the 5'-flanking DNA of *Em* identified a 50-bp DNA sequence that is sufficient for ABA induction of GUS activity [107]. A DNA element, CACGTGGC, was conserved among other ABA-induced genes, including *rab16* [108]. However, this element is also related to G-box motifs, which are involved in light-induced gene expression [109]. This conserved element, which has been referred to as an ABA-responsive element (ABRE), has been shown to bind nuclear proteins [83,108]. A leucine zipper protein, EmBP-1, was identified and a cDNA cloned whose gene product binds the ABRE of *Em* in vitro [83]. This protein, EmBP-1, contains a leucine-zipper DNA binding motif, thought to be responsible for dimer formation, adjacent to a basic domain which is a cluster of positively charged amino acids responsible for sequence-specific recognition. The combination of the basic domain and the leucine zipper domain has been termed the bZIP domain. The basic domain of EmBP-1 is similar to that found in other transcription factors that bind the DNA sequence element T/CACGTGGC, including TAF-1 from tobacco, which binds the ABRE conserved in *rab16* genes [110], and GBF, which binds the G-box found in *rbcS* genes of tomato, *Arabidopsis*, and pea [109,111] (Figure 2). It is similar to other transcription factors that bind a similar DNA sequence element, TCCACGTAGA [112–114]. An element has been identified that confers ABA inducibility in a transient assay system, and a factor has been identified that can bind this DNA element. However, the identification of several similar transcription factors that bind similar DNA elements indicates that additional aspects of ABA inducibility are yet to be understood.

Additional specificity of the DNA elements may be derived from nucleotides surrounding the core DNA sequence. Williams et al. [115] characterized the sequences flanking the G-box or ABRE core CACGTG to determine how those DNA sequences affected binding. Based on the flanking sequence, the G-box elements have been divided into two different classes to which bind two distinct classes of G-box binding proteins. Further characterizations were made by using ACGT as the core and defining three different types of boxes by the nucleotide surrounding the core. EmBP-1, TAF-1, and HBP-1a were found to have the greatest affinity for the CACGTGC sequence [116].

TAF-1 from tobacco [110] and EmBP-1 [83] from wheat both bind DNA elements with the core sequence CACGTGGC and have the same DNA-binding basic motif. However, neither motif I, TACGTGGC [108], nor the hex tetramer, GGTGACGTGGC, can confer ABA responsiveness on a GUS reporter gene in transgenic tobacco [117]. But a hex mutant, GGACGCGTGGC, with greatly reduced affinity for TAF-1 can confer ABA responsiveness on a GUS reporter gene in transgenic tobacco [117]. These results indicate that there are multiple factors that can bind similar DNA motifs and that the exact DNA sequence determines which factors will recognize it. In addition, although factors have similar amino acid sequences within the DNA-binding domain, this information cannot be used to predict the exact DNA elements these factors will bind in vivo. Other regions of the transcription factor besides the DNA binding domain must also be important for determining DNA-binding specificity.

Since the studies on *Em*, other ABA-regulated genes have also been investigated. Another *lea* gene, *rab28*, is known to be regulated by endogenous ABA in maize [87]. This gene is expressed in vegetative organs during periods of water deficit, and it has been demonstrated that an ABRE is involved in the regulation of *rab28* by ABA and water stress [118]. Chimeric genes, with a portion of the *rab28* 5'-flanking DNA containing the ABRE and GUS, transfected into rice protoplasts were ABA responsive. An in vitro dimethyl sulfate footprinting experiment identified guanine residues within the ABRE that are involved in binding nuclear proteins [118]. Interestingly, when electrophoretic mobility shift assays were completed with proteins isolated from seeds or from drought-stressed leaves, complexes of two different sizes were found. Both complexes were shown to bind the ABRE [118]. Therefore, it is proposed that the same DNA sequence element or ABRE is involved in regulation of the expression of *rab28* in the seed during development and in the leaf during stress, but the transcription factors and activators that are involved in these two different types of regulation are not identical. Similarly, it has been shown that factors that bind

```
EmBP-1    MDERELKRERRKQSNRESARRSRLRKQQ        5'-GACACGTGGC-3'
TAF-1     QNERELKREKRKQSNRESARRSRLRKQA        5'-GCCACGTGGC-3'
O2        KMPTEERVRKRKESNRESARRSRYRKAA        5'-TCCACGTAGA-3'
```

**Figure 2**  Basic domain motif of three bZIP proteins: EmBP-1 [83], TAF-1 [110], and O2 [112]. Identical amino acids are shown in boldface type. The DNA-binding site for each of the bZIP is shown to the right of the amino acid sequence.

the G-box are complexes made up of at least two different proteins [119]. A complex array of events is required for ABA-regulated expression to occur in response to stress and/or developmental cues.

In transgenic tobacco plants it has been shown that expression of drought-regulated genes is properly regulated in seeds, but not in response to stress [120,121]. When −482 to +184 of *rab16B* from rice was translationally fused to GUS, expression was limited to developing seeds and was not induced by ABA or water stress in vegetative tissues. Expression of a related gene from rice, *rab16A*, was not detected in seeds or vegetative tissues [120]. These results may indicate that elements recognized in rice for ABA-regulated expression cannot be used in tobacco. However, another family member derived from maize, *rab17*, is correctly expressed in transgenic tobacco plants when −1330 to +29 is included in the reporter gene fusion constructs [122]. This gene also contains an ABRE, and only when that sequence is present in deletion constructs is the reporter gene responsive to ABA [122].

Promoter deletion analyses for several additional ABA-regulated genes are in progress [121]. In the resurrection plant, *Craterostigma plantagineum*, DNA elements required for regulation of the gene CDeT27-45, which is similar in amino acid sequence to the cotton gene *lea14*, were characterized using promoter deletion analyses. Using a transient expression system and *C. plantagineum* protoplasts, a region between −282 and −197 of the promoter was demonstrated to be required for ABA-regulated expression. Similar studies were completed on transgenic tobacco plants carrying this gene and it was found that there was not ABA-induced expression in tobacco leaves, although the genes were expressed during development in seeds and anthers. In the region required for ABA induction of CDet27-45 expression, there are no ABRE-like elements. Therefore, although the ABRE is found in many genes that are expressed during seed development and in response to ABA application, it is not found in all genes that are in the ABA-requiring category. As another example, *le16*, a gene expressed in wild-type tomato but not in the ABA-deficient mutant, does not contain a consensus ABRE [64]. Therefore, it is expected that there are multiple DNA elements involved in ABA-regulated expression. In addition, genes that are expressed during drought are also expressed during specific developmental stages and in specific cell types. Therefore, additional elements are required to control tissue- and organ-specific expression and other specific aspects of the expression pattern during water deficit.

## D. Recognition of ABA at the Cellular Level

Although the responsiveness of the gene is controlled by the DNA elements within the gene, the pathways that lead to transcription factor binding are also important to understand. For the ABA-requiring and ABA-responsive genes, there must be an ABA recognition event followed by the activation of a pathway that leads to gene induction. Cellular conditions that are required for any of these events to occur are not understood. The response of the cell to ABA may be altered by the physiological state of the cell. For example, sensitivity to ABA is altered by the osmotic potential of the cells; there is increased sensitivity to ABA with increased osmotic stress. In some cases, osmoticum can completely replace exogenous ABA. For *Em* mRNA accumulation in rice cell cultures, increasing concentrations of NaCl increased the accumulation of *Em* mRNA in response to suboptimal concentrations of ABA [123]. Therefore, the cells' response to ABA may be altered by the water potential or water content of the cell. However, another possibility should be considered. Because the cells are induced to accumulate ABA in response to osmotic stress, it becomes difficult to determine if the newly synthesized ABA is contributing to the induction of genes. The ABA that is synthesized in the cell may not be located in the same compartment within the cell as ABA that is applied to the cell [124]. Therefore, the plant may be more sensitive to endogenous ABA than to applied ABA. It is known that high levels of ABA must be applied to elicit a response similar to that stimulated by endogenous ABA concentrations.

In addition to understanding the mechanism of gene induction at the gene level, it must also be understood how the cell recognizes ABA and what signal transduction pathway is taken to gene induction. It is important to understand the aspects of the ABA molecule that are required for gene induction (Figure 3). The strategy has been taken to use ABA analogues to identify parts of the ABA molecule that are required for gene induction. Walker-Simmons et al. [125] compared optically pure ABA analogues in the induction of *rab*, *Em*, and *lea* group 3. The induction of *rab* and *lea* group 3 was similar with similar analogues; however, *Em* induction differed. These results support the conclusion that there is more than one mechanism for ABA regulation of gene expression. In the induction of *rab16* and *basi* in barley aleurone protoplasts, methylation of the carboxyl group had the least effect on the level of gene expression [126].

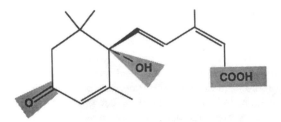

**Figure 3** Structure of (+)-abscisic acid. Modifications made to the molecule to study the molecular structure required for gene regulation are shown shaded. (From Ref. 126.)

Removal of the carboxyl group, the 1'-hydroxyl, and the 4'-carbonyl had the greatest reduction in gene expression [126]. The ABA molecule is still recognized for gene induction if the 1'-hydroxyl is removed (Figure 3).

In another attempt to determine what is required for ABA regulation of gene expression, protein synthesis has been inhibited by the application of cycloheximide to determine if proteins must be synthesized for ABA action. Interestingly, the requirement for protein synthesis in the response to ABA is dependent on the gene studied. Protein synthesis is required for ABA induction of *rd22* but not for *rd29* [88]. It was also found that ABA induction did not require protein synthesis for *rab16* [9]. Therefore, there are at least two different pathways of gene induction in response to ABA.

## V. FUTURE DIRECTIONS AND POSSIBLE AGRICULTURAL BENEFITS

Much progress has been made in the identification, isolation, and characterization of genes induced by different abiotic stresses. Studies on the isolation of genes and the regulation of specific genes have indicated that there are many similarities between stresses that result in cellular dehydration. Many of the genes that are induced by these stresses are also induced by ABA application, and several of these have been shown to require elevated levels of ABA for expression. Many of the genes induced during stress are predicted to play a protective role through direct protection of cellular contents or by altering the cellular water content.

The major challenge of the future is to obtain biochemical and genetic evidence that these gene products function in stress tolerance, improving the adaptability of plants to the environment. If adaptive gene products are characterized, these may have promise for use in the development of crop plants with increased stress tolerance. The use of different protective traits in combination, such as enhanced osmotic adjustment and overproduction of a hydrophilic gene product, may improve the chances of developing transgenic crop plants with an agricultural benefit.

## REFERENCES

1. JAD Zeevaart, RA Creelman. Annu Rev Plant Physiol Plant Mol Biol 39:439, 1988.
2. M Pierce, K Raschke. Planta 148:174, 1980.
3. JC Baker, C Steele, L Dure III. Plant Mol Biol 11:277, 1988.
4. L Dure III, M Crouch, J Harada, T-HD Ho, J Mundy, R Quatrano, T Thomas, ZR Sung. Plant Mol Biol 12:475, 1989.
5. JC Litts, GW Colwell, RL Chakerian, RS Quatrano. Nucleic Acids Res 15:3607, 1987.
6. M Espelund, S Saebe-Larssen, DW Hughes, GA Galau, F Larssen, KS Jakobsen. Plant J 2:241, 1992.
7. GA Galau, HY-C Wang, DW Hughes. Plant Physiol 99:783, 1992.
8. TJ Close, AA Kortt, PM Chandler. Plant Mol Biol 13:95, 1989.
9. J Mundy, N-H Chua. EMBO J 7:2279, 1988.
10. TJ Close, RD Fenton, A Yang, R Asghar, DA DeMason, DE Crone, NC Meyer, F Moonan. In: TJ Close, EA Bray, eds. Plant Responses to Cellular Dehydration During Environmental Stress. Current Topics in Plant Physiology, Vol 10. Rockville, MD: American Society of Plant Physiologists, 1993, p 104.
11. A Cohen, AL Plant, MS Moses, EA Bray. Plant Physiol 97:1367, 1991.
12. JA Godoy, JM Pardo, JA Pintor-Toro. Plant Mol Biol 15:695, 1990.
13. M Pla, A Goday, J Vilardell, J Gómez, M Pagès. Plant Mol Biol 13:385, 1989.
14. A Cohen, EA Bray. Planta 182:27, 1990.

15.  V Lång, ET Palva. Plant Mol Biol 20:951, 1992.
16.  K Schneider, B Wells, E Schmelzer, F Salamini, D Bartels. Planta 189:120, 1993.
17.  TJ Close, PJ Lammers. Plant Physiol 101:773, 1993.
18.  J Curry, MK Walker-Simmons. In: TJ Close, EA Bray, eds. Plant Responses to Cellular Dehydration During Environmental Stress. Current Topics in Plant Physiology, Vol 10. Rockville, MD: American Society of Plant Physiologists, 1993, p 128.
19.  L. Dure III. In: TJ Close, EA Bray, eds. Plant Responses to Cellular Dehydration During Environmental Stress. Current Topics in Plant Physiology, Vol 10. Rockville, MD: American Society of Plant Physiologists, 1993, p 91.
20.  J Vilardell, A Goday, MA Freire, M Torrent, MC Martínez, JM Torné, M Pagès. Plant Mol Biol 14:423, 1990.
21.  M Plana, E Itarte, R Eritja, A Goday, M Pagès, MC Martínez. J Biol Chem 266:22510, 1991.
22.  M Houde, J Danyluk, J-F Laliberté, E Rassart, RS Dhindsa, F Sarhan. Plant Physiol 99:1381, 1992.
23.  W Guo, RW Ward, MF Thomashow. Plant Physiol 100:915, 1993.
24.  LA Wolfraim, R Langis, H Tyson, RS Dhindsa. Plant Physiol 101:1275, 1993.
25.  L Dure III. Plant J 3:363, 1993.
26.  JK Roberts, NA DeSimone, WL Lingle, L Dure III. Plant Cell 5:769, 1993.
27.  A Cohen, EA Bray. Plant Mol Biol 18:411, 1992.
28.  K Nordin, T Vahala, ET Palva. Plant Mol Biol 21:641, 1993.
29.  K Yamaguchi-Shinozaki, K Shinozaki. Plant Physiol 101:1119, 1993.
30.  K Yamaguchi-Shinozaki, K Shinozaki. Mol Gen Genet 236:331, 1993.
31.  M Luo, L Lin, RD Hill, SS Mohapatra. Plant Mol Biol 17:1267, 1991.
32.  M Luo, J-H Liu, S Mohapatra, RD Hill, SS Mohapatra. J Biol Chem 267:15367, 1992.
33.  S Laberge, Y Castonguay, L-P Vezina. Plant Physiol 101:1411, 1993.
34.  SJ Gilmour, NN Artus, MJ Thomashow. Plant Mol Biol 18:13, 1992.
35.  S Kurkela, M Franck. Plant Mol Biol 15:137, 1990.
36.  S Kurkela, M Borg-Franck. Plant Mol Biol 19:689, 1992.
37.  AD Hanson, WD Hitz. Annu Rev Plant Physiol 33:163, 1982.
38.  TC Hsiao. Annu Rev Plant Physiol 24:519, 1973.
39.  AJ Delauney, DPS Verma. Mol Gen Genet 221:299, 1990.
40.  C-AA Hu, AJ Delauney, DPS Verma. Proc Natl Acad Sci U S A 89:9354, 1992.
41.  CL Williamson, RD Slocum. Plant Physiol 100:1464, 1992.
42.  EA Weretilnyk, AD Hanson. Proc Natl Acad Sci U S A 87:2745, 1990.
43.  D Bartels, K Engelhardt, R Roncarati, K Schneider, M Rotter, F Salamini. EMBO J 10:1037, 1991.
44.  DM Vernon, H Bohnert. EMBO J 11:2079, 1992.
45.  DM Vernon, JA Ostrem, HJ Bohnert. Plant Cell Environ 16:437, 1993.
46.  DM Vernon, MC Tarczynski, RG Jensen, HJ Bohnert. Plant J 4:199, 1993.
47.  SP Lee, TH-H Chen. Plant Physiol 101:1089, 1993.
48.  DB DeWald, HS Mason, JE Mullet. J Biol Chem 267:15958, 1992.
49.  PE Staswick, C Papa, J-F Huang. Plant Physiol 102S:27, 1993.
50.  FD Guerrero, JT Jones, JE Mullet. Plant Mol Biol 15:11, 1990.
51.  M Koizumi, K Yamaguchi-Shinozaki, H Tsuiji, K Shinozaki. Gene 129:175, 1993.
52.  MA Schaffer, RL Fischer. Plant Physiol 87:431, 1988.
53.  JC Cushman, G Meyer, CB Michalowski, JM Schmitt, HJ Bohnert. Plant Cell 1:715, 1989.
54.  JC Thomas, RL De Armond, HJ Bohnert. Plant Physiol 98:626, 1992.
55.  H Sze. Annu Rev Plant Physiol 36:175, 1985.
56.  TK Surowy, JS Boyer. Plant Mol Biol 16:251, 1991.
57.  X Niu, J-K Zhu, ML Narasimhan, RA Bressan, PM Hasegawa. Planta 190:433, 1993.
58.  RJ Anderberg, MK Walker-Simmons. Proc Natl Acad Sci U S A 89:10183, 1992.
59.  NK Singh, AK Handa, PM Hasegawa, RA Bressan. Plant Physiol 79:126, 1985.
60.  AK Konononwicz, KG Ragothama, AM Casas, M Reuveni, A-E Watad, D Liu, R Bressan, PM Hasegawa. In: TJ Close, EA Bray, eds. Plant Responses to Cellular Dehydration During Environmental Stress. Current Topics in Plant Physiology, Vol 10. Rockville, MD: American Society of Plant Physiologists, 1993, p 144.
61.  WK Roberts, CP Selitrennikoff. J Gen Microbiol 136:1771, 1990.
62.  AJ Vigers, S Weidemann, WK Roberts, M Legrand, CP Selitrennikoff, B Fritig. Plant Sci 83:155, 1992.
63.  M Yamada. Plant Cell Phyiol 33:1, 1992.
64.  AL Plant, A Cohen, MS Moses, EA Bray. Plant Physiol 97:900, 1991.
65.  S Torres-Schumann, JA Godoy, JA Pintor-Toro. Plant Mol Biol 18:749, 1992.
66.  MA Hughes, MA Dunn, RS Pearce, AJ White, L Zhang. Plant Cell Environ 15:861, 1992.
67.  P Sterk, H Booij, GA Schellekens, A Van Kammen, SC De Vries. Plant Cell 3:907, 1991.
68.  FR Terras, IJ Goderis, F Van Leuven, J Vanderleyden, BPA Cammue, WF Broekaert. Plant Physiol 100:1055, 1992.
69.  RK Hajela, DP Horvath, SJ Gilmour, MF Thomashow. Plant Physiol 93:1246, 1990.
70.  C Lin, MF Thomashow. Plant Physiol 99:519, 1992.

71. MF Thomashow. In: TJ Close, EA Bray, eds. Plant Responses to Cellular Dehydration During Environmental Stress. Current Topics in Plant Physiology, Vol 10. Rockville, MD: American Society of Plant Physiologists, 1993, p 137.

72. C Borkird, B Claes, A Caplan, C Simoens, M Van Montagu. J Plant Physiol 138:591, 1991.

73. C Almoguera, J Jordano. Plant Mol Biol 19:781, 1992.

74. E Vierling. Annu Rev Plant Physiol Plant Mol Biol 42:579, 1991.

75. WL Downing, F Mauxion, M-O Fauvarque, R-P Reviron, D de Vienne, N Vartanian, J Giraudat. Plant J 2:685, 1992.

76. K Yamaguchi-Shinozaki, M Koizumi, S Urao, K Shinozaki. Plant Cell Physiol 33:217, 1992.

77. H Höfte, L Hubbard, J Reizer, D Ludevid, EM Herman, MJ Chrispeels. Plant Physiol 99:561, 1992.

78. D Bar-Zvi, T Shagan. Plant Physiol 101:1397, 1993.

79. T Shagan, D Meraro, D Bar-Zvi. Plant Physiol 102:689, 1993.

80. C Maurel, J Reizer, JI Schroeder, MJ Chrispeels. EMBO J 12:2241, 1993.

81. J Gómez, D Sanchez-Martínez, V Stiefel, J Rigau, P Puigdomènech, M Pagès. Nature 334:262, 1988.

82. MD Ludevid, MA Freire, J Gómez, CG Burd, F Alberico, E Giralt, G Dreyfuss, M Pagès. Plant J 2:999, 1992.

83. MJ Guiltinan, WR Marcotte Jr, RS Quatrano. Science 250:267, 1990.

84. I Winicov. Plant Physiol 102:681, 1993.

85. EA Bray, MS Moses, R Imai, A Cohen, AL Plant. In: TJ Close, EA Bray, eds. Plant Responses to Cellular Dehydration During Environmental Stress. Current Topics in Plant Physiology, Vol 10. Rockville, MD: American Society of Plant Physiologists, 1993, p 167.

86. ND Iusem, DM Bartholomew, WD Hitz, PA Scolnik. Plant Physiol 102:1353, 1993.

87. M Pla, J Gómez, A Goday, M Pagès. Mol Gen. Genet 230:394, 1991.

88. K Yamaguchi-Shinozaki, K Shinozaki. Mol Gen Genet 238:17, 1993.

89. R Bassuner, H Bäumlein, A Huth, R Jung, U Wobus, TA Rapoport, G Saalbach, K Müntz. Plant Mol Biol 11:321, 1988.

90. WJ Hurkman, HP Tao, CK Tanaka. Plant Physiol 97:366, 1991.

91. E Dratewka-Kos, S Rahman, ZF Grzelczak, TD Kennedy, RK Murray, BG Lane. J Biol Chem 264:4896, 1989.

92. CB Michalowski, HJ Bohnert. Plant Physiol 100:537, 1992.

93. D Bartels, C Hanke, K Schneider, D Michel, F Salamini. EMBO J 8:2771, 1992.

94. W Kolanus, C Scharnhorst, U Kühne, F Herzfeld. Mol Gen Genet 209:234, 1987.

95. B Grimm, E Kruse, K Kloppstech. Plant Mol Biol 13:583, 1989.

96. PC LaRosa, Z Chen, DE Nelson, NK Singh, PM Hasegawa, RA Bressan. Plant Cell 4:513, 1992.

97. G Iturriaga, K Schneider, F Salamini, D Bartels. Plant Mol Biol 20:555, 1992.

98. MC Tarczynski, RG Jensen, HJ Bohnert. Proc Natl Acad Sci U S A 89:2600, 1992.

99. MC Tarczynski, RG Jensen, HJ Bohnert. Science 259:508, 1993.

100. AD Parry. Methods Plant Biochem 9:381, 1993.

101. FD Guerrero, JE Mullet. Plant Physiol 80:588, 1986.

102. IB Taylor. In: WJ Davies, HG Jones, eds. Abscisic Acid: Physiology and Biochemistry. Environmental Plant Biology Series Oxford: Bios Scientific Publishers, 1991, p 23.

103. EA Bray. Plant Physiol 88:1210, 1988.

104. SJ Gilmour, MF Thomashow. Plant Mol Biol 16:1233, 1991.

105. K Nordin, P Heino, ET Palva. Plant Mol Biol 16:1061, 1991.

106. WR Marcotte Jr, CC Bayley, RS Quatrano. Nature 335:454, 1988.

107. WR Marcotte Jr, SH Russell, RS Quatrano. Plant Cell 1:969, 1989.

108. J Mundy, K Yamaguchi-Shinozaki, N-H Chua. Proc Natl Acad Sci U S A 87:1406, 1990.

109. G Giuliano, E Pichersky, VS Malik, MP Timko, PA Skolnik, AR Cashmore. Proc Natl Acad Sci U S A 85:7089, 1988.

110. K Oeda, J Salinas, N-H Chua. EMBO J 10:1793, 1991.

111. U Schindler, AE Menkens, H Beckmann, JR Ecker, AR Cashmore. EMBO J 11:1261, 1992.

112. RJ Schmidt, FA Burr, MJ Aukerman, B Burr. Proc Natl Acad Sci U S A 87:46, 1990.

113. RJ Schmidt, M Ketuadat, MJ Aukerman, G Hoschek. Plant Cell 4:689, 1992.

114. MJ Varagona, RJ Schmidt, NV Raikhel. Plant Cell 4:1213, 1993.

115. ME Williams, R Foster, N-H Chua. Plant Cell 4:485, 1992.

116. T Izawa, R Foster, N-H Chua. J Mol Biol 230:1131, 1993.

117. E Lam, N-H Chua. J Biol Chem 266:17131, 1991.

118. M Pla, J Vilardell, MJ Guiltinan, WR Marcotte Jr, MF Niogret, RS Quatrano, M Pagès. Plant Mol Biol 21:259, 1993.

119. NC de Vetten, G Lu, RJ Ferl. Plant Cell 4:1295, 1992.

120. K Yamaguchi-Shinozaki, M Mino, J Mundy, N-H Chua. Plant Mol Biol 15:905, 1990.

121. D Michel, F Salamini, D Bartels, P Dale, M Bagga, A Szalay. Plant J 4:29, 1993.

122. J Vilardell, J Mundy, B Stilling, B Leroux, M Pla, G Freyssinet, M Pagès. Plant Mol Biol 17:985, 1991.

123. RM Bostock, RS Quatrano. Plant Physiol 98:1356, 1992.

124. EA Bray, JAD Zeevaart. Plant Physiol 80:105, 1986.

125. MK Walker-Simmons, RJ Anderberg, PA Rose, SR Abrams. Plant Physiol 99:501, 1992.

126. RM Van der Meulen, F Heidekamp, B Jastorff, R Horgan, M Wang, J Plant Growth Regul 12:13, 1993.

# 37

# How Plants Adapt Their Physiology to an Excess of Metals

## Martine Bertrand and Jean-Claude Guary

*National Institute for Marine Sciences and Techniques, Conservatoire National des Arts et Métiers, Cherbourg, France*

## Benoît Schoefs

*University of South Bohemia, Budejovice, Czech Republic\**

## I. INTRODUCTION

It is well established that trace metals are indispensable for physiological and biochemical processes in plants. For instance, in plants growing in a Zn-deficient medium the stem fails to elongate and there is a marked reduction of shoot fresh weight [1,2]. Resupply with Zn stimulates growth [1]. At the biochemical level, various metals are involved in the structure and the biological activity of many proteins (Table 1). Other molecules such as chlorophylls require Mg for their structure and function.

However, when metal levels increase in the environment, they reach concentrations that plants can no longer tolerate [3]; their ions rapidly become highly toxic. This is especially true for heavy metals such as Cd, Ag, Hg, and Pb,† which are not known to play a physiological role in organisms. Plants have developed defense mechanisms against metal pollution. These include the control of metal influx, active metal efflux as well as intracellular, extracellular metal ion sequestration, and exclusion [4]. In this chapter, we first describe adverse effects of metals on plant physiology and then focus more specifically on the plant's defense mechanisms. Both aquatic and terrestrial plants are considered.

The following abbreviations are used in this chapter: HM, heavy metal; MT, metallothionein; PC, phytochelatin.

## II. ADVERSE EFFECTS OF METAL IONS ON PLANT PHYSIOLOGY AND BIOCHEMISTRY

The benefit that a plant can usually obtain from metal assimilation can turn to disaster when the metallic concentration increases or when nonessential metals such as Cd, Hg, or Pb are absorbed. One can tentatively define metal toxicity by stating that a metal becomes toxic when there is no free specific site for it. Then it binds to any molecule that cannot chemically refuse it, modifying significantly the functional properties of the host molecule. Consequently, the general metabolism is modified. In the worse case, metal binding to functional proteins is irreversible, causing death of the cell.

---

\* *Current affiliation:* Université Joseph Fourier, Grenoble, France.
† When no indication of the oxidation state is given, the symbol refers to the element and not to a particular chemical form of it.

**TABLE 1**   Some Proteins That Require Metals for Either Structural or Activity Purposes

| Proteins | Type of metal(s) | Role of metal(s) |
|---|---|---|
| Aldehyde oxidase | Mo, Fe | Catalysis |
| Carbonic anhydrase | Zn | Structure and catalysis |
| Cytochrome oxydase | Cu | Electron carrier |
| Formiate dehydrogenase | Fe, Mo, Se, W | Catalysis |
| Nitrogenase | Fe, Mo | Catalysis |
| Peroxidase | Fe | Catalysis |
| Plastocyanin | Cu | Electron carrier |
| Sulfite oxidase | Fe, Mo | Catalysis |
| Superoxide dismutases | Cu/Zn; Fe; Mn | Structure and catalysis |
| Water oxidase | Mn | Catalysis |
| Xanthine oxidase | Fe, Mo | Electron carrier |

Making a generalization about metal toxicity is difficult. This is due to the multidimensional variations of the parameters chosen for experiments, including concentrations, biospeciation, duration of incubation, type of plant material (e.g., whole plant, algae, isolated organelles), and type of the affected target(s) (e.g., organ, cell, molecule).

A dysfunction due to an excess of metal(s) can be seen at the morphological level; e.g., roots or leaves grow more slowly [5]. Plants growing in soil containing high concentrations of Al produce a shallow root system and are sensitive to drought. In addition, they can use other soil nutrients poorly [6–8]. Under metal excess, a general chlorosis of young leaves is observed [5,9,10], reflecting a weaker chlorophyll synthesis capacity [11]. Reduction of cell growth has been reported with green algae [12,13]. In microalgae, metals can also affect cell division and separation [12].

At the cellular level, the primary target of metals is the plasma membrane, leading to loss of K by leakage into the extracellular spaces [14]. In lichens, Cu triggers larger K efflux than Pb and Zn [15]. Metal permeation in plant cells can be facilitated by chelating agents such as by some fungicides (dithiocarbamate derivatives), which are pronounced lipophilic compounds. In contrast, hydrophilic metal chelates are less available for plants and are toxic for microorganisms [16–18]. Once inside the cell, the hydrophilic metal chelates can release the metal ions [19]. Cu was found to inhibit the enzyme responsible for the destruction of hydrogen peroxide in cells; this remaining toxic compound oxidizes lipids that in turn damage membranes [20]. In addition, harmful activated oxygen species such as hydroxyl radicals and hydrogen peroxide are triggered in the presence of excess $Fe^{2+}$, $Cd^{2+}$, or $Cu^{2+}$ [21,22], finally damaging DNA. Chloroplast ultrastructure is also damaged by an excess of metals [23].

At the biochemical level, heavy metals (HMs) have been found on pigments. Cd, Cu, Hg, Ni, Pb, and Zn can substitute for the chlorophyll Mg [24,25] and lead to a breakdown of photosynthesis and even to death of the cells. When dead, the plant remains green if Cu- or Zn-chlorophylls are present because of the high stability of these substituted pigments. This was observed for plants stressed in shade [25]. Under strong light, HM-treated plants bleach almost completely because of chlorophyll decay; a structural feature was suggested by H. Küpper (personal communication) to explain that HMs cannot reach most of the chlorophylls in high-light conditions. Generally speaking, photosynthesis is less efficient when HMs have entered the chloroplasts; the pollutants interact at several steps [11], modifying, for instance, the structure of pigment-protein complexes [26]. Interactions between toxic metals and proteins alone have also been reported [21,26,27]. HMs may interact with thiol or histidyl groups of proteins, whose functions are then inhibited. In this way, several steps of the chlorophyll biosynthetic pathway [reviewed in 28] are altered [29]. In contrast, some toxic metal ions may increase enzyme activity or induce synthesis of specific proteins. For instance, Cd increases the activity of arginine carboxylase (EC 4.1.1.19) [30], an enzyme involved in the putrescine biosynthesis pathway, whereas Al increases the putrescine level in a different way.

## III.   MECHANISMS OF METAL ION UPTAKE BY CELLS

Free cations usually constitute the most available form of metals for living organisms [31]. The components associated with metals in the plant environment also have to be considered; for instance, the metal uptake by *Ulva lactuca* increases when salinity decreases [32]. The primary active transport system in algae has been identified as an $Mg^{2+}$ adenosinetriphosphatase (ATPase)–driven, V-sensitive electrogenic $H^+$ efflux pump [33]. Metallic ion accumulation by cells, such as microalgae, consists of two phases: a rapid phase of metal binding to the cell wall (i.e., biosorption) followed by a slower phase due to the simultaneous effects of growth and surface adsorption, active and passive transport [34,35]. Depending on the metal ions and on the algal species, the proportion of metals during the first phase can account for up to 50% [35]. The first phase can be described using the Freundlich adsorption isotherm [36], but a slight deviation can be observed for high HM concentrations [35,37]. The deviation can be explained by a competition between metal ions for available binding sites. A convenient way to characterize the adsorption of metal ions on algae is to use the Scatchard plot [38], from which the maximal binding capacity and the binding constant of metals can be estimated.

## IV.  METAL SUBCELLULAR LOCALIZATION

Metals are localized in different compartments of the cells; Table 2 shows the proportions measured in various organisms. A study of the accumulation of Co, Zn, and Mn by *Chlorella* indicated, for instance, that the concentration of these elements is higher in vacuoles than in cytoplasm [35]. With excess of Ni, Molas [39] observed vacuolization in leaf mesophyll cells of *Brassica oleracea*. In natural metal-rich habitats, some hyperaccumulator plants can accumulate very high amounts of a metal (Ni, most often) without showing any toxicity symptoms or reduction in growth [40,41]. *Thlaspi caerulescens* can have 20,000 $\mu$g Zn $g^{-1}$ dry weight of shoots [42]. Küpper et al. [42] estimated that more than 60% of the metal accumulated by leaves was present in the epidermal vacuoles. The mechanism involved in this preferential accumulation is not known. The Zn was found in soluble form and not in deposits as globular crystals as described by Vazquez et al. [43]. Anyway, a metal in excess changes the microelement balance and photosynthesis of a plant [44].

## V.  DEFENSE MECHANISMS AGAINST METAL TOXICITY

To avoid undesirable metal penetration, plants are able to extrude material that can chelate free metallic cations in the extracellular space. Toxic metals can also be trapped once they are inside the cells. Then they are either rapidly excluded from the cells or stored in vacuoles. Metallic sequestration often involves the formation of complexes between a metal cation and functional groups (e.g., carboxyl, carbonyl, sulfonate, phosphate) present on the surface or inside the porous structure of the biological material [45].

### A.  Extracellular Metal Sequestration

Differences in Al tolerance between several bean species have been attributed to the capacity of roots to exude citric acid, a strong Al chelator [46]. A similar conclusion was drawn for monocotyledons (barley,

**TABLE 2**  Subcellular Localization of Some Metals in Different Plants

| Metal | Localization | Organism | Reference |
|---|---|---|---|
| Co | Cell wall (38%) Cytosol (9%) Vacuoles and organelles (10%) Insoluble (43%) | *Chlorella* | 35 |
| Cu | Vacuole, chloroplast, nuclei | *Armeria maritima* | 37 |
| Mn | Cell wall (43%) Cytosol (21%) Vacuoles and organelles (15%) Insoluble (21%) | *Chlorella* | 35 |
| Pb | Dictyosome, cell wall | *Zea mays* | 39 |
| Zn | Cell wall (52%) Cytosol (16%) Vacuoles and organelles (10%) Insoluble (22%) | *Chlorella* | 35 |

wheat, maize), for which better resistance to Al toxicity is associated with root exudation of citric acid, succinic acid, and other organic acids [47–50].

It has been reported that in seawater, Cu, Cd, and Zn can also be strongly complexed by organic ligands which could have originated from organisms [51,52]. The ligands can be phytochelatins (PCs) (formerly metallothionein class III [53]; according to Zenk [54], the term PC as used here includes cadystins), usualy described as occasional intracellular low-molecular-weight proteins containing a high proportion of SH groups [55]. However, the role of PCs outside the cell remains to be conclusively demonstrated because up to the present the only example is the release of Cd-PC complexes from phytoplankton. In addition, these complexes are not very stable and as a consequence Cd is reabsorbed by the cells [56]. Nevertheless, Cd extrusion may be an important adaptative mechanism for lowering the concentration of free Cd in the cells.

Metals can also be trapped by polysaccharides, alginates, and other slimes secreted by algae [57–59]. Metal ion captured by algal polysaccharides is governed by ion-exchange selectivity [60] and is proportional to the total carboxylic acid content [45]. Again, the capacity to bind these metal ions depends on the chemical composition of secreted compounds, which unfortunately remains too often not determined. Usually, Pb is not chelated [61]. However exceptions have been described [62,63]. According to these authors, the capacity to bind Pb could be due to the presence of large amounts of proteinaceous substances in the compounds extruded by the algae.

## B.  Metal Excretion and Volatilization

When comparing a Cu-tolerant and a Cu-nontolerant strain of the green alga *Chlorella vulgaris*, Foster [12] found the first strain capable of growing in a medium containing 1 mg Cu $L^{-1}$, whereas the growth of the sensitive strain was already completely inhibited at 0.3 mg Cu $L^{-1}$ This was explained by excretion of Cu by the tolerant strain. *Chlorella* is also able to exclude Zn [64].

On transferring the brown macroalgae *Ascophyllum nodosum* from a Zn-polluted zone to a nonpolluted one, Eide et al. [65] noted a decrease in the Zn concentration in the algal thallus. This decrease was interpreted as the result of Zn excretion, possibly bound to phenols.

The mechanism of Hg and phenyl-Hg-acetate resistance in *Chlorella* appears to be similar to that observed with bacteria and yeast; i.e., they are volatilized [64]. The enzymatic system partially purified from Hg-resistant cells [64] is able to volatilize both Hg and phenyl-Hg-acetate in vitro and is very similar to that isolated from an Hg-resistant strain of *Escherichia coli*.

## C.  Intracellular Metal Sequestration

### 1.  Proteins

Metallothioneins (MTs) and phytochelatins (PCs) are two protein families capable of sequestering metals. They are both cystein-rich polypeptides having the ability to form metal-thiolate clusters. MTs have been characterized as gene-encoded proteins [66,67] with a molecular weight of 5000–20,000, whereas PCs are smaller enzymically synthesized polypeptides (molecular weight in the range 500–2300) [53]. Initial analysis of the organism responses to HMs has identified MTs as proteins rather typical of vertebrates [68,69] and PCs as rather typical of plants [54,55]. Actually, MTs have also been found in lower organisms. For fungi, Kneer and Zenk [70] generalized the original finding of Mehra et al. [71] that HMs induce both MT and PC synthesis. However, the role of MTs does not appear to be restricted to metal detoxification [66], and the complete roles of both metalloproteins are not fully understood.

METALLOTHIONEINS   The relatively high expression of MTs in diverse plant tissues indicates a fundamental role of these proteins. They are involved in homeostasis, i.e., the mechanism regulating the availability of metal ions in cells. This statement is supported by the fact that MTs and MT-like proteins are also expressed in plants growing in the absence of metal excess. In 1996, Zenk [54] noted that there is no experimental evidence that these "plant MTs" are involved in the detoxification of HMs. Consequently, plant MT-like proteins will not be described further here. The interested reader may read the comprehensive reviews on MTs [68,69].

PHYTOCHELATINS   The ability to synthesize PCs in response to HM pollution is a general feature of the plant kingdom, including algae [72]. A few organisms such as *Viola calaminaria*, *Thlaspi caerulescens*, or *Brassica juncea* are naturally capable of growing in an environment extremely enriched in metals (e.g., mine soils and deposits) or even directly on metal veins [41,73]. This adaptation of plants is not well understood. It was shown, however, that they are particularly enriched in PCs. These polypeptides are found in roots and stems of higher plants but not in leaves or fruits [74].

PCs are characterized by the general structure $(\gamma\text{-Glu-Cys})_n\text{-Gly}$, where $n = 2\text{--}11$ (Figure 1). They can, however, differ in their C-terminal amino acid [3,54]. The cystein residues ensure metal coordination via their thiol group. What were called cadystins for Cd-containing complexes in the fission yeast *Schizosaccharomyces pombe* [75] correspond in fact to $PC_2$ and $PC_3$. The high percentage of glutamic acid residues makes PCs extremely water soluble, explaining why they are usually localized in the cytoplasm. Optimized structures of different Cd-PC complexes have been proposed [76].

The fact that amino acids of PCs are linked by the $\gamma$-carbon of the carboxylic acid residue of glutamate excludes the usual protein synthesis via translation in ribosomes [77]. Actually, the PC production starts with two molecules of glutathione $\gamma$-Glu-Cys-Gly with the loss of one Gly to form the smallest PC ($n = 2$) [78,79]. To increase $n$, more free glutathione (a major intracellular reductant) is required; every

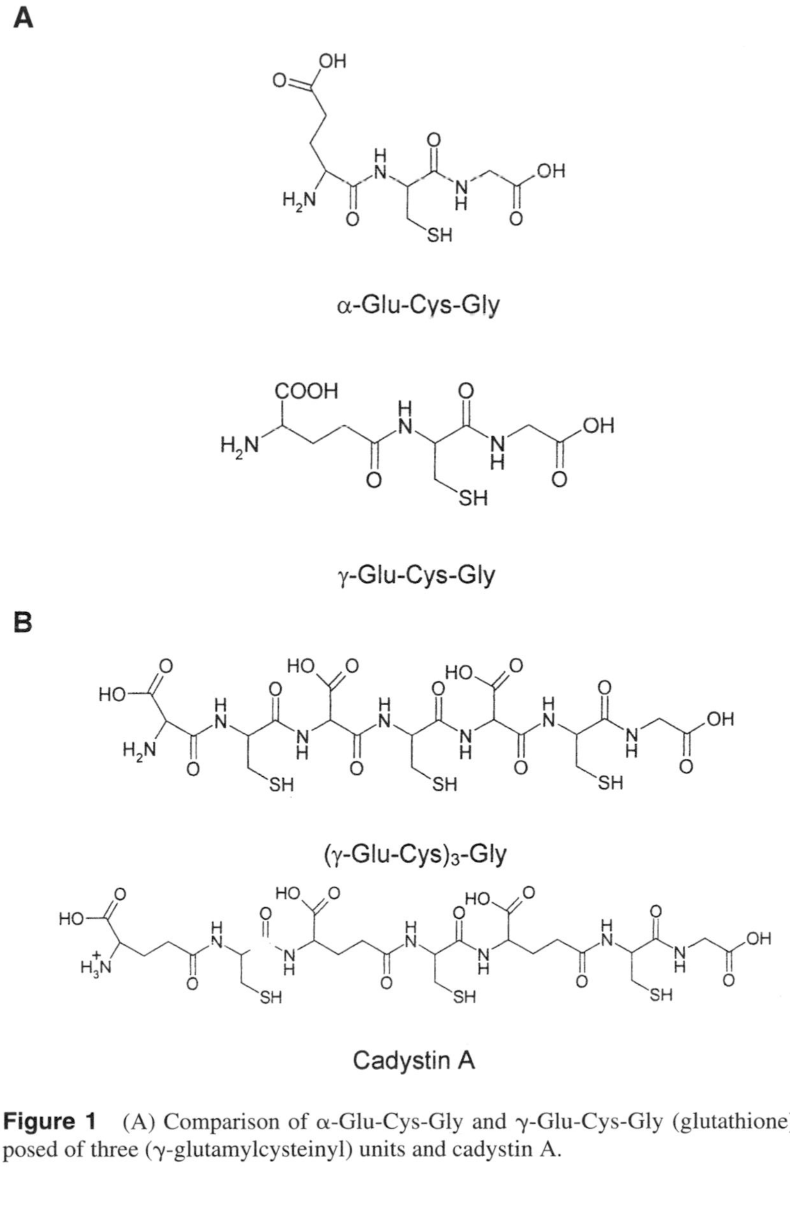

**A**

α-Glu-Cys-Gly

γ-Glu-Cys-Gly

**B**

$(\gamma\text{-Glu-Cys})_3\text{-Gly}$

Cadystin A

**Figure 1**  (A) Comparison of α-Glu-Cys-Gly and γ-Glu-Cys-Gly (glutathione). (B) Structure of PC composed of three (γ-glutamylcysteinyl) units and cadystin A.

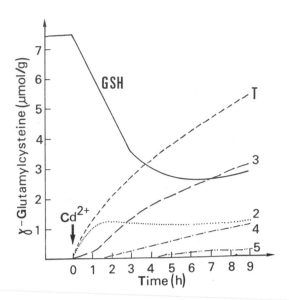

**Figure 2** Time course of PC induction and glutathione (GSH) consumption after administration of 200 μM $Cd(NO_3)_2$ to a cell suspension culture of *Rauvolfia serpentina*. Quantities of glutathione (GSH,—), total PC (-----), and individual PCs with *n* (number of γ-glutamylcysteine units per molecule) = 2 (······), 3 (———), 4 (—·—), or 5 (–··–) are expressed as μmol of γ-glutamylcysteine per g of cell dry weight. (From Ref. 79.) With permission, from the Annual Review of Plant Physiology and Plant Molecular Biology, Volume 41, © 1990, by Annual Reviews.

tripeptide loses its Gly, and the two other remaining amino acids are bound to the PC. Formation of longer PC chains increases the possibility for metal ions to be sequestrated promptly and protect HM-sensitive enzymes. The enzyme phytochelatin synthase (γ-glutamylcysteinyltransferase, EC 2.3.2.15) was found to ensure PC synthesis [78]. Consequently, as PCs become longer, the glutathione concentration drops drastically (Figure 2). Prolonged PC synthesis requires de novo synthesis of the glutathione-synthesizing enzymes [80]. When PC concentrations are sufficient to chelate the metals, the enzyme synthesis is turned off [81]. The appearance of PC only 10 to 15 min after the entry of HM in cells is well correlated with the fact that the enzymes involved in PC synthesis are constitutive in nature [54]. Chen et al. [74] demonstrated that PC synthase requires free metal ions such as $Cd^{2+}$ or $Ag^+$ for activity.

When HM pollution occurs, plants synthesize metal-glutathione complexes [53,54], which are exported to the vacuoles [82] or transfer the metal ($Cd^{2+}$, $Pb^{2+}$, $Cu^+$, and $Hg^{2+}$) to PCs that have a higher affinity for metals than glutathione itself [83–85]. PC-metal complexes accumulate in vacuoles, where inorganic sulfide and sulfite are incorporated, making the complexes more stable [86–88]. Increasing sulfide incorporation into PC leads to a substantial increase in $Cd^{2+}$/PC stoichiometry [88,89]. Mehra et al. [90] suggested that the sulfate reduction pathway, leading to the production of sulfide, may allow the formation of PC-Cd-S complexes in *Candida glabrata* cells before internalization in the vacuole. Others have suggested that the formation of PC-Cd-S complexes in fission yeast may take place in vacuoles [91]. An ATP-dependent Cd-PC active transport has been described in the tonoplast of oat root cells [92]. Once in the vacuoles, the metals are liberated and the PC moiety is degraded [93]. The metals are complexed with vacuolar organic acids while the individual amino acids can enter the cytosol again [54].

Although PC production by algae is known from laboratory experiments [56,94,95], until recently there was no evidence for metal toxicity in phytoplankton communities in situ. In 1997, Ahner et al. [96] indicated that PCs were synthesized in situ, suggesting that they can be used to detect trace metal pollution because the PC concentration increases well before other physiological parameters such as growth rate or chlorophyll content are affected.

All the studies, except Ref. 97, report that PC synthase is activated by metals [78,80]. The efficiency of various metals for PC synthase activation has been tested with various plant material [53,54,74,98]. Although the capacity of different metals to induce PC synthesis varies from one study to another, all designate $Cd^{2+}$ as the strongest inducer. Despite the fact that many metals are able to trigger PC synthesis, only some of them (i.e., $Cd^{2+}$, $Hg^{2+}$, $Pb^{2+}$, $Cu^+$, $Ag^+$) are able to form complexes with PC [98–101]. An alternative role for PC in metal tolerance could be a shuttle activity for the transfer of metals from the cytoplasm to the vacuole [102] and vice versa because metals used for growth can be stored in vacuoles [103,104]. Interestingly, $Mn^{2+}$ and $Zn^{2+}$ concentrations are higher in the vacuoles of the green alga *Chlorella* than in the cytoplasm [35]. In higher plants, Zn accumulates in vacuoles [105], usually chelated by organic acids [106,107] or precipitated as Zn-phytate [108]. In fact, PC and PC synthase could be involved in homeostasis [93].

## 2. *Proline*

Cu, Cd, and Na induce proline accumulation in some freshwater algae [13,109] and higher plants [110–112]. In *Chlorella*, proline accumulation reduces the internalization of Cu. Although the exact mechanism is still unknown, it is hypothesized that proline decreases Cu absorption [13], probably through inhibition of $K^+$ leakage as demonstrated with the cyanobacterium *Anacystis nidulans* [109]. It is interesting to note that proline may stabilize membranes [113].

In higher plants, proline can also be involved in the chelation of excess cytoplasmic metal ions that have a preference for nitrogen or oxygen coordination [114]. These metals, e.g., Ni or Zn, are usually poor inducers of PC [53]. Although there is no conclusive evidence for a direct contribution of proline in cellular HM detoxification, it is interesting to note that constitutive proline levels are higher in metal-tolerant ecotypes of *Silene vulgaris* [115] and *Armeria maritima* [116]. Proline production could not be a direct effect of HM stress but rather a consequence of the water stress induced by metals.

## VI. CAN METAL RESISTANCE OF PLANTS BE IMPROVED?

From the preceding sections, it is apparent that much work remains to be done to unveil fully the mechanisms involved in plant metal resistance. However, using the data already available, some remediation of inorganic contaminants has been successfully achieved thanks to genetic assays.

The insertion of the animal MT (class I) gene into the genome of either higher plants [117] or cyanobacteria [118] confers a stimulation of HM tolerance in the transformants.

We reported earlier that an Al resistance is linked to the secretion of citric acid [46]. One way to increase the resistance is to make the plant produce more citric acid. This was achieved for two Al-sensitive species (tobacco and papaya) after insertion in their genome of the bacterial gene coding for the citric acid synthase. Consequently, roots of the transgenic plants secreted 5–6 times more citric acid and were in turn 10 times more metal tolerant than the wild type [119,120] as they absorbed less Al.

Bacteria can reduce a number of HM ions and oxyanions to less toxic oxidation states [121]. For instance, Hg resistance in gram-negative bacteria is located on an operon that encodes different kinds of proteins: (1) transport proteins that bind and transfer Hg into the cell; (2) an organomercury lyase that catalyzes the protonolysis of C—Hg bonds, releasing $Hg^+$; and (3) a mercuric ion reductase that reduces $Hg^+$ to $Hg^0$ that is in turn volatilized from the cell [122,123]. The gene encoding the mercuric ion reductase was slightly modified and cloned successfully in *Arabidopsis*, which became resistant to mercury [124].

Only these examples let us hope that other current trials in genetic engineering [125] will provide transformants with improved resistance to toxic metals.

## ACKNOWLEDGMENTS

The authors thank Dr. R. K. Mehra (University of California, San Diego) for his constructive help and H. Küpper (University of South Bohemia, Czech Republic) for revealing his latest results. B. Schoefs and M. Bertrand thank the Ministry of Education, Youth and Sports of the Czech Republic (grant VS-96085) for their financial support.

## REFERENCES

1.  IH Cakmak, H Marschner, F Bangerth. Effect of zinc nutritional status on growth, protein metabolism and levels of indole-3-acetic acid and other phytohormones in bean (*Phaseolus vulgaris* L.). J Exp Bot 40:405–412, 1989.

2.  B Hossain, N Hirata, Y Nagatomo, M Suiko, H Takaki. Zinc nutrition and levels of endogenous indole-3-acetic acid in radish shoots. J Plant Nutr 21:1113–1128, 1998.

3.  WE Rauser. Phytochelatins and related peptides: structure, biosynthesis and function. Plant Physiol 109:1141–1149, 1995.

4.  S Silver. Plasmid-determined metal resistance mechanisms: range and overview. Plasmid 27:1–3, 1992.

5.  S Samantaray, GR Rout, P Das. Manganese toxicity in *Echinochloa colona*: effects of divalent manganese on growth and development. Isr J Plant Sci 45:9–12, 1997.

6.  MD Kauffman, EH Gardner. Segmental liming of soil and its effects on the growth of wheat. Agron J 70:331–336, 1978.

7.  CD Foy. Physiological effects of hydrogen, aluminium and manganese toxicities in acid soils. In: F Adams, ed. Soil Acidity and Liming. Monograph 12. Madison, WI: American Society of Agronomy, 1984, pp 57–97.

8.  DC Foy, JJ Murray. Developing aluminium-tolerant strains of tall fescue for acid soils. J Plant Nutr 21:1301–1325, 1998.

9.  HW Woolhouse. Toxicity and tolerance in the response of plants to metals. In: OL Lange, PS Nobel, CB Osmond, H Ziegler, eds. Encyclopedia of Plant Physiology, New Series, Vol 12 C: Plant Ecology III: Responses to the Chemical and Biological Environment. Berlin: Springer Verlag, 1983, pp 246–300.

10. Z Shen, F Zhang, F Zhang. Toxicity of copper and zinc in seedlings of mung bean and inducing accumulation of polyamine. J Plant Nutr 21:1153–1162, 1998.

11. G Horváth, M Droppa, A Oravecz, VL Raskin, JB Marder. Formation of the photosynthetic apparatus during greening of cadmium-poisoned barley leaves. Planta 199:238–243, 1996.

12. PL Foster. Copper exclusion as a mechanism of heavy metal tolerance in a green alga. Nature 269:322–323, 1977.

13. JT Wu, MT Hsieh, LC Kow. Role of proline accumulation in response to toxic copper in *Chlorella* sp. (Chlorophyceae) cells. J Phycol 34:113–117, 1998.

14. LF De Filippis. The effect of heavy metals compounds on the permeability of *Chlorella* cells. Z Planzenphysiol 92:39–49, 1979.

15. MK Chettri, T Sawidis. Impact of heavy metals on water loss from lichen thalli. Ecotoxicol Environ Saf 37:103–111, 1997.

16. A Kaar Sijpesteijn, MJ Janssen. On the mode of action of dialkyldithiocarbamates in moulds and bacteria. Antonie Leeuwenhoek J Microbiol Serol 25:422–438, 1959.

17. M Ahsanullah, TM Florence. Toxicity of copper to the marine amphipod *Allorchestes compressa* in the presence of water and lipid-soluble ligands. Mar Biol 84:41–45, 1984.

18. PGC Campbell. Interactions between trace metals and aquatic organisms: a critique of the free-ion activity model. In: A Tessier, DR Turner, eds. Metal Speciation and Bioavailability in Aquatic Systems. New York: Wiley, 1995, pp 45–101.

19. JT Phinney, KW Bruland. Trace metal exchange in solution by the fungicides ziram and maned (dithiocarbamates) and subsequent uptake of lipophilic organic zinc, copper and lead complexes into phytoplankton cells. Environ Toxicol Chem 16:2046–2053, 1997.

20. G Sandmann, O Böger. Copper-mediated lipid peroxidation processes in photosynthetic membranes. Plant Physiol 66:797–800, 1980.

21. A Stoinski, M Kozlowska. Cadmium-induced oxidative stress in potato tuber. Acta Soc Bot Pol 66:189–195, 1997.

22. W Maksymiec. Effect of copper on cellular processes in higher plants. Photosynthetica 34:321–342, 1997.

23. F Van Assche, H Clijsters. Effects of metals on enzyme activity in plants. Plant Cell Environ 13:195–206, 1990.

24. MK Chettri, CM Cook, E Vardaka, T Sawidis, T Lanaras. The effect of Cu, Zn and Pb on the chlorophyll content of the lichens *Cladonia convoluta* and *Cladonia rangiformis*. Environ Exp Bot 39:1–10, 1998.

25. H Küpper, F Küpper, M Spiller. In situ detection of heavy metal substituted chlorophylls in water plants. Photosynth Res 58:123–133, 1998.

26. S Nahar, HA Tajmir-Riahi. Do metal ions alter the proteins secondary structure of a light-harvesting complex of thylakoid membranes? J Inorg Biochem 58:223–234, 1995.

27. M Jain, R Gadre R. Inhibition of 5-amino levulinic acid dehydratase activity by selenium in excised etiolated maize leaf segments during greening. Indian J Exp Biol 32:804–806, 1994.

28. B Schoefs, M Bertrand. Chlorophyll biosynthesis. In: M Pessarakli, ed. Handbook of Photosynthesis. New York: Marcel Dekker, 1997, pp 49–69.

29. KL Hill, S Merchant. Coordinate expression of coproporphyrinogen oxidase and cytochrome c6 in the green alga *Chlamydomonas reinhardtii* in response to changes in copper availability. EMBO J 14:857–865, 1995.

30. LH Weinstein, R Kaur-Stawhney, MV Rajam, SH Wettlaufer, AW Galston. Cadmium-induced accumulation of putrescine in oat and bean leaves. Plant Physiol 82:641–645, 1986.

31. SN Luoma. Bioavailability of trace metals to aquatic organisms: a review. Sci Total Environ 128:1–22, 1993.

32. WX Wang, RCH Dei. Kinetic measurements of metal accumulation in two marine macroalgae. Mar Biol 135:11–23, 1999.

33. JA Raven. Transport systems in algae and bryophytes: an overview. Methods Enzymol 174:366–390, 1989.

34. D Kummongkol, GS Canterford, C Fryer. Accumulation of heavy metals in unicellular algae. Biotechnol Bioeng 4:2643–2660, 1982.

35. GW Garnham, GA Codd, GM Gadd. Kinetics of uptake and intracellular location of cobalt, manganese and zinc in the estuarine green alga *Chlorella salina*. Appl Microbiol Biotechnol 37:270–276, 1992.

36. H Freundlich. Colloid and Capillary Chemistry. London: Methuen, 1929.

37. D Neumann, UZ Nieder, O Lichtenberger, I Leopold. How does *Armeria maritima* tolerate high heavy metal concentration? J Plant Physiol 146:704–717, 1995.

38. G Scatchard. The attraction of proteins for small molecules and ions. Ann N Y Acad Sci 51:660–672, 1949.

39. J Molas. Ultrastructural response of cabbage outer leaf mesophyll cells (*Brassica oleracea* L.) to excess of nickel. Acta Soc Bot Pol 66:307–317, 1997.

40. RR Brooks, J Lee, RD Reeves, T Jaffre. Detection of nickeliferous rocks by analysis of herbarium specimens of indicator plants. J Geochem Explor 7:49–77, 1977.

41. SD Cunningham, DW Ow. Promises and prospects of phytoremediation. Plant Physiol 110:715–719, 1996.

42. H Küpper, FJ Zhao, SP McGrath. Cellular compartmentation of zinc in leaves of the hyperaccumulator *Thlaspi caerulescens*. Plant Physiol 119:305–311, 1999.

43. MD Vazquez, C Poschenrieder, J Barcelo, AJM Baker, P Hatton, GH Cope. Compartmentation of zinc in roots and leaves of the zinc hyperaccumulator *Thlaspi caerulescens*. J C Presl Bot Acta 107:243–250, 1994.

44. EY Zolotukhina. Change of microelement balance and photosynthesis of seaweeds under the action of heavy metals. Vestn Mosk Univ Ser XVI Biol (1) 71:46–54, 1995.

45. E Fourest, B Volesky. Alginate properties and heavy metal biosorption by marine algae. Appl Biochem Biotechnol 67:33–44, 1997.

46. SC Miyasaka, RK Buta, RK Howell, CD Foy. Mechanisms of aluminium tolerance in snapbeans: root exudation of citric acid. Plant Physiol 96:737–743, 1991.

47. DC Foy, EH Lee, SB Wilding. Differential aluminium tolerance of two barley cultivars related to organic acids in their roots. J Plant Nutr 10:1089–1101, 1987.

48. PR Ryan, E Delhaize, PJ Randall. Characterization of Al-stimulated efflux of malate from the apices of Al-tolerant wheat roots. Planta 196:103–110, 1995.

49. LRM de Andrade, M Ikeda, J Ishizuka. Stimulation of organic acid excretion by roots of aluminium-tolerant and aluminium sensitive wheat varieties under aluminium stress. Rev Bras Fisiol Veg 9:27–34, 1997.

50. DM Pellet, DL Grunes, LV Kochian. Organic exudation as an aluminium tolerance mechanism in maize. Planta 196.788–795, 1995.

51. JW Moffett, RG Zika, LE Brand. Distribution and potential sources and sinks of copper chelators in the Sargasso Sea. Deep Sea Res 37:27–36, 1990.

52. JW Moffett, LE Brand. Production of some extracellular Cu chelator by marine cyanobacteria in response to Cu stress. Limnol Oceanogr 41:388–395, 1996.

53. E Grill, EL Winnacker, MH Zenk. Phytochelatins, a class of heavy-metal-binding peptides from plants, are functionally analogous to metallothioneins. Proc Natl Acad Sci U S A 84:439–443, 1987.

54. MH Zenk. Heavy-metal detoxification in higher plants. A review. Gene 179:21–30, 1996.

55. E Grill, EL Winnacker, MH Zenk. Occurrence of heavy metal binding phytochelatins in plants growing in a mining refuse area. Experentia 44:539–540, 1988.

56. JG Lee, BA Ahner, FMM Morel. Export of cadmium and phytochelatin by the marine diatom *Thalassiosira weissflogii*. Environ Sci Technol 30:1814–1821, 1996.

57. P Foster. Concentrations and concentration factors of heavy metals in brown algae. Environ Pollut 10:45–53, 1976.

58. AAH Vieira, OR Nascimento. An EPR determination of copper complexation by excreted high molecular compounds of *Ankistrodesmus densus* (Chlorophyceae). J Plankton Res 10:1313–1315, 1988.

59. HS Lee, B Volesky. Interaction of light metals and protons with seaweed biosorbent. Water Res 31:3082–3088, 1997.

60. O Skipnes, T Roald, A Huang. Uptake of zinc and strontium by brown algae. Physiol Plant 43:314–320, 1975.

61. AT Lombardi, AAH Vieira. Lead and copper toxicity to *Nephrocytium lunatum* (Chlorophyceae) and their complexation with excreted material. Rev Microbiol 29:44–48, 1988.

62. AT Lombardi, AAH Vieira. Copper and lead complexation by high molecular weight compounds produced by *Synura* sp. (Chrysophyceae). Phycologia 37:34–39, 1988.

63. AT Lombardi, AAH Vieira. Lead- and copper-complexing extracellular ligands released by *Kirchneriella aperta* (Chloroccocales, Chlorophyta). Phycologia 38:283–288, 1999.

64. LF De Filippis LF, CK Pallaghy. The effect of sub-lethal concentrations of mercury and zinc on *Chlorella*. III. Development and possible mechanisms of resistance to metals. Z Pflanzenphysiol 79:323–335, 1976.

65. IE Eide, S Myklestad, S Melsom. Long-term uptake and release of heavy metals by *Ascophyllum nodosum* (L.) Le Jol. (Phaeophyceae) in situ. Environ Pollut (Ser A) 23:19–28, 1980.

66. CA Whitelaw, JA Le Huquet, DA Thurman, AB Tomsett. The isolation and characterisation of type II metallothionein-like genes from tomato (*Lycopersicon esculentum* L.). Plant Mol Biol 33:503–511, 1997.

67. CA Morris, B Nicolaus, V Sampson, JL Harwood, P Kille. Identification and characterization of a recombinant metallothionein protein from a marine alga, *Fucus vesiculosus*. Biochem J 338:553–560, 1999.

68.  JHR Kägi, A Schaffer. Biochemistry of metallothionein. Biochemistry 27:8509–8515, 1988.
69.  EH Fischer, EW Davie. Recent excitement regarding metallothionein. Proc Natl Acad Sci U S A
     95:3333–3334, 1998.
70.  R Kneer, MH Zenk. Phytochelatins protect plant enzymes from heavy metal poisoning. Phytochemistry
     31:2663–2667, 1992.
71.  RK Mehra, EB Tabet, WR Gray, DR Winge. Metal-specific synthesis of two metallothioneins and γ-glutamyl
     peptides in *Candida glabrata*. Proc Natl Acad Sci U S A 85:8815–8819, 1988.
72.  W Gekeler, E Grill, EL Winnacker, MH Zenk. Survey of the plant kingdom for the ability to bind heavy met-
     als through phytochelatins. Z Naturforsch 44c:361–369, 1989.
73.  I Raskin. Plant genetic-engineering may help with environmental cleanup. Proc Natl Acad Sci U S A
     93:3164–3166, 1996.
74.  JJ Chen, JM Zhou, PG Goldsbrough. Characterization of phytochelatin synthase from tomato. Physiol Plant
     101:165–172, 1997.
75.  N Kondo, M Isobe, K Imai, T Goto. Structure of cadystin the unit-peptide of cadmium-binding peptides in-
     duced in a fission yeast, *Schizosaccharomyces pombe*. Tetrahedron Lett 24:925–928, 1983.
76.  B Manunza, S Deiana, M Pintore, V Solinas, C Gessa. The complex of cadmium with phytochelatins. A quan-
     tum mechanics study. *http://antas.agraria.uniss.it/electronic_papers/eccc4/phytoc/welcome.html*, June 1998.
77.  NJ Robinson, AM Tommey, C Kuske, PJ Jackson. Plant metallothioneins. Biochem J 295:1–10, 1993.
78.  E Grill, S Loffler, EL Winnacker MH Zenk. Phytochelatins, the heavy-metal binding peptides of plants, are
     synthesized from glutathione by a specific γ-glutamylcysteine dipeptidyl transpeptidase (phytochelatin syn-
     thase). Proc Natl Acad Sci U S A 86:6838–6842, 1989.
79.  JC Steffens. The heavy metal-binding peptides of plants. Annu Rev Plant Physiol Plant Mol Biol 41:553–575,
     1990.
80.  A Rüegsegger, D Schnutz, C Brunold. Regulation of glutathione synthesis by cadmium in *Pisum sativum* L.
     Plant Physiol 93:1579–1584, 1990.
81.  S Loeffler, A Hochberger, E Grill, EL Winnacker, MH Zenk. Termination of the phytochelatin synthase reac-
     tion through sequestration of heavy metals by the reaction product. FEBS Lett 258:42–46, 1989.
82.  ZS Li, YP Lu, M Zhen, M Szcypka, DJ Thiele, PA Rea. A new pathway for vacuolar cadmium sequestration
     in *Saccharomyces cerevisiae—ycf1*-catalyzed transport of bis(glutathionato)cadmium. Proc Natl Acad Sci U S
     A 94:42–47, 1997.
83.  RK Mehra, P Mulchandani. Glutathione-mediated transfer of Cu(I) into phytochelatins. Biochem J
     307:697–705, 1995.
84.  RK Mehra, J Miclat, VR Kodati, R Abdullah, TC Hunter, P Mulchandani. Optical spectroscopic and reverse-
     phase HPLC analyses of Hg(II) binding to phytochelatins. Biochem J 314:73–82, 1996.
85.  W Bae, RK Mehra. Metal-binding characteristics of a phytochelatin analog (Gly-Cys)$_2$ Gly. J Inorg Biochem
     68:201–210, 1997.
86.  RN Reese, DR Winge. Sulfide stabilization of the cadmium-γ-glutamyl peptide complex of *Schizosaccha-
     romyces pombe*. J Biol Chem 263:12832–12835, 1988.
87.  R Vogeli-Lange, GJ Wagner. Subcellular localization of cadmium-binding peptides in tobacco leaves: impli-
     cations of a transport function for cadmium-binding peptides. Plant Physiol 92:1086–1093, 1990.
88.  CT Dameron, RN Reese, RK Mehra, AR Kortan, PJ Carroll, ML Steigerwald, LE Brus, DR Winge. Biosyn-
     thesis of cadmium sulphide quantum semiconductor crystallites. Nature 338:596–598, 1989.
89.  CT Dameron, DR Winge. Characterization of peptide-coated cadmium-sulfide crystallites. Inorg Chem
     29:1343–1348, 1990.
90.  RK Mehra, T Mulchandani, TC Hunter. Role of quantum crystallites in cadmium resistance in *Candida
     glabrata*. Biochem Biophys Res Commun 200:1193–1200, 1994.
91.  DF Ortiz, LK Kreppel, DM Speiser, G Scheel, G McDonald, DW Ow. Heavy-metal tolerance in the fission
     yeast requires an ATP-binding cassette-type vacuolar membrane transport. EMBO J 11:3491–3499, 1992.
92.  DE Salt, WE Rauser. MgATP-dependent transport of phytochelatins across the tonoplast of oat roots. Plant
     Physiol 107:1293–1301, 1995.
93.  E Grill, J Thumann, EL Winnacker, MH Zenk. Induction of heavy metal binding phytochelatins by inoculation
     of cell cultures in standard media. Plant Cell Rep 7:375–378, 1988.
94.  BA Ahner, NM Price, FMM Morel. Phytochelatin production by marine phytoplankton at low free metal ion
     concentrations: laboratory studies and field data from Massachusetts Bay. Proc Narl Acad Sci U S A
     91:8433–8436, 1994.
95.  E Morelli, E Pratesi. Production of phytochelatins in the marine diatom *Phaeodactylum tricornutum* in re-
     sponse to copper and cadmium exposure. Bull Environ Contam Toxicol 59:657–664, 1997.
96.  BA Ahner, FMM Morel, JW Moffett. Trace metal control phytochelatin production in coastal waters. Limnol
     Oceanogr 42:601–608, 1997.
97.  Y Hayashi, CW Nakagawa, N Mutoh, M Isobe, T Goto. Two pathways in the biosynthesis of cadystins (γ-
     EC)$_n$G in the cell-free system of the fission yeast. Biochem Cell Biol 69:115–121, 1991.
98.  T Maitani, H Kubota, K Sato, T Yamada. The composition of metals bound to class III metallothionein (phy-
     tochelatin and its desglycyl peptide) induced by various metals in root cultures of *Rubia tinctorium*. Plant Phys-
     iol 110:1145–1150, 1996.

99.  RK Mehra, DR Winge. Cu(I) binding to the *Schizosaccharomyces pombe* γ-glutamyltransferase peptides varying in chain lengths. Arch Biochem Biophys 265:381–389, 1988.

100. RK Mehra, VR Kodati, R Abdullah. Chain length–dependent Pb(II)-coordination in phytochelatins. Biochem Biophys Res Commun 215:730–736, 1995.

101. RK Mehra, K Tran, GW Scott, P Mulchandani, SS Saini. Ag(I)-binding to phytochelatins. J Inorg Biochem 61:125–142, 1996.

102. GJ Wagner. Accumulation of cadmium in crop plants and its consequences to human health. Adv Agron 51:173–212, 1993.

103. S Kannan. An in vitro determination of the transport of $^{59}$Fe and $^{54}$Mn to different leaves of young corn seedlings. Z Pflanzenphysiol 83:375–378, 1977.

104. F Raguzzi, E Lesuisse, RR Crichton. Iron storage in *Saccharomyces cerevisae*. FEBS Lett 231:253–258, 1988.

105. A Brune, W Urbach, KJ Dietz. Compartmentation and transport of zinc in barley primary leaves as basic mechanism involved in zinc tolerance. Plant Cell Environ 17:153–162, 1994.

106. A Brookes, JC Collins, DA Thurman. The mechanism of zinc tolerance in grasses. J Plant Nutr 3:695–705, 1981.

107. W Mathys. The role of malate, oxalate, and mustard oil glucosides in the evolution of zinc-resistance in herbage plants. Physiol Plant 40:130–136, 1977.

108. RFM Van Steveninck, ME Van Steveninck, AJ Wells, DR Fernando. Zinc tolerance and the binding of zinc as zinc phytate in *Lemna minor*. X-ray microanalytical evidence. J Plant Physiol 137:140–146, 1990.

109. JT Wu, SJ Chang, TL Chou. Intracellular proline accumulation in some algae exposed to copper and cadmium. Bot Bull Acad Sin 36:89–93, 1995.

110. R Bassi, SS Sharma. Changes in proline content accompanying the uptake of zinc and copper in *Lemna minor*. Ann Bot 72:151–154, 1993.

111. R Bassi, SS Sharma. Proline accumulation in wheat seedlings exposed to zinc and copper. Phytochemistry 33:1339–1342, 1993.

112. G Costa, JL Morel. Water relations, gas exchange and amino acid content in Cd-treated lettuce. Plant Physiol Biochem 32:561–570, 1994.

113. Y Jolivet, J Hamelin, F Larher. Osmoregulation in halophytic higher plants: the protective effect of glycine betaine and other related solutes against the oxalate destabilization of membranes in beet root-cells. Z Pflanzenphysiol 109:171–180, 1982.

114. ME Farago, WA Mullen. Plants which accumulate metals. IV. A possible copper-proline complex from the roots of *Armeria maritima*. Inorg Chim Acta 32:L93–L94, 1979.

115. H Schat, SS Sharma, R Vooijs. Heavy metal–induced accumulation of free proline in a metal tolerant and a nontolerant ecotype of *Silene vulgaris*. Physiol Plant 101:477–482, 1997.

116. ME Farago. Metal tolerant plants. Coord Chem Rev 36:155–182, 1981.

117. S Misra, L Gedamu. Heavy metal tolerant *Brassica napus* L. and *Nicotiana tabacum* L. plants. Theor Appl Genet 78:161–168, 1989.

118. L Ren, D Shi, J Dai, B Ru. Expression of the mouse metallothionein-I gene conferring cadmium resistance in a transgenic cyanobacterium. FEMS Microbiol Lett 158:127–132, 1998.

119. M Barinaga. Making plants aluminum tolerant. Science 276:1497, 1997.

120. JM De la Fuente, V Ramirez-Rodriguez, JL Cabrera-Ponce, J Herra-Estrella. Aluminum tolerance in transgenic plants by alteration of citrate synthesis. Science 276:1566–1568, 1997.

121. MD Moore, S Kaplan. Identification of intrinsic high-level resistance to rare-earth-oxides and oxyanions in members of the class Proteobacteria. Characterization of tellurite, selenite, and rhodium sequioxide in *Rhodobacter sphaeroides*. J Bacteriol 74:1505–1514, 1992.

122. AO Summers. Organization, expression, and evolution of genes for mercury resistance. Annu Rev Microbiol 40:607–634, 1986.

123. B Fox, CT Walsh. Mercuric-reductase. Purification and characterization of a transposon-encoded flavoprotein containing an oxidation-reduction-active disulfide. J Biol Chem 257:2498–2503, 1982.

124. CL Rugh, HD Wilde, NM Stack, DM Thompson, AO Summers, RB Meagher. Mercuric ion reduction and resistance in transgenic *Arabidopsis thaliana* plants expressing a modified bacterial *merA* gene. Proc Natl Acad Sci U S A 93:3182–3187, 1996.

125. L Herrera-Estrella. Transgenic plants for tropical regions: some considerations about their development and their transfer to the small farmer. Proc Natl Acad Sci U S A 96:5978–5981, 1999.

# 38

# The Negative Action of Toxic Divalent Cations on the Photosynthetic Apparatus

**Robert Carpentier**

*Université du Québec à Trois-Rivières, Trois-Rivières, Québec, Canada*

## I. INTRODUCTION

Heavy metal and other toxic metal cations are widespread pollutants that are phytotoxic. Most of them are absorbed by the plant roots, where they can accumulate. When the root tolerance is overloaded, the metals are translocated toward the leaves and affect the photosynthetic apparatus. Lower plants are also affected. In algae and cyanobacteria, the sensitivity to metal cations depends on their plasma membrane permeability and their capacity to bioaccumulate and is thus variable between species. Similarly, higher plants can be classified in three classes depending on their resistance to excess toxic divalent cations [1]. The first class of plants absorbs and translocates the ions as a function of metal concentration, the second class includes plants that are tolerant and that can exclude the metals at the root level, and the third class is represented by plants that can bioaccumulate heavy metals in their roots. The responses of the different plant species at the level of the photosynthetic electron transport system will thus be widely variable depending on species.

In this chapter, the inhibitory action of copper, mercury, cadmium, lead, zinc, and nickel will be reviewed. The sites and mechanisms of action will be discussed. The metals were studied in various plant materials ranging from whole plants and plant seedlings to isolated chloroplasts or photosynthetic membranes. The action during exposure of whole plants is often weaker than in isolated materials because of the translocation process. However, the toxicity often increases and becomes more significant with prolonged exposure periods. Thus, the more precise studies of metal action are reported in isolated thylakoid membranes and photosystem submembrane fractions, where the electron transfer components are more readily accessible to metal cations and other reagents such as artificial electron acceptors and donors that are used to localize the inhibitory active sites. Accurate knowledge of the inhibitory site and mode of action of the toxic metal cations may lead to interesting applications of these inhibitors in studies of the structure-function relationship of the photosynthetic apparatus [2].

## II. ACTIVE SITES AND MODE OF ACTION OF TOXIC METAL CATIONS

### A. Copper

Copper is an essential microelement in higher plants and algae as it occurs as part of the prosthetic groups of several enzymes. It is involved in photosynthesis as part of plastocyanine, the nearest electron donor to photosystem I. Apart from the stoichiometric amount ligated to plastocyanine, endogenous copper was also found in thylakoid membrane preparations and in isolated photosystem II submembrane fractions [3,4] but it was shown to be associated with proteins or nuclear contaminants that can be removed together with starch from Triton X-100 preparations using centrifugation at $10,000 \times g$ [5].

Addition of exogenous copper at concentrations greater than 1 μM can cause a toxic response in most photosynthetic organisms. Hence, it has been used extensively as an algaecide and herbicide. Prolonged exposure of whole plants to relatively high copper concentrations can lead to complete disintegration of the chloroplast lamellar system [6], whereas low concentrations have relatively minor effects on the photosynthetic apparatus [7]. In isolated chloroplasts or thylakoid membranes, copper is a relatively strong inhibitor of photosynthesis as a micromolar range of its chloride or sulfate salt was shown to inhibit electron transport from water to artificial electron acceptors of photosystem II or photosystem I.

Copper was found to inhibit both photosystem I and II. A direct interaction with ferredoxin was inferred to cause the inhibition on the acceptor side of photosystem I [8]. However, its action is much stronger on photosystem II [9,10]. Energy storage measured by photoacoustic spectroscopy and P700 turnover measured under red light from absorbance changes at 820 nm in intact leaves were less affected by copper than oxygen evolution [11–13]. This observation was attributed to the remaining energy storage activity during cyclic electron transport in photosystem I, which is less inhibited than photosystem II.

The several proposed sites of copper inhibition in photosystem II are reviewed by Barón et al. [14]. It was readily found that variable chlorophyll fluorescence declines in the presence of copper. Fluorescence parameters such as Fv/Fm, representing the photochemical quantum yield of photosystem II, and qP, which denotes the portion of absorbed energy that is trapped by open photosystem II centers, were shown to decline in the presence of copper [15–19]. Further, diphenylcarbazide (DPC), an artificial electron donor to photosystem II, could not restore the inhibition in the green algae *Ankistrodesmus falcatus*, indicating that the inhibitory site was located at the donor side of the photosystem near the DPC electron donation site [20]. Similar conclusions were deduced from fluorescence measurements in microalgae [21]. However other lines of evidence suggested an inhibition on the acceptor side beyond $Q_B$, the secondary quinone acceptor of photosystem II, and the inhibitory site of diuron in the $Q_B$ pocket [22]. In further studies by Hsu and Lee [23] using fluorescence induction experiments in pea thylakoids, the reduced variable chlorophyll fluorescence in the presence of copper was interpreted as an inhibitory action at the reaction center level affecting primary charge separation. This interpretation was challenged by Mohanty et al. [24], whose thermoluminescence and delayed luminescence studies supported the idea of an inhibitory action directly at the $Q_B$ site or at the nonheme iron located between $Q_A$, the primary quinone acceptor of photosystem II, and $Q_B$. This site was also supported by thermodynamic and kinetic studies of the electron transfer between $Q_A^-$ and $Q_B$ that indicated a reduced affinity for atrazine binding in the $Q_B$ pocket [25]. Yet other studies involving variable fluorescence measurements also indicated a heterogeneous inhibition of photosystem II populations inhibiting the primary photochemistry in the $Q_B$ reducing photosystem II centers but not in the non-$Q_B$ reducing centers [26–29].

Precise kinetic studies of electron transfer from $Q_A^-$ to $Q_B$, from the manganese-containing water oxidizing complex to the redox-active $Tyr_Z^{OX}$, and from $Tyr_Z^{OX}$ to $P680^+$, together with an analysis of the extent of charge separation between $P680^+$ and $Q_A^-$ using flash-induced absorption and fluorescence changes [25,30], provided evidence that copper does not affect charge separation but rather modifies $Tyr_Z$. In copper-binding proteins, copper is coordinated with the imidazole nitrogen atoms of histidines. It has been proposed that copper could bind His190 of the reaction center protein D1 [31]. This histidine is located near $Tyr_Z$ and has also been proposed to take part in proton transfer from $Tyr_Z$ during oxygen evolution. However, thermoluminescence measurements have shown that recombination of the $[His^+Q_A^-]$ couple was not affected by copper and only $Tyr_Z$ seems to be inactivated [32].

The preceding discussion indicates inhibition on both donor and acceptor sides of photosystem II. Confirmation of this proposal came from electron paramagnetic resonance (EPR) experiments showing that EPR signal II that reflects oxidation of $Tyr_Z$ could not be induced in photosystem II preparations treated with copper and further the EPR signal from the $[Q_A^--Fe^{2+}]$ couple was also lost, indicating that the nonheme iron located between $Q_A$ and $Q_B$ may be displaced by copper [31]. It can be proposed that $Cu^{2+}$ interferes with the histidine ligands to which the nonheme iron is bound. To this effect, it is interesting to note that added $Fe^{2+}$ can partly prevent copper inhibition in photosystem II [22]. Alternatively, Yruela et al. [33] proposed from picosecond time-resolved fluorescence experiments measuring charge separation in photosystem II preparations that copper may be involved in a close attractive interaction with Pheo$^-$ and Q$_A^-$ that eliminates the repulsive interaction between Pheo$^-$ and Q$_A^-$ without affecting primary charge separation.

Inhibition of the donor side of photosystem II close to the water oxidation complex was supported by the loss of the extrinsic 17-kDa polypeptide associated with the oxygen-evolving complex [31]. Furthermore, it was shown that photosystem II preparations depleted of the 17- and 23-kDa extrinsic polypeptides were more sensitive to copper and that Ca$^{2+}$, a cofactor of the water oxidation complex, specifically prevented the inhibitory action [34]. Ca$^{2+}$ also partially prevented the deleterious effects of copper in whole plants [35]. This might indicate that Cu$^{2+}$ ions are competing for the Ca$^{2+}$ binding site and may replace Ca$^{2+}$. Inhibition by copper is also competitive in respect to protons [28,36] and is thus pH sensitive. It has been suggested that it may bind an unprotonated residue close to the water-oxidizing system [9].

In the cyanobacterium *Spirulina platensis*, it was shown that copper had a much more significant inhibitory effect on photosystem II photochemistry under illumination in comparison with dark incubation. It was hypothesized that light may stimulate copper binding at its inhibitory site [29]. On the other hand, in bean plants, copper was also shown to accelerate the photoinhibition process in photosystem II by increasing the photoinhibition quantum yield and thus decreasing the steady-state concentration of active photosystem II centers [37]. It was previously proposed that copper might reduce the repair cycle of the D1 protein of the reaction center of photosystem II [38]. Copper is a relatively good catalyst of free radicals, and its presence is assumed to stimulate oxidative damage in plants. Copper was shown to increase significantly the concentration of superoxide in both thylakoid membranes and photosystem II preparations isolated from wheat seedlings exposed to copper [39]. This coincided with the induction of antioxidative enzymes even though it was also observed that in the cyanobacterium *Anabeana doliolum*, the antioxidant system could not significantly protect against copper-induced oxidative damage [40]. From studies using photosystem II preparations, Yruela et al. [41] proposed that copper strongly interacts with the acceptor side of the photosystem, near the pheophytin-Q$_A$ domain, where it catalyzes the generation of superoxide and hydroxyl radicals instead of the formation of singlet oxygen usually involved in donor-side photoinhibition. This results in greater yields of photodamage because hydroxyl radicals are more deleterious than singlet oxygen.

## B. Cadmium

Cadmium is a major environmental contaminant for which no biological function has been described. It has a number of toxic effects in plants, although the photosynthetic apparatus is particularly susceptible to this metal. Micromolar concentrations of cadmium were shown to inhibit oxygen evolution and CO$_2$ fixation in cyanobacteria [42,43]. Cadmium transport into the cells of *Synechocystis aquatilis* was pH dependent and was optimal at pH 7.5 [42]. In higher plants, it was suggested that the primary targets of cadmium during short exposures of bean plants were more at the Calvin cycle enzyme than at the electron transport reactions [44,45]. The inhibitory action also depended on leaf maturity [45]. The ultrastructure of developing chloroplasts in several plant species was shown to be greatly affected by cadmium and large destruction of the granal structure was observed under illumination [46–48]. The accumulation of chlorophylls and carotenoids was retarded by cadmium and some changes in the photosystem II light-harvesting complexes were also reported in radish seedlings, showing that the monomeric content increased following exposure to cadmium at the expense of the oligomeric form [49,50].

In the electron transport system, both photosystems were shown to be affected together with the ATP synthase/adenosinetriphosphatase (ATPase) [51–53]. The inhibition in photosystem I was proposed to affect the acceptor side of the photosystem at the level of the ferredoxin:NADP-reductase [54]. However, photosystem II was reported to be much more sensitive than photosystem I [51,52]. A cadmium-tolerant mutant of *Chlamydomonas reinhardtii* was exclusively affected by the mutation at the level of photosystem II, which also points to the inhibitory action of cadmium in photosystem II [55].

Studies using isolated chloroplasts indicated that the inhibition of photosystem II by cadmium could be assigned to the donor side of photosystem II [56–58]. However, a location within the reaction center of the photosystem has also been suggested [59]. In clover and lucerne plants the inhibition was reported to be removed by artificial electron donors specific for the photosystem such as hydroxylamine and MnCl$_2$, indicating an inhibitory site on the donor side [51]. However, contradictory results concerning the electron donors were obtained by Atal et al. [52] in wheat seedlings. Still, these authors also postulated an inhibitory site at the donor side of photosystem II but they indicated its location near the hydroxy-lamine electron donation site [52]. A decrease in the concentration of active photosystem II centers ($\alpha$ type) was also observed [52].

Cadmium was supposed to replace the Mn ions in the water-oxidizing system [60], or alternatively, an alteration of the lipid environment around the photosystem may be responsible for the inhibition [50]. The partial removal of the 16- and 24-kDa extrinsic polypeptides and the complete removal of the 33-kDa polypeptide from the oxygen-evolving complex of photosystem II during cadmium treatment also indicated inhibition at the level of the water-oxidizing system [31,61]. Cadmium has been shown to interact directly with calcium metabolism and the toxic effects of cadmium are similar in many ways to calcium deficit symptoms [62]. Thus, it can be postulated that this metal may interact with the site where calcium plays its cofactor role in the oxygen-evolving complex. Accordingly, calcium was shown to relieve the negative effect of cadmium on the primary photochemistry of bean plants [63]. Cadmium also causes effects similar to those of Fe deficiency, and an increase in Fe supply relieves the negative effects of cadmium on photosynthetic pigment accumulation and on the light phase of photosynthesis [64].

Cadmium was also shown to reduce the turnover rate of the D1 protein of the reaction center of photosystem II [48]. This action was proposed to originate from the modification of SH groups by cadmium resulting in the inhibition of messenger RNA (mRNA)-binding protein complex formation involved in D1 synthesis [48]. However, it was shown by Fourier transform infrared spectroscopy in isolated photosystem II submembrane fractions that this metal cation forms metal-protein complexes ligating $C=O$ and $C-N$ groups of amino acids but not SH groups [65].

## C. Zinc

Zinc is an important micronutrient associated with several enzymatic activities in all photosynthetic organisms [66]. However, this metal can inhibit $CO_2$ assimilation at relatively low concentrations. Higher concentrations initiate the loss of photosynthetic pigments and a decline in the chlorophyll $a/b$ ratio [67] and also result in inhibition of photosynthetic electron transport and photophosphorylation [68].

Zinc has been shown to affect the water-oxidizing complex by releasing the manganese ions involved in the oxygen evolution mechanism [69]. The Mn atoms released from the oxygen-evolving complex by millimolar concentrations of zinc are, however, sequestered within the thylakoid interior [70,71]. The extrinsic polypeptides of 16 and 24 kDa associated with the oxygen-evolving complex were significantly dissociated from photosystem II submembrane fractions treated with zinc but not the 33-kDa polypeptide [72]. Miller [73] has shown that the four Mn atoms were depleted from the thylakoid membranes as zinc concentration was increased with the concomitant release of the 16- and 24-kDa extrinsic polypeptides. The inhibitory action is thus due to the loss of manganese atoms rather than to the release of polypeptides. Although zinc noncompetitively inhibited $Ca^{2+}$ and $Mn^{2+}$ binding [74], $Ca^{2+}$ did not prevent the release of the extrinsic polypeptides [72]. Zinc has also been proposed to bind photosystem II on the acceptor side near the $Q_B$ binding site in *Rhodobacter sphaeroides*. This binding at a site distinct from the nonheme iron was shown to reduce significantly the rate of electron transfer between $Q_A$ and $Q_B$ [75].

## D. Mercury

Mercury is an environmental contaminant that is highly toxic to photosynthetic organisms at micromolar concentrations. Several sites of inhibition have been described in the photosynthetic electron transport chain, and both photosystems are affected [76–78]. Mercury was proposed to bind proteins through modification of SH groups [79]. Fourier transform infrared spectroscopy experiments indicated a strong interaction between photosystem II submembrane fractions and mercury causing some alteration of the protein structure that originated from the formation of metal-protein binding through peptide SH, $C=O$, and $C-N$ groups [65]. The binding of mercury to thylakoid membrane proteins was also illustrated by the quenching of the fluorescence emission associated with aromatic amino acids [80].

In photosystem I, the inhibition was reported at the donor side beyond the cytochrome $b/f$ complex [15]. More precisely, mercury was shown to react directly with plastocyanine, replacing copper [77,78,81]. An inhibition was also reported at the acceptor side of photosystem I, where it was proposed to alter the enzyme ferredoxin:NADP-reductase in which SH groups may be modified by mercury

[54,76,82] and the iron-sulfur center $F_B$ [83]. Electron paramagnetic resonance (EPR) experiments also indicated that the reaction center of photosystem I is oxidized by mercury in the dark [80].

Photosystem II is also affected on both donor and acceptor sides by mercury [19]. The inhibition on the acceptor side was proposed between the quinone acceptors $Q_A$ and $Q_B$ [84]. Most reports concentrated on the inhibition on the donor side of the photosystem [54,85,86]. Accordingly, mercury was shown to decrease variable fluorescence in cyanobacteria and isolated photosynthetic membranes [19,86–88]. The inhibition on the donor side was studied in more detail using photosystem II submembrane fractions. It was shown that the inhibition could be reversed by chloride ions that act as a cofactor for the oxygen-evolving complex [88] and that mercury selectively removed the 33-kDa extrinsic polypeptide associated with the oxygen-evolving complex [89]. Mercury also releases the Mn ions from the manganese cluster of the water-oxidizing complex [80]. Those results associate an inhibitory active site of mercury more closely with the oxygen-evolving complex. In an earlier report, mercury was shown to be an electron acceptor for photosystem II [90]. Those studies were performed in the presence of relatively high chloride concentrations (30 mM NaCl and 2.5 mM $MgCl_2$) that prevented the inhibitory action at the oxygen-evolving complex. Mercury was also shown to form metal-protein complexes in isolated photosystem II submembrane fractions involving SH, C=O, and C—N amino acids groups [65]; this type of interaction may be involved in the inhibitory process.

Studies have shown that the mercury-induced decline of variable fluorescence is enhanced under illumination [91]. Inhibition by mercury was associated with an increased nonphotochemical component of fluorescence quenching [15]. In the flagellate green alga *Haematococcus lacustris*, the rise of non-photochemical quenching was comparable to the rise caused by photoinhibition [91]. Mercury was thus proposed to increase the pH-independent rise of photoinhibitory quenching of chlorophyll fluorescence. The alga could recover from mercury-enhanced photoinhibition only after a few days of illumination. The recovery from the inhibition by mercury was not possible in the dark or in the presence of a chloroplast protein synthesis inhibitor, indicating a possible connection of the inhibitory mode of action of mercury with the turnover mechanism of the D1 polypeptide in the reaction center of photosystem II [92].

An action of mercury on the light-harvesting complex of photosystem II in the cyanobacteria *Spirulina platensis* was also suggested from the decline of the initial fluorescence Fo as mercury concentration was raised above 3 μM [87]. A further increase in concentration above 18 μM caused a strong quenching of chlorophyll fluorescence. Fluorescence is also quenched in isolated chloroplasts [80]. Direct interaction was also supported by Fourier transform infrared spectroscopy showing the formation of organometallic complexes between mercury and photosystem II light-harvesting complexes isolated from spinach, although those results were obtained at relatively high mercury concentrations [93]. Absorption and fluorescence spectroscopy experiments in *Synechococcus* have also shown significant changes in the chlorophyll spectral properties indicating a modification in the chlorophyll-protein complexes [94]. Mercury also affected the spectral characteristics of the phycobilisomes in *Spirulina platensis* [95].

# E.  Lead

Lead is considered another important phytotoxic pollutant that is not essential for growth. In several studies in intact plants or leaves, lead was found to affect only minimally photosynthetic electron transport [96–98]. However, some other reports indicated that the accumulation of lead in several plant species reduces the rate of photosynthetic reactions [99–101]. A reduction in the photosynthetic pigment composition has been observed [51,102,103]. Lead was also reported to disturb the granal structure of the chloroplasts [104]. A study of Parys and coauthors [105] indicated that the photochemical efficiency of photosystem II is reduced by about 10% after a 2-h exposure of detached pea leaves to lead. However, the primary inhibitory site is supposed to be at the level of the Calvin cycle enzymes [106]. It must be emphasized that lead is not very well translocated in plants and that its deleterious effects on photosynthesis are seen only after prolonged exposure.

An early study of Miles et al. [107] indicated that lead inhibited photosystem II electron transport without any effect on photosystem I. However, Wong and Govindjee [108] found that this metal directly inhibited the reaction center of photosystem I in isolated maize chloroplasts. Lead is now considered to influence both photosystems, although photosystem II is more sensitive [51]. This metal, assayed either with intact detached leaves or with isolated photosynthetic membranes, produces a decline of variable chlorophyll fluorescence indicating an inhibition on the donor side of photosystem II [19,105,109]. Accordingly, Becerril et al. [51] reported that the inhibition was partly restored by photosystem II–specific electron donors such as hydroxylamine and $MnCl_2$ but not by diphenylcarbazide.

A site at the level of the water-oxidizing complex was confirmed using photosystem II submembrane fractions isolated from spinach. It was shown that $Ca^{2+}$ and $Cl^-$, which are essential cofactors for oxygen evolution, could protect against lead inhibition, indicating that the metal competes for binding near the calcium and chloride binding sites in the water-oxidizing complex [109]. In those preparations, diphenylcarbazide was able to restore the inhibition presumably because of the loss of the Mn cluster of the oxygen-evolving complex as this electron donor was reported to be efficient only when the water-oxidizing complex is depleted of its Mn [109]. Another argument in favor of an inhibitory site at the oxygen-evolving complex is the loss of the extrinsic polypeptides of 17 and 24 kDa in lead-treated photosystem II submembrane fractions, a loss that can be prevented by added $Ca^{2+}$ [72]. A study using a combination of fluorescence and photoacoustic spectroscopy in isolated photosystem II submembrane fractions indicated that this metal may affect only the donor side of the photosystem with no effect on the acceptor side [19]. Lead was also shown to form metal-protein complexes by association with C=O and C—N groups of amino acids in photosystem II submembrane fractions [65]. These bindings may be involved in the inhibitory mode of action.

## F.  Nickel

Nickel is readily absorbed by plant roots and translocated to the leaves, where it clearly affects electron transport [97,110,111]. A significant reduction of grana structures has been observed in nickel-treated cabbage plants [112,113]. The pigment content was also shown to be reduced following exposure to this metal in various photosynthetic organisms, also depending on growth stage [97,114–116]. A possible mode of action would be the peroxidation of membrane lipids due to the induction of free radical reactions by nickel [112]. As with other metals, nickel was reported to affect both photosystems. It induced a decrease of variable chlorophyll fluorescence and an inhibition of the reduction of photosystem II artificial electron acceptors [15], indicating an inhibitory site on the donor side of photosystem II. More biochemical studies will be appropriate to clarify further the mode(s) and site(s) of action of this metal.

## III.  CONCLUSIONS

When toxic metal cations reach the photosynthetic apparatus, they clearly affect photosynthetic electron transport at the level of photosystems I and II. Except for mercury, whose inhibitory action at the level of photosystem I is well documented to be significant, the metals mostly affect photosystem II. Multiple inhibitory active sites were reported for most of the metals. However, a general consensus has been reached indicating that they mainly affect the oxygen-evolving complex with the loss of all or part of the manganese cluster together with some of the extrinsic polypeptides associated with the water oxidation mechanism. Their mode of action probably includes binding or modification of some membrane protein groups such as SH groups. Their presence in the environment makes them part of the several stresses imposed on plants. It can thus be expected that their action involves an enhancement of photoinhibition together with an increased turnover of the D1 protein, which are now though to be part of a more general protective response against environmental stresses [117,118]. Further research will probably confirm this hypothesis.

## REFERENCES

1.  AJM Baker. Accumulators and excluders: strategies in the response of plants to heavy metals. J Plant Nutr 3:643–654, 1981.
2.  A Trebst. Inhibitors in electron flow: tools for the functional and structural localization of carriers and energy conservation sites. Methods Enzymol 69:675–715, 1980.
3.  FS Holdsworth, JH Arshad. A manganese-copper-pigment-protein complex isolated from the photosystem II of *Phaeodactylum tricornutum*. Arch Biochem Biophy 183:361–373, 1977.
4.  JB Arellano, M Baron, A Chueca, M Lachica. Determination of copper in different chloroplast preparations. Plant Soil 154:7–11, 1993.
5.  M Baron, JB Arellano, W Schröder, M Lachica, A Chueca. Copper binding sites associated with photosystem 2 preparations. Photosynthetica 28:195–204, 1993.

6. M Angelov, T Tsonev, A Uzunova, K Gaidardjieva. $Cu^{2+}$ effect upon photosynthesis, chloroplast structure, RNA and protein synthesis of pea plants. Photosynthetica 28:341–350, 1993.

7. M Ciscato, R Valcke, K Van Loven, J Clijsters, F Navari-Izzo. Effects of in vivo copper treatment on the photosynthetic apparatus of two *Triticum durum* cultivars with different stress sensitivity. Physiol Plant 100:901–908, 1997.

8. Y Shioi, H Tamai, T Sasa. Effects of copper on photosynthetic electron transport systems in spinach chloroplasts. Plant Cell Physiol 19:203–209, 1978.

9. G Vierke, P Struckmeir. Inhibition of millisecond luminescence by copper(II) in spinach chloroplasts. Z Naturforsch 33c:266–270, 1978.

10. T Baszyński, A Tukendorf, M Ruszkowska, E Skórzyńska, W Maksymiec. Characteristics of the photosynthetic apparatus of copper non-tolerant spinach exposed to excess copper. J Plant Physiol 132:708–713, 1988.

11. G Ouzounidou, R Lannoye, S Karataglis. Photoacoustic measurements of photosynthetic activities in intact leaves under copper stress. Plant Sci 89:221–226, 1993.

12. G Ouzounidou. The use of photoacoustic spectroscopy in assessing leaf photosynthesis under copper stress: correlation of energy storage to photosystem II fluorescence parameters and redox change of $P_{700}$. Plant Sci 113:229–237, 1996.

13. G Ouzounidou, M Moustakas, RJ Strasser. Sites of action of copper in the photosynthetic apparatus of maize leaves: kinetic analysis of chlorophyll fluorescence, oxygen evolution, absorption changes and thermal dissipation as monitored by photoacoustic signals. Aust J Plant Physiol 24:81–90, 1997.

14. M Barón, JB Arellano, JL Gorgé. Copper and photosystem II: a controversial relationship. Physiol Plant 94:174–180, 1995.

15. DP Singh, P Khare, PS Bisen. Effect of $Ni^{2+}$, $Hg^{2+}$ and $Cu^{2+}$ on growth, oxygen evolution and photosynthetic electron transport in *Cylindrospermum* IU 942. J Plant Physiol 134:406–412, 1989.

16. FC Lidon, JC Ramalho, FS Henriques. Copper inhibition of rice photosynthesis. J Plant Physiol 142:12–17, 1993.

17. W Maksymiec, R Russa, T Urbanik-Sypniewska, T Baszyński. Effect of excess Cu on the photosynthetic apparatus of runner bean leaves treated at two different growth stages. Physiol Plant 91:715–721, 1994.

18. G Ouzounidou, M Moustakas, R Lannoye. Chlorophyll fluorescence and photoacoustic characteristics in relationship to changes in chlorophyll and $Ca^{2+}$ content of a Cu-tolerant *Silene compacta* ecotype under Cu treatment. Physiol Plant 93:551–557, 1995.

19. N Boucher, R Carpentier. $Hg^{2+}$, $Cu^{2+}$, and $Pb^{2+}$-induced changes in photosystem II photochemical yield and energy storage in isolated thylakoid membranes: a study using simultaneous fluorescence and photoacoustic measurements. Photosynth Res 59:167–174, 1999.

20. Y Shioi, H Tamai, T Sasa. Inhibition of photosystem II in the green alga *Ankistrodesmus falcatus* by copper. Physiol Plant 44:434–438, 1978.

21. A Cid, C Herrero, E Torres, J Abalde. Copper toxicity on the marine microalga *Phaeodactylum tricornutum*: effects on the photosynthesis and related parameters. Aquat Toxicol 31:165–174, 1995.

22. DP Singh, SP Singh. Action of heavy metals on Hill activity and oxygen evolution in *Anacystis nidulans*. Plant Physiol 83:12–14, 1987.

23. B-D Hsu, J-Y Lee. Toxic effects of copper on photosystem II of spinach chloroplasts. Plant Physiol 87:116–119, 1988.

24. N Mohanty, I Vass, S Demeter. Copper toxicity affects photosystem II electron transport at the secondary quinone acceptor $Q_b$. Plant Physiol 90:175–179, 1989.

25. G Renger, HM Gleiter, E Haag, F Reifarth. Photosystem II: thermodynamics and kinetics of electron transport from $QA^-$ to $Qb(Qb^-)$ and deleterious effects of copper(II). Z Naturforsch 48c:234–240, 1993.

26. G Samson, JC Morissette, R Popovic. Copper quenching of the variable fluorescence in *Dunaliella tertiolecta*. New evidence for a copper inhibition effect on PSII photochemistry. Photochem Photobiol 48:329–332, 1988.

27. M Renganathan, S Bose. Inhibition of primary photochemistry of photosystem II by copper in isolated pea chloroplasts. Biochim Biophys Acta 974:247–253, 1989.

28. JB Arellano, JJ Lázaro, J López-Gorgé, M Barón. The donor side of photosystem II as the copper-inhibitory binding site. Photosynth Res 45:127–143, 1995.

29. C-M Lu, J-H Zhang. Copper-induced inhibition of PSII photochemistry in cyanobacterium *Spirulina platensis* is stimulated by light. J Plant Physiol 154:173–178, 1999.

30. WP Schröder, JB Arellano, T Bittner, M Barón, H-J Eckert, G Renger. Flash-induced absorption spectroscopy studies of copper interaction with photosystem II in higher plants. J Biol Chem 269:32865–32870, 1994.

31. C Jegerschöld, JB Arellano, WP Schröder, PJM van Kan, S Baron, S Styring. Copper(II) inhibition of electron transfer through photosystem II studied by EPR spectroscopy. Biochemistry 34:12747–12754, 1995.

32. G Horváth, JB Arellano, M Droppa, M Barón. Alterations in photosystem II electron transport as revealed by thermoluminescence of Cu-poisoned chloroplasts. Photosynth Res 57:175–182, 1998.

33. I Yruela, G Gatzen, R Picorel, AR Holzwarth. Cu(II)-inhibitory effect on photosystem II from higher plants. A picosecond time-resolved fluorescence study. Biochemistry 35:9469–9474, 1996.

34. SC Sabat. Copper ion inhibition of electron transport activity in sodium chloride washed photosystem II particle is partially prevented by calcium ion. Z Naturforsch 51c:179–184, 1996.

35. W Maksymiec, T Baszyński. Are calcium ions and calcium channels involved in the mechanisms of $Cu^{2+}$ toxicity in bean plants? The influence of leaf age. Photosynthetica 36:267–278, 1999.

36. I Yruela, G Montoya, R Picorel. The inhibitory mechanism of Cu(II) on the photosystem II electron transport from higher plants. Photosynth Res 33:227–233, 1992.

37. E Pätsikkä, E-M Aro, E Tyystjärvi. Increase in the quantum yield of photoinhibition contributes to copper toxicity in vivo. Plant Physiol 117:619–627, 1998.

38. DV Vavilin, VA Polynov, DN Matorin, PS Venediktov. Sublethal concentrations of copper stimulate photostsystem II photoinhibition in *Chlorella pyrenoidosa*. J Plant Physiol 146:609–614, 1995.

39. F Navari-Izzo, MF Quartacci, C Pinzino, F Dalla Vecchia, CLM Sgherri. Thylakoid-bound and stromal antioxidative enzymes in wheat treated with excess copper. Physiol Plant 104:630–638, 1998.

40. N Mallick, LC Rai. Response of the antioxidant systems of the nitrogen fixing cyanobacterium *Anabaena doliolum* to copper. J Plant Physiol 155:146–149, 1999.

41. I Yruela, JJ Pueyo, PJ Alonso, R Picorel. Photoinhibition of photosystem II from higher plants. J Biol Chem 271:27408–27415, 1996.

42. B Pawlik, T Skowroński, Z Ramazanow, P Gardeström, G Samuelsson. pH dependent cadmium transport inhibits photosynthesis in the cyanobacterium *Synechocystis aquatilis*. Environ Exp Bot 33:331–337, 1993.

43. JL Moya, R Ros, I Picazo. Influence of cadmium and nickel on growth, net photosynthesis and carbohydrate distribution in rice plants. Photosynth Res 36:75–80, 1993.

44. Z Krupa, G Öquist, NPA Huner. The effect of cadmium on photosynthetis of *Phaseolus vulgaris*—a fluorescence analysis. Physiol Plant 88:626–630, 1993.

45. Z Krupa, M Moniak. The stage of leaf maturity implicates the response of the photosynthetic apparatus to cadmium toxicity. Plant Sci 138:149–156, 1998.

46. S Ghoshroy, MJ Nadakavukaren. Influence of cadmium on the ultrastructure of developing chloroplasts in soybean and corn. Environ Exp Bot 30:187–192, 1990.

47. A Siedlecka, Z Krupa. Interaction between cadmium and iron and its effects on photosynthetic capacity of primary leaves of *Phaseolus vulgaris*. Plant Physiol Biochem 34:833–841, 1996.

48. B Geiken, J Masojídek, M Rizzuto, ML Pompili, MT Giardi. Incorporation of [$^{35}$S]methionine in higher plants reveals that stimulation of the D1 reaction centre II protein turnover accompanies tolerance to heavy metal stress. Plant Cell Environ 21:1265–1273, 1998.

49. Z Krupa, E Skórzyńska, W Maksymiec, T Baszynski. Effect of cadmium treatment on the photosynthetic apparatus and its photochemical activities in greening radish seedlings. Photosynthetica 21:156–164, 1987.

50. Z Krupa. Cadmium-induced changes in the composition and structure of the light-harvesting chlorophyll *a/b* protein complex II in radish cotyledons. Physiol Plant 73:518–524, 1988.

51. JM Becerril, A Munoz-Rueda, P Aparicio-Tejo, C Gonzalez-Murua. The effects of cadmium and lead on photosynthetic electron transport in clover and lucerne. Plant Physiol Biochem 26:357–363, 1988.

52. N Atal, PP Saradhi, P Mohanty. Inhibition of the chloroplast photochemical reactions by treatment of wheat seedlings with low concentrations of cadmium: analysis of electron transport activities and changes in fluorescence yield. Plant Cell Physiol 32:943–951, 1991.

53. M Teige, B Huchzermeyer, G Schulz. Inhibition of chloroplast ATP synthase/Atpase is a primary effect of heavy metal toxicity in spinach plants. Biochem Physiol Pflanz 186:165–168, 1990.

54. LF De Filipis, R Hampp, H Ziegler. The effects of sublethal concentrations of zinc, cadmium and mercury on *Euglena*. II. Respiration, photosynthesis and photochemical activities. Arch Microbiol 128:407–411, 1881.

55. J Voigt, K Nagel, D Wrann. A cadmium-tolerant *Chlamydomonas* mutant strain impaired in photosystem II activity. J Plant Physiol 153:566–573, 1998.

56. MB Bazzaz, Govindjee. Effects of cadmium nitrate on spectral characteristics and light reactions of chloroplasts. Environ Lett 6:1–12, 1974.

57. MA Van Duijvendijk-Matteoli, GM Desmet. On the inhibitory action of cadmium on the donor side of photosystem II in isolated chloroplats. Biochim Biophys Acta 408:164–169, 1975.

58. R Hampp, K Beulich, H Ziegler. Effects of zinc and cadmium on photosynthetic $CO_2$-fixation and Hill activity of isolated spinach chloroplasts. Z Pflanzenphysiol 77:336–344, 1976.

59. EH Li, CD Miles. Effects of cadmium of photoreaction II of chloroplasts. Plant Sci Lett 5:33–40, 1975.

60. T Baszyński, L Wajda, M Krol, D Wolinska, Z Krupa, A Tukendorf. Photosynthetic activities of cadmium-treated tomato plants. Plant Physiol 98:365–370, 1980.

61. W Maksymiec, T Baszyn´ski. The effect of $Cd^{2+}$ on the release of proteins from thylakoid membranes of tamato leaves. Acta Soc Bot Pol 57:465–474, 1988.

62. M Greger, S Lindberg. Effects of $Cd^{2+}$ and EDTA on young sugar beets (*Beta vulgaris*). II. Net uptake and distribution of $Mg^{2+}$, $Ca^{2+}$ and $Fe^{2+}/Fe^{3+}$. Physiol Plant 69:81–86, 1987.

63. E Skórzyńska-Polit, A Tukendorf, E Selstam, T Baszyński. Calcium modifies Cd effect on runner bean plants. Environ Exp Bot 40:275–286, 1998.

64. A Siedlecka, Z Krupa. Cd/Fe interaction in higher plants—its consequences for the photosynthetic apparatus. Photosynthetica 36:321–331, 1999.

65. S Nahar, HA Tajmir-Riahi. Complexation of heavy metal cations Hg, Cd and Pb with proteins of PSII: evidence for metal-sulfur binding and protein conformational transition by FTIR spectroscopy. J Coll Interface Sci 178:648–656, 1996.

66. EJ Hewitt. The role of mineral elements in the activity of plant enzyme systems. In: W Ruthland, ed. Encyclopedia of Plant Physiology, Vol 4. Berlin: Springer-Verlag, 1958, pp 427–448.

67. AG Davies, JA Sleep. Photosynthesis in some British coastal water may be inhibited by zinc pollution. Nature 277:292–293, 1979.

68. CK Shrotri, VS Rathore, P Mohanty. Studies on photosynthetic electron transport; phosphorylation and $CO_2$ fixation in zinc deficient leaf cells of *Zea mays*. J Plant Nutr 3:345–354, 1981.

69. BC Tripathy, P Mohanty. Zinc inhibited electron transport of photosynthesis in isolated barley chloroplasts. Plant Physiol 66:1174–1178, 1980.

70. M Miller, RP Cox. Effect of $Zn^{2+}$ on photosynthetic ooxygen evolution and chloroplast manganese. FEBS Lett 155:331–333, 1983.

71. M Miller, RP Cox. Inhibition of photosynthetic oxygen evolution in non-vesicular preparations releases $Mn^{2+}$ into a restricted aqueous compartment. Biochem Biophys Res Commun 119:168–172, 1984.

72. A Rashid, EL Camm, AKM Ekramoddoulah. Molecular mechanism of action of $Pb^{2+}$ and $Zn^{2+}$ on water oxidizing complex of photosystem II. FEBS Lett 350:296–298, 1994.

73. M Miller. The release of polypeptides and manganese from oxygen-evolving photosystem II preparations following zinc-treatment. FEBS Lett 189:355–360, 1985.

74. A Rashid, M Bernier, L Pazdernick, R Carpentier. Interaction of $Zn^{2+}$ with the donor side of photosystem II. Photosynth Res 30:123–130, 1991.

75. LM Utschig, Y Ohigashi, MC Thurnauer, DM Tiede. A new metal-binding site in photosynthetic bacterial reaction centers that modulates $Q_A$ to $Q_B$ electron transfer. Biochemistry 37:8278–82281, 1998.

76. RC Honeycutt, DW Krogmann. Inhibition of chloroplast reactions with phenylmercuric acetate. Plant Physiol 49:376–380, 1972.

77. R Radmer, B Kok. Kinetic observation of the system II electron acceptor pool isolated by mercuric ion. Biochim Biophys Acta 357:177–180, 1974.

78. LC Rai, AK Singh, N Mallick. Studies on photosynthesis, the associated electron transport system and some physiological variables of *Chlorella vulgaris* under heavy metal stress. J Plant Physiol 137:419–424, 1991.

79. M Bernier, R Popovic, R Carpentier. Mercury inhibition at the donor side of photosystem II is reversed by chloride. FEBS Lett 321:19–23, 1993.

80. F Šeršeň, K Král'ová, A Bumbálová. Action of mercury on the photosynthetic apparatus of spinach chloroplasts. Photosynthetica 35:551–559, 1998.

81. M Kimimura, S Katoh. Studies on electron transport associated with photosystem I. I. Functional site of plastocyanin: inhibitory effects of $HgCl_2$ on electron transport and plastocyanin in chloroplasts. Biochim Biophys Acta 283:279–292, 1972.

82. H Clijsters, F Van Assche. Inhibition of photosynthesis by heavy metals. Photosynth Res 7:31–40, 1985.

83. Y-S Jung, L Yu, JH Golbeck. Reconstitution of iron-sulfur center $F_b$ results in complete restoration of $NADP^+$ photoreduction in Hg-treated photosystem I complexes from *Synechococcus* sp. (Pcc 6301). Photosynth Res 46:249–255, 1995.

84. Z Prokowski. Effects of $HgCl_2$ on long-lived delayed luminescence in *Scenedesmus quadricauda*. Photosynthetica 28:563–566, 1993.

85. CB Singh, SP Singh. Effect of mercury on photosynthesis in *Nostoc calcicola*: role of ATP and interacting heavy metal ions. J Plant Physiol 129:49–58, 1987.

86. G Samson, J-C Morissette, R Popovic. Determination of four apparent mercury interaction sites in photosystem II by using a new modification of the Stern-Voltmer analysis. Biochem Biophy Res Commun 166:873–878, 1990.

87. SDS Murthy, NG Bukhov, P Mohanty. Mercury-induced alterations of chlorophyll *a* fluorescence kinetics in cyanobacteria: multiple effects of mercury on electron transport. J Photochem Photobiol B 6:373–380, 1990.

88. M Bernier, R Popovic, R Carpentier. Mercury inhibition at the donor side of photosystem II is reversed by chloride. FEBS Lett 321:19–23, 1993.

89. M Bernier, R Carpentier. The action of mercury on the binding of the extrinsic polypeptides associated with the water oxidizing complex of photosystem II. FEBS Lett 360:251–254, 1995.

90. D Miles, P Bolen, S Farag, R Goodin, J Lutz, A Moustafa, B Rodrigez, C Weil. $Hg^{++}$—a DCMU independent electron acceptor of photosystem II. Biochem Biophys Res Commun 50:1113–1119, 1973.

91. M Xyländer, C Hagen, W Braune. Mercury increases light susceptibility in the green alga *Haematococcus lacustris*. Bot Acta 109:222–228, 1996.

92. M Xyländer, W Fisher, W Braune. Influence of mercury on the green alga *Haematococcus lacutris*. Inhibition effects and recovery of impact. Bot Acta 111:467–473, 1998.

93. S Nahar, HA Tajmir-Riahi. Do metal ions alter the protein secondary structure of a light-harvesting complex of thylacoid membranes? J Inorg Biochem 58:223–234, 1995.

94. SDS Murthy, N Mohanty, P Mohanty. Prolonged incubation with low concentrations of mercury alters energy transfer and chlorophyll (Chl) *a* protein complexes in *Synechococcus* 6301: changes in Chl *a* absorption and emission characteristics and loss of the F695 emission band. Bio Metals 8:237–242, 1995.

95. SDS Murthy, P Mohanty. Action of selected heavy metal ions on the photosystem 2 activity of the cyanobacterium *Spirulina platensis*. Biol Plant 37:79–84, 1995.

96.  Z Krupa, G Öquist, NPA Huner. The influence of cadmium on primary photosystem II photochemistry in bean as revealed by chlorophyll *a* fluorescence—a preliminary study. Acta Physiol Plant 14:71–76, 1992.

97.  Z Krupa, A Siedlecka, W Maksymiec, T Baszyński. In vivo response of photosynthetic apparatus of *Phaseolus vulgaris* L. to nickel toxicity. J Plant Physiol 142:664–668, 1993.

98.  JW Poskuta, E Waclawczyk-Lach. In vivo responses of primary photochemistry of photosystem II and $CO_2$ exchange in light and in darkness of tall fescue genotypes to lead toxicity. Acta Physiol Plant 17:233–240, 1995.

99.  FA Bazzaz, GL Rolfe, P Windle. Differing sensitivity of corn and soybean photosynthesis and transpiration to lead contamination. J Environ Qual 3:156–158, 1974.

100. RW Carlson, FA Bazzaz, GL Rolfe. The effect of heavy metals on plants. Environ Res 10:113–120, 1975.

101. JW Poskuta, E Parys, E Romanowska. The effects of lead on the gaseous exchange and photosynthetic carbon metabolism of pea seedlings. Acta Soc Bot Pol 56:127–137, 1987.

102. R Hampp, K Lendzian. Effect of lead ions on chlorophyll synthesis. Naturwissenschaften 61:218–219, 1974.

103. M Burzyński. Influence of lead on the chlorophyll content and on initial steps of its biosynthesis in greening cucumber seedlings. Acta Soc Bot Pol 54:95–105, 1985.

104. HM Rebechini, L Hanzley. Lead induced ultrastructural changes in the chloroplasts of the hydrophyte *Ceratophylum demersum*. Z Pflanzenphysiol 73:377–386, 1974.

105. E Parys, E Romanowska, M Siedlecka, JW Poskuta. The effect of lead on photosynthesis and respiration in detached leaves and in mesophyll protoplasts of *Pisum sativum*. Acta Physiol Plant 20:313–322, 1998.

106. HJ Weigel. Inhibition of photosynthetic reactions of isolated intact chloroplasts by cadmium. J Plant Physiol 119:179–189, 1985.

107. CD Miles, JR Brandle, DJ Daniel, O Chu-Der, PD Schnore, DJ Uhlik. Inhibition of photosystem II in isolated chloroplasts by lead. Plant Physiol 49:820–825, 1972.

108. D Wong, Govindjee. Effects of lead ions on photoststem I in isolated chloroplasts: studies on the reaction center. Photosynthetica 10:241–254, 1976.

109. A Rashid, R Popovic. Protective role of $CaCl_2$ against $Pb^{2+}$ inhibition in photosystem II. FEBS Lett 271:181–184, 1990.

110. BC Tripathy, B Bhatia, P Mohanty. Inactivation of chloroplast photosynthetic electron transport activity by $Ni^{2+}$. Biochim Biophys Acta 638:217–224, 1981.

111. N Mohanty, J Vass, S Demeter. Impairment of photosystem 2 activity at the level of secondary quinone electron acceptor in chloroplasts treated with cobalt, nickel and zinc ions. Physiol Plant 76:386–390, 1989.

112. J Molas. Changes in morphological and anatomical structure of cabbage (*Brassica oleracea* L.) outer leaves and in ultrastructure of their chloroplasts caused by an in vitro excess of nickel. Photosynthetica 34:513–522, 1997.

113. J Molas. Ultrastructural response of cabbage outer leaf mesophyll cells (*Brassica oleracea* L.) to excess of nickel. Acta Soc Bot Pol 66:307–317, 1997.

114. K Veeranjaneyulu, VSR Das. Intrachloroplast localization of $^{65}Zn$ and $^{63}Ni$ in Zn-tolerant plant (*Ocimum basilicum* Benth.). J Exp Bot 137:1161–1165, 1982.

115. IS Sheoran, HR Singal, R Singh. Effect of cadmium and nickel on photosynthesis and the enzymes of the photosynthetic carbon reduction cycle in pigeon pea (*Cajanus cajan* L.) Photosynth Res 23:345–351, 1990.

116. M Xyländer, W Braune. Influence of nickel on the green alga *Haematococcus lacustris* Rostafinski in phases of its life cycle. J Plant Physiol 144:86–93, 1994.

117. MT Giardi, J Masojidek, D Godde. Effects of abiotic stresses on the turnover of the $D_1$ reaction centre II protein. Physiol Plant 101:635–642, 1997.

118. E Franco, S Alessandrelli, J Masojídek, A Margonelli, MT Giardi. Modulation of $D_1$ protein turnover under cadmium and heat stresses monitored by [$^{35}S$]methionine incorporation. Plant Sci 144:53–61, 1999.

# 39
# Physiological Mechanisms of Herbicide Actions

## Francisco F. de la Rosa

*University of Seville, Seville, Spain*

## I.  GENERAL STATEMENT

The goal of this work was to study different types of herbicides as well as their diverse mechanisms of action. A short generalization on the selectivity of herbicides, forms of application, and resistance mechanisms will be given, followed by a specific study of each type of herbicide.

Herbicides are substances used to combat the development of weeds. For a herbicide to be effective, it has to have a series of characteristics:

1.  Adequate contact with plants
2.  Absorbed with facility
3.  Movement within the plant, arriving at its action site without being deactivated
4.  Reach levels with sufficient toxicity at the action site.

Finding the suitable moment at which to apply each herbicide is also important. For example, herbicides that break the cellular membrane, such as acifluorfen or paraquat, have to be applied when the plants are forming leaf materials. Inhibitors of seed germination, such as trifluralin or alachlor, need to be applied in soil before germination begins.

### A.  Herbicide Selectivity

Herbicides must have potent biological activity against a broad spectrum of weeds and at the same time be nontoxic to crop plants. Plants that quickly degrade or deactivate a herbicide can escape the toxic effects of this product. For example, corn is tolerant to herbicides derived from triazines because it quickly deactivates the herbicides, joining them to several products of its metabolism. In a similar way, soybean deactivates metribuzin upon conjugating it with sugar molecules.

On some occasions, a crop plant can be damaged by a herbicide to which it is normally tolerant. This is due to environmental stress, such as extreme temperatures, high relative humidity, or a strong hailstorm. All these situations can affect the ability of plants to prevent entry or to deactivate herbicides; cyanazine, for example, to which corn is naturally tolerant, causes substantial damages to corn plants when atmospheric conditions are cold and rainy.

Excessive herbicide treatment can cause a great deal of damage to a tolerant plant because deactivation or degradation systems can become saturated.

## B.  Herbicide Application

### 1.  Herbicides of Preemergence Application

Within this category are the herbicides that are generally applied to the soil before crops are sown and always before the plants reach the surface.

Because seeds of many species are quite small and germinate at a depth of 0.7 to 1.5 cm, it is important that herbicides are applied in the upper 2 or 3 cm of the soil.

Once herbicides come into contact with the plant, they are absorbed through roots or apical meristems. Herbicides that are absorbed by roots will be absorbed while the top of the root is in the layer of soil that contains the herbicide. Upon continuing growth, the root at a greater depth will decline in absorption. If the tops of the roots leave the contaminated layer soon, the plant will be likely to survive.

Many herbicides, upon application to the soil, are absorbed through apical meristems and can kill many scions before leaving the soil. Among of these herbicides, some enter the apical meristems as gases (e.g., thiocarbamates) or as a liquid (e.g., alachlor). There are environmental and physical factors that promote rapid growth in crop plants and therefore their time in touch with the herbicides is reduced.

Some herbicides have different abilities to move in the plant. Thus, some herbicides are absorbed by roots and move to the leaves. Others do not have movement capacity; therefore, their symptoms are expressed only at the absorption sites. Generally, symptoms are more prominent at sites where the mobile herbicides are concentrated.

### 2.  Herbicides of Postemergence Application

These herbicides are applied when the plant has emerged from the soil. Generally, they are absorbed by leaves. The relationship between the quantity of herbicide and the surface of the leaf, size of the plant, age, hydric stress, air temperature, relative humidity, etc. are factors that can influence the quantity of absorbed herbicide; for example, in dry conditions and at high temperatures, absorption is less.

Herbicides also differ in their capacity to move in the plant. Those that are nonmobile (e.g., diphenyl ether, bipyridine) should cover more of the plant. The mobile herbicides can move from their site of application to the site at which they act.

## C.  Herbicide Resistance

A great many weeds that at one time were susceptible to certain herbicides and were easily treated but at present they have now developed resistance to them. In fact, currently not less than 53 kinds of weeds are known to be resistant to at least five different herbicide families [1,2].

Resistance to herbicides is probably developed by selection of the resistant biotypes after extended exposure of a species of weed to a herbicide (we define biotypes as plants within the same species that are characterized by having a given feature in common). The resistant plants will survive the herbicide, produce seeds, and create new resistant weed generations.

Resistance mechanisms vary depending on the herbicide family. The resistant biotypes can have a slightly different biochemistry in their susceptible reactions, which can determine the resistance. For example, although photosynthesis is inhibited in the biotypes susceptible to triazine, a small change in protein D1 makes resistant biotypes that accomplish perfect photosynthesis in the presence of the herbicide. The appearance of the resistant biotypes is much easier if the herbicide has only one site of action.

In spite of resistance mechanisms, knowing the mode of action of herbicides would help develop prevention and extinction programs for resistant biotypes. The use of a single strategy with a herbicide family would increase the number of resistant biotypes, which would cause serious problems in the near future.

There is a series of recommendations to prevent and to treat resistance to herbicides:

Practice crop rotation.
Rotate herbicide families.
Mix herbicides with different modes of action.

Use correct sanitation to avoid the expansion of resistant biotypes.
Integrate different methods (chemical, mechanical, or cultural) of weed control.

## II. PHOTOSYNTHESIS INHIBITORS

The herbicides of this group bind to specific sites of chloroplasts and generally lead to slow death of the plant; however, on occasion, the plant experiences a more rapid death because of the production of toxic secondary products. Photosynthesis inhibitors can be classified according to their mode of action into the following groups.

### A. Electron Transport Inhibitors

Photosynthetic electron transport can be inhibited by inactivation of one of the redox mediators or by competition at the action site of this mediator.

Most of the inhibitors classified here have their action site near photosystem II [3–5] (Figure 1). This is observed in the inhibition of noncyclic electron transport; on the contrary, the cyclic transportation around photosystem I is maintained as well as the reduction of $NADP^+$ as long as an artificial electron donor is added with a redox potential between those of the two photosystems (for example, DPIP + ascorbate) [6].

The reaction center of photosystem II (PSII) is formed by two proteins of 32 kDa (D1 and D2), both of which are found in the interior of the thylakoid membrane. Near them are localized the transducer pigment P680, a molecule of pheophytin, the complex acceptor of electrons of PSII consisting of two quinones ($Q_A$ and $Q_B$), and a nonheme iron atom (Figure 2). When pigment P680 is excited by the arrival of a photon, the electron of higher energetic level passes to the pheophytin, which immediately reduces to $Q_A$; then $Q_A^-$ is reoxidized in a tenth of a millisecond by $Q_B$, which has the capacity to accept two electrons. Once $Q_B$ receives the two electrons, captures two protons and loses affinity to its union site, being substituted for an oxidized quinone $Q_B$ [7–9].

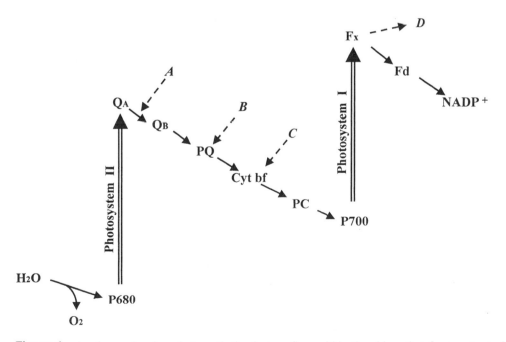

**Figure 1** A scheme showing photosynthetic electron flow within the chloroplast from water to $NADP^+$. Open arrows indicate light reactions, solid arrows indicate dark reactions. A, Site of action of ureas, triazines, and phenol-type herbicides. B, Site of inhibition by analogues of plastoquinone. C, Site of action for compounds that interfere with cytochrome *bf* complex. D, Site of diversion of electron flow by bipyridyls.

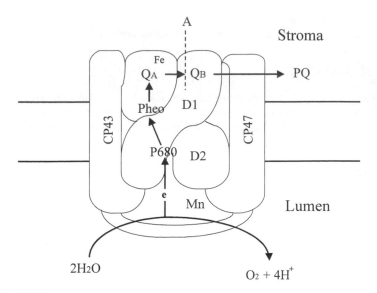

**Figure 2**  Schematic model of the PSII core complex. Proteins D1 and D2 bind the redox-active prostetic groups that are involved in electron flow. CP43 and CP47 are proteins that bind chlorophyll molecules for light absorption. A is the site of action of herbicides binding quinone.

It has been discovered that most inhibitors belonging to this group act in competition for the binding site of $Q_B$ located in D1; therefore they block the reoxidation of $Q_A^-$, inhibiting electron flow [10–12]. Protein D1 is encoded by the gene *psbA*, which has been sequenced in cyanobacteria, algae, and plants and has been confirmed to be conserved in most cases [13].

There are several families of herbicides that act in a similar way:

Ureas
Triazines
Uracils
Phenolics

Among these families, the first three consist of mobile and the last of immobile herbicides.

### 1.  Mobile Herbicides

Herbicides derived from urea and triazine have behaviors as well as action sites in common; therefore, a superfamily of herbicides including ureas and triazines together can be considered. Within this group, the two most important herbicides are $N'$-(3,4-dichlorophenyl)-$N,N$-dimethylurea (DCMU; diuron) (without doubt the most used and studied herbicide) and atrazine, both acting to displace $Q_B$ from its binding site [10–12]. The herbicides belonging to these three families tend to act in a noncompetitive way [14]. It is believed that ureas and triazines have two binding sites, one of them in common [15,16].

### 2.  Immobile Herbicides

Within the phenol-type herbicides, dinitrocresol (DNOC), bromoxynil, and ioxinil are the most used. It is believed that these compounds interact in the same region of PSII as ureas and triazines, although their mechanisms of action are different because the phenolic herbicides have a special binding site and act in a competitive way [17,18]. Interestingly, mutants showing resistance to atrazine and diuron often exhibit increased sensitivity to phenolic herbicides [6]. These compounds have a pK of about 4; their binding to the thylakoid membrane is extremely slow at physiological pH but can be markedly accelerated by a lower pH [19].

Electron flow can also be inhibited by analogues of benzoquinone, the best known of which is DB-MIB (2,5-dibromo-3-methyl-6-isopropylbenzoquinone) which prevents the oxidation of plastoquinone [20,21].

Although not well studied, the function of plastocyanin (the electron donor for PSI) is affected by potassium cyanide as well as by mercuric chloride. These compounds probably interfere with the cytochrome $b_6f$ complex [5,22] (see Figure 1).

RESISTANCE    Triazines are metabolized in naturally resistant crops by several pathways, including 2-hydroxylation, conjugation with glutathione, and $N$-dealkylation of lateral chains, followed by oxidation. For example, corn uses these three routes. The resistance to substituted ureas is due to limited absorption, rapid degradation by cytochrome P-450 monooxygenase, or oxidation [23–25].

Currently, transgenic plants are being obtained by the incorporation of a bacterial gene that codifies an enzyme that specifically produces the hydrolysis of bromoxynil.

Resistance to triazine should be due to mutations in the gene *psbA* that produces a change of serine to glycine and causes D1 not to be susceptible to triazines [26]. This gene is codified by chloroplast DNA; therefore the resistance to triazine is maternally inherited [27]. Other mutations confer different resistance spectra on triazines, ureas, and uracils. Thus, some weed biotypes have been made resistant upon increasing the expression of glutathione $S$-transferase and detoxify atrazine upon conjugating it with glutation [28]. Other biotypes are resistant to all PSII inhibitors because of the activity of cytochrome P-450 monooxygenase [29,30].

## B.  Uncouplers

Uncouplers act by dissociating ATP synthesis of electron transport by dissipating the energized state of the thylakoid membrane. There are many uncouplers, such as ammonium, CCCP (carbonyl cyanide m-chlorophenyl-hydrazone), FCCP [carbonyl cyanide p(trifluorometoxy) phenyl-hydrazone] and gramicidin, that are not herbicides. The only herbicide known that acts purely as an uncoupler in photophosphorylation is perfluidone, which has this capacity when the pH is about 8. This herbicide is protonated in the interior of the thylakoid and deprotonated in the stroma, thus breaking the proton gradient and preventing ATP synthesis [31,32].

## C.  Energy Transfer Inhibitors

Herbicides of this group inhibit electron transport as well as ATP formation in coupled systems. However, if an uncoupler that dissipates the proton gradient is added, inhibition of electron flow is relieved but without ATP formation. The inhibition by these compounds is on the phosphorylation but in a step after the action of uncouplers.

Several types of phenylureas have been reported to act as energy transfer inhibitors of photophosphorylation under certain conditions [5,33]. Nitrophen and related diphenyl ethers may inhibit by binding directly to the coupling factor and preventing the ADP exchange [5,34]. Nonherbicidal compounds that behave as energy transfer inhibitors include the antibiotic Dio-9, phlorizin, DCCD, TBT, etc. [4,35].

## D.  Inhibitory Uncouplers

The herbicides affecting electron transport as well as the proton gradient are found in this group. Like DCMU, they inhibit electron transport from water and do not inhibit reduction of $NADP^+$ with ascorbate plus DPIP, but unlike DCMU they inhibit noncyclic phosphorylation. Cyclic phosphorylation assayed in the absence of oxygen is also inhibited by treatment with herbicides of this group. With compounds such as dinitrophenol, an uncoupler action is observed at approximately pH 6; however, electron transport is inhibited at a pH higher than 8 [36]. Herbicides of this group have been provided that affect the permeability of thylakoid and mitochondrial membranes [37]. To this group belong: acylanilides, bromofenoxim, dinitrophenols, imidazoles, pyridinols, etc. [32,38].

## E.  Electron Acceptors

Compounds classified within this group are able to compete by electrons with some components of the electron transport chain. Certain bipyridyliums can compete by electrons with the acceptor of PSI and have herbicidal activity (site D in Figure 1). They are bivalent cations that intercept the electrons between ferredoxin and $NADP^+$ and then reduce oxygen to superoxide. The superoxide reacts with a wide range of molecules in the chloroplast, causing substantial damage to photosynthetic activity [39]. Phytotoxicity is associated with quaternary nitrogen salts in which the nitrogens are in positions 2,2'-, 2,4'-, and 4,4'-. Maximum herbicide activity is produced when the two rings of pyridine adopt a planar configuration. The two most important herbicides of the bipyridilium family are diquat (2,2'-bipyridyl) and paraquat, also known as methyl viologen (4,4'-bipyridyl), the latter being used more. These herbicides inhibit cyclic as well as noncyclic phosphorylation and the reduction of $NADP^+$ even upon the addition of ascorbate plus DPIP [4,12,40].

These herbicides are absorbed through the leaves. They are not selective and they do not present residual activity in soil because they quickly adhere to the colloids of the soil. Paraquat is more active on monocotyledons, diquat on dicotyledons.

In our laboratory, the action of methyl viologen was used in artificial photosystems for the production of hydrogen peroxide by free and immobilized microalgae and chloroplasts. Methyl viologen acting on PSI is reduced by electron flow and reoxidized by oxygen, producing $O_2^-$, with a concurrent increase of hydrogen peroxide present in the system [39]. Hydrogen peroxide is an energetic compound that can be used as engine and rocket fuel [41–45].

RESISTANCE    There are no known species of paraquat-resistant annual crops. About 18 species of weeds have been found with resistance, but the resistance mechanisms are not clearly understood [1,46]. A plant having the capacity to metabolize these herbicides has not been found in susceptible species or in evolved resistant biotypes [47,48]. Resistance can be conferred by increasing the ability to eliminate the toxic species of oxygen; thus, $O_2^-$ can be converted to hydrogen peroxide by superoxide dismutase and hydrogen peroxide decomposed to oxygen and water by catalase. Other enzymes, such as ascorbate peroxidase and glutathione reductase, can produce detoxification of the active oxygen species. Higher activity of these enzymes has been found in crude leaf extracts and chloroplasts [49–52]. Therefore, the increasing activity of the enzymes in transgenic plants should enable the use of paraquat in a more profitable way [53,54]. Probably another resistance mechanism involves restriction of the mobility of these herbicides, but no evidence has been found for a molecular mechanism restricting mobility or site of compartmentalization.

## III.  AMINO ACID SYNTHESIS INHIBITORS

Among herbicides included in this group are sulfonylureas, imidazolinones, sulfonamides, isopropylamines, and some substituted amino acids. These chemicals inhibit enzymes that participate in the synthesis of the amino acids necessary for normal development and plant growth.

## A.  ALS Inhibitors

Sulfonylureas, imidazolinones, and sulfonamides act on the enzyme acetolactate synthetase (ALS). This enzyme catalyzes the first reaction of the metabolic pathway that synthesizes isoleucine, leucine, and valine (Figure 3). It has been shown that these herbicides inhibit specifically and potentially the ALS of plants and bacteria in vitro [55–57].

The ALS inhibitors are absorbed by leaves and by roots, moving easily via the phloem as well as the xylem to the growing zones. These herbicides affect the leaves of both annual and perennial plants.

RESISTANCE    In crops, resistance to this type of herbicide is based on the capacity for metabolic inactivation rather than on a difference in herbicide uptake or on modification of the affinity of ALS [58–60]. For example, wheat and corn produce hydroxylation followed by conjugation of substituted sulfonylurea [61,62]. Genetic engineering has been used to produce transgenic species with less susceptible ALS. Genes from resistant mutants have been introduced into tobacco by transformation and have conferred useful levels of herbicide resistance in transgenic plants [63,64].

Multiple resistance forms have emerged in weeds because of the extended use of sulfonylureas. Many species of *Lolium*, *Lactuca*, *Salsola*, etc. have evolved resistance to ALS inhibitors in different countries [65–67]. The resistance in these cases is based fundamentally on metabolic inactivation rather than on a less susceptible ALS [68]. The resistance appears to be a quantitative (polygenic) characteristic, although in some cases, such as in *Lactuca* spp., a single mutation in ALS prevents its recognition by inhibitors [69].

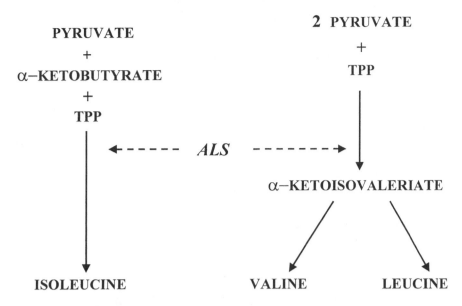

**Figure 3** Scheme of the biosynthesis of the amino acids isoleucine, leucine, and valine. The enzyme aceto-lactate synthetase (ALS) is the target site for the sulfonylurea, imidazolinone, and triazolopyrimidine herbicide classes. (TPP, coenzyme thiamine pyrophosphate)

## B. EPSPS Inhibitors

Herbicides that belong to this group are derivatives of amino acids and inhibit the enzyme 5-enolpyruvyl-shikimate-3P synthetase (EPSPS). This enzyme participates in the synthesis of shikimic acid, which is a precursor of phenylalanine, tyrosine, and tryptophan (Figure 4). Glyphosate, the isopropylamine salt of phosphonic acid, is a noncompetitive inhibitor of shikimate-3P. Glyphosate binds effectively and reversibly provided the enzyme has previously bound shikimate-3P, thereby forming a "dead-end complex" that cannot catalyze the reaction producing 5-enolpyruvyl-shikimate-3P [70–72].

The uptake of herbicides derived from amino acids occurs in leaves, and they are distributed by phloem to all parts of the plant. They are postemergence nonselective but especially useful against most annual and many perennial weeds. They have little residual soil activity, are not toxic to mammals, and are toxic to almost all annual plants.

RESISTANCE *Salmonella typhimurium* has a biotype tolerant to glyphosate thanks to a mutation in the gene *aroA* that codes for EPSPS [73,74]. This gene has been inserted by genetic engineering into various crop plants (e.g., tomato, tobacco, petunia) and its expression has been amplified with satisfactory results [75–78]. Also, resistance to this herbicide has been broadened by selection of some species of plants with enhanced EPSPS activity [79]. Weeds resistant to glyphosate with mutations in EPSPS have not been found, although some crop and weed species have low resistance levels [80].

## C. Glutamine Synthetase Inhibitors

Phosphinothricin (PPT) is a new herbicide, an analogue of glutamate, that inhibits the enzyme glutamine synthetase (GS) [1,2,4,12]. This enzyme participates in the assimilation of ammonium and in the regulation of nitrogen metabolism (Figure 5) [81,82]. PPT was originally discovered as a component of bialaphos, an antibiotic produced by *Streptomyces* species, and is synthesized chemically as the herbicide glufosinate (2-amino-methylphosphinyl-butanoic acid) [83]. Upon applying PPT to the plant, a rapid buildup of intracellular ammonium levels and an associated disruption of chloroplast structure are caused, resulting in the inhibition of photosynthesis and plant death [84,85].

In addition to the effect on glutamine synthetase, phosphinothricin produces an inhibition of photosynthesis that may be due to inhibition of protein synthesis (especially some proteins involved in electron

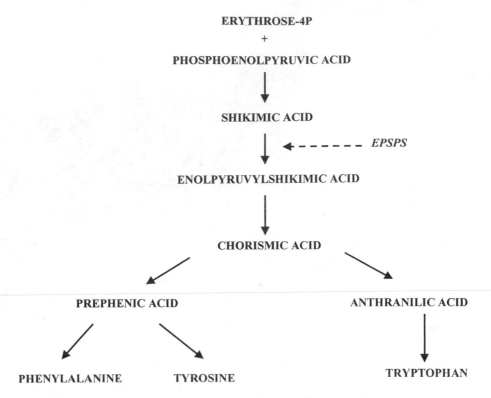

**Figure 4**   Scheme of the biosynthesis of aromatic amino acids. The enzyme enolpyruvyl-shikimate-3-phosphate synthetase (EPSPS) is the target site for glyphosate.

transport), to toxic glyoxylate accumulation, or to insufficient regeneration of intermediates of the Calvin cycle [86,87].

Glufosinate is a nonselective, postemergence contact herbicide that is highly effective and rapidly biodegraded.

RESISTANCE    Several approaches have been used to produce plants tolerant to phosphinothricin. An alfalfa cell line tolerant to PPT with an elevated level of GS was used to transform tobacco. Enhanced GS expression in tobacco was obtained, allowing sufficient enzyme activity even in the presence of the her-

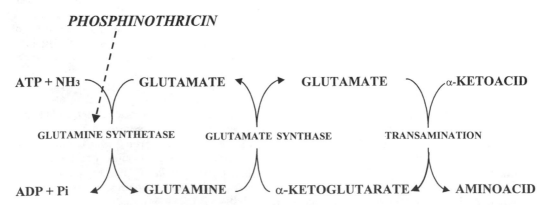

**Figure 5**   Scheme of ammonium incorporation on carbon skeletons for amino acid biosynthesis. Glutamine synthetase is the target site for the herbicide phosphinothricin and analogues.

bicide [2,88]. In *Chlamydomonas reinhardtii*, resistance is due to lack of phosphinothricin transport into the cells [89]. An alternative approach was to introduce the gene *bar*, which produces a detoxifying enzyme, into plants. The *bar* gene is from *Streptomyces hygroscopicus*, which produces bialaphos, the tripeptide precursor of PPT. The protein encoded by the *bar* gene protects these bacteria from the action of their own antibiotics by metabolizing PPT to an inactive derivative [90,91].

## IV. LIPID SYNTHESIS INHIBITORS

These compounds are formed by two herbicide families: phenoxypropionates (diclofop, haloxyfop, and trifop) and cyclohexanediones (alloxydim, sethoxydim, and dethodim). These herbicides inhibit fatty acid synthesis, which is fundamental for the biosynthesis of lipids. The lipids are also fundamental for the integrity of the cellular membranes and for the growth of plants.

All of these herbicides inhibit a key enzyme, the acetyl–coenzyme A (CoA) carboxylase, in the pathway of fatty acid synthesis [92–94]. This enzyme produces the carboxylation of acetyl-CoA, using ATP and $HCO_3^-$, giving malonyl-CoA (Figure 6). Cyclohexanediones and phenoxypropionates probably bind the same region, but in different sites, of the target enzyme [94–96].

Also, it is believed that these herbicides dissipate the transmembrane proton gradient [97,98]. They are absorbed by leaves and move by phloem to the rest of the plant.

RESISTANCE    Perennial as well as annual weeds are susceptible to these inhibitors, but most of the broadleaf plants are tolerant. Dicotyledon and monocotyledon (except graminaceous) crops are resistant to these herbicides. Dicotyledons have an acetyl-CoA carboxylase resistant to these herbicides, and many of them also have the capacity for cyclohexanedione and phenoxypropionate detoxification. Most grass crops are susceptible, but wheat is tolerant to diclofop. Wheat rapidly detoxifies diclofop with a demethylation produced by an active esterase [99,100].

Resistance to acetyl-CoA carboxylase inhibitors in crop plants was investigated [101]. In one case, mutants with increased expression of susceptible acetyl-CoA carboxylase were obtained [102]; in another case plants with a resistant acetyl-CoA carboxylase were found [103].

There are examples of weeds that are naturally tolerant to diclofop [104]. Many weeds are developing resistance to these herbicides by evolving a resistant acetyl-CoA carboxylase [105,106] or by induction of this protein to compensate for the inactivated enzyme [103,107].

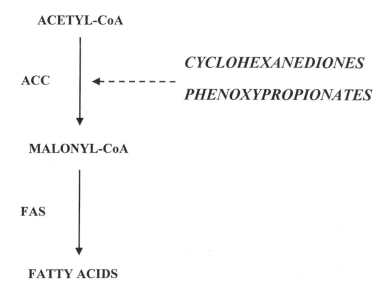

**Figure 6**    Scheme of fatty acid metabolism with the enzymes that are targets for herbicides. Dashed arrow indicates site of action of main herbicides. ACC, acetyl-CoA carboxylase; FAS, fatty acid synthetase cycle.

## V.  PIGMENT SYNTHESIS INHIBITORS

These herbicides affect the synthesis of photosynthetic pigments. As a result, the affected parts of the plant become white or translucent. There are only two families in this group of herbicides: pyridazinones with norflurazon as the main representative of the family and isoxazolidinones in which clomazone is most used at present.

Norflurazon inhibits the desaturation of phytoene [108,109], which is the first step in the synthesis of carotenes from geranylgeranylpyrophosphate (GGPP) (see Figure 7) [110]. The primary action of this family of herbicides is interference with carotenoid biosynthesis at the level of phytoene desaturase [6,111,112]. The inhibition of this enzyme results in a decrease in carotenoids together with chlorophyll and a concurrent accumulation of phytoene [110,111,113].

It is not known where clomazone acts, but it is certainly different fromt the site of action of norflurazon [114,115].

Herbicides of these two families are absorbed by roots and move via the xylem to the leaves. Developing leaves tend to be so sensitive to these herbicides that very small quantities are sufficient for their bleaching. These herbicides do not have significant agronomic interest because they have not been extensively studied and their resistance mechanisms are still unknown. A norflurazon-resistant mutant has been isolated from the alga *Chlamydomonas reinhardtii* with alterations in the target enzyme phytoene desaturase [116]. Resistance to bleaching herbicides affecting phytoene desaturase has been obtained in cyanobacteria by genetic engineering [117].

## VI.  GROWTH REGULATORS

A series of herbicide families affect the growth of crop plants. Among these families are found phenoxyacetic acids, benzoic acids, and pyridines.

Their action sites are not exactly known, but they can act at several sites in a plant. For example, they disrupt hormone balance and protein synthesis and thereby produce a variety of growth abnormalities.

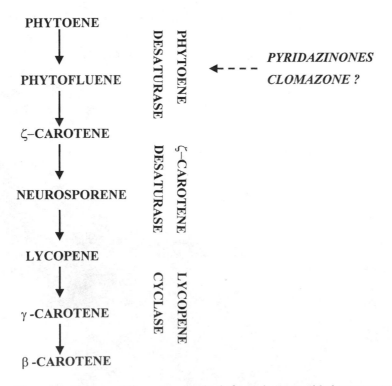

**Figure 7**   Scheme of β-carotene synthesis from phytoene with desaturases and cyclase. Dashed arrows indicate the target for herbicides pyridazinones and (probably) isoxazolidinones.

These herbicides, similar to auxin hormones, cause rapid changes in cell walls as a result of increased activity of plasma membrane adenosinetriphosphatase (ATPase) [118,119].

The herbicides belonging to this group selectively kill broadleaf plants and weeds. They tend to be absorbed by leaves, although to a lesser extent they can enter through the roots. They can move via either phloem or xylem toward the zones of the plant that are growing; therefore they are as effective in perennial plants as in annual broadleaf plants.

RESISTANCE   Many monocotyledon crops and weeds are naturally resistant to these herbicides. It has been shown that arylhydroxylation by some monooxygenases confers resistance to these herbicides [120]. Some of these genes from bacteria are being used to make resistant crop plants [121].

## VII.  SEEDLING GROWTH INHIBITORS

Within this group there are three herbicide families: dinitroanilines, acetanilides, and thiocarbamates. These herbicides reduce the ability of seedlings to develop normally in soil. Plants take up these herbicides after germinating until the seedling emerges from the soil.

Dinitroanilines act as root inhibitors, affecting division, elongation, and cellular differentiation. They bind to the heterodimers of tubulin, inhibiting polymerization, and stop mitosis in prometaphase. Thus, chromosomes are found to be condensed and cannot migrate to the poles. This causes an abnormal nucleus because the nuclear membrane is reconstituted with condensed chromosomes [122–125]. As a rule, it is believed that these compounds affect multiple sites, fundamentally lipid and protein synthesis.

Acctanilidcs and thiocarbamatcs arc inhibitors of shoots. Thcir action sitc is not known, but they probably have multiple action sites.

Dinitroanilines are effective against annual weeds, less against annual dicotyledons, but they have little effect on perennial species. In cultivation, seeds must be located below the contaminated soil layer or planted after the herbicide degradation. Susceptibility is associated with especially small seeds, low lipid content, or a high proportion of meristematic tissue with low protection in the soil where the herbicide acts.

Acetanilide derivatives are generally used for preemergence control of annual grasses and broadleaf weeds in agronomic and vegetable crop production [126].

RESISTANCE   The principal resistance mechanisms are related to decreased absorption of the herbicide and poor movement of the herbicide through the most superficial caps of cells. Some species can degrade dinitroanilines, and in others the resistance may be associated with high levels of lipids. In other cases, resistance has been associated with an increase of free amino acid levels [127]. Several biotypes that have acquired great resistance to these herbicides produce a higher quantity of tubulin and tubulin $\beta$ of greater molecular weight [122]. Researchers have found that in a biotype of goosegrass, resistance to dinitroaniline is inherited in a single recessive nuclear gene [123].

Acetanilide herbicides are detoxified in biological systems by the formation of glutathione-acetanilide conjugates. This conjugation is mediated by glutathione-$S$-transferase, which is present in microorganisms, plants, and animals [128].

## ACKNOWLEDGMENTS

We are thankful for the financial help of Ministerio de Educación y Cultura (PB96-1358) and Plan Andaluz de Investigación.

## REFERENCES

1.  JS Holt. Mechanisms and agronomic aspects of herbicide resistance. Annu Rev Plant Physiol Plant Mol Biol. 44:203–229, 1993.
2.  A Díaz, M Lacuesta, A Muñoz-Rueda. Comparative effects of phosphinothricin on nitrate and ammonium assimilation and on anaplerotic $CO_2$ fixation in N-deprived barley plants. J Plant Physiol 149:9–13, 1996.
3.  BJ Mazur, SC Falco. The development of herbicide resistant crops. Annu Rev Plant Mol Biol 40:441–470, 1989.
4.  DE Moreland. Mechanisms of action of herbicides. Annu Rev Plant Physiol 31:597–638, 1980.
5.  G Sandmann, P Böger. Sites of herbicide inhibition at the photosynthetic apparatus. In: LA Staehelin, CJ Arntzen, eds. Encyclopedia of Plant Physiology. Vol 19. Berlin: Springer-Verlag, 1986, pp 596–602.

6.  P Böger, G Sandmann. Modern herbicides affecting typical plant processes. In: WS Bower, W Ebing, D Martin, R Wegler, eds. Chemistry of Plant Protection. Vol 6. Berlin/Heidelberg: Springer, 1990, pp 173–216.
7.  C Critchley. The structure and function of photosystem II. In: M Pessarakli, ed. Handbook of Photosynthesis. New York: Marcel Dekker, 1997, pp 231–239.
8.  A Wild, R Ball. Photosynthetic unit and photosystems. Leiden: Backhuys Publishers, 1997, pp 131–153.
9.  DO Hall, KK Rao. Photosynthesis, 5th ed. Cambridge: University Press, 1994, pp 67–74.
10.  DO Hall, KK Rao. Photosynthesis, 5th ed. Cambridge: University Press, 1994, pp 97–98.
11.  I Sinning. Herbicide binding in the bacterial photosynthetic reaction center. Trends Biochem Sci 17:150–154, 1992.
12.  LA Kleczkowski. Inhibitors of photosynthetic enzymes/carriers and metabolism. Annu Rev Plant Physiol Plant Mol Biol 45:339–367, 1993.
13.  G Zurawski, HJ Böhnert, PR Whitfiel, W Bottomley. Nucleotide sequence of the gene for the Mr 32,000 thylakoid membrane protein from *Spinacia oleracea* and *Nicotiana debneyi* predicts a totally conserved primary translation protein of Mr 38,950. Proc Natl Acad Sci U S A 79:7699–7703, 1982.
14.  W Oettmeier, K Masson. Synthesis and thylakoid membrane binding of the radioactively labeled herbicide dinoseb. Pestic Biochem Physiol 14:86–97, 1980.
15.  JJS Van Rensen. Herbicides interacting with photosystem II. In. AD Dodge, ed. Herbicides and Plant Metabolism. Cambridge: University Press, 1989, pp 21–36.
16.  A Schultz, F Wengenmayer, HM Goodman. Genetic engineering of herbicide resistance in higher plants. CRC Crit Rev Plant Sci 9:1–15, 1990.
17.  W Oettmeier, K Masson, U Johanningmeier. Evidence for two herbicide-binding proteins at the reducing site of photosystem II. Biochim Biophys Acta 679:376–383, 1982.
18.  U Johanningmeier, E Neumann, W Oettmeier. Interaction of a phenolic inhibitor with photosystem II particles. J Bioenerg Biomembr 15:43–66, 1983.
19.  A Thiel, P Böhger. Binding of ioxynil to photosynthetic membranes. Pestic Biochem Physiol 25:270–278, 1986.
20.  S Izawa. Inhibitors of electron transport. In: A Trebs, M Avron, eds. Encyclopedia of Plant Physiology. New Ser. Berlin: Springer, 1977, pp 266–282.
21.  A Trebs, H Wietoska, W Draber, HJ Knops. The inhibition of photosynthetic electron flow in chloroplasts by the dinitrophenylether of bromo or iodo-nitrothymol. Z Naturforsch Teil C 33:919–927, 1978.
22.  MP Percival, NR Baker. Herbicides and photosynthesis. In: NR Baker, MP Percival eds. Herbicides. New York: Elsevier, 1991.
23.  FM Ashton, AS Crafts. Mode of Action of Herbicides, 2nd ed. New York: Wiley, 1981.
24.  KK Hatzios. Metribuzin. In: PC Kearney, DD Kaufman. Herbicides: Chemistry, Degradation and Mode of Action. New York: Marcel Dekker, 1988, pp 191–243.
25.  KK Hatzios, D Penner. Metabolism of Herbicides in Higher Plants. Minneapolis: Burgess, 1982, pp 83–92.
26.  J Hirschberg, A Bleecker, DJ Kyle, L McIntosh, CJ Arntzen. The molecular basis of triazine-resistance in higher plant chloroplast. Z Naturforsch 39c:412–419, 1984.
27.  V Souza-Machado, JD Bandeen, GR Stephenson, P Lavigne. Uniparental inheritance of chloroplast atrazine tolerance in *Brassica campestris*. Can J Plant Sci 58:977–981, 1978.
28.  RN Anderson, JW Gronwald. Atrazine resistance in a velvetleaf (*Abutilon theophrasti*) biotype due to enhanced glutathione *S*-transferase activity. Plant Physiol 96:104–109, 1991.
29.  MS Kemp, JC Caseley. Synergists to combat herbicide resistance. In: JC Caseley, GW Gussans, RK Atkin, eds. Herbicide Resistance in Weeds and Crops. Oxford: Butterworth-Heinemann, 1991, pp 279–292.
30.  MS Kemp, SR Moss, TH Thomas. Herbicide resistance in *Alopecurus myosuroides*. In: MB Green, HM Lebaron, WK Moberg, eds. Managing Resistance to Agrochemicals from Fundamental Research to Practical Strategies. ACS Symp Ser No 421. Washington, DC: ACS Books, 1990, pp 376–393.
31.  WR Alsop, DE Moreland. Effects of herbicides in the light-activated, magnesium-dependent ATPase of isolated spinach (*Spinacia oleraceae* L.) chloroplasts. Pestic Biochem Physiol 5:163–170, 1975.
32.  DE Moreland, JL Hilton. Actions on photosynthetic systems. In: LJ Audus, ed. Herbicides: Physiology, Biochemistry, Ecology. Vol 1. London: Academic Press, 1976, pp 493–523.
33.  G Hauska, A Trebst, C Kötter, H Schultz. 1,2,3-Thadiazolyphenyl-ureas, new inhibitors of photosynthetic and respiratory energy conservation. Z Naturforsch Teil C 30:505–510, 1975.
34.  B Huchzermeyer, A Loehr. Effects of nitrophen on chloroplast coupling factor–dependent reactions. Biochim Biophys Acta 724:224–229, 1983.
35.  RE McCarty. Energy transfer inhibitors of photophosphorylation in chloroplasts. In: A Trebs, M Avron, eds. Encyclopedia of Plant Physiology. New Ser. Berlin: Springer, 1977, pp 437–447.
36.  NE Good, S Izawa. Inhibition of photosynthesis. In: RM Hochster, M Kates, JH Quastel. Metabolic Inhibitors. Vol 4. New York: Academy Press, 1973, pp 179–214.
37.  DE Moreland, SC Huber. Inhibition of photosynthesis and respiration by substituted 2,6-dinitroaniline herbicides. Pestic Biochem Physiol 11:247–257, 1979.
38.  DE Moreland. Mode of action of herbicides. In: JR Plimmer. Pesticide Chemistry in the 20th Century. ACS Symp Ser No 37. Washington, DC: American Chem Society, 1977, pp 56–75.

39.  C Bowler, M Van Montagu, D Inzé. Superoxide dismutase and stress tolerance. Annu Rev Plant Physiol Plant Mol Biol 43:83–116, 1992.
40.  AD Dodge. Herbicides interacting with photosystem I. In: AD Dodge, ed. Herbicides and Plant Metabolism. Cambridge: University Press, 1989, pp 37–50.
41.  F Galván, FF de la Rosa. Inmobilized photosythetic system, utilization of solar energy for production of chemicals and fuels. In: M Pessarakli, ed. Handbook of Photosynthesis. New York: Marcel Dekker, 1997, pp 739–749.
42.  I Morales, S Batuecas. FF de la Rosa. Storage of solar energy by production of hydrogen peroxide by the blue-green alga *Anacystis nidulans* R2: stimulation by azide. Biotechnol Bioeng 40:147–150, 1992.
43.  I Morales, FF de la Rosa, Hydrogen peroxide photoproduction by immobilized cells of the blue-green alga *Anabaena variabilis*: a way to solar energy conversion. Sol Energy 49:41–46, 1992.
44.  I Morales, FF de la Rosa. Continuous photosynthetic production of hydrogen peroxide by the blue-green algae *Anacystis nidulans* R2 as a way to solar energy conversion. Sol Energy 43:373–377, 1989.
45.  W Scholz, F Galván, FF de la Rosa. The microalga *Chlamydomonas reinhardtii* CW-15 as a solar cell for hydrogen peroxide photoproduction: comparison between free and immobilized cells and thylakoids for energy conversion. Sol Energy Mater Sol Cells 39:61–69, 1995.
46.  P Fuerst, K Vaughn. Mechanisms of paraquat resistance. Weed Technol 4:150–156, 1990.
47.  BMR Harvey, J Muldoon, DB Harper. Mechanisms of paraquat tolerance in perennial ryegrass. Plant Cell Environ 1:203–209, 1978.
48.  C Preston, JAM Holtum, SB Powles. On the mechanisms of resistance to paraquat in *Hordeum glaucum* and *H. leporinum*. Plant Physiol 100:630–636, 1992.
49.  DB Herper, BMR Harveys. Mechanism of paraquat tolerance in perennial ryegrass. Role of superoxide dismutase, catalase and peroxidase. Plant Cell Environ 1:211–215, 1978.
50.  RJ Youngman, AD Dodge. On the mechanism of paraquat resistance in *Conyza* sp. In: G Akoyunoglou, ed. Photosynthesis and Environment. Philadelphia: Balalban, 1981, pp 537–544.
51.  Y Shaaltiel, J Gressel. Multienzyme oxygen radical detoxifying system correlated with paraquat resistance in *Conyza bonariensis*. Pestic Biochem Physiol 26:22–28, 1986.
52.  S Matsunaka, K Ito. Paraquat resistance in Japan. In: JC Caseley, GW Cussans, RK Atkin, eds. Herbicide Resistance in Weeds and Crops. Oxford-Butterworth-Heinemann, 1991, pp 77–86.
53.  AKMR Islam, SB Powles. Inheritance of resistance to paraquat in barley grass *Hordeum glucum*. Weed Res 28:393–397, 1988.
54.  Y Shaatiel, NH Chua, S Gepstein, J Gressel. Dominant pleitropy controls enzymes co-segregating with paraquat resistance in *Conyza bonariensis*. Theor Appl Genet 75:850–856, 1988.
55.  TB Ray. Site of action of chlorsulfuron. Inhibition of valine and isoleucine biosynthesis in plants. Plant Physiol 75:827–831, 1984.
56.  DL Shaner, PC Anderson, MA Stidham. Imidazolinones. Potent inhibitor of acetohydroxyacid synthase. Plant Physiol 76:545–546, 1984.
57.  MV Subramaniam, HY Hung, JM Dias, JH Miner, JH Butler, et al. Properties of mutant acetolactate synthase resistant to triazolopyrimidine sulfonilide. Plant Physiol 94:239–244, 1990.
58.  EM Beyer, MJ Duffy, JV Hey, DD Schlueter. Sulfonylureas. In: PC Kearney, DD Kaufman. Herbicides: Chemistry, Degradation and Mode of Action. Vol 3. New York: Marcel Dekker, 1988, pp 117–189.
59.  HM Brown, VA Witenbach, DR Forney, SD Strachan. Basis for soybean tolerance to thifensulfuron-methyl. Pestic Biochem Physiol 37:303–313, 1990.
60.  CV Eberlein, KM Rosow, JL Geadelmann, SJ Openshaw. Differential tolerance of corn genotypes to DPX-M6316. Weed Sci 37:651–657, 1989.
61.  R Fonné-Pfister, J Gaudin, K Kreuz, K Ramsteiner, E Ebert. Hydroxylation of primisulfuron by an inducible cytochrome P-450–dependent monooxygenase system from maize. Pestic Biochem Physiol 37:165–173, 1990.
62.  A Zimmerlin, F Durst. Xenobiotic metabolism in plants: aryl hydroxylation of diclofop by a cytochrome P-450 enzyme from wheat. Phytochemistry 29:1729–1732, 1990.
63.  GW Haughn, J Smith, B Mazur, C Somerville. Transformation with a mutant *Arabidopsis* acetolactate synthase gene renders tobacco resistant to sulfonylurea herbicides. Mol Gen Genet 211:266–271, 1988.
64.  KY Lee, J Townsend, J Tepperman, M Black, CF Chui, et al. The molecular basis of sulfonylurea herbicide resistance in higher plants. EMBO J 7:1241–1248, 1988.
65.  JT Christopher, SB Powles, DR Lijegren, JAM Holtum. Cross resistance to herbicides in annual ryegrass (*Lolium rigidum*). Plant Physiol 95:1036–1043, 1991.
66.  IM Heap, R Knight. The occurrence of herbicide cross-resistance in a population of annual ryegrass, *Lolium rigidum*, resistant to diclofop-methyl. Aust J Agric Res 37:149–156, 1986.
67.  CA Mallory-Smith, DC Thill, MJ Dial. Identification of sulfonylurea herbicide-resistance prickly lettuce (*Lactuca serriola*). Weed Technol 4:163–168, 1990.
68.  JC Cotterman, LL Saari. Rapid metabolic inactivation is the basis for cross-resistance to chlorsuforon in diclofop-methyl–resistant rigid ryegrass (*Lolium rigidum*) SR4/84. Pestic Biochem Physiol 43:182–192, 1992.
69.  CA Mallory-Smith, DC Thill, MJ Dial, RS Zemetra. Inheritance of sulfonylurea resistance in *Lactuca* spp. Weed Technol 4:787–790, 1990.

70.  GM Kishore, D Shah. Amino acid biosynthesis inhibitors as herbicides. Annu Rev Biochem 57:627–663, 1988.
71.  SO Duke. Glyphosate. In: PC Kearney, DD Kaufman. Herbicides: Chemistry, Degradation and Mode of Action. Vol 3. New York: Marcel Dekker, 1988, pp 1–70.
72.  E Grossbard, D Atkinson, eds. The Herbicide Glyphosate. London: Butterworth, 1985.
73.  DM Stalker, WR Hiatt, L Comai. A single amino acid substitution in the enzyme 5-enolpyruvylshikimate-5-phosphate synthase confers resistance to the herbicide glyphosate. J Biol Chem 260:4724–4728, 1985.
74.  L Comai, LC Sen, DM Stalker. An altered *aroA* gene product confers resistance to the herbicide glyphosate. Science 221:370–371, 1983.
75.  SR Padgette, G Della-Ciopa, DM Shah, RT Fraley, GM Kishore. Selective herbicide tolerance through protein engineering. In: J Schell, I Vasil, eds. Cell Culture and Somatic Cell Genetics of Plants. Vol 6. New York: Academic Press, 1989, pp 441–476.
76.  L Comai, D Facciotti, WR Hiatt, G Thompson, RE Rose, DM Stalker. Expression in plants of a mutant *aroA* gene from *Salmonella typhimurium* confers tolerance to glyphosate. Nature 317:741–744, 1985.
77.  G Della-Cioppa, SC Bauer, ML Taylor, DE Rochster, BK Klein, et al. Targeting a herbicide-resistant enzyme from *Escherichia coli* to chloroplasts of higher plants. Biotechnology 5:579–584, 1987.
78.  JJ Fillati, J Kiser, R Rose, L Comai. Efficient transfer of a glyphosate tolerance gene into tomato using a binary *Agrobacterium tumefaciens* vector. Biotechnology 5:726–730, 1987.
79.  CM Boerboom, DL Wyse, DA Somers. Mechanism of glyphosate tolerance in birdsfoot trefoil (*Lotus corniculatus*). Weed Sci 38:463–467, 1990.
80.  FP DeGennaro, SC Weller. Differential susceptibility of field bindweed (*Convolvulus arvensis*) biotypes to glyphosate. Weed Sci 32:472–476, 1984.
81.  CV Givan, KW Joy, LA Kleczkowski. A decade of photorespiratory nitrogen cycling. Trends Biochem Sci 13:433–437, 1988.
82.  AK Keys, IF Bird, MJ Cornelius, PJ Lea, RM Wallsgrove, BJ Miflin. Photorespiratory nitrogen cycle. Nature 275:741–742, 1978.
83.  K Tachibana, T Watanabe, T Sekizawa, T Takematsu. Action mechanism of bialaphos. J Pestic Sci 11:33–37, 1986.
84.  R Manderscheid, A Wild. Studies on the mechanism of inhibition by phosphinothricin of glutamine synthetase isolated from *Triticum aestivum* L. J Plant Physiol 123:135–142, 1986.
85.  R Altenburger, R Callies, LH Grimme, D Leibfritz, A Mayer. The mode of action of glufosinate in algae: the role of uptake and nitrogen assimilation pathways. Pestic Sci 45:305–310, 1995.
86.  CE Palmer, M Oelck. The relationship of phosphinothricin to growth and metabolism in cell cultures of *Brassica napus* L. J Plant Physiol 141:105–110, 1993.
87.  H Sauer, A Wild, W Rühle. The effect of phosphinothricin (glufosinate) on photosynthesis, Z Naturforsch Teil C 42:270–278, 1987.
88.  G Donn, E Tischer, JA Smith, HM Goodman. Herbicide-resistant alfalfa cells: an example of gene amplification in plants. J Mol Appl Genet 2:621–635, 1984.
89.  AR Franco, FJ López-Siles, J Cárdenas. Resistance to phosphinothricin (glufosinate) and its utilization as a nitrogen source in *Chlamydomonas reinhardtii*. Appl Environ Microbiol 62:3834–3839, 1996.
90.  C Thompson, N Movva, R Tizard, R Crameri, J Davies, et al. Characterization of the herbicide-resistance gene *bar* from *Streptomyces higroscopicus*. EMBO J 6:2519–2523, 1987.
91.  W DeGreef, R Delon, M De Block, J Leemans, J Botterman. Evaluation of herbicide resistance in transgenic crops under field conditions. Biotechnology 7:61–64, 1989.
92.  JD Burton, JW Gronwald, DA Somers, JA Connelly, BG Gengenbach, DL Wyse. Inhibition of plant acetyl–coenzyme A carboxylase by the herbicides sethoxydim and haloxyfop. Biochem Biophys Res Commun 148:1039–1044, 1987.
93.  HY Cho, JM Widholm, FW Slife. Effects of haloxyfop on corn and soybean cell suspension cultures. Weed Sci 34:496–501, 1986.
94.  BJ Incledon, C Hall. Acetyl–coenzyme A carboxylase: quaternary structure and inhibition by graminicidal herbicides. Pestic Biochem Physiol 57:255–271, 1997.
95.  HK Lichtenthaler. Mode of action of herbicides affecting acetyl-CoA carboxylase and fatty acid biosynthesis. Z Naturforsch Teil C 45:521–528, 1990.
96.  RH Shimabukuro, BL Hoffer. Induction of ethylene as an indicator of senescence in the mode of action of diclofop-methyl. Pestic Biochem Physiol 54:146–158, 1996.
97.  JM DiTomaso, PH Brown, AE Stowe, DL Linscott, LV Kochian. Effects of diclofop and diclofop-methyl on membrane potentials in roots of intact oat, maize and pea seedlings. Plant Physiol 95:1063–1069, 1991.
98.  RH Shimabukuro, BL Hoffer. Effect of diclofop on the membrane potentials of herbicide-resistant and susceptible annuals ryegrass root tips. Plant Physiol 98:1415–1422, 1992.
99.  SG Gorbach, K Kuenzler, J Asshauer, On the metabolism of Hoe 234OhOH in wheat. J Agric Food Chem 25:507–511, 1977.
100. FS Tanaka, BL Hoffer, RH Shimabukuro, RG Wien, WC Walsh. Identification of the isomeric hydroxylated metabolites of methyl 2[-4-(2,4-dichlorophenoxy)phenoxy]propanoate (diclofop-methyl) in wheat. J Agric Food Chem 38:559–565, 1990.

101. RH Shimabukuro, BL Hoffer. Perturbation of the transmembrane proton gradient and resistance to AOPP herbicides. In: R de Prado, J Jorrin, L García-Torres. Weed Crop Resistance Herbicides. Dordrecht: Kluwer, 1997, pp 71–79.

102. WB Parker, LC Marshall, JD Burton, DA Somers, DL Wyse. Dominat mutations causing alterations in acetyl–coenzyme A carboxylase confer tolerance to cyclohexanedione and aryloxyphenoxypropionate herbicides in maize. Proc Natl Acad Sci U S A 87:7175–7179, 1990.

103. WB Parker, DA Somers, DL Wise, RA Keith, JD Burton, JW Gronwald, BG Gengenbach. Selection and characterization of sethoxydim-tolerant maize tissue cultures. Plant Physiol 92:1220–1225, 1990.

104. MD Devine, SA Maclssac, ML Romano, JC Hall. Investigation of the mechanism of diclofop resistance in two biotypes of *Avena fatua*. Pestic Biochem Physiol 42:88–96, 1992.

105. LC Marshall, DA Somers, PD Dotray, BG Gengenbach, DL Wise, JW Granwald. Allelic mutations in acetyl-CoA carboxylase confer herbicide tolerance in maize. Theor Appl Genet 83:435–442, 1992.

106. J Menedez, R DePrado. Diclofop-methyl cross-resistance in a chlorotoluron-resistant biotype of *Alopecurus myosuroides*. Pestic Biochem Physiol 56:123–133, 1996.

107. AR Rendina, JM Felts, JD Beaudoin, AC Craig-Kennard, LL Look, SL Paraskos, JA Hagenah. Kinetic characterization, stereoselectivity, and species selectivity of the inhibition of plant acetyl-CoA carboxylase by the aryloxyphenoxypropionic acid grass herbicides. Arch Biochem Biophys 265:219–225, 1988.

108. AD Pardo, JA Schiff. Plastid and seedling development in SAN-9789 [4-chloro-5-(methylamine)-2-(α,α,α-trifluoro-*m*-toly)-3-(2 *H*)-pyridasinone]–treated etiolated bean seddlings. Can J Bot 58:25–35, 1980.

109. O Abrous, G Benhassaine-Kesri, A Tremolieres, P Mazliak. Effect of norflurazon on lipid metabolism in soy seedling. Phytochemistry 49:979–985, 1998.

110. FX Cunningham Jr, E Grantt. Genes and enzymes of carotenoid biosynthesis in plants. Annu Rev Plant Physiol Plant Mol Biol 49:557–583, 1998.

111. D Urbach, M Suchanka, W Urbach. Effect of substituted pyridazinone herbicides and of difunone (EMD-IT 5914) on carotenoid biosynthesis in green algae. Z Naturforsch 31C:652–655, 1976.

112. G Sandmann, IE Clarke, PM Bramley, P Böger. Inhibition of phytoene desaturase—the mode of action of certain bleaching herbicides. Z Naturforsch 39c:443–449, 1984.

113. CM Wetzel, SR Rodermel. Regulation of phytoene desaturase expression is independent of leaf pigment content in *Arabidopsis thaliana*. Plant Mol Biol 37:1045–1053, 1998.

114. G Sandman, P Böger. Interference of dimethazone with formation of terpenoid compounds. Z Naturforsch 41c:729–732, 1986.

115. G Sandmann, P Böger. Interconversion of prenyl pyrophosphates and subsequent reactions in the presence of FMC 57020. Z Naturforsch 42c:803–807, 1987.

116. V Vartak, S Bhargava. Characterization of a norflurazon-resistant mutant of *Chlamydomonas reinhardtii*. Wedd Sci 45:374–377, 1997.

117. U Windhovel, G Sandmann, P Böeger. Genetic engineering of resistance to bleaching herbicides affecting phytoene desaturase and lycopene cyclase in cyanobacterial carotenogenesis. Pestic Biochem Physiol 57:68–78, 1997.

118. D Coupland, DT Cooke, CS James. Effects of 4-chloro-2-methylphenoxypropionate (an auxin analogue) on plasma membrane ATPase activity in herbicide-resistance and herbicide-susceptible biotypes of *Stellaria media* L. J Exp Bot 42:1065–1071, 1991.

119. JB Pillmoor, JK Gaunt. The behaviour and mode of action of the phenoxyacetic acids in plants. In: DH Hutson, TR Roberts. Progress in Pesticide Biochemistry. Vol 1. Chichester, UK: John Wiley & Sons, 1981, pp 147–218.

120. NA Broadhurst, ML Montgomery, VH Freed. Metabolism of 2-methoxy-3,6-diclorobenzoic acid (dicamba) by wheat and bluegrass plants. J Agric Food Chem 14:585–588, 1966.

121. BR Lyon, DJ Llewellyn, JL Huppatz, ES Dennis, WJ Peacock. Expression of a bacterial gene in transgenic tobacco plants confers resistance to the herbicide 2,4-dichlorophenoxyacetic acid. Plant Mol Biol 13:533–540, 1989.

122. E Yamamoto, L Zeng, W Baird. Alfa-tubulin missense mutations correlate with antimicrotubule drug resistance in *Eleusine indica*. Plant Cell 10:297–308, 1998.

123. L Zeng, W Baird. Genetic basis of dinitroaniline herbicide resistance in a highly resistant biotype of goosegrass (*Eleusine indica*). J Hered 88:427–432, 1997.

124. FD Hess. Herbicide interference with cell division in plants. In: P Böger, G Sandmann, eds. Target Sites of Herbicide Action. Boca Raton, FL: CRC Press, 1989, pp 85–103.

125. FD Hess, D Bayer. Binding of the herbicide trifluralin to *Chlamydomonas* flagellas tubulin. J Cell Sci 24:351–360, 1977.

126. NE Humburg, SR Colby, RE Hill, LM Kitchen, RG Lym WJ McAvoy, R Prasad. Herbicide Handbook of the Weed Science Society of America, 6th ed. Champaign, IL: Weed Science Society of America, 1989.

127. AN Starrat, G Lazarovits. Herbicide-induced disease resistance and associated increases in free aminoacid levels in melon plants. Can J Plant Pathol 21:33–36, 1999.

128. DM Stamper, OH Tuovinen. Biodegradation of the acetanilide herbicides alachlor, metalachlor and propachlor. Crit Rev Microbiol 24:1–22, 1998.

# 40

# Parasitic Flowering Plants of Genus *Orobanche:* DNA Markers, Molecular Evolution, and Physiological Relations with the Host Plants

**Ivan N. Minkov and Antoaneta Ljubenova***

*University of Plovdiv, Plovdiv, Bulgaria*

## I. INTRODUCTION

Interactions between different living organisms are a rule rather than an exception. These interactions are called symbiosis when the species engaged live permanently together and are bound by nutritional links. The symbiosis is mutualistic when both organisms benefit from the relationship and parasitic when only one of the partners uses the resources of the system, thus damaging the development of the other one [1].

In ancient Greek the word *parasitos* means one who eats at the table of another. Parasitism is usually associated with insects, fungi, and microorganisms, but it also occurs among flowering plants. Although parasitic plants were first mentioned as botanical curiosities [2], about 4000 species in 22 dicot families are currently recognized as parasitic [3]. They can be placed into 11 independent phylogenetic clades, indicating that parasitism originated several times during the evolution of angiosperms [3].

Parasitic plants represent an extraordinary adaptation in which modified roots, haustoria, are used to transfer water, minerals and a diverse collection of carbon compounds from a host plant to the parasite [4–9].

Evolution has associated the parasitic life cycle with profound morphological, physiological, and biochemical changes [1,10–19]. On several occasions this has proceeded to an extreme form known as holoparasitism [18,19], in which the parasite obtains virtually all of its reduced carbon from the host. In addition to the loss of photosynthesis and associated pigments [20], holoparasites exhibit reduction or loss of leaves and loss of nonhaustorial roots, and in one extreme group, Rafflesiaceae [21], vegetative tissues have been reduced to a mycelium-like mass of feeding cells that reside solely within the photosynthetic host plant. There are also hemiparasites, which contain chlorophyll but in which the photosynthetic efficiency is reduced [22–24].

The molecular mechanisms of the expression of genetic material of parasitic plants are not well understood, although they can provide valuable evidence for the evolution of flowering plants as well as for the nature of metabolism during the parasite-host interaction. This review is focused on root parasites from the genus *Orobanche*, belonging to Orobanchaceae, closely related to Scrophulariaceae. We will discuss three areas of plant parasitism: the use of DNA molecular markers in solving taxonomy problems within the genus; the evolutionary fate of the parasitic plant genome, especially the plastome and chondriome and its phylogenetic consequences; and the physiologic relations between different species of hosts and parasites.

---

**Current affiliation:* University of the North, Sovenga, South Africa

## II. GENUS OROBANCHE

Theophrastus Eresius (371–286 BC) mentions parasitic seed plants such as *Cuscuta* and *Orbanche* in his book on the history of plants (cited in Ref. 18). The name *Orobanche* derives, according to Dioscorides (first century AD), from the Greek *Orobos* (= pea) and *angchein* (= to strangle).

The majority of Orobanchaceae occur in the Northern Hemisphere. There, the largest species diversity is found in the warm and temperate zones.

The genus *Orobanche* contains more than 100 species of total root parasites, all of which are achlorophyllous [25–27]. Most of the parasitic species and especially *O. ramosa* and *O. aegyptiaca*, included in the section Trionychon Wallr., and *O. cernua*, *O. cumana*, and *O. crenata*, included in the section Osproleon Wallr., have a broad host range including plant families such as Solanaceae, Brassicaceae, and Fabaceae. They are widespread in the Mediterranean countries, southeastern Europe, Middle East, and North Africa [28]. They cause tremendous agricultural problems there, especially with tomato, tobacco, and sunflower, on which they can grow rapidly, resulting in a serious reduction of crop yield [19,26,29–36].

These plants have a complicated life cycle. Their seeds are dormant after ripening and in order to germinate, as well as those of *Striga* [37–41] after contact with a germination stimulant, they require a postripening phase under warm and dry storage conditions [42–45]. Also necessary is a conditioning phase in a warm and humid surrounding [46–48]. It has been shown that not only the germination but also the penetration and the development of the haustorium are subject to chemical signaling from the host [49–52].

## III. RAPD MOLECULAR MARKERS FOR SOLVING TAXONOMY PROBLEMS

### A. *O. ramosa* L. *O. aegyptiaca* Pers.—Two Species or a Species Complex?

*O. ramosa* L. and *O. aegyptiaca* Pers. are two species of special interest because of the controversial data on their taxonomy. It is a problem for several reasons [26]:

1. There is an inherent morphological variability within populations of the plant, which is reflected in several aspects of their biology: weediness, chromosome aberrations, and reproductive strategies.
2. Their holoparasitism results in a reduced number of characters of taxonomy: they are nonphotosynthetic, they have no leaves, and they produce only short abnormal roots.
3. The host plant may influence the morphology of the parasite.

These plants are especially difficult to distinguish in the fields because too few characters are used in keys [26].

In spite of the economic importance and scientific interest of the *Orobanche* problem, relatively little is known about the *genetic variability* of the species within the genus. It is possible that genetic differences may contribute to variation in host preferences, and knowledge of the genetic diversity of parasite weed populations within and across species boundaries is important in evolving control strategies. Genetic variation has been reported with regard to susceptibility of the host and the virulence of the parasite [53]. Some cytological studies reveal that the genus seems to contain several polyploid complexes, which can produce new interspecific hybrids with unpredictable consequences concerning the range of hosts [54]. Some information has also been gained from isoenzyme analysis [55]. However, isoenzyme markers may be affected by environmental conditions and are expressed differentially at different stages of development.

DNA polymorphism is a different physical property at the same locus [56–58]. If such a polymorphic DNA sequence is detected, one can distinguish the two genomic DNAs. DNA-based markers are not dependent on environmental and developmental factors and have been applied successfully to discriminate between individual genotypes [59–74]. Random amplified polymorphic DNA (RAPD) is a technique used to detect polymorphisms in a DNA sequence [75–78]. The method relies on amplification of polymorphic DNA fragments by the polymerase chain reaction (PCR) using a single oligonucleotide primer of arbitrary sequence [79]. RAPD markers have already been used to differentiate between important species of the genus *Orobanche* in Israel [31,32,80].

## B. Establishing of *Orobanche Ecotypes*

So far, all molecular studies of *Orobanche* parasitic plants have been performed with plant material from one generation and usually so-called origins have been used, presumably representing a mixture of parasitic plant seeds as no inbred lines have been prepared for the parasite. Because of their mode of life, parasitic plants can reproduce by self-pollination [1,18,19], which makes it easy to produce plants with a relatively high degree of homozygosity and homogeneity. Such ecotypes are appropriate to trace inheritance of molecular markers and give more precise answers on species discrimination questions. They can also be used for other types of genetic research.

We produced ecotypes for three species from section Trionycon Warll.—*O. ramosa, O. aegyptiaca,* and *O. oxyloba*—and studied representatives of the last two generations by RAPD analysis for the amount of genetic variability among the ecotypes and between the generations. We examined representatives of section Trionycon of the genus *Orobanche* because it is considered the "ancestral stock" on the basis of its chromosome number, $2n = 24$ [54].

Three *O. ramosa* genotypes with different geographic origins (Spain, Bulgaria, and North America) were chosen for the study as they represent highly isolated populations of the species. In this way there was a possibility to trace the partition of variation between populations as a function of the geographical distance between the regions of collection.

An *O. aegyptiaca* genotype was used as a closely related species in order to obtain more information about the disputable species complex *O. ramosa/O. aegyptiaca.*

*O. oxyloba* is a species not yet reported to damage the crop yield, so it is most interesting to compare it with the other two in order to see the degree of relatedness between the three representatives of the same section of the genus *Orobanche*. Besides, there are no data about its chromosome number and an assumption has been made that a comparison could throw light on the evolutionary pathways within the genus [54].

Four generations of plants have been obtained, thus showing that all three species can reproduce by self-pollination. As the lives most suitable for RAPD mapping are inbred lines derived from repeated selfing of individual plants of an $F_2$ population [81], the study was performed on those and also on plants from $F_3$. Further selection was ceased to prevent inbreeding depression.

## C. Reproducible Different DNA Fingerprints for All Ecotypes

The pattern of DNA polymorphisms (Figures 1 and 2) and the genetic distances that we observed (Table 1, Figure 3) demonstrate that RAPD analysis of the five *Orobanche* ecotypes studied provides possibilities to distinguish clearly between them, as single primers and primer combinations generated reproducible different DNA fingerprints. These primers were selected because they had been used for similar purposes in Israel [80] and Spain [82]. Thus, it is possible to claim that a set of primers is already established that can be successfully used in different laboratories for identification of *Orobanche* species of economic importance.

The results of our study indicate that the genetic distance (GD) between the *O. ramosa* genotype cluster and the *O. aegyptiaca/O. oxyloba* cluster is big enough (GD = 0.72) to consider *O. ramosa* and *O. aegyptiaca* unambiguously different species. This corresponds to the findings about populations of the same species in Israel [80] and Egypt [31].

There are significant differences between the three *O. ramosa* genotypes. The most closely related proved to be the populations from America and Bulgaria (GD = 0.15). There is evidence that the American and European genera of Orobanchaceae do not share a common Orobanchaceae ancestor and thereby have evolved at slightly different rates [83]. The reason for the closer link between the two populations exhibited in our study could probably be the relatively recent introduction of this particular American genotype in the New World habitat.

Spain is the westernmost site of *Orobanche* distribution in the Mediterranean region. The reason for the pronounced genetic distance of this genotype from the other two *O. ramosa* genotypes may be that the

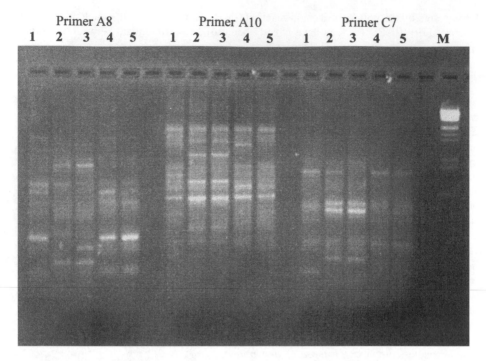

**Figure 1**  Amplification patterns obtained from five *Orobanche* ecotypes by primers A8, A10, and C7. The last line contains a lambda *Eco*RI marker. 1, *O. ramosa*, Spain; 2, *O. aegyptiaca*, Egypt; 3, *O. oxyloba*, Egypt; 4, *O. ramosa*, Bulgaria; 5, *O. ramosa*, North America.

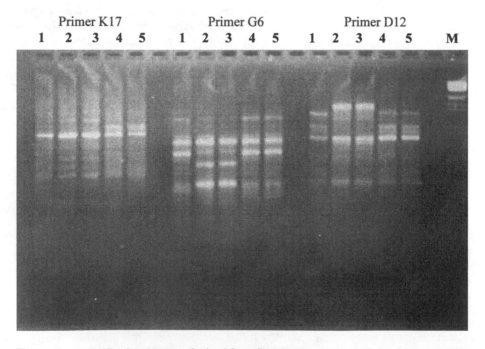

**Figure 2**  Amplification patterns obtained from five *Orobanche* ecotypes by primers K17, G6, and D12. The last line contains a lambda *Eco*RI marker. 1, *O. ramosa*, Spain; 2 *O. aegyptiaca*, Egypt; 3, *O. oxyloba*, Egypt; 4, *O. ramosa*, Bulgaria; 5, *O. ramosa*, North America.

**TABLE 1**  Genetic Distances Between Five *Orobanche* Genotypes, Calculated from the Presence and Absence of 968 Amplification Products of six RAPD Primers

|  | *O. ramosa* Spain (S) | *O. aegyptiaca* | *O. oxyloba* | *O. ramosa* Bulgaria (BG) |
|---|---|---|---|---|
| *O. aegyptiaca* | 0.717 | | | |
| *O. oxyloba* | 0.753 | 0.133 | | |
| *O. ramosa*, BG | 0.246 | 0.687 | 0.746 | |
| *O. ramosa*, S | 0.321 | 0.707 | 0.730 | 0.150 |

population has been founded by only a few individuals, and the geographic isolation may have increased the distinction.

A real surprise is the second cluster from the dendrogram, *O. aegyptiaca/O. oxyloba*. These species are the most closely related (GD = 0.13). This is similar to the findings of Zeid et al. [31], who calculated a genetic similarity of 68% between the populations of the same species in the same habitat. The results may be due to the fact that we examined representatives of the second and third generations of the parasitic plants produced by autogamous reproduction, which by decreasing the amount of heterogeneity may have contributed to the lesser difference. Yet another possibility is that because of the relatively close geographic habitats and easy seed dispersal some hybridizations between the species have occurred. That hypothesis can be proved with further molecular genetic studies with the inbred lines that have been obtained. However, the great similarity of *O. oxyloba* to *O. aegyptiaca* at the DNA level is a good reason to consider this species an important "new player" in the game. This is also supported by the knowledge that the mode of evolution of parasitic plants is by passing free-living stage, wild host, and crop host stages [1,10,84].

## IV. SOME ASPECTS OF THE EVOLUTION OF PARASITIC PLANTS

Within angiosperms, haustorial parasitism has evolved independently at least eight or nine times [3,85–87].

Atsatt [10] considers it an answer to competition for limited resources. In nutrient-poor and arid habitats parasites have competitive advantage over autotrophic plants because the host plant derives nutrients not otherwise available.

Searcy [88] offers a *molecular evolutionary hypothesis* related to parasitic plants. He proposes that parasitic plants would undergo distinct phases of evolution, each of which is composed of a multitude of individual evolutionary events. During the *first* phase, a free-living organism is involved in an initial parasitic or symbiotic relationship with the host. There would be a requirement for new or modification of the existing genetic information because of the evolution of a functional haustorium and the crucial structural and physiological innovations in the parasites. At the *second* phase, after the parasitic relationship is established, natural selection would be relaxed on functions required of a free-living organism and the genes for these functions would be free to accumulate random mutations. They would eventually be lost from the genome, which would lead to reduction of its total size and locking the plant irreversibly into an obligate parasitic relationship. A *third* phase of evolution would be possible as the obligate parasite

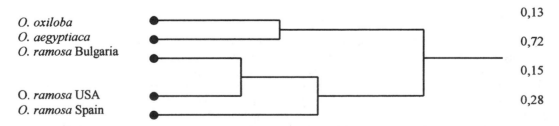

**Figure 3**  A dendrogram of five *Orobanche* genotypes based on average linkage cluster analysis using RAPD markers. Genetic distances between the species are indicated on the right of the dendrogram.

evolved more complex adaptations specific to the parasitic life cycle, such as complex life cycles or mechanisms to overcome host defenses.

Recent studies of the nuclear and chloroplast genomes of parasitic plants have proved to some extent the main ideas in Searcy's hypothesis.

## A.   State of the Nuclear and Plastid Genome of Parasitic Plants

A profound examination of the nuclear rDNA in parasitic plants shows that the ribosomal cistron is not invariant over its entire length but is a mosaic of slowly and rapidly evolving regions [86,87]. Variable and conserved domains exist in both 18S and 26S rDNA. High substitution rates have been detected in nuclear 18S rDNA, mainly in the lineages that exhibit a reduction in photosynthesis and an advanced state of nutritional dependence on the host.

The plastid genome is by far the most studied in parasitic plants: *Cuscuta*-Cuscutaceae [89–91], *Epifagus*-Orobancheceae [15,92], *Lathraea*-Scrophulariaceae [14,93], *Conopholis*-Orobancheceae [94,95], *Orobanche*-Orobancheceae [96–98].

De Pamphilis and Palmer [15] demonstrated that the root parasite *Epifagus virginiana* (beechdrops) has a plastome that lacks all the genes for photosynthesis found in the chloroplast genomes of green plants [83]. It has undergone a big reduction in size and with its 70 kb is the smallest plastid genome among plants. The *Orobanche* plastid genome has also undergone an important reduction in size, resulting in a plastid chromosome of approximately one half the length of a typical plastid genome of an autotrophic plant, such as tobacco or *Digitalis* [86,87].

It has been shown that evolution of *rbc*L within species of *Orobanche* has proceeded along divergent pathways [97]. Intact open reading frames are present in *O. corymbosa* and *O. fasciculata*, whereas *O. cernua* and *O. ramosa* have *rbc*L pseudogenes. Rubisco function is lost, and the differences suggest that this happened after the adaptation to heterotrophy.

## B.   Correlation between Nuclear, Plastid, and Mitochondrial Genomes

The genetic material of plants is compartmentalized: it is divided between the nucleus/cytosol, the plastids, and the mitochondria [99]. Plants therefore not only have to control the many genes in each of these compartments but also must coordinate the expression of genes between the three genetic compartments. Sufficient data exist in support of the interaction between the different genetic compartments [100–106]. Nuclear genes control the expression of both plastid and mitochondrial genes, and plastome and chondriome can affect nuclear gene expression [107–113].

Because holoparasitic plants are valuable model organisms that can increase our understanding of the mode and tempo of evolutionary change at the molecular level, it may be worth checking on the state of their mitochondrial genome too, as transfer of genes is a significant evolutionary event [114].

We examined the mitochondrial genes coxI, coxIII, atp6, atp9, atpA, 18S + 5S rRNA, rrn18, as well as the chloroplast probes pRp7-1 and pRp9-1 by Southern blot hybridization of total *Orobanche* DNA. The objective was to determine whether they are present in the investigated genotypes and whether there are differences between the genotypes and the different generations that can be used as molecular and genetic markers to distinguish them unambiguously.

Figures 4, 5, and 6 represent the hybridization pattern obtained with *Eco*RI-digested *Orobanche* DNA probed with the coxIII mitochondrial gene, pRp7-1 chloroplast probe, and rrn18 mitochondrial gene.

## C.   Conservative Mitochondrial Spots—Why These Genes?

Research on the characteristics of the mitochondria of *O. cernua* showed that there is no evidence to suggest any basic abnormality in their biochemical properties in relation to the obligatory parasitic form of life. The mitochondria are as functionally active as similar preparations of mitochondria from nonparasitic plants insofar as the Krebs cycle, electron transport, and accessory pathways are concerned [115,116]. However, no study of the state of different genes in the mitochondrial genome of parasitic plants has been performed. Higher plant mitochondrial genomes have characteristics of particular interest such as high frequency of recombination (by far the most unique one), large coding capacity, the encoding of genes present in the nuclear genome of other eukaryotes, the existence of genes that disrupt growth and pollen development, the phenomenon of RNA editing, *trans*-splicing, the import of tRNAs

1   2   3   4   5   6   7   8   9   10   11   12   13   14   15   16   17   18  19   20  21

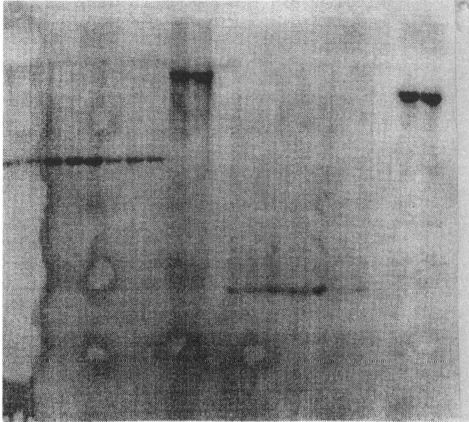

**Figure 4**   Southern blot of *Orobanche* sp. DNA digested by *Eco*RI, probed with mitochondrial gene cox III. 1, Lambda *Hind*III; 2,3, *O. ramosa*, Spain; 4,5, *O. aegyptiaca*, Egypt; 6,7, *O. oxyloba*, Egypt; 8,9, *O. ramosa*, Bulgaria; 11, *O. ramosa*, North America; 12,13, *L. esculentum*.

from the cytoplasm, and the rapid alternation in genome organization [100]. These peculiar features are most probably involved in some way in the evolution and development of parasitic plants. The chondriome contains genes for the translation apparatus of the mitochondrion as well as those coding for subunits of the respiratory chain complexes. In our study we included representatives of both groups.

The rrn5, rrn18, and tRNA[fMet] usually represent one cistron. In our investigation its sequence was present in all the investigated genotypes. Hybridization data show the same situation for coxIII, atp9, and atpA sequences. The way of life of parasitic plants allows the dropout of a large amount of biochemical pathways. Nevertheless, changes in the preceding genes detectable by Southern blot analysis have not accumulated, which could be due to the direct need for their products (cytochrome *c* oxidaze subunit III, subunits 9 and A of the ATPase complex) in the life cycle of the plants.

## D.   Polymorphism in 18S Ribosomal DNA—a Genetic and Molecular Marker

*rrn18* unambiguously distinguishes between three *O. ramosa* genotypes (seeds collected from Spain, Nothern America, Bulgaria) on the one hand and *O. aegyptiaca* and *O. oxyloba* on the other with all four restriction endonucleases and in the three generations of plants used in the study. Therefore it can be used as a molecular and genetic marker to discriminate between these species. The three distinct genomes of plants each contain a complement of ribosomal RNA genes. It has been shown that the nuclear 18S rDNA

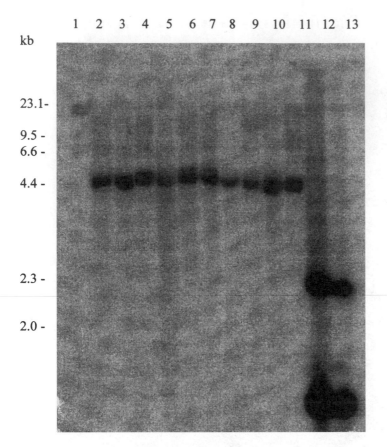

**Figure 5** Southern blot of *Orobanche* sp. DNA digested by *Eco*RI, probed with the chloroplast probe pRp 7-1. 1, Lambda *Hind*III; 2,3, *O. ramosa*, Spain; 4,5, *O. aegyptiaca*, Egypt; 6,7, *O. oxyloba*, Egypt; 8,9, *O. ramosa*, Bulgaria; 11, *O. ramosa*, North America; 12,13, *L. esculentum*.

of parasitic plants from the same genus exhibits a high level of variability [86]. We can speculate that for some reason this sequence is a kind of "hot spot" that also occurs in the mitochondrial genome of these plants. Because of the high copy number of mitochondrial chromosomes, a mitochodrial probe could be an easy applicable tool in identifying *Orobanche* species.

## E.   Conserved Chloroplast Spots

It was already mentioned that the plastome of parasitic plants (and also that of *Orobanche*) is relatively well studied. Its dynamic structure has already been shown [84,92,95,97,98]. It was tempting to compare the three *O. ramosa* genotypes from these highly separated geographic isolates (North America, Bulgaria, and Spain) and also the very closely related *O. aegyptiaca* and *O. oxyloba* by hybridization with chloroplast probes. However, the chloroplast probes we used did not show any differences in the hybridization pattern. We have probably been investigating the regions within the inverted repeats considered most conserved [84].

## V.   PHYSIOLOGIC RELATIONS WITH THE HOST PLANTS—A CASE STUDY

The influence of two species of the parasitic angiosperm *Orobanche*, *O. ramosa* L. and *O. aegyptiaca* Pers., on the physiology of its tomato host was studied in a trays-growth system.

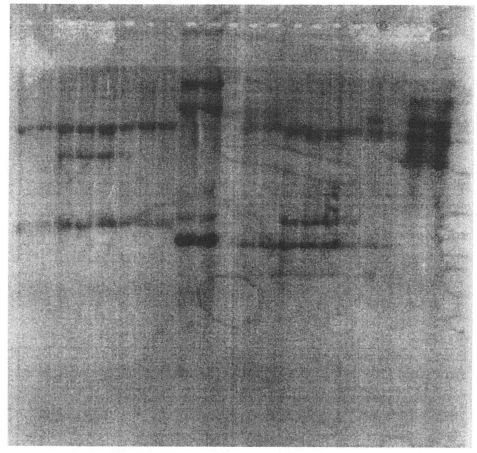

**Figure 6**  Southern blot of *Orobanche* sp. DNA digested by *Eco*RI, probed with mitochondrial gene 18 rrn. 1, Lambda *Hind*III; 2,3, *O. ramosa*, Spain; 4,5, *O. aegyptiaca*, Egypt; 6,7, *O. oxyloba*, Egypt; 8,9, *O. ramosa*, Bulgaria; 11, *O. ramosa*, North America; 12,13, *L. esculentum*.

Nine different *Lycopersicon* species—two *L. esculentum* genotypes, one *L. pennellii*, and six *L. peruvianum*—were used in a comparative study conducted on 15 host plants per species. Six weeks after inoculation of 20 germinated *Orobanche* seeds per host plant, all host plants were harvested. Measurements of the photosynthesis of the host, leaf area of the tomato leaves, stem length of the tomatoes, and fresh and dry weight of leaves, stems, and roots of the host plants as well as of the whole *Orobanche* plants were performed. Several physiological parameters—shoot/root ratio, specific leaf area, leaf area ratio, leaf weight ratio, stem weight ratio, root weight ratio—were calculated to get an idea of the effect of the interaction.

Data show that there are significant differences between the tomato host reactions at every level of the physiological interaction with *Orobanche* [117]. Differences were also found with respect to the two different species of the parasite, *O. aegyptiaca* being more virulent than *O. ramosa*. As both parasitic species are quite similar in their habitat, morphology, and host range, this study is good proof that they should not be treated as one species.

# REFERENCES

1.  E Kuiper. Comparative Studies on the Parasitism of *Striga aspera* and *Striga hermontica* on Tropical Grasses. The Netherlands: Edam, 1997.

2.  LJ Musselman. Taxonomy and spread of *Orobanche*. In: AH Pieterse, JAC Verkleij, SJ ter Borg, eds. Biology and Management of *Orobanche*. Amsterdam: Royal Tropical Institute, 1994, pp 27–35.
3.  DL Nickrent, R Joel Duff. Molecular studies of parasitic plants using ribosomal RNA. In: MT Moreno, J Cubero, eds. Advances in Parasitic Plants Research. Proceedings 6th International Symposium Parasitic Plants, Cordoba, Spain, 1996, pp 28–52.
4.  JH Visser, I Dorr. The haustorium. In: LJ Musselman ed. Parastitic Weeds in Agriculture. Vol I. Srtiga. Boca Raton, FL: CRC Press, 1987, pp 91–106.
5.  MC Press, N Shah, GR Stewart. The parasitic habit: trends in metabolic reductionism. In: SJ ter Borg, ed. Biology and Control of *Orobanche*. Wageningen, The Netherlands: LH/VPO, 1986, pp 96–106.
6.  MC Press, N Shah, JM Tuohy, GR Stewart. Carbon isotope ratios demonstrate carbon flux from $C_4$ host to $C_3$ parasite. Plant Physiol 85:1143–1145, 1987.
7.  MC Press, JD Graves. Carbon relations of angiosperm parasites and their hosts. In: SJ ter Borg, ed. Proceedings of a Workshop on the Biology and Control of *Orobanche*. Wageningen, The Netherlands: LH/VPO, 1989, pp 55–65.
8.  MC Press, S Smith, GR Stewart. Carbon acquisition and assimilation in parasitic plants. Funct Ecol 5:278–283, 1991.
9.  I Cechin, MC Press. Nitrogen relations of the sorghum–*Striga hermontica* host-parasite association: growth and photosynthesis. Plant Cell Environ 16:237–247, 1993.
10. PR Atsatt. Parasitic flowering plants: how did they evolve? Natur 107:502–510, 1973.
11. KW Cummings, TM Szaro, TD Burns. Evolution of extreme specialization within a lineage of ectomycorrhizal epiparasites. Nature 379, 63–66, 1996.
12. TE Dawson, JR Ehleringer. Ecological correlates of seed mass variation in *Phoradendron juniperum*, a xylem-tapping mistletoe. Oecologia 85:322–342, 1991.
13. AC de la Harpe, JH Visser, N Grobbelaar. Photosyntethic characteristics of some South African parasitic flowering plants. Z Pflazenphysiol 103:226–275, 1981.
14. P Delavault, V Sakanyan, P Thalouarn. Divergent evolution of two plastid genes, *rbc*L and *atp*B, in a non-photosynthetic parasitic plant. Plant Mol Biol 29:1071–1079, 1995.
15. CW de Pamphilis, JD Palmer. Loss of photosynthetic and chlororespiratory genes from the plastid genome of a parasitic flowering plant. Nature 348:337–339, 1990.
16. CW de Pamphilis, ND Young, AD Wolfe. Evolution of plastid gene *rps*2 in a lineage of hemiparasitic and holoparasitic plants: many losses of photosynthesis and complex patterns of rate variation. Proc Natl Acad Sci U S A 94:7367–7372, 1977.
17. GR Stewart, MC Press. The physiology and biochemistry of parasitic angio-sperms. Annu Rev Plant Physiol Plant Mol Biol 41:127–151, 1990.
18. J Sauerborn. Parasitic flowering plants: ecology and management. PhD dissertation, Verlag Josef Margraf FR Germany, 1991.
19. C Parker, CR Riches. Parasitic Weeds of the World: Biology and Control. Wallingford, UK: CAB International, 1993.
20. MC Press. Carbon and nitrogen relations. In: MC Press, JD Graves, eds. Parasitic Plants. London: Chapman & Hall, 1995, pp 103–124.
21. JS Pate. Mineral relationship of parasites and their hosts. In: MC Press, JD Graves, eds. Parasitic Plants. London: Chapman & Hall, 1995, pp 80–102.
22. MC Press, JM Tuohy, GR Stewart. Gas exchange characteristics of the sorghum-*Striga* host-parasite association. Plant Physiol 84:814–819, 1987.
23. MC Press, JD Graves, GR Stewart. Transpiration and carbon acquisition in root hemiparasitic angiosperms. 39:1009–1014, 1988.
24. ED Schulze, OL Lange, H Ziegler, G Gebauer. Carbon and nitrogen isotope ratios of mistletoes growing on nitrogen and non-nitrogen fixing hosts and opn CAM species in the Namib desert confirm partial heterotrophy. Oecologia 88:457–462, 1991.
25. LJ Musselman. The biology of *Striga*, *Orobanche*, and other root-parasitic weeds. Annu Rev Phytopathol 18:463–489, 1980.
26. LJ Musselman. Taxonomy and spread of *Orobanche*. In: AH Pieterse, JAC Verkleij, SJ ter Borg, eds. Biology and Management of *Orobanche*. Amsterdam: Royal Tropical Institute, 1994, pp 27–35.
27. OA Charter, DA Webb. Flora Europaea. 10:286–293, 1972.
28. AH Pieterse. The broomrapes (Orobanchaceae)—a review. Abstr Trop Agric 5(3):9–35, 1979.
29. C Parker. The present state of the *Orobanche* problem. In: AH Pieterse, JAC Verkleij, SJ ter Borg, eds. Biology and Management of *Orobanche*. Amsterdam: Royal Tropical Institute, 1994, pp 17–26.
30. JI Cubero, MT Moreno. Parasitic weed science: a quarter century. In: MT Moreno, JL Cubero, eds. Advances in Parasitic Plants Research. Proceedings 6th International Symposium Parasitic Plants, Cordoba, Spain, 1996, pp 16–21.
31. M Zeid, M Madkour, Y Koraiem, A Nawar, M Soliman, F Zaitoun. Molecular studies on *Orobanche*. Phytopathol 141:351–355, 1997.
32. I Paran, D Gidoni, R Jacobsohn. Variation between and within broomrape (*Orobanche*) species revealed by RAPD markers. Heredity 78:68–74, 1997.

33. DM Joel. Key developmental processes in parasitic weeds as potential targets for novel control methods. Proceedings 6th EWRS Mediterranean Symposium, Monpellier, France, 1998, pp 135–140.

34. H Chalakov. Present situation and prospects for solving the tobacco broomrape problem in Bulgaria. In: K Wegmann, LJ Musselman, DM Joel, eds. Current Problems of Orobanche Researches. Proceedings 4th International Orobanche Workshop, Albena, Bulgaria, 1998, pp 401–404.

35. RE Eplee, R Norris. Control of parasitic weeds. In: MC Press, JD Graves, eds. Parasitic Plants. London: Chapman & Hall, 1995, pp 256–278.

36. GB Khattril, SC Srivastawa, R Jacobsohn. Agronomic problems and control of broomrape (*Orobanche* spp) in Nepal. In: K Wegmann, LJ Musselman, DM Joel, eds. Current Problems of *Orobanche* Researches. Proceedings 4th International *Orobanche* Workshop, Albena, Bulgaria, 1998, pp 447–448.

37. CE Cook, LP Whichard, B Turner, ME Wall, GH Egley. Germination of witchweed (*Striga lutea* Lour.): isolation and properties of a potent stimulant. Science 154:1189–1190, 1966.

38. C Hauck, S Muller, H Schidklnecht. A germination stimulant for parasitic flowering plants from *Sorghum bicolor*, a genuine host plant. J Plant Physiol 139:474–478, 1992.

39. R Herb, JH Visser, H Schildknecht. Recovery, isolation and preliminary investigation of germination stimulants produced by *Vigna unguiculata* Warp. Cv Saunders Upfight. In: HC Weber, W Forstereuter, eds. Proceedings of the Fourth International Symposium on Parasitic Flowering Plants, Marburg, Germany, Philipps-Universitat, 1987, pp 315–366.

40. DC Logan, GR Stewart. Role of ethylene in the germination of the hemiparasite *Striga hermontica*. Plant Physiol 97:1435–1438, 1991.

41. BA Siame, Y Weerasuriya, K Wood, G Ejeta, LG Butler. Isolation of Srigol, a germination stimulant for *Striga asiatica*, from host plants. J Agric Food Chem 41:1486–1496, 1993.

42. LS Boone, G Fate, M Chang, DG Lynn. Seed germination. In: MC Press, JD Graves, eds. Parasitic Plants. London: Chapman & Hall, 1995, pp 14–38.

43. M Chang, DG Lynn. Haustoria and the chemistry of host recognition in parasitic plants. J Chem Ecol 12, 561–579, 1986.

44. R Jain, L Foy. Influence of various nutrients and growth regulators on germination and parasitism of *Orobanche aegyptiaca* Pers. seeds. In: HC Weber, W Forstereuter, eds. Proceedings 4th International Symposium on Parasitic Flowering Plants, Marburg, Germany, 1987, pp 427–436.

45. S Slavov, R Batchvarova. Stimulants for *Orobanche* spp. seeds germination. In: K Wegmann, LJ Musselman, DM Joel, eds. Current Problems of *Orobanche* Researches. Proceedings 4th International *Orobanche* Workshop, Albena, Bulgaria, 1998, pp 98–99.

46. GN Dhanapal, PC Struik, SJ ter Borg. Effect of natural stimulants with and without GR24 on broomrape germination. In: K Wegmann, LJ Musselman, DM Joel, eds. Current Problems of *Orobanche* Researches. Proceedings 4th International *Orobanche* Workshop, Albena, Bulgaria, 1998, pp 51–58.

47. JR Ehleringer, JD Marshall. Water relations. In: MC Press, JD Graves, eds. Parasitic Plants. London: Chapman & Hall, 1995, pp 125–140.

48. EM Estabrook, JI Yoder. Plant-plant communications: rhizosphere signaling between parasitic angiospersms and their hosts. Plant Physiol 116:1–7, 1998.

49. JD Graves. Host-plant responses to parasitism. In: MC Press, JD Graves, eds. Parasitic Plants. London: Chapman & Hall, 1995, pp 206–225.

50. DM Joel, D Losner-Goshen, T Goldman-Gues, VH Portnoy. The haustorium of *Orobanche*. In: K Wegmann, LJ Musselman, DM Joel, eds. Current Problems of *Orobanche* Researches. Proceedings 4th International *Orobanche* Workshop, Albena, Bulgaria, 1998, pp 101–106.

51. DM Joel, D Losner-Goshen, J Hershenhorn, Y Goldwasser, M Assayag. The haustorium and its development in compatible and resistant hosts. In: MT Moreno, JI Cubero, eds. Advances in Parasitic Plants Research. Proceedings 6th International Symposium Parasitic Plants, Cordoba, Spain, 1996, pp 531–542.

52. D Losner-Goshen, VH Portnoy, A Mayer, DM Joel. Pectolytic activity by the haustorium of the parasitic plant *Orobanche* L. (Orobancheceae) in host roots. Ann Bot 81:319–326, 1998.

53. JAC Verkleij, AH Pieterse. Genetic variability in *Orobanche* (broomrape) and *Striga* (witchweed) in relation to host specificity. In: AH Pieterse, JAC Verkleij, SJ ter Borg, eds. Biology and Management of *Orobanche*. Amsterdam: Royal Tropical Institute, 1994, pp 67–79.

54. JI Cubero. Cytogenetics in Orobanchaceae: a review. In: MT Moreno, JI Cubero, eds. Advances in Parasitic Plants Research. Proceedings 6th International Symposium Parasitic Plants, Cordoba, Spain, 1996, pp 76–96.

55. B Schuchardt, K Wegman. Characterization and differentiation of *Orobanche* species and races by isoenzyme analysis. In: MT Moreno, JI Cubero, eds. Advances in Parasitic Plants Research. Proceedings 6th International Symposium Parasitic Plants, Cordoba, Spain, 1996, pp 167–174.

56. BR Glick, JJ Pasternak. Molecular Biotechnology: Principles and Applications of Recombinant DNA. Washington, DC: ASM Press, 1994.

57. RJ Henry. Practical Applications of Plant Molecular Biology. London: Chapman & Hall, 1997.

58. BH Taylor, JR Manhart, RM Amasino. Isolation and characterization of plant DNA. In: BR Glick, JE Thompson, eds. Methods in Plant Molecular Biology and Biotechnology. Boca Raton, FL: CRC Press, 1993, pp 37–48.

59. G Zurawski, MT Clegg. Evolution of higher plants chloroplast DNA-encoded genes: implications for structure-function and phylogenetic studies. Annu Rev Plant Physiol 38:391–418, 1987.

60. SI Warlick, LD Black. Evaluation of the subtribes Moricandidae, Savignyinae, Vellinae, and Zillinae (Brassicaceae, tribe Brassiceae) using chloroplast DNA restriction site variation. Can J Bot 72:1692–1701, 1994.

61. RK Saiki, S Scharf, E Faloona, KB Mullis, GT Horn, HA Erlich, N Arnheim. Enzymatic amplification of B* globin genomic sequences and restriction site analysis for diagnosis of sickle cell anemia. Science 230:1350, 1985.

62. OP Rajora, BP Dancik. Chloroplast DNA variation in *Populus*. I. Intraspecific restriction fragment diversity within *Populus deltoides*, *P. nigra* and *P. maximowiczii*. Theor Appl Genet 90:317–323, 1995.

63. OP Rajora, BP Dancik. Chloroplast DNA variation in *Populus*. II. Interspecific restriction fragment polymorphisms and genetic relationships among *Populus deltoides*, *P. maximowiczii* and *P. X canadensis*. Theor Appl Genet 90:324–330, 1995.

64. OP Rajora, BP Dancik. Chloroplast DNA variation in *Populus*. III. Novel chloroplast DNA variants in natural *P. X canadensis* hybrids. Theor Appl Genet 90:331–334, 1995.

65. FR Muza, DJ Lee, DJ Andrews, SC Gupta. Mitochondrial DNA variation in finger millet (*Eleusine coracana* L. Gaertn). Euphytica 81:199–205, 1995.

66. H Luo, B Van Coppenolle, M Sequin, M Boutry. Mithochondrial DNA polymorphism and phylogenic relationships in *Hevea brasiliensis*. Mol Breed 1:51–63, 1995.

67. AJ Lohan, KH Wolfe. A subset of conserved tRNA genes in plastid DNA of nongreen plants. Genetics 150:425–433, 1998.

68. B Beyermann, P Nurnberg, A Weihe, M Meixner, JT Epplen, T Borner. Fingerprinting plant genomes with oligonucleotide probes specific for some repetitive DNA sequences. Theor Appl Genet 83:691–694, 1992.

69. NA Tinker, MG Fortin, DE Mather. Random amplified polymorphic DNA and pedigree relationships in spring barley. Theor Appl Genet 85:976–984, 1993.

70. JI Stiles, C Lemme, S Sondur, MB Morshidi, R Manshardt. Using randomly amplified polymorphic DNA for evaluating genetic relationships among papaya cultivars. Theor Appl Genet 85:697–701, 1993.

71. B Koller, A Lehmenn, JM McDermott, C Gessler. Identification of apple cultivars using RAPD markers. Theor Appl Genet 85:901–904, 1993.

72. LX Yu, HT Nguyen. Genetic variation detected with RAPD markers among upland and lowland rice cultivars (*Oryza sativa* L.). Theor Appl Genet 87:668–672, 1994.

73. RJ Mailer, R Scarth, B Fristensky. Discrimination among cultivars of rape-seed (*Brassica napus* L.) using DNA polymorphisms amplified from arbitrary primers. Theor Appl Genet 87:697–704, 1994.

74. K Sossey-Alaoui, H Serieys, M Tersac, P Lambert, E Schilling, E Griveau, F Kaan, A Berville. Evidence for several genomes in *Helianthus*. Theor Appl Genet 97:422–430, 1998.

75. JGK Williams, AR Kubelik, KJ Livak, JA Rafalski, SV Tingey. DNA polymorphisms amplified by arbitrary primers are useful as genetic markers. Nucleic Acids Res 18:6531, 1990.

76. JGK Williams, MK Hanafey, JA Rafalski, SV Tingey. 1993 Genetic analysis using random amplified polymorphic DNA markers. Methods Enzymol 218:704–740, 1993.

77. P Hedrick. Shooting the RAPDs. Nature 355:679–680, 1992.

78. G Penner, et al. Reproducibility of random amplified polymorphic DNA (RAPD) analysis among laboratories. PCR Methods Appl 2:341–345, 1993.

79. FK Yu, VA Deynze, PK Pauls. Random amplified polymorphic DNA. In: BR Glick, JE Thompson, eds. Methods in Plant Molecular Biology and Biotechnology. Boca Raton, FL: CRC Press, 1993, pp 287–301.

80. N Katzir, V Portnoy, G Tzuri, M Castejon-Munoz, DM Joel. Use of random amplified polymorphic DNA (RAPD) markers in the study of the parasitic weed *Orobanche*. Theor Appl Genet 93:367–372, 1996.

81. SB Landry. DNA mapping in plants. In: BR Glick, JE Thompson, eds. Methods in Plant Molecular Biology and Biotechnology. Boca Raton, FL: CRC Press, 1993, pp 269–285.

82. T Millan, S Cobos, AM Torres. The use of molecular markers in *Orobanche*. In: MT Moreno, JI Cubero, eds. Advances in Parasitic Plants Research. Proceedings 6th International Symposium Parasitic Plants Cordoba, Spain, 1996, pp 161–166.

83. P Delavault. Plastid genome evolution in Orobanchaceae and other holo-parasites. In: AH Pieterse, JAC Verkleij, SJ ter Borg, eds. Biology and Management of *Orobanche*. Amsterdam: Royal Tropical Institute, 1994, pp 80–93.

84. CW dePamphilis. Genes and genomes. In: MC Press, JD Graves, eds. Parasitic plants. London: Chapman & Hall, 1995, pp 177–205.

85. DL Nickerent, EM Starr. High rates of nucleotide substitution in nuclear small subunit (18 S) rDNA from holoparastitic flowering plants J Mol Evol 39:62–70, 1993.

86. DL Nickrent, Y Ouyang, R Joel Duff, CW dePamphilis. Do nonasterid holoparasitic flowering plants have plastid genomes? Plant Mol Biol 34:717–729, 1997.

87. DL Nickrent, R Joel Duff. Molecular studies of parasitic plants using ribosomal RNA. In: MT Moreno, JI Cubero, eds. Advances in Parasitic Plants Research. Proceedings 6th International Symposium Parasitic Plants, Cordoba, Spain, 1996, pp 28–52.

88. DG Searcy. Measurements by DNA hybridization in vitro of the genetic basis of parasitic reduction. Evolution 24:207–219, 1970.

89. G Haberhausen, K Valentin, K Zetsche. Organization and sequence of photosynthetic genes from the plastid genome of the holoparasitic flowering plant *Cuscuta reflexa*. Mol Genet 232:154–161, 1992

90. G Haberhausen, K Zetsche. Functional loss of all *ndh* genes in an otherwise relatively unaltered plastid genome of the holoparasitic plant *Cuscuta reflexa*. Plant Mol Biol 24:217–222, 1994.
91. MA Machado, K Zetshe. A structural, functional and molecular analysis of plastids of the holoparasites *Cuscuta reflexa* and *Cuscuta europaea*. Planta 181:91–96, 1990.
92. AD Wolfe, CW Morden, JD Palmer. Function and evolution of a minimal plastid genome from a non-photosynthetic parasitic plant. Proc Natl Acad Sci U S A 98:10648–10652, 1992.
93. P Thalouarn, S Renaudin. Polymerase chain reaction evidence of the *rbsL* gene in the Scrophulariaceae holoparasite *Lathrea cladestina* L. Comparison with the autotroph *Digitalis purpurea* L. and hemiparasite *Melampyrum pratense* L. C R Acad Sci Paris 309:381–387, 1991.
94. CF Wimpee, R Morgan, RL Wrobel. An aberrant plastid ribosomal RNA gene cluster in the root parasite *Conopholis americana*. Plant Mol Biol 18:275–285, 1992.
95. CF Wimpee, R Morgan, RL Wrobel. Loss of transfer RNA genes from the plastid 16S–23S ribosomal RNA gene spacer in a parasitic plant. Curr Genet 21:417–422, 1992.
96. P Thalouarn, C Theodet, N Russo, P Delavault. The reduced plastid genome of a nonphotosynthetic angiosperm *Orobanche hederae* has retained the *rbsL* gene. Plant Physiol Biochem 32:233–242, 1994.
97. AD Wolfe, CW dePamphillis. Alternate paths of evolution for the photosynthetic gene *rbcL* in four nonphotosynthetic species of *Orobanche*. Plant Mol Biol 33:965–977, 1997.
98. AD Wolfe, CW dePamphillis. The effect of relaxed functional constraints on the photosynthetic gene *rbcL* in photosynthetic and nonphotosynthetic parasitic plants. Mol Biol Evol 15:1243–1258, 1998.
99. P Westhoff. The genetic material of plant cells is compartmentalized: structure and expression of the subgenomes. In: P Westhoff, H Jeske, G Jurgens, K Kloppstech, G Link, eds. Molecular Plant Development, from Gene to Plant. Oxford: Oxford University Press, 1998, pp 67–109.
100. MR Hanson, O Folkerts. Structure and function of the higher plant mitochondrial genome. Int Rev Cytol 141:129–172, 1992.
101. W Schuster, A Brennicke. The plant mitochondrial genome: physical structure, information content, RNA editing, and gene migration to the nucleus. Annu Rev Plant Physiol 45:61–78, 1994.
102. C Dean, R Schmidt. Plant genomes: a current molecular description. Annu Rev Plant Physiol 46:395–418, 1995.
103. R Bock, H Koop. Extraplastidic site-specific factors mediate RNA editing in chloroplasts. EMBO 16:3282–3288, 1997.
104. R Bock. Analysis of RNA editing. In: Plastids Methods: A Companion to Methods Enzymol 15:75–83, 1998.
105. TM Clegg, MP Cummings, ML Durbin. The evolution of plant nuclear genes. Proc Natl Acad Sci U S A 94:7791–7798, 1997.
106. A Hirai, M Nakasono. Six percent of the mitochondrial genome of race came from chloroplast DNA. Plant Mol Biol Rep 11:98–100, 1993.
107. MW Gray. The endosymbiont hypothesis revised. Int Rev Cytol 141:233–358, 1992.
108. J Doebly. Genetics, development and plant evolution. Curr Opin Genet Dev 3:865–872, 1993.
109. G Moore. Cereal genome evolution: pastoral pursuits with "Lego" genomes. Curr Opin Genet Dev 5:717–724, 1995.
110. PE Thorsness, ER Weber. Escape and migration of nucleic acids between chloroplasts, mitochondria and the nucleus. Int Rev Cytol 165:207–234, 1996.
111. M Sugita, M Sugiura. Regulation of gene expression in chloroplast of higher plants. Int Rev Cytol 32:315–326, 1996.
112. S Binder, A Marchfelder, A Brennicke. Regulation of gene expression in plant mitochondria. Plant Mol Biol 32:303–314, 1996.
113. M Goldschmidt-Clermont. Coordination of nuclear and chloroplast gene expression in plant cells. Int Rev Cytol 177:115–180, 1998.
114. MW Gray. Origin and evolution of organelle genomes. Curr Opin Genet Dev 3:884–890, 1993.
115. M Singh, PS Krishnan. Isolation of mitochondria from *Orobanche*, a "total" root angiospermic parasite. Physiol Plant 39(2):179–184, 1977.
116. M Singh, PS Krishnan. Enzymatic activity in mitochondria from *Orobanche*. Physiol Plant 40(2):145–152, 1997.
117. A Ljubenova. Comparative investigation of genotypes from genus *Orobanche*—broomrape using DNA molecular markers. PhD thesis, Plovdiv, Bulgaria, 1999.

# 41

# Developmental Genetics in Lower Plants

**John C. Wallace**

*University of New Hampshire, Durham, New Hampshire*

## I. INTRODUCTION

A chapter on lower plant developmental genetics may seem out of place in a handbook largely devoted to crop plants, but as the true complexity of physiological and developmental processes in higher plants becomes ever more apparent, simpler model systems that can aid in the discovery of such processes are worthy of inclusion. The relative simplicity of lower plants and ease of genetic analysis in the haploid state can make them a model system to rival even *Arabidopsis thaliana* for some aspects of flowering plant development. As genetic engineering of crop plants progresses beyond the addition of single, entirely foreign genes to the manipulation and addition of entire pathways, knowledge of the truly fundamental aspects of plant development, which can best be discovered in lower plants, is more important than ever.

This chapter focuses on the two lower plants (a volvox and a moss) for which there exist substantial developmental genetic data; i.e., a variety of mutants whose normal ontogeny is altered have been isolated and characterized. Recent advances in molecular approaches to the analysis of these organisms, which offer even further advantages over higher plants, are already being exploited. A brief section on a third plant (a fern) is also included.

Recent studies have shown that the simplest of the three, *Volvox carteri*, is almost as closely related to animals as to plants [1], and one of the most fundamental questions that one hopes to answer by studying *Volvox*, that of the origin of the germ-soma dichotomy, is not normally even relevant to plants. But it remains classically defined as a plant, it is perhaps the simplest example in biology of how a single cell differentiates into two cell types (and thus has fascinated biologists for a very long time, e.g., Ref. 2), and work on it is beginning to answer some of these fundamental questions at the molecular level.

The moss *Physcomitrella patens* is much more obvious as a model for crop plants, and a healthy array of developmental mutants from this species have been isolated [3,4]. The ease with which it can be transformed [5,6] and the transformants grown to maturity are a great boon to its utility, and the discovery that transgenes introduced into *Physcomitrella* homologously recombine with their genomic counterparts at high frequency [7] has engendered some excitement in the plant research community [8,9].

The most complex of the three, the fern *Ceratopteris richardii*, has emerged as a model organism for plant developmental genetics [10,11]. Its special appeal lies in the fact that both gametophyte and sporophyte generations are free living, multicellular, and readily accessible for study, so the ease of genetic manipulation and analysis in the former can also be utilized to analyze traits in the latter. Also, it clearly represents the best model for the examination of the mechanisms involved in that most fundamental aspect of plant biology, alternation of generations.

## II.  VOLVOX

### A.  Volvocales—a Variety of Developmental Potentials

Although probably only distantly related to other plants [1], the order Volvocales contains organisms that display an interesting variety of developmental potentials. These range from single-celled, free-living organisms (such as those in the genus *Chlamydomonas*), through colonial organisms consisting of a single cell type (genera *Gonium, Pandorina, Eudorina*) that develops from a somatic into a reproductive cell, to organisms in which different cells specialize for separate somatic and reproductive functions (genus *Volvox*). Thus, within this order are organisms ideal for study of the genetic changes involved in the evolution of multicellularity and of cell specialization. Because the volvox are the most studied in terms of developmental genetics, this review concentrates on that group and mentions others only for purposes of comparison. Several reviews [12–14] and the comprehensive book by David Kirk [1] give more detailed accounts of volvox development than are included here.

### B.  Asexual Life Cycle and Mutants

The adult of the most studied volvox, *V. carteri f. nagariensis*, comprises a spheroid of about 2000–4000 terminally differentiated somatic cells that contains, within an extracellular glycoprotein matrix, about 16 developing embryos. Reproduction can be either asexual or sexual. In the asexual cycle (Figure 1), large reproductive cells (the gonidia), housed within the spheroid, undergo six cycles of mitosis to produce small embryonic spheres of 32 morphologically indistinguishable cells. At the next division the cells in the anterior half of each embryo divide asymmetrically to produce 16 larger cells that will give rise to the next generation of gonidia and 16 smaller cells that will, along with the remainder of the organism, form the somatic cells. The former divide asymmetrically twice more, releasing more small somatic cell precursors, and the latter undergo several more rounds of cell division until about 2000 cells are formed, including the 16 larger gonidia. Initially, the gonidia are located on the outside of the sphere and the cilia of the somatic cells point inward, but in the process known as inversion the adult form of internal gonidia and external, ciliated somatic cells is produced. After a period of expansion and cytodifferentiation, the somatic cells of the parental spheroid self-destruct, and the now juvenile *Volvox* are dispersed and live independently.

The haploid nature of all metabolically active cells in *Volvox* is a great aid to mutant isolation, and many have been described that affect the asexual cycle just outlined [15–17]. Their properties suggest several intriguing possibilities for how the initial split between somatic and germ cell lines is controlled. Mutants at the *pcd* ("premature cessation of division," Ref. 15) and the several *mul* ("multiple gonidia," Ref. 16) loci both display altered cleavage patterns and result in a greater than normal number of gonidia being formed. This led to the hypothesis that gonidial determination is a direct consequence of the larger cell size resulting from the asymmetrical cell division [15]. The competing idea [18] is that a cytoplasmic determinant that is partitioned into the larger cells is responsible for initiating reproductive cell development—a phenomenon similar to the pole plasm found to determine germ cell development in *Drosophila* [19]. Several lines of evidence favor the former hypothesis [17,20]. For example, if heat shock or microsurgery is used to alter the cleavage of early gonidial cells that normally differentiate only into somatic cells, the larger than normal offspring, which would normally become somatic, differentiate as extra gonidia. Detailed studies of normal cleavage patterns and how they give rise to the larger cells that become gonidia are reviewed in Ref. 1.

In addition to the *pcd* and *mul* loci, several other mutant phenotypes have been isolated that alter the normal developmental sequence. In the *lag* ("late gonidia") mutants, which comprise four genetic loci, the asymmetric cleavages that lead to gonidia occur as usual, but the resultant large cells at first differentiate somatically, producing cilia. Only later do they redifferentiate and follow the germ cell pathway. A similar phenotype is seen in mutants in the single *regA* locus. In this type, gonidial cell development is normal, but the somatic cells first differentiate normally, then redifferentiate to form small gonidia.

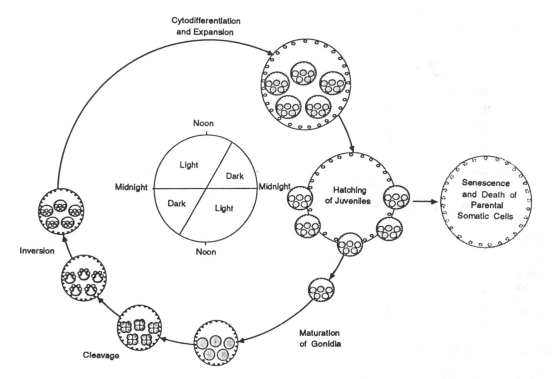

**Figure 1**   The asexual life cycle of *Volvox carteri* as synchronized by a light-dark cycle. Mature gonidia (asexual reproductive cells) undergo a rapid series of cleavage divisions, certain of which are asymmetric. The larger cells resulting from these unequal divisions will become the gonidia of the next generation, while the small cells will become part of the somatic cell population. At the end of cleavage, all cells that will be present in the adult are present in undifferentiated form, but the embryo is inside-out with respect to the adult configuration. The adult orientation is achieved through the process called inversion. Following inversion, both the parental spheroid and the juveniles contained within it expand by deposition of extracellular matrix. Midway through expansion, the juveniles hatch and swim away, leaving a "hulk" of parental somatic cells that will undergo programmed cell death. The juvenile spheroids continue to expand while their gonidia mature, preparing for a new round of embryogenesis. Under the synchronizing influence of the light-dark cycle, one asexual life cycle is completed every 48 hr, cleavage (which takes about 7 hr) begins near the end of a light period, and inversion (which takes less than an hour) occurs in the dark period. (From Ref. 13.)

Clearly the *lag* and *regA* genes are critically important to regulating the germ-soma dichotomy. Finally, mutants of a third type, *gls* ("gonidialess"), undergo no asymmetric cleavages and all cells develop somatically; clearly, this mutant cannot reproduce and to be maintained must be carried in a *regA*⁻ background. Interestingly, *regA/gls* double mutants resemble more primitive colonial volvocaleans, such as those in the genus *Eudorina*.

Based on the nature of the mutant phenotypes described here, a model for how these genes control *Volvox* development has been proposed by Kirk and colleagues (Figure 2). In the model the *gls* gene is directly responsible for the asymmetric cleavages; the *mul* genes specify the exact times and places for these divisions to occur. In the larger cells the *lag* genes become active, and their products keep those cells from undergoing somatic differentiation. In the smaller cells the *regA* gene product prevents any development as a reproductive cell, and in the absence of *lag* gene products the pathway of somatic development is followed. The complementary nature of the *lag* and *regA* genes is particularly apparent here: the former prevent expression of somatic cell genes, and the latter prevent expression of germ cell genes. The cloning of some of these genes (see later) is allowing the model to be examined and tested at an entirely new level of detail.

An especially noteworthy feature of the *regA* locus is its hypermutability [21]. It was found that *regA* mutants appeared at an exceptionally high frequency after treatment with agents that interfere with DNA

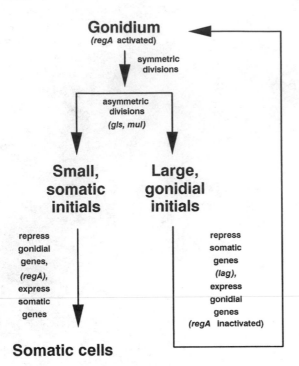

**Figure 2**  Diagrammatic representations of the asexual life cycle of *Volvox carteri*. Vegetative and reproductive functions are divided between somatic cells and gonidia, respectively. After a number of symmetric divisions have occurred, the *gls* gene acts (at times and places specified by the various *mul* genes) to cause a set of asymmetric cleavage divisions that generate large-small sister cell pairs. In the small somatic initials, the *regA* gene is expressed and acts to suppress expression of genes required for gonidial differentiation. These cells thus become terminally differentiated as somatic cells. In the large gonidial initials, meanwhile, the *lag* genes are expressed and the *regA* gene is inactivated. This leads to repression of the genes required for differentiation of somatic cell features and expression of the genes required for gonidial differentiation. The *regA* gene becomes reactivated in mature gonidia some time prior to the first cleavage. (Modified from Ref. 14.)

recombination or repair functions. The timing of treatment was also critical and the hypermutability appeared only at two times: a few hours before cleavage began in the gonidial cells and then shortly after the asymmetric cleavage. The hypermutability of the locus is suggested to be a consequence of DNA rearrangement: to ensure that the *regA* gene cannot be expressed in gonidia, it is proposed to be inactivated by an actual physical rearrangement of the gene itself. The gene is inactivated in pregonidial cells so that they can express the genes for reproductive function, then reactivated in mature gonidia prior to the first cleavage as one-celled embryos [21].

## C.  Sexual Life Cycle and Mutants

A good description of the sexual cycle in *Volvox* remained elusive until Darden [22] finally determined the conditions for its sexual propagation in culture. The cycle is summarized in Figure 3. During asexual development, males and females are morphologically indistinguishable from one another. After exposure to the exceedingly potent sexual inducer (active at concentrations lower than $10^{-16}$ M), however, both sexes undergo one more round of asexual cell division and then initiate gamete formation. For female development, the gonidia divide symmetrically up to the 64-cell stage, and then up to 48 of the cells divide asymmetrically and the larger daughter cells form eggs. If fertilization does not occur, the eggs have the capability of developing into gonidia and continuing development in an asexual manner. After induction, a male gonidium divides symmetrically up to the 256-cell stage, and then an unequal cleavage of all the cells produces 256 sperm initials along with somatic cells; the former undergo six or seven further divi-

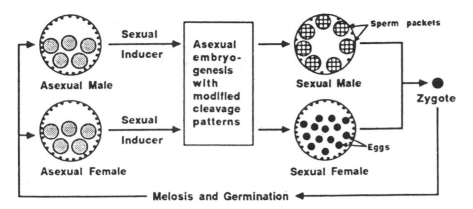

**Figure 3** The sexual cycle of *Volvox carteri*. Asexual males and females (which are morphologically indistinguishable from one another) respond to the sexual inducer by undergoing another round of asexual embryogenesis in which the patterns of asymmetric division are modified and in which the germ cells that are formed develop not as gonidia but as sperm packets or eggs. Sperm-egg fusion results in formation of a dormant, resistant zygote. When dormancy is broken by washing with fresh medium, each zygote undergoes meiosis to form one viable germling and three polar bodies. (From Ref. 13.)

sions to produce packets containing 64 or 128 mature sperm. Soon after they form, the sperm packets are released into the surrounding medium and attach to the somatic cells of female spheroids. They subsequently make a hole in the spheroid wall and the individual sperm are released into the interior, where they fertilize the eggs. The resulting zygote develops into a cold- and drought-resistant dormant zygospore that, in culture, becomes activated only when fresh medium is added. In the meiotic division that follows activation only one new germling is produced; the other meiotic products are polar bodies.

As with the asexual cycle, many mutants in sexual development have been isolated and characterized [16,23]. Somewhat surprisingly, many of these mutants mix and match the asexual, male, and female patterns of cell division with the production of gonidia, sperm packets, and eggs: gonidia can result from cleavages in either the male or female pattern, and sperm or eggs can result from the pattern normally seen in the opposite sex. Thus, it is clear that there is no tight linkage between the specific cleavage pattern and the specific type of cellular differentiation that ensues. Explanation and understanding of the controlling factors in these processes will have to await the cloning and characterization of the relevant genes. Given the newly available tools of molecular developmental analysis of *Volvox* (see later), that understanding will probably come soon.

## D. Molecular Approaches

In 1994, *Volvox* finally yielded itself up to the techniques of molecular transformation [24], and the preceding models are now being directly tested at the molecular level (see later). Early difficulties in transformation in *Volvox* were probably due to the poor performance of promoters from higher eukaryotes, to an extensive DNA methylation system, and to a very biased codon usage: these render standard selectable markers and reporter genes unusable. Only when an endogenous gene that can be used as a selectable marker was cloned [25] and employed did transformation succeed [24]. The gene is that for nitrate reductase (*nitA*), which has the very useful property that one can select for both the presence and absence of the functional gene: $nitA^-$ strains can be selected on the basis of their resistance to chlorate, and only $nitA^+$ strains will grow on nitrate as the sole nitrogen source. Subsequently, the *HUP-1* gene from *Chlorella*, which encodes a glucose/$H^+$ symporter, was also found to be useful as a selectable marker in *Volvox* [26], and a construct containing a bacterial *ble* gene under the control of a *Volvox* tubulin promoter has worked to confer stable resistance to the antibiotic zeocin [27]. Transforming DNA is easily introduced via particle bombardment and appears to integrate randomly into the host genome, although homologous recombination with the endogenous gene has been reported [28]. Nonselectable genes can be introduced on separate plasmids and cotransformed with the selectable DNA. Finally, developmental studies can now be aided by the use of the endogenous aryl sulfatase (*Ars*) gene, which can serve as a reporter gene and has an inducible promoter [29].

Another molecular tool added to the box for *Volvox* research is the characterization of a transposable element, waggishly named *Jordan* for its first-class jumping ability [30]. This allows the techniques of transposon tagging, which have been of great utility in *Arabidopsis* and other plants [31,32], to be employed in *Volvox*. Transposition of *Jordan* can be induced by temperature shock [1,30], and activity of a tagged gene is frequently restored when it is excised [30]. Its disadvantage is that it is present in roughly 50 copies/genome, so identification of a gene that has been tagged by *Jordan* is somewhat problematic. However, this problem can be overcome, and both the *regA* and a *gls* gene have been cloned via transposon tagging with *Jordan* [33,34].

The cloned *gls* gene is now referred to as *glsA*; different letters will be assigned to other genes whose disruption causes the gonidialess phenotype as they are cloned [34]. Its identity as a *gls* gene was convincingly demonstrated by its ability to rescue well-established mutants when introduced via the transformation system outlined before. Consistent with the proposed role of *gls* genes in controlling the asymmetric division that gives rise to the differentiation of germ and somatic cells (Figure 2), *glsA* messenger RNA (mRNA) is expressed during cleavage, with the highest level seen at the asymmetric cleavage itself. It is not expressed at all in mature somatic cells. The glsA protein has been localized to the mitotic spindle, and a model has been proposed for its involvement in the asymmetric cleavages that lead to the germ-soma dichotomy.

Much to the gratification of all, the sequence and expression pattern of the *regA* gene also turn out to be consistent with its proposed role (Figure 2) as an inhibitor of gonidia-specific genes [33]. *RegA* mRNA can first be detected in very young somatic cells, where it increases, and is not detectable in gonidial cells or precursors. Although its amino acid sequence does not show direct homology to any known protein, it contains several features (helix-loop-helix domain, nuclear localization signal, a region very rich in Glu, Ala, and Pro) that are compatible with its being a transcriptional repressor. Whether or not the *regA* locus' hypermutability is a consequence of programmed DNA rearrangement [21] is a target of current investigation.

A surprising finding related to the *regA* gene, however, has been the genes that are the presumed targets of its repression: the cloning and identification of complementary DNAs (cDNAs) that are specifically expressed in gonidial cells [35] have revealed that a substantial majority are nuclear-encoded genes that are utilized in the chloroplast for photosynthesis [33,36]. This has led to the hypothesis that somatic cells are unable to enter into a reproductive program largely due to an inability to photosynthesize; they senesce at least partly because they run out of metabolic reserves [33,37]. An appealing feature of this hypothesis is that it can fairly simply explain the evolutionary origin of the germ-soma dichotomy: because many photosynthetic genes in higher plants are coordinately regulated via common *cis*-acting elements in their promoters [38], it is not difficult to imagine that the *regA* gene could have evolved from a preexisting gene that encodes a transcription factor that carries out this regulation. Further examination of *regA* and its homologues in different volvocalean species will probably shed light on this hypothesis. Its relevance to basic mechanisms of cell differentiation in higher organisms should also prove very interesting.

## III.  MOSSES

Bryophytes are clearly much more closely related to higher plants than are the Volvocales, both biochemically and functionally. Mosses respond to many of the same growth regulators as higher plants (reviewed in Ref. 39)—in fact, one of the first direct demonstrations of the effect of a cytokinin on plant development was its stimulation of bud development in a moss [40]. Like higher plants, bryophytes have a multicellular, differentiated sporophyte, and there is no separation early in development of discrete germ line and somatic tissue. Thus, the study of developmental and physiological processes in mosses is quite relevant to higher plants.

As with *Volvox*, genetic analysis of bryophytes, especially the generation of mutants, is greatly aided by the fact that the dominant stage of the life cycle is haploid. Unlike the situation for *Volvox*, however, most differentiated cells of mosses, like those of many higher plants, are totipotent, remaining capable of redifferentiating and thence giving rise to an entire new plant. Thus, mutations that produce a sterile phenotype can still be maintained and studied. They can even be further mutagenized, because techniques of somatic mutagenesis and regeneration from protoplasts are well developed [41]. Complementation assays and dominance relationships can also be studied in sterile mutants by the production, via protoplast fusion, of somatic hybrids that behave as diploids [42].

Techniques for the analysis and manipulation of moss using the tools of molecular biology have advanced very rapidly in the past 10 years (see later), and with these advances has come increasing recognition of its value as a model organism for basic features of plant development [43]. In no small part this is due to the discovery that in moss a transgene will, with high frequency, homologously recombine with its genomic counterpart [7]. To date, the moss *Physcomitrella patens* is the only land plant for which the resulting analysis of gene function via "knockout" and other, more subtle manipulations is routinely possible.

## A. Basic Life Cycle and Development Program

Most of the genetic analysis that has been performed on bryophytes has been in the moss *Physcomitrella patens*; therefore much of what follows will be based on studies of that organism (see reviews in Refs. 3 and 4). *Physcomitrella* is a relatively simple moss, but its basic developmental pathways appear to be quite similar to those of all mosses [44,45]. Other moss species whose development and physiology have been extensively studied, but for which genetic analyses are less well developed, are *Ceratodon purpureus* and *Funaria hygrometrica* [46].

Although mature mosses certainly contain many cell types, in the early stages of the life cycle moss development is rather simple, involving only a few types of differentiated cells. Following germination of a haploid spore, the initial growth pattern consists of a two-dimensional filamentous network of cells called a protonema (Figure 4). In most mosses, including *Physcomitrella*, the initial protonemal cell type is the chloronema. These are chloroplast-rich cells about 115 μm in length that divide about every 20 hr. As with many other filamentous systems, growth of a protonemal filament occurs solely at the apical cell. After three or four cell divisions, an apical chloronemal cell begins to differentiate into a second protonemal cell type, the caulonema. Caulonemal cells contain fewer chloroplasts, are about 160 μm in length, and divide every 8 hr. Thus, after several days of growth, caulonemal filaments and their derivatives dominate the culture. Essentially all subapical caulonemal cells will divide to produce side branches, and it is these side branches that can give rise to the next stage of development. Most (~90%) become new chloronemal filaments, a few give rise to new caulonemata, but about 3% begin more complex two- and three-dimensional growth and become buds, the bryophyte equivalent of apical meristems. A bud then develops into a gametophore, which consists primarily of the small leafy shoot that is the most conspicuous part of a moss gametophyte. A young *Physcomitrella* gametophore comprises only a few cell types: the leaf, which is only one cell thick and contains no vascular tissue; its supporting stem; and the rhizoid, a filament resembling a caulonemal cell that extends from the base. Other mosses may contain more complex structures and even primitive conducting cells. Figure 5 shows a cell lineage chart of the basic early developmental pattern for *P. patens*.

As shown in Figure 5, light, calcium, auxins, and cytokinins are all involved in the control of these early developmental pathways. Most dramatically, exogenously added cytokinin can cause essentially all side branch initials to become buds. In addition, abscisic acid is believed to mediate stress responses [47,48], cyclic adenosine monophosphate (cAMP) and calcium have been implicated in playing important roles at the subcellular level, and there have been detailed descriptions of the movements of organelles prior to and during each cell division (reviewed in Ref. 46).

When gametophores have sufficiently developed and been exposed to the right environmental conditions (commonly, cool temperatures), they are induced to make antheridia and archegonia, the organs that produce sperm and eggs, respectively. After the presence of water allows the motile sperm to effect fertilization, the diploid sporophyte grows out of the archegonium that housed the egg. The sporophyte remains largely dependent on the gametophyte for most of its nutritional requirements, although it appears that there is at least some metabolic separation between the two stages (cited in Ref. 49). It differentiates a sporangium, or capsule, in which spore mother cells undergo meiosis and generate spores. Very little is known about the factors influencing the development of moss sporophytes. In *Physcomitrella* the sporophyte is all but invisible; only the mature spore capsule can easily be seen, sitting among the gametophores. A review of sexual reproduction in *Physcomitrella* has been published [49].

**Figure 4** Caulonemal filaments from a 3-week-old culture of *Physcomitrella patens*, which has grown under standard conditions. The main filament axes are composed of caulonemal cells. Almost every subapical
caulonemal cell has divided to produce one or more side branches. Most of these have developed into filaments
of secondary chloronema. A few have produced buds, which may be seen at various stages of development. Bar
1 mm. (From Ref. 3.)

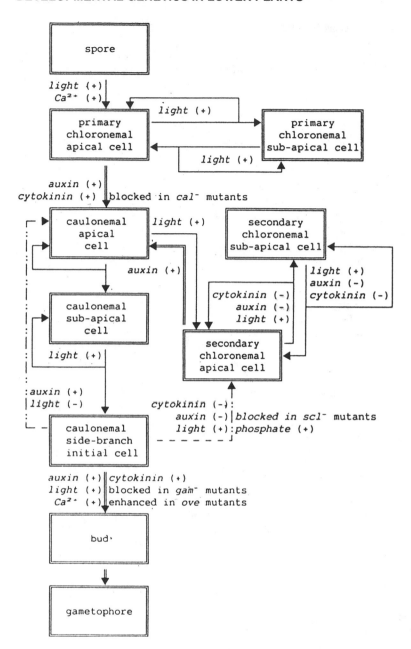

**Figure 5** Cell lineages in the development of the gametophyte of *Physcomitrella patens*. Transitions between stages that are connected by a broken arrow do not require cell division. The branched arrows with single lines represent developmental steps involving a cell division; the arrows indicate the two products of the division. Stages connected by a double-lined arrow require more than one cell division. The $(+)$ sign beside a signal indicates that it is required for or enhances the frequency of the transition, the $(-)$ sign, that it decreases the frequency of the transition. (From Ref. 3.)

## B.  Hormone Responses and Mutants

As was the case with *Volvox*, many mutants have been isolated that affect the moss developmental pathway. Because the moss can, unlike *Volvox*, be propagated from single, nonreproductive cells, mutants that block the normal reproductive pathway can be readily maintained. *Cal*⁻ mutants, for example, are unable to undergo the first differentiation from chloronema to caulonema. For some isolates with this phenotype, the defect can be overcome by the addition of auxins or cytokinins to the medium, showing that these substances are involved in the differentiation. In these cases the lesions are most likely in genes involved in the biosynthesis of these compounds. Other *cal*⁻ isolates, however, are not responsive to the addition of hormones and thus are likely to be more directly related to the developmental response itself. Similarly, *bud*⁻ mutants are unable to form buds, and *gad*⁻ mutants form buds that do not differentiate into gametophores. Some, but not all, *bud*⁻ mutants can be rescued by the exogenous addition of cytokinins. The existence of separate cytokinin-responsive *cal*⁻ and *bud*⁻ mutants suggests that the cytokinin response threshold is different for the two types of differentiations: in *bud*⁻ mutants of this type, some cytokinin must be present for caulonemal cells to form, but it apparently is not enough to trigger the development of buds. A similar two-tiered response to cytokinin has been found for induction of caulonemal branching and of bud formation in the moss *Funaria hygrometrica* [50].

Reutter et al. [51] transformed various *Physcomitrella* mutants with a construct containing the *ipt* gene from *Agrobacterium*, resulting in the endogenous production of excess cytokinin. The bacterial gene was able to rescue a *bud*⁻ mutant and another cytokinin-related mutant with defective plastid division, but not a *gad*⁻ mutant, suggesting that cytokinins are needed for bud formation but that other triggers are involved in further gametophore development. The results were rather different from those when cytokinins were added to the medium, demonstrating a difference between exogenously added and endogenously produced plant hormones.

Finally, a third class of hormone-response mutants are given the name *ove*. These mutants overproduce buds as a direct consequence of supernormal levels of cytokinin. Somewhat surprisingly, mutations in at least three separate loci give this phenotype, indicating that cytokinin concentrations must be very carefully controlled by the moss [52].

## C.  Light and Gravity Responses and Mutants

Tropisms in a moss protonema occur in the apical cells only; i.e., they involve a change in the direction of growth of a single cell (Figure 6). This makes the moss particularly attractive for the study of the intracellular signaling events involved in tropisms because the effects of treatments or manipulations that affect an individual cell can be directly monitored via time-lapse video microscopy. Similarly, the establishment of cell polarity in response to light can be readily measured in regenerating moss protoplasts [53].

As to phototropism, chloronemata, caulonemata, and gametophores all show phototropic and polarotropic responses, and the different cell types vary in their reactions to different wavelengths and light intensities [54,55]. In *Physcomitrella*, mutants in at least three loci have been described that have lost phototropism in gametophores (*ptr* mutants). Caulonemal cells are similarly affected in these mutants, but the chloronemal response is more complex (reviewed in Ref. 56). These mutations do not alter the photomorphic effects on the normal developmental pathways (Figure 5), so they are apparently not involved in light perception per se. Phytochrome is clearly involved in most of these responses: in *Ceratodon*, *ptr* mutants with both normal and altered phytochromes have been characterized [57,58]. The variation in the tropisms of the different cell types in both wild-type and mutant strains makes it clear that no simple controls of phototropism operate in mosses, and more work is needed to determine the mechanisms, both genetic and biochemical, involved.

Caulonemal cells and gametophores also show a negative gravitropic response, but this can be observed only in darkness or infrared light because the phototropic response will override it [59,60]. As is true of the phototropic response, it is only the apical cell of a caulonemal filament that responds to gravity (Figure 6), and there is increasingly good evidence that migrating amyloplasts interacting with cytoskeletal components are the statoliths [61,62]. Several mutants with altered or reduced caulonemal gravitropism (*gtr* mutants) have been isolated [59]. Some of these show no gravitropism whatsoever, and,

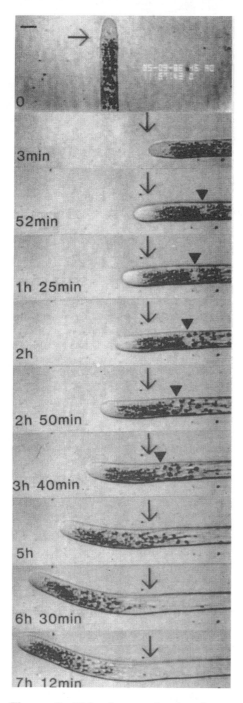

**Figure 6** Video images showing the response to 90° reorientation of an apical protonemal cell of *Physcomitrella patens* grown in infrared light. Times after reorientation are shown. The large arrow provides a reference point for the measurement of growth rate and was repositioned immediately after reorientation. The arrowhead denotes the position of the nucleus. (From Ref. 60.)

surprisingly, some show a positive gravitropism, the opposite of normal. Unlike the phototropic mutants, however, gametophores in the *gtr* mutants isolated thus far display a normal gravitropic response. Interactions between the pathways for light- and gravity-mediated responses are beginning to be sorted out [63].

## D.  Molecular Approaches

As with *Volvox*, many of the tools and techniques of molecular genetics have become readily available for moss research only in the last 10 years or so (reviewed in Ref. 64). However, they have done so with such success that *Physcomitrella* is now sometimes referred to as a "green yeast," some of its properties paralleling those of *Saccharomyces cerevisiae* [4]. It is very easily and efficiently transformed, sometimes generating hundreds of transformed plants from a microgram of DNA [5,6] (J. Wallace, unpublished results). Its genome size is estimated as a tractable 480 Mbp, about three times that of *Arabidopsis* [65]. Higher plant promoters function in moss (e.g., Ref. 47), and codon usage is very similar in both groups [48,66]. The tetracycline induction-repression system also functions well in moss, so introduced transgenes can be turned on and off at will [67]. The moss *Ceratodon purpureus* has also been successfully transformed [68], so *Physcomitrella* may soon have some competition on the molecular front.

Perhaps of most importance to the utility of *Physcomitrella* is that it is the only known land plant in which a transgene will homologously recombine with its genomic counterpart with high efficiency [7]. This is allowing the approaches of targeted gene knockout and allele replacement, often referred to as "reverse genetics," to be exploited in a plant as they have been so profitably exploited in fungal and animal systems. Among the genes whose targeted disruption have been reported so far are a Cab gene [69]; a gene encoding a delta 6-acyl-group desaturase [70]; FtsZ, an ancestral tubulin gene involved in plastid division [71]; and the multiubiquitin chain binding subunit of the 26S proteasome [72].

The disruption of the proteasome gene [72] is especially interesting, as it affects the early moss developmental pathway described earlier. The proteasome recognizes and degrades proteins that are ubiquinated, either because they are aberrant or because their degradation is programmed (reviewed in Ref. 73). Moss plants in which this gene has been knocked out behave as Bud$^-$ mutants; that is, they are unable to make the switch from protonemal (filamentous) to three-dimensional growth. Thus, it appears as though specific protein turnover is part of this basic differentiation process. Surprisingly, however, the developmental block can be overcome by treatment with auxin and cytokinin. This suggests that the switch in cellular development is mediated by at least these two processes—targeted proteolysis and changes due to hormone signal transduction—and that increasing the signal from one can overcome a deficit in the other. It seems likely that basic cell differentiations are usually effected by several things: targeted proteolysis, hormones, mRNA degradation, phosphorylations, calcium, cAMP, etc. These are probably all involved and interrelated in changing cell fate. If the example from *Physcomitrella* can be generalized, a deficiency in one of these mechanisms (e.g., specific protein degradation) can be overcome if one of the others (e.g., hormone response) is increased.

A final feature of *Physcomitrella* that makes it somewhat like a green yeast is the phenomenon of unstable transformants. When moss is transformed and placed on selective medium, three types of resistant colonies emerge: transient expressers of the transgene, which die after a few weeks; stable transformants; and unstables, which continue to grow (albeit slowly) and express the transgene as long as they are on selective medium [74,75]. If an unstable transformant is grown without selection pressure for a couple of weeks, it loses its ability to express the transgene and dies if replaced on selection. It is believed that in unstable transformants the transforming plasmid is maintained and replicates extrachromosomally, and in many cases it can be recovered from the moss by transforming the moss DNA into *E. coli*, where the plasmid will still confer prokaryotic antibiotic resistance (C. Knight and J. Wallace, unpublished results). Thus, the possibility exists of generating a "shuttle vector" that can be transferred from moss to bacteria and back.

In this era of comparative genomics and proteomics, the discovery that so many genes and proteins in higher organisms have homologues in simpler ones has proved to be of great benefit because properties and functions of the genes are generally much more easily explored in the latter. Thus, it is no surprise that studies to generate expressed sequence tag (EST) libraries in *Physcomitrella* have already begun [48,66]. One can expect that sequencing of the entire moss genome is not too far off.

## IV.  FERNS

### A.  General Considerations

Fern development, especially of the gametophyte, has been a favorite object of study for many years [76]. When compared with those of *Volvox* and *Physcomitrella*, however, genetic analysis of fern development is relatively new. Studies with the fern *Ceratopteris richardii*, however, are showing promise for the understanding of some fundamental features of its developmental genetics [10,11,77,78]. Like mosses, ferns have a free-living, independent gametophyte stage that greatly aids mutant isolation and analysis. Diploid gametophytes can be generated via apospory for assessment of dominance relationships [79]. Like *Physcomitrella*, *Ceratopteris* is self-fertile, so homozygous sporophytes can easily be obtained, and its life cycle can be completed in about 3 months. The highly developed vascular system and the independence of the sporophyte, however, clearly make the study of ferns more relevant to higher plants than that of mosses.

Fern gametophyte development has been reviewed [77]. Like that of mosses, it begins with spore germination producing a protonema. Protonemal development in ferns, however, is one dimensional and very limited: after only a few cell divisions have resulted in a linear filament, the apical cell commences two-dimensional divisions and becomes the meristem of the flat, heart-shaped prothallus. The sexual archegonia and antheridia eventually develop on the prothallus (see later). After fertilization is effected, the sporophyte rapidly outgrows the gametophyte and assumes an independent existence.

As already mentioned, the genetic study of ferns is still young, so relatively few mutants have been isolated and characterized. A number of mutants resistant to metabolic inhibitors and herbicides, very useful for traditional genetic analysis, have been isolated by Hickok and colleagues (reviewed in Ref. 10). In addition, salt-tolerant mutants [78], mutants in photomorphogenesis of germinating spores [80], and mutants with altered sex determination [81,82] have all been described. As the last subject is especially relevant to the scope of this chapter, it will be described further.

### B.  Sex Determination

Unlike most higher plants, *Ceratopteris* is homosporous; i.e., it produces only one type of spore. Based on environmental and other signals, the resulting gametophyte then develops into a male that produces antheridia or a hermaphrodite that produces both antheridia and archegonia. By contrast, in higher plants the decision as to what type of gametophyte will be produced is determined in the sporophyte, i.e., whether a stamen that produces microspores or a pistil that produces megaspores develops. There have been some advances in the genetics of sex determination in the fern sporophyte (reviewed in Refs. 77, 81, and 82), which are summarized in the following.

Male and hermaphrodite fern gametophytes differ by more than the types of gametes they produce. The hermaphrodite consists of a heart-shaped leaflike prothallus about 2 mm in diameter that contains a notched meristem. Archegonia develop below the notch and antheridia above it. The male gametophyte lacks a meristem and is consequently only about one quarter the size of the hermaphrodite, which aids in the screening for developmentally altered mutants. Essentially all cells of the male develop into antheridia.

If no outside signals are received, the default pathway for gametophyte development is the hermaphrodite. Once the meristem develops, a pheromone (antheridiogen *Ceratopteris* or $A_{CE}$) is produced, and this induces younger gametophytes to follow the male developmental pathway. Thus, in dense populations of *Ceratopteris* there is a high concentration of $A_{CE}$, and essentially all gametophytes develop as males, encouraging outcrossing and discouraging overcrowding. Abscisic acid acts as an antagonist to $A_{CE}$ [83], which is not too surprising because $A_{CE}$ is very likely a gibberellin [84]. In hermaphrodites it seems that there is nearly simultaneous development of (1) commitment to the hermaphrodite developmental pathway, (2) development of the meristem, (3) production and secretion of $A_{CE}$, and (4) loss of sensitivity to the pheromone. One suspects that these traits are all controlled by the same "master gene," perhaps the *tra* gene described in the following.

Several mutants have been described that affect the normal pathway [85–87]. *Her* mutants, in at least five loci, develop as hermaphrodites regardless of the presence of $A_{CE}$ and are thus likely to be impaired in the perception and/or signal transduction of the pheromonal signal. *Tra* mutants (at least two loci) always develop as males, and *man* mutants produce supernormal numbers of antheridia in hermaphrodites. Interestingly, some *tra* and *man* mutants also display an altered sporophyte phenotype. Finally, *fem* mutants always develop as females, which resemble hermaphrodites but lack antheridia. Based on the phe-

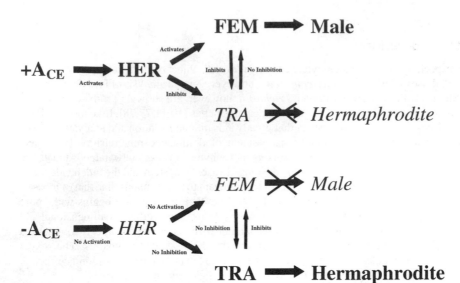

**Figure 7** Diagram of the genetic interactions proposed to control sex determination in *Ceratopteris richardii*. In the presence of the antheridiogen $A_{CE}$, the signal transduction cascade composed of the *her* genes results in activation of the *fem* genes and repression of the *tra* genes. The gametophyte develops as a male. In the absence of $A_{CE}$, the *tra* genes are active: they inhibit the *fem* genes, and a hermaphrodite develops. Active genes or processes are represented in bold, inactive ones in italics. (Modified from Ref. 82.)

notypes of double and triple mutants, a model for the control of sex determination in *Ceratopteris* has been proposed [77,81,82,88], shown in Figure 7.

Unfortunately, a lack of modern molecular tools (e.g., transformation) available to *Ceratopteris* researchers has thus far prevented the cloning of the preceding genes. However, progress is being made at the molecular level, as witnessed by the recent cloning of a gene whose expression is induced by $A_{CE}$ [88]. The gene, called *ani1*, is also constitutively expressed in a *tra* mutant, and, interestingly, is induced even in the absence of $A_{CE}$ by the protein synthesis inhibitor cycloheximide. Also, homologues of the MADS-box genes that control higher plant flowering [89,90] and of homeodomain-leucine zipper genes [91] are being examined in *Ceratopteris*, and they are certainly candidates for involvement in the sex determination process.

Perhaps the designation of *Ceratopteris* as "a model plant for the 90s" [11], was a bit premature, but surely it will catch up to the others soon, and the plant research community at large will become more aware of its unique advantages.

## V. OUTLOOK

The great excitement of discovery in plant molecular genetics that occurred in the 1980s due to the advent of the new tools of molecular cloning, sequencing, and transformation originally left lower plants a bit by the wayside. Most likely this was largely due to the relative paucity of groups working with lower plants when compared with angiosperms, so that the techniques were slower to develop. However, now that most of the molecular techniques are developed for use in these model systems, their unique advantages are once again becoming apparent: the ease of genetic analysis because of haploidy and rapid generation time, coupled with the relative simplicity of their development, make lower plants excellent model systems both for some practical aspects of crop plant development and for the study of very fundamental biological processes. It is well worth noting, in this era of bioinformatics and comparative genomics, that "lower" animals such as *C. elegans*, *Drosophila*, and zebrafish are being found to be of ever-increasing relevance to research in mammals in general and to medical research specifically. Almost certainly the same will be found true of lower plants.

# REFERENCES

1. DL Kirk. Volvox: Molecular-Genetic Origins of Multicellularity and Cellular Differentiation. Cambridge: Cambridge University Press, 1999, p 23.
2. JH Powers. Further studies in *Volvox*, with descriptions of three new species. Trans Am Microsc Soc 28:141–175, 1908.
3. DJ Cove. Regulation of development in the moss, *Physcomitrella patens*. In: VEA Russo, S Brody, D Cove, S Ottolonghi, eds. Development, the Molecular Genetic Approach. Berlin: Springer-Verlag, 1992, pp 179–193.
4. R Reski. Development, genetics and molecular biology of mosses. Bot Acta 111:1–15, 1998.
5. D Schaefer, JP Zryd, CD Knight, DJ Cove. Stable transformation of the moss *Physcomitrella patens*. Mol Gen Genet 226:418–424, 1991.
6. W Sawahel, S Onde, C Knight, D Cove. Transfer of foreign DNA into *Physcomitrella patens* protonemal tissue by using the gene gun. Plant Mol Biol Rep 10:314–315, 1992.
7. DG Schaefer, JP Zryd. Efficient gene targeting in the moss *Physcomitrella patens*. Plant J 11:1195–1206, 1997.
8. H Puchta. Towards targeted transformation in plants. Trends Plant Sci 3:77–78, 1998.
9. R Reski. *Physcomitrella* and *Arabidopsis*: the David and Goliath of reverse genetics. Trends Plant Sci 3:209–210, 1998.
10. LG Hickok, TR Warne, RS Fribourg. The biology of the fern *Ceratopteris* and its use as a model system. Int J Plant Sci 156:332–345, 1995.
11. R Chasan. *Ceratopteris*: a model plant for the 90s. Plant Cell 4:113–115, 1992.
12. DL Kirk, JF Harper. Genetic, biochemical, and molecular approaches to *Volvox* development and evolution. Int Rev Cytol 99:217–293, 1986.
13. DL Kirk. The ontogeny and phylogeny of cellular differentiation in *Volvox*. Trends Genet 4:32–36, 1988.
14. R Schmitt, S Fabry, DL Kirk. In search of the molecular origins of cellular differentiation in *Volvox* and its relatives. Int Rev Cytol 139:189–265, 1992.
15. ML Pall. Mutants of volvox showing premature cessation of division: evidence for a relationship between cell size and reproductive cell differentiation. In: D McMahon, CF Fox, eds. Developmental Biology: Pattern Formation, Genetic Regulation. Vol 2. Menlo Park, CA: Benjamin, 1975, pp 148–156.
16. RJ Huskey, BE Griffin, PO Cecil, AM Callahan. A preliminary genetic investigation of *Volvox carteri*. Genetics 91:229–244, 1979.
17. DL Kirk, MR Kaufman, RM Keeling, DA Stamer. Genetic and cytological control of the asymmetric divisions that pattern the *Volvox* embryo. Development Suppl 1:67–82, 1991.
18. G Kochert. Developmental mechanisms in volvox reproduction. In: CI Markert, J Papaconstantinou, eds. The Developmental Biology of Reproduction. New York: Academic Press, 1975, pp 55–90.
19. SF Gilbert. Developmental Biology. Sunderland, MA: Sinauer Associates, 1991, 276.
20. MM Kirk, A Ransick, SE McRae, DL Kirk. The relationship between cell size and cell fate in *Volvox carteri*. J. Cell Biol 123:191–208, 1993.
21. DL Kirk, GJ Baran, JF Harper, RJ Huskey, KS Huson, N Zagris. Stage-specific hypermutability of the regA locus of *Volvox*, a gene regulating the germ-soma dichotomy. Cell 48:11–24, 1987.
22. WH Darden. Sexual differentiation in *Volvox aureus*. J. Protozool 13:239–255, 1966.
23. AM Callahan, RJ Huskey. Genetic control of sexual development in *Volvox*. Dev Biol 80:419–435, 1980.
24. B Schiedlmeier, R Schmitt, W Muller, MM Kirk, H Gruber, W Mages, DL Kirk. Nuclear transformation of *Volvox carteri*. Proc Natl Acad Sci U S A 91:5080–5084, 1994.
25. H Gruber, SD Goetinck, DL Kirk, R Schmitt. The nitrate reductase–encoding gene of *Volvox carteri*: map location, sequence and induction kinetics. Gene 120:75–83, 1992.
26. A Hallmann, M Sumper. The *Chlorella* hexose/$H^+$ symporter is a useful selectable marker and biochemical reagent when expressed in *Volvox*. Proc Natl Acad Sci U S A 93:669–673, 1996.
27. A Hallmann, A Rappel. Genetic engineering of the multicellular green alga *Volvox*: a modified and multiplied antibiotic resistance gene as a dominant selectable marker. Plant J 17:99–109, 1999.
28. A Hallmann, A Rappel, M Sumper. Gene replacement by homologous recombination in the multicellular green alga *Volvox carteri*. Proc Natl Acad Sci U S A 94:7469–7474, 1997.
29. A Hallmann, M Sumper. Reporter genes and highly regulated promoters as tools for transformation experiments in *Volvox carteri*. Proc Natl Acad Sci U S A 91:11562–11566, 1994.
30. SM Miller, R Schmitt, DL Kirk. Jordan, an active *Volvox* transposable element similar to higher plant transposons. Plant Cell 5:1125–1138, 1993.
31. KA Feldmann. T-DNA insertion mutagenesis in *Arabidopsis*: mutational spectrum. Plant J 1:71–82, 1991.
32. A Lonneborg, C Jansson. Isolation of plant genes by T-DNA and transposon mutagenesis—gene tagging. Prog Bot 54:295–305, 1993.
33. MM Kirk, W Muller, B Taillon, SM Miller, H Gruber, K Stark, R Schmitt, DL Kirk. regA, a *Volvox* gene that plays a central role in germ-soma differentiation, encodes a novel regulatory protein. Development 126:639–647, 1999.
34. SM Miller DL Kirk. glsA, a *Volvox* gene required for asymmetric division and germ cell specification, encodes a chaperone-like protein. Development 126:649–658, 1999.

35. LW Tam, DL Kirk. Identification of cell-type-specific genes of *Volvox carteri* and characterization of their expression during the asexual life cycle. Dev Biol 145:51–66, 1991.

36. G Choi, M Przybylska, D Straus. Three abundant germ line–specific transcripts in *Volvox carteri* encode photosynthetic proteins. Curr Genet 30:347–355, 1996.

37. DL Kirk. *Volvox*: Molecular-Genetic Origins of Multicellularity and Cellular Differentiation. Cambridge: Cambridge University Press, 1999, p 321.

38. P Puente, N Wei, XW Deng. Combinatorial interplay of promoter elements constitutes the minimal determinants for light and developmental control of gene expression in *Arabidopsis*. EMBO J 15:3732–3743, 1996.

39. DJ Cove, NW Ashton. In: AF Dyer, JG Duckett, eds. The Experimental Biology of Bryophytes. London: Academic Press, 1984, pp 177–202.

40. BS Gorton, RE Eakin. Development of the gametophyte in the moss *Tortella caespitosa*. Bot Gaz 119:31–38, 1957.

41. PJ Boyd, NH Grimsley, DJ Cove. Somatic mutagenesis of the moss, *Physcomitrella patens*. Mol Gen Genet 211:545–546, 1988.

42. NH Grimsley, NW Ashton, DJ Cove. The production of somatic hybrids by protoplast fusion in the moss, *Physcomitrella patens*. Mol Gen Genet 154:97–100, 1977.

43. DJ Cove, CD Knight, T Lamparter. Mosses as model systems. Trends Plant Sci 2:99–105, 1997.

44. M Lal. The culture of bryophytes including apogamy, apospory, parthenogenesis and protoplasts. In: AF Dyer, JG Duckett, eds. The Experimental Biology of Bryophytes. London: Academic Press, 1984, pp 97–116.

45. B Knoop. Development of bryophytes. In: AF Dyer, JG Duckett, eds. The Experimental Biology of Bryophytes. London: Academic Press, 1984, pp 143–176.

46. KS Schumaker, MA Dietrich. Programmed changes in form during moss development. Plant Cell 9:1099–1107, 1997.

47. CD Knight, A Sehgal, K Atwal, J Wallace, DJ Cove, D Coates, RS Quatrano, S Bahadur, P Stockely, AC Cuming. Molecular responses to abscisic acid and stress are conserved between mosses and cereals. Plant Cell 7:499–506, 1995.

48. J Machuka, S Bashiardes, E Ruben, K Spooner, A Cuming, C Knight, D Cove. Sequence analysis of expressed sequence tags from an ABA-treated cDNA library identifies stress response genes in the moss *Physcomitrella patens*. Plant Cell Physiol 40:378–387, 1999.

49. DJ Cove, CD Knight. The moss *Physcomitrella patens*, a model system with potential for the study of plant reproduction. Plant Cell 5:1483–1488, 1993.

50. M Bopp, HJ Jacob. Cytokinin effect on branching and bud formation in *Funaria*. Planta 169:462–464, 1986.

51. K Reutter, R Atzorn, B Hadeler, T Schmulling, R Reski. Expression of the bacterial ipt gene in *Physcomitrella* rescues mutations in budding and in plastid division. Planta 206:196–203, 1998.

52. DR Featherstone, DJ Cove, NW Ashton. Genetic analysis by somatic hybridization of cytokinin overproducing developmental mutants of the moss, *Physcomitrella patens*. Mol Gen Genet 222:217–224, 1990.

53. DJ Cove, RS Quatrano, E Hartmann. The alignment of the axis of asymmetry in regenerating protoplasts of the moss, *Ceratodon purpureus*, is determined independently of axis polarity. Development 122:371–379, 1996.

54. E Hartmann, GI Jenkins. Photomorphogenesis of mosses and liverworts. In: AF Dyer, JG Duckett, eds. The Experimental Biology of Bryophytes. London: Academic Press, 1984, pp 203–228.

55. DJ Cove, A Schild, NW Ashton, E Hartmann. Genetic and physiological studies of the effect of light on the development of the moss *Physcomitrella patens*. Photochem Photobiol 27:249–254, 1978.

56. DJ Cove, CD Knight. Gravitropism and phototropism in the moss *Physcomitrella patens*. In: H Thomas, D Grierson, eds. Developmental Mutants in Higher Plants. Cambridge: Cambridge University Press, 1987, pp 181–197.

57. T Lamparter, H Esch, D Cove, J Hughes, E Hartmann. Aphototropic mutants of the moss *Ceratodon purpureus* with spectrally normal and with spectrally dysfunctional phytochrome. Plant Cell Environ 19:560–568, 1996.

58. H Esch, E Hartmann, D Cove, M Wada, T Lamparter. Phytochrome-controlled phototropism of the moss *Ceratodon purpureus*: physiology of the wild type and class 2 ptr-mutants. Planta 209:290–298, 1999.

59. GI Jenkins, GRM Courtice, DJ Cove. Gravitropic responses of wild-type and mutant strains of the moss *Physcomitrella patens*. Plant Cell Environ 9:637–644, 1986.

60. CD Knight, DJ Cove. The polarity of gravitropism in the moss *Physcomitrella patens* is reversed during mitosis and after growth on a clinostat. Plant Cell Environ 14:995–1001, 1991.

61. FD Sack. Gravitropism in protonemata of the moss *Ceratodon*. Mem Torrey Bot Club 25:36–44, 1993.

62. OA Kuznetsov, J Schwuchow, FD Sack. Curvature induced by amyloplast magnetophoresis in protonemata of the moss *Ceratodon purpureus*. Plant Physiol 119:645–650, 1999.

63. T Lamparter, J Hughes, E Hartmann. Blue light– and genetically-reversed gravitropic response in protonemata of the moss *Ceratodon purpureus*. Planta 206:95–102, 1999.

64. R Reski. Molecular genetics of *Physcomitrella*. Planta 208:301–309, 1999.

65. R Reski, M Faust, XH Wang, M Wehe, WO Abel. Genome analysis of the moss *Physcomitrella patens*. Mol Gen Genet 244:352–359, 1994.

66. R Reski, S Reynolds, M Wehe, T Kleber-Janke, S Kruse. Moss (*Physcomitrella patens*) ESTs include several sequences which are novel for plants. Bot Acta 111:145–151, 1998.

67. M Zeidler, C Gatz, E Hartmann, J Hughes. Tetracycline-regulated reporter gene expression in the moss *Physcomitrella patens*. Plant Mol Biol 30:199–205, 1996.
68. M Zeidler, E Hartmann, J Hughes. Transgene expression in the moss *Ceratodon purpureus*. J Plant Physiol 154:641–650, 1999.
69. AH Hofmann, AC Codon, C Ivascu, VEA Russo, C Knight, D Cove, DG Schaefer, M Chakhparonian, JP Zryd. A specific member of the Cab multigene family can be efficiently targeted and disrupted in the moss *Physcomitrella patens*. Mol Gen Genet 261:92–99, 1999.
70. T Girke, H Schmidt, U Zahringer, R Reski, E Heinz. Identification of a novel delta 6-acyl-group desaturase by targeted gene disruption in *Physcomitrella patens*. Plant J 15:39–48, 1999.
71. R Strepp, S Scholz, S Kruse, V Speth, R Reski. Plant nuclear gene knockout reveals a role in plastid division for the homolog of the bacterial cell division protein FtsZ, an ancestral tubulin. Proc Natl Acad Sci USA 95:4368–4373, 1998.
72. PA Girod, H Fu, JP Zryd, RD Vierstra. Multiubiquitin chain binding subunit MCB1 (RPN10) of the 26S proteasome is essential for developmental progression in *Physcomitrella patens*. Plant Cell 11:1457–1471, 1999.
73. RD Vierstra. Proteolysis in plants: mechanisms and functions. Plant Mol Biol 32:275–302, 1996.
74. CD Knight. Studying plant development in mosses: the transgenic route. Plant Cell Environ 17:669–674, 1994.
75. DG Schaefer, G Bisztray, JP Zryd. Genetic transformation of the moss *Physcomitrella patens*. Biotechnol Agric For 29:349–364, 1994.
76. V Raghavan. Developmental Biology of Fern Gametophytes. Cambridge, Cambridge University Press, 1989.
77. JA Banks. Gametophyte development in ferns. Annu Rev Plant Physiol Plant Mol Biol 50:163–186, 1999.
78. TR Warne, DL Vogelien, LG Hickok. The analysis of genetically and physiologically complex traits using *Ceratopteris*: a case study of NaCl-tolerant mutants. Int J Plant Sci 156:374–384, 1995.
79. B DeYoung, T Weber, B Hass, JA Banks. Generating autotetraploid sporophytes and their use in analyzing mutations affecting gametophyte development in the fern *Ceratopteris*. Genetics 147:809–814, 1997.
80. TJ Cooke, LG Hickok, M Sugai. The fern *Ceratopteris richardii* as a lower plant model system for studying the genetic regulation of plant photomorphogenesis. Int J Plant Sci 156:367–373, 1995.
81. J Eberle, J Nemacheck, CK Wen, M Hasebe, JA Banks. *Ceratopteris*: a model system for studying sex-determining mechanisms in plants. Int J Plant Sci 156:359–366, 1995.
82. JA Banks. Sex determination in the fern *Ceratopteris*. Trends Plant Sci 2:175–180, 1997.
83. TR Warne, LG Hickok. Control of sexual development in gametophytes of *Ceratopteris richardii*: antheridiogen and abscisic acid. Bot Gaz 152:148–153, 1991.
84. TR Warne, LG Hickok. Evidence for a gibberellin biosynthetic origin of *Ceratopteris* antheridiogen. Plant Physiol 89:535–538, 1989.
85. JA Banks. Sex-determining genes in the homosporous fern *Ceratopteris*. Development 120:1949–1958, 1994.
86. JR Eberle, JA Banks. Genetic interactions among sex-determining genes in the fern *Ceratopteris richardii*. Genetics 142:973–985, 1996.
87. JA Banks. The TRANSFORMER genes of the fern *Ceratopteris* simultaneously promote meristem and archegonia development and repress antheridia development. Genetics 147:1885–1897, 1997.
88. CK Wen, R Smith JA Banks. ANI1: a sex pheromone–induced gene in *Ceratopteris* gametophytes and its possible role in sex determination. Plant Cell 11:1307–1317, 1999.
89. T Munster, J Pahnke, A Di Rosa, JT Kim, W Martin, H Saedler, G Theissen. Floral homeotic genes were recruited from homologous MADS-box genes preexisting in the common ancestor of ferns and seed plants. Proc Natl Acad Sci USA 94:2415–2420, 1997.
90. M Hasebe, CK Wen, M Kato, JA Banks. Characterization of MADS homeotic genes in the fern *Ceratopteris richardii*. Proc Natl Acad Sci USA 95:6222–6227, 1998.
91. K Aso, M Kato, JA Banks, M Hasebe. Characterization of homeodomain-leucine zipper genes in the fern *Ceratopteris richardii* and the evolution of the homeodomain-leucine zipper gene family in vascular plants. Mol Biol Evol 16:544–552, 1999.

# 42
# Photosynthetic Efficiency and Crop Yield

## Da-Quan Xu and Yun-Kang Shen

*Shanghai Institute of Plant Physiology, Chinese Academy of Sciences, Shanghai, People's Republic of China*

## I. INTRODUCTION

Photosynthesis, the most important biochemical process on the earth, is of such vital importance that no plant, animal, or human can live without it because they all depend on the energy, organic matter, and oxygen provided by it. Photosystem II (PSII) of the photosynthetic apparatus has been regarded as the engine of life [1]. However, considering both the cooperative relation between it and photosystem I (PSI) in photosynthesis and the key role of photosynthesis in the biosphere, we prefer to consider the two photosystems together, even the whole photosynthetic apparatus, which includes the carbon assimilation enzyme systems, as the engine of life driven by the energy from sunlight.

Photosynthesis is the cornerstone of all crop production practices, and the aim of crop production is to maximize it [2]. Agriculture is basically a system of exploiting solar energy to synthesize organic matter through photosynthesis. The yield of crop plants ultimately depends on the size and efficiency of their photosynthetic system [3]. Two important determinants of biomass production of any crop are the quantity of radiation intercepted by the crop and the efficiency of using the radiation in dry matter production [4]. As the economic yield of a crop is related not only to the dry matter production but also to the harvest index, crop productivity depends primarily on how efficiently incident light is used for assimilating carbon dioxide and how efficiently this assimilated carbon is partitioned among plant parts [5].

The notion of photosynthetic efficiency in the literature involves some different terms including photosynthetic rate; quantum yield of carbon assimilation; photochemical efficiency of PSII, which is often expressed as a ratio of variable to maximal fluorescence, Fv/Fm; light utilization efficiency; etc. These terms are different but linked to each other. From the light response curve of photosynthesis it may be understood that the limiting factors of photosynthesis are different at different light intensities. In weak light, photosynthetic rate increases linearly with an increase in light intensity because radiation energy is the main limiting factor. In stronger light with an increase in light intensity, its increase lowers gradually and finally ceases because the main limiting factor has become the capacity to use light energy of the photosynthetic apparatus. In weak light one is concerned mainly with quantum yield, whereas photosynthetic rate is more noted in strong light. Both photosynthetic rate and quantum yield are related to characteristics of the leaf, cell, and chloroplast itself and environmental conditions. Photosynthetic rate is often expressed as number of molecules of $CO_2$ fixed or $O_2$ evolved per unit leaf area per unit time (for example, $\mu mol\ CO_2\ m^{-2}\ s^{-1}$), while quantum yield is expressed as number of molecules of $CO_2$ fixed or $O_2$

evolved per photon absorbed. The light utilization efficiency of a canopy depends not only on the factors already mentioned but also on the total leaf area of the canopy and the canopy architecture. It is often expressed as a ratio of the energy stored in dry matter of crop plants formed in photosynthesis to the energy received per unit ground area where the crop plants grow. For crop yield and light utilization efficiency, both photosynthetic rate in strong light and quantum yield of carbon assimilation in weak light are important under field conditions.

In this chapter, first of all, the limiting factors of photosynthetic rate and the photosynthetic rate–crop yield relation are reviewed with an emphasis that the positive correlation is the reflection of the essence of the relation. Then the significance of quantum yield in yield formation and factors affecting quantum yield are discussed. Lastly, the main characteristics of the coming new green revolution are predicted on the basis of analysis of the limitations of the first green revolution.

## II.  PHOTOSYNTHETIC RATE

Photosynthetic rate in strong light is an important parameter characterizing the photosynthetic capacity of the photosynthetic apparatus. Apparently, it is not a notion of efficiency because the efficiency is the ratio of the output to input energy [6]. In fact, it is also a notion of the efficiency because it is a determinant of crop yield and light use efficiency. Under same light intensity, especially saturating light for photosynthesis, leaves with a higher photosynthetic rate necessarily have a higher photosynthetic capacity and higher light use efficiency compared with leaves with a lower one.

### A.  Factors Limiting Photosynthetic Rate

Many external environment factors such as low or high temperature, deficiency of water or nutrient supply, low $CO_2$ or high $O_2$ concentration, and low light intensity may limit photosynthesis, leading to a decreased photosynthetic rate. Meanwhile, many plant internal factors including development, hormones, respiration, etc. may also have a significant effect on net photosynthetic rate, but the main limitation site of net photosynthetic rate in $C_3$ plants is often in the reaction catalyzed by the enzyme ribulose-1,5-bis-phosphate (RuBP) carboxylase/oxygenase (Rubisco). Therefore, reducing or eliminating its oxygenase function or photorespiration or increasing the affinity of the enzyme for $CO_2$ is a long-term goal to increase productivity [7]. Nevertheless, we found that the photosynthetic rate in some plants such as wheat and rice was increased when the ATP supply was enhanced by spraying leaves with PMS (N-methylphenazonium methosulfate) to induce cyclic photophosphorylation (PSP) or with coupling efficiency improvers, for instance, polybasic acids. These findings indicate that the ATP supply from PSP may also be a limiting factor of the photosynthetic rate [8]. Recently, it was demonstrated that photosynthetic assimilation of $CO_2$ in water-stressed leaves of sunflower is not limited by $CO_2$ diffusion but by inhibition of RuBP synthesis, related to a lower ATP content resulting from loss of ATP synthase [9].

### B.  Diurnal Variation in Leaf Photosynthetic Rate

Leaf photosynthetic rate is not a constant parameter. It often varies with development of the leaf itself and changes in the environment. So it often displays ontogenetic, seasonal, and diurnal variations.

There are two typical patterns of the diurnal course of photosynthesis under natural conditions. One is the one-peaked pattern with the maximum net photosynthetic rate around noon on cloudy days; the other is two peaked, with one of the peaks in the late morning and the other in early afternoon with a depression around noon (11:00–14:00) on clear days, the so-called midday depression of photosynthesis. When the depression is severe, the peak in the afternoon may fail to appear. The possible mechanisms of the phenomenon have been reviewed in detail [10]. For wheat, the main cause of the midday depression observed by us is attributed to the partial closure of stomata [11], although photoinhibition of photosynthesis occurs simultaneously [12].

Midday depression of photosynthesis, as a regulatory process of the plant itself, is advantageous for the survival of plants under stress conditions but is at the expense of plant productivity, as it may decrease productivity. Alleviating it by some measures, therefore, may increase crop yield significantly. For example, mist irrigation at the grain-filling stage increased stomatal conductance and thereby net photosynthetic rate in flag leaves of wheat, thus increasing grain yield by about 18% [13].

## C.  Stomatal and Nonstomatal Limitations to Leaf Photosynthesis

Under stress conditions such as water deficiency and low temperature, declines in both leaf photosynthetic rate and stomatal conductance are often observed. To determine correctly the cause-effect relation between the two declines, an analysis of the stomatal limitation of photosynthesis must be made according to criteria suggested by Farquhar and Sharkey [14]. A decline in intercellular $CO_2$ concentration (Ci) indicates that the main cause of decline in leaf photosynthetic rate is a decrease in stomatal conductance. In contrast, an increase in Ci suggests that a decrease in photosynthetic activity of mesophyll cells, namely a nonstomatal factor, is the main cause of the decline in leaf photosynthetic rate. It appears that the direction of Ci change is the most important criterion for analysis of the stomatal limitation of photosynthesis.

If some unreliable criteria are used in the analysis, an incorrect conclusion may be reached about the cause-effect relation between changes in leaf photosynthetic rate and stomatal conductance. It has been emphasized in a review [15] that (1) a necessary criterion of predominantly stomatal limitation is a decreased Ci rather than a positive correlation between leaf photosynthetic rate and stomatal conductance; (2) an important criterion of predominantly stomatal limitation is the direction rather than the extent of Ci decrease; (3) a reliable criterion of predominantly nonstomatal limitation is an increased rather than a constant Ci; and (4) a boundary line of predominantly stomatal and nonstomatal limitation is the direction of change in stomatal limitation value rather than the relative magnitudes of stomatal and nonstomatal limitation values.

## D.  Relationship between Leaf Photosynthetic Rate and Crop Yield

Except for the mineral nutrient elements, accounting for about 5% of the total, all of the dry matter of crop plants is derived from photosynthetic $CO_2$ assimilation. From this fact it is naturally expected that a high photosynthetic rate will lead to a high yield or that there is a positive correlation between leaf photosynthesis and crop yield. However, a positive relation is not often observed, and it has been stated that in most cases there is no association between them and in some cases even a negative correlation between leaf photosynthetic rate and yield [16]. This paradox puzzled many plant physiologists and agronomists for quite a long time [17]. In fact, the apparent lack of a positive correlation is not surprising because this is a very complex problem and leaf photosynthetic rate is an important but not the sole factor determining crop yield.

### 1.  Reflection of the Essence—Positive Correlation

It is well known that about 95% of the dry matter of plants comes from photosynthesis. This fundamental fact determines that the essence of the relationship between leaf photosynthetic rate and crop yield is positive but not negative or no correlation.

The economic yield (Y) of crops is a function of photosynthetic production, respiratory consumption (R), and harvest index (HI). The amount of photosynthetic production depends on photosynthetic rate (P), leaf area (A), and photosynthetic duration (T). Their relationship can be very roughly expressed in the following equation:

$$Y = HI (P \times A \times T - R)$$

From this equation, it is very clear that Y must increase when P increases and Y must decrease when P decreases provided that HI, A, T, and R remain constant. A positive relationship between P and Y is intrinsic. Therefore, an increase in P caused by some treatment such as $CO_2$ enrichment in soybean and rice [18–20], spraying water in wheat [13], spraying sodium bisulfide solution upon leaves of wheat and rice [21], and improving N nutrient in soybean [22] always leads to an increase in crop yield. On the contrary, shading treatment leads to a decline in yield due to decreased photosynthesis [23].

Positive relationships between leaf photosynthetic rates and plant productivity, indeed, have been reported for dry bean [24], wheat [25,26], soybean [27], blackgram [28], green gram [29], pea [30], cassava [31], grain sorghum [32], upland cotton [33], and asparagus [34]. It appears that a significant positive correlation between light-saturated photosynthetic rate and yield among the cultivars of many crops is a reflection of the rule rather than the exception [35]. In addition, there have been some reports indicating that cultivars with higher yields have higher photosynthetic rates in soybean [36,37], oats [38], and rice [39].

## 2.  False Appearance—Negative Correlation

By making a comparative study of wild and cultivated wheat, it was found that leaf photosynthetic rate decreased while yield increased in wheat in the development of agriculture; that is, wild wheat has a lower yield and a higher leaf photosynthetic rate than modern cultivated wheat [40]. It was then assumed that there are counterproductive associations with high photosynthetic rate [41]. This finding is often cited as evidence that there is no positive association and there is even a negative association between leaf photosynthesis and crop yield. However, wild wheat with a higher leaf photosynthetic rate also had a smaller leaf area, shorter photosynthetic functional duration, and lower harvest index. Therefore, the so-called negative association is actually a false appearance. Obviously, for wild wheat the lower yield is due to smaller leaf area, shorter functional duration, and lower harvest index rather than higher leaf photosynthesis.

A negative correlation between leaf photosynthetic rate and biological yield was also reported for three tall fescue genotypes [42]. The negative correlation is also a false appearance. For the genotype having a higher yield, the high yield is owing to its bigger leaf area rather than its lower leaf photosynthetic rate.

It appears that although the analysis of correlation is a useful method used frequently in biological studies, a positive or negative correlation between two variables is only a phenomenon, and it is not necessarily a correct reflection of an intrinsic relation, especially a cause-effect one, between them. The correlation may be altered or masked by complex changes in other variables related to the two being considered. Therefore, one should take care when a conclusion is based on data from a correlation analysis.

## 3.  Factors Masking the Intrinsic Relation

Yield formation of a crop plant is a complex process involving many factors. In other words, crop yield is influenced and determined by several factors, external and internal. Among internal factors, leaf photosynthetic rate is the basic but not the sole factor. Thus, the intrinsic relation between leaf photosynthetic rate and crop yield may be masked by other factors changed in the opposite direction.

LEAF AREA   Leaf area is a very important factor for crop yield. A close association between leaf area and yield is often observed, especially when the planting density is not high. Many species have a significant negative correlation between leaf size and photosynthetic rate calculated on a unit leaf area basis. The "dilution" effect [43] frequently masks the intrinsic relation between leaf photosynthesis and yield, resulting in the paradox of no positive or even a negative association. By examining the differences in photosynthetic rates between soybean cultivars, it was found that two cultivars had lower yields although their photosynthetic rates were 25% higher than that of another one. It was due to the fact that the latter had a bigger leaf area (20% higher) and longer leaf functional duration (8 to 10 days longer) than the former ones [44].

LEAF FUNCTIONAL DURATION   From the results mentioned above [40,44], it may be seen that leaf functional duration is also an important factor affecting yield, and a lower yield is often related to a shorter leaf functional duration.

HARVEST INDEX   When the photosynthesis-yield relation is analyzed, the partitioning of photosynthates to different organs, expressed as a partitioning coefficient or harvest index, must be considered. A change in the harvest index opposite to that in leaf photosynthesis may mask the intrinsic relation between leaf photosynthetic rate and yield. It was reported that there was a significant positive correlation between leaf photosynthetic rate and biomass and seed yield for eight of nine cultivars of dry bean grown in the field. The only exception had a lower photosynthetic rate but a higher yield because of its very high harvest index [24].

RESPIRATORY LOSS   The size of respiration loss is closely related to crop yield. Selection for low leaf respiration has led to yield increases in perennial ryegrass [45,46]. There were considerable varietal differences in leaf photosynthetic rate during flowering among 11 cultivars of rice, but the low-yielding tall varieties were not all less efficient in photosynthesis than the high-yielding dwarf varieties. The lower yield might result from higher photorespiratory activity and inferior ability to transport the postflowering photosynthates to the developing grains [47].

DEVELOPMENTAL STAGE   For many crops, more than half of the economic yield derives from photosynthesis after flowering. Therefore, photosynthesis at the reproductive stage is more directly related to yield size [48]. The positive correlation between leaf photosynthesis and yield mentioned earlier is observed mostly at this stage. Obviously, one should not expect to find the correlation at all stages of crop development.

In addition, it is likely that the apparent lack of a positive relationship between leaf photosynthesis and crop yield may also originate from unsuitable methods and techniques used in the measurement of leaf photosynthetic rate. First, in some studies the equipment for photosynthetic measurement could not carry out rapid and accurate determinations of the difference in leaf photosynthetic rates under field conditions. Second, the number of leaves for photosynthetic measurement is sometimes too small to detect significant differences between cultivars. It has been calculated that 34, 16, and 8 measurements are needed for detecting a 7–9%, 15%, and 20% difference, respectively, between different soybean genotypes in the field [16]. Third, representative leaves should be used in photosynthetic measurements. Crop yield comes from a canopy, not several leaves. Therefore, differences in leaf photosynthesis among crop genotypes have not been shown to be correlated with seed yield, probably because leaves selected for measurements have not been representative of the crop canopy [49]. Moreover, conclusions about the absence of relationship between net photosynthetic rate and crop yield are often drawn from instantaneous photosynthetic measurements conducted under standardized conditions rather than seasonal measurements conducted under field conditions [50].

It was pointed out that there are several factors that may account for the apparent lack of correlation between photosynthetic rates and yield [23]. Much of the photosynthetic data is collected on individual leaves, whereas yield is measured on the entire plant or canopy. Many photosynthetic measurements are point-in-time determinations made at varying developmental stages of the plants and do not take into consideration the entire growing season. The relationship could be masked by any of a number of biochemical and physiological events that occur between the production of photosynthates and their utilization in the accumulation of final yield. Therefore, the concept of canopy seasonal photosynthesis was proposed and may be estimated by integrating the area under the time course curve of canopy net photosynthetic rate. It was found that the grain yield of soybean is strongly dependent on seasonal photosynthesis [23]. Similarly, there have been many other reports indicating positive relationships between photosynthetic rate at the canopy level and plant productivity in barley [51], soybean [49], cotton [52], wheat [53], and maize [54]. These may also be considered as strong evidence for an intrinsic relation between photosynthetic rate and crop yield.

From the findings discussed, it is concluded that the lack of correlation or negative correlation between photosynthetic rate and crop yield is only apparent whereas the positive correlation is a reflection of the intrinsic relation but is often masked by some factors. There has been evidence indicating that genetic selection for higher photosynthetic rates could lead to increases in yield. For example, pima cotton bred for increased yield has enhanced photosynthesis. Among the cultivated types of pima cotton, genetic advances are closely associated with an increase in leaf photosynthetic rate and stomatal conductance, especially in the morning [55]. Thus, we could improve crop yield by selecting a higher photosynthetic rate. However, higher photosynthetic rate is not the sole parameter that should be considered. When a higher photosynthetic rate is selected, some unfavorable changes such as a decline in leaf area, leaf functional duration, and harvest index and an increase in respiratory rate should be avoided. Meanwhile, the quantum yield of photosynthesis in weak light is also an important parameter to be considered in crop breeding [56].

## III. QUANTUM YIELD

For a high yield of crop canopy, not only a high photosynthetic rate in strong light but also a high quantum yield in weak light is important because in a canopy not all leaves are in strong light. Apart from leaves in the upper layer, those of the middle and bottom layers in a canopy are often under light-limiting conditions even on clear days. On cloudy days or in the early morning and late afternoon on clear days, all leaves of a canopy are under weak light. So under field conditions a significant part of crop photosynthesis occurs at nonsaturating light. Therefore, Ort and Baker [57] believed that future research efforts aimed at improving crop production through improved photosynthetic performance should have a major focus on the efficiency of operation under nonsaturating light conditions. Similarly, it was considered that increasing net photosynthetic rate (Pn) and productivity in low light would require an increase in apparent quantum yield [7].

## A.  Theoretical and Actual Values

It is well known that according to the Calvin-Benson cycle in photosynthesis, assimilating one molecule of $CO_2$ into carbohydrate requires 2NADPH and 3ATP. The production of 2NADPH is the result of transporting four electrons from $2H_2O$ to $2NADP^+$ along an electron transport chain (Z-scheme). Because the chain includes two photosystems in series, two photons are needed for one electron transport. Thus, at least eight photons are required for the production of 2NADPH. Therefore, the maximal or theoretical quantum yield for photosynthetic carbon assimilation is 0.125 mole $CO_2$/mole photons. With respect to the amount of ATP produced through PSP coupled with the photosynthetic electron transport mentioned above, i.e., the number of molecules of ATP produced by coupling with the transport of two electrons or the evolution of one-half molecule $O_2$, or P/O ratio, there are several values, namely 1, 1.33, and 2 in different laboratories [58]. If P/O is lower than 1.5 or part of the ATP produced by PSP is used in the biosynthesis of compounds other than carbohydrates, cyclic or pseudocyclic PSP is required to meet the demand of carbohydrate synthesis for ATP. In such cases, the quantum requirement (the reciprocal of quantum yield) must be higher than 8. An accurate value of the minimum quantum requirement of photosynthetic carbon assimilation, in fact, is still uncertain. Values in a range of 8–12 are acceptable to most scientists in the area [59]. The uncertainty may be related to the complex regulatory mechanisms of photosynthesis and the variability of environmental conditions.

It should be noted that the values of 8–12 for minimum quantum requirements are obtained under the most suitable conditions. If an inevitable loss such as photorespiration occurs, the value will be about 17 [60]. However, such a value is seldom obtained under field conditions. It is often much higher than 25 even if under normal conditions without any environmental stress [61]. The causes leading to the difference between theoretical and actual values of the quantum requirement are worth studying.

## B.  Factors Affecting Quantum Yield

### 1.  Environmental Factors

Emerson and Lewis [62] showed that the values of quantum yield were related to the quality of light. A high quantum yield was measured at red light around 680 nm. The quantum yields of sun and shade leaves grown under different light intensities were similar, although there was a significant difference in light-saturated photosynthetic rate between them [63,64]. At 21% $O_2$ and a temperature range of 15–35°C the quantum yield decreased gradually with temperature increase in $C_3$ plants but not in $C_4$ plants [60,65]. Water deficiency and excessive water or flooding could lead to a decline in quantum yield [66,67]. After several rainy days, the photosynthetic quantum efficiency became lower in spinach leaves [68]. The reason may be that the reduction of $NADP^+$ is severely hindered in swollen chloroplasts under hypotonic conditions [69]. Decreasing $O_2$ concentration or increasing $CO_2$ concentration in air could increase quantum yield in $C_3$ plants but not in $C_4$ plants [60,70]. We found that reduced atmospheric pressure had an adverse effect on photosynthetic quantum efficiency [71]. The difference in apparent quantum yield calculated on the basis of incident photon flux density under different nitrogen nutrition levels could be attributed to decreased light absorption induced by a low nitrogen level [72]. Phosphate deficiency in nutrient solution could lead to a declined quantum yield in spinach leaves [73]. This may be due to decreased excitation energy transport from antenna pigments to PSII reaction centers and enhanced excitation energy dissipation as heat under phosphate deficiency conditions [74].

### 2.  Plant Factors

Among all internal factors, photorespiration has the most significant effect on quantum yield. The effects of air temperature and $CO_2$ or $O_2$ concentration on quantum yield mentioned earlier, in fact, are related to the changes in photorespiratory rate caused by these factors. In normal air and at 20–25°C, the quantum yields of $C_3$ and $C_4$ plants were similar. However, when the air temperature was over 30°C, the quantum yield in $C_4$ plants was slightly higher than that in $C_3$ plants [75]. When photorespiration was inhibited by high $CO_2$ and/or low $O_2$, $C_4$ plants had about 30% lower quantum yields than $C_3$ plants because they used two additional ATP molecules in the $C_4$ pathway for fixation of onc molecule of $CO_2$ to form carbohydrate [76]. So $C_4$ plants are not more efficient than $C_3$ plants in weak light. Quantum yield was lower in younger leaves than in mature leaves [77,78]. This may be because more ATP is used in the biosynthesis of components other than carbohydrates in younger leaves growing luxuriantly. The chlorophyll content of leaves affects only the apparent quantum yield calculated on the basis of the quantity of light incident on the leaf surface but not that based on the quantity of light absorbed by leaves [79]. When ATP was insufficient or NADPH was excessive, cyclic PSP would be enhanced, leading to a decline in quantum yield [80]. The dark respiratory rate does not affect the quantum yield under conditions in which dark respiratory rate is constant.

## C. Diurnal Variation of Quantum Yield

In field studies we found that the apparent quantum yield of photosynthetic carbon assimilation often displayed a significant midday decline in many $C_3$ plants such as soybean and wheat but not in $C_4$ plants such as maize and sorghum on clear days [81]. It was deduced that photoinhibition may be a cause of the midday decline of the photosynthetic efficiency [61]. The molecular mechanism of photoinhibition is still not fully clear. For more than a decade photoinhibition has been considered almost synonymous with photodamage to the photosynthetic apparatus [82], mainly the loss of D1 protein, a central component of the PSII reaction center complex. However, no evident change in D1 protein content in the leaves of sweet vibrium, wheat, and soybean was observed when photoinhibition occurred in strong light [83–85]. These results indicate that under normal conditions without other environmental stress photoinhibition is a reflection of enhanced operation of protective mechanisms rather than a result of damage to the photosynthetic apparatus [86]. DTT (dithiothreitol), an inhibitor of the xanthophyll cycle, could exacerbate photoinhibition and result in a substantial loss of D1 protein in wheat leaves after exposure to midday strong light, indicating that the xanthophyll cycle–dependent heat dissipation plays an important protective role against photodamage to the photosynthetic apparatus in strong light [84]. Nevertheless, our studies also demonstrated that the xanthophyll cycle–dependent heat dissipation is a predominantly protective mechanism only in some plant species such as wheat and barley but not in other plant species such as soybean and cotton [85,87,88]. In soybean leaves, the predominantly protective mechanisms is likely the reversibly inactivated PSII reaction center–dependent heat dissipation [85,87,88]. The mechanisms of reversible inactivation and heat dissipation of PSII reaction centers are still not clear. There have been experimental results showing that the reversible inactivation is related to the dissociation of light-harvesting complex II (LHCII) from the PSII reaction center complex [89].

In addition to photoinhibition, enhanced photorespiration is another cause of the midday decline in the photosynthetic efficiency of $C_3$ plants [90]. For a long time photorespiration has been considered a wasteful process. Many efforts have been made to eliminate it, but no success has been reported. Extensive screening programs involving several species (wheat, barley, oats, soybean, potato, tall fescue) failed to identify genotypes with a low $CO_2$ compensation point [91]. Attempts to select $C_3$ plants with low rates of photorespiration and high rates of net photosynthesis have had little success. Some mutant genotypes of tobacco with increased productivity have been selected at low $CO_2$ concentrations, but this increased productivity is related to a greater leaf area per plant and higher photosynthetic rates rather than reduced photorespiratory rate or $CO_2$ compensation point or improved Rubisco properties [92]. It is likely that it is not simply a wasteful process but a protective one for plants, at least under some stress conditions. Our study has demonstrated that it can protect the photosynthetic apparatus against photodamage through accelerating phosphate recycling during photosynthesis [93].

## IV. GREEN REVOLUTION

As mentioned before, the economic yield of crop is a function of photosynthetic production, respiration consumption, and harvest index. Thus, it is related not only to leaf photosynthetic performance but also to plant type and canopy structure. Dwarfing the stem of a crop may lead to higher yield through increasing the harvest index. Erect leaves are also favorable for an increase in crop yield because the leaves of middle and bottom layers in the canopy may receive more light energy, thus improving the light use efficiency of the canopy [94]. The breeding of high-yielding varieties with dwarf stems and erect leaves brought about a great revolution in agriculture.

## A. First Green Revolution

In the 1950s and 1960s, some agricultural scientists developed a package of high-yielding crop varieties and agricultural management techniques. The package brought about an unprecedented boom (more than double) in world grain yields since 1960. This is the so-called green revolution, one of the 20th century's greatest technological achievements [95].

The phrase green revolution was introduced by U.S. Agency for International Development administrator W.S. Gaud in 1968 [96]. The green revolution originated from the success of the American breeder Dr. Norman Borlaug's work on wheat in Mexico in the period 1961–1965 [97]. He was awarded the Nobel Peace Prize in 1970 owing to his great contribution [98]. The release of the first high-yielding modern rice cultivar IR8, a semiwarf one with erect leaves and high harvest index, by the International Rice Research Institute in 1966 marked the start of the green revolution in Asia [99]. Around the green revolution two major breakthroughs in rice breeding occurred in China. One was the wide distribution of fertilizer-responsive, lodging-resistant dwarf rice varieties with high-yielding potential 2 years before the release of IR8. Another was the commercialization of hybrid rice production in 1976 [100]. The great contribution to rice breeding of academician L.-P. Yuan et al. [101] made China the first country to commercialize the production of hybrid rice. Rice hybrids have a yield advantage of about 15% over the best inbred varieties, and approximately 50% of the rice area has been devoted to plant rice hybrids in China.

Apart from seeds of high-yielding varieties, the green revolution also needs the support of adequate amounts of chemical fertilizers, water, pesticides, improved farm equipment, etc. If all or most of them are not available, there is no guarantee of the revolution [102]. The green revolution, in fact, is only a cereal revolution. Other agricultural crops have not shown any boost increase in yield [103]. In the revolution, agricultural scientists concentrated their efforts only on yield so that those high-yielding varieties were often susceptible to various diseases and pests and had a low protein content [96,104].

## B.  Second Green Revolution

The achievement of the first green revolution is great, yet its shortcomings are also obvious. In addition to those already mentioned, one of the more important ones is that only the short stalk, erect leaves increasing the light utilization of the canopy and a high harvest index were emphasized, but the photosynthetic efficiency of the leaf was not considered an important selection criterion for high-yielding varieties. The potential for improving stalk height, leaf angle, and harvest index has been exploited to a fuller extent; thus, the rest room has been very limited (mainly for wheat and rice). The measures adopted in the first green revolution, such as improved crop management and increased inputs of water, chemical fertilizers, and pesticides, have lost their edge in increasing crop yield. Agricultural scientists have begun to seek a new revolution [95].

It was considered that to usher in a second green revolution, the following research topics should be urgently addressed: increase of potential leaf photosynthesis and canopy photosynthesis, enlargement of sink capacity for assimilates, and knowledge of photosynthetic criteria for environmental stress tolerance [105]. It appears that improving photosynthesis is a great hope of the future of agriculture [95]. On the basis of an analysis of rice production constraints in China, it was pointed out that of the 11 plant-related factors, the most important ones are plant structure, photosynthetic efficiency, and growth duration; therefore, research should concentrate on improving them [100]. It is increasingly realized that the yield potential of varieties will be increased by improving their photosynthetic efficiency. This can be possible mainly by way of DNA transfers through genetic engineering and exploitation of hybrid vigor [106]. Large benefits would result from concentrating research funds on increasing the biological efficiency of crops. This would come from success in pursuing hybrid crops with desirable traits such as improved plant capacity to generate photosynthates and to store them in the grain. Moreover, scientists clearly expect biotechnology—the modern techniques for genetic transfers—to provide large gains [107].

Indeed, some genetic engineers have aimed at enhancing crop photosynthesis [108]. In genetic engineering, Rubisco is the most important target because it is a key enzyme in photosynthetic carbon assimilation. Also, it is an enzyme with two functions, catalyzing both carboxylation and oxygenation of RuBP. It is not only the world's most abundant protein but also the world's most incompetent enzyme. Under normal air conditions, the carboxylation reaction catalyzed by the enzyme is the main rate-limiting step in the whole photosynthesis process [109]. Natural variation in the kinetic properties of the enzyme suggests that it is possible to alter the enzyme to favor the carboxylation activity relative to oxygenation [110]. Ultimately, the desire is to engineer higher plant Rubisco to increase specificity, i.e., to favor carboxylation and increase photosynthetic rate and thus plant yield [111]. Exploitation of the natural biodiversity of Rubisco molecules may be an important part of future strategies to solve the molecular basis of $CO_2/O_2$ specificity [112]. There has been no report so far that engineered Rubisco leads to higher efficiency with respect to $CO_2$ fixation [113], although it has been a prime focus for genetically engineering an increase in photosynthetic productivity [114].

In genetic engineering aimed at improving photosynthesis, another important target is the enzyme phosphoenolpyruvate carboxylase (PEPC), which catalyzes the first reaction of the $C_4$ cycle of photosynthetic carbon fixation. High-level expression of maize PEPC in transgenic rice plants has been successful. Some transgenic rice leaves have an enhanced PEPC activity two- to threefold higher than that in maize, and they showed a reduced $O_2$ inhibition of photosynthesis but no increased photosynthetic rates compared with those of untransformed plants [115]. Higher photosynthetic efficiency in $C_4$ plants depends not only on PEPC and the other enzymes of $C_4$ pathway but also on the special anatomic structure of their leaves. Thus, overexpression of PEPC alone is not likely to result in an improvement of crop yield [116].

Genetic engineering targeted at other enzymes of photosynthetic carbon metabolism such as sucrose phosphate synthase (SPS) and ADP-glucose pyrophosphorylase (AGPase), which participate in sucrose and starch biosynthesis, respectively, has been performed. These efforts are undoubtedly useful in understanding the regulatory role of these enzymes but not necessarily in increasing the yield of crop plants. SPS has been considered to be a key enzyme in the regulation of carbon assimilation and export from the leaf [117]. Although increased photosynthetic rates were observed in transformed tomato plants expressing a maize SPS gene in addition to the native enzyme, total dry matter production and fruit yield were not significantly increased [118,119]. Up- and down-regulation of SPS activity may lead to expected changes with respect to carbon partitioning. However, it remains questionable whether the actual rate of photosynthetic sucrose formation does determine final crop yield [120].

In addition, there has been a report that genetic engineering–transformed potato plants with reduced accumulation of a protein located in chloroplasts showed stunted growth, decreased tuber yield, and reduced values of nonphotochemical quenching of chlorophyll *a* fluorescence. These results indicate a preferential association of the protein with the light-harvesting complex of PSII (LHCII) and its functional role of modulating photosynthetic efficiency and dissipating excessive absorbed light energy within the antenna complex [121].

Regardless of success or failure in increasing crop yield, such efforts themselves have implied that the central object will be the improvement of photosynthetic efficiency of crops, and the sharpest tool will be genetic engineering for a new green revolution. Transforming crops through genetic engineering to get good varieties with high photosynthetic efficiency seems to be the major hope for the new revolution. However, although the study of photosynthesis has benefited from the techniques of molecular biology, these techniques alone rarely permit a mechanistic understanding of the process. Cooperative efforts for integrated experimental approaches that combine the strategies used in physiology, biochemistry, and other relevant fields will be imperative to evaluating the process fully [122]. In the process many basic details still remain to be understood, especially about the regulation of photosynthetic efficiency, including the step most seriously limiting photosynthetic efficiency, which the physiologists and biochemists of photosynthesis are exploring. A comprehensive understanding of this regulatory mechanism will be the basis of success of the molecular biologists in engineering crops with increased photosynthetic efficiency. It is impossible to find a successful target(s) in engineering crops without a clear understanding of the mechanism. Moreover, even allowing for the enormous advances in molecular biology that are being made, results are not expected to be obtained soon in genetic improvement of crop productivity [123].

Recently, it has been reported that it is now possible to insert a single, genetically dominant, potentially yield-enhancing, dwarfing gene into the genome of any transformable crop without the need for long-term conventional breeding programs and with minimal disruption of genetic background [124]. This study indicates that plant height reduction associated with yield increase is still an important aim for high-yield breeding, especially in high-stalk crops such as maize and sorghum. It appears that the two green revolutions cannot be totally separated. The second revolution will be a continuation of the first one, but it will have new characteristics. If it may be said that the first revolution was characterized by improving plant type, then the second one will be characterized mainly by improving photosynthetic efficiency.

Besides the central object of improving photosynthetic efficiency, of course, the new revolution may involve more aims such as improving grain quality, manipulating plant nutrients for human health, achieving herbicide resistance, etc. by engineering crops genetically. The initial phase of a new revolution in agriculture has already occurred. Worldwide, in 1999, about 28 million hectares of transgenic plants were being grown. It has been predicted that this area will be tripled in the next 5 years [125].

## V. CONCLUDING REMARKS

Photosynthesis is the engine of life for almost all organisms on earth because it provides food, energy, and oxygen for them and is also the basis of yield formation in crop plants. High photosynthetic efficiency, including high photosynthetic rate in strong light and high quantum yield in weak light, should be used as a selection criterion for high-yielding varieties.

The first green revolution has achieved great success in increasing crop yield since the 1960s, but it is losing its edge. Scientists are seeking a new revolution in order to feed the increasing population of the world. For the new green revolution, the central objective is improvement of the photosynthetic efficiency of crops and the sharp tool is gene engineering. The tool is powerful not only in understanding molecular mechanisms for the regulation of photosynthetic efficiency but also in engineering crops with desirable characteristics. Cooperative efforts made by scientists of many disciplines are absolutely necessary to the success of the new revolution. The proper choice of targets of engineering crops depends on a breakthrough in the comprehensive study on the regulatory mechanisms of the photosynthetic efficiency.

## ACKNOWLEDGMENTS

Some studies of the authors cited in this chapter were supported by the State Key Basic Research and Development Plan (No. G1998010100) and the National Natural Science Foundation of China (No. 39730040). We thank Professor Tian-Duo Wang for critically reading the manuscript and giving useful suggestions.

## REFERENCES

1.  J Barber. Molecular basis of photoinhibition. In: P Mathis, ed. Photosynthesis: from Light to Biosphere. Vol IV. Dordrecht, The Netherlands: Kluwer Academic Publishers, 1995, pp 159–164.
2.  NC Stoskopf. Understanding Crop Production. Reston, VA: Reston Publishing Company, 1981, pp 1–12.
3.  FP Gardner, RB Pearce, RL Mitchell. Physiology of Crop Plants. Ames: Iowa State University Press, 1985, pp 3–30.
4.  NR Baker, DR Ort. Light and crop photosynthesis performance. In: NR Baker, H Thomas, eds. Crop Photosynthesis: Spatial and Temporal Determinants. Amsterdam: Elsevier Science Publishers, 1992, pp 289–312.
5.  DR Geiger, JC Servaites, W-J Shieh. Balance in the source-sink system: a factor in crop productivity. In: NR Baker, H Thomas, eds. Crop Photosynthesis: Spatial and Temporal Determinants. Amsterdam: Elsevier Science Publishers, 1992, pp 155–176.
6.  OS Ksenzhek, AG Volkov. Plant Energetics. San Diego: Academic Press, 1998, p 120.
7.  DW Lawlor. Photosynthesis, productivity and environment. J Exp Bot 46:1449–1461, 1995.
8.  Y-K Shen. Coupling problem of photophosphorylation. In: CL Tsou, ed. Current Biochemical Research in China. San Diego: Academic Press, 1989, pp 137–147.
9.  W Tezara, VJ Mitchell, SD Driscoll, DW Lawlor. Water stress inhibits plant photosynthesis by decreasing coupling factor and ATP. Nature 401:914–917, 1999.
10. D-Q Xu, Y-K Shen. Midday depression of photosynthesis. In: M Pessarakli, ed. Handbook of Photosynthesis. New York: Marcel Dekker, 1997, pp 451–459.
11. D-Q Xu, D-Y Li, Y-G Shen, G-A Liang. On midday depression of photosynthesis of wheat leaf under field conditions. Acta Phytophysiol Sin 10:269–276, 1984.
12. D-Q Xu, Y Ding, H Wu. Relationship between diurnal variations of photosynthetic efficiency and midday depression of photosynthetic rate in wheat leaves under field conditions. Acta Phytophysiol Sin 18:279–284, 1992.
13. D-Q Xu, D-Y Li, Y-G Shen, J-Y Yan, Y-G Zhang, Y-S Zheng. On the midday depression of photosynthesis of wheat leaf under field conditions. II. The effects of spraying water on the photosynthetic rate and the grain yield of wheat. Acta Agron Sin 13:111–115, 1987.
14. GD Farquhar, TD Sharkey. Stomatal conductance and photosynthesis. Annu Rev Plant Physiol 33:317–345, 1982.
15. D-Q Xu. Some problems in stomatal limitation analysis of photosynthesis. Plant Physiol Commun Sin 33:241–244, 1997.

16. CJ Nelson. Genetic associations between photosynthetic characteristics and yield: review of the evidence. Plant Physiol Biochem 26:543–554, 1988.
17. CD Elmore. The paradox of no correlation between leaf photosynthetic rates and crop yields. In: JD Hesketh, JW Jones, eds. Predicting Photosynthesis for Ecosystem Models. Vol II. Boca Raton, FL: CRC Press, 1980, pp 155–167.
18. RL Cooper, WA Brun. Response of soybeans to a carbon dioxide–enriched atmosphere. Crop Sci 7:455–457, 1967.
19. PJ Kramer. Carbon dioxide concentration, photosynthesis, and dry matter production. Bioscience 31:29–33, 1981.
20. BA Kimball. Carbon dioxide and agricultural yield: an assemblage and analysis of 430 prior observations. Agron J 75:779–788, 1983.
21. S Tan, Y-G Shen. The effects of sodium bisulfite on photosynthetic apparatus and its operation. Acta Phytophysiol Sin 13:42–50, 1987.
22. D-Q Xu, Y-G Shen, S-J Wang, X-W Zhang. Studies on the relationship between photosynthesis and nitrogen fixation in the symbiotic system of soybean and nodule bacteria (*Rhizobium*). Acta Bot Sin 31:103–109, 1989.
23. AL Christy, CA Porter. Canopy photosynthesis and yield in soybean. In: Govindjee, ed. Photosynthesis: Development, Carbon Metabolism, and Plant Productivity. Vol II. New York: Academic Press, 1982, pp 499–511.
24. MM Peet, A Bravo, DH Wallace, JL Ozbun. Photosynthesis, stomatal resistance, and enzyme activities in relation to yield of field-grown dry bean varieties. Crop Sci 17:287–293, 1977.
25. RA Fischer, F Bidinger, JR Syme, PC Wall. Leaf photosynthesis, leaf permeability, crop growth, and yield of short spring wheat genotypes under irrigation. Crop Sci 21:367–373, 1981.
26. RA Fischer, D Rees, KD Sayre, Z-M Lu, AG Condon, A Larque Saavedra. Wheat yield progress associated with higher stomatal conductance and photosynthetic rate, and cooler canopies. Crop Sci 38:1476–1475, 1998.
27. BR Buttery, RI Buzzell, WI Findlay. Relationships among photosynthetic rate, bean yield and other characters in field-grown cultivars of soybean. Can J Plant Sci 61:191–198, 1981.
28. R Chandra Babu, PS Srinivasan, N Natarajaratnam, SR Sree Rangasamy. Relationship between leaf photosynthetic rate and yield in blackgram (*Vigna mungo* L. Hepper) genotypes. Photosynthetica 19:159–163, 1985.
29. PS Srinivasan, R Chandrababu, N Natarajaratnam, SR Sree Rangaswamy. Leaf photosynthesis and yield potential in green gram [*Vigna radiata* (L.) Wilczek] cultivars. Trop Agric 62:222–224, 1985.
30. SLA Hobbs. Relationships between carbon dioxide exchange rate, photosynthetic area and biomass in pea. Can J Plant Sci 66:465–472, 1986.
31. MA El-Sharkawy, JH Cock, JK Lynam, AdP Hernandez, LLF Cadavid. Relationships between biomass, root-yield and single-leaf photosynthesis in field-grown cassava. Field Crop Res 25:183–201, 1990.
32. S Peng, DR Krieg, FS Girma. Leaf photosynthetic rate is correlated with biomass and grain production in grain sorghum lines. Photosynth Res 28:1–7, 1991.
33. WT Pettigrew, WR Meredith. Leaf gas exchange parameters vary among cotton genotypes. Crop Sci 34:700–705, 1994.
34. MJ Faville, WB Silvester, TG Allan Green, WA Jermyn. Photosynthetic characteristics of three asparagus cultivars differing in yield. Crop Sci 39:1070–1077, 1999.
35. LT Evans. From leaf photosynthesis to crop productivity. In: N Murata, ed. Research in Photosynthesis. Vol IV. Dordrecht, The Netherlands: Kluwer Academic Publishers, 1992, pp 587–594.
36. GM Dornhoff, RM Shibles. Variety differences in net photosynthesis of soybean leaves. Crop Sci 10:42–45, 1970.
37. AK Sharma, BB Singh, SP Singh. Relationship among net assimilation rate, leaf area index and yield in soybean [*Glycine max* (L.) Merrill] genotypes. Photosynthetica 16:115–118, 1982.
38. MA Brinkman, KJ Frey. Flag leaf physiological analysis of oat isolines that differ in grain yield from their recurrent parents. Crop Sci 18:69–73, 1978.
39. S Chakrabarti, S Saha. Photosynthetic behavior of some high- and low-yielding cultivars of rice. Plant Physiol Biochem 9:119–129, 1982.
40. LT Evans, RL Dunstone. Some physiological aspects of evolution in wheat. Aust J Biol Sci 23:725–741, 1970.
41. LT Evans. The physiological basis of crop yield. In: LT Evans, ed. Crop Physiology. Cambridge: Cambridge University Press, 1975, pp 327–355.
42. JW Poskuta, CJ Nelson. Role of photosynthesis and photorespiration and of leaf area in determining yield of tall fescue genotypes. Photosynthetica 20:94–101, 1986.
43. JD Hesketh, WL Ogren, RE Hageman, DB Peters. Correlations among leaf $CO_2$-exchange rates, areas and enzyme activities among soybean cultivars. Photosynth Res 2:21–30, 1981.
44. PE Curtis, WL Ogren, RH Hageman. Varietal effect of soybean photosynthesis and photorespiration. Crop Sci 9:323–327, 1969.
45. MJ Robson. The growth and carbon economy of selection lines of *Lolium perenne* cv. S23 with differing rates of dark respiration. 2. Grown as young plants from seed. Ann Bot 49:331–339, 1982.
46. D Wilson. Response to selection for dark respiration rate of mature leaves in *Lolium perenne* and its effects on growth of young plants and simulated swards. Ann Bot 49:303–312, 1982.
47. P Palit, A Kundu, RK Mandal, SM Sircan. Productivity of rice plant in relation to photosynthesis, photorespiration and translocation. Indian J Plant Physiol 22:66–74, 1979.

48. H-C Yin, Y-K Shen, Y Chen, C-H Yu, P-C Li. Accumulation and distribution of dry matter in rice after flowering. Acta Bot Sin 5:177–184, 1956.

49. R Wells, LL Schulze, DA Ashley, HR Boerma, RH Brown. Cultivar differences in canopy apparent photosynthesis and their relationship to seed yield in soybeans. Crop Sci 22:886–890, 1982.

50. I Zelitch. The close relationship between net photosynthesis and crop yield. Bioscience 32:796–802, 1982.

51. PV Biscoe, RK Scott, JL Monteith. Barley and its environment. III. Carbon budget of the stand. J Appl Ecol 12:269–293, 1975.

52. R Wells, WR Meredith Jr, JR Williford. Canopy photosynthesis and its relationship to plant productivity in near-isogenic cotton lines differing in leaf morphology. Plant Physiol 82:635–640, 1986.

53. S-T Dong. Studies on the relationship between canopy apparent photosynthesis and grain yield in high-yielding winter wheat. Acta Agron Sin 17:461–469, 1991.

54. C-H Hu, S-T Dong, S-S Yue, Q-Y Wang, R-Q Gao, Z-L Pan. Studies on the relationship between canopy apparent photosynthesis rate and grain yield in high yielding summer corn (*Zea mays* L.). Acta Agron Sin 19:63–69, 1993.

55. K Cornish, JW Radin, EL Turcotte, Z Lu, E Zeiger. Enhanced photosynthesis and stomatal conductance of pima cotton (*Gossypium barbadense* L.) bred for increased yield. Plant Physiol 97:484–489, 1991.

56. D-Q Xu, Y-K Shen. Photosynthesis and crop yield. In: Q Zhou, X-C Wang, eds. The Physiological Basis of High Yield, High Efficiency and Resistance against Stress in Main Crops. Beijing: Science Press. 1996, pp 17–24.

57. DR Ort, NR Baker. Consideration of photosynthetic efficiency at low light as a major determinant of crop photosynthetic performance. Plant Physiol Biochem 26:555–565, 1988.

58. Y-K Shen. The physiological approach to the mechanism of photophosphorylation. Physiol Veg 23:725–729, 1985.

59. Govindjee. On the requirement of minimum number of four versus eight quanta of light for the evolution of one molecule of oxygen in photosynthesis: a historical note. Photosynth Res 59:249–254, 1999.

60. J Ehleringer, O Bjorkman. Quantum yield for $CO_2$ uptake in $C_3$ and $C_4$ plants. Plant Physiol 59:86–90, 1977.

61. D-Q Xu, B-J Xu, Y-G Shen. Diurnal variation of photosynthetic efficiency in $C_3$ plants. Acta Phytophysiol Sin 16:1–5, 1990.

62. R Emerson, CM Lewis. The dependence of the quantum yield of chlorella photosynthesis on wave length of light. Am J Bot 30:165–178, 1943.

63. O Bjorkman. Response to different quantum flux densities. In: OL Lange, PS Nobel, CB Osmond, H Ziegler, eds. Encyclopedia of Plant Physiol. Vol 12A. Responses to the Physical Environment. Berlin: Springer-Verlag, 1981, pp 57–107.

64. G Oquist, L Brunes, J-E Hallgren. Photosynthetic efficiency of *Betula pendula* acclimated to different quantum flux densities. Plant Cell Environ 5:9–15, 1982.

65. SB Ku, GE Edwards. Oxygen inhibition of photosynthesis. III. Temperature dependence of quantum yield and relation to $O_2/CO_2$ solubility ratio. Planta 140:1–6, 1978.

66. P Mohanty, JS Boyer. Chloroplast response to low leaf water potentials. IV. Quantum yield is reduced. Plant Physiol 57:704–709, 1976.

67. FS Davies, JA Flore. Short-term flooding effects on gas exchange and quantum yield of rabbiteye blueberry (*Vaccinium ashei* Reade). Plant Physiol 81:289–292, 1986.

68. D-Y Li, J-Y Ye, Y-K Shen. Effect of rainy weather on the photosynthetic efficiency in spinach. Plant Physiol Commun Sin 27:413–415, 1991.

69. J-Y Ye, D-Y Li, Y-G Shen. Effect of hypotonic swelling on photosynthesis in spinach intact chloroplasts. Acta Phytophysiol Sin 21:73–79, 1995.

70. RK Monson, JRO Littlejohn, GJ Williams III. The quantum yield for $CO_2$ uptake in $C_3$ and $C_4$ grasses. Photosynth Res 3:153–159, 1982.

71. S-Y Zhang, G-Q Lu, H Wu, Z-X Shen, H-M Zhong, Y-G Shen, D-Q Xu, H-G Ding, W-X Hu. Photosynthesis of major $C_3$ plants on Qinghai plateau. Acta Bot Sin 34:176–184, 1992.

72. BA Osborne, MK Garrett. Quantum yield for $CO_2$ uptake in some diploid and tetraploid plant species. Plant Cell Environ 6:135–144, 1983.

73. A Brooks. Effects of phosphorus nutrition on ribulose-1,5-bisphosphate carboxylase activation, photosynthetic quantum yield and amounts of some Calvin-cycle metabolites in spinch leaves. Aust J Plant Physiol 13:221–237, 1986.

74. J Jacob. Phosphate deficency increases the rate constant of thermal dissipation of excitation energy by photosystem II in intact leaves of sunflower and maize. Aust J Plant Physiol 22:417–424, 1995.

75. J Ehleringer, RW Pearcy. Variation in quantum yield for $CO_2$ uptake among $C_3$ and $C_4$ plants. Plant Physiol 73:555–559, 1983.

76. CB Osmond, O Bjorkman, DJ Anderson. Physiological Processes in Plant Ecology. Berlin: Springer-Verlag, 1980, pp 291–377.

77. G Oquist, L Brunes, J-E Hallgren. Photosynthetic efficiency during ontogenesis of leaves of *Betula pendula*. Plant Cell Environ 5:17–21, 1982.

78. Q-F Yang, H Jiang, D-Q Xu. Changes in the photosynthetic efficency of the flag leaves of wheat during development. Acta Phytophysiol Sin 25:408–412, 1999.
79. O Bjorkman, B Demmig. Photon yield of $O_2$ evolution and chlorophyll fluorescence characteristics at 77 K among vascular plants of diverse origins. Planta 170:489–504, 1987.
80. JK Hoober. Chloroplasts. New York: Plenum, 1984, pp 111–145.
81. D-Q Xu. Photosynthetic efficiency. Plant Physiol Commun Sin 24:1–6, 1988.
82. KK Niyogi. Photoprotection revisited: genetic and molecular approaches. Annu Rev Plant Physiol Plant Mol Biol 50:333–359, 1999.
83. L-W Guo, D-Q Xu. Photoinhibition of photosynthesis in sweet viburnum leaves under natural conditions. Acta Phytophysiol Sin 20:46–54, 1994.
84. L-W Guo, D-Q Xu, Shen Y-K. Photoinhibition of photosynthesis without net loss of D1 protein in wheat leaves under field conditions. Acta Bot Sin 38:196–202, 1996.
85. SS Hong, D-Q Xu. Light-induced increase in initial chlorophyll fluorescence Fo level and its possible mechanism in soybean leaves. In: G Garab, ed. Photosynthesis: Mechanisms and Effects. Vol III. Dordrecht, The Netherlands: Kluwer Academic Publishers, 1998, pp 2179–2182.
86. D-Q Xu, Y-K Shen. Light stress: photoinhibition of photosynthesis in plants under natural conditions. In: M Pessarakli, ed. Handbook of Plant and Crop Stress. New York: Marcel Dekker, 1999, pp 483–497.
87. SS Hong, D-Q Xu. Difference in response of chlorophyll fluorescence parameters to strong light between wheat and soybean leaves. Chin Sci Bulletin 42:684–688, 1997.
88. SS Hong, D-Q Xu. Light-induced increase in initial chlorophyll fluorescence Fo level and reversible inactivation of PS II reaction centers in soybean leaves. Photosynth Res 61:269–280, 1999.
89. SS Hong, D-Q Xu. Reversible inactivation of PS II reaction centers and the dissociation of LHC II from PS II complex in soybean leaves. Plant Sci 147:111–118, 1999.
90. L-W Guo, D-Q Xu, Y-K Shen. The causes of midday decline of photosynthetic efficiency in cotton leaves under field conditions. Acta Phytophysiol Sin 20:360–366, 1994.
91. RKM Hay, AJ Walker. An Introduction to the Physiology of Crop Yield. New York: Longman Scientific & Technical copublished in the United States with John Wiley & Sons, 1989, pp 31–86.
92. H Medrano, AJ Keys, DW Lawlor, MAJ Parry, J Azcon-Bieto, E Delgado. Improving plant production by selection for survival at low $CO_2$ concentrations. J Exp Bot 46:1389–1396, 1995.
93. L-W Guo, D-Q Xu, Y-K Shen. Relation between photorespiration and photoinhibition in cotton leaves. Chin Sci Bull 41:415–420, 1996.
94. H-C Yin, T-D Wang, Y-K Shen, G-X Qiu, Y-Z Li, G-M Shen, S-Y Yang. Community structure and light utilization of wheat fields. Acta Agric Sin 10:381–397, 1959.
95. CC Mann. Crop scientists seek a new revolution. Science 283:310–314, 1999.
96. LT Evans. Crop Evolution, Adaptation and Yield. Cambridge: Cambridge University Press, 1993, pp 32–61.
97. SK Mukherjee. Green revolution—outlook of a horticulturist. In: NR Dhar, Convened. Symposium on Green Revolution. Allahabad, India: National Academy of Sciences, 1974, pp 141–145.
98. NR Dhar. Green revolution and vital factors in steady increase in crop production and atmospheric nitrogen fixation in the slow oxidation of organic matter by air and the protein problem. In: NR Dhar, Convened. Symposium on Green Revolution. Allahabad, India: National Academy of Sciences, 1974, pp 1–124.
99. S Peng, GS Khush, KG Cassman. Evolution of the new plant ideotype for increased yield potential. In: KG Cassman, ed. Breaking the Yield Barrier. Los Banos, Philippines: International Rice Research Institute, 1994, pp 5–20.
100. JY Lin, M Shen. Rice production constraints in China. In: RE Evenson, RW Herdt, M Hossain, eds. Rice Research in Asia. Progress and Priorities. Cambridge: CAB International, 1996, pp 161–178.
101. L-P Yuan, Z-Y Yang, J-B Yang. Hybrid rice in China. Paper presented to the Second International Symposium on Hybrid Rice. Los Banos: IRRI. 1992, pp 21–25.
102. SK Mukherjee. Pros and cons of green revolution. In: NR Dhar, Convened. Symposium on Green Revolution. Allahabad, India: National Academy of Sciences, 1974, pp 137–140.
103. SK Majumdar. Green revolution under limited nitrogenous fertilizer availability. In: NR Dhar, Convened. Symposium on Green Revolution. Allahabad, India: National Academy of Sciences, 1974, pp 193–197.
104. SP Raychaudhri. Role of plant diseases in green revolution. In: NR Dhar, Convened. Symposium on Green Revolution. Allahabad, India: National Academy of Sciences, 1974, pp 198–201.
105. R Ishii. Leaf/canopy photosynthesis and crop productivity. In: AS Raghavendra, ed. Photosynthesis: A Comprehensive Treatise. Cambridge: Cambridge University Press, 1998, pp 215–225.
106. C Ramasamy, TR Shanmugam, D Suresh. Constraints to higher rice yields in different rice production environments and prioritization of rice research in southern India. In: RE Evenson, RW Herdt, M Hossain, eds, Rice Research in Asia. Progress and Priorities. Cambridge: CAB International, 1996, pp 145–160.
107. RW Herdt. Summary, conclusions and implications. In: RE Evenson, RW Herdt, M Hossain, eds. Rice Research in Asia: Progress and Priorities. Cambridge: CAB International, 1996, pp 393–405.
108. CC Mann. Genetic engineers aim to soup up crop photosynthesis. Science 283:314–316, 1999.

109.  IE Woodrow, JA Berry. Enzymitic regulation of photosynthetic $CO_2$ fixation in $C_3$ plants. Annu Rev Plant Physiol Plant Mol Biol 39:533–594, 1988.

110.  G Bainbridge, P Madgwick, S Parmar, R Mitchell, M Paul, J Pitts, AJ Keys, MAJ Parry. Engineering Rubisco to change its catalytic properties. J Exp Bot 46:1269–1276, 1995.

111.  S Gutteridge, J Newman, C Herrmann, D Rhoades. The crystal structures of Rubisco and opportunities for manipulating photosynthesis. J Exp Bot 46:1261–1267, 1995.

112.  FR Tabita. Microbial ribulose 1,5-bisphosphate carboxylase/oxygenase: a different perspective. Photosynth Res 60:1–28, 1999.

113.  JW Riesmeier, J Kossmann, B Muller-Rober, L Willmitzer. Transgeneic plants with special emphasis on photosynthetic carbon metabolism. In: B Andersson, AH Salter, J Barber, eds. Molecular Genetics of Photosynthesis. Oxford: IRL Press, Oxford University Press, 1996, pp 160–179.

114.  RJ Spreitzer. Questions about the complexity of chloroplast ribulose-1,5-bisphosphate carboxylase/oxygenase. Photosynth Res 60:29–42, 1999.

115.  MSB Ku, S Agarie, M Nomura, H Fukayama, H Tsuchida, K Ono, S Hirose, S Toki, M Miyao, M Matsuoka. High-level expression of maize phosphoenolpyruvate carboxylase in transgenic rice plants. Nat Biotechnol 17:76–80, 1999.

116.  RE Hausler, M Kleines, H Ubrig, H-J Hirsch, H Smets. Overexpression of phosphoenolpyruvate carboxylase from *Corynebacterium glutamicum* lowers the $CO_2$ compensation point ($\Gamma*$) and enhances dark and light respiration in transgenic potato. J Exp Bot 50:1231–1242, 1999.

117.  S Ferrario-Mery, E Murchie, B Hirel, N Galtier, WP Quick, CH Foyer. Manipulation of the pathway of sucrose biosynthesis and nitrogen assimilation in transformed plants to improve photosynthesis and productivity. In: CH Foyer, WP Quick, eds. A Molecular Approach to Primary Metabolism in Higher Plants. London: Taylor & Francis, 1997, pp 125–153.

118.  N Galtier, CH Foyer, J Huber, TA Voelker, SC Huber. Effects of elevated sucrose-phosphate synthase activity on photosynthesis, assimilate partitioning, and growth in tomato (*Lycopersicon esculentum* var UC82B). Plant Physiol 101:535–543, 1993.

119.  N Galtier, CH Foyer, E Murchie, R Alred, P Quick, TA Voelker, C Thepenier, G Lasceve, T Betsche. Effects of light and atmospheric carbon dioxide enrichment on photosynthesis and carbon partitioning in leaves of tomato (*Lycopersicon esculentum* L.) plants over-expressing sucrose phosphate synthase. J Exp Bot 46:1335–1344, 1995.

120.  U Sonnewald. Modulation of sucrose metabolism. In: CH Foyer, WP Quick, eds. A Molecular Approach to Primary Metabolism in Higher Plants. London: Taylor & Francis, 1997, pp 63–79.

121.  E Monte, D Ludevid, S Part. Leaf C40.4: a carotenoid-associated protein involved in the modulation of photosynthetic efficiency? Plant J 19:399–410, 1999.

122.  RG Herrmann. Photosynthesis research: aspects and perspectives. In: B Andersson, AH Salter, J Barber eds. Molecular Genetics of Photosynthesis. New York: Oxford University Press, 1996, pp 1–44.

123.  M Vivekanandan, VC Saralabai. The use of transgenic plants to manipulate photosynthetic processes and crop yield. In: M Pessarakli, ed. Handbook of Photosynthesis. New York: Marcel Dekker, 1997, pp 661–669.

124.  J Peng, DE Richards, NM Hartley, GP Murphy, KM Devos, JE Flintham, J Beales, LJ Fish, AJ Worland, F Pelica, D Sudhakar, P Christou, JW Snape, MD Gale, NP Harberd. 'Green revolution' genes encode mutant gibberellin response modulators. Nature 400:256–261, 1999.

125.  PH Abelson, PJ Hines. The plant revolution. Science 285:367–368, 1999.

# 43

# Transpiration Efficiency: Avenues for Genetic Improvement

**G. V. Subbarao***

*Dynamac Corporation, Kennedy Space Center, Florida*

**Chris Johansen†**

*International Crops Research Institute for the Semi-Arid Tropics (ICRISAT), Andhra Pradesh, India*

## I. INTRODUCTION

In view of the increasing demand for water for nonagricultural uses (such as for urban and industrial uses), and also to rationally redeploy available water resources for more areas of crop production, it is important to optimize the use of water for crop production [1]. Agricultural research has a major responsibility to develop and use techniques and practices that will result in more effective use of water in farming systems. This involves improvement of water use efficiency (WUE), defined here as aerial dry matter production of a crop per unit of evapotranspiration (ET). Transpiration efficiency (TE) is a component of WUE, being aerial dry matter production per unit of water transpired by the crop. The difference between WUE and TE is important, as suppression of soil evaporation and transpiration by weeds can improve WUE without improving TE, which is a direct measure of the crop species performance. Plant attributes (canopy structure, rate of canopy development, etc.) and management means (manipulating plant population, optimizing planting dates, fertilizer management, etc.) can modify soil evaporative losses ($E_s$) relative to transpiration ($T$) and can therefore affect WUE to a greater extent than TE.

Generally, any means (either genetic or management) that promotes early canopy development and radiation interception will reduce $E_s$ and increase $T$ (as evaporational losses would be negligible once the canopy closes), often with little or no increase in total ET [2,3]. For example, in Syria, erect chickpea lines intercepted less solar radiation, thus permitting greater evaporative water losses during early growth, and consequently they had a lower WUE value than chickpea lines with a prostrate habit [4]. Similarly, leafless pea had a lower WUE than either semileafless or conventionally leafed types [5]. Leafless pea intercepts less radiation than semileafless or conventionally leafed pea and therefore the crop suffers greater $E_s$ losses. Fertilizer application can increase WUE [6], as it promotes greater leaf area development and reduces $E_s$ relative to $T$. In many legumes, a basal dose of nitrogen and phosphorus promotes the early growth rate and thus minimizes $E_s$ [3]. Other management options such as improving water delivery systems, nutrient management approaches, and improved cultural practices could enhance WUE by minimizing $E_s$.

Also, vapor pressure deficit (vpd) during the growing season plays a major role in determining the WUE. When other factors are nonlimiting, the cost of producing dry matter (in terms of water) would be

---

* *Current affiliation:* Japan International Research Center for Agricultural Sciences, Ibaraki, Japan.
† *Current affiliation:* Consultant in Agricultural Research and Development, Dhaka, Bangladesh.

much higher under high vpd (i.e., results in low WUE) compared with low vpd (i.e., results in high WUE) conditions. For instance, in Mediterranean environments, the seasonal WUE varies from 8.5 g/kg (g dry matter produced per kilogram of water evaporated or transpired) in midwinter to only 2.5 g/kg in midsummer [7]. Thus, management (by early planting, optimizing the plant population and fertility requirements, etc.) and genetic means (such as early vigor, rapid canopy development, cold tolerance, and tolerance to diseases such as *Ascochyta*) that would permit full canopy development and rapid dry matter accumulation during periods when the vpd is low would maximize WUE for the growing season. Early planting (i.e., winter planting) in Mediterranean climates usually allows rapid canopy development and dry matter production when the vpd is low and thus results in higher WUE of both dry matter production and grain yield [3,8].

However, once options for minimizing $E_s$ relative to $T$ are exhausted, further improvements in WUE are possible for a given crop only by genetically improving TE value of that crop. In water-limited environments, yield is a function of $T$, TE, and harvest index (HI) [9]. Increased production may result from increased TE if other components (i.e., $T$ and HI) are independent [10] and not affected. By reducing $T$ or by allowing more efficient use of transpirational water in photosynthesis, available soil moisture could be better rationed during the cropping period, which should increase productivity [9].

Plants lose water as they fix carbon dioxide ($CO_2$) from the air. The loss is inevitable because it is necessary for $CO_2$ to dissolve in water in order to become available for photosynthesis [11]. This would lead to evaporation as the wet cell surface inside the leaf is exposed to the atmosphere. $CO_2$ diffuses down a concentration gradient to the leaf interior and water diffuses outward along a decreasing humidity gradient [11]. The lower the external humidity, the higher will be the evaporation when all the other factors are constant. This two-way diffusion of $CO_2$ and water forms the basis of improving TE [11]. Cultivars with improved TE are those with inherent characteristics that will allow increased production of dry matter per unit of water transpired [12]. This chapter focuses on exploring the opportunities for genetic improvement of the various morphological, physiological, and biochemical factors that determine TE in $C_3$ crop plants and assesses the scope for exploiting this trait in plant breeding programs.

## II.  FACTORS AFFECTING TE

Transpiration efficiency is a function of both environmental and plant attributes related to resistances to $CO_2$ fixation by leaves. Under some circumstances, the environment can have a significant influence on TE. Variation in humidity and temperature can influence TE [13]. TE is governed by three factors: (1) the vpd between air and leaf, (2) the $CO_2$ gradient from the air to the leaf, and (3) the diffusion resistances for both $CO_2$ and water [14]. The first factor is mainly abiotic, although the surface temperature of the leaf will actually respond to the atmosphere (e.g., radiation and vpd). The last two factors are largely plant-controlled factors. Also, incident irradiance has an important effect on TE [15]. There is an optimum irradiance for maximum efficiency of water use that is usually less than the irradiance incident upon a leaf [16] (see Sec. II.C for further discussion of this aspect).

A variety of morphological, anatomical, physiological, phenological, and biochemical processes enable crop plants to regulate and ration water for production of dry matter and yield in a given agroecological production system. These are discussed in the following.

### A.  Stomatal Behavior

Stomata may exert relatively greater control on water loss than that exerted by $CO_2$ uptake. This is because the rate of biochemical reactions involved in $CO_2$ assimilation ($A$) influences removal of $CO_2$ from cell solutions and thereby affects $CO_2$ gradients [17]. This is in addition to resistances faced by $CO_2$ in its transport, with stomatal resistance perhaps being a smaller component of the total resistance for $CO_2$ than for water [17]. Stomatal aperture plays a key role in maintaining the balance between taking up $CO_2$ and losing water [18]. Stomatal movements are the most rapid means by which plants can adjust to changes in the environment [18]. In particular, stomata respond directly to ambient humidity [19], thereby strongly influencing plant TE.

For $C_3$ crop plants, optimization of TE normally requires midday stomatal closure [13]. Such behavior has been observed frequently and is at least partly attributable to the effect of water deficit [20] or is a direct stomatal response to vpd [21]. If diurnal variation in a natural environment were regular and predictable, optimization would require only an appropriate circadian rhythm for stomatal movement [18]. However, this is usually not the case, and therefore optimization requires that the plant respond directly to the changing environment [18]. This demands that stomata respond to changes in external environmental conditions, which in turn influences rates of $T$ and $A$. Thus, stomata should be capable of controlling gas exchange by a feed-forward process, making it possible for $T$ to decrease when environmental changes tend to enhance the rate of $T$ (e.g., under high vpd) or for intercellular partial pressure of $CO_2$ ($P_i$) to increase when environmental changes would tend to enhance $A$ [22].

Reduced stomatal aperture increases TE because the rate of $A$ is reduced proportionately less than $T$ [23–25]. This often happens when plants are subjected to moderate levels of water stress. Factors such as osmotic adjustment (OA) can significantly influence stomatal aperture and thus determine TE under moisture stress. For example, the critical leaf water potential for stomatal closure varies with the level of OA [26,27]. Crop plants show genetic variation for stomatal characteristics such as stomatal density, aperture size, opening patterns, and sensitivity to changes in internal plant water status and soil water status [28–31]. This, in turn, affects their ability to regulate and optimize water use [32,33]. The existence of genetic variation in stomatal characteristics suggests that it may be possible to develop cultivars that utilize water more efficiently, thus contributing to their adaptation to moisture limiting environments [34,35].

## B. Canopy Structure

The aerodynamic resistance of a crop can play a role in determining the relative importance of stomatal conductance ($g_s$) to TE. If the canopy resistance to heat and water vapor diffusion is large, an increase in $g_s$ would tend to cool and humidify the air in the boundary layer, thus lowering the leaf-air vpd; TE would then increase [36,37]. Thus, cultivars with greater $g_s$ could assimilate more at the same level of TE [22,38]. Under field conditions, the boundary layer that forms over crop canopies could cause gas exchange to be less dependent on $g_s$ and is thus one of the important factors affecting TE [39].

A plant with high TE may be able to decrease the aerodynamic conductance of its canopy boundary layer through greater rigidity of the canopy while maintaining a high $g_s$ [40]. This provides it with ready access to $CO_2$ within the canopy, which is not depleted compared with the bulk atmosphere, while retaining water vapor within the canopy. Boundary layer resistance is a function of the thickness of the unstirred air boundary layer adjacent to the leaf, which in turn is determined by the leaf size [41]. Smaller leaves have a thinner unstirred boundary layer [41]. Thus, boundary layer resistance at the canopy level depends on canopy architecture, which is determined by leaf size, leaf arrangement, growth habit (i.e., prostrate versus erect), and height of the canopy. With a low canopy conductance, leaf water equilibrates with an adjacent air space of higher humidity than the bulk atmosphere [40]. However, such canopy structure may create sufficiently high levels of humidity within the canopy to be conducive to fungal disease development, thus negating the positive effects of higher TE on biomass production or yield. For instance, in chickpea the closed canopy types, which have greater WUE than open canopy types [4], also provide a conducive microenvironment for the development of *Botrytis* and *Ascochyta* blight diseases [42]. Thus, the positive effects of such closed canopies on improving TE of a crop and its production would depend on the availability of sources of resistance to such diseases, which could be incorporated into cultivars forming closed canopies if they lack disease resistance.

## C. Leaf Movements and Surface Reflectance

Incident radiation is completely absorbed by the canopy once 100% ground cover is achieved and the incident energy is partitioned between $T$ and $A$ [11]. The proportional allocation differs between species and climates and from year to year [43]. The optimum irradiance for maximum TE is usually less than the irradiance incident upon a leaf oriented normal to the sun's rays [16,44,45]. This is mainly because $T$ normally shows a positive relationship (linear or curvilinear) with increasing irradiance (due to rising leaf temperature and falling stomatal resistance), while $A$ shows a downward curvilinearity with increased irradiance [7]. Leaf movements and surface reflectance provide a means of optimizing this radiation load on the leaf for the maximization of TE. This can be particularly advantageous in water deficit environments, to dissipate the energy as latent heat, minimize heat damage, and optimize TE and radiation use efficiency (RUE) [46–49]. The main advantage of leaf movements is that they would allow maximum exposure of leaf area to direct radiation when evaporative demand is low and thus improve TE. Almost all crop plants show some degree of leaf movement in response to radiation, soil, and plant water status. However, the degree of leaf movement and the threshold soil and plant water status that triggers these movements vary among and within crop species, which could contribute to their growth performance in water-limited environments [32,50–52].

Leaf pubescence and surface reflectance can provide additional means of controlling leaf temperature and water balance, apart from stomatal control and leaf movement [53–55]. In near-isogenic lines of soybean, it was shown that lines with pubescent leaves had significantly lower $T$ than either normal or glabrous isolines [53,56]. Leaf pubescence in *Encelia farinosa* reduced absorbance of irradiance as much as 56% compared with the nonpubescent plant *E. californica* [57]. This reduced absorbance can result in lower leaf temperatures and lower $T$ [58]. However, leaf hairs can reflect radiation, which may reduce $A$. Nevertheless, it appears that in climates with high irradiance and temperatures, beneficial effects of reduced leaf temperature would more than counterbalance the effect of decreased light on $A$ [59]. Other morphological features such as cuticle thickness and wax deposits on the leaf surface can to some extent control evaporational losses from the leaf surface [60–63]. There is genetic variability in a number of crop species for leaf surface wax levels and cuticle thickness [61–63].

## D. Specific Leaf Area

Variation in TE in crop plants can result from changes in water vapor flux through stomata or changes in photosynthetic capacity [29,64]. In wheat, variation in TE is caused by stomatal mechanisms [29,65], whereas in groundnut it appears to be caused by variation in photosynthetic capacity [64,66]. Genotypic variation in photosynthetic capacity on a unit leaf area basis has been observed in many crops [67,68], and a significant negative correlation has been shown between photosynthetic capacity and specific leaf area [69]. This evidence suggests indirectly that the basis of variation in TE through specific leaf area (i.e., leaf thickness) may result from differences in photosynthetic capacity on a unit leaf area basis (see Sec. V.B for more discussion of this).

## E. Root Systems

Root distribution, density, and resistance can influence water use in space and time. Thus, WUE can be affected by the rate of growth and spread of roots, particularly during early stages of crop growth. In receding residual moisture situations, profligate water use during early crop growth might lead to water deficit conditions during reproductive growth stages. In such circumstances, induction of a large resistance within the plant to the flow of water through selection for smaller metaxylem vessel diameters in the seminal roots should change the pattern of water use for different growth phases [70,71]. Thus, the same amount of water can be transpired to produce more grain yield. Selection for increased root resistance has been shown to be amenable to genetic manipulation in cereals [72,73]. Differences in root radial resistance to water flux have been suggested to occur among groundnut genotypes [74].

## III. ASSESSMENT OF GENOTYPIC DIFFERENCES IN TE

Measurement of $T$ in the field is quite complex [75]. Even the field measurement of ET is difficult in many situations where drainage from the root zone, water uptake from saturated zones, and runon and runoff from the area are difficult to measure both temporally and spatially. Transpiration is usually estimated from evapotranspiration measurements such as by (1) subtraction of an estimate of soil evaporation ($E_s$), which is often a seasonal constant, from the measured seasonal ET [76]; (2) daily water balance simulation using empirical functions to calculate $T$ separately from daily calculations of ET, using measured plant parameters such as leaf area index (LAI) or ground cover [77,78]; or (3) measuring $E_s$ and subtracting it from measurements of ET [79]. All of these measurement techniques, however, result in indirect estimates of $T$. Direct long-term estimates of TE require accurate measurements of the water used. Rates of water movement through plants can be measured by using heat-pulse velocity techniques [80], but difficulties in volume calibrations have limited the accurate estimation of transpiration flux. However, improvements in heat-pulse instrumentation have reduced the calibration problems [81,82]. Technical problems related to data collection limit the number of plants that can be measured using this technique.

This limits its use in genetic improvement programs where large numbers of plants and genotypes need to be characterized. Pot experiments can give reliable estimations of TE as they allow accurate measurement of $T$ and dry matter production, including roots. However, these experiments are extremely laborious and not realistically applicable to screening germ plasm or to genetic studies associated with cultivar improvement [83].

Assessment of genetic variation in TE has often been based on instantaneous measurements of $CO_2$ fixation and $T$ from single leaves [84]. However, both of these processes vary markedly during the day and according to leaf and plant age. Thus, these instantaneous measurements do not integrate performance throughout the life of the plant. Also, these instantaneous measurements of TE cannot assess the impact of morphological or physiological adaptations to drought that may influence season-long TE and plant performance under water-limited conditions [85,86]. Further, these measurements have large coefficients of variation and are thus usually not suitable for screening and selection studies [87]. It is therefore apparent that breeding for improved TE has been constrained by difficulties in measuring TE on a large number of plants under field conditions [88]. Selection criteria and methods are therefore needed that are efficient and can be used at least indirectly to select genotypes with high TE from large populations in the field.

# IV. CARBON ISOTOPE DISCRIMINATION (D) AND ITS RELATION TO TE

## A. Theoretical Background

Carbon occurs naturally as two stable isotopes, $^{12}C$ and $^{13}C$. Most of the carbon is $^{12}C$ (98.9%), with 1.1% being $^{13}C$. As the $^{12}C$ isotope is lighter than $^{13}C$, $^{12}CO_2$ diffuses faster than $^{13}CO_2$. Ribulose 1,5-bisphosphate carboxylase (Rubisco) fixes the lighter isotope faster, thus discriminating against the heavier isotope $^{13}C$ [89]; these two effects cause the $^{13}C/^{12}C$ ratio to be lower in plants than in the ambient atmosphere. The link between TE and $^{13}C/^{12}C$ discrimination ($\Delta$) is via the gas exchange characteristics of the leaves [90]. Because the isotopes are stable, the information inherent in the ratio of abundance of the carbon isotopes ($^{13}C/^{12}C$) is invariant [90]. The extent of discrimination against the naturally occurring stable isotope $^{13}C$ during photosynthetic $CO_2$ fixation in $C_3$ plants is determined largely by the ratio of the intercellular to atmospheric partial pressure ($P_i/P_a$) of $CO_2$ [83,90]. As Rubisco actively discriminates against $^{13}CO_2$ [36], $^{13}CO_2$ is concentrated relative to $^{12}CO_2$ in the intercellular spaces as $P_i$ decreases. This concentrating effect results in Rubisco fixing an increased proportion of $^{13}C$ relative to $^{12}C$, and $\Delta$ decreases. This is reflected in the carbon isotope ratio of $C_3$ plants, which shows a $^{13}C$ value of around $-25\%$ [38]. Therefore, $\Delta$ normally correlates positively with $P_i/P_a$ in $C_3$ plants and not in $C_4$ plants (Figure 1), where Rubisco plays a relatively minor role in overall $CO_2$ fixation. Thus, according to theory, in $C_3$ plants a lower $^{13}C$ discrimination is associated with a higher TE. Variation exists among $C_3$ crop species in their photosynthetic rates ($A$). This leads to variation in $P_i/P_a$, and is reflected in $^{13}C$ discrimination values ranging from $-22$ to $-40\%$, depending on the crop species [91]. For $C_4$ crops, which have a higher TE than $C_3$ crops, $^{13}C$ discrimination values range from $-9$ to $-19\%$; however, these lower values are due mainly to the alternative pathways of $CO_2$ fixation in $C_4$ crops, such as phosphoenolpyruvate (PEP) carboxylase, which does not discriminate between $C_{13}$ and $C_{12}$ [91].

The carbon isotope ratio ($\delta^{13}C$) can be calculated by comparing the $^{13}C$ to $^{12}C$ composition of a sample ($R_{sample}$) relative to the Pee Dee belemnite (PDB) standard ($R_{PDB}$).

$$\delta^{13}C_{sample} = \left( \frac{R_{sample}}{R_{PDB}} - 1 \right) \times 1000 \tag{1}$$

These $\delta^{13}C$ values can be used to calculate isotope discrimination ($\Delta$), as described by Farquhar and Richards [29] and Hubick et al. [64].

$$\Delta = \frac{\delta^{13}C_{air} - \delta^{13}C_{sample}}{1 + \delta^{13}C_{sample}/1000} \tag{2}$$

The absolute isotopic composition of a sample is not easy to measure directly; the mass spectrometer measures the deviation of the isotopic composition of the material from the standard.

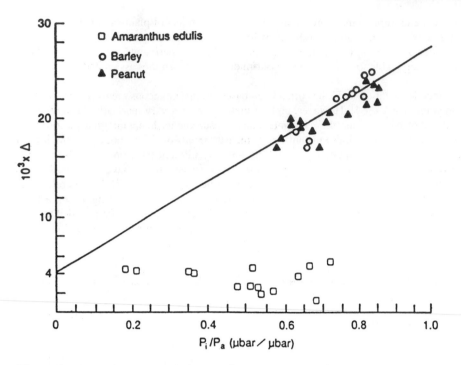

**Figure 1**   Carbon isotope discrimination, $\Delta$, versus the ratio of intercellular and ambient partial pressures of $CO_2$ ($P_i/P_a$) when both are measured simultaneously in a gas exchange system. Peanut and barley are $C_3$ species and *Amaranthus edulis* is a $C_4$ species. (From Ref. 38.)

$$\delta P = \frac{R_p - R_s}{R_s} = \frac{R_p}{R_s} - 1 \tag{3}$$

$$\delta a = \frac{R_a - R_s}{R_s} = \frac{R_a}{R_s} - 1 \tag{4}$$

where $\delta P$ is the carbon isotope composition of the plant sample, $\delta a$ the carbon isotope composition of air, $R_s$ the molar abundance ratio of $^{13}C/^{12}C$ of the standard, and $R_p$ and $R_a$ the molar abundance ratios of $^{13}C/^{12}C$ of the plant sample and air, respectively.

The reference material in determinations of carbon isotope ratios has traditionally been in $CO_2$ generated from a fossil PDB. The carbon isotope composition ($\delta$) is standardized against PDB; atmospheric $CO_2$ has a value of $-8‰$ relative to PDB [92].

The carbon isotopic technique can also be used to quantify internal $CO_2$ levels of leaves on a long-term basis. Internal $CO_2$ levels ($C_i$) represent a balance between $A$ and $T$. The existence of variation in $C_i$ confirms the existence of genotypic differences in TE. Carbon isotope discrimination and TE are related through independent relationships with $P_i/P_a$ [10]. This depends to different extents on the way in which plants coordinate leaf conductance to water vapor with the capacity for photosynthetic $CO_2$ uptake. Variation in coordination of leaf $g_s$ and $A$ can give rise to variation in $P_i/P_a$ [10]. This, in turn, results in variation in TE and carbon isotope discrimination. It has been stated that if plant breeding is to affect detectable changes in TE of dry matter production, $(1 - P_i/P_a)$ needs to be modified substantially [93]. In theory, greater TE will be associated with low $\Delta$ if the leaf-to-air vpd remains constant [10].

Farquhar et al. [90] have suggested that $\Delta$ can be expressed based on gas exchange as follows:

$$\Delta = a \frac{P_a - P_i}{P_a} + b \frac{P_i}{P_a} = a + (b - a)\frac{P_i}{P_a} \tag{5}$$

where $a$ is the fractionation due to diffusion in air, which is about $-4.4‰$ [94]; $b$ the net fractionation caused by carboxylation, which is about $-27‰$ [29]; and $P_a$ and $P_i$ are the ambient and intercellular partial pressures of $CO_2$, respectively.

The significance of $b$ in Eq. (5) is that when $g_s$ is small in relation to $CO_2$ fixation, $P_i$ is small and $\Delta$ tends toward $a$ ($-4.4‰$); when conductance is comparatively large, $P_i$ approaches $P_a$, and $\Delta$ approaches $b$ ($-27‰$ to $-30‰$, i.e., becomes more negative) [90]. Thus, $^{13}C$ discrimination measurements should be useful in studying the genetic control of $g_s$ in relation to $A$. Measurements of $\Delta$ in $C_3$ crops may contribute to selection for TE. Theory [90] and supporting empirical evidence have shown that differences in intrinsic TE were associated with $\Delta$ in a range of crops [10,29,64,66,86,95,96].

The instantaneous ratio of $CO_2$ assimilation rate of a leaf ($A$) to its $T$ is given approximately by

$$\frac{A}{T} = \frac{P_a - P_i}{1.6v} \tag{6}$$

where $v$ is the difference in partial pressure of water vapor between the intercellular spaces and the surrounding air. The factor 1.6 is the ratio of the diffusivity of water vapor and $CO_2$ in air [36].

Farquhar et al. [36] suggested that Eq. (6) may be rewritten as

$$\frac{A}{T} = \frac{P_a\left(1 - \dfrac{P_i}{P_a}\right)}{1.6v} \tag{7}$$

Equation (7) emphasizes that a small value of $P_i/P_a$ would result in an increase in TE for a constant vpd. Selecting for lower $P_i/P_a$ thus should equate with selecting for greater TE [36]. Therefore, the carbon isotope composition ($^{13}C/^{12}C$) of $C_3$ plant tissues provides a long-term integrated measure of photosynthetic capacity [97].

To account for losses of carbon and water due to metabolic and physical processes, Farquhar et al. [36] modified Eq. (7) to describe the molar ratio, $W$, of carbon gain by a plant to water loss:

$$W = \frac{P_a\left(1 - \dfrac{P_i}{P_a}\right)(1 - \phi c)}{\phantom{xxxxxxx}} \tag{8}$$

where $\phi c$ is the proportion of carbon lost due to respiration and $\phi w$ is the proportion of water lost other than through stomata (cuticular transpiration, etc.).

The presence of vpd ($v$) in Eq. (8) suggests that TE is affected by environment as well as by physiological responses of the plant [38]. Thus, $v$ can vary because of alterations in canopy interception and absorption of radiation via changing leaf angle and surface reflection properties (see Sec. II.C for more details) and increases or decreases in their coupling to ambient temperature by decreasing or increasing leaf size, respectively.

Equation (8) also explains that TE is likely to be more affected than $\Delta$ by processes independent of those resulting in variation in $P_i/P_a$ [10]. For example, genetic differences in respiratory losses of carbon and nonstomatal water losses such as cuticular transpiration may affect TE independently of $P_i/P_a$ [10]. Thus, Eqs. (8) and (5) can be combined to show that $\Delta$ is largely dependent of $P_i$ and vpd. Plants with higher TE will therefore show less negative $^{13}C$ values or lower $\Delta$ values, giving a negative correlation between TE and $\Delta$ [36]. This theoretical relationship between $\Delta$ and TE in plants with a $C_3$ photosynthetic pathway has been confirmed for several crops in pot [10,29,64,66,83,98–100] and field experiments [74,96,101] (Figure 2).

## B. Water Deficit and TE

The degree of stomatal closure induced by water stress depends on the level of stress and the ability of the crop to meet evapotranspirational demands [102]. Direct measurements of TE using whole plant carbon and water balances have shown that moderate drought can cause an increase in TE of up to 100%, whereas extreme drought could substantially decrease TE [103]. A common response to water stress is a simultaneous decrease in $A$ and $T$ and an increase in leaf temperature [104]. If $T$ decreases faster than $A$, then $P_i$ will decrease [24,105]. This response results in water savings to the plant and a subsequent increase in TE. As Rubisco discriminates against $^{13}CO_2$, the proportion of $^{13}CO_2$ to $^{12}CO_2$ also increases within the leaf. Thus $^{13}CO_2$ discrimination decreases as stress becomes more pronounced [106]. In long-term observations in both growth chamber and field conditions, plants under water deficit had lower $P_i$ as indicated by $^{13}C$

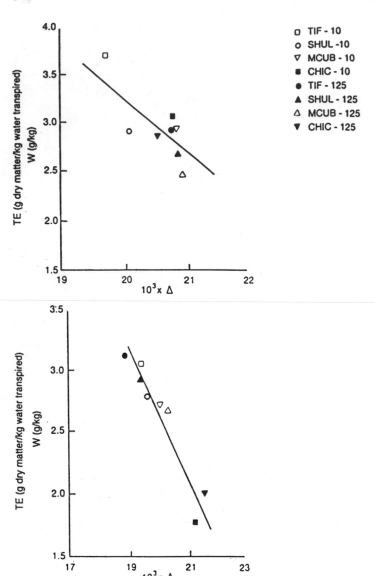

**Figure 2**   Relationship between transpiration efficiency (g dry matter per kilogram of water transpired) and carbon isotope discrimination ($\Delta$) under well-watered and moisture-deficient conditions for a range of peanut cultivars grown in field conditions. (From Ref. 74.)

discrimination analysis [29,107–110]. Several studies with a number of crop species have shown that moderate water stress leads to an increase in TE as indicated by their level of $^{13}C$ discrimination ($\Delta$) [88,98,111,112]. Water stress resulted in about 2‰ lower $\Delta$ compared with well-irrigated plants of chickpea [107]. Similarly, for cowpea (*Vigna unguiculata*), it was shown that leaves sampled from field-grown plants in a dry environment had about 1.5‰ lower $\Delta$ than plants from irrigated conditions [113].

Under severe water deficit, TE is reported to decrease [103]. This is because leaves become less efficient with respect to water and $CO_2$ exchange; water can still be lost through the cuticle but $CO_2$ entry through stomata is severely restricted, thus causing reduced TE [18]. In groundnut, the relation between $\Delta$ and TE can break down under severe drought conditions, which could be related to increased respiratory losses of carbon [74]. A similar response has been reported for sunflower [100]. Respiratory losses of carbon can be as much as 40% under severe drought conditions [103].

## C.  Influence of Crop Canopy on Δ and TE

The negative relationship between Δ and TE might hold for individual plants in pots [64] or for small plots in the field [66,74] or field-grown crops [96] but might become inconsistent when results are extended to a large area, depending on the crop and microclimate [36]. First, the microclimate in field canopies is usually different from that of isolated plants in pots. This could lead to potential differences in stomatal control of $T$ as influenced by environmental factors and thus to a breakdown in the relationship between TE and Δ. This emphasizes the problem in the field, where the aerodynamic resistance of the crop has to be taken into account if the canopy and leaf boundary layer resistances to energy flux are very large [38,74]. Because of this, it is possible that under high atmospheric evaporative demands, plants can have a high $g_s$, and thus a high Δ, but also high TE because of complete closure of the canopy [114]. However, this is less likely to occur when crops have small leaf area index (LAIs), as would be the case under conditions where stress occurs early in the cropping season, because under these conditions the crop is more closely coupled to the atmosphere [39,114]. However, if the source of variation in Δ is the capacity for photosynthesis, the effects of boundary layers are unimportant [114]. as seems to be the case for groundnut [10,74]. Therefore, at the crop level, identification of the causes underlying differences in Δ may become important.

Second, the nonstomatal loss of water (i.e., cuticular transpiration, soil evaporation) ($\varphi w$) could vary with leaf area development and the level of wax deposition on the cuticle and thus is not an independent fixed proportion of transpiration. This could influence the Δ as $\varphi w$ is an important component of WUE [Eq. (8)]. Also, because vpd is an important component of Eq. (8), any fluctuation in vpd during the growing season and the growth rate of a given variety during the growing season could influence TE. For example, the genotypes that grow faster when vpd is small because of their adaptation to low temperatures could show a greater TE for the same Δ.

## V.  SCOPE FOR GENETIC IMPROVEMENT OF TE IN C₃ CROP PLANTS

## A.  Relation Between Transpiration and Photosynthesis

Because the stomatal diffusion pathway is the same for both water vapor and $CO_2$ exchange, water is inevitably lost when stomata open and $CO_2$ is absorbed. Stomatal conductance is believed to adjust according to the assimilatory capacity of the mesophyll tissue [115]. That is, other factors being similar (i.e., nonlimiting), stomata open to the extent required to provide $CO_2$ at rates sufficient to meet the $CO_2$ fixation requirements of the metabolic pathway [116]. Close coupling between $A$ and $T$ is expected because $CO_2$ and $H_2O$ simultaneously move through the stomata [117]. The diffusive conductance of the stomatal opening imposes a major control on the rates of both processes, although the $C_i$ concentration and the external water vapor concentration determine the magnitude of the respective gradients [117]. However, changes in $g_s$ may not necessarily affect $T$ and $A$ similarly [24].

There is a strong correlation between $A$ and $g_s$ over a wide variety of plant species and under a diversity of environmental conditions [116,118]. This implies some level of regulation between $CO_2$ demand by chloroplasts and $CO_2$ supply, via stomatal control. Generally, leaf conductance and photosynthesis are correlated at low conductance levels but are uncoupled at high conductance levels [119]. If there is no deviation from the slope of photosynthesis versus conductance relationships, and if the intercept is zero (as is assumed initially), then $P_i$ values of all crop plants should be constant, dependent only on photosynthetic pathway [85]. Although many studies have shown a significant tendency for photosynthesis and conductance to be correlated [116,120], many of these data sets exhibit some deviation from a linear relationship or nonzero intercept [121,122].

Genotypic variation in TE can result from variation in $g_s$ but with the genotypes having the same level of photosynthetic capacity [57]. The slopes of the regression line of $g_{max}$ (stomatal conductance maximum) versus $A_{max}$ vary substantially among C₃ plants [57,123]. For high evaporative environments, it has been shown that genotypic differences in $P_i$, based on long-term gas exchange studies as well as on ¹³C discrimination analysis, offer the possibility of genetically modifying TE [57]. However, for low evaporative environments, it appears that $A$ is highly dependent on leaf $g_s$, suggesting little possibility of improvement of TE [57].

## B. Mechanisms by Which Genotypes Differ in TE

Any factor that influences genetic variation in either $g_s$ or $A$ in a disproportionate manner would influence $\Delta$ and thus TE [38]. If variation in $A$ was the only cause of variation in $P_i$, increasing photosynthetic capacity should lower $P_i/P_a$ and therefore lower $\Delta$. In this situation, TE would increase and the relationship between $\Delta$ and plant biomass should be negative [124]. In groundnut, differences in $A$ are reported to be largely responsible for TE variation, as dry matter production is negatively correlated with $\Delta$ in pots [10,74] and at the canopy level [66,96]. Significant variation in $A$ per unit leaf area has been reported in groundnut genotypes and there is also heterosis for this trait [68,125–127]. In cowpea, genotypic means for TE were positively correlated with $A$ but only weakly correlated with $g_s$, indicating that genotypic differences in TE were mainly due to differences in $A$ [112]. Similarly, in sunflower, tomato, and wheat genotypic differences in TE were due to differences in $A$ [128,129].

A strong positive correlation has been observed between $\Delta$ and specific leaf area (SLA) among groundnut genotypes [101,130,131]. This is consistent with the foregoing hypothesis that high TE genotypes have higher $A$. Indeed, the genotypes with thicker leaves (low SLA) had significantly higher leaf nitrogen contents, again indicative of higher photosynthetic capacity. The significant application of these observations is that breeders could use the inexpensively measured SLA, in lieu of $\Delta$, to screen for high TE among groundnut genotypes within specific environments [74].

However, if $g_s$ is the main source of variation in $P_i/P_a$, greater $g_s$ should increase $P_i/P_a$ and therefore increase $\Delta$. In adequately irrigated coffee, higher TE values of some of the genotypes tested were associated with reduced stomatal aperture rather than increased $A$ at a given $g_s$ [99]. This suggests that high TE may restrict yield when water supply is not limiting. Thus, in this case, as in wheat, selection for higher $\Delta$ could lead to increased biomass production but with decreased TE [132]. For example, in crested wheatgrass, greater TE in low $\Delta$ clones resulted from a proportionately greater decline in $g_s$ than in $A$ [106]. Similar results were reported for chickpea [107]. However, variation in $P_i/P_a$ among wheat genotypes is approximately equal to variation in leaf $g_s$ and in $A$ [65,133–135]. In wheat, it was reported that $g_s$ covaried with $A$, with the change in $g_s$ being relatively greater [135]. This means that there could be a positive correlation between $A$ and $P_i/P_a$. The effect of this on growth may be compounded if genotypes with large $P_i/P_a$ partition more carbon into shoots [136].

Cultivar differences in $\Delta$ may also result indirectly from genetic variation in root characteristics affecting the level of water stress experienced by the canopy [98,137]. Differences in root growth affect the degree of dehydration postponement, and this could prolong gas exchange activity and the maintenance of relatively high $P_i$ and thus $\Delta$ [137].

## C. Genetic Variation and Genetics of TE and $\Delta$

Genetic variation in TE and $\Delta$ has been reported in wheat [65,124,132,138], barley [95], tomato [86], sunflower [100], chickpea [107], groundnut [64,66,74,131], cowpea [88], alfalfa [139], and coffee [99]. In wheat, variation in $\Delta$ among genotypes is typically around $2 \times 10^{-3}$ [138]. This is equivalent to a variation in TE of 59% [138]. In groundnut, genotypic variation in TE is estimated as about 65% [64]. Based on extreme cases of genotypes that differ in TE, it was reported that cowpea genotypes such as vita 7 and 8049 had nearly 67% higher TE values than those of other genotypes tested [111]. Also, earliness is generally associated with low TE in cowpea; however, significant genotypic differences were noticed within any given maturity group, suggesting that these two traits are not necessarily linked [111]. Similarly, tall landrace genotypes of wheat, which are also late maturing, had higher TE than did the modern dwarf and semidwarf genotypes [124]. However, among Australian wheats, low values of $\Delta$ and thus high TE have been found to be strongly associated with the WW15 genetic background, which was introduced into Australia from CIMMYT as a major source of the dwarfing gene in Australian wheat.

The utility of a trait for selection in plant breeding programs is strongly enhanced by the consistency of genotypic ranking across environments [112]. Based on studies with wheat, cowpea, crested wheat grass, groundnut, and beans, it was found that genotypic ranking for $\Delta$ across environments is consistent [37,38,101,109,111,112,132,138,140]. For crops such as groundnut, it was shown that genotypic ranking for $\Delta$ was maintained during ontogeny [74] (Figure 3). However, in crops such as wheat, genotypic ranking could change between the early vegetative stage and the heading and grain filling

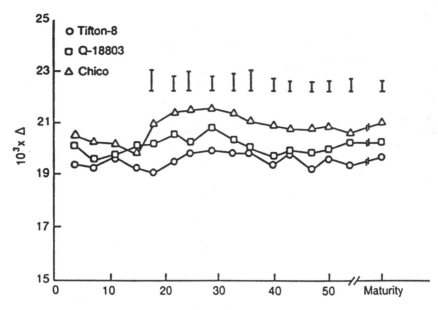

**Figure 3**  Change in carbon isotope discrimination in leaves and error variation versus time for well-watered groundnut cultivars of Tifton-8, Q 18803, and Chico grown in a greenhouse. (From Ref. 74.)

stages [138]. This could be due to a number of factors, including hormonal imbalance, causing loss of stomatal control on water loss after heading. Also, the plant material used for $\Delta$ analysis could determine the level of heritability [138]. It was shown in a number of crops that the $\Delta$ of leaf material is a better indicator of differences in TE than that of grains [10,37,64,109,111,124]. One of the main reasons could be genotypic differences in the ability to translocate preanthesis-stored carbohydrate reserves for grain filling [141].

The effectiveness of indirect selection for TE using $\Delta$ will partly depend on the magnitude of the heritabilities for TE and $\Delta$ and the genotypic correlation between these characters [142]. Broad-sense heritability, which is the proportion of total phenotypic variance that is attributable to genotypic differences, is a measure of the repeatability of the expression of those genotypic differences [138]. In many crops, heritabilities for $\Delta$ are above 80% [10,38,109,111,124,138,139].

## D.  Advantages of Using $\Delta$ for TE Evaluations

Breeding for improved TE has been limited by the lack of screening tools for identifying desirable genotypes under field conditions [112]. The $^{13}C$ discrimination technique makes it possible to survey a large number of plants with a simple, albeit expensive, analysis of the leaf tissue [11]. As $\Delta$ provides an integrated estimate of TE, it has been suggested that measurement of $\Delta$ may better differentiate among genotypes than most instantaneous physiological assays [124]. Genotypic ranking based on $\Delta$ is much more consistent than that based on gas exchange measurements [112] and thus should be easier to select for in breeding programs. Also, as $\Delta$ remains reasonably constant throughout crop ontogeny, selection could be made during crop development [74].

Further, $\Delta$ is faster and easier to measure than total growth relative to total water use [29]. It is readily determined on field-grown plants because it does not require the plant to be sheltered from rain or that any other special experimental treatment be maintained. Measurements can be made on small plant samples collected at maturity with minimal problems of storage and handling. The material can be either leaf, stem, or grain. Leaves and stems are easier to grind, and use of vegetative material has the potential advantage that selection can be made early in the crop growth cycle and thus could assist in improving selection efficiency and reducing the time and maintenance costs [29,138].

## E.   Limitations of Using Δ to Select for TE

Carbon partitioning and Δ would not be expected to be stable across all environments and with changes in plant hormonal balance. For example, cytokinins and abscisic acid (ABA) can affect both leaf gas exchange and carbon allocation [106]. Also, there are some problems of assessment of TE through carbon isotope estimations: (1) it is a ratio and not directly correlated with yield or productivity, (2) the small sample size may introduce subsampling errors and careful grinding is required, and (3) the technique requires considerable capital investment in equipment and technical expertise [106].

Also, there are a number of potential sources of nongenetic variability in the measurement of Δ. Some can be readily overcome by technical or sampling precautions, as they are associated with the composition of plant dry matter [143] and the size and storage of the dry matter sample used in the measurement [38]. Other sources of variation in Δ among plant organs result from temporal variation in the growth environment. Increased salinity [144,145], decreased soil water availability [29,66,108], soil compaction [136], and a decrease in vpd [146] could all result in lower values of Δ.

Genotypic variation for Δ measured under field conditions could be complicated by inherent differences in root growth [137]. This would affect the degree of dehydration postponement that could allow prolonged maintenance of relatively large $g_s$, thus decreasing TE but increasing growth and yield. Positive correlations between root length density and Δ have been reported in crops such as beans [137,147] (Fig. 4), and thus selection for low Δ (high TE) may lead to selection of genotypes with poor root attributes, such as shallow rooting and low root densities. Bean genotypes that had a deeper root system had high Δ values compared with the shallow-rooted genotypes [137]. Thus, leaf physiology (as measured by

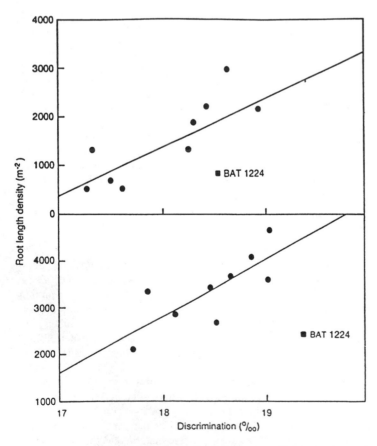

**Figure 4**   Relationship between leaf carbon isotope discrimination and root length density for rain-fed bean genotypes at two locations: Palmira (upper graph) and Quilichao (lower graph). (From Ref. 137.)

$\Delta$) is not independent of root activity, and it seems that there is a close correlation between gas exchange in water-deficit environments and root attributes [137]. One way to overcome this problem of differences in root attributes is to evaluate germ plasm lines under irrigated conditions, where differences in root growth do not affect the leaf gas exchange characteristics and thus $\Delta$. In many crop species, variation in $P_i/P_a$ and $\Delta$ has been reported among genotypes under irrigated conditions, indicating the existence of genetic variation in the "baseline $C_i$" that is expressed under nonstress conditions [137].

In crops such as groundnut, there is a moderately positive correlation ($r = 0.55$) between $\Delta$ and HI, and thus selecting for low $\Delta$ (high TE) could lead to selection of genotypes with low partitioning [10,66,96]. This indicates that selection for high TE and HI, and thus yield potential, could be difficult because of this negative association. However, the possibility of combining high HI and high TE requires further research [10,96]. This highlights the need for physiologists and breeders to be aware of the potential for negative associations between traits such as TE, partitioning of biomass, and root water uptake attributes of roots.

As several factors can alter plant dry weight independently of $\Delta$, there may not always be a direct association between $\Delta$ and productivity [36]. However in many crops, the general trend in relationship between $\Delta$ and dry matter productivity is negative; that is, higher productivity under optimum conditions (e.g., irrigated) is associated with lower $\Delta$ [140]. Thus, in crops where there is a positive association between $\Delta$ and dry matter production, it may be that high TE and potential for dry matter productivity are incompatible. For crops such as wheat, barley, and beans, where differences in TE are due mainly to differences in $g_s$, there appears to be a positive correlation between $\Delta$ and dry matter production [132]. This indicates that selection for low $\Delta$ could lead to selection of genotypes with low dry matter accumulation capability and thus potential productivity. It was suggested that selection for low $\Delta$ will improve adaptation to drought [29], whereas selection for high $\Delta$ should improve yield potential [132]. However, it should still be possible to identify genotypes that do not comply with this general relationship. For example, in barley, although there is generally a negative relationship between TE and dry matter accumulation among the genotypes tested, certain genotypes deviate from this relationship (Figure 5) [148].

For crops such as groundnut, and in cool-season grasses, where photosynthetic rates are the main source of variation in TE, selection for low $\Delta$ should lead to genotypes with high dry matter production

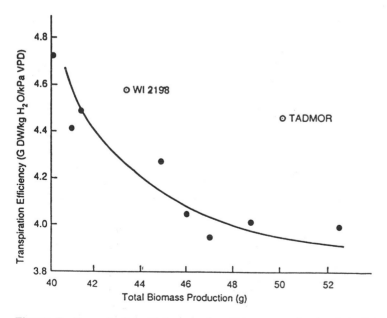

**Figure 5** Transpiration efficiency and total biomass production in barley genotypes grown in a greenhouse. (From Ref. 148.)

capabilities [10,37,66,140]. Thus, it is interesting to note that the usefulness of $\Delta$ in selection for high TE could vary depending on the crop species and the target environment; in one case it could lead to improving productivity, and in other cases it could be detrimental to productivity.

## F.   Role of TE in Improving Drought Resistance of Crops

Crop plants have evolved a variety of strategies to cope with water deficit conditions [146,149,150]. The seasonal progression of temperature, the distribution and intensity of rainfall, and the availability of soil moisture will largely determine the plant attributes that need to be altered beneficially to improve the efficiency of water use [151]. Transpiration efficiency is one of the components involved in adaptation to drought by potentially extending the period of soil moisture availability and is thus expected to contribute to improving adaptation to drought-prone environments. This is particularly so if the crop is raised on finite amounts of stored moisture. A drought-resistant groundnut genotype (drought resistance defined here as relative total dry matter production under drought conditions), Tifton-8, was found to be very efficient in its water use compared with a sensitive A. villosa [152]. Chico, a short-season groundnut variety, had the lowest TE value compared with long-season groundnut varieties [98], which are also found to be more drought resistant than the short-season varieties. In wheat, barley, cowpea, and groundnut, TE is positively correlated with days to heading, which indicates that selection for early maturity might result in decreased TE values [64,96,111,124,132,153]. However, in groundnut, there is still considerable variation in TE/$\Delta$ within similar maturity groups, indicating that the variation in TE could be located in any given maturity group [64,96]. Thus, simultaneous selection for TE and phenological characteristics should be practiced to improve TE within an optimum maturity group. Tall landrace wheat genotypes had greater total dry matter and TE but were later in maturity than the modern dwarf and semidwarf genotypes [124].

In many cropping systems where irrigation water is not readily available, yield stability can be affected by intermittent droughts [10]. Ideally, maximum growth with the water available is a goal. One possibility for improving productivity in low-rainfall and drought-prone areas is to select and breed plants that require less water for growth without losing their yield potential (i.e., to improve their TE value). However, there is a distinction between TE and drought resistance as a whole, and it needs to be recognized that the development and use of drought-resistant plants can lead to the effective use of limited soil water that would otherwise be unavailable. In effect, WUE would be increased for the entire land area even if the drought-resistant crops grown actually transpire more water per unit of dry matter than nonresistant crops.

In rain-fed environments, TE alone may not play a key role in determining the level of drought resistance of a given cultivar. The negative correlations between reduced $\Delta$, biomass, yield, and LAI indicate that greatest growth under rain-fed conditions would occur in cultivars best able to postpone desiccation and maintain relatively large stomatal conductance (i.e., mostly to deal with the efficiency with which the water is extracted rather than utilized), thus showing less reduction in $C_i$ than occurs in irrigated treatments [137]. However, high levels of TE and efficient root systems (deep root system, uniform root length distribution through the soil profile, efficient water uptake from low soil water potentials, etc.) are independent attributes of a plant; therefore, they need not be incompatible. Thus, one could improve TE of a given variety through breeding even if it is found to have a more efficient root system but a low TE. In groundnut, some of the genotypes that have deep rooting attributes and are more efficient in water uptake also had higher levels of TE than the genotypes poor in both attributes [154].

Assuming that the traits contributing to drought resistance are independent attributes, it would be necessary to develop ideotypes to suit the requirements of specific target production environments [155,156]. Then genetic improvement would depend largely on the local variety that needs to be improved, which can be guided by using the ideotype as a basis for the evaluation of traits that need to be incorporated [155]. Thus, genetic improvement for better adaptation to moisture deficit environments could be focused on a few selected traits rather than considering adaptation as a single component of improvement. This would assist in quantifying progress and devising appropriate strategies for further improvement, apart from being able to use genetic stocks developed during the process in related breeding programs in other production environments.

## VI.　FUTURE OUTLOOK

Large sums of money have been spent to develop irrigated cropping systems throughout the world, but relatively little attention has been paid to research on improving WUE, let alone genetically improving TE values of crop species [157]. Although differences among and within crop species in their TE values (thus in their total water requirements to produce a given amount of yield) were demonstrated more than 80 years ago [158], very little progress has been made since in initiating breeding programs specifically targeted at improving TE values in any crop species. This is mainly due to the lack of appropriate means of characterizing and quantifying genotypic variation in TE and the inability to handle the large number of samples required in a breeding program. The finding that TE is negatively related to $^{13}C$ discrimination ($\Delta$) has led to renewed interest in TE as a potentially exploitable trait, and thus $\Delta$ has been proposed as a selection criterion for improving TE in plant breeding programs [29]. It has now been shown that genetic variation in TE exists for many crop species in both well-watered and moisture deficit environments. The high levels of heritability for $\Delta$ have further strengthened the argument that $\Delta$ is amenable to genetic improvement. This opens the way for developing crop varieties that require less water to produce the same amount of yield according to their present potential. This also provides scope for much more rational deployment of irrigation water.

However, $^{13}C$ discrimination analysis of plant samples requires mass spectrometer facilities, and it is beyond the ability of many breeding programs to acquire and maintain such highly expensive and sensitive equipment. This is particularly so in developing countries, which are located mostly in semiarid regions, where improving crop TE could play a crucial role in improving and stabilizing crop production. Thus, this would presently be the limiting factor for the use of this technology in breeding programs focused specifically toward genetic improvement of TE. Nevertheless, it could still be handled by having centralized facilities in selected institutes where analyses could be done. Also, once the equipment is installed and maintained, the actual analysis costs may be within the capability of many breeding programs. Correlated traits such as specific leaf area, which has been shown to be related to $\Delta$, could thus be used as a surrogate to $^{13}C$ discrimination analysis [130,131]. Measuring specific leaf area could be relatively inexpensive and requires no special equipment. However, it needs to be proved that selection programs based on specific leaf area could lead to genetic enhancement of TE, and its heritability needs to be established clearly before proposing this as a surrogate to $\Delta$ in a selection program. There are indications in groundnut that it could be used effectively as an alternative to $\Delta$ in selecting for TE [101,130], but this needs to be proved convincingly. Also, recent reports indicate that molecular markers (such as restriction fragment length polymorphisms, RFLPs) could be linked to water use efficiency (see Chapter 44 for further discussion of molecular markers) and other physiological traits such as osmotic adjustment [159–167]. This could lead to better integration of physiological traits into crop breeding programs for the development of cultivars that are better adapted to moisture deficit environments without a loss in yield potential.

## ACKNOWLEDGMENTS

We wish to acknowledge editorial assistance from the ICRISAT Editorial Committee in improving the structure and presentation of the manuscript.

## REFERENCES

1.　SL Postel. Water for food production: will there be enough in 2025? Bioscience 48:629–637, 1998.
2.　PJM Cooper, JDH Keatinge, G Hughes. Crop evaporation—a technique for calculation of its components by field measurements. Field Crops Res 7:299–312, 1983.
3.　PJM Cooper, GS Campbell, MC Heath, PD Hebblethwaite. Factors which affect water use efficiency in rainfed production of food legumes, and their management. In: RJ Summerfield, ed. World Crops: Cool Season Food Legumes. London: Kluwer Academic Publishers, 1988, pp 813–829.
4.　G Hughes, JDH Keatinge, PJM Cooper, NF Dee. Solar radiation interception and utilzation by chickpea (*Cicer arietenum* L.) crops in northern Syria. J Agric Sci (Camb) 108:419–424, 1987.
5.　MC Heath, PD Hebblethwaite, Solar radiation interception by leafless, semi-leafless and leafed peas (*Pisum sativum*) under contrasting conditions. Ann Appl Biol 107:309–318, 1985.
6.　FG Viets Jr. Fertilizers and the efficient use of water. Adv Agron 14:223–264, 1962.

7.   RA Fischer, NC Turner. Plant productivity in the arid and semi-arid zones. Annu Rev Plant Physiol 29:277–317, 1978.
8.   MC Saxena. Agronomy of chickpea. In: MC Saxena, KB Singh, eds. The Chickpea. Wallingford, UK: CAB International, 1987, pp 207–232.
9.   JB Passioura. Grain yield, harvest index and water use of wheat. J Aust Inst Agric Sci 43:117–121, 1977.
10.  KT Hubick, R Shorter, GD Farquhar. Heritability and genotype X environment interactions of carbon isotope discrimination and transpiration efficiency in peanut (*Arachis hypogaea*). Aust J Plant Physiol 15:799–813, 1988.
11.  JS Boyer. Mechanisms for obtaining water use efficiency and drought resistance. In: HT Stalker, JP Murphy, eds. Plant Breeding in the 1990s. Wallingford, UK: CAB International, 1992, pp 181–200.
12.  DK Barnes. Managing root systems for efficient water use: breeding plants for efficient water use. In: HM Taylor, WR Jordan, TR Sinclair, eds. Limitations to Efficient Water Use in Crop Plants. Madison, WI: American Society of Agronomy, 1983, pp 127–136.
13.  WJ Davies. Transpiration and the water balance of plants. In: FC Steward, JF Sutcliffe, JE Dale, eds. Plant Physiology: A Treatise. Vol. IX. Water and Solutes in Plants. New York: Academic Press, 1986, pp 49–154.
14.  D Hillel. Role of irrigation in agricultural systems. In: BA Stewart, DR Nielson, eds. Irrigation of Agricultural Crops. Madison, WI: American Society of Agronomy, 1990, pp 5–30.
15.  SMA Faiz, PE Weatherley. Further investigations into the location and magnitude of the hydraulic resistances in the soil: plant system. New Phytol 81:19–28, 1978.
16.  HG Jones. Crop characteristics and the ratio between assimilation and transpiration. J Appl Ecol 13:605–622, 1976.
17.  DN Moss, JT Wooley, JF Stone. Plant modification for more efficient water use: the challenge. Agric Meteorol 14:311–320, 1974.
18.  W Wenkert. Water transport and balance within the plant: an overview. In: HM Taylor, WR Jordan, TR Sinclair, eds. Limitations to Efficient Water use in Crop Plants. Madison, WI: American Society of Agronomy, 1983, pp 137–172.
19.  GD Farquhar. Feed forward responses of stomata to humidity. Aust J Plant Physiol 5:787–800, 1978.
20.  H Meidner, TA Mansfield. Physiology of Stomata. London: McGraw-Hill, 1968.
21.  ED Schulze, OL Lange, M Evenari, L Kappen, M Evenari. The role of air humidity and temperature in controlling stomatal resistance of *Prunus armeniaca* L. under desert conditions. I. Assimilation of the daily course of stomatal resistance. Oecologia 17:159–170, 1974.
22.  IR Cowan, GD Farquhar. Stomatal function in relation to leaf metabolism and environment. In: DH Jennings, ed. Integration of Activity in the Higher Plants. Cambridge: Cambridge University Press, 1977, pp 471–505.
23.  KJ Bradford, TD Sharkey, GD Farquhar. Gas exchange, stomatal behavior and $^{13}$C values of the flacca tomato mutant in relation to abscisic acid. Plant Physiol 72:245–250, 1983.
24.  IR Cowan, JH Troughton. The relative role of stomata in transpiration and assimilation. Planta 97:323–336, 1971.
25.  JIL Morison. Sensitivity of stomata and water use efficiency to high $CO_2$. Plant Cell Environ 8:467–474, 1985.
26.  KW Brown, WR Jordan, JC Thomas. Water stress induced alterations of the stomatal response to decreases in leaf water potential. Physiol Plant 37:1–5, 1976.
27.  E Fereres, E Acevedo, D Henderson, TC Hsiao. Seasonal changes in water potential and turgor maintenance in sorghum and maize under water stress. Planta 44:261–267, 1978.
28.  MM Ludlow. Adaptive significance of stomatal responses to water stress. In: NC Turner, PJ Kramer, eds. Adaptations of Plants to Water and High Temperature Stress. New York: John Wiley & Sons, 1980, pp 123–138.
29.  GD Farquhar, RA Richards. Isotopic composition of plant carbon correlates with water-use efficiency of wheat genotypes. Aust J Plant Physiol 11:539–552, 1984.
30.  AH Markhart. Comparative water relations of *Phaseolus vulgaris* L. and *Phaseolus acutifolius* gray. Plant Physiol 77:113–117, 1985.
31.  D Vignes, A Djekoun, C Planchan. Responses of various soybean genotypes to water stress. Can J Plant Sci 66:247–255, 1986.
32.  RJ Lawn. Responses of four grain legumes to water stress in south-eastern Queensland. I. Physiological response mechanisms. Aust J Agric Res 33:481–496, 1982.
33.  RG Henzell, KJ McCree, CHM van Bavel, KF Schertz. Sorghum genotype variation in stomatal sensitivity to leaf water deficit. Crop Sci 16:660–662, 1976.
34.  L Riccardi, P Steduto. Leaf water potential and stomatal resistance variations in *Vicia faba* L. FABIS Newslett 20:21–24, 1988.
35.  MJ Hattendert, DW Evans, RN Peaden. Canopy temperature and stomatal conductance of water-stressed dormant and non-dormant alfalfa types. Agron J 82:873–877, 1990.
36.  GD Farquhar, JR Ehleringer, KT Hubick. Carbon isotope discrimination and photosynthesis. Annu Rev Plant Physiol 40:503–537, 1989.

37. JJ Read, RC Johnson, BF Carver, SA Quarrie. Carbon isotope discrimination, gas exchange, and yield of spring wheat selected for abscisic acid content. Crop Sci 31:139–146, 1991.

38. GD Farquhar, KT Hubick, AG Condon, RA Richards. Carbon isotope fractionation and plant water-use efficiency. In: JR Ehleringer, KA Nagy, eds. Stable Isotopes in Ecological Research. Ecological Studies Vol 68. New York: Springer-Verlag, 1988, pp 21–40.

39. PG Jarvis, KG McNaughton. Stomatal control of transpiration: scaling up from leaf to region. Adv Ecol Res 15:1–49, 1986.

40. CD Walker, RCN Lance. The fractionation of $^2H$ and $^{18}O$ in leaf water of barley. Aust J Plant Physiol 18:411–425, 1991.

41. DF Parkhurst, OL Loucks. Optimal leaf size in relation to environment. J Ecol 60:505–537, 1972.

42. C Johansen, B Baldev, JB Brouwer, W Erskine, WA Jermyn, Lang Li-Juan, BA Malik, A Ahad Miah, SN Silim. Biotic and abiotic stresses constraining productivity of cool season food legumes in Asia, Africa and Oceania, In: FJ Muehlbauer, WJ Kaiser, eds. Expanding the Production and Use of Cool Season Food Legumes. Dordrecht, The Netherlands: Kluwer, 1994, pp 175–194.

43. RJ Hanks. Yield and water-use relationships: an overview. In: H Taylor, WR Jordan, TR Sinclair, eds. Limitations to Efficient Water Use in Crop Production. Madison, WI: American Society of Agronomy, 1983, pp 393–411.

44. JF Beerhuizen, RO Slatyer. Effect of atmospheric concentration of water vapor and $CO_2$ in determining transpiration-photosynthesis relationships of cotton leaves. Agric Meteorol 2:259–270, 1965.

45. RW Downes. Effect of light intensity and leaf temperature on photosynthesis and transpiration in wheat and sorghum. Aust J Biol Sci 23:775–782, 1970.

46. KA Shackel, AE Hall. Reversible leaf movements in relation to drought adaptation of cowpeas (*Vigna unguiculata* (L.) *walp.*). Aust J Plant Physiol 6:265–276, 1979.

47. MM Ludlow, O Bjorkman. Paraheliotropic leaf movement in siratro as a protective mechanism against drought induced damage to primary photosynthetic reactions: damage by excessive light and heat. Planta 161:505–518, 1984.

48. IN Forseth, AH Teramura. Kuozu leaf energy budget and calculated transpiration: the influence of leaflet orientation. Ecology 67:564–571, 1986.

49. VS Berg, S Heuchelin. Leaf orientation of soybean seedlings. I. Effect of water potential and photosynthetic photon flux density on paraheliotropism. Crop Sci 30:631–638, 1990.

50. RC Muchow. An analysis of the effects of water deficits on grain legumes grown in a semi-arid environment in terms of radiation interception and its efficiency of use. Field Crops Res 11:309–323, 1985.

51. GR Squire. The Physiology of Tropical Crop Production. Wallingford, UK: CAB International, 1990.

52. RB Matthews, D Harris, JH Williams, RC Nagcswara Rao. The physiological basis for yield differences between four genotypes of groundnut (*Arachis hypogaea*) in response to drought. II. Solar radiation interception and leaf movement. Exp Agric 24:203–213, 1988.

53. SR Ghorashy, JW Pendleton, ME Bernard. Effect of leaf pubescence on transpiration, photosynthetic rate, and seed yield of near-isogenic lines of soybean. Crop Sci 11:426–427, 1971.

54. J Ehleringer. Leaf morphology and reflectance in relation to water and temperature stress. In: NC Turner, PJ Kramer, eds. Adaptation of Plants to Water and Temperature Stress. New York: Wiley Interscience, 1980, pp 295–308.

55. M Ashraf, F Karim. Screening of some cultivars/lines of black gram (*Vigna mungo* (L.) Alepper.) for resistance to water stress. Trop Agric (Trinidad) 68:57–62, 1991.

56. DD Baldocchi, SB Verma, NJ Rosenberg. Water use efficiency in a soybean field: influence of plant water stress. Agric For Meteorol 34:53–65, 1985.

57. ED Schulze, AE Hall. Stomatal responses, water loss and $CO_2$ assimilation rates of plants in contrasting environments. In: OL Lange, PS Nobel, CB Osmond, H Ziegler, eds. Physiological Plant Ecology II, Water Relations and Carbon Assimilation. New York: Springer-Verlag, 1982, pp 181–230.

58. JR Ehleringer, O Bjorkman, HA Mooney. Leaf pubescence: effects on absorptance and photosynthesis in a desert shrub. Science 192:376–377, 1976.

59. JR Ehleringer, HA Mooney. Leaf hairs: effects on physiological activity and adaptive value to a desert shrub. Oecologia 37:183–201, 1978.

60. JM Clarke, RA Richards. The effects of glaucousness, epicuticular wax, leaf age, plant height, and growth environment on water loss rates of excised wheat leaves. Can J Plant Sci 68:975–982, 1988.

61. MCM Paje, MM Ludlow, RJ Lawn. Variation among soybean (*Glycine max* (L.) Merr.) accessions in epidermal conductance of leaves. Aust J Agric Res 39:363–373, 1988.

62. JM Clarke, I Ramagosa, S Jana, JP Srivastava, TN McCaig. Relationship of excised leaf water loss rate and yield of durum wheat in diverse environments. Can J Plant Sci 69:1075–1089, 1989.

63. Y Castonguay, AH Markhart. Saturated rates of photosynthesis in water stressed leaves of common bean and tepary bean. Crop Sci 31:1605–1611, 1991.

64. KT Hubick, GD Farquhar, R Shorter. Correlation between water-use efficiency and carbon isotope discrimination in diverse peanut (*Arachis*) germplasm. Aust J Plant Physiol 13:803–816, 1986.

65.  AG Condon, GD Farquhar, RA Richards. Genotypic variation in carbon isotope discrimination and transpiration efficiency in wheat. Leaf gas exchange and whole plant studies. Aust J Plant Physiol 17:9–22, 1990.

66.  GC Wright, KT Hubick, GD Farquhar. Discrimination in carbon isotopes of leaves correlates with water use efficiency of field grown peanut cultivars. Aust J Plant Physiol 15:815–825, 1988.

67.  DH Wallace, JL Ozbun, HM Munger. Physiological genetics of crop yield. Adv Agron 24:97–146, 1972.

68.  AS Bhagsari, RH Brown. Photosynthesis of peanut (*Arachis*) genotypes. Peanut Sci 3:1–9, 1976.

69.  GM Dornhoff, RM Shibles. Leaf morphology and anatomy in relation to $CO_2$ exchange rate of soybean leaves. Crop Sci 16:377–381, 1976.

70.  JB Passioura. The effect of root geometry on the yield of wheat growing on stored water. Aust J Agric Res 23:745–752, 1972.

71.  M Gallardo, J Eastham, PJ Gregory, NC Turner. A comparison of plant hydraulic conductances in wheat and lupins. J Exp Bot 47:233–239, 1996.

72.  RA Richards, JB Passioura. Seminal root morphology and water use of wheat. I. Environmental effects. Crop Sci 21:249–252, 1981.

73.  RA Richards, JB Passioura. Seminal root morphology and water use of wheat. II. Genetic variation. Crop Sci 21:253–255, 1981.

74.  GC Wright, KT Hubick, GD Farquhar, RC Nageswara Rao. Genetic and environmental variations in transpiration efficiency and its correlation with carbon isotope discrimination and specific leaf area in peanut. In: JE Ehleringer, AE Hall, GD Farquhar, eds. Stable Isotopes and Plant Carbon-Water Relations. New York: Academic Press, 1993, pp 247–267.

75.  NL Klocke, DF Heermann, HR Duke. Measurement of evaporation and transpiration with lysimeters. Trans ASAE 28:183–189, 1985.

76.  RJ Hanks, HR Gardner, RL Florian. Plant growth-evapotranspiration relations for several crops in the central great plains. Agron J 61:30–34, 1969.

77.  TA Howell, KR Davis, RL McCormick, H Yamada, VT Walhood, DW Meek. Water-use efficiency of narrow leafed cotton. Irrig Sci 5:195–204, 1984.

78.  RJ Hanks. Crop coefficients for transpiration. Advances in Evapotranspiration. St. Joseph, MI: ASAE, pp 431–438, 1985.

79.  RJ Lascane, CHM van Bavel, JL Hatfield, DR Upchurch. Energy and water balance of a sparse crop: simulated and measured soil and crop evaporation. Soil Sci Am J 51:113–1121, 1987.

80.  ME Bloodworth, JB Page, WR Cowley. A thermoelectric method for determining the rate of water movement in plants. Soil Sci Soc Am Proc 19:411–414, 1955.

81.  T Sakuratini. Improvement of the probe for measuring water flow rate in intact plants with the stem heat balance method. Agric Meteorol 40:273–277, 1984.

82.  JM Baker, CHM van Bavel. Measurement of mass flow of water in stems of herbaceous plants. Plant Cell Environ 10:779–782, 1987.

83.  JR Evans, TD Sharkey, JA Berry, GD Farquhar. Carbon isotope discrimination measured concurrently with gas exchange to investigate $CO_2$ diffusion in leaves of higher plants. Aust J Plant Physiol 13:281–292, 1986.

84.  AB Frank, RE Barker, JD Berdahl. Water-use efficiency of grasses grown under controlled and field conditions. Agron J 79:541–544, 1987.

85.  TR Sinclair, CB Tanner, JM Bennett. Water use efficiency in crop production. Bioscience 34:36–40, 1983.

86.  B Martin, YR Thorstenson. Stable carbon isotope composition ($^{13}C$), water use efficiency, and biomass productivity of *Lycopersicon esculentum*, *L. pennelli*, and their $F_1$ hybrid. Plant Physiol 88:213–217, 1988.

87.  HG Jones. Breeding for stomatal characteristics. In: E Ziegler, GD Farquhar, IR Cowan, eds. Stomatal Function. Stanford: Stanford University Press, 1984 (cited in Ref. 24).

88.  AM Ismail, AE Hall. Correlation between water use efficiency and carbon isotope discrimination in diverse cowpea genotypes and isogenic lines. Crop Sci 32:7–12, 1992.

89.  RD Guy, MF Fogel, JA Berry, TC Hoering. Isotope fractionation during oxygen production and consumption by plants. In: J Biggins, ed. Progress in Photosynthetic Research. III. Dordrecht: Martinus Nijhoff, 1987, pp 597–600.

90.  GD Farquhar, MH O'Leary, JA Berry. On the relationship between carbon isotope discrimination and the intercellular carbon dioxide concentration in leaves. Aust J Plant Physiol 9:121–137, 1982.

91.  JH Troughton. $\delta^{13}C$ as an indicator of carboxylation reactions. In: M. Gibbs, E. Latzko, eds. Photosynthesis II, Photosynthetic Carbon Metabolism and Related Processes. Berlin: Springer-Verlag, 1979, pp 140–149.

92.  WG Mook, M Koopmans, AF Carter, CD Keeling. Seasonal, latitudinal and secular variations in the abundance of isotopic ratios of atmospheric carbon dioxide. I. Results from land stations. J Geophys Res 88:10915–10933, 1983.

93.  CB Tanner, TR Sinclair. Efficient water use in crop production: research or re-search? In: H Taylor, WR Jordan, TR Sinclair, eds. Limitations to Efficient Water Use in Crop Production. Madison, WI: American Society of Agronomy, 1983, pp 1–28.

94.  H Craig. Carbon-13 in plants and the relationship between carbon-13 and carbon-14 variations in nature. J Geol 62:115–149, 1954.

95.  KT Hubick, GD Farquhar. Carbon isotope discrimination and the ratio of carbon gained to water lost in barley cultivars. Plant Cell Environ 12:795–804, 1989.

96. RC Nageswara Rao, JH Williams, KDR Wadia, KT Hubick, GD Farquhar. Crop growth, water-use efficiency and carbon isotope discrimination in groundnut (*Arachis hypogaea* L.) genotypes under end-of-season drought conditions. Ann Appl Biol 122:357–367, 1993.

97. WSF Schuster, SL Philips, DR Sandquist, JR Ehleringer. Heritability of carbon isotope discrimination in *Qutierrezia microcephala* (Asteraceae). Am J Bot 79:216–221, 1992.

98. GC Wright, TA Sarwanto, A Rahmianna, D Syarefuddin. Investigation of drought tolerance traits conferring adaptation to drought stress in peanut. In: GC Wright, KJ Middleton, eds. Peanut Improvement: A Case Study in Indonesia. ACIAR Proceedings No. 40, Canberra, Australia, 1993, pp 74–84.

99. FC Meinzer, G Goldstein, DA Grantz. Carbon isotope discrimination in coffee genotypes grown under limited water supply. Plant Physiol 92:130–135, 1990.

100. JM Virgona, KT Hubick, HM Rawson, GD Farquhar, RW Downes. Genotypic variation in transpiration efficiency, carbon isotope discrimination and carbon allocation during early growth in sunflower. Aust J Plant Physiol 17:207–214, 1990.

101. RC Nageswara Rao, GC Wright. Stability of the relationship between specific leaf area and carbon isotope discrimination across environments in peanut. Crop Sci 34:98–103, 1994.

102. RD Guy, PG Warne, DM Reid. Stable carbon isotope ratio as an index of water-use efficiency in C$_3$ halophytes—possible relationship to strategies for osmotic adjustment. In: PW Rundel, JR Ehleringer, KA Nagy, eds. Ecological Studies—68, Stable Isotopes in Ecological Research, Berlin: Springer-Verlag, 1988, pp 55–75.

103. KJ McCree, SG Richardson. Salt increases the water use efficiency in water stressed plants. Crop Sci 27:543–547, 1987.

104. GD Farquhar, TD Sharkey. Stomatal conductance and photosynthesis. Annu Rev Plant Physiol 33:317–345, 1982.

105. IR Cowan. Regulation of water use in relation to carbon gain in higher plants. In: OL Lange, PS Nobel, CB Osmond, H Ziegler, eds. Physiological Plant Ecology. II. Water Relations and Carbon Assimilation. Encyclopaedia of Plant Physiology. New Series. Vol 12B. Berlin: Springer-Verlag, 1982, pp 589–613.

106. DA Johnson, KH Assay, LL Tieszen, JR Ehleringer, PG Jefferson. Carbon isotope discrimination: potential in screening cool-season grasses for water limited environments. Crop Sci 30:338–343, 1990.

107. K Winter. CO$_2$ and water vapour exchange, malate content, and $\delta^{13}$C value in *Cicer arietenum* grown under two water regimes. Z Pflanzenphysiol 101:421–430, 1981.

108. KT Hubick, GD Farquhar. Carbon isotope discrimination—selecting for water-use efficiency. Aust Cotton Grower 8:66–68, 1987.

109. JR Ehleringer. Correlations between carbon isotope ratio, water use efficiency and yield. In: JW White, G Hoogenboom, F Ibarra, SP Singh, eds. Research on Drought Tolerance in Common Bean. Working document no 41. Cali: CIAT, 1988, pp 165–191.

110. JR Ehleringer, TA Cooper. Correlations between isotope ratio and microhabitat in desert plants. Oecologia 76:562–566, 1988.

111. AE Hall, RG Mutters, KT Hubick, GD Farquhar. Genotypic differences in carbon isotope discrimination by cowpea under wet and dry field conditions. Crop Sci 30:300–305, 1990.

112. AE Hall, RG Mutters, GD Farquhar. Genotypic and drought induced differences in carbon isotope discrimination and gas exchange of cowpea. Crop Sci 32:1–6, 1992.

113. WR Kirchhoff, AE Hall, WW Thomson. Gas exchange, carbon isotope discrimination, and chloroplast ultra-structure of a chlorophyll deficient mutant of cowpea. Crop Sci 29:109–115, 1989.

114. IR Cowan. Stomatal physiology and gas exchange in the field. In: O.T. Denmead, ed. Flow and Transport in the Natural Environment: Advances and Applications. New York: Springer-Verlag, 1988, pp 160–172.

115. CB Osmond, K Winter, H Ziegler. Functional significance of different pathways of CO$_2$ in photosynthesis. In: OL Lange, PS Nobel, CB Osmond, H Ziegler, eds. Physiological Plant Ecology II, Water Relations and Carbon Assimilation. Encyclopaedia of Plant Physiology. New Series. Vol 12B. Berlin: Springer-Verlag, 1982, pp 479–547.

116. SC Wong, IR Cowan, GD Farquhar. Stomatal conductance correlates with photosynthetic capacity. Nature 282:424–426, 1979.

117. TA Howell. Relationships between crop production and transpiration, evapotranspiration and irrigation. In: BA Stewart, DR Nielson, eds. Irrigation of Agricultural Crops. Madison, WI: American Society of Agronomy, 1990, pp 391–434.

118. J Goudriaan, HH van Laar. Relations between leaf resistance, CO$_2$ concentration and K assimilation in maize, beans, lanlang grass and sunflower. Photosynthetica 12:241–249, 1978.

119. DR Krieg. Whole-plant response to water deficits: carbon assimilation and utilization. In: HM Taylor, WR Jordan, TR Sinclair, eds. Limitations to Efficient Water Use in Crop Production. Madison, WI: American Society of Agronomy, 1983, pp 319–330.

120. RA Fischer, D Rees, KD Sayre, ZM Lu, AG Condon, A Larque Saavedra. Wheat yield progress associated with higher stomatal conductance and photosynthetic rate, and cooler canopies. Crop Sci 38:1467–1475, 1998.

121. C Ramos, AE Hall. Relationships between leaf conductance, intercellular CO$_2$ uptake rate in two C$_3$ and C$_4$ plant species. Photosynthetica 16:343–355, 1982.

122. MH O'Leary, I Treichel, M Rooney. Short term measurement of carbon isotope fractionation in plants. Plant Physiol 80:578–582, 1986.

123. SD Wullschleger. Biochemical limitations of carbon assimilation in $C_3$ plants—a retrospective analysis of the $A/C_i$ curves from 109 species. J Exp Bot 44:907–920, 1993.

124. B Ehdaie, AE Hall, GD Farquhar, HT Nguyen, JG Waines. Water-use efficiency and carbon isotope discrimination in wheat. Crop Sci 31:1282–1288, 1991.

125. JE Pallas Jr, YB Samish. Photosynthetic response of peanut. Crop Sci 14:478–482, 1974.

126. JE Pallas Jr. Photosynthetis traits of selected peanut genotypes. Peanut Sci 9:14–17, 1982.

127. WD Branch, JE Pallas Jr. Heterosis of apparent photosynthesis rate in *Arachis hypogaea* L. Peanut Sci 11:56–57, 1984.

128. F Mojayad, C Planchon. Stomatal and photosynthetic adjustment to water deficit as the expression of heterosis in sunflower. Crop Sci 34:103–107, 1994.

129. B Martin, H Kebede, C Rilling. Photosynthetic differences among *Lycopersicon* species and *Triticum aestivum* cultivars. Crop Sci 34:113–118, 1994.

130. GC Wright, RC Nageswara Rao, GD Farquhar. Water-use efficiency and carbon isotope discrimination in peanut under water deficit conditions. Crop Sci 34:92–97, 1994.

131. PQ Craufurd, TR Wheeler, RH Ellis, RJ Summerfield, JH Williams. Effect of temperature and water deficit on water-use efficiency, carbon isotope discrimination, and specific leaf area in peanut. Crop Sci 39:136–142, 1999.

132. AG Condon, RA Richards, GD Farquhar. Carbon isotope discrimination is positively correlated with grain yield and dry matter production in field grown wheat. Crop Sci 27:996–1001, 1987.

133. D Shimshi, J Ephrat. Stomatal behavior of wheat cultivars in relation to their transpiration and photosynthesis and yield. Agron J 67:326–331, 1975.

134. RC Johnson, H Kebede, DW Mornhinweg, BF Carver, A Lanerayburn, HT Nguyen. Photosynthetic differences among *Triticum* accessions at tillering. Crop Sci 27:1046–1050, 1987.

135. RL Dunstone, RM Gifford, LT Evans. Photosynthetic characteristics of modern and primitive wheat species in relation to ontogeny and adaptation to light. Aust J Biol Sci 26:295–307, 1973.

136. J Masle, GD Farquhar. Effects of soil strength on the relation of water use efficiency and growth to carbon isotope discrimination in wheat seedlings. Plant Physiol 86:32–38, 1988.

137. JW White, JA Castillo, J Ehleringer. Associations between productivity, root growth, and carbon isotope discrimination in *Phaseolus vulgaris* under water deficit. Aust J Plant Physiol 17:189–198, 1990.

138. AG Condon, RA Richards. Broad sense heritability and GXE interaction for carbon isotope discrimination in field grown wheat. Aust J Agric Res 43:921–934, 1992.

139. IM Ray, MS Townsend, CM Muncy, JA Henning. Heritabilities of water-use efficiency traits and correlations with agronomic traits in water-stressed alfalfa. Crop Sci 39:494–498, 1999.

140. RC Johnson, LM Bassett. Carbon isotope discrimination and water use efficiency in four cool season grasses. Crop Sci 31:157–162, 1991.

141. PC Pheloung, KHM Siddique. Contribution of stem dry matter to grain yield in wheat cultivars. Aust J Plant Physiol 18:53–64, 1991.

142. DS Falconer. Introduction to Quantitative Genetics. New York: Longman Group, 1981.

143. MH O'Leary. Carbon isotope fractionation in plants. Phytochemistry 20:553–567, 1981.

144. WJS Downton, WJR Grant, SP Robinson. Photosynthetic and stomatal responses of spinach leaves to salt stress. Plant Physiol 78:85–88, 1985.

145. RD Guy, DM Reid. Photosynthesis and the influences of $CO_2$ enrichment on $\delta^{13}C$ values in a $C_3$ halophyte. Plant Cell Environ 9:65–72, 1986.

146. K Winter, JAM Holtum, GE Edwards, MH O'Leary. Effect of low relative humidity on $\delta^{13}C$ values in two $C_3$ grasses and in *Panicum melioides*, a $C_3$ and $C_4$ intermediate species. J Exp Bot 33:88–91, 1982.

147. BN Sponchiado, JW White, JA Castillo, PG James. Root growth of common bean cultivars in relation to drought tolerance in environments with contrasting soil types. Exp Agric 25:249–257, 1989.

148. E Acevedo. Improvement of winter cereal crops in Mediterranean environments. Use of yield, morphological and physiological traits. In: E Acevedo, AP Conesa, P Monneveux, JP Srivastava, eds. Physiology-Breeding of Winter Cereals for Stressed Mediterranean Environments. Paris: INRA, 1991.

149. NH Nam, GV Subbarao, YS Chauhan, C Johansen. Importance of canopy attributes in determining dry matter accumulation of pigeonpea under contrasting moisture regimes. Crop Sci 38:955–961, 1998.

150. TC Hsiao, E Acevedo. Plant responses to water deficits, water-use efficiency, and drought resistance. Agric Meteorol 14:59–84, 1974.

151. GH Heichel. Crop manipulation for efficient use of water: inadvertent and intended manipulations of crop water use. In: HM Taylor, WR Jordan, TR Sinclair, eds. Limitations to Efficient Water Use in Crop Production. Madison, WI: American Society of Agronomy, 1983, pp 375–380.

152. TA Coffelt, RO Hammons, WD Branch, PM Mozingo, PM Phipps, JC Smith, RE Lynch, CS Kvien, DL Ketring, DM Porter, AC Misen. Registration of Tifton-8 peanut germplasm. Crop Sci 25:203, 1985.

153. PQ Craufurd, AB Austin. Carbon isotope discrimination as an indicator of drought tolerance. Annual Report of the Agriculture and Food Research Council, Institute of Plant Science Research and John Innes Institute, John Catt Ltd, England, 1987, pp 12–14.

154. GC Wright, RC Nageswara Rao, HB So. Variation in root characteristics and their association with water uptake and drought tolerance in four peanut cultivars. Paper presented at the Australian Agronomy Conference, The University of Adelaide, Adelaide, South Australia, September 19–24, 1993.

155. GV Subbarao, C Johansen, AE Slinkard, RC Nageswara Rao, NP Saxena, YS Chauhan. Strategies for improving drought resistance in grain legumes. CRC Crit Rev Plant Sci 14:469–523, 1995.
156. S Ceccarelli, E Acevedo, S Grando. Breeding for yield stability in unpredictable environments: single traits, interaction between traits, and architecture of genotypes. Euphytica 56:169–185, 1991.
157. MN Christiansen. World environmental limitations to food and fiber culture. In: MN Christiansen, CF Lewis, eds. Breeding Plants for Less Favorable Environments. New York: John Wiley & Sons, 1982, pp 1–11.
158. LJ Briggs, HL Shantz. The water requirements of plants. II. A review of the literature. USDA Bur Plant Ind Bull 285:1–96, 1913.
159. MAR Mian, MA Bailey, DA Ashley, R Wells, TE Carter Jr, WA Parrott, HR Boerma. Molecular markers associated with water use efficiency and leaf ash in soybean. Crop Sci 36:1252–1257, 1996.
160. BP Forster, JR Russell, RP Ellis, LL Handley, D Robinson, CA Hackett, E Nevo, R Waugh, DC Gordon, R Keith, W Powell. Locating genotypes and genes for abiotic stress tolerance in barley: a strategy using maps, markers and the wild species. New Phytol 137:141–147, 1997.
161. I Jamaux, A Steinmetz, E Belhassen. Looking for molecular and physiological markers of osmotic adjustment in sunflower. New Phytol 137:117–127, 1997.
162. HT Nguyen, R Chandra Babu, A Blum. Breeding for drought resistance in rice: physiology and molecular genetic considerations. Crop Sci 37:1426–1434, 1997.
163. B Teulat, P Monneveux, J Wery, C Borries, I Souyris, A Charrier, D This. Relationship between relative water content and growth parameters under water stress in barley: a QTL study. New Phytol 137:99–107, 1997.
164. N Jones, H Ougham, H Thomas. Markers and mapping: we are genetists now. New Phytol 137:165–177, 1997.
165. MAR Mian, DA Ashley, HR Boerma. An additional QTL for water use efficiency in soybean. Crop Sci 38:390–393, 1998.
166. MR Tuinstra, G Ejeta, P Goldsbrough. Evaluation of near-isogenic sorghum lines contrasting for QTL markers associated with drought tolerance. Crop Sci 38:835–842, 1998.
167. J Zhang, HT Nguyen, A Blum. Genetic analysis of osmotic adjustment in crop plants. J Exp Bot 50:291–302, 1999.

# 44

# Physiological Mechanisms Relevant to Genetic Improvement of Salinity Tolerance in Crop Plants

**G. V. Subbarao***

*Dynamac Corporation, Kennedy Space Center, Florida*

**Chris Johansen†**

*International Crops Research Institute for the Semi-Arid Tropics (ICRISAT), Andhra Pradesh, India*

## I. INTRODUCTION

Crop species differ widely in their ability to grow and yield under saline conditions. However, almost all crop plants belong to the glycophytic category, except for a few crop species such as sugar beet, which has halophytic ancestors. By ecological definition, halophytes are the native flora of saline habitats [1,2]. From a crop improvement perspective, the variability of salinity tolerance within a crop species or among its wild relatives is important. It is also important to understand the physiological mechanisms of salinity tolerance operating within a crop species so that suitable breeding strategies can be developed for improving salinity tolerance. There are several reviews covering the general responses of plants to salinity stress and the mechanisms available in halophytes and glycophytes that allow them to cope with saline habitats [2–17]. However, little attempt has been made to integrate information on these physiological aspects into genetic improvement concepts.

Salinity creates stress by reducing the osmotic potential of the rooting medium and increasing ambient concentrations of ions such as Cl, $SO_4$, $CO_3$, $HCO_3$, Na, Ca, and Mg ions. Being glycophytes, crop species have no appendages such as salt glands, bladders, or hairs that excrete salts absorbed in excess from their shoot tissues. The limited compartmentation ability of the shoot demands strict regulation of ionic delivery to the shoot. Physiological mechanisms controlling salt absorption and distribution in crop plants and the osmotic adjustment that is essential for turgor driven water uptake are covered in this chapter. We specifically address the question of how information on these physiological mechanisms could be utilized in genetic improvement programs as an integrated approach toward improving salinity tolerance in a given crop.

## II. REGULATION OF ION TRANSPORT

Plants regulate their intracellular ionic composition to maintain a suitable ionic environment for the physiological and biochemical processes that proceed within a cell. This internal environment needs to be maintained within acceptable limits if plant growth and function are to proceed in saline environments [18]. Salinity under field conditions is characterized by a mixture of salts. However, Na and Cl are predominant in most

---

* *Current affiliation:* Japan International Research Center for Agricultural Sciences, Ibaraki, Japan.
† *Current affiliation:* Consultant in Agricultural Research and Development, Dhaka, Bangladesh.

situations. Therefore, most studies of salinity effects refer to NaCl salinity as a model system, although effects of all ions that are in excess in a saline environment on nutrient uptake are recognized [19,20]. Similarly, because of the importance of K in plant nutrition and because effects of Na on K uptake have been studied extensively, we refer mainly to this interaction in our discussion of ion uptake mechanisms.

## A.  Regulation at Root Membranes

The concept of dual mechanisms of ion transport is a useful framework for describing ion uptake [21] (see Chapter 17). At low concentrations of K in the external solution, below 1 mM, uptake of K is described by a discrete Michaelis-Menten kinetic equation and is thought to operate at the plasmalemma. We shall call this mechanism 1. At K concentrations in the range 1–50 mM, mechanism 2 operates. Mechanism 2 is thought to involve diffusive or at least nonselective ion movement across the plasmalemma with the rate limitation inward from the plasmalemma, probably at the tonoplast [21]. For mechanism 1, there is a high selectivity of the active transport mechanism for K over competing cations such as Na. For mechanism 2, this level of selectivity is not present. Mechanism 1 is not influenced by the concomitant counteranion, but mechanism 2 is. For example, compared with Cl, $SO_4$ severely depresses K absorption at K concentrations in the range of mechanism 2 but not in the range of mechanism 1. This dual phenomenon of ion uptake has been described for different plant and ionic species (see Ref. 21, p. 136).

Selective ion transport, at least in the range of mechanism 1, depends on metabolic energy derived from adenosine triphosphate (ATP). This allows charge separation across cell membranes, through primary transport of $H^+$, thus creating a localized electrochemical gradient for other ions to traverse the membrane [22–24]. Cations move in the opposite direction to $H^+$ (antiport), while anions are cotransported with it (symport) or move as antiport to $OH^-$ or $HCO_3^-$ [25].

Selectivity between ionic species is governed by the particular binding properties of cell membrane constituents. Little is known about this process because of limited knowledge of plant membrane structure and function [25–27]. Breakthroughs in this regard will allow an understanding of the molecular basis of ion transport and effects of salinity on this process. The entry of Na or other ions in excess in the ambient solution can be controlled by this selective binding. Another alternative for regulating K/Na levels inside root cells is by means of an outwardly directed Na pump at the plasmalemma [3,28–30].

In most situations, saline or otherwise, Na movement across the plasmalemma into root cells is thought to be passive down an electrochemical gradient [8]. For example, the membrane leakage of Na accounts for the cytoplasmic Na levels found in rice [31]. Jeschke [7] has proposed a model to explain K/Na exchange at the plasmalemma (Figure 1), the components of which are as follows:

1.  A proton pump powered by ATP generates an electrical potential difference and proton gradient across the plasmalemma.
2.  The electrical charge of H is compensated by an influx of K at a specific site or channel. This site has a lower affinity for Na.

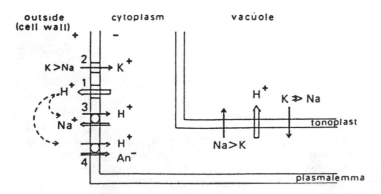

**Figure 1**  Model of the proton-mediated K/Na exchange system at the plasmalemma and Na/K exchange system at the tonoplast. 1, Proton pump; 2, K uniport (i.e., system 1 of K influx); 3, H-Na antiport; 4, $H^+$-anion symport (From Ref. 7.)

3.  The proton gradient provides energy for extrusion of Na from the cytoplasm by an H/Na antiport; this site is reported to have a lower affinity for K.

There is variation among crop species in their K/Na exchange capability [7,32]. Barley, wheat, and rye showed efficient K/Na exchange compared with sensitive species such as *Allium cepa* and *Helianthus annuus* [32]. The existence of genotypic differences in this trait within a crop species and its relation to salinity tolerance are not known. Such information is vital to an evaluation of this trait in genetic improvement programs for salinity tolerance. The relation between K/Na selectivity and salt tolerance has been reviewed [3–5,33–35]. Variation in K/Na exchange suggests at least quantitative differences in membrane properties among different crop species [7]. The general response of many crop plants to a moderate increase in external salinity is increased plant K levels and reduced Na concentrations in tolerant relative to nontolerant genotypes [36–41].

For the high-affinity system mediating K influx (Epstein's mechanism 1), a proton pump appears to be present in the plasmalemma of root cortical cells [7]. However, the graded response of Na efflux to added K suggests quantitative differences between species, and perhaps among genotypes of a crop species, in the number and efficiency of sites mediating the H/Na antiport [7]. The number of sites for the H/Na antiport needs to be quantified and the existence of genotypic variation within a crop species estimated to determine the feasibility of favorable genetic manipulation of this trait.

At K concentrations above 1 mM, in the range of mechanism 2, selectivity diminishes in the presence of competition from other ions, such as Na, in the ambient medium. Whether this is due to increased passive movement of all ambient ions across the plasmalemma, down an electrochemical gradient, or lesser selectivity in an active transport process remains unclear [21]. Eventually, however, if ambient salt concentrations reach high enough levels, membranes would become completely permeable. Information on species or genotypic differences regarding the level at which such physical disruption occurs may also provide a guide to selection for salinity tolerance [42,43].

Most of the kinetic studies just referred to were carried out on tissue previously starved of salts (low-salt status). However, as cytoplasmic concentrations of absorbed ions increase, influx rates slow down, indicating a feedback mechanism controlling active influx of ions [25,44]. For example, K concentrations in the cytoplasm of normally growing plants are maintained in the range 90–110 mM [33]. Although there is considerable speculation about the nature of such feedback mechanisms [25], their further understanding would also assist in selection of genotypes that better control their ion transport processes at the plasmalemma.

## B.  Intracellular Compartmentation in Roots

Vacuoles occupy more than 80% of a mature root cell's volume and thus provide a means of osmotic regulation for root tissue [45]. This is achieved by compartmentation of inorganic salts primarily because these are metabolically inexpensive compared with organic solutes. Salt ions move across membranes more easily than molecules of large molecular weight. There are considerable metabolic costs in transporting photosynthates from the shoots for use as osmotica in roots [46].

Inorganic ions contribute substantially to osmotic adjustment in root cells of glycophytes under saline conditions (see Chapter 17). However, the amount of osmotic adjustment varies from one species to another and could be an important factor in determining salinity tolerance. Roots of many glycophytic crop species contain substantially higher levels of Na and Cl under saline conditions than do shoots [41,47]. In pigeonpea (*Cajanus cajan*) and its wild relatives, the most tolerant genotypes retained higher levels of Na and Cl in the roots and this was associated with salinity tolerance in this crop [41,48]. Ability to retain Na and Cl in roots breaks down at a given concentration, leading to large-scale translocation of these ions to the shoot, with resultant plant mortality. This critical level varies between pigeonpea genotypes and between pigeonpea and its wild relatives and is considered a determinant of the level of salinity tolerance [41].

The cytoplasm shows a strong selectivity for K over Na, Mg over Ca, and P over Cl or $NO_3$ [39,49]. Optimal concentrations for various ions vary in the cytoplasm; thus, when ions enter the protoplast above this concentration, they may be actively transported through the tonoplast into the vacuole. However, these ions could be recovered from the vacuole, depending on the metabolic requirements in other plant parts. Retranslocation of K is one example [7].

Vacuoles play an important role in maintaining stable levels of various inorganic ions in the cytoplasm by acting as a storage reservoir for these ions [45]. Under NaCl salinity, Na and Cl are normally the predominant ions entering the protoplast of root cells. These ions are actively pumped into the vacuole after reaching a threshold concentration in the cytoplasm. This would reduce the flow into the xylem of Na and Cl and of other ions associated with salinity (e.g., Ca, Mg, $SO_4$, $CO_3$) and thus restrict their translocation to the shoot.

The general hypothesis is that Na and Cl must be excluded from the cytoplasm. This is based on the sensitivity of enzyme activities to high NaCl levels in vitro [8]. High levels of Na in the cytoplasm are reported to interfere with K metabolism, resulting in ionic toxicity, but it is not known what Na levels are biochemically compatible with other cytoplasm solutes [8]. In corn, cytoplasmic Na concentrations can reach 40–70 mM under nonsaline conditions [40] but can rise to 140 mM under 100 mM NaCl external salinity and become toxic to the plant. In roots of the halophyte *Triglochin maritima* exposed to 500 mM NaCl, the Na/K ratio was only 2 in the cytoplasm compared with 15 in the vacuole, although there was approximately 150 mM Na in both compartments [50]. Thus the tolerance of the cytoplasm to Na can vary between species. As long as tissue Na concentration is below the level acceptable for the cytoplasm, more sophisticated compartmentation may not be necessary [8].

There are several factors that could mitigate the adverse effects of excess ions in the cytoplasm. One is the type and quantity of organic solutes that could modify the tolerance level of cytoplasm to monovalent cations such as Na (see Chapter 45). Another is the existence of isoenzymes for many enzyme systems, which may have different tolerance thresholds in the cytoplasm. In *Zea mays*, although the total acid phosphatase activity was slightly reduced under salinity, certain isoenzymic forms of acid phosphatase increased in different plant parts [51]. Similarly, the relative proportions of malate dehydrogenase isoenzymes were changed during salinity stress in pea seedlings [52].

In sunflower, a plastome mutant line that has higher resistance to salinity than its parental line reportedly produced a unique isoenzyme of peroxidase under saline conditions [53]. This isoenzyme was found to be resistant to NaCl or $Na_2SO_4$ salinity up to 1.2% and 2.4%, respectively, in vitro. Cavalieri and Huang [54] reported that enzymes isolated from roots were distinctly more tolerant to Na than those from the shoots; these results might reflect other differences between shoots and roots and in compartmentation between cytoplasm and vacuole [45]. Another possibility is that certain isoenzymes exist only in certain plant parts; for example, the isoenzyme patterns of shoots could be different from those of roots [55]. Thus, the statement often made that "there are no differences in enzyme systems of halophytes and nonhalophytes in their tolerance to monovalent cations in vitro" [4,33,56–58] needs to be reexamined.

Another aspect of the adaptation of higher plants to salinity is compartmentation within the cytoplasm because the cytosol is particularly sensitive to fluctuating salt levels [47]. For cells involved in salt transport, the rough endoplasmic reticulum (RER) provides a compartment within the cytoplasm in which salt may be sequestered [47]. Substances can be transported symplastically through the RER via desmotubules. This may also provide a means of ion transfer to vacuoles without disrupting ion concentrations in the cytosol, as RER cisternae may fuse with the tonoplast, releasing their contents into the vacuole [47].

Several hypotheses have been proposed to explain the mode of ion transport from cytoplasm to vacuole through the tonoplast. Pitman and Saddler [28] located an inwardly directed Na pump at the tonoplast that would effectively deplete Na levels in the cytoplasm. Jennings [59] proposed a very similar model for transport of Na from the cytoplasm into the vacuole by means of Na/K exchange. Proton pumps powered by ATP are also thought to play a crucial role in generating the transmembrane electrochemical potential differences required to energize tonoplast ion transport [60,61]. Two types of proton pump are reported to be located in the tonoplast; they are catalyzed by functionally and physiologically distinct phosphohydralases—tp-ATPase, and tp-PPase (tonoplast pyrophosphatase) [60].

Exchange of Na and K at the tonoplast can occur only while K remains in the vacuole [7]. Thus distribution of K and Na between vacuole and cytoplasm appears to be crucial for salt tolerance [33,35], and because vacuolar K concentration represents a potential reservoir that could be removed by exchange for Na, the allocation of these ions needs to be regulated. However, the vacuole of root cortical cells is in some respects a dead end; continued selective transport across the root depends on selective transport at the point of entry of salts into the cytoplasm, which depends on the ability of the plasmamembrane to restrict passive influx of sodium and maintain high K/Na selectivity [62]. Thus, without control of the quantity of salt that is allowed into the root or that reaches the leaves, intracellular compartmentation either at root cortex or in the shoot would in any case be a very limited option [8]. The vacuole's role may be more in using Na as an osmoticum instead of K and in providing a source of stored K under salinization rather than as part of a selective system of salt transport across the root [62].

## C. Regulation of Long-Distance Transport to Shoots

Beyond the plasmalemma, there are several other possible barriers that could minimize transport of excess salts to the shoots. An important one is movement of salts from xylem parenchyma cells into the xylem stream. Evidence favors this process being mediated by active transport [21] with the possibility of further selectivity in ion transport. Xylem parenchyma cells can be differentiated as transfer cells (XPTs) with well-developed wall protuberances adjacent to the bordered pits of xylem vessels in the proximal region of roots and stems. These are reported in *Phaseolus coccineus* [47], *Glycine max* [63], maize [64,65], and squash [66]. These transfer cells accumulate K in the absence of NaCl in the growth medium and Na under saline (NaCl) conditions [63].

A salt-induced formation of wall ingrowths has been reported for xylem parenchyma cells in soybean [63,67] and for the root epidermis cells of *Phaseolus coccineus* [47]. Xylem parenchyma cells and transfer cells are both capable of restricting solutes, particularly Na, by exchange with K from the transpiration stream [43,68]. These XPTs have been reported to accumulate Na selectively from the transpiration stream and then transfer it to the phloem pathway to be extruded by the roots [69]. In *Lycopersicon*, XPTs in the leaf petiole remove Na from the xylem stream before it enters the leaf lamina [70]. It appears that the entire xylem transport pathway has a backup reabsorption system [7].

The cytoplasm of these transfer cells contains cisternae of RER that increase under NaCl or $Na_2SO_4$ salinity in *Phaseolus coccineus* hypocotyl and epicotyl [47] and in *Zea mays* [68]. RER could permit a large flow of ions through the cytoplasm of xylem parenchyma cells, assuming that ions are localized mainly in the vacuole [4]. The quantitative significance of this reabsorption process from the xylem in regulating Na ion transport to the shoot is not known.

The ability of XPTs to absorb Na is finite and could be exhausted rapidly under saline conditions [71]. Some lateral redistribution is possible, but this may not be sufficient to prevent Na from eventually reaching the shoot [72]. However, XPTs have a limited capability to store Na, and this Na needs to be removed to the lateral tissue for XPTs to continue absorbing Na from the transpiration stream. This Na could be loaded into the phloem and translocated to the roots, where it could either be further compartmentalized or extruded. Such Na extrusion has been reported in *H. vulgare* [28,30,73] and *P. vulgaris* [74]. Thus, the practical significance of XPT cells in the basal part of the stem may be limited in controlling Na flow into the shoot to a low degree or a short duration of salinity stress [75,76]. The existence of quantitative variation in XPTs among genotypes in relation to differences in salinity tolerance is not known. Such knowledge is necessary to evaluate the usefulness of this trait from a genetic improvement perspective.

## D. Apoplastic Salt Accumulation

Oertli [77] predicted that apoplastic salt load could cause water deficit and turgor loss in leaf cells and proposed it as a mechanism of salinity damage. This concept has received renewed interest [5,78–80]. Under saline conditions, Na and Cl can bypass the ion transport control mechanisms discussed earlier, be carried upward in the xylem stream, and be delivered to the apoplasts of leaf cells [81]. If shoot protoplast accumulates these ions beyond levels that are tolerated in the cytoplasm and its compartmentation capacity of the vacuole, disruption of the metabolic functions by ionic toxicity would result [82]. On the other hand, a failure to do so would lead to ion accumulation in the apoplast, which could reach very high levels in a short time as the apoplast occupies only 1% of the cell's volume [77,82]. For instance, even if 90% of the NaCl arriving in the xylem (plants grown at 50 mM NaCl external solution) is accumulated in the protoplast, the apoplastic concentrations could reach 500 mM within 7 days [82] and cause cell death, although the average tissue Na and Cl concentrations may not reach 100 mM. Because of the small apoplast volume, such ion concentrations in the apoplast could occur at overall low tissue concentrations and would thus escape detection in standard tissue analysis [82]. Excessive accumulation of salts in the leaf apoplast would cause turgor loss, stomatal closure, and cell dehydration.

Water deficits in a particular leaf, as opposed to the plant as a whole, could be an inevitable consequence of increasing apoplastic salt load [77] and will occur whenever the rate of arrival of NaCl in the xylem is greater than the rate of accumulation of these ions in leaf cells [79]. Thus arguments that plants have adjusted osmotically to external salinity, which are based on comparisons of solute concentrations in tissue water with external salinity, need to be viewed with caution [83]. The success of a crop species in surviving and reproducing under saline conditions depends considerably on its ability to regulate ion delivery into the xylem stream without causing ion toxicity in leaf protoplasts or apoplastic salt buildup [82]. Genotypes that could more effectively transfer NaCl from leaf apoplast into leaf cells would be at an advantage. Although this increases their protoplast salt concentrations because of the relative volumes of protoplast and apoplast, this is considered to be less serious than the consequences of apoplastic salt buildup [31,82].

## E. Phloem Retranslocation

When Na or Cl levels in the cytoplasm of mesophyll cells reach a tolerance threshold and their compartmentation capacity becomes saturated, additional Na or Cl ions can immediately be transported by intraveinal recycling so as to prevent apoplastic buildup of Na or Cl or ion toxicity in the cytoplasm [76]. As there is no barrier between the xylem and the leaf apoplast [84], ions can be actively loaded into phloem vessels [85]. This mechanism may play a significant role in the regulation of Na or Cl ions in the shoot [66,74,86]. Based on cytoplasmic Na concentrations, it has been estimated that nearly 25% of the Na entering the leaf can be retranslocated by the phloem [45]. However, phloem loading and retranslocation of Na or Cl is seen as metabolically expensive. Large quantities of Na or Cl in phloem reflect poor control at the root level in regulating ion flow into the xylem. This was found in studies by Lessani and Marschner [87], where phloem translocation of Na or Cl was greatest in sensitive species such as bean and least in tolerant species such as barley and sugarbeet [39].

Among a range of species, there was a significant correlation between decrease in dry matter production at 100 mM NaCl in the medium and Na retranslocation from leaves and particularly efflux from roots (Fig. 2) [87]. If incoming ions are excessive to the shoot's compartmentation ability and the phloem translocation capacity, overloading of Na or Cl ions into the phloem parenchyma transfer cells could occur. This would result in destruction of phloem transfer cells [76,88]. Although phloem retranslocation does contribute to regulation of Na or Cl levels in the shoot, it appears to have a limited role in this regard and, thus, in determining the level of salinity tolerance. Regulation of Na and Cl levels in the shoot lies primarily with the root's ability to regulate Na or Cl flow into the xylem rather than the shoot's ability to retranslocate to the root [71].

Availability of sufficient K in growing and expanding regions of the shoot and root is crucial to maintenance of K/Na selectivity and subsequent Na compartmentation in the root cortex. In addition to efficient K/Na selectivity at the plasma membrane, phloem transport of K reserves within the plant plays an important role in salinity tolerance. Potassium is remobilized from mature leaves by removal of vacuolar K through Na/K exchange at the tonoplast of mesophyll cells. This K is then retranslocated to the growing regions of the root, shoot, and expanding leaves, where there is little vacuolar space and the cytoplasm occupies a major portion of the cell. These growing zones require large quantities of K to meet their demands for osmotic adjustment in the rapidly expanding vacuolar space. Leaves develop and expand close to the shoot apex and derive their mineral nutrient supply from the phloem (which is rich in K), particularly because phloem tissue differentiates prior to xylem elements [89]. With increasing leaf age, minerals are imported mainly by the xylem, which is high in Na levels compared with the phloem supply. This Na is compartmentalized through Na/K exchange at the tonoplast; thus, K is recovered from the vacuole to provide a major source of K for retranslocation [35].

Nearly 20% of K arriving in the shoot through the xylem could be retranslocated to the growing regions of the root, where high K levels are essential [7]. Such K retranslocation has been reported in barley [90–92], tomatoes, and lupins [7]. The ability to remobilize and retranslocate K into the growing region of the root and shoot plays an important role in Na compartmentation in the root cortex and in maintaining a high K/Na ratio in shoot growing regions, thus protecting them from Na or Cl toxicity. Most tolerant crop species, such as barley and sugar beet, have a very efficient K recirculation system that is tightly linked to Na regulatory mechanisms. This mechanism may also be important in determining genotypic differences in salinity response.

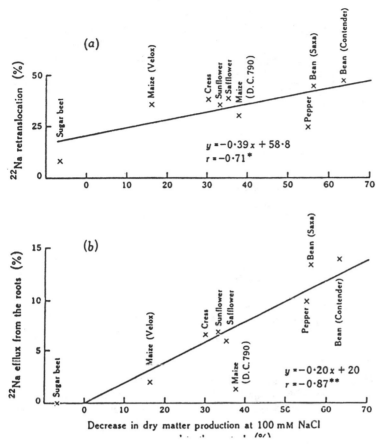

**Figure 2**  Relationship between decrease in dry matter production and (a) Na retranslocation and (b) efflux of Na from roots in species differing in their tolerance to salinity. (From Ref. 87.)

## F.  Role of Transpiration

Shoot ion concentrations are a product of transpiration rate, xylem ion concentrations, and growth rate [93]. Under high evapotranspirational demands, transpiration increases, while K/Na selectivity decreases, resulting in increased Na and Cl uptake [94,95]. Alternatively, a reduction in transpiration can decrease ion (Na and Cl) uptake [7,39,72].

A number of hypotheses have been proposed to explain increased xylem sap Na and Cl levels under high evapotranspiration rates in saline growth media. Enhanced water flow interacts with ion flow across membranes of root cells at more than one site, thus interfering with processes that regulate the balance between ion accumulation in the root cell vacuole and transport to the shoot [7]. Increased water flow due to transpiration promotes passive ion movements where there is no active transport barrier [96].

Water flow can promote the ion flow across the cortex toward a pump that secretes ions into xylem vessels [92]. Ions could either be moved by water along an apoplastic pathway at high concentrations or be coupled to water flow during symplastic passage across the root [97]. High transpiration rates increased Na transport more than K, thus shifting the selectivity toward Na [98,99]. Potassium ions absorbed in roots may be released through Na/K exchange, mainly from vacuoles, for transport to the shoot at times of high evaporative demand [7].

In halophytes, the entry of ions such as Na or Cl into the roots or their release to the xylem sap is tightly regulated at high evapotranspirative demand under saline conditions, thus regulating ion supply to the shoot. Certain morphological features, such as increased width or early development of casparian strips [100] or formation of a double endodermis [101,102], have been reported to develop under high evaporative demands, thus minimizing the passive influx and bypass flow of Na and Cl ions into the xylem. We are not aware of such anatomical changes reported in any crop species under saline conditions. It may be worthwhile to examine genotypes that show high salinity tolerance for such kinds of adaptive features. Because water use is tightly linked to ion uptake and selectivity, the morphological and physiological traits that increase water use efficiency (WUE) in a given genotype could have a role in determining salinity tolerance [93]. In rice, genotypes that showed higher WUE also had a higher level of salinity tolerance [93].

## III.  ORGANIC SOLUTE ACCUMULATION

A wide variety of organic solutes have been reported to accumulate in plant tissues during water and salt stress and are hypothesized to have functions including cytoplasmic osmotic adjustment, protecting cytoplasm and chloroplasts from sodium damage, and stabilizing proteins and membrane structure [103–105] (this volume, Chapter 45 on glycine betaine for further discussion). The chemical nature of the compatible solutes varies from one taxonomic group to another, but most are derivatives of polyols or nitrogen dipoles [39] (Table 1). Osmotic adjustment by the plant promotes turgor maintenance and is thus associated with adaptation to both high soil salinity and low soil moisture [5,106] (see Chapter 45). Compatible solutes are an important factor in the osmotic balance of the cytoplasm under salt stress [33], where sodium salts are sequestered to play a complementary osmotic role in the vacuole [4,33,35] (see Chapter 17). However, this is considered to be a halophytic mode of osmoregulation [2,107], which is energetically more efficient than overall osmoregulation by organic solutes [6,108], a common feature of glycophytes [103].

These organic solutes may comprise common metabolites such as sugars, amino acids such as proline [109–112], and organic acids such as prolinebetaine [113] and other aliphatic quaternary ammonium compounds [114] (see Chapter 45). There is evidence that solute accumulation is a regulated process and not merely the result of a discrepancy between the sensitivity of the growth process and photosynthesis to stress [115]. Nevertheless, metabolites such as glucose and sucrose accumulate in tissues whose growth has been inhibited by stress [103].

The type of stress would determine which compounds act as osmotic solutes [116]. In grain sorghum, betaine accumulates only under moderate levels of salt stress, not under water stress [116]. However, in crops such as wheat, barley, and rye, betaine accumulates under water stress as well as salinity stress [116]. Nevertheless, salinity is reported to be the more effective stimulator of betaine accumulation [117] (see Chapter 45). In barley, more glycinebetaine is accumulated under gradual stress, but proline is the predominant solute under sudden stress [118].

## A.  Role in Osmoregulation

High concentrations of organic solutes in the cytoplasm could contribute to the osmotic balance when electrolytes are lower in the cytoplasm than in the vacuole [109,119]. These compatible solutes could also act as a nitrogen source [120] or protect membranes against salt inactivation [121,122]. These proposed activities may complement each other within the integrated metabolic and ontogenic pattern of a particular species [123].

Under saline conditions, the large quantities of Na, K, and Cl and other ions that are translocated to the shoot and contribute to the osmotic adjustment are believed to accumulate mainly in the vacuole after reaching threshold levels in the cytoplasm [8]. This concentration of inorganic ions could be considered as a threshold level at which accumulation of organic solutes such as proline, betaine, or other compounds begins in the cytoplasm, thus maintaining the intracellular osmotic balance between cytoplasm and vacuole [116]. For instance, in wheat, proline began accumulating when Na+K exceeded a threshold value of 200 mol/g fresh weight [124]. Also, in grain sorghum, a moderate level of salt stress (0.4 MPa or more) is required to induce a significant betaine concentration [125,126]. Further studies are needed to determine the extent to which this threshold level varies among genotypes of a given species.

**TABLE 1**  Types of Compatible Solutes That Could Accumulate Under Salinity Stress in Various Plant Species

| Solute | Structure | Distribution |
|--------|-----------|--------------|
| D-Sorbitol | $\begin{array}{c} CH_2OH \\ H-C-OH \\ HO-C-H \\ H-C-OH \\ H-C-OH \\ CH_2OH \end{array}$ | Plantaginaceae<br>Rosaceae |
| D-Mannitol | $\begin{array}{c} CH_2OH \\ HO-C-H \\ HO-C-H \\ H-C-OH \\ H-C-OH \\ CH_2OH \end{array}$ | Combretaceae<br>Myrsinaceae<br>Rublaceae |
| D-Pinitol | *(cyclohexane ring structure with OH, OH, HO, CH₃O, OH, OH groups)* | Leguminoseae<br>Rhizophoraceae<br>Caryophyllaceae |
| L-Quebrachitol | *(cyclohexane ring structure with OH, OCH₃, OH, OH, OH, OH groups)* | Euphorblaceae |
| Glycine betaine | $(CH_3)_3N^\circ CH_3COO^-$ | Chenopodlaceae<br>Amaranthaceae<br>Asteraceae<br>Solanaceae<br>Gramineae<br>Avicenniaceae |
| β-Alanine betaine | $(CH_3)_3N^\circ CH_3CH_3COO^-$ | Plumbaginaceae |
| Proline | *(pyrrolidine ring with N, COO⁻, H₃)* | Juncaginaceae<br>Asteraceae<br>Gramineae |
| Proline betaine (stachydrine) | *(pyrrolidine ring with N, COO⁻, CH₃ CH₃)* | Lablaicae<br><br>Capparidaceae<br>Leguminosae |
| 3-Dimethylsulfonio propionate | $(CH_3)_3S^\circ CH_3CH_3COO^-$ | Asteraceae<br><br>Gramineae |

*Source:* Ref. 39.

The total quaternary ammonium compounds (QACs) in the leaf tissue in wheat species (*Triticum aestivum* and *T. durum*) shows a high positive correlation with salinity treatment [116]. The capacity to accumulate betaine in grasses has been reported to be correlated with basal levels of betaine in unstressed plants [127]. Crops such as oats and rice, which have very low betaine levels under nonsaline conditions, accumulated very little under stress conditions [116].

The relatively small increases in glycinebetaine with increasing external salinity, together with the high levels found in many halophytes at very low external salinity, imply that this solute may be redistributed between the vacuole and cytoplasm, depending on tissue electrolyte concentrations [119]. However, in crop plants such as sorghum, it is reported that betaine is relatively nonlabile compared with compounds such as proline [128,129]. A sixfold increase in glycinebetaine levels in isolated chloroplasts of spinach under saline conditions was observed, which could account for 36% of the osmotic adjustment in chloroplasts [130].

Proline levels can change quickly in response to abrupt stress, whereas other organic solutes accumulate more slowly [126]. Thus, when stress is applied slowly, less proline accumulates, but the total accumulation of organic solutes remains predictable on the basis of tissue Na and Cl levels [118]. Accumulation of free proline has been correlated with tissue Na concentration in a number of crop species [131–133]. A level of 25 mol proline/g fresh weight could produce a concentration of 280 mM if confined to cytoplasm, thus making a significant contribution to the cytoplasmic solute potential [4].

Proline concentrations were reported to be directly proportional to Na concentrations [134]; each increase in Na concentration is reported to be balanced by an increase in proline concentration equal to about 4% of the rise in Na [135]. This relationship between steady-state proline concentrations and Na levels indicates its role as a cytoplasmic solute [135]. Proline levels for various grasses (*Sorghum bicolo*, *Agrostis stolonifera*, *Cyanodendactyla*, *Paspalum vaginatum*, etc.) increased in response to Na accumulation [134]. However, overall proline levels and accumulation rates were highly variable among grasses and therefore are not reliable indicators of relative tolerance levels [134].

In pigeonpea, proline levels increased with increasing external salinity in two genotypes differing in their salt tolerance. The highest proline levels were observed at 10 dS/m, where both genotypes died subsequently [41]. Among the wild species related to pigeonpea, there is a steady increase of proline levels with increasing external salinity in only a few species (Fig. 3). There was no clear relationship between salinity tolerance and proline accumulation, as proline accumulated to higher levels in both sensitive and tolerant species [41]. Similarly, some tolerant and sensitive species did not accumulate significant levels of proline at any level of external salinity, thus defying any simple relationship between proline accumulation and salinity tolerance [41].

It is usually assumed that the cytoplasm makes up about 5% of the cell's volume, proline is accumulated in the cytoplasm, and Na is largely sequestered into the vacuole [135]. Under these conditions, proline alone could merely osmotically balance the Na. However, other ions and organic solutes are also likely to be involved, as field salinity is often caused by a mixture of salts. Thus, a variety of ions, particularly K, Mg, or Ca, can accumulate in the cytoplasm under those conditions. Given the wide range of organic solutes that can accumulate in different crop species (Table 1) or even among different genotypes within a crop species, which may have a functionally similar role, it would be unrealistic to expect any direct correlation between salinity tolerance and accumulation of any one particular organic solute, either qualitatively or quantitatively.

## B. Role in Ion Compartmentation

Compatible solutes or cytosolic solutes could play an important role in regulating intracellular ion distribution under salt stress, thus inducing Na accumulation in the vacuole [134]. Externally applied glycinebetaine was reported to increase the vacuolar Na concentration in barley roots [128]. The salt concentration required for proline accumulation could be the same as that required for salts to be sequestered into the vacuole [126]. The reported threshold of about 200 mol (Na+K)/g fresh weight is only slightly above (Na+K) levels measured in unstressed leaves [126]. In sorghum, proline accumulation seems to be related to total monovalent cation concentration whether Na or K salts were used in the salinity treatment [125]. An ion pump at the tonoplast could become active at about the same cytoplasmic salt concentration that activates the accumulation of proline or other organic solutes [126].

Proline concentration (μg/g fresh Wt.)

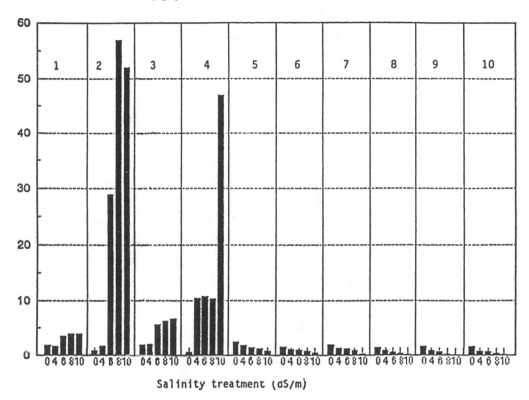

Salinity treatment (dS/m)

**Figure 3**  Proline accumulation in the wild relatives of pigeonpea (*Atylosia* sp.) at various salinity levels (leaf samples were collected 50 days after sowing). 1, *A. albicans*; 2, *A. sericea*; 3, *A. acutifolia*; 4, *A. lineata*; 5, *A. cajanifolia*; 6, *A. volubilis*; 7, *A. reticulata*; 8, *A. grandifolia*; 9, *A. goensis*; 10, *A. lanceolata*. (From Refs. 41 and 181.)

## C.  Role in Protecting Enzymes Against Monovalent Cations

Apart from the purpose of osmoregulation, organic solutes can accumulate to protect cell metabolism from the toxic effects of accumulated ions [4,132–134,136] (see Chapter 45). Pollard and Wyn Jones [137] demonstrated such protection using glycinebetaine and, in barley leaves, with the enzyme malate dehydrogenase (decarboxylating). Glycinebetaine has been reported to stabilize enzymes and membranes partially against a range of perturbations [138]. Proline levels up to 600 mM did not inhibit enzyme activity in vitro [139]. In barley, 1000 mM proline did not inhibit dehydrogenase activity [103]. Polyribosomes are stable in vitro in glycinebetaine and proline concentrations up to about 1000 mM [140].

Thus, the effect of proline and glycinebetaine on enzyme systems in the presence of inhibitory ion concentrations may be an expression of a wider role of such compounds in protein stability [103]. Most organic solutes that accumulate under stress conditions are compatible with enzyme activity and continued metabolism [103].

Osmoregulators not only can be compatible with cytoplasmic enzymes but also can either promote or inhibit enzyme activity, depending on the enzyme source [141]. The affinity of phosphoenol pyruvate carboxylase (PEPCase) (extracted from *Cynodon dactylon* and *Sporobolus pungens* grown on saline soil) for PEP was increased by betaine and proline, which resulted in full protection against NaCl inhibition [141]. However, proline did not protect PEPCase against NaCl when it was extracted from *Salsola soda*, although betaine did provide protection [141]. These differences could be due to the existence of isoenzymes.

Although organic-compatible solutes may ameliorate some of the effects of accumulated ions, it seems that ion compartmentation is of greater significance in preserving metabolic activities. In some cases, the effects of compatible solutes are apparent only under severe stress and act merely as a survival trait rather than having any beneficial effect on growth during stress [142]. But they may promote growth recovery if these solutes protect enzyme systems against stress-induced degradation so that they can recommence synthetic function rapidly [103].

## D.  Metabolic Costs of Organic Solute Accumulation

Despite active accumulation of organic osmotica, there is no evidence of an additional cost, and thus osmotic adjustment exists as an energy-efficient and physiologically effective device for alleviation of drought and salinity stress [143]. However, synthesis of organic molecules such as proline or betaine does put an additional metabolic load on the plant. When sugars are used for osmotic adjustment, they are not available for growth [143]. The accumulation of nonstructural carbon is associated with osmotic adjustment and turgor maintenance [8]. Turner [144] considered that the carbon required for osmotic adjustment would be only a small fraction of that produced by the plant. However, the metabolic cost of storing photosynthate and using it for osmotic adjustment is less than the cost of converting it to new biomass, which the nonstressed plants were better able to do [143]. This explanation was confirmed by the fact that there was a large increase in the respiration rate accompanied by a rapid increase in leaf area when stressed plants were irrigated [143].

From the preceding, it appears that a variety of organic solutes accumulate under salinity or drought stress conditions. Some of these compounds could be the result of passive accumulation (i.e., due to the general reduction in growth processes). Carbon and nitrogen compounds are simply diverted from growth-related activities to produce compounds such as proline, sucrose, or others as a way of storing them. This avoids formation of toxic compounds, such as ammonia or putricine, from excess nitrogen metabolites. However, there is evidence that solute accumulation is an active process and is very strongly regulated according to immediate plant needs as influenced by external salinity and the plant's ability to regulate ion entry into the transpiration stream. Also, apart from acting as an organic osmoticum in the cytoplasm, these compatible solutes accelerate the compartmentation of Na and Cl into the vacuole, thus playing a significant role in determining the crop species' level of salinity tolerance. However, it needs to be realized that organic solute accumulation is only one component in the overall maintenance of a stable internal ionic environment in the cytoplasm, which would ultimately determine the survival and production potential of a crop species grown in a saline environment. Thus, the ability to accumulate organic solutes would have a positive functional role only if a genotype has the "genetic knowhow" to regulate ion entry, particularly of Na and Cl, into the transpiration stream.

## IV.  ORGANISM INTEGRATION

Although various processes that play a role in ionic and osmotic regulation at the whole plant level have being discussed separately, the level of salinity tolerance of a given crop species or genotype is the collective expression of a number of processes. These are influx selectivity, K/Na exchange, and Na extrusion, Na compartmentation in the root cortex, Na and Cl regulation at the endodermis, retrieval of Na from the xylem stream by XPT, transpiration efficiency, preventing apoplastic accumulation, phloem retranslocation of Na and Cl, K retranslocation, organic solute accumulation, Na and Cl compartmentation in the leaf, and others. For this reason, it is not surprising that no single physiological mechanism or trait shows a clear-cut direct relationship to salinity tolerance. Genotypes may differ in one or many processes that regulate entry of Na or Cl ions into the plant or qualitative or quantitative differences in the organic solutes. These processes interact at the organism level to determine the ultimate level of tolerance.

## V.  CONCEPTUAL FRAMEWORK FOR INTEGRATING PHYSIOLOGICAL ASPECTS INTO GENETIC IMPROVEMENT PROGRAMS

There is a substantial amount of information on the physiological responses of crop plants to salinity (i.e., mostly NaCl) stress. A major portion of this information deals merely with the effects of excess salts on various metabolic functions of the plants. As Munns et al. [45] pointed out, most of this information describes only the consequences rather than the causes of reduced growth or injury and is thus of limited use for integration into genetic improvement programs. We believe that there is scope for more directed physiological research that would be more relevant to genetic improvement considerations. Emphasis should be given to understanding the interactions among the many possible processes involved and thus "organism integration." The two main approaches that we see for achieving this are the "black box" and "physiological ideotype" approaches.

## A. Black Box Approach

The black box approach attempts to proceed from established phenotypic differences (i.e., response to salinity) to the underlying differences in physiological mechanisms contributing to higher levels of tolerance [107,145]. Once a source of a higher level of salinity tolerance is identified in the cultivated species or its wild relatives, the next step would be to transfer this tolerance to agronomically acceptable varieties through a conventional breeding approach. Because salinity tolerance is a complex physiological trait, governed by different genes or groups of genes, the problem is how best to transfer this type of trait or ensemble of traits from the donor parent to the recipient. A black box approach is therefore enhanced by an understanding of the specific physiological traits operating in the donor parent by conducting comparative physiological studies between donor and recipient parents. This will facilitate design of the most appropriate genetic improvement procedures. In particular, simple and effective means of screening segregating populations for salinity tolerance are needed rather than having to rely on the measurement of growth or yield reduction under given levels of salinity. Identification of the predominant physiological trait or traits responsible for the genotypic differences measured is desirable.

In pigeonpea and its related wild species, there appears to be either a curvilinear or a linear relationship between dry matter and tissue Na or Cl levels [$R^2 = 0.76$, $R^2 = 0.70$ ($P < .001$), Figure 4a and b]. However, this relationship is stronger for Na than for Cl. There is a significant positive linear relationship between tissue Na and Cl levels in both shoots and roots [$R^2 = 0.66$ ($P < .001$), Figure 4e and f]. Although the overall relationship between growth reduction and tissue Na or Cl levels appears to be positive, there is considerable variation among various wild species in the level of ionic tolerance within their tissues. This is indicated by the scatter of points. For instance, for a 50% reduction in growth, tissue Cl levels ranged from <1% to about 4%, and for Na it varied from 0.02% to about 1%. For tissue K levels, we did not find any significant relationship ($R^2 = 0.008$, Figure 4c); however, there is a positive relation between K/Na in shoot and shoot growth [$R^2 = 0.73$ ($P < .001$), Figure 4d]. These data points are also very much scattered, which indicates a wide range of variation among species for their optimum K/Na requirements at a given level of growth reduction under salinity. This is not surprising given the complexity of physiological mechanisms operating in Na, K, and Cl regulation and the number of mitigating factors that could change the metabolic tolerance of Na and Cl levels in the tissues.

However, in comparing genotypes that differ in their tolerance, especially among the wild relatives of pigeonpea, we have noticed that the ability to retain higher levels of Na and Cl in the roots could be one of the crucial factors in regulating their levels in the shoot. This regulatory ability breaks down at salinity thresholds that vary across species and genotypes [41,48]. Further studies have shown that this regulatory ability is expressed in the $F_1$ hybrids of crosses between a tolerant wild relative (*Atylosia albicans*) and a sensitive pigeonpea genotype (ICP 3783) (Figure 5) [48]. Thus, this trait is heritable. Further studies are required on the segregating $F_2$ and $F_3$ generations, including the analysis of the ionic constituents, to establish the inheritance pattern of these physiological traits.

## B. Physiological Ideotype and Pyramiding Approach

An "ideotype" is defined as "a hypothetical plant described in terms of traits that are thought to enhance genetic yield potential" [146]. Thus, a physiological ideotype for salinity tolerance could be defined in terms of the specific physiological traits that are expected to contribute functionally in maintaining ionic and osmotic relations under saline conditions. As expressed on a relative yield basis, it is the collective expression of a number of physiological traits as described earlier.

Salinity stress normally varies over time within a crop cycle, from season to season and from site to site. Different landraces/genotypes/varieties that show a given level of tolerance to salinity are expected

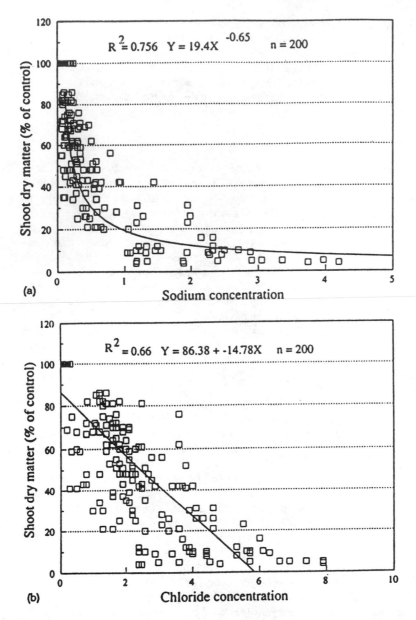

**Figure 4**   Relationships between shoot dry matter and tissue Na, Cl, K, and K/Na levels (a–d) and between Cl and Na levels in root and shoot (e and f). (Plant samples were collected for growth and chemical analysis 55 days after sowing; plants were grown at 0, 4, 6, 8, and 10 dS/m salinity levels.) (From Ref. 181.)

to have evolved a variety of mechanisms that contribute to yielding ability under those conditions. For instance, *T. aestivum*, *Secale cereale*, and *Aegilops squarrosa* have an efficient K/Na selectivity character because of the D genome but are less tolerant than crop species such as *H. vulgare* and *T. durum*, which are less efficient in K/Na selectivity but more efficient in their compartmentation ability [147,148]. Similarly, such differences can be observed among genotypes within a crop species, which is reflected in contradictory reports for various crop species either confirming or disputing direct correlation between K/Na selectivity and level of salinity tolerance [5].

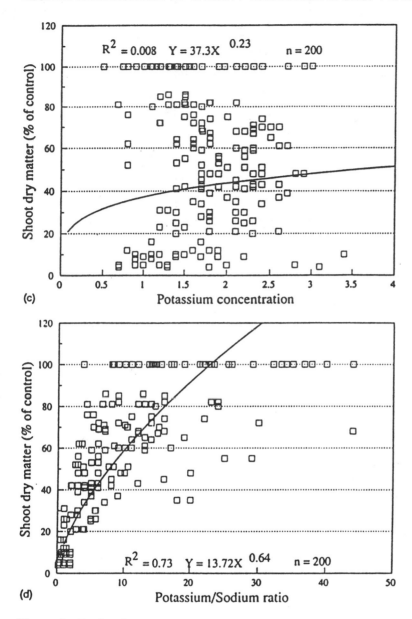

**Figure 4**  Continued.

The underlying philosophy is that, although different genotypes may show the same level of tolerance to salinity, they could attain this level of tolerance (i.e., phenotype) through different physiological mechanisms or traits. Lack of sufficient phenotypic variation for salinity tolerance is a serious problem in many crops, and this is particularly so with rice [149]. Even after screening the entire world collection of rice germ plasm, the most tolerant genotypes would still suffer about a 50% yield reduction at 5 dS/m [149–151]. Conceptually, the physiological approach for improving salinity tolerance in crop plants should be to bring together the relevant traits that would complement each other in a pyramidal manner ("building block" approach) by their selective incorporation into a single genotype or variety under improvement (i.e., optimization of several, probably independent, physiological mechanisms in a single variety) [31].

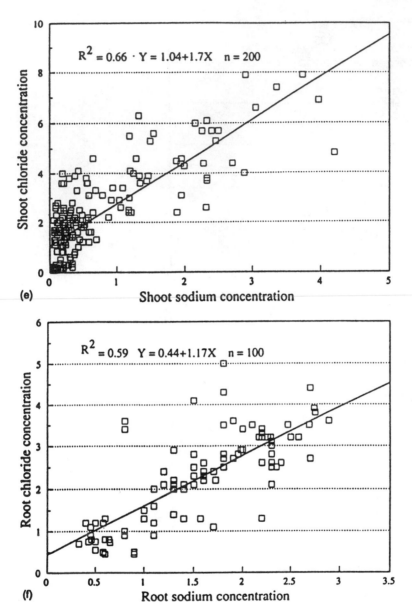

**Figure 4**  Continued.

An analogy can be drawn from disease resistance breeding. In breeding for disease resistance horizontal resistance (which can be defined as resistance to a number of physiological races of a disease) can be achieved by pyramiding different genes specifically resistant to individual physiological races. This contributes to the stability of a genotype across years in disease-prone environments. The same concept could also be applied to the genetic improvement of salinity tolerance, whereby pyramiding of genes that regulate various specific physiological traits into a single genotype or variety could provide that genotype with the necessary genetic means to respond to different types and levels of salinity stress that it is likely to experience at different locations and sites and over years. This would contribute to its stability of production as well as widening its adaptability to a greater range of saline environments.

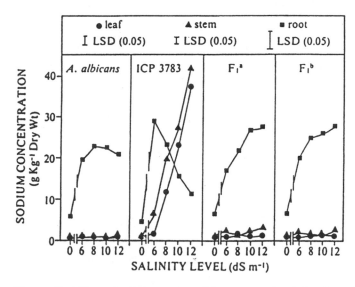

**Figure 5**   Effect of salinity on tissue Na concentration (g/kg dry weight) of *Atylosia albicans*, *Cajanus cajan* (ICP 3783), and their reciprocal $F_1$ hybrids (a and b), 75 days after transplanting. Data are means of two replications. (From Ref. 48.)

Various steps are involved in this kind of approach:

1. Define the various physiological traits having functional significance in determining the tolerance and productivity of a given crop in saline environments.

2. Establish genetic variability and locate sources of high efficiency for each physiological trait in the germ plasm. Selection should be directed toward the individual components of salinity tolerance on a trait-by-trait basis irrespective of phenotype.

3. Establish the genetic basis for each physiological trait under consideration by studying its inheritance pattern and estimating its heritability, which would determine the feasibility of using that particular trait in a breeding program.

4. Develop restriction fragment length polymorphism (RFLP) markers if easily identifiable morphological, physiological, or other markers are not readily available for each physiological trait as this would streamline the selection process of segregating materials in a breeding program.

5. Identify genotypes for each physiological trait that have good combining ability.

6. Incorporate relevant traits into an agronomically acceptable background basis.

Information generated through this exercise could be stored in a database system that would be made available to breeders interested in incorporating salinity tolerance in their breeding programs. This is similar to information databases that are available for morphological traits from the germ plasm evaluation exercises at CGIAR (Consultative Group for International Agricultural Research) centers.

Selection of traits to be introduced into a given genotype or variety under improvement depends on the target environment in which it will be grown and the specific traits a particular variety may be lacking. For instance, a variety may be very efficient in Na and Cl compartmentation in the root as well as in the shoot but may be lacking effective Na or Cl regulation at the plasmalemma. There is evidence of genotypic variation within crop species in Na compartmentation in shoots [31,152,153] and tolerance to high internal Na and Cl levels [80]. In this case, only the trait that is lacking needs to be introduced. Similarly, a given variety may be very efficient in ion regulation but lack the genetic means necessary to produce organic solutes.

Development of RFLP markers for each of these physiological components of salinity tolerance could play a crucial role in the incorporation of these physiological traits into a genotype or variety under improvement. Salinity tolerance traits are controlled by a number of genes located throughout the chromosome complement [154]. Each gene of a polygenic system may contribute only a small amount to the trait of interest. Clear dominance is not likely to be exhibited, and the phenotype (i.e., the specific trait in this case) would have a large component of environmental variance. All these characteristics conspire to make physiological traits very difficult to analyze. Thus, conventional Mendelian methods of analysis, which are suitable for traits controlled by a single or a few genes, cannot be applied to analysis of these physiological traits. This is one reason that physiological traits have not been used extensively in the genetic improvement programs for salinity or drought tolerance, although a number of them having functional significance for determining level of tolerance have been identified [7,155].

With the development of RFLP mapping techniques (for a detailed discussion of RFLP techniques see Tanksley et al. [156]), it is possible to analyze complex polygenic characters, such as physiological traits, as ensembles of single Mendelian factors. Because RFLP markers can be used to follow simultaneously the segregation of all chromosome segments during a cross, the basic idea is to look for correlations between physiological traits and specific chromosome segments marked by RFLPs. If correlations exist, the inference is that the chromosome segment must be involved in the quantitative trait. The difficult part in this procedure is to establish correlations between the trait and specific chromosome segments. The RFLP markers can be easily scored, but the physiological trait must be characterized in a conventional fashion [156]. Once this most difficult process is completed and specific chromosome segments are implicated in the trait, RFLP markers with a positive effect on a quantitative trait can be selected from a population of plants and incorporated into a single genotype. This is possible because of the ability to score for several RFLP markers simultaneously in a single plant in a manner that is free from environmental influence or gene interactions. Carbon isotope ($^{13}$C) discrimination, which is an indicator of water use efficiency, could be satisfactorily predicted from three RFLPs in tomato [157]. The K/Na discrimination trait of *Aegilops tauchii* Cosson has been linked to five RFLPs on the distal third of the long arm of chromosome 4D [158]. Also, three RFLP markers were linked to osmotic adjustment in sunflower [159]. These findings demonstrate the feasibility of using RFLP markers for physiological traits that could bridge the gap between plant physiology and breeding, to facilitate integration of these two disciplines and thus expedite development of varieties that are higher yielding and more stable across environments affected by salinity.

## VI. FUTURE OUTLOOK

The past 30 years of research (after the report of dual mechanisms of ion transport by Epstein et al. [160]) on physiological aspects of salinity tolerance has contributed substantially to an understanding of the mechanisms by which plants cope with excess salts in their habitat. In recent times, efforts have been initiated to identify genes responsible for specific physiological mechanisms [13,161–165]. Overexpression of a vacuolar $Na^+/H^+$ antiport has been linked to increased salinity tolerance in *Arabidopsis thaliana* [166]. Location of the K/Na selectivity character on the 7a chromosome of the D genome in wheat is one such example [167–169]. Similarly, Na exclusion capability and K/Na discrimination were enhanced in *T. aestivum* by the incorporation of a *Lophopyrum* genome [170]. The K/Na discriminating locus has been located on the 3E chromosome in *Lophopyrum elongatum* [154]. An association with the higher level of salinity tolerance in *Agropyron junceum* has been located in the 5J chromosome [171]. There were some efforts to link a certain ion channel type with a lower Na/K permeability ratio in salt-tolerant genotypes than in salt-sensitive genotypes of wheat [172,173]. In rice, at least three groups of genes were found to be involved in the inheritance of Na and Ca levels in the plant; Na and Ca levels in shoots and roots were reported to show additive effects with a high degree of heritability [174].

Similarly, Cl translocation is under genetic control [175,176]. Accumulation of organic solutes such as betaine has been reported to be regulated by a limited number of genes [177–180]. Our research with pigeonpea [181] has shown that the higher levels of salinity tolerance, and the associated physiological mechanisms identified in the wild relative *Atylosia albicans*, could be expressed in the reciprocal crosses of $F_1$ hybrids of this species with the cultivated species (Figure 5) [48]. Information on the genetic control of specific mechanisms is essential for proper integration of physiological research into breeding programs. Developments in biotechnology, particularly with genetic markers such as RFLPs, could accelerate this integration of disciplines. Wild relatives have been inadequately explored for their potential to contribute unique physiological mechanisms of salinity tolerance. We hope future efforts would be directed toward generating information in these areas.

# REFERENCES

1. DH Jennings. The effects of sodium chloride on higher plants. Biol Rev 51:453–486, 1976.
2. EP Glenn, J Jed Brown, E Blumwald. Salt tolerance and crop potential of halophytes. Crit Rev Plant Sci 18:227–255, 1999.
3. DW Rains. Salt transport by plants in relation to salinity. Annu Rev Plant Physiol 23:367–388, 1972.
4. TJ Flowers, PF Troke, AR Yeo. The mechanisms of salt tolerance in halohytes. Annu Rev Plant Physiol 28:89–121, 1977.
5. H Greenway, R Munns. Mechanisms of salt tolerance in nonhalophytes. Annu Rev Plant Physiol 31:149–190, 1980.
6. AR Yeo. Salinity resistance: physiologies and prices. Physiol Plant 58:214–222, 1983.
7. WD Jeschke. K-Na exchange at cellular membranes, intracellular compartmentation of cations and salt tolerance. In: RC Staples, GH Toenniessen, eds. Salinity Tolerance in Plants: Strategies for Crop Improvement. New York: John Wiley & Sons, 1984, pp 37–66.
8. JM Cheeseman. Mechanisms of salinity tolerance in plants. Plant Physiol 87:547–550, 1988.
9. R Serrano. Salt tolerance in plants and microorganisms: toxicity targets and defense responses. Int Rev Cytol 165:1–52, 1996.
10. P Neumann. Salinity resistance and plant growth revisited. Plant Cell Environ 20:1193–1198, 1997.
11. MF Robinson, A Very, D Sanders, TA Mansfield. How can stomata contribute to salt tolerance? Ann Bot 80:387–393, 1997.
12. R Romero-Aranda, JL Moya, FR Tadeo, F Legaz, E Primo-Millo, M Talon. Physiological and anatomical disturbances induced by chloride salts in sensitive and tolerant citrus: beneficial and detrimental effects of cations. Plant Cell Environ 21:1243–1253, 1998.
13. A Yeo. Molecular biology of salt tolerance in the context of whole-plant physiology. J Exp Bot 49:915–929, 1998.
14. KM Volkmar, Y Hu, H Steppuhn. Physiological responses of plants to salinity: a review. Can J Plant Sci 8:19–27, 1998.
15. A Amtmann, D Sanders. Mechanisms of Na uptake by plant cells. Adv Bot Res 29:75–112, 1999.
16. DB Lazof, N Bernstein. The NaCl induced inhibition of shoot growth: the case for disturbed nutrition with special consideration of calcium. Adv Bot Res 29:113–189, 1999.
17. R Serrano, JM Mulet, G Rios, JA Marquez, IF de Larrinoa, MP Leube, I Mendizabal, A Pascual-Ahuir, M Proft, R Ros, C Montesinos. A glimpse of the mechanisms of ion homeostasis during salt stress. J Exp Bot 50:1023–1036, 1999.
18. H Boyko. Salinity and Aridity: New Approaches to Old Problem. The Hague: Junk Publishers, 1966.
19. SR Grattan, CM Grieve. Mineral nutrient acquisition and response by plants grown in saline environments. In: M Pessarakli, ed. Handbook of Plant and Crop Stress. 2nd ed. New York: Marcel Dekker, 1999, pp 203–229.
20. TB Kinraide. Interactions among $Ca^{2+}$, $Na^+$ and $K^+$ in salinity toxicity: quantitative resolution of multiple toxic and ameliorative effects. J Exp Bot 50:1495–1505, 1999.
21. E Epstein. Mineral Nutrition of Plants: Principles and Prospectives. New York: John Wiley & Sons, 1972.
22. P Mitchell. Coupling of phosphorylation to electron and hydrogen transfer by a chemi-osmotic type mechanism. Nature 191:144–145, 1961.
23. RN Robertson. Protons, Electrons, Phosphorylation, and Active Transport. Cambridge: Cambridge University Press, 1968.
24. DP Briskin, JB Hanson. How does the plant plasma membrane $H^+$-ATPase pump protons? J Exp Bot 43:269–289, 1992.
25. DT Clarkson, C Grignon. The phosphate transport system and its regulation in roots. In: C Johansen, KK Lee, KL Sahrawat, eds. Phosphorus Nutrition of Grain Legumes in the Semi-Arid Tropics. Patancheru: ICRISAT, 1961, pp 49–62.
26. AC Leopold, RP Willing. Evidence for toxicity of effects of salt on membranes. In: RC Staples, GA Toenniessen, eds. Salt Tolerance in Plants: Strategies for Crop Improvement. New York: John Wiley & Sons, 1984, pp 67–76.
27. PJC Kuiper. Functioning of plant cell membranes under saline conditions: membrane lipid composition and ATPases. In: RC Staples, GA Toenniessen, eds. Salinity Tolerance in Plants: Strategies for Crop Improvement. New York: John Wiley & Sons, 1984, pp 77–91.
28. MG Pitman, HDW Saddler. Active sodium and potassium transport in cells of barley roots, Proc Natl Acad Sci U S A 57:44–49, 1967.
29. DH Jennings. Halophytes, succulence and sodium in plants—a unified theory. New Phytol 67:899–911, 1968.
30. WD Jeschke. K stimulated Na efflux and selective transport in barley roots. In: WP Anderson, ed. Ion Transport in Plants. London: Academic Press, 1973, pp 285–290.
31. AR Yeo, TJ Flowers. Salinity resistance in rice (*Oryza sativa*): a pyramiding approach to breeding varieties for saline fields. Aust J Plant Physiol 13:161–173, 1986.
32. WD Jeschke. Cation fluxes in excised and intact roots in relation to specific and varietal differences, In: NEL Bassam, M Dambroth, BC Loughman, eds. Genetic Aspects of Plant Nutrition. London: Kluwer Academic Publishers, 1990, pp 71–86.

33. RG Wyn Jones, CJ Brady, J Speirs. Ionic and osmotic relations in plant cells, In: DL Laidman, RG Wyn Jones, eds. Recent Advances in the Biochemistry of Cereals. London: Academic Press, 1979, pp 63–104.

34. RG Wyn Jones. An assessment of quaternary ammonium and related compounds as osmotic effects in crop plants. In: DW Rains, RC Valentine, A Hollaender, eds. Genetic Engineering of Osmoregulation: Impact on Plant Productivity for Food, Chemicals, and Energy. New York: Plenum, 1980, pp 155–170.

35. WD Jeschke. Univalent cation selectivity and compartmentation in cereals. In: DL Laidman, RG Wyn Jones, eds. Recent Advances in the Biochemistry of Cereals. London: Academic Press, 1979, pp 37–61.

36. J Rozema, ED Rozema, AHJ Freijsen, JJL Huber. Population differentiation within *Festuca rubra* (L.) with regard to soil salinity and soil water. Oecologia 34:329–341, 1978.

37. J Gorham, LL Hughes, RG Wyn Jones. Chemical composition of salt marsh plants from Ynysmon (Anglesey): the concept of physiotypes. Oecologia 3:309–318, 1980.

38. I Ahmad, SJ Wainwright, GR Stewart. The solute and water relations of *Agrostis stolonifera* ecotypes differing in their salt tolerance. New Phytol 87:615–629, 1981.

39. J Gorham, RG Wyn Jones, E McDonnel. Some mechanisms of salt tolerance in crop plants. Plant Soil 89:15–40, 1985.

40. MA Hajibagheri, DMR Harvey, TJ Flowers. Quantitative ion distribution within the root cells of salt sensitive and salt tolerant maize varieties. New Phytol 105:367–379, 1987.

41. GV Subbarao, C Johansen, MK Jana, JVDK Kumar Rao. Physiological basis of differences in salinity tolerance of pigeonpea and its related wild species. J Plant Physiol 137:64–71, 1990.

42. A Scarpa, J deGiez. Biochim Biophys Acta 241:789, 1971 (cited in Ref. 7).

43. A Lauchli. Symplasmic transport in ion release to the xylem. In: IF Wardlaw, JB Passioura, eds. Transport and Transport Processes in Plants. New York: Academic Press, 1976, pp 101–112.

44. C Johansen, DG Edwards, JF Loneragan. Potassium fluxes during potassium absorption by intact barley plants of increasing potassium content. Plant Physiol 45:601–603, 1970.

45. R Munns, H Greenway, GO Kirst. Halotolerant eukaryotes. In: OL Lange, PS Nobel, CB Osmond, H Ziegler, eds. Encyclopedia of Plant Physiology. New Series. Vol 12C. Physiological Plant Ecology. New York: Springer-Verlag, 1983, pp 59–135.

46. E Epstein. Responses of plants to saline environments. In: DW Rains, RL Valentine, A Hollaender, eds. Genetic Engineering of Osmoregulation: Impact on Plant Productivity for Food, Chemicals, and Energy. New York: Plenum, 1980, pp 7–21.

47. D Kramer, A Lauchli, AR Yeo, J Gullasch. Transfer cells in roots of *Phaseolus coccineus*: ultrastructure and possible function in exclusion of sodium from the shoot, Ann Bot 41:1031–1040, 1977.

48. GV Subbarao, C Johansen, JVDK Kumar Rao, MK Jana. Salinity tolerance in $F_1$ hybrids of pigeonpea and a tolerant wild relative. Crop Sci 30:785–788, 1990.

49. R Behl, WD Jeschke. Influence of abscisic acid on unidirectional fluxes and intracellular compartmentation of K and Na in excised barley root segments. Physiol Plant 53:95–100, 1981.

50. RL Jeffries. The ionic relations of seedlings of the halophyte *Triglochin maritima* (L.). In: WP Anderson, ed. Ion Transport in Plants. London: Academic Press, 1973.

51. SM Pan, YR Chen. The effects of salt stress on acid phosphotase activity of *Zea mays* seedlings. Bot Bull Acad Sin 29:33–38, 1988.

52. R Weinberg. Effects of sodium chloride on the activity of a soluble malate dehydrogenase from pea seeds. J Biol Chem 242:3000–3006, 1967.

53. YD Beletskii, TB Karnaukhova, NI Shevyakova. Peroxidase isoenzymes in a salt tolerant plastome mutant of sunflower and a hybrid of it. Sov Plant Physiol 33:890–895, 1986.

54. AJ Cavalieri, AHC Huang. Effect of NaCl on the in vitro activity of malate dehydrogenase in salt marsh halophytes of the USA. Physiol Plant 41:79–84, 1977.

55. SD Tanksley. Isozymes in Plant Genetics and Breeding. Vol I. Amsterdam: Elsevier, 1983.

56. JL Hall, TJ Flowers. The effect of salt on protein synthesis in the halophyte *Suaeda maritima*. Planta 110:361–368, 1973.

57. TJ Flowers. Salt tolerance in *Suaeda maritima* (L.) Dum. The effect of sodium chloride on growth, respiration and soluble enzymes in a comparative study with *Pisum sativum*. J Exp Bot 23:310–321, 1972.

58. TJ Flowers. The effect of sodium chloride on enzyme activity from four halophytic species of Chenopodiaceae. Phytochemistry 11:881–1886, 1972.

59. DH Jennings. The physiology of the uptake of ions by the growing plant cell. In: IH Rorison, ed. Ecological Aspects of the Mineral Nutrition of Plants. Oxford: Blackwell Scientific, 1969, pp 261–279.

60. PA Rea, D Sanders. Tonoplast energization: two $H^+$ pumps, one membrane. Physiol Plant 71:131–141, 1987.

61. MG Palmgren. Regulation of plant plasma membrane $H^+$-ATPase activity. Physiol Plant 83:314–323, 1991.

62. MG Pitman. Transport across the root and shoot/root interactions. In: RC Staples, GH Toenniessen, eds. Salinity Tolerance in Plants: Strategies for Crop Improvement. New York: John Wiley & Sons, 1984, pp 93–123.

63. A Lauchli, J Wieneke. Salt relations of soybean mutants differing in salt tolerance: distribution of ions and localization by x-ray microanalysis, Plant Nutrition, Proceedings of 8th International Colloquium on Plant Analysis and Fertilizer Problems, Auckland, New Zealand. Wellington: Government Printer, 1978, pp 275–282.

64. JG Johansen, JM Cheeseman. Uptake and distribution of sodium and potassium by corn seedlings. I. Role of the mesocotyl in sodium exclusion. Plant Physiol 73:153–158, 1983.

65. JG Johansen, JM Cheeseman. Uptake and distribution of sodium and potassium by corn seedlings. II. Ion transport within the mesocotyl. Plant Physiol 73:159–164, 1983.

66. BJ Cooil, RK Dela Fuente, RS Dela Pena. Absorption and transport of sodium and potassium in squash. Plant Physiol 40:625–633, 1965.

67. A Lauchli, J Wieneke. Studies on growth and distribution of Na, K, and Cl in soybean varieties differing in salt tolerance. Z Pflanzenernaehr Bodenkd 124:3–13, 1979.

68. AR Yeo, D Kramer, A Lauchli, J Gullasch. Ion distribution in salt stressed mature *Zea mays* roots in relation to ultrastructure and retention of sodium. J Exp Bot 28:17–29, 1977.

69. B Jacoby. Sodium recirculation and loss from *Phaseolus vulgaris*. Ann Bot 43:741–744, 1979.

70. RT Besford. Effect of replacing nutrient potassium by sodium on uptake and distribution of sodium in tomato plants. Plant Soil 58:399–409, 1978.

71. R Munns, DB Fisher, ML Tennet. Na and Cl transport in the phloem from leaves of NaCl treated barley. Aust J Plant Physiol 13:757–766, 1986.

72. RR Walker. Sodium exclusion and potassium-sodium selectivity in salt affected trifoliate orange (*Poncirus trifoliata*) and Cleopatra mandarin (*Citrus reticulata*) plants. Aust J Plant Physiol 13:293–303, 1986.

73. H Nassery, DA Baker. Extrusion of sodium ions by barley roots. III. The effect of high salinity on long distance sodium ion transport. Ann Bot 38:141–144, 1974.

74. H Marschner, O Ossenberg-Neuhaus. Longstrecken transport von natrium in bohnenpflanzen. Z Pflanzenernaehr Bodenkd 139:129–142, 1976.

75. U Luttge. Import and export of mineral nutrients in plant roots. In: A Lauchli, RL Bieleski, eds. Encyclopedia of Plant Physiology. New Series. Vol 15. Berlin: Springer-Verlag, 1983, pp 181–211.

76. E Winter. Salt tolerance of *Trifolium alexandrinum*. III. Effects of salt on ultrastructure of phloem and xylem transfer cells in petioles and leaves. Aust J Plant Physiol 9:239–250, 1982.

77. JJ Oertli. Extracellular salt accumulation a possible mechanism of salt injury in plants, Agrochimica 12:461–469, 1968.

78. RA Leigh, RG Wyn Jones. A hypothesis relating critical potassium concentration for growth to the distribution and functions of this ion in the plant cell. New Phytol 97:1–13, 1984.

79. R Munns, JB Passioura. Effect of prolonged exposure to NaCl on the osmotic pressure of leaf xylem sap from intact transpiring barley plants. Aust J Plant Physiol 11:479–507, 1984.

80. NIW Clipson, AD Tomos, TJ Flowers, RG Wyn Jones. Salt tolerance in the halophyte *Suaeda maritima* (L.) dum. Planta 165:392–396, 1985.

81. AR Yeo, SJM Caporn, TJ Flowers. The effect of salinity upon photosynthesis in rice (*Oryza sativa*) gas exchange by individual leaves in relation to their salt content. J Exp Bot 36:1240–1248, 1985.

82. TJ Flowers, YR Yeo. Ion relations of plants under drought and salinity. Aust J Plant Physiol 13:75–91, 1986.

83. TJ Flowers, MA Hajibagheri, AR Yeo. Ion accumulation in the cell walls of rice plants growing under saline conditions: evidence for the Oertli hypothesis. Plant Cell Environ 14:319–325, 1991.

84. JS Pate. Nutrients and metabolites of fluids recovered from xylem and phloem: significance in relation to long distance transport in plants. In: IF Wardlaw, JB Passioura, eds. Transport and Transfer Processes in Plants. New York: Academic Press, 1976, pp 253–281.

85. DR Geiger. Phloem loading in source leaves. In: JF Wardlaw, JB Passioura, eds. Transport and Transfer Processes in Plants. New York: Acadamic Press, 1976, pp 167–183.

86. E Levi. The distribution of mineral elements following leaf and root uptake. Physiol Plant 21:213–226, 1968.

87. H Lessani, H Marschner. Relation between salt tolerance and long distance transport of sodium and chloride in various crop species. Aust J Plant Physiol 5:27–37, 1978.

88. E Winter. Intraveinal recycling of sodium in the leaves of salt tolerant *T. alexandrinum*, L. Micron 2:519–520, 1980.

89. JA Webb, PR Gorham. Translocation of photosynthetically assimilated $C^{14}$ in straightnecked squash. Plant Physiol 39:663–672, 1964.

90. H Greenway, MG Pitman. Potassium retranslocation in seedlings of *Hordeum vulgare*. Aust J Biol Sci 18:235–247, 1965.

91. JS Pate. Exchange of solutes between phloem and xylem and circulation in the whole plant. In: MH Zimmerman, JA Milburn, eds. Encyclopedia of Plant Physiology. New Series. Vol 1. Berlin: Springer-Verlag, 1975, pp 451–473.

92. MG Pitman. Whole plants. In: DA Baker, JL Hall, eds. Ion Transport in Plant Cells and Tissues. Amsterdam: North Holland, 1975, pp 267–308.

93. TJ Flowers, FM Salama, YR Yeo. Water-use efficiency in rice *Oryza sativa* in relation to resistance to salinity. Plant Cell Environ 11:453–459, 1988.

94. H Greenway. Plant responses to saline substrates. IV. Chloride uptake by *Hordeum vulgare* as affected by inhibitors, transpiration, and nutrients in the medium. Aust J Biol Sci 18:249–268, 1965.

95. MG Pitman. Transpiration and the selective uptake of potassium by barley seedlings (*H. vulgare* cv. Bolivia). Aust J Biol Sci 18:987–988, 1965.

96. DTF Bowling, Weatherley. The relationship between transpiration and potassium uptake in *Ricinus communis*. J Exp Bot 16:732–741, 1965.

97.   MG Pitman. Ion transport into the xylem. Annu Rev Plant Physiol 28:71–88, 1977.

98.   F Malek, DA Baker. Proton co-transport of sugars in phloem loading. Planta 135:297–299, 1977.

99.   MG Pitman. Uptake of potassium and sodium by seedlings of *Sinapsis alba*. Aust J Biol Sci 19:257–259, 1966.

100.  A Poljakoff-Mayber. Morphological and anatomical changes in plants as a response to salinity stress. In: A Poljakoff-Mayber, J Gale, eds. Plants in Saline Environments. Berlin: Springer-Verlag, 1975, pp 97–117.

101.  R Stelzer, A Lauchli. Salt and flooding tolerance of *Puccinellia peisonis*. II. Structural differentiation of the root in relation to function. Z Pflanzenphysiol 84:95–108, 1977.

102.  MA Hajibagheri, JL Hall, TJ Flowers. The structure of the cuticle in relation to cuticular transport in leaves of the halophyte *Suaeda maritima* (L.) dum. New Phytol 94:25–131, 1983.

103.  D Aspinall. Metabolic effects of water and salinity stress in relation to expansion of the leaf surface. Aust J Plant Physiol 13:59–73, 1986.

104.  LR Rajasekaran, PE Kriedemann, D Aspinall, LG Paleg. Physiological significance of proline and glycinebetaine: maintaining photosynthesis during NaCl stress in wheat. Photosynthetica 34:357–366, 1997.

105.  CSV Rajendrakumar, T Suryanarayana, AR Reddy. DNA helix destabilization by proline and betaine: possible role in the salinity tolerance process. FEBS Lett 410:201–205, 1997.

106.  NC Turner, MM Jones. Turgor maintenance by osmotic adjustment: a review and evaluation. In: NC Turner, PJ Kramer, eds. Adaptation of Plants to Water and High Temperature Stress. New York: John Wiley & Sons, 1980, pp 87–103.

107.  M Tal. Genetics of salt tolerance in higher plants: theoretical and practical considerations. Plant Soil 89:199–226, 1985.

108.  AR Yeo. Salt tolerance in the halophyte *Suaeda maritima* (L.) dum: intracellular compartmentation of ions. J Exp Bot 32:487–497, 1981.

109.  GR Stewart, JA Lee. Role of proline accumulation in halophytes. Planta 120:279–289, 1974.

110.  GR Stewart, AD Hanson. Proline accumulation as a metabolic response to water stress. In: NC Turner, PJ Kramer, eds. Adaptation of Plants to Water and High Temperature Stress. New York: John Wiley & Sons, 1980, pp 173–189.

111.  S Treichel. The effect of NaCl on the concentration of proline in different halophytes. Z Pflanzenphysiol 76:56–58, 1975.

112.  AD Hanson, WD Hitz. Metabolic responses of mesophytes to plant water deficits. Annu Rev Plant Physiol 33:163–203, 1982.

113.  G Parameshwara. Stress effects in alfalfa (*Medicago sativa*) seedlings in relation to proline and betaines. PhD dissertation, University of Adelaide, Adelaide, Australia, 1984.

114.  R Storey, RG Wyn Jones. Quaternary ammonium compounds in plants in relation to salt resistance. Phytochemistry 16:447–453, 1977.

115.  H Greenway, R Munns, J Gibbs. Effects of accumulation of 3-$O$-methylglucose on levels of endogenous osmotic solutes in *Chlorella emersonii*. Plant Cell Environ 5:405–412, 1982.

116.  CM Grieve, EV Maas. Betaine accumulation in salt stressed sorghum. Physiol Plant 61:167–171, 1984.

117.  RG Wyn Jones, R Storey. In: LG Paleg, D Aspinall, eds. Betaines, Physiology and Biochemistry of Drought Resistance in Plants. Sydney: Academic Press, 1981, pp 171–203.

118.  RG Wyn Jones, R Storey. Salt stress and comparative physiology in the graminae. II. Glycinebetaine and proline accumulation in two salt and water stressed barley cultivars. Aust J Plant Physiol 5:817–829, 1978.

119.  Z Kefu, R Munns, RW King. Abscisic acid levels in NaCl treated barley, cotton, and saltbush. Aust J Plant Physiol 18:17–24, 1991.

120.  JF Thomson, CR Stewart, CJ Morris. Changes in aminoacid content of excised leaves during incubation. I. The effect of water content of leaves and atmospheric oxygen level. Plant Physiol 41:1578–1584, 1966.

121.  I Ahmad, F Larher, AF Mann, SF McNally, GR Stewart. Nitrogen metabolism of halophytes. IV. Characteristics of glutamine synthetase from *Triglochin maritima*. New Phytol 91:585–595, 1982.

122.  LG Paleg, TJ Douglas, A vonDaal, DB Kuch. Proline, betaine and other organic solutes protect enzymes against heat inactivation. Aust J Plant Physiol 8:107–114, 1981.

123.  RL Jeffries, T Rudmik. The responses of halophytes to salinity: an ecological perspective. In: RC Staples, GH Toenniessen, eds. Salinity Tolerance in Crop Plants: Strategies for Crop Improvement. New York: John Wiley & Sons, 1984, pp 213–227.

124.  R Weinberg. Modification of foliar solute concentrations by calcium in two species of wheat stressed with sodium chloride and or potassium chloride. Physiol Plant 73:418–425, 1988.

125.  R Weimberg, HR Lerner, A Poljakoff-Mayber. A relationship between potassium and proline accumulation in salt stressed *Sorghum bicolor*. Physiol Plant 55:5–10, 1982.

126.  G Voetberg, CR Stewart. Steady state proline levels in salt shocked barley leaves. Plant Physiol 76:567–570, 1984.

127.  WD Hitz, AD Hanson. Determination of glycinebetaine by pyrolysis–gas chromatography in cereals and grasses. Phytochemistry 19:2371–2374, 1980.

128.  N Ahmad, RG Wyn Jones. Glycinebetaine, proline and inorganic ion levels in barley seedlings following transient stress. Plant Sci Lett 15:231–237, 1979.

129.  AD Hanson, CE Nelson. Betaine accumulation and C$^{14}$ formate metabolism in water stressed barley leaves. Plant Physiol 62:305–312, 1978.

130. SP Robinson, GP Jones. Accumulation of glycinebetaine in chloroplasts provides osmotic adjustment during salt stress. Aust J Plant Physiol 13:659–668, 1986.

131. R Weimberg, HR Lerner, A Poljakoff-Mayber. Changes in growth and water soluble solute concentration in *Sorghum bicolor* stressed with sodium and potassium salts. Physiol Plant 62:472–480, 1984.

132. J Levitt. Responses of Plants to Environmental Stresses. Vol II. Water, Salt and Other Stresses. 2nd ed. New York: Academic Press, 1980.

133. RG Wyn Jones, R Storey, RA Leigh, N Ahmad, A Pollard,. A hypothesis on cytoplasmic osmoregulation. In: E Marre, O Cieferri, eds. Regulation of Cell Membrane Activities in Plants. Amsterdam: North Holland, 1977, pp 121–135.

134. WA Torello, LA Rice. Effects of NaCl stress on proline and cation accumulation in salt sensitive and tolerant turfgrass. Plant Soil 93:241–247, 1986.

135. TJ Flowers, AR Yeo. Effects of salinity on plant growth and crop yields. In: JH Cherry, ed. Environmental Stress in Plants: Biochemical and Physiological Mechanisms. NATO, ASI Series, Series G, Ecological Sciences, Vol 19. Berlin: Springer-Verlag, 1989, pp 101–119.

136. H Greenway, AP Sims. Effects of high concentrations of KCl and NaCl on responses of malate dehydrogenase (decarboxylating) to malate and various inhibitors. Aust J Plant Physiol 1:5–29, 1974.

137. A Pollard, RJ Wyn Jones. Enzyme activities in concentrated solutions of glycinebetaine and other solutes. Planta 144:291–298, 1979.

138. RG Wyn Jones. Phytochemical aspects of osmotic adaptation. In: BN Timmerman, C Steelink, FA Loewis, eds. Recent Advances in Phytochemistry. Vol 13. Phytochemical Adaptations to Stress. New York: Plenum, 1984, pp 55–78.

139. MI Lone, JSH Kuch, RG Wyn Jones, SWJ Bright. Influence of proline and glycinebetaine on salt tolerance of cultured embryos. J Exp Bot 38:479–490, 1987.

140. CJ Brady, TS Gibson, EWR Barlow, J Speirs, RG Wyn Jones. Salt tolerance in plants. I. Ions, compatible organic solutes and the stability of plant ribosomes. Plant Cell Environ 7:371–378, 1984.

141. Y Manetas, Y Petropoulore, G Karabourniotis. Compatible solutes and their effects on phosphoenolpyruvate carboxylase of $C_4$ halophytes. Plant Cell Environ 9:145–151, 1986.

142. C Itai, LG Paleg. Responses of water stressed *Hordeum distichum* (L.) and *Cucumis sativus* to proline and betaine. Plant Sci Lett 25:329–335, 1982.

143. KJ McCree. Whole plant carbon balance during osmotic adjustment to drought and salinity stress. Aust J Plant Physiol 13:33–43, 1986.

144. NC Turner. Drought resistance and adaptation to water deficits in crop plants. In: H Mussell, RC Staples, eds. Stress Physiology in Crop Plants. New York: John Wiley & Sons, 1979, pp 344–372.

145. RA Fischer. Optimizing the use of water and nitrogen through breeding of crops. Plant Soil 58:249–278, 1981.

146. DC Rasmusson. An evaluation of ideotype breeding. Crop Sci 27:1140–1146, 1987.

147. J Gorham. Salt tolerance in the Triticeae: K/Na discrimination in synthetic hexaploid wheats. J Exp Bot 41:623–627, 1990.

148. J Gorham, A Bristol, EM Young, RGW Jones, G Kashour. Salt tolerance in the Triticeae: K/Na discrimination in barley. J Exp Bot 41:1095–1101, 1990.

149. AR Yeo, ME Yeo, SA Flowers, TJ Flowers. Screening of rice (*Oryza sativa*) genotypes for physiological characters contributing to salinity resistance and their relationship to overall performance. Theor Appl Genet 79:377–384, 1990.

150. S Yoshida. Fundamentals of Rice Crop Science. Los Banos, Philippines: International Rice Research Institute, 1981.

151. AR Yeo, KS Lee, P Izard, PJ Boursier, TJ Flowers. Short and long-term effects of salinity on leaf growth in rice (*O. sativa*). J Exp Bot 42:881–889, 1991.

152. TJ Flowers, E Dugue, MA Hajibagheri, TP McGenigle, AR Yeo. The effect of salinity on the ultrastructure and net photosynthesis of two varieties of rice: further evidence for a cellular component of salt resistance. New Phytol 100:37–43, 1985.

153. Y Saranga, J Rudich, D Zamir. Salt tolerance of cultivated tomato, its wild relatives, and interspecific segregating populations. Acta Hortic 200:203, 1987.

154. J Dvorak, E Epstein, A Galvez, P Gulick, JA Omielan. Genetic basis of plant tolerance to soil toxicity. In: HT Stalker, JP Murphy, eds. Plant Breeding in the 1990s. Proceedings of the Symposium on Plant Breeding in the 1990s held at North Carolina State University, March 1991. CAB International, 1992, pp 201–217.

155. MM Ludlow, JM Muchow. A critical evaluation of traits for improving crop yields in water-limited environments. Adv Agron 43:107–153, 1990.

156. SD Tanksley, ND Young, AH Paterson, MW Bonierbale. RFLP mapping in plant breeding: new tools for an old science. Biotechnology 7:257–264, 1989.

157. B Martin, J Nienhuis, G King, A Schaeffer. Restriction fragment length polymorphisms associated with water use efficiency in tomato. Science 243:1725–1728, 1989.

158. J Gorham, J Bridges, J Dubcovsky, J Dvorak, PA Hollington, MC Luo, JA Khan. Genetic analysis and physiology of a trait for enhanced $K^+$/Na discrimination in wheat. New Phytol 137:109–116, 1997.

159. I Jamaux, A Steinmetz, E Belhassen. Looking for molecular and physiological markers of osmotic adjustment in sunflower. New Phytol 137:117–127, 1997.

160. E Epstein, DW Rains, OE Elzam. Resolution of dual mechanisms of potassium absorption by barley roots. Proc Natl Acad Sci USA 49:684–692, 1963.

161. ML Binzel. NaCl-induced accumulation of tonoplast and plasma membrane H-ATPase message in tomato. Physiol Plant 94:722–728, 1995.

162. I Winicov. New molecular approaches to improving salt tolerance in crop plants. Ann Bot 82:703–710, 1998.

163. J Zhu, J Liu, L Xiong. Genetic analysis of salt tolerance in *Arabidopsis*: evidence for a critical role of potassium nutrition. Plant Cell 10:1181–1191, 1998.

164. FJM Maathuis, A Amtmann. $K^+$ nutrition and Na toxicity: the basis of cellular K/Na ratios. Ann Bot 84:123–133, 1999.

165. TR Warne, LG Hickok, CE Sams, DL Vogelien. Sodium/potassium selectivity and pleiotropy in stl2, a highly salt tolerant mutation in *Ceratopteris richardii*. Plant Cell Environ 22:1027–1034, 1999.

166. MP Apse, GS Aharon, WA Snedden, E Blumwald. Salt tolerance conferred by overexpression of a vacuolar $Na^+/H^+$ antiport in *Arabidopsis*. Science 285:1256–1258, 1999.

167. R Storey, RG Graham, KW Shepherd. Modification of the salinity response of wheat by the genome of *Elytrigia elongatum*. Plant Soil 83:323–330, 1985.

168. SH Shah, J Gorham, BP Forster, RGW Jones, RG Wyn Jones. Salt tolerance in the Triticeae: the contribution of 'D' genome to cation selectivity in hexaploid wheat. J Exp Bot 38:254–269, 1987.

169. J Gorham, C Hardy, RCW Jones, LR Joppa, CN Law, RG Wyn Jones. Chromosomal location of a K/Na discrimination character in the D genome of wheat. Theor Appl Genet 74:584–588, 1987.

170. J Gorham. Genetics of sodium uptake in wheat. Proceedings of 7th International Wheat Genetics Symposium, 1988, pp 817–821.

171. DP Schachtman, AJ Bloom, J Dvorak. Salt tolerant *Triticum* × *Lophopyrum* derivatives limit the accumulation of sodium and chloride ions under saline stress. Plant Cell Environ 12:47–55, 1989.

172. DP Schachtman, SD Tyerman, BR Terry. The $K^+/Na^+$ selectivity of a cation channel in the plasma membrane of root cells does not differ in salt-tolerant and salt-sensitive wheat species. Plant Physiol 97:598–605, 1991.

173. GP Findlay, SD Tyerman, A Garrill, M Skerrett. Pump and $K^+$ inward rectifiers in the plasmalemma of wheat root protoplasts. J Membr Biol 139:103–116, 1994.

174. M Akbar, GS Khush, D Hilleristambers. Genetics of salt tolerance in rice. In: Rice Genetics. Manila: IRRI, 1986, pp 399–409.

175. GH Abel. Inheritance of the capacity for chloride inclusion and chloride exclusion by soybeans. Crop Sci 9:697–698, 1969.

176. AV Venables, DA Wilkins. Salt tolerance in pasture grasses. New Phytol 80:613–622, 1978.

177. R Grumet, TG Isleib, AD Hansen. Genetic control of glycinebetaine level in barley. Crop Sci 25:618–622, 1985.

178. R Grumet, AD Hansen. Genetic evidence for an osmoregulatory function of glycinebetaine accumulation in barley. Aust J Plant Physiol 13:353–364, 1986.

179. SA Quarrie. Genetic variability and heritability of drought induced abscisic acid accumulation in spring wheat. Plant Cell Environ 4:147–151, 1981.

180. SA Quarrie. Genetic differences in abscisic acid physiology and their potential uses in agriculture. In: FT Addicott, ed. Abscisic Acid. New York: Praeger, 1983, pp 365–419.

181. GV Subbarao. Salinity tolerance in pigeonpea (*Cajanus cajan*) and its wild relatives. PhD dissertation, Indian Institute of Technology, Kharagpur, India, 1989.

# 45

# Glycine Betaine Accumulation: Its Role in Stress Resistance in Crop Plants

**G. V. Subbarao\*, Lanfang He Levine, and Gray W. Stutte**

*Dynamac Corporation, Kennedy Space Center, Florida*

**Raymond M. Wheeler**

*National Aeronautics and Space Administration, Kennedy Space Center, Florida*

## I. INTRODUCTION

The concept of compatible solutes was first introduced by Brown and Simpson [1] to define substances that accumulate in the cytoplasm that are noninhibitory to metabolism when subjected to low external water potentials. Compatible solutes that accumulate in higher plants include glycerol, sucrose, trehalose, pinitol, proline, and betaines [2–8]. Plants accumulate these solutes as an adaptive mechanism to stresses such as salinity, water deficit, and temperature extremes [5,9]. Compatible solutes provide a cellular environment that maintains the macromolecular structure and function of proteins [5]. They are hypothesized to have functions including cytoplasmic osmotic adjustment [2,10], protecting cytoplasm and chloroplasts from Na damage, hydroxyl radical scavenging [11], stabilization of proteins [12–16], protecting membrane structure [17], and general maintenance of physiological stability under stressful conditions [5,8,18]. Information on the adaptive role of glycine betaine (GB) in plant stress resistance comes from studies on

> Foliar application of GB to plants
> Enhancing GB levels through traditional breeding or genetic engineering

This chapter summarizes the current state of knowledge and understanding of GB accumulation and distribution in plants. Areas covered include distribution among plant species, biosynthetic pathway, and GB's adaptive significance to stress environments. Analytical methodologies for GB are also reviewed. The potential for introducing the GB biosynthetic pathway into crop plants (that do not naturally accumulate) and the limitations associated with this approach are discussed.

## A. Glycine Betaine Accumulation in Higher Plants

Several families of flowering plants have the ability to synthesize GB (Table 1). However, osmotically significant amounts are detected in only a few families. Glycine betaine synthesis appears to be constitutive as significant amounts are produced under nonstress growing conditions [21]. In GB-accumulating species, GB synthesis can increase severalfold under stress (Table 2). Several GB-nonaccumulating species have the ability to synthesize this compound but at very low levels ($<1$ $\mu$mol g$^{-1}$ dwt), about 100 to 1000 times less than GB accumulators [8,36]. Accumulators and nonaccumulators are often found

---

\* Current affiliation: Japan International Research Center for Agricultural Sciences, Ibaraki, Japan

**TABLE 1**    Glycine Betaine Accumulation in Plants

| Family | References |
|---|---|
| Chenopodiaceae | |
| *Atriplex, Beta* | [3,4,14,19–25] |
| *Halosarcia* | |
| *Maireana, Rhagodia* | |
| *Salicornia, Salsola* | |
| *Sarcocornia* | |
| *Sclerolaena* | |
| *Spinacia, Threlkeldia* | |
| *Suaeda fruticosa* | |
| Amaranthaceae | |
| *Amaranthus* | [4,24] |
| Avicenniaceae | |
| *Avicennia* | [22] |
| Gramineae | |
| *Cymbopogon* | [14,22,26–34] |
| *Distichlis* | |
| *Enneapogon* | |
| *Eragrostis, Hordeum* | |
| *Sorghum, Spartina* | |
| *Themeda, Triodia* | |
| *Triticum, Zea* | |
| Compositae | |
| *Aster, Cassinia* | [14,22–24] |
| *Cratystylis* | |
| Gnaphalium | |
| *Helianthus, Olearia* | |
| Convolvulaceae | |
| *Convolvulus* | [22–24] |
| *Cuscuta, Evolvulus* | |
| *Wilsonia* | |
| Plumbaginaceae | |
| *Plumbago, Limonium* | [24] |
| Solanaceae | |
| *Lycium* | |
| Leguminoseae | |
| *Medicago, Trifolium* | [7,22,24,35] |
| Asteraceae | [4,24] |
| Malvaceae | [4,14,24,36] |
| Poaceae | [4,24,26,29,31,32,37–39] |
| Portulacaceae | [4,24] |
| Caryophyllaceae | [24] |

Note: In accumulating genera, the glycine betaine levels were reported to vary between 5 to 100 $\mu$mol g$^{-1}$ dwt under nonstressed conditions or 40 to 400 $\mu$mol g$^{-1}$ under saline or dry conditions. For nonaccumulating genera, glycine betaine levels were <1 $\mu$mol g$^{-1}$ (less than detectable limits).
*Source:* Adapted and modified from Ref. 8.

within the same family, genus, and even species. For example, most members of Poaceae accumulate GB, but cultivated and wild rice species do not [33,46,47]. Both GB accumulators and nonaccumulators are reported in the genus *Limonium* in the family Plumbaginaceae [48] and *Wollastonia* in Asteraceae [49,50]. Many food crops, including rice, maize, wheat, and sorghum, do not accumulate GB in appreciable amounts [31–33,51,52], even though some GB-positive lines of maize [53,54], wheat [30], and sorghum [39] have been identified.

**TABLE 2**  Glycine Betaine Levels Under Stress in Various Crop Plants

| Crop species | Stress type | Betaine levels ($\mu$mol g$^{-1}$ dwt) Control | Stress | References |
|---|---|---|---|---|
| Wheat | Salinity | 20[a] | 170 | [40] |
|  | Salinity |  | 40 | [41] |
| *Leymus sabulosus* | Salinity (250 mM NaCl) | 8.0[b] | 28[b] | [42] |
| Barley | Salinity | 20[a] | 170 | [40] |
| Sugarbeet |  |  |  |  |
| Spinach | Salinity |  | 150[a] | [43] |
| *Halophyrum mucronatum* | Salinity |  | 100 | [44] |
| *Suaeda fruticosa* | Salinity |  | 200 | [44] |
| *Haloxylon recurvum* | Salinity |  | 600 | [44] |
| *Amaranthus tricolor* | Salinity | 80 | 260 | [45] |

[a] Converted from fresh weight values and expressed as dry weight values assuming that fresh weight/dry weight ratio is about 10.
[b] Concentration in mM of leaf sap.

## B.  Glycine Betaine Accumulation in Microorganisms

Glycine betaine is one of the most widespread osmotic solutes in microorganisms, particularly in halophilic eubacteria. Glycine betaine functions as a compatible solute, enabling them to withstand the high osmotic potential (OP) of their surrounding medium [55]. In cyanobacteria GB prevents the inhibitory effects of high salt levels on enzyme activity [56]. Only three bacterial groups are reported to synthesize betaine de novo; these are cyanobacteria [57–59], anaerobic photosynthetic bacteria (*Ectohiorhodospira*), and *Actinopolyspora halophila* [60]. Transport systems for GB are reported for *Escherichia coli*, *Salmonella typhimurium*, and *Klebsiella pneumoniae* [61]; *Rhizobium meliloti* [62]; *Rhodobacter sphaeroides* [63]; and *Azospirillum brasilense* [64].

Osmoprotective effects of endogenously synthesized and exogenously supplied GB are reported for bacteria [65–67] and cyanobacteria [68]. Addition of GB to the plating medium restored the ability of *E. coli* to form colonies under otherwise toxic osmotic conditions [69]. A number of halophilic bacteria accumulate GB from the medium [70]. Under salt stress, cyanobacteria [71], halophilic bacteria [18,19,72,73], nonhalophilic eubacteria [74,75], *E. coli* [76], halotolerant microorganisms [77], and green algae [8,66,78,79] accumulate GB. Intracellular GB concentrations can vary from 0.5 to 2.0 M in halophilic eubacteria [58,73]. Glycine betaine protects the respiratory system against salt damage in moderately halophilic bacteria [80].

## II.  ANALYTICAL METHODOLOGY

The quantitative determination of betaines (fully *N*-methylated amino acids) has research implications for both plant and animal systems. Both chromatographic and spectrometric methods (see Refs. 4 and 8 for earlier reviews on these methods) are discussed in this chapter. This review is intended to provide a quick reference to researchers in finding suitable tools for their specific research. We focus only on GB analysis during this review. Glycine betaine analysis involves three important steps: sample extraction, purification, and final quantification (Figure 1).

## A.  Extraction

Quantification of GB in vivo is possible [43,81]; however, most studies of GB are performed on extracts. Glycine betaine is soluble in water (160 g/100 g water), methanol (55 g/100 g solvent), and ethanol (8.7 g/100 g solvent) but sparingly soluble in ether or chloroform. Thus, betaines are extracted with hot aqueous alcohol [82], methanol [35,83], or methanol-chloroform-water (MCW) mixtures [84–86]. The proportion of methanol-chloroform-water used in previous studies varied with plant materials. A ratio of 12:5:3 has been widely used. An optimal solvent system is one that renders the best recovery and highest

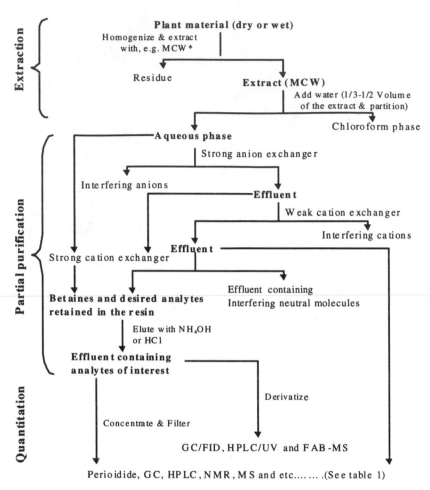

*MCW : Methanol:Chloroform:Water

**Figure 1**    Analytical approaches for the determination of glycine betaine in plant material.

reproducibility. The extract is then partitioned against chloroform. The methanol-water phase containing betaines is collected for further purification. Both fresh and dried plant materials are suitable for betaine analysis [87]. Care should be taken if plant material is homogenized in MCW mixture because the heat generated by homogenization could result in the breakdown of chloroform and acidification of the extract. The performance of the ion-exchange chromatography step (described in the following) will be very low if the extract has a pH below 5.0. It is important to homogenize tissue in extraction mixture over an ice bath or otherwise adjust the pH between 5 to 7 with 1 M NaOH.

## B.    Purification

Most of the available quantification methods require removal of interfering compounds. The most widely used purification protocols are based on the unique charge properties of betaines, a permanent positive charge on the quaternary ammonium group with a carboxyl group of low pKa. Betaines are not retained on either strong anion- or weak cation-exchange resin. This allows excellent separation from impurities by passing through strong anion- and weak cation-exchange resins in series. Strong anionic resin removes all anions, amino acids, and zwitterions except $OH^-$ and betaines. All cations are retained on the weak cation resin. The effluent of this column combination contains betaines, where amines are retained on the weak cation resin exchanger. With a strong cation exchanger in combination with weak cation exchanger, betaines or beatines plus amines are retained and are subsequently eluted with HCl or $NH_4OH$, depending on the analytes of interest. While GB may be eluted with HCl or $NH_4OH$, choline, betaine aldehyde, β-alaninebetaine, γ-butyrobetaine, and β-dimethylsulfoniopropionate are not recovered quantitatively under alkaline conditions [8]. In some quantification techniques, one-step purification using strong cation-exchange resin is sufficient. Detailed information on purification using three-resin, two-resin, and one-resin systems can be found elsewhere [23,88,89].

Recovery of quaternary ammonium compounds (QACs) can be determined by adding known amounts of betaine (either radiolabeled or not) as an internal standard to the sample. Hitz and Hanson [46] reported an average recovery of 78% of radioactive label following extraction and ion-exchange chromatography. Studies conducted in our laboratory (unpublished) indicate 70–74% recovery after passing the aqueous phase of the extract directly through a mixture of two resins. The recovery is improved by washing the resin several times with water. Guy et al. [87] did not find noticeable differences in chromatography between the two- and three-resin purification systems in 27 species examined (mostly halophytes). It is recommended that an internal standard is added during extraction and purification steps. Dimethylsulfonioacetate (the sulfur analogue of glycine betaine) is a good candidate.

## C. Separation and Detection

### 1. Spectrophotometric Method

Early analyses of QACs were based on nonspecific precipitation with periodide [90,91], reineckate salts [92], or modified Dragendorff reagent [93]. These methods are not specific, and quantification of individual compounds in mixtures requires prior separation. The periodide method was further advanced to enable determination of both betaine and choline [19]. In this modified method, choline and betaine are selectively precipitated at different pHs. The betaine or choline-periodide complex is extracted with 1,2-dichloroethane and absorbance is measured at 365 nm. Betaine can be subsequently derived from QACs (pH 2.0) after subtracting choline (pH 8.0). When these methods are employed, care should be taken to minimize the background absorbance and remove other naturally occurring nitrogenous substances. These deficiencies have promoted the development of quantification methods based on chromatographic separation in conjunction with more sophisticated detection techniques.

### 2. Chromatographic Method

TLC AND TLE.   The advancement of betaine analysis is largely dependent on the analytical tools available at the time. Thin-layer chromatography (TLC) or paper chromatography followed by visualization of separated compounds with Dragendorff's reagent was used widely before high-performance liquid chromatography (HPLC). Paper chromatography or TLC is insufficient for complete separation of all QACs [94,95]. Separation is further improved by adopting high-voltage electrophoresis [51,96]. In the early 1980s, thin-layer electrophoresis (TLE) in combination with scanning densitometry was widely used for the determination of a number of ammonium compounds [97]. Muller and Eckert [89] improved sensitivity by using methyl orange as the visualization reagent. Although reasonably rapid, adaptable, and free from interference, this approach is not particularly sensitive. Also, a standard curve is required for every plate.

GAS CHROMATOGRAPHY (GC).   Glycine betaine is a zwitterion and nonvolatile and thus cannot be directly analyzed by gas chromatography. Esterification of its carboxylic group [98], pyrolysis of glycine betaine [46], is necessary for the volatilization of GB. The esterification product of GB with N-O-bis-trimethylsilyl-trifluoracetamide is not well characterized. Ranfft and Gerstl [98] reported a detection limit of 0.42 nmol and a reproducibility of 1.79% (relative standard deviation, RSD) for a concentration of 41 ppm. The low-temperature pyrolysis–gas chromatography technique introduced by Hitz and Hanson [46] is much more sensitive (minimum detection limit, MDL, is about 2 nmol). This technique is based on pyrolytic dealkylation and deamination of the ammonium group in QACs. Low-temperature pyrolysis of the $OH^-$ form of GB favors deamination, giving trimethylamine (TMA) as the major product that can be separated by gas chromatography (GC) and determined by flame ionization detection (FID). The absolute TMA yields varied with pyrolysis probes, temperature, and concentration ranges. Hitz and Hanson [46] reported 42–51% of the theoretical yield over a concentration range of 20–150 nmol GB. Because trimethyl QACs yield TMA in addition to GB upon pyrolysis and the absolute yields of individual trimethyl QACs differ, the method measures trimethyl QACs in only a semiquantitative manner. The identity of betaines present in plant extract must be verified through other techniques. In addition, the expense of pyrolysis probes, which require frequent replacement (an average of 300 analyses), makes this technique less attractive and thus not readily adapted by the researchers working on GB.

HIGH-PERFORMANCE LIQUID CHROMATOGRAPHY.   HPLC is by far the most widely used method because of its low cost, sensitivity, specificity, accuracy, and versatility. Separation of QACs and their esters can be accomplished on strong cation [85,86,99,100], weak cation [88], amino-bonded silica gel [87,101], and reverse-phase columns [37,82]. Most betaines do not have significant ultraviolet (UV) absorbance above about 210 nm and thus require either low wavelengths (190–200 nm) or a differential refractometer for direct quantification. The differential refractometer requires a larger sample (about 20 times higher betaine content) than UV detectors. The estimated minimum detection limit for GB by low-wavelength UV varies from 0.5 to 4.4 nmol, and that is dependent on the wavelength and instrument used for analysis [85,100]. Optimum peak sensitivity is achieved at 195 nm, and increasing the wavelength above 195 nm decreases the peak area; e.g., peaks are three times larger at 195 nm than at 200 nm [85]. The sensitivity is better than that reported-for pyrolysis-GC [46] and TLE with scanning densitometry [97].

Derivatization increases selectivity and sensitivity by conferring particular physical or chemical properties to the compounds. Methyl esters of α-alaninebetaine, GB, trigonelline, and β-aminobutyric acid betaine are quantified by mobile phase ion chromatography using suppressed conductivity detection having a detection limit less than 0.42 nmol for GB [102]. Glycine betaine can also be esterified with α,p-dibromoacetophenone, p-nitrobenzyl bromide, or α-bromo-p-tolunitrile. These aromatic derivatives are then quantified either by directly measuring the absorbance at high wavelength (e.g. 262 nm for the p-bromophenacyl ester) in aqueous phase [103] or by HPLC separation on a strong cation-exchange column followed by UV detection. Complete esterification (>99%) of several betaines and betaine analogues with α, p-dibromoacetophenone is observed. Generally the p-bromophenacyl esters are preferred because of their higher molar absorptivity ($e = 1.968 \times 10^4 \, mol^{-1} \, cm^{-1}$ for GB phenacyl ester). The HPLC-UV analysis of betaine phenacyl esters can detect <1 nmol of betaines. A combination of capillary electrophoresis and UV detection of betaine phenacyl esters was introduced by Zhang et al. [84]. Estimation of betaines as either the methyl or p-phenacyl esters has several advantages over low-wavelength UV detection. It is more reliable because of less interference at high wavelength and more sensitive; thus it is able to detect smaller concentrations in plant tissues.

### 3.  Spectrometric Method

[1]H NMR.   This method utilizes the proton nuclear magnetic resonance signal of the three N-methyl groups in GB [81]. The signal from the N-methyl resonance of GB (arising from nine protons) occurs in a region of the [1]H nuclear magnetic resonance (NMR) spectrum that is generally free from interfering signals and allows the detection of <85 nmol (16 or 32 data acquisitions, a total data acquisition time of 2 min) of the compound. Greater sensitivity can be achieved by using a superconducting NMR spectrometer of high field strength [104]. This method can be used for simultaneous determination of other N-methyl compounds with different sensitivities depending on the number of protons and coupling with neighboring protons (mutiplet or singlet). Another feature of NMR spectroscopy is that the radio-frequency radiation used in the method can penetrate "opaque" objects such as seed coat, cell walls, and cellular membranes, and thus this method can be used for determination of GB levels of intact organisms and organelles [43].

NATURAL ABUNDANCE [13]C NMR.   As with [1]H NMR, the resonance signal of a selected carbon isotope ([13]C) in the compound of interest can be used for quantification. The three carbons in the trimethylated quaternary ammonium group of GB give rise to a very strong and distinct signal at 56.4 ppm. Signals from methylene groups on proline (26.54, 31.76, 48.83 ppm) and the trimethylated ammonium group on betaine (56 ppm) are distinguishable [105]. Crude extract and plant tissue in vivo can be analyzed because of the specificity of the chemical shift of a chosen carbon in the compound of interest. Also, this method has potential for nondestructive analysis of solutes extracted from the plant. It allows identification of solute molecules (concentrations of 1 μmol/g fw) relevant to osmotic adjustment in vivo.

The $^{13}$C NMR method is not suitable for routine analysis of a large number of samples because of lack of sensitivity (a total final concentration of 10 mM is required), lack of resolution (signals from GB and betaine homologues overlap), and the fact that it is also time consuming. For quantification, care should be taken to ensure that the intensity of analyte specific signals for a given specific resonance be directly proportional to analyte concentration. This is pertinent for every method but more relevant for NMR as a complete relaxation of the resonance used between data acquisition and maximum expression of the Overhauser effect are required.

MASS SPECTROMETRY    Mass spectrometric analysis of QACs of synthetic and actual origin has been approached by desorption methods. Fast atom bombardment mass spectrometry (FAB-MS), Cf fission fragment and laser desorption, thermal ionization, field desorption, direct chemical ionization, and secondary ion mass spectrometry have all been applied to determine ionization and fragmentation common to QACs. Although all of these techniques produce intact molecular cations, FAB-MS produces molecular cations of QACs with the greatest relative abundance and allows monitoring of the long-lived signal derived from stable ion emission.

Rhodes et al. [23] developed an application of FAB-MS in quantitative determination of betaines of plant extracts. The method entails converting QACs to aliphatic alcohol esters followed by FAB-MS analysis. Removal of free amino acids is essential because in the derivatization process, amino acids also yield esters that can complicate the mass spectra. Also, it is important to remove valine, which gives a molecular ion of mass identical to that of GB. The lower limit of detection for GB as the *n*-propyl ester is 0.05 nmol $\mu L^{-1}$ glycerol. Accurate quantification of QACs is accomplished by the use of deuterium-labeled internal standards or QAC homologues of distinct mass. A linear correlation between the molar ratio of GB to standard and the signal ratio of $d_0$ (*m/z* 174)/$d_9$ (*m/z* 183) is maintained over a wide range of 0.01 to 1 (correlation coefficient = 1.00). It is by far the most sensitive method (MDL < 0.05 nmol) and can discriminate the same molecules of stable isotope. Because of these attributes, this method is particularly suited to stable isotope tracer studies of the GB biosynthetic pathway and also can reliably and selectively quantify low nanomolar amounts of betaines in complex mixtures of QACs in plant extracts. The method has been applied to quantification of QACs and determination of the stable isotope abundance of QACs. This method has potential for the identification of genotypes that lack certain QACs [31,35,53].

Plasma desorption–MS (PD-MS) is also developed for quantification of GB and choline [83,106]. However, this method is different from FAB-MS as it does not require prior derivatization of GB and thus is particularly useful for choline determination. Glycine betaine is quantified from the signal intensity at *m/z* 118 relative to the stable isotope labeled [$d_9$] glycine betaine internal standard (*m/z* 127). The ion intensity ratio of *m/z* 118:127 is correlated with molar ratio [$d_0$] glycine betaine/[$d_9$] glycine betaine. A useful feature of the PD-MS method is that the signal from choline (*m/z* 104 and 113 for [$d_0$] choline and [$d_9$] choline internal standard, respectively) is more intense than the signal from GB. This method has advantages over alternative methods such as DCI-MS and FAB-MS as derivatization of the QACs is not required and choline can be readily detected and quantified even in the presence of large amounts of GB.

### 4.  *Remarks on Methodology*

The relative advantages and disadvantages associated with various analytical methods for GB are summarized in Table 3. There are several other techniques, such as radioisotope dilution in conjunction with micro-Kjeldahl analysis and methyltransferase-catalyzed reaction [107,108], that are not covered in this review. Depending on the research objectives and facilities available, researchers need to choose the appropriate analytical techniques that meet their needs. It is highly recommended that when a new plant species is under investigation, a full evaluation of recovery, reproducibility, and accuracy of the analytical methodology should be conducted as part of the standardization procedure.

## III.  BIOSYNTHETIC PATHWAY IN HIGHER PLANTS

## A.  Biosynthetic Pathway of Glycine Betaine from Choline

The biosynthetic pathway of GB is studied mostly in species of Chenopodiaceae (spinach, sugarbeet), and to some extent in Amaranthaceae and Gramineae (barley, maize). Glycine betaine is synthesized via a two-step oxidation of choline catalyzed by choline monooxygenase (CMO) and betaine aldehyde dehy-

**TABLE 3**   Advantages and Disadvantages Associated with Various Analytical Techniques for Glycine Betaine Analysis

| Method | Advantages | Disadvantages |
|---|---|---|
| Precipitation with either periodide or reineckate or Dragendorff reagent | Moderate Sensitivity | Nonspecific |
| TLC and TLE | Specific | Sensitivity varies (2.5 nmol using methyl orange reagent, 85 nmol using Dragendroff reagent) |
| | | Lack of precision, semiquantitative |
| | | Limited separation efficiency |
| | | Tedious, a complete standard curve being required for every plate |
| Pyrolysis-GC/FID | Sensitive (2 nmol limit) | Not very specific |
| | | Costly (pyrolysis probe) |
| GC/FID upon derivatization | Moderate sensitivity | Derivatization is required |
| | | Not well characterized for betaine analysis |
| HPLC/UV | Versatile | |
| | Selective and specific | |
| | Convenient | |
| | Sensitive (sensitivity equal to or better than pyrolysis GC using low UV, greater when derivatized) | |
| | Easy to automate and suitable for large number of samples | |
| $^1$H NMR and $^{13}$C NMR | Specific | Not very sensitive (MDL—85 nmol of 90 Hz $^1$H NMR) |
| | Nondestructive, suitable for analysis in vivo | Costly (instrument) |
| FAB-MS | Extremely sensitive (MDL < 0.05 nmol) | Expensive (equipment and stable isotope labeled GB) |
| | Specific | |
| | Rapid | |
| PD-MS | Simple sample preparation | Less sensitive for GB than FAB-MS |
| | Sensitive for choline analysis | Expensive (equipment and stable isotope labeled GB) |

drogenase (BADH, EC 1.2.1.8) (Figure 2) [8]. Immunological studies suggest that the pathway is common in angiosperms. Glycine betaine–accumulating species from several distantly related families of dicotyledons (*Spinacia oleracea*, *Amaranthus caudatus*, *Convolvulus arvensis*, and *Lycium ferocissimum*) have expressed immunologically related BADH enzymes [24]. The nonaccumulating species in *Magnolia* and *Soulangiana* have small amounts of GB and express a 63-kDa protein that cross-reacted with antibodies of BADH from spinach (*S. oleracea*).

The first step in GB synthesis is catalyzed by ferredoxin-dependent choline monooxygenase [109,110] that converts choline to the hydrate form, which is the dominant form (>99%) in aquous extracts. CMO is an unusual ferrodoxin-dependent enzyme that is unique to plants and has been purified only from spinach [111,112]. CMO expression is reported only in Amaranthaceae and Chenopodiaceae [113,114] and has not been reported in monocots [36]. Sodium dodecyl sulfate–polyacrylamide gel electrophoresis (PAGE) and matrix-assisted laser desorption ionization of MS showed the presence of a polypeptide with a subunit molecular mass of 45 kDa, suggesting that CMO is a homodimer of identical subunits [111,112], each of which has a Rieske-type [2Fe-2S] cluster [113,114]. CMO is located in the chloroplast stroma [110], has a pH optimum close to 8, and is strongly stimulated by light and $Mg^{2+}$ [109,115].

The second step in GB synthesis is catalyzed by a pyridine nucleotide–dependent betaine aldehyde dehydrogenase and has a strong preference for $NAD^+$ [8,24,29,116]. BADH enzyme activity is found in

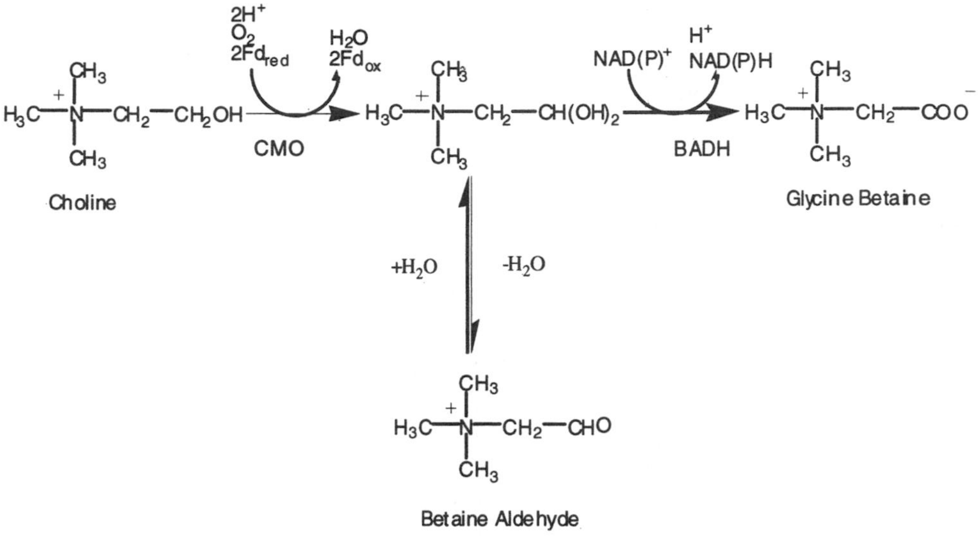

**Figure 2** The biosynthesis of glycine betaine from choline in higher plants.

members of Chenopodiaceae [117,118], Amaranthaceae [119,120], and Gramineae [32,47]. In spinach leaves, the majority of the BADH activity is located in the chloroplast stroma [8,121]; however, in members of Gramineae, BADH may be peroxisomal [47]. Substantial levels of BADH were reported in roots of sugar beet [117] and in the etiolated leaves of barley [29]. Nevertheless, GB synthesis has not been detected in sugar beet roots [21]. Thus it is not certain that GB synthesis can occur outside chloroplasts and organs other than leaves. BADH enzyme has been purified to homogeneity from spinach [122,123] and amaranth [119,120]. BADH is a dimer with subunits of 60 kDa and has a native molecular mass of 125 kDa. The spinach and amaranth BADH has a pH optimum of about 8.6 and optimum temperature of around 50°C. BADH is activated by relatively low concentrations of $K^+$, sucrose, and proline but is inhibited by $NH_4^+$, $Na^+$, and high concentrations of GB [120].

## B. Biosynthetic Pathway(s) of Choline (GB Precursor)

The biosynthetic pathway of choline has been studied in GB-accumulating (e.g., sugar beet, spinach, and barley) and GB-nonaccumulating species (e.g., carrot, *Lemna*, soybean cell cultures, and castor bean endosperm). In higher plants, choline is synthesized from serine via ethanolamine [36]. Choline biosynthesis involves three parallel, interconnected series of N-methylation reactions at the free base, phospho base, or phosphytidyl base level (Figure 3). The predominant routes have been reviewed by Rhodes and Hanson [8]. Ethanolamine kinase and three S-adenosylmethionine phospho bases, N-methyl transferases, are reported to catalyze the methylation of phosphoethanolamine [124]. The regulatory step for choline synthesis is the enzyme catalyzing the first N-methylation of phosphoethanolamine, which is stimulated by light and also is triggered by osmotic stress [125,126].

## C. Regulation of Glycine Betaine Synthesis

Glycine betaine synthesis is regulated at the biosynthetic level and is proportional to the severity of salinity or water deficits [21,127]. In vivo radiotracer studies indicate that enhanced GB synthesis is accompanied by a higher rate of choline synthesis followed by oxidation of choline to GB [8]. Consistent with these in vivo data, activities of both CMO and BADH increase under salinity and water stress. Choline monooxygenase in sugar beet and *Amaranthus caudatus* increased severalfold in response to osmotic stress [114]. Similarly, BADH levels increased two- to fourfold in leaves of sugar beet when plants were exposed to NaCl (500 mM) salinity [117,128]. The increase of BADH activity in sugar beet leaves and roots under salinity is paralleled by an increase in levels of translatable BADH messenger RNA (mRNA).

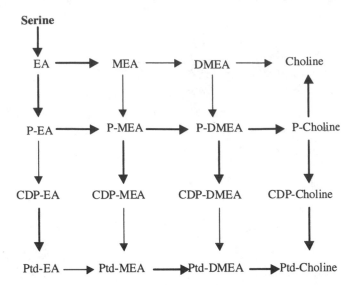

**Figure 3** Pathways of choline synthesis in higher plants. Faint arrows show steps for which there is only radiotracer evidence. The bold arrows show steps for which respective enzyme activity is found (EA, ethanolamine; p-EA, phosphoryl ethanolamine; CDP-EA, cytidine diphosphate ethanolamine ester; Ptd-EA, phosphatidyl ethanolamine; MEA, methyl ethanolamine; DMEA, dimethyl ethanolamine). (Adapted from Ref. 8.)

## IV.  PHYSIOLOGICAL SIGNIFICANCE

Glycine betaine is widely perceived as a compatible solute accumulated in the cytoplasm to reduce the intracellular water activity and $\Psi$s between the cell and its surroundings [2,5,8,19,24,129–131]. This is required to control the turgor pressure, which is the driving force for cell expansion, division, and thus growth [132]. The principal characteristic of compatible solutes is their ability to reduce $\Psi$s of cytoplasm without negative effects on metabolism. Glycine betaine preferentially excludes inorganic ions (such as Na) from the hydration sphere of proteins and thus protects enzymes from denaturation [15,16, 79,133,134]. Also, compatible solutes are believed to stabilize freeze-thaw cycles and thus act as cryoprotectants [135,136].

In general, inorganic ions in the cytoplasm are maintained at relatively constant levels for normal metabolic functioning [72,137,138]. High concentrations of inorganic solutes can be harmful in the cytoplasm as they cause protein denaturation and thus disruption of metabolic functions [25,43,139–141]. In contrast, vacuoles can tolerate higher concentrations of inorganic ions where they serve in a role of osmotic adjustment (OA). Potassium and in some cases Na are used for this function [86,142]. For example, K is the major cation contributing to OA in sorghum [143], and nearly 78% of the OA in wheat could be attributed to K accumulation [144,145].

A number of organic solutes such as betaines, tertiary sulfonium compounds such as dimethyl sulfoniopropionate (DMSP), amino acids such as proline, and polyols such as sucrose, mannitol, and trehalose have similar OA functions in the cytoplasm [10,48,50,145,146]. However, GB is widely believed to be the most effective among compatible solutes in protecting the cytoplasm from dehydration, ion toxicity (particularly from Na), and temperature stresses [8]. Also, GB can reverse damage to proteins and membranes from high levels of Na [79]. High concentrations of organic solutes are reported to stabilize macromolecules or molecular assemblies, thus decreasing the loss of either enzyme activity or membrane integrity when water is limiting [147,148]. Nevertheless, each of the structurally distinct osmoprotectants could differentially benefit the osmotically sensitive classes of molecules or structures within the cell [147,148]. The following section will discuss GB's role in protecting the metabolic functions of the cytoplasm and thus its direct physiological significance in improving stress resistance of plants.

## A.  Intracellular Osmotic Adjustment

Plants rely on inorganic solutes (such as K or Na in vacuoles) to maintain $\Psi$s when subjected to water deficits or high salts in the root zone [143,144]. Using organic solutes (such as sugars, betaines, or amino acids) to regulate $\Psi$s in vacuolar volumes is too expensive metabolically as an effective strategy of adaptation [141,149]. Equivalent molar concentrations of inorganic solutes, such as NaCl or KCl, have twice the $\Psi$s compared with GB (Figure 4a–c). For example, NaCl, KCl, and GB at 100 mM reduce $\Psi$s by 0.41, 0.42, and 0.23 MPa, respectively (Figure 4a–c). But, as noted earlier, inorganic solutes are potentially damaging to the cytoplasm where organic solutes must be used for $\Psi$s maintenance [141,149]. Glycine betaine is used in this capacity to maintain osmotic balance between the cytoplasm and vacuole [19]. Typical GB levels are between 50 and 200 $\mu$mol g$^{-1}$ dwt and account for 2 to 3% of the $\Psi$s of the leaf sap [2,86,150,151].

Spinach leaves with a $\Psi$s of $-2.0$ MPa had a GB content of 320 $\mu$mol g$^{-1}$ dwt [127]; these GB levels could contribute $-0.07$ MPa to the total $\Psi$s if distributed uniformly throughout the leaf. Similarly, in red beet, leaf GB levels reached 104 $\mu$mol g$^{-1}$ dwt under moderately saline conditions at a $\Psi$s of 1.39 MPa, accounting for 2.5% of the total $\Psi$s (G.V. Subbarao et al., unpublished results). Using histochemical techniques, Hall et al. [152] demonstrated cytoplasmic localization of GB in leaf cells of salt-grown *Suaeda maritima*. In wheat, GB accounted for 4.5% of the $\Psi$s of the leaf sap, but in chloroplasts, the GB concentrations were about 20 times higher and accounted for a major portion of the chloroplast's $\Psi$s [153]. For beets, concentrations of GB in cytoplasm ranged from 45 to 470 mM [154]. In *Atriplex gmelini*, nearly 320 mM GB levels were reported in cytoplasm, with only 0.24 mM GB in the vacuole [155]. For many halophytic chenopods, GB is predominantly localized in the cytoplasm of the leaf cells [152]. Concentrations of GB up to 300 mM were measured in isolated chloroplasts of salinized spinach leaves [28,43]. At these concentrations, GB contributes significantly to chloroplast OA, volume maintenance, and photosynthetic capacity at low leaf water potentials [43]. Many hypotheses about GB as an intracellular osmoticum are valid only if it is localized in the cytoplasmic compartment of the cell [9,156].

In contrast, Leigh et al. [157] reported that 26 to 84% of the total tissue GB is localized in the vacuole for red beet. The physiological significance of this finding is not clear. High vacuolar concentrations might result from passive diffusion through the tonoplast as cytoplasmic concentrations increase [43], or GB may be actively, and reversibly, transported to the vacuole in response to the level of osmotic stress. In highly stressed cells, most of the GB can be located in the cytoplasm, but after removal of stress, GB could possibly be transferred to the vacuole instead of being degraded.

### 1.  Threshold Stress for Glycine Betaine Accumulation

Glycine betaine accumulates in the cytoplasm only after a threshold turgor is reached [2,40, 137,156,158,159], suggesting that cytoplasmic $\Psi$s is initially regulated using inorganic solutes. For example, if K salts and other solutes maintain a basal cytoplasmic osmotic pressure of 300 to 400 mOsm

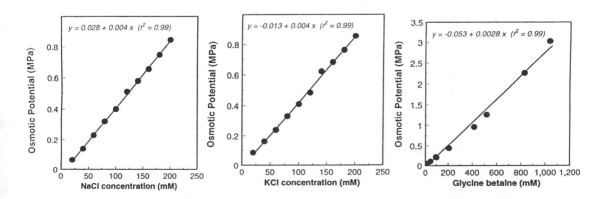

**Figure 4**   Osmotic potential of NaCl, KCl, and glycine betaine.

$kg^{-1}$, then GB accumulation is necessary only if vacuolar osmotic pressures rise above this level [156,157]. Also, the threshold K level for GB may vary depending on the external K concentration, the type of stress, and genetic factors. In yeast (*Saccharomyces cerevisiae*), two plasma membrane proteins, Sln1 and Sho1, operate as sensors for turgor loss under mild osmotic stress and these proteins activate several defense mechanisms that include Na efflux pumps [160]. It is not known whether such sensor proteins are involved in the activation or acceleration of GB synthesis in plants. Similarly, abscisic acid (ABA) synthesis, stomatal closure, and subsequent physiological responses do not occur until this threshold is met [158,161].

## 2. What Triggers Glycine Betaine Production? (Osmotic Stress or Internal Sodium Levels?)

Internal Na levels alone can trigger GB production in red beet when tissue Na levels are increased without subjecting the plants to osmotic stress [86]. In contrast, tissue Na levels alone do not trigger GB production in spinach under similar conditions (G.V. Subbarao and R.M. Wheeler, unpublished). Several reports indicate that spinach accumulates GB when subjected to high levels of NaCl salinity, suggesting that osmotic stress may be the physiological trigger [43]. Glycine betaine concentration and BADH activity increased severalfold in spinach leaves under salt stress [29,128,162]. Also, GB accumulation appears to be dependent on the rate at which water stress develops; when water stress develops gradually, GB accumulates, but when water stress develops abruptly, it does not [30].

## 3. The Relationship Between Glycine Betaine and the Solute Potential ($\Psi s$) of the Leaf Sap

Glycine betaine accumulation is linearly related to $\Psi s$ in sugar beet ($r^2 = 0.99$), although GB's contribution to the total leaf $\Psi s$ is <5% [27]. In other crops, nearly linear relationships between leaf $\Psi s$ and GB levels were observed when plants were exposed to NaCl salinity or water deficits [21,163–165]. In barley, isopopulations that differ nearly twofold in unstressed GB content, it was shown that the high-GB population maintained 0.1 MPa lower $\Psi s$ at 300 mM NaCl. For several halophytic species belonging to the family Chenopodiaceae, GB levels were correlated with sap $\Psi s$ under salinity [163]. In red beet, turgid leaf $\Psi s$ is linearly related to GB levels (Figure 5). Collectively, GB levels appear to be a genetic phenomenon and modulated by the environment. In addition, GB accumulation might be tightly linked to other loci that regulate the accumulation of other organic and inorganic solutes, which together determine the water relations of plants [43].

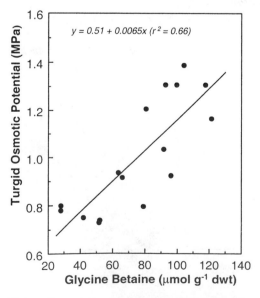

**Figure 5**  Relationship between glycine betaine and turgid osmotic potential of leaf sap in red beet.

## B.  Protection Against Na Ions in Cytoplasm

High intracellular concentrations of both Na and Cl can be deleterious to cellular systems. High salt concentrations (greater than 400 mM) inhibit most enzymes because they perturb the hydrophobic-electrostatic balance needed to maintain protein structure [139]. However, toxic effects of Na occur at much lower concentrations (such as 100 mM Na), suggesting specific Na toxicity targets in the cell [141]. Sodium interferes with cationic sites involved in binding of K, Ca, and Mg [139,141]. For example, 20 mM Na in cytoplasm inhibited several nucleotidases and ribonucleases by displacing the essential Mg from the protein complexes [166]. Chloride may interfere with anionic binding of RNA and anionic metabolites such as bicarbonate, carboxylates, and sugar phosphates.

Glycine betaine is widely believed to protect the cytoplasm from Na toxicity [167]. It is hypothesized that the dipole character neutralize Na and Cl during salt stress, and the hydrophobic methyl groups stabilize the hydrophobic domains of proteins [15,16,133,167]. In vitro studies showed that GB (200 to 500 mM) protected enzyme activity from Na toxicity [147,168–175]. Glycine betaine also protects PEP-carboxylase against excessive concentrations of Na ions [176,177]. Also, GB can alter the thermodynamic properties of membranes by indirectly interacting with phosphatidylcholine moieties [178,179].

Glycine betaine is less inhibitory to enzyme activity [180] and mRNA translation than equivalent concentrations of other organic solutes in vitro [42,181]. In wheat and sugar beet, protein synthesis (translation of mRNA) was maintained at GB concentrations up to 500 mM [42], whereas sucrose levels above 100 mM and proline levels above 300 mM were inhibitory to protein synthesis [42]. In sugar beet, GB protects the root membranes from heat destabilization [178,182]. These properties are of interest because crop plants transformed with enzymes involved in osmolyte biosynthesis may also exhibit increased tolerance to heat stress [15,16].

### 1.  *Positive Correlation Between Tissue Salt and Glycine Betaine Levels*

As mentioned earlier, if GB is involved in the osmotic adjustment of the cytoplasm, then its production should be regulated at the biosynthetic level with its production proportional to the severity of salinity or water stress [21,127]. A positive correlation between salt concentration and GB accumulation was reported in *Atriplex semibaccata*, *A. halimus* [183], *Spartina alterniflora* [184], *Sporobolus virginicus* [185], *Limonium* sp. [48,50], and *Suaeda monoica* [163]. In red beet, GB accumulation increased linearly with leaf Na levels (Figure 6).

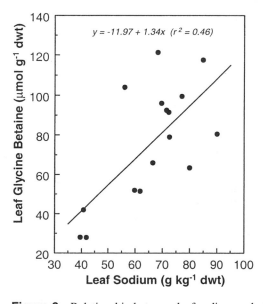

**Figure 6**  Relationship between leaf sodium and glycine betaine accumulation in red beet.

## C. Protection of Chloroplasts and Photosystem II from Na Damage

Glycine betaine is hypothesized to stabilize photosynthetic reactions, the structure of extrinsic proteins of photosystem II (PSII) complex, and ATP synthesis under Na stress conditions [186]. Chloroplasts of sugar beet can accumulate high levels of Na but still retain structural and functional integrity [142]. In contrast, chloroplasts of bean leaves showed pronounced swelling when 75% of their K was replaced by Na and consequently were unable to function [142]. This is consistent with reports of sensitivity of leaf photosynthesis to salinity stress in beans [187]. Glycine betaine can prevent chlorophyll loss and proteolysis and may improve resistance to drought or salinity [188,189]. In sugar beet leaves, most of the GB is localized in the chloroplasts [2,43,190,191]. In spinach [43] and *Suaeda* [148], GB concentrations in chloroplast reached 300 mM under salt stress. Glycine betaine may also protect the oxygen-evolving PSII complex against the inhibitory effects of NaCl [148,175,192,193] and freezing stress [194].

In red beet, leaf photosynthetic rates are highly tolerant to internal Na. For example, leaf photosynthetic rates showing no trend across internal Na levels ranging from 40 to 100 g kg$^{-1}$ dwt (G. V. Subbarao and R. M. Wheeler, unpublished results). Stress may cause lesions in the reaction centers of PSII [195,196]. Unlike the case of beets, when tissue K levels were replaced with Na in spinach, GB levels decreased linearly as tissue Na increased (Figure 7). Chlorophyll levels and photosynthetic rates are sensitive to leaf Na in spinach (G. V. Subbarao and R. M. Wheeler, unpublished results).

In wheat, pretreatment with GB alleviated NaCl-induced stomatal and nonstomatal inhibition of photosynthesis completely [197]. In vitro studies have shown that GB can protect thylakoid membranes from freezing stress [194]. Glycine betaine–deficient maize lines are more sensitive to high-temperature stresses than GB-producing lines. Membrane stability, resistance to photoinhibition, and steady-state yield of electron transport over PSII are adversely affected by high temperatures in GB-deficient lines [198]. Williams et al. [199] reported that GB can increase the thermal stability of PSII. Several studies have indicated that GB protects the PSII against high Na in vitro [175]. Also, transgenic *Arabidopsis thaliana* lines that overexpress GB maintain normal PSII function under high salt levels [200] or cold stresses [201]. Betaine-treated spring and winter wheat seedlings exhibited a greater capacity to prevent the closure of PSII reaction centers than control plants when subjected to cold stress [153]. It is hypothesized that GB protects PSII from denaturation and inactivation (from Na) by forming layers of preferentially oriented dipoles on protein surfaces that shield them both sterically and electrostatically from chaotropic solutes (such as Na) [192]. Incharoensakdi et al. [172] reported that GB may protect ribulose-1,5-bisphosphate carboxylase/oxygenase (Rubisco) activity from high salts by acting at the protein-water interface, decreasing the effects of excess salts on the enzymes and other macromolecules [182,202].

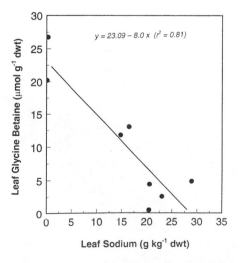

**Figure 7**   Relationship between leaf sodium and glycine betaine accumulation in spinach.

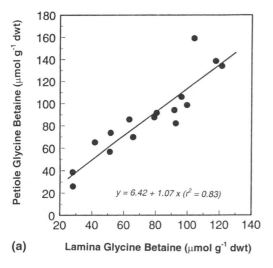

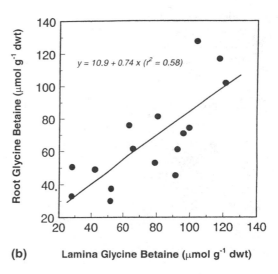

(a)    Lamina Glycine Betaine ($\mu$mol g$^{-1}$ dwt)

(b)    Lamina Glycine Betaine ($\mu$mol g$^{-1}$ dwt)

**Figure 8**    (a) Relationship between lamina and petiole glycine betaine levels of red beet. (b) Relationship between lamina and root glycine betaine of red beet.

## D.    Distribution of Glycine Betaine in the Plant

Although GB synthesis is largely confined to chloroplasts, GB is readily translocated to other tissues [21,203]. In sugar beet, there is a linear relationship between GB levels in leaves and roots [21]. In red beet, leaf GB levels are linearly related to petiole and root GB, indicating that GB is readily translocated from leaves to other parts of the plant (Figure 8a and b). Thus, the distribution of GB in red beet appears to be fairly uniform in various plant parts (leaves, storage root, and fibrous roots), although slightly higher concentrations are found in leaf lamina (G. V. Subbarao and R. M. Wheeler, unpublished results). In other chenopods, GB accumulates primarily in mature leaves under salt stress, but under nonstress conditions, GB is found mostly in young expanding leaves [34,127,203]. In barley under water stress, GB accumulates mostly in mature leaves but is then translocated to the expanding leaves upon rewatering [104,203].

## E.    Stability of Glycine Betaine in Plants

Glycine betaine constitutes essentially an inert end product that is not metabolized by the plant [4,204,205]. Glycine betaine is not degraded upon relieving the stress in wheat [206,207], barley [29,128,203], sugar beet [21], alfalfa [208], tobacco [118], or spinach [191]. But GB levels may decline once stress is removed due to growth in the absence of subsequent synthesis [204,207]. Because of this stability, GB content may serve as a cumulative index of the internal water status [204], with potential applications in both plant breeding and crop management.

## F.    Physiological Costs Associated with Glycine Betaine Accumulation

Energetically, synthesis of GB is expensive for plants as it contains N and its biosynthesis requires NADPH [14]. In barley leaves, GB can represent nearly 2% of the Kjeldahl nitrogen of the tissue [204,209]. In GB accumulators (such as halophytes), where concentrations can reach 500 to 1000 mM, betaine represents 20 to 30% of the total N of the plant [2,15,16].

Conventional genetic approaches provide a glimpse of possible benefits and costs of GB accumulation. Homozygous GB-accumulating (*Bet1/Bet1*) maize plants exhibited less growth inhibition than near-isogenic GB-deficient (*bet1/bet1*) plants under salinity stress. However, high GB accumulation in maize is associated with a 5% reduction in grain yield under well-irrigated field conditions. The molar concentration of GB in sugar beet can reach 5 to 10% that of sucrose and conceivably reduce sugar yields indirectly as the synthesis of GB requires about the same energy input as that of sucrose [164,210]. Photosynthate diverted to GB represents an appreciable cost in energy and carbon that is available neither for storage as sucrose nor for plant processes that contribute indirectly to economic yield.

## V. ROLE IN ADAPTATION TO DROUGHT AND SALINITY

Mechanisms that either permit resistance to cellular dehydration or minimize water loss contribute to productivity in water-limited environments [211,212]. Osmotic adjustment, a well-defined adaptation to water deficit, is associated with maintenance of protoplast volume, cell turgor [211], and avoidance of lethal relative water content [213–215]. In some plant species, osmotic adjustment is partially achieved by accumulation of GB [4,130]. Increase in cellular osmolarity, particularly in the cytoplasm, results from the accumulation of nontoxic, osmotically active solutes and is accompanied by either water influx or reduced efflux from cells to provide the turgor necessary for cell expansion and growth [9].

Salt resistance may be correlated with the accumulation of GB and choline in a number of species [3,216]. Sugar beet, which has higher resistance to salinity than spinach, also accumulates higher levels of betaine [21,121,217]. There is some evidence that GB accumulation contributes to salt resistance in *Lophopyrum elongatum* [218]. In sorghum, drought-resistant genotypes accumulated nearly three times more GB than sensitive genotypes under well-watered and drought-stressed conditions [219]. Nevertheless, the adaptive value of betaine accumulation in salt-stressed and water-stressed plants is still speculative and needs further investigation [107,164,168].

### A. Exogenous Glycine Betaine Application for Improving Drought and Salt Resistance

Although traces of GB are detected in legumes, e.g., bean (*Phaseolus vulgaris* L), pea, soybean, lupins, and alfalfa [52], and other crops such as tomato, potato, rice, maize, and rape (*Brassica napus* L.), no significant amounts of GB accumulate in these crops [4,220]. It is hypothesized that foliar application of GB in nonaccumulating species can ameliorate the negative effects of drought and salinity on productivity [221,222]. Foliar-applied GB rapidly penetrates leaf tissue and is quickly translocated to other plant organs [223]. Also, surfactants (such as 'kinetic', 'lus-50', and 'sito+') can enhance the penetration and uptake of GB in leaf tissue [223]. Foliar application of GB to potato and tomato is reported to reduce crop failures in arid climates [221,223–228]. Fruit yield of tomato increased up to 40% by the foliar application of GB when grown in saline soils or exposed to high temperatures in California [223,226].

Exogenous GB application is reported to improve growth and yield of tobacco under drought conditions [221,222]. Field experiments with wheat and barley have indicated that foliar application of GB at 18 kg ha$^{-1}$ increased yield [227,229]. Rice growth was improved under salinity by foliar application of GB [230]. Foliar application of GB to sorghum, barley, wheat, and soybean has improved growth under field drought conditions [221–223,226,229]. Exogenous GB application has also resulted in improvement of growth under drought and saline conditions for lupins (cited in Agboma et al. [222]), alfalfa [62,208,231], cotton [232], and maize [221]. Thus, application of GB, which is a by-product of sugar beet [226], environmentally safe, nontoxic, and water soluble, should be explored further for its potential use as a management strategy in mitigating the negative impact of water stress and salinity on crop production [226].

However, for some crops that do not naturally accumulate it, GB application had no effect or reduced the yield and hence may not be compatible in crops that do not naturally accumulate it [233]. Foliar application of GB on turnip, rapeseed, and spring cereals did not improve yield in drought studies in Finland [227]. In rape (*Brassica napus* L. var. *oleifera* cv. Samourai), the viability of GB-treated leaf disks was substantially reduced when subjected to PEG stress. This was attributed to toxic effects of GB on Rubisco and protein synthesis [234]. In contrast, GB application enhanced the viability of leaf disks under PEG stress in spinach. This suggests that GB is not a compatible organic osmoticum for all plants [234].

## VI.  GENETICS OF GLYCINE BETAINE ACCUMULATION

### A.  Genetic Control

Analysis of the progeny from high-GB × low-GB crosses suggests that GB levels are controlled by a small number of nuclear-coded genes [235]. The precise function of these genes is unknown. It is suggested that these GB genes may have a role in the overall control of osmoregulation rather than the control of GB alone [236]. In maize, a single gene *bet1* located on chromosome 3 [54] confers the presence or absence of GB. Glycine betaine deficiency in homozygous *bet1* plants is associated with an inability to oxidize choline to betaine aldehyde [8,31,83,237]. A number of maize inbreds lack GB completely [31,53,54], and this is controlled by recessive alleles of a single nuclear gene [31,54,238]. Maize lines that are deficient in GB synthesis (homozygous *bet1*) are more susceptible to salinity stress than maize lines that synthesize GB (homozygous *Bet1*) [83,236]. In sorghum, GB synthesis is also governed by a single nuclear gene [39].

### B.  Genetic Variation

In cultivated barley and its wild progenitor (*H. spontaneum*), the natural variability for GB levels among genotypes tested (71 cultivated and 268 wild types) ranged from 19 to 40 $\mu$mol g$^{-1}$ dwt under irrigated conditions and 15 to 90 $\mu$mol g$^{-1}$ dwt under stress conditions [239]. Salt-tolerant rice cultivars CSC 1, AU1, and Co 43 accumulated GB levels up to 16.4 to 21.3 $\mu$mol g$^{-1}$ dwt, whereas salt-sensitive cultivars TKM 9, TKM 4, CSC 2, Co 36, IR 20, and GR 3 accumulated little GB during salinization [240]. The GB concentration in Co 43, CSC 1, and AU1 was consistently higher than in the other cultivars during the entire growing period of the salinization [240]. Glycine betaine–accumulating maize lines showed higher stomatal conductance than GB-deficient lines under drought [223,236].

### C.  Genetic Engineering

Genetic engineering of GB synthesis in higher plants requires at least two genes: choline monooxygenase (CMO) and betaine aldehyde dehydrogenase (BADH). Also, free choline pools in the metabolic pathway are required, although their size is not known [36,241]. Manipulation of foliate-mediated methyl group metabolism may be required to ensure appropriate reserves of these choline pools [242,243]. Also, transport systems for choline and GB are necessary to increase salt resistance through in situ GB production [230]. The BADH genes from *E. coli* [244], spinach [118], or barley [162] have been introduced into tobacco plants. But the BADH transgenic tobacco lines did not exhibit increased stress tolerance. In contrast, tobacco plants engineered to accumulate trehalose, mannitol, proline, fructan, sorbitol, inositol, and ononitol are reported to show improved performance under drought or salinity conditions (Table 4). Enhanced osmotic tolerance in transgenic tobacco plants engineered to accumulate proline [247], fructans [248], and mannitol [245] is associated with increased root/shoot ratio. In sorghum and wheat, increased OA leads to improved root growth, larger shoots, and improved yield under drought conditions [214,215]. Tobacco plants expressing a gene for trehalose synthesis lose water more slowly than nontransformed plants [244]. However, the levels of trehalose expressed in these transgenic plants are not sufficiently high to reduce the leaf $\Psi$s. It is hypothesized that trehalose preserves protein and membrane integrity under drought and salinity [169,170,174].

## VII.  CONCLUDING REMARKS

Early concepts of GB as being a useless metabolic waste product (Winterstein, 1910; cited in Ref. 3) have evolved into defining a biochemically significant role for GB in stress adaptation of plants [8,18]. However, much additional research is required to establish the importance of GB in conferring drought or salinity resistance. Given the inherent complexity of drought and/or salinity stresses and the extensive mechanisms evolved in plants for adaptation to these stresses [9,16,65,66,214,215] (also see Chapter 44), it is unlikely that a single biochemical trait such as GB accumulation can have an overriding effect. However, GB accumulation could be part of multiple mechanisms that are needed to improve crop adaptation to stresses such as drought and salinity, which defy easy genetic solutions. Salinity resistance is a combi-

**TABLE 4**  Transgenic Plants Where Osmotically Active Substances Were Introduced to Improve Osmotic Adjustment[a]

| Gene product | Source | Product accumulated | Increased resistance | References |
|---|---|---|---|---|
| Mannitol-1-phosphate dehydrogenase | *Escherichia coli* | Mannitol | Yes | [245] [246] |
| p5C synthetase | *Vigna aconitifolia* | Proline | Yes | [247] |
| Fructosyltransferase (levan-sucrase) | *Bacillus subtilis* | Fructan | Yes | [248] |
| Betaine aldehyde dehydrogenase (BADH) | *Hordeum vulgare* | Betaine | No | [118] [162] |
| Sorbitol-6-phosphate dehydrogenase | *Malus domestica* | Sorbitol | ND[c] | [249] |
| Choline dehydrogenase | *Escherichia coli* | Glycine betaine? | Yes | [250] |
| Trehalose-6-phosphate synthetase | *Saccharomyces cerevisiae* | Trehalose | Yes | [244] [251] |
| Trehalose-6-phosphate synthetase | *E. coli* | Trehalose | No | [252] |
| *myo*-Inositol-3-phosphate synthase | *Spirodela polyrrhiza* | Inositol | No | [253] |
| *myo*-Inositol-3-methyl transferase | *Mesembryanthemum crystallinum* | Ononitol | Yes | [254] |
| Choline oxidase[b] | *Arthrobacter globiformis* | Glycine betaine | Yes | [200] |

[a] All these studies involved transformation of tobacco (*Nicotiana tabacum*).
[b] Introduced into *Arbidopsis thaliana*.
[c] ND, stress resistance was not determined.
*Source:* Adapted from Ref. 9.

nation of a regulated uptake of ions in the roots, the cell-, tissue-specific compartmentation of ions, and osmotic adjustment [133,137,255] (also see Chapter 44).

Most information available on the functional role of GB is based on in vitro studies. Glycine betaine has only a limited capacity to regulate total leaf $\Psi$s, as its concentrations rarely exceed 100 to 200 $\mu$mol $g^{-1}$ dwt or about 5% of total leaf $\Psi$s. Thus, GB can have significant impact in $\Psi$s only if localized in the cytoplasm. Also, the functional relationship between GB accumulation and stress resistance in the field has received limited attention. Characterization of genetic stocks that differ in GB accumulation and their adaptive potential to stressful environments has not been well established. Development of near-isogenic lines that differ in GB accumulating capability is essential for evaluating GB as an adaptive trait for crop improvement programs. But surprisingly little progress has been made in this direction.

Glycine betaine concentrations are tightly correlated with leaf osmotic adjustment under saline and drought conditions, although a mechanism for this response is not known. Also, to have a positive effect on plant water relations over extended period of time under water deficits, osmotic adjustment should stimulate root growth to permit better water extraction. However, to our knowledge, mechanisms that could link GB accumulation or osmotic adjustment to the root growth stimulation have not been established.

Since GB is not metabolized when stress is removed, this represents a permanent cost to plants that synthesize the compound. Remobilization of C and N compounds plays an important role in limiting water stress effects on the seed-filling phase of legumes [256] (G.V. Subbarao et al. unpublished). Negative osmotic adjustment during seed filling of pigeonpea is observed in genetic stocks that have the highest yield potential under moisture deficits (G.V. Subbarao et al. unpublished data). Thus, osmotic adjustment as a mechanism of stress resistance may not limit yield if carbon and nitrogen from the organic solutes can be utilized subsequently for reproductive growth. Because GB cannot be metabolized like other organic solutes, a potential exists that GB accumulation can reduce yield. Nevertheless, GB may confer metabolic stability as it can be redistributed when plants are subjected to intermittent and transient stresses.

Effects of exogenous application of GB as a management strategy for improving stress resistance in drought-prone and saline environments need to be evaluated. Potential impacts of introducing the GB biosynthetic pathway into crops (that do not accumulate) are not fully understood. There are concerns that introducing genes for GB synthesis may not be sufficient if precursor compounds such as choline are still limiting. Introducing genes that promote synthesis of these precursor compounds (such as choline from serine) would also need to be evaluated. Also, transport mechanisms for GB between cell compartments and tissues are not well understood and need further study [8].

Successful development of stress-tolerant crops will require large-scale metabolic engineering of not only genes encoding osmolytes but also those for antioxidants, water channel proteins, etc. because osmotic adjustment is just part of a complex web of adaptive strategies in plants [215,257]. The complexity and range of physiological mechanisms that have similar adaptive roles for drought and salinity make it a challenging task to establish the functional role of GB in crop adaptation to stressful environments.

## ACKNOWLEDGMENTS

We would like to thank Drs. Wade Berry (University of California, Los Angeles) and M. Shannon (USDA Salinity, Riverside, CA), who went through the earlier versions of the manuscript and made several suggestions to improve it. Also, we would like to thank Dr. William Knott, NASA Biomedical Office, Kennedy Space Center, for his support and encouragement for this project. This research was supported through the National Research Council and NASA's Life Support Contract (NAS10-12180).

## REFERENCES

1.  AD Brown, JR Simpson. Water relations of sugar tolerant yeasts: the role of intracellular polyols. J Gen Microbiol 72:589–591, 1972.
2.  GR Stewart, JA Lee. The role of proline accumulation in halophytes. Planta 120:279–289, 1974.
3.  R Storey, RG Wyn Jones. Betaine and choline levels in plants and their relationship to NaCl Stress. Plant Sci Lett 4:161–168, 1975.
4.  RG Wyn Jones, R Storey. Betaines. In: LG Paleg, D Aspinall, eds. The Physiology and Biochemistry of Drought Resistance in Plants. Sydney: Academic Press, 1981, pp 171–204.
5.  PH Yancey, ME Clark, SC Hand, RD Bowlus, GN Somero. Living with water stress: evolution of osmolyte systems. Science 217:1214–1222, 1982.
6.  RG Wyn Jones. An assessment of quaternary ammonium and related compounds as omotic effectors in crop plants (phytochemical aspects of osmotic adaptation). Recent Adv Phytochem 18:55–58, 1984.
7.  BP Naidu, LG Paleg, GP Jones. Nitrogenous compatible solutes in drought stressed *Medicago* spp. Phytochemistry 31:1195–1197, 1992.
8.  D Rhodes, AD Hanson. Quaternary ammonium and tertiary sulfonium in higher plants. Annu Rev Plant Physiol Plant Mol Biol 44:357–384, 1993.
9.  PD Hare, WA Cress, J Van Staden. Dissecting the roles of osmolytes accumulation during stress. Plant Cell Environ 21:535–553, 1998.
10.  AJ Delauney, DPS Verma. Proline biosynthesis and osmoregulation in plants. Plant J 4:215–223, 1993.
11.  N Smirnoff, QJ Cumber. Hydroxyl radical scavenging activity of compatible solutes. Phytochemistry 28:1057–1060, 1989.
12.  B Schobert. Is there an osmotic regulatory mechanism in algae and higher plants? J Theor Biol 68:17–26, 1977.
13.  SN Timasheff, T Arakawa. Stabilization of protein structure by solvents. In: TE Creighton, ed. Protein Structure: A Practical Approach. Oxford: IRL Press, 1989, pp 331–345.
14.  J Gorham. Betaines in higher plants—biosynthesis and role in stress metabolism. In: RM Wallsgrove, ed. Aminoacids and Their Derivatives in Higher Plants. Cambridge: University Press, 1995, pp 172–203.
15.  HJ Bohnert, RG Jensen. Metabolic engineering for increased salt tolerance—the next step. Aust J Plant Physiol 23:661–667, 1996.
16.  HJ Bohnert, RJ Jensen. Strategies for engineering water-stress tolerance in plants. TIBTECH 14:89–97, 1996.
17.  JH Crowe, FA Hoekstra, LM Crowe. Anhydrobiosis. Annu Rev Physiol 54:579–599, 1992.
18.  EA Galinski, HG Truper. Betaine, a compatible solute in the extremely halophilic phototrophic bacterium *Ectothiorhodospira halochloris*. FEMS Microbiol Lett 13:357–360, 1982.
19.  R Storey, RG Wyn Jones. Quaternary ammonium compounds in plants in relation to salt resistance. Phytochemistry 16:447–453, 1977.
20.  M Briens, F Larher. Osmoregulation in halophytic higher plants: a comparative study of soluble carbohydrates, polyols, betaines and free proline. Plant Cell Environ 5:287–292, 1982.

21. ADH Hanson, R Wyse. Biosynthesis, translocation and accumulation of betaine in sugar-beet and its progenitors in relation to salinity. Plant Physiol 70:1191–1198, 1982.

22. A Poljakoff-Mayber, DE Symon, GP Jones, BP Naidu, LG Paleg. Nitrogenous compatible solutes in native South Australian plants. Aust J Plant Physiol 14:341–350, 1987.

23. D Rhodes, PJ Rich, AC Myers, CC Reuter, GC Jamieson. Determination of betaines by fast atom bombardment mass spectrometry: identification of glycinebetaine deficient genotypes of *Zea mays*. Plant Physiol 84:781–788, 1987.

24. EA Weretilnyk, S Bednarek, KF McCue, D Rhodes, AD Hanson. Comparative biochemical and immunological studies of the glycinebetaine synthesis pathway in diverse families of dicotyledons. Planta 178:342–352, 1989.

25. KF McCue, AD Hanson. Drought and salt tolerance: towards understanding and application. Trends Biotech 8:358–362, 1990.

26. AD Hanson. Interpreting the metabolic responses of plants to water stress. HortScience 15:623–629, 1980.

27. R Grumet, AD Hanson. Genetic evidence for an osmoregulatory function of glycinebetaine accumulation in barley. Aust J Plant Physiol 13:353–364, 1986.

28. G Schroppel-Meier, WM Kaiser. Ion homeostasis in chloroplasts under salinity and mineral deficiency. I. Solute concentrations in leaves and chloroplasts from spinach plants under NaCl or $NaNO_3$ salinity. Plant Physiol 87:822–827, 1988.

29. K Arakawa, M Katayama, T Takabe. Levels of betaine in green leaves and etiolated leaves and roots of barley. Plant Cell Physiol 31:797–803, 1990.

30. BP Naidu, LG Paleg, D Aspinall, AC Jennings, GP Jones. Rate of imposition of water stress alters the accumulation of nitrogen-containing solutes by wheat seedlings. Aust J Plant Physiol 17:653–664, 1990.

31. C Lerma, PJ Rich, GC Ju, WJ Yang, AD Hanson, D Rhodes. Betaine deficiency in maize: complementation tests and metabolic basis. Plant Physiol 95:1113–1119, 1991.

32. M Ishitani, K Arakawa, K Mizuno, S Kishitani, T Takabe. Betaine aldehyde dehydrogenase in the Gramineae: levels in leaves of both betaine-accumulating and nonaccumulating cereal plants. Plant Cell Physiol 34:493–495, 1993.

33. B Rathinasabapathi, DA Gage, DJ Mackill, AD Hanson. Cultivated and wild rices do not accumulate glycinebetaine due to deficiencies in two biosynthetic steps. Crop Sci 33:534–538, 1993.

34. T Nakamura, M Ishitani, P Harinasut, M Nomura, T Takabe, T Takabe. Distribution of glycinebetaine in old and yound leaf blades of salt stressed barley plants. Plant Cell Environ 37:873–877, 1996.

35. KV Wood, KJ Stringham, DL Smith, JJ Volenec, KL Hendershot. Betaines of alfalfa: characterization by fast atom bombardment and desorption chemical ionization mass spectrometry. Plant Physiol 96:892–897, 1991.

36. DA Gage, B Rathinasabapathi. Role of glycine betaine and dimethylsulfoniopropionate in water-stress tolerance. In: K Shinozaki, ed. Cold, Drought, Heat and Salt Stress: Molecular Responses in Higher Plants. Boca Raton, FL: RG Landes Co, 1999, pp 125–152.

37. A Al-Amoudi, AY Ali. Some practical aspects of measurements of betaines and their sulphur analogues by the use of HPLC. J Microbiol Methods 10:289–296, 1989.

38. U Anthoni, C Christophersen, L Hougaard, PH Nielsen. Quaternary ammonium compounds in the biosphere: an example of a versatile adaptive strategy. Comp Biochem Physiol 99B:1–18, 1991.

39. EM Grote, G Ejeta, D Rhodes. Inheritance of glycinebetaine deficiency in sorghum. Crop Sci 34:1217–1220, 1994.

40. F Araya, O Abarca, GE Zuniga, LJ Corcuera. Effects of NaCl on glycinebetaine and on aphids in cereal seedlings. Phytochemistry 30:1793–1795, 1991.

41. R Krishnamurthy, KA Bhagwat. Accumulation of choline and glycinebetaine in salt-stressed wheat seedlings. Curr Sci 59:111–112, 1990.

42. TS Gibson, J Speirs, CJ Brady. Salt tolerance in plants. II. In vitro translation of mRNAs from salt-tolerant and salt-sensitive plants on wheat germ ribosomes. Responses to ions and compatible organic solutes. Plant Cell Environ 7:579–587, 1984.

43. SP Robinson, GP Jones. Accumulation of glycinebetaine in chloroplasts provide osmotic adjustment during salt stress. Aust J Plant Physiol 13:659–668, 1986.

44. M Ajmal Khan, IA Ungar, AM Showalter, HD Dewald. NaCl-induced accumulation of glycinebetaine in four subtropical halophytes from Pakistan. Physiol Plant 102:487–492, 1998.

45. Y Wang, Y Meng, H Ishikawa, T Hibino, Y Tanaka, N Nii, T Takabe. Photosynthetic adaptation to salt stress in three-color leaves of a $C_4$ plant *Amaranthus tricolor*. Plant Cell Physiol 40:668–674, 1999.

46. WD Hitz, AD Hanson. Determination of glycine betaine by pyrolysis–gas chromatography in cereals and grasses. Phytochemistry 19:2371–2374, 1980.

47. T Nakamura, S Yokota, Y Muramoto, K Tsutsui, Y Oguri, K Kukui, T Takabe. Expression of a betaine aldehyde dehydrogenase gene in rice, a glycinebetaine nonaccumulator and possible localization of its protein in peroxisomes. Plant J 11:1115–1120, 1997.

48. AD Hanson, B Rathinasabapathi, J Rivoal, M Burnet, MO Dillon, DA Gage. Osmoprotectant compounds in the Plumbaginaceae: a natural experiment in metabolic engineering. Proc Natl Acad Sci U S A 91:306–310, 1994.

49. R Storey, J Gorham, MG Pitman, AD Hanson, DA Gage. Responses of *Melanthera biflora* to salinity and water stress. J Exp Bot 44:1551–1560, 1993.

50. AD Hanson, J Rivoal, L Paquet, DA Gage. Biosynthesis of 3-dimethylsulfoniopropionate in *Wollastonia biflora* (L.) DC. Evidence that *S*-methylmethionine is a intermediate. Plant Physiol 105:103–110, 1994.

51. AD Hanson, NA Scott. Betaine synthesis from radioactive precursors in attached water stressed leaves. Plant Physiol 66:342–348, 1980.

52. AL Takhtajan. Outline of the classification of flowering plants (Magnoliophyta). Bot Rev 46:225–359, 1980.

53. DG Brunk, PJ Rich, D Rhodes. Genotypic variation for glycinebetaine among public inbreds of maize. Plant Physiol 91:1122–1125, 1989.

54. D Rhodes, PJ Rich, DG Brunk, GC Ju, JC Rhodes, MH Pauly, LA Hansen. Development of two isogenic sweet corn hybrids differing for glycinebetaine content. Plant Physiol 91:1112–1121, 1989.

55. HG Truper, EA Galinski. Compatible solutes in halophilic phototrophic procaryotes. In: Y Cohen, E Rosenberg, eds. Microbial Mats. Physiological Econogy of Benthic Microbial Communities. Washington, DC: American Society of Microbiology, 1989, pp 342–348.

56. SRC Warr, RH Reed, WDP Stewart. The compatibility of osmotica in cyanobacteria. Plant Cell Environ 11:137–142, 1988.

57. LJ Borowitzka. Solute accumulation and regulation of cell water activity. In: LB Paleg, D Aspinall, eds. The Physiology and Biochemistry of Drought Resistance in Plants. Sydney: Academic Press, 1980, pp 11841–11846.

58. FAA Mohammad, RH Reed, WDP Stewart. The halophilic cyanobacterium *Synechocystis* DUN52 and its osmotic responses. FEMS Microbiol Lett 16:287–290, 1983.

59. Gabbay-Azaria, E Tel-Or, M Schonfeld. Glycinebetaine as an osmoregulant and compatible solute in the marine cyanobacterium *Spirulina subsalsa*. Arch Biochem Biophys 264:333–339, 1988.

60. J Severin, A Wohlfarth, EA Galinski. The predominant role of recently discovered tetrahydropyrimidines for the osmoadaptation of halophilic eubacteria. J Gen Microbiol 138:1629–1638, 1992.

61. D Le Rudulier, L Bouillard. Glycinebetaine, an osmotic effector of *Klebsiella pneumoniae* and other members of the Enterobacteriaceae. Appl Environ Microbiol 46:152–159, 1983.

62. F Fougere, D Le Ruddlier. Uptake of glycine betaine and its analogues by bacterioids of *Rhizobium meliloti*. J Gen Microbiol 136:157–163, 1990.

63. T Abee, R Palmen, KJ Hellingwerf, NW Konings. Osmoregulation in *Rhodobacter sphaeroides*. J Bacteriol 173:229–233, 1990.

64. N Riou, MC Poggi, D Le Rudulier. Characterization of an osmoregulated periplasmic glycinebetaine-binding protein in *Azospirillum brasilense* sp7. Biochimie 73:1187–1193, 1991.

65. LN Csonka. Physiological and genetic responses of bacteria to osmotic stress. Microbiol Rev 53:121–147, 1989.

66. LN Csonka, AD Hanson. Prokaryotic osmoregulation: genetics and physiology. Annu Rev Microbiol 45:569–606, 1991.

67. KR Sowers, RP Gunsalus. Halotolerance in *Methanosarcina* spp.: role of $N^{\varepsilon}$-acetyl-β-lycine, α-glutamate, glycine betaine, and K as compatible solutes for osmotic adaptation. Appl Environ Microbiol 61:4382–4388, 1995.

68. LJ Borowitzka. Osmoregulation in blue-green algae. Prog Phycol Res 4:243–256, 1986.

69. WG Roth, MP Leckie, DN Dietzler. Restoration of colony-forming activity in osmotically stressed *Escherichia coli* by betaine. Appl Environ Microbiol 54:3142–3146, 1988.

70. DE Robertson, D Noll, MF Roberts, JAGF Menaia, DR Boone. Detection of the osmoregulator betaine in methanogens. Appl Environ Microbiol 56:563–565, 1990.

71. RH Reed, LJ Borowitzka, MA Mackay, JA Chudek, R Foster, SRC Warr, DJ Moore, WDP Stewart. Organic solute accumulation in osmotically stressed cyanobacteria. FEMS Microbiol Rev 39:51–56, 1986.

72. JA Hellebust. Osmoregulation. Annu Rev Plant Physiol 27:485–505, 1976.

73. JF Imhoff, F Rodriguez-Valera. Betaine is the main compatible solute of halophilic eubacteria. J Bacteriol 160:478–479, 1984.

74. J Cairney, IR Booth, CF Higgins. Osmoregulation of gene expression in *Salmonella typhimurium*: *proU* encodes an osmotically induced betaine transport system. J Bacteriol 164:1224–1232, 1985.

75. J Cairney, IR Booth, CF Higgins. *Salmonella typhimurium* encodes a transport system for the osmoprotectant beatine. J Bacteriol 164:1218–1223, 1885.

76. S Cayley, BA Lewis, MT Record Jr. Origins of the osmoprotective properties of betaine and proline in *Escherichia coli* K-12. J Bacteriol 174:1586–1595, 1992.

77. S Ken-Dror, R Preger, Y Avi-Dor. Role of betaine in the control of respiration and osmoregulation of a halotolerant bacterium. FEMS Microbiol Rev 39:115–120, 1986.

78. D Le Rudulier. Elucidation of the role of osmoprotective compounds and osmoregulatory genes: the key role of bacteria. In: H Leith, A Al Masoom, eds. Towards the Rational Use of High Saline Tolerant Plants. Vol 1. Dordrecht: Kluwer Academic, 1993, pp 313–322.

79. PH Yancey. Compatible and counteracting solutes. In: K Strange, ed. Cellular and Molecular Physiology of Cell Volume Regulation. Boca Raton, FL: CRC Press, 1994, pp 81–109.

80. D Rafaeli-Eshkol, Y Avi-Dor. Studies in halotolerance in a moderately halophilic bacterium. Effect of betaine on salt resistance of the respiratory system. Biochem J 109:687–691, 1968.

81. GP Jones, BP Naidu, RK Starr, LG Paleg. Estimates of solutes accumulating in plants by $^1$H nuclear magnetic resonance spectroscopy. Aust J Plant Physiol 13:649–658, 1986.

82. CW Ford. Simultaneous determination of proline and betaines by high performance liquid chromatography. J Sci Food Agric 35:881–886, 1984.

83. WJ Yang, A Nadolska-Orczyk, KV Wood, DT Hahn, PJ Rich, AJ Wood, H Saneoka, GS Premachandra, CC Bonham, JC Rhodes, RJ Joly, Y Samaras, PB Goldsbrough, D Rhodes. Near-isogenic lines of maize differing for glycinebetaine. Plant Physiol 107:621–630, 1995.

84. JH Zhang, A Okubo, S Yamazaki. Determination of betaine in plants by low-pH capillary electrophoresis as their phenacyl esters. Bunseki Kagaku (Japanese) 46:275–279, 1997.

85. BP Naidu. Separation of sugars, polyols, proline analogues, and betaines in stressed plant extracts by high performance liquid chromatography and quantification by ultra violet detection. Aust J Plant Physiol 25:793–800, 1998.

86. GV Subbarao, RM Wheeler, GW Stutte, LH Levine. How far can sodium substitute for potassium in red beet? J Plant Nutr 22:1745–1761, 1999.

87. RD Guy, PG Warne, DM Reid. Glycine betaine content of halophytes: improved analysis by liquid chromatography and interpretation of results. Physiol Plant 61:195–202, 1984.

88. J Gorham. Separation of plant betaines and their sulphur analogues by cation-exchange high performance liquid chromatography. J Chromatogr 287:345–351, 1984.

89. H Muller, H Eckert. Simultaneous determination of monoethanolamine and glycine betaine in plants. J Chromatogr 479:452–458, 1989.

90. JS Wall, DD Christinanson, RJ Dimler, FR Senti. Spectrophotometric determination of betaines and other quaternary nitrogen compounds as their periodides. Anal Chem 32:870–873, 1960.

91. CM Grieve, SR Grattam. Rapid assay for determination of water soluble quaternary ammonium compounds. Plant Soil 70:303–307, 1983.

92. RC Greene. Biosynthesis of dimethyl-β-propiothetin. J Biol Chem 237:2251–2254, 1962.

93. DK Stumpf. Quantification and purification of quaternary ammonium compounds from halophyte tissue. Plant Physiol 75:273–274, 1984.

94. GM Blunden, M ElBarouni, SM Gordon, WFH McLean, DJ Rogers. Extraction, purification and characterization of Dragendorff-positive compounds from some marine algae. Bot Mar 24:451–456, 1981.

95. GM Blunden, SM Gordon. Betaines and their sulphonio analogues in marine algae. Prog Phycol Res 4:39–80, 1986.

96. N Ahmad, RG Wyn Jones. Glycinebetaine, proline and inorganic ion levels in barley seedlings following transient stress. Plant Sci Lett 15:231–237, 1979.

97. J Gorham, SJ Coughlan, R Storey, RG Wyn Jones. Estimation of quaternary ammonium compounds and tertiary sulphonium compounds by thin-layer electrophoresis and scanning densitometry. J Chromatogr 210:550–554, 1981.

98. K Ranfft, G Gerstl. Gas-chromatographic determination of glycocoll betaine in feeding-stuffs. Z Anal Chem 276:51–54, 1975.

99. H Eckert, H Bergmann, G Eckert, H Mueller. Influence of amino alcohols on the glycine betaine accumulation in barley plants at moderate drought stress. Angew Bot 66:124–129, 1992.

100. A Zamarreno, RG Cantera, M Garcia-Mina. Extraction and determination of glycine betaine in liquid fertilizers. J Agric Food Chem 45:774–776, 1997.

101. J Vialle, M Kolosky, JL Rocca. Determination of betaine in sugar and wine by liquid chromatography. J Chromatogr 204:429–435, 1981.

102. J Gorham. Separation and quantitative estimation of betaine ester by high-performance liquid chromatography. J Chromatorg 361:301–310, 1986.

103. J Gorham, E McDonnell, RG Wyn Jones. Determination of betaines as ultraviolet-absorbing esters. Anal Chim Acta 138:277–283, 1982.

104. PD Erskine, GR Stewart, S Schmidt, MH Turnbull, M Unkovich, JS Pate. Water availability—a physiological constraint on nitrate utilization in plants of Australian semi-arid mulga woodlands. Plant Cell Environ 19:1149–1159, 1996.

105. F Larher. Natural abundance carbon-13 NMR studies on the comatible solutes of halophytic higher plants. Plant Physiol Biochem (Paris) 1149–1159, 1996.

106. AG Harrison, RJ Cotter. General techniques. 1. Methods of ionization. Methods Enzymol 193:3–37, 1990.

107. AD Hanson, CE Nelson. Betaine accumulation and [c$^{14}$C]formate metabolism in water-stressed barley leaves. Plant Physiol 62:305–312, 1978.

108. JJ Martin, JD Finkelstein. Enzymatic determination of betaine in a rat tissues. Anal Biochem 111:72–76, 1981.

109. C Lerma, AD Hanson, D Rhodes. Oxygen-18 and deuterium labeling studies of choline oxidation by spinach and sugar beet. Plant Physiol 88:695–702, 1988.

110. R Brouquisse, P Weigel, D Rhodes, CF Yocum, AD Hanson. Evidence for a ferrodoxin dependent choline monooxygenase from spinach chloroplast stroma. Plant Physiol 90:322–329, 1989.

111. M Burnet, PJ Lafontaine, AD Hanson. Assay, purification, and partial characterization of choline monooxygenase from spinach. Plant Physiol 108:581–588, 1995.

112. B Rathinasabapathi, M Burnet, BL Russell, DA Gage, PC Liao, GJ Nye, P Scott, JH Golbeck, AD Hanson. Choline monooxygenase, an unusual iron-sulfur enzyme catalyzing the first step of glycine betaine synthesis in plants: prosthetic group characterization and cDNA cloning. Proc Natl Acad Sci U S A 94:3454–3458, 1997.

113. B Rathinasabapathi, M Burnet, B Russell, DA Gage, P Scott, JH Golbeck, AD Hanson. Metabolic engineering of glycine betaine synthesis in plants: molecular cloning and characterization of choline monooxygenase from spinach. Plant Physiol 114:109, 1997.

114. BL Russell, B Rathinasabapathi, AD Hanson. Osmotic stress induces expression of choline monooxygenase in sugar beet and amaranth. Plant Physiol 116:859–865, 1998.

115. P Weigel, C Lerma, AD Hanson. Choline oxidation by intact spinach chloroplasts. Plant Physiol 86:54–60, 1988.

116. SM Pan. Betaine aldehyde dehydrogenase in spinach. Bot Bull Acad Sin 29:255–263, 1988.

117. KF McCue, AD Hanson. Salt inducible betaine aldehyde dehydrogenase from sugar beet: cDNA cloning and expression. Plant Mol Biol 18:1–11, 1992.

118. B Rathinasabapathi, KF McCue, DA Gage, AD Hanson. Metabolic engineering of glycine betaine synthesis: plant betaine aldehyde dehydrogenases lacking typical transit peptides are targeted to tobacco chloroplasts where they confer betaine aldehyde resistance. Planta 193:155–162, 1994.

119. EM Valenzuela-Soto, RA Munoz-Clares. Betaine-aldehyde dehydrogenase from leaves of *Amaranthus hypochondriacus* L. exhibits an iso ordered bi bi steady state mechanism. J Biol Chem 268:23818–23824, 1994.

120. EM Valenzuela-Sol, RA Munoz-Clares. Purification and properties of betaine aldehyde dehydrogenas extracted from detached leaves of *Amaranthus hypochondriacus* L. subjected to water deficit. J Plant Physiol 143:145–152, 1994.

121. P Weigel, EA Weretilnyk, AD Hanson. Betaine aldehyde oxidation by spinach chloroplasts. Plant Physiol 82:753–759, 1986.

122. K Arakawa, T Takabe, T Sugiyama, T Akazawa. Purification of betaine-aldehyde dehydrogenase from spinach leaves and preparation of its antibody. J Biochem 101:1485–1488, 1987.

123. EA Weretilnyk, AD Hanson. Betaine aldehyde dehydrogenase from spinach leaves: purification in vitro translation of the mRNA, and regulation by salinity. Arch Biochem Biophys 271:56–63, 1989.

124. PS Summers, EA Weretilyn. Choline synthesis in spinach in relation to salt stress. Plant Physiol 103:1269–1276, 1993.

125. CJ Coughlan, RG Wyn Jones. Glycinebetaine biosynthesis and its control in detached secondary leaves of spinach. Planta 154:6–17, 1982.

126. EA Weretilnyk, DD Smith, GA Wilch, PS Summers. Enzymes of choline synthesis in spinach. Response of phospho-base *N*-methyltransferase activities to light and salinity. Plant Physiol 109:1085–1091, 1995.

127. SJ Coughlan, RG Wyn Jones. Some responses of *Spinacia oleracea* to salt stress. J Exp Bot 31:883–893, 1980.

128. K Arakawa, K Mizuno, S Kishitani, T Takabe. Immunological studies of betaine aldehyde dehydrogenase in barley. Plant Cell Physiol 33:833–840, 1992.

129. DJ Kushner. Life in high salt and solute concentrations: halophilic bacteria. In: DJ Kushner, ed. Microbial Life in Extreme Environments. New York: Academic Press, 1978, pp 317–368.

130. LJ Borowitzka. Solute accumulation and regulation of cell water. In: LG Paleg, D Aspinall, eds. The Physiology and Biochemistry of Drought Resistance in Plants. Sydney: Academic Press, 1981, pp 97–130.

131. Y Jolivet, J Haelin, F Larher. Osmoregulation in halophytic higher plant: the protective effects of glycinebetaine and other related sources against oxalate destabilization of membranes in beet root cells. Z Pflanzenphysiol 109:171–230, 1983.

132. WJ Cram. The regulation of concentration and hydrostatic pressure in cells in relation to growth. Bull R Soc N Z 12:183–189, 1975.

133. JH Crowe, LM Crowe, JF Carpenter, AS Rudolph, CA Wistrom, BJ Spargo, TJ Anchordoguy. Interactions of sugars with membranes. Biochim Biophys Acta 974:367–372, 1988.

134. B Jacoby. Mechanisms involved in salt tolerance by plants. In: M. Pessarakli, ed. Handbook of Plant and Crop Stress. New York: Marcel Dekker, 1994, pp 97–124.

135. JF Carpenter, JH Crowe. The mechanism of cryoprotection of proteins by solutes. Cryobiology 25:244–255, 1988.

136. GO Kist. Osmotic adjustment in phytoplankton and macroalgae. The use of dimethylsulfoniopropionate. In: RP Kiene, PT Visscher, MD Keller, GO Kirst, eds. Biological and Environmental Chemistry of DMSP and Related Sulfonium Compounds. New York: Plenum, 1996, pp 121–129.

137. TJ Flowers, PF Troke, AR Yeo. The mechanism of salt tolerance in halophytes. Annu Rev Plant Physiol 28:89–121, 1977.

138. H Greenway, R Munns. Mechanisms of salt tolerance in non-halophytes. Annu Rev Plant Physiol 31:149–190, 1980.

139. RG Wyn Jones, A Pollard. Proteins, enzymes and inorganic ions. In: A Lauchli, A Pirson, eds. Encyclopedia of Plant Physiology. New Series, Vol 15B. Berlin: Springer, 1983, pp 528–562.

140. PS Low. Molecular basis of the biological compatibility of nature's osmolytes. In: R Gilles, M GillesBaillien, eds. Transport Processes, Iono and Osmoregulation. Berlin: Springer-Verlag, 1985, pp 469–477.

141. R Serrano. Salt tolerance in plants and microorganisms: toxicity targets and defence responses. Int Rev Cytol 165:1–52, 1996.

142. H Marschner. Why can sodium replace potassium in plants? Proceedings of 8th Colloquium, International Potash Institute, Berlin, 1971, pp 50–63.

143. MM Jones, CB Osmond, NC Turner. Accumulation of solutes in leaves of sorghum and sunflower in response to water deficits. Aust J Plant Physiol 7:193–205, 1980.

144. JM Morgan. Osmotic components and properties associated with genotypic differences in osmoregulation in wheat. Aust J Plant Physiol 19:67–76, 1992.

145. R. Weimberg, HR Lerner, A Poljakoff-Mayber. A relationship between potassium and proline accumulation in salt-stressed *Sorghum bicolor*. Physiol Plant 55:5–10, 1982.

146. JMH Stoop, JD Williamson, DM Pharr. Mannitol metabolism in plants: a method for coping with stress. Trends Plant Sci 1:139–144, 1996.

147. KB Schwab, DF Gaff. Influence of compatible solutes on soluble enzymes from desiccation-tolerant *Sporobolus stapfianus* and desiccation-sensitive *Sporobolus pyramidalis*. J Plant Physiol 137:208–215, 1990.

148. H Genard, J Le Saos, JP Billard, A Tremolieres, J Boucaud. Effect of salinity on lipid composition, glycinebetaine content and photosynthetic activity in chloroplasts of *Suaeda maritima*. Plant Physiol Biochem 29:421–427, 1991.

149. PD Hare, WA Cress. Metabolic implications of stress induced proline accumulation in plants. Plant Growth Regul 21:79–102, 1997.

150. LR Rajasekaran. The mechanism of accumulation of proline and quaternary ammonium compounds and their adaptive significance in tomato (*Lycopersicon esculentum* Mill) and wheat (*Triticum aestivum*) exposed to salt stress. PhD dissertation, University of Adelaide, Glen Osmond, 1988.

151. IS Fedina, AV Popova, Photosynthesis, photorespiration and proline accumulation in water stressed pea leaves. Photosynthetica 32:213–220, 1996.

152. JL Hall, DMR Harvey, TJ Flowers, B Kent. Evidence for the cytoplasmic localization of betaine in the leaf cells of *Suaeda maritima*. Planta 140:59–62, 1978.

153. F Allard, M Houde, M Krol, A Ivanov, NPA Huner, F Sarhan. Betaine improves freezing tolerance in wheat. Plant Cell Physiol 39:1194–1202, 1998.

154. RA Leigh, N Ahmad, RG Wyn Jones. Assessment of glycinebetaine and proline compartmentation by analysis of isolated beet vacuoles. Planta 153:34–41, 1981.

155. T Matoh, J Watanabe, E Takahashi. Sodium, potassium, chloride and betaine concentrations in isolated vacuoles from salt-grown *Atriplex gmelini* leaves. Plant Physiol 84:173–177, 1987.

156. RG Wyn Jones, R Storey, RA Leigh, N Ahmad, A Pollard. A hypothesis of cell membrane activity in plants. In: E Marre, O Ciferri, eds. Regulation of Cell Membrane Activity in Plants. Amsterdam: Elsevier, 1977, pp 121–136.

157. RA Leigh, N Ahmad, RG Wyn Jones. Assessment of glycinebetaine and proline compartment by analysis of isolated beet vacuoles. Planta 153:34–41, 1981.

158. KJ Bradford, TC Hsiao. Physiological responses to moderate water stress. In: OL Lange, PS Nobel, CB Osmond, H Ziegler, eds. Encyclopedia of Plant Physiology. New Series, Vol 12B. Physiological Plant Ecology II. Heidelberg: Springer-Verlag, 1982, pp 236–324.

159. LB Turner, GR Stewart. The effect of water stress upon polyamine levels in barley (*Hordeum vulgare* L.) leaves. J Exp Bot 37:170–177, 1986.

160. R Serrano, JM Mulet, G Rios, JA Marquez, IF de Larrinoa, MP Leube, I Mendizabal, A Pascual-Ahuir, M Proft, R Ros, C Montesinos. A glimpse of the mechanisms of ion homeostasis during salt stress. J Exp Bot 50:1023–1036, 1999.

161. WJ Davies, JA Wilson, RE Sharp, O Osonubi. Control of stomatal behaviour in water-stressed plants. In: PG Jarvis, TA Mansfield, eds. Stomatal Physiology. Society for Experimental Biology Seminar Series, 8. Cambridge: Cambridge University Press, 1981, pp 163–185.

162. M Ishitani, T Nakamura, SY Han, T Takabe. Expression of betaine aldehyde dehydrogenase gene in barley in response to osmotic stress and abscisic acid. Plant Mol Biol 27:307–315, 1995.

163. R Storey, RG Wyn Jones. Response of *Atriplex spongiosa* and *Suaeda monoica* to salinity. Plant Physiol 63:156–162, 1979.

164. WD Hitz, JAR Ladyman, AD Hanson. Betaine synthesis and accumulation in barley during field water-stress. Crop Sci 22:47–54, 1982.

165. CM Grieve, EV Maas. Betaine accumulation in salt-stressed sorghum. Physiol Plant 61:167–171, 1984.

166. JR Atack. Structure and mechanism of inositol monophosphatase. FEBS Lett 361:1–7, 1995.

167. M Nomura, T Hibino, T Takabe, T Sugiyama, A Yokota, H Miyake, T Takabe. Transgenically produced glycinebetaine protects ribulose 1,5-biphosphate carboxylase/oxygenase from inactivation in *Synechococcus* sp. PCC 7942 under salt stress. Plant Cell Physiol 39:425–432, 1998.

168. A Pollard, RG Wyn Jones. Enzyme activities in concentrated solutions of glycinebetaine and other solutes. Planta 144:291–298, 1979.

169. JH Crow, LM Crow, D Chapman. Preservation of membranes in anhydrobiotic organisms: the role of trehalose. Science 223:701–703, 1984.

170. JH Crow, LM Crow, D Chapman, LM Crowe. Interactions of phospholipid monolayers with carbohydrates. Biochim Biophys Acta 769:151–159, 1984.

171. K Arakawa, N Timasheff. The stabilization of proteins by osmolytes. Biophys J 47:411–414, 1985.

172. A Incharoensakdi, T Takabe, T Akasawa. Effect of betaine on enzyme activity and subunit interaction of ribu-lose-1,5-biphosphate carboxylase/oxygenase from *Aphanothece halophytica*. Plant Physiol 81:1044–1049, 1986.

173. T Matoh, S Yasuoka, T Ishikawa, E Takahashi. Potassium requirement of pyruvate kinase extracted from leaves of halophytes. Physiol Plant 74:675–678, 1988.

174. C Colaco, S Sen, M Thangavelu, S Pinder, B Roser. Extraordinary stability of enzymes dried in trehalose: sim-plified molecular biology. Biotech 10:1007–1011, 1992.

175. N Murata, PS Mohanty, H Hayashi, GC Papageorgiou. Glycinebetaine stabilizes the association of extrinsic proteins with the photosynthetic oxygen-evolving complex. FEBS Lett 296:187–189, 1992.

176. E Selinioti, D Nikolopoulus, Y Manetas. Organic cosolutes as stabilizers of phosphoenolpyruvate carboxylase in storage: an interpretation of their action. Aust J Plant Physiol 14:203–210, 1987.

177. D Nikolopoulos, Y Manetas. Compatible solute and in vitro stability of *Salsola soda* enzymes: proline incom-patibility. Phytochemistry 30:411–413, 1991.

178. Y Jolivet, F Larher, J Hamelin. Osmoregulation in halophytic higher plants: the protective effect of glycinebe-taine against the heat destabilization of membranes. Plant Sci Lett 25:193–201, 1982.

179. AS Rudolph, JH Crowe, LM Crowe. Effects of three stabilizing agents—proline, betaine and trehalose—on membrane phospholipids. Arch Biochem Biophys 245:134–143, 1986.

180. TJ Flowers, JL Hall, ME Ward. Salt tolerance in the halophyte, *Suaeda maritima* (L.) Dum: properties of malic enzyme and PEP carboxylase. Ann Bot 42:1065–1074, 1978.

181. CJ Brady, TS Gibson, EW Barlow, R Speirs, RG Wyn Jones. Salt tolerance in plants. I. Ions, compatible or-ganic solutes and the stability of plant ribosomes. Plant Cell Environ 7:571–578, 1984.

182. LG Paleg, GR Stewart, JW Bradbeer. Proline and glycinebetaine influence protein solvation. Plant Physiol 75:974–978, 1984.

183. MAH Koheil, SH Hilal, TS El-Alfy, E Leistner. Quaternary ammonium compounds in intact plants and cell suspension cultures of *Atriplex semibaccata* and *A. halimus* during osmotic stress. Phytochemistry 31:2003–2008, 1992.

184. AJ Cavalieri. Proline and glycine-betaine accumulation by *Spartina alterniflora* Loisel in response to NaCl and nitrogen in a controlled environment. Oecologia (Berl) 57:20–24, 1983.

185. KB Marcum, CL Murdoch. Salt tolerance of coastal salt marsh grass, *Sporobolus virginicus* (L.) kunth. New Phytol 120:281–288, 1992.

186. MD Mamedov, H Hayashi, H Wada, PS Mohanty, GC Papageorgiou, N Murata. Glycinebetaine enhances and stabilizes the evolution of oxygen and the synthesis of ATP by cyanobacterial thylakoid membranes. FEBS Lett 294:271–274, 1991.

187. AR Yeo, TJ Flowers. Accumulation and localization of sodium ions within the shoots of rice (*Oryza sativa*) varieties differing in salinity resistance. Physiol Plant 56:343–348, 1982.

188. RJ Cooke, J Oliver, DD Davies. Stress and protein turnover in *Lemna minor*. Plant Physiol 64:1109–1113, 1979.

189. A Kyparissis, Y Petropoulos, Y Manetas. Summer survival of leaves in a soft-leaved shrub (*Phlomis fruticosa* L. Labiatae) under Mediterranean conditions: avoidance of photoinhibitory damage through decreased chloro-phyll contents. J Exp Bot 46:1825–1831, 1995.

190. M Noguchi, A Kowai, E Tanaki. Studies on nitrogen metabolismin tobacco plants. VIII. $\Delta'$-pyrroline-5-car-boxylate reductase from tobacco leaves. Agric Biol Chem 30:492–496, 1966.

191. AD Hanson, AM May, R Grumet, J Bode, GJ Jamieson, D Rhodes. Betaine synthesis in chenopods: localiza-tion in chloroplasts. Proc Natl Acad Sci U S A 82:3678–3682, 1985.

192. GC Papageorgiou, Y Fujimura, N Murata. Protection of the oxygen-evolving photosystem II complex by glycinebetaine. Biochim Biophys Acta 1057:361, 1991.

193. K Kalosaka, GC Papageorgiou. Glycinebetaine, sucrose, or chloride ions protect isolated photosystem-2 parti-ciles from denaturation and inactivation. In: JH Argyroudi-Akoyunoglou, ed. Regulation of Chloroplast Bio-genesis. New York: Plenum, 1992, pp 391–395.

194. SJ Coughlan, U Heber. The role of glycinebetaine in the protection of thylakoids against the freezing stress. Planta 156:62–69, 1982.

195. J Cao, Govindjee. Chlorophyll *a* fluorescence transient as an indicator of active and inactive photosystem II in thylakoid membranes. Biochim Biophys Acta 1015:180–188, 1990.

196. J Berry, O Bjorkman. Photosynthetic response and adaptation to temperature in higher plants. Annu Rev Plant Physiol 31:491–543, 1996.

197. LR Rajasekaran, PE Kriedemann, D Aspinall, LG Paleg. Physiological significance of proline and glycinebe-taine: maintaining photosynthesis during NaCl stress in wheat. Photosynthetica 34:357–366, 1997.

198. G Yang, D Rhodes, RJ Joly. Effects of high temperature on membrane stability and chlorophyll fluorescence in glycinebetaine-deficient and -containing maize lines. Aust J Plant Physiol 23:437–443, 1996.

199. WP Williams, APR Brian, PJ Dominy. Induction of non-bilayer lipid phase separations in chloroplast thylakoid membranes by compatible co-solutes and its relation to the thermal stability of photosystem II. Biochim Biophys Acta 1099:137–144, 1992.

200. H Hayashi, ML Alia, P Deshnium, M Ida, N Murata. Transformation of *Arabidopsis thaliana* with the *codA* gene for choline oxidase: accumulation of glycinebetaine and enhanced tolerance to salt and cold stress. Plant J 12:133–142, 1997.

201. P Deshnium, Z Gombos, Y Nishiyama, N Murata. The action in vivo of glycinebetaine in enhancement of tolerance of *Synechococcus* sp. strain PCC7942 to low temperature. J Bacteriol 179:339–344, 1997.

202. LG Paleg, TJ Douglas, A Van Dall, DB Keech. Proline and betaine protect enzymes against heat inactivation. Aust J Plant Physiol 8:107–114, 1981.

203. JAR Ladyman, WD Hitz, AD Hanson. Translocation and metabolism of glycinebetaine by barley plants in relation to water stress. Planta 150:191–196, 1980.

204. AD Hanson, CE Nelsen, JAR Ladyman. Betaine accumulation in water-stressed barley leaves (Abstract). Plant Physiol (Suppl) 61:81, 1978.

205. ADH Hanson, WD Hitz. Metabolic responses of mesophytes to plant water deficits. Annu Rev Plant Physiol 33:163–203, 1982.

206. MS Bowman, R Rohringer. Formate metabolism and betaine formation in healthy and rust-affected wheat. Can J Bot 48:803–811, 1970.

207. SR Grattan, CM Grieve. Betaine status in wheat in relation to nitrogen stress and to transient salinity stress. Plant Soil 85:3–9, 1985.

208. JA Pocard, T Bernard, D Le Rudulier. Translocation and metabolism of glycinebetaine in nodulated alfalfa plants subjected to salt stress. Physiol Plant 81:95–102, 1991.

209. RE Tully, AD Hanson, CE Nelsen. Proline accumulation in water-stressed barley leaves in relation to translocation and the nitrogen budget. Plant Physiol 63:518–523, 1979.

210. GA Smith, SS Martin. Effects of plant density and nitrogen fertility on purity components of sugarbeet. Crop Sci 17:469–472, 1977.

211. JM Morgan. Osmoregulation and water stress in higher plants. Annu Rev Plant Physiol 35:299–319, 1984.

212. J Zhang, HT Nguyen, A Blum. Genetic analysis of osmotic adjustment in crop plants. J Exp Bot 50:291–302, 1999.

213. DJ Flower, MM Ludlow. Contribution of osmotic adjustment to the dehydration tolerance of water stressed pigeonpea (*Cajanus cajan* L. Mill sp.) leaves. Plant Cell Environ 9:33–40, 1986.

214. MM Ludlow, RC Muchow. A critical examination of traits for improving crop yields in water-limited environments. Adv Agron 43:107–153, 1990.

215. GV Subbarao, C Johansen, AE Slinkard, RC Nageswara Rao, NP Saxena, YS Chauhan. Strategies for improving drought resistance in grain legumes. Crit Rev Plant Sci 14:469–523, 1995.

216. R Storey, N Ahmad, RG Wyn Jones. Taxonomic and ecological aspects of the distribution of glycinebetaine and related compounds in plants. Oecologia 27:319–332, 1977.

217. JC Papp, MC Ball, N Terry. A comparative study of the effects of NaCl salinity on respiration, photosynthesis, and leaf extension growth in *Beta vulgaris* L. (sugar beet). Plant Cell Environ 6:675–677, 1983.

218. TD Colmer, E Epstein, J Dvorak. Differential solute regulation in leaf blades of various ages in salt-sensitive wheat and a salt-tolerant wheat × *Lophopyrum elongatum* (Host) A. Love amphiploid. Plant Physiol 108:1715–1724, 1995.

219. GS Premachandra, DT Hahn, D Rhodes, RJ Joly. Leaf water relations and solute accumulation in two grain sorghum lines exhibiting contrasting drought tolerance. J Exp Bot 46:1833–1841, 1995.

220. G Selvaraj, RK Jain, DJ Olson, R Hirja, S Jana, LR Hogge. Glycinebetaine in oilseed rape and flax leaves: detection by liquid chromatography/continuous flow secondary ion–mass spectrometry. Phytochemistry 38:1143–1146, 1995.

221. PC Agboma, MGK Jones, P Peltonen-Sainio, H Rita, E Pehu. Exogenous glycinebetaine enhances grain yield of maize, sorghum and wheat grown under two supplementary watering regimes. J Agron Crop Sci 178:29–37, 1997.

222. PC Agboma, TR Sinclair, K Jokinen, P Peltonen-Sainio, E Pehu. An evaluation of the effect of exogenous glycinebetaine on the growth and yield of soybean: timing of application, watering regimes and cultivars. Field Crops Res 54:51–64, 1997.

223. P Makela, P Peltonen-Sainio, K Jokinen, E Pehu, H Setala, R Hinkkanen, S Somersalo. Effect of foliar applications of glycinebetaine on stomatal conductance, abscisic acid and solute concentrations in leaves of salt- or drought-stresed tomato. Aust J Plant Physiol 25:655–663, 1998.

224. DL Ehret, LC Ho. The effects of salinity on dry matter partitioning and fruit growth of tomatoes grown in nutrient film culture. J Hortic Sci 61:361–367, 1986.

225. S El-Amin. Effect of glycinebetaine (betaine) on water stress in potato and tomato crops in Sudan (cited in Makela et al., 1998, 226), 1993.

226. P Makela, K Jokinen, M Kontturi, P Peltonen-Sainio, E Pehu, S Somersalo. Foliar application of glycinebetaine—a novel product from sugarbeet—as an approach to increase tomato yield. Ind Crops Prod 7:139–148, 1998.

227. P Makela, J Mantila, R Hinkkanen, E Pehu, P Peltonen-Sainio. Effects of foliar applications of glycinebetaine on stress tolerance, growth, and yield of spring cereals and summer turnip rape in Finland. J Agric Crop Sci 176:223–234, 1996.

228. P Makela, P Peltonen-Sainio, K Jokinen, E Pehu, H Setala, R Hinkkanen, S Somersalo. Uptake and translocation of foliar-applied glycinebetaine in crop plants. Plant Sci 121:221–230, 1996.

229. S Borojevic, T Cupina, M Krsmanovic. Green area parameters in relation to grain yield of different wheat genotypes. Z Pflanzenzuecht 84:265–283, 1980.

230. P Harinasut, K Tsutsui, T Takabe, M Nomura, T Takabe, S Kishitani. Exogenous glycinebetaine accumulation and increased salt-tolerance in rice seedlings. Biosci Biotechnol Biochem 60:366–368, 1996.

231. JA Pocard, T Bernard, G Goas, D Le Rudulier. Restauration partielle, par la glycine betaine et la proline betaine, de l'activité fixatrice de jeunes plantes de *Medicago sativa* L. soumises a un stress hydrique. C R Acad Sci Ser III 298:477–480, 1984.

232. BP Naidu, DF Cameron, SV Konduri. Improving drought tolerance of cotton by glycinebetaine application and selection. Proceeding of Australian Agronomy Society Meetings, 1999.

233. R Sulpice, Y Gibon, A Bouchereau, F Larher. Exogenously supplied glycinebetaine in spinach and rapeseed leaf discs: compatibility or non-compatibility? Plant Cell Environ 21:1285–1292, 1998.

234. Y Gibon, MA Bessieres, F Larher. Is glycine betaine a non-compatible solute in higher plants that do not accumulate it? Plant Cell Environ 20:329–340, 1997.

235. AD Hanson, R Grumet. Betaine accumulation: metabolic pathways and genetics. In: JL Key, T Kosuge, eds. Cellular and Molecular Biology of Plant Stress. New York: Alan R Liss, 1985, pp 71–92.

236. H Saneoka, C Nagasaka, DT Hahn, W Yang, GP Premachandra, RJ Joly, D Rhodes. Salt tolerance of glycinebetaine-deficient and -containing maize lines. Plant Physiol 107:631–638, 1995.

237. GC Papageorgiou, N Murata. The unusually strong stabilizing effects of glycinebetaine on the structure and function of the oxygen-evolving photosystem II complex. Photosynth Res 44:243–252, 1995.

238. D Rhodes, PJ Rich. Preliminary genetic studies of the phenotype of betaine deficiency in *Zea mays* L. Plant Physiol 88:102–108, 1988.

239. JAR Ladyman, KM Ditz, R Grumet, AD Hanson. Genotypic variation for glycinebetaine accumulation by cultivated and wild barley in relation to water stress. Crop Sci 23:465–468, 1983.

240. R Krishnamurthy, M Anbazhagan, KA Bhagwat. Glycinebetaine accumulation and varietal adaptability to salinity as a potential metabolic measure of salt tolerance in rice. Curr Sci 57:259–261, 1988.

241. SD McNeil, ML Nuccio, AD Hanson. Betaines and related osmoprotectants. Targets for metabolic engineering of stress resistance. Plant Physiol 120:945–949, 1999.

242. JE Bailey. Toward a science of metabolic engineering. Science 252:1668–1675, 1991.

243. EA Cossins, L Chen. Folates and one carbon metabolism in plants and fungi. Phytochemistry 45:437–452, 1997.

244. KO Holmstrom, E Mantyla, B Welin, A Mandal, ET Palva, OE Tunnela, J Londesborough. Drought tolerance in tobacco. Nature 379:683–684, 1996.

245. MC Tarczynski, RG Jensen, HJ Bohnert. Stress protection of transgenic tobacco by production of the osmolyte mannitol. Science 259:508–510, 1993.

246. B Karakas, P Ozias-Akins, C Stushnoff, M Suefferheld, M Rieger. Salinity and drought tolerance of mannitol-accumulating transgenic tobacco. Plant Cell Environ 20:609–616, 1997.

247. PB Kavi Kishore, Z Hong, G Mio, CA Hu, DPS Verma. Overexpression of $\Delta^1$-pyrroline-5-carboxylase synthetase increases proline production and confers osmotolerance in transgenic plants. Plant Physiol 108:1387–1394, 1995.

248. EAH Pilon-Smits, MJM Ebskamp, MJJ Paul, MJW Jeuken, PJ Weisbeek, SCM Smeekens. Improved performance of transgenic fructan-accumulating tobacco under drought stress. Plant Physiol 107:125–130, 1995.

249. R Tao, SL Uratsu, AM Dandekar. Sorbitol synthesis in transgenic tobacco with apple cDNA encoding NADP-dependent sorbitol-6-phosphate dehydrogenase. Plant Cell Physiol 36:525–532, 1995.

250. G Lilius, N Holmberg, L Bulow. Enhanced NaCl stress tolerance in transgenic tobacco expressing bacterial choline dehydrogenase. Biotechnology 14:177–180, 1996.

251. C Romero, JM Belles, JL Vaya, R Serrano, FA Culianez-Macia. Expression of the yeast trehalose 6-phosphate synthase gene in transgenic tobacco plants: pleiotropic phenotypes include drought tolerance. Planta 201:293–297, 1997.

252. OJM Goddijn, TC Verwoerd, E Voogd, RWHH Krutwagen, PTHM de Graaf, J Poels, K van Dun, AS Ponstein, B Damm, J Pen. Inhibition of trehalase activity enhances trehalose accumulation in transgenic plants. Plant Physiol 113:181–190, 1997.

253. CC Smart, S Flores. Overexpression of *d*-myo-inositol-3-phosphate synthase leads to elevated elvels of inositol in *Arabidopsis*. Plant Mol Biol 33:811–820, 1997.

254. E Sheveleva, W Chmara, HJ Bohnert, RG Jensen. Increased salt and drought tolerance by D-ononitol production in transgenic *Nicotiana tabacum* L. Plant Physiol 115:1211–1219, 1997.

255. R Munns. Physiological processes limiting plant growth in saline soils: some dogmas and hypothesis. Plant Cell Environ 16:15–24, 1993.

256. H Schnyder. The role of carbohydrate storage and redistribution in the source-sink relations of wheat and barley during grain filling—a review. New Phytol 123:233–245, 1993.

257. EA Bray. Plant responses to water deficit. Trends Plant Sci 2:48–54, 1997.

# 46

# Computer Simulation of Plant and Crop Allocation Processes

**Donna M. Dubay and Monica A. Madore**

*University of California, Riverside, California*

## I. INTRODUCTION

Plant growth is balanced between growth above ground (shoot growth) and below ground (root growth). This balance provides adequate surface area to intercept light for photosynthesis as well as sufficient area to acquire needed water and nutrients in the soil. In this way, the requirements for carbon (C) and mineral nutrients such as nitrogen (N) within the plant are satisfied to maximize plant growth or productivity. Each plant species appears to have a preferred balance, which is apparently genetically controlled but which can be altered by environmental conditions. For example, if a needed resource such as water or a nutrient is limiting, a plant often tends to grow a proportionally greater amount of root [1,2]. This gives the plant more surface area within the soil to absorb the limited supply of resources and also to search out pockets of resources deeper within the soil. In a similar sense, if the shoot is exposed to a limit in a resource (most commonly light, which is necessary to acquire C through photosynthesis), growth will shift to favor shoot growth over root growth. This supplies the plant with greater surface area to intercept light and absorb carbon dioxide and brings balance back to the plant.

Although these growth responses have been repeatedly demonstrated experimentally, no physiological mechanism has yet been found that regulates these processes. In fact, how the balance of root and shoot growth in a plant is attained even without a limited resource is still a mystery [3,4].

## II. MODELING PLANT GROWTH

Allometrically, the relationship between the mass of the shoot and root can be explained by the formula [5]:

$$y = bx^k \qquad \text{(or: } \log y = \log b + k \log x) \tag{1}$$

where  $y$ = root dry weight
 $x$ = shoot dry weight
 $b$ and $k$ are constants

This allometric depiction is purely empirical and does not explain to any degree the plasticity of growth responses observed in nature in response to limitations of substrate during the growth period.

In the 1960s, Brouwer and others developed the concept of functional equilibrium [6,7] to attempt to describe mathematically the observed dependence of root/shoot ratios on the availability of substrates. This model can be depicted mathematically as follows [7]:

$$W_r S_r \propto W_s S_s \tag{2}$$

where   $W_r$ = root mass (g)
         $S_r$ = specific absorption rate of the root for a particular nutrient (g g$^{-1}$ day$^{-1}$)
         $W_s$ = shoot mass (g)
         $S_s$ = specific photosythesis rate of the shoot (g g$^{-1}$ day$^{-1}$)

The functional equilibrium concept is based upon the idea that substrates required for growth are composed of elements obtained either above or below ground and that growth of any part of the plant is dependent upon the availability of all the required growth substrates. Furthermore, it assumes that each substrate is first available to the plant parts in the region in which the substrate is first acquired. For example, N, acquired from the soil by uptake processes occurring in the roots, would first be available for root shoot growth. Likewise, C, obtained through photosynthesis in the shoots, would first be available for shoot growth. Thus, for a given substrate, limitation of that substrate would reduce the growth of regions away from the acquisition site. For instance, if N is limited, root growth would continue and the shoot growth would decrease because N is first available to the root through uptake from the soil. Conversely, if C was limiting, shoot growth would continue and the root growth would decrease because C is first available to the shoot through photosynthesis.

This concept was further refined by Thornley and colleagues [8–10], who added the assumption that dry matter distribution between root and shoot was only indirectly regulated by the uptake activities of the shoot and root, as indicated in Eq. (1) but rather was controlled by the *availability* of those substrates in the form of labile storage pools. Pool sizes in turn would depend not only on supply (i.e., $W_s S_s$ and $W_r S_r$) but also on utilization and, more important, transport of the substrates between shoot and root. Thus, Thornley's model takes a much more mechanistic approach than simple utilization of the functional equilibrium equation. The Thornley model has since been extended to the point of being an ecosystem model for grassland crops and was extensively characterized in a recent volume [11]. In this latest rendition, the detail for the plant submodel alone points to the complexity of the models that must be invoked to simulate plant growth effectively [11].

## III.  COMPUTER SIMULATION OF PLANT GROWTH

The ability to express plant growth responses as mathematical functions as predicted by these models makes it highly attractive to input these equations into computer programs to use as predictors of plant growth under various environmental conditions. These computer simulations have been used by biologists to predict plant growth for about 30 years [2,12], but until recently their use has been limited to a small group of researchers. With the advent of more powerful and inexpensive computational equipment, the past few decades have seen a huge increase in the number of simulations published for horticultural and agronomic purposes [13,14]. As personal computers continue to become more accessible and more powerful, and coupled with vast improvements in equipment for field data collection, simulations for plant growth are becoming more practical and have a much wider range of applications than when these models were first developed. Users of a simulation now simply supply input to a computer program, which may include information defining the growing environment to which the plants are expected to respond, such as weather, soil, and water, to name a few. Input also generally includes a set of initial conditions, the starting point for growth to be simulated, and information regarding what is being simulated. The computer program then simulates the growing process of the plants using a combination of models within the simulation that model the process of growth by way of mathematical equations and relationships. The output of the program is then the growth data based on the input provided by the user. The growth data can be in many forms depending on what is required by the user.

### A.  Methods Used to Model Allocation

So how do simulations model partitioning? We can first consider some methods used to model partitioning in an optimum situation. Scientists have been able to make generalized observations of plant growth patterns and apply them to partitioning models to make them more mechanistically representative of the process within the plant [15]. One way to address the problem, as described previously, is simply by scaling or allometry. It has been shown that ratios such as relative growth rate of root and shoot generally remain constant even as the age and size of a plant increase [16,17]. For many simple growth models, these allometric relations are directly applied [18] and usually do not include modification by environmental factors, although there are some that do [3].

In addition to growing root and shoot, there are generally other destinations (sinks) for newly produced photosynthate. Models often use a priority system, assigning priorities to destinations and uses such as respiration and fruit development as well as structural growth of various plant organs [12,19]. This is often coupled with the functional equilibrium concept, as it is used to establish the basis for production of new photosynthate by balancing the aboveground and belowground growth. The concept models growth of a region as substrate limited and the availability of each substrate, generally C and N, as a direct function of the regions above and below ground to supply a substrate [8,20]. The ability to supply substrate is usually a function of size and the environment. In addition to simply assigned priorities, multiple sinks are assigned priorities based on proximity. This is simply the assumption that the closest sink to the source gets first delivery and essentially has the best access. Sinks farther from the source have reduced priority as they get what is left over. This is done by representing the plant structure generalized as multiple sinks, with each sink supplied photosynthate on the basis of a combination of assigned priority and proximity. Proximity is often modeled in terms of resistance along the pathway from source to sink. The difference in activity of the source and activity in the sink regions provides a potential gradient for flow of substrate through assigned resistances throughout the plant. Again, often the very basis of this model is functional equilibrium, used to balance the supply of available substrates to the source.

Thus, the models that make up a simulation can be described as either mechanistic or empirical. A mechanistic model uses equations to model what we understand of the actual physiological processes. Empirical models use previous data to predict future performance. For example, once a response has been characterized over a range of values, that response can be statistically curve fit using regression techniques, resulting in an equation to model the response. One of the drawbacks to this includes the fact that the response needs to be previously characterized over the expected range. Furthermore, if the model is used to predicted responses that are outside the previous range, extrapolation may be required [3,21]. Mechanistic models tend to be more robust, giving a more reliable prediction over a wider range of conditions; however, they can be developed only for processes in which we understand the mechanisms well enough to apply equations [22].

## IV. USES FOR COMPUTER SIMULATIONS

Partitioning new growth is an integral part of any plant growth simulation. Because no mechanism has been identified to regulate the process, modelers have had to rely on essentially empirical models [23]. These models are often based on the performance of plants experiencing optimum growth such as a plant would experience with an abundance of everything it needed from the environment. Then the effects of something limited in the environment are added in to reduce the optimum growth. Often, the result is both a reduction in overall growth rate and a shift in partitioning. This shift can be difficult to predict but can ultimately be an important aspect of what is being predicted. Growth partitioned above ground is often the area of most interest to users of a simulation. For example, more vegetative mass above ground can translate to more fruit or yield to a farmer [3]. But growth below ground is also important; such is the case in ecological simulations when making predictions about the organic storage matter in soils [24].

Currently, uses for these simulations generally fall into two categories, ecological applications and agriculture. Some of the ecological applications include predictions about how future climates such as increases in carbon dioxide and temperature may affect forests [25–28] and grasslands [27,28] or how years of grazing may affect pasture lands [11]. They are used in agriculture to maximize the effectiveness of farmer input by using artificial neural networks [29] or other decision-making methods such as COMAX in the simulation for cotton [30] or the Penn State Apple Orchard Consultant (PSAOC) expert system used in apple orchards [31]. They are also used to aid in managing tree crops for activities such as fruit thinning and canopy management [32]. Farmers can determine such things as optimum N applications [33], irrigation plans [34], and pesticide applications based on how they affect the final yield [21].

Simulations for predicting the effects of future climates on forests and grasslands have various levels of complexity depending upon the scale of what is being modeled, such as BACROS, BIOMASS, FORGRO, and MAESTRO as reviewed by Ågren [35]. The most basic of these simulations model plant physiology and how plants respond to the environmental changes [35]. These simulations include models that predict aspects such as conversion of light within the canopy to assimilated carbon, water and nutrient uptake, and partitioning of dry weight in the form of growth. Other processes within the plant that they model generally include respiration, partitioning, reproduction, and senescence. All of these processes are affected by environmental factors such as temperature; herbivory (grazing and insects); availability of water, nitrogen, and phosphorus; and atmospheric levels of $CO_2$. All these factors are integrated over time to make future predictions based on the combination of initial conditions and ongoing environmental changes. Many simulations also include feedback from the growing plants to the environment in such ways as litter accumulation and degradation of fallen matter. These models give a physiological basis upon which other models, which encompass larger scales, can be used to predict such things as how much C can be stored in the earth's forest over time given changes in temperature and the atmospheric $CO_2$ levels.

Agricultural simulations tend to be oriented to a smaller scale than the ecological simulations, putting more emphasis on plants within a single crop. Examples of some of these include SOYGRO for soybeans, PNUTGRO for peanuts, BEANGRO for dry bean [36], and GOSSYM for cotton [34]. However, they are not limited to that and are gaining use for estimating potential yields in a marketing application as well as predicting fertilizer losses to the environment [13,21] and changing climate effects on crops [37]. They have more variability in input because many factors are controlled by the farmer such as water, nutrients, and, for those designed for glasshouses, climate to some extent. The physiological models within the simulation are generally the same as those in ecological applications, conversion of light to assimilated carbon, uptake of water and nutrients, partitioning, respiration, and reproduction. The environmental factors affecting the process tend to be more local and smaller in scale, such as day-to-day climate, plant spacing, and specific pests. The predictions (output) are generally dry matter accumulation specific to the crop—for instance, fruit size and quantity—and estimates for timing of harvest. But they are certainly not limited to this, as demonstrated by the use of the Erosion-Productivity Impact Calculator (EPIC) crop model used to predict the relationship between soil erosion and soil productivity [38].

## V. CONCLUDING REMARKS

Unfortunately, even with all the marvels of new computing power, computer simulations are only as good as the programs written for them. The programs, in turn, are ultimately a reflection of our biological understanding of plant growth. Some processes are more clearly understood than others, but a simulation relies on the combination of the models used to predict the outcome. If one or more of the models is not reliable over the range of input, results may be misleading [11]. Thus, it is necessary that all plant processes be modeled as accurately as possible. However, it is usually the case that we do not have a deep enough understanding of all of the myriad of biological processes occurring in a growing plant to even begin to adequately quantify the outcome with mathematical equations.

## REFERENCES

1. R Brouwer. In: FL Milthorpe, JD Ivins, eds. The Growth of Cereals and Grasses. London: Butterworths, 1965, pp 153–166.
2. R Brouwer, CT De Wit. In: WJ Whittington, ed. Root Growth. New York: Plenum, 1968, pp 224–244.
3. LFM Marcelis, E Heuvelink, J Goudriaan. Sci Hortic 74:83, 1998.
4. IF Wardlaw. New Phytol 116:341, 1990.
5. A Troughton. J Br Grassland Soc 6:56, 1956.
6. R Brouwer. Neth J Agric Sci 10:399, 1962.
7. RL Davidson. Ann Bot 40:561, 1969.
8. JHM Thornley. Ann Bot 36:431, 1972.
9. JF Reynolds, JHM Thornley. Ann Bot 49:585, 1982.
10. IR Johnson, JHM Thornley. Ann Bot 60:133, 1987.
11. JHM Thornley. Grassland Dynamics: An Ecosystem Simulation Model. Wallingford, UK: CAB International, 1998.

12. GW Fick, RS Loomis, WA Williams. In: LT Evans, ed. Crop Physiology: Some Case Histories. Cambridge, UK: Cambridge University Press, 1975, pp 259–295.
13. C Gary, JW Jones, M Tchamichian. Sci Hortic 74:3, 1998.
14. TR Sinclair, NG Seligman. Agron J 88:698, 1996.
15. JB Wilson. Ann Bot 61:433, 1987.
16. J Farrar, S Gunn. In: H Lambers, H Poorter, MMI Van Vuuren, eds. Inherent Variation in Plant Growth: Physiological Mechanisms and Ecological Consequences. Leiden: Backhuys Publishers, 1998, pp 183–198.
17. KJ Niklas. Plant Allometry: The Scaling of Form and Process. Chicago: University of Chicago Press, 1994.
18. E Veneklaas, L Poorter. In: H Lambers, H Poorter, MMI Van Vuuren, eds. Inherent Variation in Plant Growth: Physiological Mechanisms and Ecological Consequences. Leiden: Backhuys Publishers, 1998, pp 337–361.
19. GW Fick, WA Williams, RS Loomis. Crop Sci 13:413, 1973.
20. AA Makela, RP Sievanen. Ann Bot 59:129, 1987.
21. KJ Boote, JW Jones, NB Pickering. Agron J 88:704, 1996.
22. HG Jones. Plants and Microclimate: A Quantitative Approach to Environmental Plant Physiology. Cambridge, UK: Cambridge University Press, 1992, p 428.
23. LFM Marcelis. Acta Hortic 328:49, 1993.
24. GR Shaver, JD Aber. In: AI Breyer, DO Hall, JM Melillo, GI Agren, eds. Global Change: Effects on Coniferous Forests and Grasslands. New York: Wiley, 1996, pp 183–198.
25. MGR Cannell, RC Dewar. Adv Ecol Res 25:59, 1994.
26. JHM Thornley, MGR Cannell. Plant Cell Environ 19:1331, 1996.
27. JR Melillo, DO Hall, GI Agren. In: AI Breyer, DO Hall, JM Melillo, GI Agren, eds. Global Change: Effects on Coniferous Forests and Grasslands. New York: Wiley, 1996, pp 1–16.
28. JHM Thornley, MGR Cannell. Ann Bot 80:205, 1997.
29. N McRoberts, GN Foster, S Wale, K Davies, RG McKinlay, A Hunter. Acta Hortic 476:243, 1998.
30. AC Gertsis, FD Whisler. Acta Hortic 476:213, 1998.
31. JW Travis, E Rajotte, R Bankhert, KD Hickey, LA Hull, V Eby, PH Heinemann, R Crassweller, J McClure. Plant Dis 76:545, 1992.
32. TM DeJong, YL Grossman. Acta Hortic 313:21, 1992.
33. WD Batchelor, JW Jones, KJ Boote, HO Pinnschmidt. Trans ASAE 36:551, 1993.
34. SA Staggenborg, RJ Lascano, DR Krieg. Agron J 88:740, 1996.
35. GI Agren, RE McMurtrie, WJ Parton, J Pastor, HH Shugart. Ecol Appl 1:118, 1991.
36. G Hoogenboom, JW Jones, KJ Boote. Trans ASAE 35:2043, 1992.
37. FN Tubiello, C Rosenzweig, BA Kimball, PJ Pinter, GW Wall, DJ Hunsaker, RL LaMorte, RL Garcia. Agron J 91:247, 1999.
38. JR Williams, CA Jones, JR Kiniry, DA Spanel. Trans ASAE 32:497, 1989.

# 47

# Composite Lighting for Controlled-Environment Plant Factories

**Joel L. Cuello**

*The University of Arizona, Tucson, Arizona*

## I. INTRODUCTION

One of the consequences of employing artificial lighting to supplement solar irradiance either in a greenhouse [1–3] or in a controlled-environment plant growth chamber, wherein solar irradiance is transmitted through optical cables from solar concentrating systems [4–6], is the subjection of the growing crops to lighting profiles that differ from the conventional lighting profile. The daily lighting profile of a conventional electric-based plant-lighting system can generally be represented by a rectangular wave (Figure 1A) whose height represents the magnitude of the instantaneous photosynthetic photon flux (PPF, in $\mu$mol m$^{-2}$ sec$^{-1}$), whose length represents the daily photoperiod ($P$, in hr), and whose area represents the daily integrated PPF ($Q$, in mol m$^{-2}$ day$^{-1}$). For a hybrid solar and artificial lighting system, the daily lighting profile that results is a composite lighting profile (Figure 1B), typically consisting of an approximately bell-shaped curve, representing the solar component, that is superimposed over a rectangular wave, representing the artificial lighting component. The total of the area under the solar curve and the area of the rectangular wave represents the daily integrated PPF.

This chapter would show that composite lighting could significantly influence the physiological responses of crops, particularly photosynthesis and respiration, and thus could be harnessed as a practical strategy for improving crop growth and productivity. The adoption of composite lighting for controlled-environment crop production rests on the principal premise that, depending on the lighting profile employed, equal moles of photons delivered to two crop treatments do not necessarily result in equal growths for the two treatments.

## II. FEATURES, TYPES, AND PARAMETERS

### A. Features of Composite Lighting

Composite lighting is a lighting profile that possesses the following essential features:

1. Two significantly distinct instantaneous PPF levels (one high and one low), each applied with its own photoperiod (Figure 2A). It should be noted that a solar component's nonrectangular curve could be represented by an equivalent rectangular wave whose area is equal to that of the original curve and whose height is the average instantaneous PPF in the original curve.

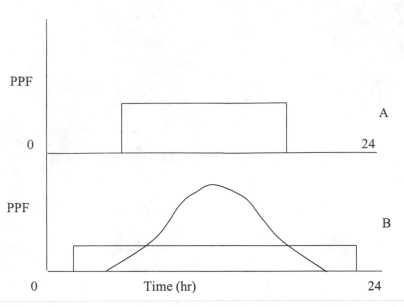

**Figure 1** Daily profiles for conventional lighting (A) as a rectangular wave, whose height is the magnitude of the photosynthetic photon flux (PPF) in $\mu$mol m$^{-2}$ sec$^{-1}$, whose length is the daily photoperiod in hr, and whose area is the daily integrated PPF in mol m$^{-2}$ day$^{-1}$, and for composite lighting (B) with a bell-shaped solar component and a rectangular-wave artificial-lighting component.

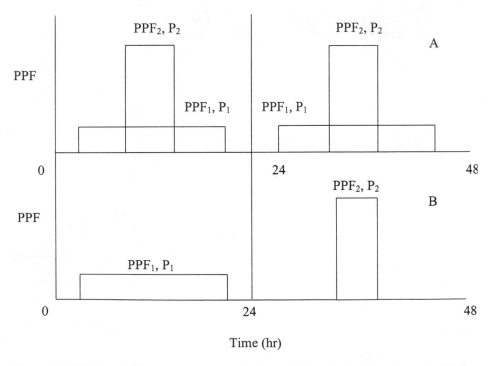

**Figure 2** Lighting profiles over two consecutive days: (A) for simultaneous composite lighting consisting of a component with low PPF (PPF$_1$) and long photoperiod ($P_1$) and a second component with high PPF (PPF$_2$) and short photoperiod ($P_2$) and (B) for alternating composite lighting with each component alternating between days.

2. The photoperiod of the lower instantaneous PPF level is usually significantly longer than that of the higher instantaneous PPF level (Figure 2A), although the two photoperiods could be made different in other ways or even be made equal.

3. The two instantaneous PPF levels with their respective photoperiods may be applied either simultaneously or alternately (Figure 2A and B).

4. The two instantaneous PPF levels may be generated by the same kind of light source (homogeneous lighting) or different kinds of light source (heterogeneous or hybrid lighting), the latter being either artificial or solar.

## B. Types of Composite Lighting

Based on the enumerated essential features, the six types of composite lighting (Figure 3) are as follows:

1. Simultaneous hybrid solar-artificial composite lighting—combination of instantaneous PPF from solar radiation and instantaneous PPF from an artificial light source, both applied concurrently generally according to diurnal (or daily) cycles

2. Alternating hybrid solar-artificial composite lighting—combination of instantaneous PPF from solar radiation and instantaneous PPF from an artificial light source, applied in sequence or alternately generally according to diurnal (or daily) cycles

3. Simultaneous hybrid artificial composite lighting—combination of instantaneous PPF from one artificial light source and instantaneous PPF from a second artificial light source, both applied concurrently generally according to diurnal (or daily) cycles

4. Alternating hybrid artificial composite lighting—combination of instantaneous PPF from one artificial light source and instantaneous PPF from a second artificial light source, applied in sequence or alternately generally according to diurnal (or daily) cycles

5. Simultaneous homogeneous composite lighting—combination of two instantaneous PPF levels from one artificial light source, both applied concurrently generally according to diurnal (or daily) cycles

6. Alternating homogeneous composite lighting—combination of two instantaneous PPF levels from one artificial light source, applied in sequence or alternately generally according to diurnal (or daily) cycles

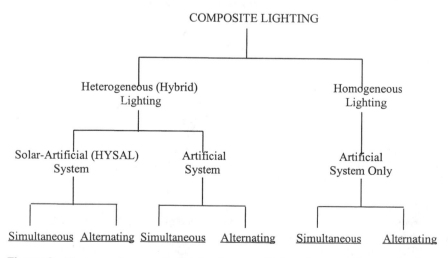

**Figure 3** Six types of composite lighting for controlled-environment plant production.

## C. Parameters of Composite Lighting

Using the symbols given in Figure 2A for the general case of composite lighting, the two basic parameters of composite lighting could be calculated as follows. The daily integrated PPF is given by

$$Q = \text{PPF}_1 P_1 + \text{PPF}_2 P_2 \tag{1}$$

where $Q$ is daily integrated PPF in mol m$^{-2}$ day$^{-1}$, PPF$_1$ and PPF$_2$ are component instantaneous PPF in μmol m$^{-2}$ sec$^{-1}$, and $P_1$ and $P_2$ are photoperiods in hr of the component instantaneous PPF.

The average instantaneous PPF is given by

$$\text{PPF}_{\text{ave}} = \text{PPF}_1 \left( \frac{P_1}{P_1 + P_2} \right) + \text{PPF}_2 \left( \frac{P_2}{P_1 + P_2} \right) \tag{2}$$

where PPF$_{\text{ave}}$ is average instantaneous PPF in μmol m$^{-2}$ sec$^{-1}$, PPF$_1$ and PPF$_2$ are component instantaneous PPF in μmol m$^{-2}$ sec$^{-1}$, and $P_1$ and $P_2$ are photoperiods in hr of the component instantaneous PPF.

Note that, given Eq. (1), Eq. (2) may also be expressed as

$$\text{PPF}_{\text{ave}} = \left( \frac{Q}{P_1 + P_2} \right) \tag{3}$$

## III. PHYSIOLOGICAL BASES

There are two major physiological bases for the premise that, for a given constant daily integrated PPF, a given lighting profile can significantly affect crop growth or yield: (1) the duration of the dark period implemented by the lighting profile, which affects the extent of the crop's dark respiration, and (2) the average instantaneous PPF implemented by the lighting profile, which determines in large measure the crop's light compensation point (LCP) at a given air temperature and $CO_2$ concentration.

## A. For a Constant Daily Integrated PPF, a Longer Daily Photoperiod Results in Lower Maintenance Respiration, Which Translates into Greater Growth

Studies have shown that reduction in the maintenance component of dark respiration could improve a crop's carbon balance and ultimately its yield [7–11]. Indeed, there is a preponderance of evidence in the literature showing that respiration rates and growth are negatively correlated. Heichel [12], working with two varieties of maize seedlings (*Zea mays*) in a growth chamber environment, found that the leaf specific respiration rate was about 40% higher in the slower growing variety. And while the stem specific respiration rate was about the same, the root specific respiration rate was also higher by about 35% in the slower growing variety. Because the photosynthetic rates were significantly indistinguishable between the two varieties, a slow rate of respiration resulted in greater carbon accumulation. Investigating tall fescue (*Festuca arundinaceae*) genotypes, Volenec et al. [13] similarly found significant negative correlations between dark respiration and yield per tiller and between dark respiration and specific leaf weight. A reduction in respiration by 47% at 20°C resulted in an increase in specific leaf weight by 52% and a rise in yield per tiller by 90%. Results of Winzeler et al. [14] showed that among genotypes of winter wheat (*Triticum aestivum*), a reduction in respiration by 16% translated into an increase in area per leaf by 22% and a rise in dry weight per leaf by 19%. And Wilson and Jones [15] observed a 10% increase in the annual productivity of field-grown ryegrass swards (*Lolium perenne*) for a 20% reduction in mature tissue respiration. Hence, minimizing respiratory carbon loss is one approach to increasing rates of dry matter accumulation [14,16].

One avenue of manipulating the maintenance respiration of a given crop is through regulation of its daily photoperiod. Logendra and Janes [17] investigated the influence of light duration on carbon partitioning and translocation (references) in young tomato plants (*Lycopersicon esculentum*) growing under similar daily integrated PPF. Using incandescent and fluorescent lamps as light sources, the plants were grown inside controlled-environment growth chambers under daily photoperiods of 8 hr (and PPF of 300 μmol m$^{-2}$ sec$^{-1}$) and 16 hr (and PPF of 150 μmol m$^{-2}$ sec$^{-1}$) at a constant daily integrated PPF of 8.64

mol m$^{-2}$ day$^{-1}$. Although the plants in both treatments fixed statistically indistinguishable amounts of carbohydrates per day, the plants in the 8-hr treatment respired a total amount of carbohydrates that was 184% greater than that for the plants in the 16-hr treatment. In addition, the amount of carbon translocated during the light period in the 16-hr treatment was 157% greater than that in the 8-hr treatment. Over 24 hr, the amount of carbon translocated in the 16-hr treatment remained greater by 41% than that in the 8-hr treatment. Thus, despite the plants in both treatments fixing comparable amounts of carbohydrates, more carbon was partitioned into sucrose and translocated out of the leaves in the plants grown under the 16-hr treatment. These results agreed with the findings that plants grown under short light periods have greater starch accumulation rates [18–22], starch content [20,21,23], and low sucrose content [24] in contrast with those grown under long light periods. The high starch accumulation rates observed under short light periods are associated with low translocation rates [19,22] and decreased amounts of carbon translocated [24]. Under short light periods, plants accumulate higher amounts of starch, most likely to satisfy the carbohydrate requirements of the subsequent longer nights [18].

Similar results were obtained earlier by Jiao et al. [25,26], who investigated the influence of radiation on whole-plant net $CO_2$ exchanges in rose plants (*Rosa hybrida*). They found that rose plants exposed to a daily photoperiod of 24 hr (and PPF of 204 μmol m$^{-2}$ sec$^{-1}$) retained 80% more carbon than those exposed to a shorter daily photoperiod of 12 hr (and higher PPF of 410 μmol m$^{-2}$ sec$^{-1}$), despite the two treatments receiving the same daily integrated PPF of 17.6 mol m$^{-2}$ day$^{-1}$ as supplied by high-pressure sodium lamps.

The results of Grange [24] on pepper plants (*Capsicum annuum*) demonstrated the same trends. Using controlled-environment growth chambers equipped with warm-white fluorescent and tungsten lamps as light sources, three treatments with comparable daily integrated PPF values of 9.27, 9.94, and 8.94 mol m$^{-2}$ day$^{-1}$ were given daily photoperiods of 14 hr (and a PPF of 184 μmol m$^{-2}$ sec$^{-1}$), 10 hr (and a PPF of 276 μmol m$^{-2}$ sec$^{-1}$), and 6 hr (and a PPF of 414 μmol m$^{-2}$ sec$^{-1}$), respectively. Indeed, the treatment with the longest photoperiod of 14 hr (and lowest PPF of 184 μmol m$^{-2}$ sec$^{-1}$) yielded the greatest dry weight per plant of 20.3 g, and the treatment with the shortest photoperiod of 6 hr (and highest PPF of 414 μmol m$^{-2}$ sec$^{-1}$) produced the smallest dry weight per plant of 6.2 g. The treatment with the intermediate photoperiod of 10 hr (and intermediate PPF of 276 μmol m$^{-2}$ sec$^{-1}$) yielded an intermediate dry weight per plant of 16.5 g. Note that although the maximum daily integrated PPF level among the treatments was greater than the minimum by only 11%, the resulting maximum dry weight per plant among the treatments exceeded the resulting minimum by 227%.

Citing the results of Hurd and Thornly [27] for tomato plants as evidence, Moe [28] concluded that if long photoperiods do not cause adverse effects, such as leaf damage or prevention of flowering in short-day plants, prolonged low instantaneous PPF should be more effective than providing the same daily integrated PPF at a higher instantaneous PPF for a shorter period. But although Moe [28] and the foregoing authors correctly established the significant correlations between protracted photoperiod, reduced maintenance respiration, and increased growth or yield, it should also be pointed out that the concomitant decline in the instantaneous PPF when the photoperiod is prolonged—while keeping the daily integrated PPF constant—contributes as well to the reduction in maintenance respiration. The latter constitutes the second physiological basis for composite lighting.

## B. For a Constant Daily Integrated PPF, a Lower Average Instantaneous PPF Results in Lower LCP, Which in Turn Results in Lower Maintenance Respiration, Which Translates into Greater Growth

Fonteno and McWilliams [29] found that a 15-week acclimatization of four tropical foliage species to 27 μmol m$^{-2}$ sec$^{-1}$, using cool-white fluorescent lamps as light source for 12 hr per day, resulted in a significant reduction in LCP accompanied by a significant decline in dark respiration for each species. Light compensation points decreased between week 1 and week 15 as follows: 33 to 7 μmol m$^{-2}$ sec$^{-1}$ in *Philodendron scandens* subsp. *oxycardium*, 38 to 6 μmol m$^{-2}$ sec$^{-1}$ in *Epipremnum aureum*, 14 to 4 μmol m$^{-2}$ sec$^{-1}$ in *Brassaia actinophylla*, and 119 to 15 μmol m$^{-2}$ sec$^{-1}$ in *Dracaena sanderana*. Concomitantly, dark respiration decreased 63% in *P. scandens* subsp. *oxycardium*, 71% in *E. aureum*, 53% in *B. actinophylla*, and 64% in *D. sanderana* during the acclimatization period. Fonteno and McWilliams

[29] concluded that, by slowly lowering the conditioning irradiance in the acclimatization area, the LCPs of these shade-tolerant species could be lowered. In general, leaves that have low LCP are such not because they photosynthesize better but because they respire less [30]. Consequently, they frequently achieve more net photosynthesis because they respire less [30]. Thus, a lowered instantaneous PPF, which goes hand in hand with a prolonged photoperiod for a given daily integrated PPF, by itself contributes to the reduction in the crop's maintenance respiration.

## IV.  INDUSTRY APPLICATIONS

### A.  Greenhouse Application

The term "composite lighting" was first used by Cuello et al. [5] in describing hybrid solar and artificial lighting in a controlled-environment plant growth chamber. Although composite lighting had in practice been used or demonstrated earlier in greenhouses, it had not been differentiated from what was merely supplemental lighting and had not been recognized and adopted as a practical strategy for influencing crop growth or yield through active regulation of both the crop's photoperiod and light compensation point. In supplemental lighting for greenhouses, the main interest had simply been in augmenting, using artificial-light sources, whatever solar irradiance was available in order to increase the level of the daily integrated PPF.

The results obtained by Gislerod et al. [2] in a greenhouse study, however, were quite telling. Investigating the effects of photoperiod and instantaneous PPF on the growth of four greenhouse plants, they determined how plants responded when they were exposed to the same level of daily integrated PPF but at different average instantaneous PPF and photoperiod levels. *Begonia* × *hiemalis* cultivar 'Schwabeland', *Kalanchoe blossfeldiana* cultivar 'Pollux', *Hedera helix* cultivars 'Svendborg' and 'Gloire de Marengo', and *Pelargonium* × *hortorum* cultivar 'Alex' were all supplied with the same daily integrated PPF, using high-pressure sodium lamps, but under two different scenarios: (1) at a PPF of 85 $\mu$mol m$^{-2}$ sec$^{-1}$ and photoperiod of 16 hr and (2) at a lower PPF of 68 $\mu$mol m$^{-2}$ sec$^{-1}$ and a longer photoperiod of 20 hr. The solar integrated PPF on average was approximately 39% of the total integrated PPF, that is, of the combined solar and artificial-lighting components. The results showed that, for all four species, the resulting dry weight per plant and percent dry matter were consistently and significantly greater in the treatment with lower instantaneous PPF and longer photoperiod than with the treatment with higher instantaneous PPF and shorter photoperiod. The dry weight per plant in the first treatment exceeded that in the second treatment by 20% for *Begonia*, 35% for *Kalanchoe*, 42% for *Pelargonium*, and 42% for *Hedera*. For percent dry matter, the first treatment exceeded the second treatment by 14% for *Begonia*, 11% for *Kalanchoe*, 9% for *Pelargonium*, and 8% for *Hedera*. These results made clear that the specific design of a composite-lighting profile could significantly affect a crop's growth performance even without changing the total integrated PPF delivered to the crop. This is a restatement of the principal premise for composite lighting enunciated in the Introduction.

### B.  Growth Chamber Application

The study conducted by Cuello et al. [5] on composite lighting appears to be the first performed in a growth chamber environment. The hybrid solar and artificial lighting (HYSAL) system used in this study consisted of a mirror-based optical waveguide (OW) solar lighting system as the solar component and four 60-W xenon–metal halide illuminators as the artificial-light component. A reference (or control) system consisted of a conventional 250-W high-pressure sodium (HPS) lamp. Solar irradiance was harnessed whenever available for the HYSAL treatment. During the course of the 30-day growth period for lettuce (*Lactuca sativa*), the HYSAL's instantaneous solar PPF varied with the natural fluctuations of terrestrial solar irradiance, which changed dramatically within each day and between days. When averaged over the entire growth period, the average instantaneous solar PPF for the HYSAL treatment turned out to be 322 $\mu$mol m$^{-2}$ sec$^{-1}$ for an average daily photoperiod of only 3.86 hr owing to numerous cloudy days.

Over the whole growth period, the xenon–metal halide lamps provided an average instantaneous PPF of 30 $\mu$mol m$^{-2}$ sec$^{-1}$ continuously for 24 hr each day. The resulting total moles of photons received with the HYSAL treatment for 30 days were 199 moles/m$^2$, being 60.6% solar and 39.4% artificial. The HPS reference was made to receive the same daily moles of photons as in the HYSAL treatment throughout the growth period, resulting in both HPS reference and HYSAL treatments having the same total number of moles (199 moles/m$^2$) at the end of the growth period. Over the entire growth period, the HPS reference had an average instantaneous PPF of 194 $\mu$mol m$^{-2}$ sec$^{-1}$ and an average daily photoperiod of 9.5 hr. The resulting average total dry weight per plant for the HYSAL treatment of 1.37 $\pm$ 0.38 g exceeded significantly by 76% ($\alpha = 0.05$) that for the HPS reference of only 0.78 $\pm$ 0.17 g. This significant discrepancy could be explained physiologically by the HPS reference having both a significantly longer dark period and a higher light compensation point than the HYSAL treatment. Whereas the HYSAL treatment had no dark period at all, the HPS reference had 14.5 hr of dark period each day or a total of 435 hr (18.1 days) over the 30-day growth period. The resulting average light compensation point for the HYSAL treatment of 55.3 $\mu$mol m$^2$ sec$^{-1}$ was also significantly lower ($\alpha = 0.05$) than the average LCP for the HPS reference of 169.1 $\mu$mol m$^{-2}$ sec$^{-1}$. Further experimentation showed that it was indeed the composite lighting profile of the HYSAL treatment, not the light quality factor, that caused the biomass discrepancy. In the same experiment, Cuello et al. [5] also successfully demonstrated that, with respect to crop response, a simultaneous composite lighting profile (as that shown in Figure 2A) was significantly indistinguishable from an alternating composite lighting profile (as that depicted in Figure 2B). Although needing further study, these results provide strong preliminary evidence that composite lighting could be employed in growth-chamber settings as a practical strategy to optimize crop performance at a given total integrated PPF.

## V.  DESIGN OF COMPOSITE PROFILES

The practical usefulness of composite lighting is that it makes possible the optimization of a given crop's performance by making allowance for its daily photoperiod to be extended as much as permissible with respect to the crop and for its average instantaneous PPF to be maintained at a desired relatively low level even when the daily integrated PPF is raised significantly. This important flexibility is simply lacking in conventional lighting. For instance, given a conventional-lighting case where the instantaneous PPF is set at 100 $\mu$mol m$^{-2}$ sec$^{-1}$ and the daily photoperiod is stretched to the maximum possible value of 24 hr, the resulting daily integrated PPF is 8.64 mol m$^{-2}$, which is the maximum daily integrated PPF corresponding to the set instantaneous PPF of 100 $\mu$mol m$^{-2}$ sec$^{-1}$. This means that conventional lighting can only be used for implementing an instantaneous PPF of 100 $\mu$mol m$^{-2}$ sec$^{-1}$ if, and only if, the specified daily integrated PPF does not exceed 8.64 mol m$^{-2}$. Conversely, this means that conventional lighting cannot be employed to implement an instantaneous PPF of 100 $\mu$mol m$^{-2}$ sec$^{-1}$ when the specified daily integrated PPF exceeds 8.64 mol m$^{-2}$.

Considering a second conventional-lighting case where the instantaneous PPF is set at 100 $\mu$mol m$^{-2}$ sec$^{-1}$ and the daily photoperiod is set at 16 hr, yielding a daily integrated PPF of 5.76 mol m$^{-2}$, there are only three possible ways by which the daily integrated PPF can be raised significantly above 5.76 mol m$^{-2}$. First is by keeping the instantaneous PPF at 100 $\mu$mol m$^{-2}$ sec$^{-1}$ while lengthening the photoperiod beyond 16 hr, which may not be an issue for some crops but may be for others. Second is by keeping the photoperiod at 16 hr but raising the instantaneous PPF above 100 $\mu$mol m$^{-2}$ sec$^{-1}$, with the likely undesirable consequence of raising the crop's light compensation point. And third is by both lengthening the photoperiod beyond 16 hr and raising the instantaneous PPF above 100 $\mu$mol m$^{-2}$ sec$^{-1}$. Composite lighting circumvents this difficulty by providing for the flexibility of raising the daily integrated PPF significantly without changing either the effective daily photoperiod or the average instantaneous PPF.

Thus, given the desired values for the daily integrated PPF ($Q$), average instantaneous PPF (PPF$_{ave}$), and the long photoperiod ($P_1$), the composite lighting profile is designed as follows. Expressing the long photoperiod as a multiple of the short photoperiod,

$$P_1 = kP_2 \tag{4}$$

where $P_1$ is the long photoperiod, $P_2$ is the short photoperiod, and $k$ is a constant. Similarly, expressing the low instantaneous PPF as a fraction of the high instantaneous PPF, then,

$$PPF_1 = mPPF_2 \tag{5}$$

where $PPF_1$ is the low instantaneous PPF, $PPF_2$ is the high instantaneous PPF, and $m$ is a constant. Combining Eqs. (2), (4), and (5) yields

$$PPF_{ave} = \left( \frac{mk + 1}{k + 1} \right) PPF_2 \tag{6}$$

Because the desired value for the long photoperiod $P_1$ is set, the short photoperiod $P_2$ may be readily calculated from Eq. (3) as

$$P_2 = \left( \frac{Q}{PPF_{ave}} - P_1 \right) \tag{7}$$

where $Q$ and $PPF_{ave}$ are the known desired daily integrated PPF and average instantaneous PPF, respectively. Hence, from Eq. (4), $k$ is calculated as

$$k = P_1/P_2 \tag{8}$$

Letting

$$z = \left( \frac{mk + 1}{k + 1} \right) \tag{9}$$

and understanding that $z$ is the ratio of the average instantaneous PPF ($PPF_{ave}$) to the high instantaneous PPF ($PPF_2$), a desired value for $z$, which must be less than 1.0, is assigned (e.g., 0.20). With the values of $k$ and $z$ now known, $m$ can be calculated from Eq. (9) as

$$m = \frac{z(k + 1) - 1}{k} \tag{10}$$

Therefore, the high instantaneous PPF ($PPF_2$) may now be calculated as

$$PPF_2 = \frac{PPF_{ave}}{\left( \dfrac{mk + 1}{k + 1} \right)} \tag{11}$$

Subsequently, the low PPF ($PPF_1$) is calculated as

$$PPF_1 = mPPF_2 \tag{12}$$

As an illustration, recall the second conventional-lighting case where the instantaneous PPF is set at 100 $\mu$mol m$^{-2}$ sec$^{-1}$ and the daily photoperiod is set at 16 hr, yielding a daily integrated PPF of 5.76 mol m$^{-2}$. Assume that it is desired that the daily integrated PPF be increased by over 80% to 10.44 mol m$^{-2}$ while keeping the daily photoperiod at 16 hr and the average instantaneous PPF at 100 $\mu$mol m$^{-2}$ sec$^{-1}$. Thus, $Q = 10.44$ mol m$^{-2}$, $P_1 = 16$ hr, and $PPF_{ave} = 100$ $\mu$mol m$^{-2}$ sec$^{-1}$. From Eq. (7), $P_2$ is calculated to be 13 hr and, from Eq. (8), $k$ is calculated to be 1.23. Letting $z = 0.75$, $m$ is calculated from Eq. (10) to be 0.55. Hence, $PPF_2$, from Eq. (11) is calculated to be 133.3 $\mu$mol m$^{-2}$ sec$^{-1}$ and $PPF_1$, from Eq. (12), is calculated to be 73.3 $\mu$mol m$^{-2}$ sec$^{-1}$. Using the foregoing calculated values for Eqs. (1) and (2), note that the resulting $Q$ and $PPF_{ave}$ are 10.44 mol m$^{-2}$ and 100 $\mu$mol m$^{-2}$ sec$^{-1}$, respectively, as desired. Thus, the daily integrated PPF could be increased significantly without necessarily changing the values of the average instantaneous PPF of 100 $\mu$mol m$^{-2}$ sec$^{-1}$ and the effective photoperiod of 16 hr.

## VI.  CONCLUDING REMARKS

Composite lighting allows an increase in the daily integrated PPF delivered to a given crop to ensure that the lighting requirement by the crop is properly met, while also providing a high degree of freedom in allowing both the daily photoperiod and the average instantaneous PPF to be regulated as desired so as to minimize maintenance respiration and improve crop growth or yield. It is desirable for the daily photoperiod to be extended as much as possible. Whereas crops such as lettuce, sweet pepper, roses, and chrysanthemum can tolerate long photoperiods, other crops such as tomato plants require 4–6 hr of darkness [28]. Gislerod et al. [2] found that, when using the same daily integrated PPF with *Begonia, Chrysanthemum, Hedera, Kalanchoe,* and *Pelargonium,* the growth was best when the plants were allowed to have 4–6 hr in darkness during a 24-hr cycle. Also, for any given daily integrated PPF, a lower average instantaneous PPF is preferable to decrease the crop's light compensation point. The combination of a longer photoperiod and a lower average instantaneous PPF, for any given daily integrated PPF, constitutes the ideal scenario for lowering maintenance respiration and improving the crop's growth or yield.

Because proper design of a composite lighting profile could result in significant crop growth for a given total integrated PPF, an important implication of composite lighting is that it could potentially cut down significantly the electrical power required to achieve a given desired level of crop biomass. It is entirely probable that employing an appropriate composite lighting profile could obviate the delivery of additional moles of photons to a given crop to increase its growth or yield because composite lighting, relative to certain conventional lighting profiles, could increase crop growth or yield without increasing the total integrated PPF. The capacity of composite lighting to conserve electrical power by its ability to optimize crop yield requires further investigation. Its benefits would be of significant importance not only in terrestrial greenhouse and growth-chamber applications but also in extraterrestrial advanced life support systems for the human exploration and development of space.

# REFERENCES

1. GT Bruggink, E Heuvelink. Influence of light on growth of young tomato, cucumber, and sweet pepper plants in the greenhouse: effects on relative growth rate, net assimilation rate and leaf area ratio. Sci Hortic 31:161–174, 1987.
2. HR Gislerod, IM Eidstein, LM Mortensen. The interaction of daily lighting period and light intensity on growth of some greenhouse plants. Sci Hortic 38:295–304, 1989.
3. LM Mortensen, HR Gislerod, H Mikkelsen. Maximizing the yield of greenhouse roses with respect to artificial lighting. Norw J Agric Sci 6:27–34, 1992.
4. JL Cuello, D Jack, P Sadler, T Nakamura. Hybrid solar and artificial lighting (HYSAL): next-generation lighting strategy for bioregenerative advanced life support. Proceedings of the 29th International Conference on Environmental Systems, Denver, 1999.
5. JL Cuello, Y Yang, E Ono, K Jordan, T Nakamura. Hybrid solar and xenon–metal halide lighting for lunar and martian bioregenerative life support. Proceedings of the 30th International Conference on Environmental Systems, Toulouse, 2000.
6. JL Cuello, D Jack, E Ono, T Nakamura. Supplemental terrestrial solar lighting for an experimental subterranean biomass production chamber. Proceedings of the 30th International Conference on Environmental Systems, Toulouse, 2000.
7. FWT Penning de Vries. Substrate utilization and respiration in relation to growth and maintenance of higher plants. Neth J Agric Sci 22:40–44, 1974.
8. FWT Penning de Vries. The cost of maintenance processes in plant cells. Ann Bot 39:77–92, 1975.
9. D Wilson. Variation in leaf respiration in relation to growth and photosynthesis in *Lolium*. Ann Appl Biol 80:323–338, 1975.
10. JE Sheehy, JM Cobby, GJA Ryle. The growth of perennial rye grass: a model. Ann Bot 43:335–354, 1979.
11. JE Sheehy, JM Cobby, GJA Ryle. The use of a model to investigate the influence of some environmental factors on the growth of perennial rye grass. Ann Bot 46:343–365, 1980.
12. GH Heichel. Confirming measurements of respiration and photosynthesis with dry matter accumulation. Photosynthetica 5:93–98, 1971.
13. JJ Volenec, CJ Nelson, DA Sleper. Influence of temperature on leaf dark respiration of diverse tall fescue genotypes. Crop Sci 24:907–912, 1984.
14. M Winzeler, DE McCullough, LA Hunt. Genotypic differences in dark respiration of mature leaves in winter wheat (*Triticum aestivum* L.). Can J Plant Sci 68:669–675, 1988.
15. D Wilson, JG Jones. Effect of selection for dark respiration rate of mature leaves on crop yields of *Lolium perenne* cv. S23. Ann Bot 49:313–320, 1982.
16. JS Amthor. Respiration and Crop Productivity. New York: Springer-Verlag, 1989, pp 164–169.
17. S Logendra, HW Janes. Light duration effects on carbon partitioning and translocation in tomato. Sci Hortic 52:19–25, 1992.
18. NJ Chatterton, JE Silvius. Photosynthate partitioning into starch in soybean leaves. 1. Effects of photoperiod versus photosynthetic period duration. Plant Physiol 64:749–753, 1979.
19. NJ Chatterton, JE Silvius. Acclimation of photosynthate partitioning and photosynthetic rates to changes in length of the daily photosynthetic period. Ann Bot 46:739–745, 1980.
20. RC Sicher, WG Harris, DF Kremer, NJ Chatterton. Effects of shortened daylength upon translocation and starch accumulation by maize, wheat, and pangola grass leaves. Can J Bot 60:1304–1309, 1982.
21. JM Robinson. Photosynthetic carbon metabolism in leaves and isolated chloroplasts from spinach plants grown under short and intermediate photosynthetic periods. Plant Physiol 75:397–409, 1984.

22. SC Huber, TW Rufty, PS Kerr. Effects of photoperiod on photosynthate partitioning and diurnal rhythms in sucrose phosphate synthase activity in leaves of soybean (*Glycine max*) and tobacco (*Nicotiana tabacum*). Plant Physiol 75:1080–1084, 1984.
23. NJ Chatterton, JE Silvius. Photosynthate partitioning into leaf starch as affected by daily photosynthetic period duration in six species. Physiol Plant 49:141–144, 1980.
24. RI Grange. Carbon partitioning in mature leaves of pepper: effects of daylength. J Exp Bot 36:1749–1759, 1985.
25. J Jiao, MJ Tsujita, B Grodzinski. Influence of radiation and $CO_2$ enrichment on whole plant net $CO_2$ exchange in roses. Can J Plant Sci 71:245–252, 1991.
26. J Jiao, MJ Tsujita, B Grodzinski. Influence of temperature on net $CO_2$ exchanges in roses. Can J Plant Sci 71:235–243, 1991.
27. RG Hurd, JHM Thornly. An analysis of the growth of young tomato plants in water culture at different light integrals and $CO_2$ concentrations. I. Physiological aspects. Ann Bot 38:375–388, 1974.
28. R Moe. Physiological aspects of supplementary lighting in horticulture. Third International Symposium on Artificial Lighting in Horticulture, Noordwijkerhout, 1994, pp 17–24.
29. WC Fonteno, EL McWilliams. Light compensation points and acclimatization of four tropical foliage plants. J Am Soc Hortic Sci 103:52–56, 1978.
30. D Walker. Excited leaves. Tansley review No 36. New Phytol 121:325–345, 1992.

# 48

# Plant Growth and Human Life Support for Space Travel

**Raymond M. Wheeler**

*National Aeronautics and Space Administration, Kennedy Space Center, Florida*

**Gary W. Stutte, G. V. Subbarao\*, and Neil C. Yorio**

*Dynamac Corporation, Kennedy Space Center, Florida*

## I. BIOREGENERATIVE SYSTEMS

## A. Background

The balance of the carbon dioxide ($CO_2$), oxygen ($O_2$), and water in Earth's biosphere is largely dependent on photosynthetic and transpiration processes of green plants. Indeed, it is photosynthesis that ultimately provides the food and energy that humans and other animals depend on. As technology levels advance, humans will eventually be able to travel away from Earth's sustaining biosphere on long-term space travel. One approach to providing consumables for space travel would be to grow plants (crops) as part of a bioregenerative life support system.

The concept of using bioregenerative life support for space has been studied since the 1950s and 1960s, with most of these early studies centered on the use of algae (e.g., *Chlorella*) for $O_2$ production and $CO_2$ removal [1–3]. Testing was expanded in the late 1960s and 1970s by Russian researchers to include higher plants [4,5] and in the late 1970s by the U.S. National Aeronautics and Space Administration (NASA) under the Controlled Ecological Life Support System, or CELSS program [6]. The CELSS plant research consisted primarily of laboratory-scale studies carried out at U.S. universities and by several European and Japanese investigators [6–8]. These studies were coupled with large-scale, closed-system tests conducted at NASA's Kennedy Space Center and Johnson Space Center to assess the performance of plants in prototype life support systems [9–11] (Figure 1). The CELSS or bioregenerative life support research is currently consolidated under NASA's Advanced Life Support (ALS) program.

## B. Mission Constraints: When Would Plants Have a Role?

To date, human space travel has been short in duration and close to the Earth; for example, NASA space shuttle missions are typically 7 to 14 days. Even when missions have been longer, such as with the orbiting Mir Space Station, the distance from Earth is still relatively short. This has allowed most of the life support consumables to be stowed or replenished from Earth on a continuing basis. But supply-line economics dictate that as mission distances increase, so will the costs of stowage and resupply [12,13]. This has forced spaceflight engineers to explore regenerative approaches for providing human life support [14]. Physicochemical regenerative technologies have already been used for water recycling on Mir and are planned for use along with $CO_2$ reduction and $O_2$ production systems on the International Space Station [14].

---

\* *Current affiliation:* Japan International Research Center for Agricultural Sciences, Ibaraki, Japan

**Figure 1** NASA's Biomass Production Chamber located at Kennedy Space Center, Florida. The chamber provides 20 m$^2$ of growing area in a closed volume of 113 m$^3$ for testing crop growth and yield for life support applications.

**TABLE 1**  Possible Crops for Life Support Systems in Space

| Tibbitts and Alford | Hoff, Howe, and Mitchell | Salisbury and Clark | BIOS-3 tests |
|---|---|---|---|
| Wheat | Wheat | Wheat | Wheat |
| Soybean | Potato | Rice | Potato |
| Potato | Soybean | Sweetpotato | Carrot |
| Lettuce | Rice | Broccoli | Radish |
| Sweetpotato | Peanut | Kale | Beet |
| Peanut | Dry bean | Lettuce | Nut sedge (chuffa) |
| Rice | Tomato | Carrot | Onion |
| Sugar beet | Carrot | Rape seed (canola) | Cabbage |
| Pea | Chard | Soybean | Tomato |
| Taro | Cabbage | Peanut | Pea |
| Winged bean | | Chickpea | Dill |
| Broccoli | | Lentil | Cucumber |
| Onion | | Tomato | Salad spp. |
| Strawberry | | Onion | |
| | | Chili pepper | |

*Sources:* Tibbitts and Alford [17]; Hoff, Howe, and Mitchell [18]; Salisbury and Clark [19]; Gitelson and Okladnikov [4]—diet also included supplemental animal protein and sugar.

When one considers travel to other planets, resupply costs will force even more recycling and eventually some in situ food production on these missions. At this point, plants and bioregenerative systems may be an option. Mars would likely be the first planet visited by humans, and early Mars missions would rely heavily on stowage and resupply [13]. But as mission durations increase, so will the need for greater autonomy and closure, not only to reduce costs but also to provide contingencies for mission delays or failures [9]. Even prior to the establishment of surface colonies, plants could play an important role for Earth-orbiting space stations or planetary transit vehicles by providing fresh food for the diet. This concept has often been referred to as a "salad machine," where a small amount of vegetables and fruits could be provided to the crew [15]. Although this would not have a large impact on total dietary needs, the fresh food supplements could have a positive psychological effect on the humans living in confined space habitats; moreover, the presence of plants, their lighting systems, and the humidities and aromas associated with the plants could have positive effects on the crew [16].

## C.  Crop Selection

Criteria for choosing life support crops include obvious attributes, such as crop yield, nutritional value, horticulture, and processing requirements [17–20]. For space applications, it will be important to optimize yield as well as minimize the area requirements and associated infrastructure costs for growing plants (e.g., lighting and watering system components). Thus high yield per unit area per unit time (i.e., $g \, m^{-2} \, day^{-1}$) is especially important [9,17,21]. In addition, characteristics such as high harvest index (ratio of edible to total biomass) and short stature are important to minimize inedible wastes and allow the crops to fit in volume-limited systems [9,17,22]. Processing requirements for converting the harvested biomass into useful foods must also be considered, and crops that require extensive processing may be too costly to include, especially for early missions where processing equipment may not be available [13,17,18].

Table 1 lists some crop species that have been suggested for bioregenerative life support. These lists were based largely on human nutritional requirements, but meeting all these nutritional requirements will be difficult with such short lists. It is more economical to provide minor nutrients (e.g., vitamin $B_{12}$) with supplements from Earth rather than producing them on site [13,17–19].

## II.  GROWING PLANTS FOR LIFE SUPPORT

## A.  Environmental Management

Because of the harsh environment of space, growing crops for life support will require protected environments. Light, $CO_2$, temperature, humidity, and mineral nutrition will all need to be managed carefully to optimize crop performance. The gravitational environment must also be considered: for example, orbiting space stations and transit missions to planets would be nearly weightless, and surface settlements on the moon would have about $\frac{1}{6}g_n$ and on Mars about $\frac{1}{3}g_n$ [21,23,24]. Evidence to date suggests that even under near weightlessness, acceptable plant growth and development should be possible provided the other environmental needs are satisfied [25–28]. Yet providing all these environmental needs in weightlessness can be difficult. For example, watering plants in weightlessness will require closed plumbing systems to prevent water from escaping and a distribution system to maintain both adequate water and oxygen throughout root zones [29,30]. On the moon or Mars, the gravitational fields should provide sufficient mechanical advantage for moving water and nutrients and allow use of such conventional recirculating hydroponic approaches. Plants might be grown on centrifuges in spaceflight to impose an artificial gravity and potentially alleviate root-zone drainage problems, but this would present additional engineering challenges, especially for large-scale systems.

## 1. Light

PHOTOSYNTHETICALLY ACTIVE RADIATION   Of all the environmental factors for growing crops for life support, light (photosynthetically active radiation, PAR) is perhaps the most important with regard to crop yields and system costs. At low to moderate light levels, crop yields are near-linear functions of total PAR [10,31,32], and for some species (e.g., wheat), yields continue to increase even at very high PAR levels [31] (Figure 2). These findings are encouraging for life support and suggest that growing areas could be reduced significantly with higher light intensities. Whether productivities of broad leaf crops would continue to increase at such high PAR levels needs further study, but results with potato and soybean suggest that maximum yields might be achieved near 800 to 1000 $\mu$mol m$^{-2}$ sec$^{-1}$ [33,34]. For some crops, high PAR levels may be undesirable because of injuries such as tipburn (e.g., lettuce) [35] and leaf chlorosis, particularly under high-intensity discharge lamps [36].

Most of the testing to date with plants for life support has used electric lighting. The results from these studies should be applicable to space habitats where sufficient electrical power is available. Alternatively, incident solar lighting might be used to grow plants to reduce large electrical power requirements, but this would require transparent materials with the appropriate pressure and thermal integrity, or light collection/conduit systems [37,38]. Solar light collection systems would also be subject to the local photoperiods and the native solar intensity. For example, the solar radiation constant for the moon is similar to that just outside the Earth's atmosphere (1370 W m$^{-2}$ total radiation with ~600 W m$^{-2}$ PAR), but the light cycle at most latitudes is ~14.7 days with a corresponding ~14.7 day night cy-

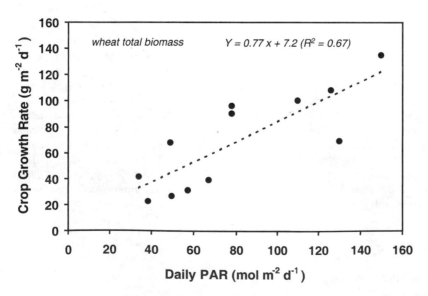

**Figure 2**   Crop growth rate of wheat versus photosynthetically active radiation (PAR). (Data from Refs. 10, 11, and 31.)

cle [21,24]. An Earth-orbiting space station would typically have a 90-min orbital cycle, providing about 60 min of light and 30 min of dark [39]; a planetary transit vehicle could capture uninterrupted solar light, but the intensity would continually drop off with distance. Mars receives about 45% of the solar radiation that Earth does (~600 W m$^{-2}$ total) and has a relatively favorable photoperiod of ~25 hr [21,23]. But large dust storms occur at some latitudes on Mars, which would affect solar light collection systems [24,40].

Depending on the crop growing area, electric lighting could represent the single largest energy demand in the life support system. This includes not only making the light from electricity but also removing the heat generated by the lamps. If 40 m$^2$ of crop area and a PAR input of 1000 μmol m$^{-2}$ sec$^{-1}$ (~200 W m$^{-2}$) is needed to support one human (see later), then 200 W m$^{-2}$ PAR/20% (electrical conversion efficiency for lamping systems) or about 1 kW of electrical power would be required for each m$^2$ of crop area. Then, 1 kW m$^{-2}$ × 40 m$^{-2}$ person$^{-1}$, or 40 kW would be required per person for the electric lighting. This power requirement might be doubled to accommodate heat rejection, water pumping, air circulation, etc., indicating that perhaps 100 kW of electrical power would be required per person to grow the food just to meet caloric needs.

Thus if electric lighting is used, it will be critical to use efficient electric lamps to minimize the power requirements. High-intensity discharge (HID) lamps, such as high-pressure sodium and metal halide lamps, have high electrical conversion efficiencies and have been used extensively for plant research [41]. Yet these are relatively hot, point sources of light that require a sufficient distance from the plants to achieve adequate distribution and avoid plant damage. Fluorescent lamps have a larger radiating surface but are less efficient and shorter lived than HID lamps. Innovative lighting technologies, such as light-emitting diodes (LEDs), are relatively cool light sources that can be positioned much closer to the plants [42,43]. In addition, LEDs can have very long life spans, which could provide substantial savings for system maintenance [42,43]. Recently developed microwave lamps have very high electrical conversion efficiencies (35–40%) and hold promise as sources for plant lighting [44]. Light collection and delivery technologies are also being studied in which either sunlight or light from a bright, artificial source is collected and delivered to plants via conduits or optical fibers [37,38]. These light conduits might even be used for intracanopy lighting to increase system efficiency [45,46].

PHOTOPERIOD   Long photoperiods have worked well for growing wheat and lettuce in life support studies [31,47–49], whereas short photoperiods have been best for soybean, potato, rice, and sweetpotato [50–55]. Thus, there will probably be a range of optimal photoperiods for different crops in life support systems. Tomatoes and some cultivars of potatoes are intolerant (show injury) to very long photoperiods [51,52,56], and some short-day crops are sensitive to night breaks or dim light during the dark periods. For example, dim day length extensions of 5 μmol m$^{-2}$ sec$^{-1}$ PAR effectively blocked tuber development in potatoes [51], and light leakage even as low as 0.4 μmol m$^{-2}$ sec$^{-1}$ during a dark period can delay or inhibit tuber initiation [57]. Thus, sensitive short-day species might require light barriers to separate them from areas on different lighting cycles, or crops might be grown in separate long- and short-day chambers.

As an alternative to adjusting the lighting environment, day-neutral species and cultivars might be selected to avoid photoperiod complications. Despite being short-day plants, some potato cultivars grow and tuberize under continuous light, provided temperatures are kept sufficiently cool or cycled on a diurnal basis [51,58]. Some sweetpotato and soybean cultivars also tolerate continuous light [34,59]. Yet even with day-neutral cultivars, growth can still be affected by photoperiod: for example, early-season potato cultivars and day-neutral rice cultivars can still show reductions in yield and harvest index under long photoperiods [52,55].

SPECTRAL QUALITY   The spectral balance of the light also needs to be considered for achieving optimal plant growth. If solar light is used, this would provide a broad spectrum to which the plants are well adapted. If electric lamps are used, an acceptable photomorphological spectrum must be provided [41,60]. For example, the lack of blue light in red LEDs or low-pressure sodium lamps (monochromatic 589 nm) could affect phototropic orientation and stomatal functions in many species [60,61]. Even with high-pressure sodium lamps, which have a broader spectrum, there still might be insufficient blue light to prevent excessive stem growth [34,61,62]. In studies with red LEDs, it is usually necessary to supplement the red light with small amount of blue to get acceptable growth [43]; this can be done with broader spectrum lamps or by including blue LEDs in the lighting arrays.

## 2. Carbon Dioxide

With the assistance of physicochemical control systems, adequate $CO_2$ concentrations (i.e., $CO_2$ partial pressures) should be achievable for plant production systems in space. Studies of canopy gas exchange typically show increased photosynthesis as $CO_2$ is increased from 0.035 to 0.07 kPa (350 up to ~700 ppm at 101 kPa total pressure) [63], with maximum rates for $C_3$ species occurring near 0.10–0.15 kPa [10,47,64]. Likewise, biomass yields of $C_3$ crops usually reach a maximum at 0.10–0.15 kPa [33,54,65]. The effects of $CO_2$ partial pressures greater than this are less clear, however [54,65]. Total biomass production in radish and lettuce was reduced by superelevated $CO_2$ (e.g., 1.00 kPa), whereas no effects were observed on biomass yields of soybean, potato, and wheat [65–68] (Figure 3). Others have reported no effects on biomass but decreased seed yield in wheat and rice [54,69]. For potato, soybean, radish, and sweetpotato, water use increased substantially at superelevated $CO_2$ because of increased stomatal conductance [66,68]. This finding was surprising in light of $CO_2$'s tendency to reduce conductance as it increases across lower concentrations (e.g., 0.04 to 0.10 kPa) [63]. If plant growth systems are closely linked with atmospheres in human habitats in space, then $CO_2$ pressures may become superelevated as a result of human respiration, ranging from 0.2 to 0.6 kPa [14,70]. In addition, if plants were grown in enclosures on Mars where local $CO_2$ was used as a pressurizing gas, then $CO_2$ partial pressures would probably exceed 1 kPa.

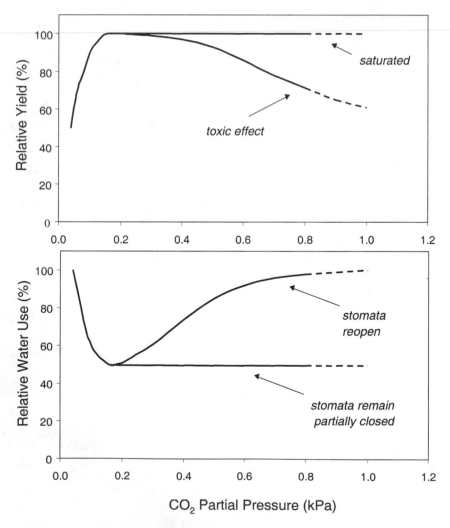

**Figure 3**   Possible effects of carbon dioxide ($CO_2$) on plant growth and water use. Superelevated $CO_2$ levels (>0.2 kPa) could be toxic for some species and/or result in increased water use [54,65,68].

## 3. Temperature

Many of the crops considered for life support (rice, peanut, soybean, and sweetpotato) prefer warmer temperatures (e.g., 25 to 30°C), whereas potatoes and wheat do well at cooler temperatures (e.g., 15 to 20°C) [47,50,53,71]. Wheat is tolerant of warmer temperatures (20–25°C), but the life cycle decreases and yields tend to drop [22,47]. Potatoes also tolerate warm temperatures but do not tuberize well in controlled environments when temperatures are >20°C [71]. Lettuce and tomatoes do well at intermediate temperatures, e.g., 23°C [72,73], whereas other species considered for life support studies, such as cabbage, chard, and carrot, typically grow better at cooler temperatures (e.g., 16 to 18°C) [74]. This range of temperature optima suggests that it may be most efficient to partition plant production systems into at least warm and cool growth areas. In all cases, freezing temperatures would be lethal for most actively growing crops and must be avoided.

## 4. Volatile Organic Compounds

Life support habitats for space must be tightly closed to minimize atmospheric gas losses and associated resupply costs. This tight closure could cause volatile organic compounds (VOCs) to accumulate, which could affect both humans and plants in these systems [14,70,75]. These VOCs emanate from a variety of sources, including plastics, paints, glues, caulking, and the plants themselves [75–77]. Closed-system tests with plants have documented the presence of compounds known to be associated with plant metabolism (Table 2), and the effects that these have on plants and humans is not well studied [75–79]. In addition to VOCs, volatile nitrogenous compounds such as $N_2O$ can accumulate in closed systems with plants as a result of bacterial denitrification in root zones [80].

With plants in closed systems, there is a special concern with ethylene gas [81]. Closed-system studies conducted at NASA's Kennedy Space Center showed that ethylene is a natural product of plant stands, with the rates of production varying with species and developmental stage [78]. Somewhat surprisingly, the highest ethylene production typically occurred during rapid vegetative growth [78]. Environmental stresses, such as using continuous light with potato, or periods of fruit ripening with tomato also resulted in rapid production of ethylene (B. V. Peterson et al., unpublished). High ethylene can result in leaf epinasty, flower abortion, reduced stem elongation, and reduced seed set [81]. Current approaches for controlling volatile organics in closed atmospheres of space habitats include the use of carbon filtration, chemical oxidants such as potassium permanganate, or catalytic burners [14].

**TABLE 2** Volatile Organic Compounds (VOCs) from Humans and Plants

| Humans[a] | Plants[b] |
|---|---|
| Acetaldehyde | Benzaldehyde |
| Acetone | 2-Butanone |
| Ammonia | Carbon disulfide |
| n-Butyl alcohol | Ethylene |
| Carbon monoxide | 2-Ethyl-1-hexanol |
| Caprylic acid | Heptanal |
| Ethanol | Hexanal |
| Ethyl mercaptan | 2-Hexen-1-ol acetate |
| Hydrogen | Isoprene |
| Hydrogen sulfide | Limonene |
| Indole | 2-Methylfuran |
| Methanol | Nonanal |
| Methane | Ocimene |
| Methyl mercaptan | α-Pinene |
| Propyl mercaptan | β-Pinene |
| Pyruvic acid | α-Terpinene |
| Skatole | Tetrahydrofuran |
| Valeraldehyde | Tetramethylurea |
| Valeric acid | Thiobismethane |

[a] Reed and Coulter [70].
[b] Stutte and Wheeler [77]; Stutte [75].

## B.  Horticultural Considerations

Plant tests for bioregenerative technologies have utilized a range of nutrient/water delivery concepts, including solution cultures [82,83], solid media [45,84], and nutrient film technique (NFT) [85,86]. For space applications, it will be necessary to recycle (recirculate) nutrient solutions to conserve water and nutrients [82,85]. NFT has been used to grow a wide range of species, including wheat, soybean, potato, lettuce, tomato, peanut, sweetpotato, spinach, beet, and rice [85–88] and has the advantage of low mass (low water volume) and relatively simple harvesting for root-zone crops. On the other hand, low water volumes make the system more susceptible to failure if circulation is interrupted (e.g., pump failures). An additional concern is that solution culture systems typically lack buffering capacity unless buffering agents are added [89]. The lack of solution buffering requires close management of nutrient concentrations and pH to avoid periods of nutrient depletion or pH variations [85]. In studies in which plants were grown in recirculating NFT with nitrate-nitrogen, acid requirements for nutrient solution pH control could exceed 1.0 mmol $H^+$ per gram of dry mass produced or over 40 mmol $m^{-2}$ crop growing area $day^{-1}$ [85]. These acid requirements could add substantial costs to operating hydroponic systems for life support but might be reduced by manipulating the nutrient solution composition, e.g., substituting some $NH_4^+$ for $NO_3^-$ salts [90]. For early missions with relatively simple control systems, it would seem most efficient to select a universal nutrient solution formulation that could be used with a range of crops [85].

Most of the plant testing for life support has involved single plantings, after which the crops were harvested and the system was cleaned [10,47,85]. Yet for actual life support applications, crop production would have to be sustained on a continuous basis. Controlled environment tests with NFT-grown potatoes showed that productivities could be maintained through four production cycles (418 days) by planting directly back into harvested spots and continually managing the nutrient solution [91,92]. But these tests also revealed that tuber "promoting" compound(s) built up in the nutrient solution over time, which resulted in reduced shoot growth and early tuberization in successive plantings [92,93]. These inductive effects could be removed by placing activated-carbon filters in the nutrient solution, but this illustrates the challenges that can arise for sustaining crop productivities over long periods of time.

An alternative to hydroponic approaches for space applications would be to use media that are preloaded with the essential nutrients [84,94]. These systems would not require the monitoring and control of a recirculated hydroponic culture, and condensed water could be returned directly to the rooting media. A disadvantage is that the medium would eventually become nutrient depleted and require recharge or disposal. In addition, the nutrient loading may have to be tailored for each species to optimize growth. Clearly, the choice of culture system will depend on mission constraints and costs. Early efforts to grow plants on planetary surfaces might involve simple, deployable systems that are relatively autonomous and might be used only for one crop cycle. Later missions might employ more sophisticated, human-tended systems where optimizing crop growth and continuous production are required.

## C.  Crop Improvements

With the exception of lettuce and tomato, most of the crop cultivars studied for life support testing were developed for field settings [52,65,87]. These cultivars were then screened for their performance under controlled environment conditions [47,52,65,87]. For wheat, however, breeding lines were established specifically for controlled environment performance [22]. This led to the development of cv. 'Apogee', which grows only to about 50 cm in height and produces exceptionally high yields in controlled environments [22]. Tests are also being conducted with dwarf soybean and rice cultivars that are well adapted for space systems [34] (B. Bugbee, personal communication). Molecular techniques have been employed to improve crops for life support applications (e.g., sweetpotato protein content) [95], and additional use of molecular approaches could accelerate development of short-stature, high-yielding crops for space agriculture.

## III.  WASTE RECYCLING WITH PLANTS

The advantages of using plants for $CO_2$ removal while producing $O_2$ and food are obvious, but another possible function of plant systems for life support is their capability for waste recycling, especially wastewater. The phrase "plant growing system" is key here, in that the microbial communities associated with the plant roots serve a critical role in degrading many of the soluble organics [96,97]. The driving process for purifying water is transpiration, where water is taken up by the plants and then evaporated from the leaves. This resultant water vapor (humidity) can then be condensed as a source of potable water [10]. The condensate quality is dependent on the type of condenser used and may contain some dissolved volatile compounds from the atmosphere, but the water is essentially distilled.

Gray water (i.e., soap-containing water from laundry, dishes, and showers) represents the largest waste stream mass in closed life support system, with an expected production close to 25 L person$^{-1}$ day$^{-1}$ [70]. Studies have shown that gray water containing Igepon soap from shower and laundry water can be added directly to plant hydroponic systems [97]. Microbial communities in the rhizosphere of the hydroponic systems were able to degrade the organic soap rapidly and prevent any buildup and damage to the plants. Related studies showed that higher concentrations of soap can be toxic to some species [98], hence having stable microbial communities in the rhizosphere along with carefully controlled additions of gray water would be critical if plant systems were used for wastewater purification [96].

An additional concern for recycling wastewater directly would be the possibility of human pathogen buildup in the plant systems. Studies in which four different human-associated bacteria were added directly to plant hydroponic systems showed that three of the four species dropped below detectable limits within several days. The fourth species, *Pseudomonas aeruginosa*, also dropped sharply from the inoculated levels but was still detectable [99]. The findings suggest that most human-associated organisms will not be able to compete effectively in an ecologically diverse rhizosphere, but further studies are needed.

Another large liquid waste source in life support systems would be urine, with outputs ranging from 1.3 to 2.1 L (kg) person$^{-1}$ day$^{-1}$ [4,70]. Urine can contain substantial amounts of nitrogen, which could be recycled to the plants, but it can also contain large amounts of NaCl, depending on the crew diet. If urine were recycled to plants to retrieve the N while purifying the water, the challenge would be to prevent Na from reaching to toxic levels in the system. This might be accomplished by separating the Na and N in some pretreatment step, e.g., electrodialysis [100], yet this represents an additional energy and mass requirement for the system. Another approach would be to use plants capable of removing Na from the wastewater; these plants would have to partition the Na in edible tissues, thereby recycling to the human diet and avoiding its buildup in the plant production system [101].

Studies at NASA's Kennedy Space Center have shown that most of the nutrients in inedible portions of crops (e.g., leaves and stems) can be retrieved by processing the biomass in liquid stirred-tank reactors [91,102,103]. These bioreactors can effectively release up to 80% of inorganic nutrients contained in inedible plant biomass, which can subsequently be recycled to grow more plants [91,102,103]. Composting offers another approach for processing the waste biomass, where the nutrients could be leached from the compost and returned to plants, or the compost might be used directly or in combinations with local regolith to generate soils for growing plants [104].

## IV.  PLANTS AS LIFE SUPPORT MACHINES: WHAT DO WE KNOW?

### A.  Biomass Yields

Studies by Bruce Bugbee and colleagues at Utah State University have demonstrated that wheat yields increased in a near-linear fashion with light, even up to irradiances of 2000 μmol m$^{-2}$ sec$^{-1}$ PAR provided on a continuous basis, or ~170 mol m$^{-2}$ day$^{-1}$ (Figure 2). At these high PAR levels, wheat stands produced remarkably high yields, e.g., ~4 kg m$^{-2}$ seed dry mass, far surpassing yields recorded from field settings [22,31,47]. Related studies with lettuce showed that high yields were possible with high light, provided plants were harvested prior to heading and onset of tipburn injury [35,48,49]. Likewise, potatoes grown in controlled environments produced yields up to 20 kg m$^{-2}$ fresh mass [45], which is nearly double the best reported field yields [105].

Results from large-scale (20 m$^2$) tests with several species considered for life support testing are shown in Table 3. These tests were conducted in a closed atmosphere to simulate a situation that might be faced in space settings [10,85]. The best edible biomass productivities from these tests were achieved with potatoes—18.4 g m$^{-2}$ day$^{-1}$ at a PAR input of 42.2 mol m$^{-2}$ day$^{-1}$. This equated to a radiation use efficiency of 0.44 g edible biomass mol$^{-1}$ (Table 3), which compares favorably with or exceeds values reported in the literature [32,106], and could be improved even further by more effective horticultural techniques. For example, plants could have been spaced more closely for the first ~15 days and then trans-

**TABLE 3**   Productivities of Some Crops Grown in NASA's Biomass Production Chamber[a]

| Crop | PAR[b] (mol m$^{-2}$ day$^{-1}$) | Edible biomass (g m$^{-2}$ day$^{-1}$) | Total biomass (g m$^{-2}$ day$^{-1}$) | Radiation use efficiency (edible) (g mol$^{-1}$) | Radiation use efficiency (total) (g mol$^{-1}$) | Edible biomass, optimized spacing[c] (g m$^{-2}$ day$^{-1}$) |
|---|---|---|---|---|---|---|
| Wheat | 57.5 | 12.6 | 31.6 | 0.22 | 0.55 | — |
| Soybean | 36.9 | 6.0 | 15.7 | 0.16 | 0.43 | 6.7 |
| Potato | 42.2 | 18.4 | 27.2 | 0.44 | 0.64 | 20.3 |
| Lettuce | 16.8 | 7.1 | 7.7 | 0.42 | 0.46 | 11.0 |
| Tomato | 38.6 | 9.8 | 19.6 | 0.25 | 0.51 | 11.1 |

[a] Plants grown in 20 m$^2$ area in closed chamber with $CO_2$ enriched to 1000 to 1200 μmol mol$^{-1}$.
[b] PAR = photosynthetically active radiation. Total PAR varied between crops depending on photoperiod requirements and types of lamps used.
[c] Estimated yield if 10 days were eliminated from growth cycle by using transplanted seedlings. Transplanting of wheat seedlings would be impractical.

planted to their final spacing, thereby improving space use efficiency and productivities [107]. Thus edible biomass productivities of ~20 g m$^{-2}$ day$^{-1}$ and conversion rates of ~0.5 g mol$^{-1}$ appear achievable with careful horticultural techniques and $CO_2$ enrichment.

By using the potato productivities of ~20 g m$^{-2}$ day$^{-1}$ (Table 3) and a tuber caloric value of about 3.7 kcal g$^{-1}$ [10], this productivity would equate to about 75 kcal m$^{-2}$ day$^{-1}$. To provide a conservative requirement of 3000 kcal person$^{-1}$ day$^{-1}$ would then require about 40 m$^2$ per person (3000 kcal person$^{-1}$ day$^{-1}$/75 kcal m$^{-2}$ day$^{-1}$). Doing a similar calculation but using the higher productivities achieved with wheat suggests that area requirements could be reduced to ~15 m$^2$ per person, assuming the high irradiances could be provided [21,31,47]. Although these estimates consider only caloric yield of the crops, the results suggest that much of the food production could be met with reasonably small areas, provided the lighting energy is available (Figure 4). Alternatively, if high light cannot be provided, more growing area

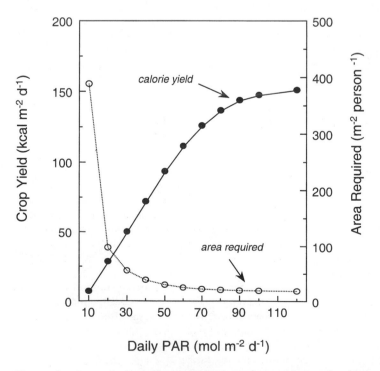

**Figure 4**   Crop caloric yield and crop area requirement to sustain one person as a function of total photosynthetically active radiation (PAR) provided to plants.

would be needed (Figure 4). Depending on mission constraints, this might be acceptable if energy is limited but growing volume is available.

Crops grown in controlled environments for life support testing showed no major changes in proximate or elemental composition when compared with field-grown crops, but ash and protein can be higher in controlled environment–grown crops [108–110]. Much of this may be the result of luxuriant uptake of nutrients, especially nitrogen from hydroponic culture, which can artificially increase protein estimates [108]. Hydroponically grown crops can also have a high P content in their tissues, which can interfere with Ca nutrition in humans. Also, leafy vegetables can accumulate high levels of nitrate, which may pose a health concern [108]. But these concerns might be managed through cultivar selection and careful control of the plant nutrient solutions, e.g., reducing nitrate concentrations prior to harvest.

## B.  Atmospheric Regeneration

Several investigators have studied photosynthetic gas exchange rates of crops using specially built or modified chambers [8,9,11,47,64,111,112]. These chambers allowed direct measurements of $CO_2$ removal by plant stands and could be used to estimate standing biomass throughout growth and development [9,64,112]. In nearly all cases, $CO_2$ uptake was strongly affected by canopy cover, PAR, and $CO_2$ concentration and could be used to detect stresses or perturbations to the crop stands [9,64,112]. Oxygen production was also tracked in some studies, which provided information on assimilation ratios of the crop canopies [9]. In studies where canopy gas exchange was not measured, $CO_2$ uptake and $O_2$ production could still be estimated from elemental analysis of biomass following crop harvests [10,111].

Although total $O_2$ production and $CO_2$ removal by plants is a function of the total biomass, only a fraction of this biomass is edible—perhaps 50% for mix of species [9]. Thus, if enough edible biomass is produced to meet human dietary needs, the $O_2$ requirements and $CO_2$ removal should also be met [9]. The balance of $O_2$ and $CO_2$ in closed life support systems will also depend on waste recycling strategies; if waste (inedible) biomass is oxidized, this would consume some of the $O_2$ produced in photosynthesis. Alternatively, if waste biomass is discarded following nutrient extraction, then an equivalent amount of carbon would need to be replaced, such as with stowed food. By using crops with a high harvest index (i.e., percent edible biomass) the proportion of waste biomass can be reduced, which in turn can reduce $O_2$ requirements for waste processing [9]. The composition of stowed foods must also be considered, where, for example, consumption of foods containing fat would reduce the respiration quotient ($CO_2$ produced/$O_2$ consumed) of humans [2,3]. Likewise, production of fat by plants could change the assimilation quotients, as could nitrate reduction requirements, which utilize energy from photosynthesis [2,3,9]. Precisely balancing $O_2$ and $CO_2$ through photosynthesis may be difficult, but these imbalances could be managed with the use of supplemental physicochemical gas control technologies [2,4,5].

## C.  Spaceflight Testing

Because of the high cost of spaceflight testing, most plant studies for life support have been conducted in "ground-based" settings, but spaceflight testing would provide a good first step for demonstrating the ultimate use of plants for life support. Numerous plant experiments have been carried out in space, but these studies were generally focused on fundamental biological questions [25]. Nonetheless, results from these studies suggest that plant growth and development can proceed in space if a good growing environment is maintained—adequate light, water, nutrients, etc. [25]. A key step toward achieving this will be the development of a reliable water and nutrient delivery system that operates in weightlessness [29,30]. Limited electrical power for lighting and cooling and relatively small growing volumes most likely will continue to impose constraints for spaceflight testing in the near future.

Of the food crops discussed for life support, wheat has been studied the most in space. These studies were carried out both on the U.S. Space Shuttle and the Russian Mir Space Station [26,113]. Initial efforts to produce seeds from the wheat in the Mir studies were unsuccessful, and it now appears that this was due to high background levels (~1 ppm) of ethylene in the growing environment [26,113]. Steps were taken to reduce ethylene in subsequent studies, and wheat plants produced viable seeds in these tests [113].

Spaceflight testing has also been carried out with potato but using explants (leaf cuttings) to study early tuber development. Results from a 16-day Space Shuttle experiment in 1995 using leaf cuttings showed that potato tubers could form and accumulate starch in spaceflight [27,28]. In addition, gas monitoring equipment in the spaceflight growth chamber used for this study showed active photosynthesis ($CO_2$ uptake during light cycles) and respiration ($CO_2$ production during the dark cycles) by the potato leaves in space [28].

Twelve sweetpotato stem cuttings were flown aboard the Space Shuttle in 1999 to study adventitious root development in spaceflight. Results showed that all the cuttings rooted well, regardless of whether the basal or apical end of the cutting was placed in agar (D. Mortley et al., unpublished), and suggest that vegetative propagation of sweetpotato should be possible under weightless conditions.

## V.   CONSIDERATIONS FOR MARS "GREENHOUSES"

Perhaps some of the first opportunities for testing bioregenerative life support approaches will occur when human missions reach Mars. With current propulsion technologies, minimal durations for Mars missions would be in the range of 3 years. This would account for travel to Mars, time for orbital realignment, and travel back to Earth. Such missions might be preceded by unmanned missions to stockpile consumables and possibly establish in situ propellant production systems for returning to Earth [114]. Conceivably, plant production systems or "greenhouses" might also be deployed prior to human arrival to provide some food and $O_2$. As a human presence on Mars increases, larger and more sophisticated plant growing systems may be feasible.

### A.   What Would It Take to Set Up A Plant Growing System on Mars?

The Martian surface environment is cold in comparison with Earth, with an average temperature of 210 K (Earth average 275 K), although temperatures can rise above 273 K (0°C) at some locations during the day [23]. Because of this, any plant growing enclosure would have to be well insulated, particularly at night. Achieving this insulation may be difficult with transparent structures, but supplemental nighttime covers or enclosures might be considered [115]. Temperature-sensitive hinges might be used to open insulating covers in the morning and then close them at night. These covers could then also be designed to reflect additional light at the transparent structure. Day lengths on Mars are 24.6 hr; hence the natural photoperiod is close to the circadian cycle on Earth. But the incident solar radiation is only about 45% that of Earth's, and dust storms can reduce the amount reaching the surface even further [40].

The Martian atmosphere is tenuous (~0.6 kPa) compared with Earth's (~100 kPa) and is composed primarily of $CO_2$ [23,24] (Table 4). This Martian $CO_2$ could be used for sustaining photosynthesis, but the ambient pressure is too low to support plants. Thus a Mars greenhouse would have to be pressurized to some minimum level to sustain acceptable plant growth. The saturation pressure of water does not change much with total pressure (across range of 5 to 100 kPa) [116], and this factor must be considered in assessing low-pressure thresholds for plants. If adequate water vapor pressure cannot be maintained, this combined with increased gas diffusion coefficients at low pressures could increase transpiration and result in water stress to the plants [117]. In addition to $CO_2$ and water vapor, some minimal level of $O_2$ would be required to sustain respiration [118,119], particularly in root zones and during dark cycles.

Determining the minimum pressures acceptable for plant enclosures is critical because lower pressures could reduce structural mass and gas leakage, which would reduce system costs [115,117]. This in-

**TABLE 4**   Gases and Elements Available on Mars

| Atmospheric composition[a] | Regolith composition[b] |
|---|---|
| $CO_2$ (95.3%) | $SiO_2$ (40–60%) |
| $N_2$ (2.7%) | FeO or $Fe_2O_3$ (12–17%) |
| Ar (1.6%) | $Al_2O_3$ (7–11%) |
| $O_2$ (0.13%) | $SO_3$ (5–8%) |
| CO (0.08%) | MgO (2–7%) |
| $H_2O$ (~0.03%) | CaO (6–7%) |
| | $K_2O$ (0–1%) |

[a] Total pressure of ~0.6 kPa [23,24].
[b] Data from Viking and Pathfinder landings [24,114].

creases the probability of finding acceptable, transparent materials that could be used for low-mass, inflatable greenhouses [115].

If an initial charge of water and $O_2$ were added to the greenhouse (i.e., brought from Earth to start the system), could local $CO_2$ be used as a pressurizing gas? Water in the greenhouse could be recycled, although there would be some leakage and incorporation into biomass, and $O_2$ produced from photosynthesis would have to be removed and stored. If, for example, a minimum pressure of 10 kPa (0.1 atm) is required for good plant growth, and 2 kPa of water vapor and 5 kPa of $O_2$ were needed [115–117], then the balance (3 kPa) might be $CO_2$. Studies with $CO_2$ levels >0.5 kPa indicate that these high levels can cause unexpected results in some plants (Figure 3) [54,65,68]. Obviously, much more research is needed to further define the potential for growing plants at low pressures and at vastly different partial pressures than have been studied before.

Some essential elements for plant growth are available on the Martian surface [114] (Table 4) but whether these would be useful for early efforts to establish plant systems is unknown. A more likely scenario would be to supply the necessary nutrients initially and then incorporate recycling approaches with the inedible biomass from previous plantings and even human wastes following human arrival [91,97,101,103]. Eventually, residual biomass might be composted and incorporated with local regolith to generate soils for supporting plant growth as systems expand [104].

## VI.  CONCLUDING REMARKS

As humankind advances and technologies improve, the exploration and colonization of other planets in our solar system seem inevitable. This exploration will require safe and reliable life support technologies that minimize stowage and resupply costs. Green plants (crops) could serve a vital role for these regenerative life support systems, where photosynthesis is used to provide oxygen and food while removing waste carbon dioxide. In addition, plant transpiration in combination with root-zone microbes could be used to process and purify wastewater. For bioregenerative systems to succeed, the growing environment and horticultural approaches must be carefully managed to optimize crop outputs. In some settings, e.g., Mars "greenhouse," this might involve the use of low atmospheric pressures and/or gas partial pressures vastly different from the terrestrial environment. A key factor for implementing crop production systems will be the development of energy-efficient lighting approaches. For early missions, stowage and physicochemical technologies will provide most of the consumables, with plants possibly grown to provide a modest supply of fresh foods. As mission distances and durations increase, the role for plants could expand, where crops are then used for most of the atmospheric regeneration and provide major portions of carbohydrate, protein, and oil for the crew. The self-regenerating nature of biological systems could also provide a degree of autonomy for surviving potential system failures or mission delays.

## REFERENCES

1.  J Myers. Basic remarks on the use of plants as a biological gas exchangers in a closed system. J Aviat Med 25:407–411, 1954.
2.  JH Eley, J Myers. Study of a photosynthetic gas exchanger. A quantitative repetition of the Priestley experiment. Tex J Sci 16:296–333, 1964.
3.  RL Miller, CH Ward. Algal bioregenerative systems. In: E Kammermeyer, ed. Atmosphere in Space Cabins and Closed Environments. New York: Appleton-Century-Croft, 1966, pp 186–221.
4.  JI Gitelson, YuN Okladnikov. Man as a component of a closed ecological life support system. Life Support Biosphere Sci 1:73–81, 1994.
5.  FB Salisbury, JI Gitelson, GM Lisovsky. Bios-3: Siberian experiments in bioregenerative life support. BioScience 47:575–585, 1997.
6.  RD MacElroy, NV Martello, DT Smernoff. Controlled ecological life support systems: CELSS '85 workshop. NASA Technical Memorandum 88215, NASA Ames Research Center, Moffett Field, CA, 1986.
7.  K Nitta, M Oguchi, S Kanda. CELSS experiment model and design concept of gas recycle system. In: RD MacElroy, NV Martello, DT Smernoff, eds. Controlled Ecological Life Support Systems: CELSS '85 Workshop. NASA Technical Memorandum 88215, Ames Research Center, Moffett Field, CA, 1986, pp 35–46.
8.  M Andre, H Du Cloux, Ch Richaud. Wheat response to carbon dioxide enrichment: carbon dioxide exchange rates, transpiration, and mineral uptake. In: RD MacElroy, NV Martello, DT Smernoff, eds. Controlled Ecological Life Support Systems: CELSS '85 Workshop. NASA Technical Memorandum 88215, NASA Ames Research Center, Moffett Field, CA, 1986, pp 405–428.

9. RM Wheeler. Gas balance in a plant-based CELSS. In: H Suge, ed. Plants in Space Biology. Tohoku: Institute of Genetic Ecology, Tohoku University Press, 1996, pp 207–216.

10. RM Wheeler, CL Mackowiak, GW Stutte, JC Sager, NC Yorio. LM Ruffe, RE Fortson, TW Dreschel, WM Knott, KA Corey. NASA's Biomass Production Chamber: a testbed for bioregenerative life support studies. Adv Space Res 18:215–224, 1996.

11. DJ Barta, K Henderson. Performance of wheat for air revitalization and food production during the lunar-Mars life support test project phase III test. SAE Technical Paper Series, 981704, 1998.

12. AE Drysdale. The effect of resource cost on life support selection. SAE Technical Paper 951492, 1995.

13. J Hunter. Engineering concepts for food processing in bioregenerative life support systems. Life Support Biosphere Sci 6:53–60, 1999.

14. KL Mitchell, RM Bagdigian, RL Carrasquillo, DL Carter, GD Franks, DW Holder, CF Hutchens, KY Ogle, JL Perry, CD Ray. Technical assessment of Mir-1 life support hardware for the International Space Station. NASA Technical Memorandum 108441, Marshall Space Flight Center, 1994.

15. RD MacElroy, M Kliss, C Straight. Life support systems for Mars transit. Adv Space Res 12:159–166, 1992.

16. VI Lohr. Quantifying the intangible. Interior Landscape. August:32–39, 1992.

17. TW Tibbitts, DK Alford. Controlled ecological life support system. Use of higher plants. Ames Research Center, Moffett Field, CA, NASA Conference Publication 2231, 1982.

18. JE Hoff, JM Howe, CA Mitchell. Nutritional and cultural aspects of plant species selection for a regenerative life support system. Report to NASA Ames Research Center, NSG2401 and NSG 2404, 1982.

19. FB Salisbury, MAZ Clark. Choosing plants to be grown in a controlled environment life support system (CELSS) based upon attractive vegetarian diets. Life Support Biosphere Sci 2:169–179, 1996.

20. CA Mitchell, TAO Dougher, SS Nielsen, MA Belury, RM Wheeler. Costs of providing edible biomass for a balanced vegetarian diet in a controlled ecological life support system. In: H Suge, ed. Plants in Space Biology. Tohoku: Institute of Genetic Ecology, Tohoku University, 1996, pp 245–254.

21. FB Salisbury. Some challenges in designing a lunar, Martian, or microgravity CELSS. Acta Astronaut 27:211–217, 1992.

22. B Bugbee, G Koerner. Yield comparisons and unique characteristics of the dwarf wheat cultivar 'USU Apogee'. Adv Space Res 20:1891–1894, 1997.

23. C McKay. A short guide to Mars. In: PJ Boston, ed. The Case for Mars. Vol 57, Science and Technology Series, San Diego: American Astronautical Society Publishers, 1984, pp 303–310.

24. W Mendell, J Plescia, AC Tribble. Surface environments. Physiology of spaceflight. In: WJ Larson, LK Pranke, eds. Human Spaceflight: Mission Analysis and Design. New York: McGraw-Hill, 1999, pp 77–101.

25. AD Krikorian, HG Levine. Development and growth in space. In: RGS Bidwell, ed. Plant Physiology: A Treatise. Orlando, FL: Academic Press, 1991, pp 491–555.

26. FB Salisbury. Growing super-dwarf wheat in space station Mir. Life Support Biosphere Sci 4:155–166, 1997.

27. J Croxdale, M Cook, TW Tibbitts, CS Brown, RM Wheeler. Structure of potato tubers formed during spaceflight. J Exp Bot 48:2037–2043, 1997.

28. CS Brown, TW Tibbitts, JG Croxdale, RM Wheeler. Potato tuber formation in the spaceflight environment. Life Support Biosphere Sci 4:71–76, 1997.

29. BD Wright, WC Bausch, WM Knott. A hydroponic system for microgravity plant experiments. Tran ASAE 31:440–446, 1988.

30. TW Dreschel, JC Sager. Control of water and nutrient using a porous tube: A method for growth plants in space. HortScience 24:944–947, 1989.

31. BG Bubgee, FB Salisbury. Exploring the limits of crop productivity. Photosynthetic efficiency of wheat in high irradiance environments. Plant Physiol 88:869–878, 1988.

32. E Heuvelink. Growth, development and yield of a tomato crop: periodic destructive measurements in a greenhouse. Sci Hortic 61:77–99, 1995.

33. RM Wheeler, TW Tibbitts, AH Fitzpatrick. Carbon dioxide effects on potato growth under different photoperiods and irradiance. Crop Sci 31:1209–1213, 1991.

34. TAO Dougher, BG Bugbee. Blue light and temperature effects on internode elongation, growth and yield. Adv Space Res 20:1895–1899, 1997.

35. GF Collier, TW Tibbitts. Tipburn of lettuce. Hortic Rev 4:49–65, 1982.

36. AV Barker, KA Corey, LE Craker. Nutritional stresses in tomato genotypes grown under high-pressure sodium vapor lamps. HortScience 24:255–258, 1989.

37. K Mori, H Ohya, K Matsumoto, H Furune. Sunlight supply and gas exchange systems in microalgal bioreactor. In: RD MacElroy, DT Smernoff, eds. Controlled Ecological Life Support Systems, Regenerative Life Support System in Space. NASA Conf Publ 2480, Ames Research Center, Moffett Field, CA, 1987, pp 45–50.

38. JL Cuello, P Sadler, D Jack, E Ono, KA Jordan. Evaluation of light transmission and distribution materials for lunar and Martian bioregenerative life support. Life Support Biosphere Sci 5:389–402, 1998.

39. RC Morrow. RJ Bula, TW Tibbitts. Orbital light:dark cycle effects on potato productivity. American Society for Gravitational and Space Biology, 3rd Annual Meeting, Logan, UT, 1997, p 30.

40. GA Landis. Dust obscuration of Mars solar arrays. Acta Astronaut 38:885–891, 1996.

41. JC Sager, JL Edwards, WH Klein. Light energy utilization efficiency for photosynthesis. Trans ASAE 25:1737–1746, 1982.

42. RJ Bula, RC Morrow, TW Tibbitts, DJ Barta, RW Ignatius, TS Martin. Light-emitting diodes as a radiation source for plants. HortScience 26:203–205, 1992.

43. GD Goins, NC Yorio, MM Sanwo, CS Brown. Photomorphogenesis, photosynthesis, and seed yield of wheat plants grown under red light-emitting diodes (LEDs) with and without supplemental blue lighting. J Exp Bot 48:1407–1413, 1997.

44. DA MacLennan, BP Turner, JT Dolan, MG Ury, P Gustafson. Efficient, full-spectrum, long-lived, non-toxic microwave lamp for plant growth. In: TW Tibbitts, ed. International Lighting in Controlled Environments Workshop, Madison, WI, NASA-CP-95-3309, Kennedy Space Center, FL, 1994, pp 243–254.

45. TW Tibbitts, W Cao, RM Wheeler. Growth of potatoes for CELSS. Ames Research Center, Moffett Field, CA, NASA Contractor Report 177646, 1994.

46. MA Stasiak, R Cote, M Dixon, B Grodzinski. Increasing plant productivity in closed environments with inner canopy illumination. Life Support Biosphere Sci 5:175–182, 1998.

47. B Bugbee. Determining the potential productivity of food crops in controlled environments. Adv Space Res 12:85–95, 1992.

48. SL Knight, CA Mitchell. Enhancement of lettuce yield by manipulation of light and nitrogen nutrition. J Am Soc Hortic Sci 108:750–754, 1983.

49. SL Knight, CA Mitchell. Growth and yield characteristics of 'Waldmann's Green' leaf lettuce under different photon fluxes from metal halide or incandescent + fluorescent radiation. Sci Hortic 35:51–61, 1988.

50. JF Thomas, CD Raper. Photoperiod and temperature regulation of floral initiation and anthesis in soya bean. Ann Bot 51:481–489, 1983.

51. RM Wheeler, TW Tibbitts. Growth and tuberization of potato (*Solanum tuberosum* L) under continuous light. Plant Physiol 80:801–804, 1986.

52. RM Wheeler, TW Tibbitts. Utilization of potatoes for life support systems in space. I. Cultivar-photoperiod interaction. Am Potato J 63:315–323, 1986.

53. WA Hill, PA Loretan, CK Bonsi, CE Morris, JY Lu, C Ogbuehi. Utilization of sweetpotatoes in controlled ecological life support systems. Adv Space Res 9:29–41, 1989.

54. B Bugbee, B Spanarkel, S Johnson, O Monje, G Koerner. $CO_2$ crop growth enhancement and toxicity in wheat and rice. Adv Space Res 14:257–267, 1994.

55. KR Goldman. CA Mitchell. Transfer from long to short photoperiod affects production efficiency of day-neutral rice. HortScience 34:875–877, 1999.

56. WS Hillman. Injury of tomato plants by continuous light and unfavorable photoperiodic cycles. Am J Bot 43:89–96, 1956.

57. NC Yorio, GW Stutte, RM Wheeler, LM Ruffe. Low level irradiance during the dark period prevents photoperiodic-induced tuberization of potato. HortScience 33:542, 1998.

58. TW Tibbitts, SM Bennett, W Cao. Control of continuous irradiation injury on potatoes with daily temperature cycling. Plant Physiol 93:409–411, 1990.

59. CK Bonsi, PA Loretan, WA Hill, DG Mortley. Response of sweetpotatoes to continuous light. HortScience 27:471, 1992.

60. H Smith. Plants and the Daylight Spectrum. London: Academic Press, 1981.

61. SJ Britz, JC Sager. Photomorphogenesis and photoassimilation in soybean and sorghum grown under broad spectrum and blue-deficient light sources. Plant Physiol 94:448–454, 1991.

62. RM Wheeler, CL Mackowiak, JC Sager. Soybean stem growth under high-pressure sodium with supplemental blue lighting. Agron J 83:903–906, 1991.

63. BG Drake, MA Gonzalez-Meler, SP Long. More efficient plants: a consequence of rising atmospheric $CO_2$? Annu Rev Plant Physiol Plant Mol Biol 48:609–639, 1996.

64. RM Wheeler, KA Corey, JC Sager, WM Knott. Gas exchange rates of wheat stands grown in a sealed chamber. Crop Sci 33:161–168, 1993.

65. RM Wheeler, CL Mackowiak, LM Siegriest, JC Sager. Supraoptimal carbon dioxide effects on growth of soybean (*Glycine max* (L.) Merr.). J Plant Physiol 142:173–178, 1993.

66. CL Mackowiak, LM Ruffe, NC Yorio, RM Wheeler. Effect of carbon dioxide enrichment on radish production using nutrient film technique (NFT). NASA Technical Memorandum 109198, 1994.

67. CL Mackowiak. RM Wheeler. Growth and stomatal behavior of hydroponically cultured potato (*Solanum tuberosum* L.) at elevated and super-elevated $CO_2$. J Plant Physiol 149:205–210, 1996.

68. RM Wheeler, CL Mackowiak, NC Yorio, JC Sager. Effects of $CO_2$ on stomatal conductance: do stomata open at very high $CO_2$ concentrations? Ann Bot 83:243–251, 1999.

69. TP Grotenhuis, B Bugbee. Super-optimal $CO_2$ reduces seed yield but not vegetative growth in wheat. Crop Sci 37:1215–1222, 1997.

70. RD Reed, GR Coulter. Physiology of spaceflight. In: WJ Larson, LK Pranke, eds. Human Spaceflight: Mission Analysis and Design. New York: McGraw-Hill, 1999, pp 103–132.

71. RM Wheeler, KL Steffen, TW Tibbitts, JP Palta. Utilization of potato for life support systems. II. The effects of temperature under 24-h and 12-h photoperiod. Am Potato J 63:639–647, 1986.

72. PR Hicklenton, MS Wolynetz. Influence of light- and dark-period air temperatures and root temperature on growth of lettuce in nutrient flow systems. J Am Soc Hortic Sci 112:932–935, 1987.

73. H Janes, R McAvoy. Environmental control of a single-cluster greenhouse tomato crop. HortTech 1:110–114, 1991.
74. M Yamaguchi. World Vegetables. Principles, Production and Nutritive Values. New York: Van Nostrand Reinhold, 1983.
75. GW Stutte. Photochemicals: implications for long duration space missions. In: HC Cutler, SS Culter, eds. Biologically Active Natural Products: Agrochemicals. New York: CRC Press, 1999, pp 275–286.
76. JH Batten, GW Stutte, RM Wheeler. Effect of crop development on biogenic emissions from plant populations grown in a closed plant growth chambers. Phytochemistry 39:1351–1357, 1995.
77. GW Stutte, RM Wheeler. Accumulation and effects of volatile organic compounds in closed life support systems. Adv Space Res 20:1913–1922, 1997.
78. RM Wheeler, BV Peterson, JC Sager, WM Knott. Ethylene production by plants in a closed environment. Adv Space Res 18:193–196, 1996.
79. CS Charron, DJ Cantliffe, RM Wheeler. Photosynthetic photon flux, photoperiod, and temperature effects on volatile emission from lettuce. J Am Soc Hortic Sci 121:488–493, 1996.
80. GW Stutte. Nitrogen dynamics in the CELSS Breadboard Facility at Kennedy Space Center. Life Support Biosphere Sci 3:67–74, 1996.
81. FB Abeles, PW Morgan, ME Saltveit. Ethylene in Plant Biology. San Diego: Academic Press, 1992.
82. BG Bugbee, FB Salisbury. Controlled environment crop production: hydroponic vs. lunar regolith. In: DW Ming, DL Henninger, eds. Lunar Base Agriculture: Soils for Plant Growth. Madison, WI: American Society of Agronomy, 1989, pp 107–129.
83. JD Vessey, CD Raper, L Tolley Henry. Cyclic variations in nitrogen uptake rate in soybean plants: uptake during reproductive growth. J Exp Bot 41:1579–1584, 1990.
84. ER Allen, DW Ming, LR Hossner, DL Henninger, C Galindo. Growth and nutrient uptake of wheat in clinoptilolite-phosphate rock substrates. Agron J 87:1052–1059, 1995.
85. RM Wheeler, CL Mackowiak, WL Berry, GW Stutte, NC Yorio, LM Ruffe, JC Sager. Nutrient, acid, and water budgets of hydroponically grown crops. Acta Hortic 481:655–661, 1999.
86. CL Mackowiak, RM Wheeler, GW Stutte, NC Yorio, LM Ruffe. A recirculating hydroponic system for studying peanut (Arachis hypogaea L.). HortScience 33:650–651, 1998.
87. DG Mortley, CK Bonsi, PA Loretan, CE Morris, WA Hill, CR Ogbuehi. Evaluation of sweet potato genotypes for adaptability to hydroponic systems. Crop Sci 31:84–847, 1991.
88. GV Subbarao, RM Wheeler, GW Stutte, LH Levine. How far can sodium substitute for potassium in red beet. J Plant Nutr 2:1745–1761, 1999.
89. B Bugbee, FB Salisbury. An evaluation of MES [2(N-morpholino)-ethanesulfonic acid] and amberlite IRC-50 as pH buffers for nutrient solution studies. J Plant Nutr 8:567–583, 1985.
90. H Marschner. Mineral Nutrition of Higher Plants, New York: Academic Press, 1995.
91. CL Mackowiak, RM Wheeler, GW Stutte, NC Yorio, JC Sager. Use of biologically reclaimed minerals for continuous hydroponic potato production in a CELSS. Adv Space Res 20:1815–1820, 1997.
92. GW Stutte, CM Mackowiak, NC Yorio, RM Wheeler. Theoretical and practical considerations of staggered crop production in a BLSS. Life Support Biosphere Sci 6:287–291, 1999.
93. RM Wheeler, GW Stutte, CL Mackowiak, NC Yorio, LM Ruffe. Accumulation of possible potato tuber-inducing factor in continuous use recirculating NFT systems. HortScience 30:790, 1995.
94. DW Ming. Manufactured soils for plant growth at a lunar base. In: DW Ming, DL Henninger, eds. Lunar Base Agriculture. Soils for Plant Growth. Madison, WI: American Society of Agronomy, 1989, pp 93–105.
95. G He, CS Prakash, RL Jarret. Analysis of genetic diversity in a sweetpotato (Ipomoea batatas) germplasm collection using DNA amplification fingerprinting. Genome 38:938–945, 1985.
96. JL Garland. The structure and function of microbial communities in recirculating hydroponic systems. Adv Space Res 14:383–386, 1994.
97. CA Loader, JL Garland, LH Levine, KL Cook, CL Mackowiak, HE Vivenzio. Direct recycling of human hygiene water into hydroponic plant growth systems. Life Support Biosphere Sci 6:141–152, 1999.
98. D Bubenheim, K Wignarajah, W Berry, T Wydeven. Phytotoxic effects of gray water due to surfactants. J Am Soc Hortic Sci 122:792–796, 1997.
99. A Morales, JL Garland, DV Lim. Survival of potentially pathogenic human-associated bacteria in the rhizosphere of hydroponically grown wheat. FEMS Microb Ecol 20:155–162, 1996.
100. H Kurokawa, K Funabashi, M Oda, S Sugawara, A Ashida, K Nitta. Mineral recovery system for closed ecology experiment facility (CEEF). Life Support Biosphere Sci 5:249–253, 1998.
101. GV Subbarao, RM Wheeler, CL Mackowiak. Recycling of Na in advanced life support: strategies based on crop production systems. Life Support Biosphere Sci 6:153–160, 1999.
102. BW Finger, RF Strayer. Development of an intermediate-scale aerobic bioreactor to regenerate nutrients from inedible crop residues. SAE Tech Paper 941501, 1994.
103. CL Mackowiak, JL Garland, JC Sager. Recycling crop residues for use in recirculating hydroponic crop production. Acta Hortic 440:19–24, 1996.
104. RF Strayer, CF Atkinson. Recycling nutrient from crop residues for space applications. Compost Sci Util 5:25–31, 1997.

105. L Sibma. Maximization of arable crop yields in the Netherlands. Neth J Agric Sci 25:278–287, 1977.
106. TR Sinclair. Canopy carbon assimilation and crop radiation-use efficiency dependence on leaf nitrogen content. In: KJ Boote, RS Loomis, eds. Modeling Crop Photosynthesis—From Biochemistry to Canopy. Madison, WI: Crop Science Society of America, 1991, pp 95–107.
107. RP Prince, JW Bartok. Plant spacing for controlled environment plant growth. Trans ASAE 21:332–336, 1978.
108. JD McKeehen, CA Mitchell, RM Wheeler, B Bugbee, SS Nielsen. Excess nutrients in hydroponic solutions alter nutrient content of rice, wheat, and potato. Adv Space Res 18:73–83, 1996.
109. RM Wheeler, CL Mackowiak, JC Sager, WM Knott, WL Berry. Proximate composition of CELSS crops grown in NASA's biomass production chamber. Adv Space Res 18:43–47, 1996.
110. AM Almazan, X Zhou. Biomass yield and composition of sweetpotato grown in a nutrient film technique system. Plant Foods Hum Nutr 50:259–268, 1997.
111. SL Knight, CA Mitchell. Effects of $CO_2$ and photosynthetic photon flux on yield, gas exchange and growth rate of *Lactuca sativa* L. 'Waldmann's Green. J Exp Bot 39:317–328, 1988.
112. B Bugbee, O Monje. The limits of crop productivity. BioScience 42:494–502, 1992.
113. GE Bingham, VN Sytchev, MA Levinskikh, IG Podolsky. Final plant experiments on Mir provide second generation wheat and seeds. Gravitational Space Biol Bull 13:48, 1999.
114. CC Allen, R. Zubrin. In-situ resources. In: WJ Larson, LK Pranke, eds. Human Spaceflight: Mission Analysis and Design. New York: McGraw-Hill, 1999, pp 477–512.
115. PJ Boston. Low-pressure greenhouses and plants for a manned research station on Mars. J Br Interplanetary Soc 34:189–192, 1981.
116. JH Kennan, FG Keyes. Thermodynamic Properties of Steam. New York: John Wiley & Sons, 1962.
117. M Andre, D Massimino. Growth of plants at reduced pressures: experiments in wheat—technological advantages and constraints. Adv Space Res 12:97–106, 1992.
118. ME Musgrave, WA Gerth, HW Scheld, BR Strain. Growth and mitochondrial respiration of mungbeans (*Phaseolus aureus* Roxb.) germinated at low pressure. Plant Physiol 86:19–22, 1988.
119. SH Schwartzkopf, RL Mancinelli. Germination and growth of wheat in simulated Martian atmospheres. Acta Astronaut 25:245–247, 1991.

# Index